FIFTH EDITION

FUNDAMENTALS OF
PHYSICS

FUNDAMENTALS OF
PHYSICS
EXTENDED

DAVID HALLIDAY
University of Pittsburgh

ROBERT RESNICK
Rensselaer Polytechnic Institute

JEARL WALKER
Cleveland State University

JOHN WILEY & SONS, INC.

New York • Chichester • Brisbane • Toronto • Singapore

ACQUISITIONS EDITOR Stuart Johnson
DEVELOPMENTAL EDITOR Rachel Nelson
SENIOR PRODUCTION SUPERVISOR Cathy Ronda
PRODUCTION ASSISTANT Raymond Alvarez
MARKETING MANAGER Catherine Faduska
ASSISTANT MARKETING MANAGER Ethan Goodman
DESIGNER Dawn L. Stanley
MANUFACTURING MANAGER Mark Cirillo
PHOTO EDITOR Hilary Newman
COVER PHOTO William Warren/Westlight
ILLUSTRATION EDITOR Edward Starr
ILLUSTRATION Radiant/Precision Graphics

This book was set in Times Roman by Progressive Information Technologies, and printed and bound by Von Hofmann Press. The cover was printed by Phoenix Color Corp.

Hello There!

You are about to begin your first college level physics course. You may have heard from friends and fellow students that physics is a difficult course, especially if you don't plan to go on to a career in the hard sciences. But that doesn't mean it has to be difficult for you. The key to success in this course is to have a good understanding of each chapter before moving on to the next. When learned a little bit at a time, physics is straightforward and simple. Here are some ideas that can make this text and this class work for you:

- Read through the **Sample Problems** and solutions carefully. These problems are similar to many of the end-of-chapter exercises, so reading them will help you solve homework problems. In addition, they offer a look at how an experienced physicist would approach solving the problem.

- Try to answer the **Checkpoint** questions as you read through the chapter. Most of these can be answered by thinking through the problem, but it helps if you have some scratch paper and a pencil nearby to work out some of the harder questions. Hold the answer page at the back of the book with your thumb (or a tab or bookmark) so you can refer to it easily. The end-of-chapter **Questions** are very similar— you can use them to quiz yourself after you've read the whole chapter.

- Use the **Review & Summary** sections at the end of each chapter as the first place to look for formulas you might neeed to solve homework problems. These sections are also helpful in making study sheets for exams.

- The biggest tip I can give you is pretty obvious. Do your homework! Understanding the homework problems is the best way to master the material and do well in the course. Doing a lot of homework problems is also the best way to review for exams. And by all means, consult your classmates whenever you are stuck. Working in groups will make your studying more effective.

The study of basic physics is required for degrees in Engineering, Physics, Biology, Chemistry, Medicine, and many other sciences because the fundamentals of physics are the framework on which every other science is built. Therefore, a solid grasp of basic physical principles will help you understand upper-level science courses and make your study of these courses easier.

I took introductory physics because it was a prerequisite for my B.S. in Mechanical Engineering. Even though engineering and pure physics are worlds apart, I find myself using this book as a reference almost every day. I urge you to keep it after you have finished your course. The knowledge you will gain from this book and your introductory physics course is the foundation for all other sciences. This is the primary reason to take this class seriously and be successful in it.

Best of luck!

Josh Kane

Josh Kane

Preface

For four editions, *Fundamentals of Physics* has been successful in preparing physics students for careers in science and engineering. The first three editions were coauthored by the highly regarded team of David Halliday and Robert Resnick, who developed a groundbreaking text replete with conceptual structure and applications. In the fourth edition, the insights provided by new coauthor, Jearl Walker, took the text into the 1990s and met the challenge of guiding students through a time of tremendous advances and a ferment of activity in the science of physics. Now, in the fifth edition, we have expanded on the conventional strengths of the earlier editions and enhanced the applications that help students forge a bridge between concepts and reasoning. We not only *tell* students how physics works, we *show* them, and we give them the opportunity to show us what they have learned by testing their understanding of the concepts and applying them to real-world scenarios. Concept checkpoints, problem solving tactics, sample problems, electronic computations, exercises and problems—all of these skill-building signposts have been developed to help students establish a connection between conceptual theories and application. The students reading this text today are the scientists and engineers of tomorrow. It is our hope that the fifth edition of *Fundamentals of Physics* will help prepare these students for future endeavors by contributing to the enhancement of physics education.

CHANGES IN THE FIFTH EDITION

Although we have retained the basic framework of the fourth edition of *Fundamentals of Physics,* we have made extensive changes in portions of the book. Each chapter and element has been scrutinized to ensure clarity, currency, and accuracy, reflecting the needs of today's science and engineering students.

Content Changes

Mindful that textbooks have grown large and that they tend to increase in length from edition to edition, we have reduced the length of the fifth edition by combining several chapters and pruning their contents. In doing so, six chapters have been rewritten completely, while the remaining chapters have been carefully edited and revised, often ex-

tensively, to enhance their clarity, incorporating ideas and suggestions from dozens of reviewers.

- ***Chapters 7 and 8 on energy*** (and sections of later chapters dealing with energy) have been rewritten to provide a more careful treatment of energy, work, and the work–kinetic energy theorem. As the same time, the text material and problems at the end of each chapter still allow the instructor to present the more traditional treatment of these subjects.

- ***Temperature, heat, and the first law of thermodynamics*** have been condensed from two chapters to one chapter (*Chapter 19*).

- ***Chapter 21 on entropy*** now includes a statistical mechanical presentation of entropy that is tied to the traditional thermodynamical presentation.

- ***Chapters on Faraday's law and inductance*** have been combined into one new chapter (*Chapter 31*).

- ***Treatment of Maxwell's equations*** has been streamlined and moved up earlier into the chapter on magnetism and matter (*Chapter 32*).

- ***Coverage of electromagnetic oscillations and alternating currents*** has been combined into one chapter (*Chapter 33*).

- ***Chapters 39, 40, and 41 on quantum mechanics*** have been rewritten to modernize the subject. They now include experimental and theoretical results of the last few years. In addition, quantum physics and special relativity are introduced in some of the early chapters in short sections that can be covered quickly. These early sections lay some of the groundwork for the ''modern physics'' topics that appear later in the extended version of the text and add an element of suspense about the subject.

New Pedagogy

In the interest of addressing the needs of science and engineering students, we have added a number of new pedagogical features intended to help students forge a bridge between concepts and reasoning and to marry theory with practice. These new features are designed to help students test their understanding of the material. They were also developed to help students prepare to apply the information to exam questions and real-world scenarios.

- To provide opportunities for students to check their understanding of the physics concepts they have just read, we have placed **Checkpoint** questions within the chapter presentations. Nearly 300 Checkpoints have been added to help guide the student away from common errors and misconceptions. All of the Checkpoints require decision making and reasoning on the part of the student (rather than computations requiring calculators) and focus on the key points of the physics that students need to understand in order to tackle the exercises and problems at the end of each chapter. Answers to all of the Checkpoints are found in the back of the book, sometimes with extra guidance to the student.

- Continuing our focus on the key points of the physics, we have included additional **Checkpoint-type questions** in the Questions section at the end of each chapter. These new questions require decision making and reasoning on the part of the student; they ask the student to organize the physics concepts rather than just plug numbers into equations. Answers to the odd-numbered questions are now provided in the back of the book.

- To encourage the use of computer math packages and graphing calculators, we have added an **Electronic Computation problem section** to the Exercises and Problems sections of many of the chapters.

These new features are just a few of the pedagogical elements available to enhance the student's study of physics. A number of tried-and-true features of the previous edition have been retained and refined in the fifth edition, as described below.

CHAPTER FEATURES

The pedagogical elements that have been retained from previous editions have been carefully planned and crafted to motivate students and guide their reasoning process.

- ***Puzzlers*** Each chapter opens with an intriguing photograph and a ''puzzler'' that is designed to motivate the student to read the chapter. The answer to each puzzler is provided within the chapter, but it is not identified as such to ensure that the student reads the entire chapter.

- ***Sample Problems*** Throughout each chapter, sample problems provide a bridge from the concepts of the chapter to the exercises and problems at the end of the chapter. Many of the nearly 400 sample problems featured in the text have been replaced with new ones that more sharply focus on the common difficulties students experience in solving the exercises and problems. We have been especially mindful of the mathematical difficulties students face. The sample problems also provide

an opportunity for the student to see how a physicist thinks through a problem.

- ***Problem Solving Tactics*** To help further bridge concepts and applications and to add focus to the key physics concepts, we have refined and expanded the number of problem solving tactics that are placed within the chapters, particularly in the earlier chapters. These tactics provide guidance to the students about how to organize the physics concepts, how to tackle mathematical requirements in the exercises and problems, and how to prepare for exams.

- ***Illustrations*** Because the illustrations in a physics textbook are so important to an understanding of the concepts, we have altered nearly 30 percent of the illustrations to improve their clarity. We have also removed some of the less effective illustrations and added many new ones.

- ***Review & Summary*** A review and summary section is found at the end of each chapter, providing a quick review of the key definitions and physics concepts *without* being a replacement for reading the chapter.

- ***Questions*** Approximately 700 thought-provoking questions emphasizing the conceptual aspects of physics appear at the ends of the chapters. Many of these questions relate back to the checkpoints found throughout the chapters, requiring decision making and reasoning on the part of the student. Answers to the odd-numbered questions are provided in the back of the book.

- ***Exercises & Problems*** There are approximately 3400 end-of-chapter exercises and problems in the text, arranged in order of difficulty, starting with the exercises (labeled ''E''), followed by the problems (labeled ''P''). Particularly challenging problems are identified with an asterisk (*). Those exercises and problems that have been retained from previous editions have been edited for greater clarity; many have been replaced. Answers to the odd-numbered exercises and problems are provided in the back of the book.

VERSIONS OF THE TEXT

The fifth edition of *Fundamentals of Physics* is available in a number of different versions, to accommodate the individual needs of instructors and students alike. The Regular Edition consists of Chapters 1 through 38 (ISBN 0-471-10558-9). The Extended Edition contains seven additional chapters on quantum physics and cosmology (Chapters 1–45) (ISBN 0-471-10559-7). Both editions are available as single, hardcover books, or in the alternative versions listed on page ix:

- Volume 1—Chapters 1–21 (Mechanics/Thermodynamics), cloth, 0-471-15662-0
- Volume 2—Chapters 22–45 (E&M and Modern Physics), cloth, 0-471-15663-9
- Part 1—Chapters 1–12, paperback, 0-471-14561-0
- Part 2—Chapters 13–21, paperback, 0-471-14854-7
- Part 3—Chapters 22–33, paperback, 0-471-14855-5
- Part 4—Chapters 34–38, paperback, 0-471-14856-3
- Part 5—Chapters 39–45, paperback, 0-471-15719-8

The Extended edition of the text is also available on CD ROM.

SUPPLEMENTS

The fifth edition of *Fundamentals of Physics* is supplemented by a comprehensive ancillary package carefully developed to help teachers teach and students learn.

Instructor's Supplements

- **Instructor's Manual** by J. RICHARD CHRISTMAN, U.S. Coast Guard Academy. This manual contains lecture notes outlining the most important topics of each chapter, as well as demonstration experiments, and laboratory and computer exercises; film and video sources are also included. Separate sections contain articles that have appeared recently in the *American Journal of Physics* and *The Physics Teacher.*
- **Instructor's Solutions Manual** by JERRY J. SHI, Pasadena City College. This manual provides worked-out solutions for all the exercises and problems found at the end of each chapter within the text. *This supplement is available only to instructors.*
- **Solutions Disk.** An electronic version of the Instructor's Solutions Manual, for instructors only, available in TeX for Macintosh and Windows™.
- **Test Bank** by J. RICHARD CHRISTMAN, U.S. Coast Guard Academy. More than 2200 multiple-choice questions are included in the Test Bank for *Fundamentals of Physics.*

- **Computerized Test Bank.** IBM and Macintosh versions of the entire Test Bank are available with full editing features to help you customize tests.
- **Animated Illustrations.** Approximately 85 text illustrations are animated for enhanced lecture demonstrations.
- **Transparencies.** More than 200 four-color illustrations from the text are provided in a form suitable for projection in the classroom.

Student's Supplements

- **A Student's Companion** by J. RICHARD CHRISTMAN, U.S. Coast Guard Academy. Much more than a traditional study guide, this student manual is designed to be used in close conjunction with the text. The Student's Companion is divided into four parts, each of which corresponds to a major section of the text, beginning with an overview ''chapter.'' These overviews are designed to help students understand how the important topics are integrated and how the text is organized. For each chapter of the text, the corresponding Companion chapter offers: Basic Concepts, Problem Solving, Notes, Mathematical Skills, and Computer Projects and Notes.
- **Solutions Manual** by J. RICHARD CHRISTMAN, U.S. Coast Guard Academy and EDWARD DERRINGH, Wentworth Institute. This manual provides students with complete worked-out solutions to 30 percent of the exercises and problems found at the end of each chapter within the text.
- **Interactive Learningware** by JAMES TANNER, Georgia Institute of Technology, with the assistance of GARY LEWIS, Kennesaw State College. This software contains 200 problems from the end-of-chapter exercises and problems, presented in an interactive format, providing detailed feedback for the student. Problems from Chapter 1 to 21 are included in Part 1, from Chapters 22 to 38 in Part 2. The accompanying workbooks allow the student to keep a record of the worked-out problems. The Learningware is available in IBM 3.5″ and Macintosh formats.
- **CD Physics.** The entire Extended Version of the text (Chapters 1–45) is available on CD ROM, along with the student solutions manual, study guide, animated illustrations, and Interactive Learningware.

Acknowledgments

A textbook contains far more contributions to the elucidation of a subject than those made by the authors alone. J. Richard Christman, of the U.S. Coast Guard Academy, has once again created many fine supplements for us; his knowledge of our book and his recommendations to students and faculty are invaluable. James Tanner, of Georgia Institute of Technology, and Gary Lewis, of Kennesaw State College, have provided us with innovative software, closely tied to the text's exercises and problems. J. Richard Christman, of the U.S. Coast Guard Academy, and Glen Terrell, of the University of Texas at Arlington, contributed problems to the Electronic Computation sections of the text. Jerry Shi, of Pasadena City College, performed the Herculean task of working out solutions for every one of the Exercises and Problems in the text. We thank John Merrill, of Brigham Young University, and Edward Derringh, of the Wentworth Institute of Technology for their many contributions in the past. We also thank George W. Hukle of Oxnard, California, for his check of the answers at the back of the book.

At John Wiley, publishers, we have been fortunate to receive strong coordination and support from our former editor, Cliff Mills. Cliff guided our efforts and encouraged us along the way. When Cliff moved on to other responsibilities at Wiley, we were ably guided to completion by his successor, Stuart Johnson. Rachel Nelson has coordinated the developmental editing and multilayered preproduction process. Catherine Faduska, our senior marketing manager, and Ethan Goodman, assistant marketing manager, have been tireless in their efforts on behalf of this edition. Jennifer Bruer has built a fine supporting package of ancillary materials. Monica Stipanov and Julia Salsbury managed the review and administrative duties admirably.

We thank Lucille Buonocore, our able production manager, and Cathy Ronda, our production editor, for pulling all the pieces together and guiding us through the complex production process. We also thank Dawn Stanley, for her design; Brenda Griffing, for her copy editing; Edward Starr, for managing the line art program; Lilian Brady, for her proofreading; and all other members of the production team.

Stella Kupferburg and her team of photo researchers, particularly Hilary Newman and Pat Cadley, were inspired in their search for unusual and interesting photographs that communicate physics principles beautifully. We thank Boris Starosta and Irene Nunes for their careful development of a full-color line art program, for which they scrutinized and suggested revisions of every piece. We also owe a debt of gratitude for the line art to the late John Balbalis, whose careful hand and understanding of physics can still be seen in every diagram.

We especially thank Edward Millman for his developmental work on the manuscript. With us, he has read every word, asking many questions from the point of view of a student. Many of his questions and suggested changes have added to the clarity of this volume. Irene Nunes added a final, valuable developmental check in the last stages of the book.

We owe a particular debt of gratitude to the numerous students who used the fourth edition of *Fundamentals of Physics* and took the time to fill out the response cards and return them to us. As the ultimate consumers of this text, students are extremely important to us. By sharing their opinions with us, your students help us ensure that we are providing the best possible product and the most value for their textbook dollars. We encourage the users of this book to contact us with their thoughts and concerns so that we can continue to improve this text in the years to come. In particular, we owe a special debt of gratitude to the students who participated in a final focus group at Union College in Schenecdaty, New York: Matthew Glogowski, Josh Kane, Lauren Papa, Phil Tavernier, Suzanne Weldon, and Rebecca Willis.

Finally, our external reviewers have been outstanding and we acknowledge here our debt to each member of that team:

MARIS A. ABOLINS
Michigan State University

BARBARA ANDERECK
Ohio Wesleyan University

ALBERT BARTLETT
University of Colorado

MICHAEL E. BROWNE
University of Idaho

TIMOTHY J. BURNS
Leeward Community College

JOSEPH BUSCHI
Manhattan College

PHILIP A. CASABELLA
Rensselaer Polytechnic Institute

RANDALL CATON
Christopher Newport College

J. RICHARD CHRISTMAN
U.S. Coast Guard Academy

ROGER CLAPP
University of South Florida

W. R. CONKIE
Queen's University

PETER CROOKER
University of Hawaii at Manoa

WILLIAM P. CRUMMETT
Montana College of Mineral Science and Technology

EUGENE DUNNAM
University of Florida

ROBERT ENDORF
University of Cincinnati

F. PAUL ESPOSITO
University of Cincinnati

JERRY FINKELSTEIN
San Jose State University

ALEXANDER FIRESTONE
Iowa State University

ALEXANDER GARDNER
Howard University

ANDREW L. GARDNER
Brigham Young University

JOHN GIENIEC
Central Missouri State University

JOHN B. GRUBER
San Jose State University

ANN HANKS
American River College

SAMUEL HARRIS
Purdue University

EMILY HAUGHT
Georgia Institute of Technology

LAURENT HODGES
Iowa State University

JOHN HUBISZ
North Carolina State University

JOEY HUSTON
Michigan State University

DARRELL HUWE
Ohio University

CLAUDE KACSER
University of Maryland

LEONARD KLEINMAN
University of Texas at Austin

EARL KOLLER
Stevens Institute of Technology

ARTHUR Z. KOVACS
Rochester Institute of Technology

KENNETH KRANE
Oregon State University

SOL KRASNER
University of Illinois at Chicago

PETER LOLY
University of Manitoba

ROBERT R. MARCHINI
Memphis State University

DAVID MARKOWITZ
University of Connecticut

HOWARD C. MCALLISTER
University of Hawaii at Manoa

W. SCOTT MCCULLOUGH
Oklahoma State University

JAMES H. MCGUIRE
Tulane University

DAVID M. MCKINSTRY
Eastern Washington University

JOE P. MEYER
Georgia Institute of Technology

ROY MIDDLETON
University of Pennsylvania

IRVIN A. MILLER
Drexel University

EUGENE MOSCA
United States Naval Academy

MICHAEL O'SHEA
Kansas State University

PATRICK PAPIN
San Diego State University

GEORGE PARKER
North Carolina State University

ROBERT PELCOVITS
Brown University

OREN P. QUIST
South Dakota State University

JONATHAN REICHART
SUNY—Buffalo

MANUEL SCHWARTZ
University of Louisville

DARRELL SEELEY
Milwaukee School of Engineering

BRUCE ARNE SHERWOOD
Carnegie Mellon University

JOHN SPANGLER
St. Norbert College

ROSS L. SPENCER
Brigham Young University

HAROLD STOKES
Brigham Young University

JAY D. STRIEB
Villanova University

DAVID TOOT
Alfred University

J. S. TURNER
University of Texas at Austin

T. S. VENKATARAMAN
Drexel University

GIANFRANCO VIDALI
Syracuse University

FRED WANG
Prairie View A&M

ROBERT C. WEBB
Texas A&M University

GEORGE WILLIAMS
University of Utah

DAVID WOLFE
University of New Mexico

We hope that our words here reveal at least some of the wonder of physics, the fundamental clockwork of the universe. And, hopefully, those words might also reveal some of our awe of that clockwork.

DAVID HALLIDAY
6563 NE Windermere Road
Seattle, WA 98105

ROBERT RESNICK
Rensselaer Polytechnic Institute
Troy, NY 12181

JEARL WALKER
Cleveland State University
Cleveland, OH 44115

Chapter Opening Puzzlers

Each chapter opens with an intriguing example of physics in action. By presenting high-interest applications of each chapters concepts, the puzzlers are intended to peak your interest and motivate you to read the chapter.

In 1977, Kitty O'Neil set a dragster record by reaching 392.54 mi/h in a sizzling time of 3.72 s. In 1958, Eli Beeding Jr. rode a rocket sled from a standstill to a speed of 72.5 mi/h in an elapsed time of 0.04 s (less than an eye blink). How can we compare these two rides to see which was more exciting (or more frightening)—by final speeds, by elapsed times, or by some other quantity?

SAMPLE PROBLEM 2-6

(a) When Kitty O'Neil set the dragster records for the greatest speed and least elapsed time, she reached 392.54 mi/h in 3.72 s. What was her average acceleration?

SOLUTION: From Eq. 2-7, O'Neil's average acceleration was

$$\bar{a} = \frac{\Delta v}{\Delta t} = \frac{392.54 \text{ mi/h} - 0}{3.72 \text{ s} - 0}$$

$$= +106 \frac{\text{mi}}{\text{h} \cdot \text{s}}, \qquad \text{(Answer)}$$

where the motion is taken to be in the positive x direction. In

Answers to Puzzlers

All chapter-opening puzzlers are answered later in the chapter, either in text discussion or in a sample problem.

If the car
r, the bob
oninertial

w wish to
accelera-
the stan-
use) the
been as-

ictionless
at by trial
accelera-
lefinition,
ody has a

body by

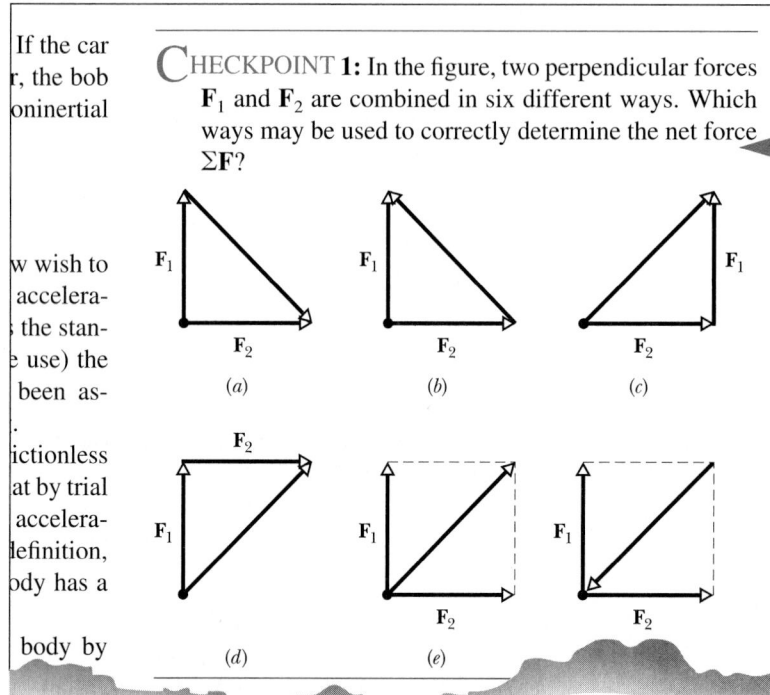

CHECKPOINT **1:** In the figure, two perpendicular forces F_1 and F_2 are combined in six different ways. Which ways may be used to correctly determine the net force ΣF?

(a) (b) (c)

(d) (e)

Checkpoints

Checkpoints appear throughout the text, focusing on the key points of physics you will need to tackle the exercises and problems found at the end of each chapter. These checkpoints help guide you away from common errors and misconceptions.

Checkpoint Questions

Checkpoint-type questions at the end of each chapter ask you to organize the physics concepts rather than plug numbers into equations. Answers to the odd-numbered questions are provided in the back of the book.

ey puck in

$v = -2t\mathbf{i}$

ponents of
ion vector
nd and t is
-2 and 3?

al projec-
t identical
n the same
nal speeds

(c)

peed (a) a

$4.9\mathbf{j}$ (x is
). Has the

9. Figure 4-25 shows three paths for a kicked football. Ignoring the effects of air on the flight, rank the paths according to (a) time of flight, (b) initial vertical velocity component, (c) initial horizontal velocity component, and (d) initial speed. Place the greatest first in each part.

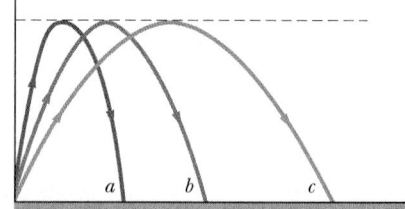

FIGURE 4-25 Question 9.

10. Figure 4-26 shows the velocity and acceleration of a particle at a particular instant in three situations. In which situation, and at that instant, is (a) the speed increasing, (b) the speed decreasing, (c) the speed not changing, (d) $\mathbf{v} \cdot \mathbf{a}$ positive, (e) $\mathbf{v} \cdot \mathbf{a}$ negative, and (f) $\mathbf{v} \cdot \mathbf{a} = 0$?

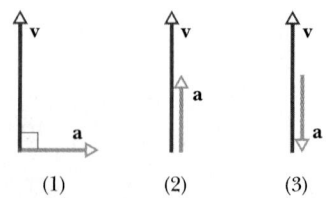

(1) (2) (3)

FIGURE 4-26 Question 10.

Sample Problems

The sample problems offer you the opportunity to work through the physics concepts just presented. Often built around real-world applications, they are closely coordinated with the end-of-chapter Questions, Exercises, and Problems.

SAMPLE PROBLEM 4-1

The position vector for a particle is initially

$$\mathbf{r}_1 = -3\mathbf{i} + 2\mathbf{j} + 5\mathbf{k}$$

and then later is

$$\mathbf{r}_2 = 9\mathbf{i} + 2\mathbf{j} + 8\mathbf{k}$$

(see Fig. 4-2). What is the displacement from $\mathbf{r}_1$ to $\mathbf{r}_2$?

SOLUTION: Recall from Chapter 3 that we add (or subtract) two vectors in unit-vector notation by combining the components, axis by axis. So Eq. 4-2 becomes

$$\Delta\mathbf{r} = (9\mathbf{i} + 2\mathbf{j} + 8\mathbf{k}) - (-3\mathbf{i} + 2\mathbf{j} + 5\mathbf{k})$$
$$= 12\mathbf{i} + 3\mathbf{k}. \qquad \text{(Answer)}$$

The displacement vector is parallel to the xz plane, because it lacks any y component, a fact that is easier to pick out in the numerical result than in Fig. 4-2.

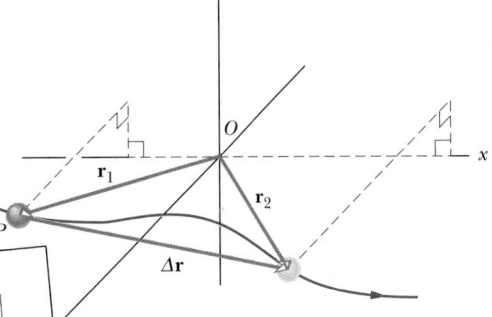

Sample Problem 4-1. The displacement $\Delta\mathbf{r} =$
d of $\mathbf{r}_1$ to the head of $\mathbf{r}_2$.

PROBLEM SOLVING TACTICS

TACTIC 1: *Reading Force Problems*

Read the problem statement several times until you have a good mental picture of what the situation is, what data are given, and what is requested. In Sample Problem 5-1, for example, you should tell yourself: "Someone is pushing a sled. Its speed changes, so acceleration is involved. The motion is along a straight line. A force is given in one part and asked for in the other, and so the situation looks like Newton's second law applied to one-dimensional motion."

If you know what the problem is about but don't know what to do next, put the problem aside and reread the text. If you are hazy about Newton's second law, reread that section. Study the sample problems. The one-dimensional-motion parts of Sample Problem 5-1 and the constant acceleration should send you back to Chapter 2 and especially to Table 2-1, which displays all the equations you are likely to need.

TACTIC 2: *Draw Two Types of Figures*

You may need two figures. One is a rough sketch of the actual real-world situation. When you draw the forces on it, place the tail of each force vector either on the boundary of or within the body feeling that force. The other figure is a free-body diagram in which the forces on a *single* body are drawn, with the body represented with a dot or a sketch. Place the tail of each force vector on the dot or sketch.

TACTIC 3: *What I*

Problem Solving Tactics

Careful attention has been paid to helping you develop your problem-solving skills. Problem-solving tactics are closely related to the sample problems and can be found throughout the text, though most fall within the first half. The tactics are designed to help you work through assigned homework problems and prepare for exams. Collectively, they represent the stock in trade of experienced problem solvers and practicing scientists and engineers.

Review and Summary

Review & Summary sections at the end of each chapter review the most important concepts and equations.

REVIEW & SUMMARY

Conservative Forces

A force is a **conservative force** if the net work it does on a particle moving along a closed path from an initial point and then back to that point is zero. Or, equivalently, it is conservative if its work on a particle moving between two points does not depend on the path taken by the particle. The gravitational force (weight) and the spring force are conservative forces; the kinetic frictional force is a **nonconservative force.**

Potential Energy

A **potential energy** is energy that is associated with the configuration of a system in which a conservative force acts. When the conservative force does work W on a particle within the system, the change ΔU in the potential energy of the system is

$$\Delta U = -W. \qquad (8\text{-}1)$$

If the particle moves from point x_i to point x_f, the change in the potential energy of the system is

$$\Delta U = -\int_{x_i}^{x_f} F(x)\, dx. \qquad (8\text{-}6)$$

Gravitational Potential Energy

The potential energy associated with a system consisting of the Earth and a nearby particle is the **gravitational potential energy.** If the particle moves from height y_i to height y_f, the change in the gravitational potential energy of the particle–Earth system is

$$\Delta U = mg(y_f - y_i) = mg\,\Delta y. \qquad (8\text{-}7)$$

If the **reference position** of the particle is set as $y_i = 0$ and the corresponding gravitational potential energy of the system is set as $U_i = 0$, then the gravitational potential energy U when the particle is at any position y is

$$U = mgy. \qquad (8\text{-}9)$$

in which the subscripts refer to different instants during an transfer process. This conservation can also be written as

$$\Delta E = \Delta K + \Delta U = 0.$$

Potential Energy Curves

If we know the **potential energy function** $U(x)$ for a sy which a force F acts on a particle, we can find the force

$$F(x) = -\frac{dU(x)}{dx}.$$

If $U(x)$ is given on a graph, then at any value of x, the fo the negative of the slope of the curve there and the kinetic of the particle is given by

$$K(x) = E - U(x),$$

where E is the mechanical energy of the system. A **turnin** is a point x where the particle reverses its motion (there, $\blacksquare$ The particle is in **equilibrium** at points where the slope $U(x)$ curve is zero (there, $F(x) = 0$).

Work by Nonconservative Forces

If a nonconservative applied force F does work on particle part of a system having a potential energy, then the wo done on the system by F is equal to the change ΔE in the m ical energy of the system:

$$W_{\text{app}} = \Delta K + \Delta U = \Delta E. \qquad (8\text{-}24$$

If a kinetic frictional force $\mathbf{f}_k$ does work on an obj change ΔE in the total mechanical energy of the object a system containing it is given by

$$\Delta E = -f_k d,$$

in which d is the displacement of the object during the wo

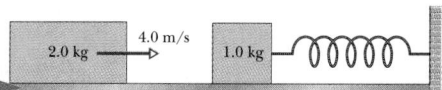

FIGURE 10-44 Problem 56.

Exercises and Problems

A hallmark of this text, nearly 3400 end-of-chapter exercises and problems are arranged in order of difficulty, starting with the exercises (labeled "E"), followed by the problems (labeled "P"). Particularly difficult problems are identified with an asterisk (*). Answers to all the odd-numbered exercises and problems are provided in the back of the book. New electronic computation problems, which require the use of math packages and graphing calculators, have been added to many of the chapters.

57P. Two 22.7 kg ice sleds are placed a short distance apart, one directly behind the other, as shown in Fig. 10-45. A 3.63 kg cat, standing on one sled, jumps across to the other and immediately back to the first. Both jumps are made at a speed of 3.05 m/s relative to the ice. Find the final speeds of the two sleds.

FIGURE 10-45 Problem 57.

58P. The bumper of a 1200 kg car is designed so that it can just absorb all the energy when the car runs head-on into a solid wall at 5.00 km/h. The car is involved in a collision in which it runs at 70.0 km/h into the rear of a 900 kg car moving at 60.0 km/h in the same direction. The 900 kg car is accelerated to 70.0 km/h as a result of the collision. (a) What is the speed of the 1200 kg car immediately after impact? (b) What is the ratio of the kinetic energy absorbed in the collision to that which can be absorbed by the bumper of the 1200 kg car?

59P. A railroad freight car weighing 32 tons and traveling at 5.0

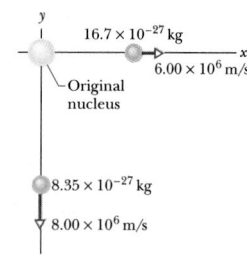

FIGURE 10-46 Exercise 62.

63E. In a game of pool, the cue ball strikes anothe at rest. After the collision, the cue ball moves at 3.5 line making an angle of 22.0° with its original dir tion, and the second ball has a speed of 2.00 m/s angle between the direction of motion of the secon original direction of motion of the cue ball and (b speed of the cue ball. (c) Is kinetic energy conserve

64E. Two vehicles A and B are traveling west and tively, toward the same intersection, where they co together. Before the collision, A (total weight 2700 with a speed of 40 mi/h and B (total weight 3600 lb of 60 mi/h. Find the magnitude and direction of the v (interlocked) vehicles immediately after the collisi

65E. In a game of billiards, the cue ball is given a V and strikes the pack of 15 stationary balls. All engage in nume $\quad$ and ball–cushion co time lat $\quad$ ne accident) all

Brief Contents

Contents

CHAPTER 10

COLLISIONS *214*

Is a board or a concrete block easier to break in karate?

CHAPTER 11

ROTATION *238*

What advantages does physics offer in judo throws?

CHAPTER 12

ROLLING, TORQUE, AND ANGULAR MOMENTUM *268*

Why is a quadruple somersault so difficult in trapeze acts?

CHAPTER 13

EQUILIBRIUM AND ELASTICITY *297*

Can you safely rest in a fissure during a chimney climb?

CHAPTER 14

GRAVITATION *322*

How can a black hole be detected?

How does a bat detect a moth in total darkness?

CHAPTER 19

TEMPERATURE, HEAT, AND THE FIRST LAW OF THERMODYNAMICS *453*

Why are black robes worn in extremely hot climates?

CHAPTER 20

THE KINETIC THEORY OF GASES *484*

When a room's air is heated, does the internal energy of the air increase?

CHAPTER 21

ENTROPY AND THE SECOND LAW OF THERMODYNAMICS *509*

What in the world gives direction to time?

CHAPTER 28

CIRCUITS   *673*

*How does an electric eel
produce a large current?*

CHAPTER 29

MAGNETIC FIELDS   *700*

Why is an aurora so thin yet so tall and wide?

CHAPTER 30

MAGNETIC FIELDS DUE TO CURRENTS   *728*

How can cargo be shot into space?

CHAPTER 31

INDUCTION AND INDUCTANCE   *752*

How did the electric guitar revolutionize rock?

CHAPTER 37

DIFFRACTION   929

Why do the colors in a pointillism painting change with viewing distance?

CHAPTER 38

RELATIVITY   958

Why is special relativity so important in modern navigation?

CHAPTER 39

PHOTONS AND MATTER WAVES   985

How can a particle such as an electron be a wave?

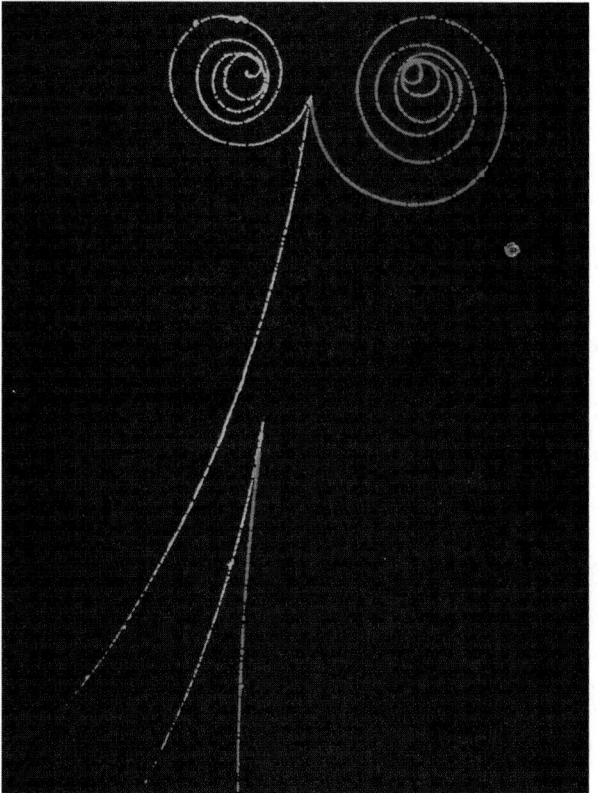

CHAPTER 45

QUARKS, LEPTONS, AND THE BIG BANG   *1118*

How can a photograph of the early universe be taken?

APPENDICES   *A1*

ANSWERS TO CHECKPOINTS AND ODD-NUMBERED QUESTIONS, EXERCISES AND PROBLEMS   *AN1*

INDEX   *I1*

FUNDAMENTALS OF
PHYSICS

1
Measurement

You can watch the Sun set and disappear over a calm ocean once while you lie on a beach, and once again if you stand up. Surprisingly, by measuring the time between the two sunsets, you can approximate the radius of the Earth. How can such a simple observation be used to measure the Earth?

1-1 MEASURING THINGS

Physics is based on measurement. We discover physics by learning how to measure the quantities that are involved in physics. Among these quantities are length, time, mass, temperature, pressure, and electrical resistance.

To describe a physical quantity we first define a **unit,** that is, a measure of the quantity that is defined to be exactly 1.0. Then we define a **standard,** that is, a reference to which all other examples of the quantity are compared. For example, the unit of length is the meter, and, as you will see, the standard for the meter is defined to be the distance traveled by light in vacuum during a certain fraction of a second. We are free to define a unit and its standard in any way we care to. The important thing is to do so in such a way that scientists around the world will agree that our definitions are both sensible and practical.

Once we have set up a standard, say, for length, we must work out procedures by which any length whatever, be it the radius of a hydrogen atom, the wheelbase of a skateboard, or the distance to a star, can be expressed in terms of the standard. Rulers, which approximate our length standard, give us one such procedure for measuring length. But many of our comparisons must be indirect. You cannot use a ruler, for example, to measure either the radius of an atom or the distance to a star.

There are so many physical quantities that it is a problem to organize them. Fortunately, they are not all independent. Speed, for example, is the ratio of a length to a time. So what we do is pick out—by international agreement—a small number of physical quantities, such as length and time, and assign standards to them alone. We then define all other physical quantities in terms of these *base quantities* and their standards. Speed, for example, is defined in terms of the base quantities length and time and the associated base standards.

The base standards must be both accessible and invariable. If we define the length standard as the distance between one's nose and the index finger on an outstretched arm, we certainly have an accessible standard—but it will, of course, vary from person to person. The demand for precision in science and engineering pushes us to aim first for invariability. We then exert great effort to make duplicates of the base standards that are accessible to those who need them.

1-2 THE INTERNATIONAL SYSTEM OF UNITS

In 1971, the 14th General Conference on Weights and Measures picked seven quantities as base quantities, thereby forming the basis of the International System of Units, abbreviated SI from its French name and popularly known as the *metric system.* Table 1-1 shows the units for the three base quantities—length, mass, and time—that we use in the early chapters of this book. The units for the quantities were chosen to be on a "human scale."

Many SI *derived units* are defined in terms of these base units. For example, the SI unit for power, called the **watt** (abbreviated as W), is defined in terms of the base units for mass, length, and time. Thus, as you will see in Chapter 7,

$$1 \text{ watt} = 1 \text{ W} = 1 \text{ kg} \cdot \text{m}^2/\text{s}^3. \tag{1-1}$$

To express the very large and very small quantities that we often run into in physics, we use the so-called *scientific notation,* which employs powers of 10. In this notation,

$$3{,}560{,}000{,}000 \text{ m} = 3.56 \times 10^9 \text{ m} \tag{1-2}$$

and
$$0.000\,000\,492 \text{ s} = 4.92 \times 10^{-7} \text{ s}. \tag{1-3}$$

Since the advent of computers, scientific notation sometimes takes on an even briefer look, as in 3.56 E9 m and 4.92 E −7 s, where E stands for "exponent of ten." It is briefer still on some calculators, where E is replaced with an empty space.

As a further convenience when dealing with very large or very small measurements, we use the prefixes listed in Table 1-2. As you can see, each prefix represents a certain power of 10, as a factor. Attaching a prefix to an SI unit has the effect of multiplying by the associated factor. Thus we can express a particular electric power as

$$1.27 \times 10^9 \text{ watts} = 1.27 \text{ gigawatts} = 1.27 \text{ GW} \tag{1-4}$$

or a particular time interval as

$$2.35 \times 10^{-9} \text{ s} = 2.35 \text{ nanoseconds} = 2.35 \text{ ns}. \tag{1-5}$$

Some prefixes, as used in milliliter, centimeter, kilogram, and megabyte, are already familiar to you.

1-3 CHANGING UNITS

We often need to change the units in which a physical quantity is expressed. We do so by a method called *chain-link conversion.* In this method, we multiply the original measurement by a **conversion factor** (a ratio of units that is equal to unity). For example, because 1 min and 60 s are identical time intervals, we can write

$$\frac{1 \text{ min}}{60 \text{ s}} = 1 \quad \text{and} \quad \frac{60 \text{ s}}{1 \text{ min}} = 1.$$

This is *not* the same as writing $\frac{1}{60} = 1$ or $60 = 1$; the *number* and its *unit* must be treated together.

TABLE 1-1 SOME SI BASE UNITS

QUANTITY	UNIT NAME	UNIT SYMBOL
Length	meter	m
Time	second	s
Mass	kilogram	kg

Because multiplying any quantity by unity leaves it unchanged, we can introduce such conversion factors wherever we find them useful. In chain-link conversion, we use the factors to cancel the unwanted units. For example, to convert 2 min to seconds, we have

$$2 \text{ min} = (2 \text{ min})(1) = (2 \text{ min})\left(\frac{60 \text{ s}}{1 \text{ min}}\right)$$
$$= 120 \text{ s}. \qquad (1\text{-}6)$$

If you introduce the conversion factor in a way that the units do *not* cancel, invert the factor and try again. Units obey the same algebraic rules as variables and numbers.

Appendix D and the inside back cover give conversion factors between SI and other systems of units. The United States is the only major country (in fact, almost the only country) that has not officially adopted the International System of Units.

SAMPLE PROBLEM 1-1

The research submersible ALVIN is diving at a speed of 36.5 fathoms per minute.

(a) Express this speed in meters per second. A *fathom* (fath) is precisely 6 ft.

SOLUTION: To find the speed in meters per second, we write

$$36.5 \frac{\text{fath}}{\text{min}} = \left(36.5 \frac{\text{fath}}{\text{min}}\right)\left(\frac{1 \text{ min}}{60 \text{ s}}\right)\left(\frac{6 \text{ ft}}{1 \text{ fath}}\right)\left(\frac{1 \text{ m}}{3.28 \text{ ft}}\right)$$
$$= 1.11 \text{ m/s}. \qquad \text{(Answer)}$$

(b) What is this speed in miles per hour?

SOLUTION: As above, we have

$$36.5 \frac{\text{fath}}{\text{min}} = \left(36.5 \frac{\text{fath}}{\text{min}}\right)\left(\frac{60 \text{ min}}{1 \text{ h}}\right)\left(\frac{6 \text{ ft}}{1 \text{ fath}}\right)\left(\frac{1 \text{ mi}}{5280 \text{ ft}}\right)$$
$$= 2.49 \text{ mi/h}. \qquad \text{(Answer)}$$

(c) What is this speed in light-years per year?

SOLUTION: A light-year (ly) is the distance that light travels in 1 year, 9.46×10^{12} km.

We start from the result of (a) above:

$$1.11 \frac{\text{m}}{\text{s}} = \left(1.11 \frac{\text{m}}{\text{s}}\right)\left(\frac{1 \text{ ly}}{9.46 \times 10^{12} \text{ km}}\right)$$
$$\times \left(\frac{1 \text{ km}}{1000 \text{ m}}\right)\left(\frac{3.16 \times 10^7 \text{ s}}{1 \text{ y}}\right)$$
$$= 3.71 \times 10^{-9} \text{ ly/y}. \qquad \text{(Answer)}$$

We can write this in the even more unlikely form of 3.71 nly/y, where ''nly'' is an abbreviation for nanolight-year.

SAMPLE PROBLEM 1-2

How many square meters are in an area of 6.0 km²?

SOLUTION: Each kilometer in the original measure must be converted. The surest way is to separate them:

TABLE 1-2 PREFIXES FOR SI UNITS[a]

FACTOR	PREFIX	SYMBOL	FACTOR	PREFIX	SYMBOL
10^{24}	yotta-	Y	10^{-24}	yocto-	y
10^{21}	zetta-	Z	10^{-21}	zepto-	z
10^{18}	exa-	E	10^{-18}	atto-	a
10^{15}	peta-	P	10^{-15}	femto-	f
10^{12}	tera-	T	**10^{-12}**	**pico-**	**p**
10^9	**giga-**	**G**	**10^{-9}**	**nano-**	**n**
10^6	**mega-**	**M**	**10^{-6}**	**micro-**	**μ**
10^3	**kilo-**	**k**	**10^{-3}**	**milli-**	**m**
10^2	hecto-	h	**10^{-2}**	**centi-**	**c**
10^1	deka-	da	10^{-1}	deci-	d

[a]The most commonly used prefixes are shown in bold type.

$$6.0 \text{ km}^2 = 6.0 \text{ (km)(km)}$$

$$= 6.0 \text{ (km)(km)} \left(\frac{1000 \text{ m}}{1 \text{ km}} \right) \left(\frac{1000 \text{ m}}{1 \text{ km}} \right)$$

$$= 6.0 \times 10^6 \text{ m}^2.$$

PROBLEM SOLVING TACTICS

TACTIC 1: *Significant Figures and Decimal Places*

If you calculated the answer to (a) in Sample Problem 1-1 without your calculator automatically rounding it off, the number 1.112804878 might have appeared in the display. The precision implied by this number is meaningless. We have properly rounded the number to 1.11 m/s, which is the precision that is justified by the precision of the original data: The given speed of 36.5 fath/min consists of three digits, called **significant figures.** Any fourth digit that might appear to the right of the 5 is not known, and so the result of the conversion is not reliable beyond three digits, or three significant figures.

In general, no final result should have more significant figures than the original data from which it was derived.

If multiple steps of calculation are involved, you should retain more significant figures than the original data have. However, when you come to the final result, you should round off according to the original data with the least significant figures. (The answers to sample problems in this book are usually presented with the symbol = instead of ≈ even if rounding off is involved.)

When a number such as 3.15 or 3.15×10^3 is provided in a problem, the number of significant figures is apparent. But how about the number 3000? Is it known to only one significant figure (it could be written as 3×10^3)? Or is it known to as many as four significant figures (it could be written as 3.000×10^3)? In this book, we assume that all the zeros in such provided numbers as 3000 are significant, but you had better not make that assumption elsewhere.

Don't confuse *significant figures* with *decimal places*. Consider the lengths 35.6 mm, 3.56 m, and 0.00356 m. They all have three significant figures but, in sequence, they have one, two, and five decimal places.

1-4 LENGTH

In 1792, the newborn Republic of France established a new system of weights and measures. Its cornerstone was the meter, defined to be one ten-millionth of the distance from the north pole to the equator. Eventually, for practical reasons, this Earth standard was abandoned and the meter came to be defined as the distance between two fine lines engraved near the ends of a platinum–iridium bar, the **standard meter bar,** which was kept at the International Bureau of Weights and Measures near Paris. Accurate copies of the bar were sent to standardizing laboratories throughout the world. These **secondary standards** were used to produce other, still more accessible standards so that ultimately every measuring device derived its authority from the standard meter bar through a complicated chain of comparisons.

In 1959, the yard was legally defined to be

$$1 \text{ yard} = 0.9144 \text{ meter (exactly)}, \tag{1-7}$$

which is equivalent to

$$1 \text{ inch} = 2.54 \text{ centimeters (exactly)}. \tag{1-8}$$

Table 1-3 lists some interesting lengths. One corresponds to a virus; an example is shown in Fig. 1-1.

Eventually, modern science and technology required a more precise standard than the distance between two fine scratches on a metal bar. In 1960, a new standard for the meter, based on the wavelength of light, was adopted. Specifically, the meter was redefined to be 1,650,763.73 wavelengths of a particular orange-red light emitted by atoms of krypton-86 in a gas discharge tube.* This awkward number of wavelengths was chosen so that the new standard would be as consistent as possible with the old meter-bar standard.

TABLE 1-3 SOME APPROXIMATE LENGTHS

LENGTH	METERS
Distance to the farthest observed quasar (1996)	2×10^{26}
Distance to the Andromeda galaxy	2×10^{22}
Distance to the nearest star (Proxima Centauri)	4×10^{16}
Distance to the farthest planet (Pluto)	6×10^{12}
Radius of the Earth	6×10^6
Height of Mt. Everest	9×10^3
Thickness of this page	1×10^{-4}
Wavelength of light	5×10^{-7}
Length of a typical virus	1×10^{-8}
Radius of a hydrogen atom	5×10^{-11}
Radius of a proton	1×10^{-15}

---

*The number 86 in the notation krypton-86 identifies a particular one of the five stable isotopes (or types) of this element. An equivalent notation is ^{86}Kr. The number is called the *mass number* of the isotope in question.

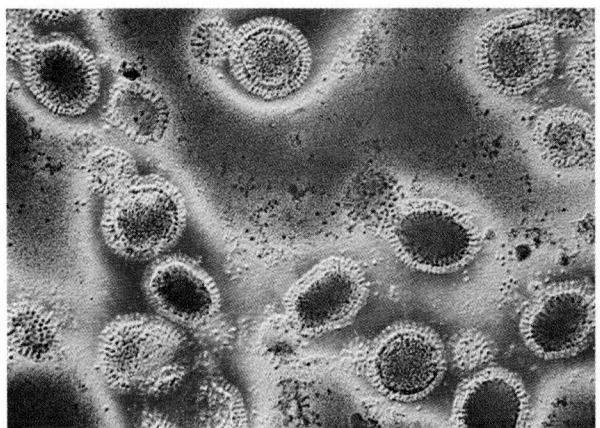

FIGURE 1-1 A "false-color" electron micrograph of influenza virus particles. Lipoproteins (colored yellow) taken from the host surround the core (colored green) of the virus. Together they are less than 50 nm in diameter.

The krypton-86 atoms of the atomic length standard are available everywhere, are identical, and emit light of precisely the same wavelength. As Philip Morrison of MIT has pointed out, every atom is a storehouse of natural standards, more secure than the International Bureau of Weights and Measures.

By 1983, the demand for higher precision had reached such a point that even the krypton-86 standard could not meet it, and in that year a bold step was taken. The meter was redefined as the distance traveled by a light wave in a specified time interval. In the words of the 17th General Conference on Weights and Measures:

The meter is the length of the path traveled by light in vacuum during a time interval of 1/299,792,458 of a second.

This number was chosen so that the speed of light c could be exactly

$$c = 299,792,458 \text{ m/s}.$$

Measurements of the speed of light had become extremely precise, so it made sense to adopt the speed of light as a defined quantity and to use it to redefine the meter.

SAMPLE PROBLEM 1-3

Both 100 yd and 100 m are used as distances for dashes in track meets.

(a) Which is longer?

SOLUTION: From Eq. 1-7, 100 yd is equal to 91.44 m, so that 100 m is longer than 100 yd.

(b) By how many meters is it longer?

SOLUTION: We represent the difference as ΔL, where Δ is the capital Greek letter delta. Thus

$$\Delta L = 100 \text{ m} - 100 \text{ yd}$$
$$= 100 \text{ m} - 91.44 \text{ m} = 8.56 \text{ m}. \quad \text{(Answer)}$$

(c) By how many feet is it longer?

SOLUTION: We can express this difference by the method used in Sample Problem 1-1,

$$\Delta L = (8.56 \text{ m})\left(\frac{3.28 \text{ ft}}{1 \text{ m}}\right) = 28.1 \text{ ft}. \quad \text{(Answer)}$$

1-5 TIME

Time has two aspects. For civil and some scientific purposes, we want to know the time of day (see Fig. 1-2) so that we can order events in sequence. And in much scientific work, we want to know how long an event lasts. Thus any time standard must be able to answer two questions: "*When* did it happen?" and "What was its *duration*?" Table 1-4 shows some time intervals.

Any phenomenon that repeats itself is a possible time standard. The rotation of the Earth, which determines the length of the day, has been used in this way for centuries. A quartz clock, in which a quartz ring continuously vibrates, can be calibrated against the rotating Earth via astronomical observations and used to measure time intervals in the laboratory. However, the calibration cannot be carried out

FIGURE 1-2 When the metric system was proposed in 1792, the hour was redefined to provide a 10-hour day. The idea did not catch on. The maker of this 10-hour watch wisely provided a small dial that kept conventional 12-hour time. Do the two dials indicate the same time?

TABLE 1-4 **SOME APPROXIMATE TIME INTERVALS**

TIME INTERVAL	SECONDS
Lifetime of the proton (predicted)	1×10^{39}
Age of the universe	5×10^{17}
Age of the pyramid of Cheops	1×10^{11}
Human life expectancy (U.S.)	2×10^{9}
Length of a day	9×10^{4}
Time between human heartbeats	8×10^{-1}
Lifetime of the muon	2×10^{-6}
Shortest laboratory light pulse (1989)	6×10^{-15}
Lifetime of most unstable particle	1×10^{-23}
The Planck time[a]	1×10^{-43}

[a]This is the earliest time after the ''Big Bang'' at which the laws of physics as we know them can be applied.

with the accuracy called for by modern scientific and engineering technology.

To meet the need for a better time standard, atomic clocks have been developed. Figure 1-3 shows such a clock, based on a characteristic frequency of the isotope cesium-133, at the National Institute of Standards and Technology (NIST). It forms the basis in this country for Coordinated Universal Time (UTC), for which time signals are available by shortwave radio (stations WWV and WWVH) and by telephone. (To set a clock extremely accurately at your particular location, you would have to account for the travel time that is required for these signals to reach you.)

Figure 1-4 shows variations in the length of one day on Earth over a 4-year period, as determined by comparison with a cesium clock. Because of the seasonal variation displayed by Fig. 1-4, we suspect the rotating Earth when there is a difference between Earth and atom as timekeepers. The variation is probably due to tidal effects caused by the Moon and to large-scale atmospheric winds.

The 13th General Conference on Weights and Measures in 1967 adopted a standard second based on the cesium clock:

One second is the time taken by 9,192,631,770 vibrations of the light (of a specified wavelength) emitted by a cesium-133 atom.

In principle, two cesium clocks would have to run for 6000 years before their readings would differ by more than 1 s. Even such accuracy pales in comparison to that of clocks currently being developed; their precision may be as fine as 1 part in 10^{18}, that is, 1 s in 1×10^{18} s (about 3×10^{10} y).

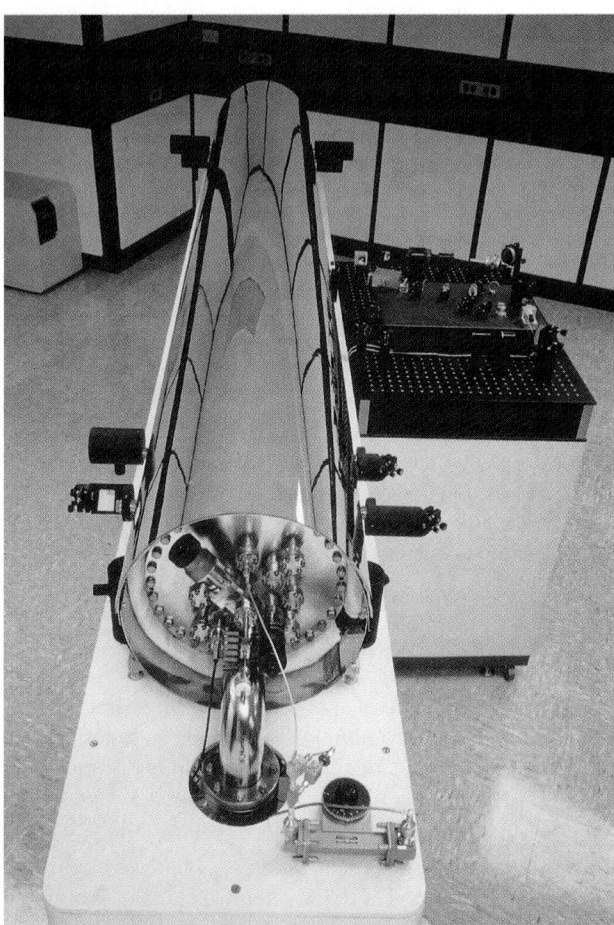

FIGURE 1-3 The cesium atomic frequency standard at the National Institute of Standards and Technology in Boulder, Colorado. It is the primary standard for the unit of time in the United States. Dial (303) 499-7111 and set your watch by it. Call (900) 410-8463 or http://tycho.usno.navy.mil/time.html for Naval Observatory time signals.

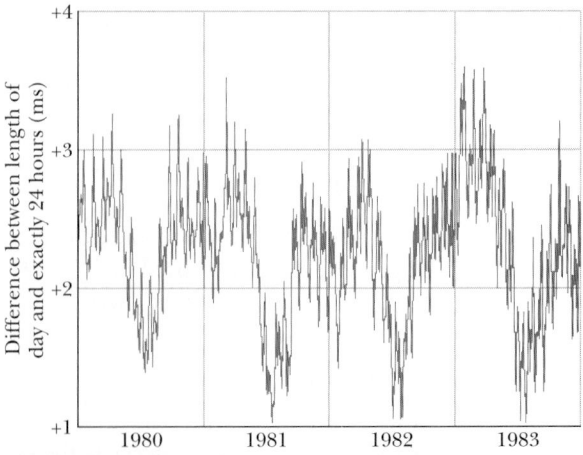

FIGURE 1-4 Variations in the length of the day over a 4-year period. Note that the entire vertical scale amounts to only 3 ms (= 0.003 s).

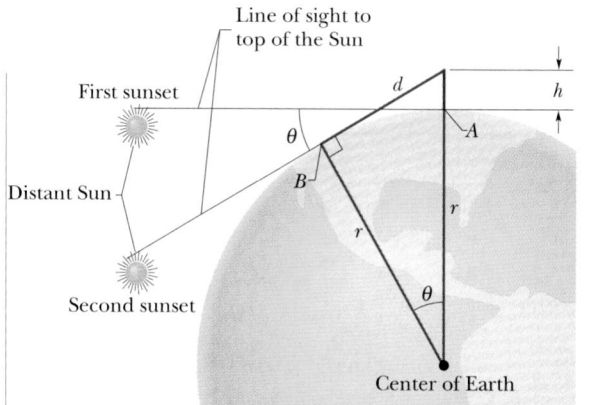

FIGURE 1-5 Sample Problem 1-4. Your line of sight to the top of the setting Sun rotates through the angle θ when you move from a prone position at point A, elevating your eyes by a distance h. (Angle θ and distance h are exaggerated here for clarity.)

SAMPLE PROBLEM 1-4*

Suppose that you watch the Sun set over a calm ocean while lying on the beach, starting a stopwatch just as the top of the Sun disappears. You then stand, elevating your eyes by a height $h = 1.70$ m, and stop the watch when the top of the Sun again disappears. If the elapsed time on the watch is $t = 11.1$ s, what is the radius r of the Earth?

SOLUTION: As shown in Fig. 1-5, your line of sight to the top of the Sun, as the top first disappears, is tangent to the Earth's surface at your location, at point A. The figure also shows that your line of sight to the top of the Sun as the top again disappears is tangent to the Earth's surface at point B. Let d represent the distance between point B and the location of your eyes when you are standing, and draw radii r as shown in Fig. 1-5. From the Pythagorean theorem, we then have

$$d^2 + r^2 = (r + h)^2 = r^2 + 2rh + h^2,$$

or $$d^2 = 2rh + h^2. \qquad (1\text{-}9)$$

Because the height h is so much smaller than the Earth's radius r, the term h^2 is negligible compared to the term $2rh$, and we can rewrite Eq. 1-9 as

$$d^2 = 2rh. \qquad (1\text{-}10)$$

In Fig. 1-5, the angle between the two tangent points A and B is θ, which is also the angle through which the Sun moves about the Earth during the measured time $t = 11.1$ s. During a full day, which is approximately 24 h, the Sun moves through an angle of $360°$ about the Earth. This allows us to write

$$\frac{\theta}{360°} = \frac{t}{24 \text{ h}},$$

which, with $t = 11.1$ s, gives us

$$\theta = \frac{(360°)(11.1 \text{ s})}{(24 \text{ h})(60 \text{ min/h})(60 \text{ s/min})} = 0.04625°.$$

From Fig. 1-5 we see that $d = r \tan \theta$. Substituting this for d in Eq. 1-10 gives us

$$r^2 \tan^2\theta = 2rh$$

or $$r = \frac{2h}{\tan^2\theta}.$$

Substituting $\theta = 0.04625°$ and $h = 1.70$ m, we find

$$r = \frac{(2)\,(1.70 \text{ m})}{\tan^2 0.04625°} = 5.22 \times 10^6 \text{ m}, \quad \text{(Answer)}$$

which is within 20% of the accepted value (6.37×10^6 m) for the (mean) radius of the Earth.

*Adapted from "Doubling Your Sunsets, or How Anyone Can Measure the Earth's Size with a Wristwatch and Meter Stick," by Dennis Rawlins, *American Journal of Physics,* Feb. 1979, Vol. 47, pp. 126–128. This technique works best at the equator.

1-6 MASS

The Standard Kilogram

The SI standard of mass is a platinum–iridium cylinder (Fig. 1-6) kept at the International Bureau of Weights and Measures near Paris and assigned, by international agreement, a mass of 1 kilogram. Accurate copies have been sent to standardizing laboratories in other countries, and the masses of other bodies can be determined by balancing them against a copy. Table 1-5 shows some masses expressed in kilograms.

The U.S. copy of the standard kilogram is housed in a vault at NIST. It is removed, no more than once a year, for the purpose of checking duplicate copies that are used elsewhere. Since 1889, it has been taken to France twice for recomparison with the primary standard. The procedure of comparison with the standard kilogram will someday be

FIGURE 1-6 The international 1 kg standard of mass, a cylinder 39 cm in height and in diameter.

TABLE 1-5 **SOME APPROXIMATE MASSES**

OBJECT	KILOGRAMS
Known universe	1×10^{53}
Our galaxy	2×10^{41}
Sun	2×10^{30}
Moon	7×10^{22}
Asteroid Eros	5×10^{15}
Small mountain	1×10^{12}
Ocean liner	7×10^{7}
Elephant	5×10^{3}
Grape	3×10^{-3}
Speck of dust	7×10^{-10}
Penicillin molecule	5×10^{-17}
Uranium atom	4×10^{-25}
Proton	2×10^{-27}
Electron	9×10^{-31}

replaced with a more reliable and accessible procedure involving a base unit of mass that is the mass of an atom.

A Second Mass Standard

The masses of atoms can be compared with each other more precisely than they can be compared with the standard kilogram. For this reason, we have a second mass standard. It is the carbon-12 atom, which, by international agreement, has been assigned a mass of 12 **atomic mass units** (u). The relation between the two units is

$$1 \text{ u} = 1.6605402 \times 10^{-27} \text{ kg}, \qquad (1\text{-}11)$$

with an uncertainty of ± 10 in the last two decimal places. Scientists can, with reasonable precision, experimentally determine the masses of other atoms relative to the mass of carbon-12. What we presently lack is a reliable means of extending that precision to more common units of mass, such as a kilogram.

REVIEW & SUMMARY

Measurement in Physics

Physics is based on measurement of physical quantities. Certain physical quantities have been chosen as **base quantities** (such as length, time, and mass); each has been defined in terms of a **standard** and given a **unit** measure (such as meter, second, and kilogram). Other physical quantities (such as speed) are defined in terms of the base quantities and their standards and units.

SI Units

The unit system emphasized in this book is the International System of Units (SI). The three physical quantities displayed in Table 1-1 are used in the early chapters of this book. Standards, which must be both accessible and invariable, have been established for these base quantities by international agreement. These standards are used in all physical measurement, for both the base quantities and the quantities derived from them. Scientific notation and the prefixes of Table 1-2 can be used to simplify measurement notation in many cases.

Changing Units

Conversion of units from one system to another (for example, from miles per hour to kilometers per second) may be performed by using *chain-link conversions* in which the units are treated as algebraic quantities and the original data are multiplied successively, by *conversion factors* written as unity, until the desired units are obtained.

Length

The unit of length—the meter—is defined as the distance traveled by light during a precisely specified time interval. The yard, together with its multiples and submultiples, is legally defined in this country in terms of the meter.

Time

The unit of time—the second—was formerly defined in terms of the rotation of the Earth. It is now defined in terms of the vibrations of the light emitted by an atomic (cesium-133) source. Accurate time signals are sent worldwide by radio signals keyed to atomic clocks in standardizing laboratories.

Mass

The unit of mass—the kilogram—is defined in terms of a particular platinum–iridium prototype kept near Paris, France. For measurements on an atomic scale, the atomic mass unit, defined in terms of the atom carbon-12, is usually used.

EXERCISES & PROBLEMS

SECTION 1-2 The International System of Units

1E. Use the prefixes in Table 1-2 to express (a) 10^6 phones; (b) 10^{-6} phone; (c) 10^1 cards; (d) 10^9 lows; (e) 10^{12} bulls; (f) 10^{-1} mate; (g) 10^{-2} pede; (h) 10^{-9} Nannette; (i) 10^{-12} boo; (j) 10^{-18} boy; (k) 2×10^2 withits; (l) 2×10^3 mockingbirds. Now that you have the idea, invent a few similar expressions. (See, in this connection, p. 61 of *A Random Walk in Science*, compiled by R.

L. Weber; Crane, Russak & Co., New York, 1974.)

2E. Some of the prefixes of the SI units have crept into everyday language. (a) What is the weekly equivalent of an annual salary of $36 K (= 36 kilobucks)? (b) A lottery awards 10 megabucks as the top prize, payable over 20 years. How much is in each monthly check? (c) The hard disk of a computer has a capacity of 30 MB (= 30 megabytes). At 8 bytes per word, how many words can it store? [In computerese, *mega* means 1,048,576 (= 2^{20}), not 1,000,000.]

SECTION 1-4 Length

3E. A space shuttle orbits the Earth at an altitude of 300 km. What is this altitude in (a) miles and (b) millimeters?

4E. What is your height in meters

5E. The micrometer (1 μm) is often called the *micron*. (a) How many microns make up 1.0 km? (b) What fraction of a centimeter equals 1.0 μm? (c) How many microns are in 1.0 yd?

6E. The Earth is approximately a sphere of radius 6.37 $\times$ 10^6 m. (a) What is its circumference in kilometers? (b) What is its surface area in square kilometers? (c) What is its volume in cubic kilometers?

7E. Calculate the number of kilometers in 20.0 mi using only the following conversion factors: 1 mi = 5280 ft, 1 ft = 12 in., 1 in. = 2.54 cm, 1 m = 100 cm, and 1 km = 1000 m.

8E. Give the relation between (a) a square yard and a square foot; (b) a square inch and a square centimeter; (c) a square mile and a square kilometer; and (d) a cubic meter and a cubic centimeter.

9P. A unit of area, often used in measuring land areas, is the *hectare,* defined as 10^4 m^2. An open-pit coal mine consumes 75 hectares of land, down to a depth of 26 m, each year. What volume of earth, in cubic kilometers, is removed in this time?

10P. The *cord* is a volume of cut wood equal to a stack 8 ft long, 4 ft wide, and 4 ft high. How many cords of wood are in 1.0 m^3?

11P. A room is 20 ft, 2 in. long and 12 ft, 5 in. wide. What is the floor area in (a) square feet and (b) square meters? If the ceiling is 12 ft, 2$\frac{1}{2}$ in. above the floor, what is the volume of the room in (c) cubic feet and (d) cubic meters?

12P. Antarctica is roughly semicircular, with a radius of 2000 km (Fig. 1-7). The average thickness of its ice cover is 3000 m. How many cubic centimeters of ice does Antarctica contain? (Ignore the curvature of the Earth.)

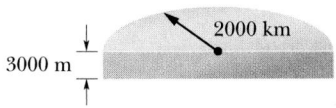

FIGURE 1-7 Problem 12.

13P. A typical sugar cube has an edge length of 1 cm. If you had a cubical box that contained a mole of sugar cubes, what would its edge length be? (One mole = 6.02 $\times$ 10^{23} units.)

14P. Hydraulic engineers often use, as a unit of volume of water, the *acre-foot,* defined as the volume of water that will cover 1 acre of land to a depth of 1 ft. A severe thunderstorm dumps 2.0 in. of rain in 30 min on a town of area 26 km^2. What volume of water, in acre-feet, fell on the town?

15P. A certain brand of house paint claims a coverage of 460 ft^2/gal. (a) Express this quantity in square meters per liter. (b) Express this quantity in SI base units (see Appendixes A and D). (c) What is the inverse of the original quantity, and what is its physical significance?

16P. Astronomical distances are so large compared to terrestrial ones that much larger units of length are needed. An *astronomical unit* (AU) is equal to the average distance from the Earth to the Sun, about 92.9 $\times$ 10^6 mi. A *parsec* (pc) is the distance at which 1 AU would subtend an angle of exactly 1 second of arc (Fig. 1-8). A *light-year* (ly) is the distance that light, traveling through a vacuum with a speed of 186,000 mi/s, would cover in 1.0 year. (a) Express the distance from the Earth to the Sun in parsecs and in light-years. (b) Express 1 ly and 1 pc in miles. Although ''light-year'' appears frequently in popular writing, the parsec is preferred by astronomers.

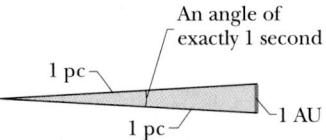

FIGURE 1-8 Problem 16.

17P. During a total solar eclipse, your view of the Sun is almost exactly replaced by your view of the moon. Assuming that the distance from you to the Sun is about 400 times the distance from you to the moon, (a) find the ratio of the Sun's diameter to the moon's diameter. (b) What is the ratio of their volumes? (c) Hold up a dime (or another small coin) so that it would just eclipse the full moon, and measure the angle it subtends at the eye. From this experimental result and the given distance between the moon and the Earth (= 3.8 $\times$ 10^5 km), estimate the diameter of the moon.

18P*. The standard kilogram (see Fig. 1-6) is in the shape of a circular cylinder with its height equal to its diameter. Show that, for a circular cylinder of fixed volume, this equality gives the smallest surface area, thus minimizing the effects of surface contamination and wear.

SECTION 1-5 Time

19E. Express the speed of light, 3.0 $\times$ 10^8 m/s, in (a) feet per nanosecond and (b) millimeters per picosecond.

20E. Enrico Fermi once pointed out that a standard lecture period (50 min) is close to 1 microcentury. How long is a microcentury in minutes, and what is the percent difference from Fermi's approximation?

21E. How many seconds are in 1 year (= 365.25 days)?

22E. A certain pendulum clock (with a 12-hour dial) happens to gain 1.0 min/day. After setting the clock to the correct time, how long must one wait until it again indicates the correct time?

23E. What is the age of the universe (see Table 1-4) in days?

24E. (a) A unit of time sometimes used in microscopic physics is the *shake*. One shake equals 10^{-8} s. Are there more shakes in a second than there are seconds in a year? (b) Humans have existed for about 10^6 years, whereas the universe is about 10^{10} years old. If the age of the universe is taken to be 1 "day," for how many "seconds" have humans existed?

25E. The maximum speeds of various animals are roughly as follows, in miles per hour: (a) snail, 3.0×10^{-2}; (b) spider, 1.2; (c) human, 23; and (d) cheetah, 70. Convert these data to meters per second. (All four calculations involve the identical conversion factor. You might want to determine that factor first and store it in the memory of your calculator, to use when needed.)

26P. An astronomical unit (AU) is the average distance of the Earth from the Sun, approximately 1.50×10^8 km. The speed of light is about 3.0×10^8 m/s. Express the speed of light in terms of astronomical units per minute.

27P. Until 1883, every city and town in the United States kept its own local time. Today, travelers reset their watches only when the time change equals 1 h. How far, on the average, must you travel in degrees of longitude until your watch must be reset by 1.0 h? *Hint:* The Earth rotates 360° in about 24 h.

28P. Assuming the length of the day uniformly increases by 0.001 s per century, calculate the cumulative effect on the measure of time over 20 centuries. Such slowing of the Earth's rotation is indicated by observations of the occurrences of solar eclipses during this period.

29P. On two *different* tracks, the winners of the mile race ran their races in 3 min, 58.05 s and 3 min, 58.20 s. In order to conclude that the runner with the shorter time was indeed faster, what is the maximum error, in feet, that can be permitted in laying out the mile distances?

30P. Five clocks are being tested in a laboratory. Exactly at noon, as determined by the WWV time signal, on successive days of a week the clocks read as in the following table. Rank the five clocks according to their relative value as good timekeepers, best to worst. Justify your choice.

CLOCK	SUN.	MON.	TUES.	WED.	THURS.	FRI.	SAT.
A	12:36:40	12:36:56	12:37:12	12:37:27	12:37:44	12:37:59	12:38:14
B	11:59:59	12:00:02	11:59:57	12:00:07	12:00:02	11:59:56	12:00:03
C	15:50:45	15:51:43	15:52:41	15:53:39	15:54:37	15:55:35	15:56:33
D	12:03:59	12:02:52	12:01:45	12:00:38	11:59:31	11:58:24	11:57:17
E	12:03:59	12:02:49	12:01:54	12:01:52	12:01:32	12:01:22	12:01:12

31P*. The time it takes the Moon to return to a given position as seen against the background of the fixed stars is called a *sidereal* month. The time interval between identical phases of the Moon is called a *lunar* month. The lunar month is longer than the sidereal month. Why, and by how much?

SECTION 1-6 Mass

32E. Using conversions and data in the chapter, determine the number of hydrogen atoms required to obtain 1.0 kg of hydrogen. A hydrogen atom has a mass of 1.0 u.

33E. One molecule of water (H_2O) contains two atoms of hydrogen and one atom of oxygen. A hydrogen atom has a mass of 1.0 u and an atom of oxygen has a mass of 16 u, approximately. (a) What is the mass in kilograms of one molecule of water? (b) How many molecules of water are in the world's oceans, which have an estimated total mass of 1.4×10^{21} kg?

34E. The Earth has a mass of 5.98×10^{24} kg. The average mass of the atoms that make up the Earth is 40 u. How many atoms are in the Earth?

35P. What mass of water fell on the town in Problem 14 during the thunderstorm? One cubic meter of water has a mass of 10^3 kg.

36P. A person on a diet might lose 2.3 kg (about 5 lb) per week. Express the mass loss rate in milligrams per second.

37P. (a) Assuming that the density (mass/volume) of water is exactly 1 g/cm³, express the density of water in kilograms per cubic meter (kg/m³). (b) Suppose that it takes 10 h to drain a container of 5700 m³ of water. What is the "mass flow rate," in kilograms per second, of water from the container?

38P. Grains of fine California beach sand are approximately spheres with an average radius of 50 μm; they are made of silicon dioxide, 1 m³ of which has a mass of 2600 kg. What mass of sand grains would have a total surface area equal to the surface area of a cube 1 m on an edge?

39P. The density of iron is 7.87 g/cm³, and the mass of an iron atom is 9.27×10^{-26} kg. If the atoms are spherical and tightly packed, (a) what is the volume of an iron atom and (b) what is the distance between the centers of adjacent atoms?

2
Motion Along a Straight Line

In 1977, Kitty O'Neil set a dragster record by reaching 392.54 mi/h in a sizzling time of 3.72 s. In 1958, Eli Beeding Jr. rode a rocket sled from a standstill to a speed of 72.5 mi/h in an elapsed time of 0.04 s (less than an eye blink). How can we compare these two rides to see which was more exciting (or more frightening)—by final speeds, by elapsed times, or by some other quantity?

2-1 MOTION

The world, and everything in it, moves. Even seemingly stationary things, such as a roadway, move with the Earth's rotation, the Earth's orbit around the Sun, the Sun's orbit around the center of the Milky Way galaxy, and that galaxy's migration relative to other galaxies. The classification and comparison of motions (called **kinematics**) is often challenging. What exactly do you measure, and how do you compare?

Before we attempt an answer, we shall examine some general properties of motion that is restricted in three ways.

1. The motion is along a straight line only. The line may be vertical (that of a falling stone), horizontal (that of a car on a level highway), or slanted, but it must be straight.

2. The cause of the motion will not be specified until Chapter 5. In this chapter you study only the motion itself. Does the object speed up, slow down, stop, or reverse direction; and, if the motion does change, how is time involved in the change?

3. The moving object is either a **particle** (by which we mean a pointlike object such as an electron) or an object that moves like a particle (such that every portion moves in the same direction and at the same rate). A pig slipping down a straight playground slide might be considered to be moving like a particle; however, a rotating merry-go-round would not, because different points around its rim move in different directions.

2-2 POSITION AND DISPLACEMENT

To locate an object means to find its position relative to some reference point, often the **origin** (or zero point) of an axis such as the x axis in Fig. 2-1. The **positive direction** of the axis is in the direction of increasing numbers (coordinates), which is toward the right as Fig. 2-1 is drawn. The opposite direction is the **negative direction.**

For example, a particle might be located at $x = 5$ m, which means that it is 5 m in the positive direction from the origin. If it were at $x = -5$ m, it would be just as far from the origin but in the opposite direction. On the axis, a coor-

dinate of -5 m is less than one of -1 m, and both coordinates are less than a coordinate of $+5$ m.

A change from one position x_1 to another position x_2 is called a **displacement** Δx, where

$$\Delta x = x_2 - x_1. \tag{2-1}$$

(The symbol Δ, which represents a change in a quantity, means the final value of that quantity less the initial value.) When numbers are inserted for the position values x_1 and x_2, a displacement in the positive direction (toward the right in Fig. 2-1) always comes out positive, and one in the opposite direction (left in the figure) negative. For example, if the particle moves from $x_1 = 5$ m to $x_2 = 12$ m, then $\Delta x = (12$ m$) - (5$ m$) = +7$ m. The positive result indicates that the motion is in the positive direction. If the particle then returns to $x = 5$ m, the displacement for the full trip is zero. The actual number of meters covered for the full trip is immaterial; displacement involves only the original and final positions.

If we ignore the sign of a displacement (and thus the direction), we are left with the **magnitude** of the displacement Δx, which is always positive.

Displacement is an example of a **vector quantity,** which is a quantity that has both a direction and a magnitude. We explore vectors more fully in Chapter 3 (in fact, some of you may have already read that chapter), but here all we need is the idea that displacement has two features: (1) its magnitude is the distance (such as the number of meters) between the original and final positions, and (2) its direction on an axis, from an original position to a final position, is represented by a plus sign or a minus sign.

What follows is the first of many checkpoints that you will see in this book. Each consists of one or more questions whose answers require some reasoning or a mental calculation, and each gives you a quick check of your understanding. The answers are listed in the back of the book.

$\mathbb{C}$HECKPOINT **1:** Here are three pairs of initial and final positions, respectively, along an x axis. Which pairs give a negative displacement: (a) -3 m, $+5$ m; (b) -3 m, -7 m; (c) 7 m, -3 m?

2-3 AVERAGE VELOCITY AND AVERAGE SPEED

A compact way to describe position is with a graph of position x plotted as a function of time t—a graph of $x(t)$. As a simple example, Fig. 2-2 shows $x(t)$ for a stationary jack rabbit (which we treat as a particle) at $x = -2$ m.

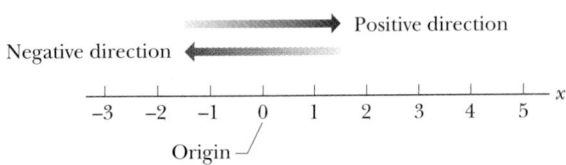

FIGURE 2-1 Position is determined on an axis that is marked in units of length and that extends indefinitely in opposite directions.

Figure 2-3*a*, also for a rabbit, is more interesting, because it involves motion. The rabbit is apparently first noticed at $t = 0$ when it is at the position $x = -5$ m. It moves toward $x = 0$, passes through that point at $t = 3$ s, and then moves on to increasingly larger positive values of x.

Figure 2-3*b* depicts the actual straight-line motion of the rabbit and is something like what you would see. The graph in Fig. 2-3*a* is more abstract and quite unlike what you would see, but it is richer in information. It also reveals how fast the rabbit moves. Actually, several quantities are associated with the phrase "how fast." One of them is the **average velocity** $\bar{v}$, which is the ratio of the displacement Δx that occurs during a particular time interval Δt to that interval:*

$$\bar{v} = \frac{\Delta x}{\Delta t} = \frac{x_2 - x_1}{t_2 - t_1}. \qquad (2\text{-}2)$$

On a graph of x versus t, $\bar{v}$ is the **slope** of the straight line that connects two particular points on the $x(t)$ curve: one is the point that corresponds to x_2 and t_2, and the other is the point that corresponds to x_1 and t_1. Like displacement, $\bar{v}$ has both magnitude and direction. (Average velocity is another vector quantity.) Its magnitude is the magnitude of the line's slope. A positive $\bar{v}$ (and slope) tells us that the line slants upward toward the right; a negative $\bar{v}$ (and slope), that the line slants upward to the left. The average velocity $\bar{v}$ always has the same sign as the displacement Δx because Δt in Eq. 2-2 is always positive.

Figure 2-4 shows how to find $\bar{v}$ for the rabbit of Fig. 2-3 for the time interval $t = 1$ s to $t = 4$ s. The average velocity during that time interval is $\bar{v} = 6$ m/3 s $= +2$ m/s; it is computed as the slope of the straight line that connects the point on the curve at the beginning of the interval and the point on the curve at the end of the interval.

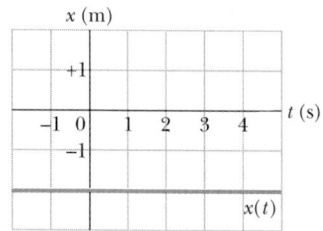

FIGURE 2-2 The graph of $x(t)$ for a jack rabbit that is stationary at $x = -2$ m. The value of x is -2 m for all times t.

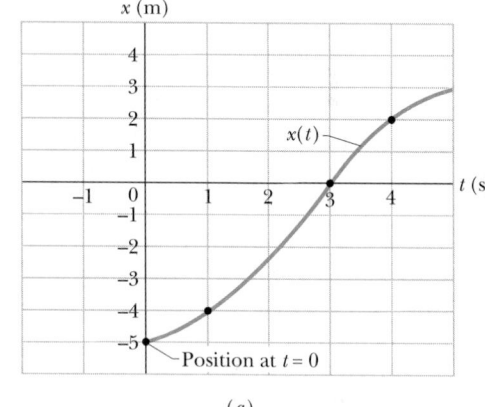

(*a*)

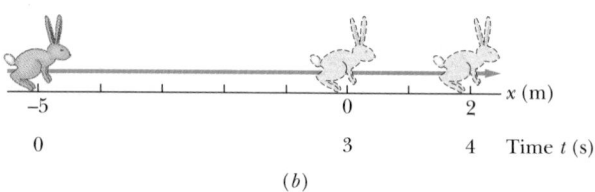

(*b*)

FIGURE 2-3 (*a*) The graph of $x(t)$ for a moving jack rabbit. (*b*) The path associated with the graph. The scale below the x axis shows the times at which the rabbit reaches various x values.

SAMPLE PROBLEM 2-1

You drive a beat-up pickup truck down a straight road for 5.2 mi at 43 mi/h, at which point you run out of fuel. You walk 1.2 mi farther, to the nearest gas station, in 27 min (= 0.450 h). What is your average velocity from the time you started your truck to the time you arrived at the station? Find the answer both numerically and graphically.

SOLUTION: To calculate $\bar{v}$ we need your displacement Δx, from start to finish, and the elapsed time Δt. Assume, for convenience, that your starting point is at the origin of an x axis (so $x_1 = 0$) and that you move in the positive direction.

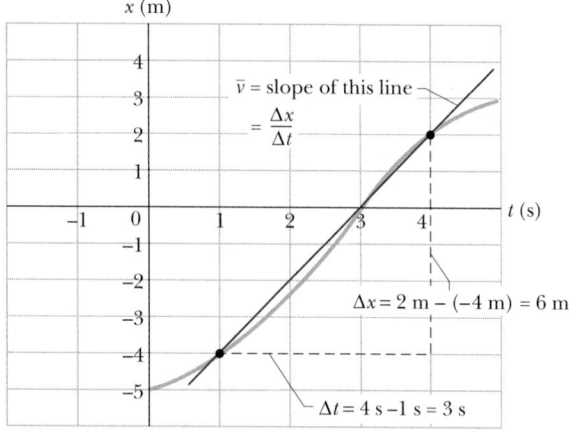

FIGURE 2-4 Calculation of the average velocity between $t = 1$ s and $t = 4$ s as the slope of the line that connects the points (on the curve) representing those times.

---

*In this book, a bar over a symbol means an average value of the quantity that the symbol represents.

You end up at $x_2 = 5.2$ mi $+ 1.2$ mi $= +6.4$ mi, and so, $\Delta x = x_2 - x_1 = +6.4$ mi. To get the driving time, we rearrange Eq. 2-2 and insert the data about the driving:

$$\Delta t = \frac{\Delta x}{\bar{v}} = \frac{5.2 \text{ mi}}{43 \text{ mi/h}} = 0.121 \text{ h},$$

or about 7.3 min. So the total time, start to finish, is

$$\Delta t = 0.121 \text{ h} + 0.450 \text{ h} = 0.571 \text{ h}.$$

Finally, we insert Δx and Δt into Eq. 2-2:

$$\bar{v} = \frac{\Delta x}{\Delta t} = \frac{6.4 \text{ mi}}{0.571 \text{ h}} \approx +11 \text{ mi/h}. \quad \text{(Answer)}$$

To find $\bar{v}$ graphically, we must first plot $x(t)$, as in Fig. 2-5 where the start and finish points on the curve are the origin and P, respectively. Your average velocity is the slope of the straight line connecting those points. The dashed lines show that the slope also gives $\bar{v} = 6.4$ mi/0.57 h $= +11$ mi/h.

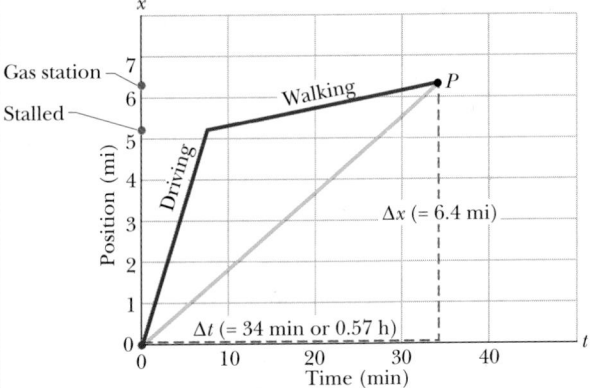

FIGURE 2-5 Sample Problem 2-1. The lines marked "Driving" and "Walking" are the position–time plots for the driver–walker in Sample Problem 2-1. The slope of the straight line joining the origin and point P is the average velocity for the trip.

SAMPLE PROBLEM 2-2

Suppose that you next carry the fuel back to the truck, making the return trip in 35 min. What is your average velocity for the full journey, from the start of your driving to your arrival back at the truck with the fuel?

SOLUTION: As previously, we must find your displacement Δx from start to finish and then divide it by the time interval Δt between start and finish. In this problem, however, the finish is back at the truck. You started at $x_1 = 0$. Back at the truck you are at position $x_2 = 5.2$ mi. And so Δx is $5.2 - 0 = 5.2$ mi. The total time Δt you take in going from start to finish is

$$\Delta t = \frac{5.2 \text{ mi}}{43 \text{ mi/h}} + 27 \text{ min} + 35 \text{ min}$$

$$= 0.121 \text{ h} + 0.450 \text{ h} + 0.583 \text{ h} = 1.15 \text{ h}.$$

So $\quad \bar{v} = \dfrac{\Delta x}{\Delta t} = \dfrac{5.2 \text{ mi}}{1.15 \text{ h}} \approx +4.5 \text{ mi/h}.$ (Answer)

This is slower than the average velocity computed in Sample Problem 2-1 because here the displacement is smaller and the time interval longer.

CHECKPOINT **2:** In Sample Problems 2-1 and 2-2, suppose that after refueling the truck, you drive back to x_1 at 52 mi/h. What is your average velocity for the entire trip?

Average speed $\bar{s}$ is a different way of describing "how fast" a particle moves. Whereas the average velocity involves the particle's displacement Δx, the average speed involves the total distance covered (for example, the number of meters run), independent of direction. That is,

$$\bar{s} = \frac{\text{total distance}}{\Delta t}. \quad (2\text{-}3)$$

Average speed also differs from average velocity in that average speed does *not* include direction and thus lacks any algebraic sign. Sometimes $\bar{s}$ is the same (except for the absence of a sign) as $\bar{v}$. But, as demonstrated in Sample Problem 2-3, when an object doubles back on its path, the results can be quite different.

SAMPLE PROBLEM 2-3

In Sample Problem 2-2, what is your average speed?

SOLUTION: From the beginning of your drive to your return to the truck with the fuel, you covered a total of 5.2 mi $+$ 1.2 mi $+$ 1.2 mi $= 7.6$ mi, taking 1.15 h; so

$$\bar{s} = \frac{7.6 \text{ mi}}{1.15 \text{ h}} \approx 6.6 \text{ mi/h}. \quad \text{(Answer)}$$

PROBLEM SOLVING TACTICS

TACTIC 1: *Do You Understand the Problem?*

For beginning problem solvers, no difficulty is more common than simply not understanding the problem. The best test of understanding is this: Can you explain the problem, in your own words, to a friend? Give it a try.

Write down the given data, with units, using the symbols of the chapter. (In Sample Problems 2-1 and 2-2, the given data allow you to find your net displacement Δx and the corresponding time interval Δt.) Identify the unknown and its symbol. (In these sample problems, the unknown is your average velocity, symbol $\bar{v}$.) Then find the connection between the unknown and the data. (The connection is Eq. 2-2, the definition of average velocity.)

TACTIC 2: *Are the Units OK?*

Be sure to use a consistent set of units when putting numbers into the equations. In Sample Problems 2-1 and 2-2, which involve a truck, the logical units in terms of the given data are miles for distances, hours for time intervals, and miles per hour for velocities. You may need to make conversions.

TACTIC 3: *Is Your Answer Reasonable?*

Look at your answer and ask yourself whether it makes sense. Is it far too large or far too small? Is the sign correct? Are the units appropriate? In Sample Problem 2-1, for example, the correct answer is 11 mi/h. If you find 0.00011 mi/h, -11 mi/h, 11 mi/s, or 11,000 mi/h, you should realize at once that you have done something wrong. The error may lie in your method, in your algebra, or in your arithmetic.

TACTIC 4: *Reading a Graph*

Figures 2-2, 2-3a, 2-4, and 2-5 are graphs that you should be able to read easily. In each graph, the variable on the horizontal axis is the time t, the direction of increasing time being to the right. In each, the variable on the vertical axis is the position x of the moving particle with respect to the origin, the direction of increasing x being upward.

   Always note the units (seconds or minutes; meters, kilometers, or miles) in which the variables are expressed, and note whether the variables are positive or negative.

2-4 INSTANTANEOUS VELOCITY AND SPEED

You have now seen two ways to describe how fast something moves: average velocity and average speed, both of which are measured over a time interval Δt. But the phrase "how fast" more commonly refers to how fast a particle is moving at a given instant—and that is its **instantaneous velocity** (or simply **velocity**) v.

   The velocity at any instant is obtained from the average velocity by shrinking the time interval Δt closer and closer to 0. As Δt dwindles, the average velocity approaches a limiting value, which is the velocity at that instant:

$$v = \lim_{\Delta t \to 0} \frac{\Delta x}{\Delta t} = \frac{dx}{dt}. \qquad (2-4)$$

Velocity is another vector quantity and thus has an associated direction.

In the language of calculus, the instantaneous velocity is the rate at which a particle's position x is changing with time at a given instant. According to Eq. 2-4, the velocity of a particle at any instant is the slope of its position-time curve at the point representing that instant.

Speed is the magnitude of velocity; that is, speed is velocity that has been stripped of any indication of direction, either in words or via an algebraic sign.* A velocity of $+5$ m/s and one of -5 m/s both have an associated speed of 5 m/s. The speedometer in a car measures the speed, not the velocity, because it cannot ascertain anything about the direction of motion.

SAMPLE PROBLEM 2-4

Figure 2-6a is an $x(t)$ plot for an elevator cab that is initially stationary, then moves upward (which we take to be the positive direction), and then stops. Plot v as a function of time.

SOLUTION: The slope, and so also the velocity, is zero in the intervals containing points a and d, when the cab is stationary. During the interval bc the slope is constant and nonzero; so the cab moves with a constant velocity. We calculate the slope of $x(t)$ as

$$\frac{\Delta x}{\Delta t} = v = \frac{24 \text{ m} - 4.0 \text{ m}}{8.0 \text{ s} - 3.0 \text{ s}} = +4.0 \text{ m/s}.$$

The plus sign indicates that the cab is moving in the positive x direction. These values ($v = 0$ and $v = 4$ m/s) are plotted in Fig. 2-6b. In addition, as the cab initially begins to move and then later slows to a stop, v might vary as indicated in the intervals 1 s to 3 s and 8 s to 9 s. (Figure 2-6c is considered in Section 2-5.)

   Given a $v(t)$ graph such as Fig. 2-6b, we could "work backward" to produce the shape of the associated $x(t)$ graph (Fig. 2-6a). However, we would not know the actual values for x at various times, because the $v(t)$ graph indicates only *changes* in x. To find such a change in x during any interval, we must, in the language of calculus, calculate the area "under the curve" on the $v(t)$ graph for the same interval. For example, during the interval 3 s to 8 s in which the cab has a velocity of 4.0 m/s, the change in x is given by the "area" under the $v(t)$ curve:

$$\text{area} = (4.0 \text{ m/s})(8.0 \text{ s} - 3.0 \text{ s}) = +20 \text{ m}.$$

(This area is positive because the $v(t)$ curve is above the t axis.) Figure 2-6a shows that x does indeed increase by 20 m in the interval. But Fig. 2-6b does not tell us the *values* of x at the beginning and end of the interval. For that we need additional information.

---

*Speed and average speed can be quite different, so you must be careful solving problems that involve either quantity.

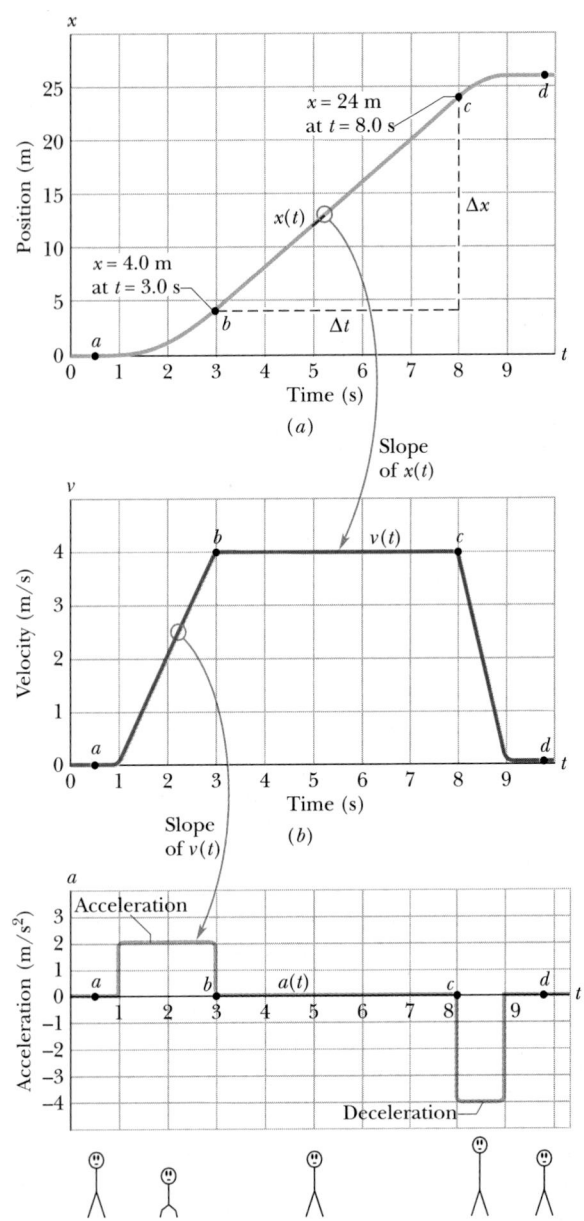

FIGURE 2-6   Sample Problem 2-4. (*a*) The $x(t)$ curve for an elevator cab that moves upward along an x axis. (*b*) The $v(t)$ curve for the cab. Note that it is the derivative of the $x(t)$ curve ($v = dx/dt$). (*c*) The $a(t)$ curve for the cab. It is the derivative of the $v(t)$ curve ($a = dv/dt$). The stick figures along the bottom suggest how a passenger's body might feel during the accelerations.

SAMPLE PROBLEM 2-5

The position of a particle moving on the x axis is given by

$$x = 7.8 + 9.2t - 2.1t^3. \tag{2-5}$$

What is its velocity at $t = 3.5$ s? Is the velocity constant, or is it continuously changing?

SOLUTION: For simplicity, the units have been omitted but you can insert them if you like by changing the coefficients to 7.8 m, 9.2 m/s, and -2.1 m/s³. To find the velocity, we use Eq. 2-4 with the right side of Eq. 2-5 substituted for x:

$$v = \frac{dx}{dt} = \frac{d}{dt}(7.8 + 9.2t - 2.1t^3),$$

which becomes

$$v = 0 + 9.2 - (3)(2.1)t^2 = 9.2 - 6.3t^2. \tag{2-6}$$

At $t = 3.5$ s,

$$v = 9.2 - (6.3)(3.5)^2 = -68 \text{ m/s.} \quad \text{(Answer)}$$

At $t = 3.5$ s, the particle is moving toward decreasing x (note the minus sign) with a speed of 68 m/s. Since the quantity t appears in Eq. 2-6, the velocity v depends on t and so is continuously changing.

$\mathbb{C}$HECKPOINT **3:** The following equations give the position $x(t)$ of a particle in four situations (in each equation, x is in meters, t is in seconds, and $t > 0$): (1) $x = 3t - 2$; (2) $x = -4t^2 - 2$; (3) $x = 2/t^2$; and (4) $x = -2$. (a) In which situation is the velocity v of the particle constant? (b) In which is v in the negative x direction? (c) In which is the particle slowing?

PROBLEM SOLVING TACTICS

TACTIC 5: *Derivatives and Slopes*

Every derivative can be interpreted as the slope of a curve at a point. In Sample Problem 2-4, for example, the velocity of the cab at any instant (a derivative; see Eq. 2-4) is the slope of the $x(t)$ curve of Fig. 2-6a at that instant. Here's how you can find a slope at a point (and thus a derivative) graphically.

Figure 2-7 shows an $x(t)$ plot for a moving particle. To find the velocity of the particle at $t = 1$ s, put a dot on the curve at the point that represents $t = 1$ s. Then draw a line tangent to the curve through the dot (*tangent* means *touching;* the tangent line touches the curve at a single point, the dot), judging carefully by eye. Then construct a right triangle ABC with sides parallel to the axes. (Although the slope is the same no matter what the size of this triangle, the larger the triangle, the more precise will be your measurement of the slope.) Find Δx and Δt, using the vertical and horizontal scales. The slope (derivative) is the quotient $\Delta x/\Delta t$. In Fig. 2-7,

$$\text{slope} = \frac{\Delta x}{\Delta t} = \frac{5.5 \text{ m} - 2.3 \text{ m}}{1.8 \text{ s} - 0.3 \text{ s}} = \frac{3.2 \text{ m}}{1.5 \text{ s}} = +2.1 \text{ m/s.}$$

As Eq. 2-4 tells you, this slope is the velocity of the particle at $t = 1$ s. If you change the scale on either axis of Fig. 2-7, the appearance of the curve and the angle θ will change, but the value you find for the velocity at $t = 1$ s will not.

If you have a mathematical expression for the function $x(t)$, as in Sample Problem 2-5, you can find the derivative dx/dt by the methods of calculus and avoid this graphical method.

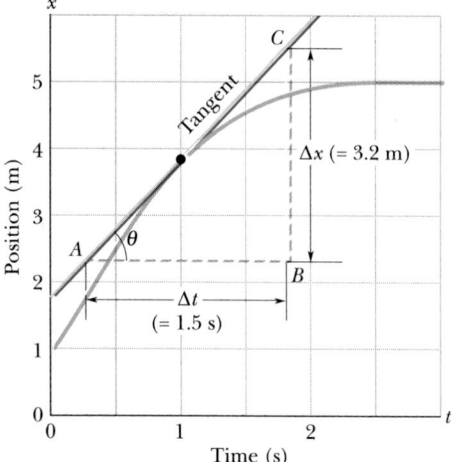

FIGURE 2-7 The derivative of a curve at any point is the slope of its tangent line at that point. At $t = 1.0$ s, the slope of the tangent line (and thus dx/dt, the instantaneous velocity) is $\Delta x / \Delta t = +2.1$ m/s.

2-5 ACCELERATION

When a particle's velocity changes, the particle is said to undergo **acceleration** (or to accelerate). The **average acceleration** $\bar{a}$ over an interval Δt is computed as

$$\bar{a} = \frac{v_2 - v_1}{t_2 - t_1} = \frac{\Delta v}{\Delta t}. \qquad (2\text{-}7)$$

The **instantaneous acceleration** (or simply **acceleration**) is the derivative of the velocity:

$$a = \frac{dv}{dt}. \qquad (2\text{-}8)$$

In words, the acceleration of a particle at any instant is the rate at which its velocity is changing at that instant. According to Eq. 2-8, the acceleration at any point is the slope of the curve of $v(t)$ at that point.

We can combine Eq. 2-8 with Eq. 2-4 to write

$$a = \frac{dv}{dt} = \frac{d}{dt}\left(\frac{dx}{dt}\right) = \frac{d^2x}{dt^2}. \qquad (2\text{-}9)$$

In words, the acceleration of a particle at any instant is the second derivative of its position $x(t)$ with respect to time.

A common unit of acceleration is the meter per second per second: m/(s·s) or m/s². You will see other units in the problems, but they will each be in the form of distance/(time·time) or distance/time². Acceleration has both magnitude and direction (it is yet another vector quantity). The algebraic sign represents the direction on an axis just as it does for displacement and velocity.

Figure 2-6c is a plot of the acceleration of the elevator cab discussed in Sample Problem 2-4. Compare this $a(t)$ curve with the $v(t)$ curve—each point on the $a(t)$ curve is the derivative (slope) of the corresponding point on the $v(t)$ curve. When v is constant (at either 0 or 4 m/s), the derivative is zero and so also is the acceleration. When the cab first begins to move, the $v(t)$ curve has a positive derivative (the slope is positive), which means that $a(t)$ is positive. When the cab slows to a stop, the derivative and slope of the $v(t)$ curve are negative; that is, $a(t)$ is negative.

Next compare the slopes of the $v(t)$ curve during the two acceleration periods. The slope associated with the cab's stopping (commonly called "deceleration") is steeper, because the cab stops in half the time it took to get up to speed. The steeper slope means that the magnitude of the deceleration is larger than that of the acceleration, as indicated in Fig. 2-6c.

The sensations you would feel while riding in the cab of Fig. 2-6 are indicated by the sketched figures. When the car first accelerates, you feel as though you are pressed downward; when later the cab is braked to a stop, you seem to be stretched upward. In between, you feel nothing special. Your body reacts to accelerations (it is an accelerometer) but not to velocities (it is not a speedometer). When you are in a car traveling at 60 mi/h or an airplane traveling at 600 mi/h, you have no bodily awareness of the motion. But if the car or plane quickly changes velocity, you may become keenly aware of the change, perhaps even frightened by it. Part of the thrill of an amusement park ride is due to the quick changes of velocity that you undergo. A more extreme example is shown in the photographs of Fig. 2-8, which were taken while a rocket sled was rapidly accelerated and then rapidly braked to a stop.

Large accelerations are sometimes expressed in terms of "g" units, with

$$1g = 9.8 \text{ m/s}^2 \qquad (g \text{ unit}). \qquad (2\text{-}10)$$

(As we shall discuss in Section 2-8, g is the magnitude of the acceleration of a falling object near the Earth's surface.) On a roller coaster, you have brief accelerations up to $3g$, which is $(3)(9.8 \text{ m/s}^2)$ or about 29 m/s².

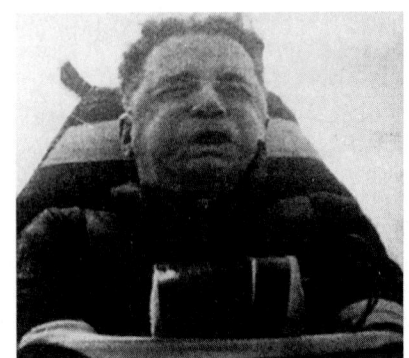

FIGURE 2-8 Colonel J. P. Stapp in a rocket sled as it is brought up to high speed (acceleration out of the page) and then very rapidly braked (acceleration into the page).

SAMPLE PROBLEM 2-6

(a) When Kitty O'Neil set the dragster records for the greatest speed and least elapsed time, she reached 392.54 mi/h in 3.72 s. What was her average acceleration?

SOLUTION: From Eq. 2-7, O'Neil's average acceleration was

$$\bar{a} = \frac{\Delta v}{\Delta t} = \frac{392.54 \text{ mi/h} - 0}{3.72 \text{ s} - 0}$$

$$= +106 \frac{\text{mi}}{\text{h} \cdot \text{s}}, \qquad \text{(Answer)}$$

where the motion is taken to be in the positive x direction. In more conventional units, her acceleration was 47.1 m/s², which is 4.8g.

(b) What was the average acceleration when Eli Beeding Jr. reached 72.5 mi/h in 0.04 s on a rocket sled?

SOLUTION: Again from Eq. 2-7,

$$\bar{a} = \frac{\Delta v}{\Delta t} = \frac{72.5 \text{ mi/h} - 0}{0.04 \text{ s} - 0}$$

$$= +1.8 \times 10^3 \frac{\text{mi}}{\text{h} \cdot \text{s}} \approx +800 \text{ m/s}^2, \quad \text{(Answer)}$$

or about 80g.

Recall our question in the chapter opener, where O'Neil and Beeding were introduced. How can we tell who had the more exciting ride? Should we compare final speeds, elapsed times, or some other quantity? You now can answer that question. Because the human body senses acceleration rather than speed, you should compare accelerations, and so Beeding wins out, even though his final speed was considerably slower than O'Neil's. In fact, Beeding's acceleration could have been lethal, had it continued for much longer.

PROBLEM SOLVING TACTICS

TACTIC 6: *An Acceleration's Sign*

Look again at the algebraic sign for the accelerations that are calculated in Sample Problem 2-6. In many common examples of acceleration, the sign has its nonscientific meaning: positive acceleration means that the speed of an object is increasing, and negative acceleration means that the speed is decreasing (the object is decelerating).

Such meanings cannot be interpreted without some thought, however. For example, if a car with an initial velocity $v = -27$ m/s ($= -60$ mi/h) is braked to a stop in 5.0 s, $\bar{a} = +5.4$ m/s². The acceleration is *positive,* but the car has slowed. The reason is the difference in signs: the direction of the acceleration is opposite that of the velocity.

Here then is the proper way to interpret the signs:

If the signs of the velocity and acceleration of a particle are the same, the speed of the particle increases. If the signs are opposite, the speed decreases.

The interpretation will have more meaning in Chapter 4, where we explore the vector nature of velocity and acceleration.

CHECKPOINT 4: A wombat moves along an x axis. What is the sign of its acceleration if it is moving (a) in the positive direction with increasing speed, (b) in the positive direction with decreasing speed, (c) in the negative direction with increasing speed, and (d) in the negative direction with decreasing speed?

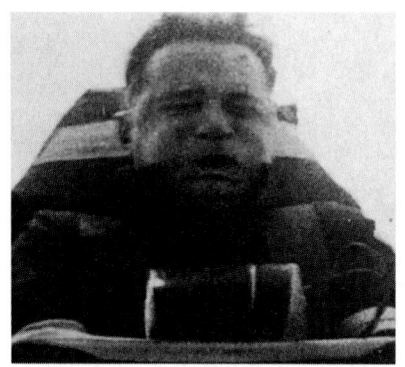

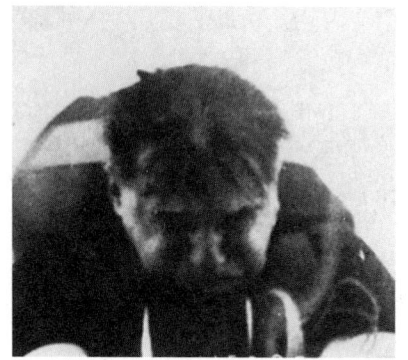

FIGURE 2-8 *Continued*

SAMPLE PROBLEM 2-7

A particle's position is given by

$$x = 4 - 27t + t^3,$$

where the units of the coefficients are meters, meters per second, and meters per second cubed, respectively, and the x axis is shown in Fig. 2-1.

(a) Find $v(t)$ and $a(t)$.

SOLUTION: To get $v(t)$, we differentiate $x(t)$ with respect to t:

$$v = -27 + 3t^2.$$ (Answer)

To get $a(t)$, we differentiate $v(t)$ with respect to t:

$$a = +6t.$$ (Answer)

(b) Is there ever a time when $v = 0$?

SOLUTION: Setting $v(t) = 0$ yields

$$0 = -27 + 3t^2,$$

which has the solution $t = \pm 3$ s. (Answer)

(c) Describe the particle's motion for $t \geq 0$.

SOLUTION: To answer, we examine the expressions for $x(t)$, $v(t)$, and $a(t)$.

At $t = 0$ the particle is at $x = +4$ m, is moving leftward with a velocity of -27 m/s, and is not accelerating.

For $0 < t < 3$ s, the particle continues to move to the left, but at decreasing speed, because it is now accelerating to the right. (Check $v(t)$ and $a(t)$ for, say, $t = 2$ s.) The rate of the acceleration is increasing.

At $t = 3$ s, the particle stops momentarily ($v = 0$) and is as far to the left as it will ever get ($x = -50$ m). It continues to accelerate to the right at an increasing rate.

For $t > 3$ s, its acceleration to the right continues to increase, and its velocity, which is now also to the right, increases rapidly. (Note that now the signs of v and a match.) The particle moves continuously to the right.

2-6 CONSTANT ACCELERATION: A SPECIAL CASE

In many common types of motion, the acceleration is either constant or approximately so. For example, you might accelerate a car at an approximately constant rate when a traffic light turns from red to green. (Graphs of your position, velocity, and acceleration would resemble those in Fig. 2-9.) If you later had to brake the car to a stop, the deceleration during the braking might also be approximately constant.

Such cases are so ubiquitous that a special set of equations has been derived for dealing with them. One approach to the derivation of the equations is given in this section. A second approach is given in the next section.

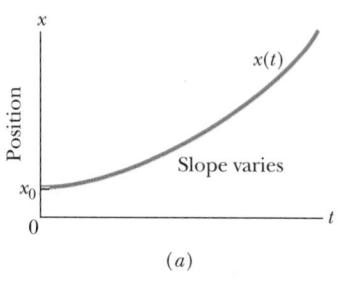
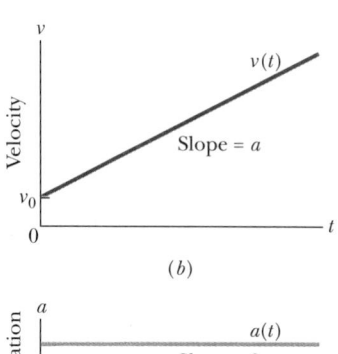

FIGURE 2-9 (*a*) The position $x(t)$ of a particle moving with constant acceleration. (*b*) Its velocity $v(t)$, given at each point by the slope of the curve in (*a*). (*c*) Its (constant) acceleration, equal to the (constant) slope of $v(t)$.

Throughout both sections and later when you work on the homework problems, keep in mind that *the equations are valid only for constant acceleration (or situations in which you can approximate the acceleration as being constant).*

When the acceleration is constant, the average acceleration and instantaneous acceleration are equal and we can write Eq. 2-7, with some changes in notation, as

$$a = \frac{v - v_0}{t - 0}.$$

Here v_0 is the velocity at time $t = 0$, and v is the velocity at any later time t. We can recast this equation as

$$v = v_0 + at. \tag{2-11}$$

As a check, note that this equation reduces to $v = v_0$ for $t = 0$, as it must. As a further check, take the derivative of Eq. 2-11. Doing so yields $dv/dt = a$, which is the definition of a. Figure 2-9b shows a plot of Eq. 2-11, the $v(t)$ function.

In similar manner we can rewrite Eq. 2-2 (with a few changes in notation) as

$$\bar{v} = \frac{x - x_0}{t - 0}$$

and then as

$$x = x_0 + \bar{v}t, \tag{2-12}$$

in which x_0 is the position of the particle at $t = 0$, and $\bar{v}$ is the average velocity between $t = 0$ and a later time t.

If you plot v against t using Eq. 2-11, a straight line results. Under these conditions, the *average* velocity over any time interval (say, $t = 0$ to a later time t) is the average of the velocity at the beginning of the interval ($= v_0$) and the velocity at the end of the interval ($= v$). For the interval $t = 0$ to the later time t then, the average velocity is

$$\bar{v} = \frac{1}{2}(v_0 + v). \tag{2-13}$$

Substituting the right side of Eq. 2-11 for v yields, after a little rearrangement,

$$\bar{v} = v_0 + \frac{1}{2}at. \tag{2-14}$$

Finally, substituting Eq. 2-14 into Eq. 2-12 yields

$$x - x_0 = v_0 t + \frac{1}{2}at^2. \tag{2-15}$$

As a check, note that putting $t = 0$ yields $x = x_0$, as it must. As a further check, taking the derivative of Eq. 2-15 yields Eq. 2-11, again as it must. Figure 2-9a is a plot of Eq. 2-15.

Five quantities can possibly be involved in any given problem regarding constant acceleration, namely, $x - x_0$, v, t, a, and v_0. Usually, one of these quantities is *not* involved in the problem, *either as a given or as an unknown.* We are then presented with three of the remaining quantities and asked to find the fourth.

Equations 2-11 and 2-15 each contain four of these quantities, but not the same four. In Eq. 2-11, the "missing ingredient" is the displacement, $x - x_0$. In Eq. 2-15, it is the velocity v. These two equations can also be combined in three ways to yield three additional equations, each of which involves a different "missing ingredient." First, we can eliminate t to obtain

$$v^2 = v_0^2 + 2a(x - x_0). \tag{2-16}$$

This equation is useful if we do not know t and are not required to find it. Second, we can eliminate the acceleration a between Eqs. 2-11 and 2-15 to produce an equation in which a does not appear:

$$x - x_0 = \frac{1}{2}(v_0 + v)t. \tag{2-17}$$

Finally, we can eliminate v_0, obtaining

$$x - x_0 = vt - \frac{1}{2}at^2. \tag{2-18}$$

Note the subtle difference between this equation and Eq. 2-15. One involves the initial velocity v_0; the other involves the velocity v at time t.

Table 2-1 lists Eqs. 2-11, 2-15, 2-16, 2-17, and 2-18 and shows which one of the five possible quantities is missing from each. To solve a constant acceleration problem, you must decide which of the five quantities is *not* involved in the problem, either as a given or as an unknown. Select the correct equation from Table 2-1 and substitute for the three given quantities to find the unknown. Or, instead of using the table, you might find it easier to remember only Eqs. 2-11 and 2-15, and solve them as simultaneous equations when needed.

TABLE 2-1 EQUATIONS FOR MOTION WITH CONSTANT ACCELERATION[a]

EQUATION NUMBER	EQUATION	MISSING QUANTITY
2-11	$v = v_0 + at$	$x - x_0$
2-15	$x - x_0 = v_0 t + \frac{1}{2}at^2$	v
2-16	$v^2 = v_0^2 + 2a(x - x_0)$	t
2-17	$x - x_0 = \frac{1}{2}(v_0 + v)t$	a
2-18	$x - x_0 = vt - \frac{1}{2}at^2$	v_0

[a]Make sure that the acceleration is indeed constant before using the equations in this table. Note that if you differentiate Eq. 2-15 you get Eq. 2-11. The other three equations are found by eliminating one or another of the variables between Eqs. 2-11 and 2-15.

CHECKPOINT **5:** The following equations give the position $x(t)$ of a particle in four situations: (1) $x = 3t - 4$; (2) $x = -5t^3 + 4t^2 + 6$; (3) $x = 2/t^2 - 4/t$; (4) $x = 5t^2 - 3$. To which of these situations do the equations of Table 2-1 apply?

SAMPLE PROBLEM 2-8

Spotting a police car, you brake a Porsche from 75 km/h to 45 km/h over a displacement of 88 m.

(a) What is the acceleration, assumed to be constant?

SOLUTION: In this problem we are given v_0, v, and $x - x_0$ and need to find a. The time is not involved, being neither given nor required. Table 2-1 then leads us to Eq. 2-16. Solving this equation for a yields

$$a = \frac{v^2 - v_0^2}{2(x - x_0)} = \frac{(45 \text{ km/h})^2 - (75 \text{ km/h})^2}{(2)(0.088 \text{ km})}$$

$$= -2.05 \times 10^4 \text{ km/h}^2 \approx -1.6 \text{ m/s}^2. \quad \text{(Answer)}$$

(In converting hours squared to seconds squared in the last step, we must convert *both* the hour units.) Note that the velocities are positive and the acceleration is negative, which is consistent with a slowing of the car.

(b) What is the elapsed time?

SOLUTION: Now time is involved in the problem, but the acceleration is not. Table 2-1 suggests Eq. 2-17. Solving that equation for t, we obtain

$$t = \frac{2(x - x_0)}{v_0 + v} = \frac{(2)(0.088 \text{ km})}{(75 + 45) \text{ km/h}}$$

$$= 1.5 \times 10^{-3} \text{ h} = 5.4 \text{ s}. \quad \text{(Answer)}$$

(c) If you continue to slow down with the acceleration calculated in (a), how much time will elapse in bringing the car to rest from 75 km/h?

SOLUTION: The quantity not given or asked for here is the displacement, $x - x_0$. Table 2-1 then suggests that we use Eq. 2-11. Solving for t gives

$$t = \frac{v - v_0}{a} = \frac{0 - (75 \text{ km/h})}{(-2.05 \times 10^4 \text{ km/h}^2)}$$

$$= 3.7 \times 10^{-3} \text{ h} = 13 \text{ s}. \quad \text{(Answer)}$$

(d) In (c), what distance will be covered?

SOLUTION: From Eq. 2-15, we have, for the displacement of the car,

$$x - x_0 = v_0 t + \tfrac{1}{2}at^2$$

$$= (75 \text{ km/h})(3.7 \times 10^{-3} \text{ h})$$

$$+ \tfrac{1}{2}(-2.05 \times 10^4 \text{ km/h}^2)(3.7 \times 10^{-3} \text{ h})^2$$

$$= 0.137 \text{ km} \approx 140 \text{ m}. \quad \text{(Answer)}$$

(Neglecting the sign on the acceleration would give a wrong result—you should always be alert to signs.)

(e) Suppose that later, using the acceleration calculated in (a) but a different initial velocity, you bring your car to rest after traversing 200 m. What is the total braking time?

SOLUTION: The missing quantity here is the initial velocity, so we use Eq. 2-18. Setting v (the final velocity) equal to zero (''rest'') and solving this equation for t, we obtain

$$t = \left(-\frac{(2)(x - x_0)}{a}\right)^{1/2} = \left(-\frac{(2)(200 \text{ m})}{-1.6 \text{ m/s}^2}\right)^{1/2}$$

$$= 16 \text{ s}. \quad \text{(Answer)}$$

PROBLEM SOLVING TACTICS

TACTIC 7: *Check the Dimensions*

The dimension of a velocity is L/T, that is, length L divided by time T. The dimension of an acceleration is L/T^2; and so on. In any equation, the dimensions of all terms must be the same. If you are in doubt about an equation, check its dimensions.

To check the dimensions of Eq. 2-15 ($x - x_0 = v_0 t + \tfrac{1}{2}at^2$), we note that every term must be a length, because that is the dimension of x and of x_0. The dimension of the term $v_0 t$ is $(L/T)(T)$, which is L. The dimension of $\tfrac{1}{2}at^2$ is $(L/T^2)(T^2)$, which is also L. This equation checks out. A pure number such as $\tfrac{1}{2}$ or π has no dimension.

2-7 ANOTHER LOOK AT CONSTANT ACCELERATION*

The first two equations in Table 2-1 are the basic equations from which the others are derived. Those two can be obtained by integration of the acceleration with the condition that a is constant. The definition of a (in Eq. 2-8) is

$$a = \frac{dv}{dt},$$

which can be rewritten as

$$dv = a \, dt.$$

To take the *indefinite integral* (or *antiderivative*) of both sides, we write

$$\int dv = \int a \, dt.$$

---

*This section is intended for students who have had integral calculus.

Since acceleration a is a constant, it can be taken outside the integration. Then we obtain

$$\int dv = a \int dt$$

or
$$v = at + C. \qquad (2\text{-}19)$$

To evaluate the constant of integration C, we let $t = 0$, at which time $v = v_0$. Substituting these values into Eq. 2-19 (which must hold for all values of t, including $t = 0$) yields

$$v_0 = (a)(0) + C = C.$$

Substituting this into Eq. 2-19 gives us Eq. 2-11.

To derive the other basic equation in Table 2-1, we rewrite the definition of velocity (Eq. 2-4) as

$$dx = v \, dt$$

and then take the indefinite integral of both sides to obtain

$$\int dx = \int v \, dt.$$

There is no reason to believe that v is constant, so we cannot move it outside the integration. But we can substitute for v with Eq. 2-11:

$$\int dx = \int (v_0 + at) dt.$$

Since v_0 is a constant, this can be rewritten as

$$\int dx = v_0 \int dt + a \int t \, dt.$$

Integration now yields

$$x = v_0 t + \tfrac{1}{2}at^2 + C', \qquad (2\text{-}20)$$

where C' is another constant of integration. At time $t = 0$, we have $x = x_0$. Substituting these values in Eq. 2-20 yields $x_0 = C'$. Replacing C' with x_0 in Eq. 2-20 gives us Eq. 2-15.

2-8 FREE-FALL ACCELERATION

If you tossed an object either up or down and could somehow eliminate the effects of air on its flight, you would find that the object accelerates downward at a certain rate. That rate is called the **free-fall acceleration** g. The acceleration is independent of the object's characteristics, such as mass, density, or shape; it is the same for all objects.

Two examples of free-fall acceleration are shown in Fig. 2-10, which is a series of stroboscopic photos of a feather and an apple. As these objects descend, they accelerate downward—both at the same rate g. Their speeds

FIGURE 2-10 A feather and an apple, undergoing free fall in a vacuum, move downward at the same acceleration g. The acceleration causes the increase in distance between images during the fall.

increase together. The value of g varies slightly with latitude and with elevation. At sea level in the midlatitudes the value is 9.8 m/s² (or 32 ft/s²), which is what you should use for the problems in this chapter.

The equations of motion in Table 2-1 for constant acceleration apply to free fall near the Earth's surface. That is, they apply to an object in vertical flight, either up or down, when the effects of the air can be neglected. However, we can make them simpler to use with two minor changes. (1) The directions of motion are now along the vertical y axis instead of the x axis, with the positive direction of y upward. (This change will reduce confusion in later chapters when combined horizontal and vertical motions are examined.) (2) The free-fall acceleration is now negative, that is, downward on the y axis, and so we replace a with $-g$ in the equations.

With these small changes, Eqs. 2-11 and 2-15 to 2-18 of Table 2-1 become, for free fall, Eqs. 2-21 to 2-25 in Table 2-2. Note carefully that:

The free-fall acceleration near the Earth's surface is $a = -g = -9.8$ m/s², and the *magnitude* of the acceleration is $g = 9.8$ m/s². Do not substitute -9.8 m/s² for g.

TABLE 2-2 EQUATIONS FOR FREE FALL

EQUATION NUMBER	EQUATION	MISSING QUANTITY
2-21	$v = v_0 - gt$	$y - y_0$
2-22	$y - y_0 = v_0 t - \frac{1}{2}gt^2$	v
2-23	$v^2 = v_0^2 - 2g(y - y_0)$	t
2-24	$y - y_0 = \frac{1}{2}(v_0 + v)t$	g
2-25	$y - y_0 = vt + \frac{1}{2}gt^2$	v_0

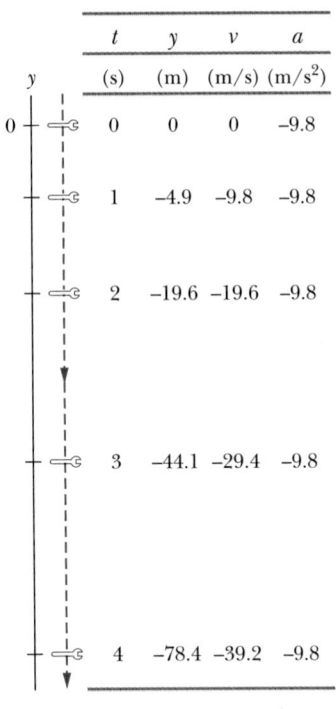

t	y	v	a
(s)	(m)	(m/s)	(m/s^2)
0	0	0	−9.8
1	−4.9	−9.8	−9.8
2	−19.6	−19.6	−9.8
3	−44.1	−29.4	−9.8
4	−78.4	−39.2	−9.8

FIGURE 2-11 Sample Problem 2-9. The position, velocity, and acceleration of a freely falling object.

Suppose that you toss a tomato directly upward with an initial velocity v_0 and then catch it when it returns to the level of release. During its *free-fall flight* (just after its release and just before it is caught), the equations of Table 2-2 apply to its motion. The acceleration is always $a = -g = -9.8$ m/s^2 and thus downward. The velocity, however, changes, as indicated by Eqs. 2-21 and 2-23: during the ascent, the magnitude of the (positive) velocity decreases, until it is momentarily zero. Because the tomato has then stopped, it is at its maximum height. During the descent, the magnitude of the (negative) velocity increases.

SAMPLE PROBLEM 2-9

A worker drops a wrench down the elevator shaft of a tall building.

(a) Where is the wrench 1.5 s later?

SOLUTION: Here we know the time t and the acceleration magnitude g, and we can assume the wrench is initially at rest ($v_0 = 0$). We want the displacement. So the missing quantity is the velocity v, which is neither given nor required. This suggests Eq. 2-22 of Table 2-2. Choose the release point of the wrench to be at the origin of the y axis. Setting $y_0 = 0$, $v_0 = 0$, and $t = 1.5$ s in Eq. 2-22 gives

$$y = (0)(1.5 \text{ s}) - \tfrac{1}{2}(9.8 \text{ m/s}^2)(1.5 \text{ s})^2$$
$$= -11 \text{ m.} \qquad \text{(Answer)}$$

The minus sign means that after falling 1.5 s the wrench is below its release point, which we certainly expect.

(b) How fast is the wrench falling just then?

SOLUTION: The wrench's velocity is, by Eq. 2-21,

$$v = v_0 - gt = 0 - (9.8 \text{ m/s}^2)(1.5 \text{ s})$$
$$= -15 \text{ m/s.} \qquad \text{(Answer)}$$

Here the minus sign means that the wrench is falling downward. Again, no great surprise. Figure 2-11 displays the important features of the motion up to $t = 4$ s.

SAMPLE PROBLEM 2-10

In 1939, Joe Sprinz of the San Francisco Baseball Club attempted to break the record for catching a baseball dropped from the greatest height. Members of the Cleveland Indians had set the record the preceding year when they caught baseballs dropped about 700 ft from atop a building. Sprinz used a blimp at 800 ft. Ignore the effects of air on the ball and assume that the ball falls 800 ft.

(a) Find its time of fall.

SOLUTION: Mentally erect a vertical y axis with its origin at the point of the ball's release in the blimp, which means that $y_0 = 0$. The initial velocity v_0 is zero. The missing ingredient is v, and so Eq. 2-22 is required:

$$y - y_0 = v_0 t - \tfrac{1}{2}gt^2$$
$$-800 \text{ ft} = 0t - \tfrac{1}{2}(32 \text{ ft/s}^2)t^2$$
$$16t^2 = 800$$
$$t = 7.1 \text{ s.} \qquad \text{(Answer)}$$

When taking a square root, we have the option of attaching a plus or minus sign to the square root. Here we choose the plus sign, since the ball reaches the ground *after* it is released.

(b) What is the velocity of the ball just before it is caught?

SOLUTION: To get the velocity from the original data, rather than from the result of (a), we use Eq. 2-23:

$$v^2 = v_0^2 - 2g(y - y_0)$$
$$= 0 - (2)(32 \text{ ft/s}^2)(-800 \text{ ft})$$
$$= 5.12 \times 10^4 \text{ ft}^2/\text{s}^2$$
$$v = -226 \text{ ft/s} \ (\approx -154 \text{ mi/h}). \quad \text{(Answer)}$$

Since the ball is moving downward, we choose the minus sign in our option of signs in taking the square root.

Neglecting the effects of air is actually unwarranted in such a fall. If you included them, you would find that the fall time was longer and the final speed was smaller than the values calculated above. Still, the speed must have been considerable, because when Sprinz finally managed to get a ball in his glove (on his fifth attempt) the impact slammed the glove and hand into his face, fracturing the upper jaw in 12 places, breaking five teeth, and knocking him unconscious. And he dropped the ball.

SAMPLE PROBLEM 2-11

A pitcher tosses a baseball straight up, with an initial speed of 12 m/s. See Fig. 2-12.

(a) How long does the ball take to reach its highest point?

SOLUTION: The ball is at its highest point when its velocity v becomes zero. From Eq. 2-21, we have

$$t = \frac{v_0 - v}{g} = \frac{12 \text{ m/s} - 0}{9.8 \text{ m/s}^2} = 1.2 \text{ s}. \quad \text{(Answer)}$$

(b) How high does the ball rise above its release point?

SOLUTION: We take the release point of the ball to be the origin of the y axis. Putting $y_0 = 0$ in Eq. 2-23 and solving for y, we obtain

$$y = \frac{v_0^2 - v^2}{2g} = \frac{(12 \text{ m/s})^2 - (0)^2}{(2)(9.8 \text{ m/s}^2)}$$
$$= 7.3 \text{ m}. \quad \text{(Answer)}$$

If we wanted to take advantage of the fact that we also know the time of flight, having found it in (a), we could also calculate the height of rise from Eq. 2-25. Check it out.

(c) How long will it take for the ball to reach a point 5.0 m above its release point?

SOLUTION: Inspection of Eqs. 2-21 to 2-25 suggests that we try Eq. 2-22. With $y_0 = 0$, we have

$$y = v_0 t - \tfrac{1}{2}gt^2,$$

so $5.0 \text{ m} = (12 \text{ m/s})t - (\tfrac{1}{2})(9.8 \text{ m/s}^2)t^2.$

If we temporarily omit the units (having noted that they are consistent), we can rewrite this as

$$4.9t^2 - 12t + 5.0 = 0.$$

Solving this quadratic equation for t yields*

$$t = 0.53 \text{ s} \quad \text{and} \quad t = 1.9 \text{ s}. \quad \text{(Answer)}$$

There are two such times! This is not really surprising because the ball passes twice through $y = 5.0$ m, once on the way up and once on the way down.

We can check our findings because the time at which the ball reaches its maximum height should lie halfway between these two times, or at

$$t = \tfrac{1}{2}(0.53 \text{ s} + 1.9 \text{ s}) = 1.2 \text{ s}.$$

This is exactly what we found in (a) for the time to reach maximum height.

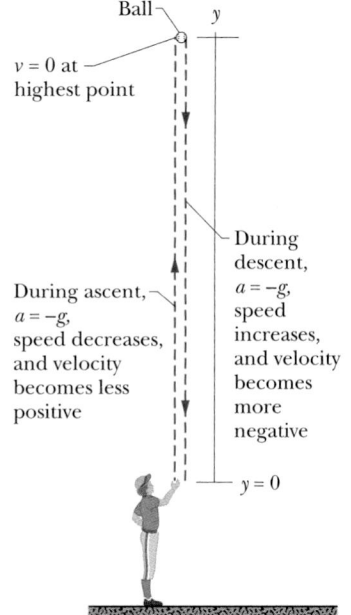

FIGURE 2-12 Sample Problem 2-11. A pitcher tosses a baseball straight up into the air. The equations of free fall apply for rising as well as for falling objects, provided any effects from the air can be neglected.

CHECKPOINT 6: In Sample Problem 2-11, what is the sign of the ball's displacement (a) during its ascent and (b) during its descent? (c) What is the ball's acceleration at its highest point?

PROBLEM SOLVING TACTICS

TACTIC 8: *Meanings of Minus Signs*
In Sample Problems 2-9, 2-10, and 2-11, many answers emerged automatically with minus signs. It is important to know what these signs mean. For falling-body problems, we established a vertical axis (the y axis) and we chose—quite arbitrarily—its upward direction to be positive.

---

*See Appendix E for a general solution to quadratic equations.

We then chose the origin of the y axis (that is, the $y = 0$ position) to suit the problem. In Sample Problem 2-9, the origin was the worker's hand; in Sample Problem 2-10 it was at the blimp; in Sample Problem 2-11 it was the pitcher's hand. A negative value of y means that the body is below the chosen origin.

A negative velocity means that the body is moving in the direction of decreasing y, that is, downward. This is true no matter where the body is located.

We have taken the acceleration to be negative ($= -9.8$ m/s^2) in all problems dealing with falling bodies. A negative acceleration means that, as time goes on, the velocity of the body becomes either less positive or more negative. This is true no matter where the body is located and no matter how fast or in what direction it is moving. In Sample Problem 2-11, the acceleration of the ball is negative throughout its flight, whether the ball is rising or falling.

TACTIC 9: *Unexpected Answers*

Mathematics often generates answers that you might not have thought of as possibilities, as in Sample Problem 2-11c. If you get more answers than you expect, do not discard out of hand the ones that do not seem to fit. Examine them carefully for physical meaning; it is often there.

If time is your variable, even a negative value can mean something; negative time simply refers to time before $t = 0$, the (arbitrary) time at which you decided to start your stopwatch.

2-9 THE PARTICLES OF PHYSICS

As we progress through the book, we plan to step aside occasionally from the familiar world of large, tangible objects and look at nature on a much finer scale. The "particles" that we have dealt with in this chapter, for example, have included pigs, baseballs, and dragsters. In the spirit of our plan, we ask: How small can a particle be? What are the *ultimate* particles of nature? **Particle physics**—for so the field that relates to our inquiry is called—attracts the attention of many of the best of today's physicists.

The realization that matter, on its finest scale, is not continuous but is made up of lumps of material called atoms was the beginning of understanding for physics and chemistry. With modern microscopes, we can "photograph" these atoms, as Fig. 2-13 makes clear. We describe the "lumpiness" of matter by saying that matter is *quantized,* using a word that comes from the Latin word *quantus,* meaning "how much." Quantization is a central feature of nature, and as we go along you will see other physical quantities that are quantized when looked at on a fine enough scale. This pervasiveness of quantization is reflected in the name we give to physics at the atomic level—**quantum physics.**

There is no sharp discontinuity between the quantum world and the world of large-scale objects. The quantum world and the laws that govern it are universal but, as we move from atoms to baseballs and automobiles, the fact of quantization becomes less noticeable and finally totally undetectable. The "graininess" effectively disappears, and the laws of **classical physics** that govern the motions of large objects emerge as special limiting forms of the more general laws of quantum physics.

The Structure of Atoms

An **atom** consists of a central, almost unimaginably compact and dense **nucleus** that is surrounded by one or more lightweight **electrons.** An atom is usually considered to be spherical; so is the nucleus. The radius of a typical atom is on the order of 10^{-10} m; the radius of a nucleus is 100,000 times smaller, about 10^{-15} m. An atom is held together by electrical attraction between the electrons, which are electrically negative, and **protons,** which are electrically positive and reside within the nucleus. The nature of that attraction is explored later in this book, but for now you might realize that were it not present, atoms could not exist, and so neither could you.

The Structure of Nuclei

The simplest nucleus, that of common hydrogen, has a single proton. There are two other, rare, versions of hydrogen: they differ from the common version by the presence

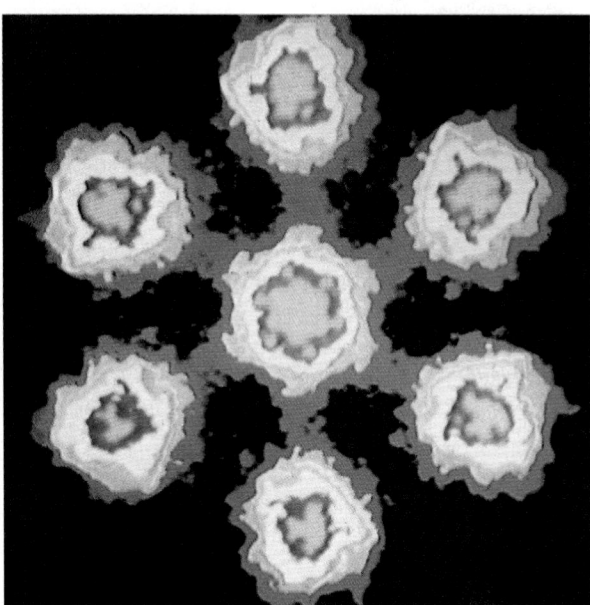

FIGURE 2-13 A hexagonal array of uranium atoms is revealed in this image from a scanning transmission electron microscope. The color has been added by a computer.

of one or two **neutrons** (electrically neutral particles) inside the nucleus. Hydrogen, in any of its versions, is an example of an **element;** each element is distinguished from all the others by the number of protons in the nucleus. When there is only one proton, the element is hydrogen. When, instead, there are six, the element is carbon. The various versions of each element are called **isotopes;** they are distinguished by the number of neutrons.

Roughly speaking, the purpose of the neutrons is to glue together the protons, which, being all electrically positive and closely packed, strongly repel one another. If the neutrons did not provide the glue, the only type of atom that could exist would be common hydrogen; all others would blow apart.

Such instability can be found in many isotopes of common elements, but thankfully not the elements on which your existence depends. For example, 19 of the 21 isotopes of copper are unstable and undergo transformations to become other elements. The two stable isotopes are the ones used in electronics and other technology.

The Structure of the Particles Within Atoms

The electron is simple but perplexingly so. When detected, it appears to be infinitesimal in size; that is, it has no size and no structure. It is a member of a family of pointlike particles called **leptons;** there are six basic types, each with an **antiparticle** version.

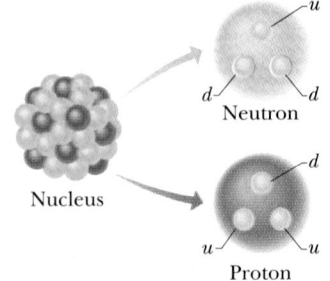

FIGURE 2-14 A representation of the nucleus of an atom, showing the neutrons and protons that make it up. These particles, in turn, are composed of ''up'' (*u*) and ''down'' (*d*) quarks.

Protons and neutrons are believed to be different from electrons and the other leptons, because each of the former appears to be a bundle of three simpler particles called **quarks.*** A proton consists of two ''up'' quarks and one ''down'' quark, and a neutron two ''down'' quarks and one ''up'' quark (Fig. 2-14). Other particles that were first thought to be fundamental appear to be similar bundles.

Provocatively, quarks come in six basic types† (each with its antiparticle) just as do leptons. Here physicists wonder: Is there a basic reason for the match in number of types? Or is the match simply coincidence? We do not know.

---

*On a quirk, the word ''quark'' was lifted from James Joyce's *Finnegans Wake*:

Three quarks for Muster Mark.
Sure he hasn't got much of a bark
And sure any he has it's all beside the mark.

†The other types are called charm, strange, top, and bottom (even physicists have their moments).

REVIEW & SUMMARY

Position

The *position x* of a particle on an axis locates the particle with respect to the **origin,** or zero point, of the axis. The position is either positive or negative, according to which side of the origin the particle is on, or zero if the particle is at the origin. The **positive direction** on an axis is the direction of increasing positive numbers; the opposite direction is the **negative direction.**

Displacement

The *displacement* Δx of a particle is the change in its position:

$$\Delta x = x_2 - x_1. \qquad (2\text{-}1)$$

Displacement is a vector quantity. It is positive if the particle has moved in the positive direction of the *x* axis, and negative if it has moved in the negative direction.

Average Velocity

When a particle has moved from position x_1 to x_2 during a time

interval $\Delta t = t_2 - t_1$, its *average velocity* is

$$\bar{v} = \frac{\Delta x}{\Delta t} = \frac{x_2 - x_1}{t_2 - t_1}. \qquad (2\text{-}2)$$

The algebraic sign of $\bar{v}$ indicates the direction of motion ($\bar{v}$ is a vector quantity). Average velocity does not depend on the actual distance a particle covers, but instead depends on its original and final positions.

On a graph of *x* versus *t*, the average velocity for a time interval Δt is the slope of the straight line connecting the points on the curve that represent the ends of the interval.

Average Speed

The *average speed* $\bar{s}$ of a particle depends on the full distance it covers in a time interval Δt:

$$\bar{s} = \frac{\text{total distance}}{\Delta t}. \qquad (2\text{-}3)$$

Instantaneous Velocity

If we allow Δt to approach zero in Eq. 2-2, then $\bar{v}$ will approach a limiting value v, the *instantaneous velocity* (or simply **velocity**) of the particle at the time in question; that is,

$$v = \lim_{\Delta t \to 0} \frac{\Delta x}{\Delta t} = \frac{dx}{dt}. \tag{2-4}$$

The instantaneous velocity (at a particular time) may be found as the slope (at that particular time) of the graph of x versus t. **Speed** is the magnitude of instantaneous velocity.

Average Acceleration

Average acceleration is the ratio of a change in velocity Δv to the time interval Δt in which the change occurs:

$$\bar{a} = \frac{\Delta v}{\Delta t}. \tag{2-7}$$

The algebraic sign indicates the direction of $\bar{a}$.

Instantaneous Acceleration

Instantaneous acceleration (or simply **acceleration**) is the rate of change of velocity with time:

$$a = \frac{dv}{dt}. \tag{2-8}$$

The acceleration can also be written as the second derivative of the position $x(t)$ with respect to time:

$$a = \frac{d^2 x}{dt^2}. \tag{2-9}$$

On a graph of v versus t, the acceleration a at any time t is the slope of the curve at the point that represents t.

Constant Acceleration

Figure 2-9 shows $x(t)$, $v(t)$, and $a(t)$ for the important case in which a is constant. In this circumstance, the five equations in Table 2-1 describe the motion:

$$v = v_0 + at, \tag{2-11}$$

$$x - x_0 = v_0 t + \tfrac{1}{2} at^2, \tag{2-15}$$

$$v^2 = v_0^2 + 2a(x - x_0), \tag{2-16}$$

$$x - x_0 = \tfrac{1}{2}(v_0 + v)t, \tag{2-17}$$

$$x - x_0 = vt - \tfrac{1}{2} at^2. \tag{2-18}$$

These are *not* valid when the acceleration is not constant.

Free-Fall Acceleration

An important example of straight-line motion with constant acceleration is that of an object rising or falling freely near the Earth's surface. The constant acceleration equations describe this motion, but we make two changes in notation: (1) we refer the motion to the vertical y axis with $+y$ vertically *up*; (2) we replace a with $-g$, where g is the magnitude of the free-fall acceleration. Near the Earth's surface, $g = 9.8$ m/s^2 ($= 32$ ft/s^2). The free-fall equations are those shown as Eqs. 2-21 to 2-25.

The Structure of Atoms

All ordinary matter is composed of **atoms,** which, in a simple model, consist of **electrons** that surround a highly compact central core, the **nucleus. Neutrons** and **protons** reside inside nuclei. Each **element** is distinguished by the number of protons in its nucleus. Variations of an element, differing in the number of neutrons, are **isotopes** of the element.

Quarks and Leptons

Electrons appear to be pointlike particles with no size or internal structure, but protons and neutrons appear to have size and to contain more elementary particles, called **quarks.** There are six basic types of quarks, each with an antiparticle version. Electrons are members of a family of particles, **leptons,** that also come in six basic types, each with an antiparticle version.

QUESTIONS

1. (a) Can an object have zero velocity and still be accelerating? (b) Can an object have constant velocity and still have a varying speed? (c) Can the velocity of an object reverse direction when the object's acceleration is constant? (d) Can an object be increasing in speed as its acceleration decreases?

2. Figure 2-15 is a graph of a particle's position along an x axis versus time. (a) At time $t = 0$, what is the sign of the particle's position? Is the particle's velocity positive, negative, or zero at (b) $t = 1$ s, (c) $t = 2$ s, and (d) $t = 3$ s? (e) How many times does the particle go through the point $x = 0$?

3. At $t = 0$, a particle moving along an x axis is at position $x_0 = -20$ m. The signs of the particle's initial velocity v_0 (at time t_0) and constant acceleration a are, respectively, for four situations: (1) +, +; (2) +, −; (3) −, +; (4) −, −. In which situation will

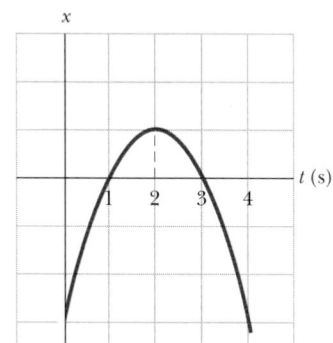

FIGURE 2-15 Question 2.

the particle (a) undergo a momentary stop, (b) definitely pass through the origin (given enough time), and (c) definitely not pass through the origin?

4. Figure 2-16 gives the velocity of a particle moving on an x axis. What are (a) the initial and (b) the final directions of travel? (c) Does the particle stop momentarily? (d) Is the acceleration positive or negative? (e) Is it constant or varying?

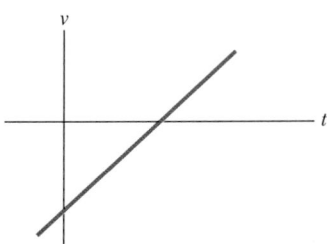

FIGURE 2-16   Question 4.

5. The initial and final velocities, respectively, of a particle in four situations are: (a) 2 m/s, 3 m/s; (b) −2 m/s, 3 m/s; (c) −2 m/s, −3 m/s; (d) 2 m/s, −3 m/s. The magnitude of the particle's constant acceleration is the same in all four situations. Rank the situations according to the magnitude of the particle's displacement, greatest first, during the change from initial to final velocity.

6. The following equations give the velocity $v(t)$ of a particle in four situations: (a) $v = 3$; (b) $v = 4t^2 + 2t − 6$; (c) $v = 3t − 4$; (d) $v = 5t^2 − 3$. To which of these situations do the equations of Table 2-1 apply?

7. Suppose that a hot-air balloonist drops an apple over the side while the balloon is accelerating upward at 4.0 m/s² during liftoff. (a) What is the apple's acceleration once it has been released? (b) If the velocity of the balloon is 2 m/s upward at the instant of release, what is the apple's velocity just then?

8. (a) Graph y, v, and a versus t for a cream tangerine that is thrown straight up from a cliff but which, upon falling back down, barely passes by the cliff edge. (b) On the same figures, graph the same quantities for a cream tangerine that is dropped (released with no initial speed) from the cliff edge.

9. You throw a ball straight up from the edge of a cliff, and it lands on the ground below the cliff. If you had, instead, thrown the ball down from the cliff edge with the same speed, would the ball's speed just before landing be larger than, smaller than, or the same as previously? (*Hint:* Consider Eq. 2-23.)

10. The driver of a blue car, moving at a speed of 60 mi/h, suddenly realizes that she is about to rear-end a red car, moving at a speed of 40 mi/h. To avoid a collision, what is the maximum speed the blue car can have just as it reaches the red car? (Warmup for Problem 57)

11. At $t = 0$ and $x = 0$, an initially stationary blue car begins to accelerate at the constant rate of 2.0 m/s² in the positive direction of the x axis. At $t = 2$ s, a red car, traveling in an adjacent lane and in the same direction, passes $x = 0$ with a speed of 8.0 m/s and a constant acceleration of 3.0 m/s². What pair of simultaneous equations should be solved to find when the red car passes the blue car? (Warmup for Problem 56)

12. Figure 2-17 shows that a particle moving along an x axis undergoes three periods of acceleration. Without written computation, rank the acceleration periods according to the increases they produce in the particle's velocity, greatest first.

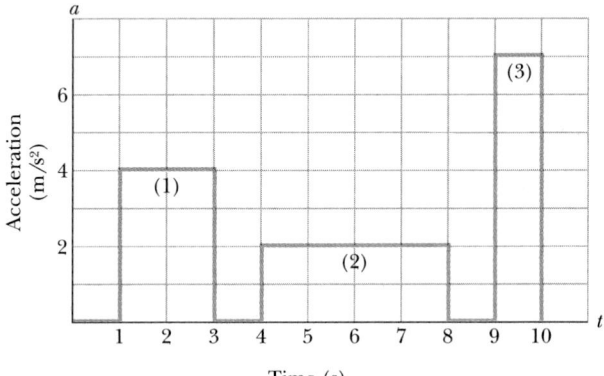

FIGURE 2-17   Question 12.

13. A second ball is dropped from a cliff 1 s after a first ball was dropped. As both fall, does the distance between them increase, decrease, or stay the same?

EXERCISES & PROBLEMS

In several of the problems that follow you are asked to graph position, velocity, and acceleration versus time. Usually a sketch will suffice, appropriately labeled and with straight and curved portions apparent. If you have a computer or programmable calculator, you might use it to produce the graph.

SECTION 2-3 Average Velocity and Average Speed

1E. Carl Lewis ran the 100 m dash in about 10 s, and Bill Rodgers ran the marathon (26 mi, 385 yd) in about 2 h 10 min. (a) What are their average speeds? (b) If Lewis could have maintained his sprint speed during a marathon, how long would he have taken to finish?

2E. During a hard sneeze, your eyes might shut for 0.50 s. If you are driving a car at 90 km/h, how far does it move during that time?

3E. On average, a blink lasts about 100 ms. How far does a MiG-25 ''Foxbat'' fighter travel during a pilot's blink if the plane's average velocity is 2110 mi/h?

4E. Boston Red Sox pitcher Roger Clemens could routinely throw a fastball at a horizontal speed of 160 km/h. How long did the ball take to reach home plate 18.4 m away?

5E. Figure 2-18 is a plot of the age of ancient seafloor material, in millions of years, against the distance from a particular ocean ridge. Seafloor material extruded from this ridge moves away from it at approximately uniform speed. Find that speed in centimeters per year.

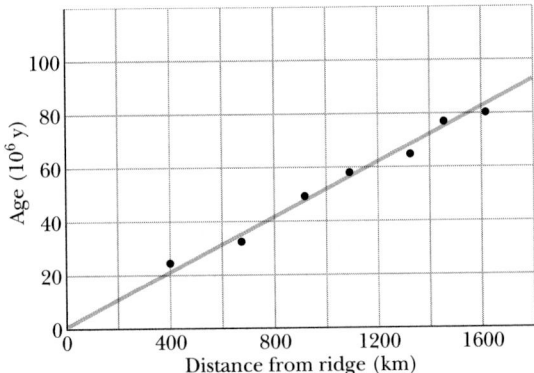

FIGURE 2-18 Exercise 5.

6E. When the legal speed limit for the New York Thruway was increased from 55 mi/h (= 88.5 km/h) to 65 mi/h (= 105 km/h), how much time was saved by a motorist who drove the 435 mi (= 700 km) between the Buffalo entrance and the New York City exit at the legal speed limit?

7E. Using the tables in Appendix D, find the speed of light (= 3×10^8 m/s) in miles per hour, feet per second, and light-years per year.

8E. An automobile travels on a straight road for 40 km at 30 km/h. It then continues in the same direction for another 40 km at 60 km/h. (a) What is the average velocity of the car during this 80 km trip? (Assume that it moves in the positive x direction.) (b) What is its average speed? (c) Graph x versus t and indicate how the average velocity is found on the graph.

9P. Compute your average velocity in the following two cases. (a) You walk 240 ft at a speed of 4.0 ft/s and then run 240 ft at a speed of 10 ft/s along a straight track. (b) You walk for 1.0 min at a speed of 4.0 ft/s and then run for 1.0 min at 10 ft/s along a straight track. (c) Graph x versus t for both cases and indicate how the average velocity is found on the graph.

10P. A car travels up a hill at a constant speed of 40 km/h and returns down the hill at a constant speed of 60 km/h. Calculate the average speed for the round trip.

11P. You drive on Interstate 10 from San Antonio to Houston, half the *time* at 35 mi/h (= 56 km/h) and the other half at 55 mi/h (= 89 km/h). On the way back you travel half the *distance* at 35 mi/h and the other half at 55 mi/h. What is your average speed (a) from San Antonio to Houston, (b) from Houston back to San Antonio, and (c) for the entire trip? (d) What is your average velocity for the entire trip? (e) Graph x versus t for (a), assuming the motion is all in the positive x direction. Indicate how the average velocity can be found on the graph.

12P. The position of an object moving in a straight line is given by $x = 3t - 4t^2 + t^3$, where x is in meters and t in seconds. (a) What is the position of the object at $t = 1, 2, 3$, and 4 s? (b) What is the object's displacement between $t = 0$ and $t = 4$ s? (c) What is the average velocity for the time interval from $t = 2$ s to $t = 4$ s? (d) Graph x versus t for $0 \le t \le 4$ s and indicate how the answer for (c) can be found on the graph.

13P. The position of a particle moving along the x axis is given in centimeters by $x = 9.75 + 1.50t^3$, where t is in seconds. Consider the time interval $t = 2.00$ s to $t = 3.00$ s and calculate (a) the average velocity; (b) the instantaneous velocity at $t = 2.00$ s; (c) the instantaneous velocity at $t = 3.00$ s; (d) the instantaneous velocity at $t = 2.50$ s; and (e) the instantaneous velocity when the particle is midway between its positions at $t = 2.00$ s and $t = 3.00$ s. (f) Graph x versus t and indicate your answers graphically.

14P. A high-performance jet plane, practicing radar avoidance maneuvers, is in horizontal flight 35 m above the level ground. Suddenly, the plane encounters terrain that slopes gently upward at 4.3°, an amount difficult to detect (see Fig. 2-19). How much time does the pilot have to make a correction to avoid flying into the ground? The speed of the plane is 1300 km/h.

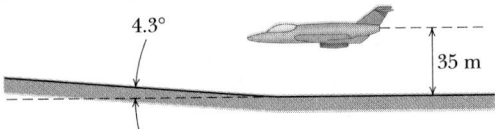

FIGURE 2-19 Problem 14.

15P. Two trains, each having a speed of 30 km/h, are headed at each other on the same straight track. A bird that can fly 60 km/h flies off the front of one train when they are 60 km apart and heads directly for the other train. On reaching the other train it flies directly back to the first train, and so forth. (We have no idea *why* a bird would behave in this way.) (a) How many trips can the bird make from one train to the other before they crash? (b) What is the total distance the bird travels?

SECTION 2-4 Instantaneous Velocity and Speed

16E. (a) If a particle's position is given by $x = 4 - 12t + 3t^2$ (where t is in seconds and x is in meters), what is its velocity at $t = 1$ s? (b) Is it moving toward increasing or decreasing x just then? (c) What is its speed just then? (d) Is the speed larger or smaller at later times? (Try answering the next two questions without further calculation.) (e) Is there ever an instant when the velocity is zero? (f) Is there a time after $t = 3$ s when the particle is moving toward decreasing x?

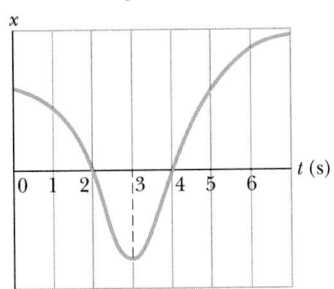

FIGURE 2-20 Exercise 17.

17E. The graph in Fig. 2-20 pertains to an armadillo that scampers left (direction of decreasing x) and right along an x axis. (a) When, if ever, is the animal to the left of the origin on the axis? When, if ever, is its velocity (b) negative, (c) positive, or (d) zero?

18E. Sketch a graph of x versus t for a mouse that is constrained in a narrow corridor (the x axis) and that scurries in the following sequence: (1) runs leftward (the direction of decreasing x) with a constant speed of 1.2 m/s, (2) gradually slows to 0.6 m/s toward the left, (3) gradually speeds up to 2.0 m/s toward the left, (4) gradually slows to a stop and then speeds up to 1.2 m/s toward the right. Where is the curve steepest? The least steep?

19P. How far does the runner whose velocity–time graph is shown in Fig. 2-21 travel in 16 s?

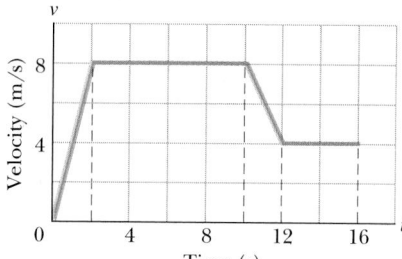

FIGURE 2-21
Problem 19.

SECTION 2-5   Acceleration

20E. A particle had a velocity of 18 m/s and 2.4 s later its velocity was 30 m/s in the opposite direction. (a) What was the magnitude of the average acceleration of the particle during this 2.4 s interval? (b) Graph v versus t and indicate how the average acceleration can be found on the graph.

21E. An object moves in a straight line as described by the velocity–time graph of Fig. 2-22. Sketch a graph that represents the acceleration of the object as a function of time.

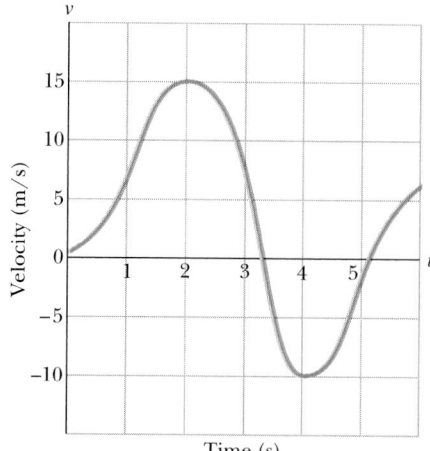

FIGURE 2-22
Exercise 21.

22E. The graph of x versus t in Fig. 2-23a is for a particle in straight-line motion. (a) State, for each of the intervals AB, BC, CD, and DE, whether the velocity v is positive, negative, or 0 and whether the acceleration a is positive, negative, or 0. (Ignore the end points of the intervals.) (b) From the curve, is there any inter-

val over which the acceleration is obviously not constant? (c) If the axes are shifted upward together such that the time axis ends up running along the dashed line, do any of your answers change?

23E. Repeat Exercise 22 for the motion described by the graph of Fig. 2-23b.

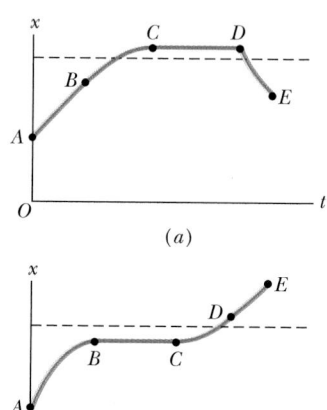

FIGURE 2-23
Exercises 22 and 23.

24E. A particle moves along the x axis with $x(t)$ as shown in Fig. 2-24. Make rough sketches of velocity versus time and acceleration versus time for this motion.

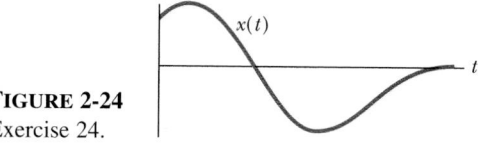

FIGURE 2-24
Exercise 24.

25E. Sketch a graph that is a possible description of position as a function of time for a particle that moves along the x axis and, at $t = 1$ s, has (a) zero velocity and positive acceleration; (b) zero velocity and negative acceleration; (c) negative velocity and positive acceleration; (d) negative velocity and negative acceleration. (e) For which of these situations is the speed of the particle increasing at $t = 1$ s?

26E. Consider the two quantities $(dx/dt)^2$ and d^2x/dt^2. (a) Are these two equivalent expressions for the same thing? (b) What are the SI units of these two quantities?

27E. A particle moves along the x axis according to the equation $x = 50t + 10t^2$, where x is in meters and t is in seconds. Calculate (a) the average velocity of the particle during the first 3.0 s of its motion, (b) the instantaneous velocity of the particle at $t = 3.0$ s, and (c) the instantaneous acceleration of the particle at $t = 3.0$ s. (d) Graph x versus t and indicate how the answer to (a) can be obtained from the plot. (e) Indicate the answer to (b) on the graph. (f) Plot v versus t and indicate on it the answer to (c).

28E. (a) If the position of a particle is given by $x = 20t - 5t^3$, where x is in meters and t is in seconds, when, if ever, is the particle's velocity zero? (b) When is its acceleration a zero? (c) When is a negative? Positive? (d) Graph $x(t)$, $v(t)$, and $a(t)$.

29P. A man stands still from $t = 0$ to $t = 5.00$ min; from $t = 5.00$ min to $t = 10.0$ min he walks briskly in a straight line at a constant speed of 2.20 m/s. What are his average velocity and average acceleration during the time intervals (a) 2.00 min to 8.00 min and (b) 3.00 min to 9.00 min? (c) Sketch x versus t and v versus t, and indicate how the answers to (a) and (b) can be obtained from the graphs.

30P. If the position of an object is given by $x = 2.0t^3$, where x is measured in meters and t in seconds, find (a) the average velocity and (b) the average acceleration between $t = 1.0$ s and $t = 2.0$ s. Then find (c) the instantaneous velocities and (d) the instantaneous accelerations at $t = 1.0$ s and $t = 2.0$ s. (e) Compare the average and instantaneous quantities and in each case explain why the larger one is larger. (f) Graph x versus t and v versus t, and indicate on the graphs your answers to (a) through (d).

31P. In an arcade video game, a spot is programmed to move across the screen according to $x = 9.00t - 0.750t^3$, where x is distance in centimeters measured from the left edge of the screen and t is time in seconds. When the spot reaches a screen edge, at either $x = 0$ or $x = 15.0$ cm, t is reset to 0 and the spot starts moving again according to $x(t)$. (a) At what time after starting is the spot instantaneously at rest? (b) Where does this occur? (c) What is its acceleration when this occurs? (d) In what direction is it moving just prior to coming to rest? (e) Just after? (f) When does it first reach an edge of the screen after $t = 0$?

32P. The position of a particle moving along the x axis depends on the time according to the equation $x = et^2 - bt^3$, where x is in feet and t in seconds. (a) What dimensions and units must e and b have? For the following, let their numerical values be 3.0 and 1.0, respectively. (b) At what time does the particle reach its maximum positive x position? (c) What distance does the particle cover in the first 4.0 s? (d) What is its displacement from $t = 0$ to $t = 4.0$ s? (e) What is its velocity at $t = 1.0, 2.0, 3.0,$ and 4.0 s? (f) What is its acceleration at these times?

SECTION 2-6 Constant Acceleration: A Special Case

33E. The head of a rattlesnake can accelerate 50 m/s² in striking a victim. If a car could do as well, how long would it take to reach a speed of 100 km/h from rest?

34E. An object has a constant acceleration of +3.2 m/s². At a certain instant its velocity is +9.6 m/s. What is its velocity (a) 2.5 s earlier and (b) 2.5 s later?

35E. An automobile increases its speed uniformly from 25 km/h to 55 km/h in 0.50 min. A bicycle rider uniformly speeds up to 30 km/h from rest in 0.50 min. Calculate their accelerations.

36E. Suppose a rocketship in deep space moves with constant acceleration equal to 9.8 m/s², which gives the illusion of normal gravity during the flight. (a) If it starts from rest, how long will it take to acquire a speed one-tenth that of light, which travels at 3.0×10^8 m/s? (b) How far will it travel in so doing?

37E. A jumbo jet must reach a speed of 360 km/h (= 225 mi/h) on the runway for takeoff. What is the least constant acceleration needed for takeoff from a 1.80 km runway?

38E. A muon (an elementary particle) enters an electric field with a speed of 5.00×10^6 m/s, whereupon the field slows it at the rate of 1.25×10^{14} m/s². (a) How far does the muon take to stop? (b) Graph x versus t and v versus t for the muon.

39E. An electron with initial velocity $v_0 = 1.50 \times 10^5$ m/s enters a region 1.0 cm long where it is electrically accelerated (Fig. 2-25). It emerges with velocity $v = 5.70 \times 10^6$ m/s. What was its acceleration, assumed constant? (Such a process occurs in the electron gun in a cathode-ray tube, used in television receivers and oscilloscopes.)

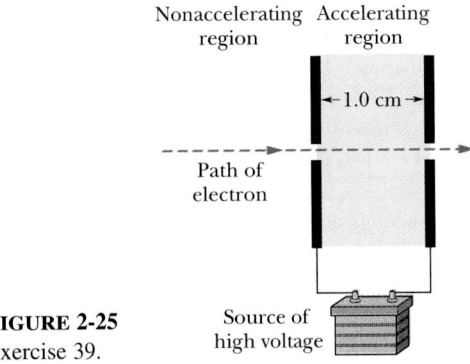

Nonaccelerating region   Accelerating region

1.0 cm

Path of electron

FIGURE 2-25
Exercise 39.

Source of high voltage

40E. A car can be braked to a stop from 60 mi/h in 43 m. (a) What is the magnitude of the acceleration in SI units and g units? (Assume that a is constant.) (b) What is the stopping time? If your reaction time T for braking is 400 ms, to how many "reaction times" does the stopping time correspond?

41E. A world's land speed record was set by Colonel John P. Stapp when, on March 19, 1954, he rode a rocket-propelled sled that moved down a track at 1020 km/h. He and the sled were brought to a stop in 1.4 s. (See Fig. 2-8.) What acceleration did he experience? Express your answer in g units.

42E. On a dry road a car with good tires may be able to brake with a deceleration of 4.92 m/s² (assume that it is constant). (a) How long does such a car, initially traveling at 24.6 m/s, take to come to rest? (b) How far does it travel in this time? (c) Graph x versus t and v versus t for the deceleration.

43E. A rocket-driven sled running on a straight, level track is used to investigate the physiological effects of large accelerations on humans. One such sled can attain a speed of 1600 km/h in 1.8 s starting from rest. Find (a) the acceleration (assumed constant) in g units and (b) the distance traveled.

44E. The brakes on your automobile are capable of creating a deceleration of 17 ft/s². (a) If you are going 85 mi/h and suddenly see a state trooper, what is the minimum time in which you can get your car under the 55 mi/h speed limit? (The answer reveals the futility of braking to keep your high speed from being detected with a radar gun.) (b) Graph x versus t and v versus t for such a deceleration.

45E. A motorcycle is moving at 30 m/s when the rider applies the brakes, giving the motorcycle a constant deceleration. During the 3.0 s interval immediately after braking begins, the speed decreases to 15 m/s. What distance does the motorcycle travel from the instant braking begins until it comes to rest?

46P. A hot rod can accelerate from 0 to 60 km/h in 5.4 s. (a) What is its average acceleration, in m/s², during this time? (b) How far will it travel during the 5.4 s, assuming its acceleration is constant? (c) How much time would it require to go a distance of 0.25 km if its acceleration could be maintained at the value in (a)?

47P. A train started from rest and moved with constant acceleration. At one time it was traveling 30 m/s, and 160 m farther on it was traveling 50 m/s. Calculate (a) the acceleration, (b) the time required to travel the 160 m mentioned, (c) the time required to attain the speed of 30 m/s, and (d) the distance moved from rest to the time the train had a speed of 30 m/s. (e) Graph x versus t and v versus t for the train, from rest.

48P. A car traveling 56.0 km/h is 24.0 m from a barrier when the driver slams on the brakes. The car hits the barrier 2.00 s later. (a) What was the car's constant deceleration before impact? (b) How fast was the car traveling at impact?

49P. A car moving with constant acceleration covers the distance between two points 60.0 m apart in 6.00 s. Its speed as it passes the second point is 15.0 m/s. (a) What is the speed at the first point? (b) What is the acceleration? (c) At what prior distance from the first point was the car at rest? (d) Graph x versus t and v versus t for the car, from rest.

50P. Two subway stops are separated by 1100 m. If a subway train accelerates at +1.2 m/s² from rest through the first half of the distance and decelerates at −1.2 m/s² through the second half, what are (a) its travel time and (b) its maximum speed? (c) Graph x, v, and a versus t for the trip.

51P. To stop a car, you require first a certain reaction time to begin braking; then the car slows under the constant braking deceleration. Suppose that the total distance moved by your car during these two phases is 186 ft when its initial speed is 50 mi/h, and 80 ft when the initial speed is 30 mi/h. What are (a) your reaction time and (b) the magnitude of the deceleration?

52P. You are driving toward a traffic signal when it turns yellow. Your speed is the legal speed limit of $v_0 = 35$ mi/h; your best deceleration rate is $a = 17$ ft/s². Your best reaction time to begin braking is $T = 0.75$ s. To avoid having the front of your car enter the intersection after the light turns red, should you brake to a stop or continue to move at 35 mi/h if the distance to the intersection and the duration of the yellow light are (a) 40 m and 2.8 s, and (b) 32 m and 1.8 s?

53P. When a driver brings a car to a stop by braking as hard as possible, the stopping distance can be regarded as the sum of a "reaction distance," which is initial speed times the driver's reaction time, and a "braking distance," which is the distance covered during braking. The following table gives typical values:

INITIAL SPEED (m/s)	REACTION DISTANCE (m)	BRAKING DISTANCE (m)	STOPPING DISTANCE (m)
10	7.5	5.0	12.5
20	15	20	35
30	22.5	45	67.5

(a) What reaction time is the driver assumed to have? (b) What is the car's stopping distance if the initial speed is 25 m/s?

54P. (a) If the maximum acceleration that is tolerable for passengers in a subway train is 1.34 m/s², and subway stations are located 806 m apart, what is the maximum speed a subway train can attain between stations? (b) What is the travel time between stations? (c) If the subway train stops for 20 s at each station, what is the maximum average speed of a subway train, from one start-up to the next? (d) Graph x, v, and a versus t.

55P. An elevator cab in the New York Marquis Marriott has a total run of 624 ft. Its maximum speed is 1000 ft/min. Its acceleration and deceleration both have a magnitude of 4.0 ft/s². (a) How far does the cab move while accelerating to full speed from rest? (b) How long does it take to make the nonstop 624 ft run, starting and ending at rest?

56P. At the instant the traffic light turns green, an automobile starts with a constant acceleration a of 2.2 m/s². At the same instant a truck, traveling with a constant speed of 9.5 m/s, overtakes and passes the automobile. (a) How far beyond the traffic signal will the automobile overtake the truck? (b) How fast will the car be traveling at that instant?

57P. When a high-speed passenger train traveling at 100 mi/h rounds a bend, the engineer is shocked to see that a locomotive has improperly entered onto the track from a siding 0.42 mi ahead; see Fig. 2-26. The locomotive is moving at 18 mi/h. The engineer of the passenger train immediately applies the brakes. (a) What must be the magnitude of the resulting constant acceleration if a collision is to be just avoided? (b) Assume that the

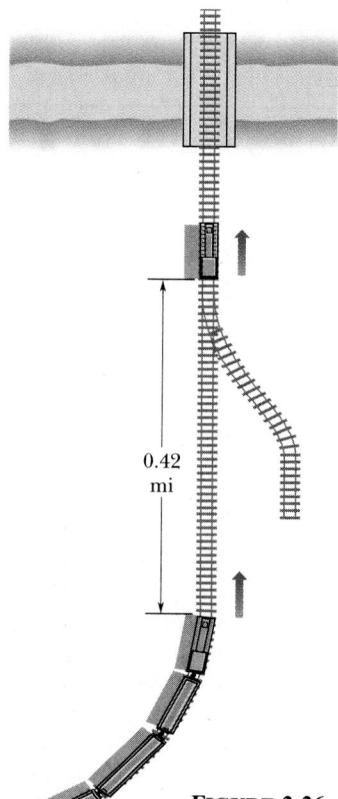

FIGURE 2-26 Problem 57.

engineer is at $x = 0$ when, at $t = 0$, he first spots the locomotive. Sketch the $x(t)$ curves representing the locomotive and passenger train for the situations in which a collision is just avoided and not quite avoided.

58P. Two trains, one traveling at 72 km/h and the other at 144 km/h, are headed toward one another along a straight, level track. When they are 950 m apart, each engineer sees the other's train and applies the brakes. The brakes decelerate each train at the rate of 1.0 m/s². Is there a collision?

59P. Sketch a $v(t)$ graph that would be associated with the $a(t)$ graph shown in Fig. 2-27.

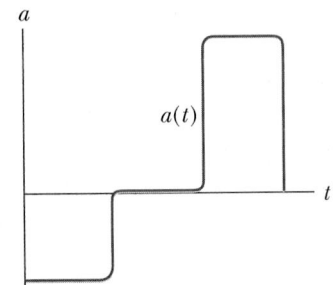

FIGURE 2-27
Problem 59.

SECTION 2-8 Free-Fall Acceleration

60E. At a construction site a pipe wrench strikes the ground with a speed of 24 m/s. (a) From what height was it inadvertently dropped? (b) How long was it falling? (c) Sketch graphs of y, v, and a versus t for the wrench.

61E. (a) With what speed must a ball be thrown vertically from ground level to rise to a maximum height of 50 m? (b) How long will it be in the air? (c) Sketch graphs of y, v, and a versus t for the ball. On the first two graphs, indicate the time at which 50 m is reached.

62E. Raindrops fall to Earth from a cloud 1700 m above the Earth's surface. If they were not slowed by air resistance, how fast would the drops be moving when they struck the ground? Would it be safe to walk outside during a rainstorm?

63E. The single cable supporting an unoccupied construction elevator breaks when the elevator is at rest at the top of a 120-m-high building. (a) With what speed does the elevator strike the ground? (b) How long was it falling? (c) What was its speed when it passed the halfway point on the way down? (d) How long had it been falling when it passed the halfway point?

64E. A hoodlum throws a stone vertically downward with an initial speed of 12.0 m/s from the roof of a building, 30.0 m above the ground. (a) How long does it take the stone to reach the ground? (b) What is the speed of the stone at impact?

65E. The Zero Gravity Research Facility at the NASA Lewis Research Center includes a 145 m drop tower. This is an evacuated vertical tower through which, among other possibilities, a 1 m diameter sphere containing an experimental package can be dropped. (a) How long is the sphere in free fall? (b) What is its speed just as it reaches a catching device at the bottom of the tower? (c) When caught, the sphere experiences an average de-

celeration of 25g as its speed is reduced to zero. Through what distance does it travel during the deceleration?

66E. A model rocket, propelled by burning fuel, takes off vertically. Plot qualitatively (numbers not required) graphs of y, v, and a versus t for the rocket's flight. Indicate when the fuel is exhausted, when the rocket reaches maximum height, and when it returns to the ground.

67E. A rock is dropped from a 100 m high cliff. How long does it take to fall (a) the first 50 m and (b) the second 50 m?

68P. A startled armadillo leaps upward (Fig. 2-28), rising 0.544 m in 0.200 s. (a) What was its initial speed? (b) What is its speed at this height? (c) How much higher does it go?

FIGURE 2-28 Problem 68.

69P. An object falls from a bridge that is 45 m above the water. It falls directly into a model boat, moving with constant velocity, that was 12 m from the point of impact when the object was released. What was the speed of the boat?

70P. A model rocket is fired vertically and ascends with a constant vertical acceleration of 4.00 m/s² for 6.00 s. Its fuel is then exhausted and it continues as a free-fall particle. (a) What is the maximum altitude reached? (b) What is the total time elapsed from takeoff until the rocket strikes the Earth?

71P. A basketball player, standing near the basket to grab a rebound, jumps 76.0 cm vertically. How much (total) time does the player spend (a) in the top 15.0 cm of this jump and (b) in the bottom 15.0 cm? Does this help explain why such players seem to hang in the air at the tops of their jumps?

72P. At the National Physical Laboratory in England, a measurement of the free-fall acceleration g was made by throwing a glass ball straight up in an evacuated tube and letting it return. Let ΔT_L in Fig. 2-29 be the time interval between the two passages of the ball across a certain lower level, ΔT_U the time interval between the two passages across an upper level, and H the distance between the two levels. Show that

$$g = \frac{8H}{\Delta T_L^2 - \Delta T_U^2}.$$

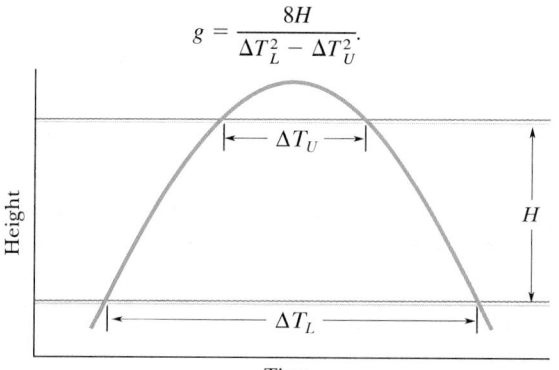

FIGURE 2-29 Problem 72.

73P. A ball of moist clay falls to the ground from a height of 15.0 m. It is in contact with the ground for 20.0 ms before coming to rest. What is the average acceleration of the clay during the time it is in contact with the ground? (Treat the ball as a particle.)

74P. A ball is thrown *down* vertically with an initial *speed* of v_0 from a height of h. (a) What is its speed just before it strikes the ground? (b) How long does the ball take to reach the ground? (c) What would be the answers to (a) and (b) if the ball were thrown *upward* from the same height and with the same initial speed? Before solving any equations, decide if the answers here should be greater than, less than, or the same as in (a) and (b).

75P. Figure 2-30 shows a simple device for measuring your reaction time. It consists of a cardboard strip marked with a scale and two large dots. A friend holds the strip *vertically,* with thumb and forefinger at the dot on the right in Fig. 2-30. You then position your thumb and forefinger at the other dot (on the left in Fig. 2-30), being careful not to touch the strip. Your friend releases the strip, and you try to pinch it as soon as possible after you see it begin to fall. The mark at the place where you pinch the strip gives your reaction time. (a) How far from the lower dot should you place the 50.0 ms mark? (b) How much higher should the marks for 100, 150, 200, and 250 ms be? (For example, should the 100 ms marker be two times as far from the dot as the 50 ms marker? Can you find any pattern in the answers?)

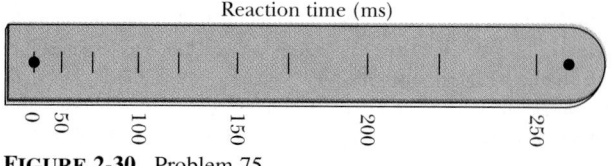

FIGURE 2-30 Problem 75.

76P. A juggler tosses balls vertically a certain distance into the air. How much higher must they be tossed if they are to spend twice as much time in the air?

77P. A stone is thrown vertically upward. On its way up it passes point A with speed v, and point B, 3.00 m higher than A, with speed $\frac{1}{2}v$. Calculate (a) the speed v and (b) the maximum height reached by the stone above point B.

78P. To test the quality of a tennis ball, you drop it onto the floor from a height of 4.00 m. It rebounds to a height of 3.00 m. If the ball was in contact with the floor for 10.0 ms, what was its average acceleration during contact

79P. Water drips from the nozzle of a shower onto the floor 200 cm below. The drops fall at regular (equal) intervals of time, the first drop striking the floor at the instant the fourth drop begins to fall. Find the locations of the second and third drops when the first strikes the floor.

80P. A lead ball is dropped into a lake from a diving board 5.20 m above the water. It hits the water with a certain velocity and then sinks to the bottom with this same constant velocity. It reaches the bottom 4.80 s after it is dropped. (a) How deep is the lake? (b) What is the average velocity of the ball? (c) Suppose that all the water is drained from the lake. The ball is now thrown from the diving board so that it again reaches the bottom in 4.80 s. What is the initial velocity of the ball?

81P. An object falls from height h from rest. If it travels $0.50h$ in the last 1.00 s, find (a) the time and (b) the height of its fall. Explain the physically unacceptable solution of the quadratic equation in t that you obtain.

82P. A woman fell 144 ft from the top of a building, landing on the top of a metal ventilator box, which she crushed to a depth of 18 in. She survived without serious injury. What acceleration (assumed uniform) did she experience during the collision? Express your answer in terms of g units.

83P. A stone is dropped into a river from a bridge 144 ft above the water. Another stone is thrown vertically down 1.00 s after the first is dropped. Both stones strike the water at the same time. (a) What was the initial speed of the second stone? (b) Plot velocity versus time on a graph for each stone, taking zero time as the instant the first stone is released.

84P. A parachutist bails out and freely falls 50 m. Then the parachute opens, and thereafter she decelerates at 2.0 m/s². She reaches the ground with a speed of 3.0 m/s. (a) How long was the parachutist in the air? (b) At what height did the fall begin?

85P. Two objects begin a free fall from rest from the same height 1.0 s apart. How long after the first object begins to fall will the two objects be 10 m apart?

86P. As Fig. 2-31 shows, Clara jumps from a bridge, followed closely by Jim. How long did Jim wait after Clara jumped? Assume that Jim is 170 cm tall and that the jumping-off level is at the top of the figure. Make scale measurements directly on the figure.

87P. A hot-air balloon is ascending at the rate of 12 m/s and is 80 m above the ground when a package is dropped over the side. (a) How long does the package take to reach the ground? (b) With what speed does it hit the ground?

88P. An elevator without a ceiling is ascending with a constant speed of 10 m/s. A boy on the elevator throws a ball directly

FIGURE 2-31 Problem 86.

upward, from a height of 2.0 m above the elevator floor, just as the elevator floor is 28 m above the ground. The initial speed of the ball with respect to the elevator is 20 m/s. (a) What is the maximum height attained by the ball? (b) How long does it take for the ball to return to the elevator floor?

89P. A steel ball is dropped from a building's roof and passes a window, taking 0.125 s to fall from the top to the bottom of the window, a distance of 1.20 m. It then falls to a sidewalk and bounces "perfectly" back past the window, moving from bottom to top in 0.125 s. (The upward flight is a reverse of the fall.) The time spent below the bottom of the window is 2.00 s. How tall is the building?

90P. A drowsy cat spots a flowerpot that sails first up and then down past an open window. The pot was in view for a total of 0.50 s, and the top-to-bottom height of the window is 2.00 m. How high above the window top did the flowerpot go?

ELECTRONIC COMPUTATION

Problems in this section are to be solved with a graphing calculator or with a mathematics package on a computer.

91. A driver is to test the braking of a car equipped with antilock brakes, which can decelerate the car at the maximum possible rate without the tires skidding. The test is made on a dry straight section of a test track. In response to a visual signal, the driver stops the car in the shortest possible distance. The stopping distances for six initial speeds are given in the following table. (a) Find an expression for the stopping distance d in terms of the initial speed v_i, the response time T_R of the driver to the visual signal, and the magnitude a' (assumed constant) of the car's deceleration. Using a least squares fit for the data, find (b) a' and (c) T_R.

v_i (m/s)	5.00	10.0	15.0	20.0	25.0	30.0
d (m)	4.6	12.1	22.3	35.2	51.0	69.5

92. A motorcyclist starts from rest and accelerates along a horizontal straight track. Photogates are attached to six posts evenly spaced every 10.0 m along the track; the first post is at the starting point. The photogates on each post measure the time required by the motorcyclist to reach that post. The following table gives the results for one test. (a) Find an expression for the distance d to each post in terms of the time t to reach that post and the acceleration a' of the motorcyclist (assumed constant). (b) Using the data of the table, graph d versus t^2. (c) Using a linear regression fit of the data, find the acceleration of the motorcyclist.

Post number	1	2	3	4	5	6
Distance traveled (m)	0	10.0	20.0	30.0	40.0	50.0
Time required (s)	0	1.63	2.33	2.83	3.31	3.79

93. The motorcyclist of Problem 92 next starts from rest and accelerates uniformly along another straight test track. A series of six posts equally spaced by a distance $d_0 = 10.0$ m line the track, beginning an unknown distance d from the motorcyclist's starting point. The following table gives the speed v at each post. (a) Find an expression for the square of the speed v_j at post j in terms of the speed v_1 at post 1, the distance d_0, the constant acceleration a', and the post number j. (b) Graph v_j^2 versus $(j - 1)$. Using a linear regression fit of the data, find (c) acceleration a' and (d) distance d.

Post number j	1	2	3	4	5	6
Speed v (m/s)	14.0	18.3	21.7	24.6	27.5	30.0

94. Suppose the coordinate of a particle moving along the x axis is given as a function of time t by

$$x(t) = -32.0 + 24.0t^2 e^{-0.0300t},$$

where x is in meters and t is in seconds. (a) Write expressions for the velocity v and acceleration a as functions of the time. (b) Plot graphs of the coordinate, velocity, and acceleration as functions of time from $t = 0$ to $t = 100$ s. (c) Find the time when the coordinate of the particle is zero; then find the velocity and acceleration at that time. (d) Find the time when the velocity of the particle is zero; then find the coordinate and acceleration at that time.

3
Vectors

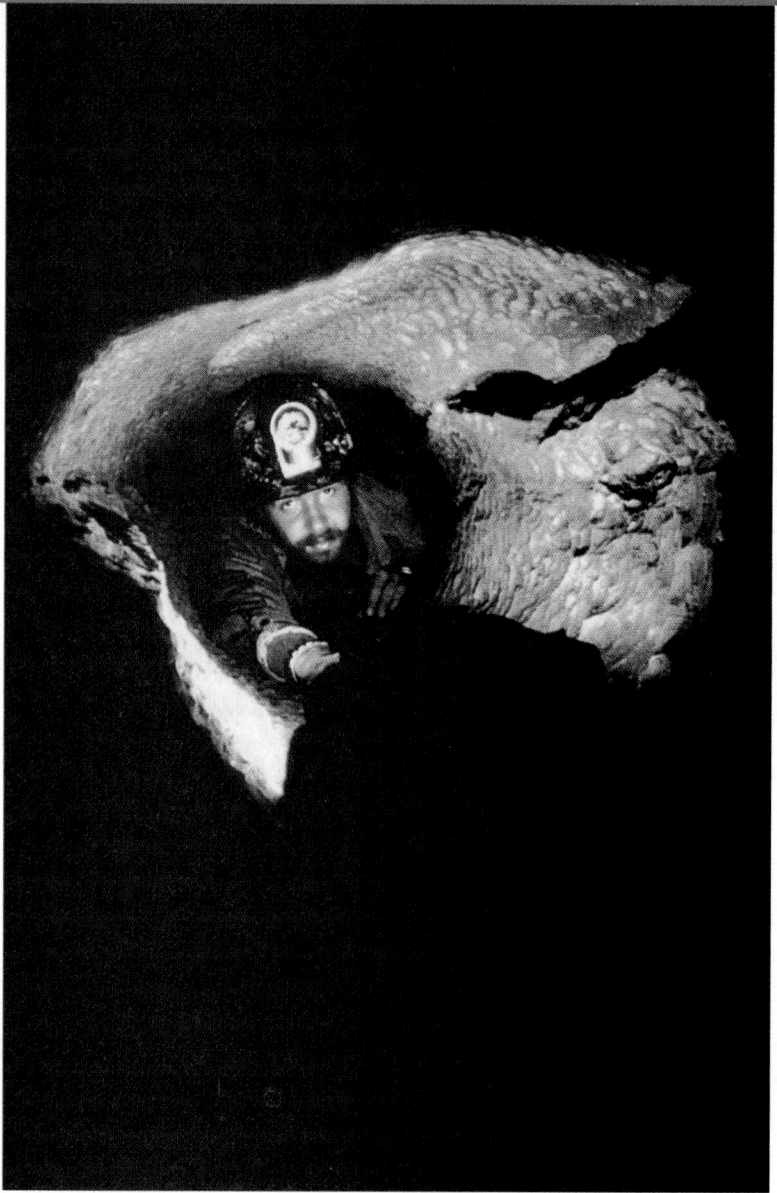

For two decades spelunking teams crawled, climbed, and squirmed through 200 km of Mammoth Cave and the Flint Ridge cave system, seeking a connection. The photograph shows Richard Zopf pushing his pack through the Tight Tube, far inside the Flint Ridge system. After 12 hours of "caving" along a labyrinthine route, Zopf and six others waded through a stretch of chilling water and found themselves in Mammoth Cave. Their breakthrough established the Mammoth–Flint cave system as being the longest cave in the world. How can their final point be related to their initial point other than in terms of the actual route they covered?

3-1 VECTORS AND SCALARS

A particle confined to a straight line can move in only two directions. We can take its motion to be positive in one of these directions and negative in the other. For a particle moving in three dimensions, however, a plus sign or minus sign is no longer enough to indicate the direction of the motion. Instead, we must use a *vector.*

A **vector** has magnitude as well as direction, and vectors follow certain rules of combination, which we examine below; a **vector quantity** is a quantity that has both a magnitude and a direction. Some physical quantities that can be represented by vectors are displacement, velocity, and acceleration.

Not all physical quantities involve a direction. Temperature, pressure, energy, mass, and time, for example, do not ''point'' in the spatial sense. We call such quantities **scalars,** and we deal with them by the rules of ordinary algebra. A single value, with a sign (as in $-40°F$), specifies a scalar.

The simplest vector quantity is displacement, or change of position. A vector that represents a displacement is called, reasonably, a **displacement vector.** (Similarly, we have velocity vectors and acceleration vectors.) If a particle changes its position by moving from A to B in Fig. 3-1a, we say that it undergoes a displacement from A to B, which we represent with an arrow pointing from A to B. The arrow specifies the vector graphically. To distinguish vector symbols from other kinds of arrows, we use the outline of a triangle as the arrowhead.

The arrows from A to B, from A' to B', and from A'' to B'' in Fig. 3-1a represent the same *change of position* for the particle and we make no distinction among them. All three arrows have the same magnitude and direction and thus specify identical displacement vectors.

The displacement vector tells us nothing about the actual path that the particle takes. In Fig. 3-1b, for example, all three paths connecting points A and B correspond to the same displacement vector, that of Fig. 3-1a. Displacement vectors represent only the overall effect of the motion, not the motion itself.

Another way to specify a vector is to give its magnitude and the angles it makes with two reference lines. We do that in Sample Problem 3-1.

SAMPLE PROBLEM 3-1

The 1972 team that connected the Mammoth–Flint cave system went from the Austin Entrance in the Flint Ridge system to Echo River in Mammoth Cave (see Fig. 3-2a), traveling a net 2.6 km westward, 3.9 km southward, and 25 m upward. What was their displacement vector?

SOLUTION: We first take an overhead view (Fig. 3-2b) to find their horizontal displacement d_h. The magnitude of d_h is given by the Pythagorean theorem:

$$d_h = \sqrt{(2.6 \text{ km})^2 + (3.9 \text{ km})^2} = 4.69 \text{ km}.$$

The angle θ of their displacement vector relative to the west is given by

$$\tan \theta = \frac{3.9 \text{ km}}{2.6 \text{ km}} = 1.5,$$

or $\qquad \theta = \tan^{-1} 1.5 = 56°.$

We next take a side view (Fig. 3-2c), to find the magnitude of the overall displacement d,

$$d = \sqrt{(4.69 \text{ km})^2 + (0.025 \text{ km})^2} = 4.69 \text{ km} \approx 4.7 \text{ km},$$

and the angle ϕ,

$$\phi = \tan^{-1} \frac{0.025 \text{ km}}{4.69 \text{ km}} = 0.3°.$$

So the team's displacement vector had a magnitude of 4.7 km and was at an angle of 56° south of west and at an angle of 0.3° upward. The net vertical motion was, of course, insignificant compared to the horizontal motion, but that fact would have been no comfort to the team as they climbed up and down countless times. The route they actually covered was quite different from the displacement vector, which merely points from start to finish.

3-2 ADDING VECTORS: GRAPHICAL METHOD

Suppose that, as in the vector diagram of Fig. 3-3a, a particle moves from A to B and then later from B to C. We can represent its overall displacement (no matter what its actual path) with two successive displacement vectors, AB and BC. The net effect of these two displacements is a single displacement from A to C. We call AC the **vector sum** (or **resultant**) of the vectors AB and BC. This sum is

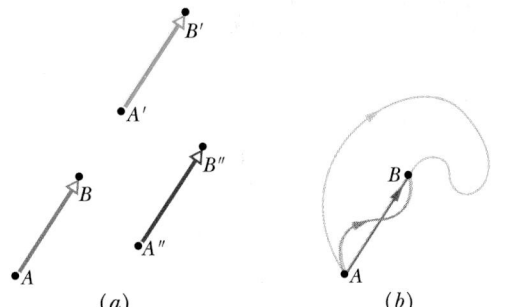

(a)   (b)

FIGURE 3-1 (a) All three arrows represent the same displacement. (b) All three paths connecting the two points correspond to the same displacement vector.

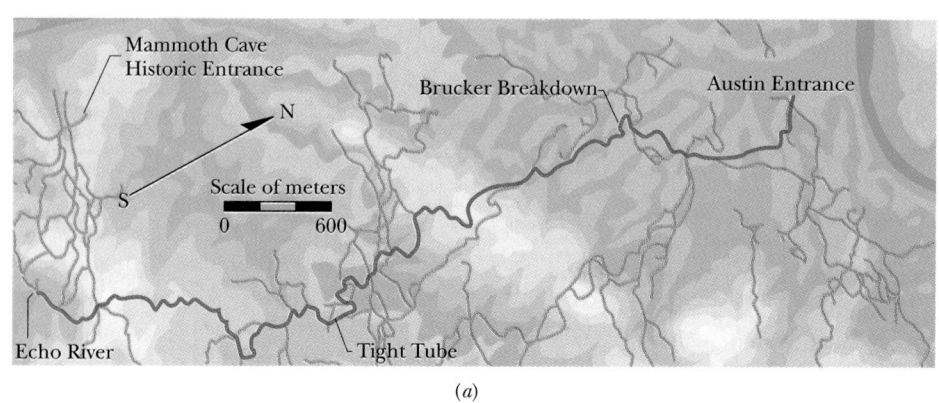

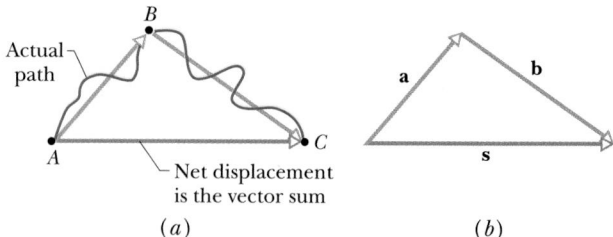

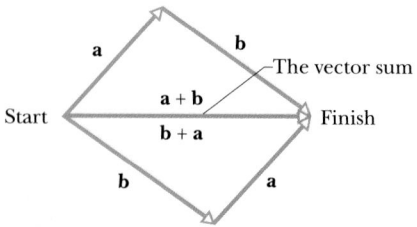

FIGURE 3-2 Sample Problem 3-1. (*a*) Part of the Mammoth–Flint cave system, with the spelunking team's route from the Austin Entrance to Echo River indicated. (*b*) An overhead view of the team's displacement. (*c*) A side view. (Adapted from map by Cave Research Foundation.)

not the usual algebraic sum, and we need more than simple numbers to specify it.

In Fig. 3-3*b*, we redraw the vectors of Fig. 3-3*a* and relabel them in the way that we shall use from now on, namely, with boldface symbols such as **a**, **b**, and **s**. In handwriting, you can place an arrow over the symbol, as in $\vec{a}$. If we want to indicate only the magnitude of the vector (a quantity that is always positive), we shall use the italic symbol, such as *a*, *b*, and *s*. (You can use just a handwritten symbol.) The boldface symbol always implies both properties of the vector, magnitude and direction.

We can represent the relation among the three vectors in Fig. 3-3*b* with the vector equation

$$\mathbf{s} = \mathbf{a} + \mathbf{b}, \tag{3-1}$$

which says that the vector **s** is the vector sum of vectors **a** and **b**. The symbol + in Eq. 3-1 and the words ''sum'' and ''add'' have different meanings for vectors than they do in the usual algebra.

Figure 3-3 suggests a procedure for adding two-dimensional vectors **a** and **b** graphically. (1) On a sheet of paper, lay out vector **a** to some convenient scale and at the proper angle. (2) Lay out vector **b** to the same scale, with its tail at the head of vector **a,** again at the proper angle. (3) Construct the vector sum **s** by drawing an arrow from the tail of **a** to the head of **b**. Note that this procedure takes into account both the magnitudes and the directions of the vectors; you can easily generalize it to add more than two vectors.

Vector addition, defined in this way, has two important properties. First, the order of addition does not matter. That is,

$$\mathbf{a} + \mathbf{b} = \mathbf{b} + \mathbf{a} \quad \text{(commutative law).} \tag{3-2}$$

Figure 3-4 should convince you that this is the case.

Second, if there are more than two vectors, it does not matter how we group them as we add them. Thus if we want to add vectors **a**, **b**, and **c**, we can add **a** and **b** first and then add their vector sum to **c**. On the other hand, we can add **b** and **c** first, and then add *that* sum to **a**. We get the same result either way. In equation form,

$$(\mathbf{a} + \mathbf{b}) + \mathbf{c} = \mathbf{a} + (\mathbf{b} + \mathbf{c}) \quad \text{(associative law).} \tag{3-3}$$

Figure 3-5 should convince you that Eq. 3-3 is correct.

FIGURE 3-3 (*a*) *AC* is the vector sum of the vectors *AB* and *BC*. (*b*) The same vectors relabeled.

FIGURE 3-4 The two vectors **a** and **b** can be added in either order; see Eq. 3-2.

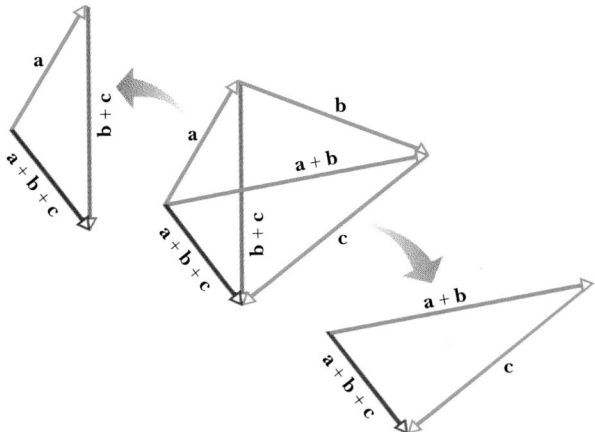

FIGURE 3-5 The three vectors **a**, **b**, and **c** can be grouped in any way as they are added; see Eq. 3-3.

The vector −**b** is a vector with the same magnitude as **b** but the opposite direction (see Fig. 3-6). If you try to add the two vectors in Fig. 3-6, you will see that

$$\mathbf{b} + (-\mathbf{b}) = 0.$$

Adding −**b** has the effect of subtracting **b**. We use this property to define the difference between two vectors: let **d** = **a** − **b**. Then

$$\mathbf{d} = \mathbf{a} - \mathbf{b} = \mathbf{a} + (-\mathbf{b}) \qquad \text{(subtraction).} \quad (3\text{-}4)$$

That is, we find the difference vector **d** by adding the vector −**b** to the vector **a**. Figure 3-7 shows how this is done graphically.

Remember, although we have used displacement vectors as a prototype, the rules for addition and subtraction hold for vectors of all kinds, whether they represent forces, velocities, or anything else. However, as in ordinary arithmetic, it is still true that we can add only vectors of the same kind. We can add two displacements, for example, or

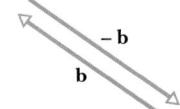

FIGURE 3-6 The vectors **b** and −**b**.

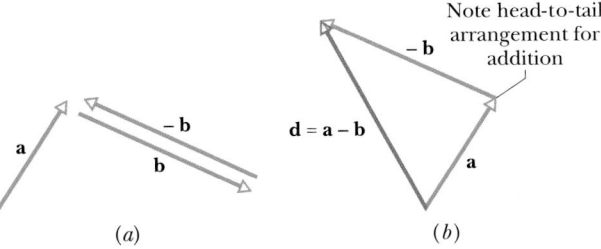

FIGURE 3-7 (*a*) Vectors **a**, **b**, and −**b**. (*b*) To subtract vector **b** from vector **a**, add vector −**b** to vector **a**.

two velocities, but it makes no sense to add a displacement and a velocity. In the world of scalars, that would be like trying to add 21 s and 12 m.

CHECKPOINT **1:** The magnitudes of displacements **a** and **b** are 3 m and 4 m, respectively, and **c** = **a** + **b**. Considering various orientations of **a** and **b**, what is (a) the maximum possible magnitude for **c** and (b) the minimum possible magnitude?

SAMPLE PROBLEM 3-2

In an orienteering class, you have the goal of moving as far (straight-line distance) from base camp as possible by making three straight-line moves. You may use the following displacements in any order: (a) **a**, 2.0 km due east; (b) **b**, 2.0 km 30° north of east; (c) **c**, 1.0 km due west. Or you may substitute either −**b** for **b** or −**c** for **c**. What is the greatest distance you can be from base camp at the end of the third displacement?

SOLUTION: Using a convenient scale, we draw vectors **a**, **b**, **c**, −**b**, and −**c** as in Fig. 3-8a. We then mentally slide the vectors over the page, connecting three of them in head-to-tail arrangements to find their vector sum **d**. The tail of the first vector represents base camp. The head of the third vector represents the point at which you stop. The vector sum **d** extends from the tail of the first vector to the head of the third vector. Its magnitude *d* is your distance from base camp.

We find that distance *d* is greatest for a head-to-tail arrangement of vectors **a**, **b**, and −**c**. They can be in any order, because their vector sum is the same for any order. The order shown in Fig. 3-8b is for the vector sum

$$\mathbf{d} = \mathbf{b} + \mathbf{a} + (-\mathbf{c}) = \mathbf{b} + \mathbf{a} - \mathbf{c}.$$

Using the scale given in Fig. 3-8a, we measure the length *d* of this vector sum, finding

$$d = 4.8 \text{ m.} \qquad \text{(Answer)}$$

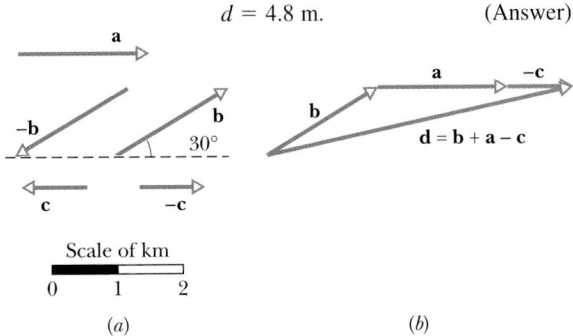

FIGURE 3-8 Sample Problem 3-2. (*a*) Displacement vectors; three are to be used. (*b*) Your distance from base camp is greatest if you undergo displacements **a**, **b**, and −**c**, in any order. One choice of order is shown; it gives vector sum **d** = **b** + **a** − **c**.

3-3 VECTORS AND THEIR COMPONENTS

Adding vectors graphically can be tedious. A neater and easier technique involves algebra but requires that the vectors be placed on a rectangular coordinate system. The x and y axes are usually drawn in the plane of the page, as in Fig. 3-9a. The z axis comes directly out of the page at the origin; we ignore it for now and deal only with two-dimensional vectors.

Vector **a** in Fig. 3-9 is in the xy plane. If we drop perpendicular lines from the ends of **a** to the coordinate axes, the quantities a_x and a_y so formed are called the **components** of the vector **a** in the x and y directions. The process of forming them is called **resolving the vector**. In general, a vector will have three components, although for the case of Fig. 3-9a the component along the z axis is zero. As Fig. 3-9b shows, if you move a vector in such a way that it remains parallel to its original direction at all times, the values of its components remain unchanged.

We can easily find the components of **a** in Fig. 3-9a from the right triangle there:

$$a_x = a \cos \theta \quad \text{and} \quad a_y = a \sin \theta, \quad (3\text{-}5)$$

where θ is the angle that the vector **a** makes with the direction of increasing x, and a is its magnitude. Figure 3-9c shows that the vector and its x and y components form a right triangle. It also shows how we can reconstruct a vector from its components: we arrange the components head to tail. Then we complete a right triangle with the vector being along the hypotenuse, from the tail of one component to the head of the other component.

Depending on the value of θ, the components of a vector may be positive, negative, or zero. In a figure, we use small, solid triangles as arrowheads to indicate the signs of the components, according to the usual convention: positive in the direction of increasing coordinate values, and negative in the opposite direction. Figure 3-10 shows a vector **b** for which b_y is negative and b_x is positive.

Once a vector has been resolved into its components, the components themselves can be used in place of the vector. For example, the two numbers a and θ specify the magnitude and direction of the two-dimensional vector **a** in Fig. 3-9a. But instead of a and θ, we can specify the vector with the two other numbers a_x and a_y. Both sets of numbers contain exactly the same information, and we can pass back and forth readily between the two descriptions. To obtain a and θ if we are given a_x and a_y, we note (see Fig. 3-9c) that

$$a = \sqrt{a_x^2 + a_y^2} \quad \text{and} \quad \tan \theta = \frac{a_y}{a_x}. \quad (3\text{-}6)$$

In solving problems, you may use *either* the a_x, a_y notation *or* the a, θ notation.

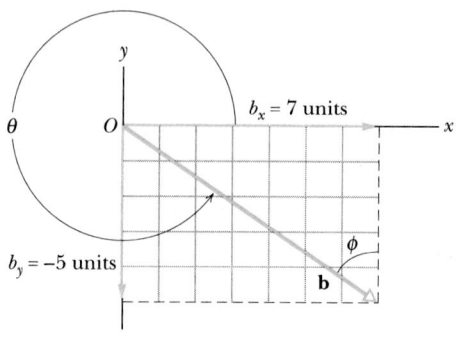

FIGURE 3-10   The components of **b** are positive on the x axis and negative on the y axis.

Cʜᴇᴄᴋᴘᴏɪɴᴛ **2:** In the figure, which of the ways indicated for combining the x and y components of vector **a** are proper to determine that vector?

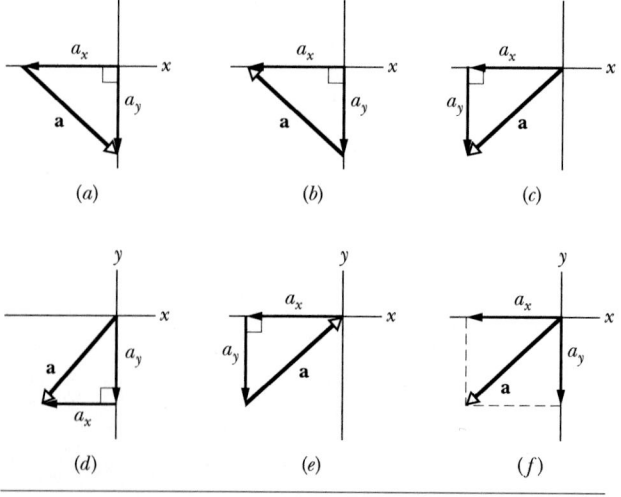

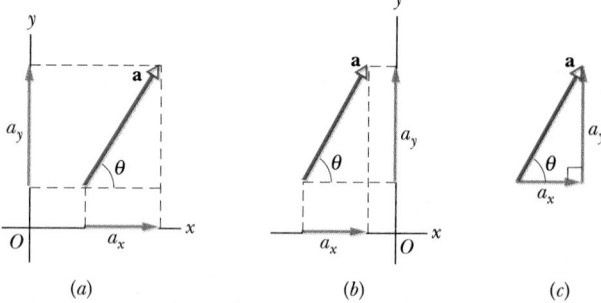

FIGURE 3-9   (a) The components of vector **a**. (b) The components are unchanged if the vector is shifted, as long as the magnitude and orientation are maintained. (c) The components form the legs of a right triangle whose hypotenuse is the magnitude of the vector.

SAMPLE PROBLEM 3-3

A small airplane leaves an airport on an overcast day and is later sighted 215 km away, in a direction making an angle of 22° east of north. How far east and north is the airplane from the airport when sighted?

SOLUTION: On an xy coordinate system the situation is as shown in Fig. 3-11, where for convenience the origin of the system has been placed at the airport. The airplane's displacement vector **d** points from the origin to where the airplane is sighted.

To answer the question, we find the components of **d**. With Eq. 3-5 and the angle $\theta = 68°$ ($= 90° - 22°$), we have

$$d_x = d \cos \theta = (215 \text{ km})(\cos 68°)$$
$$= 81 \text{ km} \qquad \text{(Answer)}$$

and
$$d_y = d \sin \theta = (215 \text{ km})(\sin 68°)$$
$$= 199 \text{ km}. \qquad \text{(Answer)}$$

So the airplane was spotted 199 km north and 81 km east of the airport.

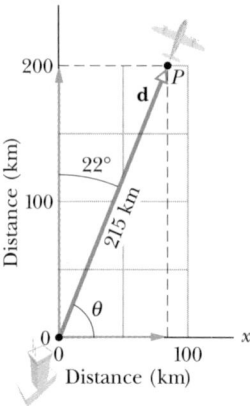

FIGURE 3-11 Sample Problem 3-3. A plane takes off from an airport at the origin and is later sighted at P.

PROBLEM SOLVING TACTICS

TACTIC 1: *Angles—Degrees and Radians*

Angles that are measured relative to the positive direction of the x axis are positive if they are measured in the counterclockwise direction, and negative if clockwise. For example, 210° and −150° are the same angle. Most calculators (try yours) will accept angles in either form when a trig function is taken.

Angles may be measured in degrees or radians (rad). You can relate the two measures by remembering that one full circle is equivalent to 360° and to 2π rad. So if you needed to convert, say, 40° to radians, you would write

$$40° \frac{2\pi \text{ rad}}{360°} = 0.70 \text{ rad}.$$

Is the answer reasonable? Quickly check by realizing that 40° is one-ninth of a full circle, and with a full circle equivalent to 2π rad, which is about 6.3 rad, the angle should be $\frac{1}{9}(6.3)$, which it is. Or, check by remembering that 1 rad $\approx 57°$.

Most calculators are in the degree mode when they are turned on, so that angles must be entered in degrees. However, you may be able to change yours to the radian mode. Check it before your first exam.

TACTIC 2: *Trig Functions*

You need to know the definitions of the common trig functions—sine, cosine, and tangent—because they are part of the language of science and engineering. They are given in Fig. 3-12 in a form that does not depend on how the triangle is labeled.

$$\sin \theta = \frac{\text{leg opposite } \theta}{\text{hypotenuse}}$$

$$\cos \theta = \frac{\text{leg adjacent to } \theta}{\text{hypotenuse}}$$

$$\tan \theta = \frac{\text{leg opposite } \theta}{\text{leg adjacent to } \theta}$$

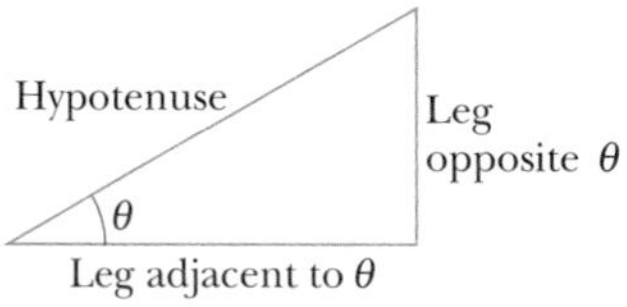

FIGURE 3-12 A triangle used to define the trigonometric functions. See also Appendix E.

You should also be able to sketch how the trig functions vary with angle, as in Fig. 3-13, in order to be able to judge whether a calculator result is reasonable. Even knowing the signs of the functions in the various quadrants can be of help.

TACTIC 3: *Inverse Trig Functions*

The most important of the inverse trig functions are $\sin^{-1}$, $\cos^{-1}$, and $\tan^{-1}$. When they are taken on a calculator, you must consider the reasonableness of the answer you get, because there is usually another possible answer that the calculator does not give. The range of operation for a calculator in taking each inverse trig function is indicated in Fig. 3-13. As an example, $\sin^{-1} 0.5$ has associated angles of 30° (which is displayed by the calculator, since 30° falls within its range of operation) and 150°. To see both angles, draw a horizontal line through 0.5 in Fig. 3-13a and note where it cuts the sine curve.

How do you distinguish a correct answer? As an example, reconsider the calculation of θ in Sample Problem 3-1, where $\tan \theta = 1.5$. Taking $\tan^{-1} 1.5$ on your calculator tells you that $\theta = 56°$, but $\theta = 236°$ ($= 180° + 56°$) also has a tangent of 1.5. Which is correct? From the physical situation (Fig. 3-2b), 56° is reasonable and 236° is clearly not.

TACTIC 4: *Measuring Vector Angles*

The equations for $\cos \theta$ and $\sin \theta$ in Eq. 3-5 and the equation for $\tan \theta$ in Eq. 3-6 are valid only if the angle is measured relative to the positive direction of the x axis. If it is measured relative to some other direction, then the trig functions in Eq. 3-5 may have to be interchanged, and the ratio in Eq. 3-6 may have to be inverted. A safer method is to convert the given angle into one that is measured from the positive direction of the x axis, as shown in Sample Problem 3-3.

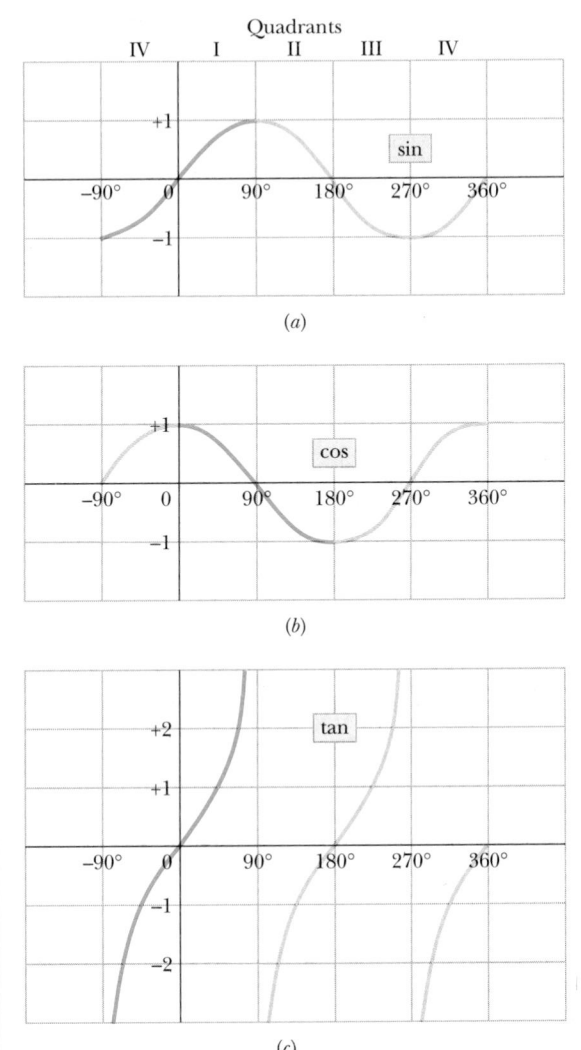

(a)

(b)

(c)

FIGURE 3-13 Three useful curves to remember. A calculator's range of operation for taking *inverse* trig functions is indicated by the darker portions of the colored curves.

3-4 UNIT VECTORS

A **unit vector** is a vector that has a magnitude of exactly 1 and points in a particular direction. It lacks both dimension and unit. Its sole purpose is to point, that is, to specify a direction. The unit vectors in the positive directions of the x, y, and z axes are labeled **i**, **j**, and **k**, as shown in Fig. 3-14.* The arrangement of axes in Fig. 3-14 is said to be a **right-handed coordinate system.** The system remains right-handed if it is rotated rigidly to a new orientation. We use such coordinate systems exclusively in this text.

*To distinguish handwritten unit vectors from other symbols, including other vectors, you might top them with a "hat": $\hat{i}, \hat{j}, \hat{k}$.

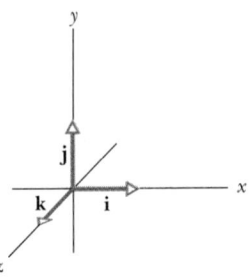

FIGURE 3-14 Unit vectors **i**, **j**, and **k** define the directions of a right-handed rectangular coordinate system.

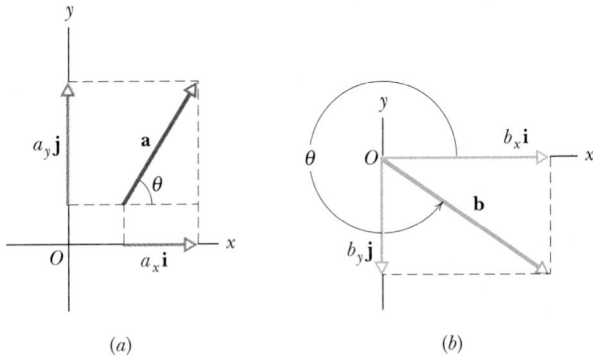

(a)   (b)

FIGURE 3-15 (a) The vector components of vector **a**. (b) The vector components of vector **b**.

Unit vectors are very useful for expressing other vectors; for example, we can express **a** and **b** of Figs. 3-9 and 3-10 as

$$\mathbf{a} = a_x\mathbf{i} + a_y\mathbf{j} \tag{3-7}$$

and

$$\mathbf{b} = b_x\mathbf{i} + b_y\mathbf{j}. \tag{3-8}$$

These two equations are illustrated in Fig. 3-15. The quantities $a_x\mathbf{i}$ and $a_y\mathbf{j}$ are the **vector components** of **a**, in contrast to a_x and a_y, which are its **scalar components** (or, as before, simply its **components**).

Look back for a moment at Sample Problem 3-1. If you superimpose the coordinate system of Fig. 3-14 at the Austin Entrance of Fig. 3-2a, with **i** eastward, **j** northward, and **k** upward, then the displacement **d** to Echo River is neatly expressed as

$$\mathbf{d} = -(2.6 \text{ km})\mathbf{i} - (3.9 \text{ km})\mathbf{j} + (0.025 \text{ km})\mathbf{k}.$$

3-5 ADDING VECTORS BY COMPONENTS

As we have noted, adding vectors graphically is tedious; it also has limited accuracy and is challenging in three dimensions. Here we find a more direct technique—adding vectors by combining their components, axis by axis.

To start, consider the statement

$$\mathbf{r} = \mathbf{a} + \mathbf{b}, \tag{3-9}$$

which says that the vector **r** is the same as the vector (**a** + **b**). If that is so, then each component of **r** must be the same as the corresponding component of (**a** + **b**):

$$r_x = a_x + b_x, \qquad (3\text{-}10)$$

$$r_y = a_y + b_y, \qquad (3\text{-}11)$$

$$r_z = a_z + b_z. \qquad (3\text{-}12)$$

In other words, two vectors are equal if their corresponding components are equal. Equations 3-10 to 3-12 tell us that to add vectors **a** and **b**, we must (1) resolve the vectors into their scalar components; (2) axis by axis, combine these scalar components to get the components of the sum **r**; and (3) if necessary, combine the components of **r** to get **r** itself. We have a choice in step 3. We can express **r** in unit-vector notation, or we can give the magnitude of **r** and its direction (by stating one angle when we are working in two dimensions or two angles when we are working in three dimensions).

CHECKPOINT **3:** (a) In the figure, what are the signs of the x components of $\mathbf{d}_1$ and $\mathbf{d}_2$? (b) What are the signs of the y components of $\mathbf{d}_1$ and $\mathbf{d}_2$? (c) What are the signs of the x and y components of $\mathbf{d}_1 + \mathbf{d}_2$?

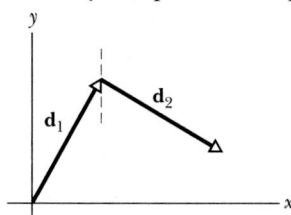

SAMPLE PROBLEM 3-4

In a road rally, you are given the following instructions: from the starting point use available roads to drive 36 km due east to checkpoint "Able," then 42 km due north to "Baker," and then 25 km northwest to "Charlie." (The roadway and checkpoints are shown in Fig. 3-16.)

(a) At "Charlie," what are the magnitude and orientation of your displacement **d** from the starting point?

SOLUTION: Figure 3-16 also shows a convenient orientation for an xy coordinate system, as well as vectors representing the three displacements you have undergone. The scalar components of **d** are

$$d_x = a_x + b_x + c_x = 36 \text{ km} + 0 + (25 \text{ km})(\cos 135°)$$
$$= (36 + 0 - 17.7) \text{ km} = 18.3 \text{ km}$$
and
$$d_y = a_y + b_y + c_y = 0 + 42 \text{ km} + (25 \text{ km})(\sin 135°)$$
$$= (0 + 42 + 17.7) \text{ km} = 59.7 \text{ km}.$$

Next, we use Eq. 3-6 to find the magnitude and direction of **d**:

$$d = \sqrt{d_x^2 + d_y^2} = \sqrt{(18.3 \text{ km})^2 + (59.7 \text{ km})^2}$$
$$= 62 \text{ km} \qquad \text{(Answer)}$$
and
$$\theta = \tan^{-1} \frac{d_y}{d_x} = \tan^{-1} \frac{59.7 \text{ km}}{18.3 \text{ km}} = 73°, \qquad \text{(Answer)}$$

where θ is the angle shown in Fig. 3-16.

(b) Write **d** in unit-vector notation.

SOLUTION: We simply write **d** as

$$\mathbf{d} = (x \text{ component})\mathbf{i} + (y \text{ component})\mathbf{j}$$
$$= (18.3 \text{ km})\mathbf{i} + (59.7 \text{ km})\mathbf{j}. \qquad \text{(Answer)}$$

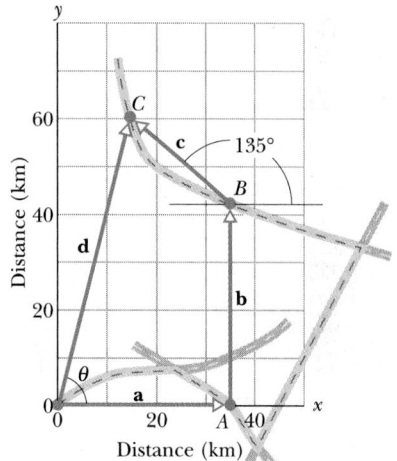

FIGURE 3-16 Sample Problem 3-4. A rally route, showing the origin, checkpoints Able (*A*), Baker (*B*), and Charlie (*C*), and the road network.

SAMPLE PROBLEM 3-5

Here are three vectors, each expressed in unit-vector notation:

$$\mathbf{a} = 4.2\mathbf{i} - 1.6\mathbf{j},$$
$$\mathbf{b} = -1.6\mathbf{i} + 2.9\mathbf{j},$$
$$\mathbf{c} = -3.7\mathbf{j}.$$

All three lie in the xy plane, and none of them has a z component. Find the vector **r** that is the sum of these three vectors. For convenience, the units have been omitted from these vector expressions; you may take them to be meters.

SOLUTION: From Eqs. 3-10 and 3-11 we have

$$r_x = a_x + b_x + c_x = 4.2 - 1.6 + 0 = 2.6$$
and
$$r_y = a_y + b_y + c_y = -1.6 + 2.9 - 3.7 = -2.4.$$

Thus

$$\mathbf{r} = 2.6\mathbf{i} - 2.4\mathbf{j}. \qquad \text{(Answer)}$$

Figure 3-17a shows the three vectors and their sum. Figure 3-17b shows **r** and its vector components.

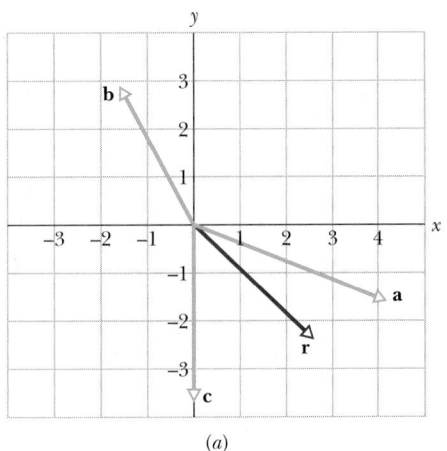

(a)

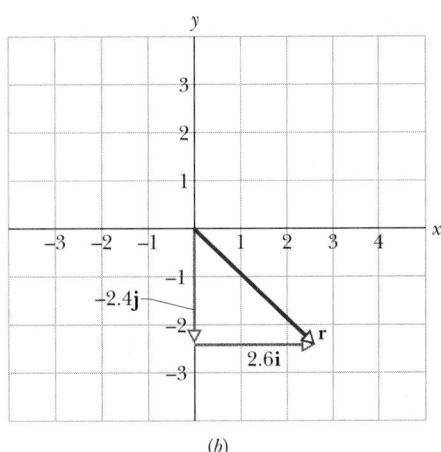

(b)

FIGURE 3-17 Sample Problem 3-5. Vector **r** is the vector sum of the other three vectors.

3-6 VECTORS AND THE LAWS OF PHYSICS

So far, in every figure that includes a coordinate system, the x and y axes are parallel to the edges of the book page. And so, when a vector **a** is included, its components a_x and a_y are also parallel to the edges (as in Fig. 3-18a). The only reason for that orientation of the axes is that it looks proper: there is no deeper reason. We could, instead, rotate the axes (but not the vector **a**) through an angle ϕ as in Fig. 3-18b, in which case the components would have new values, call them a'_x and a'_y. Since there are an infinite number of choices of ϕ, there are an infinite number of different pairs of components for **a**.

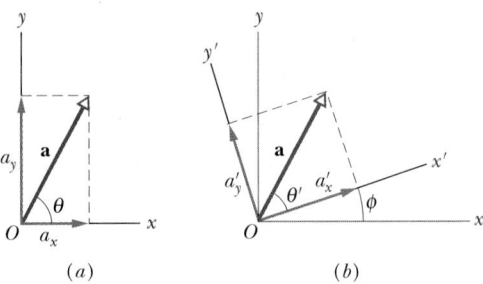

(a) (b)

FIGURE 3-18 (a) The vector **a** and its components. (b) The same vector, with the axes of the coordinate system rotated through an angle ϕ.

Which then is the "right" pair of components? The answer is that they are all equally valid because each pair (with its axes) just gives us a different way of describing the same vector **a**; all produce the same magnitude and direction for the same vector. In Fig. 3-18 we have

$$a = \sqrt{a_x^2 + a_y^2} = \sqrt{a_x'^2 + a_y'^2} \qquad (3\text{-}13)$$

and

$$\theta = \theta' + \phi. \qquad (3\text{-}14)$$

The point is that we have great freedom in choosing a coordinate system, because the relations among vectors (including, for example, the vector addition of Eq. 3-1) do not depend on the location of the origin of the coordinate system or on the orientation of the axes. This is also true of the relations of physics; they are all independent of the choice of coordinate system. Add to that the simplicity and richness of the language of vectors and you can see why the laws of physics are almost always presented in that language: one equation, like Eq. 3-9, can represent three (or even more) relations, like Eqs. 3-10, 3-11, and 3-12.

3-7 MULTIPLYING VECTORS*

There are three ways in which vectors can be multiplied. None is exactly like the usual algebraic multiplication.

Multiplying a Vector by a Scalar

If we multiply a vector **a** by a scalar s, we get a new vector. Its magnitude is the product of the magnitude of **a** and the absolute value of s. Its direction is the direction of **a** if s is positive, but the opposite direction if s is negative. To divide **a** by s, we multiply **a** by 1/s.

In either multiplication or division, the scalar may be a pure number or a physical quantity; in the latter case, the

---

*This material will not be employed until later (Chapter 7 for scalar products and Chapter 12 for vector products), and so your instructor may wish to postpone assignment of the section.

physical nature of the product differs from that of the original vector **a**.

The Scalar Product

There are two ways to multiply a vector by a vector: one way produces a scalar and the other produces a new vector. Students commonly confuse the two ways, and so starting now, you should carefully distinguish between them.

The **scalar product** of the vectors **a** and **b** in Fig. 3-19a is written as **a · b** and defined to be

$$\mathbf{a} \cdot \mathbf{b} = ab \cos \phi, \qquad (3\text{-}15)$$

where a is the magnitude of **a**, b is the magnitude of **b**, and ϕ is the angle* between **a** and **b**. Note that there are only scalars on the right side of Eq. 3-15 (including the value of $\cos \phi$). Thus **a · b** on the left side represents a *scalar* quantity; because of the notation, it is also known as the **dot product** and is spoken as "a dot b."

A dot product can be regarded as the product of two quantities: (1) the magnitude of one of the vectors and (2) the scalar component of the second vector along the direction of the first vector. For example, in Fig. 3-19b, **a** has a scalar component $a \cos \phi$ along the direction of **b**; note that a perpendicular dropped from the head of **a** to **b** determines that component. Similarly, **b** has a scalar component $b \cos \phi$ along the direction of **a**. If ϕ is 0°, the component of one vector along the other is maximum, and so also is

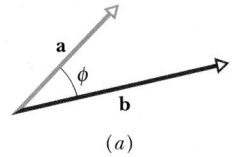

(a)

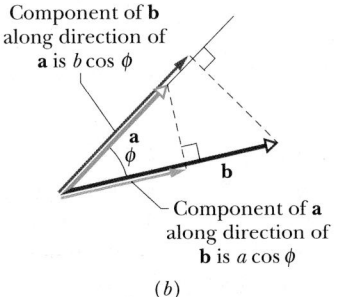

Component of **b** along direction of **a** is $b \cos \phi$

Component of **a** along direction of **b** is $a \cos \phi$

(b)

FIGURE 3-19 (a) Two vectors **a** and **b**, with an angle ϕ between them. (b) Each vector has a component along the direction of the other vector.

*In Fig. 3-19a, there are actually two angles between the vectors: ϕ and $360° - \phi$. Either can be used in Eq. 3-15, because their cosines are the same.

the dot product. If, instead, ϕ is 90°, the component of one vector along the other is zero, and so is the dot product.

Equation 3-15 can be rewritten as follows to emphasize the components:

$$\mathbf{a} \cdot \mathbf{b} = (a \cos \phi)(b) = (a)(b \cos \phi). \qquad (3\text{-}16)$$

The commutative law applies to a scalar product, so we can write

$$\mathbf{a} \cdot \mathbf{b} = \mathbf{b} \cdot \mathbf{a}.$$

When two vectors are in unit-vector notation, we write their dot product as

$$\mathbf{a} \cdot \mathbf{b} = (a_x\mathbf{i} + a_y\mathbf{j} + a_z\mathbf{k}) \cdot (b_x\mathbf{i} + b_y\mathbf{j} + b_z\mathbf{k}), \qquad (3\text{-}17)$$

for which the distributive law applies: each component of the first vector is to be dotted into each component of the second vector.

CHECKPOINT 4: Vectors **C** and **D** have magnitudes of 3 units and 4 units, respectively. What is the angle between **C** and **D** if **C · D** equals (a) zero, (b) 12 units, and (c) −12 units?

> **SAMPLE PROBLEM 3-6**
>
> What is the angle ϕ between $\mathbf{a} = 3.0\mathbf{i} - 4.0\mathbf{j}$ and $\mathbf{b} = -2.0\mathbf{i} + 3.0\mathbf{k}$?
>
> **SOLUTION:** By Eq. 3-15, the dot product is
>
> $$\mathbf{a} \cdot \mathbf{b} = ab \cos \phi = \sqrt{3.0^2 + 4.0^2}\,\sqrt{2.0^2 + 3.0^2}\,\cos \phi$$
> $$= 18.0 \cos \phi. \qquad (3\text{-}18)$$
>
> We next find the dot product with Eq. 3-17:
>
> $$\mathbf{a} \cdot \mathbf{b} = (3.0\mathbf{i} - 4.0\mathbf{j}) \cdot (-2.0\mathbf{i} + 3.0\mathbf{k}).$$
>
> With the distributive law, this yields
>
> $$\mathbf{a} \cdot \mathbf{b} = (3.0\mathbf{i}) \cdot (-2.0\mathbf{i}) + (3.0\mathbf{i}) \cdot (3.0\mathbf{k})$$
> $$+ (-4.0\mathbf{j}) \cdot (-2.0\mathbf{i}) + (-4.0\mathbf{j}) \cdot (3.0\mathbf{k}).$$
>
> We next apply Eq. 3-15 to each term. The angle between vectors for the first term is 0°, and for the other three terms it is 90°. We then have
>
> $$\mathbf{a} \cdot \mathbf{b} = -(6.0)(1) + (9.0)(0) + (8.0)(0) - (12)(0)$$
> $$= -6.0. \qquad (3\text{-}19)$$
>
> Setting the results of Eqs. 3-18 and 3-19 equal to each other, we find
>
> $$18.0 \cos \phi = -6.0$$
>
> or
>
> $$\phi = \cos^{-1} \frac{-6.0}{18.0} = 109° \approx 110°. \qquad \text{(Answer)}$$

The Vector Product

The **vector product** of **a** and **b**, written **a** × **b**, produces a third vector **c** whose magnitude is

$$c = ab \sin \phi, \qquad (3\text{-}20)$$

where ϕ is the *smaller* of the two angles between **a** and **b**.* Because of the notation, **a** × **b** is also known as the **cross product** and in speech it is "a cross b." If **a** and **b** are parallel or antiparallel, **a** × **b** = 0. The magnitude of **a** × **b**, which can be written as |**a** × **b**|, is maximum when **a** and **b** are perpendicular to each other.

The direction of **c** is perpendicular to the plane that contains **a** and **b**. Figure 3-20a shows how to determine the direction of **c** = **a** × **b** with what is known as the **right-hand rule**. Place the vectors **a** and **b** tail to tail without altering their orientations and imagine a line that is perpendicular to their plane where they meet. Pretend to grasp the line with your *right* hand in such a way that your fingers would sweep **a** into **b** through the smaller angle between them. Your outstretched thumb points in the direction of **c**.

The order of the vector multiplication is important. In Fig. 3-20b, we are determining the direction of **c'** = **b** × **a**, so the fingers are placed to sweep **b** into **a** through the smaller angle. The thumb ends up in the opposite direction from previously, and so it must be that **c'** = −**c**; that is,

$$\mathbf{b} \times \mathbf{a} = -(\mathbf{a} \times \mathbf{b}). \qquad (3\text{-}21)$$

In other words, the commutative law does not apply to a vector product.

In unit-vector notation, we write

$$\mathbf{a} \times \mathbf{b} = (a_x\mathbf{i} + a_y\mathbf{j} + a_z\mathbf{k})$$
$$\times (b_x\mathbf{i} + b_y\mathbf{j} + b_z\mathbf{k}), \qquad (3\text{-}22)$$

and the distributive law applies: each component of the first vector is to be crossed into each component of the second vector. You should use that law to expand Eq. 3-22; check your result against the expansion given in Appendix E. You can also evaluate a cross product by setting up and evaluating a determinant, as shown in Appendix E.

We can use a cross product of unit vectors to see if any given rectangular coordinate system is a right-handed coordinate system: if **i** × **j** = **k**, it is. Check the system shown in Fig. 3-14.

$\mathbb{C}$HECKPOINT 5: Vectors **C** and **D** have magnitudes of 3 units and 4 units, respectively. What is the angle between **C** and **D** if the magnitude of the vector product **C** × **D** equals (a) zero and (b) 12 units?

---

*You must use the smaller of the two angles between the vectors because $\sin \phi$ and $\sin(360° - \phi)$ differ in algebraic sign.

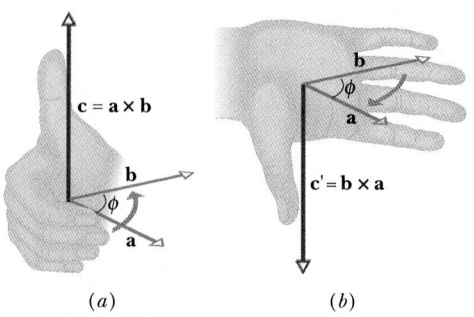

(a)   (b)

FIGURE 3-20 Illustration of the right-hand rule for vector products. (a) Swing vector **a** into vector **b** with the fingers of your right hand. Your thumb shows the direction of vector **c** = **a** × **b**. (b) Showing that (**a** × **b**) = −(**b** × **a**).

SAMPLE PROBLEM 3-7

Vector **a** lies in the xy plane in Fig. 3-21. It has a magnitude of 18 units and points in a direction 250° from the direction of increasing x. Vector **b** has a magnitude of 12 units and points along the direction of increasing z.

(a) What is the scalar product of these two vectors?

SOLUTION: The angle ϕ between these two vectors is 90° so that, from Eq. 3-15,

$$\mathbf{a} \cdot \mathbf{b} = ab \cos \phi = (18)(12)(\cos 90°) = 0. \quad \text{(Answer)}$$

The scalar product of any two vectors that are at right angles to each other is zero. This is consistent with the fact that neither of these vectors has a component along the direction of the other vector.

(b) What is the vector product **c** = **a** × **b**?

SOLUTION: The magnitude of the vector product is, from Eq. 3-20,

$$ab \sin \phi = (18)(12)(\sin 90°) = 216. \quad \text{(Answer)}$$

The direction of **c** is perpendicular to the plane formed by **a** and **b**. It must then be perpendicular to **b**, which means that it is perpendicular to the z axis. This in turn means that **c** must lie in the xy plane. The right-hand rule of Fig. 3-20 shows that **c** points as shown in Fig. 3-21. Because **c** is also perpendicular to **a**, the direction of **c** makes an angle of 250° − 90° = 160° with the direction of increasing x.

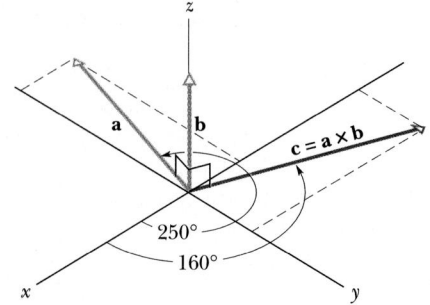

FIGURE 3-21 Sample Problem 3-7. A problem in vector multiplication.

SAMPLE PROBLEM 3-8

If $\mathbf{c} = \mathbf{a} \times \mathbf{b}$, where $\mathbf{a} = 3\mathbf{i} - 4\mathbf{j}$ and $\mathbf{b} = -2\mathbf{i} + 3\mathbf{k}$, what is $\mathbf{c}$?

SOLUTION: From Eq. 3-22 we have

$$\mathbf{c} = (3\mathbf{i} - 4\mathbf{j}) \times (-2\mathbf{i} + 3\mathbf{k}),$$

which, by the distributive law, becomes

$$\mathbf{c} = -(3\mathbf{i} \times 2\mathbf{i}) + (3\mathbf{i} \times 3\mathbf{k}) + (4\mathbf{j} \times 2\mathbf{i}) - (4\mathbf{j} \times 3\mathbf{k}).$$

We next evaluate each term with Eq. 3-20, determining the direction with the right-hand rule. (For the first term, $\phi = 0°$; for the other terms, $\phi = 90°$.) We find

$$\mathbf{c} = 0 - 9\mathbf{j} - 8\mathbf{k} - 12\mathbf{i} = -12\mathbf{i} - 9\mathbf{j} - 8\mathbf{k}. \quad \text{(Answer)}$$

The vector $\mathbf{c}$ is perpendicular to both $\mathbf{a}$ and $\mathbf{b}$, a fact that you can check by showing $\mathbf{c} \cdot \mathbf{a} = 0$ and $\mathbf{c} \cdot \mathbf{b} = 0$; that is, there is no component of $\mathbf{c}$ along the direction of either $\mathbf{a}$ or $\mathbf{b}$.

PROBLEM SOLVING TACTICS

TACTIC 5: *Common Errors with Cross Products*

Several errors are common in finding a cross product. (1) Failure to arrange vectors tail to tail is tempting when an illustration presents them head to tail: you must mentally shift (or better, redraw) one vector to the proper arrangement without changing its orientation. (2) Failing to use the right hand in applying the right-hand rule is easy when the right hand is occupied with a calculator or pencil. (3) Failure to sweep the first vector of the product into the second vector can occur when the orientations of the vectors require an awkward twisting of your hand to apply the right-hand rule. And sometimes that happens when you try to make the sweep mentally rather than actually using your hand. (4) Failure to work with a right-handed coordinate system results when you forget how to draw such a system (see Fig. 3-14).

REVIEW & SUMMARY

Scalars and Vectors

Scalars, such as temperature, have magnitude only. They are specified by a number with a unit (82°F) and obey the rules of arithmetic and ordinary algebra. *Vectors,* such as displacement, have both magnitude and direction (5 m, north) and obey the special rules of vector algebra.

Adding Vectors Geometrically

Two vectors $\mathbf{a}$ and $\mathbf{b}$ may be added geometrically by drawing them to a common scale and placing them head to tail. The vector connecting the tail of the first to the head of the second is the sum vector $\mathbf{s}$, as Fig. 3-3 shows. To subtract $\mathbf{b}$ from $\mathbf{a}$, reverse the direction of $\mathbf{b}$ to get $-\mathbf{b}$; then add $-\mathbf{b}$ to $\mathbf{a}$: see Fig. 3-7. Vector addition is commutative and obeys the associative law.

Components of a Vector

The (scalar) *components* a_x and a_y of any two-dimensional vector $\mathbf{a}$ are found by dropping perpendicular lines from the ends of $\mathbf{a}$ onto the coordinate axes. The components are given by

$$a_x = a \cos \theta \quad \text{and} \quad a_y = a \sin \theta, \quad (3\text{-}5)$$

where θ is the angle from the positive direction of the x axis to the direction of $\mathbf{a}$. The algebraic sign of a component indicates its direction along the associated axis. Given the components, we can reconstruct the vector from

$$a = \sqrt{a_x^2 + a_y^2} \quad \text{and} \quad \tan \theta = \frac{a_y}{a_x}. \quad (3\text{-}6)$$

Unit-Vector Notation

Often we find it useful to introduce *unit vectors* $\mathbf{i}$, $\mathbf{j}$, and $\mathbf{k}$, whose magnitudes are unity and whose directions are those of the x, y,

and z axes, respectively, in a right-handed coordinate system. We can write a vector $\mathbf{a}$ in terms of unit vectors as

$$\mathbf{a} = a_x\mathbf{i} + a_y\mathbf{j} + a_z\mathbf{k}, \quad (3\text{-}7)$$

in which $a_x\mathbf{i}$, $a_y\mathbf{j}$, and $a_z\mathbf{k}$ are the **vector components** and a_x, a_y, and a_z are the **scalar components** of $\mathbf{a}$.

Adding Vectors in Component Form

To add vectors in component form, we use the rules

$$r_x = a_x + b_x; \quad r_y = a_y + b_y; \quad r_z = a_z + b_z. \quad (3\text{-}10 \text{ to } 3\text{-}12)$$

Here $\mathbf{a}$ and $\mathbf{b}$ are the vectors to be added, and $\mathbf{r}$ is the vector sum.

Vectors and Physical Laws

Any physical situation involving vectors can be described using many possible coordinate systems. We usually choose the one that simplifies the work. However, the relation between the vector quantities does not depend on our choice. The laws of physics are also independent of our choice of coordinates.

Scalar Times Vector

The product of a scalar s and a vector $\mathbf{v}$ is a new vector whose magnitude is sv and whose direction is the same as that of $\mathbf{v}$ if s is positive, and opposite that of $\mathbf{v}$ if s is negative. To divide $\mathbf{v}$ by s, multiply $\mathbf{v}$ by $(1/s)$.

The Scalar Product

The scalar (or **dot**) product of two vectors $\mathbf{a}$ and $\mathbf{b}$ is written $\mathbf{a} \cdot \mathbf{b}$ and is the *scalar* quantity given by

$$\mathbf{a} \cdot \mathbf{b} = ab \cos \phi, \quad (3\text{-}15)$$

in which ϕ is the angle between the directions of **a** and **b**. The scalar product may be positive, zero, or negative, depending on the value of ϕ. Figure 3-19*b* indicates that a scalar product is the product of the magnitude of one vector and the component of the second vector along the direction of the first vector. In unit-vector notation, we have

$$\mathbf{a} \cdot \mathbf{b} = (a_x\mathbf{i} + a_y\mathbf{j} + a_z\mathbf{k}) \cdot (b_x\mathbf{i} + b_y\mathbf{j} + b_z\mathbf{k}), \quad (3\text{-}17)$$

which obeys the distributive law. Note that $\mathbf{a} \cdot \mathbf{b} = \mathbf{b} \cdot \mathbf{a}$.

The Vector Product

The vector (or **cross**) product of two vectors **a** and **b** is written

$\mathbf{a} \times \mathbf{b}$ and is a *vector* **c** whose magnitude c is given by

$$c = ab \sin \phi, \quad (3\text{-}20)$$

in which ϕ is the smaller of the angles between the directions of **a** and **b**. The direction of **c** is perpendicular to the plane defined by **a** and **b** and is given by a right-hand rule, as shown in Fig. 3-20. Note that $\mathbf{a} \times \mathbf{b} = -(\mathbf{b} \times \mathbf{a})$. In unit-vector notation we have

$$\mathbf{a} \times \mathbf{b} = (a_x\mathbf{i} + a_y\mathbf{j} + a_z\mathbf{k}) \times (b_x\mathbf{i} + b_y\mathbf{j} + b_z\mathbf{k}), \quad (3\text{-}22)$$

to which we may apply the distributive law.

QUESTIONS

1. Displacement vector **N** points from coordinates (5 m, 3 m) to coordinates (7 m, 6 m) in the *xy* plane. Which of the following displacement vectors are equivalent to **N**: vector **A**, which points from (−6 m, −5 m) to (−4 m, −2 m); vector **B**, which points from (−6 m, 1 m) to (−4 m, 4 m); vector **C**, which points from (−8 m, −6 m) to (−10 m, −9 m)?

2. If $\mathbf{d} = \mathbf{a} + \mathbf{b} + (-\mathbf{c})$, does (a) $\mathbf{a} + (-\mathbf{d}) = \mathbf{c} + (-\mathbf{b})$, (b) $\mathbf{a} = (-\mathbf{b}) + \mathbf{d} + \mathbf{c}$, and (c) $\mathbf{c} + (-\mathbf{d}) = \mathbf{a} + \mathbf{b}$?

3. Equation 3-2 shows that the addition of two vectors **a** and **b** is commutative. Does that mean subtraction is commutative, so that $\mathbf{a} - \mathbf{b} = \mathbf{b} - \mathbf{a}$?

4. In a game within a three-dimensional maze, you must move your game piece from *start,* at *xyz* coordinates (0, 0, 0), to *finish,* at coordinates (−2 cm, 4 cm, −4 cm). The game piece can undergo only the displacements (in centimeters) given below. If, along the way, the game piece lands at coordinates (−5 cm, −1 cm, −1 cm) or (5 cm, 2 cm, −1 cm), you lose the game. Which vectors and in what sequence will get your game piece to *finish*?

$$\mathbf{p} = -7\mathbf{i} + 2\mathbf{j} - 3\mathbf{k} \quad \mathbf{r} = 2\mathbf{i} - 3\mathbf{j} + 2\mathbf{k}$$
$$\mathbf{q} = 2\mathbf{i} - \mathbf{j} + 4\mathbf{k} \quad \mathbf{s} = 3\mathbf{i} + 5\mathbf{j} - 3\mathbf{k}.$$

5. Describe two vectors **a** and **b** such that
(a) $\mathbf{a} + \mathbf{b} = \mathbf{c}$ and $a + b = c$;
(b) $\mathbf{a} + \mathbf{b} = \mathbf{a} - \mathbf{b}$;
(c) $\mathbf{a} + \mathbf{b} = \mathbf{c}$ and $a^2 + b^2 = c^2$.

6. (a) Can a vector have zero magnitude if one of its components is not zero? (b) Can two vectors having different magnitudes be combined to give a vector sum of zero? Can three vectors give a vector sum of zero if they (c) are in the same plane, and (d) are not in the same plane?

7. Can the magnitude of the difference between two vectors ever be greater than (a) the magnitude of one of the vectors, (b) the magnitudes of both vectors, and (c) the magnitude of their sum?

8. Can the sum of the magnitudes of two vectors ever be equal to the magnitude of the sum of the same two vectors? If no, why not? If yes, when?

9. Which of the arrangements of axes in Fig. 3-22 can be labeled "right-handed coordinate system"? As usual, each axis label indicates the positive side of the axis.

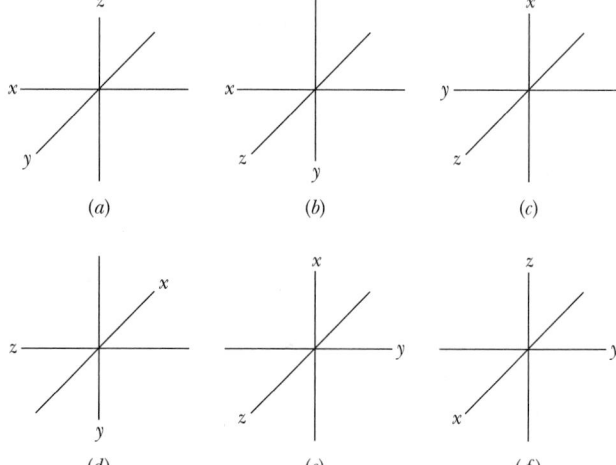

FIGURE 3-22 Question 9.

10. The *x* and *y* components of four vectors, **a, b, c,** and **d** are given below. For which vectors will your calculator give you the correct angle θ when you use it to find θ with Eq. 3-6? Answer first by examining Fig. 3-13, and then check your answers with your calculator.

$$a_x = 3 \quad a_y = 3 \quad c_x = -3 \quad c_y = -3$$
$$b_x = -3 \quad b_y = 3 \quad d_x = 3 \quad d_y = -3.$$

11. In Fig. 3-23, what are the signs of the *x* and *y* components of (a) **A**, and (b) **D**, where $\mathbf{D} = \mathbf{A} - \mathbf{B}$?

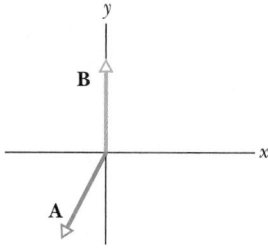

FIGURE 3-23 Question 11.

12. Figure 3-24 shows vector **A** and four choices for vector **B** that differ only in orientation. Rank the choices according to the absolute value of **A·B** that they give, greatest first.

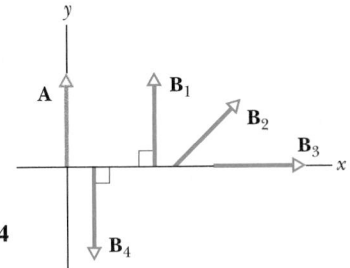

FIGURE 3-24
Question 12.

13. Figure 3-25 shows vector **A** and four other vectors that have the same magnitude but differ in orientation. (a) Which of those other four vectors have the same dot product with **A**? (b) Which have a negative dot product with **A**?

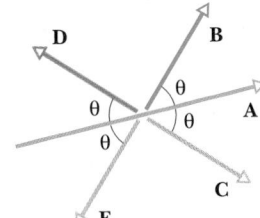

FIGURE 3-25 Question 13.

14. If **A** = 2**i** + 4**j** and **B** = 7**k**, what is **A·B**?

15. If **a·b** = **a·c**, must **b** equal **c**?

16. In Fig. 3-26, what is the direction of (a) **E** × **F**, (b) **F** × **E**, and (c) **G** × **E**, where **G** is in the *xy* plane? (d) Does the answer to (c) change if **G** is shifted parallel to the *z* axis without any change in its direction?

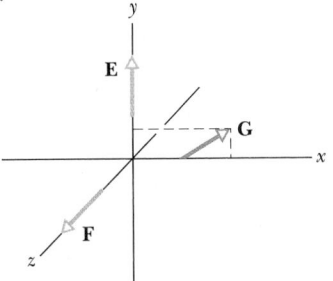

FIGURE 3-26 Question 16.

17. If **A** = 2**i** + 4**j**, what is **A** × **B** when (a) **B** = 8**i** + 16**j** and (b) **B** = − 8**i** − 16**j**? (This question can be answered without computation.)

EXERCISES & PROBLEMS

SECTION 3-2 Adding Vectors: Graphical Method

1E. Consider two displacements, one of magnitude 3 m and another of magnitude 4 m. Show how the displacement vectors may be combined to get a resultant displacement of magnitude (a) 7 m, (b) 1 m, and (c) 5 m.

2E. A woman walks 250 m in the direction 30° east of north, then 175 m directly east. (a) Using graphical methods, find her final displacement from the starting point. (b) Compare the magnitude of her displacement with the distance she walked.

3E. A person walks in the following pattern: 3.1 km north, then 2.4 km west, and finally 5.2 km south. (a) Construct the vector diagram that represents this motion. (b) How far and in what direction would a bird fly in a straight line from the same starting point to the same final point?

4E. A car is driven east for a distance of 50 km, then north for 30 km, and then in a direction 30° east of north for 25 km. Draw the vector diagram and determine the total displacement of the car from its starting point.

5P. Vector **a** has a magnitude of 5.0 units and is directed east. Vector **b** is directed 35° west of north and has a magnitude of 4.0 units. Construct vector diagrams for calculating **a** + **b** and **b** − **a**. Estimate the magnitudes and directions of **a** + **b** and **b** − **a** from your diagrams.

6P. A bank in downtown Boston is robbed (see the map in Fig. 3-27). To elude police, the robbers escape by helicopter, making three successive flights described by the following displacements: 20 miles, 45° south of east; 33 miles, 26° north of west; 16

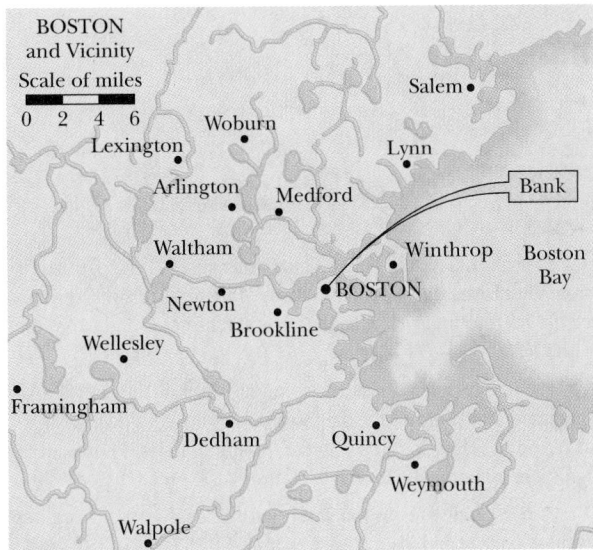

FIGURE 3-27 Problem 6.

miles, 18° east of south. At the end of the third flight they are captured. In what town are they apprehended? (Use the graphical method to add these displacements on the map.)

7P. Three vectors **a**, **b**, and **c**, each having a magnitude of 50 units, lie in the *xy* plane and make angles of 30°, 195°, and 315° with the positive *x* axis, respectively. Find graphically the magnitudes and directions of the vectors (a) **a** + **b** + **c**, (b) **a** − **b** + **c**, and (c) a vector **d** such that (**a** + **b**) − (**c** + **d**) = 0.

SECTION 3-3   Vectors and Their Components

8E. (a) Express the following angles in radians: 20.0°, 50.0°, 100°. (b) Convert the following angles to degrees: 0.330 rad, 2.10 rad, 7.70 rad.

9E. What are the *x* and *y* components of a vector **a** in the *xy* plane if its direction is 250° counterclockwise from the positive *x* axis and its magnitude is 7.3 units?

10E. The *x* component of a certain vector is −25.0 units and the *y* component is +40.0 units. (a) What is the magnitude of the vector? (b) What is the angle between the direction of the vector and the positive direction of *x*?

11E. A displacement vector **r** in the *xy* plane is 15 m long and directed as shown in Fig. 3-28. Determine the *x* and *y* components of the vector.

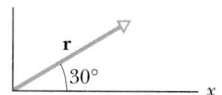

FIGURE 3-28
Exercise 11.

12E. A heavy piece of machinery is raised by sliding it 12.5 m along a plank oriented at 20.0° to the horizontal, as shown in Fig. 3-29. (a) How high above its original position is it raised? (b) How far is it moved horizontally?

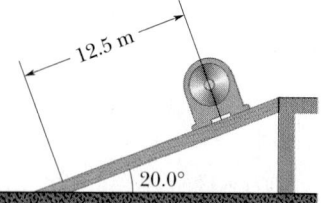

FIGURE 3-29
Exercise 12.

13E. The minute hand of a wall clock measures 10 cm from axis to tip. What is the displacement vector of its tip (a) from a quarter after the hour to half past, (b) in the next half hour, and (c) in the next hour?

14E. A ship sets out to sail to a point 120 km due north. An unexpected storm blows the ship to a point 100 km due east of its starting point. How far, and in what direction, must it now sail to reach its original destination?

15P. A person desires to reach a point that is 3.40 km from her present location and in a direction that is 35.0° north of east. However, she must travel along streets that are oriented either

north–south or east–west. What is the minimum distance she could travel to reach her destination?

16P. Rock *faults* are ruptures along which opposite faces of rock have slid past each other. In Fig. 3-30, points *A* and *B* coincided before the rock in the foreground slid down to the right. The net displacement *AB* is along the plane of the fault. The horizontal component of *AB* is the *strike-slip AC*. The component of *AB* that is directly down the plane of the fault is the *dip-slip AD*. (a) What is the net displacement *AB* if the strike-slip is 22.0 m and the dip-slip is 17.0 m? (b) If the plane of the fault is inclined 52.0° to the horizontal, what is the vertical component of *AB*?

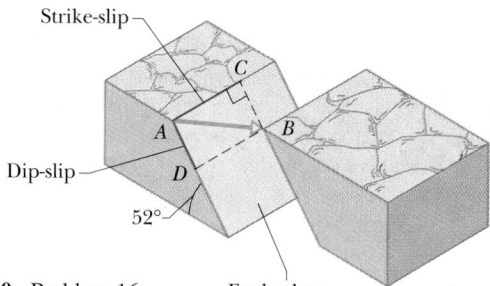

FIGURE 3-30   Problem 16.   Fault plane

17P. A wheel with a radius of 45.0 cm rolls without slipping along a horizontal floor (Fig. 3-31). At time t_1, the dot *P* painted on the rim of the wheel is at the point of contact between the wheel and the floor. At a later time t_2, the wheel has rolled through one-half of a revolution. What is the displacement of *P* during this interval?

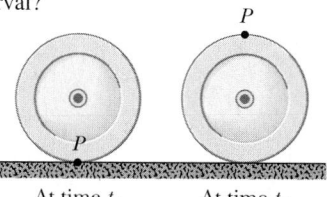

At time t_1 At time t_2

FIGURE 3-31   Problem 17.

18P. A room has dimensions 10.0 ft (height) × 12.0 ft × 14.0 ft. A fly starting at one corner flies around, ending up at the diagonally opposite corner. (a) What is the magnitude of its displacement? (b) Could the length of its path be less than this distance? Greater than this distance? Equal to this distance? (c) Choose a suitable coordinate system and find the components of the displacement vector in that system. (d) If the fly walks rather than flies, what is the length of the shortest path it can take? (This can be answered without calculus.)

SECTION 3-5   Adding Vectors by Components

19E. Find the vector components of the sum **r** of the vector displacements **c** and **d** whose components in meters along three perpendicular directions are $c_x = 7.4$, $c_y = -3.8$, $c_z = -6.1$; $d_x = 4.4$, $d_y = -2.0$, $d_z = 3.3$.

20E. (a) What is the sum in unit-vector notation of the two vectors **a** = 4.0**i** + 3.0**j** and **b** = −13**i** + 7.0**j**? (b) What are the magnitude and direction of **a** + **b**?

21E. Calculate the x and y components, magnitudes, and directions of (a) $\mathbf{a} + \mathbf{b}$ and (b) $\mathbf{b} - \mathbf{a}$ if $\mathbf{a} = 3.0\mathbf{i} + 4.0\mathbf{j}$ and $\mathbf{b} = 5.0\mathbf{i} - 2.0\mathbf{j}$.

22E. Two vectors are given by $\mathbf{a} = 4\mathbf{i} - 3\mathbf{j} + \mathbf{k}$ and $\mathbf{b} = -\mathbf{i} + \mathbf{j} + 4\mathbf{k}$. Find (a) $\mathbf{a} + \mathbf{b}$, (b) $\mathbf{a} - \mathbf{b}$, and (c) a vector $\mathbf{c}$ such that $\mathbf{a} - \mathbf{b} + \mathbf{c} = 0$.

23E. Given two vectors $\mathbf{a} = 4.0\mathbf{i} - 3.0\mathbf{j}$ and $\mathbf{b} = 6.0\mathbf{i} + 8.0\mathbf{j}$, find the magnitudes and directions of (a) $\mathbf{a}$, (b) $\mathbf{b}$, (c) $\mathbf{a} + \mathbf{b}$, (d) $\mathbf{b} - \mathbf{a}$, and (e) $\mathbf{a} - \mathbf{b}$. How do the orientations of the last two compare?

24P. If $\mathbf{a} - \mathbf{b} = 2\mathbf{c}$, $\mathbf{a} + \mathbf{b} = 4\mathbf{c}$, and $\mathbf{c} = 3\mathbf{i} + 4\mathbf{j}$, then what are $\mathbf{a}$ and $\mathbf{b}$?

25P. If vector $\mathbf{B}$ is added to vector $\mathbf{A}$, the result is $6.0\mathbf{i} + 1.0\mathbf{j}$. If $\mathbf{B}$ is subtracted from $\mathbf{A}$, the result is $-4.0\mathbf{i} + 7.0\mathbf{j}$. What is the magnitude of $\mathbf{A}$?

26P. A vector $\mathbf{B}$, when added to the vector $\mathbf{C} = 3.0\mathbf{i} + 4.0\mathbf{j}$, yields a resultant vector that is in the positive y direction and has a magnitude equal to that of $\mathbf{C}$. What is the magnitude of $\mathbf{B}$?

27P. Two vectors $\mathbf{a}$ and $\mathbf{b}$ have equal magnitudes of 10.0 units. They are oriented as shown in Fig. 3-32 and their vector sum is $\mathbf{r}$. Find (a) the x and y components of $\mathbf{r}$, (b) the magnitude of $\mathbf{r}$, and (c) the angle $\mathbf{r}$ makes with the positive x axis.

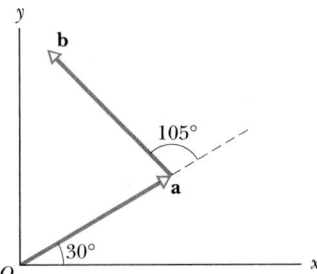

FIGURE 3-32 Problem 27.

28P. A golfer takes three putts to get the ball into the hole. The first putt displaces the ball 12 ft north, the second 6.0 ft southeast, and the third 3.0 ft southwest. What displacement was needed to get the ball into the hole on the first putt?

29P. A radar station detects an airplane approaching directly from the east. At first observation, the range to the plane is 1200 ft at 40° above the horizon. The airplane is tracked for another 123° in the vertical east–west plane, the range at final contact being 2580 ft. See Fig. 3-33. Find the displacement of the airplane during the period of observation.

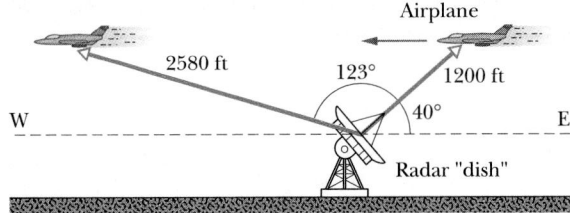

FIGURE 3-33 Problem 29.

30P. (a) A man leaves his front door, walks 1000 m east, 2000 m north, and then takes a penny from his pocket and drops it from a cliff 500 m high. Set up a coordinate system and write down an expression, using unit vectors, for the displacement of the penny from the house to its landing point. (b) The man then returns to his front door, following a different path. What is his resultant displacement for the round trip?

31P. A particle undergoes three successive displacements in a plane, as follows: 4.00 m southwest, 5.00 m east, and 6.00 m in a direction 60.0° north of east. Choose the y axis pointing north and the x axis pointing east and find (a) the components of each displacement, (b) the components of the resultant displacement, (c) the magnitude and direction of the resultant displacement, and (d) the displacement that would be required to bring the particle back to the starting point.

32P. Prove that two vectors must have equal magnitudes if their sum is perpendicular to their difference.

33P. Two vectors of lengths a and b make an angle θ with each other when placed tail to tail. Prove, by taking components along two perpendicular axes, that the length of their sum is

$$r = \sqrt{a^2 + b^2 + 2ab\cos\theta}.$$

34P. (a) Using unit vectors, write expressions for the body diagonals (the straight lines from one corner to another through the center) of a cube in terms of its edges, which have length a. (b) Determine the angles that the body diagonals make with the adjacent edges. (c) Determine the length of the body diagonals.

SECTION 3-6 Vectors and the Laws of Physics

35E. A vector $\mathbf{a}$ with a magnitude of 17.0 m is directed 56.0° counterclockwise from the $+x$ axis, as shown in Fig. 3-34. (a) What are the components a_x and a_y of the vector? (b) A second coordinate system is inclined by 18.0° with respect to the first. What are the components a_x' and a_y' in this primed coordinate system?

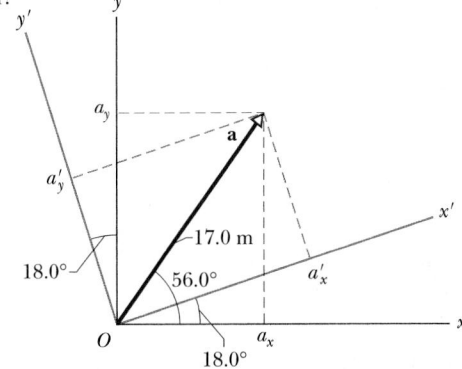

FIGURE 3-34 Exercise 35.

SECTION 3-7 Multiplying Vectors

36E. A vector $\mathbf{d}$ has a magnitude of 2.5 m and points north. What are the magnitudes and directions of the vectors (a) $4.0\mathbf{d}$ and (b) $-3.0\mathbf{d}$?

37E. Consider $\mathbf{a}$ in the positive direction of x, $\mathbf{b}$ in the positive direction of y, and a scalar d. What is the direction of $\mathbf{b}/d$ if d is (a) positive and (b) negative? What is the magnitude of (c) $\mathbf{a} \cdot \mathbf{b}$ and

(d) $\mathbf{a} \cdot \mathbf{b}/d$? What is the direction of (e) $\mathbf{a} \times \mathbf{b}$ and (f) $\mathbf{b} \times \mathbf{a}$? (g) What are the magnitudes of the cross products in (e) and (f)? (h) What are the magnitude and direction of $\mathbf{a} \times \mathbf{b}/d$?

38E. In a right-handed coordinate system show that

$$\mathbf{i} \cdot \mathbf{i} = \mathbf{j} \cdot \mathbf{j} = \mathbf{k} \cdot \mathbf{k} = 1 \quad \text{and} \quad \mathbf{i} \cdot \mathbf{j} = \mathbf{j} \cdot \mathbf{k} = \mathbf{k} \cdot \mathbf{i} = 0.$$

If the coordinate system is rectangular but not right-handed, do the results change?

39E. In a right-handed coordinate system show that

$$\mathbf{i} \times \mathbf{i} = \mathbf{j} \times \mathbf{j} = \mathbf{k} \times \mathbf{k} = 0$$

and $\qquad \mathbf{i} \times \mathbf{j} = \mathbf{k}; \quad \mathbf{k} \times \mathbf{i} = \mathbf{j}; \quad \mathbf{j} \times \mathbf{k} = \mathbf{i}.$

If the coordinate system is rectangular but not right-handed, do the results change?

40E. Show for any vector $\mathbf{a}$ that $\mathbf{a} \cdot \mathbf{a} = a^2$ and that $\mathbf{a} \times \mathbf{a} = 0$.

41E. Find (a) "north cross west," (b) "down dot south," (c) "east cross up," (d) "west dot west," and (e) "south cross south." Let each vector have unit magnitude.

42E. A vector $\mathbf{a}$ of magnitude 10 units and another vector $\mathbf{b}$ of magnitude 6.0 units point in directions differing by 60°. Find (a) the scalar product of the two vectors and (b) the magnitude of the vector product $\mathbf{a} \times \mathbf{b}$.

43E. Two vectors, $\mathbf{r}$ and $\mathbf{s}$, lie in the xy plane. Their magnitudes are 4.50 and 7.30 units, respectively, and their directions are 320° and 85.0°, respectively, as measured counterclockwise from the positive x axis. What are the values of (a) $\mathbf{r} \cdot \mathbf{s}$ and (b) $\mathbf{r} \times \mathbf{s}$?

44E. For the vectors in Fig. 3-35, calculate (a) $\mathbf{a} \cdot \mathbf{b}$, (b) $\mathbf{a} \cdot \mathbf{c}$, and (c) $\mathbf{b} \cdot \mathbf{c}$.

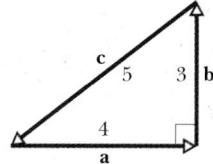

FIGURE 3-35 Exercises 44 and 45.

45E. For the vectors in Fig. 3-35, calculate (a) $\mathbf{a} \times \mathbf{b}$, (b) $\mathbf{a} \times \mathbf{c}$, and (c) $\mathbf{b} \times \mathbf{c}$.

46P. *Scalar Product in Unit-Vector Notation.* Let two vectors be represented in terms of their components as

$$\mathbf{a} = a_x\mathbf{i} + a_y\mathbf{j} + a_z\mathbf{k} \quad \text{and} \quad \mathbf{b} = b_x\mathbf{i} + b_y\mathbf{j} + b_z\mathbf{k}.$$

Show that

$$\mathbf{a} \cdot \mathbf{b} = a_xb_x + a_yb_y + a_zb_z.$$

47P. (a) Determine the components and magnitude of $\mathbf{r} = \mathbf{a} - \mathbf{b} + \mathbf{c}$ if $\mathbf{a} = 5.0\mathbf{i} + 4.0\mathbf{j} - 6.0\mathbf{k}$, $\mathbf{b} = -2.0\mathbf{i} + 2.0\mathbf{j} + 3.0\mathbf{k}$, and $\mathbf{c} = 4.0\mathbf{i} + 3.0\mathbf{j} + 2.0\mathbf{k}$. (b) Calculate the angle between $\mathbf{r}$ and the positive z axis.

48P. Use the definition of scalar product, $\mathbf{a} \cdot \mathbf{b} = ab \cos \theta$, and the fact that $\mathbf{a} \cdot \mathbf{b} = a_xb_x + a_yb_y + a_zb_z$ (see Problem 46) to calculate the angle between the two vectors given by $\mathbf{a} = 3.0\mathbf{i} + 3.0\mathbf{j} + 3.0\mathbf{k}$ and $\mathbf{b} = 2.0\mathbf{i} + 1.0\mathbf{j} + 3.0\mathbf{k}$.

49P. *Vector Product in Unit-Vector Notation.* Show that for the vectors $\mathbf{a}$ and $\mathbf{b}$ of Problem 46,

$$\mathbf{a} \times \mathbf{b} = \mathbf{i}(a_yb_z - a_zb_y) + \mathbf{j}(a_zb_x - a_xb_z) + \mathbf{k}(a_xb_y - a_yb_x).$$

50P. Two vectors are given by $\mathbf{a} = 3.0\mathbf{i} + 5.0\mathbf{j}$ and $\mathbf{b} = 2.0\mathbf{i} + 4.0\mathbf{j}$. Find (a) $\mathbf{a} \times \mathbf{b}$, (b) $\mathbf{a} \cdot \mathbf{b}$, and (c) $(\mathbf{a} + \mathbf{b}) \cdot \mathbf{b}$.

51P. Two vectors $\mathbf{a}$ and $\mathbf{b}$ have the components, in arbitrary units, $a_x = 3.2$, $a_y = 1.6$, $b_x = 0.50$, $b_y = 4.5$. (a) Find the angle between the directions of $\mathbf{a}$ and $\mathbf{b}$. (b) Find the components of a vector $\mathbf{c}$ that is perpendicular to $\mathbf{a}$, is in the xy plane, and has a magnitude of 5.0 units.

52P. Vector $\mathbf{a}$ lies in the yz plane 63° from the $+y$ axis, has a positive z component, and has magnitude 3.20 units. Vector $\mathbf{b}$ lies in the xz plane 48° from the $+x$ axis, has a positive z component, and has magnitude 1.40 units. Find (a) $\mathbf{a} \cdot \mathbf{b}$, (b) $\mathbf{a} \times \mathbf{b}$, and (c) the angle between $\mathbf{a}$ and $\mathbf{b}$.

53P. Three vectors are given by $\mathbf{a} = 3.0\mathbf{i} + 3.0\mathbf{j} - 2.0\mathbf{k}$, $\mathbf{b} = -1.0\mathbf{i} - 4.0\mathbf{j} + 2.0\mathbf{k}$, and $\mathbf{c} = 2.0\mathbf{i} + 2.0\mathbf{j} + 1.0\mathbf{k}$. Find (a) $\mathbf{a} \cdot (\mathbf{b} \times \mathbf{c})$, (b) $\mathbf{a} \cdot (\mathbf{b} + \mathbf{c})$, and (c) $\mathbf{a} \times (\mathbf{b} + \mathbf{c})$.

54P. Find the angles between the body diagonals of a cube with edge length a. See Problem 34.

55P. Show that the area of the triangle contained between the vectors $\mathbf{a}$ and $\mathbf{b}$ and the red line in Fig. 3-36 is $\frac{1}{2}|\mathbf{a} \times \mathbf{b}|$.

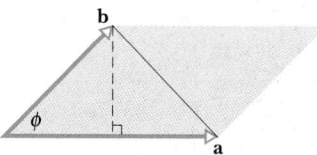

FIGURE 3-36 Problem 55.

56P. (a) Show that $\mathbf{a} \cdot (\mathbf{b} \times \mathbf{a})$ is zero for all vectors $\mathbf{a}$ and $\mathbf{b}$. (b) What is the value of $\mathbf{a} \times (\mathbf{b} \times \mathbf{a})$ if there is an angle ϕ between the directions of $\mathbf{a}$ and $\mathbf{b}$?

57P. Show that $\mathbf{a} \cdot (\mathbf{b} \times \mathbf{c})$ is equal in magnitude to the volume of the parallelepiped formed on the three vectors $\mathbf{a}$, $\mathbf{b}$, and $\mathbf{c}$ as shown in Fig. 3-37.

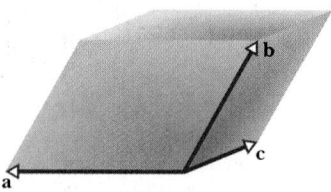

FIGURE 3-37 Problem 57.

58P. The three vectors shown in Fig. 3-38 have magnitudes $a = 3.00$, $b = 4.00$, and $c = 10.0$. (a) Calculate the x and y components of these vectors. (b) Find the numbers p and q such that $\mathbf{c} = p\mathbf{a} + q\mathbf{b}$.

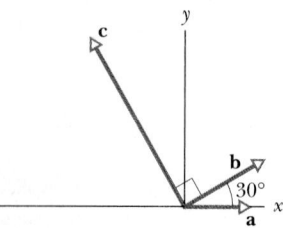

FIGURE 3-38 Problem 58.

In 1922, one of the Zacchinis, a famous family of circus performers, was the first human cannonball to be shot across an arena and into a net. To increase the excitement, the family gradually increased the height and distance of the flight until, in 1939 or 1940, Emanuel Zacchini soared over three Ferris wheels and through a horizontal distance of 225 feet. How could he know where to place the net? And how could he be certain he would clear the Ferris wheels?

4-1 MOVING IN TWO OR THREE DIMENSIONS

This chapter extends the material of the preceding two chapters to two and three dimensions. Many of the ideas of Chapter 2, such as position, velocity, and acceleration, are used here, but they are now a little more complex because of the extra dimensions. To keep the notation manageable, we use the vector algebra of Chapter 3. As you read this chapter, you might want to thumb back to those previous chapters to refresh your memory.

4-2 POSITION AND DISPLACEMENT

One general way of locating a particle-like object is with a **position vector r**, which is a vector that extends from a reference point (usually the origin of a coordinate system) to the object. In the unit-vector notation of Section 3-4, **r** can be written

$$\mathbf{r} = x\mathbf{i} + y\mathbf{j} + z\mathbf{k}, \qquad (4\text{-}1)$$

where $x\mathbf{i}$, $y\mathbf{j}$, and $z\mathbf{k}$ are the vector components of **r** and the coefficients x, y, and z are its scalar components. (This notation is slightly different from the notation we used in Chapter 3. Take a minute to convince yourself that the two are comparable.)

The coefficients x, y, and z give the object's location along the axes and relative to the origin. That is, the object has the rectangular coordinates (x, y, z). For instance, Fig. 4-1 shows an object P whose position vector is

$$\mathbf{r} = -3\mathbf{i} + 2\mathbf{j} + 5\mathbf{k}$$

and whose rectangular coordinates are $(-3, 2, 5)$. Along the x axis P is 3 units from the origin, in the $-\mathbf{i}$ direction. Along the y axis it is 2 units from the origin, in the $+\mathbf{j}$

direction. And along the z axis it is 5 units from the origin, in the $+\mathbf{k}$ direction.

As an object moves, its position vector changes in such a way that the vector always extends to the object from the origin. If the object has position vector $\mathbf{r}_1$ at time t_1 and position vector $\mathbf{r}_2$ at a later time $t_1 + \Delta t$, then its *displacement* $\Delta\mathbf{r}$ during the time interval Δt is

$$\Delta\mathbf{r} = \mathbf{r}_2 - \mathbf{r}_1. \qquad (4\text{-}2)$$

Using the unit-vector notation of Eq. 4-1, we can rewrite this displacement as

$$\Delta\mathbf{r} = (x_2\mathbf{i} + y_2\mathbf{j} + z_2\mathbf{k}) - (x_1\mathbf{i} + y_1\mathbf{j} + z_1\mathbf{k})$$

or

$$\Delta\mathbf{r} = (x_2 - x_1)\mathbf{i} + (y_2 - y_1)\mathbf{j} + (z_2 - z_1)\mathbf{k}, \qquad (4\text{-}3)$$

where coordinates (x_1, y_1, z_1) correspond to position vector $\mathbf{r}_1$ and coordinates (x_2, y_2, z_2) correspond to position vector $\mathbf{r}_2$. We can also rewrite the displacement by substituting Δx for $(x_2 - x_1)$, Δy for $(y_2 - y_1)$, and Δz for $(z_2 - z_1)$.

SAMPLE PROBLEM 4-1

The position vector for a particle is initially

$$\mathbf{r}_1 = -3\mathbf{i} + 2\mathbf{j} + 5\mathbf{k}$$

and then later is

$$\mathbf{r}_2 = 9\mathbf{i} + 2\mathbf{j} + 8\mathbf{k}$$

(see Fig. 4-2). What is the displacement from $\mathbf{r}_1$ to $\mathbf{r}_2$?

SOLUTION: Recall from Chapter 3 that we add (or subtract) two vectors in unit-vector notation by combining the components, axis by axis. So Eq. 4-2 becomes

$$\Delta\mathbf{r} = (9\mathbf{i} + 2\mathbf{j} + 8\mathbf{k}) - (-3\mathbf{i} + 2\mathbf{j} + 5\mathbf{k})$$

$$= 12\mathbf{i} + 3\mathbf{k}. \qquad \text{(Answer)}$$

The displacement vector is parallel to the xz plane, because it lacks any y component, a fact that is easier to pick out in the numerical result than in Fig. 4-2.

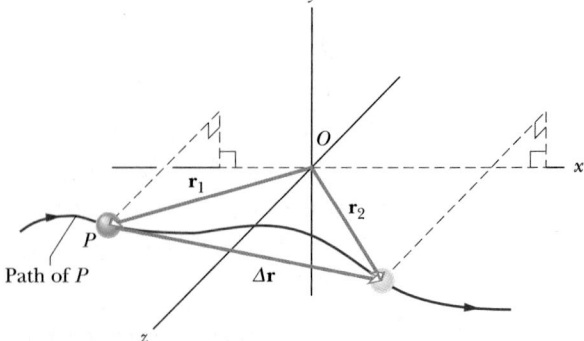

FIGURE 4-2 Sample Problem 4-1. The displacement $\Delta\mathbf{r} = \mathbf{r}_2 - \mathbf{r}_1$ extends from the head of $\mathbf{r}_1$ to the head of $\mathbf{r}_2$.

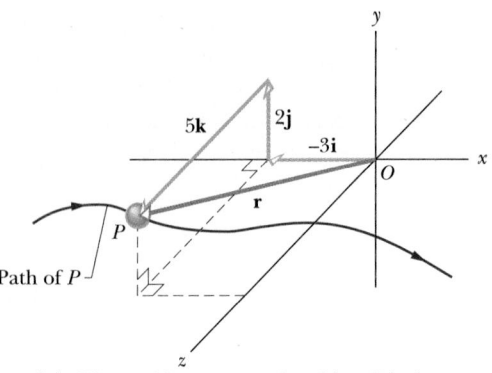

FIGURE 4-1 The position vector **r** for object P is the vector sum of its vector components.

CHECKPOINT **1:** (a) If a bat flies from *xyz* coordinates $(-2 \text{ m}, 4 \text{ m}, -3 \text{ m})$ to coordinates $(6 \text{ m}, -2 \text{ m}, -3 \text{ m})$, what is its displacement $\Delta \mathbf{r}$ in unit-vector notation? (b) Is $\Delta \mathbf{r}$ parallel to one of the three coordinate planes? If so, which plane?

4-3 VELOCITY AND AVERAGE VELOCITY

If a particle moves through a displacement $\Delta \mathbf{r}$ in a time interval Δt, then its *average velocity* is

$$\overline{\mathbf{v}} = \frac{\Delta \mathbf{r}}{\Delta t}, \tag{4-4}$$

which can be written in expanded form as

$$\overline{\mathbf{v}} = \frac{\Delta x \mathbf{i} + \Delta y \mathbf{j} + \Delta z \mathbf{k}}{\Delta t}$$

$$= \frac{\Delta x}{\Delta t} \mathbf{i} + \frac{\Delta y}{\Delta t} \mathbf{j} + \frac{\Delta z}{\Delta t} \mathbf{k}. \tag{4-5}$$

The (instantaneous) *velocity* $\mathbf{v}$ is the value that $\overline{\mathbf{v}}$ approaches in the limit as we shrink Δt to 0. It can be written as the derivative

$$\mathbf{v} = \frac{d\mathbf{r}}{dt}. \tag{4-6}$$

Substituting for $\mathbf{r}$ from Eq. 4-1 yields

$$\mathbf{v} = \frac{d}{dt}(x\mathbf{i} + y\mathbf{j} + z\mathbf{k}) = \frac{dx}{dt}\mathbf{i} + \frac{dy}{dt}\mathbf{j} + \frac{dz}{dt}\mathbf{k},$$

which can be rewritten as

$$\mathbf{v} = v_x\mathbf{i} + v_y\mathbf{j} + v_z\mathbf{k}, \tag{4-7}$$

where

$$v_x = \frac{dx}{dt}, \quad v_y = \frac{dy}{dt}, \quad \text{and} \quad v_z = \frac{dz}{dt} \tag{4-8}$$

are the scalar components of $\mathbf{v}$.

Figure 4-3 shows the path of a particle P that is restricted to the *xy* plane. As the particle travels to the right along the curve, its position vector sweeps to the right. At t_1 the position vector is $\mathbf{r}_1$, and at an arbitrary later time $t_1 + \Delta t$ the position vector is $\mathbf{r}_2$. The particle's displacement during Δt is $\Delta \mathbf{r}$. The particle's average velocity $\overline{\mathbf{v}}$ during Δt is, by Eq. 4-4, in the same direction as $\Delta \mathbf{r}$.

Three things happen as we shrink interval Δt toward zero: (1) the vector $\mathbf{r}_2$ in Fig. 4-3 moves toward $\mathbf{r}_1$ so that

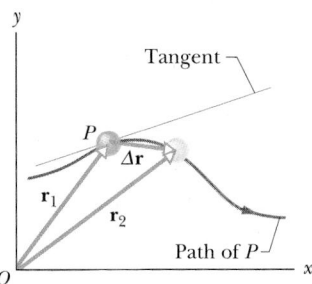

FIGURE 4-3 The position of particle P along its path, both at time t_1 and at a later time $t_1 + \Delta t$. The vector $\Delta \mathbf{r}$ is the displacement of the particle during Δt. The tangent to the particle's path at t_1 is shown.

$\Delta \mathbf{r}$ shrinks toward zero; (2) the direction of $\Delta \mathbf{r}$ (and thus the direction of $\overline{\mathbf{v}}$) approaches the direction of the tangent line in Fig. 4-3; and (3) the average velocity $\overline{\mathbf{v}}$ approaches the instantaneous velocity $\mathbf{v}$.

In the limit as $\Delta t \to 0$, we have $\overline{\mathbf{v}} \to \mathbf{v}$ and, most important here, $\overline{\mathbf{v}}$ takes on the direction of the tangent line. Hence $\mathbf{v}$ has that direction as well. That is:

> The instantaneous velocity $\mathbf{v}$ of a particle is always tangent to the path of the particle.

This is shown in Fig. 4-4, where both $\mathbf{v}$ and its scalar x and y components are included. The result is the same in three dimensions: $\mathbf{v}$ is always tangent to the particle's path.

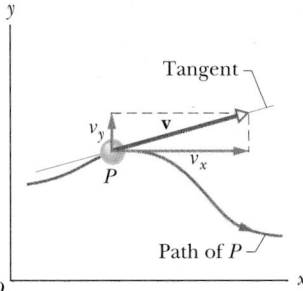

FIGURE 4-4 The velocity $\mathbf{v}$ of particle P along with the scalar components of $\mathbf{v}$. Note that $\mathbf{v}$ lies along the tangent to the path.

CHECKPOINT **2:** The figure shows a circular path taken by a particle. If the instantaneous velocity of the particle is $\mathbf{v} = (2 \text{ m/s})\mathbf{i} - (2 \text{ m/s})\mathbf{j}$, through which quadrant is the particle moving when it is traveling (a) clockwise and (b) counterclockwise around the circle?

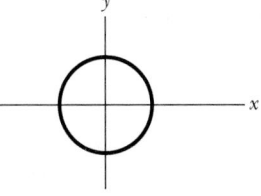

4-4 ACCELERATION AND AVERAGE ACCELERATION

When a particle's velocity changes from $\mathbf{v}_1$ to $\mathbf{v}_2$ in a time period Δt, its average acceleration $\bar{\mathbf{a}}$ during Δt is

$$\bar{\mathbf{a}} = \frac{\mathbf{v}_2 - \mathbf{v}_1}{\Delta t} = \frac{\Delta \mathbf{v}}{\Delta t}. \tag{4-9}$$

If we shrink Δt to 0, then in the limit $\bar{\mathbf{a}}$ approaches the (instantaneous) *acceleration* $\mathbf{a}$; that is,

$$\mathbf{a} = \frac{d\mathbf{v}}{dt}. \tag{4-10}$$

If the velocity changes in *either* magnitude *or* direction (or both), there is an acceleration.

Substituting $\mathbf{v}$ from Eq. 4-7 into Eq. 4-10 yields

$$\mathbf{a} = \frac{d}{dt} (v_x \mathbf{i} + v_y \mathbf{j} + v_z \mathbf{k})$$

$$= \frac{dv_x}{dt} \mathbf{i} + \frac{dv_y}{dt} \mathbf{j} + \frac{dv_z}{dt} \mathbf{k}$$

or

$$\mathbf{a} = a_x \mathbf{i} + a_y \mathbf{j} + a_z \mathbf{k}, \tag{4-11}$$

in which the three scalar components of the acceleration vector are given by

$$a_x = \frac{dv_x}{dt}, \quad a_y = \frac{dv_y}{dt}, \quad \text{and} \quad a_z = \frac{dv_z}{dt}. \tag{4-12}$$

Figure 4-5 shows an acceleration vector $\mathbf{a}$ and its scalar components for a particle P moving in two dimensions.

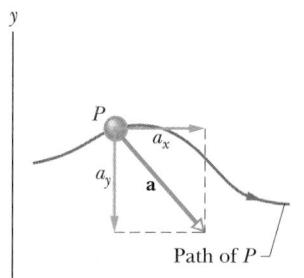

FIGURE 4-5 The acceleration $\mathbf{a}$ of particle P along with the scalar components of $\mathbf{a}$.

SAMPLE PROBLEM 4-2

A rabbit runs across a parking lot on which a set of coordinate axes has, strangely enough, been drawn. The rabbit's path is such that the components of its position with respect to an origin of coordinates are given as functions of time by

$$x = -0.31t^2 + 7.2t + 28$$

and

$$y = 0.22t^2 - 9.1t + 30,$$

with t in seconds and x and y in meters. The rabbit's position vector $\mathbf{r}$ is

$$\mathbf{r}(t) = x(t)\mathbf{i} + y(t)\mathbf{j}.$$

(a) What are the magnitude and direction of the rabbit's position vector at $t = 15$ s?

SOLUTION: At $t = 15$ s, the components of $\mathbf{r}$ are

$$x = (-0.31)(15)^2 + (7.2)(15) + 28 = 66 \text{ m}$$

and

$$y = (0.22)(15)^2 - (9.1)(15) + 30 = -57 \text{ m}.$$

The components and $\mathbf{r}$ itself are shown in Fig. 4-6a.
The magnitude of $\mathbf{r}$ is

$$r = \sqrt{x^2 + y^2} = \sqrt{(66 \text{ m})^2 + (-57 \text{ m})^2}$$

$$= 87 \text{ m}. \qquad \text{(Answer)}$$

The angle θ between $\mathbf{r}$ and the direction of increasing x is

$$\theta = \tan^{-1} \frac{y}{x} = \tan^{-1} \left(\frac{-57 \text{ m}}{66 \text{ m}} \right) = -41°. \quad \text{(Answer)}$$

(Although $\theta = 139°$ has the same tangent as $-41°$, study of the signs of the components of $\mathbf{r}$ rules out $139°$.)

(b) Also calculate the position of the rabbit at $t = 0, 5, 10, 20$, and 25 s and sketch the rabbit's path.

SOLUTION: Proceeding as in (a) leads to the following values of r and θ.

t (s)	x (m)	y (m)	r (m)	θ
0	28	30	41	$+47°$
5	56	-10	57	$-10°$
10	69	-39	79	$-29°$
15	66	-57	87	$-41°$
20	48	-64	80	$-53°$
25	14	-60	62	$-77°$

Figure 4-6b shows a plot of the rabbit's path, drawn with the x and y values.

SAMPLE PROBLEM 4-3

In Sample Problem 4-2, find the magnitude and direction of the rabbit's velocity vector at $t = 15$ s.

SOLUTION: The velocity component along the x axis (see Eq. 4-8) is

$$v_x = \frac{dx}{dt} = \frac{d}{dt}(-0.31t^2 + 7.2t + 28) = -0.62t + 7.2.$$

At $t = 15$ s, this becomes

$$v_x = (-0.62)(15) + 7.2 = -2.1 \text{ m/s}.$$

Similarly,

$$v_y = \frac{dy}{dt} = \frac{d}{dt}(0.22t^2 - 9.1t + 30) = 0.44t - 9.1.$$

At $t = 15$ s, this becomes

$$v_y = (0.44)(15) - 9.1 = -2.5 \text{ m/s}.$$

The vector **v** and its components are shown in Fig. 4-6c. The magnitude and direction of **v** are

$$v = \sqrt{v_x^2 + v_y^2} = \sqrt{(-2.1 \text{ m/s})^2 + (-2.5 \text{ m/s})^2}$$

$$= 3.3 \text{ m/s} \qquad \text{(Answer)}$$

and

$$\theta = \tan^{-1} \frac{v_y}{v_x} = \tan^{-1}\left(\frac{-2.5 \text{ m/s}}{-2.1 \text{ m/s}}\right)$$

$$= \tan^{-1} 1.19 = -130°. \qquad \text{(Answer)}$$

(Although 50° has the same tangent, inspection of the signs of the velocity components indicates that the desired angle is in the third quadrant, given by 50° − 180° = −130°.) The velocity vector in Fig. 4-6c is tangent to the path of the rabbit and points in the direction in which the rabbit is running at $t = 15$ s.

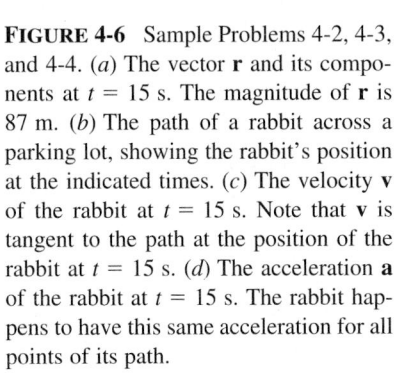

FIGURE 4-6 Sample Problems 4-2, 4-3, and 4-4. (a) The vector **r** and its components at $t = 15$ s. The magnitude of **r** is 87 m. (b) The path of a rabbit across a parking lot, showing the rabbit's position at the indicated times. (c) The velocity **v** of the rabbit at $t = 15$ s. Note that **v** is tangent to the path at the position of the rabbit at $t = 15$ s. (d) The acceleration **a** of the rabbit at $t = 15$ s. The rabbit happens to have this same acceleration for all points of its path.

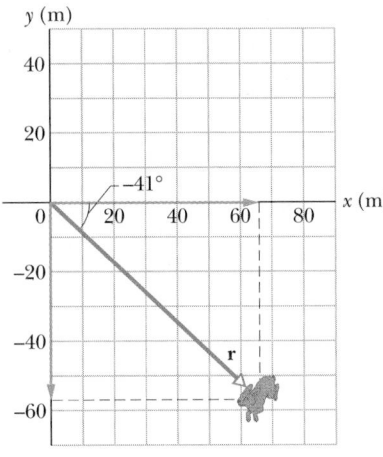

(a)

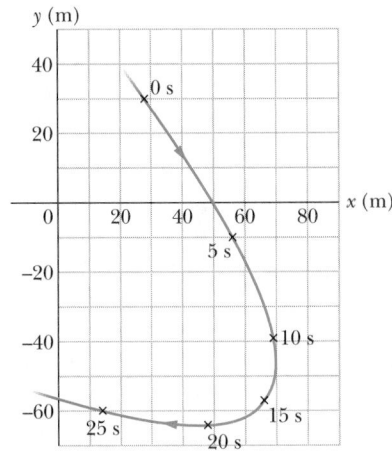

(b)

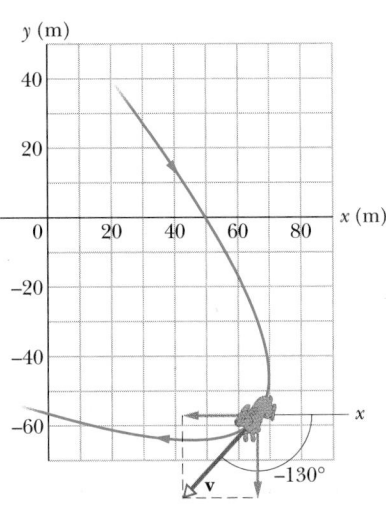

(c)

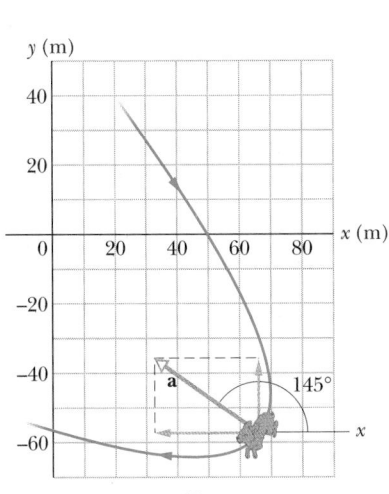

(d)

SAMPLE PROBLEM 4-4

In Sample Problem 4-2, determine the magnitude and direction of the rabbit's acceleration vector **a** at $t = 15$ s.

SOLUTION: The acceleration components (see Eq. 4-12) are

$$a_x = \frac{dv_x}{dt} = \frac{d}{dt}(-0.62t + 7.2) = -0.62 \text{ m/s}^2$$

and

$$a_y = \frac{dv_y}{dt} = \frac{d}{dt}(0.44t - 9.1) = 0.44 \text{ m/s}^2.$$

We see that the acceleration does not vary with time; it is a constant. In fact, we have differentiated the time variable completely away. The components of **a**, along with the vector itself, are shown in Fig. 4-6d.

The magnitude and direction of **a** are

$$a = \sqrt{a_x^2 + a_y^2} = \sqrt{(-0.62 \text{ m/s}^2)^2 + (0.44 \text{ m/s}^2)^2}$$

$$= 0.76 \text{ m/s}^2 \qquad \text{(Answer)}$$

and

$$\theta = \tan^{-1}\frac{a_y}{a_x} = \tan^{-1}\left(\frac{0.44 \text{ m/s}^2}{-0.62 \text{ m/s}^2}\right)$$

$$= 145°. \qquad \text{(Answer)}$$

The acceleration vector has the same magnitude and direction for all parts of the rabbit's path. Perhaps a strong southeast wind was blowing across the parking lot, causing the rabbit's acceleration toward the northwest.

CHECKPOINT **3:** Here are four descriptions of the position (in meters) of a hockey puck as it moves in the xy plane:

(1) $x = -3t^2 + 4t - 2$ and $y = 6t^2 - 4t$
(2) $x = -3t^3 - 4t$ and $y = -5t^2 + 6$
(3) $\mathbf{r} = 2t^2\mathbf{i} - (4t + 3)\mathbf{j}$
(4) $\mathbf{r} = (4t^3 - 2t)\mathbf{i} + 3\mathbf{j}$

For each description, determine whether the x and y components of the puck's acceleration are constant, and whether the acceleration **a** is constant.

SAMPLE PROBLEM 4-5

A particle with velocity $\mathbf{v}_0 = -2.0\mathbf{i} + 4.0\mathbf{j}$ (in meters per second) at $t = 0$ undergoes a constant acceleration **a** of magnitude $a = 3.0$ m/s^2 at an angle $\theta = 130°$ from the positive direction of the x axis. What is the particle's velocity **v** at $t = 2.0$ s, in unit-vector notation and as a magnitude and direction (with respect to the positive direction of the x axis)?

SOLUTION: Since **a** is constant, Eq. 2-11 ($v = v_0 + at$) applies; it should, however, be used separately to find v_x and v_y

(the x and y components of velocity **v**) because they vary independently of each other. So we write

$$v_x = v_{0x} + a_x t \quad \text{and} \quad v_y = v_{0y} + a_y t.$$

Here v_{0x} ($= -2.0$ m/s) and v_{0y} ($= 4.0$ m/s) are the x and y components of $\mathbf{v}_0$, and a_x and a_y are the x and y components of **a**. To find a_x and a_y, we resolve **a** with Eqs. 3-5:

$$a_x = a \cos \theta = (3.0 \text{ m/s}^2)(\cos 130°) = -1.93 \text{ m/s}^2,$$

$$a_y = a \sin \theta = (3.0 \text{ m/s}^2)(\sin 130°) = +2.30 \text{ m/s}^2.$$

When these values are inserted into the equations for v_x and v_y, we find that

$$v_x = -2.0 \text{ m/s} + (-1.93 \text{ m/s}^2)(2.0 \text{ s}) = -5.9 \text{ m/s},$$

$$v_y = 4.0 \text{ m/s} + (2.30 \text{ m/s}^2)(2.0 \text{ s}) = 8.6 \text{ m/s}.$$

So at $t = 2.0$ s, we have, from Eq. 4-7,

$$\mathbf{v} = (-5.9 \text{ m/s})\mathbf{i} + (8.6 \text{ m/s})\mathbf{j}. \qquad \text{(Answer)}$$

The magnitude of **v** is

$$v = \sqrt{(-5.9 \text{ m/s})^2 + (8.6 \text{ m/s})^2}$$

$$= 10 \text{ m/s}. \qquad \text{(Answer)}$$

The angle of **v** is

$$\theta = \tan^{-1}\left(\frac{8.6 \text{ m/s}}{-5.9 \text{ m/s}}\right) = 124° \approx 120°. \qquad \text{(Answer)}$$

Check the last line with your calculator. Does 124° appear on the display, or does $-55.5°$ appear? Now sketch the vector **v** with its components to see which angle is reasonable. To see why some calculators give a mathematically possible but unreasonable result here, reread Tactic 3 in Chapter 3.

PROBLEM SOLVING TACTICS

TACTIC 1: *Trig Functions and Angles*

In Sample Problem 4-3, we computed $\theta = \tan^{-1} 1.19$ and had to find θ. Your calculator might tell you $\theta = 50°$. However, Fig. 3-13c shows that $\theta = 230°$ ($= 50° + 180°$) has the same tangent. Inspection of the signs of the velocity components v_x and v_y in Fig. 4-6c tells us that this latter angle is the correct one. Some calculators can choose the correct angle for you.

There is still another decision to make. We can stick with 230° or we can relabel it as $-130°$. They are exactly the same angle, as is pointed out in Tactic 1 of Chapter 3. Here, we chose $\theta = -130°$.

TACTIC 2: *Drawing Vectors*

The vectors in Fig. 4-6 were oriented in the following way. (1) Choose the point at which you wish the tail of the vector to be. (2) From that point, draw a line in the direction of increasing x. (3) Using a protractor, mark off the appropriate angle θ, counterclockwise from this line if θ is positive or clockwise if θ is negative.

The vector **r** in Fig. 4-6*a* should be drawn to the same scale as the two axes because it is a length. The velocity vector **v** in Fig. 4-6*c* and the acceleration vector **a** in Fig. 4-6*d*, however, have no established scale and you may make them as long or as short as you wish.

It makes no sense to ask whether, for example, a velocity vector should be longer or shorter than a displacement vector. They are different physical quantities, expressed in different units, and they have no common scale.

CHECKPOINT **4:** If a marble's position is given by $\mathbf{r} = (4t^3 - 2t)\mathbf{i} + 3\mathbf{j}$, with **r** in meters and t in seconds, what must be the units of the coefficients 4, -2, and 3?

4-5 PROJECTILE MOTION

We next consider a particle that moves in a vertical plane during free fall, with its only acceleration being that of the free-fall acceleration **g**, which is downward. Such a **projectile** might be a golf ball (as in Fig. 4-7), a baseball, or any of a variety of other objects. Throughout, we shall assume that the air has no effect on the motion of the projectile.

Figure 4-8, which is analyzed in the next section, shows the path followed by a projectile under such ideal

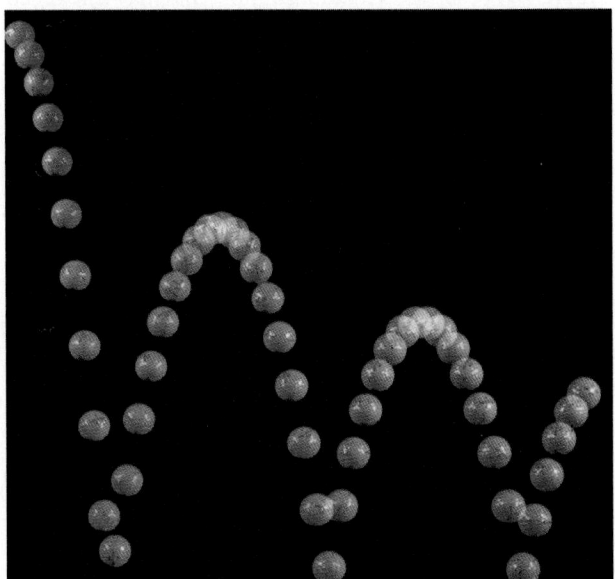

FIGURE 4-7 A stroboscopic photograph of an orange golf ball bouncing off a hard surface. Between impacts, the ball has projectile motion.

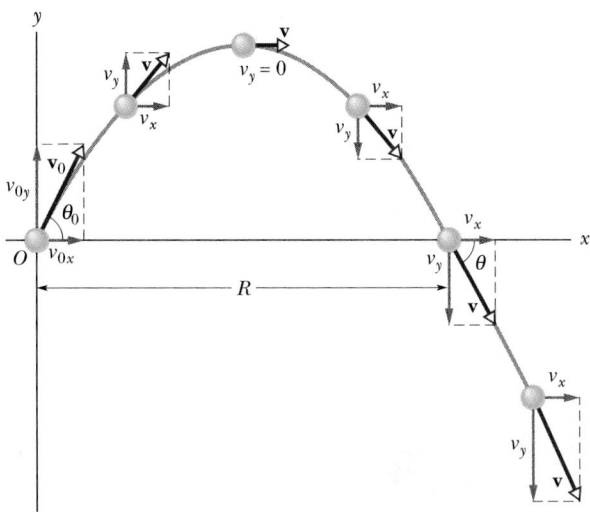

FIGURE 4-8 The path of a projectile that is launched at $x_0 = 0$ and $y_0 = 0$, with an initial velocity $\mathbf{v}_0$. The initial velocity and the velocities at various points along its path are shown, along with their components. Note that the horizontal velocity component remains constant but the vertical velocity component changes continuously. The *range R* is the horizontal distance the projectile has traveled *when it returns to its launch height.*

conditions. The projectile is launched with some initial velocity $\mathbf{v}_0$, which can be written

$$\mathbf{v}_0 = v_{0x}\mathbf{i} + v_{0y}\mathbf{j}. \qquad (4\text{-}13)$$

The components v_{0x} and v_{0y} can then be found if we know the angle θ_0 between $\mathbf{v}_0$ and the positive x direction:

$$v_{0x} = v_0 \cos \theta_0 \quad \text{and} \quad v_{0y} = v_0 \sin \theta_0. \qquad (4\text{-}14)$$

During its two-dimensional motion, the projectile's position vector **r** and velocity vector **v** change continuously, but its acceleration vector **a** is constant and *always* directed vertically downward. (The projectile has *no* horizontal acceleration.) As shown in Fig. 4-9, the angle between the direction of the velocity vector and that of the acceleration vector is not constant but varies during the motion.

Projectile motion, like that in Figs. 4-7 through 4-9, looks complicated, but we have the following simplifying feature (known from experiment):

The horizontal motion and the vertical motion are independent of each other; that is, neither motion affects the other.

This feature allows us to break up a problem involving two-dimensional motion into two separate and easier one-dimensional problems, one for the horizontal motion and

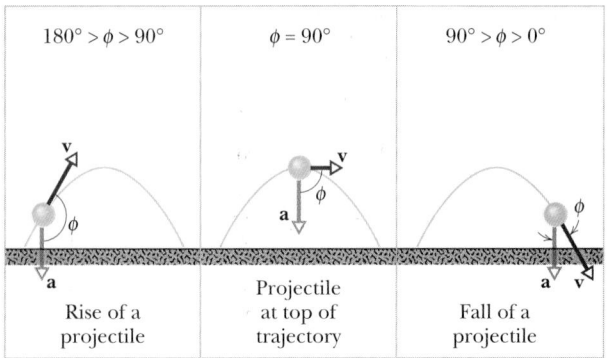

FIGURE 4-9 The velocity and acceleration vectors of a projectile for various motions. Note that the acceleration and velocity vectors do not have any fixed directional relation to each other.

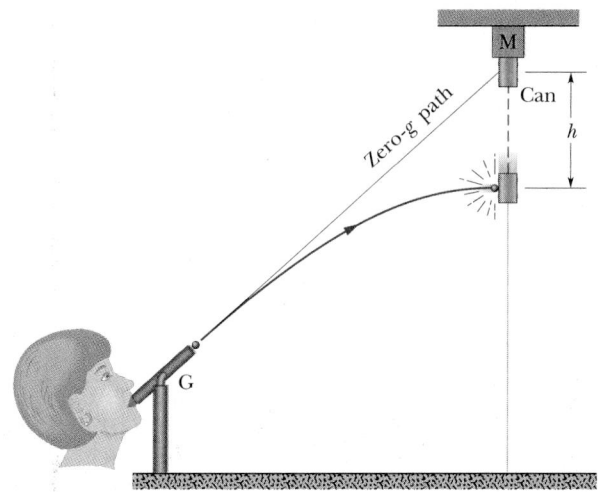

FIGURE 4-11 The projectile ball always hits the falling can. Each falls a distance h from where it would be were there no free-fall acceleration.

one for the vertical motion. Here are two experiments that show that the horizontal motion and the vertical motion are independent.

Two Golf Balls

Figure 4-10 is a stroboscopic photograph of two golf balls, one simply dropped and the other fired horizontally by a spring mechanism. They have the same vertical motion, each ball falling through the same vertical distance in the same interval of time. *The fact that one ball is moving horizontally while it is falling has no effect on its vertical motion.* To push the experiment to a limit, we say that if you were to fire a rifle horizontally and at the same time drop a bullet, in the absence of air resistance the two bullets would reach level ground at the same time.

A Great Student Rouser

Figure 4-11 shows a demonstration that has enlivened many a physics lecture. It involves a blow gun G, using a ball as a projectile. The target is a can suspended from a magnet M, and the tube of the blow gun is aimed directly at the can. The experiment is arranged so that the magnet releases the can just as the ball leaves the blow gun.

If g (the magnitude of the free-fall acceleration) were zero, the ball would follow the straight line shown in Fig. 4-11 and the can would float in place after the magnet released it. The ball would certainly hit the can.

However, g is *not* zero. The ball *still* hits the can! As Fig. 4-11 shows, during the time of flight of the ball, both ball and can fall the same distance h from their zero-g locations. The harder the demonstrator blows, the greater the ball's initial speed, the shorter the time of flight, and the smaller the value of h.

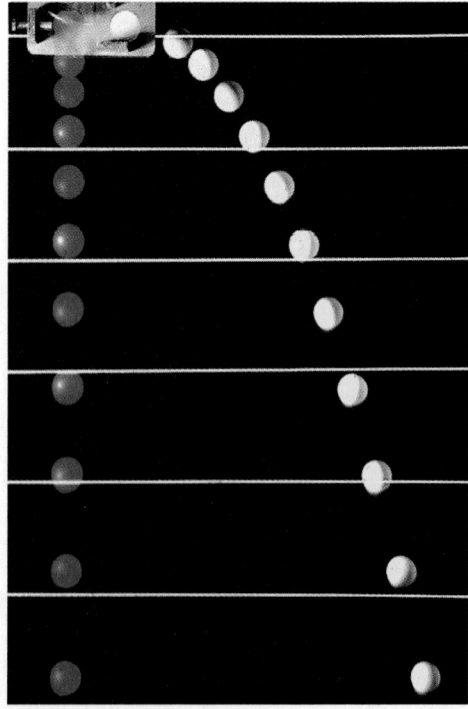

FIGURE 4-10 One ball is released from rest at the same instant that another ball is shot horizontally to the right. Their vertical motions are identical.

4-6 PROJECTILE MOTION ANALYZED

Now we are ready to analyze projectile motion, horizontally and vertically:

FIGURE 4-12 The vertical component of this skateboarder's velocity is changing, but not the horizontal component, which matches the skateboard's velocity. As a result, the skateboard stays underneath him, allowing him to land on it.

The Horizontal Motion

Because there is *no acceleration* in the horizontal direction, the horizontal component v_{0x} of the projectile's initial velocity remains unchanged throughout the motion, as demonstrated in Fig. 4-12. The horizontal displacement $x - x_0$ from an initial position x_0 is given by Eq. 2-15 with $a = 0$, which we write as

$$x - x_0 = v_{0x}t.$$

Because $v_{0x} = v_0 \cos \theta_0$, this becomes

$$x - x_0 = (v_0 \cos \theta_0)t. \qquad (4\text{-}15)$$

The Vertical Motion

The vertical motion is the motion we discussed in Section 2-8 for a particle in free fall. Equations 2-21 to 2-25 apply. Equation 2-22, for example, becomes

$$\begin{aligned} y - y_0 &= v_{0y}t - \tfrac{1}{2}gt^2 \\ &= (v_0 \sin \theta_0)t - \tfrac{1}{2}gt^2, \end{aligned} \qquad (4\text{-}16)$$

where the initial vertical velocity component v_{0y} is replaced with the equivalent $v_0 \sin \theta_0$. Equations 2-21 and 2-23 are also useful in analyzing projectile motion.

Adapted to our purpose, they are

$$v_y = v_0 \sin \theta_0 - gt \qquad (4\text{-}17)$$

and $\qquad v_y^2 = (v_0 \sin \theta_0)^2 - 2g(y - y_0). \qquad (4\text{-}18)$

As is illustrated in Fig. 4-8 and Eq. 4-17, the vertical velocity component behaves just as for a ball thrown vertically upward. It is directed upward initially, its magnitude steadily decreasing to zero, *which marks the maximum height of the path*. The vertical component then reverses direction, and its magnitude becomes larger with time.

The Equation of the Path

We can find the equation of the path (the **trajectory**) of the projectile by eliminating t between Eqs. 4-15 and 4-16. Solving Eq. 4-15 for t and substituting into Eq. 4-16, we obtain, after a little rearrangement,

$$y = (\tan \theta_0)x - \frac{gx^2}{2(v_0 \cos \theta_0)^2} \qquad \text{(trajectory)}. \quad (4\text{-}19)$$

This is the equation of the path shown in Fig. 4-8. In deriving it, for simplicity we let $x_0 = 0$ and $y_0 = 0$ in Eqs. 4-15 and 4-16, respectively. Because g, θ_0, and v_0 are constants, Eq. 4-19 is of the form $y = ax + bx^2$, in which a and b are constants. This is the equation of a parabola, so the path is *parabolic*.

The Horizontal Range

The *horizontal range R* of the projectile, as Fig. 4-8 shows, is the *horizontal* distance the projectile has traveled when it returns to its initial (launch) height. To find range R, let us put $x - x_0 = R$ in Eq. 4-15 and $y - y_0 = 0$ in Eq. 4-16, obtaining

$$x - x_0 = (v_0 \cos \theta_0)t = R$$

and $\qquad y - y_0 = (v_0 \sin \theta_0)t - \tfrac{1}{2}gt^2 = 0.$

Eliminating t between these two equations yields

$$R = \frac{2v_0^2}{g} \sin \theta_0 \cos \theta_0.$$

Using the identity $\sin 2\theta_0 = 2 \sin \theta_0 \cos \theta_0$ (see Appendix E), we obtain

$$R = \frac{v_0^2}{g} \sin 2\theta_0. \qquad (4\text{-}20)$$

Note that R has its maximum value when $\sin 2\theta_0 = 1$, which corresponds to $2\theta_0 = 90°$ or $\theta_0 = 45°$.

The horizontal range R is maximum for a launch angle of $45°$.

The Effects of the Air

We have assumed that the air through which the projectile moves has no effect on its motion, a reasonable assumption at low speeds. However, for greater speeds, the disagreement between our calculations and the actual motion of the projectile can be large because the air resists (or opposes) the motion. Figure 4-13, for example, shows two paths for a fly ball that leaves the bat at an angle of 60° with the horizontal and an initial speed of 100 mi/h (data adapted from "The Trajectory of a Fly Ball," by Peter J. Brancazio, *The Physics Teacher,* January 1985). Path I (the baseball player's fly ball) is a calculated path that approximates normal conditions of play, in air. Path II (the physics professor's fly ball) is the path that the ball would follow in a vacuum. Table 4-1 gives some data for the two cases. We shall discuss details of the effect of the air on motion in Chapter 6.

$\mathbb{C}$HECKPOINT 5: A fly ball is hit to the outfield. During its flight (ignore the effects of air), what happens to its (a) horizontal and (b) vertical components of velocity? What are the (c) horizontal and (d) vertical components of its acceleration during its ascent and its descent, and at the topmost point of its flight?

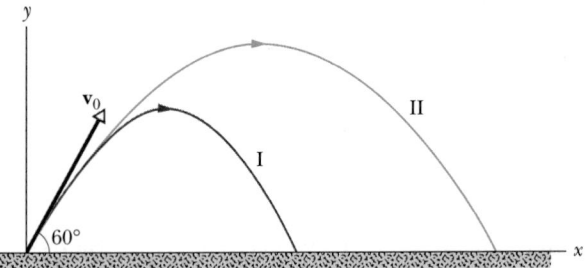

FIGURE 4-13   (I) The path of a fly ball, calculated (using a computer) by taking air resistance into account. (II) The path the ball would follow in a vacuum, calculated by the methods of this chapter. Also see Table 4-1.

TABLE 4-1 TWO FLY BALLS[a]

	PATH I (AIR)	PATH II (VACUUM)
Range	323 ft	581 ft
Maximum height	174 ft	252 ft
Time of flight	6.6 s	7.9 s

[a]See Fig. 4-13. The launch angle is 60° and the launch speed is 100 mi/h.

SAMPLE PROBLEM 4-6

A rescue plane is flying at a constant elevation of 1200 m with a speed of 430 km/h toward a point directly over a person struggling in the water (see Fig. 4-14). At what angle of sight ϕ should the pilot release a rescue capsule if it is to strike (very close to) the person in the water?

SOLUTION: The initial velocity of the capsule is the same as the velocity of the plane at the moment of release. That is, the initial capsule velocity v_0 is horizontal and has a magnitude of 430 km/h. We know the vertical distance the capsule falls, so we can find its time of flight with Eq. 4-16,

$$y - y_0 = (v_0 \sin \theta_0)t - \tfrac{1}{2}gt^2.$$

Putting $y - y_0 = -1200$ m (we use the minus sign because the person is below the origin) and $\theta_0 = 0$, we obtain

$$-1200 \text{ m} = 0 - \tfrac{1}{2}(9.8 \text{ m/s}^2)t^2.$$

Solving for t yields

$$t = \sqrt{\frac{(2)(1200 \text{ m})}{9.8 \text{ m/s}^2}} = 15.65 \text{ s}.$$

The horizontal distance covered by the capsule (and by the plane) in that time is given by Eq. 4-15:

$$
\begin{aligned}
x - x_0 &= (v_0 \cos \theta_0)t \\
&= (430 \text{ km/h})(\cos 0°)(15.65 \text{ s})(1 \text{ h}/3600 \text{ s}) \\
&= 1.869 \text{ km} = 1869 \text{ m}.
\end{aligned}
$$

If $x_0 = 0$, then $x = 1869$ m. The angle of sight is then (see Fig. 4-14)

$$\phi = \tan^{-1} \frac{x}{h} = \tan^{-1} \frac{1869 \text{ m}}{1200 \text{ m}} = 57°. \quad \text{(Answer)}$$

Because the plane and the capsule have the same horizontal velocity, the plane remains vertically over the capsule while the capsule is in flight.

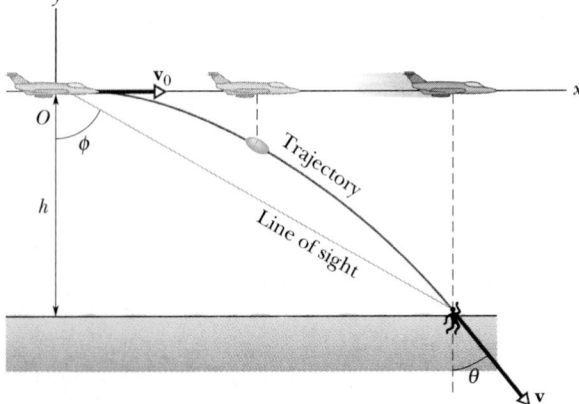

FIGURE 4-14 Sample Problem 4-6. A plane drops a rescue capsule and then continues in level flight. While the capsule is falling, its horizontal velocity component remains equal to the velocity of the plane. The capsule hits the water with velocity **v**, at angle θ from the vertical.

SAMPLE PROBLEM 4-7

A movie stuntman is to run across a rooftop and jump horizontally off it, to land on the roof of the next building (Fig. 4-15). Before he attempts the jump, he wisely asks you to determine whether it is possible. Can he make the jump if his maximum rooftop speed is 4.5 m/s?

SOLUTION: The fall of 4.8 m will take a time t, which can be obtained from Eq. 4-16. Letting $y - y_0 = -4.8$ m (note the sign) and $\theta_0 = 0$, you rearrange Eq. 4-16 to obtain

$$t = \sqrt{-\frac{2(y - y_0)}{g}} = \sqrt{-\frac{(2)(-4.8 \text{ m})}{9.8 \text{ m/s}^2}}$$

$$= 0.990 \text{ s}.$$

You now ask: "How far would he move horizontally in this time?" The answer, from Eq. 4-15, is

$$x - x_0 = (v_0 \cos \theta_0)t$$

$$= (4.5 \text{ m/s})(\cos 0°)(0.990 \text{ s}) = 4.5 \text{ m}.$$

To reach the next building, the stuntman has to move 6.2 m horizontally. Your advice: "Don't jump."

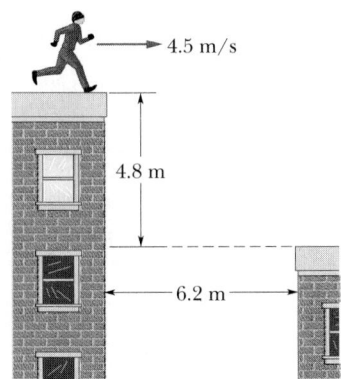

FIGURE 4-15 Sample Problem 4-7. Can the stuntman make the jump?

SAMPLE PROBLEM 4-8

Figure 4-16 shows a pirate ship, moored 560 m from a fort defending the harbor entrance of an island. The harbor defense cannon, located at sea level, has a muzzle velocity of 82 m/s.

(a) To what angle must the cannon be elevated to hit the pirate ship?

SOLUTION: Solving Eq. 4-20 for $2\theta_0$ yields

$$2\theta_0 = \sin^{-1} \frac{gR}{v_0^2} = \sin^{-1} \frac{(9.8 \text{ m/s}^2)(560 \text{ m})}{(82 \text{ m/s})^2}$$

$$= \sin^{-1} 0.816.$$

There are two angles whose sine is 0.816, namely, 54.7° and 125.3°. Thus we find

$$\theta_0 = \tfrac{1}{2}(54.7°) \approx 27° \qquad \text{(Answer)}$$

and

$$\theta_0 = \tfrac{1}{2}(125.3°) \approx 63°. \qquad \text{(Answer)}$$

The commandant of the fort can elevate the cannon to either of these two angles and (if only there were no intervening air!) hit the pirate ship.

(b) What are the times of flight for the two elevation angles calculated above?

SOLUTION: Solving Eq. 4-15 for t gives, for $\theta_0 = 27°$,

$$t = \frac{x - x_0}{v_0 \cos \theta_0} = \frac{560 \text{ m}}{(82 \text{ m/s})(\cos 27°)}$$

$$= 7.7 \text{ s}. \qquad \text{(Answer)}$$

Repeating the calculation for $\theta_0 = 63°$ yields $t = 15$ s. It is reasonable that the time of flight for the higher elevation angle should be larger.

(c) How far should the pirate ship be from the fort if it is to be beyond range of the cannon?

SOLUTION: We have seen that maximum range corresponds to an elevation angle θ_0 of 45°. Thus, from Eq. 4-20 with $\theta_0 = 45°$,

$$R = \frac{v_0^2}{g} \sin 2\theta_0 = \frac{(82 \text{ m/s})^2}{9.8 \text{ m/s}^2} \sin(2 \times 45°)$$

$$= 690 \text{ m}. \qquad \text{(Answer)}$$

As the pirate ship sails away, the two elevation angles at which the ship can be hit draw closer together, eventually merging at $\theta_0 = 45°$ when the ship is 690 m away. Beyond that distance the ship is safe.

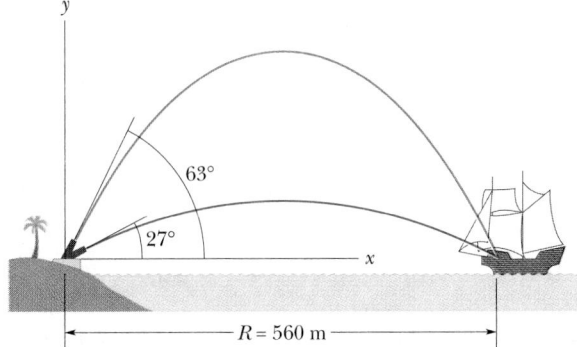

FIGURE 4-16 Sample Problem 4-8. At this range, the harbor defense cannon can hit the pirate ship at two elevation angles of the cannon.

SAMPLE PROBLEM 4-9

Figure 4-17 illustrates the flight of Emanuel Zacchini over three Ferris wheels, located as shown and each 18 m high. Zacchini is launched with speed $v_0 = 26.5$ m/s, at an angle $\theta_0 = 53°$ from the horizontal and with an initial height of 3.0 m above the ground. The net in which he is to land is at the same height.

(a) Does he clear the first Ferris wheel?

SOLUTION: We place the origin on the cannon muzzle so that $x_0 = 0$ and $y_0 = 0$. To find his height y when $x = 23$ m, we use Eq. 4-19:

$$y = (\tan \theta_0)x - \frac{gx^2}{2(v_0 \cos \theta_0)^2}$$

$$= (\tan 53°)(23 \text{ m}) - \frac{(9.8 \text{ m/s}^2)(23 \text{ m})^2}{2(26.5 \text{ m/s})^2(\cos 53°)^2}$$

$$= 20.3 \text{ m.} \qquad \text{(Answer)}$$

Since he begins 3.0 m off the ground, he clears the Ferris wheel by about 5.3 m.

(b) If he reaches his maximum height when he is over the middle Ferris wheel, what is his clearance above it?

SOLUTION: At maximum height, $v_y = 0$ and Eq. 4-18 becomes

$$v_y^2 = (v_0 \sin \theta_0)^2 - 2gy = 0.$$

Solving for y gives

$$y = \frac{(v_0 \sin \theta_0)^2}{2g} = \frac{(26.5 \text{ m/s})^2(\sin 53°)^2}{(2)(9.8 \text{ m/s}^2)} = 22.9 \text{ m,}$$

which means that he clears the middle Ferris wheel by 7.9 m.

(c) What is his time of flight t?

SOLUTION: Of the several ways to find t, we could use Eq. 4-16 and the fact that $y = 0$ when he lands. Then we have

$$y = (v_0 \sin \theta_0)t - \tfrac{1}{2}gt^2 = 0,$$

or $\qquad t = \dfrac{2v_0 \sin \theta_0}{g} = \dfrac{(2)(26.5 \text{ m/s})(\sin 53°)}{9.8 \text{ m/s}^2}$

$$= 4.3 \text{ s.} \qquad \text{(Answer)}$$

(d) How far from the cannon should the center of the net be positioned?

SOLUTION: One way to answer is with Eq. 4-15, with $x_0 = 0$:

$$x = (v_0 \cos \theta_0)t$$
$$= (26.5 \text{ m/s})(\cos 53°)(4.3 \text{ s})$$
$$= 69 \text{ m,} \qquad \text{(Answer)}$$

which is the range R of the flight.

We can now answer the questions that opened this chapter: How could Zacchini know where to place the net? And how could he be certain he would clear the Ferris wheels? He (or someone) did the calculations as we have here. Although he could not take into account the complicated effects of the air on his flight, Zacchini knew that the air would slow him and thus decrease his range from the calculated value. So he used a wide net and biased it toward the cannon. He was then relatively safe whether the effects of the air in a particular flight happened to slow him considerably or very little. Still, the variability of this factor of air effects must have played on his imagination before each flight.

The circus performer still faced a subtle danger: even for shorter flights, his propulsion in the muzzle of the cannon was so severe that he underwent a momentary blackout. If he landed during the blackout, he could break his neck. So, he had trained himself to awake quickly. Indeed, not waking up in time presents the only real danger to a human cannonball in a circus of today, where a flight is much shorter and safer than Zacchini's.

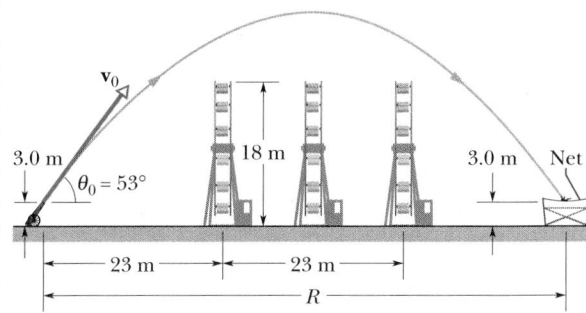

FIGURE 14-17 Sample Problem 4-9. The flight of a human cannonball over three Ferris wheels, and the desired placement of the net.

PROBLEM SOLVING TACTICS

TACTIC 3: *Numbers Versus Algebra*

One way to avoid numerical rounding errors is to solve problems algebraically, substituting numbers only in the final step. That is easy to do in Sample Problems 4-6 to 4-9, and that is the way experienced problem solvers operate. In these early chapters, however, we prefer to solve most problems in pieces, to give you a firmer numerical grasp of what is going on. Later we shall stick more to the algebra.

4-7 UNIFORM CIRCULAR MOTION

A particle is in **uniform circular motion** if it travels around a circle or circular arc at constant speed. Although the speed does not vary, *the particle is accelerating*. That fact may be surprising because we usually think of acceleration as an increase or decrease in speed. But actually **v** is a vector, not a scalar. If **v** changes, even only in direction, there is an acceleration, and that is what happens in uniform circular motion.

We use Fig. 4-18 to find the magnitude and direction of the acceleration. That figure represents a particle in uniform circular motion with speed v in a circle of radius r. Velocity vectors are drawn for two points, p and q, that are symmetric with respect to the y axis. These vectors, $\mathbf{v}_p$ and $\mathbf{v}_q$, have the same magnitude v but—because they point in different directions—they are different vectors. Their x and y components are

$$v_{px} = +v \cos \theta, \qquad v_{py} = +v \sin \theta$$

and
$$v_{qx} = +v \cos \theta, \qquad v_{qy} = -v \sin \theta.$$

The time required for the particle to move from p to q at constant speed v is

$$\Delta t = \frac{\text{arc}(pq)}{v} = \frac{r(2\theta)}{v}, \qquad (4\text{-}21)$$

in which $\text{arc}(pq)$ is the length of the arc from p to q.

We now have enough information to calculate the components of the average acceleration $\bar{\mathbf{a}}$ experienced by the particle as it moves from p to q in Fig. 4-18. For the x component, we have

$$\bar{a}_x = \frac{v_{qx} - v_{px}}{\Delta t} = \frac{v \cos \theta - v \cos \theta}{\Delta t} = 0.$$

This result is not surprising because it is clear from symmetry in Fig. 4-18 that the x component of velocity has the same value at q and at p.

For the y component of the average acceleration, we find, making use of Eq. 4-21,

$$\bar{a}_y = \frac{v_{qy} - v_{py}}{\Delta t} = \frac{-v \sin \theta - v \sin \theta}{\Delta t}$$

$$= -\frac{2v \sin \theta}{2r\theta/v} = -\left(\frac{v^2}{r}\right)\left(\frac{\sin \theta}{\theta}\right).$$

The minus sign tells us that this acceleration component points vertically downward in Fig. 4-18.

Now let us allow the angle θ in Fig. 4-18 to shrink, approaching zero as a limit. This requires that points p and q in that figure approach their midpoint, shown as point P at the top of the circle. The average acceleration $\bar{\mathbf{a}}$, whose components we have just found, then approaches the instantaneous acceleration $\mathbf{a}$ at point P.

The *direction* of this instantaneous acceleration vector at point P in Fig. 4-18 is downward, toward the center of the circle at O, because the direction of the average acceleration does not change as θ becomes smaller. To find the *magnitude* a of the instantaneous acceleration vector, we need only the mathematical fact that as θ shrinks, the ratio $(\sin \theta)/\theta$ approaches unity. From the relation given above for $\bar{a}_y$, we then have

$$a = \frac{v^2}{r} \qquad \text{(centripetal acceleration).} \qquad (4\text{-}22)$$

We conclude:

When a particle moves at constant speed v in a circle (or a circular arc) of radius r, the acceleration of the particle is directed toward the center of the circle and has a constant magnitude v^2/r.

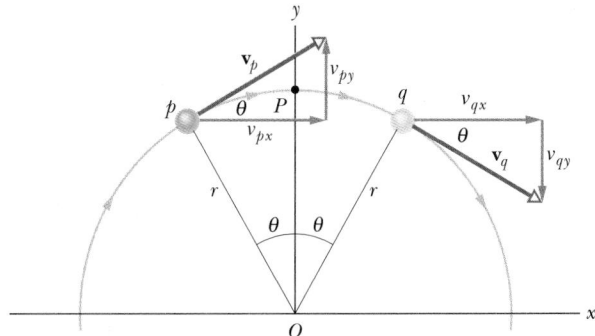

FIGURE 4-18 A particle moves in uniform circular motion at constant speed v in a circle of radius r. Its velocities $\mathbf{v}_p$ and $\mathbf{v}_q$ at points p and q, equidistant from the y axis, are shown, along with its velocity components at those points. The instantaneous acceleration of the particle at any point is directed toward the center of the circle and has a magnitude v^2/r.

In addition, during this acceleration at constant speed, the particle travels the circumference of the circle (a distance of $2\pi r$) in time

$$T = \frac{2\pi r}{v} \qquad \text{(period).} \qquad (4\text{-}23)$$

T is called the *period of revolution,* or simply the *period,* of the motion. It is, in general, the time for a particle to go around a closed path exactly once.

Figure 4-19 shows the relation between the velocity and acceleration vectors at various stages during uniform circular motion. Both vectors have constant magnitude as the motion progresses, but their directions change continuously. The velocity is always tangent to the circle in the direction of motion; the acceleration is always directed radially inward. Because of this, the acceleration associated with uniform circular motion is called a **centripetal** (meaning "center seeking") **acceleration.**

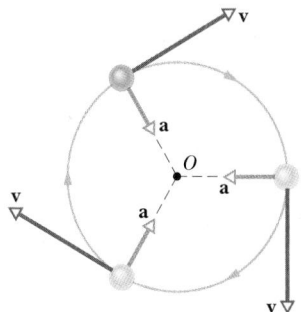

FIGURE 4-19 Velocity and acceleration vectors for a particle in uniform circular motion. Both have constant magnitude but vary continuously in direction.

The acceleration resulting from a change in the direction of a velocity is just as real as one resulting from a change in the magnitude of a velocity. In Fig. 2-8, for example, we saw Colonel John P. Stapp while his rocket sled was braked to rest. He was experiencing a velocity that was constant in direction but changing rapidly in magnitude. On the other hand, an astronaut whirling in a human centrifuge experiences a velocity that is constant in magnitude but changing rapidly in direction. The accelerations that these two people feel are indistinguishable.

CHECKPOINT **6:** An object moves in a circular path in the xy plane, with the center at the origin. When the object is at $x = -2$ m, its velocity is $-(4$ m/s$)\mathbf{j}$. Give the object's (a) velocity and (b) centripetal acceleration when it is at $y = 2$ m.

SAMPLE PROBLEM 4-10

''Top gun'' pilots have long worried about taking a turn too tightly. As a pilot's body undergoes centripetal acceleration, with the head toward the center of curvature, the blood pressure in the brain decreases, leading to loss of brain function.

There are several warning signs to signal a pilot to ease up: when the centripetal acceleration is $2g$ or $3g$, the pilot feels heavy. At about $4g$, the pilot's vision switches to black and white and narrows down to ''tunnel vision.'' If that acceleration is sustained or increased, vision ceases and soon after, the pilot is unconscious—a condition known as g-LOC for ''g-induced loss of consciousness.''

What is the centripetal acceleration, in g units, of a pilot flying an F-22 at speed $v = 1600$ mi/h (716 m/s) through a circular arc with radius of curvature $r = 3.60$ mi (5.80 km)?

SOLUTION: Substituting the given data into Eq. 4-22, we have

$$a = \frac{v^2}{r} = \frac{(716 \text{ m/s})^2}{5800 \text{ m}}$$

$$= 88.39 \text{ m/s}^2 = 9.0g. \qquad \text{(Answer)}$$

If an unwary pilot caught in a dogfight puts the aircraft into such a tight turn, the pilot goes into g-LOC almost immediately, with no warning signs to signal the danger.

SAMPLE PROBLEM 4-11

A satellite is in circular Earth orbit, at an altitude $h = 200$ km above the Earth's surface. There the free-fall acceleration g is 9.20 m/s^2. What is the orbital speed v of the satellite?

SOLUTION: We have uniform circular motion around the

Earth. The satellite's centripetal acceleration is then the free-fall acceleration g. So we can find v from Eq. 4-22, with $a = g$ and with $r = R_E + h$, where R_E is the Earth's radius (see inside front cover or Appendix C):

$$g = \frac{v^2}{R_E + h}.$$

Solving for v gives

$$v = \sqrt{g(R_E + h)}$$
$$= \sqrt{(9.20 \text{ m/s}^2)(6.37 \times 10^6 \text{ m} + 200 \times 10^3 \text{ m})}$$
$$= 7770 \text{ m/s} = 7.77 \text{ km/s}. \qquad \text{(Answer)}$$

You can show that this is equivalent to 17,400 mi/h and that the satellite would take 1.47 h to complete one orbital revolution; that is, the period T of the motion is 1.47 h.

4-8 RELATIVE MOTION IN ONE DIMENSION

Suppose you see a duck flying north at, say, 20 mi/h. To another duck flying alongside, the first duck is at rest. In other words, the velocity of a particle depends on the **reference frame** of whoever is doing the measuring. For our purposes, a reference frame is the physical object to which you attach your coordinate system.

The reference frame that seems most natural to us in our daily comings and goings is the ground beneath our feet. When a traffic officer tells you that you have been driving at 70 mi/h, the unspoken qualification, ''in a coordinate system attached to the ground,'' is always taken for granted by each of you.

If you are traveling in an airplane or a spaceship, the reference frame of the Earth may not be the most convenient one. You are free to choose any reference frame that you wish. Once having made it, however, you must always be aware of your choice and be careful to make all your measurements with respect to that reference frame.

Suppose that Alex (frame A) is parked by the side of a highway, watching car P (the ''particle'') speed past. Barbara (frame B), driving along the highway at constant speed, is also watching car P. Suppose that, as in Fig. 4-20, they both measure the position of the car at a given moment. From the figure we see that

$$x_{PA} = x_{PB} + x_{BA}. \qquad (4\text{-}24)$$

The terms in Eq. 4-24 are scalars and may be of either sign. The equation is read: ''The coordinate of P as measured by A *is equal to* the coordinate of P as measured by B plus the coordinate of B as measured by A.'' Note how this reading is supported by the sequence of the subscripts.

Taking the time derivative of Eq. 4-24, we obtain

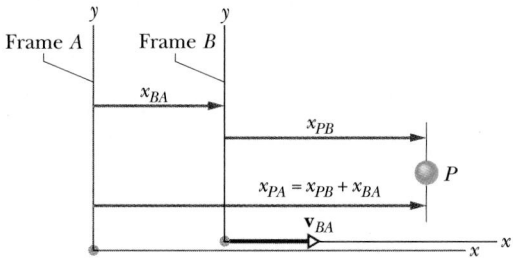

FIGURE 4-20 Alex (frame A) and Barbara (frame B) watch car P. All motion is along the common x axis of the two frames. The vector $\mathbf{v}_{BA}$ shows the relative separation velocity of the two frames. The three position measurements shown are all made at the same instant of time.

$$\frac{d}{dt}(x_{PA}) = \frac{d}{dt}(x_{PB}) + \frac{d}{dt}(x_{BA}),$$

or (because $v = dx/dt$)

$$v_{PA} = v_{PB} + v_{BA}. \qquad (4\text{-}25)$$

This scalar equation is the relation between the velocities of the same object (car P) as measured in the two frames; those measured velocities are different. In words, Eq. 4-25 says: "The velocity of P as measured by A *is equal to* the velocity of P as measured by B *plus* the velocity of B as measured by A." The term v_{BA} is the separation velocity of frame B with respect to frame A; see Fig. 4-20.

We consider only frames that move at constant velocity; such frames are called **inertial reference frames.** In our example, this means that Barbara (frame B) will drive always at constant speed with respect to Alex (frame A). Car P (the moving particle), however, may speed up, slow down, come to rest, or reverse direction.

The time derivative of the velocity equation (Eq. 4-25) yields the acceleration equation,

$$a_{PA} = a_{PB}. \qquad (4\text{-}26)$$

Note that because v_{BA} is a constant, its time derivative is zero. Equation 4-26 tells us:

Observers on different inertial frames of reference (their velocity of separation is constant) will measure the same acceleration for a moving particle.

CHECKPOINT **7**: The table gives velocities (km/h) for Barbara and car P of Fig. 4-20 for three situations. For each, what is the missing value and how is the distance between Barbara and car P changing?

SITUATION	v_{BA}	v_{PA}	v_{PB}
1	+50	+50	
2	+30		+40
3		+60	−20

SAMPLE PROBLEM 4-12

Alex, parked by the side of an east–west road, is watching car P, which is moving in a westerly direction. Barbara, driving east at a speed $v_{BA} = 52$ km/h, watches the same car. Take the easterly direction as positive.

(a) If Alex measures a speed of 78 km/h for car P, what velocity will Barbara measure?

SOLUTION: Equation 4-25 may be rearranged to yield

$$v_{PB} = v_{PA} - v_{BA}.$$

We have $v_{PA} = -78$ km/h, the minus sign telling us that car P is moving west, in the negative direction. We also have $v_{BA} = 52$ km/h, so that

$$v_{PB} = (-78 \text{ km/h}) - (52 \text{ km/h})$$
$$= -130 \text{ km/h}. \qquad \text{(Answer)}$$

If car P were connected to Barbara's car by a string wound up on a spool, the string would be unwinding at this speed as the two cars separated.

(b) If Alex sees car P brake to a halt in 10 s, what acceleration (assumed constant) will he measure for it?

SOLUTION: From Eq. 2-11 ($v = v_0 + at$) we have

$$a = \frac{v - v_0}{t} = \frac{0 - (-78 \text{ km/h})}{10 \text{ s}}$$
$$= \left(\frac{78 \text{ km/h}}{10 \text{ s}}\right)\left(\frac{1 \text{ m/s}}{3.6 \text{ km/h}}\right)$$
$$= 2.2 \text{ m/s}^2. \qquad \text{(Answer)}$$

(c) What acceleration would Barbara measure for the braking car?

SOLUTION: Barbara sees the initial speed of the car as −130 km/h, as we calculated in (a). Although the car has braked to rest, it is at rest only in Alex's reference frame. To Barbara, car P is not at rest at all but is receding from her at 52 km/h so that its final velocity in her reference frame is −52 km/h. Thus, from the relation $v = v_0 + at$,

$$a = \frac{v - v_0}{t} = \frac{(-52 \text{ km/h}) - (-130 \text{ km/h})}{10 \text{ s}}$$
$$= 2.2 \text{ m/s}^2. \qquad \text{(Answer)}$$

This is exactly the same acceleration that Alex measured, which reassures us that we made no mistakes.

4-9 RELATIVE MOTION IN TWO DIMENSIONS

Now we move from the scalar world of relative motion in one dimension to the vector world of relative motion in two (and, by extension, in three) dimensions. Figure 4-21 shows reference frames A and B, now two-dimensional. Our two observers are again watching a moving particle P. We once more assume that the two frames are separating at a constant velocity $\mathbf{v}_{BA}$ (the frames are inertial) and, for simplicity, we further assume that their x and y axes remain parallel to each other as they do so.

Let observers on frames A and B each measure the position of particle P at a certain instant. From the vector triangle in Fig. 4-21, we have the vector equation

$$\mathbf{r}_{PA} = \mathbf{r}_{PB} + \mathbf{r}_{BA}. \tag{4-27}$$

This relation is the vector equivalent of the scalar Eq. 4-24.

If we take the time derivative of Eq. 4-27, we find a connection between the (vector) velocities of the particle as measured by our two observers; namely,

$$\mathbf{v}_{PA} = \mathbf{v}_{PB} + \mathbf{v}_{BA}. \tag{4-28}$$

This is the vector equivalent of the scalar Eq. 4-25. Note that the order of the subscripts is the same as in that equation, and that again $\mathbf{v}_{BA}$ is the constant relative velocity of frame B as observed by the observer on frame A.

If we take the time derivative of Eq. 4-28, we obtain a connection between the two measured accelerations; namely,

$$\mathbf{a}_{PA} = \mathbf{a}_{PB}. \tag{4-29}$$

It remains true for motion in three dimensions that all observers on inertial reference frames will measure the same acceleration for a moving particle.

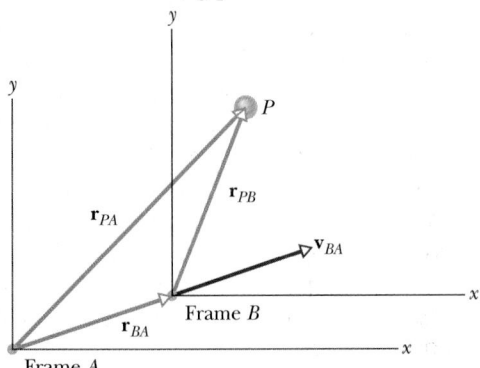

FIGURE 4-21   Reference frames in two dimensions. The vectors $\mathbf{r}_{PA}$ and $\mathbf{r}_{PB}$ show the positions of particle P in frames A and B, respectively. Vector $\mathbf{r}_{BA}$ shows the position of frame B with respect to frame A. Vector $\mathbf{v}_{BA}$ shows the (constant) separation velocity of the two frames.

SAMPLE PROBLEM 4-13

A bat detects an insect (lunch) while the two are flying with velocities $\mathbf{v}_{BG}$ and $\mathbf{v}_{IG}$, respectively, measured with respect to the ground. See Fig. 4-22a. What is the velocity $\mathbf{v}_{IB}$ of the insect with respect to the bat, in unit-vector notation?

SOLUTION: From Fig. 4-22a, the velocities of the insect and the bat, relative to the ground, are given by

$$\mathbf{v}_{IG} = (5.0 \text{ m/s})(\cos 50°)\mathbf{i} + (5.0 \text{ m/s})(\sin 50°)\mathbf{j}$$

and

$$\mathbf{v}_{BG} = (4.0 \text{ m/s})(\cos 150°)\mathbf{i} + (4.0 \text{ m/s})(\sin 150°)\mathbf{j},$$

where in each term, the angle is relative to the positive direction of the x axis. Now here is the key idea: the velocity $\mathbf{v}_{IB}$ of the *insect relative to the bat* is given by the vector sum of the velocity $\mathbf{v}_{IG}$ of the *insect relative to the ground* and the velocity $\mathbf{v}_{GB}$ of the *ground relative to the bat;* that is,

$$\mathbf{v}_{IB} = \mathbf{v}_{IG} + \mathbf{v}_{GB},$$

as shown in Fig. 4-22b. (Note that on the right-hand side of the equation, the two inner subscripts are the same. Also note that the two outer ones are the same as those on the left-hand side and appear in the same order.)

By definition the vector $\mathbf{v}_{GB}$ is in the direction opposite that of the vector $\mathbf{v}_{BG}$. Hence $\mathbf{v}_{GB} = -\mathbf{v}_{BG}$, and so

$$\mathbf{v}_{IB} = \mathbf{v}_{IG} + (-\mathbf{v}_{BG}).$$

Substituting the unit-vector expressions for $\mathbf{v}_{IG}$ and $\mathbf{v}_{BG}$ into this expression, we find

$$
\begin{aligned}
\mathbf{v}_{IB} &= (5.0 \text{ m/s})(\cos 50°)\mathbf{i} + (5.0 \text{ m/s})(\sin 50°)\mathbf{j} \\
&\quad - (4.0 \text{ m/s})(\cos 150°)\mathbf{i} - (4.0 \text{ m/s})(\sin 150°)\mathbf{j} \\
&= 3.21\mathbf{i} + 3.83\mathbf{j} + 3.46\mathbf{i} - 2.0\mathbf{j} \\
&\approx (6.7 \text{ m/s})\mathbf{i} + (1.8 \text{ m/s})\mathbf{j}. \quad\text{(Answer)}
\end{aligned}
$$

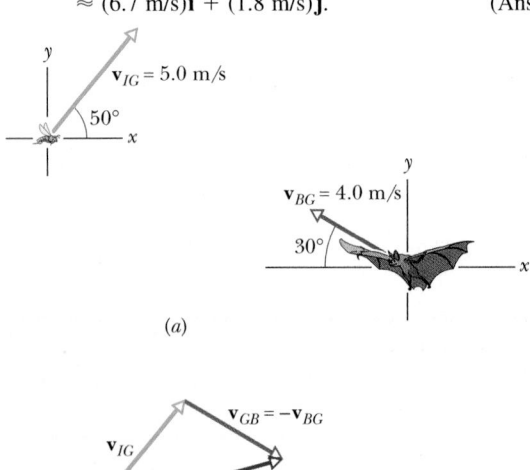

(a)

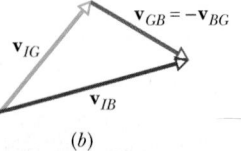

(b)

FIGURE 4-22   Sample Problem 4-13. (a) A bat detects an insect. (b) Velocity vectors of insect and bat.

SAMPLE PROBLEM 4-14

The compass in a plane indicates that the plane is headed (pointed) due east; its airspeed indicator reads 215 km/h. (Airspeed is speed relative to the air.) A steady wind of 65.0 km/h is blowing due north.

(a) What is the velocity of the plane with respect to the ground?

SOLUTION: The moving "particle" in this problem is the plane P. There are two reference frames, the ground (G) and the air mass (M). By a simple change of notation, we can rewrite Eq. 4-28 as

$$\mathbf{v}_{PG} = \mathbf{v}_{PM} + \mathbf{v}_{MG}. \qquad (4\text{-}30)$$

Figure 4-23a shows these vectors, which form a right triangle. The terms in Eq. 4-30 are, in sequence, the velocity of the plane with respect to the ground, the velocity of the plane with respect to the air, and the velocity of the air with respect to the ground (that is, the wind velocity). Note that the orientation of the plane in Fig. 4-23a is consistent with a due east reading on its compass. The plane is pointed due east but may not actually be moving in that direction.

The magnitude of the plane's velocity relative to the ground is found from the triangle in Fig. 4-23a:

$$\begin{aligned} v_{PG} &= \sqrt{v_{PM}^2 + v_{MG}^2} \\ &= \sqrt{(215 \text{ km/h})^2 + (65.0 \text{ km/h})^2} \\ &= 225 \text{ km/h}. \qquad \text{(Answer)} \end{aligned}$$

The angle α in Fig. 4-23a follows from

$$\begin{aligned} \alpha &= \tan^{-1} \frac{v_{MG}}{v_{PM}} = \tan^{-1} \frac{65.0 \text{ km/h}}{215 \text{ km/h}} \\ &= 16.8°. \qquad \text{(Answer)} \end{aligned}$$

Thus, with respect to the ground, the plane is flying at 225 km/h in a direction 16.8° north of east. Note that its speed relative to the ground (the "ground speed") is greater than its airspeed.

(b) If the pilot wishes to fly due east, what must be the heading? That is, what must the compass read?

SOLUTION: In this case the pilot must head into the wind somewhat, so that the wind velocity will help ensure that the velocity of the plane with respect to the ground points east. The wind velocity is unchanged from the preceding situation (a), and the vector diagram representing the new situation is as shown in Fig. 4-23b. Note that the three vectors still form a right triangle, as they did in Fig. 4-23a, and Eq. 4-30 still holds.

The pilot's ground speed is now, from Fig. 4-23b,

$$\begin{aligned} v_{PG} &= \sqrt{v_{PM}^2 - v_{MG}^2} \\ &= \sqrt{(215 \text{ km/h})^2 - (65.0 \text{ km/h})^2} = 205 \text{ km/h}. \end{aligned}$$

As the orientation of the plane in Fig. 4-23b indicates, the pilot must head into the wind by an angle θ given by

$$\theta = \sin^{-1} \frac{v_{MG}}{v_{PM}} = \sin^{-1} \frac{65.0 \text{ km/h}}{215 \text{ km/h}} = 17.6°. \quad \text{(Answer)}$$

Note that the ground speed is now less than the airspeed because the plane is headed into the wind.

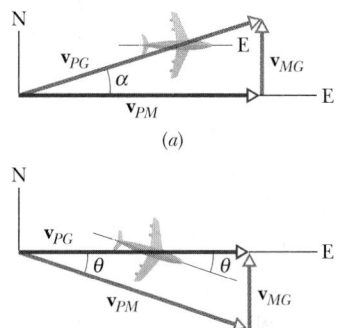

(a)

(b)

FIGURE 4-23 Sample Problem 4-14. (a) An airplane, heading due east, is blown to the north. (b) To travel due east, the plane must head into the wind.

4-10 RELATIVE MOTION AT HIGH SPEEDS (OPTIONAL)

An orbiting satellite has a speed of 17,000 mi/h. Before you call that a high speed, you must answer this question: "High compared to what?" Nature has given us a standard; it is the speed of light c, where

$$c = 299{,}792{,}458 \text{ m/s} \qquad \text{(speed of light)}. \quad (4\text{-}31)$$

As we shall examine later, no entity—be it particle or wave—can move faster than the speed of light, no matter what reference frame is used for observation. By this standard, all tangible large-scale objects—no matter how high their speeds seem by ordinary standards—are very slow indeed. The speed of the orbiting satellite, for example, is only 0.0025% of the speed of light. Subatomic particles such as electrons or protons, however, can acquire speeds very close to (but never equal to or greater than) the speed of light. Experiment shows, for example, that an electron accelerated through 10 million volts acquires a speed of 0.9988c; if you accelerate it through 20 million volts, its speed increases, but only to 0.9997c. The speed of light is a barrier that objects can approach but never reach. (Alas, the faster-than-light speeds utilized in science fiction, such as during "warp drive" in *Star Trek,* when the speed is $c2^n$, where n is the warp number, are only fictional.)

How can we be sure that the kinematics we have examined so far, which was developed by studying very slow, ordinary objects, also holds for very fast objects, such as high-speed electrons or protons? The answer, which we can find only from experiment, is that the kinematics for slow objects does *not* hold at speeds that approach the speed of light. Einstein's **theory of special relativity,** however, agrees with experiment at *all* speeds.

At "slow" speeds—all speeds that can be acquired by ordinary objects—the kinematic equations of Einstein's theory reduce to those of the kinematics we have examined. The failure of the "slow kinematics" is gradual, its predictions agreeing less and less well with experiment as the speed increases. Here is an example: Eq. 4-25,

$$v_{PA} = v_{PB} + v_{BA} \quad \text{(slow speeds),}$$

relates the speed of particle P as seen by an observer in frame B to that seen by an observer in frame A. The corresponding equation in Einstein's theory is

$$v_{PA} = \frac{v_{PB} + v_{BA}}{1 + v_{PB}v_{BA}/c^2} \quad \text{(all speeds).} \quad (4\text{-}32)$$

If $v_{PB} \ll c$ and $v_{BA} \ll c$ (which is always the case for ordinary objects), then the denominator in Eq. 4-32 is very close to unity and Eq. 4-32 reduces to Eq. 4-25, as we know it must.

The speed of light c is the central constant of Einstein's theory, and every relativistic equation contains it. A way to test the validity of such equations is to allow c to become infinitely large. Under those conditions, *all* speeds would be "slow" and "slow kinematics" would never fail. Putting $c \to \infty$ in Eq. 4-32 does indeed reduce that equation to Eq. 4-25.

SAMPLE PROBLEM 4-15

(Slow speeds)   For the case where $v_{PB} = v_{BA} = 0.0001c$ ($= 67,000$ mi/h!), what do Eqs. 4-25 and 4-32 predict for v_{PA}?

SOLUTION: From Eq. 4-25,

$$v_{PA} = v_{PB} + v_{BA} = 0.0001c + 0.0001c$$
$$= 0.0002c. \quad \text{(Answer)}$$

From Eq. 4-32,

$$v_{PA} = \frac{v_{PB} + v_{BA}}{1 + v_{PB}v_{BA}/c^2} = \frac{0.0001c + 0.0001c}{1 + (0.0001c)^2/c^2}$$
$$= \frac{0.0002c}{1.00000001} \approx 0.0002c. \quad \text{(Answer)}$$

Conclusion: At any speed acquired by ordinary tangible objects, Eqs. 4-25 and 4-32 yield essentially the same answer. For such speeds, we can use Eq. 4-25 ("slow kinematics") without a second thought.

SAMPLE PROBLEM 4-16

(High speeds)   For $v_{PB} = v_{BA} = 0.65c$, what do Eqs. 4-25 and 4-32 predict for v_{PA}?

SOLUTION: From Eq. 4-25,

$$v_{PA} = v_{PB} + v_{BA} = 0.65c + 0.65c$$
$$= 1.30c. \quad \text{(Answer)}$$

From Eq. 4-32,

$$v_{PA} = \frac{v_{PB} + v_{BA}}{1 + v_{PB}v_{BA}/c^2} = \frac{0.65c + 0.65c}{1 + (0.65c)(0.65c)/c^2}$$
$$= \frac{1.30c}{1.423} = 0.91c. \quad \text{(Answer)}$$

Conclusion: At high speeds, "slow kinematics" and special relativity predict very different results. "Slow kinematics" involves no upper limit on speed and can easily (as in this case) predict a speed greater than the speed of light. Special relativity, on the other hand, *never* predicts a speed that exceeds c, no matter how high the combining speeds. Experiment agrees with special relativity on all counts.

REVIEW & SUMMARY

Position Vector

The location of a particle relative to the origin of a coordinate system is given by a *position vector* $\mathbf{r}$, which in unit-vector notation is

$$\mathbf{r} = x\mathbf{i} + y\mathbf{j} + z\mathbf{k}. \quad (4\text{-}1)$$

Here $x\mathbf{i}$, $y\mathbf{j}$, and $z\mathbf{k}$ are the *vector components* of position vector $\mathbf{r}$, and x, y, and z are its *scalar components*. A position vector is described by a magnitude and one or two angles for orientation, or by its vector or scalar components.

Displacement

If a particle moves so that its position vector changes from $\mathbf{r}_1$ to $\mathbf{r}_2$, then the particle's *displacement* $\Delta\mathbf{r}$ is

$$\Delta\mathbf{r} = \mathbf{r}_2 - \mathbf{r}_1. \quad (4\text{-}2)$$

The displacement can also be written as

$$\Delta\mathbf{r} = (x_2 - x_1)\mathbf{i} + (y_2 - y_1)\mathbf{j} + (z_2 - z_1)\mathbf{k}, \quad (4\text{-}3)$$

where coordinates (x_1, y_1, z_1) correspond to position vector $\mathbf{r}_1$ and coordinates (x_2, y_2, z_2) correspond to position vector $\mathbf{r}_2$.

Average Velocity

If a particle undergoes a displacement $\Delta \mathbf{r}$ in time Δt, its *average velocity* $\bar{\mathbf{v}}$ for that time interval is

$$\bar{\mathbf{v}} = \frac{\Delta \mathbf{r}}{\Delta t}. \tag{4-4}$$

Velocity

As Δt in Eq. 4-4 is shrunk to 0, $\bar{\mathbf{v}}$ reaches a limit called the (instantaneous) *velocity:*

$$\mathbf{v} = \frac{d\mathbf{r}}{dt}, \tag{4-6}$$

which can be rewritten in unit-vector notation as

$$\mathbf{v} = v_x \mathbf{i} + v_y \mathbf{j} + v_z \mathbf{k}, \tag{4-7}$$

where $v_x = dx/dt$, $v_y = dy/dt$, and $v_z = dz/dt$. When the position of a moving particle is plotted on a coordinate system, $\mathbf{v}$ is always tangent to the curve representing the particle's path.

Average Acceleration

If a particle's velocity changes from $\mathbf{v}_1$ to $\mathbf{v}_2$ in time interval Δt, its *average acceleration* during Δt is

$$\bar{\mathbf{a}} = \frac{\mathbf{v}_2 - \mathbf{v}_1}{\Delta t} = \frac{\Delta \mathbf{v}}{\Delta t}. \tag{4-9}$$

Acceleration

As Δt in Eq. 4-9 is shrunk to 0, $\bar{\mathbf{a}}$ reaches a limiting value called the (instantaneous) *acceleration,* which is

$$\mathbf{a} = \frac{d\mathbf{v}}{dt}. \tag{4-10}$$

In unit-vector notation,

$$\mathbf{a} = a_x \mathbf{i} + a_y \mathbf{j} + a_z \mathbf{k}, \tag{4-11}$$

where $a_x = dv_x/dt$, $a_y = dv_y/dt$, and $a_z = dv_z/dt$.

When $\mathbf{a}$ is constant, the components of $\mathbf{a}$, $\mathbf{v}$, and $\mathbf{r}$ along any axis can be treated as in the one-dimensional motion of Chapter 2.

Projectile Motion

Projectile motion is the motion of a particle that is launched with an initial velocity $\mathbf{v}_0$ and then has the free-fall acceleration $\mathbf{g}$. If $\mathbf{v}_0$ is expressed as a magnitude (the speed v_0) and an angle θ_0, the equations of motion along the horizontal x axis and vertical y axis are

$$x - x_0 = (v_0 \cos \theta_0)t, \tag{4-15}$$

$$y - y_0 = (v_0 \sin \theta_0)t - \tfrac{1}{2}gt^2, \tag{4-16}$$

$$v_y = v_0 \sin \theta_0 - gt, \tag{4-17}$$

$$v_y^2 = (v_0 \sin \theta_0)^2 - 2g(y - y_0). \tag{4-18}$$

The path of a particle in projectile motion is parabolic and is given by

$$y = (\tan \theta_0)x - \frac{gx^2}{2(v_0 \cos \theta_0)^2}, \tag{4-19}$$

where the origin has been chosen so that x_0 and y_0 are both zero. The particle's **range** R, which is the horizontal distance from the launch point to the point at which the particle returns to the launch height, is

$$R = \frac{v_0^2}{g} \sin 2\theta_0. \tag{4-20}$$

Uniform Circular Motion

If a particle travels along a circle or circular arc with radius r at constant speed v, it is in *uniform circular motion* and has an acceleration $\mathbf{a}$ of magnitude

$$a = \frac{v^2}{r}. \tag{4-22}$$

The direction of $\mathbf{a}$ is toward the center of the circle or circular arc, and $\mathbf{a}$ is said to be centripetal. The time for the particle to complete a circle is

$$T = \frac{2\pi r}{v}. \tag{4-23}$$

T is called the *period of revolution,* or simply the *period,* of the motion.

Relative Motion

When two frames of reference A and B are moving relative to each other at constant velocity, they are said to be **inertial reference frames.** The velocity of a moving particle as measured by an observer in frame A, in general, differs from that measured from frame B. The two measured velocities are related by

$$\mathbf{v}_{PA} = \mathbf{v}_{PB} + \mathbf{v}_{BA}, \tag{4-28}$$

in which $\mathbf{v}_{BA}$ is the velocity of B with respect to A. Both observers measure the same acceleration for the particle; that is,

$$\mathbf{a}_{PA} = \mathbf{a}_{PB}. \tag{4-29}$$

If speeds near the speed of light are involved, Eq. 4-25 ($v_{PA} = v_{PB} + v_{BA}$) must be replaced with an equation derived using the **special theory of relativity.** For one-dimensional motion, the correct result is

$$v_{PA} = \frac{v_{PB} + v_{BA}}{1 + v_{PB}v_{BA}/c^2}, \tag{4-32}$$

which becomes identical to Eq. 4-25 if the speeds are all negligible compared to the speed of light c.

QUESTIONS

1. Here are four descriptions for the velocity of a hockey puck in the xy plane, all in meters per second:

(1) $v_x = -3t^2 + 4t - 2$ and $v_y = 6t - 4$
(2) $v_x = -3$ and $v_y = -5t^2 + 6$
(3) $\mathbf{v} = 2t^2\mathbf{i} - (4t + 3)\mathbf{j}$
(4) $\mathbf{v} = -2t\mathbf{i} + 3\mathbf{j}$

For each description, determine whether the x and y components of the acceleration are constant, and whether the acceleration vector $\mathbf{a}$ is constant. In description 4, if $\mathbf{v}$ is in meters per second and t is in seconds, what must be the units of the coefficients -2 and 3?

2. Figure 4-24 shows three situations in which identical projectiles are launched from the ground (at the same levels) at identical initial speeds and angles. The projectiles do not land on the same terrain, however. Rank the situations according to the final speeds of the projectiles just before they land, greatest first.

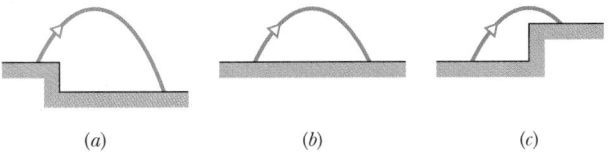

(a) (b) (c)

FIGURE 4-24 Question 2.

3. At what point in the path of a projectile is the speed (a) a minimum and (b) a maximum?

4. At a certain instant, a fly ball has velocity $\mathbf{v} = 25\mathbf{i} - 4.9\mathbf{j}$ (x is horizontal, y is upward, and $\mathbf{v}$ is in meters per second). Has the ball passed the highest point of its trajectory?

5. You are to launch a rocket, from just above the ground, with one of the following initial velocity vectors: (1) $\mathbf{v}_0 = 20\mathbf{i} + 70\mathbf{j}$, (2) $\mathbf{v}_0 = -20\mathbf{i} + 70\mathbf{j}$, (3) $\mathbf{v}_0 = 20\mathbf{i} - 70\mathbf{j}$, (4) $\mathbf{v}_0 = -20\mathbf{i} - 70\mathbf{j}$. In your coordinate system, x runs along level ground and y increases upward. (a) Rank the vectors according to the launch speed of the projectile, greatest first. (b) Rank the vectors according to the time of flight of the projectile, greatest first.

6. A snowball is thrown from ground level (by someone in a hole) with initial speed v_0 at an angle of $45°$ relative to the (level) ground, on which the snowball later lands. If the launch angle is increased, do (a) the range and (b) the flight time increase, decrease, or stay the same?

7. A mud ball is launched 2 m above the ground with initial velocity $\mathbf{v}_0 = (2\mathbf{i} + 4\mathbf{j})$ m/s. What is its velocity just before it lands on a surface that is 2 m above the ground?

8. An airplane flying horizontally at a constant speed of 350 km/h over level ground releases a bundle of food supplies. Ignore the effect of the air on the bundle. What are the bundle's initial (a) vertical and (b) horizontal components of velocity? (c) What is its horizontal component of velocity just before hitting the ground? (d) If the airplane's speed were, instead, 450 km/h, would the time of fall be larger, smaller, or the same?

9. Figure 4-25 shows three paths for a kicked football. Ignoring the effects of air on the flight, rank the paths according to (a) time of flight, (b) initial vertical velocity component, (c) initial horizontal velocity component, and (d) initial speed. Place the greatest first in each part.

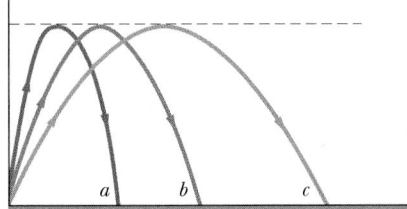

FIGURE 4-25 Question 9.

10. Figure 4-26 shows the velocity and acceleration of a particle at a particular instant in three situations. In which situation, and at that instant, is (a) the speed increasing, (b) the speed decreasing, (c) the speed not changing, (d) $\mathbf{v} \cdot \mathbf{a}$ positive, (e) $\mathbf{v} \cdot \mathbf{a}$ negative, and (f) $\mathbf{v} \cdot \mathbf{a} = 0$?

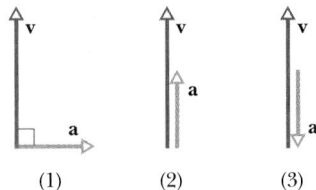

(1) (2) (3)

FIGURE 4-26 Question 10.

11. You are driving directly behind a pickup truck, going at the same speed as the truck. A crate falls from the bed of the truck to the road. (a) Will your car hit the crate before the crate hits the road if you neither brake nor swerve? (b) During the fall, is the horizontal speed of the crate more than, less than, or the same as that of the truck?

12. (a) Is it possible to be accelerating while traveling at constant speed? Is it possible to round a curve with (b) zero acceleration and (c) a constant magnitude of acceleration?

13. While riding in a moving car, you toss an egg directly upward. Does the egg tend to land behind you, in front of you, or back in your hands if the car is (a) traveling at a constant speed, (b) increasing in speed, and (c) decreasing in speed?

14. A passenger in an elevator drops a coin while the elevator descends at constant speed. According to (a) the passenger and (b) someone waiting at the next floor, is the acceleration of the coin equal to, less than, or more than g?

15. A pickpocket standing on the rear observation platform of a train moving with constant velocity drops a wallet over the rear guardrail. Describe the path of the wallet as seen by (a) the pickpocket, (b) her accomplice waiting alongside the track at the drop site, and (c) a police officer on another train moving at constant speed in the direction opposite that of the pickpocket's train, on a parallel track.

16. When the Germans shelled Paris from 70 mi away with the WWI long-range artillery piece nicknamed "Big Bertha," the shells were fired at an angle greater than 45°; the Germans had discovered that a greater angle gave their gun a greater range, possibly even twice as long as that with a 45° angle. Does that result mean that the density of the air at high altitudes increases with altitude or decreases with altitude?

17. Figure 4-27 shows one of four star cruisers that are in a race. As each cruiser passes the starting line, a shuttle craft leaves the cruiser and races toward the finish line. You are the judge and are stationary relative to the start and finish lines. The speeds v_c of the four cruisers relative to you and the speeds v_s of the shuttles relative to their corresponding cruisers are, respectively: (1)

$0.70c$, $0.40c$; (2) $0.40c$, $0.70c$; (3) $0.20c$, $0.90c$; (4) $0.50c$, $0.60c$. Without written calculation, determine (a) which shuttle wins the race and (b) which is last.

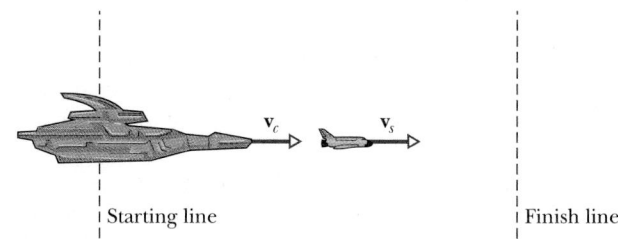

FIGURE 4-27 Question 17.

EXERCISES & PROBLEMS

SECTION 4-2 Position and Displacement

1E. A watermelon has the following coordinates: $x = -5.0$ m, $y = 8.0$ m, and $z = 0$ m. Find its position vector (a) in unit-vector notation and (b) in terms of the vector's magnitude and orientation. (c) Sketch the vector on a right-handed coordinate system. If the watermelon is moved to the xyz coordinates (3.00 m, 0 m, 0 m), what is its displacement (d) in unit-vector notation and (e) in terms of a magnitude and an orientation?

2E. The position vector for an electron is $\mathbf{r} = 5.0\mathbf{i} - 3.0\mathbf{j} + 2.0\mathbf{k}$, in meters. (a) Find the magnitude of $\mathbf{r}$. (b) Sketch the vector on a right-handed coordinate system.

3E. The position vector for a proton is initially $\mathbf{r} = 5.0\mathbf{i} - 6.0\mathbf{j} + 2.0\mathbf{k}$ and then later is $\mathbf{r} = -2.0\mathbf{i} + 6.0\mathbf{j} + 2.0\mathbf{k}$, all in meters. (a) What is the proton's displacement vector, and (b) to what plane is it parallel?

4E. A positron undergoes a displacement $\Delta\mathbf{r} = 2.0\mathbf{i} - 3.0\mathbf{j} + 6.0\mathbf{k}$, ending with the position vector $\mathbf{r} = 3.0\mathbf{j} - 4.0\mathbf{k}$, in meters. What was the positron's former position vector?

SECTION 4-3 Velocity and Average Velocity

5E. A plane flies 300 mi east from city A to city B in 45.0 min and then 600 mi south from city B to city C in 1.50 h. (a) What displacement vector represents the total trip? What are (b) the average velocity (a vector) and (c) the average speed for the trip?

6E. A train moving at a constant speed of 60.0 km/h moves east for 40.0 min, then in a direction 50.0° east of north for 20.0 min, and finally west for 50.0 min. What is the average velocity of the train during this run?

7E. In 3.50 h, a balloon drifts 21.5 km north, 9.70 km east, and 2.88 km in upward elevation from its release point on the ground. Find (a) the magnitude of its average velocity and (b) the angle its average velocity makes with the horizontal.

8E. An ion's position vector is initially $\mathbf{r} = 5.0\mathbf{i} - 6.0\mathbf{j} + 2.0\mathbf{k}$,

and 10 s later it is $\mathbf{r} = -2.0\mathbf{i} + 8.0\mathbf{j} - 2.0\mathbf{k}$, all in meters. What was its average velocity during the 10 s?

9E. The position of an electron is given by $\mathbf{r} = 3.0t\mathbf{i} - 4.0t^2\mathbf{j} + 2.0\mathbf{k}$ (where t is in seconds and the coefficients have the proper units for $\mathbf{r}$ to be in meters). (a) What is $\mathbf{v}(t)$ for the electron? (b) In unit-vector notation, what is $\mathbf{v}$ at $t = 2.0$ s? (c) What are the magnitude and direction of $\mathbf{v}$ just then?

SECTION 4-4 Acceleration and Average Acceleration

10E. A proton initially has $\mathbf{v} = 4.0\mathbf{i} - 2.0\mathbf{j} + 3.0\mathbf{k}$ and then 4.0 s later has $\mathbf{v} = -2.0\mathbf{i} - 2.0\mathbf{j} + 5.0\mathbf{k}$ (in meters per second). (a) In unit-vector notation, what is the average acceleration $\bar{\mathbf{a}}$ over the 4.0 s? (b) What are the magnitude and orientation of $\bar{\mathbf{a}}$?

11E. A particle moves so that its position as a function of time in SI units is $\mathbf{r} = \mathbf{i} + 4t^2\mathbf{j} + t\mathbf{k}$. Write expressions for (a) its velocity and (b) its acceleration as functions of time.

12E. The position $\mathbf{r}$ of a particle moving in an xy plane is given by $\mathbf{r} = (2.00t^3 - 5.00t)\mathbf{i} + (6.00 - 7.00t^4)\mathbf{j}$. Here $\mathbf{r}$ is in meters and t in seconds. Calculate (a) $\mathbf{r}$, (b) $\mathbf{v}$, and (c) $\mathbf{a}$ when $t = 2.00$ s. (d) What is the orientation of a line that is tangent to the particle's path at $t = 2.00$ s?

13E. An ice boat sails across the surface of a frozen lake with constant acceleration produced by the wind. At a certain instant the boat's velocity is $6.30\mathbf{i} - 8.42\mathbf{j}$ in meters per second. Three seconds later, because of a wind shift, the boat is instantaneously at rest. What is its average acceleration during this 3 s interval?

14P. A particle starts from the origin at $t = 0$ with a velocity of $8.0\mathbf{j}$ m/s and moves in the xy plane with a constant acceleration of $(4.0\mathbf{i} + 2.0\mathbf{j})$ m/s². (a) At the instant the x coordinate of the particle is 29 m, what is its y coordinate? (b) What is the speed of the particle at this time?

15P. A particle leaves the origin with an initial velocity $\mathbf{v} =$

3.00i, in meters per second. It experiences a constant acceleration **a** = −1.00i − 0.500j, in meters per second squared. (a) What is the velocity of the particle when it reaches its maximum *x* coordinate? (b) Where is the particle at this time?

16P. The velocity **v** of a particle moving in the *xy* plane is given by **v** = (6.0*t* − 4.0*t*²)**i** + 8.0**j**. Here **v** is in meters per second and *t* (> 0) is in seconds. (a) What is the acceleration when *t* = 3.0 s? (b) When (if ever) is the acceleration zero? (c) When (if ever) is the velocity zero? (d) When (if ever) does the speed equal 10 m/s?

17P. A particle *A* moves along the line *y* = 30 m with a constant velocity **v** (*v* = 3.0 m/s) directed parallel to the positive *x* axis (Fig. 4-28). A second particle *B* starts at the origin with zero speed and constant acceleration **a** (*a* = 0.40 m/s²) at the same instant that particle A passes the *y* axis. What angle θ between **a** and the positive *y* axis would result in a collision between these two particles? (If your computation involves an equation with a term such as *t*⁴, substitute *u* = *t*² and then consider solving the resulting quadratic equation to get *u*.)

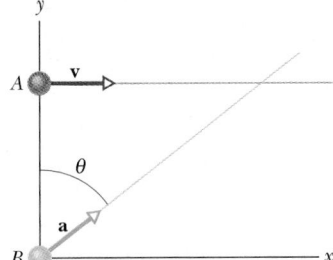

FIGURE 4-28
Problem 17.

SECTION 4-6 Projectile Motion Analyzed

In some of these problems, exclusion of the effects of the air is unwarranted but helps simplify the calculations.

18E. A dart is thrown horizontally toward the bull's-eye, point *P* on the dart board of Fig. 4-29, with an initial speed of 10 m/s. It hits at point *Q* on the rim, vertically below *P*, 0.19 s later. (a) What is the distance *PQ*? (b) How far away from the dart board did the dart thrower stand?

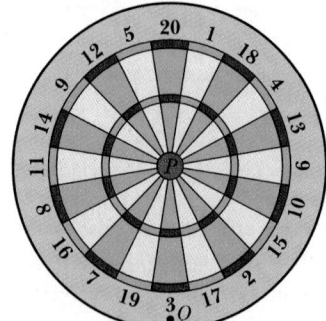

FIGURE 4-29
Exercise 18.

19E. A rifle is aimed horizontally at a target 100 ft away. The bullet hits the target 0.75 in. below the aiming point. (a) What is the bullet's time of flight? (b) What is its muzzle velocity?

20E. Electrons, like all other forms of matter, can undergo free fall. (a) If an electron is projected horizontally with a speed of

3.0 × 10⁶ m/s, how far will it fall in traversing 1.0 m of horizontal distance? (b) Does the answer increase or decrease if the initial speed is increased?

21E. In a cathode-ray tube, a beam of electrons is projected horizontally with a speed of 1.0 × 10⁹ cm/s into the region between a pair of horizontal plates 2.0 cm square. An electric field between the plates causes a constant downward acceleration of the electrons of magnitude 1.0 × 10¹⁷ cm/s². Find (a) the time required for an electron to pass through the plates, (b) the vertical displacement of the beam in passing through the plates (it does not run into a plate), and (c) the velocity of the beam as it emerges from the plates.

22E. A ball rolls horizontally off the edge of a tabletop that is 4.0 ft high. It strikes the floor at a point 5.0 ft horizontally away from the edge of the table. (a) How long was the ball in the air? (b) What was its speed at the instant it left the table?

23E. A projectile is fired horizontally from a gun that is 45.0 m above flat ground. The muzzle velocity is 250 m/s. (a) How long does the projectile remain in the air? (b) At what horizontal distance from the firing point does it strike the ground? (c) What is the magnitude of the vertical component of its velocity as it strikes the ground?

24E. A baseball leaves a pitcher's hand horizontally at a speed of 100 mi/h. The distance to the batter is 60 ft. (a) How long does it take for the ball to travel the first 30 ft horizontally? The second 30 ft? (b) How far does the ball fall under gravity during the first 30 ft of its horizontal travel? (c) During the second 30 ft? (d) Why aren't the quantities in (b) and (c) equal? (Ignore the effect of air resistance.)

25E. A projectile is launched with an initial speed of 30 m/s at an angle of 60° above the horizontal. Calculate the magnitude and direction of its velocity (a) 2.0 s and (b) 5.0 s after launch.

26E. A stone is catapulted rightward with an initial velocity of 20.0 m/s at an angle of 40.0° above level ground. Find its horizontal and vertical displacements (a) 1.10 s, (b) 1.80 s, and (c) 5.00 s after launch.

27E. You throw a ball from a cliff with an initial velocity of 15.0 m/s at an angle of 20.0° below the horizontal. Find (a) its horizontal displacement and (b) its vertical displacement 2.30 s later.

28E. You throw a ball with a speed of 25.0 m/s at an angle of 40.0° above the horizontal directly toward a wall (Fig. 4-30). The wall is 22.0 m from the release point of the ball. (a) How long does the ball take to reach the wall? (b) How far above the release point does the ball hit the wall? (c) What are the horizontal and vertical components of its velocity as it hits the wall? (d) When it hits, has it passed the highest point on its trajectory?

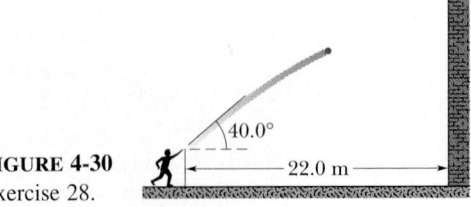

FIGURE 4-30
Exercise 28.

29E. (a) Prove that for a projectile fired from level ground at an angle θ_0 above the horizontal, the ratio of the maximum height H to the range R is given by $H/R = \frac{1}{4} \tan \theta_0$. See Fig. 4-31. (b) For what angle θ_0 does $H = R$?

30E. A projectile is fired from level ground at an angle θ_0 above the horizontal. (a) Show that the elevation angle ϕ of the highest point as seen from the launch point is related to θ_0, the elevation angle of projection, by $\tan \phi = \frac{1}{2} \tan \theta_0$. See Fig. 4-31 and Exercise 29. (b) Calculate ϕ for $\theta_0 = 45°$.

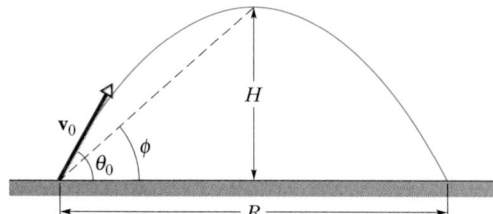

FIGURE 4-31 Exercises 29 and 30.

31E. A stone is projected at a cliff of height h with an initial speed of 42.0 m/s directed 60.0° above the horizontal, as shown in Fig. 4-32. The stone strikes at A, 5.50 s after launching. Find (a) the height h of the cliff, (b) the speed of the stone just before impact at A, and (c) the maximum height H reached above the ground.

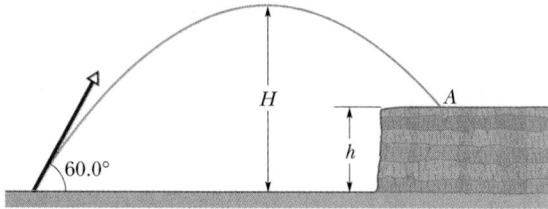

FIGURE 4-32 Exercise 31.

32P. The launching speed of a certain projectile is five times the speed it has at its maximum height. Calculate the elevation angle at launching.

33P. In Sample Problem 4-6, find (a) the speed of the capsule when it hits the water and (b) the angle θ shown in Fig. 4-14.

34P. In the 1991 World Track and Field Championships in Tokyo, Mike Powell (Fig. 4-33) jumped 8.95 m, breaking the 23-year long-jump record set by Bob Beamon by a full 5 cm. Assume that Powell's speed on takeoff was 9.5 m/s (about equal to that of a sprinter) and that $g = 9.80$ in Tokyo. How close did Powell come to the maximum possible range in the absence of air resistance?

35P. A rifle with a muzzle velocity of 1500 ft/s shoots a bullet at a target 150 ft away. How high above the target must the rifle barrel be pointed to ensure that the bullet will hit the target?

36P. Show that the maximum height reached by a projectile is $y_{max} = (v_0 \sin \theta_0)^2/2g$.

37P. A ball is shot from the ground into the air. At a height of 9.1 m, the velocity is observed to be $\mathbf{v} = 7.6\mathbf{i} + 6.1\mathbf{j}$ in meters per second ($\mathbf{i}$ horizontal, $\mathbf{j}$ upward). (a) To what maximum height will the ball rise? (b) What will be the total horizontal distance

FIGURE 4-33 Problem 34. Mike Powell's jump.

traveled by the ball? (c) What is the velocity of the ball (magnitude and direction) the instant before it hits the ground?

38P. In a detective story, a body is found 15 ft from the base of a building and 80 ft below an open window. Would you guess the death to be accidental. Explain your answer.

39P. In Galileo's *Two New Sciences,* the author states that "for elevations [angles of projection] which exceed or fall short of 45° by equal amounts, the ranges are equal. . . ." Prove this statement. (See Fig. 4-34.)

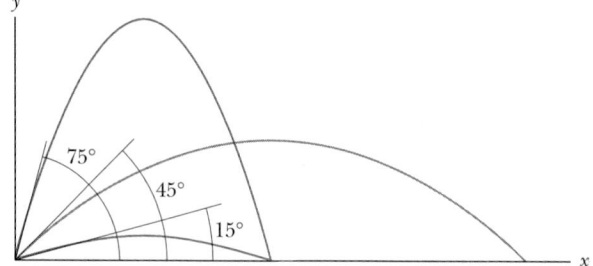

FIGURE 4-34 Problem 39.

40P. The range of a projectile depends not only on v_0 and θ_0 but also on the value g of the free-fall acceleration, which varies from place to place. In 1936, Jesse Owens established a world's running broad jump record of 8.09 m at the Olympic Games at Berlin (g = 9.8128 m/s²). Assuming the same values of v_0 and θ_0, by how much would his record have differed if he had competed instead in 1956 at Melbourne (g = 9.7999 m/s²)?

41P. A third baseman wishes to throw to first base, 127 ft distant. His best throwing speed is 85 mi/h. (a) If he throws the ball horizontally 3.0 ft above the ground, how far from first base will it hit the ground? (b) At what upward angle must the third baseman throw the ball if the first baseman is to catch it 3.0 ft above the ground? (c) What will be the time of flight in that case?

42P. During volcanic eruptions, chunks of solid rock can be blasted out of the volcano; these projectiles are called *volcanic bombs*. Figure 4-35 shows a cross section of Mt. Fuji, in Japan. (a) At what initial speed would a bomb have to be ejected, at 35° to the horizontal, from the vent at A in order to fall at the foot of the volcano at B? Ignore, for the moment, the effects of air on the bomb's travel. (b) What would be the time of flight? (c) Would the effect of the air increase or decrease your answer in (a)?

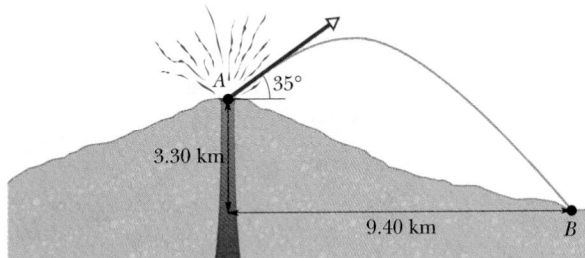

FIGURE 4-35 Problem 42.

43P. At what initial speed must the basketball player throw the ball, at 55° above the horizontal, to make the foul shot, as shown in Fig. 4-36?

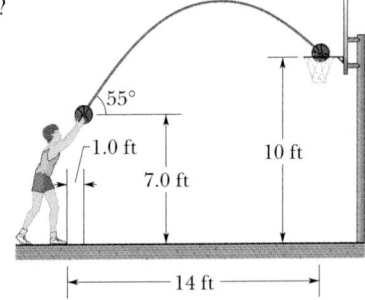

FIGURE 4-36 Problem 43.

44P. A football player punts the football so that it will have a "hang time" (time of flight) of 4.5 s and land 50 yd away. If the ball leaves the player's foot 5.0 ft above the ground, what initial velocity (magnitude and direction) must the ball have?

45P. A golfer tees off from the top of a rise, giving the golf ball an initial velocity of 43 m/s at an angle of 30° above the horizontal. The ball strikes the fairway a horizontal distance of 180 m from the tee. Assume the fairway is level. (a) How high is the rise above the fairway? (b) What is the speed of the ball as it strikes the fairway?

46P. A projectile is fired with an initial speed v_0 = 30.0 m/s from the level ground at a target on the ground a distance R = 20.0 m away (Fig. 4-37). Find the two projection angles.

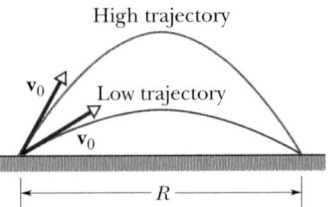

FIGURE 4-37 Problem 46.

47P. What is the maximum vertical height to which a baseball player can throw a ball if his maximum throwing *range* is 60 m?

48P. A certain airplane has a speed of 180 mi/h and is diving at an angle of 30.0° below the horizontal when a radar decoy is released. (See Fig. 4-38.) The horizontal distance between the release point and the point where the decoy strikes the ground is 2300 ft. (a) How high was the plane when the decoy was released? (b) How long was the decoy in the air?

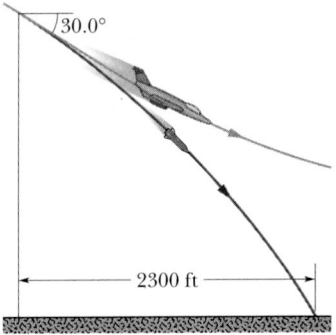

FIGURE 4-38 Problem 48.

49P. A football is kicked off with an initial speed of 64 ft/s at a projection angle of 45°. A receiver 60 yd away in the direction of the kick starts running to meet the ball at that instant. What must be his average speed if he is to catch the ball just before it hits the ground? Neglect air resistance.

50P. A ball rolls horizontally off the top of a stairway with a speed of 5.0 ft/s. The steps are 8.0 in. high and 8.0 in. wide. Which step will the ball hit first?

51P. An airplane, diving at an angle of 53.0° with the vertical, releases a projectile at an altitude of 730 m. The projectile hits the ground 5.00 s after being released. (a) What is the speed of the aircraft? (b) How far did the projectile travel horizontally during its flight? (c) What were the horizontal and vertical components of its velocity just before striking the ground?

52P. A ball is thrown horizontally from a height of 20 m and hits the ground with a speed that is three times its initial speed. What was the initial speed?

53P. (a) During a tennis match, a player serves at 23.6 m/s, the ball leaving the racquet horizontally 2.37 m above the court surface. By how much does the ball clear the net, which is 12 m away and 0.90 m high? (b) Suppose the player serves the ball as before except that the ball leaves the racquet at 5.00° below the horizontal. Does the ball clear the net now?

54P. In Sample Problem 4-8, suppose that a second identical harbor defense cannon is emplaced 30 m above sea level, rather that at sea level. How much longer is the horizontal distance from launch to impact of the second cannon than that of the first, which was found to be 690 m, if the elevation angle of fire is 45°?

55P. A batter hits a pitched ball whose center is 4.0 ft above the ground so that its angle of projection is 45° and its *range* is 350 ft. The ball will be a home run if it clears a 24 ft high fence that is 320 ft from home plate. Will the ball clear the fence? If so, by how much?

56P*. A football kicker can give the ball an initial speed of 25 m/s. Within what two elevation angles must he kick the ball to score a field goal from a point 50 m in front of goalposts whose horizontal bar is 3.44 m above the ground? (You might want to use $\sin^2 \theta + \cos^2 \theta = 1$ to get a relation between $\tan^2 \theta$ and $1/\cos^2 \theta$, and then solve the resulting quadratic equation.)

SECTION 4-7 Uniform Circular Motion

57E. In one model of the hydrogen atom, an electron orbits a proton in a circle of radius 5.28×10^{-11} m with a speed of 2.18×10^6 m/s. (a) What is the acceleration of the electron in this model? (b) What is the period of the motion?

58E. (a) What is the acceleration of a sprinter running at 10 m/s when rounding a bend with a turn radius of 25 m? (b) In what direction does the acceleration vector point?

59E. A magnetic field can force a charged particle to move in a circular path. Suppose that an electron experiences a radial acceleration of 3.0×10^{14} m/s² in a particular magnetic field. (a) What is the speed of the electron if the radius of its circular path is 15 cm? (b) What is the period of the motion?

60E. A sprinter runs at 9.2 m/s around a circular track with a centripetal acceleration of 3.8 m/s². (a) What is the track radius? (b) What is the period of the motion?

61E. An Earth satellite moves in a circular orbit 640 km above the Earth's surface. The period of the motion is 98.0 min. (a) What is the speed of the satellite? (b) What is the free-fall acceleration at the orbit height?

62E. Suppose a space probe can withstand the stresses of a $20g$ acceleration. (a) What is the minimum turning radius of such a craft moving at a speed of one-tenth the speed of light? (b) How long would it take to complete a 90° turn at this speed?

63E. A rotating fan completes 1200 revolutions every minute. Consider a point on the tip of a blade, at a radius of 0.15 m. (a) Through what distance does the point move in one revolution? (b) What is the speed of the point? (c) What is its acceleration? (d) What is the period of the motion?

64E. The fast train known as the TGV (Train à Grande Vitesse) that runs south from Paris, France, has a scheduled average speed of 216 km/h. (a) If the train goes around a curve at that speed and the acceleration experienced by the passengers is to be limited to $0.050g$, what is the smallest radius of curvature for the track that can be tolerated? (b) If there is a curve with a 1.00 km radius, to what speed must the train be slowed to keep the acceleration below the limit?

65E. When a large star becomes a *supernova*, its core may be compressed so tightly that it becomes a *neutron star*, with a radius

of about 20 km (about the size of the San Francisco area). If a neutron star rotates once every second, (a) what is the speed of a particle on the star's equator and (b) what is the particle's centripetal acceleration in meters per second squared and in g units? (c) If the neutron star rotates even faster, what happens to the answers to (a) and (b)?

66E. An astronaut is rotated in a horizontal centrifuge at a radius of 5.0 m. (a) What is the astronaut's speed if the centripetal acceleration is $7.0g$? (b) How many revolutions per minute are required to produce this acceleration? (c) What is the period of the motion?

67P. (a) What is the centripetal acceleration of an object on the Earth's equator owing to the rotation of the Earth? (b) What would the period of rotation of the Earth have to be for objects on the equator to have a centripetal acceleration equal to 9.8 m/s²?

68P. A carnival Ferris wheel has a 15 m radius and completes five turns about its horizontal axis every minute. (a) What is the period of the motion? (b) What is the centripetal acceleration of a passenger at the highest point? (c) What is the centripetal acceleration at the lowest point?

69P. Calculate the acceleration of a person at latitude 40° owing to the rotation of the Earth. (See Fig. 4-39.)

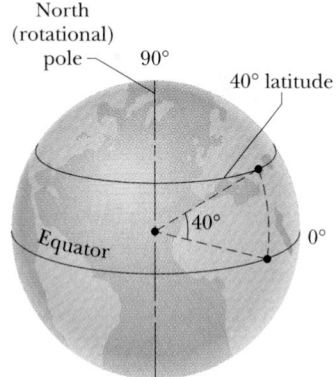

FIGURE 4-39 Problem 69.

70P. A particle P travels with constant speed on a circle of radius $r = 3.00$ m (Fig. 4-40) and completes one revolution in 20.0 s. The particle passes through O at $t = 0$. Find the magnitude and direction of each of the following vectors. (a) With respect to O, find the particle's position vector at $t = 5.00$ s, 7.50 s, and 10.0 s. For the 5.00 s interval from the end of the fifth second to the end of the tenth second, find the particle's (b) displacement and (c) average velocity. Find its (d) velocity and (e) acceleration at the beginning and end of that 5.00 s interval.

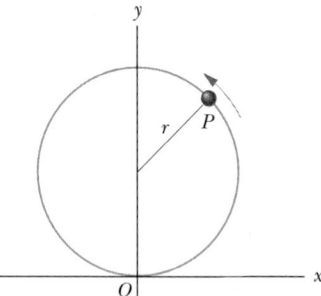

FIGURE 4-40 Problem 70.

71P. A boy whirls a stone in a horizontal circle 2.0 m above the ground by means of a string 1.5 m long. The string breaks, and the stone flies off horizontally and strikes the ground 10 m away. What was the centripetal acceleration of the stone while in circular motion?

SECTION 4-8 Relative Motion in One Dimension

72E. A boat is traveling upstream at 14 km/h with respect to the water of a river. The water itself is flowing at 9 km/h with respect to the ground. (a) What is the velocity of the boat with respect to the ground? (b) A child on the boat walks from front to rear at 6 km/h with respect to the boat. What is the child's velocity with respect to the ground?

73E. A person walks up a stalled 15 m long escalator in 90 s. When standing on the same escalator, now moving, the person is carried up in 60 s. How much time would it take that person to walk up the moving escalator? Does the answer depend on the length of the escalator?

74E. A transcontinental flight of 2700 mi is scheduled to take 50 min longer westward than eastward. The airspeed of the airplane is 600 mi/h, and the jet stream it will fly through is presumed to be moving either due east or due west. What assumptions about the jet stream wind velocity are made in preparing the schedule?

75E. A cameraman on a pickup truck is traveling westward at 40 mi/h while he videotapes a cheetah that is moving westward 30 mi/h faster than the truck. Suddenly, the cheetah stops, turns, and then runs at 60 mi/h eastward, as measured by a suddenly nervous crew member who stands alongside the cheetah's path. The change in the animal's velocity took 2.0 s. What was its acceleration from the perspective of the cameraman? From the perspective of the nervous crew member?

76P. The airport terminal in Geneva, Switzerland, has a "moving sidewalk" to speed passengers through a long corridor. Peter does not use the moving sidewalk; he takes 150 s to walk through the corridor. Paul, who simply stands on the moving sidewalk, covers the same distance in 70 s. Mary boards the sidewalk and walks along it. How long does Mary take to move through the corridor? Assume that Peter and Mary walk at the same speed.

SECTION 4-9 Relative Motion in Two Dimensions

77E. In rugby (Fig. 4-41) a player can legally pass the ball to a teammate as long as the pass is not "forward" (it must not have a velocity component parallel to the length of the field and directed toward the other team's goal). Suppose a player runs parallel to the field's length with a speed of 4.0 m/s while he passes the ball with a speed of 6.0 m/s relative to himself. What is the smallest angle from the forward direction that keeps the pass legal?

78E. Two highways intersect as shown in Fig. 4-42. At the instant shown, a police car P is 800 m from the intersection and moving at 80 km/h. Motorist M is 600 m from the intersection and moving at 60 km/h. (a) In unit-vector notation, what is the velocity of the motorist with respect to the police car? (b) For the

FIGURE 4-41 Exercise 77.

instant shown in Fig. 4-42, how does the direction of the velocity found in (a) compare to the line of sight between the two cars? (c) If the cars maintain their velocities, do the answers to (a) and (b) change as the cars move nearer the intersection?

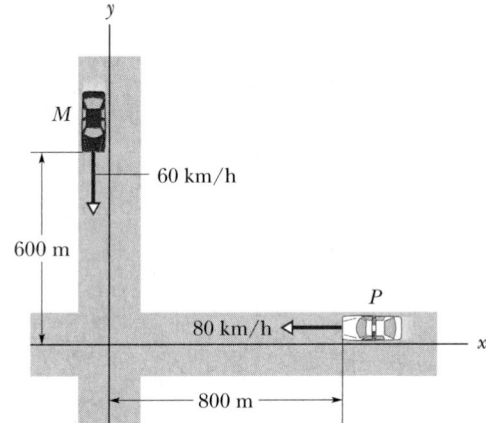

FIGURE 4-42 Exercise 78.

79E. Snow is falling vertically at a constant speed of 8.0 m/s. At what angle from the vertical do the snowflakes appear to be falling as viewed by the driver of a car traveling on a straight road with a speed of 50 km/h?

80E. In a large department store, a shopper is standing on the "up" escalator, which is traveling at an angle of 40° above the horizontal and at a speed of 0.75 m/s. He passes his daughter, who is standing on the identical, adjacent "down" escalator. (See Fig. 4-43.) Find the velocity of the shopper relative to his daughter in unit-vector notation.

81P. A helicopter is flying in a straight line over a level field at a constant speed of 6.2 m/s and at a constant altitude of 9.5 m. A package is ejected horizontally from the helicopter with an initial velocity of 12 m/s relative to the helicopter, and in a direction opposite the helicopter's motion. (a) Find the initial speed of the package relative to the ground. (b) What is the horizontal distance between the helicopter and the package at the instant the package strikes the ground? (c) What angle does the velocity vector of the

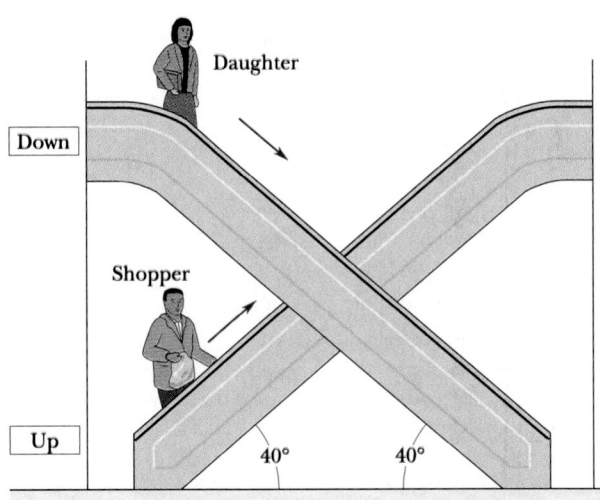

FIGURE 4-43 Exercise 80.

package make with the ground at the instant before impact, as seen from the ground?

82P. A train travels due south at 30 m/s (relative to the ground) in a rain that is blown toward the south by the wind. The path of each raindrop makes an angle of 70° with the vertical, as measured by an observer stationary on the Earth. An observer on the train, however, sees the drops fall perfectly vertically. Determine the speed of the raindrops relative to the Earth.

83P. A light plane attains an airspeed of 500 km/h. The pilot sets out for a destination 800 km to the north but discovers that the plane must be headed 20.0° east of north to fly there directly. The plane arrives in 2.00 h. What was the wind velocity vector?

84P. Two ships, A and B, leave port at the same time. Ship A travels northwest at 24 knots and ship B travels at 28 knots in a direction 40° west of south. (1 knot = 1 nautical mile per hour; see Appendix D.) (a) What are the magnitude and direction of the velocity of ship A relative to B? (b) After what time will they be 160 nautical miles apart? (c) What will be the bearing (direction) of B relative to A at that time?

85P. The New Hampshire State Police use aircraft to enforce highway speed limits. Suppose that one of the airplanes has a speed of 135 mi/h in still air. It is flying straight north so that it is at all times directly above a north–south highway. A ground observer tells the pilot by radio that a 70.0 mi/h wind is blowing, but neglects to give the wind direction. The pilot observes that in spite of the wind the plane can travel 135 mi along the highway in 1.00 h. In other words, the ground speed is the same as if there were no wind. (a) What is the direction of the wind? (b) What is the heading of the plane, that is, the angle between an axis along its length and the highway?

86P. A wooden boxcar is moving along a straight railroad track at speed v_1. A sniper fires a bullet (initial speed v_2) at it from a high-powered rifle. The bullet passes through both walls of the car, its entrance and exit holes being exactly opposite each other as viewed from within the car. From what direction, relative to the track, was the bullet fired? Assume that the bullet was not

deflected upon entering the car, but that its speed decreased by 20%. Take $v_1 = 85$ km/h and $v_2 = 650$ m/s. (Why don't you need to know the width of the boxcar?)

87P. A woman can row a boat at 4.0 mi/h in still water. (a) If she is crossing a river where the current is 2.0 mi/h, in what direction must her boat be headed if she wants to reach a point directly opposite her starting point? (b) If the river is 4.0 mi wide, how long will it take her to cross the river? (c) Suppose that instead of crossing the river she rows 2.0 mi *down* the river and then back to her starting point. How long will she take? (d) How long will she take to row 2.0 mi *up* the river and then back to her starting point? (e) In what direction should she head the boat if she wants to cross in the shortest possible time, and what is that time?

SECTION 4-10 Relative Motion at High Speeds

88E. As we head toward the center of our galaxy, we detect a burst of light traveling past us at speed c relative to us, toward that center. An observer on a second ship, traveling at a speed of $0.98c$ relative to us, also detects the light. What is the speed of the light as measured by the observer on the second ship if that ship is traveling (a) in the same direction as us and (b) in the opposite direction?

89E. An electron moves at speed $0.42c$ with respect to observer B. Observer B moves at speed $0.63c$ with respect to observer A, in the same direction as the electron. What does observer A measure for the speed of the electron?

90P. While traveling in a ship toward the star Betelgeuse, we detect a burst of protons traveling past us toward that star, with a velocity of $0.9800c$ relative to us. Detectors on a second ship, also traveling toward Betelgeuse along our line of travel, measure the velocity of the protons to be $-0.9800c$ relative to them. What is the relative speed between us and the second ship?

91P. Galaxy Alpha is observed to be receding from us with a speed of $0.35c$. Galaxy Beta, located in precisely the opposite direction, is also found to be receding from us at this same speed. What recessional speed would an observer on Galaxy Alpha find (a) for our galaxy and (b) for Galaxy Beta?

ELECTRONIC COMPUTATION

92. If the launch site of a projectile is above the landing place, a launch angle of 45° may not produce the greatest horizontal distance. Suppose a shot putter releases the shot from a point that is a distance h above a horizontal playing field. The initial speed of the shot is v_0 and the launch angle is θ. (a) Show that the horizontal distance from the putter's feet to the landing point is given by

$$d = \frac{v_0 \cos \theta}{g} [v_0 \sin \theta + \sqrt{v_0^2 \sin^2 \theta + 2gh}].$$

(b) Set up a program to calculate d for a series of values of θ, given v_0 and h. (c) Take $v_0 = 9.0$ m/s and $h = 2.1$ m and find to the nearest half degree the launch angle that produces the greatest horizontal distance. Also find that distance. (d) Does the result depend on the initial speed of the shot? Try $v_0 = 5.0$ m/s and $v_0 = 15$ m/s. (The last value is much greater than the speed generated by a champion shot putter.)

93. Immediately after launch, all projectiles move away from the launch site but later some move closer before moving further away again. It depends on the launch angle. Set up a program to compute the distance from the launch site to the projectile every 0.5 s from the time of launch to a few seconds after it is below the launch site. Take the initial speed to be 100 m/s and run the program for launch angles from 5° to 90°, with an interval of 5°. For each launch angle, search the list of distances to find when the projectile is moving away from and when it is moving toward the launch site. Classify the motions according to whether or not the projectile moves toward the launch site during any time interval and, if it does, estimate the interval. (See the article "Projectiles: Are They Coming or Going?" by James S. Walker in *The Physics Teacher,* May 1995.)

94. A batter strikes a baseball at a point 1.00 m above home plate, giving the ball an initial velocity of magnitude v_0, at angle θ above the horizontal. The ball is to barely clear a 2.40 m tall fence at a distance $R = 110$ m from home plate. (a) Find the minimum and maximum values of θ that allow this clearance for $v_0 = 35.0$ m/s. (*Hint:* rather than explicitly solving for θ in the equations of motion, find two expressions of θ, plot them, and then find their intersections.) (b) Find the minimum value of v_0 that allows the clearance for $\theta = 40.0°$. (c) Now find the minimum value of v_0 considering all values of θ and give the corresponding angle. (d) Repeat (c) for Fenway Park, with a 12.2 m tall "wall" at 96.0 m down the foul line from home plate.

95. A golfer chips balls toward a vertical wall 20.0 m straight ahead, trying to hit a 30.0 cm diameter red circle painted on the wall. The target is centered about a point 1.20 m above the point where the wall intersects the horizontal ground. On one try, the ball leaves the ground with a speed of 15.0 m/s and at an angle of 35.0° above the horizontal. (a) How long does the ball take to reach the wall? (b) Does the ball hit the red circle? (c) What is the speed of the ball just before it hits? (d) Has the ball passed the highest point of its trajectory when it hits?

96. At one instant a bicyclist is 40.0 m due east of a park's flagpole, going due south with a speed of 10.0 m/s. Then 30.0 s later, the cyclist is 40.0 m due north of the flagpole, going due east with a speed of 10.0 m/s. Find (a) the displacement, (b) the average velocity, and (c) the average acceleration of the cyclist during the 30.0 s interval. (d) Compute $(\mathbf{v}_f - \mathbf{v}_i)/2$, where $\mathbf{v}_f$ and $\mathbf{v}_i$ are the velocities at the end and at the beginning of the 30.0 s interval, respectively.

97. At one instant a butterfly has the position vector $\mathbf{D}_i = (2.00\text{ m})\mathbf{i} + (3.00\text{ m})\mathbf{j} + (1.00\text{ m})\mathbf{k}$ relative to the base of a birdbath. Then 40 s later, the butterfly has the position vector $\mathbf{D}_f = (3.00\text{ m})\mathbf{i} + (1.00\text{ m})\mathbf{j} + (2.00\text{ m})\mathbf{k}$. Find (a) the displacement (in unit-vector notation), (b) the magnitude of the displacement, (c) the average velocity, and (d) the average speed of the butterfly during the 40.0 s interval.

On April 4, 1974, John Massis of Belgium managed to move two passenger cars belonging to New York's Long Island Railroad. He did so by clamping his teeth down on a bit that was attached to the cars with a rope and then leaning backward while pressing his feet against the railway ties. The cars weighed about 80 tons. Did Massis have to pull with superhuman force to accelerate them?

5-1 WHAT CAUSES AN ACCELERATION?

If you see the velocity of a particle-like body change in either magnitude or direction, you know that something must have *caused* that change (that acceleration). Indeed, out of common experience, you know that the change in velocity must be due to an interaction between the body and something in its surroundings. For example, if you see a hockey puck that is sliding across an ice rink suddenly stop or suddenly change direction, you will suspect that the puck bumped into a slight hill on the ice surface.

An interaction that causes an acceleration of a body is called a **force**, which is, loosely speaking, a push or pull. For example, the bump on the hockey puck by the hill is a push on the puck, causing an acceleration. The relationship between a force and the acceleration it causes was first understood by Isaac Newton (1642–1727) and is the subject of this chapter. The study of that relationship, as Newton presented it, is called *Newtonian mechanics*. We will focus on its three primary laws of motion.

Newtonian mechanics does not apply to all situations. As we discussed in Chapter 4, if the speeds of the interacting bodies are an appreciable fraction of the speed of light, we must replace Newtonian mechanics with Einstein's special theory of relativity, which holds at any speed, including those near the speed of light. If the interacting bodies are on the scale of atomic structure (for example, they might be electrons within an atom), we must replace Newtonian mechanics with quantum mechanics. Physicists now view Newtonian mechanics as a special case of these two more comprehensive theories. Still, it is a very important special case because it applies to the motion of objects ranging in size from the very small (almost on the scale of atomic structure) to astronomical (objects such as galaxies and clusters of galaxies). Let us now examine the first law of motion in Newtonian mechanics.

5-2 NEWTON'S FIRST LAW

Before Newton formulated his mechanics, it was thought that some influence, a "force," was needed to keep a body moving at constant velocity. Similarly, a body was thought to be in its "natural state" when it was at rest. For it to move with constant velocity, it seemingly had to be propelled in some way, by a push or a pull. Otherwise, it would "naturally" stop moving.

These ideas were reasonable. If you send a book sliding across a wooden floor, it does indeed slow and then stop. If you want to make it move across the floor with constant velocity, you have to continuously pull or push it.

Slide the book over the ice of a skating rink, however,

and it goes a lot farther. You can imagine longer and more slippery surfaces, over which the book would slide farther and farther. In the limit you can think of a long, extremely slippery surface (said to be a **frictionless surface**), over which the book would hardly slow. (We can in fact come close to this situation in the laboratory, by sending a book sliding over a horizontal air table, across which it moves on a film of air.)

We are led to conclude that you do *not* need a force to keep a body moving with constant velocity. And that leads us to the first of Newton's three laws of motion:

Newton's First Law: Consider a body on which no force acts. If the body is at rest, it will remain at rest. If the body is moving with constant velocity, it will continue to do so.

This law fits in nicely with what we discussed in Section 4-8 about reference frames: a body on which no force acts may be stationary in one frame and moving at constant velocity with respect to another. (*Rest* and *moving with constant velocity* are not all that different.)

Newton's first law can be interpreted as a statement about reference frames, in that it defines the kinds of reference frames in which the laws of Newtonian mechanics hold: frames with constant separation velocity. From this point of view the first law is expressed as follows:

Newton's First Law: If no force acts on a body, we can always find a reference frame in which that body has no acceleration.

Newton's first law is sometimes called the *law of inertia,* and the reference frames that it defines are called *inertial reference frames* or just *inertial frames.*

Figure 5-1 shows how you can test a particular frame to see whether it is an inertial frame. With the railroad car at rest, mark the position of the stationary pendulum bob on the table. With the car in motion, the bob remains over the mark *only* if the car is moving in a straight line at

FIGURE 5-1 Testing a railroad car to see whether it is an inertial reference frame.

constant speed. The car is then an inertial frame. If the car is gaining or losing speed or is rounding a corner, the bob moves from its mark, and the car is then a noninertial reference frame.

5-3 FORCE

A force causes the acceleration of a body. We now wish to define the unit of force carefully, in terms of the acceleration that it gives to a standard reference body. As the standard body, we use (or rather we imagine that we use) the standard kilogram of Fig. 1-6. This body has been assigned, exactly and by definition, a mass of 1 kg.

We put the standard body on a horizontal frictionless table and pull the body to the right (Fig. 5-2) so that by trial and error, it eventually experiences a measured acceleration of 1 m/s². We then declare, as a matter of definition, that the force we are exerting on the standard body has a magnitude of 1 newton (abbreviated N).

We can exert a 2 N force on our standard body by pulling it so that its measured acceleration is 2 m/s², and so on. Thus in general, if our standard body of 1 kg mass has an acceleration a, we know that force F must be acting on it and that the magnitude of the force (in newtons) is equal to the magnitude of the acceleration (in meters per second per second).

Thus, a force is measured by the acceleration it produces. But acceleration is a vector quantity, with both magnitude and direction. Is force also a vector quantity? We can easily assign a direction to a force (just assign the direction of the acceleration), but that is not sufficient. We must prove by experiment that forces are vector quantities. Actually, that has been done: forces are indeed vector quantities; they have magnitudes and directions and they combine according to the vector rules of Chapter 3.

Henceforth, we shall use boldface letters, most often **F**, to represent forces. And we shall use the symbol Σ**F** for the vector sum of several forces, which we call the **resultant force** or the **net force**. As with other vectors, a force or a net force can have components along coordinate axes. Finally, we note that Newton's first law holds not only when there is no force on a body, but also when the net force is equal to zero.

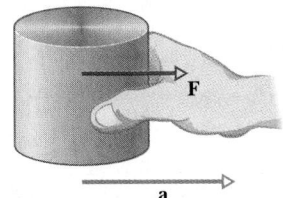

FIGURE 5-2 A force **F** on the standard kilogram gives that body an acceleration **a**.

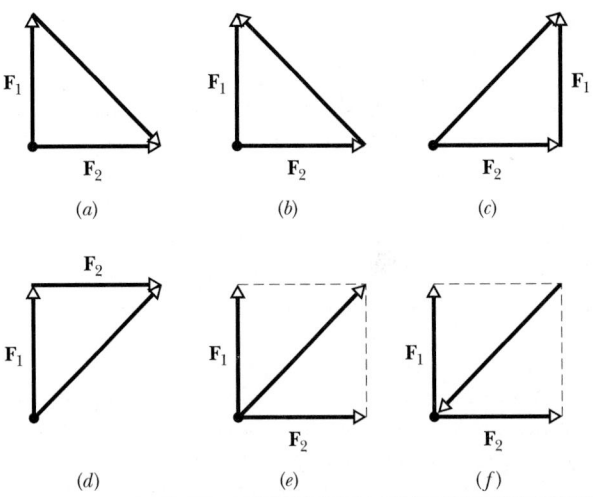

5-4 MASS

Everyday experience tells us that a given force produces different magnitudes of acceleration for different bodies. Put a baseball and a bowling ball on the floor and give both the same sharp kick. Even if you don't actually do this, you know the result: the baseball receives a noticeably larger acceleration than the bowling ball. The two accelerations differ because the mass of the baseball differs from the mass of the bowling ball. But what, exactly, is mass?

We can explain how to measure mass by imagining a series of experiments. In the first experiment we exert a force on our standard body, whose mass m_0 is defined to be 1.0 kg. Suppose that the standard body accelerates at 1.0 m/s². We can then say the force on that body is 1.0 N.

We next apply that same force (we would need some way of being certain it is the same force) to a second body, body X, whose mass is not known. Suppose we find that this body X accelerates at 0.25 m/s². We know that a *less massive* baseball receives a *greater acceleration* than a more massive bowling ball when the same force (kick) is applied to both. Let us then make the following conjecture: the ratio of the masses of two bodies is equal to the inverse of the ratio of their accelerations when the same force is applied to both. For body X and the standard body, this tells us that

$$\frac{m_X}{m_0} = \frac{a_0}{a_X}.$$

Solving for m_X yields

$$m_X = m_0 \frac{a_0}{a_X} = (1.0 \text{ kg}) \frac{1.0 \text{ m/s}^2}{0.25 \text{ m/s}^2} = 4.0 \text{ kg}.$$

Our conjecture will be useful, of course, only if it continues to hold when we change the applied force to other values. For example, if we apply an 8.0 N force to the standard body, we obtain an acceleration of 8.0 m/s². And when the 8.0 N force is applied to body X, we obtain an acceleration of 2.0 m/s². Our conjecture then gives us

$$m_X = m_0 \frac{a_0}{a_X} = (1.0 \text{ kg}) \frac{8.0 \text{ m/s}^2}{2.0 \text{ m/s}^2} = 4.0 \text{ kg},$$

consistent with our first experiment. Many experiments yielding similar results indicate that our conjecture provides a consistent and reliable means of assigning a mass to any given body.

Our measurement experiments indicate that mass is an *intrinsic* characteristic of a body—that is, a characteristic that automatically comes with the existence of the body. They also indicate that mass is a scalar quantity. However, the nagging question remains: What, exactly, is mass?

Since the word *mass* is used in everyday English, we should have some intuitive understanding of it, maybe something that we can physically sense. Is it a body's size, weight, or density? The answer is no, although those characteristics are sometimes confused with mass. We can say only that *the mass of a body is the characteristic that relates a force on the body to the resulting acceleration.* Mass has no more familiar definition than that; you can have a physical sensation of mass only when you attempt to accelerate a body, as in the kicking of a baseball or a bowling ball.

5-5 NEWTON'S SECOND LAW

All the definitions, experiments, and observations that we have described so far can be summarized in a simple vector equation, which is called Newton's second law of motion:

$$\sum \mathbf{F} = m\mathbf{a} \qquad \text{(Newton's second law).} \qquad (5\text{-}1)$$

In using Eq. 5-1, we must first be quite certain what body we are applying it to. Then $\sum \mathbf{F}$ in Eq. 5-1 is the vector sum, or net force, of *all* the forces that act *on* that body. Only forces that act *on* the body are to be included, not forces acting on other bodies that might be involved in a given problem. Finally, $\sum \mathbf{F}$ includes only *external* forces, that is, forces exerted on the body by other bodies. We do not include internal forces, in which one part of the body exerts a force on another part.

Like other vector equations, Eq. 5-1 is equivalent to three scalar equations:

$$\sum F_x = ma_x, \qquad \sum F_y = ma_y, \qquad \sum F_z = ma_z. \qquad (5\text{-}2)$$

These equations relate the three components of the net force acting on a body to the three components of the acceleration of that body.

You should note that Newton's second law includes the formal statement of Newton's first law as a special case. That is, if no force acts on a body, Eq. 5-1 tells us that the body will not be accelerated.

For SI units, Eqs. 5-2 tell us that

$$1 \text{ N} = (1 \text{ kg})(1 \text{ m/s}^2) = 1 \text{ kg} \cdot \text{m/s}^2, \qquad (5\text{-}3)$$

consistent with our discussion in Section 5-3. Although we shall use SI units almost exclusively from now on, other systems of units are still in use. Chief among these are the British system and the CGS (centimeter–gram–second) system. Table 5-1 shows the units in which Eqs. 5-1 and 5-2 are expressed in these systems. (See also Appendix D.)

To solve problems with Newton's second law, we often draw a **free-body diagram,** representing the body by a dot and each external force (or the net force $\sum \mathbf{F}$) that acts on the body by a vector with its tail on the dot. (Instead of the dot, we can sketch the body.) A set of coordinate axes is included, and sometimes a vector representing the acceleration of the body is also included.

We then should start the solution with the vector relation of Eq. 5-1. However, we shall usually quickly switch our attention to one or more of the scalar relations of Eqs. 5-2 and work along one axis at a time. The first relation of Eqs. 5-2 tells us that the sum of all the force components along the x axis causes the x component a_x of the body's acceleration but causes no acceleration in the y and z directions. Or, turned around, it says that the acceleration component a_x is caused only by the sum of the force components along the x axis. Similarly, the acceleration component a_y along the y axis is caused only by the sum of the force components along the y axis; and the acceleration component a_z along the z axis is caused only by the sum of the force components along the z axis. In general we have:

The acceleration component along a given axis is caused only by the sum of the force components along that *same* axis and not by force components along some other axis.

In Sample Problem 5-1 we shall be concerned with a single force that acts along an x axis. So, we shall not need to find several force components. However, in Sample Problem 5-2, three forces act on a body and two make

TABLE 5-1 UNITS IN NEWTON'S SECOND LAW
(Eqs. 5-1 and 5-2)

SYSTEM	FORCE	MASS	ACCELERATION
SI	newton (N)	kilogram (kg)	m/s^2
CGS[a]	dyne	gram (g)	cm/s^2
British[b]	pound (lb)	slug	ft/s^2

[a] 1 dyne = 1 g·cm/s^2.   [b] 1 lb = 1 slug·ft/s^2.

nonzero angles with an x axis and a y axis. In this two-dimensional situation, we must find the force components along the x axis and the y axis, and then use, separately, the first two relations of Eqs. 5-2.

Sample Problem 5-2 is also an example of a general type of problem: the body does not accelerate ($\mathbf{a} = 0$) in spite of the external forces on it. In such a situation, Eq. 5-1 tells us that $\Sigma\mathbf{F} = 0$. That is, the net force is zero, the forces acting on the body *balance* each other, and the body is said to be in *equilibrium*.

However, if we switch our attention to Eqs. 5-2, we see something that is more useful to problem solving. The lack of acceleration means that $a_x = 0$ and thus that $\Sigma F_x = 0$. That is, the x components of the forces also balance each other. If we replace ΣF_x with the actual x components of the forces involved, we then have an algebraic relation we can use in the solution. Similarly, $a_y = 0$ and thus $\Sigma F_y = 0$; after replacing ΣF_y with the actual y components of the forces involved, we have another algebraic relation.

In some problems you might find that the force components along one axis balance, but those along a second axis do not. So, there must be acceleration along that second axis only.

How will you know what to do in any given situation? Experience helps, and that is why this chapter contains many Sample Problems.

CHECKPOINT 2: The figure shows two horizontal forces moving a block along a frictionless floor. Assume that a third horizontal force $\mathbf{F}_3$ also acts on the block. What are the magnitude and direction of $\mathbf{F}_3$ when the block is (a) stationary and (b) moving to the left with a constant speed of 5 m/s?

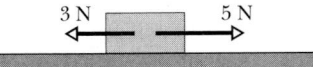

3 N   5 N

SAMPLE PROBLEM 5-1

A student (with cleated boots) pushes a loaded sled whose mass m is 240 kg for a distance d of 2.3 m over the frictionless surface of a frozen lake. He exerts a constant horizontal force

$\mathbf{F}$, with magnitude $F = 130$ N, as he does so (see Fig. 5-3a).

(a) If the sled starts from rest, what is its final velocity?

SOLUTION: Figure 5-3b is a free-body diagram for the situation. We lay out a horizontal x axis, we take the direction of increasing x to be to the right, and we treat the sled as a particle, represented by a dot. We assume that the x component F_x of the force $\mathbf{F}$ exerted by the student is the only horizontal force acting on the sled. We can then find the magnitude of the acceleration a_x of the sled from Newton's second law:

$$a_x = \frac{F_x}{m} = \frac{130 \text{ N}}{240 \text{ kg}} = 0.542 \text{ m/s}^2.$$

Because the acceleration is constant, we can use Eq. 2-16, $v^2 = v_0^2 + 2a(x - x_0)$, to find the final velocity. Putting $v_0 = 0$ and $x - x_0 = d$, and identifying a_x as a, we solve for v:

$$v = \sqrt{2ad}$$
$$= \sqrt{(2)(0.542 \text{ m/s}^2)(2.3 \text{ m})} = 1.6 \text{ m/s}. \quad \text{(Answer)}$$

The force, the acceleration, the displacement, and thus the final velocity of the sled are all positive, which means that they all point to the right in Fig. 5-3b.

(b) The student now wants to reverse the direction of the velocity of the sled in 4.5 s. With what constant force must he push on the sled to do so?

SOLUTION: Let us first find the constant acceleration required to reverse the sled's velocity in 4.5 s, using Eq. 2-11 ($v = v_0 + at$). Solving for a gives

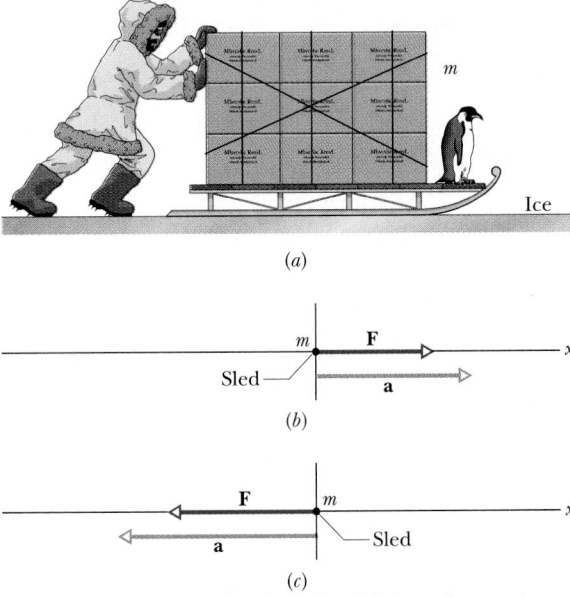

(a)

(b)

(c)

FIGURE 5-3 Sample Problem 5-1. (*a*) A student pushes a loaded sled over a frictionless surface. (*b*) A free-body diagram for part (a) in the problem, showing the net force acting on the sled and the acceleration the net force produces. (*c*) A free-body diagram for part (b). The student now pushes in the opposite direction on the sled, reversing its acceleration.

$$a = \frac{v - v_0}{t} = \frac{(-1.6 \text{ m/s}) - (1.6 \text{ m/s})}{4.5 \text{ s}}$$

$$= -0.711 \text{ m/s}^2.$$

This is larger in magnitude than the acceleration in (a), namely 0.542 m/s², so it stands to reason that the student must push with a greater force this time. We find this greater force from Eqs. 5-2, with a_x being a:

$$F_x = ma_x = (240 \text{ kg})(-0.711 \text{ m/s}^2)$$

$$= -171 \text{ N.} \qquad \text{(Answer)}$$

The minus sign shows that the student must push the sled in the direction of decreasing x, that is, to the left in Fig. 5-3c, the free-body diagram for this situation.

SAMPLE PROBLEM 5-2

In a two-dimensional tug-of-war, Alex, Betty, and Charles pull on an automobile tire, at angles as shown in Fig. 5-4a, which is an overhead view. The tire remains stationary in spite of the three pulls. Alex pulls with force $\mathbf{F}_A$ of magnitude 220 N, and Charles pulls with force $\mathbf{F}_C$ of magnitude 170 N. The direction of $\mathbf{F}_C$ is not given. What is the magnitude of Betty's force $\mathbf{F}_B$?

SOLUTION: Figure 5-4b is a free-body diagram for the tire. Because the acceleration of the tire is zero, Eq. 5-1 tells us that the net force on the tire must also be zero:

$$\Sigma \mathbf{F} = \mathbf{F}_A + \mathbf{F}_B + \mathbf{F}_C = m\mathbf{a} = 0.$$

This vector relation is equivalent to the first two scalar relations of Eqs. 5-2. Along the x axis we have

$$\Sigma F_x = F_{Ax} + F_{Bx} + F_{Cx} = 0, \qquad (5\text{-}4)$$

and along the y axis we have

$$\Sigma F_y = F_{Ay} + F_{By} + F_{Cy} = 0. \qquad (5\text{-}5)$$

Using the given data and the angles in Fig. 5-4b, we now substitute into Eqs. 5-4 and 5-5, using signs to indicate the directions of the vector components. Equation 5-4 becomes

$$\Sigma F_x = -F_A \cos 47.0° + 0 + F_C \cos \phi = 0.$$

Substituting known values yields

$$-(220 \text{ N})(\cos 47.0°) + 0 + (170 \text{ N})(\cos \phi) = 0,$$

which gives us

$$\phi = \cos^{-1} \frac{(220 \text{ N})(\cos 47.0°)}{170 \text{ N}} = 28.0°.$$

Similarly, Eq. 5-5 becomes

$$\Sigma F_y = F_A \sin 47.0° - F_B + F_C \sin \phi = 0,$$

in which we have substituted $-F_B$ for F_{By} because Betty's pull is entirely along the negative direction of y. Substituting known values here yields

$$(220 \text{ N})(\sin 47.0°) - F_B + (170 \text{ N})(\sin \phi) = 0.$$

We next substitute 28.0° for ϕ and solve the equation for F_B, finding that

$$F_B = (220 \text{ N})(\sin 47.0°) + (170 \text{ N})(\sin 28.0°)$$

$$= 241 \text{ N.} \qquad \text{(Answer)}$$

Note that we first had to solve the equations along the x axis to use $\phi = 28.0°$ in the equations along the y axis. If we had started with the equations along the y axis, we would have just gotten "stuck." So, if you get stuck on one axis, try another.

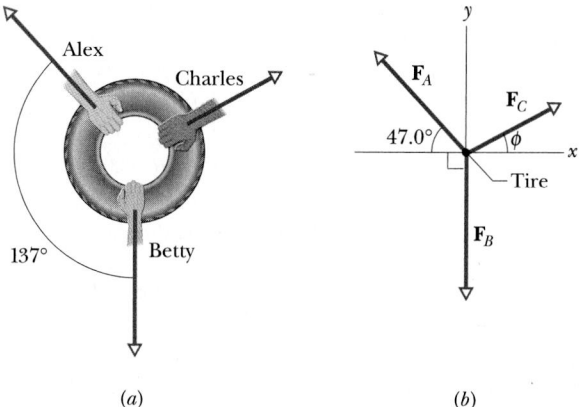

FIGURE 5-4 Sample Problem 5-2. (a) An overhead view of three people pulling on a tire. (b) A free-body diagram for the tire.

SAMPLE PROBLEM 5-3

Figure 5-5a shows an overhead view of a 2 kg cookie tin being accelerated at 8 m/s² across a frictionless surface by three horizontal forces. Forces $\mathbf{F}_1$ and $\mathbf{F}_2$ have magnitudes of 10 N and 12 N, respectively. Figure 5-5b is an incomplete free-body diagram for the situation, with the acceleration $\mathbf{a}$ included. What is the third force $\mathbf{F}_3$ in unit vector notation?

SOLUTION: The net force of the three horizontal forces causes the horizontal acceleration, and Eq. 5-1 gives us

$$\Sigma \mathbf{F} = \mathbf{F}_1 + \mathbf{F}_2 + \mathbf{F}_3 = m\mathbf{a}.$$

From Eqs. 5-2, we have, along the x axis,

$$\Sigma F_x = F_{1x} + F_{2x} + F_{3x} = ma_x, \qquad (5\text{-}6)$$

and along the y axis

$$\Sigma F_y = F_{1y} + F_{2y} + F_{3y} = ma_y. \qquad (5\text{-}7)$$

By rewriting Eq. 5-6 in terms of magnitudes and angles, and including signs to indicate directions, we obtain

$$-F_1 \cos 60° + 0 + F_{3x} = ma \sin 30°.$$

Substitution of the given data yields

$-(10 \text{ N}) \cos 60° + 0 + F_{3x} = (2 \text{ kg})(8 \text{ m/s}^2) \sin 30°,$

which tells us that

$$F_{3x} = (10 \text{ N}) \cos 60° + (2 \text{ kg})(8 \text{ m/s}^2) \sin 30°$$
$$= 13 \text{ N}.$$

Similarly, Eq. 5-7 becomes

$$-F_1 \sin 60° + F_2 + F_{3y} = -ma \cos 30°$$

and then

$$-(10 \text{ N}) \sin 60° + 12 \text{ N} + F_{3y} = -(2 \text{ kg})(8 \text{ m/s}^2) \cos 30°,$$

which tells us that

$$F_{3y} = -17.2 \text{ N} \approx 17 \text{ N}.$$

Thus, the third force is

$$\mathbf{F}_3 = (13 \text{ N})\mathbf{i} - (17 \text{ N})\mathbf{j}. \qquad \text{(Answer)}$$

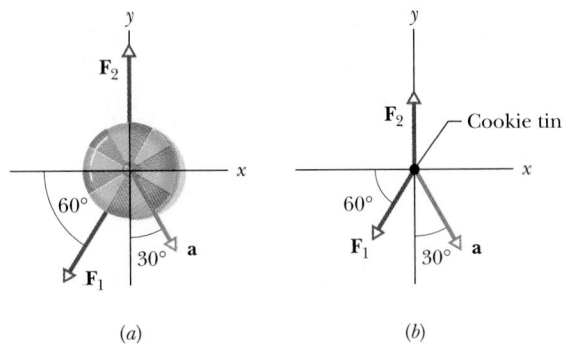

(a)   (b)

FIGURE 5-5 Sample Problem 5-3. (a) An overhead view of a cookie tin that is being accelerated by three horizontal forces, two of which are shown. (b) A free-body diagram for the cookie tin.

CHECKPOINT **3:** The figure shows overhead views of four situations in which two forces accelerate the same block across a frictionless floor. Rank the situations according to the magnitudes of (a) the net force on the block and (b) the acceleration of the block, greatest first.

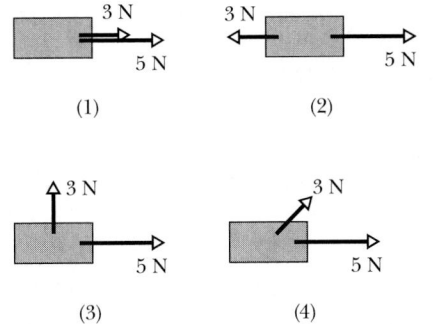

5-6 SOME PARTICULAR FORCES

Weight

The **weight W** of a body is a force that pulls the body directly toward a nearby astronomical body; in everyday circumstances that astronomical body is the Earth. The force is primarily due to an attraction—called a **gravitational attraction**—between the two bodies; we discuss it in detail in Chapter 14. For now we consider only situations in which a body with mass m is located at a point where the free-fall acceleration has magnitude g. Then the magnitude W of the weight (force) vector acting on the body is

$$W = mg. \tag{5-8}$$

The *weight vector* itself can be written either as

$$\mathbf{W} = -mg\mathbf{j} = -W\mathbf{j} \tag{5-9}$$

(where $+\mathbf{j}$ points upward, away from the Earth), or as

$$\mathbf{W} = m\mathbf{g}, \tag{5-10}$$

where $\mathbf{g}$ represents the free-fall acceleration vector. In many cases, the choice of notation is up to you, but you need to understand what you mean by it and not get tripped up by, say, writing Eq. 5-8 when you mean Eq. 5-10.

Since weight is a force, its SI unit is the newton. *It is not mass,* and its magnitude at any given location depends on the value of g there. A bowling ball might weigh 71 N on the Earth, but only 12 N on the Moon, where the free-fall acceleration is different. The ball's mass, 7.2 kg, is the same in either place, because mass is an intrinsic property of the ball alone. (If you want to lose weight, climb a mountain. Not only will the exercise reduce your mass, but the increased elevation means you are farther from the center of the Earth, and that means the value of g is less. So your weight will be less.)

Normally we assume that weight is measured from an inertial frame. If it is, instead, measured from a noninertial frame (an example comes up in Sample Problems 5-11*b* and *c*), the measurement gives an **apparent weight** instead of the actual weight.

We can *weigh* a body by placing it on one of the pans of an equal-arm balance (Fig. 5-6) and then adding reference bodies (whose masses are known) on the other pan until we strike a balance. The masses on the pans then match, and we know the mass m of the body. If we know the value of g for the location of the balance, we can find the weight of the body with Eq. 5-8.

We can also weigh a body with a spring scale (Fig. 5-7). The body stretches a spring, moving a pointer along a scale that has been calibrated and marked in either mass or weight units. (Most bathroom scales in the United States work this way and read in pounds.) If the scale is marked in mass units, it is accurate only where the value of g is the same as where the scale was calibrated.

The Normal Force

When a body is pressed against a surface, the body experiences a force that is perpendicular to the surface. The force is called the **normal force N,** the name coming from the mathematical term *normal,* meaning "perpendicular."

If a body rests on a horizontal surface as in Fig. 5-8*a,* **N** is directed upward and the body's weight $\mathbf{W} = m\mathbf{g}$ is

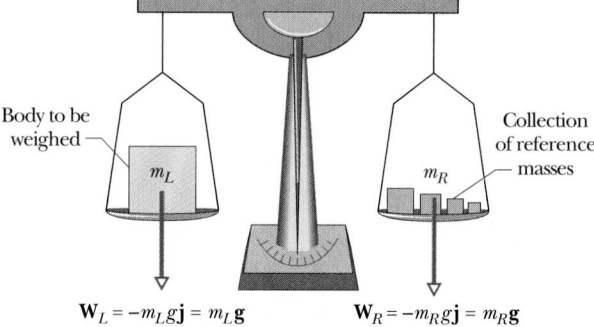

$$\mathbf{W}_L = -m_L g\mathbf{j} = m_L \mathbf{g} \qquad \mathbf{W}_R = -m_R g\mathbf{j} = m_R \mathbf{g}$$

FIGURE 5-6   An equal-arm balance. When the device is in balance, the masses on the left (L) and right (R) pans are equal.

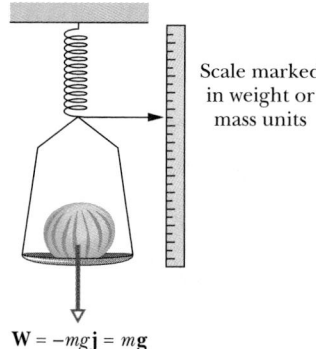

FIGURE 5-7   A spring scale. The reading is proportional to the *weight* of the object placed on the pan, and the scale gives that weight if marked in weight units. If, instead, it is marked in mass units, the reading is accurate only if the free-fall acceleration g is the same as where the scale was calibrated.

Scale marked in weight or mass units

$$\mathbf{W} = -mg\mathbf{j} = m\mathbf{g}$$

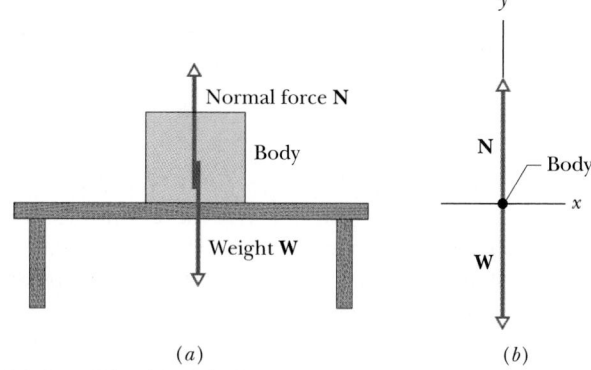

FIGURE 5-8   (*a*) A body resting on a tabletop experiences a normal force **N** perpendicular to the tabletop. (*b*) The corresponding free-body diagram.

directed downward. For this *particular* arrangement, we find the magnitude of **N** from the second of Eqs. 5-2:

$$\sum F_y = N - mg = ma_y, \tag{5-11}$$

and so, with $a_y = 0$,

$$N = mg. \tag{5-12}$$

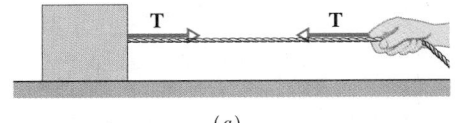

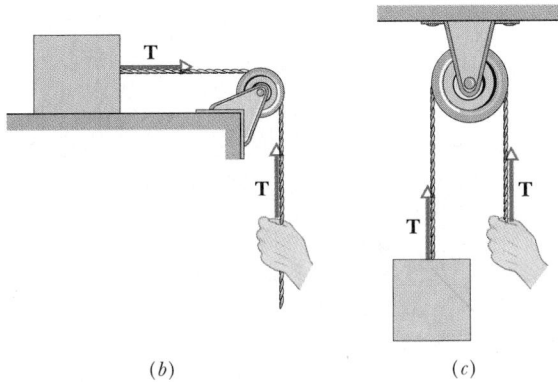

(b)   (c)

FIGURE 5-10 (a) The cord, pulled taut, is under tension. It pulls on the body and the hand with force **T**, even if it runs around a massless, frictionless pulley as in (b) and (c).

PROBLEM SOLVING TACTICS

TACTIC 5: *Normal Force*

Equation 5-12 for the normal force holds only when **N** is directed upward and the vertical acceleration is zero. So we do *not* apply it for other orientations of **N** or when the vertical acceleration is not zero. Instead, we must use Newton's second law in its component form.

   We are free to move **N** around in a figure as long as we maintain its orientation. For example, in Fig. 5-8a we can slide it downward so that its head is at the boundary of the body and the tabletop. However, **N** is least likely to be misinterpreted if its tail is at that boundary or somewhere within the body (as shown). An even better technique is to draw a free-body diagram like that in Fig. 5-8b, with the tail of **N** directly on the dot or sketch representing the body.

CHECKPOINT **4:** In Fig. 5-8, is the magnitude of the normal force **N** greater than, less than, or equal to the weight magnitude *mg* if the body and table are in an elevator that is moving upward (a) at a constant speed and (b) at an increasing speed?

Friction

If we slide or attempt to slide a body over a surface, the motion is resisted by a bonding between the body and surface. (We discuss this more in the next chapter.) The resistance is considered to be a single force **f**, called the **frictional force,** or simply **friction.** This force is directed along the surface, opposite the direction of the intended motion (Fig. 5-9). Sometimes, to simplify a situation, friction is assumed to be negligible, and the surface is said to be *frictionless.*

Tension

When a cord (or a rope, cable, or other such object) is attached to a body and pulled taut, the cord is said to be under **tension.** It pulls on the body with a force **T**, whose direction is away from the body and along the cord at the point of attachment (Fig. 5-10a).

   A cord is often said to be *massless* (meaning its mass is negligible compared to the body's mass) and unstretchable. The cord exists only as a connection between two bodies. It pulls on both bodies with the same magnitude *T*, even if the bodies and the cord are accelerating and even if the cord runs around a *massless, frictionless pulley* (Figs. 5-10b and c). Such a pulley has negligible mass compared to the bodies and has negligible friction on its axle opposing its rotation.

CHECKPOINT **5:** The body that is suspended by a rope in Fig. 5-10c has a weight of 75 N. Is *T* equal to, greater than, or less than 75 N when the body is moving upward (a) at constant speed, (b) at increasing speed, and (c) at decreasing speed?

SAMPLE PROBLEM 5-4

Let us return to John Massis and the railroad cars, and assume that Massis pulled (with his teeth) on his end of the rope with a constant force that was 2.5 times his body weight, at an angle θ of 30° from the horizontal. His mass *m* was 80 kg. The weight *W* of the passenger cars was 7.0×10^5 N (about 80 tons), and he moved them 1.0 m along the rails. Assume that the rolling wheels encountered no retarding force from the rails. What was the train's speed at the end of the pull?

SOLUTION: Figure 5-11 is a free-body diagram for the cars, which are represented by a dot. The *x* axis runs along the rails. From Eqs. 5-2, we have

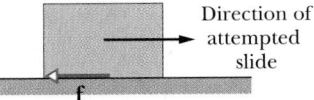

FIGURE 5-9 A frictional force **f** opposes the attempted slide of a body over a surface.

$$\sum F_x = T \cos \theta = M a_x, \qquad (5\text{-}13)$$

in which M is the mass of the cars. We can find T and M from the given data.

With our assumptions, the pull from Massis is

$$T = 2.5mg = (2.5)(80 \text{ kg})(9.8 \text{ m/s}^2) = 1960 \text{ N}$$

(or 440 lb), which is about what a good middle-weight weight lifter can lift—and far from a superhuman force.

The weight W of the cars is, by Eq. 5-8,

$$W = Mg,$$

so their mass M must be

$$M = \frac{W}{g} = \frac{7.0 \times 10^5 \text{ N}}{9.8 \text{ m/s}^2} = 7.143 \times 10^4 \text{ kg}.$$

Now, from Eq. 5-13, we find their acceleration to be

$$a_x = \frac{T \cos \theta}{M} = \frac{(1960 \text{ N})(\cos 30°)}{7.143 \times 10^4 \text{ kg}}$$

$$= 2.376 \times 10^{-2} \text{ m/s}^2.$$

To solve for the speed of the cars at the end of the pull, we use Eq. 2-16, with subscripts for the x axis and with $v_0 = 0$ and $x - x_0 = 1.0$ m:

$$v_x^2 = v_{0x}^2 + 2a_x(x - x_0),$$

or

$$v_x = \sqrt{0 + (2)(2.376 \times 10^{-2} \text{ m/s}^2)(1.0 \text{ m})}$$

$$= 0.22 \text{ m/s}. \qquad \text{(Answer)}$$

Massis would have done better if the rope had been attached higher on the car, so that it was horizontal. Can you see why?

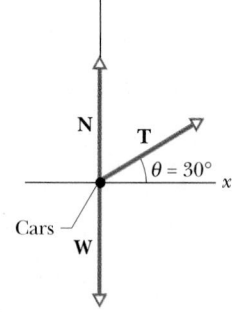

FIGURE 5-11 Sample Problem 5-4. Free-body diagram for the passenger cars pulled by Massis. The vectors are not drawn to scale; the tension in the rope is *much* smaller than the weight and normal force.

5-7 NEWTON'S THIRD LAW

Forces come in pairs. If a hammer exerts a force on a nail, the nail exerts a force of equal magnitude but opposite direction on the hammer. If you lean against a brick wall, the wall pushes back on you.

Let body A in Fig. 5-12 exert a force $\mathbf{F}_{BA}$ on body B; experiment shows that body B then exerts a force $\mathbf{F}_{AB}$ on body A. These two forces are equal in magnitude and op-

FIGURE 5-12 Newton's third law. Body A exerts a force $\mathbf{F}_{BA}$ on body B, while body B exerts a force $\mathbf{F}_{AB}$ on body A, where $\mathbf{F}_{AB} = -\mathbf{F}_{BA}$.

positely directed. That is,

$$\mathbf{F}_{AB} = -\mathbf{F}_{BA} \qquad \text{(Newton's third law).} \qquad (5\text{-}14)$$

Note the order of the subscripts. $\mathbf{F}_{AB}$, for example, is the force exerted *on* body A *by* body B. Equation 5-14 holds regardless of whether the bodies move or remain stationary.

Equation 5-14 sums up Newton's third law of motion. Commonly, one of these forces (it does not matter which) is called the **action force.** The other member of the pair is then called the **reaction force.** Every time you find a force, a good question is: Where is its reaction force?

The words "To every action there is always an equal and opposite reaction" have become enshrined in the popular language and mean various things to various speakers. In physics, however, these words mean Eq. 5-14 and nothing else. In particular, cause and effect are not involved; either force can be the action force.

You may think: "If every force has an associated force that is equal in magnitude and opposite in direction, why don't they cancel each other? How can anything ever get moving?" The answer is simple. As Fig. 5-12 shows:

The forces of an action–reaction pair *always* act on *different* bodies; thus they do not combine to give a net force and cannot cancel each other.

Two forces that act on the *same* body are *not* an action–reaction pair, even though they may be equal in magnitude

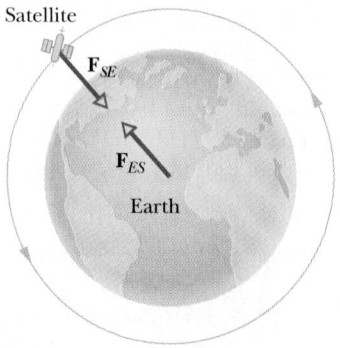

FIGURE 5-13 A satellite in Earth orbit. The forces shown are an action–reaction pair. Note that they act on different bodies.

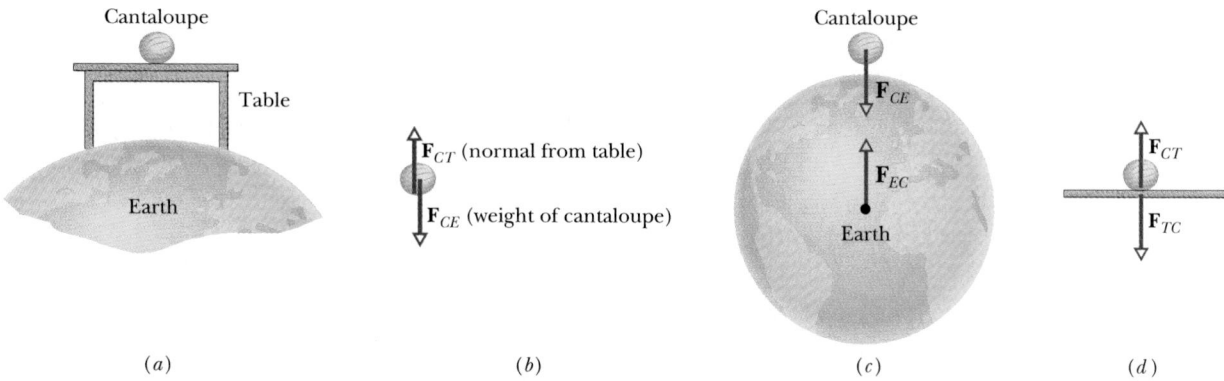

FIGURE 5-14 (*a*) A cantaloupe rests on a tabletop that rests on the Earth. (*b*) The forces *on the cantaloupe*, are $\mathbf{F}_{CT}$ and $\mathbf{F}_{CE}$. The cantaloupe is stationary because these two forces balance. (*c*) The action–reaction pair for the cantaloupe–Earth forces. (*d*) The action–reaction pair for the cantaloupe–table forces.

and opposite in direction. Let's identify the action–reaction pairs in two examples.

An Orbiting Satellite

Figure 5-13 shows an orbiting satellite. The only force that acts on it is $\mathbf{F}_{SE}$, the force exerted *on* the satellite *by* the gravitational pull of the Earth. Where is the corresponding reaction force? It is $\mathbf{F}_{ES}$, the force acting on the Earth due to the gravitational pull of the satellite; the pull is taken to be at the center of the Earth.

You may think that the tiny satellite cannot exert much of a gravitational pull on the Earth but it does, exactly as Newton's third law requires. That is, considering magnitudes only, $F_{ES} = F_{SE}$. The force $\mathbf{F}_{ES}$ causes the Earth to accelerate, but, because of the Earth's large mass, its acceleration is too small to be detected.

A Cantaloupe Resting on a Table

Figure 5-14*a* shows a cantaloupe at rest on a table.* The Earth pulls downward on the cantaloupe with a force $\mathbf{F}_{CE}$, the cantaloupe's weight. The cantaloupe does not accelerate because this force is canceled by an equal and opposite normal force $\mathbf{F}_{CT}$ exerted on the cantaloupe by the table. (See Fig. 5-14*b*.) However, $\mathbf{F}_{CE}$ and $\mathbf{F}_{CT}$ do *not* form an action–reaction pair *because they act on the same body, the cantaloupe.*

The reaction force to $\mathbf{F}_{CE}$ is $\mathbf{F}_{EC}$, the (gravitational) force with which the cantaloupe attracts the Earth. This action–reaction pair is shown in Fig. 5-14*c*.

The reaction force to $\mathbf{F}_{CT}$ is $\mathbf{F}_{TC}$, the force exerted on the table by the cantaloupe. This action–reaction pair is shown in Fig. 5-14*d*. The action–reaction pairs in this

problem, and the bodies on which they act, are then

first pair: $\mathbf{F}_{CE} = -\mathbf{F}_{EC}$ (cantaloupe and Earth)

and

second pair: $\mathbf{F}_{CT} = -\mathbf{F}_{TC}$ (cantaloupe and table).

CHECKPOINT **6:** Suppose that the cantaloupe and table of Fig. 5-14 are in an elevator cab that begins to accelerate upward. (a) Do the magnitudes of forces $\mathbf{F}_{TC}$ and $\mathbf{F}_{CT}$ increase, decrease, or stay the same? (b) Are those two forces still equal in magnitude but opposite in direction? (c) Do the magnitudes of forces $\mathbf{F}_{CE}$ and $\mathbf{F}_{EC}$ increase, decrease, or stay the same? (d) Are those two forces still equal in magnitude but opposite in direction?

5-8 APPLYING NEWTON'S LAWS

The rest of this chapter consists of sample problems. You should pore over them, learning not just their particular answers but, instead, the procedures for attacking a problem. Especially important is knowing how to translate a sketch of a situation into a free-body diagram with appropriate axes, so that Newton's laws can be applied. We begin with Sample Problem 5-5, which is worked out in exhaustive detail, using a question-and-answer format.

SAMPLE PROBLEM 5-5

Figure 5-15 shows a block (the *sliding block*) whose mass M is 3.3 kg. It is free to move along a horizontal frictionless surface such as an air table. The sliding block is connected by a cord that extends around a massless, frictionless pulley to a second block (the *hanging block*), whose mass m is 2.1 kg. The hang-

---

*We ignore small complications caused by the rotation of the Earth.

ing block falls and the sliding block accelerates to the right. Find (a) the acceleration of the sliding block, (b) the acceleration of the hanging block, and (c) the tension in the cord.

Q *What is this problem all about?*

You are given two massive objects, the sliding block and the hanging block. It might not occur to you, but you are also given the Earth, which pulls on each of these objects; without the Earth, nothing would happen. A total of five forces act on the blocks, as shown in Fig. 5-16:

1. The cord pulls to the right on the sliding block with a force of magnitude *T*.

2. The cord pulls upward on the hanging block with a force of the same magnitude *T*. This upward force keeps the hanging block from falling freely, which it would otherwise do. We assume that the cord has the same tension throughout its length; the pulley just serves to change the direction of this force, without changing its magnitude.

3. The Earth pulls down on the sliding block with a force *M***g**, the weight of the sliding block.

4. The Earth pulls down on the hanging block with a force *m***g**, the weight of the hanging block.

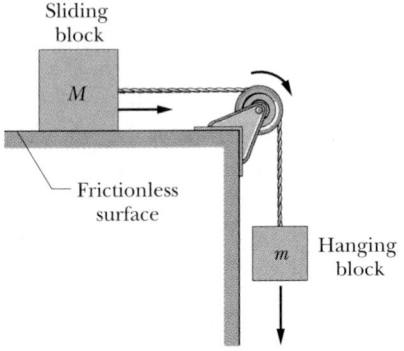

FIGURE 5-15 Sample Problem 5-5. A block of mass *M* on a horizontal frictionless surface is connected to a block of mass *m* by a cord that wraps over a pulley. Both cord and pulley are massless. The pulley is frictionless. The arrows indicate the motion when the system is released from rest.

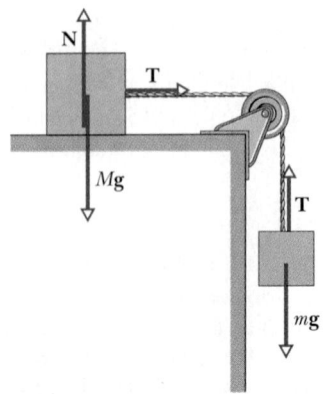

FIGURE 5-16 The forces acting on the two blocks.

5. The table pushes up on the sliding block with a normal force **N**.

There is another thing that you should note. We assume that the cord does not stretch, so that if the hanging block falls 1 mm in a certain time, the sliding block moves 1 mm to the right in that same interval. The blocks move together and their accelerations have the same magnitude *a*.

Q *How do I classify this problem? Should it suggest a particular law of physics to me?*

Yes, it should. Forces, masses, and accelerations are involved, and that should suggest Newton's second law of motion, Σ**F** = *m***a**.

Q *If I apply that law to this problem, to what body should I apply it?*

We focus on two bodies in this problem, the sliding block and the hanging block. Although they are extended objects, we can treat each block as a particle because every small part of it (every atom, say) moves in exactly the same way. Apply Newton's second law separately to each block.

Q *What about the pulley?*

We cannot represent the pulley as a particle because different parts of it move in different ways. When we discuss rotation, we shall deal with pulleys in detail. Meanwhile, we get around the problem by using a pulley whose mass is negligible compared with the masses of the two blocks.

Q *OK. Now how do I apply Σ**F** = m**a** to the sliding block?*

Represent the sliding block as a particle of mass *M* and draw *all* the forces that act *on* it, as in Fig. 5-17. This is the block's *free-body diagram*. There are three forces. Next, locate a set of axes. It makes sense to draw the *x* axis parallel to the table, in the direction in which the block moves.

Q *Thanks, but you still haven't told me how to apply Σ**F** = m**a** to the sliding block. All you have done is explain how to draw a free-body diagram.*

Right you are. The expression Σ**F** = *m***a** is a vector equation and you can write it as three scalar equations:

$$\sum F_x = Ma_x, \quad \sum F_y = Ma_y, \quad \sum F_z = Ma_z, \quad (5\text{-}15)$$

in which $\sum F_x$, $\sum F_y$, and $\sum F_z$ are the components of the net force. Because the sliding block does not move vertically, we know there is no net force in the *y* direction: the weight **W** = *M***g** of the sliding block is balanced by the upward-acting normal force **N** on the block. No force acts in the *z* direction, which is perpendicular to the page. We can, however, apply the first of Eqs. 5-15.

In the *x* direction, there is only one force component, so $\sum F_x = Ma_x$ becomes

$$T = Ma. \quad (5\text{-}16)$$

This equation contains two unknowns, *T* and *a*, so we cannot yet solve it. Recall, however, that we have not said anything about the hanging block.

Q *I agree. How do I apply Σ**F** = m**a** to the hanging block?*

Draw a free-body diagram for the block, as in Fig. 5-18. This time, we use the second of Eqs. 5-15, finding

$$\sum F_y = T - mg = -ma, \qquad (5\text{-}17)$$

where the minus sign on the right side of the equation indicates that the hanging block accelerates downward, in the negative direction of the y axis. Equation 5-17 yields

$$mg - T = ma. \qquad (5\text{-}18)$$

This contains the same two unknowns as Eq. 5-16 does. If you add these equations, T will cancel out. Solving for a then yields

$$a = \frac{m}{M + m} g. \qquad (5\text{-}19)$$

Substituting this result into Eq. 5-16 yields

$$T = \frac{Mm}{M + m} g. \qquad (5\text{-}20)$$

Putting in the numbers gives, for these two quantities,

$$a = \frac{m}{M + m} g = \frac{2.1 \text{ kg}}{3.3 \text{ kg} + 2.1 \text{ kg}} (9.8 \text{ m/s}^2)$$

$$= 3.8 \text{ m/s}^2 \qquad \text{(Answer)}$$

and

$$T = \frac{Mm}{M + m} g = \frac{(3.3 \text{ kg})(2.1 \text{ kg})}{3.3 \text{ kg} + 2.1 \text{ kg}} (9.8 \text{ m/s}^2)$$

$$= 13 \text{ N.} \qquad \text{(Answer)}$$

Q *The problem is now solved, right?*

That's a fair question, but we are here not only to solve problems but also to learn physics. This problem is not really finished until we have studied the results to see if they make sense. This is often a much more confidence-building experience than simply getting the right answer.

Look first at Eq. 5-19. Note that it is dimensionally correct and also that the acceleration a will always be less than g. This is as it must be, because the hanging block is not in free fall. The cord pulls upward on it.

Look now at Eq. 5-20, which we rewrite in the form

$$T = \frac{M}{M + m} mg. \qquad (5\text{-}21)$$

In this form, it is easier to see that this equation is also dimensionally correct, because both T and mg are forces. Equation 5-21 also lets us see that the tension in the cord is always less than mg, the weight of the hanging block. That is a comforting thought because, if T were *greater* than mg, the hanging block would accelerate upward.

We can also check the results by studying special cases, in which we can guess what the answers must be. A simple example is to put $g = 0$, as if the experiment were carried out in interstellar space. We know that in that case, the blocks would not move from rest and there would be no tension in the cord. Do the formulas predict this? Yes, they do. If you put $g = 0$ in Eqs. 5-19 and 5-20, you find $a = 0$ and $T = 0$. Two more special cases that you might try are $M = 0$ and $m \rightarrow \infty$.

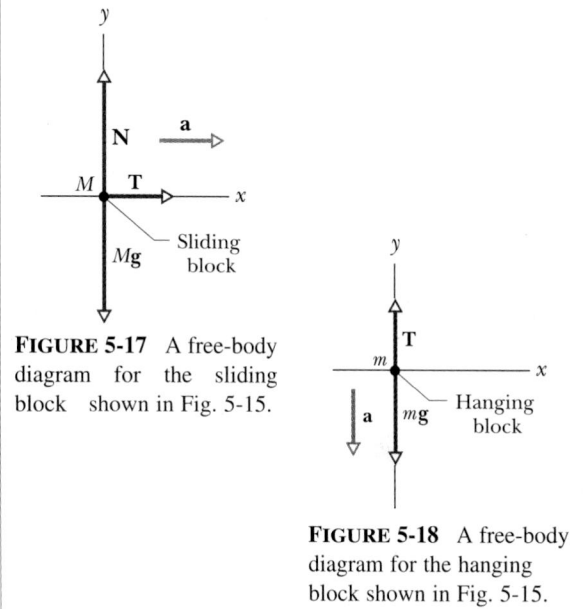

FIGURE 5-17 A free-body diagram for the sliding block shown in Fig. 5-15.

FIGURE 5-18 A free-body diagram for the hanging block shown in Fig. 5-15.

SAMPLE PROBLEM 5-5—ANOTHER WAY

The acceleration a of the blocks of Fig. 5-15 can be found in two lines of algebra if we (a) employ an unconventional axis, call it u, that runs through *both* blocks and along the cord as shown in Fig. 5-19a, and then (b) mentally straighten out the u axis as in Fig. 5-19b and treat the blocks as being portions of a single composite body with mass $M + m$. A free-body diagram for the two-block system is shown in Fig. 5-19c.

SOLUTION: Note that there is only one force acting on the composite body along the u axis, and that is the force mg in the positive direction of the axis. The tension $\mathbf{T}$ of Fig. 5-16 is now internal to the composite body and so does not enter into Newton's second law. The force exerted by the pulley on the cord is perpendicular to the u axis; so it too does not enter.

Using Eqs. 5-2 as a guide, we write a component equation for the acceleration along the u axis:

$$\sum F_u = (M + m)a_u,$$

where the mass of the body is $M + m$. The acceleration of the composite body along the u axis (and of the individual blocks, since they are connected) has magnitude a. The only force on the composite body along the u axis has a magnitude of mg. So our equation becomes

$$mg = (M + m)a,$$

or

$$a = \frac{m}{M + m} g, \qquad (5\text{-}22)$$

which matches Eq. 5-19.

To find T, we apply Newton's second law to either block, obtaining either Eq. 5-16 or Eq. 5-18. We then substitute for a from Eq. 5-22 and solve for T, getting Eq. 5-20.

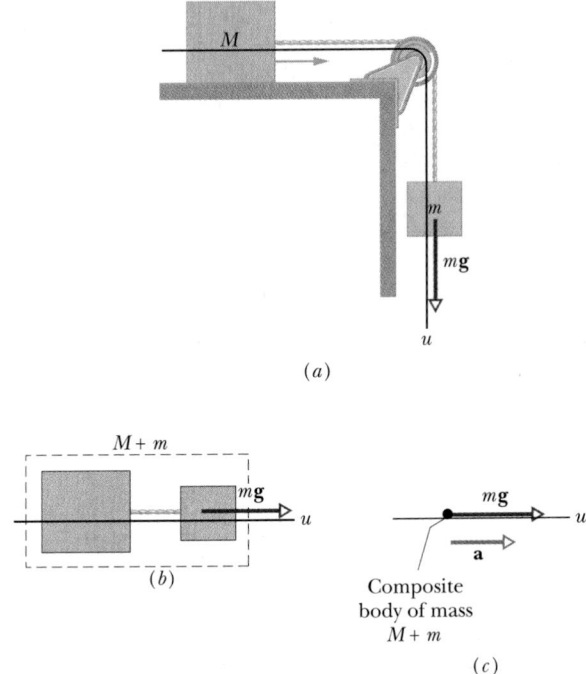

(a)

(b)

Composite
body of mass
$M + m$

(c)

FIGURE 5-19   (a) An "axis" u runs through the blocks-and-cord system of Fig. 5-15. (b) The blocks are rearranged to straighten u and then are treated as a single body with mass $M + m$. (c) The associated free-body diagram, considering only forces along u. There is one such force.

SAMPLE PROBLEM 5-6

A block whose mass M is 33 kg is pushed across a frictionless surface by a stick whose mass m is 3.2 kg, as in Fig. 5-20a. The block is moved (from rest) a distance $d = 77$ cm in 1.7 s at constant acceleration.

(a) Identify all horizontal action–reaction force pairs in this problem.

SOLUTION: As the exploded view of Fig. 5-20b shows, there are two action–reaction pairs:

first pair: $\mathbf{F}_{HS} = -\mathbf{F}_{SH}$ (hand and stick)

second pair: $\mathbf{F}_{SB} = -\mathbf{F}_{BS}$ (stick and block).

The force $\mathbf{F}_{HS}$ on the hand from the stick is the force that you would feel if the hand in Fig. 5-20 were yours.

(b) What force must the hand apply to the stick?

SOLUTION: This is the force that accelerates the block and stick. To find it, we must first find the constant acceleration a, using Eq. 2-15:

$$x - x_0 = v_0 t + \tfrac{1}{2}at^2.$$

Setting $v_0 = 0$ and $x - x_0 = d$, and solving for a, give

$$a = \frac{2d}{t^2} = \frac{(2)(0.77 \text{ m})}{(1.7 \text{ s})^2} = 0.533 \text{ m/s}^2.$$

To find the force that the hand exerts, we apply Newton's second law to a system consisting of the stick and the block taken together. Thus,

$$F_{SH} = (M + m)a = (33 \text{ kg} + 3.2 \text{ kg})(0.533 \text{ m/s}^2)$$
$$= 19.3 \text{ N} \approx 19 \text{ N}. \qquad \text{(Answer)}$$

(c) With what force does the stick push on the block?

SOLUTION: To find this force, we apply Newton's second law to the block alone:

$$F_{BS} = Ma = (33 \text{ kg})(0.533 \text{ m/s}^2)$$
$$= 17.6 \text{ N} \approx 18 \text{ N}. \qquad \text{(Answer)}$$

(d) What is the net force on the stick?

SOLUTION: We can find the magnitude F of this force in two ways. First, using results from (b) and (c), we have

$$F = F_{SH} - F_{SB} = 19.3 \text{ N} - 17.6 \text{ N}$$
$$= 1.7 \text{ N}. \qquad \text{(Answer)}$$

Note that we have used Newton's third law here, in assuming that $\mathbf{F}_{SB}$, the force on the stick from the block, has the same magnitude (17.6 N before rounding) as $\mathbf{F}_{BS}$.

The second way to arrive at an answer is to apply Newton's second law to the stick directly. We have

$$F = ma = (3.2 \text{ kg})(0.533 \text{ m/s}^2) = 1.7 \text{ N}, \quad \text{(Answer)}$$

in agreement with our first result. This is as it must be because the two methods are algebraically identical; check it out.

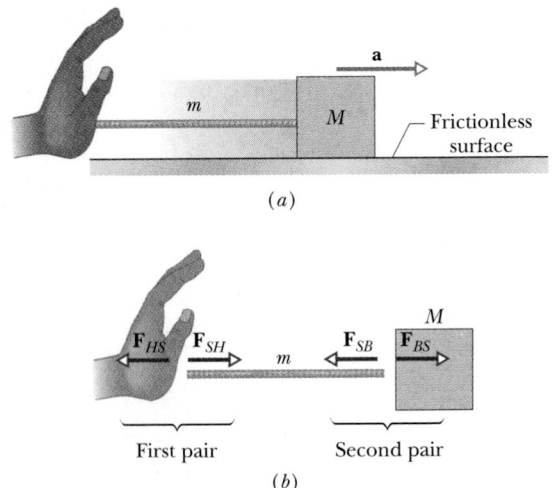

(a)

First pair   Second pair

(b)

FIGURE 5-20   Sample Problem 5-6. (a) A block of mass M is pushed over a frictionless surface by a stick of mass m. (b) An exploded view, showing the action–reaction pairs between the hand and the stick (first pair) and between the stick and the block (second pair).

SAMPLE PROBLEM 5-7

Figure 5-21a shows a block of mass $m = 15$ kg hanging from three cords. What are the tensions in the cords?

SOLUTION: In the free-body diagram for the block (Fig. 5-21b), tension $\mathbf{T}_C$ from cord C pulls upward while the block's weight $m\mathbf{g}$ is directed downward. Since the system is at rest, Newton's second law for the block yields

$$\sum \mathbf{F} = \mathbf{T}_C + m\mathbf{g} = 0.$$

Because forces $\mathbf{T}_C$ and $m\mathbf{g}$ are only vertical, this equation gives us one scalar equation:

$$\sum F_y = T_C - mg = 0.$$

Substituting known values, we find

$$T_C = mg = (15 \text{ kg})(9.8 \text{ m/s}^2)$$
$$= 147 \text{ N} \approx 150 \text{ N}. \qquad \text{(Answer)}$$

The clue to our next step is to realize that the knot where the three cords join is the only point at which all three forces act, and it is this knot to which we should apply Newton's second law. Figure 5-21c is the free-body diagram for the knot. Since the knot is not accelerated, the net force acting on it must be zero. Thus

$$\sum \mathbf{F} = \mathbf{T}_A + \mathbf{T}_B + \mathbf{T}_C = 0.$$

This vector equation is equivalent to the two scalar equations

$$\sum F_y = T_A \sin 28° + T_B \sin 47° - T_C = 0 \quad \text{(5-23)}$$

and

$$\sum F_x = -T_A \cos 28° + T_B \cos 47° = 0. \quad \text{(5-24)}$$

Note carefully that, when we write the x component of $\mathbf{T}_A$ as $T_A \cos 28°$, we must include a minus sign to show that it extends in the negative direction of the x axis.

Substituting numerical values into Eqs. 5-23 and 5-24 leads to

$$T_A(0.469) + T_B(0.731) = 147 \text{ N} \quad \text{(5-25)}$$

and

$$T_B(0.682) = T_A(0.883). \quad \text{(5-26)}$$

From Eq. 5-26, we have

$$T_B = \frac{0.883}{0.682} T_A = 1.29 T_A.$$

Substituting this into Eq. 5-25 and solving for T_A, we obtain

$$T_A = \frac{147 \text{ N}}{0.469 + (1.29)(0.731)}$$
$$= 104 \text{ N} \approx 100 \text{ N}. \qquad \text{(Answer)}$$

Finally, T_B is found from

$$T_B = 1.29 T_A = (1.29)(104 \text{ N})$$
$$= 134 \text{ N} \approx 130 \text{ N}. \qquad \text{(Answer)}$$

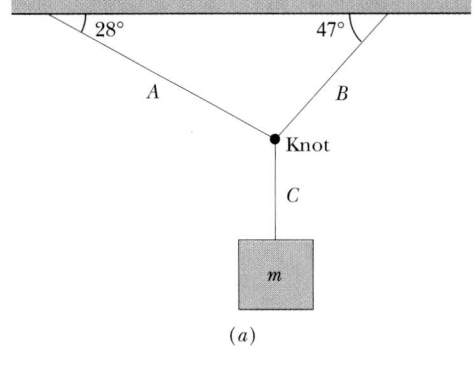

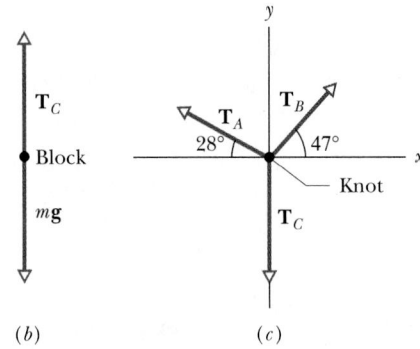

FIGURE 5-21 Sample Problem 5-7. (a) A block of mass m hangs from three cords. (b) A free-body diagram for the block. (c) A free-body diagram for the knot at the intersection of the three cords.

SAMPLE PROBLEM 5-8

Figure 5-22a shows a block of mass $m = 15$ kg held by a cord on a frictionless inclined plane. What is the tension in the cord if $\theta = 27°$? What force does the plane exert on the block?

SOLUTION: Figure 5-22b is the free-body diagram for the block. The following forces act on it: (1) a normal force $\mathbf{N}$, exerted outward on the block by the plane on which it rests, (2) the tension $\mathbf{T}$ in the cord, and (3) the weight $\mathbf{W}$ ($= m\mathbf{g}$) of the block. Because the acceleration of the block is zero, the net force acting on the block must also be zero by Newton's second law:

$$\sum \mathbf{F} = \mathbf{T} + \mathbf{N} + m\mathbf{g} = 0. \quad \text{(5-27)}$$

We choose a coordinate system with the x axis parallel to the plane. With this choice, not one but two forces ($\mathbf{N}$ and $\mathbf{T}$) line up with coordinate axes (a bonus). Note that the angle between the weight vector and the negative direction of the y axis equals the slant angle of the plane. The x and y components of that vector are found with the triangle of Fig. 5-22c. The component versions of Eq. 5-27 are

$$\sum F_x = T - mg \sin \theta = 0$$

and

$$\sum F_y = N - mg \cos \theta = 0.$$

Thus

$$T = mg \sin \theta$$
$$= (15 \text{ kg})(9.8 \text{ m/s}^2)(\sin 27°)$$
$$= 67 \text{ N} \qquad \text{(Answer)}$$

and

$$N = mg \cos \theta$$
$$= (15 \text{ kg})(9.8 \text{ m/s}^2)(\cos 27°)$$
$$= 131 \text{ N} \approx 130 \text{ N}. \qquad \text{(Answer)}$$

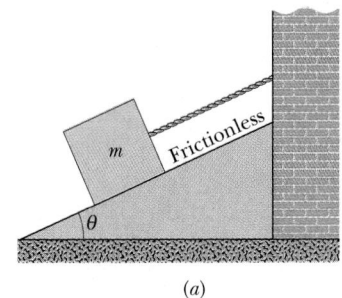

(a)

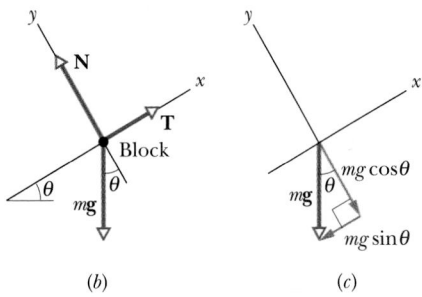

(b)   (c)

FIGURE 5-22 Sample Problems 5-8 and 5-9. (a) A block of mass m rests on a smooth plane, held there by a cord. (b) A free-body diagram for the block. Note how the coordinate axes are placed. (c) Finding the x and y components of mg.

CHECKPOINT 7: In the figure, horizontal force **F** is applied to the block. (a) Is the component of **F** that is perpendicular to the ramp $F \cos \theta$ or $F \sin \theta$? (b) Does the presence of **F** increase or decrease the magnitude of the normal force on the block from the ramp?

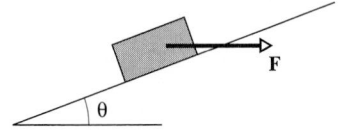

SAMPLE PROBLEM 5-9

Suppose you cut the cord holding the block on the plane in Fig. 5-22a. With what acceleration will the block move?

SOLUTION: Cutting the cord removes the tension **T** in Fig. 5-22b. The two remaining forces do not cancel and, indeed, they cannot, because they do not act along the same line. Applying Newton's second law to the x components of the forces **N** and m**g** in Fig. 5-22b now yields

$$\sum F_x = 0 - mg \sin \theta = ma,$$

so

$$a = -g \sin \theta. \qquad (5\text{-}28)$$

Note that the normal force **N** plays no role in producing the acceleration because its x component is zero.

Equation 5-28 yields

$$a = -(9.8 \text{ m/s}^2)(\sin 27°) = -4.4 \text{ m/s}^2. \qquad \text{(Answer)}$$

The minus sign indicates that the acceleration is in the direction of decreasing x, that is, down the plane.

Equation 5-28 reveals that the acceleration of the block is independent of its mass, just as the acceleration of a freely falling body is independent of the mass of the falling body. Indeed, Eq. 5-28 shows that an inclined plane can be used to "dilute" gravitation—to "slow down" free fall. For $\theta = 90°$, Eq. 5-28 yields $a = -g$; for $\theta = 0°$, it yields $a = 0$. Both are expected results.

SAMPLE PROBLEM 5-10

Figure 5-23a shows two blocks connected by a cord that passes over a massless, frictionless pulley (the arrangement is known as *Atwood's machine*). Let $m = 1.3$ kg and $M = 2.8$ kg. Find the tension in the cord and the (common) magnitude of the acceleration of the two blocks.

SOLUTION: Figures 5-23b and 5-23c are free-body diagrams for the blocks. We are given $M > m$, so we expect M to fall and m to rise. That information allows us to assign the proper algebraic signs to the accelerations of the blocks.

Before we begin the calculations, we note that the tension in the cord must be less than the weight of block M (otherwise, that block would not fall from rest) and greater than the weight of block m (otherwise, that block would not rise). The vectors in the two free-body diagrams of Fig. 5-23 are drawn to represent these facts.

Applying Newton's second law to the block with mass m, which has an acceleration of magnitude a in the positive direction of the y axis, we find

$$T - mg = ma. \qquad (5\text{-}29)$$

For the block with mass M, which has an acceleration of magnitude a in the negative direction of the y axis, we have

$$T - Mg = -Ma, \qquad (5\text{-}30)$$

or

$$-T + Mg = Ma. \qquad (5\text{-}31)$$

By adding Eqs. 5-29 and 5-31 (or eliminating T through substitution), we obtain

$$a = \frac{M - m}{M + m} g. \qquad (5\text{-}32)$$

Substituting this result in either Eq. 5-29 or Eq. 5-31 and solving for T, we obtain

$$T = \frac{2mM}{M + m} g. \qquad (5\text{-}33)$$

Inserting the given data, we obtain

$$a = \frac{M - m}{M + m} g = \frac{2.8 \text{ kg} - 1.3 \text{ kg}}{2.8 \text{ kg} + 1.3 \text{ kg}} (9.8 \text{ m/s}^2)$$

$$= 3.6 \text{ m/s}^2 \qquad \text{(Answer)}$$

and

$$T = \frac{2Mm}{M + m} g = \frac{(2)(2.8 \text{ kg})(1.3 \text{ kg})}{2.8 \text{ kg} + 1.3 \text{ kg}} (9.8 \text{ m/s}^2)$$

$$= 17 \text{ N}. \qquad \text{(Answer)}$$

You can show that the weights of the two blocks are 13 N ($= mg$) and 27 N ($= Mg$). So the tension ($= 17$ N) does indeed lie between these two values.

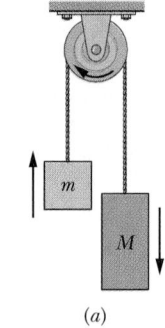

(a)

FIGURE 5-23 Sample Problem 5-10. (a) A block of mass M and one of smaller mass m are connected by a cord that passes over a pulley. The directions in which the system will accelerate from rest are shown by the arrows. Free-body diagrams for (b) block m and (c) block M.

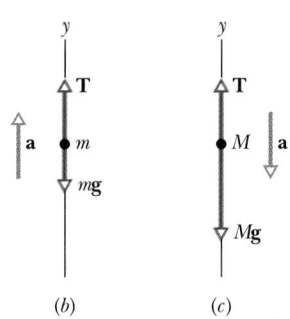

(b)   (c)

SAMPLE PROBLEM 5-10—ANOTHER WAY

Just as we redid Sample Problem 5-5 with an unconventional axis u, we can redo Sample Problem 5-10 with u.

SOLUTION: Run the u axis through the system as shown in Fig. 5-24a. Straighten out the axis as in Fig. 5-24b, and treat the blocks as a single body with a mass of $M + m$. Then draw a free-body diagram as in Fig. 5-24c. Note that along the u

axis there are two forces on the composite of the blocks: mg in the negative direction of the u axis and Mg in the positive direction. (The force exerted on the cord by the pulley is perpendicular to the u axis and so doesn't enter our calculation.) The two forces along the u axis give the composite body (and each block) an acceleration **a**. Newton's second law in component form for motion along u is

$$\sum F_u = Mg - mg = (M + m)a, \qquad (5\text{-}34)$$

which yields

$$a = \frac{M - m}{M + m} g,$$

as previously. To get T we apply Newton's second law to either block, using a conventional axis y. For the block with mass m, we obtain Eq. 5-29. With the above result for a substituted into Eq. 5-29, we obtain Eq. 5-33.

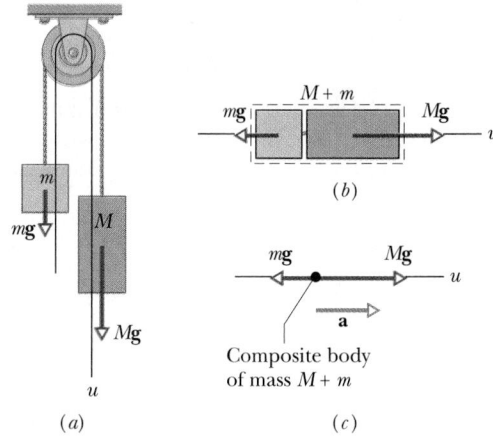

FIGURE 5-24 (a) An "axis" u runs through the system of Fig. 5-23. (b) The blocks are rearranged to straighten u and then treated as a single body with mass $M + m$. (c) The associated free-body diagram, considering only forces along u. There are two such forces.

SAMPLE PROBLEM 5-11

A passenger of mass $m = 72.2$ kg stands on a platform scale in an elevator cab (Fig. 5-25). What is the scale reading for the acceleration values given in the figure?

SOLUTION: We consider this problem from the point of view of an observer in an (inertial) reference frame fixed with respect to the Earth. Let this observer apply Newton's second law to the accelerating passenger. Figure 5-25a–c presents free-body diagrams for the passenger, treated as a particle (some particle!), for various accelerations of the cab.

Regardless of the acceleration of the cab, the Earth pulls downward on the passenger with a force having magnitude mg, in which $g = 9.80$ m/s^2 is the free-fall acceleration in the reference frame of the Earth. The elevator pushes upward on the scale platform. The scale platform, in turn, pushes upward on the passenger with a normal force whose magnitude N equals the reading of the scale. The weight that the accelerating passenger would judge himself to have is what he reads on the scale. This quantity is often called the *apparent weight;* the term *weight* (or *true weight*) is reserved for the quantity mg.

Newton's second law yields

$$N - mg = ma,$$

or $$N = m(g + a). \qquad (5\text{-}35)$$

(a) What does the scale read if the cab is at rest or moving with constant speed? (See Fig. 5-25a.)

SOLUTION: Here $a = 0$ and we have

$$N = m(g + a) = (72.2 \text{ kg})(9.80 \text{ m/s}^2 + 0)$$
$$= 708 \text{ N}. \qquad \text{(Answer)}$$

This is the weight of the passenger.

(b) What does the scale read if the cab has an upward acceleration of magnitude 3.20 m/s²? (See Fig. 5-25b.)

SOLUTION: An upward acceleration means that the cab is either moving up with increasing speed or down with decreasing speed. In either case, the cab is a noninertial frame and Eq. 5-35 yields

$$N = m(g + a) = (72.2 \text{ kg})(9.80 \text{ m/s}^2 + 3.20 \text{ m/s}^2)$$
$$= 939 \text{ N}. \qquad \text{(Answer)}$$

The passenger presses down on the scale with a greater force than if he were at rest. The passenger—reading the scale—might conclude that he has gained 231 N!

(c) What does the scale read if the cab has a downward acceleration of magnitude 3.20 m/s²? (See Fig. 5-25c.)

SOLUTION: A downward acceleration means that the cab is either moving up with decreasing speed or down with increas-

ing speed. Again the cab is a noninertial frame. Equation 5-35 now yields

$$N = m(g + a) = (72.2 \text{ kg})(9.80 \text{ m/s}^2 - 3.20 \text{ m/s}^2)$$
$$= 477 \text{ N}. \qquad \text{(Answer)}$$

The passenger presses down on the scale with less force than if the cab were at rest. He seems to have lost 231 N.

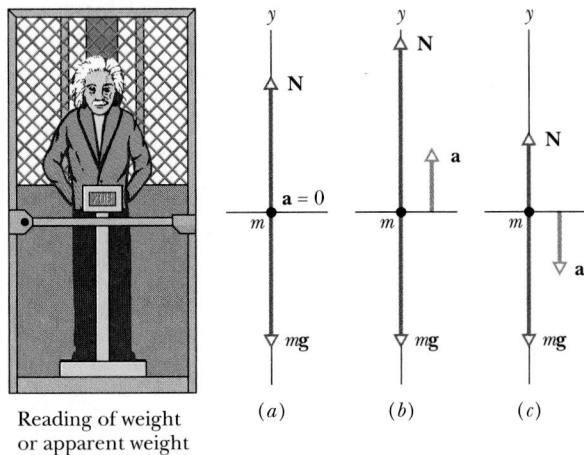

Reading of weight or apparent weight

FIGURE 5-25 Sample Problem 5-11. A passenger with mass m in an elevator, standing on a spring scale which indicates his weight or apparent weight. The free-body diagrams are for the cases in which (a) the acceleration of the elevator cab is zero, (b) $a = +3.20$ m/s², and (c) $a = -3.20$ m/s².

CHECKPOINT **8:** What does the scale read if the cable breaks, so that the cab falls freely? That is, what is the apparent weight of the passenger in free fall?

REVIEW & SUMMARY

Newtonian Mechanics
The velocity of a particle or a particle-like body can change—that is, the particle can accelerate—when the particle is acted on by one or more **forces**—pushes or pulls—exerted by other objects. *Newtonian mechanics* relates accelerations and forces.

Force
The magnitudes of forces are defined in terms of the acceleration they give the standard kilogram. A force that accelerates that standard body by exactly 1 m/s² is defined to have a magnitude of 1 N. The direction of the force is the direction of the acceleration. Forces are experimentally found to be vector quantities, so they are combined according to the rules of vector algebra. The **net force** on a body is the vector sum of all the forces acting on it.

Mass
The **mass** of a body is the characteristic of that body that relates the body's acceleration to the force (or net force) causing the acceleration. Masses are scalar quantities.

Newton's First Law
If there is no net force on a body, the body must remain at rest if it is initially at rest, or move in a straight line at constant speed if it is in motion. For such a body, there are reference frames, called *inertial frames,* from which the body's acceleration $\mathbf{a}$ will be measured as being zero.

Newton's Second Law
The net force $\Sigma \mathbf{F}$ on a body with mass m is related to the body's

acceleration **a** by

$$\sum \mathbf{F} = m\mathbf{a}, \tag{5-1}$$

which may be written in its scalar component version:

$$\sum F_x = ma_x, \quad \sum F_y = ma_y, \quad \text{and} \quad \sum F_z = ma_z. \tag{5-2}$$

The second law indicates that in SI units

$$1 \text{ N} = 1 \text{ kg} \cdot \text{m/s}^2. \tag{5-3}$$

A **free-body diagram** is helpful in solving problems with the second law: it is a stripped-down diagram in which only *one* body is considered. That body is represented by a dot. The external forces on the body are drawn as vectors, and a coordinate system is superimposed, oriented so as to simplify the solution.

Some Particular Forces

A body's **weight W** is the force on the body from a nearby astronomical body:

$$\mathbf{W} = m\mathbf{g}, \tag{5-10}$$

where **g** is the free-fall acceleration. Usually the astronomical body is the Earth.

A **normal force N** is the force exerted on a body by a surface against which the body is pressed. The normal force is always perpendicular to the surface.

A **frictional force f** is the force on a body when the body slides or attempts to slide along a surface. The force is parallel to the surface and directed so as to oppose the motion of the body. A **frictionless surface** is one where the frictional force is negligible.

A **tension T** is the force on a body from a taut cord at its point of attachment. The force points along the cord, away from the body. For **massless** cords (their mass is negligible), the pulls at both ends of the cord have the same magnitude T, even if the cord runs around a **massless, frictionless pulley** (the pulley's mass is negligible and the pulley has negligible friction on its axle opposing its rotation).

Newton's Third Law

If body A exerts a force $\mathbf{F}_{BA}$ on body B, then B must exert a force $\mathbf{F}_{AB}$ on body A. The forces are equal in magnitude and opposite in direction:

$$\mathbf{F}_{AB} = -\mathbf{F}_{BA}. \tag{5-14}$$

These forces act on *different* bodies.

QUESTIONS

1. If two forces act on a moving body, could the body possibly move (a) at constant speed or (b) at constant velocity? Could the body's velocity be zero (c) for an instant or (d) continuously?

2. Figure 5-26 shows four forces of equal magnitude. Do any three of them look as if they will leave a body's velocity unchanged when all three act on the body? If so, which three?

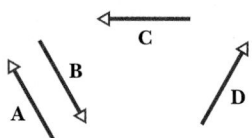

FIGURE 5-26   Question 2.

3. Figure 5-27 shows overhead views of four situations in which forces act on a block that lies on a frictionless floor. If the force magnitudes are chosen properly, in which situations is it possible that the block is (a) stationary and (b) moving with a constant velocity?

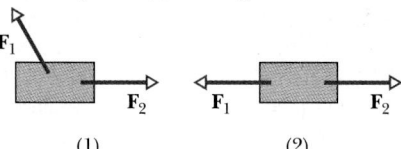

(1)   (2)

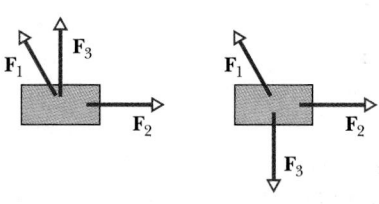

FIGURE 5-27   Question 3.   (3)   (4)

4. A vertical force **F** is applied to a block of mass m that lies on a floor. What happens to the magnitude of the normal force **N** on the block from the floor as magnitude F is increased from zero if force **F** is (a) downward and (b) upward?

5. If the body in Fig. 5-8 has a weight of 100 N, and if a body with a weight of 50 N lies on top of it, what is the normal force on (a) the upper body from the lower body and (b) the lower body from the table?

6. (a) In Fig. 5-28, does the vertical component of force **F** help support the box, or does it press the box against the floor? (b) Suppose the mass of the box is m. Then is the magnitude of the normal force on the box equal to, greater than, or less than mg? (c) Is the vertical component of the force $F \sin \theta$ or $F \cos \theta$?

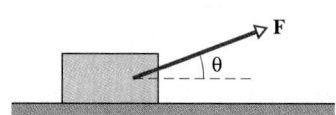

FIGURE 5-28   Question 6.

7. The body that is suspended by a rope in Fig. 5-10c has a weight of 75 N. Is T equal to, greater than, or less than 75 N when the body is moving downward at (a) increasing speed and (b) decreasing speed?

8. An elevator is supported by a single cable and there is no counterweight. The elevator receives passengers at the ground floor and takes them to the top floor, where they disembark. New passengers enter and are taken down to the ground floor. During this round trip, (a) when is the tension in the cable equal to the weight of the elevator plus passengers, (b) when is it greater, and (c) when is it less?

9. In Fig. 5-29, a massless rope is strung over a frictionless pulley. A monkey holds onto the rope, and a mirror, having the same weight as the monkey, is attached to the other side of the rope at the monkey's level. Can the monkey get away from the image it sees in the mirror by (a) climbing up the rope, (b) climbing down the rope, or (c) releasing the rope? Explain.

FIGURE 5-29 Question 9.

10. July 17, 1981, Kansas City: the newly opened Hyatt Regency is packed with people listening and dancing to a band playing favorites from the 1940s. Many of the people are crowded onto the walkways that hang like bridges across the wide atrium. Suddenly two of the walkways collapse, falling onto the merrymakers on the main floor.

The walkways were suspended one above another on vertical rods and held in place by nuts threaded onto the rods. In the original design, only two long rods were to be used, each extending through all three walkways (Fig. 5-30a). If each walkway and the merrymakers on it have a combined weight of W, what is the total weight supported by the threads and two nuts on (a) the lowest walkway and (b) the highest walkway?

Threading nuts on a rod is impossible except at the ends, so the design was changed: instead, six rods were used, each connecting two walkways (Fig. 5-30b). What now is the total weight supported by the threads and two nuts on (c) the lowest walkway and (d) the highest walkway? It was this design that failed.

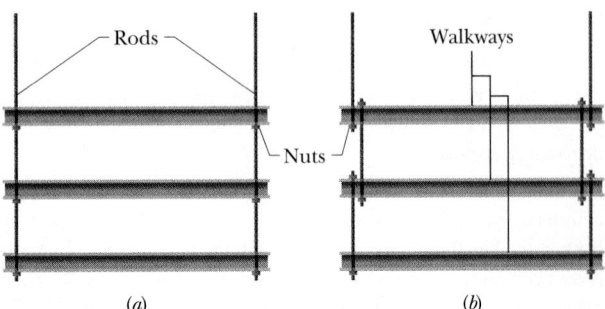

(a)   (b)

FIGURE 5-30 Question 10.

11. In Fig. 5-31, a block is attached by a rope to a bar that is itself rigidly attached to a ramp. Tell whether the magnitudes of the following increase, decrease, or remain the same as the angle θ of the ramp is increased from zero: (a) the block's weight component along the ramp, (b) the tension in the cord, (c) the block's weight component perpendicular to the ramp, and (d) the normal force on the block from the ramp.

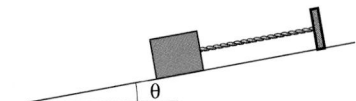

FIGURE 5-31 Question 11.

12. The box of donuts in Fig. 5-32 has a weight component of 5 N along the frictionless ramp. The force on the box from the cord has magnitude T. When the box is (a) stationary, (b) moving up the ramp at constant speed, (c) moving down the ramp at constant speed, (d) moving up the ramp at decreasing speed, and (e) moving down the ramp at decreasing speed, is T equal to, greater than, or less than 5 N?

FIGURE 5-32 Question 12.

13. Figure 5-33 shows four blocks being pulled across a frictionless floor by force $\mathbf{F}$. What total mass is accelerated to the right by (a) force $\mathbf{F}$, (b) cord 3, and (c) cord 1? (d) Rank the blocks according to their accelerations, greatest first. (e) Rank the cords according to their tension, greatest first. (Warmup for Problems 48 and 49)

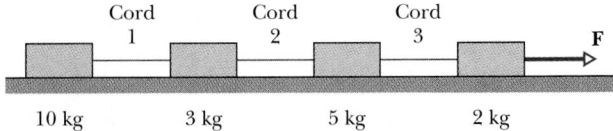

FIGURE 5-33 Question 13.

14. Figure 5-34 shows three blocks being pushed across a frictionless floor by horizontal force $\mathbf{F}$. What total mass is accelerated to the right by (a) force $\mathbf{F}$, (b) force $\mathbf{F}_{21}$ of block 1 on block 2, and (c) force $\mathbf{F}_{32}$ of block 2 on block 3? (d) Rank the blocks according to their accelerations, greatest first. (e) Rank forces $\mathbf{F}$, $\mathbf{F}_{21}$, and $\mathbf{F}_{32}$ according to their magnitude, greatest first. (Warmup for Problem 40)

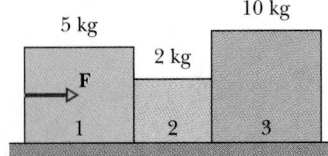

FIGURE 5-34 Question 14.

15. Figure 5-35 shows four choices for the direction of a force of magnitude F to be applied to a block on an inclined plane. The directions are either horizontal or vertical. (For choices a and b, the force is not enough to lift the block off the plane.) Rank the choices according to the magnitude of the normal force on the block from the plane, greatest first.

FIGURE 5-35
Question 15.

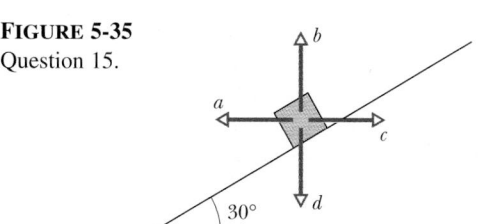

EXERCISES & PROBLEMS

SECTION 5-3 Force

1E. If the 1 kg standard body has an acceleration of 2.00 m/s² at 20° to the positive direction of the x axis, then (a) what are the x and y components of the net force on it, and (b) what is the net force in unit-vector notation?

2E. If the 1 kg standard body is accelerated by $\mathbf{F}_1 = (3.0\ \text{N})\mathbf{i} + (4.0\ \text{N})\mathbf{j}$ and $\mathbf{F}_2 = (-2.0\ \text{N})\mathbf{i} + (-6.0\ \text{N})\mathbf{j}$, then (a) what is the net force in unit-vector notation, and what are the magnitude and direction of (b) the net force and (c) the acceleration?

3P. Suppose that the 1 kg standard body accelerates at 4.00 m/s² at 160° from the positive direction of the x axis, owing to two forces, one of which is $\mathbf{F}_1 = (2.50\ \text{N})\mathbf{i} + (4.60\ \text{N})\mathbf{j}$. What is the other force (a) in unit-vector notation and (b) as a magnitude and direction?

SECTION 5-5 Newton's Second Law

4E. Two horizontal forces act on a 2.0 kg chopping block that can slide over a frictionless kitchen counter, which lies in an xy plane. One force is $\mathbf{F}_1 = (3.0\ \text{N})\mathbf{i} + (4.0\ \text{N})\mathbf{j}$. Find the acceleration of the chopping block in unit-vector notation when the other force is (a) $\mathbf{F}_2 = (-3.0\ \text{N})\mathbf{i} + (-4.0\ \text{N})\mathbf{j}$, (b) $\mathbf{F}_2 = (-3.0\ \text{N})\mathbf{i} + (4.0\ \text{N})\mathbf{j}$, and (c) $\mathbf{F}_2 = (3.0\ \text{N})\mathbf{i} + (-4.0\ \text{N})\mathbf{j}$.

5E. While two forces act on it, a particle is to move continuously with $\mathbf{v} = (3\ \text{m/s})\mathbf{i} - (4\ \text{m/s})\mathbf{j}$. One of the forces is $\mathbf{F}_1 = (2\ \text{N})\mathbf{i} + (-6\ \text{N})\mathbf{j}$. What is the other force?

6E. Three forces act on a particle that moves with an unchanging velocity of $\mathbf{v} = (2\ \text{m/s})\mathbf{i} - (7\ \text{m/s})\mathbf{j}$. Two of the forces are $\mathbf{F}_1 = (2\ \text{N})\mathbf{i} + (3\ \text{N})\mathbf{j} + (-2\ \text{N})\mathbf{k}$ and $\mathbf{F}_2 = (-5\ \text{N})\mathbf{i} + (8\ \text{N})\mathbf{j} + (-2\ \text{N})\mathbf{k}$. What is the third force?

7E. There are two horizontal forces on the 2.0 kg box in the overhead view of Fig. 5-36 but only one is shown. The box moves strictly along the x axis. For each of the following values for the acceleration a_x of the box, find the second force: (a) 10 m/s², (b) 20 m/s², (c) 0, (d) −10 m/s², and (e) −20 m/s².

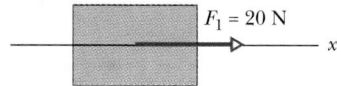

FIGURE 5-36 Exercise 7.

8E. There are two forces on the 2.0 kg box in the overhead view of Fig. 5-37 but only one is shown. The figure also shows the acceleration of the box. Find the second force (a) in unit-vector notation and (b) as a magnitude and direction.

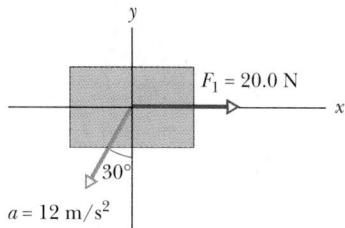

FIGURE 5-37 Exercise 8.

9E. Five forces pull on the 4.0 kg box in Fig. 5-38. Find the box's acceleration (a) in unit-vector notation and (b) as a magnitude and direction.

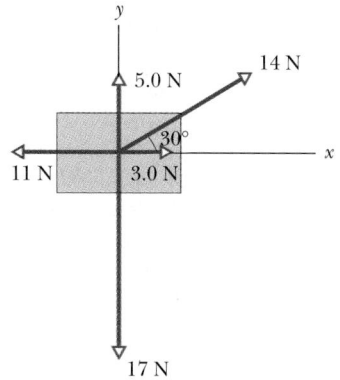

FIGURE 5-38 Exercise 9.

10P. Three astronauts, propelled by jet backpacks, push and guide a 120 kg asteroid toward a processing dock, exerting the forces shown in Fig. 5-39. What is the asteroid's acceleration (a) in unit-vector notation and (b) as a magnitude and direction?

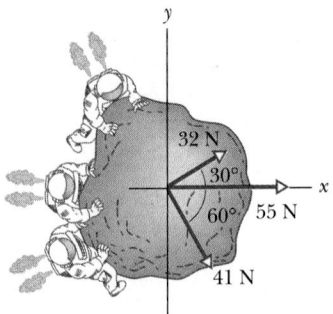

FIGURE 5-39 Problem 10.

11P. Figure 5-40 is an overhead view of a 12 kg tire that is to be pulled by three ropes. One force ($\mathbf{F}_1$, with magnitude 50 N) is indicated. Orient the other two forces $\mathbf{F}_2$ and $\mathbf{F}_3$ so that the magnitude of the resulting acceleration of the tire is least, and find that magnitude if (a) $F_2 = 30$ N, $F_3 = 20$ N; (b) $F_2 = 30$ N, $F_3 = 10$ N; and (c) $F_2 = F_3 = 30$ N.

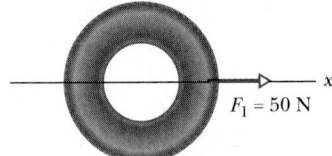

FIGURE 5-40 Problem 11.

SECTION 5-6 Some Particular Forces

12E. What are the mass and weight of (a) a 1400 lb snowmobile and (b) a 421 kg heat pump?

13E. What are the weight in newtons and the mass in kilograms of (a) a 5.0 lb bag of sugar, (b) a 240 lb fullback, and (c) a 1.8 ton automobile? (1 ton = 2000 lb.)

14E. A space traveler whose mass is 75 kg leaves the Earth. Compute his weight (a) on Earth, (b) on Mars, where $g = 3.8$ m/s², and (c) in interplanetary space, where $g = 0$. (d) What is his mass at each of these locations?

15E. A certain particle has a weight of 22 N at a point where the free-fall acceleration is 9.8 m/s². (a) What are the weight and mass of the particle at a point where the free-fall acceleration is 4.9 m/s²? (b) What are the weight and mass of the particle if it is moved to a point in space where the free-fall acceleration is zero?

16E. A weight-conscious penguin with a mass of 15.0 kg rests on a bathroom scale (Fig. 5-41). What are (a) the penguin's weight **W** and (b) the normal force **N** on the penguin? (c) What is the reading on the scale, assuming it is calibrated in weight units?

FIGURE 5-41 Exercise 16.

17E. Figure 5-42a shows a crude mobile hanging from a ceiling, with two metal pieces strung by cords of negligible mass. The masses of the pieces are given. What is the tension in (a) the bottom cord and (b) the top cord? Figure 5-42b shows a similar mobile consisting of three metal pieces. The masses of the highest and lowest pieces are given. The tension in the top cord is 199 N. What is the tension in (c) the lowest cord and (d) the middle cord?

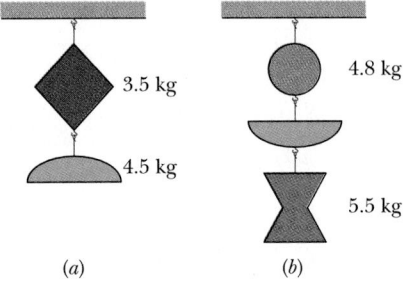

FIGURE 5-42 Exercise 17.

18E. (a) An 11.0 kg salami is supported by a cord that runs to a spring scale, which is supported by another cord from the ceiling (Fig. 5-43a). What is the reading on the scale? (b) In Fig. 5-43b the salami is supported by a cord that runs around a pulley and to a scale. The opposite end of the scale is attached by a cord to a wall. What is the reading on the scale? (c) In Fig. 5-43c the wall has been replaced with a second 11.0 kg salami on the left, and the assembly is stationary. What is the reading on the scale now?

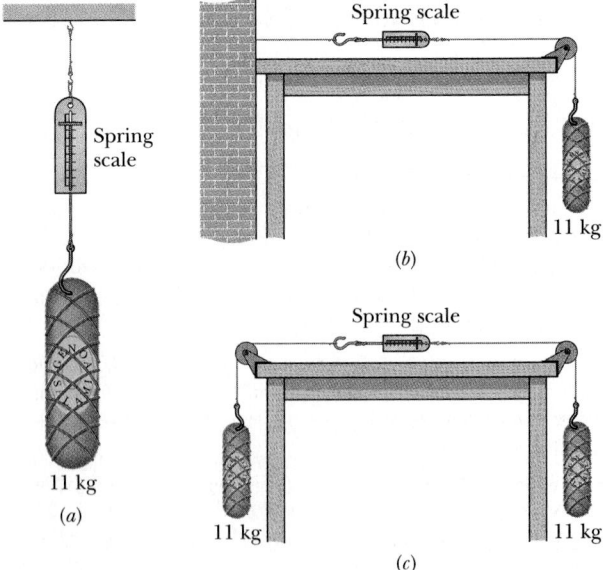

FIGURE 5-43 Exercise 18.

SECTION 5-8 Applying Newton's Laws

19E. When an airplane is in level flight, its weight is balanced by a vertical "lift," which is a force exerted by the air. How large is the lift on an airplane in such flight if the mass of the airplane is 1.20×10^3 kg?

20E. What is the magnitude of the net force acting on a 3800 lb automobile accelerating at 12 ft/s²?

21E. An experimental rocket sled can be accelerated at a constant rate from rest to 1600 km/h in 1.8 s. What is the magnitude of the required average force if the sled has a mass of 500 kg?

22E. A car traveling at 53 km/h hits a bridge abutment. A passenger in the car moves forward a distance of 65 cm (with respect to the road) while being brought to rest by an inflated air bag. What magnitude of force (assumed constant) acts on the passenger's upper torso, which has a mass of 41 kg?

23E. If a nucleus captures a stray neutron, it must bring the neutron to a stop within the diameter of the nucleus by means of the *strong force*. That force, which "glues" the nucleus together, is essentially zero outside the nucleus. Suppose that a stray neutron with an initial speed of 1.4×10^7 m/s is just barely captured by a nucleus with diameter $d = 1.0 \times 10^{-14}$ m. Assuming that the force on the neutron is constant, find the magnitude of that force. The neutron's mass is 1.67×10^{-27} kg.

24E. In a modified tug-of-war game, two people pull in opposite directions, not on a rope, but on a 25 kg sled resting on (frictionless) ice. If the participants exert forces of 90 N and 92 N, what will be the magnitude of the sled's acceleration?

25E. A 450 lb motorcycle accelerates from 0 to 55 mi/h in 6.0 s. (a) What is the magnitude of the motorcycle's constant acceleration? (b) What is the magnitude of the net force causing the acceleration?

26E. Refer to Fig. 5-15 and suppose the two masses are $m = 2.0$ kg and $M = 4.0$ kg. (a) Decide without any calculations which of them should be hanging if the magnitude of the acceleration is to be largest. What then are (b) the magnitude of the acceleration and (c) the associated tension in the cord?

27E. Refer to Fig. 5-22. Let the mass of the block be 8.5 kg and the angle θ equal 30°. Find (a) the tension in the cord and (b) the normal force acting on the block. (c) If the cord is cut, find the magnitude of the block's acceleration.

28E. A jet airplane starts from rest on the runway and accelerates for takeoff at 2.3 m/s². It has two jet engines, each of which exerts a force (thrust) on the airplane of 1.4×10^5 N. What is the weight of the airplane?

29E. *Sunjamming.* A "sun yacht" is a spacecraft with a large sail that is pushed by sunlight. Although such a push is tiny in everyday circumstances, it can be large enough to send the spacecraft outward from the Sun on a cost-free but slow trip. Suppose that the spacecraft has a mass of 900 kg and receives a push of 20 N. (a) What is the magnitude of the resulting acceleration? If the craft starts from rest, (b) how far will it travel in 1 day and (c) how fast will it then be moving?

30E. The tension at which a fishing line snaps is commonly called the line's "strength." What minimum strength is needed for a line that is to stop a 19 lb salmon in 4.4 in. if the fish is initially drifting at 9.2 ft/s? Assume a constant deceleration.

31E. In a laboratory experiment, an initially stationary electron (mass $= 9.11 \times 10^{-31}$ kg) undergoes a constant acceleration through 1.5 cm, reaching a speed of 6.0×10^6 m/s at the end of that distance. (a) What is the magnitude of the force accelerating the electron? (b) What is the weight of the electron?

32E. An electron is projected horizontally at a speed of 1.2×10^7 m/s into an electric field that exerts a constant vertical force of 4.5×10^{-16} N on it. The mass of the electron is 9.11×10^{-31} kg. Determine the vertical distance the electron is deflected during the time it has moved 30 mm horizontally.

33E. A car that weighs 1.30×10^4 N is initially moving at a speed of 40 km/h when the brakes are applied and the car is brought to a stop in 15 m. Assuming that the force that stops the car is constant, find (a) the magnitude of that force and (b) the time required for the change in speed. If the initial speed is, instead, twice as great, and the car experiences the same force during the braking, how are (c) the stopping distance and (d) the stopping time changed? (There could be a lesson here about the danger of driving at high speeds.)

34E. Compute the initial upward acceleration of a rocket of mass 1.3×10^4 kg if the initial upward force produced by its engine (the thrust) is 2.6×10^5 N. Do not neglect the weight of the rocket.

35E. A rocket and its payload have a total mass of 5.0×10^4 kg. How large is the force produced by the engine (the thrust) when (a) the rocket is "hovering" over the launchpad just after ignition, and (b) when the rocket is accelerating upward at 20 m/s²?

36P. A 40 kg girl and an 8.4 kg sled are on the surface of a frozen lake, 15 m apart. By means of a rope, the girl exerts a horizontal 5.2 N force on the sled, pulling it toward her. (a) What is the acceleration of the sled? (b) What is the acceleration of the girl? (c) How far from the girl's initial position do they meet, assuming that no frictional forces act?

37P. A 160 lb firefighter slides down a vertical pole with an acceleration of 10 ft/s², directed downward. What are the magnitudes and directions of the vertical forces (a) exerted by the pole on the firefighter and (b) exerted by the firefighter on the pole?

38P. A sphere of mass 3.0×10^{-4} kg is suspended from a cord. A steady horizontal breeze pushes the sphere so that the cord makes an angle of 37° with the vertical when at rest. Find (a) the magnitude of that push and (b) the tension in the cord.

39P. A worker drags a crate across a factory floor by pulling on a rope tied to the crate (Fig. 5-44). The worker exerts a force of 450 N on the rope, which is inclined at 38° to the horizontal, and the floor exerts a horizontal force of 125 N that opposes the motion. Calculate the acceleration of the crate (a) if its mass is 310 kg and (b) if its weight is 310 N.

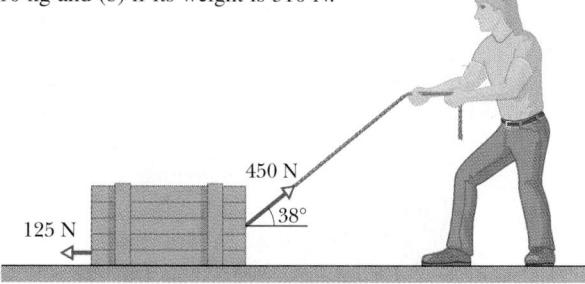

FIGURE 5-44 Problem 39.

40P. Two blocks are in contact on a frictionless table. A horizontal force is applied to one block, as shown in Fig. 5-45. (a) If

$m_1 = 2.3$ kg, $m_2 = 1.2$ kg, and $F = 3.2$ N, find the force between the two blocks. (b) Show that if a force of the same magnitude F is applied to m_2 but in the opposite direction, the force between the blocks is 2.1 N, which is not the same value calculated in (a). Explain the difference.

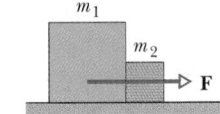

FIGURE 5-45   Problem 40.

41P. You pull a short refrigerator with a constant force F across a greased (frictionless) floor, either with F horizontal (case 1) or with F tilted upward at an angle θ (case 2). (a) What is the ratio of the refrigerator's speed in case 2 to its speed in case 1 if you pull for a certain time t? (b) What is this ratio if you pull for a certain distance d?

42P. For sport, a 12-kg armadillo runs onto a large pond of level, frictionless ice with an initial velocity of 5.0 m/s along the positive direction of the x axis. Take its initial position on the ice as being the origin. It slips over the ice while being pushed by a wind with a force of 17 N in the positive direction of the y axis. In unit-vector notation, what are the animal's (a) velocity and (b) position vectors when it has slid for 3.0 s?

43P. An elevator and its load have a combined mass of 1600 kg. Find the tension in the supporting cable when the elevator, originally moving downward at 12 m/s, is brought to rest with constant acceleration in a distance of 42 m.

44P. An object is hung from a spring balance attached to the ceiling of an elevator. The balance reads 65 N when the elevator is standing still. What is the reading when the elevator is moving upward (a) with a constant speed of 7.6 m/s and (b) with a speed of 7.6 m/s while decelerating at a rate of 2.4 m/s²?

45P. A 1400 kg jet engine is fastened to the fuselage of a passenger jet by just three bolts (this is the usual practice). Assume that each bolt supports one-third of the load. (a) Calculate the force on each bolt as the plane waits in line for clearance to take off. (b) During flight, the plane encounters turbulence, which suddenly imparts an upward vertical acceleration of 2.6 m/s² to the plane. Calculate the force on each bolt now.

46P. In Fig. 5-46 a 15,000 kg helicopter is lifting a 4500 kg truck with an upward acceleration of 1.4 m/s². Calculate (a) the force the air exerts on the helicopter blades and (b) the tension in the upper supporting cable.

FIGURE 5-46   Problem 46.

47P. An 80 kg man jumps down to a concrete patio from a window ledge only 0.50 m above the ground. He neglects to bend his knees on landing, so that his motion is arrested in a distance of 2.0 cm. (a) What is the average acceleration of the man from the time his feet first touch the patio to the time he is brought fully to rest? (b) With what force does this jump jar his bone structure?

48P. Three blocks are connected, as shown in Fig. 5-47, on a horizontal frictionless table and pulled to the right with a force $T_3 = 65.0$ N. If $m_1 = 12.0$ kg, $m_2 = 24.0$ kg, and $m_3 = 31.0$ kg, calculate (a) the acceleration of the system and (b) the tensions T_1 and T_2 in the interconnecting cords.

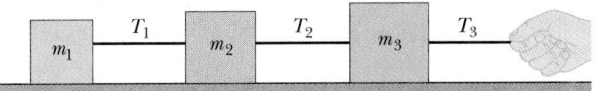

FIGURE 5-47   Problem 48.

49P. Figure 5-48 shows four penguins that are being playfully pulled along very slippery (frictionless) ice by a curator. The masses of three penguins and the tension in two of the cords are given. Find the penguin mass that is not given.

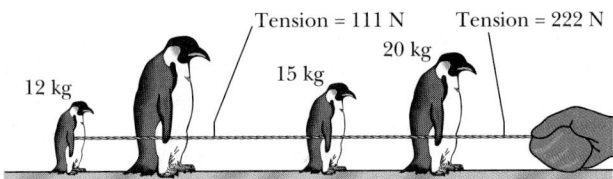

FIGURE 5-48   Problem 49.

50P. An elevator weighing 6240 lb is pulled upward by a cable with an acceleration of 4.00 ft/s². (a) Calculate the tension in the cable. (b) What is the tension when the elevator is decelerating at the rate of 4.00 ft/s² but is still moving upward?

51P. An 80-kg person is parachuting and experiencing a downward acceleration of 2.5 m/s². The mass of the parachute is 5.0 kg. (a) What upward force is exerted on the open parachute by the air? (b) What downward force is exerted by the person on the parachute?

52P. An 85-kg man lowers himself to the ground from a height of 10.0 m by holding onto a rope that runs over a frictionless pulley to a 65-kg sandbag. (a) With what speed does the man hit the ground if he started from rest? (b) Is there anything he could do to reduce the speed with which he hits the ground?

53P. A new 26-ton Navy jet (Fig. 5-49) requires an airspeed of 280 ft/s for liftoff (1 ton = 2000 lb). The engine develops a maximum force of 24,000 lb, but that is insufficient for reaching takeoff speed in the 300 ft available on an aircraft carrier. What minimum force (assumed constant) is needed from the catapult that is used to help launch the jet? Assume that the catapult and the jet's engine each exert a constant force over the 300 ft distance used for takeoff.

FIGURE 5-49   Problem 53.

54P. Imagine a landing craft approaching the surface of Callisto, one of Jupiter's moons. If the engine provides an upward force (thrust) of 3260 N, the craft descends at constant speed; if the engine provides only 2200 N, the craft accelerates downward at 0.39 m/s². (a) What is the weight of the landing craft in the vicinity of Callisto's surface? (b) What is the mass of the craft? (c) What is the free-fall acceleration near the surface of Callisto?

55P. A 52 kg circus performer is to slide down a rope that will snap if the tension exceeds 425 N. (a) What happens if the performer hangs stationary on the rope? (b) At what magnitude of acceleration does the performer just avoid breaking the rope?

56P. A chain consisting of five links, each of mass 0.100 kg, is lifted vertically with a constant acceleration of 2.50 m/s², as shown in Fig. 5-50. Find (a) the forces acting between adjacent links, (b) the force **F** exerted on the top link by the person lifting the chain, and (c) the *net* force accelerating each link.

FIGURE 5-50   Problem 56.

57P. A 1.0 kg mass on a 37° incline is connected to a 3.0 kg mass on a horizontal surface (Fig. 5-51). The surfaces and the pulley are frictionless. If $F = 12$ N, what is the tension in the connecting cord?

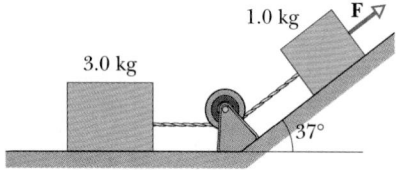

FIGURE 5-51   Problem 57.

58P. A block of mass $m_1 = 3.70$ kg on a frictionless inclined plane of angle 30.0° is connected by a cord over a massless, frictionless pulley to a second block of mass $m_2 = 2.30$ kg hanging vertically (Fig. 5-52). What are (a) the magnitude of the acceleration of each block and (b) the direction of the acceleration of m_2? (c) What is the tension in the cord?

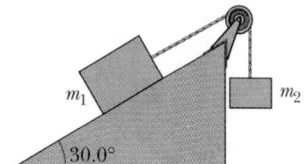

FIGURE 5-52   Problem 58.

59P. You need to lower a bundle of old roofing material weighing 100 lb to the ground with a rope that will snap if the tension in it exceeds 87 lb. (a) How can you avoid snapping the rope during the descent? (b) If the descent is 20 ft and you just barely avoid snapping the rope, with what speed will the bundle hit the ground?

60P. A block is projected up a frictionless inclined plane with initial speed v_0. The angle of incline is θ. (a) How far up the plane does it go? (b) How long does it take to get there? (c) What is its speed when it gets back to the bottom? Find numerical answers for $\theta = 32.0°$ and $v_0 = 3.50$ m/s.

61P. A spaceship lifts off vertically from the moon, where the free-fall acceleration is 1.6 m/s². If the spaceship has an upward acceleration of 1.0 m/s² as it lifts off, what is the force of the spaceship on an astronaut who weighs 735 N on Earth?

62P. A lamp hangs vertically from a cord in a descending elevator that decelerates at 2.4 m/s². (a) If the tension in the cord is 89 N, what is the lamp's mass? (b) What is the cord's tension when the elevator ascends with an upward acceleration of 2.4 m/s²?

63P. A 100 kg crate is pushed at constant speed up the frictionless 30.0° ramp shown in Fig. 5-53. (a) What horizontal force **F** is required? (b) What force is exerted by the ramp on the crate?

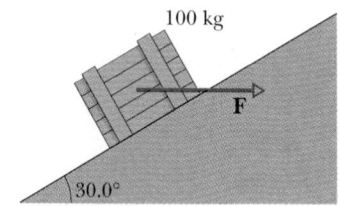

FIGURE 5-53   Problem 63.

64P. A 10 kg monkey climbs up a massless rope that runs over a frictionless tree limb and back down to a 15 kg package on the ground (Fig. 5-54). (a) What is the magnitude of the least acceleration the monkey must have if it is to lift the package off the ground? If, after the package has been lifted, the monkey stops its climb and holds onto the rope, what are (b) the monkey's acceleration and (c) the tension in the rope?

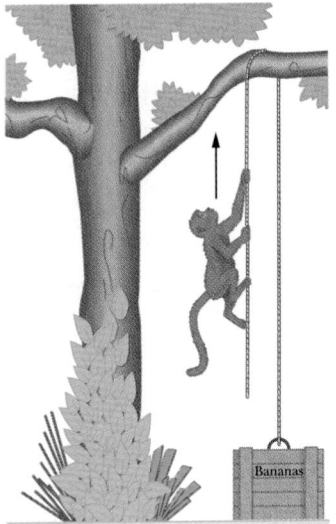

FIGURE 5-54 Problem 64.

65P. Figure 5-55 shows a section of an alpine cable-car system. The maximum permissible mass of each car with occupants is 2800 kg. The cars, riding on a support cable, are pulled by a second cable attached to each pylon. What is the difference in tension between adjacent sections of pull cable if the cars are at the maximum permissible mass and are being accelerated up the 35° incline at 0.81 m/s²?

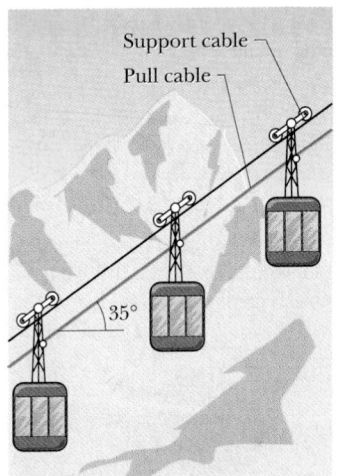

FIGURE 5-55 Problem 65.

66P. An interstellar ship has a mass of 1.20×10^6 kg and is initially at rest relative to a star system. (a) What constant acceleration is needed to bring the ship up to a speed of $0.10c$ (where c is the speed of light) relative to the star system in 3.0 days? (You need not consider Einstein's special relativity.) (b) What is that acceleration in g units? (c) What force is required for the acceleration? (d) If the engines are shut down when $0.10c$ is reached, how long does the ship take (start to finish) to journey 5.0 light-months, the distance that light travels in 5.0 months?

67P. The elevator in Fig. 5-56 consists of a 1150 kg cage (A), a 1400 kg counterweight (B), a driving mechanism (C), a cable, and two pulleys. In operation, mechanism C grabs the cable, either forcing it along or retarding its motion. This process means that tension T_1 in the cable on one side of C differs from tension T_2 on the other side. Suppose that the upward acceleration of A and the downward acceleration of B have the magnitude $a = 2.0$ m/s². Neglecting the pulleys and the mass of the cable, find (a) T_1, (b) T_2, and (c) the force on the cable produced by C.

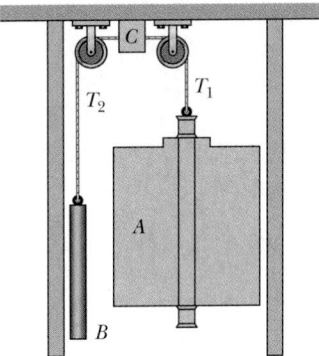

FIGURE 5-56 Problem 67.

68P. A 5.00 kg block is pulled along a horizontal frictionless floor by a cord that exerts a force $F = 12.0$ N at an angle $\theta = 25.0°$ above the horizontal, as shown in Fig. 5-57. (a) What is the acceleration of the block? (b) The force F is slowly increased. What is its value just before the block is lifted (completely) off the floor? (c) What is the acceleration of the block just before it is lifted (completely) off the floor?

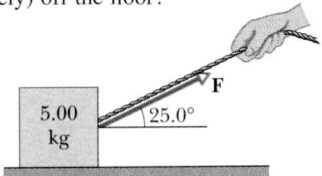

FIGURE 5-57 Problem 68.

69P. In earlier days, horses pulled barges down canals in the manner shown in Fig. 5-58. Suppose that the horse pulls on the rope with a force of 7900 N at an angle of 18° to the direction of motion of the barge, which is headed straight along the canal. The mass of the barge is 9500 kg, and its acceleration is 0.12 m/s². Calculate the force exerted by the water on the barge.

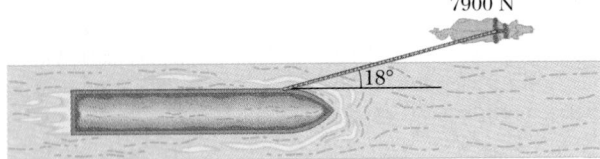

FIGURE 5-58 Problem 69.

70P. A hot-air balloon of mass M is descending vertically with downward acceleration a (Fig. 5-59). How much mass must be thrown out to give the balloon an upward acceleration a (same magnitude but opposite direction)? Assume that the upward force from the air (the lift) does not change because of the mass (the ballast) that is lost.

Ballast

FIGURE 5-59 Problem 70.

71P. A certain force gives mass m_1 an acceleration of 12.0 m/s² and mass m_2 an acceleration of 3.30 m/s². What acceleration would the force give to an object with a mass of (a) $m_2 - m_1$ and (b) $m_2 + m_1$?

72P. A rocket with mass 3000 kg is fired from the ground at an angle of elevation of 60°. The motor creates a force (thrust) on the rocket of 6.0×10^4 N at a constant angle of 60° to the horizontal for 50 s and then cuts out. As a rough approximation, ignore the mass of fuel consumed and neglect forces from the air. Calculate (a) the altitude of the rocket at motor cutout and (b) the total horizontal distance from firing point to eventual impact with the ground (assumed to be level).

73P. A block of mass M is pulled along a horizontal frictionless surface by a rope of mass m, as shown in Fig. 5-60. A horizontal force **F** is applied to one end of the rope. (a) Show that the rope *must* sag, even if only by an imperceptible amount. Then, assuming that the sag is negligible, find (b) the acceleration of rope and block, (c) the force that the rope exerts on the block, and (d) the tension in the rope at its midpoint.

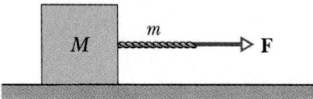

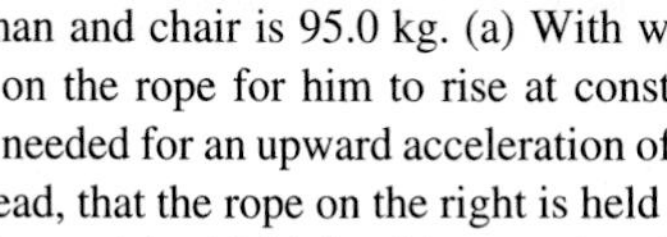

FIGURE 5-60 Problem 73.

74P. Figure 5-61 shows a man sitting in a bosun's chair that dangles from a massless rope, which runs over a massless, frictionless pulley and back down to the man's hand. The combined mass of the man and chair is 95.0 kg. (a) With what force must the man pull on the rope for him to rise at constant speed? (b) What force is needed for an upward acceleration of 1.30 m/s²? (c) Suppose, instead, that the rope on the right is held by a person on the ground. Repeat (a) and (b) for this new situation. (d) In each of the four cases, what is the force exerted on the ceiling by the pulley system?

Electronic Computation

75. Fig. 5-62 shows two blocks on a frictionless plane, with inclination angle θ, and a third block. Blocks m_1 and m_2 are con-

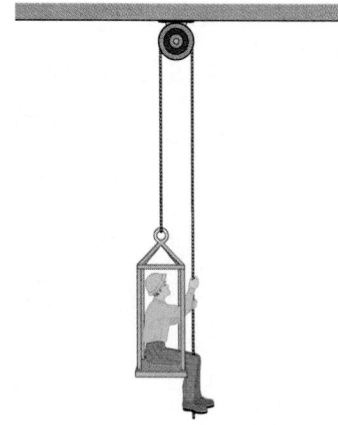

FIGURE 5-61 Problem 74.

nected by a cord with tension T_1; blocks m_2 and m_3 are connected by a cord (around a frictionless, massless pulley) with tension T_2. If $\theta = 20°$, $m_1 = 2.00$ kg, $m_2 = 1.00$ kg, and $m_3 = 3.00$ kg, find T_1, T_2, and the acceleration of the blocks. (*Hint:* Use Newton's second law for each block to write three simultaneous equations, which can then be solved electronically.)

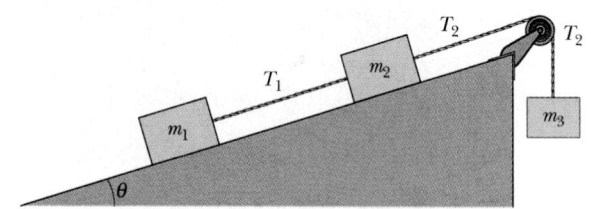

FIGURE 5-62 Problem 75.

76. A 2.00 kg object is subjected to three forces that give it an acceleration $\mathbf{a} = -(8.00 \text{ m/s}^2)\mathbf{i} + (6.00 \text{ m/s}^2)\mathbf{j}$. If two of the three forces are $\mathbf{F}_1 = (30.0 \text{ N})\mathbf{i} + (16.0 \text{ N})\mathbf{j}$ and $\mathbf{F}_2 = -(12.0 \text{ N})\mathbf{i} + (8.00 \text{ N})\mathbf{j}$, find the third force.

6
Force and Motion—II

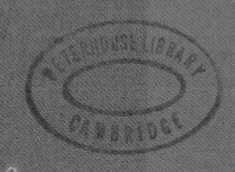

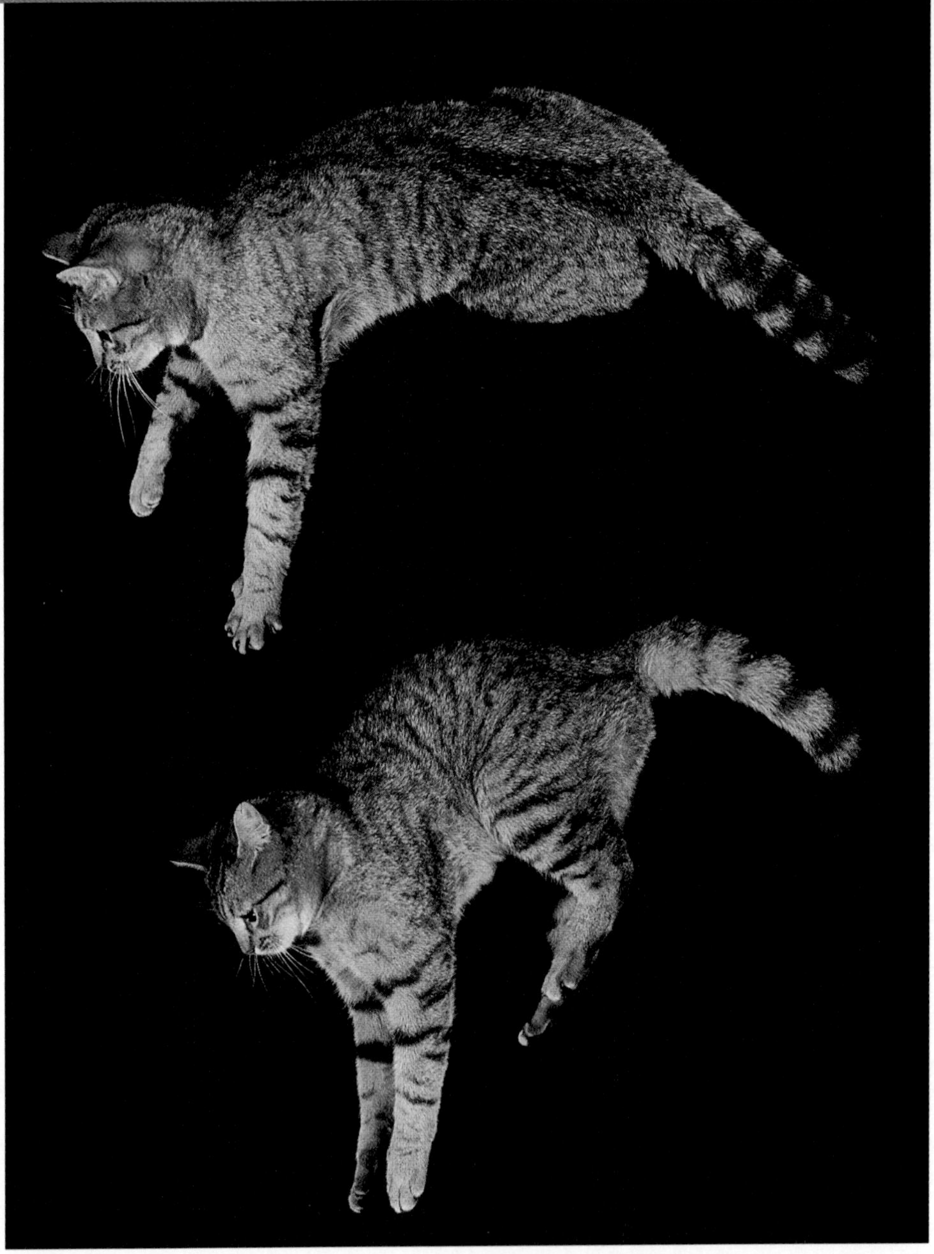

Cats, who enjoy sleeping on window sills, are often kept in apartment buildings. When a cat accidentally falls out of a window and onto a sidewalk, the extent of injury (such as the number of fractured bones or the certainty of death) decreases with height if the fall is more than seven or eight floors. (There is even a record of a cat who fell 32 floors and suffered only slight damage to its thorax and one tooth.) How can that be?

6-1 FRICTION

Frictional forces are unavoidable in our daily lives. Left to act alone, they would stop every moving object and bring to a halt every rotating shaft. In an automobile, about 20% of the gasoline is used to counteract friction in the engine and in the drive train. On the other hand, if friction were totally absent, we could not get an automobile to go anywhere, and we could not walk or ride a bicycle. We could not hold a pencil and, if we could, it would not write. Nails and screws would be useless, woven cloth would fall apart, and knots would come undone.

Here we deal with the frictional forces that exist between dry solid surfaces, moving across each other at slow speeds. Consider two simple experiments:

1. **First experiment.** Send a book sliding across a tabletop. A frictional force, exerted by the tabletop on the bottom of the sliding book, slows the book and eventually stops it. If you want to make the book move across the table with constant velocity, you must push or pull it with a steady force of equal magnitude and opposite direction to that of the frictional force, which opposes the motion.

2. **Second experiment.** A heavy crate is resting on the floor of a warehouse. You push on it horizontally with a steady force but it does not move. That is because the force that you apply is balanced by a frictional force, exerted horizontally on the bottom of the crate by the floor and opposite the direction of your push. Remarkably, this frictional force automatically adjusts itself, in both magnitude and direction, to cancel exactly whatever force you decide to apply. Of course, if you can push hard enough, you will be able to move the crate (see the first experiment).

Figure 6-1 shows a similar situation in detail. In Fig. 6.1a, a block rests on a tabletop, its weight **W** balanced by an equal but opposite normal force **N**. In Fig. 6-1b, you exert a force **F** on the block, attempting to pull it to the left. In response, a frictional force $\mathbf{f}_s$ arises, pointing to the right, exactly matching the force that you have applied. The force $\mathbf{f}_s$ is called the **static frictional force.**

Figures 6-1c and 6-1d show that as you increase your applied force, the static frictional force $\mathbf{f}_s$ increases also and the block remains at rest. When the applied force reaches a certain value, however, the block ''breaks away'' from its intimate contact with the tabletop and accelerates leftward (Fig. 6-1e). The frictional force that then opposes the motion is called the **kinetic frictional force** $\mathbf{f}_k$.

Usually, the kinetic frictional force, which acts when there is motion, is less than the maximum value of the static frictional force, which acts when there is no motion. So, if you wish the block to move across the surface with a

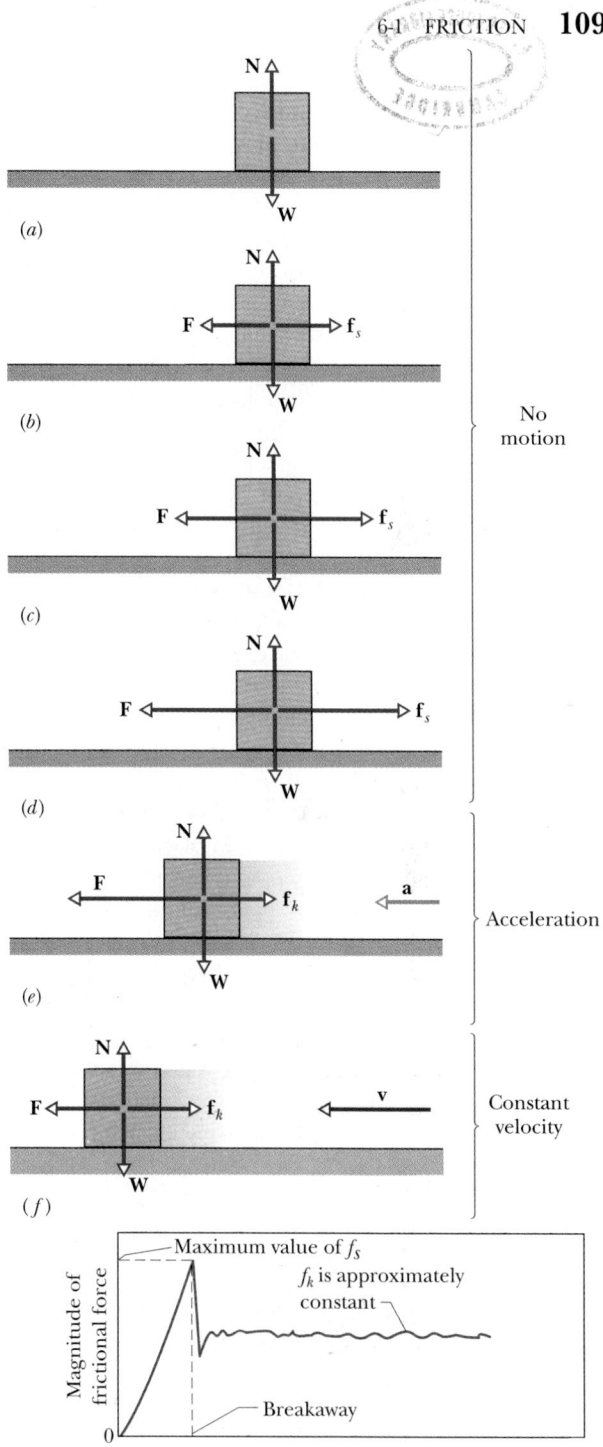

FIGURE 6-1 (a) The forces on a stationary block. (b–d) An external force **F**, applied to the block, is counterbalanced by an equal but opposite static frictional force $\mathbf{f}_s$. As **F** is increased, $\mathbf{f}_s$ also increases, until $\mathbf{f}_s$ reaches a certain maximum value. (e) The block then ''breaks away,'' accelerating suddenly to the left. (f) If the block is now to move with constant velocity, the applied force **F** must be reduced from the maximum value it had just before the block broke away. (g) Some experimental results for the sequence (a) through (f).

constant speed, you must usually decrease the applied force once the block begins to move, as in Fig. 6-1f. For example, Fig. 6-1g shows the results of an experiment in which the force on a block was slowly increased until breakaway occurred. Note the reduced force needed to keep the block moving at constant speed after breakaway.

Basically, the frictional force is a force acting between the surface atoms of one body and those of another. If two highly polished and carefully cleaned metal surfaces are brought together in a very good vacuum, they cannot be made to slide over each other. Instead, they *cold-weld* together instantly, forming a single piece of metal. If a machinist's specially polished gage blocks are brought together in air, they stick almost as firmly to each other and can be separated only by means of a wrenching motion. Usually, however, such intimate atom-to-atom contact is not possible. Even a highly polished metal surface is far from being flat on the atomic scale. Moreover, the surfaces of everyday objects have layers of oxides and other contaminants that reduce cold-welding.

When two such surfaces are placed together, only the high points touch each other. (It is like having the Alps of Switzerland turned over and placed down on the Alps of Austria.) The actual *micro*scopic area of contact is much less than the apparent *macro*scopic contact area, perhaps by a factor of 10^4. Nonetheless, many contact points cold-weld together. These welds produce static friction when an applied force attempts to slide the surfaces.

When the surfaces are pulled across each other, there is first a tearing of welds (breakaway) and then a continuous tearing apart and re-forming of welds as additional chance contacts are made (Fig. 6-2). The kinetic friction $\mathbf{f}_k$ is the vector sum of forces at those many chance contacts. Often, the motion is "jerky," because the two surfaces

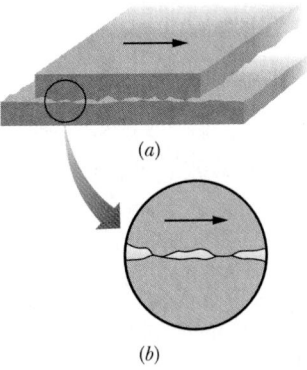

(a)

(b)

FIGURE 6-2 The mechanism of sliding friction. (a) The upper surface is sliding to the right over the lower surface in this enlarged view. (b) A detail, showing two spots where cold-welding has occurred. Force is required to break these welds and maintain the motion.

briefly stick together and then slip. Such repetitive *stick-and-slip* can produce squeaking or squealing, as when tires skid on dry pavement, fingernails scratch along a chalkboard, a rusty hinge is opened, and a bow is drawn across a violin string.

6-2 PROPERTIES OF FRICTION

Experiment shows that when a dry and unlubricated body is pressed against a surface in the same condition, and an applied force $\mathbf{F}$ attempts to slide the body along the surface, the resulting frictional force has three properties:

PROPERTY 1. If the body does not move, then the static frictional force $\mathbf{f}_s$ and the component of $\mathbf{F}$ that is parallel to the surface are equal in magnitude, and $\mathbf{f}_s$ is directed opposite that component of $\mathbf{F}$.

PROPERTY 2. The magnitude of $\mathbf{f}_s$ has a maximum value $f_{s,\mathrm{max}}$ that is given by

$$f_{s,\mathrm{max}} = \mu_s N, \tag{6-1}$$

where μ_s is the **coefficient of static friction** and N is the magnitude of the normal force. If the magnitude of the component of $\mathbf{F}$ that is parallel to the surface exceeds $f_{s,\mathrm{max}}$, then the body begins to slide along the surface.

PROPERTY 3. If the body begins to slide along the surface, the magnitude of the frictional force rapidly decreases to a value f_k given by

$$f_k = \mu_k N, \tag{6-2}$$

where μ_k is the **coefficient of kinetic friction.** Thereafter during the sliding, the kinetic frictional force $\mathbf{f}_k$ has a magnitude given by Eq. 6-2.

Properties 1 and 2 are worded in terms of a single applied force $\mathbf{F}$, but they also hold for the resultant of several applied forces acting on the body. Equations 6-1 and 6-2 are *not* vector equations; the direction of $\mathbf{f}_s$ or $\mathbf{f}_k$ is always parallel to the surface and opposite the intended motion, and $\mathbf{N}$ is perpendicular to the surface.

The coefficients μ_s and μ_k are dimensionless and must be determined experimentally. Their values depend on certain properties of both the body and the surface; hence, they are usually referred to with the preposition "between," as in "the value of μ_s *between* a sled and asphalt is 0.5." We assume that the value of μ_k does not depend on the speed at which the body slides along the surface.

CHECKPOINT 1: A block lies on a floor. (a) What is the magnitude of the frictional force on it from the floor? (b) If a horizontal force of 5 N is now applied to the block, but the block does not move, what is the magnitude of the frictional force on it? (c) If the maximum value $f_{s,max}$ of the static frictional force on the block is 10 N, will the block move if the horizontally applied force is 8 N? (d) If the force is 12 N?

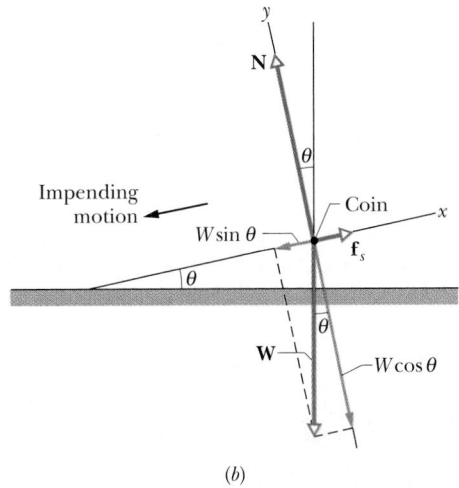

(b)

FIGURE 6-3 Sample Problem 6-1. (a) A coin is about to slide from rest down the cover of a book. (b) A free-body diagram for the coin, showing the three forces (drawn to scale) that act on it. The weight **W** is shown resolved into its components along the x and the y axes, whose orientations are chosen to simplify the problem.

SAMPLE PROBLEM 6-1

Figure 6-3a shows a coin resting on a book that has been tilted at an angle θ with the horizontal. By trial and error you find that when θ is increased to 13°, the coin begins to slide down the book. What is the coefficient of static friction μ_s between the coin and the book?

SOLUTION: Figure 6-3b is a free-body diagram for the coin when it is on the verge of sliding. The forces on the coin are the normal force **N**, pushing outward from the plane of the book, the weight **W** of the coin, and the frictional force $\mathbf{f}_s$, which points up the plane because the impending motion is down the plane. Since the coin is in equilibrium, the net force on it must be zero. From Newton's second law, we have

$$\sum \mathbf{F} = \mathbf{f}_s + \mathbf{W} + \mathbf{N} = 0. \qquad (6\text{-}3)$$

For the x components, this vector equation gives us

$$\sum F_x = f_s - W \sin \theta = 0, \quad \text{or} \quad f_s = W \sin \theta. \qquad (6\text{-}4)$$

For the y components, we have

$$\sum F_y = N - W \cos \theta = 0, \quad \text{or} \quad N = W \cos \theta. \qquad (6\text{-}5)$$

When the coin is on the verge of sliding (*and only then*) the magnitude of the static frictional force acting on it has the maximum value $\mu_s N$. Substituting $\mu_s N$ into Eq. 6-4 and dividing by Eq. 6-5, we obtain

$$\frac{f_s}{N} = \frac{\mu_s N}{N} = \frac{W \sin \theta}{W \cos \theta} = \tan \theta,$$

or

$$\mu_s = \tan \theta = \tan 13° = 0.23. \quad \text{(Answer)} \qquad (6\text{-}6)$$

You can easily measure μ_s for a coin and this text; you do not need a protractor. You measure the two lengths shown in Fig. 6-3a with a ruler, and their ratio h/d is tan θ.

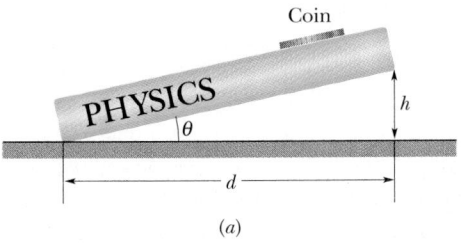

(a)

SAMPLE PROBLEM 6-2

If a car's wheels are "locked" (kept from rolling) during emergency braking, the car slides along the road. Ripped-off bits of tire and small melted sections of road form the "skid marks" that reveal the cold-welding during the slide. The record for the longest skid marks on a public road was reportedly set in 1960 by a Jaguar on the M1 highway in England— the marks were 290 m (about 950 ft) long! Assuming that $\mu_k = 0.60$, how fast was the car going when the wheels became locked?

SOLUTION: Figure 6-4a depicts the car's travel; Fig. 6-4b is the car's free-body diagram during deceleration, showing the car's weight **W**, the normal force **N**, and the kinetic frictional force $\mathbf{f}_k$ acting on the car. We can use Eq. 2-16,

$$v^2 = v_0^2 + 2a_x(x - x_0),$$

with $v = 0$ and $x - x_0 = d$, to find the car's initial speed v_0. Substituting these values and rearranging yield

$$v_0 = \sqrt{-2a_x d}. \qquad (6\text{-}7)$$

To find a_x, we apply Newton's second law along the x axis. If we ignore the effects of the air on the car, the only force with a component along the x axis is $\mathbf{f}_k$, and that component is $-f_k$. So we have

$$-f_k = ma_x, \quad \text{or} \quad a_x = -\frac{f_k}{m} = -\frac{\mu_k N}{m}, \qquad (6\text{-}8)$$

where m is the car's mass, and Eq. 6-2 tells us that $f_k = \mu_k N$.

The normal force **N** has magnitude $N = W = mg$. Substituting this result into Eq. 6-8 yields

$$a_x = -\frac{\mu_k mg}{m} = -\mu_k g. \qquad (6\text{-}9)$$

Substituting Eq. 6-9 into Eq. 6-7, we find

$$v_0 = \sqrt{2\mu_k gd} = \sqrt{(2)(0.60)(9.8 \text{ m/s}^2)(290 \text{ m})}$$

$$= 58 \text{ m/s} = 210 \text{ km/h} \qquad \text{(Answer)}$$

(about 130 mi/h). In obtaining the answer, we implicitly assumed that $v = 0$ at the far end of the skid marks. Actually, the marks ended only because the Jaguar left the road after 290 m. So v_0 was at least 210 km/h, and possibly much more.

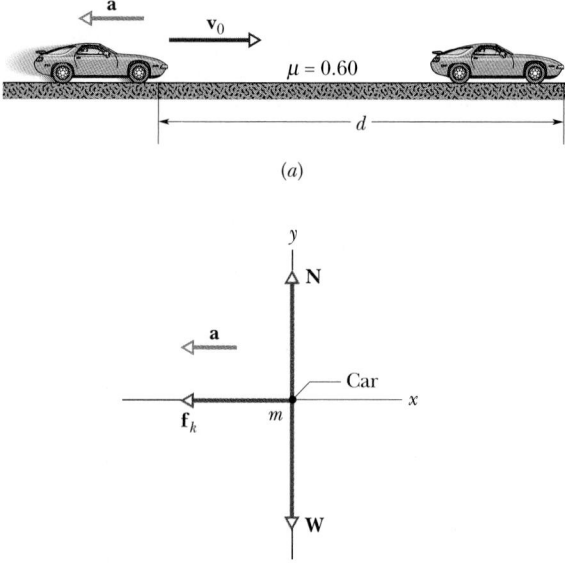

(a)

(b)

FIGURE 6-4 Sample Problem 6-2. (a) A car, sliding to the right and finally stopping after a displacement d. (b) A free-body diagram for the decelerating car. The acceleration vector points to the left, in the direction of the frictional force $\mathbf{f}_k$.

SAMPLE PROBLEM 6-3

A woman pulls a loaded sled of mass $m = 75$ kg along a horizontal surface at constant velocity. The coefficient of kinetic friction μ_k between the runners and the snow is 0.10, and the angle ϕ in Fig. 6-5 is 42°.

(a) What is the tension T in the rope?

SOLUTION: Figure 6-5b is the free-body diagram for the sled. Applying Newton's second law in the horizontal direction yields

$$T \cos \phi - f_k = ma_x = 0, \qquad (6\text{-}10)$$

where a_x is zero because the velocity is constant. In the vertical direction, we have

$$T \sin \phi + N - mg = ma_y = 0, \qquad (6\text{-}11)$$

in which mg is the weight of the sled. From Eq. 6-2,

$$f_k = \mu_k N. \qquad (6\text{-}12)$$

These three equations contain T, N, and f_k as unknowns. Eliminating N and f_k will allow us to find the remaining variable T. We start by adding Eqs. 6-10 and 6-12, obtaining

$$T \cos \phi = \mu_k N,$$

or

$$N = \frac{T \cos \phi}{\mu_k}. \qquad (6\text{-}13)$$

Substituting this into Eq. 6-11 and solving for T, we obtain

$$T = \frac{\mu_k mg}{\cos \phi + \mu_k \sin \phi} \qquad (6\text{-}14)$$

$$= \frac{(0.10)(75 \text{ kg})(9.8 \text{ m/s}^2)}{\cos 42° + (0.10)(\sin 42°)}$$

$$= 91 \text{ N}, \qquad \text{(Answer)}$$

which is considerably less than the weight of the sled.

(b) What is the normal force with which the snow pushes vertically upward on the sled?

SOLUTION: We can find N by substituting $T = 91$ N and given data into either Eq. 6-11 or 6-13. With Eq. 6-11 we find

$$N = mg - T \sin \phi$$

$$= (75 \text{ kg}) (9.8 \text{ m/s}^2) - (91 \text{ N}) \sin 42°$$

$$= 670 \text{ N}. \qquad \text{(Answer)}$$

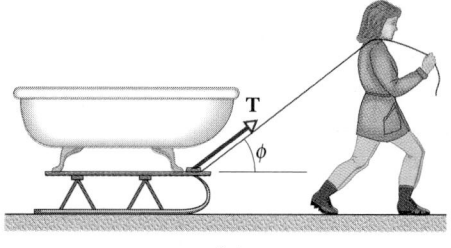

(a)

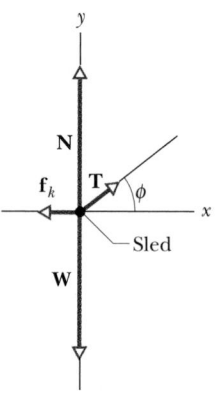

(b)

FIGURE 6-5 Sample Problem 6-3. (a) A woman exerts a force $\mathbf{T}$ on a sled, pulling it at a constant velocity. (b) A free-body diagram for the sled and its load.

CHECKPOINT **2:** In the figure, horizontal force $\mathbf{F}_1$ of magnitude 10 N is applied to a box on a floor, but the box does not slide. Then, as the magnitude of vertically applied force $\mathbf{F}_2$ is increased from zero, but before the box begins to slide, do the following quantities increase, decrease, or stay the same: (a) the magnitude of the frictional force on the box; (b) the magnitude of the normal force on the box from the floor; (c) the maximum value $f_{s,\max}$ of the static frictional force on the box?

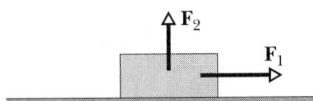

SAMPLE PROBLEM 6-4

In Fig. 6-6a, a crate of dilled pickles with mass $m_1 = 14$ kg moves along a plane that makes an angle of $\theta = 30°$ with the horizontal. That crate is connected to a crate of pickled dills with mass $m_2 = 14$ kg by a taut, massless cord that runs around a frictionless, massless pulley. The hanging crate of dills descends with constant velocity.

(a) What are the magnitude and direction of the frictional force exerted on m_1 by the plane?

SOLUTION: The fact that m_2 descends indicates that m_1 moves up the plane, and so a *kinetic* frictional force $\mathbf{f}_k$ must point down the plane.

We cannot use Eq. 6-2 to find the magnitude of $\mathbf{f}_k$, because we do not know the coefficient of kinetic friction μ_k between m_1 and the plane. However, we can use the techniques of Chapter 5. To start, we draw the free-body diagrams for m_1 and m_2 in Figs. 6-6b and 6-6c, where $\mathbf{T}$ is the pull from the tension in the cord and the weight vectors are $\mathbf{W}_1 = m_1\mathbf{g}$ and $\mathbf{W}_2 = m_2\mathbf{g}$.

With $\mathbf{W}_1$ resolved into x and y components we have, from Newton's second law applied to the x axis in Fig. 6-6b,

$$\sum F_x = T - f_k - m_1g \sin\theta = m_1a_x = 0, \quad (6\text{-}15)$$

where $a_x = 0$ because m_1 must move at constant velocity. Next, for m_2 we apply Newton's second law along the y axis in Fig. 6-6c and use the fact that m_2 moves at constant velocity. We find

$$\sum F_y = T - m_2g = m_2a_y = 0,$$

or

$$T = m_2g. \quad (6\text{-}16)$$

We then substitute T from Eq. 6-16 into Eq. 6-15 and solve for f_k, obtaining

$$f_k = m_2g - m_1g \sin\theta$$
$$= (14 \text{ kg})(9.8 \text{ m/s}^2) - (14 \text{ kg})(9.8 \text{ m/s}^2)(\sin 30°)$$
$$= 68.6 \text{ N} \approx 69 \text{ N}. \quad \text{(Answer)}$$

(b) What is μ_k?

SOLUTION: We can use Eq. 6-2 to find μ_k, but first we need the magnitude of the normal force $\mathbf{N}$ acting on m_1. To find N, we apply Newton's second law for m_1 along the y axis in Fig. 6-6b:

$$\sum F_y = N - m_1g \cos\theta = m_1a_y = 0,$$

or

$$N = m_1g \cos\theta.$$

From Eq. 6-2 we now have

$$\mu_k = \frac{f_k}{N} = \frac{f_k}{m_1g \cos\theta}$$

$$= \frac{68.6 \text{ N}}{(14 \text{ kg})(9.8 \text{ m/s}^2)(\cos 30°)} = 0.58. \quad \text{(Answer)}$$

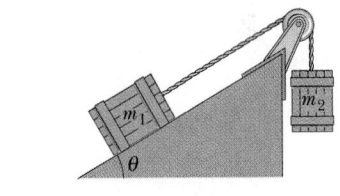

(a)

FIGURE 6-6 Sample Problem 6-4. (a) Mass m_1 moves up the plane while mass m_2 descends at constant velocity. (b) A free-body diagram for mass m_1. (c) A free-body diagram for mass m_2.

6-3 THE DRAG FORCE AND TERMINAL SPEED

A **fluid** is anything that can flow—generally either a gas or a liquid. When there is a relative velocity between a fluid and a body (either because the body moves through the fluid or because the fluid moves past the body), the body experiences a **drag force D** that opposes the relative motion and points in the direction in which the fluid flows relative to the body.

Here we examine only cases in which air is the fluid, the body is blunt (like a baseball) rather than slender (like a javelin), and the relative motion is fast enough so that the air becomes turbulent (breaks up into swirls) behind the body. In such cases, the magnitude of the drag force **D** is related to the relative speed v by an experimentally determined **drag coefficient** C according to

$$D = \tfrac{1}{2}C\rho Av^2, \qquad (6\text{-}17)$$

where ρ is the air density (mass per volume) and A is the **effective cross-sectional area** of the body (the area of a cross section taken perpendicular to the velocity **v**). The drag coefficient C (typical values range from 0.4 to 1.0) is not truly a constant for a given body, because if v varies significantly, the value of C can vary as well. Here, we ignore such complications.

Downhill speed skiers know well that drag depends on A and v^2. To reach high speeds a skier must reduce D as much as possible by, for example, riding the skis in the "egg position" (Fig. 6-7) to minimize A.

Equation 6-17 indicates that when a blunt object falls from rest through air, D gradually increases from zero as the speed of the body increases. As suggested in Fig. 6-8, if the body falls far enough, D eventually equals the body's weight W ($= mg$), and the net vertical force on the body is then zero. By Newton's second law, the acceleration must also be zero then, and so the body's speed no longer increases. The body then falls at a constant **terminal speed** v_t, which we find by setting $D = mg$ in Eq. 6-17, obtaining

$$\tfrac{1}{2}C\rho Av_t^2 = mg,$$

which gives

$$v_t = \sqrt{\frac{2mg}{C\rho A}}. \qquad (6\text{-}18)$$

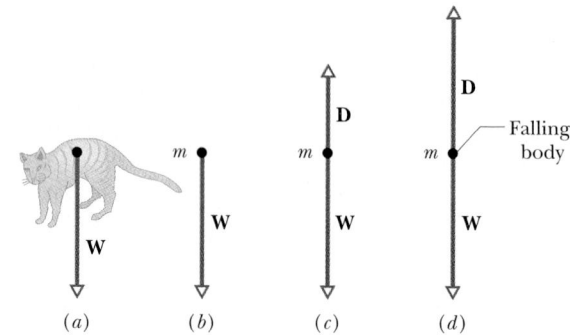

FIGURE 6-8   The forces that act on a body falling through air: (*a*) the body when it has just begun to fall, (*b*) the free-body diagram just then, and (*c*) the free-body diagram a little later, after a drag force has developed. (*d*) The drag force has increased until it balances the weight of the body. The body now falls at its constant terminal speed.

Table 6-1 gives values of v_t for some common objects.

According to calculations* based on Eq. 6-17, a cat must fall about six floors to reach terminal speed. Until it does so, $W > D$ and the cat accelerates downward because of the net downward force. Recall from Chapter 2 that your body is an accelerometer, not a speedometer. Because the cat too senses the acceleration, it is frightened and keeps its feet underneath its body, its head tucked in, and its spine bent upward, making A small, v_t large, and injury on landing likely.

However, if the cat does reach v_t, the acceleration vanishes and the cat relaxes somewhat, stretching its legs

TABLE 6-1 SOME TERMINAL SPEEDS IN AIR

OBJECT	TERMINAL SPEED (m/s)	95% DISTANCE[a] (m)
16 lb Shot	145	2500
Sky diver (typical)	60	430
Baseball	42	210
Tennis ball	31	115
Basketball	20	47
Ping-Pong ball	9	10
Raindrop (radius = 1.5 mm)	7	6
Parachutist (typical)	5	3

[a]This is the distance through which the body must fall from rest to reach 95% of its terminal speed.

Source: Adapted from Peter J. Brancazio, *Sport Science,* Simon & Schuster, New York, 1984.

FIGURE 6-7   This skier crouches in an "egg position" so as to minimize her effective cross-sectional area and thus the air drag acting on her.

*W. O. Whitney and C. J. Mehlhaff, "High-rise syndrome in cats," *The Journal of the American Veterinary Medical Association,* Vol. 191, pages 1399–1403 (1987).

and neck horizontally outward and straightening its spine (it then resembles a flying squirrel). These actions increase A and D, and the cat begins to slow because now $D > W$ —the net force is upward—until a new, smaller v_t is reached. The decrease in v_t reduces the possibility of serious injury on landing. Just before the end of the fall, when it sees it is nearing the ground, the cat pulls its legs back beneath its body to prepare for the landing.

SAMPLE PROBLEM 6-5

If a falling cat reaches a first terminal speed of 60 mi/h while it is tucked in and then stretches out, doubling A, how fast is it falling when it reaches a new terminal speed?

SOLUTION: We let v_{to} and v_{tn} represent the original and new terminal speeds, and A_o and A_n the original and new areas. We then use Eq. 6-18 to set up a ratio of speeds:

$$\frac{v_{tn}}{v_{to}} = \frac{\sqrt{2mg/C\rho A_n}}{\sqrt{2mg/C\rho A_o}} = \sqrt{\frac{A_o}{A_n}} = \sqrt{\frac{A_o}{2A_o}} = \sqrt{0.5} \approx 0.7,$$

which means that $v_{tn} \approx 0.7 v_{to}$, or about 40 mi/h.

In April 1987, during a jump, parachutist Gregory Robertson noticed that fellow parachutist Debbie Williams had been knocked unconscious in a collision with a third sky diver and was unable to open her parachute. Robertson, who was well above Williams at the time and who had not yet opened his parachute for the 13,500-ft plunge, reoriented his body head-down so as to minimize A and maximize his downward speed. Reaching an estimated v_t of 200 mi/h, he caught up with Williams and then went into a horizontal "spread eagle" (as in Fig. 6-9) to increase D so that he could grab her. He opened her parachute and then, after releasing her, his own, with a scant 10 s before impact. Williams received extensive internal injuries due to her lack of control on landing but survived.

FIGURE 6-9 A parachutist in a horizontal "spread eagle" maximizes the air drag.

SAMPLE PROBLEM 6-6

A raindrop with radius $R = 1.5$ mm falls from a cloud that is at height $h = 1200$ m above the ground. The drag coefficient C for the drop is 0.60. Assume that the drop is spherical throughout its fall. The density of water ρ_w is 1000 kg/m³, and the density of air ρ_a is 1.2 kg/m³.

(a) What is the terminal speed of the drop?

SOLUTION: The volume of a sphere is $\frac{4}{3}\pi R^3$, and its effective area A is that of a circle with radius R. So, for the drop:
$$m = \frac{4}{3}\pi R^3 \rho_w \quad \text{and} \quad A = \pi R^2.$$

Then, from Eq. 6-18, we find
$$v_t = \sqrt{\frac{2mg}{C\rho_a A}} = \sqrt{\frac{8\pi R^3 \rho_w g}{3C\rho_a \pi R^2}} = \sqrt{\frac{8R\rho_w g}{3C\rho_a}}$$
$$= \sqrt{\frac{(8)(1.5 \times 10^{-3} \text{ m})(1000 \text{ kg/m}^3)(9.8 \text{ m/s}^2)}{(3)(0.60)(1.2 \text{ kg/m}^3)}}$$
$$= 7.4 \text{ m/s} (= 17 \text{ mi/h}). \quad \text{(Answer)}$$

Note that the height of the cloud does not enter into the calculation. The raindrop (see Table 6-1) reaches terminal speed after falling just a few meters.

(b) What would have been the speed just before impact if there had been no drag force?

SOLUTION: From Eq. 2-23, with $h = -(y - y_0)$ and $v_0 = 0$, we find
$$v = \sqrt{2gh} = \sqrt{(2)(9.8 \text{ m/s}^2)(1200 \text{ m})}$$
$$= 150 \text{ m/s} (= 340 \text{ mi/h}). \quad \text{(Answer)}$$

Under these conditions, Shakespeare would scarcely have written, "it droppeth as the gentle rain from heaven, upon the place beneath."

CHECKPOINT 3: Near the ground, is the speed of large raindrops greater than, less then, or the same as that of small raindrops, assuming that both large and small raindrops are spherical?

6-4 UNIFORM CIRCULAR MOTION

Recall that when a body moves in a circle (or a circular arc) at constant speed v, it is said to be in uniform circular motion. Also recall that the body has a centripetal acceleration (directed toward the center of the circle), of constant magnitude given by Eq. 4-22:

$$a = \frac{v^2}{r} \quad \text{(centripetal acceleration),} \quad (6\text{-}19)$$

where r is the radius of the circle.

This centripetal acceleration is caused by a **centripetal force** that acts on the body and that is directed toward the center of the circle. The magnitude F of this force is constant and is given by Newton's second law as

$$F = ma = \frac{mv^2}{r} \qquad \text{(centripetal force).} \quad (6\text{-}20)$$

If this force is not present, the body cannot undergo uniform circular motion. Both the centripetal acceleration and the centripetal force are vector quantities whose magnitudes are constant but whose directions are always changing so as to point toward the center of the circle.

If the body in uniform circular motion is, say, a hockey puck whirled around on the end of a string as in Fig. 6-10, the centripetal force is provided by the tension in the string. For the Moon in its (nearly) uniform circular motion around the Earth, the centripetal force is the gravitational attraction of the Earth. Thus a centripetal force is not a new kind of force; it can be a tension force, a gravitational force, or any other force.

Let us compare two familiar examples of uniform circular motion:

1. *Rounding a curve in a car.* You are sitting in the center of the rear seat of a car moving at high speed over flat road. When the driver suddenly turns left, rounding a corner in a circular arc, you are slid across the seat toward the right and jammed against the interior wall of the car. What is going on?

While the car is moving in the circular arc, it is in uniform circular motion. The centripetal force that causes the motion is a frictional force exerted by the roadway on the tires. This force points radially inward and—spread over the four tires—has a magnitude given by Eq. 6-20.

You would have been in uniform circular motion in the center of the seat if the frictional force exerted on you

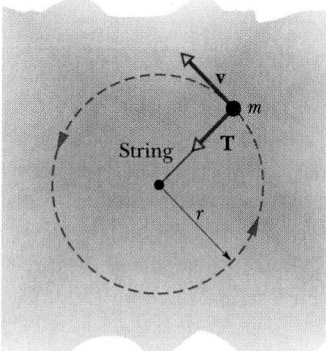

FIGURE 6-10   A hockey puck of mass m moves with constant speed v in a circular path on a horizontal frictionless surface. The centripetal force acting on the puck is **T**, the pull from the string.

by the seat had been great enough. However, it was not, and so you slid across the seat. Viewed from the reference frame of the ground, you actually continued moving in a straight line while the seat slid beneath you, until you met the car's wall. Its push on you was a centripetal force, and you then joined the car's uniform circular motion.

2. *Orbiting the Earth.* This time you are a passenger in the space shuttle *Atlantis*, orbiting the Earth and experiencing "weightlessness." What is going on in this case?

The centripetal force that keeps you and the shuttle in uniform circular motion is the gravitational attraction of the Earth for you and the shuttle. This force is directed radially inward, toward the center of the Earth, and has a magnitude given by Eq. 6-20.

In both car and shuttle you are in uniform circular motion, acted on by a centripetal force. Yet your experiences in the two situations are quite different. In the car, jammed up against the wall, you are aware of being compressed by the wall. In the orbiting shuttle, on the other hand, you are floating around with no sensation of any force acting on you. Why this great difference?

The difference is due to the nature of the two centripetal forces. In the car, the centripetal force is a *contact force*, exerted by the wall externally on the part of your body touching the wall. In the shuttle, the centripetal force is a *volume force*, due to the Earth's gravitational pull on every atom of your body and on every atom of the shuttle, in proportion to the mass of that atom. Thus there is no compression on any one part of your body, nor even any sensation of a force acting on you.

SAMPLE PROBLEM 6-7

Igor is a cosmonaut-engineer in the spacecraft *Vostok II*, orbiting the Earth at an altitude h of 520 km with a speed v of 7.6 km/s. Igor's mass m is 79 kg.

(a) What is his acceleration?

SOLUTION: Igor is in uniform circular motion in a circle of radius $R_E + h$, where R_E is the radius of the Earth. His centripetal acceleration is given by Eq. 6-19:

$$a = \frac{v^2}{r} = \frac{v^2}{R_E + h}$$

$$= \frac{(7.6 \times 10^3 \text{ m/s})^2}{6.37 \times 10^6 \text{ m} + 0.52 \times 10^6 \text{ m}}$$

$$= 8.38 \text{ m/s}^2 \approx 8.4 \text{ m/s}^2, \qquad \text{(Answer)}$$

which is the value of the free-fall acceleration at Igor's altitude. If he were lifted to that altitude and released, instead of being put into orbit there, he would fall toward the Earth's center, starting out with that value for the acceleration. The difference in the two situations is that when he orbits the Earth, he always has a "sideways" motion as well: as he falls, he also moves to the side, so that he ends up moving along a curved path around the Earth.

(b) What (centripetal) gravitational force does the Earth exert on Igor?

SOLUTION: The centripetal force is

$$F = ma = (79 \text{ kg})(8.38 \text{ m/s}^2)$$
$$= 660 \text{ N} \approx 150 \text{ lb}. \qquad \text{(Answer)}$$

If Igor were to stand on a scale placed on the top of a tower with height $h = 520$ km, the scale would read 660 N or 150 lb. In orbit, the scale (if Igor could "stand" on it) would read zero because he and the scale are in free fall together, and therefore his feet do not actually press against it.

PROBLEM SOLVING TACTICS

TACTIC 1: *Looking Things Up*

In Sample Problem 6-7, we had to know the radius of the Earth, which was not given in the problem statement. You need to become familiar with sources of this kind of information, starting with this book. Many useful data are given in the inside covers, in the various appendixes, and in tables. The *Handbook of Chemistry and Physics,* updated every year by the publisher, CRC Press, is an invaluable resource.

For practice, see if you can track down the density of iron, the series expansion of e^x, the number of centimeters in a mile, the mean distance of Saturn from the Sun, the mass of the proton, the speed of light, and the atomic number of samarium. You can find them all in this book.

CHECKPOINT 4: When you ride in a Ferris wheel at constant speed, what are the directions of your acceleration **a** and the normal force **N** on you (from the seat) as you pass through (a) the highest point and (b) the lowest point of the ride?

SAMPLE PROBLEM 6-8

In a 1901 circus performance, Allo "Dare Devil" Diavolo introduced the stunt of riding a bicycle in a loop-the-loop (Fig. 6-11*a*). Assuming that the loop is a circle with radius $R = 2.7$ m, what is the least speed v Diavolo could have at the top of the loop if he was to remain in contact with it there?

SOLUTION: Figure 6-11*b* is a free-body diagram for Diavolo and the bicycle (taken together as being a single particle) at the top of the loop, showing the normal force **N** exerted on them by the loop, and their weight **W** $= m$**g**. Their accelera-

tion **a** points downward, toward the center of the loop and, according to Eq. 6-19, has magnitude $a = v^2/R$. Applying Newton's second law along the y axis, we have

$$\sum F = -N - mg = -ma = -m\frac{v^2}{R}.$$

If he is on the verge of losing contact with the loop, $N = 0$, and so we then have

$$mg = m\frac{v^2}{R},$$

or
$$v = \sqrt{gR} = \sqrt{(9.8 \text{ m/s}^2)(2.7 \text{ m})}$$
$$= 5.1 \text{ m/s}. \qquad \text{(Answer)}$$

To avoid losing contact, Diavolo made certain that his speed at the top of the loop was greater than 5.1 m/s. Then $N > 0$ and, more important to him, he and the bicycle exerted a force of magnitude N on the track.

FIGURE 6-11 Sample Problem 6-8. (*a*) Contemporary advertisement for Diavolo and (*b*) free-body diagram for the performer at the top of the loop.

SAMPLE PROBLEM 6-9

Figure 6-12*a* shows a *conical pendulum*. Its bob, whose mass m is 1.5 kg, whirls around in a horizontal circle at constant speed v at the end of a cord whose length L, measured to the center of the bob, is 1.7 m. The cord makes an angle θ of 37° with the vertical. As the bob swings around in a circle, the cord sweeps out the surface of a cone. Find the period τ of the pendulum bob.

SOLUTION: Figure 6-12*b*, the free-body diagram for the bob, shows the forces on the bob: the pull **T** from the cord due to the cord's tension, and the bob's weight **W** ($= m$**g**). We place the origin of axes at the center of the bob, as shown. Instead of the usual x axis (which is stationary), we use a radial axis r that always points from the bob toward the center of the circle.

The y and r components of **T** are $T \cos \theta$ and $T \sin \theta$, respectively. Since $a_y = 0$, Newton's second law gives

$$T \cos \theta - mg = ma_y = 0, \quad \text{or} \quad T \cos \theta = mg. \quad (6\text{-}21)$$

There must be a net force along the r axis to provide the centripetal acceleration for the bob. The only force component in that direction is $T \sin \theta$. So along the r axis, Eq. 6-20 gives

$$T \sin \theta = ma_r = \frac{mv^2}{R}, \quad (6\text{-}22)$$

where R is the radius of the bob's circular path. Dividing Eq. 6-22 by Eq. 6-21 and solving for v, we obtain

$$v = \sqrt{\frac{gR \sin \theta}{\cos \theta}}.$$

We can substitute $2\pi R/\tau$ (the distance around the circle divided by the period) for the speed v of the bob. Doing so and solving for τ, we obtain

$$\tau = 2\pi \sqrt{\frac{R \cos \theta}{g \sin \theta}}. \quad (6\text{-}23)$$

Now, from Fig. 6-12a we see that $R = L \sin \theta$. Making this substitution in Eq. 6-23 yields

$$\tau = 2\pi \sqrt{\frac{L \cos \theta}{g}} \quad (6\text{-}24)$$

$$= 2\pi \sqrt{\frac{(1.7 \text{ m})(\cos 37°)}{9.8 \text{ m/s}^2}} = 2.3 \text{ s}. \quad \text{(Answer)}$$

From Eq. 6-24 we see that the period τ does not depend on the mass of the bob, but only on $L \cos \theta$, the vertical distance of the bob from its point of support. Thus, as shown in Fig. 6-12c, if several conical pendulums of different lengths are swung from the same support *with the same period*, their bobs will all lie in the same horizontal plane.

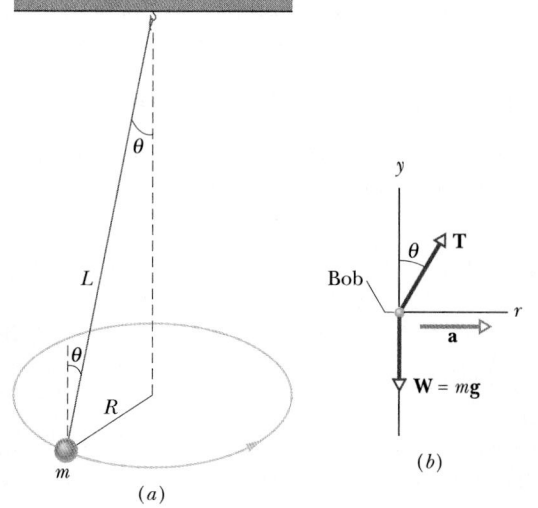

(a)

(b)

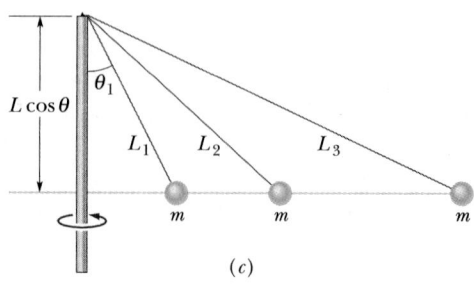

(c)

FIGURE 6-12   Sample Problem 6-9. (a) A conical pendulum, its cord making an angle θ with the vertical. (b) A free-body diagram for the pendulum bob. The axes point in the vertical and radial directions. The resultant force (and thus the acceleration) points radially inward toward the center of the circle. (c) Three pendulums, of different lengths, are whirled around by a rotating shaft; their bobs circulate in the same horizontal plane, as Eq. 6-24 predicts.

SAMPLE PROBLEM 6-10

Figure 6-13a represents a car of mass $m = 1600$ kg traveling at a constant speed $v = 20$ m/s along a flat, circular road of radius $R = 190$ m. What is the minimum value of μ_s between the tires of the car and the road that will prevent the car from slipping?

SOLUTION: The centripetal force that causes the car to travel in a circle is the radial frictional force $\mathbf{f}_s$ exerted on the tires by the road. (Although the car is moving, it is not sliding radially, and so the frictional force is $\mathbf{f}_s$ and not $\mathbf{f}_k$.)

Figure 6-13b, the free-body diagram for the car, shows the forces on the car: $\mathbf{f}_s$, **N**, and $\mathbf{W} = m\mathbf{g}$. Since the car is not accelerating vertically, $a_y = 0$, and Newton's second law leads to the now familiar result $N = W = mg$.

In the radial direction, however, there must be a net force $\Sigma \mathbf{F}_r$ to give the car a centripetal acceleration $\mathbf{a}_r$. (Otherwise the car would move off the road along a straight line.) According to Eq. 6-20, $\Sigma F_r = mv^2/R$. Because the only radial force we have is f_s, it must be true that

$$f_s = \frac{mv^2}{R}. \quad (6\text{-}25)$$

Next, recall that a body is on the verge of slipping when f_s reaches its maximum value, $\mu_s N$. Since the problem concerns just such a critical situation, we set $f_s = \mu_s N$ in Eq. 6-25 and then substitute mg for N. We get

$$\mu_s mg = \frac{mv^2}{R},$$

or

$$\mu_s = \frac{v^2}{gR} \quad (6\text{-}26)$$

$$= \frac{(20 \text{ m/s})^2}{(9.8 \text{ m/s}^2)(190 \text{ m})} = 0.21. \quad \text{(Answer)}$$

If $\mu_s \geq 0.21$, the car will be held in a circle by $\mathbf{f}_s$. But if $\mu_s < 0.21$, the car will slide out of the circle.

Note two additional features of Eq. 6-26. First, the value

of μ_s depends on the square of v. That means that much more friction is required as the turning speed increases. You may have noted this effect if you have ever taken a flat turn too fast and suddenly felt the tires slip. Second, the mass m dropped out in our derivation of Eq. 6-26. That means Eq. 6-26 holds for a vehicle of any mass, from a kiddy car to a bicycle to a heavy truck.

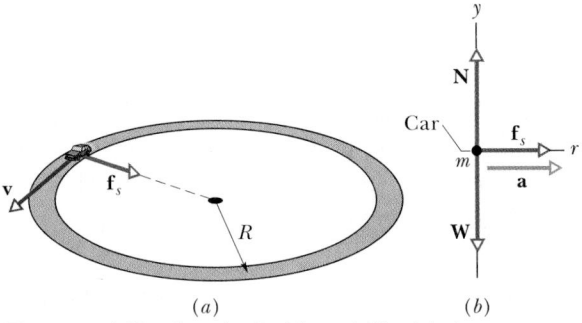

(a)   (b)

FIGURE 6-13 Sample Problem 6-10. (*a*) A car moves around a flat curved road at constant speed. The frictional force $\mathbf{f}_s$ provides the necessary centripetal force. (*b*) A free-body diagram (not to scale) for the car, in the vertical plane.

ÇHECKPOINT **5:** In Fig. 6-13, suppose the car is on the verge of sliding when the radius of the circle is R_1. (a) If we double the car's speed, what is the least radius that would now keep the car from sliding? (b) If we also double the weight of the car (say by adding sandbags), what is the least radius that would now keep the car from sliding?

$$N_r = N \sin \theta = \frac{mv^2}{R}. \tag{6-28}$$

Dividing Eq. 6-28 by Eq. 6-27 gives

$$\tan \theta = \frac{v^2}{gR}.$$

Thus we have

$$\theta = \tan^{-1} \frac{v^2}{gR} \tag{6-29}$$

$$= \tan^{-1} \frac{(20 \text{ m/s})^2}{(9.8 \text{ m/s}^2)(190 \text{ m})} = 12°. \quad \text{(Answer)}$$

Equations 6-26 and 6-29 tell us that the critical coefficient of friction for an unbanked road is the same as the tangent of the bank angle for a banked road. The road must produce a certain centripetal force one way or the other—either with friction or by being banked.

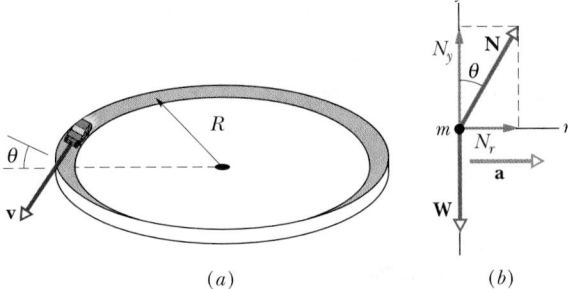

(a)   (b)

FIGURE 6-14 Sample Problem 6-11. (*a*) A car moves around a curved banked road at constant speed. The bank angle is exaggerated for clarity. (*b*) A free-body diagram for the car, assuming that friction between tires and road is zero. The radially inward component of the normal force provides the necessary centripetal force. The resulting acceleration also points radially inward.

SAMPLE PROBLEM 6-11

You cannot always count on friction to get your car around a curve, especially if the road is icy or wet. That is why highway curves are banked. As in Sample Problem 6-10, suppose that a car of mass m moves at a constant speed v of 20 m/s around a curve, now banked, whose radius R is 190 m (Fig. 6-14*a*). What bank angle θ makes reliance on friction unnecessary?

SOLUTION: The centripetal acceleration and the required centripetal force ΣF_r are the same as in the preceding sample problem. The effect of banking is to tilt the normal force $\mathbf{N}$ toward the center of curvature of the road, so that now it has an inward radial component N_r that can supply the required centripetal force.

Since there is no acceleration in the vertical direction,

$$N_y = N \cos \theta = W = mg. \tag{6-27}$$

In the radial direction the only force component is N_r (we assume a frictional force is unnecessary). So, by Eq. 6-20,

SAMPLE PROBLEM 6-12

Even some seasoned roller coaster riders blanch at the thought of riding the Rotor, which is essentially a large hollow cylinder that is rotated rapidly around its central axis (Fig. 6-15). Before the ride begins, a rider enters the cylinder through a door on the side and stands on a floor, up against a canvas-covered wall. The door is closed, and as the cylinder begins to turn, the rider, wall, and floor move in unison. When the rider's speed reaches some predetermined value, the floor abruptly and alarmingly falls away. The rider does not fall with it but instead is pinned to the wall while the cylinder rotates, as if an unseen (and somewhat unfriendly) agent is pressing the body to the wall. Later, the floor is eased back to the rider's feet, the cylinder slows, and the rider sinks a few centimeters to regain footing on the floor. (Some riders consider all this to be fun.)

Suppose that the coefficient of static friction μ_s between the rider's clothing and the canvas is 0.40 and that the cylinder's radius R is 2.1 m.

(a) What minimum speed v must the cylinder and rider have if the rider is not to fall when the floor drops?

SOLUTION: The rider will not be pulled down by her weight $\mathbf{W}$ provided the magnitudes of $\mathbf{W}$ and the frictional force $\mathbf{f}_s$, which is exerted upward on her by the wall, are equal. At the minimum required speed, she is on the verge of slipping, which means that f_s must be at its maximum value of $\mu_s N$. So in this critical situation

$$\mu_s N = mg, \tag{6-30}$$

where m is her mass.

The normal force $\mathbf{N}$ is, as usual, perpendicular to the surface against which the body (here, the woman) is pressed, but note that now it points horizontally toward the central axis. That force accounts for the centripetal force that provides the woman's centripetal acceleration a_r and keeps her moving in a circle. Thus, by Eq. 6-20,

$$N = \frac{mv^2}{R}. \tag{6-31}$$

Substituting this expression for N in Eq. 6-30 and solving for v, we have

$$v = \sqrt{\frac{gR}{\mu_s}} = \sqrt{\frac{(9.8 \text{ m/s}^2)(2.1 \text{ m})}{0.40}}$$

$$= 7.17 \text{ m/s} \approx 7.2 \text{ m/s}. \qquad \text{(Answer)}$$

Note that the result is independent of the rider's mass; it holds for anyone riding the Rotor, from a child to a sumo wrestler.

(b) If the rider's mass is 49 kg, what is the magnitude of the centripetal force on her?

SOLUTION: According to Eq. 6-31,

$$N = \frac{mv^2}{R} = \frac{(49 \text{ kg})(7.17 \text{ m/s})^2}{2.1 \text{ m}}$$

$$\approx 1200 \text{ N}. \qquad \text{(Answer)}$$

FIGURE 6-15 Sample Problem 6-12. A Rotor in an amusement park, showing the forces on a rider. The centripetal force is the normal force with which the wall pushes inward on the rider.

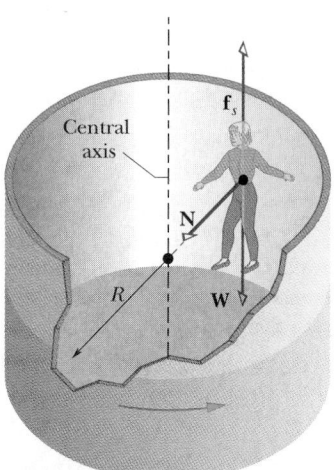

Although this force is directed toward the central axis, the rider has an overwhelming sensation that the force pinning her against the wall is directed radially outward. Her sensation stems from the fact that she is in a noninertial frame (she and it are accelerating). As measured from such frames, forces can be illusionary. The illusion is part of the Rotor's attraction.

CHECKPOINT **6:** If the Rotor of Sample Problem 6-12 initially moves at the minimum required speed for the rider not to fall and then its speed is increased in steps, do the following increase, decrease, or remain the same: (a) the magnitude of $\mathbf{f}_s$; (b) the magnitude of $\mathbf{N}$; (c) the value of $f_{s,\max}$?

6-5 THE FORCES OF NATURE

We have used $\mathbf{F}$ as a generic symbol for force. We have also used other symbols: $\mathbf{W}$ for the weight of a body, $\mathbf{T}$ for the pull from a cord under tension, $\mathbf{f}$ for a frictional force, $\mathbf{N}$ for a normal force, and $\mathbf{D}$ for the drag force exerted, for example, by air on a sky diver. At the fundamental level, all these forces fall into two types: (1) the **gravitational force,** of which weight is our only example, and (2) the **electromagnetic force,** which includes—without exception—all the others. The electromagnetic force is the combination of electrical forces and magnetic forces. The force that makes an electrically charged balloon stick to a wall and the force with which a magnet picks up an iron nail are other examples of it. In fact, aside from the gravitational force, *all* forces that we can experience directly as a push or pull are electromagnetic in nature. That is, all such forces, including frictional forces, normal forces, contact forces, and tension forces, involve, fundamentally, electromagnetic forces exerted by one atom on another. The tension in a taut rope, for example, exists only because the atoms of the rope attract one another.

Only two other fundamental forces are known, and they both act over such short distances that we cannot ex-

Banked tracks are needed for turns that are taken so quickly that friction alone cannot provide enough centripetal force.

TABLE 6-2 THE QUEST FOR THE SUPERFORCE—A PROGRESS REPORT

DATE	RESEARCHER	ACHIEVEMENT
1687	Newton	Showed that the same laws apply to astronomical bodies and to objects on Earth. Unified celestial and terrestrial mechanics.
1820 1830s	Oersted Faraday	Showed, by brilliant experiments, that the then separate sciences of electricity and magnetism are intimately linked.
1873	Maxwell	Unified the sciences of electricity, magnetism, and optics into the single subject of electromagnetism.
1979	Glashow, Salam, Weinberg	Received the Nobel prize for showing that the weak force and the electromagnetic force could be viewed as different aspects of a single *electroweak force*. This combination of forces reduced the number of fundamental forces from four to three.
1984	Rubbia, van der Meer	Received the Nobel prize for verifying experimentally the predictions of the theory of the electroweak force.

Work in Progress

Grand unification theories (GUTs), seek to unify the electroweak force and the strong force.
Supersymmetry theories: seek to unify all forces, including the gravitational force, within a single framework.
Superstring theories: interpret pointlike particles, such as electrons, as being unimaginably tiny, closed loops. Strangely, extra dimensions beyond the familiar four dimensions of spacetime appear to be required.

perience them directly through our senses. They are the **weak force,** which is involved in certain kinds of radioactive decay, and the **strong force,** which binds together the quarks that make up protons and neutrons and is the "glue" that holds together an atomic nucleus.

Physicists have long believed that nature has an underlying simplicity and that the number of fundamental forces can be reduced. Einstein spent most of his working life trying to interpret these forces as different aspects of a single *superforce*. He failed, but in the 1960s and 1970s, other physicists showed that the weak force and the electromagnetic force are different aspects of a single **electroweak force.** The quest for further reduction continues today, at the very forefront of physics. Table 6-2 lists the progress that has been made toward **unification** (as the goal is called) and gives some hints about the future.

REVIEW & SUMMARY

Friction

When a force $\mathbf{F}$ attempts to slide a body along a surface, a **frictional force** is exerted on the body by the surface. The frictional force is parallel to the surface and directed so as to oppose the sliding. It is due to bonding between the body and the surface.

If the body does not slide, the frictional force is a **static frictional force** $\mathbf{f}_s$. If there is sliding, the frictional force is a **kinetic frictional force** $\mathbf{f}_k$.

Properties of Friction

Property 1. If the body does not move, then the static frictional force $\mathbf{f}_s$ and the component of $\mathbf{F}$ that is parallel to the surface are equal in magnitude, and $\mathbf{f}_s$ is directed opposite that component. If that parallel component increases, $\mathbf{f}_s$ also increases.

Property 2. The magnitude of $\mathbf{f}_s$ has a maximum value $f_{s,\max}$ that is given by

$$f_{s,\max} = \mu_s N, \tag{6-1}$$

where μ_s is the **coefficient of static friction** and N is the magnitude of the normal force. If the component of $\mathbf{F}$ that is parallel to the surface exceeds $f_{s,\max}$, then the body slides on the surface.

Property 3. If the body begins to slide along the surface, the

magnitude of the frictional force rapidly decreases to a constant value f_k given by

$$f_k = \mu_k N, \tag{6-2}$$

where μ_k is the **coefficient of kinetic friction.**

Drag Force

When there is a relative velocity between air (or some other fluid) and a body, the body experiences a **drag force D** that opposes the relative motion and points in the direction in which the fluid flows relative to the body. The magnitude of **D** is related to the relative speed v by an experimentally determined **drag coefficient** C according to

$$D = \tfrac{1}{2} C \rho A v^2, \tag{6-17}$$

where ρ is the fluid density (mass per volume) and A is the **effective cross-sectional area** of the body (the area of a cross section taken perpendicular to the relative velocity $\mathbf{v}$).

Terminal Speed

When a blunt object falls far enough through air, the magnitudes of the drag force and the object's weight are equal. The body then falls at a constant **terminal speed** v_t given by

$$v_t = \sqrt{\frac{2mg}{C\rho A}}, \qquad (6\text{-}18)$$

where m is the body's mass.

Uniform Circular Motion

If a particle moves in a circle or a circular arc with radius r at constant speed v, it is said to be in **uniform circular motion.** It then has a **centripetal acceleration** with a magnitude given by

$$a = \frac{v^2}{r}, \qquad (6\text{-}19)$$

which is due to a **centripetal force** with a magnitude given by

$$F = \frac{mv^2}{r}, \qquad (6\text{-}20)$$

where m is the particle's mass. The vectors **a** and **F** point toward the center of curvature of the particle's path.

Fundamental Forces

The myriad examples of forces can be reduced to three fundamental types: **gravitational, electroweak** (a combination of the historic grouping of **electric** and **magnetic** forces with the **weak** force), and **strong**. Only the gravitational, electric, and magnetic forces are readily apparent in the everyday world. Physicists hope to reduce the list of three fundamental forces to a single force, the elusive *superforce* that would include all others.

QUESTIONS

1. Figure 6-16 shows four blocks arranged on a board. The board will be lifted by its right end (like the book in Fig. 6-3a) until the blocks begin to slide down it. The blocks are made of the same material and their masses are:

block 1, 5 kg	block 3, 10 kg
block 2, 10 kg	block 4, 5 kg

In which order, left to right, should the blocks be placed for them to begin sliding at the smallest possible angle between the board and the horizontal?

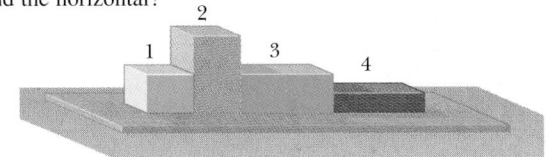

FIGURE 6-16 Question 1.

2. In Fig. 6-17, horizontal force F_1 of magnitude 10 N is applied to a box on a floor, but the box does not slide. Then, as the magnitude of vertical force F_2 is increased from zero, do the following quantities increase, decrease, or stay the same: (a) the magnitude of the frictional force f_s on the box; (b) the magnitude of the normal force N on the box from the floor; (c) the maximum value $f_{s,\max}$ of the static frictional force on the box? (d) Does the box eventually slide?

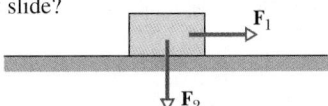

FIGURE 6-17 Question 2.

3. If you press an apple crate against a wall so hard that the crate cannot slide down the wall, what is the direction of (a) the static frictional force f_s on the crate from the wall and (b) the normal force N on the crate from the wall? If you increase your push, what happens to (c) f_s, (d) N, and (e) $f_{s,\max}$?

4. A box is on a ramp that is at angle θ to the horizontal. As θ is increased from zero, and before the box slips, do the following increase, decrease, or remain the same: (a) the weight component of the box along the ramp, (b) the magnitude of the static frictional force on the box from the ramp, (c) the weight component of the box perpendicular to the ramp, (d) the normal force on the box from the ramp, and (e) the maximum value $f_{s,\max}$ of the static frictional force?

5. In Fig. 6-18, a block is held stationary on a ramp by the frictional force on it from the ramp. A force **F**, directed up the ramp, is then applied to the block and gradually increased in magnitude from zero. During the increase, what happens to the direction and magnitude of the frictional force on the block?

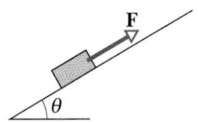

FIGURE 6-18 Question 5.

6. Reconsider Question 5 but with the force **F** now directed down the ramp. As the magnitude of **F** is increased from zero, what happens to the direction and magnitude of the frictional force on the block?

7. In Fig. 6-19, if the angle θ of force **F** on the stationary box is increased, do the following quantities increase, decrease, or remain the same: (a) F_x; (b) f_s; (c) N; (d) $f_{s,\max}$?

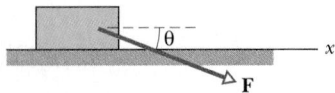

FIGURE 6-19 Question 7.

8. Repeat Question 7 for force **F** angled upward instead of downward.

9. In the conical pendulum of Sample Problem 6-9, what (a) period and (b) speed are associated with $\theta = 90°$?

10. A particle is made to move around three circular arcs, with the following speeds and radii of curvature:

ARC	SPEED	RADIUS
1	$2v_0$	r_0
2	$3v_0$	$3r_0$
3	$2v_0$	$4r_0$

Rank the arcs according to the magnitude of the centripetal force acting on the particle, greatest first.

11. Figure 6-20 shows an overhead view of a amusement park ride that travels at constant speed through five circular arcs of radii R_0, $2R_0$, and $3R_0$. Rank the arcs according to the magnitude of the centripetal force acting on a rider while traveling in the arcs, greatest first.

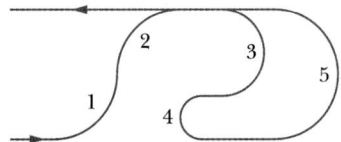

FIGURE 6-20 Question 11.

12. Figure 6-21 shows a section of a circular space station that rotates about its center so as to give an apparent weight to the crew. One of the crew is shown at the outer wall of the station, which has velocity $\mathbf{v}_s$. (a) If the astronaut moves to a point closer to the center of the station (say by taking an elevator), does his apparent weight increase, decrease, or remain the same? (b) If, instead, the astronaut runs along the outer wall in the direction opposite $\mathbf{v}_s$ (with a speed less than the magnitude of $\mathbf{v}_s$), does his apparent weight increase, decrease, or remain the same?

FIGURE 6-21 Question 12.

13. Figure 6-22 shows overhead views of two stones that travel in circles over a frictionless surface. Each stone is tied to a cord whose opposite end is anchored at the center of the circle. Is the tension in the longer cord greater than, less than, or the same as that in the shorter cord if the stones travel (a) at the same speed and (b) with the same period of motion?

FIGURE 6-22 Question 13.

14. A coin lies on a turntable whose speed can be gradually increased from zero in steps. What happens to the magnitude of the frictional force on the coin from the turntable as the speed is increased to a large value?

EXERCISES & PROBLEMS

SECTION 6-2 Properties of Friction

1E. A bedroom bureau with a mass of 45 kg, including drawers and clothing, rests on the floor. (a) If the coefficient of static friction between the bureau and the floor is 0.45, what is the minimum horizontal force a person must apply to start the bureau moving? (b) If the drawers and clothing, with 17 kg mass, are removed before the bureau is pushed, what is the new minimum magnitude?

2E. A baseball player with mass $m = 79$ kg, sliding into second base, is retarded by a force of friction $f = 470$ N. What is the coefficient of kinetic friction μ_k between the player and the ground?

3E. The coefficient of static friction between Teflon and scrambled eggs is about 0.04. What is the smallest angle from the horizontal that will cause the eggs to slide across the bottom of a Teflon-coated skillet?

4E. A 100 N force, directed at an angle θ above the horizontal, is applied to a 25.0 kg chair sitting on the floor. (a) For each of the following angles θ, calculate the magnitude of the normal force of the floor on the chair and the horizontal component of the applied force: (i) 0°, (ii) 30.0°, (iii) 60.0°. (b) Take the coefficient of static friction between the chair and the floor to be 0.420 and, for each of the values of θ, decide if the chair remains at rest or slides.

5E. In Nevada and southern California, stones leave trails in the hard-baked desert floor, as if they had been migrating (Fig. 6-23). For years curiosity mounted about the unseen motion that caused the trails. The answer finally came in the 1970s: when an occasional storm hits the desert, a thin layer of mud may form over a still-firm base, greatly reducing the coefficient of friction between the stones and ground. If a strong wind accompanies the storm, it pushes the stones, leaving trails that are later baked hard by the Sun. Suppose a stone's mass is 300 kg (about the greatest stone mass that has left a trail) and the coefficient of static friction is reduced to 0.15. Of what magnitude is the force from a horizontal gust that is needed to move the stone?

FIGURE 6-23   Exercise 5.

6E. What is the greatest acceleration that can be generated by a runner if the coefficient of static friction between shoes and track is 0.95? (Only one foot is on the track during the acceleration.)

7E. A worker pushes horizontally on a 35 kg crate with a 110 N force. The coefficient of static friction between the crate and the floor is 0.37. (a) What is the frictional force exerted on the crate by the floor? (b) What is the maximum magnitude $f_{s,max}$ of the static frictional force under the circumstances? (c) Does the crate move? (d) Suppose, next, that a second worker pulls directly upward on the crate to help out. What is the least pull she can exert that will allow the first worker's 110 N push to move the crate? (e) If, instead, the second worker pulls horizontally to help out, what is the least pull she can exert to get the crate moving?

8E. A person pushes horizontally with a force of 220 N on a 55 kg crate to move it across a level floor. The coefficient of kinetic friction is 0.35. (a) What is the magnitude of the frictional force? (b) What is the acceleration of the crate?

9E. A trunk with a weight of 220 N rests on the floor. The coefficient of static friction between the trunk and the floor is 0.41, while the coefficient of kinetic friction is 0.32. (a) What is the minimum magnitude for a horizontal force with which a person must push on the trunk to start it moving? (b) Once the trunk is moving, what magnitude of horizontal force must the person apply to keep it moving with constant velocity? (c) If the person continued to push with the force used to start the motion, what would be the acceleration of the trunk?

10E. A filing cabinet with a weight of 556 N rests on the floor. The coefficient of static friction between it and the floor is 0.68, and the coefficient of kinetic friction is 0.56. In four different attempts to move it, it is pushed with horizontal forces of (a) 222 N, (b) 334 N, (c) 445 N, and (d) 556 N. For each attempt, determine whether the cabinet moves, and calculate the magnitude of the frictional force the floor exerts on it. The cabinet is initially at rest for each attempt.

11E. A horizontal force F of 12 N pushes a block weighing 5.0 N against a vertical wall (Fig. 6-24). The coefficient of static

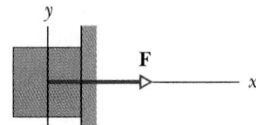

FIGURE 6-24   Exercise 11.

friction between the wall and the block is 0.60, and the coefficient of kinetic friction is 0.40. Assume that the block is not moving initially. (a) Will the block start moving? (b) In unit-vector notation, what is the force exerted on the block by the wall?

12E. A 49 kg rock climber is climbing a "chimney" between two rock slabs as shown in Fig. 6-25. The static coefficient of friction between her shoes and the rock is 1.2; between her back and the rock it is 0.80. She has reduced her push against the rock until her back and her shoes are on the verge of slipping. (a) What is her push against the rock? (b) What fraction of her weight is supported by the frictional force on her shoes?

FIGURE 6-25   Exercise 12.

13E. A house is built on the top of a hill with a nearby 45° slope (Fig. 6-26). An engineering study indicates that the slope angle should be reduced because the top layers of soil along the slope might slip past the lower layers. If the static coefficient of friction between two such layers is 0.5, what is the least angle ϕ through which the present slope should be reduced to prevent slippage?

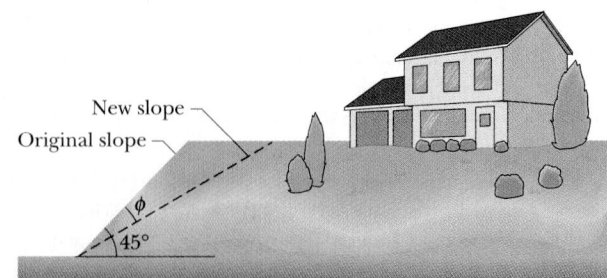

FIGURE 6-26   Exercise 13.

14E. The coefficient of kinetic friction in Fig. 6-27 is 0.20. What is the acceleration of the block if (a) it is sliding down the slope and (b) it has been given an upward shove and is still sliding up the slope?

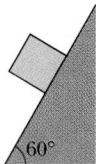

FIGURE 6-27 Exercise 14.

15E. A 110 g hockey puck slides on the ice for 15 m before it stops. (a) If its initial speed was 6.0 m/s, what was the magnitude of the frictional force on the puck during the sliding? (b) What was the coefficient of friction between the puck and the ice?

16P. A student, crazed with final exams, uses a force **P** of magnitude 80 N to push a 5.0 kg block across the ceiling of his room, as shown in Fig. 6-28. If the coefficient of kinetic friction between the block and surface is 0.40, what is the magnitude of the acceleration of the block?

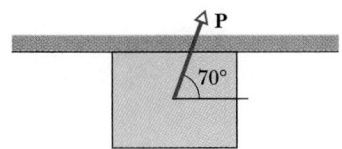

FIGURE 6-28 Problem 16.

17P. A student wants to determine the coefficients of static friction and kinetic friction between a box and a plank. She places the box on the plank and gradually raises one end of the plank. When the angle of inclination with the horizontal reaches 30°, the box starts to slip, and it slides 2.5 m down the plank in 4.0 s. What are the coefficients of friction?

18P. A worker wishes to pile a cone of sand onto a circular area in his yard. The radius of the circle is R, and no sand is to spill onto the surrounding area (Fig. 6-29). If μ_s is the static coefficient of friction between each layer of sand along the slope and the sand beneath it (along which it might slip), show that the greatest volume of sand that can be stored in this manner is $\pi \mu_s R^3 / 3$. (The volume of a cone is $Ah/3$, where A is the base area and h is the cone's height.)

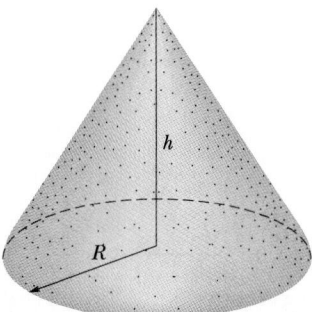

FIGURE 6-29 Problem 18.

19P. A ski that is placed on snow will stick to the snow. However, when the ski is moved along the snow, the rubbing warms and partially melts the snow, reducing the coefficient of friction and promoting sliding. Waxing the ski makes it water repellent

and reduces friction with the resulting layer of water. A magazine reports that a new type of plastic ski is especially water repellent and that, on a gentle 200 m slope in the Alps, a skier reduced his top-to-bottom time from 61 s with standard skis to 42 s with the new skis. (a) Determine the magnitudes of his average acceleration with each pair of skis. (b) Assuming a 3.0° slope, compute the coefficient of kinetic friction for each case.

20P. An 11 kg block of steel is at rest on a horizontal table. The coefficient of static friction between block and table is 0.52. (a) What is the magnitude of the horizontal force that will just start the block moving? (b) What is the magnitude of a force acting upward 60° from the horizontal that will just start the block moving? (c) If the force acts down at 60° from the horizontal, how large can its magnitude be without causing the block to move?

21P. A railroad flatcar is loaded with crates having a coefficient of static friction of 0.25 with the floor. If the train is moving at 48 km/h, in how short a distance can the train be stopped at constant deceleration without causing the crates to slide?

22P. A block slides down an inclined plane of slope angle θ with constant velocity. It is then projected up the same plane with an initial speed v_0. (a) How far up the incline will it move before coming to rest? (b) Will it slide down again? Give an argument to back your answer.

23P. A 68 kg crate is dragged across a floor by pulling on a rope inclined 15° above the horizontal. (a) If the coefficient of static friction is 0.50, what minimum tension in the rope is required to start the crate moving? (b) If $\mu_k = 0.35$, what is the magnitude of the initial acceleration of the crate?

24P. A pig slides down a 35° incline (Fig. 6-30) in twice the time it would take to slide down a frictionless 35° incline. What is the coefficient of kinetic friction between the pig and the incline?

FIGURE 6-30 Problem 24.

25P. In Fig. 6-31, A and B are blocks with weights of 44 N and 22 N, respectively. (a) Determine the minimum weight (block C) that must be placed on A to keep it from sliding, if μ_s between A

and the table is 0.20. (b) Block *C* suddenly is lifted off *A*. What is the acceleration of block *A*, if μ_k between *A* and the table is 0.15?

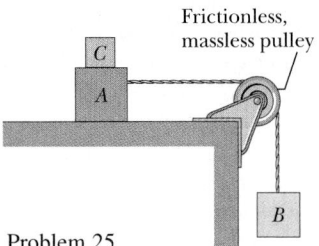

FIGURE 6-31 Problem 25.

26P. A 3.5 kg block is pushed along a horizontal floor by a force $F = 15$ N that makes an angle $\theta = 40°$ with the horizontal (Fig. 6-32). The coefficient of kinetic friction between the block and floor is 0.25. Calculate (a) the magnitude of the frictional force exerted on the block and (b) the acceleration of the block.

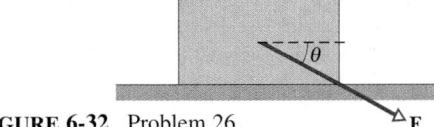

FIGURE 6-32 Problem 26.

27P. In Fig. 6-33 a fastidious worker pushes directly along the handle of a mop with a force **F**. The handle is at an angle θ with the vertical, and μ_s and μ_k are the coefficients of static and kinetic friction between the head of the mop and the floor. Ignore the mass of the handle and assume that all the mop's mass *m* is in its head. (a) If the mop head moves along the floor with a constant velocity, then what is *F*? (b) Show that if θ is less than a certain value θ_0, then **F** (still directed along the handle) is unable to move the mop head. Find θ_0.

FIGURE 6-33 Problem 27.

28P. A 5.0 kg block on an inclined plane is acted on by a horizontal force **F** with magnitude 50 N (Fig. 6-34). The coefficient of kinetic friction between block and plane is 0.30. The coefficient of static friction is not given (but you might still know something about it). (a) What is the acceleration of the block if it is moving up the plane? (b) With the horizontal force still acting, how far up the plane will the block go if it has an initial upward speed of 4.0 m/s? (c) What happens to the block after it reaches the highest point? Give an argument to back your answer.

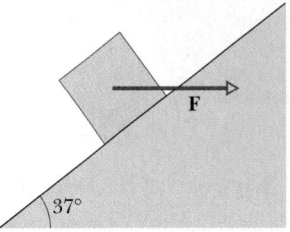

FIGURE 6-34 Problem 28.

29P. Figure 6-35 shows the cross section of a road cut into the side of a mountain. The solid line *AA'* represents a weak bedding plane along which sliding is possible. Block *B* directly above the highway is separated from uphill rock by a large crack (called a *joint*), so that only the force of friction between the block and the bedding plane prevents sliding. The mass of the block is 1.8×10^7 kg, the *dip angle* θ of the failure plane is 24°, and the coefficient of static friction between block and plane is 0.63. (a) Show that the block will not slide. (b) Water seeps into the joint and expands upon freezing, exerting on the block a force **F** parallel to *AA'*. What minimum value of *F* will trigger a slide?

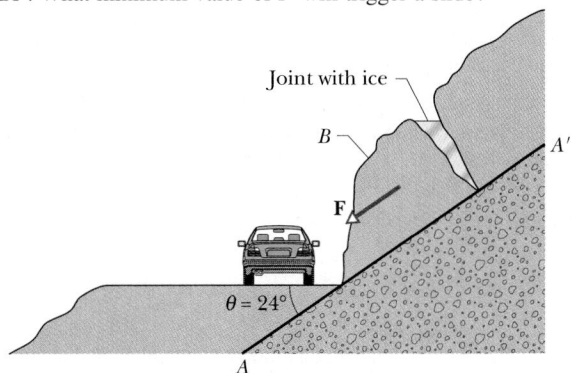

FIGURE 6-35 Problem 29.

30P. A block weighing 80 N rests on a plane inclined at 20° to the horizontal (Fig. 6-36). The coefficient of static friction is 0.25, and the coefficient of kinetic friction is 0.15. (a) What is the minimum magnitude of the force **F**, parallel to the plane, that will prevent the block from slipping down the plane? (b) What is the minimum magnitude *F* that will start the block moving up the plane? (c) What value of *F* is required to move the block up the plane at constant velocity?

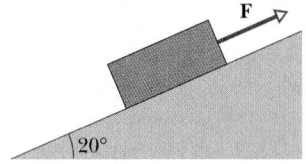

FIGURE 6-36 Problem 30.

31P. Block *B* in Fig. 6-37 weighs 711 N. The coefficient of static friction between block and horizontal surface is 0.25. Find the maximum weight of block *A* for which the system will be stationary.

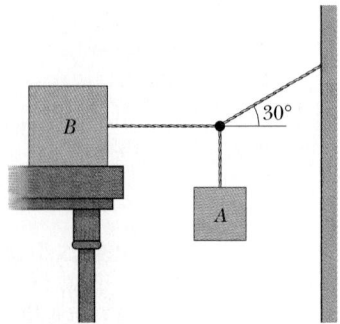

FIGURE 6-37 Problem 31.

32P. Body *A* in Fig. 6-38 weighs 102 N, and body *B* weighs 32 N. The coefficients of friction between *A* and the incline are $\mu_s = 0.56$ and $\mu_k = 0.25$. Angle θ is 40°. Find the acceleration of the system if (a) *A* is initially at rest, (b) *A* is moving up the incline, and (c) *A* is moving down the incline.

33P. Two blocks are connected over a pulley as shown in Fig. 6-38. The mass of block *A* is 10 kg and the coefficient of kinetic friction is 0.20. Angle θ is 30°. Block *A* slides down the incline at constant speed. What is the mass of block *B*?

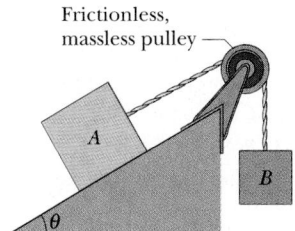

FIGURE 6-38   Problems 32 and 33.

34P. Block m_1 in Fig. 6-39 has a mass of 4.0 kg and m_2 has a mass of 2.0 kg. The coefficient of kinetic friction between m_2 and the horizontal plane is 0.50. The inclined plane is frictionless. Find (a) the tension in the cord and (b) the acceleration of the blocks.

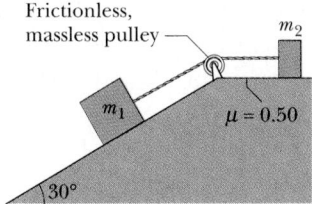

FIGURE 6-39   Problem 34.

35P. Two blocks, of weights 8.0 lb and 16 lb, are connected by a massless string and slide down a 30° inclined plane. The coefficient of kinetic friction between the 8.0 lb block and the plane is 0.10; that between the 16 lb block and the plane is 0.20. Assuming that the 8.0 lb block leads, find (a) the acceleration of the blocks and (b) the tension in the string. (c) Describe the motion if the blocks are reversed.

36P. Two masses, $m_1 = 1.65$ kg and $m_2 = 3.30$ kg, attached by a massless rod parallel to the inclined plane on which both slide (Fig. 6-40), travel down along the plane with m_1 trailing m_2. The angle of incline is $\theta = 30°$. The coefficient of kinetic friction between m_1 and the incline is $\mu_1 = 0.226$; that between m_2 and the incline is $\mu_2 = 0.113$. Compute (a) the tension in the rod and (b) the common acceleration of the two masses. (c) How would the answers to (a) and (b) change if m_2 trailed m_1?

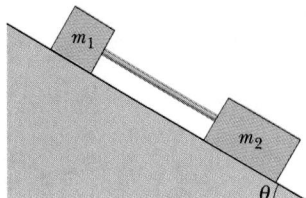

FIGURE 6-40   Problem 36.

37P. A 4.0 kg block is put on top of a 5.0 kg block. To cause the top block to slip on the bottom one, while the bottom one is held fixed, a horizontal force of at least 12 N must be applied to the top block. The assembly of blocks is now placed on a horizontal, frictionless table (Fig. 6-41). Find (a) the magnitude of the maximum horizontal force **F** that can be applied to the lower block so that the blocks will move together and (b) the resulting acceleration of the blocks.

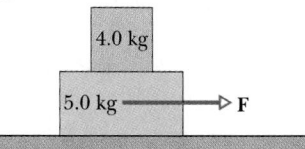

FIGURE 6-41   Problem 37.

38P. The two blocks (with $m = 16$ kg and $M = 88$ kg) shown in Fig. 6-42 are not attached. The coefficient of static friction between the blocks is $\mu_s = 0.38$, but the surface beneath *M* is frictionless. What is the minimum magnitude of the horizontal force **F** required to hold *m* against *M*?

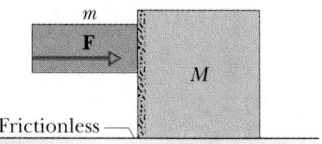

FIGURE 6-42   Problem 38.

39P. A 40 kg slab rests on a frictionless floor. A 10 kg block rests on top of the slab (Fig. 6-43). The coefficient of static friction μ_s between the block and the slab is 0.60, whereas the kinetic coefficient μ_k is 0.40. The 10 kg block is pulled by a horizontal force with a magnitude of 100 N. What are the resulting acceleration magnitudes of (a) the block and (b) the slab?

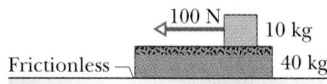

FIGURE 6-43   Problem 39.

40P. A crate slides down an inclined right-angled trough as in Fig. 6-44. The coefficient of kinetic friction between the crate and the trough is μ_k. What is the acceleration of the crate in terms of μ_k, θ, and g?

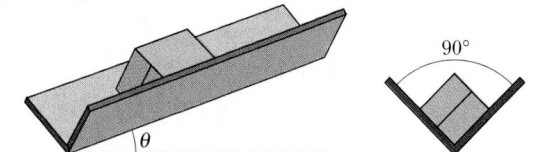

FIGURE 6-44   Problem 40.

41P. A locomotive accelerates a 25-car train along a level track. Every car has a mass of 50 metric tons and is subject to a frictional force $f = 250v$, where the speed v is in meters per second and the force f is in newtons. At the instant when the speed of the train is 30 km/h, the acceleration is 0.20 m/s². (a) What is the tension in the coupling between the first car and the locomotive? (b) If this tension is the maximum force the locomotive can exert on the train, what is the steepest grade up which the locomotive can pull the train at 30 km/h?

42P. An initially stationary box of sand is to be pulled across a floor by means of a cord in which the tension should not exceed 1100 N. The coefficient of static friction between the box and floor is 0.35. (a) What should be the angle between the cord and the horizontal in order to pull the greatest possible amount of sand, and (b) what is the weight of the sand and box in that situation?

43P*. A 1000 kg boat is traveling at 90 km/h when its engine is shut off. The magnitude of the frictional force $\mathbf{f}_k$ between boat and water is proportional to the speed v of the boat: $f_k = 70v$, where v is in meters per second and f_k is in newtons. Find the time required for the boat to slow down to 45 km/h.

SECTION 6-3 The Drag Force and Terminal Speed

44E. Calculate the drag force on a missile 53 cm in diameter cruising with a speed of 250 m/s at low altitude, where the density of air is 1.2 kg/m^3. Assume $C = 0.75$.

45E. The terminal speed of a sky diver in the spread-eagle position is 160 km/h. In the nosedive position, the terminal speed is 310 km/h. Assuming that C does not change from one position to the other, find the ratio of the effective cross-sectional area A in the slower position to that in the faster position.

46E. Calculate the ratio of the drag force on a passenger jet flying with a speed of 1000 km/h at an altitude of 10 km to the drag force on a prop-driven transport flying at half the speed and half the altitude of the jet. At 10 km the density of air is 0.38 kg/m^3 and at 5.0 km it is 0.67 kg/m^3. Assume that the airplanes have the same effective cross-sectional area and the same drag coefficient C.

47P. From the data in Table 6-1, deduce the diameter of the 16 lb shot. Assume that $C = 0.49$ and the density of air is 1.2 kg/m^3.

SECTION 6-4 Uniform Circular Motion

48E. If the coefficient of static friction for tires on a road is 0.25, at what maximum speed can a car round a level curve of 47.5 m radius without slipping?

49E. What is the smallest radius of an unbanked curve around which a bicyclist can travel if her speed is 18 mi/h and the coefficient of static friction between the tires and the road is 0.32?

50E. During an Olympic bobsled run, a European team takes a turn of radius 25 ft at a speed of 60 mi/h. How many g's do the riders experience during the turn?

51E. A car weighing 10.7 kN and traveling at 13.4 m/s attempts to round an unbanked curve with a radius of 61.0 m. (a) What force of friction is required to keep the car on its circular path? (b) If the coefficient of static friction between the tires and road is 0.35, is the attempt at taking the curve successful?

52E. A circular curve of highway is designed for traffic moving at 60 km/h. (a) If the radius of the curve is 150 m, what is the correct angle of banking of the road? (b) If the curve were not banked, what would be the minimum coefficient of friction between tires and road that would keep traffic from skidding at this speed?

53E. A banked circular highway curve is designed for traffic moving at 60 km/h. The radius of the curve is 200 m. Traffic is moving along the highway at 40 km/h on a rainy day. What is the minimum coefficient of friction between tires and road that will allow cars to negotiate the turn without sliding off the road?

54E. A child places a picnic basket on the outer rim of a merry-go-round that has a radius of 4.6 m and revolves once every 30 s. (a) What is the speed of a point on that rim? (b) How large must the coefficient of static friction between the basket and the merry-go-round be for the basket to stay on the ride?

55E. A conical pendulum is formed by attaching a 50 g mass to a 1.2 m string. The mass swings around a horizontal circle of radius 25 cm. (a) What is the speed of the mass? (b) What is the acceleration of the mass? (c) What is the tension in the string?

56E. In the Bohr model of the hydrogen atom, the electron revolves in a circular orbit around the nucleus. If the radius is 5.3×10^{-11} m and the electron circles 6.6×10^{15} times per second, find (a) the speed of the electron, (b) the acceleration (magnitude and direction) of the electron, and (c) the centripetal force acting on the electron. (This force is the result of the attraction between the positively charged nucleus and the negatively charged electron.) The electron's mass is 9.11×10^{-31} kg.

57E. A mass m on a frictionless table is attached to a hanging mass M by a cord through a hole in the table (Fig. 6-45). Find the speed with which m must move in order for M to stay at rest.

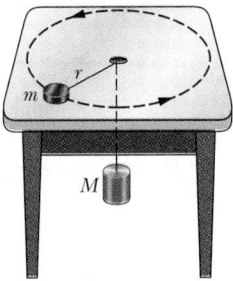

FIGURE 6-45 Exercise 57.

58E. A stuntman drives a car over the top of a hill, the cross section of which can be approximated by a circle of radius 250 m, as in Fig. 6-46. What is the greatest speed at which he can drive without the car leaving the road at the top of the hill?

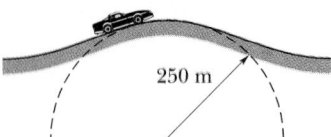

250 m

FIGURE 6-46 Exercise 58.

59P. A small coin is placed on a flat, horizontal turntable. The turntable is observed to make three revolutions in 3.14 s. (a) What is the speed of the coin when it rides without slipping at a distance of 5.0 cm from the center of the turntable? (b) What is the acceleration (magnitude and direction) of the coin? (c) What is the magnitude of the frictional force acting on the coin if the coin has a mass of 2.0 g? (d) What is the coefficient of static friction between the coin and the turntable if the coin is observed to slide off the turntable when it is more than 10 cm from the center of the turntable?

60P. A small object is placed 10 cm from the center of a phonograph turntable. It remains in place when the table rotates at $33\frac{1}{3}$ rev/min but slides off when the table rotates at 45 rev/min. Between what limits must the coefficient of static friction between the object and the surface of the turntable lie?

61P. A bicyclist travels in a circle of radius 25.0 m at a constant speed of 9.00 m/s. The combined mass of the bicycle and rider is 85.0 kg. Calculate the magnitudes of (a) the force of friction exerted by the road on the bicycle and (b) the total force exerted by the road on the bicycle.

62P. A 150 lb student on a steadily rotating Ferris wheel has an apparent weight of 125 lb at the highest point. (a) What is the student's apparent weight at the lowest point? (b) What is the student's apparent weight at the highest point if the wheel's speed is doubled?

63P. A car is rounding a flat curve of radius $R = 220$ m at the curve's maximum design speed $v = 94.0$ km/h. What *total* force does a passenger with mass $m = 85.0$ kg exert on the seat cushion?

64P. A stone tied to the end of a string is whirled around in a vertical circle of radius R. Find the critical speed below which the string would become slack at the highest point.

65P. A certain string can withstand a maximum tension of 9.0 lb without breaking. A child ties a 0.82 lb stone to one end and, holding the other end, whirls the stone in a vertical circle of radius 3.0 ft, slowly increasing the speed until the string breaks. (a) Where is the stone on its path when the string breaks? (b) What is the speed of the stone as the string breaks?

66P. An airplane is flying in a horizontal circle at a speed of 480 km/h. If the wings of the plane are tilted 40° to the horizontal, what is the radius of the circle in which the plane is flying? (See Fig. 6-47.) Assume that the required force is provided entirely by an "aerodynamic lift" that is perpendicular to the wing surface.

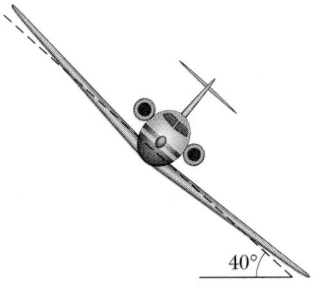

40°

FIGURE 6-47 Problem 66.

67P. A frigate bird is soaring in a horizontal circular path. Its bank angle (relative to the horizontal) is estimated to be 25° and the bird takes 13 s to complete one circle. (a) How fast is the bird flying? (b) What is the radius of the circle?

68P. A model airplane of mass 0.75 kg is flying at constant speed in a horizontal circle at one end of a 30 m cord and at a height of 18 m. The other end of the cord is tethered to the ground. The airplane circles 4.4 times per minute and has its wings horizontal so that the air is pushing vertically upward. (a) What is the acceleration of the plane? (b) What is the tension in the cord? (c) What is the total upward force (lift) on the plane's wings?

69P. An old streetcar rounds a corner on unbanked tracks. If the radius of the tracks is 30 ft and the car's speed is 10 mi/h, what angle with the vertical will be made by the loosely hanging hand straps?

70P. As shown in Fig. 6-48, a 1.34 kg ball is connected by means of two massless strings to a vertical, rotating rod. The strings are tied to the rod, are taut, and form two sides of an equilateral triangle. The tension in the upper string is 35 N. (a) Draw the free-body diagram for the ball. (b) What is the tension in the lower string? (c) What is the net force on the ball at the instant shown in Fig. 6-48? (d) What is the speed of the ball?

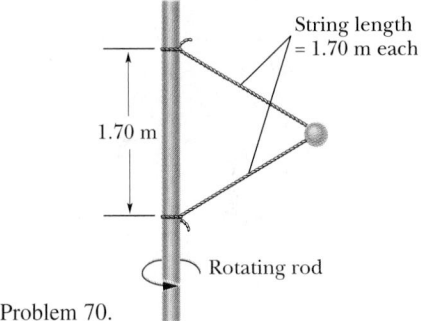

String length = 1.70 m each

1.70 m

Rotating rod

FIGURE 6-48 Problem 70.

71P. Assume that the standard kilogram mass would weigh exactly 9.80 N at sea level on the Earth's equator if the Earth did not rotate. Then take into account the fact that the Earth does rotate, so that this mass moves in a circle of radius 6.40×10^6 m (the Earth's radius) at a constant speed of 465 m/s. (a) Determine the centripetal force needed to keep the standard mass moving in its circular path. (b) Determine the force exerted by the standard mass on a spring balance from which it is suspended at the equator (that force is its "apparent weight").

72P. Suppose that the space station in Question 12 has a radius of 500 m. (a) If a crew member weighs 600 N on Earth, what must be the speed v_s of the outer wall of the station if that crew member is to have an apparent weight of 300 N when standing near the outer wall (as in Fig. 6-21)? (b) What is the apparent weight if the crew member sprints along the outer wall at 10 m/s (relative to the outer wall) in the same direction as $\mathbf{v}_s$?

In the weight-lifting competition of the 1976 Olympics, Vasili Alexeev astounded the world by lifting a record-breaking 562 lb (2500 N) from the floor to over his head (about 2 m). In 1957 Paul Anderson stooped beneath a reinforced wood platform, placed his hands on a short stool to brace himself, and then pushed upward on the platform with his back, lifting the platform and its load about a centimeter. On the platform were auto parts and a safe filled with lead; the composite weight of the load was 6270 lb (27,900 N)! Who, Alexeev or Anderson, did more work on the objects they lifted?

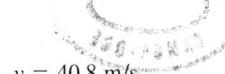

7-1 KINETIC ENERGY

We begin this chapter with a definition: **energy** is a scalar quantity that is associated with a state of one or more objects. The term *state* here has its common meaning: it is the condition of an object. In this chapter we shall focus on one form of energy, **kinetic energy** K, which is associated with the *state of motion* of an object. The faster the object moves, the greater is its kinetic energy. And when the object is stationary, its kinetic energy is zero.

For an object of mass m and whose speed v is well below the speed of light, we define kinetic energy as

$$K = \tfrac{1}{2}mv^2 \qquad \text{(kinetic energy).} \qquad (7\text{-}1)$$

Kinetic energy can never be negative because m and v^2 can never be negative.

The SI unit of kinetic energy (and every other type of energy) is the **joule** (J), named for James Prescott Joule, an English scientist of the 1800s. It is derived directly from the units for mass and velocity:

$$1 \text{ joule} = 1 \text{ J} = 1 \text{ kg} \cdot \text{m}^2/\text{s}^2. \qquad (7\text{-}2)$$

A convenient unit of energy for dealing with atoms or with subatomic particles is the **electron-volt** (eV):

$$1 \text{ electron-volt} = 1 \text{ eV} = 1.60 \times 10^{-19} \text{ J}. \qquad (7\text{-}3)$$

Three common multiples of this unit are the kiloelectron-volt ($1 \text{ keV} = 10^3 \text{ eV}$), the megaelectron-volt ($1 \text{ MeV} = 10^6 \text{ eV}$), and the gigaelectron-volt ($1 \text{ GeV} = 10^9 \text{ eV}$).

SAMPLE PROBLEM 7-1

In 1896 in Waco, Texas, William Crush of the ''Katy'' railroad parked two locomotives at opposite ends of a 6.4 km track, fired them up, tied their throttles open, and then allowed them to crash head-on at full speed (Fig. 7-1), in front of 30,000 spectators. Hundreds of people were hurt by flying debris; several were killed. Assuming the weight of each locomotive was 1.2×10^6 N and its acceleration prior to the collision was a constant 0.26 m/s^2, what was the total kinetic energy of the two locomotives just before the collision?

SOLUTION: To find the kinetic energy of each locomotive, we need its mass and its speed just before the collision. To find the speed v, we use Eq. 2-16,

$$v^2 = v_0^2 + 2a(x - x_0).$$

With $v_0 = 0$ and $x - x_0 = 3.2 \times 10^3$ m (half the initial separation), this yields

$$v^2 = 0 + 2(0.26 \text{ m/s}^2)(3.2 \times 10^3 \text{ m}),$$

or

$$v = 40.8 \text{ m/s}$$

(about 90 mi/h). To find the mass m of each locomotive, we divide its weight by g:

$$m = \frac{1.2 \times 10^6 \text{ N}}{9.8 \text{ m/s}^2} = 1.22 \times 10^5 \text{ kg}.$$

Now, using Eq. 7-1, we find the total kinetic energy of both locomotives just before the collision as:

$$K = 2(\tfrac{1}{2}mv^2) = (1.22 \times 10^5 \text{ kg})(40.8 \text{ m/s})^2$$
$$= 2.0 \times 10^8 \text{ J}. \qquad \text{(Answer)}$$

This is equivalent to a detonation of about 100 lb of TNT.

FIGURE 7-1 Sample Problem 7-1. The aftermath of an 1896 crash of two locomotives.

7-2 WORK

The energy of an object changes if an exchange of energy occurs between the object and its environment. Such a transfer can occur due to a force or due to an exchange of heat. We discuss the exchange of heat in Chapter 19. Here, we discuss the transfer of energy via a force, a process known as doing **work.**

If you accelerate an object to a greater speed by applying a force to the object, you increase the kinetic energy K ($= \tfrac{1}{2}mv^2$) of the object. Similarly, if you decelerate the object to a lesser speed by applying a force, you decrease the kinetic energy of the object. We account for these changes in kinetic energy by saying that your force has transferred energy *to* the object from yourself or *from* the object to yourself.

In such a transfer of energy via a force, *work* W is said to be *done on the object by the force.* More formally, we define work as follows:

Work W is energy transferred to or from an object by means of a force acting on the object. Energy transferred to the object is positive work, and energy transferred from the object is negative work.

"Work," then, is transferred energy; "doing work" is the act of transferring the energy. Work has the same units as energy and is a scalar quantity.

The term *transfer* can be misleading; it does not mean that anything material flows into or out of the object. That is, the transfer is not like a flow of water. Rather it is like the electronic transfer of money between two bank accounts: the number in one account goes up while the number in the other account goes down, with nothing material passing between the two accounts.

Note that we are not concerned here with the common meaning of the word "work," which implies that *any* physical or mental labor is work. For example, if you push hard against a wall, you tire owing to the continuously repeated contractions of your muscles that are required and you are, in the common sense, working. But such effort does not cause an energy transfer to or from the wall and thus is not work done on the wall as defined here.

To avoid confusion in this chapter, we shall use the symbol W only for work and shall represent weight with $m\mathbf{g}$ or mg.

7-3 WORK AND KINETIC ENERGY

Let us now relate the work done on an object by a force and the corresponding change in the kinetic energy of the object. If the force changes the speed of the object, it also changes the kinetic energy of the object. If the kinetic energy is the only type of energy of the object being changed by the force, then the change ΔK in kinetic energy is equal to the work W done by the force:

$$\Delta K = K_f - K_i = W \qquad \text{(work–kinetic energy theorem).} \qquad (7\text{-}4)$$

Here K_i is the initial kinetic energy ($= \frac{1}{2}mv_0^2$) and K_f is the kinetic energy ($+ \frac{1}{2}mv^2$) after the work—the energy transfer—is done.

The right-hand equality of Eq. 7-4 can also be written

$$K_f = K_i + W. \qquad (7\text{-}5)$$

Equations 7-4 and 7-5 are equivalent statements of the **work–kinetic energy theorem.**

If the object's energy other than kinetic energy is being changed by the force, then Eqs. 7-4 and 7-5 do not apply. A kinetic frictional force, for example, can change

both the kinetic energy and the *thermal energy* of an object. (Thermal energy is associated with the random motions of atoms and molecules within an object.) Equations 7-4 and 7-5 then do not apply.

Some forces cause a transfer of energy within the object itself, and again Eqs. 7-4 and 7-5 do not apply. For example, suppose you push yourself away from a wall while ice skating; then the push (a force) transfers energy internally from a biological type of energy in your muscles to kinetic energy of your body as a whole. We shall consider kinetic frictional forces in Chapter 8 and internal energy transfers in Chapter 9.

CHECKPOINT 1: A particle moves along an x axis. Does the kinetic energy of the particle increase, decrease, or remain the same if the particle's velocity changes (a) from -3 m/s to -2 m/s and (b) from -2 m/s to 2 m/s? (c) In each situation, is the work done on the particle positive, negative, or zero?

Work Done by a Single Force

We would now like to relate the change ΔK in kinetic energy of an object to the magnitude F of the force causing the change. We start with a particle; the only type of energy this simplest type of object can have is kinetic energy. In Fig. 7-2, a particle moves along an x axis on a horizontal frictionless floor while a constant force $\mathbf{F}$ acts on it at a constant angle ϕ to the particle's path. Because the horizontal force component $F \cos \phi$ gives the particle an acceleration a_x along the path, the force changes the particle's velocity from its initial value $\mathbf{v}_0$. Thus the force also changes the particle's kinetic energy.

Suppose the force acts on the particle through a displacement $\mathbf{d}$, giving it a velocity $\mathbf{v}$ whose magnitude is, by Eq. 2-16,

$$v^2 = v_0^2 + 2a_x d. \qquad (7\text{-}6)$$

Multiplying both sides of Eq. 7-6 by the mass m of the particle and rearranging yield

$$\tfrac{1}{2}mv^2 - \tfrac{1}{2}mv_0^2 = ma_x d. \qquad (7\text{-}7)$$

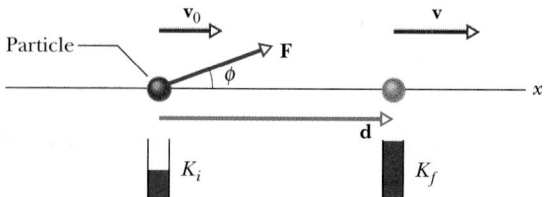

FIGURE 7-2 A constant force $\mathbf{F}$ directed at angle ϕ to the displacement $\mathbf{d}$ of a particle accelerates the particle along that path, changing the velocity of the particle from $\mathbf{v}_0$ to $\mathbf{v}$. A "kinetic energy gauge" indicates the resulting change in the kinetic energy of the particle, from the value K_i to K_f.

The left side of Eq. 7-7 is the difference between the initial kinetic energy K_i $(= \frac{1}{2}mv_0^2)$ of the particle and the final kinetic energy K_f $(= \frac{1}{2}mv^2)$. This difference is the change ΔK in the kinetic energy of the particle due to the force **F**. Making that substitution, and substituting $F \cos \phi$ for the product ma_x according to Newton's second law, we get

$$\Delta K = Fd \cos \phi. \qquad (7\text{-}8)$$

Now by comparing Eq. 7-8 with Eq. 7-4, we see that the right side of Eq. 7-8 gives the work W done on the particle by force **F**. Thus, we may write

$$W = Fd \cos \phi \qquad \text{(work done by a constant force).} \qquad (7\text{-}9)$$

If the angle ϕ is less than 90°, then the work W is positive, which means that energy is transferred *to* the particle and the kinetic energy of the particle *increases*. If ϕ is greater than 90° (up to 180°), then the work W is negative, which means that energy is transferred *from* the particle and the kinetic energy of the particle *decreases*.

From Eq. 7-9 we see that in addition to the joule, another SI unit of work is the newton-meter (N·m). The corresponding unit in the British system is the foot-pound (ft·lb). Thus, we can extend Eq. 7-2 to

$$1 \text{ J} = 1 \text{ kg}\cdot\text{m}^2/\text{s}^2 = 1 \text{ N}\cdot\text{m} = 0.738 \text{ ft}\cdot\text{lb}. \quad (7\text{-}10)$$

The right side of Eq. 7-9 is equivalent to the scalar (or dot) product **F**·**d**. So in vector form Eq. 7-9 is

$$W = \mathbf{F}\cdot\mathbf{d} \qquad \text{(work by a constant force).} \qquad (7\text{-}11)$$

(This is our first application of a scalar product; you may wish to review its discussion in Section 3-7.) Equation 7-11 is especially useful when **F** and **d** are given in unit-vector notation.

As derived, Eqs. 7-9 and 7-11 give the work done by a constant force that changes the kinetic energy of a particle. However, in certain cases we can extend them to objects that are clearly not particles:

If a force acting alone on an object changes only the kinetic energy of the object (and no other energy of the object), then the work done by the force is given by Eqs. 7-9 and 7-11.

Such objects are said to be *particle-like*.

For example, Fig. 7-3 shows a student propelling a bed in an intramural bed race. If his force on the bed is constant, do Eqs. 7-9 and 7-11 give the work done by the force? We can easily see that most but not all of the trans-

FIGURE 7-3 A bed race. We can approximate the bed as being a particle for the purpose of calculating the work done on the bed by the force applied by the student.

ferred energy goes into the kinetic energy of the bed and penguin-rider moving along the street. Some small amount goes into the rotation of the wheels. However, if we choose to neglect that small amount, then Eqs. 7-9 and Eq. 7-11 do give the work done by the force applied by the student, and the bed and rider are particle-like.

CHECKPOINT 2: The figure shows four situations in which a force acts on a box while the box slides rightward a distance d across a frictionless floor. The magnitudes of the forces are identical; their orientations are as shown. Rank the situations according to the work done on the box during the displacement, from most positive to most negative.

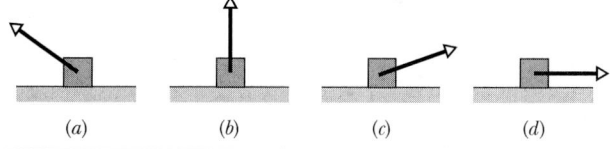

Work Done by Multiple Forces

If several forces act on a particle, we can replace **F** in Eq. 7-11 with the net force $\Sigma\mathbf{F}$, where

$$\sum \mathbf{F} = \mathbf{F}_1 + \mathbf{F}_2 + \mathbf{F}_3 + \cdots, \qquad (7\text{-}12)$$

where $\mathbf{F}_j$ are the individual forces. Then

$$W = \left(\sum \mathbf{F}\right)\cdot\mathbf{d} \qquad (7\text{-}13)$$

is the work done by the net force during a displacement **d** of the particle. With Eq. 7-12, we can rewrite Eq. 7-13 as

$$W = \mathbf{F}_1\cdot\mathbf{d} + \mathbf{F}_2\cdot\mathbf{d} + \mathbf{F}_3\cdot\mathbf{d} + \cdots$$
$$= W_1 + W_2 + W_3 + \cdots \qquad \text{(total work).} \quad (7\text{-}14)$$

This equation tells us that the total work done on the particle is the sum of the work done by all the forces acting on the particle.

We can now rewrite Eq. 7-4 (the work–kinetic energy theorem) as

$$\Delta K = K_f - K_i = W_1 + W_2 + W_3 + \cdots. \quad (7\text{-}15)$$

This tells us that the change ΔK in the kinetic energy of a particle is equal to the total work done by all the forces acting on the particle.

SAMPLE PROBLEM 7-2

Figure 7-4a shows two industrial spies sliding an initially stationary 225 kg floor safe a distance of 8.50 m along a straight line toward their truck. The push $\mathbf{F}_1$ of Spy 001 is 12.0 N, directed at an angle of 30° downward from the horizontal; the pull $\mathbf{F}_2$ of Spy 002 is 10.0 N, directed at 40° above the horizontal. The floor and safe make frictionless contact.

(a) What is the total work done on the safe by forces $\mathbf{F}_1$ and $\mathbf{F}_2$ during the 8.50 m displacement $\mathbf{d}$?

SOLUTION: Figure 7-4b is a free-body diagram for the safe, considered to be a particle. We can find the total work done on the safe by finding the work done by each force and then adding the results. From Eq. 7-9, the work done by $\mathbf{F}_1$ is

$$W_1 = F_1 d \cos \phi_1 = (12.0 \text{ N})(8.50 \text{ m})(\cos 30°)$$
$$= 88.33 \text{ J},$$

and the work done by $\mathbf{F}_2$ is

$$W_2 = F_2 d \cos \phi_2 = (10.0 \text{ N})(8.50 \text{ m})(\cos 40°)$$
$$= 65.11 \text{ J}.$$

So, from Eq. 7-14, the total work W is

$$W = W_1 + W_2 = 88.33 \text{ J} + 65.11 \text{ J}$$
$$= 153.4 \text{ J} \approx 153 \text{ J}. \qquad \text{(Answer)}$$

Thus during the 8.50 m displacement, the spies transfer 153 J of energy to the kinetic energy of the safe.

(b) During the displacement, what is the work W_g done on the safe by its weight mg and what is the work W_N done on the safe by the normal force $\mathbf{N}$ due to the floor?

SOLUTION: Both these forces are perpendicular to the displacement. Thus Eq. 7-9 tells us that

$$W_g = mgd \cos 90° = mgd(0) = 0 \qquad \text{(Answer)}$$

and

$$W_N = Nd \cos 90° = Nd(0) = 0. \qquad \text{(Answer)}$$

These forces do not transfer any energy to or from the safe.

(c) The safe is initially stationary. What is its speed v at the end of the 8.50 m displacement?

SOLUTION: The speed of the safe changes because its kinetic energy is changed when energy is transferred to it by the forces. We relate the speed to the work done by combining Eqs. 7-4 and 7-1:

$$W = K_f - K_i = \tfrac{1}{2}mv^2 - \tfrac{1}{2}mv_0^2.$$

The initial speed v_0 is zero, and we now know that the work done is 153.4 J. Solving for v and then substituting the known data, we find that

$$v = \sqrt{\frac{2W}{m}} = \sqrt{\frac{2(153.4 \text{ J})}{225 \text{ kg}}}$$
$$= 1.17 \text{ m/s}. \qquad \text{(Answer)}$$

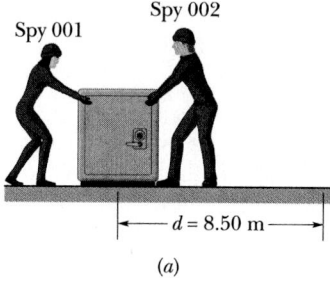

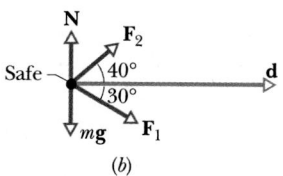

FIGURE 7-4 Sample Problem 7-2. (a) Two spies move a floor safe. (b) A free-body diagram for the safe, with the displacement $\mathbf{d}$ of the safe included.

SAMPLE PROBLEM 7-3

A runaway crate of prunes slides over a floor toward you. To slow the crate you push against it with a force $\mathbf{F} = (2.0 \text{ N})\mathbf{i} + (-6.0 \text{ N})\mathbf{j}$ while running backward (Fig. 7-5). During your pushing, the crate goes through a displacement $\mathbf{d} = (-3.0 \text{ m})\mathbf{i}$.

(a) How much work has your force done on the crate during the displacement?

SOLUTION: From Eq. 7-11, the work is

$$W = \mathbf{F} \cdot \mathbf{d} = [(2.0 \text{ N})\mathbf{i} + (-6.0 \text{ N})\mathbf{j}] \cdot [(-3.0 \text{ m})\mathbf{i}].$$

Of the possible unit-vector dot products, only $\mathbf{i} \cdot \mathbf{i}$, $\mathbf{j} \cdot \mathbf{j}$, and $\mathbf{k} \cdot \mathbf{k}$ are nonzero (see Section 3-7). Here we have

$$W = (2.0)(-3.0 \text{ m})\mathbf{i} \cdot \mathbf{i} + (-6.0 \text{ N})(-3.0 \text{ m})\mathbf{j} \cdot \mathbf{i}$$
$$= (-6.0 \text{ J})(1) + 0 = -6.0 \text{ J}. \qquad \text{(Answer)}$$

Thus the force transfers 6.0 J of energy from the kinetic energy of the crate.

(b) If the crate has a kinetic energy of 10 J at the beginning of the displacement $\mathbf{d}$, what is its kinetic energy at the end of the displacement?

SOLUTION: Using Eq. 7-5 with $K_i = 10$ J and $W = -6$ J, we find

$$K_f = K_i + W = 10 \text{ J} + (-6.0 \text{ J}) = 4.0 \text{ J}. \qquad \text{(Answer)}$$

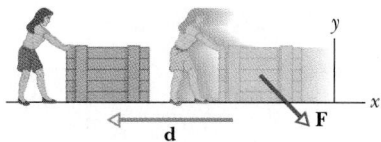

FIGURE 7-5 Sample Problem 7-3. Slowing a runaway prune crate with force **F** applied during displacement **d**.

7-4 WORK DONE BY WEIGHT

We next examine the work done on an object by a particular type of force, namely, its weight. Figure 7-6 shows a particle-like tomato of mass m that is thrown upward with initial speed v_0 and thus with initial kinetic energy $K_i = \frac{1}{2}mv_0^2$. As the tomato rises, it slows because a constant downward force, its weight mg, acts on it.

Because the tomato slows, its kinetic energy decreases. We know from experimental observation that if the weight mg is the only force acting on the tomato (air drag is somehow eliminated), then kinetic energy is the only energy of the tomato that changes. Thus, to find the work W_g done on the tomato by its weight, we substitute mg for F in Eq. 7-9, finding

$$W_g = mgd \cos \phi \quad \begin{array}{l}\text{(work done}\\ \text{by weight).}\end{array} \quad (7\text{-}16)$$

For such a rising object, the weight mg is directed opposite the displacement $\mathbf{d}$, as indicated in Fig. 7-6. Then $\phi = 180°$ and

$$W_g = mgd \cos (180°) = mgd(-1) = -mgd. \quad (7\text{-}17)$$

The minus sign tells us that during the object's rise, the weight of the object transfers energy in the amount mgd from the kinetic energy of the object. This is consistent with the slowing of the object as it rises.

After the object has reached its maximum height and

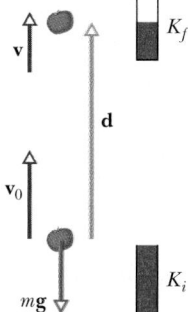

FIGURE 7-6 A particle-like tomato of mass m thrown upward slows from velocity $\mathbf{v}_0$ to velocity $\mathbf{v}$ during displacement $\mathbf{d}$ because its weight mg acts on it. A kinetic energy gauge indicates the resulting change in the kinetic energy of the object, from K_i ($= \frac{1}{2}mv_0^2$) to K_f ($= \frac{1}{2}mv^2$).

is falling back down, the angle ϕ between the weight mg and the displacement $\mathbf{d}$ is zero. Thus,

$$W_g = mgd \cos (0°) = mgd(+1) = +mgd. \quad (7\text{-}18)$$

The plus sign tells us that the weight now transfers energy in the amount mgd to the kinetic energy of the object. This is consistent with the speeding up of the object as it falls. (Actually, as we shall see in Chapter 8, energy transfers associated with lifting and lowering an object involve not just the object, but the full object-Earth system. Without the Earth, of course, "lifting" would be meaningless.)

Work Done in Lifting and Lowering an Object

Now suppose we lift a particle-like object by applying a force **F** to it. During the upward displacement, our applied force does positive work W_a on the object while the object's weight also does negative work W_g on it. That is, our force tends to transfer energy to the object while its weight tends to transfer energy from it. By Eq. 7-15, the change ΔK in the kinetic energy of the object due to these two energy transfers is

$$\Delta K = K_f - K_i = W_a + W_g, \quad (7\text{-}19)$$

in which K_f is the kinetic energy at the end of the displacement and K_i is that at the start of the displacement. This equation also applies if we lower the object; but then the weight tends to transfer energy *to* the object while our force tends to transfer energy *from* it.

In one common situation the object is stationary before and after the lift — for example, when you lift a book from the floor to a shelf. Then K_f and K_i are both zero, and Eq. 7-19 reduces to

$$W_a + W_g = 0$$

or

$$W_a = -W_g. \quad (7\text{-}20)$$

Note that we get the same result if K_f and K_i are not zero but are still equal. Either way, the result means that the work done by the applied force is the negative of the work done by the weight. That is, the applied force transfers the same amount of energy to the object as its weight transfers from the object. Using Eq. 7-16, we can rewrite Eq. 7-20 as

$$W_a = -mgd \cos \phi \quad \begin{array}{l}\text{(work in lifting and}\\ \text{lowering; } K_f = K_i),\end{array} \quad (7\text{-}21)$$

with ϕ being the angle between mg and $\mathbf{d}$. If the displacement is vertically upward (Fig. 7-7a), then $\phi = 180°$ and the work done by our force equals mgd. If the displacement is vertically downward (Fig. 7-7b), then $\phi = 0°$ and the work done by the applied force equals $-mgd$.

Equations 7-20 and 7-21 apply to any situation in

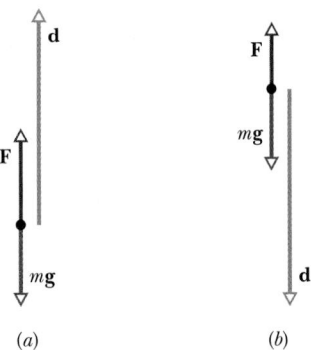

(a)   (b)

FIGURE 7-7 (*a*) An applied force **F** lifts an object. The displacement **d** of the object makes an angle $\phi = 180°$ with the weight vector *mg* of the object. The applied force does positive work on the object. (*b*) An applied force **F** lowers an object. The displacement **d** of the object makes an angle $\phi = 0°$ with weight vector *mg*. The applied force does negative work on the object.

which an object is lifted or lowered, with the object stationary before and after the lift. They are independent of the magnitude of the force used. For example, when Alexeev lifted the record-breaking weight of 2500 N, his force on the weights varied considerably during the lift. Still, because the weights were stationary before and after the lift, the work he did is given by Eqs. 7-20 and 7-21, where, in Eq. 7-21, *mg* is the weight he lifted and *d* is the distance he lifted that weight.

SAMPLE PROBLEM 7-4

Let us return to the weight-lifting feats of Vasili Alexeev and Paul Anderson.

(a) When Alexeev lifted a weight of 2500 N a distance of 2.0 m, how much work was done on the weights by their weight *mg*?

SOLUTION: The magnitude of the weight vector *mg* is *mg*. The angle ϕ between that vector and the displacement vector **d** is 180°. From Eq. 7-16, the work done by *mg* is

$$W_g = mgd \cos \phi = (2500 \text{ N})(2.0 \text{ m})(\cos 180°)$$
$$= -5000 \text{ J}. \qquad \text{(Answer)}$$

(b) How much work was done by Alexeev's force during the lift?

SOLUTION: Because the weights were stationary at the start and end of the lift, we can use Eq. 7-20, finding

$$W_{VA} = -W_g = +5000 \text{ J}. \qquad \text{(Answer)}$$

(c) While Alexeev held the weights stationary above his head, how much work was done by his force on the weights?

SOLUTION: When he supported the weights, they were stationary. Thus their displacement $d = 0$ and, by Eq. 7-9, the

Using a harness, Paul Anderson lifts 30 people having a combined weight of about 2400 lb.

work done on the weights was zero (even though supporting the weights was a very tiring task).

(d) How much work was done by the force Paul Anderson applied to lift a weight of 27,900 N a distance of 1.0 cm?

SOLUTION: From Eq. 7-21, with $mg = 27,900$ N and $d = 1.0$ cm, we find

$$W_{PA} = -mgd \cos \phi = -mgd \cos 180°$$
$$= -(27,900 \text{ N})(0.010 \text{ m})(-1) = 280 \text{ J}. \qquad \text{(Answer)}$$

Anderson's lift required a tremendous upward force but only a small energy transfer of 280 J, owing to the short displacement involved.

SAMPLE PROBLEM 7-5

An initially stationary 15.0 kg crate of cheese is pulled, via a cable, a distance $L = 5.70$ m up a frictionless ramp, to a height *h* of 2.50 m, where it stops (Fig. 7-8*a*).

(a) How much work is done on the crate by its weight *mg* during the lift?

SOLUTION: We calculate this work with Eq. 7-16, using *L* for the magnitude of the displacement. The angle between *mg* and the displacement is $\theta + 90°$ (see the free-body diagram of Fig. 7-8*b*). We have

$$W_g = mg \, L \cos (\theta + 90°) = -mgL \sin \theta.$$

From Fig. 7-8*a*, we see that $L \sin \theta$ is the height *h* moved by the crate. So we have

$$W_g = -mgh. \qquad (7\text{-}22)$$

This means that the work done by the weight depends on the

vertical displacement of the crate and not on the horizontal displacement. (We shall return to this point in Chapter 8.) Inserting the given data in Eq. 7-22 yields

$$W_g = -(15.0 \text{ kg})(9.8 \text{ m/s}^2)(2.50 \text{ m})$$
$$= -368 \text{ J}. \quad \text{(Answer)}$$

(b) How much work is done on the crate by the force **T** applied by the cable, which pulls the crate up the ramp?

SOLUTION: Because the crate is stationary before and after the lift, the change ΔK in its kinetic energy must be zero. Then from Eq. 7-15,

$$\Delta K = W_1 + W_2 + W_3 + \cdots, \quad (7\text{-}23)$$

we know that the total work done by all the forces acting on it must be zero. Besides the weight of the crate, there are only two other forces acting on the crate: the normal force **N** due to the ramp and the force **T** from the cable. Because **N** is perpendicular to the displacement of the crate along the ramp, the normal force does zero work on the crate. So, with W_T representing the work done by the force **T**, Eq. 7-23 becomes

$$0 = W_g + W_T.$$

Substituting -368 J for W_g, we find

$$W_T = 368 \text{ J}. \quad \text{(Answer)}$$

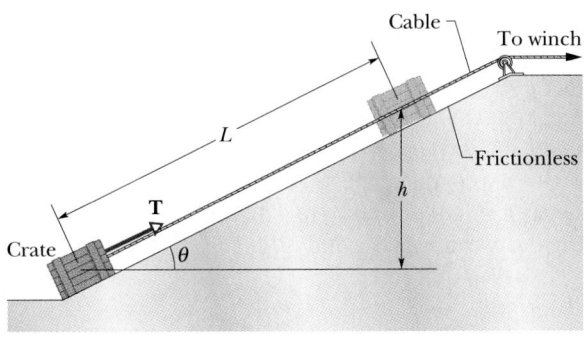

(a)

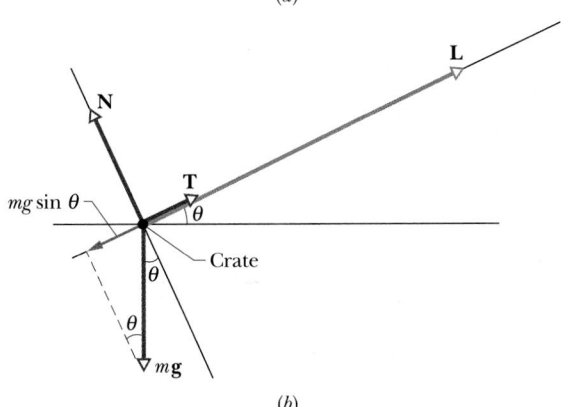

(b)

FIGURE 7-8 Sample Problem 7-5. (a) A crate is pulled up a frictionless ramp by a force parallel to the ramp. (b) A free-body diagram for the crate, showing all the forces that act on it. Its displacement **L** is also shown.

CHECKPOINT **3:** Suppose we raise the crate in Sample Problem 7-5 by the same height h but with a longer ramp. (a) Is the work done by force **T** now greater, smaller, or the same as before? (b) Is the magnitude of **T** needed to move the crate now greater, smaller, or the same as before?

SAMPLE PROBLEM 7-6

A 500 kg elevator cab is descending with speed $v_i = 4.0$ m/s when the cable that controls it begins to slip, allowing it to fall with constant acceleration $\mathbf{a} = \mathbf{g}/5$ (Fig. 7-9a).

(a) During its fall through a distance $d = 12$ m, what is the work W_1 done on the cab by its weight $m\mathbf{g}$?

SOLUTION: The cab's free-body diagram during the 12-m fall is shown in Fig. 7-9b. Note that the angle between the cab's displacement **d** and its weight $m\mathbf{g}$ is 0°. With Eq. 7-16 we find

$$W_1 = mgd \cos 0° = (500 \text{ kg})(9.8 \text{ m/s}^2)(12 \text{ m})(1)$$
$$= 5.88 \times 10^4 \text{ J} \approx 5.9 \times 10^4 \text{ J}. \quad \text{(Answer)}$$

(b) During the 12 m fall, what is the work W_2 done on the cab by the upward pull **T** exerted by the elevator cable?

SOLUTION: The situation here differs from those in Sample Problems 7-4 and 7-5, because the kinetic energy of the cab at the beginning of the displacement does not equal that at the end of the displacement. So Eqs. 7-20 and 7-21 do not apply; the work done by force **T** on the cab *is not* the negative of the

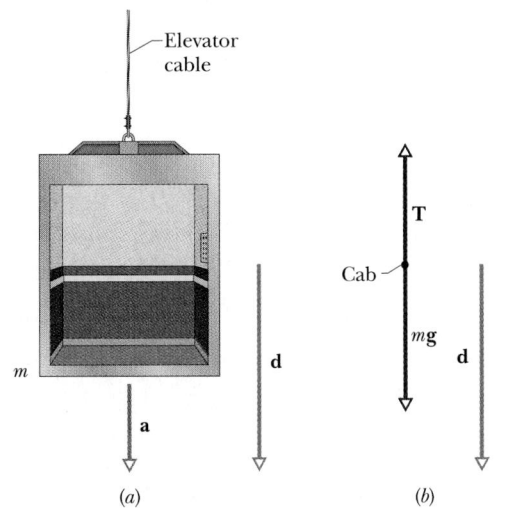

FIGURE 7-9 Sample Problem 7-6. An elevator cab, descending with speed v_i, suddenly begins to accelerate downward. (a) It moves through a displacement **d** with constant acceleration $\mathbf{a} = \mathbf{g}/5$. (b) A free-body diagram for the cab, with the displacement included.

work done by the weight of the cab. We must calculate the work W_2 done by **T** with Eq. 7-9 ($W = Fd \cos \phi$). To do so, we first must find the magnitude of **T**. Applying Newton's second law to the cab yields

$$\sum F = T - mg = ma.$$

Since **a** has magnitude $g/5$ and is directed downward, we get

$$T = m(g + a) = m(g - g/5)$$
$$= (500 \text{ kg})(\tfrac{4}{5})(9.8 \text{ m/s}^2) = 3920 \text{ N}.$$

We can now use Eq. 7-9 to find the work done by **T**. The angle between **T** and the cab's displacement **d** is 180°. The magnitude of **T** is 3920 N. And the magnitude of the displacement is 12 m. So

$$W_2 = Td \cos 180° = (3920 \text{ N})(12 \text{ m})(-1)$$
$$= -4.70 \times 10^4 \text{ J}. \qquad \text{(Answer)}$$

(c) What is the total work W done on the cab in the 12 m fall?

SOLUTION: The total work is the algebraic sum of the work done by the two forces (by Eq. 7-14):

$$W = W_1 + W_2 = 5.88 \times 10^4 \text{ J} - 4.70 \times 10^4 \text{ J}$$
$$= 1.18 \times 10^4 \text{ J} \approx 1.2 \times 10^4 \text{ J}. \qquad \text{(Answer)}$$

Thus, during the 12 m fall, a net energy of 1.2×10^4 J is transferred to the cab.

We can also find W a different way. We first find the net force on the cab, using Newton's second law:

$$\sum F = ma = (500 \text{ kg}) \left(-\frac{9.8 \text{ m/s}^2}{5} \right) = -980 \text{ N}.$$

We next find the work done on the cab by that net force, which acts downward and thus at an angle of 0° with **d**:

$$W = (980 \text{ N})(12 \text{ m}) \cos 0°$$
$$= 1.18 \times 10^4 \text{ J} \approx 1.2 \times 10^4 \text{ J}. \qquad \text{(Answer)}$$

(d) What is the cab's kinetic energy at the end of the 12 m fall?

SOLUTION: The kinetic energy K_i at the start of the fall, when the speed is $v_i = 4.0$ m/s, is

$$K_i = \tfrac{1}{2}mv_i^2 = \tfrac{1}{2}(500 \text{ kg})(4.0 \text{ m/s})^2 = 4000 \text{ J}.$$

The kinetic energy K_f at the end of the fall is given by Eq. 7-5,

$$K_f = K_i + W = 4000 \text{ J} + 1.18 \times 10^4 \text{ J}$$
$$= 1.58 \times 10^4 \text{ J} \approx 1.6 \times 10^4 \text{ J}. \qquad \text{(Answer)}$$

(e) What is the speed v_f of the cab at the end of the 12 m fall?

SOLUTION: From Eq. 7-1, we have

$$K_f = \tfrac{1}{2}mv_f^2,$$

which we solve for v_f:

$$v_f = \sqrt{\frac{2K_f}{m}} = \sqrt{\frac{(2)(1.58 \times 10^4 \text{ J})}{500 \text{ kg}}}$$
$$= 7.9 \text{ m/s}. \qquad \text{(Answer)}$$

7-5 WORK DONE BY A VARIABLE FORCE

One-Dimensional Analysis

Let us return to the situation of Fig. 7-2 but now consider the force to be directed along the x axis and the force magnitude to vary with position x. Thus, as the particle moves, the magnitude of the force doing work on it changes. Only the magnitude of this **variable force** changes, not its direction. Moreover, its magnitude changes with the position of the particle, but not over time.

Figure 7-10a shows a plot of such a one-dimensional variable force. How do we find the work done on the particle by this force as the particle moves from an initial point x_i to a final point x_f? We cannot use Eq. 7-9, because it applies only for a constant force **F**. To develop a new approach, let us divide the total displacement of the particle into a number of intervals of width Δx. We choose Δx small enough to permit us to take the force $F(x)$ as being reasonably constant over that interval. We let $\overline{F_j(x)}$ be the average value of $F(x)$ within the jth interval.

The increment (small amount) of work ΔW_j done by the force in the jth interval is now given by Eq. 7-9 and is

$$\Delta W_j = \overline{F_j(x)} \, \Delta x. \qquad (7\text{-}24)$$

On the graph of Fig. 7-10b, $\overline{F_j(x)}$ is the height of the jth strip, and Δx is its width; ΔW_j is then equal in magnitude to the area of the strip.

To approximate the total work W done by the force as the particle moves from x_i to x_f, we add the areas of all the strips between x_i and x_f in Fig. 7-10b. That is,

$$W = \sum \Delta W_j = \sum \overline{F_j(x)} \, \Delta x. \qquad (7\text{-}25)$$

Equation 7-25 is an approximation because the broken "skyline" formed by the tops of the rectangular strips in Fig. 7-10b only approximates the actual curve.

We can make the approximation better by reducing the strip width Δx and using more strips, as in Fig. 7-10c. In the limit, we let the strip width approach zero; the number of strips then becomes infinitely large and we have, as an exact result,

$$W = \lim_{\Delta x \to 0} \sum \overline{F_j(x)} \, \Delta x. \qquad (7\text{-}26)$$

This limit is exactly what we mean by the integral of the function $F(x)$ between the limits x_i and x_f. Thus Eq. 7-26 becomes

$$W = \int_{x_i}^{x_f} F(x) \, dx \qquad \text{(work: variable force).} \qquad (7\text{-}27)$$

If we know the function $F(x)$, we can substitute it into

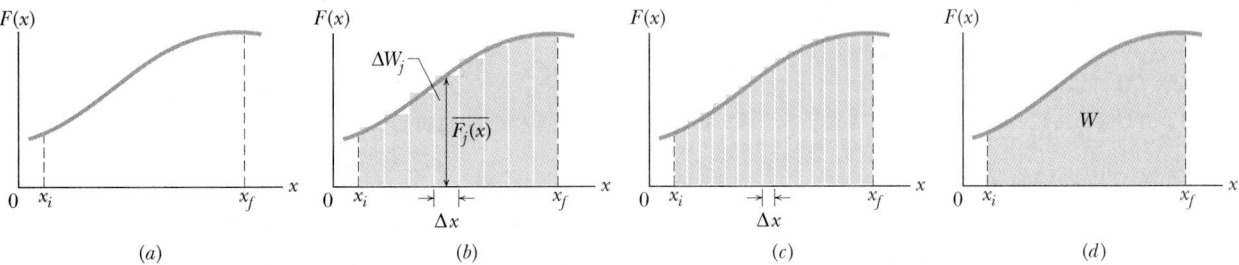

FIGURE 7-10 (*a*) A generalized one-dimensional force plotted against the displacement x of a particle on which it acts. The particle moves from x_i to x_f. (*b*) Same as (*a*) but with the area under the curve divided into narrow strips. (*c*) Same as (*b*) but with the area divided into narrower strips. (*d*) The limiting case. The work done by the force is given by Eq. 7-27 and is represented geometrically by the shaded area between the curve and the x axis and between x_i and x_f.

Eq. 7-27, introduce the proper limits of integration, carry out the integration, and thus find the work. (Appendix E contains a list of common integrals.) Geometrically, the work is equal to the area between the $F(x)$ curve and the x axis, between the limits x_i and x_f (shaded in Fig. 7-10*d*).

Three-Dimensional Analysis

Consider now a particle that is acted on by a three-dimensional force

$$\mathbf{F} = F_x\mathbf{i} + F_y\mathbf{j} + F_z\mathbf{k}, \qquad (7\text{-}28)$$

and let some or all of the components F_x, F_y, and F_z depend on the position of the particle, which means that they are functions of that position. Furthermore, let the particle move through an incremental displacement

$$d\mathbf{r} = dx\mathbf{i} + dy\mathbf{j} + dz\mathbf{k}. \qquad (7\text{-}29)$$

The increment of work dW done on the particle by $\mathbf{F}$ during the displacement $d\mathbf{r}$ is, by Eq. 7-11,

$$dW = \mathbf{F} \cdot d\mathbf{r} = F_x\,dx + F_y\,dy + F_z\,dz. \qquad (7\text{-}30)$$

The work W done by $\mathbf{F}$ while the particle moves from an initial position r_i with coordinates (x_i, y_i, z_i) to a final position r_f with coordinates (x_f, y_f, z_f) is then

$$W = \int_{r_i}^{r_f} dW = \int_{x_i}^{x_f} F_x\,dx + \int_{y_i}^{y_f} F_y\,dy + \int_{z_i}^{z_f} F_z\,dz. \qquad (7\text{-}31)$$

If $\mathbf{F}$ has only an x component, then the y and z terms in Eq. 7-31 are zero and the equation reduces to Eq. 7-27.

Work–Kinetic Energy Theorem with a Variable Force

Equation 7-27 gives the work done by a variable force on a particle in a one-dimensional situation. Let us now make certain that the work calculated with Eq. 7-27 is indeed equal to the change in kinetic energy of the particle, as the work–kinetic energy theorem states.

Consider a particle of mass m, moving along the x axis and acted on by a net force $F(x)$ that points along that axis. The work done on the particle by this force as the particle moves from an initial position x_i to a final position x_f is given by Eq. 7-27 as

$$W = \int_{x_i}^{x_f} F(x)\,dx = \int_{x_i}^{x_f} ma\,dx, \qquad (7\text{-}32)$$

in which we use Newton's second law to replace $F(x)$ with ma. We can write the quantity $ma\,dx$ in Eq. 7-32 as

$$ma\,dx = m\frac{dv}{dt}\,dx. \qquad (7\text{-}33)$$

From the "chain rule" of calculus, we have

$$\frac{dv}{dt} = \frac{dv}{dx}\frac{dx}{dt} = \frac{dv}{dx}v, \qquad (7\text{-}34)$$

and Eq. 7-33 becomes

$$ma\,dx = m\frac{dv}{dx}v\,dx = mv\,dv. \qquad (7\text{-}35)$$

Substituting Eq. 7-35 into Eq. 7-32 yields

$$\begin{aligned} W &= \int_{v_i}^{v_f} mv\,dv = m\int_{v_i}^{v_f} v\,dv \\ &= \tfrac{1}{2}mv_f^2 - \tfrac{1}{2}mv_i^2. \end{aligned} \qquad (7\text{-}36)$$

Note that when we change the variable from x to v we are required to express the limits on the integral in terms of the new variable. Note also that because the mass m is a constant, we are able to move it outside the integral.

Recognizing the terms on the right of Eq. 7-36 as kinetic energies allows us to write this equation as

$$W = K_f - K_i = \Delta K,$$

which is the work–kinetic energy theorem.

SAMPLE PROBLEM 7-7

Force $\mathbf{F} = (3x \text{ N})\mathbf{i} + (4 \text{ N})\mathbf{j}$, with x in meters, acts on a particle, changing only the kinetic energy of the particle. How much work is done on the particle as it moves from coordinates (2 m, 3 m) to (3 m, 0 m)? Does the speed of the particle increase, decrease, or remain the same?

SOLUTION: From Eq. 7-31 we have

$$W = \int_2^3 3x \, dx + \int_3^0 4 \, dy = 3 \int_2^3 x \, dx + 4 \int_3^0 dy.$$

Using the list of integrals in Appendix E, we obtain

$$W = 3[\tfrac{1}{2}x^2]_2^3 + 4[y]_3^0$$
$$= \tfrac{3}{2}[3^2 - 2^2] + 4[0 - 3]$$
$$= -4.5 \text{ J} \approx -5 \text{ J}. \qquad \text{(Answer)}$$

The negative result tells us that energy is transferred from the particle by force $\mathbf{F}$. Because the kinetic energy of the particle decreases, its speed must decrease.

7-6 WORK DONE BY A SPRING FORCE

We next want to examine the work done on a particle-like object by a particular type of variable force, namely, a **spring force**—the force exerted by a spring. Many forces in nature have the same mathematical form as the spring force. So, by examining this one force, you can gain an understanding of many others.

The Force Exerted by a Spring

Figure 7-11a shows a spring in its **relaxed state**, that is, neither compressed nor extended. One end is fixed, and a particle-like object, say, a block, is attached to the other, free end. In Fig. 7-11b, we stretch the spring by pulling the block to the right. In reaction, the spring pulls on the block toward the left, in the direction that will restore the relaxed state. (A spring's force is sometimes said to be a *restoring force.*) In Fig. 7-11c, we compress the spring by pushing the block to the left. The spring now pushes on the block toward the right, again so as to restore the relaxed state.

To a good approximation for many springs, the force $\mathbf{F}$ exerted by the spring is proportional to the displacement $\mathbf{d}$ of the free end from its position when the spring is in the relaxed state. That is, the *spring force* is given by

$$\mathbf{F} = -k\mathbf{d} \qquad \text{(Hooke's law)}, \qquad (7\text{-}37)$$

which is known as **Hooke's law** after Robert Hooke, an English scientist of the late 1600s. The minus sign in Eq. 7-37 indicates that the spring force is always opposite in direction from the displacement of its free end. The con-

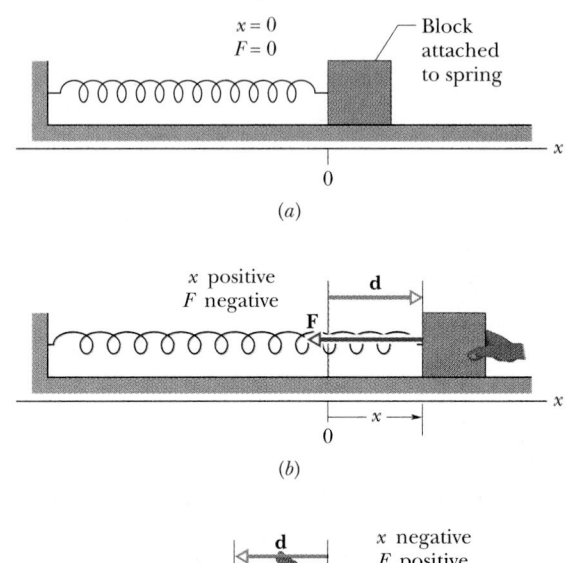

FIGURE 7-11 (a) A spring in its relaxed state. The origin of an x axis has been placed at the end of the spring that is attached to a block. (b) The block is displaced by $\mathbf{d}$, and the spring is stretched by an amount x. Note the restoring force $\mathbf{F}$ exerted by the spring. (c) The spring is compressed by an amount x. Again, note the restoring force.

stant k is called the **spring constant** (or **force constant**) and is a measure of the stiffness of the spring. The larger k is, the stiffer the spring; that is, the stronger the spring will pull or push for a given displacement. The SI unit for k is the newton per meter.

In Fig. 7-11 an x axis has been placed parallel to the length of the spring, with the origin ($x = 0$) at the position of the free end when the spring is in its relaxed state. For this common arrangement, Eq. 7-37 becomes

$$F = -kx \qquad \text{(Hooke's law)}. \qquad (7\text{-}38)$$

Note that a spring force is a variable force because it depends on the position of the free end: F can be symbolized as $F(x)$, as in Section 7-5. Also note that Hooke's law is a linear relationship between the force magnitude F and the position x of the free end.

Work Done by a Spring Force

Suppose we give the block in Fig. 7-11a an abrupt rightward jerk to provide the block with kinetic energy. And let us assume that the contact between block and floor is frictionless, that the spring has negligible mass compared to the block (the spring is *massless*), and that the spring obeys Hooke's law exactly (it is *ideal*).

As the block moves rightward, the spring force **F** slows it and thus decreases its kinetic energy. Experimentally we find that when our assumptions about the spring and floor hold, the only energy of the block that is changed by **F** is the kinetic energy. Thus the work–kinetic energy theorem applies to this situation. To find the work, we must use Eq. 7-27 because the spring force is a variable force, with F given by Eq. 7-38. So, as the block moves from a position x_i to x_f, the work done on the block by the spring force is

$$W_s = \int_{x_i}^{x_f} F \, dx = \int_{x_i}^{x_f} (-kx) \, dx = -k \int_{x_i}^{x_f} x \, dx$$

$$= (-\tfrac{1}{2}k) \left[x^2 \right]_{x_i}^{x_f} = (-\tfrac{1}{2}k)(x_f^2 - x_i^2). \qquad (7\text{-}39)$$

Multiplied out this yields

$$W_s = \tfrac{1}{2}kx_i^2 - \tfrac{1}{2}kx_f^2 \qquad \begin{array}{l}\text{(work by a}\\ \text{spring force).}\end{array} \quad (7\text{-}40)$$

This work W_s done by the spring force can have a positive or negative value, depending on whether the *net* transfer of energy is to or from the block as the block moves from x_i to x_f. If $x_i = 0$ and if we call the final position x, then Eq. 7-40 becomes

$$W_s = -\tfrac{1}{2}kx^2 \qquad \begin{array}{l}\text{(work by a}\\ \text{spring force).}\end{array} \quad (7\text{-}41)$$

Now suppose that we displace the block along x while continuing to apply a force $\mathbf{F}_a$ to it. During the displacement, our applied force does work W_a on the block while the spring force does work W_s. By Eq. 7-15, the change ΔK in the kinetic energy of the block due to these two energy transfers is

$$\Delta K = K_f - K_i = W_a + W_s, \qquad (7\text{-}42)$$

in which K_f is the kinetic energy at the end of the displacement and K_i is that at the start of the displacement. If the block is stationary before and after the displacement, then K_f and K_i are both zero and Eq. 7-42 reduces to

$$W_a + W_s = 0,$$

or

$$W_a = -W_s. \qquad (7\text{-}43)$$

The result means that the work done on the block by the applied force is the negative of the work done on the block by the spring force.

Note that the length of the spring does not appear explicitly in the expressions for the spring force (Eqs. 7-37 and 7-38) and for the work done by the spring force (Eqs. 7-40 and 7-41). The length of the spring is one of several factors that determine the spring constant k; thus, the length is in those equations implicitly.

SAMPLE PROBLEM 7-8

You apply a 4.9 N force F_a to a block attached to the free end of a spring to keep the spring stretched from its relaxed length by 12 mm, as in Fig. 7-11*b*.

(a) What is the spring constant of the spring?

SOLUTION: The stretched spring pulls with a force of -4.9 N. From Eq. 7-38, with $x = 12$ mm, we have

$$k = -\frac{F}{x} = -\frac{-4.9 \text{ N}}{12 \times 10^{-3} \text{ m}}$$

$$= 408 \text{ N/m} \approx 410 \text{ N/m}. \qquad \text{(Answer)}$$

Note that we do not need to know the length of the spring to find k. The plot of Eq. 7-38 in Fig. 7-12 refers to this spring. The slope of the line is -410 N/m.

(b) What force does the spring exert on the block if you stretch the spring by 17 mm?

SOLUTION: From Eq. 7-38 we have

$$F = -kx = -(408 \text{ N/m})(17 \times 10^{-3} \text{ m})$$

$$= -6.9 \text{ N}. \qquad \text{(Answer)}$$

The dot on the curve of Fig. 7-12 represents this force and the corresponding displacement. Note that x is positive and F is negative, as required by Eq. 7-38.

(c) How much work does the spring force do on the block as the spring is stretched 17 mm as in (b)?

SOLUTION: Because the spring is initially in its relaxed state, we can use Eq. 7-41:

$$W_s = -\tfrac{1}{2}kx^2 = -(\tfrac{1}{2})(408 \text{ N/m})(17 \times 10^{-3} \text{ m})^2$$

$$= -5.9 \times 10^{-2} \text{ J} = -59 \text{ mJ}. \qquad \text{(Answer)}$$

The shaded area in Fig. 7-12 represents this work. The work is negative because the spring force and the displacement of the block are in opposite directions. Note that the amount of work done by the spring force would be the same if the spring had been compressed (rather than stretched) by 17 mm.

(d) With the spring initially stretched by 17 mm, you allow the block to return to $x = 0$ (the spring returns to its relaxed state); you then compress the spring by 12 mm. How much work does the spring force do on the block during this total displacement of the block?

SOLUTION: For this situation, we have $x_i = +17$ mm (the spring is initially stretched) and $x_f = -12$ mm (the spring is finally compressed). Equation 7-40 becomes

$$W_s = \tfrac{1}{2}kx_i^2 - \tfrac{1}{2}kx_f^2 = \tfrac{1}{2}k(x_i^2 - x_f^2)$$

$$= \tfrac{1}{2}(408 \text{ N/m})[(17 \times 10^{-3} \text{ m})^2 - (-12 \times 10^{-3} \text{ m})^2]$$

$$= 0.030 \text{ J} = 30 \text{ mJ}. \qquad \text{(Answer)}$$

This work done on the block by the spring force is positive because the spring force does more positive work as the block moves from $x_i = +17$ mm to $x = 0$ than it does negative work as the block moves from $x = 0$ to $x_f = -12$ mm.

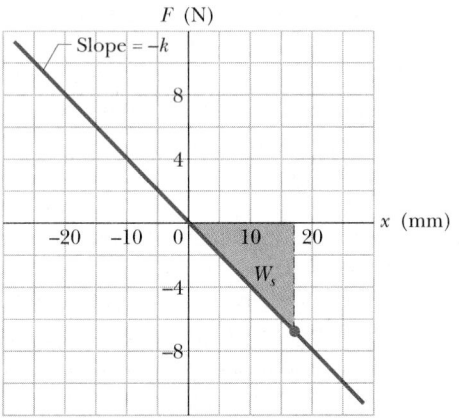

FIGURE 7-12   The force–distance plot for the spring of Sample Problem 7-8. The spring obeys Hooke's law (Eqs. 7-37 and 7-38) and has a spring constant $k = 410$ N/m. For the significance of the dot and the shaded area marked W_s, see Sample Problem 7-8(b) and 7-8(c), respectively.

$\mathbf{C}$HECKPOINT **4:** For three situations, the initial and final positions, respectively, along the x axis for the block in Fig. 7-11 are (a) -3 cm, 2 cm; (b) 2 cm, 3 cm; and (c) -2 cm, 2 cm. In each situation is the work done by the spring force on the block positive, negative, or zero?

SAMPLE PROBLEM 7-9

A block whose mass m is 5.7 kg slides on a horizontal frictionless tabletop with a constant speed v of 1.2 m/s. It is brought momentarily to rest as it compresses a spring in its path (Fig. 7-13). By what distance d is the spring compressed? The spring constant k is 1500 N/m.

SOLUTION: From Eq. 7-41, the work done *by* the spring force *on* the block as the spring is compressed a distance d from its rest state is given by

$$W_s = -\tfrac{1}{2}kd^2.$$

The change in the kinetic energy of the block as it is stopped is

$$\Delta K = K_f - K_i = 0 - \tfrac{1}{2}mv^2.$$

The work–kinetic energy theorem (Eq. 7-4) requires that these two quantities be equal. Setting them so and solving for d, we obtain

$$d = v\sqrt{\frac{m}{k}} = (1.2 \text{ m/s})\sqrt{\frac{5.7 \text{ kg}}{1500 \text{ N/m}}}$$

$$= 7.4 \times 10^{-2} \text{ m} = 7.4 \text{ cm}. \qquad \text{(Answer)}$$

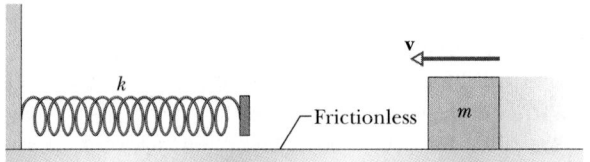

FIGURE 7-13   Sample Problem 7-9. A block moves toward a spring with velocity $\mathbf{v}$. When it is momentarily stopped by the spring, it will have compressed the spring by a distance d.

PROBLEM SOLVING TACTICS

TACTIC 1: *Derivatives and Integrals; Slopes and Areas*

If you know a function $y = F(x)$, you can find the value of its derivative (for any value of x) or of its integral (between any two values of x) from the rules of calculus. If you do not know the function analytically but have a plot of it, you can find the values of both its derivative and its integral graphically. Finding a derivative graphically is shown in Tactic 5 of Chapter 2. Here we evaluate an integral graphically.

Figure 7-14 is a plot of a particular force function $F(x)$. Let us calculate graphically the work W done by this force as the particle on which it acts moves from $x_i = 2.0$ cm to $x_f = 5.0$ cm. According to Eq. 7-27, the work is

$$W = \int_{x_i}^{x_f} F(x)\, dx,$$

which is equal to the (total) shaded area shown under the curve between the two points.

You can approximate this area with a rectangle formed by drawing a horizontal line across Fig. 7-14. Draw it at a level such that the areas marked "1" and "2" appear to be equal. A line at $F = 44$ N is about right, and the area of the equivalent rectangle ($= W$) is then

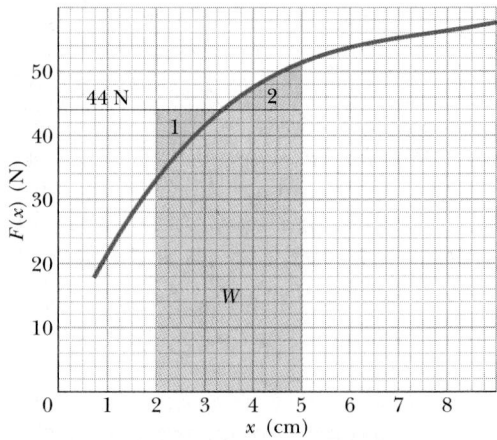

FIGURE 7-14   The graph of a one-dimensional force $F(x)$. The shaded area under the curve (which represents the work done by F) is approximated by a rectangle formed by excluding area 2 and including (approximately equal) area 1.

$$W = \text{height} \times \text{base} = (44\ \text{N})(5.0\ \text{cm} - 2.0\ \text{cm})$$
$$= 132\ \text{N} \cdot \text{cm} \approx 1.3\ \text{N} \cdot \text{m} = 1.3\ \text{J}.$$

You can also find the area by counting the small squares underneath the curve. The shaded area contains about 260 squares, and each square represents $(2\ \text{N})(0.25\ \text{cm}) = 0.5$ N·cm. The work is then

$$W = (260\ \text{squares})\left(\frac{0.5\ \text{N} \cdot \text{cm}}{1\ \text{square}}\right) = 130\ \text{N} \cdot \text{cm}$$
$$= 1.3\ \text{J},$$

just as above. *Remember:* On a two-dimensional graph, a derivative is represented by a slope, and an integral by an area.

7-7 POWER

A contractor wishes to lift a load of bricks from the sidewalk to the top of a building by means of a winch. We can now calculate how much work the force applied by the winch must do on the load to make the lift. The contractor, however, is much more interested in the *rate* at which that work is done. Will the job take 5 minutes (acceptable) or a week (unacceptable)?

The rate at which work is done by a force is said to be the **power** due to the force. If an amount of work W is done in an amount of time Δt by a force, the **average power** due to the force during that time interval is

$$\overline{P} = \frac{W}{\Delta t} \qquad \text{(average power).} \qquad (7\text{-}44)$$

The **instantaneous power** P is the instantaneous rate of doing work, which we can write as

$$P = \frac{dW}{dt} \qquad \text{(instantaneous power).} \qquad (7\text{-}45)$$

The SI unit of power is the joule per second. This unit is used so often that it has a special name, the **watt** (W), after James Watt, who greatly improved the rate at which steam engines could do work. In the British system, the unit of power is the foot-pound per second. Often the horsepower is used. Some relations among these units are

$$1\ \text{watt} = 1\ \text{W} = 1\ \text{J/s} = 0.738\ \text{ft} \cdot \text{lb/s} \qquad (7\text{-}46)$$

and

$$1\ \text{horsepower} = 1\ \text{hp} = 550\ \text{ft} \cdot \text{lb/s}$$
$$= 746\ \text{W}. \qquad (7\text{-}47)$$

Inspection of Eq. 7-44 shows that work can be expressed as power multiplied by time, as in the common unit, the kilowatt-hour. Thus

$$1\ \text{kilowatt-hour} = 1\ \text{kW} \cdot \text{h} = (10^3\ \text{W})(3600\ \text{s})$$
$$= 3.60 \times 10^6\ \text{J} = 3.60\ \text{MJ}. \qquad (7\text{-}48)$$

Perhaps because of our utility bills, the watt and the kilowatt-hour have become identified as electrical units. They can be used equally well as units for other examples of power and energy. Thus if you pick up this book from the floor and put it on a tabletop, you are free to report the work that you have done as $4 \times 10^{-6}\ \text{kW} \cdot \text{h}$ (or more conveniently as $4\ \text{mW} \cdot \text{h}$).

We can also express the rate at which a force does work on a particle (or particle-like object) in terms of that force and the body's velocity. For a particle moving along a straight line (say, the x axis) and acted on by a constant force $\mathbf{F}$ directed at some angle ϕ to that line, Eq. 7-45 becomes

$$P = \frac{dW}{dt} = \frac{F\cos\phi\ dx}{dt} = F\cos\phi\left(\frac{dx}{dt}\right),$$

or

$$P = Fv\cos\phi. \qquad (7\text{-}49)$$

Reorganizing the right side of Eq. 7-49 as the dot product $\mathbf{F} \cdot \mathbf{v}$, we may also write Eq. 7-49 as

$$P = \mathbf{F} \cdot \mathbf{v} \qquad \text{(instantaneous power).} \qquad (7\text{-}50)$$

For example, the truck in Fig. 7-15 exerts a force $\mathbf{F}$ on the trailing load, which has velocity $\mathbf{v}$ at some instant. The instantaneous power due to $\mathbf{F}$ is the rate at which $\mathbf{F}$ transfers energy to the load at that instant and is given by Eqs. 7-49 and 7-50. Saying that this power is "the power of the truck" is often acceptable, but we should keep in mind what is meant: power is the rate at which the applied *force* transfers energy.

FIGURE 7-15 The power due to the truck's applied force on the trailing load is the rate at which that force does work on the load.

SAMPLE PROBLEM 7-10

Figure 7-16 shows forces $\mathbf{F}_1$ and $\mathbf{F}_2$ acting on a box as the box slides rightward across a frictionless floor. Force $\mathbf{F}_1$ is horizontal, with magnitude 2.0 N; force $\mathbf{F}_2$ is angled upward by 60° to the floor and has magnitude 4.0 N. The speed v of the box at a certain instant is 3.0 m/s.

(a) What is the power due to each force acting on the box at that instant, and what is the net power? Is the net power changing at that instant?

SOLUTION: We use Eq. 7-49 to find the power due to each force. For force $\mathbf{F}_1$, at angle $\phi_1 = 180°$ to velocity $\mathbf{v}$, we have

$$P_1 = F_1 v \cos \phi_1 = (2.0 \text{ N})(3.0 \text{ m/s}) \cos 180°$$
$$= -6.0 \text{ W}. \qquad \text{(Answer)}$$

This result tells us that force $\mathbf{F}_1$ is transferring energy from the box at the rate of 6.0 J/s.

For force $\mathbf{F}_2$, at angle $\phi_2 = 60°$ to velocity $\mathbf{v}$, we have

$$P_2 = F_2 v \cos \phi_2 = (4.0 \text{ N})(3.0 \text{ m/s}) \cos 60°$$
$$= 6.0 \text{ W}. \qquad \text{(Answer)}$$

This result tells us that force $\mathbf{F}_2$ is transferring energy to the box at the rate of 6.0 J/s.

The net power is the sum of the individual powers:

$$P_{\text{net}} = P_1 + P_2$$
$$= -6.0 \text{ W} + 6.0 \text{ W} = 0, \qquad \text{(Answer)}$$

which tells us that the net rate of transfer of energy to or from the box is zero. Thus, the kinetic energy ($K = \frac{1}{2}mv^2$) of the box is not changing, and so the speed of the box will remain at 3.0 m/s. With neither the forces $\mathbf{F}_1$ and $\mathbf{F}_2$ nor the velocity $\mathbf{v}$ changing, we see from Eq. 7-50 that P_1 and P_2 are not changing and thus neither is P_{net}.

(b) If the magnitude of $\mathbf{F}_2$ is, instead, 6.0 N, what now is the power due to each force acting on the box at the given instant, and what is the net power? Is the net power changing?

SOLUTION: For force $\mathbf{F}_2$, we now have

$$P_2 = F_2 v \cos \theta_2 = (6.0 \text{ N})(3.0 \text{ m/s}) \cos 60°$$
$$= 9.0 \text{ W}. \qquad \text{(Answer)}$$

The power of force $\mathbf{F}_1$ is still

$$P_1 = -6.0 \text{ W}. \qquad \text{(Answer)}$$

So, the net power is now

$$P_{\text{net}} = P_1 + P_2 = -6.0 \text{ W} + 9.0 \text{ W}$$
$$= 3.0 \text{ W}, \qquad \text{(Answer)}$$

which tells us that the net rate of transfer of energy to the box has a positive value. Thus, the kinetic energy of the box is increasing, and so also is the speed of the box. With the speed increasing, we see from Eq. 7-50 that the values of P_1 and P_2, and thus also of P_{net}, will be changing. Hence, this net power of 3.0 W is the net power only at the instant the speed is the given 3.0 m/s.

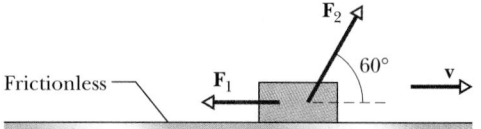

FIGURE 7-16 Sample Problem 7-10. Two forces $\mathbf{F}_1$ and $\mathbf{F}_2$ act on a box that slides rightward across a frictionless floor. The velocity of the box is $\mathbf{v}$.

C**HECKPOINT 5:** A block moves with uniform circular motion because a cord tied to the block is anchored at the center of a circle. Is the power of the force exerted on the block by the cord positive, negative, or zero?

7-8 KINETIC ENERGY AT HIGH SPEEDS (OPTIONAL)

In Section 4-10 we saw that for particles moving at speeds near the speed of light, Newtonian mechanics fails and must be replaced by Einstein's theory of special relativity. One consequence is that we can no longer use the expression $K = \frac{1}{2}mv^2$ for the kinetic energy of a particle. Instead, we must use

$$K = mc^2 \left(\frac{1}{\sqrt{1 - (v/c)^2}} - 1 \right), \qquad (7-51)$$

in which c is the speed of light.

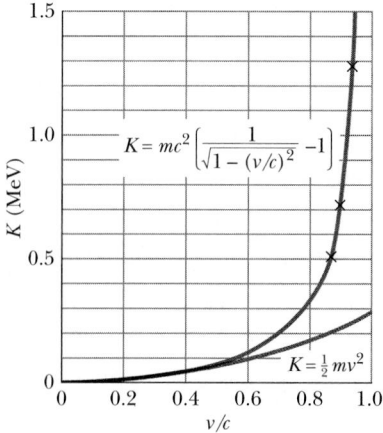

FIGURE 7-17 The relativistic (Eq. 7-51) and classical (Eq. 7-1) equations for the kinetic energy of an electron, plotted as a function of v/c, where v is the speed of the electron and c is the speed of light. Note that the two curves blend together at low speeds and diverge widely at high speeds. Experimental data (at the × marks) show that at high speeds the relativistic curve agrees with experiment but the classical curve does not.

Figure 7-17 shows that these two formulas, which look so different, do indeed give widely different results at high speeds. Experiment shows that—beyond any doubt—the relativistic expression (Eq. 7-51) is correct and the classical expression (Eq. 7-1) is not. At low speeds, however, the two formulas merge, yielding the same result. In particular, both formulas yield $K = 0$ for $v = 0$.

All relativistic formulas must reduce to their classical counterparts at low speeds. To see how this comes about for Eq. 7-51, let us write that equation in the form

$$K = mc^2[(1 - \beta^2)^{-1/2} - 1]. \qquad (7\text{-}52)$$

Here, for convenience, we have substituted the dimensionless *speed parameter* β for the speed ratio v/c.

At very low speeds, $v \ll c$ and therefore $\beta \ll 1$. At low speeds, we can then expand the quantity $(1 - \beta^2)^{-1/2}$ by the binomial theorem, obtaining (see Tactic 2, below)

$$(1 - \beta^2)^{-1/2} = 1 + \tfrac{1}{2}\beta^2 + \cdots. \qquad (7\text{-}53)$$

Substituting Eq. 7-53 into Eq. 7-52 leads to

$$K = mc^2[(1 + \tfrac{1}{2}\beta^2 + \cdots) - 1]. \qquad (7\text{-}54)$$

For very small β, the terms represented by the dots in Eq. 7-54 decrease rapidly in size. We can then, with little error, replace the entire sum in parentheses with its first two terms, obtaining

$$K \approx (mc^2)[(1 + \tfrac{1}{2}\beta^2) - 1]$$

or, with $\beta = v/c$,

$$K \approx (mc^2)(\tfrac{1}{2}\beta^2) = \tfrac{1}{2}mv^2,$$

which is exactly what we set out to show.

PROBLEM SOLVING TACTICS

TACTIC 2: *Approximations with the Binomial Theorem*

Often we have a quantity written in the form $(a + b)^n$ and we want to find its approximate value for the case in which $b \ll a$. To do so, it is simplest to recast the quantity in the form $(1 + x)^n$, where x is dimensionless and is much less than unity. Thus we can put

$$(a + b)^n = a^n(1 + b/a)^n = (a^n)(1 + x)^n.$$

This is of the desired form, with $x \,(= b/a)$ being dimensionless. We can then evaluate $(1 + x)^n$ with the binomial theorem, keeping only as many terms as are appropriate to the problem. (That decision takes some experience.)

The binomial theorem (given in Appendix E) is

$$(1 + x)^n = 1 + \frac{n}{1!}x + \frac{n(n-1)}{2!}x^2 + \cdots. \qquad (7\text{-}55)$$

The exclamation marks in Eq. 7-55 identify factorials, or products of all the whole numbers from the given number

down to 1. As an example, $4! = 4 \times 3 \times 2 \times 1 = 24$; you probably have a key for factorials on your calculator.

We applied the binomial theorem in Eq. 7-53, letting $x = -\beta^2$ and $n = -\tfrac{1}{2}$.

As an exercise, evaluate $(1 + 0.045)^{-2.3}$ both on your calculator and by expansion, using Eq. 7-55 with $x = 0.045$ and $n = -2.3$. Check the various terms in the binomial sum to see how rapidly they decrease.

7-9   REFERENCE FRAMES

We apply Newton's laws of mechanics only in inertial reference frames. Recall that these frames move at constant velocity.

For some physical quantities, observers in different inertial reference frames will measure the exact same values. In Newtonian mechanics, these *invariant* quantities (as they are called) are force, mass, acceleration, and time. As an example, if an observer in one inertial frame finds that a certain particle has a mass of 3.15 kg, observers in all other inertial frames will measure the same mass for the particle. For other physical quantities, such as the displacement and velocity of a particle, observers in different inertial frames will measure different values; these quantities are *not* invariant.

If the displacement of a particle depends on the reference frame of the observer, so must the work done on the particle because work ($W = \mathbf{F} \cdot \mathbf{d}$) is defined in terms of displacement. If the displacement of a particle during a given interval is measured to be $+2.47$ m in one reference frame, it might be measured to be zero in another and -3.64 m in a third. Because the force $\mathbf{F}$ does not change (it is invariant), work that is positive in one reference frame can be zero in another and negative in a third.

What about the kinetic energy of the particle? If its velocity depends on the reference frame of the observer, then so must its kinetic energy because kinetic energy ($K = \tfrac{1}{2}mv^2$) is defined in terms of velocity. Does that mean that the work–kinetic energy theorem is in trouble?

From Galileo to Einstein, physicists have come to believe in the **principle of invariance:**

The laws of physics must have the same form in all inertial reference frames.

That is, even though some *physical quantities* have different values in different reference frames, the *laws of physics* must hold true in all frames. Behind this formal statement of invariance is a feeling that—in some deep sense—if different observers look at the same event, they must see nature working in the same way.

Among the laws to which this invariance principle applies is the work–kinetic energy theorem. Thus, even though different observers watching the same moving particle might measure different values for work and for kinetic energy, they would all find that the work–kinetic energy theorem holds in their respective frames. Let us look at a simple example.

In Fig. 7-18, Sally rides up in an elevator cab at constant speed v, holding a book. Steve, standing on a facing balcony, observes her as the cab rises through a height h. From these two references frames, what are the work–kinetic energy relations that apply to the book?

1. Sally's report. "My reference frame is the elevator cab. I am exerting an upward force on the book, but this force does no work because the book is not moving in my reference frame. The weight of the book, acting downward, does no work, for the same reason. Thus the total work done on the book by all the forces that act on it is zero. According to the work–kinetic energy theorem, the kinetic energy of the book should not change. That is what I observe; the kinetic energy of the book is zero in my reference frame and remains so. Everything fits together."

2. Steve's report. "My reference frame is the balcony. I see that Sally exerts a force $\mathbf{F}$ on her book. In my frame, the point of application of $\mathbf{F}$ is moving and the work done by $\mathbf{F}$ as the book moves upward through a height h is $+mgh$. I also know that the book's weight does work on the book, in the amount $-mgh$. Thus the total work done on the book during its rise is zero. According to the work–kinetic energy theorem, the kinetic energy of the book

FIGURE 7-18 Sally rides up in an elevator, holding a book. Steve watches her. Both check the work–kinetic energy theorem in their respective reference frames, as it applies to the motion of the book.

should not change. That is what I observe. In my frame, the kinetic energy is a constant $\frac{1}{2}mv^2$. Everything fits."

Although Steve and Sally do not agree about the displacement of the book and its kinetic energy, they do agree that the work–kinetic energy theorem holds in their respective reference frames.

It does not matter what (inertial) reference frame you pick in which to solve a problem as long as you (1) make sure you know what that frame is and (2) use that frame consistently throughout the problem.

REVIEW & SUMMARY

Kinetic Energy

The **kinetic energy** K associated with the motion of a particle of mass m and speed v, where v is well below the speed of light, is

$$K = \tfrac{1}{2}mv^2 \quad \text{(kinetic energy).} \tag{7-1}$$

Work

Work W is energy transferred to or from an object via a force acting on the object. Energy transferred to the object is positive work, and that transferred from the object is negative work.

Work and Kinetic Energy

If the only type of energy of an object being changed by work W is kinetic energy, then we can write the change ΔK in kinetic energy of the object as

$$\Delta K = K_f - K_i = W \quad \begin{array}{l}\text{(work–kinetic}\\\text{energy theorem),}\end{array} \tag{7-4}$$

in which K_i is the initial kinetic energy of the object and K_f is the kinetic energy after the work is done. Equation 7-4 rearranged gives us

$$K_f = K_i + W. \tag{7-5}$$

Work Done by a Constant Force

The work done on a particle by a constant force $\mathbf{F}$ during displacement $\mathbf{d}$ of the particle is

$$W = Fd \cos \phi = \mathbf{F} \cdot \mathbf{d} \quad \begin{array}{l}\text{(work,}\\\text{constant force),}\end{array} \tag{7-9, 7-11}$$

in which ϕ is the constant angle between $\mathbf{F}$ and $\mathbf{d}$. When several forces $\mathbf{F}_1, \mathbf{F}_2, \mathbf{F}_3, \ldots$ act on a particle, the work done by the net force $\Sigma\mathbf{F}$ is

$$W = \left(\sum \mathbf{F}\right) \cdot \mathbf{d} \quad \begin{array}{l}\text{(work by constant}\\\text{net force),}\end{array} \tag{7-13}$$

This work is equal to the total work done by all the forces:

$$W = \mathbf{F}_1 \cdot \mathbf{d} + \mathbf{F}_2 \cdot \mathbf{d} + \mathbf{F}_3 \cdot \mathbf{d} + \cdots$$
$$= W_1 + W_2 + W_3 + \cdots \quad \text{(total work)}. \quad (7\text{-}14)$$

Substituting this total work into the work–kinetic energy theorem, we have

$$\Delta K = K_f - K_i = W_1 + W_2 + W_3 + \cdots, \quad (7\text{-}15)$$

which gives us the change in the kinetic energy of the particle due to the total work.

Work Done by Weight

The work W_g done by the weight mg of a particle-like object during a displacement $\mathbf{d}$ of the object is given by

$$W_g = mgd \cos \phi, \quad (7\text{-}16)$$

in which ϕ is the angle between mg and $\mathbf{d}$.

Work Done in Lifting and Lowering an Object

The work W_a done by an applied force during a lifting or lowering of a particle-like object is related to the work W_g done by the object's weight and the change ΔK in the object's kinetic energy by

$$\Delta K = K_f - K_i = W_a + W_g. \quad (7\text{-}19)$$

If the kinetic energy at the beginning of a lift equals that at the end of the lift, then Eq. 7-19 reduces to

$$W_a = -W_g, \quad (7\text{-}20)$$

which tells us that the applied force transfers as much energy to the object as the weight of the object transfers from the object.

Work Done by a Variable Force

When the force $\mathbf{F}$ on a particle-like object depends on the position of the object, the work done by $\mathbf{F}$ on the object while the object moves from an initial position r_i with coordinates (x_i, y_i, z_i) to a final position r_f with coordinates (x_f, y_f, z_f) is

$$W = \int_{x_i}^{x_f} F_x \, dx + \int_{y_i}^{y_f} F_y \, dy + \int_{z_i}^{z_f} F_z \, dz. \quad (7\text{-}31)$$

If $\mathbf{F}$ has only an x component, then Eq. 7-31 reduces to

$$W = \int_{x_i}^{x_f} F(x) \, dx. \quad (7\text{-}27)$$

Spring Force

The force $\mathbf{F}$ exerted by a spring is

$$\mathbf{F} = -k\mathbf{d} \quad \text{(Hooke's law)}, \quad (7\text{-}37)$$

where $\mathbf{d}$ is the displacement of its free end from the position when the spring is in its **relaxed state** (neither compressed or extended)

and k is the **spring constant** (a measure of the spring's stiffness). If an x axis lies along the spring, with the origin at the location of the spring's free end when the spring is in its relaxed state, Eq. 7-37 can be written as

$$F = -kx \quad \text{(Hooke's law)}. \quad (7\text{-}38)$$

Work Done by a Spring Force

If an object is attached to the spring's free end, the work W_s done on the object by the spring force when the object is moved from an initial position x_i to a final position x_f is

$$W_s = \tfrac{1}{2}kx_i^2 - \tfrac{1}{2}kx_f^2. \quad (7\text{-}40)$$

If $x_i = 0$ and $x_f = x$, then Eq. 7-40 becomes

$$W_s = -\tfrac{1}{2}kx^2. \quad (7\text{-}41)$$

Power

The **power** due to a force is the *rate* at which that force does work on an object. If the force does work W during a time interval Δt, the *average power* due to the force over that time interval is

$$\overline{P} = \frac{W}{\Delta t}. \quad (7\text{-}44)$$

Instantaneous power is the instantaneous rate of doing work:

$$P = \frac{dW}{dt}. \quad (7\text{-}45)$$

If a force $\mathbf{F}$ is at an angle ϕ to the direction of travel of the object, the instantaneous power is

$$P = Fv \cos \phi = \mathbf{F} \cdot \mathbf{v}, \quad (7\text{-}49, 7\text{-}50)$$

in which $\mathbf{v}$ is the instantaneous velocity of the object.

Relativistic Kinetic Energy

The kinetic energy of objects moving at speeds near the speed of light c must be calculated using the relativistic expression

$$K = mc^2 \left(\frac{1}{\sqrt{1 - (v/c)^2}} - 1 \right). \quad (7\text{-}51)$$

This equation reduces to Eq. 7-1 when v is much less than c.

The Principle of Invariance

Some quantities (such as mass, force, acceleration, and time in Newtonian mechanics) are *invariant;* that is, they have the same numerical values when measured in different inertial reference frames. Others (for example, velocity, kinetic energy, and work) have different values in different frames. However, the *laws* of physics have the same *form* in all inertial reference frames. This is called the **principle of invariance.**

QUESTIONS

1. Rank the following velocities according to the kinetic energy a particle will have with each velocity, greatest first: (a) $\mathbf{v} = 4\mathbf{i} + 3\mathbf{j}$, (b) $\mathbf{v} = -4\mathbf{i} + 3\mathbf{j}$, (c) $\mathbf{v} = -3\mathbf{i} + 4\mathbf{j}$, (d) $\mathbf{v} = 3\mathbf{i} - 4\mathbf{j}$, (e) $\mathbf{v} = 5\mathbf{i}$, and (f) $v = 5$ m/s at 30° to the horizontal.

2. For $t > 0$, is the kinetic energy of a particle increasing, decreasing, or remaining the same if the particle's velocity is given by (a) $\mathbf{v} = 3\mathbf{i}$ and (b) $v = -4t$?

3. For $t > 0$, is the kinetic energy of a particle increasing, decreasing, or remaining the same if the particle's position is given by (a) $x = 4t^2 - 2$, (b) $x = -3t + 14$, (c) $\mathbf{r} = 2\mathbf{i} - 3t\mathbf{j}$, and (d) $\mathbf{r} = (2t^2 - 3)\mathbf{i} + (4t - 2)\mathbf{j}$?

4. A force $\mathbf{F}$ that is directed along an x axis moves a particle 5 m along that axis. Can we use Eqs. 7-9 and 7-11 to find the work done if the magnitude (in newtons) of $\mathbf{F}$ is given by (a) $F = 3$; (b) $F = 2x$; (c) $F = 2t$?

5. Is positive or negative work done by a constant force $\mathbf{F}$ on a particle during a straight-line displacement $\mathbf{d}$ if (a) the angle between $\mathbf{F}$ and $\mathbf{d}$ is 30°; (b) the angle is 100°; (c) $\mathbf{F} = 2\mathbf{i} - 3\mathbf{j}$ and $\mathbf{d} = -4\mathbf{i}$?

6. If a particle-like object of mass 5 kg travels in uniform circular motion at speed 4 m/s, how much work is done on the particle by the centripetal force acting on it (a) at any instant and (b) during one complete revolution?

7. Figure 7-19 shows six situations in which two forces act simultaneously on a box after the box has been sent sliding over a frictionless surface, either to the left or to the right. The forces are either 1 N or 2 N in magnitude, as indicated by the vector lengths. For each situation, is the work done on the box by the net force during the indicated displacement $\mathbf{d}$ positive, negative, or zero?

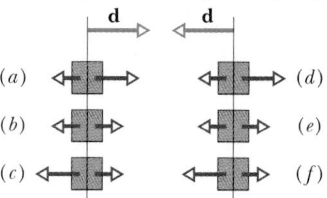

FIGURE 7-19 Question 7.

8. In overhead views, Fig. 7-20 shows three situations in which a box is acted on by two forces ($\mathbf{F}_1$ and $\mathbf{F}_2$) of equal magnitude. As the box moves, the forces maintain their orientation relative to the velocity $\mathbf{v}$. For each situation, is the total work done on the box positive, negative, or zero?

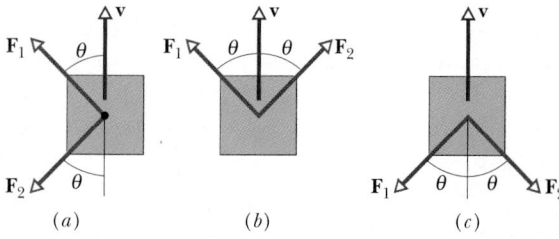

FIGURE 7-20 Question 8.

9. In Fig. 7-21, a greased pig has a choice of three frictionless slides along which to slide to the ground. Rank the slides according to how much work the pig's weight does on the pig during the descent, greatest first.

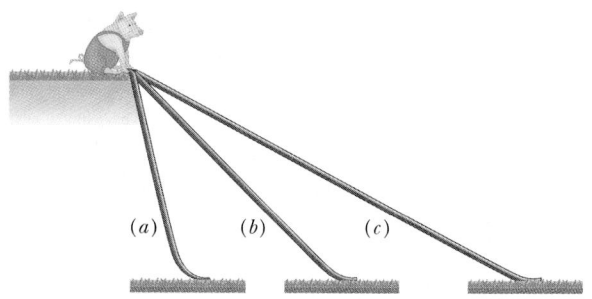

FIGURE 7-21 Question 9.

10. You lift an armadillo to a shelf. Does the work done by your force on the armadillo depend on (a) the mass of the armadillo, (b) the weight of the armadillo, (c) the height of the shelf, (d) the time you take, or (e) whether you move the armadillo sideways or directly upward?

11. Figure 7-22 shows a bundle of magazines that is lifted by a cord through distance d. The table shows six pairs of values of the initial speed v_0 and final speed v (in meters per second) of the bundle at the beginning and end of distance d. Rank the pairs according to the work done by the cord's force on the bundle over distance d, most positive first, most negative last.

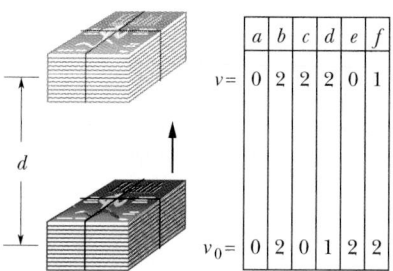

	a	b	c	d	e	f
$v=$	0	2	2	2	0	1
$v_0=$	0	2	0	1	2	2

FIGURE 7-22 Question 11.

12. Figure 7-23 shows four graphs (drawn to the same scale) of the values of a variable force $\mathbf{F}$ (directed along an x axis) versus the position x of a particle on which the force acts. Rank the graphs according to the work done by $\mathbf{F}$ on the particle from $x = 0$ to x_1, from most positive work to most negative work.

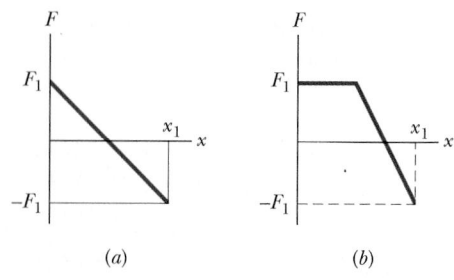

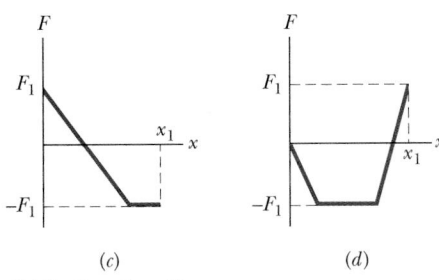

(c)   (d)

FIGURE 7-23 Question 12.

13. Figure 7-24 shows the values of a force **F**, directed along an x axis, that will act on a particle at the corresponding values of x. If the particle begins at rest at $x = 0$, what is the particle's coordinate when it has (a) its greatest kinetic energy, (b) its greatest speed, and (c) zero speed? (d) What is its direction of travel after it reaches $x = 6$ m?

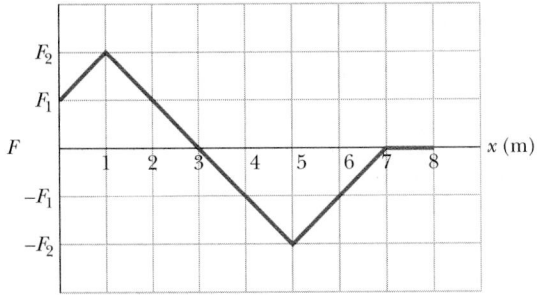

FIGURE 7-24 Question 13.

14. A block is attached to a relaxed spring as in Fig. 7-25a. The spring constant k of the spring is such that for a certain rightward displacement **d** of the block, the spring force acting on the block has magnitude F_1 and has done work W_1 on the block. A second, identical spring is then attached on the opposite side of the block,

as shown in Fig. 7-25b; both springs are in their relaxed state in the figure. If the block is again displaced by **d**, (a) what is the magnitude of the net force on it from both springs and (b) how much work has been done on it by the spring forces?

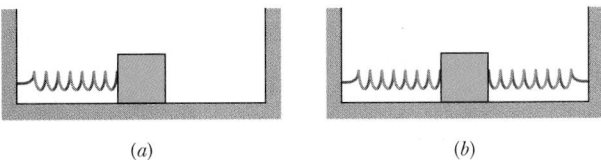

(a)   (b)

FIGURE 7-25 Question 14.

15. Spring A is stiffer than spring B, that is, $k_A > k_B$. The spring force of which spring does more work if the springs are compressed (a) the same distance and (b) by the same applied force?

16. Figure 7-26 shows three arrangements of a block attached to identical springs that are in their relaxed state when the block is centered as shown. Rank the arrangements according to the magnitude of the net force on the block, largest first, when the block is displaced by distance d (a) to the right and (b) to the left. Rank the arrangements according to the work done on the block by the spring forces, greatest first, when the block is displaced by d (c) to the right and (d) to the left.

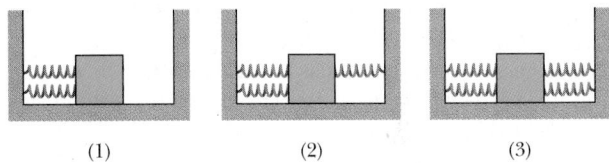

(1)   (2)   (3)

FIGURE 7-26 Question 16.

17. If you cut a spring in half, the spring constant of either half is what multiple of the spring constant of the original spring? (*Hint:* Consider the amount by which each half stretches for a given value of force.)

EXERCISES & PROBLEMS

SECTION 7-1 Kinetic Energy

1E. If a Saturn V rocket with an Apollo spacecraft attached has a combined mass of 2.9×10^5 kg and is to reach a speed of 11.2 km/s, how much kinetic energy will it then have?

2E. If an electron (mass $m = 9.11 \times 10^{-31}$ kg) in copper near the lowest possible temperature has a kinetic energy of 6.7×10^{-19} J, what is the speed of the electron?

3E. Calculate the kinetic energies of the following objects moving at the given speeds: (a) a 110 kg football linebacker running at 8.1 m/s; (b) a 4.2 g bullet at 950 m/s; (c) the aircraft carrier *Nimitz,* 91,400 tons at 32 knots.

4E. On August 10, 1972, a large meteorite skipped across the atmosphere above western United States and Canada, much like a stone skipped across water. The accompanying fireball was so bright that it could be seen in the daytime sky (Fig. 7-27). The meteorite's mass was about 4×10^6 kg; its speed was about 15

FIGURE 7-27 Exercise 4 A large meteorite skips across the atmosphere in the sky above the mountains (upper right).

km/s. Had it entered the atmosphere vertically, it would have hit the Earth's surface with about the same speed. (a) Calculate the meteorite's loss of kinetic energy (in joules) that would have been associated with the vertical impact. (b) Express the energy as a multiple of the explosive energy of 1 megaton of TNT, which is 4.2×10^{15} J. (c) The energy associated with the atomic bomb explosion over Hiroshima was equivalent to 13 kilotons of TNT. To how many "Hiroshima bombs" would the meteorite impact have been equivalent?

5E. An explosion at ground level leaves a crater with a diameter that is proportional to the energy of the explosion raised to the $\frac{1}{3}$ power; an explosion of 1 megaton of TNT leaves a crater with a 1 km diameter. Below Lake Huron in Michigan there appears to be an ancient impact crater with a 50 km diameter. What was the kinetic energy associated with that impact, in terms of (a) megatons of TNT and (b) Hiroshima bomb equivalents (see Exercise 4)? (Such meteorite or comet impacts may have significantly altered the Earth's climate and contributed to the extinction of the dinosaurs and other life-forms.)

6P. A proton (mass $m = 1.67 \times 10^{-27}$ kg) is being accelerated along a straight line at 3.6×10^{15} m/s^2 in a machine called a linear accelerator. If the proton has an initial speed of 2.4×10^7 m/s and travels 3.5 cm, what then is (a) its speed and (b) the increase in its kinetic energy in electron-volts?

7P. A man racing his son has half the kinetic energy of the son, who has half the mass of the father. The man speeds up by 1.0 m/s and then has the same kinetic energy as the son. What were the original speeds of man and son?

8P. What is the kinetic energy associated with the Earth's orbiting of the Sun? (See Appendix C for numerical data.)

SECTION 7-3 Work and Kinetic Energy

9E. A 102 kg object moves in a horizontal straight line with an initial speed of 53 m/s. If it is stopped along that line with a deceleration of 2.0 m/s^2, (a) what magnitude of force is required, (b) what distance does it travel while decelerating, and (c) what work is done by the decelerating force on the object? (d) Answer questions (a) to (c) for a deceleration of 4.0 m/s^2.

10E. To pull a 50 kg crate across a frictionless floor, a worker applies a force of 210 N, directed 20° above the horizontal. As the crate moves 3.0 m, what work is done on the crate by (a) the worker's force, (b) the weight of the crate, and (c) the normal force exerted by the floor on the crate? (d) What is the total work done on the crate?

11E. A floating ice block is pushed through a displacement $\mathbf{d} = (15 \text{ m})\mathbf{i} - (12 \text{ m})\mathbf{j}$ along a straight embankment by rushing water, which exerts a force $\mathbf{F} = (210 \text{ N})\mathbf{i} - (150 \text{ N})\mathbf{j}$ on the block. How much work does the force do on the block during the displacement?

12E. A particle moves along a straight path through a displacement $\mathbf{d} = (8 \text{ m})\mathbf{i} + c\mathbf{j}$ while a force $\mathbf{F} = (2 \text{ N})\mathbf{i} - (4 \text{ N})\mathbf{j}$ acts on it. (Other forces also act on the particle.) What is the value of c if the work done by $\mathbf{F}$ on the particle is (a) zero, (b) positive, and (c) negative?

13E. An initially stationary proton is accelerated in a cyclotron to a final speed of 3.0×10^6 m/s. How much work, in electron-volts, is done on the proton by the (electric) force accelerating it? (Although the speed is 1% the speed of light, relativity theory is not required here.)

14E. A single force acts on a body that moves along a straight line. Figure 7-28 shows the velocity versus time for the body. Find the sign (plus or minus) of the work done by the force on the body in each of the intervals *AB, BC, CD,* and *DE.*

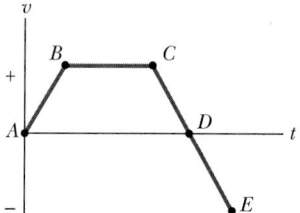

FIGURE 7-28 Exercise 14.

15E. A firehose 12 m long (Fig. 7-29) is uncoiled by pulling the nozzle end horizontally along a frictionless surface at the steady speed of 2.3 m/s. The mass of 1.0 m of the hose is 0.25 kg. How much work has been done on the hose by the applied force when the entire hose is moving?

FIGURE 7-29 Exercise 15.

16P. Figure 7-30 shows three forces applied to a greased trunk that moves leftward by 3.00 m over a frictionless floor. The force magnitudes are $F_1 = 5.00$ N, $F_2 = 9.00$ N, and $F_3 = 3.00$ N. During the displacement, (a) what is the net work done on the trunk by the three forces and (b) does the kinetic energy of the trunk increase or decrease?

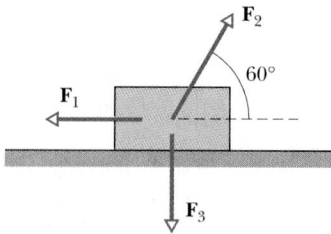

FIGURE 7-30 Problem 16.

17P. A force acts on a 3.0 kg particle-like object in such a way that the position of the object as a function of time is given by $x = 3.0t - 4.0t^2 + 1.0t^3$, with x in meters and t in seconds. Find the work done on the object by the force from $t = 0$ to $t = 4.0$ s. (*Hint:* What are the speeds at those times?)

18P. Figure 7-31 shows an overhead view of three horizontal forces acting on a cargo canister that is initially stationary but which now moves across a frictionless floor. The force magnitudes are $F_1 = 3.00$ N, $F_2 = 4.00$ N, and $F_3 = 10.0$ N. What is the net work done on the canister by the three forces during the first 4.00 m of displacement?

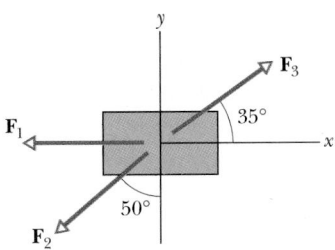

FIGURE 7-31   Problem 18.

19P. A constant force of magnitude 10 N makes an angle of 150° (measured counterclockwise) with the direction of increasing x as the force acts on a 2.0 kg object. How much work is done on the object by the force as the object moves from the origin to the point with position vector $(2.0 \text{ m})\mathbf{i} - (4.0 \text{ m})\mathbf{j}$?

SECTION 7-4 Work Done by Weight

20E. (a) In 1975 the roof of Montreal's Velodrome, with a weight of 41,000 tons, was lifted by 4.0 in. so that it could be centered. How much work was done on the roof by the forces making the lift? (b) In 1960 Mrs. Maxwell Rogers of Tampa, Florida, reportedly raised one end of a car that had fallen onto her son when a jack failed. If her panic lift effectively raised 900 lb of the 3600 lb car by 2 in., how much work was done by her force acting on the car?

21E. To push a 25.0 kg crate up a frictionless incline, angled at 25.0° to the horizontal, a worker exerts a force of 209 N, parallel to the incline. As the crate slides 1.50 m, how much work is done on the crate by (a) the worker's applied force, (b) the weight of the crate, and (c) the normal force exerted by the incline on the crate? (d) What is the total work done on the crate?

22E. A 45 kg block of ice slides down a frictionless incline 1.5 m long and 0.91 m high. A worker pushes up against the ice, parallel to the incline, so that the block slides down at constant speed. (a) Find the force exerted by the worker. How much work is done on the block by (b) the worker's force, (c) the weight of the block, (d) the normal force exerted by the surface of the incline on the block, and (e) the resultant force on the block?

23E. In Fig. 7-32, a cord runs around two massless, frictionless pulleys; a canister with mass $m = 20$ kg hangs from one pulley; and you exert a force $\mathbf{F}$ on the free end of the cord. (a) What must

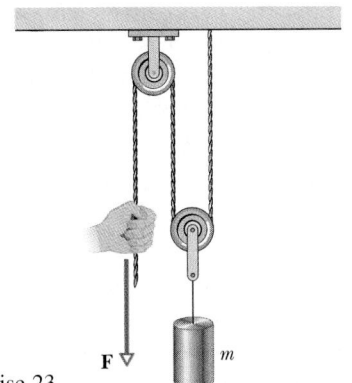

FIGURE 7-32   Exercise 23.

be the magnitude of $\mathbf{F}$ if you are to lift the canister at a constant speed? (b) To lift the canister by 2.0 cm, how far must you pull the free end of the cord? During that lift, what is the work done on the canister by (c) your force (via the cord) and (d) the weight $m\mathbf{g}$ of the canister? (*Hint:* When a cord loops around a pulley as shown, it pulls on the pulley with a net force that is twice the tension in the cord.)

24P. A helicopter lifts a 72 kg astronaut 15 m vertically from the ocean by means of a cable. The acceleration of the astronaut is $g/10$. How much work is done on the astronaut by (a) the force from the helicopter and (b) her weight? What are (c) the kinetic energy and (d) the speed of the astronaut just before she reaches the helicopter?

25P. A cord is used to vertically lower an initially stationary block of mass M at a constant downward acceleration of $g/4$. When the block has fallen a distance d, find (a) the work done by the cord's force on the block, (b) the work done by the weight of the block, (c) the kinetic energy of the block, and (d) the speed of the block.

26P. A cave rescue team lifts an injured spelunker directly upward and out of a sinkhole by means of a motor-driven cable. The lift is performed in three stages, each requiring a vertical distance of 10.0 m: (1) the initially stationary spelunker is accelerated to a speed of 5.00 m/s; (2) he is then lifted at the constant speed of 5.00 m/s; (3) finally he is decelerated to zero speed. How much work is done on the 80.0 kg rescuee by the force lifting him during each stage?

SECTION 7-5 Work Done by a Variable Force

27E. A 5.0 kg block moves in a straight line on a horizontal frictionless surface under the influence of a force that varies with position as shown in Fig. 7-33. How much work is done by the force as the block moves from the origin to $x = 8.0$ m?

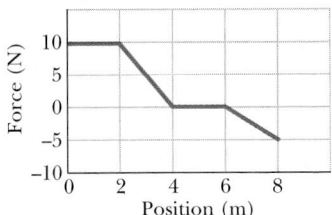

FIGURE 7-33   Exercise 27.

28E. A 10 kg mass moves along an x axis. Its acceleration as a function of its position is shown in Fig. 7-34. What is the net work performed on the mass by the force causing the acceleration as the mass moves from $x = 0$ to $x = 8.0$ m?

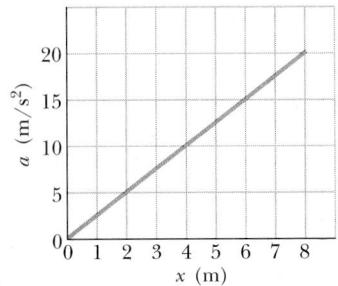

FIGURE 7-34   Exercise 28.

29P. (a) Estimate the work done by the force represented by the graph of Fig. 7-35 in displacing a particle from $x = 1$ m to $x = 3$ m. Refine your method to see how close you can come to the exact answer of 6 J. (b) The curve is given by $F = a/x^2$, with $a = 9$ N·m². Calculate the work using integration.

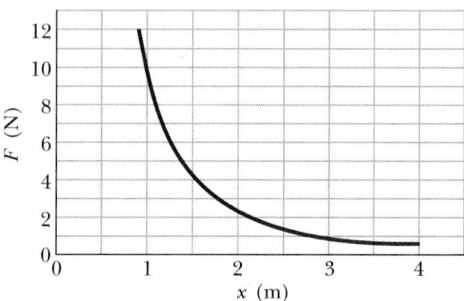

FIGURE 7-35   Problem 29.

30P. The force on a particle is $F = F_0(x/x_0 - 1)$. Find the work done by the force in moving the particle from $x = 0$ to $x = 2x_0$ by (a) plotting $F(x)$ and measuring the work from the graph and (b) integrating $F(x)$.

31P. What work is done by a force $\mathbf{F} = (2x \text{ N})\mathbf{i} + (3 \text{ N})\mathbf{j}$, with x in meters, that moves a particle from a position $\mathbf{r}_i = (2 \text{ m})\mathbf{i} + (3 \text{ m})\mathbf{j}$ to a position $\mathbf{r}_f = -(4 \text{ m})\mathbf{i} - (3 \text{ m})\mathbf{j}$?

32P. A force $\mathbf{F}$ in the direction of increasing x acts on an object moving along the x axis. If the magnitude of the force is $F = 10e^{-x/2.0}$ N, where x is in meters, find the work done by $\mathbf{F}$ as the object moves from $x = 0$ to $x = 2.0$ m (a) by plotting $F(x)$ and finding the area under the curve and (b) by integrating to find the work analytically.

33P. The only force acting on a 2.0 kg body as it moves along the x axis varies as shown in Fig. 7-36. The velocity of the body at $x = 0$ is 4.0 m/s. (a) What is the kinetic energy of the body at $x = 3.0$ m? (b) At what value of x will the body have a kinetic energy of 8.0 J? (c) What is the maximum kinetic energy attained by the body between $x = 0$ and $x = 5.0$ m?

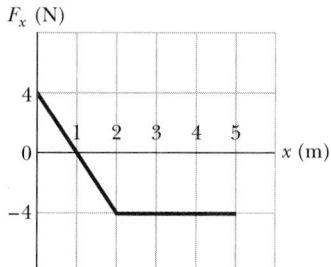

FIGURE 7-36   Problem 33.

34P. A 230 kg crate hangs from the end of a 12.0 m rope. You push horizontally on the crate with a varying force $\mathbf{F}$ to move it 4.00 m to the side (Fig. 7-37). (a) What is the magnitude of $\mathbf{F}$ when the crate is in this final position? During the crate's displacement, what are (b) the total work done on it, (c) the work done by the weight of the crate, and (d) the work done by the pull on the crate from the rope? (e) Knowing that the crate is motion-

less before and after its displacement, use the answers to (b), (c), and (d) to find the work your force $\mathbf{F}$ does on the crate. (f) Why is the work of your force not equal to the product of the horizontal displacement and the answer to (a)?

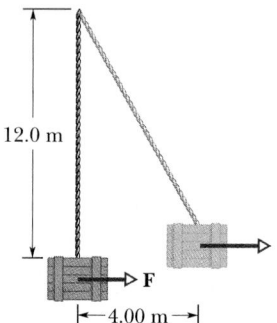

FIGURE 7-37   Problem 34.

SECTION 7-6   Work Done by a Spring Force

35E. A spring with a spring constant of 15 N/cm has a cage attached to one end (Fig. 7-38). (a) How much work does the spring force do on the cage if the spring is stretched from its relaxed length by 7.6 mm? (b) How much additional work is done by the spring force if it is stretched by an additional 7.6 mm?

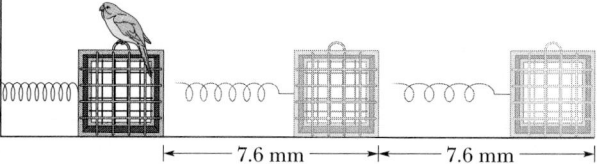

FIGURE 7-38   Exercise 35.

36E. During spring semester at MIT, residents of the parallel buildings of the East Campus dorms battle one another with large catapults that are made with surgical hose mounted on a window frame. A balloon filled with dyed water is placed in a pouch attached to the hose, which is then stretched through the width of the room. Assume that the stretching of the hose obeys Hooke's law and has a spring constant of 100 N/m. If the hose is stretched by 5.00 m and then released, how much work does the force from the hose do on the balloon in the pouch by the time the hose reaches its relaxed length?

37P. The only force acting on a 2.0 kg body as it moves along the positive x axis has an x component $F_x = -6x$ N, where x is in meters. The velocity of the body at $x = 3.0$ m is 8.0 m/s. (a) What is the velocity of the body at $x = 4.0$ m? (b) At what positive value of x will the body have a velocity of 5.0 m/s?

38P. A spring with a pointer attached is hanging next to a scale marked in millimeters. Three different weights are hung from the spring, in turn, as shown in Fig. 7-39. (a) If all weight is removed from the spring, which mark on the scale will the pointer indicate? (b) What is the weight W?

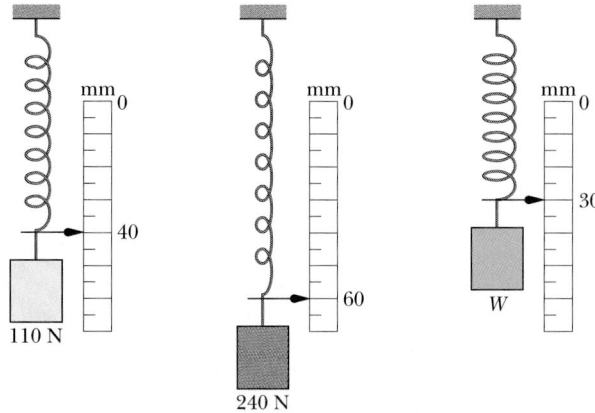

FIGURE 7-39 Problem 38.

39P. In Fig. 7-40, two identical springs, each with a relaxed length of 50 cm and a spring constant of 500 N/m, are connected by a short cord of length 10 cm. The upper spring is attached to the ceiling; a box that weighs 100 N hangs from the lower spring. Two additional cords, each 85 cm long, are also tied to the assembly; they are limp. (a) If the short cord is cut, so that the box then hangs from the springs and the two longer cords, does the box move up or down? (b) How far does the box move? (c) How much total work do the two spring forces (one directly, the other via a cord) do on the box during that move?

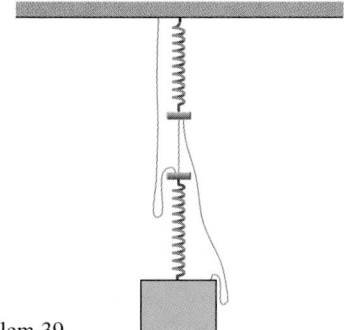

FIGURE 7-40 Problem 39.

40P. A 250 g block is dropped onto a vertical spring with spring constant $k = 2.5$ N/cm (Fig. 7-41). The block becomes attached to the spring, and the spring compresses 12 cm before momentarily stopping. While the spring is being compressed, what work is

done on the block (a) by its weight and (b) by the spring force? (c) What is the speed of the block just before it hits the spring? (Assume that friction is negligible.) (d) If the speed at impact is doubled, what is the maximum compression of the spring?

SECTION 7-7 Power

41E. The loaded cab of an elevator has a mass of 3.0×10^3 kg and moves 210 m up the shaft in 23 s at constant speed. At what average rate does the force from the cable do work on the cab?

42E. If a ski lift raises 100 passengers averaging 150 lb a height of 500 ft in 60 s, at constant speed, what average power is required of the force making the lift?

43E. An elevator cab in the New York Marriott Marquis has a mass of 4500 kg and can carry a maximum load of 1800 kg. If the cab is moving upward at full load at 3.80 m/s, what power is required of the force moving the cab to maintain that speed?

44E. (a) At a certain instant, a particle experiences a force $\mathbf{F} = (4.0$ N$)\mathbf{i} - (2.0$ N$)\mathbf{j} + (9.0$ N$)\mathbf{k}$ while having a velocity $\mathbf{v} = -(2.0$ m/s$)\mathbf{i} + (4.0$ m/s$)\mathbf{k}$. What is the instantaneous rate at which the force does work on the particle? (b) At some other time, the velocity consists of only a $\mathbf{j}$ component. If the force is unchanged, and the instantaneous power is -12 W, what is the velocity of the particle just then?

45P. A 100 kg block is pulled at a constant speed of 5.0 m/s across a horizontal floor by an applied force of 122 N directed 37° above the horizontal. What is the rate at which the force does work on the block?

46P. A horse pulls a cart with a force of 40 lb at an angle of 30° above the horizontal and moves along at a speed of 6.0 mi/h. (a) How much work does the force do in 10 min? (b) What is the average power (in horsepower) of the force?

47P. An initially stationary 2.0 kg object accelerates horizontally and uniformly to a speed of 10 m/s in 3.0 s. (a) In that 3.0 s interval, how much work is done on the object by the force accelerating it? What is the instantaneous power due to that force (b) at the end of the interval and (c) at the end of the first half of the interval?

48P. A force of 5.0 N acts on a 15 kg body initially at rest. Compute (a) the work done by the force in the first, second, and third seconds and (b) the instantaneous power due to the force at the end of the third second.

49P. A fully loaded, slow-moving freight elevator has a cab with a total mass of 1200 kg, which is required to travel upward 54 m in 3.0 min. The elevator's counterweight has a mass of only 950 kg. So, the elevator motor must help pull the cab upward. What average power (in horsepower) is required of the force that motor exerts on the cab via the cable?

50P. The force (but not the power) required to tow a boat at constant velocity is proportional to the speed. If a speed of 2.5 mi/h requires 10 hp, how much horsepower does a speed of 7.5 mi/h require?

51P. A 0.30 kg mass sliding on a horizontal frictionless surface is attached to one end of a horizontal spring (with $k = 500$ N/m)

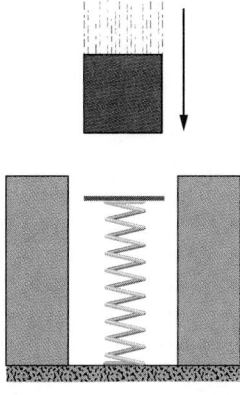

FIGURE 7-41 Problem 40.

whose other end is fixed. The mass has a kinetic energy of 10 J as it passes through its equilibrium position (the point at which the spring force is zero). (a) At what rate is the spring doing work on the mass as the mass passes through its equilibrium position? (b) At what rate is the spring doing work on the mass when the spring is compressed 0.10 m and the mass is moving away from the equilibrium position?

SECTION 7-8 Kinetic Energy at High Speeds

52E. An electron moves through 5.1 cm in 0.25 ns. (a) What is its speed in terms of the speed of light? (b) What is its kinetic energy in electron-volts? (c) What percentage error do you make if you use the classical formula to compute the kinetic energy?

53E. The work–kinetic energy theorem applies to particles at all speeds. How much work, in electron-volts, must be done to increase the speed of an electron from rest to (a) $0.500c$, (b) $0.990c$, and (c) $0.999c$?

54P. An electron has a speed of $0.999c$. (a) What is its kinetic energy? (b) If its speed is increased by 0.05%, by what percentage does its kinetic energy increase?

Electronic Computation

55P. A force $\mathbf{F} = (3.00 \text{ N})\mathbf{i} + (7.00 \text{ N})\mathbf{j} + (7.00 \text{ N})\mathbf{k}$ acts on a 2.00 kg object which moves from an initial position of $\mathbf{d}_i = (3.00 \text{ m})\mathbf{i} - (2.00 \text{ m})\mathbf{j} + (5.00 \text{ m})\mathbf{k}$ to a final position of $\mathbf{d}_f = -(5.00 \text{ m})\mathbf{i} + (4.00 \text{ m})\mathbf{j} + (7.00 \text{ m})\mathbf{k}$ in 4.00 s. Find (a) the work done on the object by the force in the 4.00 s interval, (b) the average power due to the force during the 4.00 s interval, and (c) the angle between vectors $\mathbf{d}_i$ and $\mathbf{d}_f$.

8
Potential Energy and Conservation of Energy

In the dangerous "sport" of bungee-jumping, a participant attaches a specially designed elastic cord to the ankles and leaps from a tall structure. How can we tell just how far down a jumper will go? The answer is, of course, important to the jumper, who is keenly aware of what lies below.❓

8-1 POTENTIAL ENERGY

In this chapter we continue the discussion of energy that we began in Chapter 7. To do so, we define a second form of energy: **potential energy** U is energy that can be associated with the configuration (or arrangement) of a system of objects that exert a force on one another. If the configuration of the system changes, then the potential energy of the system also changes.

One type of potential energy is the **gravitational potential energy** that is associated with the state of separation between objects, which attract one another via the gravitational force. For example, when Vasili Alexeev lifted 562 lb above his head in the 1976 Olympics, he increased the separation between the weights and the Earth. His effort changed the gravitational potential energy of the weights–Earth system by changing the configuration of the system, that is, by shifting the relative locations of the Earth and the weights (Fig. 8-1).

Another type of potential energy is **elastic potential energy,** which is associated with the state of compression or extension of an elastic (springlike) object. The force involved here is the spring force. If you compress or extend a spring, you change the relative locations of the coils within the spring. A spring force resists that change, and the result of your effort is an increase in the elastic potential energy of the spring.

Work and Potential Energy

In Chapter 7 we discussed the relation between work and a change in kinetic energy. Here we discuss the relation between work and a change in potential energy.

Let us throw a tomato upward (Fig. 8-2). We already know that as the tomato rises, the work W_g done on the tomato by its weight (the gravitational force acting on it) is negative because the weight transfers energy *from* the ki-

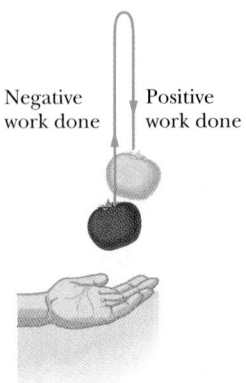

FIGURE 8-2 A tomato is thrown upward. As it rises, its weight does negative work on it, decreasing its kinetic energy by transferring energy to the gravitational potential energy of the tomato–Earth system. As the tomato descends, its weight does positive work on it, increasing its kinetic energy by transferring energy from the gravitational potential energy of the system.

netic energy of the tomato. We can now finish the story by saying that this energy is transferred by the weight *to* the gravitational potential energy of the tomato–Earth system.

The tomato slows, stops, and then begins to fall back down because of its weight. During the fall, the transfer is reversed: the work W_g done on the tomato by its weight is now positive—the weight transfers energy *from* the gravitational potential energy of the tomato–Earth system *to* the kinetic energy of the tomato.

For either rise or fall, the change ΔU in the gravitational potential energy is defined to equal the negative of the work done on the tomato by its weight. Using the general symbol W for work, we write this as

$$\Delta U = -W. \tag{8-1}$$

The same relationship applies to a block–spring system, as in Fig. 8-3. If we abruptly shove the block to send it rightward, the spring force does negative work on the block during the rightward motion, transferring energy from the kinetic energy of the block to the elastic potential energy of the spring. The block slows, stops, and then begins to move in the opposite direction because of the spring force. The transfer of energy is then reversed.

Conservative and Nonconservative Forces

Let us list the key elements of the two situations we just discussed:

1. The *system* consists of two or more objects.

2. A *force* acts between a particle-like object (tomato or block) in the system and the rest of the system.

3. When the system configuration changes, the force does *work* (call it W_1) on the particle-like object, transferring

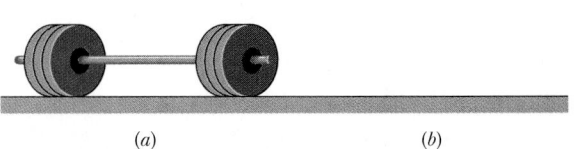

(a)   (b)

FIGURE 8-1 When Alexeev lifted the weights above his head, he increased the separation between the weights and the Earth and thus changed the configuration of the weights–Earth system from that in (a) to that in (b).

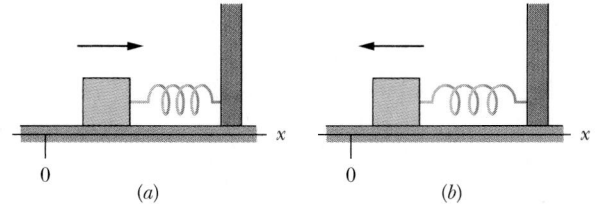

FIGURE 8-3 A block, attached to a spring and initially at rest at $x = 0$, is set in motion toward the right. (*a*) As the block moves rightward (as indicated by the arrow), the spring force does negative work on it, decreasing its kinetic energy by transferring energy to the elastic potential energy of the spring. The block momentarily stops when it has zero kinetic energy. (*b*) Then, as the block moves back toward $x = 0$, the spring force does positive work on it, increasing its kinetic energy by transferring energy from the elastic potential energy.

energy between the kinetic energy K of the object and some other form of energy of the system.

4. When the configuration change is reversed, the force reverses the energy transfer, doing work W_2 in the process.

If $W_1 = -W_2$ is always true, then the other form of energy is a potential energy, and the force is said to be a **conservative force.** As you might suspect, the gravitational force and the spring force are both conservative (since otherwise we could not have spoken of gravitational potential energy and elastic potential energy, as above.)

Not all forces are conservative. For an example, let us send a block sliding across a floor that is not frictionless. During the sliding, a kinetic frictional force from the floor does negative work on the block, slowing the block by transferring energy from its kinetic energy to thermal energy of the block–floor system. We know from experiment that this energy transfer cannot be reversed (thermal energy cannot be transferred back to kinetic energy of the block by the kinetic frictional force). So, although we have a system, a force that acts between parts of the system, and a transfer of energy by the force, the force is not conservative (it is **nonconservative**). Thus the thermal energy of the block–floor system is not a potential energy.

When only conservative forces act on a particle-like object, we can greatly simplify otherwise difficult problems involving motion of the object. The next section, in which we develop a test for identifying conservative forces, provides one means for simplifying such problems.

8-2 PATH INDEPENDENCE OF CONSERVATIVE FORCES

The primary test for determining whether a force is conservative or nonconservative is this: let the force act (alone) on a particle that moves along a *closed path* begin-

ning at some initial position and eventually returning to that position (so that the particle makes a *round trip* beginning and ending at the initial position). The force is conservative only if the total energy it transfers to and from the particle during the round trip is zero. In other words:

The net work done by a conservative force on a particle moving around any closed path is zero.

We know from experiment that the gravitational force passes this test. An example is the tossed tomato of Fig. 8-2. The tomato leaves the launch point with a speed v_0 and kinetic energy $\frac{1}{2}mv_0^2$. The gravitational force acting on the tomato slows it, stops it, and then makes it fall back down. When the tomato returns to the launch point, it again has speed v_0 and kinetic energy $\frac{1}{2}mv_0^2$. Thus the gravitational force transfers as much energy *from* the tomato during the ascent as it transfers *to* the tomato during the descent back to the launch point. The net work done on the tomato by the gravitational force during the round trip is zero.

Figure 8-4*a* shows an arbitrary round trip for a particle acted upon by a single force. The particle moves from an initial point *a* to point *b* along path 1 and then back to point *a* along path 2. The force does work on the particle as the particle moves along each path. Without worrying about where positive work is done and where negative work is done, let us just represent the work done from *a* to *b* along path 1 as $W_{ab,1}$ and the work done from *b* back to *a* along path 2 as $W_{ba,2}$. If the force is conservative, then the net work done during the round trip must be zero:

$$W_{ab,1} + W_{ba,2} = 0,$$

and thus

$$W_{ab,1} = -W_{ba,2}. \tag{8-2}$$

In words, the work done along the outward path must be the negative of the work done along the path back.

Let us now consider the work $W_{ab,2}$ done on the particle by the force as the particle moves from *a* to *b* along path 2, as indicated in Fig. 8-4*b*. If the force is conservative, that

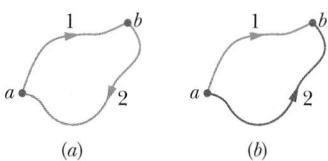

FIGURE 8-4 (*a*) A particle, acted on by a conservative force, moves in a round trip from point *a* to point *b* along path 1 and then back to point *a* along path 2. (*b*) The particle can move from point *a* to point *b* by following either path 1 or path 2.

work is the negative of $W_{ba,2}$:

$$W_{ab,2} = -W_{ba,2}. \tag{8-3}$$

Substituting $W_{ab,2}$ for $-W_{ba,2}$ in Eq. 8-2, we obtain

$$W_{ab,1} = W_{ab,2}. \tag{8-4}$$

This simple equation is a powerful one, because it allows us to simplify difficult problems when only a conservative force is involved. It tells us that the work done on a particle by a conservative force as the particle moves from some initial point a to some other point b does not depend on *which* path is taken.

> The work done by a conservative force on a particle moving between two points does not depend on the path taken by the particle.

Suppose you need to calculate the work done by a conservative force along a given path between two points, and the calculation is difficult. You can find the work by using another path between those two points for which the calculation is easier. Sample Problem 8-1 gives an example.

SAMPLE PROBLEM 8-1

Figure 8-5a shows a 2.0 kg block of slippery cheese that slides along a frictionless track from point a to point b. The cheese travels through a total distance of 2.0 m along the track, and a net vertical distance of 0.80 m. How much work is done on the cheese by its weight during the slide?

SOLUTION: From Eq. 7-16, we know that the work done by a weight force $m\mathbf{g}$, over a displacement $\mathbf{d}$ at angle ϕ to $m\mathbf{g}$, is given by $W = mgd \cos \phi$. But using that equation here appears to be impossible because we do not know the actual shape of the track. And even if we did, the calculation would be difficult because the angle ϕ is not constant over the path taken by the cheese.

However, because weight is a conservative force, we can find the work by choosing another path between a and b—one that makes the calculation easy. Let us choose the dashed path in Fig. 8-5b; it consists of two straight segments. Along the horizontal segment, the angle ϕ is a constant 90°. Even though we do not know the displacement along that horizontal segment, Eq. 7-16 tells us that the work W_h done there is

$$W_h = mgd \cos 90° = 0.$$

Along the vertical segment, the displacement d is 0.80 m and, with $m\mathbf{g}$ and $\mathbf{d}$ both downward, the angle ϕ is a constant 0°. So Eq. 7-16 gives us, for the work W_v done along the vertical part of the dashed path,

$$W_v = mgd \cos 0°$$
$$= (2.0 \text{ kg})(9.8 \text{ m/s}^2)(0.80 \text{ m})(1) = 15.7 \text{ J}.$$

The total work done on the cheese by its weight as the cheese moves from point a to point b along the dashed path is then

$$W = W_h + W_v = 0 + 15.7 \text{ J} \approx 16 \text{ J}. \quad \text{(Answer)}$$

This is also the work done as the cheese moves along the track from a to b.

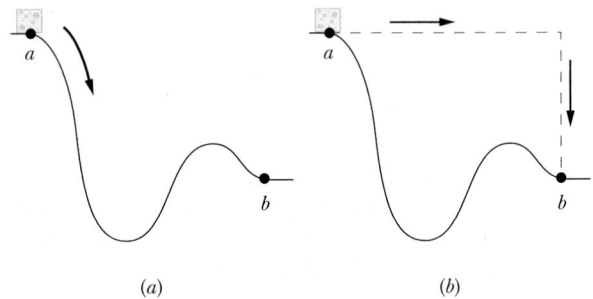

(a)   (b)

FIGURE 8-5 Sample Problem 8-1. (a) A block of cheese slides along a frictionless track from point a to point b. (b) Finding the work done on the cheese by its weight is easier along the dashed path than along the actual path taken by the cheese; the result is the same for both paths.

CHECKPOINT 1: The figure shows three paths connecting points a and b. A single force **F** does the indicated work on a particle moving along each path in the indicated direction. On the basis of this information, is force **F** conservative?

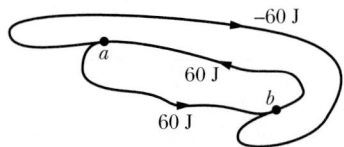

8-3 DETERMINING POTENTIAL ENERGY VALUES

Consider now a particle-like object (say, a melon) that is part of a system (say, a melon–Earth system) in which a conservative force **F** acts. When that force does work W on the object, the change ΔU in the potential energy associated with the system is the negative of the work done. We wrote this fact as Eq. 8-1 ($\Delta U = -W$). For the most general case, in which the force may vary with position, we may write the work W as in Eq. 7-27:

$$W = \int_{x_i}^{x_f} F(x) \, dx. \tag{8-5}$$

This equation gives the work done by the force when the object moves from point x_i to point x_f, changing the configuration of the system. (Because F is a conservative force, the work is the same for all paths between those two points.)

Substituting Eq. 8-5 into Eq. 8-1, we find that the change in potential energy due to the change in configuration is

$$\Delta U = -\int_{x_i}^{x_f} F(x)\, dx. \qquad (8\text{-}6)$$

Gravitational Potential Energy

Now consider a particle with mass m moving vertically along a y axis. As the particle moves from point y_i to point y_f, its weight mg does work on it. To find the corresponding change in the gravitational potential energy of the particle–Earth system, we change the integration in Eq. 8-6 to be along the y axis and we substitute $-mg$ for force F. We then have

$$\Delta U = -\int_{y_i}^{y_f} (-mg)\, dy = mg \int_{y_i}^{y_f} dy = mg \left[y \right]_{y_i}^{y_f},$$

which yields

$$\Delta U = mg(y_f - y_i) = mg\, \Delta y. \qquad (8\text{-}7)$$

Only a change ΔU in gravitational potential energy (or any other type of potential energy) is physically important. However, to simplify a calculation or a discussion, we can say that a certain gravitational potential energy U is associated with any given configuration of the system, with the particle at a given position y. To do so, we rewrite Eq. 8-7 as

$$U - U_i = mg(y - y_i). \qquad (8\text{-}8)$$

Then we take U_i to be the gravitational potential energy of the system when it is in a **reference configuration,** with the particle at a **reference point** y_i. Usually, we set $U_i = 0$ and $y_i = 0$. Doing this changes Eq. 8-8 to

$$U - 0 = mg(y - 0),$$

which gives us

$$U = mgy \qquad \text{(gravitational potential energy).} \qquad (8\text{-}9)$$

Equation 8-9 tells us that the gravitational potential energy associated with a particle–Earth system depends on the vertical position y (or height) of the particle relative to the reference position of $y = 0$, not on the horizontal position.

Elastic Potential Energy

We next consider the block–spring system shown in Fig. 8-3, with the block moving on the end of a spring of spring constant k. As the block moves from point x_i to point x_f, the spring force $F = -kx$ does work on the block. To find the

corresponding change in the elastic potential energy of the block–spring system, we substitute $-kx$ for $F(x)$ in Eq. 8-6. We then have

$$\Delta U = -\int_{x_i}^{x_f} (-kx)\, dx = k \int_{x_i}^{x_f} x\, dx = \tfrac{1}{2}k \left[x^2 \right]_{x_i}^{x_f},$$

and

$$\Delta U = \tfrac{1}{2}kx_f^2 - \tfrac{1}{2}kx_i^2. \qquad (8\text{-}10)$$

To associate a potential energy U with any given configuration of the system, with the block at position x, we set the reference point for the block as $x_i = 0$, which is always at the equilibrium position of the block. And we set the corresponding elastic potential energy of the system as $U_i = 0$. Thus Eq. 8-10 becomes

$$U - 0 = \tfrac{1}{2}kx^2 - 0,$$

which gives us

$$U(x) = \tfrac{1}{2}kx^2 \qquad \text{(elastic potential energy).} \qquad (8\text{-}11)$$

PROBLEM SOLVING TACTICS

TACTIC 1: *Using the Term "Potential Energy"*

A potential energy is associated with a system as a whole. However, you might see statements that associate it with only part of the system. For example, you might read, "An apple hanging in a tree has a gravitational potential energy of 30 J." Such statements are often acceptable, but you should always keep in mind that the potential energy is actually associated with a system—here the apple–Earth system. Also keep in mind that assigning a particular potential energy value, such as 30 J here, to an object or even a system makes sense *only* if the reference potential energy value is known.

SAMPLE PROBLEM 8-2

A 2.0 kg sloth clings to a limb that is 5.0 m above the ground (Fig. 8-6).

(a) What is the gravitational potential energy U of the sloth–Earth system if we take the reference point $y = 0$ to be (1) at the ground, (2) at the bottom of a balcony that is 3.0 m above the ground, (3) at the limb, and (4) 1.0 m above the limb? Take the gravitational potential energy to be zero at $y = 0$.

SOLUTION: With Eq. 8-9, we can calculate U for each choice of $y = 0$. For example, for (1) the sloth is initially at $y = 5.0$ m, and

$$U = mgy = (2.0 \text{ kg})(9.8 \text{ m/s}^2)(5.0 \text{ m})$$

= 98 J   (Answer)

For the other choices, the values of U are

    (2) $U = mgy = mg(2.0 \text{ m}) = 39$ J,

    (3) $U = mgy = mg(0) = 0$ J,

    (4) $U = mgy = mg(-1.0 \text{ m}) = -19.6$ J

                ≈ -20 J.   (Answer)

(b) The sloth drops to the ground. For each choice of reference point, what is the change in the potential energy of the sloth–Earth system due to the fall?

SOLUTION: For all four situations, we have $\Delta y = -5.0$ m. So for (1) to (4), Eq. 8-7 tells us that

$$\Delta U = mg\,\Delta y = (2.0 \text{ kg})(9.8 \text{ m/s}^2)(-5.0 \text{ m})$$

$$= -98 \text{ J}. \qquad \text{(Answer)}$$

Thus, although the value of U depends on the choice of where we let $y = 0$, the *change* in potential energy does not. Remember, only the change ΔU in potential energy is physically important—not the value of U, which depends on arbitrary choices about the reference configuration.

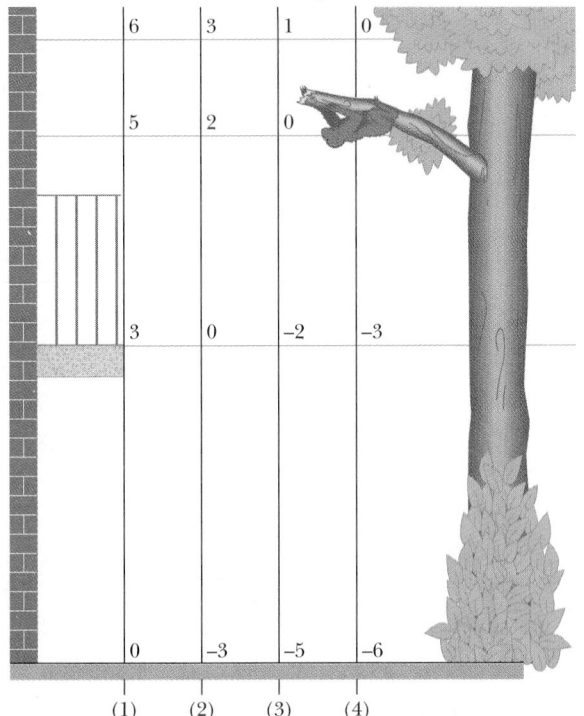

FIGURE 8-6 Sample Problem 8-2. Four choices of reference point $y = 0$. Each y axis is marked in units of meters.

C**HECKPOINT 2:** A particle is to move along the x axis from $x = 0$ to x_1 while a conservative force, directed along the x axis, acts on the particle. The figure shows three situations in which the magnitude of that force varies with x. The force has the same maximum magnitude F_1 in all three situations. Rank the situations

according to the change in the associated potential energy during the particle's motion, most positive first, most negative last.

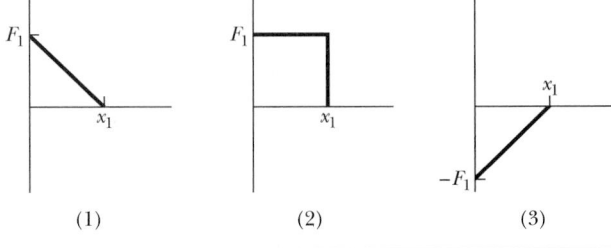

8-4 CONSERVATION OF MECHANICAL ENERGY

The **mechanical energy** E of a system is the sum of its potential energy U and the kinetic energy K of the objects within it:

$$E = K + U \qquad \text{(mechanical energy).} \qquad (8\text{-}12)$$

In this section, we examine what happens to this mechanical energy when a conservative force acts within the system. (Later we shall consider the effects of nonconservative forces.)

    When a conservative force does work W on an object within the system, it transfers energy between kinetic en-

In olden days, a native Alaskan would be tossed via a blanket to be able to see farther over the flat terrain. Nowadays, it is done just for fun. During the ascent of the child in the photograph, energy is transferred from kinetic energy to gravitational potential energy. The maximum height is reached when that transfer is complete. Then the transfer is reversed during the fall.

ergy K of the object and potential energy U of the system. From Eq. 7-4, the change ΔK in kinetic energy is

$$\Delta K = W \qquad (8\text{-}13)$$

and from Eq. 8-1, the change ΔU in potential energy is

$$\Delta U = -W. \qquad (8\text{-}14)$$

Combining Eqs. 8-13 and Eq. 8-14, we find that

$$\Delta K = -\Delta U. \qquad (8\text{-}15)$$

In other words, one of these forms of energy increases exactly as much as the other decreases.

We can rewrite Eq. 8-15 as

$$K_2 - K_1 = -(U_2 - U_1), \qquad (8\text{-}16)$$

where the subscripts 1 and 2 refer to two different instants during the particle's motion, and thus to two different configurations of the system. Rearranging Eq. 8-16 yields

$$K_2 + U_2 = K_1 + U_1 \quad \begin{matrix}\text{(conservation of}\\ \text{mechanical energy).}\end{matrix} \quad (8\text{-}17)$$

The left side of Eq. 8-17 is the mechanical energy of the system at one instant; the right side is the mechanical energy at another instant and in a different configuration. The equation tells us that these two energies are equal:

When only a conservative force acts within a system, the kinetic energy and potential energy can change. However, their sum, the mechanical energy E of the system, does not change.

This result is called the **principle of conservation of mechanical energy.** (Now you can see where *conservative* forces got their name.) With the aid of Eq. 8-15, we can write this principle in one more form, as

$$\Delta E = \Delta K + \Delta U = 0. \qquad (8\text{-}18)$$

The principle of conservation of mechanical energy allows us to solve problems that would be quite difficult to solve using only Newton's laws:

When the mechanical energy of a system is conserved, we can relate the total of kinetic energy and potential energy at one instant to that at another instant *without considering the intermediate motion* and *without finding the work done by the forces involved.*

Figure 8-7 shows an example in which the principle of conservation of mechanical energy can be applied: as a

pendulum swings, the energy of the pendulum–Earth system is transferred back and forth between kinetic energy K and gravitational potential energy U, with the sum $K + U$ being constant. If we are given the gravitational potential energy when the pendulum bob is at its highest point (Fig. 8-7c), we can find the kinetic energy of the bob at the lowest point (Fig. 8-7e) by using Eq. 8-17.

For example, let us choose the lowest point as the reference point and set the corresponding gravitational potential energy $U_2 = 0$. Suppose then that the potential energy at the highest point is $U_1 = 20$ J relative to the reference point. Because the bob momentarily stops at its highest point, the kinetic energy there is $K_1 = 0$. Substituting these values into Eq. 8-17 gives us the kinetic energy K_2 at the lowest point:

$$K_2 + 0 = 0 + 20\,\text{J} \quad \text{or} \quad K_2 = 20\,\text{J}.$$

Note that we get this result without considering the motion between the highest and lowest points (such as in Fig. 8-7d) and without finding the work done by any forces involved in the motion.

CHECKPOINT **3:** The figure shows four situations— one in which an initially stationary block is dropped and three in which the block is allowed to slide down frictionless ramps. (a) Rank the situations according to the kinetic energy of the block at point B, greatest first. (b) Rank them according to the speed of the block at point B, greatest first.

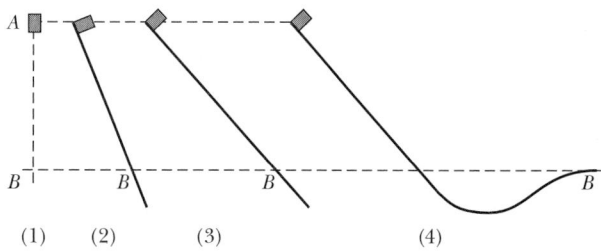

SAMPLE PROBLEM 8-3

In Fig. 8-8, a child of mass m is released from rest at the top of a water slide, at height $h = 8.5$ m above the bottom of the slide. Assuming that the slide is frictionless because of the water on it, find the child's speed at the bottom of the slide.

SOLUTION: If we use only the physics of Chapters 2 through 6, this problem is impossible to solve because we are given no information about the shape of the slide. However, with the physics of Chapters 7 and 8, it is easy. We first note that, in the absence of friction, the only force exerted on the child by the slide is a normal force, which is always perpendicular to the surface of the slide. Since the child always moves *along* the slide, this force is always perpendicular to

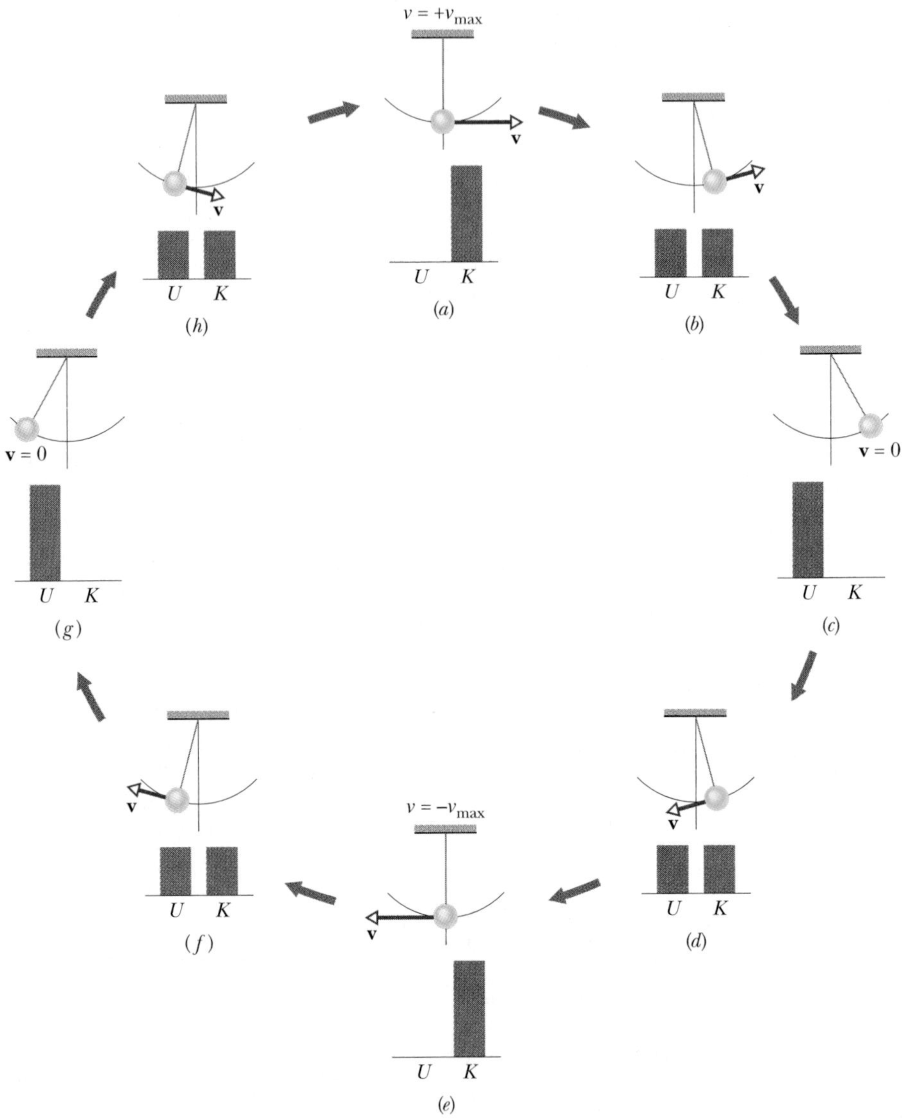

FIGURE 8-7 A pendulum, with its mass concentrated in a bob at the lower end, swings back and forth. One full cycle of the motion is shown. During the cycle the values of the potential and kinetic energies of the pendulum–Earth system vary as the bob rises and falls, but the mechanical energy E of the system remains constant. The energy E can be described as continuously shifting between the kinetic and potential forms. In stages (a) and (e), all the energy is kinetic energy. The bob then has its greatest speed and is at its lowest point. In stages (c) and (g), all the energy is potential energy. The bob then has zero speed and is at its highest point. In stages (b), (d), (f), and (h), half the energy is kinetic energy and half is potential energy. If the swinging involved a frictional force at the point where the pendulum is attached to the ceiling, or a drag force due to the air, then E would not be conserved, and eventually the pendulum would stop.

the child's displacement and thus cannot do work on the child. The only force that does work is the weight $m\mathbf{g}$ of the child, a conservative force. We therefore have a child–Earth system for which the mechanical energy E is conserved throughout the child's motion.

Thus, the mechanical energy E_b at the bottom of the slide

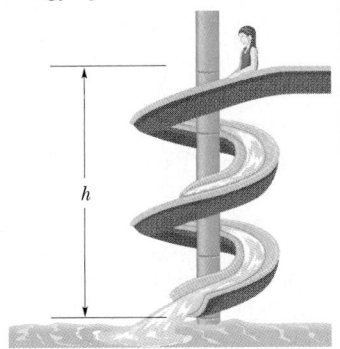

FIGURE 8-8 Sample Problem 8-3. A child slides down a water slide as she descends a height h.

and the mechanical energy E_t at the top are equal: $E_b = E_t$. Writing this conservation of energy equation in the form of Eq. 8-17 gives us

$$K_b + U_b = K_t + U_t,$$

or

$$\tfrac{1}{2}mv_b^2 + mgy_b = \tfrac{1}{2}mv_t^2 + mgy_t.$$

Dividing by m and rearranging yield

$$v_b^2 = v_t^2 + 2g(y_t - y_b).$$

Putting $v_t = 0$ and $y_t - y_b = h$ leads to

$$v_b = \sqrt{2gh} = \sqrt{(2)(9.8 \text{ m/s}^2)(8.5 \text{ m})}$$
$$= 13 \text{ m/s.} \qquad \text{(Answer)}$$

This is the same speed that the child would reach if she fell 8.5 m. On an actual slide, some frictional forces would act and the child would not be moving quite so fast.

This problem is hard to solve directly with Newton's laws. Using conservation of mechanical energy makes the solution much easier. However, if you were asked to find the time taken for the child to reach the bottom of the slide, energy methods would be of no use; you would need to know the shape of the slide and you would have a difficult problem.

SAMPLE PROBLEM 8-4

The spring of a spring gun is compressed a distance d of 3.2 cm from its relaxed state, and a ball of mass $m = 12$ g is put in the barrel. With what speed will the ball leave the barrel when the gun is fired? The spring constant k is 7.5 N/cm. Assume no friction and a horizontal gun barrel. Also assume that the ball leaves the spring and the spring stops when the spring reaches its relaxed length.

SOLUTION: Let E_i be the mechanical energy of the gun–ball system in the initial state (before the gun has been fired), and E_f be the system's mechanical energy in the final state (as the ball leaves the barrel). Initially, the mechanical energy is the spring's potential energy $U_i = \frac{1}{2}kd^2$. In the final state, the mechanical energy is the ball's kinetic energy $K_f = \frac{1}{2}mv^2$. Because mechanical energy is conserved, we have

$$E_i = E_f,$$
$$U_i + K_i = U_f + K_f,$$

and
$$\tfrac{1}{2}kd^2 + 0 = 0 + \tfrac{1}{2}mv^2.$$

Solving for v yields

$$v = d\sqrt{\frac{k}{m}} = (0.032 \text{ m})\sqrt{\frac{750 \text{ N/m}}{12 \times 10^{-3} \text{ kg}}}$$
$$= 8.0 \text{ m/s.} \qquad \text{(Answer)}$$

SAMPLE PROBLEM 8-5

A 61.0 kg bungee-cord jumper is on a bridge 45.0 m above a river. In its relaxed state, the elastic bungee cord has length $L = 25.0$ m. Assume that the cord obeys Hooke's law, with a spring constant of 160 N/m.

(a) If the jumper stops before reaching the water, what is the height h of her feet above the water at her lowest point?

SOLUTION: As shown in Fig. 8-9, let d be the amount by which the cord has stretched when she momentarily stops at her lowest point. As a result of her jump, the change ΔU_g in her gravitational potential energy is

$$\Delta U_g = mg\,\Delta y = -mg(L + d),$$

where m is her mass. The change in the elastic potential energy U_e of the cord is

$$\Delta U_e = \tfrac{1}{2}kd^2.$$

Her kinetic energy K is zero both initially and when she stops.

From Eq. 8-18 and the above, we have for the cord–jumper–Earth system,

$$\Delta K + \Delta U_e + \Delta U_g = 0$$
$$0 + \tfrac{1}{2}kd^2 - mg(L + d) = 0$$
$$\tfrac{1}{2}kd^2 - mgL - mgd = 0.$$

Inserting the given data, we obtain

$$\tfrac{1}{2}(160 \text{ N/m})d^2 - (61.0 \text{ kg})(9.8 \text{ m/s}^2)(25.0 \text{ m})$$
$$- (61.0 \text{ kg})(9.8 \text{ m/s}^2)d = 0,$$

which we solve for d with the quadratic formula, getting

$$d = 17.9 \text{ m.}$$

(The quadratic formula also gives a negative value for d, which has no meaning here.) Her feet are then a distance of $(L + d) = 42.9$ m below their initial height. Thus

$$h = 45.0 \text{ m} - 42.9 \text{ m} = 2.1 \text{ m.} \qquad \text{(Answer)}$$

If she happens to be especially tall, the jumper could be in for a dunking.

(b) What is the net force on her at her lowest point (in particular, is it zero)?

SOLUTION: Her weight mg acts downward and has magnitude $mg = 597.8$ N. The upward force exerted on her by the cord at the moment of stopping is given by Hooke's law, $F = -kx$, where x is the displacement of the free end of the cord. Here, the displacement is downward, so $x = -d$ and

$$F = -kx = -(160 \text{ N/m})(-17.9 \text{ m}) = 2864 \text{ N.}$$

The net force on the jumper is then

$$2864 \text{ N} - 597.8 \text{ N} \approx 2270 \text{ N.} \qquad \text{(Answer)}$$

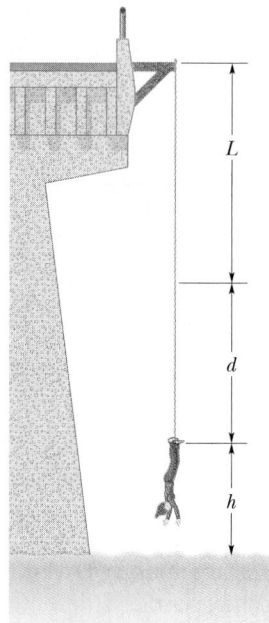

FIGURE 8-9 Sample Problem 8-5. A bungee-cord jumper at the lowest point of the jump.

Thus at the lowest point, where she momentarily stops, there is a net upward force of 2270 N (almost four times her weight) acting on her. This force will yank her back upward.

PROBLEM SOLVING TACTICS

TACTIC 2: *Conservation of Mechanical Energy*

Asking the following questions will help you to solve problems involving the conservation of mechanical energy.

For what system is mechanical energy conserved? You should be able to separate your system from its environment. Imagine drawing a closed surface such that whatever is inside is your system and whatever is outside is the environment of that system. In Sample Problem 8-3 the system is the *child + Earth*. In Sample Problem 8-4, it is the *ball + gun*. In Sample Problem 8-5, it is the *cord + jumper + Earth*.

Is friction or drag present? If friction or drag is present, mechanical energy is not conserved.

Is your system isolated? Conservation of mechanical energy applies only to isolated systems. That means that no *external forces* (forces exerted by objects outside the system) should do work on the objects in the system. In Sample Problem 8-4, if you chose the ball alone as your system, it is not isolated because the spring force does work on it. Thus the mechanical energy of the ball is not conserved.

What are the initial and final states of your system? The system changes from some initial state (or configuration) to some final state. You apply the principle of conservation of mechanical energy by saying that *E* has the same value in both these states. Be very clear about what these two states are.

8-5 READING A POTENTIAL ENERGY CURVE

Once again consider a particle that is part of a system in which a conservative force acts. This time suppose that the particle is constrained to move along an *x* axis while the conservative force does work on it. We can learn a lot about the motion of the particle from a plot of the system's potential energy $U(x)$. However, before we discuss such plots, we need one more relationship.

Finding the Force Analytically

Equation 8-6 tells us how to find the potential energy $U(x)$ in a one-dimensional situation if we know the force $F(x)$. But now we want to go the other way. That is, we know the potential energy function $U(x)$ and want to find the force.

For one-dimensional motion, the work W done by a force that acts on a particle as the particle moves through a distance Δx is $F(x)\, \Delta x$. We can then write Eq. 8-1 as

$$\Delta U(x) = -W = -F(x)\, \Delta x.$$

Solving for $F(x)$ and passing to the differential limit yield

$$F(x) = -\frac{dU(x)}{dx} \qquad \text{(one-dimensional motion)}. \qquad (8\text{-}19)$$

We can check this result by putting $U(x) = \frac{1}{2}kx^2$, which is the elastic potential energy function for a spring force. Equation 8-19 then yields, as expected, $F(x) = -kx$, which is Hooke's law. Similarly, we can substitute $U(x) = mgx$, which is the gravitational potential energy function for a particle of mass m at height x above the Earth's surface. Equation 8-19 then yields $F = -mg$, which is the weight that acts on the particle.

The Potential Energy Curve

Figure 8-10*a* is a plot of a potential energy function $U(x)$ for a system in which a particle is in one-dimensional motion while a conservative force $F(x)$ does work on it. We can easily find $F(x)$ by (graphically) taking the slope of the $U(x)$ curve at various points, as Eq. 8-19 tells us to do. Figure 8-10*b* is a plot of $F(x)$ found in this way.

Turning Points

In the absence of a nonconservative force, the mechanical energy E of the system has a constant value given by

$$U(x) + K(x) = E. \qquad (8\text{-}20)$$

Here $K(x)$ is the *kinetic energy function* of the particle (this $K(x)$ gives the kinetic energy as a function of the particle's location x). We may rewrite Eq. 8-20 as

$$K(x) = E - U(x). \qquad (8\text{-}21)$$

Suppose that E (which has a constant value, remember) happens to be 5.0 J. It would be represented in Fig. 8-10 by a horizontal line that runs through the value 5.0 J on the energy axis. (It is shown in Fig. 8-10*a*.)

Equation 8-21 tells us how to determine the kinetic energy K for any location x of the particle: on the $U(x)$ curve, find U for that location x and then subtract U from E. For example, if the particle is at any point to the right of x_5, then $K = 1.0$ J. The value of K is greatest (5.0 J) when the particle is at x_2, and least (0 J) when the particle is at x_1.

Since K can never be negative (because v^2 is always positive), the particle can never move to the left of x_1, where $E - U$ is negative. As the particle moves toward x_1 from x_2, K decreases (the particle slows) until $K = 0$ at x_1 (the particle stops).

Note that when the particle reaches x_1, the force on the particle, given by Eq. 8-19, is positive (because the slope dU/dx is negative). This means that the particle does not remain at x_1 but instead begins to move to the right, opposite its earlier motion. Hence x_1 is a **turning point,** a place where $K = 0$ and the particle changes direction. There is

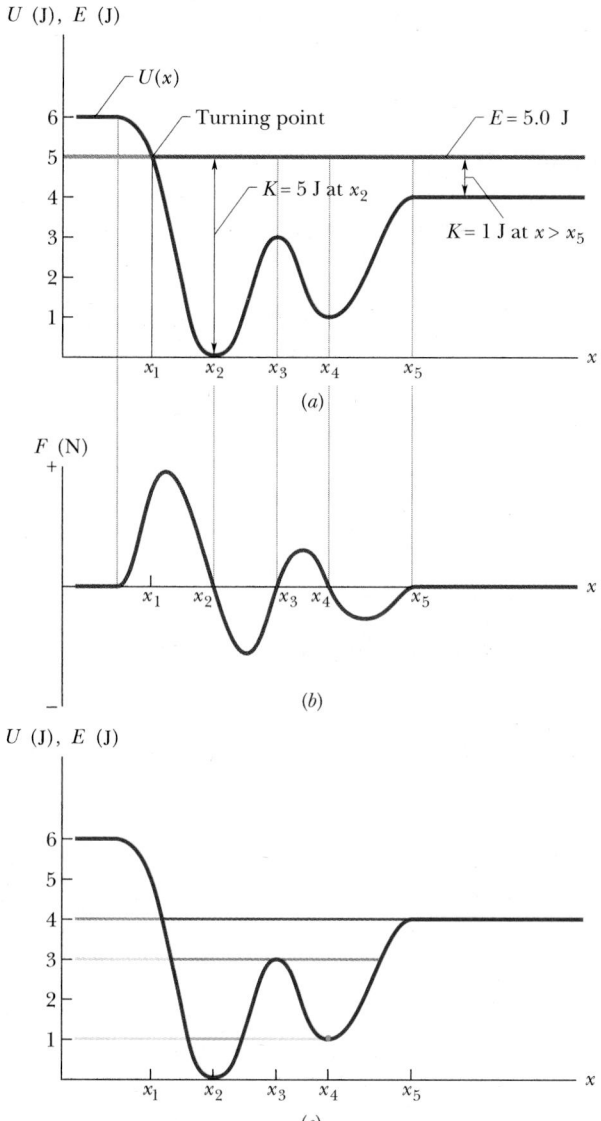

FIGURE 8-10 (a) A plot of $U(x)$, the potential energy function of a system containing a particle confined to move along the x axis. There is no friction, so mechanical energy is conserved. (b) A plot of the force $F(x)$ acting on the particle, derived from the potential energy plot by taking its slope at various points. (c) The $U(x)$ plot of (a) with three possible values of E shown.

no turning point (where $K = 0$) on the right side of the graph. When the particle heads toward increasing values of x, it will continue indefinitely.

Equilibrium Points

Figure 8-10c shows three additional values for E superposed on the plot of the potential energy $U(x)$. Let us see how they would change the situation. If $E = 4.0$ J (purple line), the turning point shifts from x_1 to a point between x_1 and x_2. Also, at any point to the right of x_5, the system's

mechanical energy is equal to its potential energy; thus, the particle has no kinetic energy and (by Eq. 8-19) no force acts on it, and so it must be stationary there. A particle at such a position is said to be in **neutral equilibrium.** (A marble placed on a horizontal tabletop is in that state.)

If $E = 3.0$ J (pink line), there are two turning points: one is between x_1 and x_2, and the other is between x_4 and x_5. In addition, x_3 is a point at which $K = 0$. If the particle is located exactly there, the force on it is also zero, and the particle remains stationary. However, if it is displaced even slightly in either direction, a nonzero force pushes it farther in the same direction, and the particle continues to move. A particle at such a position is said to be in **unstable equilibrium.** (A marble balanced on top of a bowling ball is an example.)

Next consider the particle's behavior if $E = 1.0$ J (green line). If we place it at x_4, it is stuck there. It cannot move left or right on its own because to do so would require a negative kinetic energy. If we push it slightly left or right, a restoring force appears that moves it back to x_4. A particle at such a position is said to be in **stable equilibrium.** (A marble placed at the bottom of a hemispherical bowl is an example.) If we place the particle in the cuplike *potential well* centered at x_2, it can still move left and right somewhat, but only partway to x_1 or x_3.

CHECKPOINT 4: The figure gives the potential energy function $U(x)$ for a system in which a particle is in one-dimensional motion. (a) Rank regions AB, BC, and CD according to the magnitude of the force on the particle, greatest first. (b) What is the direction of the force when the particle is in region AB?

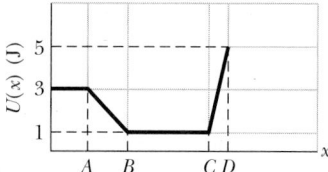

8-6 WORK DONE BY NONCONSERVATIVE FORCES

When only a conservative force does work on an object within a system, the mechanical energy E of the system stays constant. However, when a nonconservative force does work on an object, the mechanical energy of the system changes. In that situation we *cannot* use Eqs. 8-17 and 8-18. What, then, can we say about the energy changes? We answer this question by examining nonconservative forces of two types: an applied force (applied by, say, you) and a kinetic frictional force.

Work Done by an Applied Force

Suppose you push a bowling ball directly upward. During the upward motion and before the ball leaves your hand, your applied force does work W_{app} on the ball while the weight of the ball does work W_g on the ball. From observation we find that the ball can be treated as a particle. Thus, we can relate the work done on the ball to the change ΔK in its kinetic energy by using the work–kinetic energy theorem (Eq. 7-15):

$$W_{app} + W_g = \Delta K. \qquad (8\text{-}22)$$

Because the weight of the ball is a conservative force, we can use Eq. 8-1 to relate W_g to the corresponding change ΔU in the gravitational potential energy of the ball–Earth system:

$$W_g = -\Delta U. \qquad (8\text{-}23)$$

Substituting for W_g in Eq. 8-22, we get

$$W_{app} = \Delta K + \Delta U. \qquad (8\text{-}24)$$

The right side of this equation is the change ΔE in the mechanical energy of the ball–Earth system, so

$$W_{app} = \Delta E. \qquad (8\text{-}25)$$

Equation 8-25 says that the work done by your applied force changes the mechanical energy of the ball–Earth system. Specifically, your force transfers energy to the kinetic energy of the ball (you make it move faster) and to the gravitational potential energy of ball–Earth system (you increase the separation between the ball and the Earth).

Previously we defined work as being energy transferred to or from an object via a force. Now we can expand that definition to include situations in which the object is part of a system:

Work is energy transferred to or from a system via a force.

In some situations, the system consists of just an object, as in Chapter 7; in other situations, the system consists of two (or more) objects, as in the ball–Earth system here.

Work Done by a Kinetic Frictional Force

Figure 8-11 shows a block of mass m and initial speed v_0 sliding across a floor that is not frictionless. A kinetic frictional force $\mathbf{f}_k$ stops the block during displacement $\mathbf{d}$. From experiment we know that this force reduces the kinetic energy K of the block by transferring energy to thermal energy E_{int} of the block and the portion of the floor along which the block slides. Thermal energy is said to be an

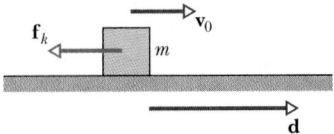

FIGURE 8-11 A block of mass m and initial speed v_0 slides across a floor. A kinetic frictional force $\mathbf{f}_k$ does work on the block, stopping it during displacement $\mathbf{d}$.

internal energy of an object because it involves random motions of the atoms and molecules within the object; hence, we represent it with the symbol E_{int}.

Because we also know from experiment that the transfer to thermal energy is not reversible, we say that the block's kinetic energy is **dissipated** by force $\mathbf{f}_k$. The work W_f done on the block by $\mathbf{f}_k$ is not the entire amount of dissipated energy, but only the part that is transferred *from* the block to the floor. The rest of the dissipated energy remains within the block as thermal energy.

As an example, suppose we find experimentally that the change in kinetic energy of the block is $\Delta K = -100$ J. Then 100 J is dissipated by $\mathbf{f}_k$, and the change in thermal energy of the block and floor is $\Delta E_{int} = +100$ J. Suppose we also find experimentally that 40 J is transferred (internally) to thermal energy of the block and 60 J is transferred (externally) to thermal energy of the floor. Then the work W_f done on the block by $\mathbf{f}_k$ is $W_f = -60$ J.

The situation in Fig. 8-11 is comparable to that in Fig. 7-2 (Section 7-3), so Eq. 7-8 applies here with two changes. The force magnitude is now f_k instead of F, and the angle ϕ between the displacement and the force is $180°$. With those changes, Eq. 7-8 becomes

$$\Delta K = f_k d \cos 180° = -f_k d. \qquad (8\text{-}26)$$

Thus the product $-f_k d$ tells us the amount by which $\mathbf{f}_k$ decreases the kinetic energy K of the block in Fig. 8-11. If a potential energy were also involved (say, the block were sliding along a ramp), the product $-f_k d$ would tell us the amount by which $\mathbf{f}_k$ decreases (by dissipating) the mechanical energy E of the system:

$$\Delta E = -f_k d \qquad \begin{array}{c}\text{(dissipated mechanical} \\ \text{energy).}\end{array} \qquad (8\text{-}27)$$

Only a portion of this dissipated energy is transferred from the block to the floor, and only that portion is equal to the work W_f done by $\mathbf{f}_k$. *Without additional information, we cannot determine W_f.*

SAMPLE PROBLEM 8-6

Figure 8-12 shows a disabled robot of mass $m = 40$ kg being dragged by a cable up the $30°$ inclined wall inside a volcano crater. The applied force $\mathbf{F}$ exerted on the robot by the cable has a magnitude of 380 N. The kinetic frictional force $\mathbf{f}_k$ acting on the robot has a magnitude of 140 N. The robot moves through a displacement $\mathbf{d}$ of magnitude 0.50 m along the crater wall.

(a) How much of the mechanical energy of the robot–Earth system is dissipated by the kinetic frictional force $\mathbf{f}_k$ during displacement $\mathbf{d}$?

SOLUTION: From Eq. 8-27 and the given data, we have

$$\Delta E = -f_k d = -(140 \text{ N})(0.50 \text{ m}) = -70 \text{ J}. \quad \text{(Answer)}$$

This result means that during the displacement, 70 J is transferred to thermal energy of the robot and the wall along which it slides.

(b) What is the work W_g done on the robot by its weight during the displacement?

SOLUTION: From Eqs. 8-1 and 8-7, we know that

$$W_g = -\Delta U = -mg \, \Delta y. \quad (8\text{-}28)$$

In Fig. 8-12, the change Δy in the height of the robot is $h = d \sin 30°$. Substituting this and given data into Eq. 8-28, we find

$$
\begin{aligned}
W_g &= -mgh = -mgd \sin 30° \\
&= -(40 \text{ kg})(9.8 \text{ m/s}^2)(0.50 \text{ m})(0.5) \\
&= -98 \text{ J}. \quad \text{(Answer)}
\end{aligned}
$$

This result means that during the displacement, 98 J is transferred to gravitational potential energy of the robot–Earth system.

(c) What is the work W_{app} done by the applied force $\mathbf{F}$?

SOLUTION: We use Eq. 7-9 ($W = Fd \cos \phi$) to find the work W_{app} done by $\mathbf{F}$ in Fig. 8-12. The angle ϕ between $\mathbf{F}$ and $\mathbf{d}$ is 0°. Substituting this and given data into Eq. 7-9, we find

$$
\begin{aligned}
W_{app} &= Fd \cos \phi = (380 \text{ N})(0.50 \text{ m}) \cos 0° \\
&= 190 \text{ J}. \quad \text{(Answer)}
\end{aligned}
$$

This result means that during the displacement, 190 J is transferred by the applied force $\mathbf{F}$ to the robot–Earth system.

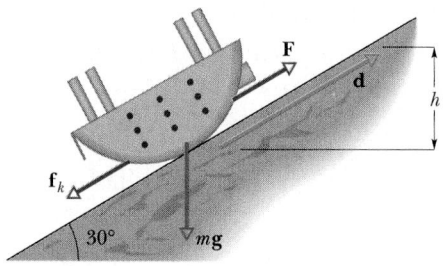

FIGURE 8-12 Sample Problem 8-6. A disabled robot is dragged up the wall inside a volcano crater, through a displacement $\mathbf{d}$ and vertical distance h, by force $\mathbf{F}$. A kinetic frictional force $\mathbf{f}_k$ opposes the motion.

CHECKPOINT 5: The figure shows three choices for the orientation of a plane that is not frictionless and for the direction in which a block slides along the plane. The block begins with the same speed in all three choices and slides until the kinetic frictional force has stopped it. Rank the choices according to the amount of mechanical energy that is dissipated, greatest first.

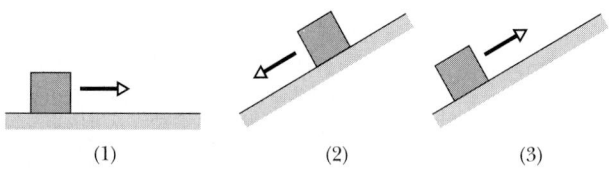

(1)   (2)   (3)

8-7 CONSERVATION OF ENERGY

Picture a block sliding over a frictionless floor. We can consider the block to be an isolated system in the sense that no external force transfers energy into or out of the system. So, a conservation principle applies to our system (albeit a simple one): the energy of the block is conserved.

Now suppose the block slides over a floor that is not frictionless. If we again take the block to be our system, we must say that the system is no longer isolated: Now an external force—the kinetic frictional force—acts on the block, and part of the energy transferred by that force is transferred to the floor. So now the energy of the system (the block) is not conserved.

However, we can expand the system to include both the block and the floor. For this new system, the kinetic frictional force is no longer an external force, and so it no longer transfers energy from the system, only within the system. So again we have an isolated system within which energy is conserved. Expanding what we mean by *the system* allows us to write a new conservation principle, one that helps us account for the energy transfers.

To find this new conservation principle, we use Eq. 8-27 to write the amount of mechanical energy dissipated by $\mathbf{f}_k$. With the block sliding across the floor, the only mechanical energy involved is kinetic energy, so we write Eq. 8-27 as

$$\Delta K = -f_k d, \quad (8\text{-}29)$$

in which ΔK is the change (decrease) in kinetic energy. This entire amount of energy is transferred to thermal energy of the block and floor. Thus, the change ΔE_{int} in thermal energy is

$$\Delta E_{int} = -\Delta K, \quad (8\text{-}30)$$

which gives us

$$\Delta K + \Delta E_{int} = 0. \quad (8\text{-}31)$$

So, although the mechanical energy of the block is not conserved, the sum of the mechanical energy of the block and the thermal energy of the block and floor *is* conserved. That sum is called the **total energy** E_{tot} of the block–floor system, and our new conservation principle is called the **law of conservation of energy.** Using Eq. 8-31, we can write this law as

$$\Delta E_{tot} = \Delta K + \Delta E_{int} = 0. \qquad (8\text{-}32)$$

If, in addition, conservative forces act on parts of our system, we can again expand the meaning of *the system* to include any transfers to and from potential energy by those forces. Then the total energy of this new, expanded isolated system does not change and our conservation law is written as

$$\Delta E_{tot} = \Delta K + \Delta U + \Delta E_{int} = 0 \quad \begin{array}{l}\text{(isolated system,}\\ \text{conservation of}\\ \text{total energy).}\end{array}$$
$$(8\text{-}33)$$

If we discover other forms of energy, we can always include them in our conservation law by rewriting Eq. 8-33 as

$$\Delta E_{tot} = \Delta K + \Delta U + \Delta E_{int}$$
$$+ \left(\begin{array}{c}\text{changes in other}\\ \text{forms of energy}\end{array}\right) = 0. \qquad (8\text{-}34)$$

Note that this law of conservation of energy is *not* something we derived. Instead, it is a law based on countless experiments. Scientists and engineers have never found an exception to it. The law is extremely powerful in analyzing difficult situations because it tells us something that does

FIGURE 8-13 To descend, the rock climber must transfer energy from the gravitational potential energy of a system consisting of her, her gear, and the Earth. She has wrapped the rope around metal rings so that the rope rubs against the rings. This allows most of the transferred energy to go to the thermal energy of the rope and bars rather than to her kinetic energy.

not change when various types of energies *are* changing. In words, it says:

> In an isolated system, energy can be transferred from one type to another, but the total energy of the system remains constant.

In less formal wording, it says: energy cannot magically appear or disappear.

For example, in Fig. 8-13, we can consider the Earth and the climber and her gear to be one system. As the climber rappels down the rock face, changing the configuration of the system, she needs to control the transfer of energy from the gravitational potential energy of the system (it cannot just magically disappear). Some of it is transferred to her kinetic energy, but she obviously does not want much transferred to that form of energy. So she has wrapped the rope around metal rings. The resulting frictional force between the rope and the rings dissipates much of the mechanical energy of the system in a controlled way by slowly transferring energy to thermal energy of the rope and the rings.

If a system is not isolated and external forces transfer energy to or from the system, then Eqs. 8-33 and 8-34 do not apply. Instead we account for those transfers by writing

$$W = \Delta E_{tot} = \Delta K + \Delta U + \Delta E_{int}, \qquad (8\text{-}35)$$

in which W is the work done on the system by the external forces. For example, in Fig. 8-13, if we consider the rope to be external to the system, then the frictional force exerted by the rope on the metal rings of the system does work W on the system, transferring energy from the system to thermal energy of the rope while, within the system, the values of K, U, and E_{int} change.

Power

Now that you have seen how energy can be transferred from one form to another and have been given a hint that additional forms of energy exist, we can expand the definition of power given in Section 7-7. There it is the rate at which work is done by a force. In a more general sense, power P is the rate at which energy is transferred by a force from one form to another. If an amount of energy ΔE is transferred in an amount of time Δt, the **average power** due to the force is

$$\overline{P} = \frac{\Delta E}{\Delta t}. \qquad (8\text{-}36)$$

Similarly, the **instantaneous power** due to the force is

$$P = \frac{dE}{dt}. \qquad (8\text{-}37)$$

SAMPLE PROBLEM 8-7

In Fig. 8-14, a circus beagle of mass 6.0 kg runs onto the left end of a curved ramp with speed $v_0 = 7.8$ m/s at height $y_0 = 8.5$ m above the floor. It then slides to the right and comes to a momentary stop when it reaches a height $y = 11.1$ m from the floor. What is the increase in thermal energy of the beagle and ramp due to the sliding?

SOLUTION: The isolated system to which Eq. 8-33 applies is the beagle–Earth–ramp system, because the only forces acting on the beagle are its weight $m\mathbf{g}$ (due to the Earth) and frictional and normal forces (due to the ramp). The weight is, of course, a conservative force with which we associate a change ΔU in gravitational potential energy. The frictional force dissipates that potential energy and the beagle's kinetic energy during the slide, increasing the thermal energy of the beagle and ramp by an amount ΔE_{int}. The normal force causes no energy transfers because it is always perpendicular to the path of the beagle.

At the stopping point, the speed of the beagle is zero, and thus so is its kinetic energy. Applying Eq. 8-33 to the beagle–Earth–ramp system, we find that

$$\Delta K + \Delta U + \Delta E_{int} = 0$$

and $(0 - \frac{1}{2}mv_0^2) + mg(y - y_0) + \Delta E_{int} = 0.$

Solving for ΔE_{int}, we get

$$\Delta E_{int} = \frac{1}{2}mv_0^2 - mg(y - y_0)$$
$$= \frac{1}{2}(6.0 \text{ kg})(7.8 \text{ m/s})^2$$
$$-(6.0 \text{ kg})(9.8 \text{ m/s}^2)(11.1 \text{ m} - 8.5 \text{ m})$$
$$\approx 30 \text{ J}. \qquad \text{(Answer)}$$

FIGURE 8-14 Sample Problem 8-7. A beagle slides along a curved ramp, starting with speed v_0 at height y_0, and reaching a height y where it comes to a momentary stop.

SAMPLE PROBLEM 8-8

A steel ball whose mass m is 5.2 g is fired vertically downward from a height h_1 of 18 m with an initial speed v_0 of 14 m/s (Fig. 8-15a). It buries itself in sand to a depth h_2 of 21 cm.

(a) What is the change in the mechanical energy of the ball?

SOLUTION: At the stopping depth h_2, the speed of the ball is zero, and thus so is its kinetic energy. The change in the mechanical energy of the ball is given by

$$\Delta E = \Delta K + \Delta U, \qquad (8-38)$$

or, with Eq. 8-8 ($\Delta U = mg \, \Delta y$),

$$\Delta E = (0 - \frac{1}{2}mv_0^2) - mg(h_1 + h_2),$$

where $-(h_1 + h_2)$ is the total *downward* displacement of the ball. Inserting the given data, we find

$$\Delta E = -\frac{1}{2}(5.2 \times 10^{-3} \text{ kg})(14 \text{ m/s})^2$$
$$-(5.2 \times 10^{-3} \text{ kg})(9.8 \text{ m/s}^2)(18 \text{ m} + 0.21 \text{ m})$$
$$= -1.437 \text{ J} \approx -1.4 \text{ J}. \qquad \text{(Answer)}$$

(Note that here we assigned the potential energy to the ball alone rather than, more properly, to the ball–Earth system—recall Problem Solving Tactic 1.)

(b) What is the change in the internal energy of the ball–Earth–sand system?

SOLUTION: This system is an isolated system because once the ball has been fired, the only forces that act on it are its weight $m\mathbf{g}$ (due to the Earth) and the average upward force $\mathbf{F}$ exerted on it by the sand (Fig. 8-15b). So Eq. 8-33 applies. Substituting Eq. 8-38 into Eq. 8-33, we find that for the ball–Earth–sand system

$$\Delta E + \Delta E_{int} = 0,$$

or

$$\Delta E_{int} = -\Delta E = -(-1.437 \text{ J}) \approx 1.4 \text{ J}. \qquad \text{(Answer)}$$

As the ball moves through the sand, $\mathbf{F}$ dissipates all the ball's mechanical energy, transferring that energy to thermal energy of the ball and sand.

(c) What is the magnitude of the average force $\mathbf{F}$ exerted on the ball by the sand?

SOLUTION: The mechanical energy of the ball is conserved until it reaches the sand. Then, as the ball moves through a distance h_2 in the sand, its mechanical energy is changed by ΔE. So Eq. 8-27 ($-f_k d = \Delta E$) can be rewritten as

$$-Fh_2 = \Delta E.$$

Solving this for F, we find

$$F = \frac{\Delta E}{-h_2} = \frac{-1.437 \text{ J}}{-0.21 \text{ m}} = 6.84 \text{ N} \approx 6.8 \text{ N}. \qquad \text{(Answer)}$$

We could also find F by first using the techniques of Chapter 2 to find the ball's speed at the surface of the sand and its average deceleration within the sand. Then, using Newton's second law, we would have F. Obviously, more algebraic steps would be required.

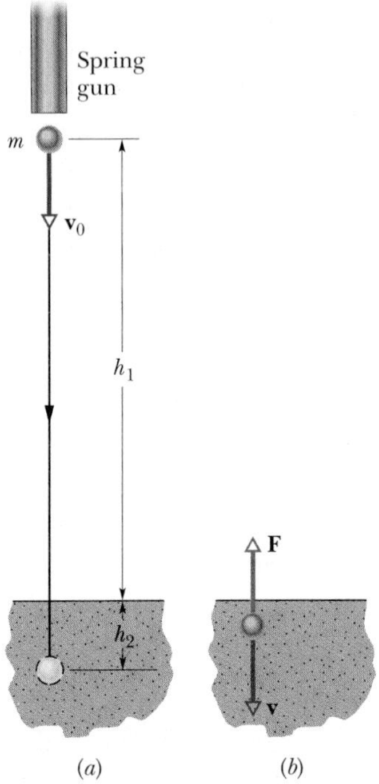

Spring gun

m

$\triangledown\ \mathbf{v}_0$

h_1

h_2

$\triangle\ \mathbf{F}$

$\triangledown\ \mathbf{v}$

(a)   (b)

FIGURE 8-15 Sample Problem 8-8. (a) A ball is fired downward, coming to rest in sand. Mechanical energy is conserved over path h_1 but not over path h_2, where a nonconservative force $\mathbf{F}$ acts on the ball. (b) Force $\mathbf{F}$, due to the sand, transfers energy primarily to thermal energy of the ball and the sand through which the ball travels.

8-8 MASS AND ENERGY (OPTIONAL)

The science of chemistry was built up by assuming that in chemical reactions, energy and mass are each conserved. In 1905, Einstein showed that as a consequence of his theory of special relativity, mass can be considered to be another form of energy. Thus the law of conservation of energy is really the law of conservation of mass–energy.

In a chemical reaction, the amount of mass that is transferred into other forms of energy (or vice versa) is such a tiny fraction of the total mass involved that there is no hope of measuring the mass change with even the best laboratory balances. Mass and energy truly *seem* to be separately conserved. In a nuclear reaction, however, the energy released is often about a million times greater than in a chemical reaction, and the change in mass can easily be measured. Taking mass–energy transfers into account where nuclear reactions are involved becomes a matter of necessary laboratory routine.

Mass and energy are related by what is certainly the best-known equation in physics (see Fig. 8-16), namely,

FIGURE 8-16 In 1979, students of Shenandoah Junior High School, in Miami, Florida, honored Albert Einstein on the 100th anniversary of his birth by spelling out his famous formula with their bodies.

$$E = mc^2, \qquad (8\text{-}39)$$

in which E is the energy equivalent (called the **mass energy**) of mass m, and c is the speed of light. (If you continue your study of physics beyond this book, you will see more refined discussions of the relation of mass and energy. You might even encounter disagreements about just what that relation is and means.)

Table 8-1 shows the energy equivalents of the masses of a few objects. The amount of energy lying dormant in ordinary objects is enormous. The energy equivalent of the mass of a penny, for example, would cost over a million dollars to purchase from your local utility company. The mass equivalents of some energies are equally striking. The entire annual U.S. electrical energy production, for example, corresponds to a mass of only a few hundred kilograms of matter (stones, potatoes, anything!).

In applying Eq. 8-39 to nuclear or chemical reactions between particles, we can rewrite it as

$$Q = -\Delta m\ c^2, \qquad (8\text{-}40)$$

TABLE 8-1 THE ENERGY EQUIVALENTS OF A FEW OBJECTS

OBJECT	MASS (kg)	ENERGY EQUIVALENT
Electron	9.11×10^{-31}	8.2×10^{-14} J ($= 511$ keV)
Proton	1.67×10^{-27}	1.5×10^{-10} J ($= 938$ MeV)
Uranium atom	4.0×10^{-25}	3.6×10^{-8} J ($= 225$ GeV)
Dust particle	1×10^{-13}	1×10^{4} J ($= 2$ kcal)
Penny	3.1×10^{-3}	2.8×10^{14} J ($= 78$ GW·h)

in which Q (called simply the *Q of the reaction*) is the energy released (positive value) or absorbed (negative value) in the reaction, and Δm is the corresponding decrease or increase in the mass of the particles as a result of the reaction. If the reaction is nuclear fission (a nucleus splits into smaller nuclei), less than 0.1% of the mass initially present is transformed into other forms of energy. If the reaction is a chemical one, the percentage is much less, typically a million times less.

In practice, SI units are rarely used with Eq. 8-40, because they are too large to be convenient. Masses are usually measured in atomic mass units (abbreviated u; see Section 1-6), where

$$1 \text{ u} = 1.66 \times 10^{-27} \text{ kg}, \tag{8-41}$$

and energies are usually measured in electron-volts or multiples thereof. Equation 7-3 tells us that

$$1 \text{ eV} = 1.60 \times 10^{-19} \text{ J}. \tag{8-42}$$

In the units of Eqs. 8-41 and 8-42, the multiplying constant c^2 has the values

$$c^2 = 9.315 \times 10^8 \text{ eV/u} = 9.315 \times 10^5 \text{ keV/u}$$
$$= 931.5 \text{ MeV/u}. \tag{8-43}$$

SAMPLE PROBLEM 8-9

In the *nuclear fission* reaction

$$\text{n} + {}^{235}\text{U} \rightarrow {}^{140}\text{Ce} + {}^{94}\text{Zr} + \text{n} + \text{n},$$

a neutron (n) combines with a uranium nucleus (^{235}U), making the nucleus unstable and causing it to split (to fission) into two smaller nuclei (^{140}Ce and ^{94}Zr) and to release two neutrons. The masses involved are

mass(^{235}U) = 235.04 u   mass(^{94}Zr) = 93.91 u

mass(^{140}Ce) = 139.91 u   mass(n) = 1.00867 u

(a) What is the fractional change in the mass of the two interacting particles?

SOLUTION: To find the mass change Δm, we subtract the mass of the reacting particles from the mass of the product particles:

$$\Delta m = (139.91 + 93.91 + 2 \times 1.00867)$$
$$- (235.04 + 1.00867)$$
$$= -0.211 \text{ u}.$$

The mass of the reacting particles is

$$M = 235.04 \text{ u} + 1.00867 \text{ u} = 236.05 \text{ u},$$

so the fractional decrease in mass is

$$\frac{|\Delta m|}{M} = \frac{0.211 \text{ u}}{236.05 \text{ u}}$$
$$= 0.00089, \text{ or about } 0.1\%. \quad \text{(Answer)}$$

Although this is small, it is easily measurable.

(b) How much energy is released during each fission reaction?

SOLUTION: From Eq. 8-40 we have

$$Q = -\Delta m\, c^2 = -(-0.211 \text{ u})(931.5 \text{ MeV/u})$$
$$= 197 \text{ MeV}, \quad \text{(Answer)}$$

where the value for c^2 is taken from Eq. 8-43. The energy release of 197 MeV per reaction is much larger than the energy releases of a few electron-volts per reaction that are typical of chemical reactions.

SAMPLE PROBLEM 8-10

The nucleus of an atom of deuterium (or heavy hydrogen) is called a **deuteron.** It is composed of a proton and a neutron. If a deuteron is torn apart, how much energy is involved in the process? Is energy absorbed or released by the reaction? The masses involved are

deuteron: $m_d = 2.01355$ u

proton: $m_p = 1.00728$ u

neutron: $m_n = 1.00867$ u $\Big\}$ 2.01595 u

SOLUTION: Since the total mass of the separated proton and neutron is greater than the deuteron mass, energy must be added to the deuteron to cause the reaction. The increase in mass due to the reaction is

$$\Delta m = (m_p + m_n) - m_d$$
$$= (1.00728 + 1.00867) - (2.01355) = 0.00240 \text{ u}.$$

The corresponding energy is then, from Eq. 8-40,

$$Q = -\Delta m\, c^2 = -(0.00240 \text{ u})(931.5 \text{ MeV/u})$$
$$= -2.24 \text{ MeV}. \quad \text{(Answer)}$$

The minus sign means that energy must be added to the deuteron. The quantity 2.24 MeV is called the *binding energy* of the deuteron. The binding energy of any nucleus is the amount of energy that must be supplied to separate the nucleus into its basic units of protons and neutrons. Conversely, it is the amount of energy that would be released if the nucleus could somehow be constructed from those basic units.

CHECKPOINT **6:** If we could somehow bring together six protons and six neutrons to form the nucleus of a ^{12}C atom, would the total mass of those 12 particles decrease or increase?

8-9 QUANTIZED ENERGY

So far, we have assumed that the energy within a system can take on *any* value. That seems a reasonable assumption for everyday systems involving springs and weights and such. However, it is not true for such microscopic systems as an atom and a *quantum dot* (a laboratory-produced confinement of electrons that resembles an atom). For these microscopic systems, the energy internal to the system is **quantized;** that is, the energy is restricted to certain values. When the system has a particular energy value, it is said to be in the **quantum state** that is characterized by that value.

Figure 8-17 is a typical diagram of the energy values (or *energy levels*) for an atom: energy is plotted vertically and each of the five lowest possible energy values E_0, E_1, . . . , E_4 is represented by a horizontal line (hence the term *level*). The atom cannot have any intermediate value of energy. The quantum state associated with the lowest energy, labeled E_0 in Fig. 8-17 and assigned the arbitrary value of zero, is called the *ground state* of the atom. The quantum states with greater energies are called *excited states* of the atom: the state with energy E_1 is the first excited state, the state with energy E_2 is the second excited state, and so on.

The atom tends to be in its ground state so as to have the lowest allowed energy (much as a ball tends to roll downhill). The atom can move to an excited state, with greater energy, only if an external source provides energy equal to the energy difference between the ground state and the excited state. The atom can gain energy in a collision with another atom or a free electron. Or it can gain energy by absorbing light. However, in either process the atom must gain enough energy to put it into one of the higher energy levels. Thus, not only the energy values of an atom are quantized, but also the amounts of energy that an atom can gain (or lose) as it makes *quantum jumps* (or transitions) from one energy level to another are quantized.

The upward arrow in Fig. 8-17 represents a quantum jump from the ground state to the second excited state, which (according to the diagram) requires that the atom gain 2.5 eV of energy. If it gains the energy by absorbing light, that light ceases to exist and its energy is fully transferred to the energy of the atom. Thus, that light must have an energy of 2.5 eV if it is to be absorbed by the atom. If the light has slightly more or slightly less energy, it cannot be absorbed.

If an atom is to absorb light, the energy of that light must equal the energy difference between the initial energy level of the atom and a higher level.

When the atom reaches an excited state, it does not stay there but quickly *de-excites* by decreasing its energy, either in a collision or by emitting light. In the latter process, the emitted light is actually created by the atom (the

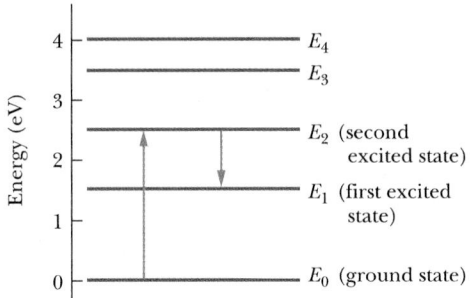

FIGURE 8-17   An energy level diagram for an atom showing the quantized energy values (or levels). Each energy level corresponds to a quantum state of the atom. The lowest energy, marked E_0, is associated with the ground state. The upward arrow represents a quantum jump of the atom from the ground state to the second excited state, which has energy E_2. The downward arrow represents a quantum jump from there to the first excited state, which has energy E_1.

light did not exist prior to de-excitation). In either process, the atom must lose enough energy to reach one of its lower energy levels.

If an atom emits light, the energy of that light equals the energy difference between the initial energy level of the atom and a lower level.

As an example, the downward arrow in Fig. 8-17 represents a quantum jump from the second excited state to the first excited state. The light that is emitted in the jump has an energy of 1.0 eV. The atom must next jump to the ground state with a second emission of light having an energy of 1.5 eV. Instead of this two-jump process, the atom could have jumped from the second excited state directly to the ground state with a single emission of light having an energy of 2.5 eV.

CHECKPOINT **7:** An atom that is being monitored emits light four times without being re-excited. Three of the emissions are detected at energies of 1.1 eV, 1.4 eV, and 2.8 eV. The sequence of the emissions is not known. Here are the energy levels of the atom:

$E_8 = 7.7$ eV	$E_5 = 4.8$ eV	$E_2 = 2.7$ eV
$E_7 = 6.6$	$E_4 = 4.2$	$E_1 = 1.3$
$E_6 = 5.5$	$E_3 = 3.9$	$E_0 = 0$

(a) What was the initial state of the atom and (b) what was the energy of the fourth emission?

REVIEW & SUMMARY

Conservative Forces

A force is a **conservative force** if the net work it does on a particle moving along a closed path from an initial point and then back to that point is zero. Or, equivalently, it is conservative if its work on a particle moving between two points does not depend on the path taken by the particle. The gravitational force (weight) and the spring force are conservative forces; the kinetic frictional force is a **nonconservative force.**

Potential Energy

A **potential energy** is energy that is associated with the configuration of a system in which a conservative force acts. When the conservative force does work W on a particle within the system, the change ΔU in the potential energy of the system is

$$\Delta U = -W. \tag{8-1}$$

If the particle moves from point x_i to point x_f, the change in the potential energy of the system is

$$\Delta U = -\int_{x_i}^{x_f} F(x)\, dx. \tag{8-6}$$

Gravitational Potential Energy

The potential energy associated with a system consisting of the Earth and a nearby particle is the **gravitational potential energy.** If the particle moves from height y_i to height y_f, the change in the gravitational potential energy of the particle–Earth system is

$$\Delta U = mg(y_f - y_i) = mg\,\Delta y. \tag{8-7}$$

If the **reference position** of the particle is set as $y_i = 0$ and the corresponding gravitational potential energy of the system is set as $U_i = 0$, then the gravitational potential energy U when the particle is at any position y is

$$U = mgy. \tag{8-9}$$

Elastic Potential Energy

Elastic potential energy is the energy associated with the state of compression or extension of an elastic object. For a spring that exerts a spring force $F = -kx$, the elastic potential energy is

$$U(x) = \tfrac{1}{2}kx^2. \tag{8-11}$$

The reference configuration is with the spring at its relaxed length, with $x = 0$ and $U = 0$.

Mechanical Energy

The **mechanical energy** E of a system is the sum of its kinetic energy K and its potential energy U:

$$E = K + U. \tag{8-12}$$

If only a conservative force within the system does work, then the mechanical energy E of the system cannot change. This **conservation of mechanical energy** is written as

$$K_2 + U_2 = K_1 + U_1, \tag{8-17}$$

in which the subscripts refer to different instants during an energy transfer process. This conservation can also be written as

$$\Delta E = \Delta K + \Delta U = 0. \tag{8-18}$$

Potential Energy Curves

If we know the **potential energy function** $U(x)$ for a system in which a force F acts on a particle, we can find the force as

$$F(x) = -\frac{dU(x)}{dx}. \tag{8-19}$$

If $U(x)$ is given on a graph, then at any value of x, the force F is the negative of the slope of the curve there and the kinetic energy of the particle is given by

$$K(x) = E - U(x), \tag{8-21}$$

where E is the mechanical energy of the system. A **turning point** is a point x where the particle reverses its motion (there, $K = 0$). The particle is in **equilibrium** at points where the slope of the $U(x)$ curve is zero (there, $F(x) = 0$).

Work by Nonconservative Forces

If a nonconservative applied force **F** does work on particle that is part of a system having a potential energy, then the work W_{app} done on the system by **F** is equal to the change ΔE in the mechanical energy of the system:

$$W_{app} = \Delta K + \Delta U = \Delta E. \tag{8-24, 8-25}$$

If a kinetic frictional force $\mathbf{f}_k$ does work on an object, the change ΔE in the total mechanical energy of the object and any system containing it is given by

$$\Delta E = -f_k d, \tag{8-27}$$

in which d is the displacement of the object during the work. The mechanical energy lost by this transfer is said to be **dissipated** by $\mathbf{f}_k$. A portion of the dissipated energy is transferred from the object; that amount is equal to the work done by $\mathbf{f}_k$.

Principle of Conservation of Energy

In an isolated system, energy may be transferred from one type to another, but the total energy E_{tot} of the system always remains constant. This conservation law is written as

$$\Delta E_{tot} = \Delta K + \Delta U + \Delta E_{int}$$
$$+ \left(\begin{array}{c}\text{changes in other}\\\text{forms of energy}\end{array}\right) = 0. \tag{8-34}$$

Here ΔE_{int} is the change in the internal energy of the bodies within the system.

If, instead, the system is not isolated, then an external force can change the total energy of the system by doing work W. In that case

$$W = \Delta E_{tot} = \Delta K + \Delta U + \Delta E_{int}. \tag{8-35}$$

Power

The **power** of a force is the *rate* at which that force transfers energy. If an amount of energy ΔE is transferred by a force in an amount of time Δt, the **average power** of the force is

$$\overline{P} = \frac{\Delta E}{\Delta t}. \tag{8-36}$$

The **instantaneous power** of a force is

$$P = \frac{dE}{dt}. \tag{8-37}$$

Mass and Energy

The **mass energy** E of an object is the energy equivalent of its mass m:

$$E = mc^2, \tag{8-39}$$

in which c is the speed of light. To convert a mass given in atomic mass units (u) to an energy equivalent in megaelectron-volts, multiply by 931.5 MeV/u. In a nuclear or chemical reaction, the energy released or absorbed is the Q of the reaction, given as

$$Q = -\Delta m\, c^2, \tag{8-40}$$

in which Δm is the corresponding decrease or increase in mass. (Q is positive for an energy release, corresponding to a mass decrease.)

Quantized Energy

The energy within a microscopic system, such as an atom, is **quantized** (restricted to certain values) and the system is said to be in a **quantum state** that is characterized by its energy value. The system cannot have any other values of energy. The lowest energy corresponds to the **ground state** of the system. Greater values of energy correspond to **excited states** of the system. If the system gains or loses energy via light, the energy of the light must equal the difference between two of the allowed energy values of the system.

QUESTIONS

1. When a particle moves from f to i and from j to i along the paths shown in Fig. 8-18, and in the indicated directions, a conservative force **F** does the indicated amounts of work on it. How much work is done by **F** when the particle moves from f to j?

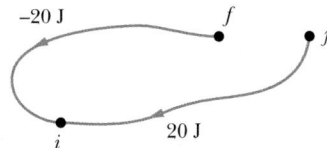

FIGURE 8-18 Question 1.

2. Figure 8-19 shows one direct path and four indirect paths from point i to point f. Along the direct path and three of the indirect paths, only a conservative force F_c acts on a certain particle. Along the fourth indirect path, both F_c and a nonconservative force F_{nc} act on the particle. The work (in joules) done on a particle in going from i to f is indicated along each straight line segment of the indirect paths. (a) How much work is done on the particle in moving from i to f along the direct path? (b) How much work is done on the particle by F_{nc} along the one path where it acts?

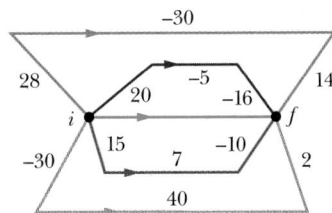

FIGURE 8-19 Question 2.

3. A spring is initially stretched by 3.0 cm from its relaxed length. Here are four choices: the initial stretch is changed to (a) a stretch of 2.0 cm, (b) a compression of 2.0 cm, (c) a compression of 4.0 cm, and (d) a stretch of 4.0 cm. Rank the choices according to the change in the elastic potential energy of the spring, most positive first, most negative last.

4. In Fig. 8-20, a brave skater slides down three slopes of frictionless ice whose vertical heights d are identical. Rank the slopes according to (a) the work done on the skater by her weight during the descent on each slope and (b) the change in her kinetic energy produced along the slope, greatest first.

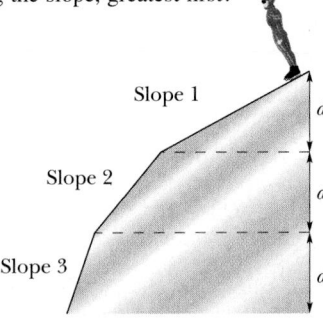

FIGURE 8-20 Question 4.

5. A coconut is thrown from a cliff edge toward a wide, flat valley, with initial speed 8 m/s. Rank the following choices for the launch direction according to (a) the initial kinetic energy of the coconut and (b) its kinetic energy just before hitting the valley bottom, greatest first: (1) **v** almost vertically upward, (2) **v** angled upward by 45°, (3) **v** horizontal, (4) **v** angled downward by 45°, and (5) **v** almost vertically downward.

6. Figure 8-21 shows three plums that are launched from the same level with the same speed. One moves straight upward, one is launched at a small angle to the vertical, and one is launched along a frictionless incline. Rank the plums according to their speed when they reach the level of the dashed line, greatest first.

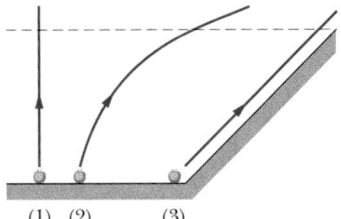

FIGURE 8-21 Question 6. (1) (2) (3)

7. In Fig. 8-22, a horizontally moving block can take three frictionless routes, differing in elevation, to reach the dashed finish line. Rank the routes according to (a) the speed of the block at the finish line and (b) the travel time of the block to the finish line, greatest first.

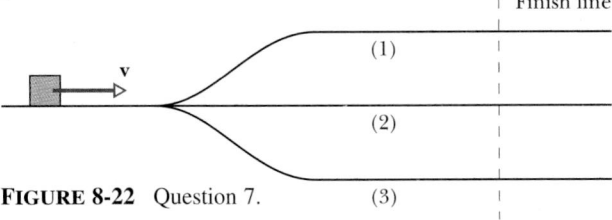

FIGURE 8-22 Question 7. (3)

8. In Fig. 8-23, a small, initially stationary block is released on a frictionless ramp at a height of 3.0 m. Hill heights along the ramp are as shown. The hills have identical circular tops (assume that the block does not fly off any hill). (a) Which hill is the first the block cannot cross? (b) What does it do after failing to cross that hill? On which hilltop is (c) the centripetal acceleration of the block greatest? (d) On which hilltop is the normal force on the block least?

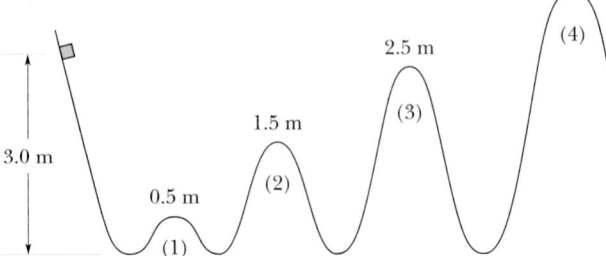

FIGURE 8-23 Question 8.

9. Figure 8-24 shows two arrangements of the same two blocks on a frictionless plane; the blocks are connected with a taut cord that runs over a massless, frictionless pulley. In each arrangement, the hanging block descends when the blocks are released. Consider the total kinetic energy of the two blocks when that block has descended by a certain distance d. Is the total kinetic energy in arrangement (a) more than, less than, or the same as that in arrangement (b)?

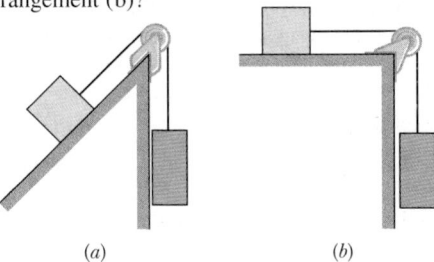

(a) (b)

FIGURE 8-24 Question 9.

10. In Fig. 8-25, an initially stationary block is released at time t_1 to slide down a frictionless ramp to a massless spring, which it reaches at time t_2 and then compresses until the maximum compression is reached at time t_3. From t_1 to t_3, what happens to (a) the kinetic energy of the block, (b) the gravitational potential energy of the block–Earth system, (c) the elastic potential energy of the spring, (d) the mechanical energy of the block–Earth system, (e) the mechanical energy of the spring, and (f) the mechanical energy of the block–Earth–spring system?

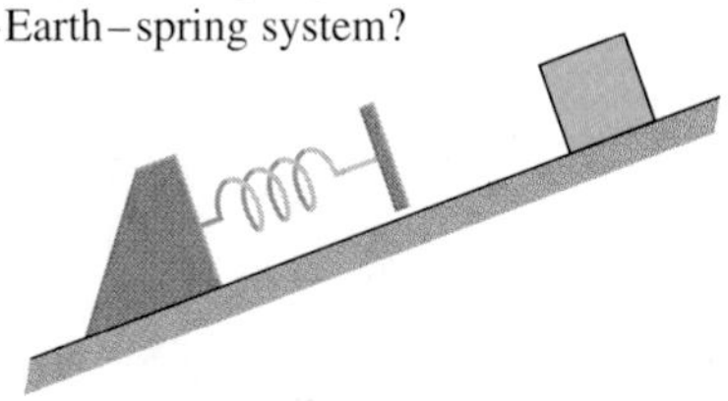

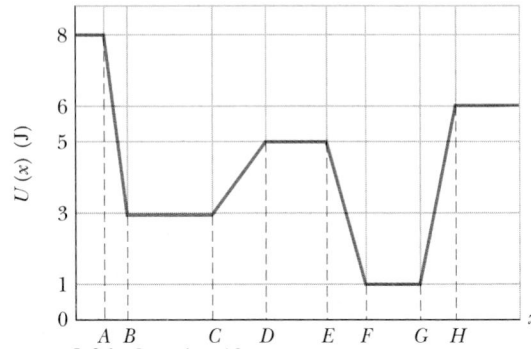

FIGURE 8-25
Question 10.

11. In Checkpoint 4, what values must the mechanical energy E and kinetic energy K of the particle not exceed if the particle is to be (a) trapped in the potential well and (b) able to move to the left of point D but not to the right of it.

12. Figure 8-26 gives the potential energy function of a particle. (a) Rank regions AB, BC, CD, and DE according to the magnitude of the force on the particle, greatest first. What value must the energy E of the particle not exceed if the particle is to be (b) trapped in the potential well at the left, (c) trapped in the potential well at the right, and (d) able to move between the two potential wells but not to the right of point H? For the situation of (d), in what region will the particle have (e) the greatest kinetic energy and (f) the least speed?

FIGURE 8-26 Question 12.

13. In Fig. 8-27, a block slides from A to C along a frictionless ramp, and then it passes through horizontal region CD, where a frictional force acts on it. Is the block's kinetic energy increasing, decreasing, or constant in (a) region AB, (b) region BC, and (c) region CD? (d) Is the block's mechanical energy increasing, decreasing, or constant in those regions?

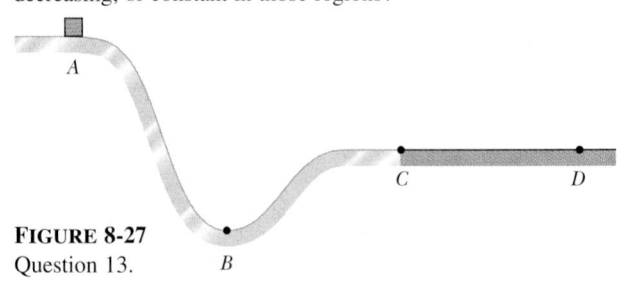

FIGURE 8-27
Question 13.

EXERCISES & PROBLEMS

SECTION 8-3 Determining Potential Energy Values

1E. The only force acting on a particle is conservative force **F**. If the particle is at point A, the potential energy of the system associated with **F** and the particle is 40 J. If the particle moves from point A to point B, the work done on the particle by **F** is +25 J. What is the potential energy of the system with the particle at B?

2E. What is the spring constant of a spring that stores 25 J of elastic potential energy when compressed by 7.5 cm from its relaxed length?

3E. You drop a 2.00 kg textbook to a friend who stands on the ground 10.0 m below the textbook with outstretched hands 1.50 m above the ground (Fig. 8-28). (a) How much work is done on the textbook by its weight as it drops to your friend's hands? (b) What is the change in the gravitational potential energy of the textbook–Earth system during the drop? If the gravitational potential energy of that system is taken to be zero at ground level, what is its potential energy when the textbook (c) is released and (d) reaches the hands?

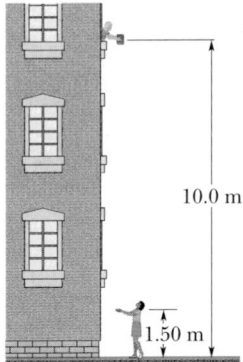

10.0 m

1.50 m

FIGURE 8-28 Exercises 3 and 12.

4E. In Fig. 8-29, a 2.00 g ice flake is released from the edge of a hemispherical bowl whose radius r is 22.0 cm. The flake–bowl contact is frictionless. (a) How much work is done on the flake by its weight during the flake's descent to the bottom of the bowl? (b) What is the change in the potential energy of the flake–Earth system during that descent? (c) If that potential energy is taken to be zero at the bottom of the bowl, what is its value when the flake is released? (d) If, instead, the potential energy is taken to be zero at the release point, what is its value when the flake reaches the bottom of the bowl?

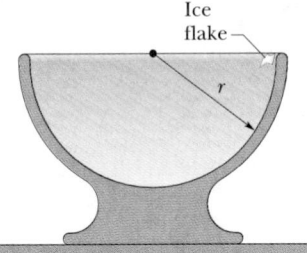

Ice
flake

r

FIGURE 8-29 Exercises 4 and 13.

5E. In Fig. 8-30, a frictionless roller coaster of mass m tops the first hill with speed v_0. How much work does its weight do on it from that point to (a) point A, (b) point B, and (c) point C? If the gravitational potential energy of the coaster–Earth system is taken to be zero at point C, what is its value when the coaster is at (d) point B and (e) point A?

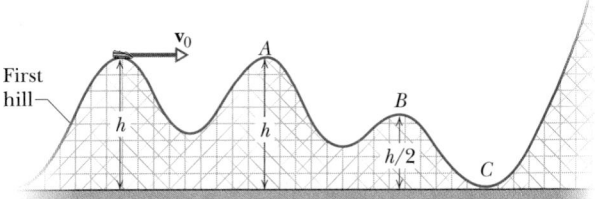

v_0

First
hill

A

h

h

B

$h/2$

C

FIGURE 8-30 Exercises 5 and 14.

6E. Figure 8-31 shows a ball with mass m attached to the end of a thin rod with length L and negligible mass. The other end of the rod is pivoted so that the ball can move in a vertical circle. The rod is held in the horizontal position as shown and then given enough of a downward push to cause the ball to swing down and around and just reach the vertically upward position, with zero speed there. How much work is done on the ball by its weight from the initial point to (a) the lowest point, (b) the highest point, and (c) the point on the right at which the ball is level with the initial point? (d) If the gravitational potential energy of the ball–Earth system is taken to be zero at the initial point, what is its value when the ball reaches those three other points, respectively?

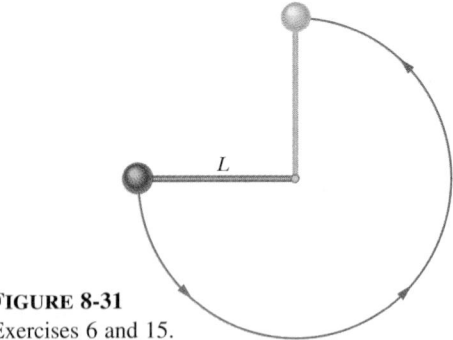

L

FIGURE 8-31
Exercises 6 and 15.

7P. A spring with a spring constant of 3200 N/m is initially stretched until the elastic potential energy is 1.44 J. ($U = 0$ for no stretch.) What is the change in the elastic potential energy if the initial stretch is changed to (a) a stretch of 2.0 cm, (b) a compression of 2.0 cm, and (c) a compression of 4.0 cm?

8P. A 1.50 kg snowball is fired from a cliff 12.5 m high with an initial velocity of 14.0 m/s, directed 41.0° above the horizontal. (a) How much work is done on the snowball by its weight during its flight to the ground below the cliff? (b) What is the change in the gravitational potential energy of the snowball–Earth system during the flight? (c) If that gravitational potential energy is taken to be zero at the height of the cliff, what is its value when the snowball reaches the ground?

9P. Figure 8-32 shows a thin rod, of length L and negligible mass, that can pivot about one end to rotate in a vertical circle. A heavy ball of mass m is attached to the other end. The rod is pulled aside through an angle θ and released. As the ball descends to its lowest point, (a) how much work does its weight do on it and (b) what is the change in the gravitational potential energy of the ball–Earth system? (c) If the gravitational potential energy is taken to be zero at the lowest point, what is its value just as the ball is released?

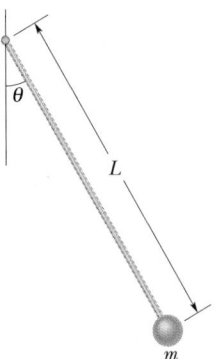

FIGURE 8-32 Problems 9 and 20.

10P. In Fig. 8-33, a small block of mass m can slide along the frictionless loop-the-loop. The block is released from rest at point P, at height $h = 5R$ above the bottom of the loop. How much work does the weight of the block do on the block as the block travels from point P to (a) point Q and (b) the top of the loop? If the gravitational potential energy of the block–Earth system is taken to be zero at the bottom of the loop, what is that potential energy when the block is (c) at point P, (d) at point Q, and (e) at the top of the loop?

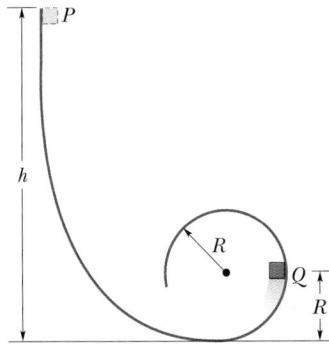

FIGURE 8-33 Problems 10 and 39.

11P. A force $\mathbf{F}$ that is directed along an x axis acts on a particle as the particle moves from $x = 1.0$ m to $x = 4.0$ m and then back to $x = 1.0$ m. What is the net work done on the particle by $\mathbf{F}$ for the round trip if the values of the force during the outward and the return trips are (a) 3.0 N and -3.0 N; (b) 5.0 N and 5.0 N; (c) $2.0x$ and $-2.0x$; (d) $3.0x^2$ and $3.0x^2$? Here, x is in meters and F is in newtons. (e) In which situation(s) is the force conservative?

SECTION 8-4 Conservation of Mechanical Energy

12E. (a) In Exercise 3, what is the speed of the textbook when it reaches the hands? (b) If we substituted a second textbook with twice the mass, what would its speed be?

13E. (a) In Exercise 4, what is the speed of the flake when it reaches the bottom of the bowl? (b) If we substituted a second flake with twice the mass, what would its speed be?

14E. In Exercise 5, what is the speed of the coaster at (a) point A, (b) point B, and (c) point C? (d) How high will it go on the last hill, which is too high for it to cross? (e) If we substitute a second coaster with twice the mass, what then are the answers (a) through (d)?

15E. (a) In Exercise 6, what initial speed is given the ball? What is its speed at (b) the lowest point and (c) the point on the right at which the ball is level with the initial point?

16E. A 70.0 kg man jumping from a window lands in an elevated fire rescue net 11.0 m below the window. He momentarily stops when he has stretched the net by 1.50 m. Assuming that mechanical energy is conserved during this process and that the net functions like an ideal spring, find the elastic potential energy of the net when it is stretched by 1.50 m.

17E. In Fig. 8-34, a runaway truck with failed brakes is moving downgrade at 80 mi/h just before the driver has the truck travel up an emergency escape ramp with an inclination of 15°. What minimum length L must the ramp have if the truck is to stop (momentarily) along it? Why are real escape ramps often covered with a thick layer of sand or gravel?

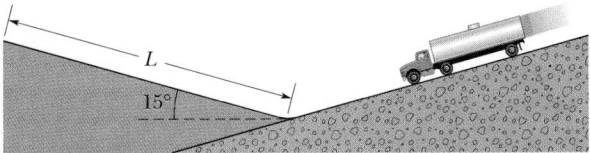

FIGURE 8-34 Exercise 17.

18E. A volcanic ash flow is moving across horizontal ground when it encounters a 10° upslope. The front of the flow then travels 920 m on the upslope before stopping. Assume that the gases entrapped in the flow lift the flow and thus make the frictional force from the ground negligible; assume also that mechanical energy of the front of the flow is conserved. What was the initial speed of the front of the flow?

19P. A 1.50 kg water balloon is shot straight up with an initial speed of 3.00 m/s. (a) What is the kinetic energy of the balloon just as it is launched? (b) How much work does the weight of the balloon do on the balloon during the balloon's full ascent? (c) What is the change in the gravitational potential energy of the balloon–Earth system during the full ascent? (d) If the gravitational potential energy is taken to be zero at the launch point, what is its value when the balloon reaches its maximum height? (e) If, instead, the gravitational potential energy is taken to be zero at the maximum height, what is its value at the launch point? (f) What is the maximum height of the balloon?

20P. In Problem 9, what is the speed of the ball at the lowest point if $L = 2.00$ m and $\theta = 30.0°$?

21P. (a) In Problem 8, using energy techniques, rather than techniques of Chapter 4, find the speed of the snowball as it reaches the ground below the cliff. (b) What is that speed if, instead, the launch angle is 41.0° *below* the horizontal?

22P. Figure 8-35 shows an 8.00 kg stone resting on a spring. The spring is compressed 10.0 cm by the stone. (a) What is the spring constant? (b) The stone is pushed down an additional 30.0 cm and released. What is the elastic potential energy of the compressed spring just before that release? (c) What is the change in the gravitational potential energy of the stone–Earth system when the stone moves from the release point to its maximum height? (d) What is that maximum height, measured from the release point?

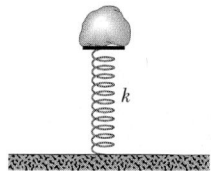

FIGURE 8-35 Problem 22.

23P. A 5.0 g marble is fired vertically upward using a spring gun. The spring must be compressed 8.0 cm if the marble is to just reach a target 20 m above the marble's position on the compressed spring. (a) What is the change in the gravitational potential energy of the marble–Earth system during the 20 m ascent? (b) What is the change in the elastic potential energy of the spring during its launch of the marble? (c) What is the spring constant of the spring?

24P. Figure 8-36a applies to the spring in a cork gun (Fig. 8-36b): it shows the spring force as a function of the stretch or compression of the spring. The spring is compressed by 5.5 cm and used to propel a 3.8 g cork from the gun. (a) What is the speed of the cork if it is released as the spring passes through its relaxed position? (b) Suppose, instead, that the cork sticks to the spring and stretches it 1.5 cm before separation occurs. What now is the speed of the cork at the time of release?

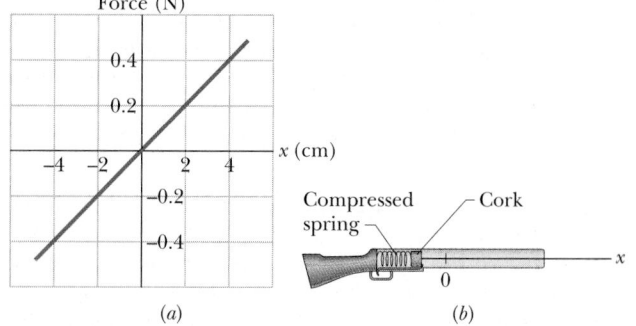

FIGURE 8-36 Problem 24.

25P. A 2.00 kg block is placed against a spring on a frictionless 30.0° incline (Fig. 8-37). The spring, whose spring constant is 19.6 N/cm, is compressed 20.0 cm and then released. (a) What is the elastic potential energy of the compressed spring? (b) What is the change in the gravitational potential energy of the block–Earth system as the block moves from the release point to its

highest point on the incline? (c) How far along the incline is the highest point from the release point?

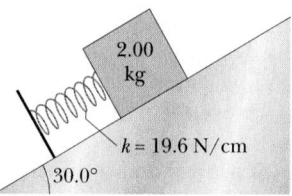

FIGURE 8-37 Problem 25.

26P. In Fig. 8-38, a 12 kg block is released from rest on an incline angled at $\theta = 30°$. Below the block is a spring that can be compressed 2.0 cm by a force of 270 N. The block momentarily stops when it compresses the spring by 5.5 cm. (a) How far has the block moved down the incline to this stopping point? (b) What is the speed of the block just as it touches the spring?

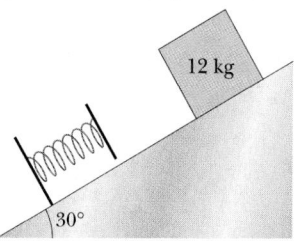

FIGURE 8-38 Problem 26.

27P. A 0.55 kg projectile is launched from the edge of a cliff with an initial kinetic energy of 1550 J and at its highest point is 140 m above the launch point. (a) What is the horizontal component of its velocity? (b) What was the vertical component of its velocity just after launch? (c) At one instant during its flight the vertical component of its velocity is 65 m/s. At that time, how far is it above or below the launch point?

28P. A 50 g ball is thrown from a window with an initial velocity of 8.0 m/s at an angle of 30° above the horizontal. Using energy methods, determine (a) the kinetic energy of the ball at the top of its flight and (b) its speed when it is 3.0 m below the window. Does the answer to (b) depend on either (c) the mass of the ball or (d) the initial angle?

29P. The spring of a child's spring gun has a spring constant of 4.0 lb/in. When the gun is inclined upward by 30° to the horizontal, a 2.0 oz ball is shot to a height of 6.0 ft above the muzzle of the gun. (a) What was the muzzle speed of the ball? (b) By how much must the spring have been compressed initially?

30P. In Fig. 8-39, the pulley is massless, and both it and the inclined plane are frictionless. If the masses are released from rest with the connecting cord taut, what is their total kinetic energy when the 2.0 kg mass has fallen 25 cm?

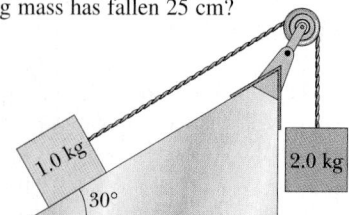

FIGURE 8-39 Problem 30.

31P. A 1.50 kg snowball is shot upward at an angle of 34.0° to the horizontal with an initial speed of 20.0 m/s. (a) What is its initial kinetic energy? (b) By how much does the gravitational potential energy of the snowball–Earth system change as the snowball moves from the launch point to the point of maximum height? (c) What is that maximum height?

32P. A pendulum consists of a 2.0 kg stone swinging on a 4.0 m string of negligible mass. The stone has a speed of 8.0 m/s when it passes its lowest point. (a) What is the speed when the string is at 60° to the vertical? (b) What is the greatest angle with the vertical that the string will reach during the stone's motion? (c) If the potential energy of the pendulum–Earth system is taken to be zero at the stone's lowest point, what is the total mechanical energy of the system?

33P. The string in Fig. 8-40 is $L = 120$ cm long, and the distance d to the fixed peg at point P is 75.0 cm. When the initially stationary ball is released with the string horizontal as shown, it will swing along the dashed arc. What is its speed when it reaches (a) its lowest point and (b) its highest point after the string catches on the peg?

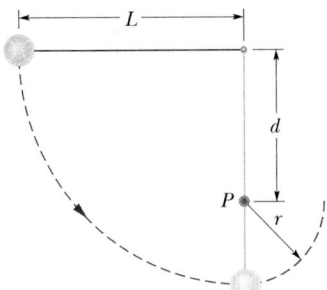

FIGURE 8-40 Problems 33 and 41.

34P. A 2.0 kg block is dropped from a height of 40 cm onto a spring of spring constant $k = 1960$ N/m (Fig. 8-41). Find the maximum distance the spring is compressed.

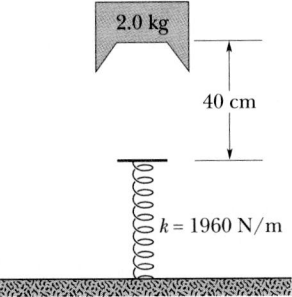

FIGURE 8-41 Problem 34.

35P. Figure 8-42 shows a pendulum of length L. Its bob (which effectively has all the mass) has speed v_0 when the cord makes an angle θ_0 with the vertical. (a) Derive an expression for the speed of the bob when it is in its lowest position. What is the least value that v_0 can have if the pendulum is to swing down and then up (b) to a horizontal position, and (c) to a vertical position with the cord remaining straight?

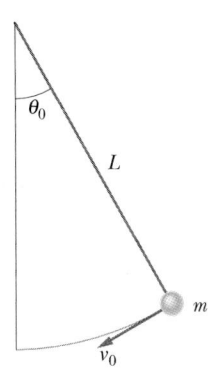

FIGURE 8-42 Problem 35.

36P. Two children are playing a game in which they try to hit a small box on the floor with a marble fired from a spring-loaded gun that is mounted on a table. The target box is 2.20 m horizontally from the edge of the table; see Fig. 8-43. Bobby compresses the spring 1.10 cm, but the center of the marble falls 27.0 cm short of the center of the box. How far should Rhoda compress the spring to score a direct hit?

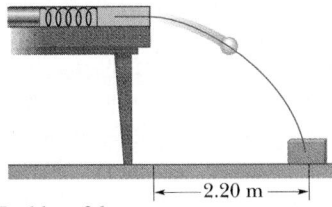

FIGURE 8-43 Problem 36.

37P. The magnitude of the gravitational force between a particle of mass m_1 and one of mass m_2 is given by

$$F(x) = G\frac{m_1 m_2}{x^2},$$

where G is a constant and x is the distance between the particles. (a) What is the corresponding potential energy function $U(x)$? Assume that $U(x) \to 0$ as $x \to \infty$. (b) How much work is required to increase the separation of the particles from $x = x_1$ to $x = x_1 + d$?

38P. A 20 kg object is acted on by a conservative force given by $F = -3.0x - 5.0x^2$, with F in newtons and x in meters. Take the potential energy associated with the force to be zero when the object is at $x = 0$. (a) What is the potential energy of the system associated with the force when the object is at $x = 2.0$ m? (b) If the object has a velocity of 4.0 m/s in the negative direction of the x axis when it is at $x = 5.0$ m, what is its speed when it passes through the origin? (c) What are the answers to (a) and (b) if the potential energy of the system is taken to be -8.0 J when the object is at $x = 0$?

39P. (a) In Problem 10, what is the *net* force acting on the block when it reaches point Q? (b) At what height h should the block be released from rest so that it is on the verge of losing contact with the track at the top of the loop? (*On the verge of losing contact* means that the normal force on the block from the track has just then become zero.)

40P. Tarzan, who weighs 688 N, swings from a cliff at the end of a convenient vine that is 18 m long (Fig. 8-44). From the top of the cliff to the bottom of the swing, he descends by 3.2 m. The vine will break if the force on it exceeds 950 N. (a) Does the vine break? (b) If no, what is the greatest force on it during the swing? If yes, at what angle does it break?

FIGURE 8-44   Problem 40.

41P. In Fig. 8-40 show that, if the ball is to swing completely around the fixed peg, then $d > 3L/5$. (*Hint:* The ball must still be moving at the top of its swing. Do you see why?)

42P. An effectively massless rigid rod of length L has a ball with mass m attached to its end, forming a pendulum. The pendulum is inverted, with the rod straight up, and then released. What are (a) the ball's speed at the lowest point and (b) the tension in the rod at that point? (c) The same pendulum is next put in a horizontal position and released from rest. At what angle from the vertical do the magnitudes of the tension in the rod and the weight of the ball match?

43P*. A chain is held on a frictionless table with one-fourth of its length hanging over the edge, as shown in Fig. 8-45. If the chain has length L and mass m, how much work is required to pull the hanging part back onto the table?

FIGURE 8-45   Problem 43.

44P*. A 3.20 kg block starts at rest and slides a distance d down a frictionless 30.0° incline, where it runs into a spring (Fig. 8-46). The block slides an additional 21.0 cm before it is brought to rest momentarily by compressing the spring, whose spring constant k is 431 N/m. (a) What is the value of d? (b) What is the distance between the point of first contact and the point where the block's speed is greatest?

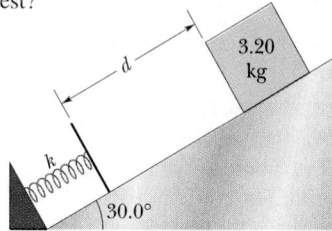

FIGURE 8-46   Problem 44.

45P*. A boy is seated on the top of a hemispherical mound of ice (Fig. 8-47). He is given a very small push and starts sliding down the ice. Show that he leaves the ice at a point whose height is $2R/3$ if the ice is frictionless. (*Hint:* The normal force vanishes as he leaves the ice.)

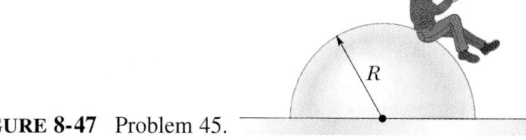

FIGURE 8-47   Problem 45.

SECTION 8-5   Reading a Potential Energy Curve

46E. A conservative force $F(x)$ acts on a particle that moves along the x axis. Figure 8-48 shows how the potential energy $U(x)$ associated with $F(x)$ varies with the particle's position. (a) Plot $F(x)$, using the same x scale as in Fig. 8-48. (b) The mechanical energy E of the system is 4.0 J. Plot the particle's kinetic energy $K(x)$ on Fig. 8-48.

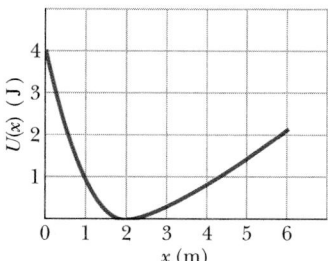

FIGURE 8-48   Exercise 46.

47P. The potential energy of a diatomic molecule (a two-atom system like H_2 or O_2) is given by

$$U = \frac{A}{r^{12}} - \frac{B}{r^6},$$

with r being the separation of the two atoms of the molecule and A and B being positive constants. This potential energy is associated with the force that binds the two atoms together. (a) Find the *equilibrium separation,* that is, the distance between the atoms at which the force on each atom is zero. Is the force repulsive (the atoms are pushed apart) or attractive (they are pulled together) if their separation is (b) smaller and (c) larger than the equilibrium separation?

48P. A conservative force $F(x)$ acts on a 2.0 kg particle that

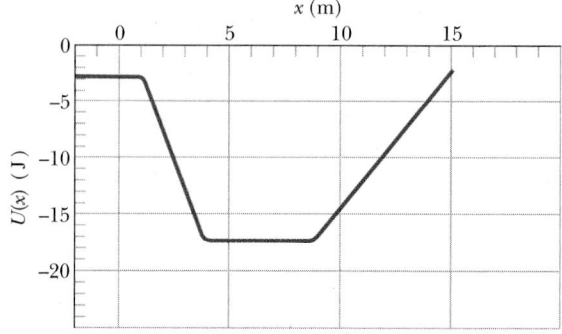

FIGURE 8-49   Problem 48.

moves along the x axis. The potential energy $U(x)$ associated with $F(x)$ is graphed in Fig. 8-49. When the particle is at $x = 2.0$ m, its velocity is -1.5 m/s. (a) What are the magnitude and direction of $F(x)$ at this position? (b) Between what limits of x does the particle move? (c) What is its speed at $x = 7.0$ m?

49P. Figure 8-50a shows a molecule consisting of two atoms of masses m and M (with $m \ll M$) and separation r. Figure 8-50b shows the potential energy $U(r)$ of the molecule as a function of r. Describe the motion of the atoms (a) if the total mechanical energy E of the two-atom system is greater than zero (as is E_1), and (b) if E is less than zero (as is E_2). For $E_1 = 1 \times 10^{-19}$ J and $r = 0.3$ nm, find (c) the potential energy of the system, (d) the total kinetic energy of the atoms, and (e) the force (magnitude and direction) acting on each atom. For what values of r is the force (f) repulsive, (g) attractive, and (h) zero?

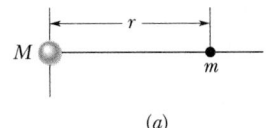

(a)

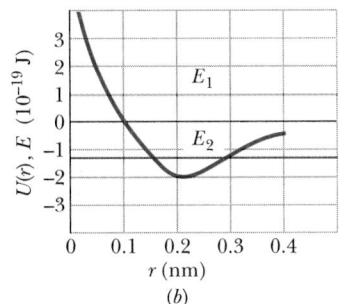

(b)

FIGURE 8-50   Problem 49.

SECTION 8-6 Work Done by Nonconservative Forces

50E. A collie drags its bed box across a floor by applying a horizontal force of 8.0 N. The kinetic frictional force acting on the box has magnitude 5.0 N. As the box is dragged through 0.70 m along the way, (a) what is the work done by the collie's applied force and (b) how much mechanical energy is dissipated by the frictional force?

51E. The temperature of a plastic cube is monitored while the cube is pushed 3.0 m across a floor at constant speed by a horizontal force of 15 N. The monitoring reveals that the thermal energy of the cube increases by 20 J. How much work is done on the cube by the kinetic frictional force acting on it?

52P. A worker pushed a 27 kg block 9.2 m along a level floor at constant speed with a force directed 32° below the horizontal. (a) If the coefficient of kinetic friction is 0.20, how much work was done by the worker's force? (b) How much energy was dissipated by the frictional force?

53P. A 50 kg trunk is pushed 6.0 m at constant speed up a 30° incline by a constant horizontal force. The coefficient of kinetic friction between the trunk and the incline is 0.20. Calculate the work done by (a) the applied force and (b) the weight of the trunk. (c) How much energy was dissipated by the frictional force acting on the trunk?

54P. A 3.57 kg block is drawn at constant speed 4.06 m along a horizontal floor by a rope exerting a 7.68 N force at an angle of 15.0° above the horizontal. Compute (a) the work done by the rope's force and (b) the coefficient of kinetic friction between block and floor. (c) How much energy is dissipated by the frictional force?

55P. A 1400 kg block of granite is pulled up an incline at a constant speed of 1.34 m/s by a cable and winch (Fig. 8-51). The coefficient of kinetic friction between the block and the incline is 0.40. What is the power due to the force applied to the block by the cable?

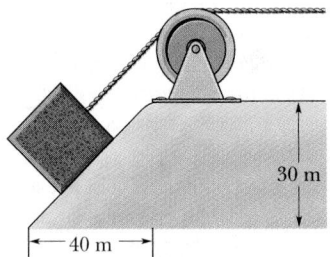

FIGURE 8-51   Problem 55.

SECTION 8-7 Conservation of Energy

56E. If a 70 kg baseball player steals home by sliding into the plate with an initial speed of 10 m/s, (a) how much kinetic energy is dissipated by the frictional force stopping him and (b) what is the change in the thermal energy of his body and the ground along which he slides?

57E. A 75 g Frisbee is thrown from a point 1.1 m above the ground with a speed of 12 m/s. When it has reached a height of 2.1 m, its speed is 10.5 m/s. (a) How much work was done on the Frisbee by its weight? (b) How much of the Frisbee's mechanical energy was dissipated by air drag?

58E. An outfielder throws a baseball with an initial speed of 120 ft/s. Just before an infielder catches the ball at the same level, its speed is 110 ft/s. How much of the ball's mechanical energy is dissipated by air drag? The weight of a baseball is 9.0 oz.

59E. A 0.63 kg ball is thrown up with an initial speed of 14 m/s and reaches a maximum height of 8.1 m. How much energy is dissipated by the air drag acting on the ball during the ascent?

60E. A 9.4 kg projectile is fired vertically upward. Air drag dissipates 68 kJ during its ascent. How much higher would it have gone were air drag negligible?

61E. A river descends 15 m through rapids. The speed of the water is 3.2 m/s upon entering the rapids and 13 m/s upon leaving. What percentage of the gravitational potential energy of the water–Earth system is transferred to kinetic energy during the descent? (*Hint:* Consider the descent of, say, 10 kg of water.)

62E. Approximately 5.5×10^6 kg of water falls 50 m over Niagara Falls each second. (a) What is the decrease in the gravitational potential energy of the water–Earth system each second?

(b) If all this energy could be converted to electrical energy by an electric generating plant (it cannot be), at what rate would electrical energy by supplied? (The mass of 1 m³ of water is 1000 kg.) (c) If the electrical energy were sold at 1 cent/kW · h, what would be the yearly cost?

63E. Each second, 1200 m³ of water passes over a waterfall 100 m high. Three-fourths of the kinetic energy gained by the water in falling is transferred to electrical energy by a hydroelectric generator. At what rate does the generator produce electrical energy? (The mass of 1 m³ of water is 1000 kg.)

64E. The area of the continental United States is about 8 × 10⁶ km², and the average elevation of its land surface is about 500 m (above sea level). The average yearly rainfall is 75 cm. Two-thirds of this rainwater returns to the atmosphere by evaporation, but the rest eventually flows into the ocean. If the associated decrease in gravitational potential energy of the water–Earth system could be fully converted to electrical energy, what would be the average power of the conversion? (The mass of 1 m³ of water is 1000 kg.)

65E. A 68 kg sky diver falls at a constant terminal speed of 59 m/s. (a) At what rate is the gravitational potential energy of the Earth–sky diver system being reduced? (b) At what rate is mechanical energy being dissipated?

66E. A 25 kg bear slides, from rest, 12 m down a lodgepole pine tree, moving with a speed of 5.6 m/s just before hitting the ground. (a) What change occurs in the gravitational potential energy of the bear–Earth system during the slide? (b) What is the kinetic energy of the bear just before hitting the ground? (c) What is the average frictional force that acts on the bear?

67E. During a rockslide, a 520 kg rock slides from rest down a hillside that is 500 m long and 300 m high. The coefficient of kinetic friction between the rock and the hill surface is 0.25. (a) If the gravitational potential energy U of the rock–Earth system is set to zero at the bottom of the hill, what is the value of U just before the slide? (b) How much mechanical energy is dissipated by frictional forces during the slide? (c) What is the kinetic energy of the rock as it reaches the bottom of the hill? (d) What is its speed then?

68E. A 30 g bullet, with a horizontal velocity of 500 m/s, stops 12 cm within a solid wall. (a) What is the change in its mechanical energy? (b) What is the magnitude of the average force from the wall stopping it?

69P. As Fig. 8-52 shows, a 3.5 kg block is accelerated by a compressed spring whose spring constant is 640 N/m. After leaving the spring at the spring's relaxed length, the block travels over a horizontal surface, with a coefficient of kinetic friction of 0.25, for a distance of 7.8 m before stopping. (a) How much mechanical energy was dissipated by the frictional force in stopping the block? (b) What was the maximum kinetic energy of the block? (c) Through what distance was the spring compressed before the block began to move?

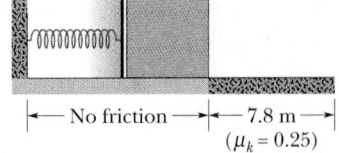

FIGURE 8-52 Problem 69.

70P. In Fig. 8-53, a block is moved down an incline a distance of 5.0 m from point A to point B by a force $\mathbf{F}$ that is parallel to the incline and has a magnitude of 2.0 N. The magnitude of the frictional force acting on the block is 10 N. If the kinetic energy of the block increases by 35 J between A and B, how much work is done on the block by its weight as the block moves from A to B?

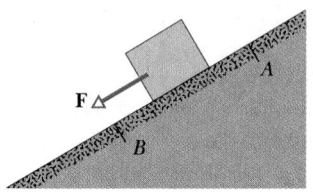

FIGURE 8-53 Problem 70.

71P. Fasten one end of a vertical spring to a ceiling, attach a cabbage to the other end, and then slowly lower the cabbage until the upward force exerted by the spring on the cabbage balances the weight of the cabbage. Show that the loss of gravitational potential energy of the cabbage–Earth system equals twice the gain in the spring's potential energy. Why are these two quantities not equal?

72P. You push a 2.0 kg block against a horizontal spring, compressing the spring by 15 cm. When you release the block, the spring forces it to slide across a tabletop. It stops 75 cm from where you released it. The spring constant is 200 N/m. What is the coefficient of kinetic friction between the block and the table?

73P. A moving 2.5 kg block (Fig. 8-54) collides with a horizontal spring whose spring constant is 320 N/m. The block compresses the spring a maximum distance of 7.5 cm from its rest position. The coefficient of kinetic friction between the block and the horizontal surface is 0.25. (a) How much work is done by the spring in bringing the block to rest? (b) How much mechanical energy is dissipated by the force of friction while the block is being brought to rest by the spring? (c) What was the speed of the block when it hit the spring?

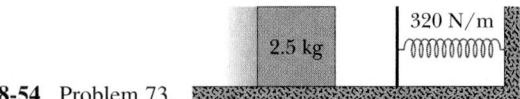

FIGURE 8-54 Problem 73.

74P. Two snowy peaks are 850 m and 750 m above the valley between them. A ski run extends down from the top of the higher peak and then back up to the top of the lower one, with a total length of 3.2 km and an average slope of 30° (Fig. 8-55). (a) A skier starts from rest on the higher peak. At what speed will he arrive at the top of the lower peak if he just coasts without using the poles? Ignore friction. (b) Approximately what coefficient of kinetic friction between snow and skis would make him stop just at the top of the lower peak?

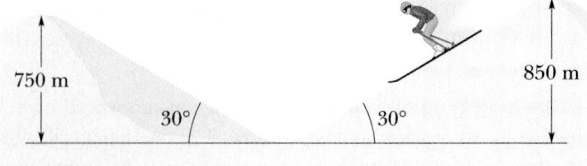

FIGURE 8-55 Problem 74.

75P. A factory worker accidentally releases a 400 lb crate that was being held at rest at the top of a 12 ft long ramp inclined at 39° to the horizontal. The coefficient of kinetic friction between the crate and the ramp, and between the crate and the horizontal factory floor, is 0.28. (a) How fast is the crate moving as it reaches the bottom of the ramp? (b) How far will it subsequently slide across the factory floor? (Assume that the crate's kinetic energy does not change as it moves from the ramp onto the floor.) (c) Why don't the answers to (a) and (b) depend on the mass of the crate?

76P. Two blocks are connected by a string, as shown in Fig. 8-56. They are released from rest. Show that, after they have moved a distance L, their common speed is given by

$$v = \sqrt{\frac{2(m_2 - \mu m_1)gL}{m_1 + m_2}},$$

in which μ is the coefficient of kinetic friction between the upper block and the surface. Assume that the pulley is massless and frictionless.

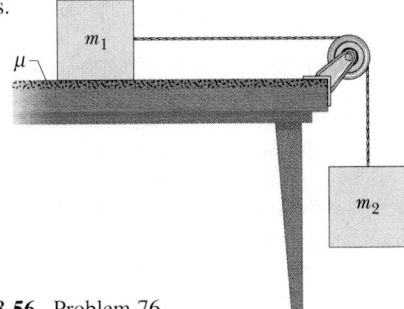

FIGURE 8-56 Problem 76.

77P. A 4.0 kg bundle starts up a 30° incline with 128 J of kinetic energy. How far will it slide up the plane if the coefficient of friction is 0.30?

78P. A cookie jar is moving up a 40° incline. At a point 1.8 ft from the bottom of the incline (measured along the incline), it has a speed of 4.5 ft/s. The coefficient of kinetic friction between jar and incline is 0.15. (a) How much farther up the incline will the jar move? (b) How fast will it be going when it slides back to the bottom of the incline?

79P. A certain spring is found *not* to conform to Hooke's law. The force (in newtons) it exerts when stretched a distance x (in meters) is found to have magnitude $52.8x + 38.4x^2$ in the direction opposing the stretch. (a) Compute the work required to stretch the spring from $x = 0.500$ m to $x = 1.00$ m. (b) With one end of the spring fixed, a particle of mass 2.17 kg is attached to the other end of the spring when it is extended by an amount $x = 1.00$ m. If the particle is then released from rest, compute its speed at the instant the spring has returned to the configuration in which the extension is $x = 0.500$ m. (c) Is the force exerted by the spring conservative or nonconservative? Explain.

80P. A girl whose weight is 267 N slides down a 6.1 m playground slide that makes an angle of 20° with the horizontal. The coefficient of kinetic friction is 0.10. (a) Find the work done on her by her weight. (b) Find the amount of energy dissipated by the frictional force. (c) If she starts at the top with a speed of 0.457 m/s, what is her speed at the bottom?

81P. In Fig. 8-57, a block slides along a track from one level to a higher level, by moving through an intermediate valley. The track is frictionless until the block reaches the higher level. There a frictional force stops the block in a distance d. The block's initial speed v_0 is 6.0 m/s; the height difference h is 1.1 m; and the coefficient of kinetic friction μ is 0.60. Find d.

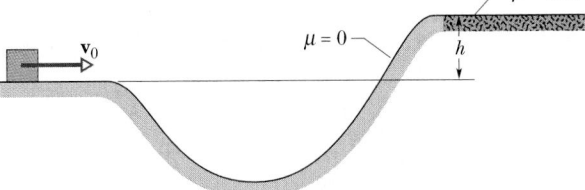

FIGURE 8-57 Problem 81.

82P. In the hydrogen atom, the magnitude of the force of attraction between the positively charged nucleus (a proton) and the negatively charged electron is given by

$$F = k\frac{e^2}{r^2},$$

where e is the magnitude of the charge of the electron and the proton, k is a constant, and r is the separation between electron and nucleus. Assume that the nucleus is fixed in place. Imagine that the electron, which is initially moving in a circle of radius r_1 about the nucleus, suddenly "jumps" into a circular orbit of smaller radius r_2 (Fig. 8-58). (a) Calculate the change in kinetic energy of the electron, using Newton's second law. (b) Using the relation between force and potential energy, calculate the change in potential energy of the atom. (c) By how much has the atom's total energy decreased in this process? (The total energy of the atom must decrease to provide the energy of the light that the atom emits because of the electron's jump.)

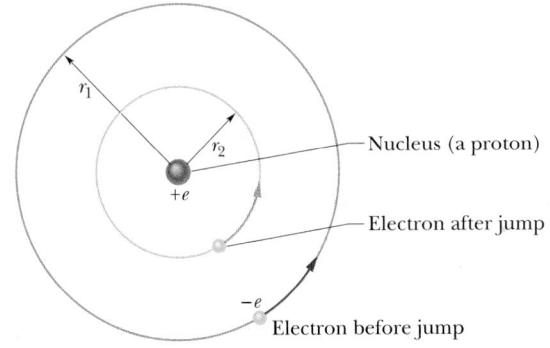

FIGURE 8-58 Problem 82.

83P. A stone with weight w is thrown vertically upward into the air from ground level with initial speed v_0. If a constant force f due to air drag acts on the stone throughout its flight, (a) show that the maximum height reached by the stone is

$$h = \frac{v_0^2}{2g(1 + f/w)}.$$

(b) Show that the speed of the stone just before impact with the ground is

$$v = v_0\left(\frac{w - f}{w + f}\right)^{1/2}.$$

84P. A playground slide is in the form of an arc of a circle with a maximum height of 4.0 m, with a radius of 12 m, and with the ground tangent to the circle (Fig. 8-59). A 25 kg child starts from rest at the top of the slide and is observed to have a speed of 6.2 m/s at the bottom. (a) What is the length of the slide? (b) What average frictional force acts on the child over this distance? If, instead of the ground, a vertical line through the *top of the slide* is tangent to the circle, what are (c) the length of the slide and (d) the average frictional force on the child?

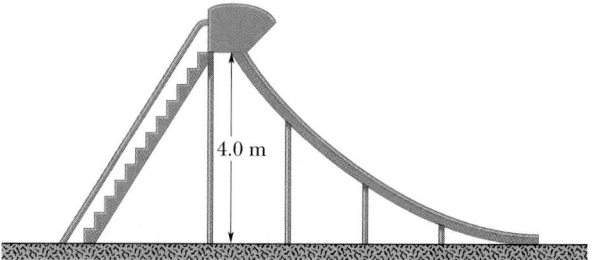

FIGURE 8-59 Problem 84.

85P. A particle can slide along a track with elevated ends and a flat central part, as shown in Fig. 8-60. The flat part has length L. The curved portions of the track are frictionless, but for the flat part the coefficient of kinetic friction is $\mu_k = 0.20$. The particle is released from rest at point A, which is a height $h = L/2$ above the flat part of the track. Where does the particle finally come to rest?

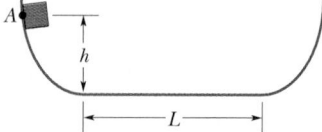

FIGURE 8-60 Problem 85.

86P. A massless rigid rod of length L has a ball of mass m attached to one end (Fig. 8-61). The other end is pivoted in such a way that the ball will move in a vertical circle. The system is launched downward from the horizontal position A with initial speed v_0. The ball just reaches point D and then stops. (a) Derive an expression for v_0 in terms of L, m, and g. (b) What is the tension in the rod when the ball is at B? (c) A little grit is placed on the pivot, after which the ball just reaches C when launched from A with the same speed as before. How much mechanical energy is dissipated by friction during this motion? (d) How much total mechanical energy has been dissipated by friction when the ball finally comes to rest at B after several oscillations?

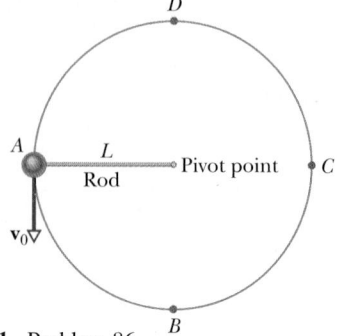

FIGURE 8-61 Problem 86.

87P. The cable of the 4000 lb elevator in Fig. 8-62 snaps when the elevator is at rest at the first floor, where the bottom is a distance $d = 12$ ft above a cushioning spring whose spring constant is $k = 10,000$ lb/ft. A safety device clamps the elevator against guide rails so that a constant frictional force of 1000 lb opposes the motion of the elevator. (a) Find the speed of the elevator just before it hits the spring. (b) Find the maximum distance x that the spring is compressed. (c) Find the distance that the elevator will bounce back up the shaft. (d) Using conservation of energy, find the approximate total distance that the elevator will move before coming to rest. Why is the answer not exact?

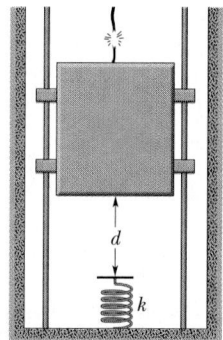

FIGURE 8-62 Problem 87.

88P. A metal tool is sharpened by being held against the rim of a wheel on a grinding machine by a force of 180 N. The wheel has a radius of 20 cm and rotates at 2.5 rev/s. The coefficient of kinetic friction between the wheel and the tool is 0.32. At what rate is energy being transferred from the motor driving the wheel to the thermal energy of the wheel and tool and to the kinetic energy of the material thrown from the tool?

89P*. At a factory, a 300 kg crate is dropped vertically from a packing machine onto a conveyor belt moving at 1.20 m/s (Fig. 8-63). (A motor maintains the belt's constant speed.) The coefficient of kinetic friction between belt and crate is 0.400. After a short time, slipping between the belt and the crate ceases and the crate then moves along with the belt. For the period of time during which the crate is being brought to rest relative to the belt, calculate, for a coordinate system at rest in the factory, (a) the kinetic energy supplied to the crate, (b) the magnitude of the kinetic frictional force acting on the crate, and (c) the energy supplied by the motor. (d) Explain why the answers to (a) and (c) are different.

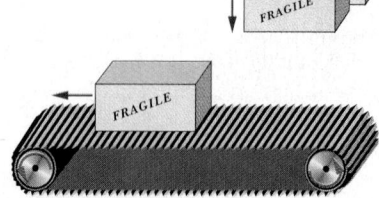

FIGURE 8-63 Problem 89.

SECTION 8-8 Mass and Energy

90E. (a) How much energy in joules is represented by a mass of 102 g? (b) For how many years would this supply the energy needs of a one-family home consuming energy at the average rate of 1.00 kW?

91E. The magnitude M of an earthquake on the so-called Richter scale is related to the released energy E in joules by the equation

$$\log E = 5.24 + 1.44M.$$

(a) The 1906 San Francisco earthquake was of magnitude 8.2 (Fig. 8-64). How much energy was released? (b) How much mass is equivalent to this amount of energy?

92E. The United States generated about 2.31×10^{12} kW·h of electrical energy in 1983. What is the mass equivalent of this electrical energy?

93P. What is the minimum energy that is required to break a nucleus of ^{12}C (of mass 11.99671 u) into three nuclei of ^{4}He (of mass 4.00151 u each)?

94P. The nucleus of a gold atom contains 79 protons and 118 neutrons and has a mass of 196.9232 u. What is its binding energy? See Sample Problem 8-10 for additional data.

95P. In the nuclear fusion reaction $d + t \rightarrow {}^{4}He + n$, a deuteron (d) combines with the nucleus of a tritium atom (t, with one proton and two neutrons) to produce the nucleus of a helium atom (with two protons and two neutrons) and a free neutron (n). The masses involved are

d: 2.01355 u ^{4}He: 4.00151 u

t: 3.01550 u   n: 1.00867 u

(a) Does the reaction release or absorb energy? (b) How much energy is absorbed or released?

SECTION 8-9 Quantized Energy

96E. Here are the lowest five energy levels, in electron-volts, of atoms of three types:

FIGURE 8-64 Exercise 91. Destruction on Nob Hill in San Francisco due to the earthquake of 1906. Over 400 km of the San Andreas fault line ruptured during the earthquake.

TYPE A	TYPE B	TYPE C
3.8	3.2	3.1
3.0	2.7	2.9
2.4	2.0	2.2
1.4	1.2	1.5
0	0	0

Two emissions of light from atoms are detected at the energies of (a) 1.4 eV and (b) 1.5 eV. For each emission, identify all transitions that could have produced the light.

97E. An atom that is being monitored emits light five times without being re-excited. The energies detected for four of those emissions are 0.7 eV, 0.8 eV, 0.9 eV, and 2.0 eV. The energy information for the other emission was lost by the computer controlling the detectors, as was information about the sequence of detection. The lowest 12 energy levels of the atom, in electron-volts, are

6.5	4.1	2.6
5.3	3.8	2.0
4.9	3.4	1.5
4.5	2.9	0

(a) From which energy level did the atom begin its quantum jumping downward in energy? (b) What value of energy was lost by the computer?

ELECTRONIC COMPUTATION

98. A nerfball™ has a mass of 9.8 g and a terminal velocity in air of 7.3 m/s. It is thrown straight up with an initial speed of 15 m/s. Use numerical integration to calculate the coordinate and velocity of the ball every 0.1 s from the time it is thrown to the time it returns to its initial position. For each of those times calculate the potential energy, kinetic energy, and total mechanical energy. On the way up how much of the mechanical energy of the ball-Earth system is dissipated? How much is dissipated on the way down? Take the potential energy to be zero initially.

99. A 700 g block is released from rest at height h_0 above a vertical spring with negligible mass and a spring constant $k = 400$ N/m. The block sticks to the spring and momentarily comes to rest after compressing the spring 19.0 cm. How much work is done (a) by the block on the spring and (b) by the spring on the block? (c) What was the value of h_0? (d) If the block was released from a height $2h_0$ above the spring, what would be the maximum compression of the spring?

100. To make a pendulum, a 300 g mass is tied to the end of a string 1.4 m long. The mass is pulled to one side until the string makes an angle of 30.0° with the vertical; then (with the string taut) the mass is released from rest. Find (a) the speed of the mass when the string makes an angle of 20.0° with the vertical and (b) the maximum speed of the mass. (c) What is the angle between the string and the vertical when the speed of the mass is one third its maximum value?

9
Systems of Particles

If you leap forward, chances are that your head and torso will follow a parabolic path, like a baseball thrown in from the outfield. But when a skilled ballet dancer leaps across the stage in a <u>grand jeté</u> as shown in the photograph, the path taken by her head and torso is nearly horizontal during much of the jump. She seems to be floating across the stage. The audience may not know about gravitational attraction, but they still sense that something unusual has happened. How does the ballerina seemingly "turn off" gravity?

9-1 A SPECIAL POINT

Physicists love to look at something complicated and find in it something simple and familiar. Here is an example. If you flip a baseball bat into the air, its motion as it turns is clearly more complicated than that of, say, a nonspinning tossed ball (Fig. 9-1a), which moves like a particle. Every part of the bat moves in a different way from every other part, so you cannot represent the bat as a particle that is tossed into the air; instead, it is a system of particles.

However, if you look closely, you will find that one special point of the bat moves in a simple parabolic path, just as a particle would if tossed into the air (Fig. 9-1b). In fact, that special point moves as though (1) the bat's total mass were concentrated there and (2) the weight of the bat acted only there. This special point is said to be the **center of mass** of the bat. In general:

The center of mass of a body or a system of bodies is the point that moves as though all of the mass were concentrated there and all external forces were applied there.

The center of mass of a baseball bat lies along the central axis. You can locate it by balancing the bat horizontally on an outstretched finger: the center of mass is on the bat's axis just above your finger.

9-2 THE CENTER OF MASS

We shall now spend some time determining how to find the center of mass in various systems. We start with a system of a few particles, and then we consider a system of a great many particles (as in a baseball bat).

Systems of Particles

Figure 9-2a shows two particles of masses m_1 and m_2 separated by a distance d. We have arbitrarily chosen the origin of the x axis to coincide with m_1. We *define* the position of the center of mass of this two-particle system to be

$$x_{cm} = \frac{m_2}{m_1 + m_2} d. \qquad (9-1)$$

Suppose, as an example, that $m_2 = 0$. Then there is only one particle (m_1), and the center of mass must lie at the position of that particle; Eq. 9-1 dutifully reduces to $x_{cm} = 0$. If $m_1 = 0$, there is again only one particle (m_2), and we have, as we expect, $x_{cm} = d$. If $m_1 = m_2$, the masses of the particles are equal and the center of mass

(a)

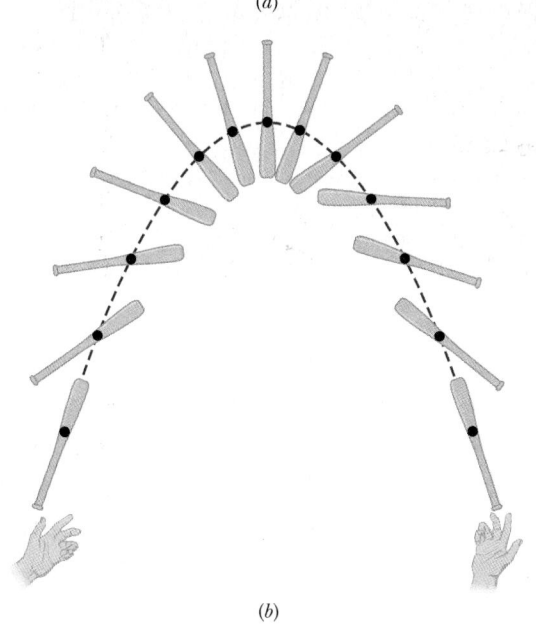

(b)

FIGURE 9-1 (a) A ball tossed into the air follows a parabolic path. (b) The center of mass (the black dot) of a baseball bat that is flipped into the air does also, but all other points of the bat follow more complicated curved paths.

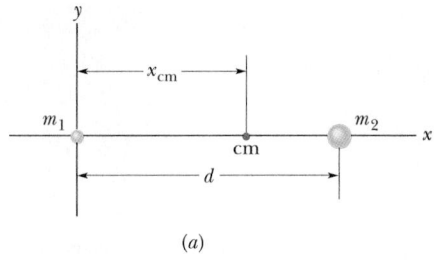

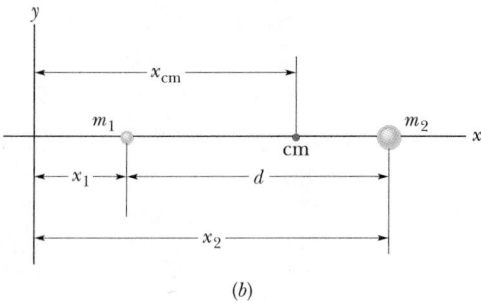

FIGURE 9-2 (a) Two particles of masses m_1 and m_2 are separated by a distance d. The dot labeled cm shows the position of the center of mass, calculated from Eq. 9-1. (b) The same as (a) except that the origin is located farther from the particles. The position of the center of mass is calculated from Eq. 9-2. The relative location of the center of mass (with respect to the particles) is the same in both cases.

should be halfway between them; Eq. 9-1 reduces to $x_{cm} = \frac{1}{2}d$, again as we expect. Finally, Eq. 9-1 tells us that if neither m_1 nor m_2 is zero, x_{cm} can have only values that lie between zero and d; that is, the center of mass must lie somewhere between the two particles.

Figure 9-2b shows a more generalized situation, in which the coordinate system has been shifted leftward. The position of the center of mass is now defined as

$$x_{cm} = \frac{m_1 x_1 + m_2 x_2}{m_1 + m_2}. \qquad (9\text{-}2)$$

Note that if we put $x_1 = 0$, then x_2 becomes d and Eq. 9-2 reduces to Eq. 9-1, as it must. Note also that in spite of the shift of the coordinate system, the center of mass is still the same distance from each particle.

We can rewrite Eq. 9-2 as

$$x_{cm} = \frac{m_1 x_1 + m_2 x_2}{M}, \qquad (9\text{-}3)$$

in which M is the total mass of the system. Here, $M = m_1 + m_2$. We can extend this equation to a more general situation in which n particles are strung out along the x axis. Then the total mass is $M = m_1 + m_2 + \cdots + m_n$,

and the location of the center of mass is

$$x_{cm} = \frac{m_1 x_1 + m_2 x_2 + m_3 x_3 + \cdots + m_n x_n}{M}$$

$$= \frac{1}{M} \sum_{i=1}^{n} m_i x_i. \qquad (9\text{-}4)$$

Here the subscript i is a running number, or index, that takes on all integer values from 1 to n. It identifies the various particles, their masses, and their x coordinates.

If the particles are distributed in three dimensions, the center of mass must be identified by three coordinates. By extension of Eq. 9-4, they are

$$x_{cm} = \frac{1}{M} \sum_{i=1}^{n} m_i x_i,$$

$$y_{cm} = \frac{1}{M} \sum_{i=1}^{n} m_i y_i, \qquad (9\text{-}5)$$

$$z_{cm} = \frac{1}{M} \sum_{i=1}^{n} m_i z_i.$$

We can also define the center of mass with the language of vectors. The position of a particle whose coordinates are x_i, y_i, and z_i is given by a position vector:

$$\mathbf{r}_i = x_i \mathbf{i} + y_i \mathbf{j} + z_i \mathbf{k}. \qquad (9\text{-}6)$$

Here the index identifies the particle, and $\mathbf{i}$, $\mathbf{j}$, and $\mathbf{k}$ are unit vectors pointing, respectively, in the direction of the x, y, and z axes. Similarly, the position of the center of mass of a system of particles is given by a position vector:

$$\mathbf{r}_{cm} = x_{cm} \mathbf{i} + y_{cm} \mathbf{j} + z_{cm} \mathbf{k}. \qquad (9\text{-}7)$$

The three scalar equations of Eq. 9-5 can now be replaced by a single vector equation,

$$\mathbf{r}_{cm} = \frac{1}{M} \sum_{i=1}^{n} m_i \mathbf{r}_i, \qquad (9\text{-}8)$$

where again M is the total mass of the system. You can check that this equation is correct by substituting Eqs. 9-6 and 9-7 into it, and then separating out the x, y, and z components. The scalar relations of Eq. 9-5 result.

Rigid Bodies

An ordinary object, such as a baseball bat, contains so many particles (atoms) that we can best treat it as a continuous distribution of matter. The "particles" then become differential mass elements dm, the sums of Eq. 9-5 become integrals, and the coordinates of the center of mass are defined as

$$x_{cm} = \frac{1}{M} \int x \, dm,$$

$$y_{cm} = \frac{1}{M} \int y \, dm, \qquad (9\text{-}9)$$

$$z_{cm} = \frac{1}{M} \int z \, dm,$$

where M is now the mass of the object. The integrals are to be evaluated over all the mass elements in the object. In practice, however, we rewrite them in terms of the coordinates of the mass elements. If the object has uniform density ρ (mass per volume), then we can write

$$\rho = \frac{dm}{dV} = \frac{M}{V}, \qquad (9\text{-}10)$$

where dV is the volume occupied by a mass element dm, and V is the total volume of the object (ρ is "rho"). We next substitute dm from Eq. 9-10 into Eq. 9-9, finding

$$x_{cm} = \frac{1}{V} \int x \, dV,$$

$$y_{cm} = \frac{1}{V} \int y \, dV, \qquad (9\text{-}11)$$

$$z_{cm} = \frac{1}{V} \int z \, dV.$$

These integrals are evaluated over the volume of the object. An example is given in Sample Problem 9-4.

Many objects have a point, a line, or a plane of symmetry. The center of mass of such an object then lies at that point, on that line, or in that plane. For example, the center of mass of a homogeneous sphere (which has a point of symmetry) is at the center of the sphere. The center of mass of a homogeneous cone (whose axis is a line of symmetry) lies on the axis of the cone. The center of mass of a banana (which has a plane of symmetry that splits it into two equal parts) lies somewhere in that plane.

The center of mass of an object need not lie within the object. There is no dough at the center of mass of a doughnut, and no iron at the center of mass of a horseshoe.

SAMPLE PROBLEM 9-1

Figure 9-3 shows three particles of masses $m_1 = 1.2$ kg, $m_2 = 2.5$ kg, and $m_3 = 3.4$ kg located at the corners of an equilateral triangle of edge $a = 140$ cm. Where is the center of mass?

SOLUTION: We choose our x and y coordinate axes so that one of the particles is located at the origin and the x axis coincides with one of the sides of the triangle. The three particles then have the following coordinates:

PARTICLE	MASS (kg)	x (cm)	y (cm)
m_1	1.2	0	0
m_2	2.5	140	0
m_3	3.4	70	121

Because of our wise choice of coordinate axes, three of the coordinates in the table are zero, simplifying the calculations. The total mass M of the system is 7.1 kg.

From Eq. 9-5, the coordinates of the center of mass are

$$x_{cm} = \frac{1}{M} \sum_{i=1}^{3} m_i x_i = \frac{m_1 x_1 + m_2 x_2 + m_3 x_3}{M}$$

$$= \frac{(1.2 \text{ kg})(0) + (2.5 \text{ kg})(140 \text{ cm}) + (3.4 \text{ kg})(70 \text{ cm})}{7.1 \text{ kg}}$$

$$= 83 \text{ cm} \qquad \text{(Answer)}$$

and

$$y_{cm} = \frac{1}{M} \sum_{i=1}^{3} m_i y_i = \frac{m_1 y_1 + m_2 y_2 + m_3 y_3}{M}$$

$$= \frac{(1.2 \text{ kg})(0) + (2.5 \text{ kg})(0) + (3.4 \text{ kg})(121 \text{ cm})}{7.1 \text{ kg}}$$

$$= 58 \text{ cm}. \qquad \text{(Answer)}$$

In Fig. 9-3, the center of mass is located by the position vector $\mathbf{r}_{cm}$, with components x_{cm} and y_{cm}.

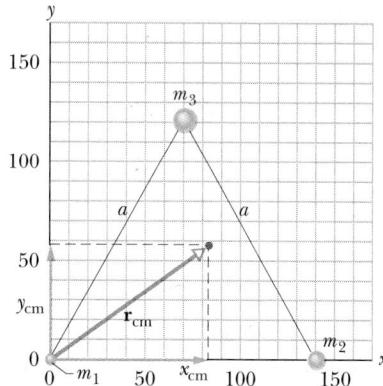

FIGURE 9-3 Sample Problem 9-1. Three particles having different masses form an equilateral triangle of side a. The center of mass is located by the position vector $\mathbf{r}_{cm}$.

SAMPLE PROBLEM 9-2

Find the center of mass of the uniform triangular plate that is shown in each part of Fig. 9-4.

SOLUTION: Figure 9-4*a* shows the plate divided into thin slats, parallel to one side of the triangle. From symmetry, the center of mass of a thin, uniform slat is at its midpoint. The center of mass of the triangular plate must then lie somewhere along the line that connects the midpoints of all the slats. That bisecting line also connects the upper vertex with the midpoint of the opposite side. The plate would balance if it were placed on a knife-edge coinciding with this line of symmetry.

In Figs. 9-4*b* and 9-4*c*, we subdivide the plate into slats parallel to the other two sides of the triangle. Again, the center of mass must lie somewhere along each of the bisecting lines shown. Hence the center of mass of the plate must lie at the intersection of these three symmetry lines, as Fig. 9-4*d* shows. It is the only point that the three lines have in common.

You can check this conclusion experimentally by taking advantage of the (correct) intuitive notion that an object suspended from a point will orient itself so that its center of mass lies vertically below that point. Suspend the triangle from each vertex in turn, and draw a line vertically downward from the suspension point, as in Fig. 9-4*e*. The center of mass of the triangle will be at the intersection of the three lines.

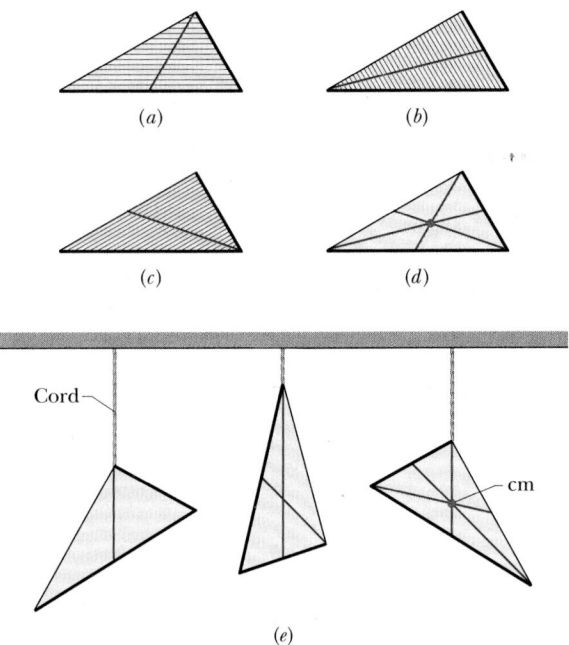

(a)   *(b)*

(c)   *(d)*

(e)

FIGURE 9-4 Sample Problem 9-2. In (*a*), (*b*), and (*c*), the triangular plate is divided into thin slats, parallel to a side. The center of mass must lie along the bisecting lines shown. (*d*) The dot, the only point common to all three lines, is the position of the center of mass. (*e*) Finding the center of mass by suspending the triangle from each vertex in turn.

Cʜᴇᴄᴋᴘᴏɪɴᴛ **1:** The figure shows a uniform square plate from which four identical squares at the corners will be removed. (a) Where is the center of mass of the

plate originally? Where is it after the removal of (b) square 1; (c) squares 1 and 2; (d) squares 1 and 3; (e) squares 1, 2, and 3; (f) all four squares? Answer in terms of quadrants, axes, or points (without calculation, of course).

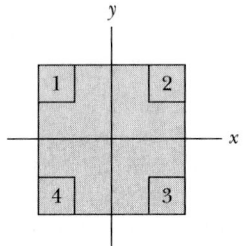

SAMPLE PROBLEM 9-3

Figure 9-5*a* shows a uniform circular metal plate of radius $2R$ from which a disk of radius R has been removed. Let us call this plate with a hole object X. Its center of mass is shown as a dot on the x axis. Locate this point.

SOLUTION: Figure 9-5*b* shows object X before the disk was removed. Call the disk object D, and the original composite plate object C. From symmetry, the center of mass of object C is at the origin of a coordinate system placed as shown.

In finding the center of mass of a composite object, we can assume that the masses of its components are concentrated at their individual centers of mass. Thus object C can be treated as equivalent to two point masses, representing objects X and D. Figure 9-5*c* shows the positions of the centers of mass of these three objects.

The position of the center of mass of object C is given, from Eq. 9-2, as

$$x_C = \frac{m_D x_D + m_X x_X}{m_D + m_X},$$

in which x_D and x_X are the positions of the centers of mass of objects D and X, respectively. Noting that $x_C = 0$ and solving for x_X, we obtain

$$x_X = -\frac{x_D m_D}{m_X}. \tag{9-12}$$

If ρ is the density (mass per volume) of the plate material and t is the uniform thickness of the plate, we have

$$m_D = \pi R^2 \rho t \quad \text{and} \quad m_X = \pi (2R)^2 \rho t - \pi R^2 \rho t.$$

With these substitutions and with $x_D = -R$, Eq. 9-12 becomes

$$x_X = -\frac{(-R)(\pi R^2 \rho t)}{\pi (2R)^2 \rho t - \pi R^2 \rho t} = \tfrac{1}{3} R. \quad \text{(Answer)}$$

Note that the uniform density and the uniform thickness of the plate cancel out and thus do not determine x_X.

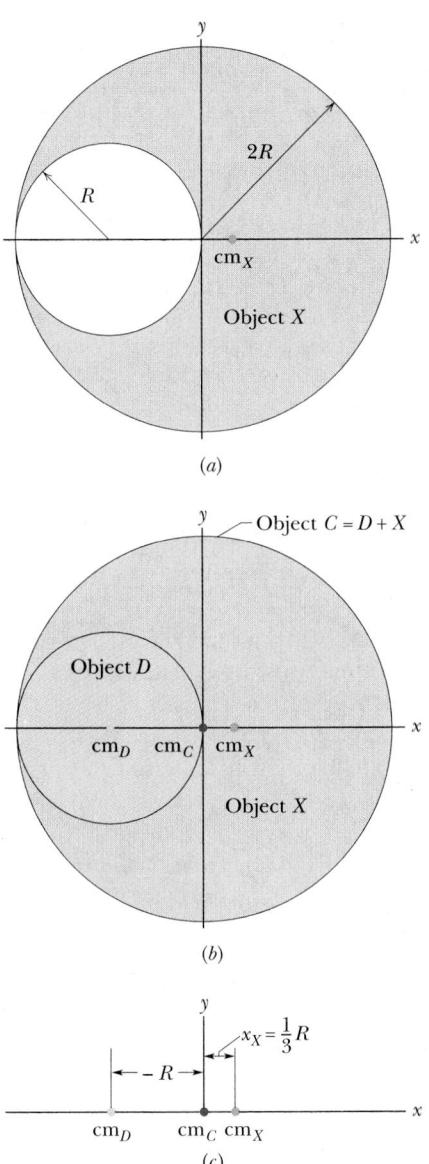

FIGURE 9-5 Sample Problem 9-3. (a) Object X is a metal disk of radius $2R$ with a hole of radius R cut in it; its center of mass is at point cm_X. (b) Object D is a metal disk that fills the hole in object X; its center of mass is at point cm_D, at $x = -R$. Object C is the composite object made up of objects X and D; its center of mass is at the origin of coordinates. (c) The centers of mass of the three objects.

SAMPLE PROBLEM 9-4

Silbury Hill (Fig. 9-6a), a mound on the plains near Stonehenge, was built 4600 years ago for unknown reasons, perhaps as a burial site. It is an incomplete right-circular cone (see Fig. 9-6b), with a flattened top of radius $r_2 = 16$ m, a base radius $r_1 = 88$ m, a height $h = 40$ m, and a volume V of $4.09 \times$

(a)

(b)

FIGURE 9-6 Sample Problem 9-4. (a) Silbury Hill in England, built by Neolithic people, required an estimated 1.8×10^7 hours of labor. (b) An incomplete, right-circular cone that resembles Silbury Hill. A "wafer" of radius r and thickness dz is shown at height z from the base.

10^5 m^3. The sides of the cone make an angle $\theta = 30°$ with the horizontal.

(a) Where is the center of mass of the mound?

SOLUTION: Because of the circular symmetry of the mound, the center of mass lies on the central vertical axis of the cone, at height z_{cm} above the base. To find z_{cm}, we use the last part of Eq. 9-11. We can simplify the integral by using the symmetry of the mound. To do this, we consider a thin, horizontal "wafer," as shown in Fig. 9-6b. The wafer has radius r, thickness dz, and horizontal area πr^2 and is at height z from the base. Its volume dV is

$$dV = \pi r^2 \, dz. \qquad (9\text{-}13)$$

The mound consists of a stack of such wafers, with radii ranging from r_1 at the bottom of the stack to r_2 at the top. If the cone were complete, it would have a height that we call H in

Fig. 9-6*b*. The radius *r* of each wafer is related to *H* by

$$\tan \theta = \frac{H}{r_1} = \frac{H - z}{r},$$

or

$$r = (H - z)\frac{r_1}{H}. \qquad (9\text{-}14)$$

Substituting Eqs. 9-13 and 9-14 into the last part of Eq. 9-11, we have

$$z_{cm} = \frac{1}{V} \int z \, dV = \frac{\pi r_1^2}{VH^2} \int_0^h z(H - z)^2 \, dz$$

$$= \frac{\pi r_1^2}{VH^2} \int_0^h (z^3 - 2z^2 H + zH^2) \, dz$$

$$= \frac{\pi r_1^2}{VH^2} \left[\frac{z^4}{4} - \frac{2z^3 H}{3} + \frac{z^2 H^2}{2} \right]_0^h$$

$$= \frac{\pi r_1^2 h^4}{VH^2} \left[\frac{1}{4} - \frac{2H}{3h} + \frac{H^2}{2h^2} \right].$$

Substituting known values, we find

$$z_{cm} = \frac{\pi (88 \text{ m})^2 (40 \text{ m})^4}{(4.09 \times 10^5 \text{ m}^3)(50.8 \text{ m})^2}$$

$$\times \left[\frac{1}{4} - \frac{2(50.8 \text{ m})}{3(40 \text{ m})} + \frac{(50.8 \text{ m})^2}{2(40 \text{ m})^2} \right]$$

$$= 12.37 \text{ m} \approx 12 \text{ m}. \qquad \text{(Answer)}$$

(b) If Silbury Hill has density $\rho = 1.5 \times 10^3$ kg/m³, then how much work was required to lift the dirt from the level of the base to build the mound?

SOLUTION: To find the work dW required to lift a mass element dm to height z, we use Eq. 7-21, with $\phi = 180°$:

$$dW = -dm \, gz \cos 180° = dm \, gz.$$

With Eq. 9-10, we substitute $\rho \, dV$ for dm, finding

$$dW = \rho gz \, dV.$$

To find the total work required to lift all the mass of Silbury Hill into place, we sum, via integration, the work dW associated with each volume element dV:

$$W = \int dW = \int \rho gz \, dV = \rho g \int z \, dV.$$

Using Eq. 9-11, we replace the integral with Vz_{cm}, finding

$$W = \rho V g z_{cm}. \qquad (9\text{-}15)$$

This tells us that the work required to lift all the mass of Silbury Hill into place is the same as if all the mass were lifted to (and somehow concentrated at) the center of mass of the hill. Substituting the known data in Eq. 9-15 yields

$$W = (1.5 \times 10^3 \text{ kg/m}^3)(4.09 \times 10^5 \text{ m}^3)$$

$$\times (9.8 \text{ m/s}^2)(12.37 \text{ m})$$

$$= 7.4 \times 10^{10} \text{ J}. \qquad \text{(Answer)}$$

PROBLEM SOLVING TACTICS

TACTIC 1: *Center-of-Mass Problems*
Sample Problems 9-1 to 9-3 provide three strategies for simplifying center-of-mass problems. (1) Make full use of the symmetry of the object, be it point, line, or plane. (2) If the object can be divided into several parts, treat each of these parts as a particle, located at its own center of mass. (3) Choose your axes wisely: if your system is a group of particles, choose one of the particles as your origin. If your system is a body with a line of symmetry, there is your *x* axis! The choice of origin is completely arbitrary; the location of the center of mass is the same regardless of the origin from which it is measured.

9-3 NEWTON'S SECOND LAW FOR A SYSTEM OF PARTICLES

If you roll a cue ball at a second billiard ball that is at rest, you expect that the two-ball system will continue to have some forward motion after impact. You would be surprised, for example, if both balls came back toward you or if both moved to the right or to the left.

What continues to move forward, its steady motion completely unaffected by the collision, is the center of mass of the two balls. If you focus on this point—which is always halfway between these bodies of identical mass—you can easily convince yourself by trial at a pool table that this is so. No matter whether the collision is glancing, head on, or somewhere in between, the center of mass moves majestically forward, as if the collision had never occurred. Let us look into this in more detail.

To do so, we replace the pair of billiard balls with an assemblage of *n* particles of (possibly) different masses. We are interested not in the individual motions of these particles but *only* in the motion of their center of mass. Although the center of mass is just a point, it moves like a particle whose mass is equal to the total mass of the system; we can assign a position, a velocity, and an acceleration to it. We state (and shall prove below) that the (vector) equation that governs the motion of the center of mass of such a system of particles is

$$\sum \mathbf{F}_{ext} = M\mathbf{a}_{cm} \quad \text{(system of particles)}. \qquad (9\text{-}16)$$

Equation 9-16 is Newton's second law governing the motion of the center of mass of a system of particles. Remarkably, it retains the same form ($\Sigma \mathbf{F} = m\mathbf{a}$) that holds for the motion of a single particle. In using Eq. 9-16, the three quantities that appear in it must be evaluated with some care:

1. $\Sigma\mathbf{F}_{\text{ext}}$ is the *vector* sum of *all* the *external forces* that act on the system. Forces exerted by one part of the system on another are called *internal forces,* and you must be careful not to include them when using Eq. 9-16.

2. *M* is the *total mass* of the system. We assume that no mass enters or leaves the system as it moves, so that *M* remains constant. The system is said to be **closed.**

3. $\mathbf{a}_{\text{cm}}$ is the acceleration of the *center of mass* of the system. Equation 9-16 gives no information about the acceleration of any other point of the system.

Equation 9-16, like all other vector equations, is equivalent to three equations involving the components of $\Sigma\mathbf{F}_{\text{ext}}$ and $\mathbf{a}_{\text{cm}}$ along the three coordinate axes. These equations are

$$\sum F_{\text{ext},x} = Ma_{\text{cm},x},$$
$$\sum F_{\text{ext},y} = Ma_{\text{cm},y}, \qquad (9\text{-}17)$$
$$\sum F_{\text{ext},z} = Ma_{\text{cm},z}.$$

Now we can go back and examine the behavior of the billiard balls. Once the cue ball has begun to roll, no net external force acts on the (two-ball) system. And thus, because $\Sigma\mathbf{F}_{\text{ext}} = 0$, Eq. 9-16 tells us that $\mathbf{a}_{\text{cm}} = 0$ also. Because acceleration is the rate of change of velocity, we conclude that the velocity of the center of mass of the system of two balls does not change. When the two balls collide, the forces that come into play are *internal* forces, exerted by one ball on the other. Such forces do not contribute to $\Sigma\mathbf{F}_{\text{ext}}$, which remains zero. Thus the center of mass of the system, which was moving forward before the collision, must continue to move forward after the collision, with the same speed and in the same direction.

Equation 9-16 applies not only to a system of particles but also to a solid body, such as the bat of Fig. 9-1*b*. In that case, *M* in Eq. 9-16 is the mass of the bat and $\Sigma\mathbf{F}_{\text{ext}}$ is the weight *M***g** of the bat. Equation 9-16 then tells us that $\mathbf{a}_{\text{cm}} = \mathbf{g}$. In other words, the center of mass of the bat moves as if the bat were a single particle of mass *M*, with force *M***g** acting on it.

Figure 9-7 shows another interesting case. Suppose that at a fireworks display, a rocket is launched on a parabolic path. At a certain point, it explodes into fragments. If the explosion had not occurred, the rocket would have continued along the trajectory shown in the figure. The forces of the explosion are *internal* to the system (the rocket or its fragments); that is, they are forces exerted by one part of the system on another part. If we ignore air resistance, the total *external* force $\Sigma\mathbf{F}_{\text{ext}}$ acting on the system is the weight *M***g** of the system, regardless of whether the rocket explodes. Thus, from Eq. 9-16, the acceleration

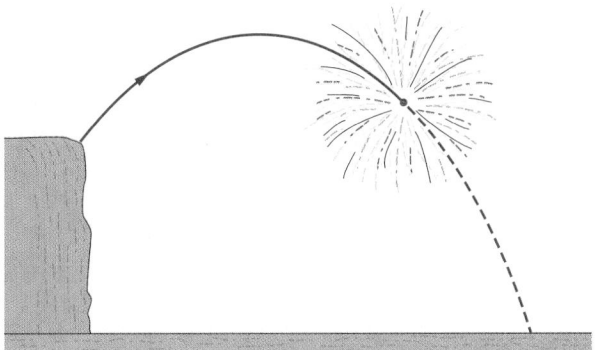

FIGURE 9-7 A fireworks rocket explodes in flight. In the absence of air drag, the center of mass of the fragments would continue to follow the original parabolic path, until fragments began to hit the ground.

of the center of mass of the fragments (while they are in flight) remains equal to **g**, and the center of mass of the fragments follows the same parabolic trajectory that the unexploded rocket would have followed.

When a ballet dancer leaps across the stage in a grand jeté, she raises her arms and stretches her legs out horizontally as soon as her feet leave the stage (Fig. 9-8). These actions shift her center of mass upward through her body. Although the shifting center of mass faithfully follows a parabolic path across the stage, its movement relative to the body decreases the height that would be attained by the head and torso in a normal jump. The result is that the head and torso follow a nearly horizontal path.

Proof of Equation 9-16

Now let us prove this important equation. From Eq. 9-8 we have, for a system of *n* particles,

$$M\mathbf{r}_{\text{cm}} = m_1\mathbf{r}_1 + m_2\mathbf{r}_2 + m_3\mathbf{r}_3 + \cdots + m_n\mathbf{r}_n, \quad (9\text{-}18)$$

in which *M* is the system's total mass and $\mathbf{r}_{\text{cm}}$ is the vector locating the position of the system's center of mass.

Differentiating Eq. 9-18 with respect to time gives

$$M\mathbf{v}_{\text{cm}} = m_1\mathbf{v}_1 + m_2\mathbf{v}_2 + m_3\mathbf{v}_3 + \cdots + m_n\mathbf{v}_n. \quad (9\text{-}19)$$

Here $\mathbf{v}_i \ (= d\mathbf{r}_i/dt)$ is the velocity of the *i*th particle, and $\mathbf{v}_{\text{cm}}$ $(= d\mathbf{r}_{\text{cm}}/dt)$ is the velocity of the center of mass.

Differentiating Eq. 9-19 with respect to time leads to

$$M\mathbf{a}_{\text{cm}} = m_1\mathbf{a}_1 + m_2\mathbf{a}_2 + m_3\mathbf{a}_3 + \cdots + m_n\mathbf{a}_n. \quad (9\text{-}20)$$

Here $a_i \ (= d\mathbf{v}_i/dt)$ is the acceleration of the *i*th particle, and $\mathbf{a}_{\text{cm}} \ (= d\mathbf{v}_{\text{cm}}/dt)$ is the acceleration of the center of mass. Although the center of mass is just a geometrical point, it has a position, a velocity, and an acceleration, as if it were a particle.

From Newton's second law, $m_i\mathbf{a}_i$ is equal to the result-

FIGURE 9-8 A grand jeté. (Adapted from *The Physics of Dance,* by Kenneth Laws, Schirmer Books, 1984.)

ant force $\mathbf{F}_i$ that acts on the *i*th particle. Thus we can rewrite Eq. 9-20 as

$$M\mathbf{a}_{cm} = \mathbf{F}_1 + \mathbf{F}_2 + \mathbf{F}_3 + \cdots + \mathbf{F}_n. \quad (9\text{-}21)$$

Among the forces that contribute to the right side of Eq. 9-21 will be forces that the particles of the system exert on each other (internal forces) and forces exerted on the particles from outside the system (external forces). By Newton's third law, the internal forces form action–reaction pairs and cancel out in the sum that appears on the right side of Eq. 9-21. What remains is the vector sum of all the *external* forces that act on the system. Equation 9-21 then reduces to Eq. 9-16, the relation that we set out to prove.

CHECKPOINT **2:** Two skaters on frictionless ice hold opposite ends of a pole of negligible mass. An axis runs along the pole, and the origin of the axis is at the center of mass of the two-skater system. One skater, Fred, weighs twice a much as the other skater, Ethel. Where do the skaters meet if (a) Fred pulls hand over hand along the pole so as to draw himself to Ethel, (b) Ethel pulls hand over hand to draw herself to Fred, and (c) both skaters pull hand over hand?

SAMPLE PROBLEM 9-5

Figure 9-9*a* shows a system of three particles, each particle acted on by a different external force and all initially at rest. What is the acceleration of the center of mass of this system?

SOLUTION: The position of the center of mass, calculated by the method of Sample Problem 9-1, is marked by a dot in the figure. As Fig. 9-9*b* suggests, we treat this point as if it were a real particle, assigning to it a mass M equal to the total mass of the system (16 kg) and assuming that all external forces are applied at that point.

The *x* component of the net external force $\Sigma\mathbf{F}_{ext}$ acting on the center of mass is

$$\sum F_{ext,x} = 14\ \text{N} - 6.0\ \text{N} + (12\ \text{N})(\cos 45°)$$
$$= 16.5\ \text{N},$$

and the *y* component is

$$\sum F_{ext,y} = (12\ \text{N})(\sin 45°) = 8.49\ \text{N}.$$

The net external force thus has the magnitude

$$\sum F_{ext} = \sqrt{(16.5\ \text{N})^2 + (8.49\ \text{N})^2} = 18.6\ \text{N}$$

and makes an angle with the *x* axis given by

$$\theta = \tan^{-1}\frac{8.49\ \text{N}}{16.5\ \text{N}} = \tan^{-1} 0.515$$
$$= 27°. \qquad \text{(Answer)}$$

This is also the direction of the acceleration $\mathbf{a}_{cm}$ of the center of mass. From Eq. 9-16, the magnitude of $\mathbf{a}_{cm}$ is given by

$$a_{cm} = \frac{\sum F_{ext}}{M} = \frac{18.6\ \text{N}}{16\ \text{kg}} = 1.16\ \text{m/s}^2$$
$$\approx 1.2\ \text{m/s}^2. \qquad \text{(Answer)}$$

The three particles of Fig. 9-9*a* and their center of mass move with (different) constant accelerations. Since the particles start from rest, each will move, with ever-increasing speed, along a straight line in the direction of the force acting on it. The center of mass will move in the directon of $\mathbf{a}_{cm}$.

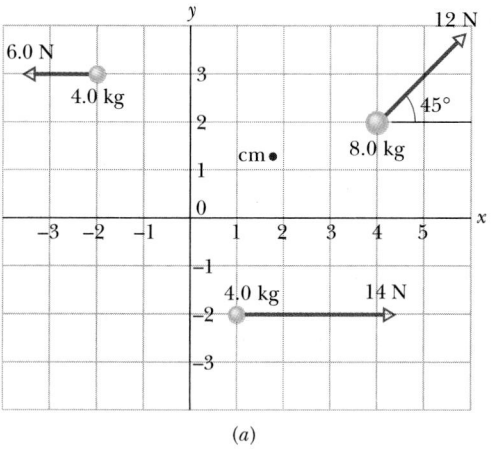

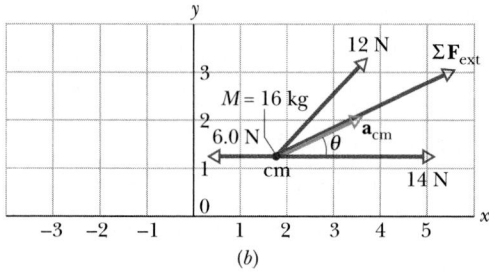

FIGURE 9-9 Sample Problem 9-5. (*a*) Three particles, placed at rest in the positions shown, are acted on by the external forces shown. The center of mass of the system is marked. (*b*) The forces are now transferred to the center of mass of the system, which behaves like a particle whose mass *M* is equal to the total mass of the system. The net force and the acceleration of the center of mass are shown.

9-4 LINEAR MOMENTUM

Momentum is another word that has several meanings in everyday language but only a single precise meaning in physics. The **linear momentum** of a particle is a vector **p**, defined as

$$\mathbf{p} = m\mathbf{v} \qquad \begin{array}{l}\text{(linear momentum}\\ \text{of a particle),}\end{array} \qquad (9\text{-}22)$$

in which *m* is the mass of the particle and **v** is its velocity. (The adjective *linear* is often dropped, but it serves to distinguish **p** from *angular* momentum, which is introduced in Chapter 12 and which is associated with rotation.) Note that since *m* is always a positive scalar quantity, Eq. 9-22 tells us that **p** and **v** are in the same direction. Equation 9-22 also tells us that the SI unit for momentum is the kilogram-meter per second.

Newton actually expressed his second law of motion in terms of momentum:

The rate of change of the momentum of a particle is proportional to the net force acting on the particle and is in the direction of that force.

In equation form this becomes

$$\sum \mathbf{F} = \frac{d\mathbf{p}}{dt}. \qquad (9\text{-}23)$$

Substituting for **p** from Eq. 9-22 gives

$$\sum \mathbf{F} = \frac{d\mathbf{p}}{dt} = \frac{d}{dt}(m\mathbf{v}) = m\frac{d\mathbf{v}}{dt} = m\mathbf{a}.$$

Thus the relations $\sum \mathbf{F} = d\mathbf{p}/dt$ and $\sum \mathbf{F} = m\mathbf{a}$ are completely equivalent expressions of Newton's second law of motion as it applies to the motion of single particles in classical mechanics.

Momentum at Very High Speeds

For particles moving with speeds that are near the speed of light, Newtonian mechanics predicts results that do not agree with experiment. In such cases, we must use Einstein's theory of special relativity. In relativity, the formulation $\mathbf{F} = d\mathbf{p}/dt$ is correct, *provided* we define the momentum of a particle not as $m\mathbf{v}$ but as

$$\mathbf{p} = \frac{m\mathbf{v}}{\sqrt{1 - (v/c)^2}}, \qquad (9\text{-}24)$$

in which *c*, the speed of light, is a sure indicator of a relativistic equation.

The speeds of common macroscopic objects such as baseballs, bullets, or space probes are so much lower than the speed of light that the quantity $(v/c)^2$ in Eq. 9-24 is very close to zero. Under these conditions, Eq. 9-24 reduces to Eq. 9-22 and Einstein's relativity theory reduces to Newtonian mechanics. For electrons and other subatomic particles, however, speeds very close to that of light are easily obtained and the definition in Eq. 9-24 *must* be used, often as a matter of routine engineering practice.

9-5 THE LINEAR MOMENTUM OF A SYSTEM OF PARTICLES

Consider now a system of *n* particles, each with its own mass, velocity, and linear momentum. The particles may interact with each other, and external forces may act on them as well. The system as a whole has a total linear momentum **P**, which is defined to be the vector sum of the individual particles' linear momenta. Thus

TABLE 9-1 **SOME DEFINITIONS AND LAWS IN CLASSICAL MECHANICS**

LAW OR DEFINITION	SINGLE PARTICLE		SYSTEM OF PARTICLES	
Newton's second law	$\Sigma \mathbf{F} = m\mathbf{a}$		$\Sigma \mathbf{F}_{ext} = M\mathbf{a}_{cm}$	(9-16)
Linear momentum	$\mathbf{p} = m\mathbf{v}$	(9-22)	$\mathbf{P} = M\mathbf{v}_{cm}$	(9-26)
Newton's second law	$\Sigma \mathbf{F} = \dfrac{d\mathbf{p}}{dt}$	(9-23)	$\Sigma \mathbf{F}_{ext} = \dfrac{d\mathbf{P}}{dt}$	(9-28)
Work–kinetic energy theorem	$W = \Delta K$			

$$\mathbf{P} = \mathbf{p}_1 + \mathbf{p}_2 + \mathbf{p}_3 + \cdots + \mathbf{p}_n$$
$$= m_1\mathbf{v}_1 + m_2\mathbf{v}_2 + m_3\mathbf{v}_3 + \cdots + m_n\mathbf{v}_n. \quad (9\text{-}25)$$

If we compare this equation with Eq. 9-19, we see that

$$\mathbf{P} = M\mathbf{v}_{cm} \quad \begin{matrix}\text{(linear momentum,}\\ \text{system of particles),}\end{matrix} \quad (9\text{-}26)$$

which gives us another way to define the linear momentum of a system of particles:

The linear momentum of a system of particles is equal to the product of the total mass M of the system and the velocity of the center of mass.

If we take the time derivative of Eq. 9-26, we find

$$\frac{d\mathbf{P}}{dt} = M\frac{d\mathbf{v}_{cm}}{dt} = M\mathbf{a}_{cm}. \quad (9\text{-}27)$$

Comparing Eqs. 9-16 and 9-27 allows us to write Newton's second law for a system of particles in the equivalent form

$$\sum \mathbf{F}_{ext} = \frac{d\mathbf{P}}{dt}. \quad (9\text{-}28)$$

This equation is the generalization of the single-particle equation $\Sigma \mathbf{F} = d\mathbf{p}/dt$ to a system of many particles. Table 9-1 displays the important relations that we have derived for single particles and for systems of particles.

CHECKPOINT **3:** The figure gives the linear momentum versus time for a particle moving along an axis. A force directed along the axis acts on the particle. (a) Rank the four regions indicated according to the magnitude of the force, greatest first. (b) In which region is the particle slowing?

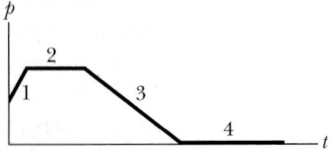

SAMPLE PROBLEM 9-6

Figure 9-10a shows a 2.0 kg toy race car before and after taking a turn on a track. Its speed is 0.50 m/s before the turn and 0.40 m/s after the turn. What is the change $\Delta\mathbf{P}$ in the linear momentum of the car?

SOLUTION: Before we can use Eq. 9-26 to find the car's linear momenta before and after the turn, we need its velocity $\mathbf{v}_i$ before the turn and its velocity $\mathbf{v}_f$ after the turn. Using the coordinate system of Fig. 9-10a, we write $\mathbf{v}_i$ and $\mathbf{v}_f$ as

$$\mathbf{v}_i = -(0.50 \text{ m/s})\mathbf{j} \quad \text{and} \quad \mathbf{v}_f = (0.40 \text{ m/s})\mathbf{i}.$$

Equation 9-26 then gives us the linear momentum $\mathbf{P}_i$ before the turn and the linear momentum $\mathbf{P}_f$ after the turn:

$$\mathbf{P}_i = M\mathbf{v}_i = (2.0 \text{ kg})(-0.50 \text{ m/s})\mathbf{j} = (-1.0 \text{ kg} \cdot \text{m/s})\mathbf{j}$$

and

$$\mathbf{P}_f = M\mathbf{v}_f = (2.0 \text{ kg})(0.40 \text{ m/s})\mathbf{i} = (0.80 \text{ kg} \cdot \text{m/s})\mathbf{i}.$$

Because these linear momenta are not along the same axis, we *cannot* find the change in linear momentum $\Delta\mathbf{P}$ by merely subtracting the magnitude of $\mathbf{P}_i$ from the magnitude of $\mathbf{P}_f$. Instead, we write the change in linear momentum as

$$\Delta\mathbf{P} = \mathbf{P}_f - \mathbf{P}_i \quad (9\text{-}29)$$

and then as

$$\Delta\mathbf{P} = (0.80 \text{ kg} \cdot \text{m/s})\mathbf{i} - (-1.0 \text{ kg} \cdot \text{m/s})\mathbf{j}$$
$$= (0.8\mathbf{i} + 1.0\mathbf{j}) \text{ kg} \cdot \text{m/s}. \quad \text{(Answer)}$$

Figure 9-10b shows $\Delta\mathbf{P}$, $\mathbf{P}_f$, and $-\mathbf{P}_i$. Note that we subtract $\mathbf{P}_i$ from $\mathbf{P}_f$ by adding $-\mathbf{P}_i$ to $\mathbf{P}_f$.

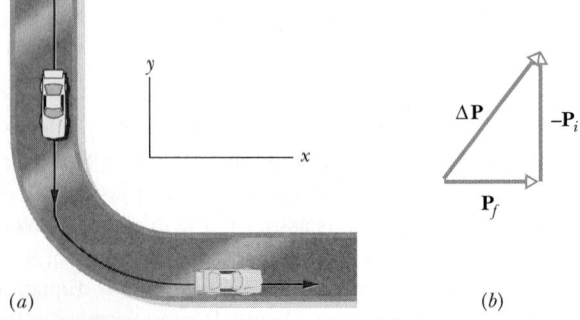

FIGURE 9-10 Sample Problem 9-6. (*a*) A toy car takes a turn. (*b*) The change $\Delta\mathbf{P}$ in the car's linear momentum is the vector sum of its final linear momentum $\mathbf{P}_f$ and the negative of its initial linear momentum $\mathbf{P}_i$.

9-6 CONSERVATION OF LINEAR MOMENTUM

Suppose that the sum of the external forces acting on a system of particles is zero (the system is isolated) and that no particles leave or enter the system (the system is closed). Putting $\Sigma\mathbf{F}_{ext} = 0$ in Eq. 9-28 then yields $d\mathbf{P}/dt = 0$, or

$$\mathbf{P} = \text{constant} \qquad \text{(closed, isolated system).} \qquad (9\text{-}30)$$

This important result, called the **law of conservation of linear momentum,** can also be written as

$$\mathbf{P}_i = \mathbf{P}_f \qquad \text{(closed, isolated system),} \qquad (9\text{-}31)$$

where the subscripts refer to the values of $\mathbf{P}$ at some initial time i and some later time f. Equations 9-30 and 9-31 tell us that if no net external force acts on a system of particles, the total linear momentum of the system remains constant.

Like the law of conservation of energy that you met in Chapter 8, the law of conservation of linear momentum is a more general law than Newtonian mechanics itself. It holds in the subatomic realm, where Newton's laws fail. It holds for the highest particle speeds, where Einstein's relativity theory prevails, that is, where Eq. 9-24, rather than Eq. 9-22, must be used to find the linear momentum.

From Eq. 9-26 ($\mathbf{P} = M\mathbf{v}_{cm}$) we see that if $\mathbf{P}$ is constant, then $\mathbf{v}_{cm}$, the velocity of the center of mass, is also constant. This in turn means that $\mathbf{a}_{cm}$, the acceleration of the center of mass, must be zero, which is consistent with Eq. 9-16 ($\Sigma\mathbf{F}_{ext} = M\mathbf{a}_{cm}$) when $\Sigma\mathbf{F}_{ext} = 0$. Thus the law of conservation of linear momentum is consistent with both forms of Newton's second law for systems of particles, as displayed in Table 9-1.

Equations 9-30 and 9-31 are vector equations and, as such, both are equivalent to three equations corresponding to the conservation of linear momentum in three mutually perpendicular directions. Depending on the forces acting on a system, linear momentum might be conserved in one or two directions but not in all directions:

If a component of the net *external* force on a closed system is zero along an axis, then the component of the linear momentum of the system along that axis cannot change.

As an example, suppose that you toss a grapefruit across a room. During its flight, the only external force acting on the grapefruit (which we take as the system) is its weight $m\mathbf{g}$, which is directed vertically downward. Thus, the vertical component of the linear momentum of the grapefruit changes, but since no horizontal external force acts on the grapefruit, the horizontal component of the linear momentum cannot change.

Note we focus on external forces acting on a closed system. Although internal forces can change the linear momentum of portions of the system, they cannot change the total linear momentum of the system itself.

CHECKPOINT 4: An initially stationary device lying on a frictionless floor explodes into two pieces, which then slide across the floor. One piece slides in the positive direction of an x axis. (a) What is the sum of the momenta of the two pieces after the explosion? (b) Can the second piece move at an angle to the x axis? (c) What is the direction of the momentum of the second piece?

SAMPLE PROBLEM 9-7

A box with mass $m = 6.0$ kg slides with speed $v = 4.0$ m/s across a frictionless floor in the positive direction of an x axis. It suddenly explodes into two pieces: one piece, with mass $m_1 = 2.0$ kg, moves in the positive direction of the x axis with speed $v_1 = 8.0$ m/s. What is the velocity of the second piece, with mass m_2?

SOLUTION: Our reference frame will be that of the floor. Our system, which initially consists of the original box and then of the two pieces, is closed but is not isolated, since the box and pieces have weight and each experiences a normal force from the floor. However, those forces are all vertical and thus cannot change the horizontal component of the momentum of the system. Neither can the forces produced by the explosion, because those forces are internal to the system. Thus, the horizontal component of the momentum of the system is conserved, and we can apply Eq. 9-31 along the x axis.

The initial momentum of the system was that of the box:

$$\mathbf{P}_i = m\mathbf{v}.$$

Similarly, we can write the final momenta of the two pieces as

$$\mathbf{P}_{f1} = m_1\mathbf{v}_1 \quad \text{and} \quad \mathbf{P}_{f2} = m_2\mathbf{v}_2.$$

The final total momentum $\mathbf{P}_f$ of the system is the vector sum of the momenta of the two pieces:

$$\mathbf{P}_f = \mathbf{P}_{f1} + \mathbf{P}_{f2} = m_1\mathbf{v}_1 + m_2\mathbf{v}_2.$$

Since all the velocities and momenta in this problem are vectors along a single axis, we can represent them by their magnitudes, along with an implied plus sign or an explicit minus sign. Doing so while applying Eq. 9-31, we now obtain

$$P_i = P_f$$

or
$$mv = m_1v_1 + m_2v_2.$$

Inserting the given data, and using the fact that the mass of the second piece is $m_2 = m - m_1 = 4.0$ kg, we find

$$(6.0 \text{ kg})(4.0 \text{ m/s}) = (2.0 \text{ kg})(8.0 \text{ m/s}) + (4.0 \text{ kg})v_2$$

and thus

$$v_2 = 2.0 \text{ m/s.} \qquad \text{(Answer)}$$

Since the result is positive, the second piece moves in the positive direction of the x axis.

SAMPLE PROBLEM 9-8

As Fig. 9-11 shows, a cannon whose mass M is 1300 kg fires a ball whose mass m is 72 kg in a horizontal direction with a velocity $\mathbf{v}$ relative to the cannon, which recoils (freely) with velocity $\mathbf{V}$ relative to the Earth. The magnitude of $\mathbf{v}$ is 55 m/s.

(a) What is $\mathbf{V}$?

SOLUTION: We choose the cannon plus the ball as our system. By doing so, we ensure that the forces involved in the firing of the cannon are internal to the system, and we do not have to deal with them. The external forces acting on the system have no components in the horizontal direction. Thus the horizontal component of the total linear momentum of the system must remain unchanged as the cannon is fired.

We choose the Earth as a reference frame. Because all the velocities here are horizontal, they are positive if they point rightward in Fig. 9-11 and negative if they point leftward. Although $\mathbf{V}$ is shown pointing leftward, we actually do not yet know its direction.

The velocity of the ball relative to the Earth is $\mathbf{v}_E$, represented by v_E. The ball's velocity relative to the cannon is the difference between its velocity relative to the Earth and the cannon's velocity relative to the Earth. We represent this as

$$v = v_E - V,$$

and thus

$$v_E = v + V. \qquad \text{(9-32)}$$

Before the cannon is fired, the system has an initial linear momentum P_i of zero. While the ball is in flight, the system has a horizontal linear momentum P_f which, with the aid of Eq. 9-32, is given by

$$P_f = MV + mv_E = MV + m(v + V),$$

in which the first term on the right is the linear momentum of the cannon and the second term that of the ball.

Conservation of linear momentum in the horizontal direction requires that $P_i = P_f$, or

$$0 = MV + m(v + V).$$

Solving for V yields

$$V = -\frac{mv}{M + m} = -\frac{(72 \text{ kg})(55 \text{ m/s})}{1300 \text{ kg} + 72 \text{ kg}}$$

$$= -2.9 \text{ m/s.} \qquad \text{(Answer)}$$

The minus sign tells us that the cannon recoils to the left in Fig. 9-11, as we know it does.

(b) What is v_E?

SOLUTION: From Eq. 9-32, we find

$$v_E = v + V = 55 \text{ m/s} + (-2.9 \text{ m/s})$$

$$= 52 \text{ m/s.} \qquad \text{(Answer)}$$

Because of the recoil, the ball is moving a little slower relative to the Earth than it would were there no recoil.

Note the importance in this problem of choosing the system (*cannon + ball*) wisely and of being clear about the reference frame (Earth or recoiling cannon) to which the various measurements are referenced.

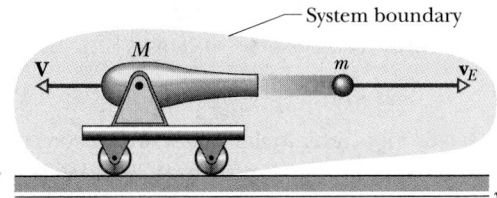

FIGURE 9-11 Sample Problem 9-8. A cannon of mass M fires a ball of mass m. The ball has velocity $\mathbf{v}_E$ relative to the Earth and velocity $\mathbf{v}$ relative to the cannon. The recoiling cannon has velocity $\mathbf{V}$ relative to the Earth.

SAMPLE PROBLEM 9-9

Imagine a spaceship and cargo module, of total mass M, traveling in deep space with velocity $v_i = 2100$ km/h relative to the Sun (Fig. 9-12a). With a small explosion, the ship ejects the cargo module, of mass $0.20M$ (Fig. 9-12b). The ship then travels 500 km/h faster than the module; that is, the relative speed v_{rel} between the module and the ship is 500 km/h. What then is the velocity v_f of the ship relative to the Sun?

SOLUTION: Because the ship–module system is closed and isolated, its total linear momentum is conserved; that is,

$$P_i = P_f, \qquad \text{(9-33)}$$

where the subscripts i and f refer to values before and after the ejection, respectively. Before the ejection, we have

$$P_i = Mv_i. \qquad \text{(9-34)}$$

Let U be the velocity of the ejected module relative to the Sun. The total linear momentum of the system after the ejection is then

$$P_f = (0.20M)U + (0.80M)v_f, \qquad \text{(9-35)}$$

where the first term on the right is the linear momentum of the module and the second term is that of the ship.

The relative speed v_{rel} between the module and the ship is the difference in their velocities

$$v_{rel} = v_f - U,$$

from which we write

$$U = v_f - v_{rel}.$$

Substituting this expression for U into Eq. 9-35, and then substituting Eqs. 9-34 and 9-35 into Eq. 9-33, we find

$$Mv_i = 0.20M(v_f - v_{rel}) + 0.80Mv_f,$$

which gives us

$$v_f = v_i + 0.20v_{rel},$$

or

$$v_f = 2100 \text{ km/h} + (0.20)(500 \text{ km/h})$$
$$= 2200 \text{ km/h}. \qquad \text{(Answer)}$$

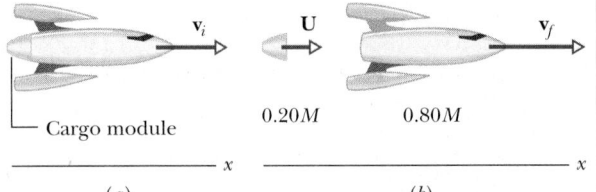

x

(a) $\qquad\qquad$ (b)

FIGURE 9-12 Sample Problem 9-9. (a) A spaceship, with a cargo module, moving at velocity $\mathbf{v}_i$. (b) The spaceship has ejected the cargo module. The spaceship now moves at velocity $\mathbf{v}_f$ and the cargo module moves at velocity $\mathbf{U}$.

$$P_f = m_1v_1 + m_2v_2.$$

Conservation of momentum requires that $P_i = P_f$, or

$$0 = m_1v_1 + m_2v_2. \qquad (9\text{-}36)$$

Thus we have

$$\frac{v_1}{v_2} = -\frac{m_2}{m_1}. \qquad (9\text{-}37)$$

The minus sign tells us that the two velocities always have opposite signs. Equation 9-37 holds at every instant after release, no matter what the actual speeds of the blocks are.

(b) What is the ratio K_1/K_2 of the kinetic energies of the blocks as their separation decreases?

SOLUTION: We can write the ratio K_1/K_2 as

$$\frac{K_1}{K_2} = \frac{\frac{1}{2}m_1v_1^2}{\frac{1}{2}m_2v_2^2} = \frac{m_1}{m_2}\left(\frac{v_1}{v_2}\right)^2.$$

Substituting for v_1/v_2 from Eq. 9-37 and simplifying, we find

$$\frac{K_1}{K_2} = \frac{m_2}{m_1}. \qquad (9\text{-}38)$$

As the blocks move toward each other and the stretch of the spring decreases, energy is transferred from the elastic potential energy of the spring to the kinetic energies K_1 and K_2 of the blocks. Although K_1 and K_2 then increase, Eq. 9-38 tells us that their ratio does not change but is preset by the ratio of the masses. After the spring reaches its rest length and begins to be compressed by the blocks, the energy transfer is reversed, but Eq. 9-38 still holds.

Equations 9-36 through 9-38 apply to other situations in which two bodies attract (or repel) each other. For example, they apply to a stone falling toward the Earth. The stone corresponds to, say, block 1 in Fig. 9-13 and the Earth to block 2. The gravitational force between the stone and the Earth corresponds to the mutual force between the blocks that is provided by the spring in Fig. 9-13. Our reference frame is the frame in which the center of mass of the stone–Earth system is stationary. In this frame, Eq. 9-36 tells us that the magnitudes of the linear momenta of the stone and the Earth remain equal throughout the fall. Equations 9-37 and 9-38 tell us that, because $m_2 \gg m_1$, the stone has much greater speed and kinetic energy than the Earth during the fall.

CHECKPOINT 5: The table gives velocities of the ship and module in Sample Problem 9-9 (after ejection and relative to the Sun), and the relative speed between the ship and the module for three situations. What are the missing values?

VELOCITIES (km/h)		RELATIVE SPEED
MODULE	SHIP	(km/h)
(a) 1500	2000	
(b)	3000	400
(c) 1000		600

SAMPLE PROBLEM 9-10

Figure 9-13 shows two blocks that are connected by an ideal spring and are free to slide on a frictionless horizontal surface. Block 1 has mass m_1 and block 2 has mass m_2. The blocks are pulled in opposite directions (stretching the spring) and then released from rest.

(a) What is the ratio v_1/v_2 of the velocity of block 1 to the velocity of block 2 as the separation between the blocks decreases?

SOLUTION: We take the two blocks and the spring as our system, and the horizontal surface on which they slide as our reference frame. The initial momentum P_i of the system before the blocks are released is zero. The momentum of the system at any instant as the blocks move toward each other is

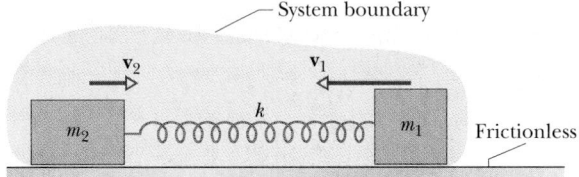

FIGURE 9-13 Sample Problem 9-10. Two blocks, resting on a frictionless surface and connected by a spring, have been pulled apart and then released from rest. The vector sum of their linear momenta remains zero during their subsequent motions. The system boundary is shown.

SAMPLE PROBLEM 9-11

A firecracker placed inside a coconut of mass M, initially at rest on a frictionless floor, blows the fruit into three pieces and sends them sliding across the floor. An overhead view is shown in Fig. 9-14a. Piece C, with mass $0.30M$, has final speed $v_{fC} = 5.0$ m/s.

(a) What is the speed of piece B, with mass $0.20M$?

SOLUTION: We superimpose an xy coordinate system as shown in Fig. 9-14b, with the direction of decreasing x coinciding with the direction of $\mathbf{v}_{fA}$. The x axis makes an angle of $80°$ with $\mathbf{v}_{fC}$ and an angle of $50°$ with $\mathbf{v}_{fB}$.

The linear momentum of the coconut (and its pieces) is conserved along both the x and y axes, because the forces involved in the explosion are internal forces, and because no external force acts on the coconut along the x or y axis. Along the y axis the conservation of linear momentum is written as

$$P_{iy} = P_{fy}, \qquad (9\text{-}39)$$

where the subscript i refers to the initial value (before the explosion), and the subscript y refers to the y component of the vector $\mathbf{P}_i$ or $\mathbf{P}_f$.

The component P_{iy} of the initial linear momentum is zero, because the coconut is initially at rest. To get an expression for P_{fy}, we find the y component of the final linear momentum of each piece:

$$p_{fA,y} = 0,$$

$$p_{fB,y} = -0.20Mv_{fB,y} = -0.20Mv_{fB}\sin 50°,$$

$$p_{fC,y} = 0.30Mv_{fC,y} = 0.30Mv_{fC}\sin 80°.$$

(Note that $p_{fA,y} = 0$ because of our choice of axes.) Equation 9-39 can now be written as

$$P_{iy} = P_{fy} = p_{fA,y} + p_{fB,y} + p_{fC,y}.$$

Then, with $v_{fC} = 5.0$ m/s, we have

$$0 = 0 - 0.20Mv_{fB}\sin 50° + (0.30M)(5.0 \text{ m/s})\sin 80°,$$

from which we find

$$v_{fB} = 9.64 \text{ m/s} \approx 9.6 \text{ m/s}. \qquad \text{(Answer)}$$

(b) What is the speed of piece A?

SOLUTION: Because linear momentum is also conserved along the x axis, we have

$$P_{ix} = P_{fx}, \qquad (9\text{-}40)$$

where $P_{ix} = 0$ because the coconut is initially at rest. To get P_{fx}, we find the x components of the final momenta, using the fact that piece A must have a mass of $0.50M$:

$$p_{fA,x} = -0.50Mv_{fA},$$

$$p_{fB,x} = 0.20Mv_{fB,x} = 0.20Mv_{fB}\cos 50°,$$

$$p_{fC,x} = 0.30Mv_{fC,x} = 0.30Mv_{fC}\cos 80°.$$

Equation 9-40 can now be written as

$$P_{ix} = P_{fx} = p_{fA,x} + p_{fB,x} + p_{fC,x}.$$

Then, with $v_{fC} = 5.0$ m/s and $v_{fB} = 9.64$ m/s, we have

$$0 = -0.50Mv_{fA} + 0.20M(9.64 \text{ m/s})\cos 50°$$
$$+ 0.30M(5.0 \text{ m/s})\cos 80°,$$

from which we find

$$v_{fA} = 3.0 \text{ m/s}. \qquad \text{(Answer)}$$

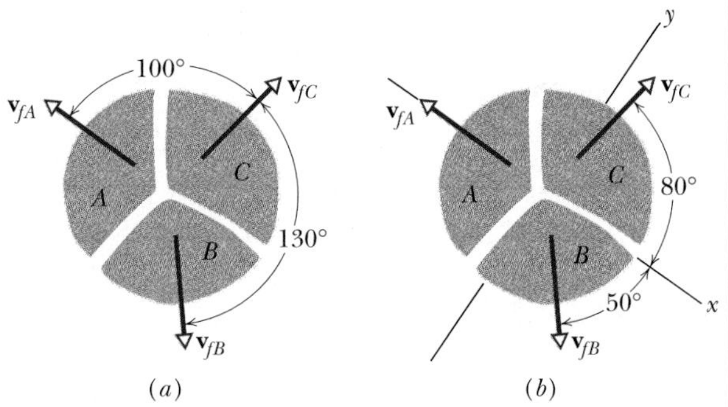

FIGURE 9-14 Sample Problem 9-11. Three pieces of an exploded coconut move off in three directions along a frictionless floor. (a) An overhead view of the event. (b) The same with a two-dimensional axis system imposed.

CHECKPOINT **6:** Suppose that the exploding coconut in Sample Problem 9-11 is accelerating in the negative direction of y in Fig. 9-14 (say it is on a ramp that slants downward in that direction). Is linear momentum conserved along (a) the x axis (as stated in Eq. 9-40) and (b) the y axis (as stated in Eq. 9-39)?

PROBLEM SOLVING TACTICS

TACTIC 2: *Conservation of Linear Momentum*

This is a good time to reread Tactic 2 of Chapter 8, which deals with problems involving the conservation of mechanical energy. The same broad suggestions hold here for problems involving the conservation of linear momentum.

First, make sure that you have chosen a closed, isolated system. *Closed* means that no matter (no particles) passes through the system boundary in any direction. *Isolated* means that the net external force acting on the system is zero. If the system is *not* isolated or *not* closed, then Eqs. 9-30 and 9-31 do not hold.

Remember that linear momentum is a vector quantity so that each component can be conserved separately, provided only that the corresponding component of the net external force is zero. In Sample Problem 9-8, no net external force acts horizontally on the *cannon + ball* system, so the horizontal component of linear momentum is conserved. However, the net external force acting vertically on this system is not zero; weight acts on the cannonball while it is in flight.

Thus the vertical component of linear momentum for this system is not conserved.

Select two appropriate states of the system (which you may choose to call the initial state and the final state) and write expressions for the linear momentum of the system in each of these two states. In writing these expressions, make sure that you know what inertial reference frame you are using, and make sure also that you include the entire system, not missing any part of it and not including objects that do not belong to your system. In Sample Problem 9-8 we must be particularly clear about the reference frame (Earth or recoiling cannon) to which the various measurements are referenced.

Finally, set your expressions for $\mathbf{P}_i$ and $\mathbf{P}_f$ equal to each other and solve for what is requested.

9-7 SYSTEMS WITH VARYING MASS: A ROCKET (OPTIONAL)

In the systems we have dealt with so far, we have assumed that the total mass of the system remains constant. Sometimes, as in a rocket (Fig. 9-15), it does not. Most of the mass of a rocket on its launching pad is fuel, all of which will eventually be burned and ejected from the nozzle of the rocket engine.

We handle the variation of the mass of the rocket as the rocket accelerates by applying Newton's second law, not to the rocket alone but to the rocket and its ejected

FIGURE 9-15 Liftoff of Project Mercury spacecraft.

combustion products taken together. The mass of *this* system does *not* change as the rocket accelerates.

Finding the Acceleration

Assume that you are in an inertial reference frame, watching a rocket accelerate through deep space with no gravitational or atmospheric drag forces acting on it. For this one-dimensional motion, let M be the mass of the rocket and v its velocity at an arbitrary time t (see Fig. 9-16a).

Figure 9-16b shows how things stand a time interval dt later. The rocket now has velocity $v + dv$ and mass $M + dM$, where the change in mass dM is a *negative quantity*. The exhaust products released by the rocket during interval dt have mass $-dM$ and velocity U relative to our inertial reference frame.

Our system consists of the rocket and the exhaust products released during interval dt. The system is closed and isolated, so the linear momentum of the system must be conserved during dt; that is,

$$P_i = P_f, \qquad (9\text{-}41)$$

where the subscripts i and f indicate the values at the beginning and end of time interval dt. We can rewrite Eq. 9-41 as

$$Mv = -dM\,U + (M + dM)(v + dv), \qquad (9\text{-}42)$$

where the first term on the right is the linear momentum of the exhaust products released during interval dt and the second term is the linear momentum of the rocket at the end of interval dt.

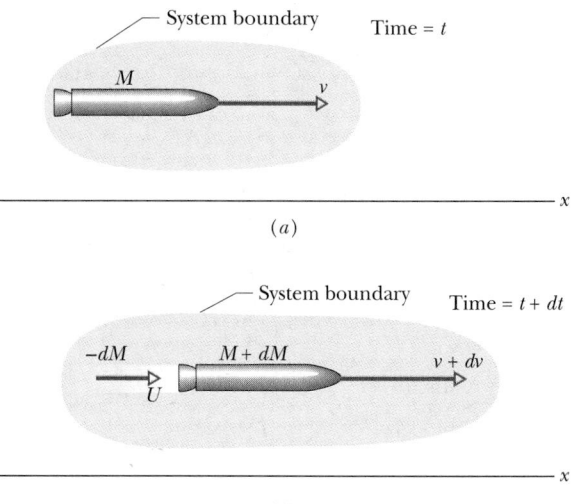

FIGURE 9-16 (a) An accelerating rocket of mass M at time t, as seen from an inertial reference frame. (b) The same but at time $t + dt$. The exhaust products released during interval dt are shown.

We can simplify Eq. 9-42 by using the speed u of the exhaust products relative to the rocket. We find this relative speed u by subtracting the velocity U of the exhaust products from the velocity $v + dv$ of the rocket at the end of interval dt:

$$u = (v + dv) - U.$$

This gives us

$$U = v + dv - u. \qquad (9\text{-}43)$$

Substituting this result for U into Eq. 9-42 yields, with a little algebra,

$$-dM\,u = M\,dv. \qquad (9\text{-}44)$$

Dividing each side by dt yields

$$-\frac{dM}{dt}u = M\frac{dv}{dt}. \qquad (9\text{-}45)$$

We replace dM/dt (the rate at which the rocket loses mass) by $-R$, where R is the (positive) rate of fuel consumption. And we recognize that dv/dt is the acceleration of the rocket. With these changes, Eq. 9-45 becomes

$$Ru = Ma \qquad \text{(first rocket equation).} \qquad (9\text{-}46)$$

Equation 9-46 holds at any instant, with the mass M, the fuel consumption rate R, and the acceleration a evaluated at that instant.

The left side of Eq. 9-46 has the dimensions of a force $(\text{kg} \cdot \text{m/s}^2 = \text{N})$ and depends only on design characteristics of the rocket engine, namely, the rate R at which it consumes fuel mass and the speed u with which that mass is ejected relative to the rocket. We call this term Ru the **thrust** of the rocket engine and represent it with T. Newton's second law emerges clearly if we write Eq. 9-46 as $T = Ma$, in which a is the acceleration of the rocket at the time that its mass is M.

Finding the Velocity

How will the velocity of a rocket change as it consumes its fuel? From Eq. 9-44 we have

$$dv = -u\frac{dM}{M}.$$

Integrating leads to

$$\int_{v_i}^{v_f} dv = -u \int_{M_i}^{M_f} \frac{dM}{M},$$

in which M_i is the initial mass of the rocket and M_f its final mass. Evaluating the integral then gives

$$v_f - v_i = u \ln \frac{M_i}{M_f} \qquad \begin{array}{c}\text{(second rocket}\\\text{equation)}\end{array} \qquad (9\text{-}47)$$

for the increase in the speed of the rocket during the change in mass from M_i to M_f.* We see here the advantage of multistage rockets, in which M_f is reduced by discarding successive stages when their fuel is depleted. An ideal rocket would reach its destination with only its payload remaining.

SAMPLE PROBLEM 9-12

A rocket whose initial mass M_i is 850 kg consumes fuel at the rate $R = 2.3$ kg/s. The speed u of the exhaust gases relative to the rocket engine is 2800 m/s.

(a) What thrust does the rocket engine provide?

SOLUTION: The thrust is

$$T = Ru = (2.3 \text{ kg/s})(2800 \text{ m/s})$$
$$= 6440 \text{ N} \approx 6400 \text{ N}. \qquad \text{(Answer)}$$

(b) What is the initial acceleration of the rocket?

SOLUTION: From Newton's second law, we have

$$a = \frac{T}{M_i} = \frac{6440 \text{ N}}{850 \text{ kg}} = 7.6 \text{ m/s}^2. \qquad \text{(Answer)}$$

To be launched from the Earth's surface, a rocket must have an initial acceleration greater than $g = 9.8$ m/s². Put another way, the thrust T of the rocket engine must exceed the initial weight of the rocket, which here is $M_i g = 850$ kg $\times$ 9.8 m/s² = 8300 N. Because the acceleration or thrust requirement is not met (here $T = 6400$ N), our rocket could not be launched from the Earth's surface. It would have to be launched into space by another, more powerful, rocket.

(c) Suppose, instead, that the rocket is launched from a spacecraft already in deep space, where we can neglect any gravitational force acting on it. The mass M_f of the rocket when its fuel is exhausted is 180 kg. What is its speed relative to the spacecraft at that time? Assume that the spacecraft is so massive that the launch does not alter its speed.

SOLUTION: The rocket begins with a speed of $v_i = 0$ relative to the spacecraft. From Eq. 9-47 we have

$$v_f = u \ln \frac{M_i}{M_f}$$
$$= (2800 \text{ m/s}) \ln \frac{850 \text{ kg}}{180 \text{ kg}}$$
$$= (2800 \text{ m/s}) \ln 4.72 \approx 4300 \text{ m/s}. \qquad \text{(Answer)}$$

Note that the ultimate speed of the rocket can exceed the exhaust speed u.

---

*The symbol "ln" in Eq. 9-47 means the *natural logarithm*, which is the logarithm taken to the base e (= 2.718 . . .); punch "ln," not "log," on your calculator.

9-8 EXTERNAL FORCES AND INTERNAL ENERGY CHANGES

As the ice skater in Fig. 9-17a pushes herself away from a railing, the railing exerts a force $\mathbf{F}_{ext}$ on her at angle ϕ to the horizontal. This force accelerates her from an initial velocity of zero to some final velocity with which she leaves the railing (Fig. 9-17b). Thus the force increases her kinetic energy. This example differs in two ways from earlier examples in which an external force changes an object's kinetic energy:

1. Previously, each part of an object moved rigidly in the same direction; here the skater's arm does not move like the rest of her body.

2. Previously, the external force transferred energy between the object and its environment; here the external force transfers energy from an internal biochemical energy of the skater's muscles to the kinetic energy of her body as a whole.

In spite of these two differences, we can relate the external force to the change in kinetic energy much as we did for a particle in Section 7-3. To do so, we treat the skater as a particle located at her center of mass, with all of her mass M concentrated there. Then we treat the external force $\mathbf{F}_{ext}$ as acting on the center of mass (Fig. 9-17c). The horizontal force component $F_{ext} \cos \phi$ gives the center of mass a horizontal acceleration $a_{cm,x}$, changing the velocity of the center of mass from an initial value $\mathbf{v}_{cm,0}$.

Suppose that with displacement $\mathbf{d}_{cm}$ of the center of mass, the velocity of the center of mass becomes $\mathbf{v}_{cm}$. Then the magnitude of $\mathbf{v}_{cm}$ is, by Eq. 2-16,

$$v_{cm}^2 = v_{cm,0}^2 + 2a_{cm,x}d_{cm}. \qquad (9\text{-}48)$$

Multiplying both sides of Eq. 9-48 by mass M and rearranging yield

$$\tfrac{1}{2}Mv_{cm}^2 - \tfrac{1}{2}Mv_{cm,0}^2 = Ma_{cm,x}d_{cm}. \qquad (9\text{-}49)$$

The left side of Eq. 9-49 is the difference between the initial kinetic energy $K_{cm,i}$ of the center of mass and the final kinetic energy $K_{cm,f}$. This difference is the change ΔK_{cm} in the kinetic energy of the center of mass due to the force $\mathbf{F}_{ext}$. Making that substitution, and substituting the product $F_{ext} \cos \phi$ for the product $Ma_{cm,x}$ according to Newton's second law, we get

$$\Delta K_{cm} = F_{ext}d_{cm} \cos \phi. \qquad (9\text{-}50)$$

Because the skater is an isolated system in the sense that no energy is transferred to or from her, the change

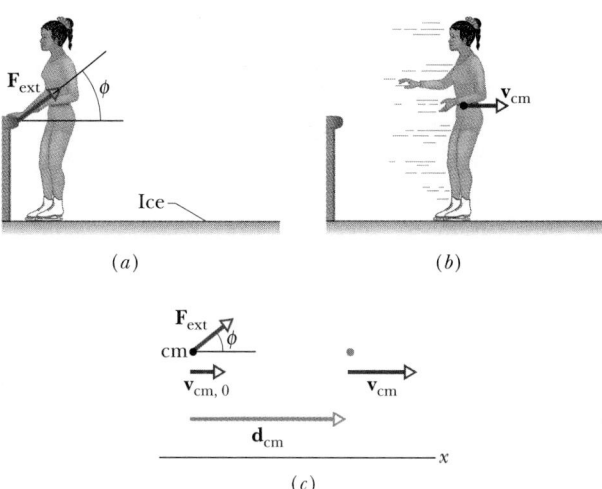

FIGURE 9-17 (a) As a skater pushes herself away from a railing, the railing exerts a force $\mathbf{F}_{ext}$ on her. (b) After the skater leaves the railing, her center of mass has velocity $\mathbf{v}_{cm}$. (c) External force $\mathbf{F}_{ext}$ is taken to act at the skater's center of mass; the vector is at angle ϕ with a horizontal x axis. When the center of mass goes through displacement $\mathbf{d}_{cm}$, its velocity is changed from $\mathbf{v}_{cm,0}$ to $\mathbf{v}_{cm}$ by the horizontal component of $\mathbf{F}_{ext}$.

ΔK_{cm} in her kinetic energy must be accompanied by a change ΔE_{int} in an internal energy. (Let us assume that the only internal energy changed is muscular biochemical energy.) From the principle of the conservation of energy, we write

$$\Delta K_{cm} + \Delta E_{int} = 0,$$

or

$$\Delta E_{int} = -\Delta K_{cm}. \qquad (9\text{-}51)$$

Substituting Eq. 9-50 into Eq. 9-51 yields the change in her internal energy:

$$\Delta E_{int} = -F_{ext}d_{cm} \cos \phi. \qquad (9\text{-}52)$$

Taken together, Eqs. 9-51 and 9-52 say that the external force transfers an amount of energy equal to $F_{ext}d_{cm} \cos \phi$ from internal energy to kinetic energy.

If the height of the skater's center of mass changes while the external force is applied, a change ΔU_{cm} in the gravitational potential energy of the skater–Earth system occurs. Then the isolated system to which the principle of conservation of energy applies is the skater–Earth system, and that principle is written as

$$\Delta K_{cm} + \Delta U_{cm} + \Delta E_{int} = 0. \qquad (9\text{-}53)$$

Substituting Eq. 9-52 into Eq. 9-53, we find

$$\Delta K_{cm} + \Delta U_{cm} - F_{ext}d_{cm} \cos \phi = 0,$$

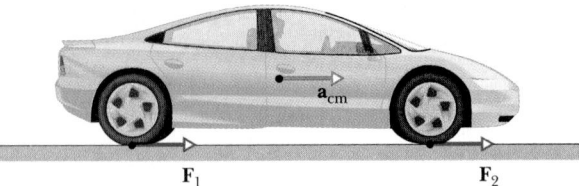

FIGURE 9-18 A car, initially at rest, accelerates to the right. The road exerts four frictional forces (two of them shown) on the bottom surfaces of the tires. Taken together, these four forces makeup the net external force $\mathbf{F}_{ext}$ acting on the car.

or

$$\Delta E = \Delta K_{cm} + \Delta U_{cm} = F_{ext}d_{cm}\cos\phi, \quad (9\text{-}54)$$

where ΔE is the change in the mechanical energy of the skater–Earth system. In words, Eq. 9-54 tells us that the external force transfers an amount of energy equal to the product $F_{ext}d_{cm}\cos\phi$ from internal energy to the mechanical energy of the system, increasing either the kinetic energy of the skater or the height of the skater, or both.

If the external force $\mathbf{F}_{ext}$ is not constant in magnitude, we can replace the symbol F_{ext} in Eqs. 9-52 and 9-54 with the symbol $\overline{F}_{ext}$ for the average magnitude.

Although we have used an ice skater to derive Eqs. 9-52 and 9-54, the equations hold for other objects on which an external force causes a change ΔE_{int} in an internal energy of the object. For example, consider a car that is increasing in speed. During the acceleration, the engine causes the tires to push backward on the road's surface. This push produces frictional forces that act on each tire in the forward direction (Fig. 9-18). The net external force $\mathbf{F}_{ext}$, which is the sum of these frictional forces, gives the car's center of mass an acceleration $\mathbf{a}_{cm}$. It also transfers energy from an internal energy (released by the combustion of fuel) to the kinetic energy of the car. If $\mathbf{F}_{ext}$ is constant, then for a given displacement $\mathbf{d}_{cm}$ of the car's center of mass along level road, we can relate the change ΔK_{cm} in the kinetic energy to the external force $\mathbf{F}_{ext}$ with Eq. 9-54, setting $\Delta U_{cm} = 0$ and $\phi = 0°$.

If the driver applies the brakes, Eq. 9-54 still applies. Now the net external force $\mathbf{F}_{ext}$ due to the frictional forces is toward the rear, and $\phi = 180°$. And now energy is transferred from kinetic energy of the car's center of mass to thermal energy of the brakes.

SAMPLE PROBLEM 9-13

When a click beetle is upside down on its back, it jumps upward by suddenly arching its back, transferring energy stored in a muscle to kinetic energy of its center of mass. This launching mechanism produces an audible click, giving the beetle its name. Videotape of a certain jump shows that the center of mass of a click beetle of mass $m = 4.0 \times 10^{-6}$ kg moved upward by displacement $\mathbf{d}_{cm} = 0.77$ mm during the launch and then to a maximum height of $h = 0.30$ m. What was the average magnitude of the external force $\mathbf{F}_{ext}$ on the beetle's back from the floor during the launch?

SOLUTION: The beetle–Earth system is an isolated system in the sense that no energy is transferred to or from it. Thus we can apply the principle of conservation of energy to the system during the time interval T from the beginning of the launch to the moment the center of mass reaches maximum height h. During interval T, the changes in energy are the following: (1) the change ΔK_{cm} in the kinetic energy of the center of mass is zero (the beetle is stationary both initially and at maximum height h); (2) the change ΔU_{cm} in the gravitational potential energy of the beetle–Earth system is equal to mgh; (3) during the launch, the external force $\mathbf{F}_{ext}$ on the beetle causes a change ΔE_{int} in an internal energy of the beetle.

We write the conservation of energy during time interval T as

$$\Delta K_{cm} + \Delta U_{cm} + \Delta E_{int} = 0.$$

Substituting for ΔK_{cm} and ΔU_{cm}, we have

$$0 + mgh + \Delta E_{int} = 0,$$

or

$$\Delta E_{int} = -mgh. \quad (9\text{-}55)$$

Substituting Eq. 9-52 into Eq. 9-55 yields

$$\overline{F}_{ext}d_{cm}\cos\phi = mgh,$$

or

$$\overline{F}_{ext} = \frac{mgh}{d_{cm}\cos\phi}. \quad (9\text{-}56)$$

The angle ϕ between upward force $\mathbf{F}_{ext}$ and upward displacement $\mathbf{d}_{cm}$ is 0°. Substituting this and given data into Eq. 9-56, we find

$$\overline{F}_{ext} = \frac{(4.0 \times 10^{-6}\text{ kg})(9.8\text{ m/s}^2)(0.30\text{ m})}{(7.7 \times 10^{-4}\text{ m})(\cos 0°)}$$

$$= 1.5 \times 10^{-2}\text{ N.} \quad (\text{Answer})$$

This force magnitude may seem small, but to the click beetle it is enormous because, as you can show, it gives the beetle an acceleration of over $380g$ during the launch.

REVIEW & SUMMARY

Center of Mass

The **center of mass** of a system of particles is defined to be the point whose coordinates are given by

$$x_{cm} = \frac{1}{M} \sum_{i=1}^{n} m_i x_i, \quad y_{cm} = \frac{1}{M} \sum_{i=1}^{n} m_i y_i, \quad z_{cm} = \frac{1}{M} \sum_{i=1}^{n} m_i z_i,$$

(9-5)

or

$$\mathbf{r}_{cm} = \frac{1}{M} \sum_{i=1}^{n} m_i \mathbf{r}_i,$$

(9-8)

where M is the total mass of the system. If the mass is continuously distributed, the center of mass is given by

$$x_{cm} = \frac{1}{M} \int x \, dm, \quad y_{cm} = \frac{1}{M} \int y \, dm, \quad z_{cm} = \frac{1}{M} \int z \, dm.$$

(9-9)

If the density (mass per volume) is uniform, then Eq. 9-9 can be rewritten as

$$x_{cm} = \frac{1}{V} \int x \, dV, \quad y_{cm} = \frac{1}{V} \int y \, dV, \quad z_{cm} = \frac{1}{V} \int z \, dV,$$

(9-11)

where V is the volume occupied by M.

Newton's Second Law for a System of Particles

The motion of the center of mass of any system of particles is governed by **Newton's second law for a system of particles**, which is written as

$$\sum \mathbf{F}_{ext} = M\mathbf{a}_{cm}.$$

(9-16)

Here $\sum \mathbf{F}_{ext}$ is the resultant of the *external* forces acting on the system, M is the total mass of the system, and $\mathbf{a}_{cm}$ is the acceleration of the system's center of mass.

Linear Momentum and Newton's Second Law

For a single particle, we define a vector quantity $\mathbf{p}$ called the **linear momentum** as

$$\mathbf{p} = m\mathbf{v},$$

(9-22)

and write Newton's second law in terms of momentum:

$$\sum \mathbf{F} = \frac{d\mathbf{p}}{dt}.$$

(9-23)

For a system of particles these relations become

$$\mathbf{P} = M\mathbf{v}_{cm} \quad \text{and} \quad \sum \mathbf{F}_{ext} = \frac{d\mathbf{P}}{dt}. \quad \text{(9-26, 9-28)}$$

Relativistic Momentum

A more complete (relativistic) definition of linear momentum is

$$\mathbf{p} = \frac{m\mathbf{v}}{\sqrt{1 - (v/c)^2}}.$$

(9-24)

Equation 9-24 must be used whenever a particle's speed is near the speed of light c. It is valid under all circumstances and is equivalent to Eq. 9-22 whenever $v \ll c$.

Conservation of Linear Momentum

If a system is isolated so that no net *external* force acts on the system, the linear momentum $\mathbf{P}$ of the system remains constant:

$$\mathbf{P} = \text{constant} \quad \text{(closed, isolated system),} \quad \text{(9-30)}$$

which can also be written as

$$\mathbf{P}_i = \mathbf{P}_f \quad \text{(closed, isolated system),} \quad \text{(9-31)}$$

where the subscripts refer to the values of $\mathbf{P}$ at some initial time and at a later time. Equations 9-30 and 9-31 are equivalent statements of the **law of conservation of linear momentum.**

Variable-Mass Systems

For a system of varying mass, we redefine the system, enlarging its boundaries until it encompasses a larger system whose mass *does* remain constant; then we apply the law of conservation of linear momentum. For a rocket, this means that the system includes both the rocket and its exhaust gases. Such analysis shows that in the absence of external forces a rocket accelerates at an instantaneous rate given by

$$Ru = Ma \quad \text{(first rocket equation),} \quad \text{(9-46)}$$

in which M is the rocket's instantaneous mass (including unexpended fuel), R is the fuel consumption rate, and u is the fuel's exhaust speed relative to the rocket. The term Ru is the **thrust** of the rocket engine. For a rocket with constant R and u, whose speed changes from v_i to v_f when its mass changes from M_i to M_f,

$$v_f - v_i = u \ln \frac{M_i}{M_f} \quad \begin{array}{l}\text{(second rocket} \\ \text{equation).}\end{array} \quad \text{(9-47)}$$

External Forces and Internal Energy Changes

An external force $\mathbf{F}_{ext}$ acting on an object can cause a change ΔE_{int} in an internal energy of the object:

$$\Delta E_{int} = -F_{ext} d_{cm} \cos \phi.$$

(9-52)

Here $\mathbf{d}_{cm}$ is the displacement of the object's center of mass and ϕ is the angle between $\mathbf{d}_{cm}$ and $\mathbf{F}_{ext}$. If the external force transfers energy from an internal energy to the kinetic energy of the object's center of mass, the change ΔK_{cm} in that kinetic energy is

$$\Delta K_{cm} = F_{ext} d_{cm} \cos \phi.$$

(9-50)

If the transfer is to the mechanical energy of a system containing the object and having potential energy U, then the change ΔE in the mechanical energy is

$$\Delta E = \Delta K_{cm} + \Delta U_{cm} = F_{ext} d_{cm} \cos \phi.$$

(9-54)

QUESTIONS

1. An amateur sculptor cuts the shape of a bird from a single sheet of metal of uniform thickness (Fig. 9-19). Of the numbered points, which is most likely to be the center of mass of the "sculpture"?

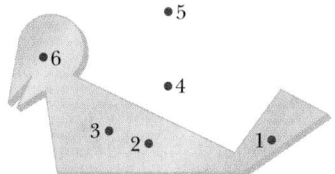

FIGURE 9-19   Question 1.

2. Figure 9-20 shows four situations in which a uniform square metal plate has had a section removed. The origin of the x and y axes is at the center of the original plate, and in each situation the center of mass of the removed section was at the origin. In each situation, where is the center of mass of the remaining section of the plate? Give quadrant, line, or point.

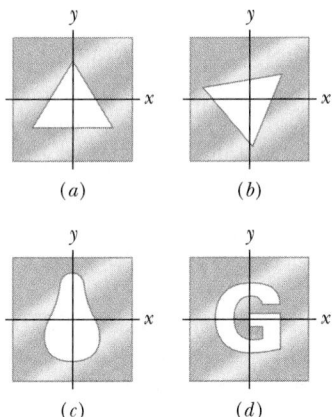

FIGURE 9-20   Question 2.

3. In Fig. 9-21, a penguin stands at the left edge of a uniform sled of length L, which lies on frictionless ice. The sled and penguin have equal masses. (a) Where is the center of mass of the sled? (b) How far and in what direction is the center of the sled from the center of mass of sled–penguin system? The penguin then waddles to the right edge of the sled and the sled slides on the ice. (c) Does the center of mass of the sled–penguin system move leftward, rightward, or not at all? (d) Now how far and in what direction is the center of the sled from the center of mass of the sled–penguin system? (e) How far does the penguin move relative to the sled? Relative to the center of mass of the sled–penguin system, how far does (f) the center of the sled move and (g) the penguin move? (Warmup for Problem 23)

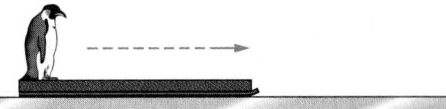

FIGURE 9-21   Questions 3 and 4.

4. In Question 3 and Fig. 9-21, suppose that the sled and penguin are initially moving rightward at speed v_0. (a) As the penguin waddles to the right edge of the sled, is the speed v of the sled less than, greater than, or equal to v_0? (b) If the penguin then waddles back to the left edge, during that motion, is the speed v of the sled less than, greater than, or equal to v_0?

5. Figure 9-22 shows an overhead view of four particles of equal mass sliding over a frictionless surface at constant velocity. The directions of the velocities are indicated; their magnitudes are equal. Consider pairing the particles. Which pairs form a system with a center of mass that (a) is stationary, (b) is stationary and at the origin, and (c) passes through the origin?

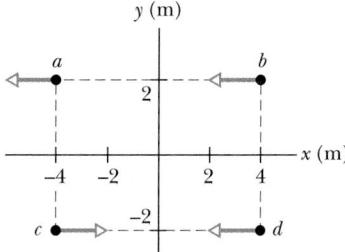

FIGURE 9-22   Question 5.

6. If you drop two watermelons side by side from a bridge, what is the acceleration of the center of mass of the two melon system? (b) If you delay dropping one of the melons, what is the acceleration of the two-melon system while both are falling?

7. Figure 9-23 shows an overhead view of three particles on which external forces act. The magnitudes and directions of the forces on two of the particles are indicated. What are the magnitude and direction of the force acting on the third particle if the center of mass of the three-particle system is (a) stationary, (b) moving at a constant velocity rightward, (c) accelerating rightward?

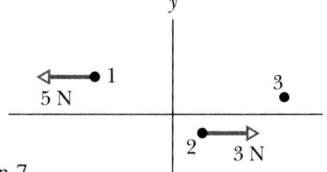

FIGURE 9-23   Question 7.

8. A container sliding along an x axis on a frictionless surface explodes into three pieces. The pieces then move along the x axis in the directions indicated in Figure 9-24. The following table gives four sets of magnitudes (in kg · m/s) for the momenta $\mathbf{p}_1$, $\mathbf{p}_2$, and $\mathbf{p}_3$ of the pieces. Rank the sets according to the initial speed of the container, greatest first.

	p_1	p_2	p_3			p_1	p_2	p_3
(a)	10	2	6		(b)	10	6	2
(c)	2	10	6		(d)	6	2	10

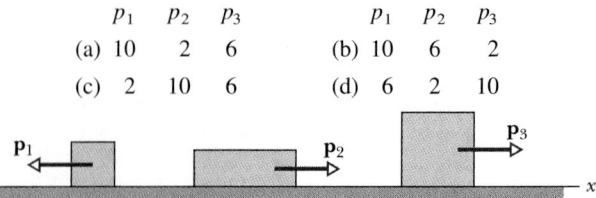

FIGURE 9-24   Question 8.

9. Consider a box, like that in Sample Problem 9-7, which explodes into two pieces while moving with a constant positive velocity along an x axis. If one piece, with mass m_1, ends up with positive velocity $\mathbf{v}_1$, then the second piece, with mass m_2, could end up with (a) a positive velocity $\mathbf{v}_2$ (Fig. 9-25a) or (b) a negative velocity $\mathbf{v}_2$ (Fig. 9-25b), or (c) zero velocity (Fig. 9-25c). Rank those three possible results for the second piece according to the corresponding magnitude of $\mathbf{v}_1$, greatest first.

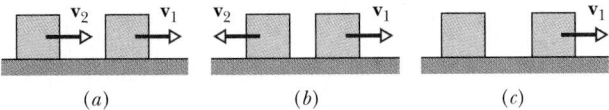

(a)   (b)   (c)

FIGURE 9-25   Question 9.

10. Figure 9-26 shows overhead views of three two-dimensional explosions in which a stationary grapefruit is blown by a firecracker into three pieces, seven pieces, and nine pieces. The pieces then slide over a frictionless floor. For each situation, Fig. 9-26 shows the momentum vectors of all but one piece; that piece has momentum $\mathbf{P}'$. The numbers next to the vectors are the magnitudes of the momenta (in kilogram-meter per second). Rank the three situations according to the magnitudes of (a) P'_x, (b) P'_y, and (c) $\mathbf{P}'$, greatest first.

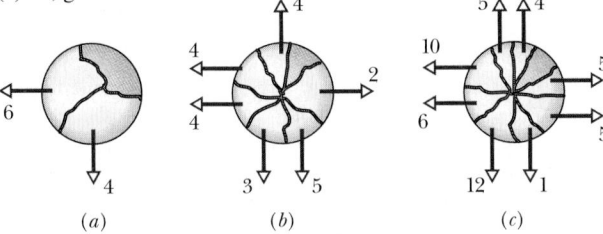

(a)   (b)   (c)

FIGURE 9-26   Question 10.

11. Figure 9-27 shows an overhead view showing six particles that have emerged from a region in which a two-dimensional explosion took place. The explosion was of an object that had been stationary on a frictionless floor. The directions of the momenta of the particles are indicated by the vectors; the numbers give the magnitudes of the momenta (in kg·m/s). (a) Will more particles be emerging from the explosion region? (b) If so, give (b) their net momentum and (c) their direction of travel.

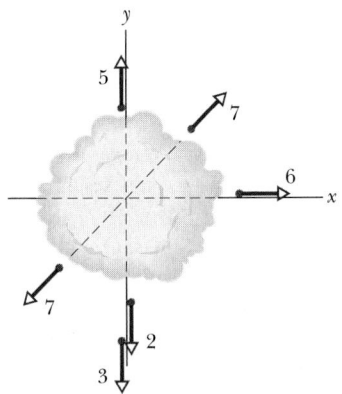

FIGURE 9-27   Question 11.

12. For three pairs of particles, here are the masses of the particles and their initial separation d:

	m_1	m_2	INITIAL SEPARATION d
Pair 1	$2m$	$8m$	1.0 m
Pair 2	$3m$	$6m$	2.0 m
Pair 3	$4m$	$9m$	0.5 m

The particles in each pair attract each other. They are released from rest at the given initial separation d. When the separation has decreased to $d/2$, the speed of m_1 is v_1 and the speed of m_2 is v_2. Without written calculation, rank the pairs according to the ratio v_1/v_2, greatest first. (*Hint:* See Sample Problem 9-10.)

EXERCISES & PROBLEMS

SECTION 9-2 The Center of Mass

1E. (a) How far is the center of mass of the Earth–Moon system from the center of the Earth? (Appendix C gives the masses of the Earth and the Moon and the distance between the two.) (b) Express the answer to (a) as a fraction of the Earth's radius.

2E. The distance between the centers of the carbon (C) and oxygen (O) atoms in a carbon monoxide (CO) gas molecule is 1.131×10^{-10} m. Locate the center of mass of a CO molecule relative to the carbon atom. (Find the masses of C and O in Appendix D.)

3E. (a) What are the coordinates of the center of mass of the three-particle system shown in Fig. 9-28? (b) What happens to the center of mass as the mass of the topmost particle is gradually increased?

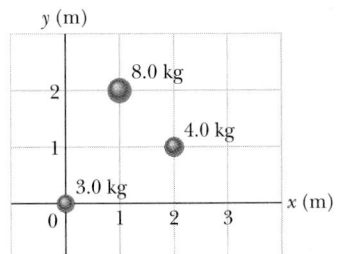

FIGURE 9-28   Exercise 3.

4E. Three thin rods each of length L are arranged in an inverted U, as shown in Fig. 9-29. The two rods on the arms of the U each have mass M; the third rod has mass $3M$. Where is the center of mass of the assembly?

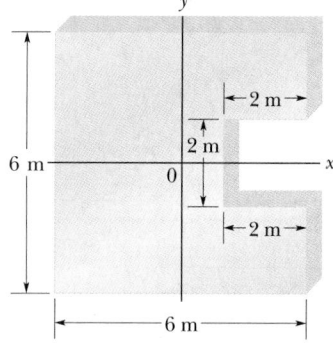

FIGURE 9-29 Exercise 4.

5E. A uniform square plate 6 m on a side has had a square piece 2 m on a side cut out of it (Fig. 9-30). The center of that piece is at $x = 2$ m, $y = 0$. The center of the square plate is at $x = y = 0$. Find the coordinates of the center of mass of the remaining piece.

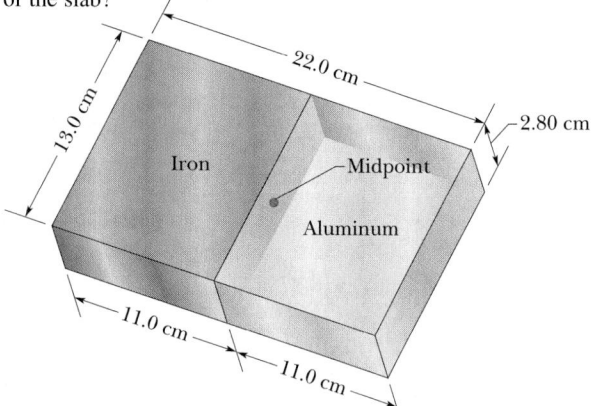

FIGURE 9-30 Exercise 5.

6P. Show that the ratio of the distances of two particles from the center of mass of the two-particle system is the inverse ratio of their masses.

7P. Figure 9-31 shows the dimensions of a composite slab; half the slab is made of aluminum (density = 2.70 g/cm³) and half is made of iron (density = 7.85 g/cm³). Where is the center of mass of the slab?

FIGURE 9-31 Problem 7.

8P. In the ammonia (NH_3) molecule (see Fig. 9-32), the three hydrogen (H) atoms form an equilateral triangle; the center of the triangle is 9.40×10^{-11} m from each hydrogen atom. The nitrogen (N) atom is at the apex of a pyramid, with the three hydrogen atoms forming the base. The nitrogen-to-hydrogen distance is 10.14×10^{-11} m, and the nitrogen-to-hydrogen atomic mass ratio is 13.9. Locate the center of mass of the molecule relative to the nitrogen atom.

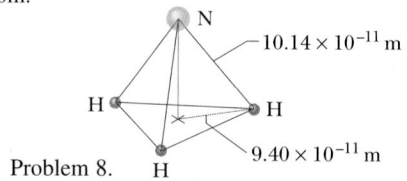

FIGURE 9-32 Problem 8.

9P. A cubical box, open at the top, with edge length 40 cm, is constructed from metal plate of uniform density and negligible thickness (Fig. 9-33). Find the coordinates of the center of mass of the box.

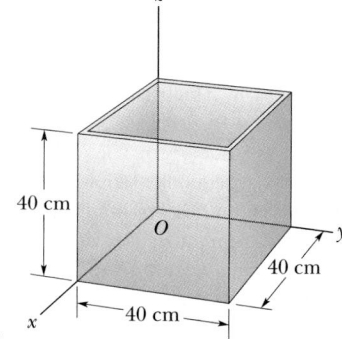

FIGURE 9-33 Problem 9.

10P. The Great Pyramid of Cheops at El Gizeh, Egypt (Fig. 9-34a), had height $H = 147$ m before its topmost stone fell. Its base is a square with edge length $L = 230$ m (see Fig. 9-34b). Its volume V is equal to $L^2 H / 3$. Assuming that it has uniform density $\rho = 1.8 \times 10^3$ kg/m³, find (a) the original height of its center of mass above the base and (b) the work required to lift the blocks into place from the base level.

(a)

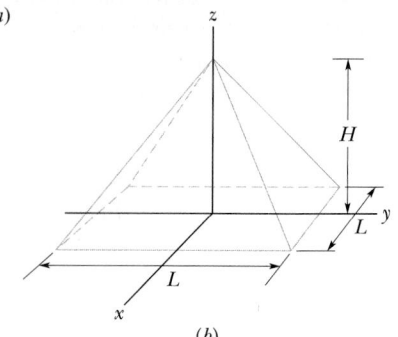

FIGURE 9-34
Problem 10.

(b)

11P*. A right cylindrical can with mass M, height H, and uniform density is initially filled with pop of mass m (Fig. 9-35). We punch small holes in the top and bottom to drain the pop; we then consider the height h of the center of mass of the can and any pop within it. What is h (a) initially, and (b) when all the pop has drained? (c) What happens to h during the draining of the pop? (d) If x is the height of the remaining pop at any given instant, find x (in terms of M, H, and m) when the center of mass reaches its lowest point.

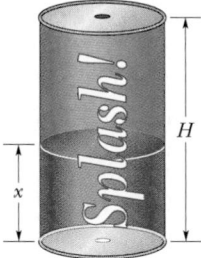

FIGURE 9-35 Problem 11.

SECTION 9-3 Newton's Second Law for a System of Particles

12E. Two skaters, one with mass 65 kg and the other with mass 40 kg, stand on an ice rink holding a pole with a length of 10 m and a negligible mass. Starting from the ends of the pole, the skaters pull themselves along the pole until they meet. How far will the 40 kg skater move?

13E. An old Chrysler with mass 2400 kg is moving along a straight stretch of road at 80 km/h. It is followed by a Ford with mass 1600 kg moving at 60 km/h. How fast is the center of mass of the two cars moving?

14E. A man of mass m clings to a rope ladder suspended below a balloon of mass M; see Fig. 9-36. The balloon is stationary with respect to the ground. (a) If the man begins to climb the ladder at speed v (with respect to the ladder), in what direction and with what speed (with respect to the Earth) will the balloon move? (b) What is the state of the motion after the man stops climbing?

FIGURE 9-36 Exercise 14.

15E. Two particles P and Q are initially at rest 1.0 m apart. P has a mass of 0.10 kg and Q a mass of 0.30 kg. P and Q attract each other with a constant force of 1.0×10^{-2} N. No external forces act on the system. (a) Describe the motion of the center of mass. (b) At what distance from P's original position do the particles collide?

16E. A cannon and a supply of cannonballs are inside a sealed railroad car of length L, as in Fig. 9-37. The cannon fires to the right; the car recoils to the left. Fired cannonballs remain in the car after hitting the far wall and landing on the floor there. (a) After all the cannonballs have been fired, what is the greatest distance the car can have moved from its original position? (b) Under what circumstance would the car move that distance? (c) What is the speed of the car just after all the cannonballs have been fired?

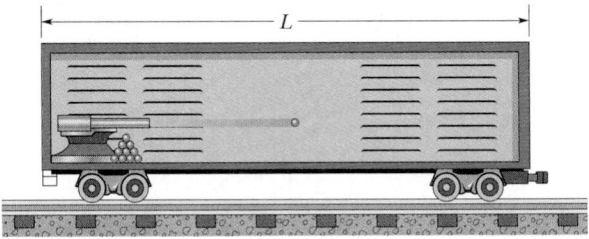

FIGURE 9-37 Exercise 16.

17P. At the instant a 3.0 kg particle has a velocity of 6.0 m/s in the direction of decreasing y, a 4.0 kg particle has a velocity of 7.0 m/s in the direction of increasing x. What is the speed of the center of mass of the two-particle system?

18P. A stone is dropped at $t = 0$. A second stone, with twice the mass of the first, is dropped from the same point at $t = 100$ ms. (a) Where is the center of mass of the two stones at $t = 300$ ms? Neither stone has yet reached the ground. (b) How fast is the center of mass of the two-stone system moving at that time?

19P. A 1000 kg automobile is at rest at a traffic signal. At the instant the light turns green, the automobile starts to move with a constant acceleration of 4.0 m/s². At the same instant a 2000 kg truck, traveling at a constant speed of 8.0 m/s, overtakes and passes the automobile. (a) How far is the center of mass of the automobile–truck system from the traffic light at $t = 3.0$ s? (b) What is the speed of the center of mass of the automobile–truck system then?

20P. Ricardo, mass 80 kg, and Carmelita, who is lighter, are enjoying Lake Merced at dusk in a 30 kg canoe. When the canoe is at rest in the placid water, they exchange seats, which are 3.0 m apart and symmetrically located with respect to the canoe's center. Ricardo notices that the canoe moved 40 cm relative to a submerged log during the exchange and calculates Carmelita's mass, which she has not told him. What is it?

21P. A shell is fired from a gun with a muzzle velocity of 20 m/s, at an angle of 60° with the horizontal. At the top of the trajectory, the shell explodes into two fragments of equal mass (Fig. 9-38). One fragment, whose speed immediately after the explosion is zero, falls vertically. How far from the gun does the other fragment land, assuming that the terrain is level and that the air drag is negligible?

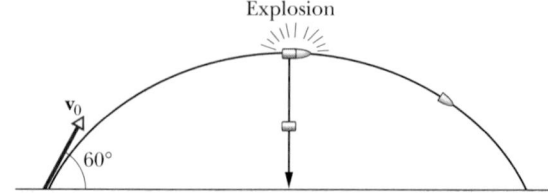

FIGURE 9-38   Problem 21.

22P. Two identical containers of sugar are connected by a mass-less cord that passes over a massless, frictionless pulley with a diameter of 50 mm (Fig. 9-39). The two containers are at the same level. Each originally has a mass of 500 g. (a) Locate the horizontal position of their center of mass. (b) Now 20 g of sugar is transferred from one container to the other, but the containers are prevented from moving. Locate the new horizontal position of their center of mass. (c) The two containers are now released. In what direction does the center of mass move? (d) What is its acceleration?

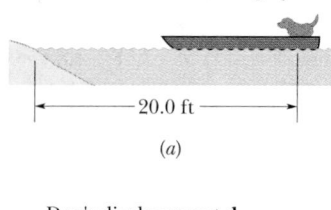

FIGURE 9-39   Problem 22.

23P. A dog, weighing 10.0 lb, is standing on a flatboat so that he is 20.0 ft from the shore (to the left in Fig. 9-40a). He walks 8.00 ft on the boat toward shore and then halts. The boat weighs 40.0 lb, and one can assume there is no friction between it and the water. How far is the dog then from the shore? (*Hint:* See Fig. 9-40b. The dog moves leftward; the boat moves rightward; but does the center of mass of the *boat + dog* system move?)

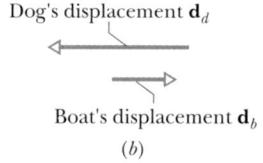

FIGURE 9-40   Problem 23.

SECTION 9-5   The Linear Momentum of a System of Particles

24E. Suppose that your mass is 80 kg. How fast would you have to run to have the same linear momentum as a 1600 kg car moving at 1.2 km/h?

25E. How fast must an 816 kg Geo travel (a) to have the same linear momentum as a 2650 kg Cadillac going 16 km/h and (b) to have the same kinetic energy?

26E. What is the linear momentum of an electron of speed 0.99c ($= 2.97 \times 10^8$ m/s)?

27E. The linear momentum of a particle moving at 1.5×10^8 m/s is measured to be 2.9×10^{-19} kg·m/s. By finding its mass, determine whether the particle is an electron or a proton.

28E. A 0.70 kg mass is moving horizontally with a speed of 5.0 m/s when it strikes a vertical wall. The mass rebounds with a speed of 2.0 m/s. What is the magnitude of the change in linear momentum of the mass?

29P. A 0.165 kg cue ball with an initial speed of 2.00 m/s bounces off the rail in a game of pool, as shown from overhead in Fig. 9-41. For x and y axes located as shown, the bounce reverses the y component of the ball's velocity but does not alter the x component. (a) What is θ in Fig. 9-41? (b) What is the change in the ball's linear momentum in unit-vector notation? (The fact that the ball rolls is not relevant to either question.)

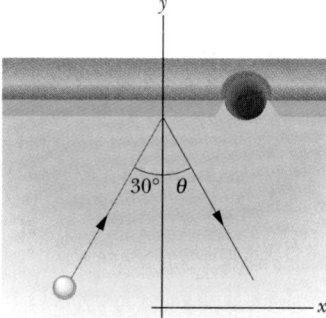

FIGURE 9-41   Problem 29.

30P. A 2100 kg truck traveling north at 41 km/h turns east and accelerates to 51 km/h. (a) What is the change in the kinetic energy of the truck? (b) What are the magnitude and direction of the change in the linear momentum of the truck?

31P. An object is tracked by a radar station and found to have a position vector given by $\mathbf{r} = (3500 - 160t)\mathbf{i} + 2700\mathbf{j} + 300\mathbf{k}$, with $\mathbf{r}$ in meters and t in seconds. The radar station's x axis points east, its y axis north, and its z axis vertically up. If the object is a 250 kg meteorological missile, what are (a) its linear momentum, (b) its direction of motion, and (c) the net force on it?

32P. A 50 g ball is thrown from ground level into the air with an initial speed of 16 m/s at an angle of 30° above the horizontal. (a) What are the values of the kinetic energy of the ball initially and just before it hits the ground? (b) Find the corresponding values of the linear momentum (magnitude and direction). (c) Show that the change in linear momentum is just equal to the weight of the ball multiplied by the time of flight.

33P. A particle with mass m has linear momentum $\mathbf{p}$ of magnitude mc. What is its speed in terms of c, the speed of light?

SECTION 9-6   Conservation of Linear Momentum

34E. A 200 lb man standing on a surface of negligible friction kicks forward a 0.15 lb stone lying at his feet, giving it a speed of 13 ft/s. What velocity does the man acquire as a result?

35E. Two blocks of masses 1.0 kg and 3.0 kg are connected by a spring and rest on a frictionless surface. They are given velocities toward each other such that the 1.0 kg block travels initially at 1.7 m/s toward the center of mass, which remains at rest. What is the initial velocity of the other block?

36E. A space vehicle is traveling at 4300 km/h relative to the Earth when the exhausted rocket motor is disengaged and sent backward with a speed of 82 km/h relative to the command module. The mass of the motor is four times the mass of the module. What is the speed of the command module relative to Earth after the separation?

37E. A 75 kg man is riding on a 39 kg cart traveling at a speed of 2.3 m/s. He jumps off with zero horizontal speed. What is the resulting change in the speed of the cart?

38E. A railroad flatcar of weight W can roll without friction along a straight horizontal track. Initially, a man of weight w is standing on the car, which is moving to the right with speed v_0; see Fig. 9-42. What is the change in velocity of the car if the man runs to the left so that his speed relative to the car is v_{rel}?

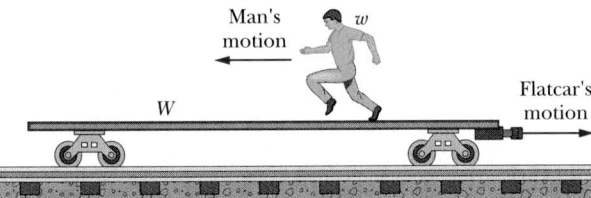

FIGURE 9-42 Exercise 38.

39P. The last stage of a rocket is traveling at a speed of 7600 m/s. This last stage is made up of two parts that are clamped together, namely, a rocket case with a mass of 290.0 kg and a payload capsule with a mass of 150.0 kg. When the clamp is released, a compressed spring causes the two parts to separate with a relative speed of 910.0 m/s. (a) What are the speeds of the two parts after they have separated? Assume that all velocities are along the same line. (b) Find the total kinetic energy of the two parts before and after they separate; account for any difference.

40P. A radioactive nucleus, initially at rest, decays by emitting an electron and a neutrino perpendicular to each other. (A *neutrino* is one of the fundamental particles of physics.) The linear momentum of the electron is 1.2×10^{-22} kg·m/s and that of the neutrino is 6.4×10^{-23} kg·m/s. (a) Find the direction and magnitude of the linear momentum of the nucleus as it recoils from the decay. (b) The mass of the residual nucleus is 5.8×10^{-26} kg. What is its kinetic energy as it recoils?

41E. An electron (mass $m_1 = 9.11 \times 10^{-31}$ kg) and a proton (mass $m_2 = 1.67 \times 10^{-27}$ kg) attract each other via an electrical force. Suppose that an electron and a proton are released from rest with an initial separation $d = 3.0 \times 10^{-6}$ m. When their separation has decreased to 1.0×10^{-6} m, what is the ratio of (a) the electron's linear momentum to the proton's linear momentum, (b) the electron's speed to the proton's speed, and (c) the electron's kinetic energy to the proton's kinetic energy? (d) As the separation continues to decrease, do the answers to (a) through (c) increase, decrease, or remain the same?

42P. A 4.0 kg mass sliding on a frictionless surface explodes into two 2.0 kg masses. After the explosion the velocities of the 2.0 kg masses are 3.0 m/s, due north, and 5.0 m/s, 30° north of east. What was the original speed of the 4.0 kg mass?

43P. A vessel at rest explodes, breaking into three pieces. Two pieces, having equal mass, fly off perpendicular to one another with the same speed of 30 m/s. The third piece has three times the mass of each other piece. What are the direction and magnitude of its velocity immediately after the explosion?

44P. A 2140 kg railroad flatcar, which can move with negligible friction, is motionless next to a platform. A 242 kg sumo wrestler runs at 5.3 m/s along the platform (parallel to the track) and then jumps onto the flatcar. What is the speed of the flatcar if he then (a) stands on it, (b) runs at 5.3 m/s relative to it in his original direction, and (c) turns and runs at 5.3 m/s relative to the flatcar opposite his original direction?

45P. A rocket sled with a mass of 2900 kg moves at 250 m/s on a set of rails. At a certain point, a scoop on the sled dips into a trough of water located between the tracks and scoops water into an empty tank on the sled. By applying the principle of conservation of linear momentum, determine the speed of the sled after 920 kg of water has been scooped up. Ignore any retarding force on the scoop.

46P. A body of mass 8 kg is traveling at 2 m/s with no external force acting. At a certain instant an internal explosion occurs, splitting the body into two chunks of 4 kg mass each. The explosion gives the chunks an additional 16 J of kinetic energy. Neither chunk leaves the line of original motion. Determine the speed and direction of motion of each of the chunks after the explosion.

47P. You are on an iceboat on frictionless, flat ice; you and the boat have a combined mass M. Along with you are two stones with masses m_1 and m_2 such that $M = 6.00m_1 = 12.0m_2$. To get the boat moving, you throw the stones rearward, either in succession or together, but in each case with a certain speed v_{rel} relative to the boat. What is the resulting speed of the boat if you throw the stones (a) simultaneously, (b) m_1 and then m_2, and (c) m_2 and then m_1?

48P. A 1400 kg cannon, which fires a 70.0 kg shell with a speed of 556 m/s relative to the muzzle, is set at an elevation angle of 39.0° above the horizontal. The cannon is mounted on frictionless rails, so that it recoils freely. (a) What is the speed of the shell with respect to the Earth? (b) At what angle with the ground is the shell projected? (*Hint:* The horizontal component of the linear momentum of the system remains unchanged as the gun is fired.)

SECTION 9-7 Systems with Varying Mass: A Rocket

49E. A rocket at rest in space, where there is virtually no gravitational force, has a mass of 2.55×10^5 kg, of which 1.81×10^5 kg is fuel. The engine consumes fuel at the rate of 480 kg/s, and the exhaust speed is 3.27 km/s. The engine is fired for 250 s. (a) Find the thrust of the rocket engine. (b) What is the mass of the rocket after the engine burn? (c) What is the final speed attained?

50E. A rocket is moving away from the solar system at a speed of 6.0×10^3 m/s. It fires its engine, which ejects exhaust with a velocity of 3.0×10^3 m/s relative to the rocket. The mass of the rocket at this time is 4.0×10^4 kg, and its acceleration is 2.0 m/s². (a) What is the thrust of the engine? (b) At what rate, in kilograms per second, was exhaust ejected during the firing?

51E. A 6090 kg space probe, moving nose-first toward Jupiter at 105 m/s relative to the Sun, fires its rocket engine, ejecting 80.0 kg of exhaust at a speed of 253 m/s relative to the space probe. What is the final velocity of the probe?

52E. Consider a rocket at rest in deep space. For the rocket's speed to be (a) equal to the exhaust speed and (b) equal to twice the exhaust speed after firing of the rocket engine, what must be the rocket's *mass ratio* (ratio of initial to final mass)?

53E. During a lunar mission, it is necessary to increase the speed of a spacecraft by 2.2 m/s when it is moving at 400 m/s. The exhaust speed of rocket engine is 1000 m/s. What fraction of the initial mass of the spacecraft must be burned and ejected for the increase?

54E. A railroad car moves at a constant speed of 3.20 m/s under a grain elevator. Grain drops into it at the rate of 540 kg/min. What force must be applied to the railroad car, in the absence of friction, to keep it moving at constant speed?

55P. A single-stage rocket, at rest in a certain inertial reference frame, has mass M when the rocket engine is ignited. Show that when the mass has decreased to $0.368M$, the gases streaming out of the rocket engine at that time will be at rest in the original reference frame.

56P. A 6100 kg rocket is set for vertical firing from the Earth's surface. If the exhaust speed is 1200 m/s, how much gas must be ejected each second if the thrust (a) is to equal the weight of the rocket and (b) is to give the rocket an initial upward acceleration of 21 m/s²?

57P. Two long barges are moving in the same direction in still water, one with a speed of 10 km/h and the other with a speed of 20 km/h. While they are passing each other, coal is shoveled from the slower to the faster one at a rate of 1000 kg/min; see Fig. 9-43. How much additional force must be provided by the driving engines of each barge if neither is to change speed? Assume that the shoveling is always perfectly sideways and that the frictional forces between the barges and the water do not depend on the weight of the barges.

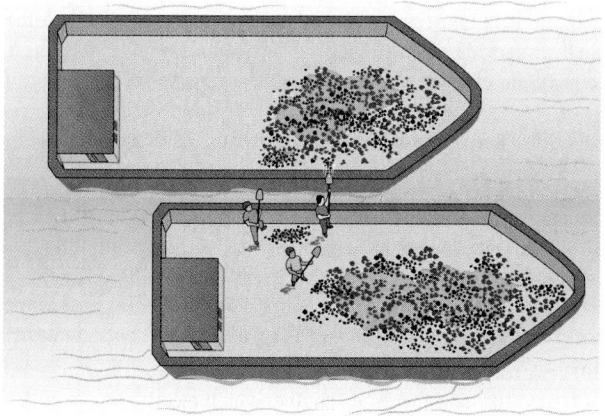

FIGURE 9-43 Problem 57.

58P. A jet airplane is traveling 180 m/s. Each second, the engine takes in 68 m³ of air, which has a mass of 70 kg. The air is used to burn 2.9 kg of fuel each second. The energy is used to compress the products of combustion and to eject them at the rear of the plane at 490 m/s relative to the plane. Find (a) the thrust of the jet engine and (b) its power in watts.

SECTION 9-8 External Forces and Internal Energy Changes

59E. The summit of Mount Everest is 8850 m above sea level. (a) How much energy would a 90 kg climber expend against the gravitational force (his weight) in climbing to the summit from sea level? (b) How many candy bars, at 300 kcal per bar, would supply an energy equivalent to this? Your answer should suggest that work done against the gravitational force is a very small part of the energy expended in climbing a mountain.

60E. A 51 kg boy climbs, with constant speed, a vertical rope 6.0 m long in 10 s. (a) What is the increase in the boy–Earth gravitational potential energy? (b) What is the boy's average power during the climb?

61E. A 55 kg woman runs up a flight of stairs with a height of 4.5 m in 3.5 s. What average power is required of her?

62E. A sprinter who weighs 670 N runs the first 7.0 m of a race in 1.6 s, starting from rest and accelerating uniformly. What are the sprinter's (a) speed and (b) kinetic energy at the end of that 1.6 s? (c) What average power does the sprinter generate during the 1.6 s interval?

63E. The luxury liner *Queen Elizabeth 2* has a diesel-electric powerplant with a maximum power of 92 MW at a cruising speed of 32.5 knots. What forward force is exerted on the ship at this highest attainable speed? (1 knot = 6076 ft/h.)

64E. What power, in horsepower, is required of the engine of a 1600 kg car moving at 25.1 m/s on a level road if the forces of resistance total 703 N?

65E. A swimmer moves through the water at an average speed of 0.22 m/s. The average drag force opposing this motion is 110 N. What average power is required of the swimmer?

66E. The energy required for a person to run is about 335 J/m, regardless of speed. What is a runner's average power during (a) a 100 m dash (time = 10 s) and (b) a marathon (distance = 26.2 mi; time = 2 h 10 min)?

67E. An automobile with passengers has weight 16,400 N and is moving at 113 km/h when the driver brakes to a stop. The road exerts a frictional force of 8230 N on the wheels. Find the stopping distance.

68E. You crouch from a standing position, lowering your center of mass 18 cm in the process. Then you jump vertically. The average force exerted on you by the floor while you jump is three times your weight. What is your upward speed as you pass through your standing position in leaving the floor?

69E. A 55-kg woman leaps vertically from a crouching position in which her center of mass is 40 cm above the ground. As her feet leave the floor her center of mass is 90 cm above the ground;

it rises to 120 cm at the top of her leap. (a) What average force was exerted on her by the ground during the jump? (b) What maximum speed does she attain?

70P. A 110 kg ice hockey player skates at 3.0 m/s toward a railing at the edge of the ice and then stops himself by grasping the railing with his outstretched arms. During this stopping process his center of mass moves 30 cm toward the railing. (a) What is the change in the kinetic energy of his center of mass during this process? (b) What average force must he exert on the railing?

71P. A 1500 kg automobile starts from rest on a horizontal road and gains a speed of 72 km/h in 30 s. (a) What is the kinetic energy of the auto at the end of the 30 s? (b) What is the average power required of the car during the 30 s interval? (c) What is the instantaneous power at the end of the 30 s interval, assuming that the acceleration was constant?

72P. While a 1710 kg automobile is moving at a constant speed of 15.0 m/s, the engine supplies 16.0 kW of power to overcome friction, wind resistance, and so on. (a) What is the effective retarding force associated with all the frictional forces combined? (b) What power must the engine supply if the car is to move up an 8.00% grade (8.00 m vertically for each 100 m horizontally) at 15.0 m/s? (c) On what downgrade, expressed as a percentage, would the car coast at 15.0 m/s?

73P. If a locomotive with a power capability of 1.5 MW can accelerate a train from a speed of 10 m/s to 25 m/s in 6.0 min, (a) calculate the mass of the train. Find (b) the speed of the train and (c) the force accelerating the train as functions of time (in seconds) during the 6.0 min interval. (d) Find the distance moved by the train during the interval.

74P. Resistance to the motion of an automobile consists of road friction, which is almost independent of speed, and air drag, which is proportional to speed squared. For a car with a weight of 12,000 N, the total resistant force F is given by $F = 300 + 1.8v^2$, where F is in newtons and v is in meters per second. Calculate the power required to accelerate the car at 0.92 m/s^2 when the speed is 80 km/h.

75P*. A dragster with mass m races another dragster through a distance d from an initial dead stop. Assume that its engine provides a constant instantaneous power P for the entire race and that the dragster can be treated as a particle. Find the elapsed time for the race.

Electronic Computation

76. The following gives the masses of three objects and, for a certain instant, the (xy) coordinates and the velocities of the objects. At that instant, what are (a) the position and (b) the velocity of the center of mass of the three particle system, and (c) what is the net linear momentum of the system?

OBJECT	MASS (kg)	COORDINATES (m)	VELOCITY (m/s)
1	4.00	(0, 0)	$1.50\mathbf{i} - 2.50\mathbf{j}$
2	3.00	(7.00, 3.00)	0
3	5.00	(3.00, 2.00)	$2.00\mathbf{i} - 1.00\mathbf{j}$

77. A 2.00 kg block is released from rest over the side of a very tall building ($y = 0$) at time $t = 0$. At time $t = 1.00$ s, a 3.00 kg block is released from rest at the same point. The first block hits the ground at $t = 5.00$ s. Plot, for the time interval $t = 0$ to $t = 6.00$ s, (a) the position and (b) the speed of the center of mass of the two block system.

10
Collisions

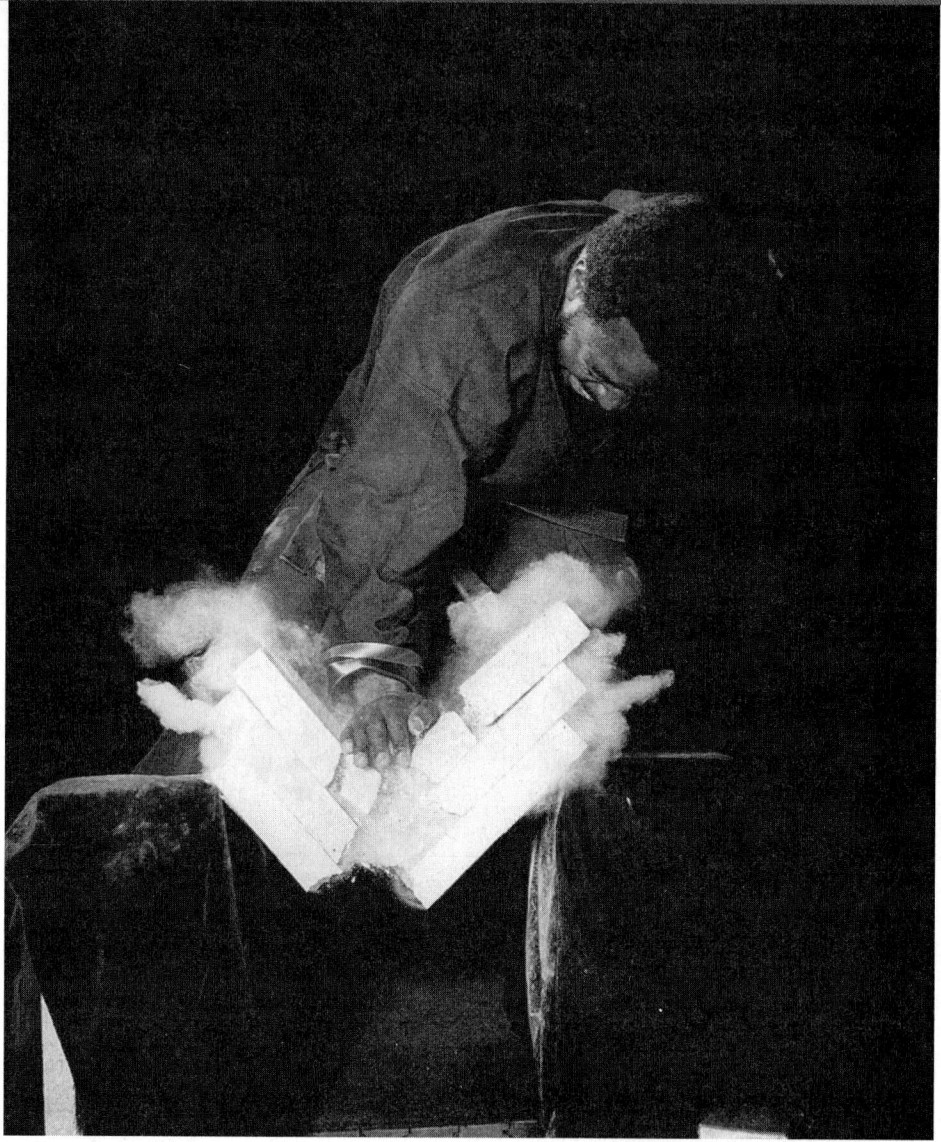

*Ronald McNair, a physicist and one of the astronauts killed in the explosion
of the* Challenger *space shuttle, held a black belt in karate. Here he breaks
several concrete slabs with one blow. In such karate demonstrations, a pine
board or a concrete "patio block" is typically used. When struck, the board
or block bends, storing energy like a stretched spring does, until a critical
energy is reached. Then the object breaks. Surprisingly, the energy
necessary to break the block is about one-third of that for the board,
yet the board is considerably easier to break. Why?*

10-1   WHAT IS A COLLISION?

In everyday language, a *collision* occurs when objects crash into each other. Although we will refine that definition, it conveys the meaning well enough and covers common collisions, such as those between billiard balls, a hammer and a nail, and—too commonly—automobiles. Figure 10-1*a* shows the lasting result of a collision (an impressive crash) that occurred about 20,000 years ago. Collisions range from the microscopic scale of subatomic particles (Fig. 10-1*b*) to the astronomic scale of colliding stars and colliding galaxies. Even when they are on a normal scale, they are often too brief to be visible, although they involve significant distortion of the *colliding bodies* (Fig. 10-1*c*).

We shall use the following more formal definition of collision:

> A **collision** is an isolated event in which two or more bodies (the colliding bodies) exert relatively strong forces on each other for a relatively short time.

We must be able to distinguish times that are *before, during,* and *after* a collision, as suggested in Fig. 10-2. In the illustration, a system boundary surrounds the two colliding bodies. The forces that the bodies exert on each other are internal to the system.

Note that our formal definition of collision does not require the "crash" of our informal definition. When a space probe swings around a large planet to pick up speed (a *slingshot* encounter), that too is a collision. The probe and planet do not actually "touch," but a collision does not require contact, and a collision force does not have to be a force of contact; it can just as easily be a gravitational force, as in this case.

Many physicists today spend their time playing what we can call "the collision game." A principal goal of this game is to find out as much as possible about the forces that act during a collision, knowing the state of the particles both before and after the collision. Virtually all our knowledge of the subatomic world—electrons, protons, neutrons, muons, quarks, and the like—comes from experiments involving collisions. The rules of the game are the laws of conservation of momentum and of energy.

(*a*)

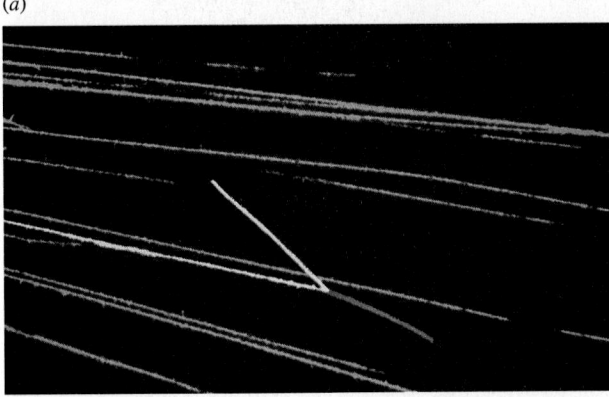

(*b*)

(*c*)

FIGURE 10-1   Collisions range widely in scale. (*a*) Meteor Crater in Arizona is about 1200 m wide and 200 m deep. (*b*) An alpha particle coming in from the left (whose trail is colored yellow in this false-color photograph) bounces off a nitrogen nucleus that had been stationary and that now moves toward the bottom right (red trail). (*c*) In a tennis match, the ball is in contact with the racquet for about 4 ms in each collision (for a cumulative time of only 1 s per set).

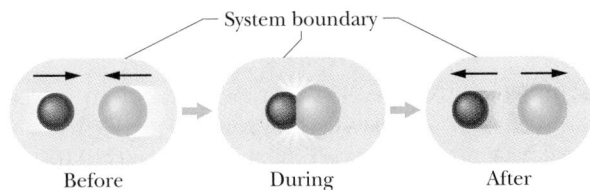

Before   During   After

FIGURE 10-2   A flowchart showing the system in which a collision occurs.

10-2 IMPULSE AND LINEAR MOMENTUM

Single Collision

Figure 10-3 shows the equal but opposite forces, $\mathbf{F}(t)$ and $-\mathbf{F}(t)$, that act during a simple head-on collision between two particle-like bodies of different masses. These forces will change the linear momentum of both bodies; the amount of the change will depend not only on the average values of the forces, but also on the time Δt during which they act. To see this quantitatively, let us apply Newton's second law in the form $\mathbf{F} = d\mathbf{p}/dt$ to body R on the right in Fig. 10-3. We have

$$d\mathbf{p} = \mathbf{F}(t)\,dt, \tag{10-1}$$

in which $\mathbf{F}(t)$ is the time-varying force, displayed as the curve in Fig. 10-4a. Let us integrate Eq. 10-1 over the collision interval Δt, that is, from an initial time t_i (just before the collision) to a final time t_f (just after the collision). We obtain

$$\int_{\mathbf{p}_i}^{\mathbf{p}_f} d\mathbf{p} = \int_{t_i}^{t_f} \mathbf{F}(t)\,dt. \tag{10-2}$$

The left side of this equation is $\mathbf{p}_f - \mathbf{p}_i$, the change in linear momentum of the body R. The right side, which is a measure of both the strength and the duration of the collision force, is called the **impulse J** of the collision. Thus

$$\mathbf{J} = \int_{t_i}^{t_f} \mathbf{F}(t)\,dt \qquad \text{(impulse defined).} \tag{10-3}$$

Equation 10-3 tells us that the impulse is equal to the area under the $F(t)$ curve of Fig. 10-4a.

From Eqs. 10-2 and 10-3 we see that the change in the linear momentum of each body in a collision is equal to the impulse that acts on that body:

$$\mathbf{p}_f - \mathbf{p}_i = \Delta\mathbf{p} = \mathbf{J} \qquad \begin{array}{l}\text{(impulse–linear}\\ \text{momentum theorem).}\end{array} \tag{10-4}$$

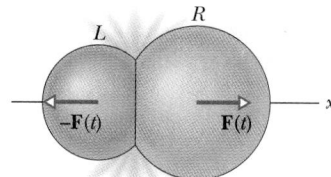

FIGURE 10-3 Two particle-like bodies L and R collide with each other. During the collision, body L exerts force $\mathbf{F}(t)$ on body R, and body R exerts force $-\mathbf{F}(t)$ on body L. Forces $\mathbf{F}(t)$ and $-\mathbf{F}(t)$ are an action–reaction pair. Their magnitudes vary with time during the collision, but at any given instant these magnitudes are equal.

(Furthermore, from the conservation of linear momentum, we know that $\Delta\mathbf{p}$ for body R in Fig. 10-3 equals $-\Delta\mathbf{p}$ for body L.) Equation 10-4 can also be written in component form as

$$p_{fx} - p_{ix} = \Delta p_x = J_x, \tag{10-5}$$

$$p_{fy} - p_{iy} = \Delta p_y = J_y, \tag{10-6}$$

and

$$p_{fz} - p_{iz} = \Delta p_z = J_z. \tag{10-7}$$

Impulse and linear momentum are both vectors, and they have the same units and dimensions. The **impulse–linear momentum theorem** of Eq. 10-4, like the work–kinetic energy theorem, is not a new and independent theorem but is a direct consequence of Newton's second law. Both theorems are special forms of this law, useful for special purposes.

If $\overline{F}$ is the average magnitude of the force in Fig. 10-4a, we can write the magnitude of the impulse as

$$J = \overline{F}\,\Delta t, \tag{10-8}$$

where Δt is the duration of the collision. The value of $\overline{F}$ is chosen so that the area within the rectangle of Fig. 10-4b is equal to the area under the $F(t)$ curve of Fig. 10-4a.

CHECKPOINT **1:** A paratrooper whose chute fails to open lands in snow; he is hurt slightly. Had he landed on bare ground, the stopping time would have been 10 times shorter and the collision lethal. Does the presence of the snow increase, decrease, or leave unchanged the values of (a) the paratrooper's change in momentum, (b) the impulse stopping the paratrooper, and (c) the force stopping the paratrooper?

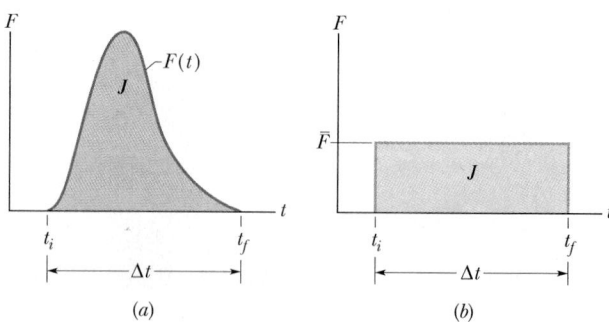

(a)   (b)

FIGURE 10-4 (a) The curve shows the magnitude of the time-varying force $F(t)$ that acts on body R during the collision of Fig. 10-3. The area under the curve for $F(t)$ is equal to the magnitude of the impulse $\mathbf{J}$ on body R in the collision. (b) The height of the rectangle represents the average force $\overline{F}$ acting on body R over the time interval Δt. The area within the rectangle is equal to the area under the curve in (a) and thus is also equal to the magnitude of the impulse $\mathbf{J}$ in the collision.

Series of Collisions

In Fig. 10-5, a steady stream of bodies, with identical linear momenta mv, collides with body R, which is fixed in place. In each collision of this one-dimensional situation, the impulse J acting on body R and the change Δp in linear momentum of the incident body have the same magnitude but opposite directions; that is, $J = -\Delta p$. If n bodies collide with R in time interval Δt, then by Eq. 10-4 the total impulse J acting on body R during Δt is

$$J = -n\,\Delta p. \qquad (10\text{-}9)$$

By rearranging Eq. 10-8 and substituting Eq. 10-9, we find the average force $\overline{F}$ acting on body R during the collisions:

$$\overline{F} = \frac{J}{\Delta t} = -\frac{n}{\Delta t}\,\Delta p = -\frac{n}{\Delta t}\,m\,\Delta v. \qquad (10\text{-}10)$$

This equation gives us $\overline{F}$ in terms of $n/\Delta t$, the rate at which the bodies collide with body R, and Δv, the change in the velocity of those bodies.

   If the colliding bodies stop upon impact, then in Eq. 10-10 we substitute

$$\Delta v = v_f - v_i = 0 - v = -v, \qquad (10\text{-}11)$$

where $v_i\,(= v)$ and $v_f\,(= 0)$ are the velocities before and after the collision, respectively. If, instead, the colliding bodies bounce directly backward from body R with no change in speed, then $v_f = -v$ and we substitute

$$\Delta v = v_f - v_i = -v - v = -2v. \qquad (10\text{-}12)$$

   In time interval Δt, an amount of mass $\Delta m = nm$ collides with body R. With this result, we can rewrite Eq. 10-10 as

$$\overline{F} = -\frac{\Delta m}{\Delta t}\,\Delta v. \qquad (10\text{-}13)$$

This equation gives the average force $\overline{F}$ in terms of $\Delta m/\Delta t$, the rate at which mass collides with body R. Here again we can substitute Eq. 10-11 or 10-12 for Δv, depending on what the colliding bodies do.

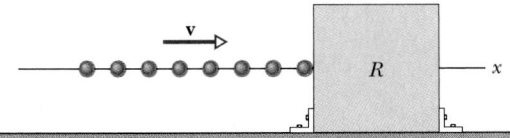

FIGURE 10-5 A steady stream of bodies, with identical linear momenta, collides with body R, which is fixed in place. The average force $\overline{F}$ on body R is to the right and has a magnitude that depends on the rate at which the bodies collide or, equivalently, the rate at which mass collides.

SAMPLE PROBLEM 10-1

A pitched 140 g baseball, in horizontal flight with a speed v_i of 39 m/s, is struck by a batter. After leaving the bat, the ball travels in the opposite direction with speed v_f, also 39 m/s.

(a) What impulse J acts on the ball while it is in contact with the bat?

SOLUTION: We can calculate the impulse from the change it produces in the ball's linear momentum, using Eq. 10-4 for one-dimensional motion. Let us choose the direction in which the bat is moving to be positive. From Eq. 10-4 we have

$$\begin{aligned}
J &= p_f - p_i = mv_f - mv_i \\
&= (0.14\ \text{kg})(39\ \text{m/s}) - (0.14\ \text{kg})(-39\ \text{m/s}) \\
&= 10.9\ \text{kg}\cdot\text{m/s} \approx 11\ \text{kg}\cdot\text{m/s}. \qquad \text{(Answer)}
\end{aligned}$$

With our sign convention, the initial velocity of the ball is negative and the final velocity is positive. The impulse turned out to be positive, which tells us that the direction of the impulse vector acting on the ball is the direction in which the bat was swinging, which makes sense.

(b) The impact time Δt for the baseball–bat collision is 1.2 ms. What average force acts on the baseball?

SOLUTION: From Eq. 10-8 we have

$$\begin{aligned}
\overline{F} &= \frac{J}{\Delta t} = \frac{10.9\ \text{kg}\cdot\text{m/s}}{0.0012\ \text{s}} \\
&= 9100\ \text{N}, \qquad \text{(Answer)}
\end{aligned}$$

which is about a ton. The *maximum* force will be larger than this. The sign of the average force exerted on the ball is positive, which means that the direction of the force vector is the same as that of the impulse vector.

(c) What is the average acceleration a of the baseball?

SOLUTION: We find this with

$$\overline{a} = \frac{\overline{F}}{m} = \frac{9100\ \text{N}}{0.14\ \text{kg}} = 6.5 \times 10^4\ \text{m/s}^2, \qquad \text{(Answer)}$$

or about $6600g$.

   In defining a collision, we assumed that no net external force acts on the colliding bodies. That is not true in this case, because the weight $m\mathbf{g}$ of the ball always acts on the ball, whether the ball is in flight or in contact with the bat. However, this force, with a magnitude of 1.4 N, is negligible compared to the average force exerted by the bat, which has a magnitude of 9100 N. We are quite safe in treating the collision as ''isolated.''

SAMPLE PROBLEM 10-2

As in Sample Problem 10-1, the baseball approaches the bat horizontally at a speed v_i of 39 m/s, but now the collision is not head-on, and the ball leaves the bat with a speed v_f of 45 m/s at an upward angle of 30° from the horizontal (Fig. 10-6). What is the average force $\overline{\mathbf{F}}$ exerted on the ball if the collision lasts 1.2 ms?

SOLUTION: We find the components J_x and J_y of the impulse from Eqs. 10-5 and 10-6:

$$J_x = p_{fx} - p_{ix} = m(v_{fx} - v_{ix})$$
$$= (0.14 \text{ kg})[(45 \text{ m/s})(\cos 30°) - (-39 \text{ m/s})]$$
$$= 10.92 \text{ kg} \cdot \text{m/s},$$

and
$$J_y = p_{fy} - p_{iy} = m(v_{fy} - v_{iy})$$
$$= (0.14 \text{ kg})[(45 \text{ m/s})(\sin 30°) - 0]$$
$$= 3.150 \text{ kg} \cdot \text{m/s}.$$

The magnitude of the impulse $\mathbf{J}$ is given by

$$J = \sqrt{J_x^2 + J_y^2}$$
$$= \sqrt{(10.92 \text{ kg} \cdot \text{m/s})^2 + (3.150 \text{ kg} \cdot \text{m/s})^2}$$
$$= 11.37 \text{ kg} \cdot \text{m/s}.$$

From Eq. 10-8, the magnitude $\overline{F}$ of the average force is

$$\overline{F} = \frac{J}{\Delta t} = \frac{11.37 \text{ kg} \cdot \text{m/s}}{0.0012 \text{ s}}$$
$$= 9475 \text{ N} \approx 9500 \text{ N}. \qquad \text{(Answer)}$$

The impulse $\mathbf{J}$ is angled upward from the horizontal by θ, where θ is given by

$$\tan \theta = \frac{J_y}{J_x} = \frac{3.150 \text{ kg} \cdot \text{m/s}}{10.92 \text{ kg} \cdot \text{m/s}} = 0.288,$$

or
$$\theta = 16°. \qquad \text{(Answer)}$$

The average force $\overline{\mathbf{F}}$ is in the same direction as $\mathbf{J}$. Note that, unlike Sample Problem 10-1, the direction of $\overline{\mathbf{F}}$ and $\mathbf{J}$ in this two-dimensional situation is different from the direction taken by the ball as it leaves the bat.

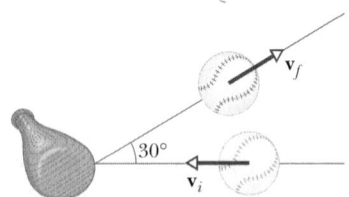

FIGURE 10-6 Sample Problem 10-2. A bat collides with a pitched baseball, sending the ball off at an angle of 30° from the horizontal.

CHECKPOINT 2: The figure shows an overhead view of a ball bouncing from a wall without any change in its speed. Consider the change $\Delta\mathbf{p}$ in the ball's linear momentum. (a) Is Δp_x positive, negative, or zero? (b) Is Δp_y positive, negative, or zero? (c) What is the direction of $\Delta\mathbf{p}$?

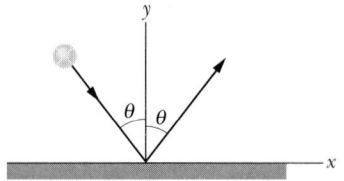

10-3 ELASTIC COLLISIONS IN ONE DIMENSION

Stationary Target

Consider a simple head-on collision of two bodies of masses m_1 and m_2 as illustrated in Fig. 10-7. For convenience, we take one of the bodies to be stationary, with velocity $\mathbf{v}_{2i} = 0$ before the collision. That body will be the "target," and the other body will be the "projectile,"* with velocity $\mathbf{v}_{1i}$ before the collision. We assume that this two-body system is closed (no mass enters or leaves it) and isolated (no net external force acts on it). Let us also make another, special assumption: the kinetic energy of the system is the same before and after the collision. The collision is then said to be of a special type called **elastic collisions.**

In an elastic collision, the kinetic energy of each colliding body can change, but the total kinetic energy of the system does not change.

Before
(a)

During
(b)

After
(c)

FIGURE 10-7 Two bodies undergo an elastic collision. One of them (the target body with mass m_2) is initially at rest before the collision. The velocities are shown (a) before, (b) during, and (c) after the collision. (The velocity during the collision is the velocity of the center of mass of the two bodies, which are momentarily touching.) The velocities are drawn to scale for the case in which $m_1 = 3m_2$.

---

*If the target body is moving with respect to our laboratory frame, we can always find another inertial reference frame in which the target body is initially stationary.

Note carefully that the linear momentum of a closed, isolated system is *always* conserved in a collision, regardless of whether the collision is elastic, because the forces involved in the collision are all internal forces.

In a closed, isolated system, the linear momentum of each colliding body can change, but the net linear momentum cannot change, regardless of whether the collision is elastic.

The conservations of linear momentum and of kinetic energy for the collision of Fig. 10-7 give us

$$m_1 v_{1i} = m_1 v_{1f} + m_2 v_{2f} \qquad \text{(linear momentum)} \qquad (10\text{-}14)$$

and

$$\tfrac{1}{2} m_1 v_{1i}^2 = \tfrac{1}{2} m_1 v_{1f}^2 + \tfrac{1}{2} m_2 v_{2f}^2 \qquad \text{(kinetic energy)}. \qquad (10\text{-}15)$$

In each of these equations, the subscript i identifies the initial velocities and the subscript f the final velocities of the bodies. If we know the masses of the bodies and if we also know v_{1i}, the initial velocity of body 1, the only unknown quantities are v_{1f} and v_{2f}, the final velocities of the two bodies. With two equations at our disposal, we should be able to find these two unknowns.

To do so, we rewrite Eq. 10-14 as

$$m_1(v_{1i} - v_{1f}) = m_2 v_{2f} \qquad (10\text{-}16)$$

and Eq. 10-15 as*

$$m_1(v_{1i} - v_{1f})(v_{1i} + v_{1f}) = m_2 v_{2f}^2. \qquad (10\text{-}17)$$

After dividing Eq. 10-17 by Eq. 10-16 and doing some more algebra, we obtain

$$v_{1f} = \frac{m_1 - m_2}{m_1 + m_2} v_{1i} \qquad (10\text{-}18)$$

and

$$v_{2f} = \frac{2m_1}{m_1 + m_2} v_{1i}. \qquad (10\text{-}19)$$

We note from Eq. 10-19 that v_{2f} is always positive (the target body with mass m_2 always moves forward). From Eq. 10-18 we see that v_{1f} may be of either sign (the projectile body with mass m_1 moves forward if $m_1 > m_2$ but rebounds if $m_1 < m_2$).

CHECKPOINT **3:** What is the final linear momentum of the target in Fig. 10-7 if the initial linear momen-

---

*In this step, we use the identity $a^2 - b^2 = (a - b)(a + b)$. It reduces the amount of algebra needed to solve the simultaneous equations, Eqs. 10-16 and 10-17.

tum of the projectile is 6 kg·m/s and the final linear momentum of the projectile is (a) 2 kg·m/s and (b) −2 kg·m/s? (c) What is the final kinetic energy of the target if the initial and final kinetic energies of the projectile are, respectively, 5 J and 2 J?

Let us look at a few special situations.

1. *Equal masses.* If $m_1 = m_2$, Eqs. 10-18 and 10-19 reduce to

$$v_{1f} = 0 \quad \text{and} \quad v_{2f} = v_{1i},$$

which we might call a pool player's result. It predicts that after a head-on collision of bodies with equal masses, body 1 (initially moving) stops dead in its tracks and body 2 (initially at rest) takes off with the initial speed of body 1. In head-on collisions, bodies of equal mass simply exchange velocities. This is true even if the target particle (body 2) is not initially at rest.

2. *A massive target.* In terms of Fig. 10-7, a massive target means that $m_2 \gg m_1$. For example, we might fire a golf ball at a cannonball. Equations 10-18 and 10-19 then reduce to

$$v_{1f} \approx -v_{1i} \quad \text{and} \quad v_{2f} \approx \left(\frac{2m_1}{m_2}\right) v_{1i}. \qquad (10\text{-}20)$$

This tells us that body 1 (the golf ball) simply bounces back in the same direction from which it came, its speed essentially unchanged. Body 2 (the cannonball) moves forward at a low speed, because the quantity in parentheses in Eq. 10-20 is much less than unity. All this is what we should expect.

3. *A massive projectile.* This is the opposite case; that is, $m_1 \gg m_2$. This time, we fire a cannonball at a golf ball. Equations 10-18 and 10-19 reduce to

$$v_{1f} \approx v_{1i} \quad \text{and} \quad v_{2f} \approx 2v_{1i}. \qquad (10\text{-}21)$$

Equation 10-21 tells us that body 1 (the cannonball) simply keeps on going, scarcely slowed by the collision. Body 2 (the golf ball) charges ahead at twice the speed of the cannonball.

You may wonder: "Why twice the speed?" As a starting point in thinking about the matter, recall the collision described by Eq. 10-20, in which the velocity of the incident light body (the golf ball) changed from $+v$ to $-v$, a velocity *change* of $2v$. The same *change* in velocity (from zero to $2v$) occurs in this example also.

4. *Motion of the center of mass.* The center of mass of two colliding bodies continues to move, totally uninfluenced by the collision. This follows from conservation

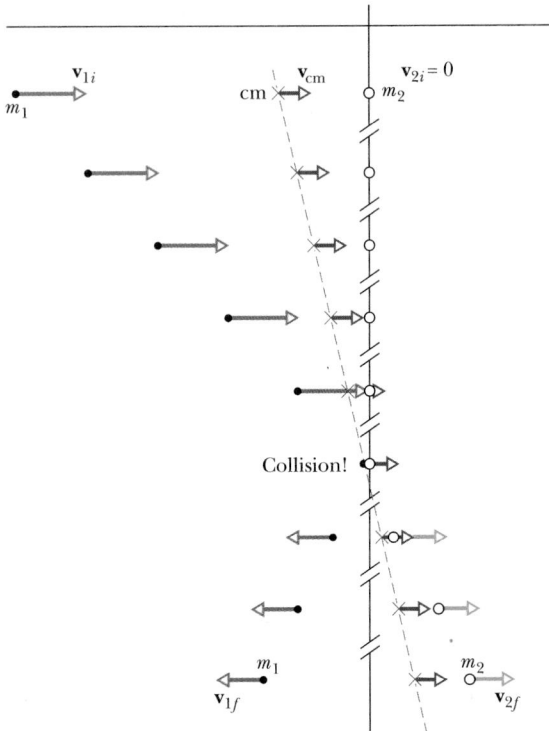

FIGURE 10-8 Some freeze-frames of two bodies undergoing an elastic collision. Body 2 is initially at rest, and $m_2 = 3m_1$. The velocity of the center of mass is also shown. Note that it is unaffected by the collision.

of linear momentum and from Eq. 9-26,

$$P = Mv_{cm} = (m_1 + m_2)v_{cm}, \quad (10\text{-}22)$$

which relates the linear momentum P of the system of two bodies to v_{cm}, the velocity of their center of mass. Because the momentum P is unchanged by the collision, v_{cm} must also be unchanged. So the center of mass continues to move in the same direction and at the same speed. From Eq. 10-22, the velocity of the center of mass for the collision shown in Fig. 10-7 (target initially at rest) is

$$v_{cm} = \frac{P}{m_1 + m_2} = \frac{m_1}{m_1 + m_2} v_{1i}. \quad (10\text{-}23)$$

Figure 10-8, a series of freeze-frames of a typical elastic collision, shows that the center of mass does indeed move steadily forward, unaffected in any way by the collision.

Moving Target

Now that we have examined the elastic collision of a projectile and a stationary target, let us examine the situation in which both bodies are moving before they undergo an elastic collision.

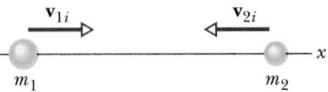

FIGURE 10-9 Two bodies headed for an elastic collision.

For the situation of Fig. 10-9, the conservation of linear momentum is written as

$$m_1 v_{1i} + m_2 v_{2i} = m_1 v_{1f} + m_2 v_{2f}, \quad (10\text{-}24)$$

and the conservation of kinetic energy is written as

$$\tfrac{1}{2}m_1 v_{1i}^2 + \tfrac{1}{2}m_2 v_{2i}^2 = \tfrac{1}{2}m_1 v_{1f}^2 + \tfrac{1}{2}m_2 v_{2f}^2. \quad (10\text{-}25)$$

To solve these simultaneous equations for v_{1f} and v_{2f}, we first rewrite Eq. 10-24 as

$$m_1(v_{1i} - v_{1f}) = -m_2(v_{2i} - v_{2f}), \quad (10\text{-}26)$$

and Eq. 10-25 as

$$m_1(v_{1i} - v_{1f})(v_{1i} + v_{1f})$$
$$= -m_2(v_{2i} - v_{2f})(v_{2i} + v_{2f}). \quad (10\text{-}27)$$

After dividing Eq. 10-27 by Eq. 10-26 and doing some more algebra, we obtain

$$v_{1f} = \frac{m_1 - m_2}{m_1 + m_2} v_{1i} + \frac{2m_2}{m_1 + m_2} v_{2i} \quad (10\text{-}28)$$

and

$$v_{2f} = \frac{2m_1}{m_1 + m_2} v_{1i} + \frac{m_2 - m_1}{m_1 + m_2} v_{2i}. \quad (10\text{-}29)$$

Note that the assignment of subscripts 1 and 2 to the bodies is arbitrary. If we exchange those subscripts in Fig. 10-9 and in Eqs. 10-28 and 10-29, we end up with the same set of equations. Note also that if we set $v_{2i} = 0$, body 2 becomes a stationary target, and Eqs. 10-28 and 10-29 reduce to Eqs. 10-18 and 10-19, respectively.

From Eq. 10-22, the constant velocity v_{cm} of the center of mass of the two-body system of Fig. 10-9 is

$$v_{cm} = \frac{P}{m_1 + m_2} = \frac{m_1 v_{1i} + m_2 v_{2i}}{m_1 + m_2}. \quad (10\text{-}30)$$

Because the two-body system is closed and isolated, its linear momentum P is unchanged by the collision. Thus, the velocity v_{cm} of its center of mass is also unchanged.

C**HECKPOINT** 4: The initial momenta of body 1 and body 2 in Fig. 10-9 are, respectively, 10 kg·m/s and −8 kg·m/s. What is the final momentum of body 2 if the final momentum of body 1 is (a) 2 kg·m/s, and (b) −2 kg·m/s?

SAMPLE PROBLEM 10-3

Two metal spheres, suspended by vertical cords, initially just touch, as shown in Fig. 10-10. Sphere 1, with mass $m_1 = 30$ g, is pulled to the left to height $h_1 = 8.0$ cm, and then released. After swinging down, it undergoes an elastic collision with sphere 2, whose mass $m_2 = 75$ g.

(a) What is the velocity v_{1f} of sphere 1 just after the collision?

SOLUTION: We let v_{1i} represent the speed of sphere 1 just before the collision. When the sphere begins its descent, its kinetic energy is zero and the gravitational potential energy is $m_1 g h_1$. Just before the collision, sphere 1 has kinetic energy $\frac{1}{2} m_1 v_{1i}^2$ and the gravitational potential energy is zero. For the descent, the conservation of mechanical energy gives us

$$\tfrac{1}{2} m_1 v_{1i}^2 = m_1 g h_1,$$

which we solve for the speed v_{1i} of sphere 1 just before the collision:

$$v_{1i} = \sqrt{2gh_1} = \sqrt{(2)(9.8 \text{ m/s}^2)(0.080 \text{ m})} = 1.252 \text{ m/s}.$$

Although sphere 1 swings down in a two-dimensional arc, its velocity is horizontal when it collides with sphere 2. Thus the collision is one-dimensional, and we can represent the velocity of sphere 1 just before that collision as v_{1i}.

To find the velocity v_{1f} of sphere 1 just after the collision, we use Eq. 10-18:

$$v_{1f} = \frac{m_1 - m_2}{m_1 + m_2} v_{1i} = \frac{0.030 \text{ kg} - 0.075 \text{ kg}}{0.030 \text{ kg} + 0.075 \text{ kg}} (1.252 \text{ m/s})$$

$$= -0.537 \text{ m/s} \approx -0.54 \text{ m/s}. \qquad \text{(Answer)}$$

The minus sign tells us that sphere 1 moves to the left just after the collision.

(b) To what height h_1' does sphere 1 swing to the left after the collision?

SOLUTION: When sphere 1 begins its upward swing, its kinetic energy is $\frac{1}{2} m_1 v_{1f}^2$ and the gravitational potential energy is zero. When it momentarily stops at height h_1', its kinetic energy is zero and the gravitational potential energy is $m_1 g h_1'$. Conserving mechanical energy for the upward swing, we find

$$m_1 g h_1' = \tfrac{1}{2} m_1 v_{1f}^2,$$

or

$$h_1' = \frac{v_{1f}^2}{2g} = \frac{(-0.537 \text{ m/s})^2}{(2)(9.8 \text{ m/s}^2)}$$

$$= 0.0147 \text{ m} \approx 1.5 \text{ cm}. \qquad \text{(Answer)}$$

(c) What is the velocity v_{2f} of sphere 2 just after the collision?

SOLUTION: From Eq. 10-19, we have

$$v_{2f} = \frac{2m_1}{m_1 + m_2} v_{1i} = \frac{(2)(0.030 \text{ kg})}{0.030 \text{ kg} + 0.075 \text{ kg}} (1.252 \text{ m/s})$$

$$= 0.715 \text{ m/s} \approx 0.72 \text{ m/s}. \qquad \text{(Answer)}$$

(d) To what height h_2 does sphere 2 swing after the collision?

SOLUTION: Sphere 2 begins its ascent with kinetic energy $\frac{1}{2} m_2 v_{2f}^2$. When it momentarily stops at height h_2, the gravitational potential energy is $m_2 g h_2$. Conservation of mechanical energy for the ascent gives us

$$m_2 g h_2 = \tfrac{1}{2} m_2 v_{2f}^2,$$

or

$$h_2 = \frac{v_{2f}^2}{2g} = \frac{(0.715 \text{ m/s})^2}{(2)(9.8 \text{ m/s}^2)}$$

$$= 0.0261 \text{ m} \approx 2.6 \text{ cm}. \qquad \text{(Answer)}$$

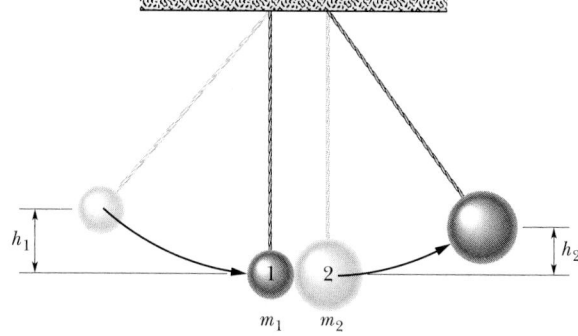

FIGURE 10-10 Sample Problem 10-3. Two metal spheres suspended by cords just touch when they are at rest. Sphere 1, with mass m_1, is pulled to the left to height h_1 and then released. The subsequent elastic collision with sphere 2 sends sphere 2 to height h_2.

SAMPLE PROBLEM 10-4

In a nuclear reactor, newly produced fast neutrons must be slowed down before they can participate effectively in the chain-reaction process. This is done by allowing them to collide with the nuclei of atoms in a *moderator*.

(a) By what fraction is the kinetic energy of a neutron (of mass m_1) reduced in a head-on elastic collision with a nucleus of mass m_2, initially at rest?

SOLUTION: The initial and final kinetic energies of the neutron are

$$K_i = \tfrac{1}{2} m_1 v_{1i}^2 \quad \text{and} \quad K_f = \tfrac{1}{2} m_1 v_{1f}^2.$$

The fraction we seek is then

$$\text{frac} = \frac{K_i - K_f}{K_i} = \frac{v_{1i}^2 - v_{1f}^2}{v_{1i}^2} = 1 - \frac{v_{1f}^2}{v_{1i}^2}. \quad (10\text{-}31)$$

For such a collision we have, from Eq. 10-18,

$$\frac{v_{1f}}{v_{1i}} = \frac{m_1 - m_2}{m_1 + m_2}. \quad (10\text{-}32)$$

Substituting Eq. 10-32 into Eq. 10-31 yields, after a little algebra,

$$\text{frac} = \frac{4m_1m_2}{(m_1 + m_2)^2}. \qquad \text{(Answer)} \qquad \text{(10-33)}$$

(b) Evaluate the fraction for lead, carbon, and hydrogen. The ratios of the mass of a nucleus to the mass of a neutron ($= m_2/m_1$) for these nuclei are 206 for lead, 12 for carbon, and about 1 for hydrogen.

SOLUTION: The following values of the fraction can be calculated with Eq. 10-33: for lead ($m_2 = 206m_1$),

$$\text{frac} = \frac{(4)(206)}{(1 + 206)^2} = 0.019 \text{ or } 1.9\%; \qquad \text{(Answer)}$$

for carbon ($m_2 = 12m_1$),

$$\text{frac} = \frac{(4)(12)}{(1 + 12)^2} = 0.28 \text{ or } 28\%; \qquad \text{(Answer)}$$

and for hydrogen ($m_2 \approx m_1$),

$$\text{frac} = \frac{(4)(1)}{(1 + 1)^2} = 1 \text{ or } 100\%. \qquad \text{(Answer)}$$

These results partially explain why water, which contains lots of hydrogen, is a much better moderator of neutrons than lead.

10-4 INELASTIC COLLISIONS IN ONE DIMENSION

An **inelastic collision** is one in which the kinetic energy of the system of colliding bodies is not conserved. If you drop a Superball onto a hard floor, it loses very little of its kinetic energy on impact and rebounds to almost its original height. If the ball did regain the original height, its collision with the floor would be elastic. However, the small loss of kinetic energy in the collision lowers the rebound height, and therefore the collision is somewhat inelastic.

A dropped golf ball will lose more of its kinetic energy and will rebound to only 60% of its original height. This collision is noticeably inelastic. If you drop a ball of wet putty onto the floor, it sticks to the floor and does not rebound at all. Because the putty sticks and does not rebound, this collision is said to be a **completely inelastic collision.**

The kinetic energy that is lost in any inelastic collision is transferred to some other form of energy, perhaps thermal energy. Nonetheless, the linear momentum of the system is always conserved (provided the system is isolated and closed). Since kinetic energy and linear momentum both involve the speeds of the colliding bodies, the conservation of linear momentum limits just how much kinetic energy is lost by a system in an inelastic collision. When the bodies stick together in a completely inelastic collision, the amount of kinetic energy that is lost is the maximum

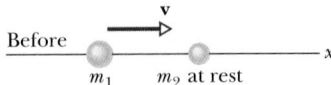

FIGURE 10-11 A completely inelastic collision between two bodies. Before the collision, the body of mass m_2 is at rest. Afterward, the bodies stick together, which is the criterion for a *completely* inelastic collision. The velocities are drawn to scale for the case in which $m_1 = 3m_2$.

allowed by the conservation of linear momentum. In some situations, this maximum loss is all the kinetic energy of the system.

In this section, we restrict ourselves to completely inelastic collisions. Figure 10-11 shows a one-dimensional inelastic collision in which one body is initially stationary. The law of conservation of linear momentum holds, so

$$m_1v = (m_1 + m_2)V, \qquad \text{(10-34)}$$

or

$$V = v\,\frac{m_1}{m_1 + m_2}, \qquad \text{(10-35)}$$

where V represents the final velocity of the stuck-together bodies. Equation 10-35 tells us that the final speed is always less than that of the incoming body.

Figure 10-12 (compare it with Fig. 10-8) shows that the motion of the center of mass in a completely inelastic collision is unaffected by the collision. Here, even though the bodies stick together, the kinetic energy associated with the motion of the center of mass is still present. Kinetic energy cannot vanish *totally* in an inelastic collision unless the reference frame is fixed with respect to the center of mass of the colliding bodies. In such cases, all motion relative to the reference frame ceases when the bodies stick together. If the target body m_2 happens to be exceedingly massive, the center of mass of the system essentially coincides with that of the target. This is the situation in which a putty ball is dropped onto the floor, the target body being the Earth. As the ball hits, all the kinetic energy is dissipated, being transferred to some other form of energy.

If both bodies are moving prior to a completely inelastic collision, we replace Eq. 10-34 with

$$m_1v_1 + m_2v_2 = (m_1 + m_2)V, \qquad \text{(10-36)}$$

where m_1v_1 is the initial linear momentum of one body, and m_2v_2 is that of the other body. Here again, all the kinetic energy is dissipated only if the reference frame is fixed with respect to the center of mass of the bodies. Figure

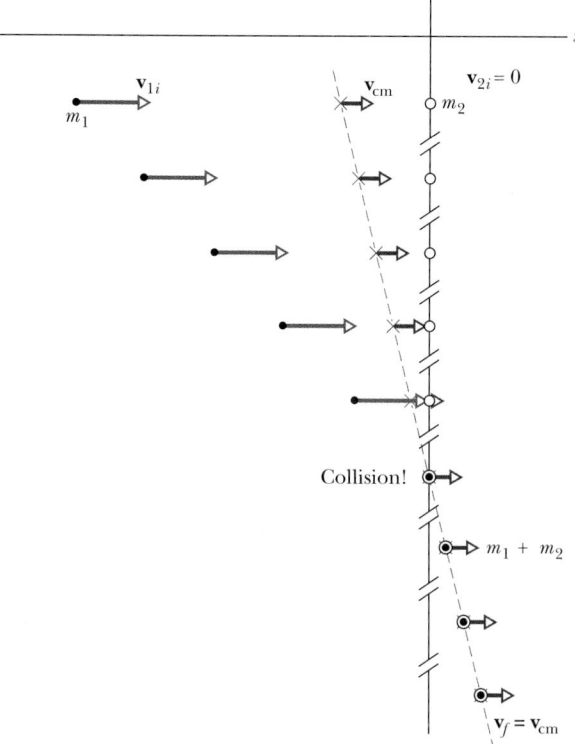

FIGURE 10-12 Some freeze-frames of two bodies undergoing a completely inelastic collision. Body 2 is initially at rest, and $m_2 = 3m_1$. The bodies stick together after the collision and move forward together. The velocity of the center of mass is also shown. Note that it is unaffected by the collision and that it is equal to the final velocity of the stuck-together bodies.

FIGURE 10-13 Two cars after an almost head-on, almost completely inelastic collision.

10-13 shows an example: identical cars driven at identical speeds underwent an almost head-on and almost completely inelastic collision. Their center of mass was stationary relative to the Earth. So a bystander, who would have been stationary relative to the Earth, would have seen these cars stop dead because of the collision, rather than move afterward in one direction or another.

CHECKPOINT **5:** Body 1 and body 2 are in a completely inelastic one-dimensional collision. What is their final momentum if their initial momenta are, respectively, (a) 10 kg·m/s and 0; (b) 10 kg·m/s and 4 kg·m/s; (c) 10 kg·m/s and −4 kg·m/s?

SAMPLE PROBLEM 10-5

The *ballistic pendulum* was used to measure the speeds of bullets before electronic timing devices were developed. The version shown in Fig. 10-14 consists of a large block of wood of mass $M = 5.4$ kg, hanging from two long cords. A bullet of mass $m = 9.5$ g is fired into the block, coming quickly to rest. The *block + bullet* then swing upward, their center of mass rising a vertical distance $h = 6.3$ cm before the pendulum comes momentarily to rest at the end of its arc.

(a) What was the speed v of the bullet just prior to the collision?

SOLUTION: Just after the collision, the *bullet + block* have speed V. Applying the conservation of linear momentum to

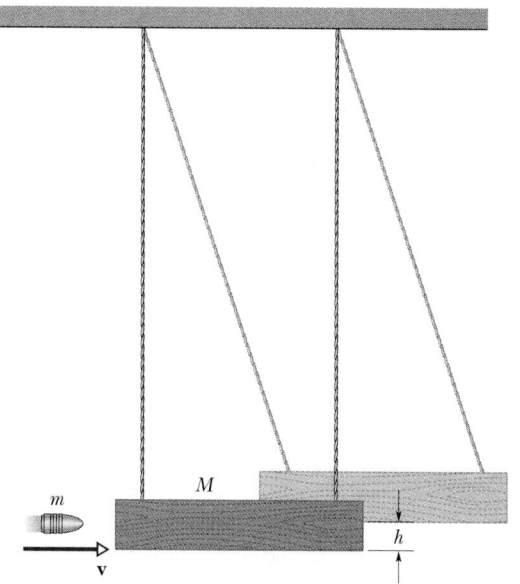

FIGURE 10-14 Sample Problem 10-5. A ballistic pendulum, used to measure the speeds of bullets.

the collision, we have

$$mv = (M + m)V.$$

Since the bullet and block stick together, the collision is completely inelastic, and kinetic energy is not conserved during it. However, *after* the collision, mechanical energy *is* conserved, because no force then acts to dissipate that energy. So the kinetic energy of the system when the block is at the bottom of its arc must equal the potential energy of the system when the block is at the top:

$$\tfrac{1}{2}(M + m)V^2 = (M + m)gh.$$

Eliminating V between these two equations leads to

$$v = \frac{M + m}{m}\sqrt{2gh}$$

$$= \left(\frac{5.4 \text{ kg} + 0.0095 \text{ kg}}{0.0095 \text{ kg}}\right)\sqrt{(2)(9.8 \text{ m/s}^2)(0.063 \text{ m})}$$

$$= 630 \text{ m/s}. \qquad \text{(Answer)}$$

The ballistic pendulum is a kind of "transformer," exchanging the high speed of a light object (the bullet) for the low—and thus more easily measurable—speed of a massive object (the block).

(b) What is the initial kinetic energy of the bullet? How much of this energy remains as mechanical energy of the swinging pendulum?

SOLUTION: The kinetic energy of the bullet is

$$K_b = \tfrac{1}{2}mv^2 = (\tfrac{1}{2})(0.0095 \text{ kg})(630 \text{ m/s})^2$$

$$= 1900 \text{ J}. \qquad \text{(Answer)}$$

The mechanical energy of the swinging pendulum is equal to the gravitational potential energy when the block is at the top of its swing:

$$E = (M + m)gh$$

$$= (5.4 \text{ kg} + 0.0095 \text{ kg})(9.8 \text{ m/s}^2)(0.063 \text{ m})$$

$$= 3.3 \text{ J}. \qquad \text{(Answer)}$$

Thus only 3.3/1900, or 0.2%, of the original kinetic energy of the bullet is transferred to mechanical energy of the pendulum. The rest is transferred to thermal energy of the block and bullet, or goes into the breaking of wood fibers as the bullet bores into the wood.

SAMPLE PROBLEM 10-6

A karate expert strikes downward with his fist (of mass $m_1 = 0.70$ kg), breaking a 0.14 kg board (Fig. 10-15a). He then does the same to a 3.2 kg concrete block. The spring constants k for bending are 4.1×10^4 N/m for the board and 2.6×10^6 N/m for the block. Breaking occurs at a deflection d of 16 mm for the board and 1.1 mm for the block (Fig. 10-15c).*

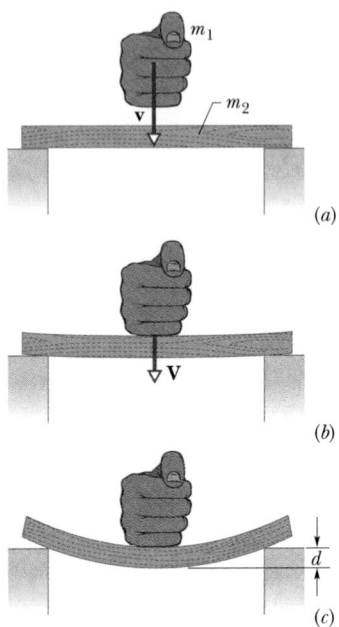

(a)

(b)

(c)

FIGURE 10-15 Sample Problem 10-6. (*a*) A karate expert strikes at a flat object with speed v. (*b*) Fist and object undergo a completely inelastic collision, and bending begins. The *fist* + *object* then have speed V. (*c*) The object breaks when its center has been deflected by an amount d.

(a) Just before the board and block break, what is the energy stored in each?

SOLUTION: We treat the bending as the compression of a spring for which Hooke's law applies. The stored energy is then, from Eq. 8-11, $U = \tfrac{1}{2}kd^2$. For the board,

$$U = \tfrac{1}{2}(4.1 \times 10^4 \text{ N/m})(0.016 \text{ m})^2$$

$$= 5.248 \text{ J} \approx 5.2 \text{ J}. \qquad \text{(Answer)}$$

For the block,

$$U = \tfrac{1}{2}(2.6 \times 10^6 \text{ N/m})(0.0011 \text{ m})^2$$

$$= 1.573 \text{ J} \approx 1.6 \text{ J}. \qquad \text{(Answer)}$$

(b) What fist speed v is required to break the board and the block? Assume that mechanical energy is conserved during the bending, that the fist and struck object stop just before the break, and that the fist–object collision at the onset of bending (Fig. 10-15b) is completely inelastic.

SOLUTION: If mechanical energy is conserved during the bending, then the kinetic energy K of the fist–object system at the onset of bending must be equal to the stored energy U just at breaking: 5.2 J for the board and 1.6 J for the block. The fist

---

*The data are taken from "The Physics of Karate," by S. R. Wilk, R. E. McNair, and M. S. Feld, *American Journal of Physics,* September 1983.

speed required to break the object is the speed required to produce that kinetic energy K. To find it, we must first find the speed V of *fist + object* at the onset of bending. We write the equality between K and U as

$$K = \tfrac{1}{2}(m_1 + m_2)V^2 = U,$$

from which we find

$$V = \sqrt{\frac{2U}{m_1 + m_2}},$$

where V is the speed of *fist + object* at the onset of bending, $m_1 = 0.70$ kg, and m_2 is 0.14 kg for the board or 3.2 kg for the block. For the board we have

$$V = \sqrt{\frac{2(5.248 \text{ J})}{0.70 \text{ kg} + 0.14 \text{ kg}}} = 3.534 \text{ m/s}.$$

For the block we have

$$V = \sqrt{\frac{2(1.573 \text{ J})}{0.70 \text{ kg} + 3.2 \text{ kg}}} = 0.8981 \text{ m/s}.$$

Now let the fist have speed v just before hitting the board or block. Then Eq. 10-35 holds for the collision. By rearranging that equation, we get

$$v = \left(\frac{m_1 + m_2}{m_1}\right)V.$$

For the board we find

$$v = \left(\frac{0.70 \text{ kg} + 0.14 \text{ kg}}{0.70 \text{ kg}}\right)(3.534 \text{ m/s})$$

$$\approx 4.2 \text{ m/s}, \qquad \text{(Answer)}$$

and for the block we find

$$v = \left(\frac{0.70 \text{ kg} + 3.2 \text{ kg}}{0.70 \text{ kg}}\right)(0.8981 \text{ m/s})$$

$$\approx 5.0 \text{ m/s}. \qquad \text{(Answer)}$$

The fist speed must be about 20% faster for the fist to break the block, because the larger mass of the block makes the transfer of energy to the block more difficult.

10-5 COLLISIONS IN TWO DIMENSIONS

Here we consider a *glancing collision* (it is not head-on) between a projectile body and a target body at rest. Figure 10-16 shows a typical situation. After the collision, the two bodies fly off at angles θ_1 and θ_2, as the figure shows.

From the conservation of linear momentum (a vector relation), we can write two scalar equations:

$$m_1v_{1i} = m_1v_{1f}\cos\theta_1 + m_2v_{2f}\cos\theta_2 \quad \text{(x component)}$$
$$\text{(10-37)}$$

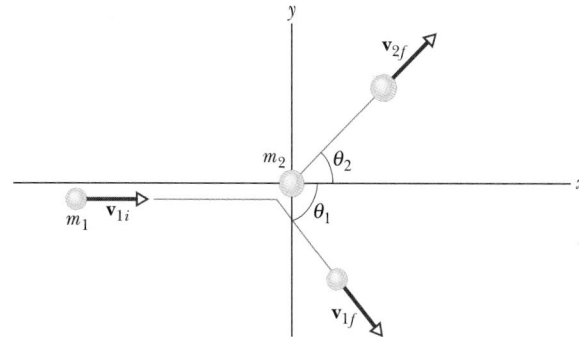

FIGURE 10-16 An elastic collision between two bodies in which the collision is not head-on. The body with mass m_2 (the target) is initially at rest.

and

$$0 = -m_1v_{1f}\sin\theta_1 + m_2v_{2f}\sin\theta_2 \quad \text{(y component)}.$$
$$\text{(10-38)}$$

If the collision is elastic, kinetic energy is also conserved:

$$\tfrac{1}{2}m_1v_{1i}^2 = \tfrac{1}{2}m_1v_{1f}^2 + \tfrac{1}{2}m_2v_{2f}^2 \quad \text{(kinetic energy)}. \quad \text{(10-39)}$$

These three equations contain seven variables: two masses, m_1 and m_2; three speeds, v_{1i}, v_{1f}, and v_{2f}; and two angles, θ_1 and θ_2. If we know any four of these quantities, we can solve the three equations for the remaining three quantities. Often the known quantities are the two masses, the initial speed, and one of the angles. The unknowns to be solved for are then the two final speeds and the remaining angle.

$\mathbb{C}$HECKPOINT **6:** In Fig. 10-16, suppose that the projectile has an initial momentum of 6 kg·m/s, a final x component of momentum of 4 kg·m/s, and a final y component of momentum of -3 kg·m/s. For the target, what then are (a) the final x component of momentum and (b) the final y component of momentum?

SAMPLE PROBLEM 10-7

Two particles of equal masses have an elastic collision, the target particle being initially at rest. Show that (unless the collision is head-on) the two particles will always move off perpendicular to each other after the collision.

SOLUTION: You might be tempted to jump into the problem and solve it in a straightforward way, by applying Eqs. 10-37, 10-38, and 10-39. There is a neater way.

Figure 10-17a shows the situation before and after the collision, each particle with its linear momentum vector attached. Because linear momentum is conserved in the collision, these vectors must form a closed triangle, as Fig. 10-17b shows. (The vector $m\mathbf{v}_{1i}$ must be the vector sum of $m\mathbf{v}_{1f}$ and

$m\mathbf{v}_{2f}$.) Because the masses of the particles are equal, the closed linear momentum triangle of Fig. 10-17b is also a closed velocity triangle, because dividing by the scalar m does not change the relation of the vectors. That is, we may draw the velocity vectors as in Fig. 10-17c, because

$$\mathbf{v}_{1i} = \mathbf{v}_{1f} + \mathbf{v}_{2f}. \qquad (10\text{-}40)$$

Because kinetic energy is conserved, Eq. 10-39 holds. With the equal terms $\frac{1}{2}m$ canceled out, this equation becomes

$$v_{1i}^2 = v_{1f}^2 + v_{2f}^2. \qquad (10\text{-}41)$$

Equation 10-41 applies to the lengths of the sides in the triangle of Fig. 10-17c. For it to hold, the triangle must be a right triangle (and Eq. 10-41 is then the Pythagorean theorem). Therefore the angle ϕ between the vectors $\mathbf{v}_{1f}$ and $\mathbf{v}_{2f}$ in Fig. 10-17 must be 90°, which is what we set out to prove.

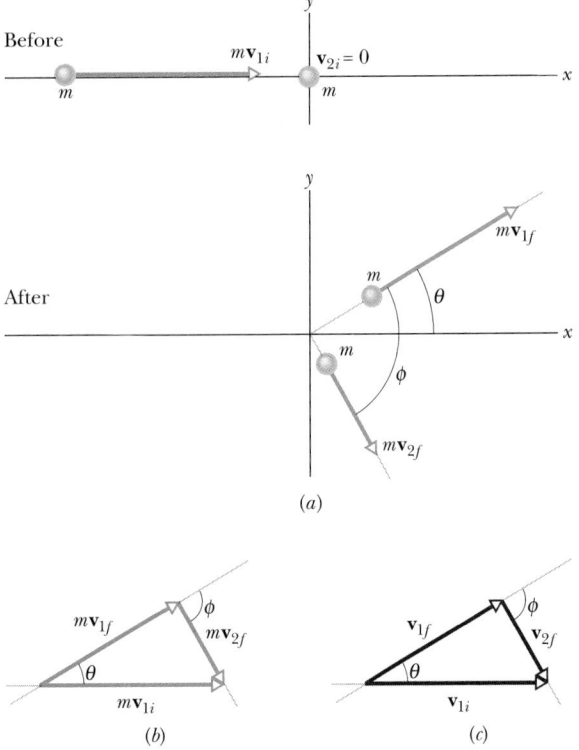

(a)

(b)   (c)

FIGURE 10-17 Sample Problem 10-7. A neat proof that in an elastic collision between two particles of the same mass, the particles fly off at 90° to each other afterward. For this to hold true, the target particle must be initially at rest and the collision must not be head-on.

SAMPLE PROBLEM 10-8

Two skaters collide and embrace, in a completely inelastic collision. That is, they stick together after impact, as suggested by Fig. 10-18, where the origin is placed at the point of collision. Alfred, whose mass m_A is 83 kg, is originally moving east with speed $v_A = 6.2$ km/h. Barbara, whose mass m_B is 55 kg, is originally moving north with speed $v_B = 7.8$ km/h.

(a) What is the velocity $\mathbf{V}$ of the couple after impact?

SOLUTION: Linear momentum is conserved during the collision. We can write, for the linear momentum components in the x and y directions,

$$m_A v_A = MV \cos\theta \qquad (x \text{ component}) \qquad (10\text{-}42)$$

and

$$m_B v_B = MV \sin\theta \qquad (y \text{ component}), \qquad (10\text{-}43)$$

in which $M = m_A + m_B$. Dividing Eq. 10-43 by Eq. 10-42 yields

$$\tan\theta = \frac{m_B v_B}{m_A v_A} = \frac{(55 \text{ kg})(7.8 \text{ km/h})}{(83 \text{ kg})(6.2 \text{ km/h})} = 0.834.$$

Thus

$$\theta = \tan^{-1} 0.834 = 39.8° \approx 40°. \qquad \text{(Answer)}$$

From Eq. 10-43 we then have

$$V = \frac{m_B v_B}{M \sin\theta} = \frac{(55 \text{ kg})(7.8 \text{ km/h})}{(83 \text{ kg} + 55 \text{ kg})(\sin 39.8°)}$$

$$= 4.86 \text{ km/h} \approx 4.9 \text{ km/h}. \qquad \text{(Answer)}$$

(b) What is the velocity of the center of mass of the two skaters before and after the collision?

SOLUTION: We can answer this without further calculation. After the collision, the velocity of the center of mass is the same as the velocity that we calculated in part (a), namely, 4.9 km/h at 40° north of east. Because the velocity of the center of mass is not changed by the collision, the same value must prevail before the collision.

(c) What is the fractional change in the kinetic energy of the skaters because of the collision?

SOLUTION: The initial kinetic energy is

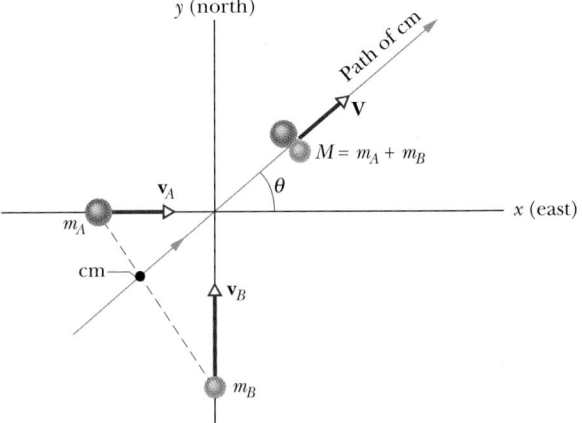

FIGURE 10-18 Sample Problem 10-8. Two skaters, Alfred (A) and Barbara (B), represented with spheres in this simplified overhead view, have a completely inelastic collision. Afterward, they move off together at angle θ, with speed V. The path of their center of mass is shown. The position of the center of mass for the indicated positions of the skaters before the collision is also shown.

$$K_i = \tfrac{1}{2}m_A v_A^2 + \tfrac{1}{2}m_B v_B^2$$
$$= (\tfrac{1}{2})(83 \text{ kg})(6.2 \text{ km/h})^2 + (\tfrac{1}{2})(55 \text{ kg})(7.8 \text{ km/h})^2$$
$$= 3270 \text{ kg} \cdot \text{km}^2/\text{h}^2.$$

The final kinetic energy is

$$K_f = \tfrac{1}{2}MV^2$$
$$= (\tfrac{1}{2})(83 \text{ kg} + 55 \text{ kg})(4.86 \text{ km/h})^2$$
$$= 1630 \text{ kg} \cdot \text{km}^2/\text{h}^2.$$

The fractional change is then

$$\text{frac} = \frac{K_f - K_i}{K_i}$$
$$= \frac{1630 \text{ kg} \cdot \text{km}^2/\text{h}^2 - 3270 \text{ kg} \cdot \text{km}^2/\text{h}^2}{3270 \text{ kg} \cdot \text{km}^2/\text{h}^2}$$
$$= -0.50. \qquad \text{(Answer)}$$

Thus 50% of the initial kinetic energy is lost as a result of the collision.

$\mathbb{C}$HECKPOINT **7**: In Sample Problem 10-8 and Fig. 10-18, does θ increase, decrease, or remain the same if we increase (a) the speed of Barbara and (b) the mass of Barbara?

PROBLEM SOLVING TACTICS

TACTIC 1: *Switch to SI Units?*

More often than not, it is wise to express all physical quantities in their basic SI units; thus all speeds in meters per second, all masses in kilograms, and so on. Sometimes, however, it is not necessary to do this. In Sample Problem 10-8a, for example, the units cancel out when we calculate the angle θ. In Sample Problem 10-8c, they cancel when we calculate the dimensionless quantity frac. In this latter case, for example, there is no need to change the kinetic energy units to joules, the basic SI energy unit; we can leave them in the units kg·km²/h² because we can look ahead and see that they are going to cancel when we compute the ratio frac.

10-6 REACTIONS AND DECAY PROCESSES (OPTIONAL)

Here we consider certain collisions (called **nuclear reactions**) in which the identity and even the number of interacting nuclear particles may change because of the collisions. We consider too the *spontaneous decay* of unstable particles, in which one particle is transformed to two other particles. For both kinds of event, there is a clear distinction between times "before the event" and times "after the event," and both linear momentum and *total* energy are conserved. In short, we can treat these events by the methods we have already developed for collisions.

SAMPLE PROBLEM 10-9

A radioactive nucleus of uranium-235 decays spontaneously to thorium-231 by emitting an alpha particle (the nucleus of a helium atom, symbolized as α, helium-4, or ^{4}He):

$$^{235}\text{U} \rightarrow \alpha + {}^{231}\text{Th}.$$

The alpha particle ($m_\alpha = 4.00$ u) has a kinetic energy K_α of 4.60 MeV. What is the kinetic energy of the thorium-231 nucleus ($m_{\text{Th}} = 231$ u)?

SOLUTION: The ^{235}U nucleus is initially at rest in the laboratory. After decay, the alpha particle flies off and the ^{231}Th moves in the opposite direction, with kinetic energies K_α and K_{Th}, respectively. Applying the law of conservation of linear momentum leads to

$$0 = m_{\text{Th}}v_{\text{Th}} + m_\alpha v_\alpha,$$

which we can recast as

$$m_{\text{Th}}v_{\text{Th}} = -m_\alpha v_\alpha. \qquad (10\text{-}44)$$

Squaring both sides of Eq. 10-44 yields

$$m_{\text{TH}}^2 v_{\text{Th}}^2 = m_\alpha^2 v_\alpha^2. \qquad (10\text{-}45)$$

Because $K = \tfrac{1}{2}mv^2$, we can write Eq. 10-45 as

$$m_{\text{Th}}K_{\text{Th}} = m_\alpha K_\alpha.$$

Thus

$$K_{\text{Th}} = K_\alpha \frac{m_\alpha}{m_{\text{Th}}} = (4.60 \text{ MeV})\frac{4.00 \text{ u}}{231 \text{ u}}$$
$$= 7.97 \times 10^{-2} \text{ MeV} = 79.7 \text{ keV}. \qquad \text{(Answer)}$$

We see that, of the total amount of kinetic energy made available during the decay (4.60 MeV + 0.0797 MeV = 4.68 MeV), the heavy thorium-231 nucleus receives only about 0.0797/4.68, or 1.7%.

SAMPLE PROBLEM 10-10

A nuclear reaction of great importance for the generation of energy by nuclear fusion is the so-called d-d reaction, one form of which is

$$d + d \rightarrow t + p. \qquad (10\text{-}46)$$

The particles represented by these letters are all isotopes of hydrogen, with the following properties:

SYMBOLS		NAME	MASS
p	^{1}H	Proton	$m_p = 1.00783$ u
d	^{2}H	Deuteron	$m_d = 2.01410$ u
t	^{3}H	Triton	$m_t = 3.01605$ u

(a) How much energy appears because of the mass change Δm that occurs in this reaction?

SOLUTION: From Eq. 8-40, the Q of a reaction (the energy released or absorbed in the reaction) is $Q = -\Delta m\, c^2$. Here the change Δm in mass is $\Delta m = m_p + m_t - 2m_d$. So, here Q is

$$Q = -\Delta m\, c^2 = (2m_d - m_p - m_t)c^2$$
$$= (2 \times 2.01410\ u - 1.00783\ u - 3.01605\ u)$$
$$\times (931.5\ \text{MeV/u})$$
$$= (0.00432\ u)(931.5\ \text{MeV/u})$$
$$= 4.02\ \text{MeV}. \qquad \text{(Answer)}$$

We have used the value of 931.5 MeV/u for c^2 (Eq. 8-43).

A positive value for Q (as in this case) means that the reaction is **exothermic;** in this reaction mass energy is transferred to kinetic energy of the products. Only $0.00432/(2 \times 2.01410)$, or about 0.1%, of the mass energy originally present has been so transferred. A negative Q signals an **endothermic** reaction, in which the transfer is from kinetic energy to mass energy. And $Q = 0$ means an *elastic encounter,* with no mass change and kinetic energy conserved.

(b) A deuteron of kinetic energy $K_d = 1.50$ MeV strikes a stationary deuteron, initiating the reaction of Eq. 10-46. A proton is observed to move off at an angle of 90° to the incident direction, with a kinetic energy of 3.39 MeV; see Fig. 10-19. What is the kinetic energy of the triton?

SOLUTION: The energy Q released due to the decrease in mass appears as an increase in the kinetic energies of the particles. Thus we can write

$$Q = \Delta K = K_p + K_t - K_d,$$

where Q is the energy computed for this reaction in (a). Solving for K_t gives us

$$K_t = Q + K_d - K_p$$
$$= 4.02\ \text{MeV} + 1.50\ \text{MeV} - 3.39\ \text{MeV}$$

$$= 2.13\ \text{MeV}. \qquad \text{(Answer)}$$

(c) At what angle ϕ with the incident direction (see Fig. 10-19) does the triton emerge?

SOLUTION: We have not yet made use of the fact that linear momentum is conserved in the reaction of Eq. 10-46. The law of conservation of linear momentum yields two scalar equations:

$$m_d v_d = m_t v_t \cos \phi \qquad (x\ \text{component}) \qquad (10\text{-}47)$$

and

$$0 = m_p v_p + m_t v_t \sin \phi \qquad (y\ \text{component}). \qquad (10\text{-}48)$$

From Eq. 10-48 we have

$$\sin \phi = -\frac{m_p v_p}{m_t v_t}. \qquad (10\text{-}49)$$

Using the relation $K = \frac{1}{2}mv^2$, we can rewrite each linear momentum mv as $\sqrt{2mK}$ and thus recast Eq. 10-49 as

$$\phi = \sin^{-1}\left(-\sqrt{\frac{m_p K_p}{m_t K_t}}\right)$$
$$= \sin^{-1}\left(-\sqrt{\frac{(1.01\ u)(3.39\ \text{MeV})}{(3.02\ u)(2.13\ \text{MeV})}}\right)$$
$$= \sin^{-1}(-0.730) = -46.9°. \qquad \text{(Answer)}$$

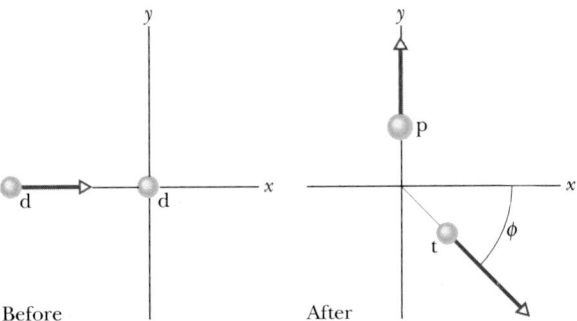

FIGURE 10-19 Sample Problem 10-10. A moving deuteron (d) strikes a stationary deuteron, producing a proton (p) and a triton (t).

REVIEW & SUMMARY

Collisions

In a **collision,** two bodies exert strong forces on each other for a relatively short time. These forces are internal to the two-body system and are significantly larger than any external force during the collision. The laws of conservation of energy and of linear momentum, applied immediately before and after a collision, allow us to predict the outcome of the collision and to understand the interactions between the colliding bodies.

Impulse and Linear Momentum

Applying Newton's second law in momentum form to a particle-like body involved in a collision leads to the **impulse–linear momentum theorem:**

$$\mathbf{p}_f - \mathbf{p}_i = \Delta \mathbf{p} = \mathbf{J}, \qquad (10\text{-}4)$$

where $\mathbf{p}_f - \mathbf{p}_i = \Delta \mathbf{p}$ is the change in the body's linear momentum, and $\mathbf{J}$ is the **impulse** due to the force $\mathbf{F}(t)$ exerted on the body by the other body in the collision:

$$\mathbf{J} = \int_{t_i}^{t_f} \mathbf{F}(t)\, dt. \qquad (10\text{-}3)$$

If $\overline{F}$ is the average of $\mathbf{F}(t)$ during the collision and Δt is the duration of the collision, then for one-dimensional motion

$$J = \overline{F}\, \Delta t. \qquad (10\text{-}8)$$

When a steady stream of bodies, each with mass m and speed v, collides with a body fixed in position, the average force on the fixed body is

$$\overline{F} = -\frac{n}{\Delta t}\Delta p = -\frac{n}{\Delta t} m \, \Delta v, \qquad (10\text{-}10)$$

where $n/\Delta t$ is the rate at which the bodies collide with the fixed body, and Δv is the change in velocity of each colliding body. This average force can also be written as

$$\overline{F} = -\frac{\Delta m}{\Delta t}\Delta v, \qquad (10\text{-}13)$$

where $\Delta m/\Delta t$ is the rate at which mass collides with the fixed body. In Eqs. 10-10 and 10-13, $\Delta v = -v$ if the bodies stop upon impact or $\Delta v = -2v$ if they bounce directly backward with no change in speed.

Elastic Collision — One Dimension

An *elastic collision* is one in which the kinetic energy of a system of two colliding bodies is conserved. For a one-dimensional situation in which one body (the target) is stationary and the other body (the projectile) is initially moving, we get the following relations from conservation of kinetic energy and linear momentum:

$$v_{1f} = \frac{m_1 - m_2}{m_1 + m_2} v_{1i} \qquad (10\text{-}18)$$

and

$$v_{2f} = \frac{2m_1}{m_1 + m_2} v_{1i}. \qquad (10\text{-}19)$$

Here subscripts i and f refer to the velocities immediately before and after the collision, respectively. If both bodies are moving prior to the collision, their velocities immediately after the collision are given by

$$v_{1f} = \frac{m_1 - m_2}{m_1 + m_2} v_{1i} + \frac{2m_2}{m_1 + m_2} v_{2i} \qquad (10\text{-}28)$$

and

$$v_{2f} = \frac{2m_1}{m_1 + m_2} v_{1i} + \frac{m_2 - m_1}{m_1 + m_2} v_{2i}. \qquad (10\text{-}29)$$

Inelastic Collision — One Dimension

An *inelastic collision* is one in which the kinetic energy of a system of two colliding bodies is not conserved. The total linear momentum of the system must, however, still be conserved. If the

colliding bodies stick together, the collision is **completely inelastic,** and the reduction in kinetic energy is maximum (but not necessarily to zero). For one-dimensional motion, with one body initially stationary and the other having initial velocity v, the velocity V of the stuck-together bodies is found by applying conservation of linear momentum to the system:

$$m_1 v = (m_1 + m_2)\, V. \qquad (10\text{-}34)$$

If both bodies are moving prior to the collision, conservation of linear momentum is written as

$$m_1 v_1 + m_2 v_2 = (m_1 + m_2)\, V. \qquad (10\text{-}36)$$

Motion of the Center of Mass

The center of mass of a closed, isolated system of colliding bodies is unaffected by the collision, whether the collision is elastic or inelastic. For a one-dimensional collision, the constant velocity v_{cm} of the center of mass is given by

$$v_{cm} = \frac{P}{m_1 + m_2} = \frac{m_1 v_{1i} + m_2 v_{2i}}{m_1 + m_2} \qquad (10\text{-}30)$$

before and after the collision.

Collisions in Two Dimensions

Collisions in two dimensions are governed by the conservation of vector linear momentum, a condition that leads to two-component equations. These determine the final motion if the collision is completely inelastic. Otherwise, the laws of conservation of linear momentum and of energy generally lead to equations that cannot be solved completely unless other data, such as the final direction of one of the velocities, are available.

Nuclear Reactions and Decay

In a *nuclear reaction* or *decay* of nuclear particles, linear momentum and *total* energy are conserved. If the mass of a system of such particles changes by Δm, the mass energy of the system changes by $\Delta m\, c^2$. The Q of the reaction or decay is defined as

$$Q = -\Delta m\, c^2.$$

This process is said to be **exothermic,** and Q is a positive quantity, if mass energy is transferred to kinetic energy of particles in the system. It is said to be **endothermic,** and Q is a negative quantity, if kinetic energy of particles in the system is transferred to mass energy.

QUESTIONS

1. Figure 10-20 shows three graphs of force magnitude versus time for a body involved in a collision. Rank the graphs according to the magnitude of the impulse on the body, greatest first.

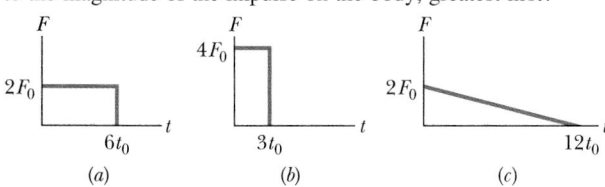

FIGURE 10-20 Question 1.

2. Figure 10-21 shows an overhead view of a golf ball bouncing off a tree trunk with no change in its speed. During the collision the force $\mathbf{F}$ on the ball by the trunk causes a change $\Delta \mathbf{p}$ in the linear momentum of the ball. If angle θ is increased (and assuming that the duration of the collision is unchanged), do the following increase, decrease, or remain the same: (a) Δp_x, (b) Δp_y, (c) magnitude of $\Delta \mathbf{p}$, (d) F_x, (e) F_y, and (f) magnitude of $\mathbf{F}$?

3. The following table gives, for three situations, the masses (kilograms) and velocities (meters per second) for the two parti-

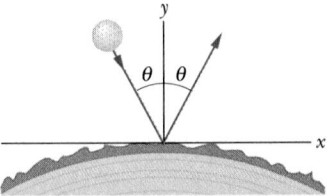

FIGURE 10-21
Question 2.

cles in Fig. 10-9. For which situations is the center of mass of the two-particle system stationary?

SITUATION	m_1	v_1	m_2	v_2
a	2	3	4	−3
b	6	2	3	−4
c	4	3	4	−3

4. (a) In Sample Problem 10-3, does the rebound height of sphere 1 increase, decrease, or remain the same if we increase the mass of sphere 2? If, instead, we set $m_2 = m_1$, (b) what is the rebound height of sphere 1, and (c) what is the height h_2 reached by sphere 2?

5. Two bodies undergo an elastic one-dimensional collision along an x axis. Figure 10-22 is a graph of position versus time for those bodies and for their center of mass. (a) Were both bodies initially moving, or was one initially stationary? Which line segment corresponds to the motion of the center of mass (b) before the collision and (c) after the collision? (d) Is the mass of the body that was moving faster before the collision greater than, less than, or equal to that of the other body?

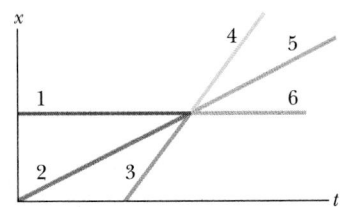

FIGURE 10-22 Question 5.

6. One body catches up with a second body and they then undergo a one-dimensional collision. Figure 10-23 is a graph of position versus time for those bodies and for their center of mass. Which line segment corresponds to (a) the faster body before the collision, (b) the center of mass before the collision, (c) the center of mass after the collision, and (d) the initially faster body after the collision? (e) Is the mass of the initially faster body greater than, less than, or equal to that of the other body?

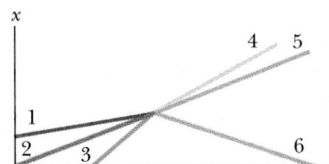

FIGURE 10-23 Question 6.

7. Figure 10-24 shows, for four situations, three identical blocks that undergo elastic collisions on a frictionless surface. In situations 1 and 2, two of the blocks are glued together. In all four situations, the initially moving blocks have the same velocity **v**. Rank the situations according to (a) the total linear momentum of the blocks after the collisions and (b) the speed of the rightmost block after the collisions, greatest first.

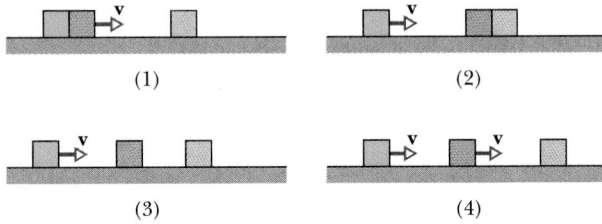

FIGURE 10-24 Question 7.

8. Figure 10-25 shows seven identical blocks on a frictionless floor. Initially, blocks a and b are moving rightward and block g is moving leftward, each with speed $v = 3$ m/s. The other blocks are stationary. A series of elastic collisions occurs. After the last collision, what are the speeds and directions of motion of each of the seven blocks?

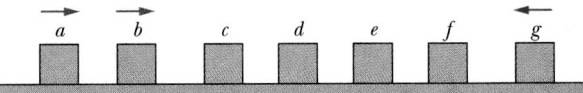

FIGURE 10-25 Question 8.

9. In Fig. 10-26, blocks A and B have linear momenta with directions as shown and with magnitudes of 9 kg·m/s and 4 kg·m/s, respectively. (a) What is the direction of motion of the center of mass of the two-block system over the frictionless floor? (b) If the blocks stick together during their collision, in what direction will they move? (c) If, instead, block A ends up moving to the left, will the magnitude of its momentum then be smaller than, more than, or the same as that of block B?

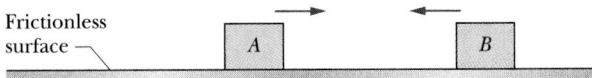

FIGURE 10-26 Question 9.

10. Figure 10-27 shows four graphs of position versus time for two bodies and their center of mass. The two bodies undergo a completely inelastic, one-dimensional collision while moving along an x axis. In graph 1, are (a) the two bodies and (b) the center of mass moving in the positive or negative direction of the x axis? (c) Which graphs correspond to a physically impossible situation?

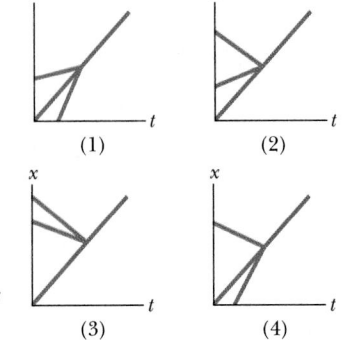

FIGURE 10-27
Question 10.

11. Body Q, with linear momentum $\mathbf{p}_Q = (2\mathbf{i} - 3\mathbf{j})$ kg·m/s, collides with and also sticks to body R, with linear momentum $\mathbf{p}_R = (8\mathbf{i} + 3\mathbf{j})$ kg·m/s. In what direction do the bodies move after the collision?

12. Drop, in succession, a baseball and a basketball from about shoulder height above a hard floor, and note how high each rebounds. Then align the baseball above the basketball (with a small separation as in Fig. 10-28a) and drop them simultaneously. (Be prepared to duck, and guard your face.) (a) Is the rebound height of the basketball now higher or lower than before (Fig. 10-28b)? (b) Is the rebound height of the baseball less than or greater than the sum of the individual baseball and basketball rebound heights? (See also Problem 37.)

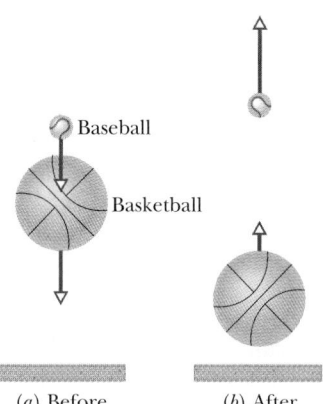

(a) Before   (b) After

FIGURE 10-28   Question 12 and Problem 37.

EXERCISES & PROBLEMS

SECTION 10-2   Impulse and Linear Momentum

1E. The linear momentum of a 1500 kg car increased by 9.0×10^3 kg·m/s in 12 s. (a) What is the magnitude of the constant force that accelerated the car? (b) By how much did the speed of the car increase?

2E. A cue stick strikes a stationary pool ball, exerting an average force of 50 N over a time of 10 ms. If the ball has mass 0.20 kg, what speed does it have after impact?

3E. The National Transportation Safety Board is testing the crash-worthiness of a new car. The 2300 kg vehicle, moving at 15 m/s, is allowed to collide with a bridge abutment, being brought to rest in a time of 0.56 s. What force, assumed constant, acted on the car during impact?

4E. A ball of mass m and speed v strikes a wall perpendicularly and rebounds in the opposite direction with the same speed. (a) If the time of collision is Δt, what is the average force exerted by the ball on the wall? (b) Evaluate this average force numerically for a rubber ball with mass 140 g moving at 7.8 m/s; the duration of the collision is 3.8 ms.

5E. A 150 g (weight ≈ 5.3 oz) baseball pitched at a speed of 40 m/s (≈ 130 ft/s) is hit straight back to the pitcher at a speed of 60 m/s (≈ 200 ft/s). What average force was exerted by the bat if it was in contact with the ball for 5.0 ms?

6E. Until he was in his seventies, Henri LaMothe excited audiences by belly-flopping from a height of 40 ft into 12 in. of water (Fig. 10-29). Assuming that he stops just as he reaches the bottom of the water, what is the average force on him from the water? Assume his weight is 160 lb.

7E. In February 1955, a paratrooper fell 1200 ft from an airplane without being able to open his chute but happened to land in snow, suffering only minor injuries. Assume that his speed at impact was 56 m/s (terminal speed), that his mass (including gear) was 85 kg, and that the force on him from the snow was at the survivable limit of 1.2×10^5 N. What is the minimum depth of snow that would have stopped him safely?

8E. A force that averages 1200 N is applied to a 0.40 kg steel ball moving at 14 m/s in a collision lasting 27 ms. If the force is in a direction opposite the initial velocity of the ball, find the final speed and direction of the ball.

9E. A 1.2 kg medicine ball drops vertically onto a floor, hitting with a speed of 25 m/s. It rebounds with an initial speed of 10 m/s. (a) What impulse acts on the ball during the contact? (b) If the ball is in contact with the floor for 0.020 s, what is the average force exerted on the floor?

10E. A golfer hits a golf ball, giving it an initial velocity of magnitude 50 m/s directed 30° above the horizontal. Assuming that the mass of the ball is 46 g and the club and ball are in contact

FIGURE 10-29   Exercise 6.

for 1.7 ms, find (a) the impulse on the ball, (b) the impulse on the club, (c) the average force exerted on the ball by the club, and (d) the work done on the ball.

11P. A 1400 kg car moving at 5.3 m/s is initially traveling north in the positive y direction. After completing a 90° right-hand turn to the positive x direction in 4.6 s, the inattentive operator drives into a tree, which stops the car in 350 ms. In unit-vector notation, what is the impulse on the car (a) during the turn and (b) during the collision? What is the magnitude of the average force that acts on the car (c) during the turn and (d) during the collision? (e) What is the angle between the average force in (c) and the positive x direction?

12P. The force on a 10 kg object increases uniformly from zero to 50 N in 4.0 s. What is the object's speed at the end of the 4.0 s interval if it started from rest?

13P. A pellet gun fires ten 2.0 g pellets per second with a speed of 500 m/s. The pellets are stopped by a rigid wall. (a) What is the momentum of each pellet? (b) What is the kinetic energy of each pellet? (c) What is the average force exerted by the stream of pellets on the wall? (d) If each pellet is in contact with the wall for 0.6 ms, what is the average force exerted on the wall by each pellet during contact? Why is this average force so different from the average force calculated in (c)?

14P. A movie set machine gun fires 50 g bullets at a speed of 1000 m/s. An actor, holding the machine gun in his hands, can exert an average force of 180 N against the gun. Determine the maximum number of bullets he can fire per minute while still holding the gun steady.

15P. It is well known that bullets and other missiles fired at Superman simply bounce off his chest (Fig. 10-30). Suppose that a gangster sprays Superman's chest with 3 g bullets at the rate of 100 bullets/min, the speed of each bullet being 500 m/s. Suppose too that the bullets rebound straight back with no change in speed. What is the average force exerted by the stream of bullets on Superman's chest?

16P. During a violent thunderstorm, hail of diameter 1.0 cm falls at a speed of 25 m/s. There are estimated to be 120 hailstones per cubic meter of air. Ignore the bounce of the hail on impact. (a) What is the mass of each hailstone (density = 0.92 g/cm³)? (b) What average force is exerted by hail on a 10 m × 20 m flat roof during the storm?

17P. A stream of water impinges on a stationary "dished" turbine blade, as shown in Fig. 10-31. The speed of the water is v, both before and after it strikes the curved surface of the blade, and the mass of water striking the blade per unit time is constant at the value μ. Find the force exerted by the water on the blade.

FIGURE 10-30 Problem 15.

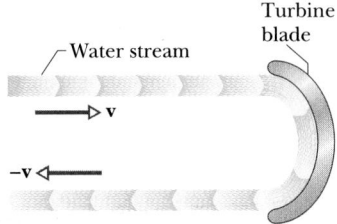

FIGURE 10-31 Problem 17.

18P. A stream of water from a hose is sprayed on a wall. If the speed of the water is 5.0 m/s and the hose sprays 300 cm³/s, what is the average force exerted on the wall by the stream of water? Assume that the water does not spatter back appreciably. Each cubic centimeter of water has a mass of 1.0 g.

19P. Figure 10-32 shows an approximate plot of force versus time during the collision of a 58 g tennis ball with a wall. The initial velocity of the ball is 34 m/s perpendicular to the wall; it rebounds with the same speed, also perpendicular to the wall. What is F_{max}, the maximum value of the contact force during the collision?

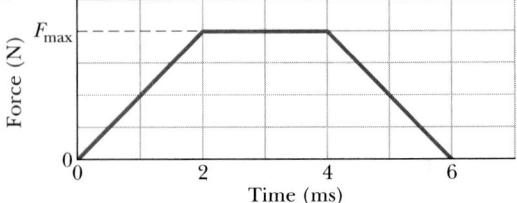

FIGURE 10-32 Problem 19.

20P. A ball having a mass of 150 g strikes a wall with a speed of 5.2 m/s and rebounds with only 50% of its initial kinetic energy.

(a) What is the speed of the ball immediately after rebounding? (b) What was the magnitude of the impulse of the ball on the wall? (c) If the ball was in contact with the wall for 7.6 ms, what was the magnitude of the average force exerted by the wall on the ball during this time interval?

21P. In the overhead view of Fig. 10-33, a 300 g ball with a speed v of 6.0 m/s strikes a wall at an angle θ of 30° and then rebounds with the same speed and angle. It is in contact with the wall for 10 ms. (a) What is the impulse on the ball? (b) What is the average force exerted by the ball on the wall?

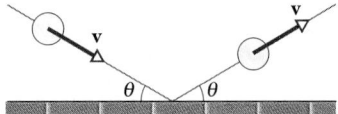

FIGURE 10-33 Problem 21.

22P. A 2500 kg unmanned space probe is moving in a straight line at a constant speed of 300 m/s. Control rockets on the space probe execute a burn in which a thrust of 3000 N acts for 65.0 s. (a) What is the change in linear momentum (magnitude only) of the probe if the thrust is backward, forward, or directly sideways? (b) What is the change in kinetic energy under the same three conditions? Assume that the mass of the ejected fuel is negligible compared to the mass of the space probe.

23P. A force exerts an impulse J on an object of mass m, changing its speed from v to u. The force and the object's motion are along the same straight line. Show that the work done by the force is $\frac{1}{2}J(u + v)$.

24P. A spacecraft is separated into two parts by detonating the explosive bolts that hold them together. The masses of the parts are 1200 kg and 1800 kg; the magnitude of the impulse on each part is 300 N·s. With what relative speed do the two parts separate because of the detonation?

25P. A stationary croquet ball with mass 0.50 kg is struck by a mallet, receiving the impulse shown in Fig. 10-34. What is the ball's velocity just after the force has become zero?

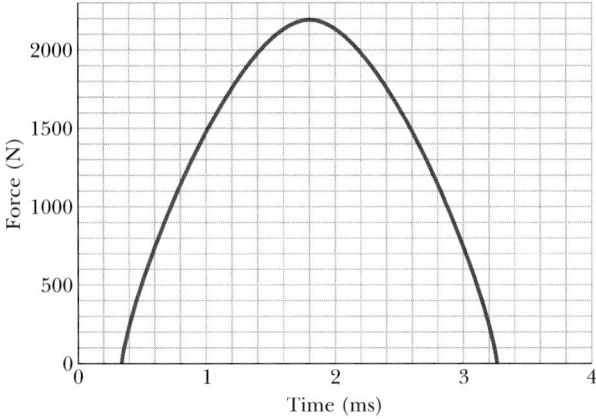

FIGURE 10-34 Problem 25.

26P. A soccer player kicks a soccer ball of mass 0.45 kg that is initially at rest. The player's foot is in contact with the ball for 3.0×10^{-3} s, and the force of the kick is given by

$$F(t) = [(6.0 \times 10^6)t - (2.0 \times 10^9)t^2] \text{ N},$$

for $0 \leq t \leq 3.0 \times 10^{-3}$ s, where t is in seconds. Find the magnitudes of the following: (a) the impulse imparted to the ball, (b) the average force exerted by the player's foot on the ball during the period of contact, (c) the maximum force exerted by the player's foot on the ball during the period of contact, and (d) the ball's velocity immediately after it loses contact with the player's foot.

27P. Spacecraft *Voyager 2* (of mass m and speed v relative to the Sun) approaches the planet Jupiter (of mass M and speed V relative to the Sun) as shown in Fig. 10-35. The spacecraft rounds the planet and departs in the opposite direction. What is its speed, relative to the Sun, after this slingshot encounter, which can be analyzed as a collision? Assume $v = 12$ km/s and $V = 13$ km/s (the orbital speed of Jupiter). The mass of Jupiter is very much greater than the mass of the spacecraft; $M \gg m$.

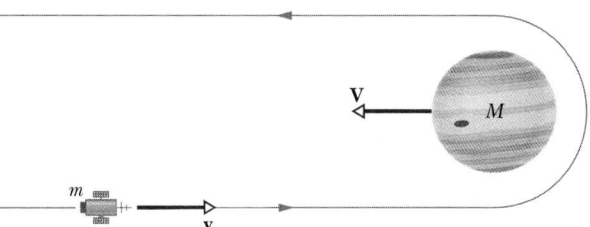

FIGURE 10-35 Problem 27.

SECTION 10-3 Elastic Collisions in One Dimension

28E. The blocks in Fig. 10-36 slide without friction. (a) What is the velocity **v** of the 1.6 kg block after the collision? (b) Is the collision elastic? (c) Suppose the initial velocity of the 2.4 kg block is the reverse of what is shown. Can the velocity **v** of the 1.6 kg block after the collision be in the direction shown?

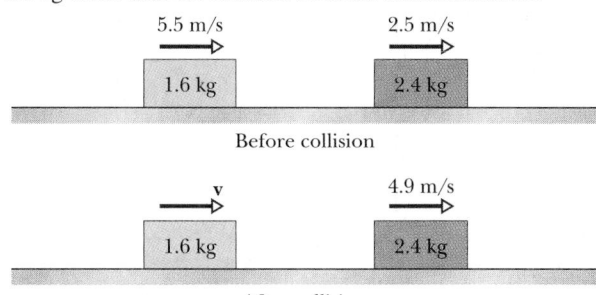

FIGURE 10-36 Exercise 28.

29E. A hovering fly is approached by an enraged elephant charging at 2.1 m/s. Assuming that the collision is elastic, at what speed does the fly rebound? Note that the projectile (the elephant) is much more massive than the stationary target (the fly).

30E. An electron collides elastically with a hydrogen atom initially at rest. (All motions are along the same straight line.) What percentage of the electron's initial kinetic energy is transferred to the hydrogen atom? The mass of the hydrogen atom is 1840 times the mass of the electron.

31E. A cart with mass 340 g moving on a frictionless linear air track at an initial speed of 1.2 m/s strikes a second cart of unknown mass at rest. The collision between the carts is elastic.

After the collision, the first cart continues in its original direction at 0.66 m/s. (a) What is the mass of the second cart? (b) What is its speed after impact? (c) What is the speed of the two-cart center of mass?

32E. An alpha particle (mass 4 u) experiences an elastic head-on collision with a gold nucleus (mass 197 u) that is originally at rest. What percentage of its original kinetic energy does the alpha particle lose?

33E. A body of mass 2.0 kg makes an elastic collision with another body at rest and continues to move in the original direction but with one-fourth of its original speed. (a) What is the mass of the struck body? (b) What is the speed of the two-body center of mass if the initial speed of the 2.0 kg body was 4.0 m/s?

34P. A steel ball of mass 0.500 kg is fastened to a cord 70.0 cm long and fixed at the far end, and is released when the cord is horizontal (Fig. 10-37). At the bottom of its path, the ball strikes a 2.50 kg steel block initially at rest on a frictionless surface. The collision is elastic. Find (a) the speed of the ball and (b) the speed of the block, both just after the collision.

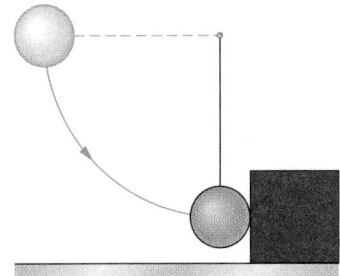

FIGURE 10-37 Problem 34.

35P. A platform scale is calibrated to indicate the mass in kilograms of an object placed on it. Particles fall from a height of 3.5 m and collide with the platform of the scale. The collisions are elastic; the particles rebound upward with the same speed they had before hitting the pan. If each particle has a mass of 110 g and collisions occur at the rate of 42 s^{-1}, what is the average scale reading?

36P. Two titanium spheres approach each other head-on with the same speed and collide elastically. After the collision, one of the spheres, whose mass is 300 g, remains at rest. (a) What is the mass of the other sphere? (b) What is the speed of the two-sphere center of mass if the initial speed of each sphere was 2.0 m/s?

37P. A ball of mass m is aligned above a ball of mass M (with a slight separation, as in Fig. 10-28a), and the two are dropped simultaneously from height h. (Assume the radius of each ball is negligible compared to h.) (a) If M rebounds elastically from the floor and then m rebounds elastically from M, what ratio m/M results in M stopping upon its collision with m? (The answer is approximately the mass ratio of a baseball to a basketball, as in Question 12.) (b) What height does m then reach?

38P. A block of mass m_1 is at rest on a long frictionless table, one end of which terminates at a wall. A block of mass m_2 is placed between the first block and the wall and set in motion to the left, toward m_1, with constant speed v_{2i}, as in Fig. 10-38. Assuming that all collisions are elastic, find the value of m_2 (in terms of m_1) for which both blocks move with the same velocity after m_2 has collided once with m_1 and once with the wall. Assume the wall to have infinite mass.

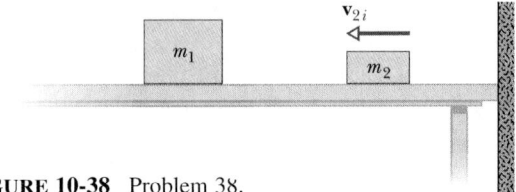

FIGURE 10-38 Problem 38.

39P. A target glider, whose mass m_2 is 350 g, is at rest on an air track, a distance $d = 53$ cm from the end of the track. A projectile glider, whose mass m_1 is 590 g, approaches the target glider with velocity $v_{1i} = -75$ cm/s and collides elastically with it (Fig. 10-39). The target glider rebounds elastically from a short spring at the end of the track and meets the projectile glider for a second time. How far from the end of the track does this second collision occur?

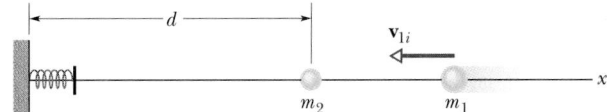

FIGURE 10-39 Problem 39.

SECTION 10-4 Inelastic Collisions in One Dimension

40E. Meteor Crater in Arizona (Fig. 10-1a) is thought to have been formed by the impact of a meteor with the Earth some 20,000 years ago. The mass of the meteor is estimated at 5×10^{10} kg, and its speed at 7200 m/s. What speed would such a meteor impart to the Earth in a head-on collision?

41E. A 6.0 kg box sled is coasting across frictionless ice at a speed of 9.0 m/s when a 12 kg package is dropped into it from above. What is the new speed of the sled?

42E. A 5.20 g bullet moving at 672 m/s strikes a 700 g wooden block at rest on a frictionless surface. The bullet emerges with its speed reduced to 428 m/s. (a) What is the resulting speed of the block? (b) What is the speed of the bullet–block center of mass?

43E. Two 2.0 kg masses, A and B, collide. The velocities before the collision are $v_A = 15i + 30j$ and $v_B = -10i + 5.0j$. After the collision, $v'_A = -5.0i + 20j$. All speeds are given in meters per second. (a) What is the final velocity of B? (b) How much kinetic energy was gained or lost in the collision?

44E. A bullet of mass 10 g strikes a ballistic pendulum of mass 2.0 kg. The center of mass of the pendulum rises a vertical distance of 12 cm. Assuming that the bullet remains embedded in the pendulum, calculate the bullet's initial speed.

45E. A bullet of mass 4.5 g is fired horizontally into a 2.4 kg wooden block at rest on a horizontal surface. The coefficient of kinetic friction between block and surface is 0.20. The bullet comes to rest in the block, which moves 1.8 m. (a) What is the speed of the block immediately after the bullet comes to rest within it? (b) At what speed is the bullet fired?

46E. A 5.0 kg block with a speed of 3.0 m/s collides with a 10 kg block that has a speed of 2.0 m/s in the same direction. After the collision, the 10 kg block is observed to be traveling in the original direction with a speed of 2.5 m/s. (a) What is the speed of the 5.0 kg block immediately after the collision? (b) By how much does the total kinetic energy of the system of two blocks change because of the collision? (c) Suppose, instead, that the 10 kg block ends up with a speed of 4.0 m/s. What then is the change in the total kinetic energy? (d) Account for the result you obtained in (c).

47P. Two cars A and B slide on an icy road as they attempt to stop at a traffic light. The mass of A is 1100 kg, and the mass of B is 1400 kg. The coefficient of kinetic friction between the locked wheels of either car and the road is 0.13. Car A succeeds in coming to rest at the light, but car B cannot stop and rear-ends car A. After the collision, A comes to rest 8.2 m ahead of the impact point and B 6.1 m ahead; see Fig. 10-40. Both drivers had their brakes locked throughout the incident. (a) From the distance each car moved after the collision, find the speed of each car immediately after impact. (b) Use conservation of linear momentum to find the speed at which car B struck car A. On what grounds can the use of linear momentum conservation be criticized here?

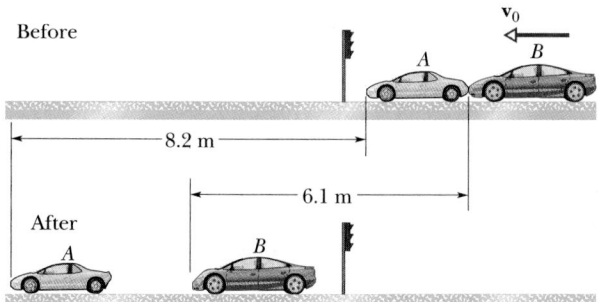

FIGURE 10-40 Problem 47.

48P. A 3000 kg weight falls vertically through 6.0 m and then collides with a 500 kg pile, driving it 3.0 cm into bedrock. Assuming that the weight–pile collision is completely inelastic, find the magnitude of the average force on the pile by the bedrock during the 3.0 cm descent.

49P. Two particles, one having twice the mass of the other, are held together with a compressed spring between them. The energy stored in the spring is 60 J. How much kinetic energy does each particle have after the two are released? Assume that all the stored energy is transferred to the particles and that neither particle is attached to the spring after the release.

50P. A 3.50 g bullet is fired horizontally at two blocks resting on a smooth tabletop, as shown in Fig. 10-41a. The bullet passes

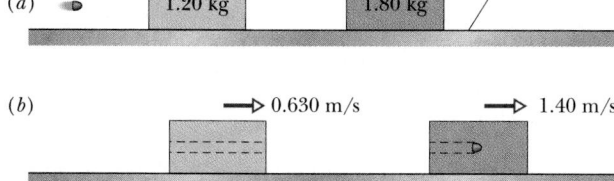

FIGURE 10-41 Problem 50.

through the first block, with mass 1.20 kg, and embeds itself in the second, with mass 1.80 kg. Speeds of 0.630 m/s and 1.40 m/s, respectively, are thereby imparted to the blocks, as shown in Fig. 10-41b. Neglecting the mass removed from the first block by the bullet, find (a) the speed of the bullet immediately after it emerges from the first block and (b) the bullet's original speed.

51P. An object with mass m and speed v explodes into two pieces, one three times as massive as the other; the explosion takes place in gravity-free space. The less massive piece comes to rest. How much kinetic energy was added to the system in the explosion?

52P. A box is put on a scale that is marked in units of mass and adjusted to read zero when the box is empty. A stream of marbles is then poured into the box from a height h above its bottom at a rate of R (marbles per second). Each marble has mass m. If the collisions between the marbles and the box are completely inelastic, find the scale reading at time t after the marbles begin to fill the box. Determine a numerical answer when $R = 100 \ \text{s}^{-1}$, $h = 7.60$ m, $m = 4.50$ g, and $t = 10.0$ s.

53P. A 35-ton railroad freight car collides with a stationary caboose car. They couple together, and 27% of the initial kinetic energy is dissipated as heat, sound, vibrations, and so on. Find the weight of the caboose.

54P. A ball of mass m is projected with speed v_i into the barrel of a spring gun of mass M initially at rest on a frictionless surface; see Fig. 10-42. The ball sticks in the barrel at the point of maximum compression of the spring. No mechanical energy is dissipated by friction. (a) What is the speed of the spring gun after the ball comes to rest in the barrel? (b) What fraction of the initial kinetic energy of the ball is stored in the spring?

FIGURE 10-42 Problem 54.

55P. A block of mass $m_1 = 2.0$ kg slides along a frictionless table with a speed of 10 m/s. Directly in front of it, and moving in the same direction, is a block of mass $m_2 = 5.0$ kg moving at 3.0 m/s. A massless spring with spring constant $k = 1120$ N/m is attached to the near side of m_2, as shown in Fig. 10-43. When the blocks collide, what is the maximum compression of the spring? (*Hint:* At the moment of maximum compression of the spring, the two blocks move as one. Find the velocity by noting that the collision is completely inelastic to this point.)

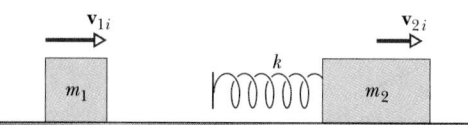

FIGURE 10-43 Problem 55.

56P. A 1.0 kg block at rest on a horizontal frictionless surface is connected to an unstretched spring ($k = 200$ N/m) whose other end is fixed (Fig. 10-44). A 2.0 kg block whose speed is 4.0 m/s collides with the 1.0 kg block. If the two blocks stick together after the one-dimensional collision, what maximum compression of the spring occurs when the blocks momentarily stop?

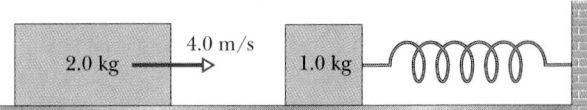

FIGURE 10-44 Problem 56.

57P. Two 22.7 kg ice sleds are placed a short distance apart, one directly behind the other, as shown in Fig. 10-45. A 3.63 kg cat, standing on one sled, jumps across to the other and immediately back to the first. Both jumps are made at a speed of 3.05 m/s relative to the ice. Find the final speeds of the two sleds.

FIGURE 10-45 Problem 57.

58P. The bumper of a 1200 kg car is designed so that it can just absorb all the energy when the car runs head-on into a solid wall at 5.00 km/h. The car is involved in a collision in which it runs at 70.0 km/h into the rear of a 900 kg car moving at 60.0 km/h in the same direction. The 900 kg car is accelerated to 70.0 km/h as a result of the collision. (a) What is the speed of the 1200 kg car immediately after impact? (b) What is the ratio of the kinetic energy absorbed in the collision to that which can be absorbed by the bumper of the 1200 kg car?

59P. A railroad freight car weighing 32 tons and traveling at 5.0 ft/s overtakes one weighing 24 tons and traveling at 3.0 ft/s in the same direction. If the cars couple together, find (a) the speed of the cars after collision and (b) the loss of kinetic energy during collision. (c) If instead, as is very unlikely, the collision is elastic, find the speeds of the cars after collision.

SECTION 10-5 Collisions in Two Dimensions

60E. An alpha particle collides with an oxygen nucleus, initially at rest. The alpha particle is scattered at an angle of 64.0° above its initial direction of motion, and the oxygen nucleus recoils at an angle of 51.0° below this initial direction. The final speed of the nucleus is 1.20×10^5 m/s. Find (a) the final speed and (b) the initial speed of the alpha particle. (The mass of an alpha particle is 4.0 u; the mass of an oxygen nucleus is 16 u.)

61E. A proton (atomic mass 1 u) with a speed of 500 m/s collides elastically with another proton at rest. The projectile proton is scattered 60° from its initial direction. (a) What is the direction of the velocity of the target proton after the collision? (b) What are the speeds of the two protons after the collision?

62E. A certain nucleus, at rest, spontaneously disintegrates into three particles. Two of them are detected; their masses and velocities are as shown in Fig. 10-46. (a) In unit-vector notation, what is the linear momentum of the third particle, which is known to have a mass of 11.7×10^{-27} kg? (b) How much kinetic energy appears in the disintegration process?

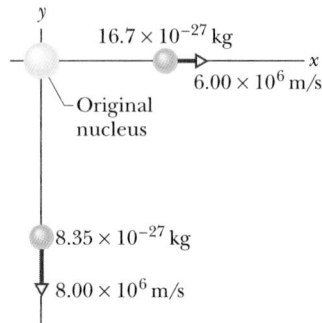

FIGURE 10-46 Exercise 62.

63E. In a game of pool, the cue ball strikes another ball initially at rest. After the collision, the cue ball moves at 3.50 m/s along a line making an angle of 22.0° with its original direction of motion, and the second ball has a speed of 2.00 m/s. Find (a) the angle between the direction of motion of the second ball and the original direction of motion of the cue ball and (b) the original speed of the cue ball. (c) Is kinetic energy conserved?

64E. Two vehicles A and B are traveling west and south, respectively, toward the same intersection, where they collide and lock together. Before the collision, A (total weight 2700 lb) is moving with a speed of 40 mi/h and B (total weight 3600 lb) has a speed of 60 mi/h. Find the magnitude and direction of the velocity of the (interlocked) vehicles immediately after the collision.

65E. In a game of billiards, the cue ball is given an initial speed V and strikes the pack of 15 stationary balls. All 16 balls then engage in numerous ball–ball and ball–cushion collisions. Some time later, it is observed that (by some accident) all 16 balls have the same speed v. Assuming that all collisions are elastic and ignoring the rotational aspect of the balls' motion, calculate v in terms of V.

66P. A 20.0 kg body is moving in the direction of the positive x axis with a speed of 200 m/s when, owing to an internal explosion, it breaks into three parts. One part, whose mass is 10.0 kg, moves away from the point of explosion with a speed of 100 m/s in the direction of the positive y axis. A second fragment, with a mass of 4.00 kg, moves in the direction of the negative x axis with a speed of 500 m/s. (a) What is the velocity of the third (6.00 kg) fragment? (b) How much energy was released in the explosion? Ignore effects due to gravity.

67P. Two balls A and B, having different but unknown masses, collide. A is initially at rest, and B has speed v. After collision, B has speed v/2 and moves perpendicularly to its original motion. (a) Find the direction in which ball A moves after collision. (b) Can you determine the speed of A from the information given? Explain.

68P. Show that if a neutron is scattered through 90° in an elastic collision with a deuteron that is initially at rest, the neutron loses two-thirds of its initial kinetic energy to the deuteron. (The mass of a neutron is 1.0 u; the mass of a deuteron is 2.0 u.)

69P. After a completely inelastic collision, two objects of the same mass and same initial speed are found to move away together at half their initial speed. Find the angle between the initial velocities of the objects.

70P. Two pendulums, both of length l, are initially situated as in Fig. 10-47. The left pendulum is released and strikes the other. Assume that the collision is completely inelastic, and neglect the mass of the strings and any frictional effects. How high does the center of mass of the pendulum system rise after the collision?

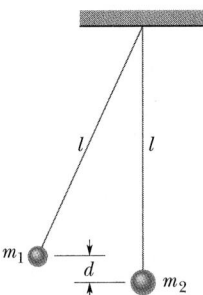

FIGURE 10-47 Problem 70.

71P. A billiard ball moving at a speed of 2.2 m/s strikes an identical stationary ball a glancing blow. After the collision, one ball is found to be moving at a speed of 1.1 m/s in a direction making a 60° angle with the original line of motion. (a) Find the velocity of the other ball. (b) Can the collision be inelastic, given these data?

72P. A ball with an initial speed of 10 m/s collides elastically with two identical stationary balls whose centers are on a line perpendicular to the initial velocity and that are initially in contact with each other (Fig. 10-48). The first ball is aimed directly at the contact point, and all motion is frictionless. Find the velocities of all three balls after the collision. (*Hint:* With friction absent, each impulse is directed along the line connecting the centers of the colliding balls, normal to the colliding surfaces.)

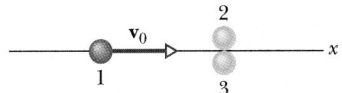

FIGURE 10-48 Problem 72.

73P. A barge with mass 1.50×10^5 kg is proceeding down river at 6.2 m/s in heavy fog when it collides broadside with a barge heading directly across the river (see Fig. 10-49). The second

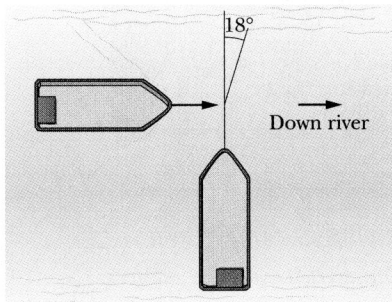

FIGURE 10-49 Problem 73.

barge has mass 2.78×10^5 kg and was moving at 4.3 m/s. Immediately after impact, the second barge finds its course deflected by 18° in the downriver direction and its speed increased to 5.1 m/s. The river current was practically zero at the time of the

accident. (a) What are the speed and direction of motion of the first barge immediately after the collision? (b) How much kinetic energy is lost in the collision?

SECTION 10-6 Reactions and Decay Processes

74E. The precise masses in the reaction

$$p + {}^{19}F \rightarrow \alpha + {}^{16}O$$

have been determined to be

$$m_p = 1.007825 \text{ u}, \qquad m_\alpha = 4.002603 \text{ u},$$
$$m_F = 18.998405 \text{ u}, \qquad m_O = 15.994915 \text{ u}.$$

Calculate the Q of the reaction from these data.

75E. A particle called Σ^- (sigma minus) is initially at rest and decays spontaneously into two other particles according to

$$\Sigma^- \rightarrow \pi^- + n.$$

The masses are

$$m_\Sigma = 2340.5 m_e, \quad m_\pi = 273.2 m_e, \quad m_n = 1838.65 m_e,$$

where m_e (9.11×10^{-31} kg) is the electron mass. (a) How much energy is transferred to kinetic energy in this process? (b) How do the linear momenta of the decay products (π^- and n) compare? (c) Which product gets the larger share of the kinetic energy?

76P*. An alpha particle with kinetic energy 7.70 MeV strikes an ^{14}N nucleus at rest. An ^{17}O nucleus and a proton are produced; the proton is emitted at 90° to the direction of the incident alpha particle and has a kinetic energy of 4.44 MeV. The masses of the various particles are: alpha particle, 4.00260 u; ^{14}N, 14.00307 u; proton, 1.007825 u; and ^{17}O, 16.99914 u. (a) What is the kinetic energy of the oxygen nucleus? (b) What is the Q of the reaction?

77P*. Consider the alpha decay of radium (Ra) to radon (Rn), according to the reaction

$$^{226}Ra \rightarrow \alpha + {}^{222}Rn.$$

The masses of the various nuclei are: ^{226}Ra, 226.0254 u; alpha, 4.0026 u; ^{222}Rn, 222.0175 u. (a) Calculate the Q of the reaction. (b) What value of Q would be obtained if the accurate masses given above were rounded off to three significant figures? What is the kinetic energy of (c) the alpha particle and (d) the radon nucleus? (For this calculation the rounded-off values of the masses *can* be used. Why?)

ELECTRONIC COMPUTATION

78. A 6.00 kg model rocket is traveling horizontally and due south with a speed of 20.0 m/s when it explodes into two pieces. The velocity of one piece, with a mass of 2.00 kg, is

$$\mathbf{v}_1 = (-12.0 \text{ m/s})\mathbf{i} + (30.0 \text{ m/s})\mathbf{j} - (15.0 \text{ m/s})\mathbf{k},$$

with **i** pointing due east, **j** pointing due north, and **k** pointing vertically upward. (a) What is the linear momentum of the other piece, in unit-vector notation? (b) What is the kinetic energy of the other piece? (c) How much kinetic energy is produced by the explosion?

11
Rotation

In judo, a weaker and smaller fighter who understands physics can defeat a stronger and larger fighter who does not. This fact is demonstrated by the basic "hip throw," in which a fighter rotates the fighter's opponent around his hip and—if the throw is successful—onto the mat. Without the proper use of physics, the throw requires considerable strength and can easily fail. *What is the advantage offered by physics❓*

11-1 TRANSLATION AND ROTATION

The graceful movement of a figure skater can be used to illustrate, in an aesthetically pleasing way, two kinds of pure, or unmixed, motion. Figure 11-1a shows a skater gliding across the ice in a straight line with constant speed. Her motion is one of pure **translation.** Figure 11-1b shows her spinning at a constant rate about a fixed vertical axis, in a motion of pure **rotation.** This second kind of motion is our focus in this chapter.

Translation is motion along a straight line, the motion we have discussed almost exclusively so far. Rotation is the motion of wheels, gears, motors, the hands of clocks, the rotors of jet engines, and the blades of helicopters. It is the motion of hurricanes, planets, stars, and galaxies.

11-2 THE ROTATIONAL VARIABLES

In this chapter, we deal with the rotation of a *rigid* body about a *fixed* axis. The first of these restrictions means that we shall not examine the rotation of such objects as the Sun, because the Sun—a ball of gas—is not a rigid body. Our second restriction rules out objects like a bowling ball rolling down a bowling lane. Such a ball is in *rolling* motion, rotating about a *moving* axis.

Figure 11-2 shows a rigid body of arbitrary shape in pure *rotation around a fixed axis,* called the **axis of rotation** or the **rotation axis.** Every point of the body moves in a circle whose center lies on the axis of rotation, and every point moves through the same angle during a particular time interval. (In pure translation, every point of the body moves in a straight line, and every point moves through the same *linear distance* during a particular time interval. Comparisons between linear and angular motion will be a constant part of what follows.)

We deal now—one at a time—with the angular equivalents of the linear quantities position, displacement, velocity, and acceleration.

FIGURE 11-1 Figure skater Kristi Yamaguchi in motion of (*a*) pure translation along a fixed direction and (*b*) pure rotation about a fixed axis.

(*a*)

(*b*)

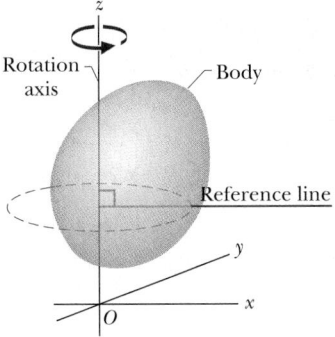

FIGURE 11-2 A rigid body of arbitrary shape in pure rotation about the *z* axis of a coordinate system. The position of the *reference line* with respect to the rigid body is arbitrary, but it is perpendicular to the rotation axis. It is fixed in the body and rotates with the body.

Angular Position

Figure 11-2 also shows a reference line, fixed in the body, perpendicular to the rotation axis, and rotating with the body. We can describe the motion of the rotating body by specifying the **angular position** of this line, that is, the angle of the line relative to a fixed direction. In Fig. 11-3, the angular position θ is measured relative to the positive direction of the x axis, and θ is given by

$$\theta = \frac{s}{r} \quad \text{(radian measure)}. \quad (11\text{-}1)$$

Here s is the length of arc (or the arc distance) along a circle and between the x axis and the reference line, and r is the radius of that circle.

An angle defined in this way is measured in **radians** (rad) rather than in revolutions (rev) or degrees. The radian, being the ratio of two lengths, is a pure number and thus has no dimension. Because the circumference of a circle of radius r is $2\pi r$, there are 2π radians in a complete circle:

$$1 \text{ rev} = 360° = \frac{2\pi r}{r} = 2\pi \text{ rad}, \quad (11\text{-}2)$$

and thus

$$1 \text{ rad} = 57.3° = 0.159 \text{ rev}. \quad (11\text{-}3)$$

We do *not* reset θ to zero with each complete rotation of the reference line about the rotation axis. If the reference line completes two revolutions, then the angular position θ is $\theta = 4\pi$ rad.

For pure translational motion along the x direction, we know all there is to know about a moving body if we know $x(t)$, its position as a function of time. Similarly, for pure rotation, we know all there is to know about a rotating body if we know $\theta(t)$, the angular position of the body's reference line as a function of time.

FIGURE 11-3 The rotating rigid body of Fig. 11-2 in cross section, viewed from above. The plane of the cross section is perpendicular to the rotation axis, which now extends out of the page, toward you. In this position of the body, the reference line makes an angle θ with the x axis.

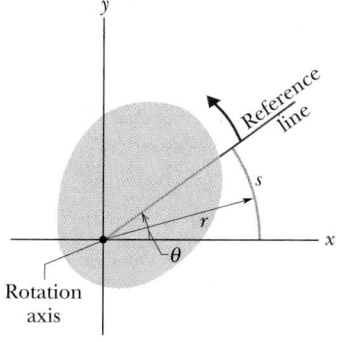

Angular Displacement

If the body of Fig. 11-3 rotates about the rotation axis as in Fig. 11-4, changing the angular position of the reference line from θ_1 to θ_2, the body undergoes an **angular displacement** $\Delta\theta$ given by

$$\Delta\theta = \theta_2 - \theta_1. \quad (11\text{-}4)$$

This definition of angular displacement holds not only for the rigid body as a whole but also for *every particle within that body*.

If a body is in translational motion along an x axis, its displacement Δx is either positive or negative, depending on whether the body is moving in the direction of increasing x or decreasing x. Similarly, the angular displacement $\Delta\theta$ of a rotating body can be either positive or negative, depending on whether the body is rotating in the direction of increasing θ (counterclockwise, as in Figs. 11-3 and 11-4) or decreasing θ (clockwise).

CHECKPOINT **1:** The disk in the figure can rotate about its central axis like a merry-go-round. Which of the following pairs of values for its initial and final angular positions, respectively, give a negative angular displacement: (a) -3 rad, $+5$ rad, (b) -3 rad, -7 rad, (c) 7 rad, -3 rad?

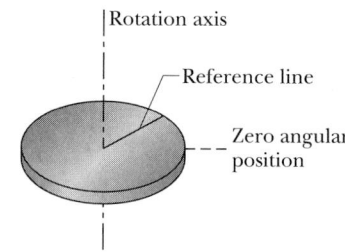

FIGURE 11-4 The reference line of the rigid body of Figs. 11-2 and 11-3 is at angular position θ_1 at time t_1 and at angular position θ_2 at a later time t_2. The quantity $\Delta\theta$ ($= \theta_2 - \theta_1$) is the angular displacement that occurs during the interval Δt ($= t_2 - t_1$). The body itself is not shown.

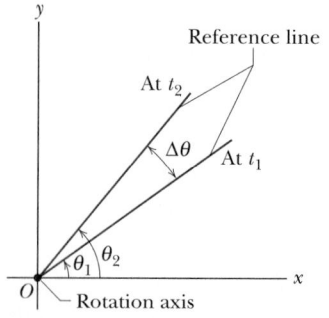

Angular Velocity

Suppose (see Fig. 11-4) that our rotating body is at angular position θ_1 at time t_1 and at angular position θ_2 at time t_2. We define the **average angular velocity** of the body in the time interval Δt from t_1 to t_2 to be

$$\overline{\omega} = \frac{\theta_2 - \theta_1}{t_2 - t_1} = \frac{\Delta\theta}{\Delta t}, \qquad (11\text{-}5)$$

in which $\Delta\theta$ is the angular displacement that occurs during Δt (ω is the lowercase Greek letter omega).

The **(instantaneous) angular velocity** ω, with which we shall be most concerned, is the limit of the ratio in Eq. 11-5 as Δt is made to approach zero. Thus

$$\omega = \lim_{\Delta t \to 0} \frac{\Delta\theta}{\Delta t} = \frac{d\theta}{dt}. \qquad (11\text{-}6)$$

If we know $\theta(t)$, we can find the angular velocity ω by differentiation.

Equations 11-5 and 11-6 hold not only for the rotating rigid body as a whole but also for *every particle of that body*. The unit of angular velocity is commonly the radian per second (rad/s) or the revolution per second (rev/s).

If a particle moves in translation along an x axis, its linear velocity v can be either positive or negative, depending on whether the particle is moving in the direction of increasing x or decreasing x. Similarly, the angular velocity ω of a rotating rigid body can be either positive or negative, depending on whether the body is rotating in the direction of increasing θ (counterclockwise) or of decreasing θ (clockwise). The magnitude of an angular velocity is called the **angular speed,** which is also represented with ω.

Angular Acceleration

If the angular velocity of a rotating body is not constant, then the body has an angular acceleration. Let ω_2 and ω_1 be the angular velocities at times t_2 and t_1, respectively. The **average angular acceleration** of the rotating body in the interval from t_1 to t_2 is defined as

$$\overline{\alpha} = \frac{\omega_2 - \omega_1}{t_2 - t_1} = \frac{\Delta\omega}{\Delta t}, \qquad (11\text{-}7)$$

in which $\Delta\omega$ is the change in the angular velocity that occurs during the time interval Δt. The **(instantaneous) angular acceleration** α, with which we shall be most concerned, is the limit of this quantity as Δt is made to approach zero. Thus

$$\alpha = \lim_{\Delta t \to 0} \frac{\Delta\omega}{\Delta t} = \frac{d\omega}{dt}. \qquad (11\text{-}8)$$

Equations 11-7 and 11-8 hold not only for the rotating rigid body as a whole but also for *every particle of that body*. The unit of angular acceleration is commonly the radian per second-squared (rad/s^2) or the revolution per second-squared (rev/s^2).

SAMPLE PROBLEM 11-1

The angular position of a reference line on a spinning wheel is given by

$$\theta = t^3 - 27t + 4,$$

where t is in seconds and θ is in radians.

(a) Find $\omega(t)$ and $\alpha(t)$.

SOLUTION: To get $\omega(t)$, we differentiate $\theta(t)$ with respect to t:

$$\omega = \frac{d\theta(t)}{dt} = 3t^2 - 27. \qquad \text{(Answer)}$$

To get $\alpha(t)$, we differentiate $\omega(t)$ with respect to t:

$$\alpha = \frac{d\omega(t)}{dt} = \frac{d(3t^2 - 27)}{dt} = 6t. \qquad \text{(Answer)}$$

(b) Do we ever find $\omega = 0$?

SOLUTION: Setting $\omega(t) = 0$ yields

$$0 = 3t^2 - 27,$$

which we solve, finding

$$t = \pm 3 \text{ s}. \qquad \text{(Answer)}$$

That is, the angular velocity is momentarily zero 3 s before and 3 s after our clock reads zero.

(c) Describe the wheel's motion for $t \geq 0$.

SOLUTION: To answer, we examine the expressions for $\theta(t)$, $\omega(t)$, and $\alpha(t)$.

At $t = 0$, the reference line on the wheel is at $\theta = 4$ rad, and the wheel is rotating with an angular velocity of -27 rad/s (that is, *clockwise* at an angular speed of 27 rad/s) and an angular acceleration of zero.

For $0 < t < 3$ s, the wheel continues to rotate clockwise, but at decreasing angular speed, because it now has a positive (counterclockwise) angular acceleration. (Check $\omega(t)$ and $\alpha(t)$ for, say, $t = 2$ s.)

At $t = 3$ s, the wheel stops momentarily ($\omega = 0$) and has rotated as far clockwise as it will ever get (the reference line is now at $\theta = -50$ rad).

For $t > 3$ s, the wheel's angular acceleration continues to increase. Its angular velocity, which is now also counterclockwise, increases rapidly because the signs of ω and α are the same.

SAMPLE PROBLEM 11-2

A child's top is spun with angular acceleration

$$\alpha = 5t^3 - 4t,$$

where the coefficients are in units compatible with seconds and radians. At $t = 0$, the top has angular velocity 5 rad/s, and a reference line on it is at angular position $\theta = 2$ rad.

(a) Obtain an expression for the angular velocity $\omega(t)$ of the top.

SOLUTION: From Eq. 11-8 we have

$$d\omega = \alpha \, dt,$$

which we integrate to get

$$\omega = \int \alpha \, dt = \int (5t^3 - 4t) \, dt$$
$$= \tfrac{5}{4}t^4 - \tfrac{4}{2}t^2 + C.$$

To evaluate the constant of integration C, we note that $\omega = 5$ rad/s at $t = 0$. Substituting these values in our expression for ω yields

$$5 \text{ rad/s} = 0 - 0 + C,$$

so $C = 5$ rad/s. Then

$$\omega = \tfrac{5}{4}t^4 - 2t^2 + 5. \qquad \text{(Answer)}$$

(b) Obtain an expression for the angular position $\theta(t)$ of the top.

SOLUTION: From Eq. 11-6 we have

$$d\theta = \omega \, dt,$$

which we integrate to get

$$\theta = \int \omega \, dt = \int (\tfrac{5}{4}t^4 - 2t^2 + 5) \, dt,$$
$$= \tfrac{1}{4}t^5 - \tfrac{2}{3}t^3 + 5t + C'$$
$$= \tfrac{1}{4}t^5 - \tfrac{2}{3}t^3 + 5t + 2, \qquad \text{(Answer)}$$

where C' has been evaluated by noting that $\theta = 2$ rad at $t = 0$.

11-3 ARE ANGULAR QUANTITIES VECTORS?

We can describe the position, velocity, and acceleration of a single particle by means of vectors. If the particle is confined to a straight line, however, we do not really need the power of vectors. Such a particle has only two directions available to it, and we can designate these with plus and minus signs.

In the same way, a rigid body rotating about a fixed

axis can rotate only clockwise or counterclockwise about this axis, and again we can select between them by means of plus and minus signs. The question arises: "Can we treat the angular displacement, velocity, and acceleration of a rotating body as vectors?" The answer is a qualified "yes" (see caution below, in connection with angular displacements).

Consider the angular velocity. Figure 11-5a shows a phonograph record rotating about a fixed spindle. The record has a fixed rotation rate ω $(= 33\tfrac{1}{3}$ rev/min) and a fixed direction of rotation (clockwise as viewed from above). By convention, we represent its angular velocity as a vector $\boldsymbol{\omega}$ pointing along the axis of rotation, as in Fig. 11-5b. We choose the length of this vector according to some convenient scale, for example, with 1 cm corresponding to 10 rev/min.

We establish a direction for the vector $\boldsymbol{\omega}$ by using a **right-hand rule,** as Fig. 11-5c shows. Curl your right hand about the rotating record, your fingers pointing *in the direction of rotation.* Your extended thumb will then point in the direction of the angular velocity vector. If the record were to rotate in the opposite sense, the right-hand rule would tell you that the angular velocity vector then points in the opposite direction.

It is not easy to get used to representing angular quantities as vectors. We instinctively expect that something should be moving *along* the direction of a vector. That is not the case here. Instead, something (the rigid body) is rotating *around* the direction of the vector. In the world of pure rotation, a vector defines an axis of rotation, not a direction in which something moves. Nonetheless, the

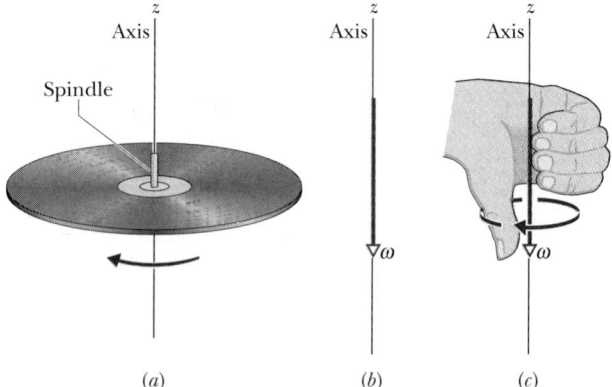

FIGURE 11-5 (a) A record rotating about a vertical axis that coincides with the axis of the spindle. (b) The angular velocity of the rotating record can be represented by the vector $\boldsymbol{\omega}$, lying along the axis and pointing down, as shown. (c) We establish the direction of the angular velocity vector as downward by using a right-hand rule. When the fingers of the right hand curl around the record and point the way it is moving, the extended thumb points in the direction of $\boldsymbol{\omega}$.

vector also defines the motion. Furthermore, it obeys all the rules for vector manipulation discussed in Chapter 3. The angular acceleration $\boldsymbol{\alpha}$ is another vector, and it too obeys those rules.

In this chapter we consider only rotations that are about a fixed axis. For such situations, we need not consider vectors—we can represent angular velocity with ω and angular acceleration with α, and we can indicate a direction with an implied plus sign for counterclockwise or an explicit minus sign for clockwise. In more complicated situations, however, we would use vectors $\boldsymbol{\omega}$ and $\boldsymbol{\alpha}$.

Now for the caution. Angular *displacements* (unless they are very small) *cannot* be treated as vectors. Why not? We can certainly give them both magnitude and direction, just as we did for the angular velocity vector in Fig. 11-5. However, that is (as the mathematicians say) a necessary condition but not sufficient. To be represented as a vector, a quantity must *also* obey the rules of vector addition, one of which says that if you add two vectors, the order in which you add them does not matter. Angular displacements fail this test.

To see this, place a book flat on a table, as in Fig. 11-6a. Now give the book two successive 90° angular displacements, *first* about the (horizontal) x axis and *then* about the (vertical) y axis, using the right-hand rule as a guide to positive rotation in each case.

Now, with a book in the same initial position (Fig. 11-6b), carry out these two angular displacements in the reverse order (that is, *first* about the y axis and *then* about the x axis). As the diagrams show, the book ends up in a very different orientation.

Thus the same two operations produce different results, depending on the order in which you carry them out. Addition of angular displacements is thus not commutative, so angular displacements are not vector quantities. With some practice, you should be able to show that the final positions of the book are much closer together if you use displacements much smaller than 90°. In the limiting case of differential angular displacements (such as $d\theta$ in Eq. 11-6), angular displacements *can* be treated as vectors.

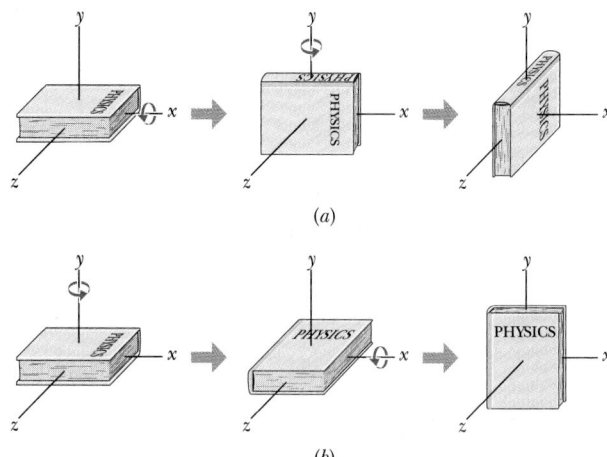

(a)

(b)

FIGURE 11-6 (a) From its initial position on the left, the book is given two successive 90° rotations, first about the (horizontal) x axis and then about the (vertical) y axis. (b) The book is given the same rotations, but in the reverse order. If angular displacement were truly a vector quantity, the order of these displacements would not matter. It clearly does matter, so (large) angular displacements are not vector quantities, even though we can assign magnitude and direction to them.

11-4 ROTATION WITH CONSTANT ANGULAR ACCELERATION

In pure translation, motion with a *constant linear acceleration* (for example, that of a falling body) is an important special case. In Table 2-1, we displayed a series of equations that hold for such motion.

In pure rotation, the case of *constant angular acceleration* is also important, and a parallel set of equations holds for this case also. We shall not derive them here, but simply write them from the corresponding linear equations, substituting equivalent angular quantities for the linear ones. This is done in Table 11-1, which displays both sets of equations (Eqs. 2-11 and 2-15 to 2-18; 11-9 to 11-13). For simplicity, we let $x_0 = 0$ and $\theta_0 = 0$ in these equa-

TABLE 11-1 EQUATIONS OF MOTION FOR CONSTANT LINEAR AND FOR CONSTANT ANGULAR ACCELERATION

EQUATION NUMBER	LINEAR FORMULA	MISSING VARIABLE		ANGULAR FORMULA	EQUATION NUMBER
(2-11)	$v = v_0 + at$	x	θ	$\omega = \omega_0 + \alpha t$	(11-9)
(2-15)	$x = v_0 t + \frac{1}{2}at^2$	v	ω	$\theta = \omega_0 t + \frac{1}{2}\alpha t^2$	(11-10)
(2-16)	$v^2 = v_0^2 + 2ax$	t	t	$\omega^2 = \omega_0^2 + 2\alpha\theta$	(11-11)
(2-17)	$x = \frac{1}{2}(v_0 + v)t$	a	α	$\theta = \frac{1}{2}(\omega_0 + \omega)t$	(11-12)
(2-18)	$x = vt - \frac{1}{2}at^2$	v_0	ω_0	$\theta = \omega t - \frac{1}{2}\alpha t^2$	(11-13)

tions. With those *initial conditions,* a linear displacement Δx ($= x - x_0$) is equal to x, and an angular displacement $\Delta\theta$ ($= \theta - \theta_0$) is equal to θ.

CHECKPOINT **2:** In four situations, a rotating body has angular position $\theta(t)$ given by (a) $\theta = 3t - 4$, (b) $\theta = -5t^3 + 4t^2 + 6$, (c) $\theta = 2/t^2 - 4/t$, and (d) $\theta = 5t^2 - 3$. To which situations do the angular equations of Table 11-1 apply?

SAMPLE PROBLEM 11-3

A grindstone (Fig. 11-7) has a constant angular acceleration $\alpha = 0.35$ rad/s². It starts from rest (that is, $\omega_0 = 0$) with an arbitrary reference line horizontal, at angular position $\theta_0 = 0$.

(a) What is the angular displacement θ of the reference line (hence of the wheel) at $t = 18$ s?

SOLUTION: From Eq. 11-10 of Table 11-1 ($\theta = \omega_0 t + \frac{1}{2}\alpha t^2$), we obtain:

$$\theta = (0)(18 \text{ s}) + (\tfrac{1}{2})(0.35 \text{ rad/s}^2)(18 \text{ s})^2$$
$$= 56.7 \text{ rad} \approx 57 \text{ rad} \approx 3200° \approx 9.0 \text{ rev.} \quad \text{(Answer)}$$

(b) What is the wheel's angular velocity at $t = 18$ s?

SOLUTION: From Eq. 11-9 of Table 11-1 ($\omega = \omega_0 + \alpha t$) we now get:

$$\omega = 0 + (0.35 \text{ rad/s}^2)(18 \text{ s})$$
$$= 6.3 \text{ rad/s} = 360°/\text{s} = 1.0 \text{ rev/s.} \quad \text{(Answer)}$$

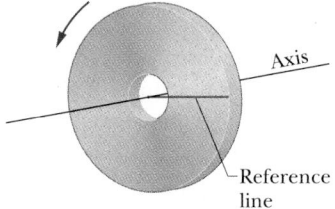

FIGURE 11-7 Sample Problems 11-3 and 11-4. A grindstone. At $t = 0$ the reference line (which we imagine to be marked on the stone) is horizontal.

SAMPLE PROBLEM 11-4

For the grindstone of Sample Problem 11-3, let us assume the same angular acceleration ($\alpha = 0.35$ rad/s²), but let us now assume that the wheel does not start from rest but has an initial angular velocity ω_0 of -4.6 rad/s; that is, the angular acceleration acts initially to slow the wheel (because the signs of α and ω_0 are opposite).

(a) At what time t will the grindstone momentarily stop?

SOLUTION: Solving Eq. 11-9 ($\omega = \omega_0 + \alpha t$) for t yields

$$t = \frac{\omega - \omega_0}{\alpha} = \frac{0 - (-4.6 \text{ rad/s})}{0.35 \text{ rad/s}^2} = 13 \text{ s.} \quad \text{(Answer)}$$

(b) At what time will the grindstone have rotated such that its angular displacement is five revolutions in the positive direction of rotation? (The angular displacement of the reference line will then be $\theta = 5$ rev.)

SOLUTION: The wheel is initially rotating in the negative (clockwise) direction with $\omega_0 = -4.6$ rad/s, but its angular acceleration α is positive (counterclockwise). This initial opposition of the signs of angular velocity and angular acceleration means that the wheel slows in its rotation in the negative direction, stops, and then reverses to rotate in the positive direction. After the reference line comes back through its initial orientation of $\theta = 0$, the wheel must turn an additional five revolutions to reach the angular displacement we want. This is all "taken care of" if we use Eq. 11-10:

$$\theta = \omega_0 t + \tfrac{1}{2}\alpha t^2.$$

Substituting known values and setting $\theta = 5$ rev $= 10\pi$ rad give us

$$10\pi \text{ rad} = (-4.6 \text{ rad/s})t + (\tfrac{1}{2})(0.35 \text{ rad/s}^2)t^2.$$

Note that for t in seconds, the units in this equation are consistent. Dropping units (for convenience) and rearranging give

$$t^2 - 26.3t - 180 = 0. \quad (11\text{-}14)$$

Solving this quadratic equation for t and discarding the negative root, we obtain

$$t = 32 \text{ s.} \quad \text{(Answer)}$$

PROBLEM SOLVING TACTICS

TACTIC 1: *Unexpected Answers*

Do not hasten to throw away one root of a quadratic equation as meaningless. Often, as in Sample Problem 11-4, a discarded root has physical meaning.

The two solutions to Eq. 11-14 are $t = 32$ s and $t = -5.6$ s. We chose the first (positive) solution and ignored the second as perhaps meaningless. But is it? A negative time in this problem simply means a time before $t = 0$, that is, a time before we started to pay attention to what was going on.

Figure 11-8 is a plot of the angular position θ of the reference line on the grindstone of Sample Problem 11-4 as a function of time, for both positive and negative times. It is a plot of Eq. 11-10 ($\theta = \omega_0 t + \frac{1}{2}\alpha t^2$), using as constants $\omega_0 = -4.6$ rad/s and $\alpha = +0.35$ rad/s². Point a corresponds to $t = 0$, at which time we arbitrarily took the angular position of the reference line to be zero. The wheel is moving in the direction of decreasing θ at that time, and continues to do so until coming to rest at point b at $t = 13$ s. It then reverses, with the reference line returning to its $\theta = 0$ position at point c and

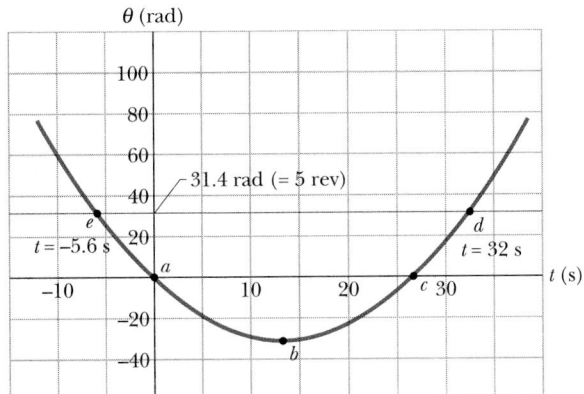

FIGURE 11-8 A plot of angular position versus time for the grindstone of Sample Problem 11-4. Negative times (that is, times before $t = 0$) have been included. The two roots of Eq. 11-14 are indicated by points d and e.

going on for five additional revolutions ($= 31.4$ rad) to point d. This latter point (with $t = 32$ s) represents the root that we accepted as the answer to our problem.

Note, however, that the reference line was at this same angular position at $t = -5.6$ s, before the "official start" of the problem. This root (point e) is just as valid as the root at point d. More important, by asking if this negative root can have any physical meaning, you learn a little more about the rotation of the wheel.

SAMPLE PROBLEM 11-5

During an analysis of a helicopter engine, you determine that the rotor's velocity changes from 320 rev/min to 225 rev/min in 1.50 min as the rotor is slowing to a stop.

(a) What is the average angular acceleration of the rotor blades during this interval?

SOLUTION: From Eq. 11-7,

$$\bar{\alpha} = \frac{\omega - \omega_0}{\Delta t} = \frac{225 \text{ rev/min} - 320 \text{ rev/min}}{1.50 \text{ min}}$$

$$= -63.3 \text{ rev/min}^2. \qquad \text{(Answer)}$$

(b) Assume the angular acceleration α is constant at this average value. How long will the rotor blades take to stop from their initial angular velocity of 320 rev/min?

SOLUTION: Solving Eq. 11-9 ($\omega = \omega_0 + \alpha t$) for t gives

$$t = \frac{\omega - \omega_0}{\alpha} = \frac{0 - 320 \text{ rev/min}}{-63.3 \text{ rev/min}^2}$$

$$= 5.1 \text{ min}. \qquad \text{(Answer)}$$

(c) How many revolutions will the rotor blades make in coming to rest from their initial angular velocity of 320 rev/min?

SOLUTION: Solving Eq. 11-11 ($\omega^2 = \omega_0^2 + 2\alpha\theta$) for θ gives us

$$\theta = \frac{\omega^2 - \omega_0^2}{2\alpha} = \frac{0 - (320 \text{ rev/min})^2}{(2)(-63.3 \text{ rev/min}^2)}$$

$$= 809 \text{ rev}. \qquad \text{(Answer)}$$

11-5 RELATING THE LINEAR AND ANGULAR VARIABLES

In Section 4-7, we discussed uniform circular motion, in which a particle travels at constant linear speed v along a circle and around an axis of rotation. When a rigid body, such as a merry-go-round, turns around an axis of rotation, each particle in the body moves in its own circle around that axis. Since the body is rigid, all the particles make one revolution in the same amount of time; that is, they all have the same angular speed ω.

However, the farther a particle is from the axis, the greater the circumference of its circle is, and so the faster its linear speed v must be. You can notice this on a merry-go-round. You turn with the same angular speed ω regardless of your distance from the center, but your linear speed v increases noticeably if you move to the outside edge of the merry-go-round.

We often need to relate the linear variables s, v, and a for a particular point in a rotating body to the angular variables θ, ω, and α for that body. The two sets of variables are related by r, the *perpendicular distance* of the point from the rotation axis. This perpendicular distance is the distance between the point and the rotation axis, measured along a perpendicular to the axis. It is also the radius r of the circle traveled by the point around the axis of rotation.

The Position

If a reference line on a rigid body is rotated through an angle θ, a point within the body is moved a distance s along a circular arc, where s is given by Eq. 11-1:

$$s = \theta r \qquad \text{(radian measure)}. \qquad (11\text{-}15)$$

This is the first of our linear–angular relations. The angle θ must be measured in radians because Eq. 11-15 is itself the definition of angular measure in radians.

The Speed

Differentiating Eq. 11-15 with respect to time—with r held constant—leads to

$$\frac{ds}{dt} = \frac{d\theta}{dt} r.$$

But ds/dt is the linear speed (the magnitude of the linear velocity) of the point in question and $d\theta/dt$ is the angular speed ω of the rotating body, so

$$v = \omega r \qquad \text{(radian measure).} \qquad (11\text{-}16)$$

Again, the angular speed ω must be expressed in radian measure. Equation 11-16 tells us that since all points within the rigid body have the same angular speed ω, points with greater radius r have greater linear speed v. Figure 11-9a shows that the linear velocity is always tangent to the circular path of the point in question.

If the angular speed ω of the rigid body is constant, then Eq. 11-16 tells us that the linear speed v for any point within it is also constant. Thus each point within the body undergoes uniform circular motion. The period of revolution T for the motion of each point and for the rigid body itself is given by Eq. 4-23:

$$T = \frac{2\pi r}{v}. \qquad (11\text{-}17)$$

This equation tells us that the time for one revolution is the distance $2\pi r$ traveled in one revolution divided by the speed at which that distance is traveled. Substituting for v from Eq. 11-16 and canceling r, we find also that

$$T = \frac{2\pi}{\omega} \qquad \text{(radian measure).} \qquad (11\text{-}18)$$

This equivalent equation says that the time for one revolution is the angle 2π rad traversed in one revolution divided by the angular speed at which that angle is traversed.

The Acceleration

Differentiating Eq. 11-16 with respect to time—again with r held constant—leads to

$$\frac{dv}{dt} = \frac{d\omega}{dt}r. \qquad (11\text{-}19)$$

Here we run up against a complication. In Eq. 11-19, dv/dt represents only the part of the linear acceleration that is responsible for changes in the *magnitude* v of the linear velocity $\mathbf{v}$. Like $\mathbf{v}$, that part of the linear acceleration is tangent to the path of the point in question. We call it the *tangential component* a_t of the linear acceleration of the point, and we write

$$a_t = \alpha r \qquad \text{(radian measure),} \qquad (11\text{-}20)$$

where $\alpha = d\omega/dt$.

In addition, as Eq. 4-22 tells us, a particle (or point) moving in a circular path has a *radial component* of linear acceleration, $a_r = v^2/r$ (radially inward), that is responsi-

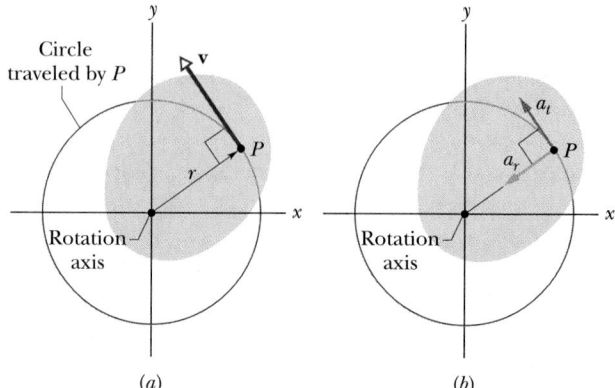

FIGURE 11-9 The rotating rigid body of Fig. 11-2, shown in cross section. Every point of the body (such as P) moves in a circle around the rotation axis. (a) The linear velocity $\mathbf{v}$ of every point is tangent to the circle in which the point moves. (b) The linear acceleration $\mathbf{a}$ of the point has (in general) two components: a tangential component a_t and a radial component a_r.

ble for changes in the *direction* of the linear velocity $\mathbf{v}$. By substituting for v from Eq. 11-16, we can write this component as

$$a_r = \frac{v^2}{r} = \omega^2 r \qquad \text{(radian measure).} \qquad (11\text{-}21)$$

Thus, as Fig. 11-9b shows, the linear acceleration of a point on a rotating rigid body has, in general, two components. The radially inward component a_r (given by Eq. 11-21) is present unless the angular velocity of the body is zero. The tangential component a_t (given by Eq. 11-20) is present unless the angular acceleration is zero.

CHECKPOINT **3:** A cockroach rides the rim of a rotating merry-go-round. If the angular speed of this system (*merry-go-round + cockroach*) is constant, does the cockroach have (a) radial acceleration and (b) tangential acceleration? If the angular speed is decreasing, does the cockroach have (c) radial and (d) tangential acceleration?

SAMPLE PROBLEM 11-6

Figure 11-10 shows a centrifuge used to accustom astronaut trainees to high accelerations. The radius r of the circle traveled by an astronaut is 15 m.

(a) At what constant angular velocity must the centrifuge rotate if the astronaut is to be subject to a linear acceleration that is $11g$?

SOLUTION: Because the angular velocity is constant, the angular acceleration α ($= d\omega/dt$) is zero and so is the tangential component of the linear acceleration (see Eq. 11-20). This leaves only the radial component. From Eq. 11-21 ($a_r = \omega^2 r$), with $a_r = 11g$, we have

$$\omega = \sqrt{\frac{a_r}{r}} = \sqrt{\frac{(11)(9.8 \text{ m/s}^2)}{15 \text{ m}}}$$

$$= 2.68 \text{ rad/s} \approx 26 \text{ rev/min}. \qquad \text{(Answer)}$$

(b) What is the tangential acceleration of the astronaut if the centrifuge accelerates uniformly from rest to the angular velocity of (a) in 120 s?

SOLUTION: Since the angular acceleration is constant during the speed-up of the centrifuge, Eq. 11-9 applies and we get

$$\alpha = \frac{\omega - \omega_0}{t} = \frac{2.68 \text{ rad/s} - 0}{120 \text{ s}} = 0.0223 \text{ rad/s}^2.$$

With Eq. 11-20, we then find

$$a_t = \alpha r = (0.0223 \text{ rad/s}^2)(15 \text{ m})$$

$$= 0.33 \text{ m/s}^2. \qquad \text{(Answer)}$$

Although the final radial acceleration a_r ($= 11g$) is large (and alarming), the tangential acceleration a_t ($= 0.034g$) during the speed-up is not.

FIGURE 11-10 Sample Problem 11-6. A centrifuge in Cologne, Germany, is used to accustom astronauts to the large acceleration experienced during a liftoff.

PROBLEM SOLVING TACTICS

TACTIC 2: *Units for Angular Variables*

In Eq. 11-1 ($\theta = s/r$), we committed ourselves to the use of radian measure for all angular variables whenever we are using equations that contain both angular and linear variables. That is, we must express angular displacements in radians, angular velocities in rad/s and rad/min, and angular accelerations in rad/s^2 and rad/min^2. Equations 11-15, 11-16, 11-18, 11-20, and 11-21 are marked to emphasize this. The only exceptions to this rule are equations that involve *only* angular

variables, such as the angular equations listed in Table 11-1. Here you are free to use any unit you wish for the angular variables. That is, you may use radians, degrees, or revolutions, as long as you use them consistently.

In equations where radian measure must be used, you need not keep track of the unit "radian" (rad) algebraically, as you must do for other units. You can add or delete it at will, to suit the context. In Sample Problem 11-6a the unit was added to the answer; in Sample Problem 11-6b it was omitted from the answer.

11-6 KINETIC ENERGY OF ROTATION

The rapidly rotating blade of a table saw certainly has kinetic energy. How can we express it? We cannot use the familiar formula $K = \frac{1}{2}mv^2$ directly because it applies only to particles, and we would not know what to use for v.

Instead, we shall treat the table saw (and any other rotating rigid body) as a collection of particles —all with different speeds. We can then add up the kinetic energies of these particles to find the kinetic energy of the body as a whole. In this way we obtain, for the kinetic energy of a rotating body,

$$K = \tfrac{1}{2}m_1 v_1^2 + \tfrac{1}{2}m_2 v_2^2 + \tfrac{1}{2}m_3 v_3^2 + \cdots$$
$$= \sum \tfrac{1}{2}m_i v_i^2, \qquad (11\text{-}22)$$

in which m_i is the mass of the ith particle and v_i is its speed. The sum is taken over all the particles in the body.

The problem with Eq. 11-22 is that v_i is not the same for all particles. We solve this problem by substituting for v from Eq. 11-16 ($v = \omega r$), so that we have

$$K = \sum \tfrac{1}{2}m_i(\omega r_i)^2 = \tfrac{1}{2}\left(\sum m_i r_i^2\right)\omega^2, \quad (11\text{-}23)$$

in which ω *is* the same for all particles.

The quantity in parentheses on the right side of Eq. 11-23 tells us how the mass of the rotating body is distributed about its axis of rotation. We call that quantity the **rotational inertia** (or **moment of inertia**) I of the body with respect to the axis of rotation. It is a constant for a particular rigid body and for a particular rotation axis. (That axis must always be specified if the value of I is to be meaningful.)

We may now write

$$I = \sum m_i r_i^2 \qquad \text{(rotational inertia)} \quad (11\text{-}24)$$

and substitute into Eq. 11-23, obtaining

$$K = \tfrac{1}{2}I\omega^2 \qquad \text{(radian measure)} \quad (11\text{-}25)$$

as the expression we seek. Because we have used the relation $v = \omega r$ in deriving Eq. 11-25, ω must be expressed in radian measure. The SI unit for I is the kilogram–square meter ($\mathrm{kg \cdot m^2}$).

Equation 11-25, which gives the kinetic energy of a rigid body in pure rotation, is the angular equivalent of the formula $K = \frac{1}{2}Mv_{cm}^2$, which gives the kinetic energy of a rigid body in pure translation. In each case, there is a factor of $\frac{1}{2}$. Where mass M (which can be called the *translational inertia*) appears in one formula, I (the *rotational inertia*) appears in the other. Finally, each equation contains as a factor the square of a speed—translational or rotational as appropriate. The kinetic energies of translation and of rotation are not different kinds of energy. They are both kinetic energy, expressed in ways that are appropriate to the motion at hand.

The rotational inertia of a rotating body depends not only on its mass but also on how that mass is distributed with respect to the rotation axis. Figure 11-11 suggests a convincing way to develop a physical feeling for rotational inertia. Figure 11-11a shows the exterior of either of two plastic rods that outwardly appear to be identical. The dimensions and the weights of the two rods are the same, and both rods balance at their midpoints. If you grasp each rod at its center and move it back and forth rapidly in translational motion, you still cannot tell them apart.

However, a truly striking difference appears if, with a twisting wrist motion, you twist the rods (like a baton) rapidly in back-and-forth angular motion. One rod twists quite easily; the other does not. As Figs. 11-11b and 11-11c show, the "easy" rod has two internal weights near its center and the "hard" rod has one at each end. Although the masses of the rods are equal, their rotational inertias about a central axis are quite different because their mass distributions are different.

CHECKPOINT **4:** The figure shows three masses that rotate about a vertical axis. The perpendicular distance between the axis and the center of each mass is given. Rank the three masses according to their rotational inertia about that axis, greatest first.

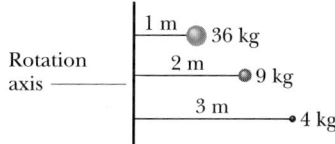

11-7 CALCULATING THE ROTATIONAL INERTIA

If a rigid body is made up of discrete particles, we can calculate its rotational inertia from Eq. 11-24. If the body is continuous, we can replace the sum in Eq. 11-24 with an integral, and the definition of rotational inertia becomes

$$I = \int r^2\, dm \qquad \text{(rotational inertia, continuous).} \qquad (11\text{-}26)$$

In the sample problems that follow this section, we calculate I for bodies of both kinds. In general, the rotational inertia of any rigid body with respect to a rotation axis depends on (1) the shape of the body, (2) the perpendicular distance from the axis to the body's center of mass, and (3) the orientation of the body with respect to the axis.

Table 11-2 gives the rotational inertias of several common bodies, about various axes. Note how the distribution of mass relative to the rotational axis affects the value of the rotational inertia I. For example, the rod in (f) has more mass far from the rotational axis than does the equally long rod in (e). So, if the rods have the same mass M, the rod in (f) has the greater rotational inertia.

The Parallel-Axis Theorem

If you know the rotational inertia of a body about any axis that passes through its center of mass, you can find its rotational inertia about any other axis parallel to that axis with the **parallel-axis theorem:**

$$I = I_{cm} + Mh^2 \qquad \text{(parallel-axis theorem).} \qquad (11\text{-}27)$$

Here M is the mass of the body and h is the perpendicular

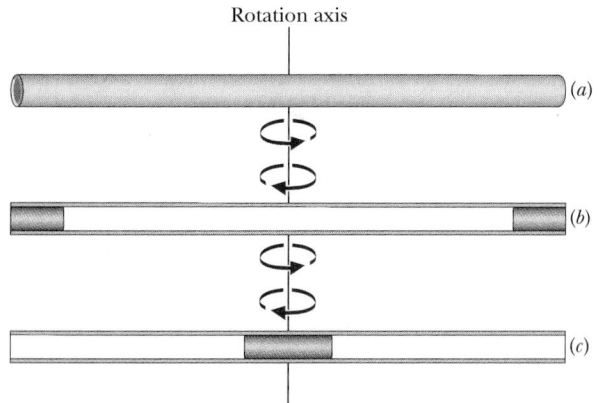

FIGURE 11-11 (*a*) Two plastic rods like this seem identical until you try to twist them back and forth rapidly about their midpoints. Rod (*c*) twists readily; rod (*b*) does not. Although both rods have the same mass, the rotational inertia of rod (*b*) about an axis through its midpoint is considerably greater than that of rod (*c*) because of the different internal distributions of mass.

TABLE 11-2 **SOME ROTATIONAL INERTIAS**

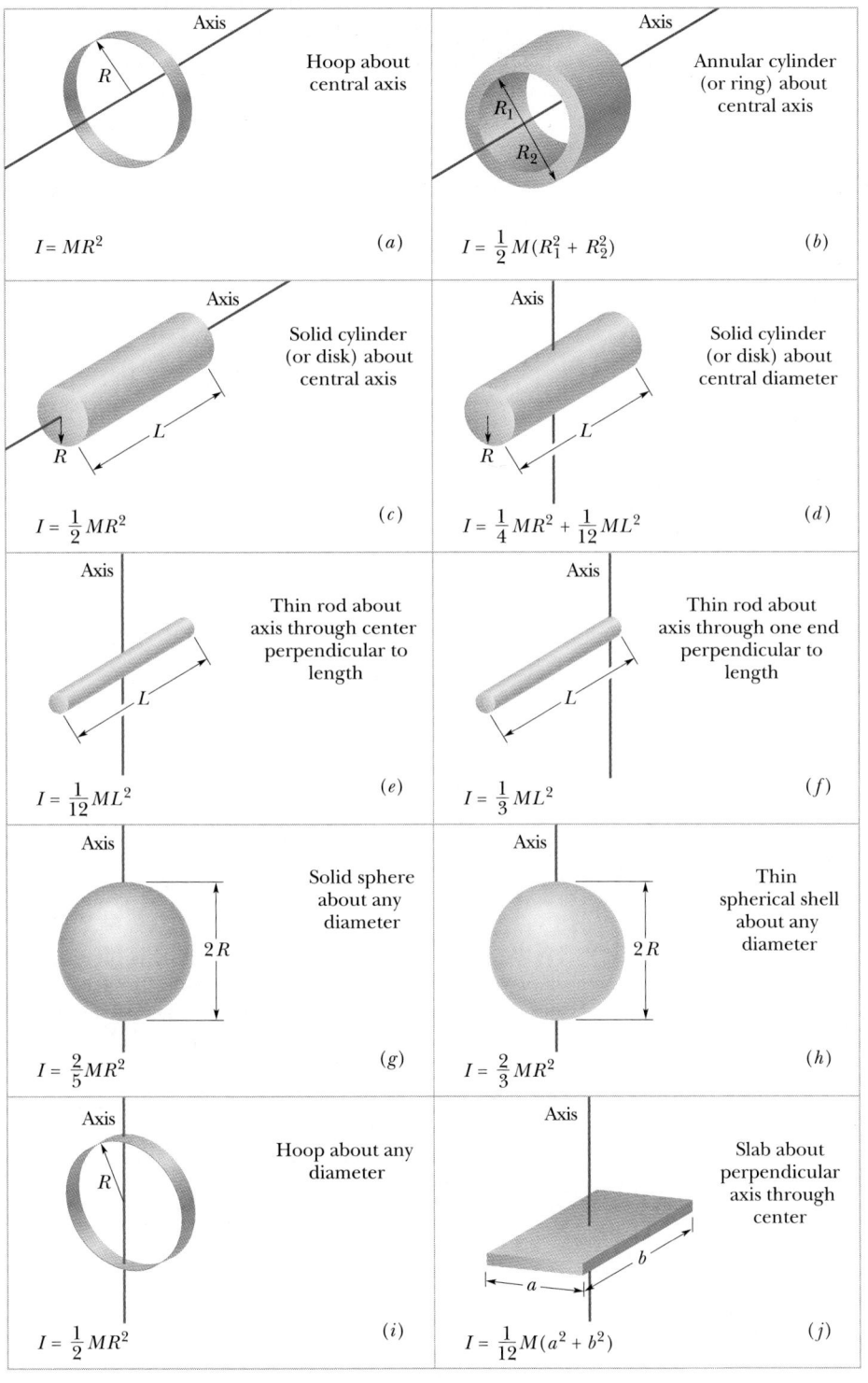

Hoop about
central axis

$I = MR^2$ (a)

Annular cylinder
(or ring) about
central axis

$I = \frac{1}{2} M (R_1^2 + R_2^2)$ (b)

Solid cylinder
(or disk) about
central axis

$I = \frac{1}{2} MR^2$ (c)

Solid cylinder
(or disk) about
central diameter

$I = \frac{1}{4} MR^2 + \frac{1}{12} ML^2$ (d)

Thin rod about
axis through center
perpendicular to
length

$I = \frac{1}{12} ML^2$ (e)

Thin rod about
axis through one end
perpendicular to
length

$I = \frac{1}{3} ML^2$ (f)

Solid sphere
about any
diameter

$I = \frac{2}{5} MR^2$ (g)

Thin
spherical shell
about any
diameter

$I = \frac{2}{3} MR^2$ (h)

Hoop about any
diameter

$I = \frac{1}{2} MR^2$ (i)

Slab about
perpendicular
axis through
center

$I = \frac{1}{12} M (a^2 + b^2)$ (j)

distance between the two (parallel) axes. In words, this theorem can be stated as follows:

The rotational inertia of a body about any axis is equal to the rotational inertia ($= Mh^2$) it would have about that axis if all its mass were concentrated at its center of mass *plus* its rotational inertia ($= I_{cm}$) about a parallel axis through its center of mass.

Proof of the Parallel-Axis Theorem

Let O be the center of mass of the arbitrarily shaped body shown in cross section in Fig. 11-12. Place the origin of coordinates at O. Consider an axis through O perpendicular to the plane of the figure, and another axis through point P parallel to the first axis. Let the coordinates of P be a and b.

Let dm be a mass element with coordinates x and y. The rotational inertia of the body about the axis through P is then, from Eq. 11-26,

$$I = \int r^2 \, dm = \int [(x-a)^2 + (y-b)^2] \, dm,$$

which we can rearrange as

$$I = \int (x^2 + y^2) \, dm - 2a \int x \, dm$$

$$- 2b \int y \, dm + \int (a^2 + b^2) \, dm. \quad (11\text{-}28)$$

From the definition of the center of mass (Eq. 9-9), the middle two integrals of Eq. 11-28 give the coordinates of the center of mass (multiplied by a constant) and thus must

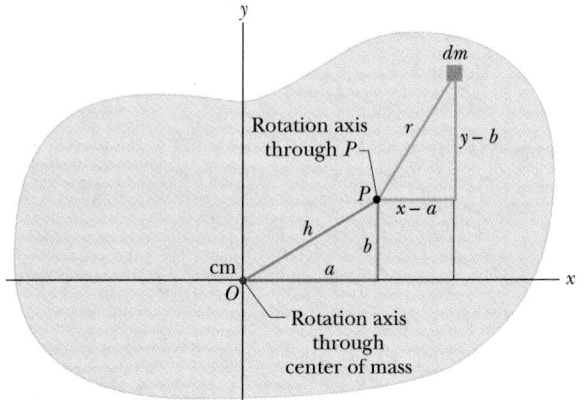

FIGURE 11-12 A rigid body in cross section, with its center of mass at O. The parallel-axis theorem (Eq. 11-27) relates the rotational inertia of the body about an axis through O to that about a parallel axis through a point such as P, a distance h from the body's center of mass. Both axes are perpendicular to the plane of the figure.

each be zero. Because $x^2 + y^2$ is equal to R^2, where R is the distance from O to dm, the first integral is simply I_{cm}, the rotational inertia of the body about an axis through its center of mass. Inspection of Fig. 11-12 shows that the last term in Eq. 11-28 is Mh^2, where M is the total mass. Thus Eq. 11-28 reduces to Eq. 11-27, which is the relation that we set out to prove.

CHECKPOINT **5:** The figure shows a booklike object (one side is longer than the other) and four choices of rotation axes, all perpendicular to the face of the object. Rank the choices according to the rotational inertia of the object about the axis, greatest first.

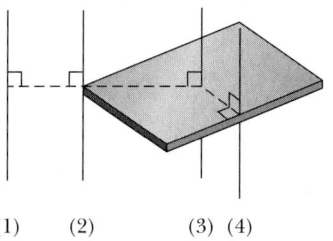

(1)　　(2)　　　　(3) (4)

SAMPLE PROBLEM 11-7

Figure 11-13 shows a rigid body consisting of two particles of mass m connected by a rod of length L and negligible mass.

(a) What is the rotational inertia of this body about an axis through its center, perpendicular to the rod (see Fig. 11-13a)?

SOLUTION: From Eq. 11-24 we have

$$I = \sum m_i r_i^2 = (m)(\tfrac{1}{2}L)^2 + (m)(\tfrac{1}{2}L)^2$$

$$= \tfrac{1}{2}mL^2. \qquad \text{(Answer)}$$

(b) What is the rotational inertia of the body about an axis through one end of the rod and parallel to the first axis, as in Fig. 11-13b?

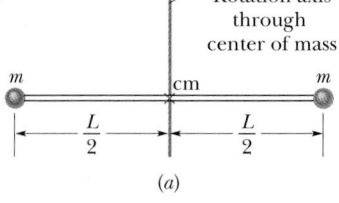

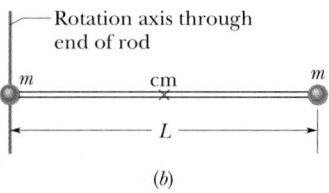

FIGURE 11-13 Sample Problem 11-7. A rigid body that consists of two particles of mass m which are joined by a rod of negligible mass.

SOLUTION: We can use the parallel-axis theorem of Eq. 11-27. We have just calculated I_{cm} in (a), and the distance h between the parallel axes is half the length of the rod. Thus, from Eq. 11-27,

$$I = I_{cm} + Mh^2 = \tfrac{1}{2}mL^2 + (2m)(\tfrac{1}{2}L)^2$$
$$= mL^2. \qquad \text{(Answer)}$$

We can check this result by direct calculation, using Eq. 11-24:

$$I = \sum m_i r_i^2 = (m)(0)^2 + (m)(L)^2$$
$$= mL^2. \qquad \text{(Answer)}$$

SAMPLE PROBLEM 11-8

Figure 11-14 shows a thin, uniform rod of mass M and length L.

(a) What is its rotational inertia about an axis perpendicular to the rod, through its center of mass?

SOLUTION: We place the rod on an x axis, with its center of mass at the origin, and we choose as a mass element a slice dx of the rod. The center of the slice is a distance x from the rotation axis. The mass per unit length of the rod is M/L, so the mass dm of the element dx is

$$dm = \left(\frac{M}{L}\right) dx.$$

From Eq. 11-26 we have

$$I = \int r^2 \, dm = \int_{x=-L/2}^{x=+L/2} x^2 \left(\frac{M}{L}\right) dx$$
$$= \frac{M}{3L}\left[x^3\right]_{-L/2}^{+L/2} = \frac{M}{3L}\left[\left(\frac{L}{2}\right)^3 - \left(-\frac{L}{2}\right)^3\right]$$
$$= \tfrac{1}{12}ML^2. \qquad \text{(Answer)}$$

This agrees with the result given in Table 11-2(e).

(b) What is the rotational inertia of the rod about an axis perpendicular to the rod through one end?

SOLUTION: We combine the result in (a) with the parallel-axis theorem (Eq. 11-27), obtaining

$$I = I_{cm} + Mh^2$$
$$= \tfrac{1}{12}ML^2 + (M)(\tfrac{1}{2}L)^2 = \tfrac{1}{3}ML^2, \qquad \text{(Answer)}$$

which agrees with the result given in Table 11-2(f).

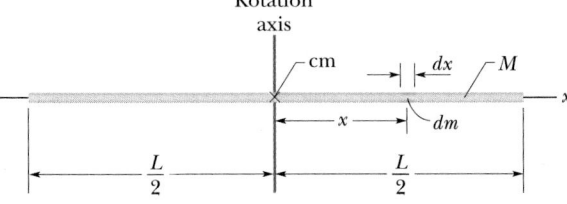

FIGURE 11-14 Sample Problem 11-8. A uniform rod of length L and mass M.

SAMPLE PROBLEM 11-9

A hydrogen chloride molecule consists of a hydrogen atom whose mass m_H is 1.01 u and a chlorine atom whose mass m_{Cl} is 35.0 u. The centers of the two atoms are a distance $d = 1.27 \times 10^{-10}$ m = 127 pm apart (Fig. 11-15). What is the rotational inertia of the molecule about an axis perpendicular to the line joining the two atoms and passing through the center of mass of the molecule?

SOLUTION: Let x be the distance from the center of mass of the molecule to the chlorine atom. Then from Fig. 11-15 and Eq. 9-3, we see that

$$0 = \frac{-m_{Cl}x + m_H(d-x)}{m_{Cl} + m_H},$$

or

$$m_{Cl}x = m_H(d - x).$$

This yields

$$x = \frac{m_H}{m_{Cl} + m_H} d. \qquad (11\text{-}29)$$

From Eq. 11-24, the rotational inertia about an axis through the center of mass is

$$I = \sum m_i r_i^2 = m_H(d - x)^2 + m_{Cl}x^2.$$

Substituting for x from Eq. 11-29 leads, after some algebra, to

$$I = d^2 \frac{m_H m_{Cl}}{m_{Cl} + m_H} = (127 \text{ pm})^2 \frac{(1.01 \text{ u})(35.0 \text{ u})}{35.0 \text{ u} + 1.01 \text{ u}}$$
$$= 15{,}800 \text{ u} \cdot \text{pm}^2. \qquad \text{(Answer)}$$

These units are convenient enough when dealing with the rotational inertia of molecules. If we use the angstrom as a length unit (1 Å = 10^{-10} m), the answer above becomes $I = 1.58$ u·Å², an even more convenient unit.

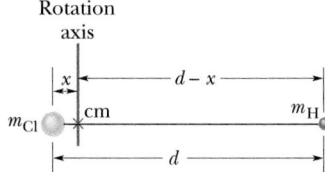

FIGURE 11-15 Sample Problem 11-9. A hydrogen chloride molecule, shown schematically. A rotation axis extends through its center of mass and is perpendicular to the line joining the two atomic centers.

SAMPLE PROBLEM 11-10

With modern technology, it is possible to construct a flywheel that stores enough energy to run an automobile. The energy is stored as rotational kinetic energy when the flywheel is initially made to spin by a machine. The stored energy is then gradually transferred to the automobile by a gear system as the automobile is being driven. Suppose that such a wheel is a solid cylinder whose mass M is 75 kg and whose radius R is

25 cm. If the wheel is spun at 85,000 rev/min, how much rotational kinetic energy can it store?

SOLUTION: The rotational inertia of the cylindrical wheel follows from Table 11-2(c):

$$I = \tfrac{1}{2}MR^2 = (\tfrac{1}{2})(75 \text{ kg})(0.25 \text{ m})^2 = 2.34 \text{ kg} \cdot \text{m}^2.$$

The angular velocity of the wheel is

$$\omega = (85{,}000 \text{ rev/min})(2\pi \text{ rad/rev})(1 \text{ min/60 s})$$
$$= 8900 \text{ rad/s}.$$

From Eq. 11-25, the kinetic energy of rotation is then

$$K = \tfrac{1}{2}I\omega^2 = (\tfrac{1}{2})(2.34 \text{ kg} \cdot \text{m}^2)(8900 \text{ rad/s})^2$$
$$= 9.3 \times 10^7 \text{ J} = 26 \text{ kW} \cdot \text{h}. \qquad \text{(Answer)}$$

This amount of energy, used with the expected efficiency, would take a small car about 200 mi.

11-8 TORQUE

A doorknob is located as far as possible from the door's hinge line for a good reason. If you want to open a heavy door you must certainly apply a force; that alone, however, is not enough. Where you apply that force and in what direction you push are also important. If you apply your force nearer to the hinge line than the knob, or at any angle other than 90° to the plane of the door, you must use a greater force to move the door than if you apply the force at the knob and perpendicular to the door's plane.

Figure 11-16a shows a cross section of a body that is free to rotate about an axis passing through O and perpendicular to the cross section. A force $\mathbf{F}$ is applied at point P, whose position relative to O is defined by a position vector $\mathbf{r}$. Vectors $\mathbf{F}$ and $\mathbf{r}$ make an angle ϕ with each other. (For simplicity, we consider only forces that have no compo-

nent parallel to the rotation axis; thus, $\mathbf{F}$ is in the plane of the page.)

To determine how $\mathbf{F}$ results in a rotation of the body around the rotation axis, we resolve $\mathbf{F}$ into two components (Fig. 11-16b). One component, called the *radial component* F_r, points along $\mathbf{r}$. This component does not cause rotation, because it acts along a line that extends through O. (If you pull on a door parallel to the plane of the door, you do not rotate the door.) The other component of $\mathbf{F}$, called the *tangential component* F_t, is perpendicular to $\mathbf{r}$ and has magnitude $F_t = F \sin \phi$. This component *does* cause rotation. (If you pull on a door perpendicular to its plane, you rotate the door.)

The ability of $\mathbf{F}$ to rotate the body depends not only on the magnitude of its tangential component F_t, but also on just how far from O it is applied. To include both these factors, we define a quantity called **torque** τ as the product of the two factors and write it as

$$\tau = (r)(F \sin \phi). \qquad (11\text{-}30)$$

Two equivalent ways of computing the torque are

$$\tau = (r)(F \sin \phi) = rF_t \qquad (11\text{-}31)$$

and

$$\tau = (r \sin \phi)(F) = r_\perp F, \qquad (11\text{-}32)$$

where $r_\perp$ is the perpendicular distance between the rotation axis at O and an extended line running through the vector $\mathbf{F}$ (Fig. 11-16c). This extended line is called the **line of action** of $\mathbf{F}$, and $r_\perp$ is called the **moment arm** of $\mathbf{F}$. Figure 11-16b shows that we can describe r, the magnitude of $\mathbf{r}$, as being the moment arm of the force component F_t.

Torque, which comes from the Latin word meaning "to twist," may be loosely identified as the turning or twisting action of the force $\mathbf{F}$. When you apply a force to an object—such as a screwdriver or pipe wrench—with the

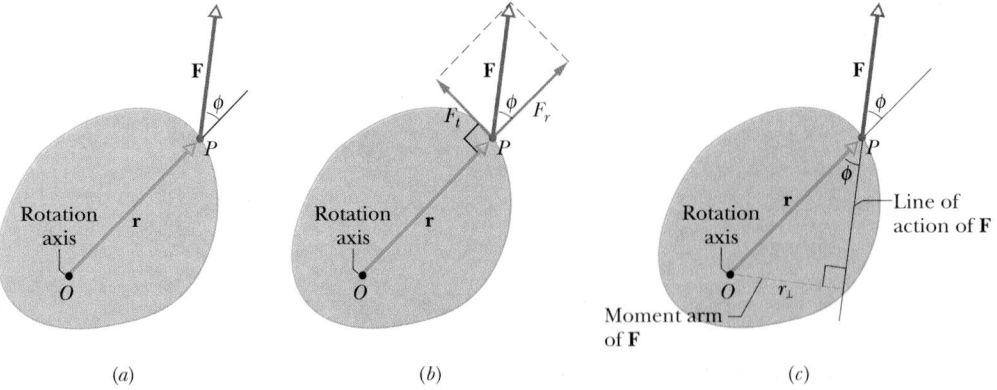

(a) $\qquad\qquad\qquad$ (b) $\qquad\qquad\qquad$ (c)

FIGURE 11-16 (a) A force $\mathbf{F}$ acts at point P on a rigid body that is free to rotate about an axis through O; the axis is perpendicular to the plane of the cross section shown here. (b) The torque exerted by this force is $(r)(F \sin \phi)$. We can also write it as rF_t, where F_t is the tangential component of $\mathbf{F}$. (c) The torque can also be written as $r_\perp F$, where $r_\perp$ is the moment arm of $\mathbf{F}$.

purpose of turning that object, you are applying a torque. The SI unit of torque is the newton-meter (N·m).* A torque τ is positive if it tends to rotate the body counterclockwise, in the direction of increasing θ, as in Fig. 11-16. It is negative if it tends to rotate the body clockwise.

The definition of torque in Eq. 11-30 can be rewritten as a *vector cross product:*

$$\tau = \mathbf{r} \times \mathbf{F}. \qquad (11\text{-}33)$$

Thus torque τ is a vector that is directed perpendicular to the plane containing $\mathbf{r}$ and $\mathbf{F}$. The magnitude of τ is given by Eqs. 11-30, 11-31, and 11-32, and its direction is given by the right-hand rule. We shall make use of Eq. 11-33 in Chapter 12.

CHECKPOINT **6:** The figure shows an overhead view of a meter stick that can pivot about the dot at the position marked 20 (for 20 cm). All five horizontal forces on the stick have the same magnitude. Rank those forces according to the magnitude of the torque that they produce, greatest first.

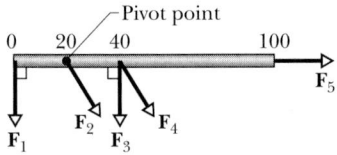

11-9 NEWTON'S SECOND LAW FOR ROTATION

Figure 11-17 shows a simple case of rotation about a fixed axis. The rotating rigid body consists of a single particle of mass m fastened to the end of a massless rod of length r. A force $\mathbf{F}$ acts as shown, causing the particle to move in a circle about the axis. The particle has a tangential component of acceleration a_t governed by Newton's second law:

$$F_t = ma_t.$$

The torque acting on the particle is, from Eq. 11-31,

$$\tau = F_t r = ma_t r.$$

From Eq. 11-20 ($a_t = \alpha r$) we can write this as

$$\tau = m(\alpha r)r = (mr^2)\alpha. \qquad (11\text{-}34)$$

The quantity in parentheses on the right side of Eq. 11-34 is the rotational inertia of the particle about the rotation axis

---

*The newton-meter is also the unit of work. Torque and work, however, are quite different quantities and must not be confused. Work is often expressed in joules (1 J = 1 N·m), but torque never is.

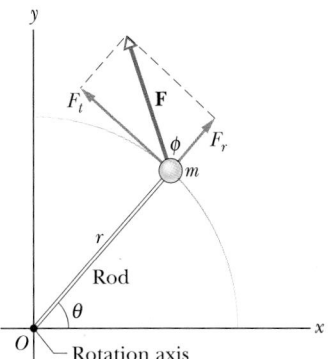

FIGURE 11-17 A simple rigid body, free to rotate about an axis through O, consists of a particle of mass m fastened to the end of a rod of length r and negligible mass. An applied force $\mathbf{F}$ causes the body to rotate.

(see Eq. 11-24). So, Eq. 11-34 reduces to

$$\tau = I\alpha \qquad \text{(radian measure)}. \qquad (11\text{-}35)$$

For the situation in which more than one force is applied to the particle, we can extend Eq. 11-35 as

$$\sum \tau = I\alpha \qquad \text{(radian measure)}, \qquad (11\text{-}36)$$

where $\Sigma\tau$ is the net torque (the sum of all external torques) acting on the particle. Equation 11-36 is the angular (or rotational) form of Newton's second law.

Although we derived Eq. 11-35 for the special case of a single particle rotating about a fixed axis, it holds for any rigid body rotating about a fixed axis, because any such body can be analyzed as an assembly of single particles.

CHECKPOINT **7:** The figure shows an overhead view of a meter stick that can pivot about the point indicated, which is to the left of the stick's midpoint. Two horizontal forces, $\mathbf{F}_1$ and $\mathbf{F}_2$, are applied to the stick. Only $\mathbf{F}_1$ is shown. $\mathbf{F}_2$ is perpendicular to the stick and is applied at the right end. If the stick is not to turn, (a) what should be the direction of $\mathbf{F}_2$, and (b) should F_2 be greater than, less than, or equal to F_1?

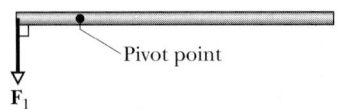

SAMPLE PROBLEM 11-11

Figure 11-18a shows a uniform disk, whose mass M is 2.5 kg and whose radius R is 20 cm, mounted on a fixed horizontal axle. A block whose mass m is 1.2 kg hangs from a massless cord that is wrapped around the rim of the disk. Find the acceleration of the falling block (assuming that it does fall), the angular acceleration of the disk, and the tension in the cord. The cord does not slip, and there is no friction at the axle.

SOLUTION: Figure 11-18b is a free-body diagram for the block. We assume the block accelerates downward, so the magnitude mg of its weight must exceed the tension T in the cord. From Newton's second law, we have

$$T - mg = ma. \qquad (11\text{-}37)$$

Figure 11-18c is a free-body diagram for the disk. The torque acting on the disk is $-TR$, negative because it rotates the disk clockwise. From Table 11-2(c), we know that the rotational inertia I of the disk is $\frac{1}{2}MR^2$. (Two other forces also act on the disk, its weight $M\mathbf{g}$ and the normal force $\mathbf{N}$ exerted on the disk by its support. Since, however, both these forces act at the axis of the disk, they exert no torque on the disk.) Applying Newton's second law in angular form ($\tau = I\alpha$) to the disk, we obtain

$$-TR = \tfrac{1}{2}MR^2\alpha.$$

Because the cord does not slip, we assume that the linear acceleration a of the block and the (tangential) linear acceleration a_t of the rim of the disk are equal. Then, by Eq. 11-20, $\alpha = a/R$, and the equation above reduces to

$$T = -\tfrac{1}{2}Ma. \qquad (11\text{-}38)$$

Combining Eqs. 11-37 and 11-38 leads to

$$a = -g\,\frac{2m}{M + 2m} = -(9.8 \text{ m/s}^2)\,\frac{(2)(1.2 \text{ kg})}{2.5 \text{ kg} + (2)(1.2 \text{ kg})}$$

$$= -4.8 \text{ m/s}^2. \qquad \text{(Answer)}$$

We then use Eq. 11-38 to find T:

$$T = -\tfrac{1}{2}Ma = -\tfrac{1}{2}(2.5 \text{ kg})(-4.8 \text{ m/s}^2)$$

$$= 6.0 \text{ N}. \qquad \text{(Answer)}$$

As we should expect, the acceleration of the falling block is

less than g, and the tension in the cord ($= 6.0$ N) is less than the weight of the hanging block ($= mg = 11.8$ N). We see also that the acceleration of the block and the tension depend on the mass of the disk but not on its radius. As a check, we note that the formulas derived above predict $a = -g$ and $T = 0$ for the case of a massless disk ($M = 0$). This is what we would expect; the block simply falls as a free body, trailing the string behind it.

From Eq. 11-20, the angular acceleration of the disk is

$$\alpha = \frac{a}{R} = \frac{-4.8 \text{ m/s}^2}{0.20 \text{ m}} = -24 \text{ rad/s}^2. \qquad \text{(Answer)}$$

SAMPLE PROBLEM 11-12

To throw an 80 kg opponent with a basic judo hip throw, you intend to pull his uniform with a force $\mathbf{F}$ and a moment arm $d_1 = 0.30$ m from a pivot point (rotation axis) on your right hip, about which you wish to rotate him with an angular acceleration of -6.0 rad/s^2, that is, with a clockwise acceleration (Fig. 11-19). Assume that his rotational inertia I relative to the pivot point is 15 kg·m^2.

(a) What must the magnitude of $\mathbf{F}$ be if you initially bend your opponent forward to bring his center of mass to your hip (Fig. 11-19a)?

SOLUTION: If your opponent's center of mass is at the rotation axis, his weight vector produces no torque about that axis. The only torque on him then is due to your pull $\mathbf{F}$. From Eqs. 11-31 and 11-35, we have, for the clockwise torque,

$$\tau = -d_1 F = I\alpha,$$

which gives

$$F = \frac{-I\alpha}{d_1} = \frac{-(15 \text{ kg·m}^2)(-6.0 \text{ rad/s}^2)}{0.30 \text{ m}}$$

$$= 300 \text{ N}. \qquad \text{(Answer)}$$

FIGURE 11-19 Sample Problem 11-12. A judo hip throw (a) correctly executed and (b) incorrectly executed.

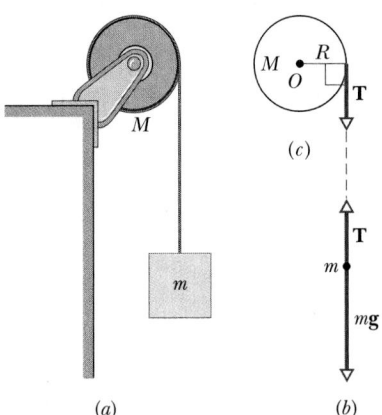

FIGURE 11-18 Sample Problems 11-11 and 11-13. (a) The falling block causes the disk to rotate. (b) A free-body diagram for the block. (c) A free-body diagram for the disk.

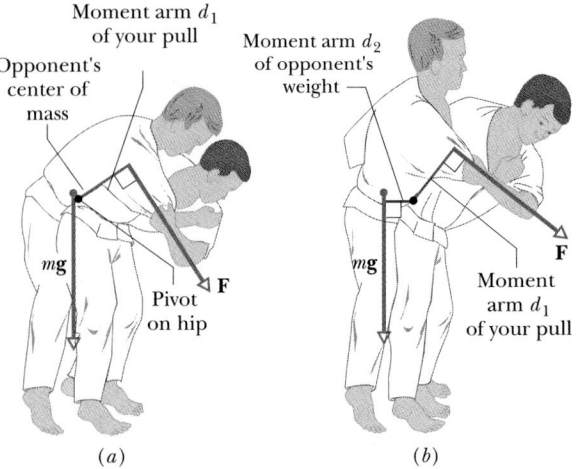

(b) What must the magnitude of **F** be if he remains upright and his weight m**g** has a moment arm $d_2 = 0.12$ m from the pivot point (Fig. 11-19b)?

SOLUTION: In this situation, your opponent's weight m**g** provides a positive (counterclockwise) torque that counters your torque. From Eqs. 11-31 and 11-36, we have

$$\sum \tau = -d_1 F + d_2 mg = I\alpha,$$

which gives

$$F = -\frac{I\alpha}{d_1} + \frac{d_2 mg}{d_1}.$$

From (a), we know that the first term on the right is equal to 300 N. Substituting this and the given data, we have

$$F = 300 \text{ N} + \frac{(0.12 \text{ m})(80 \text{ kg})(9.8 \text{ m/s}^2)}{0.30 \text{ m}}$$

$$= 613.6 \text{ N} \approx 610 \text{ N.} \qquad \text{(Answer)}$$

The results indicate that you will have to pull much harder if you do not initially bend your opponent to bring his center of mass to your hip. A good judo fighter knows this lesson from physics. (An analysis of judo and aikido is given in "The Amateur Scientist," *Scientific American*, July 1980.)

11-10 WORK AND ROTATIONAL KINETIC ENERGY

Let us again consider the situation of Fig. 11-17, in which force **F** rotates a rigid body consisting of a single particle of mass m fastened to the end of a massless rod. During the rotation, force **F** does work on the body. Let us assume that the only energy of the body that is changed by **F** is the kinetic energy. Then we can apply the work–kinetic energy theorem of Eq. 7-4:

$$\Delta K = K_f - K_i = W. \qquad (11\text{-}39)$$

Using $K = \frac{1}{2}mv^2$ and Eq. 11-16 ($v = \omega r$), we can rewrite Eq. 11-39 as

$$\Delta K = \tfrac{1}{2}mr^2\omega_f^2 - \tfrac{1}{2}mr^2\omega_i^2 = W. \qquad (11\text{-}40)$$

From Eq. 11-24, the rotational inertia for a body with one particle is $I = mr^2$. Substituting this into Eq. 11-40 yields

$$\Delta K = \tfrac{1}{2}I\omega_f^2 - \tfrac{1}{2}I\omega_i^2 = W. \qquad (11\text{-}41)$$

Equation 11-41 says that the work W done by **F** on the body of Fig. 11-17 changes the rotational kinetic energy K ($= \frac{1}{2}I\omega^2$) of the body. Equation 11-41 is the angular equivalent of the work–kinetic energy theorem for translational motion. We derived it for a rigid body with one particle, but it holds for any rigid body rotated about a fixed axis.

We next relate the work W done on the body in Fig. 11-17 to the torque τ on the body due to force **F**. If the particle in Fig. 11-17 were to move a differential distance ds along its circular path, the body would rotate through differential angle $d\theta$, with $ds = r\,d\theta$. We would use Eq. 7-11 to show the work dW done on the body by force **F** during this motion:

$$dW = \mathbf{F} \cdot d\mathbf{s} = F_t\,ds = F_t r\,d\theta. \qquad (11\text{-}42)$$

(Recall that F_t is the component of **F** along the particle's path.) From Eq. 11-31, we see that the product $F_t r$ is equal to the torque τ. So, we can rewrite Eq. 11-42 as

$$dW = \tau\,d\theta. \qquad (11\text{-}43)$$

The work done during a finite angular displacement from θ_i to θ_f is then

$$W = \int_{\theta_i}^{\theta_f} \tau\,d\theta. \qquad (11\text{-}44)$$

TABLE 11-3 SOME CORRESPONDING RELATIONS FOR TRANSLATIONAL AND ROTATIONAL MOTION

PURE TRANSLATION (FIXED DIRECTION)		PURE ROTATION (FIXED AXIS)	
Position	x	Angular position	θ
Velocity	$v = dx/dt$	Angular velocity	$\omega = d\theta/dt$
Acceleration	$a = dv/dt$	Angular acceleration	$\alpha = d\omega/dt$
Mass	m	Rotational inertia	I
Newton's second law	$F = ma$	Newton's second law	$\tau = I\alpha$
Work	$W = \int F\,dx$	Work	$W = \int \tau\,d\theta$
Kinetic energy	$K = \frac{1}{2}mv^2$	Kinetic energy	$K = \frac{1}{2}I\omega^2$
Power	$P = Fv$	Power	$P = \tau\omega$
Work–kinetic energy theorem	$W = \Delta K$	Work–kinetic energy theorem	$W = \Delta K$

Equation 11-44, which holds for any rigid body rotating about a fixed axis, is the rotational equivalent of Eq. 7-27, which is

$$W = \int_{x_i}^{x_f} F \, dx.$$

We can find the power P for rotational motion from Eq. 11-43:

$$P = \frac{dW}{dt} = \tau \frac{d\theta}{dt} = \tau\omega. \qquad (11\text{-}45)$$

This is the rotational analog of $P = Fv$ (from Eq. 7-49), which gives the rate at which a force F does work on a particle moving with speed v, with $\mathbf{F}$ and $\mathbf{v}$ parallel.

Table 11-3 summarizes the equations that apply to the rotation of a rigid body about a fixed axis and the equivalent relations for translational motion.

SAMPLE PROBLEM 11-13

(a) In the arrangement of Fig. 11-18, through what angle does the disk rotate in 2.5 s, starting from rest.

SOLUTION: From Eq. 11-10 ($\theta = \omega_0 t + \frac{1}{2}\alpha t^2$), we have, putting $\omega_0 = 0$ and using the value of α calculated in Sample Problem 11-11,

$$\theta = 0 + (\tfrac{1}{2})(-24 \text{ rad/s}^2)(2.5 \text{ s})^2$$
$$= -75 \text{ rad.} \qquad \text{(Answer)}$$

(b) What is the angular velocity of the disk at $t = 2.5$ s?

SOLUTION: We can find this with Eq. 11-9 ($\omega = \omega_0 + \alpha t$). Putting $\omega_0 = 0$ and using the value of α calculated in Sample Problem 11-11, we obtain

$$\omega = 0 + (-24 \text{ rad/s}^2)(2.5 \text{ s})$$
$$= -60 \text{ rad/s.} \qquad \text{(Answer)}$$

(c) What is the kinetic energy K of the disk at $t = 2.5$ s?

SOLUTION: From Eq. 11-25, the kinetic energy of the disk is $\frac{1}{2}I\omega^2$, in which $I = \frac{1}{2}MR^2$. Thus, using the value of ω found in (b), we have

$$K = \tfrac{1}{2}I\omega^2 = \tfrac{1}{2}(\tfrac{1}{2}MR^2)\omega^2$$
$$= (\tfrac{1}{4})(2.5 \text{ kg})(0.20 \text{ m})^2(-60 \text{ rad/s})^2$$
$$= 90 \text{ J.} \qquad \text{(Answer)}$$

Another approach to this problem is to calculate the change in the rotational kinetic energy using the work–kinetic energy theorem. To do so, we must calculate the work W done on the disk by the torque that acts on it. This torque, exerted by the tension T in the cord, is constant. Equation 11-44 then gives us,

$$W = \int_{\theta_i}^{\theta_f} \tau \, d\theta = \tau \int_{\theta_i}^{\theta_f} d\theta = \tau(\theta_f - \theta_i).$$

For the torque τ we use $-TR$, in which T is the tension in the cord ($= 6.0$ N; see Sample Problem 11-11) and R ($= 0.20$ m) is the radius of the disk. The quantity $\theta_f - \theta_i$ is just the angular displacement that we calculated in (a). Thus

$$W = \tau(\theta_f - \theta_i) = -TR(\theta_f - \theta_i)$$
$$= -(6.0 \text{ N})(0.20 \text{ m})(-75 \text{ rad})$$
$$= 90 \text{ J.} \qquad \text{(Answer)}$$

Because K_i is zero (the system starts from rest), Eq. 11-41 tells us that this answer is equal to K.

SAMPLE PROBLEM 11-14

A rigid sculpture, consisting of a thin hoop (of mass m and radius $R = 0.15$ m) and two thin rods (each of mass m and length $L = 2.0R$), is arranged as shown in Fig. 11-20. The sculpture can pivot around a horizontal axis in the plane of the hoop, passing through its center.

(a) In terms of m and R, what is the sculpture's rotational inertia I about the rotation axis?

SOLUTION: From Table 11-2(i), the hoop has rotational inertia $I_{\text{hoop}} = \frac{1}{2}mR^2$ about its diameter. From Table 11-2(e), rod A has rotational inertia $I_{\text{cm}, A} = mL^2/12$ about an axis through its center of mass and parallel to the rotation axis. To find the rotational inertia I_A of rod A about the rotation axis, we use Eq. 11-27, the parallel-axis theorem:

$$I_A = I_{\text{cm}, A} + mh_{\text{cm}, A}^2 = \frac{mL^2}{12} + m\left(R + \frac{L}{2}\right)^2$$
$$= 4.33mR^2,$$

where we have used the fact that $L = 2.0R$ and where $h_{\text{cm}, A}$ ($= R + L/2$) is the perpendicular distance between the center of rod A and the rotation axis.

We treat rod B similarly: it has zero rotational inertia $I_{\text{cm}, B}$ about an axis along its length, and its rotational inertia I_B about the rotation axis is

$$I_B = I_{\text{cm}, B} + mh_{\text{cm}, B}^2 = 0 + mR^2 = mR^2.$$

FIGURE 11-20   Sample Problem 11-14. A rigid sculpture consisting of a hoop and two rods can rotate around a horizontal axis.

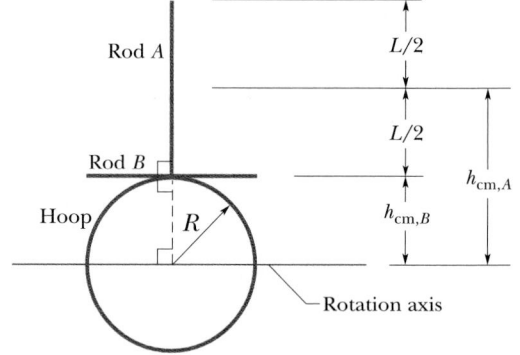

Here $h_{\text{cm}, B} (= R)$ is the perpendicular distance between rod B and the rotation axis. Thus the rotational inertia I of the sculpture about the rotation axis is

$$I = I_{\text{hoop}} + I_A + I_B = \tfrac{1}{2}mR^2 + 4.33mR^2 + mR^2$$
$$= 5.83mR^2 \approx 5.8mR^2. \qquad \text{(Answer)}$$

(b) Starting from rest, the sculpture rotates around the rotation axis from the initial upright orientation of Fig. 11-20. What is its angular speed ω about the axis when it is inverted?

SOLUTION: Before the sculpture moves, its center of mass is a distance y_{cm} above the rotation axis, with y_{cm} given by Eq. 9-5:

$$y_{\text{cm}} = \frac{m(0) + mR + m(R + L/2)}{3m} = R,$$

where $3m$ is the mass of the sculpture. As the sculpture rotates, its center of mass descends to the same distance *below* the rotation axis. Thus the center of mass undergoes a vertical displacement of $\Delta y_{\text{cm}} = -2R$. During the descent, gravita-

tional potential energy U of the sculpture is transferred to kinetic energy K of its rotation. The change ΔU in potential energy is the product of the magnitude of the sculpture's weight ($3mg$) and the vertical displacement Δy_{cm}. Thus

$$\Delta U = 3mg \, \Delta y_{\text{cm}} = 3mg(-2R) = -6mgR.$$

From Eq. 11-25, the associated change in kinetic energy is

$$\Delta K = \tfrac{1}{2}I\omega^2 - 0 = \tfrac{1}{2}I\omega^2.$$

So by conservation of mechanical energy

$$\Delta K + \Delta U = 0,$$

and

$$\tfrac{1}{2}I\omega^2 - 6mgR = 0.$$

Substituting $I = 5.83mR^2$ and solving for ω give

$$\omega = \sqrt{\frac{12g}{5.83R}} = \sqrt{\frac{(12)(9.8 \text{ m/s}^2)}{(5.83)(0.15 \text{ m})}}$$
$$= 12 \text{ rad/s}. \qquad \text{(Answer)}$$

REVIEW & SUMMARY

Angular Position

To describe the rotation of a rigid body about a fixed axis, called the **rotation axis,** we assume a **reference line** is fixed in the body, perpendicular to that axis and rotating with the body. We measure the **angular position** θ of this line relative to a fixed direction. When θ is measured in **radians,**

$$\theta = \frac{s}{r} \qquad \text{(radian measure)}, \qquad (11\text{-}1)$$

where s is the arc length of a circular path of radius r and angle θ. Radian measure is related to angle measure in revolutions and degrees by

$$1 \text{ rev} = 360° = 2\pi \text{ rad.} \qquad (11\text{-}2)$$

Angular Displacement

A body that rotates about a rotation axis, changing its angular position from θ_1 to θ_2, undergoes an **angular displacement**

$$\Delta\theta = \theta_2 - \theta_1, \qquad (11\text{-}4)$$

where $\Delta\theta$ is positive for counterclockwise rotation and negative for clockwise rotation.

Angular Velocity and Speed

If a body rotates through an angular displacement $\Delta\theta$ in a time interval Δt, its **average angular velocity** $\overline{\omega}$ is

$$\overline{\omega} = \frac{\Delta\theta}{\Delta t}. \qquad (11\text{-}5)$$

The **(instantaneous) angular velocity** ω of the body is

$$\omega = \frac{d\theta}{dt}. \qquad (11\text{-}6)$$

Both $\overline{\omega}$ and ω are vectors, with directions given by the **right-hand rule** of Fig. 11-5. They are positive for counterclockwise rotation and negative for clockwise rotation. The magnitude of the body's angular velocity is the **angular speed.**

Angular Acceleration

If the angular velocity of a body changes from ω_1 to ω_2 in a time interval $\Delta t = t_2 - t_1$, the **average angular acceleration** $\overline{\alpha}$ of the body is

$$\overline{\alpha} = \frac{\omega_2 - \omega_1}{t_2 - t_1} = \frac{\Delta\omega}{\Delta t}. \qquad (11\text{-}7)$$

The **(instantaneous) angular acceleration** α of a body is

$$\alpha = \frac{d\omega}{dt}. \qquad (11\text{-}8)$$

Both $\overline{\alpha}$ and α are vectors.

The Kinematic Equations for Constant Angular Acceleration

Constant angular acceleration ($\alpha = $ constant) is an important special case of rotational motion. The appropriate kinematic equations, given in Table 11-1, are

$$\omega = \omega_0 + \alpha t, \qquad (11\text{-}9) \qquad \theta = \tfrac{1}{2}(\omega_0 + \omega)t, \qquad (11\text{-}12)$$
$$\theta = \omega_0 t + \tfrac{1}{2}\alpha t^2, \qquad (11\text{-}10) \qquad \theta = \omega t - \tfrac{1}{2}\alpha t^2. \qquad (11\text{-}13)$$
$$\omega^2 = \omega_0^2 + 2\alpha\theta, \qquad (11\text{-}11)$$

Linear and Angular Variables Related

A point in a rigid rotating body, at a *perpendicular distance* r from the rotation axis, moves in a circle with radius r. If the body rotates through an angle θ, the point moves along an arc with length s given by

$$s = \theta r \quad \text{(radian measure)}, \quad (11\text{-}15)$$

where θ is in radians.

The linear velocity **v** of the point is tangent to the circle; the point's linear speed v is given by

$$v = \omega r \quad \text{(radian measure)}, \quad (11\text{-}16)$$

where ω is the angular speed (in radians per second) of the body.

The linear acceleration **a** of the point has both *tangential* and *radial* components. The tangential component is

$$a_t = \alpha r \quad \text{(radian measure)}, \quad (11\text{-}20)$$

where α is the magnitude of the angular acceleration (in radians per second-squared) of the body. The radial component of **a** is

$$a_r = \frac{v^2}{r} = \omega^2 r \quad \text{(radian measure)}. \quad (11\text{-}21)$$

If the point moves in uniform circular motion, the period T of the motion for the point and the body is

$$\text{(radian measure).} \quad (11\text{-}17, 11\text{-}18)$$
$$\omega$$

Rotational Kinetic Energy and Rotational Inertia

The kinetic energy K of a rigid body rotating about a fixed axis is given by

$$K = \tfrac{1}{2}I\omega^2 \quad \text{(radian measure)}, \quad (11\text{-}25)$$

in which I is the **rotational inertia** of the body, defined as

$$I = \sum m_i r_i^2 \quad (11\text{-}24)$$

for a system of discrete particles and as

$$I = \int r^2 \, dm \quad (11\text{-}26)$$

for a body with continuously distributed mass. The r and r_i in these expressions represent the perpendicular distance from the axis of rotation to each mass element in the body.

The Parallel-Axis Theorem

The *parallel-axis theorem* relates the rotational inertia I of a body about any axis to that of the same body about a parallel axis through the center of mass:

$$I = I_{cm} + Mh^2. \quad (11\text{-}27)$$

Here h is the perpendicular distance between the two axes.

Torque

Torque is a turning or twisting action on a body about a rotation axis due to a force **F**. If **F** is exerted at a point given by the position vector **r** relative to the axis, then the torque $\boldsymbol{\tau}$ (a vector quantity) is

$$\boldsymbol{\tau} = \mathbf{r} \times \mathbf{F}. \quad (11\text{-}33)$$

The magnitude of the torque is

$$\tau = rF_t = r_\perp F = rF \sin\phi, \quad (11\text{-}31, 11\text{-}32, 11\text{-}30)$$

where F_t is the component of **F** perpendicular to **r**, and ϕ is the angle between **r** and **F**. The quantity $r_\perp$ is the perpendicular distance between the rotation axis and an extended line running through the **F** vector. This line is called the **line of action** of **F**, and $r_\perp$ is called the **moment arm** of **F**. Similarly, r is the moment arm of F_t.

The SI unit of torque is the newton-meter (N·m). A torque τ is positive if it tends to rotate the body counterclockwise and negative if it tends to rotate the body in the clockwise direction.

Newton's Second Law in Angular Form

The rotational analog of Newton's second law is

$$\sum \tau = I\alpha, \quad (11\text{-}36)$$

where $\sum\tau$ is the net torque acting on a particle or rigid body, I is the rotational inertia of the particle or body about the rotation axis, and α is the resulting angular acceleration about that axis.

Work and Rotational Kinetic Energy

The equations used for calculating work and power in rotational motion are analogs of the corresponding equations governing translational motion and are

$$W = \int_{\theta_i}^{\theta_f} \tau \, d\theta \quad (11\text{-}44)$$

and

$$P = \frac{dW}{dt} = \tau \frac{d\theta}{dt} = \tau\omega. \quad (11\text{-}45)$$

The form of the work–kinetic energy theorem used for rotating bodies is

$$\Delta K = \tfrac{1}{2}I\omega_f^2 - \tfrac{1}{2}I\omega_i^2 = W. \quad (11\text{-}41)$$

QUESTIONS

1. Figure 11-21b is a graph of the angular position of the rotating disk of Fig. 11-21a. Is the angular velocity of the disk positive, negative, or zero at (a) $t = 1$ s, (b) $t = 2$ s, and (c) $t = 3$ s? (d) Is the angular acceleration positive or negative?

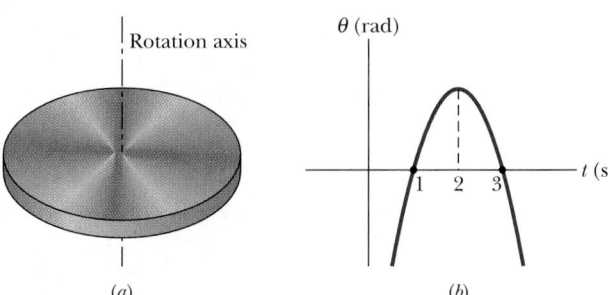

FIGURE 11-21   Question 1. (a) (b)

2. Figure 11-22 is a graph of the angular velocity of the rotating disk of Fig. 11-21*a*. What are the (a) initial and (b) final directions of rotation? (c) Does the disk momentarily stop? (d) Is the angular acceleration positive or negative? (e) Is the angular acceleration constant or varying?

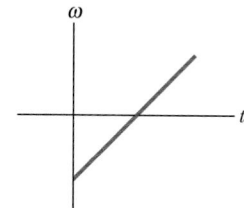

FIGURE 11-22
Question 2.

3. At $t = 0$, the rotating disk of Fig. 11-21*a* is at angular position $\theta_0 = -2$ rad. Here are the signs of the disk's initial angular velocity and constant angular acceleration, respectively, for four situations: (1) +, +; (2) +, −; (3) −, +; (4) −, −. For which situations will the disk (a) undergo a momentary stop, (b) definitely pass through angular position $\theta = 0$ (given enough time), and (c) definitely not pass through that angular position?

4. For the rotating disk of Fig. 11-21*a*, here are the initial and final angular velocities, respectively, in four situations: (a) 2 rad/s, 3 rad/s; (b) −2 rad/s, 3 rad/s; (c) −2 rad/s, −3 rad/s; (d) 2 rad/s, −3 rad/s. The magnitude of the angular acceleration of the disk has the same constant value in all four situations. Rank the situations according to the magnitude of the disk's displacement, greatest first, during the change from the initial to the final angular velocity.

5. For which of the following expressions for $\omega(t)$ of a rotating object do the angular equations of Table 11-1 apply? (a) $\omega = 3$; (b) $\omega = 4t^2 + 2t - 6$; (c) $\omega = 3t - 4$; (d) $\omega = 5t^2 - 3$.

6. Figure 11-23 gives the angular velocity versus time for the rotating disk of Fig. 11-21*a*. For a point on the rim of the disk, rank the four instants *a*, *b*, *c*, and *d* according to the magnitude of (a) the tangential acceleration and (b) the radial acceleration, greatest first.

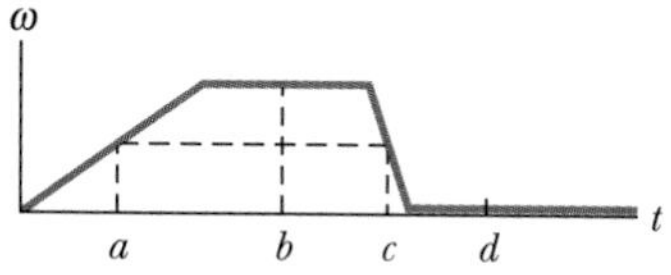

FIGURE 11-23
Question 6.

7. Figure 11-24 shows four gears that rotate together because of friction between them. Gear 2 has radius R; gear 3 has radius $2R$; and gears 1 and 4 have radius $3R$. Gear 2 is forced to rotate by a motor. Rank the four gears according to (a) the linear speed of their rims and (b) their angular speed of rotation, greatest first.

8. Circular disks A and B have the same weight and thickness, but the density of A is greater than that of B. Is the rotational inertia of A about its central axis greater than, less than, or the same as that of B?

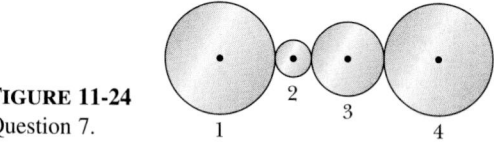

FIGURE 11-24
Question 7.

9. Figure 11-25 shows three uniform disks, along with their radii R and masses M. Rank the disks according to their rotational inertias about their central axes, greatest first.

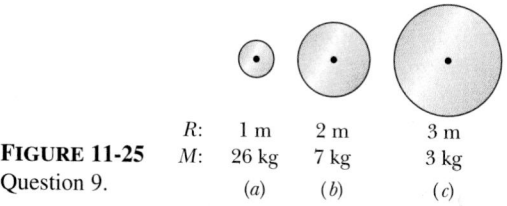

FIGURE 11-25 | R: | 1 m | 2 m | 3 m |
Question 9. | M: | 26 kg | 7 kg | 3 kg |
| | (a) | (b) | (c) |

10. Five solids with identical masses are shown in cross section in Fig. 11-26. The cross sections have equal widths at the widest part and equal heights (but not necessarily equal thicknesses). (a) Which one has the greatest rotational inertia about an axis through its center of mass and perpendicular to its cross section? (b) Which has the least?

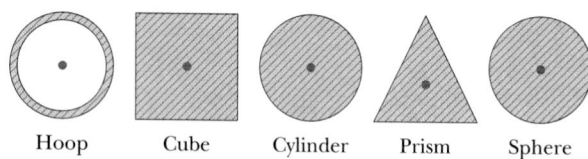

Hoop   Cube   Cylinder   Prism   Sphere
FIGURE 11-26   Question 10.

11. Figure 11-27*a* shows a meter stick, half wood and half steel, that is pivoted at the wood end at *O*. A force **F** is applied to the steel end at *a*. In Fig. 11-27*b*, the stick is reversed and pivoted at the steel end at *O'*, and the same force is applied at the wood end at *a'*. Is the resulting angular acceleration of Fig. 11-27*a* greater than, less than, or the same as that of Fig. 11-27*b*?

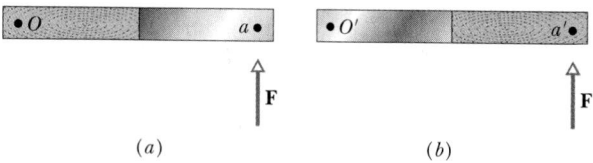

(a)   (b)
FIGURE 11-27   Question 11.

12. Figure 11-28*a* shows an overhead view of a horizontal bar that can pivot about the point indicated. Two horizontal forces act on the bar, but the bar is stationary. If the angle between the bar and force $\mathbf{F}_2$ is now decreased from the initial 90°, and the bar is still not to turn, should the magnitude of $\mathbf{F}_2$ be made larger, be made smaller, or left the same?

13. Figure 11-28*b* shows an overhead view of a horizontal bar that is rotated about the pivot point by two horizontal forces, $\mathbf{F}_1$ and $\mathbf{F}_2$, at opposite ends of the bar. The direction of $\mathbf{F}_2$ is at angle ϕ to the bar. Rank the following values of ϕ according to the magnitude of the angular acceleration of the bar, greatest first: 90°, 70°, and 110°.

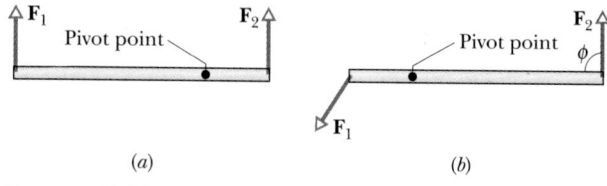

(a)   (b)
FIGURE 11-28   Questions 12 and 13.

14. A force is applied to the rim of the disk in Fig. 11-21a so as to change its angular velocity. Its initial and final angular velocities, respectively, for four situations are: (a) −2 rad/s, 5 rad/s; (b) 2 rad/s, 5 rad/s; (c) −2 rad/s, −5 rad/s; and (d) 2 rad/s, −5 rad/s. Rank the situations according to the work done by the torque due to the force, greatest first.

15. Hold your right arm downward, palm toward your thigh. Keeping your wrist rigid, (1) lift the arm until it is horizontal and forward, (2) move it horizontally until it is pointed toward the right, and (3) then bring it down to your side. Your palm faces forward. If you start over, but reverse the steps, why does your palm *not* face forward?

EXERCISES & PROBLEMS

SECTION 11-2 The Rotational Variables

1E. (a) What angle in radians is subtended by an arc that has length 1.80 m and is part of a circle of radius 1.20 m? (b) Express the same angle in degrees. (c) The angle between two radii of a circle is 0.620 rad. What arc length is subtended if the radius is 2.40 m?

2E. During a time interval t the flywheel of a generator turns through the angle $\theta = at + bt^3 - ct^4$, where a, b, and c are constants. Write expressions for the wheel's (a) angular velocity and (b) angular acceleration.

3E. What is the angular speed of (a) the second hand, (b) the minute hand, and (c) the hour hand of a watch? Answer in radians per second.

4E. Our Sun is 2.3×10^4 ly (light-years) from the center of our Milky Way galaxy and is moving in a circle around that center at a speed of 250 km/s. (a) How long does it take the Sun to make one revolution about the galactic center? (b) How many revolutions has the Sun completed since it was formed about 4.5×10^9 years ago?

5E. The angular position of a point on the rim of a rotating wheel is given by $\theta = 4.0t - 3.0t^2 + t^3$, where θ is in radians if t is given in seconds. (a) What are the angular velocities at $t = 2.0$ s and $t = 4.0$ s? (b) What is the average angular acceleration for the time interval that begins at $t = 2.0$ s and ends at $t = 4.0$ s? (c) What are the instantaneous angular accelerations at the beginning and end of this time interval?

6E. The angular position of a point on a rotating wheel is given by $\theta = 2 + 4t^2 + 2t^3$, where θ is in radians and t is in seconds. At $t = 0$, what are (a) the angular position and (b) the angular velocity? (c) What is the angular velocity at $t = 4.0$ s? (d) Calculate the angular acceleration at $t = 2.0$ s. (e) Is the angular acceleration constant?

7P. A wheel rotates with an angular acceleration α given by $\alpha = 4at^3 - 3bt^2$, where t is the time and a and b are constants. If the wheel has an initial angular speed ω_0, write equations for (a) the angular speed and (b) the angular displacement of the wheel as functions of time.

8P. A good baseball pitcher can throw a baseball toward home plate at 85 mi/h with a spin of 1800 rev/min. How many revolutions does the baseball make on its way to home plate? For simplicity, assume that the 60 ft trajectory is a straight line.

9P. A diver makes 2.5 complete revolutions on the way from a 10 m high platform to the water. Assuming zero initial vertical velocity, find the diver's average angular velocity during a dive.

10P. The wheel in Fig. 11-29 has eight spokes and a radius of 30 cm. It is mounted on a fixed axle and is spinning at 2.5 rev/s. You want to shoot a 20 cm long arrow parallel to this axle and through the wheel without hitting any of the spokes. Assume that the arrow and the spokes are very thin. (a) What minimum speed must the arrow have? (b) Does it matter where between the axle and rim of the wheel you aim? If so, where is the best location?

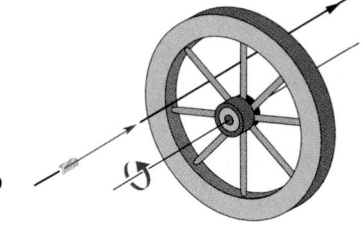

FIGURE 11-29
Problem 10.

SECTION 11-4 Rotation with Constant Angular Acceleration

11E. The angular speed of an automobile engine is increased from 1200 rev/min to 3000 rev/min in 12 s. (a) What is its angular acceleration in rev/min², assuming it to be uniform? (b) How many revolutions does the engine make during this time?

12E. A phonograph turntable rotating at $33\frac{1}{3}$ rev/min slows down and stops in 30 s after the motor is turned off. (a) Find its (uniform) angular acceleration in units of rev/min². (b) How many revolutions did it make in this time?

13E. A disk, initially rotating at 120 rad/s, is slowed down with a constant angular acceleration of magnitude 4.0 rad/s². (a) How much time elapses before the disk stops? (b) Through what angle does the disk rotate in coming to rest?

14E. A pulley wheel that is 8.0 cm in diameter has a 5.6 m long cord wrapped around its periphery. Starting from rest, the wheel is given a constant angular acceleration of 1.5 rad/s². (a) Through what angle must the wheel turn for the cord to unwind? (b) How long does the unwinding take?

15E. A heavy flywheel rotating on its central axis is slowing down because of friction in its bearings. At the end of the first minute of slowing, its angular velocity is 0.90 of its initial angular velocity of 250 rev/min. Assuming constant frictional forces, find its angular velocity at the end of the second minute.

16E. The flywheel of an engine is rotating at 25.0 rad/s. When the engine is turned off, the flywheel decelerates at a constant rate and comes to rest after 20.0 s. Calculate (a) the angular acceleration (in rad/s^2) of the flywheel, (b) the angle (in rad) through which the flywheel rotates in coming to rest, and (c) the number of revolutions made by the flywheel in coming to rest.

17E. Starting from rest, a disk rotates about its central axis with constant angular acceleration. In 5.0 s, it has rotated 25 rad. (a) What was the angular acceleration during this time? (b) What was the average angular velocity? (c) What is the instantaneous angular velocity of the disk at the end of the 5.0 s? (d) Assuming that the acceleration does not change, through what additional angle will the disk turn during the next 5.0 s?

18P. Starting from rest at $t = 0$, a wheel undergoes a constant angular acceleration. When $t = 2.0$ s, the angular velocity of the wheel is 5.0 rad/s. The acceleration continues until $t = 20$ s, when it abruptly ceases. Through what angle does the wheel rotate in the interval $t = 0$ to $t = 40$ s?

19P. A wheel, starting from rest, rotates with a constant angular acceleration of 2.00 rad/s^2. During a certain 3.00 s interval, it turns through 90.0 rad. (a) How long had the wheel been turning before the start of the 3.00 s interval? (b) What was the angular velocity of the wheel at the start of the 3.00 s interval?

20P. A wheel has a constant angular acceleration of 3.0 rad/s^2. In a 4.0 s interval, it turns through an angle of 120 rad. Assuming that the wheel started from rest, how long had it been in motion at the start of this 4.0 s interval?

21P. A flywheel completes 40 rev as it slows from an angular speed of 1.5 rad/s to a complete stop. (a) Assuming uniform acceleration, what is the time required for it to come to rest? (b) What is the angular acceleration? (c) How much time is required for it to complete the first 20 of the 40 revolutions?

22P. At $t = 0$, a flywheel has an angular velocity of 4.7 rad/s, an angular acceleration of -0.25 rad/s^2, and a reference line at $\theta_0 = 0$. (a) Through what maximum angle θ_{max} will the reference line turn in the positive direction? At what times t will the line be at (b) $\theta = \frac{1}{2}\theta_{max}$ and (c) $\theta = -10.5$ rad (consider both positive and negative values of t)? (d) Graph θ versus t, and indicate the answers to (a), (b), and (c) on the graph.

23P. A disk rotates about its central axis starting from rest and accelerates with constant angular acceleration. At one time it is rotating at 10 rev/s. After 60 more complete revolutions, its angular speed is 15 rev/s. Calculate (a) the angular acceleration, (b) the time required to complete the 60 revolutions mentioned, (c) the time required to attain the 10 rev/s angular speed, and (d) the number of revolutions from rest until the time the disk attained the 10 rev/s angular speed.

24P. A wheel turns through 90 revolutions in 15 s, its angular speed at the end of the period being 10 rev/s. (a) What was its angular speed at the beginning of the 15 s interval, assuming constant angular acceleration? (b) How much time had elapsed between the time the wheel was at rest and the beginning of the 15 s interval?

SECTION 11-5 Relating The Linear and Angular Variables

25E. What is the acceleration of a point on the rim of a 12 in. ($= 30$ cm) diameter record rotating at $33\frac{1}{3}$ rev/min?

26E. A phonograph record on a turntable rotates at $33\frac{1}{3}$ rev/min. (a) What is the angular speed in rad/s? What is the linear speed of a point on the record at the needle at (b) the beginning and (c) the end of the recording? The distances of the needle from the turntable axis are 5.9 in. and 2.9 in. at these two positions.

27E. What is the angular speed of a car traveling at 50 km/h and rounding a circular turn of radius 110 m?

28E. A flywheel with a diameter of 1.20 m is rotating at 200 rev/min. (a) What is the angular velocity of the flywheel in rad/s? (b) What is the linear velocity of a point on the rim of the flywheel? (c) What constant angular acceleration (in rev/min^2) will increase the wheel's angular speed to 1000 rev/min in 60 s? (d) How many revolutions does the wheel make during this 60 s acceleration?

29E. A point on the rim of a 0.75 m diameter grinding wheel changes speed uniformly from 12 m/s to 25 m/s in 6.2 s. What is the average angular acceleration of the wheel during this interval?

30E. The Earth's orbit about the Sun is almost a circle. With respect to the Sun, what are the Earth's (a) angular speed, (b) linear speed, and (c) acceleration?

31E. At 7:14 A.M. on June 30, 1908, a huge explosion occurred above remote central Siberia at latitude 61° N and longitude 102° E; the fireball thus created was the brightest flash seen by anyone before the advent of nuclear weapons. The *Tunguska Event,* which according to one chance witness "covered an enormous part of the sky," was probably the explosion of a *stony asteroid* about 140 m wide. (a) Considering only the Earth's rotation, determine how much later the asteroid would have had to arrive to put the explosion above Helsinki at longitude 25° E? This would have obliterated the city. (b) If the asteroid had, instead, been a *metallic asteroid,* it could have reached the Earth's surface. How much later would such an asteroid have had to arrive to put the impact in the Atlantic Ocean at longitude 20° W? (The resulting tsunamis would have wiped out coastal civilization on both sides of the Atlantic.)

32E. An astronaut is being tested in a centrifuge. The centrifuge has a radius of 10 m and, in starting, rotates according to $\theta = 0.30t^2$, where t in seconds gives θ in radians. When $t = 5.0$ s, what are the astronaut's (a) angular velocity, (b) linear speed, (c) tangential acceleration (magnitude only), and (d) radial acceleration (magnitude only)?

33E. What are (a) the angular speed, (b) the radial acceleration, and (c) the tangential acceleration of a spaceship negotiating a circular turn of radius 3220 km at a speed of 29,000 km/h?

34E. A coin of mass M is placed a distance R from the center of a phonograph turntable. The coefficient of static friction is μ_s. The angular speed of the turntable is slowly increased to a value ω_0, at which time the coin slides off. (a) Find ω_0 in terms of the quantities M, R, g, and μ_s. (b) Make a sketch showing the approximate path of the coin as it flies off the turntable.

35P. The flywheel of a steam engine runs with a constant angular speed of 150 rev/min. When steam is shut off, the friction of the bearings and of the air brings the wheel to rest in 2.2 h. (a) What is the constant angular acceleration, in rev/min², of the wheel during the slowdown? (b) How many rotations does the wheel make before coming to rest? (c) At the instant the flywheel is turning at 75 rev/min, what is the tangential component of the linear acceleration of a flywheel particle that is 50 cm from the axis of rotation? (d) What is the magnitude of the net linear acceleration of the particle in (c)?

36P. A gyroscope flywheel of radius 2.83 cm is accelerated from rest at 14.2 rad/s² until its angular speed is 2760 rev/min. (a) What is the tangential acceleration of a point on the rim of the flywheel? (b) What is the radial acceleration of this point when the flywheel is spinning at full speed? (c) Through what distance does a point on the rim move during the acceleration?

37P. If an airplane propeller rotates at 2000 rev/min while the airplane flies at a speed of 480 km/h relative to the ground, what is the speed of a point on the tip of the propeller, at radius 1.5 m, as seen by (a) the pilot and (b) an observer on the ground? The plane's velocity is parallel to the propeller's axis of rotation.

38P. An early method of measuring the speed of light makes use of a rotating slotted wheel. A beam of light passes through a slot at the outside edge of the wheel, as in Fig. 11-30, travels to a distant mirror, and returns to the wheel just in time to pass through the next slot in the wheel. One such slotted wheel has a radius of 5.0 cm and 500 slots at its edge. Measurements taken when the mirror was $l = 500$ m from the wheel indicated a speed of light of 3.0×10^5 km/s. (a) What was the (constant) angular speed of the wheel? (b) What was the linear speed of a point on the edge of the wheel?

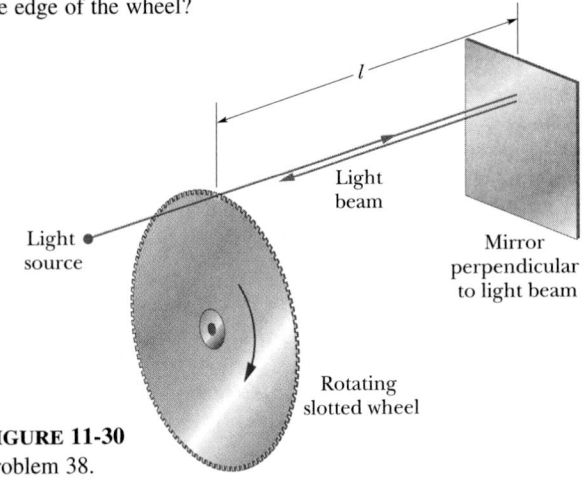

FIGURE 11-30
Problem 38.

39P. A car starts from rest and moves around a circular track of radius 30.0 m. Its speed increases at the constant rate of 0.500 m/s². (a) What is the magnitude of its *net* linear acceleration 15.0 s later? (b) What angle does this net acceleration vector make with the car's velocity at this time?

40P. (a) What is the angular speed about the polar axis of a point on the Earth's surface at a latitude of 40° N? (b) What is the linear speed of the point? (c) What are the corresponding values for a point at the equator?

41P. In Fig. 11-31, wheel A of radius $r_A = 10$ cm is coupled by belt B to wheel C of radius $r_C = 25$ cm. Wheel A increases its angular speed from rest at a uniform rate of 1.6 rad/s². Find the time for wheel C to reach a rotational speed of 100 rev/min, assuming the belt does not slip. (*Hint:* If the belt does not slip, the linear speeds at the rims of the two wheels must be equal.)

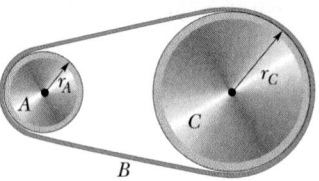

FIGURE 11-31 Problem 41.

42P. In Fig. 11-32, four pulleys are connected by two belts. Pulley A (radius 15 cm) is the drive pulley, and it rotates at 10 rad/s. Pulley B (radius 10 cm) is connected by belt 1 to pulley A. Pulley B' (radius 5 cm) is concentric with pulley B and is rigidly attached to it. Pulley C (radius 25 cm) is connected by belt 2 to pulley B'. Calculate (a) the linear speed of a point on belt 1, (b) the angular speed of pulley B, (c) the angular speed of pulley B', (d) the linear speed of a point on belt 2, and (e) the angular speed of pulley C. (See hint to Problem 41.)

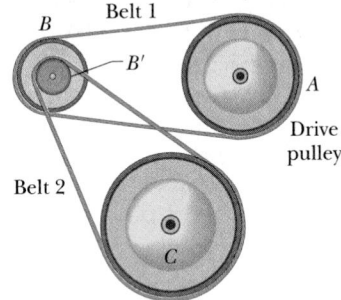

FIGURE 11-32 Problem 42.

43P. A phonograph turntable is rotating at $33\frac{1}{3}$ rev/min. A small object is on the turntable 6.0 cm from the axis of rotation. (a) Calculate the acceleration of the object, assuming that the object does not slip. (b) What is the minimum value of the coefficient of static friction between the object and the turntable? (c) Suppose that the turntable achieved its angular speed by starting from rest and undergoing a constant angular acceleration for 0.25 s. Calculate the minimum coefficient of static friction required for the object not to slip during the acceleration period.

44P. A pulsar is a rapidly rotating neutron star that emits radio pulses with precise synchronization, one such pulse for each rotation of the star. The *period T* of rotation is found by measuring the time between pulses. At present, the pulsar in the central region of the Crab nebula (see Fig. 11-33) has a period of rotation of $T = 0.033$ s, and this period is observed to be increasing at the rate of 1.26×10^{-5} s/y. (a) What is the value of the angular acceleration in rad/s²? (b) If its angular acceleration is constant, how many years from now will the pulsar stop rotating? (c) The pulsar originated in a supernova explosion seen in the year A.D. 1054. What was T for the pulsar when it was born? (Assume constant angular acceleration since then.)

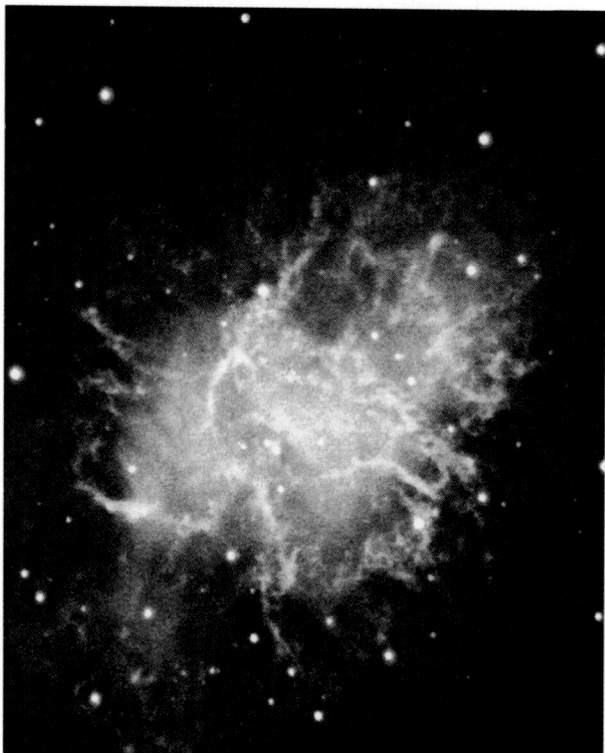

FIGURE 11-33 Problem 44. The Crab nebula resulted from a star whose explosion was seen in A.D. 1054. In addition to the gaseous debris seen here, the explosion left a spinning neutron star at its center. The star has a diameter of only 30 km.

SECTION 11-6 Kinetic Energy of Rotation

45E. Calculate the rotational inertia of a wheel that has a kinetic energy of 24,400 J when rotating at 602 rev/min.

46P. The oxygen molecule, O_2, has a total mass of 5.30×10^{-26} kg and a rotational inertia of 1.94×10^{-46} kg·m² about an axis through the center of the line joining the atoms and perpendicular to that line. Suppose that such a molecule in a gas has a speed of 500 m/s and that its rotational kinetic energy is two-thirds of its translational kinetic energy. Find its angular velocity.

SECTION 11-7 Calculating the Rotational Inertia

47E. Calculate the kinetic energies of two uniform solid cylinders, each rotating about its central axis. They have the same mass, 1.25 kg, and rotate with the same angular velocity, 235 rad/s, but the first has a radius of 0.25 m and the second a radius of 0.75 m.

48E. A molecule has a rotational inertia of 14,000 u·pm² and is spinning at an angular speed of 4.3×10^{12} rad/s. (a) Express the rotational inertia in kg·m². (b) Calculate the rotational kinetic energy in electron-volts.

49E. A communications satellite is a solid cylinder with mass 1210 kg, diameter 1.21 m, and length 1.75 m. Prior to launching from the shuttle cargo bay, it is set spinning at 1.52 rev/s about the cylinder axis (Fig. 11-34). Calculate the satellite's (a) rotational inertia about the rotation axis and (b) rotational kinetic energy.

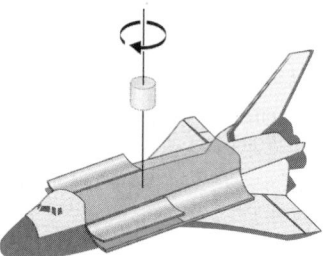

FIGURE 11-34 Exercise 49.

50E. Two particles, each with mass m, are fastened to each other, and to a rotation axis at O, by two thin rods, each with length l and mass M as shown in Fig. 11-35. The combination rotates around the rotation axis with angular velocity ω. Obtain algebraic expressions for (a) the rotational inertia of the combination about O and (b) the kinetic energy of rotation about O.

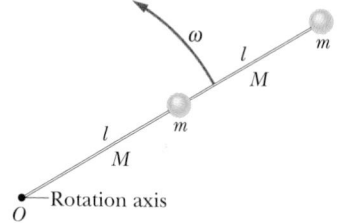

FIGURE 11-35 Exercise 50.

51E. Each of the three helicopter rotor blades shown in Fig 11-36 is 5.20 m long and has a mass of 240 kg. The rotor is rotating at 350 rev/min. (a) What is the rotational inertia of the rotor assembly about the axis of rotation? (Each blade can be considered to be a thin rod.) (b) What is the kinetic energy of rotation?

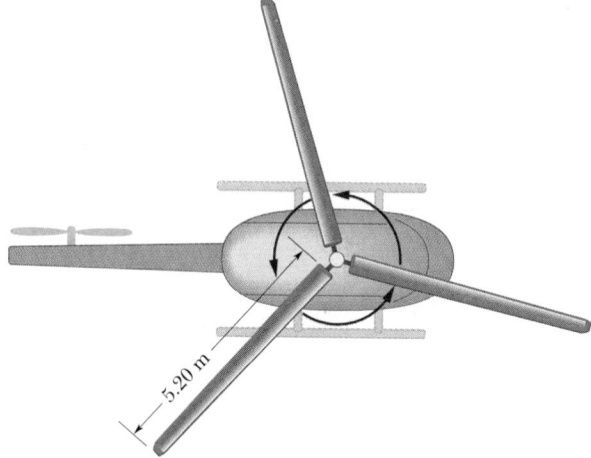

FIGURE 11-36 Exercise 51 and Problem 85.

52E. Assume the Earth to be a sphere of uniform density. Calculate (a) its rotational inertia and (b) its rotational kinetic energy. (c) Suppose that this energy could be harnessed for our use. How long could the Earth supply 1.0 kW of power to each of the 6.4×10^9 persons on Earth?

53E. Calculate the rotational inertia of a meter stick, with mass 0.56 kg, about an axis perpendicular to the stick and located at the 20 cm mark. (Treat the stick as a thin rod.)

54P. Show that the axis about which a given rigid body has its smallest rotational inertia must pass through its center of mass.

55P. Derive the expression for the rotational inertia of a hoop of mass M and radius R about its central axis; see Table 11-2(a).

56P. Figure 11-37 shows a uniform solid block of mass M and edge lengths a, b, and c. Calculate its rotational inertia about an axis through one corner and perpendicular to the large faces.

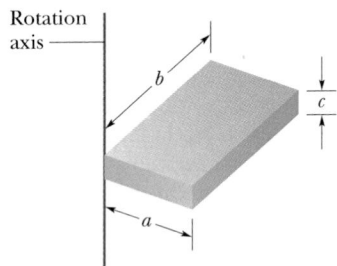

FIGURE 11-37 Problem 56.

57P. The masses and coordinates of four particles are as follows: 50 g, $x = 2.0$ cm, $y = 2.0$ cm; 25 g, $x = 0$, $y = 4.0$ cm; 25 g, $x = -3.0$ cm, $y = -3.0$ cm; 30 g, $x = -2.0$ cm, $y = 4.0$ cm. What is the rotational inertia of this collection with respect to the (a) x, (b) y, and (c) z axes? (d) If the answers to (a) and (b) are A and B, respectively, then what is the answer to (c) in terms of A and B?

58P. (a) Show that the rotational inertia of a solid cylinder of mass M and radius R about its central axis is equal to the rotational inertia of a thin hoop of mass M and radius $R/\sqrt{2}$ about its central axis. (b) Show that the rotational inertia I of any given body of mass M about any given axis is equal to the rotational inertia of an *equivalent hoop* about that axis, if the hoop has the same mass M and a radius k given by

$$k = \sqrt{\frac{I}{M}}.$$

The radius k of the equivalent hoop is called the *radius of gyration* of the given body.

59P. Delivery trucks that operate by making use of energy stored in a rotating flywheel have been used in Europe. The trucks are charged by using an electric motor to get the flywheel up to its top speed of 200π rad/s. One such flywheel is a solid, homogeneous cylinder with a mass of 500 kg and a radius of 1.0 m. (a) What is the kinetic energy of the flywheel after charging? (b) If the truck operates with an average power requirement of 8.0 kW, for how many minutes can it operate between chargings?

SECTION 11-8 Torque

60E. The length of a bicycle pedal arm is 0.152 m, and a downward force of 111 N is applied by the foot. What is the magnitude of the torque about the pivot point when the arm makes an angle of (a) 30°, (b) 90°, and (c) 180° with the vertical?

61E. A small 0.75 kg ball is attached to one end of a 1.25 m long massless rod, and the other end of the rod is hung from a pivot. When the resulting pendulum is 30° from the vertical, what is the magnitude of the torque about the pivot?

62E. A bicyclist of mass 70 kg puts all his weight on each downward-moving pedal as he pedals up a steep road. Take the diameter of the circle in which the pedals rotate to be 0.40 m, and determine the magnitude of the maximum torque he exerts.

63P. The body in Fig. 11-38 is pivoted at O, and two forces act on it as shown. (a) Find an expression for the magnitude of the net torque on the body about the pivot. (b) If $r_1 = 1.30$ m, $r_2 = 2.15$ m, $F_1 = 4.20$ N, $F_2 = 4.90$ N, $\theta_1 = 75.0°$, and $\theta_2 = 60.0°$, what is the net torque about the pivot?

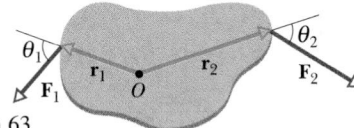

FIGURE 11-38 Problem 63.

64P. The body in Fig. 11-39 is pivoted at O. Three forces act on it in the directions shown on the figure: $F_A = 10$ N at point A, 8.0 m from O; $F_B = 16$ N at point B, 4.0 m from O; and $F_C = 19$ N at point C, 3.0 m from O. What is the net torque about O?

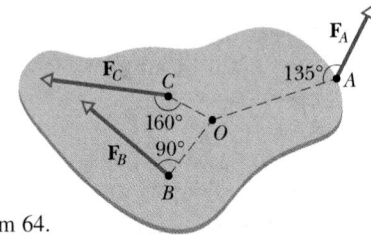

FIGURE 11-39 Problem 64.

SECTION 11-9 Newton's Second Law for Rotation

65E. When a torque of 32.0 N·m is applied to a certain wheel, the wheel acquires an angular acceleration of 25.0 rad/s². What is the rotational inertia of the wheel?

66E. Launching herself from a board, a diver changed her angular velocity from zero to 6.20 rad/s in 220 ms. Her rotational inertia is 12.0 kg·m². (a) What was her angular acceleration during the launch? (b) What external torque acted on the diver during the launch?

67E. A cylinder having a mass of 2.0 kg can rotate about its central axis through point O. Forces are applied as in Fig. 11-40:

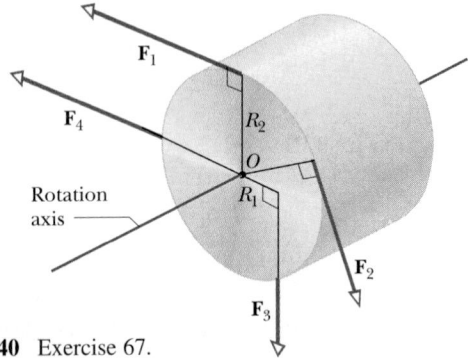

FIGURE 11-40 Exercise 67.

$F_1 = 6.0$ N, $F_2 = 4.0$ N, $F_3 = 2.0$ N, and $F_4 = 5.0$ N. Also, $R_1 = 5.0$ cm and $R_2 = 12$ cm. Find the magnitude and direction of the angular acceleration of the cylinder. (During the rotation, the forces maintain their same angles relative to the cylinder).

68E. A small object with mass 1.30 kg is mounted on one end of a rod 0.780 m long and of negligible mass. The system rotates in a horizontal circle about the other end of the rod at 5010 rev/min. (a) Calculate the rotational inertia of the system about the axis of rotation. (b) There is an air drag of 2.30×10^{-2} N on the object, directed opposite its motion. What torque must be applied to the system to keep it rotating at constant speed?

69E. A thin spherical shell has a radius of 1.90 m. An applied torque of 960 N·m imparts to the shell an angular acceleration equal to 6.20 rad/s² about an axis through the center of the shell. (a) What is the rotational inertia of the shell about the axis of rotation? (b) Calculate the mass of the shell.

70P. Figure 11-41 shows the massive shield door at a neutron test facility at Lawrence Livermore Laboratory; this is the world's heaviest hinged door. The door has a mass of 44,000 kg, a rotational inertia about an axis through its hinges of 8.7×10^4 kg·m², and a (front) face width of 2.4 m. Neglecting friction, what steady force, applied at its outer edge and perpendicular to the plane of the door, can move it from rest through an angle of 90° in 30 s? Assume no friction acts on the hinges.

FIGURE 11-41 Problem 70.

71P. A pulley, with a rotational inertia of 1.0×10^{-3} kg·m² about its axle and a radius of 10 cm, is acted on by a force applied tangentially at its rim. The force magnitude varies in time as $F = 0.50t + 0.30t^2$, where F is in newtons if t is given in seconds. The pulley is initially at rest. At $t = 3.0$ s what are (a) its angular acceleration and (b) its angular velocity?

72P. A wheel of radius 0.20 m is mounted on a frictionless horizontal axis. A massless cord is wrapped around the wheel and attached to a 2.0 kg object that slides on a frictionless surface inclined at an angle of 20° with the horizontal, as shown in Fig. 11-42. The object accelerates down the incline at 2.0 m/s². What is the rotational inertia of the wheel about its axis of rotation?

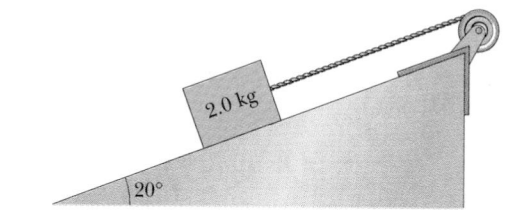

FIGURE 11-42 Problem 72.

73P. Two uniform solid spheres have the same mass, 1.65 kg, but one has a radius of 0.226 m while the other has a radius of 0.854 m. (a) For each of the spheres, find the torque required to bring the sphere from rest to an angular velocity of 317 rad/s in 15.5 s. Each sphere rotates about an axis through its center. (b) For each sphere, what force applied tangentially at the equator would provide the needed torque?

74P. In an Atwood's machine (Fig. 5-23), one block has a mass of 500 g, and the other a mass of 460 g. The pulley, which is mounted in horizontal frictionless bearings, has a radius of 5.00 cm. When released from rest, the heavier block is observed to fall 75.0 cm in 5.00 s (without the cord slipping on the pulley). (a) What is the acceleration of each block? What is the tension in the part of the cord that supports (b) the heavier block and (c) the lighter block? (d) What is the angular acceleration of the pulley? (e) What is its rotational inertia?

75P. Figure 11-43 shows two blocks, each of mass m, suspended from the ends of a rigid weightless rod of length $l_1 + l_2$, with $l_1 = 20$ cm and $l_2 = 80$ cm. The rod is held horizontally on the fulcrum shown in the figure and then released. Calculate the accelerations of the two blocks as they start to move.

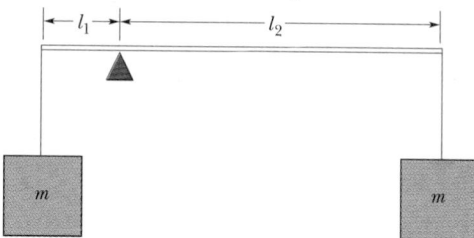

FIGURE 11-43 Problem 75.

76P. Two identical blocks, each of mass M, are connected by a massless string over a pulley of radius R and rotational inertia I (Fig. 11-44). The string does not slip on the pulley; it is not known whether there is friction between the table and the sliding block; the pulley's axis is frictionless. When this system is released, it is found that the pulley turns through an angle θ in time t

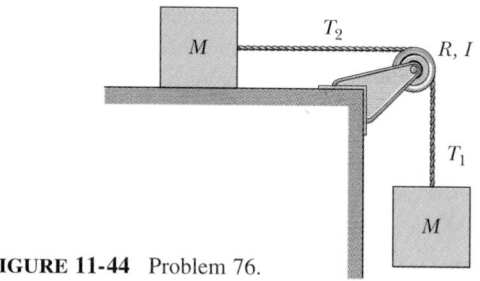

FIGURE 11-44 Problem 76.

and the acceleration of the blocks is constant. (a) What is the angular acceleration of the pulley? (b) What is the acceleration of the two blocks? (c) What are the tensions in the upper and lower sections of the string? Express the answers in terms of M, I, R, θ, g, and t.

SECTION 11-10 Work and Rotational Kinetic Energy

77E. (a) If $R = 12$ cm, $M = 400$ g, and $m = 50$ g in Fig. 11-18, find the speed of the block after it has descended 50 cm starting from rest. Solve the problem using energy conservation principles. (b) Repeat (a) with $R = 5.0$ cm.

78E. An automobile crankshaft develops 100 hp ($= 74.6$ kW) when rotating at a speed of 1800 rev/min. What torque does the crankshaft deliver?

79E. A 32.0 kg wheel, essentially a thin hoop with radius 1.20 m, is rotating at 280 rev/min. It must be brought to a stop in 15.0 s. (a) How much work must be done to stop it? (b) What is the required power?

80E. A thin rod of length l and mass m is suspended freely from one end. It is pulled to one side and then allowed to swing like a pendulum, passing through its lowest position with an angular speed ω. (a) Calculate its kinetic energy as it passes through its lowest position. (b) How high does its center of mass rise above its lowest position? Neglect friction and air resistance.

81P. Calculate (a) the torque, (b) the energy, and (c) the average power required to accelerate the Earth in 1 day from rest to its present angular speed about its axis.

82P. A meter stick is held vertically with one end on the floor and is then allowed to fall. Find the speed of the other end when it hits the floor, assuming that the end on the floor does not slip. (*Hint:* Consider the stick to be a thin rod and use the conservation of energy.)

83P. A rigid body is made of three identical thin rods, each with length l, fastened together in the form of a letter H (Fig. 11-45). The body is free to rotate about a horizontal axis that runs along the length of one of the legs of the H. The body is allowed to fall from rest from a position in which the plane of the H is horizontal. What is the angular speed of the body when the plane of the H is vertical?

FIGURE 11-45 Problem 83.

84P. A uniform cylinder of radius 10 cm and mass 20 kg is mounted so as to rotate freely about a horizontal axis that is parallel to and 5.0 cm from the central longitudinal axis of the cylinder. (a) What is the rotational inertia of the cylinder about the axis of rotation? (b) If the cylinder is released from rest with its central longitudinal axis at the same height as the axis about which the cylinder rotates, what is the angular speed of the cylinder as it passes through its lowest position? (*Hint:* Use the principle of conservation of energy.)

85P. A uniform helicopter rotor blade (see Fig. 11-36) is 7.80 m

long and has a mass of 110 kg. (a) What force is exerted on the bolt attaching the blade to the rotor axle when the rotor is turning at 320 rev/min? (*Hint:* For this calculation the blade can be considered to be a point mass at the center of mass. Why?) (b) Calculate the torque that must be applied to the rotor to bring it to full speed from rest in 6.7 s. Ignore air resistance. (The blade cannot be considered to be a point mass for this calculation. Why not? Assume the mass distribution of a uniform, thin rod.) (c) How much work did the torque do on the blade in order for the blade to reach a speed of 320 rev/min?

86P. A uniform spherical shell of mass M and radius R rotates about a vertical axis on frictionless bearings (Fig. 11-46). A massless cord passes around the equator of the shell, over a pulley of rotational inertia I and radius r, and is attached to a small object of mass m. There is no friction on the pulley's axle; the cord does not slip on the pulley. What is the speed of the object after it has fallen a distance h from rest? Use work–energy considerations.

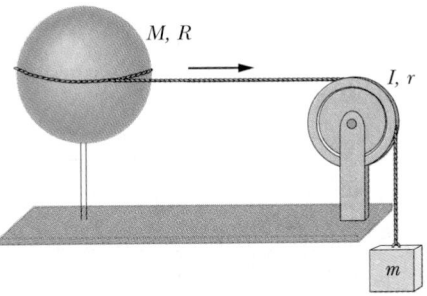

FIGURE 11-46 Problem 86.

87P. A tall, cylinder-shaped chimney falls over when its base is ruptured. Treating the chimney as a thin rod with height h, express the (a) radial and (b) tangential components of the linear acceleration of the top of the chimney as a function of the angle θ made by the chimney with the vertical. (c) At what angle θ does the linear acceleration equal g?

88P. Attached to each end of a thin steel rod of length 1.20 m and mass 6.40 kg is a small ball of mass 1.06 kg. The rod is constrained to rotate in a horizontal plane about a vertical axis through its midpoint. At a certain instant, it is observed to be rotating with an angular velocity of 39.0 rev/s. Because of friction, it comes to rest 32.0 s later. Assuming a constant frictional torque, compute (a) the angular acceleration, (b) the retarding torque exerted by the friction, (c) the total mechanical energy dissipated by the friction, and (d) the number of revolutions executed during the 32.0 s. (e) Now suppose that the frictional torque is known not to be constant. Which, if any, of the quantities (a), (b), (c), or (d) can still be computed without additional information? If such a quantity exists, give its value.

89P. A uniform rod of mass 1.5 kg is 2.0 m long (Fig. 11-47). The rod can pivot about a horizontal, frictionless pin through one end. It is released from rest at an angle of 40° above the horizontal. (a) What is the angular acceleration of the rod at the instant it is released? (The rotational inertia of the rod about the pin is 2.0 kg·m².) (b) Use the principle of conservation of energy to determine the angular speed of the rod as it passes through the horizontal position.

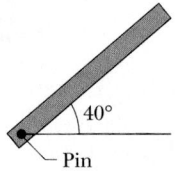

FIGURE 11-47 Problem 89.

90P. The rigid body shown in Fig. 11-48 consists of three particles connected by massless rods. It is to be rotated about an axis perpendicular to its plane through point P. If $M = 0.40$ kg, $a = 30$ cm, and $b = 50$ cm, how much work is required to take the body from rest to an angular speed of 5.0 rad/s?

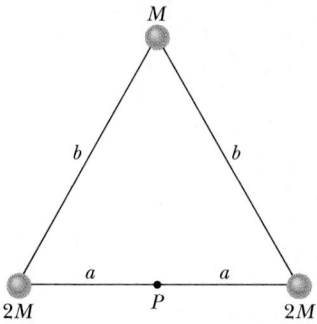

FIGURE 11-48 Problem 90.

91P*. A car is fitted with an energy-conserving flywheel, which in operation is geared to the driveshaft so that the flywheel rotates at 240 rev/s when the car is traveling at 80 km/h. The total mass of the car is 800 kg; the flywheel weighs 200 N and is a uniform disk 1.1 m in diameter. The car descends a 1500-m-long, 5° slope, starting from rest, with the flywheel engaged and no power generated from the engine. Neglecting friction and the rotational inertia of the wheels, find (a) the speed of the car at the bottom of the slope, (b) the angular acceleration of the flywheel at the bottom of the slope, and (c) the rate at which energy is stored in the rotation of the flywheel by the driveshaft as the car reaches the bottom of the slope.

ELECTRONIC COMPUTATION

92. In a certain machine two disks are free to rotate independently on parallel axles. The first has a radius of 7.0 cm and a rotational inertia of 1.5 kg·m². The second has a radius of 15 cm and a rotational inertia of 3.5 kg·m². While the first disk is at rest and the second is rotating at 175 rad/s, the axles are moved closer together so that the rims of the disks touch. Each disk exerts a normal force of 150 N on the other. (a) Take the coefficient of kinetic friction between the rims to be 0.25 and generate a table listing the values of the angular velocities for every 2 s during the first 80 s after the disks touch. Plot these values as functions of time. Also plot the total rotational kinetic energy of the disks. Is it conserved? (b) What influence does the frictional force have on the motions of the disks? Repeat the calculations for a coefficient of kinetic friction of 0.50. Is the time the disks take to reach their final angular velocities for the first μ_k value the same as for the second value? Are the final angular velocities the same for the two values? Is the final rotational kinetic energy the same?

93. Five particle-like objects, positioned in the xy plane according to the following table, form a rigidly connected body. What is the rotational inertia of the body about (a) the x axis, (b) the y axis, and (c) the z axis? (d) What is the center of mass of the body?

Object	1	2	3	4	5
Mass (grams)	500	400	300	600	450
x (cm)	15	−13	17	−4.0	−5.0
y (cm)	20	13	−6.0	−7.0	9.0

94. In Fig. 11-49, two blocks, of mass $m_1 = 400$ g and $m_2 = 600$ g, are connected by a massless cord that is wrapped around a uniform disk of mass $M = 500$ g and radius $R = 12.0$ cm. The disk can rotate without friction about a fixed horizontal axis through its center; the cord cannot slip on the disk. The system is released from rest. Find (a) the magnitude of the acceleration of the blocks, (b) the tension T_1 in the cord at the left, and (c) the tension T_2 in the cord at the right.

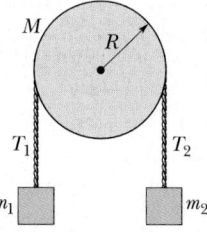

FIGURE 11-49 Problem 94.

12
Rolling, Torque, and Angular Momentum

In 1897, a European "aerialist" made the first triple somersault during the flight from a swinging trapeze to the hands of a partner. For the next 85 years aerialists attempted to complete a <u>quadruple</u> somersault, but not until 1982 was it done before an audience: Miguel Vazquez of the Ringling Bros. and Barnum & Bailey Circus rotated his body in four complete circles in midair before his brother Juan caught him. Both were stunned by their success. Why was the feat so difficult, and what feature of physics made it (finally) possible?

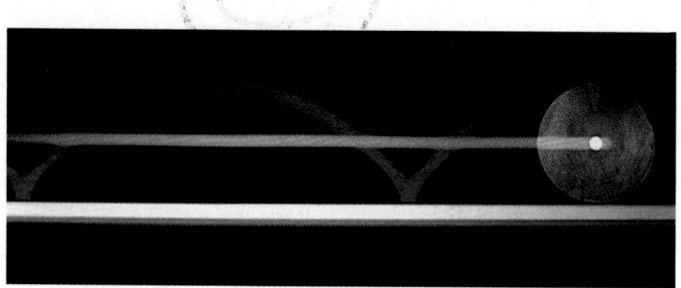

FIGURE 12-1 A time exposure photograph of a rolling disk. Small lights have been attached to the disk, one at its center and one at its edge. The latter traces out a curve called a *cycloid*.

12-1 ROLLING

When a bicycle moves along a straight track, the center of each wheel moves forward in pure translation. A point on the rim of the wheel, however, traces out a more complex path, as Fig. 12-1 shows. In what follows, we analyze the motion of a rolling wheel first by viewing it as a combination of pure translation and pure rotation, and then by viewing it as rotation alone.

Rolling as Rotation and Translation Combined

Imagine that you are watching the wheel of a bicycle, which passes you at constant speed while rolling smoothly, without slipping, along a street. As shown in Fig. 12-2, the center of mass O of the wheel moves forward at constant speed v_{cm}. The point P at which the wheel contacts the street also moves forward at speed v_{cm}, so that it always remains directly below O.

During a time interval t, you see both O and P move forward by a distance s. The bicycle rider sees the wheel rotate through an angle θ about the center of the wheel, with the point of the wheel that was touching the street at the beginning of t moving through arc length s. Equation

11-15 relates the arc length s to the rotation angle θ:

$$s = R\theta, \qquad (12\text{-}1)$$

where R is the radius of the wheel. The linear speed v_{cm} of the center of the wheel (the center of mass of this uniform wheel) is ds/dt, and the angular speed ω of the wheel about its center is $d\theta/dt$. So, differentiating Eq. 12-1 with respect to time gives us

$$v_{cm} = \omega R \qquad \text{(rolling motion)}. \qquad (12\text{-}2)$$

(Note that Eq. 12-2 holds only if the wheel rolls *smoothly*; that is, it does not slip over the street.)

Figure 12-3 shows that the rolling motion of a wheel is a combination of purely translational and purely rotational motions. Figure 12-3a shows the purely rotational motion (as if the rotation axis through the center were stationary): every point on the wheel rotates about the center with angular speed ω. (This is the type of motion we considered in Chapter 11.) Every point on the outside edge of the wheel has linear speed v_{cm} given by Eq. 12-2. Figure 12-3b shows the purely translational motion (as if the wheel did not rotate at all): every point on the wheel moves to the right with speed v_{cm}.

The combination of Figs. 12-3a and 12-3b yields the actual rolling motion of the wheel, Fig. 12-3c. Note that in this combination of motions, the portion of the wheel at the bottom (at point P) is stationary and the portion at the top (at point T) is moving at speed $2v_{cm}$, faster than any other portion of the wheel. These results are demonstrated in Fig. 12-4, which is a time exposure of a rolling bicycle wheel. The blurring of the spokes near the top of the wheel com-

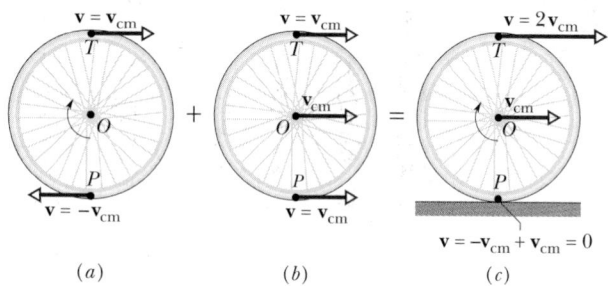

$$\mathbf{v} = -\mathbf{v}_{cm} + \mathbf{v}_{cm} = 0$$

(a)   (b)   (c)

FIGURE 12-3 Rolling motion of a wheel as a combination of purely rotational motion and purely translational motion. (a) The purely rotational motion: all points on the wheel move with the same angular speed ω. Points on the outside edge of the wheel all move with the same linear speed $v = v_{cm}$. The linear velocities $\mathbf{v}$ of two such points, at top (T) and bottom (P) of the wheel, are shown. (b) The purely translational motion: all points on the wheel move to the right with the same linear velocity $\mathbf{v}_{cm}$ as the center of the wheel. (c) The rolling motion of the wheel is the combination of (a) and (b).

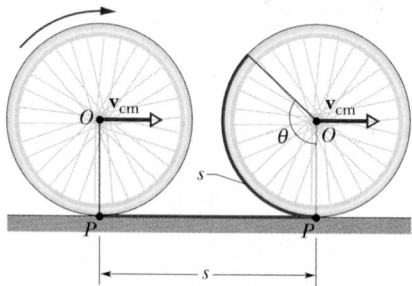

FIGURE 12-2 The center of mass O of a rolling wheel moves a distance s at velocity $\mathbf{v}_{cm}$, while the wheel rotates through angle θ. The contact point P between the wheel and the surface over which the wheel rolls also moves a distance s.

FIGURE 12-4 A photograph of a rolling bicycle wheel. The spokes near the top of the wheel are more blurred than those near the bottom of the wheel because they are moving faster, as Fig. 12-3c shows.

pared with the sharper images of the spokes near the bottom of the wheel shows that the wheel is moving faster near its top than near its bottom.

The motion of any round body rolling smoothly over a surface can be separated into purely rotational and purely translational motions, as in Figs. 12-3a and 12-3b.

Rolling as Pure Rotation

Figure 12-5 suggests another way to look at the rolling motion of a wheel, namely, as pure rotation about an axis that always extends through the point where the wheel contacts the street as the wheel moves. That is, we consider the rolling motion to be pure rotation about an axis passing through point P in Fig. 12-3c and perpendicular to the

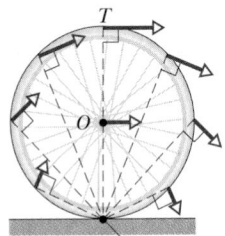

Rotational axis at P

FIGURE 12-5 Rolling can be viewed as pure rotation, with angular speed ω, about an axis that always extends through P. The vectors show the instantaneous linear velocities of selected points on the rolling wheel. You can obtain the vectors by combining the translational and rotational motions as in Fig. 12-3.

plane of the figure. The vectors in Fig. 12-5 then represent the instantaneous velocities of points on the rolling wheel.

Question. What angular speed about this new axis will a stationary observer assign to a rolling bicycle wheel?
Answer. The same angular speed ω that the rider assigns to the wheel as she or he observes it in pure rotation about an axis through its center of mass.

To verify this answer, let us use it to calculate the linear speed of the top of the rolling wheel from the point of view of a stationary observer. If we call the wheel's radius R, the top is a distance $2R$ from the axis through P in Fig. 12-5, so the linear speed at the top should be (using Eq. 12-2)

$$v_{\text{top}} = (\omega)(2R) = 2(\omega R) = 2v_{\text{cm}},$$

in exact agreement with Fig. 12-3c. You can similarly verify the linear speeds shown for the portion of the wheel at points O and P in Fig. 12-3c.

> **C**HECKPOINT **1:** The rear wheel on a clown's bicycle has twice the radius of the front wheel. (a) Is the linear speed at the very top of the rear wheel greater than, less than, or the same as that of the front wheel? (b) Is the angular speed of the rear wheel greater than, less than, or the same as that of the front wheel?

The Kinetic Energy

Let us now calculate the kinetic energy of the rolling wheel as measured by the stationary observer. If we view the

As this hoop rolls, does the dog walk as fast as the pony?

rolling as pure rotation about an axis through P in Fig. 12-5, we have

$$K = \tfrac{1}{2}I_P\omega^2, \qquad (12\text{-}3)$$

in which ω is the angular speed of the wheel and I_P is the rotational inertia of the wheel about the axis through P. From the parallel-axis theorem (Eq. 11-27) we have

$$I_P = I_{\text{cm}} + MR^2, \qquad (12\text{-}4)$$

in which M is the mass of the wheel and I_{cm} is its rotational inertia about an axis through its center of mass. Substituting Eq. 12-4 into Eq. 12-3, we obtain

$$K = \tfrac{1}{2}I_{\text{cm}}\omega^2 + \tfrac{1}{2}MR^2\omega^2,$$

and using the relation $v_{\text{cm}} = \omega R$ (Eq. 12-2) yields

$$K = \tfrac{1}{2}I_{\text{cm}}\omega^2 + \tfrac{1}{2}Mv_{\text{cm}}^2. \qquad (12\text{-}5)$$

We can interpret the first of these terms ($\tfrac{1}{2}I_{\text{cm}}\omega^2$) as the kinetic energy associated with the rotation of the wheel about an axis through its center of mass (Fig. 12-3a), and the second term ($\tfrac{1}{2}Mv_{\text{cm}}^2$) as the kinetic energy associated with the translational motion of the wheel (Fig. 12-3b).

Friction and Rolling

If the wheel rolls at constant speed, as in Fig. 12-2, it has no tendency to slide at the point of contact P, and thus there is no frictional force acting on the wheel there. However, if a force acts on the wheel, changing the speed v_{cm} of the center of the wheel or the angular speed ω about the center, then there is a tendency for the wheel to slide at P, and a frictional force acts on the wheel at P to oppose that tendency. Until the wheel actually begins to slide, the frictional force is a *static* frictional force $\mathbf{f}_s$. If the wheel begins to slide, then the force is a *kinetic* frictional force $\mathbf{f}_k$.

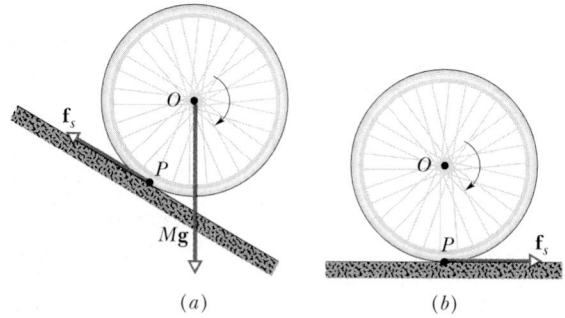

(a)   (b)

FIGURE 12-6 (a) A wheel rolls down an incline without sliding. A static frictional force $\mathbf{f}_s$ acts on the wheel at P, opposing the wheel's tendency to slide because of its weight $M\mathbf{g}$. (b) A wheel rolls horizontally without sliding while its angular speed is increased. A static frictional force $\mathbf{f}_s$ acts on the wheel at P, opposing the tendency to slide. If in (a) or (b) the wheel slides, the frictional force is a kinetic frictional force $\mathbf{f}_k$.

In Fig. 12-6a, a wheel rolls down an incline. Its weight $M\mathbf{g}$ acts on the wheel at the wheel's center of mass. Since that force $M\mathbf{g}$ lacks any moment arm relative to the center of the wheel, it cannot produce a torque about the center and cause the wheel to rotate about the center. But since $M\mathbf{g}$ tends to slide the wheel down the incline, a frictional force acts on the wheel at contact point P to oppose the tendency to slide. This force, which points up along the incline, *does* have a moment arm relative to the center; the moment arm is the radius of the wheel. Thus this frictional force produces a torque about the center and causes the wheel to rotate about the center.

In Fig. 12-6b, a wheel is being made to rotate faster while rolling along a flat surface, as on an accelerating bicycle. The increase in ω tends to slide the bottom of the wheel toward the left. A frictional force acts on the wheel toward the right at P to oppose the tendency to slide. (This frictional force is the external force acting on an accelerating bicycle–rider system that causes the acceleration.)

SAMPLE PROBLEM 12-1

A uniform solid cylindrical disk, whose mass M is 1.4 kg and whose radius R is 8.5 cm, rolls across a horizontal table at a speed v of 15 cm/s.

(a) What is the speed of the top of the rolling disk?

SOLUTION: When we speak of the speed of a rolling object, we always mean the speed of its center of mass. From Fig. 12-3c we see that the speed of the top of the disk is just twice this, or

$$v_{\text{top}} = 2v_{\text{cm}} = (2)(15 \text{ cm/s}) = 30 \text{ cm/s}. \quad \text{(Answer)}$$

(b) What is the angular speed ω of the rolling disk?

SOLUTION: From Eq. 12-2 we have

$$\omega = \frac{v_{\text{cm}}}{R} = \frac{15 \text{ cm/s}}{8.5 \text{ cm}}$$

$$= 1.8 \text{ rad/s} = 0.28 \text{ rev/s}. \quad \text{(Answer)}$$

This value applies whether the axis of rotation is taken to be an axis through point P in Fig. 12-5 or an axis through the center of mass.

(c) What is the kinetic energy K of the rolling disk?

SOLUTION: From Eq. 12-5 we have, putting $I_{\text{cm}} = \tfrac{1}{2}MR^2$ and using the relation $v_{\text{cm}} = \omega R$,

$$K = \tfrac{1}{2}I_{\text{cm}}\omega^2 + \tfrac{1}{2}Mv_{\text{cm}}^2$$

$$= (\tfrac{1}{2})(\tfrac{1}{2}MR^2)(v_{\text{cm}}/R)^2 + \tfrac{1}{2}Mv_{\text{cm}}^2 = \tfrac{3}{4}Mv_{\text{cm}}^2$$

$$= \tfrac{3}{4}(1.4 \text{ kg})(0.15 \text{ m/s})^2$$

$$= 0.024 \text{ J} = 24 \text{ mJ}. \quad \text{(Answer)}$$

TABLE 12-1 **THE RELATIVE SPLITS BETWEEN ROTATIONAL AND TRANSLATIONAL ENERGIES FOR ROLLING OBJECTS**

OBJECT	ROTATIONAL INERTIA I_{cm}	PERCENTAGE OF ENERGY IN	
		TRANSLATION	ROTATION
Hoop	$1MR^2$	50%	50%
Disk	$\frac{1}{2}MR^2$	67%	33%
Sphere	$\frac{2}{5}MR^2$	71%	29%
General[a]	βMR^2	$100\,\dfrac{1}{1+\beta}$ %	$100\,\dfrac{\beta}{1+\beta}$ %

[a]β may be computed for any rolling object as I_{cm}/MR^2.

(d) What fraction of the kinetic energy is associated with the motion of translation and what fraction with the motion of rotation about an axis through the center of mass?

SOLUTION: The kinetic energy associated with translation is the second term of Eq. 12-5, or $\frac{1}{2}Mv_{cm}^2$. The fraction we seek is then, using the expression derived in (c),

$$\text{frac} = \frac{\frac{1}{2}Mv_{cm}^2}{\frac{3}{4}Mv_{cm}^2} = \frac{2}{3} \quad \text{or} \quad 67\%. \quad \text{(Answer)}$$

The remaining 33% is associated with rotation about an axis through the center of mass.

The relative split between translational and rotational energy depends on the rotational inertia of the rolling object. As Table 12-1 shows, the rolling object (the hoop) that has its mass farthest from the central axis of rotation (and so has the largest rotational inertia) has the largest share of its kinetic energy in rotational motion. The object (the sphere) that has its mass closest to the central axis of rotation (and so has the smallest rotational inertia) has the smallest share in that form.

The formulas at the bottom of Table 12-1 apply to the generic rolling object with **rotational inertia parameter** β. This parameter has the value 1 for a hoop, $\frac{1}{2}$ for a disk, and $\frac{2}{5}$ for a sphere.

SAMPLE PROBLEM 12-2

A bowling ball, whose radius R is 11 cm and whose mass M is 7.2 kg, rolls from rest down a ramp whose length L is 2.1 m. The ramp is inclined at an angle θ of 34° to the horizontal; see the sphere in Fig. 12-7. How fast is the ball moving when it reaches the bottom of the ramp? Assume the ball is uniform in density.

SOLUTION: The center of the ball falls a vertical distance $h = L \sin \theta$; so the decrease in gravitational potential energy is $MgL \sin \theta$. This loss of potential energy equals the gain in kinetic energy. Thus we can write (see Eq. 12-5)

$$MgL \sin \theta = \frac{1}{2}I_{cm}\omega^2 + \frac{1}{2}Mv_{cm}^2. \quad (12\text{-}6)$$

From Table 11-2(g) we see that, for a solid sphere, $I_{cm} = \frac{2}{5}MR^2$. We can also replace ω with its equal, v_{cm}/R. Substituting both these quantities into Eq. 12-6 yields

$$MgL \sin \theta = (\tfrac{1}{2})(\tfrac{2}{5})(MR^2)(v_{cm}/R)^2 + \tfrac{1}{2}Mv_{cm}^2.$$

Solving for v_{cm} yields

$$v_{cm} = \sqrt{(\tfrac{10}{7})\, gL \sin \theta}$$
$$= \sqrt{(\tfrac{10}{7})(9.8 \text{ m/s}^2)(2.1 \text{ m})(\sin 34°)}$$
$$= 4.1 \text{ m/s}. \quad \text{(Answer)}$$

Note that the answer does not depend on the mass or radius of the ball.

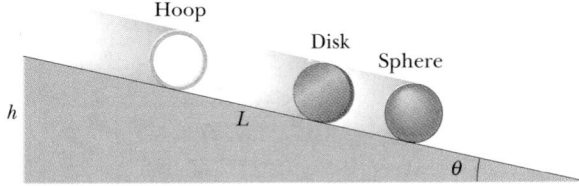

FIGURE 12-7 Sample Problems 12-2 and 12-3. A hoop, a disk, and a sphere roll from rest down a ramp of angle θ. Although released from rest at the same position and time, they arrive at the bottom in the order shown.

SAMPLE PROBLEM 12-3

Here we generalize the result of Sample Problem 12-2. A uniform hoop, disk, and sphere, having the same mass M and the same radius R, are released simultaneously from rest at the top of a ramp whose length L is 2.5 m and whose ramp angle θ is 12° (Fig. 12-7).

(a) Which object reaches the bottom first?

SOLUTION: Table 12-1 gives us the answer. The sphere puts the largest share of its kinetic energy (71%) into translational motion, so it wins the race. Next comes the disk and then the hoop.

(b) How fast are the objects moving at the ramp's bottom?

SOLUTION: The center of mass of each object rolling down the ramp falls the same vertical distance h. Like a body in free fall, the object loses potential energy in amount Mgh and thus gains this amount of kinetic energy. At the bottom of the ramp then, the total kinetic energies of all three objects are the same. How these kinetic energies are divided between the translational and rotational forms depends on each object's distribution of mass.

From Eq. 12-5 we can write (putting $\omega = v_{cm}/R$)

$$Mgh = \tfrac{1}{2}I_{cm}\omega^2 + \tfrac{1}{2}Mv_{cm}^2$$
$$= \tfrac{1}{2}I_{cm}(v_{cm}^2/R^2) + \tfrac{1}{2}Mv_{cm}^2$$
$$= \tfrac{1}{2}(I_{cm}/R^2)v_{cm}^2 + \tfrac{1}{2}Mv_{cm}^2. \quad (12\text{-}7)$$

Putting $h = L \sin \theta$ and solving for v_{cm}, we obtain

$$v_{cm} = \sqrt{\frac{2gL \sin \theta}{1 + I_{cm}/MR^2}}, \quad \text{(Answer)} \quad (12\text{-}8)$$

which is the symbolic answer to the question.

Note that the speed depends not on the mass or the radius of the rolling object, but only on the distribution of its mass about its central axis, which enters through the term I_{cm}/MR^2. A marble and a bowling ball will have the same speed at the bottom of the ramp and will thus roll down the ramp in the same time. A bowling ball will beat a disk of any mass or radius, and almost anything that rolls will beat a hoop.

For the rolling hoop (see the hoop listing in Table 12-1) we have $I_{cm}/MR^2 = 1$, so Eq. 12-8 yields

$$\begin{aligned} v_{cm} &= \sqrt{\frac{2gL \sin \theta}{1 + I_{cm}/MR^2}} \\ &= \sqrt{\frac{(2)(9.8 \text{ m/s}^2)(2.5 \text{ m})(\sin 12°)}{1 + 1}} \\ &= 2.3 \text{ m/s}. \quad \text{(Answer)} \end{aligned}$$

From a similar calculation, we obtain $v_{cm} = 2.6$ m/s for the disk ($I_{cm}/MR^2 = \frac{1}{2}$) and 2.7 m/s for the sphere ($I_{cm}/MR^2 = \frac{2}{5}$). This supports our prediction of (a) that the win, place, and show sequence in this race will be sphere, disk, and hoop.

SAMPLE PROBLEM 12-4

Figure 12-8 shows a round uniform body of mass M and radius R rolling down a ramp at an angle θ. This time let us analyze the motion directly from Newton's laws, rather than by energy methods as we did in Sample Problem 12-3.

(a) What is the linear acceleration of the rolling body?

SOLUTION: Figure 12-8 also shows the forces that act on the body: its weight $M\mathbf{g}$, a normal force $\mathbf{N}$, and a static frictional force $\mathbf{f}_s$. The weight can be considered to act at the center of mass, which is at the center of this uniform body. The normal force and the frictional force act on the portion of the body in contact with the ramp at point P. The weight and normal force have zero moment arms about an axis through the center of the body. So they cannot cause the body to rotate about that center. Clockwise rotation of the body results from a negative torque due to the frictional force; that force has a moment arm of R about the center of the body.

We now apply the linear form of Newton's second law ($\Sigma F = Ma$) along the ramp, taking the positive direction to be up the ramp. We obtain

$$\sum F = f_s - Mg \sin \theta = Ma. \quad (12\text{-}9)$$

This equation has two unknowns, f_s and a. To obtain another equation in the same two unknowns, we next apply the angular form of Newton's second law ($\Sigma \tau = I\alpha$) about the rotation axis through the center of mass. (Although we derived the

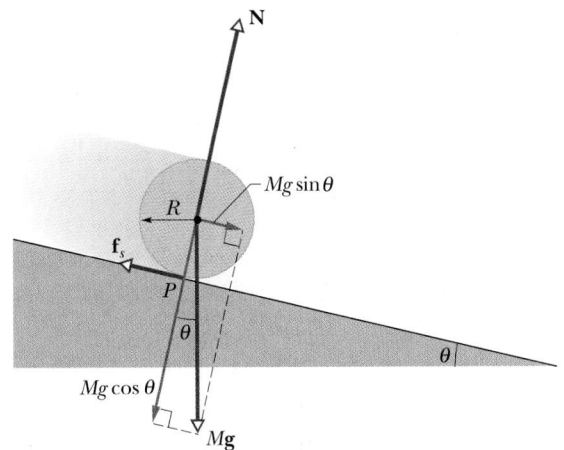

FIGURE 12-8 Sample Problem 12-4. A round uniform body of radius R rolls down a ramp. The forces that act on it are its weight $M\mathbf{g}$, a normal force $\mathbf{N}$, and a frictional force $\mathbf{f}_s$ pointing up the ramp. (For clarity, $\mathbf{N}$ has been shifted along its line of action until its tail is at the center of the body.)

relation $\Sigma \tau = I\alpha$ in Chapter 11 for an axis fixed in an inertial frame, it holds also for a rotation axis through the center of mass of this or another accelerating body, provided the axis does not change direction.) We get

$$\sum \tau = -f_s R = I_{cm}\alpha = \frac{I_{cm}a}{R}, \quad (12\text{-}10)$$

where we have used the relation $\alpha = a/R$ (Eq. 11-20).

Solving Eq. 12-10 for the frictional force f_s gives

$$f_s = -\frac{I_{cm}a}{R^2}, \quad (12\text{-}11)$$

where the minus sign reminds us that the frictional force $\mathbf{f}_s$ acts in the direction opposite that of the acceleration $\mathbf{a}$. Substituting Eq. 12-11 into Eq. 12-9 and solving for a, we have

$$a = -\frac{g \sin \theta}{1 + I_{cm}/MR^2}. \quad \text{(Answer)} \quad (12\text{-}12)$$

We could have, instead, found a second equation by summing torques and applying Newton's law in angular form about an axis through the *point of contact P*. This time, $\Sigma \tau$ would consist only of the torque due to the force component $Mg \sin \theta$, acting at the center of the body with moment arm R:

$$\sum \tau = -(Mg \sin \theta)(R) = I_P\alpha = \frac{I_P a}{R}, \quad (12\text{-}13)$$

where I_P is the rotational inertia about an axis through P. To find I_P, we would use the parallel-axis theorem:

$$I_P = I_{cm} + MR^2. \quad (12\text{-}14)$$

Substituting I_P from Eq. 12-14 in Eq. 12-13 and solving for a, we would again obtain Eq. 12-12.

(b) What is the frictional force f_s?

SOLUTION: Substituting Eq. 12-12 into Eq. 12-11 yields

$$f_s = Mg \frac{\sin \theta}{1 + MR^2/I_{cm}}. \quad \text{(Answer)} \quad (12\text{-}15)$$

Study of Eq. 12-15 shows that the frictional force is less than $Mg \sin \theta$, the component of the weight that acts parallel to the ramp. This is necessarily true if the object is to accelerate down the ramp.

Table 12-1 shows that if the rolling body is a solid disk, $I_{cm}/MR^2 = \frac{1}{2}$. The acceleration and the frictional force then follow from Eqs. 12-12 and 12-15 and are

$$a = -\tfrac{2}{3}g \sin \theta \quad \text{and} \quad f_s = \tfrac{1}{3}Mg \sin \theta.$$

(c) What is the speed of the rolling body at the bottom of the ramp if the ramp has length L?

SOLUTION: The motion is one of constant acceleration, so we can use the relation

$$v^2 = v_0^2 + 2a(x - x_0). \quad (12\text{-}16)$$

Putting $x - x_0 = -L$ and $v_0 = 0$ and introducing a from Eq. 12-12, we obtain Eq. 12-8—the result that we derived by energy methods.

$\mathbb{C}$HECKPOINT **2:** Disks A and B are identical and roll across a floor with equal speeds. Then disk A rolls up an incline, reaching a maximum height h, and disk B moves up an incline that is identical except that it is frictionless. Is the maximum height reached by disk B greater than, less than, or equal to h?

12-2 THE YO-YO

A yo-yo is a physics lab that you can fit in your pocket. If a yo-yo rolls down its string for a distance h, it loses potential energy in amount mgh but gains kinetic energy in both translational ($\tfrac{1}{2}mv_{cm}^2$) and rotational ($\tfrac{1}{2}I_{cm}\omega^2$) forms. When it is climbing back up, it loses kinetic energy and regains potential energy.

In a modern yo-yo the string is not tied to the axle but is looped around it. When the yo-yo "hits" the bottom of its string, an upward force on the axle from the string stops the descent, removing the small remaining translational kinetic energy. The yo-yo then spins, axle inside loop, with only rotational kinetic energy. The yo-yo keeps spinning ("sleeping") until you "wake it" by jerking on the string, causing the string to catch on the axle and the yo-yo to climb back up. The rotational kinetic energy of the yo-yo at the bottom of its string (and thus the sleeping time) can be considerably increased by throwing the yo-yo downward so that it starts down the string with initial speeds v_{cm} and ω instead of rolling down from rest.

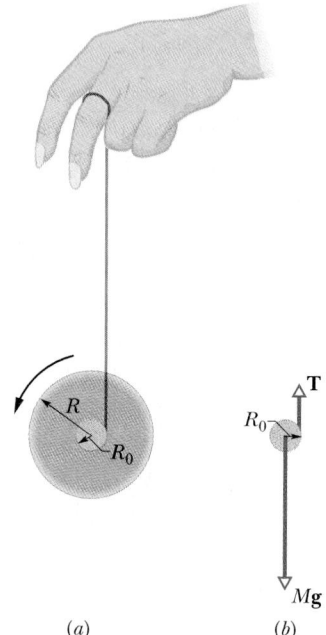

(a) $\qquad$ (b)

FIGURE 12-9 (a) A yo-yo, shown in cross section. The string, of assumed negligible thickness, is wound around an axle of radius R_0. (b) A free-body diagram for the falling yo-yo. Only the axle is shown. In a real yo-yo, the thickness of the string cannot be neglected. It changes the effective radius of the yo-yo axle, which then varies with the amount of wound-up string.

Let us analyze the motion of the yo-yo directly with Newton's second law. Figure 12-9a shows an idealized yo-yo, in which the thickness of the string can be neglected. Figure 12-9b is a free-body diagram, in which only the yo-yo axle is shown. We seek an expression for the linear acceleration a of the yo-yo. Applying Newton's second law in its linear form ($\Sigma F = ma$) yields

$$\sum F = T - Mg = Ma. \quad (12\text{-}17)$$

Here M is the mass of the yo-yo, and T is the tension in the yo-yo's string.

Applying Newton's second law in angular form ($\Sigma \tau = I\alpha$) about an axis through the center of mass yields

$$\sum \tau = TR_0 = I\alpha, \quad (12\text{-}18)$$

where R_0 is the radius of the yo-yo axle and I is the rotational inertia of the yo-yo about its central axis. The linear acceleration a of the yo-yo is downward (and thus *negative*). From the perspective of Fig. 12-9, the angular acceleration α of the yo-yo is counterclockwise (and thus *positive*) because from that perspective the torque given by Eq. 12-18 is counterclockwise. So we relate α to a with the relation $a = -\alpha R_0$. Solving this relation for $\alpha (= -a/R_0)$ and substituting into Eq. 12-18, we find

$$TR_0 = -\frac{Ia}{R_0}.$$

After eliminating T between this and Eq. 12-17, we solve for a, obtaining

$$a = -g \frac{1}{1 + I/MR_0^2}. \qquad (12\text{-}19)$$

Thus an ideal yo-yo rolls down its string with constant acceleration. For a small acceleration, you need a light yo-yo with a large rotational inertia and a small axle radius.

SAMPLE PROBLEM 12-5

A yo-yo is constructed of two brass disks whose thickness b is 8.5 mm and whose radius R is 3.5 cm, joined by a short axle whose radius R_0 is 3.2 mm.

(a) What is the yo-yo's rotational inertia about its central axis? Neglect the rotational inertia of the axle. The density ρ of brass is 8400 kg/m^3.

SOLUTION: The rotational inertia I of a disk about its central axis is $\frac{1}{2}MR^2$. In this problem, we can treat the two disks together, as a single disk. We first find its mass from its density and its volume V:

$$\begin{aligned} M &= V\rho = (2)(\pi R^2)(b)(\rho) \\ &= (2)(\pi)(0.035 \text{ m})^2(0.0085 \text{ m})(8400 \text{ kg/m}^3) \\ &= 0.550 \text{ kg}. \end{aligned}$$

The rotational inertia is then

$$\begin{aligned} I &= \tfrac{1}{2}MR^2 = (\tfrac{1}{2})(0.550 \text{ kg})(0.035 \text{ m})^2 \\ &= 3.4 \times 10^{-4} \text{ kg}\cdot\text{m}^2. \qquad \text{(Answer)} \end{aligned}$$

(b) A string of length $l = 1.1$ m and of negligible thickness is wound on the axle. What is the linear acceleration of the yo-yo as it rolls down the string from rest?

SOLUTION: From Eq. 12-19,

$$\begin{aligned} a &= -g \frac{1}{1 + I/MR_0^2} \\ &= -\frac{9.8 \text{ m/s}^2}{1 + \dfrac{3.4 \times 10^{-4} \text{ kg}\cdot\text{m}^2}{(0.550 \text{ kg})(0.0032 \text{ m})^2}} \\ &= -0.16 \text{ m/s}^2. \qquad \text{(Answer)} \end{aligned}$$

The acceleration points downward and has this value whether the yo-yo is rolling down the string or climbing it.

Note that the quantity I/MR_0^2 in Eq. 12-19 is simply the rotational inertia parameter β that was introduced in Table 12-1. For this yo-yo, we have $\beta = 60$, a much greater value than that for any of the objects listed in that table. The acceleration of our yo-yo is small, corresponding to that of a hoop rolling down a 1.9° ramp.

(c) What is the tension in the string of the yo-yo?

SOLUTION: We can find this by substituting a from Eq. 12-19 into Eq. 12-17 and solving for T. We find

$$T = \frac{Mg}{1 + MR_0^2/I}, \qquad (12\text{-}20)$$

which tells us, as it must, that the tension in the string is smaller than the weight of the yo-yo. Numerically, we have

$$\begin{aligned} T &= \frac{(0.550 \text{ kg})(9.8 \text{ m/s}^2)}{1 + (0.550 \text{ kg})(0.0032 \text{ m})^2/(3.4 \times 10^{-4} \text{ kg}\cdot\text{m}^2)} \\ &= 5.3 \text{ N}. \qquad \text{(Answer)} \end{aligned}$$

This is the tension whether the yo-yo is falling down or climbing up the string.

12-3 TORQUE REVISITED

In Chapter 11 we defined torque τ for a rigid body that can rotate around a fixed axis, with each particle in the body forced to move in a path that is a circle about that axis. We now expand the definition of torque to apply it to an individual particle that moves along any path relative to a fixed *point* (rather than a fixed axis). The path need no longer be a circle.

Figure 12-10a shows such a particle at point P in the xy plane. A single force $\mathbf{F}$ in that plane acts on the particle, and the particle's position relative to the origin O is given by position vector $\mathbf{r}$. The torque τ acting on the particle relative to the fixed point O is a vector quantity defined as

$$\boldsymbol{\tau} = \mathbf{r} \times \mathbf{F} \qquad \text{(torque defined)}. \qquad (12\text{-}21)$$

We can evaluate the vector (or cross) product in this definition of τ by using the rules for such products given in Section 3-7. To find the direction of τ, we slide the vector $\mathbf{F}$ (without changing its direction) until its tail is at origin O, so that the two vectors in the vector product are tail to tail as in Fig. 12-10b. We then use the right-hand rule for vector products in Fig. 3-20a, sweeping the fingers of the right hand from $\mathbf{r}$ (the first vector in the product) into $\mathbf{F}$ (the second vector). The outstretched right thumb then gives the direction of τ. In Fig. 12-10b, τ points along the positive direction of the z axis.

To find the magnitude of τ, we use the general result of Eq. 3-20 ($c = ab \sin \phi$), finding

$$\tau = rF \sin \phi, \qquad (12\text{-}22)$$

where ϕ is the angle between $\mathbf{r}$ and $\mathbf{F}$ when the vectors are tail to tail. From Fig. 12-10b, we see that Eq. 12-22 can be rewritten as

$$\tau = rF_\perp, \qquad (12\text{-}23)$$

where $F_\perp$ ($= F \sin \phi$) is the component of $\mathbf{F}$ perpendicular to $\mathbf{r}$. From Fig. 12-10c, we see that Eq. 12-22 can also be

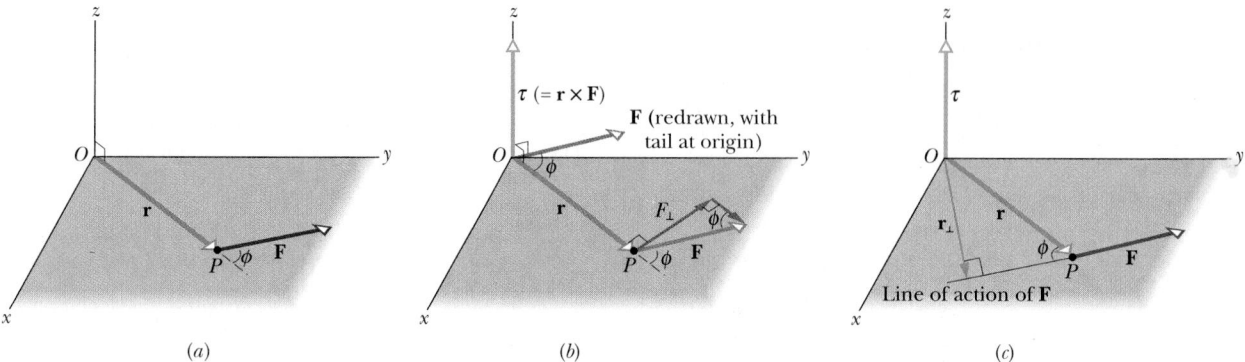

FIGURE 12-10 Defining torque. (*a*) A force **F**, lying in the *xy* plane, acts on a particle at point *P*. (*b*) This force produces a torque τ ($= \mathbf{r} \times \mathbf{F}$) on the particle with respect to the origin *O*. By the right-hand rule for vector (cross) products, the torque vector points in the direction of increasing *z*. Its magnitude is given by $rF_\perp$ in (*b*) and by $r_\perp F$ in (*c*).

rewritten as

$$\tau = r_\perp F, \qquad (12\text{-}24)$$

where $r_\perp$ ($= r \sin \phi$) is the moment arm of **F** (the perpendicular distance between *O* and the line of action of **F**).

SAMPLE PROBLEM 12-6

In Fig. 12-11*a*, three forces, each of magnitude 2.0 N, act on a particle. The particle is in the *xz* plane at point *P* given by position vector **r**, where $r = 3.0$ m and $\theta = 30°$. Force $\mathbf{F}_1$ is parallel to the *x* axis, force $\mathbf{F}_2$ is parallel to the *z* axis, and force $\mathbf{F}_3$ is parallel to the *y* axis. What is the torque, with respect to the origin *O*, due to each force?

SOLUTION: Figures 12-11*b* and 12-11*c* are direct views of the *xz* plane, redrawn with vectors $\mathbf{F}_1$ and $\mathbf{F}_2$ shifted so their tails are at the origin, to better show the angles between those vectors and vector **r**. The angle between **r** and $\mathbf{F}_3$ is 90°. Applying Eq. 12-22 for each force, we find the magnitudes of the torques to be

$$\tau_1 = rF_1 \sin \phi_1 = (3.0 \text{ m})(2.0 \text{ N})(\sin 150°)$$
$$= 3.0 \text{ N} \cdot \text{m},$$

$$\tau_2 = rF_2 \sin \phi_2 = (3.0 \text{ m})(2.0 \text{ N})(\sin 120°)$$
$$= 5.2 \text{ N} \cdot \text{m},$$

and $\quad \tau_3 = rF_3 \sin \phi_3 = (3.0 \text{ m})(2.0 \text{ N})(\sin 90°)$

$$= 6.0 \text{ N} \cdot \text{m}. \qquad \text{(Answer)}$$

To find the directions of these torques, we use the right-hand rule, placing the fingers of the right hand so as to rotate **r** into **F** through the *smaller* of the two angles they make. In Fig. 12-11*d*, we represent $\mathbf{F}_3$ with a circled cross $\otimes$, suggesting the tail of an arrow. (Were it in the opposite direction, $\mathbf{F}_3$ would be represented by a circled dot $\odot$, suggesting the tip of an arrow.) Rotating **r** into $\mathbf{F}_3$ with the fingers of the right hand yields the direction of τ_3 (in the direction of the thumb). All three of the torque vectors are shown in Fig. 12-11*e*.

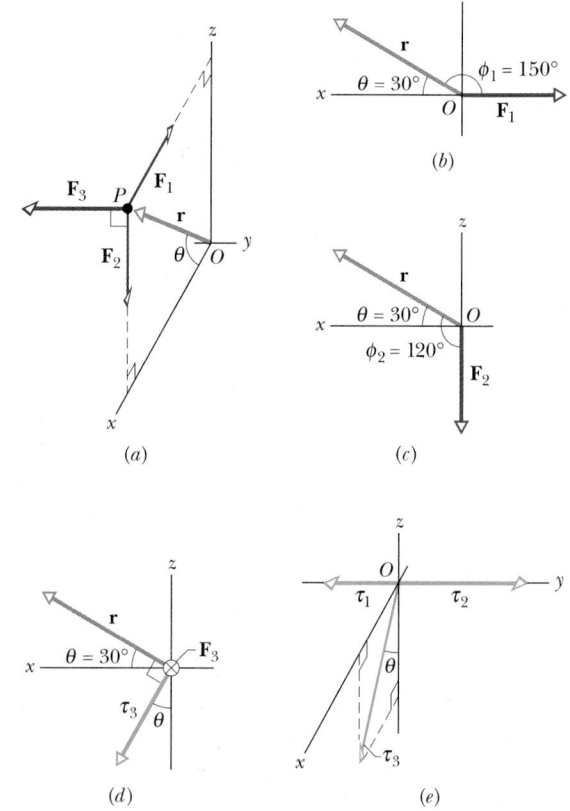

FIGURE 12-11 Sample Problem 12-6. (*a*) A particle at point *P* is acted on by three forces, each parallel to a coordinate axis. The angle ϕ (used in finding torque) is shown (*b*) for $\mathbf{F}_1$ and (*c*) for $\mathbf{F}_2$. (*d*) Torque τ_3 is perpendicular to both **r** and $\mathbf{F}_3$ (force $\mathbf{F}_3$ is directed into the plane of the figure). (*e*) The torques (relative to the origin *O*) acting on the particle.

PROBLEM SOLVING TACTICS

TACTIC 1: *Vector Products and Torques*
Equation 12-21 for torques is our first application of the vector

(or cross) product. You might want to review Section 3-7, where the rules for the vector prouct are given. In that section, Problem Solving Tactic 5 lists many common errors in finding the direction of the vector product.

Keep in mind that a torque is calculated *with respect to* (or *about*) a point, which must be known if the value of the torque is to be meaningful. Changing the point can change the torque in both magnitude and direction. For example, in Sample Problem 12-6, the torques due to the three forces are calculated about the origin O. You can show that the torques due to the same three forces are each zero if they are calculated about point P (at the position of the particle).

CHECKPOINT 3: The position vector $\mathbf{r}$ of a particle points along the positive direction of the z axis. If the torque on the particle is (a) zero, (b) in the negative direction of x, and (c) in the negative direction of y, in what direction is the force producing the torque?

12-4 ANGULAR MOMENTUM

Like all other linear quantities, linear momentum has its angular counterpart. Figure 12-12 shows a particle with linear momentum $\mathbf{p}\ (= m\mathbf{v})$ located at point P in the xy plane. The **angular momentum** $\boldsymbol{\ell}$ of this particle with respect to the origin O is a vector quantity defined as

$$\boldsymbol{\ell} = \mathbf{r} \times \mathbf{p} = m(\mathbf{r} \times \mathbf{v}) \quad \text{(angular momentum defined)},$$
$$(12\text{-}25)$$

where $\mathbf{r}$ is the position vector of the particle with respect to O. As the particle moves relative to O, in the direction of its momentum $\mathbf{p}\ (= m\mathbf{v})$, position vector $\mathbf{r}$ rotates around O. Note carefully that to have angular momentum about O, the particle does *not* itself have to rotate around O. Comparison of Eqs. 12-21 and 12-25 shows that angular momentum bears the same relation to linear momentum that torque does to force. The SI unit of angular momentum is the kilogram-meter-squared per second ($\text{kg} \cdot \text{m}^2/\text{s}$), equivalent to the joule-second ($\text{J} \cdot \text{s}$).

To find the direction of the angular momentum vector $\boldsymbol{\ell}$ in Fig. 12-12, we slide the vector $\mathbf{p}$ until its tail is at origin O. Then we use the right-hand rule for vector products, sweeping the fingers from $\mathbf{r}$ into $\mathbf{p}$. The outstretched thumb then shows that vector $\boldsymbol{\ell}$ points in the positive direction of the z axis in Fig. 12-12. This positive direction is consistent with the counterclockwise rotation of the particle's position vector $\mathbf{r}$ about the z axis, as the particle continues to move. (A negative direction of $\boldsymbol{\ell}$ would be consistent with a clockwise rotation of $\mathbf{r}$ about the z axis.)

To find the magnitude of $\boldsymbol{\ell}$, we use the general result of Eq. 3-20, to write

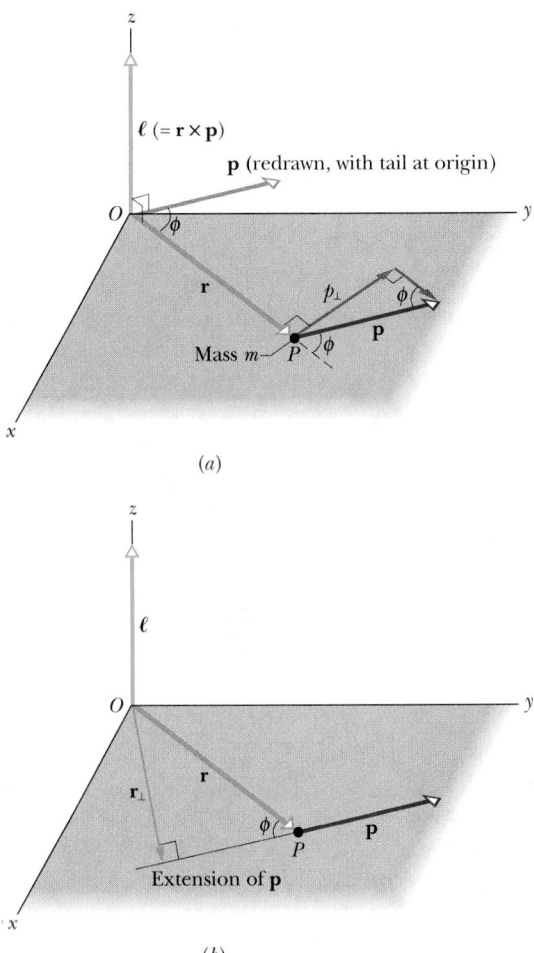

FIGURE 12-12 Defining angular momentum. A particle of mass m at point P has linear momentum $\mathbf{p}\ (= m\mathbf{v})$, assumed to lie in the xy plane. The particle has angular momentum $\boldsymbol{\ell}\ (= \mathbf{r} \times \mathbf{p})$ with respect to the origin O. By the right-hand rule, the angular momentum vector points in the direction of increasing z. (a) The magnitude of $\boldsymbol{\ell}$ is given by $\ell = rp_\perp = rmv_\perp$. (b) The magnitude of $\boldsymbol{\ell}$ is also given by $\ell = r_\perp p = r_\perp mv$.

$$\ell = rmv \sin \phi, \quad (12\text{-}26)$$

where ϕ is the angle between $\mathbf{r}$ and $\mathbf{p}$ when these two vectors are tail to tail. From Fig. 12-12a, we see that Eq. 12-26 can be rewritten as

$$\ell = rp_\perp = rmv_\perp, \quad (12\text{-}27)$$

where $p_\perp$ is the component of $\mathbf{p}$ perpendicular to $\mathbf{r}$ and $p_\perp = mv_\perp$. From Fig. 12-12b, we see that Eq. 12-26 can also be rewritten as

$$\ell = r_\perp p = r_\perp mv, \quad (12\text{-}28)$$

where $r_\perp$ is the perpendicular distance between O and an extension of $\mathbf{p}$.

Just as is true for torque, angular momentum has

meaning only with respect to a specified origin. Moreover, if the particle in Fig. 12-12 did not lie in the xy plane, or if the linear momentum **p** of the particle did not also lie in that plane, the angular momentum ℓ would not be parallel to the z axis. The direction of the angular momentum vector is always perpendicular to the plane formed by the vectors **r** and **p**.

CHECKPOINT 4: In part a of the figure, particles 1 and 2 move around point O in opposite directions, in circles with radii 2 m and 4 m. In part b, particles 3 and 4 travel along straight lines, in the same direction, with their closest distances to point O being 2 m and 4 m. Particle 5 moves directly away from O. All five particles have the same mass and the same constant speed. (a) Rank the particles according to the magnitudes of their angular momentum about point O, greatest first. (b) Which particles have negative angular momentum about point O?

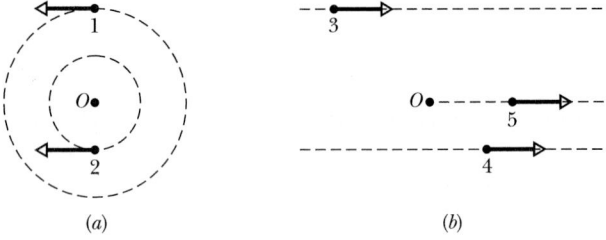

(a)   *(b)*

SAMPLE PROBLEM 12-7

Figure 12-13 shows an overhead view of two particles moving at constant momentum along horizontal paths. Particle 1, with momentum magnitude $p_1 = 5.0$ kg·m/s, has position vector r_1 and will pass 2.0 m from point O. Particle 2, with momentum magnitude $p_2 = 2.0$ kg·m/s, has position vector r_2 and will pass 4.0 m from point O. What is the net angular momentum **L** about point O of the two-particle system?

SOLUTION: To find **L**, we first must find the individual angular momentum vectors ℓ_1 and ℓ_2 of the two particles. To find the magnitude of ℓ_1, we use Eq. 12-28, substituting $r_{\perp 1} = 2.0$ m and $p_1 = 5.0$ kg·m/s:

$$\ell_1 = r_{\perp 1}p_1 = (2.0 \text{ m})(5.0 \text{ kg·m/s})$$
$$= 10 \text{ kg·m}^2/\text{s}.$$

To find the direction of vector ℓ_1, we use Eq. 12-25 and the right-hand rule for vector products. For $\mathbf{r}_1 \times \mathbf{p}_1$, the vector product is out of the page, perpendicular to the plane of Fig. 12-13. This is a positive direction, consistent with the counterclockwise rotation of the particle's position vector $\mathbf{r}_1$ around O as particle 1 moves. Thus the angular momentum vector for particle 1 is

$$\ell_1 = +10 \text{ kg·m}^2/\text{s}.$$

Similarly, the magnitude of ℓ_2 is

$$\ell_2 = r_{\perp 2}p_2 = (4.0 \text{ m})(2.0 \text{ kg·m/s})$$
$$= 8.0 \text{ kg·m}^2/\text{s},$$

and the vector product $\mathbf{r}_2 \times \mathbf{p}_2$ is into the page, which is a negative direction, consistent with the clockwise rotation of $\mathbf{r}_2$ around O as particle 2 moves. Thus, the angular momentum vector for particle 2 is

$$\ell_2 = -8.0 \text{ kg·m}^2/\text{s}.$$

The net angular momentum for the two-particle system is then

$$L = \ell_1 + \ell_2$$
$$= +10 \text{ kg·m}^2/\text{s} + (-8.0 \text{ kg·m}^2/\text{s})$$
$$= +2.0 \text{ kg·m}^2/\text{s}. \qquad \text{(Answer)}$$

The plus sign means that the system's net angular momentum about point O is out of the page.

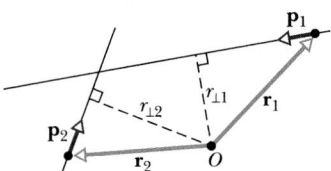

FIGURE 12-13   Sample Problem 12-7. Two particles pass near point O.

12-5 NEWTON'S SECOND LAW IN ANGULAR FORM

Newton's second law written in the form

$$\sum \mathbf{F} = \frac{d\mathbf{p}}{dt} \qquad \text{(single particle)} \qquad (12\text{-}29)$$

expresses the close relation between force and linear momentum for a single particle. We have seen enough of the parallelism between linear and angular quantities to be pretty sure that there is also a close relation between torque and angular momentum. Guided by Eq. 12-29, we can even guess that it must be

$$\sum \boldsymbol{\tau} = \frac{d\ell}{dt} \qquad \text{(single particle)} \qquad (12\text{-}30)$$

Equation 12-30 is indeed an angular form of Newton's second law for a single particle:

The (vector) sum of all the torques acting on a particle is equal to the time rate of change of the angular momentum of that particle.

Equation 12-30 has no meaning unless the torques $\boldsymbol{\tau}$ and the angular momentum $\boldsymbol{\ell}$ are defined with respect to the same origin.

Proof of Equation 12-30

We start with Eq. 12-25, the definition of the angular momentum of a particle:

$$\boldsymbol{\ell} = m(\mathbf{r} \times \mathbf{v}),$$

where $\mathbf{r}$ is the position vector of the particle and $\mathbf{v}$ is the velocity of the particle. Differentiating* each side with respect to time t yields

$$\frac{d\boldsymbol{\ell}}{dt} = m\left(\mathbf{r} \times \frac{d\mathbf{v}}{dt} + \frac{d\mathbf{r}}{dt} \times \mathbf{v}\right). \qquad (12\text{-}31)$$

But $d\mathbf{v}/dt$ is the acceleration $\mathbf{a}$ of the particle, and $d\mathbf{r}/dt$ is its velocity $\mathbf{v}$. Thus we can rewrite Eq. 12-31 as

$$\frac{d\boldsymbol{\ell}}{dt} = m(\mathbf{r} \times \mathbf{a} + \mathbf{v} \times \mathbf{v}).$$

Now $\mathbf{v} \times \mathbf{v} = 0$ (the vector product of any vector with itself is zero because the angle between the two vectors is necessarily zero). This leads to

$$\frac{d\boldsymbol{\ell}}{dt} = m(\mathbf{r} \times \mathbf{a}) = \mathbf{r} \times m\mathbf{a}.$$

We now use Newton's second law ($\Sigma\mathbf{F} = m\mathbf{a}$) to replace $m\mathbf{a}$ with its equal, the vector sum of the forces that act on the particle, obtaining

$$\frac{d\boldsymbol{\ell}}{dt} = \mathbf{r} \times \left(\sum\mathbf{F}\right) = \sum(\mathbf{r} \times \mathbf{F}). \qquad (12\text{-}32)$$

Finally, Eq. 12-21 shows us that $\mathbf{r} \times \mathbf{F}$ is the torque associated with the force $\mathbf{F}$, so Eq. 12-32 becomes

$$\sum\boldsymbol{\tau} = \frac{d\boldsymbol{\ell}}{dt}.$$

This is Eq. 12-30, the relation that we set out to prove.

$\mathbf{C}$HECKPOINT 5: The figure shows the position vector $\mathbf{r}$ of a particle at a certain instant, and four choices for the direction of a force that is to accelerate the particle. All four choices are in the xy plane. (a) Rank the choices according to the magnitude of the time rate of change ($d\boldsymbol{\ell}/dt$) they produce in the angular momentum of the particle about point O, greatest first. (b) Which choice results in a negative rate of change about O?

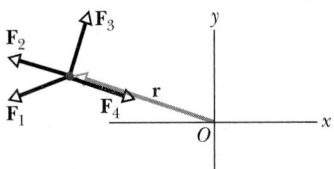

SAMPLE PROBLEM 12-8

A penguin of mass m falls from rest at point A, a horizontal distance d from the origin O in Fig. 12-14.

(a) Find an expression for the angular momentum of the falling penguin about O.

SOLUTION: The angular momentum is given by Eq. 12-25 ($\boldsymbol{\ell} = \mathbf{r} \times \mathbf{p}$); its magnitude is (from Eq. 12-26)

$$\ell = rmv \sin \phi.$$

Here, $r \sin \phi = d$ no matter how far the penguin falls, and $v = gt$. Thus ℓ has magnitude

$$\ell = mgtd. \qquad \text{(Answer)} \quad (12\text{-}33)$$

The right-hand rule shows that the angular momentum vector $\boldsymbol{\ell}$ is directed into the plane of Fig. 12-14, in the direction of decreasing z. We represent $\boldsymbol{\ell}$ with a circled cross $\otimes$ at the origin. The vector $\boldsymbol{\ell}$ changes with time in magnitude only; its direction remains unchanged.

(b) What torque does the weight $m\mathbf{g}$ acting on the penguin exert about the origin O?

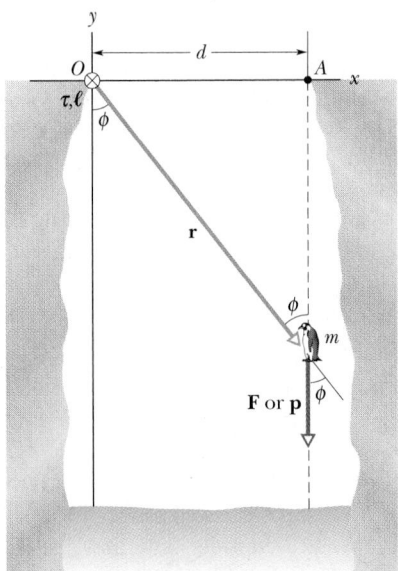

FIGURE 12-14 Sample Problem 12-8. A penguin of mass m falls vertically from point A. The torque $\boldsymbol{\tau}$ and the angular momentum $\boldsymbol{\ell}$ of the falling penguin with respect to the origin O are directed into the plane of the figure at O.

---

*In differentiating a vector product, be sure not to change the order of the two quantities (here $\mathbf{r}$ and $\mathbf{v}$) that form that product. (See Eq. 3-21.)

SOLUTION: The torque is given by Eq. 12-21 ($\boldsymbol{\tau} = \mathbf{r} \times \mathbf{F}$); its magnitude is (from Eq. 12-22)

$$\tau = rF \sin \phi.$$

Again $r \sin \phi = d$, and $F = mg$. Thus

$$\tau = mgd = \text{a constant.} \quad \text{(Answer)} \quad (12\text{-}34)$$

Note that the magnitude of the torque is simply the product of the force magnitude mg and the moment arm d. The right-hand rule shows that the torque vector $\boldsymbol{\tau}$ is directed into the plane of Fig. 12-14, in the direction of decreasing z, hence parallel to $\boldsymbol{\ell}$. (Note that we can also derive Eq. 12-34 by differentiating Eq. 12-33 with respect to t and then substituting the result into Eq. 12-30.)

We see that $\boldsymbol{\tau}$ and $\boldsymbol{\ell}$ depend very much (through d) on the location of the origin. If the penguin falls from the origin, we have $d = 0$ and thus no torque or angular momentum.

12-6 THE ANGULAR MOMENTUM OF A SYSTEM OF PARTICLES

Now we turn our attention to the motion of a system of particles with respect to an origin. Note that "a system of particles" includes a rigid body as a special case. The total angular momentum $\mathbf{L}$ of a system of particles is the (vector) sum of the angular momenta $\boldsymbol{\ell}$ of the particles:

$$\mathbf{L} = \boldsymbol{\ell}_1 + \boldsymbol{\ell}_2 + \boldsymbol{\ell}_3 + \cdots + \boldsymbol{\ell}_n = \sum_{i=1}^{n} \boldsymbol{\ell}_i, \quad (12\text{-}35)$$

in which i ($= 1, 2, 3, \ldots$) labels the particles.

With time, the angular momenta of individual particles may change, either because of interactions within the system (between the individual particles) or because of influences that may act on the system from the outside. We can find the change in $\mathbf{L}$ as these changes take place by taking the time derivative of Eq. 12-35. Thus

$$\frac{d\mathbf{L}}{dt} = \sum_{i=1}^{n} \frac{d\boldsymbol{\ell}_i}{dt}. \quad (12\text{-}36)$$

From Eq. 12-30, $d\boldsymbol{\ell}_i/dt$ is just $\Sigma\boldsymbol{\tau}_i$, the (vector) sum of the torques that act on the ith particle.

Some torques are *internal*, associated with forces that the particles within the system exert on one another; other torques are *external*, associated with forces that act from outside the system. The internal forces, because of Newton's law of action and reaction, cancel in pairs. So, to add the torques, we need consider only those associated with external forces. Equation 12-36 then becomes

$$\sum \boldsymbol{\tau}_{\text{ext}} = \frac{d\mathbf{L}}{dt} \quad \text{(system of particles).} \quad (12\text{-}37)$$

Equation 12-37 is Newton's second law for rotation in angular form, expressed for a system of particles; it is analogous to $\Sigma\mathbf{F}_{\text{ext}} = d\mathbf{P}/dt$ (Eq. 9-28). In words, Eq. 12-37 tells us that the (vector) sum of the *external torques* acting on a system of particles is equal to the time rate of change of the *angular momentum* of that system. Equation 12-37 has meaning only if the torque and angular momentum vectors are referred to the same origin. In an inertial reference frame, Eq. 12-37 can be applied with respect to any point. In an accelerating frame (such as a wheel rolling down a ramp), Eq. 12-37 can be applied *only* with respect to the *center of mass* of the system.

12-7 THE ANGULAR MOMENTUM OF A RIGID BODY ROTATING ABOUT A FIXED AXIS

We next evaluate the angular momentum of a system of particles that form a rigid body which rotates about a fixed axis. Figure 12-15a shows such a body. The fixed axis of rotation is the z axis, and the body rotates about it with constant angular speed ω. We wish to find the angular momentum of the body about the axis of rotation.

We can find the angular momentum by summing the z components of the angular momenta of the mass elements in the body. In Fig. 12-15a, a typical mass element Δm_i of the body moves around the z axis in a circular path. The position of the mass element is located relative to the origin

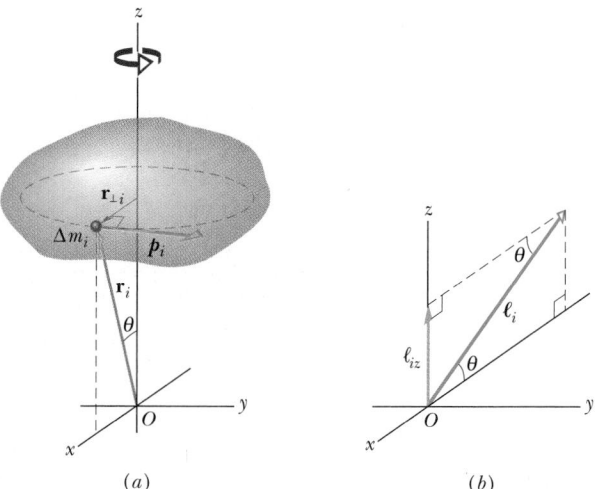

(a) $\qquad\qquad\qquad (b)$

FIGURE 12-15 (a) A rigid body rotates about the z axis with angular speed ω. A mass element Δm_i within the body moves about the z axis in a circle with radius $r_{\perp i}$. The mass element has linear momentum $\mathbf{p}_i$, and it is located relative to the origin O by position vector $\mathbf{r}_i$. Here Δm_i is shown when $r_{\perp i}$ is parallel to the x axis. (b) The angular momentum $\boldsymbol{\ell}_i$, with respect to O, of the mass element in (a). The z component ℓ_{iz} of $\boldsymbol{\ell}_i$ is also shown.

O by position vector $\mathbf{r}_i$. The radius of the mass element's circular path is $r_{\perp i}$, the perpendicular distance between the element and the z axis.

The magnitude of the angular momentum ℓ_i of this mass element, with respect to O, is given by Eq. 12-26:

$$\ell_i = (r_i)(p_i)(\sin 90°) = (r_i)(\Delta m_i\, v_i),$$

where p_i and v_i are the linear momentum and linear speed of the mass element, and 90° is the angle between $\mathbf{r}_i$ and $\mathbf{p}_i$. The angular momentum vector ℓ_i for the mass element in Fig. 12-15a is shown in Fig. 12-15b; its direction must be perpendicular to those of $\mathbf{r}_i$ and $\mathbf{p}_i$.

We are interested in the component of ℓ_i that is parallel to the rotation axis, here the z axis. That z component is

$$\ell_{iz} = \ell_i \sin \theta = (r_i \sin \theta)(\Delta m_i\, v_i) = r_{\perp i}\, \Delta m_i\, v_i.$$

The z component of the angular momentum for the rotating rigid body as a whole is found by adding up the contributions of all the mass elements that make up the body. Thus, because $v = \omega r_\perp$, we may write

$$L_z = \sum_{i=1}^{n} \ell_{iz} = \sum_{i=1}^{n} \Delta m_i\, v_i\, r_{\perp i} = \sum_{i=1}^{n} \Delta m_i (\omega r_{\perp i}) r_{\perp i}$$

$$= \omega \left(\sum_{i=1}^{n} \Delta m_i\, r_{\perp i}^2 \right). \qquad (12\text{-}38)$$

We can remove ω from the summation here because it is a constant: it has the same value for all points of the rotating rigid body.

The quantity $\Sigma \Delta m_i\, r_{\perp i}^2$ in Eq. 12-38 is the rotational inertia I of the body about the fixed axis (see Eq. 11-24). Thus Eq. 12-38 reduces to

$$L = I\omega \qquad \text{(rigid body, fixed axis).} \qquad (12\text{-}39)$$

We have dropped the subscript z, but you must remember that the angular momentum defined by Eq. 12-39 is the

angular momentum about the rotation axis. Also, I in that equation is the rotational inertia about that same axis.

Table 12-2, which supplements Table 11-3, extends our list of corresponding linear and angular relations.

SAMPLE PROBLEM 12-9

A tomahawk expert knows how to throw the instrument so that it completes an integer number of full revolutions about its center of mass during its flight, to bury its edge in the target (Fig. 12-16). Suppose that for a flight of horizontal distance $d = 5.90$ m, with a horizontal component of velocity $v_x = 20.0$ m/s, a tomahawk rotates 1.00 rev. Suppose also that the rotational inertia I of the tomahawk about its center of mass is 1.95×10^{-3} kg·m².

(a) What is the magnitude of the tomahawk's angular momentum about the center of mass during the tomahawk's flight?

SOLUTION: The tomahawk rotates at a constant angular speed ω about an axis that is through its center of mass. So, we can calculate the angular momentum of the tomahawk about that axis with Eq. 12-39 ($L = I\omega$) if we first find ω. From Eq. 11-5, we know that ω is related to an angle of rotation $\Delta\theta$ and the time interval Δt_1 for the rotation by

$$\omega = \frac{\Delta\theta}{\Delta t_1}.$$

Replacing Δt_1 with d/v_x and substituting $\Delta\theta = 1.00$ rev $= 2\pi$ rad, we have

$$\omega = \frac{v_x\, \Delta\theta}{d} = \frac{(20.0 \text{ m/s})(2\pi \text{ rad})}{5.90 \text{ m}} = 21.3 \text{ rad/s}.$$

Substituting this and the given value of I in Eq. 12-39, we find

$$L = I\omega = (1.95 \times 10^{-3} \text{ kg·m}^2)(21.3 \text{ rad/s})$$
$$= 4.15 \times 10^{-2} \text{ kg·m}^2/\text{s}. \qquad \text{(Answer)}$$

(b) The launch required a time interval $\Delta t_2 = 0.150$ s. About

---

TABLE 12-2 MORE CORRESPONDING RELATIONS FOR TRANSLATIONAL AND ROTATIONAL MOTION[a]

TRANSLATIONAL		ROTATIONAL	
Force	$\mathbf{F}$	Torque	$\boldsymbol{\tau}\ (= \mathbf{r} \times \mathbf{F})$
Linear momentum	$\mathbf{p}$	Angular momentum	$\boldsymbol{\ell}\ (= \mathbf{r} \times \mathbf{p})$
Linear momentum[b]	$\mathbf{P}\ (= \Sigma \mathbf{p}_i)$	Angular momentum[b]	$\mathbf{L}\ (= \Sigma \boldsymbol{\ell}_i)$
Linear momentum[b]	$\mathbf{P} = M\mathbf{v}_{\text{cm}}$	Angular momentum[c]	$L = I\omega$
Newton's second law[b]	$\Sigma\mathbf{F}_{\text{ext}} = \dfrac{d\mathbf{P}}{dt}$	Newton's second law[b]	$\Sigma\boldsymbol{\tau}_{\text{ext}} = \dfrac{d\mathbf{L}}{dt}$
Conservation law[d]	$\mathbf{P} = $ a constant	Conservation law[d]	$\mathbf{L} = $ a constant

[a] See also Table 11-3. [b] For systems of particles, including rigid bodies.

[c] For a rigid body about a fixed axis, with L being the component along that axis.

[d] For an isolated system.

the tomahawk's center of mass and from the perspective of Fig. 12-16, what was the average torque applied by the expert to the tomahawk during the launch?

SOLUTION: From Eq. 12-37, we can relate the average torque $\bar{\tau}$ to the change in angular momentum ΔL during the launching time interval Δt_2 with

$$\bar{\tau} = \frac{\Delta L}{\Delta t_2} = \frac{L_f - L_i}{\Delta t_2}.$$

Here, the initial and final angular momenta for the launching period are $L_i = 0$ and $L_f = -4.15 \times 10^{-2}$ kg·m²/s (negative because the tomahawk rotates clockwise in Fig. 12-16). Substituting these values and the given value for Δt_2 yields

$$\bar{\tau} = \frac{-4.15 \times 10^{-2} \text{ kg·m}^2/\text{s}}{0.150 \text{ s}} = -0.277 \text{ N·m}. \quad \text{(Answer)}$$

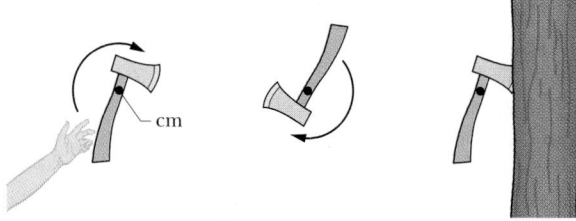

FIGURE 12-16   Sample Problem 12-9. Once thrown, a tomahawk rotates around an axis through its center of mass. (The parabolic path of the center of mass is not shown.)

CHECKPOINT 6: In the figure, a disk, a hoop, and a solid sphere are made to spin about fixed central axes (like a top) by means of strings wrapped around them, with the strings producing the same constant tangential force **F** on all three objects. The three objects have the same mass and radius, and they are initially stationary. Rank the objects according to (a) their angular momentum about their central axes and (b) their angular speed, greatest first, when the strings have been pulled for a certain time t.

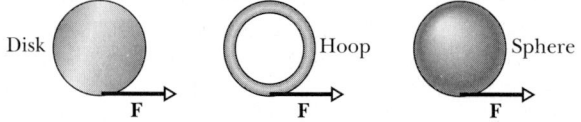

12-8 CONSERVATION OF ANGULAR MOMENTUM

So far we have discussed two powerful conservation laws, the conservation of energy and the conservation of linear momentum. Now we meet a third law of this type, the conservation of angular momentum. We start from Eq. 12-37 ($\Sigma \boldsymbol{\tau}_{\text{ext}} = d\mathbf{L}/dt$), which is Newton's second law in an-

gular form. If no net external torque acts on the system, this equation becomes $d\mathbf{L}/dt = 0$, or

$$\mathbf{L} = \text{a constant} \quad \text{(isolated system).} \quad (12\text{-}40)$$

This result, called the **law of conservation of angular momentum**, can also be written as

$$\mathbf{L}_i = \mathbf{L}_f \quad \text{(isolated system),} \quad (12\text{-}41)$$

where the subscripts refer to the values of **L** at some initial time i and later time f. Equations 12-40 and 12-41 tell us:

> If the net external torque acting on a system is zero, the angular momentum **L** of the system remains constant, no matter what changes take place within the system.

Equations 12-40 and 12-41 are vector equations; as such, they are equivalent to three scalar equations corresponding to the conservation of angular momentum in three mutually perpendicular directions. Depending on the torques acting on a system, the angular momentum of the system might be conserved in only one or two directions but not in all directions:

> If any component of the net *external* torque on a system is zero, then that component of the angular momentum of the system along that axis cannot change, no matter what changes take place within the system.

We can apply this law to the isolated body in Fig. 12-15, which rotates around the z axis. Suppose that the initially rigid body somehow redistributes its mass relative to that rotation axis, changing its rotational inertia. Equations 12-40 and 12-41 state that the angular momentum of the body cannot change. Substituting Eq. 12-39 (for the angular momentum along the rotational axis) into Eq. 12-41, we write this conservation law as

$$I_i \omega_i = I_f \omega_f. \quad (12\text{-}42)$$

Here the subscripts refer to the values of the rotational inertia I and angular speed ω before and after the redistribution of mass.

Like the other two conservation laws that we have discussed, Eqs. 12-40 and 12-41 hold beyond the limitations of Newtonian mechanics. They hold for particles whose speeds approach that of light (where the theory of relativity reigns), and they remain true in the world of subatomic particles (where quantum mechanics reigns). No exceptions to the law of conservation of angular momentum have ever been found.

We now discuss four examples involving this law.

1. *The spinning volunteer.* Figure 12-17 shows a student seated on a stool that can rotate freely about a vertical axis. The student, who has been set into rotation at a modest initial angular speed ω_i, holds two dumbbells in his outstretched hands. His angular momentum vector **L** lies along the vertical axis, pointing upward.

The instructor now asks the student to pull in his arms; this action reduces his rotational inertia from its initial value I_i to a smaller value I_f, because he moves mass closer to the rotation axis. His rate of rotation increases markedly, from ω_i to ω_f. If the student wishes to slow down, he has only to extend his arms once more.

No net external torque acts on the system consisting of the student, stool, and dumbbells. Thus the angular momentum of that system about the rotation axis must remain constant, no matter how the student maneuvers the weights. In Fig. 12-17a, the student's angular speed ω_i is relatively low and his rotational inertia I_i relatively high. According to Eq. 12-42, his angular speed in Fig. 12-17b must be greater to compensate for the decreased rotational inertia due to mass being closer to the rotational axis.

2. *The springboard diver.* Figure 12-18 shows a diver doing a forward one-and-a-half-somersault dive. As you should expect, her center of mass follows a parabolic path. She leaves the springboard with a definite angular momentum **L** about an axis through her center of mass, represented by a vector pointing into the plane of Fig. 12-18, perpendicular to the page. When she is in the air, the diver

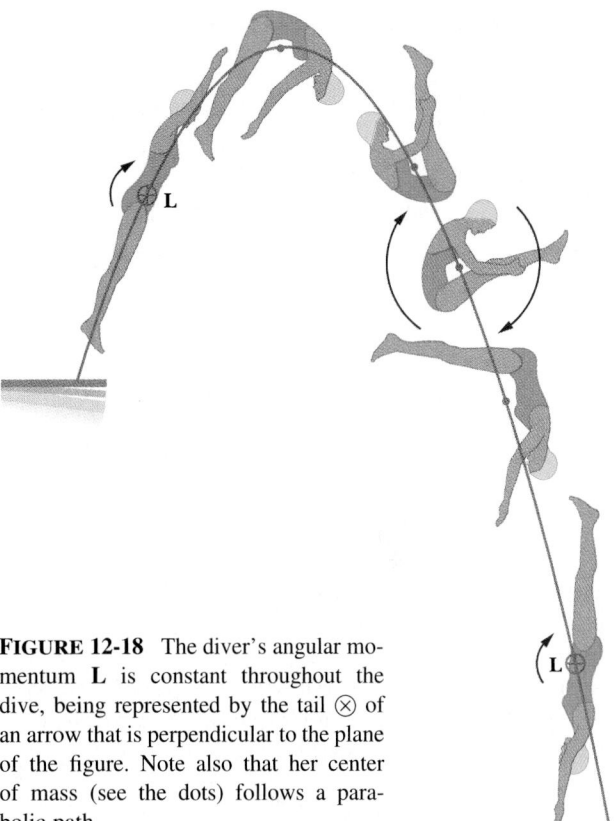

FIGURE 12-18 The diver's angular momentum **L** is constant throughout the dive, being represented by the tail ⊗ of an arrow that is perpendicular to the plane of the figure. Note also that her center of mass (see the dots) follows a parabolic path.

forms an isolated system and her angular momentum cannot change. By pulling her arms and legs into the closed *tuck position,* she can considerably reduce her rotational inertia about the same axis and thus, according to Eq. 12-42, considerably increase her angular speed. Pulling out of the tuck position (into the *open layout position*) at the end of the dive increases her rotational inertia and thus slows her rotation rate so she can enter the water with little splash. Even in a more complicated dive involving both twisting and somersaulting, the angular momentum of the diver must be conserved, in both magnitude *and* direction, throughout the dive.

3. *Spacecraft orientation.* Figure 12-19, which represents a spacecraft with a rigidly mounted flywheel, suggests a scheme (albeit crude) for orientation control. The *spacecraft + flywheel* form an isolated system. So, if the total angular momentum **L** of the system is zero because neither spacecraft nor flywheel is turning, it must remain zero (as long as the system remains isolated).

To change the orientation of the spacecraft, the flywheel is started up (Fig. 12-19a). The spacecraft will start to rotate in the opposite sense to maintain the system's angular momentum at zero. When the flywheel is then brought to rest, the spacecraft will also stop rotating but will have changed its orientation (Fig. 12-19b). Throughout, the angular momentum of the system *spacecraft + flywheel* never differs from zero.

Interestingly, the spacecraft *Voyager 2*, on its 1986

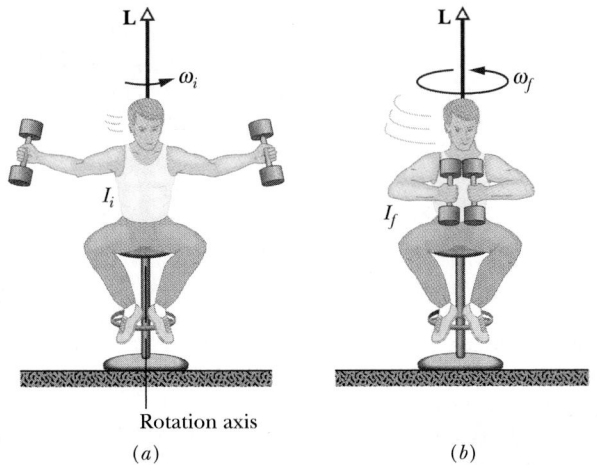

FIGURE 12-17 (*a*) The student has a relatively large rotational inertia and a relatively small angular speed. (*b*) By decreasing his rotational inertia, the student automatically increases his angular speed. The angular momentum **L** of the rotating system remains unchanged.

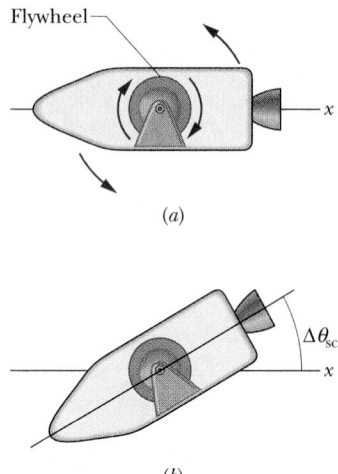

(a)

(b)

FIGURE 12-19 (a) An idealized spacecraft containing a flywheel. If the flywheel is made to rotate clockwise as shown, the spacecraft itself will rotate counterclockwise because the total angular momentum must remain zero. (b) When the flywheel is braked to rest, the spacecraft will also stop rotating but will have reoriented its axis by the angle $\Delta\theta_{sc}$.

flyby of the planet Uranus, was set into unwanted rotation by this flywheel effect every time its tape recorder was turned on at high speed. The ground staff at the Jet Propulsion Laboratory had to program the on-board computer to turn on counteracting thruster jets every time the tape recorder was turned on or off.

4. The incredible shrinking star. When the nuclear fire in the core of a star burns low, the star may eventually begin to collapse, building up pressure in its interior. The collapse may go so far as to reduce the radius of the star from something like that of the Sun to the incredibly small value of a few kilometers. The star then becomes a *neutron star*—its material has been compressed to an incredibly dense gas of neutrons.

During this shrinking process, the star is an isolated system and its angular momentum **L** cannot change. Because its rotational inertia is greatly reduced, its angular speed is correspondingly greatly increased, to as much as 600–800 revolutions per *second*. For comparison, the Sun, a typical star, rotates at about one revolution per month.

CHECKPOINT 7: A rhinoceros beetle rides the rim of a small disk that rotates like a merry-go-round. If the beetle crawls toward the center of the disk, do the following (each relative to the central axis) increase, decrease, or remain the same: (a) the rotational inertia of the beetle–disk system, (b) the angular momentum of the system, and (c) the angular speed of the beetle and disk?

SAMPLE PROBLEM 12-10

Figure 12-20a shows a student, again sitting on a stool that can rotate freely about a vertical axis. The student, initially at rest, is holding a bicycle wheel whose rim is loaded with lead and whose rotational inertia I about its central axis is 1.2 kg·m². The wheel is rotating at an angular speed ω_i of 3.9 rev/s; as seen from overhead, the rotation is counterclockwise. The axis of the wheel is vertical, and the angular momentum L_i of the wheel points vertically upward. The student now inverts the wheel (Fig. 12-20b); as a result, the student and stool rotate about the stool axis. With what angular speed and direction does the student then rotate? (The rotational inertia I_0 of the *student + stool + wheel* system about the stool axis is 6.8 kg·m².)

SOLUTION: There is no net torque acting on the *student + stool + wheel* to change the angular momentum of that system about any vertical axis. The initial angular momentum L_i of the system is that of the bicycle wheel alone. After the wheel has been inverted, the system must still have a net angular momentum of the same magnitude *and* direction.

After the inversion, the angular momentum of the wheel is $-L_i$. In addition, the *student + stool* must acquire some angular momentum; call it **L**. Then, as shown in Fig. 12-20c, we have

$$L_i = L + (-L_i),$$

or

$$L = 2L_i = I_0\omega,$$

in which ω is the angular speed acquired by the student after the wheel's inversion. This yields

$$\omega = \frac{2L_i}{I_0} = \frac{2I\omega_i}{I_0}$$

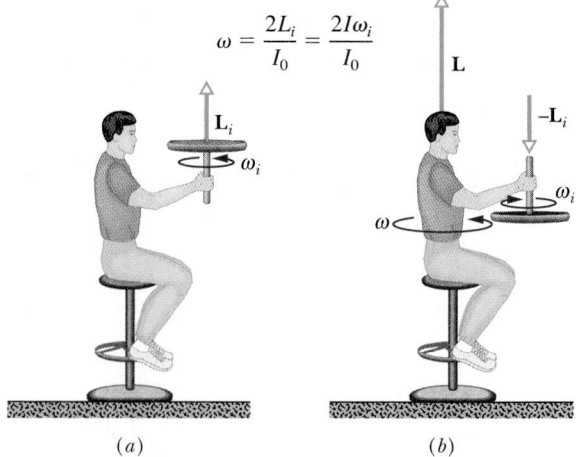

(a)   (b)

FIGURE 12-20 Sample Problem 12-10. (a) A student holds a bicycle wheel rotating around the vertical. (b) The student inverts the wheel, setting himself into rotation. (c) The net angular momentum of the system must remain the same in spite of the inversion.

$$= \frac{(2)(1.2 \text{ kg} \cdot \text{m}^2)(3.9 \text{ rev/s})}{6.8 \text{ kg} \cdot \text{m}^2}$$

$$= 1.4 \text{ rev/s}. \qquad \text{(Answer)}$$

This positive result tells us that the student rotates counterclockwise about the stool axis as seen from overhead. If the student wishes to stop rotating, he has only to invert the wheel once more.

In inverting the wheel, the student will become well aware of the need to apply a torque. However, this torque is internal to the *student + stool + wheel* system and so cannot change the total angular momentum of this system.

We could, however, decide to adopt as our system *student + stool* alone; the wheel then would be external to our system. From this point of view, as the student exerts a torque on the wheel, the wheel exerts a reaction torque on him and this torque is now an external torque. It is the action of this external torque that changes the angular momentum of the *student + stool* system, setting it spinning. Whether a torque is internal or external depends only on how you choose to define your system.

SAMPLE PROBLEM 12-11

During a jump to his partner, an aerialist is to make a triple somersault lasting a time $t = 1.87$ s. For the first and last quarter-revolution, he is in the extended orientation shown in Fig. 12-21, with rotational inertia $I_1 = 19.9 \text{ kg} \cdot \text{m}^2$ around his center of mass. During the rest of the flight he is in a moderate tuck, with rotational inertia $I_2 = 5.50 \text{ kg} \cdot \text{m}^2$.

(a) What must be his initial angular speed ω_1 around his center of mass?

SOLUTION: He rotates in the extended position for two quarter-turns, or a total angle $\theta_1 = 0.500$ rev, in total time t_1; he is in the tuck for an angle θ_2 during a time t_2. These two times are given by

$$t_1 = \frac{\theta_1}{\omega_1}, \qquad \text{and} \qquad t_2 = \frac{\theta_2}{\omega_2}, \qquad (12\text{-}43)$$

where ω_2 is his angular speed in the tuck. We can find an expression for ω_2 by noting that his angular momentum is conserved throughout the flight. Then, from Eq. 12-42,

$$I_2\omega_2 = I_1\omega_1,$$

from which

$$\omega_2 = \frac{I_1}{I_2} \omega_1. \qquad (12\text{-}44)$$

His total flight time is

$$t = t_1 + t_2,$$

which, with substitutions from Eqs. 12-43 and 12-44, becomes

$$t = \frac{\theta_1}{\omega_1} + \frac{\theta_2 I_2}{\omega_1 I_1} = \frac{1}{\omega_1} \left(\theta_1 + \theta_2 \frac{I_2}{I_1} \right). \qquad (12\text{-}45)$$

Inserting the given data, we obtain

$$1.87 \text{ s} = \frac{1}{\omega_1} \left(0.500 \text{ rev} + 2.50 \text{ rev} \frac{5.50 \text{ kg} \cdot \text{m}^2}{19.9 \text{ kg} \cdot \text{m}^2} \right),$$

where, for the triple somersault, he must spend $\theta = 2.5$ revolutions in a tuck. From this equation we find

$$\omega_1 = 0.6369 \text{ rev/s} \approx 0.637 \text{ rev/s}. \qquad \text{(Answer)}$$

(b) If he now attempts a quadruple somersault, with the same ω_1 and t, by using a tighter tuck, what must his rotational inertia I_2 be during the tuck?

SOLUTION: The angle of rotation θ_2 during the tuck is now 3.50 rev, and Eq. 12-45 becomes

$$1.87 \text{ s} = \frac{1}{0.6369 \text{ rev/s}}$$
$$\times \left(0.500 \text{ rev} + 3.50 \text{ rev} \frac{I_2}{19.9 \text{ kg} \cdot \text{m}^2} \right),$$

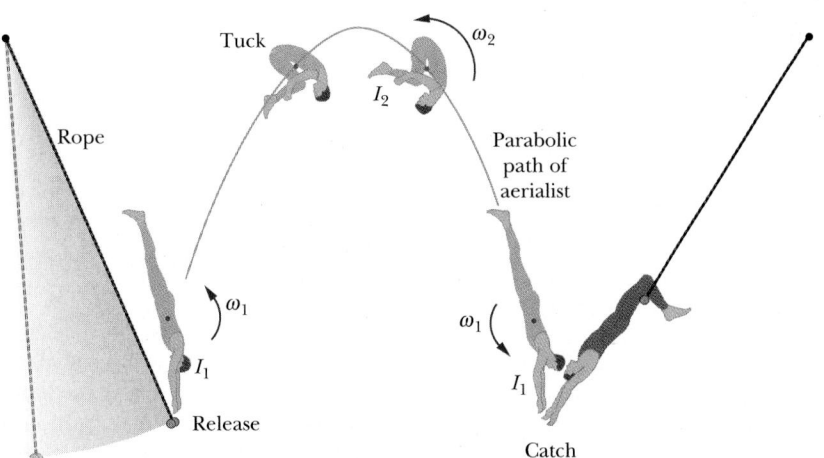

FIGURE 12-21 Sample Problem 12-11. The triple somersault.

from which we find

$$I_2 = 3.929 \text{ kg} \cdot \text{m}^2 \approx 3.93 \text{ kg} \cdot \text{m}^2. \quad \text{(Answer)}$$

This smaller value for I_2 allows faster turning in the tuck and is about the smallest value possible for an aerialist. To make a four-and-a-half somersault, an aerialist would have to increase either the time of flight or the initial angular speed, but either change would make the catch by his partner more difficult. (Can you see why?)

(c) For the quadruple somersault, what is the rotation period T during the tuck?

SOLUTION: We first find the angular speed ω_2 during the tuck from Eq. 12-44:

$$\omega_2 = \frac{I_1}{I_2}\,\omega_1 = \frac{19.9 \text{ kg} \cdot \text{m}^2}{3.929 \text{ kg} \cdot \text{m}^2}\,0.6369 \text{ rev/s}$$
$$= 3.226 \text{ rev/s}.$$

Then we find the time T for one rotation from

$$T = \frac{1 \text{ rev}}{\omega_2} = \frac{1 \text{ rev}}{3.226 \text{ rev/s}} = 0.310 \text{ s}. \quad \text{(Answer)}$$

One reason the quadruple somersault is so difficult is that rotation occurs too quickly for the aerialist to see his surroundings clearly or to "fine-tune" the angular speed by adjusting his rotational inertia during flight.

SAMPLE PROBLEM 12-12

Four thin rods, each with mass M and length $d = 1.0$ m, are rigidly connected in the form of a plus sign; the entire assembly rotates in a horizontal plane around a vertical axle at the center, with initial (clockwise) angular velocity $\omega_i = -2.0$ rad/s (see Fig. 12-22). A mud ball with mass m and initial speed $v_i = 12$ m/s is thrown at, and sticks to, the end of one rod. Let $M = 3m$. What is the final angular velocity ω_f of the *plus sign + mud ball* system if the initial path of the mud ball is each of the four paths shown in Fig. 12-22: path 1 (contact is made when the ball's velocity is perpendicular to the rod), path 2 (radial contact), path 3 (perpendicular contact), and path 4 (contact is made at 60° to the perpendicular)?

SOLUTION: The total angular momentum L of the system about the axle is conserved during the collision:

$$L_f = L_i, \quad (12\text{-}46)$$

where the subscripts f and i represent final and initial values. Let I_+ represent the rotational inertia of the plus sign about the axle. From Table 11-2(f), we have, for the four rods,

$$I_+ = 4\left(\frac{Md^2}{3}\right).$$

The rotational inertia of the mud ball about the axle as the ball rotates on the plus sign is $I_{mb} = md^2$. Let ℓ_i represent the

initial (before contact) angular momentum of the mud ball about the axle, and ω_f represent the final angular velocity of the system. Using $L = I\omega$, we may rewrite Eq. 12-46 as

$$I_+\omega_f + I_{mb}\omega_f = I_+\omega_i + \ell_i,$$

and then as

$$(\tfrac{4}{3}Md^2)\omega_f + (md^2)\omega_f = (\tfrac{4}{3}Md^2)\omega_i + \ell_i. \quad (12\text{-}47)$$

Substituting $M = 3m$ and $\omega_i = -2.0$ rad/s and solving for ω_f, we find that

$$\omega_f = \frac{1}{5md^2}\left(4md^2\,(-2 \text{ rad/s}) + \ell_i\right). \quad (12\text{-}48)$$

We evaluate the magnitude of ℓ_i for paths 1 and 3 with Eq. 12-28, where $r_\perp = d$ and $v = v_i$. For path 2, we use Eq. 12-28 with $r_\perp = 0$. For path 4 we use Eq. 12-27, where $r = d$ and $v_\perp = v_i \cos 60°$. To determine the sign of ℓ_i for an approaching mud ball, we draw a position vector from the axle of the plus sign to the mud ball. Then we note how the position vector rotates around the axle as the mud ball continues to move toward the plus sign. If the position vector rotates clockwise, ℓ_i is negative; if it rotates counterclockwise, ℓ_i is positive. The results are:

path 1: $\ell_i = -\,mdv_i$;   path 2: $\ell_i = 0$;

path 3: $\ell_i = mdv_i$;   path 4: $\ell_i = mdv_i \cos 60°$.

The speed v_i is given as 12 m/s. Substituting these values in turn into Eq. 12-48, we find ω_f to be:

path 1: -4.0 rad/s;   path 2: -1.6 rad/s;

path 3: 0.80 rad/s;   path 4: -0.40 rad/s.   (Answer)

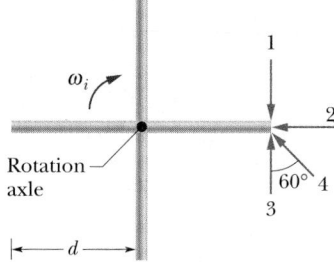

FIGURE 12-22 Sample Problem 12-12. An overhead view of four rigidly connected rods rotating freely around a central axle, and four paths a mud ball can take to stick onto one of the rods.

12-9 QUANTIZED ANGULAR MOMENTUM

A physical quantity is said to be **quantized** if it can have only certain discrete values, and all intermediate values are prohibited. We have met two examples so far, both on a microscopic level: the quantization of mass (Section 2-9) and the quantization of energy (Section 8-9). Angular momentum, on a microscopic level, is a third example.

Fundamental particles such as electrons and protons always have a certain intrinsic angular momentum called **spin angular momentum** S, as if they always spin like a top at a certain rate. (However, they do not spin. Spin angular momentum is a more abstract quantity than the angular momentum of, say, a merry-go-round, which involves actual rotation.) Spin angular momentum S is quantized; its value (along any given direction) is given by

$$S = m_s \frac{h}{2\pi}, \qquad (12\text{-}49)$$

where h is Planck's constant (6.63×10^{-34} J·s) and m_s is a quantum number. For electrons, protons, positrons, and antiprotons, the values of m_s can be only $+\frac{1}{2}$ and $-\frac{1}{2}$. If $m_s = +\frac{1}{2}$, the particle is said to be *spin up* and it has spin $S = +\frac{1}{2}(h/2\pi)$. If $m_s = -\frac{1}{2}$, the particle is said to be *spin down* and it has spin $S = -\frac{1}{2}(h/2\pi)$.

When fundamental particles collide or transform into other fundamental particles, the angular momentum of the system of those particles must be conserved. For example, it must be conserved when a proton (p) and an antiproton ($\bar{\text{p}}$) collide and *annihilate* each other. The result can be the appearance of four positive pions (π^+) and four negative pions (π^-), which are also fundamental particles:

$$\text{p} + \bar{\text{p}} \rightarrow 4\pi^+ + 4\pi^-.$$

For any pion, $m_s = 0$. So, by Eq. 12-49, for any pion, $S = 0$. Thus, after the p–$\bar{\text{p}}$ annihilation, the system has a total spin angular momentum of zero. By conservation of angular momentum, the total spin angular momentum of the colliding proton and antiproton must also be zero if the annihilation is to occur by the process above. That is, either the proton or the antiproton must be spin up and the other must be spin down.

REVIEW & SUMMARY

Rolling Bodies

For a wheel of radius R that rolls without slipping,

$$v_{\text{cm}} = \omega R, \qquad (12\text{-}2)$$

where v_{cm} is the linear speed of the wheel's center and ω is the angular speed of the wheel about its center. The wheel may also be viewed as rotating instantaneously about the point P of the "road" that is in contact with the wheel. The angular speed of the wheel about this point is the same as the angular speed of the wheel about its center. The rolling wheel has kinetic energy

$$K = \tfrac{1}{2} I_{\text{cm}} \omega^2 + \tfrac{1}{2} M v_{\text{cm}}^2, \qquad (12\text{-}5)$$

where I_{cm} is the rotational moment of the wheel about its center.

Torque as a Vector

In three dimensions, *torque* $\boldsymbol{\tau}$ is a vector quantity defined relative to a fixed point (usually an origin); it is

$$\boldsymbol{\tau} = \mathbf{r} \times \mathbf{F}, \qquad (12\text{-}21)$$

where $\mathbf{F}$ is a force applied to a particle and $\mathbf{r}$ is a position vector locating the particle relative to the fixed point (or origin). The magnitude of $\boldsymbol{\tau}$ is given by

$$\tau = rF \sin \phi = rF_\perp = r_\perp F, \quad (12\text{-}22, 12\text{-}23, 12\text{-}24)$$

where ϕ is the angle between $\mathbf{F}$ and $\mathbf{r}$, $F_\perp$ is the component of $\mathbf{F}$ perpendicular to $\mathbf{r}$, and $r_\perp$ is the moment arm of $\mathbf{F}$. The direction of $\boldsymbol{\tau}$ is given by the right-hand rule for cross products.

Angular Momentum of a Particle

The **angular momentum** $\boldsymbol{\ell}$ of a particle with linear momentum $\mathbf{p}$, mass m, and linear velocity $\mathbf{v}$ is a vector quantity defined relative to a fixed point (usually an origin); it is

$$\boldsymbol{\ell} = \mathbf{r} \times \mathbf{p} = m(\mathbf{r} \times \mathbf{v}). \qquad (12\text{-}25)$$

The magnitude of $\boldsymbol{\ell}$ is given by

$$\ell = rmv \sin \phi \qquad (12\text{-}26)$$
$$= rp_\perp = rmv_\perp \qquad (12\text{-}27)$$
$$= r_\perp p = r_\perp mv, \qquad (12\text{-}28)$$

where ϕ is the angle between $\mathbf{r}$ and $\mathbf{p}$, $p_\perp$ and $v_\perp$ are the components of $\mathbf{p}$ and $\mathbf{v}$ perpendicular to $\mathbf{r}$, and $r_\perp$ is the perpendicular distance between the fixed point and an extension of $\mathbf{p}$. The direction of $\boldsymbol{\ell}$ is given by the right-hand rule.

Newton's Second Law in Angular Form

Newton's second law for a particle can be written in vector angular form as

$$\sum \boldsymbol{\tau} = \frac{d\boldsymbol{\ell}}{dt}, \qquad (12\text{-}30)$$

where $\sum \boldsymbol{\tau}$ is the net torque acting on the particle, and $\boldsymbol{\ell}$ is the angular momentum of the particle.

Angular Momentum of a System of Particles

The angular momentum $\mathbf{L}$ of a system of particles is the vector sum of the angular momenta of the individual particles:

$$\mathbf{L} = \boldsymbol{\ell}_1 + \boldsymbol{\ell}_2 + \cdots + \boldsymbol{\ell}_n = \sum_{i=1}^{n} \boldsymbol{\ell}_i. \qquad (12\text{-}35)$$

The time rate of change of this angular momentum is equal to the sum of the external torques on the system (the torques due to interactions of the particles of the system with particles external to the system):

$$\sum \boldsymbol{\tau}_{\text{ext}} = \frac{d\mathbf{L}}{dt} \qquad \text{(system of particles)}. \qquad (12\text{-}37)$$

Angular Momentum of a Rigid Body

For a rigid body rotating about a fixed axis, the component of its angular momentum parallel to the rotation axis is

$$L = I\omega \quad \text{(rigid body, fixed axis).} \quad (12\text{-}39)$$

Conservation of Angular Momentum

The angular momentum **L** of a system remains constant if the net external torque acting on the system is zero:

$$\mathbf{L} = \text{a constant} \quad \text{(isolated system)} \quad (12\text{-}40)$$

or

$$\mathbf{L}_i = \mathbf{L}_f \quad \text{(isolated system).} \quad (12\text{-}41)$$

This is the **law of conservation of angular momentum.** It is one of the fundamental conservation laws of nature, having been veri-

fied even in situations (involving high-speed particles or subatomic dimensions) in which Newton's laws are not applicable.

Quantized Angular Momentum

The spin angular momentum S of fundamental particles is quantized; its value (along any given direction) is

$$S = m_s \frac{h}{2\pi}, \quad (12\text{-}49)$$

where h is Planck's constant (6.63×10^{-34} J·s) and m_s is a quantum number. *Spin up* and *spin down* correspond to $m_s = +\frac{1}{2}$ and $m_s = -\frac{1}{2}$, respectively. The angular momentum of a system of fundamental particles is conserved during a collision or a transformation.

QUESTIONS

1. In Fig. 12-23, a block slides down a frictionless ramp and a sphere rolls without sliding down a ramp of the same angle θ. The block and sphere have the same mass, start from rest at point A, and descend through point B. (a) In that descent, is the work done by the block's weight on the block greater than, less than, or the same as the work done by the sphere's weight on the sphere? At B, which object has more (b) translational kinetic energy and (c) speed down the ramp?

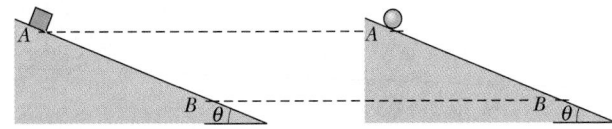

FIGURE 12-23   Question 1.

2. A cannonball and a marble roll from rest down an incline without sliding. (a) Does the cannonball take more, less, or the same time as the marble to reach the bottom? (b) Is the fraction of the cannonball's kinetic energy that is associated with translation more than, less than, or the same as that of the marble?

3. A cannonball rolls down an incline without sliding. If the roll is now repeated with an incline that is less steep but of the same height as the first incline, are (a) the ball's time to reach the bottom and (b) its translational kinetic energy at the bottom greater than, less than, or the same as previously?

4. A solid brass cylinder and a solid wood cylinder have the same radius and mass (the wood cylinder is longer). Released together from rest, they roll down an incline. (a) Which cylinder reaches the bottom first, or do they tie? (b) The wood cylinder is then shortened to match the length of the brass cylinder, and the brass cylinder is drilled out along its long axis to match the mass of the wood cylinder. Which cylinder now wins the race?

5. In Fig. 12-24, a woman rolls a cylindrical drum, by means of

a board on top, through the distance $L/2$, which is half the board's length. The drum rolls smoothly, and the board does not slide over the drum. (a) What length of board has rolled over the top of the drum? (b) How far has the woman walked?

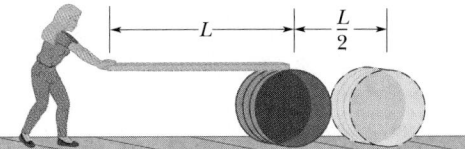

FIGURE 12-24   Question 5.

6. The position vector **r** of a particle relative to a certain point has a magnitude of 3 m, and the force **F** on the particle has a magnitude of 4 N. What is the angle between the directions of **r** and **F** if the magnitude of the associated torque equals (a) zero and (b) 12 N·m?

7. Figure 12-25 shows a particle moving at constant velocity **v** and five points with their xy coordinates. Rank the points according to the magnitude of the angular momentum of the particle measured about them, greatest first.

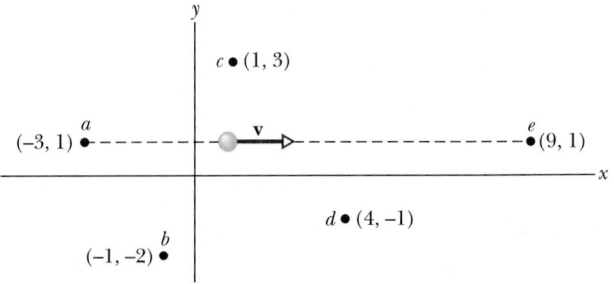

FIGURE 12-25   Question 7.

8. (a) In Checkpoint 4, what is the torque on particles 1 and 2 about point O due to the centripetal forces that cause those particles to circle at constant speed? (b) As particles 3, 4, and 5 move from the left to the right of point O, do their individual angular momenta increase, decrease, or stay the same?

9. Figure 12-26 shows three particles of the same mass and the same constant speed moving as indicated by the velocity vectors. Points a, b, c, and d form a square, with point e at the center. Rank the points according to the magnitude of the net angular momentum of the three-particle system when measured about the points, greatest first.

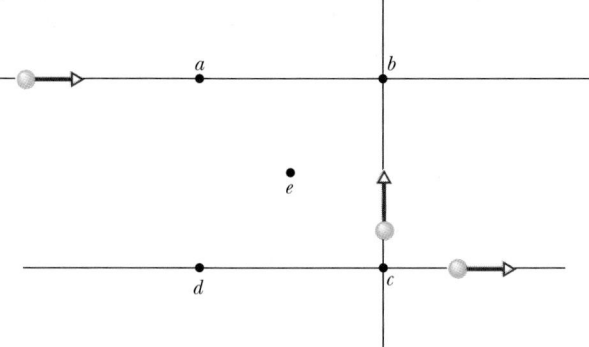

FIGURE 12-26 Question 9.

10. The following gives the angular momentum $\ell(t)$ of a particle in four situations: (1) $\ell = 3t + 4$; (2) $\ell = -6t^2$; (3) $\ell = 2$; (4) $\ell = 4/t$. In which situation is the net torque on the particle (a) zero, (b) positive and constant, (c) negative and increasing in magnitude ($t > 0$), and (d) negative and decreasing in magnitude ($t > 0$)?

11. A bola, which consists of three heavy balls connected to a common point by identical lengths of sturdy string, is readied for launch by holding one of the balls overhead and rotating the wrist, causing the other two balls to rotate in a horizontal circle about the hand. The bola is then released, and its configuration rapidly changes from that shown in the overhead view of Fig. 12-27a to that of Fig. 12-27b. During that change do its (a) angular momentum about its center and (b) angular speed increase, decrease, or stay the same?

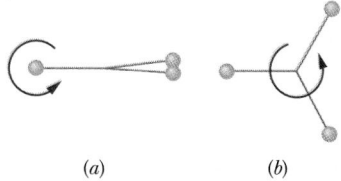

(a)   (b)

FIGURE 12-27 Question 11.

12. A rhinoceros beetle rides the rim of a disk rotating like a merry-go-round counterclockwise. If it then walks along the rim in the direction of the rotation, will the following increase, decrease, or remain the same: (a) the angular momentum of the beetle–disk system, (b) the angular momentum and angular velocity of the beetle, and (c) the angular momentum and angular

velocity of the disk? (d) What are your answers if the beetle walks in the direction opposite the rotation?

13. In Fig. 12-28, a disk is spinning freely at the bottom of an axle and with an angular momentum (about its center) of 50 units, counterclockwise. Four more spinning disks are to be dropped down the axle to land on and couple (via friction) to the first disk. Their angular momenta are: (1) 20 units clockwise, (2) 10 units counterclockwise, (3) 10 units clockwise, and (4) 60 units clockwise. (a) What is the final angular momentum of the system? (b) In what order should the five disks be added so that at one stage the disks that are then coupled on the axle are stationary?

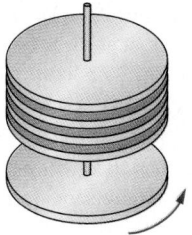

FIGURE 12-28 Question 13.

14. Figure 12-29 shows an overhead view of a rectangular slab that can spin like a merry-go-round about its center at O. Also shown are seven paths along which wads of bubble gum can be thrown (all with the same speed and mass) to stick onto the stationary slab. (a) Rank the paths according to the angular speed that the slab (and gum) will have after the gum sticks, greatest first. (b) For which paths will the angular momentum of the slab (and gum) about O be negative?

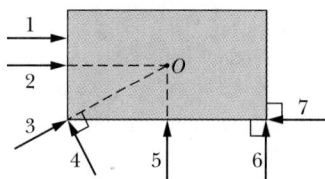

FIGURE 12-29 Question 14.

15. What happens to the initially stationary yo-yo in Fig. 12-30 if you pull it via its string with (a) force $\mathbf{F}_2$ (the line of action passes through the point of contact on the table, as indicated), (b) force $\mathbf{F}_1$ (the line of action passes above the point of contact), and (c) force $\mathbf{F}_3$ (the line of action passes to the right of the point of contact)?

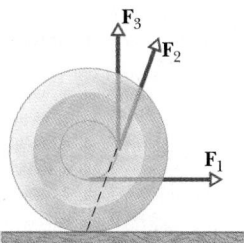

FIGURE 12-30 Question 15.

EXERCISES & PROBLEMS

SECTION 12-1 Rolling

1E. A thin-walled pipe rolls along the floor. What is the ratio of its translational kinetic energy to its rotational kinetic energy about an axis parallel to its length and through its center of mass?

2E. A 140 kg hoop rolls along a horizontal floor so that its center of mass has a speed of 0.150 m/s. How much work must be done on the hoop to stop it?

3E. An automobile traveling 80.0 km/h has tires of 75.0 cm diameter. (a) What is the angular speed of the tires about the axle? (b) If the car is brought to a stop uniformly in 30.0 turns of the tires (without skidding), what is the angular acceleration of the wheels? (c) How far does the car move during the braking?

4E. A 1000 kg car has four 10 kg wheels. When the car is moving, what fraction of the total kinetic energy of the car is due to rotation of the wheels about their axles? Assume that the wheels have the same rotational inertia as uniform disks of the same mass and size. Why do you not need the radius of the wheels?

5E. A wheel of radius 0.250 m, which is moving initially at 43.0 m/s, rolls to a stop in 225 m. Calculate (a) its linear acceleration and (b) its angular acceleration. (c) The wheel's rotational inertia is 0.155 kg·m² about its central axis. Calculate the torque exerted by the friction on the wheel, about the central axis.

6E. An automobile has a total mass of 1700 kg. It accelerates from rest to 40 km/h in 10 s. Assume each wheel is a uniform 32 kg disk. Find, for the end of the 10 s interval, (a) the rotational kinetic energy of each wheel about its axle, (b) the total kinetic energy of each wheel, and (c) the total kinetic energy of the automobile.

7E. A uniform sphere rolls down an incline. (a) What must be the incline angle if the linear acceleration of the center of the sphere is to be 0.10g? (b) For this angle, what would be the acceleration of a frictionless block sliding down the incline?

8E. A solid sphere of weight 8.00 lb rolls up an incline with an inclination angle of 30.0°. At the bottom of the incline the center of mass of the sphere has a translational speed of 16.0 ft/s. (a) What is the kinetic energy of the sphere at the bottom of the incline? (b) How far does the sphere travel up the incline? (c) Does the answer to (b) depend on the weight of the sphere?

9P. A constant horizontal force of 10 N is applied to a wheel of mass 10 kg and radius 0.30 m as shown in Fig. 12-31. The wheel

rolls without slipping on the horizontal surface, and the acceleration of its center of mass is 0.60 m/s². (a) What are the magnitude and direction of the frictional force on the wheel? (b) What is the rotational inertia of the wheel about an axis through its center of mass and perpendicular to the plane of the wheel?

10P. Consider a 66 cm diameter tire on a car traveling at 80 km/h on a level road in the direction of increasing x. As seen by a passenger in the car, what are the linear velocity and the magnitude of the linear acceleration of (a) the center of the wheel, (b) a point at the top of the tire, and (c) a point at the bottom of the tire? (d) Repeat (a) to (c), in the same order, for a stationary observer alongside the road.

11P. A body of radius R and mass m is rolling smoothly with speed v on a horizontal surface. It then rolls up a hill to a maximum height h. (a) If $h = 3v^2/4g$, what is the body's rotational inertia about the rotational axis through its center of mass? (b) What might the body be?

12P. A homogeneous sphere starts from rest at the upper end of the track shown in Fig. 12-32 and rolls without slipping until it rolls off the right-hand end. If $H = 6.0$ m and $h = 2.0$ m and the track is horizontal at the right-hand end, how far horizontally from point A does the sphere land on the floor?

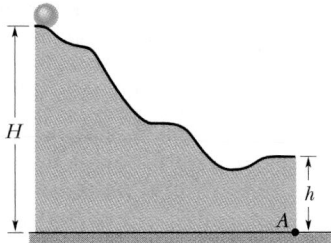

FIGURE 12-32 Problem 12.

13P. A small sphere, with radius r and mass m, rolls without slipping on the inside of a large fixed hemisphere with radius R and a vertical axis of symmetry. It starts at the top from rest. (a) What is its kinetic energy at the bottom? (b) What fraction of its kinetic energy at the bottom is associated with rotation about an axis through its center of mass? (c) What normal force does the small sphere exert on the hemisphere at the bottom if $r \ll R$?

14P. A solid cylinder of radius 10 cm and mass 12 kg starts from rest and rolls without slipping a distance of 6.0 m down a house roof that is inclined at 30°. (See Fig. 12-33.) (a) What is the angular speed of the cylinder about its center as it leaves the house roof? (b) The outside wall of the house is 5.0 m high. How far from the edge of the roof does the cylinder hit the level ground?

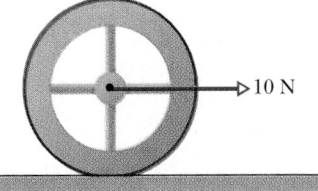

FIGURE 12-31 Problem 9.

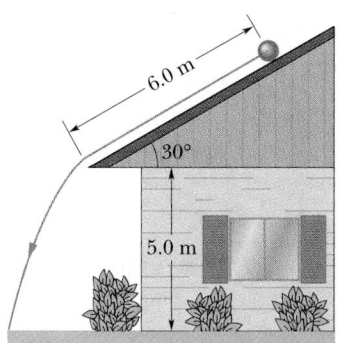

FIGURE 12-33 Problem 14.

15P. A small solid marble of mass m and radius r will roll without slipping along the loop-the-loop track shown in Fig. 12-34 if it is released from rest somewhere on the straight section of track. (a) From what minimum height h above the bottom of the track must the marble be released to ensure that it does not leave the track at the top of the loop? (The radius of the loop-the-loop is R; assume $R \gg r$.) (b) If the marble is released from height $6R$ above the bottom of the track, what is the horizontal component of the force acting on it at point Q?

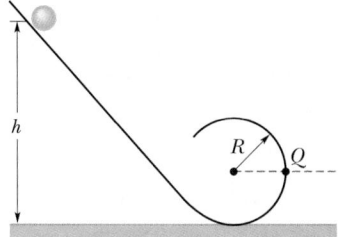

FIGURE 12-34 Problem 15.

16P. A bowler throws a bowling ball of radius $R = 11$ cm down a lane. The ball slides on the lane, with initial speed $v_{cm,0} = 8.5$ m/s and initial angular speed $\omega_0 = 0$. The coefficient of kinetic friction between the ball and the lane is 0.21. The kinetic frictional force $\mathbf{f}_k$ acting on the ball (Fig. 12-35) causes a linear acceleration of the ball while producing a torque that causes an angular acceleration of the ball. When speed v_{cm} has decreased enough and angular speed ω has increased enough, the ball stops sliding and then rolls smoothly. (a) What then is v_{cm} in terms of ω? During the sliding, what are the ball's (b) linear acceleration and (c) angular acceleration? (d) How long does the ball slide? (e) How far does the ball slide? (f) What is the speed of the ball when smooth rolling begins?

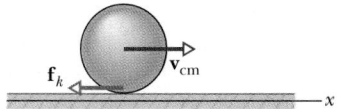

FIGURE 12-35 Problem 16.

SECTION 12-2 The Yo-Yo

17E. A yo-yo has a rotational inertia of 950 g·cm² and a mass of 120 g. Its axle radius is 3.2 mm, and its string is 120 cm long. The yo-yo rolls from rest down to the end of the string. (a) What is its linear acceleration? (b) How long does it take to reach the end of the string? As it reaches the end of the string, what are its (c) linear speed, (d) translational kinetic energy, (e) rotational kinetic energy, and (f) angular speed?

18P. Suppose that the yo-yo in Exercise 17, instead of rolling from rest, is thrown so that its initial speed down the string is 1.3 m/s. (a) How long does the yo-yo take to reach the end of the string? As it reaches the end of the string, what are its (b) total kinetic energy, (c) linear speed, (d) translational kinetic energy, (e) angular speed, and (f) rotational kinetic energy?

SECTION 12-3 Torque Revisited

19E. Given that $\mathbf{r} = x\mathbf{i} + y\mathbf{j} + z\mathbf{k}$ and $\mathbf{F} = F_x\mathbf{i} + F_y\mathbf{j} + F_z\mathbf{k}$, show that the torque $\boldsymbol{\tau} = \mathbf{r} \times \mathbf{F}$ is given by

$$\boldsymbol{\tau} = (yF_z - zF_y)\mathbf{i} + (zF_x - xF_z)\mathbf{j} + (xF_y - yF_x)\mathbf{k}.$$

20E. Show that, if $\mathbf{r}$ and $\mathbf{F}$ lie in a given plane, the torque $\boldsymbol{\tau} = \mathbf{r} \times \mathbf{F}$ has no component in that plane.

21E. What are the magnitude and direction of the torque about the origin on a plum located at coordinates $(-2.0$ m, 0, 4.0 m) due to force $\mathbf{F}$ whose only component is (a) $F_x = 6.0$ N, (b) $F_x = -6.0$ N, (c) $F_z = 6.0$ N, and (d) $F_z = -6.0$ N?

22E. What are the magnitude and direction of the torque about the origin on a particle located at coordinates $(0, -4.0$ m, 3.0 m) due to (a) force $\mathbf{F}_1$ with components $F_{1x} = 2.0$ N and $F_{1y} = F_{1z} = 0$, and (b) force $\mathbf{F}_2$ with components $F_{2x} = 0$, $F_{2y} = 2.0$ N, and $F_{2z} = 4.0$ N?

23P. Force $\mathbf{F} = (2.0$ N$)\mathbf{i} - (3.0$ N$)\mathbf{k}$ acts on a pebble with position vector $\mathbf{r} = (0.50$ m$)\mathbf{j} - (2.0$ m$)\mathbf{k}$, relative to the origin. What is the resulting torque acting on the pebble about (a) the origin and (b) a point with coordinates $(2.0$ m, 0, -3.0 m)?

24P. What is the torque about the origin on a jar of jalapeño peppers located at coordinates $(3.0$ m, -2.0 m, 4.0 m) due to (a) force $\mathbf{F}_1 = (3.0$ N$)\mathbf{i} - (4.0$ N$)\mathbf{j} + (5.0$ N$)\mathbf{k}$, (b) force $\mathbf{F}_2 = (-3.0$ N$)\mathbf{i} - (4.0$ N$)\mathbf{j} - (5.0$ N$)\mathbf{k}$, and (c) the vector sum of $\mathbf{F}_1$ and $\mathbf{F}_2$? (d) Repeat (c) for a point with coordinates $(3.0$ m, 2.0 m, 4.0 m) instead of the origin.

25P. What is the net torque about the origin on a flea located at coordinates $(0, -4.0$ m, 5.0 m) when forces $\mathbf{F}_1 = (3.0$ N$)\mathbf{k}$ and $\mathbf{F}_2 = (-2.0$ N$)\mathbf{j}$ act on the flea?

26P. Force $\mathbf{F} = (-8.0$ N$)\mathbf{i} + (6.0$ N$)\mathbf{j}$ acts on a particle with position vector $\mathbf{r} = (3.0$ m$)\mathbf{i} + (4.0$ m$)\mathbf{j}$. What are (a) the torque on the particle about the origin and (b) the angle between the directions of $\mathbf{r}$ and $\mathbf{F}$?

SECTION 12-4   Angular Momentum

27E. Two objects are moving as shown in Fig. 12-36. What is their total angular momentum about point O?

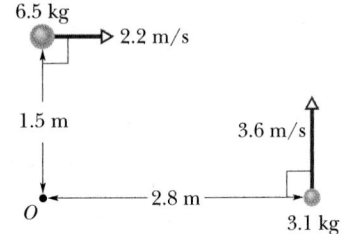

FIGURE 12-36   Exercise 27.

28E. A 1200 kg airplane is flying in a straight line at 80 m/s, 1.3 km above the ground. What is the magnitude of its angular momentum with respect to a point on the ground directly under the path of the plane?

29E. A particle P with mass 2.0 kg has position vector $\mathbf{r}$ ($r = 3.0$ m) and velocity $\mathbf{v}$ ($v = 4.0$ m/s) as shown in Fig. 12-37. It is

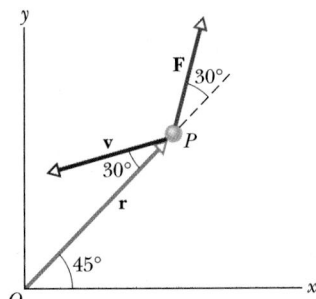

FIGURE 12-37
Exercise 29.

acted on by force $\mathbf{F}$ ($F = 2.0$ N). All three vectors lie in the xy plane. About the origin, what are (a) the angular momentum of the particle and (b) the torque acting on the particle?

30E. If we are given r, p, and ϕ, we can calculate the angular momentum of a particle from Eq. 12-26. Sometimes, however, we are given the components (x, y, z) of $\mathbf{r}$ and (v_x, v_y, v_z) of $\mathbf{v}$ instead. (a) Show that the components of ℓ along the x, y, and z axes are then given by $\ell_x = m(yv_z - zv_y)$, $\ell_y = m(zv_x - xv_z)$, and $\ell_z = m(xv_y - yv_x)$. (b) Show that if the particle moves only in the xy plane, the angular momentum vector has only a z component.

31E. At a certain time, the position vector in meters of a 0.25 kg object is $\mathbf{r} = 2.0\mathbf{i} - 2.0\mathbf{k}$. At that instant, its velocity in meters per second is $\mathbf{v} = -5.0\mathbf{i} + 5.0\mathbf{k}$, and the force in newtons acting on it is $\mathbf{F} = 4.0\mathbf{j}$. (a) What is the angular momentum of the object about the origin? (b) What torque acts on it? (*Hint:* See Exercises 19 and 30.)

32P. What is the magnitude of the angular momentum, about the Earth's center, of an 84 kg person on the equator due to the rotation of the Earth?

33P. Two particles, each of mass m and speed v, travel in opposite directions along parallel lines separated by a distance d. (a) In terms of m, v, and d, find an expression for the magnitude L of the angular momentum of the two-particle system around a point

midway between the two lines. (b) Does the expression change if we change the point about which L is calculated? (c) Now reverse the direction of travel for one of the particles and repeat (a) and (b).

34P. A 2.0 kg object moves in a plane with velocity components $v_x = 30$ m/s and $v_y = 60$ m/s as it passes through the point $(x, y) = (3.0, -4.0)$ m. (a) What is its angular momentum relative to the origin at this moment? (b) What is its angular momentum relative to the point $(-2.0, -2.0)$ m at this same moment?

35P. (a) Use the data given in the appendices to compute the total of the magnitudes of the angular momenta of all the planets owing to their revolution about the Sun. (b) What fraction of this total is associated with the planet Jupiter?

SECTION 12-5   Newton's Second Law in Angular Form

36E. A 3.0 kg particle is at $x = 3.0$ m, $y = 8.0$ m with a velocity of $\mathbf{v} = (5.0$ m/s$)\mathbf{i} - (6.0$ m/s$)\mathbf{j}$. It is acted on by a 7.0 N force in the negative x direction. (a) What is the angular momentum of the particle about the origin? (b) What torque about the origin acts on the particle? (c) At what rate is the angular momentum of the particle changing with time?

37E. A particle is acted on by two torques about the origin: τ_1 has a magnitude of 2.0 N·m and points in the direction of increasing x, and τ_2 has a magnitude of 4.0 N·m and points in the direction of decreasing y. What are the magnitude and direction of $d\ell/dt$, where ℓ is the angular momentum of the particle about the origin?

38E. What torque about the origin acts on a particle moving in the xy plane if the particle has the following values of angular momentum about the origin:

(a) -4.0 kg·m^2/s,   (c) $-4.0\sqrt{t}$ kg·m^2/s,

(b) $-4.0t^2$ kg·m^2/s,   (d) $-4.0/t^2$ kg·m^2/s?

39E. A 3.0 kg toy car on the x axis has velocity $\mathbf{v} = -2.0t^3$ m/s along that axis. About the origin and for $t > 0$, what are (a) the car's angular momentum and (b) the torque acting on the car? (c) Repeat (a) and (b) for a point with coordinates (2.0 m, 5.0 m, 0) instead of the origin. (d) Repeat (a) and (b) for a point with coordinates (2.0 m, -5.0 m, 0) instead of the origin.

40P. At $t = 0$, a 2.0 kg particle has position vector $\mathbf{r} = (4.0$ m$)\mathbf{i} - (2.0$ m$)\mathbf{j}$ relative to the origin. Its velocity is given by $\mathbf{v} = (-6.0t^4$ m/s$)\mathbf{i} + (3.0$ m/s$)\mathbf{j}$. About the origin and for $t > 0$, what are (a) the particle's angular momentum and (b) the torque acting on the particle? (c) Repeat (a) and (b) for a point with coordinates $(-2.0$ m, -3.0 m, 0) instead of the origin.

41P. A projectile of mass m is fired from the ground with an initial speed v_0 and an initial angle θ_0 above the horizontal. (a) Find an expression for the magnitude of its angular momentum about the firing point as a function of time. (b) Find the rate at which the angular momentum changes with time. (c) Evaluate the magnitude of $\mathbf{r} \times \mathbf{F}$ directly and compare the result with (b). Why should the results be identical?

SECTION 12-7 The Angular Momentum of a Rigid Body Rotating About a Fixed Axis

42E. A sanding disk with rotational inertia 1.2×10^{-3} kg·m² is attached to an electric drill whose motor delivers a torque of 16 N·m. Find (a) the angular momentum of the disk about its central axis and (b) the angular speed of the disk 33 ms after the motor is turned on.

43E. The angular momentum of a flywheel having a rotational inertia of 0.140 kg·m² about its axis decreases from 3.00 to 0.800 kg·m²/s in 1.50 s. (a) What is the average torque acting on the flywheel about its central axis during this period? (b) Assuming a uniform angular acceleration, through what angle will the flywheel have turned? (c) How much work was done on the wheel? (d) What is the average power of the flywheel?

44E. Three particles, each of mass m, are fastened to each other and to a rotation axis at O by three massless strings, each with length l as shown in Fig. 12-38. The combination rotates around the rotational axis with angular velocity ω in such a way that the particles remain in a straight line. In terms of m, l, and ω, and relative to point O, what are (a) the rotational inertia of the combination, (b) the angular momentum of the middle particle, and (c) the total angular momentum of the three particles?

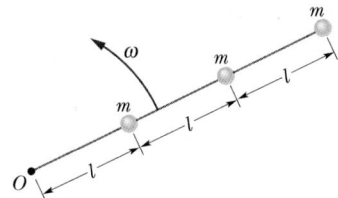

FIGURE 12-38 Exercise 44.

45E. A uniform rod rotates in a horizontal plane about a vertical axis through one end. The rod is 6.00 m long, weighs 10.0 N, and rotates at 240 rev/min clockwise when seen from above. Calculate (a) the rotational inertia of the rod about the axis of rotation and (b) the angular momentum of the rod about that axis.

46P. Figure 12-39 shows a rigid structure consisting of a circular hoop, of radius R and mass m, and a square made of four thin bars, each of length R and mass m. The rigid structure rotates at a constant speed about a vertical axis with a period of rotation of 2.5 s. Assuming $R = 0.50$ m and $m = 2.0$ kg, calculate (a) the structure's rotational inertia about the axis of rotation and (b) its angular momentum about that axis.

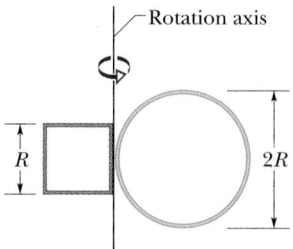

FIGURE 12-39 Problem 46.

47P. Wheels A and B in Fig. 12-40 are connected by a belt that does not slip. The radius of wheel B is three times the radius of wheel A. What would be the ratio of the rotational inertias I_A/I_B if both wheels had (a) the same angular momenta about their central axes and (b) the same rotational kinetic energies?

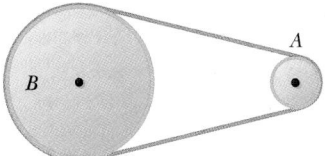

FIGURE 12-40 Problem 47.

48P. An impulsive force $F(t)$ acts for a short time Δt on a rotating rigid body with rotational inertia I. Show that

$$\int \tau \, dt = \overline{F} R \, \Delta t = I(\omega_f - \omega_i),$$

where R is the moment arm of the force, $\overline{F}$ is the average value of the force during the time it acts on the body, and ω_i and ω_f are the angular velocities of the body just before and just after the force acts. (The quantity $\int \tau \, dt = \overline{F} R \, \Delta t$ is called the *angular impulse*, in analogy with $\overline{F} \, \Delta t$, the linear impulse.)

49P*. Two cylinders having radii R_1 and R_2 and rotational inertias I_1 and I_2 about the central axis are supported by axes perpendicular to the plane of Fig. 12-41. The large cylinder is initially

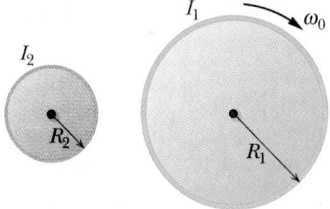

FIGURE 12-41 Problem 49.

rotating with angular velocity ω_0. The small cylinder is moved to the right until it touches the large cylinder and is caused to rotate by the frictional force between the two. Eventually, slipping ceases, and the two cylinders rotate at constant rates in opposite directions. Find the final angular velocity ω_2 of the small cylinder in terms of I_1, I_2, R_1, R_2, and ω_0. (*Hint:* Neither angular momentum nor kinetic energy is conserved. Apply the angular impulse equation of Problem 48.)

SECTION 12-8 Conservation of Angular Momentum

50E. The rotor of an electric motor has rotational inertia $I_m = 2.0 \times 10^{-3}$ kg·m² about its central axis. The motor is used to change the orientation of the space probe in which it is mounted. The motor axis is mounted parallel to the axis of the probe, which has rotational inertia $I_p = 12$ kg·m² about its axis. Calculate the number of revolutions of the rotor required to turn the probe through 30° about its axis.

51E. A man stands on a platform that is rotating (without friction) with an angular speed of 1.2 rev/s; his arms are outstretched and he holds a weight in each hand. The rotational inertia of the system of man, weights, and platform about the central axis is 6.0 kg·m². If by moving the weights the man decreases the rotational inertia of the system to 2.0 kg·m², (a) what is the resulting angular speed of the platform and (b) what is the ratio of the new kinetic energy of the system to the original kinetic energy? (c) What provided the added kinetic energy?

52E. Two disks are mounted on low-friction bearings on the same axle and can be brought together so that they couple and rotate as one unit. (a) The first disk, with rotational inertia 3.3 kg·m² about its central axis, is set spinning at 450 rev/min. The second disk, with rotational inertia 6.6 kg·m² about its central axis, is set spinning at 900 rev/min in the same direction as the first. They then couple together. What is their angular speed after coupling? (b) If instead the second disk is set spinning at 900 rev/min in the direction opposite the first disk's rotation, what is the angular speed after coupling?

53E. A wheel is rotating freely with an angular speed of 800 rev/min on a shaft whose rotational inertia is negligible. A second wheel, initially at rest and with twice the rotational inertia of the first, is suddenly coupled to the same shaft. (a) What is the angular speed of the resultant combination of the shaft and two wheels? (b) What fraction of the original rotational kinetic energy is lost?

54E. The rotational inertia of a collapsing spinning star changes to one-third its initial value. What is the ratio of the new rotational kinetic energy to the initial rotational kinetic energy?

55E. Suppose that the Sun runs out of nuclear fuel and suddenly collapses to form a white dwarf star, with a diameter equal to that of the Earth. Assuming no mass loss, what would then be the Sun's new rotation period, which currently is about 25 days? Assume that the Sun and the white dwarf are uniform, solid spheres.

56E. In a playground, there is a small merry-go-round of radius 1.20 m and mass 180 kg. Its radius of gyration (see Problem 58 of Chapter 11) is 91.0 cm. A child of mass 44.0 kg runs at a speed of 3.00 m/s along a path that is tangent to the rim of the initially stationary merry-go-round and then jumps on. Neglect friction between the bearings and the shaft of the merry-go-round. Calculate (a) the rotational inertia of the merry-go-round about its axis of rotation, (b) the angular momentum of the child, while running, about the axis of rotation of the merry-go-round, and (c) the angular speed of the merry-go-round and child after the child has jumped on.

57E. A horizontal platform in the shape of a circular disk rotates on a frictionless bearing about a vertical axle through the center of the disk. The platform has a mass of 150 kg, a radius of 2.0 m, and a rotational inertia of 300 kg·m² about the axis of rotation. A 60 kg student walks slowly from the rim of the platform toward the center. If the angular speed of the system is 1.5 rad/s when the student starts at the rim, what is the angular speed when she is 0.50 m from the center?

58E. With center and spokes of negligible mass, a certain bicycle wheel has a thin rim of radius 1.14 ft and weight 8.36 lb; it can turn on its axle with negligible friction. A man holds the wheel above his head with the axle vertical while he stands on a turntable free to rotate without friction; the wheel rotates clockwise, as seen from above, with an angular speed of 57.7 rad/s, and the

turntable is initially at rest. The rotational inertia of *wheel + man + turntable* about the common axis of rotation is 1.54 slug·ft². The man's free hand suddenly stops the rotation of the wheel (relative to the turntable). Determine the resulting angular velocity (magnitude and direction) of the system.

59P. Two skaters, each of mass 50 kg, approach each other along parallel paths separated by 3.0 m. They have equal and opposite velocities of 1.4 m/s. The first skater carries one end of a long pole with negligible mass, and the second skater grabs the other end of it as she passes; see Fig. 12-42. Assume frictionless ice. (a) Describe quantitatively the motion of the skaters after they have become connected by the pole. (b) By pulling on the pole, the skaters reduce their separation to 1.0 m. What is their angular speed then? (c) Calculate the kinetic energy of the system in (a) and (b). (d) What is the source of the added kinetic energy?

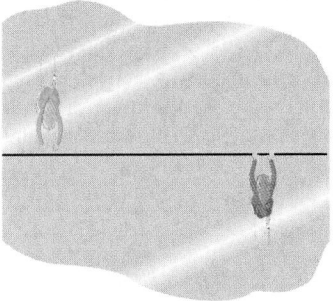

FIGURE 12-42 Problem 59.

60P. Two children, each with mass M, sit on opposite ends of a narrow board with length L and mass M (the same as the mass of each child). The board is pivoted at its center and is free to rotate in a horizontal circle without friction. (Treat it as a thin rod.) (a) What is the rotational inertia of the board plus the children about a vertical axis through the center of the board? (b) What is the angular momentum of the system if it is rotating with angular speed ω_0 in a clockwise direction as seen from above? What is the direction of the angular momentum? (c) While the system is rotating, the children pull themselves toward the center of the board until they are half as far from the center as before. What is the resulting angular speed in terms of ω_0? (d) What is the change in kinetic energy of the system as a result of the children changing their positions? (What is the source of the added kinetic energy?)

61P. A toy train track is mounted on a large wheel that is free to turn with negligible friction about a vertical axis (Fig. 12-43). A toy train of mass m is placed on the track and, with the system initially at rest, the electrical power is turned on. The train reaches a steady speed v with respect to the track. What is the angular

FIGURE 12-43 Problem 61.

velocity ω of the wheel, if its mass is M and its radius R? (Treat the wheel as a hoop, and neglect the mass of the spokes and hub.)

62P. A cockroach of mass m runs counterclockwise around the rim of a lazy Susan (a circular dish mounted on a vertical axle) of radius R and rotational inertia I and having frictionless bearings. The cockroach's speed (relative to the Earth) is v, whereas the lazy Susan turns clockwise with angular speed ω_0. The cockroach finds a bread crumb on the rim and, of course, stops. (a) What is the angular speed of the lazy Susan after the cockroach stops? (b) Is mechanical energy conserved?

63P. A girl of mass M stands on the rim of a frictionless merry-go-round of radius R and rotational inertia I that is not moving. She throws a rock of mass m horizontally in a direction that is tangent to the outer edge of the merry-go-round. The speed of the rock, relative to the ground, is v. Afterwards what are (a) the angular speed of the merry-go-round and (b) the linear speed of the girl?

64P. A phonograph record of mass 0.10 kg and radius 0.10 m rotates about a vertical axis through its center with an angular speed of 4.7 rad/s. The rotational inertia of the record about its axis of rotation is 5.0×10^{-4} kg·m². A wad of putty of mass 0.020 kg drops vertically onto the record from above and sticks to the edge of the record. What is the angular speed of the record immediately after the putty sticks to it?

65P. A uniform thin rod of length 0.50 m and mass 4.0 kg can rotate in a horizontal plane about a vertical axis through its center. The rod is at rest when a 3.0-g bullet traveling in the horizontal plane of the rod is fired into one end of the rod. As viewed from above, the direction of the bullet's velocity makes an angle of 60° with the rod (Fig. 12-44). If the bullet lodges in the rod and the angular velocity of the rod is 10 rad/s immediately after the collision, what is the magnitude of the bullet's velocity just before impact?

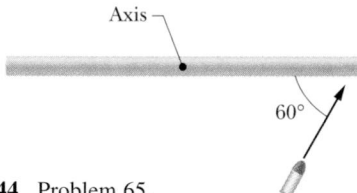

FIGURE 12-44 Problem 65.

66P. A cockroach of mass m lies on the rim of a uniform disk of mass 10.0m that can rotate freely about its center like a merry-go-round. Initially the cockroach and disk rotate together with an angular velocity of ω_0. Then the cockroach walks half way to the center of the disk. (a) What is the change $\Delta\omega$ in the angular velocity of the cockroach-disk system? (b) What is the ratio K/K_0 of the new kinetic energy of the system to its initial kinetic energy? (c) What accounts for the change in the kinetic energy?

67P. A uniform disk of mass 10m and radius 3.0r can rotate freely about its fixed center like a merry-go-round. A smaller uniform disk of mass m and radius r lies on top of the larger disk, concentric with it. Initially the two disks rotate together with an angular velocity of 20 rad/s. Then a slight disturbance causes the smaller disk to slide outward across the larger disk, until the outer edge of the smaller disk catches on the outer edge of the larger

disk. Afterwards, the two disks again rotate together (without further sliding). (a) What then is their angular velocity about the center of the larger disk? (b) What is the ratio K/K_0 of the new kinetic energy of the two-disk system to the system's initial kinetic energy?

68P. If the Earth's polar ice caps melted and the water returned to the oceans, the oceans would be made deeper by about 30 m. What effect would this have on the Earth's rotation? Make an estimate of the resulting change in the length of the day. (Concern has been expressed that warming of the atmosphere resulting from industrial pollution could cause the ice caps to melt.)

69P*. The particle of mass m in Fig. 12-45 slides down the frictionless surface and collides with the uniform vertical rod, sticking to it. The rod pivots about O through the angle θ before momentarily coming to rest. Find θ in terms of the other parameters given in the figure.

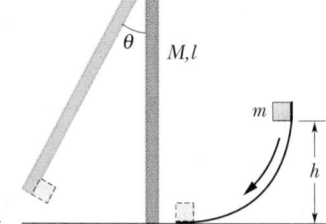

FIGURE 12-45 Problem 69.

70P*. Two 2.00 kg balls are attached to the ends of a thin rod of negligible mass, 50.0 cm long. The rod is free to rotate in a vertical plane without friction about a horizontal axis through its center. While the rod is horizontal (Fig. 12-46), a 50.0 g putty wad drops onto one of the balls with a speed of 3.00 m/s and sticks to it. (a) What is the angular speed of the system just after the putty wad hits? (b) What is the ratio of the kinetic energy of the entire system after the collision to that of the putty wad just before? (c) Through what angle will the system rotate until it momentarily stops?

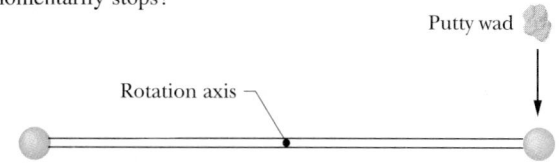

FIGURE 12-46 Problem 70.

SECTION 12-9 Quantized Angular Momentum

71E. What is the value of spin angular momentum S (along any given direction) for an electron?

72E. A positive pion (π^+) can spontaneously decay to (suddenly transform to) a positive muon (μ^+) and a neutrino (ν):

$$\pi^+ \rightarrow \mu^+ + \nu.$$

If the neutrino has $m_s = +\frac{1}{2}$, what are the values of (a) m_s and (b) S for the positive muon?

73E. Two protons colliding at high speeds can transform to three protons and one antiproton:

$$p + p \rightarrow p + p + p + \bar{p}.$$

If the two colliding protons are both spin up, what are the spin orientations (up or down) of the transformation products?

ELECTRONIC COMPUTATION

74. When a bowling ball is bowled, it usually skids for a short distance, then rolls without slipping. While it is skidding, a force of kinetic friction acts to slow the translational motion and increase the angular speed. Skidding stops when $\omega R = v$, where ω is the angular speed, R is the radius, and v is the speed of the center of mass. The net force acting on the ball has magnitude $f = \mu_k N$, where μ_k is the coefficient of kinetic friction and N is the magnitude of the normal force. Since the alley is horizontal, $N = mg$, where m is the mass of the ball. The net torque on the ball has magnitude $\tau = fR$, where R is the radius of the ball. A bowling ball has a mass of 7.25 kg, a radius of 10.9 cm, and a rotational inertia that is nearly the same as that of a uniform sphere. Set up a program to compute the angular speed and translational speed every 0.1 s from the time the ball is bowled to the time it stops skidding. Take the coefficient of kinetic friction to be 0.35 and the initial translational speed to be 20 m/s. For each of the initial conditions given below, plot a graph of the speed as a function of time. Find the time when the ball stops skidding and also the linear speed of the ball at that time. Except for a slight decrease due to air drag, this is the speed with which the ball hits the pins. (a) The ball is not rotating initially. (b) The ball is released with an angular speed of 150 rad/s, spinning in the same direction as when it stops skidding. (c) The ball is released with the same angular speed but in the "wrong" direction. (d) How should a bowling ball be released to have the greatest linear speed as it hits the pins?

75. A projectile with a mass of 0.15 kg is fired with an initial speed of 50 m/s at an angle of 35° above the horizontal. (a) Neglect air resistance and use a numerical integration program to calculate its angular momentum about the firing point every 0.1 s from the time it is fired until it again reaches the height from which it was fired. Use $\tau = \mathbf{r} \times \mathbf{F}$, where $\mathbf{r}$ is the position vector of the projectile and $\mathbf{F}$ is the gravitational force on the projectile, to calculate the torque τ on the projectile (about the firing point) at each of those times. Plot the magnitude of the angular momentum as a function of the square of the time and verify that the graph is a straight line. Plot the magnitude of the torque as a function of time and verify that the graph is a straight line. Are these graphs consistent with Newton's second law for rotation, $\tau = d\mathbf{L}/dt$? (b) Now include air resistance. Take the terminal speed to be $v_t = 75$ m/s. The acceleration can be written $\mathbf{a} = -g\mathbf{j} - bv\mathbf{v}$, where $b = g/v_t^2$. Does the magnitude of the angular momentum increase in proportion to the square of the time? Is the magnitude of the torque proportional to the time? At any position of the projectile, is the magnitude of the torque greater or less than it is in the absence of drag? Give an argument based on the directions of the forces acting on the projectile to show that your numerical result is plausible.

76. At the instant the displacement of a 2.00 kg object relative to the origin is $\mathbf{d} = (2.00 \text{ m})\mathbf{i} + (4.00 \text{ m})\mathbf{j} - (3.00 \text{ m})\mathbf{k}$, its velocity is $\mathbf{v} = -(6.00 \text{ m/s})\mathbf{i} + (3.00 \text{ m/s})\mathbf{j} + (3.00 \text{ m/s})\mathbf{k}$, and it is subject to a force $\mathbf{F} = (6.00 \text{ N})\mathbf{i} - (8.00 \text{ N})\mathbf{j} + (4.0 \text{ N})\mathbf{k}$. Find (a) the acceleration of the object, (b) the angular momentum of the object about the origin, (c) the torque about the origin acting on the object, and (d) the angle between the velocity of the object and the force acting on the object.

13
Equilibrium and Elasticity

Rock climbing may be the ultimate physics exam. Failure can mean death, and even "partial credit" can mean severe injury. For example, in a long chimney climb, in which your shoulders are pressed against one wall of a wide, vertical fissure and your feet are pressed against the opposite wall, you need to rest occasionally or you will fall due to exhaustion. Here the exam consists of a single question: What can you do to relax your push on the walls in order to rest? If you relax without considering the physics, the walls will not hold you up. So, what is the answer to this life-and-death, one-question exam?

13-1 EQUILIBRIUM

Consider these objects: (1) a book resting on a table, (2) a hockey puck sliding across a frictionless surface with constant velocity, (3) the rotating blades of a ceiling fan, and (4) the wheel of a bicycle that is traveling along a straight path at constant speed. For each of these four objects:

1. The linear momentum **P** of its center of mass is constant.

2. Its angular momentum **L** about its center of mass, or about any other point, is also constant.

We say that such objects are in **equilibrium.** The two requirements for equilibrium are then

$$\mathbf{P} = \text{a constant} \quad \text{and} \quad \mathbf{L} = \text{a constant.} \quad (13\text{-}1)$$

Our concern in this chapter is with situations in which the constants in Eq. 13-1 are in fact zero. That is, we are concerned largely with objects that are not moving in any way—either in translation or in rotation—in the reference frame from which we observe them. Such objects are in **static equilibrium.** Of the four objects mentioned at the beginning of this section, only one—the book resting on the table—is in static equilibrium.

The balancing rock of Fig. 13-1 is another example of an object that, for the present at least, is in static equilib-

FIGURE 13-1 A balancing rock near Petrified Forest National Park in Arizona. Although its perch seems precarious, the rock is in static equilibrium.

rium. It shares this property with countless other structures, such as cathedrals, houses, filing cabinets, and taco stands, that remain stationary over time.

As we discussed in Section 8-5, if a body returns to a state of static equilibrium after having been displaced from it by a force, the body is said to be in *stable* static equilibrium. A marble placed at the bottom of a hemispherical bowl is an example. If, however, a small force can displace the body and end the equilibrium, the body is in *unstable* static equilibrium.

For example, suppose we balance a domino as in Fig. 13-2a, with the domino's center of mass vertically above a supporting edge. The torque about the supporting edge due to the domino's weight **W** is zero, because the line of action of **W** is through that edge. Since the weight then cannot cause the domino to rotate about the edge, the domino is in equilibrium. Of course, even a slight force on it due to some chance disturbance ends the equilibrium: as the weight's line of action moves to one side of the supporting edge (as in Fig. 13-2b), the torque due to the weight increases the domino's rotation. Thus the domino in Fig. 13-2a is in unstable static equilibrium.

The domino in Fig. 13-2c is not quite as unstable. To topple this domino, a force would have to rotate it through and beyond the balance position of Fig. 13-2a, in which the center of mass is above a supporting edge. So, a slight force will not topple this domino, but a vigorous flick of the finger against the domino certainly will. (If we arrange a chain of such upright dominos, a finger flick against the first can cause the whole chain to fall.)

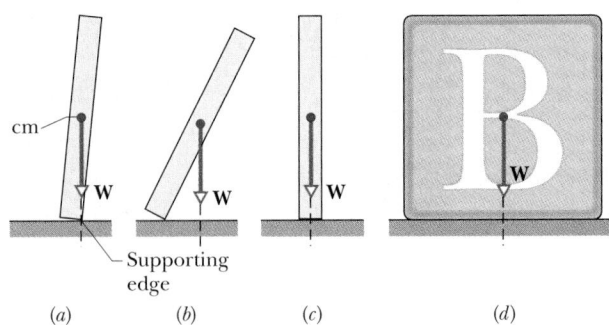

FIGURE 13-2 (a) A domino balanced on one edge, with its center of mass vertically above that edge. The domino's weight **W** is directed through the supporting edge. (b) If the domino is rotated even slightly from the balanced orientation, its weight **W** causes a torque that increases the rotation. (c) A domino upright on a narrow side is somewhat more stable than the domino in (a). (d) A square block is even more stable.

FIGURE 13-3 A construction worker balanced above New York City is in static equilibrium but is more stable parallel to the beam than perpendicular to it.

The child's square block in Fig. 13-2d is even more stable because its center of mass would have to be moved even farther to get it to pass above a supporting edge. A flick of the finger may not topple the block. (This is why you never see a chain of toppling square blocks.) The worker in Fig. 13-3 is like both the domino and the square block: parallel to the beam, his stance is wide and he is stable; perpendicular to the beam, his stance is narrow and he is unstable (and at the mercy of a chance gust of wind).

The analysis of static equilibrium is very important in engineering practice. The design engineer must isolate and identify all the external forces and torques that may act on a structure and, by good design and wise choice of materials, make sure that the structure will tolerate these loads. Such analysis is necessary to make sure, for example, that bridges do not collapse under their traffic and wind loads, and that the landing gear of aircraft will survive the shock of rough landings.

13-2 THE REQUIREMENTS OF EQUILIBRIUM

The translational motion of a body is governed by Newton's second law in its linear momentum form, given by Eq. 9-28 as

$$\sum \mathbf{F}_{ext} = \frac{d\mathbf{P}}{dt}. \qquad (13\text{-}2)$$

If the body is in translational equilibrium, that is, if **P** is a constant, then $d\mathbf{P}/dt = 0$ and we must have

$$\sum \mathbf{F}_{ext} = 0 \qquad \text{(balance of forces).} \quad (13\text{-}3)$$

The rotational motion of a body is governed by Newton's second law in its angular momentum form, given by Eq. 12-37 as

$$\sum \boldsymbol{\tau}_{ext} = \frac{d\mathbf{L}}{dt}. \qquad (13\text{-}4)$$

If the body is in rotational equilibrium, that is, if **L** is a constant, then $d\mathbf{L}/dt = 0$ and we must have

$$\sum \boldsymbol{\tau}_{ext} = 0 \qquad \text{(balance of torques).} \quad (13\text{-}5)$$

Thus the two requirements for a body to be in equilibrium are as follows:

---

1. The vector sum of all the external forces that act on the body must be zero.

2. The vector sum of all the external torques that act on the body, measured about *any* possible point, must also be zero.

These requirements obviously hold for *static* equilibrium as well as the more general equilibrium in which **P** and **L** are constant but not zero.

Equations 13-3 and 13-5, as vector equations, are each equivalent to three independent scalar equations, one for each direction of the coordinate axes:

Balance of forces	Balance of torques	
$\sum F_x = 0$	$\sum \tau_x = 0$	
$\sum F_y = 0$	$\sum \tau_y = 0$	(13-6)
$\sum F_z = 0$	$\sum \tau_z = 0$	

For convenience, we have dropped the subscript ext.

We shall simplify matters by considering only situations in which the forces that act on the body lie in the xy plane. This means that the only torques that can act on the body must tend to cause rotation around an axis parallel to the z axis. With this assumption, we eliminate one force equation and two torque equations from Eq. 13-6, leaving

$$\sum F_x = 0 \qquad \text{(balance of forces),} \qquad (13\text{-}7)$$

$$\sum F_y = 0 \qquad \text{(balance of forces),} \qquad (13\text{-}8)$$

$$\sum \tau_z = 0 \qquad \text{(balance of torques).} \qquad (13\text{-}9)$$

Here, F_x and F_y are, respectively, the x and the y components of the external forces that act on the body, and τ_z represents the torques that these forces exert either about the z axis or about *any* axis parallel to it.

A hockey puck sliding at constant velocity over ice satisfies Eqs. 13-7, 13-8, and 13-9 and is thus in equilibrium *but not in static equilibrium*. For static equilibrium, the linear momentum **P** of the puck must be not only constant but also zero; the puck must be resting on the ice. Thus there is another requirement for static equilibrium:

---

3. The linear momentum **P** of the body must be zero.

CHECKPOINT **1:** The figure gives six overhead views of a uniform rod on which two or more forces act perpendicularly to the rod. If the magnitudes of the forces are adjusted properly (but kept nonzero), in which situations can the rod be in static equilibrium?

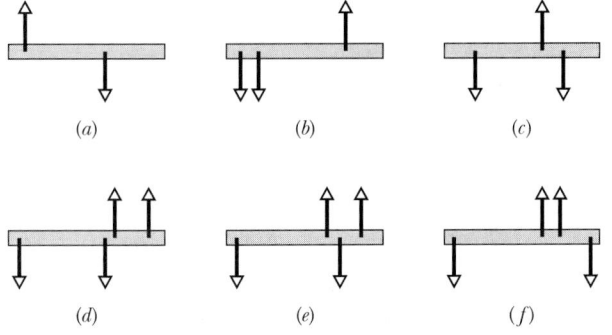

(a)   (b)   (c)

(d)   (e)   (f)

13-3 THE CENTER OF GRAVITY

The weight **W** of an extended body is the vector sum of the gravitational forces acting on the individual elements (the atoms) of the body. Instead of considering all those individual elements, we can say that a single force **W** effectively acts at a single point called the **center of gravity** (cg) of the body. By *effectively* we mean that if the forces on the individual elements were somehow turned off and force **W** at the center of gravity turned on, the net force and the net torque (about any point) acting on the body would not change.

Michel Menin walks a taut wire 10,335 ft above French farmland, adjusting the position of a heavy pole to keep the center of mass of the Menin–pole system over the wire in spite of wind gusts.

Up until now, we have assumed that the weight **W** acts at the center of mass (cm) of the body, which is equivalent to assuming that the center of gravity is at the center of mass. Here we show that this assumption is valid *provided* the acceleration **g** due to the gravitational force is constant over the body.

Figure 13-4*a* shows an extended body, of mass M, and one of its elements, of mass m_i. Each such element has weight $m_i \mathbf{g}_i$, where $\mathbf{g}_i$ is the acceleration due to the gravitational force at the location of the element. In Fig. 13-4*a*, each weight $m_i \mathbf{g}_i$ produces a torque τ_i on the element about the origin O. Using Eq. 11-32 ($\tau = r_\perp F$), we can write torque τ_i as

$$\tau_i = x_i m_i g_i, \qquad (13\text{-}10)$$

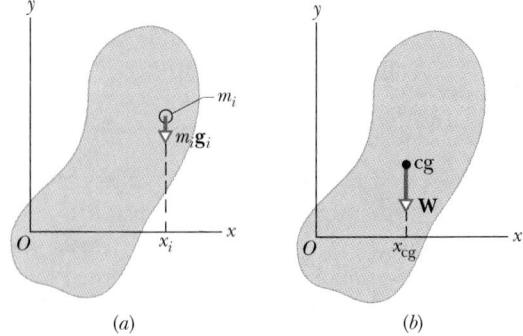

(a)   (b)

FIGURE 13-4 (*a*) An element of mass m_i in an extended body has weight $m_i \mathbf{g}_i$, with moment arm x_i about the origin O of a coordinate system. (*b*) The weight **W** of a body is said to act at the center of gravity (cg) of the body. The moment arm of **W** about O is equal to x_{cg}.

where x_i is the moment arm $r_\perp$ of force $m_i \mathbf{g}_i$. The net torque on all the elements of the body is then

$$\tau_{\text{net}} = \sum \tau_i = \sum x_i m_i g_i. \qquad (13\text{-}11)$$

Figure 13-4b shows the body's weight $\mathbf{W}$ acting at the body's center of gravity. From Eq. 11-32, the torque about O due to $\mathbf{W}$ is

$$\tau = x_{\text{cg}} W, \qquad (13\text{-}12)$$

where x_{cg} is the moment arm of $\mathbf{W}$. The body's weight $\mathbf{W}$ is equal to the sum of the weights $m_i \mathbf{g}_i$ of its elements. So we can substitute $\Sigma m_i g_i$ for W in Eq. 13-12 to write

$$\tau = x_{\text{cg}} \sum m_i g_i. \qquad (13\text{-}13)$$

By definition, the torque due to weight $\mathbf{W}$ acting at the center of gravity is equal to the net torque due to the weights of all the individual elements of the body. So, τ in Eq. 13-13 is equal to τ_{net} in Eq. 13-11, and we can write

$$x_{\text{cg}} \sum m_i g_i = \sum x_i m_i g_i \qquad (13\text{-}14)$$

If the accelerations g_i are all equal, we can cancel them from the two sides of Eq. 13-14. Substituting the body's mass M for Σm_i on the left side of Eq. 13-14 and then rearranging yield

$$x_{\text{cg}} = \frac{1}{M} \sum x_i m_i. \qquad (13\text{-}15)$$

With Eq. 9-4 we see that the right side of Eq. 13-15 gives the coordinate x_{cm} for the center of mass. We can now write

$$x_{\text{cg}} = x_{\text{cm}}. \qquad (13\text{-}16)$$

Thus, the body's center of gravity and its center of mass have the same x coordinate.

We can generalize this result for three dimensions by using vector notation. The generalized result is: If the gravitational acceleration is constant over a body, the body's center of gravity is at its center of mass. For the rest of this book, we shall assume that these points coincide.

From Eq. 13-12, we see that the torque due to the weight of a body is zero only if the moment arm x_{cg} is zero. This means that if the body is suspended from some arbitrary point S about which it can rotate, it rotates (due to the torque $\tau = x_{\text{cg}} W$ about S) until x_{cg} is zero. Its center of gravity is then vertically below the suspension point, as in Figs. 13-5a and b, and the body is in equilibrium. If the body is suspended at its center of gravity, as in Fig. 13-5c, then for any orientation of the body, x_{cg} is zero and the body is in equilibrium.

13-4 SOME EXAMPLES OF STATIC EQUILIBRIUM

In this section we examine six sample problems involving static equilibrium. In each, we select a system of one or more objects to which we apply the equations of equilibrium (Eqs. 13-7, 13-8, and 13-9). The forces involved in the equilibrium are all in the xy plane, which means that the torques involved are parallel to the z axis. Thus, in applying Eq. 13-9, the balance of torques, we select an axis parallel to the z axis about which to calculate the torques. Although Eq. 13-9 is satisfied for *any* such choice of axis, we shall see that certain choices simplify the application of Eq. 13-9 by eliminating one or more unknown force terms.

SAMPLE PROBLEM 13-1

A uniform beam of length L whose mass m is 1.8 kg rests with its ends on two digital scales, as in Fig. 13-6a. A uniform block whose mass M is 2.7 kg rests on the beam, its center a distance $L/4$ from the beam's left end. What do the scales read?

We choose as our system the beam and the block, taken together. Figure 13-6b is a free-body diagram for this system, showing all the forces that act on it. The scales push upward at the ends of the beam with forces $\mathbf{F}_l$ and $\mathbf{F}_r$. The magnitudes of these two forces are the scale readings that we seek. The weight of the beam, $m\mathbf{g}$, acts downward at the beam's center of mass. Similarly, $M\mathbf{g}$, the weight of the block, acts downward at *its* center of mass. In Fig. 13-6b, the block is represented by a dot within the boundary of the beam, and the vector $M\mathbf{g}$ is drawn with its tail on that dot. (In drawing Fig. 13-6b from 13-6a, the vector $M\mathbf{g}$ is shifted in the direction it points, that is, along its line of action. The shift does not alter $M\mathbf{g}$, or a torque due to $M\mathbf{g}$ about any axis perpendicular to the figure.)

Our system is in static equilibrium so that the balance of forces equations (Eqs. 13-7 and 13-8) and the balance of torques equation (Eq. 13-9) apply. We solve this problem in two equivalent ways.

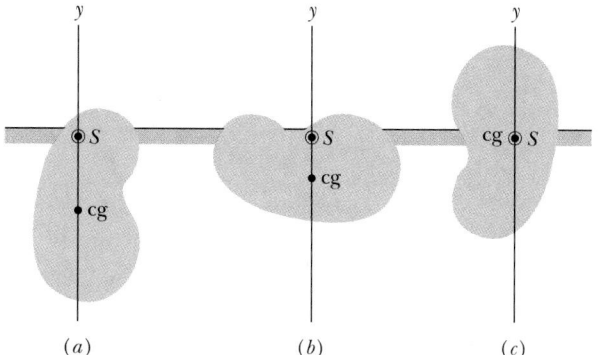

FIGURE 13-5 A body free to rotate about a suspension point S will rotate until its center of gravity is vertically below S as in (a) and (b), unless the center of gravity is at S as in (c).

FIRST SOLUTION: The forces have no x components, so Eq. 13-7, which is $\Sigma F_x = 0$, provides no information. Equation 13-8 gives, for the magnitudes of the y components,

$$\sum F_y = F_l + F_r - Mg - mg = 0. \qquad (13\text{-}17)$$

We have two unknown forces (F_l and F_r), but we cannot find them separately because we have only one equation. Fortunately, we have another equation at hand, namely, Eq. 13-9, the balance of torques equation.

We can apply Eq. 13-9 to *any* axis perpendicular to the plane of Fig. 13-6. Let us choose an axis through the left end of the beam. We take as positive those torques that—acting alone—would produce a counterclockwise rotation of the beam about our axis. We then have, from Eq. 13-9,

$$\sum \tau_z = (F_l)(0) - (Mg)(L/4) - (mg)(L/2) + (F_r)(L)$$
$$= 0,$$

or
$$F_r = (g/4)(2m + M)$$
$$= (\tfrac{1}{4})(9.8 \text{ m/s}^2)(2 \times 1.8 \text{ kg} + 2.7 \text{ kg})$$
$$= 15 \text{ N.} \qquad \text{(Answer)} \quad (13\text{-}18)$$

Note how choosing an axis that passes through the application point of one of the unknown forces, F_l, eliminates that force

from Eq. 13-9 and allows us to solve directly for the other force. *Such a choice can help simplify a problem.*

If we solve Eq. 13-17 for F_l and substitute known quantities, we find

$$F_l = (M + m)g - F_r$$
$$= (2.7 \text{ kg} + 1.8 \text{ kg})(9.8 \text{ m/s}^2) - 15 \text{ N}$$
$$= 29 \text{ N.} \qquad \text{(Answer)}$$

SECOND SOLUTION: As a check, let us solve this problem in a different way, by applying the balance of torques equation about two different axes. Choosing first an axis through the left end of the beam, as we did above, we find Eq. 13-18 and the solution $F_r = 15$ N.

For an axis passing through the right end of the beam, Eq. 13-9 yields

$$\sum \tau_z = -(F_l)(L) + (Mg)(3L/4) + (mg)(L/2) + (F_r)(0)$$
$$= 0.$$

Solving for F_l, we find

$$F_l = (g/4)(2m + 3M)$$
$$= (\tfrac{1}{4})(9.8 \text{ m/s}^2)(2 \times 1.8 \text{ kg} + 3 \times 2.7 \text{ kg})$$
$$= 29 \text{ N,} \qquad \text{(Answer)}$$

in agreement with our earlier result. Note that the length of the beam enters this problem not explicitly but only as it affects the mass of the beam.

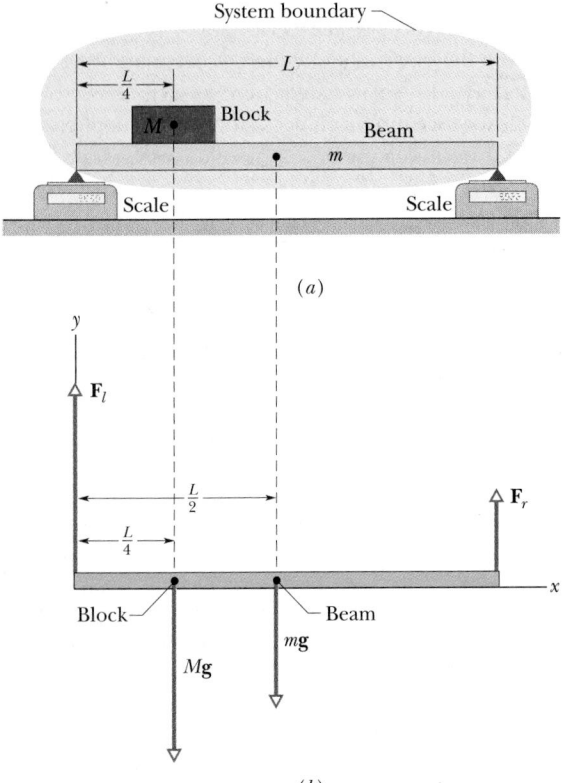

(a)

(b)

FIGURE 13-6 Sample Problem 13-1. (a) A beam of mass m supports a block of mass M. The system boundary is marked. (b) A free-body diagram, showing the forces that act on the system *beam + block*.

CHECKPOINT **2:** The figure gives an overhead view of a uniform rod in static equilibrium. (a) Can you find the magnitudes of unknown forces $\mathbf{F}_1$ and $\mathbf{F}_2$ by balancing the forces? (b) If you wish to find the magnitude of force $\mathbf{F}_2$ by using a single equation, where should you place a rotational axis? (c) The magnitude of $\mathbf{F}_2$ turns out to be 65 N. What then is the magnitude of $\mathbf{F}_1$?

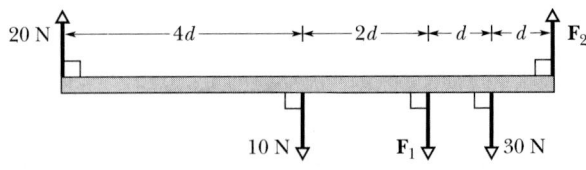

SAMPLE PROBLEM 13-2

A bowler holds a bowling ball whose mass M is 7.2 kg in the palm of his hand. As Fig. 13-7a shows, his upper arm is vertical and his lower arm is horizontal. What forces must the biceps muscle and the bony structure of the upper arm exert on the lower arm? The forearm and hand together have a mass m

of 1.8 kg; the needed dimensions are as shown in Fig. 13-7a.

SOLUTION: Our system is the lower arm and the bowling ball, taken together. Figure 13-7b shows a free-body diagram of the system. (The ball is represented by a dot within the boundary of the lower arm; the ball's weight vector $M\mathbf{g}$ has its tail on that dot. In drawing Fig. 13-7b from 13-7a, the vector $M\mathbf{g}$ is shifted along its line of action; the shift does not alter $M\mathbf{g}$, or a torque due to $M\mathbf{g}$ about any axis perpendicular to the figure.) The unknown forces are $\mathbf{T}$, the force exerted by the biceps muscle, and $\mathbf{F}$, the force exerted by the upper arm on the lower arm. The forces are all vertical.

From Eq. 13-8, which is $\Sigma F_y = 0$, we find

$$\sum F_y = T - F - mg - Mg = 0. \qquad (13\text{-}19)$$

Applying Eq. 13-9 about an axis through O at the elbow, and taking torques that would cause counterclockwise rotations as positive, we obtain

$$\sum \tau_z = (F)(0) + (T)(d) - (mg)(D) - (Mg)(L)$$
$$= 0. \qquad (13\text{-}20)$$

By choosing our axis to pass through point O, we have eliminated the variable F from Eq. 13-20. Solved for T, it yields

$$T = g\,\frac{mD + ML}{d}$$

$$= (9.8 \text{ m/s}^2)\,\frac{(1.8 \text{ kg})(15 \text{ cm}) + (7.2 \text{ kg})(33 \text{ cm})}{4.0 \text{ cm}}$$

$$= 648 \text{ N} \approx 650 \text{ N}. \qquad \text{(Answer)}$$

Thus the biceps muscle must pull up on the forearm with a force that is about nine times larger than the weight of the bowling ball—holding the ball as in Fig. 13-7a is difficult.

If we now solve Eq. 13-19 for F and substitute known quantities into it, we find

$$F = T - g(M + m)$$
$$= 648 \text{ N} - (9.8 \text{ m/s}^2)(7.2 \text{ kg} + 1.8 \text{ kg})$$
$$= 560 \text{ N}, \qquad \text{(Answer)}$$

which is about eight times the weight of the bowling ball.

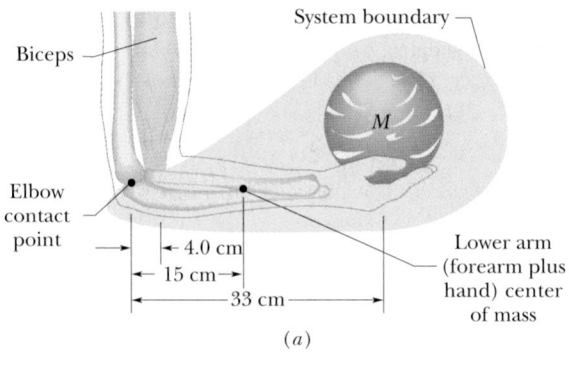

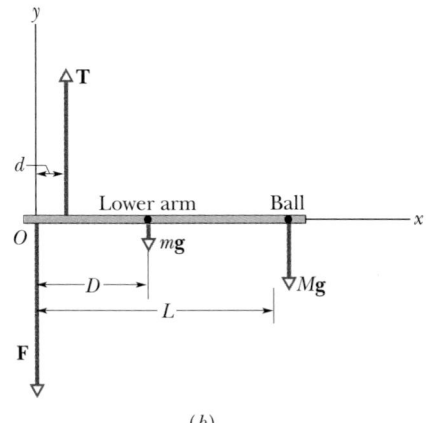

FIGURE 13-7 Sample Problem 13-2. (a) A hand holds a bowling ball. The system boundary is marked. (b) A free-body diagram of the *lower arm + ball* system, showing the forces that act. The vectors are not to scale; the powerful forces exerted by the biceps muscle and the elbow joint are many times larger than either weight.

SAMPLE PROBLEM 13-3

A ladder whose length L is 12 m and whose mass m is 45 kg rests against a wall. Its upper end is a distance h of 9.3 m above the ground, as in Fig. 13-8a. The center of mass of the ladder is one-third of the way up the ladder. A firefighter whose mass M is 72 kg climbs the ladder until her center of mass is halfway up. Assume that the wall, but not the ground, is frictionless. What forces are exerted on the ladder by the wall and by the ground?

SOLUTION: Figure 13-8b shows a free-body diagram of the *firefighter + ladder* system. (The firefighter is represented by a dot within the boundary of the ladder; her weight vector $M\mathbf{g}$ has its tail on that dot. In drawing Fig. 13-8b from 13-8a, the vector $M\mathbf{g}$ is shifted along its line of action; the shift does not alter $M\mathbf{g}$, or a torque due to $M\mathbf{g}$ about any axis perpendicular to the figure.) The wall exerts a horizontal force $\mathbf{F}_w$ on the ladder; it can exert no vertical force because the wall–ladder contact is assumed to be frictionless. The ground exerts a force $\mathbf{F}_g$ on the ladder, with horizontal component F_{gx} (due to friction) and vertical component F_{gy} (the usual normal force). We choose coordinate axes as shown, with the origin O at the point where the ladder meets the ground. The distance a from the wall to the foot of the ladder is readily found from

$$a = \sqrt{L^2 - h^2} = \sqrt{(12 \text{ m})^2 - (9.3 \text{ m})^2} = 7.58 \text{ m}.$$

From Eqs. 13-7 and 13-8, the balance of forces equations, we have for the system, respectively,

$$\sum F_x = F_w - F_{gx} = 0 \qquad (13\text{-}21)$$

and $$\sum F_y = F_{gy} - Mg - mg = 0. \qquad (13\text{-}22)$$

Equation 13-22 yields

$$F_{gy} = g(M + m) = (9.8 \text{ m/s}^2)(72 \text{ kg} + 45 \text{ kg})$$
$$= 1146.6 \text{ N} \approx 1100 \text{ N}. \qquad \text{(Answer)}$$

We next balance the torques acting on the system, choosing an axis through O, perpendicular to the plane of the figure. The moment arms about O for $\mathbf{F}_w$, $M\mathbf{g}$, $m\mathbf{g}$, $\mathbf{F}_{gx}$ and $\mathbf{F}_{gy}$ are h, $a/2$, $a/3$, zero, and zero, respectively. The zero moment arms for $\mathbf{F}_{gx}$ and $\mathbf{F}_{gy}$ mean that these forces produce zero torque about O. From Eq. 13-9, the balance of torques equation, we then have

$$\sum \tau_z = -(F_w)(h) + (Mg)(a/2) + (mg)(a/3)$$
$$= 0. \qquad \text{(13-23)}$$

Solving Eq. 13-23 for F_w, we find that

$$F_w = \frac{ga(M/2 + m/3)}{h}$$
$$= \frac{(9.8 \text{ m/s}^2)(7.58 \text{ m})(72/2 \text{ kg} + 45/3 \text{ kg})}{9.3 \text{ m}}$$
$$= 407 \text{ N} \approx 410 \text{ N}. \qquad \text{(Answer)}$$

From Eq. 13-21 we then have

$$F_{gx} = F_w = 410 \text{ N}. \qquad \text{(Answer)}$$

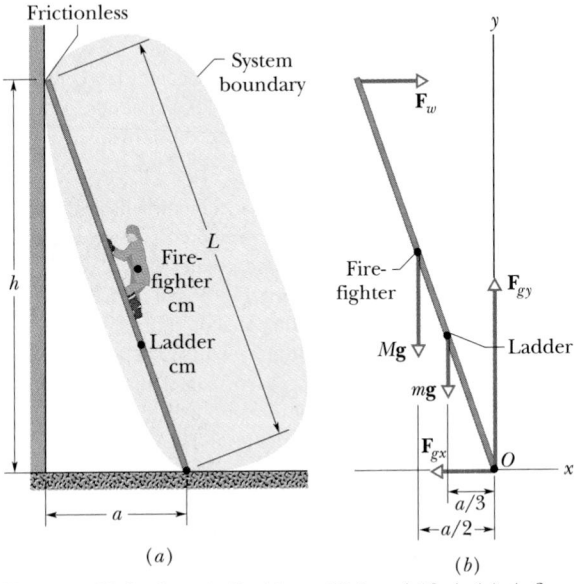

Frictionless

System boundary

L

Fire-fighter cm

Ladder cm

h

a

(a)

y

$\mathbf{F}_w$

Fire-fighter

$\mathbf{F}_{gy}$

Ladder

$M\mathbf{g}$

$m\mathbf{g}$

$\mathbf{F}_{gx}$

O

x

a/3

a/2

(b)

FIGURE 13-8 Sample Problems 13-3 and 13-4. (*a*) A firefighter climbs halfway up a ladder that is leaning against a frictionless wall. The ground is not frictionless. (*b*) A free-body diagram, showing the forces that act on the firefighter–ladder system. The origin O of a coordinate system is chosen at the point of application of the unknown force $\mathbf{F}_g$ (whose vector components $\mathbf{F}_{gx}$ and $\mathbf{F}_{gy}$ are shown). This choice simplifies the solution for the other unknown force, $\mathbf{F}_w$.

SAMPLE PROBLEM 13-4

In Sample Problem 13-3, let the coefficient of static friction μ_s between the ladder and the ground be 0.53. How far up the ladder can the firefighter go before the ladder starts to slip?

SOLUTION: The forces that act have the same labels as in Fig. 13-8. Let q be the fraction of the ladder's length the firefighter climbs before slipping occurs (she is then a horizontal distance qa from O). At the onset of slipping, we have

$$F_{gx} = \mu_s F_{gy}, \qquad \text{(13-24)}$$

in which F_{gx} is the static frictional force (usually symbolized as f_s) and F_{gy} is the normal force (usually symbolized as N).

If we apply Eq. 13-9, the balance of torques equation, about an axis through O we have, also at the onset of slipping,

$$\sum \tau_z = -(F_w)(h) + (Mg)(qa) + (mg)(a/3) = 0,$$

or

$$F_w = \frac{ga}{h} (\tfrac{1}{3}m + Mq). \qquad \text{(13-25)}$$

This equation shows us that as the firefighter climbs the ladder, that is, as q increases, the force F_w exerted by the wall must increase if equilibrium is to be maintained. To find q at the onset of slipping, we must first find F_w.

Equation 13-7, the balance of forces equation for the x direction, gives

$$\sum F_x = F_w - F_{gx} = 0.$$

If we combine this equation with Eq. 13-24 we find, for the ladder at the onset of slipping,

$$F_w = F_{gx} = \mu_s F_{gy}. \qquad \text{(13-26)}$$

From Eq. 13-8, the balance of forces equation for the y direction, we have

$$\sum F_y = F_{gy} - Mg - mg,$$

or

$$F_{gy} = (M + m)g. \qquad \text{(13-27)}$$

Combining Eqs. 13-26 and 13-27, we find

$$F_w = \mu_s g(M + m). \qquad \text{(13-28)}$$

If, finally, we combine Eqs. 13-25 and 13-28 and solve for q, we have

$$q = \frac{\mu_s h}{a} \frac{(M + m)}{M} - \frac{m}{3M} \qquad \text{(13-29)}$$

$$= \frac{(0.53)(9.3 \text{ m})}{7.6 \text{ m}} \frac{(72 \text{ kg} + 45 \text{ kg})}{72 \text{ kg}} - \frac{45 \text{ kg}}{(3)(72 \text{ kg})}$$

$$= 0.85. \qquad \text{(Answer)}$$

The firefighter can climb 85% of the way up the ladder before it starts to slip.

You can show from Eq. 13-29 that the firefighter can climb all the way up the ladder (which corresponds to $q = 1$) without slipping if $\mu_s > 0.61$. On the other hand, the ladder will slip when weight is put on the first rung (which corresponds to $q \approx 0$) if $\mu_s < 0.11$.

CHECKPOINT **3:** In the figure, a stationary 5 kg rod *AC* is held against a wall by a rope and friction between rod and wall. The uniform rod is 1 m long, and angle $\theta = 30°$. (a) If you are to find the magnitude of the force **T** on the rod from the rope with a single equation, at what labeled point should a rotational axis be placed? With that choice of axis and counterclockwise torques positive, what is the sign of (b) the torque τ_w due to the rod's weight and (c) the torque τ_r due to the pull on the rod by the rope? (d) Is τ_r greater than, less than, or equal to τ_w?

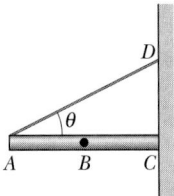

SAMPLE PROBLEM 13-5

Figure 13-9*a* shows a safe, whose mass *M* is 430 kg, hanging by a rope from a boom whose dimensions *a* and *b* are 1.9 m and 2.5 m, respectively. The boom's uniform beam has a mass *m* of 85 kg; the mass of the horizontal cable is negligible.

(a) Find the tension *T* in the cable.

SOLUTION: Figure 13-9*b* is a free-body diagram of the beam, which we take as our system. The forces acting on it are the tension **T** exerted by the cable, the weight *M***g** of the safe, the weight *m***g** of the beam itself (acting at the center of the beam), and the force components F_h and F_v exerted on the beam by the hinge that fastens the beam to the wall.

Let us apply Eq. 13-9, the balance of torques equation, to an axis through the hinge perpendicular to the plane of the figure. Taking torques that would cause counterclockwise rotations as positive, we have

$$\sum \tau_z = (T)(a) - (Mg)(b) - (mg)(\tfrac{1}{2}b) = 0.$$

By our wise choice of an axis, we have eliminated the unknown forces F_h and F_v from this equation (since they create no torque about that axis), leaving only the unknown force *T*. Solving for *T* yields

$$T = \frac{gb(M + \tfrac{1}{2}m)}{a}$$

$$= \frac{(9.8 \text{ m/s}^2)(2.5 \text{ m})(430 \text{ kg} + 85/2 \text{ kg})}{1.9 \text{ m}}$$

$$= 6090 \text{ N} \approx 6100 \text{ N}. \qquad \text{(Answer)}$$

(b) Find the force components F_h and F_v exerted on the beam by the hinge.

SOLUTION: We now apply the balance of forces equations. From Eq. 13-7 we have

$$\sum F_x = F_h - T = 0,$$

and so $\qquad F_h = T = 6090 \text{ N} \approx 6100 \text{ N}. \qquad$ (Answer)

From Eq. 13-8 we have

$$\sum F_y = F_v - mg - Mg = 0,$$

and so

$$F_v = g(m + M) = (9.8 \text{ m/s}^2)(85 \text{ kg} + 430 \text{ kg})$$

$$= 5047 \text{ N} \approx 5000 \text{ N}. \qquad \text{(Answer)}$$

(c) What is the magnitude *F* of the net force exerted by the hinge on the beam?

SOLUTION: From the figure we see that

$$F = \sqrt{F_h^2 + F_v^2}$$

$$= \sqrt{(6090 \text{ N})^2 + (5047 \text{ N})^2} \approx 7900 \text{ N}. \quad \text{(Answer)}$$

Note that *F* is substantially greater than either the combined weights of the safe and the beam, 5000 N, or the tension in the horizontal wire, 6100 N.

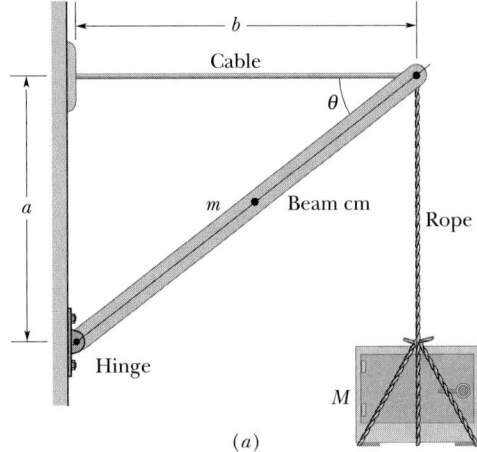

(*a*)

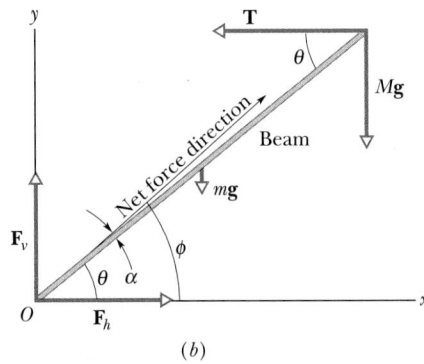

(*b*)

FIGURE 13-9 Sample Problem 13-5. (*a*) A heavy safe is hung from a boom consisting of a horizontal steel cable and a uniform beam. (*b*) A free-body diagram for the beam. Note that the net force acting on the beam due to **F**$_v$ and **F**$_h$ does not point directly along the beam axis.

(d) What is the angle α between the direction of the net force **F** exerted on the beam by the hinge and the center line of the beam?

SOLUTION: From the figure we see that

$$\theta = \tan^{-1}\frac{a}{b} = \tan^{-1}\frac{1.9\ \text{m}}{2.5\ \text{m}} = 37.2°,$$

and

$$\phi = \tan^{-1}\frac{F_v}{F_h} = \tan^{-1}\frac{5047\ \text{N}}{6090\ \text{N}} = 39.6°.$$

So

$$\alpha = \phi - \theta = 39.6° - 37.2° = 2.4°. \quad \text{(Answer)}$$

If the weight of the beam were small enough to neglect, you would find $\alpha = 0$; that is, the hinge force would point directly along the beam axis.

SAMPLE PROBLEM 13-6

In Fig. 13-10, a rock climber with mass $m = 55$ kg rests during a "chimney climb," pressing only with her shoulders and feet against the walls of a fissure of width $w = 1.0$ m. Her center of mass is a horizontal distance $d = 0.20$ m from the wall against which her shoulders are pressed. The coefficient of static friction between her shoes and the wall is $\mu_1 = 1.1$, and between her shoulders and the wall it is $\mu_2 = 0.70$.

(a) What minimum horizontal push must the climber exert on the walls to keep from falling?

SOLUTION: Her horizontal push against the wall is the same at her shoulders and at her feet. That push must be equal in magnitude to the normal force **N** on her due to the wall at each of those places. Thus there is no net horizontal force on her and Eq. 13-7, $\Sigma F_x = 0$, is satisfied.

As her weight **mg** acts to slide her down the walls, static frictional forces $\mathbf{f}_1$ and $\mathbf{f}_2$ automatically act upward at feet and shoulders, respectively, to counter the tendency to slide. As long as she does not fall, Eq. 13-8 ($\Sigma F_y = 0$) is satisfied and we have

$$f_1 + f_2 = mg. \quad (13\text{-}30)$$

Let us assume that the climber initially pushes hard against the walls, and then slowly relaxes her push. As she decreases her push, the magnitude N of the normal force decreases, and so do the products $\mu_1 N$ and $\mu_2 N$ that limit the static friction at her feet and shoulders, respectively (see Eq. 6-1).

Suppose N is reduced until the limit $\mu_1 N$ equals the magnitude f_1 of the frictional force at her feet and the limit $\mu_2 N$ equals the magnitude f_2 of the frictional force at her shoulders. She is then on the verge of slipping at both places. If she were to relax her push any more, the resulting decrease in the limits $\mu_1 N$ and $\mu_2 N$ would leave $f_1 + f_2$ less than mg, and she would fall. Thus when she is on the verge of slipping, Eq. 13-30 becomes

$$\mu_1 N + \mu_2 N = mg, \quad (13\text{-}31)$$

which yields

$$N = \frac{mg}{\mu_1 + \mu_2} = \frac{(55\ \text{kg})(9.8\ \text{m/s}^2)}{1.1 + 0.70} = 299\ \text{N} \approx 300\ \text{N}.$$

Thus her minimum horizontal push must be about 300 N.

(b) For that push, what must be the vertical distance h between her feet and shoulders if she is to be stable?

SOLUTION: To satisfy Eq. 13-9, $\Sigma\tau = 0$, the forces acting on her must not create a net torque about *any* rotation axis. Let us consider a rotational axis perpendicular to the page at her shoulders. The net torque about that axis is given by

$$\sum\tau = -f_1 w + Nh + mgd = 0. \quad (13\text{-}32)$$

Solving this for h, setting $f_1 = \mu_1 N$, and substituting $N = 299$ N and other known values, we find

$$h = \frac{f_1 w - mgd}{N} = \frac{\mu_1 Nw - mgd}{N} = \mu_1 w - \frac{mgd}{N}$$

$$= (1.1)(1.0\ \text{m}) - \frac{(55\ \text{kg})(9.8\ \text{m/s}^2)(0.20\ \text{m})}{299\ \text{N}}$$

$$= 0.739\ \text{m} \approx 0.74\ \text{m}. \quad \text{(Answer)}$$

We would find the same required value of h if we chose any other rotation axis perpendicular to the page, such as one at her feet.

(c) What are the magnitudes of the frictional forces supporting the climber?

SOLUTION: For $N = 299$ N, we have

$$f_1 = \mu_1 N = (1.1)(299\ \text{N})$$
$$= 328.9\ \text{N} \approx 330\ \text{N}, \quad \text{(Answer)}$$

and from Eq. 13-30 we then have

$$f_2 = mg - f_1 = (55\ \text{kg})(9.8\ \text{m/s}^2) - 328.9\ \text{N}$$
$$= 210.1\ \text{N} \approx 210\ \text{N}. \quad \text{(Answer)}$$

(d) Is she stable if she exerts the same force (299 N) on the walls when her feet are higher, with $h = 0.37$ m?

SOLUTION: Again $N = 299$ N, and the net torque on the climber about any rotation axis must be zero. For the axis at her shoulders, we obtain f_1 from Eq. 13-32:

$$f_1 = \frac{Nh + mgd}{w}$$

$$= \frac{(299\ \text{N})(0.37\ \text{m}) + (55\ \text{kg})(9.8\ \text{m/s}^2)(0.20\ \text{m})}{1.0\ \text{m}}$$

$$= 218\ \text{N}.$$

This is less than the limit $\mu_1 N\ (= 329$ N) and thus is possible to attain.

Next we use Eq. 13-30 to find the value of f_2 that will ensure that $\Sigma F_y = 0$:

$$f_2 = mg - f_1 = (55\ \text{kg})(9.8\ \text{m/s}^2) - 218\ \text{N} = 321\ \text{N}.$$

This exceeds the limit $\mu_2 N\ (= 209$ N) and thus is impossible to attain with a push of 299 N. The only way the climber can avoid slipping when $h = 0.37$ m (or for any value of h less than 0.74 m) is by pushing harder than 299 N on the walls so as to increase the limit $\mu_2 N$.

Similarly, if $h > 0.74$ m, she must also exert a force greater than 299 N on the walls to be stable. Here, then, is the advantage of knowing the physics before you climb a chimney. When you need to rest, you will avoid the (dire) error of novice climbers who place their feet too high or too low. Instead, you will know that there is a ''best'' distance between shoulders and feet, requiring the least push, and giving you a good chance to rest.

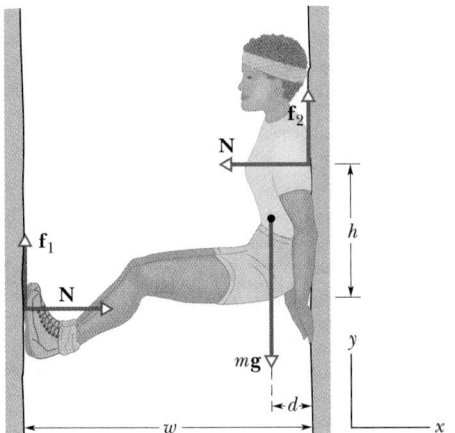

FIGURE 13-10 Sample Problem 13-6. The forces on a climber resting in a rock chimney. The push of the climber on the chimney walls gives rise to the normal forces **N** (which are equal in magnitude) and the frictional forces $\mathbf{f}_1$ and $\mathbf{f}_2$.

PROBLEM SOLVING TACTICS

TACTIC 1: *Static Equilibrium Problems*
Here is a list of steps for solving static equilibrium problems:

1. Draw a *sketch* of the problem.
2. Select the *system* to which you will apply the laws of equilibrium, drawing a closed curve around it on your sketch to fix it clearly in your mind. In some situations you can select a single object as the system; it is the object you wish to be in equilibrium (such as the rock climber in Sample Problem 13-6). In other situations, you might include additional objects in the system *if* their inclusion simplifies the calculations for equilibrium. For example, suppose in Sample Problems 13-3 and 13-4 you select only the ladder as the system. Then in Fig. 13-8b you will have to account for additional unknown forces exerted on the ladder by the hands and feet of the firefighter. These additional unknowns complicate the equilibrium calculations. The sys-

tem of Fig. 13-8 was chosen to include the firefighter so that those unknown forces are *internal* to the system and thus need not be found in order to solve Sample Problems 13-3 and 13-4.

3. Draw a *free-body diagram* of the system. Show all the forces that act on the system, labeling them clearly and making sure that their points of application and lines of action are correctly shown.
4. Draw in the *x and y axes* of a coordinate system. Choose them so that at least one axis is parallel to one or more unknown force. Resolve into components the forces that do not lie along one of the axes. In all our sample problems it made sense to choose the *x* axis horizontal and the *y* axis vertical.
5. Write the two *balance of forces equations,* using symbols throughout.
6. Choose one or more rotational axes perpendicular to the plane of the figure and write the *balance of torques equation* for each axis. If you choose an axis that passes through the line of action of an unknown force, the equation will be simplified because that force will not appear in it.
7. *Solve* your equations *algebraically* for the unknowns. Some students feel more confident in substituting numbers with units in the independent equations at this stage, especially if the algebra is particularly involved. However, experienced problem solvers prefer the algebraic approach, which reveals the dependence of solutions on the various variables.
8. Finally, *substitute numbers* with units in your algebraic solutions, obtaining numerical values for the unknowns.
9. Look at your answer—does it make sense? Is it obviously too large or too small? Is the sign correct? Are the units appropriate?

13-5 INDETERMINATE STRUCTURES

For the problems of this chapter, we have only three independent equations at our disposal, usually two balance of forces equations and one balance of torques equation about a given axis. Thus if a problem has more than three unknowns, we cannot solve it.

It is easy to find such problems. In Sample Problems 13-3 and 13-4, for example, we could have assumed that there is friction between the wall and the top of the ladder. Then there would have been a vertical frictional force acting where the ladder touches the wall, making a total of four unknown forces. With only three equations, we could not have solved this problem.

Consider also an unsymmetrically loaded car. What are the forces—all different—on the four tires? Again, we cannot find them because we have only three independent equations with which to work. Similarly, we can solve an equilibrium problem for a table with three legs but not one with four legs. Problems like these, in which there are more unknowns than equations, are called **indeterminate.**

And yet, solutions to indeterminate problems exist in the real world. If you rest the tires of the car on four platform scales, each scale will register a definite reading, the sum of the readings being the weight of the car. What is eluding us in our efforts to solve equations for the individual scale readings?

The problem is that we have assumed—without making a great point of it—that the bodies to which we apply the equations of static equilibrium are perfectly rigid. By this we mean that they do not deform when forces are applied to them. Strictly, there are no such bodies. The tires of the car, for example, deform easily under load until the car settles into a position of static equilibrium.

We have all had experience with a wobbly restaurant table, which we usually level by putting folded paper under one of the legs. If a big enough elephant sat on such a table, however, you may be sure that if the table did not collapse, it would deform just like the tires of a car. Its legs would all touch the floor, the forces acting upward on the table legs would all assume definite (and different) values as in Fig. 13-11, and the table would no longer wobble. But how do we find the values of those forces acting on the legs?

To solve such indeterminate equilibrium problems, we must supplement equilibrium equations with some knowledge of *elasticity,* the branch of physics and engineering that describes how real bodies deform when forces are applied to them. The next section provides an introduction to this subject.

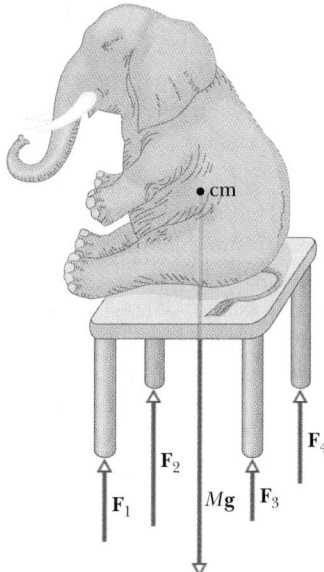

FIGURE 13-11 The table is an indeterminate structure. The four forces on the table legs are different in magnitude and cannot be found from the laws of static equilibrium alone.

CHECKPOINT **4:** A horizontal uniform bar of weight 10 N is to hang from a ceiling by two wires that exert upward forces $\mathbf{F}_1$ and $\mathbf{F}_2$. The figure shows four arrangements for the wires. Which arrangements, if any, are indeterminate (so that we cannot solve for numerical values of $\mathbf{F}_1$ and $\mathbf{F}_2$)?

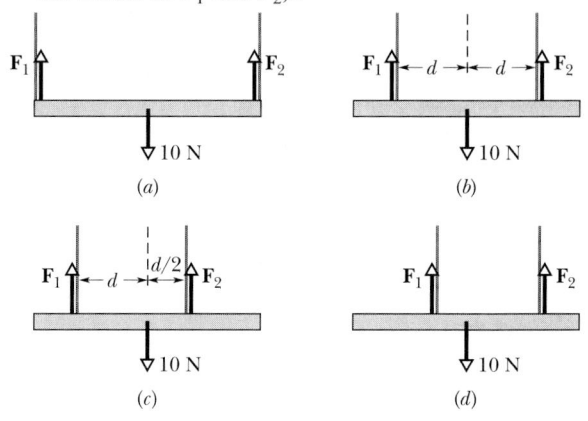

13-6 ELASTICITY

When a large number of atoms come together to form a metallic solid, such as an iron nail, they settle into equilibrium positions in a three-dimensional *lattice,* a repetitive arrangement in which each atom has a well-defined equilibrium distance from its nearest neighbors. The atoms are held together by interatomic forces that are represented by springs in Fig. 13-12.* The lattice is remarkably rigid, which is another way of saying that the "interatomic springs" are extremely stiff. It is for this reason that we perceive many ordinary objects such as metal ladders, tables, and spoons as perfectly rigid. Of course, some ordinary objects, such as garden hoses or rubber gloves, do not strike us as rigid at all. The atoms that make up these objects *do not* form a rigid lattice like that of Fig. 13-12 but are aligned in long flexible molecular chains, each chain being only loosely bound to its neighbors.

All real "rigid" bodies are to some extent **elastic,** which means that we can change their dimensions slightly by pulling, pushing, twisting, or compressing them. To get a feeling for the orders of magnitude involved, consider a vertical steel rod 1 m long and 1 cm in diameter. If you hang a subcompact car from the end of such a rod, the rod

---

*Ordinary metal objects, such as an iron nail, are made up of grains of iron, each grain formed as a more or less perfect lattice, such as that of Fig. 13-12. The forces between the grains are much weaker than the forces that hold the lattice together, so that rupture usually occurs along grain boundaries.

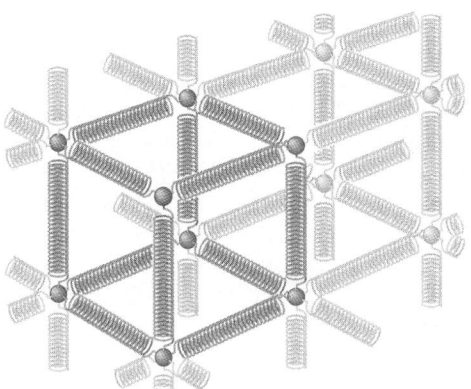

FIGURE 13-12 The atoms of a metallic solid are distributed on a repetitive three-dimensional lattice. The springs represent interatomic forces.

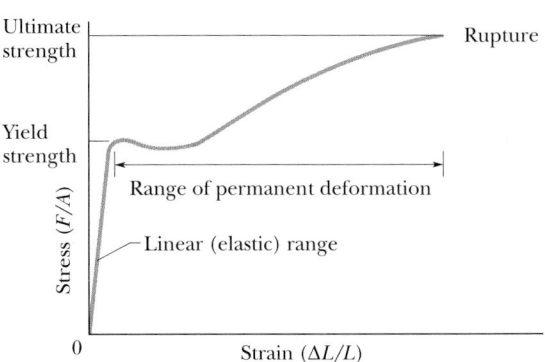

FIGURE 13-14 A stress-strain curve for a steel test specimen such as that of Fig. 13-15. The specimen deforms permanently when the stress is equal to the *yield strength* of the material. It ruptures when the stress is equal to the *ultimate strength* of the material.

will stretch, but only by about 0.5 mm, or 0.05%. Furthermore, the rod will return to its original length when the car is removed.

If you hang two cars from the rod, the rod will be permanently stretched and will not recover its original length when you remove the load. If you hang three cars from the rod, the rod will break. Just before rupture, the elongation of the rod will be less than 0.2%. Although deformations of this size seem small, they are important in engineering practice. (Whether a wing under load will stay on an airplane is obviously important.)

Figure 13-13 shows three ways that a solid might change its dimensions when forces act on it. In Fig. 13-13a, a cylinder is stretched. In Fig. 13-13b, a cylinder is deformed by a force perpendicular to its axis, much as one might deform a pack of cards or a book. In Fig. 13-13c, a

solid object, placed in a fluid under high pressure, is compressed uniformly on all sides. What the three deformation types have in common is that a **stress,** or deforming force per unit area, produces a **strain,** or unit deformation. In Fig. 13-13, *tensile stress* (associated with stretching) is in (a), *shearing stress* is in (b), and *hydraulic stress* is in (c).

The stresses and the strains take different forms in the three cases of Fig. 13-13, but—over the range of engineering usefulness—stress and strain are proportional to each other. The constant of proportionality is called a **modulus of elasticity,** so that

$$\text{stress} = \text{modulus} \times \text{strain}. \qquad (13\text{-}33)$$

Figure 13-14 shows the relation between stress and strain for a steel test cylinder such as that of Fig. 13-15. In a

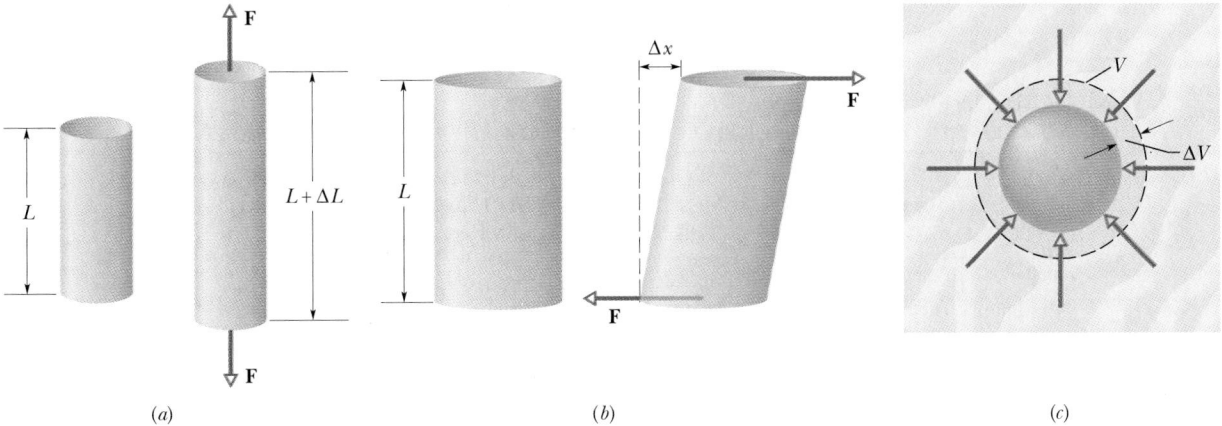

(a)　　　　　　　　　　　　(b)　　　　　　　　　　　　(c)

FIGURE 13-13 (a) A cylinder subject to *tensile stress* stretches by an amount ΔL. (b) A cylinder subject to *shearing stress* deforms by an amount Δx, somewhat like a pack of playing cards would. (c) A solid sphere subject to uniform *hydraulic stress* from a fluid shrinks in volume by an amount ΔV. All the deformations shown are greatly exaggerated.

FIGURE 13-15 A test specimen, used to determine a stress–strain curve such as that of Fig. 13-14. The change that occurs in a certain length L is measured in a tensile stress–strain test.

standard test, the tensile stress on a test cylinder is slowly increased from zero to the point where the cylinder fractures, and the strain is carefully measured and graphed. For a substantial range of applied stresses, the stress–strain relation is linear, and the specimen recovers its original dimensions when the stress is removed; it is here that Eq. 13-33 applies. If the stress is increased beyond the **yield strength** S_y of the specimen, the specimen becomes permanently deformed. If the stress continues to increase, the specimen eventually ruptures, at a stress called the **ultimate strength** S_u.

Tension and Compression

For simple tension or compression, the stress is defined as F/A, the force divided by the area over which it acts (the force is perpendicular to the area, as you can see in Fig. 13-13a). The strain, or unit deformation, is then the dimensionless quantity $\Delta L/L$, the fractional (or sometimes percentage) change in the length of the specimen. If the specimen is a long rod and the stress does not exceed the yield strength, then not only the entire rod but also every section of it experiences the same strain when a given stress is applied. Because the strain is dimensionless, the modulus in Eq. 13-33 has the same dimensions as the stress, namely, force per unit area.

The modulus for tensile and compressive stresses is called the **Young's modulus** and is represented in engi-

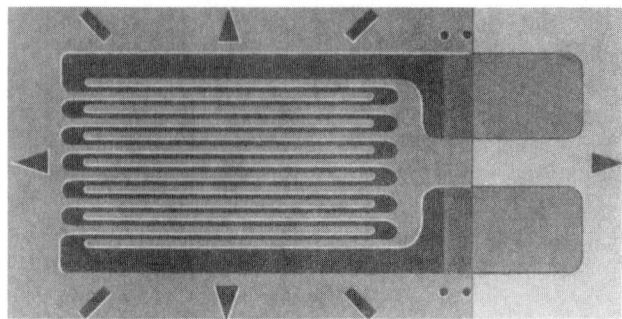

FIGURE 13-16 A strain gauge of overall dimensions 9.8 mm by 4.6 mm. The gauge is fastened with adhesive to the object whose strain is to be measured; it experiences the same strain as the object. The electrical resistance of the gauge varies with the strain, permitting strains up to about 3% to be measured.

neering practice by the symbol E. Equation 13-33 becomes

$$\frac{F}{A} = E\,\frac{\Delta L}{L}. \qquad (13\text{-}34)$$

The strain $\Delta L/L$ in a specimen can often be measured conveniently with a *strain gauge* (Fig. 13-16). These simple and useful devices, which can be attached directly to operating machinery with adhesives, are based on the principle that the electrical properties of the gauge are dependent on the strain it undergoes.

Although the Young's modulus for an object may be almost the same for tension and compression, the object's ultimate strength may well be different for the two types of stress. Concrete, for example, is very strong in compression but is so weak in tension that it is almost never used in that manner. Table 13-1 shows the Young's modulus and other elastic properties for some materials of engineering interest.

TABLE 13-1
SOME ELASTIC PROPERTIES OF SELECTED MATERIALS OF ENGINEERING INTEREST

MATERIAL	DENSITY ρ (kg/m³)	YOUNG'S MODULUS E (10^9 N/m²)	ULTIMATE STRENGTH S_u (10^6 N/m²)	YIELD STRENGTH S_y (10^6 N/m²)
Steel[a]	7860	200	400	250
Aluminum	2710	70	110	95
Glass	2190	65	50[b]	—
Concrete[c]	2320	30	40[b]	—
Wood[d]	525	13	50[b]	—
Bone	1900	9[b]	170[b]	—
Polystyrene	1050	3	48	—

[a] Structural steel (ASTM-A36).　　[b] In compression.　　[c] High strength.　　[d] Douglas fir.

Shearing

In the case of shearing, the stress is also a force per unit area, but the force vector lies in the plane of the area rather than at right angles to it. The strain is the dimensionless ratio $\Delta x/L$, with the quantities defined as shown in Fig. 13-13b. The corresponding modulus, which is given the symbol G in engineering practice, is called the **shear modulus.** For shearing, Eq. 13-33 is written as

$$\frac{F}{A} = G\frac{\Delta x}{L}. \qquad (13\text{-}35)$$

Shearing stresses play a critical role in the buckling of shafts that rotate under load and in bone fractures caused by bending.

Hydraulic Stress

In Fig. 13-13c, the stress is the fluid pressure p on the object, which, as you will see in Chapter 15, is a force per unit area. The strain is $\Delta V/V$, where V is the original volume of the specimen and ΔV is the absolute value of the change in volume. The corresponding modulus, with symbol B, is called the **bulk modulus** of the material. The object is said to be under *hydraulic compression,* and the pressure can be called the *hydraulic stress.* For this situation, we write Eq. 13-33 as

$$p = B\frac{\Delta V}{V}. \qquad (13\text{-}36)$$

The bulk modulus of water is 2.2×10^9 N/m², and that of steel is 16×10^{10} N/m². The pressure at the bottom of the Pacific Ocean, at its average depth of about 4000 m, is 4.0×10^7 N/m². The fractional compression $\Delta V/V$ of a volume of water due to this pressure is 1.8%; that for a steel object is only about 0.025%. In general, solids—with their rigid atomic lattices—are less compressible than liquids, in which the atoms or molecules are less tightly coupled to their neighbors.

SAMPLE PROBLEM 13-7

A structural steel rod has a radius R of 9.5 mm and a length L of 81 cm. A force F of 6.2×10^4 N (about 7 tons) stretches it along its length.

(a) What is the stress in the rod?

SOLUTION: From its definition, the stress is

$$\text{stress} = \frac{F}{A} = \frac{F}{\pi R^2} = \frac{6.2 \times 10^4 \text{ N}}{(\pi)(9.5 \times 10^{-3} \text{ m})^2}$$
$$= 2.2 \times 10^8 \text{ N/m}^2. \qquad \text{(Answer)}$$

The yield strength for structural steel is 2.5×10^8 N/m², so that this rod is dangerously close to its yield strength.

(b) What is the elongation of the rod under this load? What is the strain?

SOLUTION: From Eq. 13-34, using the result we have just calculated and the value of E for steel (Table 13-1), we obtain

$$\Delta L = \frac{(F/A)L}{E} = \frac{(2.2 \times 10^8 \text{ N/m}^2)(0.81 \text{ m})}{2.0 \times 10^{11} \text{ N/m}^2}$$
$$= 8.9 \times 10^{-4} \text{ m} = 0.89 \text{ mm}. \qquad \text{(Answer)}$$

Thus the strain is

$$\frac{\Delta L}{L} = \frac{8.9 \times 10^{-4} \text{ m}}{0.81 \text{ m}}$$
$$= 1.1 \times 10^{-3} = 0.11\%. \qquad \text{(Answer)}$$

SAMPLE PROBLEM 13-8

The femur, which is the principal bone of the thigh, has a minimum diameter in an adult male of about 2.8 cm, corresponding to a cross-sectional area A of 6×10^{-4} m². At what compressive load would it break?

SOLUTION: From Table 13-1 we see that the ultimate strength S_u for bone in compression is 170×10^6 N/m². The compressive force at fracture is the force F that produces the stress S_u; so

$$F = S_u A = (170 \times 10^6 \text{ N/m}^2)(6 \times 10^{-4} \text{ m}^2)$$
$$= 1.0 \times 10^5 \text{ N}. \qquad \text{(Answer)}$$

This is 23,000 lb or 11 tons. Although this is a large force, it can be encountered during, for example, an unskillful parachute landing on hard ground. The force need not be sustained to break the bone; a few milliseconds will do it.

SAMPLE PROBLEM 13-9

A table has three legs that are 1.00 m in length and a fourth leg that is longer by a distance $d = 0.50$ mm, so that the table wobbles slightly. A heavy steel cylinder whose mass M is 290 kg is placed upright on the table (whose mass is much less than M) so that all four legs compress and the table no longer wobbles. The legs are wooden cylinders whose cross-sectional area A is 1.0 cm². The Young's modulus E for the wood is 1.3×10^{10} N/m². Assume that the tabletop remains level and that the legs do not buckle. With what force does the floor push upward on each leg?

SOLUTION: We take the table plus steel cylinder as our system. The situation is like Fig. 13-11, except we now have a steel cylinder on the table. If the tabletop remains level, each of the three short legs must be compressed by the same amount (call it ΔL_3) and thus by the same force F_3. The single long leg must be compressed by a larger amount ΔL_4, by a larger force F_4, and we must have

$$\Delta L_4 = \Delta L_3 + d. \tag{13-37}$$

We can rewrite Eq. 13-34 as $\Delta L = FL/EA$. We then use this relationship to substitute for L_4 and L_3 in Eq. 13-37, letting L represent, in turn, the initial length of the three short legs (1.0 m) and the approximate initial length of the long leg (also 1.0 m). Equation 13-37 becomes

$$F_4 L = F_3 L + dAE. \tag{13-38}$$

From Eq. 13-8, the balance of forces in the vertical direction, we have for our system

$$\sum F_y = 3F_3 + F_4 - Mg = 0. \tag{13-39}$$

If we solve Eqs. 13-38 and 13-39 for the unknown force F_3, we find

$$F_3 = \frac{Mg}{4} - \frac{dAE}{4L}$$

$$= \frac{(290 \text{ kg})(9.8 \text{ m/s}^2)}{4}$$

$$- \frac{(5.0 \times 10^{-4} \text{ m})(10^{-4} \text{ m}^2)(1.3 \times 10^{10} \text{ N/m}^2)}{(4)(1.00 \text{ m})}$$

$$= 711 \text{ N} - 163 \text{ N} = 548 \text{ N} \approx 550 \text{ N}. \qquad \text{(Answer)}$$

From Eq. 13-39 we then find

$$F_4 = Mg - 3F_3 = (290 \text{ kg})(9.8 \text{ m/s}^2) - 3(548 \text{ N})$$

$$\approx 1200 \text{ N}. \qquad \text{(Answer)}$$

You can show that to reach their equilibrium configuration, the three short legs were each compressed by 0.42 mm and the single long leg by 0.92 mm, the difference being 0.50 mm.

CHECKPOINT **5:** The figure shows a horizontal block that is suspended by two wires, A and B, which are identical except for their original lengths. The center of mass of the block is closer to wire B than to wire A. (a) Measuring torques about the block's center of mass, state whether the torque due to wire A is greater than, less than, or equal to the torque due to wire B. (b) Which wire exerts more force on the block? (c) If the wires are now equal in length, which one was originally shorter?

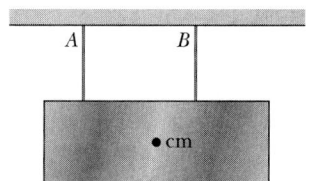

REVIEW & SUMMARY

Static Equilibrium

A rigid body at rest is said to be in **static equilibrium.** For such a body, the vector sum of the external forces acting on it is zero:

$$\sum \mathbf{F}_{\text{ext}} = 0 \qquad \text{(balance of forces).} \tag{13-3}$$

If all the forces lie in the xy plane, this vector equation is equivalent to two scalar component equations:

$$\sum F_x = 0 \quad \text{and} \quad \sum F_y = 0 \qquad \begin{array}{l}\text{(balance of}\\\text{forces).}\end{array} \tag{13-7, 13-8}$$

Static equilibrium also implies that the vector sum of the external torques acting on the body about *any* point is zero, or

$$\sum \tau_{\text{ext}} = 0 \qquad \text{(balance of torques).} \tag{13-5}$$

If the forces lie in the xy plane, all torque vectors are parallel to the z axis, and Eq. 13-5 is equivalent to the single scalar component equation

$$\sum \tau_z = 0 \qquad \text{(balance of torques).} \tag{13-9}$$

Center of Gravity

The gravitational force acts individually on all the individual elements of a body. The net effect of all the individual actions may be found by imagining an equivalent total gravitational force $M\mathbf{g}$

acting at a specific point called the **center of gravity.** If the gravitational acceleration $\mathbf{g}$ is the same for all the particles in a body, the center of gravity is at the center of mass.

Elastic Moduli

Three **elastic moduli** are used to describe the elastic behavior (deformations) of objects as they respond to the forces that act on them. The **strain** (fractional change in length) is linearly related to the applied **stress** (force per unit area) by the modulus in each case. The general relation is

$$\text{stress} = \text{modulus} \times \text{strain}. \tag{13-33}$$

Tension and Compression

When an object is under tension or compression, Eq. 13-33 is written as

$$\frac{F}{A} = E \frac{\Delta L}{L}, \tag{13-34}$$

where $\Delta L/L$ is the strain of the object, F is the magnitude of the applied force $\mathbf{F}$ causing the strain, A is the cross-sectional area over which $\mathbf{F}$ is applied (perpendicular to A, as in Fig. 13-13a), and E is the **Young's modulus** for the object. The stress is F/A.

Shearing

When an object is under a shearing stress, Eq. 13-33 is written as

$$\frac{F}{A} = G\frac{\Delta x}{L}, \qquad (13\text{-}35)$$

where $\Delta x/L$ is the strain of the object, Δx is the displacement of one end of the object in the direction of the applied force **F** (as in Fig. 13-13*b*), and *G* is the **shear modulus** of the object. The stress is *F/A*.

Hydraulic Stress

When an object undergoes *hydraulic compression* due to a stress exerted by a surrounding fluid, Eq. 13-33 is written as

$$p = B\frac{\Delta V}{V}, \qquad (13\text{-}36)$$

where *p* is the pressure (*hydraulic stress*) on the object due to the fluid, $\Delta V/V$ (the strain) is the absolute value of the fractional change in the object's volume due to that pressure, and *B* is the **bulk modulus** of the object.

QUESTIONS

1. Figure 13-17 shows an overhead view of a uniform stick on which four forces act. Suppose we choose a rotational axis through point *O*, calculate the torques about that axis due to the forces, and find that these torques balance. Will the torques balance if, instead, the rotational axis is chosen to be at (a) point *A*, (b) point *B*, or (c) point *C*? (d) Suppose, instead, that we find that the torques about point *O* do not balance. Is there another point about which the torques will balance?

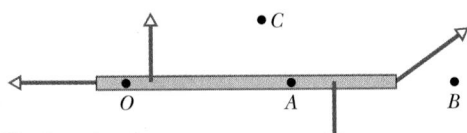

FIGURE 13-17 Question 1.

2. Figure 13-18 shows four overhead views of rotating uniform disks that are sliding across a frictionless floor. Three forces, of magnitude *F*, 2*F*, or 3*F*, act on each disk; each force is applied either at the rim, at the center, or halfway between rim and center. The force vectors rotate along with the disks, and, in the "snapshots" of Fig. 13-18, point left or right. Which disks are in equilibrium?

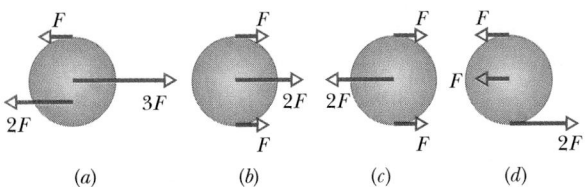

FIGURE 13-18 Question 2.

3. Figure 13-19 shows overhead views of two structures on which three forces act. The directions of the forces are as indicated. If the magnitudes of the forces are adjusted properly (but kept nonzero), which structure can be in static equilibrium?

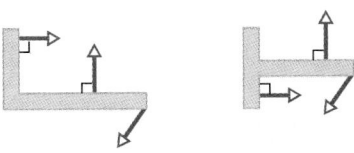

FIGURE 13-19 Question 3.   *(a)*   *(b)*

4. Figure 13-20 shows a mobile of toy penguins hanging from a ceiling. Each crossbar is horizontal, has negligible mass, and extends three times as far to the right of the wire supporting it as to the left. Penguin 1 has mass $m_1 = 48$ kg. What are the masses of the other penguins?

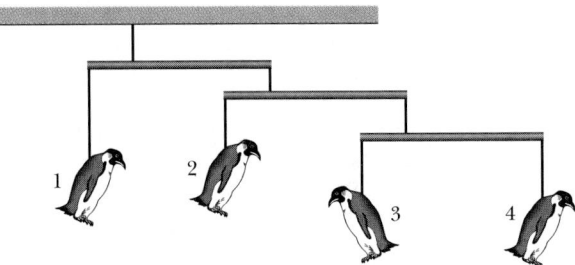

FIGURE 13-20 Question 4.

5. Figure 13-21 shows an overhead view of a metal square lying on a frictionless floor. Three forces, which are drawn to scale, act at the corners of the square. (a) Is the first requirement of equilibrium (in Eq. 13-1) satisfied? (b) Is the second requirement of equilibrium satisfied? (c) If the answer to either (a) or (b) is no, could a fourth force acting on the square then satisfy both requirements of equilibrium?

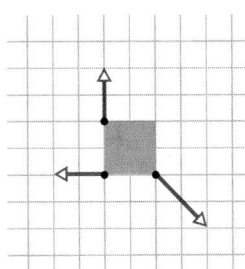

FIGURE 13-21 Question 5.

6. (a) How many different free-standing towers can you build with three of the small (four-projection) Lego blocks? (Two such blocks are stackable one directly above the other, or one displaced half a block length left or right of the other. Mirror-image arrangements count as one.) How many such towers are in (b) stable equilibrium and (c) unstable equilibrium (center of mass almost over an edge)? (d) Which arrangement is most stable (hardest to topple)? Why?

7. Figure 13-22 shows four arrangements in which a painting is suspended from a wall with two identical lengths of wire. The wires in Fig. 13-22b and c all make the same angle with the horizontal. Rank the four arrangements according to the tension in the wires, largest first.

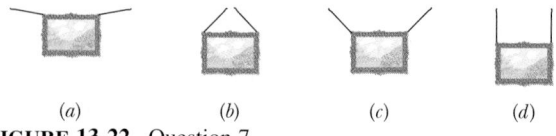

(a)   (b)   (c)   (d)

FIGURE 13-22 Question 7.

8. A ladder leans against a frictionless wall but is prevented from falling because of friction between it and the ground. Suppose you shift the base of the ladder toward the wall. Tell whether the following become larger, smaller, or stay the same: (a) the normal force on the ladder from the ground, (b) the force on the ladder from the wall, (c) the static frictional force on the ladder from the ground, and (d) the maximum value $f_{s,max}$ of the static frictional force.

9. A physical therapist gone wild has constructed the (stationary) assembly of massless pulleys and cords seen in Fig. 13-23. One long cord wraps around all the pulleys and shorter cords suspend pulleys from the ceiling or weights from the pulleys. Except for one, the weights (in newtons) are indicated. (a) What is that last weight? (*Hint:* When a cord loops halfway around a pulley as here, it pulls on the pulley with a net force that is twice the tension in the cord.) (b) What is the tension in the short cord labeled with T?

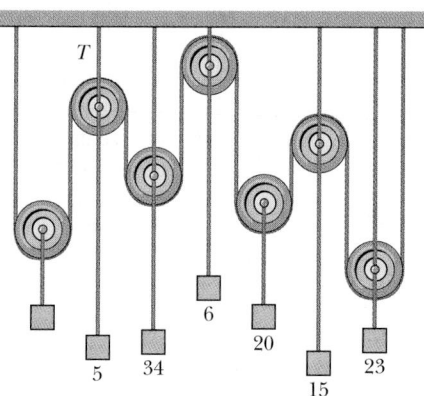

FIGURE 13-23 Question 9.

10. Three piñatas hang from the (stationary) assembly of massless pulleys and cords seen in Fig. 13-24. One long cord runs from the ceiling at the right to the lower pulley at the left. Several shorter cords suspend pulleys from the ceiling or piñatas from the pulleys. The weights (in newtons) of two piñatas are given.

(a) What is the weight of the third piñata? (*Hint:* When a cord loops around a pulley, it pulls on the pulley with a net force that is twice the tension in the cord.) (b) What is the tension in the short cord labeled with T?

FIGURE 13-24 Question 10.

11. (a) In Checkpoint 3, to express τ_r in terms of T, should you use $\sin \theta$ or $\cos \theta$? (b) If angle θ is decreased (by shortening the rope but still keeping the rod horizontal), does the torque τ_r required for equilibrium become larger, smaller, or stay the same? (c) Does the corresponding force T become larger, smaller, or stay the same?

12. The table gives the areas of three surfaces and the magnitude of a force that is applied perpendicular to the surface and uniformly across it. Rank the surfaces according to the stress on them, greatest first.

	AREA	FORCE
Surface A	$0.5A_0$	$2F_0$
Surface B	$2A_0$	$4F_0$
Surface C	$3A_0$	$6F_0$

13. The table gives the initial lengths of three rods and the changes in their length when forces are applied to their ends to put them under strain. Rank the rods according to their strain, greatest first.

	INITIAL LENGTH	CHANGE IN LENGTH
Rod A	$2L_0$	ΔL_0
Rod B	$4L_0$	$2\Delta L_0$
Rod C	$10L_0$	$4\Delta L_0$

EXERCISES & PROBLEMS

SECTION 13-4 Some Examples of Static Equilibrium

1E. An eight-member family, whose weights in pounds are indicated in Fig. 13-25, is balanced on a seesaw. What is the number of the person who causes the largest torque, about the rotation axis at *fulcrum f*, directed (a) out of the page and (b) into the page?

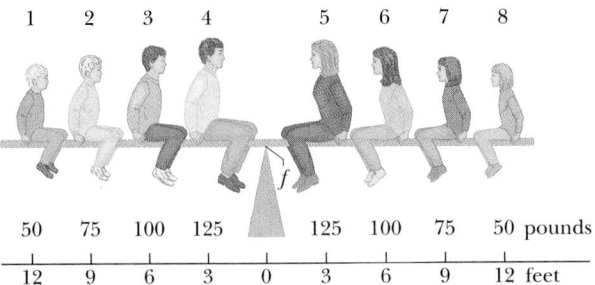

| 1 | 2 | 3 | 4 | | 5 | 6 | 7 | 8 |

| 50 | 75 | 100 | 125 | | 125 | 100 | 75 | 50 pounds |
| 12 | 9 | 6 | 3 | 0 | 3 | 6 | 9 | 12 feet |

FIGURE 13-25 Exercise 1.

2E. A certain nut is known to require forces of 40 N exerted on its shell from both sides to crack it. What force components $F_\perp$, perpendicular to the handles, will be required when the nut is placed in the nutcracker shown in Fig. 13-26?

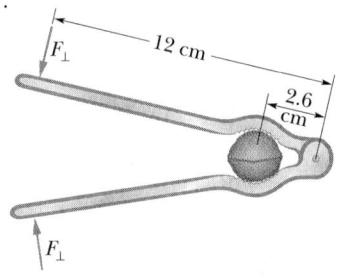

$F_\perp$ — 12 cm — 2.6 cm

$F_\perp$

FIGURE 13-26 Exercise 2.

3E. The leaning Tower of Pisa (see Fig. 13-27) is 55 m high and 7.0 m in diameter. The top of the tower is displaced 4.5 m from the vertical. Treat the tower as a uniform, circular cylinder. (a)

FIGURE 13-27 Exercise 3. The leaning Tower of Pisa (the photograph is not distorted).

What additional displacement, measured at the top, will bring the tower to the verge of toppling? (b) What angle with the vertical will the tower make at that moment?

4E. A particle is acted on by forces given, in newtons, by $F_1 = 10i - 4j$ and $F_2 = 17i + 2j$. (a) What force F_3 balances these forces? (b) What direction does F_3 have relative to the x axis?

5E. A bow is drawn until the tension in the string is equal to the force exerted by the archer. What is the angle between the two parts of the string?

6E. In Fig. 13-28, a man is trying to get his car out of mud on the shoulder of a road. He ties one end of a rope tightly around the front bumper and the other end tightly around a utility pole 60 ft away. He then pushes sideways on the rope at its midpoint with a force of 125 lb, displacing the center of the rope 1.0 ft from its previous position, and the car barely moves. What force does the rope exert on the car? (The rope stretches somewhat.)

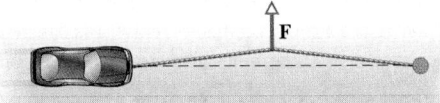

F

FIGURE 13-28 Exercise 6.

7E. A rope, assumed massless, is stretched horizontally between two supports that are 3.44 m apart. When an object of weight 3160 N is hung at the center of the rope, the rope is observed to sag by 35.0 cm. What is the tension in the rope?

8E. The system in Fig. 13-29 is in equilibrium, but it begins to slip if any additional mass is added to the 5.0 kg object. What is the coefficient of static friction between the 10 kg block and the plane on which it rests?

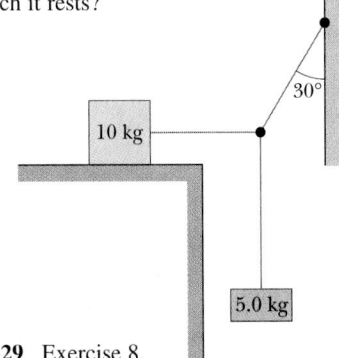

30°

10 kg

5.0 kg

FIGURE 13-29 Exercise 8.

9E. A scaffold of mass 60 kg and length 5.0 m is supported in a horizontal position by one vertical cable at each end. A window washer of mass 80 kg stands at a point 1.5 m from one end. What is the tension in the cable (a) closest to the window washer and (b) farthest away from the window washer?

10E. A beam is carried by three men, one man at one end and the other two supporting the beam between them on a crosspiece placed so that the load is equally divided among the three men.

Where is the crosspiece placed? (Neglect the mass of the crosspiece.)

11E. A uniform cubical crate is 0.750 m on each side and weighs 500 N. It rests on the floor with one edge against a very small, fixed obstruction. At what height above the floor must a horizontal force of 350 N be applied to the crate to just tip it?

12E. A uniform sphere of weight W and radius r is held in place by a rope attached to a frictionless wall a distance L above the center of the sphere, as in Fig. 13-30. Find (a) the tension in the rope and (b) the force exerted on the sphere by the wall.

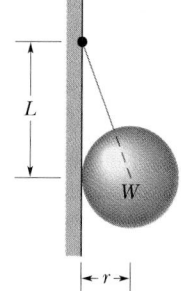

FIGURE 13-30 Exercise 12.

13E. An automobile with a mass of 1360 kg has 3.05 m between the front and rear axles. Its center of gravity is located 1.78 m behind the front axle. Determine (a) the force exerted on each of the front wheels (assumed the same) and (b) the force exerted on each of the back wheels (assumed the same) by the level ground.

14E. A 160 lb man is walking across a level bridge and stops one-fourth of the way from one end. The bridge is uniform and weighs 600 lb. What are the vertical forces exerted on the bridge by its supports at (a) the far end and (b) the near end?

15E. A diver of weight 580 N stands at the end of a 4.5 m diving board of negligible weight. The board is attached to two pedestals 1.5 m apart, as shown in Fig. 13-31. What are the magnitude and direction of the force on the board from (a) the left pedestal and (b) the right pedestal? (c) Which pedestal is being stretched, and which compressed?

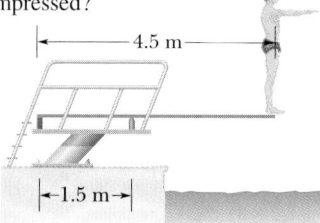

FIGURE 13-31 Exercise 15.

16E. A meter stick balances horizontally on a knife-edge at the 50.0 cm mark. With two nickels stacked over the 12.0 cm mark, the stick is found to balance at the 45.5 cm mark. A nickel has a mass of 5.0 g. What is the mass of the meter stick?

17E. A 75 kg window cleaner uses a 10 kg ladder that is 5.0 m long. He places one end on the ground 2.5 m from a wall, rests the upper end against a cracked window, and climbs the ladder. He climbs 3.0 m up the ladder when the window breaks. Neglecting friction between the ladder and window and assuming that the base of the ladder did not slip, find (a) the force exerted on the

window by the ladder just before the window breaks and (b) the magnitude and direction of the force exerted on the ladder by the ground just before the window breaks.

18E. Figure 13-32 shows the anatomical structures in the lower leg and foot that are involved in standing tiptoe with the heel raised off the floor so that the foot effectively contacts the floor at only one point, shown as P in the figure. Calculate, in terms of a person's weight W, the forces that must be exerted on the foot by (a) the calf muscle (at A) and (b) the lower-leg bones (at B) when the person stands tiptoe on one foot. Assume that $a = 5.0$ cm and $b = 15$ cm.

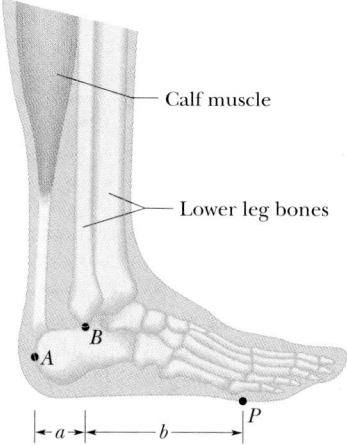

FIGURE 13-32 Exercise 18.

19P. By means of a turnbuckle G, bar AB of the square frame $ABCD$ in Fig. 13-33 is put in tension, as if its ends A and B were subject to the horizontal, outward forces **T** shown. Determine the forces on the other bars; identify those bars that are in tension and those that are being compressed. The diagonals AC and BD pass each other freely at E. Symmetry considerations can lead to considerable simplification in this and similar problems.

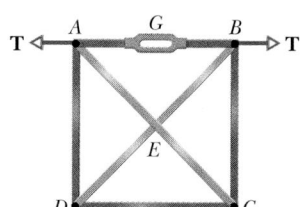

FIGURE 13-33 Problem 19.

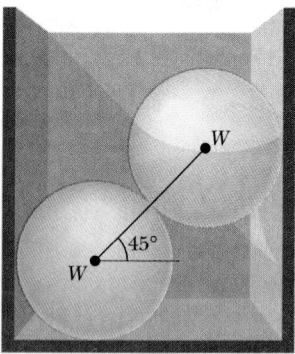

FIGURE 13-34 Problem 20.

20P. Two identical, uniform, frictionless spheres, each of weight *W*, rest in a rigid rectangular container as shown in Fig. 13-34. Find, in terms of *W*, the forces acting on the spheres due to (a) the container surfaces and (b) one another, if the line of centers of the spheres makes an angle of 45° with the horizontal.

21P. An 1800 lb construction bucket is suspended by a cable *A* that is attached at *O* to two other cables *B* and *C*, making angles of 51° and 66° with the horizontal (Fig. 13-35). Find the tension in (a) cable *A*, (b) cable *B*, and (c) cable *C*. (*Hint:* To avoid solving two equations in two unknowns, position the axes as shown in the figure.)

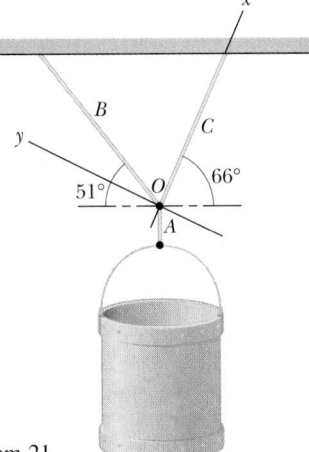

FIGURE 13-35 Problem 21.

22P. The force **F** in Fig. 13-36 is just sufficient to hold the 14 lb block and weightless pulleys in equilibrium. There is no appreciable friction. Calculate the tension *T* in the upper cable.

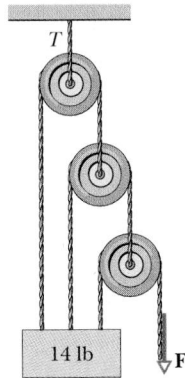

FIGURE 13-36 Problem 22.

23P. The system in Fig. 13-37 is in equilibrium with the string in the center exactly horizontal. Find (a) tension T_1, (b) tension T_2, (c) tension T_3, and (d) angle θ.

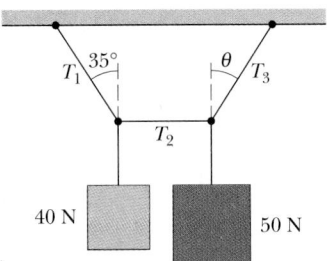

FIGURE 13-37 Problem 23.

24P. A balance is made up of a rigid, massless rod supported at and free to rotate about a point not at the center of the rod. It is balanced by unequal weights placed in the pans at each end of the rod. When an unknown mass *m* is placed in the left-hand pan, it is balanced by a mass m_1 placed in the right-hand pan; and when the mass *m* is placed in the right-hand pan, it is balanced by a mass m_2 in the left-hand pan. Show that $m = \sqrt{m_1 m_2}$.

25P. A 15 kg weight is being lifted by the pulley system shown in Fig. 13-38. The upper arm is vertical, whereas the forearm makes an angle of 30° with the horizontal. What forces are being exerted on the forearm by (a) the triceps muscle and (b) the upper-arm bone (the humerus)? The forearm and hand together have a mass of 2.0 kg with a center of mass 15 cm (measured along the arm) from the point where the forearm and upper-arm bones are in contact. The triceps muscle pulls vertically upward at a point 2.5 cm behind the contact point.

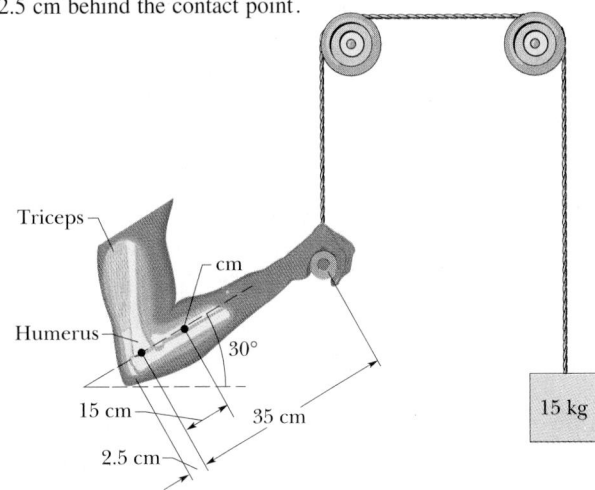

FIGURE 13-38 Problem 25.

26P. A 50.0 kg uniform square sign, 2.00 m on a side, is hung from a 3.00 m rod of negligible mass. A cable is attached to the end of the rod and to a point on the wall 4.00 m above the point where the rod is fixed to the wall (Fig. 13-39). (a) What is the tension in the cable? What are the (b) horizontal and (c) vertical components of the force exerted by the wall on the rod?

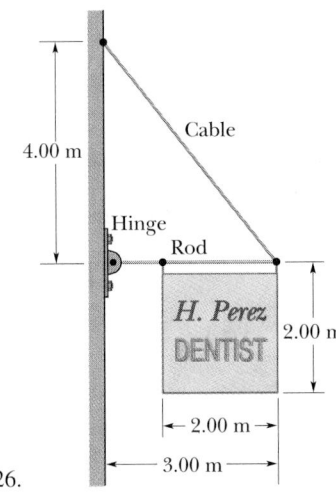

FIGURE 13-39 Problem 26.

27P. In Fig. 13-40, a 55 kg rock climber is in a lie-back climb along a fissure, with hands pulling on one side of the fissure and feet pressed against the opposite side. The fissure has width $w = 0.20$ m, and the center of mass of the climber is a horizontal distance $d = 0.40$ m from the fissure. The coefficient of static friction between hands and rock is $\mu_1 = 0.40$, and between boots and rock it is $\mu_2 = 1.2$. (a) What is the least horizontal pull by the hands and push by the feet that will keep him stable? (b) For the horizontal pull of (a), what must be the vertical distance h between hands and feet? (c) If the climber encounters wet rock, so that μ_1 and μ_2 are reduced, what happens to the answers to (a) and (b), respectively?

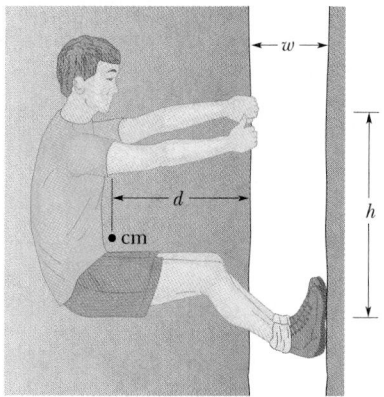

FIGURE 13-40 Problem 27.

28P. Forces $\mathbf{F}_1$, $\mathbf{F}_2$, and $\mathbf{F}_3$ act on the structure of Fig. 13-41, shown in an overhead view. We wish to put the structure in equilibrium by applying a force, at a point such as P, whose vector components are $\mathbf{F}_h$ and $\mathbf{F}_v$. We are given that $a = 2.0$ m, $b = 3.0$ m, $c = 1.0$ m, $F_1 = 20$ N, $F_2 = 10$ N, and $F_3 = 5.0$ N. Find (a) F_h, (b) F_v, and (c) d.

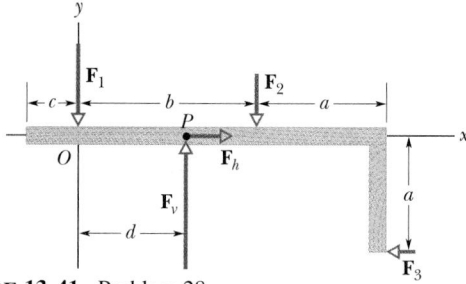

FIGURE 13-41 Problem 28.

29P. In Fig. 13-42, what magnitude of force $\mathbf{F}$ applied horizontally at the axle of the wheel is necessary to raise the wheel over an obstacle of height h? Take r as the radius of the wheel and W as its weight.

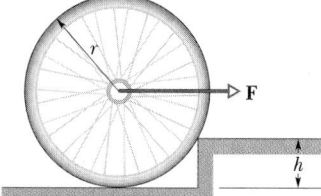

FIGURE 13-42 Problem 29.

30P. A trap door in a ceiling is 0.91 m square, has a mass of 11 kg, and is hinged along one side with a catch at the opposite side. If the center of gravity of the door is 10 cm toward the hinged side from the door's center, what forces must (a) the catch and (b) the hinge sustain?

31P. Four identical uniform bricks, each of length L, are put on top of one another (Fig. 13-43) in such a way that part of each extends beyond the one beneath. Find, in terms of L, the maximum values of (a) a_1, (b) a_2, (c) a_3, (d) a_4, and (e) h, such that the stack is in equilibrium.

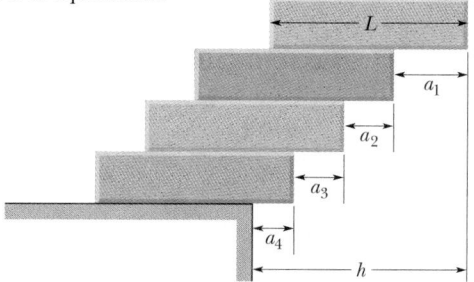

FIGURE 13-43 Problem 31.

32P. One end of a uniform beam that weighs 50.0 lb and is 3.00 ft long is attached to a wall with a hinge. The other end is supported by a wire (see Fig. 13-44). (a) Find the tension in the wire. What are the (b) horizontal and (c) vertical components of the force of the hinge on the beam?

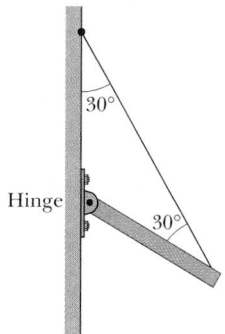

FIGURE 13-44 Problem 32.

33P. The system in Fig. 13-45 is in equilibrium. A mass of 225 kg hangs from the end of the uniform strut whose mass is 45.0 kg. Find (a) the tension T in the cable and the (b) horizontal and (c) vertical force components exerted on the strut by the hinge.

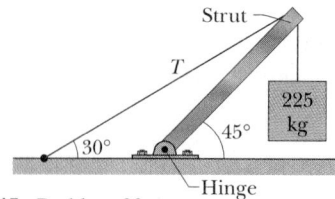

FIGURE 13-45 Problem 33.

34P. A door 2.1 m high and 0.91 m wide has a mass of 27 kg. A hinge 0.30 m from the top and another 0.30 m from the bottom each support half the door's weight. Assume that the center of gravity is at the geometrical center of the door and determine the (a) vertical and (b) horizontal force components exerted by each hinge on the door.

35P. A nonuniform bar of weight W is suspended at rest in a horizontal position by two massless cords as shown in Fig. 13-46. One cord makes the angle $\theta = 36.9°$ with the vertical; the other makes the angle $\phi = 53.1°$ with the vertical. If the length L of the bar is 6.10 m, compute the distance x from the left-hand end of the bar to its center of gravity.

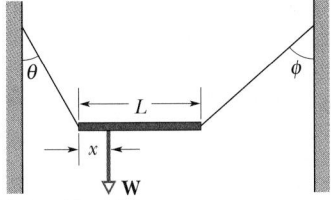

FIGURE 13-46 Problem 35.

36P. In Fig. 13-47, a thin horizontal bar AB of negligible weight and length L is pinned to a vertical wall at A and supported at B by a thin wire BC that makes an angle θ with the horizontal. A weight W can be moved anywhere along the bar; its position is defined by the distance x from the wall to its center of mass. As a function of x, find (a) the tension in the wire, and the (b) horizontal and (c) vertical components of the force exerted on the bar by the pin at A.

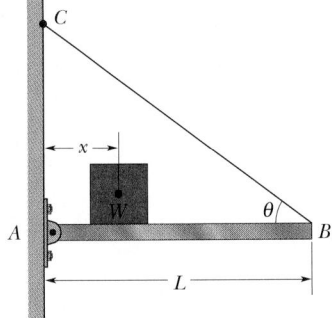

FIGURE 13-47 Problems 36 and 37.

37P. In Fig. 13-47, suppose the length L of the uniform bar is 3.0 m and its weight is 200 N. Also, let $W = 300$ N and $\theta = 30°$. The wire can withstand a maximum tension of 500 N. (a) What is the maximum possible distance x before the wire breaks? With W placed at this maximum x, what are the (b) horizontal and (c) vertical components of the force exerted on the bar by the pin at A?

38P. Two uniform beams, A and B, are attached to a wall with hinges and then loosely bolted together as in Fig. 13-48. Find the

horizontal and vertical components of the force on (a) beam A due to its hinge, (b) beam A due to the bolt, (c) beam B due to its hinge, and (d) beam B due to the bolt.

39P. Four identical, uniform bricks of length L are stacked on a table in two ways, as shown in Fig. 13-49 (compare with Problem 31). We seek to maximize the overhang distance h in both arrangements. Find the optimum distances a_1, a_2, b_1, and b_2, and calculate h for the two arrangements. (See "The Amateur Scientist," *Scientific American,* June 1985, for a discussion and an even better version of arrangement (b).)

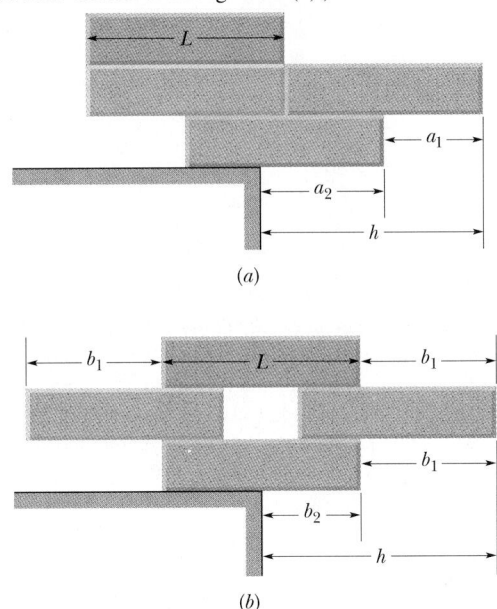

FIGURE 13-49 Problem 39.

40P. A uniform plank, with length $L = 20$ ft and weight $W = 100$ lb, rests on the ground and against a frictionless roller at the top of a wall of height $h = 10$ ft (see Fig. 13-50). The plank remains in equilibrium for any value of $\theta \geq 70°$ but slips if $\theta < 70°$. Find the coefficient of static friction between the plank and the ground.

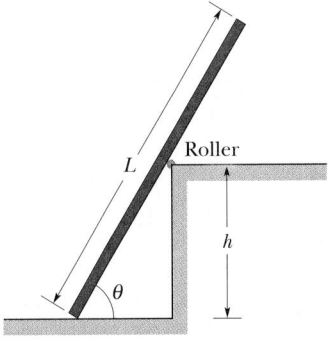

FIGURE 13-50 Problem 40.

41P. For the stepladder shown in Fig. 13-51, sides AC and CE are each 8.0 ft long and hinged at C. Bar BD is a tie-rod 2.5 ft long, halfway up. A man weighing 192 lb climbs 6.0 ft along the ladder. Assuming that the floor is frictionless and neglecting the

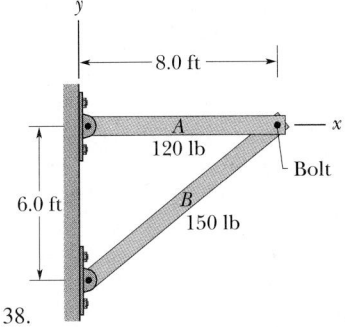

FIGURE 13-48 Problem 38.

weight of the ladder, find (a) the tension in the tie-rod and the forces exerted on the ladder by the floor at (b) *A* and (c) *E*. (*Hint:* It will help to isolate parts of the ladder in applying the equilibrium conditions.)

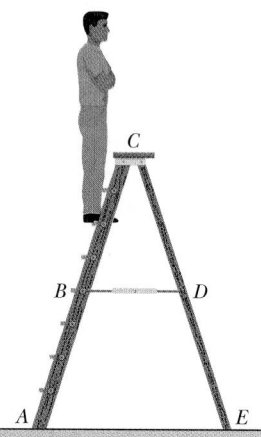

FIGURE 13-51 Problem 41.

42P. A uniform cube of side length *L* rests on a horizontal floor. The coefficient of static friction between cube and floor is μ. A horizontal pull *P* is applied perpendicular to one of the vertical faces of the cube, at a distance *h* above the floor on the vertical midline of the cube face. As *P* is slowly increased, the cube will either (a) begin to slide or (b) begin to tip. What is the condition on μ for (a) to occur? For (b)? (*Hint:* At the onset of tipping, where is the normal force located?)

43P. A cubical box is filled with sand and weighs 890 N. We wish to "roll" the box by pushing horizontally on one of the upper edges. (a) What minimum force is required? (b) What minimum coefficient of static friction between box and floor is required? (c) Is there a more efficient way to roll the box? If so, find the smallest possible force that would have to be applied directly to the box to roll it. (*Hint:* See the hint for Problem 42.)

44P. A crate, in the form of a cube with edge lengths of 4.0 ft, contains a piece of machinery whose design is such that the center of gravity of the crate and its contents is located 1.0 ft above its geometrical center. The crate rests on a ramp that makes an angle θ with the horizontal. As θ is increased from zero, an angle will be reached at which the crate will either start to slide down the ramp or tip over. Which event will occur (a) when the coefficient of static friction between ramp and crate is 0.60 and (b) when it is 0.70? In each case, give the angle at which the event occurs. (*Hint:* See the hint for Problem 42.)

45P*. A car on a horizontal road makes an emergency stop by applying the brakes so that all four wheels lock and skid along the road. The coefficient of kinetic friction between tires and road is 0.40. The separation between the front and rear axles is 4.2 m, and the center of mass of the car is located 1.8 m behind the front axle and 0.75 m above the road; see Fig. 13-52. The car weighs 11 kN. Calculate (a) the braking deceleration of the car, (b) the normal force on each wheel, and (c) the braking force on each wheel. (*Hint:* Although the car is not in translational equilibrium, it *is* in rotational equilibrium.)

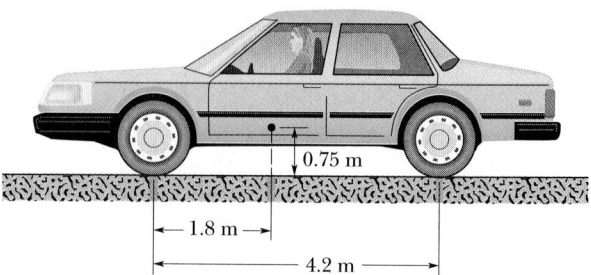

FIGURE 13-52 Problem 45.

SECTION 13-6 Elasticity

46E. Figure 13-53 shows the stress–strain curve for quartzite. What are (a) the Young's modulus and (b) the approximate yield strength for this material?

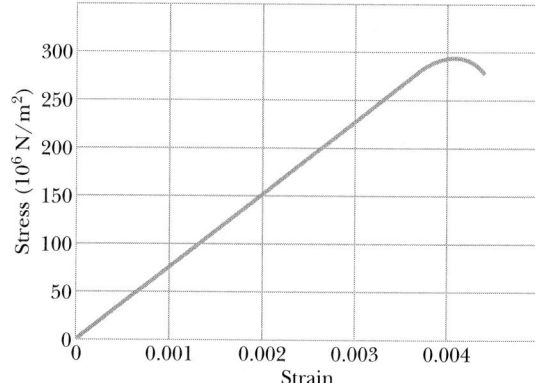

FIGURE 13-53 Exercise 46.

47E. After a fall, a 95 kg rock climber finds himself dangling from the end of a rope that had been 15 m long and 9.6 mm in diameter but which has stretched by 2.8 cm. For the rope, calculate (a) the strain, (b) the stress, and (c) the Young's modulus.

48E. A mine elevator is supported by a single steel cable 2.5 cm in diameter. The total mass of the elevator cage plus occupants is 670 kg. By how much does the cable stretch when the elevator is (a) at the surface, 12 m below the elevator motor, and (b) at the bottom of the shaft, which is 350 m deep? (Neglect the mass of the cable.)

49E. Suppose the (square) beam in Fig. 13-9a is of Douglas fir. What must be its thickness to keep the compressive stress on it to $\frac{1}{6}$ of its ultimate strength? (See Sample Problem 13-5.)

50E. A horizontal aluminum rod 4.8 cm in diameter projects 5.3 cm from a wall. A 1200 kg object is suspended from the end of the rod. The shear modulus of aluminum is 3.0×10^{10} N/m^2. Neglecting the rod's weight, find (a) the shear stress on the rod and (b) the vertical deflection of the end of the rod.

51E. A solid copper cube has an edge length of 85.5 cm. How much pressure must be applied to the cube to reduce the edge length to 85.0 cm? The bulk modulus of copper is 1.4×10^{11} N/m^2.

52P. A tunnel 150 m long, 7.2 m high, and 5.8 m wide (with a flat roof) is to be constructed 60 m beneath the ground. (See Fig. 13-54.) The tunnel roof is to be supported entirely by square steel columns, each with a cross-sectional area of 960 cm². The density of the ground material is 2.8 g/cm³. (a) What is the total weight that the columns must support? (b) How many columns are needed to keep the compressive stress on each column at one-half its ultimate strength?

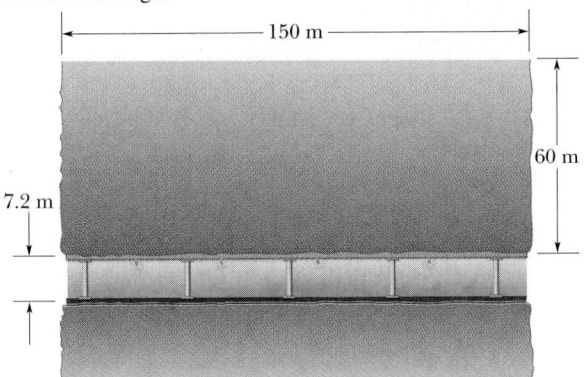

FIGURE 13-54 Problem 52.

53P. A rectangular slab of slate rests on a 26° incline; see Fig. 13-55. The slab has dimensions 43 m long, 2.5 m thick, and 12 m wide. Its density is 3.2 g/cm³. The coefficient of static friction between the slab and the underlying rock is 0.39. (a) Calculate the component of the slab's weight acting parallel to the incline. (b) Calculate the static frictional force on the slab. By comparing (a) and (b), you can see that the slab is in danger of sliding and is prevented from doing so only by chance protrusions between the slab and the underlying rock. (c) To stabilize the slab, bolts are driven perpendicular to the incline. If each bolt has a cross-sectional area of 6.4 cm² and will snap under a shearing stress of 3.6×10^8 N/m², what is the minimum number of bolts needed? Assume that the bolts do not affect the normal force.

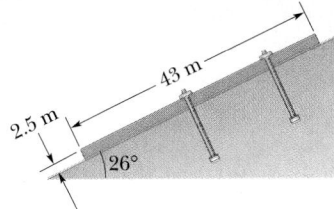

FIGURE 13-55 Problem 53.

54P. In Fig. 13-56, a lead brick rests horizontally on cylinders A and B. The areas of the top faces of the cylinders are related by $A_A = 2A_B$; the Young's moduli of the cylinders are related by $E_A = 2E_B$. The cylinders had identical lengths before the brick was placed on them. What fraction of the brick's weight is supported (a) by cylinder A and (b) by cylinder B? The horizontal distances between the center of mass of the brick and the centerlines of the cylinders are d_A for cylinder A and d_B for cylinder B. (c) What is the ratio d_A/d_B?

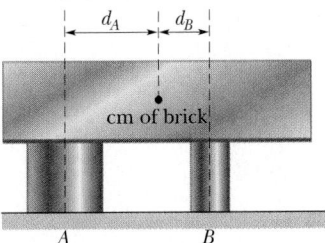

FIGURE 13-56 Problem 54.

55P. In Fig. 13-57, a 103 kg uniform log hangs by two steel wires, A and B, both of radius 1.20 mm. Initially, wire A was 2.50 m long and 2.00 mm shorter than wire B. The log is now horizontal. What forces are exerted on it by (a) wire A and (b) wire B? (c) What is the ratio d_A/d_B?

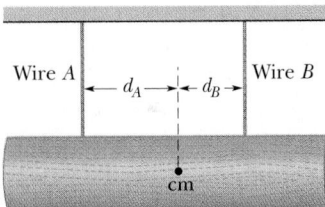

FIGURE 13-57 Problem 55.

56P. Figure 13-58 is an overhead view of a rigid rod that turns about a vertical axle until the identical rubber stoppers A and B are forced against rigid walls at distances r_A and r_B from the axle. Initially the stoppers touch the walls, without being compressed. Then force F is applied perpendicular to the rod at a distance R from the axle. Find expressions for the forces compressing (a) stopper A and (b) stopper B?

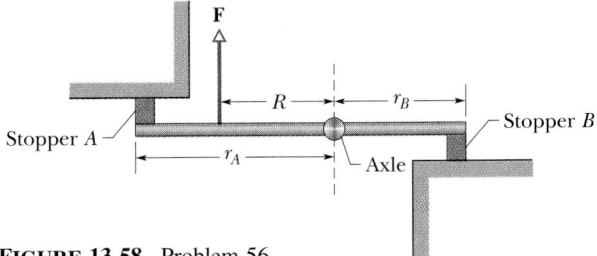

FIGURE 13-58 Problem 56.

14
Gravitation

The Milky Way galaxy is a disk-shaped collection of dust, planets, and billions of stars, including our Sun and solar system. The force that binds it or any other galaxy together is the same force that holds Earth's moon in orbit and you on Earth—the gravitational force. That force is also responsible for one of nature's strangest objects, the black hole, a star that has completely collapsed onto itself. The gravitational force near a black hole is so strong that not even light can escape it. But if that is the case, how can a black hole be detected?

14-1 THE WORLD AND THE GRAVITATIONAL FORCE

The drawing that opens this chapter shows our view of the Milky Way galaxy: we are near the edge of the disk of the galaxy, about 26,000 light-years (2.5×10^{20} m) from its center, which in the drawing lies in the star collection known as Sagittarius. Our galaxy is a member of the Local Group of galaxies, which includes the Andromeda galaxy (Fig. 14-1) at a distance of 2.3×10^6 light-years, and several closer dwarf galaxies, such as the Large Magellanic Cloud shown in the opening drawing.

The Local Group is part of the Local Supercluster of galaxies. Measurements taken during and since the 1980s suggest that the Local Supercluster and the supercluster consisting of the clusters Hydra and Centaurus are all moving toward an exceptionally massive region called the Great Attractor. This region appears to be about 150 million light-years away, on the opposite side of the Milky Way from us, past the clusters Hydra and Centaurus.

The force that binds together these progressively larger structures, from star to galaxy to supercluster, and may be drawing them all toward the Great Attractor, is the gravitational force. That force not only holds you on Earth but also reaches out across intergalactic space.

FIGURE 14-1 The Andromeda galaxy. Located 2.3×10^6 light-years from us, and faintly visible to the naked eye, it is very similar to our home galaxy, the Milky Way.

14-2 NEWTON'S LAW OF GRAVITATION

Physicists like to examine seemingly unrelated phenomena to show that a relationship can be found if they are examined closely enough. This search for unification has been going on for centuries. In 1665, the 23-year-old Isaac Newton made a basic contribution to physics when he showed that the force that holds the Moon in its orbit is the same force that makes an apple fall. We take this so much for granted now that it is not easy for us to comprehend the ancient belief that the motions of earthbound bodies and heavenly bodies were different in kind and were governed by different laws.

Newton concluded that not only does Earth attract an apple and the Moon but every body in the universe attracts every other body; this tendency of bodies to move toward each other is called **gravitation.** Newton's conclusion takes a little getting used to, because the familiar attraction of Earth for earthbound bodies is so great that it overwhelms the attraction that earthbound bodies have for each other. For example, Earth attracts an apple with a force of several ounces. You also attract a nearby apple (and it attracts you), but the force of attraction is less than the weight of a speck of dust.

Quantitatively, Newton proposed a *force law* that we call **Newton's law of gravitation:** every particle attracts any other particle with a **gravitational force** whose magnitude is given by

$$F = G \frac{m_1 m_2}{r^2}$$

(Newton's law of gravitation). (14-1)

Here m_1 and m_2 are the masses of the particles, r is the distance between them, and G is the **gravitational constant,** whose value is now known to be

$$G = 6.67 \times 10^{-11} \text{ N} \cdot \text{m}^2/\text{kg}^2$$
$$= 6.67 \times 10^{-11} \text{ m}^3/\text{kg} \cdot \text{s}^2. \quad (14-2)$$

As Fig. 14-2 shows, a particle m_2 attracts a particle m_1 with a gravitational force **F** that is directed toward particle m_2.

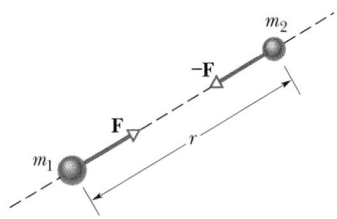

FIGURE 14-2 Two particles, of masses m_1 and m_2 and with a separation of r, attract each other according to Newton's law of gravitation, Eq. 14-1. The forces of attraction, **F** and $-$**F**, are equal in magnitude and are in opposite directions.

And particle m_1 attracts particle m_2 with a gravitational force $-\mathbf{F}$ that is directed toward m_1. The forces $\mathbf{F}$ and $-\mathbf{F}$ form an action–reaction pair and are opposite in direction but equal in magnitude. They depend on the separation of the two particles, but not on their location: the particles could be in a deep cave or in deep space. And $\mathbf{F}$ and $-\mathbf{F}$ are not altered by the presence of other bodies, even if those bodies lie between the two particles we are considering.

The strength of the gravitational force, that is, how strongly two given masses at a given separation attract each other, depends on the value of the gravitational constant G. If G—by some miracle—were suddenly multiplied by a factor of 10, you would be crushed to the floor by Earth's attraction. If it were divided by this factor, Earth's attraction would be weak enough for you to jump over a tall building.

Although Newton's law of gravitation applies strictly to particles, we can also apply it to real objects as long as the sizes of the objects are small compared to the distance between them. The Moon and Earth are far enough apart so that, to a good approximation, we can treat them both as particles. But what about an apple and Earth? From the point of view of the apple, the broad and level Earth, stretching out to the horizon beneath the apple, certainly does not look like a particle.

Newton solved the apple–Earth problem by proving an important theorem called the *shell theorem:*

A uniform spherical shell of matter attracts a particle that is outside the shell as if all the shell's mass were concentrated at its center.

Earth can be thought of as a nest of such shells, one within another, and each attracting a particle outside Earth's surface as if the mass of that shell were at the center of the shell. Thus, from the apple's point of view, Earth *does* behave like a particle, located at the center of Earth and having a mass equal to that of the planet.

Suppose, as in Fig. 14-3, that Earth pulls down on an apple with a force of 0.80 N. The apple must then pull up on Earth with a force of 0.80 N, which we take to act at the

center of Earth. Although the forces are matched in magnitude, they produce different accelerations when the apple is released. For the apple, the acceleration is about 9.8 m/s², the familiar acceleration of a falling body near Earth's surface. For Earth, the acceleration measured in a reference frame attached to the center of mass of the apple–Earth system is only about 1×10^{-25} m/s².

CHECKPOINT 1: A particle is to be placed, in turn, outside four objects, each of mass m: (1) a large uniform solid sphere, (2) a large uniform spherical shell, (3) a small uniform solid sphere, and (4) a small uniform shell. In each situation, the distance between the particle and the center of the object is d. Rank the objects according to the magnitude of the gravitational force they exert on the particle, greatest first.

14-3 GRAVITATION AND THE PRINCIPLE OF SUPERPOSITION

Given a group of particles, we find the net (or resultant) gravitational force exerted on any one of them by using the **principle of superposition.** This is a general principle that says a net effect is the sum of the individual effects. Here, the principle means that we first compute the gravitational force that acts on our selected particle due to each of the other particles, in turn. We then find the net force by adding these forces vectorially, as usual.

For n interacting particles, we can write the principle of superposition for gravitational forces as

$$\mathbf{F}_1 = \mathbf{F}_{12} + \mathbf{F}_{13} + \mathbf{F}_{14} + \mathbf{F}_{15} + \cdot \cdot \cdot + \mathbf{F}_{1n}. \quad (14\text{-}3)$$

Here $\mathbf{F}_1$ is the net force on particle 1 and, for example, $\mathbf{F}_{13}$ is the force exerted on particle 1 by particle 3. We can express this equation more compactly as a vector sum:

$$\mathbf{F}_1 = \sum_{i=2}^{n} \mathbf{F}_{1i}. \quad (14\text{-}4)$$

What about the gravitational force exerted on a particle by a real extended object? The force is found by dividing the object into units small enough to treat as particles and then using Eq. 14-4 to find the vector sum of the forces exerted on the particle by all the units. In the limiting case, we can divide the extended object into differential units of mass dm, each of which exerts only a differential force $d\mathbf{F}$ on the particle. In this limit, the sum of Eq. 14-4 becomes an integral and we have

$$\mathbf{F}_1 = \int d\mathbf{F}, \quad (14\text{-}5)$$

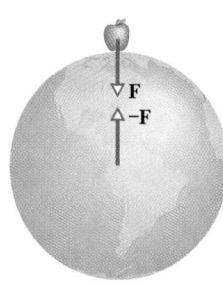

FIGURE 14-3 The apple pulls up on Earth just as hard as Earth pulls down on the apple.

in which the integral is taken over the entire extended object. If the object is a uniform sphere or a spherical shell, we can avoid the integration of Eq. 14-5 by assuming that the object's mass is concentrated at the object's center and using Eq. 14-1.

SAMPLE PROBLEM 14-1

Figure 14-4a shows an arrangement of five particles, with masses $m_1 = 8.0$ kg, $m_2 = m_3 = m_4 = m_5 = 2.0$ kg, and with $a = 2.0$ cm and $\theta = 30°$. What is the net gravitational force $\mathbf{F}_1$ acting on m_1 due to the other particles?

SOLUTION: From Eq. 14-4 we know that $\mathbf{F}_1$ is the vector sum of forces $\mathbf{F}_{12}$, $\mathbf{F}_{13}$, $\mathbf{F}_{14}$, and $\mathbf{F}_{15}$, which are the gravitational forces acting on m_1 due to the other particles. Because m_2 and m_4 are equal and because both are a distance $r = 2a$ from m_1, we have from Eq. 14-1

$$F_{12} = F_{14} = \frac{Gm_1m_2}{(2a)^2}. \qquad (14\text{-}6)$$

Similarly, since m_3 and m_5 are equal and are both a distance $r = a$ from m_1, we have

$$F_{13} = F_{15} = \frac{Gm_1m_3}{a^2}. \qquad (14\text{-}7)$$

Figure 14-4b is a free-body diagram for m_1. It and Eq. 14-6 show that $\mathbf{F}_{12}$ and $\mathbf{F}_{14}$ are equal in magnitude but opposite in direction; thus those forces cancel. Inspection of Fig. 14-4b and Eq. 14-7 reveals that the x components of $\mathbf{F}_{13}$ and $\mathbf{F}_{15}$ also cancel, and that their y components are identical in magnitude and both point toward increasing y. Thus $\mathbf{F}_1$ points toward increasing y, and its magnitude is twice the y component of $\mathbf{F}_{13}$:

$$\begin{aligned} F_1 &= 2F_{13} \cos \theta = 2 \frac{Gm_1m_3}{a^2} \cos \theta \\ &= 2 \frac{(6.67 \times 10^{-11} \text{ m}^3/\text{kg} \cdot \text{s}^2)(8.0 \text{ kg})(2.0 \text{ kg})}{(0.020 \text{ m})^2} \cos 30° \\ &= 4.6 \times 10^{-6} \text{ N}. \qquad \text{(Answer)} \end{aligned}$$

Note that the presence of m_5 along the line between m_1 and m_4 does not alter the gravitational force exerted by m_4 on m_1.

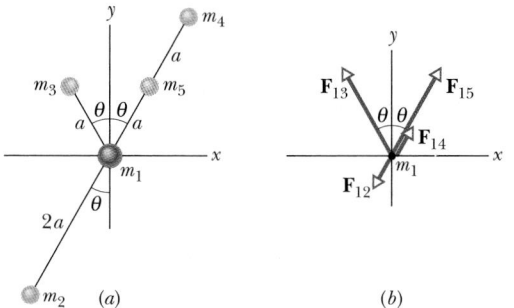

FIGURE 14-4 Sample Problem 14-1. (a) An arrangement of five particles. (b) The forces acting on the particle of mass m_1 due to the other four particles.

CHECKPOINT **2:** The figure shows four arrangements of three particles of equal masses. (a) Rank the arrangements according to the magnitude of the net gravitational force on the particle labeled m, greatest first. (b) In arrangement 2, is the direction of the net force closer to the line of length d or to the line of length D?

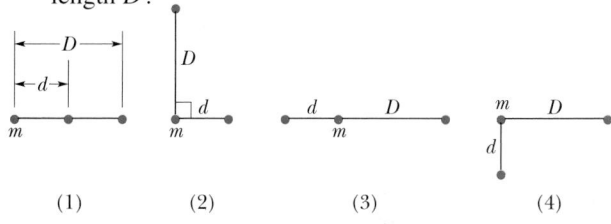

SAMPLE PROBLEM 14-2

In Fig. 14-5, a particle of mass $m_1 = 0.67$ kg is a distance $d = 23$ cm from one end of a uniform rod with length $L = 3.0$ m and mass $M = 5.0$ kg. What is the magnitude of the gravitational force $\mathbf{F}_1$ on the particle due to the rod?

SOLUTION: We consider a differential mass dm of the rod, located a distance r from m_1 and occupying a length dr along the rod. From Eq. 14-1, we may write the differential gravitational force dF_1 on m_1 due to dm as

$$dF_1 = \frac{Gm_1}{r^2} \, dm. \qquad (14\text{-}8)$$

The direction of this force is to the right in Fig. 14-5. In fact, because m_1 is located on a line through the length of the rod, each differential force dF_1 due to each differential mass dm in the rod is directed to the right. Thus, we can find the magnitude F_1 of the net force on m_1 simply by summing the magnitudes of these differential forces. We do so by integrating Eq. 14-8 along the length of the rod. (If m_1 were not on the central axis of the rod, the differential force vectors would point in different directions and the net force would have to be found by vector summation.)

The right-hand side of Eq. 14-8 contains two variables, r and dm. Before we can integrate, we must eliminate dm from that equation; we do so by using the fact that the rod is uniform in density to write

$$\frac{dm}{dr} = \frac{M}{L}. \qquad (14\text{-}9)$$

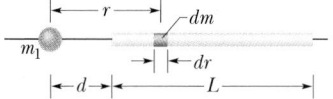

FIGURE 14-5 Sample Problem 14-2. A particle of mass m_1 is a distance d from one end of a rod of length L. A differential mass dm of the rod is a distance r from m_1.

This allows us to substitute $dm = (M/L)\,dr$ in Eq. 14-8. Then we integrate as indicated in Eq. 14-5 to get

$$F_1 = \int dF_1 = \int_d^{L+d} \frac{Gm_1}{r^2} \frac{M}{L}\,dr = \frac{Gm_1M}{L} \int_d^{L+d} \frac{dr}{r^2}$$

$$= -\frac{Gm_1M}{L} \left[\frac{1}{r}\right]_d^{L+d} = -\frac{Gm_1M}{L} \left[\frac{1}{L+d} - \frac{1}{d}\right]$$

$$= \frac{Gm_1M}{d(L+d)}$$

$$= \frac{(6.67 \times 10^{-11}\ \text{m}^3/\text{kg}\cdot\text{s}^2)(0.67\ \text{kg})(5.0\ \text{kg})}{(0.23\ \text{m})(3.0\ \text{m} + 0.23\ \text{m})}$$

$$= 3.0 \times 10^{-10}\ \text{N}. \qquad \text{(Answer)}$$

PROBLEM SOLVING TACTICS

TACTIC 1: *Drawing Gravitational Force Vectors*

When you are given a diagram of particles, such as Fig. 14-4a, and asked to find the net gravitational force on one of them, you should usually draw a free-body diagram showing only the particle of concern and the forces *it alone* experiences, as in Fig. 14-4b. If, instead, you choose to superimpose the vectors on the given diagram, be sure to draw the vectors with either their tails (preferably) or their heads on the particle experiencing those forces. If you draw the vectors elsewhere, you invite confusion. And confusion is guaranteed if you draw the vectors on the particles *causing* the forces on the particle of concern.

TACTIC 2: *Simplifying a Sum of Forces with Symmetry*

In Sample Problem 14-1 we used the symmetry of the situation to reduce the time and amount of calculation involved in the solution. By realizing that m_2 and m_4 are positioned symmetrically about m_1, and thus that $\mathbf{F}_{12}$ and $\mathbf{F}_{14}$ cancel, we avoided calculating either force. And by realizing that the x components of $\mathbf{F}_{13}$ and $\mathbf{F}_{15}$ cancel and that their y components are identical and add, we saved even more effort.

CHECKPOINT **3:** In the figure, what is the direction of the net gravitational force on the particle of mass m_1 due to the other particles, each of mass m, that are arranged symmetrically relative to the y axis?

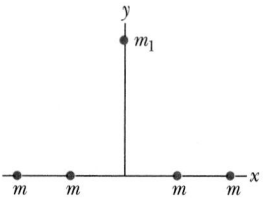

14-4 GRAVITATION NEAR EARTH'S SURFACE

Let us ignore the rotation of Earth for a moment and assume the planet to be a nonrotating uniform sphere of mass M. The magnitude of the gravitational force acting on a particle of mass m, located outside Earth a distance r from Earth's center, is then given by Eq. 14-1 as

$$F = G\frac{Mm}{r^2}. \qquad (14\text{-}10)$$

If the particle is released, it will fall toward the center of Earth, as a result of the gravitational force F, with an acceleration we shall call the **gravitational acceleration** a_g. Newton's second law tells us that F and a_g are related by

$$F = ma_g. \qquad (14\text{-}11)$$

Now, substituting F from Eq. 14-10 into Eq. 14-11 and solving for a_g, we find

$$a_g = \frac{GM}{r^2}. \qquad (14\text{-}12)$$

Table 14-1 shows values of a_g computed for various altitudes above Earth's surface.

The gravitational acceleration a_g computed with Eq. 14-12 is not the same as the free-fall acceleration g that we would measure for the falling particle (and that we have approximated as $9.80\ \text{m/s}^2$ near Earth's surface). The two accelerations differ for three reasons: (1) Earth is not uniform, (2) it is not a perfect sphere, and (3) it rotates. Moreover, because g differs from a_g, the weight mg of the particle differs from the gravitational force on the particle as given by Eq. 14-10 for the same three reasons. Let us now examine those reasons.

1. *Earth is not uniform.* The density of Earth varies radially as shown in Fig. 14-6, and the density of the crust (or outer section) of Earth varies from region to region over Earth's surface. Thus g varies from region to region.

2. *Earth is not a sphere.* Earth is approximately an ellipsoid, flattened at the poles and bulging at the equator. Its equatorial radius is greater than its polar radius by 21 km.

TABLE 14-1 **VARIATION OF a_g WITH ALTITUDE**

ALTITUDE (km)	a_g (m/s²)	ALTITUDE EXAMPLE
0	9.83	Mean Earth surface
8.8	9.80	Mt. Everest
36.6	9.71	Highest manned balloon
400	8.70	Space shuttle orbit
35,700	0.225	Communications satellite

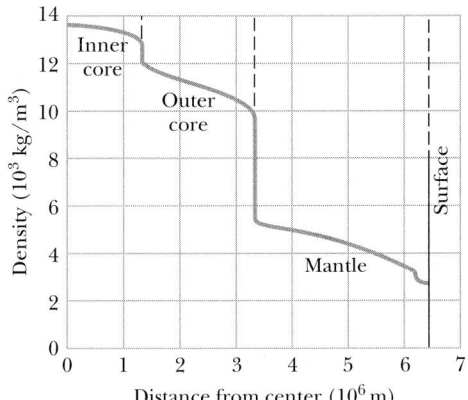

FIGURE 14-6 The density of Earth as a function of distance from the center. The limits of the solid inner core, the largely liquid outer core, and the solid mantle are shown, but the crust of Earth is too thin to show clearly on this plot.

Thus a point at the poles is closer to the dense core of Earth than is a point on the equator. This is one reason the free-fall acceleration g increases as one proceeds, at sea level, from the equator toward the poles.

3. Earth is rotating. The rotation axis runs through the north and south poles of Earth. An object located on Earth's surface anywhere except at those poles must rotate in a circle about the rotation axis and thus must have a centripetal acceleration that points toward the center of the circle. This centripetal acceleration requires a centripetal force that also points toward that center.

To see how Earth's rotation causes the free-fall acceleration g to differ from the gravitational acceleration a_g, and causes weight to differ from the gravitational force given by Eq. 14-10, we analyze a simple situation in which a crate of mass m rests on a scale at the equator. Figure 14-7a shows this situation as viewed from a point in space above the north pole.

Figure 14-7b is a free-body diagram for the crate. The centripetal acceleration **a** of the crate points toward the center of the circle it travels, which is coincident with the center of Earth (assumed spherical). Earth exerts a gravitational force on the crate with magnitude ma_g (Eq. 14-11). The scale exerts an upward normal force **N** on the crate. Applying Newton's second law to the crate, with the positive direction being radially inward toward Earth's center, we find

$$\sum F = ma_g - N = ma. \qquad (14\text{-}13)$$

The magnitude of **N** is the reading on the scale and thus is equal to the weight mg of the crate. Substituting mg for N in Eq. 14-13 yields

$$ma_g - mg = ma, \qquad (14\text{-}14)$$

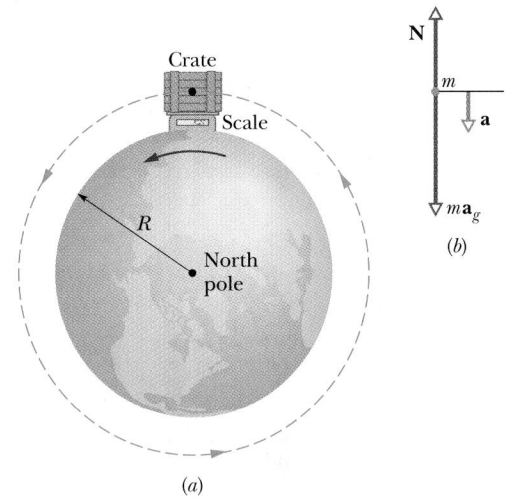

FIGURE 14-7 (a) A crate resting on a platform scale at Earth's equator as Earth rotates. The view is along Earth's rotational axis, looking down on the north pole. (b) A free-body diagram for the crate. The crate is in uniform circular motion and is thus accelerated toward the center of Earth. The gravitational force on it has magnitude ma_g. The normal force **N** from the scale has magnitude mg, where g is the free-fall acceleration.

which shows that the weight mg of the crate differs from the magnitude ma_g of the gravitational force acting on the crate. Dividing Eq. 14-14 by m shows also that g is different from a_g by an amount equal to the centripetal acceleration a.

For the centripetal acceleration a we can substitute $\omega^2 R$, where ω is the angular speed of Earth's rotation and R is the radius of the circular path taken by the crate. (R is approximately the radius of Earth.) For ω we can substitute $2\pi/T$, where $T = 24$ h is approximately the time for one rotation of Earth. When we make these substitutions and divide by m, Eq. 14-14 gives us

$$a_g - g = \omega^2 R = \left(\frac{2\pi}{T}\right)^2 R \qquad (14\text{-}15)$$

$$= 0.034 \text{ m/s}^2.$$

Thus the free-fall acceleration g (9.78 m/s$^2 \approx$ 9.8 m/s^2) measured on the equator of the real, rotating planet is slightly less than the gravitational acceleration a_g due strictly to the gravitational force. The difference between g and a_g becomes progressively smaller as the crate is moved to progressively greater latitudes, because the crate then travels in a smaller circle, making R in Eq. 14-15 smaller. So, often we can approximate the free-fall acceleration g with the gravitational acceleration a_g. And we can approximate the weight mg of an object with the gravitational force (given by Eq. 14-10).

SAMPLE PROBLEM 14-3

Consider a pulsar, a collapsed star of extremely high density, with a mass M equal to that of the Sun (1.98×10^{30} kg), a radius R of only 12 km, and a rotational period T of 0.041 s. At its equator, by what percentage does the free-fall acceleration g differ from the gravitational acceleration a_g?

SOLUTION: To find a_g on the surface of the pulsar, we use Eq. 14-12, with R replacing r and with M being the mass of the pulsar. Substituting the known values gives

$$a_g = \frac{GM}{R^2} = \frac{(6.67 \times 10^{-11} \text{ m}^3/\text{kg} \cdot \text{s}^2)(1.98 \times 10^{30} \text{ kg})}{(12{,}000 \text{ m})^2}$$

$$= 9.2 \times 10^{11} \text{ m/s}^2.$$

Dividing Eq. 14-15 (which applies to any rotating body) by a_g and substituting the known data, we now find

$$\frac{a_g - g}{a_g} = \left(\frac{2\pi}{T}\right)^2 \frac{R}{a_g} = \left(\frac{2\pi}{0.041 \text{ s}}\right)^2 \frac{12{,}000 \text{ m}}{9.2 \times 10^{11} \text{ m/s}^2}$$

$$= 3.1 \times 10^{-4} = 0.031\%. \qquad \text{(Answer)}$$

Even though a pulsar rotates extremely rapidly, its rotation reduces the free-fall acceleration from the gravitational acceleration only slightly, because its radius is so small.

SAMPLE PROBLEM 14-4

(a) An astronaut whose height h is 1.70 m floats feet ''down'' in an orbiting space shuttle at a distance $r = 6.77 \times 10^6$ m from the center of Earth. What is the difference in the gravitational acceleration at her feet and at her head?

SOLUTION: Equation 14-12 tells us the gravitational acceleration at any distance r from the center of Earth:

$$a_g = \frac{GM_E}{r^2}, \qquad (14\text{-}16)$$

where M_E is Earth's mass. We cannot simply apply Eq. 14-16 twice, first with, say, $r = 6.77 \times 10^6$ m for the feet and then with $r = 6.77 \times 10^6$ m + 1.70 m for the head. If we do, a calculator will give us the same value for a_g twice, and thus a difference of zero, because h is so small compared to r. Instead, we differentiate Eq. 14-16 with respect to r, obtaining

$$da_g = -2\frac{GM_E}{r^3}\, dr, \qquad (14\text{-}17)$$

where da_g is the differential change in the gravitational acceleration due to a differential change dr in r. For the astronaut, $dr = h$ and $r = 6.77 \times 10^6$ m. Substituting data into Eq. 14-17, we find

$$da_g = -2\frac{(6.67 \times 10^{-11} \text{ m}^3/\text{kg} \cdot \text{s}^2)(5.98 \times 10^{24} \text{ kg})}{(6.77 \times 10^6 \text{ m})^3}(1.70 \text{ m})$$

$$= -4.37 \times 10^{-6} \text{ m/s}^2. \qquad \text{(Answer)}$$

This result means that the gravitational acceleration of the astronaut's feet toward Earth is slightly greater than the gravitational acceleration of her head toward Earth. This difference in acceleration tends to stretch her body, but the difference is so small that the stretching is unnoticeable.

(b) If the astronaut is now feet ''down'' at the same orbital radius r of 6.77×10^6 m about a black hole of mass $M_h = 1.99 \times 10^{31}$ kg (which is 10 times our Sun's mass), what is the difference in the gravitational acceleration at her feet and head? The black hole has a surface (called the *horizon* of the black hole) of radius $R_h = 2.95 \times 10^4$ m. Nothing, not even light, can escape from the surface or anywhere inside it. Note that the astronaut is (wisely) well outside the surface (at $r = 229R_h$).

SOLUTION: We can again use Eq. 14-17 if we substitute $M_h = 1.99 \times 10^{31}$ kg for M_E. We find

$$da_g = -2\frac{(6.67 \times 10^{-11} \text{ m}^3/\text{kg} \cdot \text{s}^2)(1.99 \times 10^{31} \text{ kg})}{(6.77 \times 10^6 \text{ m})^3}(1.70 \text{ m})$$

$$= -14.5 \text{ m/s}^2. \qquad \text{(Answer)}$$

This means that the gravitational acceleration of the astronaut's feet toward the black hole is noticeably larger than that of her head. The resulting tendency to stretch her body would be bearable but quite painful. If she drifted closer to the black hole, the stretching tendency would increase drastically.

14-5 GRAVITATION INSIDE EARTH

Newton's shell theorem can also be applied to a situation in which a particle is located *inside* a uniform shell, to show the following:

A uniform shell of matter exerts no *net* gravitational force on a particle located inside it.

If the density of Earth were uniform, the gravitational force acting on a particle would be a maximum at Earth's surface. The force would, as we expect, decrease as the particle moved outward. If the particle were to move inward, perhaps down a deep mine shaft, the gravitational force would change for two reasons. (1) It would tend to increase because the particle would be moving closer to the center of Earth. (2) It would tend to decrease because the shell of material lying outside the particle's radial position would not exert any net force on the particle.

For a uniform Earth, the second influence prevails and the force on the particle steadily decreases to zero as the particle approaches the center of Earth. However, for the real (nonuniform) Earth, the force on the particle actually increases as the particle begins to descend. The force reaches a maximum at a certain depth; only then does it begin to decrease as the particle descends further.

SAMPLE PROBLEM 14-5

Suppose a tunnel runs through Earth from pole to pole, as in Fig. 14-8. Assume that Earth is a nonrotating, uniform sphere. Find the gravitational force on a particle of mass m dropped into the tunnel when it reaches a distance r from Earth's center.

SOLUTION: The force that acts on the particle is associated only with the mass M' of Earth that lies within a sphere of radius r. The portion of Earth that lies outside this sphere does not exert any net force on the particle.

Mass M' is given by

$$M' = \rho V' = \rho \frac{4\pi r^3}{3}, \qquad (14\text{-}18)$$

in which V' is the volume occupied by M' (which lies within the dashed line in Fig. 14-8), and ρ is the assumed uniform density of Earth.

The force acting on the particle is then, using Eqs. 14-1 and 14-18,

$$F = -\frac{GmM'}{r^2} = -\frac{Gm\rho 4\pi r^3}{3r^2} = -\left(\frac{4\pi mG\rho}{3}\right) r$$

$$= -Kr, \qquad \text{(Answer)} \quad (14\text{-}19)$$

in which K, a constant, is equal to $4\pi mG\rho/3$. We have inserted a minus sign to indicate that the force $\mathbf{F}$ and the displacement $\mathbf{r}$ are in opposite directions, the former being toward the center of Earth and the latter away from that point. Thus Eq. 14-19 tells us that the force acting on the particle is proportional to the displacement of the particle but oppositely directed, precisely the situation for Hooke's law.

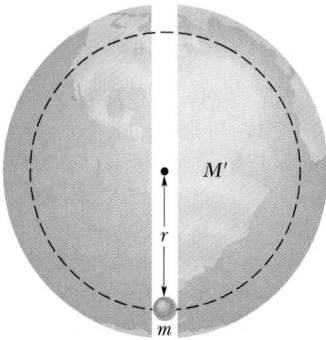

FIGURE 14-8 Sample Problem 14-5. A particle is dropped into a tunnel drilled through Earth.

14-6 GRAVITATIONAL POTENTIAL ENERGY

In Section 8-3, we discussed the gravitational potential energy of a particle–Earth system. We were careful to keep the particle near Earth's surface, so that we could regard the gravitational force as constant. We then chose some reference configuration of the system as having a gravitational potential energy of zero. Often, in this configuration the particle was on Earth's surface. For particles not on Earth's surface, the gravitational potential energy decreased when the separation between the particle and Earth decreased.

Here, we broaden our view and consider the gravitational potential energy U of two particles, of masses m and M, separated by a distance r. We again choose a reference configuration with U equal to zero. However, to simplify the equations, the separation distance r in the reference configuration is now large enough to be approximated as *infinite*. As before, the gravitational potential energy decreases when the separation decreases. Since $U = 0$ for $r = \infty$, the potential energy is negative for any finite separation and becomes progressively more negative as the particles move closer together.

With these facts in mind and as we shall justify below, we take the gravitational potential energy of the two-particle system to be

$$U = -\frac{GMm}{r} \qquad \begin{array}{l}\text{(gravitational}\\ \text{potential energy).}\end{array} \quad (14\text{-}20)$$

Note that $U(r)$ approaches zero as r approaches infinity and that for any finite value of r, the value of $U(r)$ is negative.

The potential energy given by Eq. 14-20 is a property of the system of two particles rather than of either particle alone. There is no way to divide this energy and say that so much belongs to one particle and so much to the other. Nevertheless, if $M \gg m$, as is true for Earth and a baseball, we often speak of "the potential energy of the baseball." We can get away with this because, when a baseball moves in the vicinity of Earth, changes in the potential energy of the baseball–Earth system appear almost entirely as changes in the kinetic energy of the baseball, since changes in the kinetic energy of Earth are too small to be measured. Similarly, in Section 14-8 we shall speak of "the potential energy of an artificial satellite" orbiting Earth, because the satellite's mass is so much smaller than Earth's mass. When we speak of the potential energy of bodies of comparable mass, however, we have to be careful to treat them as a system.

If our system contains more than two particles, we consider each pair of particles in turn, calculate the gravi-

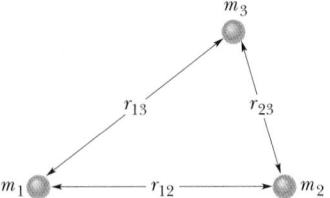

FIGURE 14-9 Three particles exert gravitational forces on one another. The gravitational potential energy of the system is the sum of the gravitational potential energies of each of the three possible pairs of the particles.

tational potential energy of that pair with Eq. 14-20 as if the other particles were not there, and then algebraically sum the results. Applying Eq. 14-20 to each of the three pairs of Fig. 14-9, for example, gives the potential energy of the system as

$$U = -\left(\frac{Gm_1m_2}{r_{12}} + \frac{Gm_1m_3}{r_{13}} + \frac{Gm_2m_3}{r_{23}} \right). \quad (14\text{-}21)$$

A *globular cluster* (Fig. 14-10) in the constellation Sagittarius is a good example of a naturally occurring system of particles. It contains about 70,000 stars, which can be paired in 2.5×10^9 different ways. Contemplation of this structure suggests the enormous amount of gravitational potential energy stored in the universe.

Proof of Eq. 14-20

Let a baseball, starting from rest at a great (infinite) distance from Earth, fall toward point P, as in Fig. 14-11. The potential energy of the baseball–Earth system is initially

FIGURE 14-10 A globular cluster, like this one in the constellation Sagittarius, contains tens of thousands of stars in an almost spherical collection. There are many such clusters in our Milky Way galaxy, some of which are visible with a small telescope.

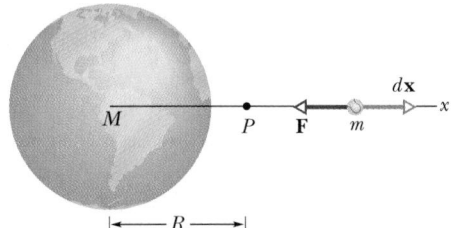

FIGURE 14-11 A baseball of mass m falls toward Earth from infinity, along a radial line (an x axis) passing through point P at a distance R from the center of Earth.

zero. When the baseball reaches P, the potential energy is the negative of the work W done by the gravitational force as the baseball moves to P from its distant position. Thus, from Eqs. 8-1 and 8-6 (generalized to vector form), we have

$$U = -W = -\int_\infty^R \mathbf{F}(x) \cdot d\mathbf{x}. \quad (14\text{-}22)$$

The limits on the integral are the initial distance of the baseball, which we have taken to be infinitely great, and its final distance R.

The vector $\mathbf{F}(x)$ in Eq. 14-22 points radially inward toward the center of Earth in Fig. 14-11, and the vector $d\mathbf{x}$ points radially outward, the angle ϕ between them being 180°. Thus

$$\mathbf{F}(x) \cdot d\mathbf{x} = F(x)(\cos 180°)(dx)$$
$$= -F(x)\,dx. \quad (14\text{-}23)$$

For $F(x)$ in Eq. 14-23, we now substitute from Newton's law of gravitation (Eq. 14-1) written as a function of x rather than r, obtaining

$$\mathbf{F}(x) \cdot d\mathbf{x} = -\frac{GMm}{x^2}\,dx.$$

Putting that result into Eq. 14-22, we obtain

$$U = \int_\infty^R \left(\frac{GMm}{x^2} \right) dx = -\left[\frac{GMm}{x} \right]_\infty^R = -\frac{GMm}{R},$$

which, with a change in symbol from R to r, is Eq. 14-20.

Equation 14-20 holds no matter what path the baseball takes in moving in toward Earth. Consider a path made up of small steps, as in Fig. 14-12. No work is done along circular steps like AB, because along them the (radial) gravitational force is perpendicular to the displacement. The total work done along all the radial steps, such as BC, is the same as the work done in moving along a single radial line, as in Fig. 14-11. Thus the work done on a particle by the gravitational force when the particle moves between any two points is independent of the path that the

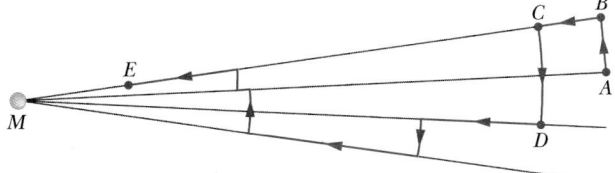

FIGURE 14-12 The work done by the gravitational force as a baseball moves from A to E is independent of the path followed.

particle follows and depends only on the particle's initial and final positions. The work may be computed simply as the negative of the difference in potential energy between the two points:

$$W = -\Delta U = -(U_f - U_i),$$ (14-24)

where U_f and U_i are the potential energies associated with the final and initial positions. This is what we mean when we say, as we did in Chapter 8, that the gravitational force is *conservative*. And, as discussed there, if the work *did* depend on the path, as it does for frictional forces, there would be no potential energy associated with the gravitational force.

Potential Energy and Force

In the proof of Eq. 14-20, we derived the potential energy function U from the force function F. We should be able to go the other way, that is, to start from the potential energy function and derive the force function. Guided by Eq. 8-19, we can write

$$F = -\frac{dU}{dr} = -\frac{d}{dr}\left(-\frac{GMm}{r}\right)$$

$$= -\frac{GMm}{r^2}.$$ (14-25)

This is just Newton's law of gravitation (Eq. 14-1). The minus sign indicates that the force on mass m points radially inward, toward mass M.

Escape Speed

If you fire a projectile upward, usually it will slow, stop momentarily, and return to Earth. There is, however, a certain minimum initial speed that will cause it to move upward forever, theoretically coming to rest only at infinity. This initial speed is called the **escape speed.**

Consider a projectile of mass m, leaving the surface of a planet (or some other astronomical body or system) with escape speed v. It has a kinetic energy K given by $\frac{1}{2}mv^2$ and a potential energy U given by Eq. 14-20:

$$U = -\frac{GMm}{R},$$

in which M is the mass of the planet and R its radius.

When the projectile reaches infinity, it stops and thus has no kinetic energy. It also has no potential energy because this is our zero-potential-energy configuration. Its total energy at infinity is therefore zero. From the principle of conservation of energy, its total energy at the planet's surface must also have been zero, so

$$K + U = \frac{1}{2}mv^2 + \left(-\frac{GMm}{R}\right) = 0.$$

This yields

$$v = \sqrt{\frac{2GM}{R}}.$$ (14-26)

The escape speed does not depend on the direction in which a projectile is fired. However, attaining that speed is easier if the projectile is fired in the direction the launch site is moving as Earth rotates about its axis. For example, rockets are launched eastward at Cape Canaveral to take advantage of the Cape's eastward speed of 1500 km/h.

Equation 14-26 can be applied to find the escape speed of a projectile from any astronomical body, provided we substitute the mass of the body for M and the radius of the body for R. Table 14-2 shows escape speeds from some astronomical bodies.

CHECKPOINT **4:** You move a ball of mass m away from a sphere of mass M. (a) Does the gravitational potential energy of the ball–sphere system increase or decrease? (b) Is positive or negative work done by the gravitational force between the ball and the sphere?

TABLE 14-2 SOME ESCAPE SPEEDS

BODY	MASS (kg)	RADIUS (m)	ESCAPE SPEED (km/s)
Ceres[a]	1.17×10^{21}	3.8×10^5	0.64
Earth's moon	7.36×10^{22}	1.74×10^6	2.38
Earth	5.98×10^{24}	6.37×10^6	11.2
Jupiter	1.90×10^{27}	7.15×10^7	59.5
Sun	1.99×10^{30}	6.96×10^8	618
Sirius B[b]	2×10^{30}	1×10^7	5200
Neutron star[c]	2×10^{30}	1×10^4	2×10^5

[a] The most massive of the asteroids.

[b] A *white dwarf* (a star in a final stage of evolution) that is a companion of the bright star Sirius.

[c] The collapsed core of a star that remains after that star has exploded in a *supernova* event.

SAMPLE PROBLEM 14-6

An asteroid, headed directly toward Earth, has a speed of 12 km/s relative to the planet when it is at a distance of 10 Earth radii from Earth's center. Ignoring the effects of the terrestrial atmosphere on the asteroid, find the asteroid's speed when it reaches Earth's surface.

SOLUTION: Because the mass of an asteroid is much less than that of Earth, we can assign the gravitational potential energy of the asteroid–Earth system to the asteroid alone, and we can neglect any change in the speed of Earth relative to the asteroid during the asteroid's fall. Because we are to ignore the effects of the atmosphere on the asteroid, the mechanical energy of the asteroid is conserved during its fall; that is,

$$K_f + U_f = K_i + U_i,$$

where K and U are the kinetic energy and gravitational potential energy of the asteroid, and the subscripts i and f refer to initial values (at a distance of 10 Earth radii) and final values (at a distance of one Earth radius).

Let m represent the mass of the asteroid, M the mass of Earth ($= 5.98 \times 10^{24}$ kg), and R the radius of Earth ($= 6.37 \times 10^6$ m). Using Eq. 14-20 for the gravitational potential energy and $\frac{1}{2}mv^2$ for K, we have

$$\tfrac{1}{2}mv_f^2 - \frac{GMm}{R} = \tfrac{1}{2}mv_i^2 - \frac{GMm}{10R}.$$

Rearranging and substituting known values, we find

$$
\begin{aligned}
v_f^2 &= v_i^2 + \frac{2GM}{R}\left(1 - \frac{1}{10}\right) \\
&= (12 \times 10^3 \text{ m/s})^2 \\
&\quad + \frac{2(6.67 \times 10^{-11} \text{ m}^3/\text{kg} \cdot \text{s}^2)(5.98 \times 10^{24} \text{ kg})}{6.37 \times 10^6 \text{ m}}\, 0.9 \\
&= 2.567 \times 10^8 \text{ m}^2/\text{s}^2,
\end{aligned}
$$

and thus

$$v_f = 1.60 \times 10^4 \text{ m/s} = 16 \text{ km/s}. \qquad \text{(Answer)}$$

At this speed, the asteroid would not have to be particularly large to do considerable damage. As an example, if it were only 5 m across, the impact could release about as much energy as the nuclear explosion at Hiroshima. Alarmingly, about 500 million asteroids of this size are near Earth's orbit, and in 1994 one of them apparently penetrated Earth's atmosphere and exploded at an altitude of 20 km near a remote South Pacific island (setting off nuclear-explosion warnings on six military satellites). An asteroid 500 m across (there may be a million of them near Earth's orbit) could end modern civilization and almost eliminate humans worldwide.

14-7 PLANETS AND SATELLITES: KEPLER'S LAWS

The motions of the planets, as they seemingly wander against the background of the stars, have been a puzzle since the dawn of history. The "loop-the-loop" motion of Mars, shown in Fig. 14-13, was particularly baffling. Johannes Kepler (1571–1630), after a lifetime of study, worked out the empirical laws that govern these motions. Tycho Brahe (1546–1601), the last of the great astronomers to make observations without the help of a telescope, compiled the extensive data from which Kepler was able to derive the three laws of planetary motion that now bear his name. Later, Newton (1642–1727) showed that his law of gravitation leads to Kepler's empirical laws.

We discuss each of Kepler's laws in turn. Although we apply the laws here to planets orbiting the Sun, they hold equally well for satellites, either natural or artificial, orbiting Earth or any other massive central body.

---

1. THE LAW OF ORBITS: All planets move in elliptical orbits, with the Sun at one focus.

Figure 14-14 shows a planet, of mass m, moving in such an orbit around the Sun, whose mass is M. We assume that $M \gg m$, so that the center of mass of the planet–Sun system is virtually at the center of the Sun.

The orbit in Fig. 14-14 is described by giving its **semimajor axis** a and its **eccentricity** e, the latter defined

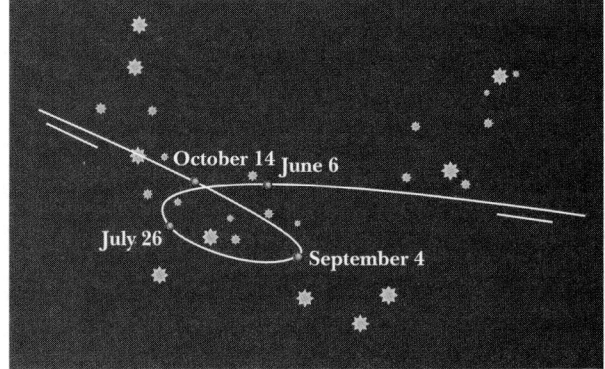

FIGURE 14-13 The path of the planet Mars as it moved against a background of the constellation Capricorn during 1971. Its position on four selected days is marked. Both Mars and Earth are moving in orbits around the Sun so that we see the position of Mars relative to us; this sometimes results in an apparent loop in the path of Mars.

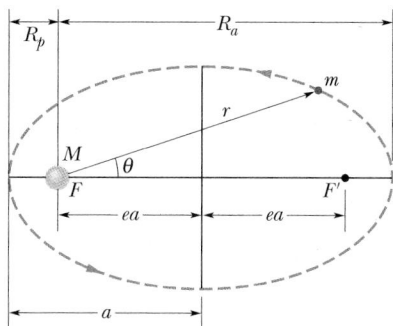

FIGURE 14-14 A planet of mass m moving in an elliptical orbit around the Sun. The Sun, of mass M, is at one focus F of the ellipse. The other, or "empty," focus is F'. Each focus is a distance ea from the center, e being the eccentricity of the ellipse. The semimajor axis a of the ellipse, the perihelion (nearest the Sun) distance R_p, and the aphelion (farthest from the Sun) distance R_a are also shown.

so that ea is the distance from the center of the ellipse to either focus F or F'. *An eccentricity of zero corresponds to a circle,* in which the two foci merge to a single central point. The eccentricities of the planetary orbits are not large, so—sketched on paper—the orbits look circular. The eccentricity of the ellipse of Fig. 14-14, which has been exaggerated for clarity, is 0.74. The eccentricity of Earth's orbit is only 0.0167.

---

2. THE LAW OF AREAS: A line that connects a planet to the Sun sweeps out equal areas in equal times.

Qualitatively, this second law tells us that the planet will move most slowly when it is farthest from the Sun and most rapidly when it is nearest to the Sun. As it turns out, Kepler's second law is totally equivalent to the law of conservation of angular momentum. Let us prove it.

The area of the shaded wedge in Fig. 14-15a closely approximates the area swept out in time Δt by a line connecting the Sun and the planet, which are separated by a distance r. The area ΔA of the wedge is approximately the area of a triangle with base $r\,\Delta\theta$ and height r. Thus $\Delta A \approx \frac{1}{2}r^2\,\Delta\theta$. This expression for ΔA becomes more exact as Δt (hence $\Delta\theta$) approaches zero. The instantaneous rate at which area is being swept out is then

$$\frac{dA}{dt} = \frac{r^2}{2}\frac{d\theta}{dt} = \frac{r^2\omega}{2}, \qquad (14\text{-}27)$$

in which ω is the angular speed of the rotating line connecting Sun and planet.

Figure 14-15b shows the linear momentum $\mathbf{p}$ of the planet, along with its components. From Eq. 12-27, the magnitude of the angular momentum $\mathbf{L}$ of the planet about the Sun is given by the product of r and the component of $\mathbf{p}$ perpendicular to r, or

$$L = rp_\perp = (r)(mv_\perp) = (r)(m\omega r)$$
$$= mr^2\omega, \qquad (14\text{-}28)$$

where we have replaced $v_\perp$ with its equivalent ωr (Eq. 11-16). Eliminating $r^2\omega$ between Eqs. 14-27 and 14-28 leads to

$$\frac{dA}{dt} = \frac{L}{2m}. \qquad (14\text{-}29)$$

If dA/dt is constant, as Kepler said it is, then Eq. 14-29 means that L must also be constant—angular momentum is conserved. So Kepler's second law is indeed equivalent to the law of conservation of angular momentum.

---

3. THE LAW OF PERIODS: The square of the period of any planet is proportional to the cube of the semimajor axis of its orbit.

To see this, consider a circular orbit with radius r (the radius is equivalent to the semimajor axis of an ellipse).

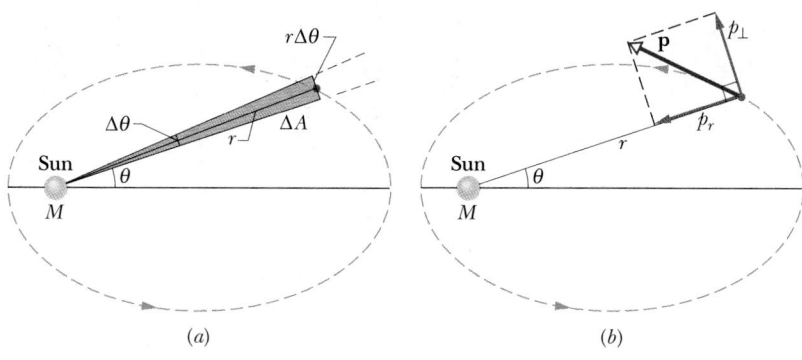

FIGURE 14-15 (a) In time Δt, the line r connecting the planet to the Sun (of mass M) sweeps through an angle $\Delta\theta$, sweeping out an area ΔA. (b) The linear momentum $\mathbf{p}$ of the planet and its components.

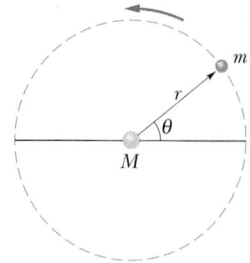

FIGURE 14-16 A planet of mass m moving around the Sun in a circular orbit of radius r.

Applying Newton's second law, $F = ma$, to the orbiting planet in Fig. 14-16 yields

$$\frac{GMm}{r^2} = (m)(\omega^2 r). \qquad (14\text{-}30)$$

Here we have substituted from Eq.14-1 for the force F and used Eq. 11-21 to substitute $\omega^2 r$ for the centripetal acceleration. If we replace ω with $2\pi/T$ (Eq. 11-18), where T is the period of the motion, we obtain Kepler's third law:

$$T^2 = \left(\frac{4\pi^2}{GM}\right) r^3 \qquad \text{(law of periods).} \quad (14\text{-}31)$$

The quantity in parentheses is a constant, its value depending only on the mass of the central body.

Equation 14-31 holds also for elliptical orbits, provided we replace r with a, the semimajor axis of the ellipse.

TABLE 14-3
KEPLER'S LAW OF PERIODS FOR THE SOLAR SYSTEM

PLANET	SEMIMAJOR AXIS a (10^{10} m)	PERIOD T (y)	T^2/a^3 (10^{-34} y²/m³)
Mercury	5.79	0.241	2.99
Venus	10.8	0.615	3.00
Earth	15.0	1.00	2.96
Mars	22.8	1.88	2.98
Jupiter	77.8	11.9	3.01
Saturn	143	29.5	2.98
Uranus	287	84.0	2.98
Neptune	450	165	2.99
Pluto	590	248	2.99

This law predicts that the ratio T^2/a^3 has essentially the same value for every planetary orbit around a given massive body. Table 14-3 shows how well it holds for the orbits of the planets of the solar system.

CHECKPOINT **5:** Satellite 1 is in a certain circular orbit about a planet, while satellite 2 is in a larger circular orbit. Which satellite has (a) the longer period and (b) the greater speed?

On February 7, 1984, at a height of 102 km above Hawaii and with a speed of about 29,000 km/h, Bruce McCandless stepped (untethered) into space from a space shuttle and became the first human satellite.

SAMPLE PROBLEM 14-7

A satellite in circular orbit at an altitude h of 230 km above Earth's surface has a period T of 89 min. What mass of Earth follows from these data?

SOLUTION: We apply Kepler's law of periods to the satellite–Earth system. Solving Eq. 14-31 for M yields

$$M = \frac{4\pi^2 r^3}{GT^2}. \qquad (14\text{-}32)$$

The radius r of the satellite orbit is

$$r = R + h = 6.37 \times 10^6 \text{ m} + 230 \times 10^3 \text{ m}$$
$$= 6.60 \times 10^6 \text{ m},$$

in which R is the radius of Earth. Substituting this value and the period T into Eq. 14-32 yields

$$M = \frac{(4\pi^2)(6.60 \times 10^6 \text{ m})^3}{(6.67 \times 10^{-11} \text{ m}^3/\text{kg} \cdot \text{s}^2)(89 \times 60 \text{ s})^2}$$
$$= 6.0 \times 10^{24} \text{ kg.} \qquad \text{(Answer)}$$

In the same way, we could find the mass of our Sun from the period and radius of Earth's orbit (assumed circular), and the mass of Jupiter from the period and orbital radius of any one of its moons (without knowing the mass of that moon).

SAMPLE PROBLEM 14-8

Comet Halley orbits about the Sun with a period of 76 years and, in 1986, had a distance of closest approach to the Sun, its *perihelion distance* R_p, of 8.9×10^{10} m. Table 14-3 shows that this is between the orbits of Mercury and Venus.

(a) What is the comet's farthest distance from the Sun, its *aphelion distance* R_a?

SOLUTION: We can find the semimajor axis of the orbit of comet Halley from Eq. 14-31. Substituting a for r in that equation and solving for a yield

$$a = \left(\frac{GMT^2}{4\pi^2}\right)^{1/3}. \qquad (14\text{-}33)$$

If we substitute the mass M of the Sun, 1.99×10^{30} kg, and the period T of the comet, 76 years or 2.4×10^9 s, into Eq. 14-33 we find that $a = 2.7 \times 10^{12}$ m.

Study of Fig. 14-14 shows that $R_a + R_p = 2a$, or

$$
\begin{aligned}
R_a &= 2a - R_p \\
&= (2)(2.7 \times 10^{12}\text{ m}) - 8.9 \times 10^{10}\text{ m} \\
&= 5.3 \times 10^{12}\text{ m}. \qquad \text{(Answer)}
\end{aligned}
$$

Table 14-3 shows that this is just a little less than the semimajor axis of the orbit of Pluto.

(b) What is the eccentricity of the orbit of comet Halley?

SOLUTION: Figure 14-14 shows that $ea = a - R_p$, or

$$
\begin{aligned}
e &= \frac{a - R_p}{a} = 1 - \frac{R_p}{a} \\
&= 1 - \frac{8.9 \times 10^{10}\text{ m}}{2.7 \times 10^{12}\text{ m}} = 0.97. \qquad \text{(Answer)}
\end{aligned}
$$

This cometary orbit, with an eccentricity approaching unity, is a long thin ellipse.

SAMPLE PROBLEM 14-9

Observations of the light from a certain star indicate that it is part of a binary (two-star) system. This visible star has orbital speed $v = 270$ km/s, orbital period $T = 1.70$ days, and approximate mass $m_1 = 6M_s$, where M_s is the Sun's mass, 1.99×10^{30} kg. Assuming that the visible star and its companion object, which is dark and unseen, are in circular orbits (see Fig. 14-17), determine the approximate mass m_2 of the dark object.

SOLUTION: As in the two particle systems of Section 9-2, the center of mass of this two-star system lies along a line connecting their centers. As with the freely rotating bodies and systems of Chapter 12, this two-star system rotates about its center of mass. In Fig. 14-17, the center of mass is labeled point O; the visible star and dark object orbit O with orbital

radii r_1 and r_2, respectively, and have a separation of $r = r_1 + r_2$. From Eq. 14-1, the gravitational force on the visible star from the dark object is

$$F = \frac{Gm_1m_2}{r^2}.$$

Applying Newton's second law, $F = ma$, to the visible star then yields

$$\frac{Gm_1m_2}{r^2} = m_1a = (m_1)(\omega^2 r_1), \qquad (14\text{-}34)$$

where ω is the angular speed of the visible star and $\omega^2 r_1$ is its centripetal acceleration toward O.

We can obtain another equation in these same variables by locating the center of mass O. Because O is a distance r_1 from the visible star, we can use Eq. 9-1 to write

$$r_1 = \frac{m_2 r}{m_1 + m_2}.$$

This yields

$$r = r_1 \frac{m_1 + m_2}{m_2}. \qquad (14\text{-}35)$$

Now substituting Eq. 14-35 for r and $2\pi/T$ for ω into Eq. 14-34 gives us, after some algebraic manipulation,

$$\frac{m_2^3}{(m_1 + m_2)^2} = \frac{4\pi^2}{GT^2} r_1^3. \qquad (14\text{-}36)$$

We still have two unknowns, m_2 and r_1. We can, however, find r_1 from the orbital motion of the visible star: the orbital period T is equal to the circumference of the orbit $(2\pi r_1)$ divided by the speed v of the planet. This gives us

$$T = \frac{2\pi r_1}{v},$$

or

$$r_1 = \frac{vT}{2\pi}. \qquad (14\text{-}37)$$

If we substitute $m_1 = 6M_s$ and Eq. 14-37 for r_1, then Eq. 14-36 becomes

$$
\begin{aligned}
\frac{m_2^3}{(6M_s + m_2)^2} &= \frac{v^3 T}{2\pi G} \\
&= \frac{(2.7 \times 10^5\text{ m/s})^3(1.70\text{ days})(86,400\text{ s/day})}{(2\pi)(6.67 \times 10^{-11}\text{ N} \cdot \text{m}^2/\text{kg}^2)} \\
&= 6.90 \times 10^{30}\text{ kg},
\end{aligned}
$$

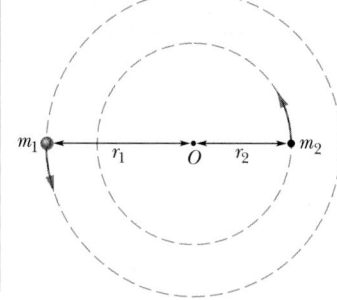

FIGURE 14-17 Sample Problem 14-9. A visible star with mass m_1 and a dark, unseen object with mass m_2 orbit around the center of mass of the two-star system at O.

or

$$\frac{m_2^3}{(6M_s + m_2)^2} = 3.47M_s. \quad (14\text{-}38)$$

We can solve this cubic equation for m_2, or, since we are working with approximate masses anyway, we can substitute integer multiples of M_s for m_2 until we find

$$m_2 \approx 9M_s. \quad (\text{Answer})$$

The data here approximate those for the binary system LMC X-3 in the Large Magellanic Cloud (shown in the figure that begins this chapter). From other data, the dark object is known to be especially compact: it may be a star that collapsed under its own gravitational pull to become a neutron star or a black hole. Since a neutron star cannot have a mass larger than about $2M_s$, the result $m_2 \approx 9M_s$ strongly suggests that the dark object is a black hole.

Thus, one can detect the presence of a black hole if it is part of a binary system with a visible star whose mass, orbital speed, and orbital period can be measured.

14-8 SATELLITES: ORBITS AND ENERGY (OPTIONAL)

As a satellite orbits Earth on its elliptical path, both its speed, which fixes its kinetic energy K, and its distance from the center of Earth, which fixes its gravitational potential energy U, fluctuate with fixed periods. However, the mechanical energy E of the satellite remains constant. (Since the satellite's mass is so much smaller than Earth's mass, we assign U and E of the Earth–satellite system to the satellite alone.)

The potential energy is given by Eq. 14-20 and is

$$U = -\frac{GMm}{r}$$

(with $U = 0$ for infinite separation). Here r is the radius of the orbit, assumed for the time being to be circular.

To find the kinetic energy of a satellite in a circular orbit, we write Newton's second law, $F = ma$, as

$$\frac{GMm}{r^2} = m\frac{v^2}{r}, \quad (14\text{-}39)$$

where v^2/r is the centripetal acceleration of the satellite. Then from Eq. 14-39, the kinetic energy is

$$K = \tfrac{1}{2}mv^2 = \frac{GMm}{2r}, \quad (14\text{-}40)$$

which shows us that for a satellite in a circular orbit,

$$K = -\frac{U}{2} \quad (\text{circular orbit}). \quad (14\text{-}41)$$

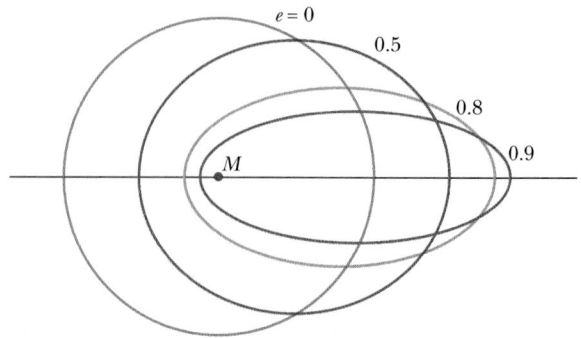

FIGURE 14-18 Four orbits about an object of mass M. All four orbits have the same semimajor axis a and thus correspond to the same total mechanical energy E. Their eccentricities e are marked.

The total mechanical energy of the orbiting satellite is

$$E = K + U = \frac{GMm}{2r} - \frac{GMm}{r}$$

or

$$E = -\frac{GMm}{2r} \quad (\text{circular orbit}). \quad (14\text{-}42)$$

This tells us that for a satellite in a circular orbit, the total energy E is the negative of the kinetic energy K:

$$E = -K \quad (\text{circular orbit}). \quad (14\text{-}43)$$

For a satellite in an elliptical orbit of semimajor axis a, we can substitute a for r in Eq. 14-42 to find the mechanical energy as

$$E = -\frac{GMm}{2a} \quad (\text{elliptical orbit}). \quad (14\text{-}44)$$

Equation 14-44 tells us that the total energy of an orbiting satellite depends only on the semimajor axis of its orbit and not on its eccentricity e. For example, four orbits with the same semimajor axis are shown in Fig. 14-18; the same satellite would have the same total mechanical energy E in all four orbits. Figure 14-19 shows the variation of K, U, and E with r for a satellite moving in a circular orbit about a massive central body.

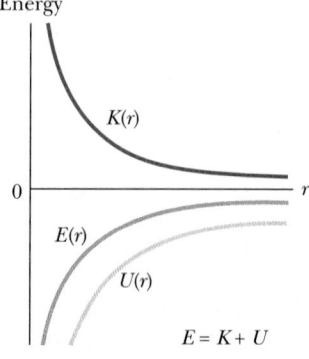

FIGURE 14-19 The variation of kinetic energy K, potential energy U, and total energy E with radius r for a satellite in a circular orbit. For any value of r, the values of U and E are negative, the value of K is positive, and $E = -K$. As r approaches infinity, all three energy curves approach a value of zero.

CHECKPOINT **6:** In the figure, a space shuttle is initially in a circular orbit of radius r about Earth. At point P, the pilot briefly fires a forward-pointing thruster to decrease the shuttle's kinetic energy K and mechanical energy E. (a) Which of the dashed elliptical orbits shown in the figure will the shuttle then take? (b) Is the orbital period T of the shuttle (the time to return to P) then greater than, less than, or the same as in the circular orbit?

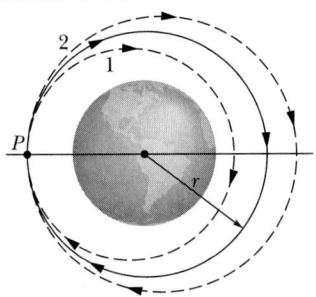

SAMPLE PROBLEM 14-10

A playful astronaut releases a bowling ball, of mass $m = 7.20$ kg, into a circular orbit about Earth at an altitude h of 350 km.

(a) What is the mechanical energy E of the ball in its orbit?

SOLUTION: The orbital radius r is given by

$$r = R + h = 6370 \text{ km} + 350 \text{ km} = 6.72 \times 10^6 \text{ m},$$

in which R is the radius of Earth. From Eq. 14-42, the mechanical energy is

$$
\begin{aligned}
E &= -\frac{GMm}{2r} \\
&= -\frac{(6.67 \times 10^{-11} \text{ N} \cdot \text{m}^2/\text{kg}^2)(5.98 \times 10^{24} \text{ kg})(7.20 \text{ kg})}{(2)(6.72 \times 10^6 \text{ m})} \\
&= -2.14 \times 10^8 \text{ J} = -214 \text{ MJ}. \quad \text{(Answer)}
\end{aligned}
$$

(b) What was the mechanical energy E_0 of the ball on the launchpad at Cape Canaveral? From there to the orbit, what was the change ΔE in the ball's mechanical energy?

SOLUTION: On the launchpad, the ball has some kinetic energy because of the rotation of Earth, but we can show that this is small enough to neglect. Thus the total energy E_0 is equal to the potential energy U_0, which is given by Eq. 14-20:

$$
\begin{aligned}
E_0 = U_0 &= -\frac{GMm}{R} \\
&= -\frac{(6.67 \times 10^{-11} \text{ N} \cdot \text{m}^2/\text{kg}^2)(5.98 \times 10^{24} \text{ kg})(7.20 \text{ kg})}{6.37 \times 10^6 \text{ m}} \\
&= -4.51 \times 10^8 \text{ J} = -451 \text{ MJ}. \quad \text{(Answer)}
\end{aligned}
$$

You might be tempted to say that the potential energy of the ball on Earth's surface is zero. Recall, however, that our zero-potential-energy configuration is one in which the ball is removed to a great distance from Earth. You also might be tempted to use Eq. 14-42 to find E_0, but recall that Eq. 14-42 applies to a satellite *in orbit*.

The *increase* in the mechanical energy of the ball from launchpad to orbit was

$$
\begin{aligned}
\Delta E = E - E_0 &= (-214 \text{ MJ}) - (-451 \text{ MJ}) \\
&= 237 \text{ MJ}. \quad \text{(Answer)}
\end{aligned}
$$

You can buy this amount of energy from your utility company for a few dollars.

14-9 EINSTEIN AND GRAVITATION

Principle of Equivalence

Einstein once said: "I was . . . in the patent office at Bern when all of a sudden a thought occurred to me: 'If a person falls freely, he will not feel his own weight.' I was startled. This simple thought made a deep impression on me. It impelled me toward a theory of gravitation."

Thus Einstein tells us how he began to form his **general theory of relativity.** The fundamental postulate of this theory about gravitation (the gravitating of objects toward each other) is called the **principle of equivalence,** which says that gravitation and acceleration are equivalent. If a physicist were locked up in a small box as in Fig. 14-20, he would not be able to tell whether the box was at

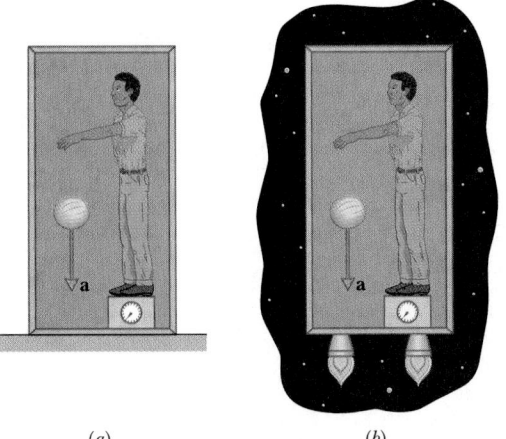

(a)   (b)

FIGURE 14-20 (a) A physicist in a box resting on Earth sees a cantaloupe falling with acceleration $a = 9.8$ m/s². (b) If he and the box accelerate in deep space at 9.8 m/s², the cantaloupe has the same acceleration relative to him. It is not possible, by doing experiments within the box, for the physicist to tell which situation he is in. For example, the platform scale on which he stands reads the same weight in both situations.

rest on Earth (and subject only to Earth's gravitational force), as in Fig. 14-20a, or accelerating through interstellar space at 9.8 m/s² (and subject only to the force producing that acceleration), as in Fig. 14-20b. In both situations he would feel the same and would read the same value for his weight on a scale. Moreover, if he watched an object fall past him, the object would have the same acceleration relative to him in both situations.

Curvature of Space

We have thus far explained gravitation as due to a force between masses. Einstein showed that, instead, gravitation is due to a curvature (or shape) of space that is caused by the masses. (As is discussed later in this book, space and time are entangled; so, the curvature of which Einstein spoke is really a curvature of *spacetime,* the combined four dimensions of our universe.)

Picturing how space (such as vacuum) can have curvature is difficult. An analogy might help: Suppose that

from orbit we watch a race in which two boats begin on the equator with a separation of 20 km and head due south (Fig. 14-21a). To the sailors, the boats travel along flat, parallel paths. However, with time the boats draw together until, nearer the south pole, they touch. The sailors in the boats can interpret this drawing together in terms of a force acting on the boats. However, we can see that the boats

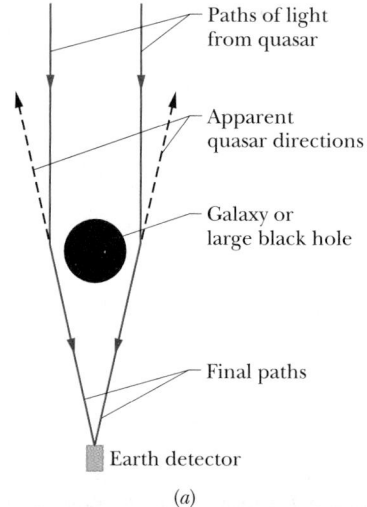

(a)

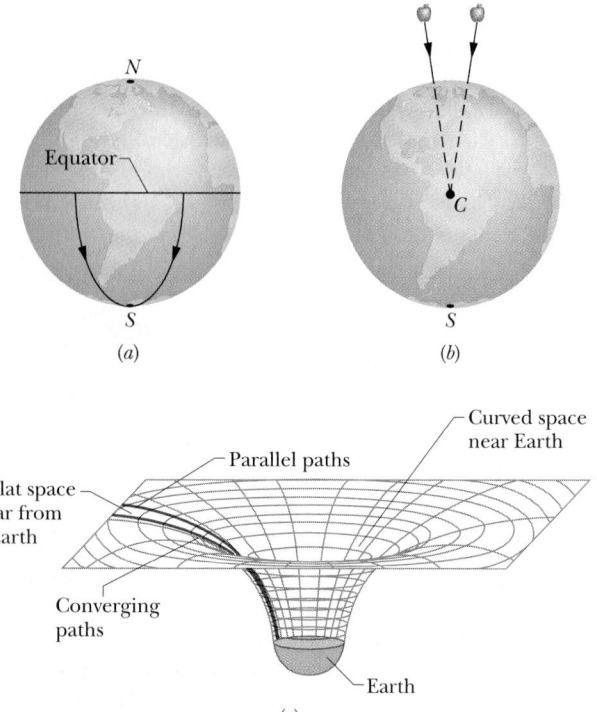

FIGURE 14-21 (a) Two objects moving along lines of longitude toward the south pole converge because of the curvature of Earth's surface. (b) Two objects falling freely near Earth move along lines that converge toward the center of Earth because of the curvature of space near Earth. (c) Far from Earth (and other masses), space is flat and parallel paths remain parallel. Close to Earth, the parallel paths begin to converge because space is curved by Earth's mass.

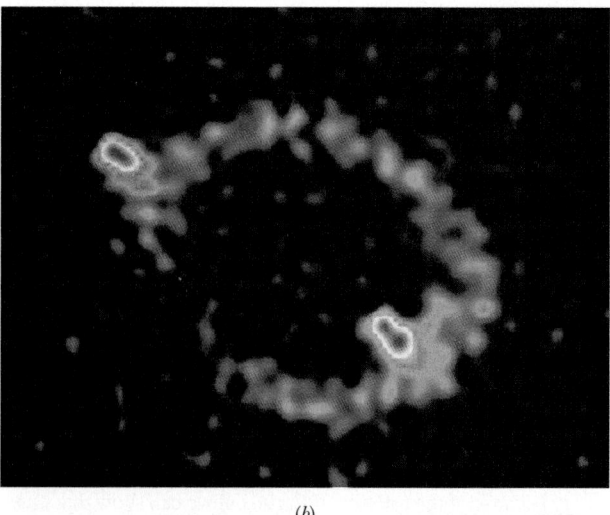

(b)

FIGURE 14-22 (a) Light from a distant quasar follows curved paths around a galaxy or a large black hole because the mass of the galaxy or black hole has curved the adjacent space. If the light is detected, it appears to have originated along the backward extensions of the final paths (dashed lines). (b) The Einstein ring known as MG1131+0456 on the computer screen of a telescope. The source of the light (actually, radio waves, which is a form of invisible light) is far behind the large, unseen galaxy that produces the ring; a portion of the source appears as the two bright spots seen along the ring.

draw together simply because of the curvature of Earth's surface. We can see this because we are viewing the race from "outside" that surface.

Figure 14-21b shows a similar race: two horizontally separated apples are dropped from the same height above Earth. Although the apples may appear to travel along parallel paths, they actually move toward each other because they both fall toward the Earth's center. We can interpret the motion of the apples in terms of the gravitational force on the apples by Earth. But we can also interpret the motion in terms of a curvature of the space near Earth, due to the presence of Earth's mass. This time we cannot see the curvature because we cannot get "outside" the curved space, as we got "outside" the curved Earth in the boat example. But we can depict the curvature with a drawing like Fig. 14–21c; there the apples would move along a surface that curves toward Earth because of Earth's mass.

When light passes near Earth, its path bends slightly because of the curvature of space there, an effect called *gravitational lensing*. When it passes a more massive structure, like a galaxy or a black hole having large mass, its path can be bent more. If such a massive structure is between us and a quasar (an extremely bright, extremely distant source of light), the light from the quasar can bend around the massive structure and toward us (Fig. 14-22a). Then, because the light seems to be coming from slightly different directions in the sky, we see the same quasar in those different directions. In some situations, the quasars we see blend together to form a giant luminous arc, which is called an *Einstein ring* (Fig. 14-22b).

Is gravitation best attributed to the curvature of space-time due to the presence of masses or to a force between masses? Or should it be attributed to the actions of a type of fundamental particle called a *graviton,* as conjectured in some modern physics theories? We do not know.

REVIEW & SUMMARY

The Law of Gravitation

Any particle in the universe attracts any other particle with a **gravitational force** whose magnitude is

$$F = G \frac{m_1 m_2}{r^2} \quad \text{(Newton's law of gravitation),} \quad (14\text{-}1)$$

where m_1 and m_2 are the masses of the particles and r is their separation. The **gravitational constant** G is a universal constant whose value is 6.67×10^{-11} N·m²/kg².

The Gravitational Behavior of Uniform Spherical Shells

Equation 14-1 holds only for particles. The gravitational force between extended bodies must generally be found by adding (integrating) the individual forces on individual particles within the bodies. However, if either of the bodies is a uniform spherical shell or a spherically symmetric solid, the net gravitational force it exerts on an *external* object may be computed as if all the mass of the shell or body were located at its center.

Superposition

Gravitational forces obey the **principle of superposition;** that is, the total force $\mathbf{F}_1$ on a particle labeled as particle 1 is the sum of the forces exerted on it by all other particles taken one at a time:

$$\mathbf{F}_1 = \sum_{i=2}^{n} \mathbf{F}_{1i}, \quad (14\text{-}4)$$

in which the sum is a vector sum of the forces $\mathbf{F}_{1i}$ exerted on particle 1 by particles 2, 3, $\cdots$, n. The gravitational force $\mathbf{F}_1$ exerted on a particle by an extended body is found by dividing the body into units of differential mass dm, each of which exerts a differential force $d\mathbf{F}$ on the particle, and then integrating to find the sum of those forces:

$$\mathbf{F}_1 = \int d\mathbf{F}. \quad (14\text{-}5)$$

Gravitational Acceleration

The *gravitational acceleration* a_g of a particle (of mass m) is due solely to the gravitational force acting on it. When the particle is at distance r from the center of a uniform, spherical body of mass M, the magnitude of the gravitational force on the particle is given by Eq. 14-1. In addition, by Newton's second law,

$$F = ma_g, \quad (14\text{-}11)$$

and thus a_g is given by

$$a_g = \frac{GM}{r^2}. \quad (14\text{-}12)$$

Free-Fall Acceleration and Weight

The actual free-fall acceleration **g** of a particle near Earth differs slightly from the gravitational acceleration $\mathbf{a}_g$, and the weight $m\mathbf{g}$ of the particle differs from the gravitational force (Eq. 14-1) acting on the particle, because Earth is not uniform or spherical and because Earth rotates.

Gravitation Within a Spherical Shell

A uniform shell of matter exerts no gravitational force on a particle located inside it. This means that if a particle is located inside a uniform solid sphere of matter at distance r from its center, the gravitational force exerted on the particle is due only to the mass

M' that lies within a sphere of radius r. This mass M' is given by

$$M' = \rho \frac{4\pi r^3}{3}, \qquad (14\text{-}18)$$

where ρ is the density of the sphere.

Gravitational Potential Energy

The gravitational potential energy $U(r)$ of a system of two particles, with masses M and m and separated by a distance r, is the negative of the work that would be done by the gravitational force of either particle acting on the other as the separation between the particles changes from infinite (very large) to r. This energy is

$$U = -\frac{GMm}{r} \qquad \begin{array}{l}\text{(gravitational}\\ \text{potential energy).}\end{array} \qquad (14\text{-}20)$$

Potential Energy of a System

If a system contains more than two particles, the total gravitational potential energy is the sum of terms representing the potential energies of all the pairs; as an example, we have for three particles, of masses m_1, m_2, and m_3,

$$U = -\left(\frac{Gm_1 m_2}{r_{12}} + \frac{Gm_1 m_3}{r_{13}} + \frac{Gm_2 m_3}{r_{23}}\right). \qquad (14\text{-}21)$$

Escape Speed

An object will escape the gravitational pull of an astronomical body of mass M and radius R if the object's speed near the body's surface is at least equal to the **escape speed,** given by

$$v = \sqrt{\frac{2GM}{R}}. \qquad (14\text{-}26)$$

Kepler's Laws

Gravitational attraction holds the solar system together and makes possible orbiting Earth satellites, both natural and artificial. Such motions are governed by Kepler's three laws of planetary motion, all of which are direct consequences of Newton's laws of motion and gravitation:

1. *The law of orbits.* All planets move in elliptical orbits with the Sun at one focus.

2. *The law of areas.* A line joining any planet to the Sun sweeps out equal areas in equal times (a statement equivalent to conservation of angular momentum).

3. *The law of periods.* The square of the period T of any planet about the Sun is proportional to the cube of the semimajor axis a of the orbit. For circular orbits with radius r, the semimajor axis a is replaced by r and the law is written as

$$T^2 = \left(\frac{4\pi^2}{GM}\right) r^3 \qquad \text{(law of periods),} \qquad (14\text{-}31)$$

where M is the mass of the attracting body—the Sun in the case of the solar system. This result is generally valid for elliptical planetary orbits, when the semimajor axis a is inserted in place of the circular radius r.

Energy in Planetary Motion

When a planet or satellite with mass m moves in a circular orbit with radius r, its potential energy U and kinetic energy K are given by

$$U = -\frac{GMm}{r} \quad \text{and} \quad K = \frac{GMm}{2r}. \qquad (14\text{-}20, 14\text{-}40)$$

The mechanical energy $E = K + U$ is

$$E = -\frac{GMm}{2r}. \qquad (14\text{-}42)$$

For an elliptical orbit of semimajor axis a,

$$E = -\frac{GMm}{2a}. \qquad (14\text{-}44)$$

Einstein's View of Gravitation

Einstein pointed out that gravitation and acceleration are equivalent. This **principle of equivalence** led him to a theory of gravitation (the **general theory of relativity**) that explains gravitational effects in terms of a curvature of space.

QUESTIONS

1. In Fig. 14-23, two particles, of masses m and $2m$, are fixed in place on an axis. (a) Where on the axis can a third particle of mass $3m$ be placed (other than at infinity) so that the net gravitational force on it from the first two particles is zero: to the left of the first two particles, to their right, between them but closer to the more massive particle, or between them but closer to the less massive particle? (b) Does the answer change if the third particle has, instead, a mass of $16m$? (c) Is there a point off the axis at which the net force on the third particle would be zero?

2. In Fig. 14-24, a central particle is surrounded by two circular rings of particles, at radii r and R, with $R > r$. All the particles have mass m. What are the magnitude and direction of the net gravitational force on the central particle due to the particles in the rings?

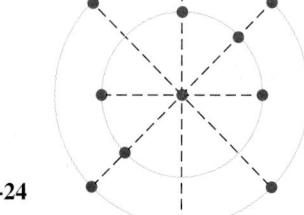

FIGURE 14-23
Question 1.

FIGURE 14-24
Question 2.

3. In Fig. 14-25, a central particle of mass M is surrounded by a square array of other particles, separated by either distance d or $d/2$ along the perimeter of the square. What are the magnitude and direction of the net gravitational force on the central particle due to the other particles?

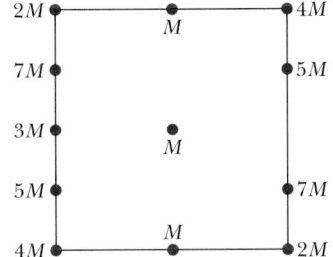

FIGURE 14-25
Question 3.

4. Figure 14-26 shows a particle of mass m that is moved from an infinite distance to the center of a ring of mass M, along the central axis of the ring. For the trip, how does the magnitude of the gravitational force on the particle due to the ring change?

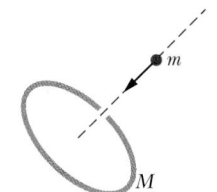

FIGURE 14-26
Question 4.

5. Figure 14-27 shows four arrangements of a particle of mass m and one or more uniform rods of mass M and length L, each a distance d from the particle. Rank the arrangements according to the magnitude of the net gravitational force on the particle from the rods, greatest first.

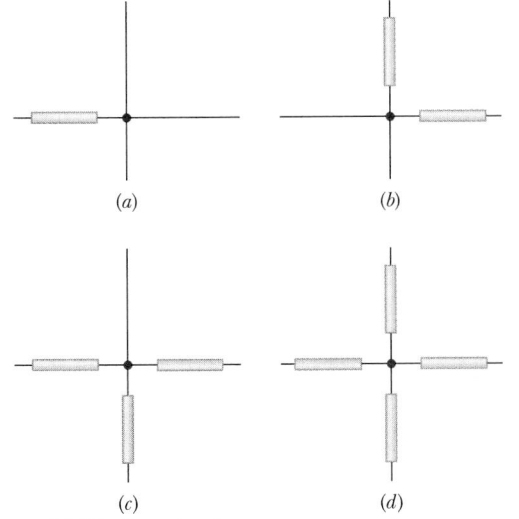

FIGURE 14-27 Question 5.

6. Rank the four systems of equal-mass particles in Checkpoint 2 according to the absolute value of the gravitational potential energy of the system, greatest first.

7. In Fig. 14-28, a particle of mass m (not shown) is to be moved from an infinite distance to one of the three possible locations a, b, and c. Two other particles, of masses m and $2m$, are fixed in place. Rank the three possible locations according to the work done by the net gravitational force on the moving particle due to the fixed particles, greatest first.

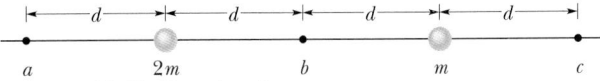

FIGURE 14-28 Question 7.

8. In Fig. 14-29, a particle of mass m is initially at point A, at distance d from the center of one uniform sphere and distance $4d$ from the center of another uniform sphere, both of mass $M \gg m$. State whether, if you moved the particle to point D, the following would be positive, negative, or zero: (a) the change in the gravitational potential energy of the particle, (b) the work done by the net gravitational force on the particle, (c) the work done by your force. (d) What are the answers if, instead, the move were from point B to point C?

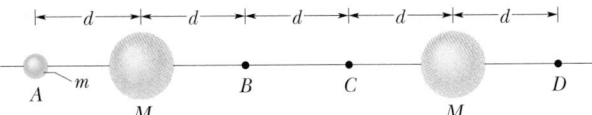

FIGURE 14-29 Question 8.

9. Reconsider the situation of Question 8. Would the work done by you be positive, negative, or zero if you moved the particle (a) from A to B, (b) from A to C, (c) from B to D? (d) Rank those moves according to the absolute value of the work done by your force, greatest first.

10. The following are the masses and radii of three planets: planet A, $2M$ and R; planet B, $3M$ and $2R$; planet C, $4M$ and $2R$. Rank the planets according to the escape speeds from their surfaces, greatest first.

11. Figure 14-30 shows six paths by which a rocket orbiting a moon might move from point a to point b. Rank the paths according to (a) the corresponding change in the gravitational potential energy of the rocket–moon system and (b) the net work done on the rocket by the gravitational force from the moon, greatest first.

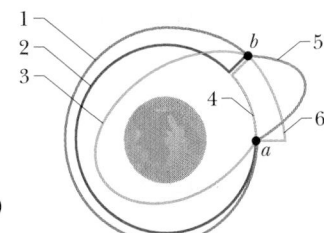

FIGURE 14-30
Question 11.

12. Which of the orbital paths (for, say, a spy satellite) in Fig. 14-31 do not require continuous adjustments by booster rockets? Orbit 1 is at latitude 60°; orbit 2 is in an equatorial plane; orbit 3 is about the center of Earth, between latitudes 60° N and 60° S.

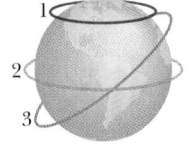

FIGURE 14-31
Question 12.

13. A satellite, with speed v_1 and mass m, is in a circular orbit about a planet of mass M_1. Another satellite, with speed v_2 and mass $2m$, is in a circular orbit of the same radius about a planet of mass M_2. Is M_2 greater than, less than, or equal to M_1 (a) if the satellites have the same period and (b) if $v_2 > v_1$?

EXERCISES & PROBLEMS

SECTION 14-2 Newton's Law of Gravitation

1E. What must the separation be between a 5.2 kg particle and a 2.4 kg particle in order for their gravitational attraction to be 2.3×10^{-12} N?

2E. Some believe that the positions of the planets at the time of birth influence the newborn. Others deride this belief and claim that the gravitational force exerted on a baby by the obstetrician is greater than that exerted by the planets. To check this claim, calculate and compare the gravitational force exerted on a 3 kg baby (a) by a 70 kg obstetrician who is 1 m away and roughly approximated as a point mass, (b) by the massive planet Jupiter ($m = 2 \times 10^{27}$ kg) at its closest approach to Earth (= 6×10^{11} m), and (c) by Jupiter at its greatest distance from Earth (= 9×10^{11} m). (d) Is the claim correct?

3E. The Sun and Earth each exert a gravitational force on the Moon. What is the ratio $F_{\text{Sun}}/F_{\text{Earth}}$ of these two forces? (The average Sun–Moon distance is equal to the Sun–Earth distance.)

4E. One of the *Echo* satellites consisted of an inflated spherical aluminum balloon 30 m in diameter and of mass 20 kg. Suppose a meteor having a mass of 7.0 kg passes within 3.0 m of the surface of the satellite. What is the gravitational force on the meteor from the satellite at the closest approach?

5P. A mass M is split into two parts, m and $M - m$, which are then separated by a certain distance. What ratio m/M maximizes the gravitational force between the parts?

SECTION 14-3 Gravitation and the Principle of Superposition

6E. How far from Earth must a space probe be along a line toward the Sun so that the Sun's gravitational pull on the probe balances Earth's pull?

7E. A spaceship is on a straight-line path between Earth and its moon. At what distance from Earth is the net gravitational force on the spaceship zero?

8P. What is the percent change in the acceleration of Earth toward the Sun when the alignment of Earth, Sun, and Moon changes from an eclipse of the Sun (with the Moon between Earth and Sun) to an eclipse of the Moon (Earth between Moon and Sun)?

9P. Four spheres, with masses $m_1 = 400$ kg, $m_2 = 350$ kg, $m_3 = 2000$ kg, and $m_4 = 500$ kg, have (x, y) coordinates of (0, 50 cm), (0, 0), (−80 cm, 0), and (40 cm, 0), respectively. What is the net gravitational force $\mathbf{F}_2$ on m_2 due to the other masses?

10P. In Fig. 14-32a, four spheres form the corners of a square whose side is 2.0 cm long. What are the magnitude and direction of the net gravitational force from them on a central sphere with mass $m_5 = 250$ kg?

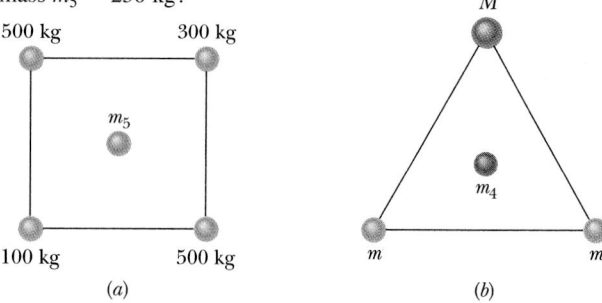

FIGURE 14-32 Problems 10 and 11.

11P. In Fig. 14-32b, two spheres of mass m and a third sphere of mass M form an equilateral triangle, and a fourth sphere of mass m_4 is at the center of the triangle. The net gravitational force on m_4 from the three other spheres is zero; what is M in terms of m?

12P. Two spheres with masses $m_1 = 800$ kg and $m_2 = 600$ kg are separated by 0.25 m. What is the net gravitational force (in both magnitude and direction) from them on a 2.0 kg sphere located 0.20 m from m_1 and 0.15 m from m_2?

13P. The masses and coordinates of three spheres are as follows: 20 kg, $x = 0.50$ m, $y = 1.0$ m; 40 kg, $x = -1.0$ m, $y = -1.0$ m; 60 kg, $x = 0$ m, $y = -0.50$ m. What is the magnitude of the gravitational force on a 20 kg sphere located at the origin due to the other spheres?

14P. Figure 14-33 shows two uniform rods of the same length L and mass M, lying on an axis with separation d. Using the result of Sample Problem 14-2, set up an integral (with integration limits and ready for integration) to find the gravitational force of one rod on the other.

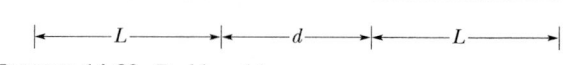

FIGURE 14-33 Problem 14.

15P. Figure 14-34 shows a spherical hollow inside a lead sphere of radius R; the surface of the hollow passes through the center of the sphere and "touches" the right side of the sphere. The mass of the sphere before hollowing was M. With what gravitational force does the hollowed-out lead sphere attract a small sphere of mass m that lies at a distance d from the center of the lead sphere, on the straight line connecting the centers of the spheres and of the hollow?

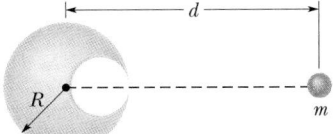

FIGURE 14-34 Problem 15.

SECTION 14-4 Gravitation Near Earth's Surface

16E. Calculate the gravitational acceleration on the surface of the Moon from values of the mass and radius of the Moon found in Appendix C.

17E. At what altitude above Earth's surface would the gravitational acceleration be 4.9 m/s²?

18E. You weigh 120 lb at sidewalk level outside the World Trade Center in New York City. Suppose that you ride from this level to the top of one of its 1350 ft towers. Ignoring Earth's rotation, how much less would you weigh there (because you are slightly farther from the center of Earth)?

19E. A typical neutron star may have a mass equal to that of the Sun but a radius of only 10 km. (a) What is the gravitational acceleration at the surface of such a star? (b) How fast would an object be moving if it fell from rest through a distance of 1.0 m on such a star? (Assume the star does not rotate.)

20E. An object lying on Earth's equator is accelerated (a) toward the center of Earth because Earth rotates, (b) toward the Sun because Earth revolves around the Sun in an almost circular orbit, and (c) toward the center of our galaxy because the Sun moves about the galactic center. For the latter, the period is 2.5×10^8 y and the radius is 2.2×10^{20} m. Calculate these three accelerations as multiples of $g = 9.8$ m/s².

21E. (a) What will an object weigh on the Moon's surface if it weighs 100 N on Earth's surface? (b) How many Earth radii must this same object be from the center of Earth if it is to weigh the same as it does on the Moon?

22P. If g is to be determined by dropping an object through a distance of exactly 10 m, what percent error in the measurement of the time of fall results in a 0.1% error in the value of g?

23P. The fastest possible rate of rotation of a planet is that for which the gravitational force on material at the equator just barely provides the centripetal force needed for the rotation. (Why?) (a) Show that the corresponding shortest period of rotation is

$$T = \sqrt{\frac{3\pi}{G\rho}},$$

where ρ is the uniform density of the spherical planet. (b) Calculate the rotation period assuming a density of 3.0 g/cm³, typical of many planets, satellites, and asteroids. No astronomical object has ever been found to be spinning with a period shorter than that determined by this analysis.

24P. In Fig. 14-35, identical masses m hang from strings on a balance at the surface of Earth. The strings have negligible mass and differ in length by h. Assume that Earth is spherical, with density $\rho = 5.5$ g/cm³. (a) Show that the difference ΔW in the weights, due to one mass being closer to Earth than the other, is $8\pi G\rho mh/3$. (b) Find the difference in length that will give a ratio $\Delta W/W = 1 \times 10^{-6}$, where W is either weight.

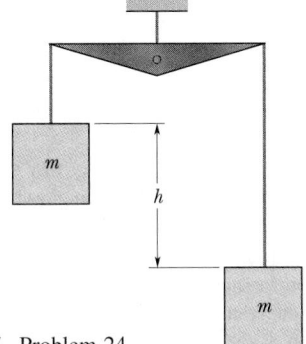

FIGURE 14-35 Problem 24.

25P. A body is suspended from a spring scale in a ship sailing along the equator with a speed v. (a) Show that the scale reading will be very close to $W_0(1 \pm 2\omega v/g)$, where ω is the angular speed of Earth and W_0 is the scale reading when the ship is at rest. (b) Explain the $\pm$ sign.

26P. Certain neutron stars (extremely dense stars) are believed to be rotating at about 1 rev/s. If such a star has a radius of 20 km, what must be its minimum mass so that material on its surface remains in place during the rapid rotation?

27P. The radius R_h and mass M_h of a black hole are related by $R_h = 2GM_h/c^2$, where c is the speed of light. Assume that the gravitational acceleration a_g of an object at a distance $r_o = 1.001R_h$ from the center of a black hole is given by Eq. 14-12 (it is, for large black holes). (a) Find an expression for a_g at r_o in terms of M_h. (b) Does a_g at r_o increase or decrease with an increase of M_h? (c) What is a_g at r_o for a very large black hole whose mass is 1.55×10^{12} times the solar mass of 1.99×10^{30} kg? (d) If the astronaut of Sample Problem 14-4 is at r_o with her feet toward this black hole, what is the difference in gravitational acceleration at head and feet? (e) Is the tendency to stretch the astronaut severe?

SECTION 14-5 Gravitation Inside Earth

28E. Two concentric shells of uniform density having masses M_1 and M_2 are situated as shown in Fig. 14-36. Find the force on a particle of mass m when the particle is located at (a) $r = a$, (b) $r = b$, and (c) $r = c$. The distance r is measured from the center of the shells.

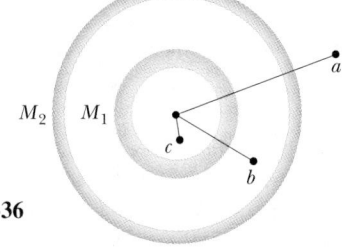

FIGURE 14-36
Exercise 28.

29E. With what speed would mail pass through the center of Earth if it were dropped down the tunnel of Sample Problem 14-5?

30E. Show that, at the bottom of a vertical mine shaft dug to depth D, the measured value of a_g will be

$$a_g = a_{gs} \left(1 - \frac{D}{R} \right),$$

a_{gs} being the surface value. Assume that Earth is a uniform sphere of radius R.

31P. A solid sphere of uniform density has a mass of 1.0×10^4 kg and a radius of 1.0 m. What is the gravitational force due to the sphere on a particle of mass m located at a distance of (a) 1.5 m and (b) 0.50 m from the center of the sphere? (c) Write a general expression for the gravitational force on m at a distance $r \le 1.0$ m from the center of the sphere.

32P. A uniform solid sphere of radius R produces a gravitational acceleration of a_g on its surface. At what two distances from the center of the sphere is the gravitational acceleration $a_g/3$? (*Hint:* Consider distances both inside and outside the sphere.)

33P. Figure 14-37 shows, not to scale, a cross section through the interior of Earth. Rather than being uniform throughout, Earth is divided into three zones: an outer *crust*, a *mantle*, and an inner *core*. The dimensions of these zones and the masses contained within them are shown on the figure. Earth has a total mass of 5.98×10^{24} kg and a radius of 6370 km. Ignore rotation and assume that Earth is spherical. (a) Calculate a_g at the surface. (b) Suppose that a bore hole (the *Mohole*) is driven to the crust–mantle interface at a depth of 25 km; what would be the value of a_g at the bottom of the hole? (c) Suppose that Earth were a uniform sphere with the same total mass and size. What would be the value of a_g at a depth of 25 km? (See Exercise 30.) (Precise measurements of a_g are sensitive probes of the interior structure of Earth, although results can be clouded by local density variations.)

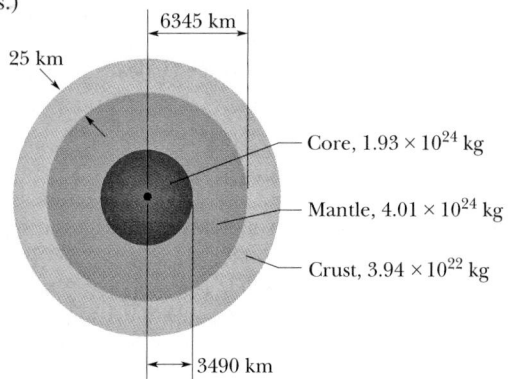

6345 km

25 km

Core, 1.93×10^{24} kg

Mantle, 4.01×10^{24} kg

Crust, 3.94×10^{22} kg

3490 km

FIGURE 14-37   Problem 33. Not to scale.

SECTION 14-6   Gravitational Potential Energy

34E. (a) What is the gravitational potential energy of the two-particle system in Exercise 1? If you triple the separation between the particles, how much work is done (b) by the gravitational force between the particles and (c) by you?

35E. (a) In Problem 9, remove m_1 and calculate the gravitational potential energy of the remaining three-particle system. (b) If m_1 is then put back in place, is the potential energy of the four-particle system more or less than that of the system in (a)? (c) In (a), is the work done by you to remove m_1 positive or negative? (d) In (b), is the work done by you to replace m_1 positive or negative?

36E. In Problem 5, what ratio m/M gives the least gravitational potential energy for the system?

37E. The mean diameters of Mars and Earth are 6.9×10^3 km and 1.3×10^4 km, respectively. The mass of Mars is 0.11 times Earth's mass. (a) What is the ratio of the mean density of Mars to that of Earth? (b) What is the value of g on Mars? (c) What is the escape speed on Mars?

38E. A spaceship is idling at the fringes of our galaxy, 80,000 light-years from the galactic center. What is the ship's escape speed from the galaxy? The mass of the galaxy is 1.4×10^{11} times that of our sun. Assume, for simplicity, that the matter forming the galaxy is distributed in a uniform sphere.

39E. Calculate the amount of energy required to escape from (a) Earth's moon and (b) Jupiter relative to that required to escape from Earth.

40E. Show that the escape speed from the Sun at Earth's distance from the Sun is $\sqrt{2}$ times the speed of the Earth in its orbit, assumed to be a circle. (This is a specific case of a general result for circular orbits: $v_{esc} = \sqrt{2}v_{orb}$.)

41E. A particle of comet dust with mass m is a distance R from Earth's center and a distance r from the Moon's center. If Earth's mass is M_E and the Moon's mass is M_m, what is the total gravitational potential energy of the particle–Earth system and the particle–Moon system?

42E. Upon "burning out," a large star can collapse under its own gravitational force to become a black hole. The star's surface then has a radius R_s such that the removal of a mass m from the surface to infinity would require work equal to the mass energy mc^2 of that mass. If M_s is the star's mass, show with Newton's law of gravitation that $R_s = GM_s/c^2$. (The correct value for R_s is actually twice this. To get it, we would have to apply Einstein's general theory of relativity rather than Newton's law.)

43P. The three spheres in Fig. 14-38, with masses $m_1 = 800$ g, $m_2 = 100$ g, and $m_3 = 200$ g, have their centers on a common line, with $L = 12$ cm and $d = 4.0$ cm. You move the middle sphere until its center-to-center separation from m_3 is $d = 4.0$ cm. How much work is done on m_2 (a) by you and (b) by the net gravitational force on m_2 due to m_1 and m_3?

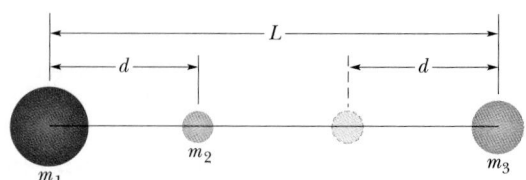

L

d

d

m_1

m_2

m_3

FIGURE 14-38   Problem 43.

44P. A rocket is accelerated to a speed of $v = 2\sqrt{gR_e}$ near Earth's surface (where Earth's radius is R_e) and then coasts upward. (a) Show that it will escape from Earth. (b) Show that very far from Earth its speed is $v = \sqrt{2gR_e}$.

45P. (a) What is the escape speed on a spherical asteroid whose radius is 500 km and whose gravitational acceleration at the surface is 3.0 m/s²? (b) How far from the surface will a particle go if it leaves the asteroid's surface with a radial speed of 1000 m/s? (c) With what speed will an object hit the asteroid if it is dropped from 1000 km above the surface?

46P. Zero, a hypothetical planet, has a mass of 5.0×10^{23} kg, a radius of 3.0×10^6 m, and no atmosphere. A 10 kg space probe is to be launched vertically from its surface. (a) If the probe is launched with an initial energy of 5.0×10^7 J, what will be its kinetic energy when it is 4.0×10^6 m from the center of Zero? (b) If the probe is to achieve a maximum distance of 8.0×10^6 m from the center of Zero, with what initial kinetic energy must it be launched from the surface of Zero?

47P. In a double-star system, two stars of mass 3.0×10^{30} kg each rotate about the system's center of mass at a radius of 1.0×10^{11} m. (a) What is their common angular speed? (b) If a meteoroid passes through this center of mass perpendicular to the orbital plane of the stars, what value must its speed exceed at that point if it is to escape to "infinity" from the star system?

48P. Two neutron stars are separated by a distance of 10^{10} m. They each have a mass of 10^{30} kg and a radius of 10^5 m. They are initially at rest with respect to each other. (a) How fast are they moving when their separation has decreased to one-half its initial value? (b) How fast are they moving just before they collide?

49P. A projectile is fired vertically from Earth's surface with an initial speed of 10 km/s. Neglecting air drag, how far above the surface of Earth will it go?

50P. A sphere of matter, of mass M and radius a, has a concentric cavity of radius b, as shown in cross section in Fig. 14-39. (a) Sketch a curve of the gravitational force F exerted by the sphere on a particle of mass m, located a distance r from the center of the sphere, as a function of r in the range $0 \le r \le \infty$. Consider $r = 0$, b, a, and ∞ in particular. (b) Sketch the corresponding curve for the potential energy $U(r)$ of the system.

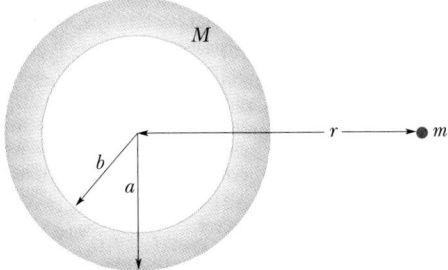

FIGURE 14-39 Problem 50.

51P. A 20 kg mass is located at the origin, and a 10 kg mass is located on the x axis at $x = 0.80$ m. The 10 kg mass is released from rest while the 20 kg mass is held in place at the origin. (a) What is the gravitational potential energy of the two-mass system immediately after the 10 kg mass is released? (b) What is the

kinetic energy of the 10 kg mass after it has moved 0.20 m toward the 20 kg mass?

52P*. Several planets (Jupiter, Saturn, Uranus) possess nearly circular surrounding rings, perhaps composed of material that failed to form a satellite. In addition, many galaxies contain ring-like structures. Consider a homogeneous ring of mass M and radius R. (a) What gravitational attraction does it exert on a particle of mass m located a distance x from the center of the ring along its axis? See Fig. 14-40. (b) Suppose the particle falls from rest as a result of the attraction of the ring of matter. Find an expression for the speed with which it passes through the center of the ring.

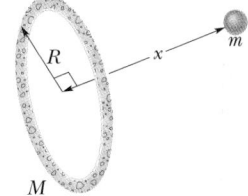

FIGURE 14-40 Problem 52.

53P*. The gravitational force between two particles with masses m and M, initially at rest at great separation, pulls them together. Show that at any instant the speed of either particle relative to the other is $\sqrt{2G(M + m)/d}$, where d is their separation at that instant. (*Hint:* Use the laws of conservation of energy and conservation of linear momentum.)

SECTION 14-7 Planets and Satellites: Kepler's Laws

54E. The mean distance of Mars from the Sun is 1.52 times that of Earth from the Sun. From Kepler's law of periods, calculate the number of years required for Mars to make one revolution about the Sun; compare your answer with the value given in Appendix C.

55E. The planet Mars has a satellite, Phobos, which travels in an orbit of radius 9.4×10^6 m with a period of 7 h 39 min. Calculate the mass of Mars from this information.

56E. Determine the mass of Earth from the period T and the radius r of the Moon's orbit about Earth: $T = 27.3$ days and $r = 3.82 \times 10^5$ km. Assume the Moon orbits the center of Earth rather than the center of mass of the Earth–Moon system.

57E. Our Sun, with mass 2.0×10^{30} kg, revolves about the center of the Milky Way galaxy, which is 2.2×10^{20} m away, once every 2.5×10^8 years. Assuming that each of the stars in the galaxy has a mass equal to that of our Sun, that the stars are distributed uniformly in a sphere about the galactic center, and that our Sun is essentially at the edge of that sphere, estimate roughly the number of stars in the galaxy.

58E. A satellite is placed in a circular orbit with a radius equal to one-half the radius of the Moon's orbit. What is its period of revolution in lunar months? (A lunar month is the period of revolution of the Moon.)

59E. (a) What linear speed must an Earth satellite have to be in a circular orbit at an altitude of 160 km? (b) What is the period of revolution?

60E. Most asteroids revolve around the Sun between Mars and Jupiter. However, several ''Apollo asteroids'' with diameters of about 30 km move in orbits that cross the orbit of Earth. The orbit of one of these asteroids is shown to scale in Fig. 14-41. By taking measurements directly from the figure, calculate the asteroid's period of revolution in years.

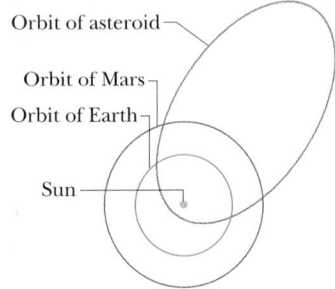

FIGURE 14-41 Exercise 60.

61E. A satellite, moving in an elliptical orbit, is 360 km above Earth's surface at its farthest point and 180 km above at its closest point. Calculate (a) the semimajor axis and (b) the eccentricity of the orbit. (*Hint:* See Sample Problem 14-8.)

62E. The Sun's center is at one focus of Earth's orbit. How far from this focus is the other focus? Express your answer in terms of the solar radius, 6.96×10^8 m. The eccentricity of Earth's orbit is 0.0167, and the semimajor axis may be taken to be 1.50×10^{11} m. See Fig. 14-14.

63E. (a) Using Kepler's third law (Eq. 14-31), express the gravitational constant G in terms of the astronomical unit as a length unit, the solar mass as a mass unit, and the year as a time unit. (One astronomical unit = 1 AU = 1.496×10^{11} m. One solar mass = $1M_s = 1.99 \times 10^{30}$ kg. One year = 1 y = 3.156×10^7 s.) (b) What form does Kepler's third law take in these units?

64E. A satellite hovers over a certain spot on the equator of (rotating) Earth. What is the altitude of its orbit (called a *geosynchronous orbit*)?

65E. A comet that was seen in April 574 by Chinese astronomers on a day known by them as the Woo Woo day was spotted again in May 1994. Assume the time between observations is the period of the Woo Woo day comet and take its eccentricity as 0.11. What are (a) the semimajor axis of the comet's orbit and (b) its greatest distance from the Sun in terms of the mean orbital radius R_P of Pluto.

66E. In 1993 the spacecraft *Galileo* sent home an image (Fig. 14-42) of asteroid 243 Ida and an orbiting tiny moon (now known as Dactyl), the first confirmed example of an asteroid–moon system. In the image, the moon, which is 1.5 km wide, is 100 km from the center of the asteroid, which is 55 km long. The shape of the moon's orbit is not well known; assume it is circular with a period of 27 h. (a) What is the mass of the asteroid? (b) The volume of the asteroid, measured from the *Galileo* images, is 14,100 km³. What is the density of the asteroid?

67P. Assume that the satellite of Exercise 64 is in orbit at the longitude of Chicago. You are in Chicago (latitude 47.5°) and

FIGURE 14-42 Exercise 66. A picture from the spacecraft *Galileo* shows a tiny moon orbiting asteroid 243 Ida.

want to pick up the signals the satellite broadcasts. In what direction should you point your antenna?

68P. In 1610, Galileo used his telescope to discover four prominent moons around Jupiter. Their mean orbital radii a and periods T are

NAME	a (10^8 m)	T (days)
Io	4.22	1.77
Europa	6.71	3.55
Ganymede	10.7	7.16
Callisto	18.8	16.7

(a) Plot log a (y axis) against log T (x axis) and show that you get a straight line. (b) Measure the slope of the line and compare it with the value that you expect from Kepler's third law. (c) Find the mass of Jupiter from the intercept of this line with the y axis.

69P. Show how, guided by Kepler's third law (Eq. 14-31), Newton could deduce that the force holding the Moon in its orbit, assumed circular, depends on the inverse square of the distance from the center of Earth.

70P. In a certain binary-star system, each star has the same mass as our Sun, and they revolve about their center of mass. The distance between them is the same as the distance between Earth and the Sun. What is their period of revolution in years?

71P. A certain triple-star system consists of two stars, each of mass m, revolving about a central star of mass M in the same circular orbit of radius r (Fig. 14-43). The two stars are always at opposite ends of a diameter of the circular orbit. Derive an expression for the period of revolution of the stars.

72P. (a) What is the escape speed from the Sun for an object in Earth's orbit (of orbital radius R) but far from Earth? (b) If an object already has a speed equal to Earth's orbital speed, what additional speed must it be given to escape as in (a)? (c) Suppose an object is launched from Earth in the direction of Earth's orbital

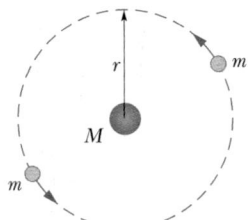

FIGURE 14-43 Problem 71.

motion. What speed must it be given during the launch so that when it is far from Earth, but still at a distance of about R from the Sun, it can escape from the Sun? (This is the speed required for an Earth-launched object to escape from the Sun.)

73P*. Three identical stars of mass M are located at the vertices of an equilateral triangle with side L. At what speed must they move if they all revolve under the influence of one another's gravitational force in a circular orbit circumscribing the triangle while still preserving the equilateral triangle?

74P*. A satellite in a circular orbit had been intended to hover over a certain spot on Earth's surface. Through error, however, the satellite's orbital radius was made 1.0 km too large for this to happen. At what rate and in what direction does the point directly below the satellite move across Earth's surface?

SECTION 14-8 Satellites: Orbits and Energy

75E. An asteroid, whose mass is 2.0×10^{-4} times the mass of Earth, revolves in a circular orbit around the Sun at a distance that is twice Earth's distance from the Sun. (a) Calculate the period of revolution of the asteroid in years. (b) What is the ratio of the kinetic energy of the asteroid to that of Earth?

76E. Consider two satellites, A and B, of equal mass m, moving in the same circular orbit of radius r around Earth, of mass M_E, but in opposite senses of rotation and therefore on a collision course (see Fig. 14-44). (a) In terms of G, M_E, m, and r, find the total mechanical energy $E_A + E_B$ of the two-satellite-plus-Earth system before collision. (b) If the collision is completely inelastic so that the wreckage remains as one piece of tangled material (mass $= 2m$), find the total mechanical energy immediately after collision. (c) Describe the subsequent motion of the wreckage.

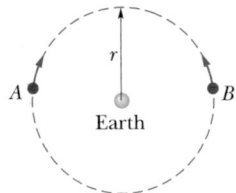

FIGURE 14-44 Exercise 76.

77P. Two Earth satellites, A and B, each of mass m, are to be launched into circular orbits about Earth's center. Satellite A is to orbit at an altitude of 4000 mi. Satellite B is to orbit at an altitude of 12,000 mi. The radius of Earth R_E is 4000 mi. (a) What is the

ratio of the potential energy of satellite B to that of satellite A, in orbit? (b) What is the ratio of the kinetic energy of satellite B to that of satellite A, in orbit? (c) Which satellite has the greater total energy if each has a mass of 14.6 kg? By how much?

78P. Use the conservation of mechanical energy and Eq. 14-44 to show that if an object is in an elliptical orbit about a planet, then its distance r from the planet and speed v are related by

$$v^2 = GM \left(\frac{2}{r} - \frac{1}{a} \right).$$

79P. Use the result of Problem 78 and data contained in Sample Problem 14-8 to calculate (a) the speed v_p of comet Halley at perihelion and (b) its speed v_a at aphelion. (c) Using conservation of angular momentum relative to the Sun, find the ratio of the comet's perihelion distance R_p to its aphelion distance R_a in terms of v_p and v_a.

80P. (a) Does it take more energy to get a satellite up to 1000 mi above Earth than to put it in circular orbit once it is there? (Take Earth's radius to be 4000 mi.) (b) What about 2000 mi? (c) What about 3000 mi?

81P. One way to attack a satellite in Earth orbit is to launch a swarm of pellets in the same orbit as the satellite but in the opposite direction. Suppose a satellite in a circular orbit 500 km above Earth's surface collides with a pellet having mass 4.0 g. (a) What is the kinetic energy of the pellet in the reference frame of the satellite? (b) What is the ratio of this kinetic energy to the kinetic energy of a 4.0 g bullet from a modern army rifle with a muzzle velocity of 950 m/s?

82P. Consider a satellite in a circular orbit about Earth. State how the following properties of the satellite depend on the radius r of its orbit: (a) period, (b) kinetic energy, (c) angular momentum, and (d) speed.

83P. What are (a) the speed and (b) the period of a 220 kg satellite in an approximately circular orbit 640 km above the surface of Earth? Suppose the satellite loses mechanical energy at the average rate of 1.4×10^5 J per orbital revolution. Adopting the reasonable approximation that the trajectory is a "circle of slowly diminishing radius," determine the satellite's (c) altitude, (d) speed, and (e) period at the end of its 1500th revolution. (f) What is the magnitude of the average retarding force? (g) Is angular momentum around Earth's center conserved for the satellite or the satellite–Earth system?

84P. The orbit of Earth about the Sun is *almost* circular: the closest and farthest distances are 1.47×10^8 km and 1.52×10^8 km, respectively. Determine the corresponding variations in (a) total energy, (b) potential energy, (c) kinetic energy, and (d) orbital speed. (*Hint:* Use the laws of conservation of energy and conservation of angular momentum.)

85P. In a shuttle craft of mass $m = 2000$ kg, Captain Janeway orbits a planet of mass $M = 5.98 \times 10^{24}$ kg, in a circular orbit of radius $r = 6.80 \times 10^6$ m. What are (a) the period of the orbit and (b) the speed of the shuttle craft? Janeway briefly fires a forward-pointing thruster, reducing her speed by 1.00%. Just then, what are (c) the speed, (d) the kinetic energy, (e) the gravitational potential energy, and (f) the mechanical energy of the shuttle craft? (g) What is the semimajor axis of the elliptical orbit now taken by the craft? (h) What is the difference between the period of the original circular orbit and that of the new elliptical orbit, and which orbit has the smaller period?

SECTION 14-9 Einstein and Gravitation

86E. In Fig. 14-20b, the scale on which the 60 kg physicist stands reads 220 N. How long will the cantaloupe take to reach the floor if the physicist drops it from rest (relative to himself), 2.1 m from the floor?

87P. Figure 14-45 shows the walls of a duct in a ship in deep space; the ship has an acceleration $\mathbf{a} = (2.5 \text{ m/s}^2)\mathbf{i}$. An electron is sent across the duct's width of 3.0 cm, from the origin shown and with an initial velocity of $\mathbf{v}_0 = (0.40 \text{ m/s})\mathbf{j}$. In unit vector notation and relative to the ship, what are (a) the displacement of the electron at the end of its flight and (b) its velocity just before impact with the far wall?

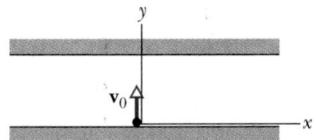

FIGURE 14-45 Problem 87.

Electronic Computation

88. A 6000 kg planetary probe is in a circular solar orbit with a radius of 108×10^6 km (the orbital radius of Venus). The space agency wants to put the probe into the same orbit as Earth, a circular orbit with a radius of 150×10^6 km. The first step is to increase the probe's speed so that the probe is in an elliptical orbit with a perihelion distance equal to the radius of its initial orbit and an aphelion distance equal to the radius of its final orbit. The mass of the Sun is 1.99×10^{30} kg. (a) Calculate the increases in speed and energy that are required. Plot the elliptical orbit. (In polar coordinates, with the origin at the center, the equation for an ellipse is

$$r^2 = \frac{r_p r_a (r_p + r_a)^2}{(r_p + r_a)^2 \sin^2 \theta + 4 r_p r_a \cos^2 \theta},$$

where r_p is the perihelion distance and r_a is the aphelion distance.) (b) When the aphelion position is reached, the speed of the probe is again changed to put it into the desired circular orbit. What changes in the speed and energy are required?

89. In your calculator, make a list of the periods T and a separate list for the semimajor axes a for the planets in Table 14-3. Multiply the list for T by an appropriate factor so that the unit of T is seconds. (a) Store values of T^2 and a^3 in two new lists. Have the calculator do a linear regression fit of T^2 versus a^3. From the parameters of the fit and using the known value of G, determine the mass of the Sun. (b) Store values of $\log T$ in a list and $\log a$ in another list. Have the calculator plot $\log T$ versus $\log a$ and then do a linear regression fit of the plot. From the parameters of this fit and using the known value of G, again determine the mass of the Sun.

15
Fluids

The force exerted by water on the body of a descending diver increases noticeably, even for a relatively shallow descent to the bottom of a swimming pool. However, in 1975, using scuba gear with a special gas mixture for breathing, William Rhodes emerged from a chamber that had been lowered 1000 ft into the Gulf of Mexico, and he then swam to a record depth of 1148 ft. Strangely, a novice scuba diver practicing in a swimming pool might be in more danger from the force exerted by the water than was Rhodes. And occasionally, novice scuba divers die because they have neglected that danger. What is this potentially lethal risk?

15-1 FLUIDS AND THE WORLD AROUND US

Fluids—which include both liquids and gases—play a central role in our daily lives. We breathe and drink them, and a rather vital fluid circulates in the human cardiovascular system. There is the fluid ocean and the fluid atmosphere.

In a car, there are fluids in the tires, the gas tank, the radiator, the combustion chambers of the engine, the exhaust manifold, the battery, the air conditioning system, the windshield wiper reservoir, the lubrication system, and the hydraulic system. (*Hydraulic* means operated via a liquid.) The next time you see a large piece of earthmoving machinery, count the hydraulic cylinders that permit the machine to do its work. Large jet planes have scores of them.

We use the kinetic energy of a moving fluid in windmills, and the potential energy of another fluid in hydroelectric power plants. Given time, fluids carve the landscape. We often travel great distances just to watch fluids move. Perhaps it is time to see what physics can tell us about fluids.

15-2 WHAT IS A FLUID?

A **fluid,** in contrast to a solid, is a substance that can flow. Fluids conform to the boundaries of any container in which we put them. They do so because a fluid cannot sustain a force that is tangential to its surface. (In the more formal language of Section 13-6, a fluid is a substance that flows because it cannot withstand a shearing stress. It can, however, exert a force in the direction perpendicular to its surface.) Some materials, such as pitch, take a long time to conform to the boundaries of a container, but they do so eventually; thus we classify them as fluids.

You may wonder why we lump liquids and gases together and call them fluids. After all (you may say), liquid water is as different from steam as it is from ice. Actually, it is not. Ice, like other crystalline solids, has its constituent atoms organized in a fairly rigid three-dimensional array called a crystalline lattice. In neither steam nor liquid water, however, is there any such orderly long-range arrangement.

15-3 DENSITY AND PRESSURE

When we discuss rigid bodies, we are concerned with particular lumps of matter, such as wooden blocks, baseballs, or metal rods. Physical quantities that we find useful, and in whose terms we express Newton's laws, are *mass* and *force*. We might speak, for example, of a 3.6 kg block acted on by a 25 N force.

With fluids, we are more interested in properties that vary from point to point in the extended substance than with properties of specific lumps of that substance. It is more useful to speak of **density** and **pressure** than of mass and force.

Density

To find the density ρ (rho) of a fluid at any point, we isolate a small volume element ΔV around that point and measure the mass Δm of the fluid contained within that element. The **density** is then

$$\rho = \frac{\Delta m}{\Delta V}. \qquad (15\text{-}1)$$

In theory, the density at any point in a fluid is the limit of this ratio as the volume element ΔV at that point is made smaller and smaller. In practice, we assume that a fluid sample is large compared to atomic dimensions and thus "smooth" (with uniform density), rather than "lumpy" with atoms. This assumption allows us to write Eq. 15-1 in the form $\rho = m/V$, where m and V are the mass and volume of the sample.

Density is a scalar property; its SI unit is the kilogram per cubic meter. Table 15-1 shows the densities of some

TABLE 15-1 SOME DENSITIES

MATERIAL OR OBJECT	DENSITY (kg/m^3)
Interstellar space	10^{-20}
Best laboratory vacuum	10^{-17}
Air: 20°C and 1 atm	1.21
20°C and 50 atm	60.5
Styrofoam	1×10^2
Water: 20°C and 1 atm	0.998×10^3
20°C and 50 atm	1.000×10^3
Seawater: 20°C and 1 atm	1.024×10^3
Whole blood	1.060×10^3
Ice	0.917×10^3
Iron	7.9×10^3
Mercury	13.6×10^3
Earth: average	5.5×10^3
core	9.5×10^3
crust	2.8×10^3
Sun: average	1.4×10^3
core	1.6×10^5
White dwarf star (core)	10^{10}
Uranium nucleus	3×10^{17}
Neutron star (core)	10^{18}
Black hole (1 solar mass)	10^{19}

substances and the average densities of some objects. Note that the density of a gas (see Air in the table) varies considerably with pressure, but the density of a liquid (see Water) does not. That is, gases are readily *compressible* but liquids are not.

Pressure

Let a small pressure-sensing device be suspended inside a fluid-filled vessel, as in Fig. 15-1a. The sensor (Fig. 15-1b) consists of a piston of area ΔA riding in a close-fitting cylinder and resting against a spring. A readout arrangement allows us to record the amount by which the (calibrated) spring is compressed by the surrounding fluid, thus indicating the magnitude ΔF of the force that acts on the piston. We define the **pressure** exerted by the fluid on the piston as

$$p = \frac{\Delta F}{\Delta A}. \tag{15-2}$$

In theory, the pressure at any point in the fluid is the limit of this ratio as the area ΔA of the piston, centered on that point, is made smaller and smaller. However, if the force is uniform over a flat area A, we can write Eq. 15-2 as $p = F/A$. (When we say a force is uniform over an area, we mean that the same force would be measured at every point of the area.)

We find by experiment that at a given point in a fluid at rest, the pressure p defined by Eq. 15-2 has the same value no matter how the pressure sensor is oriented. Pressure is a scalar, having no directional properties. It is true that the force acting on the piston of our pressure sensor is a vector, but Eq. 15-2 involves only the *magnitude* of that force, a scalar quantity.

The SI unit of pressure is the newton per square meter, which is given a special name, the **pascal** (Pa). In metric countries, tire pressure gauges are calibrated in kilopascals. The pascal is related to some other common (non-SI) pressure units as follows:

$$1 \text{ atm} = 1.01 \times 10^5 \text{ Pa} = 760 \text{ torr} = 14.7 \text{ lb/in.}^2.$$

The *atmosphere* (atm) is, as the name suggests, the approximate average pressure of the atmosphere at sea level. The *torr* (named for Evangelista Torricelli, who invented the mercury barometer in 1674) was formerly called the *millimeter of mercury* (mm Hg). The pound per square inch is often abbreviated psi. Table 15-2 shows some pressures.

TABLE 15-2 SOME PRESSURES

	PRESSURE (Pa)
Center of the Sun	2×10^{16}
Center of Earth	4×10^{11}
Highest sustained laboratory pressure	1.5×10^{10}
Deepest ocean trench (bottom)	1.1×10^8
Spike heels on a dance floor	1×10^6
Automobile tire[a]	2×10^5
Atmosphere at sea level	1.0×10^5
Normal blood pressure[a,b]	1.6×10^4
Best laboratory vacuum	10^{-12}

[a] Pressure in excess of atmospheric pressure.

[b] The systolic pressure, corresponding to 120 torr on the physician's pressure gauge.

SAMPLE PROBLEM 15-1

A living room has floor dimensions of 3.5 m and 4.2 m and a height of 2.4 m.

(a) What does the air in the room weigh?

SOLUTION: If V is the volume of the room and ρ is the density of air at 1 atm (see Table 15-1), then the weight W of the air is

$$W = mg = \rho V g$$
$$= (1.21 \text{ kg/m}^3)(3.5 \text{ m} \times 4.2 \text{ m} \times 2.4 \text{ m})(9.8 \text{ m/s}^2)$$
$$= 418 \text{ N} \approx 420 \text{ N}. \quad \text{(Answer)}$$

This is about 94 lb.

(b) What force does the atmosphere exert on the floor of the room?

SOLUTION: The force is

$$F = pA \doteq (1.0 \text{ atm}) \left(\frac{1.01 \times 10^5 \text{ N/m}^2}{1 \text{ atm}} \right) (3.5 \text{ m})(4.2 \text{ m})$$
$$= 1.5 \times 10^6 \text{ N}. \quad \text{(Answer)}$$

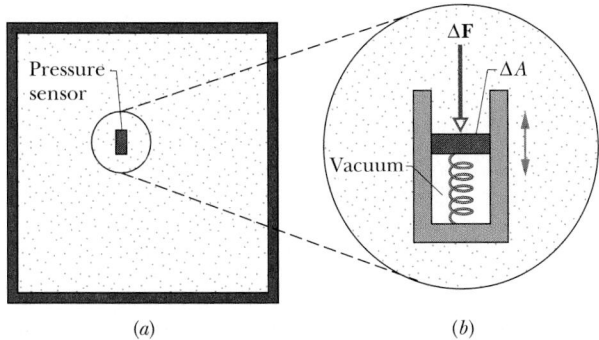

(a)   (b)

FIGURE 15-1 (a) A fluid-filled vessel containing a small pressure sensor, shown in (b). The pressure is measured by the relative position of the movable piston in the sensor.

This force ($\approx$ 170 tons) is the weight of a column of air covering the floor and extending all the way to the top of the atmosphere. It is equal to the force that would be exerted on the floor if (in the absence of the atmosphere) the room were filled with mercury to a depth of 30 in. Why doesn't this enormous force break the floor?

15-4 FLUIDS AT REST

Figure 15-2a shows a tank of water—or other liquid—open to the atmosphere. As every diver knows, the pressure *increases* with depth below the air–water interface. The diver's depth gauge, in fact, is a pressure sensor much like that of Fig. 15-1b. As every mountaineer knows, the pressure *decreases* with altitude as one ascends into the atmosphere. The pressures encountered by the diver and the mountaineer are usually called *hydrostatic pressures,* because they are due to fluids that are static (at rest).

Let us look first at the increase in pressure with depth below the water's surface. We set up a vertical y axis, its origin being at the air–water interface and the direction of increasing y being up. Consider a water sample contained in an imaginary right circular cylinder of horizontal base area A, and let y_1 and y_2 (both of which are *negative* numbers) be the depths below the surface of the upper and the lower cylinder faces, respectively.

Figure 15-2b shows a free-body diagram for the water in the cylinder. The water sample is in equilibrium, its weight (downward) being exactly balanced by the difference between the force of magnitude $F_2 = p_2A$ acting upward on its lower face and the force of magnitude $F_1 = p_1A$ acting downward on its upper face. Thus

$$F_2 = F_1 + W. \qquad (15\text{-}3)$$

The volume V of the cylinder is $A(y_1 - y_2)$. Thus the mass m of the water in the cylinder is $\rho A(y_1 - y_2)$, in which ρ is the uniform density of water. The weight W is then $mg = \rho Ag(y_1 - y_2)$. Substituting this for W in Eq. 15-3 and substituting for F_1 and F_2 yield

$$p_2A = p_1A + \rho Ag(y_1 - y_2),$$

or

$$p_2 = p_1 + \rho g(y_1 - y_2). \qquad (15\text{-}4)$$

This equation can be used to find pressures both in a liquid and in the atmosphere. For the former, suppose we seek the pressure p at a depth h below the liquid surface. Then we choose level 1 to be the surface, level 2 to be a distance h below it (as in Fig. 15-3), and p_0 to represent the atmospheric pressure on the surface. We then substitute

$$y_1 = 0, \quad p_1 = p_0 \quad \text{and} \quad y_2 = -h, \quad p_2 = p$$

into Eq. 15-4, which becomes

$$p = p_0 + \rho gh \qquad \text{(pressure at depth h)}. \quad (15\text{-}5)$$

Note that the pressure at a given depth depends on that depth but not on any horizontal dimension. Thus Eq. 15-5 holds no matter what the shape of the containing vessel. If the bottom surface of the vessel is at depth h, then Eq. 15-5 gives the pressure p on that surface.

In Eq. 15-5, p is said to be the total pressure or **absolute pressure** at level 2. To see why, note in Fig. 15-3 that the pressure p at level 2 consists of two contributions: (1) p_0, the pressure due to the atmosphere, which bears down on the liquid, and (2) ρgh, the pressure due to the liquid above level 2, which bears down on level 2. In general, the difference between an absolute pressure and an atmospheric pressure is called the **gauge pressure.** So, for the situation of Fig. 15-3, the gauge pressure is ρgh.

Equation 15-4 also holds above the liquid surface: it gives the atmospheric pressure at a given distance above level 1 in terms of the atmospheric pressure p_1 at level 1

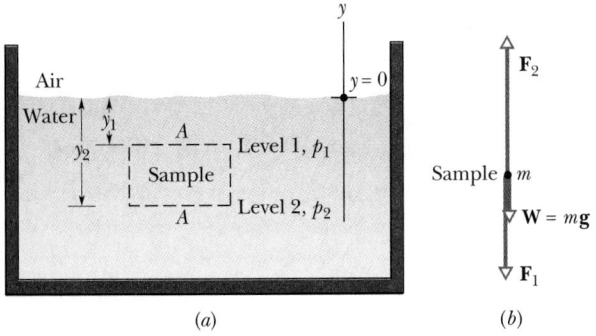

(a)   (b)

FIGURE 15-2 (a) A tank of water in which a sample of water is contained in an imaginary cylinder of horizontal base area A. (b) A free-body diagram of the water sample. The water in the sample is in static equilibrium, its weight being balanced by the net upward buoyant force that acts on it.

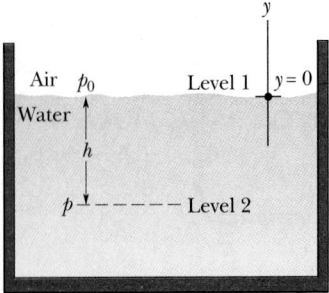

FIGURE 15-3 The pressure p increases with depth h below the water surface according to Eq. 15-5.

(*assuming* that the atmospheric density is uniform over that distance). For example, to find the atmospheric pressure at a distance d above level 1 in Fig. 15-3, we substitute

$$y_1 = 0, \quad p_1 = p_0 \quad \text{and} \quad y_2 = d, \quad p_2 = p.$$

Then with $\rho = \rho_{\text{air}}$, we obtain

$$p = p_0 - \rho_{\text{air}} g d.$$

CHECKPOINT **1:** The figure shows four containers of olive oil. Rank them according to the pressure at depth h, greatest first.

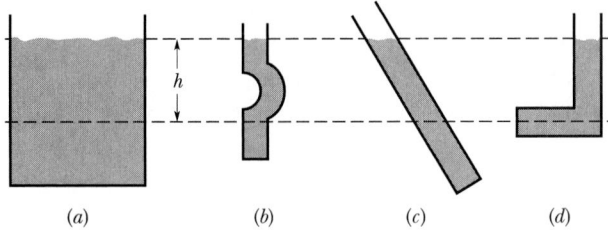

(a)   (b)   (c)   (d)

SAMPLE PROBLEM 15-2

(a) An enterprising diver reasons that if a typical snorkel tube, which is 20 cm long, works, a tube 6.0 m long should also work. If he foolishly uses such a tube (Fig. 15-4), what is the pressure difference Δp between the external pressure on him and the air pressure in his lungs? Why is he in danger?

SOLUTION: First consider the diver at depth $L = 6.0$ m without the snorkel tube. The external pressure on him is given by Eq. 15-5 as

$$p = p_0 + \rho g L.$$

His body adjusts to that pressure by contracting slightly until the internal pressures are in equilibrium with the external pressure. In particular, the average blood pressure increases, and the average air pressure in his lungs increases until it equals the external pressure p.

If he then foolishly uses the 6.0 m tube to breathe, the pressurized air in his lungs will be expelled upward through the tube to the atmosphere, and the air pressure in his lungs will rapidly drop to atmospheric pressure p_0. Assuming he is in fresh water of density 1000 kg/m^3, the pressure difference Δp between the lower pressure within his lungs and the higher pressure on his chest will then be

$$\begin{aligned}
\Delta p = p - p_0 &= \rho g L \\
&= (1000 \text{ kg/m}^3)(9.8 \text{ m/s}^2)(6.0 \text{ m}) \\
&= 5.9 \times 10^4 \text{ Pa}. \quad \text{(Answer)}
\end{aligned}$$

This pressure difference, about 0.6 atm, is sufficient to collapse the lungs and force the still-pressurized blood into them, a process known as lung squeeze.

(b) A novice scuba diver practicing in a swimming pool takes enough air from his tank to fully expand his lungs before abandoning the tank at depth L and swimming to the surface. He ignores instructions and fails to exhale during his ascent. When he reaches the surface, the pressure difference between the external pressure on him and the air in his lungs is 70 torr. From what depth did he start? What potentially lethal danger does he face?

SOLUTION: When he fills his lungs at depth L, the external pressure on him (and the air pressure within his lungs) is again given by Eq. 15-5 as

$$p = p_0 + \rho g L.$$

As he ascends, the external pressure on him decreases, until it is atmospheric pressure p_0 at the surface. His blood pressure also decreases, until it is normal. But because he does not exhale, the air pressure in his lungs remains at the value it had at depth L. So, at the surface, the pressure difference between the higher pressure in his lungs and the lower pressure on his chest is given by

$$\Delta p = p - p_0 = \rho g L,$$

from which we find

$$\begin{aligned}
L &= \frac{\Delta p}{\rho g} \\
&= \frac{70 \text{ torr}}{(1000 \text{ kg/m}^3)(9.8 \text{ m/s}^2)} \left(\frac{1.01 \times 10^5 \text{ Pa}}{760 \text{ torr}} \right) \\
&= 0.95 \text{ m}. \quad \text{(Answer)}
\end{aligned}$$

The pressure difference of 70 torr (about 9% of atmospheric pressure) is sufficient to rupture the diver's lungs and force air from them into the depressurized blood, which then carries the air to the heart, killing the diver. If the diver follows instructions and gradually exhales as he ascends, he allows the pressure in his lungs to equalize with the external pressure, and then there is no danger.

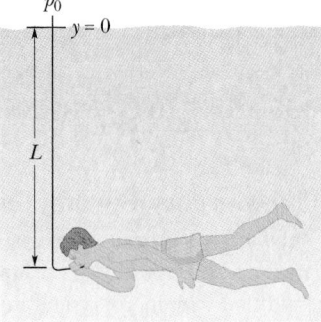

FIGURE 15-4 Sample Problem 15-2. DON'T TRY THIS with a tube that is longer than a standard snorkel tube or the attempt to breathe through the tube could kill you. The reason is that, because the external (water) pressure on your chest can be so much greater than the internal (air) pressure, you might not be able to expand your lungs to inhale.

SAMPLE PROBLEM 15-3

The U-tube in Fig. 15-5 contains two liquids in static equilibrium: water of density ρ_w is in the right arm, and oil of unknown density ρ_x is in the left. Measurement gives $l = 135$ mm and $d = 12.3$ mm. What is the density of the oil?

SOLUTION: If the pressure at the oil–water interface in the left arm is p_{int}, then the pressure in the right arm at the level of that interface must also be p_{int}, because the left and right arms are connected by water below the level of the interface. In the right arm, the interface is a distance l below the free surface of the *water* and we have, from Eq. 15-5,

$$p_{int} = p_0 + \rho_w gl \quad \text{(right arm)}.$$

In the left arm, the interface is a distance $l + d$ below the free surface of the *oil* and we have, again from Eq. 15-5,

$$p_{int} = p_0 + \rho_x g(l + d) \quad \text{(left arm)}.$$

Equating these two expressions and solving for the unknown density yield

$$\rho_x = \rho_w \frac{l}{l + d} = (1000 \text{ kg/m}^3) \frac{135 \text{ mm}}{135 \text{ mm} + 12.3 \text{ mm}}$$

$$= 916 \text{ kg/m}^3. \quad \text{(Answer)}$$

Note that the answer does not depend on the atmospheric pressure p_0 or the free-fall acceleration g.

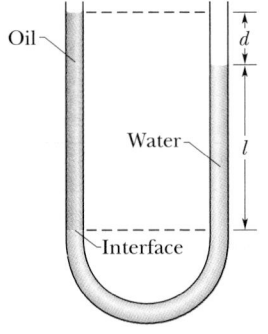

FIGURE 15-5 Sample Problem 15-3. The oil in the left arm stands higher than the water in the right arm because the oil is less dense than the water. Both fluid columns produce the same pressure P_{int} at the level of the interface.

15-5 MEASURING PRESSURE

The Mercury Barometer

Figure 15-6a shows a very basic *mercury barometer,* a device used to measure the pressure of the atmosphere. The long glass tube is filled with mercury and inverted with its open end in a dish of mercury, as the figure shows. The space above the mercury column contains only mercury vapor, whose pressure is so small at ordinary temperatures that it can be neglected.

We can use Eq. 15-4 to find the atmospheric pressure p_0 in terms of the height h of the mercury column. We choose level 1 of Fig. 15-2 to be that of the air–mercury

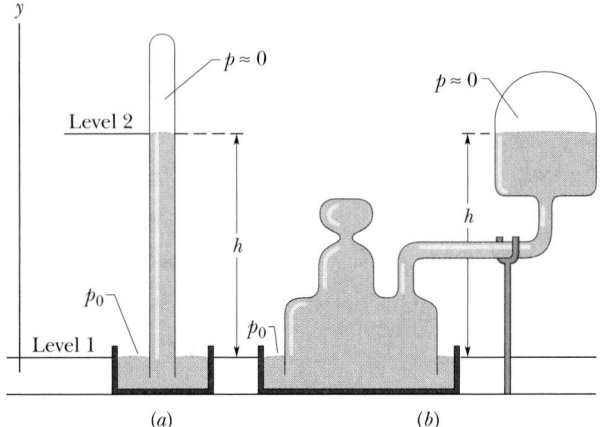

FIGURE 15-6 (a) A mercury barometer. (b) Another mercury barometer. The distance h is the same in both cases.

interface and level 2 to be that of the top of the mercury column, as labeled in Fig. 15-6a. We then substitute

$$y_1 = 0, \quad p_1 = p_0 \quad \text{and} \quad y_2 = h, \quad p_2 = 0$$

into Eq. 15-4, finding that

$$p_0 = \rho gh, \quad (15\text{-}6)$$

where ρ is the density of the mercury.

For a given pressure, the height h of the mercury column does not depend on the cross-sectional area of the vertical tube. The fanciful mercury barometer of Fig. 15-6b gives the same reading as that of Fig. 15-6a; all that counts is the vertical distance h between the mercury levels.

Equation 15-6 shows that, for a given pressure, the height of the column of mercury depends on the value of g at the location of the barometer and on the density of mercury, which varies with temperature. The column height (in millimeters) is numerically equal to the pressure (in torr) *only* if the barometer is at a place where g has its accepted standard value of 9.80665 m/s² *and* the temperature of the mercury is 0°C. If these conditions do not prevail (and they rarely do), small corrections must be made before the height of the mercury column can be transformed into a pressure.

The Open-Tube Manometer

An *open-tube manometer* (Fig. 15-7) measures the gauge pressure of a gas. It consists of a U-tube containing a liquid, with one end of the tube connected to the vessel whose gauge pressure we wish to measure and the other end open to the atmosphere. We can use Eq. 15-4 to find the gauge pressure in terms of the height h shown in Fig. 15-7. Let us choose levels 1 and 2 as shown in Fig. 15-7. We then

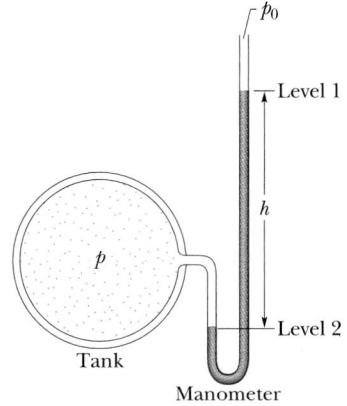

FIGURE 15-7 An open-tube manometer, connected so that the gauge pressure of the gas in the tank on the left is read. The right arm of the U-tube is open to the atmosphere.

substitute

$$y_1 = 0, \quad p_1 = p_0 \quad \text{and} \quad y_2 = -h, \quad p_2 = p$$

into Eq. 15-4, finding that

$$p_g = p - p_0 = \rho gh. \tag{15-7}$$

The gauge pressure p_g is directly proportional to h.

The gauge pressure can be positive or negative, depending on whether $p > p_0$ or $p < p_0$. In inflated tires or the human circulatory system, the (absolute) pressure is greater than atmospheric pressure, so the gauge pressure is a positive quantity, sometimes called the *overpressure*. If you suck on a straw to pull fluid up the straw, the (absolute) pressure in your lungs is actually less than atmospheric pressure. The gauge pressure in your lungs is then a negative quantity.

SAMPLE PROBLEM 15-4

The column in a mercury barometer has a measured height h of 740.35 mm. The temperature is $-5.0°C$, at which temperature the density of mercury is 1.3608×10^4 kg/m³. The free-fall acceleration g at the site of the barometer is 9.7835 m/s². What is the atmospheric pressure in pascals and in torr?

SOLUTION: From Eq. 15-6 we have

$$p_0 = \rho gh$$
$$= (1.3608 \times 10^4 \text{ kg/m}^3)(9.7835 \text{ m/s}^2)(0.74035 \text{ m})$$
$$= 9.8566 \times 10^4 \text{ Pa.} \quad \text{(Answer)}$$

Barometer readings are usually expressed in torr, where 1 torr is the pressure exerted by a column of mercury 1 mm high at a place where g has an accepted standard value of 9.80665 m/s² and at a temperature (0.0°C) at which mercury has a density of 1.35955×10^4 kg/m³. Thus from Eq. 15-6,

$$1 \text{ torr} = (1.35955 \times 10^4 \text{ kg/m}^3)(9.80665 \text{ m/s}^2)$$
$$\times (1 \times 10^{-3} \text{ m})$$
$$= 133.326 \text{ Pa.}$$

Applying this conversion factor yields, for the atmospheric pressure recorded on the barometer,

$$p_0 = 9.8566 \times 10^4 \text{ Pa} = 739.29 \text{ torr.} \quad \text{(Answer)}$$

Note that the pressure in torr (739.29 torr) is numerically close to—but otherwise differs significantly from—the height h of the mercury column expressed in mm (740.35 mm).

15-6 PASCAL'S PRINCIPLE

When you squeeze one end of a tube of toothpaste, you are watching **Pascal's principle** in action. This principle is also the basis for the Heimlich maneuver, in which a sharp pressure increase properly applied to the abdomen is transmitted to the throat, forcefully ejecting food lodged there. The principle was first stated clearly in 1652 by Blaise Pascal (for whom the unit of pressure is named):

A change in the pressure applied to an enclosed incompressible fluid is transmitted undiminished to every portion of the fluid and to the walls of the containing vessel.

Demonstrating Pascal's Principle

Consider the case in which the fluid is an incompressible liquid contained in a tall cylinder, as in Fig. 15-8. The cylinder is fitted with a piston on which a container of lead shot rests. The atmosphere, container, and shot exert pressure p_{ext} on the piston and thus on the liquid. The pressure p at any point P in the liquid is then

$$p = p_{ext} + \rho gh. \tag{15-8}$$

Let us add a little more lead shot to the container to increase p_{ext} by an amount Δp_{ext}. The quantities ρ, g, and h in Eq. 15-8 are unchanged, so the pressure change at P is

$$\Delta p = \Delta p_{ext}. \tag{15-9}$$

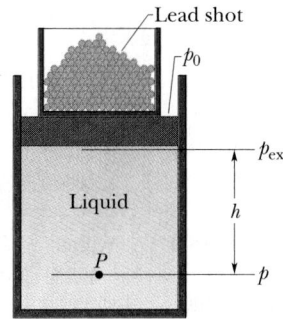

FIGURE 15-8 Weights loaded onto the piston create a pressure p_{ext} at the top of the enclosed (incompressible) liquid. If p_{ext} is increased, by adding more weights, the pressure increases by the same amount at all points within the liquid.

This pressure change is independent of h, so it must hold for all points within the liquid, as Pascal's principle states.

Pascal's Principle and the Hydraulic Lever

Figure 15-9 shows how Pascal's principle can be made the basis of a hydraulic lever. In operation, let an external force of magnitude F_i be exerted downward on the left-hand (or input) piston, whose area is A_i. An incompressible liquid in the device then exerts an upward force of magnitude F_o on the righthand (or output) piston, whose area is A_o. To keep the system in equilibrium, an external load (not shown) must exert a downward force of magnitude F_o on the output piston. The force F_i applied on the left and the downward force F_o exerted by the load on the right produce a change Δp in the pressure of the liquid that is given by

$$\Delta p = \frac{F_i}{A_i} = \frac{F_o}{A_o},$$

so

$$F_o = F_i \frac{A_o}{A_i}. \quad (15\text{-}10)$$

Equation 15-10 shows that the output force F_o exerted on the load must be larger than the input force F_i if $A_o > A_i$, as is the case in Fig. 15-9.

If we move the input piston downward a distance d_i, the output piston moves upward a distance d_o, such that the same volume V of the incompressible liquid is displaced at both pistons. Then

$$V = A_i d_i = A_o d_o,$$

which we can write as

$$d_o = d_i \frac{A_i}{A_o}. \quad (15\text{-}11)$$

This shows that, if $A_o > A_i$ (as in Fig. 15-9), the output piston moves a smaller distance than the input piston moves.

From Eqs. 15-10 and 15-11 we can write the output work as

$$W = F_o d_o = \left(F_i \frac{A_o}{A_i} \right) \left(d_i \frac{A_i}{A_o} \right) = F_i d_i, \quad (15\text{-}12)$$

which shows that the work W done *on* the input piston by the applied force is equal to the work W done *by* the output piston in lifting the load placed on it.

We see here that a given force exerted over a given distance can be transformed to a larger force exerted over a smaller distance. The product of force and distance remains unchanged so that the same work is done. However, there is often tremendous advantage in being able to exert the larger force. Most of us, for example, cannot lift an automobile and thus welcome the availability of a hydraulic jack, even though we have to pump the handle farther than the automobile rises. In this device, the displacement d_i is accomplished not in a single stroke but over a series of small strokes.

15-7 ARCHIMEDES' PRINCIPLE

Figure 15-10 shows a student in a swimming pool, manipulating a very thin plastic sack filled with water. She will find that it is in static equilibrium, tending neither to rise nor to sink. Yet the water in the sack has weight and—for that reason—should sink. The weight must be balanced by an upward force whose magnitude is equal to the weight of the water in the sack.

This upward **buoyant force** $\mathbf{F}_b$ is exerted on the water in the sack by the water that surrounds the sack. This buoyant force exists because—as you have already seen—the pressure in the water increases with depth below the surface, so the pressure near the bottom of the sack is greater than the pressure near the top.

Let us now remove the sack of water. Figure 15-11a shows the forces acting at the hole in the water formerly occupied by the sack. The buoyant force $\mathbf{F}_b$, which points up, is the vector sum of all these forces.

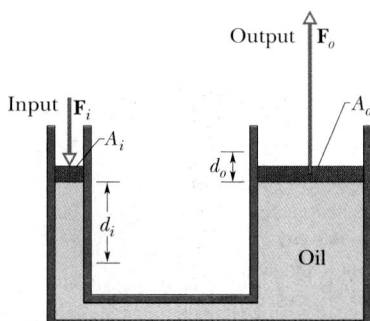

FIGURE 15-9 A hydraulic arrangement, used to magnify a force $\mathbf{F}_i$. The work done by $\mathbf{F}_i$, however, is not magnified and is the same for both the input and output forces.

FIGURE 15-10 A thin-walled plastic sack of water is in static equilibrium in the pool. Its weight must be balanced by a net upward force exerted on the sack by the water surrounding it.

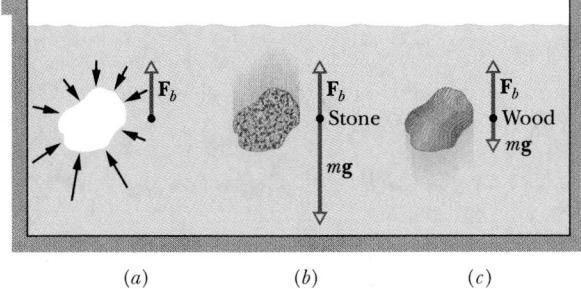

FIGURE 15-11 (*a*) The water surrounding the hole in the water exerts a resultant upward buoyant force on whatever fills the hole. (*b*) For a stone of the same volume as the hole, the weight exceeds the buoyant force. (*c*) For a lump of wood of the same volume, the weight is less than the buoyant force.

Let us fill the hole in Fig. 15-11*a* with a stone of exactly the same dimensions, as in Fig. 15-11*b*. *The same upward buoyant force that acted on the water-filled sack will act on the stone.* However, this force is too small to balance the weight of the stone, so the stone will sink. Even though the stone sinks, the water's buoyant force reduces its apparent weight, making the stone easier to lift as long as it is under water.

If we now fill the hole of Fig. 15-11*a* with a block of wood of the same dimensions, as in Fig. 15-11*c*, the same upward buoyant force acts on the wood. This time, however, the upward force is *greater* than the weight of the wood, and the wood will rise toward the surface. We summarize these facts with **Archimedes' principle**:

A body fully or partially immersed in a fluid is buoyed up by a force equal to the weight of the fluid that the body displaces.

It is this principle that explains floating. When, say, the piece of wood in Fig. 15-11*c* rises enough to break through the water's surface, it displaces less water than when it was submerged. According to Archimedes' principle, its buoyant force decreases. The wood continues to rise out of the water until the buoyant force acting on it has decreased to exactly the weight of the wood. The wood is then in static equilibrium; it is floating.

Recall that we say the weight of an object acts effectively at the object's center of mass. Similarly, we can say the buoyant force on a fully or partially immersed object acts effectively at a point called the **center of buoyancy.** This point is located where the center of mass of the displaced fluid would have been, had it not been displaced. If a uniformly dense object is fully submerged, then the center of buoyancy and the object's center of mass coincide. If the object is only partially submerged, those points are separated; then the buoyant force can produce a torque on the object about its center of mass.

CHECKPOINT **2:** A penguin floats first in a fluid of density ρ_0, then in a fluid of density $0.95\rho_0$, and then in a fluid of density $1.1\rho_0$. (a) Rank the densities according to the buoyant force on the penguin, greatest first. (b) Rank the densities according to the amount of fluid displaced by the penguin, greatest first.

SAMPLE PROBLEM 15-5

The "tip of the iceberg" in popular speech has come to mean a small visible fraction of something that is mostly hidden. For real icebergs, what is this fraction?

In the late evening of August 21, 1986, something (possibly a volcanic tremor) disturbed Cameroon's Lake Nyos, which has a high concentration of dissolved carbon dioxide. The disturbance caused that gas to form bubbles. Being lighter than the surrounding fluid (the water), those bubbles were buoyed to the surface, where they released the carbon dioxide. The gas, being heavier than the surrounding fluid (now the air), rushed down the mountainside like a river, asphyxiating 1700 persons and the scores of animals seen here.

SOLUTION: The weight of an iceberg of total volume V_i is

$$W_i = \rho_i V_i g,$$

where $\rho_i = 917$ kg/m³ is the density of ice.

The weight of the displaced seawater, which is the buoyancy force F_b, is

$$W_w = F_b = \rho_w V_w g,$$

where $\rho_w = 1024$ kg/m³ is the density of seawater and V_w is the volume of the displaced water, that is, the submerged volume of the iceberg. For the floating iceberg, these two forces are equal, or

$$\rho_i V_i g = \rho_w V_w g.$$

From this equation, we find that the fraction we seek is

$$\text{frac} = \frac{V_i - V_w}{V_i} = 1 - \frac{V_w}{V_i} = 1 - \frac{\rho_i}{\rho_w}$$

$$= 1 - \frac{917 \text{ kg/m}^3}{1024 \text{ kg/m}^3}$$

$$= 0.10 \text{ or } 10\%. \qquad \text{(Answer)}$$

SAMPLE PROBLEM 15-6

A spherical, helium-filled balloon has a radius R of 12.0 m. The balloon, support cables, and basket have a mass m of 196 kg. What maximum load M can the balloon carry? Take $\rho_{He} = 0.160$ kg/m³ and $\rho_{air} = 1.25$ kg/m³; the volume of air displaced by the load is negligible.

SOLUTION: The weight of the displaced air, which is the buoyant force, and the weight of the helium in the balloon are

$$W_{air} = \rho_{air} V g \quad \text{and} \quad W_{He} = \rho_{He} V g,$$

in which $V (= 4\pi R^3/3)$ is the volume of the balloon.

At equilibrium, from Archimedes' principle,

$$W_{air} = W_{He} + mg + Mg$$

or

$$M = (\tfrac{4}{3}\pi)(R^3)(\rho_{air} - \rho_{He}) - m$$

$$= (\tfrac{4}{3}\pi)(12.0 \text{ m})^3 (1.25 \text{ kg/m}^3 - 0.160 \text{ kg/m}^3)$$

$$\quad - 196 \text{ kg}$$

$$= 7690 \text{ kg.} \qquad \text{(Answer)}$$

An object with this mass would weigh 17,000 lb at sea level.

15-8 IDEAL FLUIDS IN MOTION

The motion of *real fluids* is very complicated and not yet fully understood. Instead, we shall discuss the motion of an **ideal fluid,** which is simpler to handle mathematically and yet provides useful results. Here are four assumptions that we make about our ideal fluid:

1. *Steady flow* In *steady* (or *laminar*) *flow* the velocity of the moving fluid at any fixed point does not change with time, either in magnitude or in direction. The gentle flow of water near the center of a quiet stream is steady; that in a chain of rapids is not. Figure 15-12 shows a transition from steady flow to *nonsteady* (or *turbulent*) flow for a rising stream of smoke. The speed of the smoke particles increases as they rise and, at a certain critical speed, the flow changes from steady to nonsteady.

2. *Incompressible flow* We assume, as we have already done for fluids at rest, that our ideal fluid is incompressible. That is, its density has a constant, uniform value.

3. *Nonviscous flow* Roughly speaking, the viscosity of a fluid is a measure of how resistive the fluid is to flow. For example, thick honey is more resistive to flow than water, and so honey is said to be more viscous than water. Viscosity is the fluid analog of friction between solids; both are mechanisms by which the kinetic energy of moving objects can be transferred to thermal energy. In the absence of friction, a block could glide at constant speed along a horizontal surface. In the same way, an object moving through a nonviscous fluid would experience no *viscous drag force,* that is, no resistive force due to viscosity. The British scientist Lord Rayleigh noted that in an ideal fluid a ship's propeller would not work but, on the other hand, a ship (once set into motion) would not need a propeller!

FIGURE 15-12   At a certain point, the rising flow of smoke and heated gas changes from steady to turbulent.

4. *Irrotational flow* Although it need not concern us further, we also assume that the flow is *irrotational.* To test for this property, let a tiny grain of dust move with the fluid. Although this test body may (or may not) move in a circular path, in irrotational flow the test body will not rotate about an axis through its own center of mass. For a loose analogy, the motion of a Ferris wheel is rotational; that of its passengers is irrotational.

15-9 STREAMLINES AND THE EQUATION OF CONTINUITY

Figure 15-13 shows streamlines traced out by dye injected into a moving fluid; Fig. 15-14 shows similar streamlines revealed by smoke. A **streamline** is the path traced out by a tiny fluid element, which we may call a fluid "particle." As the fluid particle moves, its velocity may change, both in magnitude and in direction. However, its velocity vector at any point is always tangent to the streamline at that point (Fig. 15-15). Streamlines never cross because, if they did, a fluid particle arriving at the intersection would have two velocities simultaneously, an impossibility.

In flows like that of Figs. 15-13 and 15-14, we can isolate a *tube of flow* whose boundary is made up of streamlines. Such a tube acts like a pipe because any fluid particle that enters it cannot escape through its walls; if a particle did escape, we would have a case of streamlines crossing each other.

Figure 15-16 shows two cross sections, of areas A_1 and A_2, along a thin tube of flow. Let us station ourselves at B and monitor the fluid, moving with speed v_1, for a short time interval Δt. During this interval, a fluid particle moves

FIGURE 15-14 Smoke reveals streamlines in airflow past a car in a wind-tunnel test.

a small distance $v_1 \, \Delta t$, and a volume ΔV of fluid, given by

$$\Delta V = A_1 v_1 \, \Delta t,$$

passes through area A_1.

Assume that the fluid is incompressible and cannot be created or destroyed. Thus in this same time interval, the same volume of fluid must pass point C, farther down the tube of flow. If the speed there is v_2, this means that

$$\Delta V = A_1 v_1 \, \Delta t = A_2 v_2 \, \Delta t,$$

or

$$A_1 v_1 = A_2 v_2.$$

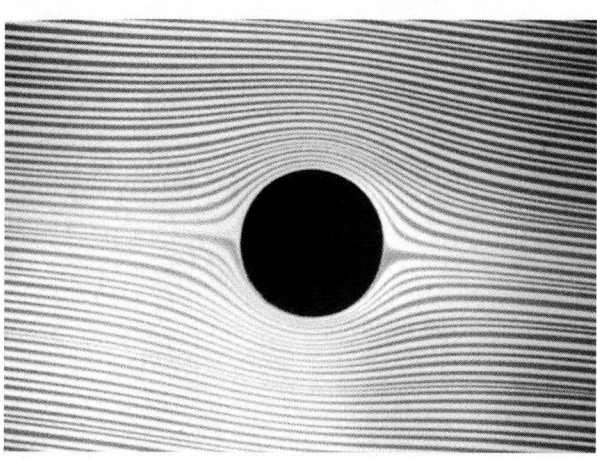

FIGURE 15-13 The steady flow of a fluid around a cylinder, as revealed by a dye tracer.

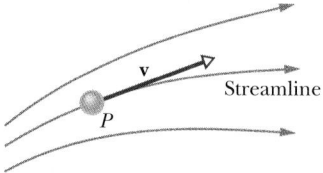

FIGURE 15-15 A fluid particle P traces out a streamline as it moves. The velocity vector of the particle is tangent to the streamline at every point.

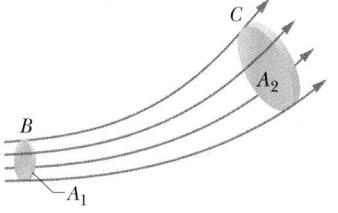

FIGURE 15-16 A tube of flow is defined by the streamlines that form its boundary. The flow rate of fluid must be the same for all cross sections of the tube of flow.

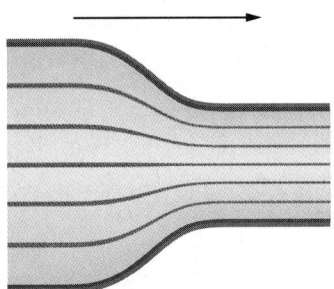

FIGURE 15-17   When a channel, such as a pipe, constricts, the streamlines draw closer together, signaling an increase in the fluid velocity. The arrow shows the direction of flow.

Thus, along the tube of flow,

$$R = Av = \text{a constant}, \qquad (15\text{-}13)$$

in which R, whose SI unit is the cubic meter per second, is the **volume flow rate** of the fluid. Equation 15-13 is called the **equation of continuity** for fluid flow. It tells us that the flow is faster in the narrower parts of a tube of flow, where the streamlines are closer together, as in Fig. 15-17.

Equation 15-13 is actually an expression of the law of conservation of mass in a form useful in fluid mechanics. In fact, if we multiply R by the (assumed uniform) density of the fluid, we get the quantity $Av\rho$, which is the **mass flow rate,** whose SI unit is the kilogram per second. Then Eq. 15-13 effectively tells us that the mass that flows through point B in Fig. 15-16 each second must be equal to the mass that flows through point C each second.

CHECKPOINT **3:** The figure shows a pipe and gives the volume flow rate (in cm³/s) and the direction of flow for all but one section. What are the volume flow rate and the direction of flow for that section?

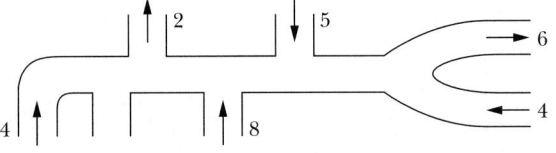

SAMPLE PROBLEM 15-7

The cross-sectional area A_0 of the aorta (the major blood vessel emerging from the heart) of a normal resting person is 3 cm², and the speed v_0 of the blood is 30 cm/s. A typical capillary (diameter $\approx$ 6 μm) has a cross-sectional area A of 3×10^{-7} cm² and a flow speed v of 0.05 cm/s. How many capillaries does such a person have?

SOLUTION: All the blood that passes through the capillaries must have passed through the aorta so that, from Eq. 15-13,

$$A_0v_0 = nAv,$$

where n is the number of capillaries. Solving for n yields

$$n = \frac{A_0v_0}{Av} = \frac{(3 \text{ cm}^2)(30 \text{ cm/s})}{(3 \times 10^{-7} \text{ cm}^2)(0.05 \text{ cm/s})}$$

$$= 6 \times 10^9 \text{ or 6 billion.} \qquad \text{(Answer)}$$

You can easily show that the combined cross-sectional area of the capillaries is about 600 times the area of the aorta.

SAMPLE PROBLEM 15-8

Figure 15-18 shows how the stream of water emerging from a faucet "necks down" as it falls. The cross-sectional area A_0 is 1.2 cm², and A is 0.35 cm². The two levels are separated by a vertical distance $h = 45$ mm. At what rate does water flow from the tap?

SOLUTION: From the equation of continuity (Eq. 15-13) we have

$$A_0v_0 = Av, \qquad (15\text{-}14)$$

where v_0 and v are the water velocities at the corresponding levels. From Eq. 2-23 we can also write, because the water is falling freely with acceleration g,

$$v^2 = v_0^2 + 2gh. \qquad (15\text{-}15)$$

Eliminating v between Eqs. 15-14 and 15-15 and solving for v_0, we obtain

$$v_0 = \sqrt{\frac{2ghA^2}{A_0^2 - A^2}}$$

$$= \sqrt{\frac{(2)(9.8 \text{ m/s}^2)(0.045 \text{ m})(0.35 \text{ cm}^2)^2}{(1.2 \text{ cm}^2)^2 - (0.35 \text{ cm}^2)^2}}$$

$$= 0.286 \text{ m/s} = 28.6 \text{ cm/s}.$$

The volume flow rate R is then

$$R = A_0v_0 = (1.2 \text{ cm}^2)(28.6 \text{ cm/s})$$

$$= 34 \text{ cm}^3/\text{s}. \qquad \text{(Answer)}$$

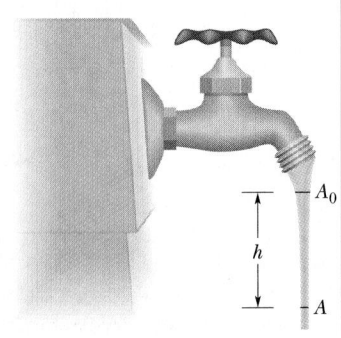

FIGURE 15-18   Sample Problem 15-8. As water falls from a tap, its speed increases. Because the flow rate must be the same at all cross sections, the stream must "neck down."

15-10 BERNOULLI'S EQUATION

Figure 15-19 represents a tube of flow (or an actual pipe, for that matter) through which an ideal fluid is flowing at a steady rate. In a time interval Δt, suppose that a volume of fluid ΔV, colored purple in Fig. 15-19a, enters the tube at the left (or input) end and an identical volume, colored green in Fig. 15-19b, emerges at the right (or output) end. The emerging volume must be the same as the entering volume because the fluid is incompressible, with an assumed constant density ρ.

Let y_1, v_1, and p_1 be the elevation, speed, and pressure of the fluid entering at the left, and y_2, v_2, and p_2 be the corresponding quantities for the fluid emerging at the right. By applying the principle of conservation of energy to the fluid, we shall show that these quantities are related by

$$p_1 + \tfrac{1}{2}\rho v_1^2 + \rho g y_1 = p_2 + \tfrac{1}{2}\rho v_2^2 + \rho g y_2. \quad (15\text{-}16)$$

We can also write this equation as

$$p + \tfrac{1}{2}\rho v^2 + \rho g y = \text{a constant}. \quad (15\text{-}17)$$

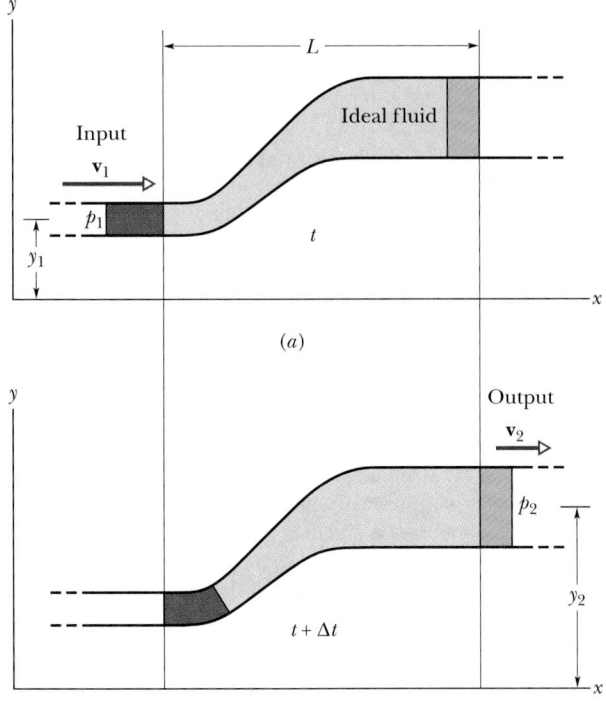

(a)

(b)

FIGURE 15-19 Fluid flows at a steady rate through a length L of a tube, from the input end at the left to the output end at the right. From time t in (a) to time $t + \Delta t$ in (b), the amount of fluid shown in purple enters the input end and the equal amount shown in green emerges from the output end.

Equations 15-16 and 15-17 are equivalent forms of **Bernoulli's equation**, after Daniel Bernoulli, who studied fluid flow in the 1700s.* Like the equation of continuity (Eq. 15-13), Bernoulli's equation is not a new principle but simply the reformulation of a familiar principle in a form more suitable to fluid mechanics. As a check, let us apply Bernoulli's equation to fluids at rest, by putting $v_1 = v_2 = 0$ in Eq. 15-16. The result is

$$p_2 = p_1 + \rho g(y_1 - y_2),$$

which is Eq. 15-4.

A major prediction of Bernoulli's equation emerges if we take y to be a constant ($y = 0$, say) so that the fluid does not change elevation as it flows. Equation 15-16 then becomes

$$p_1 + \tfrac{1}{2}\rho v_1^2 = p_2 + \tfrac{1}{2}\rho v_2^2, \quad (15\text{-}18)$$

which tells us that:

> If the speed of a fluid particle increases as it travels along a horizontal streamline, the pressure of the fluid must decrease, and conversely.

Put another way, where the streamlines are relatively close together (that is, where the velocity is relatively great), the pressure is relatively low, and conversely.

The link between a change in speed and a change in pressure makes sense if you consider a fluid particle. When the particle nears a narrow region, the higher pressure behind it accelerates it so that it then has a greater speed in the narrow region. And when it nears a wide region, the higher pressure ahead of it decelerates it so that it then has a lesser speed in the wide region.

Bernoulli's equation is strictly valid only to the extent that the fluid is ideal. If viscous forces are present, thermal energy will be involved. We take no account of this in the derivation that follows.

Proof of Bernoulli's Equation

Let us take as our system the entire volume of the (ideal) fluid shown in Fig. 15-19. We shall apply the principle of conservation of energy to this system as it moves from its initial state (Fig. 15-19a) to its final state (Fig. 15-19b). The fluid lying between the two vertical planes separated by a distance L in Fig. 15-19 does not change its properties during this process; we need be concerned only with changes that take place at the input and the output ends.

---

*For irrotational flow (which we assume), the constant in Eq. 15-17 has the same value for all points within the tube of flow; the points do not have to lie along the same streamline. Similarly, the points 1 and 2 in Eq. 15-16 can lie anywhere within the tube of flow.

We apply energy conservation in the form of the work–kinetic energy theorem,

$$W = \Delta K, \tag{15-19}$$

which tells us that the change in the kinetic energy of our system must equal the net work done on the system. The change in kinetic energy results from the change in speed between the ends of the pipe and is

$$\Delta K = \tfrac{1}{2}\Delta m \, v_2^2 - \tfrac{1}{2}\Delta m \, v_1^2$$
$$= \tfrac{1}{2}\rho \, \Delta V(v_2^2 - v_1^2), \tag{15-20}$$

in which $\Delta m \,(= \rho \, \Delta V)$ is the mass of the fluid that enters at the input end and leaves at the output end during a small time interval Δt.

The work done on the system arises from two sources. The work W_g done by the weight ($\Delta m \, \mathbf{g}$) of mass Δm during the vertical lift of the mass from the input level to the output level is

$$W_g = -\Delta m \, g(y_2 - y_1)$$
$$= -\rho g \, \Delta V(y_2 - y_1). \tag{15-21}$$

This work is negative because the upward displacement and the downward weight point in opposite directions.

Work W_p must also be done *on* the system (at the input end) to push the entering fluid into the tube and *by* the system (at the output end) to push forward the fluid ahead of the emerging fluid. In general, the work done by a force of magnitude F, acting on a fluid sample contained in a tube of area A to move the fluid through a distance Δx, is

$$F \, \Delta x = (pA)(\Delta x) = p(A \, \Delta x) = p \, \Delta V.$$

The work done on the system is then $p_1 \, \Delta V$, and the work done by the system is $-p_2 \, \Delta V$. Their sum W_p is

$$W_p = -p_2 \, \Delta V + p_1 \, \Delta V$$
$$= -(p_2 - p_1)\Delta V. \tag{15-22}$$

The work–kinetic energy theorem of Eq. 15-19 now becomes

$$W = W_g + W_p = \Delta K.$$

Substituting from Eqs. 15-20, 15-21, and 15-22 yields

$$-\rho g \, \Delta V(y_2 - y_1) - \Delta V(p_2 - p_1) = \tfrac{1}{2}\rho \, \Delta V(v_2^2 - v_1^2).$$

This, after a slight rearrangement, matches Eq. 15-16, which we set out to prove.

SAMPLE PROBLEM 15-9

Ethanol of density $\rho = 791 \text{ kg/m}^3$ flows smoothly through a horizontal pipe that tapers (as in Fig. 15-17) in cross-sectional area from $A_1 = 1.20 \times 10^{-3} \text{ m}^2$ to $A_2 = A_1/2$. The pressure difference Δp between the wide and narrow sections of pipe is 4120 Pa. What is the volume flow rate R of the ethanol?

SOLUTION: Rearranging Eq. 15-18 (Bernoulli's equation for level flow) yields

$$p_1 - p_2 = \tfrac{1}{2}\rho v_2^2 - \tfrac{1}{2}\rho v_1^2 = \tfrac{1}{2}\rho(v_2^2 - v_1^2), \tag{15-23}$$

where subscripts 1 and 2 refer to the wide and narrow sections of pipe, respectively. Equation 15-13 (the continuity equation) tells us that the flow is faster in the narrower section. Here that means $v_2 > v_1$. Equation 15-23 then tells us that $p_1 > p_2$.

Equation 15-13 also tells us the volume flow rate R is the same in the wide and narrow sections. So

$$R = v_1 A_1 = v_2 A_2.$$

These equations, with $A_2 = A_1/2$, give us

$$v_1 = \frac{R}{A_1} \quad \text{and} \quad v_2 = \frac{R}{A_2} = \frac{2R}{A_1}.$$

Substituting these expressions into Eq. 15-23, setting $p_1 - p_2 = \Delta p$, and rearranging give us

$$\Delta p = \tfrac{1}{2}\rho \left(\frac{4R^2}{A_1^2} - \frac{R^2}{A_1^2} \right) = \frac{3\rho R^2}{2A_1^2}.$$

Solving for R, we find

$$R = A_1 \sqrt{\frac{2\Delta p}{3\rho}}$$
$$= 1.20 \times 10^{-3} \text{ m}^2 \sqrt{\frac{(2)(4120 \text{ Pa})}{(3)(791 \text{ kg/m}^3)}}$$
$$= 2.24 \times 10^{-3} \text{ m}^3/\text{s}. \tag{Answer}$$

SAMPLE PROBLEM 15-10

In the old West, a desperado fires a bullet into an open water tank (Fig. 15-20), creating a hole a distance h below the water surface. What is the speed v of the water emerging from the hole?

SOLUTION: This situation is essentially that of water moving (downward) with speed V through a wide pipe (the tank) of cross-sectional area A and then moving (horizontally) with speed v through a narrow pipe (the hole) of cross-sectional area a. From Eq. 15-13, we know that

$$R = av = AV$$

and thus that

$$V = \frac{a}{A} v.$$

Because $a \ll A$, we see that $V \ll v$.

We can also relate v to V (and to h) through the Bernoulli equation (Eq. 15-16). We take the level of the hole as our reference level for measuring elevations (and thus gravitational potential energy). Noting that the pressure at the top of the tank and at the bullet hole is the atmospheric pressure p_0 (because both places are exposed to the atmosphere), we write Eq. 15-16 as

$$p_0 + \tfrac{1}{2}\rho V^2 + \rho g h = p_0 + \tfrac{1}{2}\rho v^2 + \rho g(0). \quad (15\text{-}24)$$

(Here the top of the tank is represented by the left side of the equation, and the hole by the right side. The zero on the right

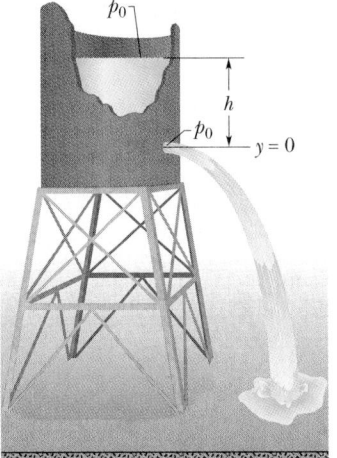

FIGURE 15-20
Sample Problem 15-10. Water pours through a hole in a water tank, at a distance h below the water surface. The water pressure at the surface and at the hole is atmospheric pressure p_0.

indicates that the hole is at our reference level.) Before we solve Eq. 15-24 for v, we can use our result that $V \ll v$ to simplify it: We assume that V^2, and thus the term $\tfrac{1}{2}\rho V^2$ in Eq. 15-24, is negligible compared to the other terms, and we drop it. Solving the remaining equation for v then yields

$$v = \sqrt{2gh}. \quad \text{(Answer)}$$

This is the same speed that an object would have when falling a height h from rest.

CHECKPOINT **4:** Water flows smoothly through the pipe shown in the figure, descending in the process. Rank the four numbered sections of pipe according to (a) the volume flow rate R through them, (b) the flow speed v through them, and (c) the water pressure p within them, greatest first.

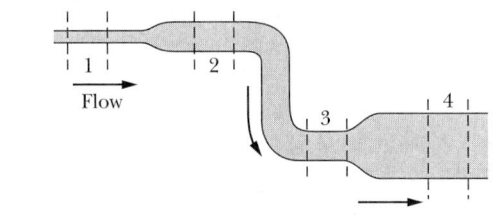

REVIEW & SUMMARY

Density

The **density** ρ of any material is defined as its mass per unit volume:

$$\rho = \frac{\Delta m}{\Delta V}. \quad (15\text{-}1)$$

Usually, where a material sample is large compared with atomic dimensions, we can write Eq. 15-1 as $\rho = m/V$.

Fluid Pressure

A **fluid** is a substance that can flow; it conforms to the boundaries of its container because it cannot withstand shearing stress. It can, however, exert a force perpendicular to its surface. That force is described in terms of **pressure** p:

$$p = \frac{\Delta F}{\Delta A}, \quad (15\text{-}2)$$

in which ΔF is the force acting on a surface element of area ΔA. If the force is uniform over a flat area, Eq. 15-2 can be written as $p = F/A$. The force resulting from fluid pressure at a particular point in a fluid has the same magnitude in all directions. *Gauge pressure* is the difference between the actual pressure (or *absolute pressure*) at a point and the atmospheric pressure.

Pressure Variation with Height and Depth

Pressure in a fluid at rest varies with vertical position y. For y measured positive upward,

$$p_2 = p_1 + \rho g(y_1 - y_2). \quad (15\text{-}4)$$

The pressure is the same for all points at the same level. If h is the *depth* of a fluid sample below some reference level at which the pressure is p_0, Eq. 15-4 becomes

$$p = p_0 + \rho g h. \quad (15\text{-}5)$$

Pascal's Principle

Pascal's principle, which can be derived from Eq. 15-4, states that a change in the pressure applied to an enclosed fluid is transmitted undiminished to every portion of the fluid and to the walls of the containing vessel.

Archimedes' Principle

The surface of an immersed object is acted on by forces associated with the fluid pressure. The vector sum of those forces (called the **buoyant force**) acts vertically upward, through the center of mass of the displaced fluid (the **center of buoyancy**). Archimedes' principle states that the magnitude of the buoyant force is equal to the weight of the fluid displaced by the object. When an object floats, its weight is equal to the buoyant force acting on it.

Flow of Ideal Fluids

An *ideal fluid* is incompressible and lacks viscosity, and its flow is steady and irrotational. A **streamline** is the path followed by an individual fluid particle. A *tube of flow* is a bundle of streamlines. The principle of conservation of mass shows that the flow within any tube of flow obeys the **equation of continuity:**

$$R = Av = \text{a constant}, \tag{15-13}$$

in which R is the **volume flow rate,** A the cross-sectional area of the tube of flow at any point, and v the speed of the fluid at that point, assumed to be constant across A. The **mass flow rate** $Av\rho$ is also constant.

Bernoulli's Equation

Applying the principle of conservation of mechanical energy to the flow of an ideal fluid leads to **Bernoulli's equation:**

$$p + \tfrac{1}{2}\rho v^2 + \rho gy = \text{a constant} \tag{15-17}$$

along any tube of flow.

QUESTIONS

1. Figure 15-21 shows a tank filled with water. Five horizontal floors and ceilings are indicated; all have the same area and are located at distances L, $2L$, or $3L$ below the top of the tank. Rank the floors and ceilings according to the force on them due to the water, greatest first.

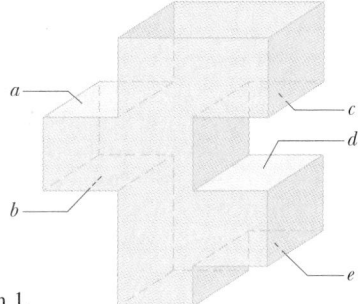

FIGURE 15-21 Question 1.

2. Figure 15-22 shows four situations in which a red liquid and a gray liquid are in a U-tube. In one situation the liquids cannot be in static equilibrium. (a) Which situation is that? (b) For the other three situations, assume static equilibrium. For each, is the density of the red liquid greater than, less than, or equal to the density of the gray liquid?

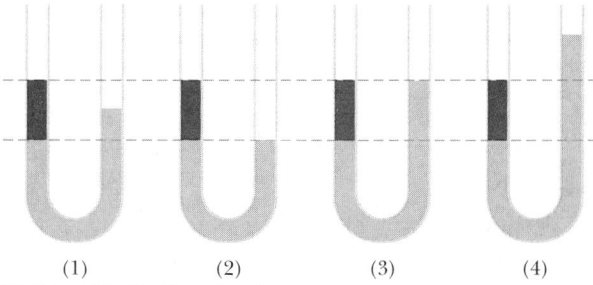

FIGURE 15-22 Question 2.

3. The containers of Fig. 15-23 have the same level of water and same base area, and are made of the same material. (a) Rank the containers (and contents) according to their weights on a scale, greatest first. (b) Rank the containers according to the pressure of the water on the base of the container. (c) Does Eq. 15-2 indicate that the answers to (a) and (b) are inconsistent? This apparent inconsistency is known as the *hydrostatic paradox*.

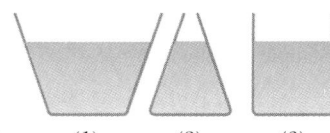

FIGURE 15-23 Question 3. (1) (2) (3)

4. We fully submerge an irregular 3 kg lump of material in a certain fluid. The fluid that would have been in the space now occupied by the lump has a mass of 2 kg. (a) When we release the lump, does it move upward, move downward, or remain in place? (b) If we next fully submerge the lump in a less dense fluid and again release it, what does it do?

5. Figure 15-24 shows four solid objects floating in corn syrup. Rank the objects according to their density, greatest first.

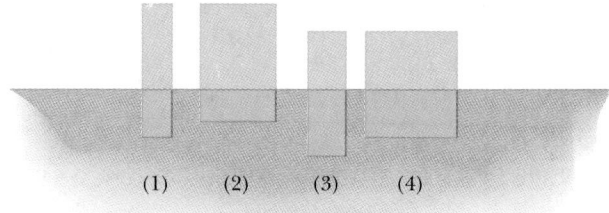

(1) (2) (3) (4)

FIGURE 15-24 Question 5.

6. Figure 15-25 shows three identical open-top containers filled to the brim with water; toy ducks float in two of them. Rank the containers and contents according to their weight, greatest first.

(a) (b) (c)

FIGURE 15-25 Question 6.

7. A boat with an anchor on board floats in a swimming pool that is somewhat wider than the boat. Does the water level in the pool move upward, move downward, or remain the same if the anchor is (a) dropped into the water or (b) thrown onto the surrounding ground? (c) Does the water level in the pool move upward, move downward, or remain the same if, instead, a cork is dropped from the boat into the water, where it floats?

8. Three gas bubbles of identical sizes are submerged in water: bubble 1 is filled with hydrogen, bubble 2 is filled with helium, and bubble 3 is filled with carbon dioxide. Rank the bubbles according to the magnitude of the buoyant force acting on them, greatest first.

9. A block of wood floats at a certain level in a pail of water in a stationary elevator. Does the block float higher, lower, or at the same level when the elevator cab (a) moves upward at constant speed, (b) moves downward at constant speed, (c) accelerates upward, and (d) accelerates downward with an acceleration magnitude less than 9.8 m/s²?

10. A container of water is suspended from a spring balance. Does the weight reading increase, decrease, or remain the same

when we, without spilling any water, (a) lower a heavy metal object via string into the water and (b) float a cork in the water?

11. Figure 15-26 shows two rectangular, uniformly dense blocks that are floating in water. Both blocks have been tipped by hand and then released. For each block, (a) does buoyancy cause a clockwise torque or a counterclockwise torque about the block's center of mass and (b) will the block right itself or tip over more?

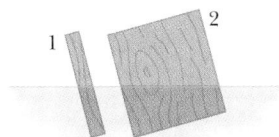

FIGURE 15-26 Question 11.

EXERCISES & PROBLEMS

SECTION 15-3 Density and Pressure

1E. Convert a density of 1.0 g/cm³ to kg/m³.

2E. Three liquids that will not mix are poured into a cylindrical container. The volumes and densities of the liquids are 0.50 L, 2.6 g/cm³; 0.25 L, 1.0 g/cm³; and 0.40 L, 0.80 g/cm³. What is the force acting on the bottom of the container due to these liquids? One liter = 1 L = 1000 cm³. (Ignore the contribution due to the atmosphere.)

3E. Find the pressure increase in the fluid in a syringe when a nurse applies a force of 42 N to the syringe's circular piston, which has a radius of 1.1 cm.

4E. You inflate the front tires on your car to 28 psi. Later, you measure your blood pressure, obtaining a reading of 120/80, the readings being in mm Hg. In metric countries (which is to say, most of the rest of the world), these pressures are customarily reported in kilopascals (kPa). In kilopascals, what are (a) your tire pressure and (b) your blood pressure?

5E. An office window has dimensions 3.4 m by 2.1 m. As a result of the passage of a storm, the outside air pressure drops to 0.96 atm, but inside the pressure is held at 1.0 atm. What net force pushes out on the window?

6E. A fish maintains its depth in fresh water by adjusting the air content of porous bone or air sacs to make its average density the same as that of the water. Suppose that with its air sacs collapsed, a fish has a density of 1.08 g/cm³. To what fraction of its expanded body volume must the fish inflate the air sacs to reduce its density to that of water?

7P. An airtight box having a lid with an area of 12 in.² is partially evacuated. If a force of 108 lb is required to pull the lid off the box and the outside atmospheric pressure is 15 lb/in.², what is the air pressure in the box?

8P. In 1654 Otto von Guericke, inventor of the air pump, gave a demonstration before the noblemen of the Holy Roman Empire in which two teams of eight horses could not pull apart two evacu-

ated brass hemispheres. (a) Assuming that the hemispheres have thin walls, so that R in Fig. 15-27 may be considered both the inside and outside radius, show that the force F required to pull apart the hemispheres is $F = \pi R^2 \Delta p$, where Δp is the difference between the pressures outside and inside the sphere. (b) Taking R equal to 1.0 ft and the inside pressure as 0.10 atm, find the force the teams of horses would have had to exert to pull apart the hemispheres. (c) Why are two teams of horses used? Would not one team have proved the point just as well if the hemispheres were attached to a sturdy wall?

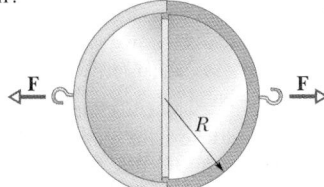

FIGURE 15-27 Problem 8.

SECTION 15-4 Fluids at Rest

9E. Calculate the hydrostatic difference in blood pressure between the brain and the foot in a person of height 1.83 m. The density of blood is 1.06 × 10³ kg/m³.

10E. Find the pressure, in pascals, 150 m below the surface of the ocean. The density of seawater is 1.03 g/cm³, and the atmospheric pressure at sea level is 1.01 × 10⁵ Pa.

11E. The sewer outlets of a house constructed on a slope are 8.2 m below street level. If the sewer is 2.1 m below street level, find the minimum pressure differential that must be created by the sewage pump to transfer waste of average density 900 kg/m³.

12E. Figure 15-28 displays the *phase diagram* of carbon, showing the ranges of temperature and pressure in which carbon will crystallize either as diamond or graphite. What is the minimum depth at which diamonds can form if the temperature at that depth is 1000°C and the subsurface rocks have density 3.1 g/cm³? Assume that, as in a fluid, the pressure is due to the weight of material lying above.

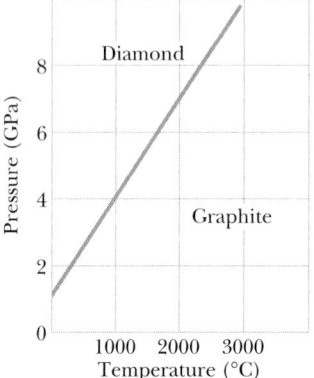

FIGURE 15-28 Exercise 12.

13E. The human lungs can operate against a pressure differential of up to about $\frac{1}{20}$ of an atmosphere. If a diver uses a snorkel for breathing, about how far below water level can she or he swim?

14E. A swimming pool has the dimensions 80 ft × 30 ft × 8.0 ft. (a) When it is filled with water, what is the force (resulting from the water alone) on the bottom, on the short sides, and on the long sides? (b) If you are concerned with the possibility that the concrete walls and floor will collapse, is it appropriate to take the atmospheric pressure into account? Why?

15E. (a) Find the total weight of water on top of a nuclear submarine at a depth of 200 m, assuming that its (horizontal cross-sectional) hull area is 3000 m². (b) What water pressure would a diver experience at this depth? Express your answer in atmospheres. Do you think that occupants of a damaged submarine at this depth could escape without special equipment? Assume the density of seawater is 1.03 g/cm³.

16E. Crew members attempt to escape from a damaged submarine 100 m below the surface. What force must they apply to a pop-out hatch, which is 1.2 m by 0.60 m, to push it out? Assume that the density of the ocean water is 1025 kg/m³.

17E. A simple open U-tube contains mercury. When 11.2 cm of water is poured into the right arm of the tube, how high above its initial level does the mercury rise in the left arm?

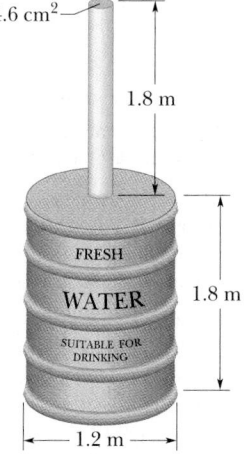

FIGURE 15-29 Exercise 18.

18E. A cylindrical barrel has a narrow tube fixed to the top, as shown with dimensions in Fig. 15-29. The vessel is filled with water to the top of the tube. Calculate the ratio of the hydrostatic force exerted on the bottom of the barrel to the weight of the water contained inside the barrel. Why is the ratio not equal to one? (Ignore the presence of the atmosphere.)

19P. Two identical cylindrical vessels with their bases at the same level each contain a liquid of density ρ. The area of either base is A, but in one vessel the liquid height is h_1, and in the other h_2. Find the work done by the gravitational force in equalizing the levels when the two vessels are connected.

20P. (a) Consider a container of fluid subject to a *vertical upward* acceleration of magnitude a. Show that the pressure variation with depth in the fluid is given by

$$p = \rho h(g + a),$$

where h is the depth and ρ is the density. (b) Show also that if the fluid as a whole undergoes a *vertical downward* acceleration of magnitude a, the pressure at a depth h is given by

$$p = \rho h(g - a).$$

(c) What is the pressure if the fluid is in free fall?

21P. In analyzing certain geological features, it is often appropriate to assume that the pressure at some horizontal *level of compensation,* deep inside Earth, is the same over a large region and is equal to the pressure exerted by the weight of the overlying material. That is, the pressure on the level of compensation is given by the fluid pressure formula. This model requires, for example, that mountains have *roots* (Fig. 15-30). Consider a mountain 6.0 km high. The continental rocks have a density of 2.9 g/cm³, and beneath the continent is the mantle, with a density of 3.3 g/cm³. Calculate the depth D of the root. (*Hint:* Set the pressure at points a and b equal; the depth y of the level of compensation will cancel out.)

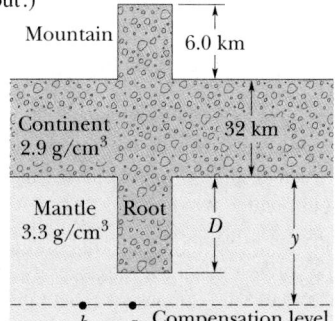

FIGURE 15-30
Problem 21.

22P. In Fig. 15-31, the ocean is about to overrun the continent. Find the depth h of the ocean using the level-of-compensation method shown in Problem 21.

23P. Water stands at a depth D behind the vertical upstream face of a dam, as shown in Fig. 15-32. Let W be the width of the dam. (a) Find the resultant horizontal force exerted on the dam by the gauge pressure of the water and (b) the net torque owing to that gauge pressure about a line through O parallel to the width of the dam. (c) Find the moment arm of the resultant horizontal force about the line through O.

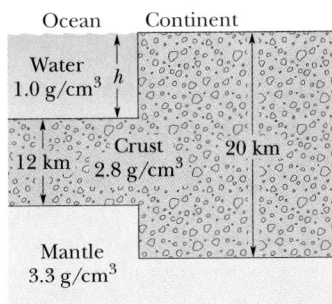

FIGURE 15-31
Problem 22.

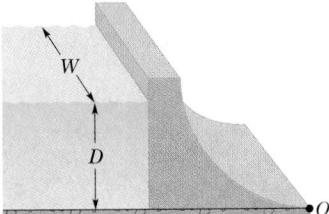

FIGURE 15-32
Problem 23.

24P. The L-shaped tank shown in Fig. 15-33 is filled with water and is open at the top. If $d = 5.0$ m, what are (a) the force on face A and (b) the force on face B due to the water?

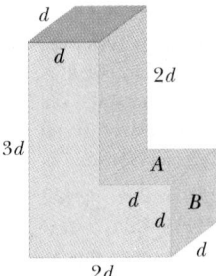

FIGURE 15-33
Problem 24.

25P. Figure 15-34 shows a dam and part of the freshwater reservoir backed up behind it. The dam is made of concrete of density 3.2 g/cm^3 and has the dimensions shown on the figure. (a) The force exerted by the water pushes horizontally on the dam face, and this is resisted by the force of static friction between the dam and the bedrock foundation on which it rests. The coefficient of friction is 0.47. Calculate the factor of safety against sliding, that is, the ratio of the maximum possible friction force to the force exerted by the water. (b) The water also tries to rotate the dam about a line running along the base of the dam through point A; see Problem 23. The torque resulting from the weight of the dam acts in the opposite sense. Calculate the factor of safety against rotation, that is, the ratio of the torque owing to the weight of the dam to the torque exerted by the water.

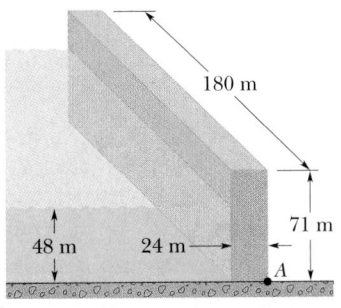

FIGURE 15-34
Problem 25.

SECTION 15-5 Measuring Pressure

26E. Calculate the height of a column of water that gives a pressure of 1 atm at the bottom. Assume that $g = 9.80$ m/s^2.

27E. To suck lemonade of density 1000 kg/m^3 up a straw to a maximum height of 4.0 cm, what minimum gauge pressure (in atmospheres) must you produce in your lungs?

28P. What would be the height of the atmosphere if the air density (a) were constant and (b) decreased linearly to zero with height? Assume a sea-level density of 1.3 kg/m^3.

SECTION 15-6 Pascal's Principle

29E. A piston of small cross-sectional area a is used in a hydraulic press to exert a small force $\mathbf{f}$ on the enclosed liquid. A connecting pipe leads to a larger piston of cross-sectional area A (Fig. 15-35). (a) What force $\mathbf{F}$ will the larger piston sustain? (b) If the small piston has a diameter of 1.5 in. and the large piston one of 21 in., what weight on the small piston will support 2.0 tons on the large piston?

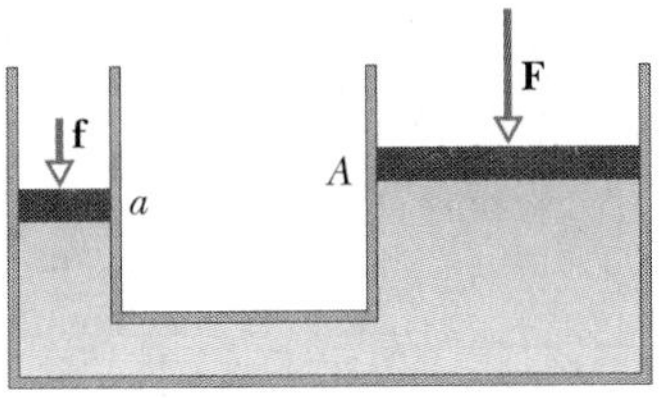

FIGURE 15-35
Exercises 29 and 30.

30E. In the hydraulic press of Exercise 29, through what distance must the large piston be moved to raise the small piston a distance of 3.5 ft?

SECTION 15-7 Archimedes' Principle

31E. A tin can has a total volume of 1200 cm^3 and a mass of 130 g. How many grams of lead shot could it carry without sinking in water? The density of lead is 11.4 g/cm^3.

32E. A boat floating in fresh water displaces 8000 lb of water. (a) How many pounds of water would this boat displace if it were floating in salt water of density 68.6 lb/ft^3? (b) Would the volume of water displaced change? If so, by how much?

33E. About one-third of the body of a physicist swimming in the Dead Sea will be above the water line. Assuming that the human body density is 0.98 g/cm^3, find the density of the water in the Dead Sea. (Why is it so much greater than 1.0 g/cm^3?)

34E. An iron anchor appears 200 N lighter in water than in air. (a) What is the volume of the anchor? (b) How much does it weigh in air? The density of iron is 7870 kg/m^3.

35E. An object hangs from a spring balance. The balance registers 30 N in air, 20 N when this object is immersed in water, and 24 N when the object is immersed in another liquid of unknown density. What is the density of that other liquid?

36E. A cubical object of dimensions $L = 2.00$ ft on a side and weight $W = 1000$ lb in a vacuum is suspended by a rope in an open tank of liquid of density $\rho = 2.00$ slugs/ft^3 as in Fig. 15-36. (a) Find the total downward force exerted by the liquid and the atmosphere on the top of the object. (b) Find the total upward force on the bottom of the object. (c) Find the tension in the rope. (d) Calculate the buoyant force on the object using Archimedes' principle. What relation exists among all these quantities?

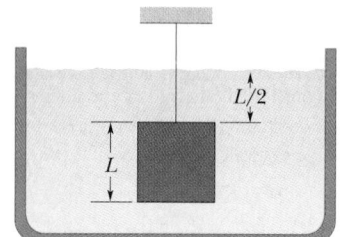

FIGURE 15-36
Exercise 36.

37E. A block of wood floats in water with two-thirds of its volume submerged. In oil the block floats with 0.90 of its volume submerged. Find the density of (a) the wood and (b) the oil.

38E. It has been proposed to move natural gas from the North Sea gas fields in huge dirigibles, using the gas itself to provide lift. Calculate the force required to tether such an airship to the ground for off-loading when it is fully loaded with 1.0×10^6 m^3 of gas at a density of 0.80 kg/m^3. (The weight of the airship is negligible by comparison.)

39E. A blimp is cruising slowly at low altitude, filled as usual with helium gas. Its maximum useful payload, including crew and cargo, is 1280 kg. How much more payload could the blimp carry if you replaced the helium with hydrogen? (Why not do it?) The volume of the helium-filled interior space is 5000 m^3. The density of helium gas is 0.16 kg/m^3, and the density of hydrogen is 0.081 kg/m^3.

40E. A helium balloon is used to lift a 40 kg payload to an altitude of 27 km, where the air density is 0.035 kg/m^3. The balloon has a mass of 15 kg, and the density of the gas in the balloon is 0.0051 kg/m^3. What is the volume of the balloon? Neglect the volume of the payload.

41P. A hollow sphere of inner radius 8.0 cm and outer radius 9.0 cm floats half-submerged in a liquid of density 800 kg/m^3. (a) What is the mass of the sphere? (b) Calculate the density of the material of which the sphere is made.

42P. A hollow spherical iron shell floats almost completely submerged in water. The outer diameter is 60.0 cm, and the density of iron is 7.87 g/cm^3. Find the inner diameter.

43P. An iron casting containing a number of cavities weighs 6000 N in air and 4000 N in water. What is the volume of the cavities in the casting? The density of iron is 7.87 g/cm^3.

44P. (a) What is the minimum area of the top surface of a slab of ice 0.30 m thick floating on fresh water that will hold up an automobile of mass 1100 kg? (b) Does it matter where the car is placed on the block of ice?

45P. Three children, each of weight 80 lb, make a log raft by lashing together logs of diameter 1.0 ft and length 6.0 ft. How many logs will be needed to keep them afloat? Take the density of wood to be 50 lb/ft^3.

46P. Assume the density of brass weights to be 8.0 g/cm^3 and that of air to be 0.0012 g/cm^3. What percent error arises from

neglecting the buoyancy of air in weighing an object of mass m and density ρ on a beam balance?

47P. A car has a total mass of 1800 kg. The volume of air space in the passenger compartment is 5.00 m^3. The volume of the motor and front wheels is 0.750 m^3, and the volume of the rear wheels, gas tank, and trunk is 0.800 m^3; water cannot enter these areas. The car is parked on a hill; the handbrake cable snaps and the car rolls down the hill into a lake (Fig. 15-37). (a) At first, no water enters the passenger compartment. How much of the car, in cubic meters, is below the water surface with the car floating as shown? (b) As water slowly enters, the car sinks. How many cubic meters of water are in the car as it disappears below the water surface? (The car remains horizontal, owing to a heavy load in the trunk.)

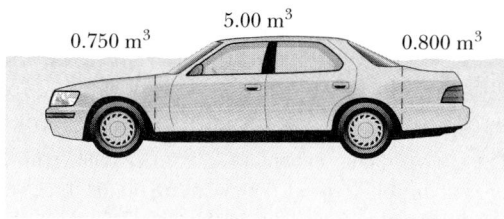

FIGURE 15-37 Problem 47.

48P. A block of wood has a mass of 3.67 kg and a density of 600 kg/m^3. It is to be loaded with lead so that it will float in water with 0.90 of its volume immersed. What mass of lead is needed (a) if the lead is on top of the wood and (b) if the lead is attached below the wood? The density of lead is 1.13×10^4 kg/m^3.

49P. You place a glass beaker, partially filled with water, in a sink (Fig. 15-38). The beaker has a mass of 390 g and an interior volume of 500 cm^3. You now start to fill the sink with water and you find, by experiment, that if the beaker is less than half full, it will float; but if it is more than half full, it remains on the bottom of the sink as the water rises to its rim. What is the density of the material of which the beaker is made?

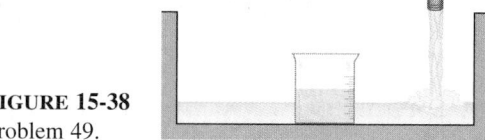

FIGURE 15-38
Problem 49.

50P. What is the acceleration of a rising hot-air balloon if the ratio of the air density outside the balloon to that inside is 1.39? Neglect the mass of the balloon fabric and the basket.

51P. A metal rod of length 80 cm and mass 1.6 kg has a uniform cross-sectional area of 6.0 cm^2. Due to a nonuniform density, the center of mass of the rod is 20 cm from one end of the rod. The rod is suspended in a horizontal position in water by ropes attached to both ends as shown in Fig. 15-39. (a) What is the ten-

FIGURE 15-39
Problem 51.

sion in the rope closer to the center of mass? (b) What is the tension in the rope farther from the center of mass?

52P*. The tension in a string holding a solid block below the surface of a liquid (of density greater than the solid) is T_0 when the containing vessel (Fig. 15-40) is at rest. Show that when the vessel has an upward vertical acceleration of magnitude a, the tension T is equal to $T_0(1 + a/g)$.

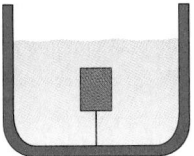

FIGURE 15-40 Problem 52.

SECTION 15-9 Streamlines and the Equation of Continuity

53E. Figure 15-41 shows the merging of two streams to form a river. One stream has a width of 8.2 m, depth of 3.4 m, and current speed of 2.3 m/s. The other stream is 6.8 m wide and 3.2 m deep, and flows at 2.6 m/s. The width of the river is 10.5 m, and the current speed is 2.9 m/s. What is its depth?

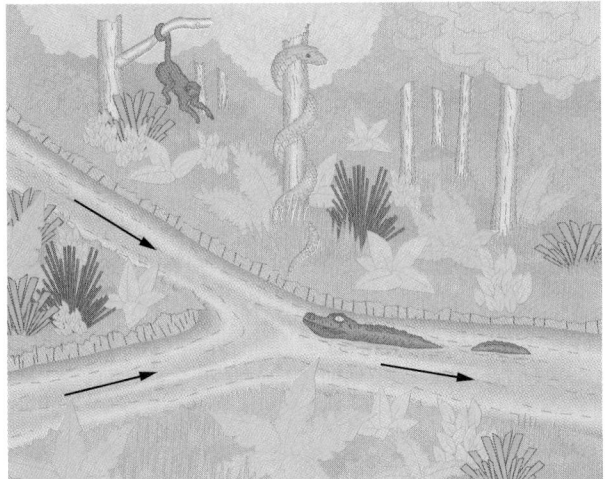

FIGURE 15-41 Exercise 53.

54E. The water flowing through a 0.75 in. (inside diameter) pipe flows out through three 0.50 in. pipes. (a) If the flow rates in the three smaller pipes are 7.0, 5.0, and 3.0 gal/min, what is the flow rate in the 0.75 in. pipe? (b) What is the ratio of the speed of water in the 0.75 in. pipe to that in the pipe carrying 7.0 gal/min?

55E. A garden hose having an internal diameter of 0.75 in. is connected to a (stationary) lawn sprinkler that consists merely of an enclosure with 24 holes, each 0.050 in. in diameter. If the water in the hose has a speed of 3.0 ft/s, at what speed does it leave the sprinkler holes?

56P. Water is pumped steadily out of a flooded basement at a speed of 5.0 m/s through a uniform hose of radius 1.0 cm. The hose passes out through a window 3.0 m above the waterline. What is the power of the pump?

57P. A river 20 m wide and 4.0 m deep drains a 3000 km² land area in which the average precipitation is 48 cm/y. One-fourth of this rainfall returns to the atmosphere by evaporation, but the remainder ultimately drains into the river. What is the average speed of the river current?

SECTION 15-10 Bernoulli's Equation

58E. Water is moving with a speed of 5.0 m/s through a pipe with a cross-sectional area of 4.0 cm². The water gradually descends 10 m as the pipe increases in area to 8.0 cm². (a) What is the speed at the lower level? (b) If the pressure at the upper level is 1.5×10^5 Pa, what is the pressure at the lower level?

59E. Models of torpedoes are sometimes tested in a horizontal pipe of flowing water, much as a wind tunnel is used to test model airplanes. Consider a circular pipe of internal diameter 25.0 cm and a torpedo model, aligned along the axis of the pipe, with a diameter of 5.00 cm. The torpedo is to be tested with water flowing past it at 2.50 m/s. (a) With what speed must the water flow in the part of the pipe that is unconstricted by the model? (b) What will the pressure difference be between the constricted and unconstricted parts of the pipe?

60E. A water intake at a pump storage reservoir (Fig. 15-42) has a cross-sectional area of 8.00 ft². The water flows in at a speed of 1.33 ft/s. At the generator building 600 ft below the intake point, the cross-sectional area is smaller than at the intake and the water flows out at 31.0 ft/s. What is the difference in pressure, in pounds per square inch, between inlet and outlet?

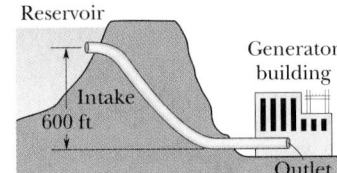

FIGURE 15-42
Exercise 60.

61E. A water pipe having a 1.0 in. inside diameter carries water into the basement of a house at a speed of 3.0 ft/s and a pressure of 25 lb/in.². If the pipe tapers to 0.50 in. and rises to the second floor 25 ft above the input point, what are (a) the speed and (b) the water pressure at the second floor?

62E. How much work is done by pressure in forcing 1.4 m³ of water through a pipe having an internal diameter of 13 mm if the difference in pressure at the two ends of the pipe is 1.0 atm?

63E. In a horizontal oil pipeline that has a constant cross-sectional area, the pressure decrease between two points 1000 ft apart is 5.0 lb/in.². What is the energy loss per cubic foot of oil per foot of pipe?

64E. A tank of large area is filled with water to a depth $D = 1.0$ ft. A hole of cross section $A = 1.0$ in.² in the bottom of the tank allows water to drain out. (a) What is the rate at which water flows out in cubic feet per second? (b) At what distance below the bottom of the tank is the cross-sectional area of the stream equal to one-half the area of the hole?

65E. Suppose that two tanks, 1 and 2, each with a large opening at the top, contain different liquids. A small hole is made in the side of each tank at the same depth h below the liquid surface, but the hole in tank 1 has half the cross-sectional area of the hole in tank 2. (a) What is the ratio ρ_1/ρ_2 of the densities of the liquids if the mass flow rate is the same for the two holes? (b) What is the ratio of the volume flow rates from the two tanks? (c) To what height above the hole in the second tank should liquid be added or drained to equalize the volume flow rates?

66E. Air flows over the top of an airplane wing of area A with speed v_t and past the underside of the wing (also of area A) with speed v_u. Show that in this simplified situation Bernoulli's equation predicts that the magnitude L of the upward lift force on the wing will be

$$L = \tfrac{1}{2}\rho A(v_t^2 - v_u^2),$$

where ρ is the density of the air.

67E. If the speed of flow past the lower surface of an airplane wing is 110 m/s, what speed of flow over the upper surface will give a pressure difference of 900 Pa between upper and lower surfaces? Take the density of air to be 1.30×10^{-3} g/cm^3, and see Exercise 66.

68E. An airplane has a wing area (each wing) of 10.0 m^2. At a certain airspeed, air flows over the upper wing surface at 48.0 m/s and over the lower wing surface at 40.0 m/s. What is the mass of the plane? Assume that the plane travels at constant velocity, that the air density is 1.20 kg/m^2, and that lift effects associated with the fuselage and tail assembly are small. Discuss the lift if the airplane, flying at the same airspeed, is (a) in level flight, (b) climbing at 15°, and (c) descending at 15°. (See Exercise 66.)

69P. Water flows through a horizontal pipe and is delivered into the atmosphere at a speed of 15 m/s as shown in Fig. 15-43. The diameters of the left and right sections of the pipe are 5.0 cm and 3.0 cm, respectively. (a) What volume of water is delivered into the atmosphere during a 10 min period? (b) What is the flow speed of the water in the left section of the pipe? (c) What is the gauge pressure in the left section of the pipe?

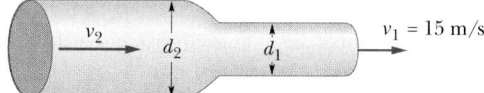

FIGURE 15-43   Problem 69.

70P. An opening of area 0.25 cm^2 in an otherwise closed beverage keg is 50 cm below the level of the liquid (of density 1.0 g/cm^3) in the keg. What is the speed of the liquid flowing through the opening if the gauge pressure in the air space above the liquid is (a) zero and (b) 0.40 atm?

71P. In a hurricane, the air (density 1.2 kg/m^3) is blowing over the roof of a house at a speed of 110 km/h. (a) What is the pressure difference between inside and outside that tends to lift the roof? (b) What would be the lifting force on a roof of area 90 m^2?

72P. The windows in an office building are of dimensions 4.00 m by 5.00 m. On a stormy day, air is blowing at 30.0 m/s past a window on the top floor. Calculate the net force on the window. The air density is 1.23 kg/m^3 there.

73P. A sniper fires a rifle bullet into a gasoline tank, making a hole 50 m below the surface of the gasoline. The tank was sealed and is under 3.0 atm absolute pressure, as shown in Fig. 15-44. The stored gasoline has a density of 660 kg/m^3. At what speed v does the gasoline begin to shoot out of the hole?

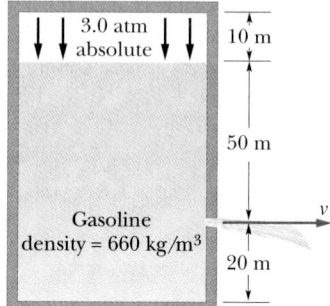

FIGURE 15-44
Problem 73.

74P. The fresh water behind a reservoir dam is 15 m deep. A horizontal pipe 4.0 cm in diameter passes through the dam 6.0 m below the water surface, as shown in Fig. 15-45. A plug secures the pipe opening. (a) Find the friction force between plug and pipe wall. (b) The plug is removed. What volume of water flows out of the pipe in 3.0 h?

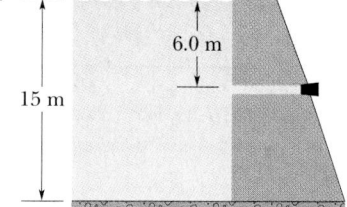

FIGURE 15-45
Problem 74.

75P. A tank is filled with water to a height H. A hole is punched in one of the walls at a depth h below the water surface (Fig. 15-46). (a) Show that the distance x from the base of the tank to the point at which the resulting stream strikes the floor is given by $x = 2\sqrt{h(H - h)}$. (b) Could a hole be punched at another depth to produce a second stream that would have the same range? If so, at what depth? (c) At what depth should the hole be placed to make the emerging stream strike the ground at the maximum distance from the base of the tank?

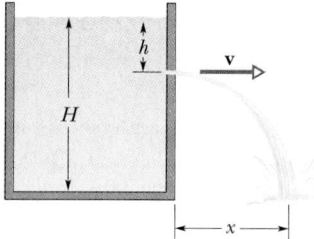

FIGURE 15-46   Problem 75.

76P. A *siphon* is a device for removing liquid from a container. It operates as shown in Fig. 15-47. Tube ABC must initially be filled, but once this has been done, liquid will flow through the tube until the liquid surface in the container is level with the tube opening at A. The liquid has density ρ and negligible viscosity. (a) With what speed does the liquid emerge from the tube at C? (b) What is the pressure in the liquid at the topmost point B? (c) Theoretically, what is the greatest possible height h_1 that a siphon can lift water?

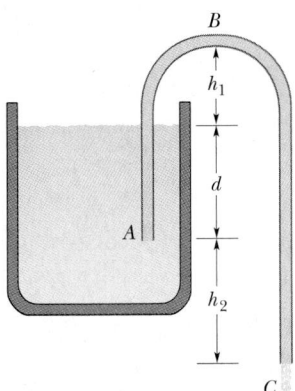

FIGURE 15-47 Problem 76.

77P. A *Venturi meter* is used to measure the flow speed of a fluid in a pipe. The meter is connected between two sections of the pipe (Fig. 15-48); the cross-sectional area A of the entrance and exit of the meter matches the pipe's cross-sectional area. Between the entrance and exit, the fluid flows from the pipe with speed v and then through a narrow "throat" of cross-sectional area a with speed V. A manometer connects the wider portion of the meter to the narrower portion. The change in the fluid's speed is accompanied by a change Δp in the fluid's pressure, which causes a height difference h of the liquid in the two arms of the manometer. (a) By applying Bernoulli's equation and the equation of continuity to points 1 and 2 in Fig. 15-48, show that

$$v = \sqrt{\frac{2a^2\,\Delta p}{\rho(A^2 - a^2)}},$$

where ρ is the density of the fluid. (b) Suppose that the fluid is water, that the cross-sectional areas are 10 in. in the pipe and 5.0 in. in the throat, and that the pressure is 8.0 lb/in.² in the pipe and 6.0 lb/in.² in the throat. What is the rate of water flow in cubic feet per second?

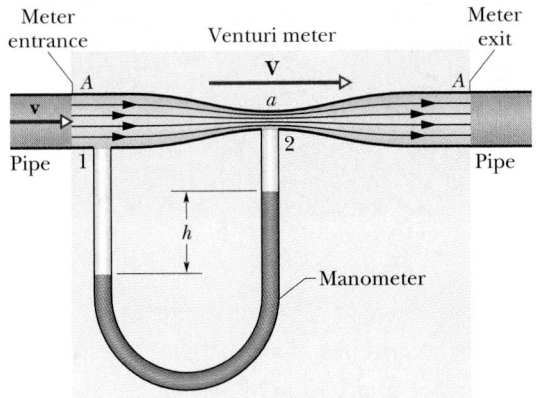

FIGURE 15-48 Problems 77 and 78.

78P. Consider the Venturi tube of Problem 77 and Fig. 15-48 without the manometer. Let A equal $5a$. Suppose that the pressure p_1 at A is 2.0 atm. (a) Compute the values of v at A and V at a that would make the pressure p_2 at a equal to zero. (b) Compute the corresponding volume flow rate if the diameter at A is 5.0 cm. The phenomenon that occurs at a when p_2 falls to nearly zero is known as cavitation. The water vaporizes into small bubbles.

79P. A pitot tube (Fig. 15-49) is used to determine the airspeed of an airplane. It consists of an outer tube with a number of small holes B (four are shown); the tube is connected to one arm of a U-tube. The other arm of the U-tube is connected to a hole, A, at the front end of the device, which points in the direction the plane is headed. At A the air becomes stagnant so that $v_A = 0$. At B, however, the speed of the air presumably equals the airspeed of the aircraft. (a) Use Bernoulli's equation to show that

$$v = \sqrt{\frac{2\rho gh}{\rho_{air}}},$$

where v is the airspeed of the plane and ρ is the density of the liquid in the U-tube. (b) The tube contains alcohol and indicates a level difference h of 26.0 cm. What is the plane's speed relative to the air? The density of the air is 1.03 kg/m³ and that of alcohol is 810 kg/m³.

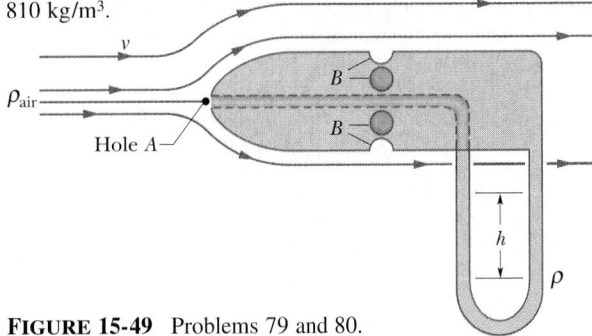

FIGURE 15-49 Problems 79 and 80.

80P. A pitot tube (see Problem 79) on a high-altitude aircraft measures a differential pressure of 180 Pa. What is the airspeed if the density of the air is 0.031 kg/m³?

16
Oscillations

Just as the third game of the 1989 World Series was about to begin near San Francisco, seismic waves from a magnitude 7.1 earthquake near Loma Prieta, 100 km distant, hit the area, causing extensive damage and killing 67 people. The photograph shows part of a 1.4 km stretch of the Nimitz Freeway, where dozens died when an upper deck collapsed onto a lower deck, trapping motorists. Obviously, the collapse was due to violent shaking by the seismic waves. But why was that particular stretch so severely damaged when the rest of the freeway, almost identical in construction, escaped collapse?

16-1 OSCILLATIONS

We are surrounded by oscillations—motions that repeat themselves. There are swinging chandeliers, boats bobbing at anchor, and the surging pistons in the engines of cars. There are oscillating guitar strings, drums, bells, diaphragms in telephones and speaker systems, and quartz crystals in wristwatches. Less evident are the oscillations of the air molecules that transmit the sensation of sound, the oscillations of the atoms in a solid that convey the sensation of temperature, and the oscillations of the electrons in the antennas of radio and TV transmitters.

Oscillations are not confined to material objects such as violin strings and electrons. Light, radio waves, x rays, and gamma rays are also oscillatory phenomena. You will study such oscillations in later chapters and will be helped greatly there by analogy with the mechanical oscillations that you are about to study here.

Oscillations in the real world are usually *damped;* that is, the motion dies out gradually, transferring mechanical energy to thermal energy by the action of frictional forces. Although we cannot totally eliminate such loss of mechanical energy, we can replenish the energy from some source. The children in Fig. 16-1, for example, know that by swinging their legs or torsos they can "pump" the swing and maintain or enhance the oscillations. In doing this, they transfer biochemical energy to mechanical energy of the oscillating systems.

16-2 SIMPLE HARMONIC MOTION

Figure 16-2 shows a sequence of "snapshots" of a simple oscillating system, a particle moving repeatedly back and forth about the origin of the x axis. In this section we simply describe the motion. Later, we shall discuss how to attain such motion.

One important property of oscillatory motion is its **frequency,** or number of oscillations that are completed each second. The symbol for frequency is f, and its SI unit is the **hertz** (abbreviated Hz), where

$$1 \text{ hertz} = 1 \text{ Hz} = 1 \text{ oscillation per second}$$
$$= 1 \text{ s}^{-1}. \tag{16-1}$$

Related to the frequency is the **period** T of the motion, which is the time for one complete oscillation (or **cycle**). That is,

$$T = \frac{1}{f}. \tag{16-2}$$

FIGURE 16-1 A child soon learns how to maintain the oscillations of a swing by transferring energy into the swing's motion.

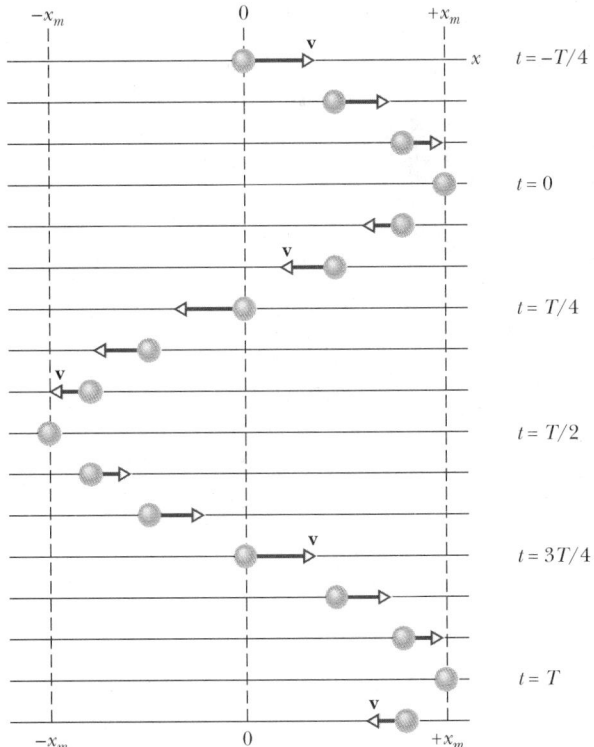

FIGURE 16-2 A sequence of "snapshots" (taken at equal time intervals) showing the position of a particle as it oscillates back and forth about the origin along the x axis, between the limits $+ x_m$ and $- x_m$. The vector arrows are scaled to indicate the speed of the particle. The speed is maximum when the particle is at the origin and zero when it is at $\pm x_m$. If the time t is chosen to be zero when the particle is at $+ x_m$, then the particle returns to $+ x_m$ at $t = T$, where T is the period of the motion. The motion is then repeated.

Any motion that repeats itself at regular intervals is called **periodic motion** or **harmonic motion.** We are interested here in motion that repeats itself in a particular way, namely, like that in Fig. 16-2. It turns out that for such motion the displacement x of the particle from the origin is given as a function of time by

$$x(t) = x_m \cos(\omega t + \phi) \quad \text{(displacement)}, \quad (16\text{-}3)$$

in which x_m, ω, and ϕ are constants. This motion is called **simple harmonic motion** (SHM), a term that means that the periodic motion is a sinusoidal function of time.

The quantity x_m in Eq. 16-3, a positive constant whose value depends on how the motion was started, is called the **amplitude** of the motion; the subscript m stands for *maximum* because the amplitude is the magnitude of the maximum displacement of the particle in either direction. The cosine function in Eq. 16-3 varies between the limits ± 1, so the displacement $x(t)$ varies between the limits $\pm x_m$, as Fig. 16-2 shows.

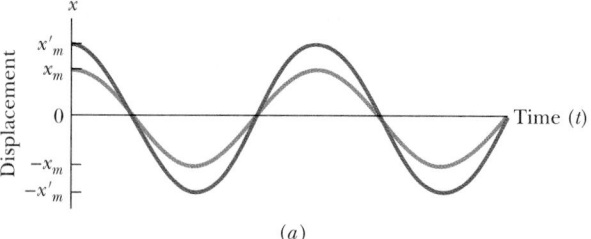

(a)

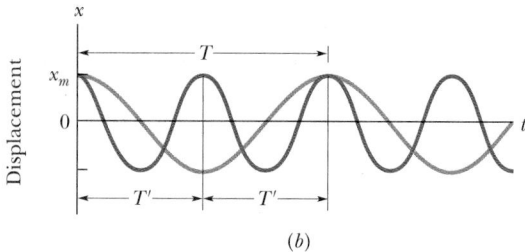

(b)

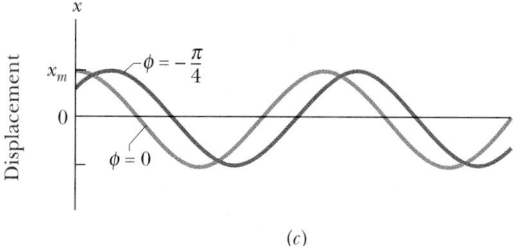

(c)

FIGURE 16-3 In all three cases, the blue curve is obtained from Eq. 16-3 with $\phi = 0$. (a) The red curve differs from the blue curve *only* in that its amplitude x'_m is greater. (b) The red curve differs from the blue curve *only* in that its period is $T' = T/2$. (c) The red curve differs from the blue curve *only* in that $\phi = -\pi/4$ rad rather than zero.

The time-varying quantity $(\omega t + \phi)$ in Eq. 16-3 is called the **phase** of the motion, and the constant ϕ is called the **phase constant** (or **phase angle**). The value of ϕ depends on the displacement and velocity of the particle at $t = 0$. For the $x(t)$ plots of Fig. 16-3a, the phase constant ϕ is zero (compare the plots and Eq. 16-3 for $t = 0$).

It remains to interpret the constant ω. The displacement $x(t)$ must return to its initial value after one period T of the motion. That is, $x(t)$ must equal $x(t + T)$ for all t. To simplify our analysis, let us put $\phi = 0$ in Eq. 16-3. From that equation we then have

$$x_m \cos \omega t = x_m \cos [\omega(t + T)].$$

The cosine function first repeats itself when its argument (the phase) has increased by 2π rad, so that we must have, in the equation above,

$$\omega(t + T) = \omega t + 2\pi$$

or

$$\omega T = 2\pi.$$

Thus from Eq. 16-2,

$$\omega = \frac{2\pi}{T} = 2\pi f. \quad (16\text{-}4)$$

The quantity ω is called the **angular frequency** of the motion; its SI unit is the radian per second. (To be consistent, then, ϕ must be in radians.) Figure 16-3 compares $x(t)$ for two simple harmonic motions that differ either in amplitude, in period (and thus in frequency and angular frequency), or in phase constant.

$\mathbf{C}$HECKPOINT **1:** A particle undergoing simple harmonic oscillation of period T (like that in Fig. 16-2) is at $-x_m$ at time $t = 0$. Is it at $-x_m$, at $+x_m$, at 0, between $-x_m$ and 0, or between 0 and $+x_m$ when (a) $t = 2.00T$, (b) $t = 3.50T$, and (c) $t = 5.25T$?

The Velocity of SHM

By differentiating Eq. 16-3, we can find an expression for the velocity of a particle moving with simple harmonic motion. That is,

$$v(t) = \frac{dx}{dt} = \frac{d}{dt} [x_m \cos(\omega t + \phi)]$$

or

$$v(t) = -\omega x_m \sin(\omega t + \phi) \quad \text{(velocity)}. \quad (16\text{-}5)$$

Figure 16-4a is a plot of Eq. 16-3 with $\phi = 0$. Figure 16-4b shows Eq. 16-5, also with $\phi = 0$. Analogous to the amplitude x_m in Eq. 16-3, the positive quantity ωx_m in Eq.

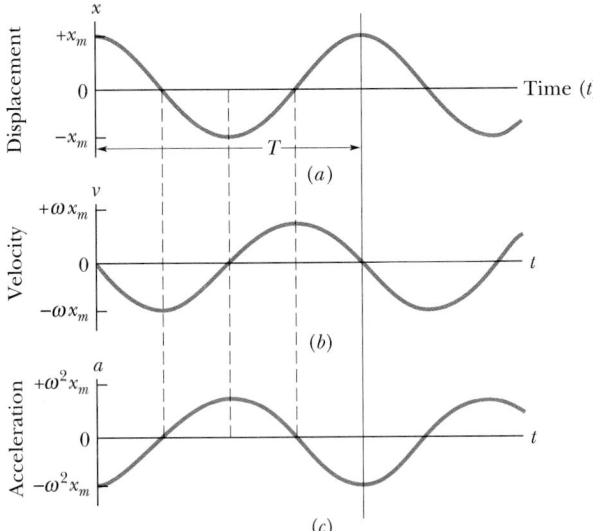

FIGURE 16-4 (a) The displacement $x(t)$ of a particle oscillating in SHM with phase angle ϕ equal to zero. The period T marks one complete oscillation. (b) The velocity $v(t)$ of the particle. (c) The acceleration $a(t)$ of the particle.

16-5 is called the **velocity amplitude** v_m. As you can see in Fig. 16-4b, the velocity of the oscillating particle varies between the limits $\pm v_m = \pm \omega x_m$. Note also in that figure that the curve of $v(t)$ is *shifted* (to the left) from the curve of $x(t)$ by one-quarter period: when the magnitude of the displacement is greatest (that is, $x(t) = x_m$), the magnitude of the velocity is least (that is, $v(t) = 0$). And when the magnitude of the displacement is least (that is, zero), the magnitude of the velocity is greatest (that is, $v_m = \omega x_m$).

The Acceleration of SHM

Knowing the velocity $v(t)$ for simple harmonic motion, we can find an expression for the acceleration of the oscillating particle by differentiating once more. Thus we have, from Eq. 16-5,

$$a(t) = \frac{dv}{dt} = \frac{d}{dt}[-\omega x_m \sin(\omega t + \phi)]$$

or

$$a(t) = -\omega^2 x_m \cos(\omega t + \phi) \quad \text{(acceleration)}. \quad (16\text{-}6)$$

Figure 16-4c is a plot of Eq. 16-6 for the case $\phi = 0$. The positive quantity $\omega^2 x_m$ in Eq. 16-6 is called the **acceleration amplitude** a_m. That is, the acceleration of the particle varies between the limits $\pm a_m = \pm \omega^2 x_m$, as Fig. 16-4c shows. Note also that the curve of $a(t)$ is shifted (to the left) by one-quarter period relative to the curve of $v(t)$.

We can combine Eqs. 16-3 and 16-6 to yield

$$a(t) = -\omega^2 x(t), \quad (16\text{-}7)$$

which is the hallmark of simple harmonic motion: the ac-

celeration is proportional to the displacement but opposite in sign, and the two quantities are related by the square of the angular frequency. Thus, as Fig. 16-4 shows, when the displacement has its greatest positive value, the acceleration has its greatest negative value, and conversely. When the displacement is zero, the acceleration is also zero.

PROBLEM SOLVING TACTICS

TACTIC 1: *Phase Angles*

Note the effect of the phase angle ϕ on a plot of $x(t)$. When $\phi = 0$, $x(t)$ has a graph like that in Fig. 16-4a, a typical cosine curve. A *negative* value for ϕ shifts the curve *rightward* along the t axis (as in Fig. 16-3c), while a *positive* value shifts it *leftward*.

Two plots of SHM with different phase angles are said to have a *phase difference;* or, each is said to be *phase-shifted* from the other, or *out of phase* with the other. The curves in Fig. 16-3c, for example, have a phase difference of $\pi/4$ rad.

Because SHM repeats after each period T and the cosine function repeats after each 2π rad, one period T represents a phase difference of 2π rad. In Fig. 16-4, $x(t)$ is phase-shifted to the right from $v(t)$ by one-quarter period, or $-\pi/2$ rad; it is shifted to the right from $a(t)$ by one-half period, or $-\pi$ rad. A phase shift of 2π rad causes a curve of SHM to coincide with itself; that is, it looks unchanged.

16-3 THE FORCE LAW FOR SIMPLE HARMONIC MOTION

Once we know how the acceleration of a particle varies with time, we can use Newton's second law to learn what force must act on the particle to give it that acceleration. If we combine Newton's second law and Eq. 16-7, we find, for simple harmonic motion,

$$F = ma = -(m\omega^2)x. \quad (16\text{-}8)$$

This result—a force proportional to the displacement but opposite in sign—is familiar. It is Hooke's law,

$$F = -kx, \quad (16\text{-}9)$$

for a spring, the spring constant here being

$$k = m\omega^2. \quad (16\text{-}10)$$

We can in fact take Eq. 16-9 as an alternative definition of simple harmonic motion. It says:

Simple harmonic motion is the motion executed by a particle of mass m subject to a force that is proportional to the displacement of the particle but opposite in sign.

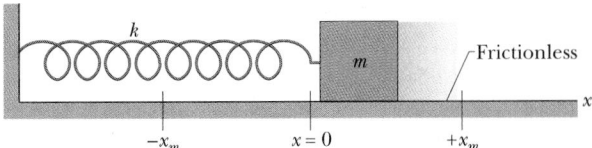

FIGURE 16-5 A linear simple harmonic oscillator. Like the particle of Fig. 16-2, the block moves in simple harmonic motion once it has been pulled to the side and released. Its displacement is then given by Eq. 16-3.

The block–spring system of Fig. 16-5 forms a **linear simple harmonic oscillator** (linear oscillator, for short), where "linear" indicates that F is proportional to x rather than to some other power of x. The angular frequency ω of the simple harmonic motion of the block is related to the spring constant k and the mass m of the block by Eq. 16-10, which yields

$$\omega = \sqrt{\frac{k}{m}} \quad \text{(angular frequency)}. \quad (16\text{-}11)$$

By combining Eqs. 16-4 and 16-11, we can write, for the **period** of the linear oscillator of Fig. 16-5,

$$T = 2\pi \sqrt{\frac{m}{k}} \quad \text{(period)}. \quad (16\text{-}12)$$

Equations 16-11 and 16-12 tell us that a large angular frequency (and thus a small period) goes with a stiff spring (large k) and a light block (small m).

Every oscillating system, be it the linear oscillator of Fig. 16-5, a diving board, or a violin string, has some element of "springiness" and some element of "inertia," or mass, and thus resembles a linear oscillator. In the linear oscillator of Fig. 16-5, these elements are located in separate parts of the system, the springiness being entirely in the spring, which we assume to be massless, and the inertia being entirely in the block, which we assume to be rigid. In a violin string, however, the two elements are both within the string itself, as you will see in Chapter 17.

CHECKPOINT **2:** Which of the following relationships between the force F on a particle and the particle's position x implies simple harmonic oscillation: (a) $F = -5x$, (b) $F = -400x^2$, (c) $F = 10x$, (d) $F = 3x^2$?

SAMPLE PROBLEM 16-1

A block whose mass m is 680 g is fastened to a spring whose spring constant k is 65 N/m. The block is pulled a distance $x = 11$ cm from its equilibrium position at $x = 0$ on a frictionless surface and released from rest at $t = 0$.

(a) What force does the spring exert on the block just before the block is released?

SOLUTION: From Hooke's law

$$F = -kx = -(65 \text{ N/m})(0.11 \text{ m})$$
$$= -7.2 \text{ N}. \qquad \text{(Answer)}$$

The minus sign reminds us that the spring force acting on the block, which points back toward the origin, is opposite the displacement of the block, which points away from the origin.

(b) What are the angular frequency, the frequency, and the period of the resulting oscillation?

SOLUTION: From Eq. 16-11 we have

$$\omega = \sqrt{\frac{k}{m}} = \sqrt{\frac{65 \text{ N/m}}{0.68 \text{ kg}}} = 9.78 \text{ rad/s}$$
$$\approx 9.8 \text{ rad/s}. \qquad \text{(Answer)}$$

The frequency follows from Eq. 16-4, which yields

$$f = \frac{\omega}{2\pi} = \frac{9.78 \text{ rad/s}}{2\pi} = 1.56 \text{ Hz} \approx 1.6 \text{ Hz}. \quad \text{(Answer)}$$

The period follows from Eq. 16-2, which yields

$$T = \frac{1}{f} = \frac{1}{1.56 \text{ Hz}} = 0.64 \text{ s} = 640 \text{ ms}. \quad \text{(Answer)}$$

(c) What is the amplitude of the oscillation?

SOLUTION: As we discussed in Section 8-4, the mechanical energy of a spring–block system like that in Fig. 16-5 is conserved because friction is not involved. Since the block is released from rest 11 cm from its equilibrium point, it has kinetic energy of zero whenever it is again 11 cm from that point. Thus its maximum displacement is 11 cm; that is,

$$x_m = 11 \text{ cm}. \qquad \text{(Answer)}$$

(d) What is the maximum speed of the oscillating block?

SOLUTION: From Eq. 16-5 we see that the velocity amplitude is

$$v_m = \omega x_m = (9.78 \text{ rad/s})(0.11 \text{ m})$$
$$= 1.1 \text{ m/s}. \qquad \text{(Answer)}$$

This maximum speed occurs when the oscillating block is rushing through the origin; compare Figs. 16-4a and 16-4b, where you can see that the speed is a maximum whenever $x = 0$.

(e) What is the magnitude of the maximum acceleration of the block?

SOLUTION: From Eq. 16-6 we see that the acceleration amplitude is

$$a_m = \omega^2 x_m = (9.78 \text{ rad/s})^2(0.11 \text{ m})$$
$$= 11 \text{ m/s}^2. \qquad \text{(Answer)}$$

This maximum acceleration occurs when the block is at the ends of its path. At those points, the force acting on the block has its maximum magnitude; compare Figs. 16-4a and 16-4c, where you can see that the magnitudes of the displacement and acceleration are maximum at the same times.

(f) What is the phase constant ϕ for the motion?

SOLUTION: At $t = 0$, the moment of release, the displacement of the block has its maximum value x_m and the velocity of the block is zero. If we put these *initial conditions,* as they are called, into Eqs. 16-3 and 16-5, we find

$$1 = \cos \phi \quad \text{and} \quad 0 = \sin \phi,$$

respectively. The smallest angle that satisfies both these requirements is

$$\phi = 0. \qquad \text{(Answer)}$$

(Any angle that is an integer multiple of 2π rad also satisfies these requirements.)

SAMPLE PROBLEM 16-2

At $t = 0$, the displacement $x(0)$ of the block in a linear oscillator like that of Fig. 16-5 is -8.50 cm. Its velocity $v(0)$ then is -0.920 m/s, and its acceleration $a(0)$ is $+47.0$ m/s^2.

(a) What are the angular frequency ω and the frequency f of this system?

SOLUTION: If we put $t = 0$ in Eqs. 16-3, 16-5, and 16-6, we find

$$x(0) = x_m \cos \phi, \qquad (16\text{-}13)$$

$$v(0) = -\omega x_m \sin \phi, \qquad (16\text{-}14)$$

and $\qquad a(0) = -\omega^2 x_m \cos \phi. \qquad (16\text{-}15)$

These three equations contain three unknowns, namely, x_m, ϕ, and ω. We should be able to find all three, but here we need only ω.

If we divide Eq. 16-15 by Eq. 16-13, the result is

$$\omega = \sqrt{-\frac{a(0)}{x(0)}} = \sqrt{-\frac{47.0 \text{ m/s}^2}{-0.0850 \text{ m}}}$$

$$= 23.5 \text{ rad/s.} \qquad \text{(Answer)}$$

The frequency f follows from Eq. 16-4 and is

$$f = \frac{\omega}{2\pi} = \frac{23.5 \text{ rad/s}}{2\pi} = 3.74 \text{ Hz.} \qquad \text{(Answer)}$$

(b) What is the phase constant ϕ?

SOLUTION: If we divide Eq. 16-14 by Eq. 16-13, we find

$$\frac{v(0)}{x(0)} = \frac{-\omega x_m \sin \phi}{x_m \cos \phi} = -\omega \tan \phi.$$

Solving for $\tan \phi$, we find

$$\tan \phi = -\frac{v(0)}{\omega x(0)} = -\frac{-0.920 \text{ m/s}}{(23.5 \text{ rad/s})(-0.0850 \text{ m})}$$

$$= -0.461.$$

This equation has two solutions:

$$\phi = -25° \quad \text{and} \quad \phi = 155°.$$

You will see in part (c) how to choose between them.

(c) What is the amplitude x_m of the motion?

SOLUTION: From Eq. 16-13 we have, provisionally putting $\phi = 155°$,

$$x_m = \frac{x(0)}{\cos \phi} = \frac{-0.0850 \text{ m}}{\cos 155°} = 0.094 \text{ m}$$

$$= 9.4 \text{ cm.} \qquad \text{(Answer)}$$

If we had chosen $\phi = -25°$, we would have found $x_m = -9.4$ cm. However, the amplitude of the motion must always be a *positive* constant, so $-25°$ cannot be the correct phase constant. We must therefore have, in part (b),

$$\phi = 155°. \qquad \text{(Answer)}$$

SAMPLE PROBLEM 16-3

In Fig. 16-6a, a uniform bar with mass m lies symmetrically across two rapidly rotating, fixed rollers, A and B, with distance $L = 2.0$ cm between the bar's center of mass and each roller. The rollers, whose directions of rotation are shown in the figure, slip against the bar with coefficient of kinetic friction $\mu_k = 0.40$. Suppose the bar is displaced horizontally by a distance x, as in Fig. 16-6b, and then released. What is the angular frequency ω of the resulting horizontal simple harmonic (back and forth) motion of the bar?

SOLUTION: To find ω, we find an expression for the horizontal acceleration a of the bar as a function of x and compare it with Eq. 16-7. We do so by applying Newton's second law vertically and horizontally; then we use the law in angular form about the contact point between the bar and roller A.

The vertical forces acting on the bar are its weight $m\mathbf{g}$ and supporting forces $\mathbf{F}_A$ due to roller A and $\mathbf{F}_B$ due to roller B. Since there is no net vertical force acting on the bar, Newton's second law gives us

$$\sum F_y = F_A + F_B - mg = 0. \qquad (16\text{-}16)$$

The horizontal forces acting on the bar are the kinetic frictional forces $f_{kA} = \mu_k F_A$ (toward the right) due to roller A

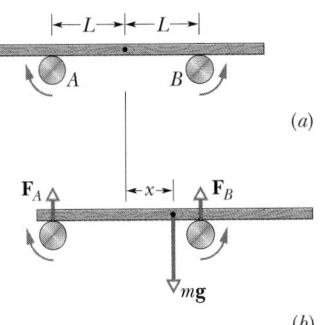

FIGURE 16-6 Sample Problem 16-3. (*a*) A bar is in equilibrium on two rotating rollers, A and B, that slip beneath it. (*b*) The bar is displaced from equilibrium by a distance x and then released.

and $f_{kB} = -\mu_k F_B$ (toward the left) due to roller B. Horizontally, Newton's second law gives us

$$\mu_k F_A - \mu_k F_B = ma,$$

or

$$a = \frac{\mu_k F_A - \mu_k F_B}{m}. \qquad (16\text{-}17)$$

The bar experiences no net torque about an axis perpendicular to the plane of Fig. 16-6 through the contact point between the bar and roller A. So Newton's second law for torques about that axis gives us

$$\sum \tau = F_A(0) + F_B 2L - mg(L + x) + f_{kA}(0) + f_{kB}(0)$$
$$= 0, \qquad (16\text{-}18)$$

where the forces $\mathbf{F}_A$, $\mathbf{F}_B$, $m\mathbf{g}$, $\mathbf{f}_{kA}$, and $\mathbf{f}_{kB}$ have moment arms about that axis of 0, $2L$, $L + x$, 0, and 0, respectively.

Solving Eqs. 16-16 and 16-18 for F_A and F_B, we find

$$F_A = \frac{mg(L - x)}{2L} \quad \text{and} \quad F_B = \frac{mg(L + x)}{2L}.$$

Substituting these results into Eq. 16-17 yields

$$a = -\frac{\mu_k g}{L} x. \qquad (16\text{-}19)$$

Comparison of Eq. 16-19 with Eq. 16-7 reveals that the bar must be undergoing simple harmonic motion with angular frequency ω given by

$$\omega^2 = \frac{\mu_k g}{L}.$$

Thus

$$\omega = \sqrt{\frac{\mu_k g}{L}} = \sqrt{\frac{(0.40)(9.8 \text{ m/s}^2)}{0.020 \text{ m}}}$$
$$= 14 \text{ rad/s}. \qquad \text{(Answer)}$$

PROBLEM SOLVING TACTICS

TACTIC 2: *Identifying SHM*

In Sample Problem 16-3 we were able to identify the motion as linear simple harmonic motion once you reached Eq. 16-19. In linear SHM the acceleration a and displacement x are related by an equation of the form

$$a = -(\text{a positive constant})x,$$

which says that the acceleration is proportional to the displacement from the equilibrium position but is in the opposite direction. Once you find such an expression, you can immediately compare it to Eq. 16-7, identify the positive constant as being equal to ω^2, and so quickly get an expression for the angular frequency of the motion. With Eq. 16-4 you then can find the period T and the frequency f.

As you will see in Sample Problem 16-8, the same technique can be used to identify angular SHM. In such motion, the angular acceleration α and angular displacement θ are re-

lated by an equation of the form

$$\alpha = -(\text{a positive constant})\theta,$$

which says that the angular acceleration is proportional to the angular displacement from the equilibrium position but is in the opposite direction. As for linear SHM, you can identify the positive constant with ω^2 to find ω, T, and f.

In some problems you might derive an expression for the force F as a function of displacement x. If the motion is linear SHM, the force and displacement are related by

$$F = -(\text{a positive constant})x,$$

which says that the force is proportional to the displacement but is in the opposite direction. Once you have found such an expression, you can immediately compare it to Eq. 16-9 and identify the positive constant as being k. If you know the mass that is involved, you can then use Eqs. 16-11, 16-12, and 16-4 to find the angular frequency ω, period T, and frequency f.

You can similarly identify angular SHM. In such motion the torque τ and angular displacement θ are related by

$$\tau = -(\text{a positive constant})\theta,$$

which says that the torque is proportional to the angular displacement from the equilibrium position but is in the opposite direction.

16-4 ENERGY IN SIMPLE HARMONIC MOTION

In Chapter 8 we saw that the energy of a linear oscillator shuttles back and forth between kinetic and potential forms, while their sum—the mechanical energy E of the oscillator—remains constant. We now consider this situation quantitatively.

The potential energy of a linear oscillator like that of Fig. 16-5 is associated entirely with the spring. Its value depends on how much the spring is stretched or compressed, that is, on $x(t)$. We can use Eqs. 8-11 and 16-3 to find

$$U(t) = \tfrac{1}{2}kx^2 = \tfrac{1}{2}kx_m^2 \cos^2(\omega t + \phi). \qquad (16\text{-}20)$$

Note carefully that a function in the form $\cos^2 A$ (as here) is the same as $(\cos A)^2$ but is *not* the same as $\cos A^2$, which means $\cos (A^2)$.

The kinetic energy of the system is associated entirely with the block. Its value depends on how fast the block is moving, that is, on $v(t)$. We can use Eq. 16-5 to find

$$K(t) = \tfrac{1}{2}mv^2 = \tfrac{1}{2}m\omega^2 x_m^2 \sin^2(\omega t + \phi). \qquad (16\text{-}21)$$

If we use Eq. 16-11 to substitute k/m for ω^2, we can write

Eq. 16-21 as

$$K(t) = \tfrac{1}{2}mv^2 = \tfrac{1}{2}kx_m^2 \sin^2(\omega t + \phi). \quad (16\text{-}22)$$

The mechanical energy follows from Eqs. 16-20 and 16-22 and is

$$E = U + K$$
$$= \tfrac{1}{2}kx_m^2 \cos^2(\omega t + \phi) + \tfrac{1}{2}kx_m^2 \sin^2(\omega t + \phi)$$
$$= \tfrac{1}{2}kx_m^2 [\cos^2(\omega t + \phi) + \sin^2(\omega t + \phi)].$$

For any angle α,

$$\cos^2 \alpha + \sin^2 \alpha = 1.$$

Thus the quantity in the square brackets above is unity and we have

$$E = U + K = \tfrac{1}{2}kx_m^2. \quad (16\text{-}23)$$

The mechanical energy of a linear oscillator is indeed a constant, independent of time. The potential energy and kinetic energy of the linear oscillator are shown as functions of time in Fig. 16-7a, and as functions of displacement in Fig. 16-7b.

You might now understand why an oscillating system normally contains an element of springiness and an ele-

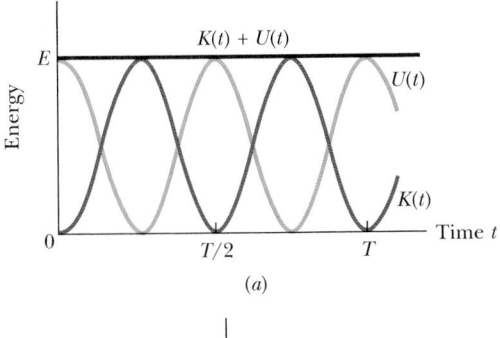

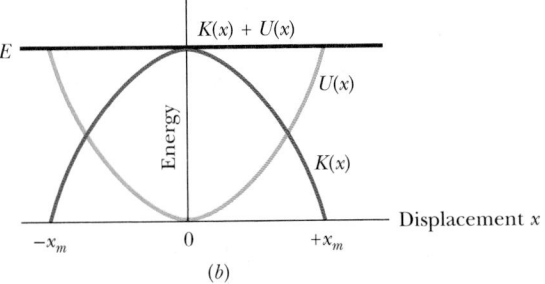

FIGURE 16-7 (a) The potential energy $U(t)$, kinetic energy $K(t)$, and mechanical energy E as functions of time, for a linear harmonic oscillator. Note that all energies are positive and that the potential energy and the kinetic energy peak twice during every period. (b) The potential energy $U(x)$, kinetic energy $K(x)$, and mechanical energy E as functions of position, for a linear harmonic oscillator with amplitude x_m. For $x = 0$ the energy is all kinetic, and for $x = \pm x_m$ it is all potential.

ment of inertia: it uses the former to store its potential energy and the latter to store its kinetic energy. The system of Sample Problem 16-3 is an exception, because no potential energy is involved. Instead, the energy transferred to the kinetic energy of the bar by the rollers actually varies over time.

CHECKPOINT **3:** In Fig. 16-5, the block has a kinetic energy of 3 J and the spring has an elastic potential energy of 2 J when the block is at $x = +2.0$ cm. (a) What is the kinetic energy at $x = 0$? What are the elastic potential energies at (b) $x = -2.0$ cm and (c) $x = -x_m$?

SAMPLE PROBLEM 16-4

(a) What is the mechanical energy of the linear oscillator of Sample Problem 16-1?

SOLUTION: We find, substituting data from Sample Problem 16-1 into Eq. 16-23,

$$E = \tfrac{1}{2}kx_m^2 = (\tfrac{1}{2})(65 \text{ N/m})(0.11 \text{ m})^2$$
$$= 0.393 \text{ J} \approx 0.39 \text{ J}. \quad \text{(Answer)}$$

This value remains constant throughout the motion.

(b) What is the potential energy of this oscillator when the block is halfway to its end point, that is, when $x = \pm\tfrac{1}{2}x_m$?

SOLUTION: For any displacement, the potential energy of a spring–block system is given by $U = \tfrac{1}{2}kx^2$. Here,

$$U = \tfrac{1}{2}kx^2 = \tfrac{1}{2}k(\tfrac{1}{2}x_m)^2 = \tfrac{1}{4}(\tfrac{1}{2}kx_m^2)$$
$$= \tfrac{1}{4}E = (\tfrac{1}{4})(0.393 \text{ J}) = 0.098 \text{ J}. \quad \text{(Answer)}$$

(c) What is the kinetic energy of the oscillator when $x = \tfrac{1}{2}x_m$?

SOLUTION: We find this from

$$K = E - U$$
$$= 0.393 \text{ J} - 0.098 \text{ J} \approx 0.30 \text{ J}. \quad \text{(Answer)}$$

Thus, at this point during the oscillation, 25% of the energy is in potential form and 75% in kinetic form.

16-5 AN ANGULAR SIMPLE HARMONIC OSCILLATOR

Figure 16-8 shows an angular version of a simple harmonic oscillator; the element of springiness or elasticity is associated with the twisting of a suspension wire rather than the extension and compression of a spring as we previously had. The device is called a **torsion pendulum,** with *torsion* referring to the twisting.

FIGURE 16-8 An angular simple harmonic oscillator, or torsion pendulum, is an angular version of the linear simple harmonic oscillator of Fig. 16-5. The disk oscillates in a horizontal plane; the reference line oscillates with angular amplitude θ_m. The twist in the suspension wire stores potential energy as a spring does and provides the restoring torque.

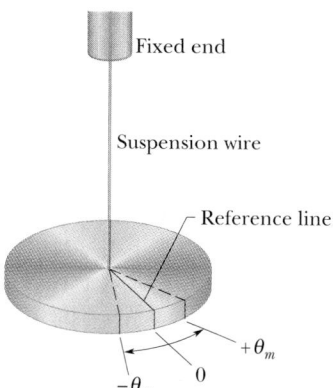

Fixed end

Suspension wire

Reference line

$+\theta_m$

$-\theta_m$

0

If we rotate the disk in Fig. 16-8 from its rest position (where the reference line is at 0) and release it, it will oscillate about that position in **angular simple harmonic motion.** Rotating the disk through an angle θ in either direction introduces a restoring torque given by

$$\tau = -\kappa\theta. \qquad (16\text{-}24)$$

Here κ (Greek *kappa*) is a constant, called the **torsion constant,** that depends on the length, diameter, and material of the suspension wire.

Comparison of Eq. 16-24 with Eq. 16-9 leads us to suspect that Eq. 16-24 is the angular form of Hooke's law, and that we can transform Eq. 16-12, which gives the period of linear SHM, into an equation for the period of angular SHM: we replace the spring constant k in Eq. 16-12 with its equivalent, the constant κ of Eq. 16-24, and we replace the mass m in Eq. 16-12 with *its* equivalent, the rotational inertia I of the oscillating disk. These replacements lead to

$$T = 2\pi\sqrt{\frac{I}{\kappa}} \qquad \text{(torsion pendulum)} \quad (16\text{-}25)$$

for the period of the angular simple harmonic oscillator, or torsion pendulum.

SAMPLE PROBLEM 16-5

As Fig. 16-9a shows, a thin rod whose length L is 12.4 cm and whose mass m is 135 g is suspended at its midpoint from a long wire. Its period T_a of angular SHM is measured to be 2.53 s. An irregularly shaped object, which we call object X, is then hung from the same wire, as in Fig. 16-9b, and its period T_b is found to be 4.76 s.

(a) What is the rotational inertia of object X about its suspension axis?

SOLUTION: In Table 11-2(e), the rotational inertia of a thin rod about a perpendicular axis through its midpoint is given as $\frac{1}{12}mL^2$. Thus we have

$$I_a = \tfrac{1}{12}mL^2 = (\tfrac{1}{12})(0.135 \text{ kg})(0.124 \text{ m})^2$$
$$= 1.73 \times 10^{-4} \text{ kg} \cdot \text{m}^2.$$

Now let us write Eq. 16-25 twice, once for the rod and once for object X:

$$T_a = 2\pi\sqrt{\frac{I_a}{\kappa}} \quad \text{and} \quad T_b = 2\pi\sqrt{\frac{I_b}{\kappa}}.$$

Here the subscripts refer to Figs. 16-9a and 16-9b. The constant κ, which is a property of the wire, is the same for both figures; only the periods and the rotational inertias differ.

Let us square each of these equations, divide the second by the first, and solve the resulting equation for I_b. The result is

$$I_b = I_a\frac{T_b^2}{T_a^2} = (1.73 \times 10^{-4} \text{ kg} \cdot \text{m}^2)\frac{(4.76 \text{ s})^2}{(2.53 \text{ s})^2}$$
$$= 6.12 \times 10^{-4} \text{ kg} \cdot \text{m}^2. \qquad \text{(Answer)}$$

(b) What would be the period of oscillation if both objects were fastened together and hung from the wire, as in Fig. 16-9c?

SOLUTION: We can again write Eq. 16-25 twice, this time as

$$T_a = 2\pi\sqrt{\frac{I_a}{\kappa}} \quad \text{and} \quad T_c = 2\pi\sqrt{\frac{I_c}{\kappa}}.$$

After dividing again and putting $I_c = I_a + I_b$, we find

$$T_c = T_a\sqrt{\frac{I_c}{I_a}} = T_a\sqrt{\frac{I_a + I_b}{I_a}} = T_a\sqrt{1 + \frac{I_b}{I_a}}$$
$$= (2.53)\sqrt{1 + \frac{6.12 \times 10^{-4} \text{ kg} \cdot \text{m}^2}{1.73 \times 10^{-4} \text{ kg} \cdot \text{m}^2}}$$
$$= 5.39 \text{ s}. \qquad \text{(Answer)}$$

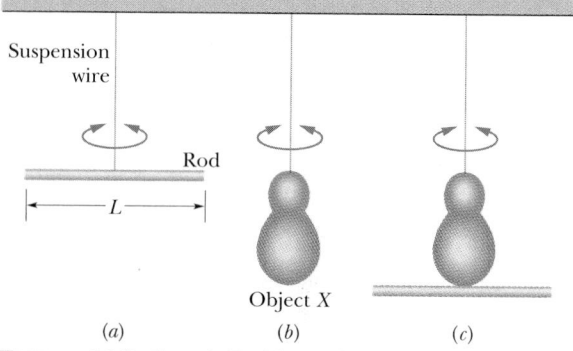

Suspension wire

Rod

L

Object X

(a)   (b)   (c)

FIGURE 16-9 Sample Problem 16-5. Three torsion pendulums consisting of a wire and (a) a rod, (b) an irregularly shaped object, and (c) the rod and the object rigidly connected.

16-6 PENDULUMS

We turn now to a class of simple harmonic oscillators in which the springiness is associated with the gravitational force rather than with the elastic properties of a twisted wire or a compressed or stretched spring.

The Simple Pendulum

If you hang an apple at the end of a long thread fixed at its upper end, and then set the apple swinging back and forth a small distance, you easily see that the apple's motion is periodic. Is it, in fact, simple harmonic motion? To idealize this situation, we consider a **simple pendulum,** which consists of a particle of mass m (called the *bob* of the pendulum) suspended from an unstretchable, massless string of length L, as in Fig. 16-10a. The bob is free to swing back and forth in the plane of the page, to the left and right of a vertical line through the point at which the upper end of the string is fixed.

The element of inertia in this pendulum is the mass m of the particle, and the element of springiness is in the gravitational attraction between the particle and Earth. Potential energy can be associated with the varying vertical distance between the swinging particle and Earth; we may view that varying distance as the varying length of a "gravitational spring."

The forces acting on the particle, shown in Fig. 16-10b, are its weight $m\mathbf{g}$ and the tension $\mathbf{T}$ in the string. We resolve $m\mathbf{g}$ into a radial component $mg \cos \theta$ and a component $mg \sin \theta$ that is tangent to the path taken by the particle. This tangential component is a restoring force, because it always acts opposite the displacement of the particle so as to bring the particle back toward its central location, the

equilibrium position $(\theta = 0)$, where it would be at rest were it not swinging. We write the restoring force as

$$F = -mg \sin \theta, \qquad (16\text{-}26)$$

where the minus sign indicates that F acts opposite the displacement.

If we assume that the angle θ in Fig. 16-10 is small, then $\sin \theta$ is very nearly equal to θ in radians. [For example, when $\theta = 5.00° (= 0.0873 \text{ rad})$, $\sin \theta = 0.0872$, a difference of only about 0.1%.] Also, the displacement s of the particle measured along its arc is equal to $L\theta$. Thus, for small θ, Eq. 16-26 becomes

$$F \approx -mg\theta = -mg \frac{s}{L} = -\left(\frac{mg}{L}\right)s. \quad (16\text{-}27)$$

A glance back at Eq. 16-9 shows that we again have Hooke's law, with the displacement now being arc length s instead of x. Thus *if a simple pendulum swings through a small angle*, it is a linear oscillator like the block–spring oscillator of Fig. 16-5; that is, it undergoes simple harmonic motion. Now the amplitude of the motion is measured as the **angular amplitude** θ_m, the maximum angle of swing. And the spring constant k is mg/L, the effective spring constant of the pendulum's gravitational spring.

By substituting mg/L for k in Eq. 16-12, we find, for the period of a simple pendulum,

$$T = 2\pi \sqrt{\frac{m}{k}} = 2\pi \sqrt{\frac{m}{mg/L}} \qquad (16\text{-}28)$$

or

$$T = 2\pi \sqrt{\frac{L}{g}} \qquad \text{(simple pendulum).} \quad (16\text{-}29)$$

Equation 16-29 holds only if the angular amplitude θ_m is small (which is what we assume in the exercises and problems for this chapter unless otherwise stated).

The element of inertia seems to be missing in Eq. 16-29 because the period is independent of the mass of the particle. This comes about because the element of springiness, which is the gravitational spring constant mg/L, is itself proportional to the mass of the particle, and the two masses cancel in Eq. 16-28. Figure 8-7 shows how energy shuttles back and forth between potential and kinetic forms during every oscillation of a simple pendulum.

The Physical Pendulum

Most pendulums in the real world are not even approximately "simple." Figure 16-11 shows a generalized **physical pendulum,** as we shall call realistic pendulums, with its weight $m\mathbf{g}$ acting at its center of mass C.

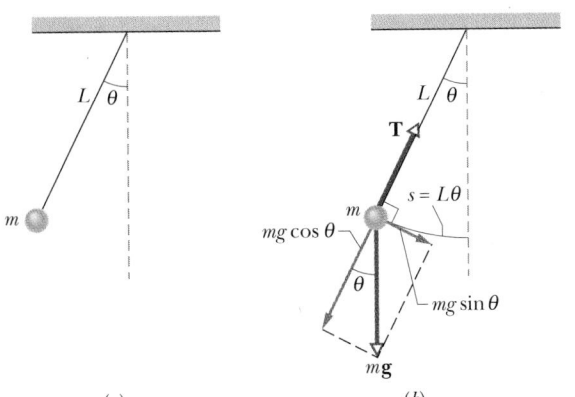

(a)   *(b)*

FIGURE 16-10 (*a*) A simple pendulum. (*b*) The forces acting on the bob are its weight $m\mathbf{g}$ and the tension $\mathbf{T}$ in the string. The tangential component $mg \sin \theta$ of the weight is a restoring force that brings the pendulum back to the central position.

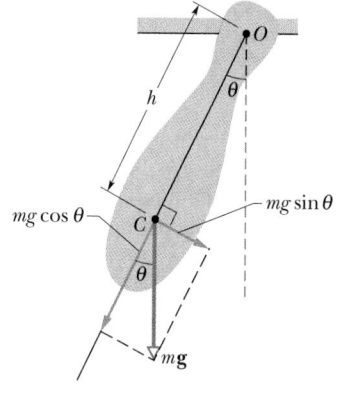

FIGURE 16-11 A physical pendulum. The restoring torque is $(mg \sin \theta)(h)$. When $\theta = 0$, the center of mass C hangs directly below the point of suspension O.

When the pendulum of Fig. 16-11 is displaced through an angle θ in either direction from its equilibrium position, a restoring torque appears. This torque acts about an axis through the suspension point O in Fig. 16-11 and has the magnitude

$$\tau = -(mg \sin \theta)(h). \qquad (16\text{-}30)$$

Here $mg \sin \theta$ is the tangential component of the weight $m\mathbf{g}$, and h (which is equal to OC) is the moment arm of this force component. The minus sign indicates that the torque is a restoring torque. That is, the torque always acts to reduce the angle θ to zero.

We once more decide to limit our interest to small amplitudes, so that $\sin \theta \approx \theta$. Then Eq. 16-30 becomes

$$\tau \approx -(mgh)\theta. \qquad (16\text{-}31)$$

Comparison with Eq. 16-24 shows that we again have Hooke's law in angular form and that the physical pendulum is an angular harmonic oscillator. Thus a physical pendulum undergoes simple harmonic motion *if* the angular amplitude θ_m of its motion is small. The term mgh of Eq. 16-31 is analogous to the torsion constant κ of Eq. 16-24. Substituting mgh for κ in Eq. 16-25, we find

$$T = 2\pi \sqrt{\frac{I}{mgh}} \quad \text{(physical pendulum)} \qquad (16\text{-}32)$$

for the period of a physical pendulum when θ_m is small. Here I is the rotational inertia of the pendulum (about an axis through its point of support and perpendicular to its plane of swing), and h is the distance between the point of support and the center of mass of the swinging pendulum.

A physical pendulum will not swing if we hang it by its center of mass. Formally, this corresponds to putting $h = 0$ in Eq. 16-32. That equation then predicts $T \to \infty$, which implies that such a pendulum will never complete one swing.

Corresponding to any physical pendulum that oscillates about a given suspension point O with period T is a simple pendulum of length L_0 with the same period T. We can find L_0 with Eq. 16-29. The point along the physical pendulum at distance L_0 from point O is called the *center of oscillation* of the physical pendulum for the given suspension point.

The physical pendulum of Fig. 16-11 includes the simple pendulum as a special case. For the latter, h is the length L of the string and I is mL^2. Making these substitutions in Eq. 16-32 leads to

$$T = 2\pi \sqrt{\frac{I}{mgh}} = 2\pi \sqrt{\frac{mL^2}{mgL}} = 2\pi \sqrt{\frac{L}{g}},$$

which is exactly Eq. 16-29, the expression for the period of a simple pendulum.

Measuring g

We can use a physical pendulum to measure the free-fall acceleration g. (Countless thousands of such measurements have been made during geophysical prospecting.)

To analyze a simple case, take the pendulum to be a uniform rod of length L, suspended from one end. For such a pendulum, h in Eq. 16-32, the distance between the suspension point and the center of mass, is $\frac{1}{2}L$. Table 11-2(f) tells us that the rotational inertia of this pendulum about a perpendicular axis through one end is $\frac{1}{3}mL^2$. If we put $h = \frac{1}{2}L$ and $I = \frac{1}{3}mL^2$ in Eq. 16-32 and solve for g, we find

$$g = \frac{8\pi^2 L}{3T^2}. \qquad (16\text{-}33)$$

Thus by measuring L and the period T, we can find the value of g. (If precise measurements are to be made, a number of refinements are needed, such as swinging the pendulum in an evacuated chamber.)

CHECKPOINT **4:** Three physical pendulums, of masses m_0, $2m_0$, and $3m_0$, have the same shape and size and are suspended at the same point. Rank the masses according to the periods of the pendulums, greatest period first.

SAMPLE PROBLEM 16-6

A meter stick, suspended from one end, swings as a physical pendulum, as in Fig. 16-12a.

(a) What is its period of oscillation?

SOLUTION: From Table 11-2(f) we see that the rotational inertia of a rod or stick of length L about a perpendicular axis through one end is $\frac{1}{3}mL^2$. The distance h from the point of suspension to the center of mass, which is point C in Fig.

16-12*a*, is $\frac{1}{2}L$. If we substitute these two quantities into Eq. 16-32, we find

$$T = 2\pi\sqrt{\frac{I}{mgh}} = 2\pi\sqrt{\frac{\frac{1}{3}mL^2}{mg(\frac{1}{2}L)}} = 2\pi\sqrt{\frac{2L}{3g}} \qquad (16\text{-}34)$$

$$= 2\pi\sqrt{\frac{(2)(1.00\text{ m})}{(3)(9.8\text{ m/s}^2)}} = 1.64\text{ s}. \qquad \text{(Answer)}$$

(b) For the stick of Fig. 16-12*a*, what is the distance L_0 between the suspension point O of the stick and the center of oscillation of the stick?

SOLUTION: To locate the center of oscillation of the meter stick, we find the length L_0 of a simple pendulum (Fig. 16-12*b*) having the same period T as the meter stick. Setting Eqs. 16-29 and 16-34 equal yields

$$T = 2\pi\sqrt{\frac{L_0}{g}} = 2\pi\sqrt{\frac{2L}{3g}}.$$

You can see by inspection that

$$L_0 = \frac{2}{3}L = (\frac{2}{3})(100\text{ cm}) = 66.7\text{ cm}. \qquad \text{(Answer)}$$

In Fig. 16-12*a*, point P marks this distance from suspension point O. Thus point P is the stick's center of oscillation for the given suspension point.

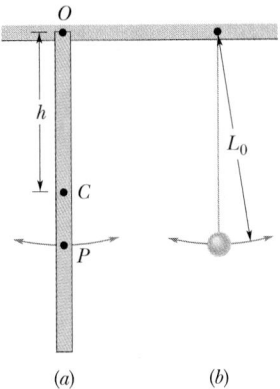

(a)   *(b)*

FIGURE 16-12 Sample Problem 16-6. (*a*) A meter stick suspended from one end as a physical pendulum. (*b*) A simple pendulum whose length L is chosen so that the periods of the two pendulums are equal. Point P on the pendulum of (*a*) marks the center of oscillation, a distance L_0 from the suspension point.

SAMPLE PROBLEM 16-7

A disk whose radius R is 12.5 cm is suspended, as a physical pendulum, from a point at distance h from its center C (Fig. 16-13). Its period T is 0.871 s when $h = R/2$. What is the free-fall acceleration g at the location of the pendulum?

SOLUTION: The rotational inertia I_{cm} of a disk about its central axis is $\frac{1}{2}mR^2$. From the parallel-axis theorem, the rotational inertia about an axis parallel to the central axis and through the point of suspension O, as in Fig. 16-13, is

$$I = I_{cm} + mh^2 = \frac{1}{2}mR^2 + m(\frac{1}{2}R)^2 = \frac{3}{4}mR^2.$$

If we put $I = \frac{3}{4}mR^2$ and $h = \frac{1}{2}R$ in Eq. 16-32, we find

$$T = 2\pi\sqrt{\frac{I}{mgh}} = 2\pi\sqrt{\frac{\frac{3}{4}mR^2}{mg(\frac{1}{2}R)}} = 2\pi\sqrt{\frac{3R}{2g}}.$$

Solving for g then gives us

$$g = \frac{6\pi^2 R}{T^2} = \frac{(6\pi^2)(0.125\text{ m})}{(0.871\text{ s})^2} = 9.76\text{ m/s}^2. \qquad \text{(Answer)}$$

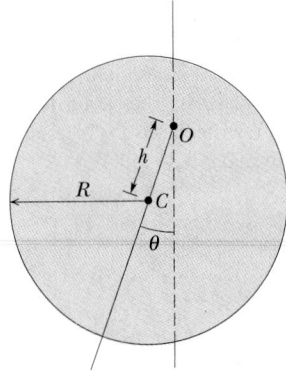

FIGURE 16-13 Sample Problem 16-7. A physical pendulum consisting of a uniform disk suspended from a point (O) that is halfway from the center of the disk to the rim.

SAMPLE PROBLEM 16-8

In Fig. 16-14, a penguin (obviously skilled in aquatic sports) dives from a uniform board that is hinged at the left and attached to a spring at the right. The board has length $L = 2.0$ m and mass $m = 12$ kg; the spring constant k is 1300 N/m. When the penguin dives, it leaves the board and spring oscillating with a small amplitude. Assume that the board is stiff enough not to bend, and find the period T of the oscillations.

SOLUTION: The spring applies a time-varying torque τ (about the hinge) to the board, which undergoes a time-varying angular acceleration α about the hinge. Since a spring is involved, we might guess that simple harmonic motion is also involved, but we will not draw that conclusion yet. Instead, we use Eqs. 11-30 and 11-35 to write

$$\tau = LF\sin 90° = I\alpha. \qquad (16\text{-}35)$$

Here I is the rotational inertia of the board about the hinge, F is the force on the right end of the board from the spring, and 90° is the angle between the length of the board and the direction of that force.

We may treat the board as being a thin rod pivoted about one end, so that from Table 11-2(f), $I = mL^2/3$. The force F from the spring is $F = -kx$, where x is the vertical displacement of the right end of the board.

Substituting these expressions for I and F into the rightmost equality of Eq. 16-35 and evaluating sin 90°, we find

$$-Lkx = \frac{mL^2\alpha}{3}. \qquad (16\text{-}36)$$

We now have a mixture of linear displacement x (vertically) and rotational acceleration α (about the hinge). We can rewrite Eq. 16-36 in terms of just the rotational motion of the right end of the board about the hinge by using Eq. 11-15,

$$s = \theta r.$$

Here θ is the angle of the board's rotation about the hinge, $r = L$ is the radius of the rotation, and s is the arc length, which is approximated by the vertical displacement x (for small θ). We can rewrite this as $x = \theta L$ and then substitute for x in Eq. 16-36 to find

$$-Lk\theta L = \frac{mL^2\alpha}{3},$$

which becomes

$$\alpha = -\frac{3k}{m}\,\theta. \qquad (16\text{-}37)$$

Equation 16-37 is an angular version of Eq. 16-7. It tells us that the board does indeed undergo simple harmonic motion, with angular acceleration α and angular displacement θ. Comparing Eqs. 16-37 and 16-7 reveals that

$$\omega^2 = \frac{3k}{m}$$

and thus that $\omega = \sqrt{3k/m}$. Using Eq. 16-4, $\omega = 2\pi/T$, we then have

$$T = 2\pi\sqrt{\frac{m}{3k}} = 2\pi\sqrt{\frac{12\text{ kg}}{3(1300\text{ N/m})}}$$

$$= 0.35\text{ s.} \qquad \text{(Answer)}$$

Perhaps surprisingly, the period is independent of the board's length L.

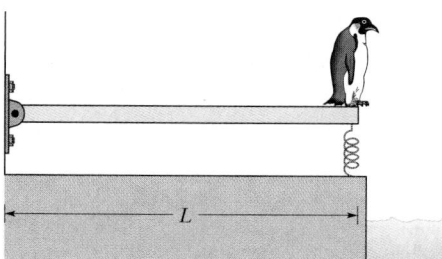

FIGURE 16-14   Sample Problem 16-8. The dive by the penguin causes the board and spring to oscillate; the board pivots about the hinge at the left.

16-7 SIMPLE HARMONIC MOTION AND UNIFORM CIRCULAR MOTION

In 1610, Galileo, using his newly constructed telescope, discovered the four principal moons of Jupiter. Over weeks of observation, each moon seemed to him to be moving back and forth relative to the planet in what today we would call simple harmonic motion; the disk of the planet was the midpoint of the motion. The record of Galileo's observations, written in his own hand, is still available. A. P. French of MIT used Galileo's data to work out the position of the moon Callisto relative to Jupiter. In the results shown in Fig. 16-15, the circles are based on Galileo's observations and the curve is a best fit to the data. The curve strongly suggests Eq. 16-3, the displacement function for SHM. A period of about 16.8 days can be measured from the plot.

Actually, Callisto moves with essentially constant speed in an essentially circular orbit around Jupiter. Its true motion—far from being simple harmonic—is uniform circular motion. What Galileo saw—and what you can see with a good pair of binoculars and a little patience—is the projection of this uniform circular motion on a line in the plane of the motion. We are led by Galileo's remarkable observations to the conclusion that simple harmonic motion is uniform circular motion viewed edge-on. In more formal language:

> Simple harmonic motion is the projection of uniform circular motion on a diameter of the circle in which the latter motion occurs.

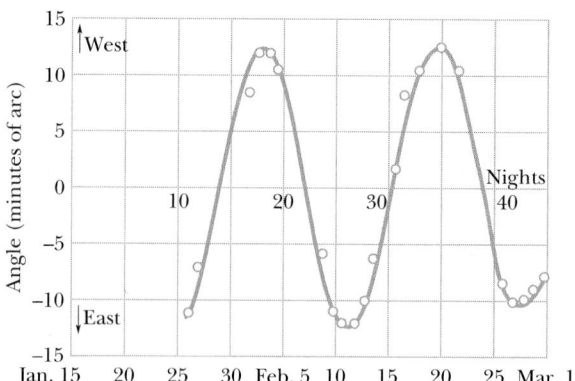

FIGURE 16-15   The angle between Jupiter and its moon Callisto as seen from Earth. The circles are based on Galileo's 1610 measurements. The curve is a best fit, strongly suggesting simple harmonic motion. At Jupiter's mean distance, 10 minutes of arc corresponds to about 2×10^6 km. (Adapted from A. P. French, *Newtonian Mechanics,* W. W. Norton & Company, New York, 1971, p. 288.)

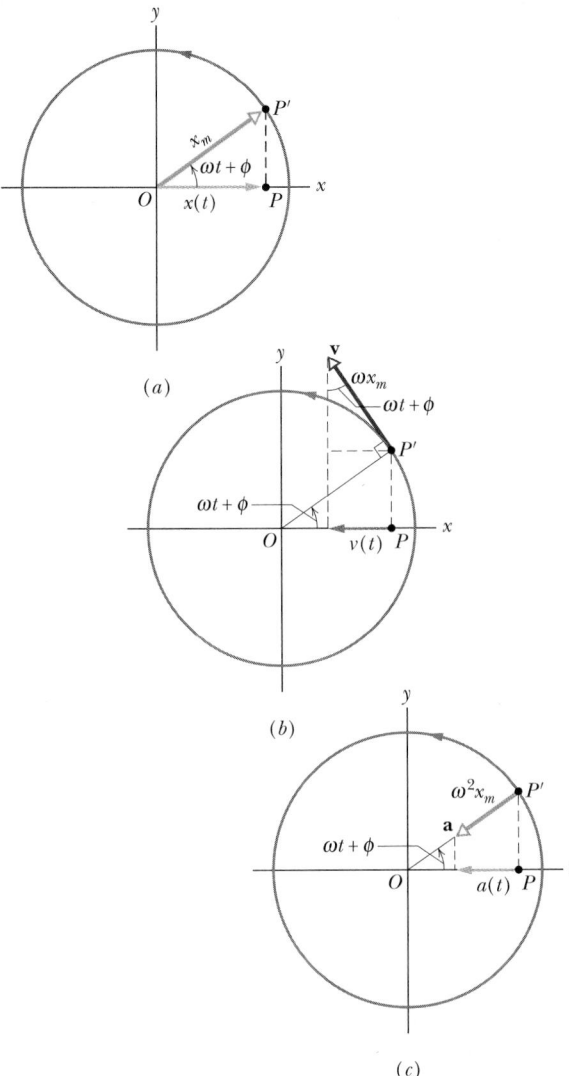

FIGURE 16-16 (a) A reference particle P' moving with uniform circular motion in a reference circle of radius x_m. Its projection P on the x axis executes simple harmonic motion. (b) The projection of the velocity **v** of the reference particle is the velocity of SHM. (c) The projection of the acceleration **a** of the reference particle is the acceleration of SHM.

Figure 16-16a gives an example. It shows a *reference particle* P' moving in uniform circular motion with (constant) angular speed ω in a *reference circle*. The radius x_m of SI circle is the magnitude of the particle's position vector. At any time t, the angular position of the particle is $\omega t + \phi$, where ϕ is the angular position at $t = 0$.

The projection of particle P' onto the x axis is a point P, which we take to be a second particle. The projection of the position vector of particle P' onto the x axis gives the location $x(t)$ of P. Thus we find

$$x(t) = x_m \cos(\omega t + \phi),$$

which is precisely Eq. 16-3. Our conclusion is correct. If reference particle P' moves in uniform circular motion, its projection particle P moves in simple harmonic motion.

This relation throws new light on the angular frequency ω of simple harmonic motion, and you can see more clearly how the "angular" originates. The quantity ω is simply the constant angular speed with which the reference particle P' moves along its reference circle. The phase constant ϕ has a value determined by the position of reference particle P' along its reference circle at $t = 0$.

Figure 16-16b shows the velocity of the reference particle. The magnitude of the velocity vector is ωx_m, and its projection on the x axis is

$$v(t) = -\omega x_m \sin(\omega t + \phi),$$

which is exactly Eq. 16-5. The minus sign appears because the velocity component of P in Fig. 16-16b points to the left, in the direction of decreasing x.

Figure 16-16c shows the acceleration of the reference particle. The magnitude of the acceleration vector is $\omega^2 x_m$ and its projection on the x axis is

$$a(t) = -\omega^2 x_m \cos(\omega t + \phi),$$

which is exactly Eq. 16-6. Thus whether we look at the displacement, the velocity, or the acceleration, the projection of uniform circular motion is indeed simple harmonic motion.

16-8 DAMPED SIMPLE HARMONIC MOTION (OPTIONAL)

A pendulum will swing hardly at all under water, because the water exerts a drag force on the pendulum that quickly eliminates the motion. A pendulum swinging in air does better, but still the motion dies out because the air exerts a drag force on the pendulum (and friction acts at its support), transferring energy from the pendulum's motion.

When the motion of an oscillator is reduced by an external force, the oscillator and its motion are said to be **damped**. An idealized example of a damped oscillator is shown in Fig. 16-17: a block with mass m oscillates on a spring with spring constant k. From the mass, a rod extends to a vane (both assumed massless) that is submerged in a liquid. As the vane moves up and down, the liquid exerts an inhibiting drag force on it and thus on the entire oscillating system. With time, the mechanical energy of the block–spring system decreases, as energy is transferred to thermal energy of the liquid and vane.

Let us assume that the liquid exerts a **damping force** $\mathbf{F}_d$ that is proportional in magnitude to the velocity **v** of the

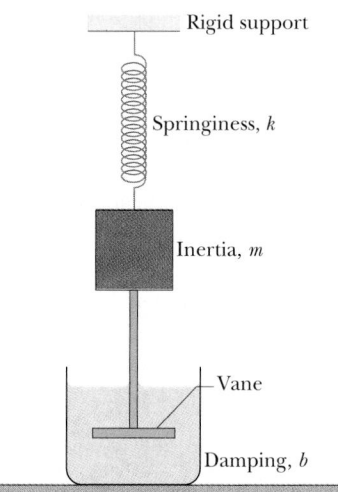

FIGURE 16-17 An idealized damped simple harmonic oscillator. A vane immersed in a liquid exerts a damping force on the oscillating block.

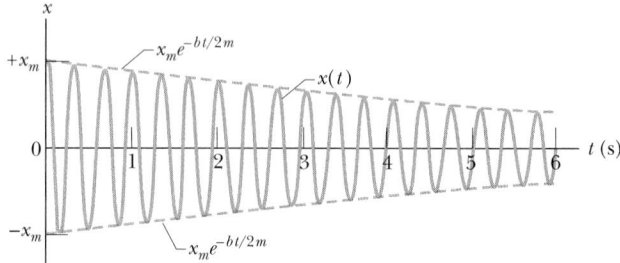

FIGURE 16-18 The displacement function $x(t)$ for the damped oscillator of Fig. 16-17. See Sample Problem 16-9. The amplitude, which is $x_m e^{-bt/2m}$, decreases exponentially with time.

vane and block (an assumption that is accurate if the vane moves slowly). Then

$$F_d = -bv, \qquad (16\text{-}38)$$

where b is a **damping constant** that depends on the characteristics of both the vane and the liquid and has the SI unit of meter per second. The minus sign indicates that $\mathbf{F}_d$ opposes the motion. The total force acting on the block is then

$$\sum F = -kx - bv,$$

or, if we set $v = dx/dt$,

$$\sum F = -kx - b\frac{dx}{dt}. \qquad (16\text{-}39)$$

Substituting this total force into Newton's second law yields the differential equation

$$m\frac{d^2x}{dt^2} + b\frac{dx}{dt} + kx = 0,$$

whose solution is

$$x(t) = x_m e^{-bt/2m} \cos(\omega' t + \phi), \qquad (16\text{-}40)$$

where ω', the angular frequency of the damped oscillator, is given by

$$\omega' = \sqrt{\frac{k}{m} - \frac{b^2}{4m^2}}. \qquad (16\text{-}41)$$

If $b = 0$ (there is no damping), then Eq. 16-41 reduces to Eq. 16-11 ($\omega = \sqrt{k/m}$) for the angular frequency of an undamped oscillator, and Eq. 16-40 reduces to Eq. 16-3 for

the displacement of an undamped oscillator. If the damping constant is not zero but small (so that $b \ll \sqrt{km}$), then $\omega' \approx \omega$.

We can regard Eq. 16-40 as a cosine function whose amplitude, which is $x_m e^{-bt/2m}$, gradually decreases with time, as Fig. 16-18 suggests. For an undamped oscillator, the mechanical energy is constant and is given by Eq. 16-23: $E = \frac{1}{2}kx_m^2$. If the oscillator is damped, the mechanical energy is not constant but decreases with time. If the damping is small, we can find $E(t)$ by replacing x_m in Eq. 16-23 with $x_m e^{-bt/2m}$, the amplitude of the damped oscillations. Doing so, we find

$$E(t) \approx \frac{1}{2}kx_m^2 e^{-bt/m}, \qquad (16\text{-}42)$$

which tells us that the mechanical energy decreases exponentially with time.

SAMPLE PROBLEM 16-9

For the damped oscillator of Fig. 16-17, $m = 250$ g, $k = 85$ N/m, and $b = 70$ g/s.

(a) What is the period of the motion?

SOLUTION: Because $b \ll \sqrt{km} = 4.6$ kg/s, the period is approximately that of the undamped oscillator. From Eq. 16-12, we then have

$$T = 2\pi\sqrt{\frac{m}{k}} = 2\pi\sqrt{\frac{0.25 \text{ kg}}{85 \text{ N/m}}} = 0.34 \text{ s}. \quad \text{(Answer)}$$

(b) How long does it take for the amplitude of the damped oscillations to drop to half its initial value?

SOLUTION: The amplitude at time t is displayed in Eq. 16-40 as $x_m e^{-bt/2m}$. It has the value x_m at $t = 0$. Thus we must find the value of t for which

$$x_m e^{-bt/2m} = \frac{1}{2}x_m.$$

Let us divide both sides by x_m and take the natural logarithm of the equation that remains. We find

$$\ln \tfrac{1}{2} = \ln(e^{-bt/2m}) = -bt/2m,$$

or $\quad t = \dfrac{-2m \ln \tfrac{1}{2}}{b} = \dfrac{-(2)(0.25 \text{ kg})(\ln \tfrac{1}{2})}{0.070 \text{ kg/s}}$

$$= 5.0 \text{ s}. \qquad \text{(Answer)}$$

Note, from the answer we found in (a), that this is about 15 periods of oscillation.

(c) How long does it take for the mechanical energy to drop to one-half its initial value?

SOLUTION: From Eq. 16-42 we see that the mechanical energy at time t is $\tfrac{1}{2}kx_m^2\, e^{-bt/m}$. It has the value $\tfrac{1}{2}kx_m^2$ at $t = 0$. Thus we must find the value of t for which

$$\tfrac{1}{2}kx_m^2\, e^{-bt/m} = \tfrac{1}{2}(\tfrac{1}{2}kx_m^2).$$

If we divide by $\tfrac{1}{2}kx_m^2$ and solve for t as we did above, we find

$$t = \dfrac{-m \ln \tfrac{1}{2}}{b} = \dfrac{-(0.25 \text{ kg})(\ln \tfrac{1}{2})}{0.070 \text{ kg/s}} = 2.5 \text{ s}. \quad \text{(Answer)}$$

This is exactly half the time calculated in (b), or about 7.5 periods of oscillation. Figure 16-18 was drawn to illustrate this sample problem.

CHECKPOINT **5:** Here are three sets of values for the spring constant, damping constant, and mass for the damped oscillator of Fig. 16-17. Rank the sets according to the time for the mechanical energy to decrease to one-fourth of its initial value, greatest first.

Set 1	$2k_0$	b_0	m_0
Set 2	k_0	$6b_0$	$4m_0$
Set 3	$3k_0$	$3b_0$	m_0

16-9 FORCED OSCILLATIONS AND RESONANCE

A person swinging passively in a swing is an example of *free oscillation.* If a kind friend pulls or pushes the swing periodically, as in Fig. 16-19, we have *forced,* or *driven, oscillations.* There are now *two* angular frequencies with which to deal: (1) the *natural* angular frequency ω of the system, which is the angular frequency at which it would oscillate if it were suddenly disturbed and then left to oscillate freely, and (2) the angular frequency ω_d of the external driving force.

We can use Fig. 16-17 to represent an idealized forced simple harmonic oscillator if we allow the structure marked "rigid support" to move up and down at a variable

FIGURE 16-19 Two frequencies are suggested in this painting by Nicholas Lancret: (1) the natural frequency at which the lady —left to herself—would swing, and (2) the frequency at which her friend tugs on the rope. If these two frequencies match, there is resonance.

angular frequency ω_d. A forced oscillator oscillates at the angular frequency ω_d of the driving force, and its displacement $x(t)$ is given by

$$x(t) = x_m \cos(\omega_d t + \phi), \qquad (16\text{-}43)$$

where x_m is the amplitude of the oscillations.

How large the displacement amplitude x_m is depends on a complicated function of ω_d and ω. The velocity amplitude v_m of the oscillations is easier to describe: it is greatest when

$$\omega_d = \omega \qquad \text{(resonance)}, \qquad (16\text{-}44)$$

a condition called **resonance.** Equation 16-44 is also *approximately* the condition at which the displacement amplitude x_m of the oscillations is greatest. Thus if you push a swing at its natural frequency, the displacement and velocity amplitudes will increase to large values, a fact that children learn quickly by trial and error. If you push at other frequencies, either higher or lower, the displacement and velocity amplitudes will be smaller.

Figure 16-20 shows how the displacement amplitude of an oscillator depends on the angular frequency ω_d of the driving force, for three values of the damping coefficient b. Note that for all three the amplitude is approximately greatest when $\omega_d/\omega = 1$, that is, when the resonance condition of Eq. 16-44 is satisfied. The curves of Fig. 16-20 show that the smaller the damping, the taller and narrower is the *resonance peak.*

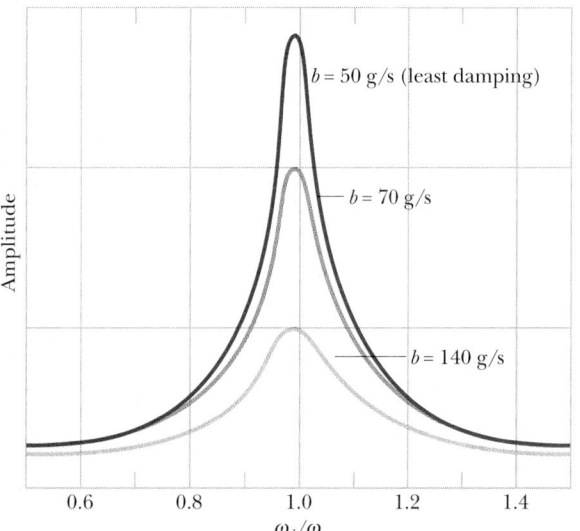

FIGURE 16-20   The displacement amplitude x_m of a forced oscillator varies as the angular frequency ω_d of the driving force is varied. The amplitude is greatest approximately at $\omega_d/\omega = 1$, the resonance condition. The curves here correspond to three values of the damping constant b.

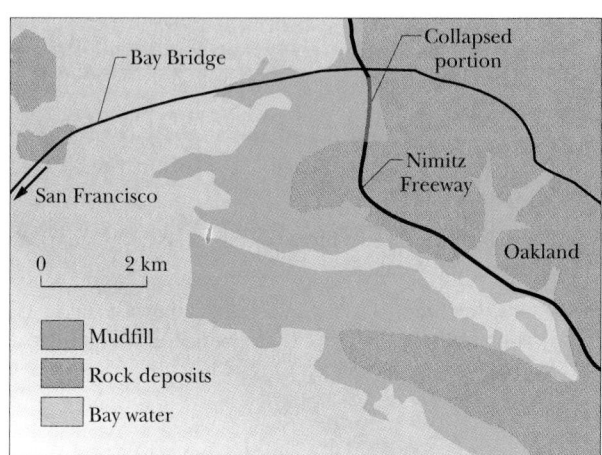

FIGURE 16-21   The geological structure in the Oakland portion of the San Francisco Bay area. The collapsed stretch of the Nimitz Freeway is indicated. (Adapted from ''Sediment-Induced Amplification and the Collapse of the Nimitz Freeway,'' by S. E. Hough et al., *Nature,* April 26, 1990.)

All mechanical structures have one or more natural frequencies, and if a structure is subjected to a strong external driving force that matches one of these frequencies, the resulting oscillations of the structure may rupture it. Thus, for example, aircraft designers must make sure that none of the natural frequencies at which a wing can vibrate matches the angular frequency of the engines at cruising speed. It obviously would be dangerous for a wing to flap violently at certain engine speeds.

The collapse of a 1.4 km stretch of the Nimitz Freeway, shown in this chapter's opening photograph, is an example of the destruction that resonance can produce. Oscillations of the ground due to the earthquake's seismic waves had their greatest velocity amplitude at an angular frequency of about 9 rad/s, which almost exactly matched a natural angular frequency of the individual horizontal sections of the freeway. The collapse was confined to a particular 1.4 km stretch because that section of the freeway was built on a loosely structured mudfill whose velocity amplitude during the quake was *at least five times larger* than that of the rock deposits under the rest of the freeway (Fig. 16-21).

REVIEW & SUMMARY

Frequency

Every oscillatory, or periodic, motion has a *frequency f* (the number of oscillations per second) measured, in the SI system, in hertz:

$$1 \text{ hertz} = 1 \text{ Hz} = 1 \text{ oscillation per second}$$
$$= 1 \text{ s}^{-1}. \qquad (16\text{-}1)$$

Period

The *period T* is the time required for one complete oscillation, or **cycle.** It is related to the frequency by

$$T = \frac{1}{f}. \qquad (16\text{-}2)$$

Simple Harmonic Motion

In *simple harmonic motion* (SHM), the displacement $x(t)$ of a particle from its equilibrium position is described by the equation

$$x = x_m \cos(\omega t + \phi) \qquad \text{(displacement)}, \qquad (16\text{-}3)$$

in which x_m is the **amplitude** of the displacement, the quantity $(\omega t + \phi)$ is the **phase** of the motion, and ϕ is the **phase constant.** The **angular frequency** ω is related to the period and frequency of the motion by

$$\omega = \frac{2\pi}{T} = 2\pi f \qquad \text{(angular frequency)}. \qquad (16\text{-}4)$$

Differentiating Eq. 16-3 leads to equations for the particle's velocity and acceleration during SHM as functions of time:

$$v = -\omega x_m \sin(\omega t + \phi) \quad \text{(velocity)} \quad (16\text{-}5)$$

and
$$a = -\omega^2 x_m \cos(\omega t + \phi) \quad \text{(acceleration).} \quad (16\text{-}6)$$

In Eq. 16-5, the positive quantity ωx_m is the **velocity amplitude** v_m of the motion. In Eq. 16-6, the positive quantity $\omega^2 x_m$ is the **acceleration amplitude** a_m of the motion.

The Linear Oscillator

A particle with mass m that moves under the influence of a Hooke's law restoring force given by $F = -kx$ exhibits simple harmonic motion with

$$\omega = \sqrt{\frac{k}{m}} \quad \text{(angular frequency)} \quad (16\text{-}11)$$

and
$$T = 2\pi \sqrt{\frac{m}{k}} \quad \text{(period).} \quad (16\text{-}12)$$

Such a system is called a **linear simple harmonic oscillator.**

Energy

A particle in simple harmonic motion has, at any time, kinetic energy $K = \frac{1}{2}mv^2$ and potential energy $U = \frac{1}{2}kx^2$. If no friction is present, the mechanical energy $E = K + U$ remains constant even though K and U change.

Pendulums

Examples of devices that undergo simple harmonic motion are the **torsion pendulum** of Fig. 16-8, the **simple pendulum** of Fig. 16-10, and the **physical pendulum** of Fig. 16-11. Their periods of oscillation for small oscillations are, respectively,

$$T = 2\pi \sqrt{I/\kappa}, \quad (16\text{-}25)$$

$$T = 2\pi \sqrt{L/g}, \quad (16\text{-}29)$$

and
$$T = 2\pi \sqrt{I/mgh}. \quad (16\text{-}32)$$

In each case, the period involves an inertial term divided by a "springiness" term that measures the strength of the oscillator's restoring force.

Simple Harmonic Motion and Uniform Circular Motion

Simple harmonic motion is the projection of uniform circular motion onto the diameter of the circle in which the latter motion occurs. Figure 16-16 shows that all parameters of circular motion (position, velocity, and acceleration) project to the corresponding values for simple harmonic motion.

Damped Harmonic Motion

The mechanical energy E in a real oscillating system decreases during the oscillations because external forces, such as a drag force, inhibit the oscillations and transfer mechanical energy to thermal energy. The real oscillator and its motion are then said to be **damped.** If the **damping force** is given by $F_d = -bv$, where v is the velocity of the oscillator and b is a **damping constant,** then the displacement of the oscillator is given by

$$x(t) = x_m e^{-bt/2m} \cos(\omega' t + \phi), \quad (16\text{-}40)$$

where ω', the angular frequency of the damped oscillator, is given by

$$\omega' = \sqrt{\frac{k}{m} - \frac{b^2}{4m^2}}. \quad (16\text{-}41)$$

If the damping constant is small ($b \ll \sqrt{km}$), then $\omega' \approx \omega$, where ω is the angular frequency of the undamped oscillator. For small b, the mechanical energy E of the oscillator is given by

$$E(t) \approx \frac{1}{2}kx_m^2 e^{-bt/m}. \quad (16\text{-}42)$$

Forced Oscillations and Resonance

If an external driving force with angular frequency ω_d acts on an oscillating system with *natural* angular frequency ω, the system oscillates with angular frequency ω_d. The velocity amplitude v_m of the system is greatest when

$$\omega_d = \omega, \quad (16\text{-}44)$$

a condition called **resonance.** The amplitude x_m of the system is (approximately) greatest under the same condition.

QUESTIONS

1. Which of the following relationships between the acceleration a and the displacement x of a particle involve SHM: (a) $a = 0.5x$, (b) $a = 400x^2$, (c) $a = -20x$, (d) $a = -3x^2$?

2. The acceleration $a(t)$ of a particle undergoing SHM is graphed in Fig. 16-22. (a) Which of the labeled points corresponds to the particle being at $-x_m$? (b) At point 4, is the velocity of the particle positive, negative, or zero? (c) At point 5, is the particle at $-x_m$, at $+x_m$, at 0, between $-x_m$ and 0, or between 0 and $+x_m$?

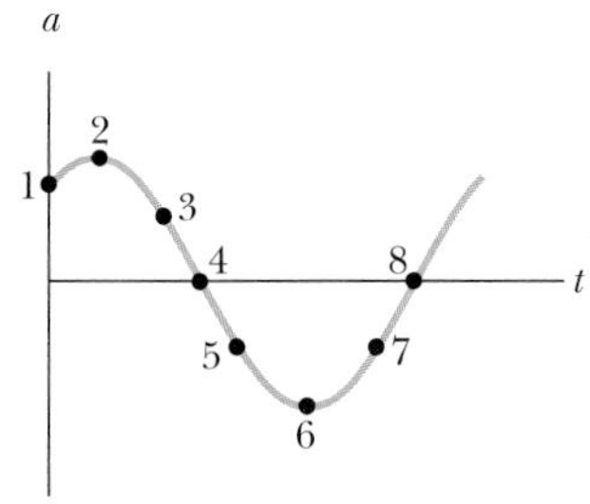

3. A particle oscillates according to

$$x = x_m \cos(\omega t + \phi).$$

FIGURE 16-22 Question 2.

At $t = 0$, is the particle at $-x_m$, at $+x_m$, at 0, between $-x_m$ and 0, or between 0 and $+x_m$ if ϕ is (a) $\pi/2$, (b) $-\pi/3$, (c) $-3\pi/4$, and (d) $3\pi/4$?

4. Which of the following describe ϕ for the SHM of Fig. 16-23a: (a) $-\pi < \phi < -\pi/2$, (b) $\pi < \phi < 3\pi/2$, (c) $-3\pi/2 < \phi < -\pi$?

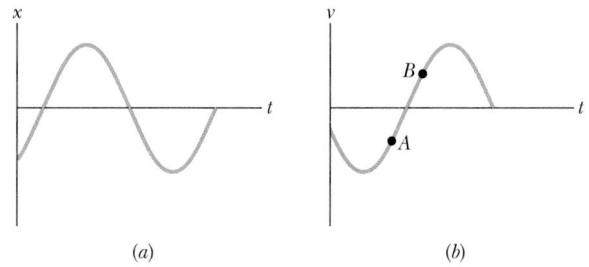

(a)   (b)

FIGURE 16-23 Questions 4 and 5.

5. The velocity $v(t)$ of a particle undergoing SHM is graphed in Fig. 16-23b. Is the particle momentarily stationary, headed toward $-x_m$, or headed toward $+x_m$ at (a) point A on the graph and (b) point B? Is the particle at $-x_m$, at $+x_m$, at 0, between $-x_m$ and 0, or between 0 and $+x_m$ when its velocity is represented by (c) point A and (d) point B? Is the speed of the particle increasing or decreasing at (e) point A and (f) point B?

6. What is the phase difference of the two linear oscillators, with identical masses and spring constants, in (a) Fig. 16-24a and (b) Fig. 16-24b? (c) What is the phase difference between the red oscillator in Fig. 16-24a and the green oscillator in Fig. 16-24b?

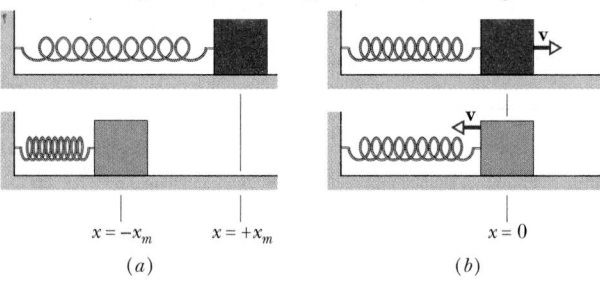

$x = -x_m$ $x = +x_m$ $x = 0$

(a)   (b)

FIGURE 16-24 Question 6.

7. A spring and a block are arranged to oscillate (1) on a frictionless horizontal surface as in Fig. 16-5, (2) on a frictionless slope of 45° (with the block at the lower end of the spring), and (3) vertically while hanging from a ceiling. Rank the arrangements according to (a) the rest length of the spring and (b) the frequency of oscillation, greatest first.

8. Three springs each hang from a ceiling with a stationary block attached to the lower end. The blocks (with $m_1 > m_2 > m_3$) stretch the springs by equal distances. Each block–spring system is then set into vertical SHM. Rank the masses according to the period of oscillation, greatest first.

9. Figure 16-25 shows three arrangements of a block attached to identical springs that are in the relaxed state when the block is centered. Rank the arrangements according to the frequency of oscillation of the block, greatest first.

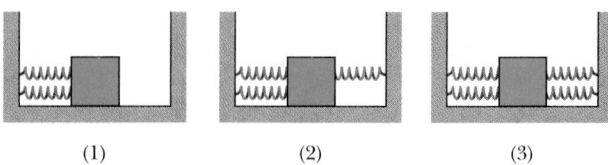

(1)   (2)   (3)

FIGURE 16-25 Question 9.

10. A mass m is suspended from a spring with spring constant k. The spring is then cut in half and the same mass is suspended from one of the halves. Which, the original spring or the half-spring, gives a greater frequency of oscillation when the mass is put into vertical SHM?

11. When mass m_1 is hung from spring A and a smaller mass m_2 is hung from spring B, the springs are stretched by the same distance. If the two spring–mass systems are then put into vertical SHM with the same amplitude, which system will have more energy?

12. Do the following increase, decrease, or remain the same if the amplitude of a simple harmonic oscillator is doubled: (a) the period, (b) the spring constant, (c) the total energy, (d) the maximum speed, (e) the maximum acceleration?

13. Figure 16-26 shows three physical pendulums consisting of identical uniform spheres of the same mass that are rigidly connected by identical rods of negligible mass. Each pendulum is vertical and can pivot about suspension point O. Rank the pendulums according to period of oscillation, greatest first.

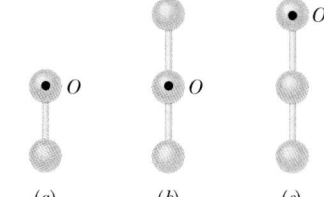

FIGURE 16-26 Question 13.   (a)   (b)   (c)

14. A pendulum suspended from the ceiling of an elevator cab has period T when the cab is stationary. Is the period larger, smaller, or the same when the cab moves (a) upward with constant speed, (b) downward with constant speed, (c) downward with constant upward acceleration, (d) upward with constant upward acceleration, (e) upward with constant downward acceleration $a = g$, and (f) downward with constant downward acceleration $a = g$?

15. A pendulum mounted in a cart has period T when the cart is stationary and on a horizontal plane. Is the period larger, smaller, or the same if the cart is on a plane inclined at angle θ (Fig. 16-27)

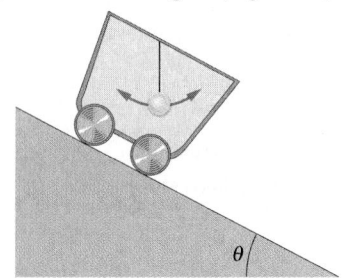

FIGURE 16-27
Question 15.

and (a) stationary, (b) moving down the plane with constant speed, (c) moving up the plane with constant speed, (d) moving up the plane with constant acceleration directed up the plane, (e) moving down the plane with constant acceleration directed up the plane, (f) moving down the plane with acceleration $a = g \sin \theta$ directed down the plane, and (g) moving up the plane with acceleration $a = g \sin \theta$ directed down the plane?

16. You are to build the oscillation transfer device shown in Fig. 16-28. It consists of two spring–block systems hanging from a flexible rod. When the spring of system 1 is stretched and then released, the resulting SHM of system 1 at frequency f_1 oscillates the rod. The rod then provides a driving force on system 2, at the same frequency f_1. For components you have to choose from four springs with spring constants k of 1600, 1500, 1400, and 1200

N/m, and four blocks with masses m of 800, 500, 400, and 200 kg. Without written calculation, determine which spring should go with which block in each of the two systems to maximize the amplitude of oscillations in system 2.

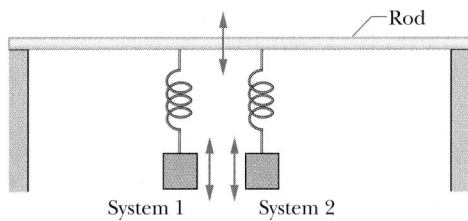

FIGURE 16-28   Question 16.

EXERCISES & PROBLEMS

SECTION 16-3 The Force Law for Simple Harmonic Motion

1E. An object undergoing simple harmonic motion takes 0.25 s to travel from one point of zero velocity to the next such point. The distance between those points is 36 cm. Calculate (a) the period, (b) the frequency, and (c) the amplitude of the motion.

2E. An oscillating mass–spring system takes 0.75 s to begin repeating its motion. Find (a) the period, (b) the frequency in hertz, and (c) the angular frequency in radians per second.

3E. A 4.00 kg block hangs from a spring, extending it 16.0 cm from its unstretched position. (a) What is the spring constant? (b) The block is removed and a 0.500 kg body is hung from the same spring. If the spring is then stretched and released, what is its period of oscillation?

4E. An oscillator consists of a block of mass 0.500 kg connected to a spring. When set into oscillation with amplitude 35.0 cm, it is observed to repeat its motion every 0.500 s. Find (a) the period, (b) the frequency, (c) the angular frequency, (d) the spring constant, (e) the maximum speed, and (f) the maximum force exerted on the block.

5E. The vibration frequencies of atoms in solids at normal temperatures are of the order of 10^{13} Hz. Imagine the atoms to be connected to one another by springs. Suppose that a single silver atom in a solid vibrates with this frequency and that all the other atoms are at rest. Compute the effective spring constant. One mole of silver (6.02×10^{23} atoms) has a mass of 108 g.

6E. What is the maximum acceleration of a platform that vibrates with an amplitude of 2.20 cm at a frequency of 6.60 Hz?

7E. A loudspeaker produces a musical sound by means of the oscillation of a diaphragm. If the amplitude of oscillation is limited to 1.0×10^{-3} mm, what frequencies will result in the acceleration of the diaphragm exceeding g?

8E. The scale of a spring balance which reads from 0 to 32.0 lb is 4.00 in. long. A package suspended from the balance is found to oscillate vertically with a frequency of 2.00 Hz. (a) What is the spring constant? (b) How much does the package weigh?

9E. A 20 N weight is hung from the bottom of a vertical spring, causing the spring to stretch 20 cm. (a) What is the spring constant? (b) This spring is now placed horizontally on a frictionless table. One end of it is held fixed, and the other end is attached to a 5.0 N weight. The weight is then moved (stretching the spring) and released from rest. What is the period of oscillation?

10E. A 50.0 g mass is attached to the bottom of a vertical spring and set vibrating. If the maximum speed of the mass is 15.0 cm/s and the period is 0.500 s, find (a) the spring constant of the spring, (b) the amplitude of the motion, and (c) the frequency of oscillation.

11E. A particle with a mass of 1.00×10^{-20} kg is vibrating with simple harmonic motion with a period of 1.00×10^{-5} s and a maximum speed of 1.00×10^3 m/s. Calculate (a) the angular frequency and (b) the maximum displacement of the particle.

12E. A small body of mass 0.12 kg is undergoing simple harmonic motion of amplitude 8.5 cm and period 0.20 s. (a) What is the maximum value of the force acting on it? (b) If the oscillations are produced by a spring, what is the spring constant?

13E. In an electric shaver, the blade moves back and forth over a distance of 2.0 mm. The motion is simple harmonic, with frequency 120 Hz. Find (a) the amplitude, (b) the maximum blade speed, and (c) the maximum blade acceleration.

14E. A loudspeaker diaphragm is vibrating in simple harmonic motion with a frequency of 440 Hz and a maximum displacement of 0.75 mm. What are (a) the angular frequency, (b) the maximum speed, and (c) the maximum acceleration?

15E. An automobile can be considered to be mounted on four identical springs as far as vertical oscillations are concerned. The springs of a certain car are adjusted so that the vibrations have a frequency of 3.00 Hz. (a) What is the spring constant of each spring if the mass of the car is 1450 kg and the weight is evenly distributed over the springs? (b) What will be the vibration frequency if five passengers, averaging 73.0 kg each, ride in the car? (Again, consider an even distribution of weight.)

16E. A body oscillates with simple harmonic motion according to the equation

$$x = (6.0 \text{ m}) \cos[(3\pi \text{ rad/s})t + \pi/3 \text{ rad}].$$

At $t = 2.0$ s, what are (a) the displacement, (b) the velocity, (c) the acceleration, and (d) the phase of the motion? Also, what are (e) the frequency and (f) the period of the motion?

17E. A particle executes linear SHM with frequency 0.25 Hz about the point $x = 0$. At $t = 0$, it has displacement $x = 0.37$ cm and zero velocity. For the motion, determine (a) the period, (b) the angular frequency, (c) the amplitude, (d) the displacement at time t, (e) the velocity at time t, (f) the maximum speed, (g) the maximum acceleration, (h) the displacement at $t = 3.0$ s, and (i) the speed at $t = 3.0$ s.

18E. The piston in the cylinder head of a locomotive has a stroke (twice the amplitude) of 0.76 m. If the piston moves with simple harmonic motion with an angular frequency of 180 rev/min, what is its maximum speed?

19P. Figure 16-29 shows an astronaut on a body-mass measuring device (BMMD). Designed for use on orbiting space vehicles, its purpose is to allow astronauts to measure their mass in the "weightless" conditions in Earth orbit. The BMMD is a spring-mounted chair; an astronaut measures his or her period of oscillation in the chair; the mass follows from the formula for the period of an oscillating block–spring system. (a) If M is the mass of the astronaut and m the effective mass of that part of the BMMD that also oscillates, show that

$$M = (k/4\pi^2)T^2 - m,$$

where T is the period of oscillation and k is the spring constant. (b) The spring constant was $k = 605.6$ N/m for the BMMD on Skylab Mission Two; the period of oscillation of the empty chair was 0.90149 s. Calculate the effective mass of the chair. (c) With

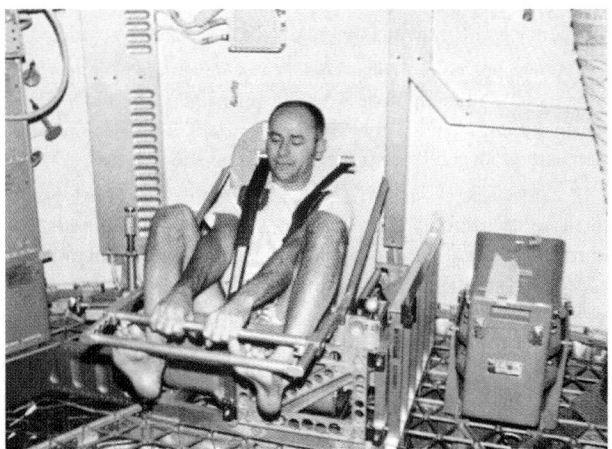

FIGURE 16-29 Problem 19.

an astronaut in the chair, the period of oscillation became 2.08832 s. Calculate the mass of the astronaut.

20P. A 2.00 kg block hangs from a spring. A 300 g body hung below the block stretches the spring 2.00 cm farther. (a) What is the spring constant? (b) If the 300 g body is removed and the block is set into oscillation, find the period of the motion.

21P. The end point of a spring vibrates with a period of 2.0 s when a mass m is attached to it. When this mass is increased by 2.0 kg, the period is found to be 3.0 s. Find the value of m.

22P. The end of one of the prongs of a tuning fork that executes simple harmonic motion of frequency 1000 Hz has an amplitude of 0.40 mm. Find (a) the maximum acceleration and (b) the maximum speed of the end of the prong. Find (c) the acceleration and (d) the speed of the end of the prong when the end has a displacement of 0.20 mm.

23P. A 0.10 kg block oscillates back and forth along a straight line on a frictionless horizontal surface. Its displacement from the origin is given by

$$x = (10 \text{ cm})\cos[(10 \text{ rad/s})t + \pi/2 \text{ rad}].$$

(a) What is the oscillation frequency? (b) What is the maximum speed acquired by the block? At what value of x does this occur? (c) What is the maximum acceleration of the block? At what value of x does this occur? (d) What force, applied to the block, results in the given oscillation?

24P. At a certain harbor, the tides cause the ocean surface to rise and fall a distance d in simple harmonic motion, with a period of 12.5 h. How long does it take for the water to fall a distance $d/4$ from its maximum height?

25P. Two blocks ($m = 1.0$ kg and $M = 10$ kg) and a spring ($k = 200$ N/m) are arranged on a horizontal, frictionless surface as shown in Fig. 16-30. The coefficient of static friction between the two blocks is 0.40. What is the maximum possible amplitude of simple harmonic motion of the spring–blocks system if no slippage is to occur between the blocks?

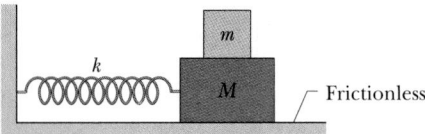

FIGURE 16-30 Problem 25.

26P. A block is on a horizontal surface (a shake table) that is moving horizontally with simple harmonic motion of frequency 2.0 Hz. The coefficient of static friction between block and surface is 0.50. How great can the amplitude of the SHM be if the block is not to slip along the surface?

27P. A block is on a piston that is moving vertically with simple harmonic motion. (a) If the SHM has period 1.0 s, at what amplitude of motion will the block and piston separate? (b) If the piston has an amplitude of 5.0 cm, what is the maximum frequency for which the block and piston will be in contact continuously?

28P. An oscillator consists of a block attached to a spring ($k = 400$ N/m). At some time t, the position (measured from the system's equilibrium location), velocity, and acceleration of the block are $x = 0.100$ m, $v = -13.6$ m/s, and $a = -123$ m/s^2. Calculate (a) the frequency of oscillation, (b) the mass of the block, and (c) the amplitude of the motion.

29P. A simple harmonic oscillator consists of a block of mass 2.00 kg attached to a spring of spring constant 100 N/m. When $t = 1.00$ s, the position and velocity of the block are $x = 0.129$ m and $v = 3.415$ m/s. (a) What is the amplitude of the oscillations? What were the (b) position and (c) velocity of the mass at $t = 0$ s?

30P. A massless spring hangs from the ceiling with a small object attached to its lower end. The object is initially held at rest in a position y_i such that the spring is at its rest length. The object is then released from y_i and oscillates up and down, with its lowest position being 10 cm below y_i. (a) What is the frequency of the oscillation? (b) What is the speed of the object when it is 8.0 cm below the initial position? (c) An object of mass 300 g is attached to the first object, after which the system oscillates with half the original frequency. What is the mass of the first object? (d) Relative to y_i, where is the new equilibrium (rest) position with both objects attached to the spring?

31P. Two particles oscillate in simple harmonic motion along a common straight line segment of length A. Each particle has a period of 1.5 s, but they differ in phase by $\pi/6$ rad. (a) How far apart are they (in terms of A) 0.50 s after the lagging particle leaves one end of the path? (b) Are they then moving in the same direction, toward each other, or away from each other?

32P. Two particles execute simple harmonic motion of the same amplitude and frequency along the same straight line. They pass each other moving in opposite directions each time their displacement is half their amplitude. What is the phase difference between them?

33P. Two identical springs are attached to a block of mass m and to fixed supports as shown in Fig. 16-31. Show that the frequency of oscillation on the frictionless surface is

$$f = \frac{1}{2\pi}\sqrt{\frac{2k}{m}}.$$

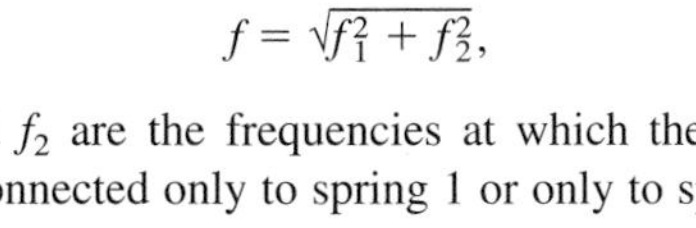

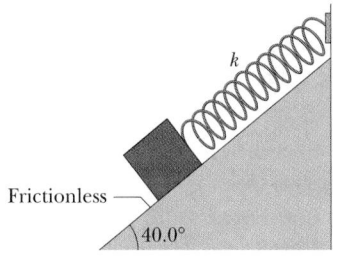

FIGURE 16-31 Problems 33 and 34.

34P. Suppose that the two springs in Fig. 16-31 have different spring constants k_1 and k_2. Show that the frequency f of oscillation of the block is then given by

$$f = \sqrt{f_1^2 + f_2^2},$$

where f_1 and f_2 are the frequencies at which the block would oscillate if connected only to spring 1 or only to spring 2.

35P. Two springs are joined and connected to a mass m as shown in Fig. 16-32. The surface is frictionless. If the springs both have force constant k, show that the frequency of oscillation of m is

$$f = \frac{1}{2\pi}\sqrt{\frac{k}{2m}}.$$

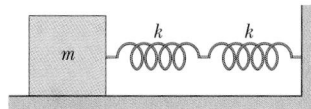

FIGURE 16-32 Problem 35.

36P. A block weighing 14.0 N, which slides without friction on a 40.0° incline, is connected to the top of the incline by a massless spring of unstretched length 0.450 m and spring constant 120 N/m, as shown in Fig. 16-33. (a) How far from the top of the incline does the block stop? (b) If the block is pulled slightly down the incline and released, what is the period of the ensuing oscillations?

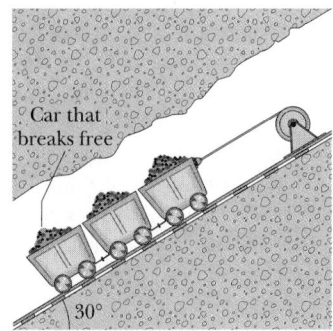

Frictionless

40.0°

FIGURE 16-33 Problem 36.

37P. A uniform spring whose unstretched length is L has a spring constant k. The spring is cut into two pieces of unstretched lengths L_1 and L_2, with $L_1 = nL_2$. (a) What are the corresponding spring constants k_1 and k_2 in terms of n and k? (b) If a block is attached to the original spring, as in Fig. 16-5, it oscillates with frequency f. If the spring is replaced with the piece L_1 or L_2, the corresponding frequency is f_1 or f_2. Find f_1 and f_2 in terms of f.

38P. Three 10,000 kg ore cars are held at rest on a 30° incline on a mine railway using a cable that is parallel to the incline (Fig. 16-34). The cable is observed to stretch 15 cm just before the coupling between the lower cars breaks, detaching the lowest car. Assuming that the cable obeys Hooke's law, find (a) the frequency and (b) the amplitude of the resulting oscillations of the remaining two cars.

Car that breaks free

30°

FIGURE 16-34 Problem 38.

39P. To alleviate the traffic congestion between two cities such as Boston and Washington, DC, engineers have proposed building a rail tunnel along a chord line connecting the cities (Fig. 16-35). A train, unpropelled by any engine and starting from rest, would fall through the first half of the tunnel and then move up

FIGURE 16-35   Problem 39.

the second half. Assuming Earth to be a uniform sphere and ignoring air drag and friction, (a) show that the travel between cities is half a cycle of simple harmonic motion and (b) find the travel time between cities.

SECTION 16-4 Energy in Simple Harmonic Motion

40E. Find the mechanical energy of a block–spring system having a spring constant of 1.3 N/cm and an amplitude of 2.4 cm.

41E. An oscillating block–spring system has a mechanical energy of 1.00 J, an amplitude of 10.0 cm, and a maximum speed of 1.20 m/s. Find (a) the spring constant, (b) the mass of the block and (c) the frequency of oscillation.

42E. A 5.00 kg object on a horizontal frictionless surface is attached to a spring with spring constant 1000 N/m. The object is displaced from equilibrium 50.0 cm horizontally and given an initial velocity of 10.0 m/s back toward the equilibrium position. (a) What is the frequency of the motion? What are (b) the initial potential energy of the block–spring system, (c) the initial kinetic energy, and (d) the amplitude of the oscillation?

43E. A vertical spring stretches 9.6 cm when a 1.3 kg block is hung from its end. (a) Calculate the spring constant. This block is then displaced an additional 5.0 cm downward and released from rest. Find (b) the period, (c) the frequency, (d) the amplitude, and (e) the maximum speed of the resulting SHM.

44E. A (hypothetical) large slingshot is stretched 1.50 m to launch a 130 g projectile with speed sufficient to escape from Earth (11.2 km/s). Assume the elastic slingshot obeys Hooke's law. (a) What is the spring constant of the device, if all the elastic potential energy is converted to kinetic energy? (b) Assume that an average person can exert a force of 220 N. How many people would be required to stretch the slingshot?

45E. When the displacement in SHM is one-half the amplitude x_m, what fraction of the total energy is (a) kinetic energy and (b) potential energy? (c) At what displacement, in terms of the amplitude, is the energy of the system half kinetic energy and half potential energy?

46E. A block of mass M, at rest on a horizontal frictionless table, is attached to a rigid support by a spring of constant k. A bullet of mass m and velocity v strikes the block as shown in Fig. 16-36. The bullet remains embedded in the block. Determine (a) the velocity of the block immediately after the collision and (b) the amplitude of the resulting simple harmonic motion.

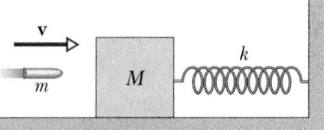

FIGURE 16-36   Exercise 46.

47P. A 3.0 kg particle is in simple harmonic motion in one dimension and moves according to the equation

$$x = (5.0 \text{ m})\cos[(\pi/3 \text{ rad/s})t - \pi/4 \text{ rad}].$$

(a) At what value of x is the potential energy of the particle equal to half the total energy? (b) How long does it take the particle to move to this position x from the equilibrium position?

48P. A 10 g particle is undergoing simple harmonic motion with an amplitude of 2.0×10^{-3} m. The maximum acceleration experienced by the particle is 8.0×10^3 m/s²; the phase constant is $-\pi/3$ rad. (a) Write an equation for the force on the particle as a function of time. (b) What is the period of the motion? (c) What is the maximum speed of the particle? (d) What is the total mechanical energy of this simple harmonic oscillator?

49P. A massless spring with spring constant 19 N/m hangs vertically. A body of mass 0.20 kg is attached to its free end and then released. Assume that the spring was unstretched before the body was released. Find (a) how far below the initial position the body descends, and (b) the frequency and (c) the amplitude of the resulting motion, assumed to be simple harmonic.

50P. A 4.0 kg block is suspended from a spring with a spring constant of 500 N/m. A 50 g bullet is fired into the block from directly below with a speed of 150 m/s and is embedded in the block. (a) Find the amplitude of the resulting simple harmonic motion. (b) What fraction of the original kinetic energy of the bullet appears as mechanical energy in the harmonic oscillator?

51P*. A solid cylinder is attached to a horizontal massless spring so that it can roll without slipping along a horizontal surface (Fig. 16-37). The spring constant k is 3.0 N/m. If the system is released from rest at a position in which the spring is stretched by 0.25 m, find (a) the translational kinetic energy and (b) the rotational kinetic energy of the cylinder as it passes through the equilibrium position. (c) Show that under these conditions the center of mass of the cylinder executes simple harmonic motion with period

$$T = 2\pi\sqrt{\frac{3M}{2k}},$$

where M is the mass of the cylinder. (*Hint:* Find the time derivative of the total mechanical energy.)

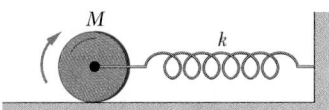

FIGURE 16-37   Problem 51.

SECTION 16-5 An Angular Simple Harmonic Oscillator

52E. A flat uniform circular disk has a mass of 3.00 kg and a radius of 70.0 cm. It is suspended in a horizontal plane by a vertical wire attached to its center. If the disk is rotated 2.50 rad about the wire, a torque of 0.0600 N·m is required to maintain the disk in this position. Calculate (a) the rotational inertia of the disk about the wire, (b) the torsion constant, and (c) the angular frequency of this torsion pendulum when it is set oscillating.

53P. A 95 kg solid sphere with a 15 cm radius is suspended by a vertical wire attached to the ceiling of a room. A torque of 0.20 N·m is required to twist the sphere through an angle of 0.85 rad. What is the period of oscillation when the sphere is released from this position?

54P. An engineer has an odd-shaped object of mass 10 kg and needs to find its rotational inertia about an axis through its center of mass. The object is supported with a wire along the desired axis. The wire has a torsion constant $\kappa = 0.50$ N·m. If this torsion pendulum oscillates through 20 complete cycles in 50 s, what is the rotational inertia of the odd-shaped object?

55P. The balance wheel of a watch oscillates with an angular amplitude of π rad and a period of 0.500 s. Find (a) the maximum angular speed of the wheel, (b) the angular speed of the wheel when its displacement is $\pi/2$ rad, and (c) the angular acceleration of the wheel when its displacement is $\pi/4$ rad.

SECTION 16-6 Pendulums

56E. What is the length of a simple pendulum whose period is 1.00 s at a point where $g = 32.2$ ft/s^2?

57E. A 2500 kg demolition ball swings from the end of a crane, as shown in Fig. 16-38. The length of the swinging segment of cable is 17 m. (a) Find the period of swing, assuming that the system can be treated as a simple pendulum. (b) Does the period depend on the ball's mass?

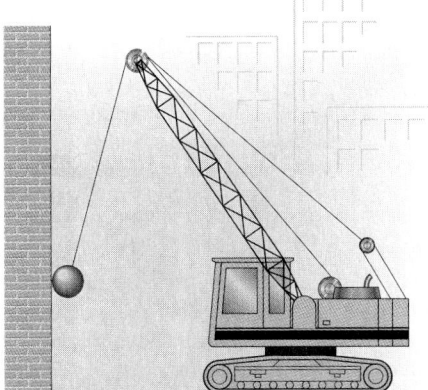

FIGURE 16-38
Exercise 57.

58E. What is the length of a simple pendulum that marks seconds by completing a full swing from left to right and then back again every 2.0 s?

59E. If a simple pendulum with length 1.50 m makes 72.0 oscillations in 180 s, what is the acceleration of gravity at its location?

60E. Two oscillating systems that you have studied are the block–spring and the simple pendulum. You can prove an interesting relation between them. Suppose that you hang a weight on the end of a spring, and when the weight is at rest, the spring is stretched a distance h as in Fig. 16-39. Show that the frequency of this block–spring system is the same as that of a simple pendulum whose length is h.

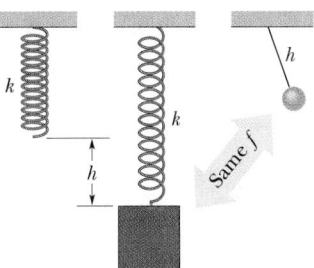

FIGURE 16-39 Exercise 60.

61E. A performer, seated on a trapeze, is swinging back and forth with a period of 8.85 s. If she stands up, thus raising the center of mass of the system *trapeze + performer* by 35.0 cm, what will be the new period of the trapeze? Treat *trapeze + performer* as a simple pendulum.

62E. A simple pendulum with length L is swinging freely with small angular amplitude. As the pendulum passes its central (or equilibrium) position, its cord is suddenly and rigidly clamped at its midpoint. In terms of the original period T of the pendulum, what will the new period be?

63E. A physical pendulum consists of a meter stick that is pivoted at a small hole drilled through the stick a distance x from the 50 cm mark. The period of oscillation is observed to be 2.5 s. Find the distance x.

64E. A pendulum is formed by pivoting a long thin rod of length L and mass m about a point on the rod that is a distance d above the center of the rod. (a) Find the period of this pendulum in terms of d, L, m, and g, assuming that it swings with small amplitude. What happens to the period if (b) d is decreased, (c) L is increased, or (d) m is increased?

65E. A physical pendulum consists of a uniform solid disk (of mass M and radius R) supported in a vertical plane by a pivot located a distance d from the center of the disk (Fig. 16-40). The disk is displaced by a small angle and released. Find an expression for the period of the resulting simple harmonic motion.

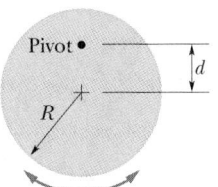

FIGURE 16-40 Exercise 65.

66E. A uniform circular disk whose radius R is 12.5 cm is suspended, as a physical pendulum, from a point on its rim. (a) What is its period of oscillation? (b) At what radial distance $r < R$ is there a point of suspension that gives the same period?

67E. A pendulum consists of a uniform disk with radius 10.0 cm and mass 500 g attached to a uniform rod with length 500 mm and mass 270 g (Fig. 16-41). (a) Calculate the rotational inertia of the pendulum about the pivot. (b) What is the distance between the pivot and the center of mass of the pendulum? (c) Calculate the period of oscillation.

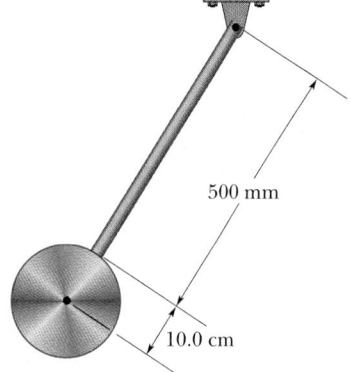

FIGURE 16-41
Exercise 67.

68E. (a) If the physical pendulum of Sample Problem 16-6 is inverted and suspended at point P, what is its period of oscillation? (b) Is the period now greater than, less than, or equal to its previous value?

69E. In Sample Problem 16-6, we saw that a physical pendulum has a center of oscillation at distance $2L/3$ from its point of suspension. Show that the distance between the point of suspension and the center of oscillation for a physical pendulum of any form is I/mh, where I and h have the meanings assigned to them in Eq. 16-32, and m is the mass of the pendulum.

70E. A meter stick swinging from one end oscillates with a frequency f_0. What would be the frequency, in terms of f_0, if the bottom half of the stick were cut off?

71P. A stick with length L oscillates as a physical pendulum, pivoted about point O in Fig. 16-42. (a) Derive an expression for the period of the pendulum in terms of L and x, the distance from the point of support to the center of mass of the pendulum. (b) For what value of x/L is the period a minimum? (c) Show that if $L = 1.00$ m and $g = 9.80$ m/s², this minimum is 1.53 s.

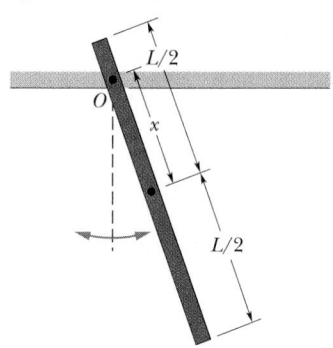

FIGURE 16-42 Problem 71.

72P. The center of oscillation of a physical pendulum has this interesting property: if an impulsive force (assumed horizontal and in the plane of oscillation) acts at the center of oscillation, no reaction is felt at the point of support. Baseball players (and players of many other sports) know that unless the ball hits the bat at this point (called the "sweet spot" by athletes), the reaction due to the impact will sting their hands. To prove this property, let the stick in Fig. 16-12a simulate a baseball bat. Suppose that a horizontal force **F** (due to impact with the ball) acts toward the right at P, the center of oscillation. The batter is assumed to hold the bat at O, the point of support of the stick. (a) What acceleration does point O undergo as a result of **F**? (b) What angular acceleration is produced by **F** about the center of mass of the stick? (c) As a result of the angular acceleration in (b), what linear acceleration does point O undergo? (d) Considering the magnitudes and directions of the accelerations in (a) and (c), convince yourself that P is indeed the "sweet spot."

73P. The fact that g varies from place to place over Earth's surface drew attention when Jean Richer in 1672 took a pendulum clock from Paris to Cayenne, French Guiana, and found that it lost 2.5 min/day. If $g = 9.81$ m/s² in Paris, what is g in Cayenne?

74P. A scientist is making a precise measurement of g at a certain point in the Indian Ocean (on the equator) by timing the swings of a pendulum of accurately known construction. To provide a stable base, the measurements are conducted in a submerged submarine. It is observed that a slightly different result for g is obtained when the submarine is moving eastward than when it is moving westward, the speed in each case being 16 km/h. Account for this difference and calculate the fractional error $\Delta g/g$ in g for either travel direction.

75P. A long uniform rod of length L and mass m is free to rotate in a horizontal plane about a vertical axis through its center. A spring with force constant k is connected horizontally between one end of the rod and a fixed wall, as shown in the overhead view of Fig. 16-43. When the rod is in equilibrium, it is parallel to the wall. What is the period of the small oscillations that result when the rod is rotated slightly and released?

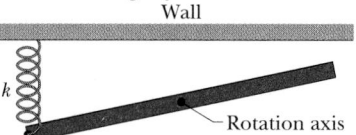

FIGURE 16-43 Problem 75.

76P. What is the frequency of a simple pendulum 2.0 m long (a) in a room, (b) in an elevator accelerating upward at a rate of 2.0 m/s², and (c) in free fall?

77P. A simple pendulum of length L and mass m is suspended in a car that is traveling with constant speed v around a circle of radius R. If the pendulum undergoes small oscillations in a radial direction about its equilibrium position, what will its frequency of oscillation be?

78P. For a simple pendulum, find the angular amplitude θ_m at which the restoring torque required for simple harmonic motion deviates from the actual restoring torque by 1.0%. (See "Trigonometric Expansions" in Appendix E.)

79P. The bob on a simple pendulum of length R moves in an arc of a circle. (a) By considering that the acceleration of the bob as it moves through its equilibrium position is that for uniform circular motion (mv^2/R), show that the tension in the string at that position is $mg(1 + \theta_m^2)$ if the angular amplitude θ_m is small. (See "Trigonometric Expansions" in Appendix E.) (b) Is the tension at other positions of the bob larger, smaller, or the same?

80P. A wheel is free to rotate about its fixed axle. A spring is attached to one of its spokes a distance r from the axle, as shown in Fig. 16-44. (a) Assuming that the wheel is a hoop of mass m and radius R, obtain the angular frequency of small oscillations of this system in terms of m, R, r, and the spring constant k. How does the result change if (b) $r = R$ and (c) $r = 0$?

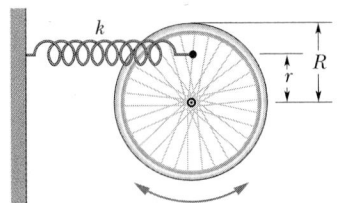

FIGURE 16-44 Problem 80.

81P. A 2.5 kg disk, 42 cm in diameter, is supported by a massless rod, 76 cm long, which is pivoted at its end, as in Fig. 16-45. (a) The massless torsion spring is initially not connected. What then is the period of oscillation? (b) The torsion spring is now connected so that, in equilibrium, the rod hangs vertically. What should be the torsional constant of the spring so that the new period of oscillation is 0.50 s shorter than before?

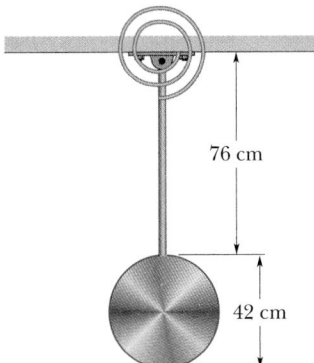

76 cm

42 cm

FIGURE 16-45 Problem 81.

82P. A physical pendulum has two possible pivot points A and B; point A has a fixed position, and B is adjustable along the length of the pendulum, as shown in Fig. 16-46. The period of the pendulum when suspended from A is found to be T. The pendulum is then reversed and suspended from B, which is moved until the pendulum again has period T. Show that the free-fall acceleration g is given by

$$g = \frac{4\pi^2 L}{T^2},$$

in which L is the distance between A and B for equal periods T. (Note that g can be measured in this way without knowing the rotational inertia of the pendulum or any of its dimensions except L.)

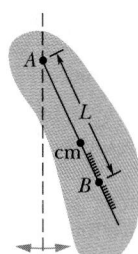

FIGURE 16-46 Problem 82.

83P*. A uniform rod with length L swings from a pivot as a physical pendulum. How far, in terms of L, should the pivot be from the center of mass to minimize the period?

SECTION 16-8 Damped Simple Harmonic Motion

84E. The amplitude of a lightly damped oscillator decreases by 3.0% during each cycle. What fraction of the energy of the oscillator is lost in each full oscillation?

85E. In Sample Problem 16-9, what is the ratio of the amplitude of the damped oscillations to the initial amplitude when 20 full oscillations have elapsed?

86E. For the system shown in Fig. 16-17, the block has a mass of 1.50 kg and the spring constant is 8.00 N/m. The damping force is given by $-b(dx/dt)$, where $b = 230$ g/s. Suppose that the block is initially pulled down a distance 12.0 cm and released. (a) Calculate the time required for the amplitude to fall to one-third of its initial value. (b) How many oscillations are made by the block in this time?

87P. A damped harmonic oscillator consists of a block ($m = 2.00$ kg), a spring ($k = 10.0$ N/m), and a damping force $F = -bv$. Initially, it oscillates with an amplitude of 25.0 cm; because of the damping, the amplitude falls to three-fourths of this initial value at the completion of four oscillations. (a) What is the value of b? (b) How much energy has been "lost" during these four oscillations?

88P. (a) In Eq. 16-39, find the ratio of the maximum damping force ($-b\,dx/dt$) to the maximum spring force ($-kx$) during the first oscillation for the data of Sample Problem 16-9. (b) Does this ratio change appreciably during later oscillations?

89P. Assume that you are examining the characteristics of the suspension system of a 2000 kg automobile. The suspension "sags" 10 cm when the weight of the entire automobile is placed on it. In addition, the amplitude of oscillation decreases by 50% during one complete oscillation. Estimate the values of k and b for the spring and shock absorber system of one wheel, assuming each wheel supports 500 kg.

SECTION 16-9 Forced Oscillations and Resonance

90E. For Eq. 16-43, suppose the amplitude x_m is given by

$$x_m = \frac{F_m}{[m^2(\omega_d^2 - \omega^2)^2 + b^2\omega_d^2]^{1/2}},$$

where F_m is the (constant) amplitude of the external oscillating force exerted on the spring by the rigid support in Fig. 16-17. At resonance, what are (a) the amplitude and (b) the velocity amplitude of the oscillating object?

91P. A 2200 lb car carrying four 180 lb people travels over a rough "washboard" dirt road with corrugations 13 ft apart which causes the car to bounce on its spring suspension.. The car bounces with maximum amplitude when its speed is 10 mi/h. The car now stops, and the four people get out. By how much does the car body rise on its suspension owing to this decrease in weight?

Electronic Computation

92. You see two carts connected by a spring and oscillating on a horizontal air track. You happen to know that the spring constant of the spring is $k = 50.0$ N/m. You use Newton's second law to determine that the period of oscillation is the same for each cart and is related to the masses of the carts by

$$T = 2\pi \sqrt{\frac{m_1 m_2}{k(m_1 + m_2)}}.$$

Using a position detector, you determine that the positions of the carts as functions of time can be expressed as $x_1(t) = 2.70[1 - \cos(18.0t)]$ and $x_2(t) = 10.70 + 1.29 \cos(18.0t)$, where the coordinates are in centimeters and the time is in seconds. (a) Use conservation of momentum to find the masses of the carts. (b) Generate a table of the values of x_1 and x_2 for every 0.01 s from $t = 0$ to $t = 35$ s. For each of these times, also calculate the position of the center of mass, the total momentum of the two-cart system, and the force of the spring on each cart. Verify that the center of mass does not move, that the total momentum is conserved, and that the forces on the carts have the same magnitude but opposite directions. (c) Use the table to find the equilibrium length of the spring.

93. A 2.0 kg block is attachd to the end of a spring with a spring constant of 350 N/m and forced to oscillate by an applied force $F = (15 \text{ N}) \sin(\omega t)$, where $\omega = 35$ rad/s. The damping constant is $b = 15$ kg/s. At $t = 0$, the block is at rest with the spring at its rest length. (a) Use numerical integration to plot the displacement of the block for the first 1.0 s of its motion. Use the motion near the end of the 1.0 s interval to estimate the amplitude, period, and angular frequency. Repeat the calculation for (b) $\omega = \sqrt{k/m}$ and (c) $\omega = 20$ rad/s.

When a beetle moves along the sand within a few tens of centimeters of this sand scorpion, the scorpion immediately turns toward the beetle and dashes to it (for lunch). The scorpion can do this without seeing (it is nocturnal) or hearing the beetle. How can it so precisely locate its prey?

17-1 WAVES AND PARTICLES

Two ways to get in touch with a friend in a distant city are to write a letter and to use the telephone.

The first choice (the letter) involves the concept of ''particle'': a material object moves from one point to another, carrying with it information and energy. Most of the preceding chapters deal with particles or with systems of particles.

The second choice (the telephone) involves the concept of ''wave,'' the subject of this chapter and the next. In a wave, information and energy move from one point to another but no material object makes that journey. In your telephone call, a sound wave carries your message from your vocal cords to the telephone. There, an electromagnetic wave takes over, passing along a copper wire or an optical fiber or through the atmosphere, possibly by way of a communications satellite. At the receiving end there is another sound wave, from a telephone to your friend's ear. Although the message is passed, nothing that you have touched reaches your friend. Leonardo da Vinci understood about waves when he wrote of water waves: ''it often happens that the wave flees the place of its creation, while the water does not; like the waves made in a field of grain by the wind, where we see the waves running across the field while the grain remains in place.''

Particle and *wave* are the two great concepts in classical physics, in the sense that we seem able to associate almost every branch of the subject with one or the other. The two concepts are quite different. The word *particle* suggests a tiny concentration of matter capable of transmitting energy. The word *wave* suggests just the opposite, namely, a broad distribution of energy, filling the space through which it passes. The job at hand is to put aside particles for a while and to learn something about waves.

17-2 TYPES OF WAVES

Waves are of three main types:

1. *Mechanical waves.* These waves are most familiar because we encounter them almost constantly; common examples include water waves, sound waves, and seismic waves. All these waves have certain central features: they are governed by Newton's laws, and they can exist only within a material medium, such as water, air, and rock.

2. *Electromagnetic waves.* These waves are less familiar, but you use them constantly; common examples include visible and ultraviolet light, radio and television waves, microwaves, x rays, and radar waves. These waves require no material medium to exist. Light waves from stars, for example, travel through the vacuum of space to reach us.

All electromagnetic waves travel through a vacuum at the same speed c, given by

$$c = 299{,}792{,}458 \text{ m/s} \quad \text{(speed of light).} \quad (17\text{-}1)$$

3. *Matter waves.* Although these waves are commonly used in modern technology, their type is probably very unfamiliar to you. Electrons, protons, and other fundamental particles, and even atoms and molecules, travel as waves. Because we commonly think of these things as constituting matter, these waves are called matter waves.

Much of what we discuss in this chapter applies to waves of all kinds. However, for specific examples we shall refer to mechanical waves.

17-3 TRANSVERSE AND LONGITUDINAL WAVES

A wave sent along a stretched, taut string is the simplest mechanical wave. If you give one end of a stretched string a single up-and-down jerk, a wave in the form of a single *pulse* travels along the string, as in Fig. 17-1a. This pulse and its motion can occur because the string is under ten-

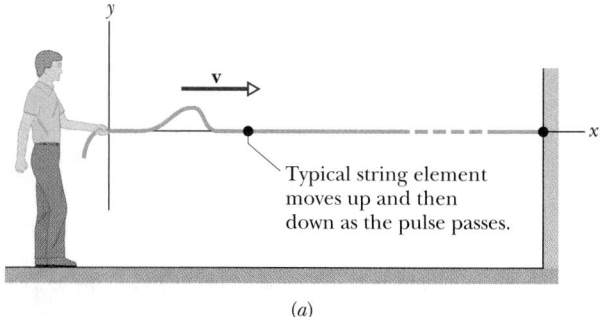

(a)

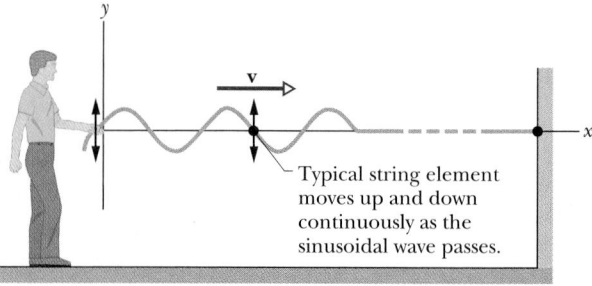

(b)

FIGURE 17-1 (a) Sending a single pulse down a long stretched string. (b) Sending a continuous sinusoidal wave down the string. Because the oscillations of any string element (represented here by a dot) are perpendicular to the direction in which the wave moves, the wave is a *transverse wave*.

sion. When you pull your end of the string upward, it begins to pull upward on the adjacent section of the string via tension between the two sections. As the adjacent section moves upward, it begins to pull the next section upward, and so on. Meanwhile, you have pulled down on your end of the string. So, as each section moves upward in turn, it begins to be pulled back downward by neighboring sections that are already on the way down. The net result is that a distortion in the string's shape (the pulse) moves along the string at some velocity **v**.

If you move your hand up and down in continuous simple harmonic motion, a continuous wave travels along the string at velocity **v**. Because the motion of your hand is a sinusoidal function of time, the wave has a sinusoidal shape at any given instant, as in Fig. 17-1b. That is, the wave has the shape of a sine curve or a cosine curve.

We consider here only an "ideal" string, in which no frictionlike forces within the string cause the wave to die out as it travels along the string. In addition, we assume that the string is so long that we need not consider a wave rebounding from the far end.

One way to study the waves of Fig. 17-1 is to monitor the **wave form** (shape of the wave) as it moves to the right. Alternatively, we can monitor the motion of an element of the string as the element oscillates up and down while the wave passes through it. We would find that the displacement of every such oscillating string element is *perpendicular* to the direction of travel of the wave, as indicated in Fig. 17-1. This motion is said to be **transverse,** and the wave is said to be a **transverse wave.**

Figure 17-2 shows how a sound wave can be produced by a piston in a long, air-filled pipe. If you suddenly move the piston rightward and then leftward, you can send a pulse of sound along the pipe. The rightward motion of the piston moves the elements of air next to it rightward, changing the air pressure there. The increased air pressure then pushes rightward on the elements of air somewhat

farther along the pipe. Once they have moved rightward, the elements move back leftward. Thus the motion of the air and the change in air pressure travel rightward along the pipe as a pulse.

If you push and pull on the piston in simple harmonic motion, as is being done in Fig. 17-2, a sinusoidal wave travels along the pipe. Because the motion of the elements of air is parallel to the direction of the wave's travel, the motion is said to be **longitudinal,** and the wave is said to be a **longitudinal wave.** In this chapter we concentrate on transverse waves, and string waves in particular; in Chapter 18 we shall concentrate on longitudinal waves, and sound waves in particular.

Both a transverse wave and a longitudinal wave are said to be **traveling waves** because the wave travels from one point to another, as from one end of the string to the other end in Fig. 17-1 or from one end of the pipe to the other end in Fig. 17-2. Note that it is the wave that moves between the two points and not the material (string or air) through which the wave moves.

The sand scorpion shown in the photograph opening this chapter uses waves of both transverse and longitudinal motion to locate its prey. When a beetle even slightly disturbs the sand, it sends pulses along the sand's surface (Fig. 17-3). One set of pulses is longitudinal, traveling with speed $v_l = 150$ m/s. A second set is transverse, traveling with speed $v_t = 50$ m/s.

The scorpion, with its eight legs spread roughly in a circle about 5 cm in diameter, intercepts the faster longitudinal pulses first and learns the direction of the beetle: it is in the direction of whichever leg is disturbed earliest by the pulses. The scorpion then senses the time interval Δt between that first interception and the interception of the slower transverse waves and uses it to determine the distance d to the beetle. This distance is given by

$$\Delta t = \frac{d}{v_t} - \frac{d}{v_l},$$

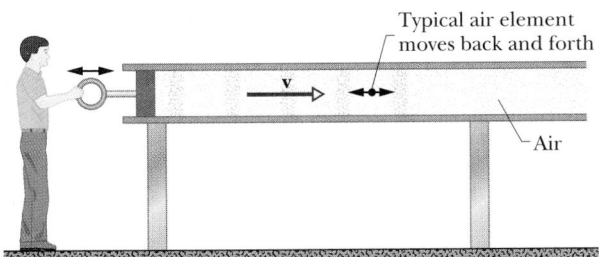

FIGURE 17-2 A sound wave is set up in an air-filled pipe by moving a piston back and forth. Because the oscillations of an element of the air (represented by the black dot) are parallel to the direction in which the wave travels, the wave is a *longitudinal wave.*

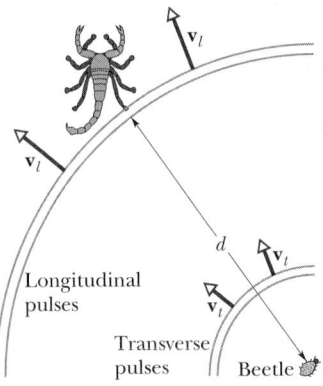

FIGURE 17-3 A beetle's motion sends fast longitudinal pulses and slower transverse pulses along the sand's surface. The sand scorpion first intercepts the longitudinal pulses; here, it is the rear-most right leg that senses the pulses earliest.

and it turns out to be

$$d = (75 \text{ m/s}) \, \Delta t.$$

For example, if $\Delta t = 4.0$ ms, then $d = 30$ cm, which gives the scorpion a perfect fix on the beetle.

17-4 WAVELENGTH AND FREQUENCY

To completely describe a wave on a string (and the motion of any element along its length), we need a function that gives the shape of the wave. This means that we need a relation in the form $y = h(x, t)$, in which y is the transverse displacement of any string element as a function h of the time t and the position x of the element along the string. In general, a sinusoidal shape like the wave in Fig. 17-1b can be described with h being either a sine function or a cosine function; both give the same general shape for the wave. In this chapter we use the sine function.

For a sinusoidal wave like that in Fig. 17-1b, traveling toward increasing values of x, the transverse displacement y of a string element at position x at time t is given by

$$y(x, t) = y_m \sin(kx - \omega t), \qquad (17\text{-}2)$$

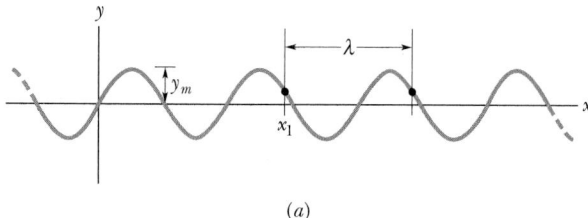

(a)

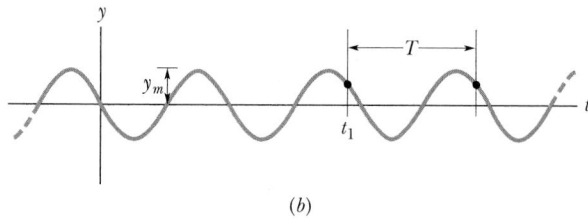

(b)

FIGURE 17-4   (a) A snapshot of a string at $t = 0$ as the sinusoidal wave of Eq. 17-2 travels along it. A typical wavelength λ is shown: it is the horizontal distance between two successive points on the string (represented by dots) with the same displacement. The amplitude y_m is also shown: it is the maximum displacement of the string. (b) A graph showing the displacement of a string element at $x = 0$ as a function of time as the sinusoidal wave passes through the element. A typical period T is shown: it is the time between identical displacements of the string element (represented by dots).

Here y_m is the **amplitude** of the wave; the subscript m stands for maximum, because the amplitude is the magnitude of the maximum displacement of the string element in either direction parallel to the y axis. (Hence, y_m is always a positive quantity.) The quantities k and ω are constants whose meanings we are about to discuss. The quantity $kx - \omega t$ is called the **phase** of the wave.

You may wonder why, of the infinite variety of wave forms that are available, we pick the sinusoidal wave of Eq. 17-2 for detailed study. This is in fact a wise choice because, as you will see in Section 17-8, *all* wave forms—including the pulse of Fig. 17-1a—can be constructed by adding sinusoidal waves of carefully selected amplitudes and values of k. Understanding sinusoidal waves is thus the key to understanding waves of any shape.

Because Eq. 17-2 has two independent variables (x and t), we cannot represent the dependent variable y on a single two-dimensional graph. We would need a videotape to show it fully and in real time. We can, however, learn a lot from plots like those of Fig. 17-4.

Wavelength and Angular Wave Number

Figure 17-4a shows how the transverse displacement y of Eq. 17-2 varies with position x at an instant, arbitrarily called $t = 0$. That is, the figure is a "snapshot" of the wave at that instant. With $t = 0$, Eq. 17-2 becomes

$$y(x, 0) = y_m \sin kx \qquad (t = 0). \qquad (17\text{-}3)$$

Figure 17-4a is a plot of this equation; it shows the shape of the actual wave at time $t = 0$.

The **wavelength** λ of a wave is the distance (along the direction of the wave) between repetitions of the wave shape. A typical wavelength is marked in Fig. 17-4a. By definition, the displacement y is the same at both ends of this wavelength, that is, at $x = x_1$ and $x = x_1 + \lambda$. Thus, by Eq. 17-3,

$$y_m \sin kx_1 = y_m \sin k(x_1 + \lambda)$$
$$= y_m \sin(kx_1 + k\lambda). \qquad (17\text{-}4)$$

A sine function begins to repeat itself when its angle (or argument) is increased by 2π rad; so Eq. 17-4 will be accurate only if $k\lambda = 2\pi$, or if

$$k = \frac{2\pi}{\lambda} \qquad \text{(angular wave number).} \qquad (17\text{-}5)$$

We call k the **angular wave number** of the wave; its SI unit is the radian per meter. (Note that the symbol k here does *not* represent a spring constant as previously.)

Period, Angular Frequency, and Frequency

Figure 17-4b shows how the displacement y of Eq. 17-2 varies with time t at a fixed position, taken to be $x = 0$. If you were to monitor the string, you would see that the single element of the string at that position moves up and down in simple harmonic motion given by Eq. 17-2 with $x = 0$:

$$y(0, t) = y_m \sin(-\omega t)$$
$$= -y_m \sin \omega t \qquad (x = 0). \qquad (17\text{-}6)$$

Here we have made use of the fact that $\sin(-\alpha) = -\sin \alpha$, where α is any angle. Figure 17-4b is a plot of this equation; it *does not* show the shape of the wave.

We define the **period** of oscillation T of a wave to be the time interval between repetitions of the motion of an oscillating string element (at any position x). A typical period is marked in Fig. 17-4b. Applying Eq. 17-6 to both ends of this time interval and equating the results yield

$$-y_m \sin \omega t_1 = -y_m \sin \omega(t_1 + T)$$
$$= -y_m \sin(\omega t_1 + \omega T). \qquad (17\text{-}7)$$

This can be true only if $\omega T = 2\pi$, or if

$$\omega = \frac{2\pi}{T} \qquad \text{(angular frequency).} \qquad (17\text{-}8)$$

We call ω the **angular frequency** of the wave; its SI unit is the radian per second.

The **frequency** f of the wave is defined as $1/T$ and is related to the angular frequency ω by

$$f = \frac{1}{T} = \frac{\omega}{2\pi} \qquad \text{(frequency).} \qquad (17\text{-}9)$$

Like the frequency of simple harmonic motion in Chapter 16, this frequency f is a number of oscillations per unit time—here, made by a string element as the wave moves through it. As in Chapter 16, f is usually measured in hertz or its multiples.

CHECKPOINT **1:** The figure shows a snapshot of three waves traveling along a string. The phases for the waves are given by (a) $2x - 4t$, (b) $4x - 8t$, and (c) $8x - 16t$. Which phase corresponds to which wave in the figure?

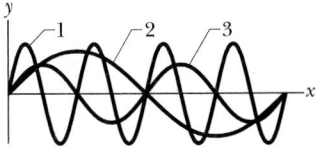

17-5 THE SPEED OF A TRAVELING WAVE

Figure 17-5 shows two snapshots of the wave of Eq. 17-2, taken a small time interval Δt apart. The wave is traveling in the direction of increasing x (to the right in Fig. 17-5), the entire wave pattern moving a distance Δx in that direction during the interval Δt. The ratio $\Delta x/\Delta t$ (or, in the differential limit, dx/dt) is the **wave speed** v. How can we find its value?

As the wave in Fig. 17-5 moves, each point of the moving wave form (such as point A) retains its displacement y. (Points on the string do not retain their displacement, but points on the wave *form* do.) For each such point, Eq. 17-2 tells us that the argument of the sine function must be a constant:

$$kx - \omega t = \text{a constant.} \qquad (17\text{-}10)$$

Note that although this argument is constant, both x and t are changing. In fact, as t increases, x must also, to keep the argument constant. This confirms that the wave pattern is moving toward increasing x.

To find the wave speed v, we take the derivative of Eq. 17-10, getting

$$k\frac{dx}{dt} - \omega = 0$$

or

$$\frac{dx}{dt} = v = \frac{\omega}{k}. \qquad (17\text{-}11)$$

Using Eq. 17-5 ($k = 2\pi/\lambda$) and Eq. 17-8 ($\omega = 2\pi/T$), we can rewrite the wave speed as

$$v = \frac{\omega}{k} = \frac{\lambda}{T} = \lambda f \qquad \text{(wave speed).} \qquad (17\text{-}12)$$

The equation $v = \lambda/T$ tells us that the wave speed is one wavelength per period: the wave moves a distance of one wavelength in one period of oscillation.

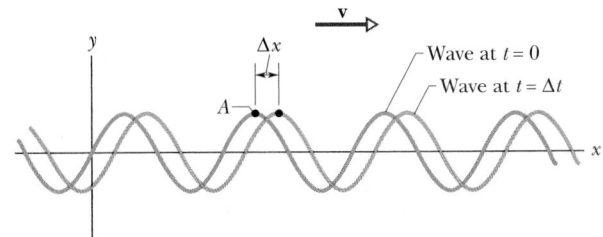

FIGURE 17-5 A snapshot of the traveling wave of Eq. 17-2 at $t = 0$ and at a later time $t = \Delta t$. During the time interval Δt, the entire curve shifts a distance Δx to the right.

Equation 17-2 describes a wave moving in the direction of increasing x. We can find the equation of a wave traveling in the opposite direction by replacing t in Eq. 17-2 with $-t$. This corresponds to the condition

$$kx + \omega t = \text{a constant,} \qquad (17\text{-}13)$$

which (compare Eq. 17-10) requires that x *decrease* with time. Thus, a wave traveling toward decreasing x is described by the equation

$$y(x, t) = y_m \sin(kx + \omega t). \qquad (17\text{-}14)$$

If you analyze the wave of Eq. 17-14 as we have just done for the wave of Eq. 17-2, you will find for its velocity

$$\frac{dx}{dt} = -\frac{\omega}{k}. \qquad (17\text{-}15)$$

The minus sign (compare Eq. 17-11) verifies that the wave is indeed moving in the direction of decreasing x and justifies our switching the sign of the time variable.

Consider now a wave of generalized shape, given by

$$y(x, t) = h(kx \pm \omega t), \qquad (17\text{-}16)$$

where h represents *any* function, the sine function being one possibility. Our analysis above shows that all waves in which the variables x and t enter in the combination $kx \pm \omega t$ are traveling waves. Furthermore, all traveling waves *must* be of the form of Eq. 17-16. Thus $y(x, t) = \sqrt{ax + bt}$ represents a possible (though perhaps physically a little bizarre) traveling wave. The function $y(x, t) = \sin(ax^2 - bt)$, on the other hand, does *not* represent a traveling wave.

SAMPLE PROBLEM 17-1

A sinusoidal wave traveling along a string is described by

$$y(x, t) = 0.00327 \sin(72.1x - 2.72t), \qquad (17\text{-}17)$$

in which the numerical constants are in SI units (0.00327 m, 72.1 rad/m, and 2.72 rad/s).

(a) What is the amplitude of this wave?

SOLUTION: By comparison with Eq. 17-2, we see that

$$y_m = 0.00327 \text{ m} = 3.27 \text{ mm.} \qquad \text{(Answer)}$$

(b) What are the wavelength, period, and frequency of this wave?

SOLUTION: By inspection of Eq. 17-17, we see that

$$k = 72.1 \text{ rad/m} \quad \text{and} \quad \omega = 2.72 \text{ rad/s.}$$

From Eq. 17-5 we have

$$\lambda = \frac{2\pi}{k} = \frac{2\pi \text{ rad}}{72.1 \text{ rad/m}} = 0.0871 \text{ m}$$

$$= 8.71 \text{ cm.} \qquad \text{(Answer)}$$

From Eq. 17-9 we have

$$T = \frac{2\pi}{\omega} = \frac{2\pi \text{ rad}}{2.72 \text{ rad/s}} = 2.31 \text{ s.} \qquad \text{(Answer)}$$

From Eq. 17-9 we have

$$f = \frac{1}{T} = \frac{1}{2.31 \text{ s}} = 0.433 \text{ Hz.} \qquad \text{(Answer)}$$

(c) What is the speed of this wave?

SOLUTION: From Eq. 17-12 we have

$$v = \frac{\omega}{k} = \frac{2.72 \text{ rad/s}}{72.1 \text{ rad/m}} = 0.0377 \text{ m/s}$$

$$= 3.77 \text{ cm/s.} \qquad \text{(Answer)}$$

Note that the quantities calculated in (b) and (c) are independent of the amplitude of the wave.

(d) What is the displacement y at $x = 22.5$ cm and $t = 18.9$ s?

SOLUTION: From Eq. 17-17 we obtain

$$y = 0.00327 \sin(72.1 \times 0.225 - 2.72 \times 18.9)$$

$$= (0.00327 \text{ m}) \sin(-35.1855 \text{ rad})$$

$$= (0.00327 \text{ m})(0.588)$$

$$= 0.00192 \text{ m} = 1.92 \text{ mm.} \qquad \text{(Answer)}$$

Thus the displacement is positive. (Be sure to change your calculator mode to radians before evaluating the sine.)

SAMPLE PROBLEM 17-2

In Sample Problem 17-1d, we showed that the transverse displacement y of an element of the string for the wave of Eq. 17-17 at $x = 0.225$ m and $t = 18.9$ s is 1.92 mm.

(a) What is u, the transverse speed of the same element of the string, at that place and at that time? (This speed, which is associated with the transverse oscillation of an element of the string, is in the y direction. Do not confuse it with v, the constant speed at which the *wave form* travels along the x axis.)

SOLUTION: The generalized equation of the wave of Eq. 17-17 is Eq. 17-2:

$$y(x, t) = y_m \sin(kx - \omega t). \qquad (17\text{-}18)$$

In this expression, let us hold x constant but allow t (for the time being) to be a variable. We then take the derivative of this equation with respect to t, obtaining*

$$u = \frac{\partial y}{\partial t} = -\omega y_m \cos(kx - \omega t). \qquad (17\text{-}19)$$

---

*A derivative taken while one (or more) of the variables is treated as a constant is called a *partial derivative* and is represented by the symbol $\partial/\partial x$, rather than d/dx.

Substituting numerical values from Sample Problem 17-1, we obtain

$$u = (-2.72 \text{ rad/s})(3.27 \text{ mm}) \cos(-35.1855 \text{ rad})$$
$$= 7.20 \text{ mm/s.} \qquad \text{(Answer)}$$

Thus, at $t = 18.9$ s, the element of string at $x = 22.5$ cm is moving in the direction of increasing y, with a speed of 7.20 mm/s.

(b) What is the transverse acceleration a_y at that position and at that time?

SOLUTION: From Eq. 17-19, treating x as a constant but allowing t (for the time being) to be a variable, we find

$$a_y = \frac{\partial u}{\partial t} = -\omega^2 y_m \sin(kx - \omega t).$$

Comparison with Eq. 17-18 shows that we can write this as

$$a_y = -\omega^2 y.$$

We see that the transverse acceleration of an oscillating string element is proportional to its transverse displacement but opposite in sign, completely consistent with the action of the element: namely, it is moving transversely in simple harmonic motion. Substituting numerical values yields

$$a_y = -(2.72 \text{ rad/s})^2 (1.92 \text{ mm})$$
$$= -14.2 \text{ mm/s}^2. \qquad \text{(Answer)}$$

Thus, at $t = 18.9$ s, the element of string at $x = 22.5$ cm is displaced from its equilibrium position by 1.92 mm in the increasing y direction and has an acceleration of magnitude 14.2 mm/s^2 in the decreasing y direction.

C HECKPOINT **2:** Here are the equations of three waves:
  (1) $y(x, t) = 2 \sin(4x - 2t)$,
  (2) $y(x, t) = \sin(3x - 4t)$,
  (3) $y(x, t) = 2 \sin(3x - 3t)$.
  Rank the waves according to their (a) wave speed and (b) maximum transverse speed, greatest first.

PROBLEM SOLVING TACTICS

TACTIC 1: *Evaluating Large Phases*
Sometimes, as in Sample Problems 17-1d and 17-2, an angle much greater than 2π rad (or 360°) crops up and you are asked to find its sine or cosine. Adding or subtracting an integral multiple of 2π rad to such an angle does not change the value of any of its trigonometric functions. In Sample Problem 17-1d, for example, the angle is -35.1855 rad. Adding $(6)(2\pi$ rad) to this angle yields

$$-35.1855 \text{ rad} + (6)(2\pi \text{ rad}) = 2.51361 \text{ rad},$$

an angle less than 2π rad that has the same trigonometric functions as -35.1855 rad (Fig. 17-6). As an example, the sine of both 2.51361 rad and -35.1855 rad is 0.588.

Your calculator will reduce such large angles for you automatically. Caution: Do not round off large angles if you intend to take their sines or cosines. In taking the sine of a very large angle, you are throwing away most of the angle and taking the sine of what is left over. If, for example, you were to round -35.1855 rad to -35 rad (a change of 0.5% and normally a reasonable step), you would be changing the sine of the angle by 27%. Also, if you change a large angle from degrees to radians, be sure to use an exact conversion factor (such as $180° = \pi$ rad) rather than an approximate one (such as $57.3° \approx 1$ rad).

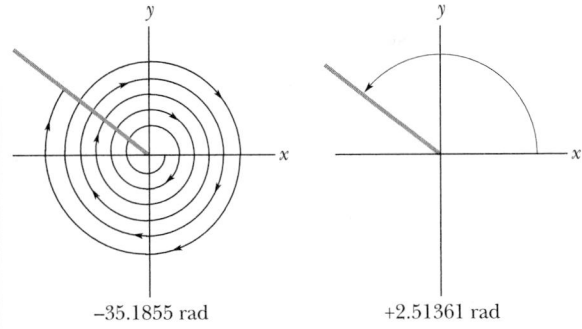

FIGURE 17-6 The two angles are different but all their trigonometric functions are identical.

17-6 WAVE SPEED ON A STRETCHED STRING

The speed of a wave is related to the wave's wavelength and frequency by Eq. 17-12, but *it is set by the medium*. If a wave is to travel through a medium such as water, air, steel, or a stretched string, it must cause the particles of that medium to oscillate as it passes. For that to happen, the medium must possess both inertia (so that kinetic energy can be stored) and elasticity (so that potential energy can be stored). These two properties determine how fast the wave can travel in the medium. And conversely, it should be possible to calculate the speed of the wave through the medium in terms of these properties. We do so now for a stretched string, in two ways.

Dimensional Analysis

In dimensional analysis we examine carefully the dimensions of all the physical quantities that enter into a given situation. In this case, we seek a speed v, which has the dimension of length divided by time, or LT^{-1}.

The inertia characteristic of a stretched string is the mass of a string element, which is represented by the mass m of the string divided by the length l of the string. We call this ratio the *linear density* μ of the string. Thus $\mu = m/l$, its dimension being mass divided by length, ML^{-1}.

You cannot send a wave along a stretched string without further stretching the string. The tension in the string does this stretching and must therefore represent the elastic characteristic of the string. The tension τ is a force and has the dimension (think about $F = ma$) MLT^{-2}. (We would normally use the symbol T for tension but we want to avoid confusion with our use of this symbol for the period of oscillation.)

The problem here is to combine μ (dimension ML^{-1}) and τ (dimension MLT^{-2}) in such a way as to generate v (dimension LT^{-1}). A little juggling of various combinations suggests

$$v = C\sqrt{\frac{\tau}{\mu}}, \qquad (17\text{-}20)$$

in which C is a dimensionless constant. The drawback of dimensional analysis is that it cannot give the value of such a dimensionless constant. In our second approach to determine the speed, you will see that Eq. 17-20 is indeed correct and that $C = 1$.

Derivation from Newton's Second Law

Instead of the sinusoidal wave of Fig. 17-1b, let us consider a single symmetrical pulse such as that of Fig. 17-7. For convenience, we choose a reference frame in which the pulse remains stationary. That is, we run along with the pulse, keeping it constantly in view. In this frame, the string appears to move past us, from right to left in Fig. 17-7, with speed v.

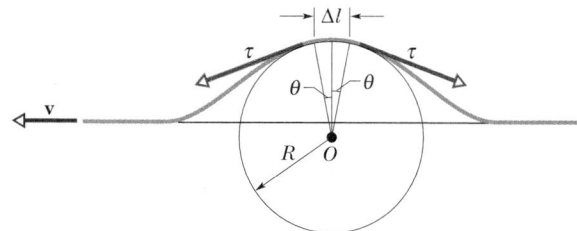

FIGURE 17-7 A symmetrical pulse, viewed from a reference frame in which the pulse is stationary and the string appears to move right to left with speed v. We find speed v by applying Newton's second law to a string element of length Δl, located at the top of the pulse.

Consider a small segment of the pulse, of length Δl, forming an arc of a circle of radius R. A force equal in magnitude to the string tension τ pulls tangentially on this segment at each end. The horizontal components of these forces cancel, but the vertical components add to form a restoring force **F**. In magnitude,

$$F = 2\tau \sin\theta \approx \tau(2\theta) = \tau\frac{\Delta l}{R} \qquad \text{(force).} \quad (17\text{-}21)$$

We have used here the approximation $\sin\theta \approx \theta$ for the small angles θ in Fig. 17-7; from that figure, we have also used $2\theta = \Delta l/R$.

The mass of the segment is given by

$$\Delta m = \mu\,\Delta l \qquad \text{(mass).} \quad (17\text{-}22)$$

At the moment shown in Fig. 17-7, the string element Δl is moving in an arc of a circle. Thus it has a centripetal acceleration toward the center of that circle, given by

$$a = \frac{v^2}{R} \qquad \text{(acceleration).} \quad (17\text{-}23)$$

Equations 17-21, 17-22, and 17-23 contain the elements of Newton's second law. Combining them in the form

$$\text{force} = \text{mass} \times \text{acceleration}$$

gives

$$\frac{\tau\,\Delta l}{R} = (\mu\,\Delta l)\frac{v^2}{R}.$$

Solving this equation for the speed v yields

$$v = \sqrt{\frac{\tau}{\mu}} \qquad \text{(speed),} \quad (17\text{-}24)$$

in exact agreement with Eq. 17-20 if the constant C in that equation is given the value unity. Equation 17-24 gives the speed of the pulse in Fig. 17-7 and the speed of any other wave on the same string under the same tension.

Equation 17-24 tells us that the speed of a wave along a stretched ideal string depends only on the characteristics of the string and not on the frequency of the wave. The *frequency* of the wave is fixed entirely by whatever generates the wave (for example, the person in Fig. 17-1b). The *wavelength* of the wave is then fixed by Eq. 17-12 in the form $\lambda = v/f$.

CHECKPOINT **3:** The figure shows two situations in which the same string is put under tension by a suspended mass of 5 kg. In which situation will the speed of waves sent along the string be greater?

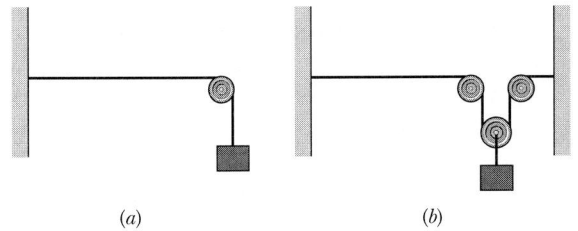

(a)   (b)

SAMPLE PROBLEM 17-3

In Fig. 17-8, a stranded climber has hooked himself onto a makeshift rope lowered by a rescuer. The rope between the climber and rescuer consists of two sections: section 1 with length l_1 and linear density μ_1, and section 2 with length $l_2 = 2l_1$ and linear density $\mu_2 = 4\mu_1$. The climber happens to pluck the bottom end of the rope (as a "ready" signal) at the same time the rescuer plucks the top end.

(a) What is the speed v_1 of the resulting pulses in section 1, in terms of their speed v_2 in section 2?

SOLUTION: We assume that the mass of the rope sections is negligible relative to the mass of the climber. Thus the tension τ in the rope is equal to the weight of the climber and is the same in the two sections. From Eq. 17-24, speeds v_1 and v_2 are given by

$$v_1 = \sqrt{\frac{\tau}{\mu_1}} \quad \text{and} \quad v_2 = \sqrt{\frac{\tau}{\mu_2}}. \qquad (17\text{-}25)$$

Dividing the first expression by the second, and substituting $\mu_2 = 4\mu_1$, we find

$$\frac{v_1}{v_2} = \sqrt{\frac{\tau}{\mu_1}} \sqrt{\frac{\mu_2}{\tau}} = \sqrt{\frac{\mu_2}{\mu_1}} = \sqrt{\frac{4\mu_1}{\mu_1}} = 2,$$

or
$$v_1 = 2v_2. \qquad \text{(Answer)} \qquad (17\text{-}26)$$

(b) In terms of l_2, at what distance below the rescuer do the two pulses pass through each other?

SOLUTION: We can simplify our calculations by first deciding whether the pulses pass each other above or below the knot joining the two sections. Let t be the time the pulses take to reach each other. From Eq. 17-26, we know that the climber's pulse moves through section 1 twice as fast as the rescuer's pulse moves through section 2. Because $l_2 = 2l_1$, we also know that the climber's pulse must travel half as far as the rescuer's pulse to reach the knot. So the climber's pulse must reach the knot first, and the point of passing must be above the knot. Let us assume that the pulses pass at a distance d below the rescuer at time t.

To reach the point of passing, the rescuer's pulse must travel downward through a distance d at speed v_2 in time t. So

$$t = \frac{d}{v_2}. \qquad (17\text{-}27)$$

To reach the point of passing, the climber's pulse must travel upward through distance l_1 at speed v_1 and then through distance $l_2 - d$ at speed v_2, all in time t. So

$$t = \frac{l_1}{v_1} + \frac{l_2 - d}{v_2}. \qquad (17\text{-}28)$$

Substituting t from Eq. 17-27 into Eq. 17-28, and setting $l_1 = l_2/2$ and $v_1 = 2v_2$, we find

$$\frac{d}{v_2} = \frac{l_1}{v_1} + \frac{l_2 - d}{v_2} = \frac{l_2/2}{2v_2} + \frac{l_2 - d}{v_2}.$$

Multiplying through by v_2 and rearranging yield

$$d = \tfrac{5}{8}l_2. \qquad \text{(Answer)}$$

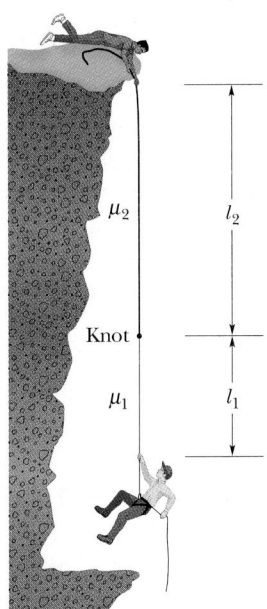

FIGURE 17-8 Sample Problem 17-3. A stranded climber hangs from a rope consisting of two sections. His rescuer has secured the rope at the top.

17-7 ENERGY AND POWER OF A TRAVELING STRING WAVE

When we set up a wave on a stretched string, we provide energy for the motion of the string. As the wave moves away from us, it transports that energy as both kinetic energy and elastic potential energy. Let us consider each form in turn.

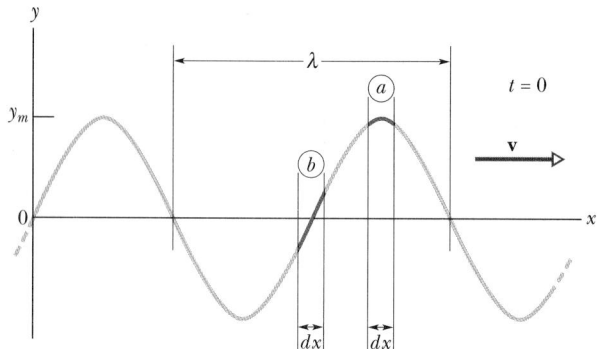

FIGURE 17-9 A snapshot of a traveling wave on a string at time $t = 0$. String element a is at displacement $y = y_m$, and string element b is at displacement $y = 0$. The kinetic energy of the string element at each position depends on the transverse velocity of the element. The potential energy depends on the amount by which the string element is stretched as the wave passes through it.

Kinetic Energy

An element of the string of mass dm, oscillating transversely in simple harmonic motion as the wave passes through it, has kinetic energy associated with its transverse velocity **u**. When the element is rushing through its $y = 0$ position (Fig. 17-9), its transverse velocity—and thus its kinetic energy—is a maximum. When the element is at its extreme position $y = y_m$, its transverse velocity—and thus again its kinetic energy—is zero.

Elastic Potential Energy

To send a sinusoidal wave along a previously straight string, the wave must necessarily stretch the string. As a string element of length dx oscillates transversely, its length must increase and decrease in a periodic way if the string element is to fit the sinusoidal wave form. Elastic potential energy is associated with these length changes, just as for a spring.

When the string element is at its $y = y_m$ position (Fig. 17-9), its length has its normal undisturbed value dx, so its elastic potential energy is zero. However, when the element is rushing through its $y = 0$ position, it is stretched to its maximum extent, and its elastic potential energy then is a maximum.

The oscillating string element thus has both its maximum kinetic energy and its maximum elastic potential energy at $y = 0$. In the snapshot of Fig. 17-9, the regions of the string at maximum displacement have no energy, and the regions at zero displacement have maximum energy. As the wave travels along the string, the tension in the string continuously does work to transfer energy from regions with energy to regions with no energy.

The Rate of Energy Transmission

The kinetic energy dK associated with a string element of mass dm is given by

$$dK = \tfrac{1}{2}\, dm\, u^2, \tag{17-29}$$

where u is the transverse speed of the oscillating string element, given by Eq. 17-19 as

$$u = \frac{\partial y}{\partial t} = -\omega y_m \cos(kx - \omega t). \tag{17-30}$$

Using this relation and putting $dm = \mu\, dx$, we rewrite Eq. 17-29 as

$$dK = \tfrac{1}{2}(\mu\, dx)(-\omega y_m)^2 \cos^2(kx - \omega t). \tag{17-31}$$

Dividing Eq. 17-31 by dt gives the rate at which the kinetic energy of a string element changes, and thus the rate at which kinetic energy is carried along by the wave. The ratio dx/dt that then appears on the right of Eq. 17-31 is the wave speed v, so we obtain

$$\frac{dK}{dt} = \tfrac{1}{2}\mu v \omega^2 y_m^2 \cos^2(kx - \omega t). \tag{17-32}$$

The *average* rate at which kinetic energy is transported is

$$\overline{\left(\frac{dK}{dt}\right)} = \tfrac{1}{2}\mu v \omega^2 y_m^2\, \overline{\cos^2(kx - \omega t)}$$

$$= \tfrac{1}{4}\mu v \omega^2 y_m^2, \tag{17-33}$$

where an overhead bar means an average value of the quantity. In Eq. 17-33 we have taken the average over an integer number of wavelengths and have used the fact that the average value of the square of a cosine function over an integer number of wavelengths is $\tfrac{1}{2}$.

Elastic potential energy is also carried along with the wave, and at the same average rate given by Eq. 17-33. Although we shall not examine the proof, you should recall that, in an oscillating system such as a pendulum or a spring–block system, the average kinetic energy and the average potential energy are indeed equal.

The **average power,** which is the average rate at which energy of both kinds is transmitted by the wave, is then

$$\overline{P} = 2\,\overline{\left(\frac{dK}{dt}\right)} \tag{17-34}$$

or, from Eq. 17-33,

$$\overline{P} = \tfrac{1}{2}\mu v \omega^2 y_m^2 \qquad \text{(average power).} \tag{17-35}$$

The factors μ and v in this equation depend on the material and tension of the string. The factors ω and y_m depend on the process that generates the wave. The dependence of the average power of a wave on the square of its amplitude and also on the square of its angular frequency is a general result, true for waves of all types.

SAMPLE PROBLEM 17-4

A string has a linear density μ of 525 g/m and is stretched with a tension τ of 45 N. A wave whose frequency f and amplitude y_m are 120 Hz and 8.5 mm, respectively, is traveling along the string. At what average rate is the wave transporting energy along the string?

SOLUTION: Before finding $\overline{P}$ from Eq. 17-35, we must calculate the angular frequency ω and the wave speed v. From Eq. 17-9,

$$\omega = 2\pi f = (2\pi)(120 \text{ Hz}) = 754 \text{ rad/s}.$$

From Eq. 17-24 we have

$$v = \sqrt{\frac{\tau}{\mu}} = \sqrt{\frac{45 \text{ N}}{0.525 \text{ kg/m}}} = 9.26 \text{ m/s}.$$

Equation 17-35 then yields

$$\overline{P} = \tfrac{1}{2}\mu v \omega^2 y_m^2$$
$$= (\tfrac{1}{2})(0.525 \text{ kg/m})(9.26 \text{ m/s})(754 \text{ rad/s})^2(0.0085 \text{ m})^2$$
$$= 100 \text{ W}. \qquad \text{(Answer)}$$

17-8 THE PRINCIPLE OF SUPERPOSITION FOR WAVES

It often happens that two or more waves pass simultaneously through the same region. When we listen to a concert, for example, sounds from many instruments fall simultaneously on our eardrums. The electrons in the antennas of our radio and TV receivers are set in motion by a whole array of signals from different broadcasting centers. The water of a lake or harbor may be churned up by the wakes of many boats.

Suppose that two waves travel simultaneously along the same stretched string. Let $y_1(x, t)$ and $y_2(x, t)$ be the displacements that the string would experience if each wave acted alone. The displacement of the string when both waves act is then

$$y'(x, t) = y_1(x, t) + y_2(x, t), \qquad (17\text{-}36)$$

the sum being an algebraic sum. This summation of displacements along the string means

---

Overlapping waves algebraically add to produce a **resultant wave.**

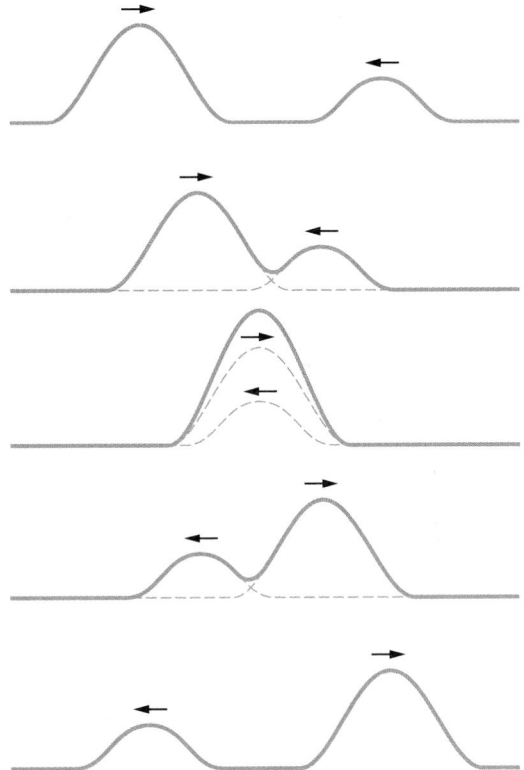

FIGURE 17-10 A series of snapshots that show two pulses traveling in opposite directions along a stretched string. The superposition principle applies as the pulses move through each other.

This is another example of the **principle of superposition,** which says that when several effects occur simultaneously, their net effect is the sum of the individual effects.

Figure 17-10 shows a sequence of snapshots of two pulses traveling in opposite directions on the same stretched string. When the pulses overlap, the resultant pulse is their sum. Moreover, each pulse moves through the other, as if the other were not present:

Overlapping waves do not in any way alter the travel of each other.

Fourier Analysis

French mathematician Jean Baptiste Fourier (1786–1830) explained how the principle of superposition can be used to analyze nonsinusoidal wave forms. He showed that any wave form can be represented as the sum of a large number of sinusoidal waves, of carefully chosen frequencies and amplitudes. English physicist Sir James Jeans expressed it well:

[Fourier's] theorem tells us that every curve, no matter what its nature may be, or in what way it was originally obtained, can be exactly reproduced by superposing a sufficient number of simple harmonic [sinusoidal] curves—in brief, every curve can be built up by piling up waves.

Figure 17-11 shows an example of a *Fourier series,* as such sums are called. The sawtooth curve in Fig. 17-11*a* shows the variation with time (at position $x = 0$) of the wave we wish to represent. The Fourier series that represents it can be shown to be

$$y(t) = -\frac{1}{\pi} \sin \omega t - \frac{1}{2\pi} \sin 2\omega t$$
$$-\frac{1}{3\pi} \sin 3\omega t \cdots, \qquad (17\text{-}37)$$

in which $\omega = 2\pi/T$, where T is the period of the sawtooth curve. The green curve of Fig. 17-11*a*, which represents the sum of the first six terms of Eq. 17-37, matches the sawtooth curve rather well. Figure 17-11*b* shows these six

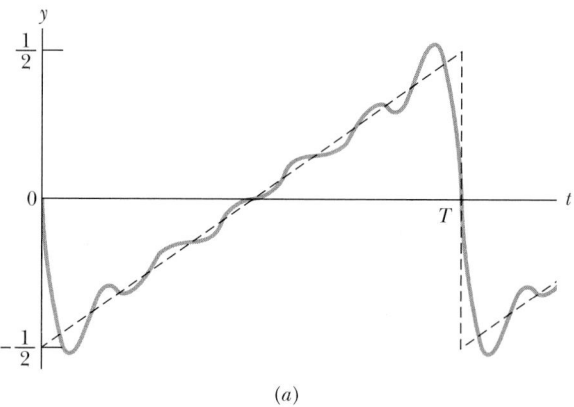

(*a*)

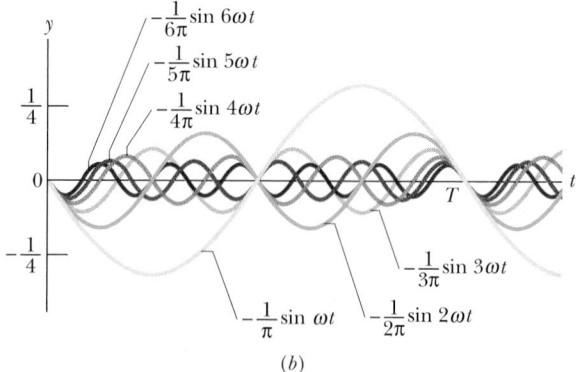

(*b*)

FIGURE 17-11 (*a*) The dashed sawtooth curve is approximated by the green curve, which is the result of evaluating the first six terms in the sum of Eq. 17-37. (Evaluating more terms would give closer approximations.) (*b*) Separate displays of the first six terms of that equation, each a sinusoidal function.

terms separately. By adding more terms, we can approximate the sawtooth curve as closely as we wish.

You can see now why we spent so much time analyzing the behavior of a sinusoidal wave. When we understand that, Fourier's theorem will open the door to all other wave shapes.

17-9 INTERFERENCE OF WAVES

Suppose we send two sinusoidal waves of the same wavelength and amplitude in the same direction along a stretched string. The superposition principle applies. What resultant wave does it predict for the string?

The resultant wave depends on the extent to which the waves are *in phase* (in step) with respect to each other, that is, how much one wave form is shifted from the other wave form. If the waves are exactly in phase (so that the peaks and valleys of one are exactly aligned with those of the other without any shift), they combine to double the displacement of either wave acting alone. If they are exactly out of phase (the peaks of one are exactly aligned with the valleys of the other), they cancel everywhere and the string remains straight. We call this phenomenon of combining and canceling of waves **interference,** and the waves are said to **interfere.** (These terms refer only to the displacements of the waves; the travel of the waves is unaffected.)

Let one wave traveling along a stretched string be given by

$$y_1(x, t) = y_m \sin(kx - \omega t) \qquad (17\text{-}38)$$

and another, shifted from the first, by

$$y_2(x, t) = y_m \sin(kx - \omega t + \phi). \qquad (17\text{-}39)$$

These waves have the same angular frequency ω (and thus the same frequency f), the same angular wave number k (and thus the same wavelength λ), and the same amplitude y_m. They travel in the same direction, that of increasing x, with the same speed, given by Eq. 17-24. They differ only by a constant angle ϕ, which we call the **phase constant.** These waves are said to be *out of phase* by ϕ or to have a *phase difference* of ϕ, or one wave is said to be *phase-shifted* from the other by ϕ.

From the principle of superposition (Eq. 17-36), the combined wave has displacement

$$y'(x, t) = y_1(x, t) + y_2(x, t)$$
$$= y_m \sin(kx - \omega t) + y_m \sin(kx - \omega t + \phi). \qquad (17\text{-}40)$$

In Appendix E we see that we can write the sum of the sines of two angles α and β as

$$\sin \alpha + \sin \beta = 2 \sin \tfrac{1}{2}(\alpha + \beta) \cos \tfrac{1}{2}(\alpha - \beta). \qquad (17\text{-}41)$$

Applying this relation to Eq. 17-40 yields

$$y'(x, t) = [2y_m \cos \tfrac{1}{2}\phi] \sin(kx - \omega t + \tfrac{1}{2}\phi). \quad (17\text{-}42)$$

The resultant wave is thus also a sinusoidal wave traveling in the direction of increasing x. It is the only wave you would actually see on the string (you would *not* see the two combining waves of Eqs. 17-38 and 17-39).

If two sinusoidal waves of the same amplitude and wavelength travel in the *same* direction along a stretched string, they interfere to produce a resultant sinusoidal wave traveling in that direction.

The resultant wave differs from the original waves in two respects: (1) its phase constant is $\tfrac{1}{2}\phi$, and (2) its amplitude y'_m is the quantity in the brackets in Eq. 17-42:

$$y'_m = 2y_m \cos \tfrac{1}{2}\phi. \quad (17\text{-}43)$$

If $\phi = 0$ rad (or 0°), the two combining waves are exactly in phase (as in Fig. 17-12a). Then Eq. 17-42 reduces to

$$y'(x, t) = 2y_m \sin(kx - \omega t) \quad (\phi = 0). \quad (17\text{-}44)$$

Note that the amplitude of the resultant wave is twice the amplitude of either combining wave. This is the greatest amplitude the resultant wave can have, because the cosine term in Eq. 17-42 and Eq. 17-43 has its greatest value (unity) when $\phi = 0$. Interference that produces the greatest possible amplitude is called *fully constructive interference*.

If $\phi = \pi$ rad (or 180°), the combining waves are exactly out of phase (as in Fig. 17-12b). Then $\cos \tfrac{1}{2}\phi$ becomes $\cos \pi/2 = 0$, and the amplitude of the resultant wave (Eq. 17-43) is zero. We then have for all values of x and t,

$$y'(x, t) = 0 \quad (\phi = \pi \text{ rad}). \quad (17\text{-}45)$$

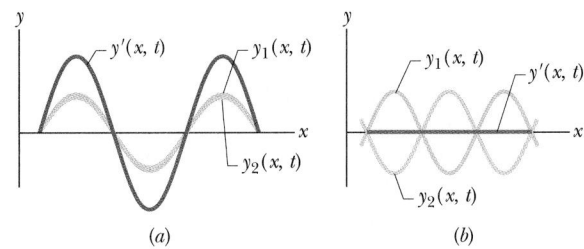

(a)   (b)

FIGURE 17-12 Two identical waves, $y_1(x, t)$ and $y_2(x, t)$, traveling in the same direction along a string, interfere to give a resultant wave $y'(x, t)$. (a) If the waves are exactly in phase, they undergo fully constructive interference and produce a resultant wave of twice their own amplitude. (b) If they are exactly out of phase, they undergo fully destructive interference and the string remains straight.

Now, although we sent two waves along the string, we see no motion of the string. This type of interference is called *fully destructive interference*.

A phase difference $\phi = 2\pi$ rad (or 360°) corresponds to a shift of one wave relative to the other wave by a distance of one wavelength. So phase differences can also be described in terms of wavelengths. For example, in Fig. 17-12b the waves may be said to be 0.50 wavelength out of phase. Table 17-1 shows some other examples of phase differences and the interference they produce. Note that when interference is neither fully constructive nor fully destructive, it is called *intermediate interference*. The amplitude of the resultant wave is then intermediate between 0 and $2y_m$.

Two waves with the same wavelength are in phase if their phase difference is zero or any integer number of wavelengths. Thus the integer part of any phase difference *expressed in wavelengths* may be discarded. For example, a phase difference of 0.40 wavelength is equivalent in every way to one of 2.40 wavelengths, and so the simpler of the two numbers can be used in computations.

TABLE 17-1 PHASE DIFFERENCES AND RESULTING INTERFERENCE TYPES[a]

PHASE DIFFERENCE, IN			AMPLITUDE OF RESULTANT WAVE	TYPE OF INTERFERENCE
DEGREES	RADIANS	WAVELENGTHS		
0	0	0	$2y_m$	Fully constructive
120	$2\pi/3$	0.33	y_m	Intermediate
180	π	0.50	0	Fully destructive
240	$4\pi/3$	0.67	y_m	Intermediate
360	2π	1.00	$2y_m$	Fully constructive
865	15.1	2.40	$0.60y_m$	Intermediate

[a]The phase difference is between two otherwise identical waves, with amplitude y_m, moving in the same direction.

SAMPLE PROBLEM 17-5

Two identical waves, moving in the same direction along a stretched string, interfere with each other. The amplitude y_m of each wave is 9.8 mm, and the phase difference ϕ between them is 100°.

(a) What is the amplitude y'_m of the resultant wave due to the interference of these two waves, and what type of interference occurs?

SOLUTION: From Eq. 17-43 we have for the amplitude

$$y'_m = 2y_m \cos \tfrac{1}{2}\phi$$
$$= (2)(9.8 \text{ mm}) \cos(100°/2)$$
$$= 13 \text{ mm.} \qquad \text{(Answer)}$$

Because the phase difference is between 0 and 180°, the interference is intermediate.

(b) What phase difference, in radians and in wavelengths, will give the resultant wave an amplitude of 4.9 mm?

SOLUTION: From Eq. 17-43 we have

$$y'_m = 2y_m \cos \tfrac{1}{2}\phi,$$

or

$$4.9 \text{ mm} = (2)(9.8 \text{ mm}) \cos \tfrac{1}{2}\phi,$$

which gives us (with a calculator in the radian mode)

$$\phi = 2 \cos^{-1} \frac{4.9 \text{ mm}}{(2)(9.8 \text{ mm})}$$
$$= \pm 2.636 \text{ rad} \approx \pm 2.6 \text{ rad.} \qquad \text{(Answer)}$$

There are two solutions because we can obtain the same resultant wave by letting the first wave *lead* (travel ahead of) or *lag* (travel behind) the second wave by 2.6 rad. In wavelengths, the phase difference is

$$\frac{\phi}{2\pi \text{ rad/wavelength}} = \frac{\pm 2.636 \text{ rad}}{2\pi \text{ rad/wavelength}}$$
$$= \pm 0.42 \text{ wavelength.} \qquad \text{(Answer)}$$

CHECKPOINT **4:** Here are four other possible phase differences between the two waves of Sample Problem 17-5, expressed in wavelengths: 0.20, 0.45, 0.60, and 0.80. Rank them according to the amplitude of the resultant wave, greatest first.

17-10 PHASORS

We can represent a string wave (or any other type of wave) vectorially with a **phasor**. In essence, a phasor is a vector that has a magnitude equal to the amplitude of the wave and that rotates around an origin; the angular speed of the phasor is equal to the angular frequency ω of the wave. For

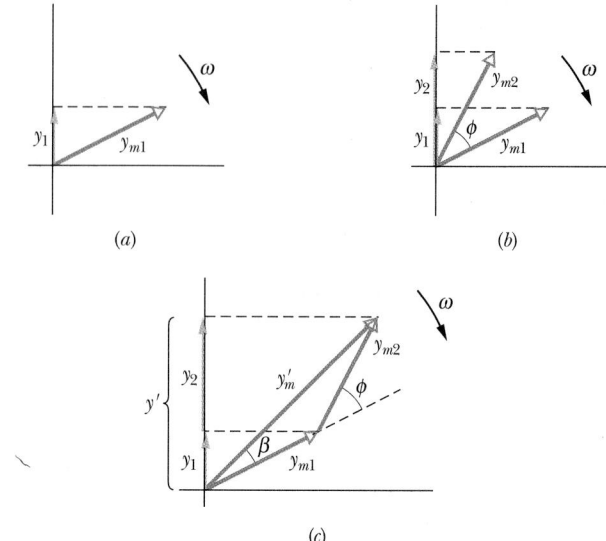

FIGURE 17-13   (*a*) A phasor of magnitude y_{m1} rotating about an origin at angular speed ω represents a sinusoidal wave. The phasor's projection y_1 on the vertical axis represents the displacement of a point through which the wave passes. (*b*) A second phasor, of magnitude y_{m2} and rotating at a constant angle ϕ to the first phasor, represents a second wave, with a phase constant ϕ. (*c*) The resultant wave of the two waves is represented by the vector sum y'_m of the two phasors. The projection y' on the vertical axis represents the displacement of a point as that resultant wave passes through it.

example, the wave

$$y_1(x, t) = y_{m1} \sin(kx - \omega t) \qquad (17\text{-}46)$$

is represented by the phasor shown in Fig. 17-13*a*. The magnitude of the phasor is the amplitude y_{m1} of the wave. As the phasor rotates around the origin at angular speed ω, its projection y_1 on the vertical axis varies sinusoidally, from a maximum of y_{m1} through zero to a minimum of $-y_{m1}$. This variation corresponds to the sinusoidal variation in the displacement y_1 of any point along the string as the wave passes through it.

When two waves travel along the same string in the same direction, we can represent them and their resultant wave in a *phasor diagram*. The phasors in Fig. 17-13*b* represent the wave of Eq. 17-46 and a second wave given by

$$y_2(x, t) = y_{m2} \sin(kx - \omega t + \phi). \qquad (17\text{-}47)$$

The angle between the two phasors in Fig. 17-13*b* is equal to the phase constant ϕ in Eq. 17-47. This angle is constant because the phasors rotate at the same angular speed ω, the angular frequency of the two waves.

Because waves y_1 and y_2 have the same angular wave number k and angular frequency ω, we know that their resultant is of the form

$$y'(x, t) = y'_m \sin(kx - \omega t + \beta), \qquad (17\text{-}48)$$

where y'_m is the amplitude of the resultant wave and β is its phase constant. To find the values of y'_m and β, we would have to sum the two combining waves, as we did to obtain Eq. 17-42.

To do this on a phasor diagram, we vectorially add the two phasors at any instant during their rotation, as in Fig. 17-13c where phasor y_{m2} has been shifted. The magnitude of the vector sum equals the amplitude y'_m in Eq. 17-48. The angle between the vector sum and the phasor for y_1 equals the phase constant β in Eq. 17-48.

Note that in contrast to the method of Section 17-9, using phasors allows us to combine waves *even if their amplitudes are different.*

SAMPLE PROBLEM 17-6

Two waves $y_1(x, t)$ and $y_2(x, t)$ have the same wavelength and travel in the same direction along a string. Their amplitudes are $y_{m1} = 4.0$ mm and $y_{m2} = 3.0$ mm, and their phase constants are 0 and $\pi/3$ rad, respectively. What are the amplitude y'_m and phase constant β of the resultant wave?

SOLUTION: Because the waves travel along the same string, Eq. 17-24 tells us they must have the same wave speed v. Then, because they also have the same wavelength (and thus the same angular wave number k), Eq. 17-12 tells us that they must have the same angular frequency ω. Thus they can be represented by two phasors rotating at the same angular speed ω about an origin, as in Fig. 17-13b. The angle ϕ between the phasors is $\pi/3$ rad.

To find the vector sum of the two phasors as in Fig. 17-13c, we are free to draw them at any instant during their rotation. To simplify the vector addition, we choose to draw them as shown in Fig. 17-14a. We now add the phasors as we would any vectors (Fig. 17-14b). The horizontal component of resultant phasor y'_m is

$$y'_{mh} = y_{m1} \cos 0 + y_{m2} \cos \pi/3$$
$$= 4.0 \text{ mm} + (3.0 \text{ mm}) \cos \pi/3 = 5.50 \text{ mm}.$$

The vertical component of y'_m is

$$y'_{mv} = y_{m1} \sin 0 + y_{m2} \sin \pi/3$$
$$= 0 + (3.0 \text{ mm}) \sin \pi/3 = 2.60 \text{ mm}.$$

Thus the resultant wave has an amplitude of

$$y'_m = \sqrt{(5.50 \text{ mm})^2 + (2.60 \text{ mm})^2}$$
$$= 6.1 \text{ mm} \qquad \text{(Answer)}$$

and a phase constant of

$$\beta = \tan^{-1} \frac{2.60 \text{ mm}}{5.50 \text{ mm}} = 0.44 \text{ rad.} \qquad \text{(Answer)}$$

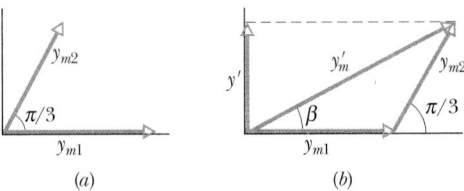

(a)   (b)

FIGURE 17-14 Sample Problem 17-6. (a) Two phasors of magnitudes y_{m1} and y_{m2} and with phase difference $\pi/3$. (b) Vector addition of these phasors at any instant during their rotation gives the magnitude y'_m of the phasor for the resultant wave.

17-11 STANDING WAVES

In the preceding two sections, we discussed two sinusoidal waves of the same wavelength and amplitude traveling *in the same direction* along a stretched string. What if they travel in opposite directions? We can again find the resultant wave by applying the superposition principle.

Figure 17-15 suggests the situation graphically. It shows the two combining waves, one traveling to the left in

(a)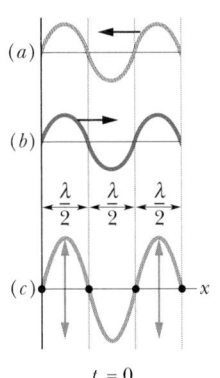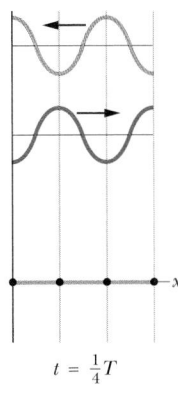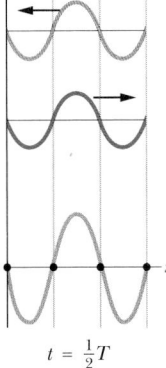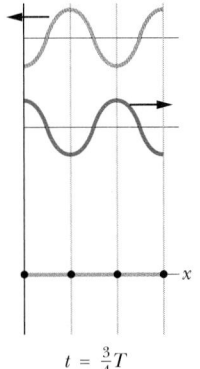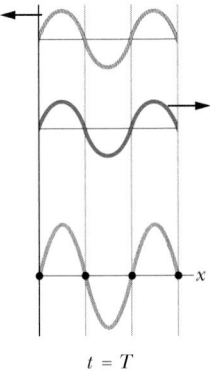

(b)

(c)

$t = 0$ $t = \frac{1}{4}T$ $t = \frac{1}{2}T$ $t = \frac{3}{4}T$ $t = T$

FIGURE 17-15 How traveling waves produce standing waves. (a) and (b) represent snapshots of two waves of the same wavelength and amplitude, traveling in opposite directions, at five different instants during a single period of oscillation. (c) Their su-

perposition at the five instants. Note the nodes and antinodes in (c), the nodes being represented by dots. There are no nodes or antinodes in the traveling waves (a) and (b).

Fig. 17-15a, the other to the right in Fig. 17-15b. Figure 17-15c shows their sum, obtained by applying the superposition principle graphically. The outstanding feature of the resultant wave is that there are places along the string, called **nodes**, where the string is permanently at rest. Four such nodes are marked by dots in Fig. 17-15c. Halfway between adjacent nodes are **antinodes,** where the amplitude of the resultant wave is a maximum. Wave patterns such as that of Fig. 17-15c are called **standing waves** because the wave patterns do not move left or right: the locations of the maxima and minima do not change.

> If two sinusoidal waves of the same amplitude and wavelength travel in *opposite* directions along a stretched string, their interference with each other produces a standing wave.

To analyze a standing wave, we represent the two combining waves with the equations

$$y_1(x, t) = y_m \sin(kx - \omega t) \qquad (17\text{-}49)$$

and
$$y_2(x, t) = y_m \sin(kx + \omega t). \qquad (17\text{-}50)$$

The principle of superposition gives, for the combined wave,

$$y'(x, t) = y_1(x, t) + y_2(x, t)$$
$$= y_m \sin(kx - \omega t) + y_m \sin(kx + \omega t).$$

Applying the trigonometric relation of Eq. 17-41 leads to

$$y'(x, t) = [2y_m \sin kx]\cos \omega t. \qquad (17\text{-}51)$$

This is *not* a traveling wave because it is not of the form of Eq. 17-16. Equation 17-51 describes a standing wave.

The quantity $2y_m \sin kx$ in the brackets of Eq. 17-51 can be viewed as the amplitude of oscillation of the string element that is located at position x. However, since an amplitude is always positive and $\sin kx$ can be negative, we take the absolute value of the quantity $2y_m \sin kx$ to be the amplitude at x.

In a traveling sinusoidal wave, the amplitude of the wave is the same for all string elements. That is not true for a standing wave, in which the amplitude *varies with position*. In the standing wave of Eq. 17-51, for example, the amplitude is zero for values of kx that give $\sin kx = 0$. Those values are

$$kx = n\pi, \qquad \text{for } n = 0, 1, 2, \cdots. \qquad (17\text{-}52)$$

Substituting $k = 2\pi/\lambda$ in this equation and rearranging, we get

$$x = n\frac{\lambda}{2}, \qquad \text{for } n = 0, 1, 2, \cdots.$$
$$\text{(nodes)}, \qquad (17\text{-}53)$$

as the positions of zero amplitude—the nodes—for the standing wave of Eq. 17-51. Note that adjacent nodes are separated by $\lambda/2$, half a wavelength.

The amplitude of the standing wave of Eq. 17-51 has a maximum value of $2y_m$, which occurs for values of kx that give $|\sin kx| = 1$. Those values are

$$kx = \tfrac{1}{2}\pi, \tfrac{3}{2}\pi, \tfrac{5}{2}\pi, \cdots$$
$$= (n + \tfrac{1}{2})\pi, \qquad \text{for } n = 0, 1, 2, \cdots. \qquad (17\text{-}54)$$

Substituting $k = 2\pi/\lambda$ in Eq. 17-54 and rearranging, we get

$$x = \left(n + \frac{1}{2}\right)\frac{\lambda}{2}, \qquad \text{for } n = 0, 1, 2, \cdots$$
$$\text{(antinodes)}, \qquad (17\text{-}55)$$

as the positions of maximum amplitude—the antinodes—of the standing wave of Eq. 17-51. The antinodes are one-half wavelength apart and are located halfway between pairs of nodes.

Reflections at a Boundary

We can set up a standing wave in a stretched string by allowing a traveling wave to be reflected from the far end of the string. The incident (original) wave and the reflected wave can then be described by Eqs. 17-49 and 17-50, respectively, and they can combine to form a pattern of standing waves.

In Fig. 17-16, we use a single pulse to show how such reflections take place. In Fig. 17-16a, the string is fixed at its left end. When the pulse arrives at that end, it exerts an upward force on the support (the wall). By Newton's third law, the support exerts an equal but opposite force on the string. This reaction force generates a pulse at the support, which travels back along the string in the direction opposite that of the incident pulse. In a "hard" reflection of this kind, there must be a node at the support because the string is fixed there. The reflected and incident pulses must have opposite signs, so as to cancel each other at that point.

In Fig. 17-16b, the left end of the string is fastened to a light ring that is free to slide without friction along a rod. When the incident pulse arrives, the ring moves up the rod. As the ring moves, it pulls on the string, stretching the string and producing a reflected pulse with the same sign and amplitude as the incident pulse. Thus in such a "soft" reflection, the incident and reflected pulses reinforce each other, creating an antinode at the end of the string; the

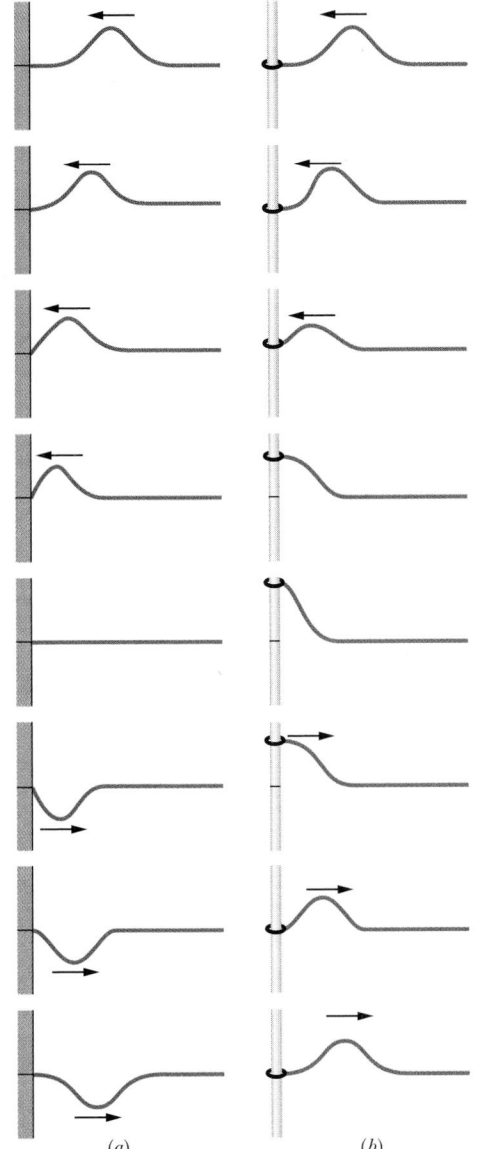

FIGURE 17-16 (*a*) A pulse incident from the right is reflected at the left end of the string, which is tied to a wall. Note that the sign of the reflected pulse is reversed. (*b*) Here the left end of the string is tied to a ring that can slide without friction up and down the rod. The pulse is reflected without a change of sign.

maximum displacement of the ring is twice the amplitude of either of these pulses.

CHECKPOINT **5:** Two waves with the same amplitude and wavelength interfere in three different situations to produce resultant waves with the following equations: (1) $y'(x, t) = 4 \sin(5x - 4t)$
(2) $y'(x, t) = 4 \sin(5x) \cos(4t)$
(3) $y'(x, t) = 4 \sin(5x + 4t)$
In which situation are the two combining waves traveling (a) toward increasing x, (b) toward decreasing x, and (c) in opposite directions?

17-12 STANDING WAVES AND RESONANCE

If (say) the left end of a stretched string is oscillated sinusoidally with the other end fixed, the oscillation sends a continuous traveling wave rightward along the string. The frequency of the wave is that of the oscillation. The wave reflects at the fixed end and travels leftward back through itself. The right-going wave and the left-going wave then interfere with each other.

For certain frequencies, the interference produces a standing wave pattern (or **oscillation mode**) with nodes and large antinodes like those in Fig. 17-17. Such a standing wave is said to be produced at **resonance,** and the string is said to *resonate* at these certain frequencies, called **resonant frequencies.** If the string is oscillated at some frequency other than a resonant frequency, a standing wave is not set up. Then the interference of the right-going wave and the left-going wave results in only small (perhaps imperceptible) oscillations of the string.

Consider a similar situation in which a string, such as a guitar string, is stretched between two clamps separated by a fixed distance L, and the string is somehow made to oscillate at a resonant frequency to set up a standing wave pattern. Since each end of the string is fixed, there must be a node at each end. The simplest pattern that meets this

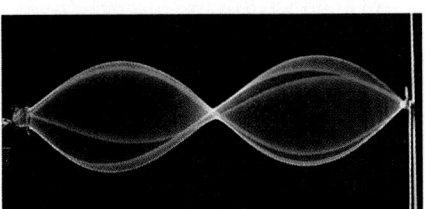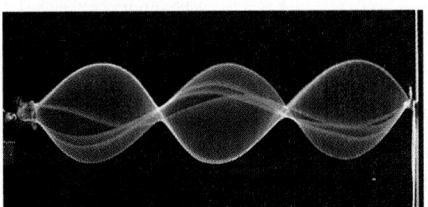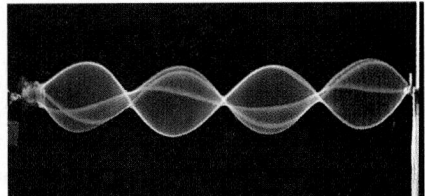

FIGURE 17-17 Stroboscopic photographs reveal (imperfect) standing wave patterns on a string being made to oscillate by a vibrator at the left end. The patterns occur at certain frequencies of oscillation.

requirement is that in Fig. 17-18a, which shows the string at both its extreme displacements (one solid and one dashed). Note that there is only one antinode, which is at the center of the string. Also note that there is half a wavelength in the length L, which we take to be the string's length. Thus, for this pattern, $\lambda/2 = L$. This condition tells us that if the left-going and right-going traveling waves are to set up this pattern by their interference, they must have the wavelength $\lambda = 2L$.

A second simple pattern meeting the requirement of fixed ends is shown in Fig. 17-18b. This pattern has three nodes and two antinodes. For the left-going and right-going waves to set it up, they must have a wavelength $\lambda = L$. A third pattern is shown in Fig. 17-18c. It has four nodes and three antinodes, and the wavelength is $\lambda = \frac{2}{3}L$. You could continue the progression by drawing increasingly more complicated patterns. In each step of the progression, the pattern would have one more node and one more antinode than the preceding step, and an additional $\lambda/2$ would be fitted into the distance L.

The relation between λ and L for a standing wave on a string can be summarized as

$$\lambda = \frac{2L}{n}, \quad \text{for } n = 1, 2, 3, \cdots. \quad (17\text{-}56)$$

The resonant frequencies that correspond to these wavelengths follow from Eq. 17-12:

$$f = \frac{v}{\lambda} = n\frac{v}{2L}, \quad \text{for } n = 1, 2, 3, \cdots. \quad (17\text{-}57)$$

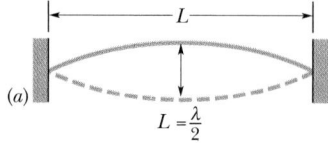
(a)
$L = \frac{\lambda}{2}$

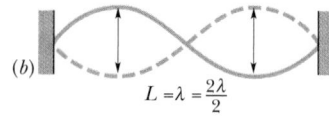
(b)
$L = \lambda = \frac{2\lambda}{2}$

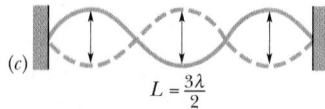
(c)
$L = \frac{3\lambda}{2}$

FIGURE 17-18 A string, stretched between two clamps, is made to oscillate in standing wave patterns. (a) The simplest possible pattern consists of one *loop*, which refers to the composite shape formed by the string in its extreme displacements (the solid and dashed lines). (b) The next simplest pattern has two loops. (c) The next pattern has three loops.

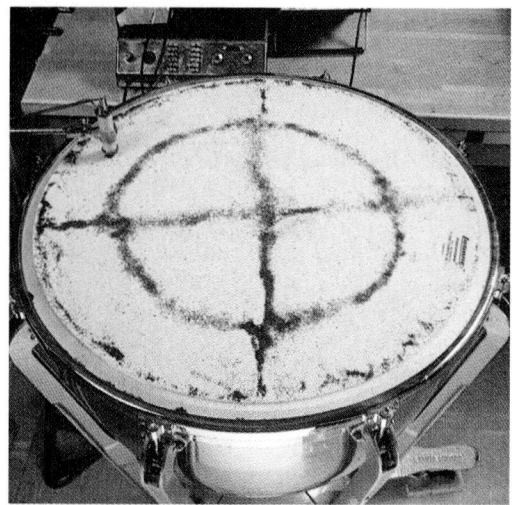

FIGURE 17-19 One of many possible standing wave patterns for a kettledrum head, made visible by dark powder sprinkled on the drumhead. As the head is set into oscillation at a single frequency by a mechanical vibrator at the upper left of the photograph, the powder collects at the nodes, which are circles and straight lines (rather than points) in this two-dimensional example.

Here v is the speed of the traveling waves on the string.

Equation 17-57 tells us that the resonant frequencies are integer multiples of the lowest resonant frequency, $f = v/2L$, which corresponds to $n = 1$. The oscillation mode with that lowest frequency is called the *fundamental mode* or the *first harmonic*. The *second harmonic* is the oscillation mode with $n = 2$; the *third harmonic* is that with $n = 3$; and so on. The frequencies associated with these modes are often labeled f_1, f_2, f_3, and so on. The collection of all possible oscillation modes is called the **harmonic series,** and n is called the **harmonic number** of the nth harmonic.

The phenomenon of resonance is common to all oscillating systems and can occur in two and three dimensions. For example, Fig. 17-19 shows a two-dimensional standing wave pattern on the oscillating head of a kettledrum.

SAMPLE PROBLEM 17-7

In Fig. 17-20, a string, tied to a sinusoidal vibrator at P and running over a support at Q, is stretched by a block of mass m. The separation L between P and Q is 1.2 m, the linear density of the string is 1.6 g/m, and the frequency f of the vibrator is fixed at 120 Hz. The amplitude of the motion at P is small enough for that point to be considered a node. A node also exists at Q.

(a) What mass m allows the vibrator to set up the fourth harmonic on the string?

SOLUTION: The resonant frequencies are given by Eq. 17-57 as

$$f = \frac{v}{2L}\, n, \qquad \text{for } n = 1, 2, 3, \cdots . \quad (17\text{-}58)$$

We need to set the tension τ in the string so that the vibrator frequency is equal to the frequency of the fourth harmonic as given by this equation.

The speed v of waves on the string is given by Eq. 17-24 as

$$v = \sqrt{\frac{\tau}{\mu}} = \sqrt{\frac{mg}{\mu}}, \quad (17\text{-}59)$$

where the tension τ in the string is equal to the weight mg of the block. Substituting v from Eq. 17-59 into Eq. 17-58, setting $n = 4$ for the fourth harmonic, and solving for m give us

$$m = \frac{4L^2 f^2 \mu}{n^2 g} \quad (17\text{-}60)$$

$$= \frac{(4)(1.2 \text{ m})^2(120 \text{ Hz})^2(0.0016 \text{ kg/m})}{(4)^2(9.8 \text{ m/s}^2)}$$

$$= 0.846 \text{ kg} \approx 0.85 \text{ kg}. \qquad \text{(Answer)}$$

(b) What standing wave mode is set up if $m = 1.00$ kg?

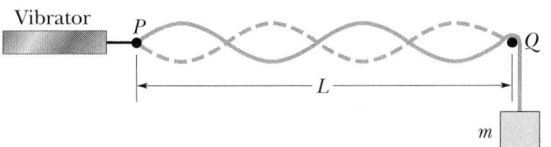

FIGURE 17-20 Sample Problem 17-7. A string under tension connected to a vibrator. For a fixed vibrator frequency, standing wave patterns will occur for certain values of the string tension.

SOLUTION: If we insert this value of m into Eq. 17-60 and solve for n, we find that $n = 3.7$, an impossibility since n must be an integer. Thus, with $m = 1.00$ kg, the vibrator cannot set up a standing wave on the string, and any oscillation of the string will be small, perhaps even imperceptible.

CHECKPOINT **6:** In the following series of resonant frequencies, one frequency (lower than 400 Hz) is missing: 150, 225, 300, 375 Hz. (a) What is the missing frequency? (b) What is the frequency of the seventh harmonic?

PROBLEM SOLVING TACTICS

TACTIC 2: *Harmonics on a String*

When you need to obtain information about a certain harmonic on a stretched string of given length L, first draw that harmonic (as in Fig. 17-18). If you are asked about, say, the fifth harmonic, you need to draw five loops between the fixed support points. That would mean that five loops, each of length $\lambda/2$, occupy the length L of the string. Thus, $5(\lambda/2) = L$, and $\lambda = 2L/5$. You can then use Eq. 17-12 ($f = v/\lambda$) to find the frequency of the harmonic.

Keep in mind that the wavelength of a harmonic is set only by the length L of the string, but the frequency depends also on the wave speed v, which is set by the tension and the linear density of the string via Eq. 17-24.

REVIEW & SUMMARY

Transverse and Longitudinal Waves

Mechanical waves can exist only in material media and are governed by Newton's laws. **Transverse** mechanical waves, like those on a stretched string, are waves in which the particles of the medium oscillate perpendicular to the wave's direction of travel. Waves in which the particles of the medium oscillate parallel to the wave's direction of travel are **longitudinal** waves.

Sinusoidal Waves

A sinusoidal wave moving in the $+x$ direction has the mathematical form

$$y(x, t) = y_m \sin(kx - \omega t), \quad (17\text{-}2)$$

where y_m is the **amplitude** of the wave, k the **angular wave**

number, ω the **angular frequency,** and $kx - \omega t$ the **phase.** The wavelength λ is related to k by

$$k = \frac{2\pi}{\lambda}. \quad (17\text{-}5)$$

The **period** T and **frequency** f of the wave are related to ω by

$$\frac{\omega}{2\pi} = f = \frac{1}{T}. \quad (17\text{-}9)$$

Finally, the **wave speed** v is related to these other parameters by

$$v = \frac{\omega}{k} = \frac{\lambda}{T} = \lambda f. \quad (17\text{-}12)$$

Equation of a Traveling Wave
In general, any function of the form

$$y(x, t) = h(kx \pm \omega t) \qquad (17\text{-}16)$$

can represent a **traveling wave** with a wave speed given by Eq. 17-12 and a wave shape given by the mathematical form of h. The plus sign is for a wave traveling in the $-x$ direction, and the minus sign for a wave in the $+x$ direction.

Wave Speed on Stretched String
The speed of a wave on a stretched string with tension τ and linear density μ is

$$v = \sqrt{\frac{\tau}{\mu}}. \qquad (17\text{-}24)$$

Power
The **average power,** or average rate at which energy is transmitted by a sinusoidal wave on a stretched string, is given by

$$\overline{P} = \tfrac{1}{2}\mu v \omega^2 y_m^2. \qquad (17\text{-}35)$$

Superposition of Waves
When two or more waves traverse the same medium, the displacement of any particle of the medium is the sum of the displacements that the individual waves would give it. This is called **superposition.**

Fourier Series
With a *Fourier series,* any wave can be constructed as the superposition of appropriate sinusoidal waves.

Interference of Waves
Two sinusoidal waves on the same string exhibit **interference,** adding or canceling according to the principle of superposition. If the two are traveling in the same direction and have the same amplitude y_m and frequency (hence the same wavelength) but differ in phase by a **phase constant** ϕ, the result is a single wave with this same frequency:

$$y'(x, t) = [2y_m \cos \tfrac{1}{2}\phi] \sin(kx - \omega t + \tfrac{1}{2}\phi). \qquad (17\text{-}42)$$

If $\phi = 0$, the waves are in phase and their interference is (fully) constructive; if $\phi = \pi$ rad, they are out of phase and their interference is destructive.

Phasors
A wave $y(x, t)$ can be represented with a *phasor.* This is a vector that has a magnitude equal to the amplitude y_m of the wave and rotates about an origin with an angular speed equal to the angular frequency ω of the wave. The projection of the rotating phasor on a vertical axis gives the displacement y of the wave.

Standing Waves
The interference of two identical sinusoidal waves moving in opposite directions produces **standing waves.** For a string with fixed ends, the standing wave is given by

$$y'(x, t) = [2y_m \sin kx]\cos \omega t. \qquad (17\text{-}51)$$

Standing waves are characterized by fixed positions of zero displacement called **nodes** and fixed positions of maximum displacement called **antinodes.**

Resonance
Standing waves on a string can be set up by reflection of traveling waves from the ends of the string. If an end is fixed, it must be the position of a node. This limits the frequencies at which standing waves will occur on a given string. Each possible frequency is a **resonant frequency,** and the corresponding standing wave pattern is an **oscillation mode.** For a stretched string of length L with fixed ends, the resonant frequencies are

$$f = \frac{v}{\lambda} = n \frac{v}{2L}, \qquad \text{for } n = 1, 2, 3, \cdots. \qquad (17\text{-}57)$$

The oscillation mode corresponding to $n = 1$ is called the *fundamental mode* or the *first harmonic*; the mode corresponding to $n = 2$ is the *second harmonic*; and so on.

QUESTIONS

1. What is the wavelength of the (strange) wave in Fig. 17-21, where each segment of the wave has length d?

FIGURE 17-21   Question 1.

2. A string is put under tension and then three different sinusoidal waves are separately sent along it. Figure 17-22 gives plots of the resulting displacement versus time for an element on the string.

Rank the three plots according to the wavelength of the corresponding wave, greatest first.

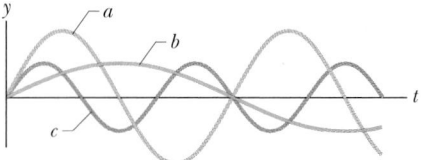

FIGURE 17-22   Question 2.

3. In Fig. 17-23, four strings are placed under tension by one or two suspended blocks, all of the same mass. Strings A, B, and C

have the same linear density; string D has a greater linear density. Rank the strings according to the speed that waves will have when sent along them, greatest first.

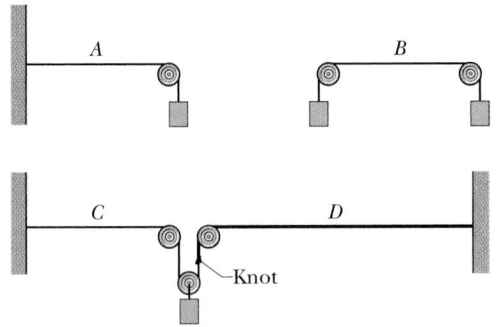

FIGURE 17-23 Question 3.

4. In Fig. 17-24, wave 1 consists of a rectangular peak of height 4 units and width d, and a rectangular valley of depth 2 units and width d. The wave travels rightward along an x axis. Waves 2, 3, and 4 are similar waves, with the same heights, depths, and widths, that will travel leftward along that axis and through wave 1. With which choice will the interference give, for an instant, (a) the deepest valley, (b) a flat line, and (c) a level peak $2d$ wide?

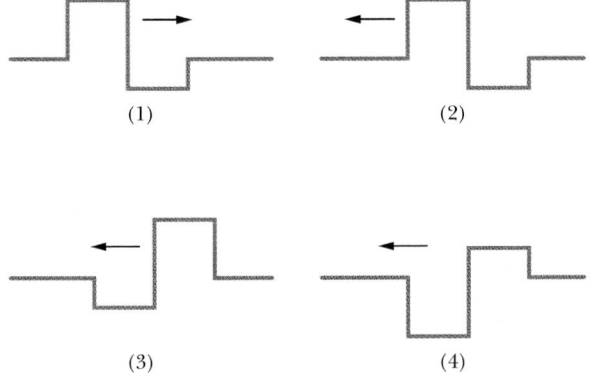

FIGURE 17-24 Question 4.

5. If you start with two waves traveling in phase on a string and then somehow phase-shift one of them by 5.4 wavelengths, what type of interference will occur on the string?

6. Here are some more phase differences between the two waves of Sample Problem 17-5, expressed in radians: $\pi/4$, $7\pi/4$, $-\pi/4$, and $-7\pi/4$. (a) Without written calculation, rank them in terms of the amplitude of the resultant wave, greatest first. (b) What type of interference occurs with each phase difference?

7. The amplitudes and phase differences for four pairs of waves of equal wavelengths are (a) 2 mm, 6 mm, and π rad; (b) 3 mm, 5 mm, and π rad; (c) 7 mm, 9 mm, and π rad; (d) 2 mm, 2 mm, and 0 rad. Each pair travels in the same direction along the same string. Without written calculation, rank the four pairs according to the amplitude of their resultant wave, greatest first. (*Hint:* Construct phasor diagrams.)

8. (a) Three waves equal in amplitude and wavelength travel along a string in the same direction. Take one to have a phase

constant of 0°. What are the phase constants of the other waves if all three together produce fully destructive interference? (*Hint:* Construct a phasor diagram.) (b) What are the phase constants if there are four such waves? Now there are two (different) answers.

9. If you set up the seventh harmonic on a string, (a) how many nodes are present, and (b) is there a node, antinode, or some intermediate state at the midpoint? If you next set up the sixth harmonic, (c) is its resonant wavelength longer or shorter than that for the seventh harmonic, and (d) is the resonant frequency higher or lower?

10. (a) In Sample Problem 17-7 and Fig. 17-20, if we gradually increase the mass of the block the (frequency remains fixed), new resonant modes appear. Do the harmonic numbers of the new resonant modes increase or decrease from one to the next? (b) Is the shift from one resonant mode to the next gradual, or does each resonant mode disappear well before the next one appears?

11. A string is stretched between two fixed supports separated by distance L. (a) For which harmonic numbers is there a node at the point located $L/3$ from one of the supports? Is there a node, antinode, or some intermediate state at a point located $2L/5$ from one support if (b) the fifth harmonic is set up and (c) the tenth harmonic is set up?

12. Strings A and B have identical lengths and linear densities, but string B is under greater tension than is string A. Figure 17-25 shows four situations, (a) through (d), in which standing wave patterns exist on the two strings. In which situations is there the possibility that strings A and B are oscillating at the same resonant frequency?

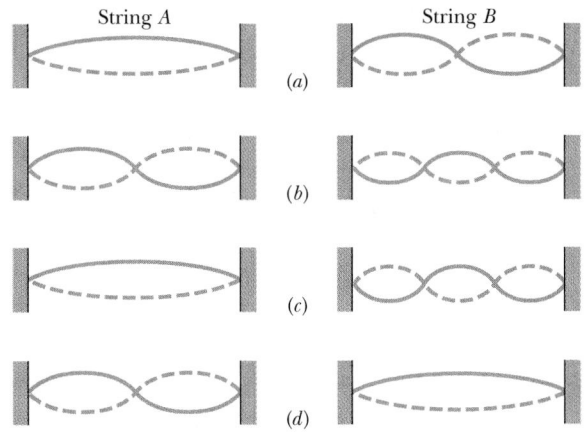

FIGURE 17-25 Question 12.

13. Two strings of equal lengths but unequal linear densities are tied together with a knot and stretched between two supports. A particular frequency happens to produce a standing wave on each length, with a node at the knot, as shown in Fig. 17-26. Which string has the greater linear density?

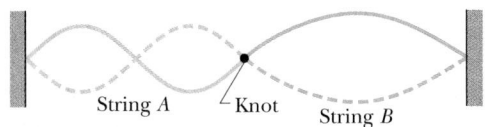

FIGURE 17-26 Question 13.

14. Do the wavelength and frequency, respectively, of the second harmonic on a string stretched between two supports increase, decrease, or remain the same if we (a) increase the distance between the supports without increasing the tension, (b) increase the tension in the string, and (c) switch to a string of greater linear density?

15. Guitar players know that prior to a show, a guitar must be played and the strings then tightened, because during the first few minutes of playing, the strings warm and loosen slightly. Does that loosening increase or decrease the resonant frequencies of the strings?

EXERCISES & PROBLEMS

SECTION 17-5 The Speed of a Traveling Wave

1E. A wave has a speed of 240 m/s and a wavelength of 3.2 m. What are the (a) frequency and (b) period of the wave?

2E. A wave has angular frequency 110 rad/s and wavelength 1.80 m. Calculate (a) the angular wave number and (b) the speed of the wave.

3E. The speed of electromagnetic waves in vacuum is 3.0×10^8 m/s. (a) Wavelengths of (visible) light waves range from about 400 nm in the violet to about 700 nm in the red. What is the range of frequencies of light waves? (b) The range of frequencies for shortwave radio (for example, FM radio and VHF television) is 1.5–300 MHz. What is the corresponding wavelength range? (c) X rays are also electromagnetic. Their wavelength range extends from about 5.0 nm to about 1.0×10^{-2} nm. What is the frequency range for x rays?

4E. A sinusoidal wave travels along a string. The time for a particular point to move from maximum displacement to zero is 0.170 s. What are the (a) period and (b) frequency? (c) The wavelength is 1.40 m; what is the wave speed?

5E. Write the equation for a wave traveling in the negative direction along the x axis and having an amplitude of 0.010 m, a frequency of 550 Hz, and a speed of 330 m/s.

6E. A traveling wave on a string is described by

$$y = 2.0 \sin \left[2\pi \left(\frac{t}{0.40} + \frac{x}{80} \right) \right],$$

where x and y are in centimeters and t is in seconds. (a) For $t = 0$, plot y as a function of x for $0 \le x \le 160$ cm. (b) Repeat (a) for $t = 0.05$ s and $t = 0.10$ s. (c) From your graphs, what is the wave speed, and in which direction (+x or −x) is the wave traveling?

7E. Show that $y = y_m \sin(kx - \omega t)$ may be written in the alternative forms

$$y = y_m \sin k(x - vt), \qquad y = y_m \sin 2\pi \left(\frac{x}{\lambda} - ft \right),$$

$$y = y_m \sin \omega \left(\frac{x}{v} - t \right), \qquad y = y_m \sin 2\pi \left(\frac{x}{\lambda} - \frac{t}{T} \right).$$

8E. A single pulse, whose wave shape is given by the function $h(x - 5t)$, is shown in Fig. 17-27 for $t = 0$. Here x is in centimeters and t is in seconds. What are the (a) speed and (b) direction of travel of the pulse? (c) Plot $h(x - 5t)$ as a function of x for $t = 2$ s. (d) Plot $h(x - 5t)$ as a function of t for $x = 10$ cm.

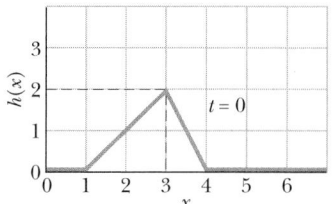

FIGURE 17-27
Exercise 8.

9E. Show (a) that the maximum transverse speed of a particle in a string owing to a traveling wave is given by $u_{max} = \omega y_m = 2\pi f y_m$ and (b) that the maximum transverse acceleration is $a_{y,max} = \omega^2 y_m = 4\pi^2 f^2 y_m$.

10E. The equation of a transverse wave traveling along a string is given by

$$y = (2.0 \text{ mm}) \sin[(20 \text{ m}^{-1})x - (600 \text{ s}^{-1})t].$$

(a) Find the amplitude, frequency, velocity, and wavelength of the wave. (b) Find the maximum transverse speed of a particle in the string.

11E. (a) Write an expression describing a sinusoidal transverse wave traveling on a cord in the +y direction with an angular wave number of 60 cm^{-1}, a period of 0.20 s, and an amplitude of 3.0 mm. Take the transverse direction to be the z direction. (b) What is the maximum transverse speed of a point on the cord?

12P. The equation of a transverse wave traveling along a very long string is given by $y = 6.0 \sin(0.020\pi x + 4.0\pi t)$, where x and y are expressed in centimeters and t is in seconds. Determine (a) the amplitude, (b) the wavelength, (c) the frequency, (d) the speed, (e) the direction of propagation of the wave, and (f) the maximum transverse speed of a particle in the string. (g) What is the transverse displacement at $x = 3.5$ cm when $t = 0.26$ s?

13P. (a) Write an equation describing a sinusoidal transverse wave traveling on a cord in the +x direction with a wavelength of 10 cm, a frequency of 400 Hz, and an amplitude of 2.0 cm. (b) What is the maximum speed of a point on the cord? (c) What is the speed of the wave?

14P. Prove that if a transverse sinusoidal wave is traveling along a string, then the slope at any point of the string is equal to the ratio of the particle speed to the wave speed at that point.

15P. A transverse sinusoidal wave of wavelength 20 cm is moving along a string toward increasing x. The transverse displacement of the string particle at $x = 0$ as a function of time is shown in Fig. 17-28. (a) Make a rough sketch of one wavelength of the wave (the portion between $x = 0$ and $x = 20$ cm) at time $t = 0$. (b) What is the velocity of propagation of the wave? (c) Write the equation for the wave with all the constants evaluated. (d) What is the transverse velocity of the particle at $x = 0$ at $t = 5.0$ s?

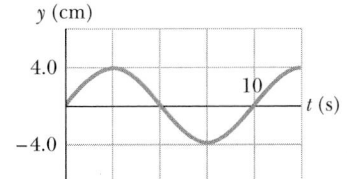

FIGURE 17-28
Problem 15.

16P. A sinusoidal wave of frequency 500 Hz has a velocity of 350 m/s. (a) How far apart are two points that differ in phase by $\pi/3$ rad? (b) What is the phase difference between two displacements at a certain point at times 1.00 ms apart?

SECTION 17-6 Wave Speed on a Stretched String

17E. What is the speed of a transverse wave in a rope of length 2.00 m and mass 60.0 g under a tension of 500 N?

18E. The heaviest and lightest strings in a certain violin have linear densities of 3.0 and 0.29 g/m. What is the ratio, heavier to lighter, of the diameters of these strings, assuming that they are made of the same material?

19E. The speed of a transverse wave on a string is 170 m/s when the string tension is 120 N. To what value must the tension be changed to raise the wave speed to 180 m/s?

20E. The tension in a wire clamped at both ends is doubled without appreciably changing its length. What is the ratio of the new to the old wave speed for transverse waves in this wire?

21E. Show that, in terms of the tensile stress S and the volume density ρ, the speed v of transverse waves in a wire is given by

$$v = \sqrt{\frac{S}{\rho}}.$$

22E. The equation of a transverse wave on a string is

$$y = (2.0 \text{ mm}) \sin[(20 \text{ m}^{-1})x - (600 \text{ s}^{-1})t].$$

The tension in the string is 15 N. (a) What is the wave speed? (b) Find the linear density of this string in grams per meter.

23E. The linear density of a string is 1.6×10^{-4} kg/m. A transverse wave is propagating on the string and is described by the following equation: $y = (0.021 \text{ m})\sin[(2.0 \text{ m}^{-1})x + (30 \text{ s}^{-1})t]$. (a) What is the wave speed? (b) What is the tension in the string?

24E. What is the fastest transverse wave that can be sent along a steel wire? Allowing for a reasonable safety factor, the maximum tensile stress to which steel wires should be subjected is 7.0×10^8 N/m². The density of steel is 7800 kg/m³. Show that your answer does not depend on the diameter of the wire.

25P. A stretched string has a mass per unit length of 5.0 g/cm and a tension of 10 N. A sinusoidal wave on this string has an amplitude of 0.12 mm and a frequency of 100 Hz and is traveling toward decreasing x. Write an equation for this wave.

26P. For a sinusoidal transverse wave on a stretched cord, find the ratio of the maximum particle speed (the speed with which a single particle in the cord moves transverse to the wave) to the wave speed. If a wave having a certain frequency and amplitude is sent along a cord, would this speed ratio depend on the material of which the cord is made, such as wire or nylon?

27P. A sinusoidal transverse wave is traveling along a string toward decreasing x. Figure 17-29 shows a plot of the displacement as a function of position at time $t = 0$. The string tension is 3.6 N, and its linear density is 25 g/m. Find (a) the amplitude, (b) the wavelength, (c) the wave speed, and (d) the period of the wave. (e) Find the maximum speed of a particle in the string. (f) Write an equation describing the traveling wave.

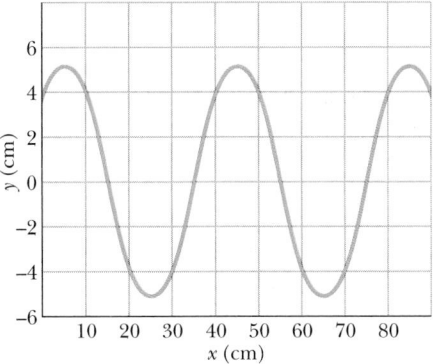

FIGURE 17-29 Problem 27.

28P. A sinusoidal wave is traveling on a string with speed 40 cm/s. The displacement of the particles of the string at $x = 10$ cm is found to vary with time according to the equation $y = (5.0 \text{ cm}) \sin[1.0 - (4.0 \text{ s}^{-1})t]$. The linear density of the string is 4.0 g/cm. What are (a) the frequency and (b) the wavelength of the wave? (c) Write the general equation giving the transverse displacement of the particles of the string as a function of position and time. (d) Calculate the tension in the string.

29P. In Fig. 17-30a, string 1 has a linear density of 3.00 g/m, and string 2 has a linear density of 5.00 g/m. They are under tension owing to the hanging block of mass $M = 500$ g. (a) Cal-

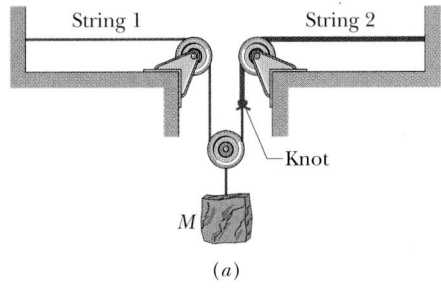

(a)

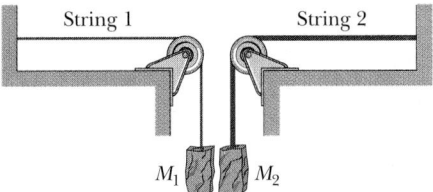

FIGURE 17-30 Problem 29.   (b)

culate the wave speed in each string. (b) The block is now divided into two blocks (with $M_1 + M_2 = M$) and the apparatus rearranged as shown in Fig. 17-30b. Find M_1 and M_2 such that the wave speeds in the two strings are equal.

30P. A wire 10.0 m long and having a mass of 100 g is stretched under a tension of 250 N. If two pulses, separated in time by 30.0 ms, are generated, one at each end of the wire, where will the pulses first meet?

31P. The type of rubber band used inside some baseballs and golf balls obeys Hooke's law over a wide range of elongation of the band. A segment of this material has an unstretched length l and a mass m. When a force F is applied, the band stretches an additional length Δl. (a) What is the speed (in terms of m, Δl, and the force constant k) of transverse waves on this stretched rubber band? (b) Using your answer to (a), show that the time required for a transverse pulse to travel the length of the rubber band is proportional to $1/\sqrt{\Delta l}$ if $\Delta l \ll l$ and is constant if $\Delta l \gg l$.

32P*. A uniform rope of mass m and length l hangs from a ceiling. (a) Show that the speed of a transverse wave on the rope is a function of y, the distance from the lower end, and is given by $v = \sqrt{gy}$. (b) Show that the time a transverse wave takes to travel the length of the rope is given by $t = 2\sqrt{l/g}$.

SECTION 17-7 Energy and Power of a Traveling String Wave

33E. Energy is transmitted at rate P_1 by a wave of frequency f_1 on a string with tension τ_1. What is the new energy transmission rate P_2 in terms of P_1 (a) if the tension of the string is increased to $\tau_2 = 4\tau_1$ and (b) if, instead, the frequency of the wave is decreased to $f_2 = f_1/2$?

34E. A string along which waves can travel is 2.70 m long and has a mass of 260 g. The tension in the string is 36.0 N. What must be the frequency of traveling waves of amplitude 7.70 mm in order that the average power be 85.0 W?

35P. A transverse sinusoidal wave is generated at one end of a long, horizontal string by a bar that moves up and down through a distance of 1.00 cm. The motion is continuous and is repeated regularly 120 times per second. The string has linear density 120 g/m and is kept under a tension of 90.0 N. Find (a) the maximum value of the transverse speed u and (b) the maximum value of the transverse component of the tension. (c) Show that the two maximum values calculated above occur at the same phase values for the wave. What is the transverse displacement y of the string at these phases? (d) What is the maximum rate of energy transfer along the string? (e) What is the transverse displacement y when this maximum transfer occurs? (f) What is the minimum rate of energy transfer along the string? (g) What is the transverse displacement y when this minimum transfer occurs?

SECTION 17-9 Interference of Waves

36E. Two identical traveling waves, moving in the same direction, are out of phase by $\pi/2$ rad. What is the amplitude of the resultant wave in terms of the common amplitude y_m of the two combining waves?

37E. What phase difference between two otherwise identical traveling waves, moving in the same direction along a stretched string, will result in the combined wave having an amplitude 1.50 times that of the common amplitude of the two combining waves? Express your answer in degrees, radians, and wavelengths.

38P. Two sinusoidal waves, identical except for phase, travel in the same direction along a string and interfere to produce a resultant wave given by $y'(x, t) = (3.0 \text{ mm}) \sin(20x - 4.0t + 0.820 \text{ rad})$, with x in meters and t in seconds. What are (a) the wavelength λ of the two waves, (b) the phase difference between them, and (c) their amplitude y_m?

SECTION 17-10 Phasors

39E. Determine the amplitude of the resultant wave when two sinusoidal string waves having the same frequency and traveling in the same direction are combined, if their amplitudes are 3.0 cm and 4.0 cm and they have phase constants of 0 and $\pi/2$ rad, respectively.

40E. The amplitudes of two sinusoidal waves traveling in the same direction along a stretched string are 3.0 and 5.0 mm; their phase constants are $0°$ and $70°$, respectively. The waves have the same wavelength. What are (a) the amplitude and (b) the phase constant of the resultant wave?

41E. Two sinusoidal waves of the same wavelength travel in the same direction along a stretched string with amplitudes of 4.0 and 7.0 mm and phase constants of 0 and 0.8π rad, respectively. What are (a) the amplitude and (b) the phase constant of the resultant wave?

42P. Two sinusoidal waves of the same period, with amplitudes of 5.0 and 7.0 mm, travel in the same direction along a stretched string; they produce a resultant wave with an amplitude of 9.0 mm. The phase constant of the 5.0 mm wave is 0. What is the phase constant of the 7.0 mm wave?

43P. Three sinusoidal waves of the same frequency travel along a string in the positive direction of an x axis. Their amplitudes are y_1, $y_1/2$, and $y_1/3$, and their phase constants are 0, $\pi/2$, and π, respectively. What are (a) the amplitude and (b) the phase constant of the resultant wave? (c) Plot the wave form of the resultant wave at $t = 0$, and discuss its behavior as t increases.

44P. Four sinusoidal waves travel in the positive x direction along the same string. Their frequencies are in the ratio $1:2:3:4$, and their amplitudes are in the ratio $1:\frac{1}{2}:\frac{1}{3}:\frac{1}{4}$, respectively. When $t = 0$, at $x = 0$, the first and third waves are $180°$ out of phase with the second and fourth. Plot the resultant wave form when $t = 0$, and discuss its behavior as t increases.

SECTION 17-12 Standing Waves and Resonance

45E. A string under tension τ_i oscillates in the third harmonic at frequency f_3, and the waves on the string have wavelength λ_3. If the tension is increased to $\tau_f = 4\tau_i$ and the string is again made to oscillate in the third harmonic, what then are (a) the frequency of oscillation in terms of f_3 and (b) the wavelength of the waves in terms of λ_3?

46E. A nylon guitar string has a linear density of 7.2 g/m and is under a tension of 150 N. The fixed supports are 90 cm apart. The string is oscillating in the standing wave pattern shown in Fig. 17-31. Calculate the (a) speed, (b) wavelength, and (c) frequency of the traveling waves whose superposition gives this standing wave.

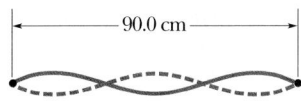

FIGURE 17-31 Exercise 46.

47E. Two sinusoidal waves with identical wavelengths and amplitudes travel in opposite directions along a string with a speed of 10 cm/s. If the time interval between instants when the string is flat is 0.50 s, what is the wavelength of the waves?

48E. When played in a certain manner, the lowest resonant frequency of a certain violin string is concert A (440 Hz). What are the frequencies of the second and third harmonics of that string?

49E. A string fixed at both ends is 8.40 m long and has a mass of 0.120 kg. It is subjected to a tension of 96.0 N and set oscillating. (a) What is the speed of the waves on the string? (b) What is the longest possible wavelength for a standing wave? (c) Give the frequency of that wave.

50E. The equation of a transverse wave traveling along a string is

$$y = 0.15 \sin(0.79x - 13t),$$

in which x and y are in meters and t is in seconds. (a) What is the displacement y at $x = 2.3$ m, $t = 0.16$ s? (b) Write the equation of a wave that, when added to the given one, would produce standing waves on the string. (c) What is the displacement of the resultant standing wave at $x = 2.3$ m, $t = 0.16$ s?

51E. A 120 cm length of string is stretched between fixed supports. What are the three longest possible wavelengths for traveling waves on the string that can produce standing waves? Sketch the corresponding standing waves.

52E. A 125 cm length of string has a mass of 2.00 g. It is stretched with a tension of 7.00 N between fixed supports. (a) What is the wave speed for this string? (b) What is the lowest resonant frequency of this string?

53E. What are the three lowest frequencies for standing waves on a wire 10.0 m long having a mass of 100 g, which is stretched under a tension of 250 N?

54E. A 1.50 m wire has a mass of 8.70 g and is held under a tension of 120 N. The wire is held rigidly at both ends and set into vibration. Calculate (a) the speed of waves on the wire, (b) the wavelengths of the waves that produce one- and two-loop standing waves on the string, and (c) the frequencies of the waves that produce one- and two-loop standing waves.

55E. String A is stretched between two clamps separated by distance l. String B, with the same linear density and under the same tension as string A, is stretched between two clamps separated by distance $4l$. Consider the first eight harmonics of string B. Which, if any, has a resonant frequency that matches a resonant frequency of string A?

56P. A string is stretched between fixed supports separated by 75.0 cm. It is observed to have resonant frequencies of 420 and 315 Hz, and no other resonant frequencies between these two. (a) What is the lowest resonant frequency for this string? (b) What is the wave speed for this string?

57P. Two waves are propagating on the same very long string. A generator at one end of the string creates a wave given by

$$y = (6.0 \text{ cm}) \cos \frac{\pi}{2} [(2.0 \text{ m}^{-1})x + (8.0 \text{ s}^{-1})t],$$

and one at the other end creates the wave

$$y = (6.0 \text{ cm}) \cos \frac{\pi}{2} [2.0 \text{ m}^{-1})x - (8.0 \text{ s}^{-1})t].$$

(a) Calculate the frequency, wavelength, and speed of each wave. (b) Find the points on the string at which there is no motion (the nodes). (c) At which points is the motion of the string a maximum (the antinodes)?

58P. A string oscillates according to the equation

$$y' = (0.50 \text{ cm}) \left[\sin \left(\frac{\pi}{3} \text{ cm}^{-1} \right) x \right] \cos[(40\pi \text{ s}^{-1})t].$$

(a) What are the amplitude and speed of the two waves (identical except for direction of travel) whose superposition gives this oscillation? (b) What is the distance between nodes? (c) What is the speed of a particle of the string at the position $x = 1.5$ cm when $t = \frac{9}{8}$ s?

59P. Two transverse sinusoidal waves travel in opposite directions along a string. Each wave has an amplitude of 0.30 cm and a wavelength of 6.0 cm. The speed of a transverse wave on the string is 1.5 m/s. Plot the shape of the string at times $t = 0$ (arbitrary), $t = 5.0$, $t = 10$, $t = 15$, and $t = 20$ ms.

60P. Two pulses travel along a string in opposite directions, as in Fig. 17-32. (a) If the wave speed v is 2.0 m/s and the pulses are 6.0 cm apart at $t = 0$, sketch the patterns when t is equal to 5.0, 10, 15, 20, and 25 ms. (b) What has happened to the energy of the pulses at $t = 15$ ms?

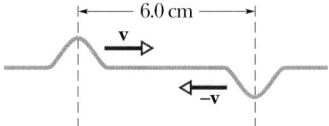

FIGURE 17-32 Problem 60.

61P. Two waves on a string are described by these equations:

$$y_1 = (0.10 \text{ m}) \sin 2\pi [(0.50 \text{ m}^{-1})x + (20 \text{ s}^{-1})t],$$
$$y_2 = (0.20 \text{ m}) \sin 2\pi [(0.50 \text{ m}^{-1})x - (20 \text{ s}^{-1})t].$$

Sketch the total response for the point on the string at $x = 3.0$ m; that is, plot y versus t for that value of x.

62P. A string 3.0 m long is oscillating as a three-loop standing wave whose amplitude is 1.0 cm. The wave speed is 100 m/s. (a) What is the frequency? (b) Write equations for two waves that, when combined, will result in this standing wave.

63P. Vibration from a 600 Hz tuning fork sets up standing waves in a string clamped at both ends. The wave speed for the string is 400 m/s. The standing wave has four loops and an amplitude of 2.0 mm. (a) What is the length of the string? (b) Write an equation for the displacement of the string as a function of position and time.

64P. In an experiment on standing waves, a string 90 cm long is attached to the prong of an electrically driven tuning fork that oscillates perpendicular to the length of the string at a frequency of 60 Hz. The mass of the string is 0.044 kg. What tension must the string be under (weights are attached to the other end) if it is to vibrate in four loops?

65P. Show that the maximum kinetic energy in each loop of a standing wave produced by two traveling waves of identical amplitudes is $2\pi^2 \mu y_m^2 f v$.

66P. An aluminum wire, of length $l_1 = 60.0$ cm, cross-sectional area 1.00×10^{-2} cm^2, and density 2.60 g/cm^3, is joined to a steel wire, of density 7.80 g/cm^3 and the same cross-sectional area. The compound wire, loaded with a block of mass $m = 10.0$ kg, is arranged as in Fig. 17-33 so that the distance l_2 from the joint to the supporting pulley is 86.6 cm. Transverse waves are set up in the wire by using an external source of variable frequency; a node is located at the pulley. (a) Find the lowest frequency of excitation for which standing waves are observed such that the joint in the wire is a node. (b) How many nodes are observed at this frequency?

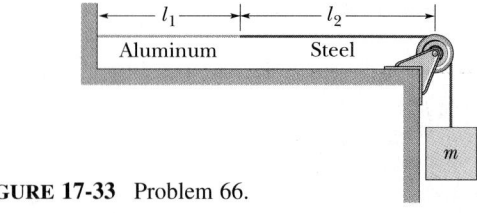

FIGURE 17-33 Problem 66.

Electronic Computation

67. Two waves,

$$y_1 = (2.50 \text{ mm})\sin[(25.1 \text{ rad/m})x - (440 \text{ rad/s})t]$$

and

$$y_2 = (1.50 \text{ mm})\sin[(25.1 \text{ rad/m})x + (440 \text{ rad/s})t],$$

travel along a stretched string. (a) Plot $y'(x, t) = y_1(x, t) + y_2(x, t)$ as a function of t for $x = 0, \lambda/8, \lambda/4, 3\lambda/8,$ and $\lambda/2$, where λ is the wavelength. The graphs should extend from $t = 0$ to a little over one period. (b) The result is the superposition of a standing wave and a traveling wave. In which direction does the traveling wave travel? How can you change the original waves so the resultant is the superposition of standing and traveling waves with the same amplitudes as before but with the traveling wave moving in the opposite direction? (c) Use your graphs to find the place at which the amplitude of vibration is the largest and the smallest. (d) How are the maximum and minimum amplitudes related to the amplitudes 2.50 mm and 1.50 mm of the original two traveling waves?

18
Waves—II

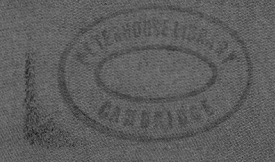

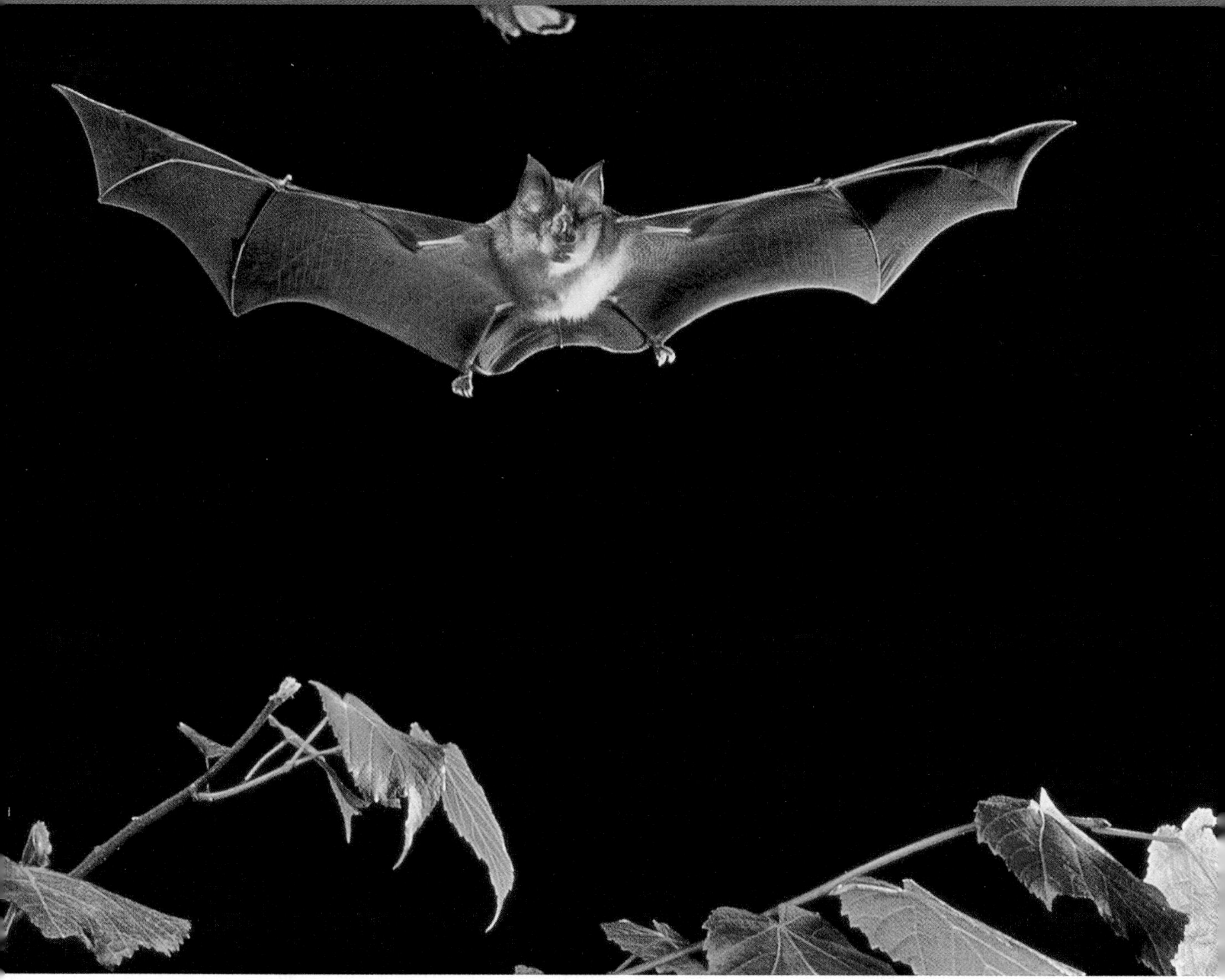

This horseshoe bat not only can locate a moth flying in total darkness but can also determine the moth's relative speed, to home in on the insect. How does the detection system work? And how can a moth "jam" the system or otherwise reduce its effectiveness?

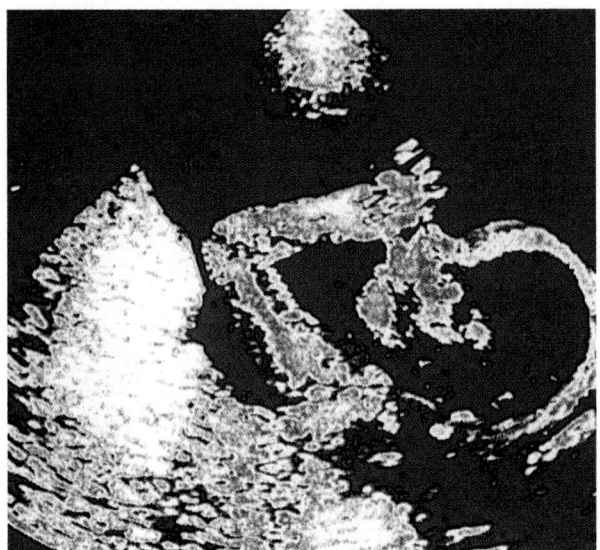

FIGURE 18-1 An ultrasound image of a fetus searching for a thumb to suck.

18-1 SOUND WAVES

As we saw in Chapter 17, mechanical waves are waves that require a material medium to exist. There are two types of mechanical waves: *transverse waves* involve oscillations perpendicular to the direction in which the wave travels; *longitudinal waves* involve oscillations parallel to the direction of wave travel.

A **sound wave** can be roughly defined as any longitudinal wave. Seismic prospecting teams use such waves to probe Earth's crust for oil. Ships carry sound ranging gear (sonar) to detect underwater obstacles. Submarines use sound waves to stalk other submarines, largely by listening for the characteristic noises produced by the propulsion system. Figure 18-1, a computer-processed image of a fetal head, shows how sound waves can be used to explore the

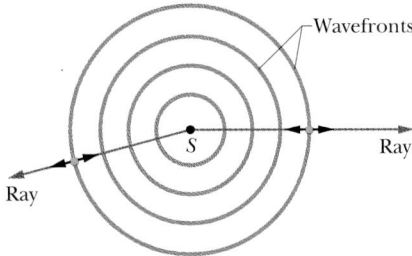

FIGURE 18-2 A sound wave travels from a point source S through a three-dimensional medium. The wavefronts form spheres centered on S; the rays are radial to S. The short, double-headed arrows indicate that elements of the medium oscillate parallel to the rays.

soft tissues of the human body. In this chapter we shall focus on sound waves that travel through the air and that are audible to people.

Figure 18-2 illustrates several ideas that we shall use in our discussions. Point S represents a tiny sound source, called a *point source*, that emits sound waves in all directions. The *wavefronts* and *rays* indicate the direction of travel and the spread of the sound waves. **Wavefronts** are surfaces over which the oscillations of the air due to the sound wave have the same value; such surfaces are represented by whole or partial circles in a two-dimensional drawing for a point source. **Rays** are directed lines perpendicular to the wavefronts that indicate the direction of travel of the wavefronts. The short double arrows superimposed on the rays of Fig. 18-2 indicate that the longitudinal oscillations of the air are parallel to the rays.

Near the point source of Fig. 18-2, the wavefronts are spherical and are spreading out in three dimensions, and the waves are said to be *spherical*. As the wavefronts move outward and their radii become larger, their curvature decreases. Far from the source, we approximate the wavefronts as straight lines and the waves are said to be *planar*.

18-2 THE SPEED OF SOUND

The speed of any mechanical wave, transverse or longitudinal, depends on both an inertial property of the medium (to store kinetic energy) and an elastic property of the medium (to store potential energy). Thus we can generalize Eq. 17-24, which gives the speed of a transverse wave along a stretched string, by writing

$$v = \sqrt{\frac{\tau}{\mu}} = \sqrt{\frac{\text{elastic property}}{\text{inertial property}}}, \quad (18\text{-}1)$$

where (for transverse waves) τ is the tension in the string and μ is the string's linear density. If the medium is air, we can guess that the inertial property, corresponding to μ, is the volume density ρ of air. What shall we put for the elastic property?

In a stretched string, potential energy is associated with the periodic stretching of the string elements as the wave passes through them. As a sound wave passes through air, potential energy is associated with periodic compressions and expansions of small volume elements of the air. The property that determines the extent to which an element of the medium changes its volume as the pressure (force per unit area) applied to it is increased or decreased is the **bulk modulus** B, as defined (from Eq. 13-36)

$$B = -\frac{\Delta p}{\Delta V/V} \quad \text{(definition of } B\text{)}. \quad (18\text{-}2)$$

TABLE 18-1 THE SPEED OF SOUND[a]

MEDIUM	SPEED (m/s)	MEDIUM	SPEED (m/s)
Gases		*Solids*	
Air (0°C)	331	Aluminum	6420
Air (20°C)	343	Steel	5941
Helium	965	Granite	6000
Hydrogen	1284		
Liquids			
Water (0°C)	1402		
Water (20°C)	1482		
Seawater[b]	1522		

[a]0°C and 1 atm pressure, except where noted.
[b]At 20°C and 3.5% salinity.

Here $\Delta V/V$ is the fractional change in volume produced by a change in pressure Δp. As explained in Section 15-3, the SI unit for pressure is the newton per square meter, which is given a special name, the *pascal* (Pa). From Eq. 18-2 we see that the unit for B is also the pascal. The signs of Δp and ΔV are always opposite: when we increase the pressure on a fluid element (Δp positive), its volume decreases (ΔV negative). We include a minus sign in Eq. 18-2 so that B will always be a positive quantity. Now substituting B for τ and ρ for μ in Eq. 18-1 yields

$$v = \sqrt{\frac{B}{\rho}} \quad \text{(speed of sound)} \quad (18\text{-}3)$$

for a medium with bulk modulus B and density ρ. Table 18-1 lists the speed of sound in various media.

The density of water is almost 1000 times greater than the density of air. If this were the only relevant factor, we would expect from Eq. 18-3 that the speed of sound in water would be considerably less than the speed of sound in air. However, Table 18-1 shows us that the reverse is true. We conclude (again from Eq. 18-3) that the bulk modulus of water must be more than 1000 times greater than that of air. This is indeed the case. Water is much more incompressible than air, which (see Eq. 18-2) is another way of saying that its bulk modulus is much greater.

Formal Derivation of Eq. 18-3

We now derive Eq. 18-3 by direct application of Newton's laws. Let a single pulse in which air is compressed travel (from right to left) with speed v through the air in a long tube, like that in Fig. 17-2. Let us run along with the pulse at that speed, so that the pulse appears to stand still in our reference frame. Figure 18-3a shows the situation as it is viewed from that frame. The pulse is standing still, and air is moving at speed v through it from left to right.

Let the pressure of the undisturbed air be p and the pressure inside the pulse be $p + \Delta p$, where Δp is positive owing to the compression. Consider a slice of air of thickness Δx and face area A, moving toward the pulse at speed v. As this element of air enters the pulse, its leading face encounters a region of higher pressure, which slows it to speed $v + \Delta v$, in which Δv is negative. This slowing is complete when the rear face reaches the pulse, which requires time interval

$$\Delta t = \frac{\Delta x}{v}. \quad (18\text{-}4)$$

Let us apply Newton's second law to the element. During Δt, the average force on the element's trailing face is pA toward the right, and the average force on the leading

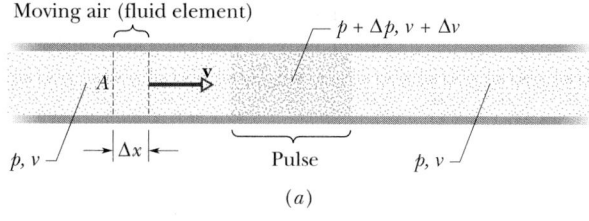

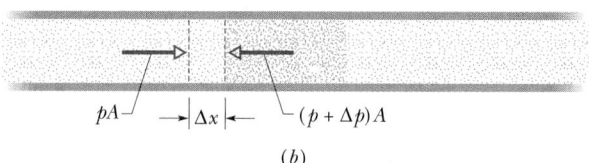

FIGURE 18-3 A pulse (compression) is sent down a long air-filled tube. The reference frame of the figure is chosen so that the pulse is at rest and the air moves from left to right. (*a*) A slice of air of width Δx moves toward the pulse with speed v. (*b*) The leading face of the slice enters the pulse. The forces acting on the leading and trailing faces (due to air pressure) are shown.

face is $(p + \Delta p)A$ toward the left (Fig. 18-3b). So the average net force on the element during Δt is

$$F = pA - (p + \Delta p)A$$
$$= -\Delta p\, A \qquad \text{(net force)}. \qquad (18\text{-}5)$$

The minus sign indicates that the net force on the fluid element points to the left in Fig. 18-3b. The volume of the element is $A\,\Delta x$, so with the aid of Eq. 18-4, we can write its mass as

$$\Delta m = \rho A\,\Delta x = \rho A v\,\Delta t \qquad \text{(mass)}. \qquad (18\text{-}6)$$

Then the average acceleration of the element during Δt is

$$a = \frac{\Delta v}{\Delta t} \qquad \text{(acceleration)}. \qquad (18\text{-}7)$$

From Newton's second law ($F = ma$), we have, from Eqs. 18-5, 18-6, and 18-7,

$$-\Delta p\, A = (\rho A v\,\Delta t)\frac{\Delta v}{\Delta t},$$

which we can write as

$$\rho v^2 = -\frac{\Delta p}{\Delta v/v}. \qquad (18\text{-}8)$$

The air that occupies a volume $V (= A v\,\Delta t)$ outside the pulse is compressed by an amount $\Delta V (= A\,\Delta v\,\Delta t)$ as it enters the pulse. Thus

$$\frac{\Delta V}{V} = \frac{A\,\Delta v\,\Delta t}{A v\,\Delta t} = \frac{\Delta v}{v}. \qquad (18\text{-}9)$$

Substituting Eq. 18-9 and then Eq. 18-2 into Eq. 18-8 leads to

$$\rho v^2 = -\frac{\Delta p}{\Delta v/v} = -\frac{\Delta p}{\Delta V/V} = B.$$

Solving for v yields Eq. 18-3 for the speed of the air toward the right in Fig. 18-3, and thus for the actual speed of the pulse toward the left.

SAMPLE PROBLEM 18-1

One clue used by your brain to determine the direction of a source of sound is the time delay Δt between the arrival of the sound at the ear closer to the source and the arrival at the farther ear. Assume that the source is distant so that a wavefront from it is approximately straight when it reaches you, and let D represent the separation between your ears.

(a) Find an expression that gives Δt in terms of D and the angle θ between the direction of the source and the forward direction.

SOLUTION: The situation is shown in Fig. 18-4 for wavefronts approaching you from a source that is located in front of

you and to your right. The time delay Δt is due to the distance d that each wavefront must travel to reach your left ear (L) after it reaches your right ear (R). From Fig. 18-4, we find

$$\Delta t = \frac{d}{v} = \frac{D \sin \theta}{v}, \qquad \text{(Answer)} \quad (18\text{-}10)$$

where v is the speed of sound in air. Based on a lifetime of experience, your brain correlates each detected value of Δt (from zero to the maximum value) with a value of θ (from zero to 90°) for the direction of the sound source.

(b) Suppose that you are submerged in water at 20°C when a wavefront arrives from directly to your right. Based on the time-delay clue, at what angle θ from the forward direction will the source seem to be?

SOLUTION: We find the time delay Δt_w in this situation with Eq. 18-10, substituting $\theta = 90°$ and using v_w, the speed of sound in water, instead of v, the speed of sound in air:

$$\Delta t_w = \frac{D \sin 90°}{v_w} = \frac{D}{v_w}. \qquad (18\text{-}11)$$

Since v_w is about four times v, Δt_w is about one-fourth the maximum time delay in air. Based on experience, your brain will process the time delay as if it occurred in air. Thus the sound source will appear to be at an angle θ smaller than 90°. To find that apparent angle, we substitute the time delay D/v_w from Eq. 18-11 for Δt in Eq. 18-10, obtaining

$$\frac{D}{v_w} = \frac{D \sin \theta}{v}. \qquad (18\text{-}12)$$

Substituting $v = 343$ m/s and $v_w = 1482$ m/s (from Table 18-1) into Eq. 18-12, we then find

$$\sin \theta = \frac{v}{v_w} = \frac{343 \text{ m/s}}{1482 \text{ m/s}} = 0.231$$

and thus

$$\theta = 13°. \qquad \text{(Answer)}$$

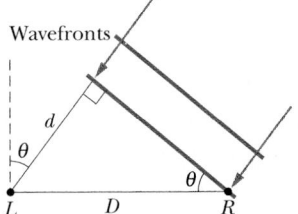

FIGURE 18-4 Sample Problem 18-1. A wavefront travels a distance d ($= D \sin \theta$) farther to reach the left ear (L) than to reach the right ear (R).

18-3 TRAVELING SOUND WAVES

Here we examine the displacements and pressure variations associated with a sinusoidal sound wave passing through air. Figure 18-5a displays such a wave traveling rightward through a long air-filled tube. Recall that we can produce such a wave by sinusoidally moving a piston at the

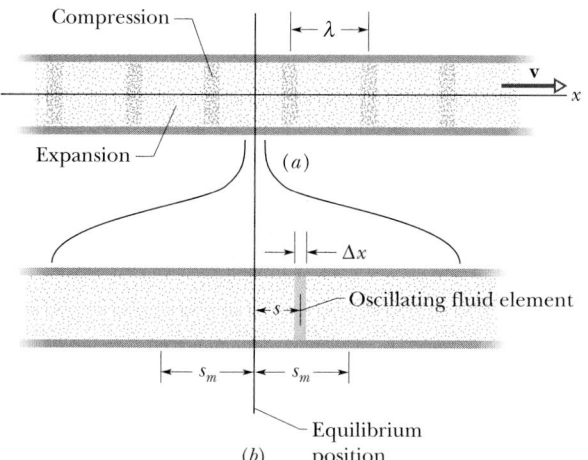

FIGURE 18-5 (a) A sound wave, traveling through a long air-filled tube with speed v, consists of a moving, periodic pattern of expansions and compressions of the air. The wave is shown at an arbitrary instant. (b) A horizontally expanded view of a short piece of the tube. As the wave passes, a fluid element of thickness Δx oscillates left and right in simple harmonic motion about its equilibrium position. At the instant shown in (b), the element happens to be displaced a distance s to the right of its equilibrium position. Its maximum displacement, either right or left, is s_m.

left end of the tube (as in Fig. 17-2). The piston's rightward motion moves the element of air next to it and compresses that air; the piston's leftward motion allows the element of air to move back to the left and the pressure to decrease. As each element of air pushes on the next element in turn, the right–left motion of the air and the change in its pressure travel along the tube as a sound wave.

Consider a thin element of air of thickness Δx, located at a position x along the tube. As the wave passes through x, the element of air oscillates left and right in simple harmonic motion about its equilibrium position (Fig. 18-5b). Thus the oscillations of each air element due to the passing sound wave are like those of a string element due to a transverse wave, except that the air element oscillates *longitudinally* rather than *transversely*. To describe the displacement $s(x, t)$ of an element of air from its equilibrium position, we can use either a sine function or a cosine function. In this chapter we use a cosine function:

$$s(x, t) = s_m \cos(kx - \omega t). \qquad (18\text{-}13)$$

Here s_m is the **displacement amplitude,** that is, the maximum displacement of the air element to either side of its equilibrium position (Fig. 18-5b).* The angular wave

number k, angular frequency ω, frequency f, wavelength λ, speed v, and period T for a sound (longitudinal) wave are defined and interrelated exactly as for a transverse wave, except that λ is now the distance (again along the direction of travel) in which the pattern of compression and expansion due to the wave begins to repeat itself (see Fig. 18-5a). (We assume s_m is much less than λ.)

As the wave moves, the air pressure at any position x in Fig. 18-5a varies sinusoidally, as we prove below. To describe this variation we write

$$\Delta p\,(x,\,t) = \Delta p_m \sin(kx - \omega t). \qquad (18\text{-}14)$$

A negative value of Δp in Eq. 18-14 corresponds to an expansion of the air, and a positive value to a compression. Here Δp_m is the **pressure amplitude,** which is the maximum increase or decrease in pressure due to the wave; Δp_m is normally very much less than the pressure p present when there is no wave. As we shall prove, the pressure amplitude Δp_m is related to the displacement amplitude s_m in Eq. 18-13 by

$$\Delta p_m = (v\rho\omega)\,s_m. \qquad (18\text{-}15)$$

Figure 18-6 shows plots of Eqs. 18-13 and 18-14 at $t = 0$; with time, the two curves would move rightward along the horizontal axes. Note that the displacement and pressure variation are $\pi/2$ rad (or 90°) out of phase. Thus, for example, the pressure variation is zero when the displacement is a maximum.

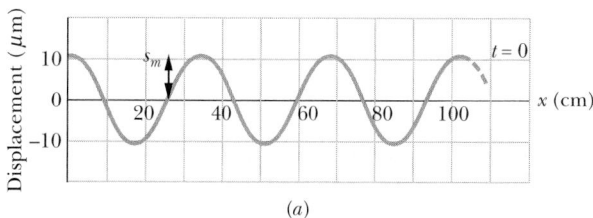

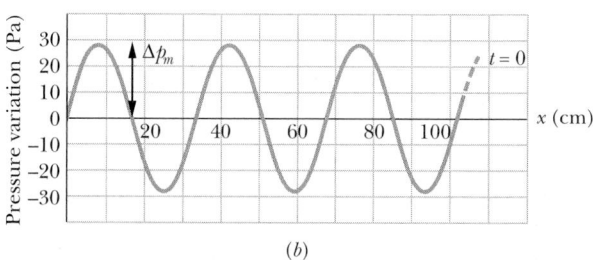

FIGURE 18-6 (a) A plot of the displacement function (Eq. 18-13) for $t = 0$. (b) A similar plot of the pressure variation function (Eq. 18-14). Both plots are for a 1000 Hz sound wave whose pressure amplitude is at the threshold of pain; see Sample Problem 18-2.

---

*For the transverse displacement of an element of a stretched string, we used the symbol $y(x, t)$. We use the notation $s(x, t)$ here to avoid writing $x(x, t)$ for the longitudinal displacement of an element of air.

CHECKPOINT **1:** When the oscillating fluid element in Fig. 18-5*b* is moving rightward through the point of zero displacement, is the pressure at its equilibrium value, just beginning to increase, or just beginning to decrease?

Derivation of Eqs. 18-14 and 18-15

Figure 18-5*b* shows an oscillating element of air of cross-sectional area A and thickness Δx, with its center displaced from its equilibrium position by distance s.

From Eq. 18-2 we can write, for the pressure variation in the displaced element,

$$\Delta p = -B \frac{\Delta V}{V}. \qquad (18\text{-}16)$$

The quantity V in Eq. 18-16 is the volume of the element, given by

$$V = A \, \Delta x. \qquad (18\text{-}17)$$

The quantity ΔV in Eq. 18-16 is the change in volume that occurs when the element is displaced. This volume change comes about because the displacements of the two faces of the element are not quite the same, differing by some amount Δs. Thus we can write the change in volume as

$$\Delta V = A \, \Delta s. \qquad (18\text{-}18)$$

Substituting Eqs. 18-17 and 18-18 into Eq. 18-16 and passing to the differential limit yield

$$\Delta p = -B \frac{\Delta s}{\Delta x} = -B \frac{\partial s}{\partial x}. \qquad (18\text{-}19)$$

The symbols ∂ indicate that the derivative in Eq. 18-19 is a *partial derivative* which tells us how s changes with x when the time t is fixed. From Eq. 18-13 we then have, treating t as a constant,

$$\frac{\partial s}{\partial x} = \frac{\partial}{\partial x} [s_m \cos(kx - \omega t)] = -ks_m \sin(kx - \omega t).$$

Substituting this quantity for the partial derivative in Eq. 18-19 yields

$$\Delta p = Bks_m \sin(kx - \omega t),$$

which is Eq. 18-14, which we set out to prove, with $\Delta p_m = Bks_m$.

Using Eq. 18-3, we can now write

$$\Delta p_m = (Bk)s_m = (v^2 \rho k)s_m.$$

Equation 18-15, which we also promised to prove, follows at once if we eliminate k by using $v = \omega/k$ (Eq. 17-12).

SAMPLE PROBLEM 18-2

The maximum pressure amplitude Δp_m that the human ear can tolerate in loud sounds is about 28 Pa (which is very much less than the normal air pressure of about 10^5 Pa). What is the displacement amplitude s_m for such a sound in air of density $\rho = 1.21$ kg/m^3, at a frequency of 1000 Hz?

SOLUTION: From Eq. 18-15 we have

$$\begin{aligned}
s_m &= \frac{\Delta p_m}{v \rho \omega} = \frac{\Delta p_m}{v \rho (2 \pi f)} \\[4pt]
&= \frac{28 \text{ Pa}}{(343 \text{ m/s})(1.21 \text{ kg/m}^3)(2\pi)(1000 \text{ Hz})} \\[4pt]
&= 1.1 \times 10^{-5} \text{ m} = 11 \ \mu\text{m}. \qquad \text{(Answer)}
\end{aligned}$$

The displacement amplitude for even the loudest sound that the ear can tolerate is obviously very small, being about one-seventh the thickness of this page.

The pressure amplitude Δp_m for the *faintest* detectable sound at 1000 Hz is 2.8×10^{-5} Pa. Proceeding as above leads to $s_m = 1.1 \times 10^{-11}$ m or 11 pm, which is about one-tenth the radius of a typical atom. The ear is indeed a sensitive detector of sound waves. The ear, in fact, can detect a pulse of sound whose total energy is as small as a few electron-volts, about the same as the energy required to remove an outer electron from a single atom.

18-4 INTERFERENCE

Figure 18-7 shows two point sources S_1 and S_2 that emit sound waves of wavelength λ in phase; that is, the emerging waves reach their maximum displacement values simultaneously. Suppose also that the sound waves pass through some common point P while traveling in (approximately) the same direction. If they have traveled along paths with identical lengths to reach point P, they will be in phase there. However, if they have traveled along paths with different lengths, as in Fig. 18-7, they may not be in phase. Their phase difference at point P depends on the **path length difference** ΔL between the two paths.

When two waves travel along different paths, the path length difference ΔL can alter their phase difference.

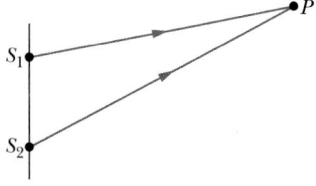

FIGURE 18-7   Two point sources S_1 and S_2 emit spherical sound waves in phase. The rays indicate that the waves pass through a common point P.

We find the phase difference ϕ between the waves by recalling (from Section 17-4) that a phase difference of 2π rad corresponds to one wavelength. Thus we can write the proportion

$$\frac{\phi}{2\pi} = \frac{\Delta L}{\lambda}, \qquad (18\text{-}20)$$

from which

$$\phi = \frac{\Delta L}{\lambda} 2\pi. \qquad (18\text{-}21)$$

Like transverse waves, sound waves undergo constructive interference and destructive interference. They undergo fully constructive interference when ϕ is zero or an integer multiple of 2π, that is, when

$$\phi = m(2\pi), \qquad m = 0, 1, 2, \cdots$$
$$\text{(fully constructive interference).} \quad (18\text{-}22)$$

They undergo fully destructive interference when ϕ is an odd multiple of π, that is, when

$$\phi = (m + \tfrac{1}{2})(2\pi), \qquad m = 0, 1, 2, \cdots$$
$$\text{(fully destructive interference).} \quad (18\text{-}23)$$

From Eq. 18-21, we see that these conditions correspond to

$$\Delta L = m\lambda, \qquad m = 0, 1, 2, \cdots$$
$$\text{(fully constructive interference).} \quad (18\text{-}24)$$

and

$$\Delta L = (m + \tfrac{1}{2})\lambda, \qquad m = 0, 1, 2, \cdots$$
$$\text{(fully destructive interference).} \quad (18\text{-}25)$$

SAMPLE PROBLEM 18-3

In Fig. 18-8a, two point sources S_1 and S_2, which are in phase and separated by a distance $D = 1.5\lambda$, emit identical sound waves of wavelength λ.

(a) What is the phase difference of the waves from S_1 and S_2 at point P_1, which lies on the perpendicular bisector of the distance D, but at a distance greater than D from the sources. What type of interference occurs at P_1?

SOLUTION: To reach P_1, waves emitted by S_1 and S_2 travel along paths that are not in the same direction. However, because P_1 is farther from the sources than the source separation D, we can approximate the directions of the waves as being the same. Furthermore, those waves travel identical distances to reach P_1. Thus the path length difference ΔL is 0, and Eq. 18-21 tells us that

$$\phi = \frac{\Delta L}{\lambda} 2\pi = 0. \qquad \text{(Answer)}$$

The value $\phi = 0$ satisfies Eq. 18-22 with $m = 0$, and so it corresponds to fully constructive interference. We get the same result with Eq. 18-24: the path length difference $\Delta L = 0$ satisfies that equation, with $m = 0$.

(b) What are the phase difference and type of interference at point P_2 in Fig. 18-8a?

SOLUTION: Point P_2 is on the line that extends through S_1 and S_2. So to reach P_2, waves emitted by S_1 must travel a distance D farther than those emitted by S_2. From Eq. 18-21 with $\Delta L = D = 1.5\lambda$, we find

$$\phi = \frac{\Delta L}{\lambda} 2\pi = \frac{1.5\lambda}{\lambda} 2\pi = 3\pi \text{ rad}. \qquad \text{(Answer)}$$

This value for ϕ satisfies Eq. 18-23 with $m = 1$, and so it corresponds to fully destructive interference. We reach the same conclusion with Eq. 18-25: because $\Delta L = 1.5\lambda$, that equation is satisfied with $m = 1$. Note that whether the phase difference is expressed in radians or wavelengths, the result is independent of the distance between P_2 and the source S_2.

(c) Figure 18-8b shows a circle with a radius much greater than D, centered on the midpoint between sources S_1 and S_2. What is the number of points N around this circle at which the interference is fully constructive?

SOLUTION: From (a), we know that the path length difference is $\Delta L = 0$ at points a and b in Fig. 18-8b, where the

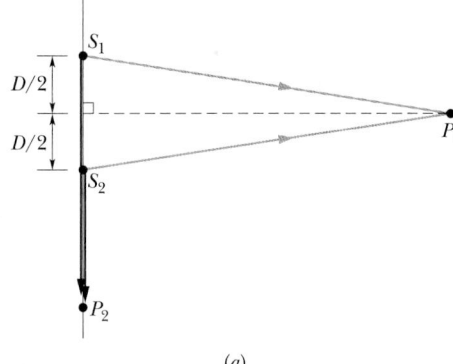

(a)

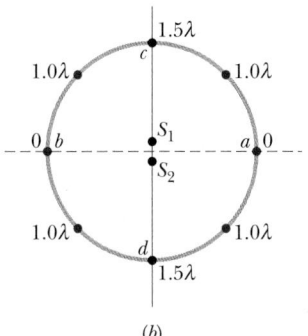

(b)

FIGURE 18-8 Sample Problem 18-3. (a) Two point sources S_1 and S_2, separated by distance D, emit spherical sound waves in phase. The waves travel equal distances to reach point P_1. Point P_2 is on the line extending through S_1 and S_2. (b) The phase difference (in terms of wavelengths) between the waves from S_1 and S_2, at eight points on a large circle around the sources.

circle intersects the perpendicular bisector of a line through the sources. From (b), we know that the path length difference is $\Delta L = 1.5\lambda$ at points c and d where the circle intersects a straight line through the sources. These results indicate that there must be intermediate points along the circle where $\Delta L = 1.0\lambda$; fully constructive interference occurs at those points. Without locating these four points exactly, we can indicate their presence as in Fig. 18-8b. We then count the points of fully constructive interference around the circle, finding

$$N = 6. \qquad \text{(Answer)}$$

CHECKPOINT **2:** In Sample Problem 18-3, if the distance D between sources S_1 and S_2 were, instead, equal to 4λ, what type of interference would occur, and what value of m would be associated with it, at (a) point P_1 and (b) point P_2?

18-5 INTENSITY AND SOUND LEVEL

If you have ever tried to sleep while someone played loud music nearby, you are well aware that there is more to sound than frequency, wavelength, and speed. There is also intensity. The **intensity** I of a sound wave at a surface is the average rate per unit area at which energy is transferred by the wave through or onto the surface. We can write this as

$$I = \frac{P}{A}, \qquad (18\text{-}26)$$

where P is the rate of energy transfer (power) of the sound wave and A is the area of the surface intercepting the sound. The intensity I is related to the displacement amplitude s_m of the sound wave by

$$I = \tfrac{1}{2}\rho v\omega^2 s_m^2, \qquad (18\text{-}27)$$

which we shall derive shortly.

Variation of Intensity with Distance

How intensity varies with distance from a real sound source is often complex: some real sources (like loudspeakers) may beam sound in particular directions, and the environment usually produces echoes (reflected sound waves) that overlap the direct sound waves. In some situations, however, we can ignore echoes and assume that the sound source is a point source that emits the sound *isotropically,* that is, with equal intensity in all directions. The wavefronts spreading from such an isotropic point source S at a particular instant are shown in Fig. 18-9.

Let us assume that the mechanical energy of the sound waves is conserved as they spread from this source. Let us

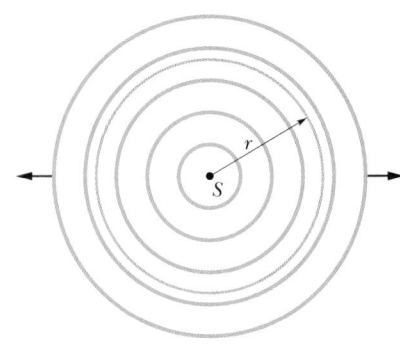

FIGURE 18-9 A point source S emits sound waves uniformly in all directions. The waves pass through an imaginary sphere of radius r that is centered on S.

also center an imaginary sphere of radius r on the source, as shown in Fig. 18-9. All the energy emitted by the source must pass through the surface of the sphere. Thus, the rate at which energy is transferred through the surface by the sound waves must equal the rate at which energy is emitted by the source (that is, the power P_s of the source). From Eq. 18-26, the intensity I at the sphere must then be

$$I = \frac{P_s}{4\pi r^2}, \qquad (18\text{-}28)$$

where $4\pi r^2$ is the area of the sphere. Equation 18-28 tells us that the intensity of sound from an isotropic point source decreases with the square of the distance r from the source.

CHECKPOINT **3:** The figure indicates three small patches 1, 2, and 3 that lie on the surfaces of two imaginary spheres; the spheres are centered on an isotropic point source of sound S. The rates at which energy is transmitted through the three patches by the sound waves are equal. Rank the patches according to (a) the intensity of the sound on them and (b) their area, greatest first.

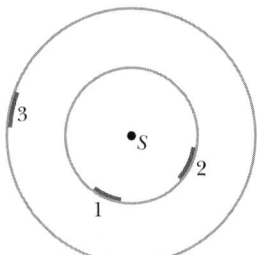

The Decibel Scale

You saw in Sample Problem 18-2 that the displacement amplitude at the human ear ranges from about 10^{-5} m for the loudest tolerable sound to about 10^{-11} m for the faintest detectable sound, a ratio of 10^6. From Eq. 18-27 we see that the intensity of a sound varies as the *square* of its amplitude, so the ratio of intensities at these two limits of the human auditory system is 10^{12}. Humans can hear over an enormous range of intensities.

Sound can cause the wall of a drinking glass to oscillate. If the sound produces a standing wave of oscillations and if the intensity of the sound is large enough, the glass will shatter.

We deal with such an enormous range of values by using logarithms. Consider the relation

$$y = \log x,$$

in which x and y are variables. It is a property of this equation that if we *multiply* x by 10, then y increases by 1. To see this, we write

$$y' = \log(10x) = \log 10 + \log x = 1 + y.$$

Similarly, if we multiply x by 10^{12}, y increases by only 12.

Thus instead of speaking of the intensity I of a sound wave, it is much more convenient to speak of its **sound level** β, defined as

$$\beta = (10 \text{ dB}) \log \frac{I}{I_0}. \qquad (18\text{-}29)$$

Here dB is the abbreviation for **decibel,** the unit of sound level, a name that was chosen to recognize the work of Alexander Graham Bell. I_0 in Eq. 18-29 is a standard reference intensity ($= 10^{-12}$ W/m^2), chosen because it is near the lower limit of the human range of hearing. For $I = I_0$, Eq. 18-29 gives $\beta = 10 \log 1 = 0$, so our standard reference level corresponds to zero decibels. Then β increases by 10 dB every time the sound intensity increases by an order of magnitude (a factor of 10). Thus, $\beta = 40$ corresponds to an intensity that is 10^4 times the standard reference level. Table 18-2 lists the sound levels for a variety of environments.

TABLE 18-2 SOME SOUND LEVELS (dB)

Hearing threshold	0	Rock concert	110
Rustle of leaves	10	Pain threshold	120
Conversation	60	Jet engine	130

Derivation of Eq. 18-27

Consider, in Fig. 18-5a, a thin slice of air of thickness dx, area A, and mass dm, oscillating back and forth as the sound wave of Eq. 18-13 passes through it. The kinetic energy dK of the slice of air is

$$dK = \tfrac{1}{2} dm\, v_s^2. \qquad (18\text{-}30)$$

Here v_s is not the speed of the wave but the speed of the oscillating element of air, obtained from Eq. 18-13 as

$$v_s = \frac{\partial s}{\partial t} = -\omega s_m \sin(kx - \omega t).$$

Using this relation and putting $dm = \rho A\, dx$ allow us to rewrite Eq. 18-30 as

$$dK = \tfrac{1}{2}(\rho A\, dx)(-\omega s_m)^2 \sin^2(kx - \omega t). \qquad (18\text{-}31)$$

Dividing Eq. 18-31 by dt gives the rate at which kinetic energy moves along with the wave. As we saw in Chapter 17 for transverse waves, the ratio dx/dt is the wave speed, so we have

$$\frac{dK}{dt} = \tfrac{1}{2}\rho A v \omega^2 s_m^2 \sin^2(kx - \omega t). \qquad (18\text{-}32)$$

The *average* rate at which kinetic energy is transported is

$$\overline{\left(\frac{dK}{dt}\right)} = \tfrac{1}{2}\rho A v \omega^2 s_m^2 \,\overline{\sin^2(kx - \omega t)}$$
$$= \tfrac{1}{4}\rho A v \omega^2 s_m^2. \qquad (18\text{-}33)$$

To obtain this equation, we have used the fact that the average value of the square of a sine (or a cosine) function over one wavelength is $\tfrac{1}{2}$.

We assume that *potential* energy is carried along with the wave at this same average rate. The wave intensity I, which is the average rate per unit area at which energy of both kinds is transmitted by the wave, is then, from Eq. 18-33,

$$I = \frac{2\overline{(dK/dt)}}{A} = \tfrac{1}{2}\rho v \omega^2 s_m^2,$$

which is Eq. 18-27, the equation we set out to derive.

SAMPLE PROBLEM 18-4

An electric spark jumps along a straight line of length $L = 10$ m, emitting a pulse of sound that travels radially outward from the spark. (The spark is said to be a *line source* of sound.) The power of the emission is $P_s = 1.6 \times 10^4$ W.

(a) What is the intensity I of the sound when it reaches a distance $r = 12$ m from the spark?

SOLUTION: Let us center an imaginary cylinder of radius $r = 12$ m and length $L = 10$ m (open at both ends) on the spark, as shown in Fig. 18-10. The rate at which energy passes through the cylinder must equal the rate P_s at which energy is emitted by the source. From Eq. 18-26, the intensity I at the cylindrical surface must then equal the power P_s divided by the area $2\pi rL$ of that surface:

$$I = \frac{P_s}{2\pi rL}. \qquad (18\text{-}34)$$

This tells us that the intensity of the sound from a line source decreases with distance r (and not with the square of distance r as for a point source). Substituting the given data, we find

$$I = \frac{1.6 \times 10^4 \text{ W}}{2\pi(12 \text{ m})(10 \text{ m})} = 21.2 \text{ W/m}^2 \approx 21 \text{ W/m}^2. \quad \text{(Answer)}$$

(b) At what rate P_d is sound energy intercepted by an acoustic detector of area $A_d = 2.0 \text{ cm}^2$, aimed at the spark and located a distance $r = 12$ m from the spark?

SOLUTION: From Eq. 18-26, we know that

$$I = \frac{P_d}{A_d}.$$

Solving for P_d and substituting the intensity I from (a) and the given value for A_d, we find

$$P_d = (21.2 \text{ W/m}^2)(2.0 \times 10^{-4} \text{ m}^2) = 4.2 \text{ mW}. \quad \text{(Answer)}$$

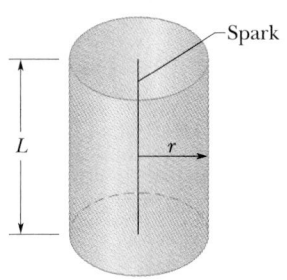

FIGURE 18-10 Sample Problem 18-4. A spark along a straight path of length L emits sound waves radially outward. The waves pass through an imaginary cylinder of radius r and length L that is centered on the spark.

SAMPLE PROBLEM 18-5

In 1976, the Who set a record for the loudest concert: the sound level 46 m in front of the speaker systems was $\beta_2 = 120$ dB. What is the ratio of the intensity I_2 of the band at that spot to the intensity I_1 of a jackhammer operating at sound level $\beta_1 = 92$ dB?

SOLUTION: Let us write the ratio of the two intensities as

$$\frac{I_2}{I_1} = \frac{I_2/I_0}{I_1/I_0}.$$

Taking logarithms on each side and multiplying through by 10 dB, we have

$$(10 \text{ dB}) \log \frac{I_2}{I_1} = (10 \text{ dB}) \log \frac{I_2}{I_0} - (10 \text{ dB}) \log \frac{I_1}{I_0}.$$

From Eq. 18-29 we then see that the terms on the right are β_2

and β_1. So

$$(10 \text{ dB}) \log \frac{I_2}{I_1} = \beta_2 - \beta_1. \qquad (18\text{-}35)$$

Note that the *ratio* of two intensities corresponds to a *difference* in their sound levels. Substituting given data yields

$$(10 \text{ dB}) \log \frac{I_2}{I_1} = 120 \text{ dB} - 92 \text{ dB} = 28 \text{ dB}$$

and

$$\log \frac{I_2}{I_1} = \frac{28 \text{ dB}}{10 \text{ dB}} = 2.8.$$

Taking the antilog of both sides gives us

$$\frac{I_2}{I_1} = 630. \qquad \text{(Answer)}$$

Thus, the Who was *very* loud.

Temporary exposure to sound intensities as great as those of a jackhammer and the 1976 Who concert results in a temporary reduction of hearing. Repeated or prolonged exposure can result in permanent reduction of hearing (Fig. 18-11). Loss of hearing is a clear risk for anyone continually listening to, say, heavy metal at high volume.

FIGURE 18-11 Sample Problem 18-5. Peter Townshend of the Who, playing in front of a speaker system. His repeated and prolonged exposure to high-intensity sound, especially when playing into a speaker to produce feedback, resulted in a permanent reduction in his hearing.

18-6 SOURCES OF MUSICAL SOUND

Musical sounds can be set up by oscillating strings (guitar, piano, violin), membranes (kettledrum, snare drum), air columns (flute, oboe, pipe organ, and the fujara of Fig. 18-12), wooden blocks or steel bars (marimba, xylophone), and many other oscillating bodies. Most instruments involve more than a single oscillating part. In the violin, for example, not only the strings but also the body of the instrument participates in producing the music.

Recall from Chapter 17 that standing waves can be set up on a stretched string that is fixed at both ends. They arise because waves traveling along the string are reflected back onto the string at each end. If the wavelength of the waves is suitably matched to the length of the string, the superposition of waves traveling in opposite directions produces a standing wave pattern (or oscillation mode). The wavelength required of the waves for such a match is one that corresponds to a *resonant frequency* of the string. The advantage of setting up standing waves is that the string then oscillates with a large, sustained amplitude, pushing back and forth against the surrounding air and thus generating a noticeable sound wave with the same fre-

quency as the oscillations of the string. This production of sound is of obvious importance to, say, a guitarist.

We can set up standing waves of sound in an air-filled pipe in a similar way. As sound waves travel through the air in the pipe, they reflect at each end and back through the pipe. (The reflection occurs even if an end is open, but the reflection is not as complete as when the end is closed.) If the wavelength of the sound waves is suitably matched to the length of the pipe, the superposition of waves traveling in opposite directions through the pipe sets up a standing wave pattern. The wavelength required of the sound waves for such a match is one that corresponds to a resonant frequency of the pipe. The advantage of such a standing wave is that the air in the pipe oscillates with a large, sustained amplitude, emitting at any open end a sound wave that has the same frequency as the oscillations in the pipe. This emission of sound is of obvious importance to, say, an organist.

Many other aspects of standing sound wave patterns are similar to those of string waves: the closed end of a pipe is like the fixed end of a string in that there must be a node (zero displacement) there. And the open end of a pipe is like the end of a string attached to a freely moving ring, as in Fig. 17-16b, in that there must be an antinode there. (Actually, the antinode for the open end of a pipe is located slightly beyond the end, but we shall not dwell on that detail here.)

The simplest standing wave pattern that can be set up in a pipe with two open ends is shown in Fig. 18-13a. There is an antinode across each open end, as required. There is also a node across the middle of the pipe. An easier way of representing this standing longitudinal sound wave is shown in Fig. 18-13b—by drawing it as a standing transverse string wave.

The standing wave pattern of Fig. 18-13a is called the *fundamental mode* or *first harmonic*. For it to be set up, the sound waves in a pipe of length L must have a wavelength

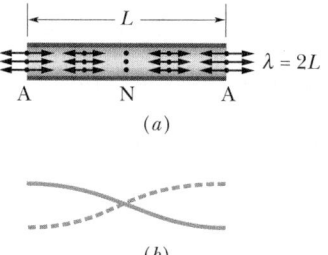

(a)

(b)

FIGURE 18-13 (a) The simplest standing wave pattern of displacement for (longitudinal) sound waves in a pipe with both ends open has an antinode (A) across each end and a node (N) across the middle. (The displacements represented by the double arrows are greatly exaggerated.) (b) The corresponding standing wave pattern for (transverse) string waves.

FIGURE 18-12 The air column within a fujara oscillates when that traditional Slovakian instrument is played.

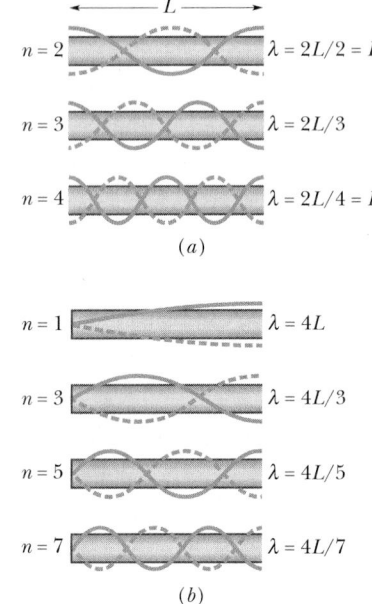

(a)

(b)

FIGURE 18-14 Standing wave patterns for string waves superimposed on pipes to represent standing sound wave patterns in the pipes. (a) With *both* ends of the pipe open, any harmonic can be set up in the pipe. (b) But with only *one* end open, only odd harmonics can be set up.

given by $L = \lambda/2$, so that $\lambda = 2L$. Several more standing sound wave patterns for a pipe with two open ends are shown in Fig. 18-14a using string wave representations. The *second harmonic* requires sound waves of wavelength $\lambda = L$. The *third harmonic* requires wavelength $\lambda = 2L/3$. And so on.

More generally, the resonant frequencies for a pipe of length L with two open ends correspond to the wavelengths

$$\lambda = \frac{2L}{n}, \qquad n = 1, 2, 3, \cdots, \qquad (18\text{-}36)$$

where n is called the *harmonic number*. The resonant frequencies are then given by

$$f = \frac{v}{\lambda} = \frac{nv}{2L}, \qquad n = 1, 2, 3, \cdots$$

$$\text{(pipe, two open ends)}, \qquad (18\text{-}37)$$

where v is the speed of sound.

Figure 18-14b shows (using string wave representations) some of the standing sound wave patterns that can be set up in a pipe with only one open end. As required, across the open end there is an antinode and across the closed end there is a node. The simplest pattern requires sound waves having a wavelength given by $L = \lambda/4$, so that $\lambda = 4L$. The next simplest pattern requires a wavelength given by $L = 3\lambda/4$, so that $\lambda = 4L/3$. And so on.

More generally, the resonant frequencies for a pipe of length L with only one open end correspond to the wavelengths

$$\lambda = \frac{4L}{n}, \qquad n = 1, 3, 5, \cdots, \qquad (18\text{-}38)$$

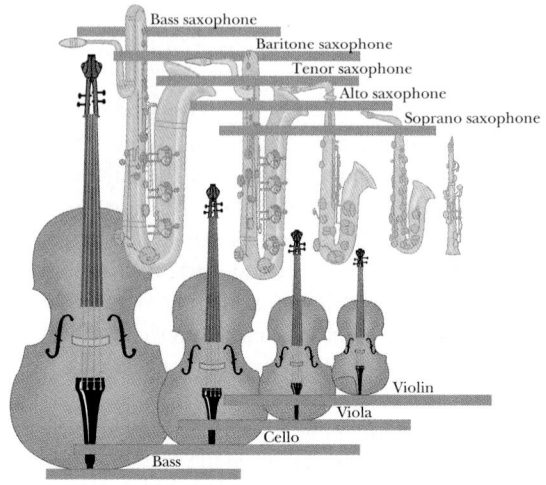

FIGURE 18-15 The saxophone and violin families, showing the relations between instrument length and frequency range. The frequency range of each instrument is indicated by a horizontal bar along a frequency scale suggested by the keyboard at the bottom; the frequency increases toward the right.

in which the harmonic number n *must be an odd number*. The resonant frequencies are then given by

$$f = \frac{v}{\lambda} = \frac{nv}{4L}, \qquad n = 1, 3, 5, \cdots$$

$$\text{(pipe, one open end)}. \qquad (18\text{-}39)$$

Note again that only odd harmonics can exist in a pipe with one open end. For example, the second harmonic, with $n = 2$, cannot be set up in such a pipe. Note also that for such a pipe the adjective in a phrase such as "the third harmonic" still refers to the harmonic number n (and not to, say, the third possible harmonic).

The length of a musical instrument reflects the range of frequencies over which the instrument is designed to function: smaller length implies higher frequencies. Figure 18-15, for example, shows the saxophone and violin families, with their frequency ranges suggested by the piano keyboard. Note that, for every instrument, there is overlap with its higher- and lower-frequency neighbors.

In any oscillating system that gives rise to a musical sound, whether it be a violin string or the air in an organ pipe, the fundamental and one or more of the higher harmonics are usually generated simultaneously. The resultant sound is the superposition of these components. For different instruments, the higher harmonics have different intensities, which accounts for the different sounds produced by different instruments playing the same note. Figure 18-16 shows, for example, the resultant wave forms when the same note, with the same fundamental frequency, is sounded on three different instruments.

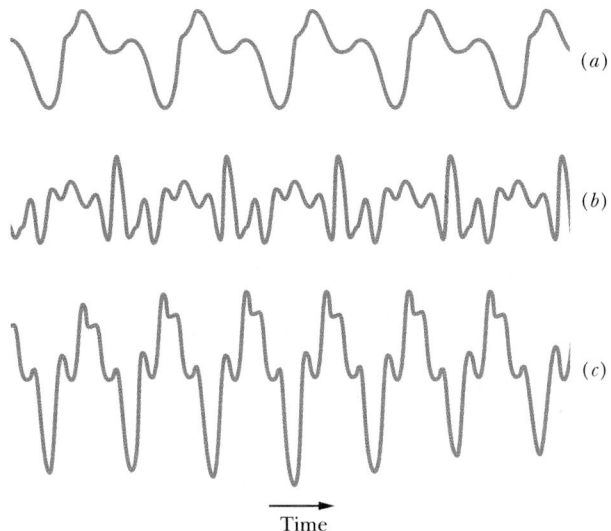

FIGURE 18-16 The wave forms produced by (a) a flute, (b) an oboe, and (c) a saxophone when they all play the same note, with the same first harmonic frequency.

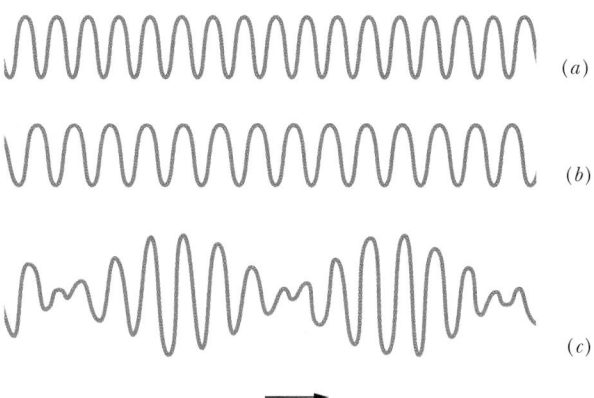

FIGURE 18-17 (a, b) The pressure variations Δp of two sound waves that are detected separately. The frequencies of the waves are nearly equal. (c) The resultant pressure variation if the two waves are detected simultaneously.

SAMPLE PROBLEM 18-6

Weak background noises from a room set up the fundamental standing wave in a cardboard tube of length $L = 67.0$ cm with two open ends. Assume that the speed of sound in the air within the tube is 343 m/s.

(a) What frequency do you hear from the tube if you jam your ear against one end?

SOLUTION: With your ear closing one end, the fundamental frequency is given by Eq. 18-39 with $n = 1$:

$$f = \frac{v}{4L} = \frac{343 \text{ m/s}}{(4)(0.670 \text{ m})} = 128 \text{ Hz}. \quad \text{(Answer)}$$

If the background noises set up any higher harmonics, such as the third harmonic, you may also hear frequencies that are *odd* multiples of 128 Hz.

(b) What frequency do you hear from the tube if you move your head away enough so that the tube has two open ends?

SOLUTION: With both ends open, the fundamental frequency is given by Eq. 18-37 with $n = 1$:

$$f = \frac{v}{2L} = \frac{343 \text{ m/s}}{(2)(0.670 \text{ m})} = 256 \text{ Hz}. \quad \text{(Answer)}$$

If the background noises set up any higher harmonics, such as the second harmonic, you may also hear frequencies that are *integer* multiples of 256 Hz. However, sound with a frequency of 128 Hz is no longer emitted by the tube.

CHECKPOINT **4:** Pipe *A*, with length *L*, and pipe *B*, with length 2*L*, both have two open ends. Which harmonic of pipe *B* has the same frequency as the fundamental of pipe *A*?

18-7 BEATS

If we listen, a few minutes apart, to two sounds whose frequencies are, say, 552 and 564 Hz, most of us cannot tell one from the other. However, if the sounds reach our ears simultaneously, what we hear is a sound whose frequency is 558 Hz, the *average* of the two combining frequencies. We also hear a striking variation in the intensity of this sound: it increases and decreases in slow, wavering *beats* that repeat at a frequency of 12 Hz, the *difference* between the two combining frequencies. Figure 18-17 shows this **beat** phenomenon.

Let the time-dependent variations of the displacements due to two sound waves at a particular location be

$$s_1 = s_m \cos \omega_1 t \quad \text{and} \quad s_2 = s_m \cos \omega_2 t. \quad \text{(18-40)}$$

We have assumed, for simplicity, that the waves have the same amplitude. According to the superposition principle, the resultant displacement is

$$s = s_1 + s_2 = s_m(\cos \omega_1 t + \cos \omega_2 t).$$

Using the trigonometric identity (see Appendix E)

$$\cos \alpha + \cos \beta = 2 \cos \tfrac{1}{2}(\alpha - \beta) \cos \tfrac{1}{2}(\alpha + \beta)$$

allows us to write the resultant displacement as

$$s = 2s_m \cos \tfrac{1}{2}(\omega_1 - \omega_2)t \cos \tfrac{1}{2}(\omega_1 + \omega_2)t. \quad \text{(18-41)}$$

If we write

$$\omega' = \tfrac{1}{2}(\omega_1 - \omega_2) \quad \text{and} \quad \omega = \tfrac{1}{2}(\omega_1 + \omega_2), \quad \text{(18-42)}$$

we can then write Eq. 18-41 as

$$s(t) = [2s_m \cos \omega' t] \cos \omega t. \quad \text{(18-43)}$$

We now assume that the angular frequencies ω_1 and ω_2 of the combining waves are almost equal, which means that $\omega \gg \omega'$ in Eq. 18-42. We can then regard Eq. 18-43 as a cosine function whose angular frequency is ω and whose amplitude (which is not constant but varies with frequency ω') is the quantity in the brackets.

A maximum amplitude will occur whenever cos $\omega' t$ in Eq. 18-43 has the value $+1$ or -1, which happens twice in each repetition of the cosine function. Because cos $\omega' t$ has angular frequency ω', the angular frequency ω_{beat} at which beats occur is $\omega_{\text{beat}} = 2\omega'$. Then, with the aid of Eq. 18-42, we can write

$$\omega_{\text{beat}} = 2\omega' = (2)(\tfrac{1}{2})(\omega_1 - \omega_2) = \omega_1 - \omega_2.$$

Because $\omega = 2\pi f$, we can recast this as

$$f_{\text{beat}} = f_1 - f_2 \quad \text{(beat frequency).} \quad (18\text{-}44)$$

Musicians use the beat phenomenon in tuning their instruments. If an instrument is sounded against a standard frequency (for example, the lead oboe's reference A) and tuned until the beat disappears, then the instrument is in tune with that standard. In musical Vienna, concert A (440 Hz) is available as a telephone service for the benefit of the city's many professional and amateur musicians.

SAMPLE PROBLEM 18-7

You wish to tune the note A_3 on a piano to its proper frequency of 220 Hz. You have available a tuning fork whose frequency is 440 Hz. How should you proceed?

SOLUTION: These two frequencies are too far apart to produce beats. Recall that in our analysis of Eq. 18-43 we assumed that the two combining frequencies were reasonably close to each other. However, note that the second harmonic of A_3 is 2×220 or 440 Hz.

Let us say that the actual piano string is mistuned such that its fundamental frequency is not exactly 220 Hz. You listen for beats between the fundamental frequency of the tuning fork and the second harmonic of A_3, and you hear a beat frequency of 6 Hz. You then change the tension in the corresponding string until the beat note disappears. The string is then in tune.

CHECKPOINT 5: In Sample Problem 18-7, you tighten the string and the beat frequency increases from 6 Hz. Should you continue to tighten the string or should you loosen the string to tune the string in tune?

18-8 THE DOPPLER EFFECT

A police car is parked by the side of the highway, sounding its 1000 Hz siren. If you are also parked by the highway, you will hear that same frequency. But if there is relative motion between you and the police car, either toward or away from each other, you will hear a different frequency. For example, if you are driving *toward* the police car at 120 km/h (about 75 mi/h), you will hear a *higher* frequency (1096 Hz, an *increase* of 96 Hz). If you are driving *away from* the police car at that same speed, you will hear a *lower* frequency (904 Hz, a *decrease* of 96 Hz).

These motion-related frequency changes are examples of the **Doppler effect.** The effect was proposed (although not fully worked out) in 1842 by Austrian physicist Johann Christian Doppler. It was tested experimentally in 1845 by Buys Ballot in Holland, "using a locomotive drawing an open car with several trumpeters."

The Doppler effect holds not only for sound waves but also for electromagnetic waves, including microwaves, radio waves, and visible light. Police use the Doppler effect with microwaves to determine the speed of a car: a radar unit beams microwaves of a certain frequency f toward the oncoming car. The microwaves that reflect from the metal portions of the car back to the radar unit have a higher frequency f' owing to the motion of the car relative to the radar unit. The radar unit captures the difference between f' and f and converts it into the speed of the car, which is then displayed for the operator to see. However, the displayed speed is accurate only if the car is moving directly toward (or away from) the radar unit. (Misalignment reduces f' and also the displayed speed.)

In the analyses that follow, we restrict ourselves to sound waves, and we take as a reference frame the body of air through which these waves travel. This means that we

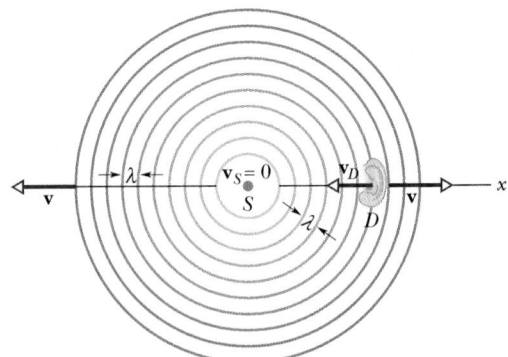

FIGURE 18-18 A stationary source of sound S emits spherical wavefronts, shown one wavelength apart, that expand outward at speed v. A sound detector D, represented by an ear, moves with velocity $\mathbf{v}_D$ toward the source. The detector senses a higher frequency because of its motion.

measure the speeds of a source S of sound waves and a detector D of those waves *relative to that body of air.* (Unless otherwise stated, the body of air is stationary relative to Earth, so the speeds can also be measured relative to Earth.) We shall assume that S and D move either directly toward or directly away from each other, at speeds less than the speed of sound.

We shall now derive equations for the Doppler effect for two particular situations: (1) for the detector moving and the source stationary and (2) for the source moving and the detector stationary. Then we shall combine the equations for these particular situations to find the *general Doppler effect equation,* which holds for both those situations and also when both detector and source are moving.

Detector Moving; Source Stationary

In Fig. 18-18, a detector D (represented by an ear) is moving at speed v_D toward a stationary source S that emits spherical wavefronts, of wavelength λ and frequency f, moving at the speed v of sound in air. The wavefronts are drawn one wavelength apart. The frequency detected by detector D is the rate at which D intercepts wavefronts (or individual wavelengths). If D were stationary, that rate would be f, but since D is moving into the wavefronts, the rate of interception is greater, and thus the detected frequency f' is greater than f.

Let us for the moment consider the situation in which D is stationary (Fig. 18-19). In time t, the wavefronts move to the right a distance vt. The number of wavelengths in that distance vt is the number of wavelengths intercepted by D in time t, and that number is vt/λ. The rate at which D intercepts wavelengths, which is the frequency f detected by D, is

$$f = \frac{vt/\lambda}{t} = \frac{v}{\lambda}. \qquad (18\text{-}45)$$

In this situation, with D stationary, there is no Doppler effect: the frequency detected by D is the frequency emitted by S.

Now let us again consider the situation in which D moves opposite the wavefronts (Fig. 18-20). In time t, the wavefronts move to the right a distance vt as previously, but now D moves to the left a distance $v_D t$. Thus in this time t, the distance moved by the wavefronts relative to D is $vt + v_D t$. The number of wavelengths in this relative distance $vt + v_D t$ is the number of wavelengths intercepted by D in time t, and is $(vt + v_D t)/\lambda$. The *rate* at which D intercepts wavelengths in this situation is the frequency f', given by

$$f' = \frac{(vt + v_D t)/\lambda}{t} = \frac{v + v_D}{\lambda}. \qquad (18\text{-}46)$$

From Eq. 18-45, we have $\lambda = v/f$. Then Eq. 18-46 becomes

$$f' = \frac{v + v_D}{v/f} = f\frac{v + v_D}{v}. \qquad (18\text{-}47)$$

Note that in Eq. 18-47, f' must be greater than f unless $v_D = 0$ (the detector is stationary).

Similarly, we can find the frequency detected by D if D moves away from the source. In this situation, the wavefronts move a distance $vt - v_D t$ relative to D in time t, and f' is given by

$$f' = f\frac{v - v_D}{v}. \qquad (18\text{-}48)$$

In Eq. 18-48, f' must be less than f unless $v_D = 0$.

We can summarize Eqs. 18-47 and 18-48 with

$$f' = f\frac{v \pm v_D}{v} \quad \begin{array}{l}\text{(detector moving;} \\ \text{source stationary).}\end{array} \qquad (18\text{-}49)$$

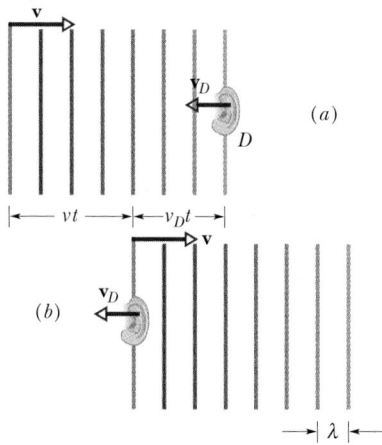

FIGURE 18-20 Wavefronts (*a*) reach and (*b*) pass detector D, which moves opposite the wavefronts. In time t, the wavefronts move a distance vt to the right and D moves a distance $v_D t$ to the left.

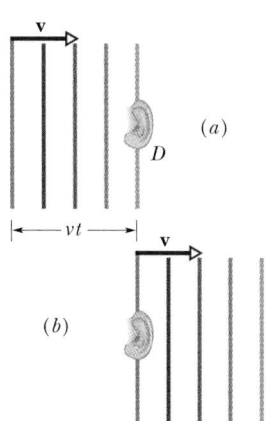

FIGURE 18-19 Wavefronts of Fig. 18-18, assumed planar, (*a*) reach and (*b*) pass a stationary detector D; they move a distance vt to the right in time t.

You can determine which sign applies in Eq. 18-49 by remembering the physical result: when the detector moves toward the source, the frequency is greater (*toward* means *greater*), which requires a plus sign in the numerator. Otherwise, a minus sign is required.

Source Moving; Detector Stationary

Let detector D be stationary with respect to the body of air, and let source S move toward D at speed v_S (Fig. 18-21). The motion of S changes the wavelength of the sound waves it emits, and thus the frequency detected by D.

To see this change, let $T \, (= 1/f)$ be the time between the emission of any pair of successive wavefronts W_1 and W_2. During T, wavefront W_1 moves a distance vT and the source moves a distance $v_S T$. At the end of T, wavefront W_2 is emitted. In the direction in which S moves, the distance between W_1 and W_2, which is the wavelength λ' of the waves moving in that direction, is $vT - v_S T$. If D detects those waves, it detects frequency f' given by

$$f' = \frac{v}{\lambda'} = \frac{v}{vT - v_S T} = \frac{v}{v/f - v_S/f}$$

$$= f \frac{v}{v - v_S}. \qquad (18\text{-}50)$$

Note that f' must be greater than f unless $v_S = 0$.

In the direction opposite that taken by S, the wavelength λ' of the waves is $vT + v_S T$. If D detects those

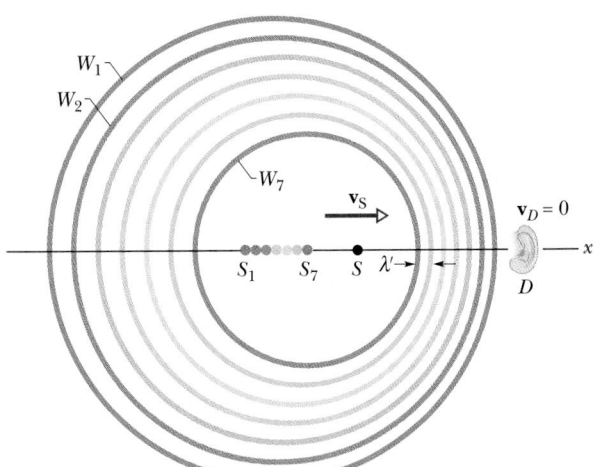

FIGURE 18-21 The detector D is stationary, with the source S moving toward it at speed v_S. Wavefront W_1 was emitted when the source was at S_1, wavefront W_7 when it was at S_7. At the moment depicted, the source is at S. The detector perceives a higher frequency because the moving source, chasing its own wavefronts, emits a reduced wavelength λ' in the direction of its motion.

waves, it detects frequency f' given by

$$f' = f \frac{v}{v + v_S}. \qquad (18\text{-}51)$$

Now f' must be less than f unless $v_S = 0$.

We can summarize Eqs. 18-50 and 18-51 with

$$f' = f \frac{v}{v \mp v_S} \qquad \begin{array}{l}\text{(source moving;}\\ \text{detector stationary).}\end{array} \qquad (18\text{-}52)$$

You can determine which sign applies in Eq. 18-52 by remembering the physical result: when the source moves toward the detector, the frequency is greater (*toward* means *greater*), which requires a minus sign in the denominator. Otherwise a plus sign should be used.

General Doppler Effect Equation

We can combine Eqs. 18-49 and 18-52 to produce the general Doppler effect equation, in which both the source and the detector can be moving with respect to the air mass. Replacing f in Eq. 18-52 (the frequency of the source) with f' of Eq. 18-49 (the frequency associated with motion of the detector) leads to

$$f' = f \frac{v \pm v_D}{v \mp v_S} \qquad \begin{array}{l}\text{(general Doppler}\\ \text{effect).}\end{array} \qquad (18\text{-}53)$$

Putting $v_S = 0$ in Eq. 18-53 reduces it to Eq. 18-49, and putting $v_D = 0$ reduces it to Eq. 18-52. The plus and minus signs are determined as in Eqs. 18-49 and 18-52 (*toward* means *greater*).

The Doppler Effect at Low Speeds

The Doppler effects for a moving detector (Eq. 18-49) and for a moving source (Eq. 18-52) are different, even though the detector and the source may be moving at the same speed. However, if the speeds are low enough (that is, if $v_D \ll v$ and $v_S \ll v$), the frequency changes produced by these two motions are essentially the same.

By using the binomial theorem (see Tactic 2 of Chapter 7), you can show that Eq. 18-53 can be written in the form

$$f' \approx f \left(1 \pm \frac{u}{v}\right) \qquad \text{(low speeds only),} \qquad (18\text{-}54)$$

in which $u \, (= |v_S \pm v_D|)$ is the *relative* speed of the source with respect to the detector. The rule for signs remains the same. If the source and the detector are moving *toward* each other, we anticipate a *greater* frequency; this requires that we choose the plus sign in Eq. 18-54. On the other hand, if the source and the detector are moving away from each other, we anticipate a frequency decrease and choose the minus sign in Eq. 18-54.

CHECKPOINT **6:** The figure indicates the motion of a sound source and a detector for six situations in stationary air. For each situation, is the detected frequency greater than or less than the emitted frequency, or can't we tell without more information?

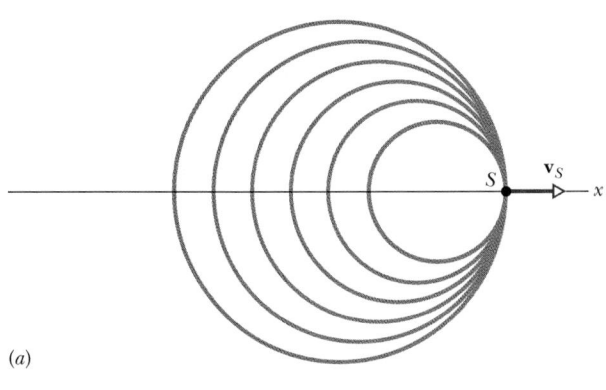

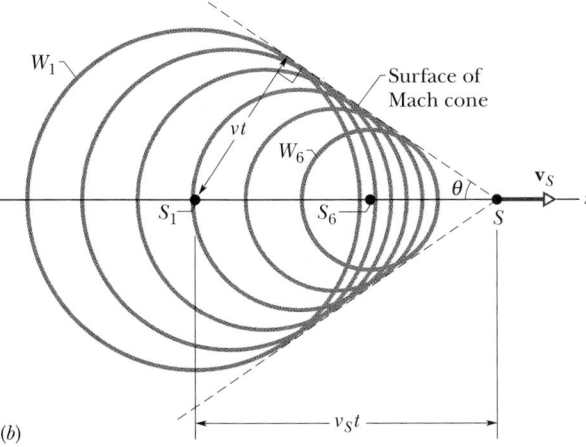

FIGURE 18-22 (*a*) A source of sound *S* moves at speed v_S equal to the speed of sound and thus as fast as the wavefronts it generates. (*b*) A source *S* moves at speed v_S faster than the speed of sound and thus faster than the wavefronts. When the source was at position S_1 it generated wavefront W_1, and at position S_6 it generated W_6. All the spherical wavefronts expand at the speed of sound *v* and bunch along the surface of a cone called the Mach cone, forming a shock wave. The surface of the cone has half-angle θ and is tangent to all the wavefronts.

Supersonic Speeds; Shock Waves

If a source is moving toward a stationary detector at a speed equal to the speed of sound, that is, if $v_S = v$, Eq. 18-52 predicts that the detected frequency f' will be infinitely great. This means that the source is moving so fast that it keeps pace with its own spherical wavefronts, as Fig. 18-22*a* suggests. What happens when the speed of the source *exceeds* the speed of sound?

For such *supersonic* speeds, Eq. 18-52 no longer applies. Figure 18-22*b* depicts the spherical wavefronts that originated at various positions of the source. The radius of any wavefront in this figure is vt, where v is the speed of sound and t is the time that has elapsed since the source emitted that wavefront. Note that all the wavefronts bunch along a V-shaped envelope in Fig. 18-22*b*, which in three dimensions is a cone called the *Mach cone*. A **shock wave** is said to exist along the surface of this cone, because the bunching of wavefronts causes an abrupt rise and fall of air pressure as the surface passes through any point. From Fig. 18-22*b*, we see that the half-angle θ of the cone, called the *Mach cone angle*, is given by

$$\sin \theta = \frac{vt}{v_S t} = \frac{v}{v_S} \qquad \text{(Mach cone angle).} \quad (18\text{-}55)$$

The ratio v_S/v is called the *Mach number.* When you hear that a particular plane has flown at Mach 2.3, it means that its speed was 2.3 times the speed of sound in the air through which the plane was flying. The shock wave generated by a supersonic aircraft or projectile (Fig. 18-23) produces a burst of sound, called a *sonic boom,* in which the air pressure first suddenly increases and then suddenly decreases below normal before returning to normal.

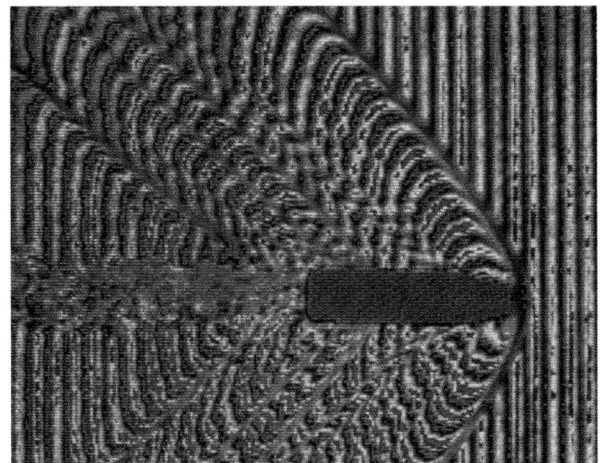

FIGURE 18-23 A false-color, high-speed image of a 20 mm bullet traveling at about Mach 1.3. Note the prominent Mach cone produced by the nose of the bullet and the secondary cones produced by irregular features along the sides.

SAMPLE PROBLEM 18-8

A toy rocket moves at a speed of 242 m/s directly toward a stationary pole (through stationary air) while emitting sound waves at frequency $f = 1250$ Hz.

(a) What frequency f' is sensed by a detector that is attached to the pole?

SOLUTION: To find f', we shall use Eq. 18-53 for the general Doppler effect. Because the detector is stationary, we set $v_D = 0$. And because the sound source (the rocket) is moving *toward* the detector, we choose the minus sign in the denominator. Then substituting $v = 343$ m/s from Table 18-1 and the given data, we find that the detected frequency is

$$f' = f \frac{v}{v - v_S} = (1250 \text{ Hz}) \frac{343 \text{ m/s}}{343 \text{ m/s} - 242 \text{ m/s}}$$

$$= 4245 \text{ Hz} \approx 4250 \text{ Hz}. \qquad \text{(Answer)}$$

As a quick check on the answer, recall the physical result: when a source moves *toward* a stationary detector, the detected frequency (here, 4245 Hz) should be *greater* than the emitted frequency (here, 1250 Hz).

(b) Some of the sound reaching the pole reflects back to the rocket, which has an onboard detector. What frequency f'' does it detect?

SOLUTION: The pole now acts as the source of sound in that it reflects sound waves, producing an echo. The frequency of the waves from the pole is the same as the frequency of the waves "sensed" by the pole, namely, $f' = 4245$ Hz. Because now the source (the pole) is stationary, we set $v_S = 0$ in Eq. 18-53. And because now the detector (the onboard detector) is moving toward the new source, we choose the plus sign in the numerator. So the frequency sensed by the onboard detector is

$$f'' = f' \frac{v + v_D}{v} = (4245 \text{ Hz}) \frac{343 \text{ m/s} + 242 \text{ m/s}}{343 \text{ m/s}}$$

$$= 7240 \text{ Hz}. \qquad \text{(Answer)}$$

Again, as a quick check on the answer, recall the physical result: when a detector moves *toward* a stationary source, the detected frequency (here, 7240 Hz) should be *greater* than the emitted frequency (here, 4245 Hz).

$\mathbb{C}$ HECKPOINT **7:** In Sample Problem 18-8, if the air is moving toward the pole at speed 20 m/s, (a) what value for the source speed v_S should be used in the solution of part a, and (b) what value for the detector speed v_D should be used in the solution of part b?

SAMPLE PROBLEM 18-9

Bats navigate and search out prey by emitting, and then detecting reflections of, ultrasonic waves, which are sound waves with frequencies greater than what can be heard by a human. Suppose a horseshoe bat flies toward a moth at speed $v_b = 9.0$ m/s, while the moth flies toward the bat with speed $v_m = 8.0$ m/s. From its nostrils, the bat emits ultrasonic waves of frequency f_{be} that reflect from the moth back to the bat with frequency f_{bd}. The bat adjusts the emitted frequency f_{be} until the returned frequency f_{bd} is 83 kHz, at which the bat's hearing is best.

(a) What is the frequency f_m of the waves heard and reflected by the moth when f_{bd} is 83 kHz?

SOLUTION: To find f_m, we use Eq. 18-53, with the moth being the source (of reflected waves with frequency f_m) and the bat being the detector (of the echo with frequency $f_{bd} = 83$ kHz). Since the detector moves toward the source (with speed v_b), we choose the plus sign in the numerator of Eq. 18-53. Since the source moves toward the detector (with speed v_m), we choose the minus sign in the denominator. We then have

$$f_{bd} = f_m \frac{v + v_b}{v - v_m},$$

or $\qquad 83 \text{ kHz} = f_m \dfrac{343 \text{ m/s} + 9.0 \text{ m/s}}{343 \text{ m/s} - 8.0 \text{ m/s}},$

from which

$$f_m = 78.99 \text{ kHz} \approx 79 \text{ kHz}. \qquad \text{(Answer)}$$

(b) What is the frequency f_{be} emitted by the bat when f_{bd} is 83 kHz?

SOLUTION: We again use Eq. 18-53, but now with the bat as source (at frequency f_{be}) and the moth as detector (of frequency f_m). Since the detector moves toward the source (with speed v_m), we choose the plus sign in the numerator of Eq. 18-53. Since the source moves toward the detector (with speed v_b), we choose the minus sign in the denominator. We then have

$$f_m = f_{be} \frac{v + v_m}{v - v_b}$$

or $\qquad 78.99 \text{ kHz} = f_{be} \dfrac{343 \text{ m/s} + 8.0 \text{ m/s}}{343 \text{ m/s} - 9.0 \text{ m/s}},$

from which

$$f_{be} = 75 \text{ kHz}. \qquad \text{(Answer)}$$

The bat determines the relative speed of the moth (17 m/s) from the 8 kHz (= 83 kHz − 75 kHz) it must lower its emitted frequency to hear an echo with a frequency of 83 kHz (at which it hears best). Some moths evade capture by flying away from the direction in which they hear ultrasonic waves. That choice of flight path reduces the frequency difference between what the bat emits and what it hears, and then the bat may not notice the echo. Some moths avoid capture by clicking to produce their own ultrasonic waves, thus "jamming" the detection system and confusing the bat.

18-9 THE DOPPLER EFFECT FOR LIGHT

It is tempting to apply to light waves the Doppler effect equation (Eq. 18-53) developed for sound waves in the preceding section, simply substituting c, the speed of light, for v, the speed of sound. We must avoid this temptation.

The reason is that sound waves require a medium through which to travel, but light waves do not. Thus, although the speed of sound is always measured with respect to the medium, the speed of light is not. Moreover, the speed of light always has the same value c, in all directions and in all inertial frames. For these reasons, Einstein's theory of special relativity shows that the Doppler effect for light depends only on the relative motion between the source of the light and a detector.

Even though the Doppler equations for light and for sound are different, at low enough speeds they reduce to the same approximate result. (In fact, *all* the predictions of special relativity theory reduce to their *classical* counterparts at low enough speeds.) Thus, Eq. 18-54, with v replaced by c, holds for light waves if $u \ll c$, where u is the relative speed of the source and the detector. That is, as a close approximation,

$$f' = f(1 \pm u/c) \qquad \text{(light waves; } u \ll c\text{)}. \quad (18\text{-}56)$$

If the source and the detector are *approaching* each other, we anticipate a frequency *increase,* and our rule for signs calls for the plus sign in Eq. 18-56.

In Doppler observations in astronomy (which involve light rather than sound), the wavelength is easier to measure than the frequency. This suggests that we replace f' and f in Eq. 18-56 with c/λ' and c/λ, respectively. Doing so, we find

$$\lambda' = \lambda(1 \pm u/c)^{-1} \approx \lambda(1 \mp u/c).$$

We can write this as

$$\frac{\lambda' - \lambda}{\lambda} = \mp \frac{u}{c}$$

or as

$$u = \frac{\Delta\lambda}{\lambda} c \qquad \text{(light waves; } u \ll c\text{)}, \quad (18\text{-}57)$$

in which $\Delta\lambda$ is the *magnitude* (no signs) of the Doppler wavelength shift.

Equation 18-57 gives us a way to determine the relative speed of a light source and a detector from a measured wavelength shift. If the wavelength decreases (called a *blue shift* because the blue portion of the visible spectrum has the shortest wavelengths), the frequency necessarily increases. That means the source and detector are approaching each other. If the wavelength increases (a *red shift*), the source and detector are moving away from each

other. Astronomers have measured the wavelength shifts of light reaching Earth from distant stars and galaxies. They have discovered that light from *all* the distant galaxies has undergone a red shift, indicating that all those galaxies are moving away from us. Moreover, the more distant a galaxy is, the faster it is moving.

SAMPLE PROBLEM 18-10

Figure 18-24a shows curves of intensity versus wavelength for light reaching us from interstellar gas on two opposite sides of galaxy M87 (Fig. 18-24b). One curve peaks at 499.8 nm; the other at 501.6 nm. The gas orbits the core of the galaxy at a radius $r = 100$ light-years, apparently moving toward us on one side of the core and moving away from us on the opposite side.

(a) Which curve corresponds to the gas moving toward us? What is the speed of the gas relative to us (and relative to the galaxy's core)?

SOLUTION: If the gas were not moving around the galaxy's core, the light from it would be detected at some wavelength λ (set by the emission process and the motion of the galaxy away from us). However, the motion of the gas changes the detected wavelength via the Doppler effect, increasing the wavelength for the gas moving away from us and decreasing it for the gas moving toward us. Thus, the curve peaking at 501.6 nm corresponds to motion away from us, and that peaking at 499.8 nm corresponds to motion toward us.

Let us assume that the increase and the decrease in wavelength due to the motion of the gas are equal in magnitude. Then the unshifted wavelength λ must be the average of the two shifted wavelengths:

$$\lambda = \frac{501.6 \text{ nm} + 499.8 \text{ nm}}{2} = 500.7 \text{ nm}.$$

The Doppler shift $\Delta\lambda$ of the light from the gas moving away from us is then

$$\Delta\lambda = 501.6 \text{ nm} - 500.7 \text{ nm} = 0.90 \text{ nm}.$$

Substituting this and $\lambda = 500.7$ nm into Eq. 18-57, we find that the speed of the gas is

$$u = \frac{\Delta\lambda}{\lambda} c = \frac{0.90 \text{ nm}}{501.6 \text{ nm}} 3.0 \times 10^8 \text{ m/s}$$

$$= 5.39 \times 10^5 \text{ m/s}. \qquad \text{(Answer)}$$

(b) The gas orbits the core of the galaxy because it experiences a gravitational force due to the mass M of the core. What is that mass in multiples of the Sun's mass M_S ($= 1.99 \times 10^{30}$ kg)?

SOLUTION: From Eq. 14-1, the gravitational force on an orbiting gas element of mass m at orbital radius r is

$$F = \frac{GMm}{r^2}.$$

Applying Newton's second law to the gas element and substituting the centripetal acceleration u^2/r for a, we have

$$\frac{GMm}{r^2} = ma = \frac{mu^2}{r}.$$

Solving this for M and substituting known data, we find

$$M = \frac{u^2 r}{G}$$

$$= \frac{(5.39 \times 10^5 \text{ m/s})^2 (100 \text{ ly})(9.46 \times 10^{15} \text{ m/ly})}{6.67 \times 10^{-11} \text{ N} \cdot \text{m}^2/\text{kg}^2}$$

$$= 4.12 \times 10^{39} \text{ kg} = (2.1 \times 10^9)M_S. \quad \text{(Answer)}$$

This result tells us that a mass equivalent to two billion suns has been compacted into the core of the galaxy, strongly suggesting that a "supermassive" black hole occupies the core.

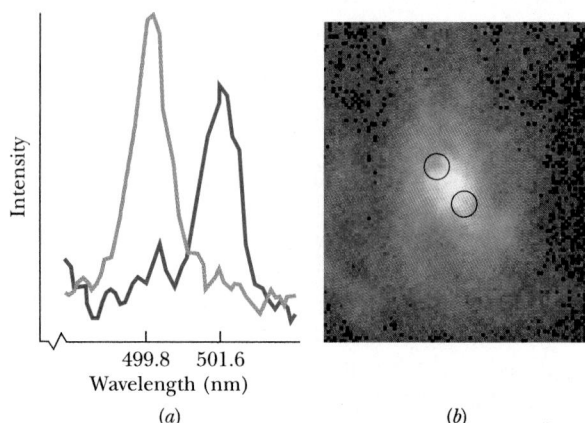

(a)　　　　　　　(b)

FIGURE 18-24 Sample Problem 18-10. (*a*) Plots of intensity versus wavelength for light emitted by gas on opposite sides of galaxy M87. (*b*) The central region of M87. The circles indicate the locations of the gas whose intensity is given in (*a*). The center of M87 is halfway between the circles.

REVIEW & SUMMARY

Sound Waves

Sound waves are longitudinal mechanical waves that can travel through solids, liquids, or gases. The speed v of a sound wave in a medium having **bulk modulus** B and density ρ is

$$v = \sqrt{\frac{B}{\rho}} \quad \text{(speed of sound).} \quad (18\text{-}3)$$

In air at 20°C, the speed of sound is 343 m/s.

A sound wave causes a longitudinal displacement s of a mass element in a medium as given by

$$s = s_m \cos(kx - \omega t), \quad (18\text{-}13)$$

where s_m is the **displacement amplitude** (maximum displacement) from equilibrium, $k = 2\pi/\lambda$, and $\omega = 2\pi f$, λ and f being the wavelength and frequency, respectively, of the sound wave. The sound wave also causes a pressure change Δp of the medium from the equilibrium pressure:

$$\Delta p = \Delta p_m \sin(kx - \omega t), \quad (18\text{-}14)$$

where the **pressure amplitude** is

$$\Delta p_m = (v\rho\omega)s_m. \quad (18\text{-}15)$$

Interference

The interference of two sound waves with identical wavelengths passing through a common point depends on their phase difference ϕ there. If the sound waves were emitted in phase and are traveling in approximately the same direction, ϕ is given by

$$\phi = \frac{\Delta L}{\lambda} 2\pi, \quad (18\text{-}21)$$

where ΔL is their **path length difference** (the difference in the distances traveled by the waves to reach the common point). Conditions for fully constructive and fully destructive interference of the waves are given by

$$\phi = m2\pi, \qquad m = 0, 1, 2, \cdots$$
$$\text{(fully constructive interference)} \quad (18\text{-}22)$$

and

$$\phi = (m + \tfrac{1}{2})2\pi, \qquad m = 0, 1, 2, \cdots$$
$$\text{(fully destructive interference).} \quad (18\text{-}23)$$

These conditions correspond to

$$\Delta L = m\lambda, \qquad m = 0, 1, 2, \cdots$$
$$\text{(fully constructive interference)} \quad (18\text{-}24)$$

and

$$\Delta L = (m + \tfrac{1}{2})\lambda, \qquad m = 0, 1, 2, \cdots$$
$$\text{(fully destructive interference).} \quad (18\text{-}25)$$

Sound Intensity

The **intensity** I of a sound wave at a surface is the average rate per unit area at which energy is transferred by the wave through or onto the surface:

$$I = \frac{P}{A}, \quad (18\text{-}26)$$

where P is the rate of energy transfer (power) of the sound wave and A is the area of the surface intercepting the sound. The intensity I is related to the displacement amplitude s_m of the sound wave by

$$I = \tfrac{1}{2}\rho v\omega^2 s_m^2. \quad (18\text{-}27)$$

The intensity at a distance r from a point source that emits sound waves of power P_s is

$$I = \frac{P_s}{4\pi r^2}. \quad (18\text{-}28)$$

Sound Level in Decibels

The *sound level* β in *decibels* (dB) is defined as

$$\beta = (10 \text{ dB}) \log \frac{I}{I_0}, \qquad (18\text{-}29)$$

where $I_0 \, (= 10^{-12} \text{ W/m}^2)$ is a reference intensity level to which all intensities are compared. For every factor-of-10 increase in intensity, 10 dB is added to the sound level.

Standing Wave Patterns in Pipes

Standing sound wave patterns can be set up in pipes. A pipe open at both ends will resonate at frequencies

$$f = \frac{v}{\lambda} = \frac{nv}{2L}, \qquad n = 1, 2, 3, \cdots$$

$$\text{(pipe, two open ends),} \quad (18\text{-}37)$$

where v is the speed of sound in the air in the pipe. For a pipe closed at one end and open at the other, the resonant frequencies are

$$f = \frac{v}{\lambda} = \frac{nv}{4L}, \qquad n = 1, 3, 5, \cdots$$

$$\text{(pipe, one open end).} \quad (18\text{-}39)$$

Beats

Beats arise when two waves having slightly different frequencies, f_1 and f_2, are detected together. The beat frequency is

$$f_{\text{beat}} = f_1 - f_2. \qquad (18\text{-}44)$$

The Doppler Effect

The *Doppler effect* is a change in the observed frequency of a wave when the source or the detector moves relative to the medium. For sound the observed frequency f' is given in terms of the source frequency f by

$$f' = f \frac{v \pm v_D}{v \mp v_S}, \qquad \begin{array}{c}\text{(general Doppler}\\ \text{effect),}\end{array} \quad (18\text{-}53)$$

where v_D is the speed of the detector relative to the medium, v_S is that of the source, and v is the speed of sound in the medium. The signs are chosen such that f' tends to be *greater* for motion (of detector or source) ''toward'' and *less* for motion ''away.''

Shock Wave

If the source speed relative to the medium exceeds the speed of sound in the medium, the Doppler equation no longer applies. In such a case, shock waves result. The half angle θ (see Fig. 18-22) of the wavefront is given by

$$\sin \theta = \frac{v}{v_S} \qquad \text{(Mach cone angle).} \quad (18\text{-}55)$$

Doppler Effect for Light

When there is a relative speed u between a light source and a detector, the detected frequency f' of the light is

$$f' = f(1 \pm u/c), \qquad (18\text{-}56)$$

where f is the frequency that would be detected were there no relative motion. The relative speed u is related to the Doppler shift in wavelength $\Delta\lambda$ by

$$u = \frac{\Delta\lambda}{\lambda} c, \qquad (18\text{-}57)$$

where λ is the wavelength that would be detected were there no relative motion. If the source and detector move toward each other, the wavelength shift $\Delta\lambda$ is negative; if they move away from each other, $\Delta\lambda$ is positive.

QUESTIONS

1. Figure 18-25 shows the paths taken by two pulses of sound that begin simultaneously and then race each other through equal distances in air. The only difference between the paths is that a region of hot (low density) air lies along path 2. Which pulse wins the race?

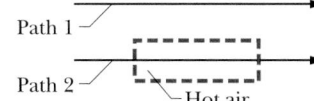

FIGURE 18-25 Question 1.

2. A sound wave of wavelength λ and displacement amplitude s_m begins to travel down a passageway (a tube, the opening of the ear, etc.). When a small device in the passageway detects this wave, it issues a second sound wave (said to be *antisound*) that is able to cancel the first wave, so that nothing is heard at the far end of the passageway. For such cancellation, what must be (a) the direction of travel, (b) the wavelength, and (c) the displacement amplitude of the second wave? (d) What must be the phase difference between the two waves? (Such antisound devices are used to eliminate unwanted sound in a noisy environment.)

3. In Fig. 18-26, two point sources S_1 and S_2, which are in phase, emit identical sound waves of wavelength 2.0 m. In terms of wavelengths, what is the phase difference between the waves arriving at point P if (a) $L_1 = 38$ m and $L_2 = 34$ m, and (b) $L_1 = 39$ m and $L_2 = 36$ m? (c) Assuming that the source separation is much smaller than L_1 and L_2, what type of interference occurs at P in situations (a) and (b), respectively?

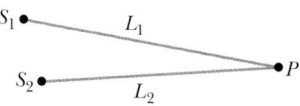

FIGURE 18-26 Question 3.

4. In Fig. 18-27, sound waves of wavelength λ are emitted by a point source S and travel to a detector D directly along path 1 and via reflection from a panel along path 2. Initially, the panel is almost along path 1 and the waves arriving at D along the two paths are almost exactly in phase. Then the panel is moved away from path 1 as shown until the waves arriving at D are exactly out of phase. What then is the path length difference Δd of the waves along the two paths?

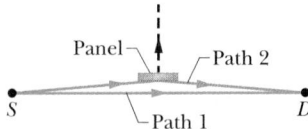

FIGURE 18-27 Question 4.

5. In Fig. 18-28, two point sources S_1 and S_2, which are in phase, emit identical sound waves of wavelength λ, and point P is at equal distances from them. Then S_2 is moved directly away from P by a distance equal to $\lambda/4$. Are the waves at P then exactly in phase, exactly out of phase, or do they have some intermediate phase relation if (a) S_1 is moved directly toward P by a distance equal to $\lambda/4$ and (b) S_1 is moved directly away from P by a distance equal to $3\lambda/4$?

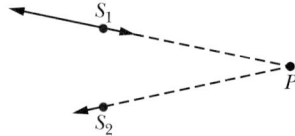

FIGURE 18-28 Question 5.

6. In Sample Problem 18-3 and Fig. 18-8a, the waves arriving at point P_1 on the perpendicular bisector are exactly in phase; that is, the waves from S_1 and S_2 always tend to move an element of air at P_1 in the same direction. Let the intersection of the perpendicular bisector and the line through S_1 and S_2 be point P_3. (a) Are the waves arriving at P_3 exactly in phase, exactly out of phase, or do they have some intermediate relation? (b) What is the answer if we increase the separation between the sources to 1.7λ?

7. Without using a calculator, determine the increase in a sound level when the intensity of the sound source becomes 10^7 times what it was previously.

8. A standing sound wave in a pipe has five nodes and five antinodes. (a) How many open ends does the pipe have? (b) What is the harmonic number n for this standing wave?

9. The sixth harmonic is set up in a pipe. (a) How many open ends does the pipe have (it has at least one)? (b) Is there a node, antinode, or some intermediate state at the midpoint?

10. (a) When an orchestra warms up, the players' warm breath increases the temperature of the air within the wind instruments (and thus decreases the density of that air). Do the resonant frequencies of those instruments increase or decrease? (b) When the slide of a slide trombone is pushed outward, do the resonant frequencies of the instrument increase or decrease?

11. Pipe A has length L and one open end. Pipe B has length $2L$ and two open ends. Which harmonics of pipe B have a frequency that matches a resonant frequency of pipe A?

12. Figure 18-29 shows a stretched string of length L and pipes a, b, c, and d of lengths L, $2L$, $L/2$, and $L/2$, respectively. The string's tension is adjusted until the speed of waves on the string equals the speed of sound waves in the air. The fundamental mode of oscillation is then set up on the string. In which pipe will the sound produced by the string cause resonance, and what oscillation mode will that sound set up?

FIGURE 18-29 Question 12.

13. You are given four tuning forks. The fork with the lowest frequency oscillates at 500 Hz. By striking two tuning forks at a time, you can produce the following beat frequencies: 1, 2, 3, 5, 7, and 8 Hz. What are the possible frequencies of the other three forks? (There are two sets of answers.)

14. A friend rides, in turn, the rims of three fast merry-go-rounds while holding a sound source that emits isotropically at a certain frequency. You stand far from each merry-go-round. The frequency you hear for each of your friend's three rides varies as the merry-go-round rotates. The variations in frequency for the three rides are given by the three curves in Fig. 18-30. Rank the curves according to (a) the linear speed v of the sound source, (b) the angular speeds ω of the merry-go-rounds, and (c) the radii r of the merry-go-rounds, greatest first.

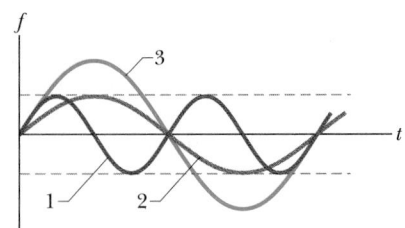

FIGURE 18-30 Question 14.

EXERCISES & PROBLEMS

Where needed in the problems, use

speed of sound in air = 343 m/s = 1125 ft/s

and

density of air = 1.21 kg/m³

unless otherwise specified.

SECTION 18-2 The Speed of Sound

1E. A rule for finding your distance in miles from a lightning flash is to count seconds from the time you see the flash until you hear the thunder and then divide the count by five. (a) Explain this rule and determine the percent error in it at 20°C, assuming that the sound travels to you along a straight path. (b) Devise a similar rule for obtaining the distance in kilometers.

2E. A column of soldiers, marching at 120 paces per minute, keep in step with the music of a band at the head of the column. It is observed that the soldiers in the rear end of the column are striding forward with the left foot when musicians are advancing with the right. What is the length of the column approximately?

3E. You are at a large outdoor concert, seated 300 m from the speaker system. The concert is also being broadcast live via satellite (at the speed of light). Consider a listener 5000 km away who receives the broadcast. Who hears the music first, you or the listener, and by what time difference?

4E. Two spectators at a soccer game in Montjuic Stadium see, and a moment later hear, the ball being kicked on the playing field. The time delay for one spectator is 0.23 s and for the other 0.12 s. Sight lines from each spectator to the player kicking the ball meet at an angle of 90°. (a) How far is each spectator from the player? (b) How far are the spectators from each other?

5E. The average density of Earth's crust 10 km beneath the continents is 2.7 g/cm³. The speed of longitudinal seismic waves at that depth, found by timing their arrival from distant earthquakes, is 5.4 km/s. Use this information to find the bulk modulus of Earth's crust at that depth. For comparison, the bulk modulus of steel is about 16×10^{10} Pa.

6E. What is the bulk modulus of oxygen at standard temperature (0°C) and pressure (1 atm) if under these conditions of temperature and pressure 1 mol (32.0 g) of oxygen occupies 22.4 L and the speed of sound in the oxygen is 317 m/s?

7P. An experimenter wishes to measure the speed of sound in an aluminum rod 10 cm long by measuring the time it takes for a sound pulse to travel the length of the rod. If results good to four significant figures are desired, how precisely must the length of the rod be known and how closely must she be able to resolve time intervals?

8P. The speed of sound in a certain metal is V. One end of a long pipe of that metal of length L is struck a hard blow. A listener at the other end hears two sounds, one from the wave that has traveled along the pipe and the other from the wave that has traveled through the air. (a) If v is the speed of sound in air, what time interval t elapses between the arrivals of the two sounds? (b) Suppose that $t = 1.00$ s and the metal is steel. Find the length L.

9P. A man strikes a long aluminum rod at one end. Another man, at the other end with his ear close to the rod, hears the sound of the blow twice (once through air and once through the rod), with a 0.120 s interval between. How long is the rod?

10P. Earthquakes generate sound waves inside Earth. Unlike a gas, Earth can experience both transverse (S) and longitudinal (P) sound waves. Typically, the speed of S waves is about 4.5 km/s and that of P waves 8.0 km/s. A seismograph records P and S waves from an earthquake. The first P waves arrive 3.0 min before the first S waves (Fig. 18-31). Assuming the waves traveled in a straight line, how far away did the earthquake occur?

11P. A stone is dropped into a well. The sound of the splash is heard 3.00 s later. What is the depth of the well?

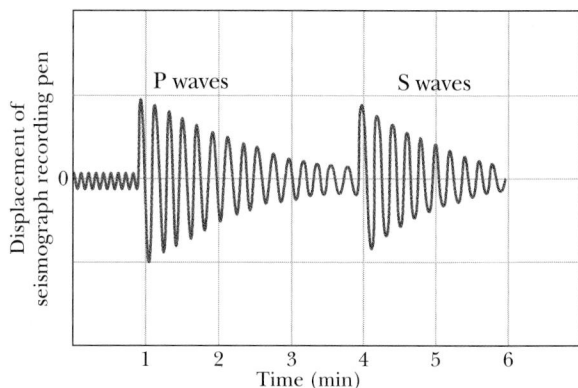

FIGURE 18-31 Problem 10.

SECTION 18-3 Traveling Sound Waves

12E. The audible frequency range for normal hearing is from about 20 Hz to 20 kHz. What are the wavelengths of sound waves at these frequencies?

13E. The shortest wavelength emitted by a bat is about 3.3 mm. What is the corresponding frequency?

14E. Diagnostic ultrasound of frequency 4.50 MHz is used to examine tumors in soft tissue. (a) What is the wavelength in air of such a sound wave? (b) If the speed of sound in tissue is 1500 m/s, what is the wavelength of this wave in tissue?

15E. (a) A conical loudspeaker has a diameter of 15.0 cm. At what frequency will the wavelength of the sound it emits in air be equal to its diameter? Be ten times its diameter? Be one-tenth its diameter? (b) Make the same calculations for a speaker of diameter 30.0 cm.

16E. Remarkably detailed images of transistors can be formed by an acoustic microscope. If the sound waves in such a microscope have a frequency of 4.2 GHz and a speed (in the liquid helium in which the specimen is immersed) of 240 m/s, what is their wavelength?

17P. (a) A continuous sinusoidal longitudinal wave is sent along a very long coiled spring from an oscillating source attached to it. The frequency of the source is 25 Hz, and the distance between successive points of maximum expansion in the spring is 24 cm. Find the wave speed. (b) If the maximum longitudinal displacement of a particle in the spring is 0.30 cm and the wave moves in the negative direction of an x axis, write the equation for the wave. Let the source be at $x = 0$ and the displacement at $x = 0$ be zero when $t = 0$.

18P. The pressure in a traveling sound wave is given by the equation

$$\Delta p = (1.5 \text{ Pa}) \sin \pi[(1.00 \text{ m}^{-1})x - (330 \text{ s}^{-1})t].$$

Find (a) the pressure amplitude, (b) the frequency, (c) the wavelength, and (d) the speed of the wave.

19P. Two sound waves, from two different sources with the same frequency, 540 Hz, travel at a speed of 330 m/s. The sources are in phase. What is the phase difference of the waves at a point that is 4.40 m from one source and 4.00 m from the other? The waves are traveling in the same direction.

20P. At a certain point in space, two waves produce pressure variations given by

$$\Delta p_1 = \Delta p_m \sin \omega t,$$

$$\Delta p_2 = \Delta p_m \sin(\omega t - \phi).$$

What is the pressure amplitude of the resultant wave at this point when $\phi = 0$, $\phi = \pi/2$, $\phi = \pi/3$, and $\phi = \pi/4$?

SECTION 18-4   Interference

21P. In Fig. 18-32, two loudspeakers, separated by a distance of 2.00 m, are in phase. Assume the amplitudes of the sound from the speakers are approximately the same at the position of a listener, who is 3.75 m directly in front of one of the speakers. (a) For what frequencies in the audible range (20–20,000 Hz) does the listener hear a minimum signal? (b) For what frequencies is the signal a maximum?

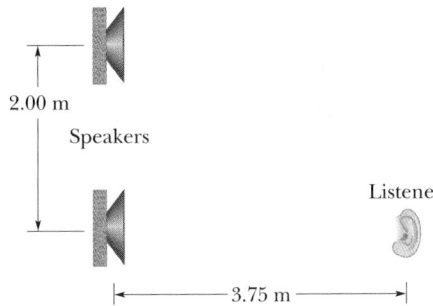

2.00 m

Speakers

Listener

|← 3.75 m →|

FIGURE 18-32   Problem 21.

22P. Two loudspeakers are located 11.0 ft apart on the stage of an auditorium. A listener is seated 60.0 ft from one and 64.0 ft from the other. A signal generator drives the two speakers in phase with the same amplitude and frequency. The transmitted frequency is swept through the audible range (20–20,000 Hz). (a) What are the three lowest frequencies at which the listener will hear a minimum signal because of destructive interference? (b) What are the three lowest frequencies at which the listener will hear a maximum signal?

23P. Two point sources of sound waves of identical wavelength λ and amplitude are separated by distance $D = 2.0\lambda$. The sources are in phase. (a) How many points of maximum signal (maximum constructive interference) lie along a large circle around the sources? (b) How many points of minimum signal (destructive interference)?

24P. A sound wave of 40.0 cm wavelength enters the tube shown in Fig. 18-33 at the source end. What must be the smallest radius r such that a minimum will be heard at the detector end?

FIGURE 18-33
Problem 24.

Source   Detector

25P. In Fig. 18-34, a point source S of sound waves lies near a reflecting wall AB. A sound detector D intercepts sound ray R_1

traveling directly from S. It also intercepts sound ray R_2 that reflects from the wall such that the *angle of incidence* θ_i is equal to the *angle of reflection* θ_r. Find two frequencies at which there is maximum constructive interference of R_1 and R_2 at D. (Reflection of sound by the wall does not alter the phase of the sound wave.)

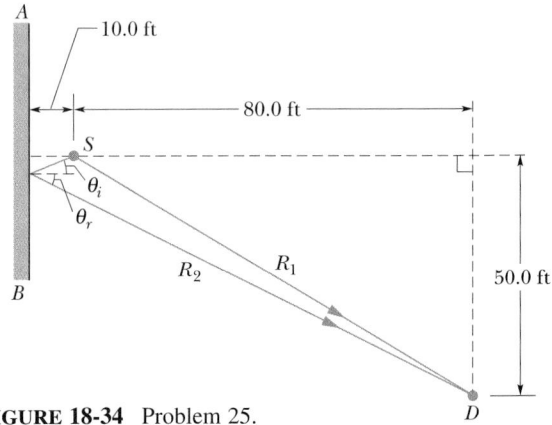

FIGURE 18-34   Problem 25.

26P*. Two point sources, separated by a distance of 5.00 m, emit sound waves at the same amplitude and frequency (300 Hz), but the sources are exactly out of phase. At what points along the line between the sources do the sound waves result in maximum oscillations of the air molecules? (*Hint:* One such point is midway between the sources.)

SECTION 18-5   Intensity and Sound Level

27E. A 1.0 W point source emits sound waves isotropically. Assuming that the energy of the waves is conserved, what is the intensity (a) 1.0 m from the source and (b) 2.5 m from the source?

28E. A source emits sound waves isotropically. The intensity of the waves 2.50 m from the source is 1.91×10^{-4} W/m². Assuming that the energy of the waves is conserved, what is the power of the source?

29E. A sound wave of frequency 300 Hz has an intensity of 1.00 μW/m². What is the amplitude of the air oscillations caused by this wave?

30E. Two sounds differ in sound level by 1.00 dB. What is the ratio of the greater intensity to the smaller intensity?

31E. A certain sound level is increased by 30 dB. By what multiple is (a) its intensity increased and (b) its pressure amplitude increased?

32E. A salesperson claimed that a stereo system had a maximum audio power of 120 W. Testing the system with several speakers set up so as to simulate a point source, the consumer noted that she could get as close as 1.2 m with the volume full on before the sound hurt her ears. Should she report the firm to the Consumer Product Safety Commission?

33E. A certain loudspeaker system emits sound isotropically with a frequency of 2000 Hz and an intensity of 0.960 mW/m² at a distance of 6.10 m. Assume that there are no reflections. (a) What is the intensity at 30.0 m? At 6.10 m, what are (b) the displacement amplitude and (c) the pressure amplitude?

34E. The source of a sound wave has a power of 1.00 μW. If it is a point source, (a) what is the intensity 3.00 m away and (b) what is the sound level in decibels at that distance?

35E. (a) If two sound waves, one in air and one in (fresh) water, are equal in intensity, what is the ratio of the pressure amplitude of the wave in water to that of the wave in air? Assume the water and the air are at 20°C. (See Table 15-1). (b) If the pressure amplitudes are equal instead, what is the ratio of the intensities of the waves?

36P. (a) Show that the intensity I of a wave is the product of the wave's energy per unit volume u and its speed v. (b) Radio waves travel at a speed of 3.00×10^8 m/s. Find u for a radio wave 480 km from a 50,000 W source, assuming the wavefronts to be spherical.

37P. Assume that a noisy freight train on a straight track emits a cylindrical, expanding sound wave. Assuming that the air absorbs no energy, how does the amplitude s_m of the wave depend on the perpendicular distance r from the source?

38P. A sound wave travels out uniformly in all directions from a point source. (a) Justify the following expression for the displacement s of the medium at any distance r from the source:

$$s = \frac{b}{r} \sin k(r - vt),$$

where b is a constant. Consider the speed, direction of propagation, periodicity, and intensity of the wave. (b) What are the dimensions of the constant b?

39P. Find the ratios of (a) the intensities, (b) the pressure amplitudes, and (c) the particle displacement amplitudes for two sounds whose sound levels differ by 37 dB.

40P. At a distance of 10 km, a 100 Hz horn, assumed to be a point source, is barely audible. At what distance would it begin to cause pain?

41P. You are standing at a distance D from a source that emits sound waves equally in all directions. You walk 50.0 m toward the source and observe that the intensity of these waves has doubled. Calculate the distance D.

42P. A point source emits 30.0 W of sound isotropically. A small microphone of effective cross-sectional area 0.750 cm² is located 200 m from the source. Calculate (a) the sound intensity there and (b) the power intercepted by the microphone.

43P. In a test, a subsonic jet flies overhead at an altitude of 100 m. The sound intensity on the ground as the jet passes overhead is 150 dB. At what altitude should the plane fly so that the ground noise is no greater than 120 dB, the threshold of pain? Ignore the finite time required for the sound to reach the ground.

44P. An audio engineer has designed a loudspeaker that is spherical and emits sound isotropically. The speaker emits 10 W of acoustic power into a room with completely absorbent walls, floor, and ceiling (an *anechoic chamber*). (a) What is the intensity (W/m²) of the sound waves 3.0 m from the center of the source? (b) How does the amplitude of the waves at 4.0 m compare with that at 3.0 m from the center of the source?

45P. Figure 18-35 shows an air-filled, acoustic interferometer, used to demonstrate the interference of sound waves. S is an oscillating diaphragm; D is a sound detector, such as the ear or a microphone. Path SBD can be varied in length, but path SAD is fixed. At D, the sound wave coming along path SBD interferes with that coming along path SAD. In one demonstration, the sound intensity at D has a minimum value of 100 units at one position of B and continuously climbs to a maximum value of 900 units when B is shifted by 1.65 cm. Find (a) the frequency of the sound emitted by the source and (b) the ratio of the amplitude at D of the SAD wave to that of the SBD wave. (c) How can it happen that these waves have different amplitudes, considering that they originate at the same source?

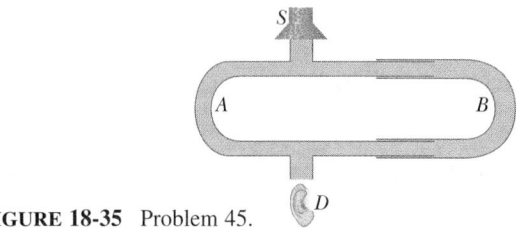

FIGURE 18-35 Problem 45.

SECTION 18-6 Sources of Musical Sound

46E. A sound wave of frequency 1000 Hz propagating through air has a pressure amplitude of 10.0 Pa. What are the (a) wavelength, (b) particle displacement amplitude, and (c) maximum particle speed? (d) An organ pipe open at both ends has this frequency as a fundamental. How long is the pipe?

47E. A sound wave in a fluid medium is reflected at a barrier so that a standing wave is formed. The distance between nodes is 3.8 cm, and the speed of propagation is 1500 m/s. Find the frequency of the sound wave.

48E. A violin string 15.0 cm long and fixed at both ends oscillates in its $n = 1$ mode. The speed of waves on the string is 250 m/s, and the speed of sound in air is 348 m/s. What are (a) the frequency and (b) the wavelength of the emitted sound wave?

49E. A violin string, oscillating in its fundamental mode, generates a sound wave with wavelength λ. By what multiple must the tension be increased if the string, still oscillating in its fundamental mode, is to generate a sound wave with wavelength $\lambda/2$?

50E. Organ pipe A, with both ends open, has a fundamental frequency of 300 Hz. The third harmonic of organ pipe B, with one end open, has the same frequency as the second harmonic of pipe A. How long are (a) pipe A and (b) pipe B?

51E. The water level in a vertical glass tube 1.00 m long can be adjusted to any position in the tube. A tuning fork vibrating at 686 Hz is held just over the open top end of the tube. At what positions of the water level will there be resonance?

52E. (a) Find the speed of waves on a violin string of mass 800 mg and length 22.0 cm if the fundamental frequency is 920 Hz. (b) What is the tension in the string? For the fundamental, what is the wavelength of (c) the waves on the string and (d) the sound waves emitted by the string?

53P. A certain violin string is 30 cm long between its fixed ends and has a mass of 2.0 g. The "open" string (no applied finger) sounds an A note (440 Hz). (a) To play a C (523 Hz), how far down the string must one place a finger? (b) What is the ratio of the wavelength of the string waves required for an A note to that required for a C note? (c) What is the ratio of the wavelength of the sound wave for an A note to that for a C note?

54P. A string on a cello has length L, for which the fundamental frequency is f. (a) By what length l must the string be shortened by fingering to change the fundamental frequency to rf? (b) What is l if L = 0.80 m and r = 1.2? (c) For r = 1.2, what is the ratio of the wavelength of the new sound wave emitted by the string to that of the wave emitted before fingering?

55P. In Fig. 18-36, S is a small loudspeaker driven by an audio oscillator and amplifier, adjustable in frequency from 1000 to 2000 Hz only. The tube D is a piece of cylindrical sheet-metal pipe 18.0 in. long and open at both ends. (a) If the speed of sound in air is 1130 ft/s at the existing temperature, at what frequencies will resonance occur in the pipe when the frequency emitted by the speaker is varied from 1000 Hz to 2000 Hz? (b) Sketch the standing waves (using the convention of Fig. 18-13b) for each resonant frequency.

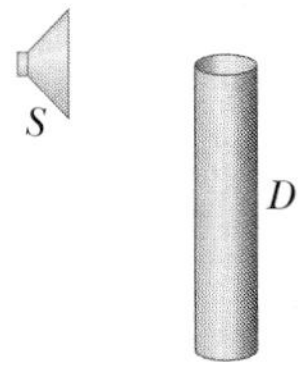

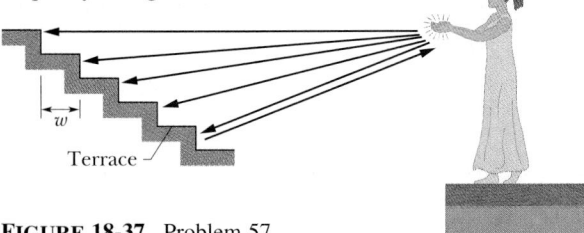

FIGURE 18-36 Problem 55.

56P. A well with vertical sides and water at the bottom resonates at 7.00 Hz and at no lower frequency. The air in the well has a density of 1.10 kg/m³ and a bulk modulus of 1.33×10^5 Pa. How deep is the well?

57P. A hand-clap on stage in an amphitheater (Fig. 18-37) sends out sound waves that scatter from terraces of width w = 0.75 m. The sound returns to the stage as a periodic series of pulses, one from each terrace; the parade of pulses sounds like a played note. (a) Assuming that all the rays are horizontal, find the frequency at which the pulses return (that is, the frequency of the perceived note). (b) If the width w of the terraces were smaller, would the frequency be higher or lower?

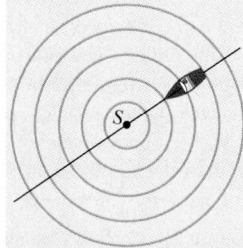

FIGURE 18-37 Problem 57.

58P. A tube 1.20 m long is closed at one end. A stretched wire is placed near the open end. The wire is 0.330 m long and has a mass of 9.60 g. It is fixed at both ends and vibrates in its fundamental mode. By resonance, it sets the air column in the tube into oscillation at that column's fundamental frequency. Find (a) that frequency and (b) the tension in the wire.

59P. The period of a pulsating variable star may be estimated by considering the star to be executing *radial* longitudinal pulsations in the fundamental standing wave mode; that is, the star's radius varies periodically with time, with a displacement antinode at the star's surface. (a) Would you expect the center of the star to be a displacement node or antinode? (b) By analogy with a pipe with one open end, show that the period of pulsation T is given by

$$T = \frac{4R}{v},$$

where R is the equilibrium radius of the star and v is the average sound speed. (c) Typical white dwarf stars are composed of material with a bulk modulus of 1.33×10^{22} Pa and a density of 10^{10} kg/m³. They have radii equal to 9.0×10^{-3} solar radius. What is the approximate pulsation period of a white dwarf?

60P. A violin string 30.0 cm long with linear density 0.650 g/m is placed near a loudspeaker that is fed by an audio oscillator of variable frequency. It is found that the string is set into oscillation only at the frequencies 880 and 1320 Hz as the frequency of the oscillator is varied over the range 500–1500 Hz. What is the tension in the string?

SECTION 18-7   Beats

61E. A tuning fork of unknown frequency makes three beats per second with a standard fork of frequency 384 Hz. The beat frequency decreases when a small piece of wax is put on a prong of the first fork. What is the frequency of this fork?

62E. The A string of a violin is a little too tightly stretched. Four beats per second are heard when the string is sounded together with a tuning fork that is oscillating accurately at concert A (440 Hz). What is the period of the violin string oscillation?

63P. Two identical piano wires have a fundamental frequency of 600 Hz when kept under the same tension. What fractional increase in the tension of one wire will lead to the occurrence of 6 beats/s when both wires oscillate simultaneously?

64P. You have five tuning forks that oscillate at different frequencies. Give (a) the maximum number and (b) the minimum number of different beat frequencies you can produce by using the forks two at a time.

SECTION 18-8   The Doppler Effect

65E. A source S generates circular waves on the surface of a lake; the pattern of wave crests is shown in Fig. 18-38. The speed

FIGURE 18-38 Exercise 65.

of the waves is 5.5 m/s, and the crest-to-crest separation is 2.3 m. You are in a small boat heading directly toward *S* at a constant speed of 3.3 m/s with respect to the shore. What frequency of the waves do you observe?

66E. Trooper *B* is chasing speeder *A* along a straight stretch of road. Both are moving at a speed of 100 mi/h. Trooper *B*, failing to catch up, sounds his siren again. Take the speed of sound in air to be 1100 ft/s and the frequency of the source to be 500 Hz. What is the Doppler shift in the frequency heard by speeder *A*?

67E. A whistle used to call a dog has a frequency of 30 kHz. The dog, however, ignores it. The owner of the dog, who cannot hear sounds above 20 kHz, wants to use the Doppler effect to make certain that the whistle is working. She asks a friend to blow the whistle from a moving car while the owner remains stationary and listens. (a) How fast would the car have to move and in what direction for the owner to hear the whistle at 20 kHz? (b) Repeat for a whistle frequency of 22 kHz instead of 30 kHz.

68E. The 16,000 Hz whine of the turbines in the jet engines of an aircraft moving with speed 200 m/s is heard at what frequency by the pilot of a second craft trying to overtake the first at a speed of 250 m/s?

69E. An ambulance with a siren emitting a whine at 1600 Hz overtakes and passes a cyclist pedaling a bike at 8.00 ft/s. After being passed, the cyclist hears a frequency of 1590 Hz. How fast is the ambulance moving?

70E. In 1845, Buys Ballot first tested the Doppler effect for sound. He put a trumpet player on a flatcar drawn by a locomotive and another player near the tracks. If each player blew a 440 Hz note and there were 4.0 beats/s as they approached each other, what was the speed of the flatcar?

71E. The speed of light in water is about three-fourths the speed of light in vacuum. A beam of high-speed electrons from a beta-tron emits Cerenkov radiation in water; the wavefront of this light forms a cone of angle 60°. Find the speed of the electrons in the water.

72E. A bullet is fired with a speed of 2200 ft/s. Find the angle made by the shock cone with the line of motion of the bullet.

73P. Two identical tuning forks can oscillate at 440 Hz. A person is located somewhere on the line between them. Calculate the beat frequency as measured by this individual if (a) she is standing still and the tuning forks both move to the right at 30.0 m/s, and (b) the tuning forks are stationary and the listener moves to the right at 30.0 m/s.

74P. A whistle of frequency 540 Hz moves in a circle of radius 2.00 ft at an angular speed of 15.0 rad/s. What are (a) the lowest and (b) the highest frequencies heard by a listener a long distance away at rest with respect to the center of the circle?

75P. A plane flies at 1.25 times the speed of sound. The sonic boom reaches a man on the ground exactly 1 min after the plane passed directly overhead. What is the altitude of the plane? Assume the speed of sound to be 330 m/s.

76P. A jet plane passes overhead at a height of 5000 m and a speed of Mach 1.5. (a) Find the Mach cone angle. (b) How long after the jet has passed directly overhead will the shock wave reach the ground? Use 331 m/s for the speed of sound.

77P. Figure 18-39 shows a transmitter and receiver of waves contained in a single instrument. It is used to measure the speed *u* of a target object (idealized as a flat plate) that is moving directly toward the unit, by analyzing the waves reflected from the target. (a) Show that the frequency f_r of the reflected waves at the receiver is related to their source frequency f_s by

$$f_r = f_s \left(\frac{v + u}{v - u} \right),$$

where *v* is the speed of the waves. (b) In a great many practical situations, $u \ll v$. In this case, show that the equation above becomes

$$\frac{f_r - f_s}{f_s} \approx \frac{2u}{v}.$$

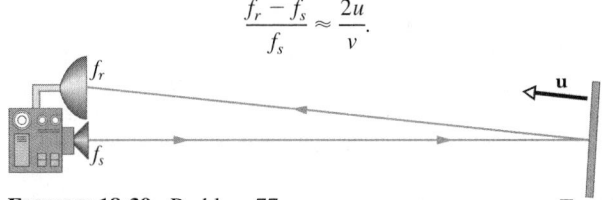

FIGURE 18-39 Problem 77.   Target

78P. A stationary motion detector sends sound waves of 0.150 MHz toward a truck approaching at a speed of 45.0 m/s. What is the frequency of the waves reflected back to the detector?

79P. A siren emitting a sound of frequency 1000 Hz moves away from you toward the face of a cliff at a speed of 10 m/s. Take the speed of sound in air as 330 m/s. (a) What is the frequency of the sound you hear coming directly from the siren? (b) What is the frequency of the sound you hear reflected off the cliff? (c) What is the beat frequency between the two sounds? Is it perceptible (it must be less than 20 Hz to be perceived)?

80P. A person on a railroad car blows a trumpet note at 440 Hz. The car is moving toward a wall at 20.0 m/s. Calculate (a) the frequency of the sound as received at the wall and (b) the frequency of the reflected sound arriving back at the source.

81P. A French submarine and a U.S. submarine move head-on during maneuvers in motionless water in the North Atlantic (Fig. 18-40). The French sub moves at 50.0 km/h, and the U.S. sub at 70.0 km/h. The French sub sends out a sonar signal (sound wave in water) at 1000 Hz. Sonar waves travel at 5470 km/h. (a) What is the signal's frequency as detected by the U.S. sub? (b) What frequency is detected by the French sub in the signal reflected back to it by the U.S. sub?

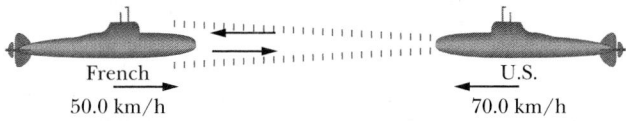

French   U.S.

50.0 km/h   70.0 km/h

FIGURE 18-40 Problem 81.

82P. A source of sound waves of frequency 1200 Hz moves to the right with a speed of 98.0 ft/s relative to the air. Ahead of it is a reflecting surface moving to the left with a speed of 216 ft/s relative to the air. Take the speed of sound in air to be 1080 ft/s and find (a) the wavelength of the sound emitted toward the reflector by the source, (b) the number of wavefronts per second arriving at the reflecting surface, (c) the speed of the reflected waves, (d) the wavelength of the reflected waves, and (e) the number of reflected wavefronts per second arriving at the source.

83P. In a discussion of Doppler shifts of ultrasonic waves used in medical diagnosis, the authors remark: "For every millimeter per second that a structure in the body moves, the frequency of the incident ultrasonic wave is shifted approximately 1.30 Hz per MHz." What speed of the ultrasonic waves in tissue do you deduce from this statement?

84P. An acoustic burglar alarm consists of a source emitting waves of frequency 28.0 kHz. What will be the beat frequency of waves reflected from an intruder walking at an average speed of 0.950 m/s directly away from the alarm?

85P. A bat is flitting about in a cave, navigating via ultrasonic bleeps. Assume that the sound emission frequency of the bat is 39,000 Hz. During one fast swoop directly toward a flat wall surface, the bat is moving at 0.025 times the speed of sound in air. What frequency does the bat hear reflected off the wall?

86P. A submarine, near the water surface, moves north at speed 75.0 km/h in a northbound current of speed 30.0 km/h, where both speeds are relative to the ocean floor. The sub emits a sonar signal (sound wave), of frequency $f = 1000$ Hz and speed 5470 km/h, that is detected by a destroyer north of the sub. What is the difference between the detected frequency and f if the destroyer (a) drifts with the current at 30.0 km/h and (b) is stationary relative to the ocean floor?

87P. A 2000 Hz siren and a civil defense official are both at rest with respect to the ground. What frequency does the official hear if the wind is blowing at 12 m/s (a) from source to observer and (b) from observer to source?

88P. Two trains are traveling toward each other at 100 ft/s relative to the ground. One train is blowing a whistle at 500 Hz. (a) What frequency will be heard on the other train in still air? (b) What frequency will be heard on the other train if the wind is blowing at 100 ft/s toward the whistle and away from the listener? (c) What frequency will be heard if the wind direction is reversed?

89P. A girl is sitting near the open window of a train that is moving at a velocity of 10.00 m/s to the east. The girl's uncle stands near the tracks and watches the train move away. The locomotive whistle emits sound at frequency 500.0 Hz. The air is still. (a) What frequency does the uncle hear? (b) What frequency does the girl hear? A wind begins to blow from the east at 10.00 m/s. (c) What frequency does the uncle now hear? (d) What frequency does the girl now hear?

SECTION 18-9   The Doppler Effect for Light

90E. Figure 18-41 is a graph of the intensity versus wavelength for light reaching Earth from galaxy NGC 7319, which is about 3×10^8 light-years away. The most intense light was emitted by the oxygen in NGC 7319. In a laboratory that emission is at wavelength $\lambda = 513$ nm, but in the light from NGC 7319 it has been shifted to 525 nm due to the Doppler effect (all the emissions from NGC 7319 have been shifted). (a) What is the speed of NGC 7319 relative to Earth? (b) Is the relative motion toward or away from our planet?

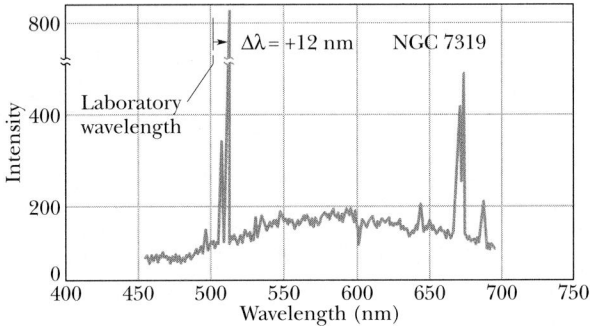

FIGURE 18-41   Exercise 90.

91E. Certain wavelengths in the light from a galaxy in the constellation Virgo are observed to be 0.4% longer than the corresponding light from terrestrial sources. What is the radial speed of this galaxy with respect to Earth? Is it approaching or receding?

92E. In the "red shift" of radiation from a distant galaxy, a certain radiation, known to have a wavelength of 434 nm when observed in the laboratory, has a wavelength of 462 nm. (a) What is the speed of the galaxy (along the line of sight) relative to Earth? (b) Is the galaxy approaching or receding?

93E. Assuming that Eq. 18-57 holds, find how fast you would have to go through a red light to have it appear green. Take 620 nm as the wavelength of red light and 540 nm as the wavelength of green light.

94P. The period of rotation of the Sun at its equator is 24.7 days; its radius is 7.00×10^5 km. What Doppler wavelength shift is expected for light with wavelength 550 nm emitted from the edge of the Sun's disk?

95P. An Earth satellite, transmitting at a frequency of 40 MHz (exactly), passes directly over a radio receiving station at an altitude of 400 km and at a speed of 3.0×10^4 km/h. Plot the change in frequency attributable to the Doppler effect as a function of time, counting $t = 0$ as the instant the satellite is over the station. (*Hint:* The speed u in the Doppler formula is not the actual speed of the satellite but its component in the direction of the station. Neglect the curvature of Earth and that of the satellite orbit.)

96P. Microwaves, which travel at the speed of light, are reflected from a distant airplane approaching the wave source. It is found that when the reflected waves are beat against the waves radiating from the source, the beat frequency is 990 Hz. If the microwaves are 0.100 m in wavelength, what is the approach speed of the airplane?

19
Temperature, Heat, and the First Law of Thermodynamics

An object with a black surface usually heats up more than one with a white surface when both are in sunlight. Such is true of the robes worn by Bedouins in the Sinai desert: black robes heat up more than white robes. Why then would a Bedouin ever wear a black robe? Wouldn't that actually decrease the chance of survival in the harsh desert environment?

19-1 THERMODYNAMICS

With this chapter, we leave the subject of mechanics and begin a new subject—*thermodynamics*. Mechanics deals with the mechanical energy of systems and is governed by Newton's laws. Thermodynamics deals with an internal energy of systems —thermal energy—and is governed by a new set of laws, which you will come to know in this and the next two chapters.

The central concept of thermodynamics is temperature. This word is so familiar that most of us—because of our built-in sense of hot and cold—tend to be overconfident in our understanding of it. Our "temperature sense" is in fact not always reliable. On a cold winter day, for example, an iron railing seems much colder to the touch than a wooden fence post, yet both are at the same temperature. This difference in our sense perception comes about because iron removes energy from our fingers more quickly than wood does. Here, we shall develop the concept of temperature from its foundations, without relying in any way on our temperature sense.

Temperature is one of the seven SI base quantities. Physicists measure temperature on the **Kelvin scale,** which is marked in units called *kelvins*. Although the temperature of a body apparently has no upper limit, it does have a lower limit; this limiting low temperature is taken as the zero of the Kelvin temperature scale. Room temperature is about 290 kelvins, or 290 K as we write it, above this

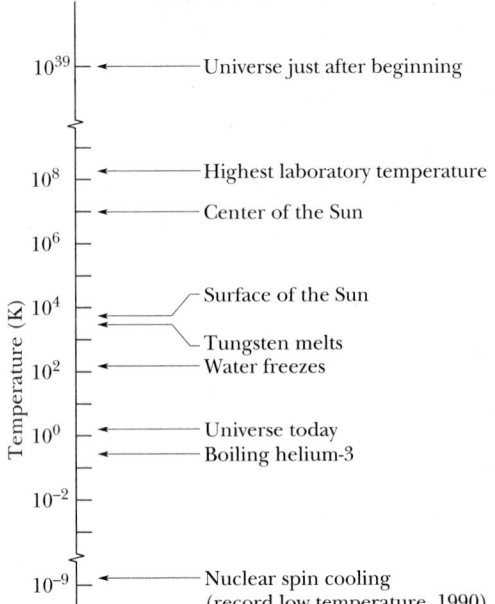

FIGURE 19-1 Some temperatures on the Kelvin scale. Temperature $T = 0$ corresponds to $10^{-\infty}$ and cannot be plotted on this logarithmic scale.

absolute zero. Figure 19-1 shows the wide range over which temperatures are determined.

When the universe began, some 10–20 billion years ago, its temperature was about 10^{39} K. As the universe expanded it cooled, and it has now reached an average temperature of about 3 K. We on Earth are a little warmer than that because we happen to live near a star. Without our Sun, we too would be at 3 K (or rather, we could not exist).

19-2 THE ZEROTH LAW OF THERMODYNAMICS

The properties of many bodies change as we alter their temperature, perhaps by moving them from a refrigerator to a warm oven. To give a few examples: as their respective temperatures increase, the volume of a liquid increases, a metal rod grows a little longer, and the electrical resistance of a wire increases, as does the pressure exerted by a confined gas. We can use any one of these properties as the basis of an instrument that will help us to pin down the concept of temperature.

Figure 19-2 shows such an instrument. Any resourceful engineer could design and construct it, using any one of the properties listed above. The instrument is fitted with a digital readout display and has the following properties: if you heat it with a Bunsen burner, the displayed number starts to increase; if you then put it into a refrigerator, the displayed number starts to decrease. The instrument is not calibrated in any way, and the numbers have (as yet) no physical meaning. The device is a *thermoscope* but not (as yet) a *thermometer.*

Suppose that, as in Fig. 19-3a, we put the thermoscope (which we shall call body T) into intimate contact with another body (body A). The entire system is confined within a thick-walled insulating box. The numbers displayed by the thermoscope roll by until, eventually, they come to rest (let us say the reading is ''137.04'') and no further change takes place. In fact, we suppose that every measurable property of body T and of body A has assumed a stable, unchanging value. Then we say that the two bodies are in *thermal equilibrium* with each other. And

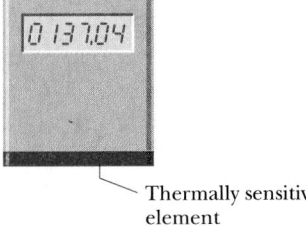

FIGURE 19-2 A thermoscope. The numbers increase when the device is heated and decrease when it is cooled. The thermally sensitive element could be —among many possibilities—a coil of wire whose electrical resistance is measured and displayed.

Thermally sensitive element

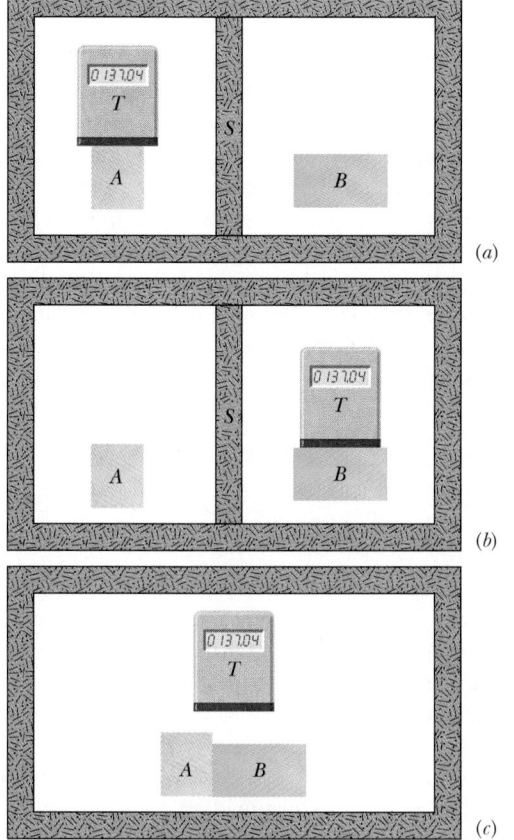

FIGURE 19-3 (*a*) Body *T* (a thermoscope) and body *A* are in thermal equilibrium. (Body *S* is a thermally insulating screen.) (*b*) Body *T* and body *B* are also in thermal equilibrium, at the same reading of the thermoscope. (*c*) If (*a*) and (*b*) are true, the zeroth law of thermodynamics states that body *A* and body *B* will also be in thermal equilibrium.

even though the displayed readings for body *T* have not been calibrated, we conclude that bodies *T* and *A* must be at the same (unknown) temperature.

Suppose that we next put body *T* in intimate contact with body *B* (Fig. 19-3*b*) and find that the two bodies come to thermal equilibrium *at the same reading of the thermoscope.* Then bodies *T* and *B* must be at the same (still unknown) temperature. If we now put bodies *A* and *B* into intimate contact (Fig. 19-3*c*), will they immediately be in thermal equilibrium with each other? Experimentally, we find that they are.

The experimental facts shown in Fig. 19-3 are summed up in the **zeroth law of thermodynamics:**

If bodies *A* and *B* are each in thermal equilibrium with a third body *T*, then they are in thermal equilibrium with each other.

In less formal language, the message of the zeroth law is: "Every body has a property called **temperature.** When

two bodies are in thermal equilibrium, their temperatures are equal." And vice versa. We can now make our thermoscope (the third body *T*) into a thermometer, confident that its readings will have physical meaning. All we have to do is calibrate it.

We use the zeroth law constantly in the laboratory. If we want to know whether the liquids in two beakers are at the same temperature, we measure the temperature of each with a thermometer. We do not need to bring the two liquids into intimate contact and observe whether they are or are not in thermal equilibrium.

The zeroth law, which has been called a logical afterthought, came to light only in the 1930s, long after the first and the second law of thermodynamics had been discovered and numbered. Because the concept of temperature is fundamental to those two laws, the law that establishes temperature as a valid concept should have the lowest number. Hence the zero.

19-3 MEASURING TEMPERATURE

Let us see how we define and measure temperatures on the Kelvin scale. Equivalently, let us see how to calibrate a thermoscope so as to make it into a usable thermometer.

The Triple Point of Water

To set up a temperature scale, we pick some reproducible thermal phenomenon and, quite arbitrarily, assign a certain Kelvin temperature to its thermal environment. That is, we select a *standard fixed point* and give it a standard fixed-point temperature. We could, for example, select the freezing point or the boiling point of water but, for various technical reasons, we do not. Instead we select the **triple point of water.**

Liquid water, solid ice, and water vapor (gaseous water) can coexist, in thermal equilibrium, at only one set of values of pressure and temperature. Figure 19-4 shows a triple-point cell, in which this so-called triple point of

FIGURE 19-4 A triple-point cell, in which solid ice, liquid water, and water vapor coexist in thermal equilibrium. By international agreement, the temperature of this mixture has been defined to be 273.16 K. The bulb of a constant-volume gas thermometer is shown inserted into the well of the cell.

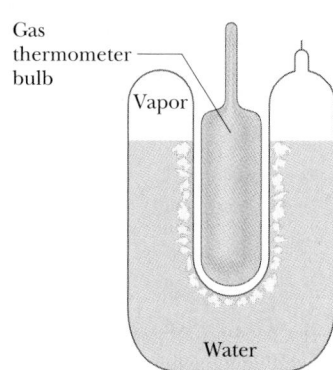

water can be achieved in the laboratory. By international agreement, the triple point of water has been assigned a value of 273.16 K as the standard fixed-point temperature for the calibration of thermometers. That is,

$$T_3 = 273.16 \text{ K} \qquad \text{(triple-point temperature),} \qquad (19\text{-}1)$$

in which the subscript 3 reminds us of the triple point. This agreement also sets the size of the kelvin as 1/273.16 of the difference between absolute zero and the triple-point temperature of water.

Note that we do not use a degree mark in reporting Kelvin temperatures. It is 300 K (not 300°K), and it is read "300 kelvins" (not "300 degrees Kelvin"). The usual SI prefixes apply. Thus 0.0035 K is 3.5 mK. No distinction in nomenclature is made between Kelvin temperatures and temperature differences. Thus we can write, "the boiling point of sulfur is 717.8 K" and "the temperature of this water bath was raised by 8.5 K."

The Constant-Volume Gas Thermometer

So far, we have not discussed the particular temperature-sensitive physical property on which, by international agreement, we shall base our thermometer. Should it be the length of a metal rod, the electrical resistance of a wire, the pressure exerted by a confined gas, or something else? The choice is important because different choices lead to different temperatures for—to give one example—the boiling point of water. For reasons that will emerge below, the standard thermometer, against which all other thermometers are to be calibrated, is based on the pressure exerted by a gas confined to a fixed volume.

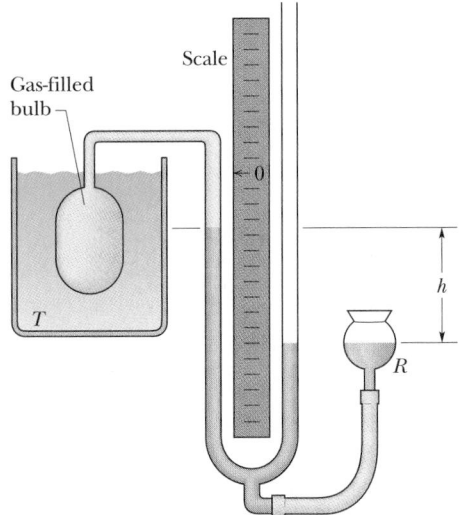

FIGURE 19-5 A constant-volume gas thermometer, its bulb immersed in a bath whose temperature T is to be measured.

Figure 19-5 shows such a **(constant-volume) gas thermometer.** It consists of a gas-filled bulb made of glass, quartz, or platinum (depending on the temperature range over which the thermometer is to be used) connected by a tube to a mercury manometer. By raising and lowering reservoir R, the mercury level on the left can always be brought to the zero of the manometer scale, thus assuring that the volume of the confined gas remains constant. The temperature of any body in thermal contact with the bulb is then defined to be

$$T = Cp, \qquad (19\text{-}2)$$

in which p is the pressure exerted by the gas and C is a constant. The pressure p is calculated from the relation

$$p = p_0 - \rho g h, \qquad (19\text{-}3)$$

in which p_0 is the atmospheric pressure, ρ is the density of the mercury in the manometer, and h is the measured level difference of mercury in the two arms of the tube.

With the bulb of the gas thermometer immersed in a triple-point cell, as in Fig. 19-4, we have

$$T_3 = Cp_3, \qquad (19\text{-}4)$$

in which p_3 is the pressure reading under this condition. Eliminating C between Eqs. 19-2 and 19-4 leads to

$$T = T_3 \left(\frac{p}{p_3} \right)$$
$$= (273.16 \text{ K}) \left(\frac{p}{p_3} \right) \qquad \text{(provisional).} \qquad (19\text{-}5)$$

Equation 19-5 is not yet our final definition of a temperature measured with a gas thermometer. We have said nothing about what gas (or how much gas) we are to place in the thermometer. If our thermometer is used to measure some temperature, such as the boiling point of water, we find that different gases in the bulb give slightly different measured temperatures. However, as we use smaller and smaller amounts of gas to fill the bulb, the readings converge nicely to a single temperature, no matter what gas we use. Figure 19-6 shows this comforting convergence.*

Thus we write, as our final recipe for measuring temperature with a gas thermometer,

$$T = (273.16 \text{ K}) \left(\lim_{m \to 0} \frac{p}{p_3} \right). \qquad (19\text{-}6)$$

---

*For pressure units, we shall use units introduced in Section 15-3. The SI unit for pressure is the newton per square meter, which is called the pascal (Pa). The pascal is related to other common pressure units by

1 atm = 1.01 × 10⁵ Pa = 760 torr = 14.7 lb/in.²

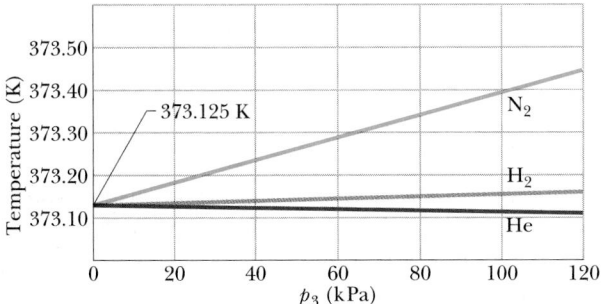

FIGURE 19-6 Temperatures calculated from Eq. 19-5 for a constant-volume gas thermometer whose bulb is immersed in boiling water. Different gases were used within the bulb, each at a variety of densities (which gave varying values for p_3). Note that all readings converge in the limit of zero density to a temperature of 373.125 K.

This instructs us to measure an unknown temperature T as follows. Fill the bulb with an arbitrary mass m of *any* gas (for example, nitrogen) and measure p_3 (using a triple-point cell) and p, the gas pressure at the temperature being measured. Calculate the ratio p/p_3. Then repeat both measurements with a smaller amount of gas in the bulb, and again calculate this ratio. Continue this way, using smaller and smaller amounts of gas, until you can extrapolate to the ratio p/p_3 that you would find if there were approximately no gas in the bulb. Calculate the temperature T by substituting that extrapolated ratio into Eq. 19-6. (The temperature measured in this way is the *ideal gas temperature*.)

If temperature is to be a truly fundamental physical quantity, one in which the laws of thermodynamics may be expressed, it is absolutely necessary that its definition be independent of the properties of specific materials. It would not do, for example, to have a quantity as basic as temperature depend on the expansivity of mercury, the electrical resistivity of platinum, or any other such "handbook" property. We chose the gas thermometer as our standard instrument precisely because no such specific properties of materials are involved in its operation. Use *any* gas—you will always get the same result.

19-4 THE CELSIUS AND FAHRENHEIT SCALES

So far, we have discussed only the Kelvin scale, used in basic scientific work. In nearly all countries of the world, the Celsius scale (formerly called the centigrade scale) is the scale of choice for popular and commercial use and much scientific use. Celsius temperatures are measured in degrees, and the Celsius degree has the same size as the kelvin. However, the zero of the Celsius scale is shifted to a more convenient value than absolute zero. If T_C represents a Celsius temperature, then

$$T_C = T - 273.15°. \qquad (19\text{-}7)$$

In expressing temperatures on the Celsius scale, the degree symbol is commonly used. Thus we write 20.00°C for a Celsius reading but 293.15 K for a Kelvin reading.

The Fahrenheit scale, used in the United States, employs a smaller degree than the Celsius scale and a different zero of temperature. You can easily verify both these differences by examining an ordinary room thermometer on which both scales are marked. The relation between the Celsius and Fahrenheit scales is

$$T_F = \tfrac{9}{5}T_C + 32°, \qquad (19\text{-}8)$$

where T_F is Fahrenheit temperature. Transferring between these two scales can be done easily by remembering a few corresponding points (such as the freezing and boiling points of water; see Table 19-1) and by making use of the fact that 9 degrees on the Fahrenheit scale equals 5 degrees on the Celsius scale. Figure 19-7 compares the Kelvin, Celsius, and Fahrenheit scales.

We use the letters C and F to distinguish measurements and degrees on the two scales. Thus,

$$0°C = 32°F$$

SAMPLE PROBLEM 19-1

The bulb of a gas thermometer is filled with nitrogen to a pressure of 120 kPa. What provisional value (see Fig. 19-6) would this thermometer yield for the boiling point of water and what is the error in this value?

SOLUTION: From Fig. 19-6, the curve for nitrogen shows that, at 120 kPa, the provisional temperature for the boiling point of water would be about 373.44 K. The actual boiling point of water (found by extrapolation on Fig. 19-6) is 373.125 K. Thus using the provisional temperature leads to an error of 0.315 K, or 315 mK.

TABLE 19-1 **SOME CORRESPONDING TEMPERATURES**

TEMPERATURE	°C	°F
Boiling point of water[a]	100	212
Normal body temperature	37.0	98.6
Accepted comfort level	20	68
Freezing point of water[a]	0	32
Zero of Fahrenheit scale	≈ −18	0
Scales coincide	−40	−40

[a]Strictly, the boiling point of water on the Celsius scale is 99.975°C, and the freezing point is 0.00°C. Thus there is slightly less than 100 C° between those two points.

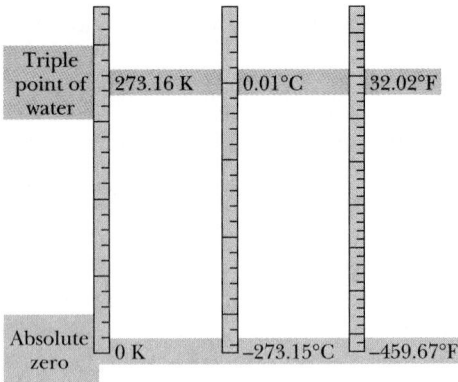

FIGURE 19-7 The Kelvin, Celsius, and Fahrenheit temperature scales compared.

means that 0° on the Celsius scale measures the same temperature as 32° on the Fahrenheit scale, whereas

$$5 \text{ C}° = 9 \text{ F}°$$

means that a temperature difference of five Celsius degrees (note the degree symbol is *after* the letter) is equivalent to a temperature difference of nine Fahrenheit degrees.

SAMPLE PROBLEM 19-2

Suppose you come across old scientific notes that describe a temperature scale called Z on which the boiling point of water is 65.0°Z and the freezing point is −14.0°Z.

(a) What temperature change ΔT on the Z scale would correspond to a change of 53.0 F°?

SOLUTION: To find a conversion factor between the two scales, we can use the boiling and freezing points of water. On the Z scale, the temperature difference between those two points is 65.0°Z − (−14.0°Z), or 79.0 Z°. On the Fahrenheit scale, it is 212°F − 32.0°F, or 180 F°. Thus a change of 79.0 Z° equals a change of 180 F°. For a change of 53.0 F°, we can now write

$$\Delta T = 53.0 \text{ F}° = 53.0 \text{ F}° \left(\frac{79.0 \text{ Z}°}{180 \text{ F}°} \right)$$

$$= 23.3 \text{ Z}°. \qquad \text{(Answer)}$$

(b) To what temperature on the Fahrenheit scale would a temperature $T = −98.0°Z$ correspond?

SOLUTION: The freezing point of water is −14.0°Z, so the difference between T and the freezing point is 84.0 Z°. To convert this difference to Fahrenheit degrees, we write

$$\Delta T = 84.0 \text{ Z}° \left(\frac{180 \text{ F}°}{79.0 \text{ Z}°} \right) = 191 \text{ F}°.$$

Thus T is 191 F° below the freezing point of water and is, on the Fahrenheit scale,

$$T = 32.0°\text{F} − 191 \text{ F}° = −159°\text{F}. \qquad \text{(Answer)}$$

CHECKPOINT **1:** The figure shows three temperature scales with the freezing and boiling points of water indicated. (a) Rank the size of a degree on these scales, greatest first. (b) Rank the following temperatures, highest first: 50°X, 50°W, and 50°Y.

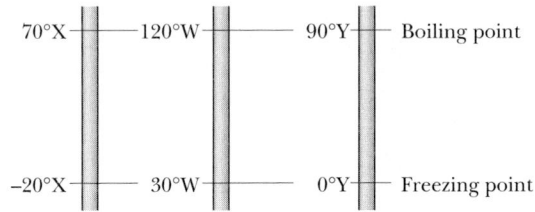

PROBLEM SOLVING TACTICS

TACTIC 1: *Temperature Changes*

Between the boiling and freezing points of water, there are (approximately) 100 kelvins and 100 Celsius degrees. Thus a kelvin is the same size as a Celsius degree. From this or from Eq. 19-7, we then know that any temperature change is the same number whether expressed in kelvins or Celsius degrees. For example, a temperature change of 10 K is equivalent to a change of 10 C°.

Between the boiling and freezing points of water, there are 180 Fahrenheit degrees. Thus 180 F° = 100 K, and a Fahrenheit degree must be 100 K/180 F°, or 5/9, the size of a kelvin or Celsius degree. From this or from Eq. 19-8, we then know that any temperature change expressed in Fahrenheit degrees must be $\frac{9}{5}$ times that same temperature change expressed in either kelvins or Celsius degrees. For example, in Fahrenheit degrees, a temperature change of 10 K is (9/5)(10 K), or 18 F°.

You should take care not to confuse a *temperature* with a temperature *change*. A temperature of 10 K is certainly not the same as one of 10°C or 18°F but, as above, a temperature *change* of 10 K is the same as one of 10 C° or 18 F°.

19-5 THERMAL EXPANSION

You can often loosen a tight metal jar lid by holding it under a stream of hot water. Both the metal of the lid and the glass of the jar expand as the hot water adds energy to their atoms. (With the added energy, the atoms can move a bit farther from each other than usual, against the spring-like interatomic forces that hold every solid together.) However, because the atoms in the metal move farther apart than those in the glass, the lid expands more than the jar and thus is loosened.

FIGURE 19-8 Railroad tracks in Asbury Park, New Jersey, distorted because of thermal expansion on a very hot July day.

TABLE 19-2 SOME COEFFICIENTS OF LINEAR EXPANSION[a]

SUBSTANCE	$\alpha\ (10^{-6}/C°)$	SUBSTANCE	$\alpha\ (10^{-6}/C°)$
Ice (at 0°C)	51	Steel	11
Lead	29	Glass (ordinary)	9
Aluminum	23	Glass (Pyrex)	3.2
Brass	19	Diamond	1.2
Copper	17	Invar[b]	0.7
Concrete	12	Fused quartz	0.5

[a]Room temperature values except for the listing for ice.

[b]This alloy was designed to have a low coefficient of expansion. The word is a shortened form of "invariable."

Such **thermal expansion** is not always desirable, as Fig. 19-8 suggests. To preclude buckling, therefore, expansion slots are placed in bridges to accommodate roadway expansion on hot days. Dental materials used for fillings must be matched in their thermal expansion properties to those of tooth enamel (otherwise consuming hot coffee or cold ice cream would be quite painful). In aircraft manufacture, however, rivets and other fasteners are often cooled in dry ice before insertion and then allowed to expand to a tight fit.

Thermometers and thermostats may be based on the differences in expansion between the components of a *bimetal strip* (Fig. 19-9). And the familiar liquid-in-glass thermometers are based on the fact that liquids such as mercury or alcohol expand to a different (greater) extent than do their glass containers.

Linear Expansion

If the temperature of a metal rod of length L is raised by an amount ΔT, its length is found to increase by an amount

$$\Delta L = L\alpha\,\Delta T, \qquad (19\text{-}9)$$

in which α is a constant called the **coefficient of linear expansion.** The coefficient α has the unit "per degree" or "per kelvin" and depends on the material. If we rewrite Eq. 19-9 as

$$\alpha = \frac{\Delta L/L}{\Delta T}, \qquad (19\text{-}10)$$

we see that α is the fractional increase in length per unit change in temperature. Although α varies somewhat with temperature, for most practical purposes it can be taken as constant for a particular material. Table 19-2 shows some coefficients of linear expansion.

The thermal expansion of a solid is like (three-dimensional) photographic enlargement. Figure 19-10*b*

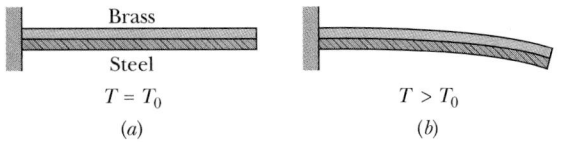

FIGURE 19-9 (*a*) A bimetal strip, consisting of a strip of brass and a strip of steel welded together, at temperature T_0. (*b*) The strip bends as shown at temperatures above this reference temperature. Below the reference temperature the strip bends the other way. Many thermostats operate on this principle, making and breaking an electrical contact as the temperature rises and falls.

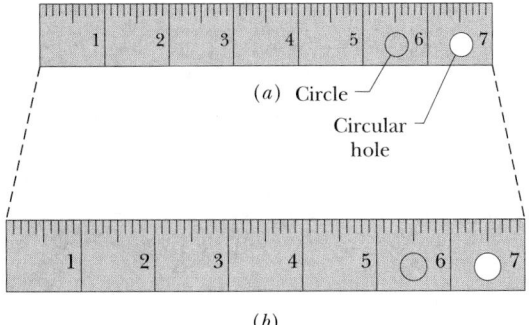

FIGURE 19-10 The same steel rule at two different temperatures. When it expands, every dimension is increased in the same proportion. The scale, the numbers, the thickness, and the diameters of the circle and circular hole are all increased by the same factor. (The expansion has been exaggerated for clarity.)

shows the (exaggerated) expansion of a steel rule after its temperature is increased from that of Fig. 19-10a. Equation 19-9 applies to every linear dimension of the rule, including its edge, thickness, diagonals, and the diameters of the circle etched on it and the circular hole cut in it. If the disk cut from that hole originally fits snugly in the hole, it will continue to fit snugly if it undergoes the same temperature increase as the rule.

Volume Expansion

If all dimensions of a solid expand with temperature, the volume of that solid must also expand. For liquids, volume expansion is the only meaningful expansion parameter. If the temperature of a solid or liquid whose volume is V is increased by an amount ΔT, the increase in volume is found to be

$$\Delta V = V\beta\,\Delta T, \qquad (19\text{-}11)$$

where β is the **coefficient of volume expansion** of the solid or liquid. The coefficients of volume expansion and linear expansion for a solid are related by

$$\beta = 3\alpha. \qquad (19\text{-}12)$$

The most common liquid, water, does not behave like other liquids. Above about 4°C, water expands as the temperature rises, as we would expect. Between 0 and about 4°C, however, water *contracts* with increasing temperature. Thus, at about 4°C, the density of water passes through a maximum. At all other temperatures, the density of water is less than this maximum value.

This behavior of water is the reason that lakes freeze from the top down rather than from the bottom up. As the water on the surface is cooled from, say, 10°C toward the freezing point, it becomes denser (''heavier'') than the lower water and sinks to the bottom. Below 4°C, however, further cooling makes the water then on the surface *less* dense (''lighter'') than the lower water, so it stays on the surface until it freezes. If lakes froze from the bottom up, the ice so formed would tend not to melt completely during the summer, being insulated by the water above. After a few years, many bodies of open water in the temperate zones of Earth would be frozen solid all year round. And aquatic life as we know it could not exist.

CHECKPOINT **2:** The figure shows four rectangular metal plates, with sides of L, $2L$, or $3L$. They are all made of the same material, and their temperature is to be increased by the same amount. Rank the plates according to the expected increase in (a) their vertical heights and (b) their areas, greatest first.

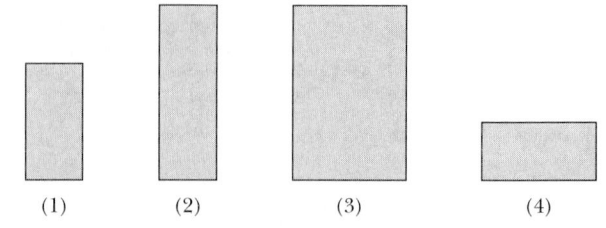

(1)   (2)   (3)   (4)

SAMPLE PROBLEM 19-3

A steel wire whose length L is 130 cm and whose diameter d is 1.1 mm is heated to an average temperature of 830°C and stretched taut between two rigid supports. What tension develops in the wire as it cools to 20°C?

SOLUTION: First let us calculate the amount by which the wire would shrink if it were allowed to cool freely. From Eq. 19-9 and Table 19-2, we can write the amount of shrinkage as

$$\Delta L = L\alpha\,\Delta T = (1.3 \text{ m})(11 \times 10^{-6}/\text{C}°)(830°\text{C} - 20°\text{C})$$
$$= 1.16 \times 10^{-2} \text{ m} = 1.16 \text{ cm}.$$

However, the wire is not permitted to shrink. We therefore calculate what force would be required to stretch the wire by this amount. From Eq. 13-34 we have

$$F = \frac{\Delta L\,EA}{L} = \frac{\Delta L\,E(\pi/4)d^2}{L},$$

in which E is Young's modulus for steel (see Table 13-1) and A is the cross-sectional area of the wire. Substitution yields

$$F = (1.16 \times 10^{-2} \text{ m})(200 \times 10^9 \text{ N/m}^2)(\pi/4)$$
$$\times \frac{(1.1 \times 10^{-3} \text{ m})^2}{1.3 \text{ m}}$$
$$= 1700 \text{ N.} \qquad \text{(Answer)}$$

Can you show that this answer is independent of the length of the wire?

Sometimes the bulging brick walls of old buildings are supported by running a steel stress rod from outside wall to outside wall, through the building. The rod is then heated, and nuts on the outside walls are tightened. As the rod cools and contracts, tension develops in it, helping to keep the walls from bulging outward.

SAMPLE PROBLEM 19-4

On a hot day in Las Vegas, an oil trucker loaded 9785 gal of diesel fuel. He encountered cold weather on the way to Payson, Utah, where the temperature was 41 F° lower than in Las Vegas, and where he delivered his entire load. How many gallons did he deliver? The coefficient of volume expansion for diesel fuel is $9.5 \times 10^{-4}/\text{C}°$, and the coefficient of linear expansion for his steel truck tank is $11 \times 10^{-6}/\text{C}°$.

SOLUTION: From Eq. 19-11,

$$\Delta V = V\beta \, \Delta T$$

$$= (9785 \text{ gal})(9.5 \times 10^{-4}/\text{C}°)(-41 \text{ F}°)\left(\frac{5 \text{ C}°}{9 \text{ F}°}\right)$$

$$= -212 \text{ gal}.$$

Thus the amount delivered was

$$V_{\text{del}} = V + \Delta V = 9785 \text{ gal} - 212 \text{ gal}$$

$$= 9573 \text{ gal} \approx 9600 \text{ gal}. \qquad \text{(Answer)}$$

Note that the thermal expansion of the steel tank has nothing to do with the problem. Question: Who paid for the "missing" diesel fuel?

PROBLEM SOLVING TACTICS

TACTIC 2: *Units for Temperature Changes*

The coefficient of linear expansion α is defined to be the fractional change in length (a dimensionless number) per unit change in temperature. In Table 19-2, that unit change in temperature is expressed in Celsius degrees (C°). Since any temperature *change* expressed in Celsius degrees is numerically the same when expressed in kelvins, the values of α in Table 19-2 could just as well be expressed in kelvins.

For example, the value of α for steel can be written as either $11 \times 10^{-6}/\text{C}°$ or $11 \times 10^{-6}/\text{K}$. This means that we could have substituted either expression for α in Sample Problem 19-3. We could have made a similar substitution for β in Sample Problem 19-4.

19-6 TEMPERATURE AND HEAT

If you take a can of cola from the refrigerator and leave it on the kitchen table, its temperature will rise—rapidly at first but then more slowly—until the temperature of the cola equals that of the room (the two are then in thermal equilibrium). In the same way, the temperature of a cup of hot coffee, left sitting on the table, will fall until it also reaches room temperature.

In generalizing this situation, we describe the cola or the coffee as a *system* (with temperature T_S) and the relevant parts of the kitchen as the *environment* (with temperature T_E) of that system. Our observation is that if T_S is not equal to T_E, then T_S will change (T_E may also change some) until the two temperatures are equal and thus thermal equilibrium is reached.

Such a change in temperature is due to the transfer of a form of energy between the system and its environment. This energy is *internal energy* (or *thermal energy*), which is the collective kinetic and potential energies associated with the random motions of the atoms, molecules, and other microscopic bodies within an object. The transferred internal energy is called **heat** and is symbolized Q. Heat is *positive* when internal energy is transferred to a system from its environment (we say that heat is absorbed). Heat is *negative* when internal energy is transferred from a system to its environment (we say that heat is released or lost).

This transfer of energy is shown in Fig. 19-11. In the situation of Fig. 19-11a, in which $T_S > T_E$, internal energy is transferred from the system to the environment, so Q is negative. In Fig. 19-11b, in which $T_S = T_E$, there is no such transfer, Q is zero, and heat is neither released nor absorbed. In Fig. 19-11c, in which $T_S < T_E$, the transfer is to the system from the environment, so Q is positive.

We are led then to this definition of heat:

> Heat is the energy that is transferred between a system and its environment because of a temperature difference that exists between them.

Recall that energy can also be transferred between a system and its environment by means of *work W*, which we always associate with a force acting on a system during a displacement of the system. Heat and work, unlike temperature, pressure, and volume, are not intrinsic properties

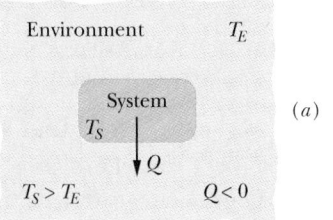

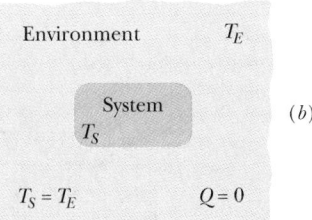

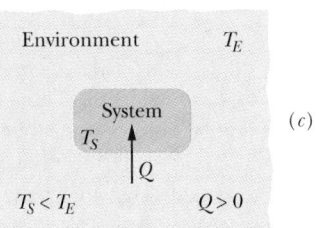

FIGURE 19-11 If the temperature of a system exceeds that of its environment as in (a), heat Q is lost by the system to the environment until thermal equilibrium (b) is established. (c) If the temperature of the system is below that of the environment, heat is absorbed by the system until thermal equilibrium is established.

of a system. They have meaning only as they describe the transfer of energy into or out of a system. Thus it is proper to say: "During the last 3 min, 15 J of heat was transferred to the system from its environment" or "During the last minute, 12 J of work was done on the system by its environment." It is meaningless to say: "This system contains 450 J of heat" or "This system contains 385 J of work."

Before scientists realized that heat is transferred energy, heat was measured in terms of its ability to raise the temperature of water. Thus the **calorie** (cal) was defined as the amount of heat that would raise the temperature of 1 g of water from 14.5°C to 15.5°C. In the British system, the corresponding unit of heat was the **British thermal unit** (Btu), defined as the amount of heat that would raise the temperature of 1 lb of water from 63°F to 64°F.

In 1948, the scientific community decided that since heat (like work) is transferred energy, the SI unit for heat should be the same as that for energy, namely, the **joule.** The calorie is now defined to be 4.1860 J (exactly) with no reference to the heating of water. (The "calorie" used in nutrition, sometimes called the Calorie (Cal), is really a kilocalorie.) The relations among the various heat units are

$$1 \text{ cal} = 3.969 \times 10^{-3} \text{ Btu} = 4.186 \text{ J}. \quad (19\text{-}13)$$

19-7 THE ABSORPTION OF HEAT BY SOLIDS AND LIQUIDS

Heat Capacity

The **heat capacity** C of an object (for example, a Pyrex coffee pot or a marble slab) is the proportionality constant between an amount of heat and the change in temperature that this heat produces in the object. Thus

$$Q = C(T_f - T_i), \quad (19\text{-}14)$$

in which T_i and T_f are the initial and final temperatures of the object. Heat capacity C has the unit of energy per degree or energy per kelvin. The heat capacity C of, say, a marble slab used in a bun warmer might be 179 cal/C°, which we can also write as 179 cal/K or as 747 J/K.

The word "capacity" in this context is really misleading in that it suggests analogy with the capacity of a bucket to hold water. *That analogy is false,* and you should not think of the object as "containing" heat or being limited in its ability to absorb heat. Heat transfer can proceed without limit as long as the necessary temperature difference is maintained. The object may, of course, melt or vaporize during the process.

Specific Heat

Two objects made of the same material, say marble, will have heat capacities proportional to their masses. It is therefore convenient to define a "heat capacity per unit mass" or **specific heat** c that refers not to an object but to a unit mass of the material of which the object is made. Equation 19-14 then becomes

$$Q = cm(T_f - T_i). \quad (19\text{-}15)$$

Through experiment we would find that although the heat capacity of the particular marble slab mentioned above is 179 cal/C° (or 747 J/K), the specific heat of marble itself (in that slab or in any other marble object) is 0.21 cal/g·C° (or 880 J/kg·K).

From the way the calorie and the British thermal unit were initially defined, the specific heat of water is

$$c = 1 \text{ cal/g} \cdot \text{C}° = 1 \text{ Btu/lb} \cdot \text{F}°$$
$$= 4190 \text{ J/kg} \cdot \text{K}. \quad (19\text{-}16)$$

Table 19-3 shows the specific heats of some substances at room temperature. Note that the value for water is relatively high. The specific heat of any substance actually depends somewhat on temperature, but the values in Table 19-3 apply reasonably well in a range of temperatures near room temperature.

TABLE 19-3 SPECIFIC HEATS OF SOME SUBSTANCES AT ROOM TEMPERATURE

SUBSTANCE	SPECIFIC HEAT cal/g·K	SPECIFIC HEAT J/kg·K	MOLAR SPECIFIC HEAT J/mol·K
Elemental Solids			
Lead	0.0305	128	26.5
Tungsten	0.0321	134	24.8
Silver	0.0564	236	25.5
Copper	0.0923	386	24.5
Aluminum	0.215	900	24.4
Other Solids			
Brass	0.092	380	
Granite	0.19	790	
Glass	0.20	840	
Ice (−10°C)	0.530	2220	
Liquids			
Mercury	0.033	140	
Ethyl alcohol	0.58	2430	
Seawater	0.93	3900	
Water	1.00	4190	

CHECKPOINT 3: A certain amount of heat Q will warm 1 g of material A by 3 C° and 1 g of material B by 4 C°. Which material has the greater specific heat?

Molar Specific Heat

In many instances the most convenient unit for specifying the amount of a substance is the mole (mol), where

$$1 \text{ mol} = 6.02 \times 10^{23} \text{ elementary units}$$

of *any* substance. Thus 1 mol of aluminum means 6.02×10^{23} atoms (the atom being the elementary unit), and 1 mol of aluminum oxide means 6.02×10^{23} molecules of the oxide (because the molecule is the elementary unit of a compound).

When quantities are expressed in moles, the specific heat must also involve moles (rather than a mass unit); it is then called a **molar specific heat.** Table 19-3 shows the values for some elemental solids (each consisting of a single element) at room temperature.

Note that the molar specific heats of all the elements listed in Table 19-3 have about the same value at room temperature, namely, 25 J/mol·K. As a matter of fact, the molar specific heats of all solids increase toward that value as the temperature increases. (Some substances, such as carbon and beryllium, do not reach this limiting value until temperatures well above room temperature. Other substances may melt or vaporize before they reach this limit.)

When we compare two substances on a molar basis, we are comparing samples that contain the same number of elementary units. The fact that, at high enough temperatures, all solid elements have the same molar specific heat tells us that atoms of all kinds—whether they be aluminum, copper, uranium, or anything else—absorb heat in the same way.

An Important Point

In determining and then using the specific heat of any substance, we need to know the conditions under which heat transfer occurs. For solids and liquids, we usually assume that the sample is under constant pressure (usually atmospheric) during the heat transfer. It is also conceivable that the sample is held at constant volume while the heat is absorbed. This means that thermal expansion of the sample is prevented by applying external pressure. For solids and liquids, this is very hard to arrange experimentally but the effect can be calculated, and it turns out that the specific heats under constant pressure and constant volume for any solid or liquid differ usually by no more than a few percent. Gases, as you will see, have quite different values for their specific heats under constant-pressure conditions and under constant-volume conditions.

Heats of Transformation

When heat is absorbed by a solid or liquid, the temperature of the sample does not necessarily rise. Instead, the sample may change from one *phase,* or *state,* (that is, solid, liquid, or gas) to another. Thus ice may melt and water may boil, absorbing heat in each case without a temperature change. In the reverse processes (water freezing, steam condensing), heat is released by the sample, again while the temperature of the sample remains constant.

The amount of heat per unit mass that must be transferred when a sample completely undergoes a phase change is called the **heat of transformation** L. So when a sample of mass m completely undergoes a phase change, the total heat transferred is

$$Q = Lm. \qquad (19\text{-}17)$$

When the phase change is from liquid to gas (then the sample must absorb heat) or from gas to liquid (then the sample must release heat), the heat of transformation is called the **heat of vaporization** L_V. For water at its normal boiling or condensation temperature,

$$L_V = 539 \text{ cal/g} = 40.7 \text{ kJ/mol}$$
$$= 2256 \text{ kJ/kg}. \qquad (19\text{-}18)$$

When the phase change is from solid to liquid (then the sample must absorb heat) or from liquid to solid (then the sample must release heat), the heat of transformation is called the **heat of fusion** L_F. For water at its normal freezing or melting temperature,

$$L_F = 79.5 \text{ cal/g} = 6.01 \text{ kJ/mol}$$
$$= 333 \text{ kJ/kg}. \qquad (19\text{-}19)$$

Table 19-4 shows the heats of transformation for some substances.

SAMPLE PROBLEM 19-5

A candy bar has a marked nutritional value of 350 Cal. How many kilowatt-hours of energy will it deliver to the body as it is digested?

SOLUTION: The Calorie in this case is a kilocalorie so that

$$\text{energy} = (350 \times 10^3 \text{ cal})(4.19 \text{ J/cal})$$
$$= (1.466 \times 10^6 \text{ J})(1 \text{ W} \cdot \text{s/J})$$
$$\times (1 \text{ h}/3600 \text{ s})(1 \text{ kW}/1000 \text{ W})$$
$$= 0.407 \text{ kW} \cdot \text{h}. \qquad \text{(Answer)}$$

This amount of energy would keep a 100 W light bulb burning

TABLE 19-4 SOME HEATS OF TRANSFORMATION

| | MELTING | | BOILING | |
SUBSTANCE	MELTING POINT (K)	HEAT OF FUSION L_F (kJ/kg)	BOILING POINT (K)	HEAT OF VAPORIZATION L_V (kJ/kg)
Hydrogen	14.0	58.0	20.3	455
Oxygen	54.8	13.9	90.2	213
Mercury	234	11.4	630	296
Water	273	333	373	2256
Lead	601	23.2	2017	858
Silver	1235	105	2323	2336
Copper	1356	207	2868	4730

for 4.1 h. To burn up this much energy by exercise, a person would have to jog about 3 or 4 mi.

A generous human diet corresponds to about 3.5 kW·h per day, which represents the absolute maximum amount of work that a human can do in one day. In an industrialized country, this amount of energy can be purchased for perhaps 35 cents.

SAMPLE PROBLEM 19-6

(a) How much heat is needed to take ice of mass $m = 720$ g at $-10°C$ to a liquid state at $15°C$?

SOLUTION: To answer, we must consider three steps. Step 1 is to raise the temperature of the ice from $-10°C$ to the melting point at $0°C$. We use Eq. 19-15, with the specific heat c_{ice} of ice given in Table 19-3. For this step, the initial temperature T_i is $-10°C$ and the final temperature T_f is $0°C$. We find

$$Q_1 = c_{ice}m(T_f - T_i)$$
$$= (2220 \text{ J/kg·K})(0.720 \text{ kg})[0°C - (-10°C)]$$
$$= 15,984 \text{ J} \approx 15.98 \text{ kJ}.$$

Step 2 is to melt the ice (there can be no change in temperature until this step is completed). We now use Eqs. 19-17 and 19-19, finding

$$Q_2 = L_Fm = (333 \text{ kJ/kg})(0.720 \text{ kg}) \approx 239.8 \text{ kJ}.$$

Step 3 is to raise the temperature of the now liquid water from $0°C$ to $15°C$. We again use Eq. 19-15, but now with the specific heat c_{liq} of liquid water given in Table 19-3. In this step the initial temperature T_i is $0°C$ and the final temperature T_f is $15°C$. We find

$$Q_3 = c_{liq}m(T_f - T_i)$$
$$= (4190 \text{ J/kg·K})(0.720 \text{ kg})(15°C - 0°C)$$
$$= 45,252 \text{ J} \approx 45.25 \text{ kJ}.$$

The total required heat Q_{tot} is the sum of the amounts required in the three steps:

$$Q_{tot} = Q_1 + Q_2 + Q_3$$
$$= 15.98 \text{ kJ} + 239.8 \text{ kJ} + 45.25 \text{ kJ}$$
$$\approx 300 \text{ kJ}. \qquad \text{(Answer)}$$

Note that the heat required to melt the ice is much larger than the heat required to raise the temperature of either the ice or the liquid water.

(b) If we supply the ice with a total heat of only 210 kJ, what then are the final state and temperature of the water?

SOLUTION: From step 1, we know that 15.98 kJ is needed to raise the temperature of the ice to the melting point. The remaining heat Q_{rem} is then 210 kJ − 15.98 kJ, or about 194 kJ. From step 2, we can see that this amount of heat is insufficient to melt all the ice. We can find the mass m of ice that is melted by the heat Q_{rem} by using Eqs. 19-17 and 19-19:

$$m = \frac{Q_{rem}}{L_F} = \frac{194 \text{ kJ}}{333 \text{ kJ/kg}} = 0.583 \text{ kg} \approx 580 \text{ g}.$$

Thus the mass of the ice that remains is 720 g − 580 g, or 140 g. Since not all the ice is melted, the temperature of the ice–liquid water must be $0°C$. Hence we have

$$580 \text{ g water} \quad \text{and} \quad 140 \text{ g ice, at } 0°C. \quad \text{(Answer)}$$

SAMPLE PROBLEM 19-7

A copper slug whose mass m_c is 75 g is heated in a laboratory oven to a temperature T of $312°C$. The slug is then dropped into a glass beaker containing a mass $m_w = 220$ g of water. The heat capacity C_b of the beaker is 45 cal/K. The initial temperature T_i of the water and the beaker is $12°C$. What is the final temperature T_f of the slug, the beaker, and the water when thermal equilibrium is reached?

SOLUTION: We take as our system *water + beaker + copper slug*. No heat enters or leaves this system, so the algebraic sum of the internal heat transfers that occur must be zero. There are three such transfers:

for the water: $Q_w = m_w c_w (T_f - T_i)$;

for the beaker: $Q_b = C_b (T_f - T_i)$;

for the copper: $Q_c = m_c c_c (T_f - T)$.

The temperature difference is written—in all three cases—as the final temperature minus the initial temperature (T_i for the water and beaker, and T for the copper). We do this even though we realize that Q_w and Q_b are positive (indicating that heat is added to the initially cool water and beaker) and that Q_c is negative (indicating that heat is released by the initially hot copper slug). Doing this allows us to write

$$Q_w + Q_b + Q_c = 0. \qquad (19\text{-}20)$$

Substituting the heat transfer expressions into Eq. 19-20 yields

$$m_w c_w (T_f - T_i) + C_b (T_f - T_i)$$
$$+ m_c c_c (T_f - T) = 0. \qquad (19\text{-}21)$$

Temperatures enter Eq. 19-21 only as differences. Thus, because the intervals on the Celsius and Kelvin scales are identical, we can use either of these scales in this equation. Solving Eq. 19-21 for T_f, we have

$$T_f = \frac{m_c c_c T + C_b T_i + m_w c_w T_i}{m_w c_w + C_b + m_c c_c}.$$

The numerator is, with Celsius temperatures,

$(75 \text{ g})(0.092 \text{ cal/g} \cdot \text{K})(312°\text{C}) + (45 \text{ cal/K})(12°\text{C})$
$\quad + (220 \text{ g})(1.00 \text{ cal/g} \cdot \text{K})(12°\text{C})$
$= 5332.8 \text{ cal},$

and the denominator is

$(220 \text{ g})(1.00 \text{ cal/g} \cdot \text{K}) + 45 \text{ cal/K}$
$\quad + (75 \text{ g})(0.092 \text{ cal/g} \cdot \text{K})$
$= 271.9 \text{ cal/C}°.$

So, we then have

$$T_f = \frac{5332.8 \text{ cal}}{271.9 \text{ cal/C}°} = 19.6°\text{C} \approx 20°\text{C}. \quad \text{(Answer)}$$

From the given data you can show that

$$Q_w \approx 1670 \text{ cal}, \quad Q_b \approx 342 \text{ cal}, \quad Q_c \approx -2020 \text{ cal}.$$

Apart from rounding errors, the algebraic sum of these three heat transfers is indeed zero, as Eq. 19-20 requires.

19-8 A CLOSER LOOK AT HEAT AND WORK

Here we look in some detail at how heat and work are exchanged between a system and its environment. Let us take as our system a gas confined to a cylinder with a movable piston, as in Fig. 19-12. The upward force on the piston due to the pressure of the confined gas is equal to the weight of lead shot loaded onto the top of the piston. The walls of the cylinder are made of insulating material that does not allow any heat transfer. The bottom of the cylinder rests on a reservoir for thermal energy, a *thermal reservoir* (perhaps a hot plate) whose temperature T you can control by turning a knob.

The system (the gas) starts from an *initial state i*, described by a pressure p_i, a volume V_i, and a temperature T_i. You want to change the system to a *final state f*, described by a pressure p_f, a volume V_f, and a temperature T_f. The procedure by which you change the system from its initial state to its final state is called a *thermodynamic process*. During such a process, heat may be transferred into the system from the thermal reservoir (positive heat) or vice versa (negative heat). And work is done by the system to raise the loaded piston (positive work) or lower it (negative work). We assume that all such changes occur slowly, with the result that the system is always in (approximate) thermal equilibrium, (that is, every part of the system is always in thermal equilibrium with every other part).

Suppose that you remove a few lead shot from the piston of Fig. 19-12, allowing the gas to push the piston

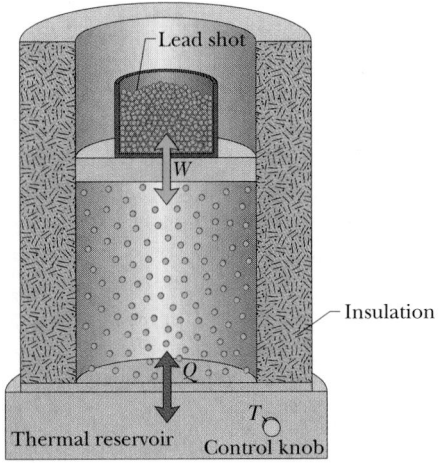

FIGURE 19-12 A gas is confined to a cylinder with a movable piston. Heat Q can be added to, or withdrawn from, the gas by regulating the temperature T of the adjustable thermal reservoir. Work W can be done by the gas by raising or lowering the piston.

and remaining shot upward through a differential displacement $d\mathbf{s}$ with an upward force $\mathbf{F}$. Since the displacement is tiny, we can assume that $\mathbf{F}$ is constant during the displacement. Then $\mathbf{F}$ has a magnitude that is equal to pA, where p is the pressure of the gas and A is the face area of the piston. The differential work dW done by the gas during the displacement is

$$dW = \mathbf{F} \cdot d\mathbf{s} = (pA)(ds) = p(A\ ds)$$
$$= p\ dV, \qquad (19\text{-}22)$$

in which dV is the differential change in the volume of the gas owing to the movement of the piston. When you have removed enough shot to allow the gas to change its volume from V_i to V_f, the total work done by the gas is

$$W = \int dW = \int_{V_i}^{V_f} p\ dV. \qquad (19\text{-}23)$$

During the change in volume, the pressure and temperature of the gas may also change. To evaluate the integral in Eq. 19-23 directly, we would need to know how pressure varies with volume for the actual process by which the system changes from state i to state f.

There are actually many ways to take the gas from state i to state f. One way is shown in Fig. 19-13a, which is a plot of the pressure of the gas versus its volume and which is called a p-V diagram. In Fig. 19-13a, the curve indicates that the pressure decreased as the volume increased. The integral of Eq. 19-23 (and thus the work W done by the gas) is represented by the shaded area under the curve between points i and f. Regardless of what exactly we did to take the gas along the curve, that work is positive, owing to the fact that the gas increased its volume by forcing the piston upward.

Another way to get from state i to state f is shown in Fig. 19-13b: there the change takes place in two steps— the first from state i to state a, and the second from state a to state f.

Step ia of this process is carried out at constant pressure, which means that you leave undisturbed the lead shot that rides on top of the piston in Fig. 19-12. You cause the volume to increase (from V_i to V_f) by slowly turning up the temperature control knob, raising the temperature of the gas to some higher value T_a. (Increasing the temperature increases the force by the gas on the piston, moving it upward.) During this process, positive work is done by the expanding gas (to lift the loaded piston) and heat is added to the system from the thermal reservoir (in response to the arbitrarily small temperature differences that you create as you turn up the temperature). This heat is positive because it is added to the system.

Step af of the process of Fig. 19-13b is carried out at

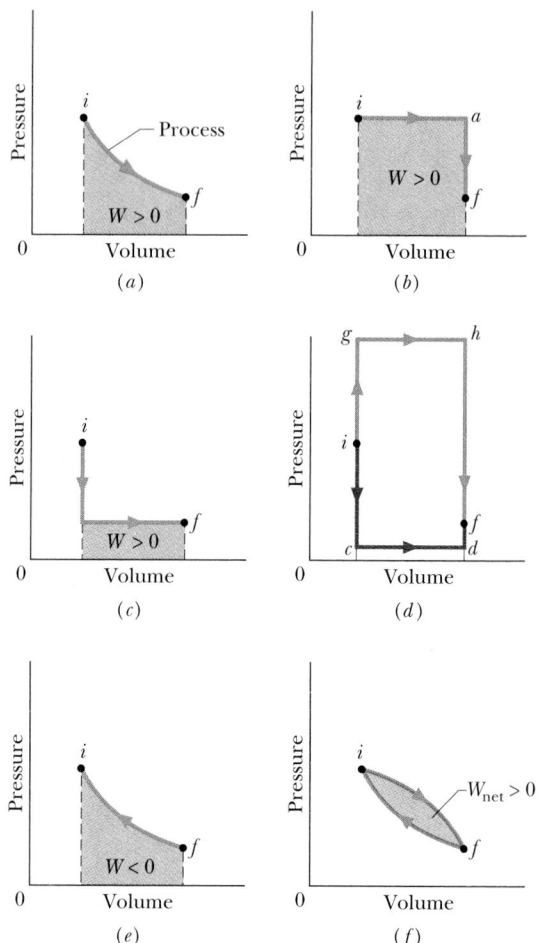

FIGURE 19-13 (a) The system of Fig. 19-12 goes from an *initial state i* to a *final state f* by means of a *thermodynamic process*. The area marked W represents the work done *by* the system during this process. The work is positive, because the process proceeds toward greater volume. (b) Another process for moving between the same two states; the work is now greater than that in (a). (c) Still another process, requiring less (positive) work. (d) The work can be made as small as you like (path $icdf$) or as large as you like (path $ighf$). (e) When the volume is reduced (by some external force), the work done *by* the system is negative. (f) The net work done by the system during a (closed) cycle is represented by the enclosed area, which is the difference in the areas beneath the two curves that make up the cycle.

constant volume, so you must wedge the piston, preventing it from moving. And then you must use the control knob to decrease the temperature so that the pressure drops from p_a to its final value p_f. During this process, heat is lost by the system to the thermal reservoir.

For the overall process iaf, the work W, which is positive and is carried out only during step ia, is represented by the shaded area under the curve. Heat is transferred during both steps ia and af, with a net heat transfer Q.

Figure 19-13c shows a process in which the two steps above are carried out in reverse order. The work W in this case is smaller than for Fig. 19-13b, as is the net heat absorbed. Figure 19-13d suggests that you can make the work done by the gas as small as you want (by following a path like $icdf$) or as large as you want (by following a path like $ighf$).

To sum up: a system can be taken from a given initial state to a given final state by an infinite number of processes. Heat may or may not be involved, and in general, the work W and the heat Q will have different values for different processes. We say that heat and work are *path-dependent* quantities.

Figure 19-13e shows an example in which negative work is done by a system as some external force compresses the system, reducing its volume. The absolute value of the work done is still equal to the area beneath the curve, but because the gas is *compressed,* the work done by the gas is negative.

Figure 19-13f shows a *thermodynamic cycle* in which the system is taken from some initial state i to some other state f and then back to i. The net work done by the system during the cycle is the sum of the positive work done during the expansion and the negative work done during the compression. In Fig. 19-13f, the net work is positive because the area under the expansion curve (i to f) is greater than the area under the compression curve (f to i).

CHECKPOINT 4: The p-V diagram shows six curved paths (connected by vertical paths) that can be followed by a gas. Which two of them should be part of a closed cycle if the net work done by the gas is to be at its maximum positive value?

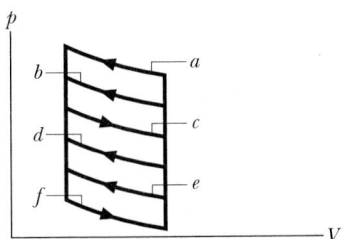

19-9 THE FIRST LAW OF THERMODYNAMICS

You have just seen that when a system changes from a given initial state to a given final state, both the work W done and the heat Q exchanged depend on the nature of the process. Experimentally, however, we find a surprising thing. *The quantity $Q - W$ is the same for all processes.* It depends only on the initial and final states and does not

depend at all on how the system gets from one to the other. All other combinations of Q and W, including Q alone, W alone, $Q + W$, and $Q - 2W$, are *path dependent;* only the quantity $Q - W$ is not.

The quantity $Q - W$ must represent a change in some intrinsic property of the system. We call this property the *internal energy* E_{int} and we write

$$\Delta E_{int} = E_{int,f} - E_{int,i}$$
$$= Q - W \quad \text{(first law).} \quad (19\text{-}24)$$

Equation 19-24 is the **first law of thermodynamics.** If the thermodynamic system undergoes only a differential change, we can write the first law as*

$$dE_{int} = dQ - dW \quad \text{(first law).} \quad (19\text{-}25)$$

The internal energy E_{int} of a system tends to increase if energy is added via heat Q and tends to decrease if energy is lost via work W done by the system.

In Chapter 8, we discussed the principle of energy conservation as it applies to isolated systems, that is, to systems in which no energy enters or leaves the system. The first law of thermodynamics is an extension of that principle to systems that are *not* isolated. In such cases, energy may be transferred into or out of the system as either work W or heat Q. In our statement of the first law of thermodynamics above, we assume that there are no changes in the kinetic energy or the potential energy of the system as a whole; that is, $\Delta K = \Delta U = 0$.

Before this chapter, the term *work* and the symbol W always meant the work done *on* a system. But starting with Eq. 19-22 and continuing through the next two chapters about thermodynamics, we focus on the work done *by* a system, such as the gas in Fig. 19-12.

The work done *on* a system is always the negative of the work done *by* the system. So if we rewrite Eq. 19-24 in terms of the work W_{on} done *on* the system, we have $\Delta E_{int} = Q + W_{on}$. This tells us the following: the internal energy of a system tends to increase if heat is absorbed by the system or if positive work is done *on* the system. Conversely, the internal energy tends to decrease if heat is lost by the system or if negative work is done *on* the system.

---

*Here dQ and dW, unlike dE_{int}, are not true differentials. That is, there are no such functions as $Q(p, V)$ and $W(p, V)$ that depend only on the state of the system. The quantities dQ and dW are called *inexact differentials* and are usually represented by the symbols đQ and đW. For our purposes, we can treat them simply as infinitesimally small energy transfers.

19-10 SOME SPECIAL CASES OF THE FIRST LAW OF THERMODYNAMICS

Here we look at four different thermodynamic processes, in each of which a certain restriction is imposed on the system. We then see what consequences follow when we apply the first law of thermodynamics to the process.

1. *Adiabatic processes.* An adiabatic process is one that occurs so rapidly or occurs in a system that is so well insulated that *no transfer of heat* occurs between the system and its environment. Putting $Q = 0$ in the first law (Eq. 19-24) then leads to

$$\Delta E_{int} = -W \qquad \text{(adiabatic process).} \qquad (19\text{-}26)$$

This tells us that if work is done *by* the system (that is, if W is positive), the internal energy of the system decreases by the amount of work. Conversely, if work is done *on* the system (that is, if W is negative), the internal energy of the system increases by that amount.

Figure 19-14 shows an idealized adiabatic process. Heat cannot enter or leave the system because of the insulation. Thus the only way energy can be transferred between the system and its environment is by work. If we remove shot from the piston and allow the gas to expand, the work done by the system (the gas) is positive and the internal energy of the gas decreases. If, instead, we add shot and compress the gas, the work done by the system is negative and the internal energy of the gas increases.

2. *Constant-volume processes.* If the volume of a system (such as a gas) is held constant, that system can do no work. Putting $W = 0$ in the first law (Eq. 19-24) yields

$$\Delta E_{int} = Q \qquad \text{(constant-volume process).} \qquad (19\text{-}27)$$

Thus if heat is added to a system (that is, if Q is positive), the internal energy of the system increases. Conversely, if heat is removed during the process (that is, if Q is negative), the internal energy of the system must decrease.

3. *Cyclical processes.* There are processes in which, after certain interchanges of heat and work, the system is restored to its initial state. In that case, no intrinsic property of the system—including its internal energy—can possibly change. Putting $\Delta E_{int} = 0$ in the first law (Eq. 19-24) yields

$$Q = W \qquad \text{(cyclical process).} \qquad (19\text{-}28)$$

Thus the net work done during the process must exactly equal the net amount of heat transferred; the store of internal energy of the system remains unchanged. Cyclical processes form a closed loop on a pressure–volume plot, as in Fig. 19-13*f*. We shall discuss such processes in some detail in Chapter 21.

4. *Free expansion.* These are adiabatic processes in which no work is done on or by the system. Thus $Q = W = 0$ and the first law requires that

$$\Delta E_{int} = 0 \qquad \text{(free expansion).} \qquad (19\text{-}29)$$

Figure 19-15 shows how such an expansion can be carried out. A gas, which is in thermal equilibrium within itself, is initially confined by a closed stopcock to one half of an insulated double chamber; the other half is evacuated. The stopcock is opened, and the gas expands freely to fill both halves of the chamber. No heat is transferred to or from the gas because of the insulation. No work is done by the gas because it rushes into a vacuum, its motion unopposed by any counteracting pressure.

A free expansion differs from all other processes we have considered because it cannot be done slowly and in a controlled way. As a result, at any given instant during the sudden expansion, the gas is not in thermal equilibrium and its pressure is not the same everywhere. So, although

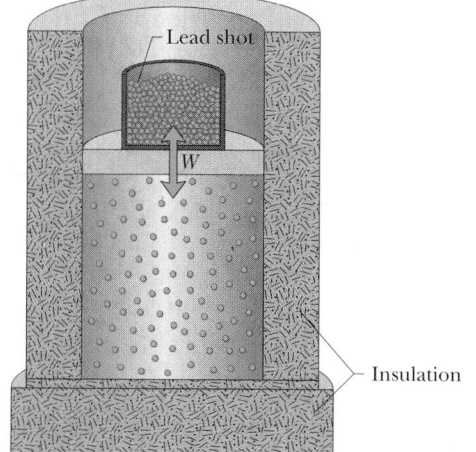

FIGURE 19-14 An adiabatic expansion can be carried out by slowly removing lead shot from the top of the piston. Adding lead shot reverses the process at any stage.

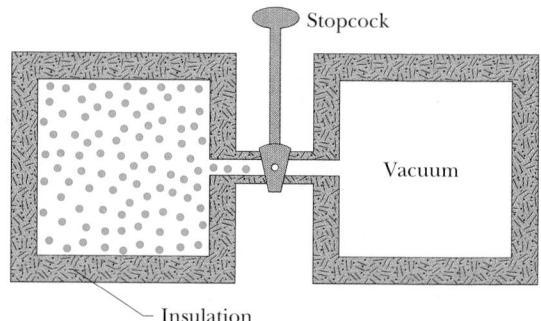

FIGURE 19-15 The initial stage of a free-expansion process. After the stopcock is opened, the gas eventually reaches an equilibrium final state, filling both chambers.

TABLE 19-5   THE FIRST LAWS OF THERMODYNAMICS: FOUR SPECIAL CASES

	The Law: $\Delta E_{int} = Q - W$ (Eq. 19-24)	
PROCESS	RESTRICTION	CONSEQUENCE
Adiabatic	$Q = 0$	$\Delta E_{int} = -W$
Constant volume	$W = 0$	$\Delta E_{int} = Q$
Closed cycle	$\Delta E_{int} = 0$	$Q = W$
Free expansion	$Q = W = 0$	$\Delta E_{int} = 0$

we can plot the initial and final states on a *p-V* diagram, we cannot plot the expansion itself.

Table 19-5 summarizes the characteristics of the processes of this section.

CHECKPOINT **5:** For one complete cycle as shown in the *p-V* diagram, are (a) ΔE_{int} for the gas and (b) the net heat transfer Q positive, negative, or zero?

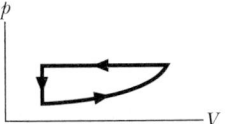

SAMPLE PROBLEM 19-8

Let 1.00 kg of liquid water at 100°C be converted to steam at 100°C by boiling at standard atmospheric pressure (1.00 atm or 1.01×10^5 Pa). The volume changes from an initial value of 1.00×10^{-3} m³ as a liquid to 1.671 m³ as steam (Fig. 19-16).

(a) How much work is done by the system this process?

SOLUTION: The work is given by Eq. 19-23. Because the pressure is constant (at 1.01×10^5 Pa) during the boiling process, we can take *p* outside the integral, obtaining

$$W = \int_{V_i}^{V_f} p\, dV = p \int_{V_i}^{V_f} dV = p(V_f - V_i)$$
$$= (1.01 \times 10^5 \text{ Pa})(1.671 \text{ m}^3 - 1.00 \times 10^{-3} \text{ m}^3)$$
$$= 1.69 \times 10^5 \text{ J} = 169 \text{ kJ}. \qquad \text{(Answer)}$$

The result is positive, indicating that work is done *by* the system on its environment in lifting the weighted piston of Fig. 19-16.

(b) How much heat must be added to the system during the process?

SOLUTION: Since there is no temperature change, but only a phase change, we use Eqs. 19-17 and 19-18:

$$Q = L_V m = (2260 \text{ kJ/kg})(1.00 \text{ kg})$$
$$= 2260 \text{ kJ}. \qquad \text{(Answer)}$$

The result is positive, which indicates that heat is *added to* the system, as we expect.

(c) What is the change in the internal energy of the system during the boiling process?

SOLUTION: We find this from the first law (Eq. 19-24):

$$\Delta E_{int} = Q - W = 2260 \text{ kJ} - 169 \text{ kJ}$$
$$\approx 2090 \text{ kJ} = 2.09 \text{ MJ}. \qquad \text{(Answer)}$$

This quantity is positive, indicating that the internal energy of the system has increased during the boiling process. This energy goes into separating the H_2O molecules, which strongly attract each other in the liquid state.

We see that, when water is boiled, about 7.5% (= 169 kJ/2260 kJ) of the added heat goes into the work of pushing back the atmosphere. The rest goes into internal energy that is added to the system.

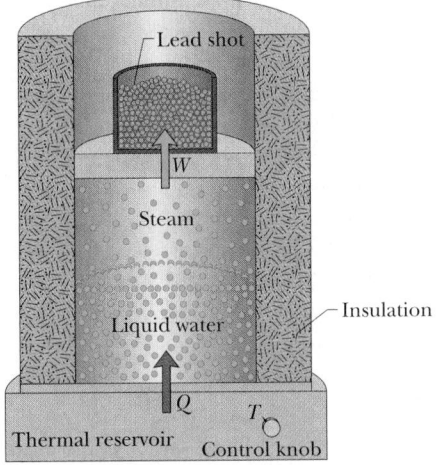

FIGURE 19-16   Sample Problem 19-8. Water boiling at constant pressure. Heat is added from the thermal reservoir until the liquid water has changed completely into steam. Work is done by the expanding gas as it lifts the loaded piston.

19-11   HEAT TRANSFER MECHANISMS

We have discussed the transfer of heat between a system and its environment, but we have not yet described how that transfer takes place. There are three transfer mechanisms: conduction, convection, and radiation.

Conduction

If you leave a metal poker in a fire for any length of time, its handle will get hot. Energy is transferred from the fire to the handle by **conduction** along the length of the poker. The vibration amplitudes of the atoms and electrons of the metal at the fire end of the poker become relatively large because of the high temperature of their environment.

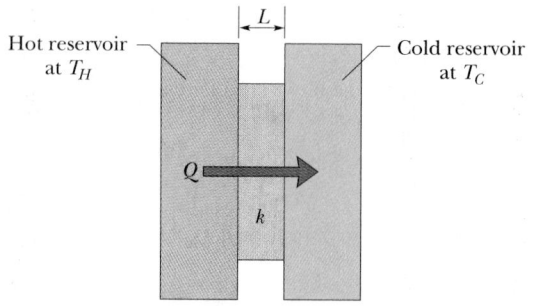

FIGURE 19-17 Thermal conduction. Heat is transferred from a reservoir at temperature T_H to a cooler reservoir at temperature T_C through a conducting slab of thickness L and thermal conductivity k.

TABLE 19-6 SOME THERMAL CONDUCTIVITIES[a]

	$k(\text{W/m}\cdot\text{K})$		$k(\text{W/m}\cdot\text{K})$
Metals		*Building Materials*	
Stainless steel	14	Polyurethane foam	0.024
Lead	35	Rock wool	0.043
Aluminum	235	Fiberglass	0.048
Copper	401	White pine	0.11
Silver	428	Window glass	1.0
Gases			
Air (dry)	0.026		
Helium	0.15		
Hydrogen	0.18		

[a]Conductivities change somewhat with temperature. The given values are at room temperature.

These increased vibrational amplitudes, and thus the associated energy, are passed along the poker, from atom to atom, during collisions between adjacent atoms. In this way, a region of rising temperature extends itself along the poker to the handle.

Consider a slab of face area A and thickness L, whose faces are maintained at temperatures T_H and T_C by a hot reservoir and a cold reservoir, as in Fig. 19-17. Let Q be the heat that is transferred through the slab, from its hot face to its cold face, in time t. Experiment shows that the rate of heat transfer H (the amount per unit time) is given by

$$H = \frac{Q}{t} = kA \frac{T_H - T_C}{L}, \qquad (19\text{-}30)$$

in which k, called the *thermal conductivity,* is a constant that depends on the material of which the slab is made. Large values of k define good heat conductors, and conversely. Table 19-6 gives thermal conductivities of some common metals, gases, and building materials.

Thermal Resistance to Conduction (*R*-Value)

If you are interested in insulating your house or in keeping cola cans cold on a picnic, you are more concerned with poor heat conductors than with good ones. For this reason, the concept of *thermal resistance R* has been introduced into engineering practice. The *R*-value of a slab of thickness L is defined as

$$R = \frac{L}{k}. \qquad (19\text{-}31)$$

Thus the lower the thermal conductivity of the material of which a slab is made, the higher the *R*-value of the slab. Note that R is a property attributed to a slab of a specified thickness, not to a material. The commonly used unit for R

(which, in the United States at least, is almost never stated) is the square foot–Fahrenheit degree-hour per British thermal unit ($\text{ft}^2 \cdot \text{F}° \cdot \text{h/Btu}$). (Now you know why the unit is rarely stated.)

Combining Eqs. 19-30 and 19-31 leads to

$$H = A \frac{T_H - T_C}{R}, \qquad (19\text{-}32)$$

which allows one to calculate the rate of heat flow through a slab if its *R*-value, its area, and the temperature difference between its faces are known.

Conduction Through a Composite Slab

Figure 19-18 shows a composite slab, consisting of two materials having different thicknesses L_1 and L_2 and different thermal conductivities k_1 and k_2. The temperatures of the outer surfaces of the slab are T_H and T_C. Each face of

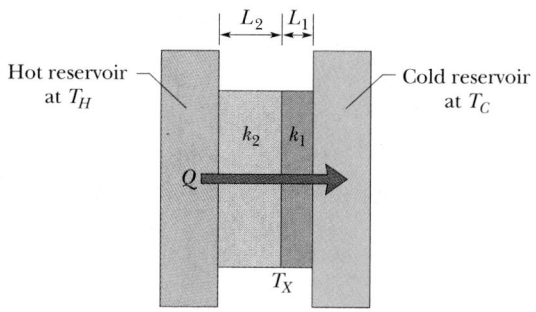

FIGURE 19-18 Heat is transferred at a steady rate through a composite slab made up of two different materials with different thicknesses and different thermal conductivities. The steady-state temperature at the interface of the two materials is T_X.

the slab has area A. Let us derive an expression for the rate of heat transfer through the slab under the assumption that the transfer is a *steady-state* process; that is, the temperatures everywhere in the slab and the rate of heat transfer do not change with time.

In the steady state, the rates of heat transfer through the two materials are equal. This is the same as saying that the heat conducted through one material in a certain time must be equal to that conducted through the other material in the same amount of time. Were this not true, temperatures in the slab would be changing and we would not have a steady-state situation. Letting T_X be the temperature of the interface between the two materials, we can now use Eq. 19-30 to write

$$H = \frac{k_2 A (T_H - T_X)}{L_2} = \frac{k_1 A (T_X - T_C)}{L_1}. \quad (19\text{-}33)$$

Solving Eq. 19-33 for T_X yields, after a little algebra,

$$T_X = \frac{k_1 L_2 T_C + k_2 L_1 T_H}{k_1 L_2 + k_2 L_1}. \quad (19\text{-}34)$$

Substituting this expression for T_X into either equality of Eq. 19-33 yields

$$H = \frac{A(T_H - T_C)}{L_1/k_1 + L_2/k_2}. \quad (19\text{-}35)$$

Equation 19-31 reminds us that $L/k = R$.

We can extend Eq. 19-35 to apply to any number n of materials making up a slab:

$$H = \frac{A(T_H - T_C)}{\Sigma (L/k)} = \frac{A(T_H - T_C)}{\Sigma R}. \quad (19\text{-}36)$$

The summation sign in the denominators tells us to add the values of L/k or R for all the materials.

CHECKPOINT **6:** The figure shows the face and interface temperatures of a composite slab consisting of four materials, of identical thicknesses, through which the heat transfer is steady. Rank the materials according to their thermal conductivities, greatest first.

25°C	15°C	10°C	−5.0°C	−10°C
	a	*b*	*c*	*d*

Convection

When you look at the flame of a candle or a match, you are watching thermal energy being transported upward by **convection.** Such heat transfer occurs when a fluid, such as

A football rally is illuminated by a fierce bonfire. Heated air and hot gases from the fire rise, and cooler air flows into the base of the fire.

air or water, is in contact with an object whose temperature is higher than that of the fluid. The temperature of the fluid that is in contact with the hot object increases, and (in most cases) the fluid expands and thus becomes less dense. Because it is now lighter than the surrounding cooler fluid, this expanded fluid rises because of buoyant forces. Some of the surrounding cooler fluid then flows so as to take the place of the rising warmer fluid and is itself warmed. The process can then continue.

Convection is part of many natural processes. Atmospheric convection plays a fundamental role in determining global climate patterns and daily weather variations. Glider pilots and birds alike seek rising thermals (currents of warm air) that keep them aloft. Huge energy transfers take place within the oceans by the same process. Finally, energy is transported to the surface of the Sun from the nuclear furnace at its core by enormous cells of convection, in which hot gas rises to the surface along the cell core and cooler gas around the core descends below the surface.

Radiation

The third method of heat transfer between an object and its environment is via electromagnetic waves (visible light is one kind of electromagnetic wave). Heat transferred in this

FIGURE 19-19 A false-color thermogram reveals the rate at which energy is radiated by houses along a street. The rates, from largest to smallest, are color coded as white, red, pink, blue, and black. You can tell where there is insulation in the walls, a heavy curtain over a window, and a higher air temperature at the ceiling on the second floor.

way is often called **thermal radiation** to distinguish it from electromagnetic *signals* (as in, say, television broadcasts) and from nuclear radiation (energy and particles emitted by nuclei). (To "radiate" generally means to emit.) When you stand in bright sunshine, you are warmed by absorbing thermal radiation from the Sun. No medium is required for heat transfer via radiation.

The rate P_r at which an object emits energy via electromagnetic radiation (that is, the power P_r of the radiating object) depends on its surface area A and the temperature T of that area in kelvins and is given by

$$P_r = \sigma \varepsilon A T^4. \qquad (19\text{-}37)$$

Here $\sigma = 5.6703 \times 10^{-8}$ W/m$^2 \cdot$K^4 is called the *Stefan–Boltzmann constant* after Josef Stefan (who discovered Eq. 19-37 experimentally in 1879) and Ludwig Boltzmann (who derived it theoretically soon after). The symbol ε represents the *emissivity* of the object's surface, which has a value between 0 and 1, depending on the composition of the surface. A surface with the maximum emissivity of 1.0 is said to be a *blackbody radiator,* but such a surface is possible only in theory. Note again that the temperature in Eq. 19-37 must be in kelvins so that a temperature of absolute zero corresponds to no radiation. Note also that every object whose temperature is above 0 K—including people—emits thermal radiation. (See Fig. 19-19.)

The rate P_a at which an object absorbs energy via thermal radiation from its environment, which we take to be at uniform temperature T_{env} (in kelvins), is

$$P_a = \sigma \varepsilon A T_{\text{env}}^4. \qquad (19\text{-}38)$$

The emissivity ε in Eq. 19-38 is the same as that in Eq. 19-37. An idealized blackbody radiator, with $\varepsilon = 1$, will absorb all the radiated energy it intercepts (rather than sending a portion back away from itself through reflection or scattering).

Because an object will radiate energy to the environment while it absorbs energy from the environment, the object's net rate P_n of energy exchange due to thermal radiation is

$$P_n = P_a - P_r$$
$$= \sigma \varepsilon A (T_{\text{env}}^4 - T^4). \qquad (19\text{-}39)$$

The emissivity of black cloth is greater than that of white cloth; hence, by Eq. 19-39 a black robe will absorb more net energy from sunshine than a white robe and thus will be at a higher temperature. In fact, research has shown that in the hot desert, a black Bedouin robe can be 6 C° higher in temperature than a similar white robe. Why, then, would someone who must avoid overheating to survive in the harsh desert wear a black robe?

FIGURE 19-20 Convection up through the hotter black robe is more vigorous than that up through the cooler white robe. (After "Why Do Bedouins Wear Black Robes in Hot Deserts?" by A. Shkolnik, C. R. Taylor, V. Finch, and A. Borut, *Nature,* Vol. 283, January 24, 1980, pp. 373–374.)

The answer is that the black robe, which is itself at a higher temperature than a comparable white robe, does indeed warm the air inside the robe. That warmer air then rises faster and leaves through the porous fabric, while external air is drawn into the robe through its open bottom (Fig. 19-20). Thus the black fabric enhances air circulation under the robe, and that keeps the Bedouin from getting any hotter than a person in a white robe. In fact, the continuous breeze blowing past his body may even make him feel more comfortable.

SAMPLE PROBLEM 19-9

A composite slab (see Fig. 19-18) whose face area A is 26 ft^2 is made up of 2.0 in. of rock wool (1.0 in. has an R-value of 3.3) and 0.75 in. of white pine (1.0 in. has an R-value of 1.3). The temperature difference between the faces of the slab is 65 F°. What is the rate of heat transfer through the slab?

SOLUTION: The R-value for the 2.0 in. of rock wool is 3.3×2.0 or 6.6 ft$^2 \cdot$°F$\cdot$h/Btu. For the 0.75 in. of wood it is 1.3×0.75 or 0.98, in the same units. The composite slab thus has an R-value of $6.6 + 0.98$ or 7.58 ft$^2 \cdot$F°$\cdot$h/Btu. Substitution into Eq. 19-36 yields

$$H = \frac{A(T_H - T_C)}{\Sigma R} = \frac{(26 \text{ ft}^2)(65 \text{ F}°)}{7.58 \text{ ft}^2 \cdot \text{F}° \cdot \text{h/Btu}}$$

$$= 223 \text{ Btu/h} \approx 220 \text{ Btu/h } (= 65 \text{ W}). \quad \text{(Answer)}$$

Thus at this temperature difference, each such insulating slab would transmit heat continuously at the rate of 65 W.

SAMPLE PROBLEM 19-10

Figure 19-21 shows the cross section of a wall made of white pine of thickness L_a and brick of thickness $L_d (= 2.0L_a)$, sandwiching two layers of unknown material with identical thicknesses and thermal conductivities. The thermal conductivity of the pine is k_a and that of the brick is $k_d (= 5.0k_a)$. The face area A of the wall is unknown. Heat conduction through the wall has reached the steady state; the only known interface temperatures are $T_1 = 25$°C, $T_2 = 20$°C, and $T_5 = -10$°C.

(a) What is interface temperature T_4?

SOLUTION: We cannot find T_4 by simply applying Eq. 19-30 layer by layer, starting with the pine and working our way rightward, because we do not know enough about the intermediate layers. However, since the heat conduction has reached the steady state, we know that the rate of conduction H_a through the pine must equal the rate of conduction H_d through the brick. From Eq. 19-30 and Fig. 19-21, we can write these rates as

$$H_a = k_a A \frac{T_1 - T_2}{L_a} \quad \text{and} \quad H_d = k_d A \frac{T_4 - T_5}{L_d}.$$

Setting $H_a = H_d$ and solving for T_4 yield

$$T_4 = \frac{k_a L_d}{k_d L_a}(T_1 - T_2) + T_5.$$

Letting $L_d = 2.0L_a$ and $k_d = 5.0k_a$, and inserting the known temperatures, we find

$$T_4 = \frac{k_a(2.0L_a)}{(5.0k_a)L_a}(25°C - 20°C) + (-10°C)$$

$$= -8.0°C. \quad \text{(Answer)}$$

(b) What is interface temperature T_3?

SOLUTION: Now that we know T_4, we can find T_3, even though we know little about the intermediate layers. (In fact, at this point you might be able to guess the answer.) Since the heat conduction process is in the steady state, the rate of conduction H_b through layer b is equal to the rate of conduction H_c through layer c. Then, from Eq. 19-30,

$$k_b A \frac{T_2 - T_3}{L_b} = k_c A \frac{T_3 - T_4}{L_c}.$$

Because the thermal conductivities k_b and k_c of the layers are equal and so are their thicknesses L_b and L_c, we have

$$T_2 - T_3 = T_3 - T_4,$$

which gives us

$$T_3 = \frac{T_2 + T_4}{2} = \frac{20°C + (-8.0°C)}{2}$$

$$= 6.0°C. \quad \text{(Answer)}$$

This shows that since the intermediate layers have identical thermal conductivities, a point midway across them has a temperature that is midway between the temperatures of their outside surfaces.

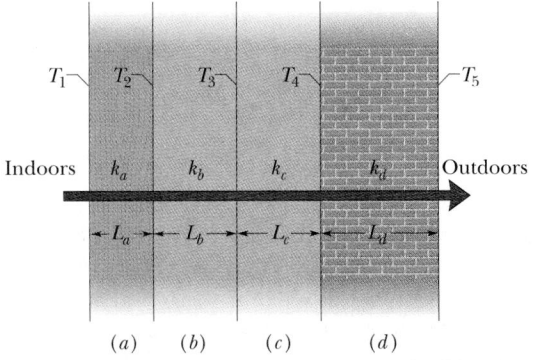

FIGURE 19-21 Sample Problem 19-10. A wall of four layers through which there is steady-state heat transfer.

SAMPLE PROBLEM 19-11

During an extended wilderness hike, you have a terrific craving for ice. Unfortunately, the air temperature drops to only 6.0°C each night—too high to freeze water. However, because a clear, moonless night sky acts like a blackbody radiator at a temperature of $T_s = -23°C$, perhaps you can make ice by letting a shallow layer of water radiate energy to such a sky. To start, you thermally insulate a container from the ground by placing a poorly conducting layer of, say, foam rubber or straw beneath it. Then you pour water into the container, forming a thin, uniform layer with mass $m = 4.5$ g, top surface area $A = 9.0$ cm^2, depth $d = 5.0$ mm, emissivity $\varepsilon = 0.90$, and initial temperature 6.0°C. Find the time required for the water to freeze via radiation. Can the freezing be accomplished during one night?

SOLUTION: If you are to make ice via thermal radiation, the radiation must first reduce the water temperature from 279 K ($= 6.0°C$) to the freezing point of 273 K. Using Eq. 19-15 and Table 19-3, we find the associated energy exchange to be

$$Q_1 = cm(T_f - T_i)$$
$$= (4190 \text{ J/kg} \cdot \text{K})(4.5 \times 10^{-3} \text{ kg})(273 \text{ K} - 279 \text{ K})$$
$$= -113 \text{ J}.$$

This energy must be radiated away by the water to drop its temperature to the freezing point.

An additional amount of energy Q_2 must be radiated away to change the phase of the water. Using Eqs. 19-17 and 19-19, and inserting a minus sign to show that the energy is lost by the water, we find

$$Q_2 = -mL_F = -(4.5 \times 10^{-3} \text{ kg})(3.33 \times 10^5 \text{ J/kg})$$
$$= -1499 \text{ J}.$$

The total amount of energy Q to be radiated away by the water is thus

$$Q = Q_1 + Q_2 = -113 \text{ J} - 1499 \text{ J} = -1612 \text{ J}.$$

The water will absorb energy from the sky while it is radiating energy to the sky. The net rate of energy exchange P_n is given by Eq. 19-39. The time t required to radiate energy Q is then

$$t = \frac{Q}{P_n} = \frac{Q}{\sigma \varepsilon A(T_s^4 - T^4)}. \qquad (19\text{-}40)$$

Although the temperature T of the water decreases slightly while the water is cooling, we can approximate T as being the freezing point, 273 K. With $T_s = 250$ K, the denominator of Eq. 19-40 is

$$(5.67 \times 10^{-8} \text{ W/m}^2 \cdot \text{K}^4)(0.90)(9.0 \times 10^{-4} \text{ m}^2)$$
$$\times [(250 \text{ K})^4 - (273 \text{ K})^4] = -7.57 \times 10^{-2} \text{ J/s},$$

and Eq. 19-40 gives us

$$t = \frac{-1612 \text{ J}}{-7.57 \times 10^{-2} \text{ J/s}} = 2.13 \times 10^4 \text{ s} = 5.9 \text{ h}. \quad \text{(Answer)}$$

Because t is less than a night, freezing water by having it radiate to the dark sky is feasible. In fact, in some parts of the world people used this technique long before the introduction of electric freezers.

REVIEW & SUMMARY

Temperature; Thermometers

Temperature is an SI base quantity related to our sense of hot and cold. It is measured with a thermometer, which contains a working substance with a measurable property, such as length or pressure, that changes in a regular way as the substance becomes hotter or colder.

Zeroth Law of Thermodynamics

When a thermometer and some other object are placed in contact with each other, they eventually reach thermal equilibrium. The reading of the thermometer is then taken to be the temperature of the other object. The process provides consistent and useful temperature measurements because of the **zeroth law of thermodynamics:** if bodies A and B are each in thermal equilibrium with a third body C (the thermometer), then A and B are in thermal equilibrium with each other.

The Kelvin Temperature Scale

Temperature is measured, in the SI system, on the **Kelvin scale.** The scale is established by defining the temperature of the *triple point* of water to be 273.16 K. Other temperatures are then defined by use of a *constant-volume gas thermometer,* in which the pressure of a constant-volume sample of a gas is proportional to its temperature. Since different gases give consistent results only at very low densities, we define the *temperature T* as measured with a gas thermometer to be

$$T = (273.16 \text{ K}) \left(\lim_{m \to 0} \frac{p}{p_3} \right). \qquad (19\text{-}6)$$

Here T is measured in kelvins, p_3 and p are the pressures of the gas at 273.16 K and the measured temperature, respectively, and m is the mass of the gas in the thermometer.

Celsius and Fahrenheit Scales

The Celsius temperature scale is defined by

$$T_C = T - 273.15°, \tag{19-7}$$

and the Fahrenheit temperature scale is defined by

$$T_F = \tfrac{9}{5}T_C + 32°. \tag{19-8}$$

Thermal Expansion

All objects change size with changes in temperature. For a temperature change ΔT, a change ΔL in any linear dimension L is given by

$$\Delta L = L\alpha\, \Delta T, \tag{19-9}$$

in which α is the **coefficient of linear expansion.** The change ΔV in the volume V of a solid or liquid is

$$\Delta V = V\beta\, \Delta T. \tag{19-11}$$

Here $\beta = 3\alpha$ is the material's **coefficient of volume expansion.**

Heat

Heat Q is energy that is transferred between a system and its environment because of a temperature difference between them. It can be measured in **joules** (J), **calories** (cal), **kilocalories** (Cal or kcal), or **British thermal units** (Btu), with

$$1 \text{ cal} = 3.969 \times 10^{-3} \text{ Btu} = 4.186 \text{ J}. \tag{19-13}$$

Heat Capacity and Specific Heat

If heat Q is added to an object, the temperature change $T_f - T_i$ is related to Q by

$$Q = C(T_f - T_i), \tag{19-14}$$

in which C is the **heat capacity** of the object. If the object has mass m, then

$$Q = cm(T_f - T_i), \tag{19-15}$$

where c is the **specific heat** of the material making up the object. The **molar specific heat** of a material is the heat capacity per mole, or per 6.02×10^{23} elementary units of the material.

Heat of Transformation

Heat supplied to a material may change the material's physical state or phase—for example, from solid to liquid or from liquid to gas. The amount of heat required per unit mass to change the phase (but not the temperature) of a particular material is its **heat of transformation** L. Thus

$$Q = Lm. \tag{19-17}$$

The **heat of vaporization** L_V is the amount of energy per unit mass that must be added to vaporize a liquid or that must be removed to condense a gas. The **heat of fusion** L_F is the amount of energy per unit mass that must be added to melt a solid or that must be removed to freeze a liquid.

Work Associated with Volume Change

A gas may exchange energy with its surroundings through work. The amount of work W done by a gas as it expands or contracts from an initial volume V_i to a final volume V_f is given by

$$W = \int dW = \int_{V_i}^{V_f} p\, dV. \tag{19-23}$$

The integration is necessary because the pressure p may vary during the volume change.

First Law of Thermodynamics

The principle of conservation of energy for a thermodynamic process is expressed in the **first law of thermodynamics**, which may assume either of the forms

$$\Delta E_{\text{int}} = E_{\text{int},f} - E_{\text{int},i} = Q - W \quad \text{(first law)} \tag{19-24}$$

or

$$dE_{\text{int}} = dQ - dW \quad \text{(first law)}. \tag{19-25}$$

E_{int} represents the internal energy of the material, which depends only on its state (temperature, pressure, and volume). Q represents the heat exchanged by the system with its surroundings; Q is positive if the system gains heat and negative if the system loses heat. W is the work done *by* the system; W is positive if the system expands against some external force exerted by the surroundings, and negative if the system contracts because of some external force. *Both Q and W are path dependent; ΔE_{int} is path independent.*

Applications of the First Law

The first law of thermodynamics finds application in several special cases:

$$
\begin{aligned}
&\textit{adiabatic processes:} && Q = 0, && \Delta E_{\text{int}} = -W \\
&\textit{constant-volume processes:} && W = 0, && \Delta E_{\text{int}} = Q \\
&\textit{cyclical processes:} && \Delta E_{\text{int}} = 0, && Q = W \\
&\textit{free expansion processes:} && Q = W = \Delta E_{\text{int}} = 0
\end{aligned}
$$

Conduction, Convection, and Radiation

The rate H at which heat is *conducted* through a slab whose faces are maintained at temperatures T_H and T_C is

$$H = \frac{Q}{t} = kA\frac{T_H - T_C}{L}, \tag{19-30}$$

in which A and L are the face area and length of the slab, and k is the thermal conductivity of the material.

Convection occurs when temperature differences cause a heat transfer by motion within a fluid. *Radiation* is heat transfer via the emission of electromagnetic energy. The rate P_r at which an object emits energy via thermal radiation is

$$P_r = \sigma\varepsilon A T^4, \tag{19-37}$$

where $\sigma\,(= 5.6703 \times 10^{-8} \text{ W/m}^2\cdot\text{K}^4)$ is the Stefan–Boltzmann constant, ε is the emissivity of the object's surface, A is its surface area, and T is its surface temperature (in kelvins). The rate P_a at which an object absorbs energy via thermal radiation from its environment, which is at the uniform temperature T_{env} (in kelvins), is

$$P_a = \sigma\varepsilon A T_{\text{env}}^4. \tag{19-38}$$

QUESTIONS

1. Figure 19-22 shows three temperature scales with the freezing and boiling points of water indicated. Rank a change of 25 R°, 25 S°, and 25 U° according to the corresponding change in temperature, greatest first.

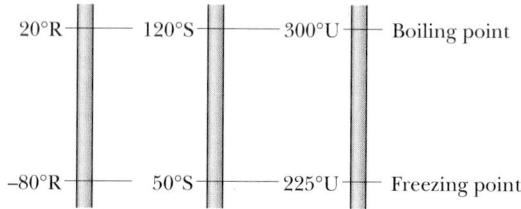

FIGURE 19-22 Question 1.

2. A rod, initially at room temperature, is to be heated and cooled in six steps. The changes in its length, expressed in terms of some basic unit, will be +7, +5, +3, −4, −6, and −4. (a) Is the final temperature of the rod equal to room temperature, above it, or below it? (b) Is there a sequence of the steps such that the rod is at room temperature after one of the intermediate steps?

3. The table gives the initial length L, change in temperature ΔT, and change in length ΔL of four rods. Rank the rods according to their coefficients of thermal expansion, greatest first.

Rod	L (m)	ΔT (C°)	ΔL (m)
a	2	10	4×10^{-4}
b	1	20	4×10^{-4}
c	2	10	8×10^{-4}
d	4	5	4×10^{-4}

4. Rank the Celsius, Kelvin, and Fahrenheit temperature scales according to the heat required to raise the temperature of one gram of water by one degree on each scale, greatest first.

5. Materials A, B, and C are solids that are at their melting temperatures. Material A requires 200 J to melt 4 kg. Material B requires 300 J to melt 5 kg. And material C requires 300 J to melt 6 kg. Rank the materials according to their heats of fusion, greatest first.

6. Figure 19-23 shows four paths on a p-V diagram along which a gas can be taken from state i to state f. Rank the paths according to (a) the change ΔE_{int}, (b) the work W done by the gas, and (c) the magnitude of the heat transfer Q, greatest first.

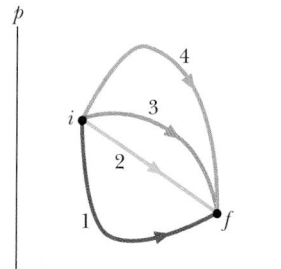

FIGURE 19-23
Question 6.

7. Figure 19-24 shows two closed cycles on p-V diagrams for a gas. The three parts of cycle 1 are of the same length and shape as those of cycle 2. For each cycle, should the cycle be traversed clockwise or counterclockwise if (a) the net work W done by the gas is to be positive and (b) the net heat transfer Q between the gas and the environment is to be positive?

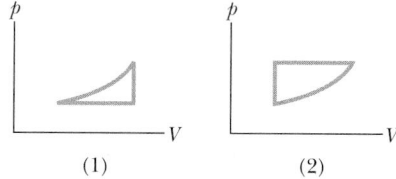

(1) (2)

FIGURE 19-24 Questions 7 and 8.

8. For which cycle in Fig. 19-24, traversed clockwise, is (a) W greater and (b) Q greater?

9. Figure 19-25 shows a composite slab of three different materials, a, b, and c, with identical thicknesses and with thermal conductivities $k_b > k_a > k_c$. The heat transfer through them is nonzero and steady. Rank the materials according to the temperature difference ΔT across them, greatest first.

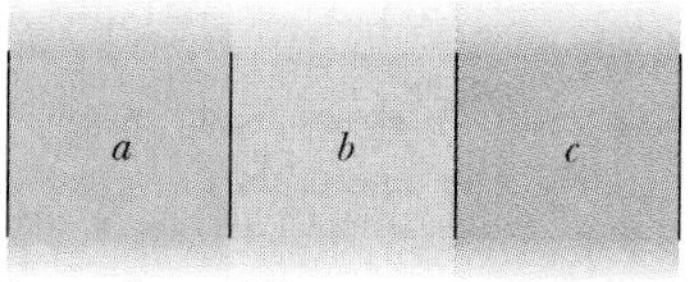

FIGURE 19-25
Question 9.

10. Figure 19-26 shows three different arrangements of materials 1, 2, and 3 to form a wall. The thermal conductivities are $k_1 > k_2 > k_3$. The left side of the wall is 20 C° higher than the right side. Rank the arrangements according to (a) the (steady state) rate of energy transfer through the wall and (b) the temperature difference across material 1, greatest first.

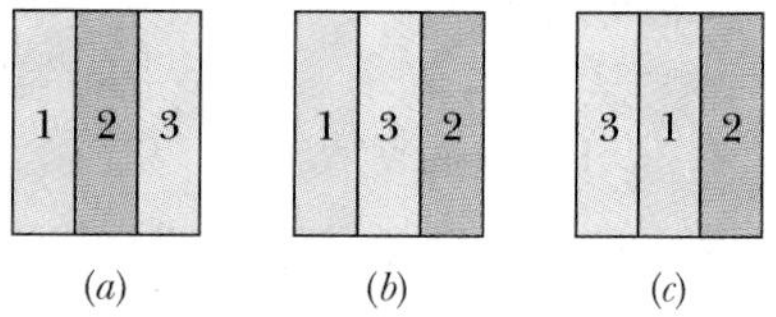

FIGURE 19-26
Question 10. (a) (b) (c)

11. During an icicle's growth, its outer surface is covered with a thin sheath of liquid water that slowly seeps downward to form drops, one at a time, that hang at the tip (Fig. 19-27). Each drop straddles a thin tube of liquid water that extends up into the icicle, toward its *root* (its top). As the water at the top of the tube gradually freezes, energy is released. Is that energy conducted radially

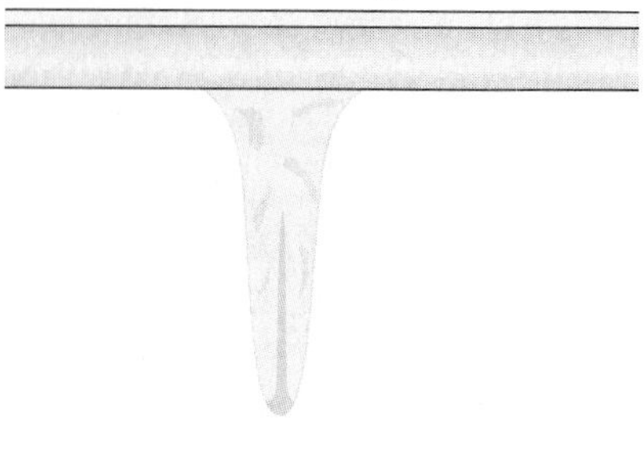

FIGURE 19-27
Question 11.

outward through the ice, downward through the water to the hanging drop, or upward toward the root? (Assume that the air temperature is below 0°C.)

12. Figure 19-28 shows a horizontal cross section (top view) of a square room surrounded on four sides by thick walls. The walls are all made of the same material and all have the same face area. They have thicknesses of either L, $2L$, or $3L$ as shown, and they are well insulated along their lengths. The faces forming the room are maintained at 5°C, and the conduction of energy outward through the walls is steady. Rank the walls according to the rate of conduction through them, greatest first.

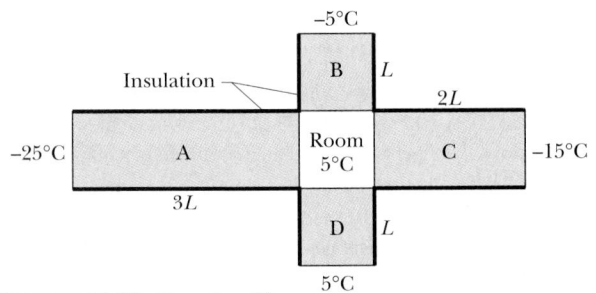

FIGURE 19-28 Question 12.

13. Suppose a block of wood and a block of metal are at the *same* temperature. When the blocks feel cold to your fingers, the metal feels colder than the wood; when the blocks feel hot to your fingers, the metal feels hotter than the wood. At what temperature will the blocks feel the same?

14. The following solid objects, made of the same material, are maintained at a temperature of 300 K in an environment whose temperature is 350 K: a cube of edge length r, a sphere of radius r, and a hemisphere of radius r. Rank the objects according to the net rate at which thermal radiation is exchanged with the environment, greatest first.

15. The following pairs give the temperatures of an object and its environment, respectively, for three situations: (1) 300 K and 350 K; (2) 350 K and 400 K; and (3) 400 K and 450 K. Without computation, rank the situations according to the object's net rate of energy exchange P_n, greatest first.

EXERCISES & PROBLEMS

SECTION 19-3 Measuring Temperature

1E. To measure temperatures, physicists and astronomers often measure how the intensity of electromagnetic radiation emitted by an object varies with wavelength. The wavelength λ_{max} at which the intensity is greatest is related to T, the temperature of the object in kelvins, by

$$\lambda_{max}T = 0.2898 \text{ cm} \cdot \text{K}.$$

In 1965, microwave radiation peaking at $\lambda_{max} = 0.107$ cm was discovered coming in all directions from space. To what temperature does this correspond? The interpretation of this *background radiation* is that it is left over from some 15 billion years ago, soon after the universe began.

2E. Suppose the temperature of a gas at the boiling point of water is 373.15 K. What then is the limiting value of the ratio of the pressure of the gas at that boiling point to its pressure at the triple point of water? (Assume the volume of the gas is the same at both temperatures.)

3P. Two constant-volume gas thermometers are assembled, one using nitrogen and the other using hydrogen. Both contain enough gas so that $p_3 = 80$ torr. What is the difference between the pressures in the two thermometers if both are inserted into a water bath at the boiling point? Which gas is at higher pressure?

4P. A particular gas thermometer is constructed of two gas-containing bulbs, each of which is put into a water bath, as shown in Fig. 19-29. The pressure difference between the two bulbs is measured by a mercury manometer as shown. Appropriate reser-

voirs, not shown in the diagram, maintain constant gas volume in the two bulbs. There is no difference in pressure when both baths are at the triple point of water. The pressure difference is 120 torr when one bath is at the triple point and the other is at the boiling point of water. Finally, the pressure difference is 90.0 torr when one bath is at the triple point and the other is at an unknown temperature to be measured. What is the unknown temperature?

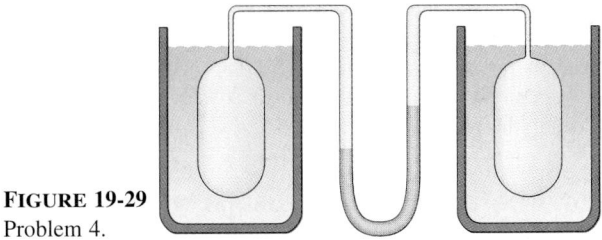

FIGURE 19-29
Problem 4.

SECTION 19-4 The Celsius and Fahrenheit Scales

5E. At what temperature is the Fahrenheit scale reading equal to (a) twice that of the Celsius and (b) half that of the Celsius?

6E. If your doctor tells you that your temperature is 310 kelvins above absolute zero, should you worry? Explain your answer.

7E. (a) In 1964, the temperature in the Siberian village of Oymyakon reached a value of −71°C. What temperature is this on the Fahrenheit scale? (b) The highest officially recorded temperature in the continental United States was 134°F in Death Valley, California. What is this temperature on the Celsius scale?

8E. (a) The temperature of the surface of the Sun is about 6000 K. Express this on the Fahrenheit scale. (b) Express normal human body temperature, 98.6°F, on the Celsius scale. (c) In the continental United States, the lowest officially recorded temperature is −70°F at Rogers Pass, Montana. Express this on the Celsius scale. (d) Express the normal boiling point of oxygen, −183°C, on the Fahrenheit scale. (e) At what Celsius temperature would you find a room to be uncomfortably warm?

9E. At what temperature do the following pairs of scales read the same: (a) Fahrenheit and Celsius (verify the listing in Table 19-1), (b) Fahrenheit and Kelvin, and (c) Celsius and Kelvin?

10P. Suppose that on a temperature scale X, water boils at −53.5°X and freezes at −170°X. What would a temperature of 340 K be on the X scale?

11P. It is an everyday observation that hot and cold objects cool down or warm up to the temperature of their surroundings. If the temperature difference ΔT between an object and its surroundings ($\Delta T = T_{obj} - T_{sur}$) is not too great, the rate of cooling or warming of the object is proportional, approximately, to this temperature difference; that is,

$$\frac{d\,\Delta T}{dt} = -A(\Delta T),$$

where A is a constant. The minus sign appears because ΔT decreases with time if ΔT is positive and increases if ΔT is negative. This is known as *Newton's law of cooling*. (a) On what factors does A depend? What are its dimensions? (b) If at some instant $t = 0$ the temperature difference is ΔT_0, show that it is

$$\Delta T = \Delta T_0 e^{-At}$$

at a later time t.

12P. The heater of a house breaks down one day when the outside temperature is 7.0°C. As a result, the inside temperature drops from 22°C to 18°C in 1.0 h. The owner fixes the heater and adds insulation to the house. Now she finds that, on a similar day, the house takes twice as long to drop from 22°C to 18°C when the heater is not operating. What is the ratio of the constant A in Newton's law of cooling (see Problem 11) after the insulation is added to the value before?

SECTION 19-5   Thermal Expansion

13E. A steel rod has a length of exactly 20 cm at 30°C. How much longer is it at 50°C?

14E. An aluminum flagpole is 33 m high. By how much does its length increase as the temperature increases by 15 C°?

15E. The Pyrex glass mirror in the telescope at the Mt. Palomar Observatory has a diameter of 200 in. The temperature ranges from −10°C to 50°C on Mt. Palomar. Determine the maximum change in the diameter of the mirror.

16E. A circular hole in an aluminum plate is 2.725 cm in diameter at 0.000°C. What is its diameter when the temperature of the plate is raised to 100.0°C?

17E. An aluminum-alloy rod has a length of 10.000 cm at

20.000°C and a length of 10.015 cm at the boiling point of water. (a) What is the length of the rod at the freezing point of water? (b) What is the temperature if the length of the rod is 10.009 cm?

18E. (a) What is the coefficient of linear expansion of aluminum per Fahrenheit degree? (b) Use your answer in (a) to calculate the change in length of a 20 ft aluminum rod if the rod is heated from 40°F to 95°F.

19E. Soon after Earth was formed, heat released by the decay of radioactive elements raised the average internal temperature from 300 to 3000 K, at about which value it remains today. Assuming an average coefficient of volume expansion of 3.0×10^{-5} K^{-1}, by how much has the radius of Earth increased since the planet was formed?

20E. The Stanford linear accelerator contains hundreds of brass disks tightly fitted into a steel tube. The system was assembled by cooling the disks in dry ice (at −57.00°C) to enable them to slide into the close-fitting tube. If the diameter of a disk is 80.00 mm at 43.00°C, what is its diameter in the dry ice?

21E. A glass window is exactly 20 cm by 30 cm at 10°C. By how much has its area increased when its temperature is 40°C?

22E. At 20°C, a brass cube has an edge length of 30 cm. What is the increase in the cube's surface area when it is heated from 20°C to 75°C?

23E. Find the change in volume of an aluminum sphere with an initial radius of 10 cm when the sphere is heated from 0.0°C to 100°C.

24E. What is the volume of a lead ball at 30°C if the ball's volume at 60°C is 50 cm³?

25E. By how much does the volume of an aluminum cube 5.00 cm on an edge increase when the cube is heated from 10.0°C to 60.0°C?

26E. An aluminum cup of 100 cm³ capacity is filled with glycerin at 22°C. How much glycerin, if any, will spill out of the cup if the temperature of the cup and glycerin is raised to 28°C? (The coefficient of volume expansion of glycerin is 5.1×10^{-4}/C°.)

27E. A steel rod at 25.0°C is bolted securely at both ends and then cooled. At what temperature will it rupture? Use Table 13-1.

28P. At 20°C, a rod is exactly 20.05 cm long on a steel ruler. Both the rod and the ruler are placed in an oven at 270°C, where the rod now measures 20.11 cm on the same ruler. What is the coefficient of thermal expansion for the material of which the rod is made?

29P. A steel rod is 3.000 cm in diameter at 25°C. A brass ring has an interior diameter of 2.992 cm at 25°C. At what common temperature will the ring just slide onto the rod?

30P. The area A of a rectangular plate is ab. Its coefficient of linear expansion is α. After a temperature rise ΔT, side a is longer by Δa and side b is longer by Δb (Fig. 19-30). Show that if we neglect the small quantity $\Delta a\,\Delta b/ab$, then $\Delta A = 2\alpha A\,\Delta T$.

31P. Density is mass divided by volume. If the volume V is temperature dependent, so is the density ρ. Show that a small change in density $\Delta\rho$ with a change in temperature ΔT is given by

$$\Delta\rho = -\beta\rho\,\Delta T,$$

where β is the coefficient of volume expansion. Explain the minus sign.

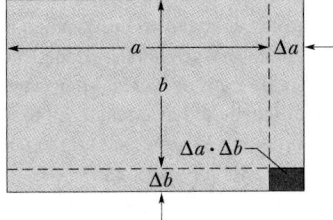

FIGURE 19-30
Problem 30.

32P. When the temperature of a metal cylinder is raised from 0.0°C to 100°C, its length increases by 0.23%. (a) Find the percent change in density. (b) What is the metal? Use Table 19-2.

33P. Show that when the temperature of a liquid in a barometer changes by ΔT, and the pressure is constant, the height h changes by $\Delta h = \beta h \, \Delta T$, where β is the coefficient of volume expansion. Neglect the expansion of the glass tube.

34P. When the temperature of a copper penny is raised by 100 C°, its diameter increases by 0.18%. To two significant figures, give the percent increase in (a) the area of a face, (b) the thickness, (c) the volume, and (d) the mass of the penny. (e) Calculate its coefficient of linear expansion.

35P. A pendulum clock with a pendulum made of brass is designed to keep accurate time at 20°C. What will be the error, in seconds per hour, if the clock operates at 0.0°C?

36P. As a result of a temperature rise of 32 C°, a bar with a crack at its center buckles upward (Fig. 19-31). If the fixed distance L_0 is 3.77 m and the coefficient of linear expansion of the bar is 25×10^{-6}/C°, find the rise x of the center.

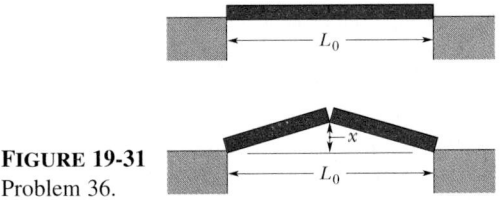

FIGURE 19-31
Problem 36.

37P. A composite bar of length $L = L_1 + L_2$ is made from a bar of material 1 and length L_1 attached to a bar of material 2 and length L_2, as shown in Fig. 19-32. (a) Show that the coefficient of linear expansion α for this composite bar is given by $\alpha = (\alpha_1 L_1 + \alpha_2 L_2)/L$. (b) Using steel and brass, design such a composite bar whose length is 52.4 cm and whose coefficient of linear expansion is 13.0×10^{-6}/C°.

FIGURE 19-32 Problem 37.

SECTION 19-7 The Absorption of Heat by Solids and Liquids

38E. It is possible to melt ice by rubbing one block of it against another. How much work, in joules, would you have to do to get 1.00 g of ice to melt?

39E. A certain substance has a mass per mole of 50 g/mol. When 314 J of heat is added to a 30.0 g sample of this material, its temperature rises from 25.0°C to 45.0°C. (a) What is the specific heat of this substance? (b) How many moles of the substance are present? (c) What is the molar specific heat of the substance?

40E. In a certain solar house, energy from the Sun is stored in barrels filled with water. In a particular winter stretch of five cloudy days, 1.00×10^6 kcal is needed to maintain the inside of the house at 22.0°C. Assuming that the water in the barrels is at 50.0°C and that the water has a density of 1.00×10^3 kg/m³, what volume of water is required?

41E. A diet doctor encourages people to diet by drinking ice water. His theory is that the body must burn off enough fat to raise the temperature of the water from 0.00°C to the body temperature of 37.0°C. How many liters of ice water would have to be consumed to burn off 454 g (about 1 lb) of fat, assuming that this much fat burning requires 3500 Cal be transferred to the ice water? Why is it not advisable to follow this diet? One liter = 10^3 cm³. The density of water is 1.00 g/cm³.

42E. Icebergs in the North Atlantic present hazards to shipping, causing the lengths of shipping routes to increase by about 30% during the iceberg season. Attempts to destroy icebergs include planting explosives, bombing, torpedoing, shelling, ramming, and coating with black soot. Suppose that direct melting of the iceberg, by placing heat sources in the ice, is tried. How much heat is required to melt 10% of an iceberg with a mass of 200,000 metric tons? (Use 1 metric ton = 1000 kg.)

43E. How much water remains unfrozen after 50.2 kJ of heat has been extracted from 260 g of liquid water initially at its freezing point?

44E. Calculate the minimum amount of heat, in joules, required to completely melt 130 g of silver initially at 15.0°C.

45E. A room is lighted by four 100 W incandescent light bulbs. (The power of 100 W is the rate at which a bulb converts electrical energy to heat and visible light.) Assuming that 90% of the energy is converted to heat, how much heat is added to the room in 1.00 h?

46E. What quantity of butter (6.0 Cal/g = 6000 cal/g) would supply the energy needed for a 160 lb man to ascend to the top of Mt. Everest, elevation 29,000 ft, from sea level?

47E. An energetic athlete dissipates all the energy in a diet of 4000 Cal/day. If he were to release this energy at a steady rate, how would this conversion of energy compare with that of a 100 W bulb? (The power of 100 W is the rate at which the bulb converts electrical energy to heat and visible light.)

48E. If the heat necessary to warm a mass m of water from 68°F to 78°F were somehow converted to translational kinetic energy of that water, what would be the speed of the water?

49E. A power of 0.400 hp is required for 2.00 min to drill a hole in a 1.60-lb copper block. (a) How much thermal energy in Btu is generated? (b) What is the rise in temperature of the copper if only 75.0% of the power warms the copper? (Use 1 ft·lb = 1.285×10^{-3} Btu.)

50E. An object of mass 6.00 kg falls through a height of 50.0 m and, by means of a mechanical linkage, rotates a paddle wheel that stirs 0.600 kg of water. The water is initially at 15.0°C. What is the maximum possible temperature rise of the water?

51E. One way to keep the contents of a garage from becoming too cold on a night when a severe subfreezing temperature is forecast is to put a tub of water in the garage. If the mass of the water is 125 kg and its initial temperature is 20°C, (a) how much energy must the water transfer to its surroundings in order to freeze completely and (b) what is the lowest possible temperature of the water and its surroundings until that happens?

52E. A small electric immersion heater is used to heat 100 g of water for a cup of instant coffee. The heater is labeled "200 watts," which means that it converts electrical energy to thermal energy at this rate. Calculate the time required to bring this water from 23°C to the boiling point, ignoring any heat losses.

53E. A pickup truck whose mass is 2200 kg is speeding along the highway at 65.0 mi/h. (a) If you could use all this kinetic energy to vaporize water already at 100°C, how much water could you vaporize? (b) If you had to buy this amount of energy from your local utility company at 12¢/kW·h, how much would it cost you? Guess at the answers before you figure them out; you may be surprised.

54E. A 150 g copper bowl contains 220 g of water, both at 20.0°C. A very hot 300 g copper cylinder is dropped into the water, causing the water to boil, with 5.00 g being converted to steam. The final temperature of the system is 100°C. (a) How much heat was transferred to the water? (b) How much to the bowl? (c) What was the original temperature of the cylinder?

55P. Calculate the specific heat of a metal from the following data. A container made of the metal has a mass of 3.6 kg and contains 14 kg of water. A 1.8 kg piece of the metal initially at a temperature of 180°C is dropped into the water. The container and water initially have a temperature of 16.0°C, and the final temperature of the entire system is 18.0°C.

56P. A thermometer of mass 0.0550 kg and of specific heat 0.837 kJ/kg·K reads 15.0°C. It is then completely immersed in 0.300 kg of water, and it comes to the same final temperature as the water. If the thermometer reads 44.4°C, what was the temperature of the water before insertion of the thermometer?

57P. How long does it take a 2.0×10^5 Btu/h water heater to raise the temperature of 40 gal of water from 70°F to 100°F?

58P. An athlete needs to lose weight and decides to do it by "pumping iron." (a) How many times must an 80.0 kg weight be lifted a distance of 1.00 m in order to burn off 1 lb of fat, assuming that that much fat is equivalent to 3500 Cal? (b) If the weight is lifted once every 2.00 s, how long does the task take?

59P. A 1500 kg Buick moving at 90 km/h brakes to rest, at uniform deceleration and without skidding, over a distance of 80 m. At what average rate is mechanical energy transferred to thermal energy in the brake system?

60P. A chef, upon finding his stove out of order, decides to boil the water for his wife's coffee by shaking it in a thermos flask. Suppose that he uses 500 cm³ of tap water at 59°F, and that the water falls 1.0 ft each shake, the chef making 30 shakes each minute. Neglecting any loss of thermal energy by the flask, how long must he shake the flask before the water boils?

61P. A block of ice, at its melting point and of initial mass 50.0 kg, slides along a horizontal surface, starting at a speed of 5.38 m/s and finally coming to rest after traveling 28.3 m. Compute the mass of ice melted as a result of the friction between the block and the surface. (Assume that all the heat generated owing to friction goes into the block of ice.)

62P. The specific heat of a substance varies with temperature according to $c = 0.20 + 0.14T + 0.023T^2$, with T in °C and c in cal/g·K. Find the heat required to raise the temperature of 2.0 g of this substance from 5.0°C to 15.0°C.

63P. In a solar water heater, energy from the Sun is gathered by water that circulates through tubes in a rooftop collector. The solar radiation enters the collector through a transparent cover and warms the water in the tubes; this water is pumped into a holding tank. Assuming that the efficiency of the overall system is 20% (that is, 80% of the incident solar energy is lost from the system), what collector area is necessary to raise the temperature of 200 L of water in the tank from 20°C to 40°C in 1.0 h? The intensity of incident sunlight is 700 W/m².

64P. An insulated thermos contains 130 cm³ of hot coffee, at a temperature of 80.0°C. You put in a 12.0 g ice cube at its melting point to cool the coffee. By how many degrees will your coffee have cooled once the ice has melted? Treat the coffee as though it were pure water.

65P. What mass of steam at 100°C must be mixed with 150 g of ice at its melting point, in a thermally insulated container, to produce liquid water at 50°C?

66P. A person makes a quantity of iced tea by mixing 500 g of hot tea (essentially water) with an equal mass of ice at its melting point. If the initial hot tea is at a temperature of (a) 90°C and (b) 70°C, what are the temperature and mass of the remaining ice when the tea and ice reach a common temperature?

67P. (a) Two 50 g ice cubes are dropped into 200 g of water in a glass. If the water was initially at 25°C, and the ice came directly from a freezer at −15°C, what will be the temperature of the drink when the ice and water reach the same temperature? (b) Suppose that only one ice cube had been used in (a); what would be the final temperature of the drink? Neglect the heat capacity of the glass.

68P. A 20.0 g copper ring has a diameter of exactly 1.00000 in. at its temperature of 0.000°C. An aluminum sphere has a diameter of exactly 1.00200 in. at its temperature of 100.0°C. The sphere is placed on top of the ring (Fig. 19-33), and the two are allowed to

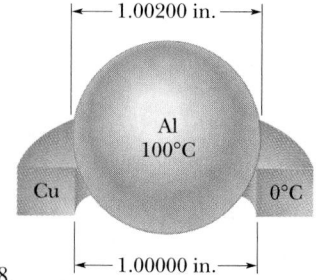

FIGURE 19-33 Problem 68.

come to thermal equilibrium, with no heat lost to the surroundings. The sphere just passes through the ring at the equilibrium temperature. What is the mass of the sphere?

69P. A *flow calorimeter* is a device used to measure the specific heat of a liquid. Heat is added at a known rate to a stream of the liquid as it passes through the calorimeter at a known rate. Measurement of the resulting temperature difference between the inflow and the outflow points of the liquid stream enables us to compute the specific heat of the liquid. Suppose a liquid of density 0.85 g/cm³ flows through a calorimeter at the rate of 8.0 cm³/s. Heat is added at the rate of 250 W by means of an electric heating coil, and a temperature difference of 15 C° is established in steady-state conditions between the inflow and the outflow points. What is the specific heat of the liquid?

70P. By means of a heating coil, energy is transferred at a constant rate to a substance in a thermally insulated container. The temperature of the substance is measured as a function of time. (a) Show how we can deduce from this information the way in which the heat capacity of the substance depends on the temperature. (b) Suppose that in a certain temperature range the temperature T is proportional to t^3, where t is the time. How does the heat capacity depend on T in this range?

SECTION 19-10 Some Special Cases of the First Law of Thermodynamics

71E. A sample of gas expands from 1.0 m³ to 4.0 m³ while its pressure decreases from 40 Pa to 10 Pa. How much work is done by the gas if its pressure changes with volume via each of the three paths shown in the p-V diagram in Fig. 19-34?

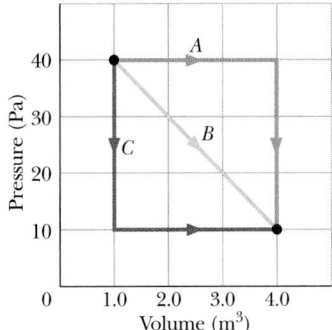

FIGURE 19-34
Exercise 71.

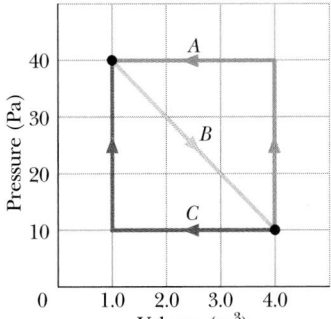

FIGURE 19-35
Exercise 72.

72E. A sample of gas expands from 1.0 m³ to 4.0 m³ along path B in the p-V diagram in Fig. 19-35. It is then compressed back to 1.0 m³ along either path A or path C. Compute the net work done by the gas for the complete cycle in each case.

73E. Consider that 200 J of work is done on a system and 70.0 cal of heat is extracted from the system. In the sense of the first law of thermodynamics, what are the values (including algebraic signs) of (a) W, (b) Q, and (c) ΔE_{int}?

74E. A thermodynamic system is taken from an initial state A to another state B and back again to A, via state C, as shown by path $ABCA$ in the p-V diagram of Fig. 19-36a. (a) Complete the table in Fig. 19-36b by filling in either $+$ or $-$ for the sign of each thermodynamic quantity associated with each process. (b) Calculate the numerical value of the work done by the system for the complete cycle $ABCA$.

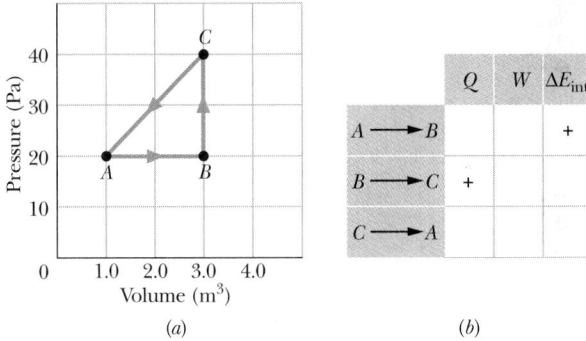

(a)   (b)

FIGURE 19-36 Exercise 74.

75E. Gas within a chamber passes through the cycle shown in Fig. 19-37. Determine the net heat added to the system during process CA if the heat Q_{AB} added during process AB is 20.0 J, no heat is transferred during process BC, and the net work done during the cycle is 15.0 J.

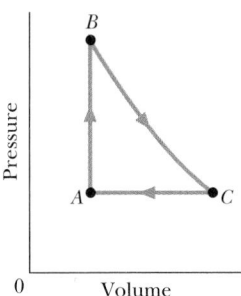

FIGURE 19-37 Exercise 75.

76E. Gas within a closed chamber undergoes the processes shown in the p-V diagram of Fig. 19-38. Calculate the net heat added to the system during one complete cycle.

77P. Figure 19-39a shows a cylinder containing gas and closed by a movable piston. The cylinder is kept submerged in an ice–water mixture. The piston is *quickly* pushed down from position 1 to position 2 and then held at position 2 until the gas is again at the temperature of the ice–water mixture; it then is *slowly* raised back to position 1. Figure 19-39b is a p-V diagram for the process.

If 100 g of ice is melted during the cycle, how much work has been done *on* the gas?

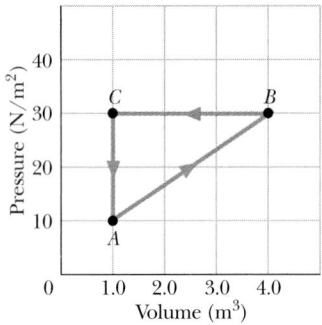

FIGURE 19-38   Exercise 76.

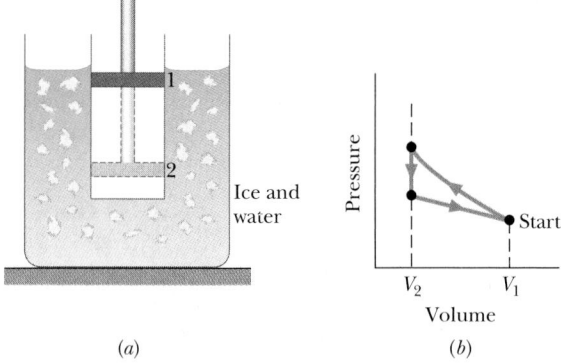

(a)   (b)

FIGURE 19-39   Problem 77.

78P.   When a system is taken from state i to state f along path iaf in Fig. 19-40, $Q = 50$ cal and $W = 20$ cal. Along path ibf, $Q = 36$ cal. (a) What is W along path ibf? (b) If $W = -13$ cal for the curved return path fi, what is Q for this path? (c) Take $E_{int,i} = 10$ cal. What is $E_{int,f}$? (d) If $E_{int,b} = 22$ cal, what are the values of Q for path ib and path bf?

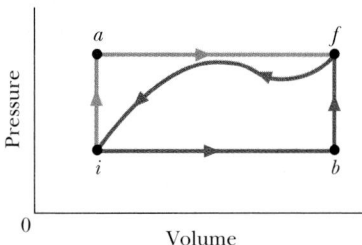

FIGURE 19-40   Problem 78.

SECTION 19-11   Heat Transfer Mechanisms

79E.   The average rate at which heat is conducted through the ground surface in North America is 54.0 mW/m², and the average thermal conductivity of the near-surface rocks is 2.50 W/m·K. Assuming a surface temperature of 10.0°C, what should be the temperature at a depth of 35.0 km (near the base of the crust)? Ignore the heat generated by the presence of radioactive elements.

80E.   The ceiling of a single-family dwelling in a cold climate should have an R-value of 30. To give such insulation, how thick

would a layer of (a) polyurethane foam and (b) silver have to be?

81E.   The thermal conductivity of Pyrex glass at 0°C is 2.9×10^{-3} cal/cm·C°·s. (a) Express the quantity in W/m·K and in Btu/ft·F°·h. (b) What is the R-value for a 0.25 in. sheet of such glass?

82E.   (a) Calculate the rate at which body heat is conducted through the clothing of a skier in a steady-state process, given the following data: the body surface area is 1.8 m² and the clothing is 1.0 cm thick; the skin surface temperature is 33°C, whereas the outer surface of the clothing is at 1.0°C; the thermal conductivity of the clothing is 0.040 W/m·K. (b) How would the answer to (a) change if, after a fall, the skier's clothes became soaked with water of thermal conductivity 0.60 W/m·K?

83E.   Consider the slab shown in Fig. 19-17. Suppose that $L = 25.0$ cm, $A = 90.0$ cm², and the material is copper. If $T_H = 125$°C, $T_C = 10.0$°C, and a steady state is reached, find the rate of heat transfer through the slab.

84E.   A cylindrical copper rod of length 1.2 m and cross-sectional area 4.8 cm² is insulated to prevent heat loss through its surface. The ends are maintained at a temperature difference of 100 C° by having one end in a water–ice mixture and the other in boiling water and steam. (a) Find the rate at which heat is conducted along the rod. (b) Find the rate at which ice melts at the cold end.

85E.   Show that the temperature T_X at the interface of a compound slab (Fig. 19-18) is given by

$$T_X = \frac{R_1 T_H + R_2 T_C}{R_1 + R_2}.$$

86E.   If you were to walk briefly in space without a spacesuit while far from the Sun (as an astronaut does in the movie *2001*), you would feel the cold of space: while you radiated thermal energy, you would absorb almost none from your environment. (a) At what rate would you lose energy? (b) How much energy would you lose in 30 s? Assume that your emissivity is 0.90, and estimate other data needed in the calculations.

87E.   Four square pieces of insulation of two different materials, all with the same thickness and area A, are available to cover an opening of area $2A$. This can be done in either of the two ways shown in Fig. 19-41. Which arrangement, (a) or (b), would give the lower heat flow if $k_2 \neq k_1$?

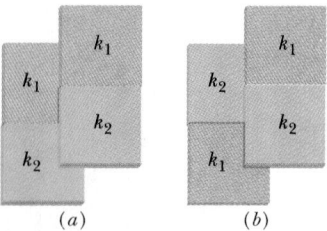

(a)   (b)

FIGURE 19-41   Exercise 87.

88P.   Two identical rectangular rods of metal are welded end to end as shown in Fig. 19-42a, and 10 J of heat is conducted (in a steady-state process) through the rods in 2.0 min. How long would it take for 10 J to be conducted through the rods if they are welded together as shown in Fig. 19-42b?

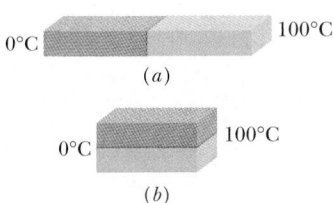

FIGURE 19-42 Problem 88.

89P. Compute the rate of heat conduction through two storm doors, both 2.0 m high and 0.75 m wide. (a) One door is made with aluminum panels 1.5 mm thick and a glass pane 3.0 mm thick; the glass covers 75% of the aluminum surface (the structural frame has a negligible area). (b) The second door is made entirely of white pine averaging 2.5 cm in thickness. Take the temperature drop across each door to be 33 C°; see Table 19-6.

90P. A large cylindrical water tank with a bottom 1.7 m in diameter is made of iron boilerplate 5.2 mm thick. Water in the tank is heated by a gas burner underneath that is able to maintain a temperature difference of 2.3 C° between the top and bottom surfaces of the bottom plate. How much heat is conducted through that plate in 5.0 min? (Iron has a thermal conductivity of 67 W/m·K.)

91P. (a) What is the rate of heat loss in watts per square meter through a glass window 3.0 mm thick if the outside temperature is −20°F and the inside temperature is +72°F? (b) A storm window is installed having the same thickness of glass but with an air gap of 7.5 cm between the two windows. What will be the corresponding rate of heat loss assuming that conduction is the only important heat-loss mechanism?

92P. A sphere of radius 0.500 m, temperature 27.0°C, and emissivity 0.850 is located in an environment of temperature 77.0°C. At what rate does the sphere (a) emit thermal radiation and (b) absorb thermal radiation? (c) What is the sphere's net rate of energy exchange?

93P. A cube of edge length 6.0×10^{-6} m, emissivity 0.75, and temperature −100°C floats in an environment of temperature −150°C. What is the cube's net thermal radiation transfer rate?

94P. A solid cylinder of radius $r_1 = 2.5$ cm, length $h_1 =$ 5.0 cm, emissivity 0.85, and temperature 30°C is suspended in an environment of temperature 50°C. (a) What is the cylinder's net thermal radiation transfer rate P_1? (b) If the cylinder is stretched until its radius is $r_2 = 0.50$ cm, its net thermal radiation transfer rate becomes P_2. What is the ratio P_2/P_1?

95P. A tank of water has been outdoors in cold weather and a slab of ice 5.0 cm thick has formed on its surface (Fig. 19-43). The air above the ice is at −10°C. Calculate the rate of formation of ice (in centimeters per hour) on the bottom surface of the ice slab. Take the thermal conductivity and density of ice to be 0.0040 cal/s·cm·C° and 0.92 g/cm³. Assume that heat is not transferred through the walls or bottom of the tank.

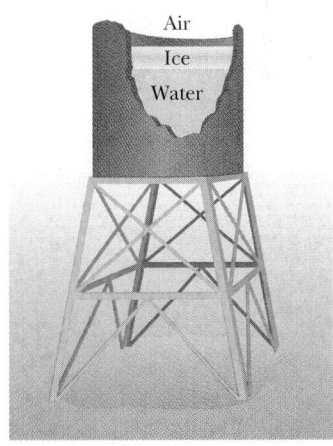

FIGURE 19-43 Problem 95.

96P. Ice has formed on a shallow pond and a steady state has been reached, with the air above the ice at −5.0°C and the bottom of the pond at 4.0°C. If the total depth of *ice + water* is 1.4 m, how thick is the ice? (Assume that the thermal conductivities of ice and water are 0.40 and 0.12 cal/m·C°·s, respectively.)

97P. Three metal rods, one copper, one aluminum, and one brass, are each 6.00 cm long and 1.00 cm in diameter. These rods are placed end to end, with the aluminum between the other two. The free ends of the copper and brass rods are maintained at the boiling point and the freezing point of water, respectively. Find the steady-state temperatures of the copper–aluminum junction and the aluminum–brass junction. The thermal conductivity of brass is 109 W/m·K.

20
The Kinetic Theory of Gases

Suppose that you return to your chilly dwelling after snowshoeing for miles on a cold winter day. Your first thought is to light a stove. But why, exactly, would you do that? Is it because the stove will increase the store of internal (thermal) energy of the air in the dwelling, until eventually the air will have enough of that internal energy to keep you comfortable? As logical as this reasoning sounds, it is flawed, because the air's store of internal energy will not be changed by the stove. How can that be? And if it is so, why would you bother to light the stove?

20-1 A NEW WAY TO LOOK AT GASES

Classical thermodynamics—the subject of the last chapter—has nothing to say about atoms. Its laws are concerned only with such macroscopic variables as pressure, volume, and temperature. However, we know that a gas is made up of atoms or molecules (groups of atoms bound together). The pressure exerted by a gas must surely be related to the steady drumbeat of its molecules on the walls of its container. The ability of a gas to take on the volume of its container must surely be due to the freedom of motion of its molecules. And the temperature and internal energy of a gas must surely be related to the kinetic energy of these molecules. Perhaps we can learn something about gases by approaching the subject from this direction. We call this molecular approach the **kinetic theory of gases.** It is the subject of this chapter.

20-2 AVOGADRO'S NUMBER

When our thinking is slanted toward molecules, it makes sense to measure the sizes of our samples in moles. If we do so, we can be certain that we are comparing samples that contain the same number of molecules. The *mole* is one of the seven SI base units and is defined as follows:

One mole is the number of atoms in a 12 g sample of carbon-12.

We speak of a ''mole of helium'' or ''a mole of water,'' meaning a certain number of the elementary units of the substance. But we could just as well speak of a mole of tennis balls, where, of course, the elementary unit would be a tennis ball.

The obvious question now is: ''Just how many atoms or molecules are there in a mole?'' The answer is determined experimentally and, as you saw in Chapter 19, is

$$N_A = 6.02 \times 10^{23} \text{ mol}^{-1} \quad \text{(Avogadro's number).} \quad (20\text{-}1)$$

This number is called **Avogadro's number** after Italian scientist Amadeo Avogadro (1776–1856), who suggested that all gases contain the same number of molecules or atoms when they occupy the same volume under the same conditions of temperature and pressure.

The number of moles n contained in a sample of any substance can be found from

$$n = \frac{N}{N_A}, \quad (20\text{-}2)$$

in which N is the number of molecules in the sample. The number of moles in a sample can also be found from the mass M_{sam} of the sample and either the *molar mass M* (the mass of 1 mole of that substance) or the mass m of one molecule:

$$n = \frac{M_{sam}}{M} = \frac{M_{sam}}{mN_A}. \quad (20\text{-}3)$$

The enormously large value of Avogadro's number suggests how tiny and how numerous atoms must be. A mole of air, for example, can easily fit into a suitcase. Yet, if these molecules were spread uniformly over the surface of Earth, there would be about 120,000 of them per square centimeter. A second example: one mole of tennis balls would fill a volume equal to that of seven Moons!

PROBLEM SOLVING TACTICS

TACTIC 1: *Avogadro's Number of What?*
In Eq. 20-1, Avogadro's number is expressed in terms of mol^{-1}, which is the inverse mole, or 1/mol. We could instead explicitly state the elementary unit involved in a given situation. For example, if the elementary unit is an atom, we might write $N_A = 6.02 \times 10^{23}$ atoms/mole. If the elementary unit is a molecule, then we might write $N_A = 6.02 \times 10^{23}$ molecules/mole. And if we are interested in a *great* many tennis balls, we can write $N_A = 6.02 \times 10^{23}$ tennis balls/mole.

20-3 IDEAL GASES

Our goal in this chapter is to explain the macroscopic properties of a gas—such as its pressure and its temperature—in terms of the behavior of the molecules that make it up. But there is an immediate problem: which gas? Should it be hydrogen or oxygen, or methane, or perhaps uranium hexafluoride? They are all different. However, experimenters have found that if we confine 1 mole samples of various gases in boxes of identical volume and hold the gases at the same temperature, then their measured pressures are nearly—though not exactly—the same. If we repeat the measurements at lower gas densities, then these small differences in the measured pressures tend to disappear. Further experiments show that, at low enough densities, all real gases tend to obey the relation

$$pV = nRT \quad \text{(ideal gas law),} \quad (20\text{-}4)$$

in which p is the absolute (not gauge) pressure, n is the number of moles of gas present, and R, the **gas constant,** has the same value for all gases, namely,

$$R = 8.31 \text{ J/mol} \cdot \text{K.} \quad (20\text{-}5)$$

The temperature T in Eq. 20-4 must be expressed in kelvins. Equation 20-4 is called the **ideal gas law.** Provided the gas density is reasonably low, Eq. 20-4 holds for any type of gas, or a mixture of different types, with n being the total number of moles present.

You may well ask, "What is an *ideal gas* and what is so 'ideal' about one?" The answer lies in the simplicity of the law (Eq. 20-4) that governs its macroscopic properties. Using this law—as you will see —we can deduce many properties of the ideal gas in a simple way. Although there is no such thing in nature as a truly ideal gas, *all* gases approach the ideal state at low enough densities, that is, under conditions in which their molecules are far enough apart. Thus the ideal gas concept allows us to gain useful insights into the limiting behavior of real gases.

Work Done by an Ideal Gas at Constant Temperature

Suppose that a sample of n moles of an ideal gas, confined to a piston–cylinder arrangement, is allowed to expand from an initial volume V_i to a final volume V_f. Suppose further that the temperature T of the gas is held constant throughout the process. Such a process is called an **isothermal expansion** (and the reverse is called an **isothermal compression**).

On a p-V diagram, an *isotherm* is a curve that connects points that have the same temperature. It is then a graph of pressure versus volume for a gas whose temperature T is held constant, that is, a graph of

$$p = nRT \frac{1}{V} = \text{(a constant)} \frac{1}{V}. \qquad (20\text{-}6)$$

Figure 20-1 shows three isotherms, each corresponding to a different (constant) value of T. Note that the values of T increase upward to the right.

The p-V diagram for an isothermal expansion or compression is also a graph of pressure versus volume for a gas whose temperature is held constant. Thus, such a p-V diagram must follow along an isotherm, as does the expansion from state i to state f in Fig. 20-1.

Let us calculate the work done by an ideal gas during an isothermal expansion. Equation 19-23,

$$W = \int_{V_i}^{V_f} p \, dV, \qquad (20\text{-}7)$$

is a general expression for the work done during any change in volume of a gas. Because we are now dealing with an ideal gas, we can use Eq. 20-4 to substitute for p, obtaining

$$W = \int_{V_i}^{V_f} \frac{nRT}{V} \, dV. \qquad (20\text{-}8)$$

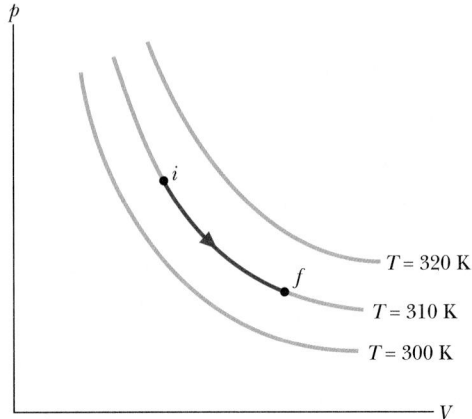

FIGURE 20-1 Three isotherms on a p-V diagram. The path shown along the middle isotherm represents an isothermal expansion of a gas from an initial state i to a final state f. The path from f to i along the isotherm would represent the reverse process, an isothermal compression.

And because we are considering an isothermal expansion, T is constant and we can move it in front of the integral sign to write

$$W = nRT \int_{V_i}^{V_f} \frac{dV}{V} = nRT \left[\ln V \right]_{V_i}^{V_f}. \qquad (20\text{-}9)$$

By evaluating at the limits and then using the relationship $\ln a - \ln b = \ln (a/b)$, we find that

$$W = nRT \ln \frac{V_f}{V_i} \qquad \begin{array}{l}\text{(ideal gas,}\\ \text{isothermal process).}\end{array} \qquad (20\text{-}10)$$

Recall that the symbol ln specifies a *natural* logarithm, that is, a logarithm to base e.

For an expansion, $V_f > V_i$ by definition, so the ratio V_f/V_i in Eq. 20-10 is greater than unity. The natural logarithm of a quantity greater than unity is positive, and so the work W done by an ideal gas during an isothermal expansion is positive, as we expect. For a compression, we have $V_f < V_i$, so the ratio of volumes in Eq. 20-10 is less than unity. The natural logarithm in that equation—hence the work W—is negative, again as we expect.

Equation 20-10 does not give the work W done by an ideal gas during *every* thermodynamic process. Instead, it gives the work only when the temperature is held constant. If the temperature varies, then the symbol T in Eq. 20-8 cannot be moved in front of the integral symbol as in Eq. 20-9, and thus we do not end up with Eq. 20-10.

However, we can go back to Eq. 20-7 to find the work W done by an ideal gas (or any other gas) during two more processes—a constant-volume process and a constant-pressure process. If the volume of the gas is constant, then Eq. 20-7 yields

$$W = 0 \qquad \text{(constant-volume process).} \qquad (20\text{-}11)$$

If, instead, the volume changes while the pressure p of the gas is held constant, then Eq. 20-7 becomes

$$W = p(V_f - V_i) = p\,\Delta V \qquad \begin{array}{c}\text{(constant-pressure}\\\text{process).}\end{array} \qquad (20\text{-}12)$$

CHECKPOINT **1:** An ideal gas has an initial pressure of 3 pressure units and an initial volume of 4 volume units. The table gives the final pressure and volume of the gas (in those same units) in five processes. Which processes start and end on the same isotherm?

	a	b	c	d	e
p	12	6	5	4	1
V	1	2	7	3	12

SAMPLE PROBLEM 20-1

A cylinder contains 12 L of oxygen at 20°C and 15 atm. The temperature is raised to 35°C, and the volume reduced to 8.5 L. What is the final pressure of the gas in atmospheres? Assume that the gas is ideal.

SOLUTION: From Eq. 20-4 we may write

$$nR = \frac{p_i V_i}{T_i} = \frac{p_f V_f}{T_f}.$$

Solving for p_f yields

$$p_f = \frac{p_i T_f V_i}{T_i V_f}. \qquad (20\text{-}13)$$

Note here that if we converted the given initial and final volumes from liters to the proper units of cubic meters, the multiplying conversion factors would cancel out of Eq. 20-13. The same would be true for conversion factors that convert the pressures from atmospheres to the proper pascals. However, to convert the given temperatures to kelvins requires the addition of an amount that would not cancel and thus must be included. Hence, we must write

$$T_i = (273 + 20)\text{ K} = 293\text{ K}$$

and $\qquad T_f = (273 + 35)\text{ K} = 308\text{ K}.$

Inserting the given data into Eq. 20-13 then yields

$$p_f = \frac{(15\text{ atm})(308\text{ K})(12\text{ L})}{(293\text{ K})(8.5\text{ L})} = 22\text{ atm.} \quad \text{(Answer)}$$

SAMPLE PROBLEM 20-2

One mole of oxygen (assume it to be an ideal gas) expands at a constant temperature T of 310 K from an initial volume V_i of 12 L to a final volume V_f of 19 L.

(a) How much work is done by the expanding gas?

SOLUTION: From Eq. 20-10 we have

$$W = nRT \ln \frac{V_f}{V_i}$$

$$= (1\text{ mol})(8.31\text{ J/mol}\cdot\text{K})(310\text{ K}) \ln \frac{19\text{ L}}{12\text{ L}}$$

$$= 1180\text{ J.} \qquad\qquad\qquad\qquad \text{(Answer)}$$

The expansion is graphed in the p-V diagram of Fig. 20-2. The work done by the gas during the expansion is represented by the area beneath the curve if.

(b) How much work is done by the gas during an isothermal *compression* from $V_i = 19$ L to $V_f = 12$ L?

SOLUTION: We proceed as in (a), finding

$$W = nRT \ln \frac{V_f}{V_i}$$

$$= (1\text{ mol})(8.31\text{ J/mol}\cdot\text{K})(310\text{ K}) \ln \frac{12\text{ L}}{19\text{ L}}$$

$$= -1180\text{ J.} \qquad\qquad\qquad\qquad \text{(Answer)}$$

This result is equal in magnitude but opposite in sign to the result found in (a) for an isothermal expansion. The minus sign tells us that an external agent must do 1180 J of work *on* the gas to compress it.

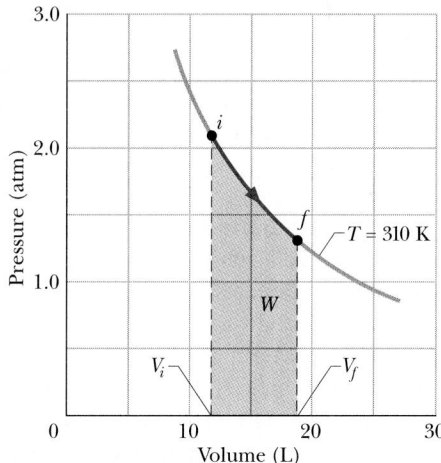

FIGURE 20-2 Sample Problem 20-2. The shaded area represents the work done by 1 mol of oxygen in expanding at a constant temperature T of 310 K.

20-4 PRESSURE, TEMPERATURE, AND RMS SPEED

Here is our first kinetic theory problem. Let n moles of an ideal gas be confined in a cubical box of volume V, as in Fig. 20-3. The walls of the box are held at temperature T. What is the connection between the pressure p exerted by the gas on the walls and the speeds of the molecules?

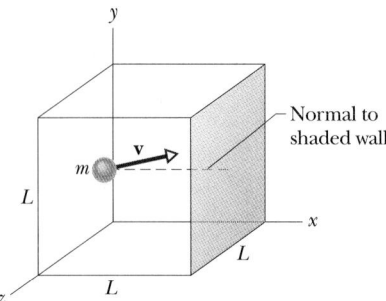

FIGURE 20-3 A cubical box of edge L, containing n moles of an ideal gas. A molecule of mass m and velocity **v** is about to collide with the shaded wall of area L^2. A normal to that wall is shown.

The molecules of gas in the box are moving in all directions and with various speeds, bumping into each other and bouncing from the walls of the box like balls in a racquetball court. We ignore (for the time being) collisions of the molecules with one another and consider only elastic collisions with the walls.

Figure 20-3 shows a typical gas molecule, of mass m and velocity **v**, that is about to collide with the shaded wall. Because we assume that any collision of a molecule with a wall is elastic, when this molecule collides with the shaded wall, the only component of its velocity that is changed is the x component, and that component is reversed. This means that the only change in the particle's momentum is then along the x axis, and that change is

$$\Delta p_x = (-mv_x) - (mv_x) = -2mv_x.$$

Hence the momentum Δp_x delivered to the wall by the molecule during the collision is $+2mv_x$. (Because in this book the symbol p represents both momentum and pressure, we must be careful to note that here p represents momentum and is a vector quantity.)

The molecule of Fig. 20-3 will hit the shaded wall repeatedly. The time Δt between collisions is the time the molecule takes to travel to the opposite wall and back again (a distance of $2L$) at speed v_x. Thus Δt is equal to $2L/v_x$. (Note that this result holds even if the molecule bounces off any of the other walls along the way, because those walls are parallel to x and so cannot change v_x.) Thus the average rate at which momentum is delivered to the shaded wall by this single molecule is

$$\frac{\Delta p_x}{\Delta t} = \frac{2mv_x}{2L/v_x} = \frac{mv_x^2}{L}.$$

From Newton's second law ($\mathbf{F} = d\mathbf{p}/dt$), the rate at which momentum is delivered to the wall is the force acting on that wall. To find the total force, we must add up the contributions of all the molecules that strike the wall, al-

lowing for the possibility that they all have different speeds. Dividing the total force F_x by the area of the wall ($= L^2$) then gives the pressure p on that wall, where now and in the rest of this discussion, p represents pressure. Thus,

$$p = \frac{F_x}{L^2} = \frac{mv_{x1}^2/L + mv_{x2}^2/L + \cdots + mv_{xN}^2/L}{L^2}$$
$$= \left(\frac{m}{L^3}\right)(v_{x1}^2 + v_{x2}^2 + \cdots + v_{xN}^2), \quad (20\text{-}14)$$

where N is the number of molecules in the box.

Since $N = nN_A$, there are nN_A terms in the second parentheses of Eq. 20-14. So we can replace that quantity by $nN_A\overline{v_x^2}$, where $\overline{v_x^2}$ is the average value of the square of the x components of all the molecular speeds. Equation 20-14 then becomes

$$p = \frac{nmN_A}{L^3}\overline{v_x^2}.$$

But mN_A is the molar mass M of the gas (that is, the mass of 1 mole of the gas). Also, L^3 is the volume of the box, so

$$p = \frac{nM\overline{v_x^2}}{V}. \quad (20\text{-}15)$$

For any molecule, $v^2 = v_x^2 + v_y^2 + v_z^2$. Because there are many molecules and because they are all moving in random directions, the average values of the squares of their velocity components are equal, so that $v_x^2 = \frac{1}{3}v^2$. Thus Eq. 20-15 becomes

$$p = \frac{nM\overline{v^2}}{3V}. \quad (20\text{-}16)$$

The square root of $\overline{v^2}$ is a kind of average speed, called the **root-mean-square speed** of the molecules and symbolized by v_{rms}. Its name describes it rather well: you *square* each speed, you find the *mean* (that is, the average) of all these squared speeds, and then you take the square *root*. With $\sqrt{\overline{v^2}} = \overline{v}_{rms}$, we can then write Eq. 20-16 as

$$p = \frac{nMv_{rms}^2}{3V}. \quad (20\text{-}17)$$

Equation 20-17 is very much in the spirit of kinetic theory. It tells us how the pressure of the gas (a purely macroscopic quantity) depends on the speed of the molecules (a purely microscopic quantity).

We can turn Eq. 20-17 around and use it to calculate v_{rms}. Combining Eq. 20-17 with the ideal gas law ($pV = nRT$) leads to

$$v_{rms} = \sqrt{\frac{3RT}{M}}. \quad (20\text{-}18)$$

TABLE 20-1 SOME MOLECULAR SPEEDS
AT ROOM TEMPERATURE ($T = 300$ K)^a

TABLE 20-1 SOME MOLECULAR SPEEDS
AT ROOM TEMPERATURE ($T = 300$ K)^a

GAS	MOLAR MASS (10^{-3} kg/mol)	v_{rms} (m/s)
Hydrogen (H_2)	2.02	1920
Helium (He)	4.0	1370
Water vapor (H_2O)	18.0	645
Nitrogen (N_2)	28.0	517
Oxygen (O_2)	32.0	483
Carbon dioxide (CO_2)	44.0	412
Sulfur dioxide (SO_2)	64.1	342

^{a}For convenience, we often set room temperature = 300 K even though (at 27°C or 81°F) that represents a fairly warm room.

Table 20-1 shows some rms speeds calculated from Eq. 20-18. The speeds are surprisingly high. For hydrogen molecules at room temperature (300 K), the rms speed is 1920 m/s or 4300 mi/h—faster than a speeding bullet! On the surface of the Sun, where the temperature is 2×10^6 K, the rms speed of hydrogen molecules would be 82 times larger than at room temperature were it not for the fact that at such high speeds, the molecules cannot survive the collisions among themselves. Remember too that the rms speed is only a kind of average speed; many molecules move much faster than this, and some much slower.

The speed of sound in a gas is closely related to the rms speed of the molecules of that gas. In a sound wave, the disturbance is passed on from molecule to molecule by means of collisions. The wave cannot move any faster than the "average" speed of the molecules. In fact, the speed of sound must be somewhat less than this "average" molecular speed because not all molecules are moving in exactly the same direction as the wave. As examples, at room temperature, the rms speeds of hydrogen and nitrogen molecules are 1920 m/s and 517 m/s, respectively. The speeds of sound in these two gases at this temperature are 1350 m/s and 350 m/s, respectively.

A question often arises: If molecules move so fast, why does it take as long as a minute or so before you can smell perfume if someone opens a bottle across a room? A glance ahead at Fig. 20-4 (Section 20-6) suggests that—although the molecules move very fast between collisions—a given molecule will wander only very slowly away from its release point.

SAMPLE PROBLEM 20-3

Here are five numbers: 5, 11, 32, 67, and 89.

(a) What is the average value $\bar{n}$ of these numbers?

SOLUTION: We find this from

$$\bar{n} = \frac{5 + 11 + 32 + 67 + 89}{5} = 40.8 \quad \text{(Answer)}$$

(b) What is the rms value n_{rms} of these numbers?

SOLUTION: We find this from

$$n_{\text{rms}} = \sqrt{\frac{5^2 + 11^2 + 32^2 + 67^2 + 89^2}{5}}$$

$$= 52.1. \quad \text{(Answer)}$$

The rms value is greater than the average value because the larger numbers—being squared—are relatively more important in forming the rms value. To test this, let us replace 89 in our set of five numbers by 300. The average value of the new set of five numbers (as you can easily show) is 2.0 times the previous average value. The rms value, however, is 2.7 times the previous rms value.

The rms values of variables occur in many branches of physics and engineering. The value 120 volts printed on an electric light bulb, for example, is an rms voltage.

20-5 TRANSLATIONAL KINETIC ENERGY

We again consider a single molecule of an ideal gas as it moves around in the box of Fig. 20-3, but we now assume that its speed changes when it collides with other molecules. Its translational kinetic energy at any instant is $\frac{1}{2}mv^2$. Its *average* translational kinetic energy over the time that we watch it is

$$\bar{K} = \overline{\tfrac{1}{2}mv^2} = \tfrac{1}{2}m\overline{v^2} = \tfrac{1}{2}mv_{\text{rms}}^2, \quad (20\text{-}19)$$

in which we make the assumption that the average speed of the molecule during our observation is the same as the average speed of all the molecules at any given time. (Provided the total energy of the gas is not changing and we observe our molecule for long enough, this assumption is appropriate.) Substituting for v_{rms} from Eq. 20-18 leads to

$$\bar{K} = (\tfrac{1}{2}m) \frac{3RT}{M}.$$

But M/m, the molar mass divided by the mass of a molecule, is simply Avogadro's number, so

$$\bar{K} = \frac{3RT}{2N_A},$$

which we can write as

$$\bar{K} = \tfrac{3}{2}kT. \quad (20\text{-}20)$$

The constant k, called the **Boltzmann constant,** is the ratio of the gas constant R to Avogadro's number N_A. It is sometimes called the gas constant for a single molecule (rather than for a mole), and its value is

$$k = \frac{R}{N_A} = \frac{8.31 \text{ J/mol} \cdot \text{K}}{6.02 \times 10^{23} \text{ mol}^{-1}} = 1.38 \times 10^{-23} \text{ J/K}$$

$$= 8.62 \times 10^{-5} \text{ eV/K}. \qquad (20\text{-}21)$$

Equation 20-20 tells us something unexpected:

At a given temperature T, all ideal gas molecules—no matter what their mass—have the same average translational kinetic energy, namely, $\frac{3}{2}kT$. When we measure the temperature of a gas, we are also measuring the average translational kinetic energy of its molecules.

CHECKPOINT **2:** A gas mixture consists of molecules of types 1, 2, and 3, with molecular masses $m_1 > m_2 > m_3$. Rank the three types according to (a) average kinetic energy and (b) rms speed, greatest first.

SAMPLE PROBLEM 20-4

What is the average translational kinetic energy (in electron-volts) of the oxygen molecules in air at room temperature (= 300 K)? Of the nitrogen molecules?

SOLUTION: The average translational kinetic energy depends only on the temperature and not on the nature of the molecule. For both oxygen and nitrogen molecules it is given by Eq. 20-20 as

$$\overline{K} = \tfrac{3}{2}kT = (\tfrac{3}{2})(8.62 \times 10^{-5} \text{ eV/K})(300 \text{ K})$$

$$= 0.039 \text{ eV.} \qquad \text{(Answer)}$$

Physicists find it useful to remember that the mean translational kinetic energy of *any* molecule at room temperature is about $\frac{1}{25}$ eV, which is essentially the above result.

From Table 20-1 we see that the rms speed of the molecules of oxygen (for which $M = 32.0$ g/mol) is 483 m/s. That for the molecules of nitrogen ($M = 28.0$ g/mol) is 517 m/s. Thus the lighter molecule has the larger rms speed, consistent with the fact that the two kinds of molecules have the same average translational kinetic energy.

20-6 MEAN FREE PATH

We continue to examine the motion of molecules in an ideal gas. Figure 20-4 shows the path of a typical molecule as it moves through the gas, changing both speed and di-

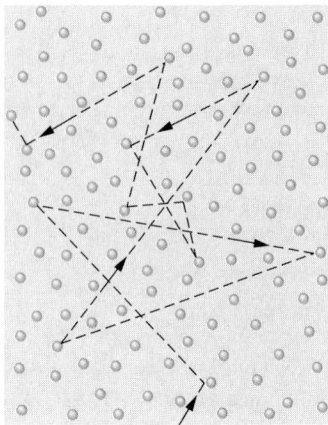

FIGURE 20-4   A molecule traveling through a gas, colliding with other gas molecules in its path. Although the other molecules are shown as stationary, they are also moving in a similar fashion.

rection abruptly as it collides elastically with other molecules. Between collisions, our typical molecule moves in a straight line at constant speed. Although the figure shows all the other molecules as stationary, they too are moving in much the same way.

One useful parameter to describe this random motion is the **mean free path** λ. As its name implies, λ is the average distance traversed by a molecule between collisions. We expect λ to vary inversely with N/V, the number of molecules per unit volume (or density of molecules). The larger N/V is, the more collisions there should be and the smaller the mean free path. We also expect λ to vary inversely with the size of the molecules. (If the molecules were true points, they would never collide and the mean free path would be infinite.) Thus the larger the molecules, the smaller the mean free path. We can even predict that λ should vary (inversely) as the *square* of the molecular diameter because the cross section of a molecule—not its diameter—determines its effective target area.

The expression for the mean free path does, in fact, turn out to be

$$\lambda = \frac{1}{\sqrt{2}\pi d^2 \, N/V} \qquad \text{(mean free path).} \quad (20\text{-}22)$$

To justify Eq. 20-22, we focus attention on a single molecule and assume—as Fig. 20-4 suggests—that our molecule is traveling with a constant speed v and that all the other molecules are at rest. Later, we shall relax this assumption.

We assume further that the molecules are spheres of diameter d. A collision will then take place if the centers of the molecules come within a distance d of each other, as in Fig. 20-5a. Another, more helpful way to look at the situation is to consider our single molecule to have a *radius* of d and all the other molecules to be *points,* as in Fig. 20-5b. This does not change our criterion for a collision.

As our single molecule zigzags through the gas, it sweeps out a short cylinder of cross-sectional area πd^2

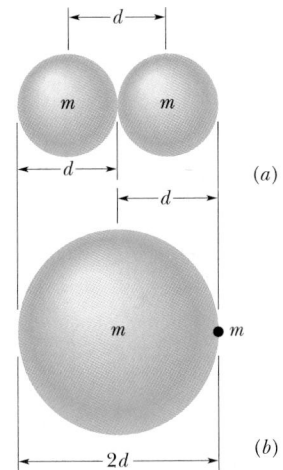

FIGURE 20-5 (*a*) A collision occurs when the centers of two molecules come within a distance *d* of each other, *d* being the molecular diameter. (*b*) An equivalent but more convenient representation is to think of the moving molecule as having a *radius d* and all other molecules as being points. The condition for a collision is unchanged.

speed of our single molecule *relative to the other molecules,* which are moving. It is this latter average speed that determines the number of collisions. A detailed calculation, taking into account the actual speed distribution of the molecules, gives $\overline{v}_{rel} = \sqrt{2}\,\overline{v}$ and thus the factor $\sqrt{2}$.

The mean free path of air molecules at sea level is about 0.1 μm. At an altitude of 100 km, the density of air has dropped to such an extent that the mean free path rises to about 16 cm. At 300 km, the mean free path is about 20 km. A problem faced by those who would study the physics and chemistry of the upper atmosphere in the laboratory is the unavailability of containers large enough to hold gas samples that simulate upper atmospheric conditions. Yet studies of the concentrations of Freon, carbon dioxide, and ozone in the upper atmosphere are of vital public concern.

between successive collisions. If we watch this molecule for a time interval Δt, it moves a distance $v\,\Delta t$, where v is its assumed speed. Thus, if we align all the short cylinders swept out in Δt, we form a composite cylinder (Fig. 20-6) of length $v\,\Delta t$ and volume $(\pi d^2)(v\,\Delta t)$. The number of collisions that occur is then equal to the number of (point) molecules that lie within this cylinder.

Since N/V is the number of molecules per unit volume, the number of collisions is N/V times the volume of the cylinder, or $(N/V)(\pi d^2 v\,\Delta t)$. The mean free path is the length of the path (and of the cylinder) divided by this number:

$$\lambda = \frac{\text{length of path}}{\text{number of collisions}} \approx \frac{v\,\Delta t}{\pi d^2 v\,\Delta t\,N/V}$$

$$= \frac{1}{\pi d^2\,N/V}. \tag{20-23}$$

This equation is only approximate because it is based on the assumption that all the molecules except one are at rest. In fact, *all* the molecules are moving; when this is taken properly into account, Eq. 20-22 results. Note that it differs from the (approximate) Eq. 20-23 only by a factor of $1/\sqrt{2}$.

We can even get a glimpse of what is "approximate" about Eq. 20-23. The v in the numerator and that in the denominator are—strictly—not the same. The v in the numerator is $\overline{v}$, the mean speed of the molecule *relative to the container.* The v in the denominator is $\overline{v}_{rel}$, the mean

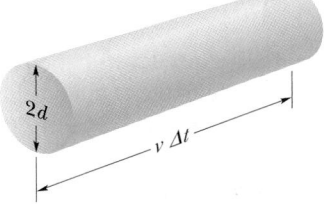

FIGURE 20-6 In time Δt the moving molecule effectively sweeps out a cylinder of length $v\,\Delta t$ and radius d.

SAMPLE PROBLEM 20-5

The molecular diameters of gas molecules of different kinds can be found experimentally by measuring the rates at which the different gases diffuse (spread) into each other. For oxygen, $d = 2.9 \times 10^{-10}$ m has been reported.

(a) What is the mean free path for oxygen at room temperature ($T = 300$ K) and at an atmospheric pressure of 1.0 atm? Assume it is an ideal gas under these conditions.

SOLUTION: Let us first find N/V, the number of molecules per unit volume under these conditions. From the ideal gas law, 1.0 mol of any ideal gas occupies a volume equal to

$$V = \frac{nRT}{p} = \frac{(1.0\ \text{mol})(8.31\ \text{J/mol}\cdot\text{K})(300\ \text{K})}{(1.0\ \text{atm})(1.01 \times 10^5\ \text{Pa/atm})}$$

$$= 2.47 \times 10^{-2}\ \text{m}^3.$$

Because we have 1.0 mol of the gas in that volume, the number of molecules per unit volume is

$$\frac{N}{V} = \frac{nN_A}{V} = \frac{(1.0\ \text{mol})(6.02 \times 10^{23}\ \text{molecules/mol})}{2.47 \times 10^{-2}\ \text{m}^3}$$

$$= 2.44 \times 10^{25}\ \text{molecules/m}^3.$$

Equation 20-22 then gives

$$\lambda = \frac{1}{\sqrt{2}\pi d^2\,N/V}$$

$$= \frac{1}{(\sqrt{2}\pi)(2.9 \times 10^{-10}\ \text{m})^2(2.44 \times 10^{25}\ \text{m}^{-3})}$$

$$= 1.1 \times 10^{-7}\ \text{m}. \qquad \text{(Answer)}$$

This is about 380 molecular diameters. On average, the molecules in such a gas are only about 11 molecular diameters apart.

(b) If the average speed of an oxygen molecule is 450 m/s, what is the average collision rate?

SOLUTION: We find this rate by dividing the average speed by the mean free path to get

$$\text{rate} = \frac{v}{\lambda} = \frac{450 \text{ m/s}}{1.1 \times 10^{-7} \text{ m}} = 4.1 \times 10^9 \text{ s}^{-1}. \quad \text{(Answer)}$$

Thus—on average—every oxygen molecule makes more than 4 billion collisions per second!

CHECKPOINT **3:** One mole of gas A, with molecular radius $2d_0$ and average molecular speed v_0, is placed inside a certain container. One mole of gas B, with molecular radius d_0 and average molecular speed $2v_0$ (the molecules of B are smaller but faster), is placed in an identical container. Which gas has the greater average collision rate within its container?

20-7 THE DISTRIBUTION OF MOLECULAR SPEEDS (OPTIONAL)

The root-mean-square speed v_{rms} gives us a general idea of the molecular speed in a gas at a given temperature. We often want to know more. For example, what fraction of the molecules have speeds greater than the rms value? Greater than twice the rms value? To answer such questions, we need to know how the possible values of speed are distributed among the molecules. Figure 20-7a shows this distribution for oxygen molecules at room temperature ($T = 300$ K); Fig. 20-7b compares it with the distribution at $T = 80$ K.

In 1852, Scottish physicist James Clerk Maxwell first solved the problem of finding the speed distribution of gas molecules. His result, known as **Maxwell's speed distribution law,** is

$$P(v) = 4\pi \left(\frac{M}{2\pi RT} \right)^{3/2} v^2 e^{-Mv^2/2RT}. \quad (20\text{-}24)$$

Here v is the molecular speed, T is the gas temperature, M is the molar mass of the gas, and R is the gas constant. It is this equation that is plotted in Fig. 20-7a,b. The quantity $P(v)$ in Eq. 20-24 and Fig. 20-7 is a *probability distribution function*, defined as follows: The product $P(v) \, dv$ (which is a dimensionless quantity) is the fraction of molecules whose speeds lie in the range v to $v + dv$.

As Fig. 20-7a shows, this fraction is equal to the area of a strip whose height is $P(v)$ and whose width is dv. The total area under the distribution curve corresponds to the fraction of the molecules whose speeds lie between zero and infinity. All molecules fall into this category, so the value of this total area is unity.

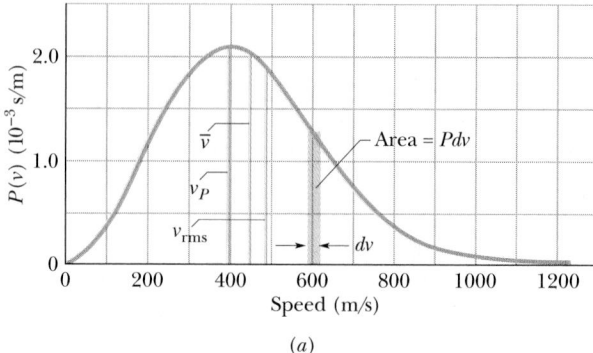

(a)

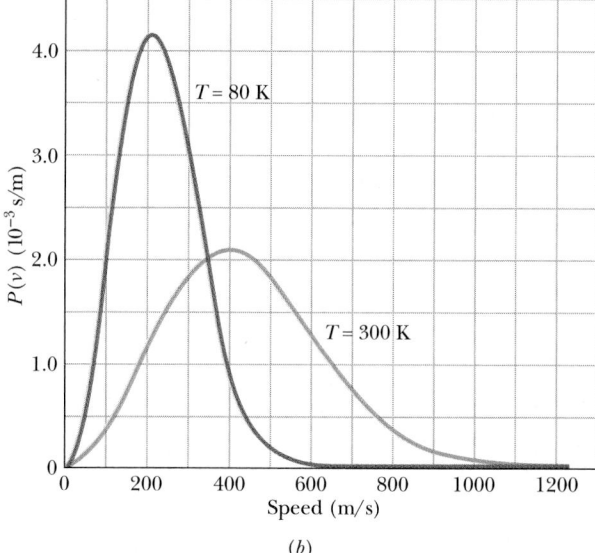

(b)

FIGURE 20-7 (a) The Maxwell speed distribution for oxygen molecules at $T = 300$ K. The three characteristic speeds are marked. (b) The curves for 300 K and 80 K. Note that the molecules move more slowly at the lower temperature. Because these are probability distributions, the area under each curve has a numerical value of unity.

Figure 20-7a also shows the root-mean-square speed v_{rms} ($= 483$ m/s) and two other measures of the speed distribution of the oxygen molecules. The *most probable speed* v_P ($= 395$ m/s) is the speed at which $P(v)$ is a maximum. The *average speed* $\bar{v}$ ($= 445$ m/s) is—as its name suggests—a simple average of the molecular speeds. A small number of molecules, lying in the extreme right-hand tail of the distribution curve, can have speeds that are several times the average speed. This simple fact, as we shall demonstrate, makes possible both rain and sunshine.

Rain: The speed distribution of water molecules in, say, a pond at summertime temperatures can be represented by a curve similar to that of Fig. 20-7a. Most of the molecules do not have nearly enough kinetic energy to escape from the water through its surface. However, small numbers of very fast molecules with speeds far out in the tail of the curve can do so. It is these water molecules that evaporate, making clouds and rain a possibility.

As the fast water molecules leave the surface, carrying energy with them, the temperature of the remaining water is maintained by heat transfer from the surroundings. Other fast molecules—produced in particularly favorable collisions—quickly take the place of those that have left, and the speed distribution is maintained.

Sunshine: Let the distribution curve of Fig. 20-7a now refer to protons in the core of the Sun. The Sun's energy is supplied by a nuclear fusion process that starts with the merging of two protons. However, protons repel each other because of their electrical charges, and protons of average speed do not have enough kinetic energy to overcome the repulsion and get close enough to merge. Very fast protons with speeds in the tail of the distribution curve can do so, however, and thus the Sun can shine.

SAMPLE PROBLEM 20-6

A container is filled with oxygen gas maintained at room temperature (300 K). What fraction of the molecules have speeds in the range 599–601 m/s? The molar mass M of oxygen is 0.0320 kg/mol.

SOLUTION: This speed interval Δv ($= 2$ m/s) is so small that we can treat it as a differential and say that the fraction frac that we seek is given very closely by $P(v)\,\Delta v$, where $P(v)$ is to be evaluated at $v = 600$ m/s, the midpoint of the interval; see the gold strip in Fig. 20-7a. Thus, using Eq. 20-24, we find

$$\text{frac} = P(v)\,\Delta v = 4\pi \left(\frac{M}{2\pi RT}\right)^{3/2} v^2 e^{-Mv^2/2RT}\,\Delta v.$$

For convenience in calculating, let us break this down into five factors, as

$$\text{frac} = (4\pi)(A)(v^2)(e^B)(\Delta v) \qquad (20\text{-}25)$$

in which A and B are

$$A = \left(\frac{M}{2\pi RT}\right)^{3/2} = \left(\frac{0.0320 \text{ kg/mol}}{(2\pi)(8.31 \text{ J/mol}\cdot\text{K})(300 \text{ K})}\right)^{3/2}$$

$$= 2.92 \times 10^{-9} \text{ s}^3/\text{m}^3$$

and

$$B = -\frac{Mv^2}{2RT} = -\frac{(0.0320 \text{ kg/mol})(600 \text{ m/s})^2}{(2)(8.31 \text{ J/mol}\cdot\text{K})(300 \text{ K})}$$

$$= -2.31.$$

Substituting A and B into Eq. 20-25 yields

$$\text{frac} = (4\pi)(A)(v^2)(e^B)(\Delta v)$$

$$= (4\pi)(2.92 \times 10^{-9} \text{ s}^3/\text{m}^3)(600 \text{ m/s})^2(e^{-2.31})(2 \text{ m/s})$$

$$= 2.62 \times 10^{-3}. \qquad \text{(Answer)}$$

Thus, at room temperature, 0.262% of the oxygen molecules will have speeds that lie in the narrow range between 599 and 601 m/s. If the gold strip of Fig. 20-7a were drawn to the scale of this problem, it would be a very thin strip indeed.

SAMPLE PROBLEM 20-7

(a) What is the average speed v of oxygen gas molecules at $T = 300$ K? The molar mass M of oxygen is 0.0320 kg/mol.

SOLUTION: To find the average speed, we weight each speed v by $P(v)\,dv$, which is the fraction of the molecules whose speeds lie in the interval v to $v + dv$. We then add up (that is, integrate) these fractions over the entire range of speeds. Thus

$$\bar{v} = \int_0^\infty v P(v)\,dv. \qquad (20\text{-}26)$$

The next step is to substitute for $P(v)$ from Eq. 20-24 and evaluate the integral that results. From a table of integrals* we find

$$\bar{v} = \sqrt{\frac{8RT}{\pi M}} \qquad \text{(average speed)}. \qquad (20\text{-}27)$$

Substituting numerical values yields

$$\bar{v} = \sqrt{\frac{(8)(8.31 \text{ J/mol}\cdot\text{K})(300 \text{ K})}{(\pi)(0.0320 \text{ kg/mol})}}$$

$$= 445 \text{ m/s}. \qquad \text{(Answer)}$$

(b) What is the root-mean-square speed v_{rms} of the oxygen molecules?

SOLUTION: We proceed as in (a) above except that we multiply v^2 (rather than simply v) by the weighting factor $P(v)\,dv$. This leads, after integration, to

$$\overline{v^2} = \int_0^\infty v^2 P(v)\,dv = \frac{3RT}{M}.$$

The rms speed is the square root of this quantity, or

$$v_{\text{rms}} = \sqrt{\overline{v^2}} = \sqrt{\frac{3RT}{M}} \qquad \text{(rms speed)}. \qquad (20\text{-}28)$$

Equation 20-28 is identical to Eq. 20-18, which we derived earlier. The numerical calculation gives

$$v_{\text{rms}} = \sqrt{\frac{(3)(8.31 \text{ J/mol}\cdot\text{K})(300 \text{ K})}{(0.0320 \text{ kg/mol})}}$$

$$= 483 \text{ m/s}. \qquad \text{(Answer)}$$

(c) What is the most probable speed v_P?

SOLUTION: The most probable speed is the speed at which $P(v)$ of Eq. 20-24 has its maximum value. We find it by re-

---

*One such table is found in Section A of the familiar *CRC Handbook of Chemistry and Physics*; see integral 667 listed under "Definite Integrals."

TABLE 20-2 SPEED PARAMETERS FOR THE MAXWELL SPEED DISTRIBUTION

PARAMETER	SYMBOL	FORMULA	FOR OXYGEN AT 300 K
Most probable speed	v_P	$\sqrt{2RT/M}$	395 m/s
Average speed	$\bar{v}$	$\sqrt{8RT/\pi M}$	445 m/s
Root-mean-square speed	v_{rms}	$\sqrt{3RT/M}$	483 m/s

quiring that $dP/dv = 0$ and solving for v. Doing so yields (as you should show)

$$v_P = \sqrt{\frac{2RT}{M}} \qquad \text{(most probable speed).} \qquad (20\text{-}29)$$

Numerically, this yields

$$v_P = \sqrt{\frac{(2)(8.31 \text{ J/mol} \cdot \text{K})(300 \text{ K})}{(0.0320 \text{ kg/mol})}}$$

$$= 395 \text{ m/s.} \qquad \text{(Answer)}$$

Table 20-2 summarizes the three measures of the Maxwell speed distribution.

20-8 THE MOLAR SPECIFIC HEATS OF AN IDEAL GAS

In this section, we want to derive from atomic or molecular considerations an expression for the internal energy E_{int} of an ideal gas. We shall then use that result to derive an expression for the molar specific heats of an ideal gas.

Internal Energy E_{int}

Let us first assume that our ideal gas is a *monatomic gas* (which has individual atoms rather than molecules), such as helium, neon, or argon. Next, recall from Chapter 8 that internal energy is the energy associated with random motions of atoms and molecules. So let us assume that the internal energy E_{int} of our ideal gas is simply the sum of the translational kinetic energies of its atoms. (Individual atoms do not have rotational kinetic energy.) The average translational kinetic energy of a single atom depends only on the gas temperature and is given by Eq. 20-20 as $\bar{K} = \frac{3}{2}kT$. A sample of n moles of such a gas contains nN_A atoms. The internal energy E_{int} of the sample is then

$$E_{int} = (nN_A)\bar{K} = (nN_A)(\tfrac{3}{2}kT)$$

or, since $N_A k = R$, the gas constant,

$$E_{int} = \tfrac{3}{2}nRT \qquad \text{(monatomic ideal gas).} \qquad (20\text{-}30)$$

Thus,

The internal energy E_{int} of an ideal gas is a function of the gas temperature *only*; it does not depend on any other variable.

With Eq. 20-30 in hand, we are now able to derive an expression for the molar specific heat of an ideal gas. Actually, we derive two expressions—one for the case in which the volume of the gas remains constant as heat is added to it, and one for the case in which the pressure of the gas remains constant as heat is added. The symbols for these two molar specific heats are C_V and C_p, respectively. (By convention, the capital letter C is used in both cases, even though C_V and C_P represent a type of specific heat.)

Molar Specific Heat at Constant Volume

Figure 20-8*a* shows n moles of an ideal gas at pressure p and temperature T, confined to a cylinder of fixed volume V. This *initial state i* of the gas is marked on the p-V diagram of Fig. 20-8*b*. Suppose now that you add a small amount of heat Q to the gas, by slowly turning up the temperature of the thermal reservoir on which the cylinder rests. The gas temperature rises a small amount to $T + \Delta T$, and its pressure to $p + \Delta p$, bringing the gas to *final state f*.

The defining equation for C_V, the molar specific heat at constant volume, is, in parallel with Eq. 19-15,

$$Q = nC_V \Delta T \qquad \text{(constant volume).} \qquad (20\text{-}31)$$

Substituting this expression for Q into the first law of thermodynamics (Eq. 19-24), we find

$$\Delta E_{int} = nC_V \Delta T - W.$$

With the volume held constant, the gas cannot do any work; so $W = 0$. Then solving for C_V we obtain

$$C_V = \frac{1}{n}\frac{\Delta E_{int}}{\Delta T}. \qquad (20\text{-}32)$$

From Eq. 20-30 we see that $\Delta E_{int}/\Delta T = \frac{3}{2}nR$. Substituting

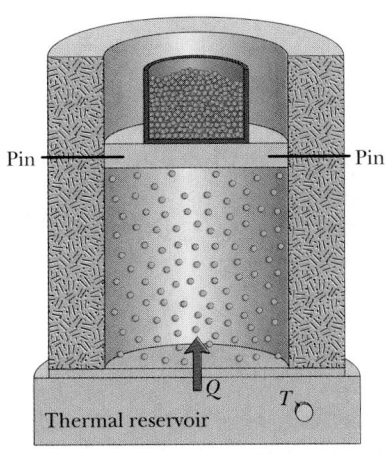

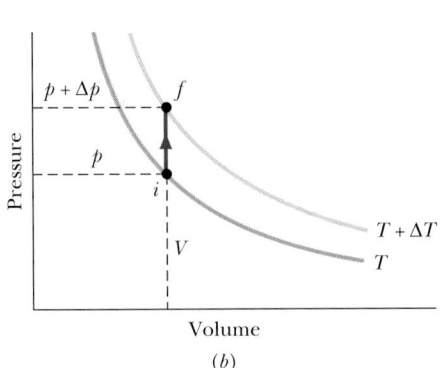

FIGURE 20-8
(*a*) The temperature of an ideal gas is raised from T to $T + \Delta T$ in a constant-volume process. Heat is added but no work is done. (*b*) The process on a *p-V* diagram.

atoms) are greater than those for monatomic gases for reasons that will be suggested in Section 20-9.

We can now generalize Eq. 20-30 for the internal energy of any ideal gas by substituting C_V for $\frac{3}{2}R$; we get

$$E_{\text{int}} = nC_V T \qquad \text{(any ideal gas).} \qquad (20\text{-}34)$$

This equation applies not only to an ideal monatomic gas but also to diatomic and polyatomic ideal gases, provided the appropriate value of C_V is used. Just as with Eq. 20-30, we see that the internal energy depends on the temperature of the gas but not on its pressure or density.

When an ideal gas that is confined to a container undergoes a temperature change ΔT, then from either Eq. 20-32 or Eq. 20-34 we can write the resulting change in its internal energy as

$$\Delta E_{\text{int}} = nC_V \, \Delta T \qquad \begin{array}{l}\text{(ideal gas,}\\\text{any process).}\end{array} \qquad (20\text{-}35)$$

This equation tells us:

A change in the internal energy E_{int} of a confined ideal gas depends on the change in the gas temperature only; it does *not* depend on what type of process produces the change in the temperature.

this result into Eq. 20-32 yields

$$C_V = \tfrac{3}{2}R = 12.5 \text{ J/mol}\cdot\text{K} \qquad \begin{array}{l}\text{(monatomic}\\\text{gas).}\end{array} \qquad (20\text{-}33)$$

As Table 20-3 shows, this prediction of the kinetic theory (for ideal gases) agrees very well with experiment for real monatomic gases, the case that we have assumed. The (predicted and) experimental values of C_V for *diatomic gases* (which have molecules with two atoms) and *polyatomic gases* (which have molecules with more than two

As examples, consider the three paths between the two isotherms in the *p-V* diagram of Fig. 20-9. Path 1 represents a constant-volume process. Path 2 represents a constant-pressure process (that we are about to examine). And path 3 represents a process in which no heat is ex-

TABLE 20-3 MOLAR SPECIFIC HEATS

MOLECULE		EXAMPLE	C_V (J/mol·K)
Monatomic	Ideal		$\frac{3}{2}R = 12.5$
	Real	He	12.5
		Ar	12.6
Diatomic	Ideal		$\frac{5}{2}R = 20.8$
	Real	N_2	20.7
		O_2	20.8
Polyatomic	Ideal		$3R = 24.9$
	Real	NH_4	29.0
		CO_2	29.7

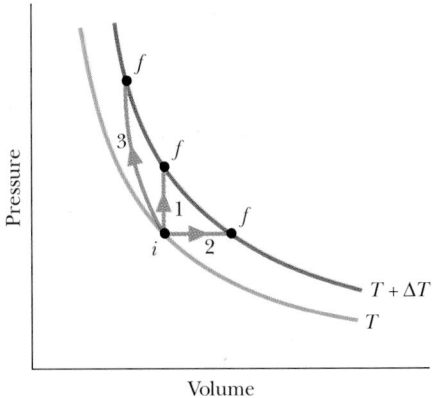

FIGURE 20-9   Three paths representing three different processes that take an ideal gas from an initial state i at temperature T to some final state f at temperature $T + \Delta T$. The change ΔE_{int} in the internal energy of the gas is the same for these three processes and for any others that result in the same change of temperature.

changed with the system's environment (we discuss this in Section 20-11). Although the values of heat Q and work W associated with these three paths differ, as do p_f and V_f, the values of ΔE_{int} associated with the three paths are identical and are all given by Eq. 20-35, because they all involve the same temperature change ΔT. So, no matter what path is actually traversed between T and $T + \Delta T$, we can always use path 1 and Eq. 20-35 to compute ΔE_{int} easily.

Molar Specific Heat at Constant Pressure

We now assume that the temperature of the ideal gas is increased by the same small amount ΔT as previously, but that the necessary heat Q is added with the gas under constant pressure. A mechanism for doing this is shown in Fig. 20-10a; the p-V diagram for the process is plotted in Fig. 20-10b. We can guess at once that the molar specific heat at constant pressure C_p, which we define with

$$Q = nC_p\, \Delta T \qquad \text{(constant pressure),} \qquad (20\text{-}36)$$

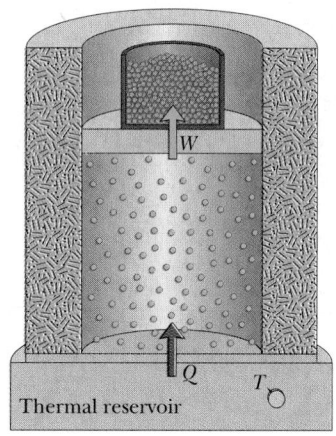

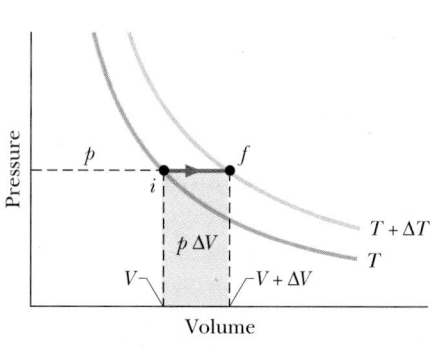

(a)

(b)

FIGURE 20-10 (a) The temperature of an ideal gas is raised from T to $T + \Delta T$ in a constant-pressure process. Heat is added and work is done in lifting the loaded piston. (b) The process on a p-V diagram. The work $p\, \Delta V$ is given by the shaded area.

is *greater* than the molar specific heat at constant volume, because energy must now be supplied not only to raise the temperature of the gas but also for the gas to do work, that is, to lift the weighted piston of Fig. 20-10a.

To relate C_p to C_V, we start with the first law of thermodynamics (Eq. 19-24):

$$\Delta E_{int} = Q - W. \qquad (20\text{-}37)$$

We next replace each term in Eq. 20-37. For ΔE_{int}, we substitute from Eq. 20-35. For Q, we substitute from Eq. 20-36. To replace W, we first note that since the pressure remains constant, Eq. 19-23 tells us that $W = p\, \Delta V$. Then we note that, using the ideal gas equation ($pV = nRT$), we can write

$$W = p\, \Delta V = nR\, \Delta T.$$

Making these substitutions in Eq. 20-37, and then dividing through by $n\, \Delta T$, we find

$$C_V = C_p - R,$$

and then

$$C_p = C_V + R. \qquad (20\text{-}38)$$

This prediction of kinetic theory agrees well with experiment, not only for monatomic gases but for gases in general, as long as their density is low enough so that we may treat them as ideal.

CHECKPOINT **4:** The figure shows five paths traversed by a gas on a p-V diagram. Rank the paths according to the change in internal energy of the gas, greatest first.

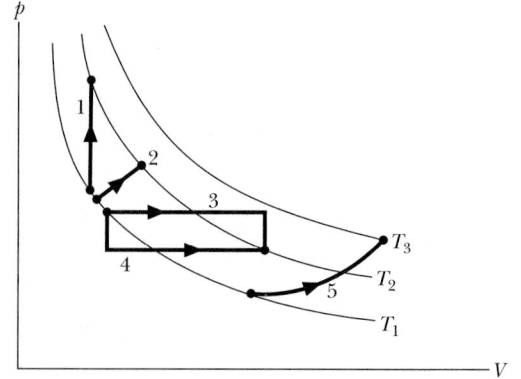

SAMPLE PROBLEM 20-8

A bubble of 5.00 mol of (monatomic) helium is submerged at a certain depth in liquid water when the water (and thus the helium) undergoes a temperature increase ΔT of 20.0 C° at constant pressure. As a result, the bubble expands.

(a) How much heat Q is added to the helium during the expansion and temperature increase?

SOLUTION: Treating the helium as an ideal gas, we use Eqs. 20-38 and 20-33 to write

$$C_p = C_V + R = \tfrac{5}{2}R.$$

We substitute this into Eq. 20-36 to obtain

$$Q = nC_p \, \Delta T = n(\tfrac{5}{2}R) \, \Delta T$$
$$= (5.00 \text{ mol})(2.5)(8.31 \text{ J/mol} \cdot \text{K})(20.0 \text{ C}°)$$
$$= 2077.5 \text{ J} \approx 2080 \text{ J}. \qquad \text{(Answer)}$$

(b) What is the change ΔE_{int} in the internal energy of the helium during the temperature increase?

SOLUTION: Even though the temperature of the helium increases at constant pressure (and *not* at constant volume), we use Eq. 20-35 to calculate the change in internal energy:

$$\Delta E_{int} = nC_V \, \Delta T$$
$$= (5.00 \text{ mol})(1.5)(8.31 \text{ J/mol} \cdot \text{K})(20.0 \text{ C}°)$$
$$= 1246.5 \text{ J} \approx 1250 \text{ J}. \qquad \text{(Answer)}$$

(c) How much work W is done by the helium as it expands against the pressure of the surrounding water during the temperature increase?

SOLUTION: From the first law of thermodynamics,

$$W = Q - \Delta E_{int} = 2077.5 \text{ J} - 1246.5 \text{ J}$$
$$= 831 \text{ J}. \qquad \text{(Answer)}$$

Note that during the temperature increase, only a portion (1250 J) of the heat (2080 J) that is transferred into the helium goes to increasing the internal energy of the helium and thus the temperature of the helium. The rest (831 J) is transferred out of the helium as work that the helium does during the expansion. If the water were frozen, it would not allow that expansion. Then the same temperature increase of 20.0 C° would require only 1250 J of heat, because no work would be done by the helium.

20-9 DEGREES OF FREEDOM AND MOLAR SPECIFIC HEATS

As Table 20-3 shows, the prediction that $C_V = \tfrac{3}{2}R$ agrees with experiment for monatomic gases but fails for diatomic and polyatomic gases. Let us try to explain the discrepancy by considering the possibility that molecules with more than one atom can store internal energy in forms other than translational motion.

Figure 20-11 shows kinetic theory models of helium (a monatomic gas), oxygen (diatomic), and methane (polyatomic). On the basis of their structure, it seems reasonable to assume that monatomic molecules—which are essentially pointlike and have only a very small rotational inertia

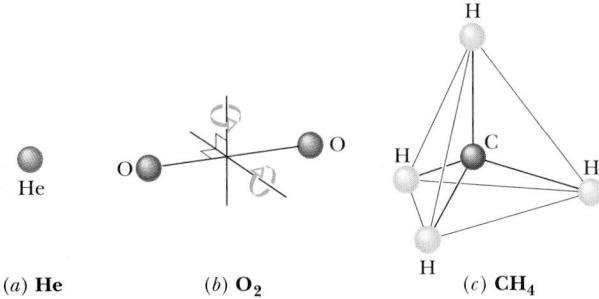

FIGURE 20-11 Models of molecules as used in kinetic theory: (*a*) helium, a typical monatomic molecule; (*b*) oxygen, a typical diatomic molecule (two rotation axes are shown); and (*c*) methane, a typical polyatomic molecule.

about any axis—can store energy only in their translational motion. Diatomic and polyatomic molecules, however, should be able to store substantial additional amounts of energy by rotating or oscillating.

To take these possibilities into account quantitatively, we use the theorem of the **equipartition of energy,** introduced by James Clerk Maxwell:

Every kind of molecule has a certain number f of *degrees of freedom*, which are independent ways in which the molecule can store energy. Each such degree of freedom has associated with it—on average—an energy of $\tfrac{1}{2}kT$ per molecule (or $\tfrac{1}{2}RT$ per mole).

For translational motion, there are three degrees of freedom, corresponding to the three perpendicular axes along which such motion can occur. For rotational motion, a monatomic molecule has no degrees of freedom. A diatomic molecule—the rigid dumbbell of Fig. 20-11*b*—has two rotational degrees of freedom, corresponding to the two perpendicular axes about which it can store rotational energy. Such a molecule cannot store rotational energy about the axis connecting the nuclei of its two constituent atoms because its rotational inertia about this axis is approximately zero. A molecule with more than two atoms has six degrees of freedom, three rotational and three translational.

To extend the treatment of Section 20-8 to ideal diatomic and polyatomic gases, it is necessary to retrace the derivations of that section in detail, replacing Eq. 20-30 ($E_{int} = \tfrac{3}{2}nRT$) by $E_{int} = (f/2)nRT$, where f is the number of degrees of freedom listed in Table 20-4. Doing so leads to the prediction

$$C_V = \left(\frac{f}{2}\right) R = 4.16f \text{ J/mol} \cdot \text{K}, \qquad (20\text{-}39)$$

TABLE 20-4 DEGREES OF FREEDOM FOR VARIOUS MOLECULES

		DEGREES OF FREEDOM			PREDICTED MOLAR SPECIFIC HEATS	
MOLECULE	EXAMPLE	TRANSLATIONAL	ROTATIONAL	TOTAL (f)	C_V (Eq. 20-39)	$C_p = C_V + R$
Monatomic	He	3	0	3	$\frac{3}{2}R$	$\frac{5}{2}R$
Diatomic	O_2	3	2	5	$\frac{5}{2}R$	$\frac{7}{2}R$
Polyatomic	CH_4	3	3	6	$3R$	$4R$

which agrees—as it must—with Eq. 20-33 for monatomic gases ($f = 3$). As Table 20-3 shows, this prediction also agrees with experiment for diatomic gases ($f = 5$), but it is too low for polyatomic gases.

SAMPLE PROBLEM 20-9

A cabin of volume V is filled with air (which we consider to be an ideal diatomic gas) at an initial low temperature T_1. After you light a wood stove, the air temperature increases to T_2. What is the resulting change in the store of internal energy of the air in the cabin?

SOLUTION: This situation differs from other situations we have examined in that the container (the cabin) is not sealed. If it were sealed, the ideal gas law ($pV = nRT$) would tell us that as the temperature of the air in the cabin increased, the air pressure would also. However, because the cabin is not airtight, air molecules leave through various openings as the air temperature increases, so that the air pressure inside the cabin always matches the air pressure outside it.

From Eq. 20-35, the change in the internal energy of the air in the room is

$$\Delta E_{int} = C_V \, \Delta(nT),$$

where both n and T can change. Using the ideal gas law, we may substitute $\Delta(pV)/R$ for $\Delta(nT)$, finding

$$\Delta E_{int} = \frac{C_V}{R} \, \Delta(pV).$$

From this we see that since neither the pressure p nor the volume V of the air within the cabin changes,

$$\Delta E_{int} = 0, \qquad \text{(Answer)}$$

even though the temperature changes.

So why does the cabin feel more comfortable at the higher temperature? There are at least two factors involved. You have a tendency to cool because (1) you emit electromagnetic radiation (thermal radiation) and (2) you exchange energy with air molecules that collide with you. If you increase the room temperature, (1) you increase the amount of thermal radiation you intercept from the surfaces within the room, replacing heat you have emitted; and (2) you increase the kinetic energy of the air molecules that collide with you, so you gain more energy from them.

20-10 A HINT OF QUANTUM THEORY

We can improve the agreement of kinetic theory with experiment by including the oscillations of the atoms in a gas of diatomic or polyatomic molecules. For example, the two atoms in the O_2 molecule of Fig. 20-11b can oscillate toward and away from each other, with the interconnecting bond acting like an oscillating spring. However, experiment shows that such oscillations occur only at relatively high temperatures of the gas—the motion is "turned on" only when the gas molecules have relatively large energies. Rotational motion is also subject to such "turning on," but at a lower gas temperature.

Figure 20-12 is of help in seeing the inception of rotational motion and oscillatory motion. The ratio C_V/R for diatomic hydrogen gas (H_2) is plotted there against temperature, with the temperature scale logarithmic to cover several orders of magnitude. Below about 80 K, we find that $C_V/R = 1.5$. This result implies that only the three translational degrees of freedom of hydrogen are involved in the specific heat. As the temperature increases, the value

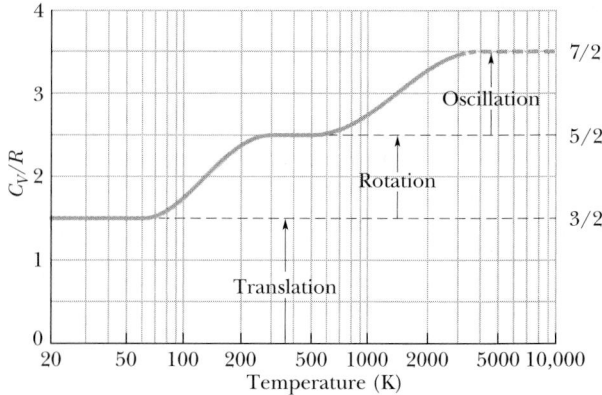

FIGURE 20-12 A plot of C_V/R versus temperature for (diatomic) hydrogen gas. Because rotational and oscillatory motions begin at certain energies, only translation is possible at very low temperatures. As the temperature increases, rotational motion can begin. At still higher temperatures, oscillatory motion can begin.

of C_V/R gradually increases to 2.5, implying that two additional degrees of freedom have become involved. Quantum theory shows that these two degrees of freedom are associated with the rotation of the hydrogen molecules and that this motion requires a certain minimum amount of energy. At very low temperatures (below 80 K), the molecules do not have enough energy to rotate. As the temperature increases from 80 K, first a few molecules and then more and more obtain enough energy to rotate, and C_V/R increases, until all of them are rotating and $C_V/R = 2.5$.

Similarly, quantum theory shows that oscillatory motion of the molecules requires a certain minimum amount of energy. This minimum amount is not met until the molecules reach a temperature of about 1000 K, as shown in Fig. 20-12. As the temperature increases beyond 1000 K, the number of molecules with enough energy to oscillate increases, and C_V/R increases, until all of them are oscillating and $C_V/R = 3.5$. (In Fig. 20-12, the plotted curve stops at 3200 K because at that temperature, the atoms of a hydrogen molecule oscillate so much that they overwhelm their common bond, and the molecule then *dissociates* into two separate atoms.)

20-11 THE ADIABATIC EXPANSION OF AN IDEAL GAS

We saw in Section 18-2 that sound waves are propagated through air and other gases as a series of compressions and expansions; these variations in the medium take place so rapidly that there is no time for heat transfer from one part of the medium to another. As we saw in Section 19-10, a process for which $Q = 0$ is an *adiabatic process*. We can ensure that $Q = 0$ either by carrying out the process very quickly (as in sound waves) or by doing it slowly in a well-insulated container. Let us see what the kinetic theory has to say about adiabatic processes.

Figure 20-13a shows an insulated cylinder containing an ideal gas and resting on an insulating stand. By removing weight from the piston, we can allow the gas to expand adiabatically. As the volume increases, both the pressure and the temperature drop. We shall prove below that the relation between the pressure and the volume during such an adiabatic process is

$$pV^\gamma = \text{a constant} \qquad \begin{array}{l}\text{(adiabatic}\\\text{process),}\end{array} \qquad (20\text{-}40)$$

in which $\gamma = C_p/C_V$, the ratio of the molar specific heats for the gas. On a p-V diagram such as that in Fig. 20-13b, the process occurs along a line (called an *adiabat*) that has the equation $p = (\text{a constant})/V^\gamma$. When the gas goes from an initial state i to a final state f, we can rewrite Eq. 20-40

as

$$p_i V_i^\gamma = p_f V_f^\gamma \qquad \text{(adiabatic process).} \quad (20\text{-}41)$$

We can also write an equation for an adiabatic process in terms of T and V. To do so, we use the ideal gas equation ($pV = nRT$) to eliminate p from Eq. 20-40, finding

$$\left(\frac{nRT}{V}\right) V^\gamma = \text{a constant.}$$

Because n and R are constants, we can rewrite this in the alternative form

$$TV^{\gamma-1} = \text{a constant} \qquad \begin{array}{l}\text{(adiabatic}\\\text{process),}\end{array} \qquad (20\text{-}42)$$

in which the constant is different from that in Eq. 20-40. When the gas goes from an initial state i to a final state f, we can rewrite Eq. 20-42 as

$$T_i V_i^{\gamma-1} = T_f V_f^{\gamma-1} \qquad \text{(adiabatic process).} \quad (20\text{-}43)$$

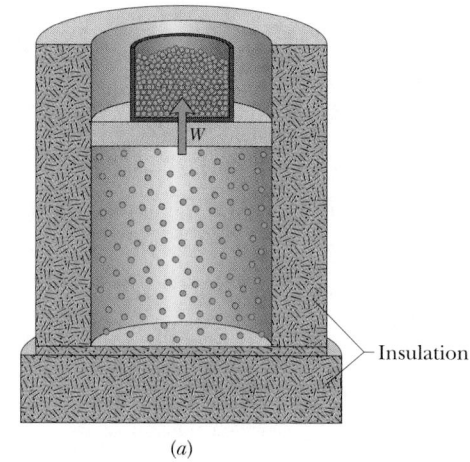

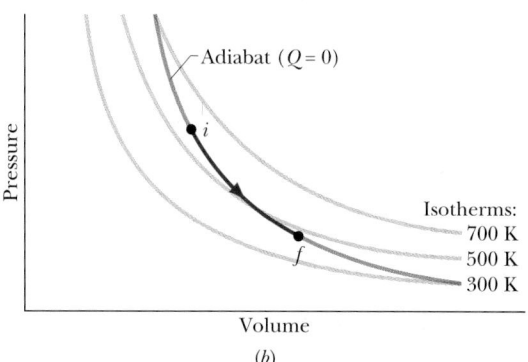

FIGURE 20-13 (a) The volume of an ideal gas is increased by removing weight from the piston. The process is adiabatic ($Q = 0$). (b) The process proceeds from i to f along an adiabat on a p-V diagram.

Proof of Eq. 20-40

Suppose that you remove some lead shot from the piston of Fig. 20-13*a*, allowing the ideal gas to push the piston and the remaining shot upward and thus to increase the volume by a differential amount dV. Since the volume change is tiny, we may assume that the pressure p of the gas on the piston is constant during the change. This assumption allows us to say that the work dW done by the gas during the volume increase is equal to $p\,dV$. From Eq. 19-25, the first law of thermodynamics can then be written as

$$dE_{int} = Q - p\,dV. \tag{20-44}$$

Since the gas is thermally insulated (and thus the expansion is adiabatic), we substitute $Q = 0$. Then we use Eq. 20-35 to substitute $nC_V\,dT$ for dE_{int}. With these substitutions, and after some rearranging, we have

$$n\,dT = -\left(\frac{p}{C_V}\right)dV. \tag{20-45}$$

Now from the ideal gas law ($pV = nRT$) we have

$$p\,dV + V\,dp = nR\,dT. \tag{20-46}$$

Replacing R with its equal, $C_p - C_V$, in Eq. 20-46 yields

$$n\,dT = \frac{p\,dV + V\,dp}{C_p - C_V}. \tag{20-47}$$

Equating Eqs. 20-45 and 20-47 and rearranging then give

$$\frac{dp}{p} + \left(\frac{C_p}{C_V}\right)\frac{dV}{V} = 0.$$

Replacing the ratio of the molar specific heats with γ and integrating (see Appendix E) yield

$$\ln p + \gamma \ln V = \text{a constant}.$$

Rewriting the left side as $\ln pV^\gamma$ and then taking the antilog of both sides, we find

$$pV^\gamma = \text{a constant}, \tag{20-48}$$

which we set out to prove.

Free Expansion

Recall from Section 19-10 that a free expansion of a gas is an adiabatic process that involves no work done on or by the gas, and no change in the internal energy of the gas. A free expansion is thus quite different from the type of adiabatic process described by Eqs. 20-40 through 20-48, in which work is done and the internal energy changes. Those equations then do *not* apply to a free expansion, even though such an expansion is adiabatic.

Also recall that in a free expansion, a gas is in equilibrium only at its initial and final points; thus we can plot only those points, but not the expansion itself, on a *p-V* diagram. In addition, because $\Delta E_{int} = 0$, the temperature of the final state must be that of the initial state. Thus, the initial and final points on a *p-V* diagram must be on the same isotherm, and instead of Eq. 20-43 we have

$$T_i = T_f \quad \text{(free expansion).} \tag{20-49}$$

If we next assume that the gas is ideal ($pV = nRT$), then with no change in temperature, there can be no change in the product pV. Thus, instead of Eq. 20-40 a free expansion involves the relation

$$p_iV_i = p_fV_f \quad \text{(free expansion).} \tag{20-50}$$

SAMPLE PROBLEM 20-10

In Sample Problem 20-2, 1 mol of oxygen (assumed to be an ideal gas) expands isothermally (at 310 K) from an initial volume of 12 L to a final volume of 19 L.

(a) What would be the final temperature if the gas had expanded adiabatically to this same final volume? Oxygen (O_2) is diatomic and here has rotation but not oscillation.

SOLUTION: For a diatomic gas whose molecules have rotation but not oscillations, $C_p = \frac{7}{2}R$ and $C_V = \frac{5}{2}R$. So, $\gamma = C_p/C_V = \frac{7}{5} = 1.40$. From Eq. 20-43, we can then write

$$T_f = \frac{T_iV_i^{\gamma-1}}{V_f^{\gamma-1}} = \frac{(310\text{ K})(12\text{ L})^{1.40-1}}{(19\text{ L})^{1.40-1}}$$
$$= 258\text{ K}. \quad \text{(Answer)}$$

The fact that the gas has cooled (from 310 K to 258 K) means that its internal energy has been correspondingly reduced. The

---

TABLE 20-5 FOUR SPECIAL PROCESSES

PATH IN FIG. 20-14	CONSTANT QUANTITY	PROCESS TYPE	SOME SPECIAL RESULTS ($\Delta E_{int} = Q - W$ and $\Delta E_{int} = nC_V\,\Delta T$ for all paths)
1	p	Isobaric	$Q = nC_p\,\Delta T$; $W = p\Delta V$
2	T	Isothermal	$Q = W = nRT\ln(V_f/V_i)$; $\Delta E_{int} = 0$
3	pV^γ, $TV^{\gamma-1}$	Adiabatic	$Q = 0$; $W = -\Delta E_{int}$
4	V	Isochoric	$Q = \Delta E_{int} = nC_V\,\Delta T$; $W = 0$

lost internal energy has gone to do the work of expansion (such as lifting the weighted piston in Fig. 20-13*a*).

(b) What would be the final temperature and pressure if, instead, the gas had expanded freely to the new volume, from an initial pressure of 2.0 Pa?

SOLUTION: Because the temperature does not change in a free expansion,

$$T_f = T_i = 310 \text{ K.} \qquad \text{(Answer)}$$

We find the new pressure using Eq. 20-50, which gives us

$$p_f = p_i \frac{V_i}{V_f} = (2.0 \text{ Pa}) \frac{12 \text{ L}}{19 \text{ L}} = 1.3 \text{ Pa.} \quad \text{(Answer)}$$

PROBLEM SOLVING TACTICS

TACTIC 2: *A Graphical Summary of Four Gas Processes*
In this chapter we have discussed four special processes that an ideal gas can undergo. An example of each is shown in Fig.

20-14, and some associated characteristics are given in Table 20-5, including two process names (isobaric and isochoric) that we have not used but which you might see in other courses.

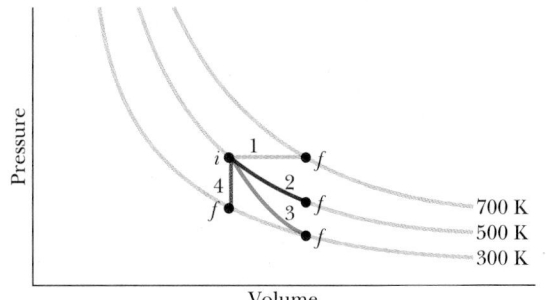

FIGURE 20-14   A *p-V* diagram representing four special processes for an ideal gas. See Table 20-5.

CHECKPOINT **5:** Rank paths 1, 2, and 3 in Fig. 20-14 according to the heat transfer to the gas, greatest first.

REVIEW & SUMMARY

Kinetic Theory of Gases

The *kinetic theory of gases* relates the *macroscopic* properties of gases (for example, pressure and temperature) to the *microscopic* properties of the gas molecules (for example, speed and kinetic energy).

Avogadro's Number

One mole of a substance contains N_A (*Avogadro's number*) elementary units (usually atoms or molecules), where N_A is found experimentally to be

$$N_A = 6.02 \times 10^{23} \text{ mol}^{-1} \qquad \begin{array}{c}\text{(Avogadro's}\\ \text{number).}\end{array} \quad (20\text{-}1)$$

One molar mass *M* of any substance is the mass of one mole of the substance.

Ideal Gas

An *ideal gas* is one for which the pressure *p*, volume *V*, and temperature *T* are related by

$$pV = nRT \qquad \text{(ideal gas law).} \quad (20\text{-}4)$$

Here *n* is the number of moles of the gas present and *R* (= 8.31 J/mol·K) is the **gas constant.**

Work in an Isothermal Volume Change

The work done *by* an ideal gas during an **isothermal** (constant-temperature) change from volume V_i to volume V_f is

$$W = nRT \ln \frac{V_f}{V_i} \qquad \begin{array}{c}\text{(ideal gas,}\\ \text{isothermal process).}\end{array} \quad (20\text{-}10)$$

Pressure, Temperature, and Molecular Speed

The pressure exerted by *n* moles of an ideal gas, in terms of the speed of its molecules, is

$$p = \frac{nMv_{\text{rms}}^2}{3V}, \quad (20\text{-}17)$$

where $v_{\text{rms}} = \sqrt{\overline{v^2}}$ is the **root-mean-square speed** of the molecules of the gas. With Eq. 20-4 this gives

$$v_{\text{rms}} = \sqrt{\frac{3RT}{M}}. \quad (20\text{-}18)$$

Temperature and Kinetic Energy

The average translational kinetic energy $\overline{K}$ per molecule of an ideal gas is

$$\overline{K} = \tfrac{3}{2}kT, \quad (20\text{-}20)$$

in which *k* (= R/N_A = 1.38×10^{-23} J/K) is the **Boltzmann constant.**

Mean Free Path

The *mean free path* λ of a gas molecule is its average path length between collisions and is given by

$$\lambda = \frac{1}{\sqrt{2}\pi d^2 \, N/V}, \quad (20\text{-}22)$$

where *N/V* is the number of molecules per unit volume and *d* is the molecular diameter.

Maxwell Speed Distribution

The *Maxwell speed distribution* $P(v)$ is a function such that

$P(v)\,dv$ gives the *fraction* of molecules with speeds between v and $v + dv$:

$$P(v) = 4\pi \left(\frac{M}{2\pi RT} \right)^{3/2} v^2 e^{-Mv^2/2RT}. \qquad (20\text{-}24)$$

Three measures of the distribution of speeds among the molecules of a gas are

$$v_P = \sqrt{\frac{2RT}{M}} \qquad \text{(most probable speed),} \qquad (20\text{-}29)$$

$$\bar{v} = \sqrt{\frac{8RT}{\pi M}} \qquad \text{(average speed),} \qquad (20\text{-}27)$$

and the rms speed defined above in Eq. 20-18.

Molar Specific Heats

The molar specific heat C_V of a gas at constant volume is defined as

$$C_V = \frac{1}{n}\frac{Q}{\Delta T} = \frac{1}{n}\frac{\Delta E_{int}}{\Delta T}, \qquad (20\text{-}31, 20\text{-}32)$$

in which Q is the heat transferred to or from a sample of n moles of the gas, ΔT is the resulting temperature change of the gas, and ΔE_{int} is the resulting change in the internal energy of the gas. For an ideal monatomic gas,

$$C_V = \tfrac{3}{2}R = 12.5 \text{ J/mol}\cdot\text{K.} \qquad (20\text{-}33)$$

The molar specific heat C_p of a gas at constant pressure is defined to be

$$C_p = \frac{1}{n}\frac{Q}{\Delta T}, \qquad (20\text{-}36)$$

in which Q, n, and ΔT are defined as above. C_p is also given by

$$C_p = C_V + R. \qquad (20\text{-}38)$$

For n moles of an ideal gas,

$$E_{int} = nC_VT \qquad \text{(ideal gas).} \qquad (20\text{-}34)$$

If n moles of a confined ideal gas undergo a temperature change ΔT due to *any* process, the change in the internal energy of the gas is

$$\Delta E_{int} = nC_V\Delta T \qquad \text{(ideal gas, any process),} \qquad (20\text{-}35)$$

in which the appropriate value of C_V must be substituted, according to the type of ideal gas.

Degrees of Freedom and C_V

We find C_V itself by using the *equipartition of energy* theorem, which states that every *degree of freedom* of a molecule (that is, every independent way it can store energy) has associated with it—on average—an energy $\tfrac{1}{2}kT$ per molecule ($= \tfrac{1}{2}RT$ per mole). If f is the number of degrees of freedom, then $E_{int} = (f/2)nRT$ and

$$C_V = \left(\frac{f}{2} \right) R = 4.16f \text{ J/mol}\cdot\text{K.} \qquad (20\text{-}39)$$

For monatomic gases $f = 3$ (three translational degrees); for diatomic gases $f = 5$ (three translational and two rotational degrees).

Adiabatic Process

When an ideal gas undergoes a slow adiabatic volume change (a change for which $Q = 0$), its pressure and volume are related by

$$pV^{\gamma} = \text{a constant} \qquad \text{(adiabatic process),} \qquad (20\text{-}40)$$

in which $\gamma \,(= C_p/C_V)$ is the ratio of molar specific heats for the gas. For a free expansion, however, $pV = $ a constant.

QUESTIONS

1. If the temperature of an ideal gas is changed from 20°C to 40°C while the volume is unchanged, is the pressure of the gas doubled, increased but less than doubled, or increased but more than doubled?

2. Two rooms of equal volume are connected by an open passageway and are maintained at different temperatures. Which room has more air molecules?

3. The molar masses and Kelvin temperatures of three ideal gases are (a) M_0 and $2T_0$, (b) $2M_0$ and T_0, and (c) $6M_0$ and $6T_0$. Rank the gases according to the rms speeds of their molecules, greatest first.

4. The volume of a gas and the number of gas molecules within that volume for four situations are (a) $2V_0$ and N_0, (b) $3V_0$ and $3N_0$, (c) $8V_0$ and $4N_0$, and (d) $3V_0$ and $9N_0$. Rank the situations according to the mean free path of the molecules, greatest first.

5. In Sample Problem 20-2, how much heat is transferred during the expansion of part (a)?

6. Figure 20-15 shows the initial state of an ideal gas and an isotherm through that state. Which of the paths shown result in a

decrease in the temperature of the gas?

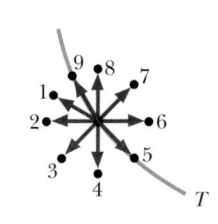

FIGURE 20-15
Question 6.

7. The table gives the heat transfer Q to or from an ideal gas and either the work W done by the gas or the work W_{on} done on the gas, all in joules, for four situations. Rank the four situations in terms of the temperature change of the gas, most positive first, most negative last.

	a	b	c	d
Q	−50	+35	−15	+20
W	−50	+35		
W_{on}			−40	+40

8. For a temperature increase of ΔT_1, a certain amount of an ideal gas requires 30 J when heated at constant volume and 50 J when heated at constant pressure. How much work is done by the gas in the second situation?

9. An ideal diatomic gas, with molecular rotation but not oscillation, loses heat Q. Is the resulting decrease in the internal energy of the gas greater if the loss occurs in a constant-volume process or in a constant-pressure process?

10. A certain amount of heat is to be transferred to 1 mol of a monatomic gas (a) at constant pressure and (b) at constant volume, and to 1 mol of a diatomic gas (c) at constant pressure and (d) at constant volume. Figure 20-16 shows four paths from an initial point to four final points on a p-V diagram. Which path goes with which process? (e) Are the molecules of the diatomic gas rotating?

FIGURE 20-16
Question 10.

11. Does the temperature of an ideal gas increase, decrease, or stay the same during (a) an isothermal expansion, (b) an expansion at constant pressure, (c) an adiabatic expansion, and (d) an increase in pressure at constant volume?

12. (a) Rank the four paths of Fig. 20-14 according to the work done by the gas, greatest first. (b) Rank paths 1, 2, and 3 according to the change in the internal energy of the gas, most positive first and most negative last.

13. In the p-V diagram of Fig. 20-17, the gas does 5 J of work along isotherm ab and 4 J along adiabat bc. What is the change in the internal energy of the gas if the gas traverses the straight path from a to c?

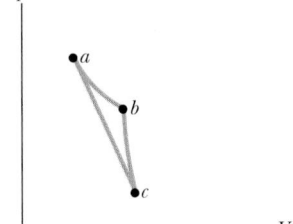

FIGURE 20-17
Question 13.

14. The p-V diagram of Fig. 20-18 (not to scale) shows eight separate transformations (lettered a to h) that an ideal gas undergoes in proceeding from an initial state to a final state. The table shows eight sets of values for Q, W, and ΔE_{int} (in joules), numbered 1 to 8. Which set of energy changes goes with which transformation? (*Hint:* Compare the paths in Fig. 20-18 to those in Fig. 20-14.)

	1	2	3	4	5	6	7	8
Q	10	5	−20		−10	−12		10
W	10		−10	10		−12	−10	
ΔE_{int}		2		−10	−10		10	10

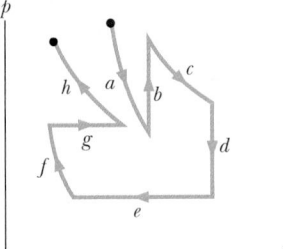

FIGURE 20-18
Question 14.

15. (a) The p-V diagram in Fig. 20-19 shows adiabatic expansions from a common point to volume V_f for a monatomic gas, a diatomic gas, and a polyatomic gas. Which path goes with which gas? (b) If the monatomic gas had, instead, freely expanded to the same final volume, would its final pressure be more than, less than, or the same as previously? (*Hint:* How would the initial and final states for the free expansion be plotted on Fig. 20-14?)

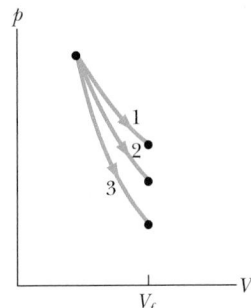

FIGURE 20-19
Question 15.

EXERCISES & PROBLEMS

SECTION 20-2 Avogadro's Number

1E. Gold has a molar mass of 197 g/mol. (a) How many moles of gold are in a 2.50 g sample of pure gold? (b) How many atoms are in the sample?

2E. Find the mass in kilograms of 7.50×10^{24} atoms of arsenic, which has a molar mass of 74.9 g/mol.

3P. If the water molecules in 1.00 g of water were distributed uniformly over the surface of Earth, how many such molecules would there be on 1.00 cm^2 of the surface?

4P. Consider this sentence: A _____ of water contains about as many molecules as there are _____ s of water in all the oceans. What single word best fits both blank spaces: drop, teaspoon, tablespoon, cup, quart, barrel, or ton? The oceans cover 75% of Earth's surface and have an average depth of about 5 km. (After Edward M. Purcell.)

5P. A distinguished scientist has written: "There are enough molecules in the ink that makes one letter of this sentence to provide not only one for every inhabitant of Earth, but one for every creature if each star of our galaxy had a planet as populous as Earth.'' Check this statement. Assume the ink sample (molar mass = 18 g/mol) to have a mass of 1 μg, the population of Earth to be 5×10^9, and the number of stars in our galaxy to be 10^{11}.

SECTION 20-3   Ideal Gases

6E. (a) What is the volume occupied by 1.00 mol of an ideal gas at standard conditions, that is, at 1.00 atm (= 1.01×10^5 Pa) and 0°C (= 273 K)? (b) Show that the number of molecules per cubic centimeter (the *Loschmidt number*) at standard conditions is 2.69×10^{19}.

7E. Compute (a) the number of moles and (b) the number of molecules in 1.00 cm³ of an ideal gas at a pressure of 100 Pa and a temperature of 220 K.

8E. The best vacuum that can be attained in the laboratory corresponds to a pressure of about 1.00×10^{-18} atm, or 1.01×10^{-13} Pa. How many gas molecules are there per cubic centimeter in such a vacuum at 293 K?

9E. A quantity of ideal gas at 10.0°C and a pressure of 100 kPa occupies a volume of 2.50 m³. (a) How many moles of the gas are present? (b) If the pressure is now raised to 300 kPa and the temperature is raised to 30.0°C, how much volume will the gas occupy? Assume no leaks.

10E. Oxygen gas having a volume of 1000 cm³ at 40.0°C and 1.01×10^5 Pa expands until its volume is 1500 cm³ and its pressure is 1.06×10^5 Pa. Find (a) the number of moles of oxygen present and (b) the final temperature of the sample.

11E. An automobile tire has a volume of 1000 in.³ and contains air at a gauge pressure of 24.0 lb/in.² when the temperature is 0.00°C. What is the gauge pressure of the air in the tires when its temperature rises to 27.0°C and its volume increases to 1020 in.³? (*Hint:* It is not necessary to convert from British units to SI units; why? Use P_{atm} = 14.7 lb/in.².)

12E. Calculate the work done by an external agent during an isothermal compression of 1.00 mol of oxygen from a volume of 22.4 L at 0°C and 1.00 atm pressure to 16.8 L.

13P. (a) What is the number of molecules per cubic meter in air at 20°C and at a pressure of 1.0 atm (= 1.01×10^5 Pa)? (b) What is the mass of this 1 m³ of air? Assume that 75% of the molecules are nitrogen (N_2) and 25% oxygen (O_2).

14P. Pressure p, volume V, and temperature T for a certain material are related by

$$p = \frac{AT - BT^2}{V},$$

where A and B are constants. Find an expression for the work done by the material if the temperature changes from T_1 to T_2 while the pressure remains constant.

15P. Air that occupies 0.14 m³ at 1.03×10^5 Pa gauge pressure is expanded isothermally to atmospheric pressure and then cooled at constant pressure until it reaches its initial volume. Compute the work done by the air.

16P. Consider a given mass of an ideal gas. Compare curves representing constant-pressure, constant-volume, and isothermal processes on (a) a p-V diagram, (b) a p-T diagram, and (c) a V-T diagram. (d) How do these curves depend on the mass of gas?

17P. A container encloses two ideal gases. Two moles of the first gas are present, with molar mass M_1. The second gas has molar mass $M_2 = 3M_1$, and 0.5 mol of this gas is present. What fraction of the total pressure on the container wall is attributable to the second gas? (The kinetic theory explanation of pressure leads to the experimentally discovered law of partial pressures for a mixture of gases that do not react chemically: *the total pressure exerted by the mixture is equal to the sum of the pressures that the several gases would exert separately if each were to occupy the vessel alone.*)

18P. A sample of an ideal gas is taken through the cyclic process *abca* shown in Fig. 20-20; at point *a*, T = 200 K. (a) How many moles of gas are in the sample? What are (b) the temperature of the gas at point *b*, (c) the temperature of the gas at point *c*, and (d) the net heat added to the gas during the cycle?

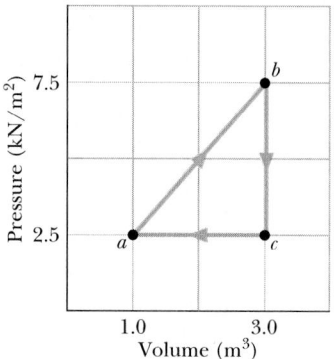

FIGURE 20-20   Problem 18.

19P. An ideal gas initially at 300 K is compressed at a constant pressure of 25 N/m² from a volume of 3.0 m³ to a volume of 1.8 m³. In the process, 75 J of heat is lost by the gas. What are (a) the change in internal energy of the gas and (b) the final temperature of the gas?

20P. A weather balloon is loosely inflated with helium at a pressure of 1.0 atm (= 760 torr) and a temperature of 20°C. The gas volume is 2.2 m³. At an elevation of 20,000 ft, the atmospheric pressure is down to 380 torr and the helium has expanded, being under no restraint from the confining bag. At this elevation the gas temperature is −48°C. What is the gas volume now?

21P. An air bubble of 20 cm³ volume is at the bottom of a lake 40 m deep where the temperature is 4.0°C. The bubble rises to the surface, which is at a temperature of 20°C. Take the temperature of the bubble to be the same as that of the surrounding water and find its volume just before it reaches the surface.

22P. A pipe of length L = 25.0 m that is open at one end contains air at atmospheric pressure. It is thrust vertically into a freshwater lake until the water rises halfway up in the pipe, as shown in Fig. 20-21. What is the depth h of the lower end of the pipe? Assume that the temperature is the same everywhere and does not change.

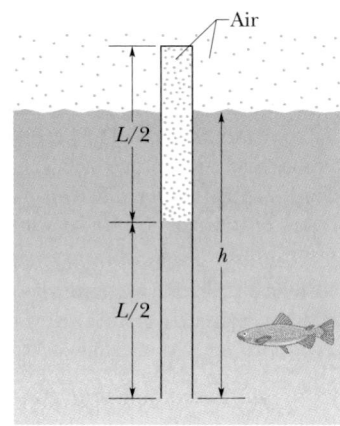

FIGURE 20-21
Problem 22.

23P. The envelope and basket of a hot-air balloon have a combined weight of 550 lb, and the envelope has a capacity of 77,000 ft³. When it is fully inflated, what should be the temperature of the enclosed air to give the balloon a lifting capacity of 600 lb (in addition to its own weight)? Assume that the surrounding air, at 20.0°C, has a weight density of 7.56×10^{-2} lb/ft³.

24P. A steel tank contains 300 g of ammonia gas (NH_3) at an absolute pressure of 1.35×10^6 Pa and temperature of 77°C. (a) What is the volume of the tank? (b) The tank is checked later when the temperature has dropped to 22°C, and the absolute pressure is found to be 8.7×10^5 Pa. How many grams of gas have leaked out of the tank?

25P. Container A in Fig. 20-22 holds an ideal gas at a pressure of 5.0×10^5 Pa and a temperature of 300 K. It is connected by a thin tube to container B with four times the volume of A. Container B holds the same ideal gas at a pressure of 1.0×10^5 Pa and a temperature of 400 K. The connecting valve is opened, and equilibrium is achieved at a common pressure while the temperature of each container is kept constant at its initial value. What is the final pressure in the system?

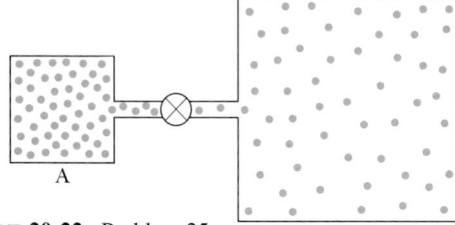

FIGURE 20-22 Problem 25.

SECTION 20-4 Pressure, Temperature, and rms Speed

26E. Calculate the root-mean-square speed of helium atoms at 1000 K. The molar mass of helium is 4.00 g/mol.

27E. The lowest possible temperature in outer space is 2.7 K. What is the root-mean-square speed of hydrogen molecules at this temperature? (Use Table 20-1.)

28E. Find the rms speed of argon atoms at 313 K. The molar mass of argon is 39.9 g/mol.

29E. The Sun is a huge ball of hot ideal gas. The temperature and pressure in the Sun's atmosphere are 2.00×10^6 K and 0.0300 Pa. Calculate the rms speed of free electrons (mass = 9.11×10^{-31} kg) there.

30E. (a) Compute the root-mean-square speed of a nitrogen molecule at 20.0°C. At what temperatures will the root-mean-square speed be (b) half that value and (c) twice that value?

31E. At what temperature do the atoms of helium gas have the same rms speed as molecules of hydrogen gas have at 20.0°C?

32P. At 273 K and 1.00×10^{-2} atm, the density of a gas is 1.24×10^{-5} g/cm³. (a) Find v_{rms} for the gas molecules. (b) Find the molar mass of the gas and identify it.

33P. The mass of the H_2 molecule is 3.3×10^{-24} g. If 10^{23} H_2 molecules per second strike 2.0 cm² of wall at an angle of 55° with the normal when moving with a speed of 1.0×10^5 cm/s, what pressure do they exert on the wall?

SECTION 20-5 Translational Kinetic Energy

34E. What is the average translational kinetic energy of nitrogen molecules at 1600 K, (a) in joules and (b) in electron-volts?

35E. (a) Determine the average value in electron-volts of the translational kinetic energy of the particles of an ideal gas at 0.00°C and at 100°C. (b) What is the translational kinetic energy per mole of an ideal gas at these temperatures, in joules?

36E. At what temperature is the average translational kinetic energy of a molecule equal to 1.00 eV?

37E. Oxygen (O_2) gas at 273 K and 1.0 atm pressure is confined to a cubical container 10 cm on a side. Calculate the ratio of (1) the change in gravitational potential energy of an oxygen molecule falling the height of the box to (2) the molecule's average translational kinetic energy.

38P. Show that the ideal gas equation, Eq. 20-4, can be written in these alternative forms: (a) $p = \rho RT/M$, where ρ is the mass density of the gas and M the molar mass; (b) $pV = NkT$, where N is the number of gas particles (atoms or molecules).

39P. Water standing in the open at 32.0°C evaporates because of the escape of some of the surface molecules. The heat of vaporization (539 cal/g) is approximately equal to ϵn, where ϵ is the average energy of the escaping molecules and n is the number of molecules per gram. (a) Find ϵ. (b) What is the ratio of ϵ to the average kinetic energy of H_2O molecules, assuming the latter is related to temperature in the same way as it is for gases?

40P. *Avogadro's law* states that under the same conditions of temperature and pressure, equal volumes of gas contain equal numbers of molecules. Is this law equivalent to the ideal gas law? Explain.

SECTION 20-6 Mean Free Path

41E. The mean free path of nitrogen molecules at 0.0°C and 1.0 atm is 0.80×10^{-5} cm. At this temperature and pressure there are 2.7×10^{19} molecules/cm³. What is the molecular diameter?

42E. At 2500 km above Earth's surface, the density of the atmosphere is about 1 molecule/cm³. (a) What mean free path is predicted by Eq. 20-22 and (b) what is its significance under these conditions? Assume a molecular diameter of 2.0×10^{-8} cm.

43E. What is the mean free path for 15 spherical jelly beans in a bag that is vigorously shaken? The volume of the bag is 1.0 L, and the diameter of a jelly bean is 1.0 cm.

44E. Derive an expression, in terms of N/V, $\bar{v}$, and d, for the collision frequency of a gas atom or molecule.

45P. In a certain particle accelerator, protons travel around a circular path of diameter 23.0 m in an evacuated chamber, whose residual gas is at 295 K and 1.00×10^{-6} torr pressure. (a) Calculate the number of gas molecules per cubic centimeter at this pressure. (b) What is the mean free path of the gas molecules if the molecular diameter is 2.00×10^{-8} cm?

46P. At what frequency would the wavelength of sound in air be equal to the mean free path of oxygen molecules at 1.0 atm pressure and 0.0°C? Take the diameter of the oxygen molecule to be 3.0×10^{-8} cm.

47P. (a) What is the molar volume (volume per mole) of an ideal gas at standard conditions (0.00°C, 1.00 atm)? (b) Calculate the ratio of the root-mean-square speed of helium atoms to that of neon atoms under these conditions. (c) What is the mean free path of helium atoms under these conditions? Assume the atomic diameter d to be 1.00×10^{-8} cm. (d) What is the mean free path of neon atoms under these conditions? Assume the same atomic diameter as for helium. (e) Comment on the results of (c) and (d) in view of the fact that the helium atoms are traveling faster than the neon atoms.

48P. The mean free path λ of the molecules of a gas may be determined from certain measurements (for example, from measurement of the viscosity of the gas). At 20°C and 750 torr pressure such measurements yield values of λ_{Ar} (argon) = 9.9×10^{-6} cm and λ_{N_2} (nitrogen) = 27.5×10^{-6} cm. (a) Find the ratio of the effective diameter of argon to that of nitrogen. What is the mean free path of argon at (b) 20°C and 150 torr, and (c) −40°C and 750 torr?

49P. Show that about 10^{13} air molecules are needed to cover the period that closes this sentence. Show that about 10^{24} air molecules collide with that period each second.

SECTION 20-7   The Distribution of Molecular Speeds

50E. The speeds of 10 molecules are 2.0, 3.0, 4.0,··, 11 km/s. (a) What is their average speed? (b) What is their root-mean-square speed?

51E. Twenty-two particles have speeds as follows (N_i represents the number of particles that have speed v_i):

N_i	2	4	6	8	2
v_i (cm/s)	1.0	2.0	3.0	4.0	5.0

(a) Compute the average speed $\bar{v}$. (b) Compute the root-mean-square speed v_{rms}. (c) Of the five speeds shown, which is the most probable speed v_P?

52E. (a) Ten particles are moving with the following speeds: four at 200 m/s, two at 500 m/s, and four at 600 m/s. Calculate the average and root-mean-square speeds. Is $v_{rms} > \bar{v}$? (b) Make up your own speed distribution for the 10 particles and show that $v_{rms} \geq \bar{v}$ for your distribution. (c) Under what condition (if any) does $v_{rms} = \bar{v}$?

53E. Consider the distribution of speeds shown in Fig. 20-23. (a) Rank v_{rms}, $\bar{v}$, and v_P, greatest first. (b) How does the ranking in (a) compare with that for a Maxwellian distribution?

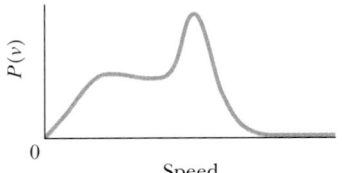

FIGURE 20-23
Exercise 53.

54E. It is found that the most probable speed of molecules in a gas at equilibrium temperature T_2 is the same as the rms speed of the molecules in this gas when its equilibrium temperature is T_1. Calculate T_2/T_1.

55P. (a) Compute the temperatures at which the rms speed is equal to the speed of escape from Earth's surface for molecular hydrogen and for molecular oxygen. (b) Do the same for the Moon, assuming the gravitational acceleration on its surface to be $0.16g$. (c) The temperature high in Earth's upper atmosphere is about 1000 K. Would you expect to find much hydrogen there? Much oxygen?

56P. A molecule of hydrogen (diameter 1.0×10^{-8} cm), traveling with the rms speed, escapes from a furnace ($T = 4000$ K) into a chamber containing atoms of cold argon (diameter 3.0×10^{-8} cm) at a density of 4.0×10^{19} atoms/cm³. (a) What is the speed of the hydrogen molecule? (b) If the H_2 molecule and an argon atom collide, what is the closest their centers can be, considering each as spherical? (c) What is the initial number of collisions per second experienced by the hydrogen molecule?

57P. Two containers are at the same temperature. The first contains gas with pressure p_1, molecular mass m_1, and root-mean-square speed v_{rms1}. The second contains gas with pressure $2p_1$, molecular mass m_2, and average speed $\bar{v}_2 = 2v_{rms1}$. Find the mass ratio m_1/m_2.

58P. For the hypothetical speed distribution for N gas particles shown in Fig. 20-24 [$P(v) = Cv^2$ for $0 < v \leq v_0$; $P(v) = 0$ for $v > v_0$], find (a) an expression for C in terms of N and v_0, (b) the average speed of the particles, and (c) their rms speed.

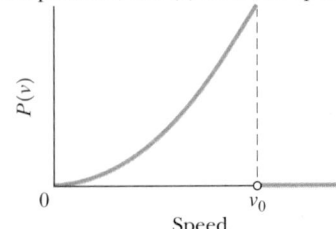

FIGURE 20-24
Problem 58.

59P. A hypothetical sample of N gas particles has the speed distribution shown in Fig. 20-25, where $P(v) = 0$ for $v > 2v_0$. (a) Express a in terms of N and v_0. (b) How many of the particles have speeds between $1.5v_0$ and $2.0v_0$? (c) Express the average speed of the particles in terms of v_0. (d) Find v_{rms}.

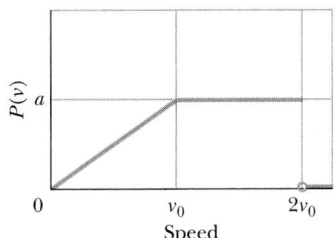

FIGURE 20-25
Problem 59.

SECTION 20-8 The Molar Specific Heats of an Ideal Gas

60E. What is the internal energy of 1.0 mol of an ideal monatomic gas at 273 K?

61E. One mole of an ideal gas undergoes an isothermal expansion. Find the heat added to the gas in terms of the initial and final volumes and the temperature. (*Hint:* Use the first law of thermodynamics.)

62E. The mass of a helium atom is 6.66×10^{-27} kg. Compute the specific heat at constant volume for (monatomic) helium gas (in $J/kg \cdot K$) from the molar specific heat at constant volume.

63P. Let 20.9 J of heat be added to a particular ideal gas. As a result, its volume changes from 50.0 cm^3 to 100 cm^3 while the pressure remains constant at 1.00 atm. (a) By how much did the internal energy of the gas change? If the quantity of gas present is 2.00×10^{-3} mol, find the molar specific heat at (b) constant pressure and (c) constant volume.

64P. A quantity of ideal monatomic gas consists of n moles initially at temperature T_1. The pressure and volume are then slowly doubled in such a manner as to trace out a straight line on a p-V diagram. In terms of n, R, and T_1, what are (a) W, (b) ΔE_{int}, and (c) Q? (d) If one were to define a molar specific heat for this process, what would be its value?

65P. A container holds a mixture of three nonreacting gases: n_1 moles of the first gas with molar specific heat at constant volume C_1, and so on. Find the molar specific heat at constant volume of the mixture, in terms of the molar specific heats and quantities of the separate gases.

66P. The mass of a gas molecule can be computed from the specific heat at constant volume c_V. Take $c_V = 0.075$ cal/g$\cdot$C° for argon and calculate (a) the mass of an argon atom and (b) the molar mass of argon.

67P. One mole of an ideal diatomic gas undergoes a transition

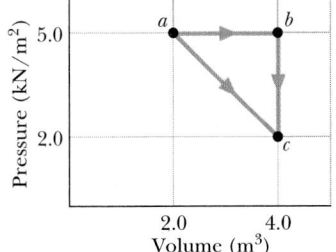

FIGURE 20-26
Problem 67.

from a to c along the diagonal path in Fig. 20-26. The temperature of the gas at point a is 1200 K. During the transition, (a) what is the change in internal energy of the gas, and (b) how much heat is added to the gas? (c) How much heat must be added to the gas if it goes from a to c along the indirect path abc?

SECTION 20-9 Degrees of Freedom and Molar Specific Heats

68E. One mole of oxygen (O_2) is heated at constant pressure starting at 0°C. How much heat must be added to the gas to double its volume? (The molecules rotate but do not oscillate.)

69E. Suppose 12.0 g of oxygen (O_2) is heated at constant atmospheric pressure from 25.0°C to 125°C. (a) How many moles of oxygen are present? (See Table 20-1.) (b) How much heat is transferred to the oxygen? (The molecules rotate but do not oscillate.) (c) What fraction of the heat is used to raise the internal energy of the oxygen?

70P. Suppose 4.00 mol of an ideal diatomic gas, with molecular rotation but not oscillation, experiences a temperature increase of 60.0 K under constant-pressure conditions. (a) How much heat was added to the gas? (b) How much did the internal energy of the gas increase? (c) How much work was done by the gas? (d) How much did the translational kinetic energy of the gas increase?

SECTION 20-11 The Adiabatic Expansion of an Ideal Gas

71E. A mass of gas occupies a volume of 4.3 L at a pressure of 1.2 atm and a temperature of 310 K. It is compressed adiabatically to a volume of 0.76 L. Determine (a) the final pressure and (b) the final temperature, assuming the gas to be an ideal gas for which $\gamma = 1.4$.

72E. (a) One liter of gas with $\gamma = 1.3$ is at 273 K and 1.0 atm pressure. It is suddenly compressed (adiabatically) to half its original volume. Find its final pressure and temperature. (b) The gas is now cooled back to 273 K at constant pressure. What is its final volume?

73E. Let n moles of an ideal gas expand adiabatically from an initial temperature T_1 to a final temperature T_2. Prove that the work done by the gas is $nC_V(T_1 - T_2)$, where C_V is the molar specific heat at constant volume. (*Hint:* Use the first law of thermodynamics.)

74E. We know that $pV^\gamma =$ a constant for an adiabatic process. Evaluate ''a constant'' for an adiabatic process involving exactly 2.0 mol of an ideal gas passing through the state having exactly $p = 1.0$ atm and $T = 300$ K. Assume a diatomic gas whose molecules have rotation but not oscillation.

75E. For adiabatic processes in an ideal gas, show that (a) the bulk modulus is given by

$$B = -V\frac{dp}{dV} = \gamma p,$$

and therefore (b) the speed of sound in the gas is

$$v_s = \sqrt{\frac{\gamma p}{\rho}} = \sqrt{\frac{\gamma RT}{M}}.$$

See Eqs. 18-2 and 18-3.

76E. Air at 0.000°C and 1.00 atm pressure has a density of 1.29×10^{-3} g/cm³, and the speed of sound in air is 331 m/s at that temperature. Compute the ratio γ of the molar specific heats of air. (*Hint:* See Exercise 75.)

77E. The speed of sound in different gases at the same temperature depends only on the molar mass of the gases. Show that $v_1/v_2 = \sqrt{M_2/M_1}$ (at constant T), where v_1 is the speed of sound in a gas of molar mass M_1 and v_2 is the speed of sound in a gas of molar mass M_2. (*Hint:* See Exercise 75.)

78P. The molar mass of iodine is 127 g/mol. A standing wave in a tube filled with iodine gas at 400 K has nodes that are 6.77 cm apart when the frequency is 1400 Hz. (a) What is γ? (b) Is iodine gas monatomic or diatomic? (*Hint:* See Exercise 75.)

79P. Knowing that C_V, the molar specific heat at constant volume, for a gas in a container is $5.0R$, calculate the ratio of the speed of sound in that gas to the rms speed of its molecules at temperature T. (*Hint:* See Exercise 75.)

80P. (a) An ideal gas initially at pressure p_0 undergoes a free expansion until its volume is 3.00 times its initial volume. What then is its pressure? (b) The gas is next slowly and adiabatically compressed back to its original volume. The pressure after compression is $(3.00)^{1/3}p_0$. Is the gas monatomic, diatomic, or polyatomic? (c) How does the average kinetic energy per molecule in this final state compare with that in the initial state?

81P. An ideal gas experiences an adiabatic compression from $p = 1.0$ atm, $V = 1.0 \times 10^6$ L, $T = 0.0°$C to $p = 1.0 \times 10^5$ atm, $V = 1.0 \times 10^3$ L. (a) Is the gas monatomic, diatomic, or polyatomic? (b) What is its final temperature? (c) How many moles of gas are present? (d) What is the total translational kinetic energy per mole before and after the compression? (e) What is the ratio of the squares of the rms speeds before and after the compression?

82P. A sample of ideal gas expands from an initial pressure and volume of 32 atm and 1.0 L to a final volume of 4.0 L. The initial temperature of the gas is 300 K. What are the final pressure and temperature of the gas and how much work is done by the gas during the expansion, if the expansion is (a) isothermal, (b) adiabatic and the gas is monatomic, and (c) adiabatic and the gas is diatomic?

83P. An ideal gas, at initial temperature T_1 and initial volume 2 m³, is expanded adiabatically to a volume of 4 m³, then expanded isothermally to a volume of 10 m³, and then compressed adiabatically until its temperature is again T_1. What is its final volume?

84P. For a certain ideal gas C_V is 6.00 cal/mol·K. The temperature of 3.0 mol of the gas is raised 50 K by each of three different processes: at constant volume, at constant pressure, and by an adiabatic compression. Complete the table, showing for each process the heat Q added (or subtracted), the work W done by the gas, the change ΔE_{int} in internal energy of the gas, and the change ΔK in total translational kinetic energy of the gas.

PROCESS	Q	W	ΔE_{int}	ΔK
Constant volume	___	___	___	___
Constant pressure	___	___	___	___
Adiabatic	___	___	___	___

85P. One mole of an ideal monatomic gas traverses the cycle shown in Fig. 20-27. Process $1 \rightarrow 2$ takes place at constant volume, process $2 \rightarrow 3$ is adiabatic, and process $3 \rightarrow 1$ takes place at constant pressure. (a) Compute the heat Q, the change in internal energy ΔE_{int}, and the work done W, for each of the three processes and for the cycle as a whole. (b) If the initial pressure at point 1 is 1.00 atm, find the pressure and the volume at points 2 and 3. Use 1.00 atm = 1.013×10^5 Pa and $R = 8.314$ J/mol·K.

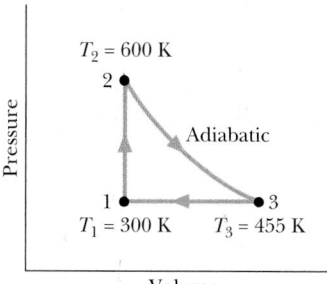

FIGURE 20-27
Problem 85.

Electronic Computation

86P. In an industrial process the volume of 25.0 moles of a monatomic ideal gas is reduced at a uniform rate from 0.616 m³ to 0.308 m³ in 2.00 h while its temperature is increased at a uniform rate from 27.0°C to 450°C. Throughout the process, the gas passes through thermodynamic equilibrium states. (a) Plot the cumulative work done on the gas and the cumulative heat absorbed by the gas as a function of time for the duration of the process. (b) What are the values of these quantities for the entire process? (c) What is the molar specific heat for the process? (*Hint:* To evaluate the integral for the work, you might use

$$\int \frac{a + bx}{A + Bx} dx = \frac{bx}{B} + \frac{aB - bA}{B^2} \ln(A + Bx),$$

an indefinite integral.) (d) Compare these values with those that would be obtained if the volume is first reduced at constant temperature, then the temperature is increased at constant volume, with the gas reaching the same final state.

21
Entropy and the Second Law of Thermodynamics

An anonymous graffito on a wall of the Pecan Street Cafe in Austin, Texas, read: "Time is God's way of keeping things from happening all at once." Time also has direction: some things happen in a certain sequence and could never happen on their own in a reverse sequence. As an example, an accidentally dropped egg splatters in a cup. The reverse process, a splattered egg re-forming into a whole egg and jumping up to an outstretched hand, will never happen on its own. But why not? Why can't that process be reversed, like a videotape run backward? What in the world gives direction to time?

21-1 SOME ONE-WAY PROCESSES

Suppose you come indoors on a very cold day and wrap your cold hands around a warm mug of cocoa. Then your hands get warmer and the mug gets cooler. But it never happens the other way around; that is, your cold hands never get still colder while the warm mug gets still warmer.

The system consisting of your hands and the mug is a *closed system*, one that is isolated from its environment. Here are some other one-way processes that occur in closed systems: (1) A crate sliding over an ordinary surface eventually stops. But you never see an initially stationary crate start to move all by itself. (2) If you drop a glob of putty, it falls to the floor. But an initially motionless glob of putty never leaps spontaneously into the air. (3) If you puncture a helium-filled balloon in a closed room, the helium gas spreads throughout the room. But the individual helium atoms will never clump up again into the shape of the balloon. We say that such one-way processes are **irreversible,** meaning that they cannot be reversed by means of only small changes in their environment.

The one-way character of such thermodynamic processes is so pervasive that we take it for granted. If these processes were to happen *spontaneously* (on their own) in the "wrong" direction, we would be astonished beyond belief. *Yet none of these "wrong-way" events would violate the law of conservation of energy.* In the cocoa mug example, that law would be obeyed even for a wrong-way transfer of heat between hands and mug. It would be obeyed even if a stationary crate or a stationary glob of putty suddenly were to transfer some of its thermal energy into kinetic energy and begin to move. And it would be obeyed even if the helium atoms released from a balloon were, on their own, to clump together again, which would require no change in their energy.

Thus changes in energy within a closed system do not set the direction of irreversible processes. Rather, that direction is set by another property that we shall discuss in this chapter—the **change in entropy** ΔS of the system. The change in entropy of a system is defined in the next section, but we can here state its central property, often called the *entropy postulate:*

If an irreversible process occurs in a closed system, the entropy S of the system always increases; it never decreases.

Entropy differs from energy in that it does *not* obey a conservation law. The *energy* of a closed system is conserved; it always remains constant. For irreversible processes, the *entropy* of a closed system always increases. Because of this property, the change in entropy is sometimes called "the arrow of time." For example, we associate the egg of our opening photograph, breaking irreversibly as it drops into a cup, with the forward direction of time and with an increase in entropy. The backward direction of time (a videotape run backward) would correspond to the broken egg re-forming into a whole egg and rising into the air. This backward process, which would result in an entropy decrease, never happens.

There are two equivalent ways to define the change in entropy of a system: (1) in terms of the system's temperature and the energy it gains or loses as heat, and (2) by counting the ways in which the atoms or molecules that make up the system can be arranged. We use the first approach in the next section, and the second in Section 21-6.

21-2 CHANGE IN ENTROPY

Let's approach the definition of *change in entropy* by looking again at a process that we described in Section 19-10: the free expansion of an ideal gas. Figure 21-1a shows the gas in its initial equilibrium state i, confined by a closed stopcock to the left half of a thermally insulated container. If we open the stopcock, the gas rushes to fill the entire container, eventually reaching the final equilibrium state f shown in Fig. 21-1b. This is an irreversible process; all the molecules of the gas will never return, by themselves, to the left half of the container.

The p-V plot of the process in Fig. 21-2 shows the pressure and volume of the gas in its initial state i and the final state f. Pressure and volume are *state properties,* properties that depend only on the state of the gas and not on how it reached that state. Other state properties are temperature and energy. We now assume that the gas has still another state property—its entropy. Furthermore, we define the **change in entropy** $S_f - S_i$ of a system during a process that takes the system from an initial state i to a final state f as

$$\Delta S = S_f - S_i = \int_i^f \frac{dQ}{T} \qquad \text{(change in entropy defined).} \qquad (21\text{-}1)$$

Here Q is the energy transferred as heat to or from the system during the process, and T is the temperature of the system in kelvins. Thus an entropy change depends not only on the heat transferred but also on the temperature at which the transfer takes place. Because T is always positive, the sign of ΔS is the same as that of Q. We see from Eq. 21-1 that the SI unit for entropy and entropy change is the joule per kelvin.

There is a problem, however, in applying Eq. 21-1 to the free expansion of Fig. 21-1. As the gas rushes to fill the entire container, the pressure, temperature, and volume of

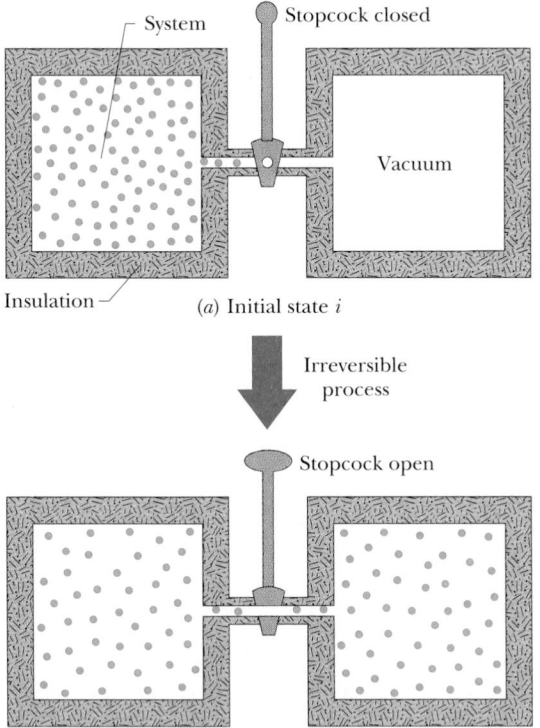

(a) Initial state *i*

(b) Final state *f*

FIGURE 21-1 The free expansion of an ideal gas. (a) The gas is confined to the left half of an insulated container by a closed stopcock. (b) When the stopcock is opened, the gas rushes to fill the entire container. This process is irreversible; that is, it does not occur in reverse, with the gas spontaneously collecting itself in the left half of the container.

the gas fluctuate unpredictably. That is, they do not have a sequence of well-defined equilibrium values during the intermediate stages of the change from initial equilibrium state *i* to final equilibrium state *f*. Thus we cannot trace a pressure–volume path for the free expansion on the *p-V* plot of Fig. 21-2 and, more important, we cannot find a relation between *Q* and *T* that allows us to integrate as Eq. 21-1 requires.

However, if entropy is truly a state property, the difference in entropy between states *i* and *f* must depend *only on those states* and not at all on the way the system went

FIGURE 21-2 A *p-V* diagram showing the initial state *i* and the final state *f* of the free expansion of Fig. 21-1. The intermediate states of the gas cannot be shown because they are not equilibrium states.

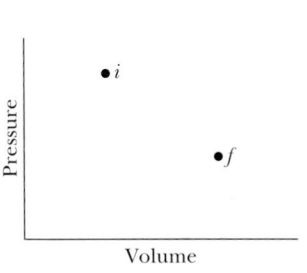

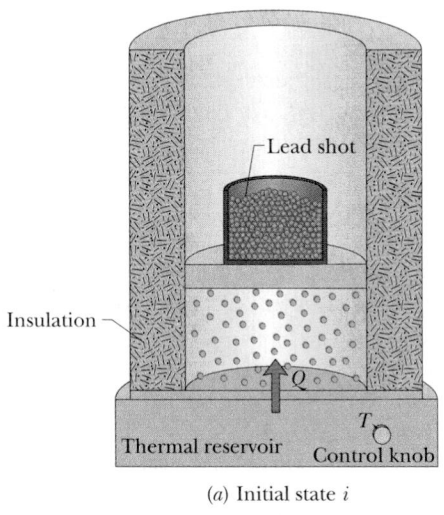

(a) Initial state *i*

FIGURE 21-3 The isothermal expansion of an ideal gas, done in a reversible way. The gas has the same initial state *i* and same final state *f* as in the irreversible process of Figs. 21-1 and 21-2.

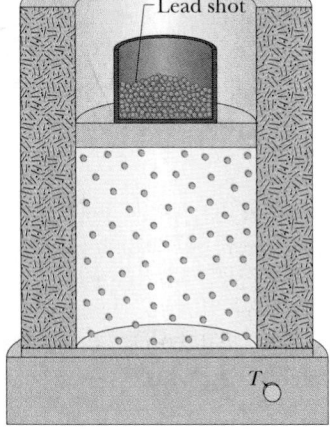

(b) Final state *f*

from one state to the other. Suppose, then, that we replace the irreversible free expansion of Fig. 21-1 with a *reversible* process that connects states *i* and *f*. With a reversible process we can trace a pressure–volume path on a *p-V* plot, and we can find a relation between *Q* and *T* that allows us to use Eq. 21-1 to obtain the entropy change.

We saw in Section 19-10 that the temperature of an ideal gas does not change during a free expansion: $T_i = T_f = T$. So points *i* and *f* in Fig. 21-2 must be on the same isotherm. A convenient replacement process is then a reversible isothermal expansion from state *i* to state *f*, which actually proceeds *along* that isotherm. Furthermore, because *T* is constant throughout a reversible isothermal expansion, the integral of Eq. 21-1 is greatly simplified.

Figure 21-3 shows how to produce such a reversible isothermal expansion. We confine the gas to an insulated cylinder that rests on a thermal reservoir maintained at the temperature T. We begin by placing just enough lead shot on the movable piston so that the pressure and volume of the gas are those of the initial state i of Fig. 21-1a. We then remove shot slowly (piece by piece) until the pressure and volume of the gas are those of the final state f of Fig. 21-1b. The temperature of the gas does not change because the gas remains in thermal contact with the reservoir throughout the process.

The reversible isothermal expansion of Fig. 21-3 is physically quite different from the irreversible free expansion of Fig. 21-1. However, *both processes have the same initial state and the same final state and thus must have the same change in entropy.* Because we removed the lead shot slowly, the intermediate states of the gas are equilibrium states, so we can plot them on a p-V diagram (Fig. 21-4).

To apply Eq. 21-1 to the isothermal expansion, we take the constant temperature T outside the integral, obtaining

$$\Delta S = \frac{1}{T} \int_i^f dQ.$$

Because $\int dQ = Q$, where Q is the total energy transferred as heat during the process, we have

$$\Delta S = S_f - S_i = \frac{Q}{T} \qquad \text{(change in entropy, isothermal process).} \qquad (21-2)$$

To keep the temperature T of the gas constant during the isothermal expansion of Fig. 21-3, heat Q must have been transferred *from* the reservoir *to* the gas. Thus Q is positive and the entropy of the gas *increases* during the isothermal process and during the free expansion of Fig. 21-1.

To summarize:

> To find the entropy change for an irreversible process occurring in a closed system, replace that process with any reversible process that connects the same initial and final states. Calculate the entropy change for this reversible process with Eq. 21-1.

When the temperature change ΔT of a system is small relative to the temperature (in kelvins) before and after the process, the entropy change can be approximated as

$$\Delta S = S_f - S_i \approx \frac{Q}{\overline{T}}, \qquad (21-3)$$

where $\overline{T}$ is the average temperature of the system during the process.

CHECKPOINT 1: Water is heated on a stove. Rank the entropy changes of the water as its temperature rises (a) from 20°C to 30°C, (b) from 30°C to 35°C, and (c) from 80°C to 85°C, greatest first.

SAMPLE PROBLEM 21-1

One mole of nitrogen gas is confined to the left side of the container of Fig. 21-1. You open the stopcock and the volume of the gas doubles. What is the entropy change of the gas for this irreversible process? Treat the gas as ideal.

SOLUTION: As we did above, we replace the irreversible process of Fig. 21-1 with the isothermal expansion of Fig. 21-3 and calculate the entropy change for this latter process.

From Table 20-5, the heat Q added to an ideal gas as it expands isothermally at temperature T from an initial volume V_i to a final volume V_f is

$$Q = nRT \ln \frac{V_f}{V_i},$$

in which n is the number of moles of gas present.

From Eq. 21-2 the entropy change for this process is

$$\Delta S = \frac{Q}{T} = \frac{nRT \ln(V_f/V_i)}{T} = nR \ln \frac{V_f}{V_i}.$$

Substituting $n = 1.00$ mol and $V_f/V_i = 2$, we find

$$\Delta S = nR \ln \frac{V_f}{V_i} = (1.00 \text{ mol})(8.31 \text{ J/mol} \cdot \text{K})(\ln 2)$$

$$= +5.76 \text{ J/K.} \qquad \text{(Answer)}$$

This is also the entropy change for the free expansion—and for all other processes that connect the initial and final states shown in Fig. 21-2. ΔS is positive, so the entropy increases, in accordance with the entropy postulate of Section 21-1.

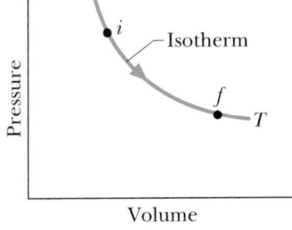

FIGURE 21-4 A p-V diagram for the reversible isothermal expansion of Fig. 21-3. The intermediate states, which are now equilibrium states, are shown.

SAMPLE PROBLEM 21-2

Figure 21-5a shows two identical copper blocks of mass $m = 1.5$ kg: block L is at temperature $T_{iL} = 60°C$ and block R is at temperature $T_{iR} = 20°C$. The blocks are in a thermally insulated box and are separated by an insulating shutter. When we lift the shutter, the blocks eventually come to the equilibrium temperature $T_f = 40°C$ (Fig. 21-5b). What is the net entropy change of the two-block system during this irreversible process? The specific heat of copper is 386 J/kg·K.

SOLUTION: To calculate the entropy change, we must find a reversible process that takes the system from the initial state of Fig. 21-5a to the final state of Fig. 21-5b. We then can calculate the net entropy change ΔS_{rev} of the reversible process using Eq. 21-1. For such a reversible process we need a thermal reservoir whose temperature can be changed slowly (say, by turning a knob). We then take the blocks through the following two steps, illustrated in Fig. 21-6:

STEP 1. With the reservoir's temperature set at 60°C, put block L on the reservoir. (Since block and reservoir are at the same temperature, they are already in thermal equilibrium.) Then slowly lower the temperature of the reservoir and the block to 40°C. As the block's temperature changes by each increment dT during this process, heat dQ is transferred *from* the block to the reservoir. Using Eq. 19-15, we can write this transferred energy as $dQ = mc\, dT$, where c is the specific heat of copper. According to Eq. 21-1, the entropy change ΔS_L of block L during the full temperature change from initial temperature T_{iL} ($= 60°C = 333$ K) to final temperature T_f ($= 40°C = 313$ K) is

$$\Delta S_L = \int_i^f \frac{dQ}{T} = \int_{T_{iL}}^{T_f} \frac{mc\, dT}{T} = mc \int_{T_{iL}}^{T_f} \frac{dT}{T}$$
$$= mc \ln \frac{T_f}{T_{iL}}.$$

Inserting the given data yields

$$\Delta S_L = (1.5 \text{ kg})(386 \text{ J/kg·K}) \ln \frac{313 \text{ K}}{333 \text{ K}}$$
$$= -35.86 \text{ J/K}.$$

STEP 2. With the reservoir's temperature now set at 20°C, put block R on the reservoir. Then slowly raise the temperature of the reservoir and the block to 40°C. With the same reasoning used to find ΔS_L, you can show that the entropy change ΔS_R of block R during this process is

$$\Delta S_R = (1.5 \text{ kg})(386 \text{ J/kg·K}) \ln \frac{313 \text{ K}}{293 \text{ K}}$$
$$= +38.23 \text{ J/K}.$$

The net entropy change ΔS_{rev} of the two-block system undergoing this two-step reversible process is then

$$\Delta S_{\text{rev}} = \Delta S_L + \Delta S_R$$
$$= -35.86 \text{ J/K} + 38.23 \text{ J/K} = 2.4 \text{ J/K}.$$

Thus, the net entropy change ΔS_{irrev} for the two-block system undergoing the actual irreversible process is

$$\Delta S_{\text{irrev}} = \Delta S_{\text{rev}} = 2.4 \text{ J/K}. \qquad \text{(Answer)}$$

This result is positive, in accordance with the entropy postulate of Section 21-1.

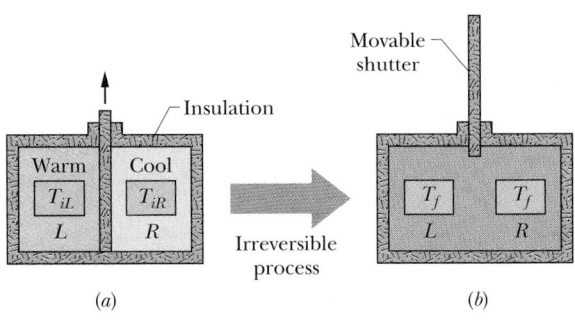

FIGURE 21-5 Sample Problem 21-2. (a) In the initial state, two copper blocks L and R, identical except for their temperatures, are in an insulating box and are separated by an insulating shutter. (b) When the shutter is removed, the blocks exchange heat and come to a final state, both with the same temperature T_f. The process is irreversible.

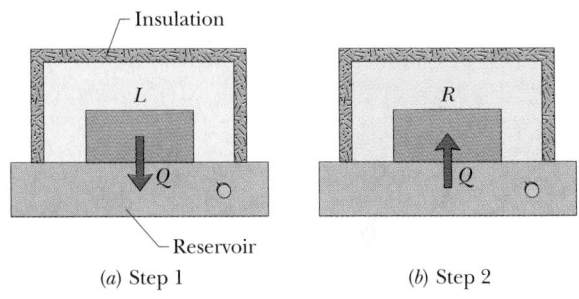

(a) Step 1   (b) Step 2

FIGURE 21-6 The blocks of Fig. 21-5 can proceed from their initial state to their final state in a reversible way if we use a reservoir with a controllable temperature (a) to extract heat reversibly from block L and (b) to add heat reversibly to block R.

CHECKPOINT 2: An ideal gas has temperature T_1 at the initial state i shown in the p-V diagram. The gas has a higher temperature T_2 at final states a and b, which it can reach along the paths shown. Is the entropy change along the path to state a larger than, smaller than, or the same as that along the path to state b?

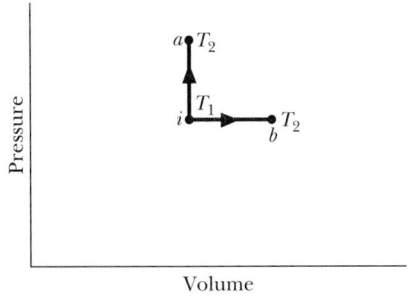

Entropy as a State Function

We have assumed that entropy, like pressure, energy, and temperature, is a property of the state of a system and independent of how that state is reached. That entropy is indeed a *state function* (as state properties are usually called) can only be deduced by experiment. However, we can prove it is a state function for the special and important case where an ideal gas is taken through a reversible process.

To make the process reversible, it is done slowly in a series of small steps, with the gas in an equilibrium state at the end of each step. For each small step, the heat transfer to or from the gas is dQ, the work done by the gas is dW, and the change in internal energy is dE_{int}. These are related by the first law of thermodynamics in differential form (Eq. 19-25):

$$dQ - dW = dE_{int}.$$

Because the steps are reversible, with the gas in equilibrium states, we can use Eq. 19-22 to replace dW with $p\,dV$ and Eq. 20-35 to replace dE_{int} with $nC_V\,dT$. Solving for dQ then leads to

$$dQ = p\,dV + nC_V\,dT.$$

Using the ideal gas law, we replace p in this equation with nRT/V. Then we divide each term in the resulting equation by T, obtaining

$$\frac{dQ}{T} = nR\frac{dV}{V} + nC_V\frac{dT}{T}.$$

Now let us integrate each term of this equation between an arbitrary initial state i and an arbitrary final state f to get

$$\int_i^f \frac{dQ}{T} = \int_i^f nR\frac{dV}{V} + \int_i^f nC_V\frac{dT}{T}.$$

The quantity on the left is the entropy change $\Delta S\ (= S_f - S_i)$ defined by Eq. 21-1. Substituting this and integrating the quantities on the right yield

$$\Delta S = S_f - S_i = nR\ln\frac{V_f}{V_i} + nC_V\ln\frac{T_f}{T_i}.$$

Note that we did not have to specify a particular reversible process when we integrated. So the integration must hold for all reversible processes that take the gas from state i to state f. Thus the change in entropy ΔS between the initial and final states of an ideal gas depends only on properties of the initial state (V_i and T_i) and properties of the final states (V_f and T_f); ΔS does not depend on how the gas changes between the two states.

21-3 THE SECOND LAW OF THERMODYNAMICS

Here is a puzzle. We saw in Sample Problem 21-1 that if we cause the reversible process of Fig. 21-3 to proceed from (*a*) to (*b*) in that figure, the change in entropy of the gas—which we take as our system—is positive. However, because the process is reversible, we can just as easily make it proceed from (*b*) to (*a*), simply by slowly adding lead shot to the piston of Fig. 21-3*b* until the original volume of the gas is restored. In this reverse process, heat must be extracted *from the gas* to keep its temperature from rising. Hence Q is negative and so, from Eq. 21-2, the entropy of the gas must decrease.

Doesn't this decrease in the entropy of the gas violate the entropy postulate of Section 21-1, which states that entropy always increases? No, because that postulate holds only for *irreversible* processes occurring in closed systems. The procedure suggested above does not meet these requirements: the process is *not* irreversible and (because energy is transferred as heat from the gas to the reservoir) the system—which is the gas alone—is *not* closed.

However, if we include the reservoir, along with the gas, as part of the system, then we do have a closed system. Let's check the change in entropy of the enlarged system *gas + reservoir* for the process that takes it from (*b*) to (*a*) in Fig. 21-3. During this reversible process, energy is transferred as heat from the gas to the reservoir, that is, from one part of the enlarged system to another. Let $|Q|$ represent the absolute value (or magnitude) of this heat. With Eq. 21-2, we can then calculate separately the entropy changes for the gas (which loses $|Q|$) and the reservoir (which gains $|Q|$). We get

$$\Delta S_{gas} = -\frac{|Q|}{T} \quad \text{and} \quad \Delta S_{res} = +\frac{|Q|}{T}.$$

The entropy change of the closed system is the sum of these two quantities, which is zero.

With this result, we can modify the entropy postulate of Section 21-1 to include both reversible and irreversible processes:

If a process occurs in a closed system, the entropy of the system increases for irreversible processes and remains constant for reversible processes. It never decreases.

Although entropy may decrease in part of a closed system, there will always be an equal or larger entropy increase in another part of the system, so that the entropy of the system as a whole never decreases. This fact is one form of the

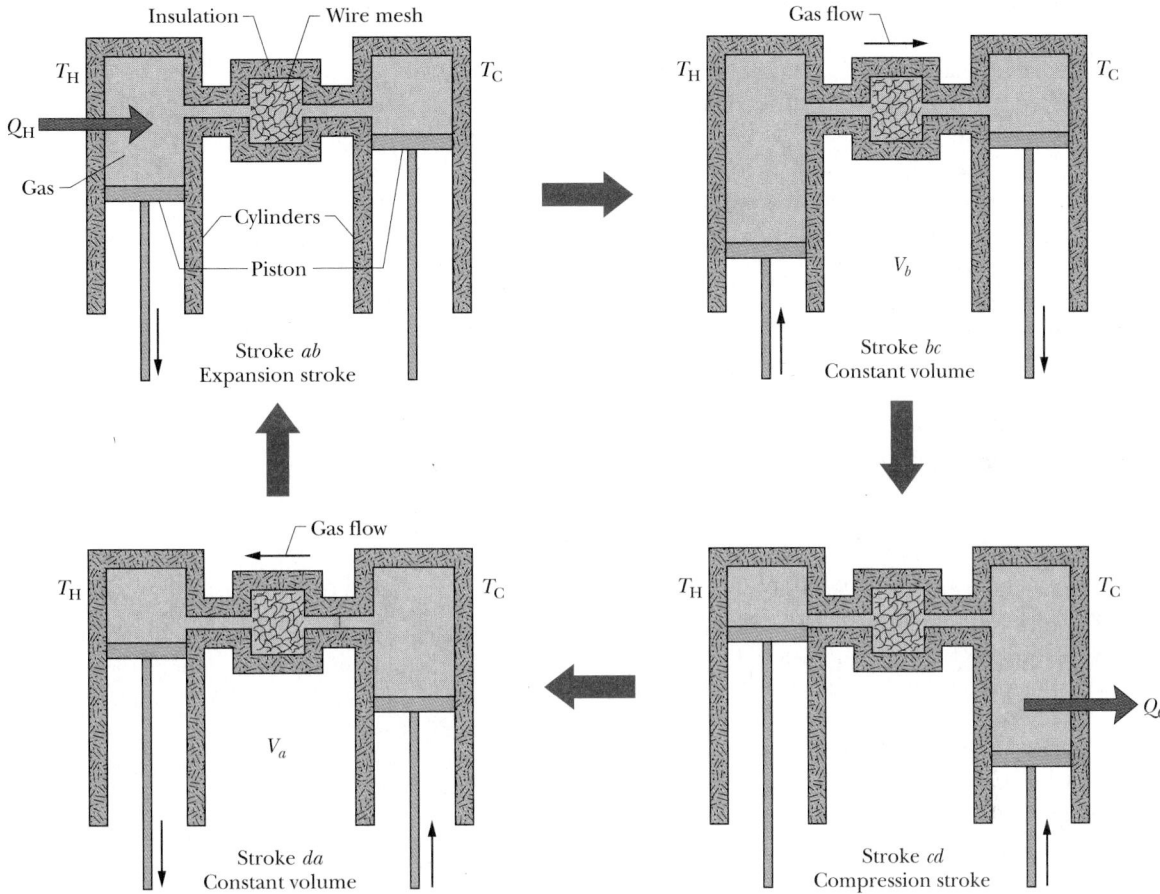

FIGURE 21-7 A schematic diagram of a Stirling engine, showing one cycle of operation, in the clockwise sequence of stroke *ab*, stroke *bc*, stroke *cd*, and then stroke *da*. Heat Q_H is added to the working substance of the engine during stroke *ab*, and heat Q_C is extracted from it during stroke *cd*. Heat is transferred to the wire mesh (to store the energy momentarily) during constant-volume stroke *bc* and then transferred back to the working substance during constant-volume stroke *da*.

second law of thermodynamics and can be written as

$$\Delta S \geq 0 \qquad \text{(second law of thermodynamics),} \qquad (21\text{-}4)$$

where the greater-than sign applies to irreversible processes and the equals sign to reversible processes. Equation 21-4 applies only to closed systems.

In the real world almost all processes are irreversible to some extent because of friction, turbulence, and other factors. So the entropy of real closed systems undergoing real processes always increases. Processes in which the system's entropy remains constant are always idealizations.

21-4 ENTROPY IN THE REAL WORLD: ENGINES

A **heat engine,** or, more simply, an **engine,** is a device that exchanges heat with its environment and does work as it continuously repeats a set sequence of processes. As a prototype, we consider the **Stirling engine,** which was first proposed in 1816 by Robert Stirling. This engine, long neglected, is now being developed for use in automobiles and in spacecraft. A Stirling engine delivering 5000 hp (3.7 MW) has been built.

Figure 21-7 shows the elements of the Stirling engine. The left-hand cylinder is maintained at a temperature T_H by a high-temperature thermal reservoir, perhaps a propane burner. The right-hand cylinder is maintained at a lower temperature T_C by a low-temperature reservoir, perhaps circulating cooling water. (Neither reservoir is shown in Fig. 21-7.) The *working substance* of the engine is a gas. As the engine operates, the working substance moves between the cylinders, transferring heat between the reservoirs and doing work. (Do not confuse the working substance of an engine with the fuel. The function of the fuel, such as propane in a burner, is to maintain the temperature of the high-temperature reservoir.)

The two cylinders in Fig. 21-7 are connected by an insulated chamber, partially stuffed with wire mesh. The mesh stores thermal energy momentarily during the operation of the engine. The two pistons, one in each cylinder, are connected by a complex mechanical linkage that permits them to move up or down as shown by the vertical arrows in Fig. 21-7. They are also connected to a crankshaft, which is not shown.

The continuously repeated sequence of processes followed by an engine is called the *cycle* of the engine. The clockwise sequence of illustrations in Fig. 21-7 is a schematic diagram of the *Stirling cycle*. The cycle, which is also represented on the p-V plot in Fig. 21-8, consists of four processes, called *strokes:*

1. An isothermal expansion (stroke *ab*) at temperature T_H, produced when the left piston moves down. As a result of this expansion, heat Q_H is transferred to the gas from the left-cylinder walls, which are maintained at temperature T_H by the high-temperature reservoir.

2. A constant-volume process (stroke *bc*) in which the temperature of the gas decreases from T_H to T_C. To accomplish this, hot gas is pushed through the wire mesh as the moving pistons increase the volume of the right cylinder, and decrease that of the left cylinder, by equal amounts. As it is pushed through, the hot gas delivers heat to the wire mesh.

3. An isothermal compression (stroke *cd*) at temperature T_C back to the original volume, produced when the right piston moves up. As a result of this compression, heat Q_C is transferred from the gas to the right-cylinder walls, which are maintained at temperature T_C by the low-temperature reservoir.

4. A constant-volume process (stroke *da*) in which the temperature of the gas increases from T_C to T_H. To accomplish this, cool gas is pushed through the hot wire mesh as the volume of the left cylinder is increased and that of the right cylinder decreased, by equal amounts.

As the pistons move during a cycle, they cause the crankshaft to rotate. The continuous cycling of the engine causes a continuous rotation of the crankshaft, which can then be put to use. That is, the engine can do work.

If we strip the Stirling engine to its essentials, we can represent it with Fig. 21-9, which actually represents any engine that operates between a high-temperature reservoir at temperature T_H and a low-temperature reservoir at temperature T_C. Once every cycle of operation, heat Q_H moves from the high-temperature reservoir to the working substance, the engine does work W, and heat Q_C moves from the working substance to the low-temperature reservoir.

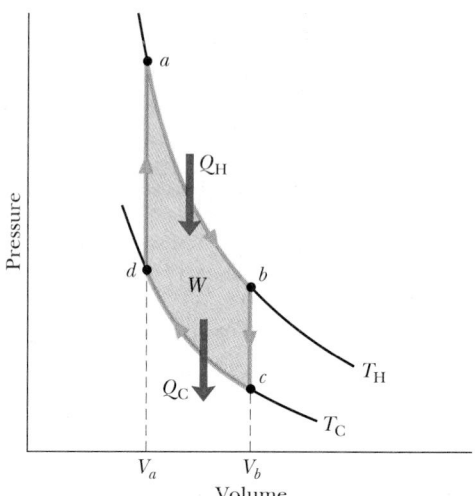

FIGURE 21-8 A p-V plot for a Stirling engine. The isothermal expansion *ab*, the isothermal compression *cd*, and the two constant-volume processes *bc* and *da* correspond to the four strokes of Fig. 21-7.

The **thermal efficiency** (or simply **efficiency**) of an engine is a measure of its ability to transform energy into work. For the engine of Fig. 21-9, heat Q_H is energy we pay for, perhaps as the cost of propane in a propane burner, and W is energy we get, to be used. So, it seems reasonable to define the efficiency of an engine as

$$\varepsilon = \frac{\text{energy we get}}{\text{energy we pay for}} = \frac{|W|}{|Q_H|} \quad \begin{array}{l}\text{(efficiency,}\\ \text{any engine).}\end{array} \quad (21\text{-}5)$$

(We use the absolute value symbols because we want a ratio of only magnitudes.)

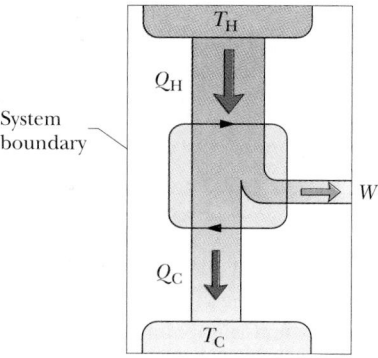

FIGURE 21-9 The elements of an engine. The two black arrowheads on the central loop suggest the working substance operating in a cycle, as if on a p-V plot. Heat Q_H is transferred from the high-temperature reservoir at temperature T_H to the working substance. Heat Q_C is transferred from the working substance to the low-temperature reservoir at temperature T_C. Work W is done by the engine (actually by the working substance) on something in the environment.

FIGURE 21-10 The North Anna nuclear power plant near Charlottesville, Virginia, which generates electrical energy at the rate of 900 MW. At the same time, by design, it discards energy into the nearby river at the rate of 2100 MW. This plant—and all others like it—throws away more energy than it delivers in useful form. It is a real counterpart to the ideal engine of Fig. 21-9.

are reversible, Eq. 21-4 tells us that the net entropy change per cycle for the entire closed system (the working substance of the engine plus the high- and low-temperature reservoirs) must be zero. The entropy change of the working substance is zero because the substance returns to its initial state at the end of each cycle. The entropy of the high-temperature reservoir *decreases* during each cycle, because heat is extracted from it. And the entropy of the low-temperature reservoir *increases,* because heat is added to it. So, for ΔS to be zero for the entire system, the magnitude of the entropy decrease of the high-temperature reservoir must equal the magnitude of the entropy increase of the low-temperature reservoir. Writing these two entropy changes in the form Q/T (via Eq. 21-2), we have

$$\frac{|Q_H|}{T_H} = \frac{|Q_C|}{T_C},$$

or

$$\frac{|Q_C|}{|Q_H|} = \frac{T_C}{T_H}. \qquad (21\text{-}8)$$

Then combining Eq. 21-8 with Eq. 21-7 yields, for the efficiency of an ideal engine operating between temperatures T_H and T_C,

$$\varepsilon = 1 - \frac{T_C}{T_H} \qquad \begin{matrix}\text{(efficiency,}\\ \text{ideal engine).}\end{matrix} \qquad (21\text{-}9)$$

Note that we have not specified either the cycle or the working substance for the ideal engine. They do not matter: any *ideal* engine whose high- and low-temperature reservoirs are at T_H and T_C has the efficiency given by Eq. 21-9. Any *real* engine, with irreversible processes and wasteful energy transfers, is less efficient. If your car engine were an ideal engine, it would have an efficiency of about 55% according to Eq. 21-9; its actual efficiency is probably about 25%. A nuclear power plant (Fig. 21-10) is a much more complex engine, drawing heat from a reactor core, doing work by means of a turbine, and discharging heat to a nearby river. If the power plant were an ideal engine, its efficiency would be about 40%; its actual efficiency is about 30%.

Inventors continually try to improve engine efficiencies by reducing the energy $|Q_C|$ that is "thrown away" during each cycle. The inventor's dream is to produce the *perfect engine,* diagrammed in Fig. 21-11, in which $|Q_C|$ is reduced to zero and $|Q_H|$ is converted completely to work $|W|$. A perfect engine on an ocean liner, for example, could extract heat from the water and use it to drive the propellers, with no fuel cost. Alas, such a perfect engine is only a dream: inspection of Eq. 21-8 shows that $|Q_C|$ can be zero only if $T_C = 0$ K or $T_H \rightarrow \infty$, requirements that are impossible to meet. Instead, decades of practical en-

Engines in the real world—*real engines*—that operate between the same high- and low-temperature reservoirs can vary in efficiency because they involve irreversible processes, with wasteful energy transfers. The wasteful transfers are often due to friction between moving parts and turbulence in the working substance. If we could make all the processes in an engine's operation reversible, eliminating wasteful energy transfers, we would have an *ideal engine.* The efficiency of this hypothetical (imaginary) engine is the upper limit for the efficiency of real engines. Let us find this limiting efficiency.

Any engine returns to its original state at the end of each cycle. This means that the internal energy of the working substance returns to its original value, and thus $\Delta E_{int} = 0$. For an ideal engine the only net energy transfers during a cycle are heat $|Q_H|$ from the high-temperature reservoir, heat $|Q_C|$ to the low-temperature reservoir, and the work $|W|$ done by the engine. The first law of thermodynamics ($\Delta E_{int} = Q - W$) then gives us

$$0 = (|Q_H| - |Q_C|) - |W|. \qquad (21\text{-}6)$$

In words, the difference between the net heat transferred per cycle $|Q_H| - |Q_C|$ and the work $|W|$ done by the engine must be zero. Solving Eq. 21-6 for $|W|$ and substituting the result into Eq. 21-5, we find that the efficiency of an ideal engine is

$$\varepsilon = \frac{|W|}{|Q_H|} = \frac{|Q_H| - |Q_C|}{|Q_H|} = 1 - \frac{|Q_C|}{|Q_H|} \qquad (21\text{-}7)$$

$$\text{(efficiency, ideal engine).}$$

We can find another expression for the efficiency of an ideal engine if we examine the entropy changes that take place during one cycle. Because all the engine processes

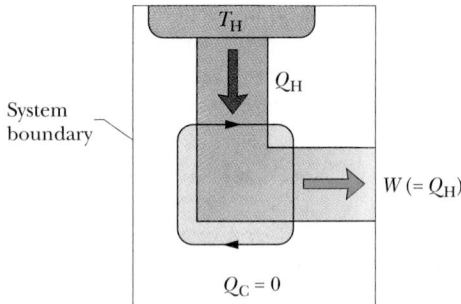

FIGURE 21-11   The elements of a perfect engine, that is, one that converts heat Q_H from a high-temperature reservoir directly to work W with 100% efficiency.

gineering experience have led to the following alternative version of the second law of thermodynamics:

> No series of processes is possible whose sole result is the absorption of heat from a thermal reservoir and the complete conversion of this energy to work.

In short, *there are no perfect engines.*

CHECKPOINT 3: Three ideal engines operate between reservoir temperatures of (a) 400 and 500 K; (b) 600 and 800 K; and (c) 400 and 600 K. Rank the engines according to their thermal efficiencies, greatest first.

SAMPLE PROBLEM 21-3

A model Stirling engine uses $n = 8.10 \times 10^{-3}$ mol of gas (assumed to be ideal) as a working substance. It operates between a high-temperature reservoir at $T_H = 95°C$ and a low-temperature reservoir at $T_C = 24°C$; the volume of its working substance doubles during the expansion stroke; and it runs at a rate of 0.70 cycle per second. Assume that the engine is ideal.

(a) How much work does the engine do per cycle?

SOLUTION: The net work per cycle is the algebraic sum of the work done by the gas during the four processes that make up the cycle shown in Figs. 21-7 and 21-8. The work done by the gas during stroke ab, an isothermal expansion from V_a to V_b, is, from Eq. 20-10,

$$W_{ab} = nRT_H \ln \frac{V_b}{V_a}.$$

The work done by the gas during stroke cd, an isothermal compression from V_b to V_a, is, similarly,

$$W_{cd} = nRT_C \ln \frac{V_a}{V_b}.$$

The work done during each of the two constant-volume processes (bc and da) is zero, so the net work for the entire cycle is

$$W = W_{ab} + W_{bc} + W_{cd} + W_{da}$$
$$= nRT_H \ln \frac{V_b}{V_a} + 0 + nRT_C \ln \frac{V_a}{V_b} + 0.$$

Because $\ln(V_a/V_b) = -\ln(V_b/V_a)$, we can write this as

$$W = nR(T_H - T_C) \ln \frac{V_b}{V_a}.$$

Substituting known data and $V_b/V_a = 2$ then yields

$$W = (8.10 \times 10^{-3} \text{ mol})(8.31 \text{ J/mol} \cdot \text{K})(95°C - 24°C)(\ln 2)$$
$$= 3.31 \text{ J} \approx 3.3 \text{ J}. \qquad \text{(Answer)}$$

Thus, during each cycle of operation, this engine does 3.3 J of work via its crankshaft.

(b) What is the power P of the engine?

SOLUTION: The power of the engine is the work W done per cycle divided by the duration t of each cycle. The engine operates at 0.70 cycle/s, so the duration of one cycle is $1/0.70 = 1.43$ s. The power is then

$$P = \frac{W}{t} = \frac{3.31 \text{ J}}{1.43 \text{ s}} \approx 2.3 \text{ W}. \qquad \text{(Answer)}$$

(c) How much heat is transferred *to* the gas *from* the high-temperature reservoir during each cycle?

SOLUTION: From Table 20-5, the magnitude $|Q_H|$ of the heat transferred to an ideal gas during an isothermal expansion (process ab in Fig. 21-8) is

$$|Q_H| = nRT_H \ln \frac{V_b}{V_a}.$$

Substituting given data and temperature $T_H = (95 + 273)$ K $= 368$ K yields

$$|Q_H| = (8.10 \times 10^{-3} \text{ mol})(8.31 \text{ J/mol} \cdot \text{K})(368 \text{ K})(\ln 2)$$
$$= 17.2 \text{ J} \approx 17 \text{ J}. \qquad \text{(Answer)}$$

(d) What is the efficiency of the engine?

SOLUTION: Substituting the temperatures of the two reservoirs into Eq. 21-9 gives us

$$\varepsilon = 1 - \frac{T_C}{T_H} = 1 - \frac{(24 + 273) \text{ K}}{(95 + 273) \text{ K}}$$
$$= 0.193 \approx 19\%. \qquad \text{(Answer)}$$

(Equations 21-5 and 21-7 give the same answer for ε.)

(e) What are the entropy changes per cycle for the gas, the high-temperature reservoir, the low-temperature reservoir, and the wire mesh?

SOLUTION: The gas operates in a cycle, so its net entropy change per cycle ΔS_{gas} is zero. For the thermal reservoirs, heat is transferred at constant temperature, so we can use Eq. 21-2 to calculate their entropy changes.

We first take the high-temperature reservoir as our system. In (c) we found that the magnitude of the heat transferred is $|Q_H| = 17.2$ J. Because the reservoir lost the heat, we can write the transfer as $-|Q_H|$. Then from Eq. 21-2, the entropy change of the reservoir is

$$\Delta S_H = \frac{-|Q_H|}{T_H} = \frac{-17.2 \text{ J}}{(95 + 273) \text{ K}} = -0.047 \text{ J/K.} \quad \text{(Answer)}$$

The entropy of the high-temperature reservoir decreases because heat is lost.

Heat enters the low-temperature reservoir from the gas, so the entropy of that reservoir increases. If you apply Eq. 21-6 or follow the example of our calculation in (c), you will find that this reservoir gains heat $|Q_C| = 13.9$ J and undergoes an entropy change of

$$\Delta S_C = \frac{+13.9 \text{ J}}{(24 + 273) \text{ K}} = +0.047 \text{ J/K.} \quad \text{(Answer)}$$

As for the mesh, the heat it gains in stroke *bc* of Fig. 21-7 it loses in stroke *da*. And these two heat transfers occur over the same range of temperatures. So from Eq. 21-2, the net entropy change per cycle ΔS_{mesh} is zero.

The entropy change per cycle for the entire system, including the reservoirs, is then

$$\Delta S = \Delta S_H + \Delta S_C + \Delta S_{gas} + \Delta S_{mesh}$$
$$= -0.047 \text{ J/K} + 0.047 \text{ J/K} + 0 + 0$$
$$= 0. \quad \text{(Answer)}$$

This is just what the second law of thermodynamics predicts for the entropy change associated with a set of reversible processes occurring in a closed system, such as the ideal engine here.

SAMPLE PROBLEM 21-4

An inventor claims to have constructed an engine that has an efficiency of 75% when operated between the boiling and freezing points of water. We shall evaluate the credibility of the inventor's claim in two (related) ways.

(a) Compare the claimed 75% efficiency to that of an ideal engine operating between the same two temperatures.

SOLUTION: With Eq. 21-9 we obtain

$$\varepsilon = 1 - \frac{T_C}{T_H} = 1 - \frac{(0 + 273) \text{ K}}{(100 + 273) \text{ K}} = 0.268 \approx 27\%.$$

An ideal engine operating between the given temperatures would have an efficiency of 27%; any real engine (with its inevitable irreversible processes and wasteful energy transfers) must be less efficient. Thus the claimed efficiency of 75% for the given temperatures is impossible.

(b) Now, starting over, assume that the engine is good enough to be approximately an ideal engine. Then determine the en-

tropy change ΔS that occurs within the closed system consisting of the engine and its reservoirs during one cycle. Compute ΔS in terms of $|Q_H|$, the magnitude of the heat transferred each cycle from the high-temperature reservoir to the working substance.

SOLUTION: Because the working substance returns to its initial state at the end of each cycle, its entropy change ΔS_{ws} during one cycle is zero. The high-temperature reservoir loses heat $|Q_H|$ at temperature T_H, and the low-temperature reservoir gains heat $|Q_C|$ at temperature T_C. Then from Eq. 21-2, their entropy changes are

$$\Delta S_H = \frac{-|Q_H|}{T_H} \quad \text{and} \quad \Delta S_C = \frac{+|Q_C|}{T_C}.$$

The total entropy change is

$$\Delta S = \Delta S_H + \Delta S_C + \Delta S_{ws}$$
$$= -\frac{|Q_H|}{T_H} + \frac{|Q_C|}{T_C} + 0. \quad (21\text{-}10)$$

We can find a substitute for $|Q_C|$ with Eq. 21-7:

$$|Q_C| = |Q_H| (1 - \varepsilon).$$

Now, combining this with Eq. 21-10, we have, after some algebra,

$$\Delta S = |Q_H| \left(\frac{1 - \varepsilon}{T_C} - \frac{1}{T_H} \right).$$

Substituting the data gives us

$$\Delta S = |Q_H| \left(\frac{1 - 0.75}{273 \text{ K}} - \frac{1}{373 \text{ K}} \right)$$
$$= -0.0018 \, |Q_H|.$$

Thus, if the inventor's claim were correct, the entropy of the system would decrease with each cycle. This is impossible. If every process that occurs in the engine were reversible, ΔS would be zero. If some were irreversible, ΔS would be greater than zero. No set of processes in a closed system can lead to ΔS less than zero.

PROBLEM SOLVING TACTICS

TACTIC 1: *The Language of Thermodynamics*

A rich, but sometimes misleading, language is used in scientific and engineering studies of thermodynamics. And regardless of your major, you will need to understand what the language means. You may see statements that say heat is absorbed, extracted, rejected, discharged, discarded, withdrawn, delivered, gained, lost, or expelled, or that it flows from one body to another (as if it were a liquid). You may also see statements that describe a body as *having* heat (as if heat can be held or possessed), or that its heat is increased or decreased. You should always keep in mind what is meant by the term *heat*:

Heat is energy that is transferred from one body to another body owing to a difference in the temperatures of the bodies.

When we identify one of the bodies as being our system of interest, any such transfer of energy into the system is positive heat Q, and any such transfer out of the system is negative Q.

The term *work* also requires close attention. You may see statements that say work is produced or generated, or combined with heat or changed from heat. Here is what is meant by the term *work*:

Work is energy that is transferred from one body to another body owing to a force that acts between them.

When we identify one of the bodies as being our system of interest, any such transfer of energy out of the system is either positive work W done *by* the system or negative work W done *on* the system. And any such transfer of energy into the system is negative work done *by* the system or positive work done *on* the system. (The preposition that is used is important.) Obviously, this can be confusing: whenever you see the term *work*, you should read carefully to determine the intent.

21-5 ENTROPY IN THE REAL WORLD: REFRIGERATORS

A **refrigerator** is a device that uses work to transfer thermal energy from a low-temperature reservoir to a high-temperature reservoir as it continuously repeats a set sequence of processes. In a household refrigerator, for example, work is done by an electrical compressor to transfer thermal energy from the food storage compartment (a low-temperature reservoir) to the room (a high-temperature reservoir). Figure 21-12 represents the basic elements of a refrigerator. If we assume that the processes involved in the refrigerator's operation are reversible, then we have an *ideal refrigerator*. Such a hypothetical refrigerator is just a (hypothetical) ideal engine operating in reverse.

Air conditioners and heat pumps are, in essence, refrigerators. The differences are only in the nature of the high- and low-temperature reservoirs. For an air conditioner, the low-temperature reservoir is the room that is to be cooled, and the high-temperature reservoir is the outdoors. A heat pump is an air conditioner that can be operated in reverse to heat a room; then the room is the high-temperature reservoir and heat is transferred to it from outdoors.

The designer of a refrigerator would like to extract as much heat $|Q_C|$ as possible from the low-temperature res-

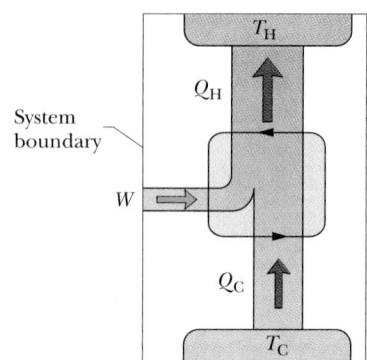

FIGURE 21-12   The elements of a refrigerator. The two black arrowheads on the central loop suggest the working substance operating in a cycle, as if on a *p-V* plot. Heat Q_C is transferred to the working substance from the low-temperature reservoir. Heat Q_H is transferred to the high-temperature reservoir from the working substance. Work W is done on the refrigerator (on the working substance) by something in the environment.

ervoir (what we want) for the least amount of work W (what we pay for). As a measure of the efficiency of a refrigerator, then, a reasonable definition is

$$K = \frac{\text{what we want}}{\text{what we pay for}} = \frac{|Q_C|}{|W|} \quad \begin{array}{l}\text{(coefficient of}\\ \text{performance)},\end{array} \quad (21\text{-}11)$$

where K is called the *coefficient of performance*. If the refrigerator is ideal, the first law of thermodynamics gives $|W| = |Q_H| - |Q_C|$, where $|Q_H|$ is the magnitude of the heat transferred to the high-temperature reservoir. Equation 21-11 then becomes

$$K = \frac{|Q_C|}{|Q_H| - |Q_C|}. \quad (21\text{-}12)$$

Because an ideal refrigerator is an ideal engine operating in reverse, we can combine Eq. 21-8 with Eq. 21-12; after some algebra we find

$$K = \frac{T_C}{T_H - T_C} \quad \begin{array}{l}\text{(coefficient of performance,}\\ \text{ideal refrigerator).}\end{array} \quad (21\text{-}13)$$

For typical room air conditioners, $K \approx 2.5$. For household refrigerators, $K \approx 5$. Perversely, the value of K is higher the closer the temperatures of the two reservoirs are to each other. That is why heat pumps are more effective in temperate climates than in climates where the outside temperature varies widely.

It would be nice to own a refrigerator that did not require some input of work, that is, one that would run without being plugged in. Figure 21-13 represents another "inventor's dream," a *perfect refrigerator* that transfers heat Q from a cold reservoir to a warm reservoir without

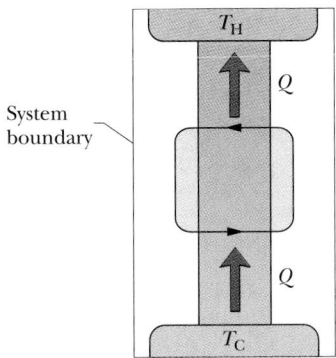

FIGURE 21-13 The elements of a perfect refrigerator, that is, one that transfers heat from a low-temperature reservoir to a high-temperature reservoir without any input of work.

the need for work. Because the unit operates in cycles, the entropy of the working substance does not change during a complete cycle. The entropies of the two reservoirs, however, do change: the entropy change for the cold reservoir is $-|Q|/T_C$, and that for the warm reservoir is $+|Q|/T_H$. So, the net entropy change for the entire system is

$$\Delta S = -\frac{|Q|}{T_C} + \frac{|Q|}{T_H}.$$

Because $T_H > T_C$, the right side of this equation is negative and thus the net change in entropy per cycle for the closed system *refrigerator + reservoirs* is also negative. Because such a decrease in entropy violates the second law of thermodynamics (Eq. 21-4), a perfect refrigerator does not exist. (If you want your refrigerator to operate, you must plug it in.)

This result leads us to another (equivalent) formulation of the second law of thermodynamics:

No series of processes is possible whose sole result is the transfer of heat from a reservoir at a given temperature to a reservoir at a higher temperature.

In short, *there are no perfect refrigerators.*

CHECKPOINT 4: You wish to increase the coefficient of performance of an ideal refrigerator. You can do so by (a) running the cold chamber at a slightly higher temperature, (b) running it at a slightly lower temperature, (c) moving the unit to a slightly warmer room, or (d) moving it to a slightly cooler room. The temperature changes are to be the same in all four cases. List the changes according to the resulting coefficients of performance, greatest first.

SAMPLE PROBLEM 21-5

An ideal refrigerator, with coefficient of performance $K = 4.7$, extracts heat from the cold chamber at the rate of 250 J/cycle.

(a) How much work per cycle is required to operate the refrigerator?

SOLUTION: From Eq. 21-11, we find

$$|W| = \frac{|Q_C|}{K} = \frac{250 \text{ J}}{4.7} = 53 \text{ J}. \qquad \text{(Answer)}$$

(b) How much heat per cycle is discharged to the room?

SOLUTION: For an ideal refrigerator, with reversible processes and no energy transfers other than $|Q_H|$, $|Q_C|$, and $|W|$, the first law of thermodynamics tells us that

$$\Delta E_{int} = (|Q_H| - |Q_C|) - |W|.$$

Here $\Delta E_{int} = 0$ because the working substance operates in a cycle. Solving for $|Q_H|$ and inserting known data then yield

$$|Q_H| = |W| + |Q_C| = 53 \text{ J} + 250 \text{ J}$$
$$= 303 \text{ J} \approx 300 \text{ J}. \qquad \text{(Answer)}$$

21-6 A STATISTICAL VIEW OF ENTROPY

In Chapter 20 we saw that macroscopic properties of gases can be explained in terms of their microscopic, or molecular, behavior; such explanations are part of the study called *statistical mechanics*. Recall, for example, that we were able to account for the pressure exerted by a gas on the walls of its container in terms of the momentum transferred to those walls by rebounding gas molecules.

Here we shall focus our attention on a single problem involving the distribution of gas molecules between two halves of a container (a box). This problem is reasonably simple to analyze, and it allows us to use statistical mechanics to calculate the entropy change for a free expansion of an ideal gas. You will see in Sample Problem 21-6 that statistical mechanics leads to the same entropy change we obtained in Sample Problem 21-1 using thermodynamics.

Consider the box of Fig. 21-14, which contains four identical (and thus indistinguishable) molecules of a gas. At any instant, a given molecule will be in either the left half of the box or the right half; because the two halves of the box have equal volumes, the molecule has the same likelihood, or probability, of being in the left half as in the right half. Table 21-1, in which L stands for a molecule in the left half of the box and R for a molecule in the right half, shows the 16 possible ways in which the four identi-

TABLE 21-1 FOUR MOLECULES IN A BOX

LOCATION OF MOLECULE a b c d				CONFIGURATION LABEL	MULTIPLICITY W (NUMBER OF MICROSTATES)	CONFIGURATION PROBABILITY	CALCULATION OF W BY EQ. 21-14	ENTROPY $(10^{-23}\,\text{J/K})$ FROM EQ. 21-15
L	L	L	L	I	1	1/16	$4!/(4!\,0!) = 1$	0.00
R	L	L	L					
L	R	L	L	II	4	4/16	$4!/(3!\,1!) = 4$	1.91
L	L	R	L					
L	L	L	R					
L	L	R	R					
L	R	L	R					
R	L	L	R	III	6	6/16	$4!/(2!\,2!) = 6$	2.47
L	R	R	L					
R	L	R	L					
R	R	L	L					
L	R	R	R					
R	L	R	R	IV	4	4/16	$4!/(1!\,3!) = 4$	1.91
R	R	L	R					
R	R	R	L					
R	R	R	R	V	1	1/16	$4!/(0!\,4!) = 1$	0.00
					Total number of microstates 16			

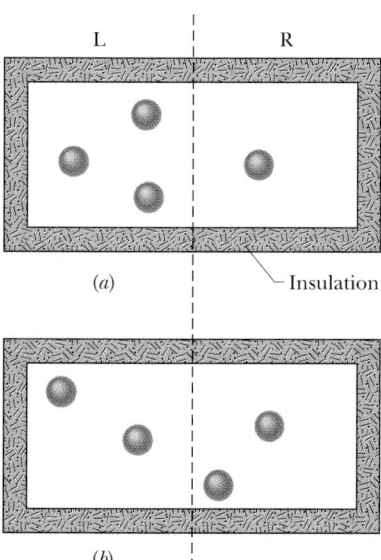

L   R

(a)   Insulation

(b)

FIGURE 21-14 An insulated box contains four gas molecules. Each molecule has the same probability of being in the left half (L) of the box as in the right half (R). The arrangement in (a) corresponds to configuration II in Table 21-1, and that in (b) corresponds to configuration III.

cal molecules, which we label a, b, c, and d, can be in the two halves of the box. We call each of these 16 arrangements a **microstate** of the system.

In Table 21-1, the 16 microstates are grouped into five **configurations**, each made up of *equivalent microstates* and labeled with a Roman numeral. Figure 21-14a shows configuration II, which we may describe as "one molecule in the right half of the box." Table 21-1 displays the four microstates that make up this configuration; they are equivalent because, since the molecules are identical, the situation in which molecule a is in the right half is physically no different from the situation in which molecule b (or c or d) is in the right half. Figure 21-14b shows configuration III, which is "two molecules in each half of the box"; Table 21-1 lists the six microstates that form this configuration. We call the number of microstates associated with a configuration the *multiplicity W* of that configuration. (We have used the same symbol for work; do not confuse these two uses.)

A basic assumption of statistical mechanics is that *all microstates are equally probable.* That is, if we were to take a great many snapshots of the molecules as they jostle around the box in random fashion, and then count the number of times each microstate occurred, we would find that all the microstates occur equally often. So, at any given instant, the probability of finding the molecules in any one microstate is equal to that of finding them in any other microstate. That is, the system will spend, on average, the same amount of time in each of the 16 microstates listed in Table 21-1.

Because the microstates are equally probable, but different configurations have *different numbers of microstates,* the configurations are *not* equally probable. The *probability (of occurrence) of a configuration* is the ratio of its multiplicity to the total number of microstates of the system. In Table 21-1, configuration III, with six microstates and thus a multiplicity of 6, is the most probable configuration, with a probability of 6/16 = 0.375. This means that the system is in configuration III 37.5% of the time. Configurations I and V, in which the gas occupies only half the box, are the least probable, with probabilities of 1/16 = 0.0625 or 6.25% each.

Now recall the free expansion process of Sample Problem 21-1; there we found that the state in which the gas fills its container has greater entropy than the state in which it occupies only half the container. Here we have found that the state (or configuration) in which the gas fills its container uniformly is more probable than a state in which the gas occupies only half the container. The state with higher entropy seems also to have the higher probability of occurrence.

Four molecules in a box ($N = 4$) are not very many on which to base a conclusion. Let us increase the number N of molecules to 100 and again compare the amount of time during which half the molecules are in each half of the box to that during which all the molecules are in the left half of the box. The ratio is not 6 to 1 (as it is for $N = 4$ in Table 21-1) but about 10^{29} to 1. Imagine what this ratio must be for the much more practical case of $N = 10^{22}$, which is

about the number of air molecules in a child's inflated balloon. The probability is then overwhelming for a uniform distribution of molecules throughout the container.

For large values of N, there are very large numbers of microstates. But nearly all the microstates correspond to an approximately equal division of the molecules between the two halves of the box, as Fig. 21-15 indicates. Even though the measured temperature and pressure of the gas remain constant, the gas is churning away endlessly as its molecules "visit" all the possible microstates with equal prob-

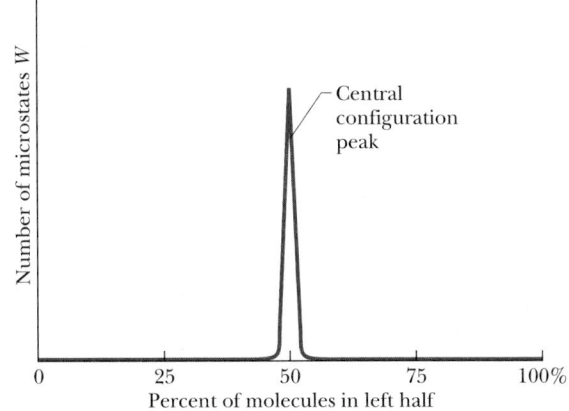

FIGURE 21-15 For a *large* number of molecules in a box, a plot of the number of microstates that require various percentages of the molecules to be in the left half of the box. Nearly all the microstates correspond to an approximately equal sharing of the molecules between the two halves of the box; those microstates form the central *configuration peak* on the plot. For $N \approx 10^{22}$, the central configuration peak would be much too narrow to be drawn on this plot.

Sometimes striking order appears naturally. (*a*) For example, Giant's Causeway in Northern Ireland consists of tall stone columns, many of which are hexagonal in cross section and appear to have been designed. The columns formed when hot magma seeped out of the ground and cooled. (*b*) Order can also be seen in

the *sorted circles* of gravel and stones that occur naturally on an island north of Norway. Just how the columns and circles were produced is still debated, but we can be certain that as the entropy of the magma and stones decreased, the entropy of their environment increased even more.

(*a*)

(*b*)

abilities. However, because so few microstates lie outside the narrow *central configuration peak* of Fig. 21-15, we might as well assume that the gas molecules are always evenly divided between the two halves of the box. And, as we saw, this is the configuration with the greatest entropy.

For the four molecules of Fig. 21-14, we can easily list the microstates associated with each of the five configurations and then find the multiplicities. For the general case of N molecules, where N can be large, we find the multiplicity of a configuration with

$$W = \frac{N!}{n_L! \, n_R!} \quad \text{(multiplicity of a configuration)}. \quad (21\text{-}14)$$

Here N is the total number of molecules, n_L is the number in the left half of the box, and n_R is the number in the right half. The symbol "!" indicates a *factorial*, which is defined for any nonnegative integer n as

$$n! = n(n-1)(n-2) \cdots (1).$$

For example, $5! = 5 \times 4 \times 3 \times 2 \times 1 = 120$. By definition, $0! = 1$. Table 21-1 includes the calculation of W for each of its configurations, using Eq. 21-14.

We have seen that configurations with higher probability of occurrence also have higher entropy. Because the multiplicity of a configuration is a measure of its probability, we may say that states with greater multiplicity have greater entropy. Austrian physicist Ludwig Boltzmann first pointed out the following relationship between these two quantities in 1877:

$$S = k \ln W \quad \text{(Boltzmann's entropy equation)}. \quad (21\text{-}15)$$

Here S is the entropy of a configuration, W is the multiplicity of the configuration, and $k \; (= 1.38 \times 10^{-23} \text{ J/K})$ is the Boltzmann constant, first encountered in Section 20-5. This famous formula is engraved on Boltzmann's tombstone.

The rightmost column of Table 21-1 displays the entropies of the configurations of the four-molecule system of Fig. 21-14, as computed with Eq. 21-15. Configuration III, which has the greatest multiplicity, also has the greatest entropy.

When you use Eq. 21-14 to calculate W, your calculator may signal an error if you try to find the factorial of a number much greater than a few hundred. Fortunately, there is a very good approximation, known as **Stirling's approximation**, not for $N!$ but for $\ln N!$—exactly what is needed in Eq. 21-15. Stirling's approximation is

$$\ln N! \approx N(\ln N) - N \quad \text{(Stirling's approximation)}. \quad (21\text{-}16)$$

(The Stirling of this approximation is not the Stirling of the Stirling engine.)

CHECKPOINT 5: A box contains one mole of a gas. Consider two configurations: (a) each half of the box contains half the molecules, and (b) each third of the box contains one-third of the molecules. Which configuration has more microstates?

SAMPLE PROBLEM 21-6

In Sample Problem 21-1 we showed that when n moles of an ideal gas doubles its volume in a free expansion, the entropy increase from the initial state i to the final state f is $S_f - S_i = nR \ln 2$. Derive this result with statistical mechanics.

SOLUTION: Let N be the number of molecules in n moles of the gas. From Eq. 21-14, the multiplicity of the initial state, in which all N molecules are in the left half of the container in Fig. 21-1, is

$$W_i = \frac{N!}{N! \, 0!} = 1.$$

The multiplicity of the final state, in which $N/2$ molecules are in each half of the container, is

$$W_f = \frac{N!}{(N/2)! \, (N/2)!}.$$

From Eq. 21-15, the initial and final entropies are

$$S_i = k \ln W_i = k \ln 1 = 0$$

and $\quad S_f = k \ln W_f = k \ln(N!) - 2k \ln[(N/2)!]. \quad (21\text{-}17)$

In writing Eq. 21-17, we have used the relation

$$\ln \frac{a}{b^2} = \ln a - 2 \ln b.$$

Now, applying Eq. 21-16 to evaluate Eq. 21-17, we find that

$$\begin{aligned} S_f &= k \ln(N!) - 2k \ln[(N/2)!] \\ &= k[N(\ln N) - N] - 2k[(N/2) \ln(N/2) - (N/2)] \\ &= k[N(\ln N) - N - N \ln(N/2) + N] \\ &= k[N(\ln N) - N(\ln N - \ln 2)] = kN \ln 2. \quad (21\text{-}18) \end{aligned}$$

We can substitute for kN by using the relations

$$k = \frac{R}{N_A} \quad \text{and} \quad N = nN_A$$

from Eqs. 20-1 and 20-2, respectively. Here N_A is the Avogadro constant and R is the universal gas constant. With these two relations we see that $kN = nR$. So, Eq. 21-18 becomes

$$S_f = nR \ln 2.$$

The change in entropy from the initial state to the final is thus

$$S_f - S_i = nR \ln 2 - 0 = nR \ln 2, \quad \text{(Answer)}$$

which is what we set out to show. In Sample Problem 21-1 we calculated this entropy increase for a free expansion with thermodynamics by finding an equivalent reversible process and calculating the entropy change for *that* process in terms of temperature and heat transfer. Here we have calculated the same increase with statistical mechanics using the fact that the system consists of molecules.

REVIEW & SUMMARY

One-Way Processes

An **irreversible process** is one that cannot be reversed by means of small changes in the environment. The direction in which an irreversible process proceeds is set by the **change in entropy** ΔS of the system undergoing the process. Entropy S is a *state property* (or *state function*) of the system; that is, it depends only on the state of the system and not on how the system reached that state. The *entropy postulate* states (in part): *If an irreversible process occurs in a closed system, the entropy of the system always increases.*

Calculating Entropy Change

The **entropy change** ΔS for an irreversible process that takes a system from an initial state i to a final state f is exactly equal to the entropy change ΔS for *any reversible process* that takes the system between those same two states. We can compute the latter (but not the former) with

$$\Delta S = S_f - S_i = \int_i^f \frac{dQ}{T}. \qquad (21\text{-}1)$$

Here Q is the energy transferred as heat to or from the system during the process, and T is the temperature of the system in kelvins.

For a reversible isothermal process, Eq. 21-1 reduces to

$$\Delta S = S_f - S_i = \frac{Q}{T}. \qquad (21\text{-}2)$$

When the temperature change ΔT of a system is small relative to the temperature (in kelvins) before and after the process, the entropy change can be approximated as

$$\Delta S = S_f - S_i \approx \frac{Q}{\overline{T}}, \qquad (21\text{-}3)$$

where $\overline{T}$ is the system's average temperature during the process.

The Second Law of Thermodynamics

This law, which is an extension of the entropy postulate, states: *If a process occurs in a closed system, the entropy of the system increases for irreversible processes and remains constant for reversible processes. It never decreases.* In equation form,

$$\Delta S \geq 0. \qquad (21\text{-}4)$$

Engines

An **engine** is a device that, operating in a cycle, extracts heat $|Q_H|$ from a high-temperature reservoir, does a certain amount of work $|W|$, and discharges heat $|Q_C|$ to a low-temperature reservoir. The

efficiency ε of an engine is defined as

$$\varepsilon = \frac{\text{energy we get}}{\text{energy we pay for}} = \frac{|W|}{|Q_H|}. \qquad (21\text{-}5)$$

An *ideal engine* is one in which all processes are reversible and the only net energy transfers in a cycle are $|Q_H|$, $|W|$, and $|Q_C|$. Then the first law of thermodynamics becomes

$$\Delta E_{int} = (|Q_H| - |Q_C|) - |W| = 0. \qquad (21\text{-}6)$$

For an ideal engine, Eq. 21-5 can then be rewritten as

$$\varepsilon = 1 - \frac{|Q_C|}{|Q_H|} = 1 - \frac{T_C}{T_H}, \qquad (21\text{-}7, 21\text{-}9)$$

in which T_H and T_C are the temperatures of the high- and low-temperature reservoirs, respectively. Real engines always have an efficiency lower than that given by Eqs. 21-7 and 21-9.

A *perfect engine* is an imaginary engine in which heat extracted from the high-temperature reservoir is converted completely to work. Because such a conversion would decrease the entropy of the system with each cycle, a perfect engine would violate the second law of thermodynamics. That law can then be restated as follows: No series of processes is possible whose sole result is the absorption of heat from a thermal reservoir and the complete conversion of this energy to work.

Refrigerators

A **refrigerator** (which may also be an air conditioner or a heat pump) is a device that, operating in a cycle, has work $|W|$ done on it, extracts heat $|Q_C|$ from a low-temperature reservoir, and discharges heat $|Q_H|$ to a high-temperature reservoir. The *coefficient of performance* K of a refrigerator is defined as

$$K = \frac{\text{what we want}}{\text{what we pay for}} = \frac{|Q_C|}{|W|}. \qquad (21\text{-}11)$$

An *ideal refrigerator* is one in which all processes are reversible and the only net energy transfers in a cycle are $|Q_H|$, $|W|$, and $|Q_C|$. For an ideal refrigerator, Eq. 21-11 becomes

$$K = \frac{|Q_C|}{|Q_H| - |Q_C|} = \frac{T_C}{T_H - T_C}, \qquad (21\text{-}12, 21\text{-}13)$$

in which T_H and T_C are the temperatures of the high- and low-temperature reservoirs, respectively.

A *perfect refrigerator* is an imaginary refrigerator in which heat extracted from the low-temperature reservoir is converted completely to heat discharged to the high-temperature reservoir, without any need for work. Because this conversion would decrease the entropy of the system with each cycle, a perfect refrigerator would violate the second law of thermodynamics. That law can then be restated as follows: No series of processes is possible whose sole result is the transfer of heat from a reservoir at a given temperature to a reservoir at a higher temperature.

Entropy from a Statistical View

The entropy of a system can be defined in terms of the possible distributions of its molecules. For identical molecules, each possible distribution of molecules is called a **microstate** of the system. All equivalent microstates are grouped into a **configuration** of the system. The number of microstates in a configuration is the **multiplicity** W of the configuration.

For a system of N molecules that may be distributed between the two halves of a box, the multiplicity is given by

$$W = \frac{N!}{n_L! \, n_R!}, \qquad (21\text{-}14)$$

in which n_L is the number of molecules in the left half of the box and n_R is the number in the right half. A basic assumption of *statistical mechanics* is that all the microstates are equally probable. Thus, configurations with a large multiplicity occur most often. When N is very large (say, $N = 10^{22}$ molecules or more), the molecules are nearly always in the configuration in which $n_L = n_R$.

The multiplicity W of a configuration of a system and the entropy S of the system in that configuration are related by Boltzmann's entropy equation:

$$S = k \ln W, \qquad (21\text{-}15)$$

where $k = 1.38 \times 10^{-23}$ J/K is the Boltzmann constant.

When N is very large (the usual case), we can approximate $\ln N!$ with *Stirling's approximation:*

$$\ln N! \approx N(\ln N) - N. \qquad (21\text{-}16)$$

QUESTIONS

1. The net heat transfer between the Sun and Earth is from Sun to Earth. Does the entropy of the Earth–Sun system increase, decrease, or remain the same during this process?

2. A gas, confined to an insulated cylinder, is compressed adiabatically to half its volume. Does the entropy of the gas increase, decrease, or remain unchanged during this process?

3. A drop of water boils away *slowly* on a hot plate whose temperature is slightly above the boiling point. (a) Does the entropy of the water increase, decrease, or remain the same during this process? (b) Does the entropy of the system *water + hot plate* increase, decrease, or remain the same?

4. Point i in Fig. 21-16 represents the initial state of an ideal gas at temperature T. Taking algebraic signs into account, rank the entropy changes as the gas moves, successively and reversibly, from point i to points a, b, c, and d, greatest first.

greater than, less than, or equal to the entropy change that would occur if you allowed the gas to expand freely from volume V directly to volume $3V$?

6. Three ideal engines operate between temperature limits of (a) 400 and 500 K; (b) 500 and 600 K; and (c) 400 and 600 K. Each engine extracts the same amount of energy per cycle from the high-temperature reservoir. Rank the magnitudes of the work done by the engines per cycle, greatest first.

7. You wish to increase the efficiency of an ideal engine. Is it better to raise the temperature of the high-temperature reservoir by a certain amount or lower the temperature of the low-temperature reservoir by the same amount?

8. The work done by an ideal engine is used to lift a heavy safe, as shown in Fig. 21-17. Does the entropy of the *engine + Earth + safe* system increase, decrease, or remain the same during this process?

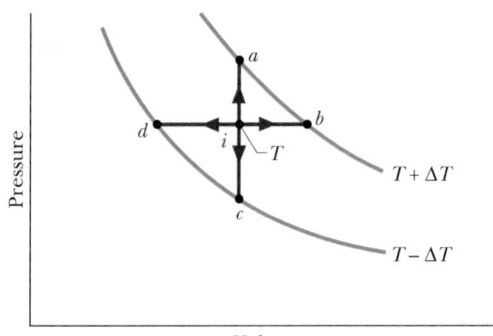

FIGURE 21-16   Question 4.

5. You allow a gas to expand freely from volume V to volume $2V$. Later you allow that gas to expand freely from volume $2V$ to volume $3V$. Is the net entropy change for these two expansions

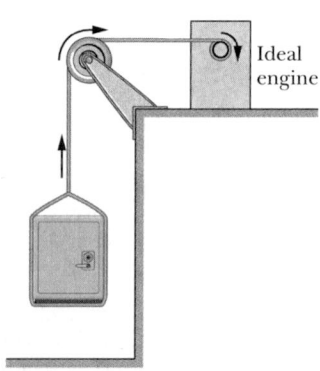

FIGURE 21-17   Question 8.

9. Does the entropy per cycle increase, decrease, or remain the same for (a) an ideal engine, (b) a real engine, and (c) a perfect engine (which is, of course, impossible to build)?

10. If you leave the door of your kitchen refrigerator open for several hours, does the temperature of the kitchen increase, decrease, or remain the same? Assume that the kitchen is closed and well insulated.

11. Does the entropy per cycle increase, decrease, or remain the same for (a) an ideal refrigerator, (b) a real refrigerator, and (c) a perfect refrigerator (which is, of course, impossible to build)?

12. In an ideal refrigerator, is the magnitude $|Q_H|$ of the heat transferred per cycle to the high-temperature reservoir always

less than, always greater than, or always equal to the magnitude of the work W done per cycle?

13. A box contains 100 atoms in a configuration, with 50 atoms in each half of the box. Suppose that you could count the different microstates associated with this configuration at the rate of 100 billion states per second, using a supercomputer. Without written calculation, guess how much computing time you would need: a day, a year, or much more than a year.

14. A basic assumption of statistical mechanics is that all microstates are equally probable. How would the air in your bedroom behave if all configurations were equally probable (they aren't)?

EXERCISES & PROBLEMS

SECTION 21-2 Change in Entropy

1E. An ideal gas undergoes a reversible isothermal expansion at 132°C. The entropy of the gas increases by 46.0 J/K. How much heat was absorbed?

2E. A 2.50 mol sample of an ideal gas expands reversibly and isothermally at 360 K until its volume is doubled. What is the increase in entropy of the gas?

3E. An ideal gas undergoes a reversible isothermal expansion at 77.0°C, increasing its volume from 1.30 L to 3.40 L. The entropy change of the gas is 22.0 J/K. How many moles of gas are present?

4E. Construct plots of T versus S, S versus E_{int}, and S versus V for the isothermal expansion process of Fig. 21-4.

5E. An ideal gas in contact with a constant-temperature reservoir undergoes a reversible isothermal expansion to twice its initial volume. Show that the reservoir's change in entropy is independent of its temperature.

6E. Four moles of an ideal gas are expanded from volume V_1 to volume $V_2 = 2V_1$. If the expansion is isothermal at temperature $T = 400$ K, find (a) the work done by the expanding gas and (b) the change in its entropy. (c) If the expansion is reversibly adiabatic instead of isothermal, what is the entropy change of the gas?

7E. Find (a) the heat absorbed and (b) the change in entropy of a 2.00 kg block of copper whose temperature is increased reversibly from 25°C to 100°C. The specific heat of copper is 386 J/kg·K.

8E. An ideal monatomic gas at initial temperature T_0 (in kelvins) expands from initial volume V_0 to volume $2V_0$ by each of the five processes indicated in the T-V diagram of Fig. 21-18. In which process is the expansion (a) isothermal, (b) isobaric (constant pressure), and (c) adiabatic? Explain your answers. (d) In which processes does the entropy of the gas decrease?

9E. The temperature of 1.0 mol of a monatomic ideal gas is raised reversibly from 300 K to 400 K, with its volume kept constant. What is the entropy change of the gas? (*Hint:* Recall that $E_{int} = \frac{3}{2}nRT$ for such a gas.)

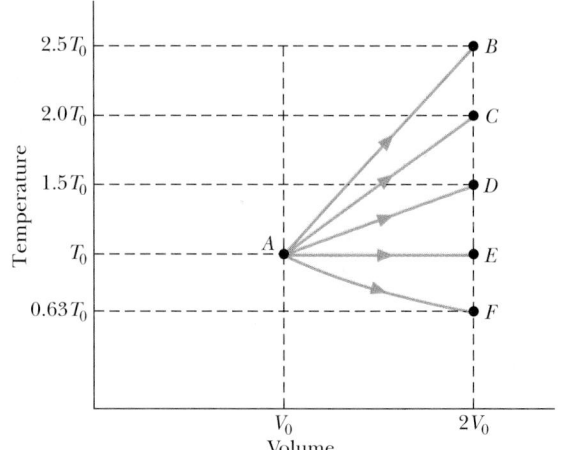

FIGURE 21-18 Exercise 8.

10E. Repeat Exercise 9, with the pressure now kept constant.

11E. (a) What is the entropy change of a 12.0 g ice cube that melts completely in a bucket of water whose temperature is just above the freezing point of water? (b) What is the entropy change of a 5.00 g spoonful of water that evaporates completely on a hot plate whose temperature is slightly above the boiling point of water?

12E. Suppose that 260 J of heat is conducted from a constant-temperature reservoir at 400 K to another reservoir at (a) 100 K, (b) 200 K, (c) 300 K, and (d) 360 K. What are the net changes in entropy of the reservoirs for those four cases? (e) What is the trend in those changes?

13E. A brass rod is in thermal contact with a constant-temperature reservoir at 130°C at one end and another reservoir at 24.0°C at the other end. (a) Compute the total change in entropy of the rod–reservoirs system that occurs when 5030 J of heat is conducted through the rod, from one reservoir to the other. (b) Does the entropy of the rod change in the process?

14E. At very low temperatures, the molar specific heat C_V of many solids is approximately $C_V = AT^3$, where A depends on the particular substance. For aluminum, $A = 3.15 \times 10^{-5}$ J/mol·K^4. Find the entropy change for 4.00 mol of aluminum when its temperature is raised from 5.00 K to 10.0 K.

15P. A 2.0 g sample of an ideal monatomic gas undergoes the reversible process shown in Fig. 21-19. (a) How much heat is absorbed by the gas? (b) What is the change in the internal energy of the gas? (c) How much work is done by the gas?

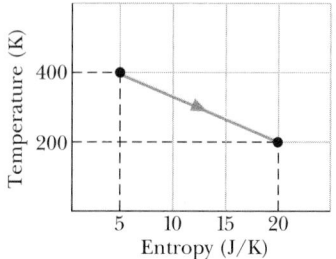

FIGURE 21-19 Problem 15.

16P. Heat can actually be removed from water at and then below the normal freezing point (0.0°C at atmospheric pressure) without causing the water to freeze; the water is said to be *supercooled*. Suppose a 1.00 g water drop is supercooled until its temperature is that of the surrounding air, which is at −5.00°C. The drop then suddenly and irreversibly freezes, transferring heat to the air until the drop is again at −5.00°C. What is the entropy change for the drop? (*Hint:* Use a three-step reversible process to take the water through the normal freezing point.) The specific heat of ice is 2220 J/kg·K.

17P. In an experiment, 200 g of aluminum (with a specific heat of 900 J/kg·K) at 100°C is mixed with 50.0 g of water at 20.0°C. (a) What is the equilibrium temperature? What are the entropy changes of (b) the aluminum, (c) the water, and (d) the aluminum−water system?

18P. An object of (constant) heat capacity C is heated from an initial temperature T_i to a final temperature T_f by a constant-temperature reservoir at T_f. (a) Represent the process on a graph of C/T versus T, and show graphically that the total change in entropy ΔS of the object−reservoir system is positive. (b) Explain how the use of reservoirs at intermediate temperatures would allow the process to be carried out in a way that makes ΔS as small as desired.

19P. In the irreversible process of Fig. 21-5, let the initial temperatures of identical blocks L and R be 305.5 and 294.5 K, respectively, and let 215 J be the heat transfer between the blocks required to reach equilibrium. Then for the reversible processes of Fig. 21-6, what are the entropy changes of (a) block L, (b) its reservoir, (c) block R, (d) its reservoir, (e) the two-block system, and (f) the system of the two blocks and the two reservoirs?

20P. Use the reversible apparatus of Fig. 21-6 to show that, if the process of Fig. 21-5 happened in reverse, the entropy of the system would decrease, a violation of the second law of thermodynamics.

21P. The entropy change ΔS between two states of a system, which is $\int (dQ/T)$, depends only on the properties of the initial and final states. That is, this integral is path independent, as is

shown for an ideal gas in Section 21-2. Show similarly that the integrals $\int dQ$, $\int (T\, dQ)$, and $\int (dQ/T^2)$ are not path independent and thus do not define new state properties for an ideal gas.

22P. An ideal diatomic gas, whose molecules are rotating but not oscillating, is taken through the cycle in Fig. 21-20. Determine for all three processes, in terms of p_1, V_1, T_1, and R: (a) p_2, p_3, and T_3 and (b) W, Q, ΔE_{int}, and ΔS per mole.

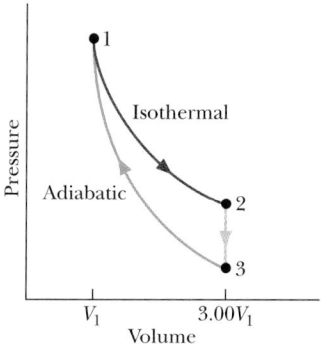

FIGURE 21-20 Problem 22.

23P. A mole of a monatomic ideal gas is taken from an initial pressure p and volume V to a final pressure $2p$ and volume $2V$ by two different processes: (I) It expands isothermally until its volume is doubled, and then its pressure is increased at constant volume to the final pressure. (II) It is compressed isothermally until its pressure is doubled, and then its volume is increased at constant pressure to the final volume. Show the path of each process on a p-V diagram. For each process calculate, in terms of p and V, (a) the heat absorbed by the gas in each part of the process, (b) the work done by the gas in each part of the process, (c) the change in internal energy of the gas, $E_{int,f} - E_{int,i}$, and (d) the change in entropy of the gas, $S_f - S_i$.

24P. A 50.0 g block of copper whose temperature is 400 K is placed in an insulating box with a 100 g block of lead whose temperature is 200 K. (a) What is the equilibrium temperature of the two-block system? (b) What is the change in the internal energy of the two-block system between the initial state and the equilibrium state? (c) What is the change in the entropy of the two-block system? See Table 19-3.

25P. A 10 g ice cube at −10°C is placed in a lake whose temperature is 15°C. Calculate the change in entropy of the cube−lake system as the ice cube comes to thermal equilibrium with the lake. The specific heat of ice is 2220 J/kg·K. (*Hint:* Will the ice cube affect the temperature of the lake?)

26P. An 8.0 g ice cube at −10°C is put into a Thermos flask containing 100 cm^3 of water at 20°C. What is the change in entropy of the cube−water system when a final equilibrium state is reached? The specific heat of ice is 2220 J/kg·K.

27P. A mixture of 1773 g of water and 227 g of ice is in an initial equilibrium state at 0.00°C. The mixture is then, in a reversible process, brought to a second equilibrium state where the water−ice ratio, by mass, is 1:1 at 0.00°C. (a) Calculate the entropy change of the system during this process. (The heat of fusion for water is 333 kJ/kg.) (b) The system is then returned to the initial equilibrium state in an irreversible process (say, by using a Bunsen burner). Calculate the entropy change of the system during this process. (c) Are your answers consistent with the second law of thermodynamics?

28P. A cylinder contains n moles of a monatomic ideal gas. If the gas undergoes a reversible isothermal expansion from initial volume V_i to final volume V_f along path I in Fig. 21-21, its change in entropy is $\Delta S = nR \ln (V_f/V_i)$. (See Sample Problem 21-1.) Now consider path II in Fig. 21-21, which takes the gas from the same initial state i to state x by a reversible adiabatic expansion, and then from state x to the same final state f by a reversible constant-volume process. (a) Describe how you would carry out the two reversible processes for path II. (b) Show that the temperature of the gas in state x is

$$T_x = T_i(V_i/V_f)^{2/3}.$$

(c) What are the heat Q_I transferred along path I and the heat Q_{II} transferred along path II? Are they equal? (d) What is the entropy change ΔS for path II? Is the entropy change for path I equal to it? (e) Evaluate T_x, Q_I, Q_{II}, and ΔS for $n = 1$, $T_i = 500$ K, and $V_f/V_i = 2$.

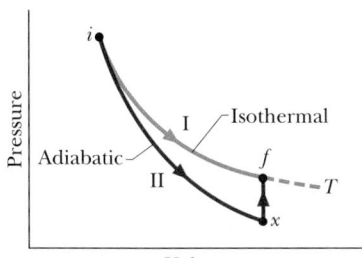

FIGURE 21-21 Problem 28.

29P. One mole of an ideal monatomic gas is taken through the cycle in Fig. 21-22. (a) How much work is done by the gas in going from state a to state c along path abc? What are the changes in internal energy and entropy in going (b) from b to c and (c) through one complete cycle? Express all answers in terms of the pressure p_0, volume V_0, and temperature T_0 of state a.

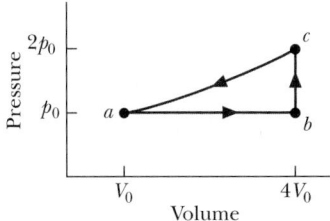

FIGURE 21-22
Problem 29.

30P. One mole of an ideal monatomic gas, at an initial pressure of 5.00 kPa and initial temperature of 600 K, expands from initial volume $V_i = 1.00$ m³ to final volume $V_f = 2.00$ m³. During the expansion, the pressure p and volume V of the gas are related by $p = 5.00 \exp[(V_i - V)/a]$, where p is in kilopascals, V_i and V are in cubic meters, and $a = 1.00$ m³. What are (a) the final pressure and (b) the final temperature of the gas? (c) How much work is done by the gas during the expansion? (d) What is the change in entropy of the gas during the expansion? (*Hint:* Use two simple reversible processes to find the entropy change.)

SECTION 21-4 Entropy in the Real World: Engines

31E. An athlete dives from a platform and comes to rest in the water; this process is irreversible, and the diver–water–Earth system is closed. Devise a reversible way of transforming the system from its initial state (diver on platform) to its final state (diver in water, with the temperature of the diver and water slightly increased due to frictional heating). You can use an ideal engine (as in Fig. 21-17) and a reservoir whose temperature can be controlled.

32E. A moving block is slowed to a stop by friction. Both the block and the surface along which it slides are in an insulated enclosure. (a) Devise a reversible process to change the system from its initial state (block moving) to its final state (block stationary; temperature of the block and surface slightly increased). Show that this reversible process results in an entropy increase for the closed block–surface system. For the process, you can use an ideal engine and a constant-temperature reservoir. (b) Show, using the same process, but in reverse, that if the temperature of the system were to decrease spontaneously and the block were to start moving again, the entropy of the system would decrease (a violation of the second law of thermodynamics).

33E. An ideal heat engine absorbs 52 kJ of heat and exhausts 36 kJ of heat in each cycle. Calculate (a) the efficiency and (b) the work done per cycle in kilojoules.

34E. Calculate the efficiency of a fossil-fuel power plant that consumes 380 metric tons of coal each hour to produce useful work at the rate of 750 MW. The heat of combustion of (the heat released by) 1.0 kg of coal is 28 MJ.

35E. An inventor claims to have invented four engines, each of which operates between constant-temperature reservoirs at 400 and 300 K. Data on each engine, per cycle of operation, are: engine A, $Q_H = 200$ J, $Q_C = -175$ J, and $W = 40$ J; engine B, $Q_H = 500$ J, $Q_C = -200$ J, and $W = 400$ J; engine C, $Q_H = 600$ J, $Q_C = -200$ J, and $W = 400$ J; engine D, $Q_H = 100$ J, $Q_C = -90$ J, and $W = 10$ J. Of the first and second laws of thermodynamics, which (if either) does each engine violate?

36E. A car engine delivers 8.2 kJ of work per cycle. (a) Before a tune-up, the efficiency is 25%. Calculate, per cycle, the heat absorbed from the combustion of fuel and the energy lost by the engine. (b) After a tune-up, the efficiency is 31%. What are the new values of the quantities calculated in (a) when 8.2 kJ of work is delivered per cycle?

37E. An ideal engine whose low-temperature reservoir is at 17°C has an efficiency of 40%. By how much should the temperature of the high-temperature reservoir be increased to increase the efficiency to 50%?

38E. An ideal heat engine operates in a Stirling cycle between 235°C and 115°C. It absorbs 6.30×10^4 J per cycle at the higher temperature. (a) What is the efficiency of the engine? (b) How much work per cycle is this engine capable of performing?

39E. In a hypothetical nuclear fusion reactor, the fuel is deuterium gas at a temperature of about 7×10^8 K. If this gas could be used to operate an ideal heat engine with $T_C = 100$°C, what would be the engine's efficiency?

40E. An ideal Stirling engine has an efficiency of 22.0%. It operates between constant-temperature reservoirs differing in temperature by 75.0 C°. What are the temperatures of the two reservoirs?

41E. Show that the heat extracted from the gas during stroke bc of the ideal Stirling engine (Fig. 21-8) is equal in magnitude to the heat added to the gas during stroke da. What is the magnitude of this heat transfer for the data given in Sample Problem 21-3? Assume that the working substance is a monatomic ideal gas.

42P. One mole of a monatomic ideal gas is taken through the reversible cycle shown in Fig. 21-23. Process bc is an adiabatic expansion, with $p_b = 10.0$ atm and $V_b = 1.00 \times 10^{-3}$ m³. Find (a) the heat added to the gas, (b) the heat leaving the gas, (c) the net work done by the gas, and (d) the efficiency of the cycle.

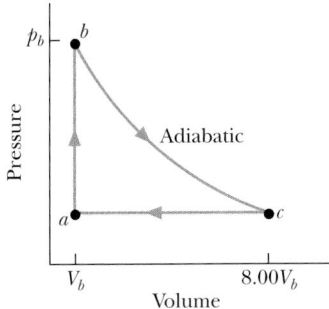

FIGURE 21-23
Problem 42.

43P. One mole of a monatomic ideal gas initially at a volume of 10 L and a temperature of 300 K is heated at constant volume to a temperature of 600 K, allowed to expand isothermally to its initial pressure, and finally compressed at constant pressure to its original volume, pressure, and temperature. (a) During the cycle, how much heat enters the system (the gas)? (b) What is the net work done by the gas? (c) What is the efficiency of the cycle?

44P. An ideal engine has a power of 500 W. It operates between constant-temperature reservoirs at 100°C and 60.0°C. What are (a) the rate of heat input and (b) the rate of exhaust heat output, in kilojoules per second?

45P. One mole of an ideal monatomic gas is taken through the cycle shown in Fig. 21-24. Assume that $p = 2p_0$, $V = 2V_0$, $p_0 = 1.01 \times 10^5$ Pa, and $V_0 = 0.0225$ m³. Calculate (a) the work done during the cycle, (b) the heat added during stroke abc, and (c) the efficiency of the cycle. (d) What is the efficiency of an ideal engine operating between the highest and lowest temperatures that occur in the cycle? How does this compare to the efficiency calculated in (c)?

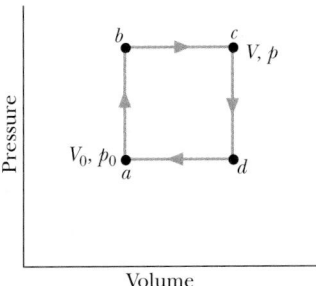

FIGURE 21-24   Problem 45.

46P. A combination mercury–steam turbine takes saturated mercury vapor from a boiler at 876°F and exhausts it to heat a steam boiler at 460°F. The steam turbine receives steam at this temperature and exhausts it to a condenser that is at 100°F. What is the maximum theoretical efficiency of the combination?

47P. In the first stage of a two-stage ideal heat engine, heat Q_1 is absorbed at temperature T_1, work W_1 is done, and heat Q_2 is expelled at a lower temperature T_2. The second stage absorbs that heat Q_2, does work W_2, and expels heat Q_3 at a still lower temperature T_3. Prove that the efficiency of the two-stage engine is $(T_1 - T_3)/T_1$.

48P. One mole of an ideal gas is used as the working substance of an engine that operates on the cycle shown in Fig. 21-25. BC and DA are reversible adiabatic processes. (a) Is the gas monatomic, diatomic, or polyatomic? (b) What is the efficiency of the engine?

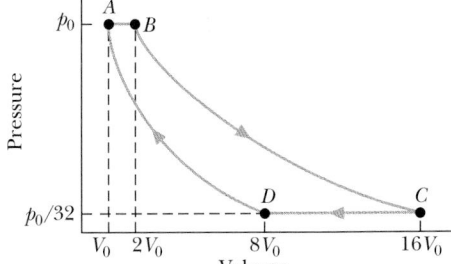

FIGURE 21-25
Problem 48.

49P. Suppose that a deep shaft were drilled in Earth's crust near one of the poles, where the surface temperature is −40°C, to a depth where the temperature is 800°C. (a) What is the theoretical limit to the efficiency of an engine operating between these temperatures? (b) If all the heat released into the low-temperature reservoir were used to melt ice that was initially at −40°C, at what rate could liquid water at 0°C be produced by a 100 MW power plant? The specific heat of ice is 2220 J/kg·K; water's heat of fusion is 333 kJ/kg. (Note that the engine can operate only between 0°C and 800°C in this case. Energy exhausted at −40°C cannot be used to raise the temperature of anything above −40°C.)

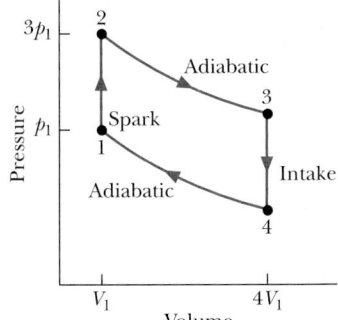

FIGURE 21-26
Problem 50.

50P. The operation of a gasoline internal combustion engine is represented by the cycle in Fig. 21-26. Assume an ideal gas and use a compression ratio of 4:1 ($V_4 = 4V_1$). Assume that $p_2 = 3p_1$. (a) Determine the pressure and temperature at each of the vertex points of the p-V diagram in terms of p_1, T_1, and the ratio γ of the molar specific heats of the gas. (b) What is the efficiency of the cycle?

SECTION 21-5 Entropy in the Real World: Refrigerators

51E. An inventor has built an engine (engine X) and claims that its efficiency ε_X is greater than the efficiency ε of an ideal engine operating between the same two temperatures. Suppose that you couple engine X to an ideal refrigerator (Fig. 21-27a) and adjust the cycle of engine X so that the work per cycle that it provides equals the work per cycle required by the ideal refrigerator. Treat this combination as a single unit and show that if the inventor's claim were true (if $\varepsilon_X > \varepsilon$), the combined unit would act as a perfect refrigerator (Fig. 21-27b), transferring heat from the low-temperature reservoir to the high-temperature reservoir without the need of work.

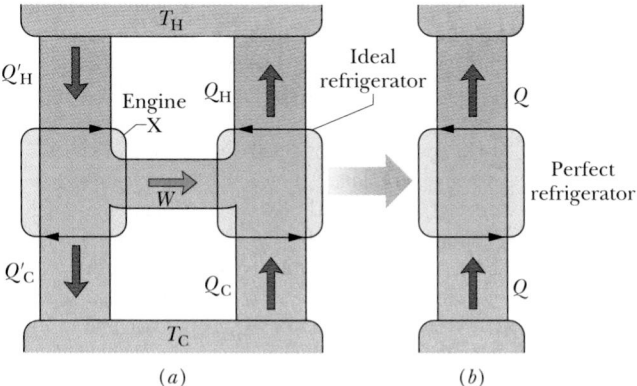

FIGURE 21-27 Exercise 51.

52E. An apparatus that liquefies helium is in a room maintained at 300 K. If the helium in the apparatus is at 4.0 K, what is the minimum ratio of heat delivered to the room to heat removed from the helium?

53E. An ideal engine has efficiency ε. Show that if you run it backward as an ideal refrigerator, the coefficient of performance will be $K = (1 - \varepsilon)/\varepsilon$.

54E. An ideal refrigerator does 150 J of work to remove 560 J of heat from its cold compartment. (a) What is the refrigerator's coefficient of performance? (b) How much heat per cycle is exhausted to the kitchen?

55E. To make ice, an ideal freezer extracts 42 kJ of heat at $-12°C$ during each cycle. The freezer has a coefficient of performance of 5.7. The room temperature is 26°C. (a) How much heat per cycle is delivered to the room? (b) How much work per cycle is required to run the freezer?

56E. (a) An ideal engine operates between a hot reservoir at 320 K and a cold reservoir at 260 K. If it absorbs 500 J of heat per cycle at the hot reservoir, how much work per cycle does it deliver? (b) If the same engine, working in reverse, functions as a refrigerator between the same two reservoirs, how much work per cycle must be supplied to remove 1000 J of heat from the cold reservoir?

57E. An ideal air conditioner takes heat from a room at 70°F and transfers it to the outdoors, which is at 96°F. For each joule of electrical energy required to operate the air conditioner, how many joules of heat are removed from the room?

58E. Heat from the outdoors at $-5.0°C$ is transferred to a room at 17°C by the electric motor of a heat pump. If the heat pump is ideal, how many joules of heat will be delivered to the room for each joule of electric energy consumed?

59E. How much work must be done by an ideal refrigerator to transfer 1.0 J of heat (a) from a reservoir at 7.0°C to one at 27°C; (b) from a reservoir at $-73°C$ to one at 27°C; (c) from a reservoir at $-173°C$ to one at 27°C; and (d) from a reservoir at $-223°C$ to one at 27°C?

60P. An ideal heat pump is used to heat a building. The outside temperature is $-5.0°C$ and the temperature inside the building is to be maintained at 22°C. The coefficient of performance is 3.8, and the heat pump delivers 7.54 MJ of heat to the building each hour. At what rate must work be done to run the heat pump?

61P. The motor in a refrigerator has a power of 200 W. If the freezing compartment is at 270 K and the outside air is at 300 K, assuming ideal efficiency, what is the maximum amount of heat that can be extracted from the freezing compartment in 10.0 min?

62P. An air conditioner operating between 93°F and 70°F is rated at 4000 Btu/h cooling capacity. Its coefficient of performance is 27% of that of an ideal refrigerator operating between the same two temperatures. What horsepower is required of the air conditioner motor?

63P. An ideal engine works between temperatures T_1 and T_2. It drives an ideal refrigerator that works between temperatures T_3 and T_4 (Fig. 21-28). Find the ratio Q_3/Q_1 in terms of T_1, T_2, T_3, and T_4.

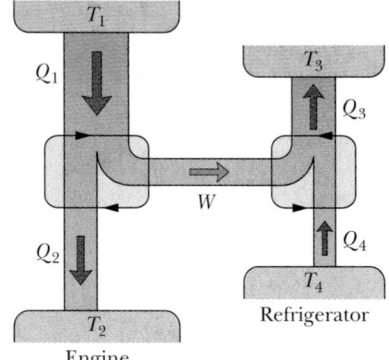

FIGURE 21-28 Problem 63.

SECTION 21-6 A Statistical View of Entropy

64E. Construct a table like Table 21-1 for six molecules rather than four.

65E. Show that for N molecules in a box, the number of possible microstates is 2^N when microstates are defined by whether a given molecule is in the left half of the box or the right half. Check this for the situation of Table 21-1.

66E. Check your calculator against Stirling's approximation, with N equal to (a) 50, (b) 100, and (c) 250. Is there a trend?

67P. Suppose you put 100 pennies in a cup, shake the cup, and then toss the pennies onto the floor. (a) How many different head–tail arrangements (microstates) are possible for the 100 pennies? (*Hint:* See Exercise 65.) What is the probability of finding exactly (b) 50 heads, (c) 48 heads, (d) 48 tails, (e) 40 heads, and (f) 30 heads?

68P. A box contains N gas molecules, equally divided between its two halves. For $N = 50$: (a) What is the multiplicity of this central configuration? (b) What is the total number of microstates for the system? (*Hint:* See Exercise 65.) (c) What percentage of the time does the system spend in its central configuration? (d) Repeat (a) through (c) for $N = 100$. (e) Repeat (a) through (c) for $N = 200$. (f) As N increases, you will find that the system spends *less* time (not more) in its central configuration. Explain why this is so.

69P. A box contains N gas molecules. Consider the box to be divided into three equal parts. (a) By extension of Eq. 21-14, write a formula for the multiplicity of any given configuration. (b) Consider two configurations: configuration A with equal numbers of molecules in all three thirds of the box, and configuration B with equal numbers of molecules in both halves of the box. What is the ratio W_A/W_B of the multiplicity of configuration A to that of configuration B? (c) Evaluate W_A/W_B for $N = 100$. (Because 100 is not evenly divisible by 3, put 34 molecules into one of the three box parts and 33 in each of the other parts for configuration A.)

70P. A box contains N molecules. Consider two configurations: configuration A with an equal division of the molecules between the two halves of the box, and configuration B with 60.0% of the molecules in the left half of the box and 40.0% in the right half. For $N = 50$, what are the multiplicities of (a) configuration A and (b) configuration B? (c) What is the ratio f of the time the system spends in configuration B to that in configuration A? (d) Repeat (a) through (c) for $N = 100$. (e) Repeat (a) through (c) for $N = 200$. (f) With an increase in N, does f increase, decrease, or remain the same?

Electronic Computation

71. The temperature of 25 mol of an ideal diatomic gas, originally 300 K, is increased by placing the gas in contact with a thermal reservoir at 800 K. The pressure remains constant throughout the increase. (a) Calculate the change in the entropy of the gas, the change in the entropy of the reservoir, and the change in the total entropy of the gas-reservoir system. (b) Suppose the same change in state is carried out by placing the gas in contact successively with n intermediate reservoirs, equally spaced in temperature between 300 K and 800 K. Calculate the total change in the entropy of the reservoirs and the change in the entropy of the closed system consisting of the gas and the reservoirs for (b) $n = 1$, (c) $n = 10$, (d) $n = 50$, and (e) $n = 100$. With each increase in n, the process more closely approximates a reversible process.

Boiling and the Leidenfrost Effect

JEARL WALKER
Cleveland State University

How does water boil? As commonplace as the event is, you may not have noticed all of its curious features. Some of the features are important in industrial applications, while others appear to be the basis for certain dangerous stunts once performed by daredevils in carnival sideshows.

Arrange for a pan of tap water to be heated from below by a flame or electric heat source. As the water warms, air molecules are driven out of solution in the water, collecting as tiny bubbles in crevices along the bottom of the pan (Fig. 1a). The air bubbles gradually inflate, and then they begin to pinch off from the crevices and rise to the top surface of the water (Fig. 1b–f). As they leave, more air bubbles form in the crevices and pinch off, until the supply of air in the water is depleted. The formation of air bubbles is a sign that the water is heating but has nothing to do with boiling.

Water that is directly exposed to the atmosphere boils at what is sometimes called its normal boiling temperature T_s. For example, T_s is about 100°C when the air pressure is 1 atm. Since the water at the bottom of your pan is not directly exposed to the atmosphere, it remains liquid even when it *superheats* above T_s by as much as a few degrees. During this process, the water is constantly mixed by convection as hot water rises and cooler water descends.

If you continue to increase the pan's temperature, the bottom layer of water begins to vaporize, with water mole- cules gathering in small vapor bubbles in the now dry crevices, as the air bubbles do in Fig. 1. This phase of boiling is signaled by pops, pings, and eventually buzzing. The water almost sings its displeasure at being heated. Every time a vapor bubble expands upward into slightly cooler water, the bubble suddenly collapses because the vapor within it condenses. Each collapse sends out a sound wave, the ping you hear. Once the temperature of the bulk water increases, the bubbles may not collapse until after they pinch off from the crevices and ascend part of the way to the top surface of the water. This phase of boiling is labeled "isolated vapor bubbles" in Fig. 2.

If you still increase the pan's temperature, the clamor of collapsing bubbles first grows louder and then disap- pears. The noise begins to soften when the bulk liquid is sufficiently hot that the vapor bubbles reach the top surface of the water. There they pop open with a light splash. The water is now in full boil.

If your heat source is a kitchen stove, the story stops at this point. However, with a laboratory burner you can con- tinue to increase the pan's temperature. The vapor bubbles

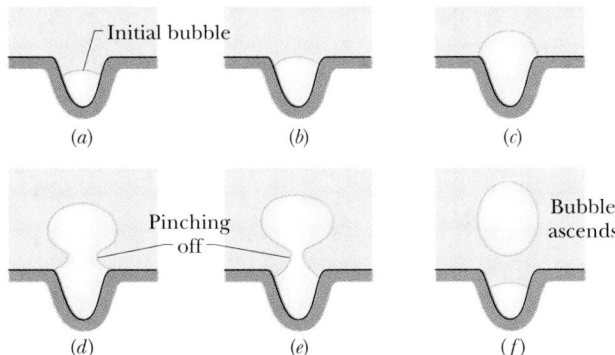

FIGURE 1 (*a*) A bubble forms in the crevice of a scratch along the bottom of a pan of water. (*b-f*) The bubble grows, pinches off, and then ascends through the water.

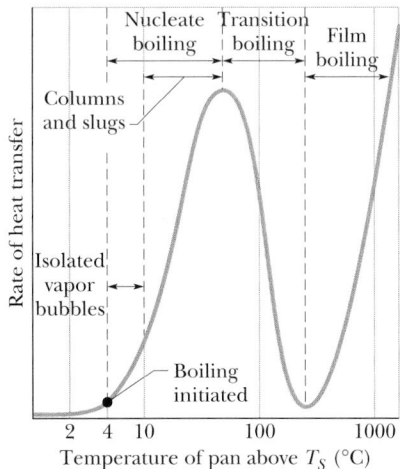

FIGURE 2 Boiling curve for water. As the temperature at the bottom of the pan is increased above the normal boiling point, the rate at which energy is transferred from the pan bottom to the water increases at first. However, above a certain temperature, the transfer almost disappears. At even higher temperatures, the transfer reappears.

533

next become so abundant and pinch off from their crevices so frequently that they coalesce, forming columns of vapor that violently and chaotically churn upward, sometimes meeting previously detached ''slugs'' of vapor.

The production of vapor bubbles and columns is called *nucleate boiling* because the formation and growth of the bubbles depend on crevices serving as *nucleating sites* (sites of formation). Whenever you increase the pan's temperature, the rate at which heat is transferred to the water increases. If you continue to raise the pan's temperature past the stage of columns and slugs, the boiling enters a new phase called the *transition regime*. Then each increase in the pan's temperature reduces the rate at which heat is transferred to the water. The decrease is not paradoxical. In the transition regime, much of the bottom of the pan is covered by a layer of vapor. Since water vapor conducts heat about an order of magnitude more poorly than does liquid water, the transfer of heat to the water is diminished. The hotter the pan becomes, the less direct contact the water has with it and the worse the transfer of heat becomes. This situation can be dangerous in a *heat exchanger,* whose purpose is to transfer heat from a heated object. If the water in the heat exchanger is allowed to enter the transition regime, the object may destructively overheat because of diminished transfer of heat from it.

Suppose you continue to increase the temperature of the pan. Eventually, the whole of the bottom surface is covered with vapor. Then heat is slowly transferred to the liquid above the vapor by radiation and gradual conduction. This phase is called *film boiling.*

Although you cannot obtain film boiling in a pan of water on a kitchen stove, it is still commonplace in the kitchen. My grandmother once demonstrated how it serves to indicate when her skillet is hot enough for pancake batter. After she heated the empty skillet for a while, she sprinkled a few drops of water into it. The drops sizzled away within seconds. Their rapid disappearance warned her that the skillet was insufficiently hot for the batter. After further heating the skillet, she repeated her test with a few more sprinkled water drops. This time they beaded up and danced over the metal, lasting well over a minute before they disappeared. The skillet was then hot enough for my grandmother's batter.

To study her demonstration, I arranged for a flat metal plate to be heated by a laboratory burner. While monitoring the temperature of the plate with a thermocouple, I carefully released a drop of distilled water from a syringe held just above the plate. The drop fell into a dent I had made in the plate with a ball-peen hammer. The syringe allowed me to release drops of uniform size. Once a drop was released, I timed how long it survived on the plate. Afterward, I plotted the survival times of the drops versus the plate temperature (Fig. 3). The graph has a curious peak. When the plate temperature was between 100 and about 200°C, each drop spread over the plate in a thin layer and rapidly vaporized. When the plate temperature was about 200°C, a drop deposited on the plate beaded up and survived for over a minute. At even higher plate temperatures, the water beads did not survive quite as long. Similar experiments with tap water generated a graph with a flatter peak, probably because suspended particles of impurities in the drops breached the vapor layer, conducting heat into the drops.

The fact that a water drop is long lived when deposited on metal that is much hotter than the boiling temperature of water was first reported by Hermann Boerhaave in 1732. It was not investigated extensively until 1756 when Johann Gottlob Leidenfrost published ''A Tract About Some Qualities of Common Water.'' Because Leidenfrost's work was not translated from the Latin until 1965, it was not widely read. Still, his name is now associated with the phenomenon. In addition, the temperature corresponding to the peak in a graph such as I made is called the Leidenfrost point.

Leidenfrost conducted his experiments with an iron spoon that was heated red-hot in a fireplace. After placing a drop of water into the spoon, he timed its duration by the swings of a pendulum. He noted that the drop seemed to suck the light and heat form the spoon, leaving a spot duller than the rest of the spoon. The first drop deposited in the spoon lasted 30 s while the next drop lasted only 10 s. Additional drops lasted only a few seconds.

Leidenfrost misunderstood his demonstrations because he did not realize that the longer-lasting drops were actually boiling. Let me explain in terms of my experi-

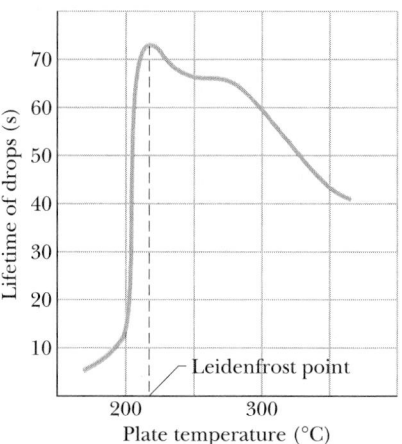

FIGURE 3 Drop lifetimes on a hot plate. Strangely, in a certain temperature range, the drops last longer when the hot plate is hotter.

ments. When the temperature of the plate is less than the Leidenfrost point, the water spreads over the plate and rapidly conducts heat from it, resulting in complete vaporization within seconds. When the temperature is at or above the Leidenfrost point, the bottom surface of a drop deposited on the plate almost immediately vaporizes. The gas pressure from this vapor layer prevents the rest of the drop from touching the plate (Fig. 4). The layer thus protects and supports the drop for the next minute or so. The layer is constantly replenished as additional water vaporizes from the bottom surface of the drop because of heat radiated and conducted through the layer from the plate. Although the layer is less than 0.1 mm thick near its outer boundary and only about 0.2 mm thick at its center, it dramatically slows the vaporization of the drop.

After reading the translation of Leidenfrost's research, I happened upon a description of a curious stunt that was performed in the sideshows of carnivals around the turn of the century. Reportedly, a performer was able to dip wet fingers into molten lead. Assuming that the stunt involved no trickery, I conjectured that it must depend on the Leidenfrost effect. As soon as the performer's wet flesh touched the hot liquid metal, part of the water vaporized, coating the fingers with a vapor layer. If the dip was brief, the flesh would not be heated significantly.

I could not resist the temptation to test my explanation. With a laboratory burner, I melted down a sizable slab of lead in a crucible. I heated the lead until its temperature was over 400°C, well above its melting temperature of 328°C. After wetting a finger in tap water, I prepared to touch the top surface of the molten lead. I must confess that I had an assistant standing ready with first-aid materials. I must also confess that my first several attempts failed because my brain refused to allow this ridiculous experiment, always directing my finger to miss the lead.

When I finally overcame my fears and briefly touched the lead, I was amazed. I felt no heat. Just as I had guessed, part of the water on the finger vaporized, forming a protective layer. Since the contact was brief, radiation and conduction of heat through the vapor layer were insufficient to raise perceptibly the temperature of my flesh. I grew braver. After wetting my hand, I dipped all my fingers into

FIGURE 5 Walker demonstrating the Leidenfrost effect with molten lead. He has just plunged his fingers into the lead, touching the bottom of the pan. The temperature of the lead is given in degrees Fahrenheit on the industrial thermometer.

the lead, touching the bottom of the container (Fig. 5). The contact with the lead was still too brief to result in a burn. Apparently, the Leidenfrost effect, or more exactly, the immediate presence of film boiling, protected my fingers.

I still questioned my explanation. Could I possibly touch the lead with a dry finger without suffering a burn? Leaving aside all rational thought, I tried it, immediately realizing my folly when pain raced through the finger. Later, I tested a dry weiner, forcing it into the molten lead for several seconds. The skin of the weiner quickly blackened. It lacked the protection of film boiling just as my dry finger had.

I must caution that dipping fingers into molten lead presents several serious dangers. If the lead is only slightly above its melting point, the loss of heat from it when the water is vaporized may solidify the lead around the fingers. If I were to pull the resulting glove of hot, solid lead up from the container, it will be in contact with my fingers so long that my fingers are certain to be badly burned. I must also contend with the possibility of splashing and spillage. In addition, there is the acute danger of having too much water on the fingers. When the surplus water rapidly va-

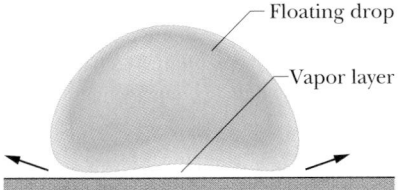

FIGURE 4 A Leidenfrost drop in cross section.

Floating drop

Vapor layer

porizes, it can blow molten lead over the surroundings and, most seriously, into the eyes. I have been scarred on my arms and face from such explosive vaporizations. *You should never repeat this demonstration.*

Film boiling can also be seen when liquid nitrogen is spilled. The drops and globs bead up as they skate over the floor. The liquid is at a temperature of about $-200°C$. When the spilled liquid nears the floor, its bottom surface vaporizes. The vapor layer then provides support for the rest of the liquid, allowing the liquid to survive for a surprisingly long time.

I was told of a stunt where a performer poured liquid nitrogen into his mouth without being hurt by its extreme cold. The liquid immediately underwent film boiling on its bottom surface and thus did not directly touch the tongue. Foolishly, I repeated this demonstration. For several dozen times the stunt went smoothly and dramatically. With a large glob of liquid nitrogen in my mouth, I concentrated on not swallowing while I breathed outward. The moisture in my cold breath condensed, creating a terrific plume that extended about a meter from my mouth. However, on my last attempt the liquid thermally contracted two of my front teeth so severely that the enamel ruptured into a "road map" of fissures. My dentist convinced me to drop the demonstration.

The Leidenfrost effect may also play a role in another foolhardy demonstration: walking over hot coals. At times the news media have carried reports of a performer striding over red-hot coals with much hoopla and mystic nonsense, perhaps claiming that protection from a bad burn is afforded by "mind over matter." Actually, physics protects the feet when the walk is successful. Particularly important is the fact that although the surface of the coals is quite hot, it contains surprisingly little energy. If the performer walks at a moderate pace, a footfall is so brief that the foot conducts little energy from the coals. Of course, a slower walk invites a burn because the longer contact allows heat to be conducted to the foot from the interior of the coals.

If the feet are wet prior to the walk, the liquid might also help protect them. To wet the feet a performer might walk over wet grass just before reaching the hot coals. Instead, the feet might just be sweaty because of the heat from the coals or the excitement of the performance. Once the performer is on the coals, some of the heat vaporizes the liquid on the feet, leaving less heat to be conducted to the flesh. In addition, there may be points of contact where the liquid undergoes film boiling, thereby providing brief protection from the coals.

I have walked over hot coals on five occasions. For four of the walks I was fearful enough that my feet were sweaty. However, on the fifth walk I took my safety so much for granted that my feet were dry. The burns I suf-

fered then were extensive and terribly painful. My feet did not heal for weeks.

My failure may have been due to a lack of film boiling on the feet, but I had also neglected an additional safety factor. On the other days I had taken the precaution of clutching an earlier edition of this book to my chest during the walks so as to bolster my belief in physics. Alas, I forgot the book on the day when I was so badly burned.

I have long argued that degree-granting programs should employ "fire-walking" as a last exam. The chairperson of the program should wait on the far side of a bed of red-hot coals while a degree candidate is forced to walk over the coals. If the candidate's belief in physics is strong enough that the feet are left undamaged, the chairperson hands the candidate a graduation certificate. The test would be more revealing than traditional final exams.

REFERENCES

Leidenfrost, Johann Gottlob, "On the Fixation of Water in Diverse Fire," *International Journal of Heat and Mass Transfer,* Vol. 9, pages 1153–1166 (1966).

Gottfried, B. S., C. J. Lee, and K. J. Bell, "The Leidenfrost Phenomenon: Film Boiling of Liquid Droplets on a Flat Plate," *International Journal of Heat and Mass Transfer,* Vol. 9, pages 1167–1187 (1966).

Hall, R. S., S. J. Board, A. J. Clare, R. B. Duffey, T. S. Playle, and D. H. Poole, "Inverse Leidenfrost Phenomenon," *Nature,* Vol. 224, pages 266–267 (1969).

Walker, Jearl, "The Amateur Scientist," *Scientific American,* Vol. 237, pages 126–131 + 140 (August 1977).

Curzon, F. L. "The Leidenfrost Phenomenon," *American Journal of Physics,* Vol. 46, pages 825–828 (1978).

Leikind, Bernard J., and William J. McCarthy, "An Investigation of Firewalking," *Skeptical Inquirer,* Vol. 10, No. 1, pages 23–34 (Fall 1985).

Bent, Henry A., "Droplet on a Hot Metal Spoon," *American Journal of Physics,* Vol. 54, page 967 (1986).

Leikind, B. J., and W. J. McCarthy, "Firewalking," *Experientia,* Vol. 44, pages 310–315 (1988).

Thimbleby, Harold, "The Leidenfrost Phenomenon," *Physics Education,* Vol. 24, pages 300–303 (1989).

Taylor, John R., "Firewalking: A Lesson in Physics," *The Physics Teacher,* Vol. 27, pages 166–168 (March 1989).

Zhang, S., and G. Gogos, "Film Evaporation of a Spherical Droplet over a Hot Surface: Fluid Mechanics and Heat/Mass Transfer Analysis," *Journal of Fluid Mechanics,* Vol. 222, pages 543–563 (1991).

Agrawal, D. C., and V. J. Menon, "Boiling and the Leidenfrost Effect in a Gravity-free Zone: A Speculation," *Physics Education,* Vol. 29, pages 39–42 (1994).

22
Electric Charge

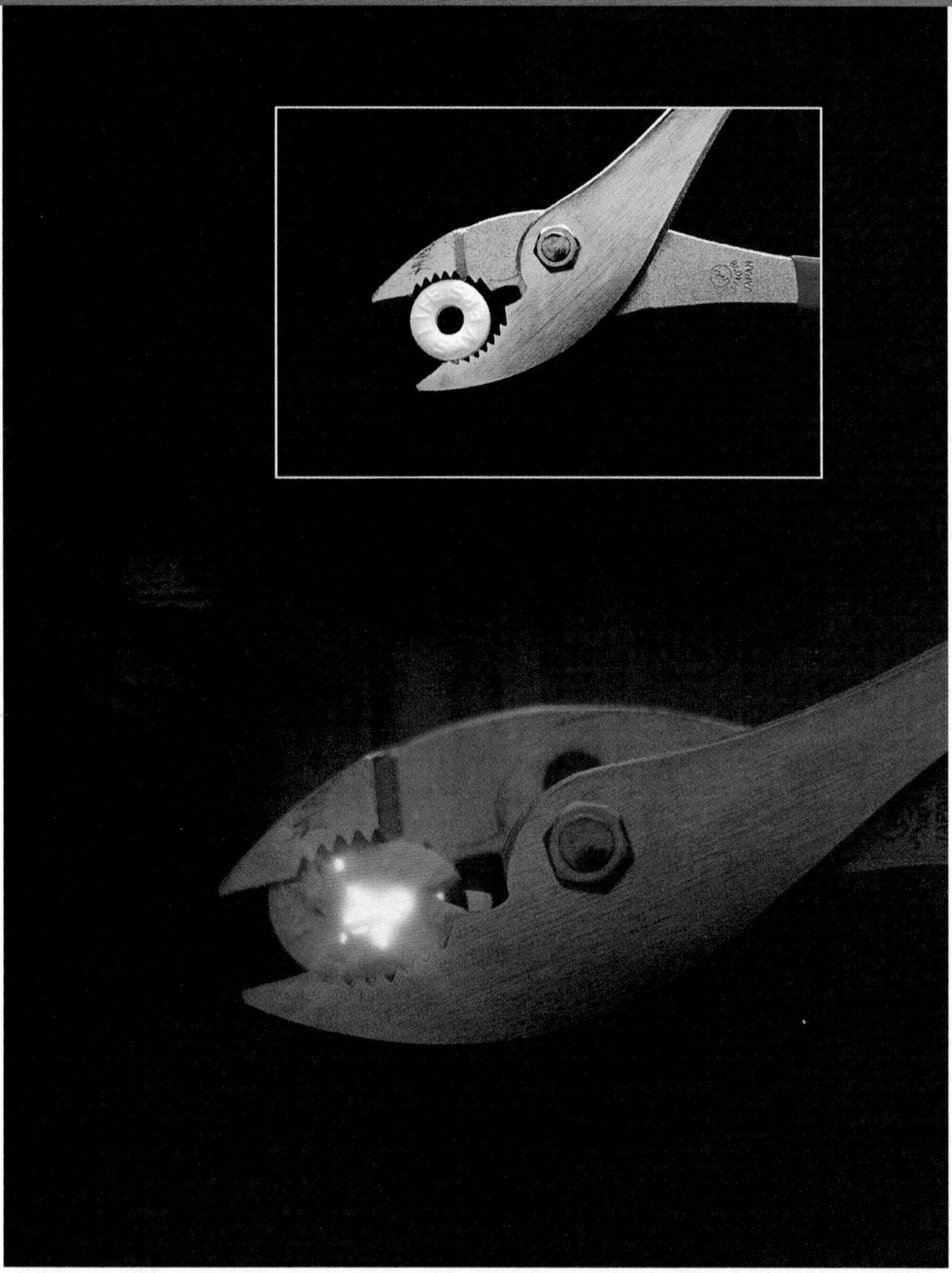

If you adapt your eyes to darkness for about 15 minutes and then have a friend chew a wintergreen LifeSaver, you will see a faint flash of blue light from your friend's mouth with each chomp. (To avoid wear on the teeth, you might crush the candy with pliers, as in the photograph.) What causes this display of light, commonly called "sparking"?

22-1 ELECTROMAGNETISM

The early Greek philosophers knew that if you rubbed a piece of amber, it would attract bits of straw. This ancient observation can be traced down directly to the electronic age in which we live. (The strength of the connection is indicated by our word *electron,* which is derived from the Greek word for amber.) The Greeks also recorded the observation that some naturally occurring "stones," known today as the mineral magnetite, would attract iron.

From these modest origins, the sciences of electricity and magnetism developed separately for centuries—until 1820, in fact, when Hans Christian Oersted found a connection between them: an electric current in a wire can deflect a magnetic compass needle. Interestingly enough, Oersted made this discovery while preparing a lecture demonstration for his physics students.

The new science of *electromagnetism* (the combination of electrical and magnetic phenomena) was developed further by workers in many countries. One of the best was Michael Faraday, a truly gifted experimenter with a talent for physical intuition and visualization. That talent is attested to by the fact that his collected laboratory notebooks do not contain a single equation. In the mid-19th century, James Clerk Maxwell put Faraday's ideas into mathematical form, introduced many new ideas of his own, and put electromagnetism on a sound theoretical basis.

Table 32-1 shows the basic laws of electromagnetism, now called Maxwell's equations. We plan to work our way through them in the chapters between here and there, but you might want to glance at them now, to see our goal.

22-2 ELECTRIC CHARGE

If you walk across a carpet in dry weather, you can produce a spark by bringing your finger close to a metal doorknob. Television advertising has alerted us to the problem of "static cling" in clothing (Fig. 22-1). On a grander scale, lightning is familiar to everyone. Each of these phenomena represents a tiny glimpse of the vast amount of *electric charge* that is stored in the familiar objects that surround us and—indeed—in our own bodies. **Electric charge** is an intrinsic characteristic of the fundamental particles making up those objects; that is, it is a characteristic that automatically accompanies those particles wherever they exist.

The vast amount of charge in an everyday object is usually hidden because the object contains equal amounts of two kinds of charge: *positive charge* and *negative charge.* With such an equality—or *balance*—of charge, the object is said to be *electrically neutral*; that is, it contains no *net* charge to interact with other objects. If the two types of charge are not in balance, then there *is* a net charge

FIGURE 22-1   Static cling, an electrical phenomenon that accompanies dry weather, causes these pieces of paper to stick to one another and to the plastic comb, and your clothing to stick to your body.

that *can* interact with other objects, and we become aware of the existence of the net charge. We say that an object is *charged* to indicate that it has a charge imbalance, or net charge. The imbalance is always very small compared to the total amounts of positive charge and negative charge contained in the object.

Charged objects interact by exerting forces on one another. To show this, we first charge a glass rod by rubbing one end with silk. At points of contact between the rod and the silk, tiny amounts of charge are transferred from one to the other, slightly upsetting the electrical neutrality of each. (We *rub* the silk over the rod to increase the number of contact points and thus the amount, still tiny, of transferred charge.)

Suppose we now suspend the charged rod from a thread to *electrically isolate* it from its surroundings so that its charge cannot change. If we bring a second, similarly charged, glass rod nearby (Fig. 22-2a), the two rods *repel*

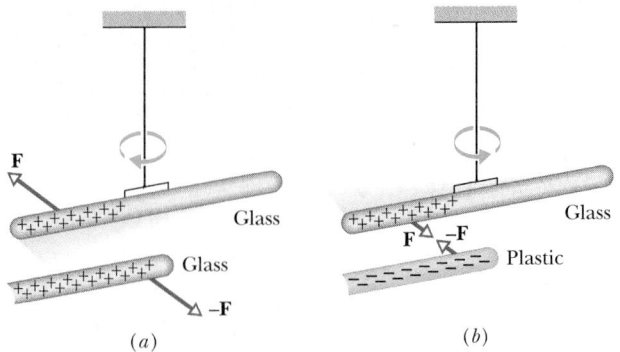

FIGURE 22-2   (*a*) Two charged rods of the same signs repel each other. (*b*) Two charged rods of opposite signs attract each other.

each other. That is, each rod experiences a force directed away from the other rod. However, if we rub a plastic rod with fur and bring it near the suspended glass rod (Fig. 22-2b), the two rods *attract* each other. That is, each rod experiences a force directed toward the other rod.

We can understand these two demonstrations in terms of positive and negative charges. When a glass rod is rubbed with silk, the glass loses some of its negative charge and then has a small unbalanced positive charge (represented by the plus signs in Fig. 22-2a). When the plastic rod is rubbed with fur, the plastic gains a small unbalanced negative charge (represented by the minus signs in Fig. 22-2b). Our two demonstrations reveal the following:

Charges with the same electrical sign repel each other, and charges with opposite electrical signs attract each other.

In Section 22-4, we shall put this rule into quantitative form as Coulomb's law of *electrostatic force* (or *electric force*) between charges. The term *electrostatic* is used to emphasize that, relative to each other, the charges are either stationary or moving only very slowly.

The "positive" and "negative" labels and signs for electric charge were chosen arbitrarily by Benjamin Franklin. He could easily have interchanged the labels or used some other pair of opposites to distinguish the two kinds of charge. (Franklin was a scientist of international reputation. It has even been said that Franklin's triumphs in diplomacy in France during the American War of Indepen-

dence were facilitated, and perhaps even made possible, because he was so highly regarded as a scientist.)

The attraction and repulsion between charged bodies have many industrial applications, including electrostatic paint spraying and powder coating, fly-ash collection in chimneys, nonimpact ink-jet printing, and photocopying. Figure 22-3 shows a tiny carrier bead in a Xerox copying machine, covered with particles of black powder called *toner,* that stick to it by means of electrostatic forces. The negatively charged toner particles are eventually attracted from the carrier bead to a rotating drum, where a positively charged image of the document being copied has formed. A charged sheet of paper then attracts the toner particles from the drum to itself, after which they are heat-fused in place to produce the copy.

22-3 CONDUCTORS AND INSULATORS

In some materials, such as metals, tap water, and the human body, some of the negative charge can move rather freely. We call such materials **conductors.** In other materials, such as glass, chemically pure water, and plastic, none of the charge can move freely. We call these materials **nonconductors** or **insulators.**

If you rub a copper rod with wool while holding the rod in your hand, you will not be able to charge the rod, because both you and the rod are conductors. The rubbing will cause a charge imbalance on the rod, but the excess charge will immediately move from the rod through you to the floor (which is connected to Earth's surface), and the rod will quickly be neutralized.

In thus setting up a pathway of conductors between an object and Earth's surface, we are said to *ground* the object. And in neutralizing the object (by eliminating an unbalanced positive or negative charge), we are said to *discharge* the object. (See Fig. 22-4 for a somewhat bizarre example of discharge.) If instead of holding the rod in your hand, you hold it via an insulating handle, you eliminate the conducting path to Earth, and the rod can then be charged by rubbing, as long as you do not touch it directly with your hand.

The properties of conductors and insulators are due to the structure and electrical nature of atoms. Atoms consist of positively charged *protons*, negatively charged *electrons*, and electrically neutral *neutrons*. The protons and neutrons are packed tightly together in a central *nucleus*; in a simple model of an atom, the electrons orbit the nucleus.

The charge of a single electron and that of a single proton have the same magnitude but are opposite in sign. Hence an electrically neutral atom contains equal numbers of electrons and protons. Electrons are held near the nu-

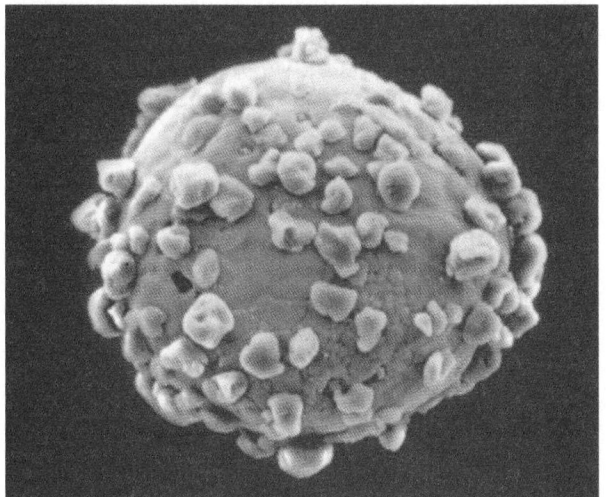

FIGURE 22-3 A carrier bead from a Xerox copying machine; it is covered with toner particles that cling to it by electrostatic attraction. The diameter of the bead is about 0.3 mm.

FIGURE 22-4 Not a parlor stunt but a serious experiment carried out in 1774 to prove that the human body is a conductor of electricity. The etching shows a person suspended by nonconducting ropes while being charged by a charged rod (which probably touched flesh instead of the trousers). When the person brought his face, left hand, or the conducting ball and rod in his right hand near one of the metallic plates, electric sparks flew through the intermediate air, discharging him.

cleus because they have the electrical sign opposite that of the protons in the nucleus and thus are attracted to the nucleus.

When atoms of a conductor like copper come together to form the solid, some of their outermost (and so most loosely held) electrons do not remain attached to the individual atoms but become free to wander about within the solid, leaving behind positively charged atoms *(positive ions)*. We call the mobile electrons *conduction electrons*. There are few (if any) free electrons in a nonconductor.

The experiment of Fig. 22-5 demonstrates the mobility of charge in a conductor. A negatively charged plastic rod will attract either end of an isolated neutral copper rod.

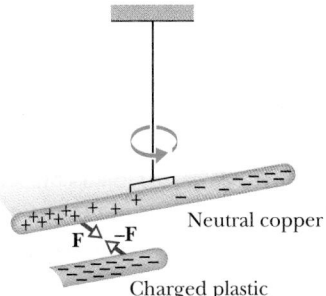

FIGURE 22-5 A neutral copper rod is electrically isolated from its surroundings by being suspended on a nonconducting thread. Either end of the copper rod will be attracted by a charged rod. Here, conduction electrons in the copper rod are repelled to the far end of that rod by the negative charge on the plastic rod. Then that negative charge attracts the remaining positive charge on the near end of the copper rod, rotating the copper rod to bring that near end closer to the plastic rod.

What happens is that many of the conduction electrons in the closer end of the copper rod are repelled by the negative charge on the plastic rod. They move to the far end of the copper rod, leaving the near end depleted in electrons and thus with an unbalanced positive charge. This positive charge is attracted to the negative charge in the plastic rod. Although the copper rod is still neutral, it is said to have an *induced charge*, which means that some of its positive and negative charges have been separated owing to the presence of a nearby charge.

Similarly, if a positively charged glass rod is brought near one end of a neutral copper rod, conduction electrons in the copper rod are attracted to that end. That end becomes negatively charged and the other end positively charged, so again an induced charge is set up in the copper rod. Although the copper rod is still neutral, it and the glass rod attract each other.

Note that it is only conduction electrons, with their negative charges, that can move; positive ions are fixed in place. Thus, an object becomes positively charged only through the *removal of negative charges*.

Semiconductors, such as silicon and germanium, are materials that are intermediate between conductors and insulators. The microelectronic revolution that has transformed our lives in so many ways is due to devices constructed of semiconducting materials.

Finally, there are **superconductors,** so called because they present no resistance to the movement of electric charge through them. When charge moves through a material, we say that an **electric current** exists in the material. Ordinary materials, even good conductors, tend to resist the flow of charge through them. In a superconductor, however, the resistance is not just small; it is precisely zero. If you set up a current in a superconducting ring, the flow of electrons persists without change for as long as you care to watch it, with no battery or other source of energy needed to maintain the current.

CHECKPOINT **1:** The figure shows five pairs of plates: *A, B,* and *D* are charged plastic plates and *C* is an electrically neutral copper plate. The electrostatic forces between the pairs of plates are shown for three of the pairs. For the remaining two pairs, do the plates repel or attract each other?

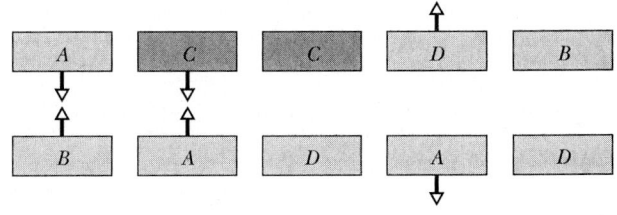

22-4 COULOMB'S LAW

Let two charged particles (also called *point charges*) have charge magnitudes q_1 and q_2 and be separated by a distance r. The **electrostatic force** of attraction or repulsion between them has the magnitude

$$F = k\frac{|q_1||q_2|}{r^2} \qquad \text{(Coulomb's law),} \quad (22\text{-}1)$$

in which k is a constant. Each particle exerts a force of this magnitude on the other particle; the two forces form an action-reaction pair. If the particles *repel* each other, the force on each particle points *away from* the other particle (as in Figs. 22-6a and b). If the particles *attract* each other, the force on each particle points *toward* the other particle (as in Fig. 22-6c).

Equation 22-1 is called **Coulomb's law** after Charles Augustin Coulomb, whose experiments in 1785 led him to it. Curiously, the form of Eq. 22-1 is the same as that of Newton's equation for the gravitational force between two particles with masses m_1 and m_2 that are separated by a distance r:

$$F = G\frac{m_1 m_2}{r^2}, \qquad (22\text{-}2)$$

in which G is the gravitational constant.

The constant k in Eq. 22-1, by analogy with the gravitational constant G in Eq. 22-2, may be called the *electrostatic constant*. Both equations describe inverse square laws that involve a property of the interacting particles—the mass in one case and the charge in the other. The laws differ in that gravitational forces are always attractive but electrostatic forces may be either attractive or repulsive,

depending on the signs of the two charges. This difference arises from the fact that, although there is only one kind of mass, there are two kinds of charge (and that is why absolute signs are needed in Eq. 22-1 but not in Eq. 22-2).

Coulomb's law has survived every experimental test; no exceptions to it have ever been found. It holds even within the atom, correctly describing the force between the positively charged nucleus and each of the negatively charged electrons, even though classical Newtonian mechanics fails in that realm and is replaced there by quantum physics. This simple law also correctly accounts for the forces that bind atoms together to form molecules, and for the forces that bind atoms and molecules together to form solids and liquids.

For practical reasons having to do with the accuracy of measurements, the SI unit of charge is derived from the SI unit of electric current, the ampere (A). The SI unit of charge is the **coulomb** (C): *One coulomb is the amount of charge that is transferred through the cross section of a wire in 1 second when there is a current of 1 ampere in the wire.* In Section 30-2 we shall describe how the ampere is defined experimentally. In general, we can write

$$dq = i\,dt, \qquad (22\text{-}3)$$

in which dq (in coulombs) is the charge transferred by a current i (in amperes) during the time interval dt (in seconds).

For historical reasons (and because doing so simplifies many other formulas), the electrostatic constant k of Eq. 22-1 is usually written $1/4\pi\epsilon_0$. Then Coulomb's law becomes

$$F = \frac{1}{4\pi\epsilon_0}\frac{|q_1||q_2|}{r^2} \qquad \text{(Coulomb's law).} \quad (22\text{-}4)$$

The constants in Eqs. 22-1 and 22-4 have the value

$$k = \frac{1}{4\pi\epsilon_0} = 8.99 \times 10^9 \text{ N}\cdot\text{m}^2/\text{C}^2. \qquad (22\text{-}5)$$

The quantity ϵ_0, called the **permittivity constant,** sometimes appears separately in equations and is

$$\epsilon_0 = 8.85 \times 10^{-12} \text{ C}^2/\text{N}\cdot\text{m}^2. \qquad (22\text{-}6)$$

Still another parallel between the gravitational force and the electrostatic force is that both obey the principle of superposition. If we have n charged particles, they interact independently in pairs, and the force on any one of them, let us say particle 1, is given by the vector sum

$$\mathbf{F}_1 = \mathbf{F}_{12} + \mathbf{F}_{13} + \mathbf{F}_{14} + \mathbf{F}_{15} + \cdots + \mathbf{F}_{1n}, \quad (22\text{-}7)$$

in which, for example, $\mathbf{F}_{14}$ is the force acting on particle 1

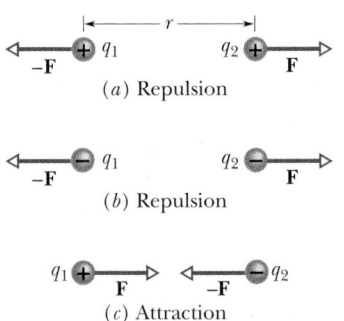

FIGURE 22-6 Two charged particles, separated by distance r, repel each other if their charges are (a) both positive and (b) both negative. (c) They attract each other if their charges are of opposite signs. In each of the three situations, the force acting on one particle is equal in magnitude to the force acting on the other particle but points in the opposite direction.

owing to the presence of particle 4. An identical formula holds for the gravitational force.

Finally, the two shell theorems that we found so useful in our study of gravitation have analogs in electrostatics:

A shell of uniform charge attracts or repels a charged particle that is outside the shell as if all the shell's charge were concentrated at its center.

A shell of uniform charge exerts no electrostatic force on a charged particle that is located inside the shell.

Spherical Conductors

If excess charge is placed on a spherical shell that is made of conducting material, the excess charge spreads uniformly over the (external) surface. For example, if we place excess electrons on a spherical metal shell, those electrons repel one another and tend to move apart, spreading over the available surface until they are uniformly distributed. That arrangement maximizes the distances between all pairs of the excess electrons. According to the first shell theorem, the shell then will attract or repel an external charge as if all the excess charge on the shell were concentrated at its center.

If we remove negative charge from a spherical metal shell, the resulting positive charge of the shell is also spread uniformly over the surface of the shell. For example, if we remove n electrons, there are then n sites of positive charge (sites missing an electron) that are spread uniformly over the shell. According to the first shell theorem, the shell will again attract or repel an external charge as if all the shell's excess charge were concentrated at its center.

PROBLEM SOLVING TACTICS

TACTIC 1: *Symbols Representing Charge*
Here is a general guide to the symbols representing charge. If the symbol q, with or without a subscript, is used in a sentence when no electrical sign has been specified, the charge can be either positive or negative. Sometimes the sign is explicitly shown, as in the notation $+q$ or $-q$.

When more than one charged object is being considered, their charges might be given as multiples of a charge magnitude. As examples, the notation $+2q$ means a positive charge with magnitude twice that of some reference charge magnitude q, and $-3q$ means a negative charge with magnitude three times that of the reference charge magnitude q.

CHECKPOINT **2:** The figure shows two protons (symbol p) and one electron (symbol e) on an axis. What are the directions of (a) the electrostatic force on the central proton due to the electron, (b) the electrostatic force on the central proton due to the other proton, and (c) the net electrostatic force on the central proton?

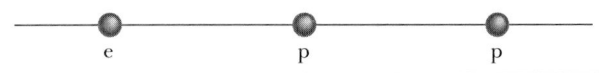

    e   p   p

SAMPLE PROBLEM 22-1

Figure 22-7a shows two particles fixed in place: a particle of charge $q_1 = +8q$ at the origin of an x axis and a particle of charge $q_2 = -2q$ at $x = L$. At what point (other than infinitely far away) can a proton be placed so that it is in *equilibrium* (meaning that the net force on it is zero)? Is that equilibrium *stable* or *unstable*?

SOLUTION: If $\mathbf{F}_1$ is the force on the proton due to charge q_1 and $\mathbf{F}_2$ is the force on the proton due to charge q_2, then the point we seek is where $\mathbf{F}_1 + \mathbf{F}_2 = 0$, which requires that

$$\mathbf{F}_1 = -\mathbf{F}_2. \tag{22-8}$$

This tell us that at the point we seek, the forces acting on the proton due to the other two particles must be of equal magnitudes,

$$F_1 = F_2, \tag{22-9}$$

and that the forces must have opposite directions.

A proton has a positive charge. Thus the proton and the particle of charge q_1 are of the same sign, and force $\mathbf{F}_1$ on the proton must point away from q_1. Also the proton and the particle of charge q_2 are of opposite signs, so force $\mathbf{F}_2$ on the proton must point toward q_2. "Away from q_1" and "toward q_2" can be in opposite directions only if the proton is located on the x axis.

If the proton is on the x axis at any point between q_1 and q_2, such as P in Fig. 22-7b, then $\mathbf{F}_1$ and $\mathbf{F}_2$ are in the same direction and not in opposite directions as required. If the proton is at any point on the x axis to the left of q_1, such as point S in Fig. 22-7b, then $\mathbf{F}_1$ and $\mathbf{F}_2$ are in opposite directions. However, Eq. 22-4 tells us that $\mathbf{F}_1$ and $\mathbf{F}_2$ cannot have equal magnitudes there: F_1 must be larger than F_2, because F_1 is produced by a closer charge (with smaller r) of larger magnitude ($8q$ versus $2q$).

Finally, if the proton is at any point on the x axis to the right of q_2, such as point R, then $\mathbf{F}_1$ and $\mathbf{F}_2$ are again in opposite directions. However, because now the charge of larger magnitude (q_1) is *farther* away from the proton than the charge of smaller magnitude, there is a point at which F_1 is equal to

F_2. Let x be the coordinate of this point, and let q_p be the charge of the proton. Then with the aid of Eq. 22-4, we can rewrite Eq. 22-9 as

$$\frac{1}{4\pi\epsilon_0}\frac{8qq_p}{x^2} = \frac{1}{4\pi\epsilon_0}\frac{2qq_p}{(x-L)^2}. \qquad (22\text{-}10)$$

(Note that only the magnitudes of the charges appear in Eq. 22-10.) Rearranging Eq. 22-10 gives us

$$\left(\frac{x-L}{x}\right)^2 = \frac{1}{4}.$$

After taking the square roots of both sides, we have

$$\frac{x-L}{x} = \frac{1}{2},$$

which gives us

$$x = 2L. \qquad \text{(Answer)}$$

The equilibrium at $x = 2L$ is unstable. That is, if the proton is displaced leftward from point R, then F_1 and F_2 both increase but F_2 increases more (because q_2 is closer than q_1), and a net force will drive the proton farther leftward. And if the proton is displaced rightward, both F_1 and F_2 decrease but F_2 decreases more, and thus a net force will then drive the proton farther rightward. In a stable equilibrium, each time the proton was displaced slightly, it would return to the equilibrium position.

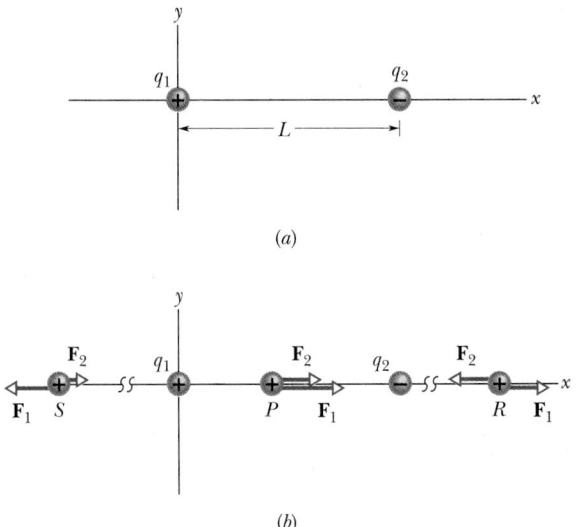

(a)

(b)

FIGURE 22-7 Sample Problem 22-1. (*a*) Two particles of charges q_1 and q_2 are fixed in place on an *x* axis, with separation *L*. (*b*) Three possible locations *S*, *P*, and *R* for a proton. At each location, the proton experiences electrostatic force $\mathbf{F}_1$ due to q_1 and electrostatic force $\mathbf{F}_2$ due to q_2.

SAMPLE PROBLEM 22-2

Figure 22-8*a* shows an arrangement of six fixed charged particles, where $a = 2.0$ cm and $\theta = 30°$. All six particles have the same magnitude of charge, $q = 3.0 \times 10^{-6}$ C; their electrical signs are as indicated. What is the net electrostatic force $\mathbf{F}_1$ acting on q_1 due to the other charges?

SOLUTION: From Eq. 22-7 we know that $\mathbf{F}_1$ is the vector sum of forces $\mathbf{F}_{12}$, $\mathbf{F}_{13}$, $\mathbf{F}_{14}$, $\mathbf{F}_{15}$, and $\mathbf{F}_{16}$, which are the electrostatic forces acting on q_1 due to the other charges. Because q_2 and q_4 are equal in magnitude and are both a distance $r = 2a$ from q_1, we have from Eq. 22-4

$$F_{12} = F_{14} = \frac{1}{4\pi\epsilon_0}\frac{|q_1||q_2|}{(2a)^2}. \qquad (22\text{-}11)$$

Similarly, since q_3, q_5, and q_6 are equal in magnitude and are each a distance $r = a$ from q_1, we have

$$F_{13} = F_{15} = F_{16} = \frac{1}{4\pi\epsilon_0}\frac{|q_1||q_3|}{a^2}. \qquad (22\text{-}12)$$

Figure 22-8*b* is a free-body diagram for q_1. It and Eq. 22-11 show that $\mathbf{F}_{12}$ and $\mathbf{F}_{14}$ are equal in magnitude but opposite in direction; thus those forces cancel. Inspection of Fig. 22-8*b* and Eq. 22-12 reveals that the *y* components of $\mathbf{F}_{13}$ and $\mathbf{F}_{15}$ also cancel, and that their *x* components are identical in magnitude and both point in the direction of decreasing *x*. Figure

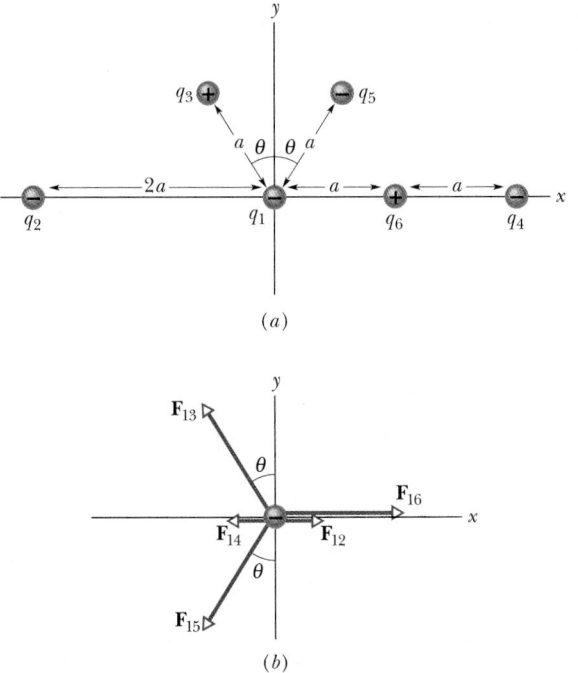

(a)

(b)

FIGURE 22-8 Sample Problem 22-2. (*a*) An arrangement of six charged particles. (*b*) The electrostatic forces acting on q_1 due to the other five charges.

22-8*b* also shows us that $\mathbf{F}_{16}$ points in the direction of increasing x. Thus $\mathbf{F}_1$ must be parallel to the x axis; its magnitude is the difference between F_{16} and twice the x component of $\mathbf{F}_{13}$:

$$F_1 = F_{16} - 2F_{13} \sin \theta$$

$$= \frac{1}{4\pi\epsilon_0}\frac{|q_1||q_6|}{a^2} - \frac{2}{4\pi\epsilon_0}\frac{|q_1||q_3|}{a^2}\sin\theta.$$

Setting $q_3 = q_6$ and $\theta = 30°$, we find

$$F_1 = \frac{1}{4\pi\epsilon_0}\frac{|q_1||q_6|}{a^2} - \frac{2}{4\pi\epsilon_0}\frac{|q_1||q_6|}{a^2}\sin 30° = 0. \quad \text{(Answer)}$$

Note that the presence of q_6 along the line between q_1 and q_4 does not alter the electrostatic force exerted by q_4 on q_1.

PROBLEM SOLVING TACTICS

TACTIC 2: *Symmetry*

In Sample Problem 22-2 we used the symmetry of the situation to reduce the time and amount of calculation involved in the solution. By realizing that q_2 and q_4 are positioned symmetrically about q_1, and thus that $\mathbf{F}_{12}$ and $\mathbf{F}_{14}$ cancel, we avoided calculating either force. And by realizing that the y components of $\mathbf{F}_{13}$ and $\mathbf{F}_{15}$ cancel and that their x components are identical and add, we saved even more effort. In fact, by using symmetry and by setting up the solution in symbols, we never had to substitute the charge magnitude 3.0×10^{-6} C given in the problem.

TACTIC 3: *Drawing Electrostatic Force Vectors*

When you are given a diagram of charged particles, such as Fig. 22-8*a*, and are asked to find the net electrostatic force on one of them, you should usually draw a free-body diagram showing only the particle of concern and the forces *it* experiences, as in Fig. 22-8*b*. If, instead, you choose to superimpose those forces on the given diagram showing all the particles, be sure to draw the force vectors with either their tails (preferably) or their heads on the particle of concern. If you draw the vectors elsewhere in the diagram, you invite confusion. And confusion is guaranteed if you draw the vectors on the particles *causing* the forces on the particle of concern.

CHECKPOINT **3:** The figure shows three arrangements of an electron e and two protons p. (a) Rank the arrangements according to the magnitude of the net electrostatic force on the electron due to the protons, largest first. (b) In situation *c*, is the angle between the

net force on the electron and the line labeled d less than or more than 45°?

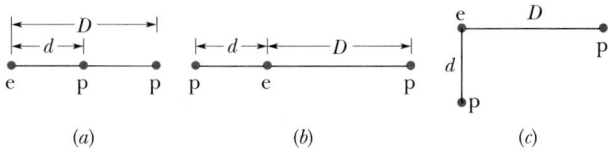

(a)   *(b)*   *(c)*

SAMPLE PROBLEM 22-3

In Fig. 22-9*a*, two identical, electrically isolated conducting spheres A and B are separated by a (center-to-center) distance a that is large compared to the spheres. Sphere A has a positive charge of $+Q$; sphere B is electrically neutral; and initially, there is no electrostatic force between the spheres.

(a) Suppose the spheres are connected for a moment by a conducting wire. The wire is thin enough so that any net charge on it is negligible. What is the electrostatic force between the spheres after the wire is removed?

SOLUTION: When the spheres are wired together, conduction electrons of sphere B are attracted to positively charged sphere A (Fig. 22-9*b*). As sphere B loses negative charge, it becomes positively charged. And as A gains negative charge, it becomes *less* positively charged. The spheres must end up with the same charge because they are identical. Thus the transfer of charge stops when the excess charge on B has increased to $+Q/2$ and the excess charge on A has decreased to $+Q/2$ (Fig. 22-9*c*). This condition occurs when a charge of $-Q/2$ has been transferred.

After the wire has been removed, we can assume that the charge on either sphere does not disturb the uniformity of the charge distribution on the other sphere, because the spheres are small relative to their separation. Thus we can apply the first shell theorem to each sphere. By Eq. 22-4 with $q_1 = q_2 = Q/2$ and $r = a$, the electrostatic force between the spheres has a magnitude of

$$F = \frac{1}{4\pi\epsilon_0}\frac{(Q/2)(Q/2)}{a^2} = \frac{1}{16\pi\epsilon_0}\left(\frac{Q}{a}\right)^2. \quad \text{(Answer)}$$

Since both spheres are now positively charged, they repel each other.

(b) Next, suppose sphere A is grounded momentarily, and then the ground connection is removed. What now is the electrostatic force between the spheres?

SOLUTION: The ground connection allows electrons, with a total charge of $-Q/2$, to move from the ground to sphere A (Fig. 22-9*d*), neutralizing that sphere (Fig. 22-9*e*). With no charge on sphere A, there is no electrostatic force between the two spheres (just as initially, in Fig. 22-9*a*).

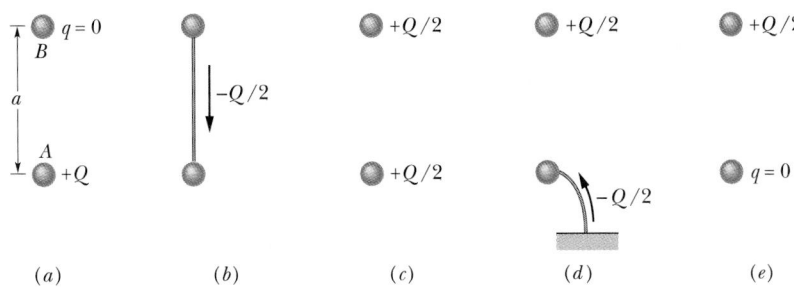

FIGURE 22-9 Sample Problem 22-3. Two small conducting spheres A and B. (*a*) To start, sphere A is charged positively. (*b*) Negative charge is transferred between the spheres through a connecting wire. (*c*) Both spheres are then charged positively. (*d*) Negative charge is transferred through a grounding wire to sphere A. (*e*) Sphere A is then neutral.

22-5 CHARGE IS QUANTIZED

In Benjamin Franklin's day, electric charge was thought to be a continuous fluid—an idea that was useful for many purposes. However, we now know that fluids themselves, such as air and water, are not continuous but are made up of atoms and molecules; matter is discrete. Experiment shows that "electrical fluid" is also not continuous but is made up of multiples of a certain elementary charge. That is, any positive or negative charge q that can be detected can be written as

$$q = ne, \qquad n = \pm 1, \pm 2, \pm 3, \cdots, \quad (22\text{-}13)$$

in which e, the **elementary charge,** has the value

$$e = 1.60 \times 10^{-19} \text{ C}. \qquad (22\text{-}14)$$

The elementary charge e is one of the important constants of nature. The electron and proton both have a charge of magnitude e (Table 22-1). (Quarks, the constituent particles of protons and neutrons, have charges of $\pm e/3$ or $\pm 2e/3$, but they apparently cannot be detected individually. Hence, we do not take their charges to be the elementary charge.)

You often see phrases—such as "the charge on a sphere," "the amount of charge transferred," and "the charge carried by the electron"—that suggest that charge is a substance. (Indeed, such statements have already appeared in this chapter.) You should, however, keep in mind what is intended: *particles* are the substance and charge happens to be one of their properties, just as mass is.

When a physical quantity such as charge can have only discrete values rather than any value, we say that the quantity is **quantized**. We have already seen that matter, energy, and angular momentum are quantized; charge adds one more important physical quantity to the list. It is possible, for example, to find a particle that has no charge at all or a charge of $+10e$ or $-6e$, but not a particle with a charge of, say, $3.57e$.

The quantum of charge is small. In an ordinary 100 W lightbulb, for example, about 10^{19} elementary charges enter the bulb every second and just as many leave. However, the graininess of electricity does not show up in such large-scale phenomena, just as you cannot feel the individual molecules of water with your hand.

The graininess of electricity is responsible for the blue glow that is emitted by a wintergreen LifeSaver while it is being crushed. When the sugar (sucrose) crystals in the candy rupture, one part of each ruptured crystal has excess electrons while the other part has excess positive ions. Almost immediately, electrons and ions jump across the gap of the rupture to neutralize the two sides. During the jumps, the electrons and positive ions collide with nitrogen molecules in the air that is then flowing into the gap. The collisions cause the nitrogen to emit ultraviolet light that you cannot see, as well as blue light (from the visible region of the spectrum) that is, however, too dim to see. Oil of wintergreen in the crystals absorbs the ultraviolet light and immediately emits enough blue light to light up a mouth or a pair of pliers. However, if the candy is wet with saliva, the demonstration fails, because the conducting saliva neutralizes the two parts of a fractured crystal before sparking can occur.

TABLE 22-1
THE CHARGES OF THREE PARTICLES

PARTICLE	SYMBOL	CHARGE
Electron	e or e$^-$	$-e$
Proton	p	$+e$
Neutron	n	0

CHECKPOINT **4:** Initially, sphere A has a charge of $-50e$ and sphere B has a charge of $+20e$. The spheres are made of conducting material and are identical in size. If the spheres then touch, what is the resulting charge on sphere A?

This is about 2×10^{12} tons! Even if the charges were separated by one Earth diameter, the attractive force would still be huge, about 120 tons. Actually, it is impossible to disturb the electrical neutrality of ordinary matter very much. If we try to remove any sizable fraction of the charge of one electrical sign from a body, a large electrostatic force appears automatically, tending to pull it back.

SAMPLE PROBLEM 22-4

An electrically neutral penny, of mass $m = 3.11$ g, contains equal amounts of positive and negative charge.

(a) Assuming that the penny is made entirely of copper, what is the magnitude q of the total positive (or negative) charge in the coin?

SOLUTION: A neutral atom has a negative charge of magnitude Ze associated with its electrons and a positive charge of the same magnitude associated with the protons in its nucleus, where Z is the *atomic number* of the element in question. For copper, Appendix F tells us that Z is 29, which means that an atom of copper has 29 protons and, when electrically neutral, 29 electrons.

The charge magnitude q we seek is equal to NZe, in which N is the number of atoms in the penny. To find N, we multiply the number of moles of copper in the penny by the number of atoms in a mole (Avogadro's number, $N_A = 6.02 \times 10^{23}$ atoms/mol). The number of moles of copper in the penny is m/M, where M is the molar mass of copper, 63.5 g/mol (from Appendix F). Thus we have

$$N = N_A \frac{m}{M} = 6.02 \times 10^{23} \text{ atoms/mol} \frac{3.11 \text{ g}}{63.5 \text{ g/mol}}$$

$$= 2.95 \times 10^{22} \text{ atoms.}$$

We then find the magnitude of the total positive or negative charge in the penny to be

$$q = NZe$$

$$= (2.95 \times 10^{22})(29)(1.60 \times 10^{-19} \text{ C})$$

$$= 137{,}000 \text{ C.} \qquad \text{(Answer)}$$

This is an enormous charge. (For comparison, if you rub a plastic rod with fur, you will be lucky to deposit any more than 10^{-9} C on the rod.)

(b) Suppose that the positive charge and the negative charge in a penny could be concentrated into two separate bundles, 100 m apart. What attractive force would act on each bundle?

SOLUTION: From Eq. 22-4 we have

$$F = \frac{1}{4\pi\epsilon_0} \frac{q^2}{r^2}$$

$$= \frac{(8.99 \times 10^9 \text{ N} \cdot \text{m}^2/\text{C}^2)(1.37 \times 10^5 \text{ C})^2}{(100 \text{ m})^2}$$

$$= 1.69 \times 10^{16} \text{ N.} \qquad \text{(Answer)}$$

SAMPLE PROBLEM 22-5

The nucleus in an iron atom has a radius of about 4.0×10^{-15} m and contains 26 protons.

(a) What is the magnitude of the repulsive electrostatic force between two of these protons that happen to be separated by 4.0×10^{-15} m?

SOLUTION: From Eq. 22-4 and Table 22-1 we can write

$$F = \frac{1}{4\pi\epsilon_0} \frac{e^2}{r^2}$$

$$= \frac{(8.99 \times 10^9 \text{ N} \cdot \text{m}^2/\text{C}^2)(1.60 \times 10^{-19} \text{ C})^2}{(4.0 \times 10^{-15} \text{ m})^2}$$

$$= 14 \text{ N.} \qquad \text{(Answer)}$$

This is a small force to be acting on a macroscopic object like a cantaloupe, but an enormous force to be acting on a proton. Such forces should blow apart the nucleus of any element but hydrogen (which has only one proton in its nucleus). But they don't, not even in nuclei with a great many protons. So there must be some attractive nuclear force to counter this enormous repulsive electrostatic force.

(b) What is the magnitude of the gravitational force between those same two protons?

SOLUTION: With m_p ($= 1.67 \times 10^{-27}$ kg) representing the mass of a proton, we write Eq. 22-2 for the gravitational force, finding

$$F = G \frac{m_p^2}{r^2}$$

$$= \frac{(6.67 \times 10^{-11} \text{ N} \cdot \text{m}^2/\text{kg}^2)(1.67 \times 10^{-27} \text{ kg})^2}{(4.0 \times 10^{-15} \text{ m})^2}$$

$$= 1.2 \times 10^{-35} \text{ N.} \qquad \text{(Answer)}$$

This result tells us that the (attractive) gravitational force is far too weak to counter the repulsive electrostatic forces between protons in a nucleus. Instead, the protons are bound together by an enormous force called (aptly) the *strong nuclear force*—a force that acts between protons (and neutrons) when they are close together, as in a nucleus.

Although the gravitational force is many, many times weaker than the electrostatic force, it is more important in large-scale situations because it is always attractive. This means that it can collect many small bodies into huge masses, such as planets and stars, that then exert large gravitational forces. The electrostatic force, on the other hand, is repulsive for charges of the same sign, so it is unable to collect either positive charge or negative charge into large concentrations that would then exert large electrostatic forces.

22-6 CHARGE IS CONSERVED

If you rub a glass rod with silk, a positive charge appears on the rod. Measurement shows that a negative charge of equal magnitude appears on the silk. This suggests that rubbing does not create charge but only transfers it from one body to another, upsetting the electrical neutrality of each body during the process. This hypothesis of **conservation of charge,** first put forward by Benjamin Franklin, has stood up under close examination, both for large-scale charged bodies and for atoms, nuclei, and elementary particles. No exceptions have ever been found. Thus we add electric charge to our list of quantities —including energy and both linear and angular momentum—that obey a conservation law.

Radioactive decay of nuclei, in which a nucleus spontaneously transforms into a different type of nucleus, gives us many instances of charge conservation at the nuclear level. For example, uranium-238, or ^{238}U, which is found in common uranium ore, can decay by emitting an alpha particle (which is a helium nucleus, ^{4}He) and transforming to thorium, ^{234}Th:

$$^{238}\text{U} \rightarrow {}^{234}\text{Th} + {}^4\text{He} \qquad \begin{array}{c}\text{(radioactive}\\\text{decay).}\end{array} \qquad (22\text{-}15)$$

The atomic number Z of the radioactive *parent* nucleus

^{238}U is 92, which tells us that this nucleus contains 92 protons and has a charge of $92e$. The emitted alpha particle has $Z = 2$, and the *daughter* nucleus ^{234}Th has $Z = 90$. Thus the amount of charge present before the decay, $92e$, is equal to the total amount present after the decay, $90e + 2e$. Charge is conserved.

Another example of charge conservation occurs when an electron e$^-$ (whose charge is $-e$) and its antiparticle, the *positron* e$^+$ (whose charge is $+e$), undergo an *annihilation process* in which they transform into two *gamma rays* (high-energy, chargeless particles of light):

$$\text{e}^- + \text{e}^+ \rightarrow \gamma + \gamma \qquad \text{(annihilation).} \qquad (22\text{-}16)$$

In applying the conservation-of-charge principle, we must add the charges algebraically, with due regard for their signs. In the annihilation process of Eq. 22-16 then, the net charge of the system is zero both before and after the event. Charge is conserved.

In *pair production,* the converse of annihilation, charge is also conserved. In this process a gamma ray transforms into an electron and a positron:

$$\gamma \rightarrow \text{e}^- + \text{e}^+ \qquad \text{(pair production).} \qquad (22\text{-}17)$$

Figure 22-10 shows such a pair-production event that occurred in a bubble chamber. A gamma ray entered the chamber directly from the left and at one point transformed into an electron and a positron. Because those new particles were charged and moving, each left a trail of tiny bubbles. (The trails were curved because a magnetic field had been set up in the chamber.) The gamma ray, being chargeless, left no trail. Still, you can tell exactly where it underwent pair production—at the tip of the curved V, where the trails of the electron and positron begin.

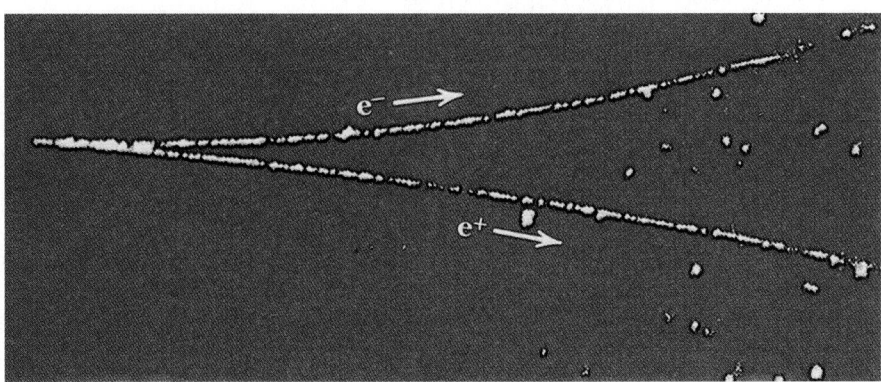

FIGURE 22-10 A photograph of trails of bubbles left in a bubble chamber by an electron and a positron. The pair of particles was produced by a gamma ray that entered the chamber directly from the left. Being chargeless, the gamma ray did not generate a telltale trail of bubbles along its path, as the electron and positron did.

REVIEW & SUMMARY

Electric Charge

The strength of a particle's electric interaction with objects around it depends on its **electric charge,** which can be either positive or negative. Charges with the same sign repel each other and charges with opposite signs attract each other. An object with equal amounts of the two kinds of charge is electrically neutral, whereas one with an imbalance is electrically charged.

Conductors are materials in which a significant number of charged particles (electrons in metals) are free to move. The charged particles in **nonconductors,** or **insulators,** are not free to move. When charge moves through a material, we say that an **electric current** exists in the material.

The Coulomb and Ampere

The SI unit of charge is the **coulomb** (C). It is defined in terms of the unit of current, the ampere (A), as the charge passing a particular point in 1 second when there is a current of 1 ampere at that point.

Coulomb's Law

Coulomb's law describes the **electrostatic force** between small (point) electric charges q_1 and q_2 at rest (or nearly at rest) and separated by a distance r:

$$F = \frac{1}{4\pi\epsilon_0} \frac{|q_1||q_2|}{r^2} \qquad \text{(Coulomb's law).} \qquad (22\text{-}4)$$

Here $\epsilon_0 = 8.85 \times 10^{-12}$ C^2/N·m^2 is the **permittivity constant;** $1/4\pi\epsilon_0 = 8.99 \times 10^9$ N·m^2/C^2.

The force of attraction or repulsion between point charges at rest acts along the line joining the two charges. If more than two charges are present, Eq. 22-4 holds for each pair of charges. The net force on each charge is then found, using the superposition principle, as the vector sum of the forces exerted on the charge by each of the others.

The two shell theorems for electrostatics are

A shell of uniform charge attracts or repels a charged particle that is outside the shell as if all the shell's charge were concentrated at its center.

A shell of uniform charge exerts no electrostatic force on a charged particle that is located inside the shell.

The Elementary Charge

Electric charge is **quantized:** any charge can be written as ne, where n is a positive or negative integer, and e is a constant of nature called the **elementary charge** (approximately 1.60×10^{-19} C). Electric charge is conserved: the (algebraic) net charge of any isolated system cannot change.

QUESTIONS

1. Does Coulomb's law hold for all charged objects?

2. A particle of charge q is to be placed, in turn, outside four metal objects, each of uniform charge Q: (1) a large solid sphere, (2) a large spherical shell, (3) a small solid sphere, and (4) a small spherical shell. The distance between the particle and the center of the object is the same in all four cases, and q is small enough not to alter significantly the uniform distribution of Q. Rank the objects according to the electrostatic force they exert on the particle, greatest first.

3. Figure 22-11 shows four situations in which charged particles are fixed in place on an axis. In which situations is there a point to the left of the particles where an electron will be in equilibrium?

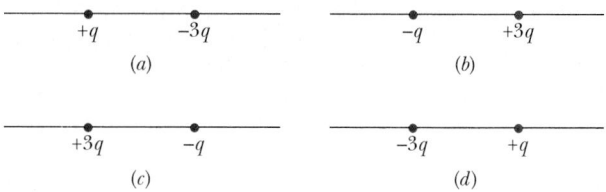

FIGURE 22-11 Question 3.

4. Figure 22-12 shows two charged particles on an axis. The charges are free to move. At one point, however, a third charged particle can be placed such that all three particles are in equilibrium. (a) Is that point to the left of the first two particles, to their right, or between them? (b) Should the third particle be positively or negatively charged? (c) Is the equilibrium stable or unstable?

FIGURE 22-12 Question 4.

5. In the figure for Checkpoint 2, two protons and one electron are fixed in place on an axis. Where on the axis could a fourth charged particle be placed so that the net electrostatic force on it due to the first three particles is zero: to the left of the first three particles; to their right; between the protons; or between the electron and the proton closer to it?

6. In Fig. 22-13, a central particle of charge $-q$ is surrounded by two circular rings of charged particles, of radii r and R, with $R > r$. What are the magnitude and direction of the net electrostatic force on the central particle due to the other particles?

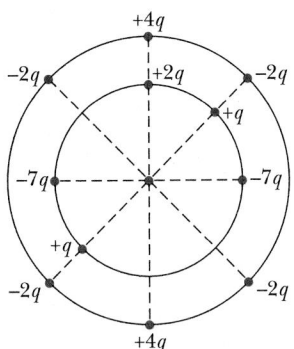

FIGURE 22-13 Question 6.

7. In Fig. 22-14, a central particle of charge $-2q$ is surrounded by a square array of charged particles, separated by either distance d or $d/2$ along the perimeter of the square. What are the magnitude and direction of the net electrostatic force on the central particle due to the other particles?

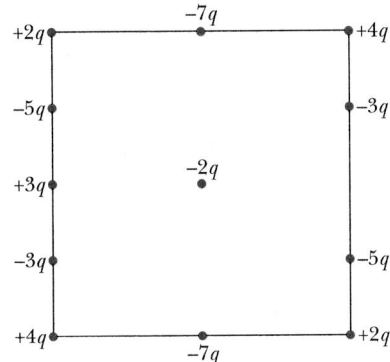

FIGURE 22-14 Question 7.

8. Figure 22-15 shows four arrangements of charged particles. Rank the arrangements according to the magnitude of the net electrostatic force on the particle with charge $+Q$, greatest first.

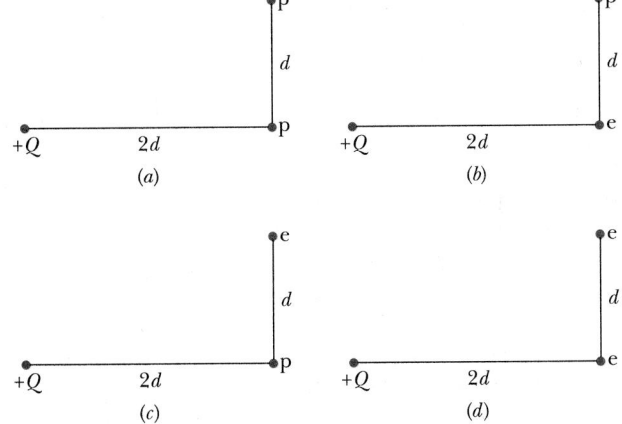

FIGURE 22-15 Question 8.

9. Figure 22-16 shows four situations in which particles of charge $+q$ or $-q$ are fixed in place. In each, the particles on the x axis are equidistant from the y axis. First, consider the middle particle in situation 1; the middle particle experiences an electrostatic force from each of the other two particles. (a) Are the magnitudes F of those forces the same or different? (b) Is the magnitude of the net force on the middle particle equal to, greater than, or less than $2F$? (c) Do the x components of the two forces add or cancel? (d) Do their y components add or cancel? (e) Is the direction of the net force on the middle particle that of the canceling components or the adding components? (f) What is the direction of that net force? Now consider the remaining situations: What is the direction of the net force on the middle particle in (g) situation 2, (h) situation 3, and (i) situation 4?

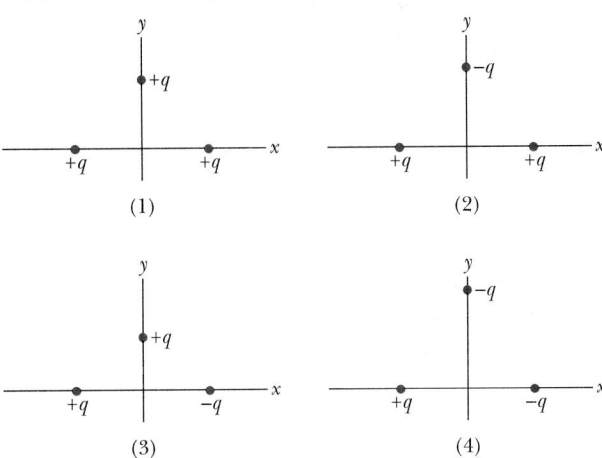

FIGURE 22-16 Question 9.

10. Figure 22-17 shows a pair of particles of charge Q and another pair of particles of charge q. The particle at the origin is free to move; the others are fixed in place. Should q be positive or negative if the net force on the free particle is to be zero and Q is (a) positive and (b) negative?

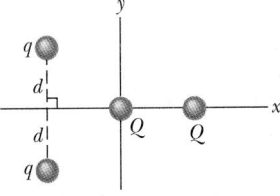

FIGURE 22-17 Question 10.

11. Four identical conducting spheres A, B, C, and D have charges of $-8.0Q$, $-6.0Q$, $-4.0Q$, and $8.0Q$, respectively. Which should be connected together (by thin wire) to produce two or more spheres with charges of (a) $-2.0Q$ and (b) $-2.5Q$? (c) What sequence of connections will produce two spheres with charges of $-3.0Q$?

12. A positively charged ball is brought close to a neutral isolated conductor. The conductor is then grounded while the ball is kept close. Is the conductor charged positively or negatively, or is it neutral, if (a) the ball is first taken away and then the ground connection is removed and (b) the ground connection is first removed and then the ball is taken away?

13. (a) A positively charged glass rod attracts an object suspended by a nonconducting thread. Is the object definitely negatively charged or only possibly negatively charged? (b) A positively charged glass rod repels a similarly suspended object. Is the object definitely positively charged or only possibly?

14. You are given two identical neutral metal spheres A and B mounted on portable insulating supports, as well as a thin conducting wire and a glass rod that you can rub with silk. You can attach the wire between the spheres or between a sphere and the ground. You cannot touch the rod to a sphere. How can you give the spheres charges of (a) equal magnitudes and the same signs and (b) equal magnitudes and opposite signs?

15. In a simple model of a helium atom, two electrons orbit a nucleus consisting of two protons. Is the magnitude of the force exerted on the nucleus by one of the electrons greater than, less than, or the same as the magnitude of the force exerted on that electron by the nucleus?

16. In Fig. 22-5, the nearby (negatively charged) plastic rod causes some of the conduction electrons in the copper rod to move to the far end of the copper rod. Why does the flow of the

conduction electrons quickly cease? After all, a huge number of them are free to move to that far end.

17. Figure 22-18 shows three small spheres that have charges of equal magnitudes and rest on a frictionless surface. Spheres y and z are fixed in place and are equally distant from sphere x. If sphere x is released from rest, which of the five paths shown will it take?

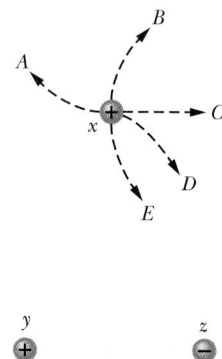

FIGURE 22-18 Question 17.

18. A person standing on an electrically insulated platform touches a charged, electrically isolated conductor. Does this discharge the conductor completely?

EXERCISES & PROBLEMS

SECTION 22-4 Coulomb's Law

1E. In the return stroke of a typical lightning bolt, a current of 2.5×10^4 A exists for 20 μs. How much charge is transferred in this event?

2E. What would be the electrostatic force between two 1.00 C charges separated by a distance of (a) 1.00 m and (b) 1.00 km if such a configuration could be set up?

3E. A point charge of $+3.00 \times 10^{-6}$ C is 12.0 cm distant from a second point charge of -1.50×10^{-6} C. Calculate the magnitude of the force on each charge.

4E. What must be the distance between point charge $q_1 = 26.0$ μC and point charge $q_2 = -47.0$ μC for the electrostatic force between them to have a magnitude of 5.70 N?

5E. Two equally charged particles, held 3.2×10^{-3} m apart, are released from rest. The initial acceleration of the first particle is observed to be 7.0 m/s^2 and that of the second to be 9.0 m/s^2. If the mass of the first particle is 6.3×10^{-7} kg, what are (a) the mass of the second particle and (b) the magnitude of the charge of each particle?

6E. In Figure 22-19, three identical conducting spheres A, B, and C form an equilateral triangle of side length d and have initial charges of $-2Q$, $-4Q$, and $8Q$, respectively. (a) What is the magnitude of the electrostatic force between spheres A and C?

The following steps are then taken: A and B are connected by a thin wire and then disconnected; B is grounded by the wire and the wire is then removed; B and C are connected by the wire and then disconnected. What now are the magnitudes of the electrostatic force (b) between spheres A and C and (c) between spheres B and C?

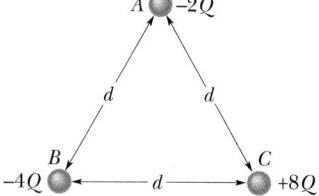

FIGURE 22-19 Exercise 6.

7E. Identical isolated conducting spheres 1 and 2 have equal amounts of charge and are separated by a distance large compared with their diameters (Fig. 22-20a). The electrostatic force acting on sphere 2 due to sphere 1 is **F**. Suppose now that a third identical sphere 3, having an insulating handle and initially neutral, is touched first to sphere 1 (Fig. 22-20b), then to sphere 2 (Fig. 22-20c), and finally removed (Fig. 22-20d). In terms of **F**, what is the electrostatic force **F'** that now acts on sphere 2?

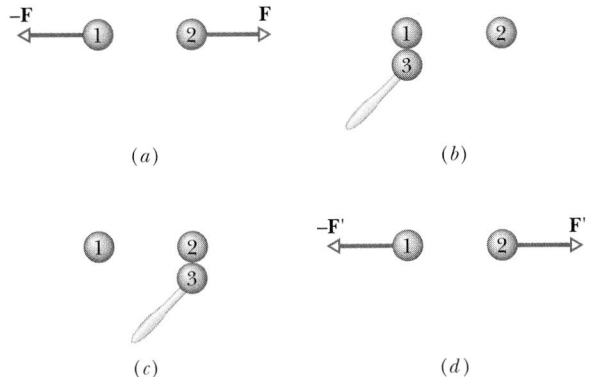

(a)   (b)

(c)   (d)

FIGURE 22-20 Exercise 7.

8P. In Fig. 22-21, three charged particles lie on a straight line and are separated by a distance d. Charges q_1 and q_2 are held fixed. Charge q_3 is free to move but happens to be in equilibrium (no net electrostatic force acts on it). Find q_1 in terms of q_2.

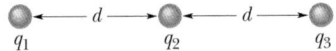

FIGURE 22-21 Problem 8.

9P. Figure 22-22a shows two charges, q_1 and q_2, held a fixed distance d apart. (a) What is the magnitude of the electrostatic force that acts on q_1? Assume that $q_1 = q_2 = 20.0 \ \mu C$ and $d = 1.50$ m. (b) A third charge $q_3 = 20.0 \ \mu C$ is brought in and placed as shown in Fig. 22-22b. What now is the magnitude of the electrostatic force on q_1?

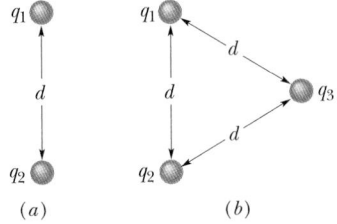

(a)   (b)

FIGURE 22-22 Problem 9.

10P. In Fig. 22-23, what are the horizontal and vertical components of the resultant electrostatic force on the charge in the lower left corner of the square if $q = 1.0 \times 10^{-7}$ C and $a = 5.0$ cm?

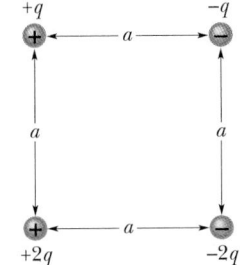

FIGURE 22-23
Problem 10.

11P. Charges q_1 and q_2 lie on the x axis at points $x = -a$ and $x = +a$, respectively. (a) How must q_1 and q_2 be related for the net electrostatic force on charge $+Q$, placed at $x = +a/2$, to

be zero? (b) Repeat (a) but with the $+Q$ charge now placed at $x = +3a/2$.

12P. Two small, positively charged spheres have a combined charge of 5.0×10^{-5} C. If each sphere is repelled from the other by an electrostatic force of 1.0 N when the spheres are 2.0 m apart, what is the charge on each sphere?

13P. Two identical conducting spheres, fixed in place, attract each other with an electrostatic force of 0.108 N when separated by 50.0 cm. The spheres are then connected by a thin conducting wire. When the wire is removed, the spheres repel each other with an electrostatic force of 0.0360 N. What were the initial charges on the spheres?

14P. Two fixed particles, of charges $q_1 = +1.0 \ \mu C$ and $q_2 = -3.0 \ \mu C$, are 10 cm apart. How far from each should a third charge be located so that no net electrostatic force acts on it?

15P. The charges and coordinates of two charged particles held fixed in the xy plane are: $q_1 = +3.0 \ \mu C$, $x_1 = 3.5$ cm, $y_1 = 0.50$ cm, and $q_2 = -4.0 \ \mu C$, $x_2 = -2.0$ cm, $y_2 = 1.5$ cm. (a) Find the magnitude and direction of the electrostatic force on q_2. (b) Where could you locate a third charge $q_3 = +4.0 \ \mu C$ such that the net electrostatic force on q_2 is zero?

16P. Two *free* point charges $+q$ and $+4q$ are a distance L apart. A third charge is placed so that the entire system is in equilibrium. (a) Find the location, magnitude, and sign of the third charge. (b) Show that the equilibrium of the system is unstable.

17P. (a) What equal positive charges would have to be placed on Earth and on the Moon to neutralize their gravitational attraction? Do you need to know the lunar distance to solve this problem? Why or why not? (b) How many thousand kilograms of hydrogen would be needed to provide the positive charge calculated in (a)?

18P. A certain charge Q is divided into two parts q and $Q - q$, which are then separated by a certain distance. What must q be in terms of Q to maximize the electrostatic repulsion between the two charges?

19P. A charge Q is fixed at each of two opposite corners of a square. A charge q is placed at each of the other two corners. (a) If the net electrostatic force on each Q is zero, what is Q in terms of q? (b) Is there any value of q that makes the net electrostatic force on each of the four charges zero? Explain.

20P. In Fig. 22-24, two tiny conducting balls of identical mass m and identical charge q hang from nonconducting threads of length L. Assume that θ is so small that $\tan \theta$ can be replaced by its

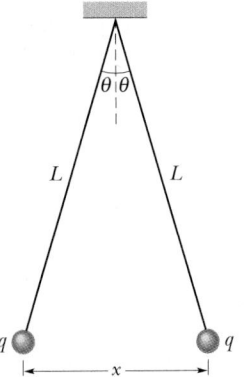

FIGURE 22-24
Problem 20.

approximate equal, sin θ. (a) Show that, for equilibrium,

$$x = \left(\frac{q^2 L}{2\pi\epsilon_0 mg}\right)^{1/3},$$

where x is the separation between the balls. (b) If $L = 120$ cm, $m = 10$ g, and $x = 5.0$ cm, what is q?

21P. Explain what happens to the balls of Problem 20b if one of them is discharged, and find the new equilibrium separation x, using the given values of L and m and the computed value of q.

22P. Figure 22-25 shows a long, nonconducting, massless rod of length L, pivoted at its center and balanced with a weight W at a distance x from the left end. At the left and right ends of the rod are attached small conducting spheres with positive charges q and $2q$, respectively. A distance h directly beneath each of these spheres is a fixed sphere with positive charge Q. (a) Find the distance x when the rod is horizontal and balanced. (b) What value should h have so that the rod exerts no vertical force on the bearing when the rod is horizontal and balanced?

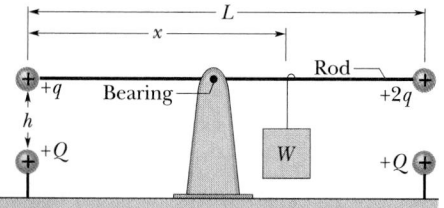

FIGURE 22-25 Problem 22.

SECTION 22-5 Charge Is Quantized

23E. What is the magnitude of the electrostatic force between a singly charged sodium ion (Na^+, of charge $+e$) and an adjacent singly charged chlorine ion (Cl^-, of charge $-e$) in a salt crystal if their separation is 2.82×10^{-10} m?

24E. A neutron consists of one "up" quark of charge $+2e/3$ and two "down" quarks each having charge $-e/3$. If the down quarks are 2.6×10^{-15} m apart inside the neutron, what is the magnitude of the electrostatic force between them?

25E. What is the total charge in coulombs of 75.0 kg of electrons?

26E. How many megacoulombs of positive (or negative) charge are in 1.00 mol of neutral molecular-hydrogen gas (H_2)?

27E. The magnitude of the electrostatic force between two identical ions that are separated by a distance of 5.0×10^{-10} m is 3.7×10^{-9} N. (a) What is the charge of each ion? (b) How many electrons are "missing" from each ion (thus giving the ion its charge imbalance)?

28E. (a) How many electrons would have to be removed from a penny to leave it with a charge of $+1.0 \times 10^{-7}$ C? (b) To what fraction of the electrons in the penny does this correspond? (See Sample Problem 22-4.)

29E. Two tiny, spherical water drops, with identical charges of -1.00×10^{-16} C, have a center-to-center separation of 1.00 cm.

(a) What is the magnitude of the electrostatic force acting between them? (b) How many excess electrons are on each drop, giving it its charge imbalance?

30E. How far apart must two protons be if the magnitude of the electrostatic force acting on either one is equal to the proton's weight at Earth's surface?

31E. An electron is in a vacuum near the surface of Earth. Where should a second electron be placed so that the electrostatic force it exerts on the first electron balances the weight of the first electron?

32P. Earth's atmosphere is constantly bombarded by *cosmic ray protons* that originate somewhere in space. If the protons were all to pass through the atmosphere, each square meter of Earth's surface would intercept protons at the average rate of 1500 protons per second. What would be the corresponding current intercepted by the total surface area of the planet?

33P. A 100 W lamp operated on a 120 V circuit has a current (assumed steady) of 0.83 A in its filament. How long does it take for 1 mol of electrons to pass through the lamp?

34P. Calculate the number of coulombs of positive charge in 250 cm³ of (neutral) water (about a glass full).

35P. In the basic CsCl (cesium chloride) crystal structure, Cs^+ ions form the corners of a cube and a Cl^- ion is at the cube's center (Fig. 22-26). The edge length of the cube is 0.40 nm. The Cs^+ ions are each deficient by one electron (and thus each has a charge of $+e$), and the Cl^- ion has one excess electron (and thus has a charge of $-e$). (a) What is the magnitude of the net electrostatic force exerted on the Cl^- ion by the eight Cs^+ ions at the corners of the cube? (b) If one of the Cs^+ ions is missing, the crystal is said to have a *defect*; what is the magnitude of the net electrostatic force exerted on the Cl^- ion by the seven remaining Cs^+ ions?

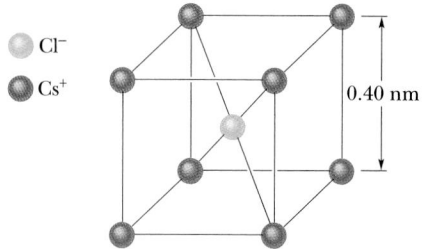

FIGURE 22-26 Problem 35.

36P. We know that, within the limits of measurement, the magnitudes of the negative charge on the electron and the positive charge on the proton are equal. Suppose, however, that these magnitudes differ from each other by 0.00010%. With what force would two copper pennies, placed 1.0 m apart, repel each other? What do you conclude? (*Hint:* See Sample Problem 22-4.)

37P. Two engineering students, John with a weight of 200 lb and Mary with a weight of 100 lb, are 100 ft apart. Suppose each has a 0.01% imbalance in the amount of positive and negative charge, one student being positive and the other negative. Estimate *roughly* the electrostatic force of attraction between them by replacing each student with a sphere of water having the same mass as the student.

SECTION 22-6 Charge Is Conserved

38E. In *beta decay* a massive fundamental particle changes to another massive particle, and either an electron or a positron is emitted. (a) If a proton undergoes beta decay to become a neutron, which particle is emitted? (b) If a neutron undergoes beta decay to become a proton, which particle is emitted?

39E. Using Appendix F, identify X in the following nuclear reactions:

(a) $^1H + ^9Be \rightarrow X + n$;

(b) $^{12}C + ^1H \rightarrow X$;

(c) $^{15}N + ^1H \rightarrow ^4He + X$.

40E. In the radioactive decay of ^{238}U (see Eq. 22-15), the center of the emerging ^4He particle is, at a certain instant, $9.0 \times$ 10^{-15} m from the center of the daughter nucleus ^{234}Th. At this instant, (a) what is the magnitude of the electrostatic force on the ^4He particle, and (b) what is that particle's acceleration?

ELECTRONIC COMPUTATION

41. In Problem 18, let $q = \alpha Q$. (a) Write an expression for the magnitude F of the force between the charges in terms of α, Q, and the charge separation d. (b) Graph F as a function of α. Graphically find the values of α that give (c) the maximum value of F and (d) half the maximum value of F.

42. Two particles, each of positive charge q, are fixed in place on an x axis, one at $x = 0$ and the other at $x = d$. A particle of charge Q is to be placed along that axis at locations given by $x = \alpha d$. (a) Write expressions, in terms of α, that give the net electrostatic force **F** acting on the third particle when it is in the three regions $x < 0$, $0 < x < d$, and $d < x$. The expressions should give a positive result when **F** is in the positive direction of the x axis and a negative result when **F** is in the negative direction. (b) Graph **F** versus α for the range $-2 < \alpha < 3$.

23
Electric Fields

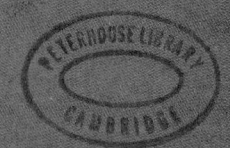

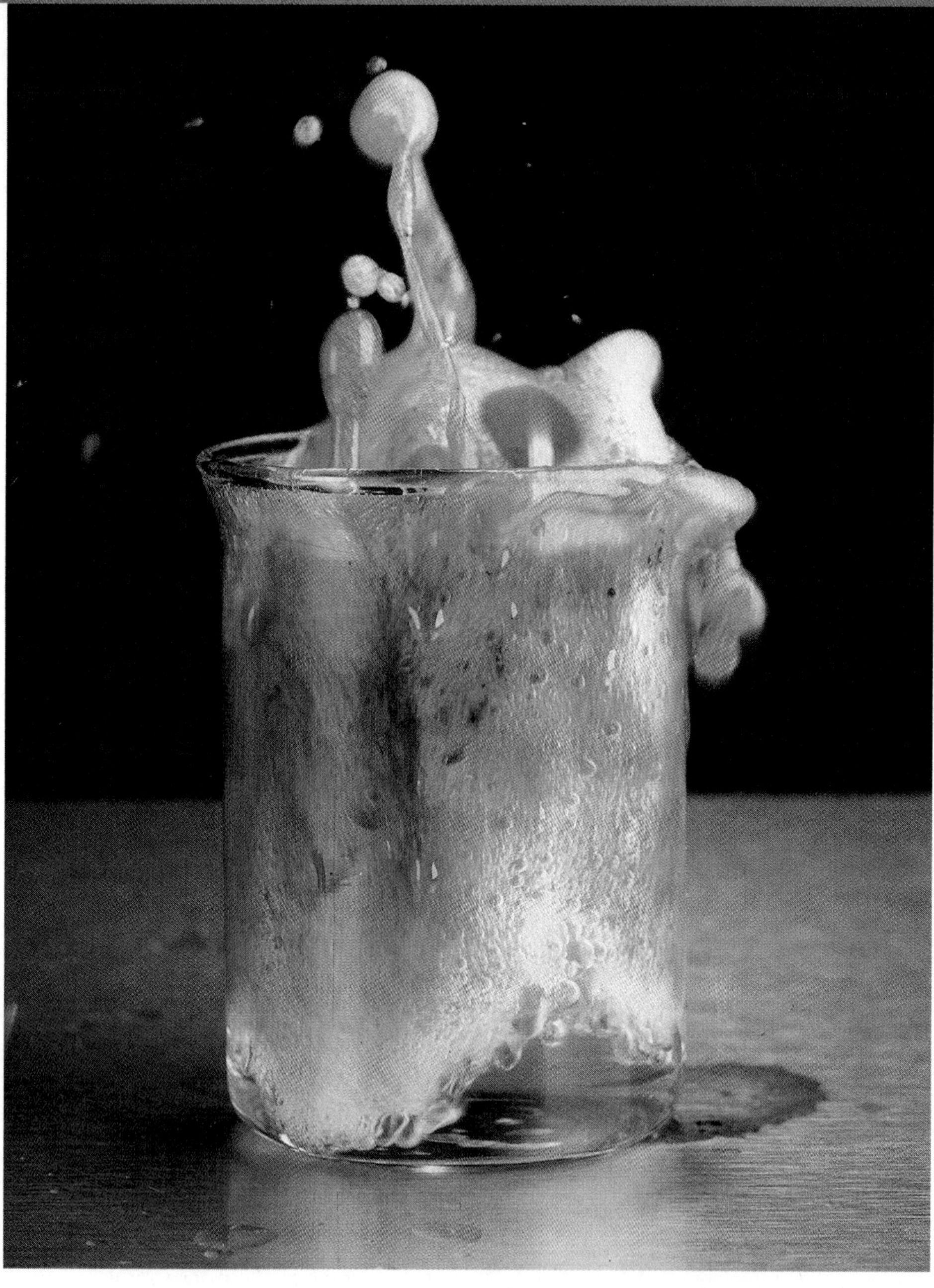

Water heats so well in a microwave oven that you might be able to heat a cup of water as much as 8 C° above the normal boiling temperature of water without causing it to boil. If you then pour coffee powder, or even chips of ice, into the water, it will erupt into a furious boil like that in the photograph, scattering water that could quickly scald you. Why do microwaves heat water?

23-1 CHARGES AND FORCES: A CLOSER LOOK

Suppose we fix a positively charged particle q_1 in place and then put a second positively charged particle q_2 near it. From Coulomb's law we know that q_1 exerts a repulsive electrostatic force on q_2 and, given enough data, we could determine the magnitude and direction of that force. Still, a nagging question remains: How does q_1 "know" of the presence of q_2? That is, since the charges do not touch, how can q_1 exert a force on q_2?

This question about *action at a distance* can be answered by saying that q_1 sets up an **electric field** in the space surrounding it. At any given point P in that space, the field has both magnitude and direction. The magnitude depends on the magnitude of q_1 and the distance between P and q_1. The direction depends on the direction from q_1 to P and the electrical sign of q_1. Thus when we place q_2 at P, q_1 interacts with q_2 through the electric field at P. The magnitude and direction of that electric field determine the magnitude and direction of the force acting on q_2.

Another action-at-a-distance problem arises if we move q_1; say, toward q_2. Coulomb's law tells us that when q_1 is closer to q_2, the repulsive electrostatic force acting on q_2 must be greater. And it is. But here the nagging question is: Does the electric field at q_2, and thus the force acting on q_2, change immediately?

The answer is no. Instead, the information about the move by q_1 travels outward from q_1 (in all directions) as an electromagnetic wave at the speed of light c. The change in the electric field at q_2, and thus the change in the force acting on q_2, occurs when the wave finally reaches q_2.

23-2 THE ELECTRIC FIELD

The temperature at every point in a room has a definite value. You can measure the temperature at any given point or combination of points by putting a thermometer there. We call the resulting distribution of temperatures a *temperature field*. In much the same way, you can imagine a *pressure field* in the atmosphere: it consists of the distribution of air pressure values, one for each point in the atmosphere. These two examples are of *scalar fields,* because temperature and air pressure are scalar quantities.

The electric field is a *vector field*: it consists of a distribution of *vectors*, one for each point in the region around a charged object, such as a charged rod. In principle, we can define the electric field at some point near the charged object, such as point P in Fig. 23-1a, by placing a *positive* charge q_0, called a *test charge*, at the point. We then measure the electrostatic force **F** that acts on the test charge.

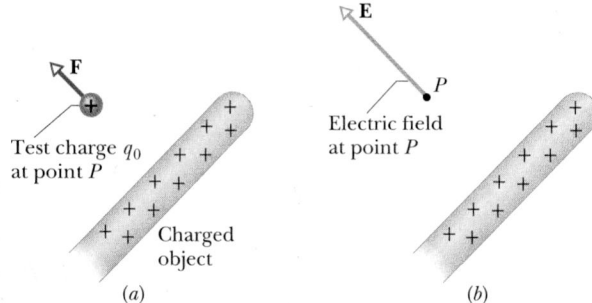

FIGURE 23-1 (*a*) A positive test charge q_0 placed at point P near a charged object. An electrostatic force **F** acts on the test charge. (*b*) The electric field **E** at point P produced by the charged object.

The electric field **E** at point P due to the charged object is defined as

$$\mathbf{E} = \frac{\mathbf{F}}{q_0} \qquad \text{(electric field)}. \qquad (23\text{-}1)$$

Thus the magnitude of the electric field **E** at point P is $E = F/q_0$, and the direction of **E** is that of the force **F** that acts on the *positive* test charge. As shown in Fig. 23-1b, we represent the electric field at P with a vector whose tail is at P. To define the electric field within some region, we must similarly measure it at all points in the region. The SI unit for the electric field is the newton per coulomb (N/C). Table 23-1 shows the electric fields that occur in a few physical situations.

Although we use a positive test charge to define the electric field of a charged object, that field exists independently of the test charge. The field at point P in Figure 23-1b existed both before and after the test charge of Fig. 23-1a was put there. (We assume that in our defining procedure, the presence of the test charge does not affect the charge distribution on the charged object, and thus does not alter the electric field we are defining.)

TABLE 23-1 SOME ELECTRIC FIELDS

FIELD LOCATION OR SITUATION	VALUE (N/C)
At the surface of a uranium nucleus	3×10^{21}
Within a hydrogen atom, at a radius of 5.29×10^{-11} m	5×10^{11}
Electric breakdown occurs in air	3×10^{6}
Near the charged drum of a photocopier	10^{5}
Near a charged plastic comb	10^{3}
In the lower atmosphere	10^{2}
Inside the copper wire of household circuits	10^{-2}

To examine the role of an electric field in the interaction between charged objects, we have two tasks: (1) calculating the electric field produced by a given distribution of charge, and (2) calculating the force that a given field exerts on a charge placed in it. We perform the first task in Sections 23-4 through 23-7 for several charge distributions. We perform the second task in Sections 23-8 and 23-9 by considering a point charge and a pair of point charges in an electric field. But first, we discuss a way to visualize electric fields.

23-3 ELECTRIC FIELD LINES

Michael Faraday, who introduced the idea of electric fields in the 19th century, thought of the space around a charged body as filled with *lines of force.* Although we no longer attach much reality to these lines, now usually called **electric field lines,** they still provide a nice way to visualize patterns in electric fields.

The relation between the field lines and electric field vectors is this: (1) at any point, the direction of a straight field line or the direction of the tangent to a curved field line gives the direction of **E** at that point, and (2) the field lines are drawn so that the number of lines per unit area, measured in a plane that is perpendicular to the lines, is proportional to the *magnitude* of **E**. This second relation means that where the field lines are close together, E is large; and where they are far apart, E is small.

Figure 23-2a shows a sphere of uniform negative charge. If we place a *positive* test charge anywhere near the sphere, an electrostatic force pointing *toward* the center of the sphere will act on the test charge as shown. In other

words, the electric field vectors at all points near the sphere are directed radially toward the sphere. This pattern of vectors is neatly displayed by the field lines in Fig. 23-2b, which point in the same directions as the force and field vectors. Moreover, the spreading of the field lines with distance from the sphere tells us that the magnitude of the electric field decreases with distance from the sphere.

If the sphere of Fig. 23-2 were of uniform *positive* charge, the electric field vectors at all points near the sphere would be directed radially *away from* the sphere. Thus the electric field lines would also extend radially away from the sphere. We then have the following rule:

Electric field lines extend away from positive charge and toward negative charge.

Figure 23-3a shows part of an infinitely large, nonconducting *sheet* (or plane) with a uniform distribution of positive charge on one side. If we were to place a positive

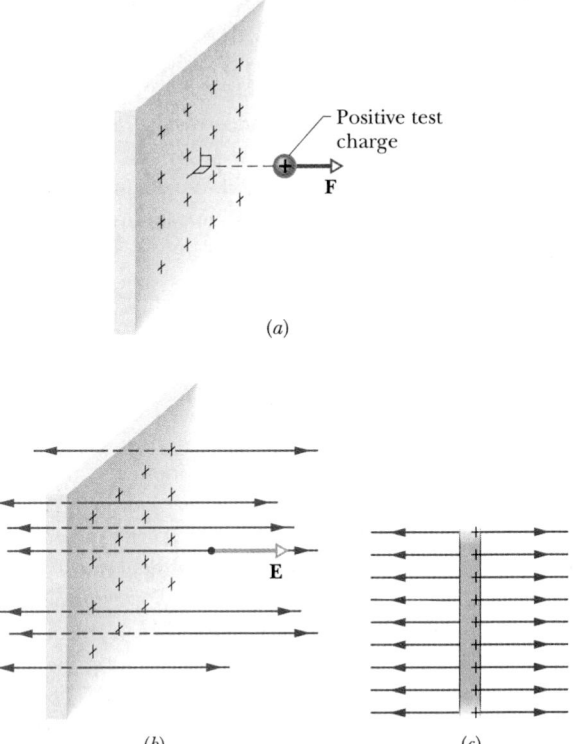

(a)

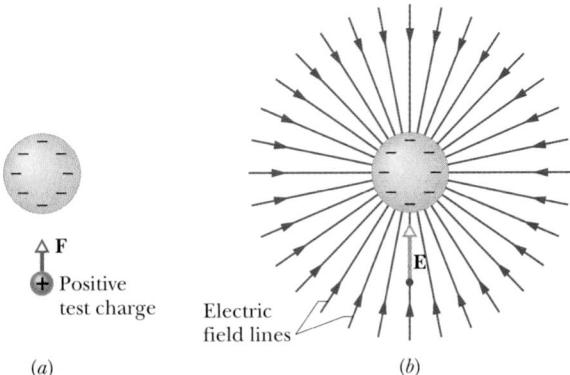

FIGURE 23-2 (a) The electrostatic force **F** acting on a positive test charge near a sphere of uniform negative charge. (b) The electric field vector **E** at the location of the test charge, and the electric field lines in the space near the sphere. The field lines extend *toward* the negatively charged sphere. (They originate on distant positive charges.)

FIGURE 23-3 (a) The electrostatic force **F** on a positive test charge near a very large, nonconducting sheet with uniformly distributed positive charge on one side. (b) The electric field vector **E** at the location of the test charge, and the electric field lines in the space near the sheet. The field lines extend *away from* the positively charged sheet. (c) Side view of (b).

test charge at any point near the sheet of Fig. 23-3a, the net electrostatic force acting on the test charge would be perpendicular to the sheet, because forces acting in all other directions would cancel one another as a result of the symmetry. Moreover, the net force on the test charge would point away from the sheet as shown. Thus the electric field vector at any point in the space on either side of the sheet is also perpendicular to the sheet and directed away from it (Figs. 23-3b and c). Since the charge is uniformly distributed along the sheet, all the field vectors have the same magnitude. Such an electric field, with the same magnitude and direction at every point, is a *uniform electric field*.

Of course, no real nonconducting sheet (such as a flat expanse of plastic) is infinitely large, but if we consider a region that is near the middle of a real sheet and not near its edges, the field lines through that region are arranged as in Fig. 23-3b and c.

Figure 23-4 shows the field lines for two equal positive charges. Figure 23-5 shows the pattern for two charges that are equal in magnitude but of opposite sign, a configuration that we call an **electric dipole.** Although we do not often use field lines quantitatively, they are very useful to visualize what is going on. Can you not almost ''see'' the charges being pushed apart in Fig. 23-4 and pulled together in Fig. 23-5?

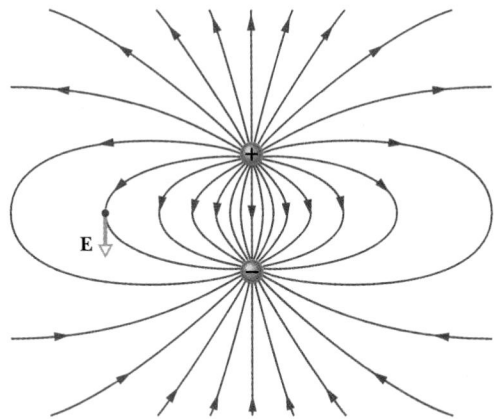

FIGURE 23-5 Field lines for a positive and a nearby negative point charge that are equal in magnitude. The charges attract each other. The pattern of field lines and the electric field it represents have rotational symmetry about an axis passing through both charges. The electric field vector at one point is shown; the vector is tangent to the field line through the point.

SAMPLE PROBLEM 23-1

In Fig. 23-2, how does the magnitude of the electric field vary with distance from the center of the uniformly charged sphere?

SOLUTION: Suppose that N field lines terminate on the sphere of Fig. 23-2. Imagine a concentric sphere of radius r surrounding the charged sphere. The number of lines per unit area on the imaginary sphere is $N/4\pi r^2$. Because E is proportional to this quantity, we can write $E \propto 1/r^2$. Thus the electric field set up by a uniform sphere of charge varies as the inverse square of the distance from the center of the sphere.

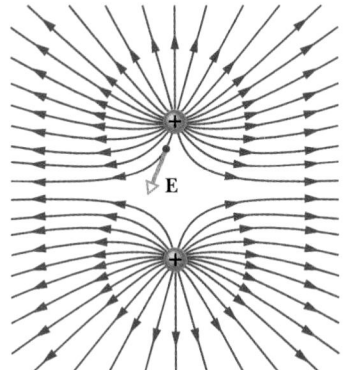

FIGURE 23-4 Field lines for two equal positive point charges. The charges repel each other. (The lines terminate on distant negative charges.) To ''see'' the actual three-dimensional pattern of field lines, mentally rotate the pattern shown here about an axis passing through both charges in the plane of the page. The three-dimensional pattern and the electric field it represents are said to have *rotational symmetry* about that axis. The electric field vector at one point is shown; note that it is tangent to the field line through that point.

23-4 THE ELECTRIC FIELD DUE TO A POINT CHARGE

To find the electric field due to a point charge q (or charged particle), we put a positive test charge q_0 at any point a distance r from the point charge. From Coulomb's law (Eq. 22-4), the magnitude of the electrostatic force acting on q_0 is

$$F = \frac{1}{4\pi\epsilon_0}\frac{|q||q_0|}{r^2}. \qquad (23\text{-}2)$$

The direction of **F** is directly away from the point charge if q is positive and directly toward the point charge if q is negative. The magnitude of the electric field vector is, from Eq. 23-1,

$$E = \frac{F}{q_0} = \frac{1}{4\pi\epsilon_0}\frac{|q|}{r^2} \qquad \text{(point charge).} \qquad (23\text{-}3)$$

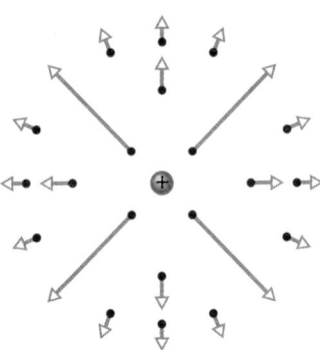

FIGURE 23-6 The electric field vectors at several points around a positive point charge.

The direction of **E** is the same as that of the force on the positive test charge: directly away from the point charge if q is positive, and toward it if q is negative.

We find the electric field in the space around a point charge by moving the test charge around in that space. The field for a positive point charge is shown in Fig. 23-6 in vector form (not as field lines).

We can find the net, or resultant, electric field due to more than one point charge with the aid of the principle of superposition. If we place a positive test charge q_0 near n point charges $q_1, q_2, \cdots, q_n$, then, from Eq. 22-7, the net force $\mathbf{F}_0$ from the n point charges acting on the test charge is

$$\mathbf{F}_0 = \mathbf{F}_{01} + \mathbf{F}_{02} + \cdots + \mathbf{F}_{0n}.$$

So, from Eq. 23-1, the net electric field at the position of the test charge is

$$\mathbf{E} = \frac{\mathbf{F}_0}{q_0} = \frac{\mathbf{F}_{01}}{q_0} + \frac{\mathbf{F}_{02}}{q_0} + \cdots + \frac{\mathbf{F}_{0n}}{q_0}$$
$$= \mathbf{E}_1 + \mathbf{E}_2 + \cdots + \mathbf{E}_n. \tag{23-4}$$

Here $\mathbf{E}_i$ is the electric field that would be set up by point charge i acting alone. Equation 23-4 shows us that the principle of superposition applies to electric fields as well as to electrostatic forces.

CHECKPOINT 1: The figure shows a proton p and an electron e on an x axis. What is the direction of the electric field due to the electron at (a) point S and (b) point R? What is the direction of the net electric field at (c) point R and (d) point S?

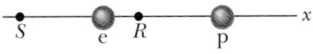

SAMPLE PROBLEM 23-2

Figure 23-7a shows three particles with charges $q_1 = +2Q$, $q_2 = -2Q$, and $q_3 = -4Q$, each a distance d from the origin. What net electric field **E** is produced at the origin?

SOLUTION: Charges q_1, q_2, and q_3 produce electric field vectors $\mathbf{E}_1$, $\mathbf{E}_2$, and $\mathbf{E}_3$, respectively, at the origin. We seek the vector sum $\mathbf{E} = \mathbf{E}_1 + \mathbf{E}_2 + \mathbf{E}_3$. For this, we first must find the magnitudes and orientations of the three field vectors. To find the magnitude of $\mathbf{E}_1$, which is due to q_1, we use Eq. 23-3, substituting d for r and $2Q$ for $|q|$ and obtaining

$$E_1 = \frac{1}{4\pi\epsilon_0} \frac{2Q}{d^2}.$$

Similarly, we find the magnitudes of the fields $\mathbf{E}_2$ and $\mathbf{E}_3$ to be

$$E_2 = \frac{1}{4\pi\epsilon_0} \frac{2Q}{d^2} \quad \text{and} \quad E_3 = \frac{1}{4\pi\epsilon_0} \frac{4Q}{d^2}.$$

We next must find the orientations of the three electric field vectors at the origin. Because q_1 is a positive charge, the field vector it produces points directly *away* from it. And because q_2 and q_3 are both negative, the field vectors they produce point directly *toward* each of them. Thus the three electric fields produced at the origin by the three charged particles are oriented as in Fig. 23-7b. (Note that we have placed the tails of the vectors at this point where the fields are to be evaluated; doing so decreases the chance of error.)

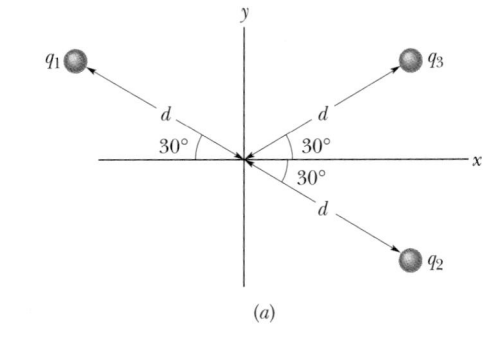

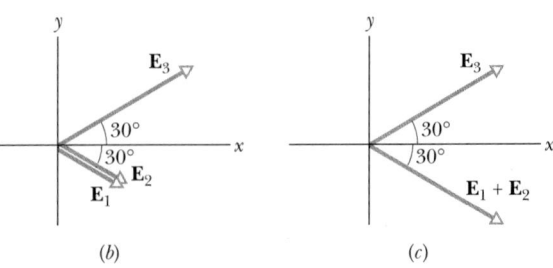

FIGURE 23-7 Sample Problem 23-2. (a) Three particles with charges q_1, q_2, and q_3 are at the same distance d from the origin. (b) The electric field vectors $\mathbf{E}_1$, $\mathbf{E}_2$, and $\mathbf{E}_3$ at the origin due to the three particles. (c) The electric field vector $\mathbf{E}_3$ and the vector sum $\mathbf{E}_1 + \mathbf{E}_2$ at the origin.

We can now add the fields vectorially as usual, by finding the x and y components of each vector, and then the net x component E_x and the net y component E_y. To get the magnitude E we would use the Pythagorean theorem, and to find the orientation of **E** we would use the definition of the tangent of an angle.

However, here we can use symmetry to simplify the procedure. From Fig. 23-7b, we see that E_1 and E_2 point in the same direction. Hence their vector sum points in that direction and has the magnitude

$$E_1 + E_2 = \frac{1}{4\pi\epsilon_0}\frac{2Q}{d^2} + \frac{1}{4\pi\epsilon_0}\frac{2Q}{d^2}$$

$$= \frac{1}{4\pi\epsilon_0}\frac{4Q}{d^2},$$

which happens to equal the magnitude of E_3.

We must now combine two vectors, E_3 and the vector sum $E_1 + E_2$, that have the same magnitude and that are oriented symmetrically about the x axis, as shown in Fig. 23-7c. From the symmetry of Fig. 23-7c, we realize that the equal y components of our two vectors cancel and the equal x components add. Thus, the net electric field **E** at the origin points along the positive direction of x and has the magnitude

$$E = 2E_{3x} = 2E_3 \cos 30°$$

$$= (2)\frac{1}{4\pi\epsilon_0}\frac{4Q}{d^2}(0.866) = \frac{6.93Q}{4\pi\epsilon_0 d^2}. \quad \text{(Answer)}$$

Since the charge of the nucleus is positive, the electric field vector **E** points outward, away from the center of the nucleus.

CHECKPOINT 2: The figure shows four situations in which charged particles are at equal distances from the origin. Rank the situations according to the magnitude of the net electric field at the origin, greatest first.

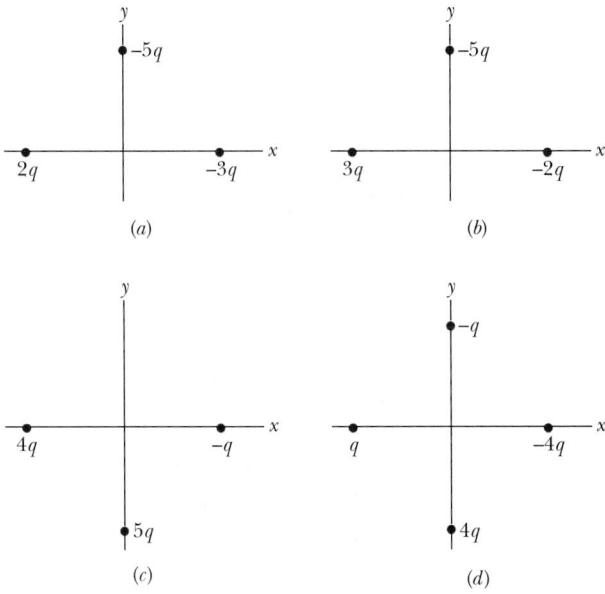

(a)   (b)

(c)   (d)

The nucleus of a uranium atom has a radius R of 6.8 fm. Assuming that the positive charge of the nucleus is distributed uniformly, determine the electric field at a point on the surface of the nucleus due to that charge.

SOLUTION: The nucleus has a positive charge of Ze, where the atomic number Z ($= 92$) is the number of protons within the nucleus, and e ($= 1.60 \times 10^{-19}$ C) is the charge of a proton. If this charge is distributed uniformly, then the first shell theorem of Chapter 22 applies. The electrostatic force on a positive test charge placed near the surface of the nucleus is the same as if the nuclear charge were concentrated at the nuclear center.

From Eq. 23-1, we then know that the electric field produced by the nucleus is also the same as if the nuclear charge were concentrated at the nuclear center. Equation 23-3 applies to such a pointlike concentration of charge, and we can write, for the magnitude of the field,

$$E = \frac{1}{4\pi\epsilon_0}\frac{Ze}{R^2}$$

$$= \frac{(8.99 \times 10^9 \text{ N} \cdot \text{m}^2/\text{C}^2)(92)(1.60 \times 10^{-19} \text{ C})}{(6.8 \times 10^{-15} \text{ m})^2}$$

$$= 2.9 \times 10^{21} \text{ N/C}. \quad \text{(Answer)}$$

23-5 THE ELECTRIC FIELD DUE TO AN ELECTRIC DIPOLE

Figure 23-8a shows two charges of magnitude q but of opposite sign, separated by a distance d. As was noted in connection with Fig. 23-5, we call this configuration an *electric dipole*. Let us find the electric field due to the dipole of Fig. 23-8a at a point P, a distance z from the midpoint of the dipole and on its central axis, which is called the *dipole axis*.

From symmetry, the electric field **E** at point P—and also the fields $E_{(+)}$ and $E_{(-)}$ due to the separate charges that make up the dipole—must lie along the dipole axis, which we take to be a z axis. Applying the superposition principle for electric fields, we find that the magnitude E of the electric field at P is

$$E = E_{(+)} - E_{(-)}$$

$$= \frac{1}{4\pi\epsilon_0}\frac{q}{r_{(+)}^2} - \frac{1}{4\pi\epsilon_0}\frac{q}{r_{(-)}^2}$$

$$= \frac{q}{4\pi\epsilon_0(z - \frac{1}{2}d)^2} - \frac{q}{4\pi\epsilon_0(z + \frac{1}{2}d)^2}. \quad (23\text{-}5)$$

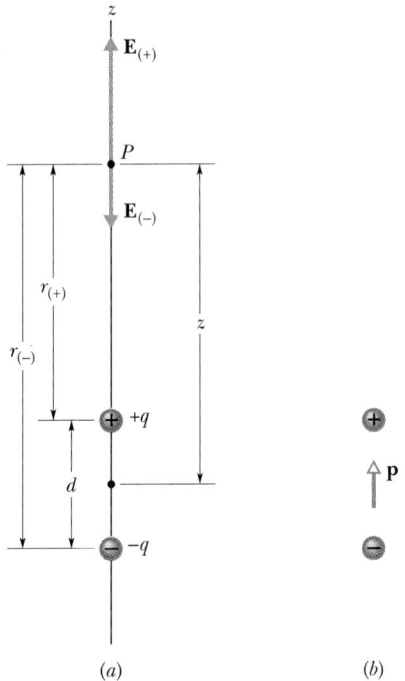

(a)   (b)

FIGURE 23-8 (a) An electric dipole. The electric field vectors $\mathbf{E}_{(+)}$ and $\mathbf{E}_{(-)}$ at point P on the dipole axis resulting from the two charges are shown. P is at distances $r_{(+)}$ and $r_{(-)}$ from the individual charges that make up the dipole. (b) The dipole moment $\mathbf{p}$ of the dipole points from the negative charge to the positive charge.

After a little algebra, we can rewrite this equation as

$$E = \frac{q}{4\pi\epsilon_0 z^2}\left[\left(1 - \frac{d}{2z}\right)^{-2} - \left(1 + \frac{d}{2z}\right)^{-2}\right]. \quad (23\text{-}6)$$

We are usually interested in the electrical effect of a dipole only at distances that are large compared with the dimensions of the dipole, that is, at distances such that $z \gg d$. At such large distances, we have $d/2z \ll 1$ in Eq. 23-6. We can then expand the two quantities in the brackets in that equation by the binomial theorem, obtaining for those quantities

$$\left[\left(1 + \frac{2d}{2z(1!)} + \cdots\right) - \left(1 - \frac{2d}{2z(1!)} + \cdots\right)\right].$$

So,

$$E = \frac{q}{4\pi\epsilon_0 z^2}\left[\left(1 + \frac{d}{z} + \cdots\right) - \left(1 - \frac{d}{z} + \cdots\right)\right].$$
$$(23\text{-}7)$$

The unwritten terms in the two expansions in Eq. 23-7 involve d/z raised to progressively higher powers. Since $d/z \ll 1$, the contributions of those terms are progressively less, and to approximate E at large distances, we can ne-

glect them. Then, in our approximation, we can rewrite Eq. 23-7 as

$$E = \frac{q}{4\pi\epsilon_0 z^2}\frac{2d}{z} = \frac{1}{2\pi\epsilon_0}\frac{qd}{z^3}. \quad (23\text{-}8)$$

The product qd, which involves the two intrinsic properties q and d of the dipole, is the magnitude p of a vector quantity known as the **electric dipole moment p** of the dipole. Thus we can write Eq. 23-8 as

$$E = \frac{1}{2\pi\epsilon_0}\frac{p}{z^3} \quad \text{(electric dipole).} \quad (23\text{-}9)$$

The direction of $\mathbf{p}$ is taken to be from the negative to the positive end of the dipole, as indicated in Fig. 23-8b. We can use $\mathbf{p}$ to specify the orientation of a dipole.

Equation 23-9 shows that, if we measure the electric field of a dipole only at distant points, we can never find q and d separately, only their product. The field at distant points would be unchanged if, for example, q were doubled and d simultaneously halved. So the dipole moment is a basic property of a dipole.

Although Eq. 23-9 holds only for distant points along the dipole axis, it turns out that E for a dipole varies as $1/r^3$ for *all* distant points, regardless of whether they lie on the dipole axis; here r is the distance between the point in question and the dipole center.

Inspection of Fig. 23-8 and of the field lines in Fig. 23-5 shows that the direction of $\mathbf{E}$ for distant points on the dipole axis is always the direction of the dipole moment vector $\mathbf{p}$. This is true whether point P in Fig. 23-8a is on the upper or the lower part of the dipole axis.

Inspection of Eq. 23-9 shows that if you double the distance of a point from a dipole, the electric field at the point drops by a factor of 8. If you double the distance from a single point charge, however (see Eq. 23-3), the electric field drops only by a factor of 4. Thus the electric field of a dipole decreases more rapidly with distance than does the electric field of a single charge. The physical reason for this rapid decrease in electric field for a dipole is that from distant points a dipole looks like two equal but opposite charges that almost—but not quite—coincide. So their electric fields at distant points almost—but not quite—cancel each other.

SAMPLE PROBLEM 23-4

A molecule of water vapor causes an electric field in the surrounding space as if it were an electric dipole like that of Fig. 23-8. Its dipole moment has a magnitude $p = 6.2 \times 10^{-30}$ C·m. What is the magnitude of the electric field at a distance $z = 1.1$ nm from the molecule on its dipole axis? (This distance is large enough for Eq. 23-9 to apply.)

SOLUTION: From Eq. 23-9

$$E = \frac{1}{2\pi\epsilon_0}\frac{p}{z^3}$$

$$= \frac{6.2 \times 10^{-30}\ \text{C}\cdot\text{m}}{(2\pi)(8.85 \times 10^{-12}\ \text{C}^2/\text{N}\cdot\text{m}^2)(1.1 \times 10^{-9}\ \text{m})^3}$$

$$= 8.4 \times 10^7\ \text{N/C}. \qquad \text{(Answer)}$$

23-6 THE ELECTRIC FIELD DUE TO A LINE OF CHARGE

So far we have considered the electric field that is produced by one or, at most, a few point charges. We now consider charge distributions that consist of a great many closely spaced point charges (perhaps billions) that are spread along a line, over a surface, or within a volume. Such distributions are said to be **continuous** rather than discrete. Since these distributions can include an enormous number of point charges, we find the electric fields that they produce by means of calculus rather than by considering the point charges one by one. In this section we discuss the electric field caused by a line of charge. We consider a charged surface in the next section. A charged volume is the subject of Sample Problem 23-3, where we found the field outside a uniformly charged sphere. In the next chapter, we shall find the field inside such a sphere.

When we deal with continuous charge distributions, it is most convenient to express the charge on an object as a *charge density* rather than as a total charge. For a line of charge, for example, we would report the linear charge density (or charge per length) λ, whose SI unit is the coulomb per meter. Table 23-2 shows the other charge densities we shall be using.

Figure 23-9 shows a thin ring of radius R with a uniform positive linear charge density λ around its circumference. We may imagine the ring to be made of plastic or some other insulator, so that the charges can be regarded as fixed in place. What is the electric field **E** at point P, a distance z from the plane of the ring along its central axis?

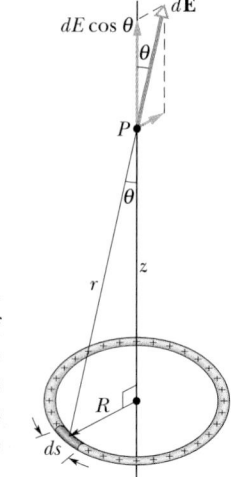

FIGURE 23-9 A ring of uniform positive charge. A differential element of charge occupies a length ds (greatly exaggerated for clarity). This element sets up an electric field $d\textbf{E}$ at point P. The component of $d\textbf{E}$ along the central axis of the ring is $dE \cos\theta$.

To answer, we cannot just apply Eq. 23-3, which gives the electric field set up by a point charge, because the ring is obviously not a point charge. However, we can mentally divide the ring into differential elements of charge that are so small that they are like point charges, and then we can apply Eq. 23-3 to each of them. Next, we can add the electric fields set up at P by all the differential elements. Their vector sum gives us the field set up at P by the ring.

Let ds be the (arc) length of any differential element of the ring. Since λ is the charge per unit length, the element has a charge of magnitude

$$dq = \lambda\ ds. \qquad (23\text{-}10)$$

This differential charge sets up a differential electric field $d\textbf{E}$ at point P, which is a distance r from the element. Treating the element as a point charge, and using Eq. 23-10, we can rewrite Eq. 23-3 to express the magnitude of $d\textbf{E}$ as

$$dE = \frac{1}{4\pi\epsilon_0}\frac{dq}{r^2} = \frac{1}{4\pi\epsilon_0}\frac{\lambda\ ds}{r^2}. \qquad (23\text{-}11)$$

From Fig. 23-9, we can rewrite Eq. 23-11 as

$$dE = \frac{1}{4\pi\epsilon_0}\frac{\lambda\ ds}{(z^2 + R^2)}. \qquad (23\text{-}12)$$

Figure 23-9 shows us that $d\textbf{E}$ is at an angle θ to the central axis (which we have taken to be a z axis) and has components perpendicular to and parallel to that axis.

Every element of charge in the ring sets up a differential field $d\textbf{E}$ at P, with magnitude given by Eq. 23-12. All these $d\textbf{E}$ vectors have identical components parallel to the central axis, in both magnitude and direction. All these $d\textbf{E}$ vectors have components perpendicular to the central axis as well; these perpendicular components are identical in

TABLE 23-2 SOME MEASURES OF ELECTRIC CHARGE

NAME	SYMBOL	SI UNIT
Charge	q	C
Linear charge density	λ	C/m
Surface charge density	σ	C/m²
Volume charge density	ρ	C/m³

magnitude but point in different directions. In fact, for any perpendicular component that points in a given direction, there is another one that points in the opposite direction. The sum of this pair of components, like the sum of all other pairs of oppositely directed components, is zero.

So the perpendicular components cancel and we need not consider them further. This leaves the parallel components; they are all in the same direction, so the net electric field at P is their sum.

From Fig. 23-9, we see that the parallel component of $d\mathbf{E}$ has magnitude $dE \cos \theta$. We also see that

$$\cos \theta = \frac{z}{r} = \frac{z}{(z^2 + R^2)^{1/2}}. \qquad (23\text{-}13)$$

Then Eqs. 23-13 and 23-12 give us, for the parallel component of $d\mathbf{E}$,

$$dE \cos \theta = \frac{z\lambda}{4\pi\epsilon_0(z^2 + R^2)^{3/2}} \, ds. \qquad (23\text{-}14)$$

To add the parallel components $dE \cos \theta$ produced by all the elements, we integrate Eq. 23-14 around the circumference of the ring, from $s = 0$ to $s = 2\pi R$. Since the only quantity in Eq. 23-14 that varies during the integration is s, the other quantities can be moved outside the integral sign. The integration then gives us

$$E = \int dE \cos \theta = \frac{z\lambda}{4\pi\epsilon_0(z^2 + R^2)^{3/2}} \int_0^{2\pi R} ds$$

$$= \frac{z\lambda(2\pi R)}{4\pi\epsilon_0(z^2 + R^2)^{3/2}}. \qquad (23\text{-}15)$$

Since λ is the charge per length of the ring, the term $\lambda(2\pi R)$ in Eq. 23-15 is q, the total charge on the ring. We then can rewrite Eq. 23-15 as

$$E = \frac{qz}{4\pi\epsilon_0(z^2 + R^2)^{3/2}} \qquad \begin{array}{c}\text{(charged}\\\text{ring).}\end{array} \quad (23\text{-}16)$$

If the charge on the ring is negative, instead of positive as we have assumed, the magnitude of the field at P is still given by Eq. 23-16. However, the electric field vector then points toward the ring instead of away from it.

Let us check Eq. 23-16 for a point on the central axis that is so far away that $z \gg R$. For such a point, the expression $z^2 + R^2$ in Eq. 23-16 can be approximated as z^2, and Eq. 23-16 becomes

$$E = \frac{1}{4\pi\epsilon_0}\frac{q}{z^2} \qquad \begin{array}{c}\text{(charged ring at}\\\text{large distance).}\end{array} \quad (23\text{-}17)$$

This is a reasonable result, because from a large distance, the ring ''looks'' like a point charge. If we replace z by r in Eq. 23-17, we indeed do have Eq. 23-3, the electric field due to a point charge.

Let us next check Eq. 23-16 for a point at the center of the ring, that is, for $z = 0$. At that point, Eq. 23-16 tells us that $E = 0$. This is a reasonable result, because if we were to place a test charge at the center of the ring, there would be no net electrostatic force acting on it: the force due to any element of the ring would be canceled by the force due to the element on the opposite side of the ring. And, by Eq. 23-1, if the force were zero, the electric field at the center of the ring would have to be zero.

SAMPLE PROBLEM 23-5

Figure 23-10a shows a plastic rod having a uniformly distributed charge $-Q$. The rod has been bent in a $120°$ circular arc of radius r. We place coordinate axes such that the axis of symmetry of the rod lies along the x axis and the origin is at the center of curvature P of the rod. In terms of Q and r, what is the electric field $\mathbf{E}$ due to the rod at point P?

SOLUTION: Consider a differential element of the rod, having arc length ds and located at an angle θ above the x axis (Fig. 23-10b). If we let λ represent the linear charge density of the rod, our element ds has a differential charge of magnitude

$$dq = \lambda \, ds. \qquad (23\text{-}18)$$

Our element produces a differential electric field $d\mathbf{E}$ at point P, which is a distance r from the element. Treating the element as a point charge, we can rewrite Eq. 23-3 to express the magnitude of $d\mathbf{E}$ as

$$dE = \frac{1}{4\pi\epsilon_0}\frac{dq}{r^2} = \frac{1}{4\pi\epsilon_0}\frac{\lambda \, ds}{r^2}. \qquad (23\text{-}19)$$

The direction of $d\mathbf{E}$ is toward ds, because charge dq is negative.

Our element has a symmetrically located (mirror image) element ds' in the bottom half of the rod. The electric field $d\mathbf{E}'$ set up at P by ds' also has the magnitude given by Eq. 23-19, but the field vector points toward ds' as shown in Fig. 23-10b. If we resolve the electric field vectors of ds and ds' into x and y components as shown in Fig. 23-10b, we see that their y components cancel (because they have equal magnitudes and are in opposite directions). We also see that their x components have equal magnitudes and are in the same direction.

Thus to find the electric field set up by the rod, we need sum (via integration) only the x components of the differential electric fields set up by all the differential elements of the rod. From Fig. 23-10b and Eq. 23-19, we can write the component dE_x set up by ds as

$$dE_x = dE \cos \theta = \frac{1}{4\pi\epsilon_0}\frac{\lambda}{r^2}\cos \theta \, ds. \quad (23\text{-}20)$$

Equation 23-20 has two variables, θ and s. Before we can integrate it, we must eliminate one variable. We do so by replacing ds, using the relation

$$ds = r \, d\theta,$$

in which $d\theta$ is the angle at P that includes arc length ds (Fig. 23-10c). With this replacement, we can integrate Eq. 23-20 over the angle made by the rod at P, from $\theta = -60°$ to $\theta = 60°$; that will give us the magnitude of the electric field at P due to the rod:

$$E = \int dE_x = \int_{-60°}^{60°} \frac{1}{4\pi\epsilon_0} \frac{\lambda}{r^2} \cos\theta \, r \, d\theta$$

$$= \frac{\lambda}{4\pi\epsilon_0 r} \int_{-60°}^{60°} \cos\theta \, d\theta = \frac{\lambda}{4\pi\epsilon_0 r} \left[\sin\theta \right]_{-60°}^{60°}$$

$$= \frac{\lambda}{4\pi\epsilon_0 r} [\sin 60° - \sin(-60°)]$$

$$= \frac{1.73\lambda}{4\pi\epsilon_0 r}. \qquad (23\text{-}21)$$

(If we had reversed the limits on the integration, we would have gotten the same result but with a minus sign. Since the integration gives only the magnitude of **E**, we would then have discarded the minus sign.)

To evaluate λ, we note that the rod has an angle of $120°$ and so is one-third of a full circle. Its arc length is then $2\pi r/3$,

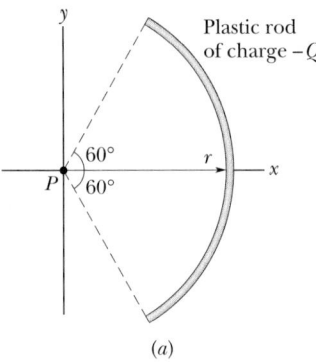

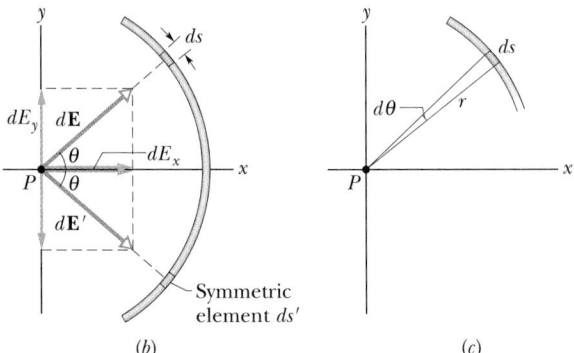

FIGURE 23-10 Sample Problem 23-5. (a) A plastic rod of charge $-Q$ is a circular section of radius r and central angle $120°$; point P is the center of curvature of the rod. (b) A differential element in the top half of the rod, at an angle θ to the x axis and of arc length ds, sets up a differential electric field $d\textbf{E}$ at P. An element ds', symmetric to ds about the x axis, sets up a field $d\textbf{E}'$ at P with the same magnitude. (c) Arc length ds makes an angle $d\theta$ about point P.

and its linear charge density must be

$$\lambda = \frac{\text{charge}}{\text{length}} = \frac{Q}{2\pi r/3} = \frac{0.477Q}{r}.$$

Substituting this into Eq. 23-21 and simplifying give us

$$E = \frac{(1.73)(0.477Q)}{4\pi\epsilon_0 r^2} = \frac{0.83Q}{4\pi\epsilon_0 r^2}. \qquad \text{(Answer)}$$

The direction of **E** is toward the rod, along the axis of symmetry of the charge distribution.

PROBLEM SOLVING TACTICS

TACTIC 1: *A Field Guide for Lines of Charge*

Here is a generic guide for finding the electric field **E** produced at a point P by a line of uniform charge, either circular or straight. The general strategy is to pick out an element dq of the charge, find $d\textbf{E}$ due to that element, and integrate $d\textbf{E}$ over the entire line of charge.

STEP 1. If the line of charge is circular, let ds be the arc length of an element of the distribution. If the line is straight, run an x axis along it and let dx be the length of an element. Mark the element on a sketch.

STEP 2. Relate the charge dq of the element to the length of the element with either $dq = \lambda \, ds$ or $dq = \lambda \, dx$. Consider dq and λ to be positive, even if the charge is actually negative. (The sign of the charge is used in the next step.)

STEP 3. Express the field $d\textbf{E}$ produced at P by dq with Eq. 23-3, replacing q in that equation with either $\lambda \, ds$ or $\lambda \, dx$. If the charge on the line is positive, then at P draw a vector $d\textbf{E}$ that points directly away from dq. If the charge is negative, draw the vector pointing directly toward dq.

STEP 4. Always look for any symmetry of the situation. If P is on an axis of symmetry of the charge distribution, resolve the field $d\textbf{E}$ produced by dq into components that are perpendicular and parallel to the axis of symmetry. Then consider a second element dq' that is located symmetrically to dq about the line of symmetry. At P draw the vector $d\textbf{E}'$ that this symmetrical element produces, and resolve it into components. One of the components produced by dq is a *canceling component*: it is canceled by the corresponding component produced by dq' and needs no further attention. The other component produced by dq is an *adding component*: it adds to the corresponding component produced by dq'. Add the adding components of all the elements via integration.

STEP 5. Here are four general types of uniform charge distributions, with strategies for simplifying the integral of step 4. Each type can be made more challenging by having the distribution consist of a line of positive charge and a line of negative charge.

Ring, with point P on (central) axis of symmetry, as in Fig. 23-9. In the expression for dE, replace r^2 with $z^2 + R^2$, as in

Eq. 23-12. Express the adding component of $d\mathbf{E}$ in terms of θ. That introduces $\cos \theta$, but θ is identical for all elements and thus is not a variable. Replace $\cos \theta$ as in Eq. 23-13. Integrate over s, around the circumference of the ring.

Circular arc, with point P at the center of curvature, as in Fig. 23-10. Express the adding component of $d\mathbf{E}$ in terms of θ. That introduces either $\sin \theta$ or $\cos \theta$. Reduce the resulting two variables s and θ to one, θ, by replacing ds with $r\, d\theta$. Integrate over θ, as in Sample Problem 23-5, from one end of the arc to the other end.

Straight line, with point P on an extension of the line, as in Fig. 23-11a. In the expression for dE, replace r with x. Integrate over x, from end to end of the line of charge.

Straight line, with point P at perpendicular distance y from the line of charge, as in Fig. 23-11b. In the expression for dE, replace r with an expression involving x and y. If P is on the perpendicular bisector of the line of charge, find an expression for the adding component of $d\mathbf{E}$. That will introduce either $\sin \theta$ or $\cos \theta$. Reduce the resulting two variables x and θ to one, x, by replacing the trigonometric function with an expression (its definition) involving x and y. Integrate over x from end to end of the line of charge. If P is not on a line of symmetry, as in Fig. 23-11c, set up an integral to sum the components dE_x, and integrate over x to find E_x. Also set up an integral to sum the components dE_y, and integrate over x again to find E_y. Use the components E_x and E_y in the usual way to find the magnitude E and the orientation of $\mathbf{E}$.

STEP 6. One arrangement of the integration limits gives a positive result. The reverse arrangement gives the same result with a minus sign; discard the minus sign. If the result is to be stated in terms of the total charge Q of the distribution, replace λ with Q/L, in which L is the length of the distribution. For a ring, L is the ring's circumference.

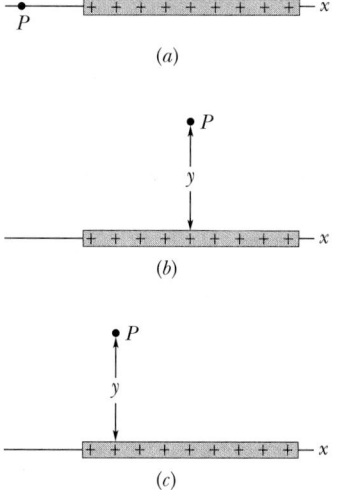

FIGURE 23-11 (a) Point P is on an extension of the line of charge. (b) P is on a line of symmetry of the line of charge, at perpendicular distance y from that line. (c) Same as (b) except that P is not on a line of symmetry.

CHECKPOINT **3:** The figure shows three nonconducting rods, one circular and two straight. Each has a uniform charge of magnitude Q along its top half and another along its bottom half. For each rod, what is the direction of the net electric field at point P?

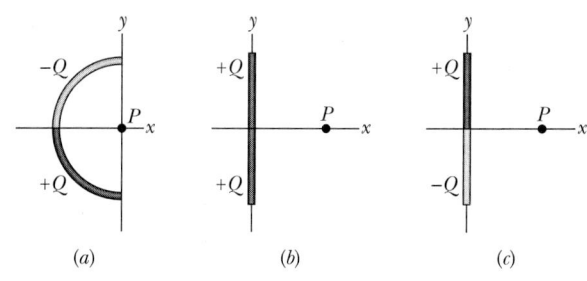

23-7 THE ELECTRIC FIELD DUE TO A CHARGED DISK

Figure 23-12 shows a circular plastic disk of radius R that has a positive surface charge of uniform density σ on its upper surface (see Table 23-2). What is the electric field at point P, a distance z from the disk along its central axis?

Our plan is to divide the disk into concentric flat rings and then to calculate the electric field at point P by adding up (that is, by integrating) the contributions of all the rings. Figure 23-12 shows one such ring, with radius r and radial width dr. Since σ is the charge per unit area, the charge on the ring is

$$dq = \sigma \, dA = \sigma(2\pi r \, dr), \tag{23-22}$$

where dA is the differential area of the ring.

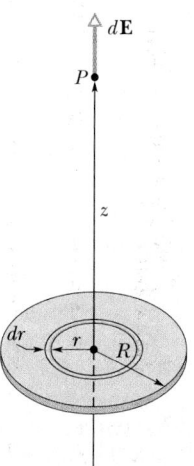

FIGURE 23-12 A disk of radius R and uniform positive charge. The ring shown has radius r and radial width dr. It sets up a differential electric field $d\mathbf{E}$ at point P on its central axis.

We have already solved the problem of the electric field due to a ring of charge. Substituting dq from Eq. 23-22 for q in Eq. 23-16, and replacing R in Eq. 23-16 with r, we obtain an expression for the electric field dE at P due to our flat ring:

$$dE = \frac{z\sigma 2\pi r\, dr}{4\pi\epsilon_0(z^2 + r^2)^{3/2}},$$

which we may write as

$$dE = \frac{\sigma z}{4\epsilon_0}\frac{2r\, dr}{(z^2 + r^2)^{3/2}}.$$

We can now find E by integrating over the surface of the disk, that is, by integrating with respect to the variable r from $r = 0$ to $r = R$. Note that z remains constant during this process. We get

$$E = \int dE = \frac{\sigma z}{4\epsilon_0}\int_0^R (z^2 + r^2)^{-3/2}(2r)\, dr. \quad (23\text{-}23)$$

To solve this integral, we cast it in the form $\int X^m\, dX$ by setting $X = (z^2 + r^2)$, $m = -\frac{3}{2}$, and $dX = (2r)\, dr$. For the recast integral we have

$$\int X^m\, dX = \frac{X^{m+1}}{m + 1},$$

so Eq. 23-23 becomes

$$E = \frac{\sigma z}{4\epsilon_0}\left[\frac{(z^2 + r^2)^{-1/2}}{-\frac{1}{2}}\right]_0^R.$$

Taking the limits and rearranging, we find

$$E = \frac{\sigma}{2\epsilon_0}\left(1 - \frac{z}{\sqrt{z^2 + R^2}}\right) \quad \begin{matrix}\text{(charged}\\ \text{disk)}\end{matrix} \quad (23\text{-}24)$$

as the magnitude of the electric field produced by a flat, circular, charged disk on its central axis. (In carrying out the integration, we assumed that $z \geq 0$.)

If we let $R \rightarrow \infty$ while keeping z finite, the second term in the parentheses in Eq. 23-24 approaches zero, and this equation reduces to

$$E = \frac{\sigma}{2\epsilon_0} \quad \text{(infinite sheet).} \quad (23\text{-}25)$$

This is the electric field produced by an infinite sheet of uniform charge located on one side of a nonconductor such as plastic. The electric field lines for such a situation are shown in Fig. 23-3.

We also get Eq. 23-25 if we let $z \rightarrow 0$ in Eq. 23-24 while keeping R finite. This shows that at points very close to the disk, the electric field set up by the disk is the same as if the disk were infinite in extent.

When the electric field surrounding a charged object (like this charged metal cap) becomes large enough, the air surrounding the object undergoes *electrical breakdown*: air molecules are ionized (electrons are removed from the molecules), and momentary conducting paths appear. The *electric sparks* you see here reveal those paths.

SAMPLE PROBLEM 23-6

The disk of Fig. 23-12 has a surface charge density σ of +5.3 μC/m^2 on its upper face. (This, incidentally, is a reasonable value for the surface charge density on the photosensitive cylinder of a photocopying machine.)

(a) What is the electric field at the surface of the disk?

SOLUTION: From Eq. 23-25 we have

$$E = \frac{\sigma}{2\epsilon_0} = \frac{5.3 \times 10^{-6}\ \text{C/m}^2}{(2)(8.85 \times 10^{-12}\ \text{C}^2/\text{N}\cdot\text{m}^2)}$$

$$= 3.0 \times 10^5\ \text{N/C}. \quad \text{(Answer)}$$

This value holds for all points that are close to the surface of the disk but not near its edge.

When the electric field in a material is large enough, the material undergoes *electrical breakdown* in which conducting paths suddenly appear in the material. Electrical breakdown occurs in air (at atmospheric pressure) when the electric field exceeds about 3×10^6 N/C. During breakdown, electrons flow along one or more conducting paths, creating *electric sparks*. Since the computed electric field in this sample problem is only 3×10^5 N/C, the charged disk will not cause sparks in the surrounding air.

(b) Using the binomial theorem, find an expression for the electric field at a point on the central axis far from the disk.

SOLUTION: The phrase *far from the disk* means that the distance z is much greater than the size of the disk, as measured by, say, its radius. As you will see, this allows us to use the binomial theorem to approximate the square-root term in Eq. 23-24.

From Appendix E, we write the general form of the binomial theorem as

$$(1 + x)^n = 1 + \frac{n}{1!}x + \frac{n(n-1)}{2!}x^2 + \cdots, \quad (23\text{-}26)$$

where $|x| \ll 1$. To get ready for the approximation, we rewrite the square-root term as

$$\frac{z}{\sqrt{z^2 + R^2}} = \frac{z}{z\sqrt{1 + \dfrac{R^2}{z^2}}} = \left(1 + \frac{R^2}{z^2}\right)^{-\frac{1}{2}},$$

which is in the proper form for use of the binomial theorem with $x = R^2/z^2$ and $n = -\frac{1}{2}$. Because z is much greater than R, the condition $|x| \ll 1$ is satisfied.

Applying Eq. 23-26, we now write

$$\left(1 + \frac{R^2}{z^2}\right)^{-\frac{1}{2}} = 1 + \frac{-\frac{1}{2}}{1!}\frac{R^2}{z^2} + \frac{-\frac{1}{2}(-\frac{1}{2} - 1)}{2!}\frac{R^4}{z^4} + \cdots.$$

Successive terms on the right side are progressively less. We can approximate the required result closely enough by discarding terms smaller than R^2/z^2, which leaves us

$$\frac{z}{\sqrt{z^2 + R^2}} = 1 - \frac{R^2}{2z^2}.$$

Substituting this expression in Eq. 23-24 gives us

$$E = \frac{\sigma}{2\epsilon_0}\left[1 - \left(1 - \frac{R^2}{2z^2}\right)\right]$$

$$= \frac{\sigma}{4\epsilon_0}\frac{R^2}{z^2}. \quad \text{(Answer)}$$

We can rewrite this in terms of the charge q on the upper face of the disk by noting that $\sigma = q/A$ and, for the disk, $A = \pi R^2$. Then

$$E = \frac{\sigma}{4\epsilon_0}\frac{R^2}{z^2} = \frac{q}{4\epsilon_0 \pi R^2}\frac{R^2}{z^2}$$

$$= \frac{1}{4\pi\epsilon_0}\frac{q}{z^2}. \quad \text{(Answer)} \quad (23\text{-}27)$$

Equation 23-27 tells us that at points on the central axis where $z \gg R$, the electric field produced by the charge q spread over the face of the disk is the same as that produced by a particle of the same charge q.

23-8 A POINT CHARGE IN AN ELECTRIC FIELD

In the preceding four sections we worked at the first of our two tasks: given a charge distribution, to find the electric field it produces in the surrounding space. Here we begin the second task: to determine what happens to a charged particle that is in an electric field that is produced by other stationary or slowly moving charges.

What happens is that an electrostatic force acts on the particle. This force, a vector quantity, is given by

$$\mathbf{F} = q\mathbf{E}, \quad (23\text{-}28)$$

in which q is the charge of the particle (including its sign)

In an electrostatic precipitator, an electric field exerts a force on charged ash as it ascends a stack, so that much of it is collected in the stack, hence does not enter and pollute the atmosphere. The precipitator is operating in the photograph at the left but not in the photograph at the right.

and **E** is the electric field that other charges have produced at the location of the particle. (The field is *not* the field set up by the particle itself; to distinguish the two fields, the field acting on the particle in Eq. 23-28 is often called the *external field*. A charged particle (or object) is not affected by its own electric field.) Equation 23-28 tells us:

The electrostatic force **F** acting on a charged particle located in an external electric field **E** points in the direction of **E** if the charge q of the particle is positive and in the opposite direction if q is negative.

CHECKPOINT **4:** (a) In the figure, what is the direction of the electrostatic force on the electron due to the electric field shown? (b) In which direction will the electron accelerate if it is moving parallel to the y axis before it encounters the electric field? (c) If, instead, the electron is initially moving rightward, will its speed increase, decrease, or remain constant?

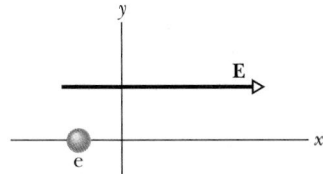

Measuring the Elementary Charge

Equation 23-28 played a role in the measurement of the elementary charge e by American physicist Robert A. Millikan in 1910–1913. Figure 23-13 is a representation of his apparatus. When tiny oil drops are sprayed into chamber A, some of them become charged, either positively or negatively, in the process. Consider a drop that drifts downward through the small hole in plate P_1 and into chamber C. Let us assume that this drop has a negative charge q.

If switch S in Fig. 23-13 is open as shown, battery B has no electrical effect on chamber C. If the switch is closed (the connection between chamber C and the positive terminal of the battery is then complete), the battery causes an excess positive charge on conducting plate P_1 and an excess negative charge on conducting plate P_2. The charged plates set up a downward-pointing electric field **E** in chamber C. According to Eq. 23-28, this field exerts an electrostatic force on any charged drop that happens to be in the chamber and affects its motion. In particular, our negatively charged drop will tend to drift upward.

By timing the motion of oil drops with the switch opened and closed and thus determining the effect of the charge q, Millikan discovered that the values of q were

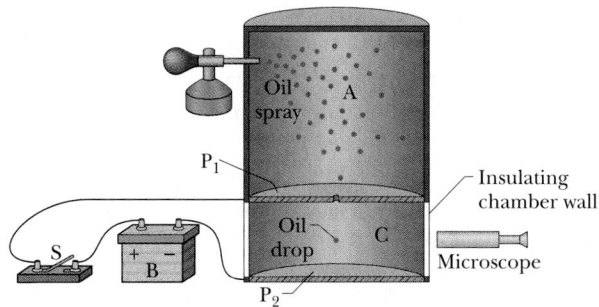

FIGURE 23-13 The Millikan oil-drop apparatus for measuring the elementary charge e. When a charged oil drop drifted into chamber C through the hole in plate P_1, its motion could be controlled by closing and opening switch S and thereby setting up or eliminating an electric field in chamber C. The microscope was used to view the drop, to permit timing of its motion.

always given by

$$q = ne, \qquad n = 0, \pm 1, \pm 2, \pm 3, \cdots, \quad (23\text{-}29)$$

in which e turned out to be the fundamental constant we call the *elementary charge,* 1.60×10^{-19} C. Millikan's experiment is convincing proof that charge is quantized, and he earned the 1923 Nobel Prize in physics in part for this work. Modern measurements of the elementary charge rely on a variety of interlocking experiments, all more precise than the pioneering experiment of Millikan.

Ink-Jet Printing

The need for high-quality, high-speed printing has caused a search for an alternative to impact printing, such as occurs in a standard typewriter. Building up letters by squirting tiny drops of ink at the paper is one such alternative.

Figure 23-14 shows a negatively charged drop moving between two conducting deflecting plates, between which

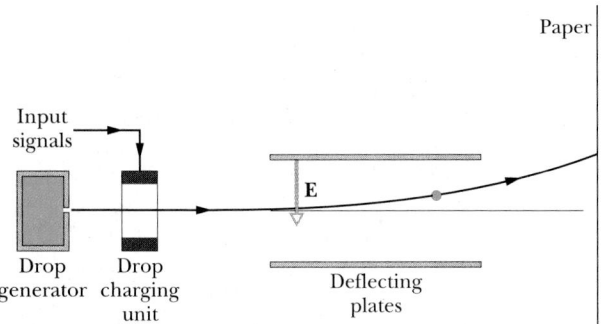

FIGURE 23-14 The essential features of an ink-jet printer. An input signal from a computer controls the charge given to each drop and thus the position on the paper at which the drop lands. About 100 tiny drops are needed to form a single character.

a uniform, downward-pointing electric field **E** has been set up. The drop is deflected upward according to Eq. 23-28 and then strikes the paper at a position that is determined by the magnitudes of **E** and the charge q of the drop.

In practice, E is held constant and the position of the drop is determined by the charge q delivered to the drop in the charging unit, through which the drop must pass before entering the deflecting system. The charging unit, in turn, is activated by electronic signals that encode the material to be printed.

SAMPLE PROBLEM 23-7

In the Millikan oil-drop apparatus of Fig. 23-13, a drop of radius $R = 2.76$ μm has an excess charge of three electrons. What are the magnitude and direction of the electric field that is required to balance the drop so it remains stationary in the apparatus? The density ρ of the oil is 920 kg/m³.

SOLUTION: To balance the drop, the electrostatic force acting on it must be upward and have a magnitude equal to the weight mg of the drop. From Eqs. 23-28 and 23-29, we can write the *magnitude* of the electrostatic force as $F = (3e)E$. We can also write the mass of the drop as the product of its volume and its density. Thus the balance of forces gives us

$$\tfrac{4}{3}\pi R^3 \rho g = (3e)E.$$

Solving for E yields

$$E = \frac{4\pi R^3 \rho g}{9e}$$

$$= \frac{(4\pi)(2.76 \times 10^{-6}\ \text{m})^3(920\ \text{kg/m}^3)(9.80\ \text{m/s}^2)}{(9)(1.60 \times 10^{-19}\ \text{C})}$$

$$= 1.65 \times 10^6\ \text{N/C}. \qquad \text{(Answer)}$$

Because the drop is negatively charged, Eq. 23-28 tells us that **E** and **F** are in opposite directions: **F** = −3e**E**. So the electric field must point downward.

SAMPLE PROBLEM 23-8

Figure 23-15 shows the deflecting plates of an ink-jet printer, with superimposed coordinate axes. An ink drop with a mass m of 1.3×10^{-10} kg and a negative charge of magnitude $Q = 1.5 \times 10^{-13}$ C enters the region between the plates, initially moving along the x axis with speed $v_x = 18$ m/s. The length L of the plates is 1.6 cm. The plates are charged and thus produce an electric field at all points between them. Assume that the downward-pointing field **E** is uniform and has a magnitude of 1.4×10^6 N/C. What is the vertical deflection of the drop at the far edge of the plates? (The weight of the drop is small relative to the electrostatic force acting on the drop and can be neglected.)

SOLUTION: Since the drop is negatively charged and the electric field is downward, Eq. 23-28 tells us that a constant electrostatic force of magnitude QE acts *upward* on the charged drop. Thus as the drop travels parallel to the x axis at constant speed v_x, it accelerates upward with constant acceleration a_y. Applying Newton's second law $F = ma$ along the y axis, we find that

$$a_y = \frac{F}{m} = \frac{QE}{m}. \qquad (23\text{-}30)$$

Let t represent the time required for the drop to pass through the region between the plates. During t the vertical and horizontal displacements of the drop are

$$y = \tfrac{1}{2}a_y t^2 \quad \text{and} \quad L = v_x t, \qquad (23\text{-}31)$$

respectively. Eliminating t between these two equations and substituting Eq. 23-30 for a_y, we find

$$y = \frac{QEL^2}{2mv_x^2}$$

$$= \frac{(1.5 \times 10^{-13}\ \text{C})(1.4 \times 10^6\ \text{N/C})(1.6 \times 10^{-2}\ \text{m})^2}{(2)(1.3 \times 10^{-10}\ \text{kg})(18\ \text{m/s})^2}$$

$$= 6.4 \times 10^{-4}\ \text{m} = 0.64\ \text{mm}. \qquad \text{(Answer)}$$

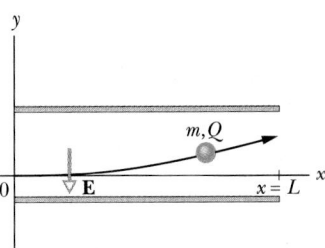

FIGURE 23-15 Sample Problem 23-8. An ink drop of mass m and charge magnitude Q is deflected in the electric field of an ink-jet printer.

23-9 A DIPOLE IN AN ELECTRIC FIELD

We have defined the electric dipole moment **p** of an electric dipole to be a vector that points from the negative to the positive end of the dipole. As you will see, the behavior of a dipole in a uniform external electric field **E** can be described completely in terms of the two vectors **E** and **p**, with no need of any details about the dipole's structure.

As was noted in Sample Problem 23-4, a molecule of water (H_2O) is an electric dipole. Figure 23-16 shows why. There the black dots represent the oxygen nucleus (having eight protons) and the two hydrogen nuclei (having one proton each). The colored enclosed areas represent the region in which the electrons orbit the nuclei.

In a water molecule, the two hydrogen atoms and the oxygen atom do not lie on a straight line but form an angle

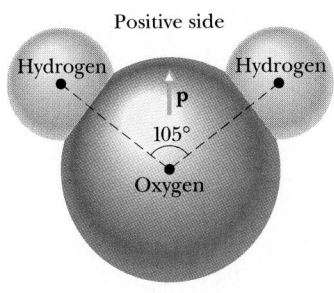

FIGURE 23-16 A molecule of H_2O, showing the three nuclei (represented by dots) and the regions in which the electrons orbit the nuclei. The electric dipole moment **p** points from the (negative) oxygen side to the (positive) hydrogen side of the molecule.

of about 105°, as shown in Fig. 23-16. As a result, the molecule has a definite "oxygen side" and "hydrogen side." Moreover, the 10 electrons of the molecule tend to remain closer to the oxygen nucleus than to the hydrogen nuclei. This makes the oxygen side of the molecule slightly more negative than the hydrogen side and creates an electric dipole moment **p** that points along the symmetry axis of the molecule as shown. If the water molecule is placed in an external electric field, it behaves as would be expected of the more abstract electric dipole of Fig. 23-8.

To examine this behavior, we now consider such an abstract dipole in a uniform external electric field **E**, as shown in Fig. 23-17a. We assume that the dipole is a rigid structure (due to internal electrostatic forces) that consists

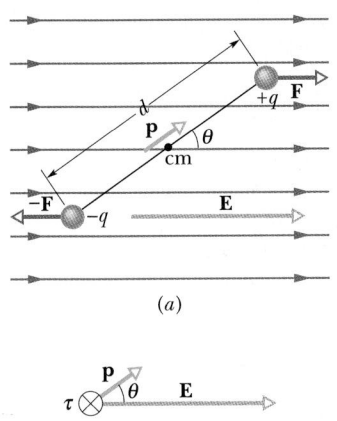

(a)

(b)

FIGURE 23-17 (a) An electric dipole in a uniform electric field **E**. Two centers of equal but opposite charge are separated by distance d. Their center of mass cm is assumed to be midway between them. The bar between them represents their rigid connection. (b) Field **E** causes a torque τ on dipole moment **p**. The direction of the torque vector τ is into the plane of the page, as represented by the symbol $\otimes$.

of two centers of opposite charge, each of magnitude q, separated by a distance d. The dipole moment **p** makes an angle θ with **E**.

At the charged ends of the dipole, electrostatic forces, **F** and $-$**F**, act in opposite directions and with the same magnitude $F = qE$. Thus the net force exerted on the dipole by the field is zero. However, these forces exert a net torque τ on the dipole about its center of mass, which we can take to be midway along the line connecting the charged ends. From Eq. 11-30, with $r = d/2$, we can write the magnitude of this net torque τ as

$$\tau = F\frac{d}{2}\sin\theta + F\frac{d}{2}\sin\theta = Fd\sin\theta. \quad (23\text{-}32)$$

We can also write the magnitude of τ in terms of the magnitudes of the electric field E and the dipole moment qd. To do so, we substitute qE for F and p/q for d in Eq. 23-32, finding that the magnitude of τ is

$$\tau = pE\sin\theta. \quad (23\text{-}33)$$

We can generalize this equation to vector form as

$$\boldsymbol{\tau} = \mathbf{p} \times \mathbf{E} \quad \text{(torque on a dipole).} \quad (23\text{-}34)$$

Vectors **p** and **E** are shown in Fig. 23-17b. The torque acting on a dipole tends to rotate **p** (hence the dipole) into the direction of **E**, thereby reducing θ. In Fig. 23-17, such rotation is clockwise. As we discussed in Chapter 11, we can represent a torque that gives rise to a clockwise rotation by including a minus sign with the magnitude of the torque. With such notation, the torque of Fig. 23-17 is

$$\tau = -pE\sin\theta. \quad (23\text{-}35)$$

Potential Energy of an Electric Dipole

Potential energy can be associated with the orientation of an electric dipole in an electric field. The dipole has its least potential energy when it is in its equilibrium orientation, which is when its moment **p** is lined up with the field **E** (then $\boldsymbol{\tau} = \mathbf{p} \times \mathbf{E} = 0$). It has greater potential energy in all other orientations. Thus the dipole is like a pendulum, which has *its* least gravitational potential energy in *its* equilibrium orientation—at its lowest point. To rotate the dipole or the pendulum to any other orientation requires work by some external agent.

In any situation involving potential energy, we are free to define the zero-potential-energy configuration in a perfectly arbitrary way, because only differences in potential energy have physical meaning. It turns out that the expression for the potential energy of an electric dipole in an external electric field is simplest if we choose the potential energy to be zero when the angle θ in Fig. 23-17 is 90°. We then can find the potential energy U of the dipole at any other value of θ with Eq. 8-1 ($\Delta U = -W$) by calculating the work W done by the field on the dipole when the dipole is rotated to that value of θ from 90°. With the aid of Eq. 11-44 ($W = \int \tau \, d\theta$) and Eq. 23-35, we find that the potential energy U at any angle θ is

$$U = -W = -\int_{90°}^{\theta} \tau \, d\theta$$

$$= \int_{90°}^{\theta} pE \sin \theta \, d\theta. \qquad (23\text{-}36)$$

Evaluating the integral leads to

$$U = -pE \cos \theta. \qquad (23\text{-}37)$$

We can generalize this equation to vector form as

$$U = -\mathbf{p} \cdot \mathbf{E} \qquad \begin{array}{l}\text{(potential energy} \\ \text{of a dipole).}\end{array} \qquad (23\text{-}38)$$

Equations 23-37 and 23-38 show us that the potential energy of the dipole is least ($U = -pE$) when $\theta = 0$, which is when $\mathbf{p}$ and $\mathbf{E}$ are in the same direction; the potential energy is greatest ($U = pE$) when $\theta = 180°$, which is when $\mathbf{p}$ and $\mathbf{E}$ are in opposite directions.

CHECKPOINT **5:** The figure shows four orientations of an electric dipole in an external electric field. Rank the orientations according to (a) the magnitude of the torque on the dipole and (b) the potential energy of the dipole, greatest first.

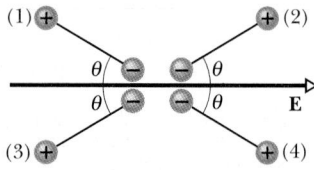

Microwave Cooking

In liquid water, where molecules are relatively free to move around, the electric field produced by each molecular dipole affects the surrounding dipoles. As a result, the molecules bond together in groups of two or three, because the negative (oxygen) end of one dipole and a positive (hydrogen) end of another dipole attract each other. Each time a group is formed, electric potential energy is transferred to the random thermal motion of the group and the surrounding molecules. And each time collisions among the molecules break up a group, the transfer is reversed. The temperature of the water (which is associated with the average thermal motion) does not change because, on the average, the net transfer of energy is zero.

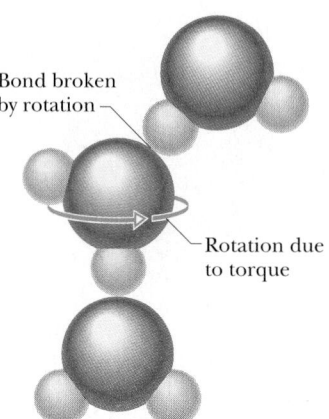

FIGURE 23-18 A group of three water molecules. A torque due to an oscillating electric field in a microwave oven breaks one of the bonds between the molecules and thus breaks up the group.

In a microwave oven, the story differs. When the oven is operated, the microwaves produce (in the oven) an electric field that rapidly oscillates back and forth in direction. If there is water in the oven, the oscillating field exerts oscillating torques on the water molecules, continually rotating them back and forth to align their dipole moments with the field direction. Molecules that are bonded as a pair can twist around their common bond to stay aligned, but molecules that are bonded in a group of three must break at least one of their two bonds (Fig. 23-18).

The energy to break these bonds comes from the electric field, that is, from the microwaves. Then molecules that have broken away from groups can form new groups, transferring the energy they just gained into thermal energy. Thus thermal energy is added to the water when the groups form but is not removed when the groups break apart, and the temperature of the water increases. Foods that contain water can be cooked in a microwave oven because of the heating of that water. If a water molecule were not an electric dipole, this would not be so and microwave ovens would be useless.

SAMPLE PROBLEM 23-9

A neutral water molecule (H_2O) in its vapor state has an electric dipole moment of 6.2×10^{-30} C·m.

(a) How far apart are the molecule's centers of positive and negative charge?

SOLUTION: There are 10 electrons and 10 protons in this molecule. So the magnitude of the dipole moment is

$$p = qd = (10e)(d),$$

in which d is the separation we are seeking and e is the elementary charge. Thus

$$d = \frac{p}{10e} = \frac{6.2 \times 10^{-30} \text{ C} \cdot \text{m}}{(10)(1.60 \times 10^{-19} \text{ C})}$$

$$= 3.9 \times 10^{-12} \text{ m} = 3.9 \text{ pm}. \qquad \text{(Answer)}$$

This distance is not only small, but it is actually smaller than the radius of a hydrogen atom.

(b) If the molecule is placed in an electric field of 1.5×10^4 N/C, what maximum torque can the field exert on it? (Such a field can easily be set up in the laboratory.)

SOLUTION: From Eq. 23-33 we know that the torque is a maximum when $\theta = 90°$. Substituting this value in that equation yields

$$\tau = pE \sin \theta$$

$$= (6.2 \times 10^{-30} \text{ C} \cdot \text{m})(1.5 \times 10^4 \text{ N/C})(\sin 90°)$$

$$= 9.3 \times 10^{-26} \text{ N} \cdot \text{m}. \qquad \text{(Answer)}$$

(c) How much work must an external agent do to turn this molecule end for end in this field, starting from its fully aligned position, for which $\theta = 0$?

SOLUTION: The work is the difference in potential energy between the positions $\theta = 180°$ and $\theta = 0$. Using Eq. 23-37, we get

$$W = U(180°) - U(0)$$

$$= (-pE \cos 180°) - (-pE \cos 0)$$

$$= 2pE = (2)(6.2 \times 10^{-30} \text{ C} \cdot \text{m})(1.5 \times 10^4 \text{ N/C})$$

$$= 1.9 \times 10^{-25} \text{ J}. \qquad \text{(Answer)}$$

REVIEW & SUMMARY

Electric Field

One way to explain the electrostatic force between charges is to assume that each charge sets up an electric field in the space around it. The electrostatic force acting on any one charge is then due to the electric field set up at its location by the other charges.

Definition of Electric Field

The *electric field* **E** at any point is defined in terms of the electrostatic force **F** that would be exerted on a positive test charge q_0 placed there:

$$\mathbf{E} = \frac{\mathbf{F}}{q_0}. \qquad (23\text{-}1)$$

Electric Field Lines

Electric field lines provide a means for visualizing the direction and magnitude of electric fields. The electric field vector at any point is tangent to a field line through that point. The density of field lines in any region is proportional to the magnitude of the electric field in that region. Field lines originate on positive charges and terminate on negative charges.

Field Due to a Point Charge

The magnitude of the electric field **E** set up by a point charge q at a distance r from the charge is

$$E = \frac{1}{4\pi\epsilon_0} \frac{|q|}{r^2}. \qquad (23\text{-}3)$$

The direction of **E** is away from the point charge if the charge is positive and toward the point charge if the charge is negative.

Field Due to an Electric Dipole

An *electric dipole* consists of two particles with charges of equal magnitude q but opposite sign, separated by a small distance d. Their **dipole moment p** has magnitude qd and points from the

negative charge to the positive charge. The magnitude of the electric field set up by the dipole at a distant point on the dipole axis (which runs through both charges) is

$$E = \frac{1}{2\pi\epsilon_0} \frac{p}{z^3}, \qquad (23\text{-}9)$$

where z is the distance between the point and the dipole center.

Field Due to a Continuous Charge Distribution

The electric field due to a *continuous charge distribution* is found by treating charge elements as point charges and then summing, via integration, the electric field vectors produced by all the charge elements.

Force on a Point Charge in an Electric Field

When a point charge q is placed in an electric field **E** set up by other charges, the electrostatic force **F** that acts on the point charge is

$$\mathbf{F} = q\mathbf{E}. \qquad (23\text{-}28)$$

Force **F** points in the direction of **E** if q is positive and opposite **E** if q is negative.

Dipole in an Electric Field

When an electric dipole of dipole moment **p** is placed in an electric field **E**, the field exerts a torque $\boldsymbol{\tau}$ on the dipole:

$$\boldsymbol{\tau} = \mathbf{p} \times \mathbf{E}. \qquad (23\text{-}34)$$

The dipole has a potential energy U associated with its orientation in the field:

$$U = -\mathbf{p} \cdot \mathbf{E}. \qquad (23\text{-}38)$$

This potential energy is defined to be zero when **p** is perpendicular to **E**; it is least ($U = -pE$) when **p** is aligned with **E**, and most ($U = pE$) when **p** is directed opposite **E**.

QUESTIONS

1. Figure 23-19 shows three electric field lines. What is the direction of the electrostatic force on a positive test charge placed at (a) point A and (b) point B? (c) At which point, A or B, will the acceleration of the test charge be greater if the charge is released?

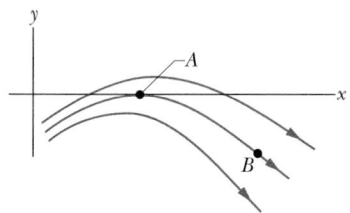

FIGURE 23-19 Question 1.

2. Figure 23-20a shows two charged particles on an axis. (a) Where on the axis (other than at an infinite distance) is there a point at which their net electric field is zero: between the charges, to their left, or to their right? (b) Is there a point of zero electric field off the axis (other than at an infinite distance)?

FIGURE 23-20 Questions 2 and 3.

3. Figure 23-20b shows two protons and an electron that are evenly spaced on an axis. Where on the axis (other than at an infinite distance) is there a point at which their net electric field is zero: to the left of the particles, to their right, between the two protons, or between the electron and the nearer proton?

4. Figure 23-21 shows two square arrays of charged particles. The squares, which are centered on point P, are misaligned. The particles are separated by either d or $d/2$ along the perimeters of the squares. What are the magnitude and direction of the net electric field at P?

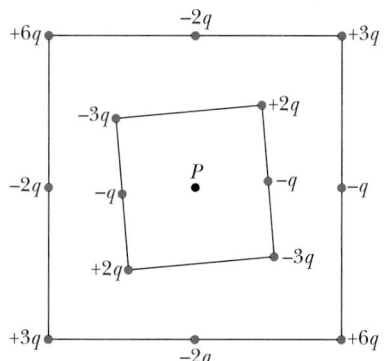

FIGURE 23-21 Question 4.

5. In Fig. 23-22, two particles of charge $-q$ are arranged symmetrically about the y axis; each produces an electric field at point P on that axis. (a) Are the magnitudes of the fields at P equal? (b) Does each electric field point toward or away from the charge producing it? (c) Is the magnitude of the net electric field at P

equal to the sum of the magnitudes of the two field vectors (is it equal to $2E$)? (d) Do the x components of those two field vectors add or cancel? (e) Do their y components add or cancel? (f) Is the direction of the net field at P that of the canceling components or the adding components? (g) What is the direction of the net field?

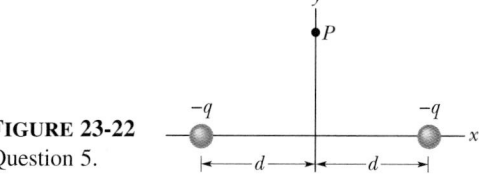

FIGURE 23-22 Question 5.

6. Three circular nonconducting rods of the same radius of curvature have uniform charges. Rod A has charge $+2Q$ and subtends an arc of 30°; rod B has charge $+6Q$ and subtends 90°; rod C has charge $+4Q$ and subtends 60°. Rank the rods according to their linear charge density, greatest first.

7. In Fig. 23-23, two identical circular nonconducting rings are centered on the same line. For three situations, the uniform charges on rings A and B are, respectively, (1) q_0 and q_0, (2) $-q_0$ and $-q_0$, (3) $-q_0$ and q_0. Rank the situations according the magnitude of the net electric field at (a) point P_1 midway between the rings, (b) point P_2 at the center of ring 2, (c) point P_3 to the right of ring 2, greatest first.

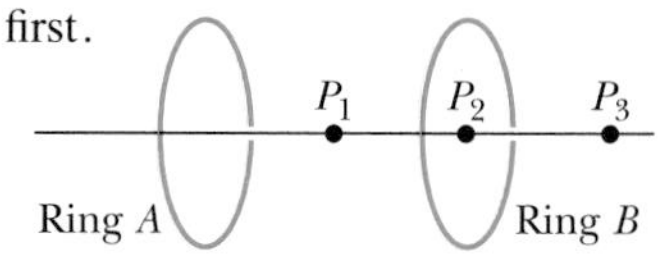

FIGURE 23-23 Question 7.

8. In Fig. 23-24a, a circular plastic rod with uniform charge $+Q$ produces an electric field of magnitude E at the center of curvature (at the origin). In Figs. 23-24b, c, and d, more circular rods with identical uniform charges $+Q$ are added until the circle is complete. A fifth arrangement (which would be labeled e) is like that in d except that the rod in the fourth quadrant has charge $-Q$.

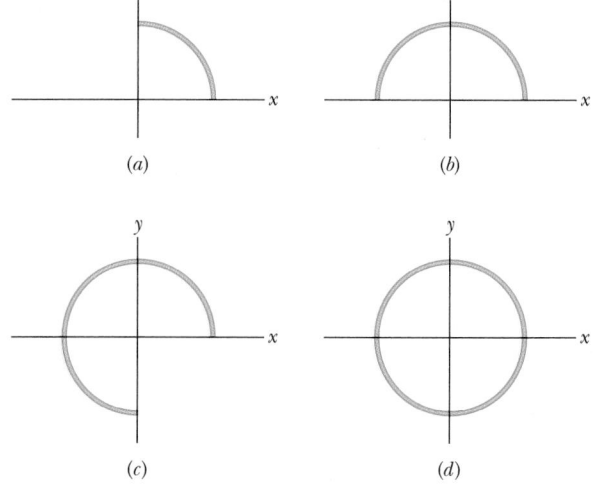

FIGURE 23-24 Question 8.

Rank the five arrangements according to the magnitude of the electric field at the center of curvature, greatest first.

9. In Fig. 23-25, an electron e travels through a small hole in plate A and then toward plate B. A uniform electric field in the region between the plates then slows the electron without deflecting it. (a) What is the direction of the field? (b) Four other particles similarly travel through small holes in either plate A or plate B and then into the region between the plates. Three have charges $+q_1$, $+q_2$, and $-q_3$. The fourth (labeled n) is a neutron, which is electrically neutral. Does the speed of each of those four other particles increase, decrease, or remain the same in the region between the plates?

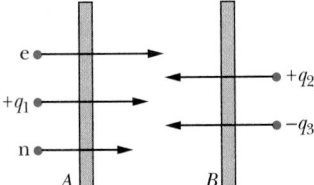

FIGURE 23-25 Question 9.

10. Figure 23-26 shows the path of negatively charged particle 1 through a rectangular region of uniform electric field; the particle is deflected toward the top of the page. (a) Is the field directed leftward, rightward, toward the top of the page, or toward the bottom? (b) Three other charged particles are shown approaching the region of electric field. Which are deflected toward the top of the page and which toward the bottom?

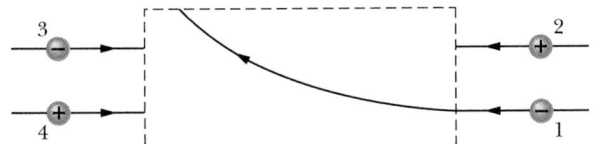

FIGURE 23-26 Question 10.

11. Figure 23-27 shows three arrangements of electric field lines. In each arrangement, a proton is released from rest at point A and is then accelerated through point B by the electric field. Points A and B have equal separations in the three arrangements. Rank the arrangements according to the linear momentum of the proton when it reaches point B, greatest first.

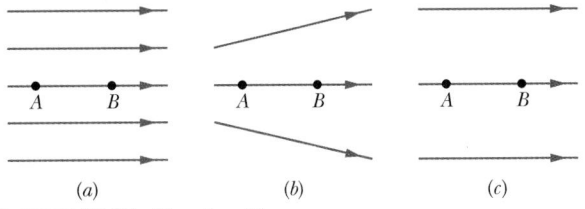

FIGURE 23-27 Question 11.

12. (a) In Checkpoint 5, if the dipole rotates from orientation 1 to orientation 2, is the work done on the dipole by the field positive, negative, or zero? (b) If, instead, the dipole rotates from orientation 1 to orientation 4, is the work done by the field more than, less than, or the same as in (a)?

13. The potential energies associated with four orientations of an electric dipole in an electric field are (1) $-5U_0$, (2) $-7U_0$, (3) $3U_0$, and (4) $5U_0$, where U_0 is positive. Rank the orientations according to (a) the angle between the electric dipole moment **p** and the electric field **E**, and (b) the magnitude of the torque on the electric dipole, greatest first.

14. If you walk across some types of carpet on a dry day and then reach for a metal doorknob or (for more fun) the back of someone's neck, you might produce a spark. Why does the spark occur? (You can increase the brightness and noise of the spark if you reach with a pointed finger or, even better, a metal key with the pointed end forward.)

EXERCISES & PROBLEMS

SECTION 23-3 Electric Field Lines

1E. In Fig. 23-28 the electric field lines on the left have twice the separation as those on the right. (a) If the magnitude of the field at A is 40 N/C, what force acts on a proton at A? (b) What is the magnitude of the field at B?

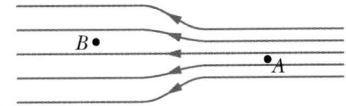

FIGURE 23-28 Exercise 1.

2E. Sketch qualitatively the electric field lines for two nearby point charges $+q$ and $-2q$.

3E. In Fig. 23-29, three point charges are arranged in an equilateral triangle. Sketch the field lines due to $+Q$ and $-Q$, and from them determine the direction of the force that acts on $+q$ because of the presence of the other two charges. (*Hint:* See Fig. 23-5.)

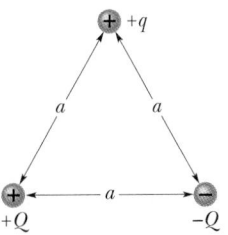

FIGURE 23-29 Exercise 3.

4E. Sketch qualitatively the electric field lines both between and outside two concentric conducting spherical shells when a uniform positive charge q_1 is on the inner shell and a uniform negative charge $-q_2$ is on the outer. Consider the cases $q_1 > q_2$, $q_1 = q_2$, and $q_1 < q_2$.

5E. Sketch qualitatively the electric field lines for a thin, circular, uniformly charged disk of radius R. (*Hint:* Consider as limiting cases points very close to the disk, where the electric field is perpendicular to the surface, and points very far from it, where the electric field is like that of a point charge.)

SECTION 23-4 The Electric Field Due to a Point Charge

6E. What is the magnitude of a point charge that would create an electric field of 1.00 N/C at points 1.00 m away?

7E. What is the magnitude of a point charge whose electric field 50 cm away has the magnitude 2.0 N/C?

8E. Two opposite charges of equal magnitude 2.0×10^{-7} C are held 15 cm apart. What are the magnitude and direction of **E** at the point midway between the charges?

9E. An atom of plutonium-239 has a nuclear radius of 6.64 fm and the atomic number $Z = 94$. Assuming that the positive charge of the nucleus is distributed uniformly, what are the magnitude and direction of the electric field at the surface of the nucleus due to the positive charge?

10P. A particle of charge $-q_1$ is located at the origin of an x axis. (a) At what location should a second particle of charge $-4q_1$ be placed so that the net electric field of the two particles is zero at $x = 2.0$ mm? (b) If, instead, a particle of charge $+4q_1$ is placed at that location, what is the direction of the net electric field at $x = 2.0$ mm?

11P. In Fig. 23-30, two point charges $q_1 = +1.0 \times 10^{-6}$ C and $q_2 = +3.0 \times 10^{-6}$ C are separated by a distance $d = 10$ cm. Plot their net electric field $E(x)$ as a function of x for both positive and negative values of x, taking E to be positive when the vector **E** points to the right and negative when **E** points to the left.

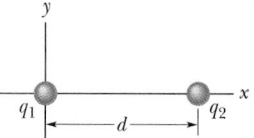

FIGURE 23-30
Problems 11 and 12.

12P. (a) In Fig. 23-30, two point charges $q_1 = -5q$ and $q_2 = +2q$ are separated by distance d. Locate the point (or points) at which the electric field due to the two charges is zero. (b) Sketch the electric field lines qualitatively.

13P. In Fig. 23-31, point charges $+1.0q$ and $-2.0q$ are fixed a distance d apart. (a) Find **E** at points A, B, and C. (b) Sketch the electric field lines.

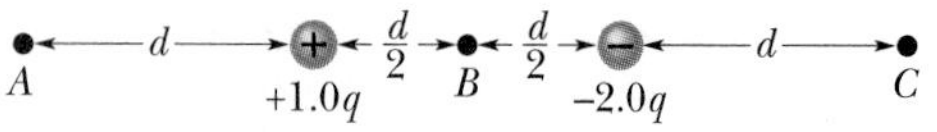

FIGURE 23-31 Problem 13.

14P. Two charges $q_1 = 2.1 \times 10^{-8}$ C and $q_2 = -4.0q_1$ are placed 50 cm apart. Find the point along the straight line passing through the two charges at which the electric field is zero.

15P. In Fig. 23-32, what is the electric field at point P due to the four point charges shown?

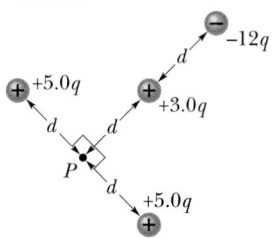

FIGURE 23-32 Problem 15.

16P. A proton and an electron form two corners of an equilateral triangle of side length 2.0×10^{-6} m. What is the magnitude of their net electric field at the third corner?

17P. A clock face has negative point charges $-q$, $-2q$, $-3q$, $\cdots$, $-12q$ fixed at the positions of the corresponding numerals. The clock hands do not perturb the net field due to the point charges. At what time does the hour hand point in the same direction as the electric field vector at the center of the dial? (*Hint:* Use symmetry.)

18P. An electron is placed at each corner of an equilateral triangle having sides 20 cm long. What is the magnitude of the electric field at the midpoint of one of the sides?

19P. Calculate the direction and magnitude of the electric field at point P in Fig. 23-33.

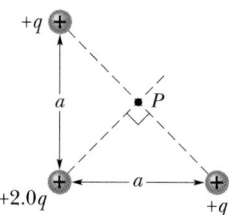

FIGURE 23-33 Problem 19.

20P. In Fig. 23-34, charges are placed at the vertices of an equilateral triangle. For what value of Q (both sign and magnitude) does the total electric field vanish at C, the center of the triangle?

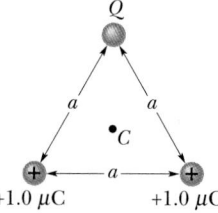

FIGURE 23-34 Problem 20.

21P. In Fig. 23-35, four charges form the corners of a square and four more charges lie at the midpoints of the sides of the square. The distance between adjacent charges on the perimeter of the square is d. What are the magnitude and direction of the electric field at the center of the square?

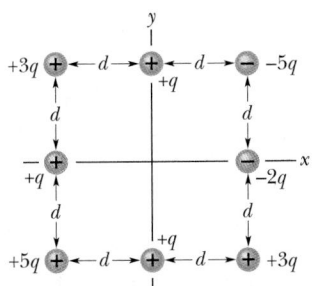

FIGURE 23-35 Problem 21.

22P. What are the magnitude and direction of the electric field at the center of the square of Fig. 23-36 if $q = 1.0 \times 10^{-8}$ C and $a = 5.0$ cm?

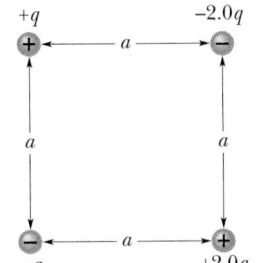

FIGURE 23-36
Problem 22.

SECTION 23-5 The Electric Field Due to an Electric Dipole

23E. Calculate the electric dipole moment of an electron and a proton 4.30 nm apart.

24E. In Fig. 23-8, let both charges be positive. Assuming $z \gg d$, show that E at point P in that figure is then given by

$$E = \frac{1}{4\pi\epsilon_0} \frac{2q}{z^2}.$$

25P. Find the magnitude and direction of the electric field at point P due to the electric dipole in Fig. 23-37. P is located at a distance $r \gg d$ along the perpendicular bisector of the line joining the charges. Express your answer in terms of the magnitude and direction of the electric dipole moment **p**.

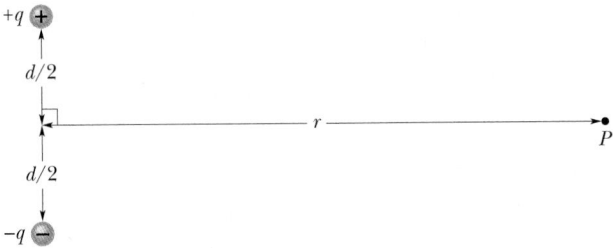

FIGURE 23-37 Problem 25.

26P*. *Electric quadrupole.* Figure 23-38 shows an electric quadrupole. It consists of two dipoles with dipole moments that are equal in magnitude but opposite in direction. Show that the

value of E on the axis of the quadrupole for points a distance z from its center (assume $z \gg d$) is given by

$$E = \frac{3Q}{4\pi\epsilon_0 z^4},$$

in which $Q \;(= 2qd^2)$ is known as the *quadrupole moment* of the charge distribution.

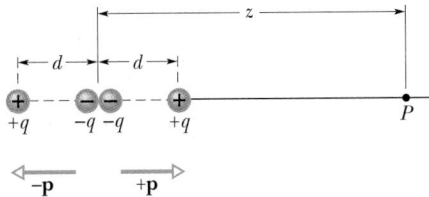

FIGURE 23-38 Problem 26.

SECTION 23-6 The Electric Field Due to a Line of Charge

27E. Make a quantitative plot of the electric field along the central axis of a charged ring having a diameter of 6.0 cm and a uniformly distributed charge of 1.0×10^{-8} C.

28E. Figure 23-39 shows two parallel nonconducting rings arranged with their central axes along a common line. Ring 1 has uniform charge q_1 and radius R; ring 2 has uniform charge q_2 and the same radius R. The rings are separated by a distance $3R$. The net electric field at point P on the common line, at distance R from ring 1, is zero. What is the ratio q_1/q_2?

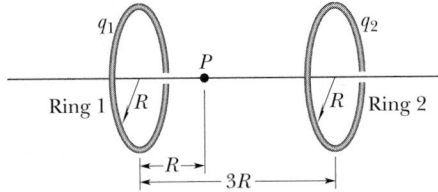

FIGURE 23-39 Exercise 28.

29P. At what distance along the central axis of a ring of radius R and uniform charge is the magnitude of the electric field due to the ring's charge maximum?

30P. An electron is constrained to the central axis of the ring of charge of radius R discussed in Section 23-6. Show that the electrostatic force exerted on the electron can cause it to oscillate through the center of the ring with an angular frequency

$$\omega = \sqrt{\frac{eq}{4\pi\epsilon_0 mR^3}},$$

where q is the ring's charge and m is the electron's mass.

31P. In Fig. 23-40a, two curved plastic rods, one of charge $+q$ and the other of charge $-q$, form a circle of radius R in an xy plane. The x axis passes through their connecting points, and the charge is distributed uniformly on both rods. What are the magnitude and direction of the electric field **E** produced at P, the center of the circle?

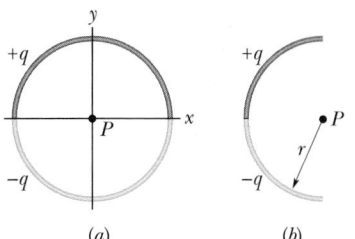

FIGURE 23-40
Problems 31 and 32.   (a)   (b)

32P. A thin glass rod is bent into a semicircle of radius r. A charge $+q$ is uniformly distributed along the upper half, and a charge $-q$ is uniformly distributed along the lower half, as shown in Fig. 23-40b. Find the magnitude and direction of the electric field **E** at P, the center of the semicircle.

33P. A thin nonconducting rod of finite length L has a charge q spread uniformly along it. Show that the magnitude E of the electric field at point P on the perpendicular bisector of the rod (Fig. 23-41) is given by

$$E = \frac{q}{2\pi\epsilon_0 y} \frac{1}{(L^2 + 4y^2)^{1/2}}.$$

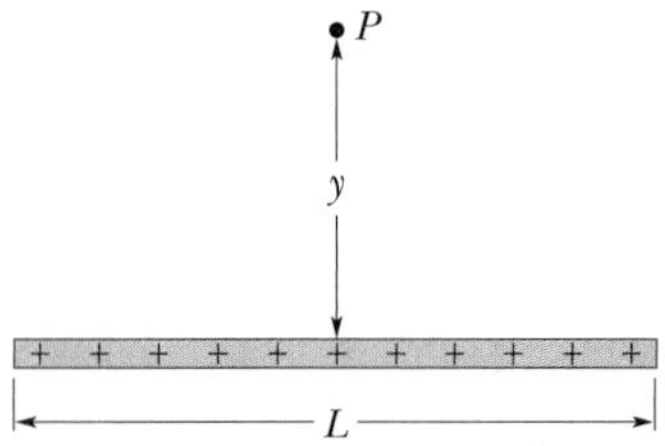

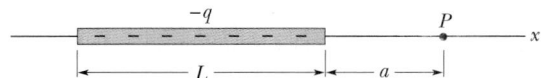

FIGURE 23-41
Problem 33.

34P. In Fig. 23-42, a nonconducting rod of length L has charge $-q$ uniformly distributed along its length. (a) What is the linear charge density of the rod? (b) What is the electric field at point P, a distance a from the end of the rod? (c) If P were very far from the rod compared to L, the rod would look like a point charge. Show that your answer to (b) reduces to the electric field of a point charge for $a \gg L$.

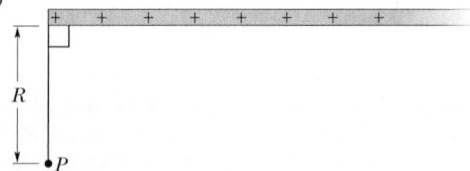

FIGURE 23-42 Problem 34.

35P*. In Fig. 23-43, a ''semi-infinite'' nonconducting rod (that is, infinite in one direction only) has uniform linear charge density λ. Show that the electric field at point P makes an angle of 45° with the rod and that this result is independent of the distance R. (*Hint:* Separately find the parallel and perpendicular (to the rod) components of the electric field at P, and then compare those components.)

FIGURE 23-43 Problem 35.

SECTION 23-7 The Electric Field Due to a Charged Disk

36E. A disk of radius 2.5 cm has a surface charge density of 5.3 $\mu C/m^2$ on its upper face. What is the magnitude of the electric field produced by the disk at a point on its central axis at distance $z = 12$ cm from the disk?

37P. (a) What total (excess) charge q must the disk in Sample Problem 23-6 (Fig. 23-12) have for the electric field on the surface of the disk at its center to equal the value at which air breaks down electrically, producing sparks? Take the disk radius as 2.5 cm, and use the listing for air in Table 23-1. (b) Suppose that each atom at the surface has an effective cross-sectional area of 0.015 nm^2. How many atoms are needed to make up the disk's surface? (c) The charge in (a) results from some of the surface atoms having one excess electron. What fraction of the surface atoms must be so charged?

38P. At what distance along the central axis of a uniformly charged plastic disk of radius R is the magnitude of the electric field equal to one-half the magnitude of the field at the center of the surface of the disk?

SECTION 23-8 A Point Charge in an Electric Field

39E. An electron is released from rest in a uniform electric field of magnitude 2.00×10^4 N/C. Calculate the acceleration of the electron. (Ignore gravitation.)

40E. An electron is accelerated eastward at 1.80×10^9 m/s² by an electric field. Determine the magnitude and direction of the electric field.

41E. Calculate the magnitude of the force, due to an electric dipole of dipole moment 3.6×10^{-29} C·m, on an electron 25 nm from the center of the dipole, along the dipole axis. Assume that this distance is large relative to the dipole's charge separation.

42E. Humid air breaks down (its molecules become ionized) in an electric field of 3.0×10^6 N/C. In that field, what is the magnitude of the electrostatic force on (a) an electron and (b) an ion with a single electron missing?

43E. An alpha particle (the nucleus of a helium atom) has a mass of 6.64×10^{-27} kg and a charge of $+2e$. What are the magnitude and direction of the electric field that will balance its weight?

44E. A charged cloud system produces an electric field in the air near the Earth's surface. A particle of charge -2.0×10^{-9} C is acted on by a downward electrostatic force of 3.0×10^{-6} N when placed in this field. (a) What is the magnitude of the electric field? (b) What are the magnitude and direction of the electrostatic force exerted on a proton placed in this field? (c) What is the gravitational force on the proton? (d) What is the ratio of the electrostatic force to the gravitational force in this case?

45E. An electric field **E** with an average magnitude of about 150 N/C points downward in the atmosphere near Earth's surface. We wish to ''float'' a sulfur sphere weighing 4.4 N in this field by charging the sphere. (a) What charge (both sign and magnitude) must be used? (b) Why is the experiment impractical?

46E. (a) What is the acceleration of an electron in a uniform electric field of 1.40×10^6 N/C? (b) How long would it take for the electron, starting from rest, to attain one-tenth the speed of light? (c) How far would it travel in that time? (Use Newtonian mechanics.)

47E. Beams of high-speed protons can be produced in "guns" using electric fields to accelerate the protons. (a) What acceleration would a proton experience if the gun's electric field were 2.00×10^4 N/C? (b) What speed would the proton attain if the field accelerated the proton through a distance of 1.00 cm?

48E. An electron with a speed of 5.00×10^8 cm/s enters an electric field of magnitude 1.00×10^3 N/C, traveling along the field in the direction that retards its motion. (a) How far will the electron travel in the field before stopping momentarily and (b) how much time will have elapsed? (c) If, instead, the region of electric field is only 8.00 mm wide (too small for the electron to stop), what fraction of the electron's initial kinetic energy will be lost in that region?

49E. A spherical water drop 1.20 μm in diameter is suspended in calm air owing to a downward-directed atmospheric electric field $E = 462$ N/C. (a) What is the weight of the drop? (b) How many excess electrons does it have?

50E. In Millikan's experiment, an oil drop of radius 1.64 μm and density 0.851 g/cm^3 is suspended in chamber C when a downward-pointing electric field of 1.92×10^5 N/C is applied. Find the charge on the drop, in terms of e.

51P. In one of his experiments, Millikan observed that the following measured charges, among others, appeared at different times on a single drop:

6.563×10^{-19} C	13.13×10^{-19} C	19.71×10^{-19} C
8.204×10^{-19} C	16.48×10^{-19} C	22.89×10^{-19} C
11.50×10^{-19} C	18.08×10^{-19} C	26.13×10^{-19} C

What value for the elementary charge e can be deduced from these data?

52P. A uniform electric field exists in a region between two oppositely charged plates. An electron is released from rest at the surface of the negatively charged plate and strikes the surface of the opposite plate, 2.0 cm away, in a time 1.5×10^{-8} s. (a) What is the speed of the electron as it strikes the second plate? (b) What is the magnitude of the electric field **E**?

53P. An object having a mass of 10.0 g and a charge of $+8.00 \times 10^{-5}$ C is placed in an electric field **E** with $E_x = 3.00 \times 10^3$ N/C, $E_y = -600$ N/C, and $E_z = 0$. (a) What are the magnitude and direction of the force on the object? (b) If the object is released from rest at the origin, what will be its coordinates after 3.00 s?

54P. At some instant the velocity components of an electron moving between two charged parallel plates are $v_x = 1.5 \times 10^5$ m/s and $v_y = 3.0 \times 10^3$ m/s. Suppose that the electric field between the plates is given by $\mathbf{E} = (120$ N/C$)\mathbf{j}$. (a) What is the acceleration of the electron? (b) What will be the velocity of the electron after its x coordinate has changed by 2.0 cm?

55P. Two large parallel copper plates are 5.0 cm apart and have

a uniform electric field between them as depicted in Fig. 23-44. An electron is released from the negative plate at the same time that a proton is released from the positive plate. Neglect the force of the particles on each other and find their distance from the positive plate when they pass each other. (Does it surprise you that you need not know the electric field to solve this problem?)

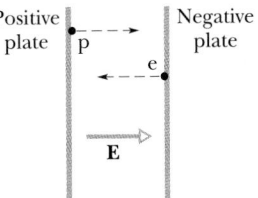

FIGURE 23-44 Problem 55.

56P. In Fig. 23-45, a pendulum is hung from the higher of two large horizontal plates. The pendulum consists of a small nonconducting sphere of mass m and charge $+q$ and an insulating thread of length l. What is the period of the pendulum if a uniform electric field **E** is set up between the plates by (a) charging the top plate negatively and the lower plate positively and (b) vice versa? In both cases, the field points away from one plate and directly toward the other plate.

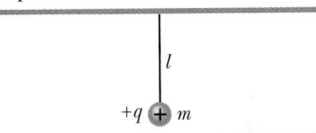

FIGURE 23-45 Problem 56.

57P. In Fig. 23-46, a uniform, upward-pointing electric field **E** of magnitude 2.00×10^3 N/C has been set up between two horizontal plates by charging the lower plate positively and the upper plate negatively. The plates have length $L = 10.0$ cm and separation $d = 2.00$ cm. An electron is then shot between the plates from the left edge of the lower plate. The initial velocity $\mathbf{v}_0$ of the electron makes an angle $\theta = 45.0°$ with the lower plate and has a magnitude of 6.00×10^6 m/s. (a) Will the electron strike one of the plates? (b) If so, which plate and how far horizontally from the left edge?

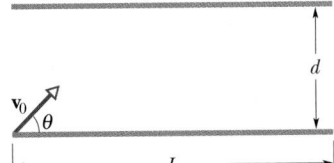

FIGURE 23-46 Problem 57.

SECTION 23-9 A Dipole in an Electric Field

58E. An electric dipole, consisting of charges of magnitude 1.50 nC separated by 6.20 μm, is in an electric field of strength 1100 N/C. (a) What is the magnitude of the electric dipole moment? (b) What is the difference in potential energy corresponding to dipole orientations parallel to and antiparallel to the field?

59E. An electric dipole consists of charges $+2e$ and $-2e$ separated by 0.78 nm. It is in an electric field of strength 3.4×10^6 N/C. Calculate the magnitude of the torque on the dipole when the dipole moment is (a) parallel to, (b) perpendicular to, and (c) antiparallel to the electric field.

60P. Find the work required to turn an electric dipole end for end in a uniform electric field $\mathbf{E}$, in terms of the magnitude p of the dipole moment, the magnitude E of the field, and the initial angle θ_0 between $\mathbf{p}$ and $\mathbf{E}$.

61P. Find the angular frequency of oscillation of an electric dipole, of dipole moment p and rotational inertia I, for small amplitudes of oscillation about its equilibrium position in a uniform electric field of magnitude E.

62P. An electric dipole with dipole moment

$$\mathbf{p} = (3.00\mathbf{i} + 4.00\mathbf{j})(1.24 \times 10^{-30}\ \text{C} \cdot \text{m})$$

is in an electric field $\mathbf{E} = (4000\ \text{N/C})\mathbf{i}$. (a) What is the potential energy of the electric dipole? (b) What is the torque acting on it?

(c) If an external agent turns the dipole until its electric dipole moment is

$$\mathbf{p} = (-4.00\mathbf{i} + 3.00\mathbf{j})(1.24 \times 10^{-30}\ \text{C} \cdot \text{m}),$$

how much work is done by the agent?

Electronic Computation

63. Two particles, each of positive charge q, are fixed in place on an y axis, one at $y = 0$ and the other at $y = -d$. (a) Write an expression that gives the magnitude E of the net electric field at points on the x axis given by $x = \alpha d$. (b) Graph E versus α for the range $0 < \alpha < 4$. From the graph, determine the values of α that give (c) the maximum value of E and (d) half the maximum value of E.

64. For the data of Problem 51, assume that the charge q on the drop is given by $q = ne$, where n is an integer and e is the elementary charge. (a) Find n for each measurement of q. (b) Do a linear regression fit of the values of q versus the values of n; from it find e.

24
Gauss' Law

Lightning strikes Manhattan in a brilliant display, each strike delivering about 10^{20} electrons from the cloud base to the ground. How wide is a lightning strike? Since it can be seen from kilometers away, is it as wide as, say, a car?

24-1 A NEW LOOK AT COULOMB'S LAW

If you want to find the center of mass of a potato, you can do so by experiment or by laborious calculation, involving the numerical evaluation of a triple integral. However, if the potato happens to be a uniform ellipsoid, you know from its symmetry exactly where the center of mass is without calculation. Such are the advantages of symmetry. Symmetrical situations arise in all areas of physics; when possible, it makes sense to cast the laws of physics in forms that take full advantage of this fact.

Coulomb's law is the governing law in electrostatics, but it is not cast in a form that particularly simplifies the work in situations involving symmetry. In this chapter we introduce a new formulation of Coulomb's law, derived by German mathematician and physicist Carl Friedrich Gauss (1777–1855). This law, called **Gauss' law,** *can* be used to take advantage of special symmetry situations. For electrostatics problems, it is the full equivalent of Coulomb's law; which of them we choose to use depends only on the problem at hand.

Central to Gauss' law is a hypothetical closed surface called a **Gaussian surface.** The Gaussian surface can be of any shape you wish to make it, but the most useful surface is one that mimics the symmetry of the problem at hand. Thus the Gaussian surface will often be a sphere, a cylinder, or some other symmetrical form. It must always be a *closed* surface, so that a clear distinction can be made between points that are inside the surface, on the surface, and outside the surface.

Imagine that you have established a Gaussian surface around a distribution of charges. Then Gauss' law comes into play:

Gauss' law relates the electric fields at points on a (closed) Gaussian surface and the net charge enclosed by that surface.

Figure 24-1 shows a simple situation in which the Gaussian surface is a sphere. Suppose you know that there is an electric field at every point on the surface and that all the fields have the same magnitude and point radially outward. Without knowing anything about Gauss' law, you can guess that some net positive charge must be inside the Gaussian surface. If you *do* know Gauss' law, you can calculate just how much net positive charge is inside the surface. To make the calculation, you need know only ''how much'' electric field is intercepted by the surface: this ''how much'' involves the *flux* of the electric field through the surface.

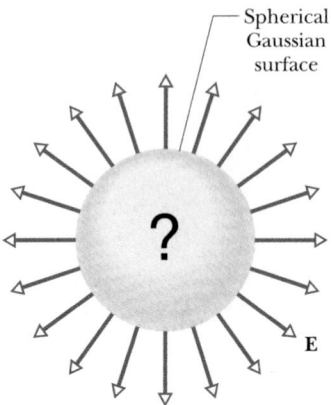

FIGURE 24-1 A spherical Gaussian surface. If the electric field vectors are of uniform magnitude and point radially outward at all surface points, you can conclude that a net positive distribution of charge must lie within the surface and have spherical symmetry.

24-2 FLUX

Suppose, as in Fig. 24-2a, that you aim a wide airstream of uniform velocity **v** at a small square loop of area A. Let Φ represent the *volume flow rate* (volume per unit time) at

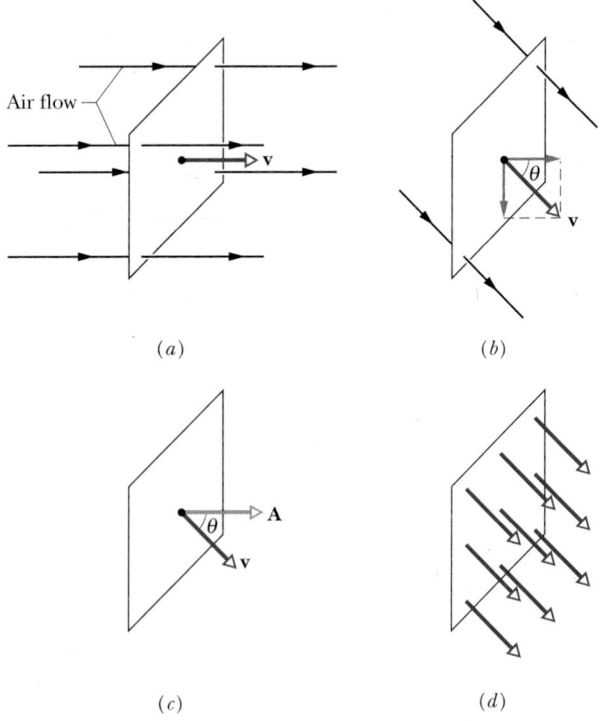

FIGURE 24-2 (a) A uniform airstream of velocity **v** is perpendicular to the plane of a square loop of area A. (b) The component of **v** perpendicular to the plane of the loop is $v \cos \theta$, where θ is the angle between **v** and a normal to the plane. (c) The area vector **A** is perpendicular to the plane of the loop and makes an angle θ with **v**. (d) The velocity field intercepted by the area of the loop.

which air flows through the loop. This rate depends on the angle between **v** and the plane of the loop. If **v** is perpendicular to the plane, the rate Φ is equal to vA.

If **v** is parallel to the plane of the loop, no air moves through the loop, so Φ is zero. For an intermediate angle θ, the rate Φ depends on the component of **v** that is normal to the plane (Fig. 24-2b). Since that component is $v \cos \theta$, the rate of volume flow through the loop is

$$\Phi = (v \cos \theta)A. \qquad (24\text{-}1)$$

This rate of flow through an area is an example of a **flux** —a *volume flux* in this situation. Before we discuss a flux that is involved in electrostatics, we need to rewrite Eq. 24-1 in terms of vectors.

To do this, we first define an *area vector* **A** as being a vector whose magnitude is equal to an area (here the area of the loop) and whose direction is normal to the plane of the area (Fig. 24-2c). We then rewrite Eq. 24-1 as the scalar (or dot) product of the velocity vector **v** of the airstream and the area vector **A** of the loop:

$$\Phi = vA \cos \theta = \mathbf{v} \cdot \mathbf{A}, \qquad (24\text{-}2)$$

where θ is the angle between **v** and **A**.

The word ''flux'' comes from the Latin word meaning ''to flow.'' That meaning makes sense if we talk about the flow of air volume through the loop. However, Eq. 24-2 can be regarded in a more abstract way. To see it, note that we can assign a velocity vector to each point in the airstream passing through the loop (Fig. 24-2d). The composite of all those vectors is a *velocity field*. So we can interpret Eq. 24-2 as giving the *flux of the velocity field through the loop*. With this interpretation, flux no longer means the actual flow of something through an area. Rather it means the product of an area and the field across that area.

24-3 FLUX OF AN ELECTRIC FIELD

To define the flux of an electric field, consider Fig. 24-3a, which shows an arbitrary (asymmetric) Gaussian surface immersed in a nonuniform electric field. Let us divide the surface into small squares of area ΔA, each square being small enough to permit us to neglect any curvature and consider the individual square to be flat. We represent each such element of area with an area vector $\Delta \mathbf{A}$, whose magnitude is the area ΔA. Each vector $\Delta \mathbf{A}$ is perpendicular to the Gaussian surface and directed away from the interior of the surface.

Because the squares have been taken to be arbitrarily small, the electric field **E** may be taken as constant over any given square. The vectors $\Delta \mathbf{A}$ and **E** for each square then make some angle θ with each other. Figure 24-3b

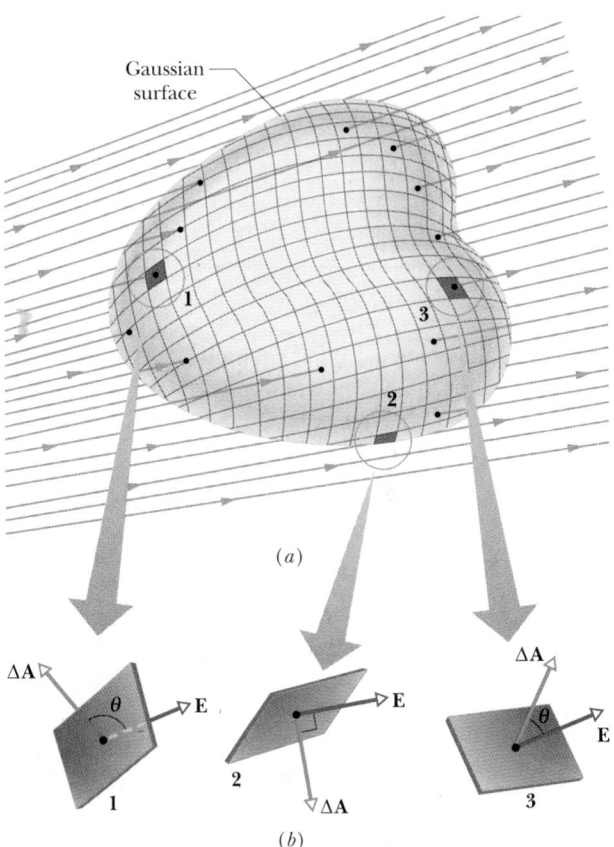

FIGURE 24-3 (a) A Gaussian surface of arbitrary shape immersed in an electric field. The surface is divided into small squares of area ΔA. (b) The electric field vectors **E** and the area vectors $\Delta \mathbf{A}$ for three representative squares, marked 1, 2, and 3.

shows an enlarged view of three squares (1, 2, and 3) on the Gaussian surface, and the angle θ for each.

A provisional definition for the flux of the electric field for the Gaussian surface of Fig. 24-3 is

$$\Phi = \sum \mathbf{E} \cdot \Delta \mathbf{A}. \qquad (24\text{-}3)$$

This equation instructs us to visit each square on the Gaussian surface, to evaluate the scalar product $\mathbf{E} \cdot \Delta \mathbf{A}$ for the two vectors **E** and $\Delta \mathbf{A}$ that we find there, and to sum the results algebraically (that is, with signs included) for all the squares that make up the surface. The sign resulting from each scalar product determines whether the flux through any given square is positive, negative, or zero. As Table 24-1 shows, squares like 1, in which **E** points inward, make a negative contribution to the sum of Eq. 24-3. Squares like 2, in which **E** lies in the surface, make zero contribution. And squares like 3, in which **E** points outward, make a positive contribution.

The exact definition of the flux of the electric field through a closed surface is found by allowing the area of the squares shown in Fig. 24-3a to become smaller and

TABLE 24-1   THREE SQUARES ON THE GAUSSIAN SURFACE OF FIG. 24-3

SQUARE	θ	DIRECTION OF E	SIGN OF $E \cdot \Delta A$
1	$> 90°$	Into the surface	Negative
2	$= 90°$	Parallel to the surface	Zero
3	$< 90°$	Out of the surface	Positive

smaller, approaching a differential limit dA. The area vectors then approach a differential limit $d\mathbf{A}$. The sum of Eq. 24-3 then becomes an integral and we have, for the definition of electric flux,

$$\Phi = \oint \mathbf{E} \cdot d\mathbf{A} \qquad \text{(electric flux through a Gaussian surface).} \qquad (24\text{-}4)$$

The circle on the integral sign indicates that the integration is to be taken over the entire (closed) surface. The flux of the electric field is a scalar, and its SI unit is the newton–square-meter per coulomb ($N \cdot m^2/C$).

We can interpret Eq. 24-4 in the following way: First recall that we can use the density of electric field lines passing through an area as a measure of an electric field $\mathbf{E}$ there. Specifically, the magnitude E is proportional to the number of electric field lines per unit area. Thus, the dot product $\mathbf{E} \cdot d\mathbf{A}$ in Eq. 24-4 is proportional to the number of electric field lines passing through area $d\mathbf{A}$. Then, because the integration in Eq. 24-4 is carried out over a Gaussian surface, which is closed, we see that

The electric flux Φ through a Gaussian surface is proportional to the net number of electric field lines passing through that surface.

SAMPLE PROBLEM 24-1

Figure 24-4 shows a Gaussian surface in the form of a cylinder of radius R immersed in a uniform electric field $\mathbf{E}$, with the cylinder axis parallel to the field. What is the flux Φ of the electric field through this closed surface?

SOLUTION: We can write the flux as the sum of three terms: integrals over the left cylinder cap a, the cylindrical surface b, and the right cap c. Thus from Eq. 24-4,

$$\Phi = \oint \mathbf{E} \cdot d\mathbf{A}$$

$$= \int_a \mathbf{E} \cdot d\mathbf{A} + \int_b \mathbf{E} \cdot d\mathbf{A} + \int_c \mathbf{E} \cdot d\mathbf{A}. \qquad (24\text{-}5)$$

For all points on the left cap, the angle θ between $\mathbf{E}$ and $d\mathbf{A}$ is 180° and the magnitude E of the field is constant. Thus,

$$\int_a \mathbf{E} \cdot d\mathbf{A} = \int E(\cos 180°)\, dA = -E \int dA = -EA,$$

where $\int dA$ gives the cap's area, $A (= \pi R^2)$. Similarly, for the right cap, where $\theta = 0$ for all points,

$$\int_c \mathbf{E} \cdot d\mathbf{A} = \int E(\cos 0)\, dA = EA.$$

Finally, for the cylindrical surface, where the angle θ is 90° at all points,

$$\int_b \mathbf{E} \cdot d\mathbf{A} = \int E(\cos 90°)\, dA = 0.$$

Substituting these results into Eq. 24-5 leads us to

$$\Phi = -EA + 0 + EA = 0. \qquad \text{(Answer)}$$

This result is perhaps not surprising because the field lines that represent the electric field all pass entirely through the Gaussian surface, entering through the left end cap, leaving through the right end cap, and giving a net flux of zero.

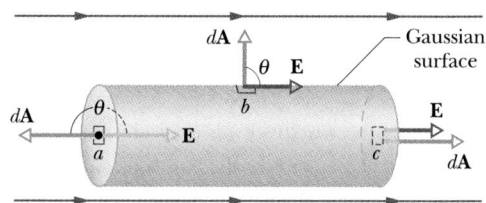

FIGURE 24-4 Sample Problem 24-1. A cylindrical Gaussian surface, closed by end caps, is immersed in a uniform electric field. The cylinder axis is parallel to the field direction.

CHECKPOINT **1:** The figure shows a Gaussian cube of face area A immersed in a uniform electric field $\mathbf{E}$ that points in the positive direction of the z axis. In terms of E and A, what is the flux through (a) the front face (which is in the xy plane), (b) the rear face, (c) the top face, and (d) the whole cube?

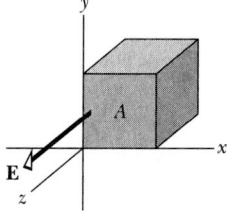

SAMPLE PROBLEM 24-2

A *nonuniform* electric field given by $\mathbf{E} = 3.0x\mathbf{i} + 4.0\mathbf{j}$ pierces the Gaussian cube shown in Fig. 24-5. (E is in newtons per coulomb and x is in meters.) What is the electric flux through the right face, the left face, and the top face?

SOLUTION: *Right face:* An area vector $\mathbf{A}$ is always perpendicular to its surface and always points away from the interior of a Gaussian surface. Thus, the vector $d\mathbf{A}$ for the right face of the cube must point in the positive x direction. In unit vector notation, then,

$$d\mathbf{A} = dA\mathbf{i}.$$

From Eq. 24-4, the flux Φ_r through the right face is then

$$\Phi_r = \int \mathbf{E} \cdot d\mathbf{A} = \int (3.0x\mathbf{i} + 4.0\mathbf{j}) \cdot (dA\mathbf{i})$$

$$= \int [(3.0x)(dA)\mathbf{i} \cdot \mathbf{i} + (4.0)(dA)\mathbf{j} \cdot \mathbf{i}]$$

$$= \int (3.0x \, dA + 0) = 3.0 \int x \, dA.$$

We are about to integrate over the right face, but we note that x has the same value everywhere on that face, namely $x = 3.0$ m. This means we can substitute that constant value for x. Then

$$\Phi_r = 3.0 \int (3.0) \, dA = 9.0 \int dA.$$

Now the integral merely gives us the area $A = 4.0$ m^2 of the right face. So,

$$\Phi_r = (9.0 \text{ N/C})(4.0 \text{ m}^2) = 36 \text{ N} \cdot \text{m}^2/\text{C}. \quad \text{(Answer)}$$

Left face: The procedure for finding the flux through the left face is the same as that for the right face. However, two factors change. (1) The differential area vector $d\mathbf{A}$ points in the negative x direction and thus $d\mathbf{A} = -dA\mathbf{i}$. (2) The term x again appears in our integration, and it is again constant over the face being considered. But on the left face, $x = 1.0$ m. With these two changes, we find that the flux Φ_l through the left face is

$$\Phi_l = -12 \text{ N} \cdot \text{m}^2/\text{C}. \quad \text{(Answer)}$$

Top face: The differential area vector $d\mathbf{A}$ points in the positive y direction and thus $d\mathbf{A} = dA\mathbf{j}$. The flux Φ_t through the top face is then

$$\Phi_t = \int (3.0x\mathbf{i} + 4.0\mathbf{j}) \cdot (dA\mathbf{j})$$

$$= \int [(3.0x)(dA)\mathbf{i} \cdot \mathbf{j} + (4.0)(dA)\mathbf{j} \cdot \mathbf{j}]$$

$$= \int (0 + 4.0 \, dA) = 4.0 \int dA$$

$$= 16 \text{ N} \cdot \text{m}^2/\text{C}. \quad \text{(Answer)}$$

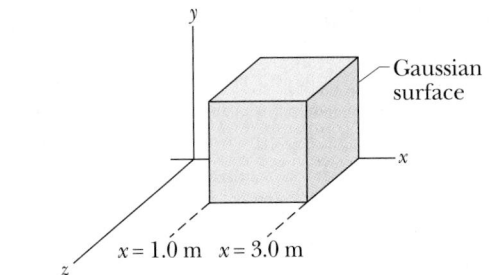

FIGURE 24-5 Sample Problem 24-2. A Gaussian cube with one edge on the x axis lies within a nonuniform electric field.

24-4 GAUSS' LAW

Gauss' law relates the net flux Φ of an electric field through a closed surface (a Gaussian surface) to the *net* charge q_{enc} that is *enclosed* by that surface. It tells us that

$$\epsilon_0 \Phi = q_{enc} \quad \text{(Gauss' law).} \quad (24\text{-}6)$$

By substituting Eq. 24-4, the definition of flux, we can also write Gauss' law as

$$\epsilon_0 \oint \mathbf{E} \cdot d\mathbf{A} = q_{enc} \quad \text{(Gauss' law).} \quad (24\text{-}7)$$

Equations 24-6 and 24-7 hold only when the net charge is located in a vacuum or (what is the same for most practical purposes) in air. In Section 26-8, we modify Gauss' law to include situations in which materials such as mica, oil, or glass are present.

In Eqs. 24-6 and 24-7, the net charge q_{enc} is the algebraic sum of all the *enclosed* positive and negative charges, and it can be positive, negative, or zero. We include the sign, rather than just use the magnitude of the charge, because the sign tells us something about the net flux through the Gaussian surface: if q_{enc} is positive, the net flux is *outward*; if q_{enc} is negative, the net flux is *inward*.

Charge outside the surface, no matter how large or how close it may be, is not included in the term q_{enc} in Gauss' law. The exact form or location of the charges inside the Gaussian surface is also of no concern; the only things that matter, on the right side of Eq. 24-7, are the magnitude and sign of the net enclosed charge. The $\mathbf{E}$ on the left side of Eq. 24-7, however, is the electric field resulting from *all* charges, both those inside and those outside the Gaussian surface. This may seem to be inconsistent, but keep in mind what we saw in Sample Problem 24-1: the electric field due to a charge outside the Gaussian surface contributes zero net flux *through* the surface, because as many field lines due to that charge enter the surface as leave it.

Let us apply these ideas to Fig. 24-6, which shows two charges, equal in magnitude but opposite in sign, and the field lines describing the electric fields that they set up in the surrounding space. Four Gaussian surfaces are also shown, in cross section. Let us consider each in turn.

SURFACE S_1. The electric field is outward for all points on this surface. Thus the flux of the electric field through this surface is positive. So is the net charge within the surface, as Gauss' law requires. (That is, in Eq. 24-6, if Φ is positive, q_{enc} must be also.)

SURFACE S_2. The electric field is inward for all points on this surface. Thus the flux of the electric field is negative and so is the enclosed charge, as Gauss' law requires.

SURFACE S_3. This surface contains no charge, and thus $q_{enc} = 0$. Gauss' law (Eq. 24-6) requires that the net flux of the electric field through this surface be zero. That is reasonable because all the field lines pass entirely through the surface, entering it at the top and leaving at the bottom.

SURFACE S_4. This surface encloses no *net* charge, because the enclosed positive and negative charges have

equal magnitudes. Gauss' law requires that the net flux of the electric field through this surface be zero. That is reasonable because there are as many field lines leaving surface S_4 as entering it.

What would happen if we were to bring an enormous charge Q up close to surface S_4 in Fig. 24-6? The pattern of the field lines would certainly change, but the net flux for the four Gaussian surfaces would not change. We can understand this because the field lines associated with the added Q would pass entirely through each of the four Gaussian surfaces, making no contribution to the net flux through any of them. The value of Q would not enter Gauss' law in any way, because Q lies outside all four of the Gaussian surfaces that we are considering.

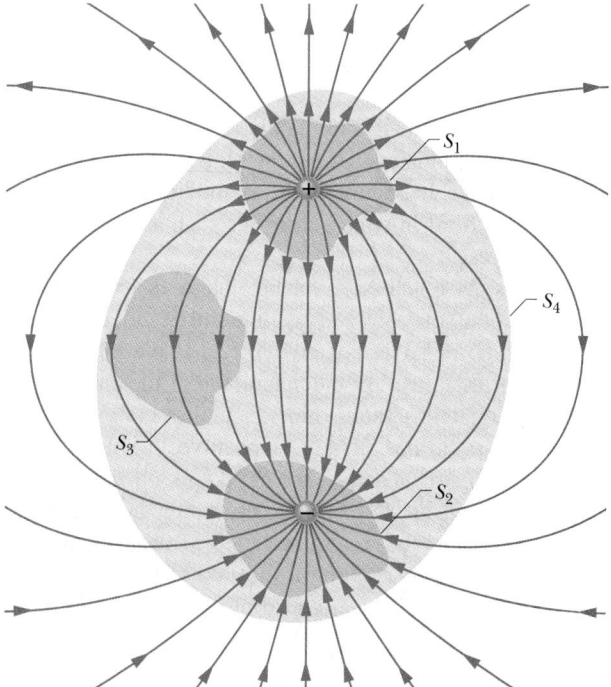

FIGURE 24-6 Two point charges, equal in magnitude but opposite in sign, and the field lines that represent their net electric field. Four Gaussian surfaces are shown in cross section. Surface S_1 encloses the positive charge. Surface S_2 encloses the negative charge. Surface S_3 encloses no charge. And surface S_4 encloses both charges, and thus no net charge.

SAMPLE PROBLEM 24-3

Figure 24-7 shows five charged lumps of plastic and an electrically neutral coin. The cross section of a Gaussian surface S is indicated. What is the net electric flux through the surface if $q_1 = q_4 = +3.1$ nC, $q_2 = q_5 = -5.9$ nC, and $q_3 = -3.1$ nC.

SOLUTION: The neutral coin makes no contribution to the net charge q_{enc} enclosed by surface S even though the positive and negative charges within the coin may be separated by the electric field in which the coin is immersed. Charges q_4 and q_5 are outside surface S and are therefore not included in q_{enc}. Thus, q_{enc} is $q_1 + q_2 + q_3$ and Eq. 24-6 gives us

$$\Phi = \frac{q_{enc}}{\epsilon_0} = \frac{q_1 + q_2 + q_3}{\epsilon_0}$$

$$= \frac{+3.1 \times 10^{-9}\text{ C} - 5.9 \times 10^{-9}\text{ C} - 3.1 \times 10^{-9}\text{ C}}{8.85 \times 10^{-12}\text{ C}^2/\text{N}\cdot\text{m}^2}$$

$$= -670\text{ N}\cdot\text{m}^2/\text{C}. \qquad \text{(Answer)}$$

The minus sign shows that the net charge within the surface is negative and that the net flux through the surface is inward.

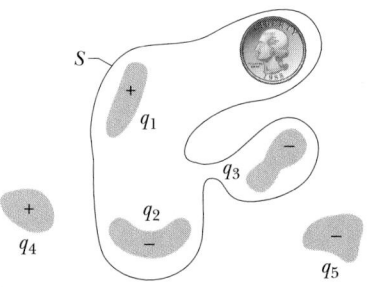

FIGURE 24-7 Sample Problem 24-3. Five plastic objects, each with an electric charge, and a coin, which has no net charge. A Gaussian surface, shown in cross section, encloses three of the plastic objects and the coin.

CHECKPOINT **2:** The figure shows three situations in which a Gaussian cube sits in an electric field. The arrows and values indicate the directions and magnitudes (in N·m²/C) of the flux through the six sides of each cube. (The lighter arrows are for the hidden faces.) In which situations does the cube enclose (a) a positive net charge, (b) a negative net charge, and (c) zero net charge?

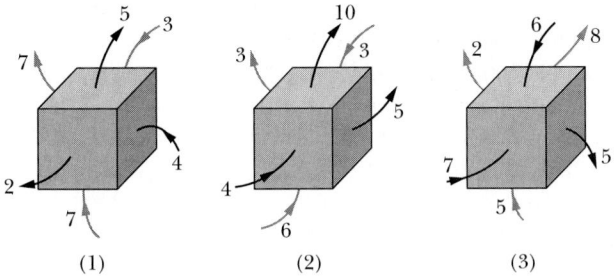

(1)   (2)   (3)

24-5 GAUSS' LAW AND COULOMB'S LAW

If Gauss' law and Coulomb's law are equivalent, we should be able to derive each from the other. Here we derive Coulomb's law from Gauss' law and some symmetry considerations.

Figure 24-8 shows a positive point charge q, around which we have drawn a concentric spherical Gaussian surface of radius r. Imagine dividing this surface into differential areas dA. By definition, the area vector $d\mathbf{A}$ at any point is perpendicular to the surface and directed outward from the interior. From the symmetry of the situation, we know that at any point the electric field $\mathbf{E}$ is also perpendicular to the surface and directed outward from the interior. Thus, since the angle θ between $\mathbf{E}$ and $d\mathbf{A}$ is zero, we can rewrite Eq. 24-7 for Gauss' law as

$$\epsilon_0 \oint \mathbf{E} \cdot d\mathbf{A} = \epsilon_0 \oint E \, dA = q_{\text{enc}}. \qquad (24\text{-}8)$$

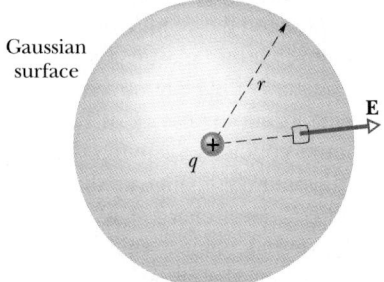

Gaussian surface

FIGURE 24-8 A spherical Gaussian surface centered on a point charge q.

Here $q_{\text{enc}} = q$. Although E varies radially with the distance from q, it has the same value everywhere on the spherical surface. Since the integral in Eq. 24-8 is taken over that surface, E is a constant in the integration and can be brought out in front of the integral sign. That gives us

$$\epsilon_0 E \oint dA = q. \qquad (24\text{-}9)$$

The integral is now merely the sum of all the differential areas dA on the sphere and thus is just the surface area, $4\pi r^2$. Substituting this, we have

$$\epsilon_0 E (4\pi r^2) = q$$

or

$$E = \frac{1}{4\pi\epsilon_0} \frac{q}{r^2}. \qquad (24\text{-}10)$$

This is exactly the electric field due to a point charge (Eq. 23-3), which we found using Coulomb's law. Thus Gauss' law is equivalent to Coulomb's law.

PROBLEM SOLVING TACTICS

TACTIC 1: *Choosing a Gaussian Surface*
The derivation of Eq. 24-10 using Gauss' law is a warm-up for derivations of electric fields produced by other charge configurations. So let us go back over the steps involved. We started with a given positive point charge q; we know that electric field lines extend radially outward from q in a spherically symmetric pattern.

To find the magnitude of the electric field E at a distance r by Gauss' law (Eq. 24-7), we had to place a hypothetical closed Gaussian surface around q, through a point that is a distance r from q. Then we had to sum via integration the values of $\mathbf{E} \cdot d\mathbf{A}$ over the full Gaussian surface. To make this integration as simple as possible, we chose a spherical Gaussian surface (to mimic the spherical symmetry of the electric field). That choice produced three simplifying features. (1) The dot product $\mathbf{E} \cdot d\mathbf{A}$ became simple, because at all points on the Gaussian surface the angle between $\mathbf{E}$ and $d\mathbf{A}$ is just zero, and so at all points we have $\mathbf{E} \cdot d\mathbf{A} = E \, dA$. (2) The electric field magnitude E is the same at all points on the spherical Gaussian surface, so E was a constant in the integration and could be brought out in front of the integral sign. (3) The result was a very simple integration—just a summation of the differential areas of the sphere, which we could immediately write as $4\pi r^2$.

Note that Gauss' law holds regardless of the shape of the Gaussian surface we choose to place around charge q_{enc}. However, if we had chosen, say, a cubical Gaussian surface, our three simplifying features would have disappeared and the integration of $\mathbf{E} \cdot d\mathbf{A}$ over the cubical surface would have been very difficult. The moral here is to choose the Gaussian surface that most simplifies the integration in Gauss' law.

CHECKPOINT 3: There is a certain net flux Φ_i through a Gaussian sphere of radius r enclosing an isolated charged particle. Suppose the enclosing Gaussian surface is changed to (a) a larger Gaussian sphere, (b) a Gaussian cube with edge length equal to r, and (c) a Gaussian cube with edge length equal to $2r$. In each case, is the net flux through the new Gaussian surface larger than, smaller than, or equal to Φ_i?

24-6 A CHARGED ISOLATED CONDUCTOR

Gauss' law permits us to prove an important theorem about isolated conductors:

If an excess charge is placed on an isolated conductor, that amount of charge will move entirely to the surface of the conductor. None of the excess charge will be found within the body of the conductor.

This might seem reasonable, considering that charges with the same sign repel each other. You might imagine that, by moving to the surface, the added charges are getting as far away from each other as they can. We turn to Gauss' law for verification of this speculation.

Figure 24-9a shows, in cross section, an isolated lump of copper hanging from an insulating thread and having an excess charge q. We place a Gaussian surface just inside the actual surface of the conductor.

The electric field inside the conductor must be zero. If this were not so, the field would exert forces on the conduction (free) electrons, which are always present in the

conductor, and thus current would always exist within the conductor. (That is, charge would flow from place to place within the conductor.) Of course, there are no such perpetual currents in an isolated conductor, and so the internal electric field is zero.

(An internal electric field *does* appear as the conductor is being charged. However, the added charge quickly distributes itself in such a way that the net internal electric field—the vector sum of the electric fields due to all the charges—is zero. The movement of charge then ceases, and the net force on each charge is zero; the charges are then in *electrostatic equilibrium.*)

If **E** is zero everywhere inside the conductor, it must be zero for all points on the Gaussian surface because that surface, though close to the surface of the conductor, is definitely inside it. This means that the flux through the Gaussian surface must be zero. Gauss' law then tells us that the net charge inside the Gaussian surface must also be zero. If the excess charge is not inside the Gaussian surface, it must be outside that surface, which means that it must lie on the actual surface of the conductor.

An Isolated Conductor with a Cavity

Figure 24-9b shows the same hanging conductor, but now with a cavity that is totally within the conductor. It is perhaps reasonable to suppose that when we scoop out the electrically neutral material to form the cavity, we should not change the distribution of charge or the pattern of the electric field that exists in Fig. 24-9a. Again, we must turn to Gauss' law for a quantitative proof.

We draw a Gaussian surface surrounding the cavity, close to its surface but inside the conducting body. Because **E** = 0 inside the conductor, there can be no flux through this new Gaussian surface. Therefore, from Gauss' law, that surface can enclose no net charge. We conclude that there is no net charge on the cavity walls; all the excess charge remains on the outer surface of the conductor, as in Fig. 24-9a.

The Conductor Removed

Suppose that, by some magic, the excess charges could be ''frozen'' into position on the conductor's surface, perhaps by embedding them in a thin plastic coating, and suppose that then the conductor could be removed completely. This is equivalent to enlarging the cavity of Fig. 24-9b until it consumes the entire conductor, leaving only the charges. The electric field would not change at all; it would remain zero inside the thin shell of charge and would remain unchanged for all external points. This shows us that the electric field is set up by the charges and not by the conductor. The conductor simply provides an initial pathway for the charges to take up their positions.

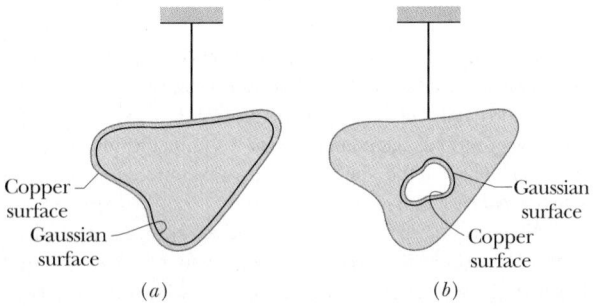

Copper surface
Gaussian surface
(a)

Gaussian surface
Copper surface
(b)

FIGURE 24-9 (a) A lump of copper with a charge q hangs from an insulating thread. A Gaussian surface is drawn within the metal, just inside the actual surface. (b) The lump of copper now has a cavity within it. A Gaussian surface lies within the metal, close to the cavity surface.

The External Electric Field

You have seen that the excess charge on an isolated conductor moves entirely to the conductor's surface. However, unless the conductor is spherical, the charge does not distribute itself uniformly. Put another way, the surface charge density σ (charge per unit area) varies over the surface of any nonspherical conductor. Generally, this variation makes the determination of the electric field set up by the surface charges very difficult.

However, the electric field just outside the surface of a conductor is easy to determine using Gauss' law. To do this, we consider a section of the surface that is small enough to permit us to neglect any curvature and take the section to be flat. We then imagine a tiny cylindrical Gaussian surface to be embedded in the section as in Fig. 24-10: one end cap is fully inside the conductor, the other is fully outside, and the cylinder is perpendicular to the conductor's surface.

The electric field **E** at and just outside the conductor's surface must also be perpendicular to that surface. If it were not, then it would have a component along the conductor's surface that would exert forces on the surface charges, causing them to move. But such motion would violate our implicit assumption that we are dealing with electrostatic equilibrium. So **E** is perpendicular to the conductor's surface.

We now sum the flux through the Gaussian surface. There is no flux through the internal end cap, because the electric field there is zero. There is no flux through the curved surface of the cylinder, because internally (in the conductor) there is no electric field and externally the electric field is parallel to the curved surface. The only flux through the Gaussian surface is that through the external end cap, where **E** is perpendicular to the plane of the cap. We assume that the cap area A is small enough that the

field magnitude E is constant over the cap. Then the flux through the cap is EA, and that is the net flux Φ through the Gaussian surface.

The charge q_{enc} enclosed by the Gaussian surface lies on the conductor's surface in an area A. If σ is the charge per unit area, then q_{enc} is equal to σA. When we substitute σA for q_{enc} and EA for Φ, Gauss' law (Eq. 24-6) becomes

$$\epsilon_0 EA = \sigma A,$$

from which we find

$$E = \frac{\sigma}{\epsilon_0} \quad \text{(conducting surface).} \quad (24\text{-}11)$$

Thus the magnitude of the electric field at a location just outside a conductor is proportional to the surface charge density at that location on the conductor. If the charge on the conductor is positive, the electric field points away from the conductor as in Fig. 24-10. It points toward the conductor if the charge is negative.

The field lines in Fig. 24-10 must terminate on negative charges somewhere in the environment. If we bring those charges near the conductor, the charge density at any given location changes and so does the magnitude of the electric field. However, the relation between σ and E is still given by Eq. 24-11.

SAMPLE PROBLEM 24-4

Figure 24-11a shows a cross section of a spherical metal shell of inner radius R. A point charge of $-5.0\ \mu C$ is located at a distance $R/2$ from the center of the shell. If the shell is electrically neutral, what are the (induced) charges on its inner and outer surfaces? Are those charges uniformly distributed? What is the field pattern inside and outside the shell?

SOLUTION: Figure 24-11b shows a cross section of a spherical Gaussian surface within the metal, just outside the inner wall of the shell. Since the electric field must be zero inside the metal (and thus on the Gaussian surface inside the metal), the electric flux through the Gaussian surface must also be zero. Gauss' law then tells us that the *net* charge enclosed by the Gaussian surface must be zero. With a point charge of $-5.0\ \mu C$ within the shell, a charge of $+5.0\ \mu C$ must lie on the inner wall of the shell.

If the point charge were centered, this positive charge would be uniformly distributed along the inner wall. However, since the point charge is off-center, the distribution of positive charge is skewed, as suggested by Fig. 24-11b, because the positive charge tends to collect on the section of the inner wall nearest the point charge.

Since the shell is electrically neutral, its inner wall can have a charge of $+5.0\ \mu C$ only if electrons, with a total charge of $-5.0\ \mu C$, leave the inner wall and move to the outer wall.

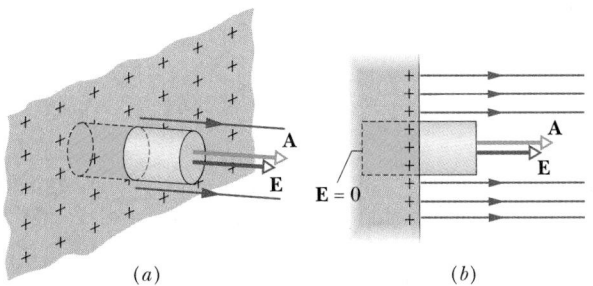

(a) $\qquad\qquad\qquad$ (b)

FIGURE 24-10 Perspective view (a) and side view (b) of a tiny portion of a large, isolated conductor with excess positive charge on its surface. A (closed) cylindrical Gaussian surface, embedded perpendicularly in the conductor, encloses some of the charge. Electric field lines pierce the external end cap of the cylinder, but not the internal end cap. The external end cap has area A and area vector **A**.

There they spread out uniformly, as is also suggested by Fig. 24-11*b*. This distribution of negative charge is uniform because the shell is spherical and because the skewed distribution of positive charge on the inner wall cannot produce an electric field in the shell to affect the distribution of charge on the outer wall.

The field lines inside and outside the shell are shown approximately in Fig. 24-11*b*. All the field lines intersect the shell and the point charge perpendicularly. Inside the shell the pattern of field lines is skewed owing to the skew of the positive charge distribution. Outside the shell the pattern is the same as if the point charge were centered and the shell were missing. In fact, this would be true no matter where inside the shell the point charge happened to be located.

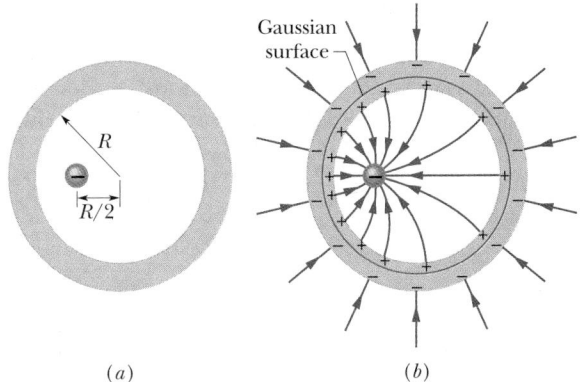

(a)   (b)

FIGURE 24-11 Sample Problem 24-4. (*a*) A negative point charge is located within a spherical metal shell that is electrically neutral. (*b*) As a result, positive charge is nonuniformly distributed on the inner wall of the shell, and an equal amount of negative charge is uniformly distributed on the outer wall. The electric field lines are shown.

CHECKPOINT 4: A ball of charge $-50e$ lies at the center of a hollow spherical metal shell that has a net charge of $-100e$. What is the charge on (a) the shell's inner surface and (b) its outer surface?

24-7 APPLYING GAUSS' LAW: CYLINDRICAL SYMMETRY

Figure 24-12 shows a section of an infinitely long cylindrical plastic rod with a uniform (positive) linear charge density λ. Let us find an expression for the magnitude of the electric field **E** at a distance r from the axis of the rod.

Our Gaussian surface should match the symmetry of the problem, which is cylindrical. We choose a circular cylinder of radius r and length h, coaxial with the rod. The Gaussian surface must be closed, so we include two end caps as part of the surface.

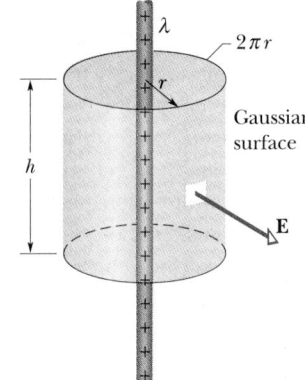

FIGURE 24-12 A Gaussian surface in the form of a closed cylinder surrounds a section of a very long, uniformly charged, cylindrical plastic rod.

Imagine now that, while you are not watching, someone rotates the plastic rod around its longitudinal axis or turns it end for end. When you look again at the rod, you will not be able to detect any change. We conclude from this symmetry that the only uniquely specified direction in this problem is along a radial line. Thus **E** must have a constant magnitude E and (for a positively charged rod) must be directed radially outward at every point on the cylindrical part of the Gaussian surface.

Since $2\pi r$ is the circumference of the cylinder and h is its height, the area of the cylindrical surface is $2\pi rh$. The flux of **E** through this cylindrical surface is then

$$\Phi = EA \cos \theta = E(2\pi rh).$$

There is no flux through the end caps because **E**, being radially directed, is parallel to the end caps at every point.

The charge enclosed by the surface is λh so that Gauss' law,

$$\epsilon_0 \Phi = q_{\text{enc}},$$

reduces to

$$\epsilon_0 E(2\pi rh) = \lambda h,$$

yielding

$$E = \frac{\lambda}{2\pi\epsilon_0 r} \qquad \text{(line of charge).} \qquad (24\text{-}12)$$

This is the electric field due to an infinitely long, straight line of charge, at a point that is a radial distance r from the line. The direction of **E** is radially outward if the charge is positive, and radially inward if it is negative.

SAMPLE PROBLEM 24-5

The visible portion of a lightning strike is preceded by an invisible stage in which a column of electrons is extended from a cloud to the ground. These electrons come from the cloud and from air molecules that are ionized within the column. The linear charge density λ along the column is typically -1×10^{-3} C/m. Once the column reaches the ground, electrons within it are rapidly dumped to the ground. During the dumping, collisions between the electrons and the air within the column result in a brilliant flash of light. If air molecules break down (ionize) in an electric field exceeding 3×10^6 N/C, what is the radius of the column?

SOLUTION: Although the column is not straight or infinitely long, we can approximate it as being a line of charge as in Fig. 24-12. (Since it contains a net negative charge, the electric field $\mathbf{E}$ points radially inward.) According to Eq. 24-12, the electric field E decreases with distance from the axis of the column of charge. The surface of the column of charge must be at a radius r where the magnitude of $\mathbf{E}$ is 3×10^6 N/C, because air molecules within that radius ionize while those farther out do not. Solving Eq. 24-12 for r and inserting the known data, we find the radius of the column to be

$$r = \frac{\lambda}{2\pi\epsilon_0 E}$$

$$= \frac{1 \times 10^{-3} \text{ C/m}}{(2\pi)(8.85 \times 10^{-12} \text{ C}^2/\text{N} \cdot \text{m}^2)(3 \times 10^6 \text{ N/C})}$$

$$= 6 \text{ m}. \qquad \text{(Answer)}$$

(The radius of the luminous portion of a lightning strike is smaller, perhaps only 0.5 m. You can get an idea of the width from Fig. 24-13.) Although the radius of the column may be

FIGURE 24-13 Lightning strikes a 20 m high sycamore. Because the tree was wet, most of the charge traveled through the water on it and the tree was unharmed.

FIGURE 24-14 Ground currents from a lightning strike have burned grass off this golf course, exposing the soil.

only 6 m, do not assume that you are safe if you are at a somewhat greater distance from the strike point, because the electrons dumped by the strike travel along the ground. Such *ground currents* are lethal. Figure 24-14 shows evidence of ground currents.

24-8 APPLYING GAUSS' LAW: PLANAR SYMMETRY

Nonconducting Sheet

Figure 24-15 shows a portion of a thin, infinite, nonconducting sheet with a uniform (positive) surface charge density σ. A sheet of thin plastic wrap, uniformly charged on one side, can serve as a simple model. Let us find the electric field $\mathbf{E}$ a distance r in front of the sheet.

A useful Gaussian surface is a closed cylinder with end caps of area A, arranged to pierce the sheet perpendicularly as shown. From symmetry, $\mathbf{E}$ must be perpendicular to the sheet, hence to the end caps. Furthermore, since the charge is positive, $\mathbf{E}$ must point *away* from the sheet, and thus the electric field lines pierce the two Gaussian end caps in an outward direction. Because the field lines do not pierce the cylinder walls, there is no flux through this portion of the Gaussian surface. Thus $\mathbf{E} \cdot d\mathbf{A}$ is simply $E \, dA$; then Gauss' law,

$$\epsilon_0 \oint \mathbf{E} \cdot d\mathbf{A} = q_{\text{enc}},$$

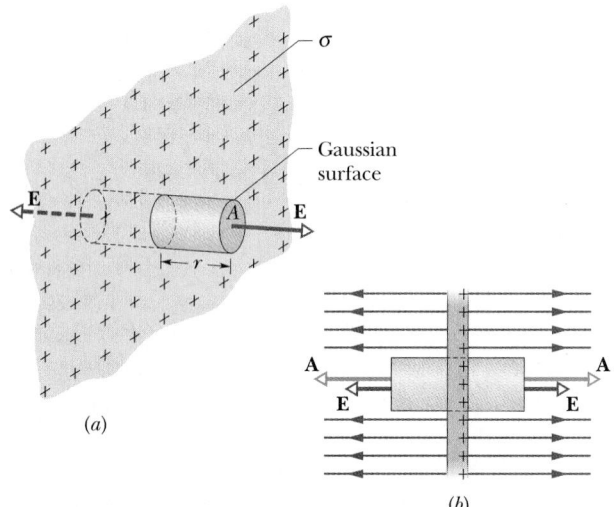

FIGURE 24-15 Perspective view (*a*) and side view (*b*) of a portion of a very large, thin plastic sheet, uniformly charged on one side to surface charge density σ. A closed cylindrical Gaussian surface passes through the sheet and is perpendicular to it.

becomes

$$\epsilon_0(EA + EA) = \sigma A,$$

where σA is the charge enclosed by the Gaussian surface. This gives

$$E = \frac{\sigma}{2\epsilon_0} \qquad \text{(sheet of charge).} \qquad (24\text{-}13)$$

Since we are considering an infinite sheet with uniform charge density, this result holds for any point at a finite distance from the sheet. Equation 24-13 agrees with Eq. 23-25, which we found by integration of the electric field components that are produced by individual charges. (Look back to that time-consuming and challenging integration, and note how much more easily we obtain the result with Gauss' law. That is one reason for devoting a whole chapter to that law: for certain symmetric arrangements of charge, it is very much easier to use than integration of field components.)

Two Conducting Plates

Figure 24-16*a* shows a cross section of a thin, infinite conducting plate with excess positive charge. From Section 24-6 we know that this excess charge lies on the surface of the plate. Since the plate is thin and very large, we can assume that essentially all the excess charge is on the two large faces of the plate.

If there is no external electric field to force the positive charge into some particular distribution, it will spread out on the two faces with a uniform surface charge density of

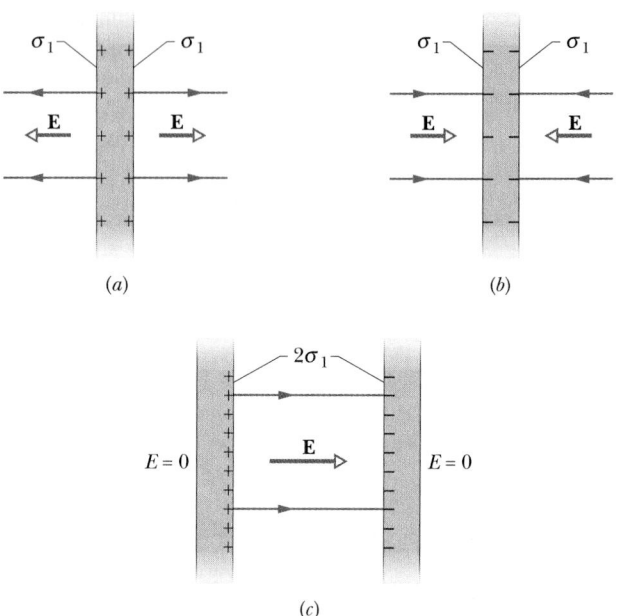

FIGURE 24-16 (*a*) A thin, very large conducting plate with excess positive charge. (*b*) An identical plate with excess negative charge. (*c*) The two plates arranged to be parallel and close.

magnitude σ_1. From Eq. 24-11 we know that just outside the plate this charge sets up an electric field of magnitude $E = \sigma_1/\epsilon_0$. Because the excess charge is positive, the field points away from the plate.

Figure 24-16*b* shows an identical plate with excess negative charge having the same magnitude of surface charge density σ_1. The only difference is that now the electric field points toward the plate.

Suppose we arrange for the plates of Figs. 24-16*a* and *b* to be close to each other and parallel (Fig. 24-16*c*). Since the plates are conductors, when we bring them into this arrangement, the excess charge on one plate attracts the excess charge on the other plate, and all the excess charge moves onto the inner faces of the plates as in Fig. 24-16*c*. With twice as much charge now on each inner face, the new surface charge density (call it σ) on each inner face is twice σ_1. Thus the electric field at any point between the plates has the magnitude

$$E = \frac{2\sigma_1}{\epsilon_0} = \frac{\sigma}{\epsilon_0}. \qquad (24\text{-}14)$$

This field points directly away from the positively charged plate and toward the negatively charged plate. Since no excess charge is left on the outer faces, the electric field to the left and right of the plates is zero.

Because the charges on the plates moved when we brought the plates close to each other, Fig. 24-16*c* is *not* the superposition of Figs. 24-16*a* and *b*; that is, the charge distribution of the two-plate system is not merely the sum of the charge distributions of the individual plates.

You may wonder why we discuss such seemingly unrealistic situations as the field set up by an infinite line of charge, an infinite sheet of charge, or a pair of infinite plates of charge. It is not enough to say that we do so because it is simple to analyze such situations with Gauss' law, although that is indeed true. The proper answer is that analyses for "infinite" situations yield good approximations to many real-world problems. Thus Eq. 24-13 holds well for a finite nonconducting sheet as long as you are close to the sheet and not too near its edges. And Eq. 24-14 holds well for a pair of finite conducting plates as long as you consider a point that is not too close to their edges.

The trouble with the edges of a sheet or a plate, and the reason we take care not to be near them, is that near an edge we can no longer use planar symmetry to find expressions for the fields. In fact, the field lines there are curved (said to be an *edge effect* or *fringing*), and the fields can be very difficult to express algebraically.

SAMPLE PROBLEM 24-6

Figure 24-17a shows portions of two large, parallel, nonconducting sheets, each with a fixed uniform charge on one side. The magnitudes of the surface charge densities are $\sigma_{(+)} = 6.8$ $\mu C/m^2$ for the positively charged sheet and $\sigma_{(-)} = 4.3$ $\mu C/m^2$ for the negatively charged sheet.

Find the electric field **E** (a) to the left of the sheets, (b) between the sheets, and (c) to the right of the sheets.

SOLUTION: Since the charges are fixed in place, we can find the electric field of the sheets in Fig. 24-17a by (1) finding the field of each sheet as if that sheet were isolated and (2) algebraically adding the fields of the isolated sheets via the superposition principle. (We can add the fields algebraically because they are parallel to each other.) From Eq. 24-13, the magnitude $E_{(+)}$ of the electric field due to the positive sheet at any point is

$$E_{(+)} = \frac{\sigma_{(+)}}{2\epsilon_0} = \frac{6.8 \times 10^{-6} \ C/m^2}{(2)(8.85 \times 10^{-12} \ C^2/N \cdot m^2)}$$
$$= 3.84 \times 10^5 \ N/C.$$

Similarly, the magnitude $E_{(-)}$ of the electric field at any point due to the negative sheet is

$$E_{(-)} = \frac{\sigma_{(-)}}{2\epsilon_0} = \frac{4.3 \times 10^{-6} \ C/m^2}{(2)(8.85 \times 10^{-12} \ C^2/N \cdot m^2)}$$
$$= 2.43 \times 10^5 \ N/C.$$

Figure 24-17b shows the fields set up by the sheets to the left of the sheets (L), between them (B), and to their right (R).

The resultant fields in these three regions follow from the superposition principle. To the left of the sheets, the field

magnitude is

$$E_L = E_{(+)} - E_{(-)}$$
$$= 3.84 \times 10^5 \ N/C - 2.43 \times 10^5 \ N/C$$
$$= 1.4 \times 10^5 \ N/C. \qquad \text{(Answer)}$$

Because $E_{(+)}$ is larger than $E_{(-)}$, the net electric field $\mathbf{E}_L$ in this region points to the left, as Fig. 24-17c shows. To the right of the sheets, the electric field $\mathbf{E}_R$ has the same magnitude but points to the right, as Fig. 24-17c shows.

Between the sheets, the two fields add and we have

$$E_B = E_{(+)} + E_{(-)}$$
$$= 3.84 \times 10^5 \ N/C + 2.43 \times 10^5 \ N/C$$
$$= 6.3 \times 10^5 \ N/C. \qquad \text{(Answer)}$$

The electric field $\mathbf{E}_B$ points to the right.

Note that outside the sheets, the electric field is the same as that from a single sheet whose surface charge density is $\sigma_{(+)} - \sigma_{(-)}$, or $+2.5 \times 10^{-6} \ C/m^2$.

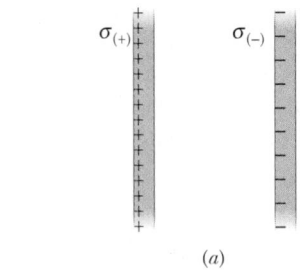

(a)

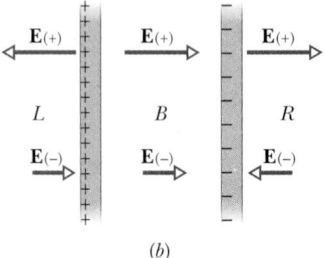

(b)

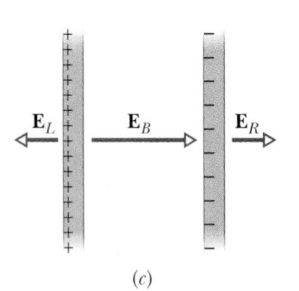

(c)

FIGURE 24-17 Sample Problem 24-6. (a) Two large parallel sheets, uniformly charged on one side. (b) The individual electric fields resulting from the two charged sheets. (c) The net field due to both charged sheets, found by superposition.

24-9 APPLYING GAUSS' LAW: SPHERICAL SYMMETRY

Here we use Gauss' law to prove the two shell theorems presented without proof in Section 22-4:

A shell of uniform charge attracts or repels a charged particle that is outside the shell as if all the shell's charge were concentrated at the center of the shell.

A shell of uniform charge exerts no electrostatic force on a charged particle that is located inside the shell.

Figure 24-18 shows a charged spherical shell of total charge q and radius R and two concentric spherical Gaussian surfaces, S_1 and S_2. Following the procedure of Section 24-5 and applying Gauss' law to surface S_2, for which $r \geq R$, we find

$$E = \frac{1}{4\pi\epsilon_0} \frac{q}{r^2} \qquad \text{(spherical shell,} \atop \text{field at } r \geq R\text{).} \qquad (24\text{-}15)$$

This is the same field that would be set up by a point charge q at the center of the shell of charge. Thus the magnitude of the force exerted by the shell on a charged particle placed outside the shell is the same as if the shell were replaced with a point charge q at the center of the shell. This proves the first shell theorem.

Applying Gauss' law to surface S_1, for which $r < R$, leads directly to

$$E = 0 \qquad \text{(spherical shell, field at } r < R\text{),} \qquad (24\text{-}16)$$

because this Gaussian surface encloses no charge. Thus if a charged particle were enclosed by the shell, the shell would exert no net electrostatic force on it. This proves the second shell theorem.

Any spherically symmetric charge distribution, such as that of Fig. 24-19, can be constructed with a nest of concentric spherical shells. For purposes of applying the two shell theorems, the volume charge density ρ should have a uniform value for each shell but need not be the same from shell to shell. That is, for the charge distribution

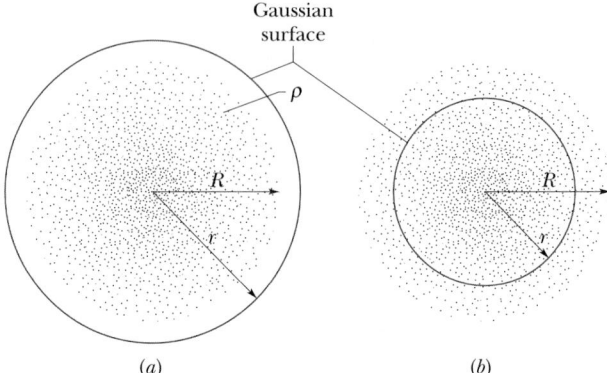

FIGURE 24-19 The dots represent a spherically symmetric distribution of charge of radius R, whose volume charge density ρ is a function only of distance from the center. The charged object is not a conductor, and the charge is assumed to be fixed in position. (a) A concentric spherical Gaussian surface with $r > R$ is included. (b) A similar Gaussian surface with $r < R$ is included.

as a whole, ρ can vary, but only with r, the radial distance from the center. We can then examine the effect of the charge distribution "shell by shell."

In Fig. 24-19a the entire charge lies within a Gaussian surface with $r > R$. The charge produces an electric field on the Gaussian surface as if the charge were a point charge located at the center, and Eq. 24-15 holds.

Figure 24-19b shows a Gaussian surface with $r < R$. To find the electric field at points on this Gaussian surface, we consider two sets of charged shells—one set inside the Gaussian surface and one set outside. Equation 24-16 says that the charge lying *outside* the Gaussian surface does not set up an electric field on the Gaussian surface. And Eq. 24-15 says that the charge *enclosed* by the surface sets up an electric field as if that enclosed charge were concentrated at the center. Letting q' represent that enclosed charge, we can then rewrite Eq. 24-15 as

$$E = \frac{1}{4\pi\epsilon_0} \frac{q'}{r^2} \qquad \text{(spherical distribution,} \atop \text{field at } r \leq R\text{).} \qquad (24\text{-}17)$$

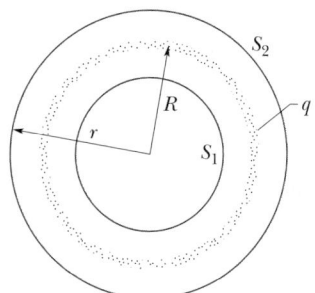

FIGURE 24-18 A thin, uniformly charged, spherical shell with total charge q, in cross section. Two Gaussian surfaces S_1 and S_2 are also shown in cross section. Surface S_2 encloses the shell, and S_1 encloses only the empty interior of the shell.

SAMPLE PROBLEM 24-7

The nucleus of an atom of gold has a radius $R = 6.2 \times 10^{-15}$ m and a positive charge $q = Ze$, where the atomic number Z of gold is 79. Plot the magnitude of the electric field from the center of the gold nucleus outward to a distance of about twice its radius. Assume that the nucleus is spherical with a uniform charge distribution.

SOLUTION: The total charge q on the nucleus is

$$q = Ze = (79)(1.60 \times 10^{-19} \text{ C}) = 1.264 \times 10^{-17} \text{ C}.$$

Outside the nucleus, the situation is represented by Fig. 24-19a and by Eq. 24-15. From this equation we have, for a point on the surface of the nucleus,

$$E = \frac{1}{4\pi\epsilon_0}\frac{q}{r^2}$$

$$= \frac{1.264 \times 10^{-17} \text{ C}}{(4\pi)(8.85 \times 10^{-12} \text{ C}^2/\text{N}\cdot\text{m}^2)(6.2 \times 10^{-15} \text{ m})^2}$$

$$= 3.0 \times 10^{21} \text{ N/C}.$$

Inside the nucleus, Fig. 24-19b and Eq. 24-17 apply. Let q' represent the charge enclosed by a Gaussian sphere of radius $r \leq R$. Since the charge is distributed uniformly throughout the volume of the nucleus, a charge enclosed by a sphere is proportional to the volume of that sphere. In particular,

$$\frac{q'}{q} = \frac{\frac{4}{3}\pi r^3}{\frac{4}{3}\pi R^3}, \qquad (24\text{-}18)$$

so

$$q' = q\frac{r^3}{R^3}. \qquad$$

If we substitute this result into Eq. 24-17, we find

$$E = \frac{1}{4\pi\epsilon_0}\frac{q'}{r^2} = \left(\frac{q}{4\pi\epsilon_0 R^3}\right)r. \qquad (24\text{-}19)$$

The quantity in parentheses is a constant, so, within the nucleus, E is directly proportional to r and is zero at the nuclear center. (Comparison of Eqs. 24-19 and 24-15 shows that they give the same result, 3.0×10^{21} N/C, at $r = R$. This simply tells us that the "inside equation" and the "outside equation," Eqs. 24-17 and 24-15, are compatible where they both apply.) Figure 24-20 shows these results graphically. To ob-

tain it, we plot Eq. 24-19 for $0 \leq r \leq 6.2 \times 10^{-15}$ m, and Eq. 24-15 for $r \geq 6.2 \times 10^{-15}$ m.

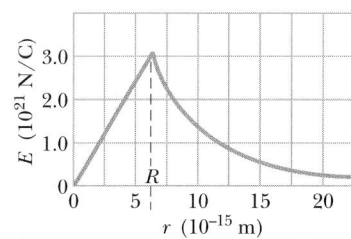

FIGURE 24-20 Sample Problem 24-7. The variation of electric field with distance from the center for the nucleus of a gold atom. The positive charge is assumed to be distributed uniformly throughout the volume of the nucleus.

CHECKPOINT **5:** The figure shows two large, parallel, nonconducting sheets with identical (positive) uniform surface charge densities, and a sphere with a uniform (positive) volume charge density. Rank the four numbered points according to the magnitude of the net electric field there, greatest first.

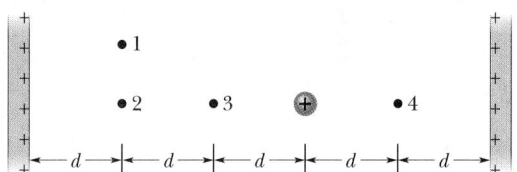

REVIEW & SUMMARY

Gauss' Law

Gauss' law and Coulomb's law, although expressed in different forms, are equivalent ways of describing the relation between charge and electric field in static situations. Gauss' law is

$$\epsilon_0\Phi = q_{\text{enc}} \qquad \text{(Gauss' law)}, \qquad (24\text{-}6)$$

in which q_{enc} is the net charge inside an imaginary closed surface (a **Gaussian surface**) and Φ is the net **flux** of the electric field through the surface:

$$\Phi = \oint \mathbf{E}\cdot d\mathbf{A} \qquad \begin{array}{l}\text{(electric flux through a}\\\text{Gaussian surface).}\end{array} \qquad (24\text{-}4)$$

Coulomb's law can readily be derived from Gauss' law.

Applications of Gauss' Law

Using Gauss' law and, in some cases, symmetry arguments, we can derive several important results in electrostatic situations. Among these are:

1. An excess charge on an *isolated conductor* is located entirely on the outer surface of the conductor.

2. The external electric field near the *surface of a charged conductor* is perpendicular to the surface and has magnitude

$$E = \frac{\sigma}{\epsilon_0} \qquad \text{(conducting surface).} \qquad (24\text{-}11)$$

Inside the conductor, $E = 0$.

3. The electric field at a point due to an infinite *line of charge* with uniform linear charge density λ is in a direction perpendicular to the line of charge and has magnitude

$$E = \frac{\lambda}{2\pi\epsilon_0 r} \qquad \text{(line of charge)}, \qquad (24\text{-}12)$$

where r is the perpendicular distance from the line of charge to the point.

4. The electric field due to an *infinite nonconducting sheet* with uniform surface charge density σ is perpendicular to the plane of the sheet and has magnitude

$$E = \frac{\sigma}{2\epsilon_0} \qquad \text{(sheet of charge)}. \qquad (24\text{-}13)$$

5. The electric field outside a *spherical shell of charge* with radius R and total charge q is directed radially and has magnitude

$$E = \frac{1}{4\pi\epsilon_0}\frac{q}{r^2} \qquad \text{(spherical shell, for } r \geq R). \quad (24\text{-}15)$$

Here r is the distance from the center of the shell to the point at

which E is measured. (The charge behaves, for external points, as if it were all at the center of the sphere.) The field *inside* a uniform spherical shell of charge is exactly zero:

$$E = 0 \qquad \text{(spherical shell, for } r < R). \qquad (24\text{-}16)$$

6. The electric field *inside a uniform sphere of charge* is directed radially and has magnitude

$$E = \left(\frac{q}{4\pi\epsilon_0 R^3}\right) r. \qquad (24\text{-}19)$$

QUESTIONS

1. A surface has the area vector $\mathbf{A} = (2\mathbf{i} + 3\mathbf{j})$ m². What is the flux of an electric field through it if the field is (a) $\mathbf{E} = 4\mathbf{i}$ N/C and (b) $\mathbf{E} = 4\mathbf{k}$ N/C?

2. What is $\int dA$ for (a) a square of edge length a, (b) a circle of radius r, and (c) the curved surface of a cylinder of length h and radius r?

3. Figure 24-21 shows four Gaussian surfaces consisting of identical cylindrical midsections but different end caps. The surfaces are in a uniform electric field $\mathbf{E}$ that is directed parallel to the central axis of the cylindrical midsections. The end caps of surface S_1 are convex hemispheres; those of surface S_2 are concave hemispheres; those of surface S_3 are cones; and those of surface S_4 are flat disks. Rank the surfaces according to (a) the net electric flux through them and (b) the electric flux through the top end caps, greatest first.

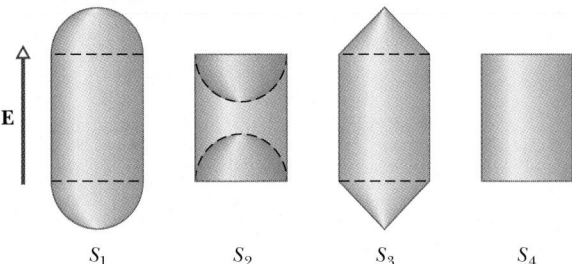

FIGURE 24-21 Question 3.

4. In Fig. 24-22, a Gaussian surface encloses two of the four positively charged particles. (a) Which of the particles contribute to the electric field at point P on the surface? (b) Which net flux of electric field through the surface is greater (if either): that due to q_1 and q_2 or that due to all four charges?

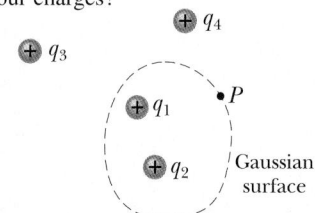

FIGURE 24-22 Question 4.

5. You are given a collection of eight particles with charges $+2q$, $+3q$, $+4q$, $+5q$, $-2q$, $-3q$, $-4q$, and $-5q$. You are also given

the goal of enclosing one or more of them with various Gaussian surfaces in turn, so that the net fluxes through the surfaces are 0, $+q/\epsilon_0$, $+2q/\epsilon_0$, $\cdots$, $+14q/\epsilon_0$. Which is impossible to produce?

6. There is a certain electric flux Φ_i through a spherical Gaussian surface of radius r when the surface encloses a proton. For the following situations, tell whether the net flux through that surface is greater than, less than, or equal to Φ_i. (a) The proton is outside the surface. (b) Two protons are inside the surface. (c) A proton is inside and another proton is outside. (d) A proton and an electron are inside.

7. Figure 24-23 shows, in cross section, a central metal ball, two spherical metal shells, and three spherical Gaussian surfaces of radii R, $2R$, and $3R$, all with the same center. The charges on the three objects are: ball, Q; smaller shell, $3Q$; larger shell, $5Q$. Rank the Gaussian surfaces according to the magnitude of the electric field at any point on the surface, greatest first.

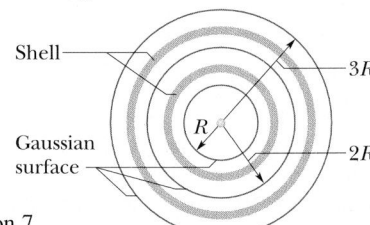

FIGURE 24-23 Question 7.

8. Figure 24-24 shows three Gaussian surfaces half-submerged in a large, thick metal plate with a uniform surface charge density. Surface S_1 is the tallest and has the smallest square end caps; surface S_3 is shortest and has the largest square end caps; and S_2 has intermediate values. Rank the surfaces according to (a) the charge they enclose, (b) the magnitude of the electric field at points on their top end cap, (c) the net electric flux through that top end cap, and (d) the net electric flux through their bottom end cap, greatest first.

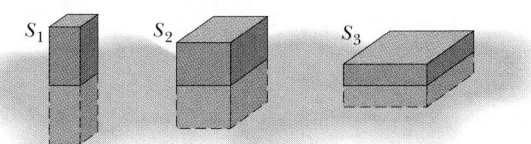

FIGURE 24-24 Question 8.

9. Figure 24-25 shows, in cross section, three cylinders, each of uniform charge Q. Concentric with each cylinder is a cylindrical Gaussian surface, all three with the same radius. Rank the Gaussian surfaces according to the electric field at any point on the surface, greatest first.

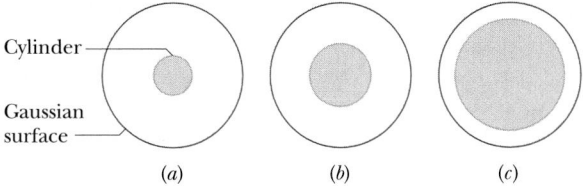

(a)   (b)   (c)

FIGURE 24-25 Question 9.

10. Figure 24-26 shows, in section, three long, uniformly charged cylinders centered on the same axis. Central cylinder A has a uniform charge of $q_A = +3q_0$. What uniform charges q_B and q_C should be on cylinders B and C so that (if possible) the net electric field is zero (a) at point 1, (b) at point 2, and (c) at point 3?

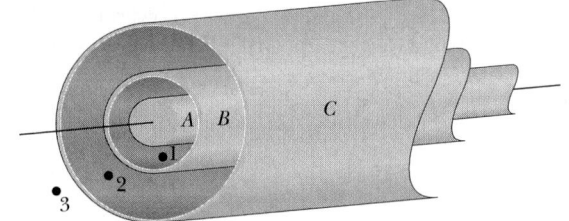

FIGURE 24-26 Question 10.

11. Three infinite nonconducting sheets, with uniform surface charge densities σ, 2σ, and 3σ, are arranged to be parallel like the two sheets in Fig. 24-17a. What is their order, from left to right, if the electric field **E** produced by the arrangement has magnitude $E = 0$ in one region and $E = 2\sigma/\epsilon_0$ in another region?

12. A small charged ball lies within the hollow of a metallic spherical shell of radius R. Here, for three situations, are the net charges on the ball and shell, respectively: (1) $+4q$, 0; (2) $-6q$, $+10q$; (3) $+16q$, $-12q$. Rank the situations according to the charge on (a) the inner surface of the shell and (b) the outer surface, most positive first.

13. Rank the situations of Question 12 according to the magnitude of the electric field (a) halfway through the shell and (b) at a point $2R$ from the center of the shell, greatest first.

14. In Checkpoint 4, what are the magnitude and direction of the electric field at a point that is a distance r from the center of the ball and spherical shell if the point is (a) between the ball and shell, (b) within the metal of the shell, (c) outside the shell?

15. A spherical nonconducting balloon has a uniform positive charge on its surface. If the balloon is expanded, does the magnitude of the electric field due to the charge increase, decrease, or remain the same at points that (a) are inside the balloon, (b) are on the balloon's surface, (c) were outside and are now inside, and (d) were and still are outside?

EXERCISES & PROBLEMS

SECTION 24-2 Flux

1E. Water in an irrigation ditch of width $w = 3.22$ m and depth $d = 1.04$ m flows with a speed of 0.207 m/s. The *mass flux* of the flowing water through an imaginary surface is the product of the water's density (1000 kg/m³) and its volume flux through that surface. Find the mass flux through the following imaginary surfaces: (a) a surface of area wd, entirely in the water, perpendicular to the flow; (b) a surface with area $3wd/2$, of which wd is in the water, perpendicular to the flow; (c) a surface of area $wd/2$, entirely in the water, perpendicular to the flow; (d) a surface of area wd, half in the water and half out, perpendicular to the flow; (e) a surface of area wd, entirely in the water, with its normal 34° from the direction of flow.

SECTION 24-3 Flux of An Electric Field

2E. The square surface shown in Fig. 24-27 measures 3.2 mm on each side. It is immersed in a uniform electric field with magnitude $E = 1800$ N/C. The field lines make an angle of 35° with a normal to the surface, as shown. Take the normal to be "outward," as though the surface were one face of a box. Calculate the electric flux through the surface.

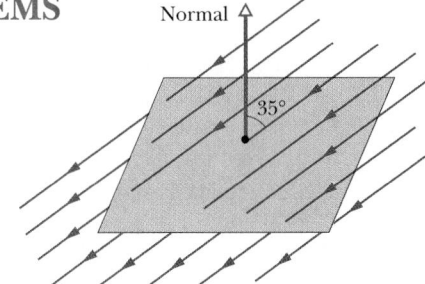

FIGURE 24-27
Exercise 2.

3E. A cube with 1.40 m edges is oriented as shown in Fig. 24-28 in a region of uniform electric field. Find the electric flux through the right face if the electric field, in newtons per coulomb, is given by (a) $6.00\mathbf{i}$, (b) $-2.00\mathbf{j}$, and (c) $-3.00\mathbf{i} + 4.00\mathbf{k}$. (d) What is the total flux through the cube for each of these fields?

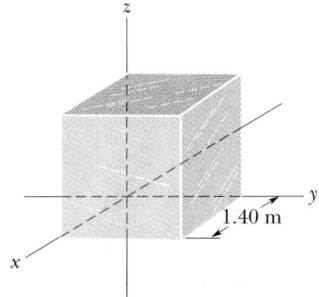

FIGURE 24-28
Exercise 3 and Problem 12.

4P. An electric field given by $\mathbf{E} = 4\mathbf{i} - 3(y^2 + 2)\mathbf{j}$ pierces the Gaussian cube of Fig. 24-5. (E is in newtons per coulomb and x is in meters.) What is the electric flux through (a) the top face, (b) the bottom face, (c) the left face, and (d) the back face? (e) What is the net electric flux through the cube?

SECTION 24-4 Gauss' Law

5E. You have four charges, $2q$, q, $-q$, and $-2q$. If possible, describe how you would place a closed surface that encloses at least the charge $2q$ and through which the net electric flux is (a) 0, (b) $+3q/\epsilon_0$, and (c) $-2q/\epsilon_0$.

6E. In Fig. 24-29, the charge on a neutral isolated conductor is separated by a nearby positively charged rod. What is the net flux through each of the five Gaussian surfaces shown in cross section? Assume that the charges enclosed by S_1, S_2, and S_3 are equal in magnitude.

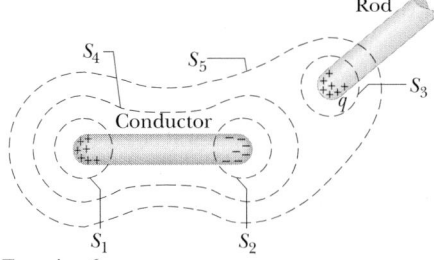

FIGURE 24-29 Exercise 6.

7E. A point charge of 1.8 μC is at the center of a cubical Gaussian surface 55 cm on edge. What is the net electric flux through the surface?

8E. The net electric flux through each face of a die (singular of dice) has a magnitude in units of 10^3 N·m²/C that is exactly equal to the number of spots N on the face (1 through 6). The flux is inward for N odd and outward for N even. What is the net charge inside the die?

9E. In Fig. 24-30, a point charge $+q$ is a distance $d/2$ directly above the center of a square of side d. What is the magnitude of the electric flux through the square? (*Hint:* Think of the square as one face of a cube with edge d.)

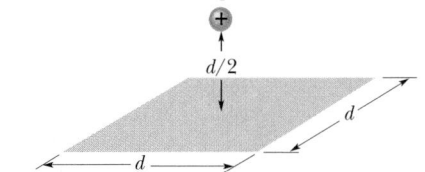

FIGURE 24-30
Exercise 9.

10E. In Fig. 24-31, a butterfly net is in a uniform electric field of magnitude E. The rim, a circle of radius a, is aligned perpendicular to the field. Find the electric flux through the netting.

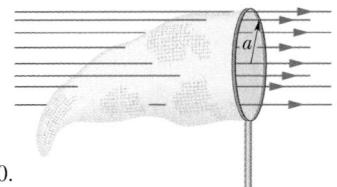

FIGURE 24-31 Exercise 10.

11E. Calculate Φ through (a) the flat base and (b) the curved surface of a hemisphere of radius R. The field $\mathbf{E}$ is uniform and perpendicular to the flat base of the hemisphere, and the field lines enter through the flat base.

12P. Find the net flux through the cube of Exercise 3 and Fig. 24-28 if the electric field is given by (a) $\mathbf{E} = 3.00y\mathbf{j}$ and (b) $\mathbf{E} = -4.00\mathbf{i} + (6.00 + 3.00y)\mathbf{j}$. E is in newtons per coulomb, and y is in meters. (c) In each case, how much charge is enclosed by the cube?

13P. What net charge is enclosed by the Gaussian cube of Problem 4 and Fig. 24-5?

14P. It is found experimentally that the electric field in a certain region of Earth's atmosphere is directed vertically down. At an altitude of 300 m the field has magnitude 60.0 N/C; at an altitude of 200 m, 100 N/C. Find the net amount of charge contained in a cube 100 m on edge, with horizontal faces at altitudes of 200 and 300 m. Neglect the curvature of Earth.

15P. A point charge q is placed at one corner of a cube of edge a. What is the flux through each of the cube faces? (*Hint:* Use Gauss' law and symmetry arguments.)

16P. "Gauss' law for gravitation" is

$$\frac{1}{4\pi G}\Phi_g = \frac{1}{4\pi G}\oint \mathbf{g}\cdot d\mathbf{A} = -m,$$

in which Φ_g is the net flux of the *gravitational field* $\mathbf{g}$ through a Gaussian surface that encloses a mass m. The field $\mathbf{g}$ is defined to be the acceleration of a test particle on which m exerts a gravitational force. Derive Newton's law of gravitation from this. What is the significance of the minus sign?

SECTION 24-6 A Charged Isolated Conductor

17E. The electric field just above the surface of the charged drum of a photocopying machine has a magnitude E of 2.3×10^5 N/C. What is the surface charge density on the drum, assuming that the drum is a conductor?

18E. A uniformly charged conducting sphere of 1.2 m diameter has a surface charge density of 8.1 μC/m². (a) Find the net charge on the sphere. (b) What is the total electric flux leaving the surface of the sphere?

19E. Space vehicles traveling through Earth's radiation belts can intercept a significant number of electrons. The resulting charge buildup can damage electronic components and disrupt operations. Suppose a spherical metallic satellite 1.3 m in diameter accumulates 2.4 μC of charge in one orbital revolution. (a) Find the resulting surface charge density. (b) Calculate the magnitude of the resulting electric field just outside the surface of the satellite due to the surface charge.

20E. A conducting sphere with positive charge Q is surrounded by a spherical conducting shell. (a) What is the net charge on the inner surface of the shell? (b) Another positive charge q is placed outside the shell. Now what is the net charge on the inner surface of the shell? (c) If q is moved to a position between the shell and the sphere, what then is the net charge on the inner surface of the shell? (d) Are your answers valid if the sphere and the shell are not concentric?

21P. An isolated conductor of arbitrary shape has a net charge of $+10 \times 10^{-6}$ C. Inside the conductor is a cavity within which is a point charge $q = +3.0 \times 10^{-6}$ C. What is the charge (a) on the cavity wall and (b) on the outer surface of the conductor?

SECTION 24-7 Applying Gauss' Law: Cylindrical Symmetry

22E. An infinite line of charge produces a field of 4.5×10^{4} N/C at a distance of 2.0 m. Calculate the linear charge density.

23E. (a) The drum of the photocopying machine in Exercise 17 has a length of 42 cm and a diameter of 12 cm. What is the total charge on the drum? (b) The manufacturer wishes to produce a desktop version of the machine. This requires reducing the size of the drum to a length of 28 cm and a diameter of 8.0 cm. The electric field at the drum surface must remain unchanged. What must be the charge on this new drum?

24P. Figure 24-32 shows a section of a long, thin-walled metal tube of radius R, carrying a charge per unit length λ on its surface. Derive expressions for E in terms of distance r from the tube axis, considering both (a) $r > R$ and (b) $r < R$. Plot your results for the range $r = 0$ to $r = 5.0$ cm, assuming that $\lambda = 2.0 \times 10^{-8}$ C/m and $R = 3.0$ cm. (*Hint:* Use cylindrical Gaussian surfaces, coaxial with the metal tube.)

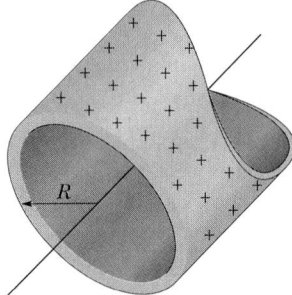

FIGURE 24-32 Problem 24.

25P. Figure 24-33 shows a section through two long thin concentric cylinders of radii a and b with $a < b$. The cylinders have equal and opposite charges per unit length λ. Using Gauss' law, prove (a) that $E = 0$ for $r < a$ and (b) that between the cylinders, where $a < r < b$,

$$E = \frac{1}{2\pi\epsilon_0} \frac{\lambda}{r}.$$

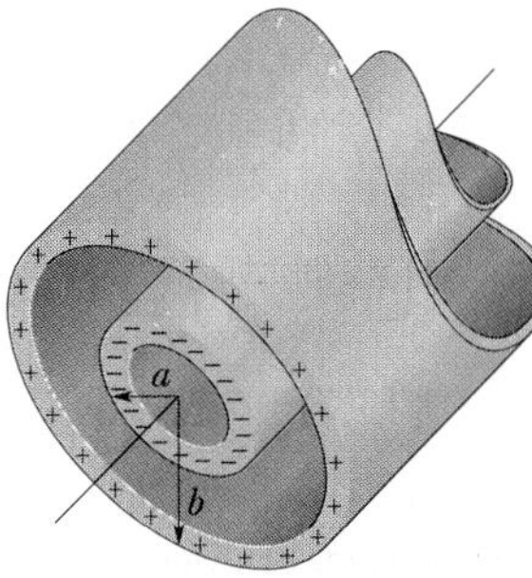

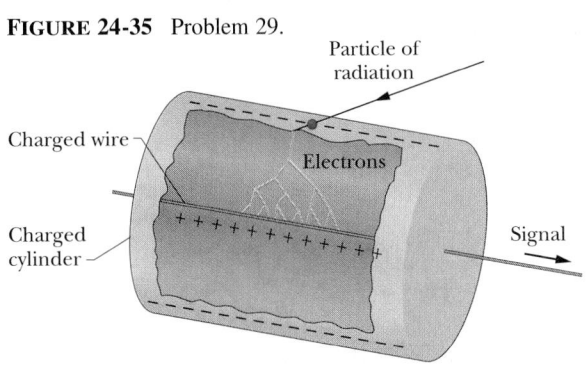

FIGURE 24-33 Problem 25.

26P. A long straight wire has fixed negative charge with a linear charge density of magnitude 3.6 nC/m. The wire is to be enclosed by a thin, nonconducting cylinder of outside radius 1.5 cm, coaxial with the wire. The cylinder is to have positive charge on its outside surface with a surface charge density σ such that the net external electric field is zero. Calculate the required σ.

27P. A very long conducting cylindrical rod of length L with a total charge $+q$ is surrounded by a conducting cylindrical shell (also of length L) with total charge $-2q$, as shown in the section in Fig. 24-34. Use Gauss' law to find (a) the electric field at points outside the conducting shell, (b) the distribution of charge on the conducting shell, and (c) the electric field in the region between the shell and rod.

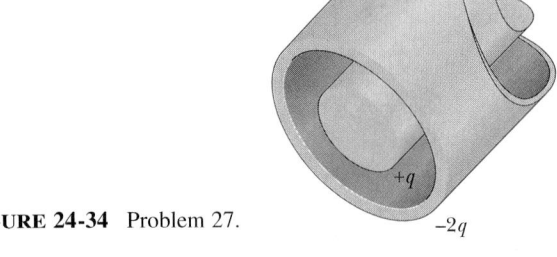

FIGURE 24-34 Problem 27.

28P. Two long, charged, concentric cylinders have radii of 3.0 and 6.0 cm. The charge per unit length is 5.0×10^{-6} C/m on the inner cylinder and -7.0×10^{-6} C/m on the outer cylinder. Find the electric field at (a) $r = 4.0$ cm and (b) $r = 8.0$ cm, where r is the radial distance from the common central axis.

29P. Figure 24-35 shows a Geiger counter, a device used to detect ionizing radiation (radiation that causes ionization of atoms). The counter consists of a thin, positively charged central wire surrounded by a concentric circular conducting cylinder with an equal negative charge. Thus a strong radial electric field is set up inside the cylinder. The cylinder contains a low-pressure inert gas. When a particle of radiation enters the device through the cylinder wall, it ionizes a few of the gas atoms. The resulting free electrons are drawn to the positive wire. However, the electric field is so intense that, between collisions with other gas atoms, the free electrons gain energy sufficient to ionize these atoms also. More free electrons are thereby created, and the process is repeated until the electrons reach the wire. The resulting

FIGURE 24-35 Problem 29.

"avalanche" of electrons is collected by the wire, generating a signal that is used to record the passage of the original particle of radiation. Suppose that the radius of the central wire is 25 μm, the radius of the cylinder 1.4 cm, and the length of the tube 16 cm. If the electric field at the cylinder's inner wall is 2.9×10^4 N/C, what is the total positive charge on the central wire?

30P. A positron, of charge 1.60×10^{-19} C, revolves in a circular path of radius r, between and concentric with the cylinders of Problem 25. What must be its kinetic energy K in electron-volts? Assume that $a = 2.0$ cm, $b = 3.0$ cm, and $\lambda = 30$ nC/m.

31P. Charge is distributed uniformly throughout the volume of an infinitely long cylinder of radius R. (a) Show that at a distance r from the cylinder axis (for $r < R$),

$$E = \frac{\rho r}{2\epsilon_0},$$

where ρ is the volume charge density. (b) Write an expression for E when $r > R$.

SECTION 24-8 Applying Gauss' Law: Planar Symmetry

32E. Figure 24-36 shows cross-sections through two large, parallel, nonconducting sheets with identical distributions of positive charge with surface charge density σ. What is **E** at points (a) above the sheets, (b) between them, and (c) below them?

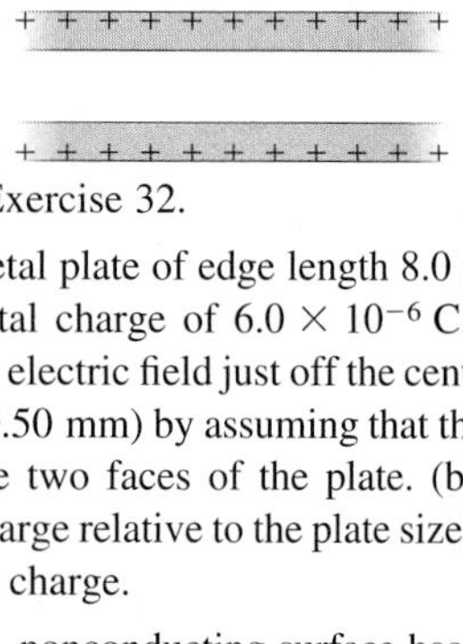

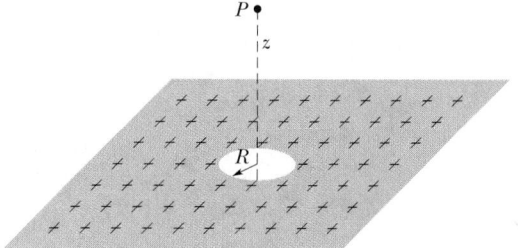

FIGURE 24-36 Exercise 32.

33E. A square metal plate of edge length 8.0 cm and negligible thickness has a total charge of 6.0×10^{-6} C. (a) Estimate the magnitude E of the electric field just off the center of the plate (at, say, a distance of 0.50 mm) by assuming that the charge is spread uniformly over the two faces of the plate. (b) Estimate E at a distance of 30 m (large relative to the plate size) by assuming that the plate is a point charge.

34E. A large, flat, nonconducting surface has a uniform charge density σ. A small circular hole of radius R has been cut in the middle of the surface, as shown in Fig. 24-37. Ignore fringing of the field lines around all edges, and calculate the electric field at point P, a distance z from the center of the hole along its axis. (*Hint:* See Eq. 23-24 and use superposition.)

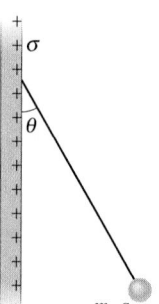

FIGURE 24-37 Exercise 34.

35P. In Fig. 24-38, a small, nonconducting ball of mass $m = 1.0$ mg and charge $q = 2.0 \times 10^{-8}$ C (distributed uniformly through its volume) hangs from an insulating thread that makes an angle $\theta = 30°$ with a vertical, uniformly charged nonconducting sheet (shown in cross section). Considering the weight of the ball and assuming that the sheet extends far vertically and into and out of the page, calculate the surface charge density σ of the sheet.

FIGURE 24-38 Problem 35.

36P. Two large thin metal plates are parallel and close to each other as in Fig. 24-16c, but with the negative plate on the left. On their inner faces, the plates have surface charge densities of opposite signs and of magnitude 7.0×10^{-22} C/m². What are the magnitude and direction of the electric field **E** (a) to the left of the plates, (b) to the right of the plates, and (c) between the plates?

37P. An electron is fired directly toward the center of a large metal plate that has excess negative charge with surface charge density 2.0×10^{-6} C/m². If the initial kinetic energy of the electron is 100 eV and if the electron is to stop (owing to electrostatic repulsion from the plate) just as it reaches the plate, how far from the plate must it be fired?

38P. Two large metal plates of area 1.0 m² face each other. They are 5.0 cm apart and have equal but opposite charges on their inner surfaces. If the magnitude E of the electric field between the plates is 55 N/C, what is the magnitude of the charge on each plate? Neglect edge effects.

39P. In a laboratory experiment, an electron's weight is just balanced by the force exerted on the electron by an electric field. If the electric field is due to charges on two large, parallel, nonconducting plates, oppositely charged and separated by 2.3 cm, (a) what is the magnitude of the surface charge density, assumed to be uniform, on the plates, and (b) in which direction does the field point?

40P*. A charge $+q$ placed a distance a from an infinite conducting plane induces negative charge on the plane with a surface charge density $\sigma = -qa/(2\pi r^3)$, where r is the distance from the charge $+q$ to a point P on the plane (Fig. 24-39). What are (a) the magnitude E of the electric field normal to the plane due to this induced charge and (b) the total negative charge induced on the plane? (c) What is the electrostatic force between charge $+q$ and the induced charge on the conducting plane? Is the force attractive or repulsive? (d) What charge, placed diametrically opposite charge $+q$ (on the other side of the plane, at the same distance from the plane) will give this same force?

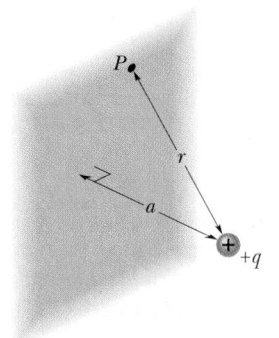

FIGURE 24-39
Problem 40.

41P*. A planar slab of thickness d has a uniform volume charge density ρ. Find the magnitude of the electric field at all points in space both (a) inside and (b) outside the slab, in terms of x, the distance measured from the central plane of the slab.

SECTION 24-9 Applying Gauss' Law: Spherical Symmetry

42E. A conducting sphere of radius 10 cm has an unknown charge. If the electric field 15 cm from the center of the sphere is 3.0×10^3 N/C and points radially inward, what is the net charge on the sphere?

43E. A point charge causes an electric flux of -750 N·m²/C to pass through a spherical Gaussian surface of 10.0 cm radius centered on the charge. (a) If the radius of the Gaussian surface were doubled, how much flux would pass through the surface? (b) What is the value of the point charge?

44E. A thin-walled metal sphere has a radius of 25 cm and a charge of 2.0×10^{-7} C. Find E for a point (a) inside the sphere, (b) just outside the sphere, and (c) 3.0 m from the center.

45E. A point charge $q = 1.0 \times 10^{-7}$ C is at the center of a spherical cavity of radius 3.0 cm in a chunk of metal. Use Gauss' law to find the electric field (a) at point P_1, halfway from the center to the surface of the cavity, and (b) at point P_2, within the metal wall.

46E. Two charged concentric spheres have radii of 10.0 and 15.0 cm. The charge on the inner sphere is 4.00×10^{-8} C and that on the outer sphere is 2.00×10^{-8} C. Find the electric field (a) at $r = 12.0$ cm and (b) at $r = 20.0$ cm.

47E. A thin, metallic, spherical shell of radius a has a charge q_a. Concentric with it is another thin, metallic, spherical shell of radius b (where $b > a$) and charge q_b. Find the electric field at radial points r where (a) $r < a$, (b) $a < r < b$, and (c) $r > b$. (d) Discuss the criterion one would use to determine how the charges are distributed on the inner and outer surfaces of the shells.

48E. In a 1911 paper, Ernest Rutherford said: "In order to form some idea of the forces required to deflect an α particle through a large angle, consider an atom containing a point positive charge Ze at its centre and surrounded by a distribution of negative electricity, $-Ze$ uniformly distributed within a sphere of radius R.

The electric field E . . . at a distance r from the center for a point *inside* the atom [is]

$$E = \frac{Ze}{4\pi\epsilon_0}\left(\frac{1}{r^2} - \frac{r}{R^3}\right)."$$

Verify this equation.

49E. Equation 24-11 ($E = \sigma/\epsilon_0$) gives the electric field at points near a charged conducting surface. Apply this equation to a conducting sphere of radius r and charge q, and show that the electric field outside the sphere is the same as the field of a point charge located at the center of the sphere.

50P. A proton with speed $v = 3.00 \times 10^5$ m/s orbits just outside a charged sphere of radius $r = 1.00$ cm. What is the charge on the sphere?

51P. A point charge $+q$ is placed at the center of an electrically neutral, spherical conducting shell with inner radius a and outer radius b. What charge appears on (a) the inner surface of the shell and (b) the outer surface? Find expressions for the net electric field at a distance r from the center of the shell if (c) $r < a$, (d) $b > r > a$, and (e) $r > b$. Sketch field lines for those three regions. For $r > b$, what is the net electric field due to (f) the central point charge and inner surface charge and (g) the outer surface charge? A point charge $-q$ is now placed outside the shell. Does this point charge change the charge distribution on (h) the outer surface and (i) the inner surface? Sketch the field lines now. (j) Is there an electrostatic force on the second point charge? (k) Is there a net electrostatic force on the first point charge? (l) Does this situation violate Newton's third law?

52P. A solid nonconducting sphere of radius R has a nonuniform charge distribution of volume charge density $\rho = \rho_s r/R$, where ρ_s is a constant and r is the distance from the center of the sphere. Show that (a) the total charge on the sphere is $Q = \pi\rho_s R^3$ and (b) the electric field inside the sphere has a magnitude given by

$$E = \frac{1}{4\pi\epsilon_0}\frac{Q}{R^4}r^2.$$

53P. In Fig. 24-40 a sphere, of radius a and charge $+q$ uniformly distributed throughout its volume, is concentric with a spherical conducting shell of inner radius b and outer radius c. This shell has a net charge of $-q$. Find expressions for the electric field, as a function of the radius r, (a) within the sphere ($r < a$); (b) between the sphere and the shell ($a < r < b$); (c) inside the shell ($b < r < c$); and (d) outside the shell ($r > c$). (e) What are the charges on the inner and outer surfaces of the shell?

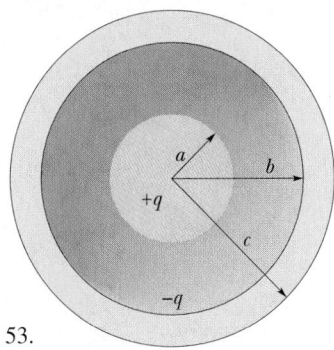

FIGURE 24-40 Problem 53.

54P. Figure 24-41*a* shows a spherical shell of charge of uniform volume charge density ρ. Plot E due to the shell for distances r from the center of the shell ranging from zero to 30 cm. Assume that $\rho = 1.0 \times 10^{-6}$ C/m^3, $a = 10$ cm, and $b = 20$ cm.

55P. In Fig. 24-41*b*, a nonconducting spherical shell, of inner radius a and outer radius b, has a volume charge density $\rho = A/r$ (within its thickness), where A is a constant and r is the distance from the center of the shell. In addition, a point charge q is located at the center. What value should A have if the electric field in the shell ($a \leq r \leq b$) is to be uniform? (*Hint:* The constant A depends on a but not on b.)

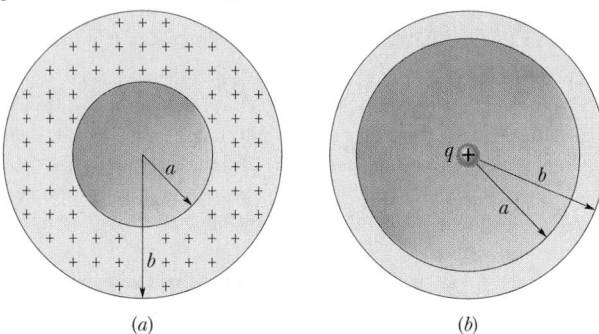

(*a*)   (*b*)

FIGURE 24-41   Problems 54 and 55.

56P. A hydrogen atom can be considered as having a central pointlike proton of positive charge $+e$ and an electron of negative charge $-e$ that is distributed about the proton according to the volume charge density $\rho = A \exp(-2r/a_0)$. Here A is a constant, $a_0 = 0.53 \times 10^{-10}$ m is the *Bohr radius,* and r is the distance from the center of the atom. (a) Using the fact that hydrogen is electrically neutral, find A. (b) Then find the electric field produced by the atom at the Bohr radius.

57P*. A nonconducting sphere has a uniform volume charge density ρ. Let **r** be the vector from the center of the sphere to a general point P within the sphere. (a) Show that the electric field at P is given by $\mathbf{E} = \rho \mathbf{r}/3\epsilon_0$. (Note that the result is independent of the radius of the sphere.) (b) A spherical cavity is hollowed out of the sphere, as shown in Fig. 24-42. Using superposition concepts, show that the electric field at all points within the cavity is $\mathbf{E} = \rho \mathbf{a}/3\epsilon_0$ (uniform field), where **a** is the position vector pointing from the center of the sphere to the center of the cavity. (Note that this result is independent of the radius of the sphere and also the radius of the cavity.)

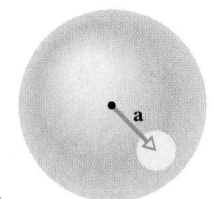

FIGURE 24-42   Problem 57.

58P*. A spherically symmetrical but nonuniform distribution of charge produces an electric field of magnitude $E = Kr^4$, directed radially outward from the center of the sphere. Here r is the radial distance from that center. What is the volume density ρ of the charge distribution?

Electronic Computation

59. From Exercise 48, rewrite Rutherford's equation for the magnitude E of the electric field inside an atom by substituting $r = \alpha R$. Use the rewritten equation to plot E versus α for the range $0 < \alpha < R$. Also plot the magnitude E' of the electric field that would be produced by the nucleus alone. From the two plots determine the value of α, for which $E = 0.500E'$.

60. A computer can be used to demonstrate Gauss' law for a situation in which the electric field is not everywhere perpendicular to the Gaussian surface. Suppose a cube with edges that are 1.00 m long is centered at the origin of a coordinate system whose axes are parallel to the edges. A charge of 1.00 μC is on the y axis at a position given below. Divide each face of the cube into a large number of small squares, calculate the electric flux through each square, and sum the results to obtain the total flux through the face. Finally, sum the fluxes through all the faces to obtain the total flux through the cube surface. Compare the result with q/ϵ_0, where q is the charge inside the cube. The more squares you use for each calculation, the better your result will be; but you should obtain accuracy to three significant figures if each square has a side that is one-thirtieth of a cube edge. If y' is the coordinate of the charge, then the electric field at the point with coordinates x, y, and z has an x component that is given by $E_x = (q/4\pi\epsilon_0)x/r^3$, a y component that is given by $E_y = (q/4\pi\epsilon_0)(y - y')/r^3$, and a z component that is given by $E_z = (q/4\pi\epsilon_0)z/r^3$, in which $r = [x^2 + (y - y')^2 + z^2]^{1/2}$. (a) Take $y' = 0$ (the charge is at the center of the cube). (b) Take $y' = 0.200$ m (the charge is inside the cube). (c) Take $y' = 0.400$ m (the charge is inside the cube). (d) Take $y' = 0.600$ m (the charge is outside the cube).

25
Electric Potential

While enjoying the Sequoia National Park from a lookout platform, this woman found her hair rising from her head. Amused, her brother took her photograph. Five minutes after they left, lightning struck the platform, killing one person and injuring seven. What had caused the woman's hair to rise? From her look, it was not fear—but she certainly should have been fearful.

25-1 ELECTRIC POTENTIAL ENERGY

Newton's law for the gravitational force and Coulomb's law for the electrostatic force are mathematically identical. Thus, the general features we have discussed for the gravitational force should apply to the electrostatic force.

In particular, we can infer that the electrostatic force is a *conservative force*. Thus when that force acts between two or more charged particles within a system of particles, we can assign an **electric potential energy** U to the system. Moreover, if the system changes its configuration from an initial state i to a different final state f, the electrostatic force does work W on the particles. From Eq. 8-1, we then know that the resulting change ΔU in the potential energy of the system is

$$\Delta U = U_f - U_i = -W. \qquad (25\text{-}1)$$

As with other conservative forces, the work done by the electrostatic force is *path independent*: Suppose a charged particle within the system moves from point i to point f while an electrostatic force between it and the rest of the system acts on it. Provided the rest of the system does not change, the work W done by the force is the same for *any* path between points i and f.

For convenience, we usually take the *reference configuration* of a system of charged particles to be that in which the particles are all infinitely separated from each other. And we usually set the corresponding *reference potential energy* to be zero. Suppose that several charged particles come together from initially infinite separations (state i) to form a system of nearby particles (state f). Let the initial potential energy U_i be zero, and let W_∞ represent the work done by the electrostatic forces between the particles during the move in from infinity. Then from Eq. 25-1, the final potential energy U of the system is

$$U = -W_\infty. \qquad (25\text{-}2)$$

As is true of other kinds of potential energy, electric potential energy is considered to be a type of mechanical energy. Recall from Chapter 8 that if only conservative forces act within a (closed) system, the mechanical energy of the system is conserved. We shall use this fact extensively in the rest of this chapter.

PROBLEM SOLVING TACTICS

TACTIC 1: *Electric Potential Energy; Work Done by a Field*

An electric potential energy is associated with a system of particles as a whole. However, you will see statements (starting with Sample Problem 25-1) that associate it with only one

particle within a system. For example, you might read, "An electron in an electric field has a potential energy of 10^{-7} J." Such statements are often acceptable, but you should always keep in mind that the potential energy is actually associated with a system—here the electron plus the charged particles that set up the electric field. Also keep in mind that it makes sense to assign a particular potential energy value, such as 10^{-7} J here, to a particle or even a system *only* if the reference potential energy value is known.

When the potential energy is associated with only one particle within a system, you often will read that the work done on the particle is *by the electric field*. What is meant is that the work is done by the force on the particle due to the charges that set up the field.

SAMPLE PROBLEM 25-1

Electrons are continually being knocked out of air molecules in the atmosphere by cosmic-ray particles coming in from space. Once released, an electron experiences an electrostatic force $\mathbf{F}$ due to the electric field $\mathbf{E}$ that is produced in the atmosphere by charged particles already on Earth. Near Earth's surface the electric field has the magnitude $E = 150$ N/C and is directed downward. What is the change ΔU in the electric potential energy of a released electron when the electrostatic force causes it to move vertically upward through a distance $d = 520$ m (Fig. 25-1)?

SOLUTION: Equation 25-1 relates the change ΔU in the electric potential energy of the electron to the work W done on the electron by the electric field. From Chapter 7 we know that the work done by a constant force $\mathbf{F}$ on a particle undergoing a displacement $\mathbf{d}$ is

$$W = \mathbf{F} \cdot \mathbf{d}. \qquad (25\text{-}3)$$

From Eq. 23-28, we know that the electrostatic force and the electric field are related by $\mathbf{F} = q\mathbf{E}$. Recall that the sign of charge q is to be used in this vector equation—here q $(= -1.6 \times 10^{-19}$ C$)$ is the charge of an electron. Substituting for $\mathbf{F}$ in Eq. 25-3 and taking the dot product yield

$$W = q\mathbf{E} \cdot \mathbf{d} = qEd \cos \theta, \qquad (25\text{-}4)$$

where θ is the angle between the directions of $\mathbf{E}$ and $\mathbf{d}$. The field $\mathbf{E}$ is directed downward and the displacement $\mathbf{d}$ is di-

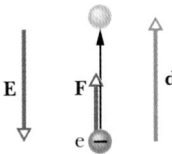

FIGURE 25-1   Sample Problem 25-1. An electron in the atmosphere is moved upward through displacement $\mathbf{d}$ by an electrostatic force $\mathbf{F}$ due to an electric field $\mathbf{E}$.

rected upward. Thus, $\theta = 180°$. Substituting this and other data into Eq. 25-4, we find

$$W = (-1.6 \times 10^{-19}\ C)(150\ N/C)(520\ m)\cos 180°$$
$$= 1.2 \times 10^{-14}\ J.$$

Equation 25-1 then yields

$$\Delta U = -W = -1.2 \times 10^{-14}\ J. \qquad \text{(Answer)}$$

This result tells us that during the 520 m ascent, the electric potential energy of the electron decreases by 1.2×10^{-14} J.

CHECKPOINT **1:** In the figure, a proton moves from point i to point f in a uniform electric field directed as shown. (a) Does the electric field do positive or negative work on the proton? (b) Does the electric potential energy of the proton increase or decrease?

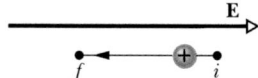

25-2 ELECTRIC POTENTIAL

As you can infer from Sample Problem 25-1, the potential energy of a charged particle in an electric field depends on the magnitude of the charge. However, the potential energy *per unit charge* has a unique value at any point in an electric field.

For example, suppose we place a test particle of positive charge 1.60×10^{-19} C at a point in an electric field where the particle has an electric potential energy of 2.40×10^{-17} J. Then the potential energy per unit charge is

$$\frac{2.40 \times 10^{-17}\ J}{1.60 \times 10^{-19}\ C} = 150\ J/C.$$

Next, suppose we replace that test particle with one having twice as much positive charge, 3.20×10^{-19} C. We would find that the second particle has an electric potential energy of 4.80×10^{-17} J, twice that of the first particle. However, the potential energy per unit charge would be the same, still 150 J/C.

Thus the potential energy per unit charge, which can be symbolized as U/q, is independent of the charge q of the particle we happen to use and is *characteristic only of the electric field* we are investigating. The potential energy per unit charge at a point in an electric field is called the **electric potential** V (or simply the **potential**) at that point. Thus,

$$V = \frac{U}{q}. \qquad (25\text{-}5)$$

Note that electric potential is a scalar, not a vector.

The *electric potential difference* ΔV between any two points i and f in an electric field is equal to the difference in potential energy per unit charge between the two points:

$$\Delta V = V_f - V_i = \frac{U_f}{q} - \frac{U_i}{q} = \frac{\Delta U}{q}. \qquad (25\text{-}6)$$

Using Eq. 25-1 to substitute $-W$ for ΔU in Eq. 25-6, we can define the potential difference between points i and f as

$$\Delta V = V_f - V_i = -\frac{W}{q} \qquad \begin{array}{c}\text{(potential}\\\text{difference defined).}\end{array} \qquad (25\text{-}7)$$

The potential difference between two points is thus the negative of the work done by the electrostatic force per unit charge that moves from one point to the other. Potential difference can be positive, negative, or zero, depending on the signs and magnitudes of q and W.

If we set $U_i = 0$ at infinity as our reference potential energy, then by Eq. 25-5, the electric potential must also be zero there. Then from Eq. 25-7, we can define the electric potential V at any point f in an electric field to be

$$V = -\frac{W_\infty}{q} \qquad \text{(potential defined),} \qquad (25\text{-}8)$$

where W_∞ is the work done by the electric field on a charged particle as that particle moves in from infinity to point f. A potential V can be positive, negative, or zero, depending on the signs and magnitudes of q and W_∞.

The SI unit for potential that follows from Eq. 25-8 is the joule per coulomb. This combination occurs so often that a special unit, the *volt* (abbreviated V) is used to represent it. That is,

$$1\ \text{volt} = 1\ \text{joule per coulomb.} \qquad (25\text{-}9)$$

This new unit allows us to adopt a more conventional unit for the electric field $\mathbf{E}$, which we have measured up to now in newtons per coulomb. With two unit conversions, we obtain

$$1\ N/C = \left(1\ \frac{N}{C}\right)\left(\frac{1\ V\cdot C}{1\ J}\right)\left(\frac{1\ J}{1\ N\cdot m}\right)$$
$$= 1\ V/m. \qquad (25\text{-}10)$$

The conversion factor in the second set of parentheses comes from Eq. 25-9; that in the third set of parentheses is derived from the definition of the joule. From now on, we shall express values of the electric field in volts per meter rather than in newtons per coulomb.

Finally, we are now in a position to define the electron-volt, the energy unit that was introduced in Section 7-1 as a convenient one for energy measurements in the atomic and subatomic domain. One *electron-volt* (eV) is the energy equal to the work required to move a single elementary charge e, such as that of the electron or the proton, through a potential difference of exactly one volt. Equation 25-7 tells us that the magnitude of this work is $q \, \Delta V$, so

$$1 \text{ eV} = e(1 \text{ V})$$
$$= (1.60 \times 10^{-19} \text{ C})(1 \text{ J/C}) = 1.60 \times 10^{-19} \text{ J}.$$

PROBLEM SOLVING TACTICS

TACTIC 2: *Electric Potential and Electric Potential Energy*
Electric potential V and electric potential energy U are quite different quantities and should not be confused.

Electric potential is a property of an electric field, regardless of whether a charged object has been placed in that field; it is measured in joules per coulomb, or volts.

Electric potential energy is an energy of a charged object in an external electric field (or more precisely, an energy of the system consisting of the object and the external electric field); it is measured in joules.

Work Done by an Applied Force

Suppose we move a particle of charge q from point i to point f in an electric field by applying a force to it. During the move, our applied force does work W_{app} on the charge while the electric field does work W on it. By the work–kinetic energy theorem of Eq. 7-15, the change ΔK in the kinetic energy of the particle is

$$\Delta K = K_f - K_i = W_{app} + W. \qquad (25\text{-}11)$$

Now suppose the particle is stationary before and after the move. Then K_f and K_i are both zero, and Eq. 25-11 reduces to

$$W_{app} = -W. \qquad (25\text{-}12)$$

In words, the work W_{app} done by our applied force during the move is equal to the negative of the work W done by the electric field.

By substituting Eq. 25-12 into Eq. 25-1, we can relate the work done by our applied force to the change in the potential energy of the particle during the move. We find

$$\Delta U = U_f - U_i = W_{app}. \qquad (25\text{-}13)$$

By similarly substituting Eq. 25-12 into Eq. 25-7, we can relate our work W_{app} to the electric potential difference ΔV

between the initial and final points of the particle. We find

$$W_{app} = q \, \Delta V. \qquad (25\text{-}14)$$

W_{app} can be positive, negative, or zero depending on the signs and magnitudes of q and ΔV. It is the work we must do to move a particle of charge q through a potential difference ΔV with no change in the particle's kinetic energy.

CHECKPOINT 2: In the figure of Checkpoint 1, we move a proton from point i to point f in a uniform electric field directed as shown. (a) Does our force do positive or negative work? (b) Does the proton move to higher or lower potential?

25-3 EQUIPOTENTIAL SURFACES

Adjacent points that have the same electric potential form an **equipotential surface,** which can be either an imaginary surface or a real, physical surface. No net work W is done on a charged particle by an electric field when the particle moves between two points i and f on the same equipotential surface. This follows from Eq. 25-7, which tells us that W must be zero if $V_f = V_i$. Because of the path independence of work (and thus of potential energy and potential), $W = 0$ for *any* path connecting points i and f, regardless of whether that path lies entirely on the equipotential surface.

Figure 25-2 shows a *family* of equipotential surfaces, associated with the electric field due to some distribution of charges. The work done by the electric field on a charged particle as the particle moves from one end to the other of paths I and II is zero because each of these paths begins and ends on the same equipotential surface. The work done as the charged particle moves from one end to the other of paths III and IV is not zero but has the same

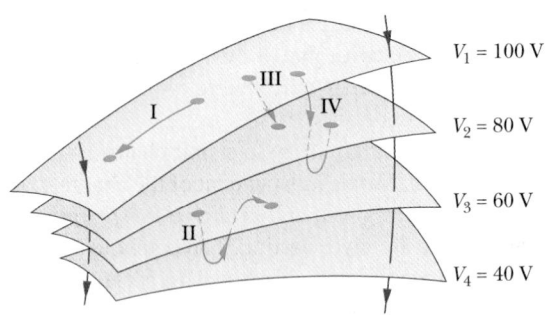

FIGURE 25-2 Portions of four equipotential surfaces. Four paths along which a test charge may move are also shown. Two electric field lines are indicated.

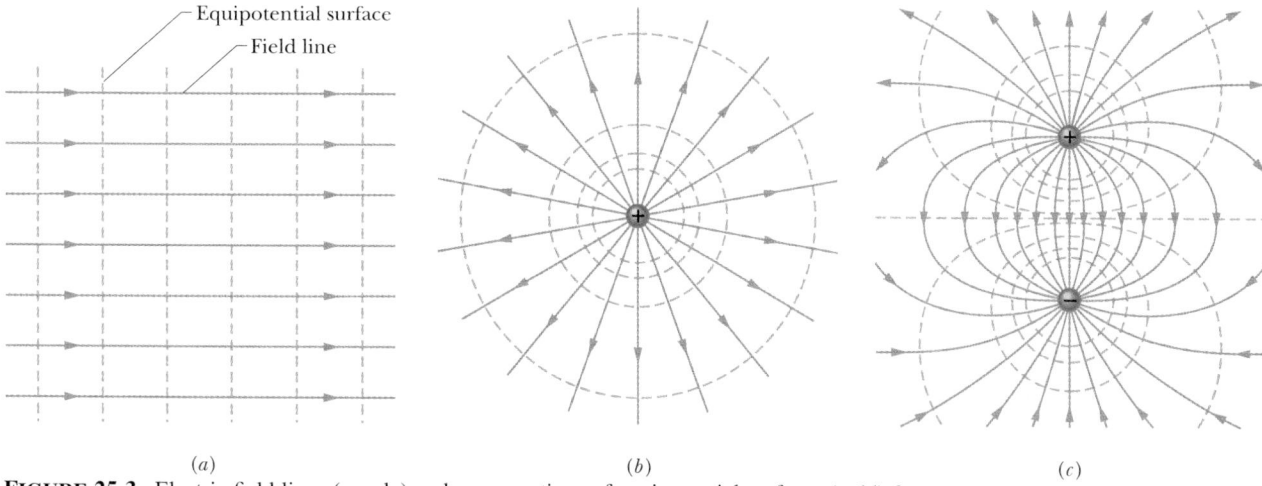

(a) (b) (c)

FIGURE 25-3 Electric field lines (purple) and cross sections of equipotential surfaces (gold) for (a) a uniform field, (b) the field of a point charge, and (c) the field of an electric dipole.

value for both these paths because the initial and final potentials are identical for the two paths. That is, paths III and IV connect the same pair of equipotential surfaces.

From symmetry, the equipotential surfaces produced by a point charge or a spherically symmetrical charge distribution are a family of concentric spheres. For a uniform field, the surfaces are a family of planes perpendicular to the field lines. In fact, equipotential surfaces are always perpendicular to electric field lines and thus to **E**, which is always tangent to these lines. If **E** were *not* perpendicular to an equipotential surface, it would have a component lying along that surface. This component would then do work on a charged particle as it moved along the surface. But by Eq. 25-7 work cannot be done if the surface is truly an equipotential surface; the only possible conclusion is that **E** must be everywhere perpendicular to the surface. Figure 25-3 shows electric field lines and cross sections of the equipotential surfaces for a uniform electric field and for the field associated with a point charge and with an electric dipole.

We now return to the woman in the opening photograph for this chapter. Because she was standing on a platform that was connected to the mountainside, she was at about the same potential as the mountainside. A highly charged cloud system created a strong electric field around her and the mountainside, with **E** pointing outward from her and the mountain. Electrostatic forces due to this field drove some of the conduction electrons in the woman downward through her body, leaving the strands of hair positively charged. The magnitude of **E** was apparently large, but less than the value of about 3×10^6 V/m that would have caused electrical breakdown of the air mole-

cules. (That value was exceeded shortly later when the lightning struck the platform.)

The equipotential surfaces surrounding the woman on the mountainside platform can be inferred from her hair: the strands are extended along the direction of **E** and thus are perpendicular to the equipotential surfaces, as drawn in Fig. 25-4. The magnitude of **E** was apparently greatest (the

FIGURE 25-4 This enhancement of the chapter's opening photograph shows the result of an overhead cloud system creating a strong electric field **E** near a woman's head. Many of the hair strands extended along the field, which was perpendicular to the equipotential surfaces and greatest where those surfaces were closest, near the top of her head.

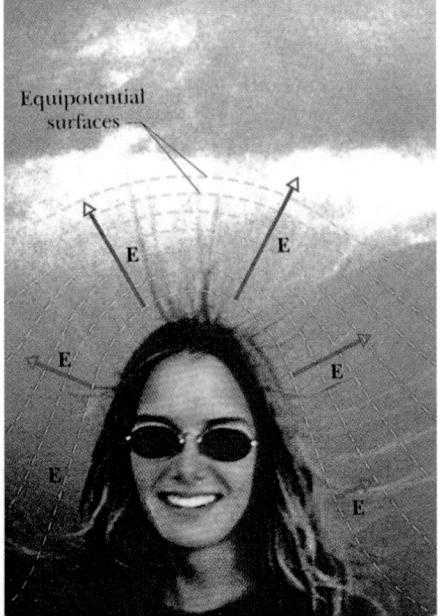

equipotential surfaces were most closely spaced) just above her head, because the hair there was extended farther than the hair around the side.

The lesson here is simple. If an electric field causes the hairs on your head to stand up, you had better run for shelter—not pose for a snapshot.

25-4 CALCULATING THE POTENTIAL FROM THE FIELD

We can calculate the potential difference between any two points i and f in an electric field if we know the field vector $\mathbf{E}$ at all positions along any path connecting those points. To make the calculation, we find the work done on a positive test charge by the field as the charge moves from i to f, and then use Eq. 25-7.

Consider an arbitrary electric field, represented by the field lines in Fig. 25-5, and a positive test charge q_0 that moves along the path shown from point i to point f. At any point on the path, an electrostatic force $q_0\mathbf{E}$ acts on the charge as it moves through a differential displacement $d\mathbf{s}$. From Chapter 7, we know that the differential work dW done on a particle by a force $\mathbf{F}$ during a displacement $d\mathbf{s}$ is

$$dW = \mathbf{F} \cdot d\mathbf{s}. \qquad (25\text{-}15)$$

For the situation of Fig. 25-5, $\mathbf{F} = q_0\mathbf{E}$ and Eq. 25-15 becomes

$$dW = q_0\mathbf{E} \cdot d\mathbf{s}. \qquad (25\text{-}16)$$

To find the total work W done on the particle by the field as the particle moves from point i to point f, we sum—via integration—the differential work done on the charge for all the differential displacements $d\mathbf{s}$ along the path:

$$W = q_0 \int_i^f \mathbf{E} \cdot d\mathbf{s}. \qquad (25\text{-}17)$$

If we substitute the total work W from Eq. 25-17 into Eq.

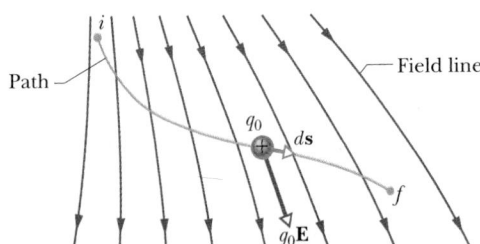

FIGURE 25-5 A test charge q_0 moves from point i to point f along the path shown in a nonuniform electric field. During a displacement $d\mathbf{s}$, an electrostatic force $q_0\mathbf{E}$ acts on the test charge. This force points in the direction of the field line at the location of the test charge.

25-7 we find

$$V_f - V_i = -\int_i^f \mathbf{E} \cdot d\mathbf{s}. \qquad (25\text{-}18)$$

Thus the potential difference $V_f - V_i$ between any two points i and f in an electric field is equal to the negative of the *line integral* (meaning the integral along the path) of $\mathbf{E} \cdot d\mathbf{s}$ from i to f. Note that this result is independent of the value of q_0 that we used to obtain it.

If the electric field is known throughout a certain region, Eq. 25-18 allows us to calculate the difference in potential between any two points in the field. Because the electrostatic force is conservative, all paths (whether easy or difficult to use) yield the same result.

If we choose the potential V_i at point i to be zero, then Eq. 25-18 becomes

$$V = -\int_i^f \mathbf{E} \cdot d\mathbf{s}, \qquad (25\text{-}19)$$

in which we have dropped the subscript f on V_f. Equation 25-19 gives us the potential V at any point f in the electric field *relative to the zero potential* at point i. If we let point i be at infinity, then Eq. 25-19 gives us the potential V at any point f relative to the zero potential at infinity.

SAMPLE PROBLEM 25-2

(a) Figure 25-6a shows two points i and f in a uniform electric field $\mathbf{E}$. The points lie on the same electric field line (not shown) and are separated by a distance d. Find the potential difference $V_f - V_i$ by moving a positive test charge q_0 from i to f along a path that is parallel to the field direction.

SOLUTION: As the test charge moves from i to f in Fig. 25-6a, its differential displacement $d\mathbf{s}$, which is always in the direction of motion, points in the same direction as the electric field $\mathbf{E}$. The angle θ between these two vectors is then zero, and Eq. 25-18 becomes

$$V_f - V_i = -\int_i^f \mathbf{E} \cdot d\mathbf{s} = -\int_i^f E(\cos 0)\, ds$$
$$= -\int_i^f E\, ds.$$

Since the field is uniform, E is constant over the path and can be moved outside the integral, giving us

$$V_f - V_i = -E \int_i^f ds = -Ed, \qquad \text{(Answer)}$$

in which the integral is simply the length d of the path. The minus sign in the result shows that the potential at point f in Fig. 25-6a is lower than the potential at point i. This is a general result: the potential always decreases along a path that extends in the direction of the electric field lines.

(b) Now find the potential difference $V_f - V_i$ by moving the positive test charge q_0 from i to f along the path icf shown in Fig. 25-6b.

SOLUTION: At all points along line ic, **E** and d**s** are perpendicular to each other. Thus $\mathbf{E} \cdot d\mathbf{s} = 0$ everywhere along this part of the path. Equation 25-18 then tells us that points i and c are at the same potential. In other words, i and c lie on the same equipotential surface.

For line cf we have $\theta = 45°$ and, from Eq. 25-18,

$$V_f - V_i = -\int_c^f \mathbf{E} \cdot d\mathbf{s} = -\int_c^f E(\cos 45°) \, ds$$

$$= -\frac{E}{\sqrt{2}} \int_c^f ds.$$

The integral in this equation is the length of line cf, which is $d/\sin 45° = \sqrt{2}d$. Thus

$$V_f - V_i = -\frac{E}{\sqrt{2}} \sqrt{2}d = -Ed. \qquad \text{(Answer)}$$

This is the same result we obtained in (a), as it must be: the potential difference between two points does not depend on the path connecting them. Moral: When you want to find the potential difference between two points by moving a test charge between them, you can save time and work by choosing a path that simplifies the use of Eq. 25-18.

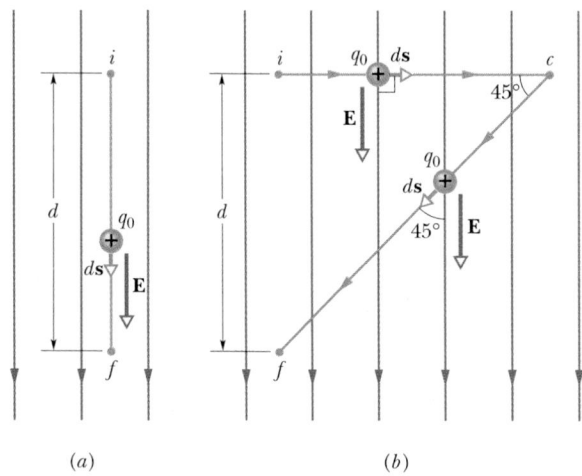

(a)   (b)

FIGURE 25-6 Sample Problem 25-2. (a) A test charge q_0 moves in a straight line from point i to point f, along the direction of a uniform electric field. (b) Charge q_0 moves along path icf in the same electric field.

CHECKPOINT 3: The figure shows a family of parallel equipotential surfaces (in cross section) and five paths along which we shall move an electron from one surface to another. (a) What is the direction of the electric field associated with the surfaces? (b) For each path, is the work we do positive, negative, or zero? (c) Rank the paths according to the work we do, greatest first.

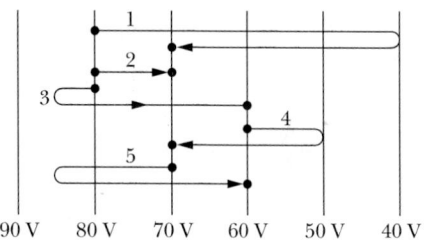

25-5 POTENTIAL DUE TO A POINT CHARGE

We are now going to use Eq. 25-19 to derive an expression for the electric potential V in the space around a point charge, relative to the zero potential at infinity.

Consider a point P at a distance r from a fixed point charge of magnitude q (Fig. 25-7). To use Eq. 25-19, we imagine that a positive test charge moves from infinity to point P. Because the path followed by the test charge does not matter, we make the simplest choice: a line that extends from infinity to P along a radius from the point charge q.

We next must evaluate the dot product $\mathbf{E} \cdot d\mathbf{s}$ in Eq. 25-19 along the path taken by the test charge. In Fig. 25-7 the test particle is at some intermediate point, at distance r' from the point charge. The electric field **E** at the location of the test particle is directed radially outward. The differential displacement $d\mathbf{s}$ of the test charge as it moves toward point P is radially inward. Thus the angle between **E** and $d\mathbf{s}$

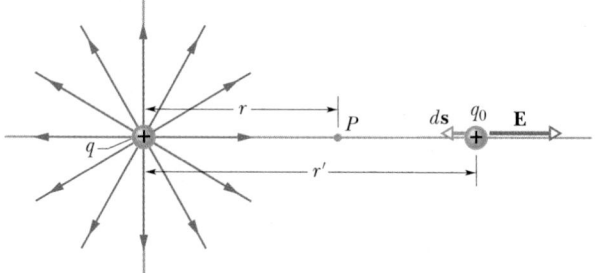

FIGURE 25-7 The positive point charge q produces an electric field **E** and an electric potential V at point P. We find the potential by moving a test charge q_0 to P from infinity. The test charge is shown at distance r' from the point charge, undergoing differential displacement $d\mathbf{s}$.

is 180°. Using this angle and writing the magnitude of the displacement ds as dr', we can write the dot product of Eq. 25-19 as

$$\mathbf{E} \cdot d\mathbf{s} = |E|\,(dr')(\cos 180°) = -|E|\,dr', \quad (25\text{-}20)$$

where $|E|$ is the absolute value of $\mathbf{E}$. (We use the absolute value sign to show explicitly that we want only the magnitude of $\mathbf{E}$ here.) Substituting Eq. 25-20 and the limits of our integration into Eq. 25-19 gives us

$$V = -\int_i^f \mathbf{E} \cdot d\mathbf{s} = \int_\infty^r |E|\,dr'. \quad (25\text{-}21)$$

The magnitude of the electric field at the site of the test charge is given by Eq. 23-3 as

$$E = \frac{1}{4\pi\epsilon_0}\frac{q}{r'^2}. \quad (25\text{-}22)$$

Substituting this result into Eq. 25-21 and integrating lead to

$$V = \frac{q}{4\pi\epsilon_0}\int_\infty^r \left|\frac{1}{r'^2}\right|\,dr' = \frac{q}{4\pi\epsilon_0}\left[\frac{1}{r'}\right]_\infty^r, \quad (25\text{-}23)$$

or

$$V = \frac{1}{4\pi\epsilon_0}\frac{q}{r} \quad \text{(point charge } +q). \quad (25\text{-}24)$$

Thus the potential V at any point around a positive point charge is positive, relative to the zero potential at infinity.

If the point charge is negative (but still of magnitude q), the electric field at P points toward q and the angle between $d\mathbf{s}$ and $\mathbf{E}$ at r' is zero. The integration in Eq. 25-23 now yields

$$V = -\frac{1}{4\pi\epsilon_0}\frac{q}{r} \quad \text{(point charge } -q). \quad (25\text{-}25)$$

Thus the potential V at any point around a negative point charge is negative, relative to the zero potential at infinity.

If we allow the symbol q to be either positive or negative instead of representing just the magnitude of the charge, we can generalize Eqs. 25-24 and 25-25 as

$$V = \frac{1}{4\pi\epsilon_0}\frac{q}{r} \quad \begin{array}{l}\text{(positive or negative}\\ \text{point charge } q).\end{array} \quad (25\text{-}26)$$

Now the sign of V is the same as the sign of q. Figure 25-8 shows a computer-generated plot of Eq. 25-26 for a positive point charge. Note also that the magnitude of V increases as $r \to 0$. In fact, according to Eq. 25-26, V for a point charge is infinite at $r = 0$, although Fig. 25-8 shows a finite, smoothed-off value there.

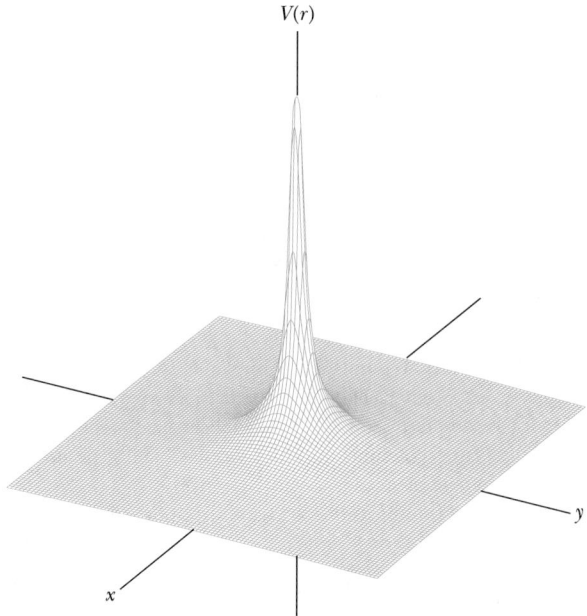

FIGURE 25-8 A computer-generated plot of electric potential $V(r)$ due to a positive point charge, which is located at the origin of an xy plane. The potentials at points in that plane are plotted vertically. (Curved lines have been added to help you visualize the plot.) The infinite value of V predicted by Eq. 25-26 for $r = 0$ is not plotted.

Equation 25-26 also gives the electric potential *outside or on the external surface of* a spherically symmetric charge distribution. We can prove this by using one of the shell theorems of Sections 22-4 and 24-9 to replace the actual charge with an equal charge concentrated at the center of the spherical distribution. Then the derivation leading to Eq. 25-26 follows, provided we do not consider a point within the actual distribution.

PROBLEM SOLVING TACTICS

TACTIC 3: *Finding a Potential Difference*
To find the potential difference ΔV between any two points in the field of an isolated point charge, we can evaluate Eq. 25-26 at each point and then subtract the results. The value of ΔV will be the same for any choice of reference potential energy because that choice is eliminated by the subtraction.

SAMPLE PROBLEM 25-3

(a) What is the electric potential V at a distance $r = 2.12 \times 10^{-10}$ m from the nucleus of a hydrogen atom (the nucleus consists of a single proton)?

SOLUTION: Substituting the given distance and the charge of a proton into Eq. 25-26 yields

$$V = \frac{1}{4\pi\epsilon_0}\frac{e}{r}$$

$$= \frac{(8.99 \times 10^9 \text{ N} \cdot \text{m}^2/\text{C}^2)(1.60 \times 10^{-19} \text{ C})}{2.12 \times 10^{-10} \text{ m}}$$

$$= 6.78 \text{ V}. \hspace{2cm} \text{(Answer)}$$

(b) What is the electric potential energy U in electron-volts of an electron at the given distance from the nucleus? (The potential energy is actually that of the electron–proton system—the hydrogen atom.)

SOLUTION: Substituting $V = 6.78$ V and the charge of an electron into Eq. 25-5 yields

$$U = qV = (-1.60 \times 10^{-19} \text{ C})(6.78 \text{ V})$$

$$= -1.09 \times 10^{-18} \text{ J} = -6.78 \text{ eV}. \hspace{0.5cm} \text{(Answer)}$$

(c) If the electron moves closer to the proton, does the electric potential energy increase or decrease?

SOLUTION: The electric potential V due to the proton at the electron's position increases. Thus the value of V in (b) increases. Because the electron is negatively charged, this means that the value of U becomes more negative. Hence, the potential energy U of the electron (that is, of the system or atom) decreases.

25-6 POTENTIAL DUE TO A GROUP OF POINT CHARGES

We can find the net potential at a point due to a group of point charges with the help of the superposition principle. We calculate the potential resulting from each charge at the given point separately, using Eq. 25-26 with the sign of the charge included. Then we sum the potentials. For n charges, the net potential is

$$V = \sum_{i=1}^{n} V_i = \frac{1}{4\pi\epsilon_0}\sum_{i=1}^{n}\frac{q_i}{r_i} \hspace{0.5cm} \begin{matrix}(n \text{ point} \\ \text{charges}).\end{matrix} \hspace{0.3cm} (25\text{-}27)$$

Here q_i is the value of the ith charge, and r_i is the radial distance of the given point from the ith charge. The sum in Eq. 25-27 is an *algebraic sum,* not a vector sum like the sum that would be used to calculate the electric field resulting from a group of point charges. Herein lies an important computational advantage of potential over electric field: it is a lot easier to sum several scalar quantities than to sum several vector quantities whose directions and components must be considered.

CHECKPOINT 4: The figure shows three arrangements of two protons. Rank the arrangements according to the net electric potential produced at point P by the protons, greatest first.

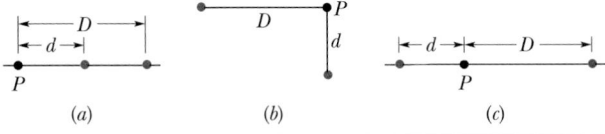

SAMPLE PROBLEM 25-4

What is the potential at point P, located at the center of the square of point charges shown in Fig. 25-9a? Assume that $d = 1.3$ m and that the charges are

$$q_1 = +12 \text{ nC}, \hspace{1cm} q_3 = +31 \text{ nC},$$

$$q_2 = -24 \text{ nC}, \hspace{1cm} q_4 = +17 \text{ nC}.$$

SOLUTION: Since each charge is the same distance r from P, Eq. 25-27 gives us

$$V = \sum_{i=1}^{4} V_i = \frac{1}{4\pi\epsilon_0}\frac{q_1 + q_2 + q_3 + q_4}{r}.$$

The distance r is $d/\sqrt{2}$, which is 0.919 m, and the sum of the charges is

$$q_1 + q_2 + q_3 + q_4 = (12 - 24 + 31 + 17) \times 10^{-9} \text{ C}$$

$$= 36 \times 10^{-9} \text{ C}.$$

So, $$V = \frac{(8.99 \times 10^9 \text{ N} \cdot \text{m}^2/\text{C}^2)(36 \times 10^{-9} \text{ C})}{0.919 \text{ m}}$$

$$\approx 350 \text{ V}. \hspace{2cm} \text{(Answer)}$$

Close to the three positive charges in Fig. 25-9a, the potential has very large positive values. Close to the single negative charge, the potential has very large negative values. Thus there must be points within the square that have the same

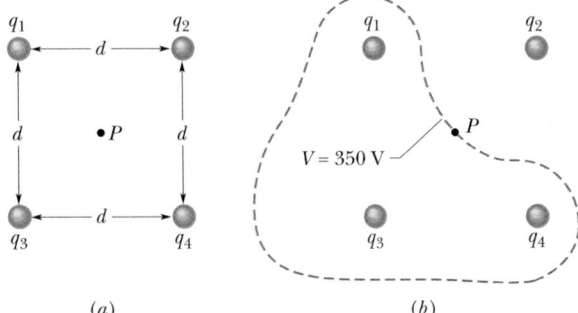

FIGURE 25-9 Sample Problem 25-4. (a) Four point charges are held fixed at the corners of a square. (b) The closed curve is a cross section, in the plane of the figure, of the equipotential surface that contains point P.

intermediate potential as that at point P. The curve in Fig. 25-9b shows the intersection of the plane of the figure with the equipotential surface that contains point P. Any point along that curve has the same potential as point P.

SAMPLE PROBLEM 25-5

(a) In Fig. 25-10a, 12 electrons (of charge $-e$) are equally spaced and fixed around a circle of radius R. Relative to $V = 0$ at infinity, what are the electric potential and electric field at the center C of the circle due to these electrons?

SOLUTION: Since electrons all have the same negative charge, and since all the electrons here are the same distance R from C, the potential at C must be, from Eq. 25-27,

$$V = -12 \frac{1}{4\pi\epsilon_0} \frac{e}{R}. \qquad \text{(Answer)} \qquad (25\text{-}28)$$

Since electric potential is a scalar, the orientation of the charges with respect to C is irrelevant to the potential V. However, since electric field is a vector, that orientation *is* important to $\mathbf{E}$. In fact, here, because of the symmetry of the arrangement, the electric field vector at C due to any given electron is canceled by the field vector due to the electron that is diametrically opposite it. Thus at C,

$$\mathbf{E} = 0. \qquad \text{(Answer)}$$

(b) If the electrons are moved along the circle until they are nonuniformly spaced over a 120° arc (Fig. 25-10b), what then is the potential at C? How does the electric field at C change (if at all)?

SOLUTION: The potential is still given by Eq. 25-28, because the distance between C and each electron is unchanged and orientation is irrelevant. The electric field is no longer zero, because the arrangement is no longer symmetric. There is now a net field that points toward the charge distribution.

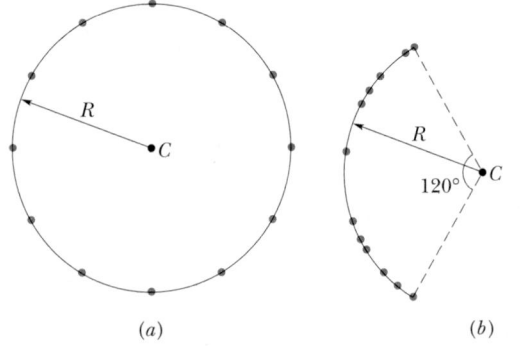

(a)   (b)

FIGURE 25-10 Sample Problem 25-5. (a) Twelve electrons uniformly spaced around a circle. (b) Those electrons are now nonuniformly spaced along an arc of the original circle.

25-7 POTENTIAL DUE TO AN ELECTRIC DIPOLE

Now let us apply Eq. 25-27 to an electric dipole to find the potential at an arbitrary point P in Fig. 25-11a. At P, the positive point charge (at distance $r_{(+)}$) sets up potential $V_{(+)}$ and the negative point charge (at distance $r_{(-)}$) sets up potential $V_{(-)}$, both potentials as given by Eq. 25-26. So the net potential at P is given by Eq. 25-27 as

$$V = \sum_{i=1}^{2} V_i = V_{(+)} + V_{(-)} = \frac{1}{4\pi\epsilon_0} \left(\frac{q}{r_{(+)}} + \frac{-q}{r_{(-)}} \right)$$

$$= \frac{q}{4\pi\epsilon_0} \frac{r_{(-)} - r_{(+)}}{r_{(-)}r_{(+)}}. \qquad (25\text{-}29)$$

Because naturally occurring dipoles—such as those possessed by many molecules—are small, we are usually interested only in points far from the dipole, such that $r \gg d$, where d is the distance between the charges. Under these conditions, the approximations that follow from Fig. 25-11b are

$$r_{(-)} - r_{(+)} \approx d \cos\theta \quad \text{and} \quad r_{(-)}r_{(+)} \approx r^2.$$

If we substitute these quantities into Eq. 25-29, we can approximate V to be

$$V = \frac{q}{4\pi\epsilon_0} \frac{d \cos\theta}{r^2},$$

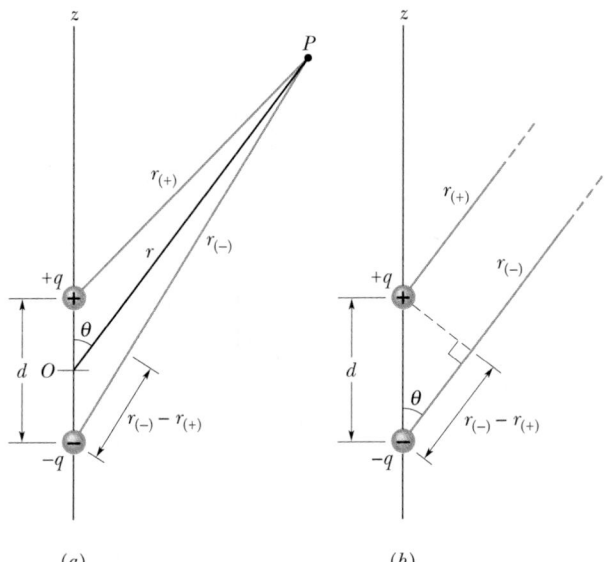

(a)   (b)

FIGURE 25-11 (a) Point P is a distance r from the midpoint O of a dipole. The line OP makes an angle θ with the dipole axis. (b) If P is far from the dipole, the lines of lengths $r_{(+)}$ and $r_{(-)}$ are approximately parallel to the line of length r, and the dashed black line is approximately perpendicular to the line of length $r_{(-)}$.

where θ is measured from the dipole axis as shown in Fig. 25-11a. We can now write V as

$$V = \frac{1}{4\pi\epsilon_0} \frac{p \cos \theta}{r^2} \qquad \text{(electric dipole),} \qquad (25\text{-}30)$$

in which p ($= qd$) is the magnitude of the electric dipole moment **p** defined in Section 23-5. The vector **p** is along the dipole axis, pointing from the negative to the positive charge. (So, θ is measured from the direction of **p**.)

CHECKPOINT **5:** Suppose that three points are set at equal (large) distances r from the center of the dipole in Fig. 25-11: point a is on the dipole axis above the positive charge, point b is on the axis below the negative charge, and point c is on a perpendicular bisector through the line connecting the two charges. Rank the points according to the electric potential of the dipole there, greatest (most positive) first.

Induced Dipole Moment

Many molecules such as water have *permanent* electric dipole moments. In other molecules (*nonpolar molecules*) and in every atom, the centers of the positive and negative charges coincide (Fig. 25-12a) and thus no dipole moment is set up. However, if we place an atom or a nonpolar molecule in an external electric field, the field distorts the electron orbits and separates the centers of positive and negative charge (Fig. 25-12b). Because the electrons are negatively charged, they tend to be shifted in a direction opposite the field. This shift sets up a dipole moment **p** that points in the direction of the field. This dipole moment is said to be *induced* by the field, and the atom or molecule is then said to be *polarized* by the field (it has a positive side and a negative side). When the field is removed, the induced dipole moment and the polarization disappear.

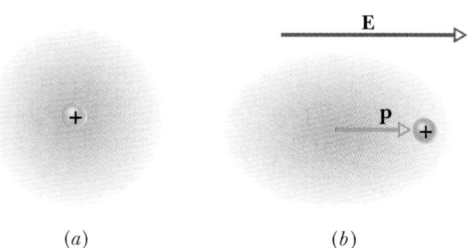

(a)   (b)

FIGURE 25-12 (a) An atom, showing the positively charged nucleus (green) and the negatively charged electrons (gold shading). The centers of positive and negative charge coincide. (b) If the atom is placed in an external electric field, the electron orbits are distorted so that the centers of positive and negative charge no longer coincide. An induced dipole moment appears. The distortion is greatly exaggerated here.

25-8 POTENTIAL DUE TO A CONTINUOUS CHARGE DISTRIBUTION

When a charge distribution q is continuous (as on a uniformly charged thin rod or disk), we cannot use the summation of Eq. 25-27 to find the potential V at a point P. Instead, we must choose a differential element of charge dq, determine the potential dV at P due to dq, and then integrate over the entire charge distribution.

Let us again take the zero of potential to be at infinity. If we treat the element of charge dq as a point charge, then we can use Eq. 25-26 to express the potential dV at point P due to dq:

$$dV = \frac{1}{4\pi\epsilon_0} \frac{dq}{r} \qquad \text{(positive or negative } dq\text{).} \qquad (25\text{-}31)$$

Here r is the distance between P and dq. To find the total potential V at P, we integrate to sum the potentials due to all the charge elements:

$$V = \int dV = \frac{1}{4\pi\epsilon_0} \int \frac{dq}{r}. \qquad (25\text{-}32)$$

The integral is to be taken over the entire charge distribution. Note that because the electric potential is a scalar, there are *no vector components* to consider in Eq. 25-32.

We now examine two continuous charge distributions, a line of charge and a charged disk.

Line of Charge

In Fig. 25-13a, a thin nonconducting rod of length L has a positive charge of uniform linear density λ. Let us determine the electric potential V due to the rod at point P, a perpendicular distance d from the left end of the rod.

We consider a differential element dx of the rod as shown in Fig. 25-13b. This (or any other) element of the rod has a differential charge of

$$dq = \lambda \, dx. \qquad (25\text{-}33)$$

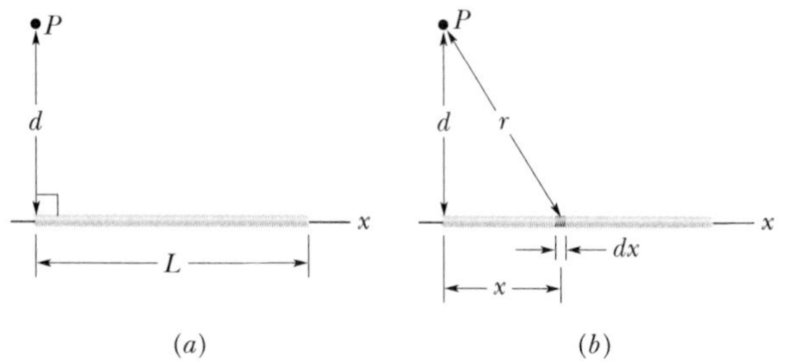

FIGURE 25-13 (a) A thin, uniformly charged rod produces an electric potential V at point P. (b) An element of charge produces a differential potential dV at P.

This element produces a potential dV at point P, which is a distance $r = (x^2 + d^2)^{1/2}$ from the element. Treating the element as a point charge, we can use Eq. 25-31 to write the potential dV as

$$dV = \frac{1}{4\pi\epsilon_0} \frac{dq}{r} = \frac{1}{4\pi\epsilon_0} \frac{\lambda\, dx}{(x^2 + d^2)^{1/2}}. \quad (25\text{-}34)$$

Since the charge on the rod is positive and we have taken $V = 0$ at infinity, we know from Section 25-5 that dV in Eq. 25-34 must be positive.

We now find the total potential V produced by the rod at point P by integrating Eq. 25-34 along the length of the rod, from $x = 0$ to $x = L$, using integral 17 in Appendix E. We find

$$V = \int dV = \int_0^L \frac{1}{4\pi\epsilon_0} \frac{\lambda}{(x^2 + d^2)^{1/2}}\, dx$$

$$= \frac{\lambda}{4\pi\epsilon_0} \int_0^L \frac{dx}{(x^2 + d^2)^{1/2}}$$

$$= \frac{\lambda}{4\pi\epsilon_0} \left[\ln\left(x + (x^2 + d^2)^{1/2} \right) \right]_0^L$$

$$= \frac{\lambda}{4\pi\epsilon_0} \left[\ln[L + (L^2 + d^2)^{1/2}] - \ln d \right].$$

We can simplify this result by using the general relation $\ln A - \ln B = \ln (A/B)$. We then find

$$V = \frac{\lambda}{4\pi\epsilon_0} \ln\left[\frac{L + (L^2 + d^2)^{1/2}}{d} \right]. \quad (25\text{-}35)$$

Because V is the sum of positive values of dV, it should be positive. But does Eq. 25-35 give a positive V? Since the argument of the logarithm is greater than one, the logarithm is a positive number and V is indeed positive.

PROBLEM SOLVING TACTICS

TACTIC 4: *Signs of Trouble with Electric Potential*

When you calculate the potential V at some point P due to a line of charge or any other continuous charge configuration, the signs can cause you trouble. Here is a generic guide to sort out the signs.

If the charge is negative, should the symbols dq and λ represent negative quantities? Or should you explicitly show the signs, using $-dq$ and $-\lambda$? You can do either as long as you remember what your notation means, so that when you get to the final step, you can correctly interpret the sign of V.

Another approach, which can be used when the entire charge distribution is of a single sign, is to let the symbols dq and λ represent magnitudes only. The result of the calculation will give you the magnitude of V at P. Then add a sign to V based on the sign of the charge. (If the zero potential is at

infinity, positive charge gives a positive potential and negative charge gives a negative potential.)

If you happen to reverse the limits on the integral used to calculate a potential, you will obtain a negative value for V. The magnitude will be correct, but discard the minus sign. Then determine the proper sign for V from the sign of the charge. As an example, we would have obtained a minus sign in Eq. 25-35 if we had reversed the limits in the integral above that equation. We would then have discarded that minus sign and noted that the potential is positive because the charge producing it is positive.

Charged Disk

In Section 23-7, we calculated the magnitude of the electric field at points on the central axis of a plastic disk of radius R that has a uniform charge density σ on one surface. Here we derive an expression for $V(z)$, the potential at any point on the central axis.

In Fig. 25-14, consider a differential element consisting of a flat ring of radius R' and radial width dR'. Its charge has magnitude

$$dq = \sigma(2\pi R')(dR'),$$

in which $(2\pi R')(dR')$ is the upper surface area of the ring. All parts of this charged element are the same distance r from point P on the disk's axis. With the aid of Fig. 25-14, we can now use Eq. 25-31 to write the contribution of this ring to the electric field at P as

$$dV = \frac{1}{4\pi\epsilon_0} \frac{dq}{r} = \frac{1}{4\pi\epsilon_0} \frac{\sigma(2\pi R')(dR')}{\sqrt{z^2 + R'^2}}. \quad (25\text{-}36)$$

We find the net potential at P by adding (via integration) the contributions of all the strips from $R' = 0$ to $R' = R$:

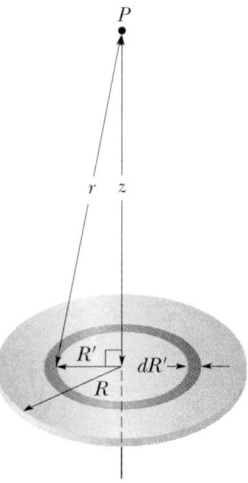

FIGURE 25-14 A plastic disk of radius R is charged on its top surface to a uniform surface charge density σ. We wish to find the potential V at point P on the central axis of the disk.

$$V = \int dV = \frac{\sigma}{2\epsilon_0} \int_0^R \frac{R'\, dR'}{\sqrt{z^2 + R'^2}}$$

$$= \frac{\sigma}{2\epsilon_0} (\sqrt{z^2 + R^2} - z). \qquad (25\text{-}37)$$

Note that the variable in the second integral of Eq. 25-37 is R' and not z, which remains constant while the integration over the surface of the disk is carried out. (Note also that, in evaluating the integral, we have assumed that $z \geq 0$.)

SAMPLE PROBLEM 25-6

The potential at the center of a uniformly charged circular disk of radius $R = 3.5$ cm is $V_0 = 550$ V.

(a) What is the total charge q on the disk?

SOLUTION: At the center of the disk, z in Eq. 25-37 is zero, so that equation reduces to

$$V_0 = \frac{\sigma R}{2\epsilon_0},$$

from which

$$\sigma = \frac{2\epsilon_0 V_0}{R}. \qquad (25\text{-}38)$$

Since σ is the surface charge density, the total charge q on the disk is $\sigma(\pi R^2)$. Using Eq. 25-38, we can now write

$$q = \sigma(\pi R^2) = 2\pi\epsilon_0 R V_0$$
$$= (2\pi)(8.85 \times 10^{-12} \text{ C}^2/\text{N} \cdot \text{m}^2)(0.035 \text{ m})(550 \text{ V})$$
$$= 1.1 \times 10^{-9} \text{ C} = 1.1 \text{ nC}, \qquad \text{(Answer)}$$

in which we use Eq. 25-9 to write 1 V = 1 J/C = 1 N·m/C.

(b) What is the potential at a point on the axis of the disk a distance $z = 5.0R$ from the center of the disk?

SOLUTION: From Eq. 25-37 we find

$$V = \frac{\sigma}{2\epsilon_0} [\sqrt{(5.0R)^2 + R^2} - 5.0R].$$

Substituting σ from Eq. 25-38 then yields

$$V = \frac{V_0}{R} (\sqrt{26R^2} - 5.0R) = V_0(\sqrt{26} - 5.0)$$
$$= (550 \text{ V})(0.099) = 54 \text{ V}. \qquad \text{(Answer)}$$

25-9 CALCULATING THE FIELD FROM THE POTENTIAL

In Section 25-4, you saw how to find the potential at a point f if you know the electric field along a path from a reference point to point f. In this section, we propose to go

the other way, that is, to find the electric field when we know the potential. As Fig. 25-3 shows, solving this problem graphically is easy: if we know the potential V at all points near an assembly of charges, we can draw in a family of equipotential surfaces. The electric field lines, sketched perpendicular to those surfaces, reveal the variation of **E**. What we are seeking here is the mathematical equivalent of this graphical procedure.

Figure 25-15 shows cross sections of a family of closely spaced equipotential surfaces, the potential difference between each pair of adjacent surfaces being dV. As the figure suggests, the field **E** at any point P is perpendicular to the equipotential surface through P.

Suppose that a positive test charge q_0 moves through a displacement $d\mathbf{s}$ from one equipotential surface to the adjacent surface. From Eq. 25-7, we see that the work that the electric field does on the test charge during the move is $-q_0\, dV$. From Eq. 25-16 and Fig. 25-15, we see that the work done by the electric field may also be written as $(q_0\mathbf{E}) \cdot d\mathbf{s}$, or $q_0 E(\cos \theta)\, ds$. Equating these two expressions for the work yields

$$-q_0\, dV = q_0 E(\cos \theta)\, ds,$$

or

$$E \cos \theta = -\frac{dV}{ds}. \qquad (25\text{-}39)$$

Since $E \cos \theta$ is the component of **E** in the direction of $d\mathbf{s}$, Eq. 25-39 becomes

$$E_s = -\frac{\partial V}{\partial s}. \qquad (25\text{-}40)$$

We have added a subscript to E and switched to the partial derivative symbols to emphasize that Eq. 25-40 involves only the variation of V along a specified axis (here called the s axis) and only the component of **E** along that axis. In

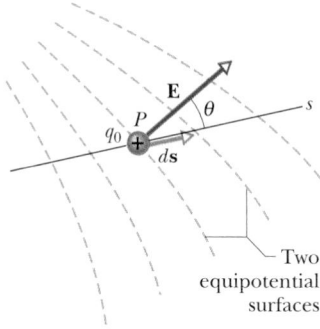

FIGURE 25-15 A test charge q_0 moves a distance $d\mathbf{s}$ from one equipotential surface to another. (The separation between the surfaces has been exaggerated for clarity.) The displacement $d\mathbf{s}$ makes an angle θ with the direction of the electric field **E**.

words, Eq. 25-40 (which is essentially the inverse of Eq. 25-18) states:

The component of **E** in any direction is the negative of the rate of change of the electric potential with distance in that direction.

If we take the s axis to be, in turn, the x, y, and z axes, we find that the x, y, and z components of **E** at any point are

$$E_x = -\frac{\partial V}{\partial x}; \quad E_y = -\frac{\partial V}{\partial y}; \quad E_z = -\frac{\partial V}{\partial z}. \quad (25\text{-}41)$$

Thus if we know V for all points in the region around a charge distribution, that is, if we know the function $V(x, y, z)$, we can find the components of **E**—and thus **E** itself—at any point by taking partial derivatives.

In the simple situation where the electric field **E** is uniform, Eq. 25-40 becomes

$$E = -\frac{\Delta V}{\Delta s}, \quad (25\text{-}42)$$

where s is perpendicular to an equipotential surface. The electric field is zero in any direction tangent to an equipotential surface.

CHECKPOINT **6:** The figure shows three pairs of parallel plates with the same separation, and the electric potential of each plate. The electric field between the plates is uniform and perpendicular to the plates. (a) Rank the pairs according to the magnitude of the electric field between the plates, greatest first. (b) For which pair is the electric field pointing rightward? (c) If an electron is released midway between the third pair of plates, does it remain there, move rightward at constant speed, move leftward at constant speed, accelerate rightward, or accelerate leftward?

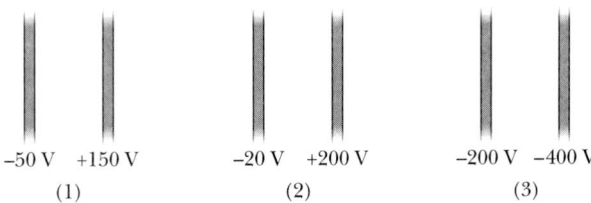

−50 V +150 V   −20 V +200 V   −200 V −400 V
  (1)   (2)   (3)

SAMPLE PROBLEM 25-7

The potential at any point on the axis of a charged disk is given by Eq. 25-37, which we can write as

$$V = \frac{\sigma}{2\epsilon_0} (\sqrt{z^2 + R^2} - z).$$

Starting with this expression, derive an expression for the electric field at any point on the axis of the disk.

SOLUTION: From symmetry, **E** must lie along the axis of the disk. If we choose the s axis to coincide with the z axis, then Eq. 25-40 gives us

$$E_z = -\frac{\partial V}{\partial z} = -\frac{\sigma}{2\epsilon_0} \frac{d}{dz} (\sqrt{z^2 + R^2} - z)$$

$$= \frac{\sigma}{2\epsilon_0} \left(1 - \frac{z}{\sqrt{z^2 + R^2}} \right). \quad \text{(Answer)}$$

This is the same expression that we derived in Section 23-7 by integration, using Coulomb's law.

25-10 ELECTRIC POTENTIAL ENERGY OF A SYSTEM OF POINT CHARGES

In Section 25-1, we discussed the electric potential energy of a test charge as a function of its position in an external electric field. In that section, we assumed that the charges that produced the field were fixed in place, so that the field could not be influenced by the presence of the test charge. In this section we can take a broader view, to find the electric potential energy of a *system* of charges due to the electric field produced *by* those charges.

For a simple example, if you push together two bodies that have charges of the same electrical sign, the work that you must do is stored as electric potential energy in the two-charge system (provided the kinetic energy of the bodies does not change). If you later release the charges, you can recover this stored energy, in whole or in part, as kinetic energy of the charged bodies as they rush away from each other.

We define the electric potential energy *of a system of point charges,* held in fixed positions by forces not specified, as follows:

The electric potential energy of a system of fixed point charges is equal to the work that must be done by an external agent to assemble the system, bringing each charge in from an infinite distance.

We assume that the charges are stationary both in their initial infinitely distant positions and in their final assembled configuration.

Figure 25-16 shows two point charges q_1 and q_2, separated by a distance r. To find the electric potential energy

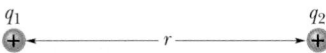

FIGURE 25-16 Two charges held a fixed distance r apart. What is the electric potential energy of the configuration?

of this two-charge system, we mentally build the system, starting with both charges infinitely far away and at rest. When we bring q_1 in from infinity and put it in place, we do no work, because no electrostatic force acts on q_1. But when we next bring q_2 in from infinity and put it in place, we must do work, because q_1 exerts an electrostatic force on q_2 during the move.

We can calculate that work with Eq. 25-8 by dropping the minus sign (so that the equation gives the work *we* do rather than the field's work) and substituting q_2 for the general charge q. Our work is then equal to q_2V, where V is the potential that has been set up by q_1 at the point where we put q_2. From Eq. 25-26, that potential is

$$V = \frac{1}{4\pi\epsilon_0} \frac{q_1}{r}.$$

Thus, from our definition, the electric potential energy of the pair of point charges of Fig. 25-16 is

$$U = W = \frac{1}{4\pi\epsilon_0} \frac{q_1 q_2}{r}. \qquad (25\text{-}43)$$

If the charges have the same sign, we have to do positive work to push them together against their mutual repulsion. Hence, as Eq. 25-43 shows, the potential energy of the system is then positive. If the charges have opposite signs, we have to do negative work against their mutual attraction to bring them together, so that they are stationary. The potential energy of the system is then negative. Sample Problem 25-8 shows how to extend this process to more than two charges.

SAMPLE PROBLEM 25-8

Figure 25-17 shows three charges held in fixed positions by forces that are not shown. What is the electric potential energy of this system of charges? Assume that $d = 12$ cm and that

$$q_1 = +q, \quad q_2 = -4q, \quad \text{and} \quad q_3 = +2q,$$

in which $q = 150$ nC.

SOLUTION: To answer, we mentally build the system of Fig. 25-17, starting with one of the charges, say q_1, in place and the others at infinity. Then we bring another one, say q_2, in from infinity and put it in place. From Eq. 25-43, with d substituted for r, the potential energy U_{12} associated with the

pair of charges q_1 and q_2 is

$$U_{12} = \frac{1}{4\pi\epsilon_0} \frac{q_1 q_2}{d}.$$

We then bring the last charge q_3 in from infinity and put it in place. The work that we must do in this last step is equal to the sum of the work we must do to bring q_3 near q_1 and the work we must do to bring it near q_2. From Eq. 25-43, with d substituted for r, that sum is

$$W_{13} + W_{23} = U_{13} + U_{23} = \frac{1}{4\pi\epsilon_0} \frac{q_1 q_3}{d} + \frac{1}{4\pi\epsilon_0} \frac{q_2 q_3}{d}.$$

The total potential energy U of the three-charge system is the sum of the potential energies associated with the three pairs of charges. This sum (which is actually independent of the order in which the charges are brought together) is

$$U = U_{12} + U_{13} + U_{23}$$

$$= \frac{1}{4\pi\epsilon_0}$$

$$\times \left(\frac{(+q)(-4q)}{d} + \frac{(+q)(+2q)}{d} + \frac{(-4q)(+2q)}{d} \right)$$

$$= -\frac{10q^2}{4\pi\epsilon_0 d}$$

$$= -\frac{(8.99 \times 10^9 \text{ N} \cdot \text{m}^2/\text{C}^2)(10)(150 \times 10^{-9} \text{ C})^2}{0.12 \text{ m}}$$

$$= -1.7 \times 10^{-2} \text{ J} = -17 \text{ mJ.} \qquad \text{(Answer)}$$

The negative potential energy means that negative work would have to be done to assemble this structure, starting with the three charges infinitely separated and at rest. Put another way, an external agent would have to do 17 mJ of work to disassemble the structure completely, ending with the three charges infinitely far apart.

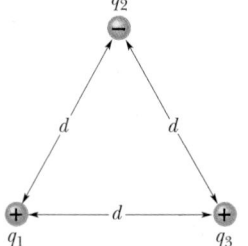

FIGURE 25-17 Sample Problem 25-8. Three charges are fixed at the vertices of an equilateral triangle. What is the electric potential energy of the configuration?

SAMPLE PROBLEM 25-9

An alpha particle (which consists of two protons and two neutrons) passes through the region of electron orbits in a gold atom, moving directly toward the gold nucleus, which has 79 protons and 118 neutrons. The alpha particle slows and then comes to a momentary stop, at a center-to-center separation r of 9.23 fm, before it begins to move back along its original path (Fig. 25-18). (Because the gold nucleus is much more massive than the alpha particle, we can assume the gold nucleus does not move.) What was the kinetic energy K of the alpha particle when it was initially far away (hence external to the gold atom)? Neglect the effect of the nuclear strong force.

SOLUTION: During the entire process, the mechanical energy of the *alpha particle + gold atom* system is conserved. When the alpha particle is outside the atom, the electric potential energy of the system is zero, because the atom has an equal number of electrons and protons, is thus electrically neutral, and so does not produce an external electric field. However, once the alpha particle has passed through the region of electron orbits on its way toward the nucleus, it is acted on by a repulsive electrostatic force due to its protons and those in the nucleus. (The neutrons, being electrically neutral, do not participate in producing this force. The electrons, now being outside the location of the alpha particle, act like a uniformly charged spherical shell, which produces no internal force.)

As the alpha particle slows because of the repulsive force, its kinetic energy is transferred to electric potential energy of the system. The transfer is complete when the alpha particle momentarily stops. Using the principle of conservation of mechanical energy, we can equate the initial kinetic energy K of the alpha particle to the electric potential energy U of the system at the instant the alpha particle stops:

$$K = U. \qquad (25-44)$$

By substituting Eq. 25-43 with $q_1 = 2e$, $q_2 = 79e$ (in which e is the elementary charge, 1.60×10^{-19} C), and $r = 9.23$ fm, we can rewrite Eq. 25-44 as

$$K = \frac{1}{4\pi\epsilon_0} \frac{(2e)(79e)}{9.23 \text{ fm}}$$

$$= \frac{(8.99 \times 10^9 \text{ N} \cdot \text{m}^2/\text{C}^2)(158)(1.60 \times 10^{-19} \text{ C})^2}{9.23 \times 10^{-15} \text{ m}}$$

$$= 3.94 \times 10^{-12} \text{ J} = 24.6 \text{ MeV}. \qquad \text{(Answer)}$$

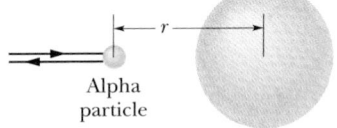

FIGURE 25-18   Sample Problem 25-9. An alpha particle, traveling head-on toward the center of a gold nucleus, has come to a momentary stop, at which time all its kinetic energy has been transferred to electric potential energy.

C HECKPOINT 7: If the alpha particle of Sample Problem 25-9 is replaced by a single proton of the same kinetic energy, will the proton momentarily stop at the same distance of 9.23 fm from the gold nucleus, farther from it, or closer to it?

25-11 POTENTIAL OF A CHARGED ISOLATED CONDUCTOR

In Section 24-6, we concluded that $\mathbf{E} = 0$ for all points inside an isolated conductor. We then used Gauss' law to prove that an excess charge placed on an isolated conductor lies entirely on its surface. (This is true even if the conductor has an empty internal cavity.) Here we use the fact that $\mathbf{E} = 0$ for all points inside an isolated conductor to prove another fact about such conductors:

An excess charge placed on an isolated conductor will distribute itself on the surface of that conductor so that all points of the conductor—whether on the surface or inside—come to the same potential. This is true regardless of whether the conductor has an internal cavity.

Our proof follows directly from Eq. 25-18, which is

$$V_f - V_i = -\int_i^f \mathbf{E} \cdot d\mathbf{s}.$$

Since $\mathbf{E} = 0$ for all points within a conductor, it follows directly that $V_f = V_i$ for all possible pairs of points i and f in the conductor.

Figure 25-19a is a plot of potential against radial distance r from the center for an isolated spherical conducting shell of 1.0 m radius, having a charge of 1.0 μC. For points outside the shell, we can calculate $V(r)$ from Eq. 25-26 because the charge q behaves for such external points as if it were concentrated at the center of the shell. That equation holds right up to the surface of the shell. Now let us push a small test charge through the shell—assuming a small hole exists—to its center. No extra work is needed to do this because no net electric force acts on the test charge once it is inside the shell. Thus the potential at all points inside the shell has the same value as that on the surface, as Fig. 25-19a shows.

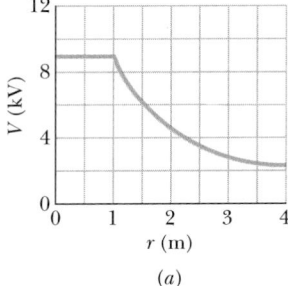

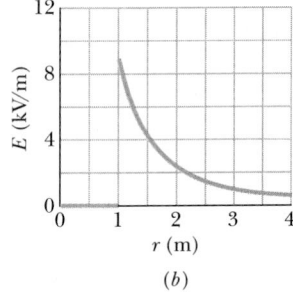

FIGURE 25-19   (a) A plot of $V(r)$ for a charged spherical shell. (b) A plot of $E(r)$ for the same shell.

FIGURE 25-20 A large spark jumps to the car's body and then exits by moving across the insulated left front tire (note the flash there), leaving the person inside unharmed.

Figure 25-19b shows the variation of electric field with radial distance for the same shell. Note that $E = 0$ everywhere inside the shell. The curves of Fig. 25-19b can be derived from the curve of Fig. 25-19a by differentiating with respect to r, using Eq. 25-40 (the derivative of a constant, recall, is zero). The curve of Fig. 25-19a can be derived from the curves of Fig. 25-19b by integrating with respect to r, using Eq. 25-19.

On nonspherical conductors, a surface charge does not distribute itself uniformly over the surface of the conductor. At sharp points or edges, the surface charge density—and thus the external electric field, which is proportional to it—may reach very high values. The air around such sharp points may become ionized, producing the corona discharge that golfers and mountaineers see on the tips of bushes, golf clubs, and rock hammers when thunderstorms

threaten. Such corona discharges, like hair that stands on end, are often the precursors of lightning strikes. In such circumstances, it is wise to enclose oneself in a cavity inside a conducting shell, where the electric field is guaranteed to be zero. A car (unless it is a convertible) is almost ideal (Fig. 25-20).

If an isolated conductor is placed in an *external electric field,* as in Fig. 25-21, all points of the conductor still come to a single potential regardless of whether the conductor has an excess charge. The free conduction electrons distribute themselves on the surface in such a way that the electric field they produce at interior points cancels the external electric field that would otherwise be there. Furthermore, the electron distribution causes the net electric field at all points on the surface to be perpendicular to the surface. If the conductor in Fig. 25-21 could be somehow removed, leaving the surface charges frozen in place, the pattern of the electric field would remain absolutely unchanged, for both exterior and interior points.

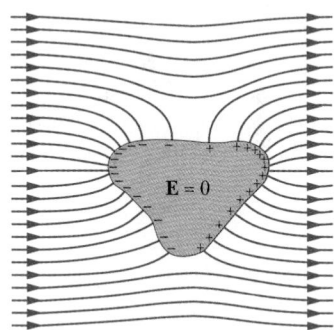

FIGURE 25-21 An uncharged conductor is suspended in an external electric field. The free electrons in the conductor distribute themselves on the surface as shown, reducing the net electric field inside the conductor to zero and making the net field at the surface perpendicular to the surface.

REVIEW & SUMMARY

Electric Potential Energy

The change ΔU in the electric potential energy U of a point charge as the charge moves from an initial point i to a final point f in an electric field is

$$\Delta U = U_f - U_i = -W, \qquad (25\text{-}1)$$

where W is the work done by the electric field on the point charge during the move from i to f. If the potential energy is defined to be zero at infinity, the **electric potential energy** U of the point charge at a particular point is

$$U = -W_\infty. \qquad (25\text{-}2)$$

Here W_∞ is the work done by the electric field on the point charge as the charge moves from infinity to the particular point.

Electric Potential Difference and Electric Potential

We define the **potential difference** ΔV between two points in an electric field as

$$\Delta V = V_f - V_i = -\frac{W}{q}, \qquad (25\text{-}7)$$

where q is the charge of a test particle on which work is done by the field. The **potential** at a point is

$$V = -\frac{W_\infty}{q}. \tag{25-8}$$

The SI unit of potential is the *volt*: 1 volt = 1 joule per coulomb.

Potential and potential difference can also be written in terms of the electric potential energy U of a particle of charge q in an electric field:

$$V = \frac{U}{q}, \tag{25-5}$$

$$\Delta V = V_f - V_i = \frac{U_f}{q} - \frac{U_i}{q} = \frac{\Delta U}{q}. \tag{25-6}$$

Equipotential Surfaces

The points on an **equipotential surface** all have the same potential. The work done on a test charge in moving it from one such surface to another is independent of the locations of the initial and final points on these surfaces and of the path that joins the points. The electric field **E** is always directed perpendicularly to equipotential surfaces.

Finding V from E

The electric potential difference between any two points is

$$V_f - V_i = -\int_i^f \mathbf{E} \cdot d\mathbf{s}, \tag{25-18}$$

where the integral is taken over any path connecting the points. If we choose $V_i = 0$ we have, for the potential at a particular point,

$$V = -\int_i^f \mathbf{E} \cdot d\mathbf{s}. \tag{25-19}$$

Potential Due to Point Charges

The electric potential due to a single point charge at a distance r from that point charge is

$$V = \frac{1}{4\pi\epsilon_0} \frac{q}{r}. \tag{25-26}$$

V has the same sign as q. The potential due to a collection of point charges is

$$V = \sum_{i=1}^n V_i = \frac{1}{4\pi\epsilon_0} \sum_{i=1}^n \frac{q_i}{r_i}. \tag{25-27}$$

Potential Due to an Electric Dipole

At a distance r from an electric dipole with dipole moment $p = qd$, the electric potential of the dipole is

$$V = \frac{1}{4\pi\epsilon_0} \frac{p \cos\theta}{r^2} \tag{25-30}$$

for $r \gg d$; the angle θ is defined in Fig. 25-11.

Potential Due to a Continuous Charge Distribution

For a continuous distribution of charge, Eq. 25-27 becomes

$$V = \frac{1}{4\pi\epsilon_0} \int \frac{dq}{r}, \tag{25-32}$$

in which the integral is taken over the entire distribution.

Calculating E from V

The component of **E** in any direction is the negative of the rate of change of the potential with distance in that direction:

$$E_s = -\frac{\partial V}{\partial s}. \tag{25-40}$$

The x, y, and z components of **E** may be found from

$$E_x = -\frac{\partial V}{\partial x}; \quad E_y = -\frac{\partial V}{\partial y}; \quad E_z = -\frac{\partial V}{\partial z}. \tag{25-41}$$

When **E** is uniform, Eq. 25-40 reduces to

$$E = -\frac{\Delta V}{\Delta s}, \tag{25-42}$$

where s is perpendicular to an equipotential surface. The electric field is zero in any direction parallel to an equipotential surface.

Electric Potential Energy of a System of Point Charges

The electric potential energy of a system of point charges is equal to the work needed to assemble the system with the charges initially at rest and infinitely distant from each other. For two charges at separation r,

$$U = W = \frac{1}{4\pi\epsilon_0} \frac{q_1 q_2}{r}. \tag{25-43}$$

Potential of a Charged Conductor

An excess charge placed on a conductor will, in the equilibrium state, be located entirely on the outer surface of the conductor. The charge distributes itself so that the entire conductor, including interior points, is at a uniform potential.

QUESTIONS

1. Figure 25-22 shows three paths along which we can move positively charged sphere A closer to positively charged sphere B, which is fixed in place. (a) Would sphere A be moved to a higher or lower electric potential? Is the work done (b) by our force and (c) by the electric field (due to the second sphere) positive, negative, or zero? (d) Rank the paths according to the work our force does, greatest first.

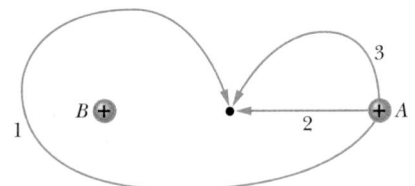

FIGURE 25-22 Question 1.

2. (a) In Fig. 25-3*a*, does the electric potential increase toward the right or toward the left? (b) If the adjacent equipotential surfaces differ by 10 V and the right-most one is at an electric potential of −100 V, what is the electric potential of the left-most one? If we move an electron toward the right, is the work done on the electron by (c) our force and (d) the electric field positive or negative?

3. Figure 25-23 shows four pairs of charged particles. Let $V = 0$ at infinity. For which pairs is there another point of zero net electric potential *on the axis* (a) between the particles and (b) to their right? (c) Where such a zero potential point exists, is the net electric field **E** due to the particles equal to zero? (d) For each pair, are there points off the axis (other than at infinity, of course) where $V = 0$?

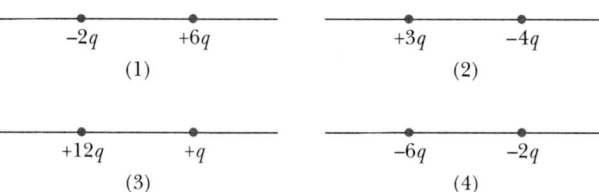

FIGURE 25-23   Questions 3 and 14.

4. Figure 25-24 shows a square array of charged particles, with distance *d* between adjacent particles. What is the electric potential at point *P* at the center of the square if the electric potential is zero at infinity?

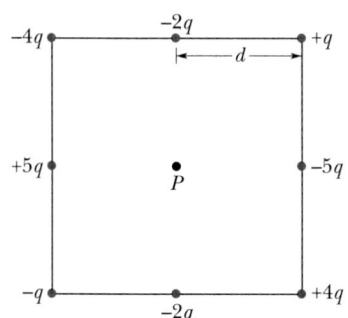

FIGURE 25-24   Question 4.

5. Figure 25-25 shows four arrangements of charged particles, all the same distance from the origin. Rank the situations according to the net electric potential at the origin, most positive first. Take the potential to be zero at infinity.

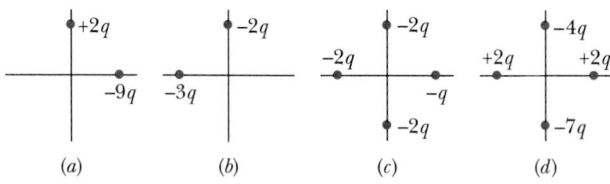

FIGURE 25-25   Question 5.

6. Figure 25-26 shows a proton at the origin, three choices for point *a* at distance *r*, and three choices for point *b* at distance 2*r*. There are nine different ways to pair the choices of *a* with the choices of *b*. Rank those nine ways according to the potential difference $V_a - V_b$ between points *a* and *b*, greatest first.

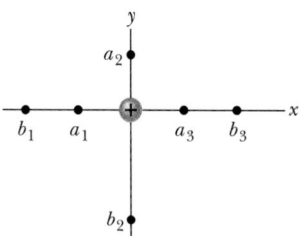

FIGURE 25-26
Question 6.

7. Figure 25-27 shows two situations in which we move an electron in from an infinite distance to a point midway between two charged particles (either proton or electron) fixed in place. In each situation, is the work done on the incoming electron *by the electric field* positive, negative, or zero?

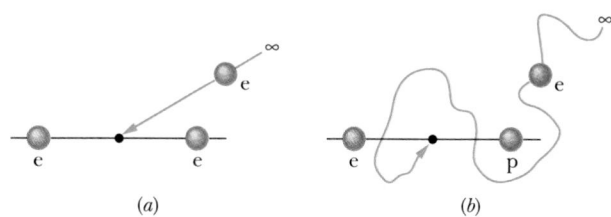

FIGURE 25-27   Question 7.

8. (a) In Fig. 25-28*a*, what is the potential at point *P* due to charge *Q* at distance *R* from *P*? Set $V = 0$ at infinity. (b) In Fig. 25-28*b*, the same charge *Q* has been spread uniformly over a circular arc of radius *R* and central angle 40°. What is the potential at point *P*, the center of curvature of the arc? (c) In Fig. 25-28*c*, the same charge *Q* has been spread uniformly over a circle of radius *R*. What is the potential at point *P*, the center of the circle? (d) Rank the three situations according to the magnitude of the electric field that is set up at *P*, greatest first.

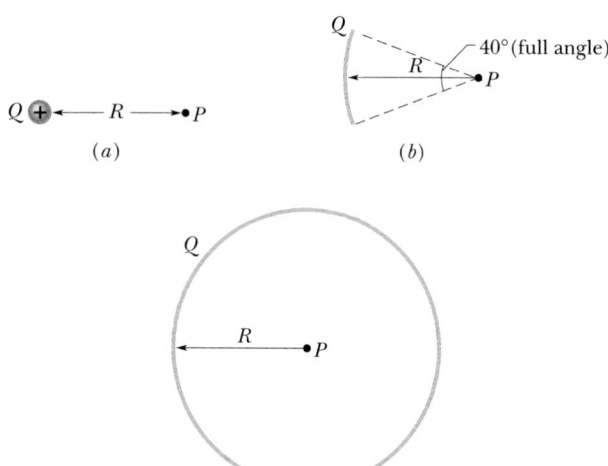

FIGURE 25-28
Question 8.

9. Figure 25-29 shows three sets of cross sections of equipotential surfaces; all three cover the same size region of space. (a) Rank the arrangements according to the magnitude of the electric field present in the region, greatest first. (b) In which is the electric field directed down the page?

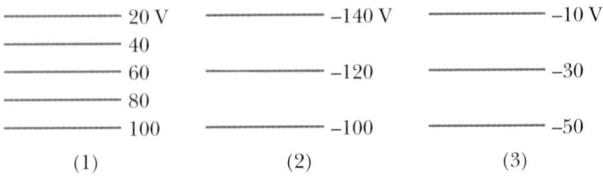

(1)   (2)   (3)

FIGURE 25-29 Question 9.

10. The electric potential at the coordinates (2 m, 0.5 m, 0.2 m) is given by $V = 2x - 3y + 4z$. Rank the magnitudes of the electric field components E_x, E_y, and E_z there, greatest first.

11. In Fig. 25-2, is the magnitude E of the electric field greater at the left or at the right?

12. Figure 25-30 gives the electric potential V as a function of distance through five regions on an x axis. (a) Rank the regions according to the magnitude of the x component of the electric field within them, greatest first. What is the direction of the field in (b) region 2 and (c) region 4?

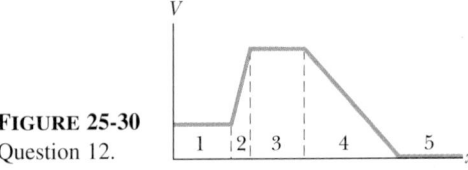

FIGURE 25-30
Question 12.

13. Rank the arrangements of Checkpoint 4 according to the electric potential energy of the system, greatest first.

14. Figure 25-23 shows four pairs of charged particles with identical separations. (a) Rank the pairs according to their electric potential energy, greatest (most positive) first. (b) For each pair, if the separation between the particles is increased, does the potential energy of the pair increase or decrease?

15. Figure 25-31 shows three systems of particles of charge $+q$ or $-q$, forming one equilateral and two isosceles triangles with edge lengths of either d, $2d$, or $d/2$. (a) Rank the systems ac-

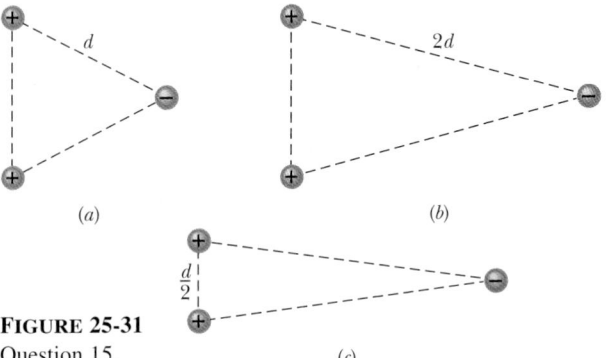

FIGURE 25-31
Question 15.

cording to their electric potential energy, most positive first. (b) How much work must we do to make the system of Fig. 25-31b if the particles are initially infinitely far apart?

16. Figure 25-32 shows a system of three charged particles. If you move the particle of charge $+q$ from point A to point D, are the following positive, negative, or zero: (a) the change in the electric potential energy of the three-particle system, (b) the work done by the net electrostatic force on the particle you moved, and (c) the work done by your force? (d) What are the answers to (a) through (c) if, instead, the move is from point B to point C?

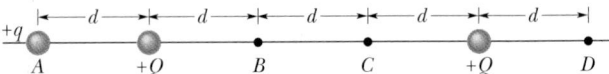

FIGURE 25-32 Questions 16 and 17.

17. Consider again the situation in Question 16. Is the work done by your force positive, negative, or zero if the move is (a) from A to B, (b) from A to C, and (c) from B to D? (d) Rank those moves according to the magnitude of the work done by your force, greatest first.

18. If the alpha particle of Sample Problem 25-9 had a smaller initial kinetic energy than the calculated 24.6 MeV, would it momentarily stop farther from, closer to, or at the same distance of 9.23 fm from the gold nucleus? (Again, neglect the strong force due to the gold nucleus.)

19. (a) If the surface of a charged conductor is an equipotential surface, does that mean that the charge is spread uniformly over the surface? (b) If the electric field is constant in magnitude over the surface of a charged conductor, does *that* mean that the charge is spread uniformly?

20. We have seen that, inside a hollow conductor, you are shielded from the fields of outside charges. If you are *outside* a hollow conductor that contains charges, are you shielded from the fields of these charges?

21. Can two equipotential surfaces of different values intersect? Explain your answer.

22. Three isolated, empty spherical shells of the same radius have the following charges: shell A, $+q$; shell B, $+2q$; shell C, $+3q$. Set the electric potential to be zero at an infinite distance from the shells. Then rank the shells, greatest first, according to (a) the electric potential at the surface of the shell, (b) the electric potential at the center of the shell, (c) the electric field magnitude at the surface of the shell, and (d) the electric field magnitude at the center of the shell.

23. Repeat Question 22 but with the electric potential now set to zero at the center of each shell.

24. Particle A of charge $+q$ and mass m and particle B of charge $-q$ and mass m are initially separated by distance d. In situation 1, we release both particles. In situation 2, we release only particle A. In which situation does particle A have greater kinetic energy when the separation between the particles has decreased to $d/2$, or does it have the same kinetic energy in the two situations?

EXERCISES & PROBLEMS

SECTION 25-2 Electric Potential

1E. The electric potential difference between the ground and a cloud in a particular thunderstorm is 1.2×10^9 V. What is the magnitude of the change in the electric potential energy (in multiples of the electron-volt) of an electron that moves between the ground and the cloud?

2E. A particular 12 V car battery can send a total charge of 84 A·h (ampere-hours) through a circuit, from one terminal to the other. (a) How many coulombs of charge does this represent? (b) If this entire charge undergoes a potential difference of 12 V, how much energy is involved?

3P. In a given lightning flash, the potential difference between a cloud and the ground is 1.0×10^9 V and the quantity of charge transferred is 30 C. (a) What is the change in energy of that transferred charge? (b) If all the energy released by the transfer could be used to accelerate a 1000 kg automobile from rest, what would be the automobile's final speed? (c) If the energy could be used to melt ice, how much ice would it melt at 0°C? The heat of fusion of ice is 3.33×10^5 J/kg.

SECTION 25-4 Calculating the Potential from the Field

4E. Two infinite lines of charge are parallel to and in the same plane with the z axis. One, of charge per unit length $+\lambda$, is a distance a to the right of this axis. The other, of charge per unit length $-\lambda$, is a distance a to the left of this axis. Sketch some of the equipotential surfaces due to this arrangement.

5E. In Fig. 25-33, three long parallel lines of charge, with the relative linear charge densities shown, extend perpendicular to the page in both directions. Sketch some electric field lines; also sketch the cross sections in the plane of the figure of some equipotential surfaces.

-2λ

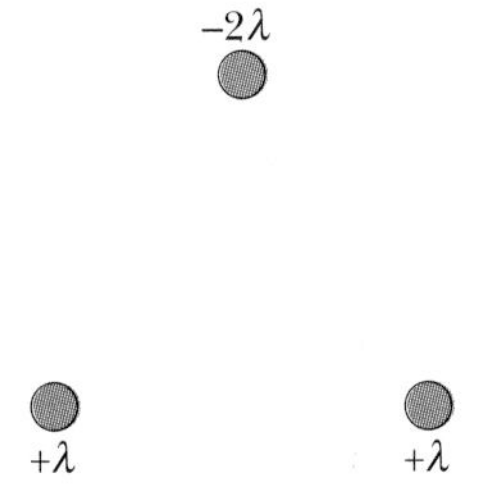

FIGURE 25-33
Exercise 5. $+\lambda$ $+\lambda$

6E. When an electron moves from A to B along an electric field line in Fig. 25-34, the electric field does 3.94×10^{-19} J of work on it. What are the electric potential differences (a) $V_B - V_A$, (b) $V_C - V_A$, and (c) $V_C - V_B$?

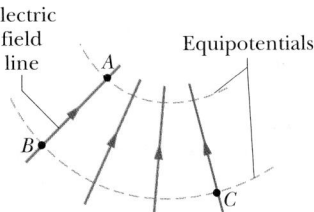

FIGURE 25-34 Exercise 6.

7E. In the Millikan oil-drop experiment (see Section 23-8), a uniform electric field of 1.92×10^5 N/C is maintained in the region between two plates separated by 1.50 cm. Find the potential difference between the plates.

8E. Two large parallel conducting plates are 12 cm apart and have charges of equal magnitude and opposite sign on their facing surfaces. An electrostatic force of 3.9×10^{-15} N acts on an electron placed anywhere between the two plates. (Neglect fringing.) (a) Find the electric field at the position of the electron. (b) What is the potential difference between the plates?

9E. An infinite nonconducting sheet has a surface charge density $\sigma = 0.10$ μC/m² on one side. How far apart are equipotential surfaces whose potentials differ by 50 V?

10P. Figure 25-35 shows, edge-on, an infinite nonconducting sheet with positive surface charge density σ on one side. (a) How much work is done by the electric field of the sheet as a small positive test charge q_0 is moved from an initial position on the sheet to a final position located a perpendicular distance z from the sheet? (b) Use Eq. 25-18 and the result from (a) to show that the electric potential of an infinite sheet of charge can be written $V = V_0 - (\sigma/2\epsilon_0)z$, where V_0 is the electric potential at the surface of the sheet.

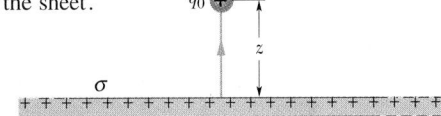

FIGURE 25-35 Problem 10.

11P. A Geiger counter has a metal cylinder 2.00 cm in diameter along whose axis is stretched a wire 1.30×10^{-4} cm in diameter. If the potential difference between them is 850 V, what is the electric field at the surface of (a) the wire and (b) the cylinder? (*Hint:* Use the result of Problem 29 of Chapter 24.)

12P. The electric field inside a nonconducting sphere of radius R, with charge spread uniformly throughout its volume, is radially directed and has magnitude

$$E(r) = \frac{qr}{4\pi\epsilon_0 R^3}.$$

Here q (positive or negative) is the total charge in the sphere, and r is the distance from the sphere center. (a) Taking $V = 0$ at the center of the sphere, find the potential $V(r)$ inside the sphere. (b) What is the difference in electric potential between a point on the surface and the sphere's center? (c) If q is positive, which of those two points is at the higher potential?

13P*. A charge q is distributed uniformly throughout a spherical volume of radius R. (a) Setting $V = 0$ at infinity, show that the potential at a distance r from the center, where $r < R$, is given by

$$V = \frac{q(3R^2 - r^2)}{8\pi\epsilon_0 R^3}.$$

(*Hint:* See Sample Problem 24-7.) (b) Why does this result differ from that in (a) of Problem 12? (c) What is the potential difference between a point on the surface and the sphere's center? (d) Why doesn't this result differ from that of (b) of Problem 12?

14P*. A thick spherical shell of charge Q and uniform volume charge density ρ is bounded by radii r_1 and r_2, where $r_2 > r_1$. With $V = 0$ at infinity, find the electric potential V as a function of the distance r from the center of the distribution, considering the regions (a) $r > r_2$, (b) $r_2 > r > r_1$, and (c) $r < r_1$. (d) Do these solutions agree at $r = r_2$ and $r = r_1$? (*Hint:* See Sample Problem 24-7.)

SECTION 25-6 Potential Due to a Group of Point Charges

15E. Consider a point charge $q = 1.0 \ \mu C$, point A at distance $d_1 = 2.0$ m from q, and point B at distance $d_2 = 1.0$ m. (a) If these points are diametrically opposite each other, as in Fig. 25-36a, what is the electric potential difference $V_A - V_B$? (b) What is that electric potential difference if points A and B are located as in Fig. 25-36b?

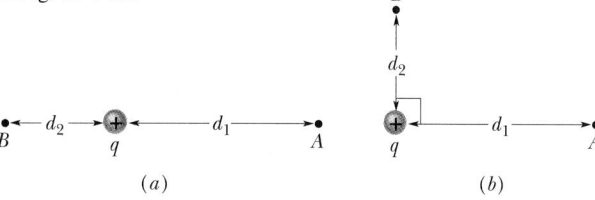

(a)   (b)

FIGURE 25-36 Exercise 15.

16E. Consider a point charge $q = 1.5 \times 10^{-8}$ C, and take $V = 0$ at infinity. (a) What are the shape and dimensions of an equipotential surface having a potential of 30 V due to q alone? (b) Are surfaces whose potentials differ by a constant amount (1.0 V, say) evenly spaced?

17E. A charge of 1.50×10^{-8} C lies on an isolated metal sphere of radius 16.0 cm. With $V = 0$ at infinity, what is the electric potential at points on the sphere's surface?

18E. As a space shuttle moves through the dilute ionized gas of Earth's ionosphere, its potential is typically changed by -1.0 V during one revolution. By assuming that the shuttle is a sphere of radius 10 m, estimate the amount of charge it collects.

19E. Much of the material making up Saturn's rings is in the form of tiny dust grains having radii on the order of 10^{-6} m. These grains are located in a region containing a dilute ionized gas, and they pick up excess electrons. As an approximation, suppose each grain is spherical, with radius $R = 1.0 \times 10^{-6}$ m. How many electrons would one grain have to pick up to have a potential of -400 V on its surface (taking $V = 0$ at infinity)?

20E. Figure 25-37 shows two charged particles on an axis. Sketch the electric field lines and the equipotential surfaces in the plane of the page for (a) $q_1 = +q$ and $q_2 = +2q$ and (b) $q_1 = +q$ and $q_2 = -3q$.

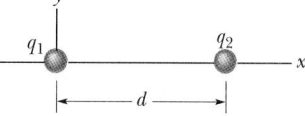

FIGURE 25-37
Exercises 20 through 23.

21E. In Fig. 25-37, set $V = 0$ at infinity and let the particles have charges $q_1 = +q$ and $q_2 = -3q$. Then locate (in terms of the separation distance d) any point on the x axis (other than at infinity) at which the net potential due to the two particles is zero.

22E. Let the separation d between the particles in Fig. 25-37 be 1.0 m; let their charges be $q_1 = +q$ and $q_2 = +2q$; and let $V = 0$ at infinity. Then locate any point on the x axis (other than at infinity) at which (a) the net electric potential due to the two particles is zero and (b) the net electric field due to them is zero.

23E. Two particles of charges q_1 and q_2 are separated by distance d in Fig. 25-37. The net electric field of the particles is zero at $x = d/4$. With $V = 0$ at infinity, locate (in terms of d) any point on the x axis (other than at infinity) at which the electric potential due to the two particles is zero.

24E. (a) If an isolated conducting sphere 10 cm in radius has a charge of 4.0 μC, and $V = 0$ at infinity, what is the potential on the surface of the sphere? (b) Can this situation actually occur, given that the air around the sphere undergoes electrical breakdown when the field exceeds 3.0 MV/m?

25P. What are (a) the charge and (b) the charge density on the surface of a conducting sphere of radius 0.15 m whose potential is 200 V (with $V = 0$ at infinity)?

26P. A spherical drop of water carrying a charge of 30 pC has a potential of 500 V at its surface (with $V = 0$ at infinity). (a) What is the radius of the drop? (b) If two such drops of the same charge and radius combine to form a single spherical drop, what is the potential at the surface of the new drop?

27P. An electric field of approximately 100 V/m is often observed near the surface of Earth. If this were the field over the entire surface, what would be the electric potential of a point on the surface? (Set $V = 0$ at infinity.)

28P. In Fig. 25-38, what is the net potential at point P due to the four point charges, if $V = 0$ at infinity?

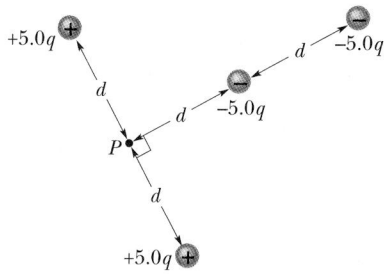

FIGURE 25-38 Problem 28.

29P. Suppose that the negative charge in a copper one-cent coin were removed to a very large distance from Earth—perhaps to a distant galaxy—and that the positive charge were distributed uniformly over Earth's surface. By how much would the electric potential at the surface change? (See Sample Problem 22-4.)

30P. In Fig. 25-39, point P is at the center of the rectangle. With $V = 0$ at infinity, what is the net electric potential at P due to the six charged particles?

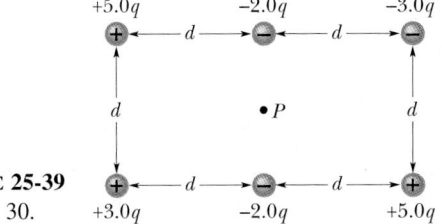

FIGURE 25-39
Problem 30.

31P. A point charge $q_1 = +6.0e$ is fixed at the origin of a rectangular coordinate system, and a second point charge $q_2 = -10e$ is fixed at $x = 8.6$ nm, $y = 0$. The locus of all points in the xy plane with $V = 0$ (other than at infinity) is a circle centered on the x axis, as shown in Fig. 25-40. Find (a) the location x_c of the center of the circle and (b) the radius R of the circle. (c) Is the xy cross section of the 5 V equipotential surface also a circle?

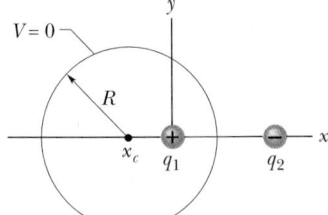

FIGURE 25-40
Problem 31.

32P. A solid copper sphere whose radius is 1.0 cm has a very thin surface coating of nickel. Some of the nickel atoms are radioactive, each atom emitting an electron as it decays. Half of these electrons enter the copper sphere, each depositing 100 keV of energy there. The other half of the electrons escape, each carrying away a charge of $-e$. The nickel coating has an activity of 10 mCi ($= 10$ millicuries $= 3.70 \times 10^8$ radioactive decays per second). The sphere is hung from a long, nonconducting string and isolated from its surroundings. (a) How long will it take for the potential of the sphere to increase by 1000 V? (b) How long will it take for the temperature of the sphere to increase by 5.0 K due to the energy deposited by the electrons? The heat capacity of the sphere is 14.3 J/K.

SECTION 25-7 Potential Due to an Electric Dipole

33E. The ammonia molecule NH_3 has a permanent electric dipole moment equal to 1.47 D, where 1 D = 1 debye unit = 3.34×10^{-30} C·m. Calculate the electric potential due to an ammonia molecule at a point 52.0 nm away along the axis of the dipole. (Set $V = 0$ at infinity.)

34P. For the charge configuration of Fig. 25-41, show that $V(r)$ for points such as P on the axis, assuming $r \gg d$, is given by

$$V = \frac{1}{4\pi\epsilon_0} \frac{q}{r}\left(1 + \frac{2d}{r}\right).$$

(*Hint:* The charge configuration can be viewed as the sum of an isolated charge and a dipole.)

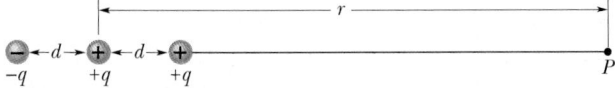

FIGURE 25-41 Problem 34.

SECTION 25-8 Potential Due to a Continuous Charge Distribution

35E. (a) Figure 25-42a shows a positively charged plastic rod of length L and uniform linear charge density λ. Setting $V = 0$ at infinity and considering Fig. 25-13 and Eq. 25-35, find the electric potential at point P without written calculation. (b) Figure 25-42b shows an identical rod, except that it is split in half and the right half is negatively charged; the left and right halves have the same magnitude λ of uniform linear charge density. With V still zero at infinity, what is the electric potential at point P in Fig. 25-42b?

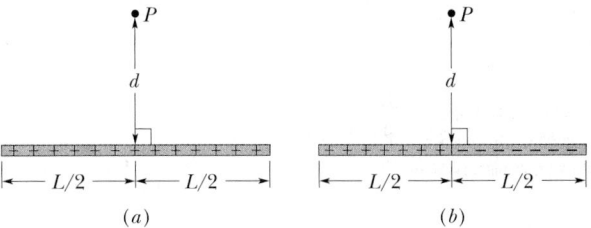

FIGURE 25-42 Exercise 35.

36E. In Fig. 25-43, a plastic rod having a uniformly distributed charge $-Q$ has been bent into a circular arc of radius R and central angle 120°. With $V = 0$ at infinity, what is the electric potential at P, the center of curvature of the rod?

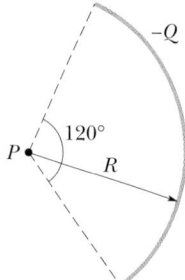

FIGURE 25-43 Exercise 36.

37E. A circular plastic rod of radius R has a positive charge $+Q$ uniformly distributed along one-quarter of its circumference and a negative charge of $-6Q$ uniformly distributed along the rest of the circumference (Fig. 25-44). With $V = 0$ at infinity, what is the electric potential (a) at the center C of the circle and (b) at point P, which is on the central axis of the circle at a distance z from the center?

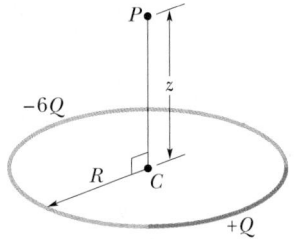

FIGURE 25-44 Exercise 37.

38E. A plastic disk is charged on one side with a uniform surface charge density λ, and then three quadrants of the disk are removed. The remaining quadrant is shown in Fig. 25-45. With $V = 0$ at infinity, what is the potential due to the remaining quadrant at point P, which is on the central axis of the original disk at a distance z from the original center?

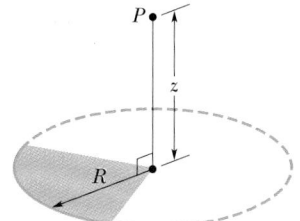

FIGURE 25-45
Exercise 38.

39P. Figure 25-46 shows a ring of outer radius R and inner radius $r = 0.200R$; the ring has a uniform surface charge density σ. With $V = 0$ at infinity, find an expression for the electric potential at point P on the central axis of the ring, at a distance $z = 2.00R$ from the center of the ring.

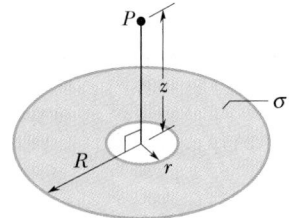

FIGURE 25-46
Problem 39.

40P. A disk like that of Fig. 25-14 has radius $R = 2.20$ cm. Its surface charge density is 1.50×10^{-6} C/m^2 from $r = 0$ to $R/2$ and 8.00×10^{-7} C/m^2 from $r = R/2$ to R. (a) What is the total charge on the disk? (b) With $V = 0$ at infinity, what is the electric potential at a point on the central axis of the disk, at a distance $z = R/2$ from the center of the disk?

41P. Figure 25-47 shows a plastic rod of length L and uniform positive charge Q lying on an x axis. With $V = 0$ at infinity, find the electric potential at point P_1 on the axis, at distance d from one end of the rod.

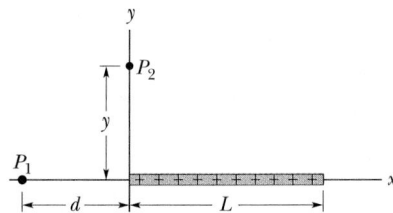

FIGURE 25-47 Problems 41, 42, 50, and 51.

42P. The plastic rod shown in Fig. 25-47 has length L and a nonuniform linear charge density $\lambda = cx$, where c is a positive constant. With $V = 0$ at infinity, find the electric potential at point P_1 on the axis, at distance d from one end.

SECTION 25-9 Calculating the Field from the Potential

43E. Two large parallel metal plates are 1.5 cm apart and have equal but opposite charges on their facing surfaces. Take the potential of the negative plate to be zero. If the potential halfway between the plates is then $+5.0$ V, what is the electric field in the region between the plates?

44E. In a certain situation, the electric potential varies along the x axis as shown in the graph of Fig. 25-48. For each of the intervals ab, bc, cd, de, ef, fg, and gh, determine the x component of the electric field, and then plot E_x versus x. (Ignore behavior at the interval end points.)

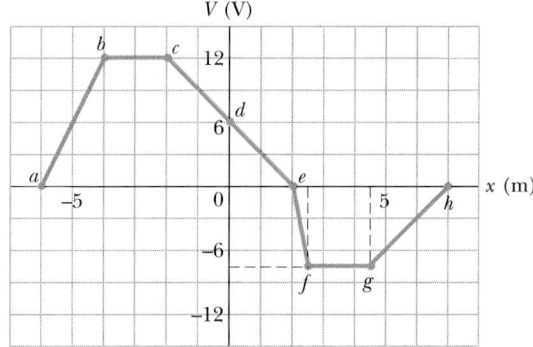

FIGURE 25-48 Exercise 44.

45E. Starting from Eq. 25-30, find the electric field due to a dipole at a point on the dipole axis.

46E. The electric potential at points in an xy plane is given by $V = (2.0 \text{ V/m}^2)x^2 - (3.0 \text{ V/m}^2)y^2$. What are the magnitude and direction of the electric field at point $(3.0 \text{ m}, 2.0 \text{ m})$?

47E. The electric potential V in the space between two flat parallel plates is given by $V = 1500x^2$, where V is in volts if x, the distance from one of the plates, is in meters. Calculate the magnitude and direction of the electric field at $x = 1.3$ cm.

48E. Exercise 48 in Chapter 24 deals with Rutherford's calculation of the electric field at a distance r from the center of an atom and inside the atom. He also gave the electric potential as

$$V = \frac{Ze}{4\pi\epsilon_0}\left(\frac{1}{r} - \frac{3}{2R} + \frac{r^2}{2R^3}\right).$$

(a) Show how the expression for the electric field given in Exercise 48 of Chapter 24 follows from the above expression for V. (b) Why does this expression for V not go to zero as $r \to \infty$?

49P. (a) Using Eq. 25-32, show that the electric potential at a point at distance z on the central axis of a thin ring of charge of radius R is

$$V = \frac{1}{4\pi\epsilon_0}\frac{q}{\sqrt{z^2 + R^2}}.$$

(b) From this result, derive an expression for E at points on the ring's axis; compare your result with the calculation of E in Section 23-6.

50P. (a) Use the result of Problem 41 to find the electric field component E_x at point P_1 in Fig. 25-47. (*Hint:* First substitute the variable x for the distance d in the result.) (b) Use symmetry to determine the electric field component E_y at P_1.

51P. The plastic rod of length L in Fig. 25-47 has the nonuniform linear charge density $\lambda = cx$, where c is a positive constant. (a) With $V = 0$ at infinity, find the electric potential at point P_2 on the y axis, a distance y from one end. (b) From that result, find the electric field component E_y at P_2. (c) Why cannot the field component E_x at P_2 be found using the result of (a)?

SECTION 25-10 Electric Potential Energy of a System of Point Charges

52E. (a) What is the electric potential energy of two electrons separated by 2.00 nm? (b) If the separation increases, does the potential energy increase or decrease?

53E. Two charges $q = +2.0\ \mu\text{C}$ are fixed in space a distance $d = 2.0$ cm apart, as shown in Fig. 25-49. (a) With $V = 0$ at infinity, what is the electric potential at point C? (b) You bring a third charge $q = +2.0\ \mu\text{C}$ from infinity to C. How much work must you do? (c) What is the potential energy U of the three-charge configuration when the third charge is in place?

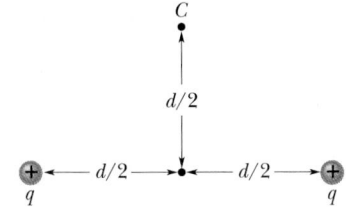

FIGURE 25-49   Exercise 53.

54E. The charges and coordinates of two point charges located in the xy plane are: $q_1 = +3.0 \times 10^{-6}$ C, $x = +3.5$ cm, $y = +0.50$ cm; and $q_2 = -4.0 \times 10^{-6}$ C, $x = -2.0$ cm, $y = +1.5$ cm. How much work must be done to locate these charges at their given positions, starting from infinite separation?

55E. A decade before Einstein published his theory of relativity, J. J. Thomson proposed that the electron might consist of small parts and attributed its mass to the electrical interaction of the parts. Furthermore, he suggested that the energy equals mc^2. Make a rough estimate of the electron mass in the following way: assume that the electron is composed of three identical parts that are brought in from infinity and placed at the vertices of an equilateral triangle having sides equal to the *classical radius* of the electron, 2.82×10^{-15} m. (a) Find the total electric potential energy of this arrangement. (b) Divide by c^2 and compare your result to the accepted electron mass (9.11×10^{-31} kg). (The result improves if more parts are assumed.)

56E. Derive an expression for the work required to set up the four-charge configuration of Fig. 25-50, assuming the charges are initially infinitely far apart.

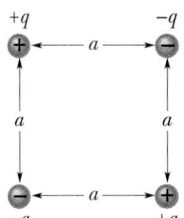

FIGURE 25-50
Exercise 56.

57E. In the quark model of fundamental particles, a proton is composed of three quarks: two "up" quarks, each having charge $+2e/3$, and one "down" quark, having charge $-e/3$. Suppose that the three quarks are equidistant from one another. Take the distance to be 1.32×10^{-15} m and calculate (a) the potential energy of the subsystem of two "up" quarks and (b) the total electric potential energy of the three-particle system.

58E. What is the electric potential energy of the charge configuration of Fig. 25-9a? Use the numerical values provided in Sample Problem 25-4.

59P. Three $+0.12$ C charges form an equilateral triangle, 1.7 m on a side. Using energy that is supplied at the rate of 0.83 kW, how many days would be required to move one of the charges to the midpoint of the line joining the other two charges?

60P. In the rectangle of Fig. 25-51, the sides have lengths 5.0 cm and 15 cm, $q_1 = -5.0\ \mu\text{C}$, and $q_2 = +2.0\ \mu\text{C}$. With $V = 0$ at infinity, what are the electric potentials (a) at corner A and (b) at corner B? (c) How much work is required to move a third charge $q_3 = +3.0\ \mu\text{C}$ from B to A along a diagonal of the rectangle? (d) Does this work increase or decrease the electric energy of the three-charge system? Is more, less, or the same work required if q_3 is moved along paths that are (e) inside the rectangle but not on a diagonal and (f) outside the rectangle?

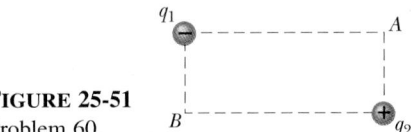

FIGURE 25-51
Problem 60.

61P. In Fig. 25-52, how much work is required to bring the charge of $+5q$ in from infinity along the dashed line and place it as shown near the two fixed charges $+4q$ and $-2q$? Take $d = 1.40$ cm and $q = 1.6 \times 10^{-19}$ C.

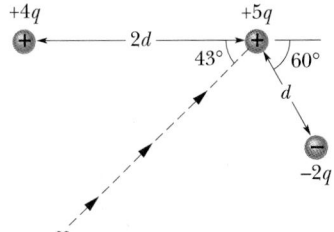

FIGURE 25-52   Problem 61.

62P. A particle of positive charge Q is fixed at point P. A second particle of mass m and negative charge $-q$ moves at constant speed in a circle of radius r_1, centered at P. Derive an expression for the work W that must be done by an external agent on the second particle to increase the radius of the circle of motion to r_2.

63P. Calculate (a) the electric potential established by the nucleus of a hydrogen atom at the average distance of the circulating electron ($r = 5.29 \times 10^{-11}$ m), (b) the electric potential energy of the atom when the electron is at this radius, and (c) the kinetic energy of the electron, assuming it to be moving in a circular orbit of this radius centered on the nucleus. (d) How much energy is required to ionize the hydrogen atom (that is, to remove the electron from the nucleus so that the separation is effectively infinite)? Express all energies in electron-volts.

64P. A particle of charge q is kept in a fixed position at a point P, and a second particle of mass m and the same charge q is initially held a distance r_1 from P. The second particle is then released. Determine its speed when it is a distance r_2 from P. Let $q = 3.1$ μC, $m = 20$ mg, $r_1 = 0.90$ mm, and $r_2 = 2.5$ mm.

65P. A charge of -9.0 nC is uniformly distributed around a ring of radius 1.5 m that lies in the yz plane with its center at the origin. A point charge of -6.0 pC is located on the x axis at $x = 3.0$ m. Calculate the work done in moving the point charge to the origin.

66P. Two tiny metal spheres A and B of mass $m_A = 5.00$ g and $m_B = 10.0$ g have equal positive charges $q = 5.00$ μC. The spheres are connected by a massless nonconducting string of length $d = 1.00$ m, a distance that is much greater than the radii of the spheres. (a) What is the electric potential energy of the system? (b) Suppose you cut the string. At that instant, what is the acceleration of each sphere? (c) A long time after you cut the string, what is the speed of each sphere?

67P. Two charged, parallel, flat conducting surfaces are spaced $d = 1.00$ cm apart and produce a potential difference $\Delta V = 625$ V between them. An electron is projected from one surface directly toward the second. What is the initial speed of the electron if it comes to rest just at the second surface?

68P. (a) A proton of kinetic energy 4.80 MeV travels head-on toward a lead nucleus. Assuming that the proton does not penetrate the nucleus and considering only electrostatic interactions, calculate the smallest center-to-center separation that occurs between the proton and the nucleus when the proton momentarily stops. (b) If the proton is replaced with an alpha particle of the same initial kinetic energy, how would the smallest center-to-center separation compare with that in (a)?

69P. A particle of mass m, positive charge q, and initial kinetic energy K is projected (from a large distance) toward a heavy nucleus of charge Q that is fixed in place. Assuming that the particle approaches head-on, how close to the center of the nucleus is the particle when it comes momentarily to rest?

70P. A thin, conducting, spherical shell of radius R is mounted on an isolating support and charged to a potential of $-V$. An electron is then fired from point P at a distance r from the center of the shell ($r \gg R$) with an initial speed v_0, directed radially inward. What value of v_0 is needed for the electron to just reach the shell before reversing direction?

71P. Two electrons are fixed 2.0 cm apart. Another electron is shot from infinity and comes to rest midway between the two. What was its initial speed?

72P. Consider an electron on the surface of a uniformly charged sphere of radius 1.0 cm and total charge 1.6×10^{-15} C. What is the *escape speed* for this electron? That is, what initial speed must it have to reach an infinite distance from the sphere and there have zero kinetic energy? (This escape speed is defined similarly to that in Chapter 14 for escaping the gravitational force, but here neglect that force.)

73P. An electron is projected with an initial speed of 3.2×10^5 m/s directly toward a proton that is fixed in place. If the electron is initially a great distance from the proton, at what distance from the proton is the speed of the electron instantaneously equal to twice the initial value?

SECTION 25-11 Potential of a Charged Isolated Conductor

74E. An empty hollow metal sphere has a potential of $+400$ V with respect to ground (defined to be at $V = 0$) and has a charge of 5.0×10^{-9} C. Find the electric potential at the center of the sphere.

75E. A thin, conducting, spherical shell of outer radius 20 cm has a charge of $+3.0$ μC. Sketch graphs of (a) the magnitude of the electric field $\mathbf{E}$ and (b) the potential V, both versus the distance r from the center of the shell. (Set $V = 0$ at infinity.)

76E. What is the excess charge on a conducting sphere of radius $r = 0.15$ m if the potential of the sphere is 1500 V and $V = 0$ at infinity?

77E. Consider two widely separated conducting spheres, 1 and 2, the second having twice the diameter of the first. The smaller sphere initially has a positive charge q, and the larger one is initially uncharged. You now connect the spheres with a long thin wire. (a) How are the final potentials V_1 and V_2 of the spheres related? (b) Find the final charges q_1 and q_2 on the spheres in terms of q. (c) What is the ratio of the final surface charge density of sphere 1 to that of sphere 2?

78P. The metal object shown in cross section in Fig. 25-53 is a figure of revolution about a horizontal axis. Suppose that it is charged negatively, and sketch a few equipotential surfaces and electric field lines. Use physical reasoning rather than mathematical analysis.

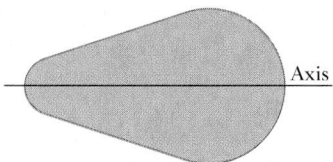

FIGURE 25-53 Problem 78.

79P. (a) If Earth had a net surface charge density of 1.0 electron per square meter (a very artificial assumption), what would its potential be? (Set $V = 0$ at infinity.) (b) What would be the electric field due to Earth just outside its surface?

80P. Two metal spheres, each of radius 3.0 cm, have a center-to-center separation of 2.0 m. One has a charge of $+1.0 \times 10^{-8}$ C; the other has a charge of -3.0×10^{-8} C. Assume that the separation is large enough relative to the size of the spheres to permit us to consider the charge on each to be uniformly distributed (the spheres are electrically isolated from each other). With $V = 0$ at infinity, calculate (a) the potential at the point halfway between their centers and (b) the potential of each sphere.

81P. A charged metal sphere of radius 15 cm has a net charge of 3.0×10^{-8} C. (a) What is the electric field at the sphere's surface? (b) If $V = 0$ at infinity, what is the electric potential at the sphere's surface? (c) At what distance from the sphere's surface has the electric potential decreased by 500 V?

82P. Two thin, isolated, concentric conducting spheres of radii R_1 and R_2 carry charges q_1 and q_2. With $V = 0$ at infinity, derive expressions for $E(r)$ and $V(r)$, where r is distance from the center of the spheres. Plot $E(r)$ and $V(r)$ from $r = 0$ to $r = 4.0$ m for $R_1 = 0.50$ m, $R_2 = 1.0$ m, $q_1 = +2.0$ μC, and $q_2 = +1.0$ μC.

Electronic Computation

83. Charge $q_1 = -1.2 \times 10^{-9}$ C is at the origin, and charge $q_2 = 2.5 \times 10^{-9}$ C is on the y axis at $y = 0.50$ m. Take the electric potential to be zero far from both charges. (a) Plot the inter-

section of the $V = 5.0$ V equipotential surface with the xy plane. It encloses one of the charges. (b) There are two equipotential surfaces corresponding to $V = 3.0$ V. One encloses one of the charges and the other encloses both charges. Plot their intersections with the xy plane. (c) Find the value of the potential for which the pattern of the electric potential switches from one to two equipotential surfaces.

84. Suppose that N electrons are to be placed in a ring of radius R and that they can be placed in either of two configurations. In the first configuration they are all placed on the circumference and are uniformly distributed so that the distance between adjacent electrons is the same everywhere; in the second configuration $N - 1$ of the electrons are placed on the circumference as before and one electron is placed in the center of the ring. (a) For which configuration is the electrostatic potential energy less? Answer for N equal to all integer values from 2 to 15. (b) What is the smallest value of N for which the second configuration is less energetic than the first? (c) For the value of N found in (b), how many rim electrons are closer to any given electron there than is the electron at the center?

26
Capacitance

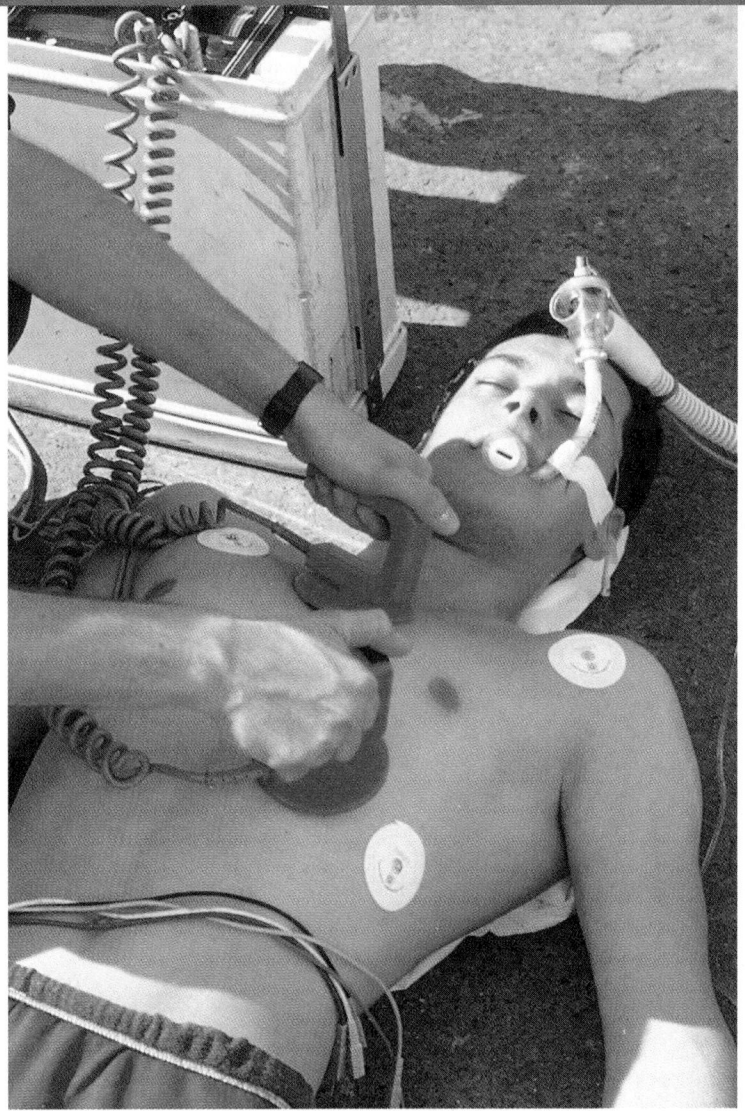

During ventricular fibrillation, a common type of heart attack, the chambers of the heart fail to pump blood because their muscle fibers randomly contract and relax. To save a victim of ventricular fibrillation, the heart muscle must be shocked to reestablish its normal rhythm. For that, 20 A of current must be sent through the chest cavity to transfer 200 J of electrical energy in about 2.0 ms. This requires about 100 kW of electrical power. The requirement may easily be met in a hospital, but what could produce that much power on, say, a remote road? Certainly not the electrical system of a car or ambulance, even if such a vehicle is available?

26-1 THE USES OF CAPACITORS

You can store energy as potential energy by pulling a bowstring, stretching a spring, compressing a gas, or lifting a book. You can also store energy as potential energy in an electric field, and a **capacitor** is a device you can use to do exactly that.

There is a capacitor in a portable battery-operated photoflash unit, for example. It accumulates charge relatively slowly during the charging process, building up an electric field as it does so. It holds this field and its energy until the energy is rapidly released during the flash.

Capacitors have many uses in our electronic and microelectronic age beyond serving as storehouses for potential energy. For one example, they are vital elements in the circuits with which we tune radio and television transmitters and receivers. For another example, microscopic capacitors form the memory banks of computers. These tiny devices are important—not so much for their stored energy as for the ON–OFF information that the presence or absence of their electric fields provides.

26-2 CAPACITANCE

Figure 26-1 shows some of the many sizes and shapes of capacitors. Figure 26-2 shows the basic elements of *any* capacitor—two isolated conductors of arbitrary shape. No matter what their geometry, flat or not, we call these conductors *plates*.

Figure 26-3a shows a less general but more conventional arrangement, called a *parallel-plate capacitor*, consisting of two parallel conducting plates of area A separated by a distance d. The symbol that we use to represent a

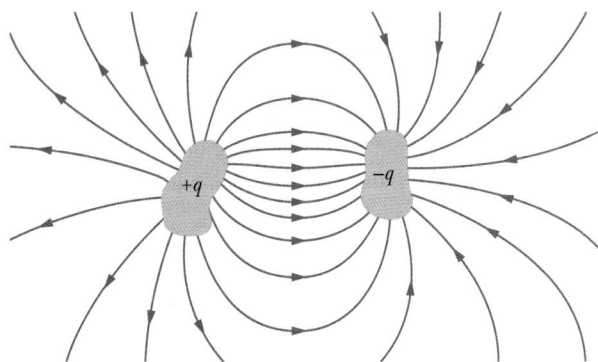

FIGURE 26-2 Two conductors, isolated electrically from each other and from their surroundings, form a *capacitor*. When the capacitor is charged, the conductors, or *plates* as they are called, carry equal but opposite charges of magnitude q.

capacitor (⊣⊢) is based on the structure of a parallel-plate capacitor but is used for capacitors of all geometries. We assume for the time being that no material medium (such as glass or plastic) is present in the region between the plates. In Section 26-6, we shall remove this restriction.

When a capacitor is *charged*, its plates have equal but opposite charges of $+q$ and $-q$. However, we refer to the

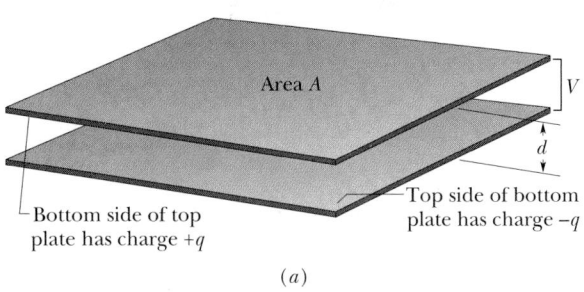

Bottom side of top plate has charge $+q$

Top side of bottom plate has charge $-q$

(a)

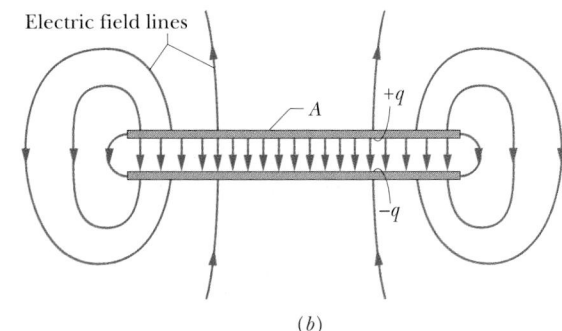

Electric field lines

(b)

FIGURE 26-3 (a) A parallel-plate capacitor, made up of two plates of area A separated by a distance d. The plates have equal and opposite charges of magnitude q on their facing surfaces. (b) As the field lines show, the electric field is uniform in the central region between the plates. The field is not uniform at the edges of the plates, as indicated by the "fringing" of the field lines there.

FIGURE 26-1 An assortment of capacitors.

charge of a capacitor as being q, the absolute value of these charges on the plates. (Note that q is not the net charge on the capacitor, which is zero.)

Because the plates are conductors, they are equipotential surfaces: all points on a plate are at the same electric potential. Moreover, there is a potential difference between the two plates. For historical reasons, we represent the absolute value of this potential difference with V rather than with ΔV as we would with previous notation.

The charge q and the potential difference V for a capacitor are proportional to each other. That is,

$$q = CV. \qquad (26\text{-}1)$$

The proportionality constant C is called the **capacitance** of the capacitor. Its value depends only on the geometry of the plates and *not* on their charge or potential difference. The capacitance is a measure of how much charge must be put on the plates to produce a certain potential difference between them: the *greater the capacitance, the more charge is required.*

The SI unit of capacitance that follows from Eq. 26-1 is the coulomb per volt. This unit occurs so often that it is given a special name, the *farad* (F):

$$1 \text{ farad} = 1 \text{ F} = 1 \text{ coulomb per volt}$$
$$= 1 \text{ C/V}. \qquad (26\text{-}2)$$

As you will see, the farad is a very large unit. Submultiples of the farad, such as the microfarad (1 μF $= 10^{-6}$ F) and the picofarad (1 pF $= 10^{-12}$ F), are more convenient units in practice.

Charging a Capacitor

One way to charge a capacitor is to place it in an electric circuit with a battery. An *electric circuit* is a path through which charge can flow. A *battery* is a device that maintains a certain potential difference between its *terminals* (points at which charge can enter or leave the battery) by means of internal electrochemical reactions.

In Fig. 26-4a, a battery B, a switch S, an uncharged capacitor C, and interconnecting wires form a circuit. The same circuit is shown in the *schematic diagram* of Fig. 26-4b, in which the symbols for a battery, a switch, and a capacitor represent those devices. The battery maintains potential difference V between its terminals. The terminal of higher potential is labeled $+$ and is often called the positive terminal; the terminal of lower potential is labeled $-$ and is often called the negative terminal.

The circuit shown in Figs. 26-4a and b is said to be incomplete because switch S is *open*; that is, it does not electrically connect the wires attached to it. When the switch is *closed*, electrically connecting those wires, the

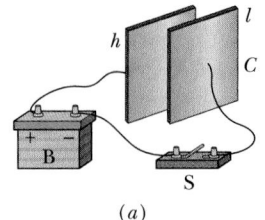

(a)

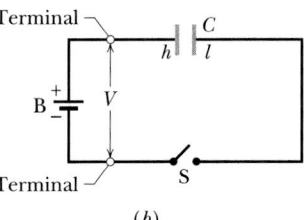

(b)

FIGURE 26-4 (a) Battery B, switch S, and plates h and l of capacitor C, connected in a circuit. (b) A schematic diagram with the *circuit elements* represented by their symbols.

circuit is complete and charge can then flow through the switch and the wires. As we discussed in Chapter 22, the charge that can flow through a conductor, such as a wire, is that of electrons. When the circuit of Fig. 26-4 is completed, electrons are driven through the wires by an electric field that the battery sets up in the wires. The field drives electrons from capacitor plate h to the positive terminal of the battery; thus plate h, losing electrons, becomes positively charged. The field drives just as many electrons from the negative terminal of the battery to capacitor plate l; thus plate l, gaining electrons, becomes negatively charged *just as much* as plate h becomes positively charged.

The potential difference between the initially uncharged plates is zero. As the plates become oppositely charged, that potential difference increases until it equals the potential difference V between the terminals of the battery. Then plate h and the positive terminal of the battery are at the same potential, and there is no longer an electric field in the wire between them. Similarly, plate l and the negative terminal reach the same potential and there is then no electric field in the wire between them. Thus, with the field zero, there is no further drive of electrons. The capacitor is then said to be *fully charged,* with a potential difference V and a charge q, which are related by Eq. 26-1.

PROBLEM SOLVING TACTICS

TACTIC 1: *The Symbol V and Potential Difference*

In previous chapters, the symbol V represents an electric potential at a point or along an equipotential surface. However, in matters concerning electrical devices, V often represents a *potential difference* between two points or two equipotential surfaces. Equation 26-1 is an example of this second use of the symbol. In Section 26-3, you will see a mixture of the two meanings of *V*. There and in later chapters, you need to be alert as to the intent of this symbol.

You will also be seeing, in this book and elsewhere, a variety of phrases regarding potential difference. A potential difference or a ''potential'' or a ''voltage'' may be *applied* to a device, or it may be *across* a device. A capacitor can be charged to a potential difference, as in ''a capacitor is charged to 12 V.'' And a battery can be characterized by the potential difference across it, as in ''a 12 V battery.'' Always keep in mind what is meant by such phrases: there is a potential difference between two points, such as two points in a circuit or at the terminals of a device such as a battery.

CHECKPOINT 1: Does the capacitance *C* of a capacitor increase, decrease, or remain the same (a) when the charge *q* on it is doubled and (b) when the potential difference *V* across it is tripled?

26-3 CALCULATING THE CAPACITANCE

Our task here is to calculate the capacitance of a capacitor once we know its geometry. Because we are going to consider a number of different geometries, it seems wise to develop a general plan to simplify the work. In brief our plan is as follows: (1) assume a charge *q* on the plates; (2) calculate the electric field **E** between the plates in terms of this charge, using Gauss' law; (3) knowing **E**, calculate the potential difference *V* between the plates from Eq. 25-18; (4) calculate *C* from Eq. 26-1.

Before we start, we can simplify the calculation of both the electric field and the potential difference by making certain assumptions. We discuss each in turn.

Calculating the Electric Field

The electric field **E** between the plates of a capacitor is related to the charge *q* on a plate by Gauss' law:

$$\epsilon_0 \oint \mathbf{E} \cdot d\mathbf{A} = q. \qquad (26\text{-}3)$$

Here *q* is the charge enclosed by a Gaussian surface, and $\oint \mathbf{E} \cdot d\mathbf{A}$ is the net electric flux through that surface. In all cases that we shall consider, the Gaussian surface will be such that whenever electric flux passes through it, **E** will have a magnitude *E* and the vectors **E** and *d***A** will be parallel. Equation 26-3 then reduces to

$$q = \epsilon_0 EA \qquad \text{(special case of Eq. 26-3),} \qquad (26\text{-}4)$$

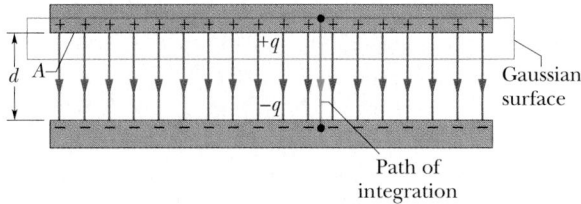

FIGURE 26-5   A charged parallel-plate capacitor. A Gaussian surface encloses the charge on the positive plate. The integration of Eq. 26-6 is taken along a path extending directly from the positive plate to the negative plate.

in which *A* is the area of that part of the Gaussian surface through which flux passes. For convenience, we shall always draw the Gaussian surface in such a way that it completely encloses the charge on the positive plate; see Fig. 26-5 for an example.

Calculating the Potential Difference

In the notation of Chapter 25 (Eq. 25-18), the potential difference between the plates is related to the electric field **E** by

$$V_f - V_i = -\int_i^f \mathbf{E} \cdot d\mathbf{s}, \qquad (26\text{-}5)$$

in which the integral is to be evaluated along any path that starts on one plate and ends on the other. We shall always choose a path that follows an electric field line from the positive plate to the negative plate. For this path, the vectors **E** and *d***s** will always point in the same direction, so the dot product **E** · *d***s** will be equal to the positive quantity *E ds*. Equation 26-5 then tells us that the quantity $V_f - V_i$ will always be negative. Since we are looking for *V*, the *absolute value* of the potential difference between the plates, we can set $V_f - V_i = -V$. Thus we can recast Eq. 26-5 as

$$V = \int_+^- E\, ds \qquad \text{(special case of Eq. 26-5),} \qquad (26\text{-}6)$$

in which the + and − remind us that our path of integration starts on the positive plate and ends on the negative plate.

We are now ready to apply Eqs. 26-4 and 26-6 to some particular cases.

A Parallel-Plate Capacitor

We assume, as Fig. 26-5 suggests, that the plates of our parallel-plate capacitor are so large and so close together that we can neglect the fringing of the electric field at the edges of the plates, taking **E** to be constant throughout the volume between the plates.

We draw a Gaussian surface that encloses just the charge q on the positive plate, as in Fig. 26-5. From Eq. 26-4 we can then write

$$q = \epsilon_0 EA, \qquad (26\text{-}7)$$

where A is the area of the plate.

Equation 26-6 yields

$$V = \int_+^- E\,ds = E\int_0^d ds = Ed. \qquad (26\text{-}8)$$

In Eq. 26-8, E can be placed outside the integral because it is a constant; the second integral then is simply the plate separation d.

If we substitute q from Eq. 26-7 and V from Eq. 26-8 into the relation $q = CV$ (Eq. 26-1), we find

$$C = \frac{\epsilon_0 A}{d} \qquad \text{(parallel-plate capacitor).} \quad (26\text{-}9)$$

So the capacitance does indeed depend only on geometrical factors, namely, the plate area A and the plate separation d. Note that C increases as we increase area A or decrease separation d.

As an aside we point out that Eq. 26-9 suggests one of our reasons for writing the electrostatic constant in Coulomb's law in the form $1/4\pi\epsilon_0$. If we had not done so, Eq. 26-9—which is used more often in engineering practice than Coulomb's law—would have been less simple in form. We note further that Eq. 26-9 permits us to express the permittivity constant ϵ_0 in a unit more appropriate for use in problems involving capacitors, namely,

$$\epsilon_0 = 8.85 \times 10^{-12}\ \text{F/m} = 8.85\ \text{pF/m}. \quad (26\text{-}10)$$

We have previously expressed this constant as

$$\epsilon_0 = 8.85 \times 10^{-12}\ \text{C}^2/\text{N}\cdot\text{m}^2, \qquad (26\text{-}11)$$

using units that are useful for problems involving Coulomb's law (see Section 22-4).

A Cylindrical Capacitor

Figure 26-6 shows, in cross section, a cylindrical capacitor of length L formed by two coaxial cylinders of radii a and b. We assume that $L \gg b$ so that we can neglect the fringing of the electric field that occurs at the ends of the cylinders. Each plate contains a charge of magnitude q.

As a Gaussian surface, we choose a cylinder of length L and radius r, closed by end caps and placed as is shown in Fig. 26-6. Equation 26-4 then yields

$$q = \epsilon_0 EA = \epsilon_0 E(2\pi rL),$$

in which $2\pi rL$ is the area of the curved part of the Gaussian surface. There is no flux through the end caps. Solving

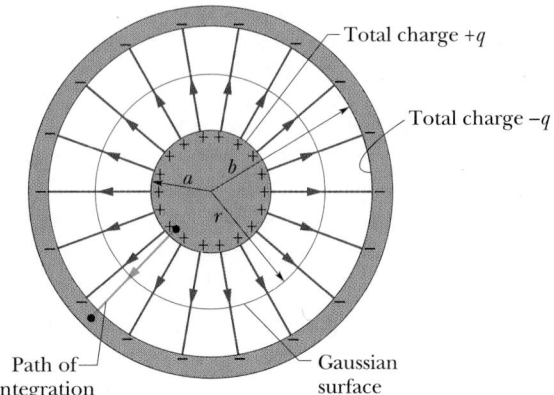

FIGURE 26-6 A cross section of a long cylindrical capacitor, showing a cylindrical Gaussian surface of radius r and the radial path of integration along which Eq. 26-6 is to be applied. This figure also serves to illustrate a spherical capacitor in a cross section through its center.

for E yields

$$E = \frac{q}{2\pi\epsilon_0 Lr}. \qquad (26\text{-}12)$$

Substitution of this result into Eq. 26-6 yields

$$V = \int_+^- E\,ds = \frac{q}{2\pi\epsilon_0 L}\int_a^b \frac{dr}{r}$$

$$= \frac{q}{2\pi\epsilon_0 L}\ln\left(\frac{b}{a}\right), \qquad (26\text{-}13)$$

where we have used the fact that here $ds = dr$. From the relation $C = q/V$, we then have

$$C = 2\pi\epsilon_0 \frac{L}{\ln(b/a)} \qquad \begin{array}{l}\text{(cylindrical} \\ \text{capacitor).}\end{array} \quad (26\text{-}14)$$

We see that the capacitance of a cylindrical capacitor, like that of a parallel-plate capacitor, depends only on geometrical factors, in this case L, b, and a.

A Spherical Capacitor

Figure 26-6 can also serve as a central cross section of a capacitor that consists of two concentric spherical shells, of radii a and b. As a Gaussian surface we draw a sphere of radius r concentric with the two shells. Applying Eq. 26-4 to this surface yields

$$q = \epsilon_0 EA = \epsilon_0 E(4\pi r^2),$$

in which $4\pi r^2$ is the area of the spherical Gaussian surface. We solve this equation for E, obtaining

$$E = \frac{1}{4\pi\epsilon_0}\frac{q}{r^2}, \qquad (26\text{-}15)$$

which we recognize as the expression for the electric field due to a uniform spherical charge distribution (Eq. 24-15).

If we substitute this expression into Eq. 26-6, we find

$$V = \int_{+}^{-} E \, ds = \frac{q}{4\pi\epsilon_0} \int_a^b \frac{dr}{r^2} = \frac{q}{4\pi\epsilon_0} \left(\frac{1}{a} - \frac{1}{b} \right)$$

$$= \frac{q}{4\pi\epsilon_0} \frac{b-a}{ab}. \tag{26-16}$$

If we substitute Eq. 26-16 into Eq. 26-1 and solve for C, we find

$$C = 4\pi\epsilon_0 \frac{ab}{b-a} \quad \begin{array}{l} \text{(spherical} \\ \text{capacitor).} \end{array} \tag{26-17}$$

An Isolated Sphere

We can assign a capacitance to a *single* isolated spherical conductor of radius R by assuming that the "missing plate" is a conducting sphere of infinite radius. After all, the field lines that leave the surface of a charged isolated conductor must end somewhere; the walls of the room in which the conductor is housed can serve effectively as our sphere of infinite radius.

To find the capacitance of the isolated conductor, we first rewrite Eq. 26-17 as

$$C = 4\pi\epsilon_0 \frac{a}{1 - a/b}.$$

If we then let $b \to \infty$ and substitute R for a, we find

$$C = 4\pi\epsilon_0 R \quad \text{(isolated sphere).} \tag{26-18}$$

Note that this formula and the others we have derived for capacitance (Eqs. 26-9, 26-14 and 26-17) involve the constant ϵ_0 multiplied by a quantity that has the dimensions of a length.

CHECKPOINT **2:** For capacitors charged by the same battery, does the charge stored by the capacitor increase, decrease, or remain the same in each of the following situations? (a) The plate separation of a parallel-plate capacitor is increased. (b) The radius of the inner cylinder of a cylindrical capacitor is increased. (c) The radius of the outer spherical shell of a spherical capacitor is increased.

SAMPLE PROBLEM 26-1

The plates of a parallel-plate capacitor are separated by a distance $d = 1.0$ mm. What must be the plate area if the capacitance is to be 1.0 F?

SOLUTION: From Eq. 26-9 we have

$$A = \frac{Cd}{\epsilon_0} = \frac{(1.0 \text{ F})(1.0 \times 10^{-3} \text{ m})}{8.85 \times 10^{-12} \text{ F/m}}$$

$$= 1.1 \times 10^8 \text{ m}^2. \quad \text{(Answer)}$$

This is the area of a square more than 10 km on edge. The farad is indeed a large unit. Modern technology, however, has permitted the construction of 1 F capacitors of very modest size. These "supercaps" are used as backup voltage sources for computers; they can maintain the computer memory for up to 30 days in case of power failure.

SAMPLE PROBLEM 26-2

A storage capacitor on a random access memory (RAM) chip has a capacitance of 55 fF. If the capacitor is charged to 5.3 V, how many excess electrons are on its negative plate?

SOLUTION: The number n of excess electrons is given by q/e, where e is the fundamental charge. Then, using Eq. 26-1, we have

$$n = \frac{q}{e} = \frac{CV}{e} = \frac{(55 \times 10^{-15} \text{ F})(5.3 \text{ V})}{1.60 \times 10^{-19} \text{ C}}$$

$$= 1.8 \times 10^6 \text{ electrons.} \quad \text{(Answer)}$$

For electrons, this is a very small number. A speck of household dust, so tiny that it essentially never settles, contains about 10^{17} electrons (and the same number of protons).

26-4 CAPACITORS IN PARALLEL AND IN SERIES

When there is a combination of capacitors in a circuit, we can sometimes replace that combination with an **equivalent capacitor,** that is, a single capacitor that has the same capacitance as the actual combination of capacitors. With such a replacement, we can simplify the circuit, affording easier solutions for unknown quantities of the circuit. Here we discuss two basic combinations of capacitors that allow such a replacement.

Capacitors in Parallel

Figure 26-7a shows three capacitors connected *in parallel* to a battery B. They are connected "in parallel" because the terminals of the battery are effectively wired directly to the plates of each of the three capacitors. Because the battery maintains a potential difference V between its terminals, it applies the same potential difference V across each capacitor.

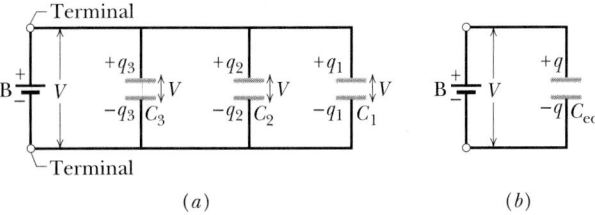

FIGURE 26-7 (a) Three capacitors connected in parallel to battery B. The battery maintains potential difference V across its terminals and thus across *each* capacitor. (b) The equivalent capacitance C_{eq} replaces the parallel combination. The charge q on C_{eq} is equal to the sum of the charges q_1, q_2, and q_3 on the capacitors of (a).

Connected capacitors are said to be in parallel when a potential difference that is applied across their combination results in that same potential difference across each capacitor.

We seek the single capacitance C_{eq} that is equivalent to this parallel combination and thus can replace the combination (as in Fig. 26-7b). By *equivalent,* we mean that when the same potential difference V is applied across it, a capacitor with capacitance C_{eq} will store the same total charge q as is stored in the combination being replaced.

For the three capacitors we can write, from Eq. 26-1,

$$q_1 = C_1 V, \quad q_2 = C_2 V, \quad \text{and} \quad q_3 = C_3 V.$$

The total charge on the parallel combination is then

$$q = q_1 + q_2 + q_3 = (C_1 + C_2 + C_3)V.$$

The equivalent capacitance, with the same total charge q and applied potential difference V as the combination, is then

$$C_{eq} = \frac{q}{V} = C_1 + C_2 + C_3,$$

a result that we can easily extend to any number n of capacitors, as

$$C_{eq} = \sum_{j=1}^{n} C_j \qquad \begin{array}{l}(n \text{ capacitors} \\ \text{in parallel}).\end{array} \qquad (26\text{-}19)$$

Thus, to find the equivalent capacitance of a parallel combination, we simply add the individual capacitances.

Capacitors in Series

Figure 26-8a shows three capacitors connected *in series* to battery B, which maintains a potential difference V across the left and right terminals of the series combination. This arrangement produces potential differences V_1, V_2, and V_3

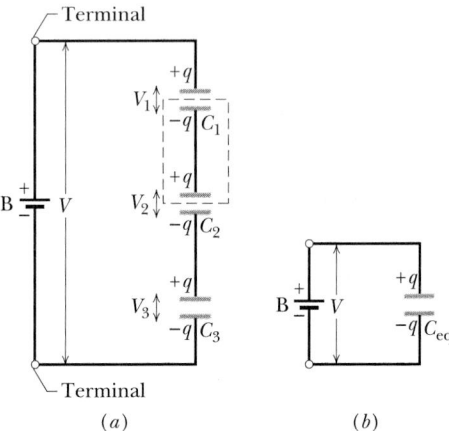

FIGURE 26-8 (a) Three capacitors connected in series to battery B. The battery maintains potential difference V between the left and right sides of the series combination. (b) The equivalent capacitance C_{eq} replaces the series combination. The potential difference V across C_{eq} is equal to the sum of the potential differences V_1, V_2, and V_3 across the capacitors of (a).

across capacitors C_1, C_2, and C_3, respectively, such that $V_1 + V_2 + V_3 = V$.

Connected capacitors are said to be in series when a potential difference that is applied across their combination is the sum of the resulting potential differences across each capacitor.

We seek the single capacitance C_{eq} that is equivalent to this series combination and thus can replace the combination (as in Fig. 26-8b). Again, *equivalent* means that the same total charge q is stored with the same applied potential difference V.

When the battery is connected, each capacitor in Fig. 26-8a must have the same charge q. This is true even though the three capacitors may be of different types and may have different capacitances. To understand this fact, note that the part of the circuit enclosed by the dashed lines in Fig. 26-8a is electrically isolated from the rest of the circuit. Thus it cannot gain or lose charge. However, the battery can *induce* charge on the isolated part, that is, redistribute the charge that is already there: When the battery produces a charge of $+q$ on the top plate of C_1, that charge attracts electrons within the isolated part, redistributing some of them. The redistribution leaves the bottom plate of C_1 with charge $-q$ and the top plate of C_2 with charge $+q$. At the same time, the battery produces a charge of $-q$ on the bottom plate of C_3, causing redistribution of charges on the connected plates of C_2 and C_3.

The net effect is that all three capacitors in the series combination have the same charge q. However, the total charge produced by the battery on the series combination is *not* the sum of these three charges. The battery has produced only charge q, using the top-most and bottom-most plates in the series. The other plates are charged only because of the redistribution of electrons already there.

Application of Eq. 26-1 to each capacitor yields

$$V_1 = \frac{q}{C_1}, \quad V_2 = \frac{q}{C_2}, \quad \text{and} \quad V_3 = \frac{q}{C_3}.$$

The potential difference across the entire series combination is then

$$V = V_1 + V_2 + V_3$$
$$= q \left(\frac{1}{C_1} + \frac{1}{C_2} + \frac{1}{C_3} \right).$$

The equivalent capacitance is then

$$C_{eq} = \frac{q}{V} = \frac{1}{1/C_1 + 1/C_2 + 1/C_3},$$

or

$$\frac{1}{C_{eq}} = \frac{1}{C_1} + \frac{1}{C_2} + \frac{1}{C_3}.$$

We can easily extend this to any number n of capacitors as

$$\frac{1}{C_{eq}} = \sum_{j=1}^{n} \frac{1}{C_j} \quad (n \text{ capacitors in series}). \quad (26\text{-}20)$$

From Eq. 26-20 you can deduce that the equivalent series capacitance is always less than the least capacitance in the series of capacitors.

As you will see in Sample Problem 26-3, some complicated combinations of capacitors can be subdivided into parallel and series combinations, which can then be replaced with equivalent capacitances. This simplifies the original combination and the analysis of circuits.

CHECKPOINT **3:** A battery of potential V stores charge q on a combination of two identical capacitors. What are the potential difference across and the charge on either capacitor if the capacitors are (a) in parallel and (b) in series?

SAMPLE PROBLEM 26-3

(a) Find the equivalent capacitance of the combination shown in Fig. 26-9a. Assume

$$C_1 = 12.0 \ \mu\text{F}, \quad C_2 = 5.30 \ \mu\text{F}, \quad \text{and} \quad C_3 = 4.50 \ \mu\text{F}.$$

SOLUTION: Capacitors C_1 and C_2 are in parallel. From Eq.

26-19, their equivalent capacitance is

$$C_{12} = C_1 + C_2 = 12.0 \ \mu\text{F} + 5.30 \ \mu\text{F} = 17.3 \ \mu\text{F}.$$

As Fig. 26-9b shows, C_{12} and C_3 now form a series combination. From Eq. 26-20, their equivalent capacitance (shown in Fig. 26-9c) is given by

$$\frac{1}{C_{123}} = \frac{1}{C_{12}} + \frac{1}{C_3} = \frac{1}{17.3 \ \mu\text{F}} + \frac{1}{4.50 \ \mu\text{F}} = 0.280 \ \mu\text{F}^{-1},$$

from which

$$C_{123} = \frac{1}{0.280 \ \mu\text{F}^{-1}} = 3.57 \ \mu\text{F}. \quad (\text{Answer})$$

(b) A potential difference $V = 12.5$ V is applied to the input terminals in Fig. 26-9a. What is the charge on C_1?

SOLUTION: We treat the equivalent capacitors C_{12} and C_{123} exactly as we would real capacitors of the same capacitances. For the charge on C_{123} in Fig. 26-9c we then have

$$q_{123} = C_{123}V = (3.57 \ \mu\text{F})(12.5 \text{ V}) = 44.6 \ \mu\text{C}.$$

This same charge exists on each capacitor in the series combination of Fig. 26-9b. Let q_{12} ($= q_{123}$) represent the charge on C_{12} in that figure. The potential difference across C_{12} is then

$$V_{12} = \frac{q_{12}}{C_{12}} = \frac{44.6 \ \mu\text{C}}{17.3 \ \mu\text{F}} = 2.58 \text{ V}.$$

This same potential difference appears across both C_1 and C_2 in Fig. 26-9a. Let V_1 ($= V_{12}$) represent the potential difference across C_1. We then have

$$q_1 = C_1V_1 = (12.0 \ \mu\text{F})(2.58 \text{ V})$$
$$= 31.0 \ \mu\text{C}. \quad (\text{Answer})$$

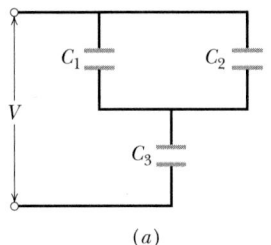

(a)

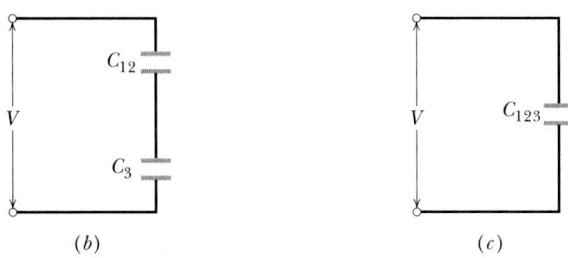

(b)   (c)

FIGURE 26-9 Sample Problem 26-3. (a) Three capacitors. (b) C_1 and C_2, a parallel combination, are replaced by C_{12}. (c) C_{12} and C_3, a series combination, are replaced by the equivalent capacitance C_{123}.

SAMPLE PROBLEM 26-4

A 3.55 μF capacitor C_1 is charged to a potential difference $V_0 = 6.30$ V, using a 6.30 V battery. The battery is then removed and the capacitor is connected as in Fig. 26-10 to an uncharged 8.95 μF capacitor C_2. When switch S is closed, charge flows from C_1 to C_2 until the capacitors have the same potential difference V. What is this common potential difference they reach?

SOLUTION: The original charge q_0 is now shared by two capacitors, so

$$q_0 = q_1 + q_2.$$

Applying the relation $q = CV$ to each term of this equation yields

$$C_1 V_0 = C_1 V + C_2 V,$$

from which

$$V = V_0 \frac{C_1}{C_1 + C_2} = \frac{(6.30 \text{ V})(3.55 \text{ } \mu\text{F})}{3.55 \text{ } \mu\text{F} + 8.95 \text{ } \mu\text{F}}$$

$$= 1.79 \text{ V}. \qquad \text{(Answer)}$$

When the capacitors reach this value of electric potential difference, the charge flow stops.

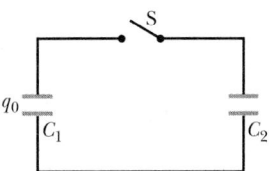

FIGURE 26-10 Sample Problems 26-4 and 26-5. A potential difference V_0 is applied to C_1 and the charging battery is removed. Switch S is then closed so that the charge on C_1 is shared with C_2.

PROBLEM SOLVING TACTICS

TACTIC 2: *Multiple Capacitor Circuits*

Let us review the procedure used in the solution of Sample Problem 26-3, in which several capacitors are connected to a battery. To find a single equivalent capacitance, we simplify the given arrangement of capacitances by replacing them, in steps, with equivalent capacitances, using Eq. 26-19 when we find capacitances in parallel and Eq. 26-20 when we find capacitances in series. Then, to find the charge stored by that single equivalent capacitance, we use Eq. 26-1 and the potential difference V imposed by the battery.

That result tells us the net charge stored on the actual arrangement of capacitors. However, to find the charge on, or the potential difference across, any particular capacitor in the actual arrangement, we need to reverse our steps of simplification. With each reversed step, we use these two rules: when capacitances are in parallel, they have the same potential difference their equivalent capacitance has, and we use Eq. 26-1 to find the charge on each capacitance; when they are in series,

they have the same charge their equivalent capacitance has, and we use Eq. 26-1 to find the potential difference across each capacitance.

TACTIC 3: *Batteries and Capacitors*

A battery maintains a certain potential difference across its terminals. So when capacitor C_1 of Sample Problem 26-4 is connected to the 6.30 V battery, charge flows between the capacitor and the battery until the capacitor has the same potential difference as the battery.

A capacitor differs from a battery in that a capacitor lacks internal electrochemical reactions to release charged particles (electrons) from internal atoms and molecules. So when the charged capacitor C_1 of Sample Problem 26-4 is disconnected from the battery and then connected to the uncharged capacitor C_2 with switch S closed, the potential difference across C_1 is not maintained. The quantity that *is* maintained is the total charge q_0 of the two-capacitor system; that is, charge obeys a conservation law, *not* electric potential.

Here is what happens to that charge. When switch S is open as shown in Fig. 26-10, charge q_0 is entirely on C_1. Charge cannot be transferred between the capacitors until there is a complete circuit, or loop, through which the charge can flow. When switch S is closed, there is such a complete circuit, and a portion of q_0 flows from C_1 to C_2, increasing the potential difference of C_2 and decreasing that of C_1, until the two capacitors have the same potential difference V. The top plates of the capacitors are then at the same electric potential, and so are the bottom plates; thus the capacitors are in equilibrium and there is no further charge flow.

CHECKPOINT **4:** In Sample Problem 26-4 and Fig. 26-10, suppose capacitor C_2 is replaced by a series combination of capacitors C_3 and C_4. (a) After the switch is closed and charge has stopped flowing, what is the relation between the initial charge q_0, the charge q_1 then on C_1, and the charge q_{34} then on the equivalent capacitance C_{34}? (b) If $C_3 > C_4$, is the charge q_3 on C_3 more than, less than, or equal to the charge q_4 on C_4?

26-5 STORING ENERGY IN AN ELECTRIC FIELD

Work must be done by an external agent to charge a capacitor. Starting with an uncharged capacitor, for example, imagine that—using "magic tweezers"—you remove electrons from one plate and transfer them one at a time to the other plate. The electric field that builds up in the space between the plates has a direction that tends to oppose further transfer. Thus, as charge accumulates on the capacitor plates, you have to do increasingly larger amounts of work to transfer additional electrons. In practice, this work is done not by "magic tweezers" but by a battery, at the expense of its store of chemical energy.

We visualize the work required to charge a capacitor as being stored in the form of **electric potential energy** U in the electric field between the plates. You can recover this energy at will, by discharging the capacitor in a circuit, just as you can recover the potential energy stored in a stretched bow by releasing the bowstring to transfer the energy to the kinetic energy of an arrow.

Suppose that, at a given instant, a charge q' has been transferred from one plate to the other. The potential difference V' between the plates at that instant will be q'/C. If an extra increment of charge dq' is then transferred, the increment of work required will be, from Eq. 25-7,

$$dW = V' \, dq' = \frac{q'}{C} \, dq'.$$

The work required to bring the total capacitor charge up to a final value q is

$$W = \int dW = \frac{1}{C} \int_0^q q' \, dq' = \frac{q^2}{2C}.$$

This work is stored as potential energy U in the capacitor, so that

$$U = \frac{q^2}{2C} \qquad \text{(potential energy).} \qquad (26\text{-}21)$$

From Eq. 26-1, we can also write this as

$$U = \tfrac{1}{2}CV^2 \qquad \text{(potential energy).} \qquad (26\text{-}22)$$

Equations 26-21 and 26-22 hold no matter what the geometry of the capacitor is.

To gain some physical insight into energy storage, consider two parallel-plate capacitors C_1 and C_2 that are identical except that C_1 has twice the plate separation of C_2. Then C_1 has twice the volume between its plates and also, from Eq. 26-9, half the capacitance of C_2. Equation 26-4 tells us that if both capacitors have the same charge q, the electric fields between their plates are identical. And Eq. 26-21 tells us that C_1 has twice the stored potential energy of C_2. Thus, of two otherwise identical capacitors with the same charge and same electric field, the one with twice the volume between its plates has twice the stored potential energy. Arguments like this tend to verify our earlier assumption:

The potential energy of a charged capacitor may be viewed as stored in the electric field between its plates.

The Medical Defibrillator

The ability of a capacitor to store potential energy is the basis of *defibrillator* devices, which are used by emergency medical teams to stop the fibrillation of heart attack victims. In the portable version, a battery charges a capacitor to a high potential difference, storing a large amount of energy in less than a minute. The battery maintains only a modest potential difference; an electronic circuit repeatedly uses that potential difference to greatly increase the potential difference of the capacitor. The power, or rate of energy transfer, during this process is also modest.

Conducting leads ("paddles") are placed on the victim's chest. When a control switch is closed, the capacitor sends a portion of its stored energy from paddle to paddle through the victim. As an example, when a 70 μF capacitor in a defibrillator is charged to 5000 V, Eq. 26-22 gives the energy stored in the capacitor as

$$U = \tfrac{1}{2}CV^2 = \tfrac{1}{2}(70 \times 10^{-6} \text{ F})(5000 \text{ V})^2 = 875 \text{ J}.$$

About 200 J of this energy is sent through the victim during a pulse of about 2.0 ms. The power of the pulse is

$$P = \frac{U}{t} = \frac{200 \text{ J}}{2.0 \times 10^{-3} \text{ s}} = 100 \text{ kW},$$

which is much greater than the power of the battery itself.

Energy Density

In a parallel-plate capacitor, neglecting fringing, the electric field has the same value for all points between the plates. Thus the **energy density** u, that is, the potential energy per unit volume between the plates, should also be uniform. We can find u by dividing the total potential energy by the volume Ad of the space between the plates. Using Eq. 26-22, we obtain

$$u = \frac{U}{Ad} = \frac{CV^2}{2Ad}.$$

With Eq. 26-9 ($C = \epsilon_0 A/d$), this result becomes

$$u = \tfrac{1}{2}\epsilon_0 \left(\frac{V}{d}\right)^2.$$

But, from Eq. 25-42, V/d equals the electric field magnitude E, so

$$u = \tfrac{1}{2}\epsilon_0 E^2 \qquad \text{(energy density).} \qquad (26\text{-}23)$$

Although we derived this result for the special case of a parallel-plate capacitor, it holds generally, whatever may be the source of the electric field. If an electric field **E** exists at any point in space, we can think of that point as the site of potential energy whose amount per unit volume is given by Eq. 26-23.

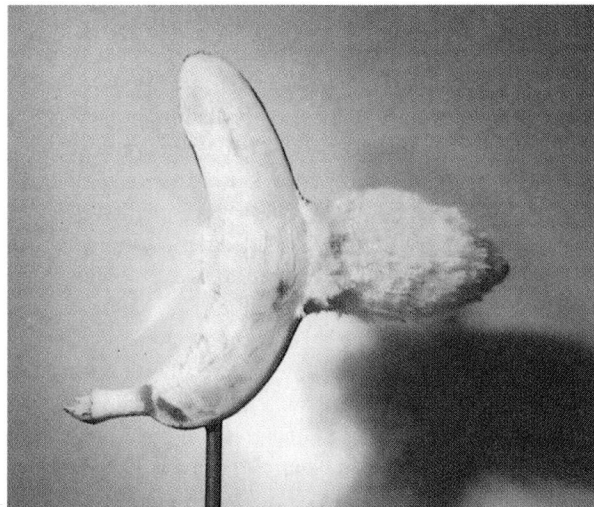

To photograph a bullet blowing apart a banana, Harold Edgerton, the inventor of the stroboscope, used a capacitor to dump electrical energy into one of his stroboscopic lamps, which then brightly illuminated the banana for only 0.3 μs.

SAMPLE PROBLEM 26-5

What is the potential energy of the two-capacitor system in Sample Problem 26-4, before and after switch S in Fig. 26-10 is closed?

SOLUTION: Initially, only capacitor C_1 is charged and has a potential energy; its potential difference is $V_0 = 6.30$ V. So from Eq. 26-22, the initial potential energy is

$$U_i = \tfrac{1}{2}C_1V_0^2 = (\tfrac{1}{2})(3.55 \times 10^{-6} \text{ F})(6.30 \text{ V})^2$$
$$= 7.04 \times 10^{-5} \text{ J} = 70.4 \ \mu\text{J.} \qquad \text{(Answer)}$$

After the switch has been closed, the capacitors come to the same final potential difference $V = 1.79$ V. The final potential energy is then

$$U_f = \tfrac{1}{2}C_1V^2 + \tfrac{1}{2}C_2V^2 = \tfrac{1}{2}(C_1 + C_2)V^2$$
$$= (\tfrac{1}{2})(3.55 \times 10^{-6} \text{ F} + 8.95 \times 10^{-6} \text{ F})(1.79 \text{ V})^2$$
$$= 2.00 \times 10^{-5} \text{ J} = 20.0 \ \mu\text{J.} \qquad \text{(Answer)}$$

Thus $U_f < U_i$, by about 72%.

  This is not a violation of the principle of energy conservation. The ''missing'' energy appears as thermal energy in the connecting wires (as we shall discuss in Chapter 27) and as radiated energy.

SAMPLE PROBLEM 26-6

An isolated conducting sphere whose radius R is 6.85 cm has a charge $q = 1.25$ nC.

(a) How much potential energy is stored in the electric field of this charged conductor?

SOLUTION: From Eqs. 26-21 and 26-18 we have

$$U = \frac{q^2}{2C} = \frac{q^2}{8\pi\epsilon_0 R}$$
$$= \frac{(1.25 \times 10^{-9} \text{ C})^2}{(8\pi)(8.85 \times 10^{-12} \text{ F/m})(0.0685 \text{ m})}$$
$$= 1.03 \times 10^{-7} \text{ J} = 103 \text{ nJ.} \qquad \text{(Answer)}$$

(b) What is the energy density at the surface of the sphere?

SOLUTION: From Eq. 26-23,

$$u = \tfrac{1}{2}\epsilon_0 E^2,$$

so we must first find E at the surface of the sphere. This is given by Eq. 24-15:

$$E = \frac{1}{4\pi\epsilon_0}\frac{q}{R^2}.$$

The energy density is then

$$u = \tfrac{1}{2}\epsilon_0 E^2 = \frac{q^2}{32\pi^2\epsilon_0 R^4} \qquad (26\text{-}24)$$
$$= \frac{(1.25 \times 10^{-9} \text{ C})^2}{(32\pi^2)(8.85 \times 10^{-12} \text{ C}^2/\text{N}\cdot\text{m}^2)(0.0685 \text{ m})^4}$$
$$= 2.54 \times 10^{-5} \text{ J/m}^3 = 25.4 \ \mu\text{J/m}^3. \qquad \text{(Answer)}$$

(c) What is the radius R_0 of an imaginary spherical surface such that half of the stored potential energy lies within it?

SOLUTION: This situation requires that

$$\int_R^{R_0} dU = \frac{1}{2}\int_R^{\infty} dU. \qquad (26\text{-}25)$$

The lower limit on the integrals is R rather than 0 because no electric field, and thus no stored potential energy, lies within the conducting sphere of radius R.

  The energy dU that lies in a spherical shell between inner and outer radii r and $r + dr$ is

$$dU = (u)(4\pi r^2)(dr), \qquad (26\text{-}26)$$

where u is again the energy density and $(4\pi r^2)(dr)$ is the volume of the spherical shell. Substituting Eq. 26-24 into Eq. 26-26 with r replacing R, we have

$$dU = \frac{q^2}{8\pi\epsilon_0}\frac{dr}{r^2}. \qquad (26\text{-}27)$$

Substituting Eq. 26-26 into both sides of Eq. 26-25 and simplifying give us

$$\int_R^{R_0}\frac{dr}{r^2} = \frac{1}{2}\int_R^{\infty}\frac{dr}{r^2},$$

which, after integration, becomes

$$\frac{1}{R} - \frac{1}{R_0} = \frac{1}{2R}.$$

Solving for R_0 yields

$$R_0 = 2R = (2)(6.85 \text{ cm}) = 13.7 \text{ cm}. \quad \text{(Answer)}$$

Thus half the stored energy is contained within a spherical surface whose radius is twice the radius of the conducting sphere.

26-6 CAPACITOR WITH A DIELECTRIC

If you fill the space between the plates of a capacitor with a *dielectric,* which is an insulating material such as mineral oil or plastic, what happens to the capacitance? Michael Faraday—to whom the whole concept of capacitance is largely due and for whom the SI unit of capacitance is named—first looked into this matter in 1837. Using simple equipment much like that shown in Fig. 26-11, he found that the capacitance *increased* by a numerical factor κ, which he called the **dielectric constant** of the introduced material. Table 26-1 shows some dielectric materials and their dielectric constants. The dielectric constant of a vacuum is unity by definition. Because air is mostly empty space, its measured dielectric constant is only slightly greater than unity.

Another effect of the introduction of a dielectric is to limit the potential difference that can be applied between the plates to a certain value $V_{\max}$, called the *breakdown potential.* If this value is substantially exceeded, the dielectric material will break down and form a conducting path between the plates. Every dielectric material has a

TABLE 26-1 SOME PROPERTIES OF DIELECTRICS[a]

MATERIAL	DIELECTRIC CONSTANT κ	DIELECTRIC STRENGTH (kV/mm)
Air (1 atm)	1.00054	3
Polystyrene	2.6	24
Paper	3.5	16
Transformer oil	4.5	
Pyrex	4.7	14
Ruby mica	5.4	
Porcelain	6.5	
Silicon	12	
Germanium	16	
Ethanol	25	
Water (20°C)	80.4	
Water (25°C)	78.5	
Titania ceramic	130	
Strontium titanate	310	8

For a vacuum, κ = unity.

[a]Measured at room temperature, except for the water.

characteristic *dielectric strength,* which is the maximum value of the electric field that it can tolerate without breakdown. A few such values are listed in Table 26-1.

As we discussed in connection with Eq. 26-18, the capacitance of any capacitor can be written in the form

$$C = \epsilon_0 \mathscr{L}, \quad (26\text{-}28)$$

in which $\mathscr{L}$ has the dimensions of a length. For example, $\mathscr{L} = A/d$ for a parallel-plate capacitor. Faraday's discovery was that, with a dielectric *completely* filling the space between the plates, Eq. 26-28 becomes

$$C = \kappa \epsilon_0 \mathscr{L} = \kappa C_{\text{air}}, \quad (26\text{-}29)$$

where C_{air} is the value of the capacitance with only air between the plates.

Figure 26-12 provides some insight into Faraday's experiments. In Fig. 26-12a the battery ensures that the potential difference V between the plates will remain constant. When a dielectric slab is inserted between the plates, the charge q increases by a factor of κ, the additional charge being delivered to the capacitor plates by the battery. In Fig. 26-12b there is no battery and therefore the charge q must remain constant as the dielectric slab is inserted; then the potential difference V between the plates decreases by a factor of κ. Both these observations are consistent (through the relation $q = CV$) with the increase in capacitance caused by the dielectric.

FIGURE 26-11 The simple electrostatic apparatus used by Faraday. An assembled apparatus (second from left) forms a spherical capacitor consisting of a central brass ball and a concentric brass shell. Faraday placed dielectric materials in the space between the sphere and shell.

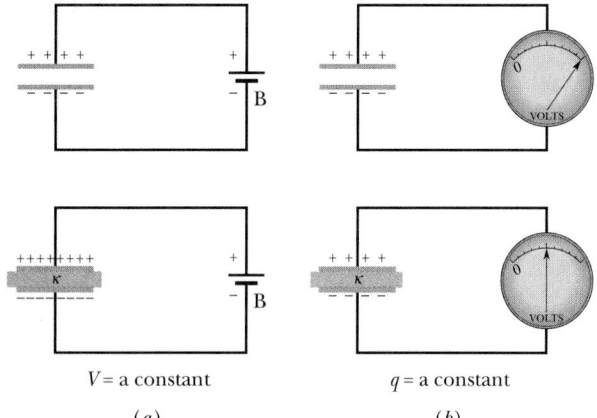

V = a constant q = a constant

        (*a*)   (*b*)

FIGURE 26-12 (*a*) If the potential difference between the plates of a capacitor is maintained, as by battery B, the effect of a dielectric is to increase the charge on the plates. (*b*) If the charge on the capacitor plates is maintained, as in this case, the effect of a dielectric is to reduce the potential difference between the plates. The scale shown is that of a *potentiometer*, a device used to measure potential difference (here, between the plates). A capacitor cannot discharge through a potentiometer.

Comparison of Eqs. 26-28 and 26-29 suggests that the effect of a dielectric can be summed up in more general terms:

In a region completely filled by a dielectric material of dielectric constant κ, all electrostatic equations containing the permittivity constant ϵ_0 are to be modified by replacing ϵ_0 with $\kappa\epsilon_0$.

Thus a point charge inside a dielectric produces an electric field that, by Coulomb's law, has magnitude

$$E = \frac{1}{4\pi\kappa\epsilon_0}\frac{q}{r^2}. \qquad (26\text{-}30)$$

Also, the expression for the electric field just outside an isolated conductor immersed in a dielectric (see Eq. 24-11) becomes

$$E = \frac{\sigma}{\kappa\epsilon_0}. \qquad (26\text{-}31)$$

Both these expressions show that *for a fixed distribution of charges, the effect of a dielectric is to weaken the electric field* that would otherwise be present.

SAMPLE PROBLEM 26-7

A parallel-plate capacitor whose capacitance C is 13.5 pF is charged to a potential difference $V = 12.5$ V between its plates. The charging battery is now disconnected and a porcelain slab ($\kappa = 6.50$) is slipped between the plates. What is the potential energy of the device, both before and after the slab is introduced?

SOLUTION: The initial potential energy is given by Eq. 26-22 as

$$U_i = \tfrac{1}{2}CV^2 = (\tfrac{1}{2})(13.5 \times 10^{-12} \text{ F})(12.5 \text{ V})^2$$
$$= 1.055 \times 10^{-9} \text{ J} = 1055 \text{ pJ} \approx 1100 \text{ pJ}. \qquad \text{(Answer)}$$

We can also write the initial potential energy, from Eq. 26-21, in the form

$$U_i = \frac{q^2}{2C}.$$

We choose to do so because, from the conditions of the problem statement, q (but not V) remains constant as the slab is introduced. After the slab is in place, C increases to κC so that

$$U_f = \frac{q^2}{2\kappa C} = \frac{U_i}{\kappa} = \frac{1055 \text{ pJ}}{6.50}$$
$$= 162 \text{ pJ} \approx 160 \text{ pJ}. \qquad \text{(Answer)}$$

When the slab is introduced, the energy decreases by a factor of $1/\kappa$.

The "missing" energy, in principle, would be apparent to the person who introduced the slab. The capacitor would exert a tiny tug on the slab and would do work on it, in amount

$$W = U_i - U_f = (1055 - 162) \text{ pJ} = 893 \text{ pJ}.$$

If the slab were allowed to slide between the plates with no restraint and if there were no friction, the slab would oscillate back and forth between the plates with a (constant) mechanical energy of 893 pJ, and this system energy would transfer back and forth between kinetic energy of the moving slab and potential energy stored in the electric field.

ℂHECKPOINT **5:** If the battery in Sample Problem 26-7 remains connected, do the following increase, decrease, or remain the same when the slab is introduced: (a) the potential difference between the capacitor plates, (b) the capacitance, (c) the charge on the capacitor, (d) the potential energy of the device, (e) the electric field between the plates? (*Hint:* For (e), note that the charge is not fixed.)

26-7 DIELECTRICS: AN ATOMIC VIEW

What happens, in atomic and molecular terms, when we put a dielectric in an electric field? There are two possibilities, depending on the nature of the molecules:

1. Polar dielectrics. The molecules of some dielectrics, like water, have permanent electric dipole moments. In

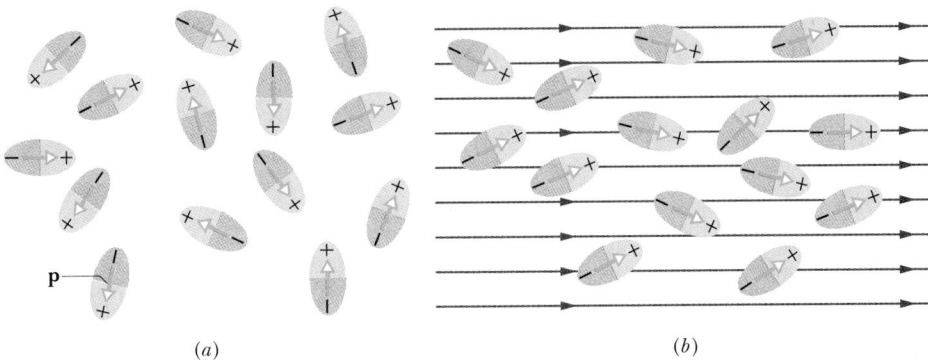

(a)   (b)

FIGURE 26-13 (a) Molecules with a permanent electric dipole moment, showing their random orientation in the absence of an external electric field. (b) An electric field is applied, producing partial alignment of the dipoles. Thermal agitation prevents complete alignment.

such materials (called *polar dielectrics*), the electric dipoles tend to line up with an external electric field as in Fig. 26-13. Because the molecules are continuously jostling each other as a result of their random thermal motion, this alignment is not complete, but it becomes more complete as the magnitude of the applied field is increased (or as the temperature, and thus the jostling, is decreased). The alignment of the electric dipoles produces an electric field that is opposite the applied field and smaller in magnitude than that field.

2. Nonpolar dielectrics. Regardless of whether they have permanent electric dipole moments, molecules acquire dipole moments by induction when placed in an external electric field. In Section 25-7 (see Fig. 25-12), we saw that this external field tends to "stretch" the molecule, separating slightly the centers of negative and positive charge.

Figure 26-14a shows a nonpolar dielectric slab with no external electric field applied. In Fig. 26-14b, an electric field $\mathbf{E}_0$ is applied via a capacitor, whose plates are charged as shown. The effect of field $\mathbf{E}_0$ is a slight separation of the centers of the positive and negative charge distributions within the slab, producing positive charge on one face of the slab (due to the positive ends of dipoles there) and negative charge on the opposite face (due to the negative ends of dipoles there). The slab as a whole remains electrically neutral and—within the slab—there is no excess charge in any volume element.

Figure 26-14c shows that the induced surface charges on the faces produce an electric field $\mathbf{E}'$ in the direction opposite that of the applied electric field $\mathbf{E}_0$. The resultant field $\mathbf{E}$ inside the dielectric (the vector sum of $\mathbf{E}_0$ and $\mathbf{E}'$) has the direction of $\mathbf{E}_0$ but is smaller in magnitude.

Both the field $\mathbf{E}'$ produced by the surface charges and the electric field produced by the permanent electric dipoles in Fig. 26-13 act in the same way: they oppose the applied field $\mathbf{E}$. Thus, the effect of both polar and nonpolar dielectrics is to weaken any applied field within them—as between the plates of a capacitor.

We can now see why the dielectric porcelain slab in

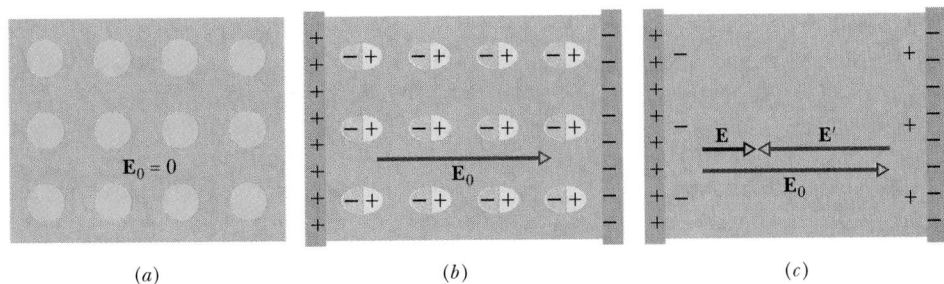

(a)   (b)   (c)

FIGURE 26-14 (a) A dielectric slab. The circles represent the electrically neutral atoms within the slab. (b) An electric field is applied via charged capacitor plates; the field slightly stretches the atoms, separating the centers of positive and negative charge. (c) The separation produces surface charges on the slab faces. These charges set up a field $\mathbf{E}'$, which opposes the applied field $\mathbf{E}_0$. The resultant field $\mathbf{E}$ inside the dielectric (the vector sum of $\mathbf{E}_0$ and $\mathbf{E}'$) has the same direction as $\mathbf{E}_0$ but less magnitude.

Sample Problem 26-7 is pulled into the capacitor: as it enters the space between the plates, the surface charge that appears on each slab face has the opposite sign as the charge on the adjacent capacitor plate. Thus, slab and plate attract each other.

26-8 DIELECTRICS AND GAUSS' LAW

In our discussion of Gauss' law in Chapter 24, we assumed that the charges existed in a vacuum. Here we shall see how to modify and generalize that law if dielectric materials, such as those listed in Table 26-1, are present. Figure 26-15 shows a parallel-plate capacitor of plate area A, both with and without a dielectric. We assume that the charge q on the plates is the same in both situations. Note that the field between the plates induces charges on the faces of the dielectric by one of the methods of Section 26-7.

For the situation of Fig. 26-15a, without a dielectric, we can find the electric field $\mathbf{E}_0$ between the plates as we did in Fig. 26-5: we enclose the charge $+q$ on the top plate with a Gaussian surface and then apply Gauss' law. Letting E_0 represent the magnitude of the field, we find

$$\epsilon_0 \oint \mathbf{E} \cdot d\mathbf{A} = \epsilon_0 E_0 A = q, \qquad (26\text{-}32)$$

or

$$E_0 = \frac{q}{\epsilon_0 A}. \qquad (26\text{-}33)$$

In Fig. 26-15b, with the dielectric in place, we can find the electric field between the plates (and within the dielectric) by using the same Gaussian surface. However, now the surface encloses two types of charge: it still encloses charge $+q$ on the top plate, but it now also encloses the induced charge $-q'$ on the top face of the dielectric. The charge on the conducting plate is said to be *free charge* because it can move if we can change the electric potential of the plate; the induced charge on the surface of the dielectric is not free charge because it cannot move from that surface.

The net charge enclosed by the Gaussian surface in Fig. 26-15b is $q - q'$. So, Gauss' law now gives

$$\epsilon_0 \oint \mathbf{E} \cdot d\mathbf{A} = \epsilon_0 E A = q - q', \qquad (26\text{-}34)$$

or

$$E = \frac{q - q'}{\epsilon_0 A}. \qquad (26\text{-}35)$$

The effect of the dielectric is to weaken the original field E_0 by a factor κ. So we may write

$$E = \frac{E_0}{\kappa} = \frac{q}{\kappa \epsilon_0 A}. \qquad (26\text{-}36)$$

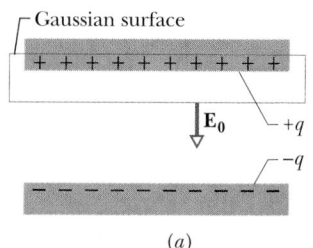

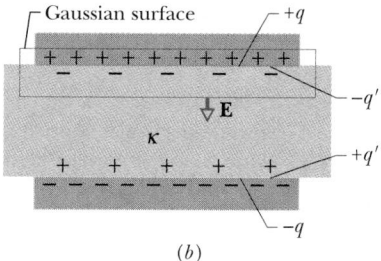

FIGURE 26-15 Parallel-plate capacitor (a) without and (b) with a dielectric slab inserted. The charge q on the plates is assumed to be the same in both cases.

Comparison of Eqs. 26-35 and 26-36 shows that

$$q - q' = \frac{q}{\kappa}. \qquad (26\text{-}37)$$

Equation 26-37 shows correctly that the magnitude q' of the induced surface charge is less than that of the free charge q and is zero if no dielectric is present, that is, if $\kappa = 1$ in Eq. 26-37.

By substituting for $q - q'$ from Eq. 26-37 in Eq. 26-34, we can write Gauss' law in the form

$$\epsilon_0 \oint \kappa \mathbf{E} \cdot d\mathbf{A} = q \qquad \begin{array}{l}\text{(Gauss' law}\\ \text{with dielectric).}\end{array} \qquad (26\text{-}38)$$

This important equation, although derived for a parallel-plate capacitor, is true generally and is the most general form in which Gauss' law can be written. Note the following:

1. The flux integral now deals with $\kappa \mathbf{E}$, not with $\mathbf{E}$. (The vector $\epsilon_0 \kappa \mathbf{E}$ is sometimes called the *electric displacement* $\mathbf{D}$, so that Eq. 26-38 can be written in the simplified form $\oint \mathbf{D} \cdot d\mathbf{A} = q$.)

2. The charge q enclosed by the Gaussian surface is now taken to be the *free charge only*. The induced surface charge is deliberately ignored on the right side of this equation, having been taken fully into account by introducing the dielectric constant κ on the left side.

3. Equation 26-38 differs from Eq. 24-7, our original statement of Gauss' law, only in that ϵ_0 in the latter equation has been replaced by $\kappa \epsilon_0$. We take κ inside the integral to allow for cases in which κ is not constant over the entire Gaussian surface.

SAMPLE PROBLEM 26-8

Figure 26-16 shows a parallel-plate capacitor of plate area A and plate separation d. A potential difference V_0 is applied between the plates. The battery is then disconnected, and a dielectric slab of thickness b and dielectric constant κ is placed between the plates as shown. Assume

$$A = 115 \text{ cm}^2, \qquad d = 1.24 \text{ cm}, \qquad V_0 = 85.5 \text{ V},$$

$$b = 0.780 \text{ cm}, \qquad \kappa = 2.61.$$

(a) What is the capacitance C_0 before the dielectric slab is inserted?

SOLUTION: From Eq. 26-9 we have

$$C_0 = \frac{\epsilon_0 A}{d} = \frac{(8.85 \times 10^{-12} \text{ F/m})(115 \times 10^{-4} \text{ m}^2)}{1.24 \times 10^{-2} \text{ m}}$$

$$= 8.21 \times 10^{-12} \text{ F} = 8.21 \text{ pF}. \qquad \text{(Answer)}$$

(b) What free charge appears on the plates?

SOLUTION: From Eq. 26-1,

$$q = C_0 V_0 = (8.21 \times 10^{-12} \text{ F})(85.5 \text{ V})$$

$$= 7.02 \times 10^{-10} \text{ C} = 702 \text{ pC}. \qquad \text{(Answer)}$$

Because the charging battery was disconnected before the slab was introduced, the free charge remains unchanged as the slab is put into place.

(c) What is the electric field E_0 in the gaps between the plates and the dielectric slab?

SOLUTION: Let us apply Gauss' law in the form given in Eq. 26-38 to Gaussian surface I in Fig. 26-16, which encloses only the free charge on the upper capacitor plate. Because the area vector $d\mathbf{A}$ and the field vector $\mathbf{E}_0$ both point downward, we have

$$\epsilon_0 \oint \kappa \mathbf{E} \cdot d\mathbf{A} = \epsilon_0(1)E_0 A = q,$$

or

$$E_0 = \frac{q}{\epsilon_0 A} = \frac{7.02 \times 10^{-10} \text{ C}}{(8.85 \times 10^{-12} \text{ F/m})(115 \times 10^{-4} \text{ m}^2)}$$

$$= 6900 \text{ V/m} = 6.90 \text{ kV/m}. \qquad \text{(Answer)}$$

Note that we put $\kappa = 1$ in this equation because the Gaussian surface over which Gauss' law was integrated does not pass through any dielectric. Note too that the value of E_0 does not change when the slab is introduced because the amount of charge enclosed by Gaussian surface I in Fig. 26-16 does not change.

(d) What is the electric field E_1 in the dielectric slab?

SOLUTION: We now apply Eq. 26-38 to Gaussian surface II in Fig. 26-16. That surface encloses free charge $-q$ and induced charge $+q'$, but we do not consider the latter when we use Eq. 26-38. We find

$$\epsilon_0 \oint \kappa \mathbf{E}_1 \cdot d\mathbf{A} = -\epsilon_0 \kappa E_1 A = -q. \qquad (26\text{-}39)$$

(The first minus sign in Eq. 26-39 comes from the dot product $\mathbf{E}_1 \cdot d\mathbf{A}$, because now the field vector $\mathbf{E}_1$ points downward and the area vector $d\mathbf{A}$ points upward.) Equation 26-39 gives us

$$E_1 = \frac{q}{\kappa \epsilon_0 A} = \frac{E_0}{\kappa} = \frac{6.90 \text{ kV/m}}{2.61}$$

$$= 2.64 \text{ kV/m}. \qquad \text{(Answer)}$$

(e) What is the potential difference V between the plates after the slab has been introduced?

SOLUTION: We answer by applying Eq. 26-6, integrating along a straight-line path extending directly from the top plate to the bottom plate. Within the dielectric, the path length is b and the electric field is E_1. Within the two gaps above and below the dielectric, the total path length is $d - b$ and the electric field is E_0. Equation 26-6 then yields

$$V = \int_+^- E \, ds = E_0(d - b) + E_1 b$$

$$= (6900 \text{ V/m})(0.0124 \text{ m} - 0.00780 \text{ m})$$

$$+ (2640 \text{ V/m})(0.00780 \text{ m})$$

$$= 52.3 \text{ V}. \qquad \text{(Answer)}$$

This contrasts with the original potential difference of 85.5 V.

(f) What is the capacitance with the slab in place?

SOLUTION: From Eq. 26-1,

$$C = \frac{q}{V} = \frac{7.02 \times 10^{-10} \text{ C}}{52.3 \text{ V}}$$

$$= 1.34 \times 10^{-11} \text{ F} = 13.4 \text{ pF}. \qquad \text{(Answer)}$$

This is more than the original capacitance of 8.21 pF.

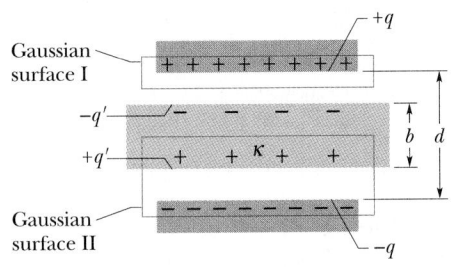

FIGURE 26-16 Sample Problem 26-8. A parallel-plate capacitor containing a dielectric slab that only partially fills the space between the plates.

CHECKPOINT 6: In Sample Problem 26-8, if the thickness b of the slab increases, do the following increase, decrease, or remain the same: (a) the electric field E_1, (b) the potential difference between the plates, and (c) the capacitance of the capacitor?

REVIEW & SUMMARY

Capacitor; Capacitance

A **capacitor** consists of two isolated conductors (the *plates*) with equal and opposite charges $+q$ and $-q$. Its **capacitance** C is defined from

$$q = CV, \qquad (26\text{-}1)$$

where V is the potential difference between the plates. The SI unit of capacitance is the farad (1 farad = 1 F = 1 coulomb per volt).

Determining Capacitance

We generally determine the capacitance of a particular capacitor configuration by (1) assuming a charge q to have been placed on the plates, (2) finding the electric field $\mathbf{E}$ due to this charge, (3) evaluating the potential difference V, and (4) calculating C from Eq. 26-1. Some specific results are the following:

A *parallel-plate capacitor* with flat parallel plates of area A and spacing d has capacitance

$$C = \frac{\epsilon_0 A}{d}. \qquad (26\text{-}9)$$

A *cylindrical capacitor* (two long coaxial cylinders) of length L and inner and outer radii a and b has capacitance

$$C = 2\pi\epsilon_0 \frac{L}{\ln(b/a)}. \qquad (26\text{-}14)$$

A *spherical capacitor* with concentric spherical plates of inner and outer radii a and b has capacitance

$$C = 4\pi\epsilon_0 \frac{ab}{b-a}. \qquad (26\text{-}17)$$

If we let $b \to \infty$ and $a = R$ in Eq. 26-17, we obtain the capacitance of an *isolated sphere* of radius R:

$$C = 4\pi\epsilon_0 R. \qquad (26\text{-}18)$$

Capacitors in Parallel and in Series

The **equivalent capacitances** C_{eq} of combinations of individual capacitors connected in **parallel** and in **series** are

$$C_{eq} = \sum_{j=1}^{n} C_j \qquad (n \text{ capacitors in parallel}) \quad (26\text{-}19)$$

and

$$\frac{1}{C_{eq}} = \sum_{j=1}^{n} \frac{1}{C_j} \qquad (n \text{ capacitors in series}). \quad (26\text{-}20)$$

These equivalent capacitances can be used to calculate the capacitances of more complicated series-parallel combinations.

Potential Energy and Energy Density

The **electric potential energy** U of a charged capacitor, given by

$$U = \frac{q^2}{2C} = \tfrac{1}{2}CV^2, \qquad (26\text{-}21, 26\text{-}22)$$

is the work required to charge it. This energy can be associated with the capacitor's electric field $\mathbf{E}$. By extension we can associate stored energy with an electric field. In vacuum, the **energy density** u, or potential energy per unit volume, is given by

$$u = \tfrac{1}{2}\epsilon_0 E^2. \qquad (26\text{-}23)$$

Capacitance with a Dielectric

If the space between the plates of a capacitor is completely filled with a dielectric material, the capacitance C is increased by a factor κ, called the **dielectric constant,** which is characteristic of the material. In a region that is completely filled by a dielectric, all electrostatic equations containing ϵ_0 must be modified by replacing ϵ_0 with $\kappa\epsilon_0$.

The effects of adding a dielectric can be understood physically in terms of the action of an electric field on the permanent or induced electric dipoles in the dielectric slab. The result is the formation of induced charges on the surfaces of the dielectric, which results in a weakening of the field within the dielectric.

Gauss' Law with a Dielectric

When a dielectric is present, Gauss' law may be generalized to

$$\epsilon_0 \oint \kappa \mathbf{E} \cdot d\mathbf{A} = q. \qquad (26\text{-}38)$$

Here q is the free charge; the induced surface charge is accounted for by including the dielectric constant κ inside the integral.

QUESTIONS

1. Figure 26-17 shows plots of charge versus potential difference for three parallel-plate capacitors, which have the plate areas and separations given in the table. Which of the plots goes with which of the capacitors?

CAPACITOR	AREA	SEPARATION
1	A	d
2	$2A$	d
3	A	$2d$

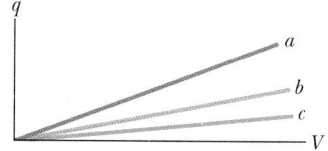

FIGURE 26-17 Question 1.

2. Figure 26-18 shows, in cross section, an isolated solid metal sphere A of radius R and two spherical capacitors B and C with inner and outer radii R and $2R$. The inner spherical ''plate'' of capacitor B is a spherical shell; that of capacitor C is a solid sphere. Rank objects A, B, and C according to their capacitance, greatest first.

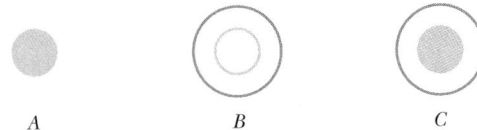

A   *B*   *C*

FIGURE 26-18 Question 2.

3. When a flat sheet of aluminum foil of negligible thickness is placed midway between the parallel plates of a capacitor, does the capacitance increase, decrease, or remain the same if the sheet is (a) electrically connected to one of the plates and (b) electrically isolated? (*Hint:* For (b), consider the equivalent capacitance.)

4. For each circuit in Fig. 26-19, are the capacitors connected in series, in parallel, or in neither mode?

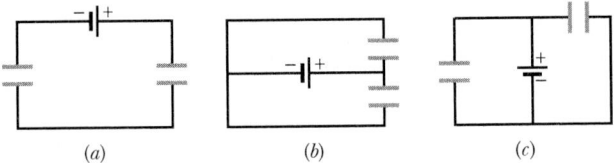

(*a*)   (*b*)   (*c*)

FIGURE 26-19 Question 4.

5. Two capacitors are wired to a battery. (a) In which arrangement, parallel or series, is the potential difference across each capacitor the same and the same as that across the equivalent capacitance? (b) In which is the charge on each capacitor the same and the same as that on the equivalent capacitance?

6. (a) In Fig. 26-20*a*, are capacitors C_1 and C_3 in series? (b) In the same figure, are capacitors C_1 and C_2 in parallel? (c) Rank the equivalent capacitances of the four circuits shown in Fig. 26-20, greatest first.

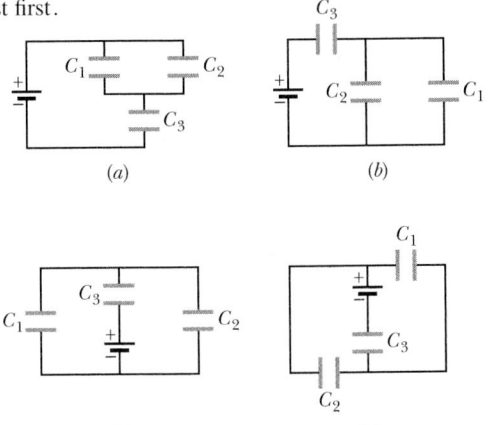

(*a*)   (*b*)

(*c*)   (*d*)

FIGURE 26-20 Question 6.

7. What is the equivalent capacitance of three capacitors, each of capacitance C, if they are connected to a battery (a) in series with one another and (b) in parallel? (c) In which arrangement is there more charge on the equivalent capacitance?

8. You are to connect capacitors C_1 and C_2, with $C_1 > C_2$, to a battery, first individually, then in series, and then in parallel. Rank those arrangements according to the amount of charge stored, greatest first.

9. (a) In Sample Problem 26-3, is the potential difference across capacitor C_2 more than, less than, or equal to that across capacitor C_1? (b) Is the charge on capacitor C_2 more than, less than, or equal to that on capacitor C_1?

10. Initially, a single capacitor C_1 is wired to a battery. Then capacitor C_2 is added in parallel. Are (a) the potential difference across C_1 and (b) the charge q_1 on C_1 now more than, less than, or the same as previously? (c) Is the equivalent capacitance C_{12} of C_1 and C_2 more than, less than, or equal to C_1? (d) Is the total charge stored on C_1 and C_2 together more than, less than, or equal to the charge stored previously on C_1?

11. Repeat Question 10 for C_2 added in series, not in parallel.

12. In Sample Problem 26-4, if we increase the capacitance of C_2, do the following increase, decrease, or remain the same: (a) the final potential difference across each capacitor, and the share of q_0 received by (b) C_1 and (c) C_2?

13. Figure 26-21 shows three circuits, each consisting of a switch and two capacitors, initially charged as indicated. After the switches have been closed, in which circuit (if any) will the charge on the left-hand capacitor (a) increase, (b) decrease, and (c) remain the same?

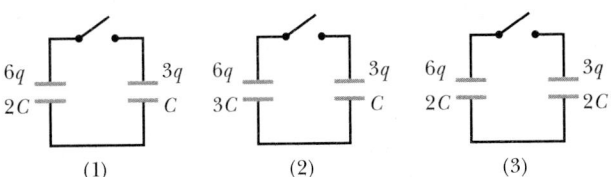

(1)   (2)   (3)

FIGURE 26-21 Question 13.

14. Two isolated metal spheres A and B have radii R and $2R$, respectively, and the same charge q. (a) Is the capacitance of A more than, less than, or equal to that of B? (b) Is the energy density just outside the surface of A more than, less than, or equal to that of B? (c) Is the energy density at radius $3R$ from the center of A more than, less than, or equal to that at the same radius from the center of B? (d) Is the total energy of the electric field due to A more than, less than, or equal to that of B?

15. An oil-filled parallel-plate capacitor was designed to have a capacitance C and to operate safely at or below a certain potential difference V_m without undergoing breakdown. However, the design is flawed and the capacitor occasionally breaks down. What can be done to redesign the capacitor, keeping C and V_m unchanged and using the same dielectric (the oil)?

16. When a dielectric slab is inserted between the plates of one of the two identical capacitors in Fig. 26-22, do the following properties of that capacitor increase, decrease, or remain the same: (a) capacitance, (b) charge, (c) potential difference, and (d) potential energy? (e) How about the same properties of the other capacitor?

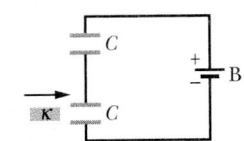

FIGURE 26-22 Question 16.

EXERCISES & PROBLEMS

SECTION 26-2 Capacitance

1E. An electrometer is a device used to measure static charge: an unknown charge is placed on the plates of the meter's capacitor, and the potential difference is measured. What minimum charge can be measured by an electrometer with a capacitance of 50 pF and a voltage sensitivity of 0.15 V?

2E. The two metal objects in Fig. 26-23 have net charges of $+70$ pC and -70 pC, which result in a 20 V potential difference between them. (a) What is the capacitance of the system? (b) If the charges are changed to $+200$ pC and -200 pC, what does the capacitance become? (c) What does the potential difference become?

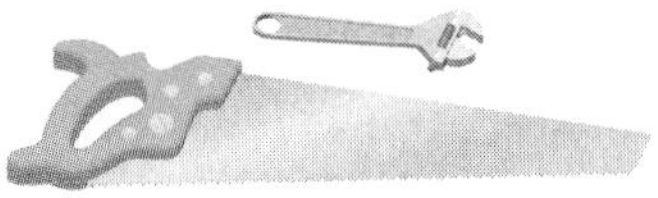

FIGURE 26-23   Exercise 2.

3E. The capacitor in Fig. 26-24 has a capacitance of 25 μF and is initially uncharged. The battery provides a potential difference of 120 V. After switch S is closed, how much charge will pass through it?

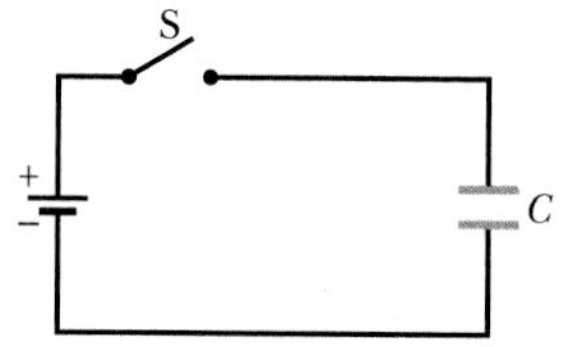

FIGURE 26-24   Exercise 3.

SECTION 26-3 Calculating the Capacitance

4E. If we solve Eq. 26-9 for ϵ_0, we see that its SI unit is the farad per meter. Show that this unit is equivalent to that obtained earlier for ϵ_0, namely, the coulomb squared per newton-meter squared.

5E. A parallel-plate capacitor has circular plates of 8.2 cm radius and 1.3 mm separation. (a) Calculate the capacitance. (b) What charge will appear on the plates if a potential difference of 120 V is applied?

6E. You have two flat metal plates, each of area 1.00 m², with which to construct a parallel-plate capacitor. If the capacitance of the device is to be 1.00 F, what must be the separation between the plates? Could this capacitor actually be constructed?

7E. The plates of a spherical capacitor have radii 38.0 and 40.0 mm. (a) Calculate the capacitance. (b) What must be the plate area of a parallel-plate capacitor with the same plate separation and capacitance?

8E. After you walk over a carpet on a dry day, your hand comes close to a metal doorknob and a 5 mm spark results. Such a spark means that there must have been a potential difference of possibly 15 kV between you and the doorknob. Assuming this potential difference, how much charge did you accumulate in walking over the carpet? For this extremely rough calculation, assume that your body can be represented by a uniformly charged conducting

sphere 25 cm in radius and electrically isolated from its surroundings.

9E. Two sheets of aluminum foil have the same area, a separation of 1.0 mm, and a capacitance of 10 pF, and are charged to 12 V. (a) Calculate the area of each sheet. The separation is now decreased by 0.10 mm with the charge held constant. (b) What is the new capacitance? (c) By how much does the potential difference change? Explain how a microphone might be constructed using this principle.

10E. A spherical drop of mercury of radius R has a capacitance given by $C = 4\pi\epsilon_0 R$. If two such drops combine to form a single larger drop, what is its capacitance?

11P. Using the approximation that $\ln(1 + x) \approx x$ when $x \ll 1$ (see Appendix E), show that the capacitance of a cylindrical capacitor approaches that of a parallel-plate capacitor when the spacing between the two cylinders is small.

12P. Suppose that the two spherical shells of a spherical capacitor have approximately equal radii. Under these conditions the device approximates a parallel-plate capacitor with $b - a = d$. Show that Eq. 26-17 does indeed reduce to Eq. 26-9 in this case.

13P. A capacitor is to be designed to operate with constant capacitance in an environment of fluctuating temperature. As shown in Fig. 26-25, the capacitor is a parallel-plate type with plastic spacers to keep the plates aligned. (a) Show that the rate of change of the capacitance C with temperature T is given by

$$\frac{dC}{dT} = C\left(\frac{1}{A}\frac{dA}{dT} - \frac{1}{x}\frac{dx}{dT}\right),$$

where A is the plate area and x the plate separation. (b) If the plates are aluminum, what should be the coefficient of thermal expansion of the spacers to ensure that the capacitance does not vary with temperature? (Ignore the effect of the spacers on the capacitance.)

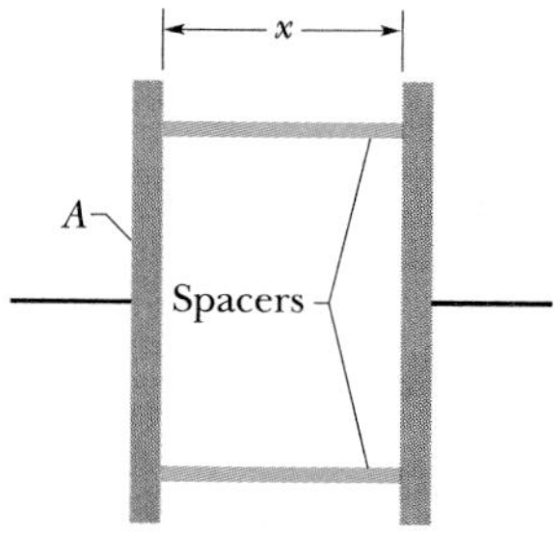

FIGURE 26-25
Problem 13.

SECTION 26-4 Capacitors in Parallel and in Series

14E. How many 1.00 μF capacitors must be connected in parallel to store a charge of 1.00 C with a potential of 110 V across the capacitors?

15E. In Fig. 26-26 find the equivalent capacitance of the combination. Assume that $C_1 = 10.0$ μF, $C_2 = 5.00$ μF, and $C_3 = 4.00$ μF.

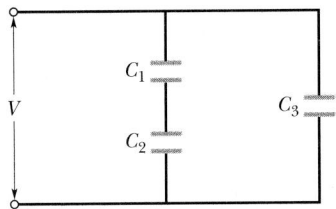

FIGURE 26-26 Exercise 15 and Problem 47.

16E. In Fig. 26-27 find the equivalent capacitance of the combination. Assume that $C_1 = 10.0$ μF, $C_2 = 5.00$ μF, and $C_3 = 4.00$ μF.

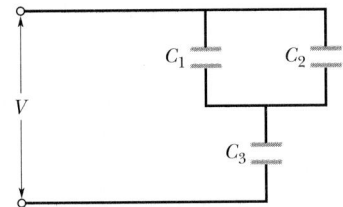

FIGURE 26-27 Exercise 16 and Problems 24 and 45.

17E. Each of the uncharged capacitors in Fig. 26-28 has a capacitance of 25.0 μF. A potential difference of 4200 V is established when the switch is closed. How many coulombs of charge then pass through meter A?

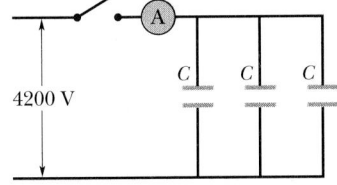

FIGURE 26-28 Exercise 17.

18E. A capacitance $C_1 = 6.00$ μF is connected in series with a capacitance $C_2 = 4.00$ μF, and a potential difference of 200 V is applied across the pair. (a) Calculate the equivalent capacitance. (b) What is the charge on each capacitor? (c) What is the potential difference across each capacitor?

19E. Repeat Exercise 18 for the same two capacitors but with them now connected in parallel.

20P. Figure 26-29 shows two capacitors in series; the center section of length b is movable vertically. Show that the equivalent capacitance of this series combination is independent of the position of the center section and is given by $C = \epsilon_0 A/(a - b)$.

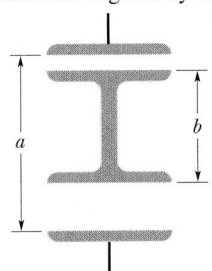

FIGURE 26-29 Problem 20.

21P. (a) Three capacitors are connected in parallel. Each has plate area A and plate spacing d. What must be the spacing of a single capacitor of plate area A if its capacitance equals that of the parallel combination? (b) What must be the spacing if the three capacitors are connected in series?

22P. (a) A potential difference of 300 V is applied to a series connection of two capacitors, of capacitance $C_1 = 2.0$ μF and capacitance $C_2 = 8.0$ μF. What are the charge on and the potential difference across each capacitor? (b) The charged capacitors are disconnected from each other and from the battery. They are then reconnected, positive plate to positive plate and negative plate to negative plate, with no external voltage being applied. What are the charge and the potential difference for each now? (c) Suppose the charged capacitors in (a) were reconnected with plates of *opposite* sign together. What then would be the steady-state charge and potential difference for each?

23P. Figure 26-30 shows a variable "air gap" capacitor of the type used in manually tuned radios. Alternate plates are connected together; one group is fixed in position and the other group is capable of rotation. Consider a pile of n plates of alternate polarity, each having an area A and separated from adjacent plates by a distance d. Show that this capacitor has a maximum capacitance of

$$C = \frac{(n - 1)\epsilon_0 A}{d}.$$

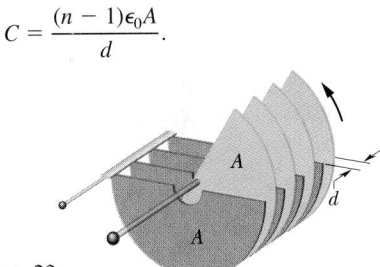

FIGURE 26-30 Problem 23.

24P. In Fig. 26-27 suppose that capacitor C_3 breaks down electrically, becoming equivalent to a conducting path. What *changes* in (a) the charge and (b) the potential difference occur for capacitor C_1? Assume that $V = 100$ V.

25P. You have several 2.0 μF capacitors, each capable of withstanding 200 V without electrical breakdown (in which they conduct charge instead of storing it). How would you assemble a combination having an equivalent capacitance of (a) 0.40 μF or (b) 1.2 μF, each capable of withstanding 1000 V?

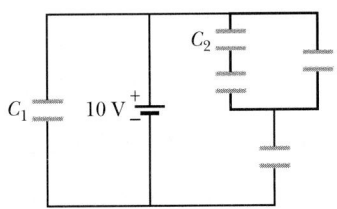

FIGURE 26-31 Problem 26.

26P. In Fig. 26-31, the battery has a potential difference of 10 V and the five capacitors each have a capacitance of 10 μF. What is the charge on (a) capacitor C_1 and (b) capacitor C_2?

27P. A 100 pF capacitor is charged to a potential difference of 50 V, and the charging battery is disconnected. The capacitor is then connected in parallel with a second (initially uncharged) capacitor. If the measured potential difference drops to 35 V, what is the capacitance of this second capacitor?

28P. In Fig. 26-32, the battery has a potential difference of 20 V. Find (a) the equivalent capacitance of all the capacitors and (b) the charge stored on that equivalent capacitance. Give the potential across and charge on (c) capacitor C_1, (d) capacitor C_2, and (e) capacitor C_3.

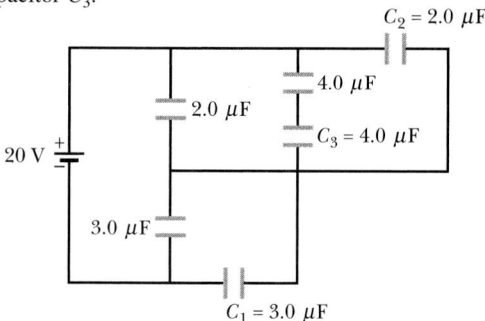

FIGURE 26-32 Problem 28.

29P. In Fig. 26-33, capacitors $C_1 = 1.0 \ \mu F$ and $C_2 = 3.0 \ \mu F$ are each charged to a potential difference of $V = 100$ V but with opposite polarity as shown. Switches S_1 and S_2 are now closed. (a) What is now the potential difference between points a and b? What are now the charges on (b) C_1 and (c) C_2?

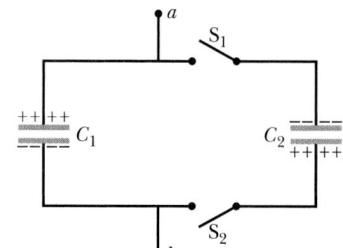

FIGURE 26-33
Problem 29.

30P. When switch S is thrown to the left in Fig. 26-34, the plates of capacitor C_1 acquire a potential difference V_0. Capacitors C_2 and C_3 are initially uncharged. The switch is now thrown to the right. What are the final charges q_1, q_2, and q_3 on the corresponding capacitors?

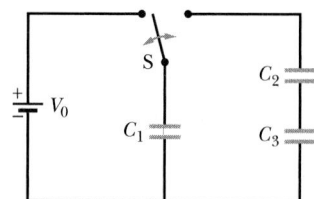

FIGURE 26-34 Problem 30.

31P. In Fig. 26-35, battery B supplies 12 V. (a) Find the charge on each capacitor first when only switch S_1 is closed and (b) later when switch S_2 is also closed. Take $C_1 = 1.0 \ \mu F$, $C_2 = 2.0 \ \mu F$, $C_3 = 3.0 \ \mu F$, and $C_4 = 4.0 \ \mu F$.

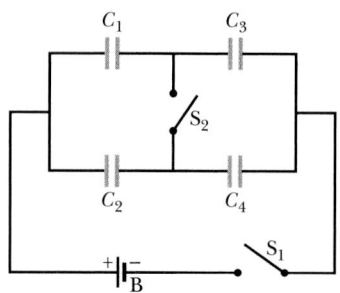

FIGURE 26-35 Problem 31.

32P. Figure 26-36 shows two identical capacitors C in a circuit with two (ideal) diodes D. (An ideal diode has the property that positive charge flows through it only in the direction of the arrow and negative charge flows through it only in the opposite direction.) A 100 V battery is connected across the input terminals, first with terminal a connected to the positive battery terminal and later with terminal b connected there. In each case, what is the potential difference across the output terminals?

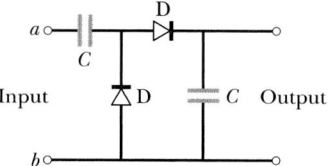

FIGURE 26-36 Problem 32.

SECTION 26-5 Storing Energy in an Electric Field

33E. How much energy is stored in one cubic meter of air due to the "fair weather" electric field of magnitude 150 V/m?

34E. Attempts to build a controlled thermonuclear fusion reactor, which, if successful, could provide the world with a vast supply of energy from heavy hydrogen in seawater, usually involve huge electric currents for short periods of time in magnetic field windings. For example, ZT-40 at Los Alamos Scientific Laboratory has rooms full of capacitors. One of the capacitor banks provides 61.0 mF at 10.0 kV. Calculate the stored energy (a) in joules and (b) in kilowatt-hours.

35E. What capacitance is required to store an energy of $10 \ kW \cdot h$ at a potential difference of 1000 V?

36E. A parallel-plate air-filled capacitor has a capacitance of 130 pF. (a) What is the stored energy if the applied potential difference is 56.0 V? (b) Can you calculate the energy density for points between the plates? Explain.

37E. A certain capacitor is charged to a potential difference V. If you wish to increase its stored energy by 10%, by what percentage should you increase V?

38E. A parallel-plate air-filled capacitor having area 40 cm² and plate spacing 1.0 mm is charged to a potential difference of 600 V. Find (a) the capacitance, (b) the magnitude of the charge on each plate, (c) the stored energy, (d) the electric field between the plates, and (e) the energy density between the plates.

39E. Two capacitors, of 2.0 and 4.0 μF capacitance, are connected in parallel across a 300 V potential difference. Calculate the total energy stored in the capacitors.

40E. (a) Calculate the energy density of the electric field at distance r from the center of an electron at rest. (b) If the electron is assumed to be an infinitesimal point, what does this calculation yield for the energy density in the limit of $r \rightarrow 0$?

41P. A charged isolated metal sphere of diameter 10 cm has a potential of 8000 V relative to $V = 0$ at infinity. Calculate the energy density in the electric field near the surface of the sphere.

42P. A parallel-connected bank of 5.00 μF capacitors is used to store electric energy. What does it cost to charge the 2000 capacitors of the bank to 50,000 V, assuming a unit cost of 3.0¢/kW·h?

43P. One capacitor is charged until its stored energy is 4.0 J. A second uncharged capacitor is then connected to it in parallel. (a) If the charge distributes equally, what is now the total energy stored in the electric fields? (b) Where did the excess energy go?

44P. Compute the energy stored for the three different connections of the capacitors of Problem 22. Compare these stored energies and explain any differences.

45P. In Fig. 26-27 find (a) the charge, (b) the potential difference, and (c) the stored energy for each capacitor. Assume the numerical values of Exercise 16, with $V = 100$ V.

46P. A parallel-plate capacitor has plates of area A and separation d and is charged to a potential difference V. The charging battery is then disconnected, and the plates are pulled apart until their separation is $2d$. Derive expressions in terms of A, d, and V for (a) the new potential difference, (b) the initial and final stored energy, and (c) the work required to separate the plates.

47P. In Fig. 26-26 find (a) the charge, (b) the potential difference, and (c) the stored energy for each capacitor. Assume the numerical values of Exercise 15, with $V = 100$ V.

48P. A cylindrical capacitor has radii a and b as in Fig. 26-6. Show that half the stored electric potential energy lies within a cylinder whose radius is $r = \sqrt{ab}$.

49P. Show that the plates of a parallel-plate capacitor attract each other with a force given by $F = q^2/2\epsilon_0 A$. Do so by calculating the work necessary to increase the plate separation from x to $x + dx$, with the charge q remaining constant.

50P. Using the result of Problem 49, show that the force per unit area (the *electrostatic stress*) acting on either capacitor plate is given by $\frac{1}{2}\epsilon_0 E^2$. (Actually, this result is true in general, for a conductor of *any* shape with an electric field **E** at its surface.)

51P*. A soap bubble of radius R_0 is slowly given a charge q. Because of mutual repulsion of the surface charges, the radius increases slightly to R. Because of the expansion, the air pressure inside the bubble drops to $p(V_0/V)$, where p is the atmospheric pressure, V_0 is the initial volume, and V is the final volume. Show that these quantities are related by

$$q^2 = 32\pi^2\epsilon_0 pR(R^3 - R_0^3).$$

(*Hint:* Consider the forces acting on a small area of the charged bubble. These forces are due to gas pressure, atmospheric pressure, and electrostatic stress; see Problem 50.)

SECTION 26-6 Capacitor with a Dielectric

52E. An air-filled parallel-plate capacitor has a capacitance of 1.3 pF. The separation of the plates is doubled and wax is inserted between them. The new capacitance is 2.6 pF. Find the dielectric constant of the wax.

53E. Given a 7.4 pF air-filled capacitor, you are asked to convert it to a capacitor that can store up to 7.4 μJ with a maximum potential difference of 652 V. What dielectric in Table 26-1 should you use to fill the gap in the air capacitor if you do not allow for a margin of error?

54E. For making a parallel-plate capacitor, you have available two plates of copper, a sheet of mica (thickness = 0.10 mm, $\kappa = 5.4$), a sheet of glass (thickness = 2.0 mm, $\kappa = 7.0$), and a slab of paraffin (thickness = 1.0 cm, $\kappa = 2.0$). To obtain the largest capacitance, which sheet should you place between the copper plates?

55E. A parallel-plate air-filled capacitor has a capacitance of 50 pF. (a) If each of its plates has an area of 0.35 m^2, what is the separation? (b) If the region between the plates is now filled with material having $\kappa = 5.6$, what is the capacitance?

56E. A coaxial cable used in a transmission line has an inner radius of 0.10 mm and an outer radius of 0.60 mm. Calculate the capacitance per meter for the cable. Assume that the space between the conductors is filled with polystyrene.

57P. A certain substance has a dielectric constant of 2.8 and a dielectric strength of 18 MV/m. If it is used as the dielectric material in a parallel-plate capacitor, what minimum area should the plates of the capacitor have to obtain a capacitance of 7.0×10^{-2} μF and to ensure that the capacitor will be able to withstand a potential difference of 4.0 kV?

58P. You are asked to construct a capacitor having a capacitance near 1 nF and a breakdown potential in excess of 10,000 V. You think of using the sides of a tall Pyrex drinking glass as a dielectric, lining the inside and outside curved surfaces with aluminum foil. The glass is 15 cm tall with an inner radius of 3.6 cm and an outer radius of 3.8 cm. What are the (a) capacitance and (b) breakdown potential?

59P. You have been assigned to design a transportable capacitor that can store 250 kJ of energy. You decide on a parallel-plate type with dielectric. (a) What is the minimum capacitor volume possible if you use a dielectric whose dielectric strength is listed in Table 26-1? (b) Modern high-performance capacitors that can store 250 kJ have volumes of 0.0870 m^3. Assuming that the dielectric used has the same dielectric strength as in (a), what must be its dielectric constant?

60P. Two parallel-plate capacitors have the same plate area A and separation d, but the dielectric constants of the materials between their plates are $\kappa + \Delta\kappa$ in one and $\kappa - \Delta\kappa$ in the other. (a) Find the equivalent capacitance when they are connected in parallel. (b) If the total charge on the parallel combination is Q, what is the charge on the capacitor with the larger capacitance?

61P. A slab of copper of thickness b is thrust into a parallel-plate capacitor of plate area A, as shown in Fig. 26-37; it is exactly halfway between the plates. (a) What is the capacitance after the slab is introduced? (b) If a charge q is maintained on the plates, what is the ratio of the stored energy before to that after the slab is inserted? (c) How much work is done on the slab as it is inserted? Is the slab sucked in or must it be pushed in?

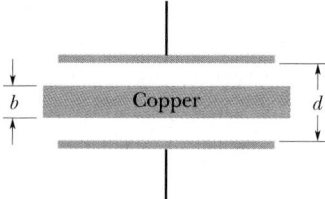

FIGURE 26-37   Problem 61.

62P. Repeat Problem 61, assuming that the potential difference rather than the charge is held constant.

63P. A parallel-plate capacitor of plate area A is filled with two dielectrics as in Fig. 26-38a. Show that the capacitance is

$$C = \frac{\epsilon_0 A}{d} \frac{\kappa_1 + \kappa_2}{2}.$$

Check this formula for limiting cases. (*Hint:* Can you justify this arrangement as being two capacitors in parallel?)

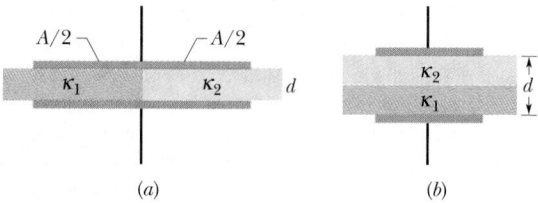

FIGURE 26-38   Problems 63 and 64.

64P. A parallel-plate capacitor of plate area A is filled with two dielectrics as in Fig. 26-38b. Show that the capacitance is

$$C = \frac{2\epsilon_0 A}{d} \frac{\kappa_1 \kappa_2}{\kappa_1 + \kappa_2}.$$

Check this formula for limiting cases. (*Hint:* Can you justify this arrangement as being two capacitors in series?)

65P. What is the capacitance of the capacitor, of plate area A, shown in Fig. 26-39? (*Hint:* See Problems 63 and 64.)

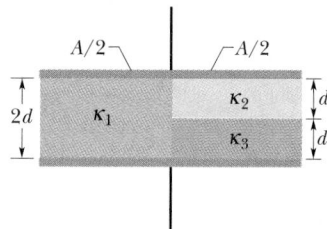

FIGURE 26-39   Problem 65.

SECTION 26-8 Dielectrics and Gauss' Law

66E. A parallel-plate capacitor has a capacitance of 100 pF, a plate area of 100 cm², and a mica dielectric ($\kappa = 5.4$). At 50 V potential difference, calculate (a) E in the mica, (b) the magnitude of the free charge on the plates, and (c) the magnitude of the induced surface charge on the mica.

67E. In Sample Problem 26-8, suppose that the battery remains connected while the dielectric slab is being introduced. Calculate (a) the capacitance, (b) the charge on the capacitor plates, (c) the electric field in the gap, and (d) the electric field in the slab, after the slab is in place.

68P. The space between two concentric conducting spherical shells of radii b and a (where $b > a$) is filled with a substance of dielectric constant κ. A potential difference V exists between the inner and outer shells. Determine (a) the capacitance of the device, (b) the free charge q on the inner shell, and (c) the charge q' induced along the surface of the inner shell.

69P. Two parallel plates of area 100 cm² are given charges of equal magnitude 8.9×10^{-7} C but opposite signs. The electric field within the dielectric material filling the space between the plates is 1.4×10^6 V/m. (a) Calculate the dielectric constant of the material. (b) Determine the magnitude of the charge induced on each dielectric surface.

70P. A parallel-plate capacitor has plates of area 0.12 m² and a separation of 1.2 cm. A battery charges the plates to a potential difference of 120 V and is then disconnected. A dielectric slab of thickness 4.0 mm and dielectric constant 4.8 is then placed symmetrically between the plates. (a) What is the capacitance before the slab is inserted? (b) What is the capacitance with the slab in place? (c) What is the free charge q before and after the slab is inserted? (d) What is the electric field in the space between the plates and dielectric? (e) What is the electric field in the dielectric? (f) With the slab in place, what is the potential difference across the plates? (g) How much external work is involved in the process of inserting the slab?

71P. In the capacitor of Sample Problem 26-8 (Fig. 26-16), (a) what fraction of the energy is stored in the air gaps? (b) What fraction is stored in the slab?

72P. A dielectric slab of thickness b is inserted between the plates of a parallel-plate capacitor of plate separation d. Show that the capacitance is then given by

$$C = \frac{\kappa \epsilon_0 A}{\kappa d - b(\kappa - 1)}.$$

(*Hint:* Derive the formula following the pattern of Sample Problem 26-8.) Does this formula predict the correct numerical result of Sample Problem 26-8? Verify that the formula gives reasonable results for the special cases of $b = 0$, $\kappa = 1$, and $b = d$.

27
Current and Resistance

The pride of Germany and a wonder of its time, the zeppelin <u>Hindenburg</u> was almost the length of three football fields—the largest flying machine that had ever been built. Although it was kept aloft by 16 cells of highly flammable hydrogen gas, it made many trans-Atlantic trips without incident. In fact, German zeppelins, which all depended on hydrogen, had never suffered an accident due to the hydrogen. But shortly after 7:21 p.m. on May 6, 1937, as the <u>Hindenburg</u> was ready to land at the U.S. Naval Air Station at Lakehurst, New Jersey, the ship burst into flames. Its crew had been waiting for a rainstorm to diminish, and handling ropes had just been let down to a navy ground crew, when ripples were sighted on the outer fabric of the ship about one-third of the way forward from the stern. Seconds later a flame erupted from that region, and a red glow illuminated the interior of the ship. Within 32 seconds the burning ship fell to the ground. Why, after so many successful flights of hydrogen-floated zeppelins, did this zeppelin burst into flames❓

27-1 MOVING CHARGES AND ELECTRIC CURRENTS

Chapters 22 through 26 deal largely with *electrostatics*, that is, with charges at rest. With this chapter we begin to focus on **electric currents,** that is, charges in motion.

Examples of electric currents abound, ranging from the large currents that constitute lightning strokes to the tiny nerve currents that regulate our muscular activity. The currents in household wiring, in lightbulbs, and in electrical appliances are familiar to all. A beam of electrons—a current—moves through an evacuated space in the picture tube of a television set. Charged particles of *both* signs flow in the ionized gases of fluorescent lamps, in the batteries of transistor radios, and in car batteries. Electric currents can also be found in the semiconductors in pocket calculators and in the chips that control microwave ovens and electric dishwashers.

On a global scale, charged particles trapped in the Van Allen radiation belts surge back and forth above the atmosphere between Earth's north and south magnetic poles. On the scale of the solar system, enormous currents of protons, electrons, and ions fly radially outward from the Sun as the *solar wind*. On the galactic scale, cosmic rays, which are largely energetic protons, stream through our Milky Way galaxy, some reaching Earth.

Although an electric current is a stream of moving charges, not all moving charges constitute an electric current. If we are to say that an electric current passes through a given surface, there must be a net flow of charge through that surface. Two examples clarify our meaning.

1. The free electrons, conduction electrons, in an isolated length of copper wire are in random motion at speeds of the order of 10^6 m/s. If you pass a hypothetical plane through such a wire, conduction electrons pass through it *in both directions* at the rate of many billions per second. Hence, there is no *net* transport of charge and thus no current through the wire. However, if you connect the ends of the wire to a battery, you slightly bias the flow in one direction, with the result that there now is a net transport of charge and thus an electric current through the wire.

2. The flow of water through a garden hose represents the directed flow of positive charge (the protons in the water molecules) at a rate of perhaps several million coulombs per second. There is no net transport of charge, however, because there is a parallel flow of negative charge (the electrons in the water molecules) of exactly the same amount moving in exactly the same direction.

In this chapter we restrict ourselves largely to the study—within the framework of classical physics—of steady currents of *conduction electrons* moving through *metallic conductors* such as copper wires.

27-2 ELECTRIC CURRENT

As Fig. 27-1a reminds us, an isolated conducting loop—regardless of whether it has an excess charge—is all at the same potential. No electric field can exist within it or parallel to its surface. Although conduction electrons are available, no net electric force acts on them and thus there is no current.

If, as in Fig. 27-1b, we insert a battery in the loop, the conducting loop is no longer at a single potential. Electric fields act inside the material making up the loop, exerting forces on the conduction electrons, causing them to move, and thus establishing a current. After a very short time, the electron flow reaches a final, constant value and the current is in its *steady state* (it is not a function of time).

Figure 27-2 shows a section of a conductor, part of a conducting loop in which current has been established. If charge dq passes through a hypothetical plane (such as aa') in time dt, then the current through that plane is defined as

$$i = \frac{dq}{dt} \qquad \text{(definition of current).} \qquad (27\text{-}1)$$

We can find the charge that passes through the plane in a time interval extending from 0 to t by integration:

$$q = \int dq = \int_0^t i \, dt, \qquad (27\text{-}2)$$

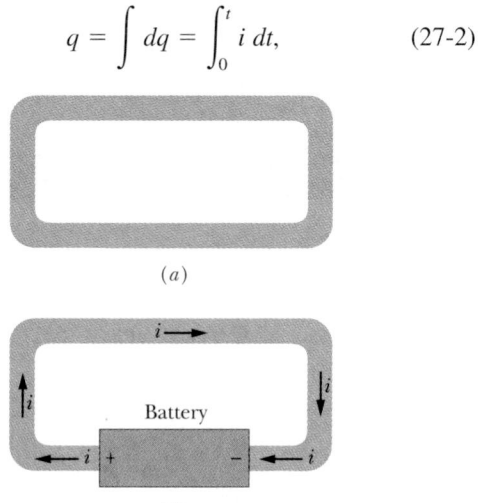

(a)

(b)

FIGURE 27-1 (a) A loop of copper in electrostatic equilibrium. The entire loop is at a single potential, and the electric field is zero at all points inside the copper. (b) Adding a battery imposes an electric potential difference between the ends of the loop that are connected to the terminals of the battery. The battery thus produces an electric field within the loop, from terminal to terminal, and the field causes charges to move around the loop. This movement of charges is a current i.

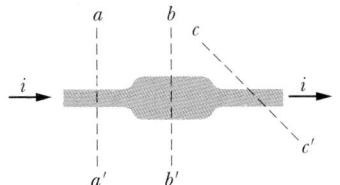

FIGURE 27-2 The current i through the conductor has the same value at planes aa', bb', and cc'.

in which the current i may be a function of time.

Under steady-state conditions, the current is the same for planes bb' and cc' and indeed for all planes that pass completely through the conductor, no matter what their location or orientation. This follows from the fact that charge is conserved. Under the steady-state conditions assumed here, an electron must enter the conductor at one end for every electron that leaves at the other. In the same way, if we have a steady flow of water through a garden hose, a drop of water must leave the nozzle for every drop that enters the hose at the other end. The amount of water in the hose is a conserved quantity.

The SI unit for current is the coulomb per second, also called the *ampere* (A):

1 ampere = 1 A = 1 coulomb per second = 1 C/s.

The ampere is an SI base unit; the coulomb is defined in terms of the ampere, as we discussed in Chapter 22. The formal definition of the ampere is presented in Chapter 30.

Current, as defined by Eq. 27-1, is a scalar because both charge and time in that equation are scalars. Yet, as in Fig. 27-1*b*, we often represent a current with an arrow to indicate the direction in which the charge is moving. Such arrows are not vectors, however, and they do not require vector addition. Figure 27-3*a* shows a conductor splitting at a junction into two branches. Because charge is conserved, the magnitudes of the currents in the branches must add to yield the magnitude of the current in the original

conductor, so that

$$i_0 = i_1 + i_2. \qquad (27\text{-}3)$$

As Fig. 27-3*b* suggests, bending or reorienting the wires in space does not change the validity of Eq. 27-3. Current arrows show only a direction (or sense) of flow along a conductor, not a direction in space.

The Directions of Currents

In Fig. 27-1*b* we drew the current arrows in the direction in which positively charged particles would be forced to move through the loop by the electric field. Such positive *charge carriers*, as they are often called, would move away from the positive battery terminal and toward the negative terminal. Actually, the charge carriers in the copper loop of Fig. 27-1*b* are electrons and thus are negatively charged. The electric field forces them to move in a direction opposite the current arrows, from the negative terminal to the positive terminal. Still, for historical reasons, we use the following convention:

A current arrow is drawn in the direction in which positive charge carriers would move, even if the actual charge carriers are negative and move in the opposite direction.

We can use this convention because in *most* situations, the assumed motion of positive charge carriers in one direction has the same effect as the actual motion of negative charge carriers in the opposite direction. (When the effect is not the same, we shall, of course, drop the convention and describe the actual motion.)

CHECKPOINT **1:** The figure shows a portion of a circuit. What are the magnitude and direction of the current i in the lower right-hand wire?

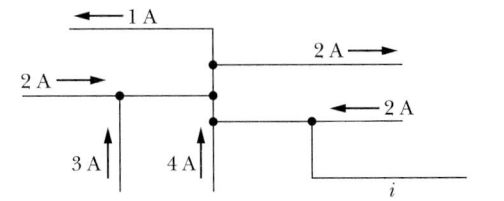

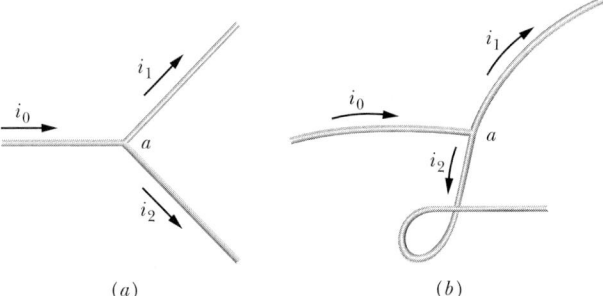

(a)   (b)

FIGURE 27-3 The relation $i_0 = i_1 + i_2$ is true at junction a no matter what the orientation in space of the three wires. Currents are scalars, not vectors.

SAMPLE PROBLEM 27-1

Water flows through a garden hose at a rate R of 450 cm³/s. To what current of negative charge does this correspond?

SOLUTION: The current of negative charge carried by the water is the rate at which water molecules pass through any plane that cuts across the hose times the amount of negative charge carried by each molecule. If ρ is the density of water and M is its molar mass, then the rate (in moles per second) at which water is flowing through the plane is $R\rho/M$. If N is the number of water molecules and N_A is Avogadro's number, the rate dN/dt at which molecules pass through the plane is

$$\frac{dN}{dt} = \frac{R\rho N_A}{M}$$

$$= (450 \times 10^{-6} \text{ m}^3/\text{s})(1000 \text{ kg/m}^3)$$

$$\times \frac{(6.02 \times 10^{23} \text{ molecules/mol})}{0.018 \text{ kg/mol}}$$

$$= 1.51 \times 10^{25} \text{ molecules/s}.$$

Each water molecule contains 10 electrons: 8 in the oxygen atom and 1 in each of the two hydrogens. Each electron has a charge of $-e$, so the current corresponding to this movement of negative charge is

$$i = \frac{dq}{dt} = 10e \frac{dN}{dt} = (10 \text{ electrons/molecule})$$

$$\times (1.60 \times 10^{-19} \text{ C/electron})$$

$$\times (1.51 \times 10^{25} \text{ molecules/s})$$

$$= 2.42 \times 10^7 \text{ C/s} = 2.42 \times 10^7 \text{ A}$$

$$= 24.2 \text{ MA}. \qquad \text{(Answer)}$$

This current of negative charge is exactly compensated by a current of positive charge associated with the nuclei of the three atoms that make up the water molecule. Thus there is no net flow of charge through the hose.

27-3 CURRENT DENSITY

Sometimes we are interested in the current i in a particular conductor. At other times we take a localized view and study the flow of charge at a particular point within a conductor. A (positive) charge carrier at a given point will flow in the direction of the electric field **E** at that point. To describe this flow, we can use the **current density J**. This vector quantity has the same direction as the electric field through a surface and has a magnitude J equal to the current per unit area through an element of that surface. We can write the amount of current through the element as **J** $\cdot d\mathbf{A}$, where $d\mathbf{A}$ is the area vector of that element, perpendicular to the element. The total current through the surface is then

$$i = \int \mathbf{J} \cdot d\mathbf{A}. \qquad (27\text{-}4)$$

If the current is uniform across the surface and parallel to $d\mathbf{A}$, then **J** is also uniform and parallel to $d\mathbf{A}$. Then Eq. 27-4 becomes

$$i = \int J \, dA = J \int dA = JA,$$

or

$$J = \frac{i}{A}, \qquad (27\text{-}5)$$

where A is the total area of the surface. From Eq. 27-4 or 27-5 we see that the SI unit for current density is the ampere per square meter (A/m^2).

In Chapter 23 we saw that we can represent an array of electric field vectors with electric field lines. Figure 27-4 shows how an array of current density vectors can be represented with a similar set of lines, which we can call *streamlines*. The current, which is toward the right in Fig. 27-4, makes a transition from the wider conductor at the left to the narrower conductor at the right. Because charge is conserved during the transition, the amount of charge and thus the amount of current cannot change. However, the current density does change—it is greater in the narrower conductor. The spacing of the streamlines suggests this increase in current density: streamlines that are closer together imply greater current density.

Drift Speed

When a conductor does not have a current through it, its conduction electrons move randomly, with no net motion in any direction. When the conductor does have a current through it, these electrons actually still move randomly, but now they tend to *drift* with a **drift speed** v_d in the direction opposite that of the applied electric field that causes the current. The drift speed is tiny compared to the speeds in the random motion. For example, in the copper conductors of household wiring, electron drift speeds are perhaps 10^{-5} or 10^{-4} m/s, while the random motion speeds are around 10^6 m/s.

We can use Fig. 27-5 to relate the drift speed v_d of the conduction electrons in a current through a wire to the magnitude J of the current density in the wire. For convenience, Fig. 27-5 shows the equivalent drift of *positive*

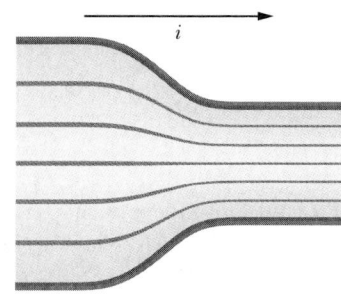

FIGURE 27-4 Streamlines representing the current density vectors in the flow of charge through a constricted conductor.

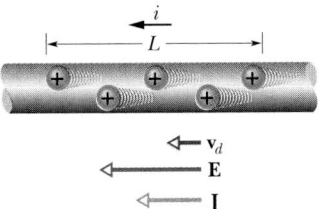

FIGURE 27-5 Positive charge carriers drift at speed v_d in the direction of the applied electric field **E**. By convention, the direction of the current density **J** and the sense of the current arrow are drawn in that same direction.

charge carriers in the direction of the applied electric field **E**. Let us assume that these charge carriers all move with the same drift speed v_d and that the current density J is uniform across the wire's cross-sectional area A. The number of charge carriers in a length L of the wire is nAL, where n is the number of carriers per unit volume. The total charge of the carriers in the length L, each with charge e, is then

$$q = (nAL)e.$$

Because the carriers all move along the wire with speed v_d, this total charge moves through any cross section of the wire in the time interval

$$t = \frac{L}{v_d}.$$

Equation 27-1 tells us that the current i is the time rate of transfer of charge across a cross section. So here we have

$$i = \frac{q}{t} = \frac{nALe}{L/v_d} = nAev_d. \qquad (27\text{-}6)$$

Solving for v_d and recalling Eq. 27-5 ($J = i/A$), we obtain

$$v_d = \frac{i}{nAe} = \frac{J}{ne}.$$

or, extended to vector form,

$$\mathbf{J} = (ne)\mathbf{v}_d. \qquad (27\text{-}7)$$

Here the product ne, whose SI unit is the coulomb per cubic meter (C/m³), is the *carrier charge density*. For positive carriers, which we always assume, ne is positive and Eq. 27-7 predicts that **J** and $\mathbf{v}_d$ point in the same direction.

CHECKPOINT **2:** The figure shows conduction electrons moving leftward through a wire. Are the following leftward or rightward: (a) the current i, (b) the current density **J**, (c) the electric field **E** in the wire?

(a) The current density in a cylindrical wire of radius $R = 2.0$ mm is uniform across a cross section of the wire and is given by $J = 2.0 \times 10^5$ A/m². What is the current through the outer portion of the wire between radial distances $R/2$ and R (Fig. 27-6a)?

SOLUTION: Because the current density is uniform across a cross section, we can use Eq. 27-5 ($J = i/A$) to find the current. However, we want only the current through a reduced cross-sectional area A' of the wire (rather than the entire area), where

$$A' = \pi R^2 - \pi \left(\frac{R}{2}\right)^2 = \pi \left(\frac{3R^2}{4}\right)$$

$$= \frac{\pi 3}{4}(0.002 \text{ m})^2 = 9.424 \times 10^{-6} \text{ m}^2.$$

We now rewrite Eq. 27-5 as

$$i = JA'$$

and then substitute the data to find

$$i = (2.0 \times 10^5 \text{ A/m}^2)(9.424 \times 10^{-6} \text{ m}^2)$$

$$= 1.9 \text{ A}. \qquad \text{(Answer)}$$

(b) Suppose, instead, that the current density through a cross section varies with radial distance r as $J = ar^2$, in which $a = 3.0 \times 10^{11}$ A/m⁴ and r is in meters. What now is the current through the same outer portion of the wire?

SOLUTION: Because the current density is not uniform across a cross section of the wire, we must use Eq. 27-4 ($i = \int \mathbf{J} \cdot d\mathbf{A}$) and integrate the current density over the portion of the wire from $r = R/2$ to $r = R$. The current density vector **J** (along the wire's length) and the differential area vector $d\mathbf{A}$ (perpendicular to a cross section of the wire) have the same direction. So,

$$\mathbf{J} \cdot d\mathbf{A} = J \, dA \cos 0 = J \, dA.$$

We need to replace the differential area dA with something we can actually integrate between the limits $r = R/2$ and $r = R$. The simplest replacement (because J is given as a

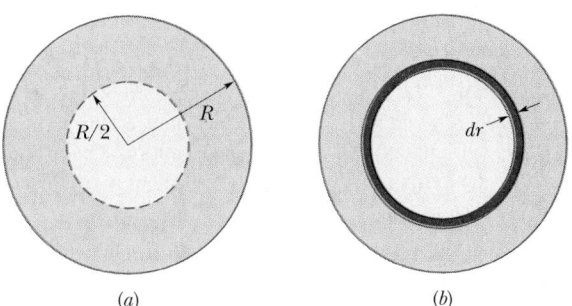

(a)   (b)

FIGURE 27-6 Sample Problem 27-2. (*a*) Cross section of a wire of radius R. (*b*) A thin ring has width dr and circumference $2\pi r$, and thus a differential area $dA = 2\pi r \, dr$.

function of r) is the area $2\pi r\, dr$ of a thin ring of circumference $2\pi r$ and width dr (Fig. 27-6b). We can then integrate with r as the variable of integration. Equation 27-4 then gives us

$$i = \int \mathbf{J} \cdot d\mathbf{A} = \int J\, dA$$

$$= \int_{R/2}^{R} ar^2\, 2\pi r\, dr = 2\pi a \int_{R/2}^{R} r^3\, dr$$

$$= 2\pi a \left[\frac{r^4}{4}\right]_{R/2}^{R} = \frac{\pi a}{2}\left[R^4 - \frac{R^4}{16}\right] = \frac{15}{32}\pi a R^4$$

$$= \frac{15}{32}\pi (3.0 \times 10^{11}\ \text{A/m}^4)(0.002\ \text{m})^4 = 7.1\ \text{A}. \quad \text{(Answer)}$$

SAMPLE PROBLEM 27-3

One end of an aluminum wire whose diameter is 2.5 mm is welded to one end of a copper wire whose diameter is 1.8 mm. The composite wire carries a steady current i of 17 mA.

(a) What is the current density in each wire?

SOLUTION: We may take the current density as constant within each wire (except near the junction, where the diameter changes). The cross-sectional area A of the aluminum wire is

$$A_{\text{Al}} = \pi\left(\frac{d}{2}\right)^2 = \frac{\pi}{4}(2.5 \times 10^{-3}\ \text{M})^2$$

$$= 4.91 \times 10^{-6}\ \text{m}^2,$$

and the current density is given by Eq. 27-5:

$$J_{\text{Al}} = \frac{i}{A_{\text{Al}}} = \frac{17 \times 10^{-3}\ \text{A}}{4.91 \times 10^{-6}\ \text{m}^2}$$

$$= 3.5 \times 10^3\ \text{A/m}^2. \quad \text{(Answer)}$$

As you can verify, the cross-sectional area of the copper wire is $2.54 \times 10^{-6}\ \text{m}^2$, so

$$J_{\text{Cu}} = \frac{i}{A_{\text{Cu}}} = \frac{17 \times 10^{-3}\ \text{A}}{2.54 \times 10^{-6}\ \text{m}^2}$$

$$= 6.7 \times 10^3\ \text{A/m}^2. \quad \text{(Answer)}$$

(b) What is the drift speed of the conduction electrons in the copper wire? Assume that, on the average, each copper atom contributes one conduction electron.

SOLUTION: We can find the drift speed from Eq. 27-7 ($\mathbf{J} = ne\mathbf{v}_d$) if we first find n, the number of electrons per unit volume. With the given assumption of about one conduction electron per atom, n is the same as the number of atoms per unit volume and can be found from

$$\frac{n}{N_A} = \frac{\rho}{M} \quad \text{or} \quad \frac{\text{atoms/m}^3}{\text{atoms/mol}} = \frac{\text{mass/m}^3}{\text{mass/mol}}$$

where ρ is the density of copper, N_A is Avogadro's number, and M is the molar mass of copper. Thus

$$n = \frac{N_A \rho}{M}$$

$$= \frac{(6.02 \times 10^{23}\ \text{mol}^{-1})(9.0 \times 10^3\ \text{kg/m}^3)}{64 \times 10^{-3}\ \text{kg/mol}}$$

$$= 8.47 \times 10^{28}\ \text{electrons/m}^3.$$

We then have from Eq. 27-7,

$$v_d = \frac{6.7 \times 10^3\ \text{A/m}^2}{\left(8.47 \times 10^{28}\ \dfrac{\text{electrons}}{\text{m}^3}\right)\left(1.6 \times 10^{-19}\ \dfrac{\text{C}}{\text{electron}}\right)}$$

$$= 4.9 \times 10^{-7}\ \text{m/s} = 1.8\ \text{mm/h}. \quad \text{(Answer)}$$

You may well ask: "If the electrons drift so slowly, why do the room lights turn on so quickly when I throw the switch?" Confusion on this point results from not distinguishing between the drift speed of the electrons and the speed at which *changes* in the electric field configuration travel along wires. This latter speed is nearly that of light; electrons everywhere in the wire begin drifting almost at once, including into the lightbulbs. Similarly, when you open the valve on your garden hose, with the hose full of water, a pressure wave travels along the hose at the speed of sound in water. The speed at which the water itself moves through the hose—measured perhaps with a dye marker—is much lower.

SAMPLE PROBLEM 27-4

Consider a strip of silicon that has a rectangular cross section with width $w = 3.2$ mm and height $h = 250\ \mu\text{m}$, and through which there is a uniform current i of 5.2 mA. The silicon is an *n-type semiconductor*, having been "doped" with a controlled phosphorus impurity. As we shall discuss in Section 27-8, the doping has the effect of greatly increasing n, the number of charge carriers per unit volume, as compared with the value for pure silicon. In this case, $n = 1.5 \times 10^{23}\ \text{m}^{-3}$.

(a) What is the current density in the strip?

SOLUTION: From Eq. 27-5,

$$J = \frac{i}{wh} = \frac{5.2 \times 10^{-3}\ \text{A}}{(3.2 \times 10^{-3}\ \text{m})(250 \times 10^{-6}\ \text{m})}$$

$$= 6500\ \text{A/m}^2. \quad \text{(Answer)}$$

(b) What is the drift speed?

SOLUTION: From Eq. 27-7,

$$v_d = \frac{J}{ne} = \frac{6500\ \text{A/m}^2}{(1.5 \times 10^{23}\ \text{m}^{-3})(1.60 \times 10^{-19}\ \text{C})}$$

$$= 0.27\ \text{m/s} = 27\ \text{cm/s}. \quad \text{(Answer)}$$

Note that the current density (6500 A/m²) for this semiconductor turns out to be comparable to the current density (6700 A/m²) for the copper conductor in Sample Problem

27-3. That is, the rate at which charge flows through a unit area is about the same for the two devices. Yet, the drift speed (0.27 m/s) in the semiconductor is *much* greater than the drift speed (4.9×10^{-7} m/s) in the copper conductor.

If you recheck the calculations, you will see that this large difference in drift speed occurs because the number n of charge carriers per unit volume is much smaller in the semiconductor. Thus, if the current densities are to be comparable, the fewer conduction electrons in the semiconductor must move much faster than the electrons in the copper conductor.

27-4 RESISTANCE AND RESISTIVITY

If we apply the same potential difference between the ends of geometrically similar rods of copper and of glass, very different currents result. The characteristic of the conductor that enters here is its **resistance.** We determine the resistance between any two points of a conductor by applying a potential difference V between those points and measuring the current i that results. The resistance R is then

$$R = \frac{V}{i} \qquad \text{(definition of } R\text{)}. \qquad (27\text{-}8)$$

The SI unit for resistance that follows from Eq. 27-8 is the volt per ampere. This combination occurs so often that we give it a special name, the **ohm** (symbol Ω). That is,

$$1 \text{ ohm} = 1\ \Omega = 1 \text{ volt per ampere}$$
$$= 1 \text{ V/A.} \qquad (27\text{-}9)$$

A conductor whose function in a circuit is to provide a specified resistance is called a **resistor** (see Fig. 27-7). We represent a resistor in a circuit diagram with the symbol ‑‑\/\/\/‑. If we write Eq. 27-8 as

$$i = \frac{V}{R},$$

we see that "resistance" is aptly named. For a given potential difference, the greater the resistance (to current), the smaller the current.

The resistance of a conductor depends on the manner in which the potential difference is applied to it. Figure

27-8, for example, shows a given potential difference applied in two different ways to the same conductor. As the current density streamlines suggest, the currents in the two cases—hence the measured resistances—will be different. Unless otherwise stated, we shall assume that any given potential difference is applied as in Fig. 27-8*b*.

As we have done several times in other connections, we often wish to take a general view and deal not with particular objects but with materials. Here we do so by focusing not on the potential difference V across a particular resistor but on the electric field **E** at a point in a resistive material. Instead of dealing with the current i through the resistor, we deal with the current density **J** at the point in question. Instead of the resistance R of an object, we deal with the **resistivity** ρ of the *material*, defined as

$$\rho = \frac{E}{J} \qquad \text{(definition of } \rho\text{)}. \qquad (27\text{-}10)$$

(Compare this equation with Eq. 27-8.)

If we combine the SI units of E and J according to Eq. 27-10, we get, for the unit of ρ, the ohm-meter ($\Omega \cdot$ m):

$$\frac{\text{unit } (E)}{\text{unit } (J)} = \frac{\text{V/m}}{\text{A/m}^2} = \frac{\text{V}}{\text{A}}\, \text{m} = \Omega \cdot \text{m}.$$

(Do not confuse the *ohm-meter,* the unit of resistivity, with the *ohmmeter,* which is an instrument that measures resistance.) Table 27-1 lists the resistivities of some materials.

We can write Eq. 27-10 in vector form as

$$\mathbf{E} = \rho\mathbf{J}. \qquad (27\text{-}11)$$

Equations 27-10 and 27-11 hold only for *isotropic* materials—materials whose electrical properties are the same in all directions.

We often speak of the **conductivity** σ of a material. This is simply the reciprocal of its resistivity, so

$$\sigma = \frac{1}{\rho} \qquad \text{(definition of } \sigma\text{)}. \qquad (27\text{-}12)$$

FIGURE 27-7 An assortment of resistors. The circular bands are color coding marks that identify the value of the resistance.

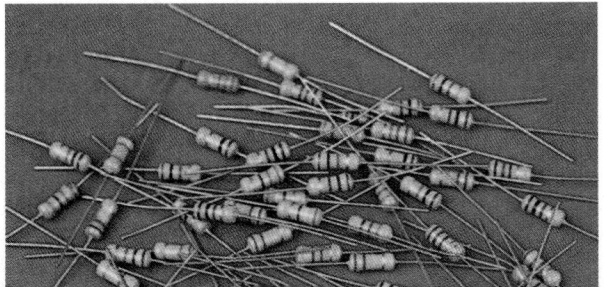

FIGURE 27-8 Two ways of applying a potential difference to a conducting rod. The heavy gray connectors are assumed to have negligible resistance. When they are arranged as in (*a*), the measured resistance is larger than when they are arranged as in (*b*).

(*a*)

(*b*)

TABLE 27-1 **RESISTIVITIES OF SOME MATERIALS AT ROOM TEMPERATURE (20°C)**

MATERIAL	RESISTIVITY, ρ ($\Omega \cdot$m)	TEMPERATURE COEFFICIENT OF RESISTIVITY, α (K^{-1})
Typical Metals		
Silver	1.62×10^{-8}	4.1×10^{-3}
Copper	1.69×10^{-8}	4.3×10^{-3}
Aluminum	2.75×10^{-8}	4.4×10^{-3}
Tungsten	5.25×10^{-8}	4.5×10^{-3}
Iron	9.68×10^{-8}	6.5×10^{-3}
Platinum	10.6×10^{-8}	3.9×10^{-3}
Manganin[a]	48.2×10^{-8}	0.002×10^{-3}
Typical Semiconductors		
Silicon, pure	2.5×10^{3}	-70×10^{-3}
Silicon, n-type[b]	8.7×10^{-4}	
Silicon, p-type[c]	2.8×10^{-3}	
Typical Insulators		
Glass	$10^{10} - 10^{14}$	
Fused quartz	$\sim 10^{16}$	

[a] An alloy specifically designed to have a small value of α.

[b] Pure silicon doped with phosphorus impurities to a charge carrier density of 10^{23} m^{-3}.

[c] Pure silicon doped with aluminum impurities to a charge carrier density of 10^{23} m^{-3}.

The SI unit of conductivity is the reciprocal ohm-meter, $(\Omega \cdot$m$)^{-1}$. The unit name mhos per meter is sometimes used (mho is ohm backward). The definition of σ allows us to write Eq. 27-11 in the alternative form

$$\mathbf{J} = \sigma\mathbf{E}. \qquad (27\text{-}13)$$

Calculating Resistance from Resistivity

We have just made an important distinction:

Resistance is a property of an object. Resistivity is a property of a material.

If we know the resistivity of a substance such as copper, we can calculate the resistance of a length of wire made of that substance. Let A be the cross-sectional area of the wire, let L be its length, and let a potential difference V exist between its ends (Fig. 27-9). If the streamlines representing the current density are uniform throughout the wire, the electric field and the current density will be con-

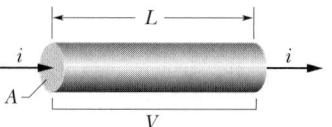

FIGURE 27-9 A potential difference V is applied between the ends of a wire of length L and cross section A, establishing a current i.

stant for all points within the wire and, from Eqs. 25-42 and 27-5, will have the values

$$E = V/L \quad \text{and} \quad J = i/A. \qquad (27\text{-}14)$$

We can then combine Eqs. 27-10 and 27-14 to write

$$\rho = \frac{E}{J} = \frac{V/L}{i/A}. \qquad (27\text{-}15)$$

But V/i is the resistance R, which allows us to recast Eq. 27-15 as

$$R = \rho \frac{L}{A}. \qquad (27\text{-}16)$$

Equation 27-16 can be applied only to a homogeneous isotropic conductor of uniform cross section, with the potential difference applied as in Fig. 27-8b.

The macroscopic quantities V, i, and R are of greatest interest when we are making electrical measurements on specific conductors. They are the quantities that we read directly on meters. We turn to the microscopic quantities E, J, and ρ when we are interested in the fundamental electrical properties of materials.

CHECKPOINT 3: The figure shows three cylindrical copper conductors along with their face areas and lengths. Rank them according to the current through them, greatest first, when the same potential difference V is placed across their lengths.

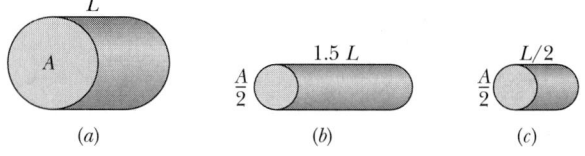

Variation with Temperature

The values of most physical properties vary with temperature, and resistivity is no exception. Figure 27-10, for example, shows the variation of this property for copper over a wide temperature range. The relation between temperature and resistivity for copper —and for metals in general —is fairly linear over a rather broad temperature range. For

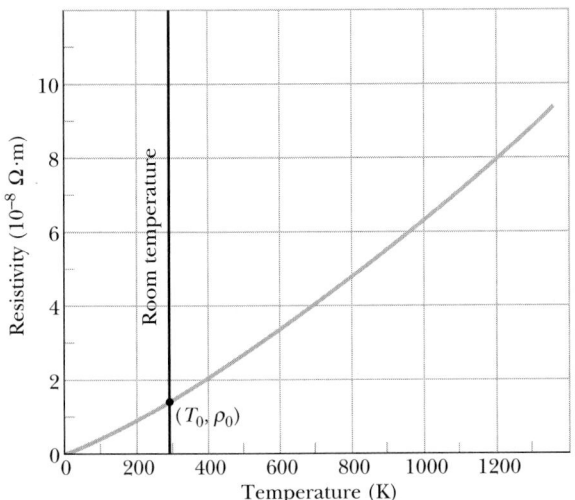

FIGURE 27-10 The resistivity of copper as a function of temperature. The dot on the curve marks a convenient reference point ($T_0 = 293$ K, $\rho_0 = 1.69 \times 10^{-8}$ $\Omega \cdot$m).

such linear relations we can write, as an empirical approximation that is good enough for most engineering purposes,

$$\rho - \rho_0 = \rho_0 \alpha (T - T_0). \quad (27\text{-}17)$$

Here T_0 is a selected reference temperature and ρ_0 is the resistivity at that temperature. Usually $T_0 = 293$ K (room temperature), for which $\rho_0 = 1.69 \times 10^{-8}$ $\Omega \cdot$m.

Because temperature enters Eq. 27-17 only as a difference, it does not matter whether you use the Celsius or Kelvin scale in that equation because the sizes of the degree on these scales are identical. The quantity α in Eq. 27-17, called the *temperature coefficient of resistivity,* is chosen so that the equation gives the best agreement with experiment for temperatures in the chosen range. Some values of α for metals are listed in Table 27-1.

The *Hindenburg*

When the zeppelin *Hindenburg* was preparing to land, the handling ropes were let down to the ground crew. Exposed to the rain, the ropes became wet (and thus able to conduct a current). In this condition, the ropes "grounded" the metal framework of the zeppelin to which they were attached. That is, the wet ropes formed a conducting path between the framework and the ground, making the electric potential of the framework the same as that of the ground. This should have also grounded the outer fabric of the zeppelin, too. The *Hindenburg*, however, had been the first zeppelin to have its outer fabric painted with a sealant of large electrical resistivity. Thus the fabric remained at the electric potential of the atmosphere at the zeppelin's

altitude of about 43 m. Owing to the rainstorm, that potential was large relative to the potential at ground level.

The handling of the ropes apparently ruptured one of the hydrogen cells and released hydrogen between that cell and the zeppelin's outer fabric, causing the reported rippling of the fabric. There was then a dangerous situation: the fabric was wet with conducting rainwater and was at a potential much different from that of the framework of the zeppelin. Apparently, charge flowed along the wet fabric and then sparked through the released hydrogen to reach the metal framework of the zeppelin, igniting the hydrogen in the process. The burning rapidly ignited the cells of hydrogen in the zeppelin and brought the ship down. If the sealant on the outer fabric of the *Hindenburg* had been of less resistivity (like that of earlier and later zeppelins), the *Hindenburg* disaster probably would not have occurred.

SAMPLE PROBLEM 27-5

(a) What is the magnitude of the electric field applied to the copper conductor of Sample Problem 27-3?

SOLUTION: In Sample Problem 27-3(a) we found the current density J to be 6.7×10^3 A/m²; from Table 27-1 we see that the resistivity ρ for copper is 1.69×10^{-8} $\Omega \cdot$m. Thus from Eq. 27-11

$$E = \rho J = (1.69 \times 10^{-8} \ \Omega \cdot \text{m})(6.7 \times 10^3 \ \text{A/m}^2)$$
$$= 1.1 \times 10^{-4} \ \text{V/m (copper)}. \quad \text{(Answer)}$$

(b) What is the magnitude of the electric field in the *n*-type silicon semiconductor of Sample Problem 27-4?

SOLUTION: In that sample problem we found that $J = 6500$ A/m², and from Table 27-1 we see that $\rho = 8.7 \times 10^{-4}$ $\Omega \cdot$m. Thus from Eq. 27-11

$$E = \rho J = (8.7 \times 10^{-4} \ \Omega \cdot \text{m})(6500 \ \text{A/m}^2)$$
$$= 5.7 \ \text{V/m (\textit{n}-type silicon)}. \quad \text{(Answer)}$$

Note that the applied electric field in the semiconductor is much greater than that in the copper conductor. If you recheck the calculations, you will find that this difference is required by the large difference in the resistivities of the two devices. The need for a much greater electric field in the semiconductor is consistent with the need for a much greater drift speed that we found in Sample Problem 27-4: if the current densities are to be comparable in the two devices, the electric field applied to the semiconductor must be greater, to make possible the acceleration of the electrons to a greater drift speed.

SAMPLE PROBLEM 27-6

A rectangular block of iron has dimensions 1.2 cm × 1.2 cm × 15 cm.

(a) What is the resistance of the block measured between the two square ends?

SOLUTION: The resistivity of iron at room temperature is $9.68 \times 10^{-8} \, \Omega \cdot m$ (Table 27-1). The area of a square end is $(1.2 \times 10^{-2} \, m)^2$, or $1.44 \times 10^{-4} \, m^2$. From Eq. 27-16,

$$R = \frac{\rho L}{A} = \frac{(9.68 \times 10^{-8} \, \Omega \cdot m)(0.15 \, m)}{1.44 \times 10^{-4} \, m^2}$$

$$= 1.0 \times 10^{-4} \, \Omega = 100 \, \mu\Omega. \qquad \text{(Answer)}$$

(b) What is the resistance between two opposite rectangular faces?

SOLUTION: The area of a rectangular face is $(1.2 \times 10^{-2} \, m)(0.15 \, m)$, or $1.80 \times 10^{-3} \, m^2$. From Eq. 27-16,

$$R = \frac{\rho L}{A} = \frac{(9.68 \times 10^{-8} \, \Omega \cdot m)(1.2 \times 10^{-2} \, m)}{1.80 \times 10^{-3} \, m^2}$$

$$= 6.5 \times 10^{-7} \, \Omega = 0.65 \, \mu\Omega. \qquad \text{(Answer)}$$

This result is much smaller than the previous result, because the distance L is smaller and the area A is larger. We assume in each part that the potential difference is applied to the block in such a way that the surfaces between which the resistance is desired are equipotential surfaces (as in Fig. 27-8b). Otherwise, Eq. 27-16 would not be valid.

27-5 OHM'S LAW

As we just discussed in Section 27-4, a resistor is a conductor with a specified resistance. It has that same resistance no matter what the magnitude and direction (*polarity*) of the applied potential difference. Other conducting devices, however, might have resistances that change with the applied potential difference.

Figure 27-11a shows how to distinguish such devices. A potential difference V is applied across the device being

FIGURE 27-11 (*a*) A device to whose terminals a potential difference V is applied, establishing a current i. (*b*) A plot of current i versus applied potential difference V when the device is a 1000 Ω resistor. (*c*) A plot when the device is a semiconducting *pn* junction diode.

tested, and the resulting current i through the device is measured as V is varied in both magnitude and polarity. The polarity of V is arbitrarily taken to be positive when the left terminal of the device is at a higher potential than the right terminal. The direction of the resulting current (from left to right) is arbitrarily assigned a plus sign. The reverse polarity of V (with the right terminal at a higher potential) is then negative; the current it causes is assigned a minus sign.

Figure 27-11b is a plot of i versus V for one device. This plot is a straight line passing through the origin, so the ratio i/V (which is the slope of the straight line) is the same for all values of V. This means that the resistance $R = V/i$ of the device is independent of the magnitude and polarity of the applied potential difference V.

Figure 27-11c is a plot for another conducting device. Current can exist in this device only when the polarity of V is positive and the applied potential difference is more than about 1.5 V. And when current does exist, the relation between i and V is not linear; it depends on the value of the applied potential difference V.

We distinguish between the two types of device by saying that one obeys Ohm's law and the other does not.

Ohm's law is an assertion that the current through a device is *always* directly proportional to the potential difference applied to the device.

(This assertion is correct only in certain situations; still, for historical reasons, the term "law" is used.) The device of Fig. 27-11b—which turns out to be a 1000 Ω resistor—obeys Ohm's law. The device of Fig. 27-11c—which turns out to be a so-called *pn* junction diode—does not.

A conducting device obeys Ohm's law when the resistance of the device is independent of the magnitude and polarity of the applied potential difference.

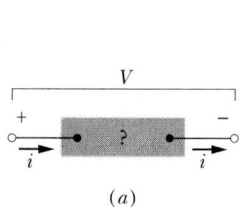

(*a*)

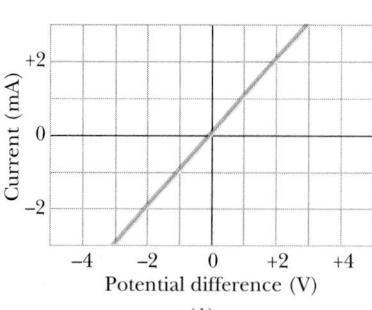

(*b*)

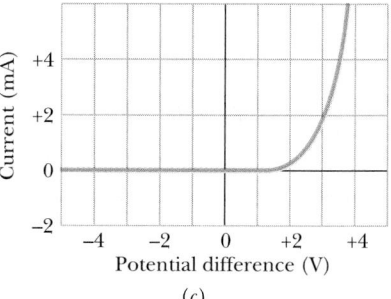

(*c*)

Modern microelectronics—and therefore much of the character of our present technological civilization—depends almost totally on devices that do *not* obey Ohm's law. Your calculator, for example, is full of them.

It is often contended that $V = iR$ is a statement of Ohm's law. That is not true! This equation is the defining equation for resistance, and it applies to all conducting devices, whether they obey Ohm's law or not. If we measure the potential difference V across, and the current i through, any device, even a *pn* junction diode, we can find its resistance *at that value of V* as $R = V/i$. The essence of Ohm's law, however, is that a plot of i versus V is linear; that is, the value of R is independent of the value of V.

We can express Ohm's law in a more general way if we focus on conducting *materials* rather than on conducting *devices*. The relevant relation is then Eq. 27-11 ($\mathbf{E} = \rho\mathbf{J}$), which is the analog of $V = iR$.

A conducting material obeys Ohm's law when the resistivity of the material is independent of the magnitude and direction of the applied electric field.

All homogeneous materials, whether they are conductors like copper or semiconductors like silicon (doped or pure), obey Ohm's law within some range of values of the electric field. If the field is too strong, however, there are departures from Ohm's law in all cases.

$\mathbf{C}$HECKPOINT 4: The following table gives the current i (in amperes) through two devices for several values of potential difference V (in volts). From these data, determine which device does not obey Ohm's law.

DEVICE 1		DEVICE 2	
V	i	V	i
2.00	4.50	2.00	1.50
3.00	6.75	3.00	2.20
4.00	9.00	4.00	2.80

27-6 A MICROSCOPIC VIEW OF OHM'S LAW

To find out *why* particular materials obey Ohm's law, we must look into the details of the conduction process at the atomic level. Here we consider only conduction in metals, such as copper. We base our analysis on the *free-electron model*, in which we assume that the conduction electrons in the metal are free to move throughout the volume of the sample, like the molecules of a gas in a closed container. We also assume that the electrons collide not with one another but only with the atoms of the metal.

According to classical physics, the electrons should have a Maxwellian speed distribution somewhat like that of the molecules in a gas. In such a distribution (see Section 20-7), the average electron speed would be proportional to the square root of the absolute temperature. The motions of the electrons, however, are governed not by the laws of classical physics but by those of quantum physics. As it turns out, an assumption that is much closer to the quantum reality is that the electrons move with a single effective speed v_{eff}, and this motion is essentially independent of the temperature. For copper, $v_{eff} \approx 1.6 \times 10^6$ m/s.

When we apply an electric field to a metal sample, the electrons modify their random motions slightly and drift very slowly—in a direction opposite that of the field—with an average drift speed v_d. As we saw in Sample Problem 27-3(b), the drift speed in a typical metallic conductor is about 4×10^{-7} m/s, less than the effective speed (1.6×10^6 m/s) by many orders of magnitude. Figure 27-12 suggests the relation between these two speeds. The gray lines show a possible random path for an electron in the absence of an applied field; the electron proceeds from A to B, making six collisions along the way. The green lines show how the same events *might* occur when an electric field $\mathbf{E}$ is applied. We see that the electron drifts steadily to the right, ending at B' rather than at B. Figure 27-12 was drawn with the assumption that $v_d \approx 0.02v_{eff}$. Since, however, the actual value is more like $v_d \approx (10^{-13})v_{eff}$, the drift displayed in the figure is greatly exaggerated.

The motion of the electrons in an electric field $\mathbf{E}$ is thus a combination of the motion due to random collisions

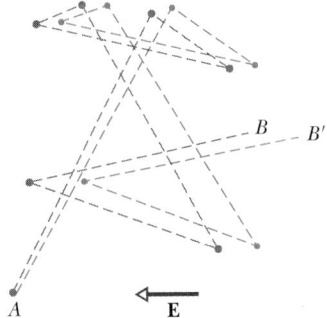

FIGURE 27-12 The gray lines show an electron moving from A to B, making six collisions en route. The green lines show what its path might be in the presence of an applied electric field $\mathbf{E}$. Note the steady drift in the direction of $-\mathbf{E}$. (Actually, the green lines should be slightly curved, to represent the parabolic paths followed by the electrons between collisions, under the influence of an electric field.)

and that due to **E**. When we consider all the free electrons, their random motions average to zero and make no contribution to the drift speed. Thus the drift speed is due only to the effect of the electric field on the electrons.

If an electron of mass m is placed in an electric field of magnitude E, the electron will experience an acceleration given by Newton's second law:

$$a = \frac{F}{m} = \frac{eE}{m}. \qquad (27\text{-}18)$$

The nature of the collisions experienced by electrons is such that, after a typical collision, the electron will—so to speak—completely lose its memory of its previous drift velocity. Each electron will then start off fresh after every encounter, moving off in a random direction. In the average time τ between collisions, the average electron will acquire a drift speed of $v_d = a\tau$. Moreover, if we measure the drift speeds of all the electrons at any instant, we will find that their average drift speed is also $a\tau$. Thus, at any instant, on average, the electrons will have drift speed $v_d = a\tau$. Then Eq. 27-18 gives us

$$v_d = a\tau = \frac{eE\tau}{m}. \qquad (27\text{-}19)$$

Combining this result with Eq. 27-7 ($J = nev_d$) yields

$$v_d = \frac{J}{ne} = \frac{eE\tau}{m},$$

which we can write as

$$E = \left(\frac{m}{e^2 n\tau}\right) J.$$

Comparing this with Eq. 27-11 ($E = \rho J$) leads to

$$\rho = \frac{m}{e^2 n\tau}. \qquad (27\text{-}20)$$

Equation 27-20 may be taken as a statement that metals obey Ohm's law if we can show that, for metals, ρ is a constant, independent of the strength of the applied electric field **E**. Because n, m, and e are constant, this reduces to convincing ourselves that τ, the average time (or *mean free time*) between collisions, is a constant, independent of the strength of the applied electric field. Indeed τ can be considered to be a constant because the drift speed v_d caused by the field is about a billion times smaller than the effective speed v_{eff}.

SAMPLE PROBLEM 27-7

(a) What is the mean free time τ between collisions for the conduction electrons in copper?

SOLUTION: From Eq. 27-20 we have

$$\tau = \frac{m}{ne^2\rho}.$$

We take the value of n, the number of conduction electrons per unit volume in copper, from Sample Problem 27-3(b). We take the value of ρ from Table 27-1. The denominator then becomes

$$(8.47 \times 10^{28} \text{ m}^{-3})(1.6 \times 10^{-19} \text{ C})^2(1.69 \times 10^{-8} \ \Omega \cdot \text{m})$$
$$= 3.66 \times 10^{-17} \text{ C}^2 \cdot \Omega/\text{m}^2 = 3.66 \times 10^{-17} \text{ kg/s},$$

where we converted units as

$$\frac{\text{C}^2 \cdot \Omega}{\text{m}^2} = \frac{\text{C}^2 \cdot \text{V}}{\text{m}^2 \cdot \text{A}} = \frac{\text{C}^2 \cdot \text{J/C}}{\text{m}^2 \cdot \text{C/s}} = \frac{\text{kg} \cdot \text{m}^2/\text{s}^2}{\text{m}^2/\text{s}} = \frac{\text{kg}}{\text{s}}.$$

For the mean free time we then have

$$\tau = \frac{9.1 \times 10^{-31} \text{ kg}}{3.66 \times 10^{-17} \text{ kg/s}} = 2.5 \times 10^{-14} \text{ s.} \quad \text{(Answer)}$$

(b) What is the mean free path λ for these collisions? Assume an effective speed v_{eff} of 1.6×10^6 m/s.

SOLUTION: In Section 20-6, we defined the mean free path as being the average distance traversed by a particle between collisions. Here the time between collisions of a free electron is τ and the speed of the electron is v_{eff}. So,

$$\lambda = \tau v_{eff} = (2.5 \times 10^{-14} \text{ s})(1.6 \times 10^6 \text{ m/s})$$
$$= 4.0 \times 10^{-8} \text{ m} = 40 \text{ nm.} \quad \text{(Answer)}$$

This is about 150 times the distance between nearest-neighbor atoms in a copper lattice.

27-7 POWER IN ELECTRIC CIRCUITS

Figure 27-13 shows a circuit consisting of a battery B that is connected by wires, which we assume to have negligible resistance, to an unspecified conducting device. The device might be a resistor, a storage battery (a rechargeable battery), a motor, or some other electrical device. The battery

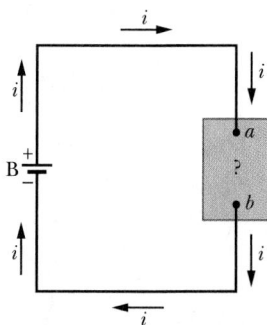

FIGURE 27-13 A battery B sets up a current i in a circuit containing an unspecified conducting device.

maintains a potential difference of magnitude V across its own terminals, and thus (because of the wires) across the terminals of the unspecified device, with a greater potential at terminal a of the device than at terminal b.

Since there is an external conducting path between the two terminals of the battery and since the potential differences set up by the battery are maintained, a steady current i is produced in the circuit, directed from terminal a to terminal b. The amount of charge dq that moves between those terminals in time interval dt is equal to $i\,dt$. This charge dq moves through a decrease in potential of magnitude V, and thus its electric potential energy decreases in magnitude by the amount

$$dU = dq\,V = i\,dt\,V.$$

The principle of conservation of energy tells us that the decrease in electric potential energy from a to b is accompanied by a transfer of energy to some other form. The power P associated with that transfer is the rate of transfer dU/dt, which is

$$P = iV \qquad \text{(rate of electrical energy transfer).} \qquad (27\text{-}21)$$

Moreover, this power P is the rate of energy transfer from the battery to the unspecified device. If that device is a

The wire coils within a toaster have appreciable resistance. When current is set up through them, electrical energy is transferred to thermal energy of the coils, increasing their temperature. The coils then emit infrared radiation and visible light that will toast (or burn) bread.

motor connected to a mechanical load, the energy is transferred as work on the load. If the device is a storage battery that is being charged, the energy is transferred to stored chemical energy in the storage battery. If the device is a resistor, the energy is transferred to internal thermal energy, tending to increase the resistor's temperature.

The unit of power that follows from Eq. 27-21 is the volt-ampere $(\text{V}\cdot\text{A})$. We can write it as

$$1\,\text{V}\cdot\text{A} = \left(1\,\frac{\text{J}}{\text{C}}\right)\left(1\,\frac{\text{C}}{\text{s}}\right) = 1\,\frac{\text{J}}{\text{s}} = 1\,\text{W}.$$

The course of an electron moving through a resistor at constant drift speed is much like that of a stone falling through water at constant terminal speed. The average kinetic energy of the electron remains constant, and its lost electric potential energy appears as thermal energy in the resistor and the surroundings. On a microscopic scale this energy transfer is due to collisions between the electron and the molecules of the resistor, which leads to an increase in the temperature of the lattice. The mechanical energy thus transferred to thermal energy is *dissipated* (lost), because the transfer cannot be reversed.

For a resistor we can combine Eqs. 27-8 ($R = V/i$) and 27-21 to obtain, for the rate of electrical energy dissipation in a resistor, either

$$P = i^2R \qquad \text{(resistive dissipation)} \qquad (27\text{-}22)$$

or

$$P = \frac{V^2}{R} \qquad \text{(resistive dissipation).} \qquad (27\text{-}23)$$

However, we must be careful to distinguish these two new equations from Eq. 27-21: $P = iV$ applies to electrical energy transfers of all kinds; $P = i^2R$ and $P = V^2/R$ apply only to the transfer of electric potential energy to thermal energy in a resistance.

CHECKPOINT **5:** A potential difference V is connected across a resistance R, causing current i through the resistance. Rank the following variations according to the change in the rate at which electrical energy is converted to thermal energy in the resistance, greatest change first: (a) V is doubled with R unchanged, (b) i is doubled with R unchanged. (c) R is doubled with V unchanged, (d) R is doubled with i unchanged.

SAMPLE PROBLEM 27-8

You are given a length of uniform heating wire made of a nickel–chromium–iron alloy called Nichrome; it has a resistance R of 72 Ω. At what rate is energy dissipated in each of the following situations? (1) A potential difference of 120 V is applied across the full length of the wire. (2) The wire is cut in half, and a potential difference of 120 V is applied across the length of each half.

SOLUTION: From Eq. 27-23, the rate of energy dissipation in situation 1 is

$$P = \frac{V^2}{R} = \frac{(120 \text{ V})^2}{72 \ \Omega} = 200 \text{ W}. \qquad \text{(Answer)}$$

In situation 2, the resistance of each half of the wire is (72 Ω)/2, or 36 Ω. Thus the dissipation rate for each half is

$$P' = \frac{(120 \text{ V})^2}{36 \ \Omega} = 400 \text{ W}. \qquad \text{(Answer)}$$

The total power of the two halves is 800 W, or four times that for the full length of wire. This would seem to suggest that you could buy a heating coil, cut it in half, and reconnect it to obtain four times the heat output. Why is this unwise? (What would happen to the amount of current in the coil?)

SAMPLE PROBLEM 27-9

A wire of length $L = 2.35$ m and diameter $d = 1.63$ mm carries a current i of 1.24 A. The wire dissipates electrical energy at the rate P of 48.5 mW. Of what is the wire made?

SOLUTION: We can identify the material by its resistivity. From Eqs. 27-16 and 27-22 we have

$$P = i^2R = \frac{i^2 \rho L}{A} = \frac{4i^2 \rho L}{\pi d^2},$$

in which $A \ (= \frac{1}{4}\pi d^2)$ is the cross-sectional area of the wire. Solving for ρ, the resistivity of the material of which the wire is made, yields

$$\rho = \frac{\pi P d^2}{4i^2L} = \frac{(\pi)(48.5 \times 10^{-3} \text{ W})(1.63 \times 10^{-3} \text{ m})^2}{(4)(1.24 \text{ A})^2(2.35 \text{ m})}$$

$$= 2.80 \times 10^{-8} \ \Omega \cdot \text{m}. \qquad \text{(Answer)}$$

Inspection of Table 27-1 tells us that the material is aluminum.

27-8 SEMICONDUCTORS

Semiconducting devices are at the heart of the microelectronic revolution that has so influenced our lives. Table 27-2 compares the properties of silicon—a typical semiconductor—and copper—a typical metallic conductor. We see that silicon has many fewer charge carriers, a much higher resistivity, and a temperature coefficient of resistivity that is both large and negative. That is, although the resistivity of copper increases with temperature, that of pure silicon decreases.

TABLE 27-2 SOME ELECTRICAL PROPERTIES OF COPPER AND SILICON[a]

PROPERTY	COPPER	SILICON
Type of material	Metal	Semiconductor
Charge carrier density, m^{-3}	9×10^{28}	1×10^{16}
Resistivity, $\Omega \cdot m$	2×10^{-8}	3×10^3
Temperature coefficient of resistivity, K^{-1}	$+4 \times 10^{-3}$	-70×10^{-3}

[a]Rounded to one significant figure for easy comparison.

The resistivity of pure silicon is so high that it is virtually an insulator and is thus not of much direct use in microelectronic circuits. The property that makes it useful is that—as Table 27-1 shows—its resistivity can be reduced in a controlled way by adding minute amounts of specific foreign "impurity" atoms, a process called *doping*.

We can explain the difference in resistivity (hence conductivity) between semiconductors and metallic conductors in terms of the energy levels of their electrons. We saw in Section 8-9 that the energies of electrons in isolated atoms are *quantized*; that is, they are restricted to certain values, *or levels*, as shown in Fig. 8-17. An electron can *occupy* (that is, have the energy of) any one of the energy levels but cannot have an intermediate energy.

Electrons in solids also occupy quantized levels, but the proximity of the atoms to each other tends to "squeeze" the many levels into a few *bands* (Fig. 27-14). An electron can occupy an energy level within a band but cannot have an energy value within the *gaps* separating the

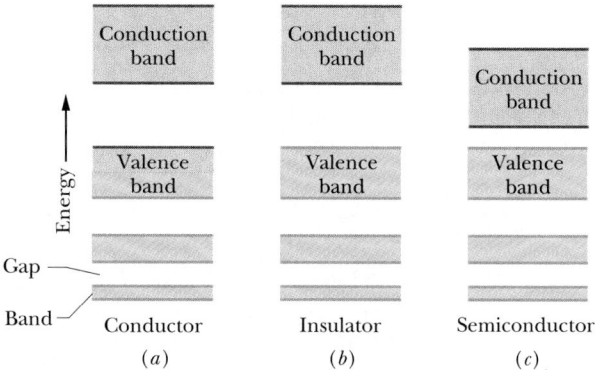

FIGURE 27-14 The allowed energy levels for the electrons in a solid form a pattern of allowed bands and forbidden gaps. Green denotes a partially or completely filled band. (*a*) In a metallic conductor, the valence band is only partially filled. (*b*) In an insulator, the valence band is completely filled and the gap between the valence band and the conduction band is relatively large. (*c*) A semiconductor resembles an insulator except that the gap between valence band and conduction band is relatively small.

bands. Moreover, the number of electrons that can occupy an energy level is limited by quantum physics. So, an electron can become more energetic *only* if it receives enough energy to reach a *vacant* higher energy level, either in the same band or in a higher band.

In a metallic conductor such as copper (Fig. 27-14*a*), the highest band that is occupied by electrons—called the *valence band*—has electrons only in its lower levels. Thus, those electrons can reach abundant vacant levels higher in the band if they receive even a modest amount of energy. We can supply such energy via an electric field applied across the conductor: the field propels some of the valence band electrons along the wire, giving them kinetic energy and thus elevating them to a higher energy level. Hence, these electrons are the conduction electrons that comprise the current through the conductor. Electrons in lower bands cannot participate in the current because the applied electric field cannot provide enough energy for them to reach vacant levels.

In an insulator (Fig. 27-14*b*), the valence band is completely filled. The next higher available vacant levels lie in an empty band (called the *conduction band*) separated from the valence band by a considerable energy gap. No current can occur when an electric field is applied because the field cannot provide enough energy for electrons to jump up to a vacant level.

A semiconductor (Fig. 27-14*c*) is like an insulator except that the energy gap between the conduction band and the valence band is small enough that the probability that electrons might ''jump the gap'' by thermal agitation is not vanishingly small. More important, controlled impurities—deliberately added—can contribute charge carriers to the conduction band. Most semiconducting devices, such as transistors and junction diodes, are fabricated by the selective doping of different regions of the silicon with impurity atoms of different kinds.

Let us now look again at Eq. 27-20, the expression for the resistivity of a conductor, with the band-gap picture in mind:

$$\rho = \frac{m}{e^2 n \tau}, \qquad (27\text{-}24)$$

where n is the number of charge carriers per unit volume and τ is the mean time between collisions of the charge carriers. (We derived this equation for conductors, but it also applies to semiconductors.) Let us consider how the variables n and τ change as the temperature is increased.

In a conductor, n is large but very nearly constant; that is, its value does not change appreciably with temperature. The increase of resistivity with temperature for metals (Fig. 27-10) is caused by an increase in the collision rate of the charge carriers, which shows up in Eq. 27-24 as a decrease in τ, the mean time between collisions.

In a semiconductor, n is small but increases very rapidly with temperature as the increased thermal agitation makes more charge carriers available. This causes the *decrease* of resistivity with increasing temperature, as indicated by the negative temperature coefficient of resistivity for silicon in Table 27-2. The same increase in collision rate that we noted for metals also occurs for semiconductors, but its effect is swamped by the rapid increase in the number of charge carriers.

27-9 SUPERCONDUCTORS

In 1911 Dutch physicist Kamerlingh Onnes discovered that the resistivity of mercury absolutely disappears at temperatures below about 4 K (Fig. 27-15). This phenomenon of **superconductivity** is of vast potential importance in technology because it would be very useful to be able to cause charge to flow through a superconducting conductor without thermal energy losses. Currents created in a superconducting ring, for example, have persisted for several years without diminution; the electrons making up the current require a force and a source of energy at start-up time, but not thereafter.

Prior to 1986, the technological development of superconductivity was throttled by the cost of producing the extremely low temperatures that were required to achieve the effect. In 1986, however, new ceramic materials were discovered that become superconducting at considerably higher (and thus cheaper to produce) temperatures. Practical application of superconducting devices at room temperature may eventually become feasible.

Superconductivity is much different from conductivity. In fact, the best of the normal conductors, such as silver and copper, cannot become superconducting at any temperature, and the new ceramic superconductors are actually insulators when they are not at low enough temperatures to be in a superconducting state.

One explanation for superconductivity is that the electrons that make up the current move in coordinated pairs. One of the electrons in a pair may electrically distort the molecular structure of the superconducting material as it moves through, creating nearby a short-lived concentration

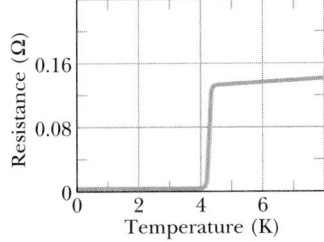

FIGURE 27-15 The resistance of mercury drops to zero at a temperature of about 4 K.

A model of the first transistor, an electronic device using semiconductor materials. Today, many thousands of these devices can be placed on a thin wafer a few millimeters wide.

A disk-shaped magnet is levitated above a superconducting material that has been cooled by liquid nitrogen. The goldfish is along for the ride.

of positive charge. The other electron in the pair may then be attracted toward this positive charge. According to the theory, such coordination between electrons would prevent them from colliding with the molecules and thus would

eliminate electrical resistance. The theory worked well to explain the pre-1986, lower temperature superconductors, but new theories appear to be needed for the newer, higher temperature superconductors.

REVIEW & SUMMARY

Current

An **electric current** i in a conductor is defined by

$$i = \frac{dq}{dt}. \tag{27-1}$$

Here dq is the amount of (positive) charge that passes in time dt through a hypothetical surface that cuts across the conductor. By convention, the direction of electric current is taken as the direction in which positive charge carriers would move. The SI unit of electric current is the **ampere** (A): 1 A = 1 C/s.

Current Density

Current (a scalar) is related to **current density J** (a vector) by

$$i = \int \mathbf{J} \cdot d\mathbf{A}, \tag{27-4}$$

where $d\mathbf{A}$ is a vector perpendicular to a surface element of area dA, and the integral is taken over any surface cutting across the conductor. The direction of $\mathbf{J}$ is that of the electric field causing the current.

Drift Speed of the Charge Carriers

When an electric field **E** is established in a conductor, the charge carriers (assumed positive) acquire a **drift speed** v_d in the direction of **E**; the velocity $\mathbf{v}_d$ is related to the current density by

$$\mathbf{J} = (ne)\mathbf{v}_d, \tag{27-7}$$

where ne is the carrier charge density.

Resistance of a Conductor

The **resistance** R of a conductor is defined as

$$R = \frac{V}{i} \quad \text{(definition of } R\text{)}, \tag{27-8}$$

where V is the potential difference across the conductor and i is the current. The SI unit of resistance is the **ohm** (Ω): 1 Ω = 1 V/A. Similar equations define the **resistivity** ρ and **conductivity** σ of a material:

$$\rho = \frac{1}{\sigma} = \frac{E}{J} \quad \begin{array}{l}\text{(definitions}\\ \text{of } \rho \text{ and } \sigma\text{)},\end{array} \tag{27-12, 27-10}$$

where E is the applied electric field. The SI unit of resistivity is the ohm-meter ($\Omega \cdot$ m). Equation 27-10 corresponds to the vector equation

$$\mathbf{E} = \rho\mathbf{J}. \qquad (27\text{-}11)$$

The resistance R of a conducting wire of length L and uniform cross section is

$$R = \frac{\rho L}{A}, \qquad (27\text{-}16)$$

where A is the cross-sectional area.

Change of ρ with Temperature

The resistivity ρ for most materials changes with temperature. For many materials, including metals, the relation between ρ and temperature T is approximated by the equation is

$$\rho - \rho_0 = \rho_0\alpha(T - T_0). \qquad (27\text{-}17)$$

Here T_0 is a reference temperature, ρ_0 is the resistivity at T_0, and α is a mean temperature coefficient of resistivity.

Ohm's Law

A given *conductor* obeys *Ohm's law* if its resistance R, defined by Eq. 27-8 as V/i, is independent of the applied potential difference V. A given *material* obeys Ohm's law if its resistivity, defined by Eq. 27-10, is independent of the magnitude and direction of the applied electric field $\mathbf{E}$.

Resistivity of a Metal

By assuming that the conduction electrons in a metal are free to move like the molecules of a gas, it is possible to derive an expression for the resistivity of a metal:

$$\rho = \frac{m}{e^2 n\tau}. \qquad (27\text{-}20)$$

Here n is the number of electrons per unit volume and τ is the mean time between the collisions of an electron with the atoms of the metal. We can explain why metals obey Ohm's law by pointing out that τ is essentially independent of E.

Power

The power P, or rate of energy transfer, in an electrical device across which a potential difference V is maintained is

$$P = iV \qquad \begin{array}{l}\text{(rate of electrical} \\ \text{energy transfer).}\end{array} \qquad (27\text{-}21)$$

Resistive Dissipation

If the device is a resistor, we can write Eq. 27-21 as

$$P = i^2R = \frac{V^2}{R} \qquad \begin{array}{l}\text{(resistive} \\ \text{dissipation).}\end{array} \qquad (27\text{-}22, 27\text{-}23)$$

In a resistor, electric potential energy is converted to internal thermal energy via collisions between charge carriers and atoms.

Semiconductors

Semiconductors are materials with few conduction electrons but with available conduction-level states that are close, in energy, to their valence bands. These materials become conductors when they are *doped* with other atoms that contribute electrons to the conduction band.

Superconductors

Superconductors are materials that lose all electrical resistance at low temperatures. Recent research has discovered materials that are superconducting at surprisingly high temperatures.

QUESTIONS

1. Figure 27-16 shows plots of the current i through the cross section of a wire over four different time periods. Rank the periods according to the net charge that passes through the cross section during each, greatest first.

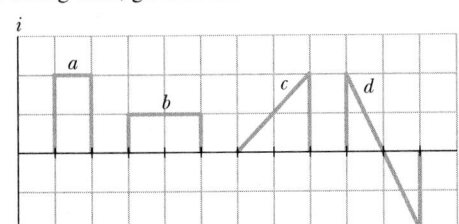

FIGURE 27-16 Question 1.

2. Figure 27-17 shows four situations in which positive and negative charges move horizontally through a region and gives the rate at which each charge moves. Rank the situations according to the effective current through the regions, greatest first.

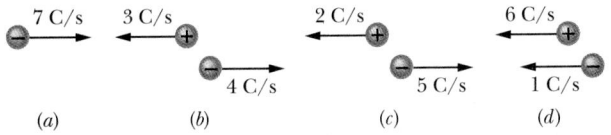

FIGURE 27-17 Question 2.

3. Figure 27-18 shows cross sections through three wires of equal length and of the same material. The figure also gives the length of each side in millimeters. Rank the wires according to their resistances (measured end to end along each wire's length), greatest first.

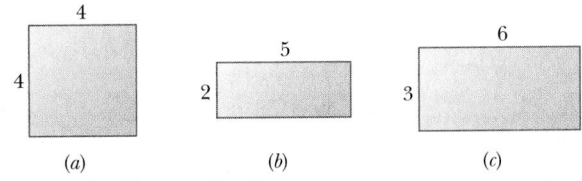

FIGURE 27-18 Question 3.

4. If you stretch a cylindrical wire and it remains cylindrical, does the resistance of the wire (measured end to end along its length) increase, decrease, or remain the same?

5. Figure 27-19 shows cross sections through three long square conductors of the same length and material, and with cross-sectional edge lengths as shown. Conductor B will fit snugly within conductor A, and conductor C will fit snugly within conductor B. Rank the following according to their end-to-end resistances, greatest first: the individual conductors and the combinations of $A + B$, $B + C$, and $A + B + C$.

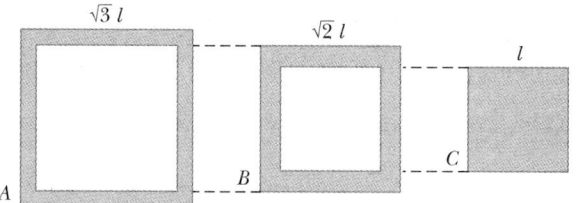

FIGURE 27-19 Question 5.

6. Figure 27-20 shows a rectangular solid conductor of edge lengths L, $2L$, and $3L$. A certain potential difference V is to be applied between pairs of opposite faces of the conductor as in Fig. 27-8b: left–right, top–bottom, and front–back. Rank those pairs according to (a) the magnitude of the electric field within the conductor, (b) the current density within the conductor, (c) the current through the conductor, and (d) the drift speed of the electrons through the conductor, greatest first.

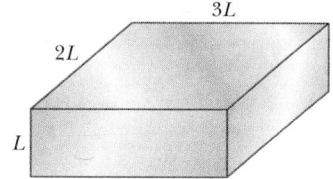

FIGURE 27-20 Question 6.

7. The following table gives the lengths of three copper rods, their diameters, and the potential differences between their ends. Rank the rods according to (a) the magnitude of the electric field within them, (b) the current density within them, and (c) the drift speed of electrons through them, greatest first.

ROD	LENGTH	DIAMETER	POTENTIAL DIFFERENCE
1	L	$3d$	V
2	$2L$	d	$2V$
3	$3L$	$2d$	$2V$

8. The following table gives the conductivity and the density of electrons for materials A, B, C, and D. Rank the materials according to the average time between collisions of the conduction electrons in the materials, greatest first.

	A	B	C	D
Conductivity	σ	2σ	2σ	σ
Electrons/m³	n	$2n$	n	$2n$

9. Three wires, of the same diameter, are connected in turn between two points maintained at a constant potential difference. Their resistivities and lengths are ρ and L (wire A), 1.2ρ and $1.2L$ (wire B), and 0.9ρ and L (wire C). Rank the wires according to the rate at which energy is transferred to thermal energy within them, greatest first.

10. In Fig. 27-21a, battery B_1 is recharging battery B_2. The current through B_2 and the potential across B_2 may be (a) 3 A and 4 V, (b) 2 A and 5 V, or (c) 6 A and 2 V. Rank these pairs of values according to the rate at which electrical energy is transferred from B_1 to B_2, greatest first.

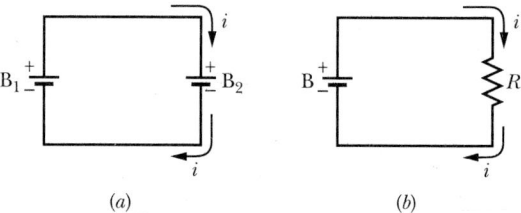

FIGURE 27-21 Questions 10 and 11.

11. In three situations, a battery B and a resistor of resistance R are connected as in Fig. 27-21b. The values of R and the current through the resistor in the three situations are (a) 4 Ω and 2 A, (b) 3 Ω and 3 A, and (c) 3 Ω and 2 A. Rank the situations according to the rate at which electrical energy is transferred to thermal energy in the resistor, greatest first.

12. Is the filament resistance lower or higher in a 500 W lightbulb than in a 100 W bulb? (The same potential difference is applied to them.)

13. Figure 27-22 gives the resistivities of four materials as a function of temperature. (a) Which materials are conductors and which semiconductors? In which materials does an increase in temperature result in (b) an increase in the number of conduction electrons per unit volume and (c) an increase in the collision rate of conduction electrons?

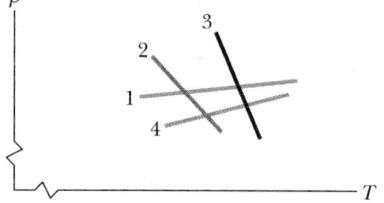

FIGURE 27-22 Question 13.

EXERCISES & PROBLEMS

SECTION 27-2 Electric Current

1E. The current in the electron beam producing a picture on a typical video display terminal is 200 μA. How many electrons strike the screen each second?

2E. A current of 5.0 A exists in a 10 Ω resistor for 4.0 min. How many (a) coulombs and (b) electrons pass through any cross section of the resistor in this time?

3P. A charged belt, 50 cm wide, travels at 30 m/s between a source of charge and a sphere. The belt carries charge into the sphere at a rate corresponding to 100 μA. Compute the surface charge density on the belt.

4P. An isolated conducting sphere has a 10 cm radius. One wire carries a current of 1.000 002 0 A into it. Another wire carries a current of 1.000 000 0 A out of it. How long would it take for the sphere to increase in potential by 1000 V?

SECTION 27-3 Current Density

5E. The (United States) National Electric Code, which sets maximum safe currents for rubber-insulated copper wires of various diameters, is given (in part) below. Plot the safe current density as a function of diameter. Which wire gauge has the maximum safe current density?

Gauge[a]	4	6	8	10	12	14	16	18
Diameter, mils[b]	204	162	129	102	81	64	51	40
Safe current, A	70	50	35	25	20	15	6	3

[a]A way of identifying the wire diameter.

[b]1 mil = 10^{-3} in.

6E. A beam contains 2.0×10^8 doubly charged positive ions per cubic centimeter, all of which are moving north with a speed of 1.0×10^5 m/s. (a) What are the magnitude and direction of the current density $\mathbf{J}$? (b) Can you calculate the total current i in this ion beam? If not, what additional information is needed?

7E. A small but measurable current of 1.2×10^{-10} A exists in a copper wire whose diameter is 2.5 mm. Assuming the current is uniform, calculate (a) the current density and (b) the electron drift speed. (See Sample Problem 27-3.)

8E. A fuse in an electric circuit is a wire that is designed to melt, and thereby open the circuit, if the current exceeds a predetermined value. Suppose that the material to be used in a fuse melts when the current density rises to 440 A/cm². What diameter of cylindrical wire should be used to limit the current to 0.50 A?

9E. A current is established in a gas discharge tube when a sufficiently high potential difference is applied across the two electrodes in the tube. The gas ionizes; electrons move toward the positive terminal and singly charged positive ions toward the negative terminal. What are the magnitude and direction of the current in a hydrogen discharge tube in which 3.1×10^{18} elec-

trons and 1.1×10^{18} protons move past a cross-sectional area of the tube each second?

10E. A *pn* junction is formed from two different semiconducting materials in the form of identical cylinders with radius 0.165 mm, as depicted in Fig. 27-23. In one application 3.50×10^{15} electrons per second flow across the junction from the *n* to the *p* side while 2.25×10^{15} holes per second flow from the *p* to the *n* side. (A hole acts like a particle with charge $+1.60 \times 10^{-19}$ C.) What are (a) the total current and (b) the current density?

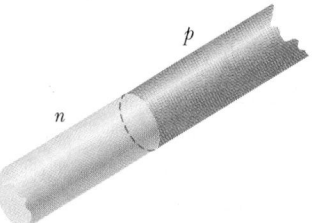

FIGURE 27-23
Exercise 10.

11P. Near Earth, the density of protons in the solar wind is 8.70 cm⁻³ and their speed is 470 km/s. (a) Find the current density of these protons. (b) If Earth's magnetic field did not deflect them, the protons would strike the planet. What total current would Earth then receive?

12P. A steady beam of alpha particles ($q = +2e$) traveling with constant kinetic energy 20 MeV carries a current of 0.25 μA. (a) If the beam is directed perpendicular to a plane surface, how many alpha particles strike the surface in 3.0 s? (b) At any instant, how many alpha particles are there in a given 20 cm length of the beam? (c) Through what potential difference was it necessary to accelerate each alpha particle from rest to bring it to an energy of 20 MeV?

13P. How long does it take electrons to get from a car battery to the starting motor? Assume the current is 300 A and the electrons travel through a copper wire with cross-sectional area 0.21 cm² and length 0.85 m. (See Sample Problem 27-3.)

14P. In a hypothetical fusion research lab, high-temperature helium gas is completely ionized, each helium atom being separated into two free electrons and the remaining positively charged nucleus (alpha particle). An applied electric field causes the alpha particles to drift to the east at 25 m/s while the electrons drift to the west at 88 m/s. The alpha particle density is 2.8×10^{15} cm⁻³. Calculate the net current density; specify the current direction.

15P. (a) The current density across a cylindrical conductor of radius R varies according to the equation

$$J = J_0\left(1 - \frac{r}{R}\right),$$

where r is the distance from the central axis. Thus the current density is a maximum J_0 at the axis ($r = 0$) and decreases linearly to zero at the surface ($r = R$). Calculate the current in terms of J_0 and the conductor's cross-sectional area $A = \pi R^2$. (b) Suppose that, instead, the current density is a maximum J_0 at the cylinder's surface and decreases linearly to zero at the axis: $J = J_0 r/R$. Calculate the current. Why is the result different from that in (a)?

SECTION 27-4   Resistance and Resistivity

16E. A steel trolley-car rail has a cross-sectional area of 56.0 cm². What is the resistance of 10.0 km of rail? The resistivity of the steel is $3.00 \times 10^{-7} \; \Omega \cdot m$.

17E. A conducting wire has a 1.0 mm diameter, a 2.0 m length, and a 50 mΩ resistance. What is the resistivity of the material?

18E. A wire of Nichrome (a nickel–chromium–iron alloy commonly used in heating elements) is 1.0 m long and 1.0 mm² in cross-sectional area. It carries a current of 4.0 A when a 2.0 V potential difference is applied between its ends. Calculate the conductivity σ of Nichrome.

19E. A human being can be electrocuted if a current as small as 50 mA passes near the heart. An electrician working with sweaty hands makes good contact with the two conductors he is holding. If his resistance is 2000 Ω, what might the fatal voltage be?

20E. A coil is formed by winding 250 turns of insulated 16-gauge copper wire (diameter = 1.3 mm) in a single layer on a cylindrical form of radius 12 cm. What is the resistance of the coil? Neglect the thickness of the insulation. (Use Table 27-1.)

21E. A wire 4.00 m long and 6.00 mm in diameter has a resistance of 15.0 mΩ. A potential difference of 23.0 V is applied between the ends. (a) What is the current in the wire? (b) What is the current density? (c) Calculate the resistivity of the wire material. Identify the material. (Use Table 27-1.)

22E. The copper windings of a motor have a resistance of 50 Ω at 20°C when the motor is idle. After the motor has run for several hours, the resistance rises to 58 Ω. What is the temperature of the windings now? Ignore changes in the dimensions of the windings. (Use Table 27-1.)

23E. (a) At what temperature would the resistance of a copper conductor be double its resistance at 20.0°C? (Use 20.0°C as the reference point in Eq. 27-17; compare your answer with Fig. 27-10.) (b) Does this same "doubling temperature" hold for all copper conductors, regardless of shape or size?

24E. Using data taken from Fig. 27-11c, plot the resistance of the *pn* junction diode as a function of applied potential difference.

25E. A 4.0 cm long caterpillar crawls in the direction of electron drift along a 5.2 mm diameter bare copper wire that carries a current of 12 A. (a) What is the potential difference between the two ends of the caterpillar? (b) Is its tail positive or negative compared to its head? (c) How much time would the caterpillar take to crawl 1.0 cm if it crawls at the drift speed of the electrons in the wire?

26E. A cylindrical copper rod of length L and cross-sectional area A is re-formed to twice its original length with no change in volume. (a) Find the new cross-sectional area. (b) The resistance between its ends was R; what is it now?

27E. A wire with a resistance of 6.0 Ω is drawn out through a die so that its new length is three times its original length. Find the resistance of the longer wire, assuming that the resistivity and density of the material are unchanged.

28E. A certain wire has a resistance R. What is the resistance of a second wire, made of the same material, that is half as long and has half the diameter?

29P. Two conductors are made of the same material and have the same length. Conductor A is a solid wire of diameter 1.0 mm. Conductor B is a hollow tube of outside diameter 2.0 mm and inside diameter 1.0 mm. What is the resistance ratio R_A/R_B, measured between their ends?

30P. A copper wire and an iron wire of the same length have the same potential difference applied to them. (a) What must be the ratio of their radii if the currents in the two wires are to be the same? (b) Can the current densities be made the same by suitable choices of the radii?

31P. An aluminum rod with a square cross section is 1.3 m long and 5.2 mm on edge. (a) What is the resistance between its ends? (b) What must be the diameter of a cylindrical copper rod of length 1.3 m if its resistance is to be the same as that of the aluminum rod?

32P. A cylindrical metal rod is 1.60 m long and 5.50 mm in diameter. The resistance between its two ends (at 20°C) is $1.09 \times 10^{-3} \; \Omega$. (a) What is the material? (b) A round disk, 2.00 cm in diameter and 1.00 mm thick, is formed of the same material. What is the resistance between the round faces, assuming that each face is an equipotential surface?

33P. An electrical cable consists of 125 strands of fine wire, each having 2.65 μΩ resistance. The same potential difference is applied between the ends of all the strands and results in a total current of 0.750 A. (a) What is the current in each strand? (b) What is the applied potential difference? (c) What is the resistance of the cable?

34P. When 115 V is applied across a wire that is 10 m long and has a 0.30 mm radius, the current density is $1.4 \times 10^4 \; A/m^2$. Find the resistivity of the wire.

35P. A common flashlight bulb is rated at 0.30 A and 2.9 V (the values of the current and voltage under operating conditions). If the resistance of the bulb filament at room temperature (20°C) is 1.1 Ω, what is the temperature of the filament when the bulb is on? The filament is made of tungsten.

36P. A block in the shape of a rectangular solid has a cross-sectional area of 3.50 cm² across its width, a front-to-rear length of 15.8 cm, and a resistance of 935 Ω. The material of which the block is made has 5.33×10^{22} conduction electrons/m³. A potential difference of 35.8 V is maintained between its front and rear. (a) What is the current in the block? (b) If the current density is uniform, what is its value? (c) What is the drift velocity of the conduction electrons? (d) What is the magnitude of the electric field in the block?

37P. Copper and aluminum are being considered for a high-voltage transmission line that must carry a current of 60.0 A. The resistance per unit length is to be 0.150 Ω/km. Compute for each choice of cable material (a) the current density and (b) the mass per meter of the cable. The densities of copper and aluminum are 8960 and 2700 kg/m^3, respectively.

38P. In Earth's lower atmosphere there are negative and positive ions, created by radioactive elements in the soil and cosmic rays from space. In a certain region, the atmospheric electric field strength is 120 V/m, directed vertically down. This field causes singly charged positive ions, 620 per cm^3, to drift downward and singly charged negative ions, 550 per cm^3, to drift upward (Fig. 27-24). The measured conductivity is $2.70 \times 10^{-14}/\Omega \cdot m$. Calculate (a) the ion drift speed, assumed to be the same for positive and negative ions, and (b) the current density.

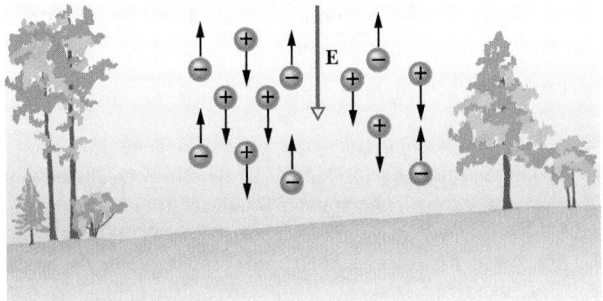

FIGURE 27-24 Problem 38.

39P. If the gauge number of a wire is increased by 6, the diameter is halved; if a gauge number is increased by 1, the diameter decreases by the factor $2^{1/6}$ (see the table in Exercise 5). Knowing this, and knowing that 1000 ft of 10-gauge copper wire has a resistance of approximately 1.00 Ω, estimate the resistance of 25 ft of 22-gauge copper wire.

40P. When a metal rod is heated, not only its resistance but also its length and its cross-sectional area change. The relation $R = \rho L/A$ suggests that all three factors should be taken into account in measuring ρ at various temperatures. (a) If the temperature changes by 1.0 C°, what percentage changes in R, L, and A occur for a copper conductor? The coefficient of linear expansion is $1.7 \times 10^{-5}/K$. (b) What conclusion do you draw?

41P. A resistor has the shape of a truncated right-circular cone (Fig. 27-25). The end radii are a and b, and the altitude is L. If the taper is small, we may assume that the current density is uniform across any cross section. (a) Calculate the resistance of this object. (b) Show that your answer reduces to $\rho(L/A)$ for the special case of zero taper (that is, for $a = b$).

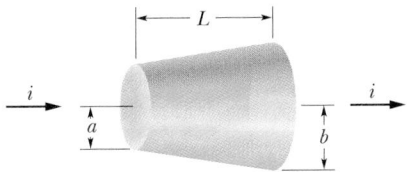

FIGURE 27-25 Problem 41.

SECTION 27-6 A Microscopic View of Ohm's Law

42P. Show that according to the free-electron model of electrical conduction in metals and classical physics, the resistivity of metals should be proportional to $\sqrt{T}$, where T is the temperature in kelvins. (See Eq. 20-27.)

SECTION 27-7 Power in Electric Circuits

43E. A student kept his 9.0 V, 7.0 W radio turned on at full volume from 9:00 p.m. until 2:00 a.m. How much charge went through it?

44E. A certain x-ray tube operates at a current of 7.0 mA and a potential difference of 80 kV. What is its power in watts?

45E. Thermal energy is developed in a resistor at a rate of 100 W when the current is 3.00 A. What is the resistance?

46E. The headlights of a moving car draw about 10 A from the 12 V alternator, which is driven by the engine. Assume the alternator is 80% efficient (its output electrical power is 80% of its input mechanical power), and calculate the horsepower the engine must supply to run the lights.

47E. A 120 V potential difference is applied to a space heater whose resistance is 14 Ω when hot. (a) At what rate is electrical energy transferred to heat? (b) At 5.0¢/kW·h, what does it cost to operate the device for 5.0 h?

48E. A 120 V potential difference is applied to a space heater that dissipates 500 W during operation. (a) What is its resistance during operation? (b) At what rate do electrons flow through any cross section of the heater element?

49E. An unknown resistor is connected between the terminals of a 3.00 V battery. Energy is dissipated in the resistor at the rate of 0.540 W. The same resistor is then connected between the terminals of a 1.50 V battery. At what rate is energy now dissipated?

50E. The National Board of Fire Underwriters has fixed safe current-carrying capacities for various sizes and types of wire. For 10-gauge rubber-coated copper wire (diameter = 0.10 in.), the maximum safe current is 25 A. At this current, find (a) the current density, (b) the electric field, (c) the potential difference across 1000 ft of wire, and (d) the rate at which thermal energy is developed in 1000 ft of wire.

51E. A potential difference of 1.20 V will be applied to a 33.0 m length of 18-gauge copper wire (diameter = 0.0400 in.). Calculate (a) the current, (b) the current density, (c) the electric field, and (d) the rate at which thermal energy will appear in the wire.

52P. A potential difference V is applied to a wire of cross section A, length L, and resistivity ρ. You want to change the applied potential difference and stretch the wire so that the energy dissipation rate is multiplied by 30 and the current is multiplied by 4. What should be the new values of L and A?

53P. A cylindrical resistor of radius 5.0 mm and length 2.0 cm is made of material that has a resistivity of 3.5×10^{-5} $\Omega \cdot m$. What are (a) the current density and (b) the potential difference when the energy dissipation rate in the resistor is 1.0 W?

54P. A heating element is made by maintaining a potential difference of 75.0 V along the length of a Nichrome wire with a $2.60 \times 10^{-6} \ m^2$ cross section and a resistivity of $5.00 \times 10^{-7} \ \Omega \cdot m$. (a) If the element dissipates 5000 W, what is its length? (b) If a potential difference of 100 V is used to obtain the same dissipation rate, what should the length be?

55P. A 100 W lightbulb is plugged into a standard 120 V outlet. (a) How much does it cost per month to leave the light turned on continuously? Assume electrical energy costs 6¢/kW·h. (b) What is the resistance of the bulb? (c) What is the current in the bulb? (d) Is the resistance different when the bulb is turned off?

56P. A 1250 W radiant heater is constructed to operate at 115 V. (a) What will be the current in the heater? (b) What is the resistance of the heating coil? (c) How much thermal energy is generated in 1.0 h by the heater?

57P. A Nichrome heater dissipates 500 W when the applied potential difference is 110 V and the wire temperature is 800°C. What would be the dissipation rate if the wire temperature were held at 200°C by immersing the wire in a bath of cooling oil? The applied potential difference remains the same, and α for Nichrome at 800°C is $4.0 \times 10^{-4}/K$.

58P. A beam of 16 MeV deuterons from a cyclotron falls on a copper block. The beam is equivalent to a current of 15 μA. (a) At what rate do deuterons strike the block? (b) At what rate is thermal energy produced in the block?

59P. A linear accelerator produces a pulsed beam of electrons. The pulse current is 0.50 A, and the pulse duration is 0.10 μs. (a) How many electrons are accelerated per pulse? (b) What is the average current for a machine operating at 500 pulses/s? (c) If the electrons are accelerated to an energy of 50 MeV, what are the average and peak powers of the accelerator?

60P. A coil of current-carrying Nichrome wire is immersed in a liquid contained in a calorimeter. When the potential difference across the coil is 12 V and the current through the coil is 5.2 A, the liquid boils at a steady rate, evaporating at the rate of 21 mg/s. Calculate the heat of vaporization of the liquid, in joules per kilogram (see Section 19-7).

61P. In Fig. 27-26, a resistance coil, wired to an external battery, is placed inside a thermally insulated cylinder fitted with a frictionless piston and containing an ideal gas. A current i = 240 mA exists in the coil, which has a resistance R = 550 Ω. At what speed v must the piston, of mass m = 12 kg, move upward to keep the temperature of the gas unchanged?

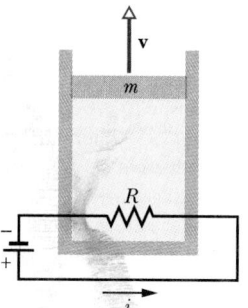

FIGURE 27-26   Problem 61.

62P. A 500 W heating unit is designed to operate with an applied potential difference of 115 V. (a) By what percentage will its heat output drop if the applied potential difference drops to 110 V? Assume no change in resistance. (b) If you took the variation of resistance with temperature into account, would the actual drop in heat output be larger or smaller than that calculated in (a)?

Electronic Computation

63. The resistance of a resistor is measured at several temperatures, as shown below. Enter the data in your graphing calculator and perform a linear regression fit of R versus T. Have your calculator graph the results of the linear regression fit; using the TRACE capability of the calculator (and perhaps the parameters of the fit), find the value of the resistance (a) at 20°C and (b) at 0°C. (c) Find the temperature coefficient *of resistance* (instead of resistivity) with a reference temperature of 20°C. (d) Find the temperature coefficient of resistance with a reference temperature of 0°C. (e) Find the resistance of the resistor at 265°C.

T, °C	50	100	150	200	250	300
R, Ω	139	171	203	234	266	298

28
Circuits

The electric eel (Electrophorus) lurks in rivers of South America, killing the fish on which it preys with pulses of current. It does so by producing a potential difference of several hundred volts along its length; the resulting current in the surrounding water, from near the eel's head to the tail region, can be as much as one ampere. If you were to brush up against this eel while swimming, you might wonder (after recovering from the very painful stun): How can the creature manage to produce a current that large without shocking itself?

28-1 "PUMPING" CHARGES

If you want to make charge carriers flow through a resistor, you must establish a potential difference between the ends of the device. One way to do this is to connect each end of the resistor to a separate conducting sphere, with one sphere charged negatively and the other positively. The trouble with this scheme is that the flow of charge acts to discharge the spheres, bringing them quickly to the same potential. When that happens, the flow of charge stops.

To produce a steady flow of charge, you need a "charge pump," a device that—by doing work on the charge carriers—maintains a potential difference between a pair of terminals. We call such a device an **emf device,** and the device is said to provide an **emf** $\mathscr{E}$, which means that it does work on charge carriers. An emf device is sometimes called a *seat of emf.* The term *emf* comes from the outdated phrase *electromotive force,* which was adopted before scientists clearly understood the function of an emf device.

A common emf device is the *battery,* used to power devices from wristwatches to submarines. The emf device that most influences our daily lives, however, is the *electric generator,* which by means of electrical lines from a generating plant, creates a potential difference in our homes and workplaces. The emf devices known as *solar cells,* long familiar as the winglike panels on spacecraft, also dot the countryside for domestic applications. Less familiar emf devices are the *fuel cells* that power the space shuttles and the *thermopiles* that provide onboard electrical power for some spacecraft and for remote stations in Antarctica and elsewhere. An emf device does not have to be an instrument: living systems, ranging from electric eels and human beings to plants, have physiological emf devices.

Although the devices we have listed differ widely in their modes of operation, they all perform the same basic function: they do work on charge carriers and thus maintain a potential difference between their terminals.

28-2 WORK, ENERGY, AND EMF

Figure 28-1 shows an emf device (consider it to be a battery) that is part of a simple circuit. The device keeps one

terminal (called the positive terminal and often labeled $+$) at a higher electric potential than the other terminal (called the negative terminal and labeled $-$). We can represent the emf of the device with an arrow that points from the negative terminal toward the positive terminal as in Fig. 28-1. This is the direction in which the device causes positive charge carriers (which make up a current) to move through itself. The device also produces a current around the circuit in the same direction (clockwise in Fig. 28-1). A small circle on the emf arrow distinguishes it from the arrows that indicate current direction.

Within the emf device, positive charge carriers move from a region of low electric potential and thus low electric potential energy (at the negative terminal) to a region of higher electric potential and higher electric potential energy (at the positive terminal). This motion is just the opposite of what the electric field between the terminals (which points from the positive terminal toward the negative terminal) would cause the charge carriers to do.

So there must be some source of energy within the device, enabling it to do work on the charges and thus

The world's largest battery, housed in Chino, California, has a power capability of 10 MW, which is put to use during peak power demands on the electric system served by Southern California Edison. Because the battery does work on charge carriers, it is an emf device.

FIGURE 28-1 A simple electric circuit, in which a device of emf $\mathscr{E}$ does work on the charge carriers and maintains a steady current i through the resistor.

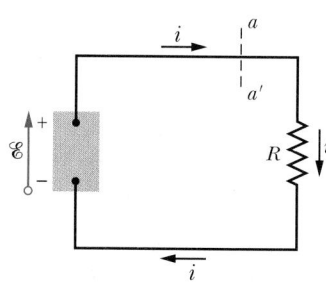

forcing them to move as they do. The energy source may be chemical, as in a battery or a fuel cell. It may involve mechanical forces, as in an electric generator. Temperature differences may supply the energy, as in a thermopile; or the Sun may supply it, as in a solar cell.

Let us now analyze the circuit of Fig. 28-1 from the point of view of work and energy transfers. In any time interval dt, a charge dq passes through any cross section of this circuit, such as aa'. This same amount of charge must enter the emf device at its low-potential end and leave at its high-potential end. The device must do an amount of work dW on the charge dq to force it to move in this way. We define the emf of the emf device in terms of this work:

$$\mathcal{E} = \frac{dW}{dq} \qquad \text{(definition of } \mathcal{E}\text{)}. \qquad (28\text{-}1)$$

In words, the emf of an emf device is the work per unit charge that the device does in moving charge from its low-potential terminal to its high-potential terminal. The SI unit for emf is the joule per coulomb; in Chapter 25 we defined that unit as the *volt*.

An **ideal emf device** is one that lacks any internal resistance to the internal movement of charge from terminal to terminal. The potential difference between the terminals of an ideal emf device is equal to the emf of the device. For example, an ideal battery with an emf of 12.0 V always has a potential difference of 12.0 V between its terminals.

A **real emf device,** such as any real battery, has internal resistance to the internal movement of charge. When a real emf device is not connected to a circuit, and thus does not have current through it, the potential difference between its terminals is equal to its emf. But when that device has current through it, the potential difference between its terminals differs from its emf. We will discuss such real batteries in Section 28-4.

When an emf device is connected to a circuit, the device transfers energy to the charge carriers passing through it. This energy can then be transferred from the charge carriers to other devices in the circuit, for example, to light a bulb. Figure 28-2a shows a circuit containing two ideal rechargeable (*storage*) batteries A and B, a resistor R, and an electric motor M that can lift an object by using energy it obtains from charge carriers in the circuit. Note that the batteries are connected so that they tend to send charges around the circuit in opposite directions. The actual direction of the current in the circuit is determined by the battery with the larger emf, which happens to be battery B. So the chemical energy within battery B is decreasing as energy is transferred to the charge carriers passing through it. But the chemical energy within battery A is increasing

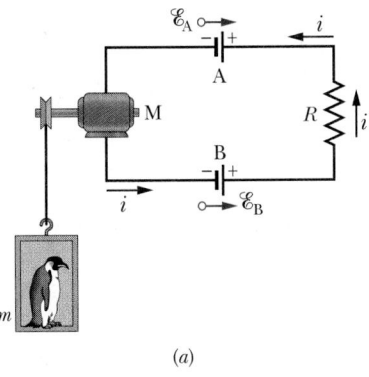

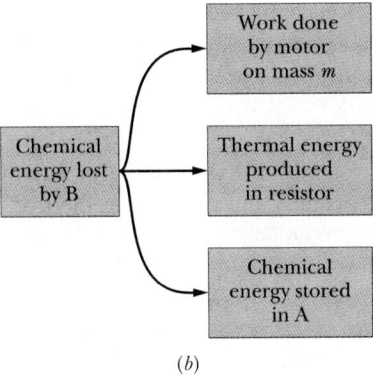

FIGURE 28-2 (a) In the circuit, $\mathcal{E}_B > \mathcal{E}_A$; so battery B determines the direction of the current. (b) The energy transfers in the circuit, assuming that no dissipation occurs in the motor.

because the current in it is directed from the positive terminal to the negative terminal. Thus battery B is charging battery A. Battery B is also providing energy to motor M and energy that is being dissipated in resistor R. Figure 28-2b shows all three energy transfers from battery B; each decreases that battery's chemical energy.

28-3 CALCULATING THE CURRENT IN A SINGLE-LOOP CIRCUIT

We discuss here two equivalent ways to calculate the current in the simple *single-loop* circuit of Fig. 28-3; one method is based on energy conservation considerations and

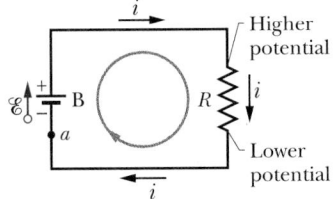

FIGURE 28-3 A single-loop circuit in which a resistance R is connected across an ideal battery B with emf $\mathcal{E}$. The resulting current i is the same throughout the circuit.

the other on the concept of potential. The circuit consists of an ideal battery B with emf $\mathscr{E}$, a resistor of resistance R, and two connecting wires. (Unless otherwise indicated, we assume that wires in circuits have negligible resistance. Their function, then, is merely to provide pathways along which charge carriers can move.)

Energy Method

Equation 27-22, $P = i^2R$, tells us that in a time interval dt an amount of energy given by $i^2R\,dt$ will appear in the resistor of Fig. 28-3 as thermal energy. (Since we assume the wires to have negligible resistance, no thermal energy will appear in them.) During the same interval, a charge $dq = i\,dt$ will have moved through battery B, and the battery will have done work on this charge, according to Eq. 28-1, equal to

$$dW = \mathscr{E}\,dq = \mathscr{E}i\,dt.$$

From the principle of conservation of energy, the work done by the battery must equal the thermal energy that appears in the resistor:

$$\mathscr{E}i\,dt = i^2R\,dt.$$

This gives us

$$\mathscr{E} = iR,$$

which in words means the following: emf $\mathscr{E}$ is the energy per unit charge transferred to the moving charges by the battery. The quantity iR is the energy per unit charge transferred *from* the moving charges to thermal energy within the resistor. The energy per unit charge transferred to the moving charges is equal to the energy per unit charge transferred from them. Solving for i, we find

$$i = \frac{\mathscr{E}}{R}. \tag{28-2}$$

Potential Method

Suppose we start at any point in the circuit of Fig. 28-3 and mentally proceed around the circuit in either direction, adding algebraically the potential differences that we encounter. When we arrive at our starting point, we must have returned to our starting potential. Before actually doing so, we shall formalize this idea in a statement that holds not only for single-loop circuits such as that of Fig. 28-3 but for any complete loop in a *multiloop* circuit, as we shall discuss in Section 28-6:

LOOP RULE: The algebraic sum of the changes in potential encountered in a complete traversal of any loop of a circuit must be zero.

This is often referred to as *Kirchhoff's loop rule* (or *Kirchhoff's voltage law*), after German physicist Gustav Robert Kirchhoff. This rule is equivalent to saying that any point on the side of a mountain must have a unique elevation above sea level. If you start from any point and return to it after walking around the mountain, the algebraic sum of the changes in elevation that you encounter must be zero.

In Fig. 28-3, let us start at point a, whose potential is V_a, and mentally walk clockwise around the circuit until we are back at a, keeping track of potential changes as we move. Our starting point is at the low-potential terminal of the battery. Since the battery is ideal, the potential difference between its terminals is equal to $\mathscr{E}$. So when we pass through the battery to the high-potential terminal, the change in potential is $+\mathscr{E}$.

As we walk along the top wire to the top end of the resistor, there is no potential change because the wire has negligible resistance: it is at the same potential as the high-potential terminal of the battery. So too is the top end of the resistor. When we pass through the resistor, however, the change in potential is $-iR$.

We return to point a along the bottom wire. Since this wire also has negligible resistance, we again find no potential change. Back at point a, the potential is again V_a. Because we traversed a complete loop, our initial potential, as modified for potential changes along the way, must be equal to our final potential; that is,

$$V_a + \mathscr{E} - iR = V_a.$$

The value of V_a cancels from this equation, which becomes

$$\mathscr{E} - iR = 0.$$

Solving this equation for i gives us the same result, $i = \mathscr{E}/R$, as the energy method (Eq. 28-2).

If we apply the loop rule to a complete *counterclockwise* walk around the circuit, the rule gives us

$$-\mathscr{E} + iR = 0$$

and we again find that $i = \mathscr{E}/R$. Thus you may mentally circle a loop in either direction to apply the loop rule.

To prepare for circuits more complex than that of Fig. 28-3, let us set down two rules for finding potential differences as we move around a loop:

RESISTANCE RULE: For a move through a resistance in the direction of the current, the change in potential is $-iR$; in the opposite direction it is $+iR$.

EMF RULE: For a move through an ideal emf device in the direction of the emf arrow, the change in potential is $+\mathscr{E}$; in the opposite direction it is $-\mathscr{E}$.

CHECKPOINT 1: The figure shows the current i in a single-loop circuit with a battery B and a resistance R (and wires of negligible resistance). (a) Should the emf arrow at B be drawn leftward or rightward? At points a, b, and c, rank (b) the magnitude of the current, (c) the electric potential, and (d) the electric potential energy of the charge carriers, greatest first.

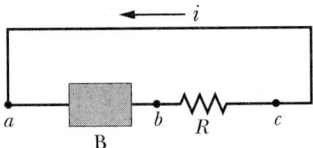

28-4 OTHER SINGLE-LOOP CIRCUITS

In this section we extend the simple circuit of Fig. 28-3 in two ways.

Internal Resistance

Figure 28-4a shows a real battery, with an internal resistance r, wired to an external resistor of resistance R. The internal resistance of the battery is the electrical resistance of the conducting materials of the battery and thus is an unremovable feature of the battery. In Fig. 28-4a, however, the battery is drawn as if it could be separated into an ideal battery with emf $\mathscr{E}$ and a resistor of resistance r. The order in which the symbols for these separated parts are drawn does not matter.

If we apply the loop rule clockwise beginning at point a, we obtain

$$\mathscr{E} - ir - iR = 0. \qquad (28\text{-}3)$$

Solving for the current, we find

$$i = \frac{\mathscr{E}}{R + r}. \qquad (28\text{-}4)$$

Note that this equation reduces to Eq. 28-2 if the battery is ideal, that is, if $r = 0$.

Figure 28-4b shows graphically the changes in electric potential around the circuit. (To better link Fig. 28-4b with the *closed circuit*, imagine curling the graph into a cylinder with point a at the left overlapping point a at the right.) Note how traversing the circuit is like walking around a (potential) mountain and returning to your starting point —you also return to the starting elevation.

In this book, when a battery is not described as real or if no internal resistance is indicated, you can generally assume that it is ideal. But, of course, in the real world, batteries are always real and have internal resistance.

Resistances in Series

Figure 28-5a shows three resistances connected **in series** to an ideal battery of emf $\mathscr{E}$. The battery applies a potential difference $V = \mathscr{E}$ across the three-resistance combination.

Connected resistances are said to be in series when a potential difference that is applied across their combination is the sum of the resulting potential differences across all the resistances.

In less formal language, this definition often means that the resistances occur one after another along a single path for the current, as in Fig. 28-5a.

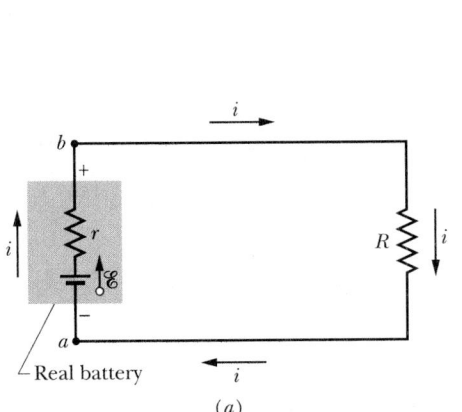

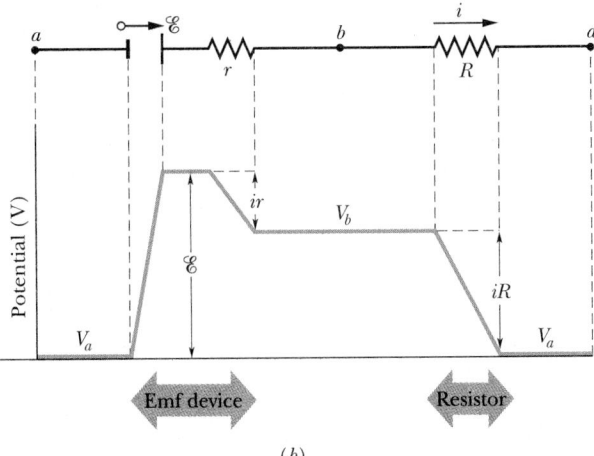

(a)

(b)

FIGURE 28-4 (a) A single-loop circuit containing a real battery having internal resistance r and emf $\mathscr{E}$. (b) The circuit is shown spread out at the top. The potentials encountered in traversing the circuit clockwise from a are shown in the graph. The potential V_a is arbitrarily assigned a value of zero, and other potentials in the circuit are graphed relative to V_a.

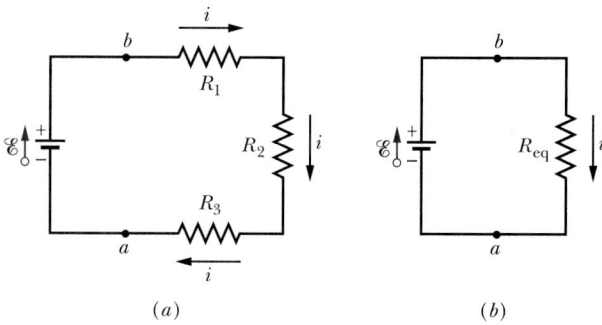

FIGURE 28-5 (a) Three resistors are connected in series between points a and b. (b) An equivalent circuit, with the three resistors replaced with their equivalent resistance R_{eq}.

We seek the single resistance R_{eq} that is equivalent to the three-resistance series combination of Fig. 28-5a. By *equivalent*, we mean that R_{eq} can replace the combination without changing the current i through the combination or the potential difference between a and b. Let us apply the loop rule, starting at terminal a and going clockwise around the circuit. We find

$$\mathscr{E} - iR_1 - iR_2 - iR_3 = 0,$$

or

$$i = \frac{\mathscr{E}}{R_1 + R_2 + R_3}. \tag{28-5}$$

If we replaced the three resistances with a single equivalent resistance R_{eq}, we would have (Fig. 28-5b)

$$i = \frac{\mathscr{E}}{R_{eq}}. \tag{28-6}$$

Comparison of Eqs. 28-5 and 28-6 shows that

$$R_{eq} = R_1 + R_2 + R_3.$$

The extension to n resistances is straightforward and is

$$R_{eq} = \sum_{j=1}^{n} R_j \qquad (n \text{ resistances in series}). \tag{28-7}$$

Note that when resistances are in series, their equivalent resistance is greater than any of the individual resistances.

28-5 POTENTIAL DIFFERENCES

We often want to find the potential difference between two points in a circuit. In Fig. 28-4a, for example, what is the potential difference between points b and a? To find out, let us start at point b and traverse the circuit clockwise to point a, passing through resistor R. If V_a and V_b are the potentials at a and b, respectively, we have

$$V_b - iR = V_a$$

because (according to our resistance rule) we experience a decrease in potential in going through a resistance in the direction of the current. We rewrite this as

$$V_b - V_a = +iR, \tag{28-8}$$

which tells us that point b is at greater potential than point a. Combining Eq. 28-8 with Eq. 28-4, we have

$$V_b - V_a = \mathscr{E} \frac{R}{R + r}, \tag{28-9}$$

where again r is the internal resistance of the emf device.

To find the potential difference between any two points in a circuit, start at one point and traverse the circuit to the other, following any path, and add algebraically the changes in potential that you encounter.

Let us again calculate $V_b - V_a$, starting again at point b but this time proceeding counterclockwise to a through the battery. We have

$$V_b + ir - \mathscr{E} = V_a$$

or

$$V_b - V_a = \mathscr{E} - ir. \tag{28-10}$$

Combining this with Eq. 28-4 again leads to Eq. 28-9.

The quantity $V_b - V_a$ in Fig. 28-4 is the potential difference of the battery across the battery terminals. As noted earlier, $V_b - V_a$ is equal to the emf $\mathscr{E}$ of the battery only if the battery has no internal resistance ($r = 0$ in Eq. 28-9) or if the circuit is open ($i = 0$ in Eq. 28-10).

Suppose that in Fig. 28-4, $\mathscr{E} = 12$ V, $R = 10$ Ω, and $r = 2.0$ Ω. Then Eq. 28-9 tells us that the potential across the battery's terminals is

$$V_b - V_a = 12 \text{ V} \frac{10 \text{ Ω}}{10 \text{ Ω} + 2.0 \text{ Ω}} = 10 \text{ V}.$$

In "pumping" charge through itself, the battery (via electrochemical reactions) does work per unit charge of $\mathscr{E} = 12$ J/C, or 12 V. However, because of the internal resistance of the battery, it produces a potential difference of only 10 J/C, or 10 V, across its terminals.

Power, Potential, and Emf

When a battery or some other type of emf device does work on the charge carriers of a current i, it transfers energy from its source of energy (such as the chemical source in a battery) to the charge carriers. Because a real emf device has an internal resistance r, it also transfers energy to internal thermal energy via resistive dissipation as discussed in Section 27-7. Let us relate these transfers.

The net rate P of energy transfer from the emf device to the charge carriers is given by Eq. 27-21:

$$P = iV, \qquad (28\text{-}11)$$

where V is the potential across the terminals of the emf device. From Eq. 28-10, we can substitute $V = \mathcal{E} - ir$ into Eq. 28-11 to find

$$P = i(\mathcal{E} - ir) = i\mathcal{E} - i^2 r. \qquad (28\text{-}12)$$

We see that the term $i^2 r$ in Eq. 28-12 is the rate P_r of energy transfer to thermal energy within the emf device:

$$P_r = i^2 r \qquad \text{(internal dissipation rate)}. \qquad (28\text{-}13)$$

Then the term $i\mathcal{E}$ in Eq. 28-12 must be the rate P_{emf} at which the emf source transfers energy to *both* the charge carriers and to internal thermal energy. Thus,

$$P_{\text{emf}} = i\mathcal{E} \qquad \text{(power of emf device)}. \qquad (28\text{-}14)$$

If a battery is being *recharged*, with the current in the "wrong way" through it, the energy transfer is then *from* the charge carriers *to* the battery—both to the battery's chemical energy and to the energy dissipated in the internal resistance r. The rate of change of the chemical energy is given by Eq. 28-14; the rate of dissipation is given by Eq. 28-13; and the rate at which the carriers supply energy is given by Eq. 28-11.

$\mathbb{C}$HECKPOINT 2: In Fig. 28-5a, if $R_1 > R_2 > R_3$, rank the three resistors according to (a) the current through them and (b) the potential difference across them, greatest first.

What is the current in the circuit of Fig. 28-6a? The emfs and the resistances have the following values:

$$\mathcal{E}_1 = 4.4 \text{ V}, \quad \mathcal{E}_2 = 2.1 \text{ V},$$
$$r_1 = 2.3 \ \Omega, \quad r_2 = 1.8 \ \Omega, \quad R = 5.5 \ \Omega.$$

SOLUTION: The two batteries are connected so that they oppose each other, but $\mathcal{E}_1$, because it is larger than $\mathcal{E}_2$, controls the direction of the current in the circuit, which is clockwise. The loop rule, applied counterclockwise from point a, yields

$$-\mathcal{E}_1 + ir_1 + iR + ir_2 + \mathcal{E}_2 = 0.$$

Check that this equation also results from applying the loop rule clockwise or from starting at some point other than a.

Also, compare this equation term by term with Fig. 28-6b, which shows the potential changes graphically (with the potential at point a arbitrarily taken to be zero).

Solving the above loop equation for the current i, we obtain

$$i = \frac{\mathcal{E}_1 - \mathcal{E}_2}{R + r_1 + r_2} = \frac{4.4 \text{ V} - 2.1 \text{ V}}{5.5 \ \Omega + 2.3 \ \Omega + 1.8 \ \Omega}$$
$$= 0.2396 \text{ A} \approx 240 \text{ mA}. \qquad \text{(Answer)}$$

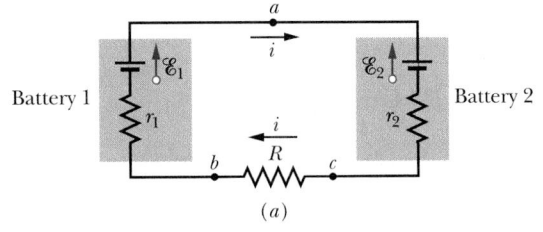

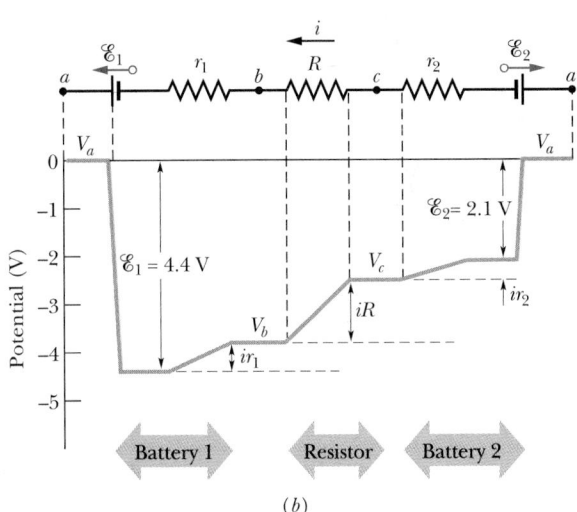

FIGURE 28-6 Sample Problems 28-1 and 28-2. (*a*) A single-loop circuit containing two real batteries and a resistor. The batteries oppose each other; that is, they tend to send current in opposite directions through the resistor. (*b*) A graph of the potentials encountered in traversing this circuit counterclockwise from point a, with the potential at a arbitrarily taken to be zero. (To better link the circuit with the graph, mentally cut the circuit at a and then unfold the left side of the circuit toward the left and the right side of the circuit toward the right.) As battery 1 is traversed from the higher-potential terminal to the lower-potential terminal against the current, the potential decreases by $\mathcal{E}_1$ and increases by ir_1. As the resistor R is traversed against the current, the potential increases by iR. As battery 2 is traversed from the lower-potential terminal to the higher-potential terminal against the current, the potential increases by ir_2 and by $\mathcal{E}_2$.

TACTIC 1: *Assuming the Direction of a Current*

In solving circuit problems, you do not need to know the direction of a current in advance. Instead, you can just assume its direction. To show this, assume the current in Fig. 28-6a is counterclockwise; that is, reverse the direction of the current arrows shown. Applying the loop rule counterclockwise from point *a* now yields

$$-\mathscr{E}_1 - ir_1 - iR - ir_2 + \mathscr{E}_2 = 0$$

or

$$i = -\frac{\mathscr{E}_1 - \mathscr{E}_2}{R + r_1 + r_2}.$$

Substituting numerical values (see above) yields $i = -240$ mA for the current. The minus sign is a signal that the current is opposite the direction we initially assumed.

SAMPLE PROBLEM 28-2

(a) What is the potential difference between the terminals of battery 1 in Fig. 28-6a?

SOLUTION: Let us start at point *b* (effectively the negative terminal of battery 1) and travel through battery 1 to point *a* (effectively the positive terminal), keeping track of potential changes. We find that

$$V_b - ir_1 + \mathscr{E}_1 = V_a,$$

which gives us

$$\begin{aligned}
V_a - V_b &= -ir_1 + \mathscr{E}_1 \\
&= -(0.2396 \text{ A})(2.3 \text{ } \Omega) + 4.4 \text{ V} \\
&= +3.84 \text{ V} \approx 3.8 \text{ V}. \quad \text{(Answer)}
\end{aligned}$$

We can verify this result by starting at point *b* in Fig. 28-6a and traversing the circuit counterclockwise to point *a*. For this different path we find

$$V_b + iR + ir_2 + \mathscr{E}_2 = V_a$$

or

$$\begin{aligned}
V_a - V_b &= i(R + r_2) + \mathscr{E}_2 \\
&= (0.2396 \text{ A})(5.5 \text{ } \Omega + 1.8 \text{ } \Omega) + 2.1 \text{ V} \\
&= +3.84 \text{ V} \approx 3.8 \text{ V}, \quad \text{(Answer)}
\end{aligned}$$

exactly as before. The potential difference between two points has the same value for all paths connecting those points.

(b) What is the potential difference between the terminals of battery 2 in Fig. 28-6a?

SOLUTION: Let us start at point *c* (the negative terminal of battery 2) and travel through battery 2 to point *a* (the positive terminal), keeping track of potential changes. We find

$$V_c + ir_2 + \mathscr{E}_2 = V_a$$

or

$$\begin{aligned}
V_a - V_c &= ir_2 + \mathscr{E}_2 \\
&= (0.2396 \text{ A})(1.8 \text{ } \Omega) + 2.1 \text{ V} \\
&= +2.5 \text{ V}. \quad \text{(Answer)}
\end{aligned}$$

Here the potential difference (2.5 V) between the terminals of the battery is *larger* than the emf (2.1 V) of the battery because charge is being forced through the battery in a direction opposite the direction in which it would normally move.

CHECKPOINT 3: A battery has an emf of 12 V and an internal resistance of 2 Ω. Is the terminal-to-terminal potential difference greater than, less than, or equal to 12 V if the current in the battery is (a) from the negative to the positive terminal, (b) from the positive terminal to the negative terminal, and (c) zero?

28-6 MULTILOOP CIRCUITS

Figure 28-7 shows a circuit containing more than one loop. For simplicity, we assume the batteries are ideal. There are two *junctions* in this circuit, at *b* and *d*, and there are three *branches* connecting these junctions. The branches are the left branch (*bad*), the right branch (*bcd*), and the central branch (*bd*). What are the currents in the three branches?

We arbitrarily label the currents, using a different symbol for each branch. Current i_1 has the same value everywhere in branch *bad*; i_2 has the same value everywhere in branch *bcd*; and i_3 is the current through branch *bd*. The directions of the currents are chosen arbitrarily.

Consider junction *d*. Charge comes into the junction via incoming currents i_1 and i_3, and it leaves via outgoing current i_2; there is no increase or decrease of charge at the junction. This condition means that

$$i_1 + i_3 = i_2. \quad (28\text{-}15)$$

You can easily check that applying this condition to junction *b* leads to exactly the same equation. Equation 28-15 suggests a general principle:

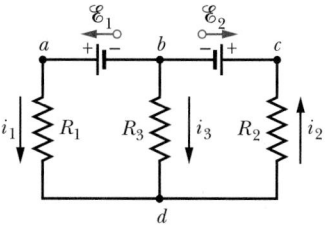

FIGURE 28-7 A multiloop circuit consisting of three branches: left-hand branch *bad*, right-hand branch *bcd*, and central branch *bd*. The circuit also consists of three loops: left-hand loop *badb*, right-hand loop *bcdb*, and big loop *badcb*.

JUNCTION RULE: The sum of the currents entering any junction must be equal to the sum of the currents leaving that junction.

This rule is often called *Kirchhoff's junction rule* (or *Kirchhoff's current law*). It is simply a statement of the conservation of charge for a steady flow of charge—there is neither a build-up nor a depletion of charge at at junction. Thus our basic tools for solving complex circuits are the *loop rule* (based on the conservation of energy) and the *junction rule* (based on the conservation of charge).

Equation 28-15 is a single equation involving three unknowns. To solve the problem completely (that is, to find all three currents), we need two more equations involving those same unknowns. We obtain them by applying the loop rule twice. In the circuit of Fig. 28-7, we have three loops from which to choose: the left-hand loop *(badb)*, the right-hand loop *(bcdb)*, and the big loop *(badcb)*. Which two loops we choose does not matter—let's choose the left-hand loop and the right-hand loop.

If we traverse the left-hand loop in a counterclockwise direction from point *b*, the loop rule gives us

$$\mathscr{E}_1 - i_1 R_1 + i_3 R_3 = 0. \qquad (28\text{-}16)$$

If we traverse the right-hand loop in a counterclockwise direction from point *b*, the loop rule gives us

$$-i_3 R_3 - i_2 R_2 - \mathscr{E}_2 = 0. \qquad (28\text{-}17)$$

We now had three equations (Eqs. 28-15, 28-16, and 28-17) in the three unknown currents, and they can be solved by a variety of techniques.

If we had applied the loop rule to the big loop, we would have obtained (moving counterclockwise from *b*) the equation

$$\mathscr{E}_1 - i_1 R_1 - i_2 R_2 - \mathscr{E}_2 = 0.$$

This equation may look like fresh information, but in fact it is only the sum of Eqs. 28-16 and 28-17. (It would, however, yield the proper results when used with Eq. 28-15 and either 28-16 or 28-17.)

Resistances in Parallel

Figure 28-8*a* shows three resistances connected **in parallel** to an ideal battery of emf $\mathscr{E}$. The battery applies a potential difference $V = \mathscr{E}$ across each resistor in this parallel combination.

Connected resistances are said to be in parallel when a potential difference that is applied across their combination results in that same potential difference across each resistance.

We seek the single resistance R_{eq} that is equivalent to this parallel combination; R_{eq} is then the resistance that can replace the combination without changing the current i through the combination or the potential difference V applied across the combination.

The currents in the three branches of Fig. 28-8*a* are

$$i_1 = \frac{V}{R_1}, \quad i_2 = \frac{V}{R_2}, \quad \text{and} \quad i_3 = \frac{V}{R_3},$$

where V is the potential difference between *a* and *b*. If we apply the junction rule at point *a* and then substitute these values, we find

$$i = i_1 + i_2 + i_3 = V\left(\frac{1}{R_1} + \frac{1}{R_2} + \frac{1}{R_3}\right). \qquad (28\text{-}18)$$

If we replaced the parallel combination with the equivalent resistance R_{eq} (Fig. 28-8*b*), we would have

$$i = \frac{V}{R_{eq}}. \qquad (2819)$$

Comparing Eqs. 28-18 and 28-19 leads to

$$\frac{1}{R_{eq}} = \frac{1}{R_1} + \frac{1}{R_2} + \frac{1}{R_3}. \qquad (28\text{-}20)$$

Extending this result to the case of *n* resistances, we have

$$\frac{1}{R_{eq}} = \sum_{j=1}^{n} \frac{1}{R_j} \qquad (n \text{ resistances in parallel}). \qquad (28\text{-}21)$$

For the case of two resistances, the equivalent resistance is their product divided by their sum. That is,

$$R_{eq} = \frac{R_1 R_2}{R_1 + R_2}. \qquad (28\text{-}22)$$

If you accidentally took the equivalent resistance to be the sum divided by the product, you would notice at once that this result would be dimensionally incorrect.

Note that when two or more resistances are connected in parallel, the equivalent resistance is smaller than any of the combining resistances. Table 28-1 summarizes the equivalence relations for resistors and capacitors in series and in parallel.

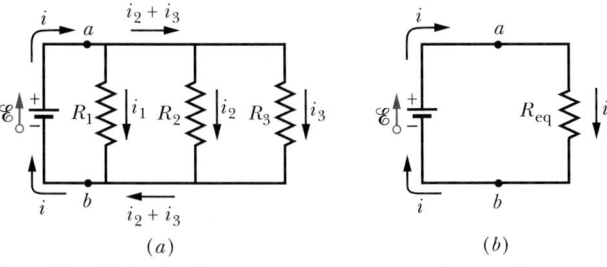

FIGURE 28-8 (*a*) Three resistors connected in parallel across points *a* and *b*. (*b*) An equivalent circuit, with the three resistors replaced with their equivalent resistance R_{eq}.

TABLE 28-1 SERIES AND PARALLEL RESISTORS AND CAPACITORS

SERIES	PARALLEL
Resistors	
$R_{eq} = \sum\limits_{j=1}^{n} R_j$ Eq. 28-7	$\dfrac{1}{R_{eq}} = \sum\limits_{j=1}^{n} \dfrac{1}{R_j}$ Eq. 28-21
Same current through all resistors	Same potential difference across all resistors
Capacitors	
$\dfrac{1}{C_{eq}} = \sum\limits_{j=1}^{n} \dfrac{1}{C_j}$ Eq. 26-20	$C_{eq} = \sum\limits_{j=1}^{n} C_j$ Eq. 26-19
Same charge on all capacitors	Same potential difference across all capacitors

CHECKPOINT **4:** A battery, with potential V across it and current i through it, is connected to a combination of two identical resistors. What are the potential difference across and the current through either resistor if the resistors are (a) in series and (b) in parallel?

SAMPLE PROBLEM 28-3

Figure 28-9a shows a multiloop circuit containing one ideal battery and four resistors with the following values:

$$R_1 = 20 \ \Omega, \quad R_2 = 20 \ \Omega, \quad \mathscr{E} = 12 \ V,$$
$$R_3 = 30 \ \Omega, \quad R_4 = 8.0 \ \Omega.$$

(a) What is the current through the battery?

SOLUTION: The current through the battery is also the current through R_1. So, to find the current, we need to write an equation for a loop through R_1; either the left-hand loop or the big loop will do. Noting that since the emf arrow of the battery points upward and the current the battery supplies is clockwise, we might consider applying the loop rule to the left-hand loop clockwise from point a, getting

$$+\mathscr{E} - iR_1 - iR_2 - iR_4 = 0 \quad \text{(incorrect)}.$$

However, this equation is incorrect because it assumes that R_1, R_2, and R_4 all have the same current i. Resistors R_1 and R_4 do have the same current, because the current passing through R_4 must pass through the battery and then through R_1 with no change in value. But that current splits at junction point b—only part passes through R_2, the rest through R_3.

To distinguish the several currents in the circuit, we must label them individually as in Fig. 28-9b. Then, circling clockwise from a, we can write the loop rule for the left-hand loop as

$$+\mathscr{E} - i_1R_1 - i_2R_2 - i_1R_4 = 0.$$

Unfortunately, this equation contains two unknown, i_1 and i_2; we would need at least one more equation to find them.

A second, much easier option is to simplify the circuit of Fig. 28-9b by finding equivalent resistances. Note carefully that R_1 and R_2 are *not* in series and thus cannot be replaced with an equivalent resistance. However, R_2 and R_3 are in parallel; so we can use either Eq. 28-21 or Eq. 28-22 to find their equivalent resistance R_{23}. From the latter,

$$R_{23} = \frac{R_2 R_3}{R_2 + R_3} = \frac{(20 \ \Omega)(30 \ \Omega)}{50 \ \Omega} = 12 \ \Omega.$$

We can now redraw the circuit as in Fig. 28-9c; note that the current through R_{23} must be i_1 because the current i_1 through R_1 and R_4 must continue through R_{23}. For this simple one-loop circuit, the loop rule (applied clockwise from point a) yields

$$+\mathscr{E} - i_1R_1 - i_1R_{23} - i_1R_4 = 0.$$

Substituting the given data, we find

$$12 \ V - i_1(20 \ \Omega) - i_1(12 \ \Omega) - i_1(8.0 \ \Omega) = 0,$$

which gives us

$$i_1 = \frac{12 \ V}{40 \ \Omega} = 0.30 \ A. \quad \text{(Answer)}$$

(b) What is the current i_2 through R_2?

SOLUTION: Look again at Fig. 28-9c. From it and the preceding answer we know that the current through R_{23} is $i_1 =$

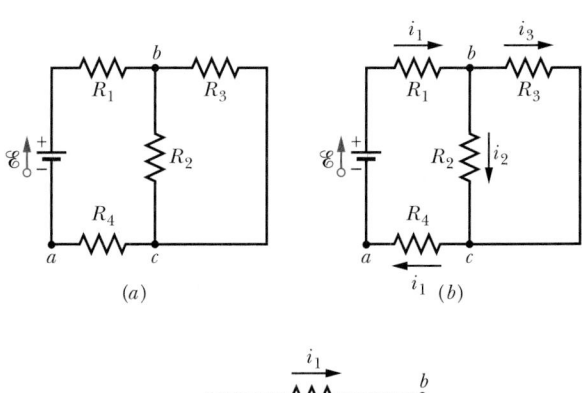

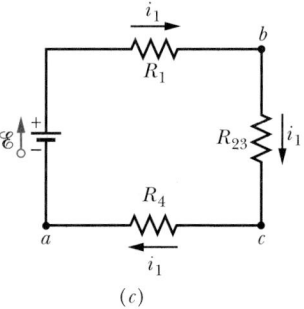

(c)

FIGURE 28-9 Sample Problem 28-3. (a) A multiloop circuit with an ideal battery of emf $\mathscr{E}$ and four resistors. (b) Assumed currents through the resistors. (c) A simplification of the circuit, with resistances R_2 and R_3 replaced with their equivalent resistance R_{23}. The current through R_{23} is equal to that through R_1 and R_4.

0.30 A. Then we can use Eq. 27-8 ($R = V/i$) to find the potential difference V_{23} across R_{23}, which is

$$V_{23} = i_1 R_{23} = (0.30 \text{ A})(12 \text{ }\Omega) = 3.6 \text{ V}.$$

This is also the potential difference across R_2 (and across R_3). Thus, applying Eq. 27-8 now to R_2, we can write

$$i_2 = \frac{V_2}{R_2} = \frac{3.6 \text{ V}}{20 \text{ }\Omega} = 0.18 \text{ A}. \qquad \text{(Answer)}$$

(c) What is the current i_3 through R_3?

SOLUTION: From Fig. 28-9b and the earlier results, application of the junction rule at point b gives us

$$i_3 = i_1 - i_2 = 0.30 \text{ A} - 0.18 \text{ A}$$
$$= 0.12 \text{ A}. \qquad \text{(Answer)}$$

SAMPLE PROBLEM 28-4

Figure 28-10 shows a circuit whose elements have the following values:

$$\mathcal{E}_1 = 3.0 \text{ V}, \quad \mathcal{E}_2 = 6.0 \text{ V},$$
$$R_1 = 2.0 \text{ }\Omega, \quad R_2 = 4.0 \text{ }\Omega.$$

The three batteries are ideal batteries. Find the magnitude and direction of the current in each of the three branches.

SOLUTION: It is not worthwhile to try to simplify this circuit, because no two resistors are in parallel, and the resistors that are in series (those in the right branch or those in the left branch) present no problem. So, we shall apply the junction and loop rules, and then solve some simultaneous equations.

Using arbitrarily chosen directions for the currents as shown in Fig. 28-10, we apply the junction rule at point a by writing

$$i_3 = i_1 + i_2. \qquad (28\text{-}23)$$

An application of the junction rule at junction b gives only the same equation. So we next apply the loop rule to any two of the three loops of the circuit. We first arbitrarily choose the left-hand loop, arbitrarily start at point a, and arbitrarily traverse the loop in the counterclockwise direction, obtaining

$$-i_1 R_1 - \mathcal{E}_1 - i_1 R_1 + \mathcal{E}_2 + i_2 R_2 = 0.$$

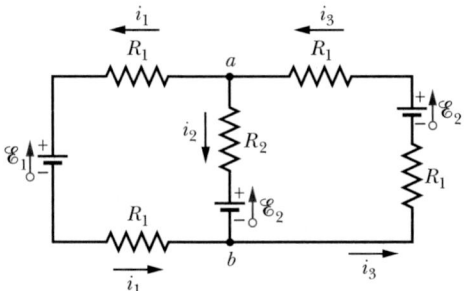

FIGURE 28-10 Sample Problem 28-4. A multiloop circuit with three ideal batteries and five resistors.

Substituting the given data and simplifying yield

$$i_1(4.0 \text{ }\Omega) - i_2(4.0 \text{ }\Omega) = 3.0 \text{ V}. \qquad (28\text{-}24)$$

For our second application of the loop rule, we arbitrarily choose to traverse the right-hand loop clockwise from point a, finding

$$+i_3 R_1 - \mathcal{E}_2 + i_3 R_1 + \mathcal{E}_2 + i_2 R_2 = 0.$$

Substituting the given data and simplifying yield

$$i_2(4.0 \text{ }\Omega) + i_3(4.0 \text{ }\Omega) = 0. \qquad (28\text{-}25)$$

Using Eq. 28-23 to eliminate i_3 from Eq. 28-25 and simplifying give us

$$i_1(4.0 \text{ }\Omega) + i_2(8.0 \text{ }\Omega) = 0. \qquad (28\text{-}26)$$

We now have a system of two equations (Eqs. 28-24 and 28-26) in two unknowns (i_1 and i_2) to solve either "by hand" (which is easy enough here) or with a "math package." (One solution technique is Cramer's rule, given in Appendix E.) We find

$$i_2 = -0.25 \text{ A}.$$

(The minus sign signals that our arbitrary choice of direction for i_2 in Fig. 28-10 is wrong; i_2 should point up through $\mathcal{E}_2$ and R_2.) Substituting $i_2 = -0.25$ A into Eq. 28-26 and solving for i_1 then give us

$$i_1 = 0.50 \text{ A}. \qquad \text{(Answer)}$$

With Eq. 28-23 we then find that

$$i_3 = i_1 + i_2 = 0.25 \text{ A}. \qquad \text{(Answer)}$$

The positive answers we obtained for i_1 and i_3 signal that our choices of directions for these currents are correct. We can now correct the direction for i_2 and write its magnitude as

$$i_2 = 0.25 \text{ A}. \qquad \text{(Answer)}$$

SAMPLE PROBLEM 28-5

Electric fish generate current with biological cells called *electroplaques*, which are physiological emf devices. The electroplaques in the South American eel shown in the photograph that opens this chapter are arranged in 140 rows, each row stretching horizontally along the body and each containing 5000 electroplaques. The arrangement is suggested in Fig. 28-11a; each electroplaque has an emf $\mathcal{E}$ of 0.15 V and an internal resistance r of 0.25 Ω.

(a) If the water surrounding the eel has resistance $R_w = 800$ Ω, how much current can the eel produce in the water, from near its head to its tail?

SOLUTION: To answer, we simplify the circuit of Fig. 28-11a, first considering a single row. The total emf $\mathcal{E}_{\text{row}}$ along a row of 5000 electroplaques is the sum of the emfs:

$$\mathcal{E}_{\text{row}} = 5000\mathcal{E} = (5000)(0.15 \text{ V}) = 750 \text{ V}.$$

The total resistance R_{row} along a row is the sum of the internal

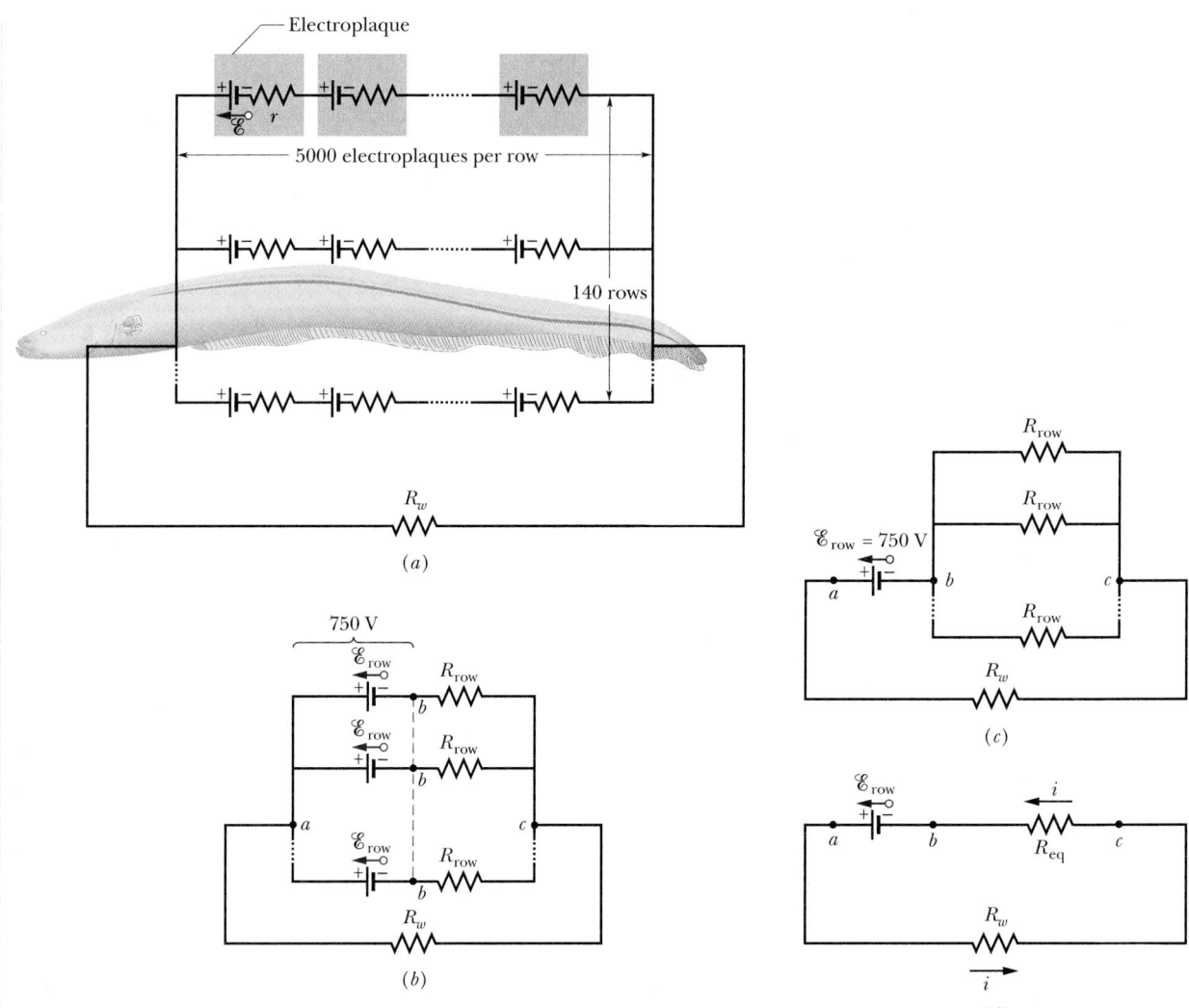

FIGURE 28-11 Sample Problem 28-5. (a) A model of the electric circuit of an eel in water. Each electroplaque of the eel has an emf $\mathscr{E}$ and internal resistance r. Along each of 140 rows extending from the head to the tail of the eel, there are 5000 electroplaques. The surrounding water has resistance R_w. (b) The emf $\mathscr{E}_{row}$ and resistance R_{row} of each row. (c) The emf between points a and b is $\mathscr{E}_{row}$. Between points b and c are 140 parallel resistances R_{row}. (d) The simplified circuit, with R_{eq} replacing the parallel combination.

resistances of the 5000 electroplaques:

$$R_{row} = 5000r = (5000)(0.25\ \Omega) = 1250\ \Omega.$$

We can now represent each of the 140 identical rows as having a single emf $\mathscr{E}_{row}$ and a single resistance R_{row}, as shown in Fig. 28-11b.

In Fig. 28-11b, the emf between point a and point b on any row is $\mathscr{E}_{row} = 750$ V. Because the rows are identical and because they are all connected together at the left in Fig. 28-11b, all points b in that figure are at the same electric potential. Thus we can consider them to be connected so that there is only a single point b. The emf between point a and this single point b is $\mathscr{E}_{row} = 750$ V, so we can draw the circuit as shown in Fig. 28-11c.

Between points b and c in Fig. 28-11c, 140 resistances of $R_{row} = 1250\ \Omega$ are in parallel. The equivalent resistance R_{eq} of this combination is given by Eq. 28-21 as

$$\frac{1}{R_{eq}} = \sum_{j=1}^{140} \frac{1}{R_j} = 140\,\frac{1}{R_{row}},$$

or

$$R_{eq} = \frac{R_{row}}{140} = \frac{1250\ \Omega}{140} = 8.93\ \Omega.$$

Replacing the parallel combination with R_{eq}, we obtain the simplified circuit of Fig. 28-11d. Applying the loop rule to this circuit counterclockwise from point b, we have

$$\mathscr{E}_{row} - iR_w - iR_{eq} = 0.$$

Solving for i and substituting the known data, we find

$$i = \frac{\mathscr{E}_{row}}{R_w + R_{eq}} = \frac{750\ \text{V}}{800\ \Omega + 8.93\ \Omega}$$

$$= 0.927\ \text{A} \approx 0.93\ \text{A}. \qquad \text{(Answer)}$$

If the head or tail of the eel is near a fish, much of this current could pass along a narrow path through the fish, stunning or killing it.

(b) How much current i_{row} travels through each row of Fig. 28-11a?

SOLUTION: Since the rows are identical, the current into and out of the eel is evenly divided among them:

$$i_{row} = \frac{i}{140} = \frac{0.927 \text{ A}}{140} = 6.6 \times 10^{-3} \text{ A}. \quad \text{(Answer)}$$

Thus the current through each row is small, about two orders of magnitude smaller than the current through the water. This means that the eel need not stun or kill itself when it stuns or kills a fish.

PROBLEM SOLVING TACTICS

TACTIC 2: *Solving Circuits of Batteries and Resistors*
Here are two general techniques for solving circuits for unknown currents or potential differences.

1. If a circuit can be simplified by replacing resistors in series or in parallel with their equivalents, do so. If you can reduce the circuit to a single loop, then you can find the current through the battery with that loop, as in Sample Problem 28-3a. You may then have to "work backward," undoing the resistor simplification process, to find the current or potential difference for any particular resistor, as in Sample Problem 28-3b.

2. If a circuit cannot be simplified to a single loop, use the junction rule and the loop rule to write a set of simultaneous equations, as in Sample Problem 28-4. You need have only as many independent equations as there are unknowns in those equations. If you have to find the current or potential difference for a particular resistor, you can ensure that its current or potential difference appears in the equations by having at least one of the loops pass through the designated resistor.

TACTIC 3: *Arbitrary Choices in Solving Circuit Problems*
In Sample Problem 28-4, we made several arbitrary choices. (1) We assumed directions for the currents in Fig. 28-10 arbitrarily. (2) We chose which of the three possible loops to write equations for arbitrarily. (3) We chose the direction in which to traverse each loop arbitrarily. (4) We chose the starting and ending point for each traversal arbitrarily.

Such arbitrariness often worries a beginning circuit solver, but an experienced circuit solver knows that it does not matter. Just keep two rules firmly in mind. First, make sure you traverse each chosen loop completely. Second, once you have chosen a direction for a current, stick with it until you get numerical values for all the currents. If you were wrong about

a direction, the algebra will signal you with a minus sign. Then you can make a correction by simply erasing the minus sign and reversing the arrow representing that current in the circuit diagram. However, *you should not make this correction until you have completed all the required calculations for the circuit*, as we did in Sample Problem 28-4.

28-7 THE AMMETER AND THE VOLTMETER

An instrument used to measure currents is called an *ammeter*. To measure the current in a wire, you usually have to break or cut the wire and insert the ammeter so that the current to be measured passes through the meter, as shown in Fig. 28-12.

It is essential that the resistance R_A of the ammeter be very small compared to other resistances in the circuit. Otherwise, the very presence of the meter will change the current to be measured.

A meter used to measure potential differences is called a *voltmeter*. To find the potential difference between any two points in the circuit, the voltmeter terminals are connected between those points, without breaking or cutting the wire (Fig. 28-12).

It is essential that the resistance R_V of a voltmeter be very large compared to the resistance of any circuit element across which the voltmeter is connected. Otherwise, the meter itself becomes an important circuit element and alters the potential difference that is to be measured.

Often a single meter is packaged so that, by means of a switch, it can be made to serve as either an ammeter or a voltmeter—and usually also as an *ohmmeter*, designed to measure the resistance of any element connected between its terminals. Such a versatile unit is called a *multimeter*.

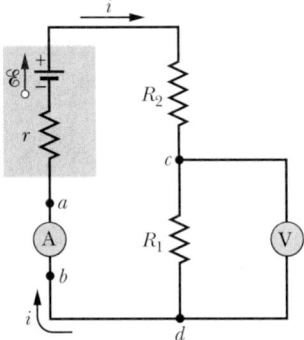

FIGURE 28-12 A single-loop circuit, showing how to connect an ammeter (A) and a voltmeter (V).

28-8 *RC* CIRCUITS

In preceding sections we dealt only with circuits in which the currents did not vary with time. Here we begin a discussion of time-varying currents.

Charging a Capacitor

The capacitor of capacitance C in Fig. 28-13 is initially uncharged. To charge it, we close switch S on point a. This completes an *RC series circuit* consisting of the capacitor, an ideal battery of emf $\mathscr{E}$, and a resistance R.

From Section 26-2, we already know that as soon as the circuit is complete, charge begins to flow (current exists) between a capacitor plate and a battery terminal on each side of the capacitor. This current increases the charge q on the plates and the potential difference V_C ($= q/C$) across the capacitor. When that potential difference equals the potential difference across the battery (which here is equal to the emf $\mathscr{E}$), the current is zero. From Eq. 26-1 ($q = CV$), the *equilibrium* (final) *charge* on the then fully charged capacitor is equal to $C\mathscr{E}$.

Here we want to examine the charging process. In particular we want to know how the charge $q(t)$ on the capacitor plates, the potential difference $V_C(t)$ across the capacitor, and the current $i(t)$ in the circuit vary with time during the charging process. We begin by applying the loop rule to the circuit, traversing it clockwise from the negative terminal of the battery. We find

$$\mathscr{E} - iR - \frac{q}{C} = 0. \qquad (28\text{-}27)$$

The last term on the left side represents the potential difference across the capacitor. The term is negative because the capacitor's top plate, which is connected to the battery's positive terminal, is at a higher potential than the lower plate. So, there is a drop in potential as we move down through the capacitor.

We cannot immediately solve Eq. 28-27 because it contains two variables, i and q. However, those variables

are not independent but are related by

$$i = \frac{dq}{dt}. \qquad (28\text{-}28)$$

Substituting this for i in Eq. 28-27 and rearranging, we find

$$R\frac{dq}{dt} + \frac{q}{C} = \mathscr{E} \qquad \text{(charging equation).} \qquad (28\text{-}29)$$

This differential equation describes the time variation of the charge q on the capacitor in Fig. 28-13. To solve it, we need to find the function $q(t)$ that satisfies this equation and also satisfies the condition that the capacitor be initially uncharged: $q = 0$ at $t = 0$.

We shall show below that the solution to Eq. 28-29 is

$$q = C\mathscr{E}(1 - e^{-t/RC}) \qquad \begin{array}{l}\text{(charging a}\\\text{capacitor).}\end{array} \qquad (28\text{-}30)$$

(Here e is the exponential base, $2.718 \cdots$, and not the elementary charge.) Note that Eq. 28-30 does indeed satisfy our required initial condition, because at $t = 0$ the term $e^{-t/RC}$ is unity; so the equation gives $q = 0$. Note also that at $t = \infty$ (that is, a long time later), the term $e^{-t/RC}$ is zero; so the equation gives the proper value for the full (equilibrium) charge on the capacitor, namely, $q = C\mathscr{E}$. A plot of $q(t)$ for the charging process is given in Fig. 28-14a.

The derivative of $q(t)$ is the current $i(t)$ charging the capacitor:

$$i = \frac{dq}{dt} = \left(\frac{\mathscr{E}}{R}\right)e^{-t/RC} \qquad \begin{array}{l}\text{(charging a}\\\text{capacitor).}\end{array} \qquad (28\text{-}31)$$

A plot of $i(t)$ for the charging process is given in Fig. 28-14b. Note that the current has the initial value $\mathscr{E}/R$ and that it decreases to zero as the capacitor becomes fully charged. Note also that the initial value $\mathscr{E}/R$ implies that at

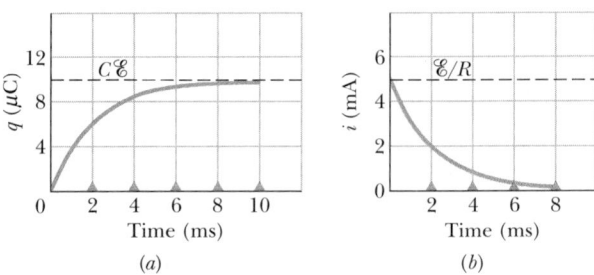

(a)   (b)

FIGURE 28-14 (a) A plot of Eq. 28-30, which shows the buildup of charge on the capacitor of Fig. 28-13. (b) A plot of Eq. 28-31, which shows the decline of the charging current in the circuit of Fig. 28-13. The curves are plotted for $R = 2000 \, \Omega$, $C = 1 \, \mu\text{F}$, and $\mathscr{E} = 10 \, \text{V}$; the small triangles represent successive intervals of one time constant.

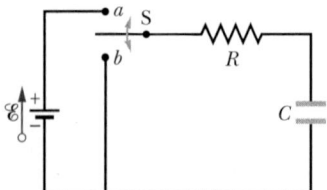

FIGURE 28-13 When switch S is closed on a, the capacitor C is *charged* through the resistor R. When the switch is afterward closed on b, the capacitor *discharges* through R.

$t = 0$, the capacitor acts as if it were a wire with negligible resistance.

By combining Eq. 26-1 ($q = CV$) and Eq. 28-30, we find that the potential difference $V_C(t)$ across the capacitor during the charging process is

$$V_C = \frac{q}{C} = \mathscr{E}(1 - e^{-t/RC}) \qquad \text{(charging a capacitor).} \qquad (28\text{-}32)$$

This tells us that $V_C = 0$ at $t = 0$ and that $V_C = \mathscr{E}$ when the capacitor is fully charged at $t = \infty$.

The Time Constant

The product RC that appears in Eqs. 28-30, 28-31, and 28-32 has the dimensions of time (because the argument of an exponential must be dimensionless); in fact, $1.0\ \Omega \times 1.0\ \text{F} = 1.0\ \text{s}$. RC is called the **capacitive time constant** of the circuit and is represented with the symbol τ:

$$\tau = RC \qquad \text{(time constant).} \qquad (28\text{-}33)$$

From Eq. 28-30, we can now see that at time $t = \tau(= RC)$, the charge on the initially uncharged capacitor of Fig. 28-13 has increased from zero to

$$q = C\mathscr{E}(1 - e^{-1}) = 0.63C\mathscr{E}. \qquad (28\text{-}34)$$

In words, during the first time constant τ the charge has increased from zero to 63% of its final value $C\mathscr{E}$. In Fig. 28-14, the small triangles along the times axes mark successive intervals of one time constant during the charging of the capacitor. The charging times for RC circuits are often stated in terms of τ: the greater τ is, the greater the charging time.

Discharging a Capacitor

Assume now that the capacitor of Fig. 28-13 is fully charged to a potential V_0 equal to the emf $\mathscr{E}$ of the battery. At a new time $t = 0$ the switch S is thrown from a to b so that the capacitor can *discharge* through resistance R. How do the charge $q(t)$ on the capacitor and the current $i(t)$ through the discharge loop of capacitor and resistance now vary with time?

The differential equation describing $q(t)$ is like Eq. 28-29 except that now, with no battery in the discharge loop, $\mathscr{E} = 0$. Thus,

$$R\frac{dq}{dt} + \frac{q}{C} = 0 \qquad \text{(discharging equation).} \qquad (28\text{-}35)$$

The solution to this differential equation is

$$q = q_0 e^{-t/RC} \qquad \text{(discharging a capacitor),} \qquad (28\text{-}36)$$

where $q_0\ (= CV_0)$ is the initial charge on the capacitor. You can verify by substitution that Eq. 28-36 is indeed a solution of Eq. 28-35.

Equation 28-36 tells us that q decreases exponentially with time, at a rate that is set by the capacitive time constant $\tau = RC$. At time $t = \tau$, the capacitor's charge has been reduced to $q_0 e^{-1}$, or about 37% of the initial value. Note that a greater τ means a greater discharge time.

Differentiating Eq. 28-36 gives us the current $i(t)$:

$$i = \frac{dq}{dt} = -\left(\frac{q_0}{RC}\right)e^{-t/RC} \qquad \text{(discharging a capacitor).} \qquad (28\text{-}37)$$

This tells us that the current also decreases exponentially with time, at a rate set by τ. The initial current i_0 is equal to q_0/RC. Note that you can find i_0 by simply applying the loop rule to the circuit at $t = 0$; just then the capacitor's initial potential V_0 is connected across the resistance R, so the current must be $i_0 = V_0/R = (q_0/C)/R = q_0/RC$. The minus sign in Eq. 38-37 can be ignored; it merely means that the capacitor's charge q is decreasing.

Derivation of Eq. 28-30

To solve Eq. 28-29, we first rewrite it as

$$\frac{dq}{dt} + \frac{q}{RC} = \frac{\mathscr{E}}{R}. \qquad (28\text{-}38)$$

The general solution to this differential equation is of the form

$$q = q_p + Ke^{-at}, \qquad (28\text{-}39)$$

where q_p is a *particular solution* of the differential equation, K is a constant to be evaluated from the initial conditions, and $a = 1/RC$ is the coefficient of q in Eq. 28-38. To find q_p, we set $dq/dt = 0$ in Eq. 28-38 (corresponding to the final condition of no further charging) and solve, obtaining

$$q_p = C\mathscr{E}. \qquad (28\text{-}40)$$

To evaluate K, we first substitute this into Eq. 28-39 to get

$$q = C\mathscr{E} + Ke^{-at}.$$

Then substituting the initial conditions $q = 0$ and $t = 0$ yields

$$0 = C\mathscr{E} + K,$$

or $K = -C\mathscr{E}$. Finally, with the values of q_p, a, and K inserted, Eq. 28-39 becomes

$$q = C\mathscr{E} - C\mathscr{E}e^{-t/RC}$$

which, with a slight modification, is Eq. 28-30.

CHECKPOINT **5:** The table gives four sets of values for the circuit elements in Fig. 28-13. Rank the sets according to (a) the initial current (as the switch is closed on a) and (b) the time required for the current to decrease to half its initial value, greatest first.

	1	2	3	4
$\mathcal{E}$ (V)	12	12	10	10
R (Ω)	2	3	10	5
C (μF)	3	2	0.5	2

SAMPLE PROBLEM 28-6

A capacitor of capacitance C is discharging through a resistor of resistance R.

(a) In terms of the time constant $\tau = RC$, when will the charge on the capacitor be half its initial value?

SOLUTION: The charge on the capacitor varies according to Eq. 28-36,

$$q = q_0 e^{-t/RC},$$

in which q_0 is the initial charge. We are asked to find the time t at which $q = \frac{1}{2}q_0$, or at which

$$\tfrac{1}{2}q_0 = q_0 e^{-t/RC}. \qquad (28\text{-}41)$$

After canceling q_0, we realize that the time t we seek is "buried" inside an exponential function. To expose the symbol t in Eq. 28-41, we take the natural logarithms of both sides of the equation. (The natural logarithm is the inverse function of the exponential function.) We find

$$\ln \tfrac{1}{2} = \ln(e^{-t/RC}) = -\frac{t}{RC}$$

or $\qquad t = (-\ln \tfrac{1}{2})RC = 0.69RC = 0.69\tau.$ (Answer)

(b) When will the energy stored in the capacitor be half its initial value?

SOLUTION: The energy stored in the capacitor is, from Eqs. 26-21 and 28-36,

$$U = \frac{q^2}{2C} = \frac{q_0^2}{2C} e^{-2t/RC} = U_0 e^{-2t/RC}, \qquad (28\text{-}42)$$

in which U_0 is the initial stored energy. We are asked to find the time at which $U = \frac{1}{2}U_0$, or at which

$$\tfrac{1}{2}U_0 = U_0 e^{-2t/RC}.$$

Canceling U_0 and taking the natural logarithms of both sides, we obtain

$$\ln \tfrac{1}{2} = -\frac{2t}{RC}$$

or $\qquad t = -RC \dfrac{\ln \tfrac{1}{2}}{2} = 0.35RC = 0.35\tau.$ (Answer)

It takes longer (0.69τ versus 0.35τ) for the *charge* to fall to half its initial value than for the *stored energy* to fall to half its initial value. Doesn't this result surprise you?

(c) At what rate P_R is thermal energy produced in the resistor during the discharging process? At what rate P_C is stored energy lost by the capacitor during the discharging process?

SOLUTION: The current through the resistor during the discharging is given by Eq. 28-37. From Eq. 27-22 ($P = i^2R$), we then have

$$P_R = i^2 R = \left[-\frac{q_0}{RC} e^{-t/RC} \right]^2 R$$

$$= \frac{q_0^2}{RC^2} e^{-2t/RC}. \qquad \text{(Answer)}$$

Stored energy is lost by the capacitor at the rate $P_C = dU/dt$, where U is the energy stored there. From Eq. 28-42, we then have

$$P_C = \frac{dU}{dt} = \frac{d}{dt}(U_0 e^{-2t/RC}) = -\frac{2U_0}{RC} e^{-2t/RC}.$$

Substituting $q_0^2/2C$ for U_0 gives us

$$P_C = -\frac{q_0^2}{RC^2} e^{-2t/RC}. \qquad \text{(Answer)}$$

Note that $P_C + P_R = 0$. In words, stored energy lost by the capacitor is transferred completely to thermal energy of the resistor.

SAMPLE PROBLEM 28-7

The circuit in Fig. 28-15 consists of an ideal battery with emf $\mathcal{E} = 12$ V, two resistors with resistances $R_1 = 4.0$ Ω and $R_2 = 6.0$ Ω, and an initially uncharged capacitor with capacitance $C = 6.0$ μC. The circuit is completed when switch S is closed at time $t = 0$.

(a) At time $t = 2.0\tau$, what is the potential difference across the capacitor?

SOLUTION: In Fig. 28-15, the capacitor is being charged through R_1 by an emf $\mathcal{E}$ connected across them, just as in Fig. 28-13. (Resistance R_2 does not change this fact.) So, we can use Eq. 28-32,

$$V_C = \mathcal{E}(1 - e^{-t/RC}),$$

to find the potential difference V_C across the capacitor, except here resistance R is R_1. Substituting $t = 2.0\tau = 2.0R_1C$ and given data, we then find

$$V_C = (12 \text{ V})(1 - e^{-2.0R_1C/R_1C})$$
$$= (12 \text{ V})(1 - e^{-2.0}) = 10 \text{ V}. \qquad \text{(Answer)}$$

(b) At time $t = 2.0\tau$, what are the potential differences V_{R_1} and V_{R_2} across the two resistors? Do those potential differences increase, decrease, or remain the same while the capacitor is being charged?

SOLUTION: If we apply the loop rule to the big loop of Fig. 28-15, clockwise from the negative terminal of the battery, we find

$$\mathscr{E} - V_C - V_{R_1} = 0. \qquad (28\text{-}43)$$

We now know that at $t = 2.0\tau$, the potential difference V_C is equal to 10 V. Substituting this and $\mathscr{E} = 12$ V into Eq. 28-43 yields

$$V_{R_1} = 2.0 \text{ V}. \qquad \text{(Answer)}$$

During the charging of the capacitor, the battery's emf $\mathscr{E}$ is constant and the potential difference V_C across the capacitor increases. By writing Eq. 28-43 as $V_{R_1} = \mathscr{E} - V_C$, we see that V_{R_1} must decrease during the charging process.

If we apply the loop rule to the left-hand loop of Fig.

28-15, again clockwise from the negative terminal, we find

$$\mathscr{E} - V_{R_2} = 0$$

and

$$V_{R_2} = \mathscr{E} = 12 \text{ V}. \qquad \text{(Answer)}$$

Thus V_{R_2} does not change during the charging process.

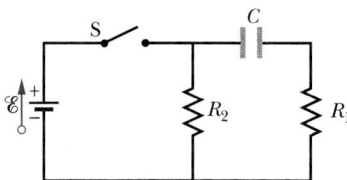

FIGURE 28-15 Sample Problem 28-7. When switch S is closed, the circuit is complete and the battery begins to charge the capacitor.

REVIEW & SUMMARY

Emf

An **emf device** does work on charges to maintain a potential difference between its output terminals. If dW is the work the device does to force positive charge dq from the negative to the positive terminal, then the **emf** (work per unit charge) of the device is

$$\mathscr{E} = \frac{dW}{dq} \qquad \text{(definition of } \mathscr{E}\text{)}. \qquad (28\text{-}1)$$

The volt is the SI unit of emf as well as of potential difference. An **ideal emf device** is one that lacks any internal resistance. The potential difference between its terminals is equal to the emf. A **real emf device** has internal resistance. The potential difference between its terminals is equal to the emf only if there is no current through the device.

Analyzing Circuits

The change in potential in traversing a resistance R in the direction of the current is $-iR$; in the opposite direction it is $+iR$. The change in potential in traversing an ideal emf device in the direction of the emf arrow is $+\mathscr{E}$; in the opposite direction it is $-\mathscr{E}$. Conservation of energy leads to the loop rule:

Loop Rule. *The algebraic sum of the changes in potential encountered in a complete traversal of any loop of a circuit must be zero.*

Conservation of charge gives us the junction rule:

Junction Rule. *The sum of the currents entering any junction must be equal to the sum of the currents leaving that junction.*

Single-Loop Circuits

The current in a single-loop circuit containing a single resistance R and an emf device with emf $\mathscr{E}$ and internal resistance r is

$$i = \frac{\mathscr{E}}{R + r}, \qquad (28\text{-}4)$$

which reduces to $i = \mathscr{E}/R$ for an ideal emf device with $r = 0$.

Power

When a real battery of emf $\mathscr{E}$ and internal resistance r does work on the charge carriers in a current i through it, the rate P of energy transfer to the charge carriers is

$$P = iV, \qquad (28\text{-}11)$$

where V is the potential across the terminals of the battery. The rate P_r of energy transfer to thermal energy within the battery is

$$P_r = i^2 r. \qquad (28\text{-}13)$$

And the rate P_{emf} at which the chemical energy within the battery changes is

$$P_{\text{emf}} = i\mathscr{E}. \qquad (28\text{-}14)$$

Series Resistances

Resistances are in **series** if the sum of their individual potential differences is equal to the potential difference applied across the combination. The equivalent resistance of the series combination is

$$R_{\text{eq}} = \sum_{j=1}^{n} R_j \qquad (n \text{ resistances in series}). \qquad (28\text{-}7)$$

Other circuit elements may also be connected in series.

Parallel Resistances

Resistances are in **parallel** if their individual potential differences are equal to the applied potential difference. The equivalent resistance of the parallel combination is

$$\frac{1}{R_{\text{eq}}} = \sum_{j=1}^{n} \frac{1}{R_j} \qquad (n \text{ resistances in parallel}). \qquad (28\text{-}21)$$

Other circuit elements may also be connected in parallel.

RC *Circuits*

When an emf $\mathscr{E}$ is applied to a resistance R and capacitance C in series, as in Fig. 28-13 with the switch at a, the charge on the capacitor increases according to

$$q = C\mathscr{E}(1 - e^{-t/RC}) \quad \text{(charging capacitor)}, \quad (28\text{-}30)$$

in which $C\mathscr{E} = q_0$ is the equilibrium (final) charge and $RC = \tau$ is the **capacitive time constant** of the circuit. During the charging, the current is

$$i = \frac{dq}{dt} = \left(\frac{\mathscr{E}}{R}\right)e^{-t/RC} \quad \begin{array}{l}\text{(charging} \\ \text{capacitor).}\end{array} \quad (28\text{-}31)$$

When a capacitor discharges through a resistance R, the charge on the capacitor decays according to

$$q = q_0 e^{-t/RC} \quad \text{(discharging capacitor)}. \quad (28\text{-}36)$$

During the discharging, the current is

$$i = \frac{dq}{dt} = -\left(\frac{q_0}{RC}\right)e^{-t/RC} \quad \begin{array}{l}\text{(discharging} \\ \text{capacitor).}\end{array} \quad (28\text{-}37)$$

QUESTIONS

1. Figure 28-16 shows current i passing through a battery. The following table gives four sets of values for i and the battery's emf $\mathscr{E}$ and internal resistance r; it also gives the *polarity* (orientation of the terminals) of the battery. Rank the sets according to the rate at which energy is transferred between the battery and the charge carriers, greatest transfer *to* the carriers first and greatest transfer *from* the carriers last.

	$\mathscr{E}$	r	i	POLARITY
(1)	$15\mathscr{E}_1$	0	i_1	+ at left
(2)	$10\mathscr{E}_1$	0	$2i_1$	+ at left
(3)	$10\mathscr{E}_1$	0	$2i_1$	− at left
(4)	$10\mathscr{E}_1$	r_1	$2i_1$	− at left

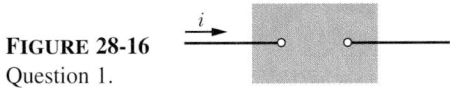

FIGURE 28-16
Question 1.

2. For each circuit in Fig. 28-17, are the resistors connected in series, in parallel, or neither?

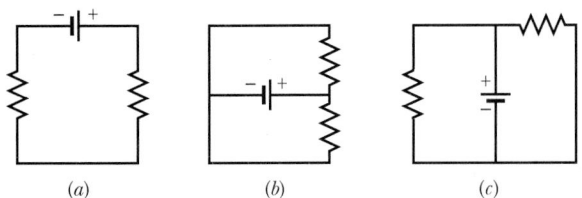

FIGURE 28-17 Question 2.

3. (a) In Fig. 28-18a, are resistors R_1 and R_3 in series? (b) Are resistors R_1 and R_2 in parallel? (c) Rank the equivalent resistances of the four circuits shown in Fig. 28-18, greatest first.

4. What is the equivalent resistance of three resistors, each of resistance R, if they are connected to an ideal battery (a) in series with one another and (b) in parallel with one another? (c) Is the potential difference across the series arrangement greater than, less than, or equal to that across the parallel arrangement?

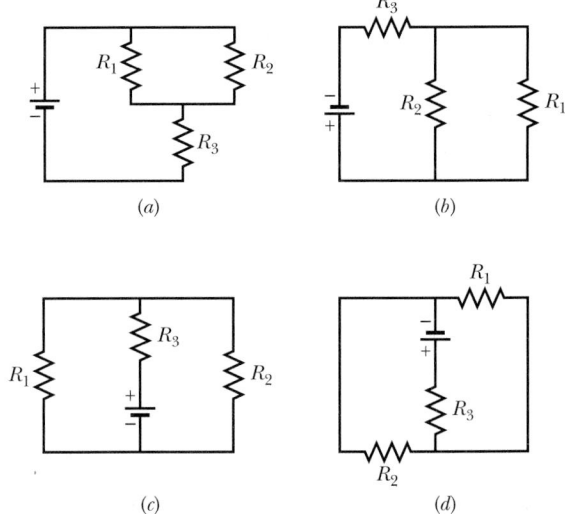

FIGURE 28-18 Questions 3 and 7.

5. You are to connect resistors R_1 and R_2, with $R_1 > R_2$, to a battery, first individually, then in series, and then in parallel. Rank those arrangements according to the amount of current through the battery, greatest first.

6. Two resistors are wired to a battery. (a) In which arrangement, parallel or series, are the potential differences across each resistor and across the equivalent resistance all equal? (b) In which arrangement are the currents through each resistor and through the equivalent resistance all equal?

7. (a) In Fig. 28-18a, with $R_1 > R_2$, is the potential difference across R_2 more than, less than, or equal to that across R_1? (b) Is the current through resistor R_2 more than, less than, or equal to that through resistor R_1?

8. Initially, a single resistor R_1 is wired to a battery. Then resistor R_2 is added in parallel. Are (a) the potential difference across R_1 and (b) the current i_1 through R_1 now more than, less than, or the same as previously? (c) Is the equivalent resistance R_{12} of R_1 and R_2 more than, less than, or equal to R_1? (d) Is the total current through R_1 and R_2 together more than, less than, or equal to the current through R_1 previously?

9. A resistor R_1 is wired to a battery; then resistor R_2 is added in series. Are (a) the potential difference across R_1 and (b) the current i_1 through R_1 now more than, less than, or the same as previously? (c) Is the equivalent resistance R_{12} of R_1 and R_2 more than, less than, or equal to R_1?

10. (a) In Fig. 28-19, when the branch with R_2 is added as indicated, does the rate at which electrical energy is transferred to thermal energy in R_1 increase, decrease, or stay the same? (b) Does the rate at which electrical energy is supplied by the battery increase, decrease, or stay the same? (c) Repeat (a) and (b) if, instead, R_2 is added in series with R_1.

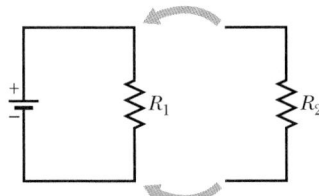

FIGURE 28-19 Question 10.

11. Without written calculation, determine the potential difference across each capacitor in Fig. 28-20.

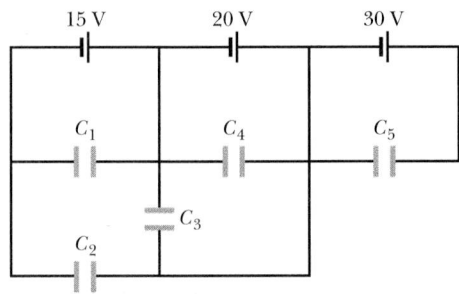

FIGURE 28-20 Question 11.

12. *Res-monster maze.* In Fig. 28-21, all the resistors have a resistance of 4.0 Ω and all the (ideal) batteries have an emf of 4.0 V. What is the current through resistor R? (If you can find the proper loop through this maze, you can answer the question with a few seconds of mental calculation.)

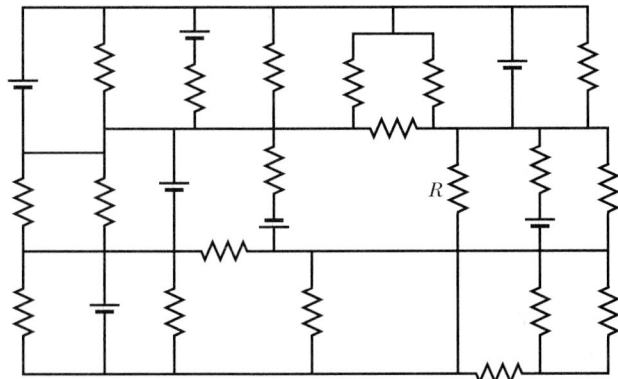

FIGURE 28-21 Question 12.

13. *Cap-monster maze.* In Fig. 28-22, all the capacitors have a capacitance of 6.0 μF, and all the batteries have an emf of 10 V. What is the charge on capacitor C? (If you can find the proper loop through this maze, you can answer the question with a few seconds of mental calculation.)

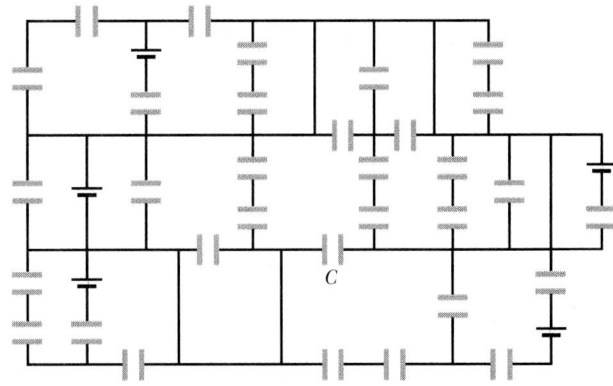

FIGURE 28-22 Question 13.

14. You are to connect n identical real batteries in series between circuit points a and b in Fig. 28-23a, and you have three choices: $n = 14$, $n = 12$, $n = 16$. Rank the choices according to (a) the total emf between a and b and (b) the total resistance between a and b, greatest first. Next, you are to connect the batteries in parallel between circuit points c and d in Fig. 28-23b. Rank the choices of n according to (c) the total emf between c and d and (d) the total resistance between c and d, greatest first.

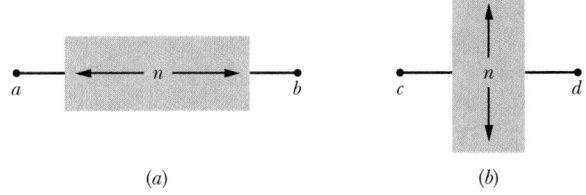

(a)   *(b)*

FIGURE 28-23 Question 14.

15. Figure 28-24 shows plots of $V(t)$ for three capacitors that discharge (separately) through the same resistor. Rank the plots according to the capacitances of the capacitors, greatest first.

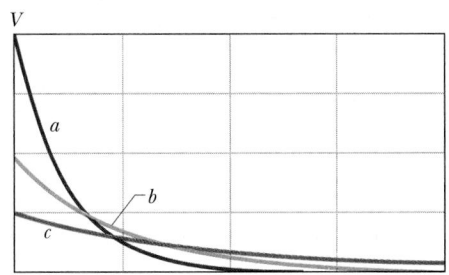

FIGURE 28-24 Question 15.

16. A capacitor is discharged from different initial charges and across different resistances in the three situations listed below. Rank the situations according to (a) the current through the resistance at the start of the discharge and (b) the time required for the current to decrease to half its starting value, greatest first.

	1	2	3
Initial charge	$12q$	$12q$	$6q$
Resistance	$2R$	$3R$	R

17. Figure 28-25 shows three sections of circuit that are to be connected in turn to the same battery via a switch as in Fig. 28-13. The resistors are all identical; so are the capacitors. Rank the sections according to (a) the final (equilibrium) charge on the capacitor and (b) the time required for the capacitor to reach 50% of its final charge, greatest first.

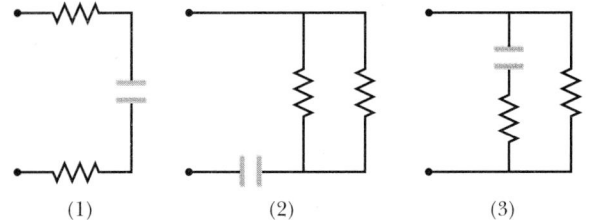

(1)   (2)   (3)

FIGURE 28-25   Question 17.

18. The five sections of circuit in Fig. 28-26 are to be connected, in turn, to the same 12 V battery via a switch as in Fig. 28-13. The resistors are all identical; so are the capacitors. Rank the sections according to the time required for the capacitors to reach 50% of their final potential, greatest first.

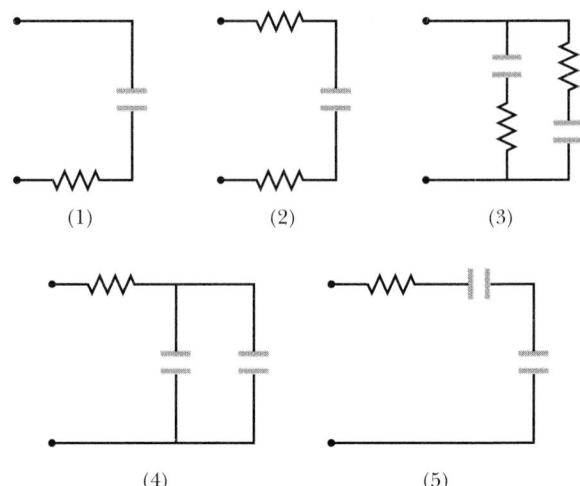

(1)   (2)   (3)

(4)   (5)

FIGURE 28-26   Questions 18 and 19.

19. Rank the five sections of Question 18 according to the potential across any resistor in the section when the potential across any capacitor in the section reaches 4 V, greatest first.

20. (a) Does the time required for the charge q on a capacitor in an RC circuit to build up to a certain fraction of its equilibrium value depend on the value of the applied emf? (b) Does the time required for the charge to change by a certain amount Δq depend on the applied emf? (c) Does the amount of charge for the fully charged capacitor depend on the internal resistance of the battery charging it?

EXERCISES & PROBLEMS

SECTION 28-5   Potential Differences

1E. A standard flashlight battery can deliver about 2.0 W·h of energy before it runs down. (a) If a battery costs 80¢, what is the cost of operating a 100 W lamp for 8.0 h using batteries? (b) What is the cost if power provided by an electric utility company, at 6¢ per kilowatt-hour, is used?

2E. (a) How much work does an ideal battery with a 12.0 V emf do on an electron that passes through the battery from the positive to the negative terminal? (b) If 3.4×10^{18} electrons pass through each second, what is the power of the battery?

3E. A 5.0 A current is set up in a circuit for 6.0 min by a rechargeable battery with a 6.0 V emf. By how much is the chemical energy of the battery reduced?

4E. A certain car battery with a 12 V emf has an initial charge of 120 A·h. Assuming that the potential across the terminals stays constant until the battery is completely discharged, for how many hours can it deliver energy at the rate of 100 W?

5E. In Fig. 28-27, $\mathscr{E}_1 = 12$ V and $\mathscr{E}_2 = 8$ V. (a) What is the direction of the current in the resistor? (b) Which battery is doing positive work? (c) Which point, A or B, is at the higher potential?

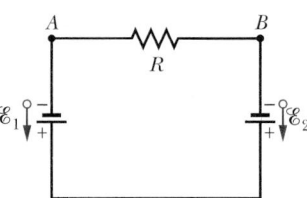

FIGURE 28-27   Exercise 5.

6E. Assume that the batteries in Fig. 28-28 have negligible internal resistance. Find (a) the current in the circuit, (b) the power dissipated in each resistor, and (c) the power of each battery, stating whether energy is supplied to or absorbed by it.

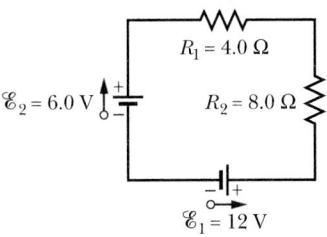

FIGURE 28-28 Exercise 6.

7E. A wire of resistance 5.0 Ω is connected to a battery whose emf $\mathscr{E}$ is 2.0 V and whose internal resistance is 1.0 Ω. In 2.0 min, (a) how much energy is transferred from chemical to electrical form? (b) How much energy appears in the wire as thermal energy? (c) Account for the difference between (a) and (b).

8E. In Fig. 28-4a, put $\mathscr{E} = 2.0$ V and $r = 100$ Ω. Plot (a) the current and (b) the potential difference across R, as functions of R over the range 0 to 500 Ω. Make both plots on the same graph. (c) Make a third plot by multiplying together, for various values of R, the corresponding values on the two plotted curves. What is the physical significance of this third plot?

9E. A car battery with a 12 V emf and an internal resistance of 0.040 Ω is being charged with a current of 50 A. (a) What is the potential difference across its terminals? (b) At what rate is energy being dissipated as thermal energy in the battery? (c) At what rate is electrical energy being converted to chemical energy? (d) What are the answers to (a) and (b) when the battery is used to supply 50 A to the starter motor?

10E. In Fig. 28-29, if the potential at point P is 100 V, what is the potential at point Q?

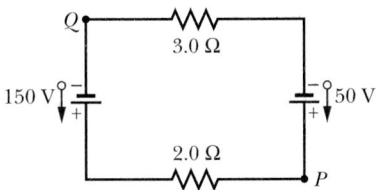

FIGURE 28-29 Exercise 10.

11E. In Fig. 28-30, the section of circuit AB absorbs 50 W of power when a current $i = 1.0$ A passes through it in the indicated direction. (a) What is the potential difference between A and B? (b) Emf device C does not have internal resistance. What is its emf? (c) What is its *polarity* (the orientation of its positive and negative terminals)?

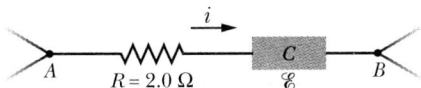

FIGURE 28-30 Exercise 11.

12E. In Fig. 28-5a calculate the potential difference across R_2, assuming $\mathscr{E} = 12$ V, $R_1 = 3.0$ Ω, $R_2 = 4.0$ Ω, and $R_3 = 5.0$ Ω.

13E. In Fig. 28-6a calculate the potential difference between a and c by considering a path that contains R, r_2, and $\mathscr{E}_2$. (See Sample Problem 28-2.)

14E. A gasoline gauge for an automobile is shown schematically in Fig. 28-31. The indicator (on the dashboard) has a resistance of 10 Ω. The tank unit is simply a float connected to a variable resistor whose resistance is 140 Ω when the tank is empty, is 20 Ω when the tank is full, and varies linearly with the volume of gasoline. Find the current in the circuit when the tank is (a) empty, (b) half-full, (c) and full.

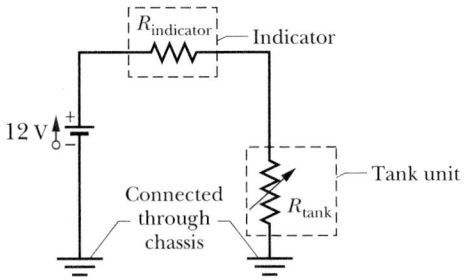

FIGURE 28-31 Exercise 14.

15P. A 10 km long underground cable extends east to west and consists of two parallel wires, each of which has resistance 13 Ω/km. A short develops at distance x from the west end when a conducting path of resistance R connects the wires (Fig. 28-32). The resistance of the wires and the short is then 100 Ω when the measurement is made from the east end, and 200 Ω when it is made from the west end. What are (a) x and (b) R?

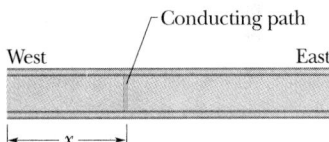

FIGURE 28-32 Problem 15.

16P. (a) In Fig. 28-33 what value must R have if the current in the circuit is to be 1.0 mA? Take $\mathscr{E}_1 = 2.0$ V, $\mathscr{E}_2 = 3.0$ V, and $r_1 = r_2 = 3.0$ Ω. (b) What is the rate at which thermal energy appears in R?

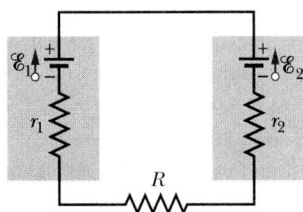

FIGURE 28-33 Problem 16.

17P. The current in a single-loop circuit with one resistance R is 5.0 A. When an additional resistance of 2.0 Ω is inserted in series with R, the current drops to 4.0 A. What is R?

18P. Thermal energy is to be generated in a 0.10 Ω resistor at the rate of 10 W by connecting the resistor to a battery whose emf is 1.5 V. (a) What potential difference must exist across the resistor? (b) What must be the internal resistance of the battery?

19P. Power is supplied by a device of emf $\mathcal{E}$ to a transmission line with resistance R. Find the ratio of the power dissipated in the line for $\mathcal{E} = 110,000$ V to that dissipated for $\mathcal{E} = 110$ V, assuming the power supplied is the same for the two cases.

20P. Wires A and B, having equal lengths of 40.0 m and equal diameters of 2.60 mm, are connected in series. A potential difference of 60.0 V is applied between the ends of the composite wire. The resistances of the wires are 0.127 and 0.729 Ω, respectively. Determine (a) the current density in each wire and (b) the potential difference across each wire. (c) Identify the wire materials. See Table 27-1.

21P. The starting motor of an automobile is turning too slowly, and the mechanic has to decide whether to replace the motor, the cable, or the battery. The manufacturer's manual says that the 12 V battery should have no more than 0.020 Ω internal resistance, the motor no more than 0.200 Ω resistance, and the cable no more than 0.040 Ω resistance. The mechanic turns on the motor and measures 11.4 V across the battery, 3.0 V across the cable, and a current of 50 A. Which part is defective?

22P. Two batteries having the same emf $\mathcal{E}$ but different internal resistances r_1 and r_2 $(r_1 > r_2)$ are connected in series to an external resistance R. (a) Find the value of R that makes the potential difference zero between the terminals of one battery. (b) Which battery is it?

23P. A solar cell generates a potential difference of 0.10 V when a 500 Ω resistor is connected across it, and a potential difference of 0.15 V when a 1000 Ω resistor is substituted. What are (a) the internal resistance and (b) the emf of the solar cell? (c) The area of the cell is 5.0 cm^2, and the rate per unit area at which it receives energy from light is 2.0 mW/cm^2. What is the efficiency of the cell for converting the light energy to thermal energy in the 1000 Ω external resistor?

24P. (a) In Fig. 28-4a, show that the rate at which energy is dissipated in R as thermal energy is a maximum when $R = r$. (b) Show that this maximum power is $P = \mathcal{E}^2/4r$.

25P. A battery of emf $\mathcal{E} = 2.00$ V and internal resistance $r = 0.500$ Ω is driving a motor. The motor is lifting a 2.00 N mass at constant speed $v = 0.500$ m/s. Assuming no energy losses, find (a) the current i in the circuit and (b) the potential difference V across the terminals of the motor. (c) Discuss the fact that there are two solutions to this problem.

26P. A temperature-stable resistor is made by connecting a resistor made of silicon in series with one made of iron. If the required total resistance is 1000 Ω in a wide temperature range around 20°C, what should be the resistances of the two resistors? See Table 27-1.

SECTION 28-6 Multiloop Circuits

27E. Four 18.0 Ω resistors are connected in parallel across a 25.0 V ideal battery. What is the current through the battery?

28E. A total resistance of 3.00 Ω is to be produced by connecting an unknown resistance to a 12.0 Ω resistance. What must be the value of the unknown resistance and should it be connected in series or in parallel?

29E. By using only two resistors—singly, in series, or in parallel—you are able to obtain resistances of 3.0, 4.0, 12, and 16 Ω. What are the two resistances?

30E. In Fig. 28-34, find the equivalent resistance between points (a) A and B, (b) A and C, and (c) B and C. (*Hint:* Imagine that a battery is connected between points A and C.)

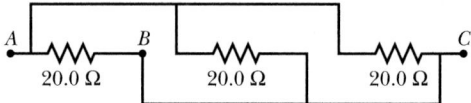

FIGURE 28-34 Exercise 30.

31E. In Fig. 28-35, find the equivalent resistance between points D and E. (*Hint:* Imagine that a battery is connected between points D and E.)

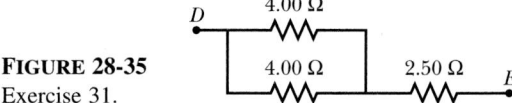

FIGURE 28-35
Exercise 31.

32E. In Fig. 28-36 find the current in each resistor and the potential difference between a and b. Put $\mathcal{E}_1 = 6.0$ V, $\mathcal{E}_2 = 5.0$ V, $\mathcal{E}_3 = 4.0$ V, $R_1 = 100$ Ω, and $R_2 = 50$ Ω.

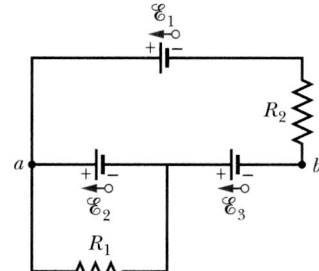

FIGURE 28-36
Exercise 32.

33E. Figure 28-37 shows a circuit containing three switches, labeled S_1, S_2, and S_3. Find the current at a for all possible combinations of switch settings. Put $\mathcal{E} = 120$ V, $R_1 = 20.0$ Ω, and $R_2 = 10.0$ Ω. Assume that the battery has no resistance.

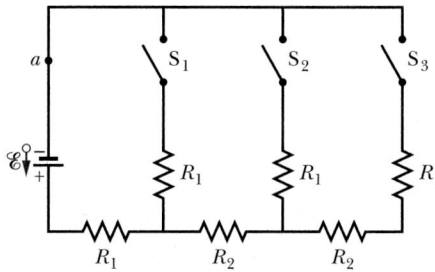

FIGURE 28-37 Exercise 33.

34E. Two lightbulbs, one of resistance R_1 and the other of resistance R_2, where $R_1 > R_2$, are connected to a battery (a) in parallel and (b) in series. Which bulb is brighter (dissipates more energy) in each case?

35E. In Fig. 28-7, calculate the potential difference between points c and d by as many paths as possible. Assume that $\mathcal{E}_1 = 4.0$ V, $\mathcal{E}_2 = 1.0$ V, $R_1 = R_2 = 10$ Ω, and $R_3 = 5.0$ Ω.

36E. Nine copper wires of length *l* and diameter *d* are connected in parallel to form a single composite conductor of resistance *R*. What must be the diameter *D* of a single copper wire of length *l* if it is to have the same resistance?

37E. A 120 V power line is protected by a 15 A fuse. What is the maximum number of 500 W lamps that can be simultaneously operated in parallel on this line without "blowing" the fuse because of an excess of current?

38E. A circuit containing five resistors connected to a battery with a 12.0 V emf is shown in Fig. 28-38. What is the potential difference across the 5.0 Ω resistor?

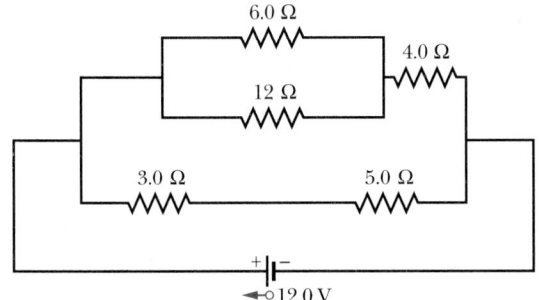

FIGURE 28-38   Exercise 38.

39P. In Fig. 28-39, find the equivalent resistance between points (a) *F* and *H* and (b) *F* and *G*. (*Hint:* For each pair of points, imagine that a battery is connected across the pair.)

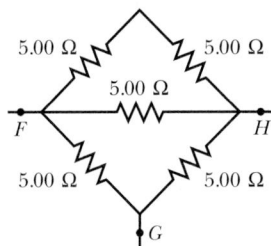

FIGURE 28-39   Problem 39.

40P. Two resistors R_1 and R_2 may be connected either in series or in parallel across an ideal battery with emf $\mathscr{E}$. We desire the rate of electrical energy dissipation of the parallel combination to be five times that of the series combination. If $R_1 = 100$ Ω, what is R_2? (*Hint:* There are two answers.)

41P. You are given a number of 10 Ω resistors, each capable of dissipating only 1.0 W without being destroyed. What is the minimum number of such resistors that you need to combine in series or parallel to make a 10 Ω resistance that is capable of dissipating at least 5.0 W?

42P. Two batteries of emf $\mathscr{E}$ and internal resistance *r* are connected in parallel across a resistor *R*, as in Fig. 28-40a. (a) For what value of *R* is the rate of electrical energy dissipation by the resistor a maximum? (b) What is the maximum energy dissipation rate?

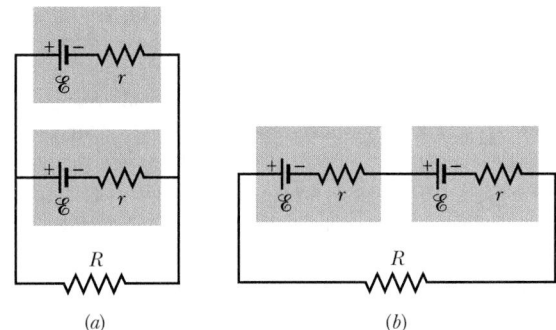

FIGURE 28-40   Problems 42 and 44.

43P. (a) Calculate the current through each ideal battery in Fig. 28-41. Assume that $R_1 = 1.0$ Ω, $R_2 = 2.0$ Ω, $\mathscr{E}_1 = 2.0$ V, and $\mathscr{E}_2 = \mathscr{E}_3 = 4.0$ V. (b) Calculate $V_a - V_b$.

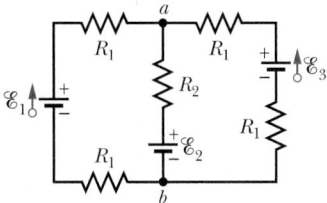

FIGURE 28-41   Problem 43.

44P. You are given two batteries of emf $\mathscr{E}$ and internal resistance *r*. They may be connected either in parallel (Fig. 28-40a) or in series (Fig. 28-40b) and are used to establish a current in a resistor *R*. (a) Derive expressions for the current in *R* for both arrangements. Which will yield the larger current (b) when $R > r$ and (c) when $R < r$?

45P. A group of *N* identical batteries of emf $\mathscr{E}$ and internal resistance *r* may be connected all in series (Fig. 28-42a) or all in parallel (Fig. 28-42b) and then across a resistor *R*. Show that both arrangements will give the same current in *R* if $R = r$.

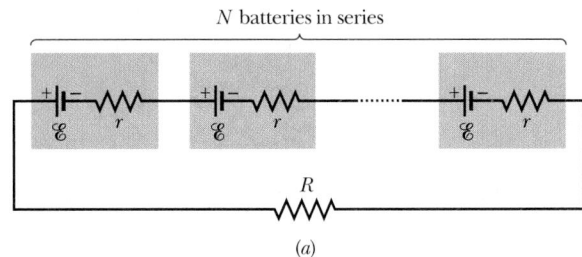

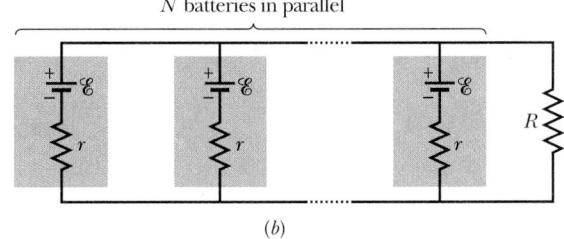

FIGURE 28-42   Problem 45.

46P. A three-way 120 V lamp bulb that contains two filaments is rated for 100-200-300 W. One filament burns out. Afterward, the bulb operates at the same intensity (dissipates energy at the same rate) on its lowest and its highest switch positions but does not operate at all on the middle position. (a) How are the two filaments wired to the three switch positions? (b) Calculate the resistances of the filaments.

47P. (a) In Fig. 28-43, what is the equivalent resistance of the network shown? (b) What is the current in each resistor? Put $R_1 = 100 \, \Omega$, $R_2 = R_3 = 50 \, \Omega$, $R_4 = 75 \, \Omega$, and $\mathcal{E} = 6.0$ V; assume the battery is ideal.

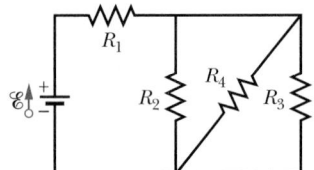

FIGURE 28-43
Problem 47.

48P. In Fig. 28-44, $\mathcal{E}_1 = 3.00$ V, $\mathcal{E}_2 = 1.00$ V, $R_1 = 5.00 \, \Omega$, $R_2 = 2.00 \, \Omega$, $R_3 = 4.00 \, \Omega$, and both batteries are ideal. (a) What is the rate at which energy is dissipated in R_1? In R_2? In R_3? (b) What is the power of battery 1? Of battery 2?

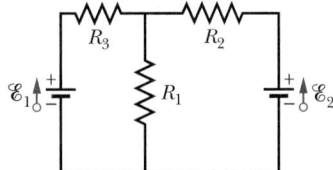

FIGURE 28-44
Problem 48.

49P. In the circuit of Fig. 28-45, for what value of R will the ideal battery transfer energy to the resistors (a) at a rate of 60.0 W, (b) at the maximum possible rate, and (c) at the minimum possible rate? (d) For (b) and (c), what are those rates?

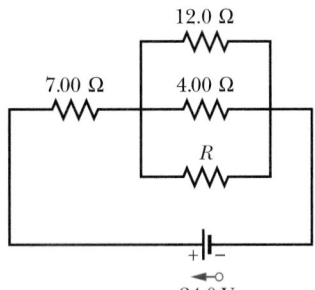

FIGURE 28-45
Problem 49.

50P. In the circuit of Fig. 28-46, $\mathcal{E}$ has a constant value but R can be varied. Find the value of R that results in the maximum heating in that resistor. The battery is ideal.

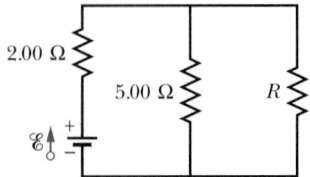

FIGURE 28-46
Problem 50.

51P. A copper wire of radius $a = 0.250$ mm has an aluminum jacket of outer radius $b = 0.380$ mm. (a) There is a current $i = $ 2.00 A in the composite wire. Using Table 27-1, calculate the current in each material. (b) What is the length of the composite wire if a potential difference $V = 12.0$ V between the ends maintains the current?

52P. Figure 28-47 shows a battery connected across a uniform resistor R_0. A sliding contact can move across the resistor from $x = 0$ at the left to $x = 10$ cm at the right. Moving the contact changes how much resistance is to the left of the contact and how much is to the right. Find an expression for the power dissipated in resistor R as a function of x. Plot the function for $\mathcal{E} = 50$ V, $R = 2000 \, \Omega$, and $R_0 = 100 \, \Omega$.

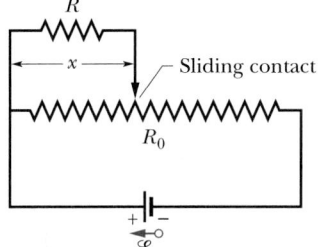

FIGURE 28-47
Problem 52.

SECTION 28-7   The Ammeter and the Voltmeter

53E. A simple ohmmeter is made by connecting a 1.50 V flashlight battery in series with a resistance R and an ammeter that reads from 0 to 1.00 mA, as shown in Fig. 28-48. Resistance R is adjusted so that when the clip leads are shorted together, the meter deflects to its full-scale value of 1.00 mA. What external resistance across the leads results in a deflection of (a) 10%, (b) 50%, and (c) 90% of full scale? (d) If the ammeter has a resistance of 20.0 Ω and the internal resistance of the battery is negligible, what is the value of R?

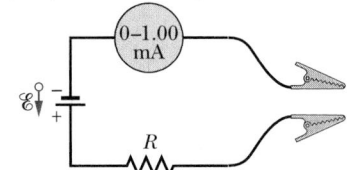

FIGURE 28-48
Exercise 53.

54E. For sensitive manual control of current in a circuit, you can use a parallel combination of variable resistors of the sliding contact type, as in Fig. 28-49. (Moving the contact changes how much resistance is in the circuit.) Suppose the full resistance R_1 of resistor A is 20 times the full resistance R_2 of resistor B. (a) What procedure should be used to adjust the current i to the desired value? (b) Why is the parallel combination better than a single variable resistor?

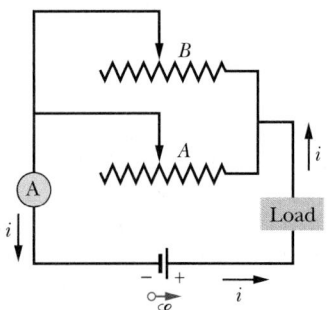

FIGURE 28-49
Exercise 54.

55P. (a) In Fig. 28-50, determine what the ammeter will read, assuming $\mathscr{E} = 5.0$ V (for the ideal battery), $R_1 = 2.0\ \Omega$, $R_2 = 4.0\ \Omega$, and $R_3 = 6.0\ \Omega$. (b) The ammeter and the source of emf are now physically interchanged. Show that the ammeter reading remains unchanged.

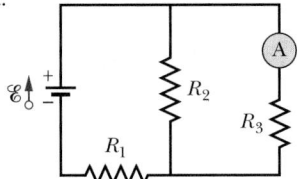

FIGURE 28-50
Problem 55.

56P. What current, in terms of $\mathscr{E}$ and R, does the ammeter in Fig. 28-51 read? Assume that it has zero resistance and that the battery is ideal.

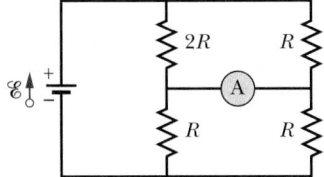

FIGURE 28-51
Problem 56.

57P. When the lights of an automobile are switched on, an ammeter in series with them reads 10 A and a voltmeter connected across them reads 12 V. See Fig. 28-52. When the electric starting motor is turned on, the ammeter reading drops to 8.0 A and the lights dim somewhat. If the internal resistance of the battery is 0.050 Ω and that of the ammeter is negligible, what are (a) the emf of the battery and (b) the current through the starting motor when the lights are on?

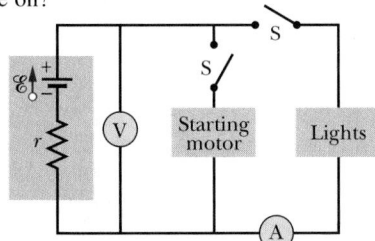

FIGURE 28-52
Problem 57.

58P. In Fig. 28-12, assume that $\mathscr{E} = 3.0$ V, $r = 100\ \Omega$, $R_1 = 250\ \Omega$, and $R_2 = 300\ \Omega$. If the voltmeter resistance $R_V = 5.0$ kΩ, what percent error is made in reading the potential difference across R_1? Ignore the presence of the ammeter.

59P. In Fig. 28-12, assume that $\mathscr{E} = 5.0$ V, $r = 2.0\ \Omega$, $R_1 = 5.0\ \Omega$, and $R_2 = 4.0\ \Omega$. If the ammeter resistance $R_A = 0.10\ \Omega$, what percent error is made in reading the current? Assume that the voltmeter is not present.

60P. A voltmeter (resistance R_V) and an ammeter (resistance R_A) are connected to measure a resistance R, as in Fig. 28-53a. The resistance is given by $R = V/i$, where V is the voltmeter reading and i is the current in the resistor R. Some of the current (i') registered by the ammeter goes through the voltmeter so that the ratio of the meter readings ($= V/i'$) gives only an *apparent* resistance reading R'. Show that R and R' are related by

$$\frac{1}{R} = \frac{1}{R'} - \frac{1}{R_V}.$$

Note that as $R_V \to \infty$, $R' \to R$.

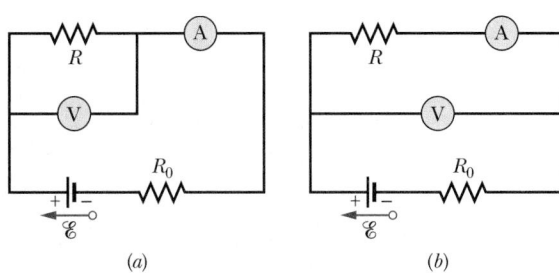

(a) (b)

FIGURE 28-53 Problems 60, 61, and 62.

61P. (See Problem 60.) If meters are used to measure resistance, they may also be connected as in Fig. 28-53b. Again the ratio of the meter readings gives only an apparent resistance R'. Show that now R' is related to R by

$$R = R' - R_A,$$

in which R_A is the ammeter resistance. Note that as $R_A \to 0$, $R' \to R$.

62P. (See Problems 60 and 61.) In Fig. 28-53 the ammeter and voltmeter resistances are 3.00 and 300 Ω, respectively. Take $\mathscr{E} = 12.0$ V for the ideal battery and $R_0 = 100\ \Omega$. If $R = 85.0\ \Omega$, (a) what will the meters read for the two different connections? (b) What apparent resistance R' will be computed in each case?

63P. In Fig. 28-54, R_s is to be adjusted in value by moving the sliding contact across it until points a and b are brought to the same potential. (One tests for this condition by momentarily connecting a sensitive ammeter between a and b; if these points are at the same potential, the ammeter will not deflect.) Show that when this adjustment is made, the following relation holds:

$$R_x = R_s \left(\frac{R_2}{R_1} \right).$$

An unknown resistance (R_x) can be measured in terms of a standard (R_s) using this device, which is called a Wheatstone bridge.

64P. (a) If points a and b in Fig. 28-54 are connected by a wire of resistance r, show that the current in the wire is

$$i = \frac{\mathscr{E}(R_s - R_x)}{(R + 2r)(R_s + R_x) + 2R_s R_x},$$

where $\mathscr{E}$ is the emf of the ideal battery and $R = R_1 = R_2$. Assume that R_0 equals zero. (b) Is this formula consistent with the result of Problem 63?

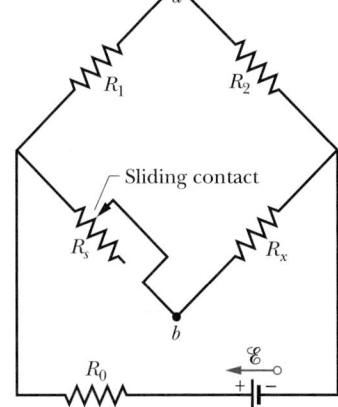

FIGURE 28-54
Problems 63 and 64.

SECTION 28-8 *RC* Circuits

65E. A capacitor with initial charge q_0 is discharged through a resistor. In terms of the time constant τ, how long is required for the capacitor to lose (a) the first one-third of its charge and (b) two-thirds of its charge?

66E. In an *RC* series circuit, $\mathscr{E} = 12.0$ V, $R = 1.40$ MΩ, and $C = 1.80$ μF. (a) Calculate the time constant. (b) Find the maximum charge that will appear on the capacitor during charging. (c) How long does it take for the charge to build up to 16.0 μC?

67E. How many time constants must elapse for an initially uncharged capacitor in an *RC* series circuit to be charged to 99.0% of its equilibrium charge?

68E. A 15.0 kΩ resistor and a capacitor are connected in series and then a 12.0 V potential difference is suddenly applied across them. The potential difference across the capacitor rises to 5.00 V in 1.30 μs. (a) Calculate the time constant of the circuit. (b) Find the capacitance of the capacitor.

69P. A 3.00 MΩ resistor and a 1.00 μF capacitor are connected in series with an ideal battery of $\mathscr{E} = 4.00$ V. At 1.00 s after the connection is made, what are the rates at which (a) the charge of the capacitor is increasing, (b) energy is being stored in the capacitor, (c) thermal energy is appearing in the resistor, and (d) energy is being delivered by the battery?

70P. A capacitor with an initial potential difference of 100 V is discharged through a resistor when a switch between them is closed at $t = 0$. At $t = 10.0$ s, the potential difference across the capacitor is 1.00 V. (a) What is the time constant of the circuit? (b) What is the potential difference across the capacitor at $t = 17.0$ s?

71P. Figure 28-55 shows the circuit of a flashing lamp, like those attached to barrels at highway construction sites. The fluorescent lamp L (of negligible capacitance) is connected in parallel across the capacitor C of an *RC* circuit. There is a current through the lamp only when the potential difference across it reaches the breakdown voltage V_L; in this event, the capacitor discharges completely through the lamp and the lamp flashes briefly. Suppose that two flashes per second are needed. For a lamp with breakdown voltage $V_L = 72.0$ V, a 95.0 V ideal battery, and a 0.150 μF capacitor, what should be the resistance R of the resistor?

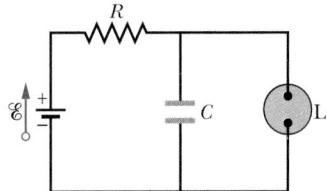

FIGURE 28-55
Problem 71.

72P. A 1.0 μF capacitor with an initial stored energy of 0.50 J is discharged through a 1.0 MΩ resistor. (a) What is the initial charge on the capacitor? (b) What is the current through the resistor when the discharge starts? (c) Determine V_C, the potential difference across the capacitor, and V_R, the potential difference across the resistor, as functions of time. (d) Express the production rate of thermal energy in the resistor as a function of time.

73P. The potential difference between the plates of a leaky (meaning that charge leaks from one plate to the other) 2.0 μF capacitor drops to one-fourth its initial value in 2.0 s. What is the equivalent resistance between the capacitor plates?

74P. An initially uncharged capacitor C is fully charged by a device of constant emf $\mathscr{E}$ connected in series with a resistor R. (a) Show that the final energy stored in the capacitor is half the energy supplied by the emf device. (b) By direct integration of i^2R over the charging time, show that the thermal energy dissipated by the resistor is also half the energy supplied by the emf device.

75P. A controller on an electronic arcade game consists of a variable resistor connected across the plates of a 0.220 μF capacitor. The capacitor is charged to 5.00 V, then discharged through the resistor. The time for the potential difference across the plates to decrease to 0.800 V is measured by a clock inside the game. If the range of discharge times that can be handled effectively is from 10.0 μs to 6.00 ms, what should be the resistance range of the resistor?

76P. The circuit of Fig. 28-56 shows a capacitor C, two ideal batteries, two resistors, and a switch S. Initially S has been open for a long time. If it is then closed for a long time, by how much does the charge on the capacitor change? Assume $C = 10$ μF, $\mathscr{E}_1 = 1.0$ V, $\mathscr{E}_2 = 3.0$ V, $R_1 = 0.20$ Ω, and $R_2 = 0.40$ Ω.

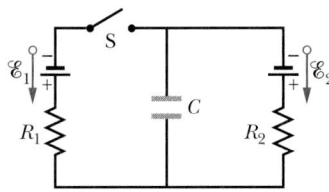

FIGURE 28-56
Problem 76.

77P*. In the circuit of Fig. 28-57, $\mathscr{E} = 1.2$ kV, $C = 6.5$ μF, $R_1 = R_2 = R_3 = 0.73$ MΩ. With C completely uncharged, switch S is suddenly closed (at $t = 0$). (a) Determine the current through each resistor for $t = 0$ and $t = \infty$. (b) Draw qualitatively a graph of the potential difference V_2 across R_2 from $t = 0$ to $t = \infty$. (c) What are the numerical values of V_2 at $t = 0$ and $t = \infty$? (d) Give the physical meaning of "$t = \infty$" in this case.

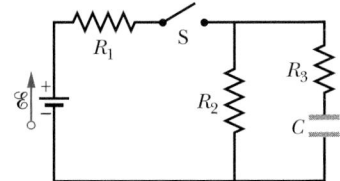

FIGURE 28-57
Problem 77.

Electronic Computation

78. Figure 28-58 shows a portion of a circuit. The rest of the circuit draws current i at the connections A and B, as indicated. Take $\mathscr{E}_1 = 10$ V, $\mathscr{E}_2 = 15$ V, $R_1 = R_2 = 5.0$ Ω, $R_3 = R_4 = 8.0$ Ω, and $R_5 = 12$ Ω. (a) For each of four values of i—0, 4.0, 8.0, and 12 A—find the current through each ideal battery and tell if the battery is charging or discharging. Also find the potential difference V_{AB}. (b) The portion of the circuit not shown consists of an emf and a resistor in series. What are their values?

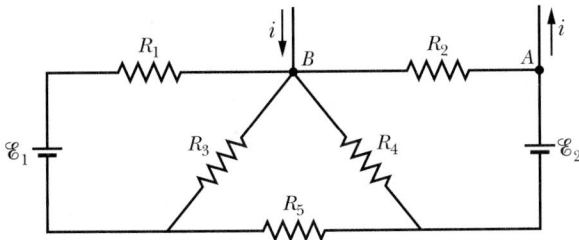

FIGURE 28-58 Problem 78.

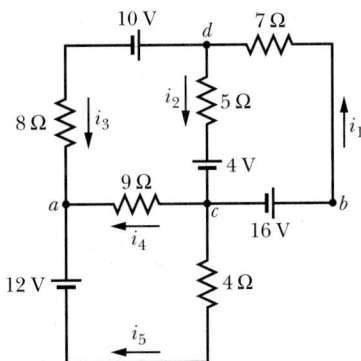

FIGURE 28-59
Problems 80 and 81.

79. The following table gives the electric potential difference V_T across the terminals of a battery as a function of current i being drawn from the battery. (a) Write an equation that represents the relationship between the terminal potential difference V_T and the current i. Enter the data into your graphing calculator and perform a linear regression fit of V_T versus i. (b) From the parameters of the fit, find (b) the battery's emf and (c) its internal resistance.

i (A):	50	75	100	125	150	175	200
V_T (V):	10.7	9.0	7.7	6.0	4.8	3.0	1.7

80. Consider the circuit in Fig. 28-59. (a) Apply the junction rule to junctions d and a and the loop rule to the three loops to produce five simultaneous, linearly independent equations. (b) Represent the five linear equations by the matrix equation $[A][B] = [C]$, where

$$[B] = \begin{bmatrix} i_1 \\ i_2 \\ i_3 \\ i_4 \\ i_5 \end{bmatrix}.$$

What are the matrices $[A]$ and $[C]$? (c) Have the calculator perform $[A]^{-1}[C]$ to find the values of i_1, i_2, i_3, i_4, and i_5.

81. For the same situation as in Problem 80 and having already solved for the five unknown currents, do the following. (a) Find the electric potential difference across the 9 Ω resistor. (b) Find the rate at which work is being done on the 7 Ω resistor. (c) Find the rate at which the 12 V battery is doing work on the circuit. (d) Find the rate at which the 4 V battery is doing work on the circuit. (e) Of the points in the circuit labeled a and c, which is at the higher electric potential?

82. A capacitor with capacitance C_0, after having been connected to a battery with emf $\mathscr{E}_0$ for a long time, is discharged through a 200,000 Ω resistor at time $t = 0$. The potential difference across the capacitor is then measured as a function of time for a brief time interval; the results are recorded below. (a) Write an equation that describes the potential difference across the capacitor as a function of time. Enter the data into your calculator and have the calculator perform a linear regression fit of $\ln V_C$ versus t. From the parameters of the fit, determine (b) the emf $\mathscr{E}_0$ of the battery and (c) the time constant τ for the circuit. (d) Finally, determine the value of C_0.

V_C (V):	9.9	7.2	5.7	4.4	3.4	2.7	2.0
t (s):	0.2	0.4	0.6	0.8	1.0	1.2	1.4

29
Magnetic Fields

If you are outside on a dark night in the middle to high latitudes, you might be able to see an aurora, a ghostly "curtain" of light that hangs down from the sky. This curtain is not just local: it may be several hundred kilometers high and several thousand kilometers long, stretching around Earth in an arc. However, it is less than 1 km thick. What produces this huge display, and what makes it so thin?

29-1 THE MAGNETIC FIELD

We have discussed how a charged plastic rod produces a vector field—the electric field **E**—at all points in the space around it. Similarly, a magnet produces a vector field—the **magnetic field B**—at all points in the space around it. You get a hint of that magnetic field whenever you attach a note to a refrigerator door with a small magnet, or accidentally erase a computer disk by bringing it near a magnet. The magnet acts on the door or disk *by means of* its magnetic field.

In a familiar type of magnet, a wire coil is wound around an iron core and a current is sent through the coil; the strength of the magnetic field is determined by the size of the current. In industry, such **electromagnets** are used for sorting scrap iron (Fig. 29-1) among many other things. You are probably more familiar with **permanent magnets** —magnets, like the refrigerator-door type, that do not need current to have a magnetic field.

FIGURE 29-1 Scrap metal collected by an electromagnet at a steel mill.

In Chapter 23 we saw that an *electric charge* sets up an electric field that can then affect other electric charges. Here, we might reasonably expect that a *magnetic charge* sets up a magnetic field that can then affect other magnetic charges. Although such magnetic charges, called *magnetic monopoles,* are predicted by certain theories, their existence has not been confirmed.

So then, how are magnetic fields set up? There are two ways. (1) Moving electrically charged particles, such as a current in a wire, create magnetic fields. (2) Elementary particles such as electrons have an intrinsic magnetic field around them; that is, these fields are a basic characteristic of the particles, just as are their mass and electric charge (or lack of charge). As we shall discuss in Chapter 32, the magnetic fields of the electrons in certain materials add together to give a net magnetic field around the material. This is true for the material in permanent magnets (which is good, because they can then hold notes to a refrigerator door). In other materials, the magnetic fields of all the electrons cancel out, giving no net magnetic field surrounding the material. This is true for the material in your body (which is also good, because otherwise you might be slammed up against a refrigerator door every time you passed one).

Experimentally we find that when a charged particle (either alone or part of a current) moves through a magnetic field, a force due to the field can act on the particle. In this chapter we focus on the relation between the magnetic field and this force.

29-2 THE DEFINITION OF B

We determined the electric field **E** at a point by putting a test particle of charge q at rest at that point and measuring the electric force $\mathbf{F}_E$ acting on the particle. We then defined **E** as

$$\mathbf{E} = \frac{\mathbf{F}_E}{q}. \tag{29-1}$$

If a magnetic monopole were available, we could define **B** in a similar way. Because such particles have not been found, we must define **B** in another way, in terms of the magnetic force $\mathbf{F}_B$ exerted on a moving electrically charged test particle.

In principle, we do this by firing a charged particle through the point where **B** is to be defined, using various directions and speeds for the particle and determining the force $\mathbf{F}_B$ that acts on the particle at that point. After many such trials we would find that when the particle's velocity **v** is along a particular axis through the point, force $\mathbf{F}_B$ is zero. For all other directions of **v,** the magnitude of $\mathbf{F}_B$ is always proportional to $v \sin \phi$, where ϕ is the angle between the zero-force axis and the direction of **v.** Furthermore, the direction of $\mathbf{F}_B$ is always perpendicular to the direction of **v**. (These results suggest that a cross product is involved.)

We can then define a magnetic field **B** to be a vector quantity that is directed along the zero-force axis. We can next measure the magnitude of $\mathbf{F}_B$ when **v** is directed perpendicular to that axis and then define the magnitude of **B** in terms of that force magnitude:

$$B = \frac{F_B}{|q|v},$$

where q is the charge of the particle.

We can summarize all these results with the following vector equation:

$$\mathbf{F}_B = q\mathbf{v} \times \mathbf{B}. \qquad (29\text{-}2)$$

That is, the force $\mathbf{F}_B$ on the particle is equal to the charge q times the cross product of its velocity **v** and the magnetic field **B**. Using Eq. 3-20 to evaluate the cross product, we can write the magnitude of $\mathbf{F}_B$ as

$$F_B = |q|vB \sin \phi, \qquad (29\text{-}3)$$

where ϕ is the angle between the directions of velocity **v** and magnetic field **B**.

Finding the Magnetic Force on a Particle

Equation 29-3 tells us that the magnitude of the force $\mathbf{F}_B$ acting on a particle in a magnetic field is proportional to the charge q and speed v of the particle. Thus, the force is equal to zero if the charge is zero or if the particle is stationary. Equation 29-3 also tells us that the magnitude of the force is zero if **v** and **B** are either parallel ($\phi = 0°$) or antiparallel ($\phi = 180°$), and the force is at its maximum when **v** and **B** are perpendicular to each other.

Equation 29-2 tells us all this plus the direction of $\mathbf{F}_B$. From Section 3-7, we know that the cross product **v** × **B** in Eq. 29-2 is a vector that is perpendicular to the two vectors **v** and **B**. The right-hand rule (Fig. 29-2a) tells us that the thumb of the right hand points in the direction of **v** × **B** when the fingers sweep **v** into **B**. If q is positive, then (by Eq. 29-2) the force $\mathbf{F}_B$ has the same sign as **v** × **B** and thus must be in the same direction. That is, for positive q, $\mathbf{F}_B$ points along the thumb as in Figs. 29-2b. If q is negative, then the force $\mathbf{F}_B$ and the cross product **v** × **B** have opposite signs and thus must be in opposite directions. So, for negative q, $\mathbf{F}_B$ points opposite the thumb as in Fig. 29-2c.

Regardless of the sign of the charge, however,

The force $\mathbf{F}_B$ acting on a charged particle moving with velocity **v** through a magnetic field **B** is *always* perpendicular to **v** and **B**.

Thus $\mathbf{F}_B$ *never* has a component parallel to **v**. This means that $\mathbf{F}_B$ cannot change the particle's speed v (and thus it cannot change the particle's kinetic energy). The force can change only the direction of **v** (and thus the direction of travel); only in this sense does $\mathbf{F}_B$ accelerate the particle.

To develop a feeling for Eq. 29-2, consider Fig. 29-3, which shows some tracks left by charged particles moving rapidly through a *bubble chamber* at the Lawrence Berkeley Laboratory. The chamber, which is filled with liquid hydrogen, is immersed in a strong uniform magnetic field that points out of the plane of the figure. At the left in Fig. 29-3 an incoming gamma ray—which leaves no track because it is uncharged—transforms into an electron (spiral track marked e⁻) and a positron (track marked e⁺) while it knocks an electron out of a hydrogen atom (long track marked e⁻). Check with Eq. 29-2 and Fig. 29-2 that the three tracks made by these two negative particles and one positive particle curve in the proper directions.

The SI unit for **B** that follows from Eqs. 29-2 and 29-3 is the newton per coulomb-meter per second. For conve-

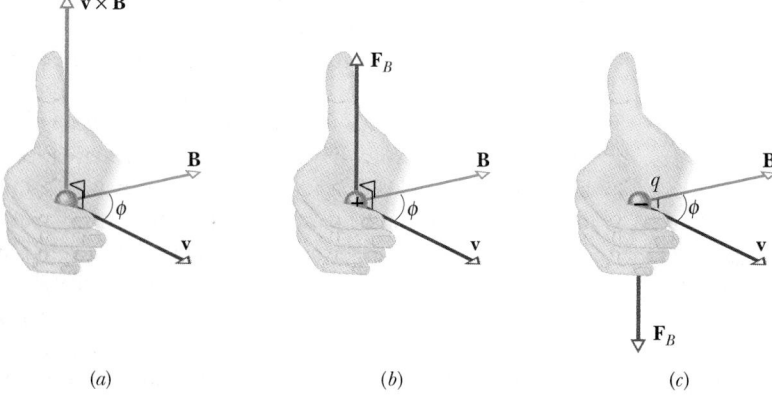

(a)   (b)   (c)

FIGURE 29-2 (a) The right-hand rule (in which **v** is swept into **B** through the smaller angle ϕ between them) gives the direction of **v** × **B** as the direction of the thumb. (b) If q is positive, then the direction of $\mathbf{F}_B = q\mathbf{v} \times \mathbf{B}$ is in the direction of **v** × **B**. (c) If q is negative, then the direction of $\mathbf{F}_B$ is opposite that of **v** × **B**.

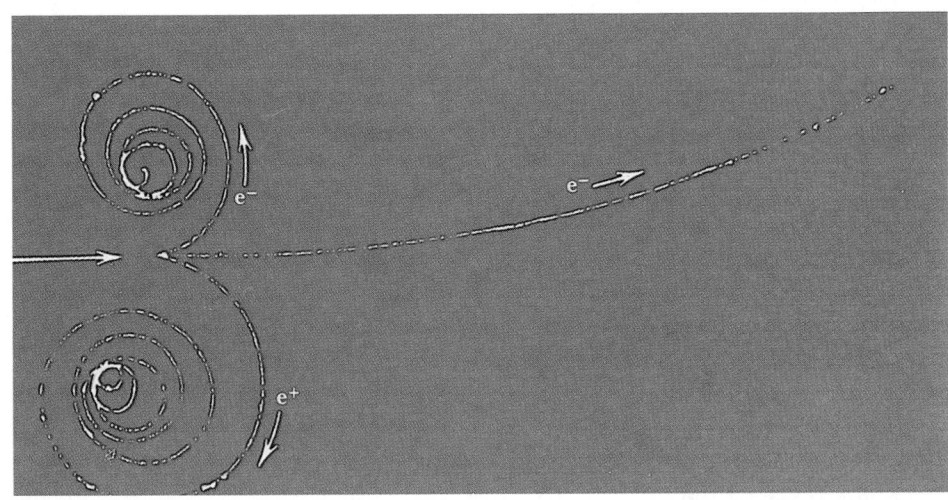

FIGURE 29-3 The tracks of two electrons (e⁻) and a positron (e⁺) in a bubble chamber that is immersed in a uniform magnetic field that points out of the plane of the page.

nience, this is called the **tesla** (T):

$$1 \text{ tesla} = 1 \text{ T} = 1 \frac{\text{newton}}{(\text{coulomb})(\text{meter/second})}.$$

Recalling that a coulomb per second is an ampere, we have

$$1 \text{ T} = 1 \frac{\text{newton}}{(\text{coulomb/second})(\text{meter})}$$

$$= 1 \frac{\text{N}}{\text{A} \cdot \text{m}}. \qquad (29\text{-}4)$$

An earlier (non-SI) unit for **B**, still in common use, is the *gauss* (G), and

$$1 \text{ tesla} = 10^4 \text{ gauss}. \qquad (29\text{-}5)$$

Table 29-1 lists the magnetic fields that occur in a few situations. Note that Earth's magnetic field near the planet's surface is about 10^{-4} T ($= 100 \ \mu$T or 1 gauss).

C HECKPOINT **1:** The figure shows three situations in which a charged particle with velocity **v** travels through a uniform magnetic field **B**. In each situation, what is the direction of the magnetic force **F**$_B$ on the particle?

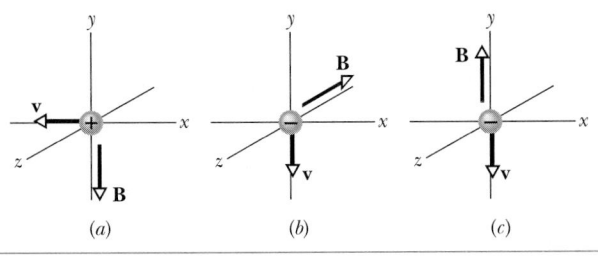

(a) (b) (c)

TABLE 29-1   SOME APPROXIMATE MAGNETIC FIELDS

At the surface of a neutron star	10^8 T
Near a big electromagnet	1.5 T
Near a small bar magnet	10^{-2} T
At Earth's surface	10^{-4} T
In interstellar space	10^{-10} T
Smallest value in a magnetically shielded room	10^{-14} T

Magnetic Field Lines

We can represent magnetic fields with field lines, just as we did for electric fields. Similar rules apply. That is, (1) the direction of the tangent to a magnetic field line at any point gives the direction of **B** at that point, and (2) the spacing of the lines represents the magnitude of **B**—the magnetic field is stronger where the lines are closer together, and conversely.

Figure 29-4a shows how the magnetic field near a *bar magnet* (a permanent magnet in the shape of a bar) can be represented by magnetic field lines. The lines all pass through the magnet, and they form closed loops (even those that are not shown closed in the figure). The external magnetic effects of a bar magnet are strongest near its ends, where the field lines are most closely spaced. Thus the magnetic field of the bar magnet in Fig. 29-4b collects the iron filings near the two ends of the magnet.

Because a magnetic field has direction, the (closed) field lines enter one end of a magnet and exit the other end. The end of a magnet from which the field lines emerge is called the *north pole* of the magnet; the other end, where field lines enter the magnet, is called the *south pole*. The magnets we use to fix notes on refrigerators are short bar

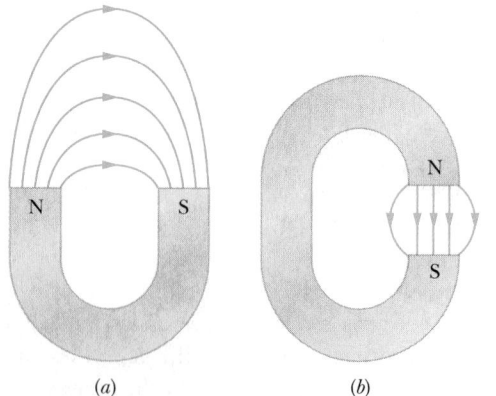

FIGURE 29-4 (a) The magnetic field lines for a bar magnet. (b) A ''cow magnet'': a bar magnet that is intended to be slipped down into the rumen of a cow to prevent accidentally ingested bits of scrap iron from reaching the cow's intestines.

magnets. Figure 29-5 shows two other common shapes for magnets: a *horseshoe magnet* and a magnet that has been bent around into the shape of a C so that the *pole faces* are facing each other. (The magnetic field between the pole faces can then be approximately uniform.) Regardless of the shape of the magnets, if we place two of them near each other we find:

FIGURE 29-5 (a) A horseshoe magnet and (b) a C-shaped magnet. (Only some of the external field lines are shown.)

Opposite magnetic poles attract each other, and like magnetic poles repel each other.

Earth has a magnetic field that is produced in its core by still unknown mechanisms. On Earth's surface, we can detect this magnetic field with a compass, which is essentially a slender bar magnet on a low-friction pivot. This bar magnet, or this needle, turns because its north-pole end is attracted toward the Arctic region of Earth. Thus, the *south pole* of Earth's magnetic field must be located toward the Arctic. Logically, we then should call the pole there a south pole. However, because we call that direction north, we are trapped into the statement that Earth has a *geomagnetic north pole* in that direction.

With more careful observation we would find that in the northern hemisphere, the magnetic field lines of Earth generally point down into Earth and toward the Arctic. And in the southern hemisphere, they generally point up out of Earth and away from the Antarctic, that is, away from Earth's *geomagnetic south pole*.

SAMPLE PROBLEM 29-1

A uniform magnetic field **B**, with magnitude 1.2 mT, points vertically upward throughout the volume of a laboratory chamber. A proton with kinetic energy 5.3 MeV enters the chamber, moving horizontally from south to north. What magnetic deflecting force acts on the proton as it enters the chamber? The proton mass is 1.67×10^{-27} kg.

SOLUTION: The magnetic deflecting force depends on the speed of the proton, which we can find from $K = \frac{1}{2}mv^2$. Solving for v, we find

$$v = \sqrt{\frac{2K}{m}} = \sqrt{\frac{(2)(5.3 \text{ MeV})(1.60 \times 10^{-13} \text{ J/MeV})}{1.67 \times 10^{-27} \text{ kg}}}$$

$$= 3.2 \times 10^7 \text{ m/s}.$$

Equation 29-3 then yields

$$F_B = |q|vB \sin \phi$$
$$= (1.60 \times 10^{-19} \text{ C})(3.2 \times 10^7 \text{ m/s})$$
$$\times (1.2 \times 10^{-3} \text{ T})(\sin 90°)$$
$$= 6.1 \times 10^{-15} \text{ N}. \qquad \text{(Answer)}$$

This may seem like a small force, but it acts on a particle of small mass, producing a large acceleration, namely,

$$a = \frac{F_B}{m} = \frac{6.1 \times 10^{-15} \text{ N}}{1.67 \times 10^{-27} \text{ kg}} = 3.7 \times 10^{12} \text{ m/s}^2.$$

It remains to find the direction of $\mathbf{F}_B$. We know that **v** points horizontally from south to north and **B** points vertically up. The right-hand rule (see Fig. 29-2b) shows us that the deflecting force $\mathbf{F}_B$ must point horizontally from west to east, as Fig. 29-6 shows. (The array of dots in the figure represents a magnetic field pointing directly out of the plane of the figure. An array of Xs would have represented a magnetic field pointing directly into that plane.)

If the charge of the particle were negative, the magnetic deflecting force would point in the opposite direction, that is, horizontally from east to west. This is predicted automatically by Eq. 29-2, if we substitute $-e$ for q.

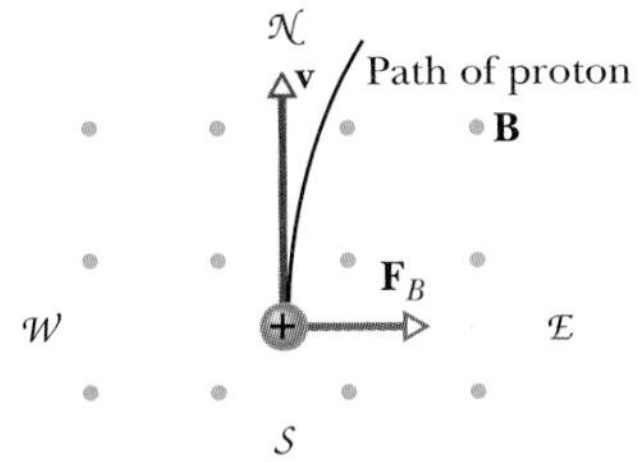

FIGURE 29-6   Sample Problem 29-1. An overhead view of a proton moving from south to north with velocity **v** in a chamber. A magnetic field points vertically upward in the chamber, as represented by the array of dots (which resemble the tips of arrows). The proton is deflected toward the east.

PROBLEM SOLVING TACTICS

TACTIC 1: *Classical and Relativistic Formulas for Kinetic Energy*

In Sample Problem 29-1, we used the (approximate) classical expression ($K = \frac{1}{2}mv^2$) for the kinetic energy of the proton rather than the (exact) relativistic expression (see Eq. 7-51). The criterion for when the classical expression may safely be used is that $K \ll mc^2$, where mc^2 is the rest energy of the particle. In this case, $K = 5.3$ MeV and the rest energy of a proton is 938 MeV. This proton passes the test and we were justified in treating it as "slow," that is, in using the classical $K = \frac{1}{2}mv^2$ formula for the kinetic energy. That is not always the case in dealing with energetic particles.

29-3 CROSSED FIELDS: DISCOVERY OF THE ELECTRON

Both an electric field **E** and a magnetic field **B** can produce a force on a charged particle. When the two fields are perpendicular to each other, they are said to be *crossed fields*. Here we shall examine what happens to charged particles, namely, electrons, as they move through crossed fields. We use as our example the experiment that led to the discovery of the electron in 1897 by J. J. Thomson at Cambridge University.

Figure 29-7 shows a modern, simplified version of Thomson's experimental apparatus—a *cathode ray tube* (which is like the "picture tube" in a standard television set). Charged particles (which we now know as electrons) are emitted by a hot filament at the rear of the evacuated tube and are accelerated by an applied potential difference V. After they pass through a slit in screen C, they form a narrow beam. They then pass through a region of crossed **E** and **B** fields, headed toward a fluorescent screen S, where they will produce a spot of light (on a television screen the spot would be part of the picture). The forces on the charged particles in the crossed-fields region can deflect them from the center of the screen. By controlling the magnitudes and directions of the fields, Thomson could thus control where the spot of light appeared on the screen. For the particular field arrangement of Fig. 29-7, electrons are forced up the page by the electric field **E** and down the page by the magnetic field **B**—that is, the forces are in *opposition*. Thomson's procedure was equivalent to the following series of steps.

1. Set $E = 0$ and $B = 0$ and note the position of the spot on screen S due to the undeflected beam.

2. Turn on **E** and measure the resulting beam deflection.

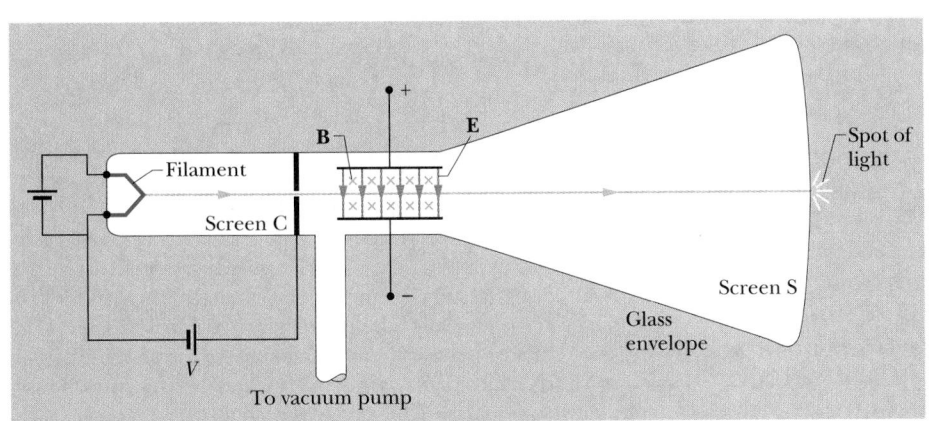

FIGURE 29-7   A modern version of J. J. Thomson's apparatus for measuring the ratio of mass to charge for the electron. The electric field **E** is established by connecting a battery across the deflecting plate terminals. The magnetic field **B** is set up by means of a current in a system of coils (not shown). The magnetic field shown is into the plane of the figure, as represented by the array of Xs (which resemble the feathered ends of arrows).

3. Maintaining **E,** now turn on **B** and adjust its value until the beam returns to the undeflected position. (With the forces in opposition, they can be made to cancel.)

We discussed the deflection of a charged particle moving through an electric field **E** between two plates (step 2 here) in Sample Problem 23-8. We found that the deflection of the particle at the far end of the plates is

$$y = \frac{qEL^2}{2mv^2}, \qquad (29\text{-}6)$$

where v is the particle's speed, m its mass, and q its charge, and L is the length of the plates. We can apply this same equation to the beam of electrons in Fig. 29-7 by measuring the deflection of the beam on screen S and then working back to calculate the deflection y at the end of the plates. (Because the direction of the deflection is set by the sign of the particle's charge, Thomson was able to show that the particles that were lighting up his screen were negatively charged.)

When the two fields in Fig. 29-7 are adjusted so that the two deflecting forces cancel (step 3), we have from Eqs. 29-1 and 29-3

$$|q|E = |q|vB \sin(90°) = |q|vB,$$

or

$$v = \frac{E}{B}. \qquad (29\text{-}7)$$

Thus the crossed fields allow us to measure the speed of the charged particles passing through them. Substituting Eq. 29-7 for v in Eq. 29-6 and rearranging yield

$$\frac{m}{q} = \frac{B^2L^2}{2yE}, \qquad (29\text{-}8)$$

in which all quantities on the right can be measured. Thus, the crossed fields allow us to measure the ratio m/q of the particles moving through Thomson's apparatus.

Thomson claimed that these particles are found in all matter. He also claimed that they are lighter than the lightest known atom (hydrogen) by a factor of more than 1000. (The exact ratio proved later to be 1836.15.) His m/q measurement, coupled with the boldness of his two claims, is considered to be the "discovery of the electron."

CHECKPOINT 2: The figure shows four directions for the velocity vector **v** of a positively charged particle moving through a uniform electric field **E** (directed out of the page and represented by an encircled dot) and a uniform magnetic field **B**. (a) Rank directions 1, 2, and 3 according to the magnitude of the net force on the particle, greatest first. (b) Of all four directions, which might result in a net force of zero?

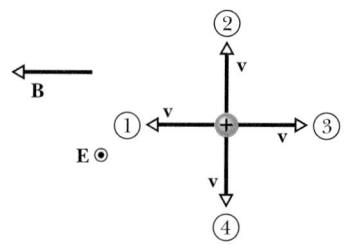

29-4 CROSSED FIELDS: THE HALL EFFECT

As we just discussed, a beam of electrons in a vacuum can be deflected by a magnetic field. Can the drifting conduction electrons in a copper wire also be deflected by a magnetic field? In 1879, Edwin H. Hall, then a 24-year-old graduate student at the Johns Hopkins University, showed that they can. This **Hall effect** allows us to find out whether the charge carriers in a conductor are positively or negatively charged. Beyond that, we can measure the number of such carriers per unit volume of the conductor.

Figure 29-8a shows a copper strip of width d, carrying a current i whose conventional direction is from the top of

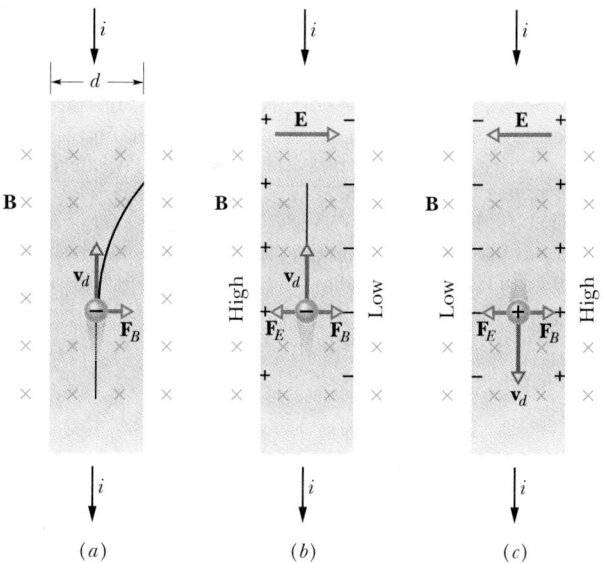

FIGURE 29-8 A strip of copper carrying a current i is immersed in a magnetic field **B**. (a) The situation immediately after the magnetic field is turned on. The curved path that will then be taken by an electron is shown. (b) The situation at equilibrium, which quickly follows. Note that negative charges pile up on the right side of the strip, leaving uncompensated positive charges on the left. Thus the left side is at a higher potential than the right side. (c) For the same current direction, if the charge carriers were positively charged, *they* would pile up on the right side, and the right side would be at the higher potential.

the figure to the bottom. The charge carriers are electrons and, as we know, they drift (with drift speed v_d) in the opposite direction, from bottom to top. At the instant shown in Fig. 29-8a, an external magnetic field **B**, pointing into the plane of the figure, has just been turned on. From Eq. 29-2 we see that a magnetic deflecting force $\mathbf{F}_B$ will act on each drifting electron, pushing it toward the right edge of the strip.

As time goes on, electrons will move to the right, mostly piling up on the right edge of the strip, leaving uncompensated positive charges in fixed positions at the left edge. The separation of positive and negative charges produces an electric field **E** within the strip, pointing from left to right in Fig. 29-8b. This field will exert an electric force $\mathbf{F}_E$ on each electron, tending to push it to the left.

An equilibrium quickly develops in which the electric force on each electron builds up until it just cancels the magnetic force. When this happens, as Fig. 29-8b shows, the force due to **B** and the force due to **E** are in balance. The drifting electrons then move along the strip toward the top of the page with no further collection of electrons on the right edge of the strip and thus no further increase in the electric field **E**.

A *Hall potential difference* V is associated with the electric field across strip width d. From Eq. 25-42, the magnitude of that potential difference is

$$V = Ed. \tag{29-9}$$

By connecting a voltmeter across the width, we can measure the potential difference between the two edges of the strip. Moreover, the voltmeter can tell us which edge is at higher potential. For the situation of Fig. 29-8a, we would find that the left edge is at higher potential, which is consistent with our assumption that the charge carriers are negatively charged.

For a moment, let us make the opposite assumption, that the charge carriers in current i are positively charged (Fig. 29-8c). Convince yourself that as these charge carriers moved from top to bottom in the strip, they would be pushed to the right edge by $\mathbf{F}_B$ and thus that the *right* edge would be at higher potential. Because that last statement is contradicted by our voltmeter reading, the charge carriers must be negatively charged.

Now for the quantitative part. When the electric and magnetic forces are in balance (Fig. 29-8b), Eqs. 29-1 and 29-3 give us

$$eE = ev_dB. \tag{29-10}$$

From Eq. 27-7, the drift speed v_d is

$$v_d = \frac{J}{ne} = \frac{i}{neA}, \tag{29-11}$$

in which $J\ (= i/A)$ is the current density in the strip, A is the cross-sectional area of the strip, and n is the *number density* of charge carriers (their number per unit volume).

In Eq. 29-10, substituting for E with Eq. 29-9 and substituting for v_d with Eq. 29-11, we obtain

$$n = \frac{Bi}{Vle}, \tag{29-12}$$

in which $l\ (= A/d)$ is the thickness of the strip. Thus we can find n in terms of quantities that we can measure.

It is also possible to use the Hall effect to measure directly the drift speed v_d of the charge carriers, which you may recall is of the order of centimeters per hour. In this clever experiment, the metal strip is moved mechanically through the magnetic field in a direction opposite that of the drift velocity of the charge carriers. The speed of the moving strip is then adjusted until the Hall potential difference vanishes. At this condition, with no Hall effect, the velocity of the charge carriers *with respect to the magnetic field* must be zero. So the velocity of the strip must be equal in magnitude but opposite in direction to the velocity of the negative charge carriers.

SAMPLE PROBLEM 29-2

Figure 29-9 shows a solid metal cube, of edge length d = 1.5 cm, moving in the positive y direction at a constant velocity **v** of magnitude 4.0 m/s. The cube moves through a uniform magnetic field **B** of magnitude 0.050 T and pointing in the positive z direction.

(a) Which cube face is at a lower electric potential and which is at a higher electric potential because of the motion through the field?

SOLUTION: When the cube first began to move through the magnetic field, the conduction electrons within the cube also began to move through the field. Because of their motion, they experienced a force $\mathbf{F}_B$ given by Eq. 29-2. In Fig. 29-9, $\mathbf{F}_B$ acts in the negative direction of the x axis. This means that some of the electrons were deflected by $\mathbf{F}_B$ to the (hidden) left cube face, making that face negatively charged and the right face positively charged. This charge separation produces an electric field **E** directed from the right face toward the left face.

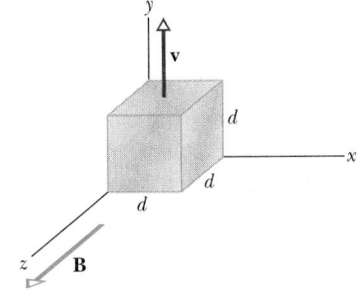

FIGURE 29-9 Sample Problem 29-2. A solid metal cube of edge length d moves at constant velocity **v** through a uniform magnetic field **B**.

Thus, the left face is at lower potential and the right face is at higher potential.

(b) What is the potential difference V between the faces of higher and lower electric potential?

SOLUTION: The electric field E that is produced by the charge separation causes a force F_E to act on the electrons; F_E is directed toward the right cube face, in the direction opposite that of force F_B. Equilibrium, in which $F_E = F_B$, is reached quickly after the cube begins to move through the magnetic field. From Eqs. 29-1 and 29-3, we then have

$$eE = evB.$$

Substituting for E with Eq. 29-9 ($V = Ed$) then yields

$$V = dvB. \qquad (29\text{-}13)$$

Substituting the given data, we now find

$$V = (0.015 \text{ m})(4.0 \text{ m/s})(0.050 \text{ T})$$
$$= 0.0030 \text{ V} = 3.0 \text{ mV}. \qquad (\text{Answer})$$

CHECKPOINT 3: The figure shows a metallic, rectangular solid that is to move at a certain speed v through the uniform magnetic field **B**. Its dimensions are multiples of d, as shown. You have six choices for the direction of the velocity of the solid: it can be parallel to x, y, or z, in either the positive or negative direction. (a) Rank the six choices according to the potential set up across the solid, greatest first. (b) For which choice is the front face at lower potential?

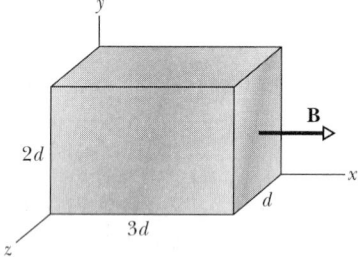

29-5 A CIRCULATING CHARGED PARTICLE

If a particle moves in a circle at constant speed, we can be sure that the net force acting on the particle is constant in magnitude and points toward the center of the circle, always perpendicular to the particle's velocity. Think of a stone tied to a string and whirled in a circle on a smooth horizontal surface, or of a satellite moving in a circular orbit around Earth. In the first case, the tension in the string provides the necessary force and centripetal acceleration. In the second case, Earth's gravitational attraction provides the force and acceleration.

Figure 29-10 shows another example: a beam of electrons is projected into a chamber by an *electron gun* G. The electrons enter in the plane of the page with velocity **v** and move in a region of uniform magnetic field **B** directed out of the plane of the figure. As a result, a magnetic force $\mathbf{F}_B = q\mathbf{v} \times \mathbf{B}$ continually deflects the electrons, and because **v** and **B** are perpendicular to each other, this deflection causes the electrons to follow a circular path. The path is visible in the photo because atoms of gas in the chamber emit light when some of the circulating electrons collide with them.

We would like to determine the parameters that characterize the circular motion of these electrons, or of any particle of charge magnitude q and mass m moving perpendicular to a uniform magnetic field **B** at speed v. From Eq. 29-3, the force acting on the particle has a magnitude of qvB. So from Newton's second law applied to uniform circular motion (Eq. 6-20),

$$F = ma = \frac{mv^2}{r}, \qquad (29\text{-}14)$$

FIGURE 29-10 Electrons circulating in a chamber containing gas at low pressure (their path is the glowing circle). A uniform magnetic field **B**, pointing directly out of the plane of the page, fills the chamber. Note the radially directed magnetic force $\mathbf{F}_B$: for circular motion to occur, $\mathbf{F}_B$ *must* point toward the center of the circle. Use the right-hand rule for cross products to confirm that $\mathbf{F}_B = q\mathbf{v} \times \mathbf{B}$ gives $\mathbf{F}_B$ the proper direction.

we have

$$qvB = \frac{mv^2}{r}. \qquad (29\text{-}15)$$

Solving for r, we find the radius of the circular path as

$$r = \frac{mv}{qB} \qquad \text{(radius).} \qquad (29\text{-}16)$$

The period T (the time for one full revolution) is equal to the circumference divided by the speed:

$$T = \frac{2\pi r}{v} = \frac{2\pi}{v}\frac{mv}{qB} = \frac{2\pi m}{qB} \qquad \text{(period).} \quad (29\text{-}17)$$

The frequency f is

$$f = \frac{1}{T} = \frac{qB}{2\pi m} \qquad \text{(frequency).} \qquad (29\text{-}18)$$

The angular frequency ω of the motion is then

$$\omega = 2\pi f = \frac{qB}{m} \qquad \text{(angular frequency).} \quad (29\text{-}19)$$

The quantities T, f, and ω do not depend on the speed of the particle (provided that speed is much less than the speed of light). Fast particles move in large circles and slow ones in small circles, but all particles with the same charge-to-mass ratio q/m take the same time T (the period) to complete one round trip. Using Eq. 29-2, you can show that if you are looking in the direction of **B**, the direction of rotation for a positive particle is always counterclockwise; that for a negative particle is always clockwise.

Helical Paths

If the velocity of a charged particle has a component parallel to the (uniform) magnetic field, the particle will move in a helical path about the direction of the field vector. Figure 29-11a, for example, shows the velocity vector **v** of such a particle resolved into two components, one parallel to **B** and one perpendicular to it:

$$v_\parallel = v \cos \phi \quad \text{and} \quad v_\perp = v \sin \phi. \qquad (29\text{-}20)$$

The parallel component determines the *pitch p* of the helix, that is, the distance between adjacent turns (Fig. 29-11b). The perpendicular component determines the radius of the helix and is the quantity to be substituted for v in Eq. 29-16.

Figure 29-11c shows a charged particle spiraling in a nonuniform magnetic field. The more closely spaced field lines at the left and right sides indicate that the magnetic field is stronger there. When the field at an end is strong enough, the particle "reflects" from that end. If the particle reflects from both ends, it is said to be trapped in a *magnetic bottle*.

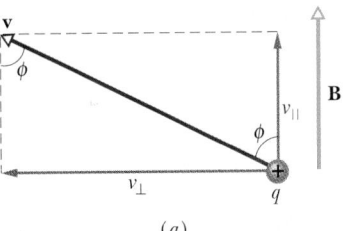

(a)

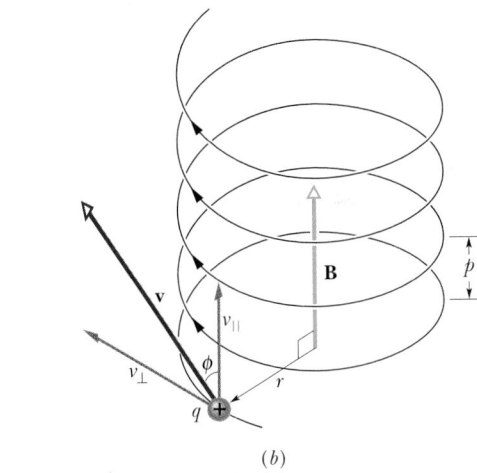

(b)

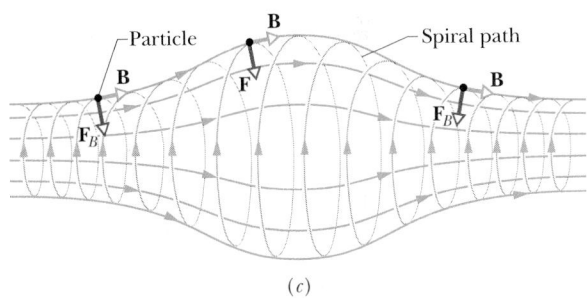

(c)

FIGURE 29-11 (a) A charged particle moves in a magnetic field, its velocity making an angle ϕ with the field direction. (b) The particle follows a helical path, of radius r and pitch p. (c) A charged particle spiraling in a nonuniform magnetic field. (The particle can become trapped, spiraling back and forth between the strong field regions at either end.) Note that the magnetic force vectors at the left and right sides have a component pointing toward the center of the figure.

Electrons and protons are trapped in this way by the terrestrial magnetic field, forming the *Van Allen radiation belts,* which loop well above Earth's atmosphere, between Earth's north and south geomagnetic poles. The trapped particles bounce back and forth, from end to end of the magnetic bottle, within a few seconds.

When a large solar flare shoots additional energetic electrons and protons into the radiation belts, an electric

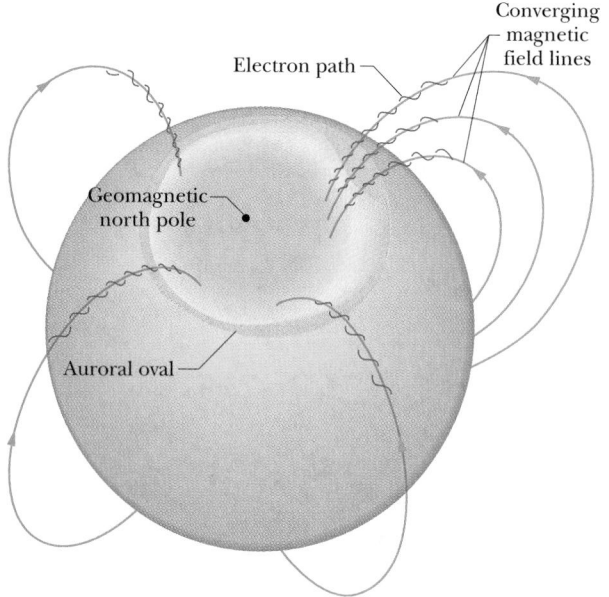

FIGURE 29-12 The auroral oval surrounding Earth's geomagnetic north pole (in northwestern Greenland). Magnetic field lines converge toward that pole. Electrons moving toward Earth are "caught by" and spiral around these field lines, entering the terrestrial atmosphere at high latitudes and producing aurora within the oval.

FIGURE 29-13 A false-color image of aurora inside the north auroral oval, recorded by the satellite *Dynamic Explorer,* using ultraviolet light emitted by oxygen atoms excited in the aurora. The sun-lit portion of Earth is the crescent at the left.

field is produced in the region where electrons normally reflect. This field eliminates the reflection and drives electrons down into the atmosphere, where they collide with atoms and molecules of air, causing that air to emit light. This light forms the aurora—a curtain of light that hangs down to an altitude of about 100 km. Green light is emitted by oxygen atoms, and pink light is emitted by nitrogen molecules, but often the light is so dim that we perceive only white light.

An auroral display extends in an arc above Earth in a region called the *auroral oval* (Figs. 29-12 and 29-13). Although the display is long, it is less than 1 km thick (north to south) because the paths of the electrons producing it converge as the electrons spiral down the converging magnetic field lines (Fig. 29-12).

CHECKPOINT 4: The figure shows the circular paths of two particles that travel at the same speed in a uniform magnetic field **B**, which points into the page. One particle is a proton; the other is an electron (which is less massive). (a) Which particle follows the smaller circle, and (b) does that particle travel clockwise or counterclockwise?

⊗
B

SAMPLE PROBLEM 29-3

Figure 29-14 shows the essentials of a *mass spectrometer,* which can be used to measure the mass of an ion: an ion of mass m (to be measured) and charge q is produced in source S. The initially stationary ion is accelerated by the electric field due to a potential difference V. The ion leaves S and enters a separator chamber in which a uniform magnetic field **B** is perpendicular to the path of the ion. The magnetic field causes the ion to move in a semicircle, striking (and thus altering) a photographic plate at distance x from the entry slit. Suppose that in a certain trial $B = 80.000$ mT and $V = 1000.0$ V and ions of charge $q = +1.6022 \times 10^{-19}$ C strike the plate at $x = 1.6254$ m. What is the mass m of the individual ions, in unified atomic mass units (1 u $= 1.6605 \times 10^{-27}$ kg)?

SOLUTION: We need to relate the ion mass m to the measured distance x in Fig. 29-14. To do so, we first note that $x = 2r$, where r is the radius of the semicircular path taken by the ion. Then we note that r is related to mass m by $r = mv/qB$ (Eq. 29-16), where v is the speed of the ion upon entering and then moving through the magnetic field.

We can relate the speed v to the accelerating potential V by applying the law of conservation of energy to the ion: its kinetic energy $\frac{1}{2}mv^2$ at the end of the acceleration is equal to its potential energy qV at the start of the acceleration. Thus

$$\tfrac{1}{2}mv^2 = qV$$

and

$$v = \sqrt{\frac{2qV}{m}}. \tag{29-21}$$

Substituting this into Eq. 29-16 gives us

$$r = \frac{mv}{qB} = \frac{m}{qB}\sqrt{\frac{2qV}{m}} = \frac{1}{B}\sqrt{\frac{2mV}{q}}.$$

Thus,

$$x = 2r = \frac{2}{B}\sqrt{\frac{2mV}{q}}.$$

Solving this for m and substituting the given data yield

$$m = \frac{B^2qx^2}{8V}$$

$$= \frac{(0.080000 \text{ T})^2(1.6022 \times 10^{-19} \text{ C})(1.6254 \text{ m})^2}{8(1000.0 \text{ V})}$$

$$= 3.3863 \times 10^{-25} \text{ kg} = 203.93 \text{ u}. \qquad \text{(Answer)}$$

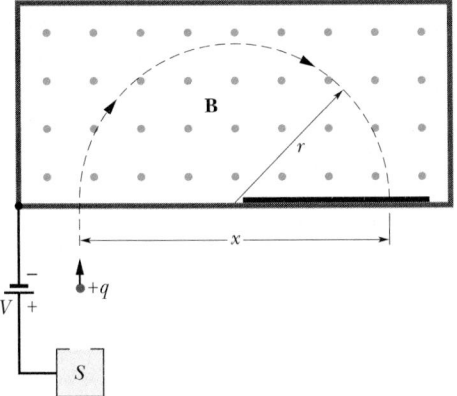

FIGURE 29-14 Sample Problem 29-3. Essentials of a mass spectrometer. A positive ion, after being accelerated from its source S by potential difference V, enters a chamber of uniform magnetic field **B**. There it travels through a semicircle of radius r and strikes a photographic plate at a distance x from where it entered the chamber.

SAMPLE PROBLEM 29-4

An electron with a kinetic energy of 22.5 eV moves into a region of uniform magnetic field **B** of magnitude 4.55×10^{-4} T. The angle between the directions of **B** and the electron's velocity **v** is 65.5°. What is the pitch of the helical path taken by the electron?

SOLUTION: The pitch p is the distance the electron travels parallel to the magnetic field **B** during one period T of revolution. That distance is $v_{\parallel}T$, where $v_{\parallel}$ is the electron's speed parallel to **B**. Using Eqs. 29-20 and 29-17, we find that

$$p = v_{\parallel}T = (v \cos \phi)\frac{2\pi m}{qB}. \qquad (29\text{-}22)$$

We can calculate the electron's speed v from its kinetic energy as we did for the proton in Sample Problem 29-1. (The kinetic energy of 22.5 eV is much less than the electron's rest energy

of 5.11×10^5 eV, so we need not use the relativistic formula for the kinetic energy.) We find that $v = 2.81 \times 10^6$ m/s. Substituting this and known data in Eq. 29-22 gives us

$$p = (2.81 \times 10^6 \text{ m/s})(\cos 65.5°)$$

$$\times \frac{2\pi(9.11 \times 10^{-31} \text{ kg})}{(1.60 \times 10^{-19} \text{ C})(4.55 \times 10^{-4} \text{ T})}$$

$$= 9.16 \text{ cm}. \qquad \text{(Answer)}$$

29-6 CYCLOTRONS AND SYNCHROTRONS

What is the structure of matter on the smallest scale? This question has always intrigued physicists. One way of getting at the answer is to allow an energetic charged particle (a proton, for example) to slam into a solid target. Better yet, allow two such energetic protons to collide head-on. Then analyze the debris from many such collisions to learn the nature of the subatomic particles of matter. The Nobel Prizes in physics for 1976 and 1984 were awarded for just such studies.

How can we give a proton enough kinetic energy for such an experiment? The direct approach is to allow the proton to "fall" through a potential difference V, thereby increasing its kinetic energy by eV. As we want higher and higher energies, however, it becomes more and more difficult to establish the necessary potential difference.

A better way is to arrange for the proton to circulate in a magnetic field, and to give it a modest electrical "kick" once per revolution. For example, if a proton circulates 100 times in a magnetic field and receives an energy boost of 100 keV every time it completes an orbit, it will end up with a kinetic energy of (100)(100 keV) or 10 MeV. Two very useful devices are based on this principle.

The Cyclotron

Figure 29-15 is a top view of the region of a *cyclotron* in which the particles (protons, say) circulate. The two hollow D-shaped objects (open on their straight edges) are made of sheet copper. These *dees*, as they are called, form part of an electrical oscillator, which establishes an alternating potential difference across the gap between them. The dees are immersed in a magnetic field ($B = 1.5$ T) whose direction is out of the plane of the page and which is set up by a large electromagnet.

Suppose that a proton, injected by source S at the center of the cyclotron in Fig. 29-15, initially moves toward a negatively charged dee. It will accelerate toward this dee and enter it. Once inside, it is shielded from electric fields by the copper walls of the dee; that is, the electric

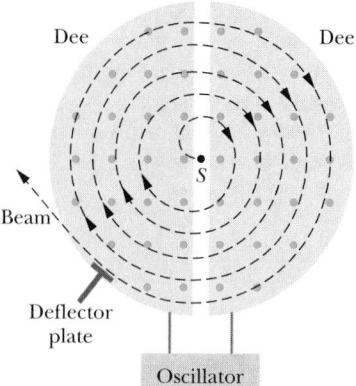

FIGURE 29-15 The elements of a cyclotron, showing the particle source S and the dees. A uniform magnetic field emerges from the plane of the figure. Circulating protons spiral outward within the hollow dees, gaining energy every time they cross the gap between the dees.

field does not enter the dee. The magnetic field, however, is not screened by the (nonmagnetic) copper dee, so the proton moves in a circular path whose radius, which depends on its speed, is given by Eq. 29-16, $r = mv/qB$.

Let us assume that at the instant the proton emerges into the center gap from the first dee, the potential difference between the dees has been reversed. Thus the proton *again* faces a negatively charged dee and is *again* accelerated. This process continues, the circulating proton always being in step with the oscillations of the dee potential, until the proton spirals out to the edge of the dee system.

The key to the operation of the cyclotron is that the frequency f at which the proton circulates in the field (and which does *not* depend on its speed) must be equal to the fixed frequency f_{osc} of the electrical oscillator, or

$$f = f_{osc} \quad \text{(resonance condition).} \quad (29\text{-}23)$$

This *resonance condition* says that, if the energy of the circulating proton is to increase, energy must be fed to it at a frequency f_{osc} that is equal to the natural frequency f at which the proton circulates in the magnetic field.

Combining Eqs. 29-18 and 29-23 allows us to write the resonance condition as

$$qB = 2\pi m f_{osc}. \quad (29\text{-}24)$$

For the proton, q and m are fixed. The oscillator (we assume) is designed to work at a single fixed frequency f_{osc}. We then "tune" the cyclotron by varying B until Eq. 29-24 is satisfied and a beam of energetic protons appears.

The Proton Synchrotron

At proton energies above 50 MeV, the conventional cyclotron begins to fail because one of the assumptions of its

design—that the frequency of revolution of a charged particle circulating in a magnetic field is independent of the particle's speed—is true only for speeds that are much less than the speed of light. At greater proton speeds, we must treat the problem relativistically.

According to relativity theory, as the speed of a circulating proton approaches that of light, the proton takes a longer and longer time to make the trip around its orbit. This means that the frequency of revolution of the circulating proton decreases steadily. Thus the protons get out of step with the cyclotron's oscillator—whose frequency remains fixed at f_{osc}—and eventually the energy of the circulating proton stops increasing.

There is another problem. For a 500 GeV proton in a magnetic field of 1.5 T, the path radius is 1.1 km. The magnet for a conventional cyclotron of the proper size would be impossibly expensive, the area of its pole faces being about 4×10^6 m^2.

The *proton synchrotron* is designed to meet these two difficulties. The magnetic field B and the oscillator frequency f_{osc}, instead of having fixed values as in the conventional cyclotron, are made to vary with time during the accelerating cycle. When this is done properly, (1) the frequency of the circulating protons remains in step with the oscillator at all times, and (2) the protons follow a circular —not a spiral—path. Thus the magnet need extend only along that circular path, not over some 4×10^6 m^2. The circular path, however, still must be large if high energies are to be achieved. The proton synchrotron at the Fermi National Accelerator Laboratory (Fermilab) in Illinois (Fig. 29-16) has a circumference of 6.3 km and can produce protons with energies of about 1 TeV (= 10^{12} eV).

FIGURE 29-16 An aerial view of Fermilab.

SAMPLE PROBLEM 29-5

Suppose a cyclotron is operated at an oscillator frequency of 12 MHz and has a dee radius $R = 53$ cm.

(a) What is the magnitude of the magnetic field needed for deuterons to be accelerated in the cyclotron?

SOLUTION: A deuteron has the same charge as a proton but approximately twice the mass ($m = 3.34 \times 10^{-27}$ kg). From Eq. 29-24,

$$B = \frac{2\pi m f_{osc}}{q} = \frac{(2\pi)(3.34 \times 10^{-27} \text{ kg})(12 \times 10^6 \text{ s}^{-1})}{1.60 \times 10^{-19} \text{ C}}$$

$$= 1.57 \text{ T} \approx 1.6 \text{ T}. \qquad \text{(Answer)}$$

Note that, to allow protons to be accelerated, B would have to be reduced by a factor of 2, assuming that the oscillator frequency remained fixed at 12 MHz.

(b) What is the resulting kinetic energy of the deuterons?

SOLUTION: From Eq. 29-16, the speed of a deuteron circulating with a radius equal to the dee radius R is given by

$$v = \frac{RqB}{m} = \frac{(0.53 \text{ m})(1.60 \times 10^{-19} \text{ C})(1.57 \text{ T})}{3.34 \times 10^{-27} \text{ kg}}$$

$$= 3.99 \times 10^7 \text{ m/s}.$$

This speed corresponds to a kinetic energy of

$$K = \tfrac{1}{2}mv^2$$

$$= \tfrac{1}{2}(3.34 \times 10^{-27} \text{ kg})(3.99 \times 10^7 \text{ m/s})^2$$

$$\times (1 \text{ MeV}/1.60 \times 10^{-13} \text{ J})$$

$$= 16.6 \text{ MeV} \approx 17 \text{ MeV}. \qquad \text{(Answer)}$$

29-7 MAGNETIC FORCE ON A CURRENT-CARRYING WIRE

We have already seen (in connection with the Hall effect) that a magnetic field exerts a sideways force on moving electrons in a wire. This force must be transmitted to the wire itself, because the conduction electrons cannot escape sideways out of the wire.

In Fig. 29-17a, a vertical wire, carrying no current and fixed in place at both ends, extends through the gap between the vertical pole faces of a magnet. The magnetic field between the faces points outward from the page. In Fig. 29-17b, a current is sent upward through the wire; the wire deflects to the right. In Fig. 29-17c, we reverse the direction of the current and the wire deflects to the left.

Figure 29-18 shows what happens inside the wire of Fig. 29-17. We see one of the conduction electrons, drifting downward with an assumed drift speed v_d. Equation 29-3, in which we must put $\phi = 90°$, tells us that a force

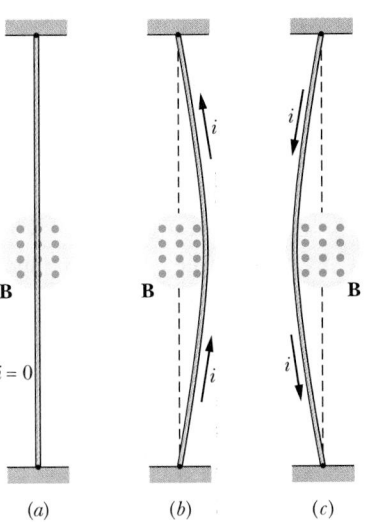

FIGURE 29-17 A flexible wire passes between the pole faces of a magnet (only the farther pole face is shown). (a) Without current in the wire, the wire is straight. (b) With upward current, the wire is deflected rightward. (c) With downward current, the deflection is leftward. The connections for getting the current into the wire at one end and out of it at the other end are not shown.

$\mathbf{F}_B$ of magnitude ev_dB must act on each such electron. From Eq. 29-2 we see that this force must point to the right. We expect then that the wire as a whole will experience a force to the right, in agreement with Fig. 29-17b.

If, in Fig. 29-18, we were to reverse *either* the direction of the magnetic field *or* the direction of the current, the force on the wire would reverse, pointing now to the left. Note too that it does not matter whether we consider negative charges drifting downward in the wire (the actual case) or positive charges drifting upward. The direction of the deflecting force on the wire is the same. We are safe then in dealing with the conventional direction of current, which assumes positive charge carriers.

Consider a length L of the wire in Fig. 29-18. The conduction electrons in this section of wire will drift past plane xx in Fig. 29-18 in a time $t = L/v_d$. Thus in that time a charge given by

$$q = it = i\frac{L}{v_d}$$

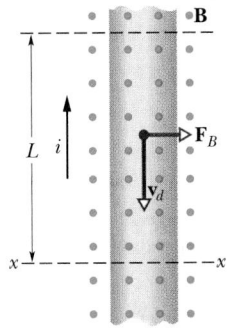

FIGURE 29-18 A close-up view of a section of the wire of Fig. 29-17b. The current direction is upward, which means that electrons drift downward. A magnetic field that emerges from the plane of the page causes the electrons and the wire to be deflected to the right.

will pass through that plane. Substituting this into Eq. 29-3 yields

$$F_B = qv_dB \sin \phi$$

$$= \frac{iL}{v_d} v_dB \sin 90°$$

or

$$F_B = iLB. \qquad (29\text{-}25)$$

This equation gives the force that acts on a segment of a straight wire of length L, carrying a current i and immersed in a magnetic field $\mathbf{B}$ that is perpendicular to the wire.

If the magnetic field is *not* perpendicular to the wire, as in Fig. 29-19, the magnetic force is given by a generalization of Eq. 29-25:

$$\mathbf{F}_B = i\mathbf{L} \times \mathbf{B} \qquad \text{(force on a current).} \qquad (29\text{-}26)$$

Here $\mathbf{L}$ is a *length vector* that points along the wire segment in the direction of the (conventional) current.

Equation 29-26 is equivalent to Eq. 29-2 in that either can be taken as the defining equation for $\mathbf{B}$. In practice, we define $\mathbf{B}$ from Eq. 29-26. It is much easier to measure the magnetic force acting on a wire than that on a single moving charge.

If a wire is not straight, we can imagine it broken up into small straight segments and apply Eq. 29-26 to each segment. The force on the wire as a whole is then the vector sum of all the forces on the segments that make it up. In the differential limit, we can write

$$d\mathbf{F}_B = i\, d\mathbf{L} \times \mathbf{B}, \qquad (29\text{-}27)$$

and we can find the resultant force on any given arrangement of currents by integrating Eq. 29-27 over that arrangement.

In using Eq. 29-27, bear in mind that there is no such thing as an isolated current-carrying wire segment of length dL. There must always be a way to introduce the current into the segment at one end and take it out at the other end.

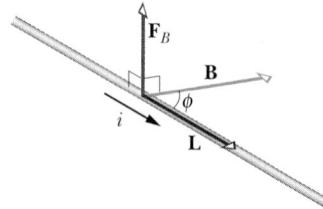

FIGURE 29-19 A wire carrying current i makes an angle ϕ with magnetic field $\mathbf{B}$. The wire has length L in the field and length vector $\mathbf{L}$ (in the direction of the current). A magnetic force $\mathbf{F}_B = i\mathbf{L} \times \mathbf{B}$ acts on the wire.

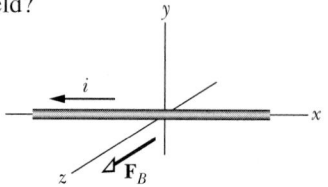

SAMPLE PROBLEM 29-6

A straight, horizontal stretch of copper wire has a current $i = 28$ A through it. What are the magnitude and direction of the minimum magnetic field $\mathbf{B}$ needed to suspend the wire, that is, to balance its weight? Its linear density is 46.6 g/m.

SOLUTION: Figure 29-20 shows the situation for a section of wire of length L, with the current out of the page. If the field is to be minimal, the force $\mathbf{F}_B$ that it exerts on the section must be upward, as shown. Equation 29-26 then requires that the field $\mathbf{B}$ be horizontal and, for the situation of Fig. 29-20, directed to the right.

In order to balance the weight of the section, the force $\mathbf{F}_B$ must have the magnitude $F_B = mg$, where m is the mass of the section. From Eq. 29-25, we then have

$$iLB = mg.$$

Solving for B and substituting known data yield

$$B = \frac{(m/L)g}{i} = \frac{(46.6 \times 10^{-3} \text{ kg/m})(9.8 \text{ m/s}^2)}{28 \text{ A}}$$

$$= 1.6 \times 10^{-2} \text{ T.} \qquad \text{(Answer)}$$

This is about 160 times the strength of Earth's magnetic field.

FIGURE 29-20 Sample Problem 29-6. A current-carrying wire (shown in cross section) can be made to "float" in a magnetic field. The current in the wire emerges from the plane of the page, and the magnetic field points to the right.

SAMPLE PROBLEM 29-7

Figure 29-21 shows a length of wire with a central semicircular arc, placed in a uniform magnetic field $\mathbf{B}$ that points out of the plane of the figure. If the wire carries a current i, what resultant magnetic force $\mathbf{F}$ acts on it?

SOLUTION: The force that acts on each straight section has the magnitude, from Eq. 29-25,

$$F_1 = F_3 = iLB$$

and points down, as shown by $\mathbf{F}_1$ and $\mathbf{F}_3$ in Figure 29-21.

A segment of the central arc of length dL has a force $d\mathbf{F}$ acting on it, whose magnitude is given by

$$dF = iB\,dL = iB(R\,d\theta)$$

and whose direction is radially toward point O, the center of the arc. Only the downward component $dF \sin\theta$ of this force element is effective. The horizontal component is canceled by an oppositely directed horizontal component associated with a symmetrically located segment on the opposite side of the arc.

Thus the total force on the central arc points down and is given by

$$F_2 = \int_0^\pi dF\sin\theta = \int_0^\pi (iBR\,d\theta)\sin\theta$$

$$= iBR\int_0^\pi \sin\theta\,d\theta = 2iBR.$$

The resultant force on the entire wire is then

$$F = F_1 + F_2 + F_3 = iLB + 2iBR + iLB$$
$$= 2iB(L + R). \qquad \text{(Answer)}$$

Note that this force is equal to the force that would act on a straight wire of length $2(L + R)$. This would be true no matter what the shape of the central segment.

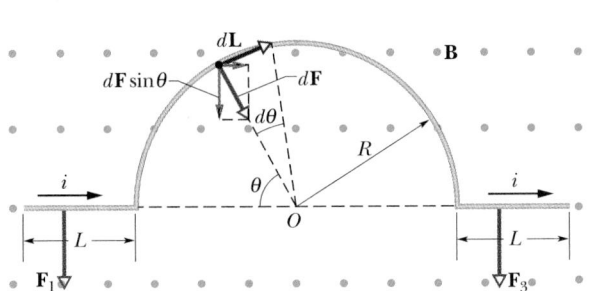

FIGURE 29-21 Sample Problem 29-7. A wire segment carrying a current i is immersed in a magnetic field. The resultant force on the wire is directed downward.

SAMPLE PROBLEM 29-8

Figure 29-22a shows a wire carrying a current $i = 6.0$ A in the positive direction of the x axis and lying in a *nonuniform* magnetic field given by $\mathbf{B} = (2.0$ T/m$)x\mathbf{i} + (2.0$ T/m$)x\mathbf{j}$, with $\mathbf{B}$ in teslas and x in meters. What is the net magnetic force $\mathbf{F}_B$ on the section of the wire between $x = 0$ and $x = 2.0$ m?

SOLUTION: Because the field varies along the section of wire, we cannot just substitute the data into Eq. 29-26, which holds only for a uniform magnetic field $\mathbf{B}$. Instead, we must

mentally divide the wire into differential lengths and then use Eq. 29-26 to find the differential force $d\mathbf{F}_B$ on each length. Then we can sum these differential forces to find the net magnetic force $\mathbf{F}_B$ on the full section of wire.

Figure 29-22b shows a differential length vector $d\mathbf{L}$ along the wire in the direction of the current; the vector has length dx and points in the positive direction of the x axis. Thus we can write this vector $d\mathbf{L}$ as

$$d\mathbf{L} = dx\,\mathbf{i}. \qquad (29\text{-}28)$$

(Be careful not to confuse the unit vector $\mathbf{i}$ with the current i.) Now, by Eq. 29-26, the differential force $d\mathbf{F}_B$ on the length dx of the wire is

$$d\mathbf{F}_B = i\,d\mathbf{L} \times \mathbf{B}$$
$$= i(dx\,\mathbf{i}) \times (2.0x\mathbf{i} + 2.0x\mathbf{j})$$
$$= i\,dx[2.0x(\mathbf{i} \times \mathbf{i}) + 2.0x(\mathbf{i} \times \mathbf{j})]$$
$$= i\,dx[0 + 2.0x\mathbf{k}] = 2.0ix\,dx\,\mathbf{k}, \qquad (29\text{-}29)$$

where the constant 2.0 has the unit teslas per meter. From this result we see that the magnetic force does not depend on the x component of $\mathbf{B}$ (because that component is along the direction of the current). We also see that the magnetic force $d\mathbf{F}_B$ on length dx of the wire is in the positive direction of the z axis (out of the page in Fig. 29-22c) and has magnitude $dF_B = (2.0$ T/m$)ix\,dx$.

Because the direction of the force $d\mathbf{F}_B$ is the same for all the differential lengths dx of the wire, we can find the magnitude of the total force by summing all the differential force magnitudes dF_B. To do so, we integrate dF_B from $x = 0$ to $x = 2.0$ m and then substitute the given data. We get

$$F_B = \int dF_B = \int_0^{2.0\text{ m}} (2.0\text{ T/m})ix\,dx$$

$$= (2.0\text{ T/m})i\left[\tfrac{1}{2}x^2\right]_0^{2.0\text{ m}} = (2.0\text{ T/m})(6.0\text{ A})(\tfrac{1}{2})(2.0\text{ m})^2$$

$$= 24\ (\text{T}\cdot\text{A}\cdot\text{m}) = 24\text{ N}. \qquad \text{(Answer)}$$

This force is directed along the positive direction of the z axis.

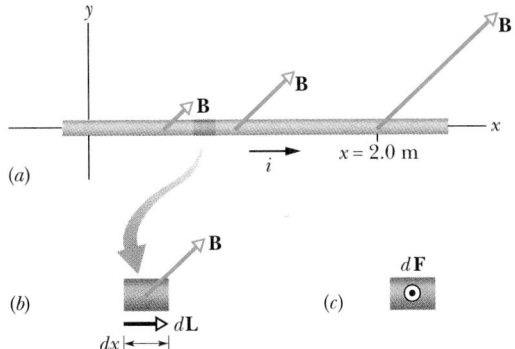

FIGURE 29-22 Sample Problem 29-8. (a) A wire with current i lies in a nonuniform magnetic field $\mathbf{B}$. (b) An element of the wire, with differential length vector $d\mathbf{L}$ and length dx. (c) The differential force $d\mathbf{F}$ acting on the element of (b) due to the magnetic field; the force is directed out of the page.

29-8 TORQUE ON A CURRENT LOOP

Much of the world's work is done by electric motors. The forces behind this work are the magnetic forces that we studied in the preceding section, that is, the forces that a magnetic field exerts on a wire that carries a current.

Figure 29-23a shows a simple motor, consisting of a single current-carrying loop immersed in a magnetic field **B**. The two magnetic forces **F** and −**F** combine to exert a torque on the loop, tending to rotate it about its central axis. Although many essential details have been omitted, the figure does suggest how the action of a magnetic field, exerting a torque on a current loop, produces the rotary motion of the electric motor. Let us analyze the action.

Figure 29-24a shows a rectangular loop of sides a and b, carrying a current i and immersed in a uniform magnetic field **B**. We place it in the field so that its long sides, labeled 1 and 3, are perpendicular to the field direction

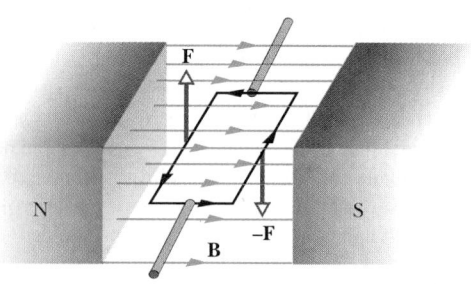

FIGURE 29-23 The elements of an electric motor. A rectangular loop of wire, carrying a current and free to rotate about a fixed axis, is placed in a magnetic field. A commutator (not shown) reverses the direction of the current every half-revolution so that the magnetic torque always acts in the same direction.

(which is into the page), but its short sides, labeled 2 and 4, are not. Wires to lead the current into and out of the loop are needed but, for simplicity, they are not shown.

To define the orientation of the loop in the magnetic field, we use a normal vector **n** that is perpendicular to the plane of the loop. Figure 29-24b shows a right-hand rule for finding the direction of **n**. Point or curl the fingers of your right hand in the direction of the current at any point on the loop. Your extended thumb then points in the direction of the normal vector **n**.

The normal vector of the loop is at an angle θ to the direction of the magnetic field **B**, as shown in Fig. 29-24c. We wish to find the net force and net torque acting on the loop in this orientation.

The net force is the vector sum of the forces acting on each of the four sides of the loop. For side 2 the vector **L** in Eq. 29-26 points in the direction of the current and has magnitude b. The angle between **L** and **B** for side 2 (see Fig. 29-24c) is $90° − \theta$. Thus the magnitude of the force acting on this side is

$$F_2 = ibB \sin(90° − \theta) = ibB \cos \theta. \quad (29\text{-}30)$$

You can show that the force **F**$_4$ acting on side 4 has the same magnitude as **F**$_2$ but points in the opposite direction. Thus **F**$_2$ and **F**$_4$ cancel out exactly. Their net force is zero and, because their common line of action is through the center of the loop, their net torque is also zero.

The situation is different for sides 1 and 3. Here the common magnitude of **F**$_1$ and **F**$_3$ is iaB, and the two forces point in opposite directions so that they do not tend to move the loop up or down. However, as Fig. 29-24c shows, these two forces do *not* share the same line of action so they *do* produce a net torque. The torque tends to rotate the loop so as to align its normal vector **n** with the direction

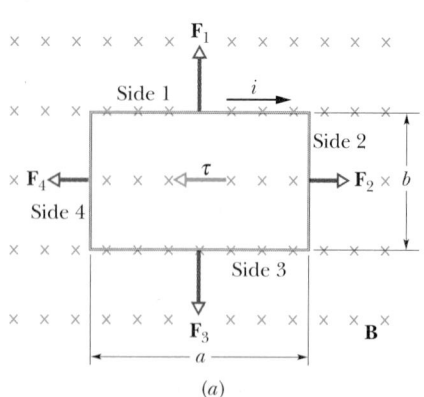
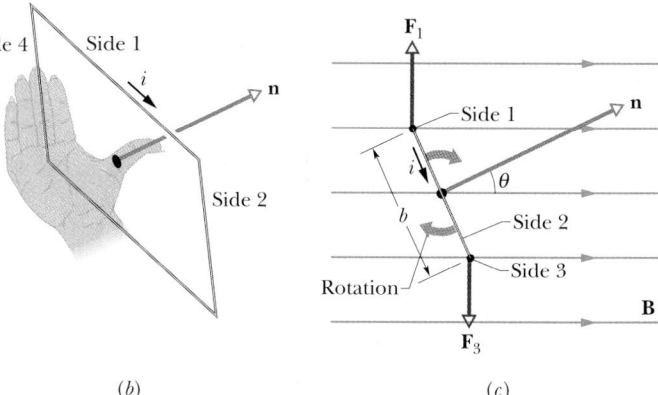

(a) (b) (c)

FIGURE 29-24 A rectangular loop, of length a and width b and carrying a current i, is placed in a uniform magnetic field. A torque τ acts to align the normal vector **n** with the direction of the field. (a) The loop as seen by looking in the direction of the

magnetic field. (b) A perspective of the loop showing how a right-hand rule gives the direction of **n**, which is perpendicular to the plane of the loop. (c) A side view of the loop, from side 2. The loop rotates as indicated.

of the magnetic field **B**. That torque has moment arm $(b/2) \sin \theta$ about the center of the loop. The magnitude τ' of the torque due to forces **F**$_1$ and **F**$_3$ is (see Fig. 29-24c)

$$\tau' = \left(iaB \frac{b}{2} \sin \theta \right) + \left(iaB \frac{b}{2} \sin \theta \right)$$
$$= iabB \sin \theta.$$

Suppose we replace the single loop of current with a *coil* of N loops, or *turns*. Further, suppose that the turns are wound tightly enough that they can be approximated as having the same dimensions and lying in a plane. Then the turns form a *flat coil* and the torque τ' derived above acts on each of them. The total torque is then

$$\tau = N\tau' = NiabB \sin \theta = (NiA)B \sin \theta, \quad (29\text{-}31)$$

in which A ($= ab$) is the area enclosed by the coil. The quantities in parentheses (NiA) are grouped together because they are all properties of the coil: its number of turns, its area, and the current it carries. This equation holds for all flat coils, no matter what their shape, provided the magnetic field is uniform.

Instead of focusing on the motion of the coil, it is simpler to keep track of the vector **n**, which is normal to the plane of the coil. Equation 29-31 tells us that a current-carrying flat coil placed in a magnetic field will tend to rotate so that **n** points in the field direction.

Analog voltmeters and ammeters work by measuring the torque exerted by a magnetic field on a current-carrying coil. The reading is displayed by means of the deflection of a pointer over a scale. Figure 29-25 shows the basic *galvanometer*, on which both analog ammeters and analog voltmeters are based. The coil is 2.1 cm high and 1.2 cm wide; it has 250 turns and is mounted so that it can rotate about an axis (into the page) in a uniform radial magnetic field with $B = 0.23$ T. For any orientation of the coil, the net magnetic field through the coil is perpendicular to the normal vector of the coil. A spring Sp provides a countertorque that balances the magnetic torque, so that a given steady current i in the coil results in a steady angular deflection ϕ. If a current of 100 μA produces an angular deflection of 28°, what must be the torsional constant κ of the spring, as used in Eq. 16-24 ($\tau = -\kappa\phi$)?

SOLUTION: Setting the magnetic torque (Eq. 29-31) equal to the spring torque and using absolute magnitudes yield

$$\tau = NiAB \sin \theta = \kappa\phi, \quad (29\text{-}32)$$

in which ϕ is the angular deflection of the coil and pointer, and A ($= 2.52 \times 10^{-4}$ m^2) is the area encircled by the coil. Since the net magnetic field through the coil is always perpen-

dicular to the normal vector of the coil, $\theta = 90°$ for any orientation of the pointer.

Solving Eq. 29-32 for κ, we find

$$\kappa = \frac{NiAB \sin \theta}{\phi}$$
$$= (250)(100 \times 10^{-6} \text{ A})(2.52 \times 10^{-4} \text{ m}^2)$$
$$\times \frac{(0.23 \text{ T})(\sin 90°)}{28°}$$
$$= 5.2 \times 10^{-8} \text{ N} \cdot \text{m/degree.} \quad \text{(Answer)}$$

Many modern ammeters and voltmeters are of the digital, direct-reading type and operate in a way that does not involve a moving coil.

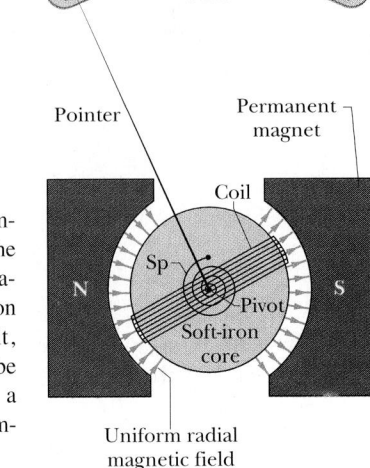

FIGURE 29-25 Sample Problem 29-9. The elements of a galvanometer. Depending on the external circuit, this device can be wired up as either a voltmeter or an ammeter.

29-9 THE MAGNETIC DIPOLE

We can describe the current-carrying coil of the preceding section with a single vector $\boldsymbol{\mu}$, its **magnetic dipole moment.** We take the direction of $\boldsymbol{\mu}$ to be that of the normal vector **n** to the plane of the coil, as in Fig. 29-24c. We define the magnitude of $\boldsymbol{\mu}$ as

$$\mu = NiA \quad \text{(magnetic moment).} \quad (29\text{-}33)$$

Thus Eq. 29-31 becomes

$$\tau = \mu B \sin \theta, \quad (29\text{-}34)$$

in which θ is the angle between the vectors $\boldsymbol{\mu}$ and **B**.

We can generalize this to the vector relation

$$\boldsymbol{\tau} = \boldsymbol{\mu} \times \mathbf{B}, \quad (29\text{-}35)$$

which reminds us very much of the corresponding equation for the torque exerted by an *electric* field on an *electric* dipole, namely Eq. 23-34:

$$\boldsymbol{\tau} = \mathbf{p} \times \mathbf{E}.$$

FIGURE 29-26 The orientations of highest and lowest energy of a magnetic dipole in an external magnetic field **B**. The direction of the current i gives the direction of the magnetic dipole moment μ via the right-hand rule shown for **n** in Fig. 29–24b.

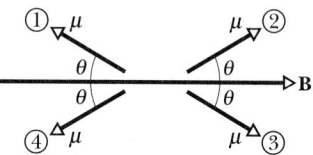

Highest energy   Lowest energy

In each case the torque exerted by the external field—either magnetic or electric—is equal to the vector product of the corresponding dipole moment and the field vector.

While an external magnetic field is exerting a torque on a magnetic dipole—such as a current-carrying coil—work must be done to change the orientation of the dipole. The magnetic dipole must then have a **magnetic potential energy** that depends on the dipole's orientation in the field. For electric dipoles we have shown (Eq. 23-38) that

$$U(\theta) = -\mathbf{p} \cdot \mathbf{E}.$$

In strict analogy, we can write for the magnetic case

$$U(\theta) = -\boldsymbol{\mu} \cdot \mathbf{B}. \qquad (29\text{-}36)$$

A magnetic dipole has its lowest energy ($= -\mu B \cos 0 = -\mu B$) when its dipole moment $\boldsymbol{\mu}$ is lined up with the magnetic field (Fig. 29-26). And it has its highest energy ($= -\mu B \cos 180° = +\mu B$) when $\boldsymbol{\mu}$ points in a direction opposite the field. The difference in energy between these two orientations is

$$\Delta U = (+\mu B) - (-\mu B) = 2\mu B. \qquad (29\text{-}37)$$

This much work must be done by an external agent (something other than the magnetic field) to turn a magnetic dipole through 180°, starting when the dipole is lined up with the magnetic field.

So far, we have identified only a current-carrying coil as a magnetic dipole. However, a simple bar magnet is also a magnetic dipole. So is a rotating sphere of charge. Earth itself is a magnetic dipole. Finally, most subatomic particles, including the electron, the proton, and the neutron, have magnetic dipole moments. As you will see in Chapter 32, all these quantities can be viewed as current loops. For comparison, some approximate magnetic dipole moments are shown in Table 29-2.

TABLE 29-2   SOME MAGNETIC DIPOLE MOMENTS

A small bar magnet	5 J/T
Earth	8.0×10^{22} J/T
A proton	1.4×10^{-26} J/T
An electron	9.3×10^{-24} J/T

CHECKPOINT **6:** The figure shows four orientations, at angle θ, of a magnetic dipole moment $\boldsymbol{\mu}$ in a magnetic field. Rank the orientations according to (a) the magnitude of the torque on the dipole and (b) the potential energy of the dipole, greatest first.

SAMPLE PROBLEM 29-10

(a) What is the magnetic dipole moment of the coil of Sample Problem 29-9, assuming that it carries a current of 100 μA?

SOLUTION: The *magnitude* of the magnetic dipole moment of the coil, whose area A is 2.52×10^{-4} m², is

$$\mu = NiA$$
$$= (250)(100 \times 10^{-6} \text{ A})(2.52 \times 10^{-4} \text{ m}^2)$$
$$= 6.3 \times 10^{-6} \text{ A} \cdot \text{m}^2 = 6.3 \times 10^{-6} \text{ J/T}. \quad \text{(Answer)}$$

You can show that these two sets of units are identical. The second set of units follows logically from Eq. 29-36.

The *direction* of $\boldsymbol{\mu}$, as inspection of Fig. 29-25 shows, is that of the pointer. You can verify this by showing that, if we assume $\boldsymbol{\mu}$ to be in the pointer direction, the torque predicted by Eq. 29-35 is such that it would indeed move the pointer clockwise across the scale.

(b) The magnetic dipole moment of the galvanometer coil is lined up with an external magnetic field whose strength is 0.85 T. How much work would be required to turn the coil end for end?

SOLUTION: The required work is equal to the increase in potential energy; that is, from Eq. 29-37,

$$W = \Delta U = 2\mu B = 2(6.3 \times 10^{-6} \text{ J/T})(0.85 \text{ T})$$
$$= 10.7 \times 10^{-6} \text{ J} \approx 11 \ \mu\text{J}. \quad \text{(Answer)}$$

This is about equal to the work needed to lift an aspirin tablet through a vertical height of 3 mm.

(c) What is the magnitude of the maximum torque τ that the external field **B** can exert on the magnetic dipole moment?

SOLUTION: From Eq. 29–34, the maximum torque occurs when the magnitude of $\sin \theta$ is 1. Thus, we have

$$\tau = \mu B \sin \theta$$
$$= (6.3 \times 10^{-6} \text{ J/T})(0.85 \text{ T})(1)$$
$$= 5.4 \times 10^{-6} \text{ N} \cdot \text{m}. \quad \text{(Answer)}$$

REVIEW & SUMMARY

Magnetic Field **B**

A **magnetic field B** is defined in terms of the force $\mathbf{F}_B$ acting on a test particle with charge q moving through the field with velocity **v**:

$$\mathbf{F}_B = q\mathbf{v} \times \mathbf{B}. \tag{29-2}$$

The SI unit for **B** is the **tesla** (T): $1\text{ T} = 1\text{ N/(A} \cdot \text{m}) = 10^4$ gauss.

The Hall Effect

When a conducting strip of thickness l carrying a current i is placed in a magnetic field **B**, some charge carriers (with charge e) build up on the sides of the conductor, creating a potential difference V across the strip. The polarity of V gives the sign of the charge carriers; the number density n of charge carriers can be calculated with

$$n = \frac{Bi}{Vle}. \tag{29-12}$$

A Charged Particle Circulating in a Magnetic Field

A charged particle with mass m and charge magnitude q moving with velocity **v** perpendicular to a magnetic field **B** will travel in a circle. Applying Newton's second law to the circular motion yields

$$qvB = \frac{mv^2}{r}, \tag{29-15}$$

from which we find the radius r of the circle to be

$$r = \frac{mv}{qB}. \tag{29-16}$$

The frequency of revolution f, the angular frequency ω, and the period of the motion T are given by

$$f = \frac{\omega}{2\pi} = \frac{1}{T} = \frac{qB}{2\pi m}. \quad \text{(29-19, 29-18, 29-17)}$$

Cyclotrons and Synchrotrons

A cyclotron is a particle accelerator that uses a magnetic field to hold a charged particle in a circular orbit of increasing radius so that a modest accelerating potential may act on the particle repeatedly, providing it with high energy. Because the moving particle gets out of step with the oscillator as its speed approaches that of light, there is an upper limit to the energy attainable with the cyclotron. A synchrotron avoids this difficulty. Here both B and the oscillator frequency f_{osc} are programmed to change cyclically so that the particle not only can go to high energies but can do so at a constant orbital radius.

Magnetic Force on a Current-Carrying Wire

A straight wire carrying a current i in a uniform magnetic field experiences a sideways force

$$\mathbf{F}_B = i\mathbf{L} \times \mathbf{B}. \tag{29-26}$$

The force acting on a current element $i\,d\mathbf{L}$ in a magnetic field is

$$d\mathbf{F}_B = i\,d\mathbf{L} \times \mathbf{B}. \tag{29-27}$$

The direction of the length vector **L** or $d\mathbf{L}$ is that of the current i.

Torque on a Current-Carrying Coil

A coil (of area A and carrying current i, with N turns) in a uniform magnetic field **B** will experience a torque $\boldsymbol{\tau}$ given by

$$\boldsymbol{\tau} = \boldsymbol{\mu} \times \mathbf{B}. \tag{29-35}$$

Here $\boldsymbol{\mu}$ is the **magnetic dipole moment** of the coil, with magnitude $\mu = NiA$ and direction given by a right-hand rule.

Orientation Energy of a Magnetic Dipole

The **magnetic potential energy** of a magnetic dipole in a magnetic field is

$$U(\theta) = -\boldsymbol{\mu} \cdot \mathbf{B}. \tag{29-36}$$

QUESTIONS

1. Figure 29-27 shows four directions for the velocity vector **v** of a negatively charged particle moving at angle θ to a uniform magnetic field **B**. (a) Rank the directions according to the magnitude of the magnetic force on the particle, greatest first. (b) Which gives a magnetic force out of the plane of the page?

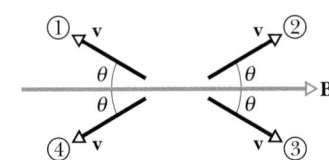

FIGURE 29-27
Question 1.

2. Here are four situations in which a proton moves with velocity **v** through a uniform magnetic field **B**:

(a) $\mathbf{v} = 2\mathbf{i} - 3\mathbf{j}$ and $\mathbf{B} = 4\mathbf{k}$

(b) $\mathbf{v} = 3\mathbf{i} + 2\mathbf{j}$ and $\mathbf{B} = -4\mathbf{k}$

(c) $\mathbf{v} = 3\mathbf{j} - 2\mathbf{k}$ and $\mathbf{B} = 4\mathbf{i}$

(d) $\mathbf{v} = 20\mathbf{i}$ and $\mathbf{B} = -4\mathbf{i}$.

Without written calculation, rank the situations according to the magnitude of the magnetic force on the proton, greatest first.

3. Figure 29-28 shows three situations in which a positive particle of velocity **v** moves through a uniform magnetic field **B** and experiences a magnetic force $\mathbf{F}_B$. In each situation, determine whether the orientations of the vectors are physically reasonable.

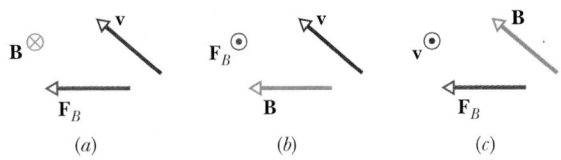

FIGURE 29-28 Question 3.

4. Figure 29-29 shows the path of an electron in a region of uniform magnetic field. The path consists of two straight sections, each between a pair of uniformly charged plates, and two half-circles. Which plate is at the higher electric potential in (a) the top pair of plates and (b) the bottom pair? (c) What is the direction of the magnetic field?

FIGURE 29-29 Question 4.

5. In Section 29-3, we discussed a charged particle moving through crossed fields with the forces $\mathbf{F}_E$ and $\mathbf{F}_B$ in opposition. We found that the particle moves in a straight line (that is, neither force dominates the motion) if its speed is given by Eq. 29-7 ($v = E/B$). Which of the two forces dominates if the speed of the particle is, instead, (a) $v < E/B$ and (b) $v > E/B$?

6. Figure 29-30 shows crossed and uniform electric and magnetic fields $\mathbf{E}$ and $\mathbf{B}$ and, at a certain instant, the velocity vectors of the 10 charged particles listed in Table 29-3. (The vectors are not drawn to scale.) The table gives the signs of the charges and the speeds of the particles; the speeds are given as either less than or greater than E/B (see Question 5). Which particles will move out of the page toward you after the instant of Fig. 29-30?

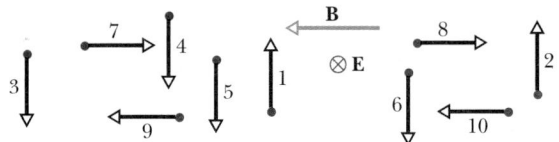

FIGURE 29-30 Question 6.

TABLE 29-3 QUESTION 6

PARTICLE	CHARGE	SPEED	PARTICLE	CHARGE	SPEED
1	+	Less	6	−	Greater
2	+	Greater	7	+	Less
3	+	Less	8	+	Greater
4	+	Greater	9	−	Less
5	−	Less	10	−	Greater

7. An airplane flies due west over Massachusetts, where Earth's magnetic field is directed downward and to the north. (a) On which wing, left or right, are some of the conduction electrons moved to the wingtip by the magnetic force on them? (b) Which wingtip gets the conduction electrons if the flight is eastward?

8. Figure 29-31 shows the cross section of a solid conductor carrying a current perpendicular to the page. (a) Which pair of the four terminals (a, b, c, d) should be used to measure the Hall voltage if the magnetic field is in the positive direction of the x axis, the charge carriers are negative, and they move out of the page? Which terminal of the pair is at the higher potential? (b) Repeat for a magnetic field in the negative direction of the y axis and positive charge carriers moving out of the page. (c) Discuss the situation if the magnetic field is in the positive z direction.

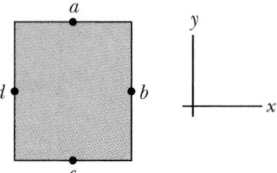

FIGURE 29-31 Question 8.

9. In Fig. 29-32, a charged particle enters a uniform magnetic field $\mathbf{B}$ with speed v_0, moves through a half-circle in time T_0, and then leaves the field. (a) Is the charge positive or negative? (b) Is the final speed of the particle greater than, smaller than, or equal to v_0? (c) If the initial speed had been $0.5v_0$, would the time spent in field $\mathbf{B}$ have been greater than, less than, or equal to T_0? (d) Would the path have been a half-circle, more than a half-circle, or less than a half-circle?

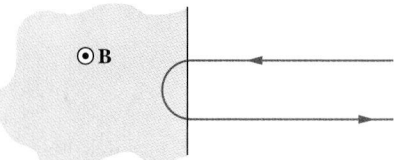

FIGURE 29-32 Question 9.

10. Figure 29-33 shows the path of a particle through six regions of uniform magnetic field, where the path is either a half-circle or a quarter-circle. Upon leaving the last region, the particle travels between two charged, parallel plates and is deflected toward the plate of higher potential. What are the directions of the magnetic fields in the six regions?

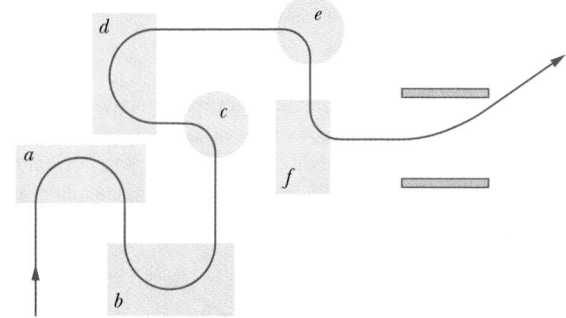

FIGURE 29-33 Question 10.

11. Figure 29-34 shows the path of an electron that passes through two regions containing uniform magnetic fields of magnitudes B_1 and B_2. Its path in each region is a half-circle. (a) Which field is stronger? (b) What are the directions of the two fields? (c) Is the time spent by the electron in the B_1 region greater than, less than, or the same as the time spent in the B_2 region?

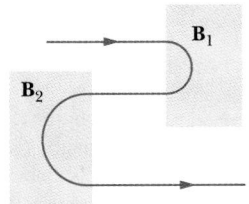

FIGURE 29-34
Question 11.

12. Particle Roundabout. Figure 29-35 shows 11 paths through a region of uniform magnetic field. One path is a straight line; the rest are half-circles. Table 29-4 gives the masses, charges, and speeds of 11 particles that take these paths through the field. Which path corresponds to which particle?

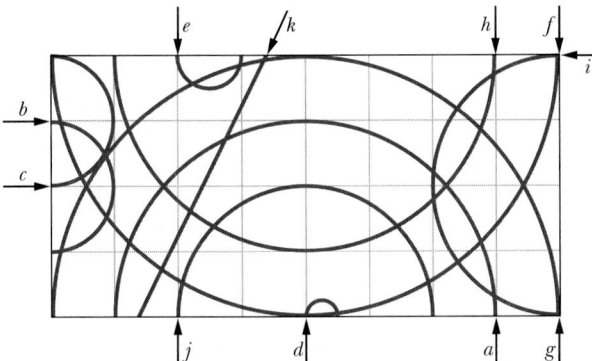

FIGURE 29-35 Question 12.

TABLE 29-4 QUESTION 12

PARTICLE	MASS	CHARGE	SPEED
1	$2m$	q	v
2	m	$2q$	v
3	$m/2$	q	$2v$
4	$3m$	$3q$	$3v$
5	$2m$	q	$2v$
6	m	$-q$	$2v$
7	m	$-4q$	v
8	m	$-q$	v
9	$2m$	$-2q$	$3v$
10	m	$-2q$	$8v$
11	$3m$	0	$3v$

13. Figure 29-36 shows three situations in which a charged particle moves in a spiral path through a uniform magnetic field. In which is the particle negatively charged?

  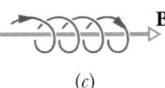

FIGURE 29-36 Question 13.

14. Figure 29-37 shows four views of a horseshoe magnet and a straight wire in which electrons are flowing out of the page, perpendicular to the plane of the magnet. In which case will the magnetic force on the wire point toward the top of the page?

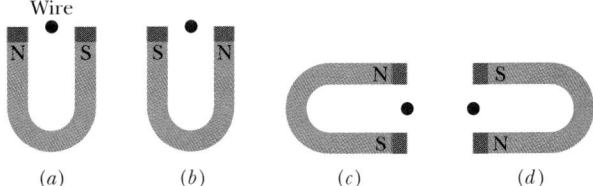

FIGURE 29-37 Question 14.

15. A wire carries a current i in the negative x direction, through a magnetic field **B**. Without written calculation, rank the following choices for **B** according to the magnitude of the magnetic forces they exert on the wire, greatest first: $\mathbf{B_1} = 2\mathbf{i} + 3\mathbf{j}$, $\mathbf{B_2} = 4\mathbf{i} - 3\mathbf{j}$, $\mathbf{B_3} = 6\mathbf{i} + 3\mathbf{k}$, and $\mathbf{B_4} = -8\mathbf{i} - 3\mathbf{k}$.

16. The dead-quiet "caterpillar drive" for submarines in the movie *The Hunt for Red October* is based on a *magnetohydrodynamic* (MHD) drive: as the ship moves forward, seawater flows through multiple channels in a structure built around the rear of the hull. Figure 29-38 shows the essentials of a channel. Magnets, positioned along opposite sides of the channel with opposite poles facing each other, create a magnetic field within the channel. Electrodes (not shown) create an electric field across the channel. The electric field drives a current across the channel and through the water; the magnetic force on the current propels the water toward the rear of the channel, thus propelling the ship forward. In Fig. 29-38, should the electric field be directed upward, downward, leftward, rightward, frontward, or rearward?

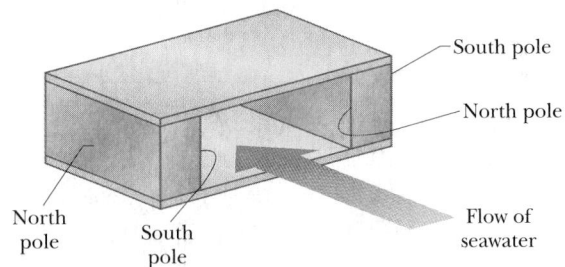

FIGURE 29-38 Question 16.

17. (a) In Checkpoint 6, if the dipole moment $\boldsymbol{\mu}$ rotates from orientation 1 to orientation 2, is the work done on the dipole *by the magnetic field* positive, negative, or zero? (b) Rank the work done on the dipole by the magnetic field for rotations from orientation 1 to (1) orientation 2, (2) orientation 3, and (3) orientation 4, greatest first.

EXERCISES & PROBLEMS

SECTION 29-2 The Definition of B

1E. Express the unit of a magnetic field B in terms of the dimensions M, L, T, and Q (mass, length, time, and charge).

2E. An alpha particle travels at a velocity $\mathbf{v}$ of magnitude 550 m/s through a uniform magnetic field $\mathbf{B}$ of magnitude 0.045 T. (An alpha particle has a charge of $+3.2 \times 10^{-19}$ C and a mass of 6.6×10^{-27} kg.) The angle between $\mathbf{v}$ and $\mathbf{B}$ is 52°. What are the magnitudes of (a) the force $\mathbf{F}_B$ acting on the particle due to the field and (b) the acceleration of the particle due to $\mathbf{F}_B$? (c) Does the speed of the particle increase, decrease, or remain equal to 550 m/s?

3E. An electron in a TV camera tube is moving at 7.20×10^6 m/s in a magnetic field of strength 83.0 mT. (a) Without knowing the direction of the field, what can you say about the greatest and least magnitudes of the force acting on the electron due to the field? (b) At one point the acceleration of the electron is 4.90×10^{14} m/s². What is the angle between the electron's velocity and the magnetic field?

4E. A proton traveling at 23.0° with respect to a magnetic field of strength 2.60 mT experiences a magnetic force of 6.50×10^{-17} N. Calculate (a) the proton's speed and (b) its kinetic energy in electron-volts.

5P. Each of the electrons in the beam of a television tube has a kinetic energy of 12.0 keV. The tube is oriented so that the electrons move horizontally from geomagnetic south to geomagnetic north. The vertical component of Earth's magnetic field points down and has a magnitude of 55.0 μT. (a) In what direction will the beam deflect? (b) What is the acceleration of a single electron due to the magnetic field? (c) How far will the beam deflect in moving 20.0 cm through the television tube?

6P. An electron that has velocity $\mathbf{v} = (2.0 \times 10^6$ m/s$)\mathbf{i} + (3.0 \times 10^6$ m/s$)\mathbf{j}$ moves through a magnetic field $\mathbf{B} = (0.030$ T$)\mathbf{i} - (0.15$ T$)\mathbf{j}$. (a) Find the force on the electron. (b) Repeat your calculation for a proton having the same velocity.

7P. An electron that is moving through a uniform magnetic field has a velocity $\mathbf{v} = (40$ km/s$)\mathbf{i} + (35$ km/s$)\mathbf{j}$ when it experiences a force $\mathbf{F} = -(4.2$ fN$)\mathbf{i} + (4.8$ fN$)\mathbf{j}$ due to the magnetic field. If $B_x = 0$, calculate the magnetic field $\mathbf{B}$.

SECTION 29-3 Crossed Fields: Discovery of the Electron

8E. A proton travels through uniform magnetic and electric fields. The magnetic field is $\mathbf{B} = -2.5\mathbf{i}$ mT. At one instant the velocity of the proton is $\mathbf{v} = 2000\mathbf{j}$ m/s. At that instant, what is the magnitude of the net force acting on the proton if the electric field is (a) $4.0\mathbf{k}$ V/m, (b) $-4.0\mathbf{k}$ V/m, and (c) $4.0\mathbf{i}$ V/m?

9E. An electron with kinetic energy 2.5 keV moves horizontally into a region of space in which there is a downward-directed electric field of magnitude 10 kV/m. (a) What are the magnitude

and direction of the (smallest) magnetic field that will cause the electron to continue to move horizontally? Ignore the gravitational force, which is rather small. (b) Is it possible for a proton to pass through this combination of fields undeflected? If so, under what circumstances?

10E. An electric field of 1.50 kV/m and a magnetic field of 0.400 T act on a moving electron to produce no net force. (a) Calculate the minimum speed v of the electron. (b) Draw the vectors $\mathbf{E}$, $\mathbf{B}$, and $\mathbf{v}$.

11P. An electron has an initial velocity of $(12.0\,\mathbf{j} + 15.0\,\mathbf{k})$ km/s and a constant acceleration of $(2.00 \times 10^{12}$ m/s²$)\mathbf{i}$ in a region in which uniform electric and magnetic fields are present. If $\mathbf{B} = (400\,\mu$T$)\mathbf{i}$, find the electric field $\mathbf{E}$.

12P. An electron is accelerated through a potential difference of 1.0 kV and directed into a region between two parallel plates separated by 20 mm with a potential difference of 100 V between them. The electron is moving perpendicular to the electric field when it enters the region between the plates. What magnetic field is necessary perpendicular to both the electron path and the electric field so that the electron travels in a straight line?

13P. An ion source is producing ions of ^{6}Li (mass = 6.0 u), each with a charge of $+e$. The ions are accelerated by a potential difference of 10 kV and pass horizontally into a region in which there is a uniform vertical magnetic field of magnitude $B = 1.2$ T. Calculate the strength of the smallest electric field, to be set up over the same region, that will allow the ^{6}Li ions to pass through undeflected.

SECTION 29-4 Crossed Fields: The Hall Effect

14E. A strip of copper 150 μm wide is placed in a uniform magnetic field $\mathbf{B}$ of magnitude 0.65 T, with $\mathbf{B}$ perpendicular to the strip. A current $i = 23$ A is then sent through the strip such that a Hall potential difference V appears across the width. Calculate V. (The number of charge carriers per unit volume for copper is 8.47×10^{28} electrons/m³.)

15E. Show that, in terms of the Hall electric field E and the current density J, the number of charge carriers per unit volume is given by $n = JB/eE$.

16P. In a Hall-effect experiment, a current of 3.0 A sent lengthwise through a conductor 1.0 cm wide, 4.0 cm long, and 10 μm thick produces a transverse (across the width) Hall voltage of 10 μV when a magnetic field of 1.5 T is passed perpendicularly through the thickness of the conductor. From these data, find (a) the drift velocity of the charge carriers and (b) the number density of charge carriers. (c) Show on a diagram the polarity of the Hall voltage with assumed current and magnetic field directions, assuming also that the charge carriers are electrons.

17P. (a) In Fig. 29-8, show that the ratio of the Hall electric field E to the electric field E_C responsible for moving charge (the current) along the length of the strip is

$$\frac{E}{E_C} = \frac{B}{ne\rho},$$

where ρ is the resistivity of the material and n is the number density of the charge carriers. (b) Compute this ratio numerically for Exercise 14. (See Table 27-1.)

18P. A metal strip 6.50 cm long, 0.850 cm wide, and 0.760 mm thick moves with constant velocity **v** through a magnetic field $B = 1.20$ mT pointing perpendicular to the strip, as shown in Fig. 29-39. A potential difference of 3.90 μV is measured between points x and y across the strip. Calculate the speed v.

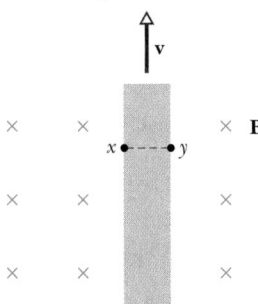

FIGURE 29-39
Problem 18.

SECTION 29-5 A Circulating Charged Particle

19E. An electron is accelerated from rest by a potential difference of 350 V. It then enters a uniform magnetic field of magnitude 200 mT with its velocity perpendicular to the field. Calculate (a) the speed of the electron and (b) the radius of its path in the magnetic field.

20E. What uniform magnetic field, applied perpendicular to a beam of electrons moving at 1.3×10^6 m/s, is required to make the electrons travel in a circular arc of radius 0.35 m?

21E. (a) In a magnetic field with $B = 0.50$ T, for what path radius will an electron circulate at 10% the speed of light? (b) What will be its kinetic energy in electron-volts? Ignore the small relativistic effects.

22E. What uniform magnetic field must be set up in space to permit a proton of speed 1.0×10^7 m/s to move in a circle the size of Earth's equator?

23E. An electron with kinetic energy 1.20 keV circles in a plane perpendicular to a uniform magnetic field. The orbit radius is 25.0 cm. Find (a) the speed of the electron, (b) the magnetic field, (c) the frequency of circling, and (d) the period of the motion.

24E. Physicist S. A. Goudsmit devised a method for measuring accurately the masses of heavy ions by timing their periods of revolution in a known magnetic field. A singly charged ion of iodine makes 7.00 rev in a field of 45.0 mT in 1.29 ms. Calculate its mass, in unified atomic mass units. Actually, the mass measurements are carried out to much greater accuracy than these approximate data suggest.

25E. An alpha particle ($q = +2e$, $m = 4.00$ u) travels in a circular path of radius 4.50 cm in a magnetic field with $B = 1.20$ T. Calculate (a) its speed, (b) its period of revolution, (c) its kinetic energy in electron-volts, and (d) the potential difference through which it would have to be accelerated to achieve this energy.

26E. (a) Find the frequency of revolution of an electron with an energy of 100 eV in a magnetic field of 35.0 μT. (b) Calculate the radius of the path of this electron if its velocity is perpendicular to the magnetic field.

27E. A beam of electrons whose kinetic energy is K emerges from a thin-foil "window" at the end of an accelerator tube. There is a metal plate a distance d from this window and perpendicular to the direction of the emerging beam (Fig. 29-40). Show that we can prevent the beam from hitting the plate if we apply a magnetic field B such that

$$B \geq \sqrt{\frac{2mK}{e^2 d^2}},$$

in which m and e are the electron mass and charge. How should **B** be oriented?

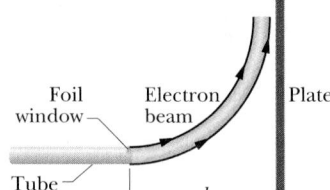

FIGURE 29-40
Exercise 27.

28P. A source injects an electron of speed $v = 1.5 \times 10^7$ m/s into a uniform magnetic field of magnitude $B = 1.0 \times 10^{-3}$ T. The velocity of the electron makes an angle $\theta = 10°$ with the direction of the magnetic field. Find the distance d from the point of injection at which the electron next crosses the field line that passes through the injection point.

29P. In a nuclear experiment a proton with kinetic energy 1.0 MeV moves in a circular path in a uniform magnetic field. What energy must (a) an alpha particle ($q = +2e$, $m = 4.0$ u) and (b) a deuteron ($q = +e$, m $= 2.0$ u) have if they are to circulate in the same orbit?

30P. A proton, a deuteron, and an alpha particle (see Problem 29), accelerated through the same potential difference, enter a region of uniform magnetic field **B**, moving perpendicular to **B**. (a) Compare their kinetic energies. If the radius of the proton's circular path is 10 cm, what are the radii of (b) the deuteron path and (c) the alpha-particle path?

31P. A proton, a deuteron, and an alpha particle (see Problem 29) with the same kinetic energies enter a region of uniform magnetic field **B,** moving perpendicular to **B**. Compare the radii of their circular paths.

32P. A proton of charge $+e$ and mass m enters a uniform magnetic field $\mathbf{B} = B\mathbf{i}$ with an initial velocity $\mathbf{v} = v_{0x}\mathbf{i} + v_{0y}\mathbf{j}$. Find an expression in unit-vector notation for its velocity **v** at any later time t.

33P. Two types of singly ionized atom having the same charge q but masses that differ by a small amount Δm are introduced into the mass spectrometer described in Sample Problem 29-3. (a) Calculate the difference in mass in terms of V, q, m (of either), B, and the distance Δx between the spots on the photographic plate. (b) Calculate Δx for a beam of singly ionized chlorine atoms of masses 35 and 37 u if $V = 7.3$ kV and $B = 0.50$ T.

34P. In a commercial mass spectrometer (see Sample Problem 29-3), uranium ions of mass 3.92×10^{-25} kg and charge 3.20×10^{-19} C are separated from related species. The ions are accelerated through a potential difference of 100 kV and then pass into a magnetic field, where they are bent in a path of radius 1.00 m. After traveling through 180° and passing through a slit of width 1.00 mm and height 1.00 cm, they are collected in a cup. (a) What is the magnitude of the (perpendicular) magnetic field in the separator? If the machine is used to separate out 100 mg of material per hour, calculate (b) the current of the desired ions in the machine and (c) the thermal energy produced in the cup in 1.00 h.

35P. Bainbridge's mass spectrometer, shown in Fig. 29-41, separates ions having the same velocity. The ions, after entering through slits S_1 and S_2, pass through a velocity selector composed of an electric field produced by the charged plates P and P′, and a magnetic field **B** perpendicular to the electric field and the ion path. The ions that pass undeviated through the crossed **E** and **B** fields enter into a region where a second magnetic field **B′** exists, where they are made to follow circular paths. A photographic plate registers their arrival. Show that, for the ions, $q/m = E/rBB'$, where r is the radius of the circular orbit.

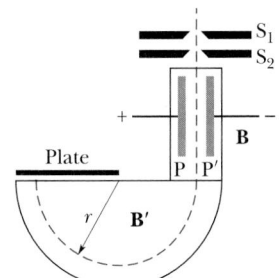

FIGURE 29-41
Problem 35.

36P. A positron with kinetic energy 2.0 keV is projected into a uniform magnetic field **B** of 0.10 T with its velocity vector making an angle of 89° with **B**. Find (a) the period, (b) the pitch p, and (c) the radius r of its helical path.

37P. A neutral particle is at rest in a uniform magnetic field of magnitude B. At time $t = 0$ it decays into two charged particles, each of mass m. (a) If the charge of one of the particles is $+q$, what is the charge of the other? (b) The two particles move off in separate paths, both of which lie in the plane perpendicular to **B**. At a later time the particles collide. Express the time from decay until collision in terms of m, B, and q.

38P. (a) What speed would a proton need to circle Earth at the equator, if Earth's magnetic field is everywhere horizontal there and directed along longitudinal lines? Relativistic effects must be taken into account. Take the magnitude of Earth's magnetic field to be 41 μT at the equator. (*Hint:* Replace the momentum mv in Eq. 29-16 with the relativistic momentum given in Eq. 9-24.) (b) Draw the velocity and magnetic field vectors corresponding to this situation.

39P. In Fig. 29-42, an electron of mass m, charge e, and negligible speed enters the region between two plates of potential difference V and plate separation d, initially headed directly toward the higher-potential top plate in the figure. A uniform magnetic field

of magnitude B is directed perpendicular to the plane of the figure. Find the minimum value of B at which the electron will not strike the top plate.

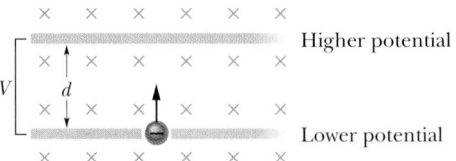

FIGURE 29-42 Problem 39.

SECTION 29-6 Cyclotrons and Synchrotrons

40E. In a certain cyclotron a proton moves in a circle of radius 0.50 m. The magnitude of the magnetic field is 1.2 T. (a) What is the cyclotron frequency? (b) What is the kinetic energy of the proton, in electron-volts?

41E. A physicist is designing a cyclotron to accelerate protons to one-tenth the speed of light. The magnet used will produce a field of 1.4 T. Calculate (a) the radius of the cyclotron and (b) the corresponding oscillator frequency. Relativity considerations are not significant.

42P. The oscillator frequency of the cyclotron in Sample Problem 29-5 has been adjusted to accelerate deuterons. (a) If protons are injected instead of deuterons, to what kinetic energy can the protons be accelerated, using the same oscillator frequency? (b) What magnetic field would be required? (c) What kinetic energy for protons could be produced if the magnetic field were left at the value used for deuterons? (d) What oscillator frequency would then be required? (e) Answer the same questions for alpha particles ($q = +2e$, $m = 4.0$ u).

43P. A deuteron in a cyclotron is moving in a magnetic field with $B = 1.5$ T and an orbit radius of 50 cm. Because of a grazing collision with a target, the deuteron breaks up, with negligible loss of kinetic energy, into a proton and a neutron. Discuss the subsequent motion of each. Assume that the deuteron energy is shared equally by the proton and neutron at breakup.

44P. Estimate the total path length traversed by a deuteron in the cyclotron of Sample Problem 29-5 during the acceleration process. Assume an accelerating potential between the dees of 80 kV.

SECTION 29-7 Magnetic Force on a Current-Carrying Wire

45E. A horizontal conductor in a power line carries a current of 5000 A from south to north. Earth's magnetic field (60.0 μT) is directed toward the north and is inclined downward at 70° to the horizontal. Find the magnitude and direction of the magnetic force on 100 m of the conductor due to Earth's field.

46E. A wire of 62.0 cm length and 13.0 g mass is suspended by a pair of flexible leads in a magnetic field of 0.440 T (Fig. 29-43). What are the magnitude and direction of the current required to remove the tension in the supporting leads?

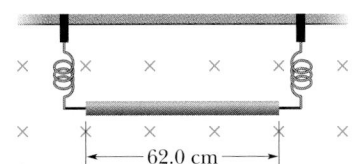

FIGURE 29-43
Exercise 46.

47E. A wire 1.80 m long carries a current of 13.0 A and makes an angle of 35.0° with a uniform magnetic field $B = 1.50$ T. Calculate the magnetic force on the wire.

48P. A wire 50 cm long lying along the x axis carries a current of 0.50 A in the positive x direction, through a magnetic field $\mathbf{B} = (0.0030$ T$)\mathbf{j} + (0.010$ T$)\mathbf{k}$. Find the force on the wire.

49P. A metal wire of mass m slides without friction on two horizontal rails spaced a distance d apart, as in Fig. 29-44. The track lies in a vertical uniform magnetic field $\mathbf{B}$. There is a constant current i through generator G, along one rail, across the wire, and back down the other rail. Find the speed and direction of the wire's motion as a function of time, assuming it to be stationary at $t = 0$.

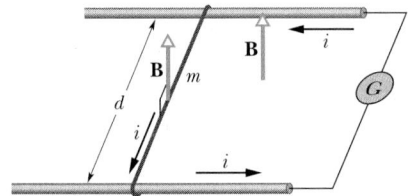

FIGURE 29-44   Problem 49.

50P. Figure 29-45 shows a wire of arbitrary shape carrying a current i between points a and b. The wire lies in a plane at right angles to a uniform magnetic field $\mathbf{B}$. (a) Prove that the force on the wire is the same as that on a straight wire carrying a current i directly from a to b. (*Hint:* Replace the wire with a series of "steps" parallel and perpendicular to the straight line joining a and b.) (b) Prove that the force on the wire becomes zero when points a and b are brought together so that the wire is a complete loop whose plane is perpendicular to $\mathbf{B}$.

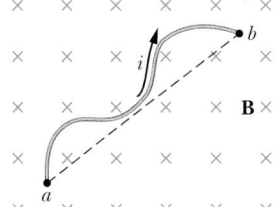

FIGURE 29-45
Problem 50.

51P. A long, rigid conductor, lying along the x axis, carries a current of 5.0 A in the negative direction. A magnetic field $\mathbf{B}$ is present, given by $\mathbf{B} = 3.0\mathbf{i} + 8.0x^2\mathbf{j}$, with x in meters and $\mathbf{B}$ in milliteslas. Calculate the force on the 2.0 m segment of the conductor that lies between $x = 1.0$ m and $x = 3.0$ m.

52P. Consider the possibility of a new design for an electric train. The engine is driven by the force on a conducting axle due to the vertical component of Earth's magnetic field. Current is down one rail, through a conducting wheel, through the axle, through another conducting wheel, and then back to the source via the other rail. (a) What current is needed to provide a modest 10 kN force? Take the vertical component of Earth's field to be 10 μT and the length of the axle to be 3.0 m. (b) How much power would be lost for each ohm of resistance in the rails? (c) Is such a train totally unrealistic or just marginally unrealistic?

53P. A 1.0 kg copper rod rests on two horizontal rails 1.0 m apart and carries a current of 50 A from one rail to the other. The coefficient of static friction between rod and rails is 0.60. What is the smallest magnetic field (not necessarily vertical) that would cause the rod to slide?

SECTION 29-8 Torque on a Current Loop

54E. A single-turn current loop, carrying a current of 4.00 A, is in the shape of a right triangle with sides 50.0, 120, and 130 cm. The loop is in a uniform magnetic field of magnitude 75.0 mT whose direction is parallel to the current in the 130 cm side of the loop. (a) Find the magnitude of the magnetic force on each of the three sides of the loop. (b) Show that the total magnetic force on the loop is zero.

55E. Figure 29-46 shows a rectangular, 20-turn coil of wire, 10 cm by 5.0 cm. It carries a current of 0.10 A and is hinged along one long side. It is mounted in the xy plane, at an angle of 30° to the direction of a uniform magnetic field of 0.50 T. Find the magnitude and direction of the torque acting on the coil about the hinge line.

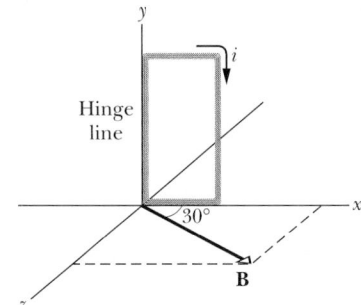

FIGURE 29-46
Exercise 55.

56E. A length L of wire carries a current i. Show that if the wire is formed into a circular coil, then the maximum torque in a given magnetic field is developed when the coil has one turn only and that maximum torque has the magnitude $\tau = (1/4\pi)L^2iB$.

57P. Prove that the relation $\tau = NiAB \sin \theta$ holds for closed loops of arbitrary shape and not only for rectangular loops as in Fig. 29-24. (*Hint:* Replace the loop of arbitrary shape with an assembly of adjacent long, thin, approximately rectangular loops that are nearly equivalent to the loop of arbitary shape as far as the distribution of current is concerned.)

58P. A closed wire loop with current i is in a uniform magnetic field $\mathbf{B}$, with the plane of the loop at angle θ to the direction of $\mathbf{B}$. Show that the total magnetic force on the loop is zero. Does your proof also hold for a nonuniform magnetic field?

59P. A particle of charge q moves in a circle of radius a with speed v. Treating the circular path as a current loop with constant current equal to its average current, find the maximum torque exerted on the loop by a uniform magnetic field of magnitude B.

60P. Figure 29-47 shows a wire ring of radius a that is perpendicular to the general direction of a radially symmetric diverging magnetic field. The magnetic field at the ring is everywhere of the same magnitude B, and its direction at the ring everywhere makes an angle θ with a normal to the plane of the ring. The twisted lead wires have no effect on the problem. Find the magnitude and direction of the force the field exerts on the ring if the ring carries a current i.

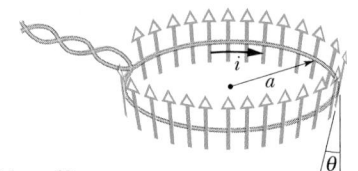

FIGURE 29-47   Problem 60.

61P. A certain galvanometer has a resistance of 75.3 Ω; its needle experiences a full-scale deflection when a current of 1.62 mA passes through its coil. (a) Determine the value of the auxiliary resistance required to convert the galvanometer to a voltmeter that reads 1.00 V at full-scale deflection. How is this resistance to be connected? (b) Determine the value of the auxiliary resistance required to convert the galvanometer to an ammeter that reads 50.0 mA at full-scale deflection. How is this resistance to be connected?

62P. Figure 29-48 shows a wooden cylinder with mass $m = 0.250$ kg and length $L = 0.100$ m, with $N = 10.0$ turns of wire wrapped around it longitudinally, so that the plane of the wire coil contains the axis of the cylinder. What is the least current i through the coil that will prevent the cylinder from rolling down a plane inclined at an angle θ to the horizontal, in the presence of a vertical, uniform magnetic field of 0.500 T, if the plane of the windings is parallel to the inclined plane?

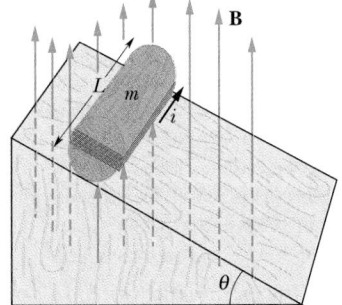

FIGURE 29-48
Problem 62.

SECTION 29-9 The Magnetic Dipole

63E. A circular coil of 160 turns has a radius of 1.90 cm. (a) Calculate the current that results in a magnetic dipole moment of 2.30 A·m². (b) Find the maximum torque that the coil, carrying this current, can experience in a uniform 35.0 mT magnetic field.

64E. The magnetic dipole moment of Earth is 8.00×10^{22} J/T. Assume that this is produced by charges flowing in Earth's mol-

ten outer core. If the radius of their circular path is 3500 km, calculate the current they produce.

65E. A circular wire loop whose radius is 15.0 cm carries a current of 2.60 A. It is placed so that the normal to its plane makes an angle of 41.0° with a uniform magnetic field of 12.0 T. (a) Calculate the magnetic dipole moment of the loop. (b) What torque acts on the loop?

66E. A current loop, carrying a current of 5.0 A, is in the shape of a right triangle with sides 30, 40, and 50 cm. The loop is in a uniform magnetic field of magnitude 80 mT whose direction is parallel to the current in the 50 cm side of the loop. Find the magnitude of (a) the magnetic dipole moment of the loop and (b) the torque on the loop.

67E. A stationary circular wall clock has a face with a radius of 15 cm. Six turns of wire are wound around its perimeter; the wire carries a current of 2.0 A in the clockwise direction. The clock is located where there is a constant, uniform external magnetic field of 70 mT (but the clock still keeps perfect time). At exactly 1:00 P.M., the hour hand of the clock points in the direction of the external magnetic field. (a) After how many minutes will the minute hand point in the direction of the torque on the winding due to the magnetic field? (b) Find the torque magnitude.

68E. Two concentric circular loops of radii 20.0 and 30.0 cm, located in the xy plane, each carry a clockwise current of 7.00 A (Fig. 29-49). (a) Find the net magnetic dipole moment of this system. (b) Repeat for reversed current in the inner loop.

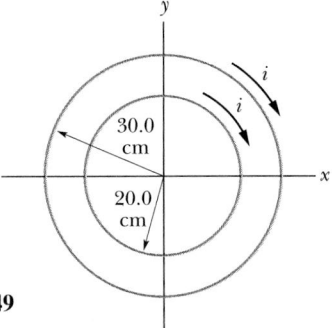

FIGURE 29-49
Exercise 68.

69P. A circular loop of wire having a radius of 8.0 cm carries a current of 0.20 A. A unit vector parallel to the dipole moment $\boldsymbol{\mu}$ of the loop is given by $0.60\mathbf{i} - 0.80\mathbf{j}$. If the loop is located in a magnetic field given by $\mathbf{B} = (0.25$ T$)\mathbf{i} + (0.30$ T$)\mathbf{k}$, find (a) the torque on the loop (in unit-vector notation) and (b) the magnetic potential energy of the loop.

70P. Figure 29-50 shows a current loop $ABCDEFA$ carrying a

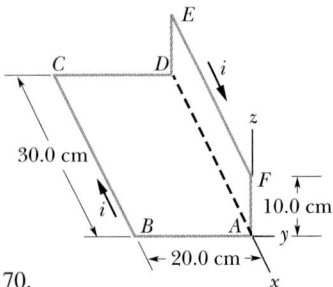

FIGURE 29-50   Problem 70.

current $i = 5.00$ A. The sides of the loop are parallel to the coordinate axes, with $AB = 20.0$ cm, $BC = 30.0$ cm, and $FA = 10.0$ cm. Calculate the magnitude and direction of the magnetic dipole moment of this loop. (*Hint:* Imagine equal and opposite currents i in the line segment AD; then treat the two rectangular loops $ABCDA$ and $ADEFA$.)

Electronic Computation

71. A particle with mass m and charge q moves in a uniform electric field **E**, in the positive y direction, and a uniform magnetic field **B**, in the positive z direction. The force on the particle is $\mathbf{F} = q[\mathbf{E} + \mathbf{v} \times \mathbf{B}]$ and the acceleration of the particle is $\mathbf{a} = (q/m)[\mathbf{E} + \mathbf{v} \times \mathbf{B}]$. If **v** is in the xy plane, then the components of the acceleration are $a_x = (qB/m)v_y$ and $a_y = (qE/m) - (qB/m)v_x$. These components can be integrated twice to obtain expressions for the coordinates of the particle. If the particle starts at the origin with an initial velocity v_0 in the positive x direction, then

$$x = \frac{E}{B} t - \frac{1}{\omega} \left[\frac{E}{B} - v_0 \right] \sin(\omega t)$$

and

$$y = -\frac{1}{\omega} \left[\frac{E}{B} - v_0 \right] [1 - \cos(\omega t)],$$

where $\omega = qB/m$. (a) By direct substitution, verify that these equations satisfy Newton's second law. Also verify that they lead to the given initial conditions. (b) Take $B = 1.2$ T, $E = 1.0 \times 10^4$ V/m, and $v_0 = 5.0 \times 10^4$ m/s and plot the trajectory of the particle for the first 4.0×10^{-6} s after it leaves the origin. The orbit can be described as a circle that translates in the positive x direction. Explain qualitatively why the motion is along the x axis when the electric field is along the y axis. Graph the trajectory for (b) $v_0 = 3.0 \times 10^4$ m/s, (c) $v_0 = 6.0 \times 10^4$ m/s, and (d) $v_0 = 9.0 \times 10^4$ m/s. (e) Why do some of the trajectories cross back over themselves while others do not? Why is one of the trajectories a straight line?

This is the way we presently launch materials into space. But when we begin mining the Moon and the asteroids, where we will not have a source of fuel for such conventional rockets, we shall need a more effective way. Electromagnetic launchers may be the answer. A small prototype, the <u>electromagnetic rail gun</u>, can presently accelerate a projectile from rest to a speed of 10 km/s (2000 mi/h) within 1 ms. How can such rapid acceleration possibly be accomplished?

30-1 CALCULATING THE MAGNETIC FIELD DUE TO A CURRENT

As we discussed in Section 29-1, one way to produce a magnetic field is with moving charges, that is, with a current. Our goal in this chapter is to calculate the magnetic field that is produced by a given distribution of currents. We shall use the same basic procedure we used in Chapter 23 to calculate the electric field produced by a given distribution of charged particles.

Let us quickly review that basic procedure. We first mentally divide the charge distribution into charge elements dq, as is done for a charge distribution of arbitrary shape in Fig. 30-1a. We then calculate the field $d\mathbf{E}$ produced by a typical charge element at some point P. Because the electric fields contributed by different elements can be superimposed, we calculate the net field $\mathbf{E}$ at P by summing, via integration, the contributions $d\mathbf{E}$ from all the elements.

Recall that we express the magnitude of $d\mathbf{E}$ as

$$dE = \frac{1}{4\pi\epsilon_0} \frac{dq}{r^2}, \qquad (30\text{-}1)$$

in which r is the distance from the charge element dq to point P. For a positively charged element, the direction of $d\mathbf{E}$ is that of $\mathbf{r}$, where $\mathbf{r}$ is the vector that extends from the charge element dq to the point P. Using $\mathbf{r}$, we can rewrite Eq. 30-1 in vector form as

$$d\mathbf{E} = \frac{1}{4\pi\epsilon_0} \frac{dq}{r^3} \mathbf{r}, \qquad (30\text{-}2)$$

which indicates that the direction of the vector $d\mathbf{E}$ produced by a positively charged element is in the direction of the vector $\mathbf{r}$. Note that Eq. 30-2 is an inverse-square law

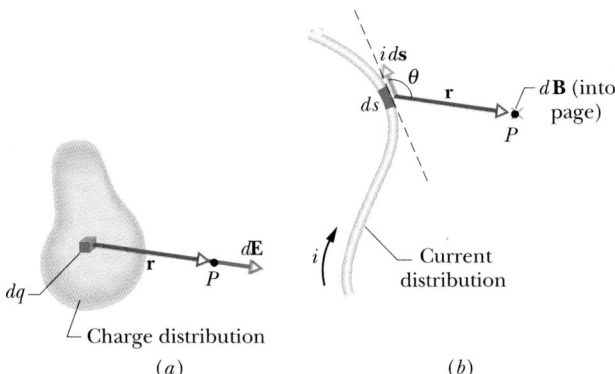

($d\mathbf{E}$ depends on inverse r^2) in spite of the exponent 3 in the denominator. That exponent is in the equation only because we added a factor of magnitude r in the numerator.

Now let us use the same basic procedure to calculate the magnetic field due to a current. Figure 30-1b shows a wire of arbitrary shape carrying a current i. We want to find the magnetic field $\mathbf{B}$ at a nearby point P. We first mentally divide the wire into differential elements $d\mathbf{s}$ that have length ds and are everywhere tangent to the wire and in the direction of the current. We can then define a differential *current-length element* to be $i\,d\mathbf{s}$; we wish to calculate the field $d\mathbf{B}$ produced at P by a typical current-length element. From experiment we find that magnetic fields, like electric fields, can be superimposed to find a net field. So, we can calculate the net field $\mathbf{B}$ at P by summing, via integration, the contributions $d\mathbf{B}$ from all the current-length elements. However, this summation is more challenging than the process associated with electric fields because of a complexity: whereas a charge element dq producing an electric field is a scalar, a current-length element $i\,d\mathbf{s}$ producing a magnetic field is the product of a scalar and a vector.

The magnitude of the field $d\mathbf{B}$ produced at point P by a current-length element $i\,d\mathbf{s}$ turns out to be

$$dB = \frac{\mu_0}{4\pi} \frac{i\,ds\,\sin\theta}{r^2}. \qquad (30\text{-}3)$$

Here μ_0 is a constant, called the *permeability constant*, whose value is defined to be exactly

$$\mu_0 = 4\pi \times 10^{-7}\ \text{T}\cdot\text{m/A}$$
$$\approx 1.26 \times 10^{-6}\ \text{T}\cdot\text{m/A}. \qquad (30\text{-}4)$$

The direction of $d\mathbf{B}$, shown as being into the page in Fig. 30-1b, is that of the cross product $d\mathbf{s} \times \mathbf{r}$, where $\mathbf{r}$ is now a vector that extends from the current element to point P. We can therefore write Eq. 30-3 in vector form as

$$d\mathbf{B} = \frac{\mu_0}{4\pi} \frac{i\,d\mathbf{s} \times \mathbf{r}}{r^3} \qquad \text{(Biot–Savart law).} \quad (30\text{-}5)$$

This vector equation and its scalar form, Eq. 30-3, are known as the **law of Biot and Savart** (rhymes with ''Leo and bazaar''). The law is an inverse-square law (the exponent in the denominator of Eq. 30-5 is 3 only because of the factor $\mathbf{r}$ in the numerator). We shall use this law to calculate the net magnetic field $\mathbf{B}$ produced at a point by various distributions of current.

Magnetic Field Due to a Current in a Long Straight Wire

Shortly we shall use the law of Biot and Savart to prove that the magnitude of the magnetic field at a perpendicular

FIGURE 30-1 (a) A charge element dq produces a differential electric field $d\mathbf{E}$ at point P. (b) A current-length element $i\,d\mathbf{s}$ produces a differential magnetic field $d\mathbf{B}$ at point P. The green $\times$ (the tail of an arrow) at the dot for point P indicates that $d\mathbf{B}$ points *into* the page there.

distance r from a long (infinite) straight wire carrying a current i is given by

$$B = \frac{\mu_0 i}{2\pi r} \qquad \text{(long straight wire).} \qquad (30\text{-}6)$$

(Note carefully that in this specialized equation r is the *perpendicular* distance between the wire and a point at which B is to be evaluated. However, in Eqs. 30-3 and 30-5—which are fundamental—r is the distance between a current-length element in the wire and that point.)

The field magnitude B in Eq. 30-6 depends only on the current and the perpendicular distance r from the wire. We shall show in our derivation that the field lines of **B** form concentric circles around the wire, as Fig. 30-2 shows and as the iron filings in Fig. 30-3 suggest. The increase in the spacing of the lines in Fig. 30-2 with increasing distance from the wire represents the $1/r$ decrease in the magnitude of **B** predicted by Eq. 30-6.

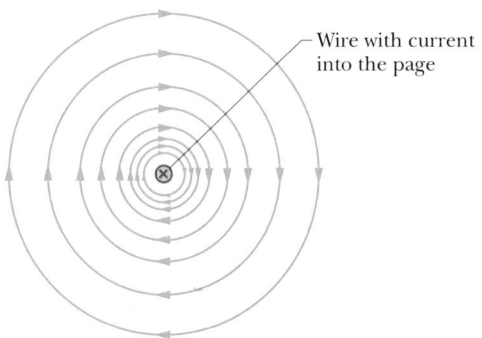

FIGURE 30-2 The magnetic field lines produced by a current in a long straight wire form concentric circles around the wire. Here the current is into the page, as indicated by the $\times$.

FIGURE 30-3 Iron filings that have been sprinkled onto cardboard collect in concentric circles when current is sent through the central wire. The alignment, which is along magnetic field lines, is caused by the magnetic field produced by the current.

Here is a simple right-hand rule for finding the direction of the magnetic field set up by a current-length element, such as a section of a long wire:

Grasp the element in your right hand with your extended thumb pointing in the direction of the current. Your fingers will then naturally curl around in the direction of the magnetic field lines due to that element.

The result of applying this right-hand rule to the current in the straight wire of Fig. 30-2 is shown in a side view in Fig. 30-4a. Note that the fingers curl around the wire as the magnetic field lines do in Fig. 30-2. To determine the direction of **B** at any particular point, place your right hand around the wire so that your fingertips pass through that point, as in Figs. 30-4a and 30-4b. The direction of the fingertips at the point then gives the direction of **B** there.

Proof of Equation 30-6

Figure 30-5, which is just like Fig. 30-1b except that now the wire is straight, illustrates the task at hand: we seek the field **B** at point P, a perpendicular distance R from the wire. The magnitude of the differential magnetic field produced at P by the current-length element $i\,d\mathbf{s}$ located a distance r from P is given by Eq. 30-3:

$$dB = \frac{\mu_0}{4\pi} \frac{i\,ds\,\sin\theta}{r^2}.$$

The direction of $d\mathbf{B}$ in Fig. 30-5 is that of the vector $d\mathbf{s} \times \mathbf{r}$, namely, directly into the page.

Note that $d\mathbf{B}$ at point P has this same direction for all the current-length elements into which the wire can be di-

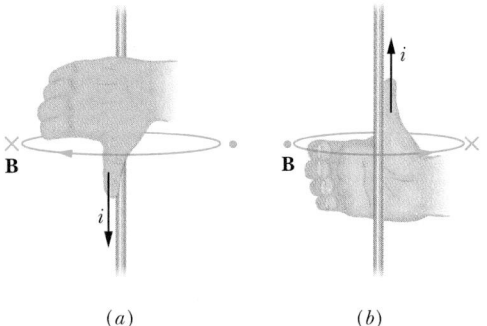

(a)   (b)

FIGURE 30-4 A right-hand rule gives the direction of the magnetic field due to a current in a wire. (a) The situation of Fig. 30-2, seen from the side. The magnetic field **B** at any point to the left of the wire points into the page, in the direction of the fingertips, as indicated by the $\times$. (b) If the current is reversed, **B** at any point to the left points out of the page, as indicated by the dot.

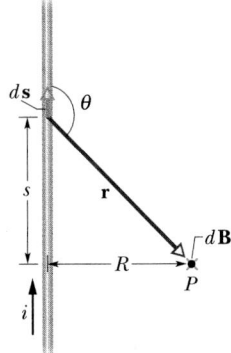

FIGURE 30-5 Calculating the magnetic field produced by a current i in a long straight wire. The field $d\mathbf{B}$ at P associated with the current-length element i $d\mathbf{s}$ points into the page, as shown.

vided. Thus we can find the magnitude of the magnetic field produced at P by the current-length elements in the upper half of the (infinitely) long wire by integrating dB in Eq. 30-3 from 0 to ∞. By Eq. 30-5, the magnetic field produced at P by a symmetrically located current-length element in the lower half of the wire has the same magnitude and direction as that from i $d\mathbf{s}$ in Fig. 30-5. Moreover, the magnetic field produced by the lower half is exactly the same as that produced by the upper half. So, to find the magnitude of the *total* magnetic field $\mathbf{B}$ at P, we need only multiply the result of our integration by 2. We get

$$B = 2 \int_0^\infty dB = \frac{\mu_0 i}{2\pi} \int_0^\infty \frac{\sin\theta\, ds}{r^2}. \qquad (30\text{-}7)$$

The variables θ, s, and r in this equation are not independent but (see Fig. 30-5) are related by

$$r = \sqrt{s^2 + R^2}$$

and
$$\sin\theta = \sin(\pi - \theta) = \frac{R}{\sqrt{s^2 + R^2}}.$$

With these substitutions and integral 19 in Appendix E, Eq. 30-7 becomes

$$B = \frac{\mu_0 i}{2\pi} \int_0^\infty \frac{R\, ds}{(s^2 + R^2)^{3/2}}$$

$$= \frac{\mu_0 i}{2\pi R}\left[\frac{s}{(s^2 + R^2)^{1/2}}\right]_0^\infty = \frac{\mu_0 i}{2\pi R}. \qquad (30\text{-}8)$$

With a small change in notation, Eq. 30-8 becomes Eq. 30-6, the relation we set out to prove. Note that the magnetic field at P due to either the lower half or the upper half of the infinite wire in Fig. 30-5 is half this value; that is,

$$B = \frac{\mu_0 i}{4\pi R} \qquad \text{(semi-infinite straight wire).} \qquad (30\text{-}9)$$

Magnetic Field Due to a Current in a Circular Arc of Wire

To find the magnetic field produced at a point by a current in a curved wire, we would again use Eq. 30-3 to write the

magnitude of the field produced by a single current-length element. And we would again integrate to find the net field produced by all the current-length elements. That integration can be difficult, depending on the shape of the wire; it is fairly straightforward, however, when the wire is a circular arc and the point is the center of curvature.

Figure 30-6a shows such an arc-shaped wire with central angle ϕ, radius R, and center C, carrying current i. At C, each current-length element i $d\mathbf{s}$ of the wire produces a magnetic field of magnitude dB given by Eq. 30-3. Moreover, as Fig. 30-6b shows, no matter where the element is located on the wire, the angle θ between the vectors $d\mathbf{s}$ and $\mathbf{r}$ is $90°$; also, $r = R$. Thus, by substituting R for r and $90°$ for θ, we obtain from Eq. 30-3,

$$dB = \frac{\mu_0}{4\pi}\frac{i\, ds \sin 90°}{R^2} = \frac{\mu_0}{4\pi}\frac{i\, ds}{R^2}. \qquad (30\text{-}10)$$

The field at C due to each current-length element in the circular arc has this same magnitude.

An application of the right-hand rule anywhere along the wire (as in Fig. 30-6c) will show that all the differential fields $d\mathbf{B}$ have the same direction at C: directly out of the page. So the total field at C is simply the sum (via integration) of all the fields $d\mathbf{B}$. We use the identity $ds = R\, d\phi$ to change the variable of integration from ds to $d\theta$ and obtain, from Eq. 30-10,

$$B = \int dB = \int_0^\phi \frac{\mu_0}{4\pi}\frac{iR\, d\phi}{R^2} = \frac{\mu_0 i}{4\pi R}\int_0^\phi d\phi.$$

Integrating, we find that

$$B = \frac{\mu_0 i\phi}{4\pi R} \qquad \text{(at center of circular arc).} \qquad (30\text{-}11)$$

Note that this equation gives us the magnetic field *only* at the center of curvature of a circular arc of current. When you insert data into the equation, you must be careful to express ϕ in radians rather than degrees.

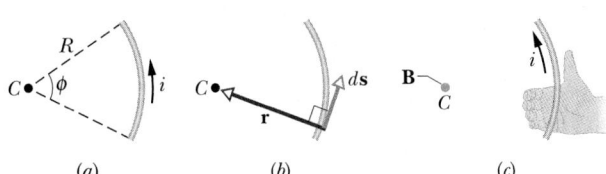

(a)   *(b)*   *(c)*

FIGURE 30-6 (a) A wire in the shape of a circular arc with center C carries current i. (b) For any element of wire along the arc, the angle between the directions of $d\mathbf{s}$ and $\mathbf{r}$ is $90°$. (c) Determining the direction of the magnetic field at the center C due to the current in the wire; the field is out of the page, in the direction of the fingertips, as indicated by the colored dot at C.

The wire in Fig. 30-7a carries a current i and consists of a circular arc of radius R and central angle $\pi/2$ rad, and two straight sections whose extensions intersect the center C of the arc. What magnetic field $\mathbf{B}$ does the current produce at C?

SOLUTION: To answer, we mentally divide the wire into three sections: (1) the straight section at left, (2) the straight section at right, and (3) the circular arc. Then we apply Eq. 30-3 to each section.

For any current-length element in section 1, the angle θ between $d\mathbf{s}$ and $\mathbf{r}$ is zero (Fig. 30-7b). So Eq. 30-3 gives us

$$dB_1 = \frac{\mu_0}{4\pi}\frac{i\,ds\,\sin\theta}{r^2} = \frac{\mu_0}{4\pi}\frac{i\,ds\,\sin 0}{r^2} = 0.$$

Thus the current along the entire length of wire in straight section 1 contributes no magnetic field at C:

$$B_1 = 0.$$

The same situation prevails in straight section 2, where the angle θ between $d\mathbf{s}$ and $\mathbf{r}$ for any current-length element is $180°$. Thus

$$B_2 = 0.$$

Because curved section 3 is a circular arc and we are to find the magnetic field at the center of curvature, we can use Eq. 30-11. Substituting $\pi/2$ rad for ϕ, we obtain

$$B_3 = \frac{\mu_0 i(\pi/2)}{4\pi R} = \frac{\mu_0 i}{8R}.$$

By applying the right-hand rule as in Fig. 30-7c, we see that $\mathbf{B}_3$ points into the page at C.

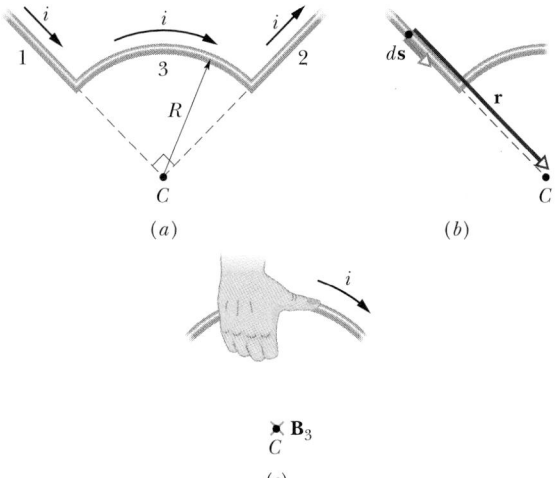

(c)

FIGURE 30-7 Sample Problem 30-1. (a) A wire consists of two straight sections (1 and 2) and a circular arc (3), and carries current i. (b) For a current-length element in section 1, the angle between $d\mathbf{s}$ and $\mathbf{r}$ is zero. (c) Determining the direction of magnetic field $\mathbf{B}_3$ at C due to the current in the circular arc; the field is into the page there.

Thus the total magnetic field $\mathbf{B}$ produced at point C by the current in the wire has magnitude

$$B = B_1 + B_2 + B_3 = 0 + 0 + \frac{\mu_0 i}{8R} = \frac{\mu_0 i}{8R} \quad \text{(Answer)}$$

and points into the plane of the page.

CHECKPOINT **1:** The figure shows three circuits consisting of concentric circular arcs (either half- or quarter-circles of radii r, $2r$, and $3r$) and radial lengths. The circuits carry the same current. Rank them according to the magnitude of the magnetic field produced at the center of curvature (the dot), greatest first.

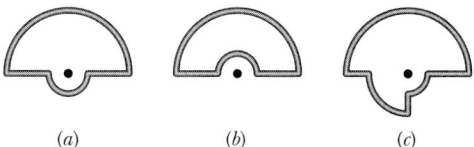

(a) (b) (c)

PROBLEM SOLVING TACTICS

TACTIC 1: *Right-Hand Rules*

To help you sort out the right-hand rules you have now seen (and the ones coming up), here is a review.

Right-Hand Rule for Cross Products. Introduced in Section 3-7, this is a way to determine the direction of the vector that results from a cross product. You point the fingers of your right hand so as to sweep the first vector expressed in the product into the second vector, through the smaller angle between the two vectors. Your outstretched thumb gives you the direction of the vector resulting from the cross product. In Chapter 12, we used this right-hand rule to find the directions of torque and angular momentum vectors; in Chapter 29, we used it to find the direction of the force on a current-carrying wire in a magnetic field.

Curled–Straight Right-Hand Rules for Magnetism. In many situations involving magnetism, you need to relate a "curled" element and a "straight" element. You can do so with the (curled) fingers and the (straight) thumb on your right hand. You have already seen an example in Section 29-8, in which we related the current around a loop (curled element) to the normal vector $\mathbf{n}$ (straight element) of the loop: you curl the fingers of your right hand around in the direction of the current along the loop; your outstretched thumb then gives the direction of $\mathbf{n}$. This is also the direction of the magnetic dipole moment $\boldsymbol{\mu}$ of the loop.

In this section you were introduced to a second curled–straight right-hand rule. To determine the direction of the magnetic field lines around a current-length element, you point the outstretched thumb of your right hand in the direction of the current. The fingers then curl around the current-length element in the direction of the field lines.

30-2 TWO PARALLEL CURRENTS

Two long parallel wires carrying currents exert forces on each other. Figure 30-8 shows two such wires, separated by a distance d and carrying currents i_a and i_b. Let us analyze the forces that these wires exert on each other.

We seek first the force on wire b in Fig. 30-8 due to the current in wire a. That current produces a magnetic field $\mathbf{B}_a$, and it is this magnetic field that actually causes the force we seek. To find the force, then, we need the magnitude and direction of the field $\mathbf{B}_a$ *at the site of wire b*. The magnitude of $\mathbf{B}_a$ at every point of wire b is, from Eq. 30-6,

$$B_a = \frac{\mu_0 i_a}{2\pi d}. \tag{30-12}$$

A (curled–straight) right-hand rule tells us that the direction of $\mathbf{B}_a$ at wire b is down, as Fig. 30-8 shows.

Now that we have the field, we can find the force it exerts on wire b. Equation 29-27 tells us that the force $\mathbf{F}_{ba}$ exerted on a length L of wire b by the external magnetic field $\mathbf{B}_a$ is

$$\mathbf{F}_{ba} = i_b \mathbf{L} \times \mathbf{B}_a. \tag{30-13}$$

In Fig. 30-8, vectors $\mathbf{L}$ and $\mathbf{B}_a$ are perpendicular. So, with Eq. 30-12, we can write

$$F_{ba} = i_b L B_a \sin 90° = \frac{\mu_0 L i_a i_b}{2\pi d}. \tag{30-14}$$

The direction of $\mathbf{F}_{ba}$ is the direction of the cross product $\mathbf{L} \times \mathbf{B}_a$. Applying the right-hand rule for cross products to $\mathbf{L}$ and $\mathbf{B}_a$ in Fig. 30-8, we find that $\mathbf{F}_{ba}$ points directly toward wire a, as shown. The general procedure for finding the force on a current-carrying wire is this:

To find the force on a current-carrying wire due to a second current-carrying wire, first find the field due to the second wire at the site of the first wire. Then find the force on the first wire due to that field.

We could now use this procedure to compute the force on wire a due to the current in wire b. We would find that the force would point directly toward wire b; hence the two wires with parallel currents attract each other. Similarly, if the two currents were antiparallel, we could show that the two wires repel each other. Thus,

Parallel currents attract, and antiparallel currents repel.

The force acting between currents in parallel wires is the basis for the definition of the ampere, which is one of the seven SI base units. The definition, adopted in 1946, is this: The ampere is that constant current which, if maintained in two straight, parallel conductors of infinite length, of negligible circular cross section, and placed 1 m apart in vacuum, would produce on each of these conductors a force of 2×10^{-7} newton per meter of length.

Rail Gun

The basics of a rail gun are shown in Fig. 30-9*a*. A large current is sent out along one of two parallel conducting rails, across a conducting "fuse" (such as a narrow piece of copper) between the rails, and then back to the current source along the second rail. The projectile to be fired lies on the far side of the fuse and fits loosely between the rails. Immediately after the current begins, the fuse element melts and vaporizes, creating a conducting gas between the rails where the fuse had been.

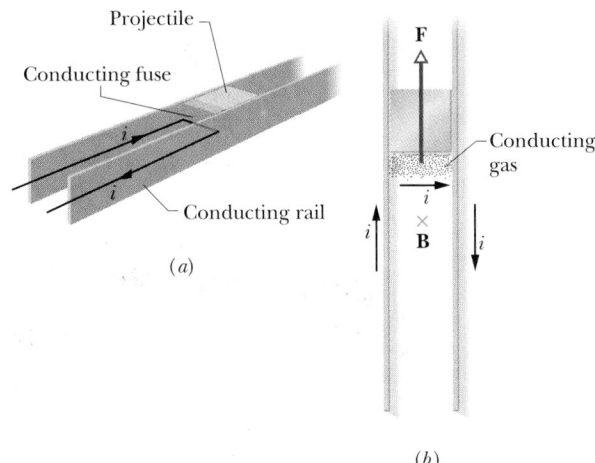

(a)

(b)

FIGURE 30-9 (*a*) A rail gun, as a current i is set up in it. The current rapidly causes the conducting fuse to vaporize. (*b*) The current produces a magnetic field $\mathbf{B}$ between the rails, and the field causes a force $\mathbf{F}$ to act on the conducting gas, which is part of the current path. The gas propels the projectile along the rails, launching it.

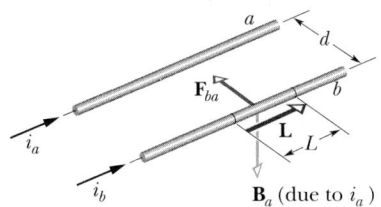

FIGURE 30-8 Two parallel wires carrying currents in the same direction attract each other. $\mathbf{B}_a$ is the magnetic field at wire b produced by the current in wire a. $\mathbf{F}_{ba}$ is the resulting force acting on wire b because it carries current in field $\mathbf{B}_a$.

The curled–straight right-hand rule of Fig. 30-4 reveals that the currents in the rails of Fig. 30-9a produce magnetic fields that are directed downward between the rails. The net magnetic field **B** exerts a force **F** on the gas due to the current i through the gas (Fig. 30-9b). With Eq. 30-13 and the right-hand rule for cross products, we find that **F** points outward along the rails. As the gas is forced outward along the rails, it pushes the projectile, accelerating it by as much as $5 \times 10^6 g$, and then launches it with a speed of 10 km/s, all within 1 ms.

SAMPLE PROBLEM 30-2

Two long parallel wires a distance $2d$ apart carry equal currents i in opposite directions, as shown in Fig. 30-10a. Derive an expression for $B(x)$, the magnitude of the resultant magnetic field for points at a distance x from the midpoint of a line joining the wires.

SOLUTION: Study of Fig. 30-10a and the use of the right-hand rule show that the fields set up by the currents in the individual wires point in the same direction for all points be-

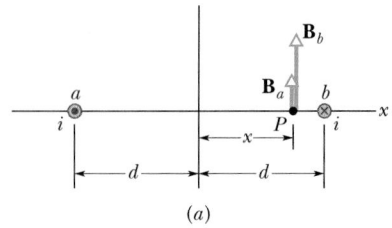

(a)

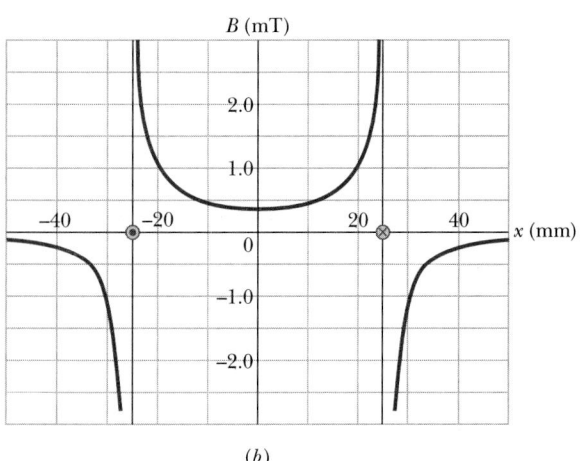

(b)

FIGURE 30-10 Sample Problem 30-2. (a) Two parallel wires carry currents of the same magnitude in opposite directions (out of and into the page). At points between the wires, such as P, the magnetic fields due to the separate currents point in the same direction. (b) A plot of $B(x)$ for $i = 25$ A and a wire separation of 50 mm.

tween the wires. From Eq. 30-6 we then have, at any point P between the wires,

$$B(x) = B_a(x) + B_b(x) = \frac{\mu_0 i}{2\pi(d+x)} + \frac{\mu_0 i}{2\pi(d-x)}$$

$$= \frac{\mu_0 i d}{\pi(d^2 - x^2)}. \qquad \text{(Answer)} \quad (30\text{-}15)$$

Inspection of this relation shows that between the wires (1) $B(x)$ is symmetric about the midpoint ($x = 0$); (2) $B(x)$ has its minimum value ($= \mu_0 i/\pi d$) at this point; and (3) $B(x) \to \infty$ as $x \to \pm d$. At $x = \pm d$, the point P in Fig. 30-10a is within the wires on their axes. Our derivation of Eq. 30-6, however, is valid only for points outside the wires, so Eq. 30-15 holds only up to the surface of the wires.

Figure 30-10b plots Eq. 30-15 for $i = 25$ A and $2d = 50$ mm. We leave it as an exercise to show what the plot suggests: that Eq. 30-15 holds also for points beyond the wires, that is, for points with $|x| > d$.

SAMPLE PROBLEM 30-3

Figure 30-11a shows two long parallel wires carrying currents i_1 and i_2 in opposite directions. What are the magnitude and direction of the resultant magnetic field at point P? Assume the following values: $i_1 = 15$ A, $i_2 = 32$ A, and $d = 5.3$ cm.

SOLUTION: Figure 30-11b shows the individual magnetic fields **B**$_1$ and **B**$_2$ set up by currents i_1 and i_2, respectively, at P. (Verify that their directions are correct, as given by the appropriate right-hand rule.) The magnitudes of these fields at P are given by Eq. 30-6 as

$$B_1 = \frac{\mu_0 i_1}{2\pi R} = \frac{\mu_0 i_1}{2\pi(d/\sqrt{2})} = \frac{\sqrt{2}\mu_0}{2\pi d} i_1$$

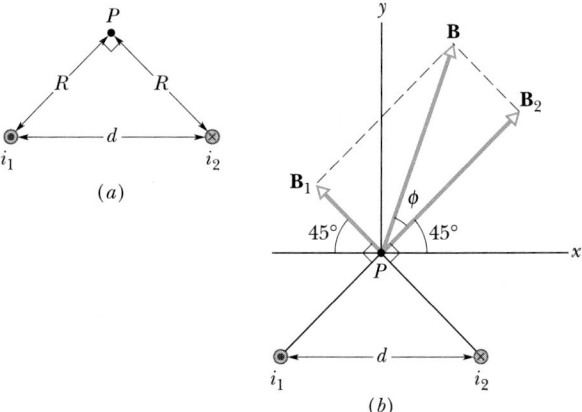

(a)

(b)

FIGURE 30-11 Sample Problem 30-3. (a) Two wires carry currents i_1 and i_2 in opposite directions (out of and into the page). Note the right angles at P. (b) The separate fields **B**$_1$ and **B**$_2$ are combined vectorially to yield the resultant field **B**.

and

$$B_2 = \frac{\mu_0 i_2}{2\pi R} = \frac{\mu_0 i_2}{2\pi(d/\sqrt{2})} = \frac{\sqrt{2}\mu_0}{2\pi d}\, i_2,$$

in which we have replaced R with its equal, $d/\sqrt{2}$, by noting that $R/d = \sin 45° = \sqrt{2}/2$.

The magnitude of the resultant magnetic field **B** is

$$B = \sqrt{B_1^2 + B_2^2} = \frac{\sqrt{2}\mu_0}{2\pi d}\sqrt{i_1^2 + i_2^2}$$

$$= \frac{(\sqrt{2})(4\pi \times 10^{-7}\text{ T}\cdot\text{m/A})\sqrt{(15\text{ A})^2 + (32\text{ A})^2}}{(2\pi)(5.3 \times 10^{-2}\text{ m})}$$

$$= 1.89 \times 10^{-4}\text{ T} \approx 190\ \mu\text{T}. \qquad \text{(Answer)}$$

The angle ϕ between **B** and $\mathbf{B}_2$ in Fig. 30-11b follows from

$$\phi = \tan^{-1}\frac{B_1}{B_2},$$

which, with B_1 and B_2 given above, yields

$$\phi = \tan^{-1}\frac{i_1}{i_2} = \tan^{-1}\frac{15\text{ A}}{32\text{ A}} = 25°.$$

The angle between **B** and the x axis is then

$$\phi + 45° = 25° + 45° = 70°. \qquad \text{(Answer)}$$

$\mathbf{C}$HECKPOINT **2:** The figure shows three long, straight, parallel, equally spaced wires with identical currents either into or out of the page. Rank the wires according to the magnitude of the force on each due to the currents in the other two wires, greatest first.

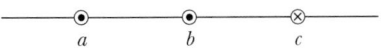

30-3 AMPERE'S LAW

We can find the net electric field due to *any* distribution of charges with the inverse-square law for the differential field $d\mathbf{E}$ (Eq. 30-2), but if the distribution is complicated, we may have to use a computer. Recall, however, that if the distribution has planar, cylindrical, or spherical symmetry, we can apply Gauss' law to find the net electric field with considerably less effort.

Similarly, we can find the net magnetic field due to *any* distribution of currents with the inverse-square law for the differential field $d\mathbf{B}$ (Eq. 30-5), but again we may have to use a computer for a complicated distribution. However, if the distribution has some symmetry, we may be able to apply **Ampere's law** to find the magnetic field with considerably less effort. This law was first advanced by André Marie Ampère (1775–1836), for whom the SI unit of current is named.

Ampere's law is

$$\oint \mathbf{B}\cdot d\mathbf{s} = \mu_0 i_{\text{enc}} \qquad \text{(Ampere's law).} \quad (30\text{-}16)$$

The circle on the integral sign means that the scalar (or dot) product $\mathbf{B}\cdot d\mathbf{s}$ is to be integrated around a *closed* loop, called an *Amperian loop*. The current i_{enc} on the right is the *net* current encircled by that loop.

To see the meaning of the scalar product $\mathbf{B}\cdot d\mathbf{s}$ and its integral, let us first apply Ampere's law to the general situation of Fig. 30-12. The figure shows the cross sections of three long straight wires that carry currents i_1, i_2, and i_3 either directly into or directly out of the page. An arbitrary Amperian loop lying in the plane of the page encircles two of the currents but not the third. The counterclockwise direction marked on the loop indicates the arbitrarily chosen direction of integration for Eq. 30-16.

To apply Ampere's law, we mentally divide the loop into differential elements $d\mathbf{s}$ that are everywhere directed along the tangent to the loop in the direction of integration. At the location of the element $d\mathbf{s}$ shown in Fig. 30-12, the net magnetic field due to the three currents is **B**. Because the wires are perpendicular to the page, we know that the magnetic field at $d\mathbf{s}$ due to each current is in the plane of Fig. 30-12; thus their net magnetic field **B** at $d\mathbf{s}$ must also be in that plane. However, we do not know the orientation of **B** in the plane. In Fig. 30-12, **B** is arbitrarily drawn at an angle θ to the direction of $d\mathbf{s}$.

The scalar product $\mathbf{B}\cdot d\mathbf{s}$ on the left of Eq. 30-16 is then equal to $B\cos\theta\, ds$. Thus Ampere's law can be written as

$$\oint \mathbf{B}\cdot d\mathbf{s} = \oint B\cos\theta\, ds = \mu_0 i_{\text{enc}}. \quad (30\text{-}17)$$

We can now interpret the scalar product $\mathbf{B}\cdot d\mathbf{s}$ as being the product of a length ds of the Amperian loop and the field component $B\cos\theta$ that is tangent to the loop. Then we can interpret the integration as being the summation of all such products around the entire loop.

When we can actually perform this integration, we do not need to know the direction of **B** before integrating.

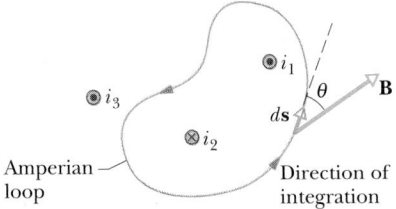

FIGURE 30-12 Ampere's law applied to an arbitrary Amperian loop that encircles two long straight wires but excludes a third wire. Note the directions of the currents.

Instead, we arbitrarily assume **B** to be generally in the direction of integration (as in Fig. 30-12). Then we use the following curled–straight right-hand rule to assign a plus sign or a minus sign to each of the currents that make up the net encircled current i_{enc}:

Curl the fingers of your right hand around the Amperian loop, with them pointing in the direction of integration. A current passing through the loop in the general direction of your outstretched thumb is assigned a plus sign, and a current moving generally in the opposite direction is assigned a minus sign.

Finally, we solve Eq. 30-17 for the magnitude of **B**. If B turns out positive, then the direction we assumed for **B** is correct. If it turns out negative, we neglect the minus sign and redraw **B** in the opposite direction.

In Fig. 30-13 we apply the curled–straight rule for Ampere's law to the situation of Fig. 30-12. With the indicated counterclockwise direction of integration, the net current encircled by the loop is

$$i_{enc} = i_1 - i_2.$$

(Current i_3 is not encircled by the loop.) So, we can rewrite Eq. 30-17 as

$$\oint B \cos \theta \, ds = \mu_0(i_1 - i_2). \quad (30\text{-}18)$$

You might wonder why, since current i_3 contributes to the magnetic-field magnitude B on the left side of Eq. 30-18, it is not needed on the right side. The answer is that the contributions of current i_3 to the magnetic field cancel out because the integration in Eq. 30-18 is made around the full loop. In contrast, the contributions of an encircled current to the magnetic field do not cancel out.

We cannot solve Eq. 30-18 for the magnitude B of the magnetic field, because for the situation of Fig. 30-12 we do not have enough information to simplify and solve the integral. However, we do know the outcome of the integration: it must be equal to the value of $\mu_0(i_1 - i_2)$, which is set by the net current passing through the loop.

We shall now apply Ampere's law to two situations in which symmetry does allow us to simplify and solve the integral, hence to find the magnetic field.

The Magnetic Field Outside a Long Straight Wire with Current

Figure 30-14 shows a long straight wire that carries current i directly out of the page. Equation 30-6 tells us that the magnetic field **B** produced by the current has the same magnitude at all points that are the same distance r from the wire. That is, the field **B** has cylindrical symmetry about the wire. We can take advantage of that symmetry to simplify the integral in Ampere's law (Eq. 30-16) if we encircle the wire with a concentric circular Amperian loop of radius r, as in Fig. 30-14. The magnetic field **B** then has the same magnitude B at every point on the loop. We shall integrate counterclockwise, and we assume that **B** points in the same direction as the element $d\mathbf{s}$ in Fig. 30-14.

We can further simplify the quantity $B \cos \theta$ in Eq. 30-17 by noting that **B** is tangent to the loop at every point along the loop. Thus at every point the angle θ between $d\mathbf{s}$ and **B** is 0°, so $\cos \theta = \cos 0° = 1$. The integral in Eq. 30-17 then becomes

$$\oint \mathbf{B} \cdot d\mathbf{s} = \oint B \cos \theta \, ds = B \oint ds = B(2\pi r).$$

Note that $\oint ds$ above is the summation of all the line segment lengths ds around the circular loop; that is, it simply gives the circumference $2\pi r$ of the loop.

Our right-hand rule gives us a plus sign for the current of Fig. 30-14. So the right side of Ampere's law becomes $+\mu_0 i$ and we then have

$$B(2\pi r) = \mu_0 i$$

or

$$B = \frac{\mu_0 i}{2\pi r}. \quad (30\text{-}19)$$

This is precisely Eq. 30-6, which we derived earlier—with considerably more effort—using the law of Biot and Savart. In addition, because the magnitude B turned out positive, we know that the correct direction of **B** must be the one shown in Fig. 30-14.

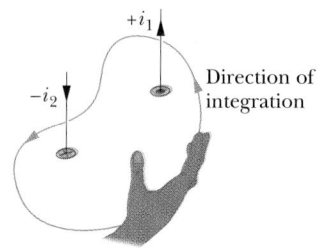

FIGURE 30-13 A right-hand rule for Ampere's law, to determine the signs for currents encircled by an Amperian loop. The situation is that of Fig. 30-12.

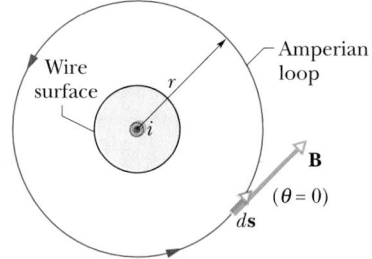

FIGURE 30-14 Using Ampere's law to find the magnetic field produced by a current i in a long straight wire. The Amperian loop is a concentric circle that lies outside the wire.

The Magnetic Field Inside a Long Straight Wire with Current

Figure 30-15 shows the cross section of a long straight wire of radius R that carries a uniformly distributed current i directly out of the page. Because the current is uniformly distributed about the center of the wire, the magnetic field **B** that it produces must be cylindrically symmetrical. So, to find the magnetic field at points inside the wire, we can again use an Amperian loop of radius r, as shown in Fig. 30-15, where now $r < R$. Symmetry again suggests that **B** is tangent to the loop, as shown. So the left side of Ampere's law again yields

$$\oint \mathbf{B} \cdot d\mathbf{s} = B \oint ds = B(2\pi r). \qquad (30\text{-}20)$$

To find the right side of Ampere's law, we note that because the current is uniformly distributed, the current i_{enc} encircled by the loop is proportional to the area encircled by the loop. That is,

$$i_{enc} = i \frac{\pi r^2}{\pi R^2}. \qquad (30\text{-}21)$$

Our right-hand rule tells us that i_{enc} gets a plus sign. Then Ampere's law gives us

$$B(2\pi r) = \mu_0 i \frac{\pi r^2}{\pi R^2}.$$

or

$$B = \left(\frac{\mu_0 i}{2\pi R^2}\right) r. \qquad (30\text{-}22)$$

Thus, inside the wire, the magnitude B of the magnetic field is proportional to r; that magnitude is zero at the center and a maximum at the surface, where $r = R$. Note that Eqs. 30-19 and 30-22 give the same (maximum) value for B at $r = R$; that is, the expressions for the magnetic field outside the wire and inside the wire yield the same result at the surface of the wire.

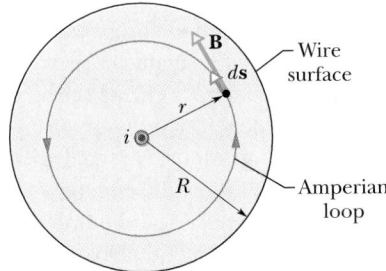

FIGURE 30-15 Using Ampere's law to find the magnetic field that a current i produces inside a long straight wire of circular cross section. The current is uniformly distributed over the cross section of the wire and emerges from the page. An Amperian loop is drawn inside the wire.

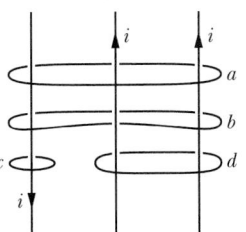

SAMPLE PROBLEM 30-4

Figure 30-16a shows the cross section of a long conducting cylinder with inner radius $a = 2.0$ cm and outer radius $b = 4.0$ cm. The cylinder carries a current out of the page, and the current density in the cross section is given by $J = cr^2$, with $c = 3.0 \times 10^6$ A/m^2 and r in meters. What is the magnetic field **B** at a point that is 3.0 cm from the central axis of the cylinder?

SOLUTION: Because the current distribution (hence the magnetic field) has cylindrical symmetry about the central axis of the cylinder, we can apply Ampere's law to the cross section to find the magnetic field. As shown in Fig. 30-16b, we draw an Amperian loop concentric with the cylinder and with radius $r = 3.0$ cm because we want to compute B at that distance from the central axis.

To apply Ampere's law, we must compute the current i_{enc} that is encircled by the Amperian loop. We cannot set up a proportionality as in Eq. 30-21, however, because here the current is not uniformly distributed. Instead, following the procedure of Sample Problem 27-2b, we must integrate the current density from the cylinder's inner radius a to the loop radius r:

$$\begin{aligned} i_{enc} &= \int J\, dA = \int_a^r cr^2\, (2\pi r\, dr) \\ &= 2\pi c \int_a^r r^3\, dr = 2\pi c \left[\frac{r^4}{4}\right]_a^r \\ &= \frac{\pi c(r^4 - a^4)}{2}. \end{aligned}$$

The direction of integration indicated in Fig. 30-16b is (arbitrarily) clockwise. Applying the right-hand rule for Ampere's law to that loop, we find that we should take i_{enc} as negative because the current is directed out of the page but our thumb is directed into the page.

We next evaluate the left side of Ampere's law exactly as we did in Fig. 30-15, and we again obtain Eq. 30-20. So, Ampere's law,

$$\oint \mathbf{B} \cdot d\mathbf{s} = \mu_0 i_{enc},$$

gives us

$$B(2\pi r) = -\frac{\mu_0 \pi c}{2} (r^4 - a^4).$$

Solving for B and substituting known data yield

$$B = -\frac{\mu_0 c}{4r} (r^4 - a^4)$$

$$= -\frac{(4\pi \times 10^{-7} \text{ T}\cdot\text{m/A})(3.0 \times 10^6 \text{ A/m}^2)}{4(0.030 \text{ m})}$$

$$\times [(0.030 \text{ m})^4 - (0.020 \text{ m})^4]$$

$$= -2.0 \times 10^{-5} \text{ T}.$$

Thus, the magnetic field **B** has the magnitude

$$B = 2.0 \times 10^{-5} \text{ T} \qquad \text{(Answer)}$$

and is directed opposite our direction of integration, hence counterclockwise in Fig. 30-16b.

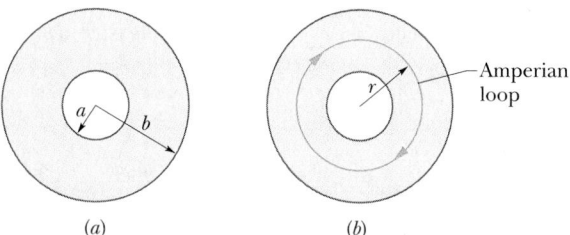

(a)   (b)

FIGURE 30-16 Sample Problem 30-4. (a) Cross section of a conducting cylinder of inner radius a and outer radius b. (b) An Amperian loop of radius r is added to compute the magnetic field at points that are a distance r from the central axis.

30-4 SOLENOIDS AND TOROIDS

Magnetic Field of a Solenoid

We now turn our attention to another situation in which Ampere's law proves useful. It concerns the magnetic field produced by the current in a long, tightly wound helical coil of wire. Such a coil is called a **solenoid** (Fig. 30-17). We assume that the length of the solenoid is much greater than the diameter.

Figure 30-18 shows a section through a portion of a "stretched-out" solenoid. The solenoid's magnetic field is the vector sum of the fields produced by the individual

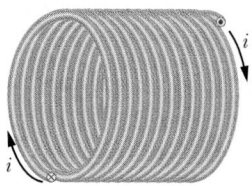

FIGURE 30-17 A solenoid carrying current i.

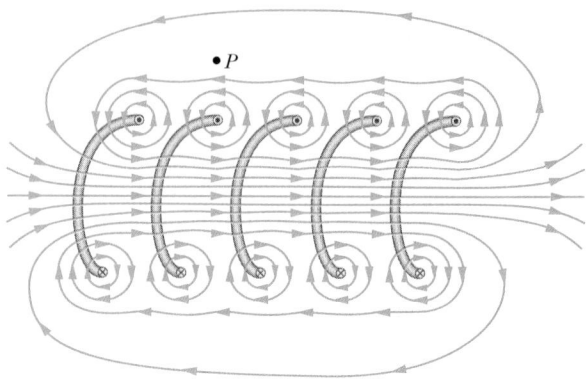

FIGURE 30-18 The magnetic field lines in a vertical cross section through the central axis of a "stretched-out" solenoid. The back portions of five turns are shown. Each turn produces circular magnetic field lines near it. Near the solenoid's axis, the field lines combine into a net magnetic field that is directed along the axis. The closely spaced field lines there indicate a strong magnetic field. Outside the solenoid the field lines are widely spaced: the field there is very weak.

turns. For points very close to each turn, the wire behaves magnetically almost like a long straight wire, and the lines of **B** there are almost concentric circles. Figure 30-18 suggests that the field tends to cancel between adjacent turns. It also suggests that, at points inside the solenoid and reasonably far from the wire, **B** is approximately parallel to the (central) solenoid axis. In the limiting case of an *ideal solenoid*, which is infinitely long and consists of tightly packed (*close-packed*) turns of square wire, the field inside the coil is uniform and parallel to the solenoid axis.

At points above the solenoid, such as P in Fig. 30-18, the field set up by the upper parts of the solenoid turns (marked $\odot$) points to the left (as drawn near P) and tends to cancel the field set up by the lower parts of the turns (marked $\otimes$), which points to the right (not drawn). In the limiting case of an ideal solenoid, the magnetic field outside the solenoid is zero. Taking the external field to be zero is an excellent assumption for a real solenoid if its length is much greater than its diameter and if we consider external points such as point P. The direction of the magnetic field along the solenoid axis is given by a curled–straight right-hand rule: grasp the solenoid with your right hand so that your fingers follow the direction of the current in the windings; your extended right thumb then points in the direction of the axial magnetic field.

Figure 30-19 shows the lines of **B** for a real solenoid. The spacing of the lines of **B** in the central region shows that the field inside the coil is fairly strong and uniform over the cross section of the coil. The external field, however, is relatively weak.

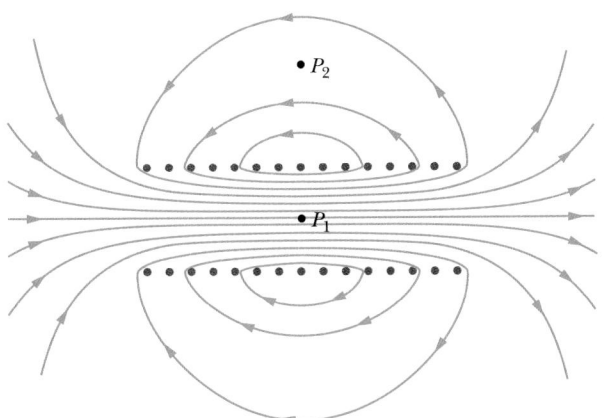

FIGURE 30-19 Magnetic field lines for a real solenoid of finite length. The field is strong and uniform at interior points such as P_1 but relatively weak at external points such as P_2.

Let us apply Ampere's law,

$$\oint \mathbf{B} \cdot d\mathbf{s} = \mu_0 i_{\text{enc}}, \qquad (30\text{-}23)$$

to the rectangular Amperian loop *abcd* in the ideal solenoid of Fig. 30-20, where **B** is uniform within the solenoid and zero outside it. We write the integral $\oint \mathbf{B} \cdot d\mathbf{s}$ as the sum of four integrals, one for each path segment:

$$\oint \mathbf{B} \cdot d\mathbf{s} = \int_a^b \mathbf{B} \cdot d\mathbf{s} + \int_b^c \mathbf{B} \cdot d\mathbf{s}$$
$$+ \int_c^d \mathbf{B} \cdot d\mathbf{s} + \int_d^a \mathbf{B} \cdot d\mathbf{s}. \qquad (30\text{-}24)$$

The first integral on the right of Eq. 30-24 is Bh, where B is the magnitude of the uniform field **B** inside the solenoid and h is the (arbitrary) length of the path from a to b. The second and fourth integrals are zero because for every element of these paths **B** either is perpendicular to the path or is zero, and thus $\mathbf{B} \cdot d\mathbf{s}$ is zero. The third integral, which is along a path that lies outside the solenoid, is zero because $B = 0$ at all external points. Thus $\oint \mathbf{B} \cdot d\mathbf{s}$ for the entire rectangular path has the value Bh.

The net current i_{enc} encircled by the rectangular Amperian loop in Fig. 30-20 is not the same as the current i in the solenoid windings because the windings pass more than once through this loop. Let n be the number of turns per unit length of the solenoid; then

$$i_{\text{enc}} = i(nh).$$

Ampere's law then gives us

$$Bh = \mu_0 inh,$$

or

$$B = \mu_0 in \qquad \text{(ideal solenoid).} \qquad (30\text{-}25)$$

Although we derived Eq. 30-25 for an infinitely long ideal solenoid, it holds quite well for actual solenoids if we apply it only at interior points, well away from the solenoid ends. Equation 30-25 is consistent with the experimental fact that B does not depend on the diameter or the length of the solenoid and that B is constant over the solenoidal cross section. A solenoid provides a practical way to set up a known uniform magnetic field for experimentation, just as a parallel-plate capacitor provides a practical way to set up a known uniform electric field.

Magnetic Field of a Toroid

Figure 30-21*a* shows a **toroid**, which we may describe as a solenoid bent into the shape of a doughnut. What magnetic field **B** is set up at its interior points (within the "tube" of the doughnut)? We can find out from Ampere's law and the symmetry.

From the symmetry, we see that the lines of **B** form concentric circles inside the toroid, directed as shown in Fig. 30-21*b*. Let us choose a concentric circle of radius r as an Amperian loop and traverse it in the clockwise direction. Ampere's law (Eq. 30-16) yields

$$(B)(2\pi r) = \mu_0 iN,$$

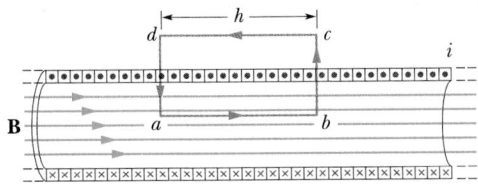

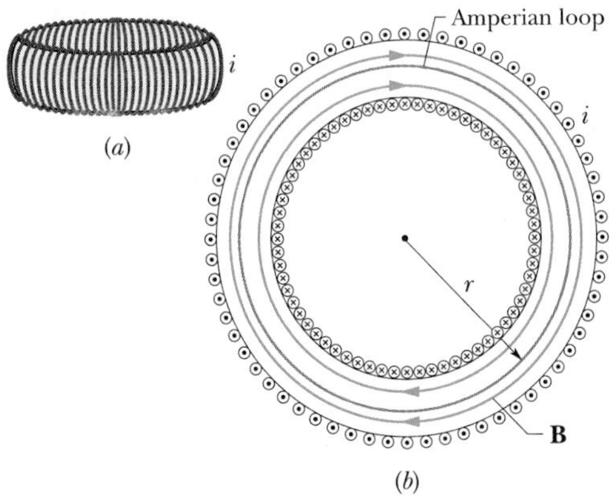

FIGURE 30-20 An application of Ampere's law to a section of a long ideal solenoid carrying a current i. The Amperian loop is the rectangle *abcd*.

FIGURE 30-21 (*a*) A toroid carrying a current i. (*b*) The toroid's cross section. The interior magnetic field (inside the doughnut-shaped tube) can be found by applying Ampere's law to the Amperian loop shown.

where i is the current in the toroid windings (and is positive) and N is the total number of turns. This gives

$$B = \frac{\mu_0 iN}{2\pi} \frac{1}{r} \qquad \text{(toroid).} \qquad (30\text{-}26)$$

In contrast to the situation for a solenoid, B is not constant over the cross section of a toroid. It is easy to show, with Ampere's law, that $B = 0$ for points outside an ideal toroid (as if the toroid were made from an ideal solenoid).

The direction of the magnetic field within a toroid follows from our curled–straight right-hand rule: grasp the toroid with the fingers of your right hand curled in the direction of the current in the windings; your extended right thumb points in the direction of the magnetic field.

SAMPLE PROBLEM 30-5

A solenoid has length $L = 1.23$ m and inner diameter $d = 3.55$ cm, and it carries a current $i = 5.57$ A. It consists of five close-packed layers, each with 850 turns along length L. What is B at its center?

SOLUTION: From Eq. 30-25

$$B = \mu_0 in = (4\pi \times 10^{-7}\ \text{T·m/A})(5.57\ \text{A})\,\frac{5 \times 850\ \text{turns}}{1.23\ \text{m}}$$

$$= 2.42 \times 10^{-2}\ \text{T} = 24.2\ \text{mT.} \qquad \text{(Answer)}$$

Note that Eq. 30-25 applies even though the solenoid has more than one layer of windings because the diameter of the windings does not enter into the equation.

30-5 A CURRENT-CARRYING COIL AS A MAGNETIC DIPOLE

So far we have examined the magnetic fields produced by a long straight wire, a solenoid, and a toroid. We turn our attention here to the field produced by a coil carrying current. You saw in Section 29-9 that such a coil behaves as a magnetic dipole in that, if we place it in an external magnetic field **B**, a torque $\boldsymbol{\tau}$ given by

$$\boldsymbol{\tau} = \boldsymbol{\mu} \times \mathbf{B} \qquad (30\text{-}27)$$

acts on it. Here $\boldsymbol{\mu}$ is the magnetic dipole moment of the coil and has the magnitude NiA, where N is the number of turns (or loops), i is the current in each turn, and A is the area enclosed by each turn.

Recall that the direction of $\boldsymbol{\mu}$ is given by a curled–straight right-hand rule: grasp the coil so that the fingers of

your right hand curl around it in the direction of the current; your extended thumb then points in the direction of the dipole moment $\boldsymbol{\mu}$.

Magnetic Field of a Coil

We turn now to the other aspect of a coil as a magnetic dipole. What magnetic field does *it* produce at a point in the surrounding space? The problem does not have enough symmetry to make Ampere's law useful, so we must turn to the law of Biot and Savart. For simplicity, we first consider only a coil with a single circular loop and only points on its central axis, which we take to be a z axis. We shall show that the magnitude of the magnetic field is

$$B(z) = \frac{\mu_0 iR^2}{2(R^2 + z^2)^{3/2}}, \qquad (30\text{-}28)$$

in which R is the radius of the circular loop and z is the distance of the point in question from the center of the loop. Furthermore, the direction of the magnetic field **B** is the same as the direction of the magnetic dipole moment $\boldsymbol{\mu}$ of the loop.

For axial points far from the loop, we have $z \gg R$ in Eq. 30-28. With that approximation, this equation reduces to

$$B(z) \approx \frac{\mu_0 iR^2}{2z^3}.$$

Recalling that πR^2 is the area A of the loop and extending our result to include a coil of N turns, we can write this equation as

$$B(z) = \frac{\mu_0}{2\pi} \frac{NiA}{z^3}$$

or, since **B** and $\boldsymbol{\mu}$ have the same direction, we can write the equation in vector form, substituting from the identity $\mu = NiA$:

$$\mathbf{B}(z) = \frac{\mu_0}{2\pi} \frac{\boldsymbol{\mu}}{z^3} \qquad \text{(current-carrying coil).} \qquad (30\text{-}29)$$

Thus we have two ways in which we can regard a current-carrying coil as a magnetic dipole: (1) it experiences a torque when we place it in an external magnetic field; (2) it generates its own intrinsic magnetic field, given, for distant points along its axis, by Eq. 30-29. Figure 30-22 shows the magnetic field of a current loop; one side of the loop acts as a north pole (in the direction of $\boldsymbol{\mu}$) and the other side as a south pole, as suggested by the ghosted magnet in the figure.

CHECKPOINT **4:** The figure shows four arrangements of circular loops of radius r or $2r$, centered on vertical axes and carrying identical currents in the directions indicated. Rank the arrangements according to the magnitude of the net magnetic field at the dot, midway between the loops on the central axis, greatest first.

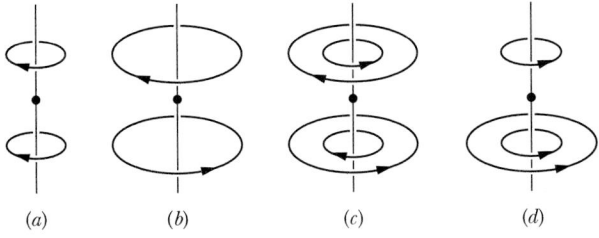

(a)   (b)   (c)   (d)

Proof of Equation 30-28

Figure 30-23 shows the back half of a circular loop of radius R carrying a current i. Consider a point P on the axis of the loop, a distance z from its plane. Let us apply the law of Biot and Savart to a length element located at the left side of the loop. The vector $d\mathbf{s}$ for this element points perpendicularly out of the page. The angle θ between $d\mathbf{s}$ and the vector $\mathbf{r}$ in Fig. 30-23 is $90°$; the plane formed by these two vectors is perpendicular to the plane of the figure and contains both $\mathbf{r}$ and $d\mathbf{s}$. From the law of Biot and Savart (and the right-hand rule), the differential field $d\mathbf{B}$ produced at point P by the current in this element is perpendicular to this plane and thus lies in the plane of the figure, perpendicular to $\mathbf{r}$, as indicated in Fig. 30-23.

Let us resolve $d\mathbf{B}$ into two components: $dB_{\|}$ along the axis of the loop, and $dB_{\perp}$ perpendicular to this axis. From the symmetry, the vector sum of all the perpendicular

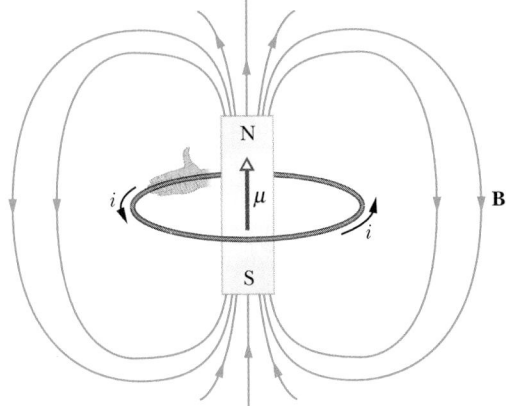

FIGURE 30-22 A current loop produces a magnetic field like that of a bar magnet and thus has associated north and south poles. The magnetic dipole moment $\boldsymbol{\mu}$ of the loop, given by a curled–straight right-hand rule, points from the south pole to the north pole, in the direction of the field $\mathbf{B}$ within the loop.

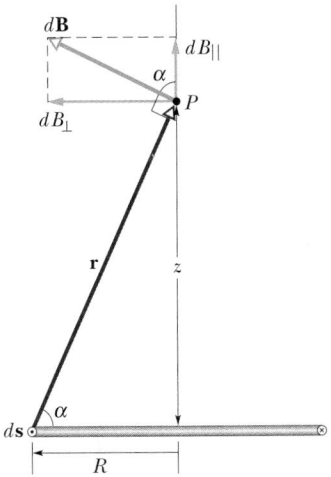

FIGURE 30-23 A current loop of radius R. The plane of the loop is perpendicular to the page and only the back half of the loop is shown. We use the law of Biot and Savart to find the magnetic field at point P on the central axis of the loop.

components $dB_{\perp}$ due to all the loop elements is zero. This leaves only the axial components $dB_{\|}$ and we have

$$B = \int dB_{\|}.$$

For the element $d\mathbf{s}$ in Fig. 30-23, the law of Biot and Savart (Eq. 30-3) gives

$$dB = \frac{\mu_0}{4\pi}\frac{i\, ds \sin 90°}{r^2}.$$

We also have

$$dB_{\|} = dB \cos \alpha.$$

Combining these two relations, we obtain

$$dB_{\|} = \frac{\mu_0 i \cos \alpha\, ds}{4\pi r^2}. \tag{30-30}$$

Figure 30-23 shows that r and α are not independent but are related to each other. Let us express each in terms of the variable z, the distance between point P and the center of the loop. The relations are

$$r = \sqrt{R^2 + z^2} \tag{30-31}$$

and

$$\cos \alpha = \frac{R}{r} = \frac{R}{\sqrt{R^2 + z^2}}. \tag{30-32}$$

Substituting Eqs. 30-31 and 30-32 into Eq. 30-30, we find

$$dB_{\|} = \frac{\mu_0 i R}{4\pi(R^2 + z^2)^{3/2}}\, ds.$$

Note that i, R, and z have the same values for all elements $d\mathbf{s}$ around the loop. So when we integrate this equation, noting that $\int ds$ is simply the circumference $2\pi r$ of the loop, we find that

$$B = \int dB_{\parallel} = \frac{\mu_0 iR}{4\pi(R^2 + z^2)^{3/2}} \int ds$$

or

$$B(z) = \frac{\mu_0 iR^2}{2(R^2 + z^2)^{3/2}},$$

which is Eq. 30-28, the relation we sought to prove.

REVIEW & SUMMARY

The Biot–Savart Law

The magnetic field set up by a current-carrying conductor can be found from the **Biot–Savart law.** This law asserts that the contribution $d\mathbf{B}$ to the field produced by a current-length element $i\,d\mathbf{s}$ at a point P, a distance r from the current element, is

$$d\mathbf{B} = \frac{\mu_0}{4\pi}\frac{i\,d\mathbf{s} \times \mathbf{r}}{r^3} \quad \text{(Biot–Savart law).} \quad (30\text{-}5)$$

Here $\mathbf{r}$ is a vector that points from the element to the point in question. The quantity μ_0, called the permeability constant, has the value $4\pi \times 10^{-7}$ T·m/A $\approx 1.26 \times 10^{-6}$ T·m/A.

Magnetic Field of a Long Straight Wire

For a long straight wire carrying a current i, the Biot–Savart law gives, for the magnetic field at a distance r from the wire,

$$B = \frac{\mu_0 i}{2\pi r} \quad \text{(long straight wire).} \quad (30\text{-}6)$$

Magnetic Field of a Circular Arc

The magnetic field at the center of a circular arc of wire of radius R that carries current i is

$$B = \frac{\mu_0 i\phi}{4\pi R} \quad \text{(at center of circular arc).} \quad (30\text{-}11)$$

The Force Between Parallel Wires Carrying Currents

Parallel wires carrying currents in the same direction attract each other, whereas parallel wires carrying currents in opposite directions repel each other. The magnitude of the force on a length L of either wire is

$$F_{ba} = i_b LB_a \sin 90° = \frac{\mu_0 L i_a i_b}{2\pi d}, \quad (30\text{-}14)$$

where d is the wire separation, and i_a and i_b are the currents in the wires.

Ampere's Law

For some current distributions, **Ampere's law,**

$$\oint \mathbf{B} \cdot d\mathbf{s} = \mu_0 i_{\text{enc}} \quad \text{(Ampere's law),} \quad (30\text{-}16)$$

can be used (instead of the Biot–Savart law) to calculate the magnetic field. The line integral in this equation is evaluated around a closed loop called an *Amperian loop.* The current i is the *net* current encircled by the loop.

Fields of a Solenoid and a Toroid

Inside a *long solenoid* carrying current i, at points not near its ends, the magnitude B of the magnetic field is

$$B = \mu_0 in \quad \text{(ideal solenoid),} \quad (30\text{-}25)$$

where n is the number of turns per unit length. At a point inside a *toroid,* the magnitude B of the magnetic field is

$$B = \frac{\mu_0 iN}{2\pi}\frac{1}{r} \quad \text{(toroid),} \quad (30\text{-}26)$$

where r is the distance from the center of the toroid to the point.

Field of a Magnetic Dipole

The magnetic field produced by a current-carrying coil, which is a *magnetic dipole,* at a point P located a distance z along the coil's central axis is parallel to the axis and is given by

$$\mathbf{B}(z) = \frac{\mu_0}{2\pi}\frac{\boldsymbol{\mu}}{z^3}, \quad (30\text{-}29)$$

where $\boldsymbol{\mu}$ is the dipole moment of the coil.

QUESTIONS

1. Figure 30-24 shows four arrangements in which long parallel wires carry equal currents directly into or out of the page at the corners of identical squares. Rank the arrangements according to the magnitude of the net magnetic field at the center of the square, greatest first.

2. Figure 30-25 shows cross sections of two long straight wires; the left-hand wire carries current i_1 directly out of the page. If the net magnetic field due to the two currents is to be zero at point P, (a) should the direction of current i_2 in the right-hand wire be directly into or out of the page and (b) should i_2 be greater than, less than, or equal to i_1?

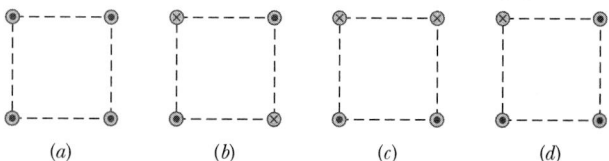

(a)   (b)   (c)   (d)

FIGURE 30-24 Question 1.

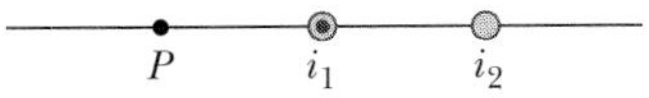

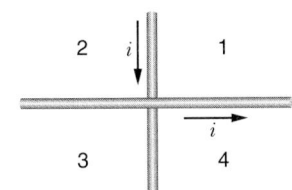

FIGURE 30-25 Question 2.

3. In Fig. 30-26, two long straight wires just pass each other perpendicularly without touching. In which quadrants are there points at which the net magnetic field is zero?

FIGURE 30-26
Question 3.

4. Figure 30-27 shows three circuits, each consisting of two concentric circular arcs, one of radius r and the other of a larger radius R, and two radial lengths. The circuits have the same current through them, and the radial lengths have the same angle between them. Rank the circuits according to the magnitude of the net magnetic field at the center, greatest first.

FIGURE 30-27
Question 4. (a) (b) (c)

5. Figure 30-28 shows three sections of circuits, each section consisting of a wire curved along a circular arc (all with the same radius) and two long straight wires that are tangential to the arc. (In a, the straight wires pass each other without touching.) The sections carry equal currents. Rank the sections according to the magnitude of the magnetic field produced at the center of curvature, greatest first.

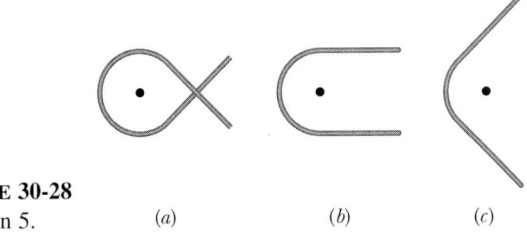

FIGURE 30-28
Question 5. (a) (b) (c)

6. Figure 30-29 shows three sections of circuits, each consisting of a curved wire along a circular arc (all with the same radius and internal angle) and two long straight wires; the straight wires are radial to the arc in circuit a and tangential to the arcs in circuits b and c (in c, the straight wires pass each other without touching). The sections carry equal currents. The net magnetic field at the center of each arc is dominated by the contribution from the arc. Rank the sections according to the magnitude of the net magnetic field at the center, greatest first.

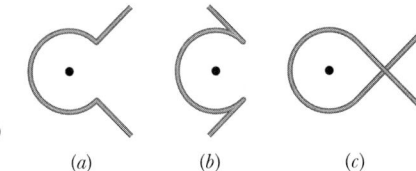

FIGURE 30-29
Question 6. (a) (b) (c)

7. Figure 30-30 shows four arrangements in which long, parallel, equally spaced wires carry equal currents directly into or out of the page. Rank the arrangements according to the magnitude of the net force on the central wire due to the currents in the other wires, greatest first.

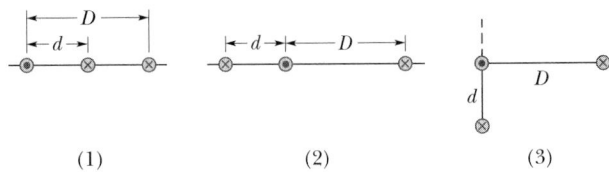

FIGURE 30-30 Question 7.

8. Figure 30-31 shows three arrangements of three long straight wires, carrying equal currents directly into or out of the page. (a) Rank the arrangements according to the magnitude of the net force on the wire with the current directed out of the page due to the currents in the other wires, greatest first. (b) In arrangement 3, is the angle between the net force on that wire and the dashed line equal to, less than, or more than 45°?

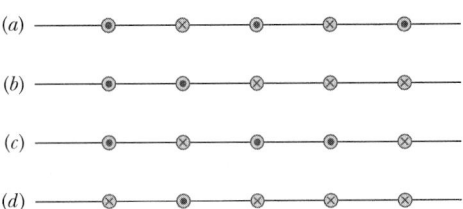

FIGURE 30-31 Question 8.

9. Figure 30-32 shows two arrangements in which two long straight wires with equal currents, directed into or out of the page, are at equal distances from a y axis. (a) For each arrangement, what is the direction of the net magnetic field at point P? (b) For each arrangement, if a third long straight wire, with a current directly out of the page, is placed at P, what is the direction of the net force on that wire due to the other currents?

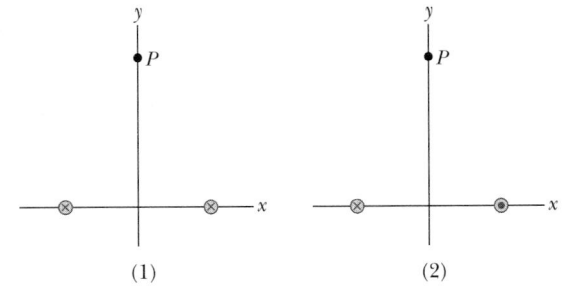

FIGURE 30-32 Question 9.

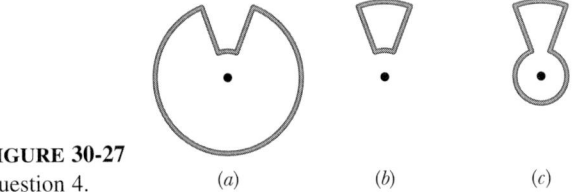

10. Figure 30-33 shows a long straight wire with current i to the right and three wire loops, all with the same clockwise current around them. The loops are all at the same distance from the long wire and have edge lengths of either L or $2L$. Rank the loops according to the magnitude of the net force on them due to the current in the long straight wire, greatest first.

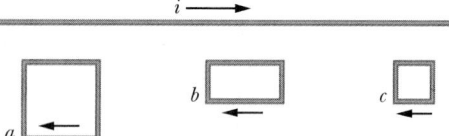

FIGURE 30-33 Question 10.

11. In Fig. 30-34, a messy loop of wire is placed on a slick table with points a and b fixed in place. If a current is then sent through the wire, will the wire be pushed outward into an arc or will it be pulled inward?

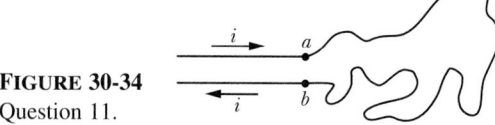

FIGURE 30-34 Question 11.

12. Figure 30-35 shows a uniform magnetic field $\mathbf{B}$ and four straight-line paths of equal lengths. Rank the paths according to the magnitude of $\int \mathbf{B} \cdot d\mathbf{s}$ taken along the paths, greatest first.

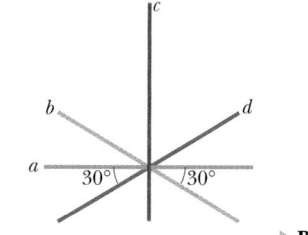

FIGURE 30-35 Question 12.

13. Figure 30-36a shows four circular Amperian loops concentric with a wire whose current is directed out of the page. The current is uniform across the wire's circular cross section. Rank the loops according to the magnitude of $\oint \mathbf{B} \cdot d\mathbf{s}$ around each, greatest first.

14. Figure 30-36b shows four circular Amperian loops and, in cross section, four long circular conductors, all of which are concentric. The currents in the conductors are, from smallest radius to largest radius, 4 A out of the page, 9 A into the page, 5 A out of the page, and 3 A into the page. Rank the loops according to the magnitude of $\oint \mathbf{B} \cdot d\mathbf{s}$ around each, greatest first.

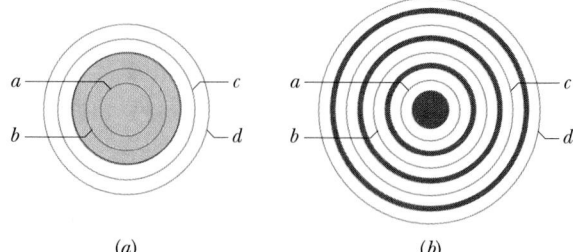

FIGURE 30-36 Questions 13 and 14.

15. Figure 30-37 shows four identical currents i and five Amperian paths encircling them. Rank the paths according to the value of $\oint \mathbf{B} \cdot d\mathbf{s}$ taken in the directions shown, most positive first and most negative last.

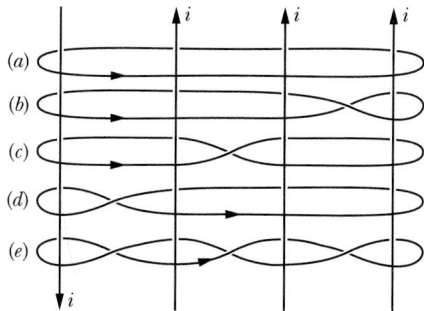

FIGURE 30-37 Question 15.

16. The following table gives the number of turns per unit length n and the current i through six ideal solenoids of different radii. You want to combine several of them concentrically to produce a net magnetic field of zero along the central axis. Can this be done with (a) two of them, (b) three of them, (c) four of them, and (d) five of them? If so, answer by listing which solenoids are to be used and indicate the directions of the currents.

Solenoid:	1	2	3	4	5	6
n:	5	4	3	2	10	8
i:	5	3	7	6	2	3

17. The position vector of a particle moving around a circle of radius r is $\mathbf{r}$. What is the value of $\oint \mathbf{r} \cdot d\mathbf{s}$ around the circle?

18. What is the value of $\oint d\mathbf{s}$ around the perimeter of (a) a square with edge length a and (b) a equilateral triangle of edge length d?

EXERCISES & PROBLEMS

SECTION 30-1 Calculating the Magnetic Field Due to a Current

1E. The magnitude of the magnetic field 88.0 cm from the axis of a long straight wire is 7.30 μT. What is the current in the wire?

2E. A 10-gauge bare copper wire (2.6 mm in diameter) can carry a current of 50 A without overheating. For this current, what is the magnetic field at the surface of the wire?

3E. A surveyor is using a magnetic compass 20 ft below a power line in which there is a steady current of 100 A. (a) What is the magnetic field at the site of the compass due to the power line? (b) Will this interfere seriously with the compass reading? The horizontal component of Earth's magnetic field at the site is 20 μT.

4E. The electron gun in a TV tube fires electrons of kinetic energy 25 keV in a beam 0.22 mm in diameter at the screen; 5.6×10^{14} electrons arrive each second. Calculate the magnetic field produced by the beam at a point 1.5 mm from the beam axis.

5E. Figure 30-38 shows a 3.0 cm segment of wire, centered at the origin, carrying a current of 2.0 A in the positive y direction (as part of some complete circuit). To calculate the magnetic field **B** at a point several meters from the origin, one may use the Biot–Savart law in the form $B = (\mu_0/4\pi)i\,\Delta s \sin\theta/r^2$, in which $\Delta s = 3.0$ cm. This is because r and θ are essentially constant over the segment of wire. Calculate **B** (magnitude and direction) at the following (x, y, z) locations: (a) (0, 0, 5.0 m), (b) (0, 6.0 m, 0), (c) (7.0 m, 7.0 m, 0), (d) (−3.0 m, −4.0 m, 0).

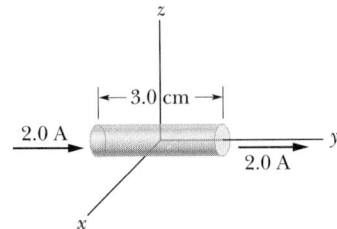

FIGURE 30-38
Exercise 5.

6E. A long wire carrying a current of 100 A is placed in a uniform external magnetic field of 5.0 mT. The wire is perpendicular to this magnetic field. Locate the points at which the resultant magnetic field is zero.

7E. At a position in the Philippines, Earth's magnetic field of 39 μT is horizontal and directed due north. Suppose the net field is zero exactly 8.0 cm above a long straight horizontal wire that carries a constant current. What are (a) the magnitude and (b) the direction of the current?

8E. A particle with positive charge q is a distance d from a long straight wire that carries a current i and is traveling with speed v perpendicular to the wire. What are the direction and magnitude of the force on the particle if it is moving (a) toward or (b) away from the wire?

9E. A straight conductor carrying a current i splits into identical semicircular arcs as shown in Fig. 30-39. What is the magnetic field at the center C of the resulting circular loop?

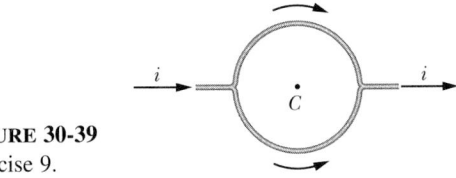

FIGURE 30-39
Exercise 9.

10E. A long straight wire carries a current of 50 A. An electron, traveling at 1.0×10^7 m/s, is 5.0 cm from the wire. What force acts on the electron if the electron velocity is directed (a) toward

the wire, (b) parallel to the wire in the direction of the current, and (c) perpendicular to both directions defined by (a) and (b)?

11P. The wire shown in Fig. 30-40 carries current i. What magnetic field **B** is produced at the center C of the semicircle by (a) each straight segment of length L, (b) the semicircular segment of radius R, and (c) the entire wire?

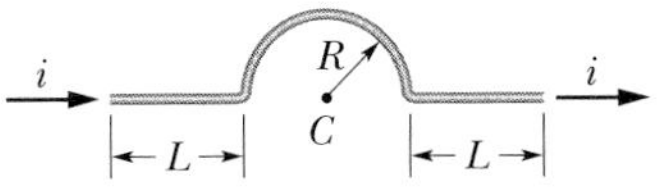

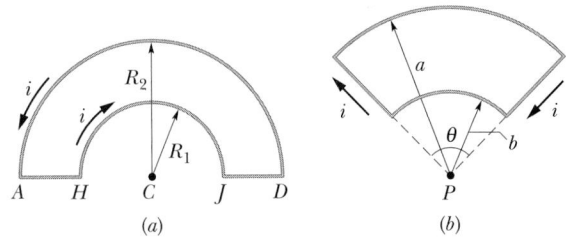

FIGURE 30-40
Problem 11.

12P. Use the Biot–Savart law to calculate the magnetic field **B** at C, the common center of the semicircular arcs AD and HJ in Fig. 30-41a. The two arcs, of radii R_2 and R_1, respectively, form part of the circuit $ADJHA$ carrying current i.

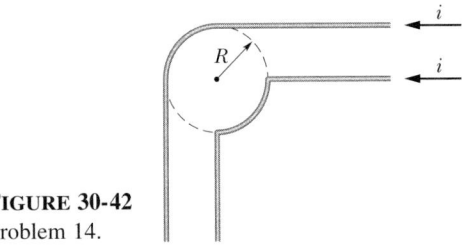

(a)   (b)

FIGURE 30-41   Problems 12 and 13.

13P. Consider the circuit of Fig. 30-41b. The curved segments are arcs of circles of radii a and b. The straight segments are along radii. Find the magnetic field **B** at point P (the center of curvature), assuming a current i in the circuit.

14P. Two infinitely long wires carry equal currents i. Each follows a 90° arc on the circumference of the same circle of radius R, in the configuration shown in Fig. 30-42. Show that **B** at the center of the circle is the same as the field **B** a distance R below an infinite straight wire carrying a current i to the left.

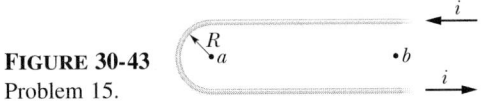

FIGURE 30-42
Problem 14.

15P. A long hairpin is formed by bending a very long wire as shown in Fig. 30-43. If the wire carries a 10 A current, what are the direction and magnitude of **B** at (a) point a and (b) midpoint b? Take $R = 5.0$ mm and the distance between a and b to be *much* larger than R (each straight section is "infinite").

FIGURE 30-43
Problem 15.

16P. A wire carrying current i has the configuration shown in Fig. 30-44. Two semi-infinite straight sections, both tangent to the same circle, are connected by a circular arc, of central angle θ, along the circumference of the circle, with all sections lying in the same plane. What must θ be in order for B to be zero at the center of the circle?

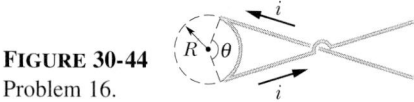

FIGURE 30-44
Problem 16.

17P. In Fig. 30-45, a straight wire of length L carries current i. Show that the magnitude of the magnetic field **B** produced by this segment at P_1, a distance R from the segment along a perpendicular bisector, is

$$B = \frac{\mu_0 i}{2\pi R} \frac{L}{(L^2 + 4R^2)^{1/2}}.$$

Show that this expression reduces to an expected result as $L \to \infty$.

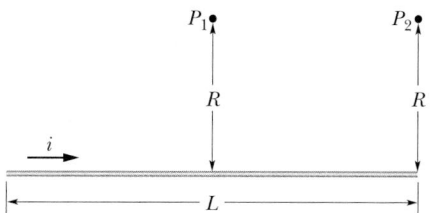

FIGURE 30-45 Problems 17 and 18.

18P. In Fig. 30-45, a straight wire of length L carries current i. Show that the magnitude of the magnetic field **B** produced by the wire at P_2, a perpendicular distance R from one end of the wire, is

$$B = \frac{\mu_0 i}{4\pi R} \frac{L}{(L^2 + R^2)^{1/2}}.$$

19P. A square loop of wire of edge length a carries current i. Using Problem 17, show that, at the center of the loop, the magnitude of the magnetic field produced by the current is

$$B = \frac{2\sqrt{2}\mu_0 i}{\pi a}.$$

20P. Using Problem 17, show that the magnitude of the magnetic field produced at the center of a rectangular loop of wire of length L and width W, carrying a current i, is

$$B = \frac{2\mu_0 i}{\pi} \frac{(L^2 + W^2)^{1/2}}{LW}.$$

Show that, for $L \gg W$, this expression reduces to a result consistent with the result of Sample Problem 30-2.

21P. A square loop of wire of edge length a carries current i. Using Problem 17, show that the magnitude of the magnetic field produced at a point on the axis of the loop and a distance x from its center is

$$B(x) = \frac{4\mu_0 i a^2}{\pi(4x^2 + a^2)(4x^2 + 2a^2)^{1/2}}.$$

Prove that this result is consistent with the result of Problem 19.

22P. Two wires, both of length L, are formed into a circle and a square, and each carries current i. Show that the square produces a greater magnetic field at its center than the circle produces at its center. (See Problem 19.)

23P. In Fig. 30-46, a current i is in a straight wire of length a. Show that the magnitude of the magnetic field produced by the current at point P is $B = \sqrt{2}\mu_0 i/8\pi a$.

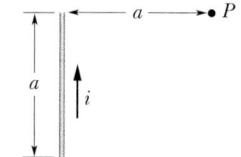

FIGURE 30-46
Problem 23.

24P. Find the magnetic field **B** at point P in Fig. 30-47 (See Problem 23.)

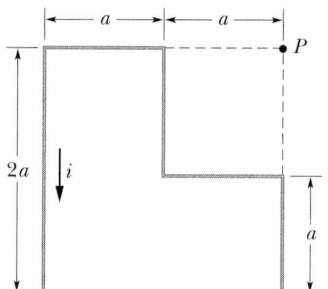

FIGURE 30-47
Problem 24.

25P. Find the magnetic field **B** at point P in Fig. 30-48 for $i = 10$ A and $a = 8.0$ cm. (See Problems 18 and 23.)

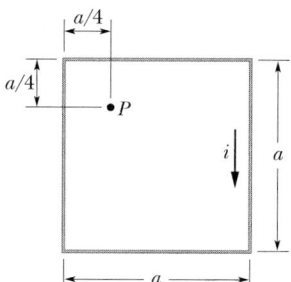

FIGURE 30-48
Problem 25.

26P. Figure 30-49 shows a cross section of a long, thin ribbon of width w that is carrying a uniformly distributed total current i into the page. Calculate the magnitude and direction of the magnetic field **B** at a point P in the plane of the ribbon at a distance d from its edge. (*Hint:* Imagine the ribbon to be constructed from many long, thin, parallel wires.)

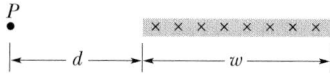

FIGURE 30-49 Problem 26.

SECTION 30-2 Two Parallel Currents

27E. Two long parallel wires are 8.0 cm apart. What equal currents must be in the wires if the magnetic field halfway between them is to have a magnitude of 300 μT? Consider both (a) parallel and (b) antiparallel currents.

28E. Two long parallel wires a distance d apart carry currents of i and $3i$ in the same direction. Locate the point or points at which their magnetic fields cancel.

29E. Two long straight parallel wires, separated by 0.75 cm, are perpendicular to the plane of the page as shown in Fig. 30-50. Wire 1 carries a current of 6.5 A into the page. What must be the current (magnitude and direction) in wire 2 for the resultant magnetic field at point P to be zero?

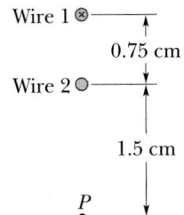

FIGURE 30-50
Exercise 29.

30E. Figure 30-51 shows five long parallel wires in the xy plane. Each wire carries a current $i = 3.00$ A in the positive x direction. The separation between adjacent wires is $d = 8.00$ cm. In unit-vector notation, what is the magnetic force per meter exerted on each of these five wires by the other wires?

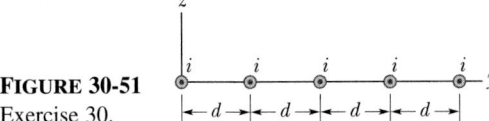

FIGURE 30-51
Exercise 30.

31E. For the wires in Sample Problem 30-2, show that Eq. 30-15 holds for points beyond the wires, that is, for points with $|x| > d$.

32E. Each of two long straight parallel wires 10 cm apart carries a current of 100 A. Figure 30-52 shows a cross section, with the wires running perpendicular to the page and point P lying on the perpendicular bisector of the line between the wires. Find the magnitude and direction of the magnetic field at P when the current in the left-hand wire is out of the page and the current in the right-hand wire is (a) out of the page and (b) into the page.

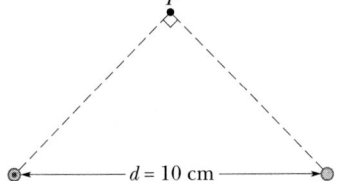

FIGURE 30-52
Exercise 32.

33P. Assume that both currents in Fig. 30-10a are in the same direction, out of the plane of the figure. Show that the magnetic field in the plane defined by the wires is

$$B(x) = \frac{\mu_0 i x}{\pi(x^2 - d^2)}.$$

Assume that $i = 10$ A and $d = 2.0$ cm in Fig. 30-10a, and plot $B(x)$ for the range -2 cm $< x < 2$ cm. Assume that the wire diameters are negligible.

34P. Four long copper wires are parallel to each other, their cross sections forming the corners of a square with sides $a = 20$ cm. A 20 A current exists in each wire in the direction shown in Fig. 30-53. What are the magnitude and direction of $\mathbf{B}$ at the center of the square?

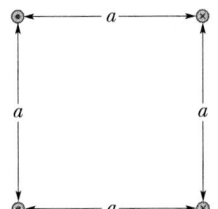

FIGURE 30-53
Problems 34, 35, and 36.

35P. Suppose, in Fig. 30-53, that the identical currents i are all out of the page. What is the force per unit length (magnitude and direction) on any one wire?

36P. In Fig. 30-53 what is the force per unit length acting on the lower left wire, in magnitude and direction, with the current directions as shown? The currents are i.

37P. Two long wires a distance d apart carry equal antiparallel currents i, as in Fig. 30-54. (a) Show that the magnitude of the magnetic field at point P, which is equidistant from the wires, is given by

$$B = \frac{2\mu_0 i d}{\pi(4R^2 + d^2)}.$$

(b) In what direction does $\mathbf{B}$ point?

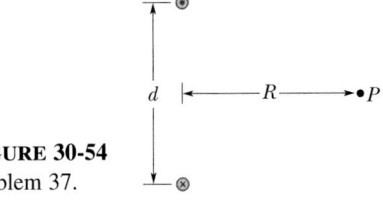

FIGURE 30-54
Problem 37.

38P. In Fig. 30-55, the long straight wire carries a current of 30 A and the rectangular loop carries a current of 20 A. Calculate the resultant force acting on the loop. Assume that $a = 1.0$ cm, $b = 8.0$ cm, and $L = 30$ cm.

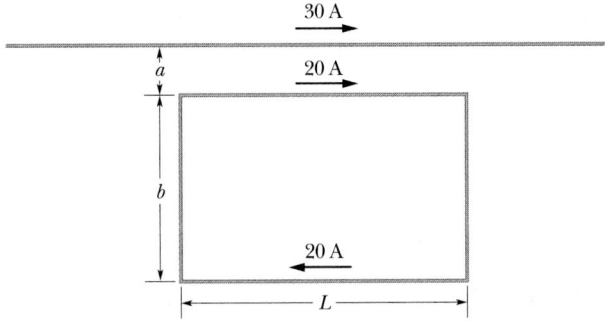

FIGURE 30-55   Problem 38.

39P. Figure 30-56 is an idealized schematic drawing of a rail gun. Projectile P sits between the two wide circular rails; a source of current sends current through the rails and through the (conducting) projectile itself (a fuse is not used). (a) Let w be the distance between the rails, R the radius of the rails, and i the current. Show that the force on the projectile is directed to the right along the rails and is given approximately by

$$F = \frac{i^2 \mu_0}{2\pi} \ln \frac{w + R}{R}.$$

(b) If the projectile starts from the left end of the rails at rest, find the speed v at which it is expelled at the right. Assume that $i = 450$ kA, $w = 12$ mm, $R = 6.7$ cm, $L = 4.0$ m, and the mass of the projectile is $m = 10$ g.

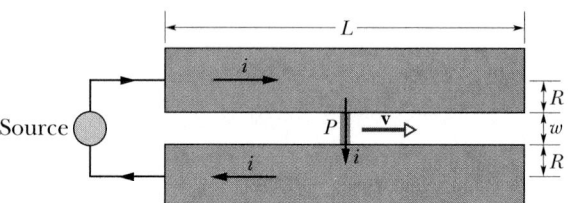

FIGURE 30-56 Problem 39.

SECTION 30-3 Ampere's Law

40E. Each of the eight conductors in Fig. 30-57 carries 2.0 A of current into or out of the page. Two paths are indicated for the line integral $\oint \mathbf{B} \cdot d\mathbf{s}$. What is the value of the integral for the path (a) at the left and (b) at the right?

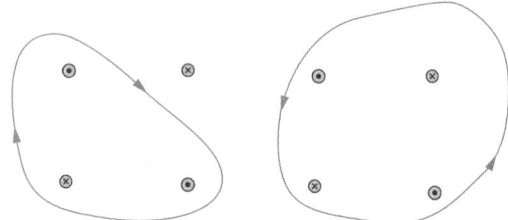

FIGURE 30-57 Exercise 40.

41E. Eight wires cut the page perpendicularly at the points shown in Fig. 30-58. A wire labeled with the integer k ($k = 1$, $2, \ldots, 8$) carries the current ki. For those with odd k, the current is out of the page; for those with even k, it is into the page. Evaluate $\oint \mathbf{B} \cdot d\mathbf{s}$ along the closed path in the direction shown.

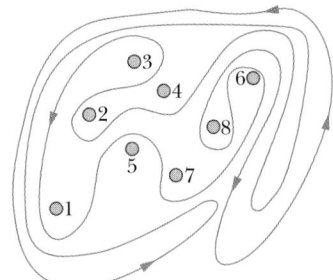

FIGURE 30-58
Exercise 41.

42E. In a certain region there is a uniform current density of 15 A/m² in the positive z direction. What is the value of $\oint \mathbf{B} \cdot d\mathbf{s}$ when the line integral is taken along the three straight-line segments from $(4d, 0, 0)$ to $(4d, 3d, 0)$ to $(0, 0, 0)$ to $(4d, 0, 0)$, where $d = 20$ cm?

43E. Figure 30-59 shows a cross section of a long cylindrical conductor of radius a, carrying a uniformly distributed current i. Assume that $a = 2.0$ cm and $i = 100$ A, and plot $B(r)$ over the range $0 < r < 6.0$ cm.

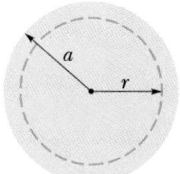

FIGURE 30-59
Exercise 43.

44P. Two square conducting loops carry currents of 5.0 and 3.0 A as shown in Fig. 30-60. What is the value of $\oint \mathbf{B} \cdot d\mathbf{s}$ for each of the two closed paths shown?

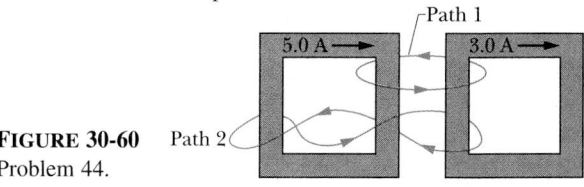

FIGURE 30-60 Path 2
Problem 44.

45P. Show that a uniform magnetic field $\mathbf{B}$ cannot drop abruptly to zero, as is suggested just to the right of point a in Fig. 30-61, as one moves perpendicular to $\mathbf{B}$, say along the horizontal arrow in the figure. (*Hint:* Apply Ampere's law to the rectangular path shown by the dashed lines.) In actual magnets "fringing" of the magnetic field lines always occurs, which means that $\mathbf{B}$ approaches zero in a gradual manner. Modify the field lines in the figure to indicate a more realistic situation.

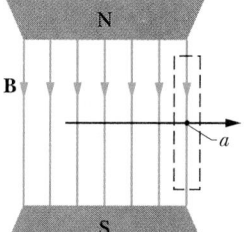

FIGURE 30-61
Problem 45.

46P. Figure 30-62a shows a cross section of a hollow cylindrical conductor of radii a and b, carrying a uniformly distributed current i. (a) Show that $B(r)$ for the range $b < r < a$ is given by

$$B = \frac{\mu_0 i}{2\pi(a^2 - b^2)} \frac{r^2 - b^2}{r}.$$

(b) Show that when $r = a$, this equation gives the magnetic field magnitude B for a long straight wire; when $r = b$, it gives zero magnetic field; and when $b = 0$, it gives the magnetic field inside a solid conductor. (c) Assume that $a = 2.0$ cm, $b = 1.8$ cm, and $i = 100$ A, and plot $B(r)$ for the range $0 < r < 6$ cm.

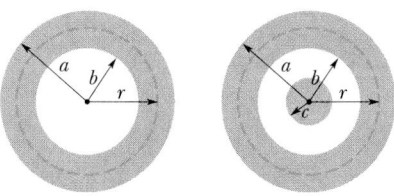

FIGURE 30-62
Problems 46 and 47.　　　(a)　　　　(b)

47P. Figure 30-62b shows a cross section of a long conductor of a type called a coaxial cable and gives its radii (a, b, c). Equal but opposite currents i are uniformly distributed in the two conductors. Derive expressions for $B(r)$ in the ranges (a) $r < c$, (b) $c < r < b$, (c) $b < r < a$, and (d) $r > a$. (e) Test these expressions for all the special cases that occur to you. (f) Assume that $a = 2.0$ cm, $b = 1.8$ cm, $c = 0.40$ cm, and $i = 120$ A and plot the function $B(r)$ over the range $0 < r < 3$ cm.

48P. The current density inside a long, solid, cylindrical wire of radius a is in the direction of the central axis and varies linearly with radial distance r from the axis according to $J = J_0 r/a$. Find the magnetic field inside the wire.

49P. A long circular pipe with outside radius R carries a (uniformly distributed) current i into the page as shown in Fig. 30-63. A wire runs parallel to the pipe at a distance of $3R$ from center to center. Find the magnitude and direction of the current in the wire such that the resultant magnetic field at point P has the same magnitude as the resultant field at the center of the pipe but is in the opposite direction.

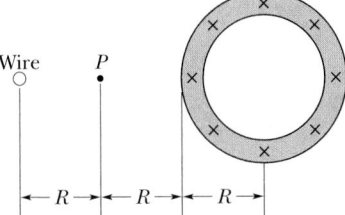

FIGURE 30-63
Problem 49.

50P. Figure 30-64 shows a cross section of a long cylindrical conductor of radius a containing a long cylindrical hole of radius b. The axes of the cylinder and hole are parallel and are a distance d apart; a current i is uniformly distributed over the tinted area. (a) Use superposition to show that the magnetic field at the center of the hole is

$$B = \frac{\mu_0 id}{2\pi(a^2 - b^2)}.$$

(b) Discuss the two special cases $b = 0$ and $d = 0$. (c) Use Ampere's law to show that the magnetic field in the hole is uniform. (*Hint:* Regard the cylindrical hole as filled with two equal currents moving in opposite directions, thus canceling each other. Assume that each of these currents has the same current density as that in the actual conductor. Then superimpose the fields due to two complete cylinders of current, of radii a and b, each cylinder having the same current density.)

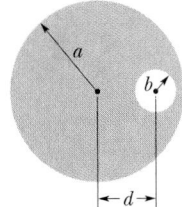

FIGURE 30-64
Problem 50.

51P. Figure 30-65 shows a cross section of an infinite conducting sheet with a current per unit x-length λ emerging perpendicu-

larly out of the page. (a) Use the Biot–Savart law and symmetry to show that for all points P above the sheet, and all points P' below it, the magnetic field **B** is parallel to the sheet and directed as shown. (b) Use Ampere's law to prove that $B = \frac{1}{2}\mu_0\lambda$ at all points P and P'.

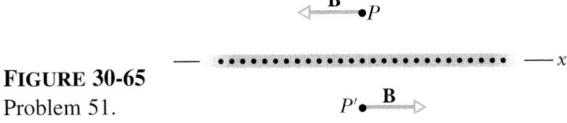

FIGURE 30-65
Problem 51.

52P*. The magnetic field in a certain region is given in milliteslas by $\mathbf{B} = 3.0\mathbf{i} + 8.0(x^2/d^2)\mathbf{j}$, where x is in meters and d is a constant with the unit of length. Some current must exist in the region to cause the specified **B** field. (a) Evaluate the integral $\oint \mathbf{B} \cdot d\mathbf{s}$ along the straight path from $(d, 0, 0)$ to $(d, d, 0)$. (b) Let $d = 0.50$ m in the expression for **B** and apply Ampere's law to determine what current flows perpendicularly through a square of side length 0.5 m that lies in the first quadrant of the xy plane, with one corner at the origin. (c) Is this current in the $+\mathbf{k}$ or $-\mathbf{k}$ direction?

SECTION 30-4 Solenoids and Toroids

53E. A 200 turn solenoid having a length of 25 cm and a diameter of 10 cm carries a current of 0.30 A. Calculate the magnitude of the magnetic field **B** inside the solenoid.

54E. A solenoid 95.0 cm long has a radius of 2.00 cm and a winding of 1200 turns; it carries a current of 3.60 A. Calculate the magnitude of the magnetic field inside the solenoid.

55E. A solenoid 1.30 m long and 2.60 cm in diameter carries a current of 18.0 A. The magnetic field inside the solenoid is 23.0 mT. Find the length of the wire forming the solenoid.

56E. A toroid having a square cross section, 5.00 cm on a side, and an inner radius of 15.0 cm has 500 turns and carries a current of 0.800 A. (It is made up of a square solenoid—instead of round as in Fig. 30-17—bent into a doughnut shape.) What is the magnetic field inside the toroid at (a) the inner radius and (b) the outer radius of the toroid?

57E. Show that if the thickness of a toroid is very small compared to its radius of curvature (a very skinny toroid), then Eq. 30-26 for the field inside a toroid reduces to Eq. 30-25 for the field inside a solenoid. Explain why this result is to be expected.

58P. Treat an ideal solenoid as a thin cylindrical conductor whose current per unit length, measured parallel to the cylinder axis, is λ. By doing so, show that the magnitude of the magnetic field inside an ideal solenoid can be written as $B = \mu_0\lambda$. This is the value of the *change* in **B** that you encounter as you move from inside the solenoid to outside, through the solenoid wall. Show that the same change occurs as you move through an infinite flat current sheet such as that of Fig. 30-65 (see Problem 51). Does this equality surprise you?

59P. In Section 30-4 we showed that the magnetic field at any radius r inside a toroid is given by

$$B = \frac{\mu_0 iN}{2\pi r}.$$

Show that as you move from a point just inside a toroid to a point just outside, the magnitude of the *change* in **B** that you encounter —at any radius r—is just $\mu_0\lambda$. Here λ is the current per unit length along a circumference of radius r within the toroid. Compare with the similar result found in Problem 58. Isn't the equality surprising?

60P. A long solenoid with 10.0 turns/cm and a radius of 7.00 cm carries a current of 20.0 mA. A current of 6.00 A exists in a straight conductor located along the axis of the solenoid. (a) At what radial distance from the axis will the direction of the resulting magnetic field be at 45.0° to the axial direction? (b) What is the magnitude of the magnetic field there?

61P. A long solenoid has 100 turns/cm and carries current i. An electron moves within the solenoid in a circle of radius 2.30 cm perpendicular to the solenoid axis. The speed of the electron is $0.0460c$ (c = speed of light). Find the current i in the solenoid.

SECTION 30-5 A Current-Carrying Coil as a Magnetic Dipole

62E. What is the magnetic dipole moment $\boldsymbol{\mu}$ of the solenoid described in Exercise 53?

63E. Figure 30-66a shows a length of wire carrying a current i and bent into a circular coil of one turn. In Fig. 30-66b the same length of wire has been bent more sharply, to give a coil of two turns, each of half the original radius. (a) If B_a and B_b are the magnitudes of the magnetic fields at the centers of the two coils, what is the ratio B_b/B_a? (b) What is the ratio of the dipole moments, μ_b/μ_a, of the coils?

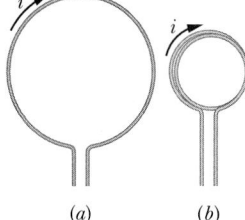

FIGURE 30-66
Exercise 63.

(a)   (b)

64E. Figure 30-67 shows an arrangement known as a Helmholtz coil. It consists of two circular coaxial coils, each of N turns and radius R, separated by a distance R. The two coils carry equal currents i in the same direction. Find the magnitude of the net magnetic field at P, midway between the coils.

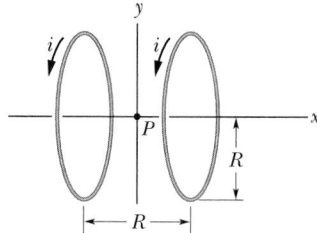

FIGURE 30-67 Exercise 64; Problems 68 and 72.

65E. A student makes a short electromagnet by winding 300 turns of wire around a wooden cylinder of diameter $d = 5.0$ cm. The coil is connected to a battery producing a current of 4.0 A in the wire. (a) What is the magnetic moment of this device? (b) At

what axial distance $z \gg d$ will the magnetic field of this dipole have the magnitude 5.0 μT (approximately one-tenth that of Earth's magnetic field)?

66E. The magnitude $B(x)$ of the magnetic field at points on the axis of a square current loop of side a is given in Problem 21. (a) Show that the axial magnetic field of this loop, for $x \gg a$, is that of a magnetic dipole (see Eq. 30-29). (b) What is the magnetic dipole moment of this loop?

67P. A length of wire is formed into a closed circuit with radii a and b, as shown in Fig. 30-68, and carries a current i. (a) What are the magnitude and direction of **B** at point P? (b) Find the magnetic dipole moment of the circuit.

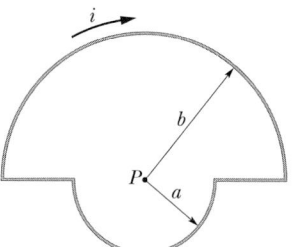

FIGURE 30-68
Problem 67.

68P. Two 300 turn coils of radius R each carry a current i. They are arranged a distance R apart, as in Fig. 30-67. For $R = 5.0$ cm and $i = 50$ A, plot the magnitude B of the net magnetic field as a function of distance x along the common axis over the range $x = -5$ cm to $x = +5$ cm, taking $x = 0$ at the midpoint P. (Such coils provide an especially uniform field **B** near point P.) (*Hint:* See Eq. 30-28.)

69P. A circular loop of radius 12 cm carries a current of 15 A. A coil of radius 0.82 cm, having 50 turns and a current of 1.3 A, is concentric with the loop. (a) What magnetic field **B** does the loop produce at its center? (b) What torque acts on the coil? Assume that the planes of the loop and coil are perpendicular and that the magnetic field due to the loop is essentially uniform throughout the volume occupied by the coil.

70P. (a) A long wire is bent into the shape shown in Fig. 30-69, without the wire actually touching itself at P. The radius of the circular section is R. Determine the magnitude and direction of **B** at the center C of the circular portion when the current i is as indicated. (b) Suppose the circular part of the wire is rotated without distortion about the indicated diameter, until the plane of the circle is perpendicular to the straight portion of the wire. The magnetic dipole moment associated with the circular loop is now in the direction of the current in the straight part of the wire. Determine **B** at C in this case.

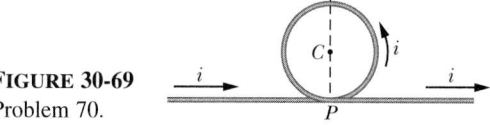

FIGURE 30-69
Problem 70.

71P. A conductor carries a current of 6.0 A along the closed path *abcdefgha* involving 8 of the 12 edges of a cube of side 10 cm as shown in Fig. 30-70. (a) Why can one regard this as the superposition of three square loops: *bcfgb*, *abgha*, and *cdefc*? (*Hint:*

Draw currents around those square loops.) (b) Use this superposition to find the magnetic dipole moment $\boldsymbol{\mu}$ (magnitude and direction) of the closed path. (c) Calculate **B** at the points $(x, y, z) = (0, 5.0 \text{ m}, 0)$ and $(5.0 \text{ m}, 0, 0)$.

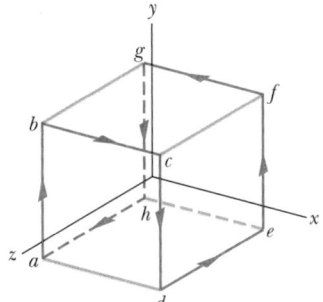

FIGURE 30-70
Problem 71.

72P. In Exercise 64 (Fig. 30-67), let the separation of the coils be a variable s (not necessarily equal to the coil radius R). (a) Show that the first derivative of the magnitude of the net magnetic field of the coils (dB/dx) vanishes at the midpoint P regardless of the value of s. Why would you expect this to be true from symmetry? (b) Show that the second derivative (d^2B/dx^2) also vanishes at P, provided $s = R$. This accounts for the uniformity of B near P for this particular coil separation.

Electronic Computation

73. The magnetic field of a circular loop with current i, at a point on the central axis through the loop, is parallel to that axis. The magnitude of the field is given by

$$B = \frac{\mu_0 i R^2}{2(R^2 + z^2)^{3/2}},$$

where R is the radius of the loop and z is the distance from the center of the loop. A solenoid can be constructed mathematically by using many such circular loops that are identical in radius and current, coaxial, and closely spaced. Suppose the solenoid has a length of 25.0 cm and a radius of 1.00 cm and consists of N equally spaced loops, each with a current of 1.00 A. For (a) $N = 11$, (b) $N = 21$, and (c) $N = 51$, compute the magnitude of the magnetic field at the center of the solenoid by summing the fields produced by the individual loops. For each value of N, compare

the result with the value found using Eq. 30-25, which holds for a long solenoid with a large number of tightly spaced loops.

74. A computer can be used to demonstrate Ampere's law for a situation in which the Amperian loop does not coincide with a magnetic field line. Suppose that a square with edges of length a is centered at the origin of a coordinate system whose x and y axes are parallel to sides of the square. A long straight wire carrying current i is perpendicular to the plane of the square and crosses the x axis at $x = x'$. Evaluate $\oint \mathbf{B} \cdot d\mathbf{s}$ numerically. Divide a side of the square into N segments of equal length Δs and for each segment evaluate $\mathbf{B} \cdot \mathbf{u} \Delta s$. Here **B** is the magnetic field at the center of the segment and **u** is a unit vector that is parallel to the segment and is in the direction of integration. For different segments, **u** might be **i**, **j**, $-\mathbf{i}$, or $-\mathbf{j}$. The magnetic field at a point with coordinates x and y is given by

$$\mathbf{B} = \frac{\mu_0 i [-y\mathbf{i} + (x - x')\mathbf{j}]}{2\pi[(x - x')^2 + y^2]}.$$

For sides that are parallel to the x axis, take $y = a/2$ or $-a/2$; for sides that are parallel to the y axis, take $x = a/2$ or $-a/2$. Suppose that the length of a side is 1.00 m and the current is 1.00 A, and, for each of the following cases, calculate the sum over segments for each side of the square separately; then add the results to find the total for the square. Compare the total to $\mu_0 i_{\text{enc}}$. The value $N = 50$ should give you three-significant-figure accuracy: (a) $x' = 0$ (the wire is at the center of the square), (b) $x' = 0.200$ m (the wire passes inside the square at an off-center point), (c) $x' = 0.400$ m (the wire passes through the square near the center of a side), and (d) $x' = 0.600$ m (the wire is outside the square).

75. Two long parallel conductors carry currents parallel to the z axis. The conductors intersect the xy plane at points along the x axis: one intersects at $x = a$ and carries a current i_1 in the $+z$ direction; the other conductor intersects at $x = 0$ and carries a current i_2 that can be varied in both magnitude and direction. The current is considered to be positive if directed in the positive z direction and negative if directed in the negative z direction. (a) Write an equation that gives the net magnetic field **B** along the x axis for $x > a$. (b) Rewrite the equation for $x = 2a$ after substituting for i_2 with $i_2 = bi_1$, where b is a variable. (c) With this rewritten equation, graph **B** versus b for the range $3 > b > -3$. Positive **B** corresponds to the magnetic field being directed in the positive y direction, negative in the negative y direction.

31
Induction and Inductance

Soon after rock began in the mid-1950s, guitarists switched from acoustic guitars to electric guitars. But it was Jimi Hendrix who first understood the electric guitar as an electronic instrument. He exploded on the scene in the 1960s, ripping his pick along the strings, positioning himself and his guitar in front of a speaker to sustain feedback, and then laying down chords on top of the feedback. He shoved rock forward from the melodies of Buddy Holly into the psychedelia of the late 1960s and into the early heavy metal of Led Zeppelin in the 1970s, and his ideas continue to influence rock today. But what is it about an electric guitar that distinguishes it from an acoustic guitar and enabled Hendrix to make so much broader use of this electronic instrument?

31-1 TWO SYMMETRIC SITUATIONS

In Section 29-8, we saw that if we put a closed conducting loop in a magnetic field and then send current through the loop, forces due to the magnetic field create a torque to turn the loop:

$$\text{current loop} + \text{magnetic field} \Rightarrow \text{torque}. \quad (31\text{-}1)$$

Suppose that, instead, with the current off, we turn the loop by hand. Will the opposite of Eq. 31-1 occur? That is, will a current now appear in the loop:

$$\text{torque} + \text{magnetic field} \Rightarrow \text{current}? \quad (31\text{-}2)$$

The answer is yes—a current does appear. The situations of Eqs. 31-1 and 31-2 are symmetric. The physical law on which Eq. 31-2 depends is called *Faraday's law of induction*. Whereas Eq. 31-1 is the basis for the electric motor, Eq. 31-2 and Faraday's law are the basis for the electric generator. This chapter is concerned with that law and the process it describes.

31-2 TWO EXPERIMENTS

Let us examine two simple experiments to prepare for our discussion of Faraday's law of induction.

First Experiment. Figure 31-1 shows a conducting loop connected to a sensitive current meter. Since there is no battery or other source of emf included, there is no current in the circuit. However, if we move a bar magnet toward the loop, a current suddenly appears in the circuit. The current disappears when the magnet stops. If we then move the magnet away, a current again suddenly appears, but now in the opposite direction. If we experimented for a while, we would discover the following:

1. A current appears only if there is relative motion between the loop and the magnet (one must move relative to the other); the current disappears when the relative motion between them ceases.

2. Faster motion produces a greater current.

3. If moving the magnet's north pole toward the loop causes, say, clockwise current, then moving the north pole away causes counterclockwise current. And moving the south pole toward or away from the loop also causes currents, but in the reversed directions.

The current produced in the loop is called an **induced current,** the work done per unit charge in producing that current (in moving the conduction electrons constituting the current) is called an **induced emf,** and the process of producing the current and emf is called **induction.**

Second Experiment. For this experiment we use the apparatus of Fig. 31-2, with the two conducting loops close to each other but not touching. If we close switch S, to turn on a current in the right-hand loop, the meter suddenly and briefly registers a current—an induced current—in the left-hand loop. If we then open the switch, another sudden and brief induced current appears in the left-hand loop, but in the opposite direction. We get an induced current (and thus an induced emf) only when the current in the right-hand loop is changing (either turning on or turning off) and not when it is constant (even if it is large).

The induced emf and induced current in these experiments are apparently caused when something changes. But what is that "something"? Faraday knew.

31-3 FARADAY'S LAW OF INDUCTION

Faraday realized that an emf and a current can be induced in a loop, as in our two experiments, by changing the *amount of magnetic field* passing through the loop. He further realized that the "amount of magnetic field" can be visualized in terms of the magnetic field lines passing through the loop. **Faraday's law of induction,** stated in terms of our experiments, is:

> An emf is induced in the left-hand loop in Figs. 31-1 and 31-2 when the number of magnetic field lines that pass through the loop is changing.

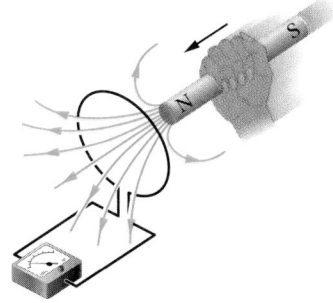

FIGURE 31-1 A current meter registers a current in the wire loop when the magnet is moving with respect to the loop.

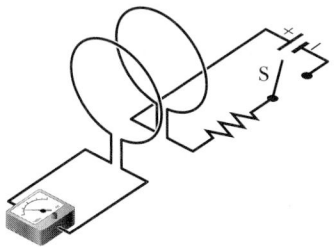

FIGURE 31-2 The current meter registers a current in the left-hand wire loop just as switch S is closed (to turn on current in the right-hand wire loop) or opened (to turn off the current in the right-hand loop). No motion of the coils is involved.

Most important, the actual number of field lines passing through the loop does not matter; the values of the induced emf and induced current are determined by the *rate* at which that number changes.

In our first experiment (Fig. 31-1), the magnetic field lines spread from the north pole of the magnet. So, as we bring the north pole toward the loop, the number of field lines passing through the loop increases. That increase apparently causes conduction electrons in the loop to move (the induced current) and provides energy (the induced emf) for their motion. When the magnet stops moving, the number of field lines through the loop no longer changes and the induced current and induced emf disappear.

In our second experiment (Fig. 31-2), when the switch is open (no current), there are no field lines. But when we turn on the current in the right-hand loop, the increasing current builds up a magnetic field around that loop and at the left-hand loop. While the field builds, the number of magnetic field lines through the left-hand loop increases. As in the first experiment, the increase in field lines through that loop apparently induces a current and an emf there. When the current in the right-hand loop reaches a final, steady value, the number of field lines through the left-hand loop no longer changes, and the induced current and induced emf disappear.

Faraday's law does not explain *why* a current and an emf are induced in either experiment; it is just a statement that helps us visualize the induction.

A Quantitative Treatment

To put Faraday's law to work, we need a way to calculate the *amount of magnetic field* that passes through a loop. In Chapter 24, in a similar situation, we needed to calculate the amount of an electric field that passes through a surface. There we defined an electric flux $\Phi_E = \int \mathbf{E} \cdot d\mathbf{A}$. So, here we define a magnetic flux. Suppose a loop enclosing an area A is placed in a magnetic field $\mathbf{B}$. Then the magnetic flux through the loop is

$$\Phi_B = \int \mathbf{B} \cdot d\mathbf{A} \qquad \begin{array}{l}\text{(magnetic flux}\\ \text{through area } A).\end{array} \qquad (31\text{-}3)$$

As in Chapter 24, $d\mathbf{A}$ is a vector of magnitude dA that is perpendicular to a differential area dA.

As a special case of Eq. 31-3, suppose that the loop lies in a plane and that the magnetic field is perpendicular to the plane of the loop. Then we can write the dot product in Eq. 31-3 as $B \, dA \cos 0° = B \, dA$. If the magnetic field is also uniform, then B can be brought out in front of the integral sign. The remaining $\int dA$ then gives just the area A of the loop. So, Eq. 31-3 reduces to

$$\Phi_B = BA \qquad (\mathbf{B} \perp \mathbf{A}, \mathbf{B} \text{ uniform}). \qquad (31\text{-}4)$$

From Eqs. 31-3 and 31-4, we see that the SI unit for magnetic flux is the tesla–square meter, which is called the *weber* (abbreviated Wb):

$$1 \text{ weber} = 1 \text{ Wb} = 1 \text{ T} \cdot \text{m}^2. \qquad (31\text{-}5)$$

With the notion of magnetic flux, we can state Faraday's law in a more quantitative and useful way:

The magnitude of the emf $\mathcal{E}$ induced in a conducting loop is equal to the rate at which the magnetic flux Φ_B through that loop changes with time.

As you will see in the next section, the induced emf $\mathcal{E}$ tends to oppose the flux change, so Faraday's law is formally written as

$$\mathcal{E} = -\frac{d\Phi_B}{dt} \qquad \text{(Faraday's law)}, \qquad (31\text{-}6)$$

with the minus sign indicating that opposition. We often neglect the minus sign in Eq. 31-6, seeking only the magnitude of the induced emf.

If we change the magnetic flux through a coil of N turns, an induced emf appears in every turn and the total emf induced in the coil is the sum of these individual induced emfs. If the coil is tightly wound *(closely packed)*, so that the same magnetic flux Φ_B passes through all the turns, the total emf induced in the coil is

$$\mathcal{E} = -N\frac{d\Phi_B}{dt} \qquad \text{(coil of } N \text{ turns)}. \qquad (31\text{-}7)$$

Here are the general means by which we can change the magnetic flux through a coil:

1. Change the magnitude B of the magnetic field within the coil.

2. Change the area of the coil, or the portion of that area that happens to lie within the magnetic field (for example, by expanding the coil or sliding it out of the field).

3. Change the angle between the direction of the magnetic field $\mathbf{B}$ and the area of the coil (for example, by rotating the coil so that $\mathbf{B}$ is first perpendicular to the plane of the coil and then is along that plane).

SAMPLE PROBLEM 31-1

The long solenoid S of Fig. 31-3 has 220 turns/cm and carries a current $i = 1.5$ A; its diameter D is 3.2 cm. At its center we place a 130-turn close-packed coil C of diameter $d = 2.1$ cm. The current in the solenoid is reduced to zero at a steady rate in 25 ms. What emf is induced in coil C while the current in the solenoid is changing?

SOLUTION: Coil C is located in the magnetic field B produced by the current in the solenoid. As the current decreases, so does B. Thus, the magnetic flux through the coil decreases. During this decrease an emf is induced in the coil via Faraday's law. To find the induced emf, we first find the initial magnetic field B_i of the solenoid by substituting given data into Eq. 30-25:

$$B_i = \mu_0 i n$$
$$= (4\pi \times 10^{-7} \text{ T} \cdot \text{m/A})$$
$$\times (1.5 \text{ A})(220 \text{ turns/cm})(100 \text{ cm/m})$$
$$= 4.15 \times 10^{-2} \text{ T}.$$

The area A of each turn of coil C is $\frac{1}{4}\pi d^2$, which is equal to 3.46×10^{-4} m². The solenoid's magnetic field is perpendicular to this area, and we assume it to be uniform across the area. So we can find the initial magnetic flux $\Phi_{B,i}$ through each turn of coil C by substituting known data into Eq. 31-4:

$$\Phi_{B,i} = BA = (4.15 \times 10^{-2} \text{ T})(3.46 \times 10^{-4} \text{ m}^2)$$
$$= 1.44 \times 10^{-5} \text{ Wb} = 14.4 \ \mu\text{Wb}.$$

The final magnetic field B_f and magnetic flux $\Phi_{B,f}$ are both zero. Thus, the change in flux through each turn of coil C is $\Delta\Phi_B = 14.4 \ \mu\text{Wb}$.

Because the solenoid's current decreases at a steady rate, so does the magnetic flux, and we can rewrite Faraday's law (Eq. 31-7) as

$$\mathcal{E} = N\frac{\Delta\Phi_B}{\Delta t},$$

where N is the number of turns (130) of the coil. (We ignore the minus sign in Eq. 31-7 because we are looking for the magnitude of $\mathcal{E}$ only.) Substituting known data then gives us

$$\mathcal{E} = (130 \text{ turns})\frac{14.4 \times 10^{-6} \text{ Wb}}{25 \times 10^{-3} \text{ s}}$$
$$= 7.5 \times 10^{-2} \text{ V} = 75 \text{ mV}. \qquad \text{(Answer)}$$

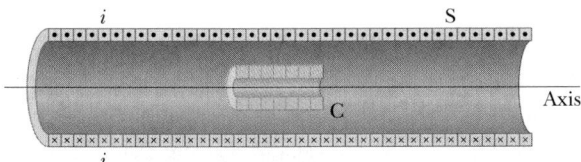

FIGURE 31-3 Sample Problem 31-1. A coil C is located inside a solenoid S, which carries current i.

CHECKPOINT **1:** The graph gives the magnitude $B(t)$ of a uniform magnetic field that exists throughout a conducting loop, perpendicular to the plane of the loop.

Rank the five regions of the graph according to the magnitude of the emf induced in the loop, greatest first.

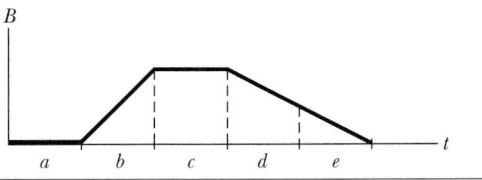

31-4 LENZ'S LAW

Soon after Faraday proposed his law of induction, Heinrich Friedrich Lenz devised a rule—now known as **Lenz's law**—for determining the direction of an induced current in a loop:

> An induced current has a direction such that the magnetic field due to *the current* opposes the change in the magnetic field that induces the current.

Furthermore, the direction of an induced emf is that of the induced current.

To get a feel for Lenz's law, let us apply it in two different but equivalent ways to Fig. 31-4, where the north pole of a magnet is being moved toward a conducting loop.

1. *Opposition to Pole Movement.* The approach of the magnet's north pole in Fig. 31-4 increases the magnetic field in the loop and thereby induces a current in the loop. From Fig. 30-22, we know that the loop then acts as a magnetic dipole with a south pole and a north pole, and that its magnetic dipole moment $\boldsymbol{\mu}$ points from south to north. To *oppose* the magnetic field increase being caused by the approaching magnet, the loop's north pole (and thus $\boldsymbol{\mu}$) must face *toward* the approaching north pole so as to repel it (Fig. 31-4). Then, the curled–straight right-hand rule for

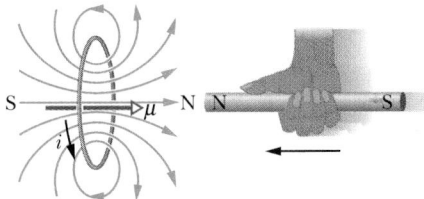

FIGURE 31-4 Lenz's law at work. As the magnet is moved toward the loop, a counterclockwise current is induced in the loop; the current produces its own magnetic field, with magnetic dipole moment $\boldsymbol{\mu}$, so as to oppose the motion of the magnet.

μ (Fig. 30-22) tells us that the current induced in the loop must be counterclockwise in Fig. 31-4.

If we next pull the magnet away from the loop, a current will again be induced in the loop. Now, however, the loop will have a south pole facing the retreating north pole of the magnet, so as to oppose the retreat. So the induced current will be clockwise.

2. *Opposition to Flux Change.* In Fig. 31-4, with the magnet initially distant, no magnetic flux passes through the loop. As the north pole of the magnet then nears the loop with its magnetic field **B** directed *toward the left,* the flux through the loop increases. To oppose this increase in flux, the induced current i must set up its own field **B**$_i$ *directed toward the right* inside the loop, as shown in Fig. 31-5a; then the rightward flux of field **B**$_i$ opposes the increasing leftward flux of field **B.** The curled–straight right-hand rule of Fig. 30-22 then tells us that i must be counterclockwise in Fig. 31-5a.

Note carefully that the flux of **B**$_i$ always opposes the *change* in the flux of **B**, but that does not always mean that **B**$_i$ points opposite **B**. For example, if we next pull the magnet away from the loop, the flux Φ_B from the magnet is still directed to the left through the loop, but it is now

decreasing. So the flux of **B**$_i$ must now be to the left inside the loop, to oppose the *decrease* in Φ_B, as shown in Fig. 31-5b. Thus, **B**$_i$ and **B** are now in the same direction.

Figures 31-5c and d show the situations in which the south pole of the magnet approaches and retreats from the loop, respectively.

Electric Guitars

Figure 31-6 shows a Fender Stratocaster, the type of electric guitar that was used by Jimi Hendrix and by many other musicians. Whereas an acoustic guitar depends for its sound on the acoustic resonance produced in the hollow body of the instrument by the oscillations of the strings, an electric guitar is a solid instrument, so there is no body resonance. Instead, the oscillations of the metal strings are sensed by electric "pickups" that send signals to an amplifier and a set of speakers.

The basic construction of a pickup is shown in Fig. 31-7. Wire connecting the instrument to the amplifier is coiled around a small magnet. The magnetic field of the magnet produces a north and south pole in the section of the metal string just above the magnet. That section then has its own magnetic field. When the string is plucked and thus made to oscillate, its motion relative to the coil changes the flux of its magnetic field through the coil, inducing a current in the coil. As the string oscillates toward and away from the coil, the induced current changes direction at the same frequency as the string's oscillations, thus relaying the frequency of oscillation to the amplifier and speaker.

On a Stratocaster, there are three groups of pickups, placed at the near end of the strings (on the wide part of the body). The group closest to the near end better detects the high-frequency oscillations of the strings; the group farthest from the near end better detects the low-frequency oscillations. By throwing a toggle switch on the guitar, the musician can select which group or which pair of groups will send signals to the amplifier and speakers.

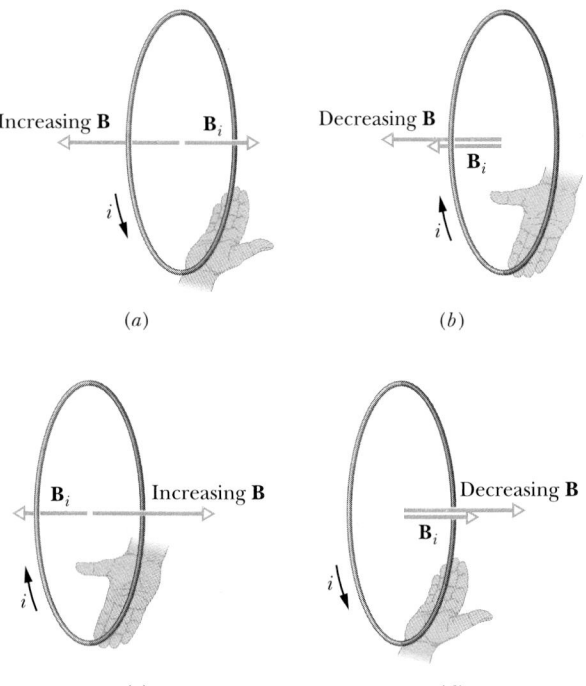

(a)   (b)

(c)   (d)

FIGURE 31-5 The current i induced in a loop has the direction such that the current's magnetic field **B**$_i$ opposes the *change* in the magnetic field **B** inducing i. The field **B**$_i$ is always directed opposite an increasing field **B** (a, c) and in the same direction as a decreasing **B** (b, d).

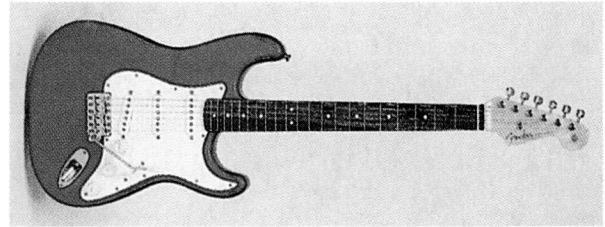

FIGURE 31-6 A Fender Stratocaster has three groups of six electric pickups each (within the wide part of the body). A toggle switch (at the bottom of the guitar) allows the musician to determine which group of pickups sends signals to an amplifier and thus to a speaker system.

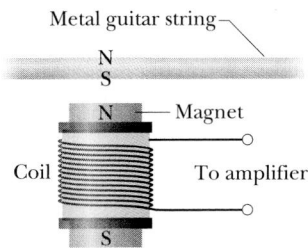

FIGURE 31-7 A side view of an electric guitar pickup. When the metal string (which acts like a magnet) is made to oscillate, the variation in magnetic flux induces a current in the coil.

To gain further control over his music, Hendrix sometimes rewrapped the wire in the pickup coils of his guitar to change the number of turns. In this way, he altered the amount of emf induced in the coils and thus their relative sensitivity. Even without this additional measure, you can see that the electric guitar offers far more control over the musical sound that is produced than can be obtained with an acoustic guitar.

CHECKPOINT **2:** The figure shows three situations in which identical circular conducting loops are in uniform magnetic fields that are either increasing (Inc) or decreasing (Dec) in magnitude at identical rates. In each, the dashed line coincides with a diameter. Rank the situations according to the magnitude of the current induced in the loops, greatest first.

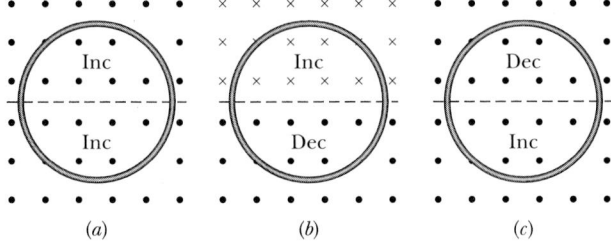

(a)   (b)   (c)

SAMPLE PROBLEM 31-2

Figure 31-8 shows a conducting loop consisting of a half-circle of radius $r = 0.20$ m and three straight sections. The half-circle lies in a uniform magnetic field **B** that is directed out of the page; the field magnitude is given by $B = 4.0t^2 + 2.0t + 3.0$, with B in teslas and t in seconds. An ideal battery with emf $\mathscr{E}_{bat} = 2.0$ V is connected to the loop. The resistance of the loop is 2.0 Ω.

(a) What are the magnitude and direction of the emf $\mathscr{E}_{ind}$ induced along the loop by **B** at $t = 10$ s?

SOLUTION: Equation 31-6 tells us that the magnitude of $\mathscr{E}_{ind}$ is equal to the rate $d\Phi_B/dt$ at which the magnetic flux through the loop changes. Because the field is uniform and perpendicular to the plane of the loop, the flux is given by Eq. 31-4: $\Phi_B = BA$. Using this equation and realizing that only the field magnitude B changes in time (not the area A), we

rewrite Eq. 31-6 as

$$\mathscr{E}_{ind} = \frac{d\Phi_B}{dt} = \frac{d(BA)}{dt} = A\frac{dB}{dt}.$$

Because the flux penetrates the loop only within the half-circle, the area A in this equation is $\frac{1}{2}\pi r^2$. Substituting this and the given expression for B yields

$$\mathscr{E}_{ind} = A\frac{dB}{dt} = \frac{\pi r^2}{2}\frac{d}{dt}(4.0t^2 + 2.0t + 3.0)$$

$$= \frac{\pi r^2}{2}(8.0t + 2.0).$$

At $t = 10$ s, then,

$$\mathscr{E}_{ind} = \frac{\pi(0.20\text{ m})^2}{2}[8.0(10) + 2.0]$$

$$= 5.152\text{ V} \approx 5.2\text{ V}. \qquad \text{(Answer)}$$

In Fig. 31-8, the flux through the loop is out of the page and increasing. Then, according to Lenz's law, the flux of the induced field B_i (due to the induced current) must be *into* the page. Using the curled–straight right-hand rule (Fig. 30-7c), we find that the induced current, must be clockwise around the loop. The induced emf $\mathscr{E}_{ind}$ must then also be clockwise.

(b) What is the current in the loop at $t = 10$ s?

SOLUTION: The induced emf $\mathscr{E}_{ind}$ tends to drive a current clockwise around the loop; the battery's emf $\mathscr{E}_{bat}$ tends to drive a current counterclockwise. Because $\mathscr{E}_{ind}$ is greater than $\mathscr{E}_{bat}$, the net emf $\mathscr{E}_{net}$ is clockwise, and thus so is the current. To find the current at $t = 10$ s, we use Eq. 28-2 ($i = \mathscr{E}/R$):

$$i = \frac{\mathscr{E}_{net}}{R} = \frac{\mathscr{E}_{ind} - \mathscr{E}_{bat}}{R} = \frac{5.152\text{ V} - 2.0\text{ V}}{2.0\ \Omega}$$

$$= 1.58\text{ A} \approx 1.6\text{ A}. \qquad \text{(Answer)}$$

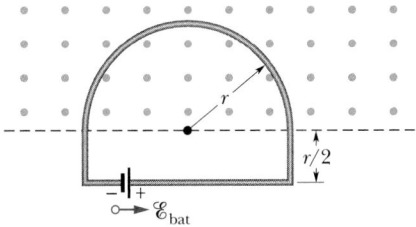

FIGURE 31-8 Sample Problem 31-2. A battery is connected to a conducting loop consisting of a half-circle of radius r that lies in a uniform magnetic field. The field is directed out of the page; its magnitude is changing.

SAMPLE PROBLEM 31-3

Figure 31-9 shows a rectangular loop of wire immersed in a nonuniform and varying magnetic field **B** that is perpendicular to and directed into the page. The field's magnitude is given by $B = 4t^2x^2$, with B in teslas, t in seconds, and x in meters. The loop has width $W = 3.0$ m and height $H = 2.0$ m. What are the magnitude and direction of the induced emf $\mathscr{E}$ along the loop at $t = 0.10$ s?

SOLUTION: The magnitude of the induced emf $\mathscr{E}$ is given by Faraday's law: $\mathscr{E} = d\Phi_B/dt$. But to use it, we need an expression for the flux Φ_B through the loop at any time t. Because B is *not* uniform over the area enclosed by the loop, we *cannot* use Eq. 31-4 ($\Phi_B = BA$) to find that expression; instead we must use Eq. 31-3 ($\Phi_B = \int \mathbf{B} \cdot d\mathbf{A}$).

In Fig. 31-9, $\mathbf{B}$ is perpendicular to the plane of the loop (hence parallel to the differential area vector $d\mathbf{A}$), so the dot product in Eq. 31-3 gives $B\, dA$. Because the magnetic field varies with the coordinate x but not with the coordinate y, we can take the differential area dA to be the area of a vertical strip of height H and width dx (as shown in Fig. 31-9). Then $dA = H\, dx$, and the flux through the loop is

$$\Phi_B = \int \mathbf{B} \cdot d\mathbf{A} = \int B\, dA = \int BH\, dx = \int 4t^2x^2H\, dx.$$

Treating t as a constant for this integration and inserting the integration limits $x = 0$ and $x = 3.0$ m, we obtain

$$\Phi_B = 4t^2H \int_0^{3.0} x^2\, dx = 4t^2H \left[\frac{x^3}{3} \right]_0^{3.0} = 72t^2,$$

where we have substituted $H = 2.0$ m and Φ_B is in webers. Now we can use Faraday's law to find the magnitude of $\mathscr{E}$ at any time t:

$$\mathscr{E} = \frac{d\Phi_B}{dt} = \frac{d(72t^2)}{dt} = 144t,$$

in which $\mathscr{E}$ is in volts. At $t = 0.10$ s,

$$\mathscr{E} = (144 \text{ V/s})(0.10 \text{ s}) \approx 14 \text{ V}. \qquad \text{(Answer)}$$

The flux of $\mathbf{B}$ through the loop is into the page in Fig. 31-9 and is increasing in magnitude because B is increasing in magnitude with time. According to Lenz's law, the field B_i of the induced current that opposes this increase is directed out of the page. The curled–straight right-hand rule of Fig. 31-5 tells us that the induced current is counterclockwise around the loop, and thus so is the induced emf $\mathscr{E}$.

31-5 INDUCTION AND ENERGY TRANSFERS

By Lenz's law, whether you move the magnet toward or away from the loop in Fig. 31-1, a force resists the motion, requiring your applied force to do positive work. At the same time, thermal energy is produced in the material of the loop because of the material's electrical resistance to the current that is induced. The energy you transfer to the closed *loop + magnet* system via your applied force ends up in this thermal energy. (For now, we neglect energy that is radiated away from the loop as electromagnetic waves during the induction.) The faster you move the magnet, the more rapidly your applied force does work, and the greater the rate of production of thermal energy in the loop.

Regardless of how current is induced in a loop, energy is always transferred to thermal energy during the process because of the electrical resistance of the loop (unless the loop is superconducting). For example, in Fig. 31-2, when switch S is closed and a current is briefly induced in the left-hand loop, energy is transferred from the battery to thermal energy in that loop.

Figure 31-10 shows another situation involving induced current. A rectangular loop of wire of width L has one end in a uniform external magnetic field that is directed perpendicularly into the plane of the loop. This field may be produced, for example, by a large electromagnet.

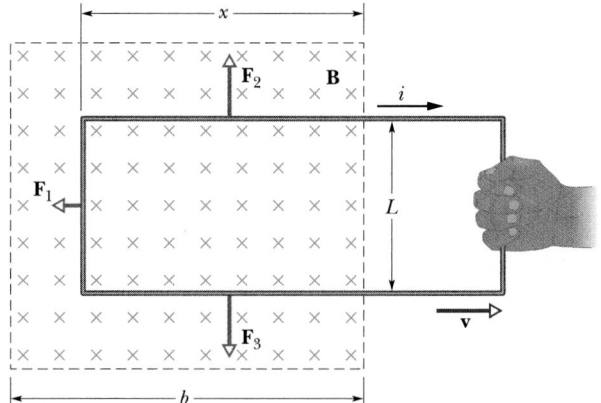

FIGURE 31-10 You pull a closed conducting loop out of a magnetic field at constant velocity **v**. While the loop is moving, a clockwise current i is induced in the loop, and the loop segments still within the magnetic field experience forces $\mathbf{F}_1$, $\mathbf{F}_2$, and $\mathbf{F}_3$.

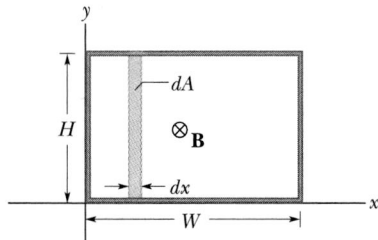

FIGURE 31-9 Sample Problem 31-3. A closed conducting loop, of width W and height H, lies in a nonuniform, varying magnetic field that points directly into the page. To apply Faraday's law, we use the vertical strip of height H, width dx, and area dA.

The dashed lines in Fig. 31-10 show the assumed limits of the magnetic field; the fringing of the field at its edges is neglected. You are asked to pull this loop to the right at a constant velocity **v**.

The situation of Fig. 31-10 does not differ in any essential way from that of Fig. 31-1. In each case a magnetic field and a conducting loop are in relative motion; in each case the flux of the field through the loop is changing with time. It is true that in Fig. 31-1 the flux is changing because **B** is changing and in Fig. 31-10 the flux is changing because the area of the loop still in the magnetic field is changing, but that difference is not important. The important difference between the two arrangements is that the arrangement of Fig. 31-10 makes calculations easier. Let us now calculate the rate at which you do mechanical work as you pull steadily on the loop in Fig. 31-10.

As you will see, if you are to pull the loop at a constant velocity **v**, you must apply a constant force **F** to the loop because a force of equal magnitude but opposite direction acts on the loop to oppose you. From Eq. 7-49, the rate at which you do work is then

$$P = Fv, \tag{31-8}$$

where F is the magnitude of your force. We wish to find an expression for P in terms of the magnitude B of the magnetic field and the characteristics of the loop, namely, its resistance R to current and its dimension L.

As you move the loop to the right in Fig. 31-10, the portion of its area within the magnetic field decreases. Thus the flux through the loop also decreases and, according to Lenz's law, a current is produced in the loop. It is the presence of this current that causes the force that opposes your pull.

To find the current, we first apply Faraday's law. When x is the length of the loop still in the magnetic field, the area of the loop still in the field is Lx. Then from Eq. 31-4, the magnitude of the flux through the loop is

$$\Phi = BA = BLx. \tag{31-9}$$

As x decreases, the flux decreases. Faraday's law tells us that with this flux decrease, an emf is induced in the loop. Dropping the minus sign in Eq. 31-6 and using Eq. 31-9, we can write the magnitude of this emf as

$$\mathcal{E} = \frac{d\Phi}{dt} = \frac{d}{dt}BLx = BL\frac{dx}{dt} = BLv, \tag{31-10}$$

in which we have replaced dx/dt with v, the speed at which the loop moves.

Figure 31-11 shows the circuit through which the charge flows: emf $\mathcal{E}$ is represented on the left, and the collective resistance R of the loop is represented on the

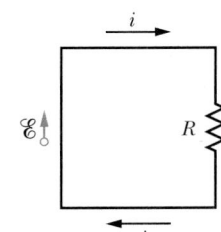

FIGURE 31-11 A circuit diagram for the loop of Fig. 31-10 while the loop is moving.

right. The direction of $\mathcal{E}$ is obtained as in Fig. 31-5b; the induced current i must have the same direction.

To find the magnitude of the induced current, we cannot apply the loop rule for potential differences in a circuit because, as you will see in Section 31-6, we cannot define a potential difference for an induced emf. However, we can apply the equation $i = \mathcal{E}/R$, as we did in Sample Problem 31-2. With Eq. 31-10, this becomes

$$i = \frac{BLv}{R}. \tag{31-11}$$

Along the three segments of the loop in Fig. 31-10 where this current is in the magnetic field, sideways deflecting forces act on the loop. From Eq. 29-26 we know that such a deflecting force is, in general notation,

$$\mathbf{F}_d = i\mathbf{L} \times \mathbf{B}. \tag{31-12}$$

In Fig. 31-10, the deflecting forces acting on the three segments of the loop are marked $\mathbf{F}_1$, $\mathbf{F}_2$, and $\mathbf{F}_3$. Note, however, that from the symmetry, $\mathbf{F}_2$ and $\mathbf{F}_3$ are equal in magnitude and cancel. This leaves only $\mathbf{F}_1$, which is directed opposite your force **F** on the loop and thus is the force that opposes you. So $\mathbf{F} = -\mathbf{F}_1$.

Using Eq. 31-12 to obtain the magnitude of $\mathbf{F}_1$ and noting that the angle between **B** and the length vector **L** for the left segment is 90°, we write

$$F = F_1 = iLB \sin 90° = iLB. \tag{31-13}$$

Substituting Eq. 31-11 for i in Eq. 31-13 then gives us

$$F = \frac{B^2L^2v}{R}. \tag{31-14}$$

Since B, L, and R are constants, the speed v at which you move the loop is constant if the magnitude F of the force you apply to the loop is also constant.

Substituting Eq. 31-14 into Eq. 31-8, we find the rate at which you do work on the loop as you pull it from the magnetic field:

$$P = Fv = \frac{B^2L^2v^2}{R} \qquad \text{(rate of doing work).} \tag{31-15}$$

To cook food on an *induction stove*, oscillating current is sent through a conducting coil that lies just below the cooking surface. The magnetic field produced by that current oscillates and induces an oscillating current in the conducting cooking pan. Because the pan has some resistance to that current, the electrical energy of the current is continuously transformed to thermal energy, resulting in a temperature increase of the pan and the food in it. The cooking surface itself might never get hot.

To complete our analysis, let us find the rate at which thermal energy appears in the loop as you pull it along at constant speed. We calculate it from Eq. 27-22,

$$P = i^2 R. \quad (31\text{-}16)$$

Substituting for i from Eq. 31-11, we find

$$P = \left(\frac{BLv}{R}\right)^2 R = \frac{B^2 L^2 v^2}{R} \quad \text{(thermal energy rate),} \quad (31\text{-}17)$$

which is exactly equal to the rate at which you are doing work on the loop (Eq. 31-15). Thus the work that you do in pulling the loop through the magnetic field appears as thermal energy in the loop, manifesting itself as a small increase in the temperature of the loop.

Eddy Currents

Suppose we replace the conducting loop of Fig. 31-10 with a solid conducting plate. If we then move the plate out of the magnetic field as we did the loop (Fig. 31-12a), the relative motion of the field and the conductor again induces a current in the conductor. Thus we again encounter an opposing force and must do work because of the induced current. With the plate, however, the conduction electrons making up the induced current do not follow one path as they do with the loop. Instead, the electrons swirl about within the plate as if they were caught in an eddy (or

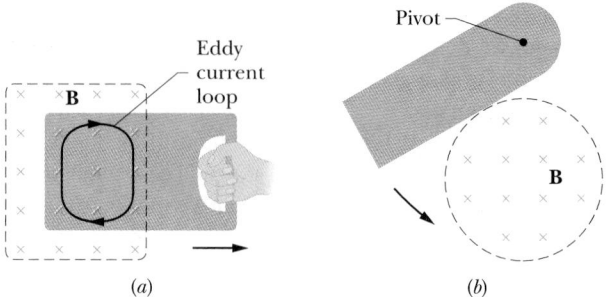

FIGURE 31-12 (*a*) As you pull a solid conducting plate out of a magnetic field, *eddy currents* are induced in the plate. A typical loop of eddy current is shown; it has the same clockwise sense of circulation as the current in the conducting loop of Fig. 31-10. (*b*) A conducting plate is allowed to swing like a pendulum about a pivot and into a region of magnetic field. As it enters and leaves the field, eddy currents are induced in the plate.

whirlpool) of water. Such a current is called an *eddy current* and can be represented as in Fig. 31-12a *as if* it followed a single path.

As with the conducting loop of Fig. 31-10, the current induced in the plate results in mechanical energy being dissipated as thermal energy. The dissipation is more apparent in the arrangement of Fig. 31-12b: a conducting plate, free to rotate about a pivot, swings down through a magnetic field like a pendulum. Each time the plate enters and leaves the field, a portion of its mechanical energy is transferred to its thermal energy. After several swings, no mechanical energy remains and the plate just hangs from its pivot.

SAMPLE PROBLEM 31-4

Figure 31-13a shows a rectangular conducting loop of resistance R, width L, and length b being pulled at constant speed v through a region of width d in which a uniform magnetic field **B** is produced by an electromagnet. Let $L = 40$ mm, $b = 10$ cm, $d = 15$ cm, $R = 1.6 \, \Omega$, $B = 2.0$ T, and $v = 1.0$ m/s.

(a) Plot the flux Φ_B through the loop as a function of the position x of the right side of the loop.

SOLUTION: The flux is zero when the loop is not in the field; it is BLb ($= 8$ mWb) when the loop is entirely in the field; it is BLx when the loop is entering the field; and then it is $BL[b - (x - d)]$ when the loop is leaving the field. These results lead to the plot of Fig. 31-13b, which you should verify.

(b) Plot the induced emf as a function of the position of the loop. Indicate the directions of the induced emf.

SOLUTION: From Eq. 31-6, the induced emf is $-d\Phi_B/dt$, which we can write as

$$\mathscr{E} = -\frac{d\Phi_B}{dt} = -\frac{d\Phi_B}{dx}\frac{dx}{dt} = -\frac{d\Phi_B}{dx}v,$$

where $d\Phi_B/dx$ is the slope of the curve of Fig. 31-13b. The emf is plotted as a function of x in Fig. 31-13c.

Lenz's law shows that when the loop is entering the field, the current and emf are counterclockwise in Fig. 31-13a; when the loop is leaving the field, the emf is clockwise in that figure. In Fig. 31-13c, a counterclockwise emf is plotted as a negative value, a clockwise emf as a positive value. There is *no* emf when the loop is either entirely out of the field or entirely in it because, in these two situations, the flux through the loop is not changing.

(c) Plot the rate of production of thermal energy in the loop as a function of the position of the loop.

SOLUTION: Substituting $i = \mathscr{E}/R$ in Eq. 31-16 gives us the rate of thermal energy production as

$$P = i^2 R = \frac{\mathscr{E}^2}{R}.$$

We can calculate P by squaring the ordinate of the curve of Fig. 31-13c and dividing by R, being careful of powers of 10

in the units. The result is plotted in Fig. 31-13d. Note that thermal energy is produced only when the loop is entering or leaving the magnetic field.

In practice, the external magnetic field **B** cannot drop sharply to zero at its boundary but must approach zero smoothly. The result would be a rounding of the corners of the curves plotted in Fig. 31-13.

CHECKPOINT **3:** The figure shows four wire loops, with edge lengths of either L or $2L$. All four loops will move through a region of uniform magnetic field **B** (directed out of the page) at the same constant velocity. Rank the four loops according to the maximum magnitude of the emf induced as they move through the field, greatest first.

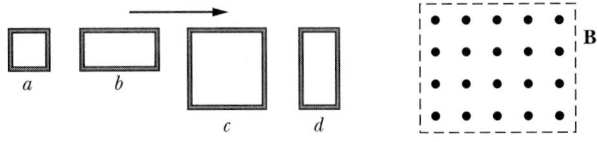

31-6 INDUCED ELECTRIC FIELDS

Let us place a copper ring of radius r in a uniform external magnetic field, as in Fig. 31-14a. The field—neglecting fringing—fills a cylindrical volume of radius R. Suppose that we increase the strength of this field at a steady rate, perhaps by increasing—in an appropriate way—the current in the windings of the electromagnet that produces the field. The magnetic flux through the ring will then change at a steady rate and—by Faraday's law—an induced emf and thus an induced current will appear in the ring. From Lenz's law we can deduce that the direction of the induced current is counterclockwise in Fig. 31-14a.

If there is a current in the copper ring, an electric field must be present along the ring; an electric field is needed to do the work of moving the conduction electrons. Moreover, the field must have been produced by the changing magnetic flux. This **induced electric field E** is just as real as an electric field produced by static charges; either field will exert a force $q_0\mathbf{E}$ on a particle of charge q_0. By this line of reasoning, we are led to a useful and informative restatement of Faraday's law of induction:

A changing magnetic field produces an electric field.

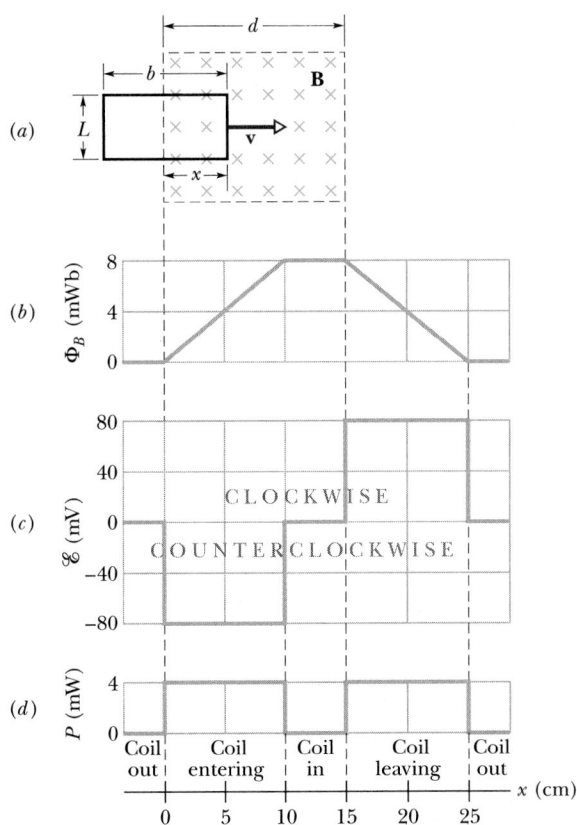

FIGURE 31-13 Sample Problem 31-4. (*a*) A closed conducting loop is pulled at constant velocity **v** completely through a magnetic field. (*b*) The flux through the loop as a function of the position x of the right side of the loop. (*c*) The induced emf as a function of x. (*d*) The rate at which thermal energy appears in the loop as a function of x.

The striking feature of this statement is that the electric field is induced even if there is no copper ring.

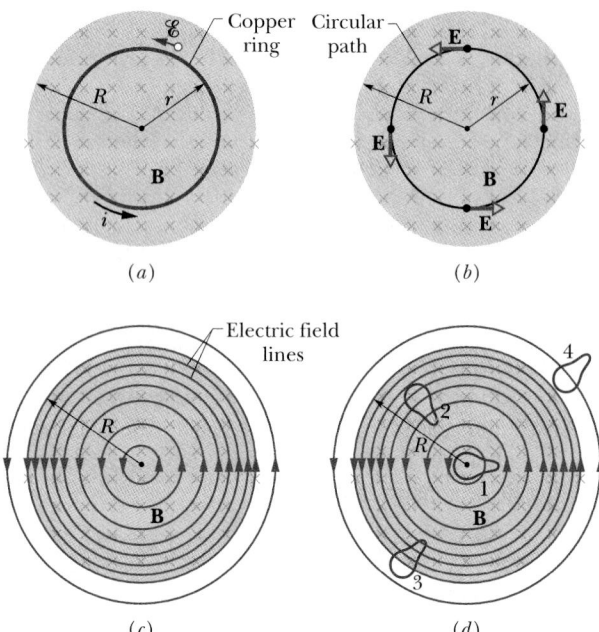

FIGURE 31-14 (a) If the magnetic field increases at a steady rate, a constant induced current appears, as shown, in the copper ring of radius r. (b) Induced electric fields appear at various points even when the ring is removed. (c) The complete picture of the induced electric fields, displayed as field lines. (d) Four similar closed paths that enclose identical areas. Equal emfs are induced around paths 1 and 2, which lie entirely within the region of changing magnetic field. A smaller emf is induced around path 3, which only partially lies in that region. No emf is induced around path 4, which lies entirely outside the magnetic field.

To fix these ideas, consider Fig. 31-14b, which is just like Fig. 31-14a except the copper ring has been replaced by a hypothetical circular path of radius r. We assume, as previously, that the magnetic field **B** is increasing in magnitude at a constant rate dB/dt. The electric field induced at various points around the circular path must—from the symmetry—be tangent to the circle, as Fig. 31-14b shows.* Hence the circular path is also an electric field line. There is nothing special about the circle of radius r, so the electric field lines produced by the changing magnetic field must be a set of concentric circles, as in Fig. 31-14c.

As long as the magnetic field is increasing with time, the electric field represented by the circular field lines in Fig. 31-14c will be present. If the magnetic field remains constant with time, there will be no induced electric field and thus no electric field lines. If the magnetic field is

*Arguments of symmetry would also permit the lines of **E** around the circular path to be *radial*, rather than tangential. However, such radial lines would imply that there are free charges, distributed symmetrically about the axis of symmetry, on which the electric field lines could begin or end; there are no such charges.

decreasing with time (at a constant rate), the electric field lines will still be concentric circles as in Fig. 31-14c, but they will now have the opposite direction. All this is what we have in mind when we say: ''A changing magnetic field produces an electric field.''

A Reformulation of Faraday's Law

Consider a particle of charge q_0 moving around the circular path of Fig. 31-14b. The work W done on it in one revolution by the induced electric field is $\mathscr{E}q_0$, where $\mathscr{E}$ is the induced emf, that is, the work done per unit charge in moving the test charge around the path. From another point of view, the work is

$$\int \mathbf{F} \cdot d\mathbf{s} = (q_0 E)(2\pi r), \quad (31\text{-}18)$$

where $q_0 E$ is the magnitude of the force acting on the test charge and $2\pi r$ is the distance over which that force acts. Setting these two expressions for W equal to each other and canceling q_0, we find that

$$\mathscr{E} = 2\pi r E. \quad (31\text{-}19)$$

More generally, we can rewrite Eq. 31-18 to give the work done on a particle of charge q_0 moving along any closed path:

$$W = \oint \mathbf{F} \cdot d\mathbf{s} = q_0 \oint E \cdot d\mathbf{s}. \quad (31\text{-}20)$$

(The circle indicates that the integral is to be taken around the closed path.) Substituting $\mathscr{E}q_0$ for W, we find that

$$\mathscr{E} = \oint \mathbf{E} \cdot d\mathbf{s}. \quad (31\text{-}21)$$

This integral reduces at once to Eq. 31-19 if we evaluate it for the special case of Fig. 31-14b.

With Eq. 31-21, we can expand the meaning of induced emf. Previously, induced emf has meant the work per unit charge done in maintaining current due to a changing magnetic flux. Or it has meant the work done per unit charge on a charged particle that moves around a closed path in a changing magnetic flux. But with Eq. 31-21, we no longer actually need a current or a particle to speak of induced emf. An induced emf is the sum—via integration—of quantities $\mathbf{E} \cdot d\mathbf{s}$ around a closed path, where **E** is the electric field induced by a changing magnetic flux and $d\mathbf{s}$ is a differential length vector along the closed path.

If we combine Eq. 31-21 with Faraday's law in Eq. 31-6 ($\mathscr{E} = -d\Phi_B/dt$), we can rewrite Faraday's law as

$$\oint \mathbf{E} \cdot d\mathbf{s} = -\frac{d\Phi_B}{dt} \quad \text{(Faraday's law).} \quad (31\text{-}22)$$

This equation says simply that a changing magnetic field induces an electric field. The changing magnetic field appears on the right side of this equation, the electric field on the left.

Faraday's law in the form of Eq. 31-22 can be applied to *any* closed path that can be drawn in a changing magnetic field. Figure 31-14*d*, for example, shows four such paths, all having the same shape and area but located in different positions in the changing field. For paths 1 and 2, the induced emfs $\mathscr{E}$ ($= \oint \mathbf{E} \cdot d\mathbf{s}$) are equal because these paths lie entirely in the magnetic field and thus have the same value of $d\Phi_B/dt$. This is true even though the electric field vectors around these paths are distributed differently, as indicated by the patterns of electric field lines. For path 3 the induced emf is smaller because the enclosed flux Φ_B (hence $d\Phi_B/dt$) is smaller, and for path 4 the induced emf is zero, even though the electric field is not zero at any point on the path.

A New Look at Electric Potential

Induced electric fields are produced not by static charges but by a changing magnetic flux. Although electric fields produced in either way exert forces on charged particles, there is an important difference between them. The simplest evidence of this difference is that the field lines of induced electric fields form closed loops, as in Fig. 31-14*c*. Field lines produced by static charges never do so but must start on positive charges and end on negative charges.

In a more formal sense, we can state the difference between electric fields produced by induction and those produced by static charges in these words:

> Electric potential has meaning only for electric fields that are produced by static charges; it has no meaning for electric fields that are produced by induction.

You can understand this statement qualitatively by considering what happens to a charged particle that makes a single journey around the circular path in Fig. 31-14*b*. It starts at a certain point and, upon its return to that same point, has experienced an emf $\mathscr{E}$ of, let us say, 5 V. Its potential should have increased by this amount. This is impossible, however, because otherwise the same point in space would have two different values of potential. We must conclude that potential has no meaning for electric fields that are set up by changing magnetic fields.

We can take a more formal look by recalling Eq. 25-18, which defines the potential difference between two points i and f:

$$V_f - V_i = -\int_i^f \mathbf{E} \cdot d\mathbf{s}. \qquad (31\text{-}23)$$

In Chapter 25 we had not yet encountered Faraday's law of induction, so the electric fields involved in the derivation of Eq. 25-18 were those due to static charges. If i and f in Eq. 31-23 are the same point, the path connecting them is a closed loop, V_i and V_f are identical, and Eq. 31-23 reduces to

$$\oint \mathbf{E} \cdot d\mathbf{s} = 0. \qquad (31\text{-}24)$$

However, when a changing magnetic flux is present, this integral is *not* zero but is $-d\Phi_B/dt$, as Eq. 31-22 asserts. Again, we conclude that electric potential has no meaning for electric fields associated with induction.

SAMPLE PROBLEM 31-5

In Fig. 31-14*b*, take $R = 8.5$ cm and $dB/dt = 0.13$ T/s.

(a) Find an expression for the magnitude E of the induced electric field at points within the magnetic field, at radius r from the center of the magnetic field. Evaluate the expression for $r = 5.2$ cm.

SOLUTION: We can find an expression for E at radius r by applying Faraday's law in the form of Eq. 31-22, with the closed integration path a circle of radius r (as in Fig. 31-14*b*). We have assumed from the symmetry that $\mathbf{E}$ in Fig. 31-14*b* is tangent to the circular path at all points. The path vector $d\mathbf{s}$ is also always tangent to the circular path, so the dot product $\mathbf{E} \cdot d\mathbf{s}$ in Eq. 31-22 must have the magnitude $E \, ds$ at all points on the path. We can also assume from the symmetry that E has the same value at all points along the circular path. Then, dropping the minus sign, Eq. 31-22 gives us

$$\oint \mathbf{E} \cdot d\mathbf{s} = \oint E \, ds = E \oint ds = E(2\pi r) = \frac{d\Phi_B}{dt}. \qquad (31\text{-}25)$$

From Eq. 31-4, the magnetic flux through the circular path of integration is

$$\Phi_B = BA = B(\pi r^2). \qquad (31\text{-}26)$$

Substituting this into Eq. 31-25, we find that

$$E(2\pi r) = (\pi r^2)\frac{dB}{dt},$$

or
$$E = \frac{r}{2}\frac{dB}{dt}. \qquad \text{(Answer)} \qquad (31\text{-}27)$$

Equation 31-27 gives the magnitude of the electric field at any point for which $r < R$ (that is, within the magnetic field). Substituting given values yields, for the magnitude of $\mathbf{E}$ at $r = 5.2$ cm,

$$E = \frac{(5.2 \times 10^{-2}\ \text{m})}{2}(0.13\ \text{T/s})$$

$$= 0.0034\ \text{V/m} = 3.4\ \text{mV/m}. \qquad \text{(Answer)}$$

(b) Find an expression for the magnitude E of the induced electric field at points that are outside the magnetic field, at radius r. Evaluate the expression for $r = 12.5$ cm.

SOLUTION: Proceeding as in (a), we again obtain Eq. 31-25. However, we do not then obtain Eq. 31-26, because the closed path is now outside the magnetic field. The magnetic flux encircled by the closed path is now the area πR^2 of the magnetic field region. So,

$$\Phi_B = BA = B(\pi R^2). \quad (31\text{-}28)$$

Substituting this into Eq. 31-25 and solving for E yield

$$E = \frac{R^2}{2r}\frac{dB}{dt}. \quad \text{(Answer)} \quad (31\text{-}29)$$

Since E is not zero here, we know that an electric field is induced even at points that are outside the changing magnetic field, an important result that (as you shall see in Section 33-11) makes transformers possible. With the given data, Eq. 31-29 yields the magnitude of **E** at $r = 12.5$ cm:

$$E = \frac{(8.5 \times 10^{-2}\ \text{m})^2}{(2)(12.5 \times 10^{-2}\ \text{m})}(0.13\ \text{T/s})$$

$$= 3.8 \times 10^{-3}\ \text{V/m} = 3.8\ \text{mV/m}. \quad \text{(Answer)}$$

Equations 31-27 and 31-29 give the same result, as they must, for $r = R$. Figure 31-15 shows a plot of $E(r)$ based on these two equations.

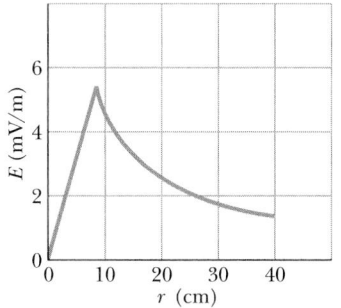

FIGURE 31-15 A plot of the induced electric field $E(r)$ for the conditions of Sample Problem 31-5.

CHECKPOINT **4:** The figure shows five lettered regions in which a uniform magnetic field extends either directly out of the page (as in region a) or into the page. The field is increasing in magnitude at the same steady rate in all five regions; the regions are identical in area. Also shown are four numbered paths along which $\oint \mathbf{E} \cdot d\mathbf{s}$ has the magnitudes given below in terms of a quantity mag. Determine whether the magnetic fields in regions b through e are directed into or out of the page.

Path:	1	2	3	4
$\oint \mathbf{E} \cdot d\mathbf{s}$:	mag	2(mag)	3(mag)	0

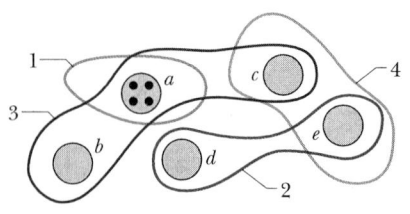

31-7 INDUCTORS AND INDUCTANCE

We found in Chapter 26 that a capacitor can be used to produce a desired electric field. We considered the parallel-plate arrangement as a basic type of capacitor. Similarly, an **inductor** (symbol ⎰⎰⎰⎰) can be used to produce a desired magnetic field. We shall consider a long solenoid (more specifically, a short length near the middle of a long solenoid) as our basic type of inductor.

If we establish a current i in the windings (or turns) of an inductor (a solenoid), the current produces a magnetic flux Φ through the central region of the inductor. The **inductance** of the inductor is then

$$L = \frac{N\Phi}{i} \quad \text{(inductance defined)}, \quad (31\text{-}30)$$

in which N is the number of turns. The windings of the inductor are said to be *linked* by the shared flux, and the product $N\Phi$ is called the *magnetic flux linkage*. The inductance **L** is thus a measure of the flux linkage produced by the inductor per unit of current.

Because the SI unit of magnetic flux is the tesla–square meter, the SI unit of inductance is the tesla–square meter per ampere (T·m²/A). We call this the **henry** (H), after American physicist Joseph Henry, the codiscoverer of the law of induction and a contemporary of Faraday. Thus

$$1\ \text{henry} = 1\ \text{H} = 1\ \text{T·m}^2/\text{A}. \quad (31\text{-}31)$$

Through the rest of this chapter we assume that all inductors, no matter what their geometric arrangement, have no magnetic materials such as iron in their vicinity. Such materials would distort the magnetic field of an inductor.

Inductance of a Solenoid

Consider a long solenoid of cross-sectional area A. What is the inductance per unit length near its middle?

To use the defining equation for inductance (Eq. 31-30), we must calculate the flux linkage set up by a given current in the solenoid windings. Consider a length l near the middle of this solenoid. The flux linkage for this section of the solenoid is

$$N\Phi = (nl)(BA)$$

The crude inductors with which Michael Faraday discovered the law of induction. In those days amenities such as insulated wire were not commercially available. It is said that Faraday insulated his wires by wrapping them with strips from one of his wife's petticoats.

in which n is the number of turns per unit length of the solenoid and B is the magnetic field within the solenoid.

The magnitude B is given by Eq. 30-25,

$$B = \mu_0 in,$$

so from Eq. 31-30,

$$L = \frac{N\Phi}{i} = \frac{(nl)(BA)}{i} = \frac{(nl)(\mu_0 in)(A)}{i}$$

$$= \mu_0 n^2 lA. \qquad (31\text{-}32)$$

Thus the inductance per unit length for a long solenoid near its center is

$$\frac{L}{l} = \mu_0 n^2 A \qquad \text{(solenoid).} \qquad (31\text{-}33)$$

Inductance—like capacitance—depends only on the geometry of the device. The dependence on the square of the number of turns per unit length is to be expected. If you, say, triple n, you triple not only the number of turns (N) but you also triple the flux ($\Phi = BA = \mu_0 inA$) through each turn, multiplying the flux linkage $N\Phi$ and thus the inductance L by a factor of 9.

If the solenoid is very much longer than its radius, then Eq. 31-32 gives its inductance to a good approximation. This approximation neglects the spreading of the magnetic field lines near the ends of the solenoid, just as the parallel-plate capacitor formula ($C = \epsilon_0 A/d$) neglects the fringing of the electric field lines near the edges of the capacitor plates.

From Eq. 31-32, and recalling that n is a number per unit length, we can see that an inductance can be written as a product of the permeability constant μ_0 and a quantity with the dimensions of a length. This means that μ_0 can be expressed in the unit henry per meter:

$$\mu_0 = 4\pi \times 10^{-7}\ \text{T} \cdot \text{m/A}$$

$$= 4\pi \times 10^{-7}\ \text{H/m.} \qquad (31\text{-}34)$$

SAMPLE PROBLEM 31-6

Figure 31-16 shows a cross section, in the plane of the page, of a toroid of N turns like that in Fig. 30-21a but of rectangular cross section; its dimensions are as indicated.

(a) What is its inductance L?

SOLUTION: To use the definition of inductance, Eq. 31-30, we need to find the magnetic flux Φ due to a current i through the toroid. From Eq. 30-26, we already know the magnitude B of the magnetic field within the toroid (also due to i):

$$B = \frac{\mu_0 iN}{2\pi r}, \qquad (31\text{-}35)$$

where r is the distance from the center of the toroid. This equation holds regardless of the shape or dimensions of the toroid's cross section.

Because B is *not* uniform over the cross section, we cannot use Eq. 31-4 ($\Phi_B = BA$) to find the flux Φ, but instead must use Eq. 31-3,

$$\Phi = \int \mathbf{B} \cdot d\mathbf{A}. \qquad (31\text{-}36)$$

The field $\mathbf{B}$ is everywhere perpendicular to the cross section, as shown in Fig. 31-16; $\mathbf{B}$ is thus parallel to the differential cross-sectional area vector $d\mathbf{A}$, so the dot product in Eq. 31-36 gives $B\,dA$. For the differential area dA, we can use the area $h\,dr$ of the strip shown in Fig. 31-16. Substituting these quantities and Eq. 31-35 into Eq. 31-36 and integrating from $r = a$ to $r = b$ yield

$$\Phi = \int_a^b Bh\,dr = \int_a^b \frac{\mu_0 iN}{2\pi r}\,h\,dr$$

$$= \frac{\mu_0 iNh}{2\pi} \int_a^b \frac{dr}{r} = \frac{\mu_0 iNh}{2\pi} \ln \frac{b}{a}.$$

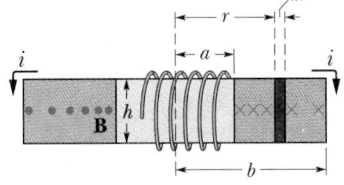

FIGURE 31-16 Sample Problem 31-6. A cross section of a toroid, showing the current in the windings and the associated magnetic field. See Fig. 30-21. The nonuniform magnetic field within the toroid is represented by nonuniformly spaced dots and $\times$s.

Equation 31-30 then gives us

$$L = \frac{N\Phi}{i} = \frac{N}{i} \frac{\mu_0 i N h}{2\pi} \ln \frac{b}{a},$$

so

$$L = \frac{\mu_0 N^2 h}{2\pi} \ln \frac{b}{a}. \quad \text{(Answer)} \quad (31\text{-}37)$$

(b) The toroid shown in Fig. 31-16 has $N = 1250$ turns, $a = 52$ mm, $b = 95$ mm, and $h = 13$ mm. What is its inductance?

SOLUTION: From Eq. 31-37

$$L = \frac{\mu_0 N^2 h}{2\pi} \ln \frac{b}{a}$$

$$= \frac{(4\pi \times 10^{-7} \text{ H/m})(1250)^2(13 \times 10^{-3} \text{ m})}{2\pi} \ln \frac{95 \text{ mm}}{52 \text{ mm}}$$

$$= 2.45 \times 10^{-3} \text{ H} \approx 2.5 \text{ mH}. \quad \text{(Answer)}$$

31-8 SELF-INDUCTION

If two coils—which we can now call inductors—are near each other, a current i in one coil produces a magnetic flux Φ through the second coil. We have seen that if we change this flux by changing the current, an induced emf appears in the second coil according to Faraday's law. An induced emf appears in the first coil as well.

An induced emf $\mathcal{E}_L$ appears in any coil in which the current is changing.

This process (see Fig. 31-17) is called **self-induction,** and the emf that appears is called a **self-induced emf.** It obeys Faraday's law of induction just as other induced emfs do.

For any inductor, Eq. 31-30 tells us that

$$N\Phi = Li. \quad (31\text{-}38)$$

Faraday's law tell us that

$$\mathcal{E}_L = -\frac{d(N\Phi)}{dt}. \quad (31\text{-}39)$$

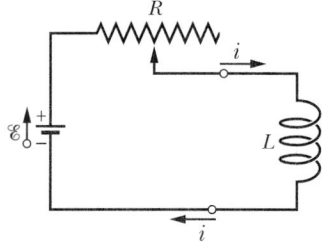

FIGURE 31-17 If the current in coil L is changed by varying the contact position on resistor R, a self-induced emf $\mathcal{E}_L$ will appear in the coil *while the current is changing.*

By combining Eqs. 31-38 and 31-39 we can write:

$$\mathcal{E}_L = -L\frac{di}{dt} \quad \text{(self-induced emf).} \quad (31\text{-}40)$$

Thus in any inductor (such as a coil, a solenoid, or a toroid) a self-induced emf appears whenever the current changes with time. The magnitude of the current has no influence on the magnitude of the induced emf; only the rate of change of the current counts.

You can find the *direction* of a self-induced emf from Lenz's law. The minus sign in Eq. 31-40 indicates that—as the law states—the self-induced emf acts to oppose the change that brings it about.

Suppose that, as in Fig. 31-18a, you set up a current i in a coil and arrange to have it increase with time at a rate di/dt. In the language of Lenz' law, this increase in the current is the "change" that the self-induction must oppose. For such opposition to occur, a self-induced emf must appear in the coil, pointing—as the figure shows—so as to oppose the increase in the current. If you cause the current to decrease with time, as in Fig. 31-18b, the self-induced emf must point in a direction that tends to oppose the decrease in the current, as the figure shows.

In Section 31-6 we saw that we cannot define an electric potential for an electric field (and thus for an emf) that is induced by a changing magnetic flux. This means that when a self-induced emf is produced in the inductor of Fig. 31-17, we cannot define an electric potential within the inductor itself, where the flux is changing. However, potentials can still be defined at points of the circuit outside this region, where the electric fields are due to charge distributions with associated electric potentials.

Moreover, we can define a potential difference V_L *across an inductor* (between its terminals, which we assume to be outside the region of changing flux). If the

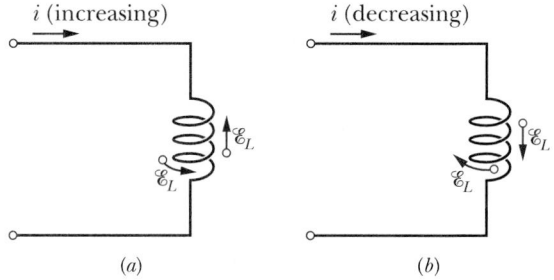

FIGURE 31-18 (a) The current i is increasing and the self-induced emf $\mathcal{E}_L$ appears along the coil in a direction such that it opposes the increase. The arrow representing $\mathcal{E}_L$ can be drawn along a turn of the coil or alongside the coil. Both are shown. (b) The current i is decreasing and the self-induced emf appears in a direction such that it opposes the decrease.

inductor is an *ideal inductor* (its wire has negligible resistance), the magnitude of V_L is equal to the magnitude of the self-induced emf $\mathscr{E}_L$.

If, instead, the wire in the inductor has resistance r, we mentally separate the inductor into a resistance r (which we take to be outside the region of changing flux) and an ideal inductor of self-induced emf $\mathscr{E}_L$. As with a real battery of emf $\mathscr{E}$ and internal resistance r, the potential difference across the terminals of a real inductor then differs from the emf. Unless otherwise indicated, we assume here that inductors are ideal.

CHECKPOINT 5: The figure shows an emf $\mathscr{E}_L$ induced in a coil. Which of the following can describe the current through the coil: (a) constant and rightward, (b) constant and leftward, (c) increasing and rightward, (d) decreasing and rightward, (e) increasing and leftward, (f) decreasing and leftward?

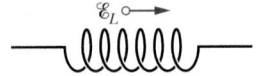

31-9 *RL* CIRCUITS

In Section 28-8 we saw that if you suddenly introduce an emf $\mathscr{E}$ into a single-loop circuit containing a resistor R and a capacitor C, the charge on the capacitor does not build up immediately to its final equilibrium value $C\mathscr{E}$ but approaches it in an exponential fashion:

$$q = C\mathscr{E}(1 - e^{-t/\tau_C}). \tag{31-41}$$

The rate at which the charge builds up is determined by the capacitive time constant τ_C, defined in Eq. 28-33 as

$$\tau_C = RC. \tag{31-42}$$

If you suddenly remove the emf from this same circuit, the charge does not immediately fall to zero but approaches zero in an exponential fashion:

$$q = q_0 e^{-t/\tau_C}. \tag{31-43}$$

The time constant τ_C describes the fall of the charge as well as its rise.

An analogous slowing of the rise (or fall) of the current occurs if we introduce an emf $\mathscr{E}$ into (or remove it from) a single-loop circuit containing a resistor R and an inductor L. When the switch S in Fig. 31-19 is closed on a, for example, the current in the resistor starts to rise. If the inductor were not present, the current would rise rapidly to a steady value $\mathscr{E}/R$. Because of the inductor, however, a self-induced emf $\mathscr{E}_L$ appears in the circuit; from Lenz's law, this emf opposes the rise of the current, which means

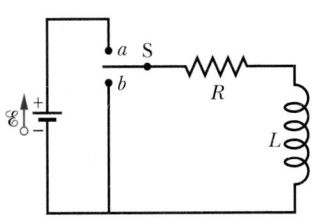

FIGURE 31-19 An *RL* circuit. When switch S is closed on a, the current rises and approaches a limiting value of $\mathscr{E}/R$.

that it opposes the battery emf $\mathscr{E}$ in polarity. Thus the resistor responds to the difference between two emfs, a constant one $\mathscr{E}$ due to the battery and a variable one $\mathscr{E}_L$ ($= -L\, di/dt$) due to self-induction. As long as $\mathscr{E}_L$ is present, the current in the resistor will be less than $\mathscr{E}/R$.

As time goes on, the rate at which the current increases becomes less rapid and the magnitude of the self-induced emf, which is proportional to di/dt, becomes smaller. Thus the current in the circuit approaches $\mathscr{E}/R$ asymptotically.

We can generalize these results as follows:

Initially, an inductor acts to oppose changes in the current through it. A long time later, it acts like ordinary connecting wire.

Now let us analyze the situation quantitatively. With the switch S in Fig. 31-19 thrown to a, the circuit is equivalent to that of Fig. 31-20. Let us apply the loop rule, starting at x in this figure and moving clockwise around the loop. For the current direction shown, x will be higher in potential than y, which means that we encounter a potential change of $-iR$ as we traverse the resistor. Point y is higher in potential than point z because, for an increasing current, the self-induced emf will oppose the rise of the current by pointing as shown. Thus, as we traverse the inductor from y to z, we encounter a potential change of $\mathscr{E}_L = -L\, di/dt$. We encounter a rise in potential of $+\mathscr{E}$ in traversing the battery from z to x. The loop rule thus gives

$$-iR - L\frac{di}{dt} + \mathscr{E} = 0$$

or

$$L\frac{di}{dt} + Ri = \mathscr{E} \qquad (RL \text{ circuit}). \tag{31-44}$$

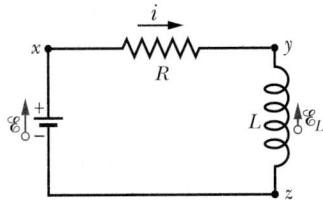

FIGURE 31-20 The circuit of Fig. 31-19 with the switch closed on a. We apply the loop rule clockwise, starting at x.

Equation 31-44 is a differential equation involving the variable i and its first derivative di/dt. We seek the function $i(t)$ such that when it and its first derivative are substituted in Eq. 31-44, the equation is satisfied and the initial condition $i(0) = 0$ is satisfied.

Equation 31-44 and its initial condition are of exactly the form of Eq. 28-29 for an RC circuit, with i replacing q, L replacing R, and R replacing $1/C$. The solution of Eq. 31-44 must then be of exactly the form of Eq. 28-30 with the same replacements. That solution is

$$i = \frac{\mathscr{E}}{R}(1 - e^{-Rt/L}), \tag{31-45}$$

which we can rewrite as

$$i = \frac{\mathscr{E}}{R}(1 - e^{-t/\tau_L}) \qquad \text{(rise of current).} \tag{31-46}$$

Here τ_L, the **inductive time constant,** is given by

$$\tau_L = \frac{L}{R} \qquad \text{(time constant).} \tag{31-47}$$

Figure 31-21 shows how the potential differences V_R ($= iR$) across the resistor and V_L ($= L\,di/dt$) across the inductor vary with time for particular values of $\mathscr{E}$, L, and R. Compare this figure carefully with the corresponding figure for an RC circuit (Fig. 28-14).

To show that the quantity τ_L ($= L/R$) has the dimensions of time, we put

$$1\,\frac{\text{H}}{\Omega} = 1\,\frac{\text{H}}{\Omega}\left(\frac{1\,\text{V}\cdot\text{s}}{1\,\text{H}\cdot\text{A}}\right)\left(\frac{1\,\Omega\cdot\text{A}}{1\,\text{V}}\right) = 1\,\text{s}.$$

The first quantity in parentheses is a conversion factor based on Eq. 31-40, and the second one is a conversion factor based on the relation $V = iR$.

The physical significance of the time constant follows from Eq. 31-46. If we put $t = \tau_L = L/R$ in this equation, it reduces to

$$i = \frac{\mathscr{E}}{R}(1 - e^{-1}) = 0.63\,\frac{\mathscr{E}}{R}.$$

Thus the time constant τ_L is the time it takes the current in the circuit to reach within $1/e$ (about 37%) of its final equilibrium value $\mathscr{E}/R$. Since the potential difference V_R across the resistor is proportional to the current i, the time dependence of the increasing current has the same shape as V_R, as plotted in Fig. 31-21a.

If the switch S in Fig. 31-19, having been closed on a long enough for the equilibrium current $\mathscr{E}/R$ to be established, is thrown to b, the effect will be to remove the battery from the circuit. (The connection to b must actually be made before the connection to a is broken. A switch that does this is called a *make-before-break* switch.)

The current through the resistor cannot drop immediately to zero but must decay to zero over time. The differential equation that governs the decay can be found by putting $\mathscr{E} = 0$ in Eq. 31-44:

$$L\frac{di}{dt} + iR = 0. \tag{31-48}$$

By analogy with Eqs. 28-35 and 28-36, the solution of this differential equation that satisfies the initial condition $i(0) = i_0 = \mathscr{E}/R$ is

$$i = \frac{\mathscr{E}}{R}e^{-t/\tau_L} = i_0 e^{-t/\tau_L} \qquad \begin{matrix}\text{(decay of}\\\text{current).}\end{matrix} \tag{31-49}$$

We see that both current rise (Eq. 31-46) and current decay (Eq. 31-49) in an RL circuit are governed by the same inductive time constant, τ_L.

We have used i_0 in Eq. 31-49 to represent the current at time $t = 0$. In our case that happened to be $\mathscr{E}/R$, but it could be any other initial value.

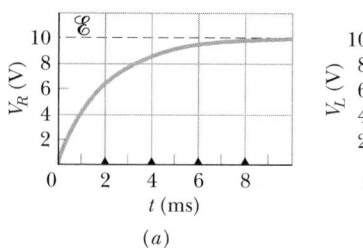
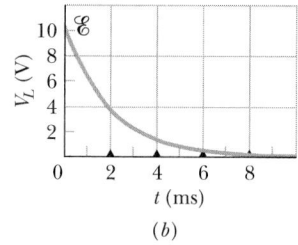

FIGURE 31-21   The variation with time of (a) V_R, the potential difference across the resistor in the circuit of Fig. 31-20, and (b) V_L, the potential difference across the inductor in that circuit. The small triangles represent successive intervals of one inductive time constant $\tau_L = L/R$. The figure is plotted for $R = 2000\,\Omega$, $L = 4.0$ H, and $\mathscr{E} = 10$ V.

SAMPLE PROBLEM 31-7

Figure 31-22a shows a circuit that contains three identical resistors with resistance $R = 9.0\,\Omega$, two identical inductors with inductance $L = 2.0$ mH, and an ideal battery with emf $\mathscr{E} = 18$ V.

(a) What is the current i through the battery just after the switch is closed?

SOLUTION: Because the current through each inductor is zero before the switch is closed, it will also be zero just afterward. So, immediately after the switch is closed, the inductors act as broken wires, as indicated in Fig. 31-22b. We then have a single-loop circuit for which the loop rule gives us

$$\mathscr{E} - iR = 0.$$

Substituting given data, we find that

$$i = \frac{\mathscr{E}}{R} = \frac{18 \text{ V}}{9.0 \ \Omega} = 2.0 \text{ A.} \qquad \text{(Answer)}$$

(b) What is the current i through the battery long after the switch has been closed?

SOLUTION: Long after the switch has been closed, the currents in the circuit have reached their equilibrium values. Then the inductors act as simple connecting wires, as indicated in Fig. 31-22c. We then have a circuit with three identical resistors in parallel; from Eq. 28-20, their equivalent resistance is $R_{\text{eq}} = R/3 = (9.0 \ \Omega)/3 = 3.0 \ \Omega$. The equivalent circuit in Fig. 31-22d then yields the loop equation $\mathscr{E} - iR_{\text{eq}} = 0$, or

$$i = \frac{\mathscr{E}}{R_{\text{eq}}} = \frac{18 \text{ V}}{3.0 \ \Omega} = 6.0 \text{ A.} \qquad \text{(Answer)}$$

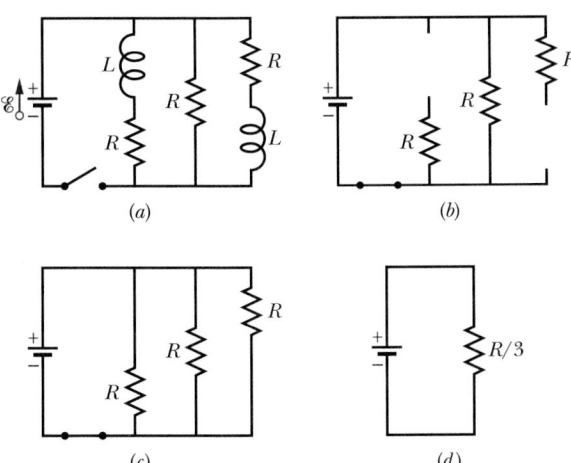

FIGURE 31-22 Sample Problem 31-7. (a) A multiloop RL circuit with an open switch. (b) The equivalent circuit just after the switch has been closed. (c) The equivalent circuit a long time later. (d) The single-loop circuit that is equivalent to circuit (c).

SAMPLE PROBLEM 31-8

A solenoid has an inductance of 53 mH and a resistance of $0.37 \ \Omega$. If it is connected to a battery, how long will the current take to reach half its final equilibrium value?

SOLUTION: The equilibrium value of the current is reached as $t \rightarrow \infty$; from Eq. 31-46 that value is $\mathscr{E}/R$. If the current has half this value at a particular time t_0, this equation becomes

$$\frac{1}{2} \frac{\mathscr{E}}{R} = \frac{\mathscr{E}}{R} (1 - e^{-t_0/\tau_L}).$$

We solve for t_0 by canceling $\mathscr{E}/R$, isolating the exponential, and taking the natural logarithm of each side. We find

$$\begin{aligned} t_0 &= \tau_L \ln 2 \\ &= \frac{L}{R} \ln 2 = \frac{53 \times 10^{-3} \text{ H}}{0.37 \ \Omega} \ln 2 \\ &= 0.10 \text{ s.} \qquad \text{(Answer)} \end{aligned}$$

CHECKPOINT **6:** The figure shows three circuits with identical batteries, inductors, and resistors. Rank the circuits according to the current through the battery (a) just after the switch is closed and (b) a long time later, greatest first.

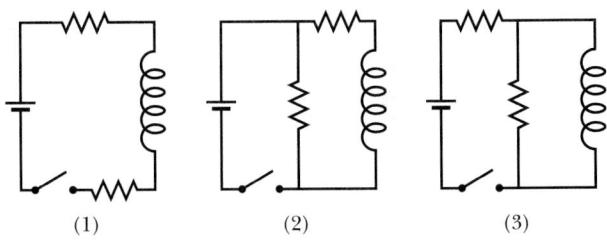

31-10 ENERGY STORED IN A MAGNETIC FIELD

When we pull two unlike charges apart, we say that the resulting electric potential energy is stored in the electric field of the charges. We get it back from the field by letting the charges move closer together again.

In the same way we can consider energy to be stored in a magnetic field. For example, two long, rigid, parallel wires carrying current in the same direction attract each other, and we must do work to pull them apart. In doing so, we store energy in the magnetic fields of the currents. We can get this stored energy back at any time by letting the wires move back to their original positions.

To derive a quantitative expression for the energy stored in a magnetic field, consider again Fig. 31-20, which shows a source of emf $\mathscr{E}$ connected to a resistor R and an inductor L. Equation 31-44, restated here for convenience,

$$\mathscr{E} = L\frac{di}{dt} + iR, \qquad (31\text{-}50)$$

is the differential equation that describes the growth of current in this circuit. We stress that this equation follows immediately from the loop rule and that the loop rule in turn is an expression of the principle of conservation of energy for single-loop circuits. If we multiply each side of Eq. 31-50 by i, we obtain

$$\mathscr{E}i = Li\frac{di}{dt} + i^2R, \qquad (31\text{-}51)$$

which has the following physical interpretation in terms of work and energy:

1. If a charge dq passes through the battery of emf $\mathcal{E}$ in Fig. 31-20 in time dt, the battery does work on it in the amount $\mathcal{E}\,dq$. The rate at which the battery does work is $(\mathcal{E}\,dq)/dt$, or $\mathcal{E}i$. Thus the left side of Eq. 31-51 represents the rate at which the emf device delivers energy to the rest of the circuit.

2. The right-most term in Eq. 31-51 represents the rate at which energy appears as thermal energy in the resistor.

3. Energy that does not appear as thermal energy must, by our conservation-of-energy hypothesis, be stored in the magnetic field of the inductor. Since Eq. 31-51 represents a statement of the conservation of energy for RL circuits, the middle term must represent the rate dU_B/dt at which energy is stored in the magnetic field; so

$$\frac{dU_B}{dt} = Li\frac{di}{dt}. \tag{31-52}$$

We can write this as

$$dU_B = Li\,di.$$

Integrating yields

$$\int_0^{U_B} dU_B = \int_0^i Li\,di$$

or

$$U_B = \tfrac{1}{2}Li^2 \quad \text{(magnetic energy),} \tag{31-53}$$

which represents the total energy stored by an inductor L carrying a current i. Note the similarity between this expression and the expression for the energy stored by a capacitor with capacitance C and charge q, namely,

$$U_E = \frac{q^2}{2C}. \tag{31-54}$$

SAMPLE PROBLEM 31-9

A coil has an inductance of 53 mH and a resistance of 0.35 Ω.

(a) If a 12 V emf is applied across the coil, how much energy is stored in the magnetic field after the current has built up to its equilibrium value?

SOLUTION: The stored energy is given by Eq. 31-53,

$$U_B = \tfrac{1}{2}Li^2.$$

To find the equilibrium stored energy, we must substitute the equilibrium current in this expression. From Eq. 31-46 the equilibrium current is

$$i_\infty = \frac{\mathcal{E}}{R} = \frac{12\text{ V}}{0.35\ \Omega} = 34.3\text{ A}.$$

The substitution yields

$$U_{B\infty} = \tfrac{1}{2}Li_\infty^2 = (\tfrac{1}{2})(53 \times 10^{-3}\text{ H})(34.3\text{ A})^2$$
$$= 31\text{ J}. \qquad \text{(Answer)}$$

(b) After how many times constants will half this equilibrium energy be stored in the magnetic field?

SOLUTION: We are asked: At what time t will the relation

$$U_B = \tfrac{1}{2}U_{B\infty}$$

be satisfied? Equation 31-53 allows us to rewrite this condition as

$$\tfrac{1}{2}Li^2 = (\tfrac{1}{2})\tfrac{1}{2}Li_\infty^2$$

or

$$i = \left(\frac{1}{\sqrt{2}}\right)i_\infty.$$

But i is given by Eq. 31-46 and i_∞ (see above) is $\mathcal{E}/R$, so this equation becomes

$$\frac{\mathcal{E}}{R}(1 - e^{-t/\tau_L}) = \frac{\mathcal{E}}{\sqrt{2}R}.$$

By canceling $\mathcal{E}/R$ and rearranging, this can be written as

$$e^{-t/\tau_L} = 1 - \frac{1}{\sqrt{2}} = 0.293,$$

which yields

$$\frac{t}{\tau_L} = -\ln 0.293 = 1.23$$

or

$$t \approx 1.2\,\tau_L. \qquad \text{(Answer)}$$

Thus the stored energy will reach half its equilibrium value after 1.2 time constants.

SAMPLE PROBLEM 31-10

A 3.56 H inductor is placed in series with a 12.8 Ω resistor, and an emf of 3.24 V is then suddenly applied across the RL combination.

(a) At 0.278 s (which is one inductive time constant) after the emf is applied, what is the rate P at which energy is being delivered by the battery?

SOLUTION: The current in the circuit is given by Eq. 31-46,

$$i = \frac{\mathcal{E}}{R}(1 - e^{-t/\tau_L}),$$

which, after one time constant, becomes

$$i = \frac{3.24\text{ V}}{12.8\ \Omega}(1 - e^{-1}) = 0.1600\text{ A}.$$

The rate at which the battery delivers energy is then given by Eq. 27-21 with $\mathcal{E}$ replacing V and is

$$P = \mathcal{E}i = (3.24 \text{ V})(0.1600 \text{ A})$$
$$= 0.5184 \text{ W} \approx 518 \text{ mW}. \qquad \text{(Answer)}$$

(b) At 0.278 s, at what rate P_R is energy appearing as thermal energy in the resistor?

SOLUTION: This is given by Eq. 27-22:

$$P_R = i^2R = (0.1600 \text{ A})^2(12.8 \text{ }\Omega)$$
$$= 0.3277 \text{ W} \approx 328 \text{ mW}. \qquad \text{(Answer)}$$

(c) At 0.278 s, at what rate P_B is energy being stored in the magnetic field?

SOLUTION: This is given by Eq. 31-52, which requires that we know di/dt. Differentiating Eq. 31-45 yields

$$\frac{di}{dt} = \frac{\mathcal{E}}{R}\frac{R}{L}(e^{-Rt/L}) = \frac{\mathcal{E}}{L}e^{-t/\tau_L}.$$

After one time constant we have

$$\frac{di}{dt} = \frac{3.24 \text{ V}}{3.56 \text{ H}}e^{-1} = 0.3348 \text{ A/s}.$$

Now from Eq. 31-52 the desired rate is

$$P_B = \frac{dU_B}{dt} = Li\frac{di}{dt}$$
$$= (3.56 \text{ H})(0.1600 \text{ A})(0.3348 \text{ A/s})$$
$$= 0.1907 \text{ W} \approx 191 \text{ mW}. \qquad \text{(Answer)}$$

Note that, as required by energy conservation,

$$P = P_R + P_B.$$

31-11 ENERGY DENSITY OF A MAGNETIC FIELD

Consider a length l near the middle of a long solenoid of cross-sectional area A; the volume associated with this length is Al. The energy stored by the length l of the solenoid must lie entirely within this volume because the magnetic field outside such a solenoid is essentially zero. Moreover, the stored energy must be uniformly distributed within the solenoid because the magnetic field is uniform everywhere inside.

Thus the energy per unit volume of the field is

$$u_B = \frac{U_B}{Al}$$

or, since

$$U_B = \tfrac{1}{2}Li^2,$$

we have

$$u_B = \frac{Li^2}{2Al} = \frac{L}{l}\frac{i^2}{2A}.$$

Substituting for L/l from Eq. 31-33, we find

$$u_B = \tfrac{1}{2}\mu_0 n^2 i^2. \qquad (31\text{-}55)$$

From Eq. 30-25 ($B = \mu_0 in$) we can write this *energy density* as

$$u_B = \frac{B^2}{2\mu_0} \qquad \text{(magnetic energy density)}. \qquad (31\text{-}56)$$

This equation gives the density of stored energy at any point where the magnetic field is B. Even though we derived it by considering a special case, the solenoid, Eq. 31-56 holds for all magnetic fields, no matter how they are generated. Equation 31-56 is comparable to Eq. 26-23, namely,

$$u_E = \tfrac{1}{2}\epsilon_0 E^2, \qquad (31\text{-}57)$$

which gives the energy density (in a vacuum) at any point in an electric field. Note that both u_B and u_E are proportional to the square of the appropriate field quantity, B or E.

$\mathcal{C}$HECKPOINT 7: The table lists the number of turns per unit length, current, and cross-sectional area for three solenoids. Rank the solenoids according to the magnetic energy density within them, greatest first.

SOLENOID	TURNS PER UNIT LENGTH	CURRENT	AREA
a	$2n_1$	i_1	$2A_1$
b	n_1	$2i_1$	A_1
c	n_1	i_1	$6A_1$

SAMPLE PROBLEM 31-11

A long coaxial cable (Fig. 31-23) consists of two thin-walled concentric conducting cylinders with radii a and b. The inner cylinder A carries a steady current i, the outer cylinder B providing the return path.

(a) Calculate the energy stored in the magnetic field between the cylinders for a length l of the cable.

SOLUTION: Consider a volume dV between the two cylinders, consisting of a cylindrical shell whose inner and outer radii are r and $r + dr$ and whose length is l. The energy dU contained within this shell is

$$dU = u_B \, dV,$$

in which u_B (the energy per unit volume) is, from Eq. 31-56, $u_B = B^2/2\mu_0$.

To find B as a function of r, we apply Ampere's law,

$$\oint \mathbf{B}\cdot d\mathbf{s} = \mu_0 i,$$

to the circle of radius r in Fig. 31-23, obtaining

$$(B)(2\pi r) = \mu_0 i,$$

or

$$B = \frac{\mu_0 i}{2\pi r}.$$

The energy density between the cylinders is then

$$u_B = \frac{1}{2\mu_0} \left(\frac{\mu_0 i}{2\pi r}\right)^2 = \frac{\mu_0 i^2}{8\pi^2 r^2}.$$

The volume dV of our shell is $(2\pi rl)(dr)$, so the energy dU contained within the shell is

$$dU = u_B \, dV = \frac{\mu_0 i^2}{8\pi^2 r^2} (2\pi rl)(dr) = \frac{\mu_0 i^2 l}{4\pi} \frac{dr}{r}.$$

The total energy is obtained by integrating this expression over the volume between the cylinders:

$$U = \int dU = \frac{\mu_0 i^2 l}{4\pi} \int_a^b \frac{dr}{r}$$

$$= \frac{\mu_0 i^2 l}{4\pi} \ln \frac{b}{a}. \qquad \text{(Answer)} \qquad (31\text{-}58)$$

No energy is stored outside the outer cylinder or inside the inner cylinder because the magnetic field is zero in both locations, as you can show with Ampere's law.

(b) What is the stored energy per unit length of the cable if $a = 1.2$ mm, $b = 3.5$ mm, and $i = 2.7$ A?

SOLUTION: From Eq. 31-58 we have

$$\frac{U}{l} = \frac{\mu_0 i^2}{4\pi} \ln \frac{b}{a}$$

$$= \frac{(4\pi \times 10^{-7} \text{ H/m})(2.7 \text{ A})^2}{4\pi} \ln \frac{3.5 \text{ mm}}{1.2 \text{ mm}}$$

$$= 7.8 \times 10^{-7} \text{ J/m} = 780 \text{ nJ/m}. \qquad \text{(Answer)}$$

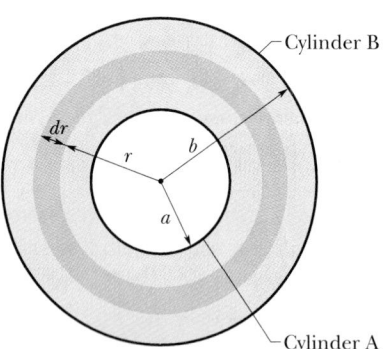

FIGURE 31-23 Sample Problem 31-11. A cross section of a long coaxial cable consisting of two thin-walled conducting cylinders, the inner cylinder of radius a and the outer cylinder of radius b.

31-12 MUTUAL INDUCTION

In this section we return to the case of two interacting coils, which we first discussed in Section 31-2, and we treat it in a somewhat more formal manner. In Fig. 31-2 we saw that if two coils are close together, a steady current i in one coil will set up a magnetic flux Φ through the other coil (*linking* the other coil). If we change i with time, an emf $\mathscr{E}$ given by Faraday's law appears in the second coil; we called this process *induction*. We could better have called it **mutual induction,** to suggest the mutual interaction of the two coils and to distinguish it from **self-induction,** in which only one coil is involved.

Let us look a little more quantitatively at mutual induction. Figure 31-24a shows two circular close-packed coils near each other and sharing a common central axis. There is a steady current i_1 in coil 1, produced by the battery in the external circuit. This current creates a magnetic field represented by the lines of $\mathbf{B}_1$ in the figure. Coil 2 is connected to a sensitive meter but contains no battery; a magnetic flux Φ_{21} (the flux through coil 2 associated with the current in coil 1) links the N_2 turns of coil 2.

We define the mutual inductance M_{21} of coil 2 with respect to coil 1 as

$$M_{21} = \frac{N_2 \Phi_{21}}{i_1}. \qquad (31\text{-}59)$$

Compare this with Eq. 31-30 ($L = N\Phi/i$), the definition of (self) inductance. We can recast Eq. 31-59 as

$$M_{21} i_1 = N_2 \Phi_{21}.$$

If, by external means, we cause i_1 to vary with time, we have

$$M_{21} \frac{di_1}{dt} = N_2 \frac{d\Phi_{21}}{dt}.$$

The right side of this equation, from Faraday's law, is, apart from a difference in sign, just the emf $\mathscr{E}_2$ appearing in coil 2 due to the changing current in coil 1. Thus

$$\mathscr{E}_2 = -M_{21} \frac{di_1}{dt}, \qquad (31\text{-}60)$$

which you should compare with Eq. 31-40 for self-induction ($\mathscr{E} = -L \, di/dt$).

Let us now interchange the roles of coils 1 and 2, as in Fig. 31-24b. That is, we set up a current i_2 in coil 2, by means of a battery, and this produces a magnetic flux Φ_{12} that links coil 1. If we change i_2 with time, we have, by the argument given above,

$$\mathscr{E}_1 = -M_{12} \frac{di_2}{dt}. \qquad (31\text{-}61)$$

Thus we see that the emf induced in either coil is

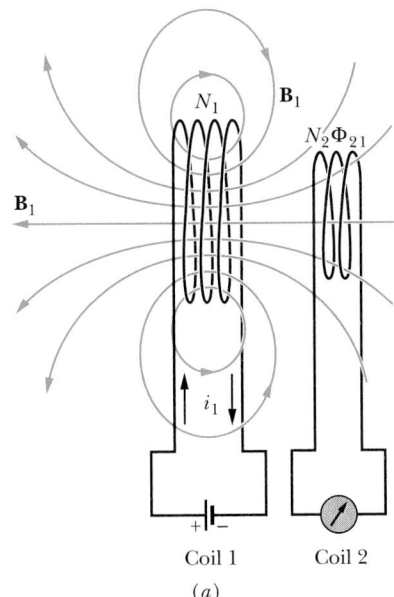

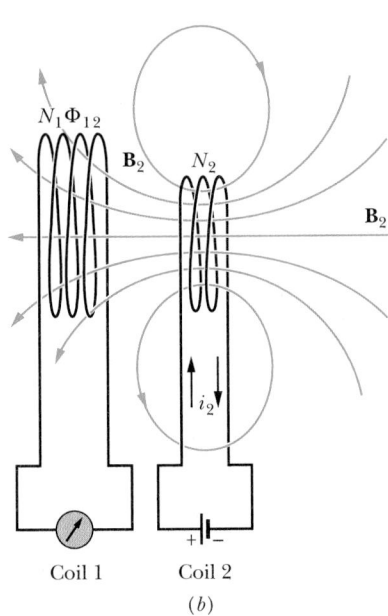

FIGURE 31-24 Mutual induction. (*a*) If the current in coil 1 changes, an emf will be induced in coil 2. (*b*) If the current in coil 2 changes, an emf will be induced in coil 1.

proportional to the rate of change of current in the other coil. The proportionality constants M_{21} and M_{12} seem to be different. We assert, without proof, that they are in fact the same so that no subscripts are needed. (This conclusion is true but is in no way obvious.) Thus we have

$$M_{21} = M_{12} = M, \qquad (31\text{-}62)$$

and we can rewrite Eqs. 31-60 and 31-61 as

$$\mathscr{E}_2 = -M \frac{di_1}{dt} \qquad (31\text{-}63)$$

and

$$\mathscr{E}_1 = -M \frac{di_2}{dt}. \qquad (31\text{-}64)$$

The induction is indeed mutual. The SI unit for M (as for L) is the henry.

SAMPLE PROBLEM 31-12

Figure 31-25 shows two circular close-packed coils, the smaller (radius R_2, with N_2 turns) being coaxial with the larger (radius R_1, with N_1 turns) and in the same plane.

(a) Derive an expression for the mutual inductance M for this arrangement of these two coils, assuming that $R_1 \gg R_2$.

SOLUTION: As Fig. 31-25 suggests, we imagine that we establish a current i_1 in the larger coil and we note the magnetic field B_1 that it sets up. The value of B_1 at the center of this coil is (from Eq. 30-28, with $z = 0$ and after multiplying the right side by N_1)

$$B_1 = \frac{\mu_0 i_1 N_1}{2R_1}.$$

Because we have assumed that $R_1 \gg R_2$, we may take B_1 to be the magnetic field at all points within the boundary of the smaller coil. The flux linkage for the smaller coil is then

$$N_2 \Phi_{21} = N_2(B_1)(\pi R_2^2) = \frac{\pi \mu_0 N_1 N_2 R_2^2 i_1}{2R_1}.$$

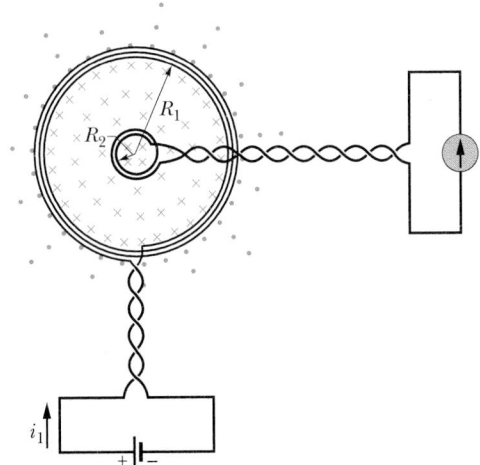

FIGURE 31-25 Sample Problem 31-12. A small coil is located at the center of a large coil. The mutual inductance of the coils can be determined by sending current i_1 through the large coil.

From Eq. 31-59 we then have

$$M = \frac{N_2 \Phi_{21}}{i_1} = \frac{\pi \mu_0 N_1 N_2 R_2^2}{2R_1}. \qquad \text{(Answer)}$$

(b) What is the value of M for $N_1 = N_2 = 1200$ turns, $R_2 = 1.1$ cm, and $R_1 = 15$ cm?

SOLUTION: The equation above yields

$$M = \frac{(\pi)(4\pi \times 10^{-7} \text{ H/m})(1200)(1200)(0.011 \text{ m})^2}{(2)(0.15 \text{ m})}$$

$$= 2.29 \times 10^{-3} \text{ H} \approx 2.3 \text{ mH}. \qquad \text{(Answer)}$$

Consider the situation if we reverse the roles of the two coils in Fig. 31-25, that is, if we produce a current i_2 in the smaller coil and try to calculate M from Eq. 31-59 in the form

$$M = \frac{N_1 \Phi_{12}}{i_2}.$$

The calculation of Φ_{12} (the flux of the smaller coil's magnetic field encompassed by the larger coil) is not simple. If we were to do the calculation numerically using a computer, we would find M to be exactly 2.3 mH, as above! This emphasizes that Eq. 31-62 ($M_{12} = M_{21} = M$) is not obvious.

REVIEW & SUMMARY

Magnetic Flux

The *magnetic flux* Φ_B through an area in a magnetic field **B** is defined as

$$\Phi_B = \int \mathbf{B} \cdot d\mathbf{A}, \qquad (31\text{-}3)$$

where the integral is taken over the area. The SI unit of magnetic flux is the weber, where 1 Wb = 1 T·m². If **B** is perpendicular to the area and uniform over it, Eq. 31-3 becomes

$$\Phi_B = BA \qquad (\mathbf{B} \perp \mathbf{A}, \mathbf{B} \text{ uniform}). \qquad (31\text{-}4)$$

Faraday's Law of Induction

If the magnetic flux Φ_B through an area bounded by a closed conducting loop changes with time, a current and an emf are produced in the loop; this process is called *induction*. The induced emf is

$$\mathscr{E} = -\frac{d\Phi_B}{dt}. \qquad \text{(Faraday's law).} \qquad (31\text{-}6)$$

If the loop is replaced by a closely packed coil of N turns, the induced emf is

$$\mathscr{E} = -N\frac{d\Phi_B}{dt}. \qquad (31\text{-}7)$$

Lenz's Law

An induced current has a direction such that the magnetic field *of the current* opposes the change in the magnetic field that produces the current.

Emf and the Induced Electric Field

An emf is induced by a changing magnetic flux even if the loop through which the flux is changing is not a physical conductor but an imaginary line. The changing flux induces an electric field **E** at every point of such a loop; the induced emf is related to **E** by

$$\mathscr{E} = \oint \mathbf{E} \cdot d\mathbf{s}, \qquad (31\text{-}21)$$

where the integration is taken around the loop. From Eq. 31-21 we can write Faraday's law in its most general form,

$$\oint \mathbf{E} \cdot d\mathbf{s} = -\frac{d\Phi_B}{dt} \qquad \text{(Faraday's law).} \qquad (31\text{-}22)$$

The essence of this law is that *a changing magnetic flux $d\Phi_B/dt$ induces an electric field* **E**.

Inductors

An **inductor** is a device that can be used to produce a known magnetic field in a specified region. If a current i is established through each of the N windings of an inductor, a magnetic flux Φ links those windings. The **inductance** L of the inductor is

$$L = \frac{N\Phi}{i} \qquad \text{(inductance defined).} \qquad (31\text{-}30)$$

The SI unit of inductance is the **henry** (H), with

$$1 \text{ henry} = 1 \text{ H} = 1 \text{ T·m}^2/\text{A}. \qquad (31\text{-}31)$$

The inductance per unit length near the middle of a long solenoid of cross-sectional area A and n turns per unit length is

$$\frac{L}{l} = \mu_0 n^2 A \qquad \text{(solenoid).} \qquad (31\text{-}33)$$

Self-Induction

If a current i in a coil changes with time, an emf is induced in the coil. This self-induced emf is

$$\mathscr{E}_L = -L\frac{di}{dt}. \qquad (31\text{-}40)$$

The direction of $\mathscr{E}_L$ is found from Lenz's law: the self-induced emf acts to oppose the change that produces it.

Series RL Circuits

If a constant emf $\mathscr{E}$ is introduced into a single-loop circuit containing a resistance R and an inductance L, the current rises to an equilibrium value of $\mathscr{E}/R$ according to

$$i = \frac{\mathscr{E}}{R}(1 - e^{-t/\tau_L}) \qquad \text{(rise of current).} \qquad (31\text{-}46)$$

Here τ_L ($= L/R$) governs the rate of rise of the current and is called the **inductive time constant** of the circuit. When the source of constant emf is removed, the current decays from a value i_0 according to

$$i = i_0 e^{-t/\tau_L} \qquad \text{(decay of current).} \qquad (31\text{-}49)$$

Magnetic Energy

If an inductor L carries a current i, the inductor's magnetic field stores an energy given by

$$U_B = \tfrac{1}{2}Li^2 \qquad \text{(magnetic energy).} \qquad (31\text{-}53)$$

If B is the magnetic field at any point (in an inductor or anywhere else), the density of stored magnetic energy at that point is

$$u_B = \frac{B^2}{2\mu_0} \qquad \text{(magnetic energy density).} \qquad (31\text{-}56)$$

Mutual Induction

If two coils (labeled 1 and 2) are near each other, a changing current in either coil can induce an emf in the other. This mutual induction is described by

$$\mathcal{E}_2 = -M\frac{di_1}{dt} \quad \text{and} \quad \mathcal{E}_1 = -M\frac{di_2}{dt}, \qquad (31\text{-}63, 31\text{-}64)$$

where M (measured in henries) is the mutual inductance for the coil arrangement.

QUESTIONS

1. Situation 1 in Fig. 31-26 shows a rectangular wire loop of length a and height b in the xy plane. Situations 2 and 3 show wire loops of the same overall dimensions but with sections brought forward parallel to the z axis. In each situation, the same uniform magnetic field pierces the loops and is increasing in magnitude at the same rate. Rank the three situations according to the magnitude of the emf induced in the loops, greatest first, if the magnetic field points in the positive direction of (a) y, (b) z, and (c) x.

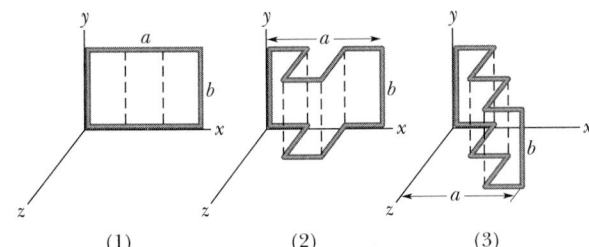

(1)   (2)   (3)

FIGURE 31-26   Question 1.

2. In Fig. 31-27, a long straight wire with current i passes (without touching) three rectangular wire loops with edge lengths L, $1.5L$, and $2L$. The loops are widely spaced (so as to not affect one another). Loops 1 and 3 are symmetric about the long wire. Rank the loops according to the size of the current induced in them if current i is (a) constant and (b) increasing, greatest first.

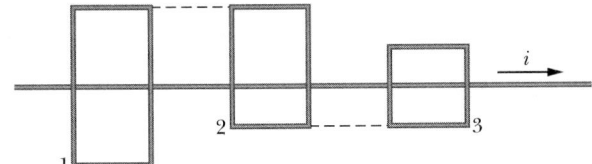

FIGURE 31-27   Question 2.

3. If the circular conductor in Fig. 31-28 undergoes thermal expansion while in a uniform magnetic field, a current will be induced clockwise around it. Is the magnetic field directed into the page or out of it?

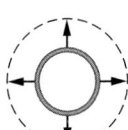

FIGURE 31-28   Question 3.

4. In Fig. 31-29, a circular loop moves at constant velocity through regions where uniform magnetic fields of the same magnitude are directed into or out of the page. (The field is zero outside the dashed lines.) At which of the seven indicated loop positions is the emf induced in the loop (a) clockwise, (b) counterclockwise, and (c) zero?

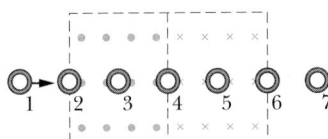

FIGURE 31-29   Question 4.

5. Figure 31-30 shows two circuits in which a conducting bar is slid at the same speed v through the same uniform magnetic field and along a U-shaped wire. The parallel lengths of the wire are separated by $2L$ in circuit 1 and by L in circuit 2. The current induced in circuit 1 is counterclockwise. (a) Is the direction of the magnetic field into or out of the page? (b) Is the direction of the current in circuit 2 clockwise or counterclockwise? (c) Is the current in circuit 1 larger than, smaller than, or the same as that in circuit 2?

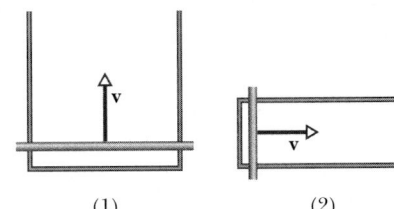

FIGURE 31-30
Question 5.   (1)   (2)

6. In Fig. 31-31, a conducting rod slides at constant velocity **v** across an incomplete conducting square, with which it makes electrical contact. The square is in a uniform magnetic field that points directly out of the page. During the slide, (a) is the induced current clockwise, counterclockwise, or does it change from one to the other midway, and (b) is the current constant, increasing, or increasing and then decreasing?

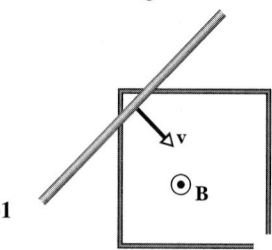

FIGURE 31-31
Question 6.

7. Figure 31-32 shows two coils wrapped around nonconducting rods. Coil X is connected to a battery and a variable resistance. What is the direction of the induced current through the current meter connected to coil Y (a) when coil Y is moved toward coil X and (b) when the current in coil X is decreased without any change in the relative positions of the coils?

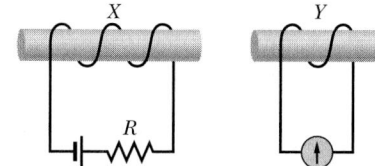

FIGURE 31-32
Question 7.

8. If the variable resistance R in the left-hand circuit of Fig. 31-33 is increased at a steady rate, is the current induced in the right-hand loop clockwise or counterclockwise?

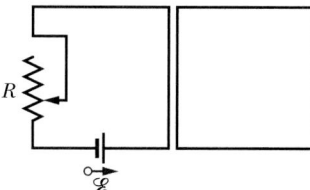

FIGURE 31-33
Question 8.

9. Figure 31-34a shows a circular region in which an increasing uniform magnetic field is directed out of the page, as well as a concentric circular path along which $\oint \mathbf{E} \cdot d\mathbf{s}$ is to be evaluated. The table gives the initial magnitude of the magnetic field, the increase in that magnitude, and the time interval for the increase, in three situations. Rank the situations according to the magnitude of the electric field induced along the path, greatest first.

SITUATION	INITIAL FIELD	INCREASE	TIME
a	B_1	ΔB_1	Δt_1
b	$2B_1$	$\Delta B_1/2$	Δt_1
c	$B_1/4$	ΔB_1	$\Delta t_1/2$

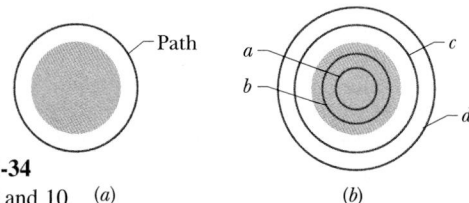

FIGURE 31-34
Questions 9 and 10. *(a)*   *(b)*

10. Figure 31-34b shows a circular region in which a decreasing uniform magnetic field is directed out of the page, as well as four concentric circular paths. Rank the paths according to the magnitude of $\oint \mathbf{E} \cdot d\mathbf{s}$ evaluated along them, greatest first.

11. The number of turns per unit length, current, and cross-sectional area for three solenoids of the same length are given in the following table. Rank the solenoids according to (a) their inductance and (b) the flux through each turn, greatest first.

SOLENOID	TURNS PER UNIT LENGTH	CURRENT	AREA
1	$2n_1$	i_1	$2A_1$
2	n_1	$2i_1$	A_1
3	n_1	i_1	$4A_1$

12. Figure 31-35 gives the variation with time of the potential difference V_R across a resistor in three circuits wired as in Fig. 31-20. The circuits contain the same resistance R and emf $\mathscr{E}$ but differ in the inductance L. Rank the circuits according to the value of L, greatest first.

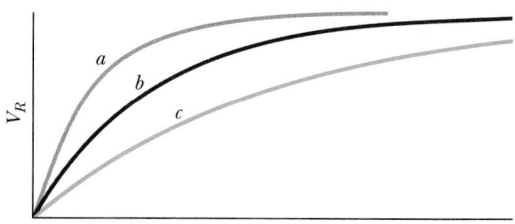

FIGURE 31-35 Question 12.

13. Here are three sets of values for the emf $\mathscr{E}$ of the battery and the potential difference V_R across the resistor in the circuit of Fig. 31-20 at some time after the current begins to increase: (a) 12 V and 3 V; (b) 24 V and 16 V; (c) 18 V and 10 V. Rank the sets according to the potential difference across the inductor at those times, greatest first.

14. Figure 31-36 shows three circuits with identical batteries, inductors, and resistors. Rank the circuits according to the time for the current to reach 50% of its equilibrium value after the switches are closed, greatest first.

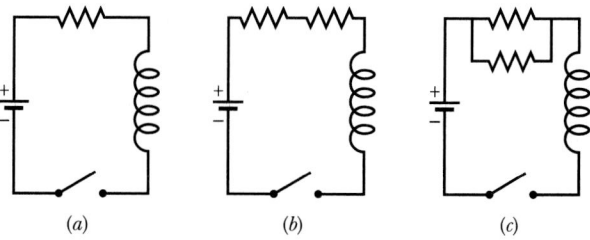

FIGURE 31-36 Question 14.

15. Figure 31-37 shows a circuit with two identical resistors and an inductor. Is the current through the central resistor more than, less than, or the same as that through the other resistor (a) just after the closing of switch S, (b) a long time after the closing of S,

(c) just after the reopening of S, a long time later, and (d) a long time after the reopening of S?

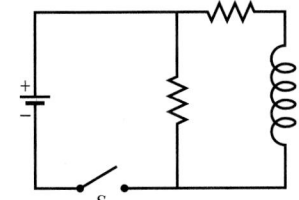

FIGURE 31-37
Question 15.

16. Figure 31-38 shows three circuits with identical batteries, inductors, and resistors. Rank the circuits, greatest first, according to the current through the resistor labeled R (a) long after the switch is closed, (b) just after the switch is reopened a long time later, and (c) long after it is reopened.

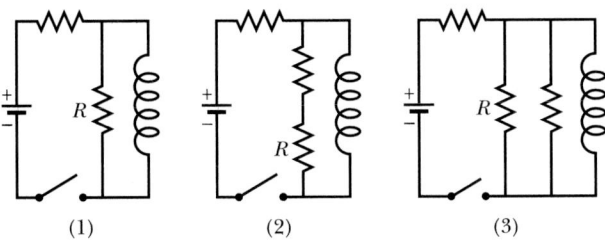

(1) (2) (3)

FIGURE 31-38 Question 16.

17. The switch in the circuit of Fig. 31-19 has been closed on a for a long time when it is then thrown to b. The resulting current through the inductor is sketched in Fig. 31-39 for four sets of values for the resistance R and inductance L: (1) R_0 and L_0; (2) $2R_0$ and L_0; (3) R_0 and $2L_0$; (4) $2R_0$ and $2L_0$. Which set goes with which curve?

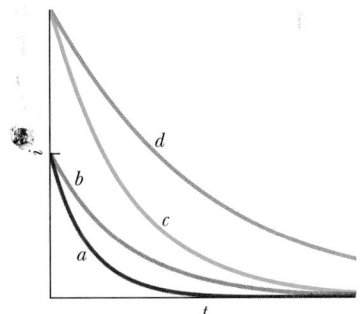

FIGURE 31-39 Question 17.

18. In Fig. 31-24, the current in coil 1 is given, in three situations, by (1) $i_1 = 3 \cos(4t)$, (2) $i_1 = 10 \cos(t)$, and (3) $i_1 = 5 \cos(2t)$, with i_1 in amperes and t in seconds. For the three situations, rank (a) the mutual inductance of the coils and (b) the magnitude of the maximum emf appearing in coil 2 due to i_1, greatest first.

EXERCISES & PROBLEMS

SECTION 31-4 Lenz's Law

1E. At a certain location in the southern hemisphere, Earth's magnetic field has a magnitude of 42 μT and points upward at 57° to the vertical. Calculate the flux through a horizontal surface of area 2.5 m^2; see Fig. 31-40, in which area vector **A** has arbitrarily been chosen to be upward.

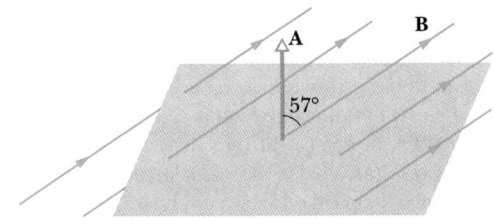

FIGURE 31-40 Exercise 1.

2E. A small loop of area A is inside of, and has its axis in the same direction as, a long solenoid of n turns per unit length and current i. If $i = i_0 \sin \omega t$, find the emf in the loop.

3E. A UHF television loop antenna has a diameter of 11 cm. The magnetic field of a TV signal is normal to the plane of the loop and, at one instant of time, its magnitude is changing at the rate 0.16 T/s. The field is uniform. What is the emf in the antenna?

4E. A uniform magnetic field **B** is perpendicular to the plane of a circular wire loop of radius r. The magnitude of the field varies with time according to $B = B_0 e^{-t/\tau}$, where B_0 and τ are constants. Find the emf in the loop as a function of time.

5E. The magnetic flux through the loop shown in Fig. 31-41 increases according to the relation $\Phi_B = 6.0t^2 + 7.0t$, where Φ_B is in milliwebers and t is in seconds. (a) What is the magnitude of the emf induced in the loop when $t = 2.0$ s? (b) What is the direction of the current through R?

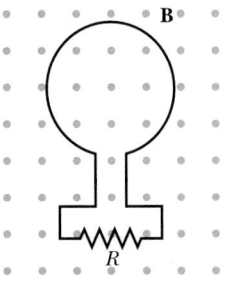

FIGURE 31-41 Exercise 5 and Problem 19.

6E. The magnetic field through a single loop of wire 12 cm in radius and of 8.5 Ω resistance changes with time as shown in Fig. 31-42. Calculate the emf in the loop as a function of time. Consider the time intervals (a) $t = 0$ to $t = 2.0$ s; (b) $t = 2.0$ s to $t = 4.0$ s; (c) $t = 4.0$ s to $t = 6.0$ s. The (uniform) magnetic field is perpendicular to the plane of the loop.

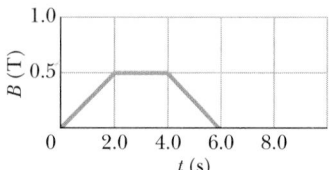

FIGURE 31-42
Exercise 6.

7E. A 20 mΩ square wire loop 20 cm on a side has its plane normal to a uniform magnetic field of magnitude $B = 2.0$ T. If you pull two opposite sides of the loop away from each other, the other two sides automatically draw toward each other, reducing the area enclosed by the loop. If the area is reduced to zero in time $\Delta t = 0.20$ s, what are (a) the average emf and (b) the average current induced in the loop during Δt?

8E. A uniform magnetic field is normal to the plane of a circular loop 10 cm in diameter and made of copper wire (of diameter 2.5 mm). (a) Calculate the resistance of the wire. (See Table 27-1.) (b) At what rate must the magnetic field change with time if an induced current of 10 A is to appear in the loop?

9P. The current in the solenoid of Sample Problem 31-1 changes, not as stated there, but according to $i = 3.0t + 1.0t^2$, where i is in amperes and t is in seconds. (a) Plot the induced emf in the coil from $t = 0$ to $t = 4.0$ s. (b) The resistance of the coil is 0.15 Ω. What is the current in the coil at $t = 2.0$ s?

10P. In Fig. 31-43 a 120-turn coil of radius 1.8 cm and resistance 5.3 Ω is placed *outside* a solenoid like that of Sample Problem 31-1. If the current in the solenoid is changed as in that sample problem, what current appears in the coil while the solenoid current is being changed?

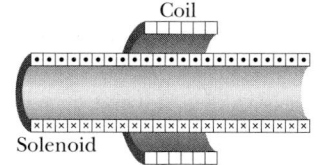

FIGURE 31-43
Problem 10.

11P. A long solenoid with a radius of 25 mm has 100 turns/cm. A single loop of wire of radius 5.0 cm is placed around the solenoid, the central axes of the loop and the solenoid coinciding. In 10 ms the current in the solenoid is reduced from 1.0 A to 0.50 A at a uniform rate. What emf appears in the loop?

12P. Derive an expression for the flux through a toroid of N turns carrying a current i. Assume that the windings have a rectangular cross section of inner radius a, outer radius b, and height h.

13P. A toroid having a 5.00 cm square cross section and an inside radius of 15.0 cm has 500 turns of wire and carries a current of 0.800 A. What is the magnetic flux through the cross section?

14P. An elastic conducting material is stretched into a circular loop of 12.0 cm radius. It is placed with its plane perpendicular to a uniform 0.800 T magnetic field. When released, the radius of the loop starts to shrink at an instantaneous rate of 75.0 cm/s. What emf is induced in the loop at that instant?

15P. A closed loop of wire consists of a pair of equal semicircles, of radius 3.7 cm, lying in mutually perpendicular planes. The loop was formed by folding a plane circular loop along a diameter until the two halves became perpendicular. A uniform magnetic field **B** of magnitude 76 mT is directed perpendicular to the fold diameter and makes equal angles ($= 45°$) with the planes of the semicircles as shown in Fig. 31-44. The magnetic field is reduced to zero at a uniform rate during a time interval of 4.5 ms. Determine the magnitude of the induced emf and the direction of the induced current in the loop during this interval.

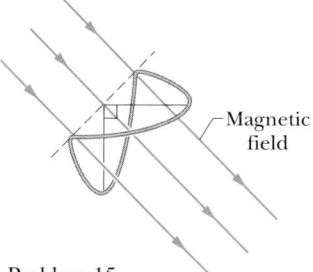

FIGURE 31-44 Problem 15.

16P. In Fig. 31-45, a circular loop of wire 10 cm in diameter (seen edge-on) is placed with its normal **N** making an angle $\theta = 30°$ with the direction of a uniform magnetic field **B** of magnitude 0.50 T. The loop is then rotated such that **N** rotates in a cone about the field direction at the constant rate of 100 rev/min; the angle θ remains unchanged during the process. What is the emf induced in the loop?

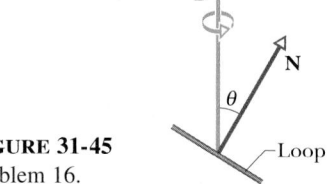

FIGURE 31-45
Problem 16.

17P. A small circular loop of area 2.00 cm² is placed in the plane of, and concentric with, a large circular loop of radius 1.00 m. The current in the large loop is changed uniformly from 200 A to -200 A (a change in direction) in a time of 1.00 s, beginning at $t = 0$. (a) What is the magnetic field at the center of the small circular loop due to the current in the large loop at $t = 0$, $t = 0.500$ s, and $t = 1.00$ s? (b) What emf is induced in the small loop at $t = 0.500$ s? (Since the inner loop is small, assume the field **B** due to the outer loop is uniform over the area of the smaller loop.)

18P. Figure 31-46 shows two parallel loops of wire having a common axis. The smaller loop (radius r) is above the larger loop (radius R), by a distance $x \gg R$. Consequently the magnetic field due to the current i in the larger loop is nearly constant throughout the smaller loop. Suppose that x is increasing at the constant rate $dx/dt = v$. (a) Determine the magnetic flux through the area bounded by the smaller loop as a function of *x*. (*Hint:* See Eq. 30-29.) In the smaller loop, find (b) the induced emf and (c) the direction of the induced current.

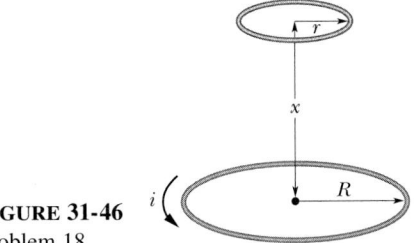

FIGURE 31-46
Problem 18.

19P. In Fig. 31-41 let the flux through the loop be $\Phi_B(0)$ at time $t = 0$. Then let the magnetic field **B** vary in a continuous but unspecified way, in both magnitude and direction, so that at time *t* the flux is represented by $\Phi_B(t)$. (a) Show that the net charge $q(t)$ that has passed through resistor *R* in time *t* is

$$q(t) = \frac{1}{R}\,[\Phi_B(0) - \Phi_B(t)]$$

and is independent of the way **B** has changed. (b) If $\Phi_B(t) = \Phi_B(0)$ in a particular case, we have $q(t) = 0$. Is the induced current necessarily zero throughout the interval from 0 to *t*?

20P. One hundred turns of insulated copper wire are wrapped around a wooden cylindrical core of cross-sectional area 1.20×10^{-3} m². The two terminals are connected to a resistor. The total resistance in the circuit is 13.0 Ω. If an externally applied uniform longitudinal magnetic field in the core changes from 1.60 T in one direction to 1.60 T in the opposite direction, how much charge flows through the circuit? (*Hint:* See Problem 19.)

21P. At a certain place, Earth's magnetic field has magnitude $B = 0.590$ gauss and is inclined downward at an angle of 70.0° to the horizontal. A flat horizontal circular coil of wire with a radius of 10.0 cm has 1000 turns and a total resistance of 85.0 Ω. It is connected to a meter with 140 Ω resistance. The coil is flipped through a half-revolution about a diameter, so that it is again horizontal. How much charge flows through the meter during the flip? (*Hint:* See Problem 19.)

22P. A square wire loop with 2.00 m sides is perpendicular to a uniform magnetic field, with half the area of the loop in the field, as shown in Fig. 31-47. The loop contains a 20.0 V battery with negligible internal resistance. If the magnitude of the field varies with time according to $B = 0.0420 - 0.870t$, with *B* in teslas and *t* in seconds, (a) what is the total emf in the circuit? (b) What is the direction of the current through the battery?

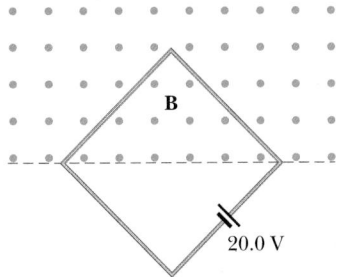

FIGURE 31-47
Problem 22.

23P. A wire is bent into three circular segments, each of radius $r = 10$ cm, as shown in Fig. 31-48. Each segment is a quadrant of a circle, *ab* lying in the *xy* plane, *bc* lying in the *yz* plane, and *ca* lying in the *zx* plane. (a) If a uniform magnetic field **B** points in the positive *x* direction, what is the magnitude of the emf developed in the wire when *B* increases at the rate of 3.0 mT/s? (b) What is the direction of the current in segment *bc*?

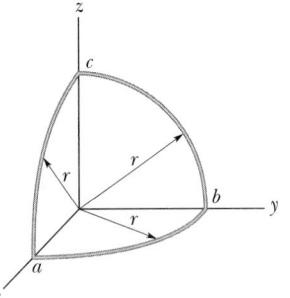

FIGURE 31-48
Problem 23.

24P. A stiff wire bent into a semicircle of radius *a* is rotated with frequency *f* in a uniform magnetic field, as suggested in Fig. 31-49. What are (a) the frequency and (b) the amplitude of the varying emf induced in the loop?

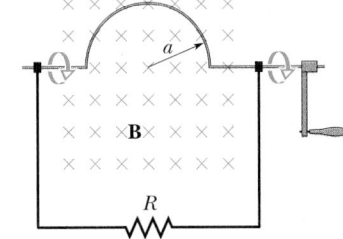

FIGURE 31-49
Problem 24.

25P. A rectangular coil of *N* turns and of length *a* and width *b* is rotated at frequency *f* in a uniform magnetic field **B**, as indicated in Fig. 31-50. The coil is connected to co-rotating cylinders, against which metal brushes slide to make contact. (a) Show that the emf induced in the coil is given (as a function of time *t*) by

$$\mathcal{E} = 2\pi fNabB \sin(2\pi ft) = \mathcal{E}_0 \sin(2\pi ft).$$

This is the principle of the commercial alternating-current generator. (b) Design a loop that will produce an emf with $\mathcal{E}_0 = 150$ V when rotated at 60.0 rev/s in a magnetic field of 0.500 T.

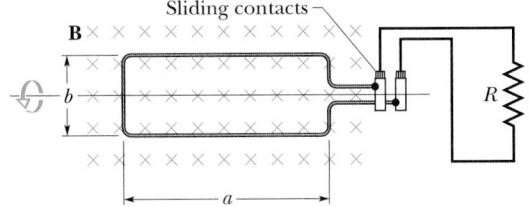

FIGURE 31-50 Problem 25.

26P. An electric generator consists of 100 turns of wire formed into a rectangular loop 50.0 cm by 30.0 cm, placed entirely in a uniform magnetic field with magnitude $B = 3.50$ T. What is the maximum value of the emf produced when the loop is spun at 1000 rev/min about an axis perpendicular to **B**?

27P. For the situation shown in Fig. 31-51, $a = 12.0$ cm and $b = 16.0$ cm. The current in the long straight wire is given by $i = 4.50t^2 - 10.0t$, where i is in amperes and t is in seconds. (a) Find the emf in the square loop at $t = 3.00$ s. (b) What is the direction of the induced current in the loop?

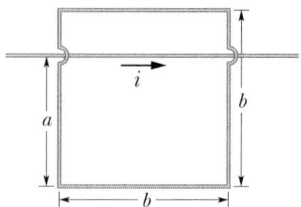

FIGURE 31-51
Problem 27.

28P. In Fig. 31-52, the square loop of wire has sides of length 2.0 cm. A magnetic field points out of the page; its magnitude is given by $B = 4.0t^2y$, where B is in teslas, t is in seconds, and y is in meters. Determine the emf around the square at $t = 2.5$ s and give its direction.

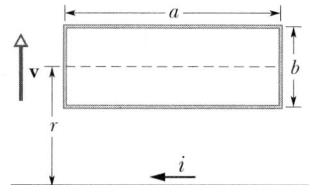

FIGURE 31-52
Problem 28.

29P. A rectangular loop of wire with length a, width b, and resistance R is placed near an infinitely long wire carrying current i, as shown in Fig. 31-53. The distance from the long wire to the center of the loop is r. Find (a) the magnitude of the magnetic flux through the loop and (b) the current in the loop as it moves away from the long wire with speed v.

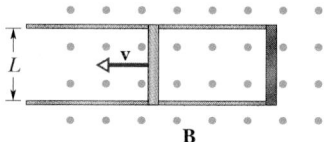

FIGURE 31-53
Problem 29.

30P*. Two long, parallel copper wires (of diameter 2.5 mm) carry currents of 10 A in opposite directions. (a) Assuming that their centers are 20 mm apart, calculate the magnetic flux per meter of wire that exists in the space between the axes of the wires. (b) What fraction of this flux lies inside the wires? (c) Repeat part (a) for parallel currents.

SECTION 31-5 Induction and Energy Transfers

31E. A loop antenna of area A and resistance R is perpendicular to a uniform magnetic field **B**. The field drops linearly to zero in a time interval Δt. Find an expression for the total thermal energy dissipated in the loop.

32E. If 50.0 cm of copper wire (diameter = 1.00 mm) is formed into a circular loop and placed perpendicular to a uniform mag-

netic field that is increasing at the constant rate of 10.0 mT/s, at what rate is thermal energy generated in the loop?

33E. A metal rod is forced to move with constant velocity **v** along two parallel metal rails, connected with a strip of metal at one end, as shown in Fig. 31-54. A magnetic field $B = 0.350$ T points out of the page. (a) If the rails are separated by 25.0 cm and the speed of the rod is 55.0 cm/s, what emf is generated? (b) If the rod has a resistance of 18.0 Ω and the rails and connector have negligible resistance, what is the current in the rod? (c) At what rate is energy being transferred to thermal energy?

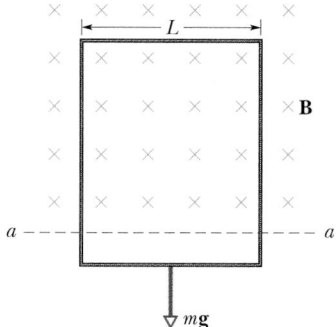

FIGURE 31-54 Exercises 33 and 34.

34E. The conducting rod shown in Fig. 31-54 has a length L and is being pulled along horizontal, frictionless, conducting rails at a constant velocity **v**. The rails are connected at one end with a metal strip. A uniform magnetic field **B**, directed out of the page, fills the region in which the rod moves. Assume that $L = 10$ cm, $v = 5.0$ m/s, and $B = 1.2$ T. (a) What is the induced emf in the rod? (b) What is the current in the conducting loop? Assume that the resistance of the rod is 0.40 Ω and that the resistance of the rails and metal strip is negligibly small. (c) At what rate is thermal energy being generated in the rod? (d) What force must be applied to the rod by an external agent to maintain its motion? (e) At what rate does this external agent do work on the rod? Compare this answer with the answer to (c).

35P. In Fig. 31-55, a long rectangular conducting loop, of width L, resistance R, and mass m, is hung in a horizontal, uniform magnetic field **B** that is directed into the page and that exists only above line aa. The loop is then dropped; during its fall, it accelerates until it reaches a certain terminal speed v_t. Ignoring air drag, find that terminal speed.

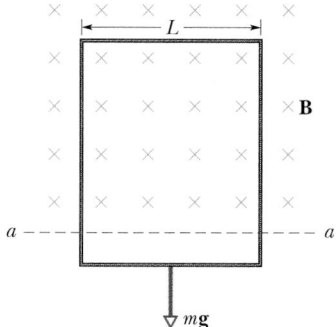

FIGURE 31-55
Problem 35.

36P. Two straight conducting rails form a right angle where their ends are joined. A conducting bar in contact with the rails starts at the vertex at time $t = 0$ and moves with a constant velocity of 5.20 m/s along them, as shown in Fig. 31-56. A 0.350 T magnetic field points out of the page. Calculate (a) the flux through the triangle formed by the rails and bar at $t = 3.00$ s and (b) the emf around the triangle at that time. (c) If we write the emf as $\mathscr{E} = at^n$, where a and n are constants, what is the value of n?

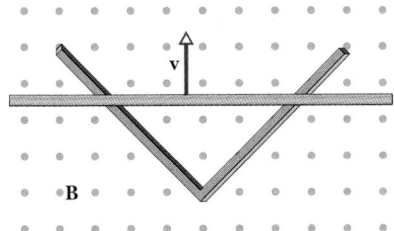

FIGURE 31-56
Problem 36.

37P. Calculate the average power supplied by the generator of Problem 25b if it is connected to a circuit of 42.0 Ω resistance. (*Hint:* The average value of $\sin^2 \theta$ over one cycle is $\frac{1}{2}$.)

38P. In Fig. 31-57 a conducting rod of mass m and length L slides without friction on two long horizontal rails. A uniform vertical magnetic field **B** fills the region in which the rod is free to move. The generator G supplies a constant current i directed as shown. (a) Find the velocity of the rod as a function of time, assuming it to be at rest at $t = 0$. The generator is now replaced by a battery that supplies a constant emf $\mathscr{E}$. (b) Show that the velocity of the rod now approaches a constant terminal value **v** and give its magnitude and direction. (c) What is the current in the rod when this terminal velocity is reached? (d) Analyze this situation and that with the generator from the point of view of energy transfers.

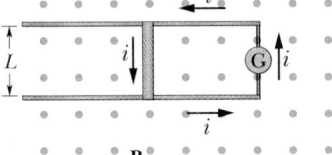

FIGURE 31-57
Problem 38.

39P. Figure 31-58 shows a rod of length L caused to move at constant speed v along horizontal conducting rails. In this case the magnetic field in which the rod moves is not uniform but is provided by a current i in a long wire parallel to the rails. Assume that $v = 5.00$ m/s, $a = 10.0$ mm, $L = 10.0$ cm, and $i = 100$ A. (a) Calculate the emf induced in the rod. (b) What is the current in the conducting loop? Assume that the resistance of the rod is 0.400 Ω and that the resistance of the rails and the strip that connects them at the right is negligible. (c) At what rate is thermal energy being generated in the rod? (d) What force must be applied to the rod by an external agent to maintain its motion? (e) At what rate does this external agent do work on the rod? Compare this answer to that for (c).

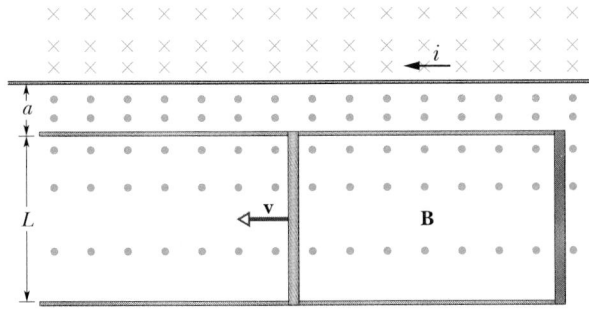

FIGURE 31-58 Problem 39.

SECTION 31-6 Induced Electric Fields

40E. A long solenoid has a diameter of 12.0 cm. When a current i exists in its windings, a uniform magnetic field $B = 30.0$ mT is produced in its interior. By decreasing i, the field is caused to decrease at the rate of 6.50 mT/s. Calculate the magnitude of the induced electric field (a) 2.20 cm and (b) 8.20 cm from the axis of the solenoid.

41E. Figure 31-59 shows two circular regions R_1 and R_2 with radii $r_1 = 20.0$ cm and $r_2 = 30.0$ cm. In R_1 there is a uniform magnetic field $B_1 = 50.0$ mT into the page and in R_2 there is a uniform magnetic field $B_2 = 75.0$ mT out of the page (ignore any fringing of these fields). Both fields are decreasing at the rate of 8.50 mT/s. Calculate the integral $\oint \mathbf{E} \cdot d\mathbf{s}$ for each of the three dashed paths.

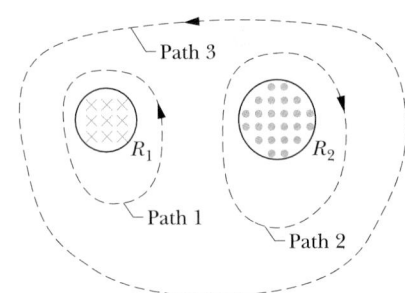

FIGURE 31-59
Exercise 41.

42P. Early in 1981 the Francis Bitter National Magnet Laboratory at M.I.T. commenced operation of a 3.3 cm diameter cylindrical magnet, which produces a 30 T field, then the world's largest steady-state field. The field can be varied sinusoidally between the limits of 29.6 and 30.0 T at a frequency of 15 Hz. When this is done, what is the maximum value of the induced electric field at a radial distance of 1.6 cm from the axis? (*Hint:* See Sample Problem 31-5.)

43P. Figure 31-60 shows a uniform magnetic field **B** confined to a cylindrical volume of radius R. The magnitude of **B** is decreasing at a constant rate of 10 mT/s. What are the instantaneous accelerations (direction and magnitude) experienced by an electron placed at a, at b, and at c? Assume $r = 5.0$ cm.

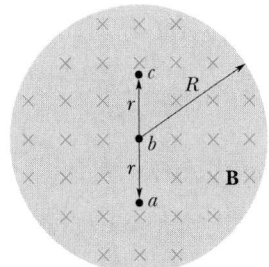

FIGURE 31-60
Problem 43.

44P. Prove that the electric field **E** in a charged parallel-plate capacitor cannot drop abruptly to zero (as is suggested at point a in Fig. 31-61), as one moves perpendicular to the field, say along the horizontal arrow in the figure. Fringing of the field lines always occurs in actual capacitors, which means that **E** approaches zero in a continuous and gradual way (see Problem 45 in

Chapter 30). (*Hint:* Apply Faraday's law to the rectangular path shown by the dashed lines.)

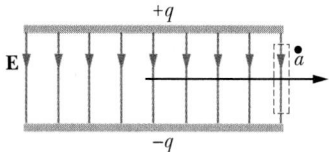

FIGURE 31-61 Problem 44.

SECTION 31-7 Inductors and Inductance

45E. The inductance of a close-packed coil of 400 turns is 8.0 mH. Calculate the magnetic flux through the coil when the current is 5.0 mA.

46E. A circular coil has a 10.0 cm radius and consists of 30.0 closely wound turns of wire. An externally produced magnetic field of 2.60 mT is perpendicular to the coil. (a) If no current is in the coil, what is the flux linkage? (b) When the current in the coil is 3.80 A in a certain direction, the net flux through the coil is found to vanish. What is the inductance of the coil?

47E. A solenoid is wound with a single layer of insulated copper wire (of diameter 2.5 mm) and is 4.0 cm in diameter and 2.0 m long. (a) How many turns are on the solenoid? (b) What is the inductance per meter for the solenoid near its center? Assume that adjacent wires touch and that insulation thickness is negligible.

48P. A long, thin solenoid can be bent into a ring to form a toroid. Show that if the solenoid is long and thin enough, the equation for the inductance of a toroid (Eq. 31-37) is equivalent to that for a solenoid of the appropriate length (Eq. 31-32).

49P. A wide copper strip of width W is bent to form a tube of radius R with two planar extensions, as shown in Fig. 31-62. There is a current i through the strip, distributed uniformly over its width. In this way a "one-turn solenoid" is formed. (a) Derive an expression for the magnitude of the magnetic field **B** in the tubular part (far away from the edges). (*Hint:* Assume that the magnetic field outside this one-turn solenoid is negligibly small.) (b) Find the inductance of this one-turn solenoid, neglecting the two planar extensions.

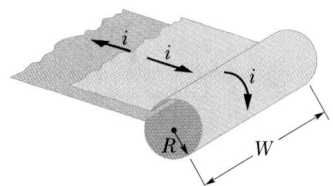

FIGURE 31-62
Problem 49.

50P. Two long parallel wires, each of radius a, whose centers are a distance d apart, carry equal currents in opposite directions. Show that, neglecting the flux within the wires, the inductance of a length l of such a pair of wires is given by

$$L = \frac{\mu_0 l}{\pi} \ln \frac{d-a}{a}.$$

See Sample Problem 30-2. (*Hint:* Calculate the flux through a rectangle of which the wires form two opposite sides.)

SECTION 31-8 Self-Induction

51E. At a given instant the current and self-induced emf in an inductor are as indicated in Fig. 31-63. (a) Is the current increasing or decreasing? (b) The induced emf is 17 V and the rate of change of the current is 25 kA/s; find the inductance.

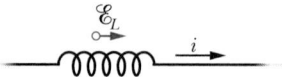

FIGURE 31-63 Exercise 51.

52E. A 12 H inductor carries a steady current of 2.0 A. How can a 60 V self-induced emf be made to appear in the inductor?

53E. A long cylindrical solenoid with 100 turns/cm has a radius of 1.6 cm. Assume that the magnetic field it produces is parallel to its axis and is uniform in its interior. (a) What is its inductance per meter of length? (b) If the current changes at the rate 13 A/s, what emf is induced per meter?

54E. The inductance of a closely wound coil is such that an emf of 3.0 mV is induced when the current changes at the rate of 5.0 A/s. A steady current of 8.0 A produces a magnetic flux of 40 μWb through each turn. (a) Calculate the inductance of the coil. (b) How many turns does the coil have?

55P. The current i through a 4.6 H inductor varies with time t as shown by the graph of Fig. 31-64. The inductor has a resistance of 12 Ω. Find the magnitude of the induced emf $\mathscr{E}$ during the time intervals (a) $t = 0$ to $t = 2$ ms; (b) $t = 2$ ms to $t = 5$ ms; (c) $t = 5$ ms to $t = 6$ ms. (Ignore the behavior at the ends of the intervals.)

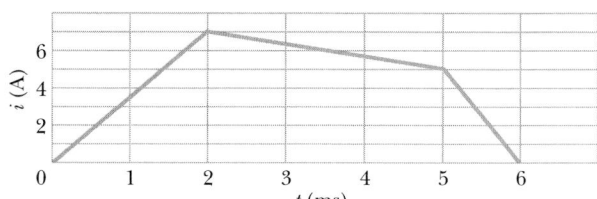

FIGURE 31-64 Problem 55.

56P. *Inductors in series.* Two inductors L_1 and L_2 are connected in series and are separated by a large distance. (a) Show that the equivalent inductance is given by

$$L_{eq} = L_1 + L_2.$$

(*Hint:* Review the derivations for resistors in series and capacitors in series. Which is similar here?) (b) Why must their separation be large for this relationship to hold? (c) What is the generalization of (a) for N inductors in series?

57P. *Inductors in parallel.* Two inductors L_1 and L_2 are connected in parallel and separated by a large distance. (a) Show that the equivalent inductance is given by

$$\frac{1}{L_{eq}} = \frac{1}{L_1} + \frac{1}{L_2}.$$

(*Hint:* Review the derivations for resistors in parallel and capacitors in parallel. Which is similar here?) (b) Why must their separation be large for this relationship to hold? (c) What is the generalization of (a) for N inductors in parallel?

SECTION 31-9 *RL* Circuits

58E. The current in an *RL* circuit builds up to one-third of its steady-state value in 5.00 s. Find the inductive time constant.

59E. In terms of τ_L, how long must we wait for the current in an *RL* circuit to build up to within 0.100% of its equilibrium value?

60E. The current in an *RL* circuit drops from 1.0 A to 10 mA in the first second following removal of the battery from the circuit. If *L* is 10 H, find the resistance *R* in the circuit.

61E. How long would it take, following the removal of the battery, for the potential difference across the resistor in an *RL* circuit (with $L = 2.00$ H, $R = 3.00$ Ω) to decay to 10.0% of its initial value?

62E. (a) Consider the *RL* circuit of Fig. 31-19. In terms of the battery emf $\mathscr{E}$, what is the self-induced emf $\mathscr{E}_L$ when the switch has just been closed on *a*? (b) What is $\mathscr{E}_L$ when $t = 2.0\tau_L$? (c) In terms of τ_L, when will $\mathscr{E}_L$ be just one-half the battery emf $\mathscr{E}$?

63E. A solenoid having an inductance of 6.30 μH is connected in series with a 1.20 kΩ resistor. (a) If a 14.0 V battery is switched across the pair, how long will it take for the current through the resistor to reach 80.0% of its final value? (b) What is the current through the resistor at time $t = 1.0\tau_L$?

64E. The flux linkage through a certain coil of 0.75 Ω resistance is 26 mWb when there is a current of 5.5 A in it. (a) Calculate the inductance of the coil. (b) If a 6.0 V battery is suddenly connected across the coil, how long will it take for the current to rise from 0 to 2.5 A?

65P. Suppose the emf of the battery in the circuit of Fig. 31-20 varies with time *t* so that the current is given by $i(t) = 3.0 + 5.0t$, where *i* is in amperes and *t* is in seconds. Take $R = 4.0$ Ω, $L = 6.0$ H, and find an expression for the battery emf as a function of time. (*Hint:* Apply the loop rule.)

66P. At $t = 0$ a battery is connected to an inductor and a resistor that are connected in series. The table below gives the measured potential difference across the inductor as a function of time following the connection of the battery. Find (a) the emf of the battery and (b) the time constant of the circuit.

t (ms)	V_L (V)	t (ms)	V_L (V)
1.0	18.2	5.0	5.98
2.0	13.8	6.0	4.53
3.0	10.4	7.0	3.43
4.0	7.90	8.0	2.60

67P. A 45.0 V potential difference is suddenly applied to a coil with $L = 50.0$ mH and $R = 180$ Ω. At what rate is the current increasing after 1.20 ms?

68P. A wooden toroidal core with a square cross section has an inner radius of 10 cm and an outer radius of 12 cm. It is wound with one layer of wire (of diameter 1.0 mm and resistance per meter 0.02 Ω/m). What are (a) the inductance and (b) the inductive time constant of the toroid? Ignore the thickness of the insulation on the wire.

69P. In Fig. 31-65, $\mathscr{E} = 100$ V, $R_1 = 10.0$ Ω, $R_2 = 20.0$ Ω, $R_3 = 30.0$ Ω, and $L = 2.00$ H. Find the values of i_1 and i_2 (a) immediately after the closing of switch S, (b) a long time later, (c) immediately after the reopening of switch S, and (d) a long time after the reopening.

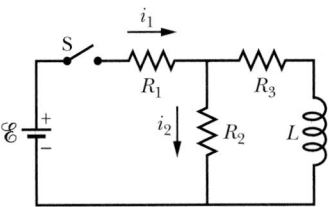

FIGURE 31-65 Problem 69.

70P. In the circuit shown in Fig. 31-66, $\mathscr{E} = 10$ V, $R_1 = 5.0$ Ω, $R_2 = 10$ Ω, and $L = 5.0$ H. For the two separate conditions (I) switch S just closed and (II) switch S closed for a long time, calculate (a) the current i_1 through R_1, (b) the current i_2 through R_2, (c) the current *i* through the switch, (d) the potential difference across R_2, (e) the potential difference across *L*, and (f) the rate of change di_2/dt.

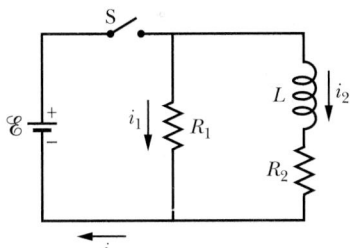

FIGURE 31-66 Problem 70.

71P. Switch S in Fig. 31-67 is closed for $t < 0$ and is opened at $t = 0$. When current i_1 through L_1 and current i_2 through L_2 are *first* equal to each other, what is their common value? (The resistors have the same resistance *R*.)

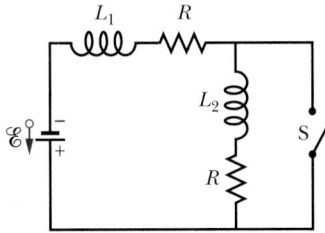

FIGURE 31-67 Problem 71.

72P. In Fig. 31-68, the component in the upper branch is an ideal 3.0 A fuse. It has zero resistance as long as the current through it remains less than 3.0 A. If the current reaches 3.0 A, it "blows" and thereafter has infinite resistance. Switch S is closed at time $t = 0$. (a) When does the fuse blow? (*Hint:* Equation 31-46 does not apply. Rethink Eq. 31-44.) (b) Sketch a graph of the current *i* through the inductor as a function of time. Mark the time at which the fuse blows.

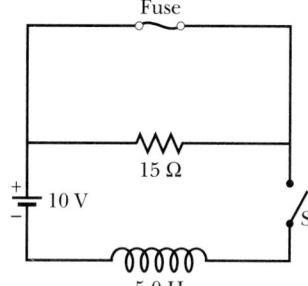

FIGURE 31-68
Problem 72.

73P*. In the circuit shown in Fig. 31-69, switch S is closed at time $t = 0$. Thereafter the constant current source, by varying its emf, maintains a constant current i out of its upper terminal. (a) Derive an expression for the current through the inductor as a function of time. (b) Show that the current through the resistor equals the current through the inductor at time $t = (L/R)\ln 2$.

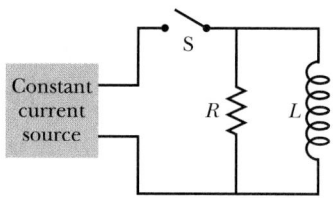

FIGURE 31-69 Problem 73.

SECTION 31-10 Energy Stored in a Magnetic Field

74E. The magnetic energy stored in a certain inductor is 25.0 mJ when the current is 60.0 mA. (a) Calculate the inductance. (b) What current is required for the stored magnetic energy to be four times as much?

75E. Consider the circuit of Fig. 31-20. In terms of the time constant, at what instant after the battery is connected will the energy stored in the magnetic field of the inductor be half its steady-state value?

76E. A coil with an inductance of 2.0 H and a resistance of 10 Ω is suddenly connected to a resistanceless battery with $\mathscr{E} = 100$ V. (a) What is the equilibrium current? (b) How much energy is stored in the magnetic field when this current exists in the coil?

77E. A coil with an inductance of 2.0 H and a resistance of 10 Ω is suddenly connected to a resistanceless battery with $\mathscr{E} = 100$ V. At 0.10 s after the connection is made, what are the rates at which (a) energy is being stored in the magnetic field, (b) thermal energy is appearing in the resistance, and (c) energy is being delivered by the battery?

78P. Suppose that the inductive time constant for the circuit of Fig. 31-20 is 37.0 ms and the current in the circuit is zero at time $t = 0$. At what time does the rate at which energy is dissipated in the resistor equal the rate at which energy is being stored in the inductor?

79P. A coil is connected in series with a 10.0 kΩ resistor. When a 50.0 V battery is applied to the two, the current reaches a value of 2.00 mA after 5.00 ms. (a) Find the inductance of the coil. (b) How much energy is stored in the coil at this same moment?

80P. For the circuit of Fig. 31-20, assume that $\mathscr{E} = 10.0$ V, $R = 6.70$ Ω, and $L = 5.50$ H. The battery is connected at time $t = 0$. (a) How much energy is delivered by the battery during the first 2.00 s? (b) How much of this energy is stored in the magnetic field of the inductor? (c) How much of this energy has been dissipated in the resistor?

81P. A solenoid, with length 80.0 cm and radius 5.00 cm, consists of 3000 turns distributed uniformly over its length. Its total resistance is 10.0 Ω. At 5.00 ms after it is connected to a 12.0 V battery, (a) how much energy is stored in its magnetic field and (b) how much energy has been supplied by the battery up to that time? (Neglect end effects.)

82P. Prove that, after switch S in Fig. 31-19 has been thrown from a to b, all the energy stored in the inductor will ultimately appear as thermal energy in the resistor.

SECTION 31-11 Energy Density of a Magnetic Field

83E. A solenoid 85.0 cm long has a cross-sectional area of 17.0 cm². There are 950 turns of wire carrying a current of 6.60 A. (a) Calculate the energy density of the magnetic field inside the solenoid. (b) Find the total energy stored in the magnetic field there (neglect end effects).

84E. A toroidal inductor with an inductance of 90.0 mH encloses a volume of 0.0200 m³. If the average energy density in the toroid is 70.0 J/m³, what is the current?

85E. What must be the magnitude of a uniform electric field if it is to have the same energy density as that possessed by a 0.50 T magnetic field?

86E. The magnetic field in the interstellar space of our galaxy has a magnitude of about 10^{-10} T. How much energy is stored in this field in a cube 10 light years on edge? (For scale, note that the nearest star is 4.3 light-years distant and the radius of our galaxy is about 8×10^4 light-years.)

87E. Use the result of Sample Problem 31-11 to obtain an expression for the inductance of a length l of the coaxial cable.

88E. Calculate the energy needed to produce, in a cube that is 10 cm on edge, (a) a uniform electric field of 100 kV/m and (b) a uniform magnetic field of 1.0 T. (Both these large fields are readily available in the laboratory.) (c) From these answers, tell which type of field can store greater amounts of energy.

89E. A circular loop of wire 50 mm in radius carries a current of 100 A. (a) Find the magnetic field strength at the center of the loop. (b) Calculate the energy density at the center of the loop.

90P. (a) For the toroid of Sample Problem 31-6b, find an expression for the energy density as a function of the radial distance r from the center. (b) By integrating the energy density over the volume of the toroid, calculate the total energy stored in the field of the toroid; assume $i = 0.500$ A. (c) Using Eq. 31-53, evaluate the energy stored in the toroid directly from the inductance and compare it with your answer in (b).

91P. A length of copper wire carries a current of 10 A, uniformly distributed. Calculate (a) the energy density of the magnetic field and (b) the energy density of the electric field at the surface of the wire. The wire diameter is 2.5 mm, and its resistance per unit length is 3.3 Ω/km.

92P. (a) What is the energy density of Earth's magnetic field, which has a magnitude of 50 μT? (b) Assuming this density to be relatively constant over distances small compared with Earth's radius and neglecting variations near the magnetic poles, how much energy would be stored between Earth's surface and a spherical shell 16 km above the surface?

SECTION 31-12 Mutual Induction

93E. Two coils are at fixed locations. When coil 1 has no current and the current in coil 2 increases at the rate 15.0 A/s, the emf in coil 1 is 25.0 mV. (a) What is their mutual inductance? (b) When coil 2 has no current and coil 1 has a current of 3.60 A, what is the flux linkage in coil 2?

94E. Coil 1 has $L_1 = 25$ mH and $N_1 = 100$ turns. Coil 2 has $L_2 = 40$ mH and $N_2 = 200$ turns. The coils are rigidly positioned with respect to each other; their mutual inductance M is 3.0 mH. A 6.0 mA current in coil 1 is changing at the rate of 4.0 A/s. (a) What flux Φ_{12} links coil 1, and what self-induced emf appears there? (b) What flux Φ_{21} links coil 2, and what mutually induced emf appears there?

95E. Two solenoids are part of the spark coil of an automobile. When the current in one solenoid falls from 6.0 A to zero in 2.5 ms, an emf of 30 kV is induced in the other solenoid. What is the mutual inductance M of the solenoids?

96P. Two coils, connected as shown in Fig. 31-70, separately have inductances L_1 and L_2. The mutual inductance is M. (a) Show that this combination can be replaced by a single coil of equivalent inductance given by

$$L_{eq} = L_1 + L_2 + 2M.$$

(b) How could the coils in Fig. 31-70 be reconnected to yield an equivalent inductance of

$$L_{eq} = L_1 + L_2 - 2M?$$

(This problem is an extension of Problem 56, but the requirement that the coils be far apart has been removed.)

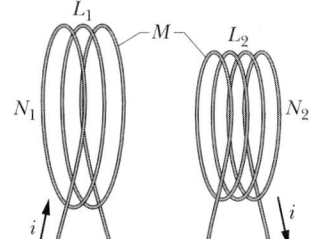

FIGURE 31-70
Problem 96.

97P. A coil C of N turns is placed around a long solenoid S of radius R and n turns per unit length, as in Fig. 31-71. Show that the mutual inductance for the coil–solenoid combination is given

by $M = \mu_0 \pi R^2 nN$. Explain why M does not depend on the shape, size, or possible lack of close-packing of the coil.

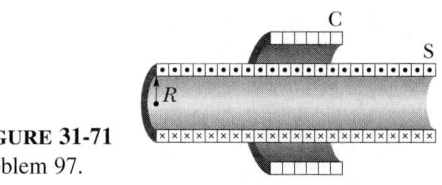

FIGURE 31-71
Problem 97.

98P. Figure 31-72 shows a coil of N_2 turns wound as shown around part of a toroid of N_1 turns. The toroid's inner radius is a, its outer radius is b, and its height is h. Show that the mutual inductance M for the toroid–coil combination is

$$M = \frac{\mu_0 N_1 N_2 h}{2\pi} \ln \frac{b}{a}.$$

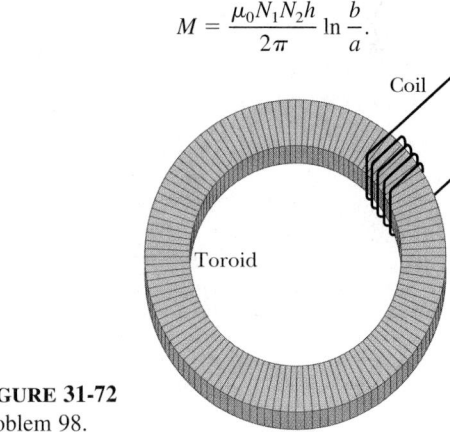

FIGURE 31-72
Problem 98.

99P. Figure 31-73 shows, in cross section, two coaxial solenoids. Show that the mutual inductance M for a length l of this solenoid–solenoid combination is given by $M = \pi R_1^2 l \mu_0 n_1 n_2$, in which n_1 and n_2 are the respective numbers of turns per unit length and R_1 is the radius of the inner solenoid. Why does M depend on R_1 and not on R_2?

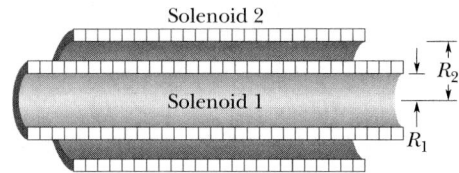

FIGURE 31-73 Problem 99.

100P. A rectangular loop of N close-packed turns is positioned near a long straight wire as in Fig. 31-74. (a) What is the mutual inductance M for the loop–wire combination? (b) Evaluate M for $N = 100$, $a = 1.0$ cm, $b = 8.0$ cm, and $l = 30$ cm.

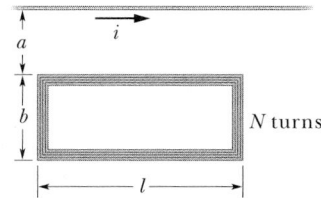

FIGURE 31-74
Problem 100.

32

Magnetism of Matter; Maxwell's Equations

The direction of Earth's magnetic field is not fixed; rather, it gradually wanders. One way we can determine the direction of the field during some period in the past is to examine a clay-walled kiln that was used to bake pottery at that time. But how and why would a clay kiln record Earth's magnetic field?

32-1 MAGNETS

The first known magnets were *lodestones*, which are stones that have been *magnetized* (made magnetic) naturally. When the ancient Greeks and ancient Chinese discovered these rare stones, they were amused by the stones' ability to attract metal over a short distance, as if by magic. Only much later did they learn to use lodestones (and artificially magnetized pieces of iron) in compasses to determine direction.

Today, magnets and magnetic materials are ubiquitous. We find them in VCRs, audio cassettes, ATM and credit cards, audio headsets, and even in the inks for paper money. In fact, some breakfast cereals that are "iron fortified" contain small bits of magnetic materials (you can collect them from a slurry of cereal and water with a magnet). More important, the modern electronics industry as we know it (including the music and information domains) would not exist without magnetic materials.

The magnetic properties of materials can be traced back to their atoms and electrons. We begin here, however, with the bar magnet in Fig. 32-1. As you have seen, iron filings sprinkled around such a magnet tend to align with the magnetic field of the magnet, and their pattern reveals the magnetic field lines. The clustering of the lines at the ends of the magnet suggests that one end is a *source* of the lines (the field diverges from it) and the other end is a *sink* of the lines (the field converges toward it). By convention, we call the source the *north pole* of the magnet and the opposite end the *south pole,* and we say that the magnet, with its two poles, is an example of a **magnetic dipole.**

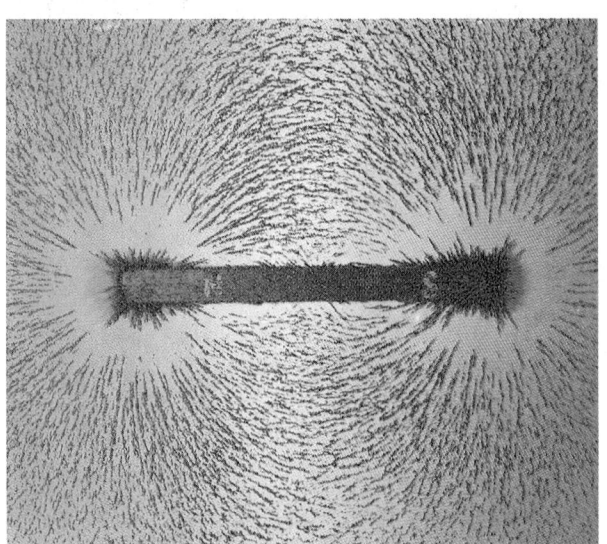

FIGURE 32-1 A bar magnet is a magnetic dipole. The iron filings suggest the magnetic field lines. (The background is illuminated with colored light.)

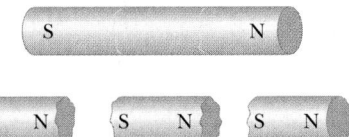

FIGURE 32-2 If you break a magnet, each fragment becomes a separate magnet, with its own north and south poles.

Suppose we break apart a bar magnet the way we break a piece of chalk (Fig. 32-2). We should, it seems, be able to isolate a single pole, or *monopole*. But surprisingly we cannot, not even if we break the magnet down to its individual atoms and then to its electrons and nuclei. Each fragment of the magnet has a north pole and a south pole. So we must conclude the following:

The simplest magnetic structure that can exist is a magnetic dipole. Magnetic monopoles do not exist (as far as we know).

32-2 GAUSS' LAW FOR MAGNETIC FIELDS

Gauss' law for magnetic fields is a formal way of saying that magnetic monopoles do not exist. The law asserts that the net magnetic flux Φ_B through any closed Gaussian surface is zero:

$$\Phi_B = \oint \mathbf{B} \cdot d\mathbf{A} = 0 \qquad \text{(Gauss' law for magnetic fields).} \qquad (32\text{-}1)$$

Contrast this with Gauss' law for electric fields,

$$\Phi_E = \oint \mathbf{E} \cdot d\mathbf{A} = \frac{1}{\epsilon_0} q \qquad \text{(Gauss' law for electric fields).}$$

In both equations, the integral is taken over a *closed* Gaussian surface. Gauss' law for electric fields says that this integral (the net electric flux through the surface) is proportional to the net electric charge q enclosed by the surface. Gauss' law for magnetic fields says that there can be no net magnetic flux through the surface because there can be no net "magnetic charge" (individual magnetic poles) enclosed by the surface. The simplest magnetic structure that can exist and thus be enclosed by a Gaussian surface is a dipole, which consists of both a source and a sink for the field lines. Thus, there must always be as much magnetic flux into the surface as out of it, and the net magnetic flux must always be zero.

Gauss' law for magnetic fields holds for more complicated structures than a magnetic dipole, and its holds even

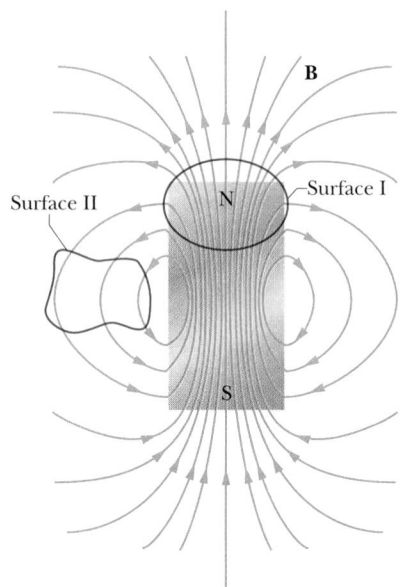

FIGURE 32-3 The field lines for the magnetic field **B** of a short bar magnet. The red curves represent cross sections of closed, three-dimensional Gaussian surfaces.

if the Gaussian surface does not enclose the entire structure. Gaussian surface II near the bar magnet of Fig. 32-3 encloses no poles, and we can easily conclude that the net magnetic flux through it is zero. But Gaussian surface I is more difficult. It may seem to enclose only the north pole of the magnet because it encloses the label N and not the label S. But a south pole must be associated with the lower boundary of the surface, because magnetic field lines enter the surface there. (The enclosed section is like one piece of the broken bar magnet in Fig. 32-2.) Thus, Gaussian surface I encloses a magnetic dipole and the net flux through the surface is zero.

$\mathbb{C}$HECKPOINT **1:** The figure shows four closed surfaces with flat top and bottom faces and curved sides. The table gives the area A of the faces and the magnitudes B of the uniform and perpendicular magnetic fields through those faces; the units of A and B are arbitrary but consistent. Rank the surfaces according to the magnitudes of the magnetic flux through their curved sides, greatest first.

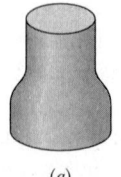  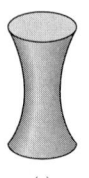 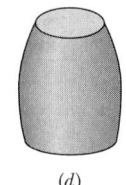

SURFACE	A_{top}	B_{top}	A_{bot}	B_{bot}
a	2	6, outward	4	3, inward
b	2	1, inward	4	2, inward
c	2	6, inward	2	8, outward
d	2	3, outward	3	2, outward

32-3 THE MAGNETISM OF EARTH

Earth is a huge magnet; for points near Earth's surface, its magnetic field can be represented as the field of a huge bar magnet—a magnetic dipole—that straddles the center of the planet. Figure 32-4 is an idealized symmetric depiction of the dipole field, without the distortion caused by passing charged particles from the Sun.

Because Earth's magnetic field is that of a magnetic dipole, a magnetic dipole moment μ is associated with the field. For the idealized field of Fig. 32-4, the magnitude of μ is 8.0×10^{22} J/T and the direction of μ makes an angle of $11.5°$ with the rotation axis (RR) of Earth. The *dipole axis* (*MM* in Fig. 32-4) lies along μ and intersects Earth's surface at the *geomagnetic north pole* in northwest Greenland and the *geomagnetic south pole* in Antarctica. The lines of the magnetic field **B** generally emerge in the southern hemisphere and reenter Earth in the northern hemisphere. Thus, the magnetic pole that is in Earth's northern hemisphere and known as a "north magnetic pole" *is really the south pole of Earth's magnetic dipole.*

The direction of the magnetic field at any location on Earth's surface is commonly specified in terms of two angles. The **field declination** is the angle (left or right)

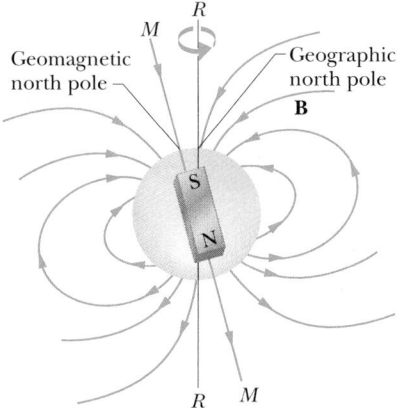

FIGURE 32-4 Earth's magnetic field represented as a dipole field. The dipole axis *MM* makes an angle of $11.5°$ with Earth's rotational axis *RR*. The south pole of the dipole is in Earth's northern hemisphere.

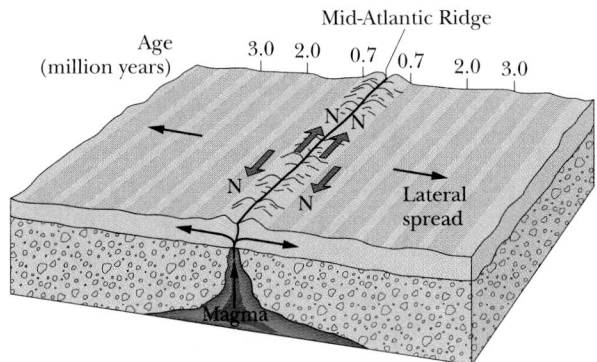

FIGURE 32-5 A magnetic profile of the seafloor on either side of the Mid-Atlantic Ridge. The seafloor, extruded through the ridge and spreading out as part of the tectonic drift system, displays a record of the past magnetic history of Earth's core. The direction of the magnetic field produced by the core reverses about every million years.

between geographic north (which is toward 90° latitude) and the horizontal component of the field. The **field inclination** is the angle (up or down) between a horizontal plane and the field's direction.

Magnetometers measure these angles and determine the field with much precision. However, you can do reasonably well with just a *compass* and a *dip meter*. A compass is simply a needle-shaped magnet that is mounted so that it can rotate freely about a vertical axis. When it is held in a horizontal plane, the north-pole end of the needle points, generally, toward the geomagnetic north pole (really a south magnetic pole, remember). The angle between the needle and geographic north is the field declination. A dip meter is a similar magnet that can rotate freely about a horizontal axis. When its vertical plane of rotation is aligned with the direction of the compass, the angle between the meter's needle and the horizontal is the field inclination.

At any point on Earth's surface, the measured magnetic field may differ appreciably, in both magnitude and direction, from the idealized dipole field of Fig. 32-4. In fact, the point where the field is actually vertical to Earth's surface and inward is not located at the geomagnetic north pole in Greenland as we would expect; instead this so-called *dip north pole* is located in the Queen Elizabeth Islands in northern Canada, far from Greenland.

In addition, the field observed at any location on the surface of Earth varies with time, by measurable amounts over a period of a few years and by substantial amounts over, say, 100 years. For example, between 1580 and 1820 the direction indicated by compass needles in London changed by 35°.

In spite of these local variations, the average dipole

field changes only slowly over such relatively short time periods. Variations over longer periods can be studied by measuring the weak magnetism of the ocean floor on either side of the Mid-Atlantic Ridge (Fig. 32-5). This floor has been formed by molten magma that oozed up through the ridge from Earth's interior, solidified, and was pulled away from the ridge (by the drift of tectonic plates) at the rate of a few centimeters per year. As the magma solidified, it became weakly magnetized with its magnetic field in the direction of Earth's magnetic field at the time of solidification. Study of this solidified magma across the ocean floor reveals that Earth's field has reversed its *polarity* (directions of the north pole and south pole) about every million years. The reason for the reversals is not known. In fact, the mechanism that produces Earth's magnetic field is only vaguely understood.

SAMPLE PROBLEM 32-1

In Tucson, Arizona, in 1964, the north pole of a compass needle pointed 13° east of geographic north, and the north pole of a dip-meter needle pointed downward, 59° below the horizontal. The horizontal component B_h of Earth's magnetic field **B** in Tucson had a magnitude of 26 μT. What was the magnitude B of the field then, in units of gauss? (Earth's field is often measured in gauss.)

SOLUTION: Figure 32-6 shows the given data; it is drawn in a plane that contains the vector **B** and so is angled 13° east of geographic north. From the figure we have

$$B = \frac{B_h}{\cos \theta} = \frac{26 \ \mu\text{T}}{\cos 59°}$$

$$= 50 \ \mu\text{T} = 0.50 \text{ gauss.} \qquad \text{(Answer)}$$

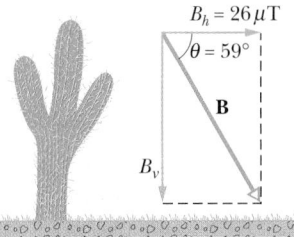

FIGURE 32-6 Sample Problem 32-1. Earth's magnetic field, along with its horizontal and vertical components, at Tucson, Arizona, in 1964.

32-4 MAGNETISM AND ELECTRONS

Magnetic materials, from lodestones to videotapes, are magnetic because of the electrons within them. We have already seen one way in which electrons can generate a magnetic field: send them through a wire as an electric current, and their motion produces a magnetic field around the wire. There are two more ways, each involving a magnetic dipole moment that produces a magnetic field in the surrounding space. However, their explanation requires quantum physics that is beyond the physics presented in this book. So here we shall only outline the results.

Spin Magnetic Dipole Moment

An electron has an intrinsic angular momentum called its **spin angular momentum** (or just **spin**) $\mathbf{S}$; associated with this spin is an intrinsic **spin magnetic dipole moment** $\boldsymbol{\mu}_s$. (By *intrinsic*, we mean that $\mathbf{S}$ and $\boldsymbol{\mu}_s$ are basic characteristics of an electron, like its mass and electric charge.) $\mathbf{S}$ and $\boldsymbol{\mu}_s$ are related by

$$\boldsymbol{\mu}_s = -\frac{e}{m}\mathbf{S}, \qquad (32\text{-}2)$$

in which e is the elementary charge (1.60×10^{-19} C) and m is the mass of an electron (9.11×10^{-31} kg). The minus sign means that $\boldsymbol{\mu}_s$ and $\mathbf{S}$ are oppositely directed.

Spin $\mathbf{S}$ is quite different from the angular momenta of Chapter 12 in two respects:

1. $\mathbf{S}$ itself cannot be measured. Instead, only its component along an axis can be measured.

2. A measured component of $\mathbf{S}$ is quantized (restricted to certain values); in fact, it always has the same magnitude (no matter which axis is chosen).

Let us assume that the component of spin $\mathbf{S}$ is measured along the z axis of a coordinate system. Then the measured component S_z can have only the two values given by

$$S_z = m_s \frac{h}{2\pi}, \qquad \text{for} \quad m_s = \pm\tfrac{1}{2}, \qquad (32\text{-}3)$$

where m_s is the *spin magnetic quantum number* and h ($= 6.63 \times 10^{-34}$ J·s) is the Planck constant, the ubiquitous constant of quantum physics. The signs given in Eq. 32-3 have to do with the direction of S_z along the z axis. When S_z is parallel to the z axis, m_s is $+\tfrac{1}{2}$ and the electron is said to be *spin up*. When S_z is antiparallel to the z axis, m_s is $-\tfrac{1}{2}$ and the electron is said to be *spin down*.

The spin magnetic dipole moment $\boldsymbol{\mu}_s$ of an electron also cannot itself be measured; only a component can be measured, and that component is quantized and always has the same magnitude. We can relate the component $\mu_{s,z}$ measured on the z axis to S_z by rewriting Eq. 32-2 in component form for the z axis as

$$\mu_{s,z} = -\frac{e}{m}S_z.$$

Substituting for S_z from Eq. 32-3 then gives us

$$\mu_{s,z} = \pm\frac{eh}{4\pi m}, \qquad (32\text{-}4)$$

where the plus and minus signs correspond to $\mu_{s,z}$ being parallel and antiparallel to the z axis, respectively.

The quantity on the right side of Eq. 32-4 is called the *Bohr magneton* μ_B:

$$\mu_B = \frac{eh}{4\pi m} = 9.27 \times 10^{-24} \text{ J/T}$$

$$\text{(Bohr magneton).} \quad (32\text{-}5)$$

Spin magnetic dipole moments of electrons and other elementary particles can be expressed in terms of μ_B. For the electron, the magnitude of the measured component of $\boldsymbol{\mu}_s$ is

$$\mu_{s,z} = 1\mu_B. \qquad (32\text{-}6)$$

(The quantum physics of the electron, called *quantum electrodynamics*, or QED, reveals that $\mu_{s,z}$ is actually slightly greater than $1\mu_B$, but we shall neglect that fact.)

When an electron is placed in an external magnetic field $\mathbf{B}_{\text{ext}}$, a potential energy U can be associated with the orientation of the electron's spin magnetic dipole moment $\boldsymbol{\mu}_s$ just as a potential energy can be associated with the orientation of the magnetic dipole moment $\boldsymbol{\mu}$ of a current loop placed in $\mathbf{B}_{\text{ext}}$. From Eq. 29-36, the potential energy for the electron is

$$U = -\boldsymbol{\mu}_s \cdot \mathbf{B}_{\text{ext}} = -\mu_{s,z}B, \qquad (32\text{-}7)$$

where the z axis is taken to be in the direction of $\mathbf{B}_{\text{ext}}$.

If we imagine an electron to be a microscopic sphere (which it is not), we can represent the spin $\mathbf{S}$, the spin magnetic dipole moment $\boldsymbol{\mu}_s$, and the associated magnetic dipole field as in Fig. 32-7. Although we use the word "spin" here, electrons do not spin like tops. How, then, can something have angular momentum without actually rotating? Again, quantum physics provides the answer.

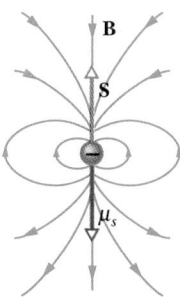

FIGURE 32-7 The spin $\mathbf{S}$, spin magnetic dipole moment $\boldsymbol{\mu}_s$, and magnetic field $\mathbf{B}$ of an electron represented as a microscopic sphere.

Protons and neutrons also have an intrinsic angular momentum called spin and an associated intrinsic spin magnetic dipole moment. For a proton those two vectors have the same direction, and for a neutron they have opposite directions. We shall not examine the contributions of these dipole moments to the magnetic fields of atoms because they are about a thousand times smaller than that due to an electron.

CHECKPOINT 2: The figure shows the spin orientations of two particles in an external magnetic field $\mathbf{B}_{ext}$. (a) If the particles are electrons, which spin orientation is at lower potential energy? (b) If, instead, the particles are protons, which spin orientation is at lower potential energy?

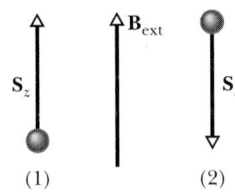

Orbital Magnetic Dipole Moment

When it is in an atom, an electron has an additional angular momentum called its **orbital angular momentum** $\mathbf{L}_{orb}$. Associated with $\mathbf{L}_{orb}$ is an **orbital magnetic dipole moment** $\boldsymbol{\mu}_{orb}$; the two are related by

$$\boldsymbol{\mu}_{orb} = -\frac{e}{2m}\mathbf{L}_{orb}. \qquad (32\text{-}8)$$

The minus sign means that $\boldsymbol{\mu}_{orb}$ and $\mathbf{L}_{orb}$ are in opposite directions.

Orbital angular momentum $\mathbf{L}_{orb}$ cannot be measured; only components along an axis can be measured, and those components are quantized. The components along, say, a z axis can have only the values given by

$$L_{orb,z} = m_l \frac{h}{2\pi}, \qquad \text{for} \quad m_l = 0, \pm 1, \pm 2, \cdots, \pm \text{ (limit)},$$
$$(32\text{-}9)$$

in which m_l is the *orbital magnetic quantum number* and "limit" refers to some largest allowed integer value for m_l. The signs in Eq. 32-9 have to do with the direction of $L_{orb,z}$ on the z axis.

The orbital magnetic dipole moment $\boldsymbol{\mu}_{orb}$ of an electron also cannot itself be measured; only a component can be measured, and that component is quantized. By writing Eq. 32-8 in component form for the z axis and then substituting for $L_{orb,z}$ from Eq. 32-9, we can write the z compo-

nent $\mu_{orb,z}$ of the orbital magnetic dipole moment as

$$\mu_{orb,z} = -m_l \frac{eh}{4\pi m} \qquad (32\text{-}10)$$

and, in terms of the Bohr magneton, as

$$\mu_{orb,z} = -m_l \mu_B. \qquad (32\text{-}11)$$

When an atom is placed in an external magnetic field $\mathbf{B}_{ext}$, a potential energy U can be associated with the orientation of the orbital magnetic dipole moment of each electron in the atom. Its value is

$$U = -\boldsymbol{\mu}_{orb} \cdot \mathbf{B}_{ext} = -\mu_{orb,z}B_{ext}, \qquad (32\text{-}12)$$

where the z axis is taken in the direction of $\mathbf{B}_{ext}$.

Although we have used the words "orbit" and "orbital" here, electrons do not orbit the nucleus of an atom like planets orbiting the Sun. How can an electron have an orbital angular momentum without orbiting in the common meaning of the term? Once more the answer requires quantum physics.

Loop Model for Electron Orbits

We can obtain Eq. 32-8 with the nonquantum derivation that follows, in which we assume that an electron moves along a circular path with a radius that is much larger than an atomic radius (hence the name "loop model"). However, the derivation does not apply to an electron within an atom (for which we need quantum physics).

We imagine an electron moving at constant speed v in a circular path of radius r, counterclockwise as shown in Fig. 32-8. The motion of the negative charge of the electron is equivalent to a conventional current i (of positive

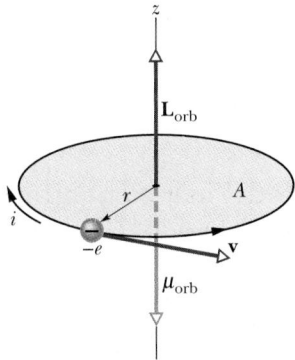

FIGURE 32-8 An electron moving at constant speed v in a circular path of radius r that encloses an area A. The electron has an orbital angular momentum $\mathbf{L}_{orb}$ and an associated orbital magnetic dipole moment $\boldsymbol{\mu}_{orb}$. A clockwise current i (of positive charge) is equivalent to the counterclockwise circulation of the negatively charged electron.

charge) that is clockwise, as also shown in Fig. 32-8. The magnitude of the orbital magnetic dipole moment of such a *current loop* is obtained from Eq. 29-33 with $N = 1$:

$$\mu_{orb} = iA, \qquad (32\text{-}13)$$

where A is the area enclosed by the loop. The direction of this magnetic dipole moment is, from the right-hand rule of Fig. 30-22, downward in Fig. 32-8.

To evaluate Eq. 32-13, we need the current i. Current is, generally, the rate at which charge passes some point in a circuit. Here, the charge of magnitude e takes a time $T = 2\pi r/v$ to circle from any point back through that point, so

$$i = \frac{charge}{time} = \frac{e}{2\pi r/v}. \qquad (32\text{-}14)$$

Substituting this and the area $A = \pi r^2$ of the loop into Eq. 32-13 gives us

$$\mu_{orb} = \frac{e}{2\pi r/v}\,\pi r^2 = \frac{evr}{2}. \qquad (32\text{-}15)$$

To get an expression for the orbital angular momentum $\mathbf{L}_{orb}$ of the electron, we use Eq. 12-25, $\ell = m(\mathbf{r} \times \mathbf{v})$. Because $\mathbf{r}$ and $\mathbf{v}$ are perpendicular, $\mathbf{L}_{orb}$ has the magnitude

$$L_{orb} = mrv \sin 90° = mrv. \qquad (32\text{-}16)$$

$\mathbf{L}_{orb}$ is directed upward in Fig. 32-8 (see Fig. 12-12). Combining Eqs. 32-15 and 32-16, generalizing to a vector formulation, and indicating the opposite directions of the vectors with a minus sign yield

$$\boldsymbol{\mu}_{orb} = -\frac{e}{2m}\mathbf{L}_{orb},$$

which is Eq. 32-8. Thus by "classical" (nonquantum) analysis we have obtained the same result, in both magnitude and direction, given by quantum physics. You might wonder, since this derivation gives the correct result for an electron within an atom, why the derivation is invalid for that situation. The answer is that this line of reasoning yields other results that are contradicted by experiments.

Loop Model in a Nonuniform Field

We continue to consider an electron orbit as a current loop, as we did in Fig. 32-8. Now, however, we draw the loop in a nonuniform magnetic field $\mathbf{B}_{ext}$ as shown in Fig. 32-9a. (This field is the diverging field near the north pole of the magnet in Fig. 32-3.) We make this change to prepare for the next several sections, in which we shall discuss the forces that act on magnetic materials when the materials are placed in a nonuniform magnetic field. We shall discuss these forces by assuming that the electron orbits in the materials are tiny current loops like that in Fig. 32-9a.

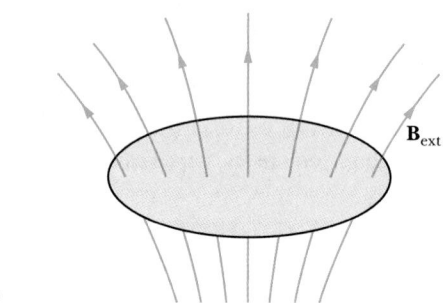

(a)

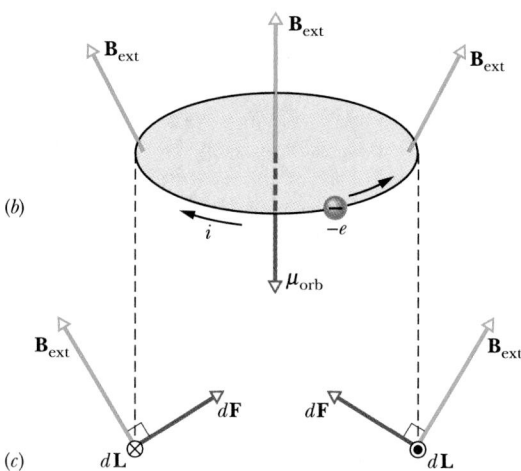

(b)

(c)

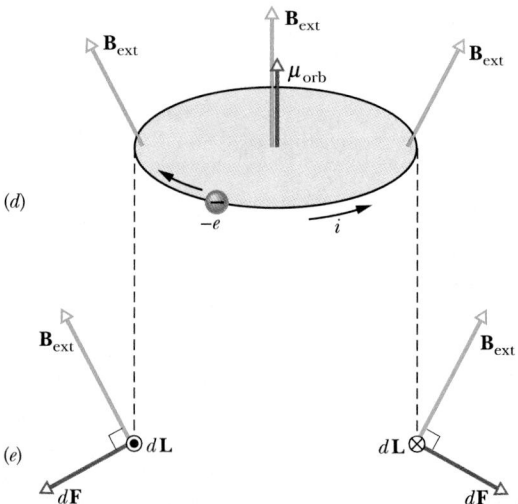

(d)

(e)

FIGURE 32-9 (a) A loop model for an electron orbiting in an atom while in a nonuniform magnetic field $\mathbf{B}_{ext}$. (b) Charge $-e$ moves counterclockwise; the associated conventional current i is clockwise. (c) The magnetic forces $d\mathbf{F}$ on the left and right sides of the loop, as seen from the plane of the loop. The net force on the loop is upward. (d) Charge $-e$ moves clockwise. (e) The net force on the loop is now downward.

Here we assume that the magnetic field vectors all around the electron's circular path have the same magnitude and form the same angle with the vertical, as shown in Figs. 32-9b and d. We also assume that all the electrons in an atom move either counterclockwise (Fig. 32-9b) or clockwise (Fig. 32-9d). The associated conventional current i around the current loop and the orbital magnetic dipole moment $\boldsymbol{\mu}_{orb}$ produced by i are shown for each of these directions of motion.

Figures 32-9c and e show diametrically opposite views of an element $d\mathbf{L}$ of the loop with the same direction as i, as seen from the plane of the orbit. Also shown are the field $\mathbf{B}_{ext}$ and the resulting magnetic force $d\mathbf{F}$ on $d\mathbf{L}$. Recall that a current along an element $d\mathbf{L}$ in a magnetic field $\mathbf{B}_{ext}$ experiences a magnetic force $d\mathbf{F}$ as given by Eq. 29-27:

$$d\mathbf{F} = i \, d\mathbf{L} \times \mathbf{B}_{ext}. \qquad (32\text{-}17)$$

On the left side of Fig. 32-9c, Eq. 32-17 tells us that the force $d\mathbf{F}$ is directed upward and rightward. On the right side, the force $d\mathbf{F}$ is just as large and is directed upward and leftward. Because their angles are the same, the horizontal components of these two forces cancel and the vertical components add. The same is true at any other two symmetric points on the loop. So, the net force on the current loop of Fig. 32-9b must be upward. The same reasoning leads to a downward net force on the loop in Fig. 32-9d. We shall use these two results shortly when we examine the behavior of magnetic materials in nonuniform magnetic fields.

32-5 MAGNETIC MATERIALS

Each electron in an atom has an orbital magnetic dipole moment and a spin magnetic dipole moment that combine vectorially. The resultant of these two vectors combines vectorially with similar resultants for all other electrons in the atom. And the resultant for each atom combines with those for all the other atoms in a sample of a material. If the combination of all these magnetic dipole moments produces a magnetic field, then the material is magnetic. There are three general types of magnetism: diamagnetism, paramagnetism, and ferromagnetism.

1. **Diamagnetism** is exhibited by all common materials but is so feeble that it is masked if the material also exhibits magnetism of either of the other two types. In diamagnetism, weak magnetic dipole moments are produced in the atoms of the material when the material is placed in an external magnetic field $\mathbf{B}_{ext}$; the combination of all those induced dipole moments gives the material as a whole only a feeble net magnetic field. The dipole moments and thus their net field disappear when $\mathbf{B}_{ext}$ is removed. The term

diamagnetic material usually refers to materials that exhibit only diamagnetism.

2. **Paramagnetism** is exhibited by materials containing transition elements, rare earth elements, and actinide elements (see Appendix G). Each atom of such a material has a permanent resultant magnetic dipole moment, but the moments are randomly oriented in the material and the material as a whole lacks a net magnetic field. However, an external magnetic field $\mathbf{B}_{ext}$ can partially align the atomic magnetic dipole moments to give the material a net magnetic field. The alignment and thus its field disappear when $\mathbf{B}_{ext}$ is removed. The term *paramagnetic material* usually refers to materials that exhibit primarily paramagnetism.

3. **Ferromagnetism** is a property of iron, nickel, and certain other elements (and of compounds and alloys of these elements). Some of the electrons in these materials align their resultant magnetic dipole moments to produce regions with strong magnetic dipole moments. An external field $\mathbf{B}_{ext}$ can then align the magnetic moments of these regions, producing a strong magnetic field for the material as a whole; the field partially persists when $\mathbf{B}_{ext}$ is removed. We usually use the term *ferromagnetic material,* and even the common term *magnetic material,* to refer to materials that exhibit primarily ferromagnetism.

The next three sections explore these three types of magnetism.

32-6 DIAMAGNETISM

We cannot yet discuss the quantum physical explanation of diamagnetism, but we can provide a classical explanation with the loop model of Figs. 32-8 and 32-9. To begin, we assume that in a diamagnetic material the electrons in an atom orbit only clockwise or counterclockwise as in Fig. 32-9. To account for the lack of magnetism in the absence of an external magnetic field $\mathbf{B}_{ext}$, we assume the atom lacks a net magnetic dipole moment. This implies that before $\mathbf{B}_{ext}$ is applied, as many electrons orbit one way as orbit the other, with the result that the net upward magnetic dipole moment of the atom equals the net downward magnetic dipole moment.

Now let's turn on the nonuniform field $\mathbf{B}_{ext}$ of Fig. 32-9 in which $\mathbf{B}_{ext}$ is directed upward but is diverging (the magnetic field lines are diverging). We could do this by increasing the current through an electromagnet or by moving the north pole of a bar magnet closer to, and below, the orbits. As the magnitude of $\mathbf{B}_{ext}$ increases from zero to its final maximum, steady-state value, a clockwise electric field is induced around the electron's orbital loop according to Faraday's law and Lenz's law. Let us see how this induced electric field affects the orbiting electrons in Figs. 32-9b and d.

In Fig. 32-9b, the counterclockwise electron is accelerated by the clockwise electric field. So as the magnetic field $\mathbf{B}_{ext}$ increases to its maximum value, the electron speed increases to a maximum value. This means that the associated conventional current i and the downward magnetic dipole moment $\boldsymbol{\mu}$ due to i also *increase*.

In Fig. 32-9d, the clockwise electron is decelerated by the clockwise electric field. So, here, the electron speed, the associated current i, and the downward magnetic dipole moment $\boldsymbol{\mu}$ due to i all *decrease*. Thus by turning on field $\mathbf{B}_{ext}$, we have given the atom a net magnetic dipole moment that is upward.

The nonuniformity of field $\mathbf{B}_{ext}$ also affects the atom. Because the current i in Fig. 32-9b increases, the upward magnetic forces $d\mathbf{F}$ in Fig. 32-9c also increase, as does the net upward force on the current loop. And because current i in Fig. 32-9d decreases, the downward magnetic forces $d\mathbf{F}$ in Fig. 32-9e also decrease, as does the net downward force on the current loop. Thus by turning on the *nonuniform* field $\mathbf{B}_{ext}$, we have produced a net force on the atom; moreover, that force is directed *away* from the region of greater magnetic field, from which the magnetic field lines diverge or to which they converge.

We have argued with fictitious electron orbits (current loops), but we have ended up with exactly what happens to a diamagnetic material: if we apply the magnetic field of Fig. 32-9, the material develops a downward magnetic dipole moment and experiences an upward force. When the field is removed, both the dipole moment and the force disappear. The external field need not be positioned as shown; similar arguments can be made for other orientations of $\mathbf{B}_{ext}$. In general:

A diamagnetic material placed in an external magnetic field $\mathbf{B}_{ext}$ develops a magnetic dipole moment directed opposite $\mathbf{B}_{ext}$. If the field is nonuniform, the diamagnetic material is repelled from a region of greater magnetic field toward a region of lesser field.

CHECKPOINT 3: The figure shows two diamagnetic spheres located near the south pole of a bar magnet. Are (a) the magnetic forces on the spheres and (b) the magnetic dipole moments of the spheres directed toward or away from the bar magnet? (c) Is the magnetic force on sphere 1 greater than, less than, or equal to that on sphere 2?

32-7 PARAMAGNETISM

In paramagnetic materials, the spin and orbital magnetic dipole moments of the electrons in each atom do not cancel but add vectorially to give the atom a net (and permanent) magnetic dipole moment $\boldsymbol{\mu}$. In the absence of an external magnetic field, these atomic dipole moments are randomly oriented, and the net magnetic dipole moment of the material is zero. However, if a sample of the material is placed in an external magnetic field $\mathbf{B}_{ext}$, the dipole moments tend to line up with the field, which gives the sample a net magnetic dipole moment. This alignment with the external field is opposite of what we saw with diamagnetic materials.

A paramagnetic material placed in an external magnetic field $\mathbf{B}_{ext}$ develops a magnetic dipole moment in the direction of $\mathbf{B}_{ext}$. If the field is nonuniform, the paramagnetic material is attracted toward a region of greater magnetic field from a region of lesser field.

A paramagnetic sample with N atoms would have a magnetic dipole moment with a magnitude of $N\mu$ if the alignment of its atomic dipoles were complete. However, random collisions of atoms due to thermal agitation transfer energy among them, disrupting their alignment and thus reducing the sample's magnetic dipole moment.

The importance of thermal agitation may be measured by comparing two energies. One, from Eq. 20-20, is the mean translational kinetic energy $K \ (= \frac{3}{2}kT)$ of an atom at temperature T, where k is the Boltzmann constant (1.38×10^{-23} J/K) and T is in kelvins (not Celsius degrees). The other, from Eq. 29-37, is the difference in energy ΔU_B

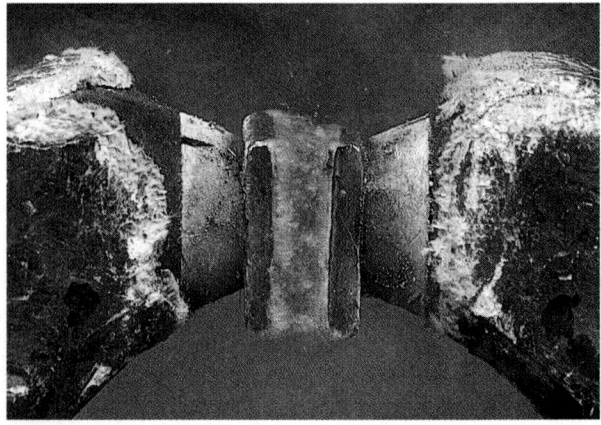

Liquid oxygen is suspended between the two pole faces of a magnet because the liquid is paramagnetic and is magnetically attracted to the magnet.

$(= 2\mu B_{\text{ext}})$ between parallel alignment and antiparallel alignment of the magnetic dipole moment of an atom and the external field. As we shall show below, $K \gg \Delta U_B$, even for ordinary temperatures and field magnitudes. Thus, energy transfers during collisions among atoms can significantly disrupt the alignment of the atomic dipole moments, keeping the magnetic dipole moment of a sample much less than $N\mu$.

We can express the extent to which a given paramagnetic sample is magnetized by finding the ratio of its magnetic dipole moment to its volume V. This vector quantity, the magnetic dipole moment per unit volume, is called the **magnetization M** of the sample, and its magnitude is

$$M = \frac{\text{measured magnetic moment}}{V}. \qquad (32\text{-}18)$$

The unit of **M** is the ampere–square meter per cubic meter, or ampere per meter (A/m). Complete alignment of the atomic dipole moments, called *saturation* of the sample, corresponds to the maximum value $M_{\text{max}} = N\mu/V$.

In 1895 Pierre Curie discovered experimentally that the magnetization of a paramagnetic sample is directly proportional to the external magnetic field $\mathbf{B}_{\text{ext}}$ and inversely proportional to the temperature T in kelvins; that is,

$$M = C\frac{B_{\text{ext}}}{T}. \qquad (32\text{-}19)$$

Equation 32-19 is known as *Curie's law,* and C is called the *Curie constant.* Curie's law is reasonable in that increasing B_{ext} tends to align the atomic dipole moments in a sample and thus to increase M, while increasing T tends to disrupt the alignment via thermal agitation and thus to decrease M. However, the law is actually an approximation that is valid only when the ratio B_{ext}/T is not too large.

Figure 32-10 shows the ratio M/M_{max} as a function of B_{ext}/T for a sample of the salt potassium chromium sulfate, in which chromium ions are the paramagnetic substance. The plot is called a *magnetization curve.* The straight line for Curie's law fits the experimental data at the left, for B_{ext}/T below about 0.5 T/K. The curve that fits all the data points is based on quantum physics. The data on the right side, near saturation, are very difficult to obtain because they require very strong magnetic fields (about 100,000 times Earth's field in Sample Problem 32-1), even at the very low temperatures noted in Fig. 32-10.

SAMPLE PROBLEM 32-2

A paramagnetic gas at room temperature ($T = 300$ K) is placed in an external uniform magnetic field of magnitude $B = 1.5$ T; the atoms of the gas have magnetic dipole moment $\mu = 1.0\mu_B$. Calculate the mean translational kinetic energy K of an atom of the gas and the energy difference ΔU_B between parallel alignment and antiparallel alignment of the atom's magnetic dipole moment with the external field.

SOLUTION: From Eq. 20-20, we have

$$K = \tfrac{3}{2}kT = \tfrac{3}{2}(1.38 \times 10^{-23}\text{ J/K})(300\text{ K})$$
$$= 6.2 \times 10^{-21}\text{ J} = 0.039\text{ eV}. \qquad \text{(Answer)}$$

From Eqs. 29-37 and 32-5, we have

$$\Delta U_B = 2\mu B = 2(9.27 \times 10^{-24}\text{ J/T})(1.5\text{ T})$$
$$= 2.8 \times 10^{-23}\text{ J} = 0.00017\text{ eV}. \qquad \text{(Answer)}$$

Here K is about 230 times ΔU_B, so energy exchanges among the atoms during their collisions with one another can easily reorient any magnetic dipole moments that might be aligned with the external magnetic field. The magnetic dipole moment exhibited by the gas must then be due to fleeting partial alignments of the atomic dipole moments.

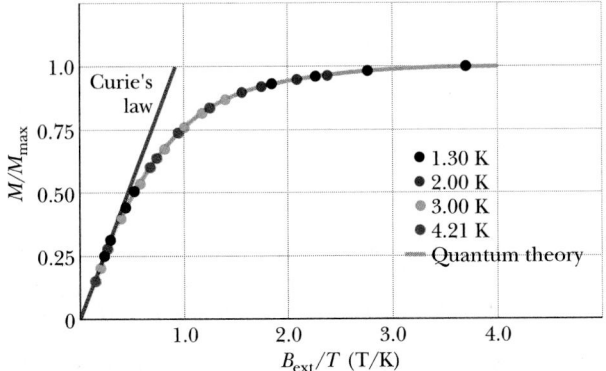

FIGURE 32-10 A *magnetization curve* for potassium chromium sulfate, a paramagnetic salt. The ratio of magnetization M of the salt to the maximum magnetization M_{max} is plotted versus the ratio of the applied magnetic field B_{ext} to the temperature T. Curie's law fits the data at the left; quantum theory fits all the data. After W. E. Henry.

CHECKPOINT 4: The figure shows two paramagnetic spheres located near the south pole of a bar magnet. Are (a) the magnetic forces on the spheres and (b) the magnetic dipole moments of the spheres directed toward or away from the bar magnet? (c) Is the magnetic force on sphere 1 greater than, less than, or equal to that on sphere 2?

32-8 FERROMAGNETISM

When we speak of magnetism in everyday conversation, we almost always have a mental picture of a bar magnet or a disk magnet (probably clinging to a refrigerator door). That is, we picture a ferromagnetic material having strong, permanent magnetism, not a diamagnetic or paramagnetic material having weak, temporary magnetism.

Iron, cobalt, nickel, gadolinium, dysprosium, and alloys of these and other elements exhibit ferromagnetism because of a quantum physical effect called *exchange coupling*. In this process the spins of the electrons in one atom interact with those of neighboring atoms. The result is an alignment of the magnetic dipole moments of the atoms, in spite of the randomizing tendency of the atomic collisions. This persistent alignment is what gives ferromagnetic materials their permanent magnetism.

If the temperature of a ferromagnetic material is raised above a certain critical value, called the *Curie temperature,* the exchange coupling ceases to be effective; most such materials then become simply paramagnetic. That is, the dipoles still tend to align with an external field but much more weakly, and thermal agitation can now more easily disrupt the alignment. The Curie temperature for iron is 1043 K (= 770°C).

The magnetization of a ferromagnetic material such as iron can be studied with an arrangement called a *Rowland ring* (Fig. 32-11). The material is formed into a thin toroidal core of circular cross section. A primary coil P having n turns per unit length is wrapped around the core and carries current i_P. (The coil is essentially a long solenoid bent into a circle.) If the iron core were not present, the magnetic field inside the coil would be, from Eq. 30-25,

$$B_0 = \mu_0 n i_P. \tag{32-20}$$

However, with the iron core present, the magnetic field B inside the coil is greater than B_0, usually by a large amount. We can write this field as

$$B = B_0 + B_M, \tag{32-21}$$

where B_M is the magnetic field contributed by the iron core. This contribution results from the alignment of the atomic dipole moments within the iron, due to exchange coupling and to the applied magnetic field B_0, and is proportional to the magnetization M of the iron. That is, the contribution B_M is proportional to the magnetic dipole moment per unit volume of the iron. To determine B_M we use a secondary coil S to measure B, compute B_0 with Eq. 32-20, and subtract as suggested by Eq. 32-21.

Figure 32-12 shows a magnetization curve for a ferromagnetic material in a Rowland ring: the ratio $B_M/B_{M,\text{max}}$, where $B_{M,\text{max}}$ is the maximum possible value of B_M, corresponding to saturation, is plotted versus B_0. The curve is similar to Fig. 32-10, the magnetization curve for a paramagnetic substance, in that both curves are measures of the extent to which an applied magnetic field can align the atomic dipole moments of a material.

For the ferromagnetic core yielding Fig. 32-12, the alignment of the dipole moments is about 70% complete for $B_0 \approx 1 \times 10^{-3}$ T. If B_0 were increased to 1 T, the alignment would be almost complete ($B_0 = 1$ T, and thus almost complete saturation, is quite difficult to obtain).

Magnetic Domains

Exchange coupling produces strong alignment of adjacent atomic dipoles in a ferromagnetic material at a temperature below the Curie temperature. Why, then, isn't the material naturally at saturation even when there is no applied mag-

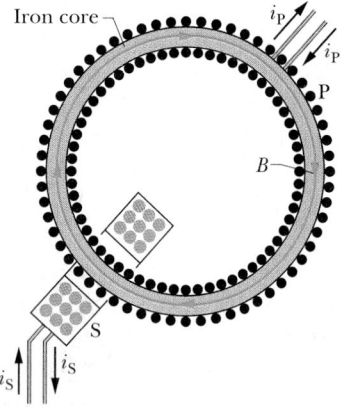

FIGURE 32-11 A Rowland ring. Current i_P is sent through a primary coil P whose core is a ferromagnetic material (here iron) that is magnetized by the current. (The turns of the coil are represented by dots.) The extent of magnetization of the core determines the total magnetic field **B** within coil P. Field **B** can be measured by means of a secondary coil S.

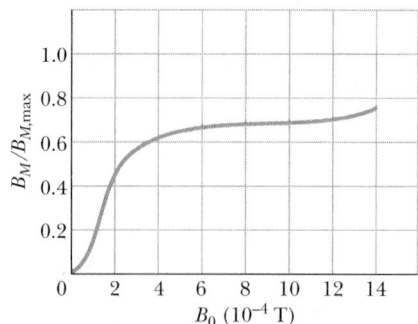

FIGURE 32-12 A magnetization curve for a ferromagnetic core material in the Rowland ring of Fig. 32-11. On the vertical axis, 1.0 corresponds to complete alignment (saturation) of the atomic dipoles within the material.

netic field B_0? That is, why isn't every piece of iron, such as an iron nail, a naturally strong magnet?

To understand this, consider a specimen of a ferromagnetic material such as iron that is in the form of a single crystal. That is, the arrangement of the atoms that make it up—its crystal lattice—extends with unbroken regularity throughout the volume of the specimen. Such a crystal will, in its normal state, be made up of a number of *magnetic domains*. These are regions of the crystal throughout which the alignment of the atomic dipoles is essentially perfect. For the crystal as a whole, however, the domains are so oriented that they largely cancel each other as far as their external magnetic effects are concerned.

Figure 32-13 is a magnified photograph of such an assembly of domains in a single crystal of nickel. It was made by sprinkling a colloidal suspension of finely powdered iron oxide on the surface of the crystal. The domain boundaries, which are thin regions in which the alignment of the elementary dipoles changes from a certain orientation in one domain to a different orientation in the other, are the sites of intense, but highly localized and nonuniform magnetic fields. The suspended colloidal particles are attracted to these boundaries and show up as the white lines. Although the atomic dipoles in each domain are completely aligned as shown by the arrows, the crystal as a whole may have a very small resultant magnetic moment.

Actually, a piece of iron as we ordinarily find it is not a single crystal but an assembly of many tiny crystals, randomly arranged; we call it a *polycrystalline solid*. Each tiny crystal, however, has its array of variously oriented domains, just as in Fig. 32-13. If we magnetize such a specimen by placing it in an external magnetic field of gradually increasing strength, we produce two effects; together they produce a magnetization curve of the shape shown in Fig. 32-12. One effect is a growth in size of the domains that are oriented along the external field at the expense of those that are not. The second effect is a shift of the orientation of the dipoles within a domain, as a unit, to become closer to the field direction.

Exchange coupling and domain shifting give us the following result:

A ferromagnetic material placed in an external magnetic field $\mathbf{B}_{ext}$ develops a strong magnetic dipole moment in the direction of $\mathbf{B}_{ext}$. If the field is nonuniform, the ferromagnetic material is attracted toward a region of greater magnetic field from a region of lesser field.

You can actually hear sound produced by shifting domains: put an audio cassette player into its play mode

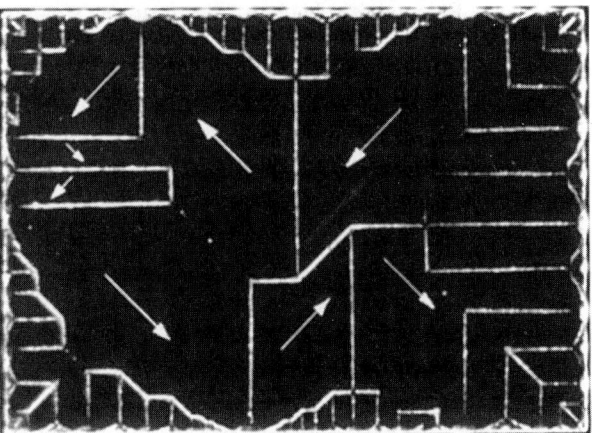

FIGURE 32-13 A photograph of domain patterns within a single crystal of nickel; white lines reveal the boundaries of the domains. The white arrows superimposed on the photograph show the orientations of the magnetic dipoles within the domains and thus the orientations of the net magnetic dipoles of the domains. The crystal as a whole is unmagnetized if the net field (the vector sum over all the domains) is zero.

without a cassette in place (or with a blank cassette) and turn the volume control to maximum. Then bring a strong magnet up to the play head (which is ferromagnetic). The magnetic field causes the domains in the play head to shift abruptly, which abruptly alters the magnetic field through a

An opaque magnetic liquid (consisting of magnetite suspended in kerosene) and a transparent nonmagnetic liquid have been placed in a thin glass cell. When the cell is upright, the slightly denser magnetic liquid (shown black here) sinks to the bottom of the cell. But when a magnetic field is then applied perpendicular to the cell face, the magnetic liquid quickly snakes its way up into the nonmagnetic liquid, forming the labyrinthine pattern shown.

coil wrapped around the play head. The resulting suddenly induced currents in the coil are amplified and fed to the speaker, producing a fizzing sound.

The Magnetism of Ancient Kilns

The clay in the walls and floor of an ancient kiln behaves similarly to iron because clay contains the iron oxides magnetite and hematite. Grains of magnetite consist of multiple domains that may be as small as 3×10^{-7} m across. Those of hematite consist of single domains that may be as large as 1 mm across. When the clay is heated to several hundred degrees Celsius (as the kiln is used), the domains in grains of both types change. In magnetite the domain walls shift so that domains that are more closely aligned with Earth's magnetic field grow, while others shrink. In hematite, the domains rotate so as to more closely align with Earth's field. For both processes, the result is that the clay then has a magnetic field that is aligned with Earth's field. When the kiln cools after use, the arrangement of the domains, and thus the magnetic field of the clay, is retained, an effect known as *thermoremanent magnetism* (TRM).

To determine the orientation Earth's field had when a kiln was last heated and cooled, an archaeologist outlines a small area of the floor, carefully measures its orientation relative to the horizontal and to geographic north, and then removes that section of the floor. By next determining the direction of the section's magnetic field relative to the section's dimensions, hence to its position in the kiln, the archaeologist then knows the direction of Earth's field (relative to the horizontal and to geographic north) when the kiln was last used. If the age of the kiln is found by radiocarbon dating or some other technique, the archaeologist also knows when Earth's field had that direction.

SAMPLE PROBLEM 32-3

A compass needle of pure iron (with density 7900 kg/m³) has a length L of 3.0 cm, a width of 1.0 mm, and a thickness of 0.50 mm. The magnitude of the magnetic dipole moment associated with an iron atom is $\mu_{Fe} = 2.1 \times 10^{-23}$ J/T.

(a) If the magnetization of the needle is equivalent to the alignment of 10% of the atoms in the needle, what is the magnitude of the needle's magnetic dipole moment μ?

SOLUTION: Alignment of all N atoms in the needle would give the needle a magnetic dipole moment of $N\mu_{Fe}$. We have only 10% alignment, so

$$\mu = 0.10 N \mu_{Fe}. \qquad (32\text{-}22)$$

The number of atoms N in the needle is

$$N = \frac{\text{needle's mass}}{\text{atomic mass of iron}}. \qquad (32\text{-}23)$$

The mass m of the needle is the product of its density, 7900 kg/m³, and its volume (3.0 cm $\times$ 1.0 mm $\times$ 0.50 mm = 1.5×10^{-8} m³) and works out to be 1.185×10^{-4} kg. The mass of an atom of iron is the ratio of the molar mass M of iron (55.847 g/mol, from Appendix F) to Avogadro's number N_A (6.02×10^{23} atoms/mol). Substituting Eq. 32-23 into Eq. 32-22 and then substituting these quantities and the given data, we find

$$\mu = 0.10 \left(\frac{m N_A}{M} \right) \mu_{Fe}$$

$$= 0.10 \frac{(1.185 \times 10^{-4} \text{ kg})(6.02 \times 10^{23} \text{ atoms/mol})}{(55.847 \text{ g/mol})(10^{-3} \text{ kg/g})}$$

$$\times (2.1 \times 10^{-23} \text{ J/T})$$

$$= 2.682 \times 10^{-3} \text{ J/T} \approx 2.7 \times 10^{-3} \text{ J/T.} \qquad \text{(Answer)}$$

(b) If the compass needle is jarred slightly from its (horizontal) north–south equilibrium position, it oscillates about that position. If the period of oscillation is 2.2 s, what is the horizontal component of the local magnetic field?

SOLUTION: The dipole moment μ of the needle is directed along its length, from its south pole to its north pole. When the needle is jarred from its equilibrium position by an angle θ, so is μ. Earth's magnetic field then exerts a torque about the needle's pivot axis, directed so as to realign μ (and the needle) with the horizontal component B_h of the magnetic field. (Remember, the needle of a compass is free to rotate only horizontally, so we are dealing only with B_h here.) From Eq. 29-34, the magnitude of this torque is

$$\tau = -\mu B_h \sin \theta, \qquad (32\text{-}24)$$

in which the minus sign indicates that τ opposes the angular displacement θ. Because the rotation angle is small, we may write $\sin \theta \approx \theta$ so that

$$\tau = -\mu B_h \theta. \qquad (32\text{-}25)$$

Because μ and B_h are both constant, Eq. 32-25 tells us that the restoring torque is proportional to the negative of the angular displacement. This kind of relation is the hallmark of angular simple harmonic motion, as we saw in Section 16-5. From Eqs. 16-24 and 16-25, the period of oscillation of the motion may then be written as

$$T = 2\pi \sqrt{\frac{I}{\mu B_h}},$$

which yields

$$B_h = \frac{I}{\mu} \left(\frac{2\pi}{T} \right)^2, \qquad (32\text{-}26)$$

where I is the rotational inertia of the needle. Approximating the needle as being a uniform thin rod, we use Table 11-2e to find

$$I = \frac{mL^2}{12} = \frac{(1.185 \times 10^{-4}\ \text{kg})(0.030\ \text{m})^2}{12}$$

$$= 8.888 \times 10^{-9}\ \text{kg} \cdot \text{m}^2.$$

Substituting this value, the value we obtained for μ, and the given value for T into Eq. 32-26, we find

$$B_h = \frac{8.888 \times 10^{-9}\ \text{kg} \cdot \text{m}^2}{2.682 \times 10^{-3}\ \text{J/T}} \left(\frac{2\pi}{2.2\ \text{s}}\right)^2$$

$$= 2.7 \times 10^{-5}\ \text{T}, \qquad \text{(Answer)}$$

which is approximately the value we used in Sample Problem 32-1 for Tucson. So, even with an inexpensive compass, we can measure a local magnetic field by jarring the needle and timing its oscillations.

Hysteresis

Magnetization curves for ferromagnetic materials do not retrace themselves as we increase and then decrease the external magnetic field B_0. Figure 32-14 is a plot of B_M versus B_0 during the following operations with a Rowland ring: (1) starting with the iron unmagnetized (point a), increase the current in the toroid until B_0 $(= \mu_0 n i)$ has the value corresponding to point b; (2) reduce the current in the toroid winding back to zero (point c); (3) reverse the toroid current and increase it in magnitude until B_0 has the value corresponding to point d; (4) reduce the current to zero again (point e); (5) reverse the current once more until point b is reached again.

The lack of retraceability shown in Fig. 32-14 is called **hysteresis,** and the curve $bcdeb$ is called a *hysteresis loop*. Note that at points c and e the iron core is magnetized, even though there is no current in the toroid windings; this is the familiar phenomenon of permanent magnetism.

Hysteresis can be understood through the concept of magnetic domains. Evidently the motions of the domain

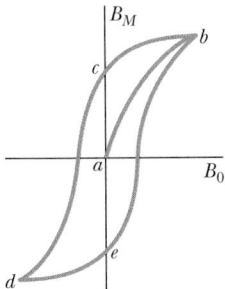

FIGURE 32-14 A magnetization curve (ab) for a ferromagnetic specimen and an associated hysteresis loop ($bcdeb$).

boundaries and the reorientations of the domain directions are not totally reversible. When the applied magnetic field B_0 is increased and then decreased back to its initial value, the domains do not return completely to their original configuration but retain some "memory" of the initial increase. This memory of magnetic materials is essential for the magnetic storage of information, as on cassette tapes and computer disks.

This memory of the alignment of domains can also occur naturally. When lightning sends currents along multiple tortuous paths through the ground, the currents produce intense magnetic fields that can suddenly magnetize any ferromagnetic material in nearby rock. Because of hysteresis, such rock material retains some of that magnetization after the lightning strike (after the currents disappear). Pieces of the rock, later exposed, broken, and loosened by weathering, are then lodestones.

32-9 INDUCED MAGNETIC FIELDS

We have to this point discussed two ways in which a magnetic field can be produced. The first, discussed in Chapter 30, is by means of an electric current; the second, discussed in this chapter, is by means of a magnetic material. There is a third way—by induction.

In Chapter 31 you saw that a changing magnetic flux induces an electric field, and we ended up with Faraday's law of induction in the form

$$\oint \mathbf{E} \cdot d\mathbf{s} = -\frac{d\Phi_B}{dt} \qquad \begin{array}{l}\text{(Faraday's law}\\ \text{of induction).}\end{array} \quad (32\text{-}27)$$

Here $\mathbf{E}$ is the electric field induced along a closed loop by the changing magnetic flux Φ_B through that loop. Because symmetry is often so powerful in physics, we should be tempted to ask whether induction can occur in the opposite sense. That is, can a changing electric flux induce a magnetic field?

The answer is that it can; furthermore, the equation governing the induction of a magnetic field is almost symmetric with Eq. 32-27. We often call it Maxwell's law of induction after James Clerk Maxwell, and we write it as

$$\oint \mathbf{B} \cdot d\mathbf{s} = \mu_0 \epsilon_0 \frac{d\Phi_E}{dt} \qquad \begin{array}{l}\text{(Maxwell's law}\\ \text{of induction).}\end{array} \quad (32\text{-}28)$$

The circle on the integral sign indicates that the integral is taken around a closed loop.

As an example of this sort of induction, we consider the charging of a parallel-plate capacitor with circular plates, as shown in Fig. 32-15a. (Although we shall focus on this particular arrangement, a changing electric flux will always induce a magnetic field whenever it occurs.) We assume that the charge on the capacitor is being increased at a steady rate by a constant current i in the connecting wires. Then the magnitude of the electric field between the plates must also be increasing at a steady rate.

Figure 32-15b is a view of the right-hand plate of Fig. 32-15a from between the plates. The electric field is directed into the page. Let us consider a circular loop through point 1 in Figs. 32-15a and b, concentric with the capacitor plates and with a radius smaller than that of the plates. Because the electric field through the loop is changing, the electric flux through the loop must also be changing. According to Eq. 32-28, this changing electric flux induces a magnetic field around the loop.

Experiment proves that a magnetic field **B** *is* indeed induced around such a loop, directed as shown. This magnetic field has the same magnitude at every point around the loop and thus has circular symmetry about the central axis of the capacitor plates.

If we now consider a larger loop, say through point 2 outside the plates in Figs. 32-15a and b, we find that a magnetic field is induced around that loop as well. Thus, while the electric field is changing, magnetic fields are induced between the plates, both inside and outside the gap. When the electric field stops changing, these induced magnetic fields disappear.

Although Eq. 32-28 is similar to Eq. 32-27, the equations differ in two ways. First, Eq. 32-28 has the two extra symbols, μ_0 and ϵ_0, but they appear only because we employ SI units. Second, Eq. 32-28 lacks the minus sign of

Eq. 32-27. That difference in sign means that the induced electric field **E** and the induced magnetic field **B** have opposite directions when they are produced in otherwise similar situations.

To see this opposition of directions, examine Fig. 32-16, in which an increasing magnetic field **B**, directed into the page, induces an electric field **E**. The induced field **E** is counterclockwise, whereas the induced magnetic field **B** in Fig. 32-15b is clockwise.

Now recall that the left side of Eq. 32-28, the integral of the dot product **B** · d**s** around a closed loop, appears in another equation, namely Ampere's law:

$$\oint \mathbf{B} \cdot d\mathbf{s} = \mu_0 i_{\text{enc}} \quad \text{(Ampere's law),} \quad (32\text{-}29)$$

where i_{enc} is the current encircled by the closed loop. Thus, our two equations that specify the magnetic field **B** produced by means other than a magnetic material (that is, by a current and by a changing electric field) give the field in exactly the same form. So we can combine the two equations into the single equation

$$\oint \mathbf{B} \cdot d\mathbf{s} = \mu_0 \epsilon_0 \frac{d\Phi_E}{dt} + \mu_0 i_{\text{enc}}$$
$$\text{(Ampere–Maxwell law).} \quad (32\text{-}30)$$

When there is a current but no change in electric flux (such as with a wire carrying a constant current), the first term on the right side of Eq. 32-30 is zero, and Eq. 32-30 reduces to Eq. 32-29, Ampere's law. When there is a change in electric flux but no current (such as inside or outside the gap of a charging capacitor), the second term on the right side of Eq. 32-30 is zero, and Eq. 32-30 reduces to Eq. 32-28, Maxwell's law of induction.

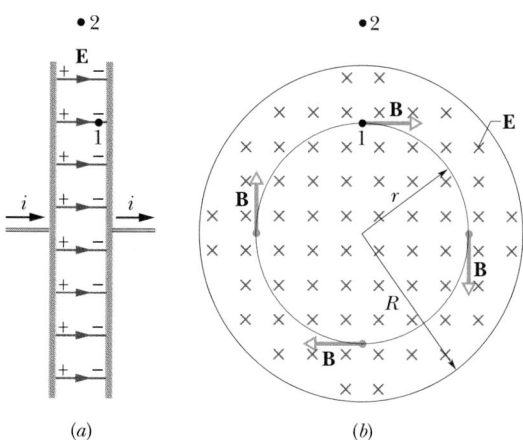

(a) (b)

FIGURE 32-15 (a) A circular parallel-plate capacitor, shown in side view, is being charged by a constant current i. (b) A view from within the capacitor, toward the plate at the right. The electric field **E** is uniform, is directed into the page (toward the plate), and grows in magnitude as the charge on the capacitor increases. The magnetic field **B** induced by this changing electric field is shown at four points on a circle with a radius r less than the plate radius R.

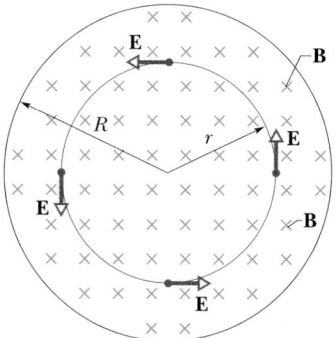

FIGURE 32-16 A uniform magnetic field **B** in a circular region. The field, directed into the page, is increasing in magnitude. The electric field **E** induced by the changing magnetic field is shown at four points on a circle concentric with the circular region. Compare this situation with that of Fig. 32-15b.

SAMPLE PROBLEM 32-4

A parallel-plate capacitor with circular plates of radius R is being charged as in Fig. 32-15a.

(a) Derive an expression for the induced magnetic field at radii r for the case of $r \leq R$.

SOLUTION: There is no current between the plates so that $i_{enc} = 0$ in Eq. 32-30, leaving

$$\oint \mathbf{B} \cdot d\mathbf{s} = \mu_0 \epsilon_0 \frac{d\Phi_E}{dt}. \qquad (32\text{-}31)$$

For an Amperian loop of radius $r \leq R$, the left side of Eq. 32-31 is $(B)(2\pi r)$. The electric flux Φ_E through that loop is $EA = \pi r^2 E$, where E is the magnitude of the electric field between the plates. So we may write Eq. 32-31 as

$$(B)(2\pi r) = \mu_0 \epsilon_0 \frac{d}{dt}(\pi r^2 E) = \mu_0 \epsilon_0 \pi r^2 \frac{dE}{dt}.$$

Solving for B, we find

$$B = \frac{\mu_0 \epsilon_0 r}{2} \frac{dE}{dt} \quad \text{(for } r \leq R\text{)}. \qquad \text{(Answer)}$$

We see that $B = 0$ at the center of the capacitor, where $r = 0$, and that B increases linearly with r out to the edge of the circular capacitor plates.

(b) Evaluate the field magnitude B for $r = R/5 = 11.0$ mm and $dE/dt = 1.50 \times 10^{12}$ V/m·s.

SOLUTION: From the answer to (a), we have

$$\begin{aligned}
B &= \frac{1}{2}\mu_0 \epsilon_0 r \frac{dE}{dt} \\
&= \tfrac{1}{2}(4\pi \times 10^{-7}\text{ T·m/A})(8.85 \times 10^{-12}\text{ C}^2/\text{N·m}^2) \\
&\quad \times (11.0 \times 10^{-3}\text{ m})(1.50 \times 10^{12}\text{ V/m·s}) \\
&= 9.18 \times 10^{-8}\text{ T}. \qquad \text{(Answer)}
\end{aligned}$$

(c) Derive an expression for the induced magnetic field for the case $r \geq R$.

SOLUTION: Outside the plate radius R, the electric field E is equal to zero, so the electric flux through an Amperian loop of radius $r \geq R$ exists only within the area πR^2 and is $\Phi_E = \pi R^2 E$. Then Eq. 32-31 becomes

$$(B)(2\pi r) = \mu_0 \epsilon_0 \frac{d}{dt}(\pi R^2 E) = \mu_0 \epsilon_0 \pi R^2 \frac{dE}{dt}.$$

Solving for B, we find

$$B = \frac{\mu_0 \epsilon_0 R^2}{2r} \frac{dE}{dt} \quad \text{(for } r \geq R\text{)}. \qquad \text{(Answer)}$$

Note that the two expressions for B that we have derived yield the same result (as we expect) for $r = R$. Furthermore, the value of B at $r = R$ is the maximum value of the induced magnetic field.

The value of B calculated in (b) is so small that it can scarcely be measured with simple apparatus. This is in sharp contrast to induced electric fields (Faraday's law), which can be demonstrated easily. This experimental difference exists partly because induced emfs can easily be multiplied by using a coil of many turns. No technique of comparable simplicity exists for multiplying induced magnetic fields. In any case, the experiment suggested by this sample problem has been done, and the presence of the induced magnetic fields has been verified quantitatively.

CHECKPOINT 5: The figure shows graphs of the electric field magnitude E versus time t for four uniform electric fields, all contained within identical circular regions as in Fig. 32-15b. Rank the fields according to the magnitude of the induced magnetic field at the edge of the region, greatest first.

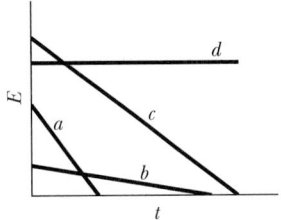

32-10 DISPLACEMENT CURRENT

If you compare the two terms on the right side of Eq. 32-30, you will see that the portion $\epsilon_0(d\Phi_E/dt)$ of the first term must have the dimension of a current. In fact, historically, that portion has been treated as being a fictitious current called the **displacement current** i_d:

$$i_d = \epsilon_0 \frac{d\Phi_E}{dt} \quad \text{(displacement current)}. \qquad (32\text{-}32)$$

''Displacement'' is poorly chosen in that nothing is being displaced, but we are stuck with the word. Nevertheless, we can now rewrite Eq. 32-30 as

$$\oint \mathbf{B} \cdot d\mathbf{s} = \mu_0 i_{d,enc} + \mu_0 i_{enc}$$

$$\text{(Ampere–Maxwell law)}, \qquad (32\text{-}33)$$

in which $i_{d,enc}$ is the displacement current that is encircled by the integration loop.

Let us again focus on a charging capacitor with circular plates, as in Fig. 32-17a. The real current i that is charging the plates changes the electric field $\mathbf{E}$ between the plates. The fictitious displacement current i_d between the plates is associated with that changing field $\mathbf{E}$. Let us relate these two currents.

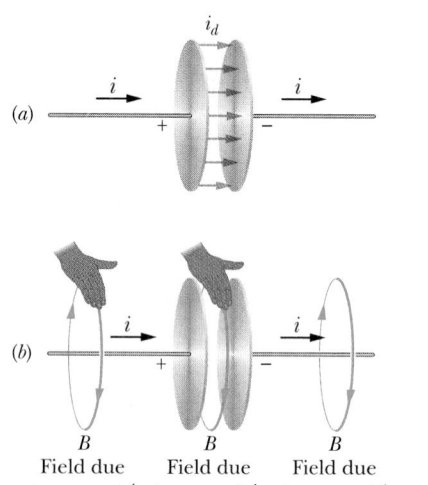

B B B
Field due   Field due   Field due
to current i to current i_d to current i

FIGURE 32-17 (a) The displacement current i_d between the plates of a capacitor that is being charged by a current i. (b) The right-hand rule for finding the direction of the magnetic field around a wire with a real current (as at the left) also gives the direction of the magnetic field around a displacement current (as in the center).

The charge q on the plates at any time is related to the magnitude E of the field between the plates at that time by Eq. 26-4:

$$q = \epsilon_0 A E, \qquad (32\text{-}34)$$

in which A is the plate area. To get the real current i, we differentiate Eq. 32-34 with respect to time, finding

$$\frac{dq}{dt} = i = \epsilon_0 A \frac{dE}{dt}. \qquad (32\text{-}35)$$

To get the displacement current i_d, we can use Eq. 32-32. Assuming that the electric field **E** between the two plates is uniform (we neglect any fringing), we can replace the electric flux Φ_E in that equation with EA. Then Eq. 32-32 becomes

$$i_d = \epsilon_0 \frac{d\Phi_E}{dt} = \epsilon_0 \frac{d(EA)}{dt} = \epsilon_0 A \frac{dE}{dt}. \qquad (32\text{-}36)$$

Comparing Eqs. 32-35 and 32-36, we see that the real current i charging the capacitor and the fictitious displacement current i_d between the plates have the same magnitude:

$$i_d = i \qquad \begin{array}{l}\text{(displacement current}\\\text{in a capacitor).}\end{array} \qquad (32\text{-}37)$$

Thus, we can consider the fictitious displacement current i_d to be simply a continuation of the real current i from one plate, across the capacitor gap, to the other plate. Because

the electric field is uniformly spread over the plates, the same is true of this fictitious displacement current i_d, as suggested by the spread of current arrows in Fig. 32-17a. Although no charge actually moves across the gap between the plates, the idea of the fictitious current i_d can help us to quickly find the direction and magnitude of an induced magnetic field, as follows.

Finding the Induced Magnetic Field

In Chapter 30 we found the direction of the magnetic field produced by a real current i by using the right-hand rule of Fig. 30-4. We can apply the same rule to find the direction of an induced magnetic field produced by a fictitious displacement current i_d, as is shown in the center of Fig. 32-17b for a capacitor.

We can also use i_d to find the magnitude of the induced magnetic field for a charging capacitor with parallel circular plates of radius R. We simply consider the space between the plates to be an imaginary circular wire of radius R carrying the imaginary current i_d. Then, from Eq. 30-22, the magnitude of the magnetic field at a point inside the capacitor at radius r from the center is

$$B = \left(\frac{\mu_0 i_d}{2\pi R^2}\right) r \qquad \text{(inside a circular capacitor).} \qquad (32\text{-}38)$$

Similarly, from Eq. 30-19, the magnitude of the magnetic field at a point outside the capacitor at radius r is

$$B = \frac{\mu_0 i_d}{2\pi r} \qquad \text{(outside a circular capacitor).} \qquad (32\text{-}39)$$

SAMPLE PROBLEM 32-5

The circular parallel-plate capacitor in Sample Problem 32-4 is being charged with a current i.

(a) Between the plates, what is the magnitude of $\oint \mathbf{B} \cdot d\mathbf{s}$, in terms of μ_0 and i, at a radius $r = R/5$?

SOLUTION: The magnetic field within the capacitor is produced by the displacement current i_d between the plates. So, we write Ampere's law for a circular loop of radius r concentric with the central axis of the capacitor as

$$\oint \mathbf{B} \cdot d\mathbf{s} = \mu_0 i_{d,\text{enc}}. \qquad (32\text{-}40)$$

We assume that the displacement current is uniformly spread over the full plate area. Thus the current $i_{d,\text{enc}}$ that is encircled by the loop is proportional to the area encircled by the loop;

that is,

$$\frac{i_{d,enc}}{i_d} = \frac{\text{encircled area}}{\text{full plate area}},$$

from which we find that

$$i_{d,enc} = i_d \frac{\pi r^2}{\pi R^2}.$$

Substituting this into Eq. 32-40, we obtain

$$\oint \mathbf{B} \cdot d\mathbf{s} = \mu_0 i_d \frac{\pi r^2}{\pi R^2}. \qquad (32\text{-}41)$$

Now substituting $i_d = i$ (from Eq. 32-37) and $r = R/5$ into Eq. 32-41 gives us

$$\oint \mathbf{B} \cdot d\mathbf{s} = \mu_0 i \frac{\pi (R/5)^2}{\pi R^2} = \frac{\mu_0 i}{25}. \qquad \text{(Answer)}$$

(b) In terms of the maximum field, what is the magnitude of the magnetic field at $r = R/5$ inside the capacitor?

SOLUTION: Because the capacitor has parallel circular plates, we can use Eq. 32-38 to find B. At $r = R/5$, that equation yields

$$B = \left(\frac{\mu_0 i_d}{2\pi R^2}\right) r = \frac{\mu_0 i_d (R/5)}{2\pi R^2} = \frac{\mu_0 i_d}{10\pi R}. \qquad (32\text{-}42)$$

The maximum field B_{max} within the capacitor occurs at $r = R$. It is

$$B_{max} = \left(\frac{\mu_0 i_d}{2\pi R^2}\right) r = \frac{\mu_0 i_d R}{2\pi R^2} = \frac{\mu_0 i_d}{2\pi R}. \qquad (32\text{-}43)$$

Dividing Eq. 32-42 by Eq. 32-43 and rearranging the result, we find

$$B = \frac{B_{max}}{5}. \qquad \text{(Answer)}$$

We should be able to obtain this result with a little reasoning and less work. Equation 32-38 tells us that inside the capacitor, B increases linearly with r. So a point $\frac{1}{5}$ the distance out to the full radius R of the plates, where B_{max} occurs, should have a field B that is $\frac{1}{5} B_{max}$.

CHECKPOINT 6: The figure is a view of one plate of a parallel-plate capacitor from within the capacitor. The dashed lines show four integration paths (path b follows the edge of the plate). Rank the paths according to the magnitude of $\oint \mathbf{B} \cdot d\mathbf{s}$ along the paths during the discharging of the capacitor, greatest first.

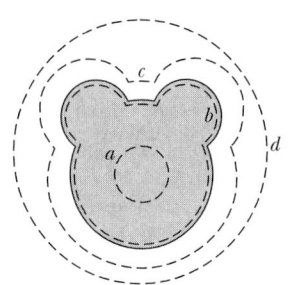

32-11 MAXWELL'S EQUATIONS

Equation 32-30 is the last of the four fundamental equations of electromagnetism, called *Maxwell's equations* and displayed in Table 32-1. These four equations explain a diverse range of phenomena, from why a compass needle points north to why a car starts when you turn the ignition key. They are the basis for the functioning of such electromagnetic devices as electric motors, cyclotrons, television transmitters and receivers, telephones, fax machines, radar, and microwave ovens.

Maxwell's equations are the basis from which many of the equations you have seen since Chapter 22 can be derived. They are also the basis of many of the equations you will see in Chapters 34 through 37, which introduce you to optics, as well as such optical devices as telescopes and eyeglasses.

TABLE 32-1 MAXWELL'S EQUATIONS[a]

NAME	EQUATION
Gauss' law for electricity	$\oint \mathbf{E} \cdot d\mathbf{A} = q/\epsilon_0$
Relates net electric flux to net enclosed electric charge	
Gauss' law for magnetism	$\oint \mathbf{B} \cdot d\mathbf{A} = 0$
Relates net magnetic flux to net enclosed magnetic charge	
Faraday's law	$\oint \mathbf{E} \cdot d\mathbf{s} = -\dfrac{d\Phi_B}{dt}$
Relates induced electric field to changing magnetic flux	
Ampere–Maxwell law	$\oint \mathbf{B} \cdot d\mathbf{s} = \mu_0 \epsilon_0 \dfrac{d\Phi_E}{dt} + \mu_0 i_{enc}$
Relates induced magnetic field to changing electric flux and to current	

[a]Written on the assumption that no dielectric or magnetic materials are present.

REVIEW & SUMMARY

Gauss' Law for Magnetic Fields

The simplest magnetic structures are magnetic dipoles. Magnetic monopoles do not exist (as far as we know). **Gauss' law** for magnetic fields,

$$\Phi_B = \oint \mathbf{B} \cdot d\mathbf{A} = 0, \qquad (32\text{-}1)$$

states that the net magnetic flux through any (closed) Gaussian surface is zero. It implies that magnetic monopoles don't exist.

Earth's Magnetic Field

Earth's magnetic field can be approximated as being that of a magnetic dipole whose dipole moment makes an angle of 11.5° with Earth's rotation axis, and with the south pole of the dipole in the northern hemisphere. The direction of the local magnetic field at any point on Earth's surface is given by the *field declination* (the angle left or right from geographic north) and the *field inclination* (the angle up or down from the horizontal).

Spin Magnetic Dipole Moment

An electron has an intrinsic angular momentum called *spin angular momentum* (or *spin*) **S**, with which an intrinsic *spin magnetic dipole moment* $\boldsymbol{\mu}_s$ is associated:

$$\boldsymbol{\mu}_s = -\frac{e}{m}\mathbf{S}. \qquad (32\text{-}2)$$

Spin **S** cannot itself be measured, but a component can be measured. Assuming that the measurement is along a z axis of a coordinate system, the component S_z can have only the values given by

$$S_z = m_s \frac{h}{2\pi}, \qquad \text{for} \qquad m_s = \pm\tfrac{1}{2}, \qquad (32\text{-}3)$$

where h (= 6.63×10^{-34} J·s) is the Planck constant. Similarly, the electron's spin magnetic dipole moment $\boldsymbol{\mu}_s$ cannot itself be measured but its component can be measured. Along a z axis, the component is

$$\mu_{s,z} = \pm\frac{eh}{4\pi m} = \pm\mu_B, \qquad (32\text{-}4, 32\text{-}6)$$

where μ_B is the *Bohr magneton*:

$$\mu_B = \frac{eh}{4\pi m} = 9.27 \times 10^{-24} \text{ J/T}. \qquad (32\text{-}5)$$

The potential energy U associated with the orientation of the spin magnetic dipole moment in an external magnetic field $\mathbf{B}_{ext}$ is

$$U = -\boldsymbol{\mu}_s \cdot \mathbf{B}_{ext} = -\mu_{s,z} B. \qquad (32\text{-}7)$$

Orbital Magnetic Dipole Moment

An electron in an atom has an additional angular momentum called its *orbital angular momentum* $\mathbf{L}_{orb}$, with which an *orbital magnetic dipole moment* $\boldsymbol{\mu}_{orb}$ is associated:

$$\boldsymbol{\mu}_{orb} = -\frac{e}{2m}\mathbf{L}_{orb}. \qquad (32\text{-}8)$$

Orbital angular momentum is quantized and can have only values given by

$$L_{orb,z} = m_l \frac{h}{2\pi}, \quad \text{for } m_l = 0, \pm 1, \pm 2, \ldots, \pm(\text{limit}). \quad (32\text{-}9)$$

So, the magnitude of the orbital angular momentum is

$$\mu_{orb,z} = -m_l \frac{eh}{4\pi m} = -m_l \mu_B. \qquad (32\text{-}10, 32\text{-}11)$$

The potential energy U associated with the orientation of the orbital magnetic dipole moment in an external magnetic field $\mathbf{B}_{ext}$ is

$$U = -\boldsymbol{\mu}_{orb} \cdot \mathbf{B}_{ext} = -\mu_{orb,z} B_{ext}. \qquad (32\text{-}12)$$

Diamagnetism

Diamagnetic materials do not exhibit magnetism until they are placed in an external magnetic field $\mathbf{B}_{ext}$. They then develop a magnetic dipole moment directed opposite $\mathbf{B}_{ext}$. If the field is nonuniform, the diamagnetic material is repelled from regions of greater magnetic field. This property is called *diamagnetism*.

Paramagnetism

In a *paramagnetic material*, each atom has a permanent magnetic dipole moment $\boldsymbol{\mu}$, but the dipole moments are randomly oriented and the material as a whole lacks a magnetic field. However, an external magnetic field $\mathbf{B}_{ext}$ can partially align the atomic dipole moments to give the material a net magnetic dipole moment in the direction of $\mathbf{B}_{ext}$. If $\mathbf{B}_{ext}$ is nonuniform, the material is attracted to regions of greater magnetic field. These properties are called *paramagnetism*.

The alignment of the atomic dipole moments increases with an increase in $\mathbf{B}_{ext}$ and decreases with an increase in temperature T. The extent to which a sample of volume V is magnetized is given by its *magnetization* **M**, whose magnitude is

$$M = \frac{\text{measured magnetic moment}}{V}. \qquad (32\text{-}18)$$

Complete alignment of all N atomic magnetic dipoles in a sample, called *saturation* of the sample, corresponds to the maximum magnetization value $M_{max} = N\mu/V$. For low values of the ratio B_{ext}/T, we have the approximation

$$M = C\frac{B_{ext}}{T} \qquad (\text{Curie's law}), \qquad (32\text{-}19)$$

where C is called the *Curie constant*.

Ferromagnetism

In the absence of an external magnetic field, some of the electrons in a ferromagnetic material have their magnetic dipole moments aligned by means of a quantum physical interaction called *exchange coupling*, producing regions (domains) within the material with strong magnetic dipole moments. An external field $\mathbf{B}_{ext}$ can align the magnetic dipole moments of those regions, producing a strong net magnetic dipole moment for the material as a whole, in the direction of $\mathbf{B}_{ext}$. This net magnetic dipole moment can partially persist when $\mathbf{B}_{ext}$ is removed. If $\mathbf{B}_{ext}$ is nonuniform, the ferromagnetic material is attracted to regions of greater magnetic field. These properties are called *ferromagnetism*. Exchange coupling disappears when a sample's temperature exceeds its *Curie temperature*, and then the sample has only paramagnetism.

Maxwell's Extension of Ampere's Law

A changing electric flux induces a magnetic field $\mathbf{B}$. Maxwell's law,

$$\oint \mathbf{B} \cdot d\mathbf{s} = \mu_0 \epsilon_0 \frac{d\Phi_E}{dt} \quad \begin{array}{l}\text{(Maxwell's law}\\ \text{of induction),}\end{array} \quad (32\text{-}28)$$

relates the magnetic field induced along a closed loop to the changing electric flux Φ_E through the loop. Ampere's law, $\oint \mathbf{B} \cdot d\mathbf{s} = \mu_0 i_{enc}$ (Eq. 32-29), gives the magnetic field generated by a current i_{enc} encircled by a loop. Maxwell's law and Ampere's law can be written as the single equation:

$$\oint \mathbf{B} \cdot d\mathbf{s} = \mu_0 \epsilon_0 \frac{d\Phi_E}{dt} + \mu_0 i_{enc}$$

(Ampere–Maxwell law). (32-30)

Displacement Current

We define the fictitious *displacement current* due to a changing electric field as

$$i_d = \epsilon_0 \frac{d\Phi_E}{dt}. \quad (32\text{-}32)$$

Equation 32-30 then becomes

$$\oint \mathbf{B} \cdot d\mathbf{s} = \mu_0 i_{d,enc} + \mu_0 i_{enc}$$

(Ampere–Maxwell law), (32-33)

where $i_{d,enc}$ is the displacement current encircled by the integration loop. The idea of a displacement current allows us to retain the notion of continuity of current through a capacitor. However, displacement current is *not* a transfer of charge.

Maxwell's Equations

Maxwell's equations, displayed in Table 32-1, summarize electromagnetism and form its foundation.

QUESTIONS

1. Figure 32-18 shows four steel bars; three are permanent magnets. One of the poles is indicated. Through experiment we find that ends a and d attract each other, ends c and f repel, ends e and h attract, and ends a and h attract. (a) Which ends are north poles? (b) Which bar is not a magnet?

FIGURE 32-18 Question 1.

2. Figure 32-19 shows four arrangements of a pair of small compass needles in a region otherwise free of magnetic field. The arrows indicate the directions of the needles and thus also the directions of the magnetic dipole moments. Which pairs are in stable equilibrium?

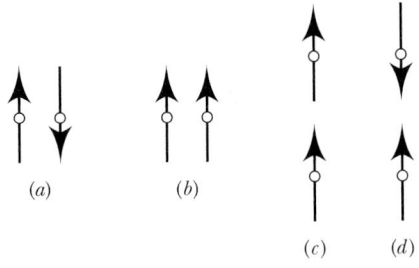

FIGURE 32-19 Question 2.

3. An electron is in an external magnetic field $\mathbf{B}_{ext}$ with spin angular momentum S_z antiparallel to $\mathbf{B}_{ext}$. If the electron undergoes a *spin-flip* so that S_z is then parallel with $\mathbf{B}_{ext}$, must energy be supplied to or lost by the electron?

4. Figure 32-20a shows the opposite spin orientations of an electron in an external magnetic field $\mathbf{B}_{ext}$. Figure 32-20b gives three choices for the graph of the potential energies associated with those orientations as a function of the magnitude of $\mathbf{B}_{ext}$. Choices b and c consist of intersecting lines, choice a of parallel lines. Which is the correct choice?

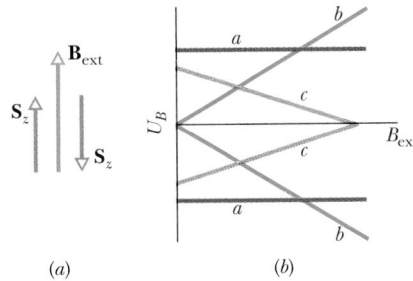

FIGURE 32-20 Question 4.

5. Figure 32-21 shows three loop models of an electron orbiting counterclockwise within a magnetic field. The fields are nonuniform for models 1 and 2 and uniform for model 3. For each model, are (a) the magnetic dipole moment of the loop and (b) the magnetic force on the loop directed up, directed down, or zero?

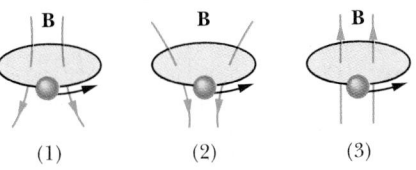

FIGURE 32-21 Questions 5, 7, and 8.

6. Does the magnitude of the net force on the loop of Figs. 32-9a and b increase, decrease, or remain the same if we increase (a) the magnitude of $\mathbf{B}_{ext}$ and (b) the divergence of $\mathbf{B}_{ext}$?

7. Replace the current loops of Question 5 and Fig. 32-21 with diamagnetic spheres. For each field, are (a) the magnetic dipole moment of the sphere and (b) the magnetic force on the sphere directed up, directed down, or zero?

8. Replace the current loops of Question 5 and Fig. 32-21 with paramagnetic spheres. For each field, are (a) the magnetic dipole moment of the sphere and (b) the magnetic force on the sphere up, down, or zero?

9. In Fig. 32-22 a nonuniform magnetic field extends between the two pole faces shown. An electron is sent through the field along a path perpendicular to the page (at the dot). Is the force on the electron due to the interaction of its intrinsic spin magnetic dipole moment with the field leftward, rightward, or zero if the electron's spin S_z is directed (a) leftward and (b) rightward? (*Hint:* Assume that the electron is a spinning ball with negative charge on its surface, resulting in a current loop like those in Fig. 32-9.)

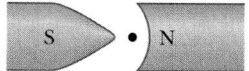

FIGURE 32-22 Question 9.

10. Figure 32-23 shows a ferromagnetic sphere that initially has no net magnetic dipole moment; it is held in place by two very fine, taut wires. When a uniform upward magnetic field **B** is switched on, the sphere acquires an upward magnetic dipole moment. As the field turns on, does the sphere rotate about the wire clockwise (as shown) or counterclockwise? (*Hint:* Consider the orientations of the spin angular momenta of the electrons and the conservation of angular momentum.)

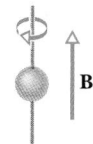

FIGURE 32-23 Question 10.

11. Figure 32-24 shows, in two situations, an electric field vector **E** and an induced magnetic field line. In each, is the magnitude of **E** increasing or decreasing?

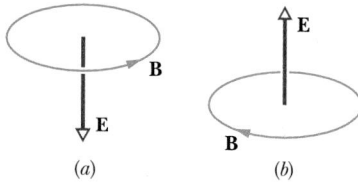

(a)   (b)

FIGURE 32-24 Question 11.

12. In Figure 32-15b, **E** is directed into the page and is increasing in magnitude. Is the direction of the magnetic field **B** clockwise or counterclockwise if, instead, **E** is out of the page and (a) increasing and (b) decreasing? (c) What is the direction of **B** if **E** is out of the page and not changing?

13. Figure 32-25 shows a face-on view of one of the two square plates of a parallel-plate capacitor, as well as four loops that are located between the plates. The capacitor is being discharged. (a) Neglecting fringing of the magnetic field, rank the loops according to the magnitude of $\oint \mathbf{B} \cdot d\mathbf{s}$ along them, greatest first. (b) Along which loop, if any, is the angle between **B** and d**s** constant (so that their dot product can easily be evaluated)? (c) Along which loop, if any, is B constant (so that B can be brought in front of the integral sign in Eq. 32-28)?

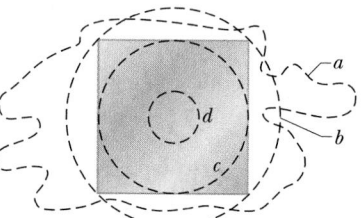

FIGURE 32-25 Question 13.

14. Figure 32-26 shows a parallel-plate capacitor and the current in the connecting wires that is discharging the capacitor. Are the directions of (a) **E** and (b) i_d leftward or rightward? (c) Is the magnetic field at point P into the page or out of the page?

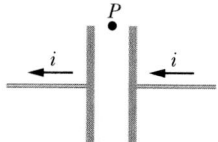

FIGURE 32-26 Question 14.

15. A parallel-plate capacitor with rectangular plates is being discharged. Consider a rectangular loop centered on the plates and between them. The loop measures L by 2L; the plate measures 2L by 4L. What fraction of the displacement current is encircled by the loop?

16. Figure 32-27a shows a capacitor that is being charged. Point a (near one of the connecting wires) and point b (inside the capacitor gap) are equidistant from the central axis, as are point c (near the wire) and point d (between the plates but outside the gap). In Fig. 32-27b, one curve gives the variation with distance r of the magnitude of the magnetic field inside and outside the wire. The other curve gives the variation with distance r of the magnitude of the magnetic field inside and outside the gap. The two curves partially overlap. Which of the three points on the curves correspond to which of the four points of Fig. 32-27a?

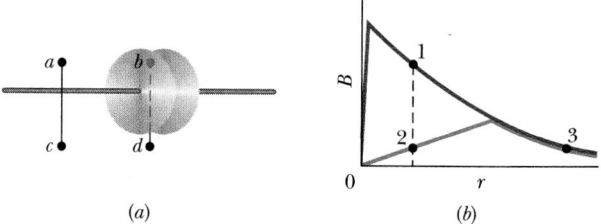

(a)   (b)

FIGURE 32-27 Question 16.

EXERCISES & PROBLEMS

SECTION 32-2 Gauss' Law for Magnetic Fields

1E. Imagine rolling a sheet of paper into a cylinder and placing a bar magnet near its end as shown in Fig. 32-28. (a) Sketch the magnetic field lines that pass through the surface of the cylinder. (b) What can you say about the sign of $\mathbf{B} \cdot d\mathbf{A}$ for every area $d\mathbf{A}$ on the surface? (c) Does this contradict Gauss' law for magnetism? Explain.

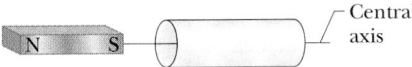

FIGURE 32-28 Exercise 1.

2E. The magnetic flux through each of five faces of a die (singular of "dice") is given by $\Phi_B = \pm N$ Wb, where N (= 1 to 5) is the number of spots on the face. The flux is positive (outward) for N even and negative (inward) for N odd. What is the flux through the sixth face of the die?

3P. A Gaussian surface in the shape of a right-circular cylinder with end caps has a radius of 12.0 cm and a length of 80.0 cm. Through one end there is an inward magnetic flux of 25.0 μWb. At the other end there is a uniform magnetic field of 1.60 mT, normal to the surface and directed outward. What is the net magnetic flux through the curved surface?

4P*. Two wires, parallel to the z axis and a distance $4r$ apart, carry equal currents i in opposite directions, as shown in Fig. 32-29. A circular cylinder of radius r and length L has its axis on the z axis, midway between the wires. Use Gauss' law for magnetism to calculate the net outward magnetic flux through the half of the cylindrical surface above the x axis. (*Hint:* Find the flux through that portion of the xz plane that is within the cylinder.)

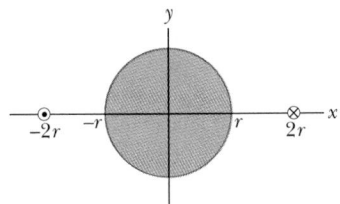

FIGURE 32-29 Problem 4.

SECTION 32-3 The Magnetism of Earth

5E. In New Hampshire the average horizontal component of Earth's magnetic field in 1912 was 16 μT and the average inclination or "dip" was 73°. What was the corresponding magnitude of Earth's magnetic field?

6E. In Sample Problem 32-1 the vertical component of Earth's magnetic field in Tucson, Arizona, was found to be 43 μT. Assume this is the average value for all of Arizona, which has an area of 295,000 km², and calculate the net magnetic flux through the rest of Earth's surface (the entire surface excluding Arizona). Is the flux outward or inward?

7E. Earth has a magnetic dipole moment of 8.0×10^{22} J/T. (a) What current would have to be produced in a single turn of wire extending around Earth at its geomagnetic equator if we wished to set up such a dipole? Could such an arrangement be used to cancel out Earth's magnetism (b) at points in space well above Earth's surface or (c) on Earth's surface?

8P. The magnetic field of Earth can be approximated as the magnetic field of a dipole, with horizontal and vertical components, at a point a distance r from Earth's center, given by

$$B_h = \frac{\mu_0 \mu}{4\pi r^3} \cos \lambda_m, \qquad B_v = \frac{\mu_0 \mu}{2\pi r^3} \sin \lambda_m,$$

where λ_m is the *magnetic latitude* (latitude measured from the geomagnetic equator toward the north or south geomagnetic pole). Assume that Earth's magnetic dipole moment is $\mu = 8.00 \times 10^{22}$ A·m². (a) Show that the magnitude of Earth's field at latitude λ_m is given by

$$B = \frac{\mu_0 \mu}{4\pi r^3} \sqrt{1 + 3 \sin^2 \lambda_m}.$$

(b) Show that the inclination ϕ_i of the magnetic field is related to the magnetic latitude λ_m by

$$\tan \phi_i = 2 \tan \lambda_m.$$

9P. Use the results displayed in Problem 8 to predict Earth's magnetic field (both magnitude and inclination) at (a) the geomagnetic equator; (b) a point at geomagnetic latitude 60°; (c) the north geomagnetic pole.

10P. Using the approximations given in Problem 8, find (a) the altitude above Earth's surface where the magnitude of its magnetic field is 50% of the surface value at the same latitude, (b) the maximum magnitude of the magnetic field at the core–mantle boundary, 2900 km below Earth's surface, and (c) the magnitude and inclination of Earth's magnetic field at the north geographic pole. Suggest why the values you calculated for (c) differ from measured values.

SECTION 32-4 Magnetism and Electrons

11E. What is the energy difference between parallel and antiparallel alignment of the z component of an electron's spin magnetic dipole moment with an external magnetic field of magnitude 0.25 T, directed parallel to the z axis?

12E. What is the measured component of the orbital magnetic dipole moment of an electron with (a) $m_l = 1$ and (b) $m_l = -2$?

13E. In the lowest energy state of the hydrogen atom, the most probable distance of the single electron from the central proton (the nucleus) is 5.2×10^{-11} m. (a) Compute the magnitude of the proton's electric field at that distance. The component $\mu_{s,z}$ of the proton's spin magnetic dipole moment measured on a z axis is 1.4×10^{-26} J/T. (b) Compute the magnitude of the proton's magnetic field at the distance 5.2×10^{-11} m on the z axis. (*Hint:*

Use Eq. 30-29.) (c) What is the ratio of the spin magnetic dipole moment of the electron to that of the proton?

14E. If an electron in an atom has an orbital angular momentum with $m_l = 0$, what are the components (a) $L_{orb,z}$ and (b) $\mu_{orb,z}$? If the atom is in an external magnetic field **B** of magnitude 35 mT and directed along the z axis, what are the potential energies associated with the orientations of (c) the electron's orbital magnetic dipole moment and (d) the electron's spin magnetic dipole moment? (e) Repeat (a) through (d) for $m_l = -3$.

15E. If an electron in an atom has orbital angular momentum with m_l values limited by ± 3, how many values of (a) $L_{orb,z}$ and (b) $\mu_{orb,z}$ can it have? In terms of h, m, and e, what are the greatest and least allowed magnitudes for (c) $L_{orb,z}$ and (d) $\mu_{orb,z}$? (e) What is the greatest allowed magnitude for the z component of its *net* angular momentum (orbital plus spin)? (f) How many values (signs included) are allowed for the z component of its net angular momentum?

16P. Figure 32-30a is a one-axis graph along which two of the allowed energy values *(levels)* of an atom are plotted (as in Fig. 8-17). When the atom is placed in a magnetic field of 0.50 T, the graph changes to that of Fig. 32-30b because of the energy associated with $\boldsymbol{\mu}_{orb} \cdot \mathbf{B}$. (We neglect $\boldsymbol{\mu}_s$.) Level E_1 is unchanged, but level E_2 splits into a (closely spaced) triplet of levels. What are the allowed values of m_l associated with (a) energy level E_1 and (b) energy level E_2? (c) On the graph, what is the spacing between the triplet levels?

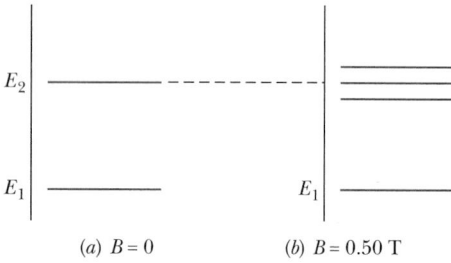

(a) $B = 0$ (b) $B = 0.50$ T

FIGURE 32-30 Problem 16.

17P. A charge q is distributed uniformly around a thin ring of radius r. The ring is rotating about an axis through its center and perpendicular to its plane, at an angular speed ω. (a) Show that the magnetic moment due to the rotating charge is

$$\mu = \tfrac{1}{2}q\omega r^2.$$

(b) What is the direction of this magnetic moment if the charge is positive?

SECTION 32-6 Diamagnetism

18E. Figure 32-31 shows a loop model (loop L) for a diamagnetic material. (a) Sketch the magnetic field lines through and about the material due to the bar magnet. (b) What are the direction of the loop's net magnetic dipole moment $\boldsymbol{\mu}$ and the direction

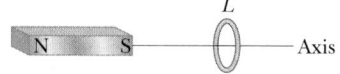

FIGURE 32-31 Exercises 18 and 22.

of the conventional current i in the loop? (c) What is the direction of the magnetic force on the loop?

19P*. Assume that an electron of mass m and charge magnitude e moves in a circular orbit of radius r about a nucleus. A uniform magnetic field **B** is then established perpendicular to the plane of the orbit. Assuming also that the radius of the orbit does not change and that the change in the speed of the electron due to field **B** is small, find an expression for the change in the orbital magnetic dipole moment of the electron.

SECTION 32-7 Paramagnetism

20E. A 0.50 T magnetic field is applied to a paramagnetic gas whose atoms have an intrinsic magnetic dipole moment of 1.0×10^{-23} J/T. At what temperature will the mean kinetic energy of translation of the gas atoms be equal to the energy required to reverse such a dipole end for end in this magnetic field?

21E. A magnet in the form of a cylindrical rod has a length of 5.00 cm and a diameter of 1.00 cm. It has a uniform magnetization of 5.30×10^3 A/m. What is its magnetic dipole moment?

22E. Repeat Exercise 18 for the case in which loop L is the model for a paramagnetic material.

23E. A sample of the paramagnetic salt to which the magnetization curve of Fig. 32-10 applies is to be tested to see whether it obeys Curie's law. The sample is placed in a uniform 0.50 T magnetic field that remains constant throughout the experiment. The magnetization M is then measured at temperatures ranging from 10 to 300 K. Will it be found that Curie's law is valid under these conditions?

24E. A sample of the paramagnetic salt to which the magnetization curve of Fig. 32-10 applies is held at room temperature (300 K). At what applied magnetic field will the degree of magnetic saturation of the sample be (a) 50% and (b) 90%? (c) Are these fields attainable in the laboratory?

25E. A sample of the paramagnetic salt to which the magnetization curve of Fig. 32-10 applies is immersed in a uniform magnetic field of 2.0 T. At what temperature will the degree of magnetic saturation of the sample be (a) 50% and (b) 90%?

26P. An electron with kinetic energy K_e travels in a circular path that is perpendicular to a uniform magnetic field, the electron's motion subject only to the force of the field. (a) Show that the magnetic dipole moment of the electron due to its orbital motion has magnitude $\mu = K_e/B$ and that it is in the direction opposite that of **B**. (b) What are the magnitude and direction of the magnetic dipole moment of a positive ion with kinetic energy K_i under the same circumstances? (c) An ionized gas consists of 5.3×10^{21} electrons/m^3 and the same number density of ions. Take the average electron kinetic energy to be 6.2×10^{-20} J and the average ion kinetic energy to be 7.6×10^{-21} J. Calculate the magnetization of the gas when it is in a magnetic field of 1.2 T.

27P. Consider a solid containing N atoms per unit volume, each atom having a magnetic dipole moment $\boldsymbol{\mu}$. Suppose the direction of $\boldsymbol{\mu}$ can be only parallel or antiparallel to an externally applied magnetic field **B** (this will be the case if $\boldsymbol{\mu}$ is due to the spin of a single electron). According to statistical mechanics, it can be shown that the probability of an atom being in a state with energy U is proportional to $e^{-U/kT}$, where T is the temperature and k is Boltzmann's constant. Thus, since $U = -\boldsymbol{\mu} \cdot \mathbf{B}$, the fraction of atoms whose dipole moment is parallel to $\mathbf{B}$ is proportional to $e^{\mu B/kT}$ and the fraction of atoms whose dipole moment is antiparallel to $\mathbf{B}$ is proportional to $e^{-\mu B/kT}$. (a) Show that the magnetization of this solid is $M = N\mu \tanh(\mu B/kT)$. Here tanh is the hyperbolic tangent function: $\tanh(x) = (e^x - e^{-x})/(e^x + e^{-x})$. (b) Show that the result given in (a) reduces to $M = N\mu^2 B/kT$ for $\mu B \ll kT$. (c) Show that the result of (a) reduces to $M = N\mu$ for $\mu B \gg kT$. (d) Show that both (b) and (c) agree qualitatively with Fig. 32-10.

SECTION 32-8 Ferromagnetism

28E. Measurements in mines and boreholes indicate that Earth's interior temperature increases with depth at the average rate of 30 C°/km. Assuming a surface temperature of 10°C, at what depth does iron cease to be ferromagnetic? (The Curie temperature of iron varies very little with pressure.)

29E. The exchange coupling mentioned in Section 32-8 as being responsible for ferromagnetism is *not* the mutual magnetic interaction between two elementary magnetic dipoles. To show this, calculate (a) the magnetic field a distance of 10 nm away, along the dipole axis, from an atom with magnetic dipole moment 1.5×10^{-23} J/T (cobalt), and (b) the minimum energy required to turn a second identical dipole end for end in this field. By comparing the latter with the results of Sample Problem 32-2, what can you conclude?

30E. The saturation magnetization M_{max} of the ferromagnetic metal nickel is 4.70×10^5 A/m. Calculate the magnetic moment of a single nickel atom. (The density of nickel is 8.90 g/cm³ and its molar mass is 58.71 g/mol.)

31E. The dipole moment associated with an atom of iron in an iron bar is 2.1×10^{-23} J/T. Assume that all the atoms in the bar, which is 5.0 cm long and has a cross-sectional area of 1.0 cm², have their dipole moments aligned. (a) What is the dipole moment of the bar? (b) What torque must be exerted to hold this magnet perpendicular to an external field of 1.5 T? The density of iron is 7.9 g/cm³.

32P. The magnetic dipole moment of Earth is 8.0×10^{22} J/T. (a) If the origin of this magnetism were a magnetized iron sphere at the center of Earth, what would be its radius? (b) What fraction of the volume of Earth would such a sphere occupy? Assume complete alignment of the dipoles. The density of Earth's inner core is 14 g/cm³. The magnetic dipole moment of an iron atom is 2.1×10^{-23} J/T. (*Note:* Earth's inner core is in fact thought to be in both liquid and solid forms and partly iron, but a permanent magnet as the source of Earth's magnetism has been ruled out by several considerations. For one, the temperature is certainly above the Curie point.)

33P. Figure 32-32 shows the apparatus used in a lecture demonstration of para- and diamagnetism. A sample of the magnetic material is suspended by a long string in the nonuniform field ($d = 2$ cm) between the poles of a powerful electromagnet. Pole

P_1 is sharply pointed and pole P_2 is rounded as indicated. Any deflection of the string from the vertical is visible to the audience by means of an optical projection system (not shown). (a) First a bismuth (highly diamagnetic) sample is used. When the electromagnet is turned on, the sample is observed to deflect slightly (about 1 mm) toward one of the poles. What is the direction of this deflection? (b) Next an aluminum (paramagnetic, conducting) sample is used. When the electromagnet is turned on, the sample is observed to deflect strongly (about 1 cm) toward one pole for about a second and then deflect moderately (a few millimeters) toward the other pole. Explain and indicate the direction of these deflections. (*Hint:* The aluminum sample is a conductor, for which Lenz's law applies.) (c) What would happen if a ferromagnetic sample were used?

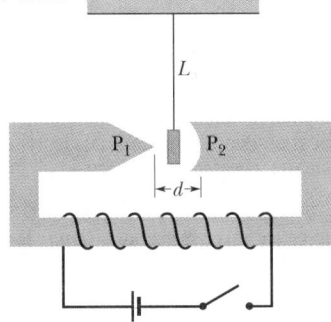

FIGURE 32-32
Problem 33.

34P. A magnetic compass has its needle of mass 0.050 kg and length 4.0 cm aligned with the horizontal component of Earth's magnetic field at a place where that component has the value $B_h = 16$ μT. After the compass is given a momentary gentle shake, the needle oscillates with angular frequency $\omega = 45$ rad/s. Assuming that the needle is a uniform thin rod mounted at its center, find its magnetic dipole moment.

35P. A Rowland ring is formed of ferromagnetic material. It is circular in cross section, with an inner radius of 5.0 cm and an outer radius of 6.0 cm, and is wound with 400 turns of wire. (a) What current must be set up in the windings to attain a toroidal field $B_0 = 0.20$ mT? (b) A secondary coil wound around the toroid has 50 turns and resistance 8.0 Ω. If, for this value of B_0, we have $B_M = 800B_0$, how much charge moves through the secondary coil when the current in the toroid windings is turned on?

SECTION 32-9 Induced Magnetic Fields

36E. For the situation of Sample Problem 32-4, at what radius r is the induced magnetic field equal to 50% of its maximum value?

37E. The induced magnetic field 6.0 mm from the central axis of a circular parallel-plate capacitor and between the plates is 2.0×10^{-7} T. The plates have radius 3.0 mm. At what rate $d\mathbf{E}/dt$ is the electric field between the plates changing?

38P. A parallel-plate capacitor, with circular plates of radius $R = 16$ mm and gap width $d = 5.0$ mm, has a uniform electric field between the plates. Starting at time $t = 0$, the potential difference between the plates is $V = (100$ V$)e^{-t/\tau}$, where the time constant $\tau = 12$ ms. At the radial distance $r = 0.80R$ from the capacitor axis within the gap, what is the magnetic field (a) as a function of time for $t \geq 0$ and (b) at time $t = 3\tau$?

39P. Suppose that a parallel-plate capacitor has circular plates with radius $R = 30$ mm and a plate separation of 5.0 mm. Suppose also that a sinusoidal potential difference with a maximum value of 150 V and a frequency of 60 Hz is applied across the plates; that is,

$$V = (150 \text{ V}) \sin[2\pi(60 \text{ Hz})t].$$

(a) Find $B_{max}(R)$, the maximum value of the induced magnetic field which occurs at $r = R$. (b) Plot $B_{max}(r)$ for $0 < r < 10$ cm.

SECTION 32-10 Displacement Current

40E. Prove that the displacement current in a parallel-plate capacitor of capacitance C can be written as $i_d = C(dV/dt)$, where V is the potential difference between the plates.

41E. At what rate must the potential difference between the plates of a parallel-plate capacitor with a 2.0 μF capacitance be changed to produce a displacement current of 1.5 A?

42E. For the situation of Sample Problem 32-4, show that the *displacement current density* is $J_d = \epsilon_0(dE/dt)$ for $r \leq R$.

43E. A parallel-plate capacitor with circular plates of radius 0.10 m is being discharged. A circular loop of radius 0.20 m is concentric with the capacitor and halfway between the plates. The displacement current through the loop is 2.0 A. At what rate is the electric field between the plates changing?

44E. A parallel-plate capacitor with circular plates of radius R is being discharged. The displacement current through a central circular area, parallel to the plates and with radius $R/2$, is 2.0 A. What is the discharge current?

45P. The magnitude of the electric field between the two circular plates of the parallel-plate capacitor in Fig. 32-33 is $E = (4.0 \times 10^5) - (6.0 \times 10^4 t)$, with E in volts per meter and t in seconds. At $t = 0$, the field is upward as shown. The plate area is 4.0×10^{-2} m². For $t \geq 0$, (a) what are the magnitude and direction of the displacement current between the plates and (b) is the direction of the induced magnetic field clockwise or counterclockwise around the plates?

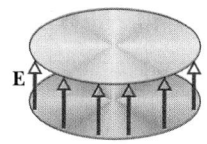

FIGURE 32-33 Problem 45.

46P. A parallel-plate capacitor with circular plates of radius R is being discharged by a current of 6.0 A. (a) At what distances from the central axis is the induced magnetic field equal to 75% of the maximum value of that field? (b) What is the maximum value of that field if $R = 0.040$ m?

47P. As a parallel-plate capacitor with circular plates 20 cm in diameter is being charged, the displacement current density throughout the region between the plates is uniform and has a magnitude of 20 A/m². (a) Calculate the magnitude B of the magnetic field at a distance $r = 50$ mm from the axis of symmetry of the region. (b) Calculate dE/dt in this region.

48P. A uniform electric field collapses to zero from an initial strength of 6.0×10^5 N/C in a time of 15 μs in the manner

shown in Fig. 32-34. Calculate the magnitude of the displacement current, through a 1.6 m² region perpendicular to the field, during each of the time intervals a, b, and c shown on the graph. (Ignore the behavior at the ends of the intervals.)

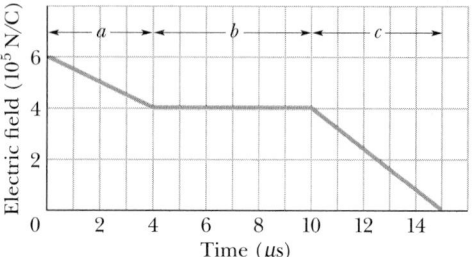

FIGURE 32-34 Problem 48.

49P. A parallel-plate capacitor has square plates 1.0 m on a side as in Fig. 32-35. A current of 2.0 A charges the capacitor, producing a uniform electric field **E** between the plates, with **E** perpendicular to the plates. (a) What is the displacement current through the region between the plates? (b) What is dE/dt in this region? (c) What is the displacement current through the square dashed path between the plates? (d) What is $\oint \mathbf{B} \cdot d\mathbf{s}$ around this square dashed path?

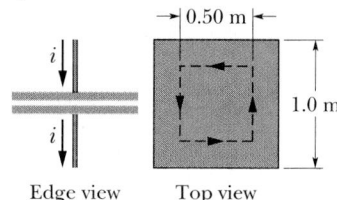

FIGURE 32-35 Problem 49.

50P. A silver wire has resistivity $\rho = 1.62 \times 10^{-8}$ $\Omega \cdot$m and a cross-sectional area of 5.00 mm². The current in the wire is uniform and changing at the rate of 2000 A/s when the current is 100 A. (a) What is the (uniform) electric field in the wire when the current in the wire is 100 A? (b) What is the displacement current in the wire at that time? (c) What is the ratio of the magnetic field due to the displacement current to that due to the current at a distance r from the wire?

51P. The capacitor in Fig. 32-36 with circular plates of radius $R = 18.0$ cm is connected to a source of emf $\mathscr{E} = \mathscr{E}_m \sin \omega t$, where $\mathscr{E}_m = 220$ V and $\omega = 130$ rad/s. The maximum value of the displacement current is $i_d = 7.60$ μA. Neglect fringing of the electric field at the edges of the plates. (a) What is the maximum value of the current i? (b) What is the maximum value of $d\Phi_E/dt$, where Φ_E is the electric flux through the region between the plates? (c) What is the separation d between the plates? (d) Find the maximum value of the magnitude of **B** between the plates at a distance $r = 11.0$ cm from the center.

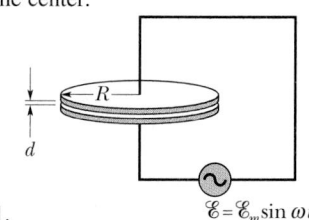

FIGURE 32-36 Problem 51.

33
Electromagnetic Oscillations and Alternating Current

When a high-voltage power transmission line requires repair, a utility company cannot just shut it down, blacking out, perhaps, an entire city. So repairs must be made while the lines are electrically "hot." The man outside the helicopter has just replaced a spacer between 500 kV lines <u>by hand</u>, a procedure that requires considerable expertise. Why, exactly, is the potential of such power transmission lines kept so high? Surprisingly, the current through them, although highly lethal, is not very large. Shouldn't it be large?

33-1 NEW PHYSICS— OLD MATHEMATICS

In this chapter you will see how the electric charge q varies with time in a circuit made up of an inductor L, a capacitor C, and a resistor R. From another point of view, we shall discuss how energy shuttles back and forth between the magnetic field of the inductor and the electric field of the capacitor, being gradually dissipated—while these oscillations continue—as thermal energy in the resistor.

We have discussed oscillations before, in another context. In Chapter 16 we saw how displacement x varies with time in a mechanical oscillating system made up of a block of mass m, a spring of spring constant k, and a viscous or frictional element such as oil; Fig. 16-17 shows such a system. You also saw how energy shuttles back and forth between the kinetic energy of the oscillating mass and the potential energy of the spring, being gradually dissipated—while the oscillations continue—as thermal energy.

The parallel between these two idealized systems is exact, and the controlling differential equations are identical. Thus there is no new mathematics to be learned; we can simply change the symbols and give our full attention to the physics of the situation.

33-2 *LC* OSCILLATIONS, QUALITATIVELY

Of the three circuit elements, resistance R, capacitance C, and inductance L, we have so far discussed the series combinations RC (in Section 28-8) and RL (in Section 31-9). In these two kinds of circuit we found that the charge, current, and potential difference grow and decay exponentially. The time scale of the growth or decay is given by a *time constant* τ, which is either capacitive or inductive.

We now examine the remaining two-element circuit combination LC. You will see that in this case the charge, current, and potential difference do not decay exponentially with time but vary sinusoidally (with period T and angular frequency ω). The circuit is said to *oscillate,* and the resulting oscillations of the capacitor's electric field

and the inductor's magnetic field are said to be **electromagnetic oscillations.**

Parts a through h of Fig. 33-1 show succeeding stages of the oscillations in a simple LC circuit. From Eq. 26-21, the energy stored in the electric field of the capacitor at any time is

$$U_E = \frac{q^2}{2C}, \qquad (33\text{-}1)$$

where q is the charge on the capacitor. From Eq. 31-53, the energy stored in the magnetic field of the inductor at any time is

$$U_B = \frac{Li^2}{2}, \qquad (33\text{-}2)$$

where i is the current through the inductor.

We now adopt the convention of representing *instantaneous values* of the electrical quantities of a sinusoidally oscillating circuit with small letters, such as q, and the *amplitudes* of those quantities with capital letters, such as Q. With this convention in mind, let us assume that initially the charge q on the capacitor in Fig. 33-1 is at its maximum value Q and that the current i through the inductor is zero. This initial state of the circuit is shown in Fig. 33-1a. The bar graphs for energy included there indicate that at this instant, with zero current through the inductor and maximum charge on the capacitor, the energy U_B of the magnetic field is zero and the energy U_E of the electric field is a maximum.

The capacitor now starts to discharge through the inductor, positive charge carriers moving counterclockwise, as shown in Fig. 33-1b. This means that a current i, given by dq/dt and pointing down in the inductor, is established. As the capacitor's charge decreases, the energy stored in the electric field within the capacitor also decreases. This energy is transferred to the magnetic field that appears around the inductor because of the current i that is building up there. Thus the electric field decreases and the magnetic field builds up as energy is transferred from the electric field to the magnetic field.

The capacitor eventually loses all its charge (Fig. 33-1c) and thus also loses its electric field and the energy stored in that field. The energy has then been fully transferred to the magnetic field of the inductor. Because the magnetic field is then at its maximum magnitude, the current through the inductor is then at its maximum value I.

Although the charge on the capacitor is now zero, the counterclockwise current must continue because the inductor does not allow it to change suddenly to zero. So, the current continues to transfer positive charge from the top

The method of repairing high-voltage lines shown in the opening photograph is patented by Scott H. Yenzer and is licensed exclusively to Haverfield Corporation of Miami, Florida. As the lineman approaches a hot line, the electric field surrounding the line brings his body to nearly the potential of the line. To match the two potentials, he then extends a conducting "wand" to the line. To avoid being electrocuted, he must be isolated from anything electrically connected to the ground. And so that his body is always at a single potential—that of the line he is working on—he wears a conducting suit, hood, and gloves, all of which are electrically connected to the line via the wand.

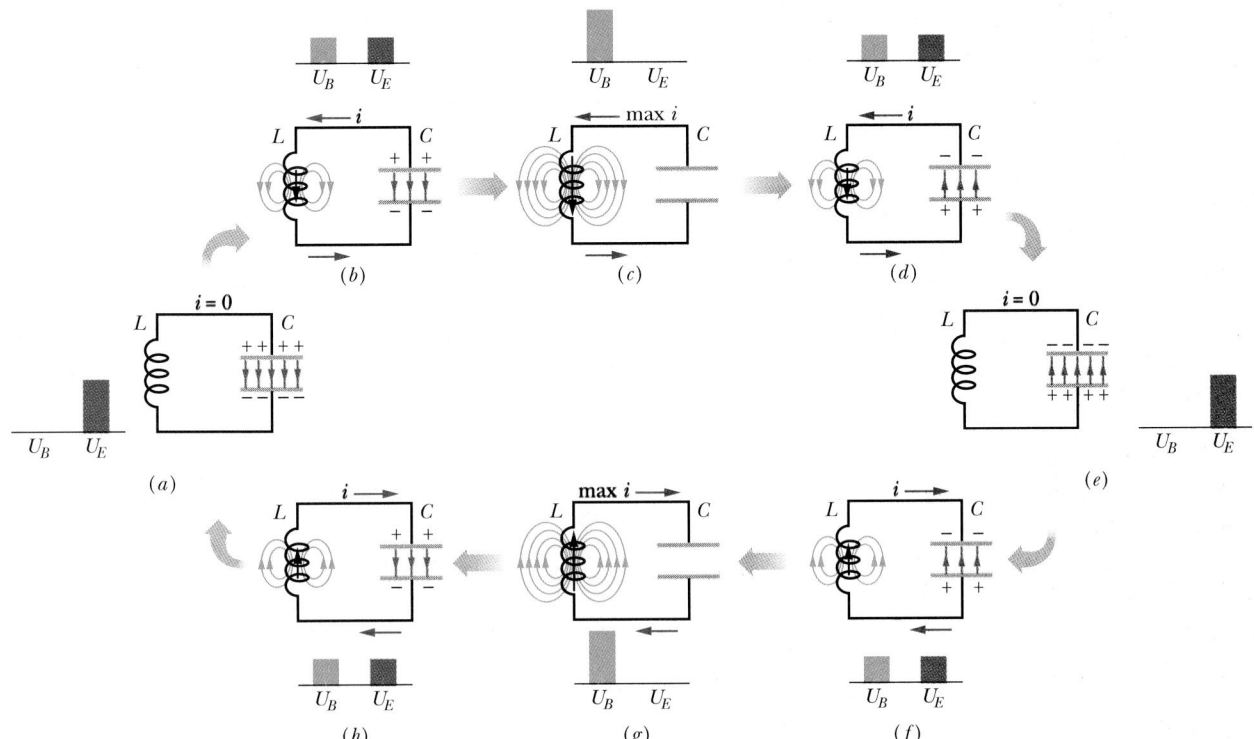

FIGURE 33-1 Eight stages in a single cycle of oscillation of a resistanceless *LC* circuit. The bar graphs by each figure show the stored magnetic and electric energies. The magnetic field lines of the inductor and the electric field lines of the capacitor are shown. (*a*) Capacitor with maximum charge, no current. (*b*) Capacitor discharging, current increasing. (*c*) Capacitor fully discharged, current maximum. (*d*) Capacitor charging but with polarity opposite that in (*a*), current decreasing. (*e*) Capacitor with maximum charge having polarity opposite that in (*a*), no current. (*f*) Capacitor discharging, current increasing with direction opposite that in (*b*). (*g*) Capacitor fully discharged, current maximum. (*h*) Capacitor charging, current decreasing.

plate to the bottom plate through the circuit (Fig. 33-1*d*). Energy now flows from the inductor back to the capacitor as the electric field within the capacitor builds up again. The current gradually decreases during this energy transfer. When, eventually, the energy has been transferred completely back to the capacitor (Fig. 33-1*e*), the current has decreased to zero (momentarily). The situation of Fig. 33-1*e* is like the initial situation, except that the capacitor is now charged oppositely.

The capacitor then starts to discharge again but now with a clockwise current (Fig. 33-1*f*). Reasoning as before, we see that the clockwise current builds to a maximum (Fig. 33-1*g*) and then decreases (Fig. 33-1*h*), until the circuit eventually returns to its initial situation (Fig. 33-1*a*). The process then repeats at a frequency *f* and thus at an angular frequency $\omega = 2\pi f$. In the ideal circuit with no resistance, there are no energy transfers other than that between the electric field of the capacitor and the magnetic field of the inductor. Owing to the conservation of energy, the oscillations continue indefinitely. The oscillations need

not begin with the energy all in the electric field; the initial situation could be any other stage of the oscillation.

To find the charge *q* on the capacitor as a function of time, we can use a voltmeter to measure the time-varying potential difference (or *voltage*) v_C that exists across the capacitor *C*. From Eq. 26-1 we can write

$$v_C = \left(\frac{1}{C}\right) q,$$

which allows us to find *q*. To measure the current, we can connect a small resistance *R* in series with the capacitor and inductor and measure the time-varying potential difference v_R across it; v_R is proportional to *i* through the relation

$$v_R = iR.$$

We assume here that *R* is so small that its effect on the behavior of the circuit is negligible. The variations in time of v_C and v_R, and thus of *q* and *i*, are shown in Fig. 33-2. All four quantities vary sinusoidally.

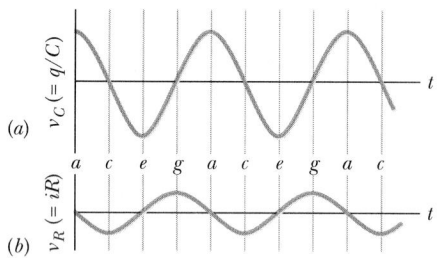

FIGURE 33-2 (*a*) The potential difference across the capacitor of the circuit of Fig. 33-1 as a function of time. This quantity is proportional to the charge on the capacitor. (*b*) A potential proportional to the current in the circuit of Fig. 33-1. The letters refer to the correspondingly labeled oscillation stages in Fig. 33-1.

In an actual *LC* circuit, the oscillations will not continue indefinitely because there is always some resistance present that will drain energy from the electric and magnetic fields and dissipate it as thermal energy; the circuit may become warmer. The oscillations, once started, will die away as Fig. 33-3 suggests. Compare this figure with Fig. 16-18, which shows the decay of mechanical oscillations caused by frictional damping in a block–spring system.

CHECKPOINT 1: A charged capacitor and an inductor are connected in series at time $t = 0$. In terms of the period T of the resulting oscillations, determine how much later the following reach their maximums: (a) the charge on the capacitor, (b) the voltage across the capacitor, with its original polarity, (c) the energy stored in the electric field, and (d) the current.

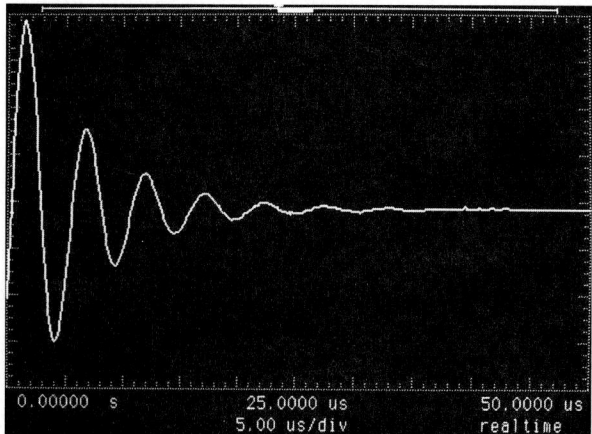

FIGURE 33-3 An oscilloscope trace showing how the oscillations in an *RLC* circuit actually die away because energy is dissipated in the resistor in thermal form.

SAMPLE PROBLEM 33-1

A 1.5 μF capacitor is charged to 57 V. The charging battery is then disconnected, and a 12 mH coil is connected in series with the capacitor so that *LC* oscillations occur. What is the maximum current in the coil? Assume that the circuit contains no resistance.

SOLUTION: From the principle of conservation of energy, the maximum stored energy in the capacitor must equal the maximum stored energy in the inductor. This leads, from Eqs. 33-1 and 33-2 to

$$\frac{Q^2}{2C} = \frac{LI^2}{2},$$

where I is the maximum current and Q is the maximum charge. (The maximum current and the maximum charge occur not at the same time but one-fourth of a cycle apart, as is evident from Figs. 33-1 and 33-2.) Solving for I and substituting CV for Q, we find

$$I = V\sqrt{\frac{C}{L}} = (57 \text{ V})\sqrt{\frac{1.5 \times 10^{-6} \text{ F}}{12 \times 10^{-3} \text{ H}}}$$

$$= 0.637 \text{ A} \approx 640 \text{ mA}. \qquad \text{(Answer)}$$

33-3 THE ELECTRICAL–MECHANICAL ANALOGY

Let us look a little closer at the analogy between the oscillating *LC* system of Fig. 33-1 and an oscillating block–spring system. Two kinds of energy are involved in the block–spring system. One is potential energy of the compressed or extended spring; the other is kinetic energy of the moving block. These two energies are given by the familiar formulas at the left in Table 33-1.

The table also shows the two kinds of energy involved in *LC* oscillations. We can see an analogy between the two pairs of energies—the mechanical energies of the block–spring system and the electromagnetic energies of the *LC* oscillator. The equations for v and i at the bottom of the table help us to improve the analogy. They tell us that q

TABLE 33-1   THE ENERGY IN TWO OSCILLATING SYSTEMS COMPARED

BLOCK–SPRING SYSTEM		*LC* OSCILLATOR	
ELEMENT	ENERGY	ELEMENT	ENERGY
Spring	Potential, $\frac{1}{2}kx^2$	Capacitor	Electric, $\frac{1}{2}(1/C)q^2$
Block	Kinetic, $\frac{1}{2}mv^2$	Inductor	Magnetic, $\frac{1}{2}Li^2$
	$v = dx/dt$		$i = dq/dt$

corresponds to x, and i corresponds to v (in both equations, the former is differentiated to obtain the latter). These correspondences lead us to pair the individual energies horizontally as shown in the table, hence suggest that $1/C$ corresponds to k and L corresponds to m. Thus

q corresponds to x, $1/C$ corresponds to k,
i corresponds to v, and L corresponds to m.

These correspondences suggest that in an *LC* oscillator, the capacitor is mathematically like the spring in a block–spring system, and the inductor is like the block.

In Section 16-3 we saw that the angular frequency of oscillation of a (frictionless) block–spring system is

$$\omega = \sqrt{\frac{k}{m}} \quad \text{(block–spring system)}. \quad (33\text{-}3)$$

The correspondences listed above suggest that to find the angular frequency of oscillation for a (resistanceless) *LC* circuit, k should be replaced by $1/C$ and m by L, yielding

$$\omega = \frac{1}{\sqrt{LC}} \quad \text{(}LC\text{ circuit)}. \quad (33\text{-}4)$$

We derive this result in the next section.

33-4 *LC* OSCILLATIONS, QUANTITATIVELY

Here we want to show explicitly that Eq. 33-4 for the angular frequency of *LC* oscillations is correct. At the same time, we want to examine even more closely the analogy between *LC* oscillations and block–spring oscillations. We start by extending somewhat our earlier treatment of the mechanical block–spring oscillator.

The Block–Spring Oscillator

We analyzed block–spring oscillations in Chapter 16 in terms of energy transfers and did not—at that early stage —derive the fundamental differential equation that governs those oscillations. We do so now.

We can write, for the total energy U of a block–spring oscillator at any instant,

$$U = U_b + U_s = \tfrac{1}{2}mv^2 + \tfrac{1}{2}kx^2, \quad (33\text{-}5)$$

where U_b and U_s are, respectively, the kinetic energy of the moving block and the potential energy of the stretched or compressed spring. If there is no friction—which we assume—the total energy U remains constant with time, even though v and x vary. In more formal language,

$dU/dt = 0$. This leads to

$$\frac{dU}{dt} = \frac{d}{dt}\left(\tfrac{1}{2}mv^2 + \tfrac{1}{2}kx^2\right)$$

$$= mv\frac{dv}{dt} + kx\frac{dx}{dt} = 0. \quad (33\text{-}6)$$

But $v = dx/dt$ and $dv/dt = d^2x/dt^2$. With these substitutions, Eq. 33-6 becomes

$$m\frac{d^2x}{dt^2} + kx = 0 \quad \begin{array}{l}\text{(block–spring}\\ \text{oscillations)}.\end{array} \quad (33\text{-}7)$$

Equation 33-7 is the fundamental *differential equation* that governs the frictionless block–spring oscillations. It involves the displacement x and its second derivative with respect to time.

The general solution to Eq. 33-7, that is, the function $x(t)$ that describes the block–spring oscillations, is (as we saw in Eq. 16-3)

$$x = X\cos(\omega t + \phi) \quad \text{(displacement)}, \quad (33\text{-}8)$$

in which X is the amplitude of the mechanical oscillations (represented by x_m in Chapter 16), ω is the angular frequency of the oscillations, and ϕ is a phase constant.

The *LC* Oscillator

Now let us analyze the oscillations of a resistanceless *LC* circuit, proceeding exactly as we just did for the block–spring oscillator. The total energy U present at any instant in an oscillating *LC* circuit is given by

$$U = U_B + U_E = \frac{Li^2}{2} + \frac{q^2}{2C}, \quad (33\text{-}9)$$

in which U_B is the energy stored in the magnetic field of the inductor and U_E is the energy stored in the electric field of the capacitor. Since we have assumed the circuit resistance to be zero, no energy is transferred to thermal energy and U remains constant with time. In more formal language, dU/dt must be zero. This leads to

$$\frac{dU}{dt} = \frac{d}{dt}\left(\frac{Li^2}{2} + \frac{q^2}{2C}\right)$$

$$= Li\frac{di}{dt} + \frac{q}{C}\frac{dq}{dt} = 0. \quad (33\text{-}10)$$

But $i = dq/dt$ and $di/dt = d^2q/dt^2$. With these substitutions, Eq. 33-10 becomes

$$L\frac{d^2q}{dt^2} + \frac{1}{C}q = 0 \quad \text{(}LC\text{ oscillations)}. \quad (33\text{-}11)$$

This is the *differential equation* that describes the oscillations of a resistanceless *LC* circuit. Careful comparison shows that Eqs. 33-11 and 33-7 are exactly of the same mathematical form, differing only in the symbols used.

Since the differential equations are mathematically identical, their solutions must also be mathematically identical. Because *q* corresponds to *x*, we can write the general solution of Eq. 33-11, giving *q* as a function of time, by analogy to Eq. 33-8 as

$$q = Q \cos(\omega t + \phi) \quad \text{(charge)}, \qquad (33\text{-}12)$$

where *Q* is the amplitude of the charge variations, ω is the angular frequency of the electromagnetic oscillations, and ϕ is the phase constant.

Taking the first derivative of Eq. 33-12 with respect to time gives us the current of the *LC* oscillator:

$$i = \frac{dq}{dt} = -\omega Q \sin(\omega t + \phi) \quad \text{(current)}. \quad (33\text{-}13)$$

The amplitude *I* of this sinusoidally varying current is

$$I = \omega Q, \qquad (33\text{-}14)$$

so we can rewrite Eq. 33-13 as

$$i = -I \sin(\omega t + \phi). \qquad (33\text{-}15)$$

We can test whether Eq. 33-12 is a solution of Eq. 33-11 by substituting it and its second derivative with respect to time into Eq. 33-11. The first derivative of Eq. 33-12 is Eq. 33-13. The second derivative is then

$$\frac{d^2q}{dt^2} = -\omega^2 Q \cos(\omega t + \phi).$$

Substituting for *q* and d^2q/dt^2 into Eq. 33-11, we obtain

$$-L\omega^2 Q \cos(\omega t + \phi) + \frac{1}{C} Q \cos(\omega t + \phi) = 0.$$

Canceling $Q \cos(\omega t + \phi)$ and rearranging lead to

$$\omega = \frac{1}{\sqrt{LC}}.$$

Thus Eq. 33-12 is indeed a solution of Eq. 33-11 if ω has the constant value $1/\sqrt{LC}$. Note that this expression for ω is exactly that given by Eq. 33-4, which we arrived at by examining correspondences.

The phase constant ϕ in Eq. 33-12 is determined by the conditions that prevail at $t = 0$. If those conditions yield $\phi = 0$, for example, then at $t = 0$, Eq. 33-12 requires that $q = Q$ and Eq. 33-13 requires that $i = 0$; these are the initial conditions represented by Fig. 33-1*a*.

The electric energy stored in the *LC* circuit at any time *t* is, from Eqs. 33-1 and 33-12,

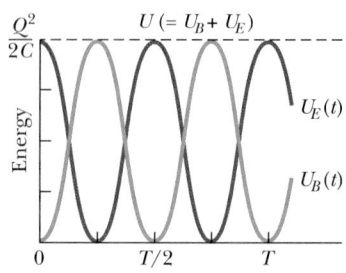

FIGURE 33-4   The stored magnetic energy and electric energy in the circuit of Fig. 33-1 as a function of time. Note that their sum remains constant. *T* is the period of oscillation.

$$U_E = \frac{q^2}{2C} = \frac{Q^2}{2C} \cos^2(\omega t + \phi), \qquad (33\text{-}16)$$

and the magnetic energy is, from Eqs. 33-2 and 33-13,

$$U_B = \tfrac{1}{2} L i^2 = \tfrac{1}{2} L \omega^2 Q^2 \sin^2(\omega t + \phi).$$

Substituting for ω from Eq. 33-4 then gives us

$$U_B = \frac{Q^2}{2C} \sin^2(\omega t + \phi). \qquad (33\text{-}17)$$

Figure 33-4 shows plots of $U_E(t)$ and $U_B(t)$ for the case of $\phi = 0$. Note that:

1. The maximum values of U_E and U_B are both $Q^2/2C$.

2. At any instant the sum of U_E and U_B is equal to $Q^2/2C$, a constant.

3. When U_E is maximum, U_B is zero, and conversely.

$\mathbb{C}$HECKPOINT 2: A capacitor in an *LC* oscillator has a maximum potential difference of 17 V and a maximum energy of 160 μJ. When the capacitor has a potential difference of 5 V and an energy of 10 μJ, what are (a) the emf across the inductor and (b) the energy stored in the magnetic field?

SAMPLE PROBLEM 33-2

(a) In an oscillating *LC* circuit, what charge *q*, expressed in terms of the maximum charge *Q*, is present on the capacitor when the energy is shared equally between the electric and magnetic fields? Assume that $L = 12$ mH and $C = 1.7$ μF.

SOLUTION: The problem requires that $U_E = \tfrac{1}{2} U_{E,\text{max}}$. The instantaneous and maximum stored energy in the capacitor are, respectively,

$$U_E = \frac{q^2}{2C} \quad \text{and} \quad U_{E,\max} = \frac{Q^2}{2C},$$

so the problem requires that

$$\frac{q^2}{2C} = \frac{1}{2}\frac{Q^2}{2C},$$

or

$$q = \frac{1}{\sqrt{2}}Q = 0.707Q. \qquad \text{(Answer)}$$

(b) When does this condition occur if the capacitor has its maximum charge at time $t = 0$?

SOLUTION: Equation 33-12 tells us how q varies with time. Because $q = Q$ at time $t = 0$, the phase constant ϕ is zero. Substituting $\phi = 0$ and $q = 0.707Q$ into Eq. 33-12 gives

$$0.707Q = Q \cos \omega t$$

from which

$$\omega t = 45° = \frac{\pi}{4} \text{ rad.}$$

This corresponds to one-eighth of a full oscillation of 2π rad. Substituting for ω from Eq. 33-4 and solving for t yield

$$t = \frac{\pi}{4\omega} = \frac{\pi\sqrt{LC}}{4} = \frac{\pi\sqrt{(12 \times 10^{-3}\text{ H})(1.7 \times 10^{-6}\text{ F})}}{4}$$

$$= 1.12 \times 10^{-4}\text{ s} \approx 110 \text{ } \mu\text{s.} \qquad \text{(Answer)}$$

33-5 DAMPED OSCILLATIONS IN AN *RLC* CIRCUIT

A circuit containing resistance, inductance, and capacitance is called an *RLC circuit*. We shall here discuss only *series RLC circuits* like that shown in Fig. 33-5. With a resistance R present, the total *electromagnetic energy U* of the circuit (the sum of the electric energy and magnetic energy) is no longer constant; instead, it decreases with time as energy is transferred to thermal energy in the resistance. Because of this loss of energy, the oscillations of charge, current, and potential difference continuously de-

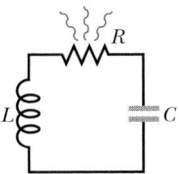

FIGURE 33-5 A series *RLC* circuit. As the charge contained in the circuit oscillates back and forth through the resistance, electromagnetic energy is dissipated as thermal energy, damping (decreasing the amplitude of) the oscillations.

crease in amplitude, and the oscillations are said to be *damped*. As you will see, they are damped in exactly the same way as those of the damped block–spring oscillator of Section 16-8.

To analyze the oscillations of this circuit, we write an equation for the total electromagnetic energy U in the circuit at any instant. Because the resistance does not store electromagnetic energy, we can use Eq. 33-9:

$$U = U_B + U_E = \frac{Li^2}{2} + \frac{q^2}{2C}. \qquad (33\text{-}18)$$

Now, however, this total energy decreases as energy is transferred to thermal energy. The rate of that transfer is, from Eq. 27-22,

$$\frac{dU}{dt} = -i^2 R, \qquad (33\text{-}19)$$

where the minus sign indicates that U decreases. By differentiating Eq. 33-18 with respect to time and then substituting the result in Eq. 33-19, we obtain

$$\frac{dU}{dt} = Li\frac{di}{dt} + \frac{q}{C}\frac{dq}{dt} = -i^2 R.$$

Substituting dq/dt for i and d^2q/dt^2 for di/dt, we obtain

$$L\frac{d^2q}{dt^2} + R\frac{dq}{dt} + \frac{1}{C}q = 0 \quad (RLC \text{ circuit}), \quad (33\text{-}20)$$

which is the differential equation that describes damped charge oscillations in an *RLC* circuit.

The solution to Eq. 33-20 is

$$q = Qe^{-Rt/2L}\cos(\omega't + \phi), \qquad (33\text{-}21)$$

in which

$$\omega' = \sqrt{\omega^2 - (R/2L)^2}, \qquad (33\text{-}22)$$

where $\omega = 1/\sqrt{LC}$, as with an undamped oscillator. Equation 33-21 tells us how the charge on the capacitor oscillates in a damped *RLC* circuit; the equation is the electromagnetic counterpart of Eq. 16-40, which gives the displacement of a damped block–spring oscillator.

Equation 33-21 describes a sinusoidal oscillation (the cosine term) with an *exponentially decaying amplitude* $Qe^{-Rt/2L}$ (the terms that multiply the cosine). The angular frequency ω' of the damped oscillations is always less than the angular frequency ω of the undamped oscillations; however, we shall here consider only situations in which R is small enough for us to replace ω' with ω.

Let us next find an expression for the total electromagnetic energy U of the circuit as a function of time. One way to do so is to monitor the energy of the electric field in the capacitor, which is given by Eq. 33-1 ($U_E = q^2/2C$).

By substituting Eq. 33-21 into Eq. 33-1, we obtain

$$U_E = \frac{q^2}{2C} = \frac{[Qe^{-Rt/2L}\cos(\omega' t + \phi)]^2}{2C}$$

$$= \frac{Q^2}{2C} e^{-Rt/L}\cos^2(\omega' t + \phi). \quad (33\text{-}23)$$

Equation 33-23 shows that the energy of the electric field oscillates according to a cosine-squared term and that the amplitude of that oscillation decreases exponentially with time. Thus, the total electromagnetic energy U must also be decreasing exponentially with time. That total energy is stored in the capacitor whenever U_E reaches a maximum, which occurs whenever $\cos^2(\omega' t + \phi)$ is 1. So we can find an expression for the total energy simply by setting the cosine-squared term in Eq. 33-23 equal to 1. We get

$$U = \frac{Q^2}{2C} e^{-Rt/L}. \quad (33\text{-}24)$$

CHECKPOINT 3: (a) The figure shows the graphs of the total electromagnetic energy U in two RLC circuits with identical capacitors and inductors. Which curve corresponds to the circuit with greater R? (b) If, instead, the circuits have the same R and C, which curve corresponds to the circuit with greater L?

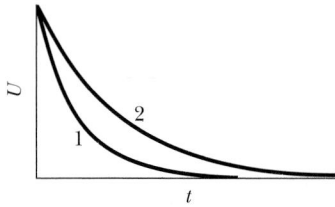

SAMPLE PROBLEM 33-3

A series RLC circuit has inductance $L = 12$ mH, capacitance $C = 1.6$ μF, and resistance $R = 1.5$ Ω.

(a) At what time t will the amplitude of the charge oscillations in the circuit be 50% of its initial value?

SOLUTION: Equation 33-21 gives the exponential decay of the charge oscillations. The amplitude of the charge oscillations has decayed to 50% of its initial value when the amplitude $Qe^{-Rt/2L}$ in Eq. 33-21 has decayed to $0.50Q$, that is, when

$$Qe^{-Rt/2L} = 0.50Q.$$

Canceling Q and taking the natural logarithms of both sides, we have

$$-\frac{Rt}{2L} = \ln 0.50.$$

Solving for t and then substituting given data yield

$$t = -\frac{2L}{R}\ln 0.50 = -\frac{(2)(12 \times 10^{-3}\text{ H})(\ln 0.50)}{1.5\ \Omega}$$

$$= 0.0111\text{ s} \approx 11\text{ ms}. \quad \text{(Answer)}$$

(b) How many oscillations are completed within this time?

SOLUTION: Since the period of oscillation is $T = 2\pi/\omega$ and the angular frequency is $\omega = 1/\sqrt{LC}$, we have $T = 2\pi\sqrt{LC}$. Each oscillation takes one period, so, in a time interval $\Delta t = 0.0111$ s, the number of complete oscillations is

$$\frac{\Delta t}{T} = \frac{\Delta t}{2\pi\sqrt{LC}}$$

$$= \frac{0.0111\text{ s}}{2\pi[(12 \times 10^{-3}\text{ H})(1.6 \times 10^{-6}\text{ F})]^{1/2}} \approx 13.$$

$$\text{(Answer)}$$

Thus the amplitude decays by 50% in about 13 complete oscillations. This damping is less severe than that shown in Fig. 33-3, where the amplitude decays by a little more than 50% in one oscillation.

33-6 ALTERNATING CURRENT

The oscillations in an RLC circuit will not damp out if an external emf device supplies enough energy to make up for the energy dissipated as thermal energy in the resistance R. Circuits in homes, offices, and factories, including countless RLC circuits, receive such energy from local power companies. In most countries the energy is supplied via oscillating emfs and currents—the current is said to be an **alternating current,** or **ac** for short. (The nonoscillating current from a battery is said to be a **direct current,** or **dc**.) These oscillating emfs and currents vary sinusoidally with time, reversing direction (in North America) 120 times per second and thus having frequency $f = 60$ Hz.

At first sight this may seem to be a strange arrangement. We have seen that the drift speed of the conduction electrons in household wiring may typically be 4×10^{-5} m/s. If we now reverse their direction every $\frac{1}{120}$ s, such electrons can move only about 3×10^{-7} m in a half-cycle. At this rate, a typical electron can drift past no more than about 10 atoms in the wiring before it is required to reverse its direction. How, you may wonder, can the electron ever get anywhere?

Although this question may be worrisome, it is a needless concern. The conduction electrons do not have to ''get anywhere.'' When we say that the current in a wire is one ampere, we mean that charge carriers pass through any plane cutting across that wire at the rate of one coulomb per second. The speed at which the carriers cross that plane does not matter directly; one ampere may correspond to many charge carriers moving very slowly or to a few moving very rapidly. Furthermore, the signal to the electrons to reverse directions—which originates in the alternating emf provided by the power company's generator—is propagated along the conductor at a speed close to that of light. All electrons, no matter where they are located, get their reversal instructions at about the same instant. Finally, we note that for many devices, such as lightbulbs and toasters, the direction of motion is unimportant as long as the electrons do move so as to transfer energy to the device via collisions with atoms in the device.

The basic advantage of alternating currents is this: *as the current alternates, so does the magnetic field that surrounds the conductor.* This makes possible the use of Faraday's law of induction, which, among other things, means that we can step up (increase) or step down (decrease) the magnitude of an alternating potential difference at will, using a device called a transformer, as you will see later in this chapter. Moreover, alternating current is more readily adaptable to rotating machinery such as generators and motors than is (nonalternating) direct current.

Figure 33-6 shows a simple model of an ac generator. As the conducting loop is forced to rotate through the external magnetic field **B**, a sinusoidally oscillating emf $\mathscr{E}$ is induced in the loop:

$$\mathscr{E} = \mathscr{E}_m \sin \omega_d t. \qquad (33\text{-}25)$$

The *angular frequency* ω_d of the emf is equal to the angular speed with which the loop rotates in the magnetic field; the *phase* of the emf is $\omega_d t$; and the *amplitude* of the emf is $\mathscr{E}_m$ (where the subscript stands for maximum). When the rotating loop is part of a closed conducting path, this emf

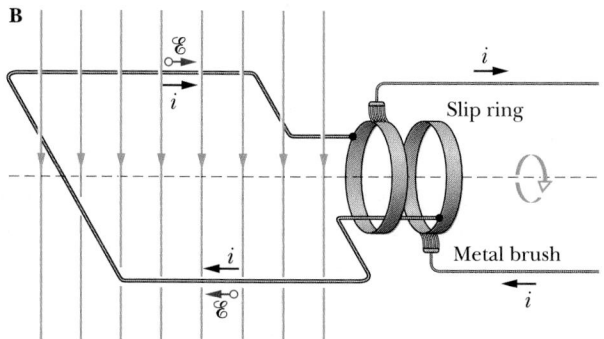

FIGURE 33-6 The basic mechanism of an alternating-current generator is a conducting loop rotated in an external magnetic field. In practice, the alternating emf induced in a coil of many turns of wire is made accessible by means of slip rings attached to the rotating loop, each ring connected to one end of the loop wire and electrically connected by a conducting brush (against which it slips) to the rest of the generator circuit.

produces *(drives)* a sinusoidal (alternating) current along the path with the same angular frequency ω_d, which then is called the **driving angular frequency.** We can write the current as

$$i = I \sin(\omega_d t - \phi), \qquad (33\text{-}26)$$

in which I is the amplitude of the driven current. (The phase $\omega_d t - \phi$ of the current is traditionally written with a minus sign instead of as $\omega_d t + \phi$.) We include a phase constant ϕ in Eq. 33-26 because the current i may not be in phase with the emf $\mathscr{E}$. (As you will see, the phase constant depends on the circuit to which the generator is connected.)

33-7 FORCED OSCILLATIONS

We have seen that once started, the charge, potential difference, and current in both undamped LC circuits and damped RLC circuits (with small enough R) oscillate at angular frequency $\omega = 1/\sqrt{LC}$. Such oscillations are said to be *free oscillations* (free of any external emf), and the angular frequency ω is said to be the circuit's **natural angular frequency.**

When the external alternating emf of Eq. 33-25 is connected to an RLC circuit, the oscillations of charge, potential difference, and current are said to be *driven oscillations* or *forced oscillations.* These oscillations always occur at the driving angular frequency ω_d:

> Whatever the natural angular frequency ω of a circuit may be, forced oscillations of charge, current, and potential difference in the circuit always occur at the driving angular frequency ω_d.

However, as you will see in Section 33-9, the amplitudes of the oscillations very much depend on how close ω_d is to ω. When the two angular frequencies match—a condition known as **resonance**—the amplitude I of the current in the circuit is maximum.

33-8 THREE SIMPLE CIRCUITS

Later, we shall connect an external alternating emf device to a series RLC circuit as in Fig. 33-7. We shall then find expressions for the amplitude I and phase constant ϕ of the sinusoidally oscillating current in terms of the amplitude $\mathscr{E}_m$ and angular frequency ω_d of the external emf. But first let us consider three simpler circuits, each having an external emf and only one other circuit element: R, C, or L. We start with a resistive element (a purely *resistive load*).

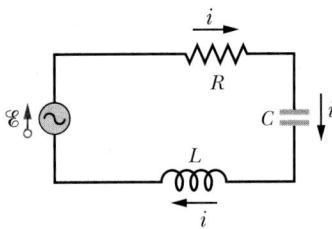

FIGURE 33-7 A single-loop circuit containing a resistor, a capacitor, and an inductor. A generator, represented by a sine wave in a circle, produces an alternating emf that establishes an alternating current; the directions of the emf and current are indicated here at only one instant.

A Resistive Load

Figure 33-8a shows a circuit containing a resistance element of value R and an ac generator with the alternating emf of Eq. 33-25. By the loop rule we have

$$\mathscr{E} - v_R = 0.$$

With Eq. 33-25, this gives us

$$v_R = \mathscr{E}_m \sin \omega_d t.$$

Because the amplitude V_R of the alternating potential difference (or voltage) across the resistance is equal to the amplitude $\mathscr{E}_m$ of the alternating emf, we can write this as

$$v_R = V_R \sin \omega_d t. \tag{33-27}$$

From the definition of resistance ($R = V/i$), we can now write the current i_R in the resistance as

$$i_R = \frac{v_R}{R} = \frac{V_R}{R} \sin \omega_d t. \tag{33-28}$$

From Eq. 33-26, we can also write this current as

$$i_R = I_R \sin(\omega_d t - \phi), \tag{33-29}$$

where I_R is the amplitude of the current i_R in the resistance. Comparing Eqs. 33-28 and 33-29, we see that for a purely resistive load the phase constant $\phi = 0°$. We also see that the voltage amplitude and current amplitude are related by

$$V_R = I_R R \quad \text{(resistor)}. \tag{33-30}$$

Although we found this relation for the circuit of Fig. 33-8a, it applies to any resistance in any ac circuit.

By comparing Eqs. 33-27 and 33-28, we see that the time-varying quantities v_R and i_R are both functions of $\sin \omega_d t$ with $\phi = 0°$. Thus these two quantities are *in phase*, which means that their corresponding maxima (and minima) occur at the same times. Figure 33-8b, which is a plot of $v_R(t)$ and $i_R(t)$, illustrates this fact. Note that v_R and i_R do not decay here, because the generator supplies energy to the circuit to make up for the energy dissipated in R.

The time-varying quantities v_R and i_R can also be represented geometrically by *phasors*. Recall from Section 17-10 that phasors are vectors that rotate around an origin. Those that represent the voltage across and current in the resistor of Fig. 33-8a are shown in Fig. 33-8c at an arbitrary time t. Such phasors have the following properties:

Angular speed: Both phasors rotate counterclockwise about the origin with an angular speed equal to the angular frequency ω_d of v_R and i_R.

Length: The length of each phasor represents the amplitude of the alternating quantity: V_R for the voltage and I_R for the current.

Projection: The projection of each phasor on the *vertical* axis represents the value of the alternating quantity at time t: v_R for the voltage and i_R for the current.

Rotation angle: The rotation angle of each phasor is equal to the phase of the alternating quantity at time t. In Fig. 33-8c, the voltage and current are in phase. So their phasors have the same phase $\omega_d t$ and the same rotation angle, and thus they rotate together.

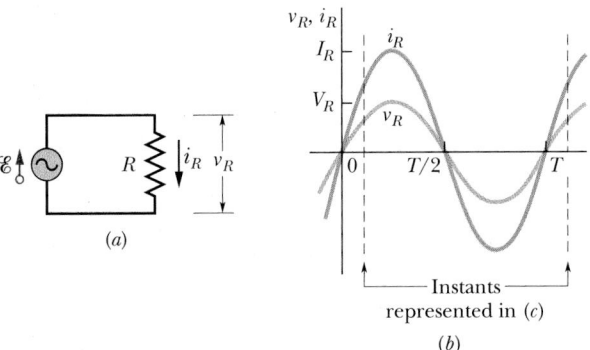

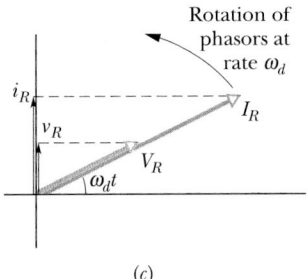

FIGURE 33-8 (a) A resistor is connected across an alternating-current generator. (b) The current and the potential difference across the resistor are in phase and complete one cycle in one period T. (c) A phasor diagram shows the same thing as (b).

Mentally follow the rotation. Can you see that when the phasors have rotated so that $\omega_d t = 90°$ (they point vertically upward), they indicate that just then $v_R = V_R$ and $i_R = I_R$? Equations 33-27 and 33-29 give the same results.

A Capacitive Load

Figure 33-9a shows a circuit containing a capacitance and a generator with the alternating emf of Eq. 33-25. Using the loop rule and proceeding as we did when we obtained Eq. 33-27, we find that the potential difference across the capacitor is

$$v_C = V_C \sin \omega_d t, \tag{33-31}$$

where V_C is the voltage amplitude across the capacitor. From the definition of capacitance we can also write

$$q_C = Cv_C = CV_C \sin \omega_d t. \tag{33-32}$$

Our concern, however, is with the current rather than the charge. Thus we differentiate Eq. 33-32 to find

$$i_C = \frac{dq_C}{dt} = \omega_d CV_C \cos \omega_d t. \tag{33-33}$$

We now recast Eq. 33-33 in two ways. First, for reasons of symmetry of notation, we introduce the quantity X_C, called the **capacitive reactance** of the capacitor, defined as

$$X_C = \frac{1}{\omega_d C} \quad \text{(capacitive reactance).} \tag{33-34}$$

Its value depends not only on the capacitance but also on the driving angular frequency ω_d. We know from the definition of the capacitive time constant ($\tau = RC$) that the SI unit for C can be expressed as seconds per ohm. Applying this to Eq. 33-34 shows that the SI unit of X_C is the *ohm*, just as for resistance R.

As our second modification of Eq. 33-33, we replace $\cos \omega_d t$ with a phase-shifted sine, namely,

$$\cos \omega_d t = \sin(\omega_d t + 90°).$$

You can verify this identity by expanding the right-hand side with the formula for $\sin(\alpha + \beta)$ listed in Appendix E.

With these two modifications, Eq. 33-33 becomes

$$i_C = \left(\frac{V_C}{X_C}\right) \sin(\omega_d t + 90°). \tag{33-35}$$

From Eq. 33-26, we can also write the current i_C in C as

$$i_C = I_C \sin(\omega_d t - \phi), \tag{33-36}$$

where I_C is the amplitude of i_C. Comparing Eqs. 33-35 and 33-36, we see that for a purely capacitive load the phase constant $\phi = -90°$. We also see that the voltage amplitude and current amplitude are related by

$$V_C = I_C X_C \quad \text{(capacitor).} \tag{33-37}$$

Although we found this relation for the circuit of Fig. 33-9a, it applies to any capacitance in any ac circuit.

Comparison of Eqs. 33-31 and 33-35, or inspection of Fig. 33-9b, shows that the quantities v_C and i_C are 90°, or one-quarter-cycle, out of phase. Furthermore, we see that i_C *leads* v_C, which means that, if you monitored the current i_C and the potential difference v_C in the circuit of Fig. 33-9a, you would find that i_C reaches its maximum *before* v_C does, by one-quarter cycle.

This relation between i_C and v_C is illustrated by the phasor diagram of Fig. 33-9c. As the phasors representing these two quantities rotate counterclockwise together, the phasor labeled I_C does indeed lead that labeled V_C, and by an angle of 90°. That is, the phasor I_C coincides with the vertical axis one-quarter cycle before the phasor V_C does. Be sure to convince yourself that the phasor diagram of Fig. 33-9c is consistent with Eqs. 33-31 and 33-35.

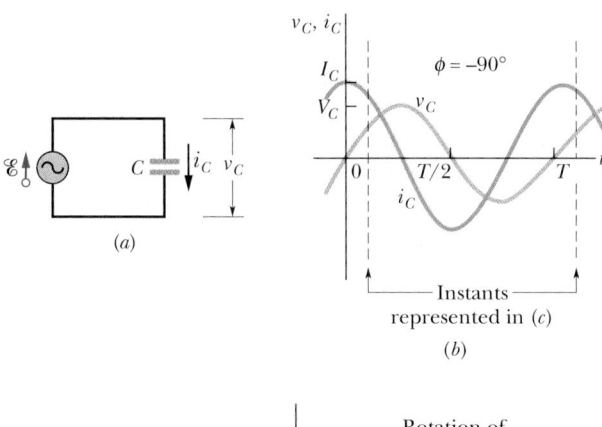

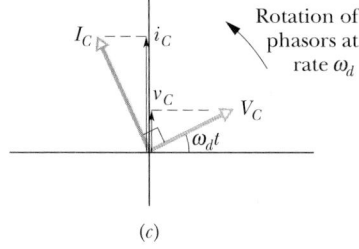

FIGURE 33-9 (a) A capacitor is connected across an alternating-current generator. (b) The current in the capacitor leads the voltage by 90°. (c) A phasor diagram shows the same thing.

CHECKPOINT 4: The figure shows, in (a), a sine curve $S(t) = \sin(\omega_d t)$ and three other sinusoidal curves $A(t)$, $B(t)$, and $C(t)$, each of the form $\sin(\omega_d t - \phi)$. (a) Rank the three other curves according to the value of ϕ, most positive first and most negative last. (b) Which curve corresponds to which phasor in (b) of the figure? (c) Which curve leads the others?

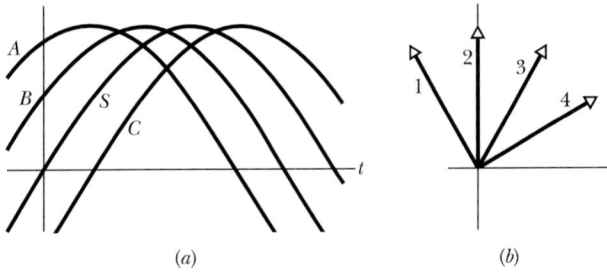

(a)   (b)

An Inductive Load

Figure 33-10a shows a circuit containing an inductance and a generator with the alternating emf of Eq. 33-25. Using the loop rule and proceeding as we did to obtain Eq. 33-27, we find that the potential difference across the inductance is

$$v_L = V_L \sin \omega_d t, \tag{33-38}$$

where V_L is the amplitude of v_L. From Eq. 31-40, we can write the potential difference across an inductance L, in which the current is changing at the rate di_L/dt, as

$$v_L = L \frac{di_L}{dt}. \tag{33-39}$$

If we combine Eqs. 33-38 and 33-39 we have

$$\frac{di_L}{dt} = \frac{V_L}{L} \sin \omega_d t. \tag{33-40}$$

Our concern, however, is with the current rather than with its time derivative. We find the former by integrating Eq. 33-40, obtaining

$$i_L = \int di_L = \frac{V_L}{L} \int \sin \omega_d t \, dt$$

$$= -\left(\frac{V_L}{\omega L}\right) \cos \omega_d t. \tag{33-41}$$

We now recast this equation in two ways. First, for reasons of symmetry of notation, we introduce the quantity X_L, called the **inductive reactance** of the inductor, which is defined as

$$X_L = \omega_d L \quad \text{(inductive reactance).} \tag{33-42}$$

The value of X_L depends on the driving angular frequency ω_d. The unit of the inductive time constant τ_L indicates that the SI unit of X_L is the *ohm*, just as it is for X_C and for R.

Second, we replace the function $-\cos \omega_d t$ in Eq. 33-

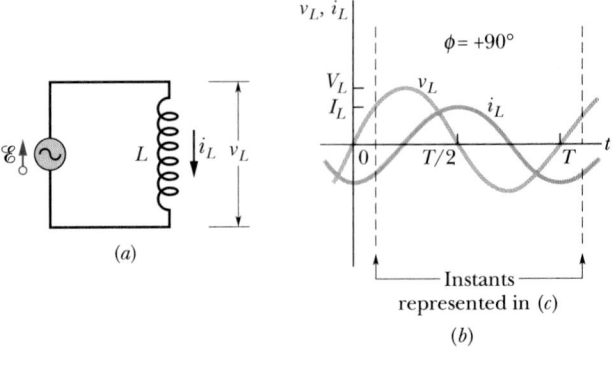

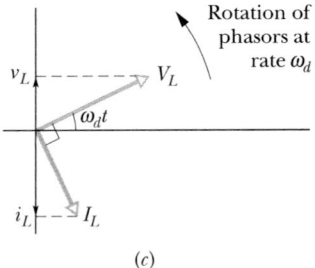

(c)

FIGURE 33-10 (a) An inductor is connected across an alternating-current generator. (b) The current in the inductor lags the voltage by 90°. (c) A phasor diagram shows the same thing.

41 with a phase-shifted sine, namely,

$$-\cos \omega_d t = \sin(\omega_d t - 90°).$$

You can verify this identity by expanding its right-hand side with the formula for $\sin(\alpha - \beta)$ given in Appendix E.

With these two changes, Eq. 33-41 becomes

$$i_L = \left(\frac{V_L}{X_L}\right) \sin(\omega_d t - 90°). \tag{33-43}$$

From Eq. 33-26, we can also write this current in the inductance as

$$i_L = I_L \sin(\omega_d t - \phi), \tag{33-44}$$

where I_L is the amplitude of the current i_L. Comparing Eqs. 33-43 and 33-44, we see that for a purely inductive load the phase constant $\phi = +90°$. We also see that the voltage amplitude and the current amplitude are related by

$$V_L = I_L X_L \quad \text{(inductor).} \tag{33-45}$$

Although we found this relation for the circuit of Fig. 33-10a, it applies to any inductance in any ac circuit.

Comparison of Eqs. 33-38 and 33-43, or inspection of Fig. 33-10b, shows that the quantities i_L and v_L are 90° out of phase. In this case, however, i_L *lags* v_L. That is, if you monitored the current i_L and the potential difference v_L in the circuit of Fig. 33-10a, you would find that i_L reaches its maximum value *after* v_L does, by one-quarter cycle.

The phasor diagram of Fig. 33-10c also contains this information. As the phasors rotate counterclockwise in the figure, the phasor labeled I_L does indeed lag that labeled V_L, and by an angle of 90°. Be sure to convince yourself that Fig. 33-10c represents Eqs. 33-38 and 33-43.

PROBLEM SOLVING TACTICS

TACTIC 1: *Leading and Lagging in ac Circuits*

Table 33-2 summarizes the relations between the current i and the voltage v for each of the three kinds of circuit elements we have considered. When an applied alternating voltage produces an alternating current in them, the current is in phase with the voltage across a resistor, leads the voltage across a capacitor, and lags the voltage across an inductor.

Many students remember these results with the mnemonic "*ELI* the *ICE* man." *ELI* contains the letter L (for inductor), and in it the letter I (for current) comes after the letter E (for emf or voltage). Thus, for an inductor, the current *lags* the voltage. Similarly *ICE* (which contains a C for capacitor) means that the current leads the voltage.

You might also use the modified mnemonic "*ELI positively* is the *ICE* man" to remember that the phase constant ϕ is positive for an inductor.

If you have difficulty in remembering whether X_C is equal to $\omega_d C$ (wrong) or $1/\omega_d C$ (right), try remembering that C is in the "cellar," that is, in the denominator.

SAMPLE PROBLEM 33-4

(a) In Fig. 33-9a, let $C = 15.0$ μF and $\mathscr{E}_m = V_C = 36.0$ V, and let the *driving frequency* f_d of the alternating emf device be 60.0 Hz. What is the current amplitude I_C?

SOLUTION: We can get this amplitude from Eq. 33-37 ($V_C = I_C X_C$) if we first find the capacitive reactance X_C. From Eq. 33-34 with $\omega_d = 2\pi f_d$, we have

$$X_C = \frac{1}{2\pi f_d C} = \frac{1}{(2\pi)(60.0 \text{ Hz})(15.0 \times 10^{-6} \text{ F})}$$
$$= 177 \ \Omega.$$

Then from Eq. 33-37,

$$I_C = \frac{V_C}{X_C} = \frac{36.0 \text{ V}}{177 \ \Omega} = 0.203 \text{ A}. \quad \text{(Answer)}$$

(b) In Fig. 33-10a, let $L = 230$ mH, $\mathscr{E}_m = V_L = 36.0$ V, and $f_d = 60.0$ Hz. What is the current amplitude I_L?

SOLUTION: We can get this amplitude from Eq. 33-45 ($I_L = V_L/X_L$) if we first find the inductive reactance X_L. From Eq. 33-42 with $\omega_d = 2\pi f_d$, we have

$$X_L = 2\pi f_d L = (2\pi)(60.0 \text{ Hz})(230 \times 10^{-3} \text{ H}) = 86.7 \ \Omega.$$

Then, from Eq. 33-45,

$$I_L = \frac{V_L}{X_L} = \frac{36.0 \text{ V}}{86.7 \ \Omega} = 0.415 \text{ A}. \quad \text{(Answer)}$$

(c) Write an expression for the time-varying current i_L in the circuit of (b).

SOLUTION: Equation 33-44 is the general equation for i_L. With I_L computed above as 0.415 A, with

$$\omega_d = 2\pi f_d = 120\pi \text{ Hz},$$

and with $\phi = 90° = \pi/2$ rad for this purely inductive circuit, we have

$$i_L = I_L \sin(\omega_d t - \phi)$$
$$= (0.415 \text{ A}) \sin\left(120\pi t - \frac{\pi}{2}\right). \quad \text{(Answer)}$$

$\mathbf{C}$HECKPOINT **5:** If we increase the driving frequency f_d in the circuits of (a) Fig. 33-9a and (b) Fig. 33-10a, does the current amplitude I increase, decrease, or stay the same?

33-9 THE SERIES *RLC* CIRCUIT

We are now ready to apply the alternating emf of Eq. 33-25,

$$\mathscr{E} = \mathscr{E}_m \sin \omega_d t \quad \text{(applied emf)}, \quad (33\text{-}46)$$

to the full *RLC* circuit of Fig. 33-7. Because R, L, and C are in series, the same current

$$i = I \sin(\omega_d t - \phi) \quad (33\text{-}47)$$

TABLE 33-2 PHASE AND AMPLITUDE RELATIONS FOR ALTERNATING CURRENTS AND VOLTAGES

CIRCUIT ELEMENT	SYMBOL	RESISTANCE OR REACTANCE	PHASE OF THE CURRENT	PHASE ANGLE ϕ	AMPLITUDE RELATION
Resistor	R	R	In phase with v_R	0°	$V_R = I_R R$
Capacitor	C	$X_C = 1/\omega_d C$	Leads v_C by 90°	−90°	$V_C = I_C X_C$
Inductor	L	$X_L = \omega_d L$	Lags v_L by 90°	+90°	$V_L = I_L X_L$

is driven in all three of them. We wish to find the current amplitude I and the phase constant ϕ.

The solution is simplified by the use of phasor diagrams. We start with Fig. 33-11a, which shows the phasor representing the current of Eq. 33-47 at an arbitrary time t. The length of the phasor is the amplitude I, the projection of the phasor on the vertical axis is the current i at time t, and the angle of rotation of the phasor is the phase $\omega_d t - \phi$ of the current at time t.

Figure 33-11b shows the phasors representing the voltages across R, L, and C at the same time t. Each phasor is oriented relative to the angle of rotation of current phasor I in Fig. 33-11a, based on the information in Table 33-2:

> ***Resistor:*** Here current and voltage are in phase; so the angle of rotation of voltage phasor V_R is the same as that of phasor I.
>
> ***Capacitor:*** Here current leads the voltage by 90°; so the angle of rotation of voltage phasor V_C is 90° less than that of phasor I.
>
> ***Inductor:*** Here current lags the voltage by 90°; so the angle of rotation of voltage phasor v_L is 90° greater than that of phasor I.

Figure 33-11b also shows the instantaneous voltages v_R, v_C, and v_L across R, C, and L at time t; those voltages are the projections of the corresponding phasors on the vertical axis of the figure.

Figure 33-11c shows the phasor representing the applied emf of Eq. 33-46. The length of the phasor is the amplitude $\mathcal{E}_m$, the projection of the phasor on the vertical axis is the emf $\mathcal{E}$ at time t, and the angle of rotation of the phasor is the phase $\omega_d t$ of the emf at time t.

Now, from the loop rule we know that at any instant the sum of the voltages v_R, v_C, and v_L is equal to the applied emf $\mathcal{E}$:

$$\mathcal{E} = v_R + v_C + v_L. \tag{33-48}$$

Thus at time t the projection $\mathcal{E}$ in Fig. 33-11c is equal to the algebraic sum of the projections v_R, v_C, and v_L in Fig. 33-11b. In fact, as the phasors rotate together, this equality always holds. This means that phasor $\mathcal{E}_m$ in Fig. 33-11c must be equal to the vector sum of the three voltage phasors V_R, V_C, and V_L in Fig. 33-11b.

We can simplify this vector sum by first noting that phasors V_C and V_L have opposite directions. We can combine them into the single phasor $V_L - V_C$ as shown in Fig. 3-11d. And we can find the vector sum of the three voltage phasors in Fig. 3-11b by finding the resultant of the two phasors V_R and $(V_L - V_C)$ in Fig. 3-11d. That resultant must coincide with phasor $\mathcal{E}_m$ as shown.

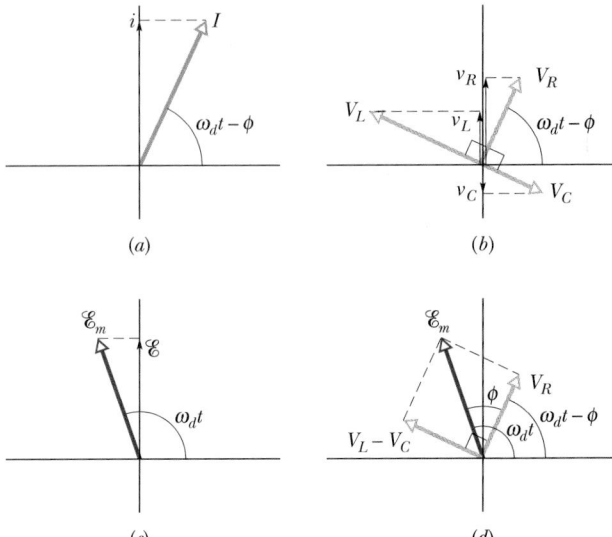

(a) (b) (c) (d)

FIGURE 33-11 (*a*) A phasor representing the alternating current in the driven *RLC* circuit of Fig. 33-7 at time *t*. The amplitude *I*, the instantaneous value *i*, and the phase ($\omega_d t - \phi$) are shown. (*b*) Phasors representing the voltages across the inductor, resistor, and capacitor, oriented with respect to the current phasor in (*a*). (*c*) A phasor representing the alternating emf that drives the current of (*a*). (*d*) The emf phasor is equal to the vector sum of the three voltage phasors of (*b*). Here, voltage phasors V_L and V_C have been added to yield their net phasor ($V_L - V_C$).

Both triangles in Fig. 3-11d are right triangles. Applying the Pythagorean theorem to either one yields

$$\mathcal{E}_m^2 = V_R^2 + (V_L - V_C)^2. \tag{33-49}$$

From the amplitude information displayed in Table 33-2 we can rewrite this as

$$\mathcal{E}_m^2 = (IR)^2 + (IX_L - IX_C)^2, \tag{33-50}$$

and then rearrange it to the form

$$I = \frac{\mathcal{E}_m}{\sqrt{R^2 + (X_L - X_C)^2}}. \tag{33-51}$$

The denominator in Eq. 33-51 is called the **impedance** Z of the circuit for the driving angular frequency ω_d:

$$Z = \sqrt{R^2 + (X_L - X_C)^2} \quad \text{(impedance defined).} \tag{33-52}$$

We can then write Eq. 33-51 as

$$I = \frac{\mathcal{E}_m}{Z}. \tag{33-53}$$

If we substitute for X_C and X_L from Eqs. 33-34 and 33-42, we can write Eq. 33-51 more explicitly as

$$I = \frac{\mathcal{E}_m}{\sqrt{R^2 + (\omega_d L - 1/\omega_d C)^2}} \quad \begin{array}{l}\text{(current}\\ \text{amplitude).}\end{array} \quad (33\text{-}54)$$

We have now accomplished half our goal: we have obtained an expression for the current amplitude I in terms of the sinusoidal driving emf and the circuit elements in a series *RLC* circuit.

The value of I depends on the difference between $\omega_d L$ and $1/\omega_d C$ in Eq. 33-54 or, equivalently, on the difference between X_L and X_C in Eq. 33-51. In either equation, it does not matter which of the two quantities is greater because the difference is always squared.

The current that we have been describing in this section is the *steady-state current* that occurs after the alternating emf has been applied for some time. When the emf is first applied to a circuit, a brief *transient current* occurs. Its duration (before settling down into the steady-state current) is determined by the time constants $\tau_L = L/R$ and $\tau_C = RC$ as the inductive and capacitive elements "turn on." This transient current can be large and can, for example, destroy a motor on start-up if it is not properly taken into account in the motor's circuit design.

The Phase Constant

We still need to find an expression for the phase constant ϕ. From the right-hand phasor triangle in Fig. 33-11*d* and from Table 33-2 we can write

$$\tan \phi = \frac{V_L - V_C}{V_R} = \frac{IX_L - IX_C}{IR}, \quad (33\text{-}55)$$

which gives us

$$\tan \phi = \frac{X_L - X_C}{R} \quad \text{(phase constant).} \quad (33\text{-}56)$$

This is the other half of our goal: an expression for the phase constant ϕ in a sinusoidally driven series *RLC* circuit. The expression gives us three different results for the phase constant, depending on the relative values of X_L and X_C:

$X_L > X_C$: The circuit is said to be *more inductive than capacitive*. Equation 33-56 tells us that ϕ is positive for such a circuit, which means that phasor $\mathcal{E}_m$ rotates ahead of phasor I (Fig. 33-12*a*). A plot of $\mathcal{E}$ and i versus time is like that in Fig. 33-12*b*. (The phasors in Figs. 33-11*c* and *d* were drawn assuming $X_L > X_C$.)

$X_C > X_L$: The circuit is said to be *more capacitive than inductive*. Equation 33-56 tells us that ϕ is nega-

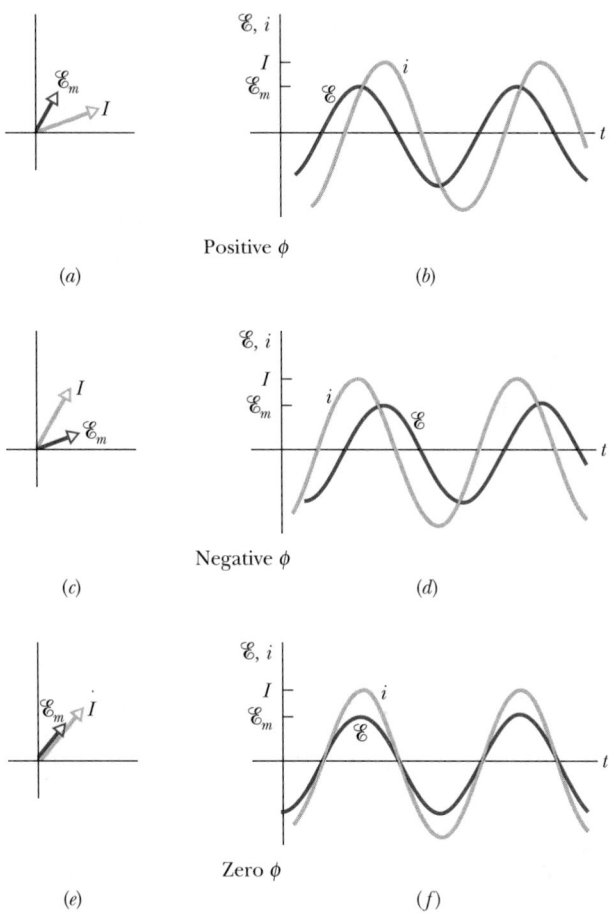

FIGURE 33-12 Phasor diagrams and plots of the alternating emf and current for the driven *RLC* circuit of Fig. 33-7, for (*a*, *b*) positive ϕ, (*c*, *d*) negative ϕ, and (*e*, *f*) zero ϕ.

tive for such a circuit, which means that phasor $\mathcal{E}_m$ rotates behind phasor I (Fig. 33-12*c*). A plot of $\mathcal{E}$ and i versus time is like that in Fig. 33-12*d*.

$X_C = X_L$: The circuit is said to be in *resonance*, a term that is explained below. Equation 33-56 tells us that $\phi = 0°$ for such a circuit, which means that phasors $\mathcal{E}_m$ and I rotate together (Fig. 33-12*e*). A plot of $\mathcal{E}$ and i versus time is like that in Fig. 33-12*f*.

As illustration, let us reconsider two extreme circuits: In the *purely inductive circuit* of Fig. 33-10*a*, where X_L is nonzero and $X_C = R = 0$, Eq. 33-56 tells us that $\phi = +90°$ (the greatest value of ϕ), consistent with Fig. 33-10*c*. In the *purely capacitive circuit* of Fig. 33-9*a*, where X_C is nonzero and $X_L = R = 0$, Eq. 33-56 tells us that $\phi = -90°$ (the least value of ϕ), consistent with Fig. 33-9*c*.

Resonance

Equation 33-54 gives the current amplitude I in an RLC circuit as a function of the driving angular frequency ω_d of the external alternating emf. For a given resistance R, that amplitude is a maximum when the quantity $\omega_d L - 1/\omega_d C$ in the denominator is zero, that is, when

$$\omega_d L = \frac{1}{\omega_d C}$$

or
$$\omega_d = \frac{1}{\sqrt{LC}} \quad \text{(maximum } I\text{).} \qquad (33\text{-}57)$$

Because the natural angular frequency ω of the RLC circuit is also equal to $1/\sqrt{LC}$, the maximum value of I occurs when the driving angular frequency matches the natural angular frequency, that is, at resonance. So in an RLC circuit, resonance and maximum current amplitude I occur when

$$\omega_d = \omega = \frac{1}{\sqrt{LC}} \quad \text{(resonance).} \qquad (33\text{-}58)$$

Figure 33-13 shows three *resonance curves* for sinusoidally driven oscillations in three series RLC circuits differing only in R. Each curve peaks at its maximum current amplitude when the ratio $\omega_d/\omega = 1.00$, but the maximum value of I decreases with increasing R. (The maximum I is always $\mathscr{E}_m/R$; to see why, combine Eqs. 33-52 and 33-53.) In addition, the curves increase in width (measured in Fig. 33-13 at half the maximum value of I) with increasing R.

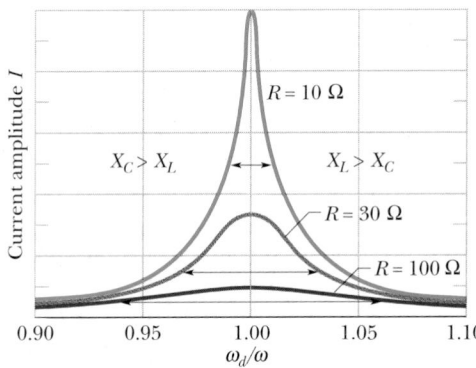

FIGURE 33-13 *Resonance curves for the driven RLC circuit of* Fig. 33-7 with $L = 100 \ \mu\text{H}$, $C = 100 \text{ pF}$, and three values of R. The current amplitude I of the alternating current depends on how close the driving angular frequency ω_d is to the natural angular frequency ω. The horizontal arrow on each curve measures the curve's width at the half-maximum level, a measure of the sharpness of the resonance. To the left of $\omega_d/\omega = 1.00$, the circuit is mainly capacitive, with $X_C > X_L$; to the right, it is mainly inductive, with $X_L > X_C$.

We can make physical sense of the resonance curves in Fig. 33-13 by considering how the reactances X_L and X_C change as we increase the driving angular frequency ω_d, starting with a value much less than the natural frequency ω. For small ω_d, reactance X_L ($= \omega_d L$) is small and reactance X_C ($= 1/\omega_d C$) is large. So, the circuit is mainly capacitive and the impedance is dominated by the large X_C, which keeps the current low.

As we increase ω_d, reactance X_C remains dominant but decreases while reactance X_L increases. The decrease in X_C decreases the impedance, allowing the current to increase, as we see on the left side of any resonance curve in Fig. 33-13. When the increasing X_L and the decreasing X_C reach equal values, the current is greatest and the circuit is in resonance, with $\omega_d = \omega$.

As we continue to increase ω_d, the increasing reactance X_L becomes progressively more dominant over the decreasing reactance X_C. So, the impedance increases because of X_L and the current decreases, as on the right side of any resonance curve in Fig. 33-13. In summary, then: the low-angular-frequency side of a resonance curve is dominated by the capacitor's reactance; the high-angular-frequency side is dominated by the inductor's reactance; and resonance occurs between the two regions.

SAMPLE PROBLEM 33-5

In Fig. 33-7 let $R = 160 \ \Omega$, $C = 15.0 \ \mu\text{F}$, $L = 230 \text{ mH}$, $f_d = 60.0 \text{ Hz}$, and $\mathscr{E}_m = 36.0 \text{ V}$. (Except for R, these parameters are the same as in Sample Problem 33-4.)

(a) What is the current amplitude I?

SOLUTION: We can get the current amplitude with Eq. 33-53 ($I = \mathscr{E}_m/Z$) if we first find the impedance Z of the circuit with Eq. 33-52. From Sample Problem 33-4a we know that the capacitive reactance X_C for the capacitor (and thus for the circuit) is 177 Ω; from Sample Problem 33-4b we know that the inductive reactance X_L for the inductor is 86.7 Ω. So, Eq. 33-52 tells us that

$$Z = \sqrt{R^2 + (X_L - X_C)^2}$$
$$= \sqrt{(160 \ \Omega)^2 + (86.7 \ \Omega - 177 \ \Omega)^2}$$
$$= 184 \ \Omega.$$

We then find

$$I = \frac{\mathscr{E}_m}{Z} = \frac{36.0 \text{ V}}{184 \ \Omega} = 0.196 \text{ A.} \qquad \text{(Answer)}$$

(b) What is the phase constant ϕ?

SOLUTION: From Eq. 33-56,

$$\tan \phi = \frac{X_L - X_C}{R} = \frac{86.7 \ \Omega - 177 \ \Omega}{160 \ \Omega} = -0.564.$$

Thus we have

$$\phi = \tan^{-1}(-0.564) = -29.4° = -0.513 \text{ rad.} \quad \text{(Answer)}$$

The negative phase angle is consistent with the fact that the load is mainly capacitive; that is, $X_C > X_L$.

CHECKPOINT **6:** Here are the capacitive reactance and inductive reactance, respectively, for three sinusoidally driven series *RLC* circuits: (1) 50 Ω, 100 Ω; (2) 100 Ω, 50 Ω; (3) 50 Ω, 50 Ω. (a) For each, does the current lead or lag the applied emf, or are the two in phase? (b) Which circuit is in resonance?

33-10 POWER IN ALTERNATING-CURRENT CIRCUITS

In the *RLC* circuit of Fig. 33-7 the source of energy is the alternating-current generator. Some of the energy that it provides is stored in the electric field in the capacitor, some is stored in the magnetic field in the inductor, and some is dissipated as thermal energy in the resistor. In steady-state operation—which we assume—the average energy stored in the capacitor and in the inductor remains constant. The net transfer of energy is thus from the generator to the resistor, where the electromagnetic energy is dissipated as thermal energy.

The instantaneous rate at which energy is dissipated in the resistor can be written, with the help of Eqs. 27-22 and

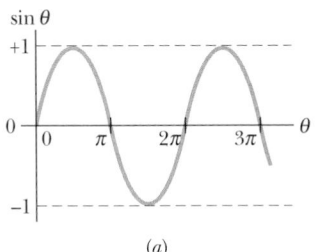

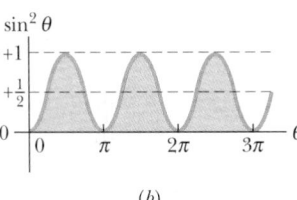

FIGURE 33-14 (*a*) A plot of sin θ versus θ. The average value over one cycle is zero. (*b*) A plot of sin² θ versus θ. The average value over one cycle is $\frac{1}{2}$.

33-26, as

$$P = i^2R = [I \sin(\omega_d t - \phi)]^2R$$
$$= I^2R \sin^2(\omega_d t - \phi). \quad (33\text{-}59)$$

The *average* rate at which energy is dissipated in the resistor, however, is the average of Eq. 33-59 over time. Although the average value of sin θ, where θ is any variable, is zero (Fig. 33-14a), the average value of sin² θ over one complete cycle is $\frac{1}{2}$ (Fig 33-14b). (Note in Fig. 33-14b how the shaded parts of the curve that lie above the horizontal line marked $+\frac{1}{2}$ exactly fill in the empty spaces below that

In August 1988, after playing only day games for 72 years, the Chicago Cubs got lamps for night games: 540 metal halide lamps, rated at 1500 W each, lit up the playing field. However, the first night game (against the Phillies) was stopped because of a thunderstorm, which some fans took as a sign that the Cubs should have stayed with day games.

At 5:17 P.M. on November 9, 1965, a faulty relay in the power system near Niagara Falls opened a circuit breaker on a transmission line, automatically causing the current to switch to other lines, which overloaded those lines and made other circuit breakers open. Within minutes the runaway shutdown blacked out much of New York, New England, and Ontario.

line.) Thus we can write, from Eq. 33-59,

$$P_{av} = \frac{I^2 R}{2} = \left(\frac{I}{\sqrt{2}}\right)^2 R. \qquad (33\text{-}60)$$

The quantity $I/\sqrt{2}$ is called the **root-mean-square, or rms,** value of the current i:

$$I_{rms} = \frac{I}{\sqrt{2}} \quad \text{(rms current)}. \qquad (33\text{-}61)$$

We can now rewrite Eq. 33-60 as

$$P_{av} = I_{rms}^2 R \quad \text{(average power)}. \qquad (33\text{-}62)$$

Equation 33-62 looks much like Eq. 27-22 ($P = i^2 R$); the message is that if we switch to the rms current, we can compute the average rate of energy dissipation for alternating-current circuits just as for direct-current circuits.

We can also define rms values of voltage and emf for alternating-current circuits:

$$V_{rms} = \frac{V}{\sqrt{2}} \quad \text{and} \quad \mathscr{E}_{rms} = \frac{\mathscr{E}_m}{\sqrt{2}} \qquad (33\text{-}63)$$

$$\text{(rms voltage; rms emf)}.$$

Alternating-current instruments, such as ammeters and voltmeters, are usually calibrated to read I_{rms}, V_{rms}, and $\mathscr{E}_{rms}$. Thus if you plug an alternating-current voltmeter into a household electric outlet and it reads 120 V, that is an rms voltage. The *maximum* value of the potential difference at the outlet is $\sqrt{2} \times (120 \text{ V})$, or 170 V.

Because the proportionality factor $1/\sqrt{2}$ in Eqs. 33-61 and 33-63 is the same for all three variables, we can write Eqs. 33-53 and 33-51 as

$$I_{rms} = \frac{\mathscr{E}_{rms}}{Z} = \frac{\mathscr{E}_{rms}}{\sqrt{R^2 + (X_L - X_C)^2}}, \qquad (33\text{-}64)$$

and, indeed, this is the form that we almost always use.

We can use the relationship $I_{rms} = \mathscr{E}_{rms}/Z$ to recast Eq. 33-62 in a useful equivalent way. We write

$$P_{av} = \frac{\mathscr{E}_{rms}}{Z} I_{rms} R = \mathscr{E}_{rms} I_{rms} \frac{R}{Z}. \qquad (33\text{-}65)$$

From Fig. 33-11d, Table 33-2, and Eq. 33-53, however, we see that R/Z is just the cosine of the phase constant ϕ:

$$\cos \phi = \frac{V_R}{\mathscr{E}_m} = \frac{IR}{IZ} = \frac{R}{Z}. \qquad (33\text{-}66)$$

Equation 33-65 then becomes

$$P_{av} = \mathscr{E}_{rms} I_{rms} \cos \phi \quad \text{(average power)}, \qquad (33\text{-}67)$$

in which the term $\cos \phi$ is called the **power factor.** Because $\cos \phi = \cos(-\phi)$, Eq. 33-67 is independent of the sign of the phase constant ϕ.

To maximize the rate at which energy is supplied to a resistive load in an *RLC* circuit, we should keep the power factor $\cos \phi$ as close to unity as possible. This is equivalent to keeping the phase constant ϕ in Eq. 33-26 as close to zero as possible. If, for example, the circuit is highly inductive, it can be made less so by adding capacitance to the circuit, thus reducing the phase constant and increasing the power factor in Eq. 33-67. Power companies place capacitors throughout their transmission systems to do this.

$\mathscr{C}$HECKPOINT **7:** (a) If the current in a sinusoidally driven series *RLC* circuit leads the emf, would we increase or decrease the capacitance to increase the rate at which energy is supplied to the resistance? (b) Will this change bring the resonant angular frequency of the circuit closer to the angular frequency of the emf or put it further away?

SAMPLE PROBLEM 33-6

A series *RLC* circuit, driven with $\mathscr{E}_{rms} = 120$ V at frequency $f_d = 60.0$ Hz, contains a resistance $R = 200 \ \Omega$, an inductance with $X_L = 80.0 \ \Omega$, and a capacitance with $X_C = 150 \ \Omega$.

(a) What are the power factor $\cos \phi$ and phase constant ϕ of the circuit?

SOLUTION: We can obtain the power factor from Eq. 33-66 ($\cos \phi = R/Z$) if we first find the impedance Z. From Eq. 33-52, we have

$$Z = \sqrt{R^2 + (X_L - X_C)^2}$$
$$= \sqrt{(200 \ \Omega)^2 + (80.0 \ \Omega - 150 \ \Omega)^2} = 211.90 \ \Omega.$$

Equation 33-66 then gives us

$$\cos \phi = \frac{R}{Z} = \frac{200 \ \Omega}{211.90 \ \Omega} = 0.944. \quad \text{(Answer)}$$

Taking inverse cosine then yields

$$\phi = \cos^{-1} 0.944 = 19.3°.$$

The implied plus sign here is wrong, however: with $X_C > X_L$, the circuit is mainly capacitive and ϕ must be negative. The problem is that the sign of the difference $X_L - X_C$ is lost in the calculation of Z. Inserting a minus sign, we then have

$$\phi = -19.3°. \quad \text{(Answer)}$$

(We could, instead, insert the known data into Eq. 33-56; we arrive at the same answer, complete with the minus sign.)

(b) What is the average rate P_{av} at which energy is dissipated in the resistance?

SOLUTION: We can calculate P_{av} with Eq. 33-67 if we first calculate I_{rms}. From Eq. 33-64 we have

$$I_{rms} = \frac{\mathscr{E}_{rms}}{Z} = \frac{120 \text{ V}}{211.90 \text{ }\Omega} = 0.5663 \text{ A}.$$

Substituting this and known data into Eq. 33-67, we find

$$P_{av} = \mathscr{E}_{rms} I_{rms} \cos \phi = (120 \text{ V})(0.5663 \text{ A})(0.944)$$
$$= 64.2 \text{ W}. \qquad \text{(Answer)}$$

(c) What change ΔC in the capacitance is needed to maximize P_{av} if the other parameters of the circuit are not changed?

SOLUTION: From Eq. 33-34 ($X_C = 1/\omega_d C$), the original capacitance is

$$C = \frac{1}{2\pi f_d X_C} = \frac{1}{(2\pi)(60.0 \text{ Hz})(150 \text{ }\Omega)} = 17.7 \text{ }\mu\text{F}.$$

The energy dissipation rate P_{av} is maximized when the circuit is in resonance, that is, when X_C is equal to X_L. Again from Eq. 33-34, now with $X_C = X_L = 80.0 \text{ }\Omega$, the capacitance would then be

$$C' = \frac{1}{2\pi f_d X_C} = \frac{1}{(2\pi)(60 \text{ Hz})(80.0 \text{ }\Omega)} = 33.2 \text{ }\mu\text{F}.$$

Thus, the required change in capacitance would be

$$\Delta C = C' - C = 33.2 \text{ }\mu\text{F} - 17.7 \text{ }\mu\text{F}$$
$$= +15.5 \text{ }\mu\text{F}. \qquad \text{(Answer)}$$

(d) With that change in capacitance, what would P_{av} be?

SOLUTION: The change would give $X_C = X_L$. Then from Eqs. 33-52 and 33-66, we would have $Z = R$ and $\cos \phi = 1$. The rms current would be, from Eq. 33-64,

$$I_{rms} = \frac{\mathscr{E}_{rms}}{Z} = \frac{120 \text{ V}}{200 \text{ }\Omega} = 0.600 \text{ A},$$

and the average power should be

$$P_{av} = \mathscr{E}_{rms} I_{rms} \cos \phi = (120\text{V})(0.600 \text{ A})(1.0)$$
$$= 72.0 \text{ W}. \qquad \text{(Answer)}$$

33-11 TRANSFORMERS

Energy Transmission Requirements

When an ac circuit has only a resistive load, the power factor in Eq. 33-67 is $\cos 0° = 1$ and the applied rms emf $\mathscr{E}$ is equal to the rms voltage V across the load. So with an rms current I in the load, energy is supplied and dissipated at the average rate of

$$P_{av} = \mathscr{E}I = IV. \qquad (33\text{-}68)$$

(In this section, we follow conventional practice and drop the subscripts identifying rms quantities. Engineers and scientists assume that all time-varying currents and voltages are reported as rms values; that is what the meters read.) Equation 33-68 tells us that, to satisfy a given power requirement, we have a range of choices, from a relatively large current I and a relatively small voltage V to just the reverse, provided only that the product IV is as required.

In electric power distribution systems it is desirable for reasons of safety and for efficient equipment design to deal with relatively low voltages at both the generating end (the electric power plant) and the receiving end (the home or factory). Nobody wants an electric toaster or a child's electric train to operate at, say, 10 kV. On the other hand, in the transmission of electric energy from the generating plant to the consumer, we want the lowest practical current (hence the largest practical voltage) to minimize I^2R losses (often called *ohmic losses*) in the transmission line.

As an example, consider the 735 kV line used to transmit electric energy from the La Grande 2 hydroelectric plant in Quebec to Montreal, 1000 km away. Suppose that the current is 500 A and the power factor is close to unity. Then from Eq. 33-67, energy is supplied at the average rate

$$P_{av} = \mathscr{E}I = (7.35 \times 10^5 \text{ V})(500 \text{ A}) = 368 \text{ MW}.$$

The resistance per kilometer is about 0.220 Ω/km for the line; thus there is a total resistance of about 220 Ω for the 1000 km stretch. Energy is dissipated owing to that resistance at a rate of about

$$P_{av} = I^2R = (500 \text{ A})^2(220 \text{ }\Omega) = 55.0 \text{ MW},$$

which is nearly 15% of the supply rate.

Imagine what would happen if we doubled the current and halved the voltage. Energy would be supplied by the plant at the same average rate of 368 MW as previously, but now energy would be dissipated at the rate of about

$$P_{av} = I^2R = (1000 \text{ A})^2(220 \text{ }\Omega) = 220 \text{ MW},$$

which is *almost 60% of the supply rate.* Hence the general energy transmission rule: transmit at the highest possible voltage and the lowest possible current.

The Ideal Transformer

The transmission rule leads to a fundamental mismatch between the requirement for efficient high-voltage transmission and the need for safe low-voltage generation and consumption. We need a device with which we can raise (for transmission) and lower (for use) the voltage in a circuit, keeping the product current × voltage essentially constant. The **transformer** is such a device. It has no moving parts, operates by Faraday's law of induction, and has no simple direct-current counterpart.

The *ideal transformer* in Fig. 33-15 consists of two coils, with different numbers of turns, wound around an iron core. (The coils are insulated from the core.) In use, the primary winding, of N_p turns, is connected to an alternating-current generator whose emf $\mathscr{E}$ is given by

$$\mathscr{E} = \mathscr{E}_m \sin \omega t. \qquad (33\text{-}69)$$

The secondary winding, of N_s turns, is connected to load resistance R, but its circuit is an open circuit as long as switch S is open (which we assume for the present). Thus there can be no current through the secondary coil. We assume further for this ideal transformer that the resistances of the primary and secondary windings are negligible, as are energy losses due to magnetic hysteresis in the iron core. Well-designed, high-capacity transformers can have energy losses as low as 1%, so our assumptions are reasonable.

For the assumed conditions, the primary winding (or *primary*) is a pure inductance, and the primary circuit is like that in Fig. 33-10a. Thus the (very small) primary current, also called the *magnetizing current* I_{mag}, lags the primary voltage V_p by 90°; the primary's power factor ($= \cos \phi$ in Eq. 33-67) is zero, and thus no power is delivered from the generator to the transformer.

However, the small alternating primary current I_{mag} induces an alternating magnetic flux Φ_B in the iron core. Because the core extends through the secondary winding (or *secondary*), this induced flux also extends through the turns of the secondary. From Faraday's law of induction (Eq. 31-6), the induced emf per turn $\mathscr{E}_{\text{turn}}$ is the same for both the primary and the secondary. Also, the voltage V_p across the primary is equal to the emf induced in the primary, and the voltage V_s across the secondary is equal to the emf induced in the secondary. Thus we can write

$$\mathscr{E}_{\text{turn}} = \frac{d\Phi_B}{dt} = \frac{V_p}{N_p} = \frac{V_s}{N_s}.$$

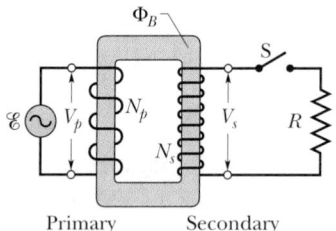

Primary   Secondary

FIGURE 33-15 An ideal transformer, two coils wound on an iron core, in a basic transformer circuit. An ac generator produces current in the coil at the left (the *primary*). The coil at the right (the *secondary*) is connected to the resistive load R when switch S is closed.

and thus

$$V_s = V_p \frac{N_s}{N_p} \quad \begin{array}{l}\text{(transformation}\\\text{of voltage).}\end{array} \qquad (33\text{-}70)$$

If $N_s > N_p$, the transformer is called a *step-up transformer* because it steps the primary's voltage V_p up to a higher voltage V_s. Similarly, if $N_s < N_p$, the device is a *step-down transformer*.

So far, with switch S open, no energy is transferred from the generator to the rest of the circuit. Now let us close S to connect the secondary to the resistive load R. (In general, the load would also contain inductive and capacitive elements, but here we consider just resistance R.) We find that now energy *is* transferred from the generator. Let us see why.

Several things happen when we close switch S. (1) An alternating current I_s appears in the secondary circuit, with corresponding energy dissipation rate I_s^2R ($= V_s^2/R$) in the resistive load. (2) This current produces its own alternating magnetic flux in the iron core, and this flux induces (from Faraday's law and Lenz's law) an opposing emf in the primary windings. (3) The voltage V_p of the primary, however, cannot change in response to this opposing emf because it must always be equal to the emf $\mathscr{E}$ that is provided by the generator; closing switch S cannot change this fact. (4) To maintain V_p, the generator now produces (in addition to I_{mag}) an alternating current I_p in the primary circuit; the magnitude and phase constant of I_p are just those required for the emf induced by I_p in the primary to exactly cancel the emf induced there by I_s. Because the phase constant of I_p is not 90° like that of I_{mag}, this current I_p can transfer energy to the primary.

We want to relate I_s to I_p. However, rather than analyze the foregoing complex process in detail, let us just apply the principle of conservation of energy. The rate at which the generator transfers energy to the primary is equal to I_pV_p. The rate at which the primary then transfers energy to the secondary (via the alternating magnetic field linking the two coils) is I_sV_s. Because we assume that no energy is lost along the way, conservation of energy requires that

$$I_pV_p = I_sV_s.$$

Substituting for V_s from Eq. 33-70, we find that

$$I_s = I_p \frac{N_p}{N_s} \quad \text{(transformation of currents).} \qquad (33\text{-}71)$$

This equation tells us that the current I_s in the secondary can be greater than, less than, or the same as the current I_p in the primary, depending on the *turns ratio* N_p/N_s.

Current I_p appears in the primary circuit because of the resistive load R in the secondary circuit. To find I_p, we substitute $I_s = V_s/R$ into Eq. 33-71 and then we substitute for V_s from Eq. 33-70. We find

$$I_p = \frac{1}{R} \left(\frac{N_s}{N_p} \right)^2 V_p. \qquad (33\text{-}72)$$

This equation has the form $I_p = V_p/R_{eq}$, where equivalent resistance R_{eq} is

$$R_{eq} = \left(\frac{N_p}{N_s} \right)^2 R. \qquad (33\text{-}73)$$

This R_{eq} is the value of the load resistance as "seen" by the generator; the generator produces the current I_p and voltage V_p as if it were connected to a resistance R_{eq}.

Impedance Matching

Equation 33-73 suggests still another function for the transformer. For maximum transfer of energy from an emf device to a resistive load, the resistance of the emf device and the resistance of the load must be equal. The same relation holds for ac circuits except that the *impedance* (rather than just the resistance) of the generator must be matched to that of the load. It often happens—as when we wish to connect a speaker set to an amplifier—that this condition is far from met, the amplifier and the speaker set being of high and low impedance, respectively. We can match the impedances of the two devices by coupling them through a transformer with a suitable turns ratio N_p/N_s.

SAMPLE PROBLEM 33-7

A transformer on a utility pole operates at $V_p = 8.5$ kV on the primary side and supplies electric energy to a number of nearby houses at $V_s = 120$ V, both quantities being rms values. Assume an ideal step-down transformer, a purely resistive load, and a power factor of unity.

(a) What is the turns ratio N_p/N_s of the transformer?

SOLUTION: From Eq. 33-70 we have

$$\frac{N_p}{N_s} = \frac{V_p}{V_s} = \frac{8.5 \times 10^3 \text{ V}}{120 \text{ V}} = 70.83 \approx 71. \quad \text{(Answer)}$$

(b) The average rate of energy consumption in the houses served by the transformer is 78 kW. What are the rms currents in the primary and secondary of the transformer?

SOLUTION: From Eq. 33-67 we have (with $\cos \phi = 1$ and $V_p = \mathcal{E}$ for a transformer connected across an emf device)

$$I_p = \frac{P_{av}}{V_p} = \frac{78 \times 10^3 \text{ W}}{8.5 \times 10^3 \text{ V}} = 9.176 \text{ A} \approx 9.2 \text{ A} \quad \text{(Answer)}$$

and

$$I_s = \frac{P_{av}}{V_s} = \frac{78 \times 10^3 \text{ W}}{120 \text{ V}} = 650 \text{ A}. \quad \text{(Answer)}$$

(c) What is the resistive load in the secondary circuit?

SOLUTION: In the secondary circuit we have

$$R_s = \frac{V_s}{I_s} = \frac{120 \text{ V}}{650 \text{ A}} = 0.1846 \ \Omega \approx 0.18 \ \Omega. \quad \text{(Answer)}$$

(d) What is the resistive load in the primary circuit?

SOLUTION: In the primary circuit we find

$$R_p = \frac{V_p}{I_p} = \frac{8.5 \times 10^3 \text{ V}}{9.176 \text{ A}} = 926 \ \Omega \approx 930 \ \Omega \quad \text{(Answer)}$$

or, using Eq. 33-73 with $R_p = R_{eq}$ and $R_s = R$,

$$R_p = \left(\frac{N_p}{N_s} \right)^2 R_s = (70.83)^2 (0.1846 \ \Omega)$$

$$= 926 \ \Omega \approx 930 \ \Omega. \quad \text{(Answer)}$$

CHECKPOINT **8:** An alternating-current emf device has a smaller resistance than that of the resistive load; to increase the transfer of energy from the device to the load, a transformer will be connected between the two. Should it be a step-up or step-down transformer?

REVIEW & SUMMARY

LC Energy Transfers

In an oscillating *LC* circuit, energy is shuttled periodically between the electric field of the capacitor and the magnetic field of the inductor; instantaneous values of the two forms of energy are

$$U_E = \frac{q^2}{2C} \quad \text{and} \quad U_B = \frac{Li^2}{2}. \qquad (33\text{-}1, 33\text{-}2)$$

The total energy $U \ (= U_E + U_B)$ remains constant.

LC Charge and Current Oscillations

The principle of conservation of energy leads to

$$L \frac{d^2 q}{dt^2} + \frac{q}{C} = 0 \quad \text{(LC oscillations)} \qquad (33\text{-}11)$$

as the differential equation of *LC* oscillations (with no resistance).

The solution of Eq. 33-11 is

$$q = Q \cos(\omega t + \phi) \quad \text{(charge)}, \qquad (33\text{-}12)$$

in which Q is the *charge amplitude* (maximum charge on the capacitor) and the angular frequency ω of the oscillations is

$$\omega = \frac{1}{\sqrt{LC}}. \qquad (33\text{-}4)$$

The phase constant ϕ in Eq. 33-12 is determined by the initial conditions (at $t = 0$) of the system.

The current i in the system is

$$i = -\omega Q \sin(\omega t + \phi) \quad \text{(current)}, \qquad (33\text{-}13)$$

in which ωQ is the *current amplitude I*.

Damped Oscillations

Oscillations in an LC circuit are damped when a dissipative element R is also present in the circuit, and then the differential equation of oscillation is

$$L \frac{d^2 q}{dt^2} + R \frac{dq}{dt} + \frac{1}{C} q = 0 \quad \text{(RLC circuit)}. \qquad (33\text{-}20)$$

Its solution is

$$q = Q e^{-Rt/2L} \cos(\omega' t + \phi), \qquad (33\text{-}21)$$

where

$$\omega' = \sqrt{\omega^2 - (R/2L)^2}. \qquad (33\text{-}22)$$

We consider only situations with small R and thus small damping; then $\omega' \approx \omega$ and the maximum energy of the electric field in the capacitor is

$$U = \frac{Q^2}{2C} e^{-Rt/L}. \qquad (33\text{-}24)$$

Alternating Currents; Forced Oscillations

A series RLC circuit may be set into *forced oscillation* at a *driving angular frequency* ω_d by an external alternating emf

$$\mathscr{E} = \mathscr{E}_m \sin \omega_d t. \qquad (33\text{-}25)$$

The current driven in the circuit by the emf is

$$i = I \sin(\omega_d t - \phi), \qquad (33\text{-}26)$$

where ϕ is a phase constant.

Resonance

The current amplitude I in a series RLC circuit driven by a sinusoidal external emf is a maximum ($I = \mathscr{E}_m/R$) when the driving angular frequency ω_d equals the natural angular frequency ω of the circuit (at *resonance*). Then $X_C = X_L$, $\phi = 0$, and the current is in phase with the emf.

Single Circuit Elements

The alternating potential difference across a resistor has amplitude $V_R = IR$; the current is in phase with the potential differ-

ence. For a *capacitor*, $V_C = IX_C$, in which $X_C = 1/\omega_d C$ is the **capacitive reactance;** the current here leads the potential difference by 90°. For an *inductor*, $V_L = IX_L$, in which $X_L = \omega_d L$ is the **inductive reactance;** the current here lags the potential difference by 90°.

Series RLC Circuits

For a series RLC circuit with external emf given by Eq. 33-25 and current given by Eq. 33-26,

$$I = \frac{\mathscr{E}_m}{\sqrt{R^2 + (X_L - X_C)^2}} = \frac{\mathscr{E}_m}{\sqrt{R^2 + (\omega_d L - 1/\omega_d C)^2}}$$
$$\text{(current amplitude)} \qquad (33\text{-}51, 33\text{-}54)$$

and

$$\tan \phi = \frac{X_L - X_C}{R} \quad \text{(phase constant)}. \qquad (33\text{-}56)$$

Defining the impedance Z of the circuit as

$$Z = \sqrt{R^2 + (X_L - X_C)^2} \quad \text{(impedance)} \qquad (33\text{-}52)$$

allows us to write Eq. 33-51 as $I = \mathscr{E}_m/Z$.

Power

In a series RLC circuit, the **average power** P_{av} of the generator is equal to the production rate of thermal energy in the resistor:

$$P_{av} = I_{rms}^2 R = \mathscr{E}_{rms} I_{rms} \cos \phi$$
$$\text{(average power)}. \qquad (33\text{-}62, 33\text{-}67)$$

Here "rms" stands for **root-mean-square;** rms quantities are related to maximum quantities by $I_{rms} = I/\sqrt{2}$, $V_{rms} = V_m/\sqrt{2}$, and $\mathscr{E}_{rms} = \mathscr{E}_m/\sqrt{2}$. The term $\cos \phi$ is called the **power factor.**

Transformers

A *transformer* (assumed to be ideal) is an iron core on which are wound a primary coil of N_p turns and a secondary coil of N_s turns. If the primary coil is connected across an alternating-current generator, the primary and secondary voltages are related by

$$V_s = V_p \frac{N_s}{N_p} \quad \begin{array}{l}\text{(transformation}\\ \text{of voltage).}\end{array} \qquad (33\text{-}70)$$

The currents are related by

$$I_s = I_p \frac{N_p}{N_s} \quad \begin{array}{l}\text{(transformation}\\ \text{of currents),}\end{array} \qquad (33\text{-}71)$$

and the equivalent resistance of the secondary circuit, as seen by the generator, is

$$R_{eq} = \left(\frac{N_p}{N_s}\right)^2 R, \qquad (33\text{-}73)$$

where R is the resistive load in the secondary circuit.

QUESTIONS

1. A charged capacitor and an inductor are connected at time $t = 0$. In terms of the period T of the resulting oscillations, what is the first later time at which the following reach a maximum: (a) U_B, (b) the magnetic flux through the inductor, (c) di/dt, and (d) the emf of the inductor?

2. What values of the phase constant ϕ in Eq. 33-12 would permit situations (a), (c), (e), and (g) of Fig. 33-1 to be the situation at $t = 0$?

3. Figure 33-16 shows three oscillating LC circuits with identical inductors and capacitors. Rank the circuits according to the time taken to fully discharge the capacitors during the oscillations, greatest first.

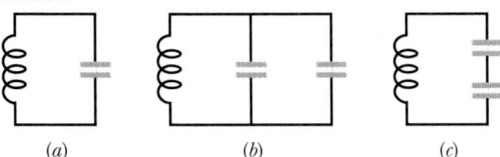

(a)   (b)   (c)

FIGURE 33-16 Question 3.

4. Figure 33-17 shows graphs of capacitor voltage v_C for LC circuits 1 and 2, which contain identical capacitances and have the same maximum charge Q. Are (a) the inductance L and (b) the maximum current I in circuit 1 greater than, less than, or the same as those in circuit 2?

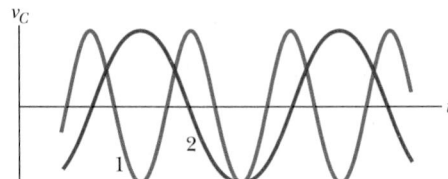

FIGURE 33-17 Question 4.

5. Charges on the capacitors in three oscillating LC circuits vary as follows: (1) $q = 2 \cos 4t$; (2) $q = 4 \cos t$; (3) $q = 3 \cos 4t$ (with q in coulombs and t in seconds). Rank the circuits according to (a) the current amplitude and (b) the period, greatest first.

6. If you increase the inductance L in an oscillating LC circuit having a given maximum charge Q, do (a) the current magnitude I and (b) the maximum magnetic energy U_B increase, decrease, or stay the same?

7. In a damped oscillating RLC circuit, does the charge decay faster than, slower than, or at the same rate as the energy?

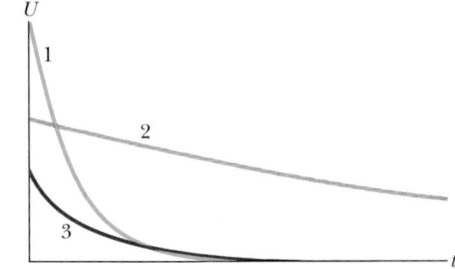

FIGURE 33-18
Question 8.

8. Figure 33-18 shows the decrease in energy with time of three damped oscillating RLC circuits with the same initial charge Q. Rank the circuits according to their (a) capacitances C and (b) values of L/R, greatest first.

9. The values of the phase constant ϕ for four sinusoidally driven series RLC circuits are (1) $-15°$, (2) $+35°$, (3) $\pi/3$ rad, and (4) $-\pi/6$ rad. (a) In which is the load primarily capacitive? (b) In which does the alternating emf lead the current?

10. Figure 33-19 shows the current i and driving emf $\mathscr{E}$ for a series RLC circuit. (a) Does the current lead or lag the emf? (b) Is the circuit's load mainly capacitive or mainly inductive? (c) Is the angular frequency of the emf greater than or less than the natural angular frequency of the circuit?

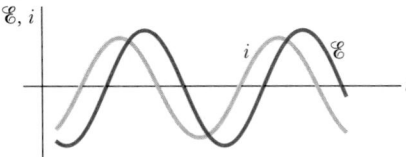

FIGURE 33-19 Questions 10 and 15.

11. The following table gives, for three series RLC circuits, the amplitude $\mathscr{E}_m$ of the driving emf and the values of R, L, and C. Without written calculation, rank the circuits according to (a) the amplitude I of the current at resonance and (b) the angular frequency at resonance, greatest first.

CIRCUIT	$\mathscr{E}_m$ (V)	R (Ω)	L (mH)	C (μF)
1	25	5.0	200	10
2	60	12	100	5.0
3	80	10	300	10

12. Suppose that for a particular driving angular frequency, the emf leads the current in a series RLC circuit. If you decrease the driving angular frequency slightly, do (a) the phase constant and (b) the current amplitude increase, decrease, or remain the same?

13. The driving angular frequency in a certain series RLC circuit is less than the natural angular frequency of the circuit. (a) Is the phase constant ϕ positive, negative, or zero? (b) Does the current lead or lag the emf?

14. Figure 33-20 shows three situations like those of Fig. 33-12. For each situation, is the driving angular frequency greater than, less than, or equal to the resonant angular frequency?

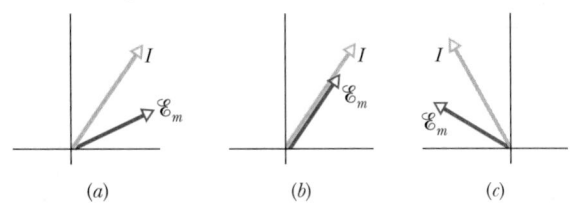

(a)   (b)   (c)

FIGURE 33-20 Question 14.

15. Figure 33-19 shows the current i and driving emf $\mathcal{E}$ for a series RLC circuit. Relative to the emf curve, does the current curve shift leftward or rightward and does the amplitude of that curve increase or decrease if we slightly increase (a) L, (b) C, and (c) the driving angular frequency of the emf?

16. Figure 33-21 shows the current i and driving emf $\mathcal{E}$ for a series RLC circuit. (a) Is the phase constant positive or negative? (b) To increase the rate at which energy is transferred to the

resistive load, should L be increased or decreased? (c) Should, instead, C be increased or decreased?

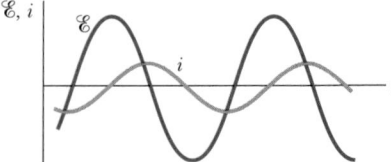

FIGURE 33-21
Question 16.

EXERCISES & PROBLEMS

SECTION 33-2 LC Oscillations, Qualitatively

1E. What is the capacitance of an oscillating LC circuit if the maximum charge on the capacitor is $1.60 \ \mu C$ and the total energy is $140 \ \mu J$?

2E. A 1.50 mH inductor in an oscillating LC circuit stores a maximum energy of $10.0 \ \mu J$. What is the maximum current?

3E. In an oscillating LC circuit $L = 1.10$ mH and $C = 4.00 \ \mu F$. The maximum charge on the capacitor is $3.00 \ \mu C$. Find the maximum current.

4E. An oscillating LC circuit consists of a 75.0 mH inductor and a 3.60 μF capacitor. If the maximum charge on the capacitor is $2.90 \ \mu C$, (a) what is the total energy in the circuit and (b) what is the maximum current?

5E. In a certain oscillating LC circuit the total energy is converted from electric energy in the capacitor to magnetic energy in the inductor in 1.50 μs. (a) What is the period of oscillation? (b) What is the frequency of oscillation? (c) How long after the magnetic energy is a maximum will it be a maximum again?

6P. The frequency of oscillation of a certain LC circuit is 200 kHz. At time $t = 0$, plate A of the capacitor has maximum positive charge. At what times $t > 0$ will (a) plate A again have maximum positive charge, (b) the other plate of the capacitor have maximum positive charge, and (c) the inductor have maximum magnetic field?

SECTION 33-3 The Electrical–Mechanical Analogy

7E. A 0.50 kg body oscillates on a spring that, when extended 2.0 mm from equilibrium, has a restoring force of 8.0 N. (a) What is the angular frequency of oscillation? (b) What is the period of oscillation? (c) What is the capacitance of the corresponding LC circuit if L is chosen to be 5.0 H?

8P. The energy in an oscillating LC circuit containing a 1.25 H inductor is 5.70 μJ. The maximum charge on the capacitor is 175 μC. Find (a) the mass, (b) the spring constant, (c) the maximum displacement, and (d) the maximum speed for the corresponding mechanical system.

SECTION 33-4 LC Oscillations, Quantitatively

9E. LC oscillators have been used in circuits connected to loudspeakers to create some of the sounds of electronic music. What inductance must be used with a 6.7 μF capacitor to produce a frequency of 10 kHz, which is near the middle of the audible range of frequencies?

10E. What capacitance would you connect across a 1.30 mH inductor to make the resulting oscillator resonate at 3.50 kHz?

11E. In an LC circuit with $L = 50$ mH and $C = 4.0 \ \mu F$, the current is initially a maximum. How long will it take before the capacitor is fully charged for the first time?

12E. Consider the circuit shown in Fig. 33-22. With switch S_1 closed and the other two switches open, the circuit has a time constant τ_C (see Section 28-8). With switch S_2 closed and the other two switches open, the circuit has a time constant τ_L (see Section 31-9). With switch S_3 closed and the other two switches open, the circuit oscillates with a period T. Show that $T = 2\pi\sqrt{\tau_C \tau_L}$.

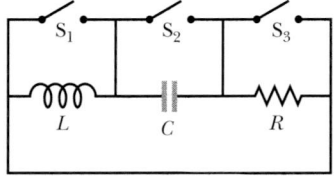

FIGURE 33-22 Exercise 12.

13E. Derive the differential equation for an LC circuit (Eq. 33-11) using the loop rule.

14E. A single loop consists of several inductors ($L_1, L_2, \ldots$), several capacitors ($C_1, C_2, \ldots$), and several resistors ($R_1, R_2, \ldots$) connected in series as shown, for example, in Fig. 33-23a. Show that regardless of the sequence of these circuit ele-

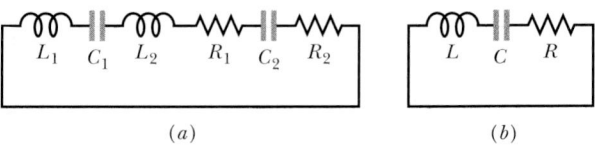

(a)   (b)

FIGURE 33-23 Exercise 14.

ments in the loop, the behavior of this circuit is identical to that of the simple LC circuit shown in Fig. 33-23b. (*Hint:* Consider the loop rule and see Problem 56 in Chapter 31)

15P. An oscillating LC circuit consisting of a 1.0 nF capacitor and a 3.0 mH coil has a maximum voltage of 3.0 V. (a) What is the maximum charge on the capacitor? (b) What is the maximum current through the circuit? (c) What is the maximum energy stored in the magnetic field of the coil?

16P. An oscillating LC circuit has an inductance of 3.00 mH and a capacitance of 10.0 μF. Calculate (a) the angular frequency and (b) the period of the oscillation. (c) At time $t = 0$ the capacitor is charged to 200 μC, and the current is zero. Sketch roughly the charge on the capacitor as a function of time.

17P. In an oscillating LC circuit in which $C = 4.00$ μF, the maximum potential difference across the capacitor during the oscillations is 1.50 V and the maximum current through the inductor is 50.0 mA. (a) What is the inductance L? (b) What is the frequency of the oscillations? (c) How much time does the charge on the capacitor take to rise from zero to its maximum value?

18P. In the circuit shown in Fig. 33-24 the switch has been in position a for a long time. It is now thrown to position b. (a) Calculate the frequency of the resulting oscillating current. (b) What is the amplitude of the current oscillations?

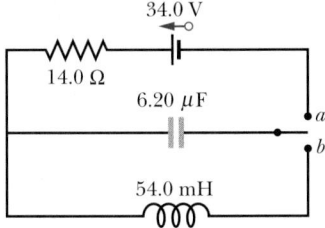

FIGURE 33-24 Problem 18.

19P. You are given a 10 mH inductor and two capacitors, of 5.0 μF and 2.0 μF capacitance. List the oscillation frequencies that can be generated by connecting these elements in various combinations.

20P. An LC circuit oscillates at a frequency of 10.4 kHz. (a) If the capacitance is 340 μF, what is the inductance? (b) If the maximum current is 7.20 mA, what is the total energy in the circuit? (c) What is the maximum charge on the capacitor?

21P. (a) In an oscillating LC circuit, in terms of the maximum charge on the capacitor, what is the charge there when the energy in the electric field is 50.0% of that in the magnetic field? (b) What fraction of a period must elapse following the time the capacitor is fully charged for this condition to arise?

22P. At some instant in an oscillating LC circuit, 75.0% of the total energy is stored in the magnetic field of the inductor. (a) In terms of the maximum charge on the capacitor, what is the charge there at this instant? (b) In terms of the maximum current in the inductor, what is the current there at this instant?

23P. An inductor is connected across a capacitor whose capacitance can be varied by turning a knob. We wish to make the frequency of the LC oscillations vary linearly with the angle of rotation of the knob, going from 2×10^5 to 4×10^5 Hz as the knob turns through 180°. If $L = 1.0$ mH, plot the required capacitance C as a function of the angle of rotation of the knob.

24P. A variable capacitor with a range from 10 to 365 pF is used with a coil to form a variable-frequency LC circuit to tune the input to a radio. (a) What ratio of maximum to minimum frequencies may be obtained with such a capacitor? (b) If this circuit is to obtain frequencies from 0.54 MHz to 1.60 MHz, the ratio computed in (a) is too large. By adding a capacitor in parallel to the variable capacitor, this range may be adjusted. How large should this added capacitor be, and what inductance should be chosen in order to obtain the desired range of frequencies?

25P. In an oscillating LC circuit, $L = 25.0$ mH and $C = 7.80$ μF. At time $t = 0$ the current is 9.20 mA, the charge on the capacitor is 3.80 μC, and the capacitor is charging. (a) What is the total energy in the circuit? (b) What is the maximum charge on the capacitor? (c) What is the maximum current? (d) If the charge on the capacitor is given by $q = Q \cos(\omega t + \phi)$, what is the phase angle ϕ? (e) Suppose the data are the same, except that the capacitor is discharging at $t = 0$. What then is ϕ?

26P. In an oscillating LC circuit, $L = 3.00$ mH and $C = 2.70$ μF. At $t = 0$ the charge on the capacitor is zero and the current is 2.00 A. (a) What is the maximum charge that will appear on the capacitor? (b) In terms of the period T of oscillation, how much time will elapse after $t = 0$ until the energy stored in the capacitor will be increasing at its greatest rate? (c) What is this greatest rate at which energy is transferred to the capacitor?

27P. Three identical inductors L and two identical capacitors C are connected in a two-loop circuit as shown in Fig. 33-25. (a) Suppose the currents are as shown in Fig. 33-25a. What is the current in the middle inductor? Write the loop equations and show that they are satisfied if the current oscillates with angular frequency $\omega = 1/\sqrt{LC}$. (b) Now suppose the currents are as shown in Fig. 33-25b. What is the current in the middle inductor? Write the loop equations and show that they are satisfied if the current oscillates with angular frequency $\omega = 1/\sqrt{3LC}$. Because the circuit can oscillate at two different frequencies, we cannot find an equivalent single-loop LC circuit to replace it.

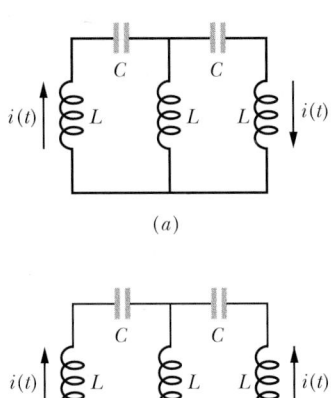

FIGURE 33-25
Problem 27.

28P. A series circuit containing inductance L_1 and capacitance C_1 oscillates at angular frequency ω. A second series circuit, containing inductance L_2 and capacitance C_2, oscillates at the same angular frequency. In terms of ω, what is the angular frequency of oscillation of a series circuit containing all four of these elements? Neglect resistance. (*Hint:* Use the formulas for equivalent capacitance and equivalent inductance; see Section 26-4 and Problem 56 in Chapter 31.

29P. In an oscillating *LC* circuit with $C = 64.0~\mu F$, the current as a function of time is given by $i = (1.60) \sin(2500t + 0.680)$, where t is in seconds, i in amperes, and the phase angle in radians. (a) How soon after $t = 0$ will the current reach its maximum value? What are (b) the inductance L and (c) the total energy?

30P*. In Fig. 33-26 the 900 μF capacitor is initially charged to 100 V and the 100 μF capacitor is uncharged. Describe in detail how one might charge the 100 μF capacitor to 300 V by manipulating switches S_1 and S_2.

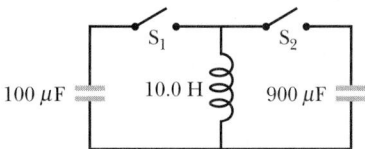

FIGURE 33-26 Problem 30.

SECTION 33-5 Damped Oscillations in an *RLC* Circuit

31E. What resistance R should be connected in series with an inductance $L = 220$ mH and capacitance $C = 12.0~\mu F$ for the maximum charge on the capacitor to decay to 99.0% of its initial value in 50.0 cycles? (Assume $\omega' \approx \omega$.)

32E. Consider a damped *LC* circuit. (a) Show that the damping term $e^{-Rt/2L}$ (which involves L but not C) can be rewritten in a more symmetric manner (involving L and C) as $e^{-\pi R\sqrt{C/L}(t/T)}$. Here T is the period of oscillation (neglecting resistance). (b) Using (a), show that the SI unit of $\sqrt{L/C}$ is the ohm. (c) Using (a), show that the condition that the fractional energy loss per cycle be small is $R \ll \sqrt{L/C}$.

33P. In an oscillating series *RLC* circuit, find the time required for the maximum energy present in the capacitor during an oscillation to fall to half its initial value. Assume $q = Q$ at $t = 0$.

34P. A single-loop circuit consists of a 7.20 Ω resistor, a 12.0 H inductor, and a 3.20 μF capacitor. Initially the capacitor has a charge of 6.20 μC and the current is zero. Calculate the charge on the capacitor N complete cycles later for $N = 5$, 10, and 100.

35P. At time $t = 0$ there is no charge on the capacitor of an *RLC* circuit but there is current I through the inductor. (a) Find the phase constant ϕ in Eq. 33-21 for the circuit. (b) Write an expression for the charge q on the capacitor as a function of time t and in terms of the current amplitude and angular frequency ω' of the oscillations.

36P. (a) By direct substitution of Eq. 33-21 into Eq. 33-20, show that $\omega' = \sqrt{(1/LC) - (R/2L)^2}$. (b) By what fraction does the fre-

quency of oscillation shift when the resistance is increased from 0 to 100 Ω in a circuit with $L = 4.40$ H and $C = 7.30~\mu F$?

37P*. In an oscillating *RLC* circuit, show that the fraction of the energy lost per cycle of oscillation, $\Delta U/U$, is given to a close approximation by $2\pi R/\omega L$. The quantity $\omega L/R$ is often called the Q of the circuit (for "quality"). A high-Q circuit has low resistance and a low fractional energy loss ($= 2\pi/Q$) per cycle.

SECTION 33-8 Three Simple Circuits

38E. A 1.50 μF capacitor is connected as in Fig. 33-9a to an ac generator with $\mathcal{E}_m = 30.0$ V. What is the amplitude of the resulting alternating current if the frequency of the emf is (a) 1.00 kHz and (b) 8.00 kHz?

39E. A 50.0 mH inductor is connected as in Fig. 33-10a to an ac generator with $\mathcal{E}_m = 30.0$ V. What is the amplitude of the resulting alternating current if the frequency of the emf is (a) 1.00 kHz and (b) 8.00 kHz?

40E. A 50 Ω resistor is connected as in Fig. 33-8a to an ac generator with $\mathcal{E}_m = 30.0$ V. What is the amplitude of the resulting alternating current if the frequency of the emf is (a) 1.00 kHz and (b) 8.00 kHz?

41E. A 45.0 mH inductor has a reactance of 1.30 kΩ. (a) What is its operating frequency? (b) What is the capacitance of a capacitor with the same reactance at that frequency? (c) If the frequency is doubled, what are the reactances of the inductor and capacitor?

42E. A 1.50 μF capacitor has a capacitive reactance of 12.0 Ω. (a) What must be its operating frequency? (b) What will be the capacitive reactance if the frequency is doubled?

43E. (a) At what frequency would a 6.0 mH inductor and a 10 μF capacitor have the same reactance? (b) What would the reactance be? (c) Show that this frequency would be the natural frequency of an oscillating circuit with the same L and C.

44P. An ac generator emf is $\mathcal{E} = \mathcal{E}_m \sin \omega_d t$, with $\mathcal{E}_m = 25.0$ V and $\omega_d = 377$ rad/s. It is connected to a 12.7 H inductor. (a) What is the maximum value of the current? (b) When the current is a maximum, what is the emf of the generator? (c) When the emf of the generator is -12.5 V and increasing in magnitude, what is the current?

45P. The ac generator of Problem 44 is connected to a 4.15 μF capacitor. (a) What is the maximum value of the current? (b) When the current is a maximum, what is the emf of the generator? (c) When the emf of the generator is -12.5 V and increasing in magnitude, what is the current?

46P. An ac generator emf is $\mathcal{E} = \mathcal{E}_m \sin(\omega_d t - \pi/4)$, where $\mathcal{E}_m = 30.0$ V and $\omega_d = 350$ rad/s. The current produced in a connected circuit is $i(t) = I \sin(\omega_d t - 3\pi/4)$, where $I = 620$ mA. (a) At what time after $t = 0$ does the generator emf first reach a maximum? (b) At what time after $t = 0$ does the current first reach a maximum? (c) The circuit contains a single element other than the generator. Is it a capacitor, an inductor, or a resistor? Justify your answer. (d) What is the value of the capacitance, inductance, or resistance, as the case may be?

47P. An ac generator emf is $\mathcal{E} = \mathcal{E}_m \sin(\omega_d t - \pi/4)$, where

$\mathcal{E}_m = 30.0$ V and $\omega_d = 350$ rad/s. The current is given by $i(t) = I \sin(\omega_d t + \pi/4)$, where $I = 620$ mA. (a) At what time after $t = 0$ does the generator emf first reach a maximum? (b) At what time after $t = 0$ does the current first reach a maximum? (c) The circuit contains a single element other than the generator. Is it a capacitor, an inductor, or a resistor? Justify your answer. (d) What is the value of the capacitance, inductance, or resistance, as the case may be?

48P. A three-phase generator G produces electrical power that is transmitted by means of three wires as shown in Fig. 33-27. The electric potentials (relative to a common reference level) of these wires are $V_1 = A \sin \omega_d t$, $V_2 = A \sin(\omega_d t - 120°)$, and $V_3 = A \sin(\omega_d t - 240°)$. Some types of heavy industrial equipment (for example, motors) have three terminals and are designed to be connected directly to these three wires. To use a more conventional two-terminal device (for example, a lightbulb), one connects it to any two of the three wires. Show that the potential difference between *any two* of the wires (a) oscillates sinusoidally with angular frequency ω_d and (b) has an amplitude of $A\sqrt{3}$.

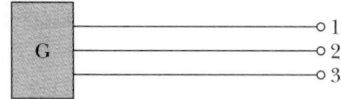

Three-wire transmission line

FIGURE 33-27   Problem 48.

SECTION 33-9 The Series *RLC* Circuit

49E. (a) Find Z, ϕ, and I for the situation of Sample Problem 33-5 with the capacitor removed from the circuit, all other parameters remaining unchanged. (b) Draw to scale a phasor diagram like that of Fig. 33-11d for this new situation.

50E. (a) Find Z, ϕ, and I for the situation of Sample Problem 33-5 with the inductor removed from the circuit, all other parameters remaining unchanged. (b) Draw to scale a phasor diagram like that of Fig. 33-11d for this new situation.

51E. (a) Find Z, ϕ, and I for the situation of Sample Problem 33-5 with $C = 70.0$ μF, the other parameters remaining unchanged. (b) Draw a phasor diagram like that of Fig. 33-11d for this new situation and compare the two diagrams closely.

52E. A generator with an adjustable frequency of oscillation is wired in series to an inductor of $L = 2.50$ mH and a capacitor of $C = 3.00$ μF. At what frequency does the generator produce the largest possible current amplitude in the circuit?

53P. In Fig. 33-28, a generator with an adjustable frequency of oscillation is connected to a variable resistance R, a capacitor of $C = 5.50$ μF, and an inductor of inductance L. With $R = 100$ Ω, the amplitude of the current produced in the circuit by the genera-

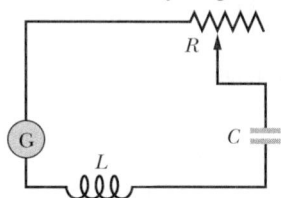

FIGURE 33-28   Problem 53.

tor is at half-maximum level when the generator's oscillations are at 1.30 and 1.50 kHz. (a) What is L? (b) If R is increased, what happens to the frequencies at which the current amplitude is at half-maximum level?

54P. Verify mathematically that the following geometric construction correctly gives both the impedance Z and the phase constant ϕ. Referring to Fig. 33-29, (i) draw an arrow in the positive y direction of magnitude X_C; (ii) draw a second arrow in the negative y direction of magnitude X_L; (iii) draw a third arrow of magnitude R in the positive x direction. Then the magnitude of the "resultant" of these arrows is Z, and the angle (measured clockwise from the positive x direction) of this resultant is ϕ.

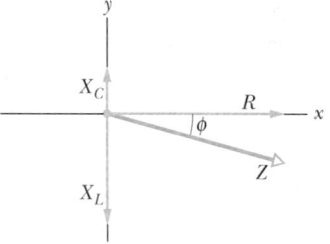

FIGURE 33-29   Problem 54.

55P. Can the amplitude of the voltage across an inductor be greater than the amplitude of the generator emf in an *RLC* circuit? Consider an *RLC* circuit with $\mathcal{E}_m = 10$ V, $R = 10$ Ω, $L = 1.0$ H, and $C = 1.0$ μF. Find the amplitude of the voltage across the inductor at resonance.

56P. A coil of inductance 88 mH and unknown resistance and a 0.94 μF capacitor are connected in series with an alternating emf of frequency 930 Hz. If the phase constant between the applied voltage and the current is 75°, what is the resistance of the coil?

57P. When the generator emf in Sample Problem 33-5 is a maximum, what is the voltage across (a) the generator, (b) the resistor, (c) the capacitor, and (d) the inductor? (e) By summing these with appropriate signs, verify that the loop rule is satisfied.

58P. An ac generator of $\mathcal{E}_m = 220$ V operating at 400 Hz causes oscillations in a series *RLC* circuit having $R = 220$ Ω, $L = 150$ mH, and $C = 24.0$ μF. Find (a) the capacitive reactance X_C, (b) the impedance Z, and (c) the current amplitude I. A second capacitor of the same capacitance is then connected in series with the other components. Determine whether the values of (d) X_C, (e) Z, and (f) I increase, decrease, or remain the same.

59P. An *RLC* circuit such as that of Fig. 33-7 has $R = 5.00$ Ω, $C = 20.0$ μF, $L = 1.00$ H, and $\mathcal{E}_m = 30.0$ V. (a) At what angular frequency ω_d will the current amplitude have its maximum value, as in the resonance curves of Fig. 33-13? (b) What is this maximum value? (c) At what two angular frequencies ω_{d1} and ω_{d2} will the current amplitude be half this maximum value? (d) What is the fractional half-width $[= (\omega_{d1} - \omega_{d2})/\omega]$ of the resonance curve for this circuit?

60P. For a certain series *RLC* circuit, the maximum generator emf is 125 V and the maximum current is 3.20 A. If the current leads the generator emf by 0.982 rad, what are (a) the impedance and (b) the resistance of the circuit? (c) Is the circuit predominantly capacitive or inductive?

61P. A series *RLC* circuit has a resonant frequency of 6.00 kHz. When it is driven at 8.00 kHz, it has an impedance of 1.00 kΩ and a phase constant of 45°. What are the values of (a) *R*, (b) *L*, and (c) *C* for this circuit?

62P. In a certain series *RLC* circuit operating at a frequency of 60.0 Hz, the maximum voltage across the inductor is 2.00 times the maximum voltage across the resistor and 2.00 times the maximum voltage across the capacitor. (a) By what phase angle does the current lag the generator emf? (b) If the maximum generator emf is 30.0 V, what should be the resistance of the circuit to obtain a maximum current of 300 mA?

63P. The circuit of Sample Problem 33-5 is not in resonance. (a) How can you tell? (b) What capacitor would you combine in parallel with the capacitor already in the circuit to bring resonance about? (c) What would the current amplitude then be?

64P. A generator is to be connected in series with an inductor of *L* = 2.00 mH and a capacitance *C*. You are to produce *C* by using capacitors of capacitances $C_1 = 4.00$ μF and $C_2 = 6.00$ μF, either singly or together. What resonant frequencies can the circuit have depending on the value of *C*?

65P. In Fig. 33-30, a generator with an adjustable frequency of oscillation is connected to resistance *R* = 100 Ω, inductances $L_1 = 1.70$ mH and $L_2 = 2.30$ mH, and capacitances $C_1 = 4.00$ μF, $C_2 = 2.50$ μF, and $C_3 = 3.50$ μF. (a) What is the resonant frequency of the circuit? (*Hint:* See Problem 56 in Chapter 31.) What happens to the resonant frequency if (b) the value of *R* is increased, (c) the value of L_1 is increased, and (d) capacitance C_3 is removed from the circuit?

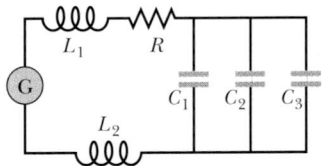

FIGURE 33-30   Problem 65.

66P. A series circuit with resistor–inductor–capacitor combination R_1, L_1, C_1 has the same resonant frequency as a second circuit with a different combination R_2, L_2, C_2. You now connect the two combinations in series. Show that this new circuit has the same resonant frequency as the separate circuits.

67P. Show that the fractional half-width (see Problem 59) of a resonance curve is given by

$$\frac{\Delta\omega_d}{\omega} = \sqrt{\frac{3C}{L}}\,R,$$

in which ω_d is the angular frequency at resonance and $\Delta\omega_d$ is the width of the resonance curve at half-amplitude. Note that $\Delta\omega_d/\omega$ decreases with *R*, as Fig. 33-13 shows. Use this formula to check the answer to Problem 59d.

68P*. The ac generator in Fig. 33-31 supplies 120 V at 60.0 Hz. With the switch open as in the diagram, the current leads the generator emf by 20.0°. With the switch in position 1 the current lags the generator emf by 10.0°. When the switch is in position 2, the current is 2.00 A. Find the values of *R*, *L*, and *C*.

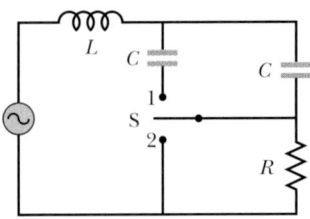

FIGURE 33-31   Problem 68.

SECTION 33-10   Power in Alternating-Current Circuits

69E. What is the maximum value of an ac voltage whose rms value is 100 V?

70E. An ac voltmeter with large impedance is connected in turn across the inductor, the capacitor, and the resistor in a series circuit having an alternating emf of 100 V (rms); it gives the same reading in volts in each case. What is this reading?

71E. (a) For the conditions in Problem 44c, is the generator supplying energy to or taking energy from the rest of the circuit? (b) Repeat for the conditions of Problem 45c.

72E. What direct current will produce the same amount of heat, in a particular resistor, as an alternating current that has a maximum value of 2.60 A?

73E. Calculate the average rate of energy dissipation in the circuits of Exercises 39, 40, 49, and 50.

74E. Show that the average rate of energy supplied to the circuit of Fig. 33-7 can also be written as $P_{av} = \mathscr{E}_{rms}^2 R/Z^2$. Show that this expression for average power gives reasonable results for a purely resistive circuit, for an *RLC* circuit at resonance, for a purely capacitive circuit, and for a purely inductive circuit.

75E. An electric motor connected to a 120 V, 60.0 Hz ac outlet does mechanical work at the rate of 0.100 hp (1 hp = 746 W). If it draws an rms current of 0.650 A, what is its effective resistance, in terms of power transfer? Is this the same as the resistance of its coils, as measured with an ohmmeter with the motor disconnected from the outlet?

76E. An air conditioner connected to a 120 V rms ac line is equivalent to a 12.0 Ω resistance and a 1.30 Ω inductive reactance in series. (a) Calculate the impedance of the air conditioner. (b) Find the average rate at which power is supplied to the appliance.

77E. An electric motor has an effective resistance of 32.0 Ω and an inductive reactance of 45.0 Ω when working under load. The rms voltage across the alternating source is 420 V. Calculate the rms current.

78P. Show mathematically, rather than graphically as in Fig. 33-14b, that the average value of $\sin^2(\omega t - \phi)$ over an integral number of half-cycles is $\frac{1}{2}$.

79P. For a sinusoidally driven series *RLC* circuit, show that over one complete cycle with period *T* (a) the energy stored in the capacitor does not change; (b) the energy stored in the inductor does not change; (c) the driving emf device supplies energy $(\frac{1}{2}T)\mathscr{E}_m I \cos \phi$; and (d) the resistor dissipates energy $(\frac{1}{2}T)RI^2$. (e) Show that the quantities found in (c) and (d) are equal.

80P. In a series oscillating *RLC* circuit, $R = 16.0 \; \Omega$, $C = 31.2$ μF, $L = 9.20$ mH, and $\mathscr{E} = \mathscr{E}_m \sin \omega_d t$ with $\mathscr{E}_m = 45.0$ V and $\omega_d = 3000$ rad/s. For time $t = 0.442$ ms find (a) the rate at which energy is being supplied by the generator, (b) the rate at which the energy in the capacitor is changing, (c) the rate at which the energy in the inductor is changing, and (d) the rate at which energy is being dissipated in the resistor. (e) What is the meaning of a negative result for any of (a), (b), and (c)? (f) Show that the results of (b), (c), and (d) sum to the result of (a).

81P. In Fig. 33-32 show that the average rate at which energy is dissipated in resistance R is a maximum when R is equal to the internal resistance r of the ac generator. (In the text we have tacitly assumed, up to this point, that $r = 0$.)

FIGURE 33-32 Problems 81 and 90.

82P. Figure 33-33 shows an ac generator connected to a "black box" through a pair of terminals. The box contains an *RLC* circuit, possibly even a multiloop circuit, whose elements and connections we do not know. Measurements outside the box reveal that

$$\mathscr{E}(t) = (75.0 \; \text{V}) \sin \omega_d t$$

and

$$i(t) = (1.20 \; \text{A}) \sin(\omega_d t + 42.0°).$$

(a) What is the power factor? (b) Does the current lead or lag the emf? (c) Is the circuit in the box largely inductive or largely capacitive? (d) Is the circuit in the box in resonance? (e) Must there be a capacitor in the box? An inductor? A resistor? (f) At what average rate is energy delivered to the box by the generator? (g) Why don't you need to know the angular frequency ω_d to answer all these questions?

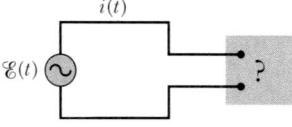

FIGURE 33-33 Problem 82.

83P. In an *RLC* circuit such as that of Fig. 33-7 assume that $R = 5.00 \; \Omega$, $L = 60.0$ mH, $f_d = 60.0$ Hz, and $\mathscr{E}_m = 30.0$ V. For what values of the capacitance would the average rate at which energy is dissipated in the resistor be (a) a maximum and (b) a minimum? (c) What are these maximum and minimum energy dissipation rates? What are (d) the corresponding phase angles and (e) the corresponding power factors?

84P. A typical "light dimmer" used to dim the stage lights in a theater consists of a variable inductor L (whose inductance is adjustable between zero and L_{max}) connected in series with the lightbulb B as shown in Fig. 33-34. The electrical supply is 120 V (rms) at 60.0 Hz; the lightbulb is rated as "120 V, 1000 W." (a)

What L_{max} is required if the rate of energy dissipation in the lightbulb is to be varied by a factor of 5 from its upper limit of 1000 W? Assume that the resistance of the lightbulb is independent of its temperature. (b) Could one use a variable resistor (adjustable between zero and R_{max}) instead of an inductor? If so, what R_{max} is required? Why isn't this done?

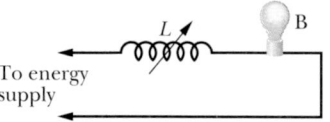

FIGURE 33-34 Problem 84.

85P. In Fig. 33-35, $R = 15.0 \; \Omega$, $C = 4.70 \; \mu$F, and $L = 25.0$ mH. The generator provides a sinusoidal voltage of 75.0 V (rms) and frequency $f = 550$ Hz. (a) Calculate the rms current. (b) Find the rms voltages V_{ab}, V_{bc}, V_{cd}, V_{bd}, V_{ad}. (c) At what average rate is energy dissipated by each of the three circuit elements?

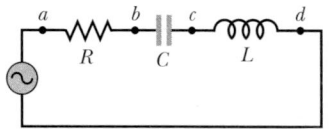

FIGURE 33-35 Problem 85.

SECTION 33-11 Transformers

86E. A generator supplies 100 V to the primary coil of a transformer of 50 turns. If the secondary coil has 500 turns, what is the secondary voltage?

87E. A transformer has 500 primary turns and 10 secondary turns. (a) If V_p is 120 V (rms), what is V_s, assuming an open circuit? (b) If the secondary is now connected to a resistive load of 15 Ω, what are the currents in the primary and secondary?

88E. Figure 33-36 shows an "autotransformer." It consists of a single coil (with an iron core). Three taps T_i are provided. Between taps T_1 and T_2 there are 200 turns, and between taps T_2 and T_3 there are 800 turns. Any two taps can be considered the "primary terminals" and any two taps can be considered the "secondary terminals." List all the ratios by which the primary voltage may be changed to a secondary voltage.

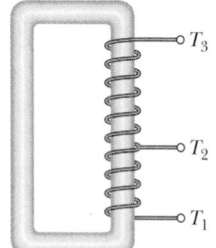

FIGURE 33-36 Exercise 88.

89P. An ac generator provides emf to a resistive load in a remote factory over a two-cable transmission line. At the factory a step-down transformer reduces the voltage from its (rms) transmission value V_t to a much lower value, safe and convenient for use in the factory. The transmission line resistance is 0.30 Ω/cable, and the power of the generator is 250 kW. Calculate the voltage drop along the transmission line and the rate at which energy is dissipated in the line as thermal energy if (a) $V_t = 80$ kV, (b) $V_t = 8.0$ kV, and (c) $V_t = 0.80$ kV. Comment on the acceptability of each choice.

90P. In Fig. 33-32 let the rectangular box on the left represent the (high-impedance) output of an audio amplifier, with $r = 1000$ Ω. Let $R = 10$ Ω represent the (low-impedance) coil of a loudspeaker. For maximum transfer of energy to the load R we must have $R = r$, and that is not true in this case. However, a transformer can be used to "transform" resistances, making them behave electrically as if they were larger or smaller than they actually are. Sketch the primary and secondary coils of a transformer that can be introduced between the amplifier and the speaker in Fig. 33-32 to match the impedances. What must be the turns ratio?

Electronic Computation

91. A 45.0 μF capacitor and a 200 Ω resistor are connected in series with an ac source with a voltage V_s of 100 V. The frequency f of the source can be varied from 0 to 100 Hz. (a) Write an equation for the capacitive reactance X_C. (b) Simultaneously plot the resistance R, the capacitive reactance X_C, and the impedance Z versus f for the range $0 < f < 100$ Hz. (c) From the plots, determine the value of f for which $X_C = R$.

92. (a) For the situation of Problem 91, simultaneously plot the voltage V_C across the capacitor, the voltage V_R across the resistor, and the (constant) voltage V_s across the source versus f for the range $0 < f < 100$ Hz. (b) From the plots, determine the value of f for which $V_C = V_R$. (c) What is V_R at that frequency? (d) Determine the value of f for which $V_R = 0.50V_s$. (e) What is V_C at that frequency? (f) Determine the value of f for which $V_C = 0.50V_s$. (g) What is V_R at that frequency?

93. A 40.0 mH inductor and a 200 Ω resistor are connected in series with an ac source with a voltage V_s of 100 V. The frequency f of the source can be varied from 0 to 2500 Hz. (a) Write an equation for the inductive reactance X_L. (b) Simultaneously plot the resistance R, the inductive reactance X_L, and the impedance Z versus f for the range $0 < f < 2500$ Hz. (c) From the plots, determine the value of f for which $X_L = R$.

94. (a) For the situation of Problem 93, simultaneously plot the voltage V_L across the inductor, the voltage V_R across the resistor, and the (constant) voltage V_s across the source versus f for the range $0 < f < 2500$ Hz. (b) From the plots, determine the value of f for which $V_L = V_R$. (c) What is V_R at that frequency? (d) Determine the value of f for which $V_R = V_s/3$. (e) What is V_L at that frequency? (f) Determine the value of f for which $V_L = V_s/3$. (g) What is V_R at that frequency?

95. A 150.0 mH inductor, a 45.0 μF capacitor, and a 90.0 Ω resistor are connected in series with an ac source with a voltage V_s of 100 V. The frequency f of the source can be varied from 0 to 1000 Hz. (a) Simultaneously plot the capacitive reactance X_C and the inductive reactance X_L versus f for the range $0 < f < 200$ Hz. (b) From the plots, determine f at which $X_C = X_L$. (c) Plot the impedance Z of the circuit versus f for the range $0 < f < 188$ Hz and, from the plot, determine the minimum value of Z and the value of f at which it occurs.

34
Electromagnetic Waves

As a comet swings around the Sun, ice on its surface vaporizes, releasing
trapped dust and charged particles. The electrically charged "solar wind" forces
the charged particles into a straight "tail" that points radially away from the
Sun. But the dust is unaffected by the solar wind and seemingly should
continue to travel along the comet's orbit. Why, instead, does much of the dust
fashion the curved lower tail seen in the photograph?

34-1 MAXWELL'S RAINBOW

James Clerk Maxwell's crowning achievement was to show that a beam of light is a traveling wave of electric and magnetic fields—an **electromagnetic wave**—and thus that optics, the study of visible light, is a branch of electromagnetism. In this chapter we move from one to the other: we conclude our discussion of strictly electric and magnetic phenomena, and we build a foundation for optics.

In Maxwell's day (mid 1800s), the visible, infrared, and ultraviolet forms of light were the only electromagnetic waves known. Spurred on by Maxwell's work, however, Heinrich Hertz discovered what we now call radio waves and verified that they move through the laboratory at the same speed as visible light.

As Fig. 34-1 shows, we now know a wide *spectrum* (or range) of electromagnetic waves, referred to by one imaginative writer as "Maxwell's rainbow." Consider the extent to which we are bathed in electromagnetic waves from across this spectrum. The Sun, whose radiations define the environment in which we as a species have evolved and adapted, is the dominant source. We are also crisscrossed by radio and television signals. Microwaves from radar systems and from telephone relay systems may reach us. There are electromagnetic waves from lightbulbs, from the heated engine blocks of automobiles, from x-ray machines, from lightning flashes, and from buried radioactive materials. Beyond this, radiation reaches us from stars and other objects in our galaxy and from other galaxies.

Electromagnetic waves also travel in the other direction. Television signals, transmitted from Earth since about 1950, have now taken news about us (along with episodes of *I Love Lucy,* albeit *very* faintly) to whatever technically sophisticated inhabitants there may be on whatever planets may encircle the nearest 400 or so stars.

In the wavelength scale in Fig. 34-1 (and similarly the corresponding frequency scale), each scale marker represents a change in wavelength (and correspondingly in frequency) by a factor of 10. The scale is open-ended; the wavelengths of electromagnetic waves have no inherent upper or lower bounds.

Certain regions of the electromagnetic spectrum in Fig. 34-1 are identified by familiar labels, such as *x rays* and *radio waves*. These labels denote roughly defined wavelength ranges within which certain kinds of sources and detectors of electromagnetic waves are in common use. Other regions of Fig. 34-1, such as those labeled television and AM radio, represent specific wavelength bands assigned by law for certain commercial or other purposes. There are no gaps in the electromagnetic spectrum. And all electromagnetic waves, no matter where they lie in the spectrum, travel through *free space* (vacuum) with the same speed c.

The visible region of the spectrum is of course of particular interest to us. Figure 34-2 shows the relative sensitivity of the human eye to light of various wavelengths. The center of the visible region is about 555 nm, which produces the sensation that we call yellow-green.

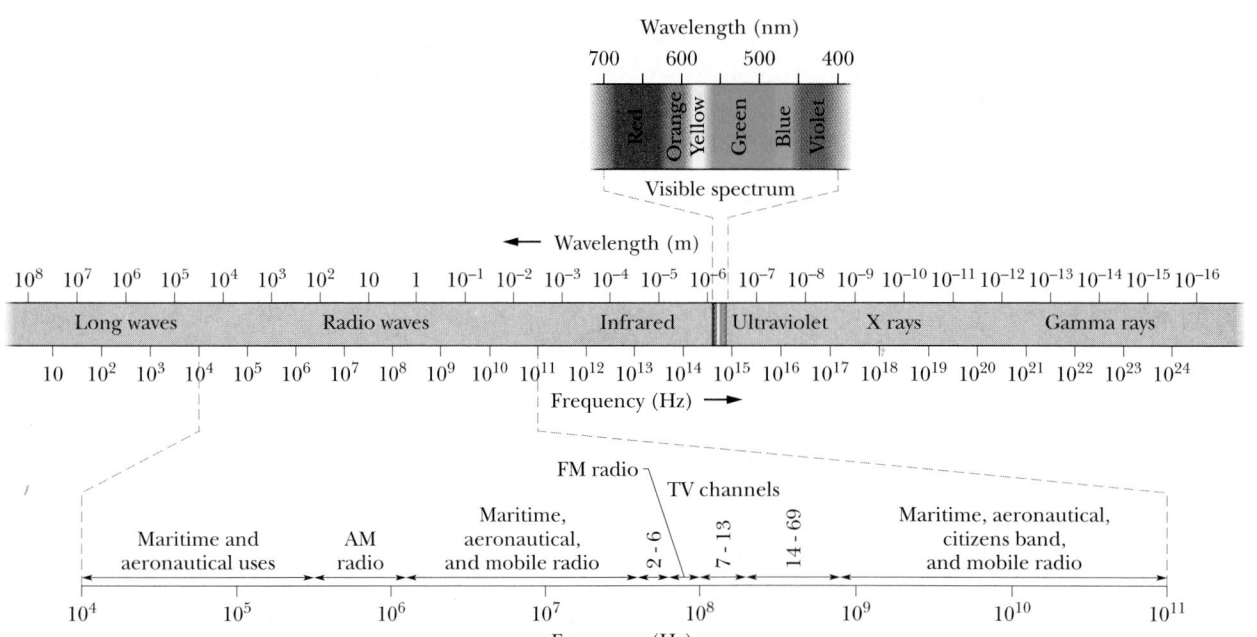

FIGURE 34-1 The electromagnetic spectrum.

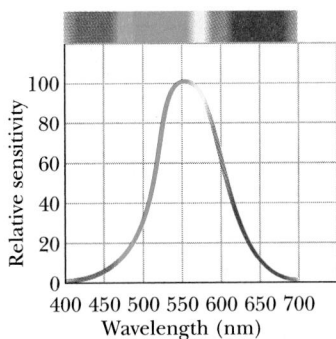

FIGURE 34-2 The relative sensitivity of the eye to electromagnetic waves at different wavelengths. This portion of the electromagnetic spectrum consists of *visible light*.

The limits of this visible spectrum are not well defined because the eye sensitivity curve approaches the zero-sensitivity line asymptotically at both long and short wavelengths. If we take the limits, arbitrarily, as the wavelengths at which eye sensitivity has dropped to 1% of its maximum value, these limits are about 430 and 690 nm; however, the eye can detect electromagnetic waves somewhat beyond these limits if they are intense enough.

34-2 THE TRAVELING ELECTROMAGNETIC WAVE, QUALITATIVELY

Some electromagnetic waves, including x rays, gamma rays, and visible light, are *radiated* (emitted) from sources that are of atomic or nuclear size, where quantum physics rules. Here we discuss how other electromagnetic waves are generated. To simplify matters, we restrict ourselves to that region of the spectrum (wavelength $\lambda \approx 1$ m) in which the source of the *radiation* (the emitted waves) is both macroscopic and of manageable dimensions.

Figure 34-3 shows, in broad outline, the generation of such waves. At its heart is an *LC oscillator*, which estab-

lishes an angular frequency ω $(= 1/\sqrt{LC})$. Charges and currents in this circuit vary sinusoidally at this frequency, as depicted in Fig. 33-1. An external source—possibly an ac generator—must be included to supply energy to compensate both for thermal losses in the circuit and for energy carried away by the radiated electromagnetic wave.

The *LC* oscillator of Fig. 34-3 is coupled by a transformer and a transmission line to an *antenna*, which consists essentially of two thin solid conducting rods. Through this coupling, the sinusoidally varying current in the oscillator causes charge to oscillate sinusoidally along the rods of the antenna at the angular frequency ω of the *LC* oscillator. The current in the rods associated with this movement of charge also varies sinusoidally, in magnitude and direction, at angular frequency ω. The antenna has the effect of an electric dipole whose electric dipole moment varies sinusoidally in magnitude and direction along the length of the antenna.

Because the dipole moment varies in magnitude and direction, the electric field produced by the dipole varies in magnitude and direction. And because the current varies, the magnetic field produced by the current varies in magnitude and direction. However, the changes in the electric and magnetic fields do not happen everywhere instantaneously; rather, the changes travel outward from the antenna at the speed of light c. Together the changing fields form an electromagnetic wave that travels away from the antenna at speed c. The angular frequency of this wave is ω, the same as that of the *LC* oscillator.

Figure 34-4 shows how the electric field **E** and the magnetic field **B** change with time as one wavelength of the wave sweeps past the distant point P of Fig. 34-3; in each part of Fig. 34-4, the wave is traveling directly out of the page. (We choose a distant point so that the curvature of the waves suggested in Fig. 34-3 is small enough to neglect. At such points, the wave is said to be a *plane wave*, and discussion of the wave is much simplified.) Note several key features in Fig. 34-4; they are present regardless of how the wave is created:

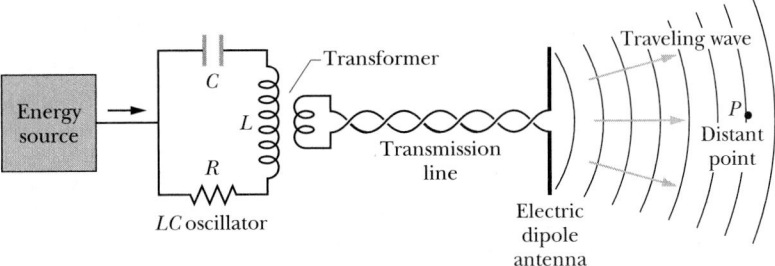

FIGURE 34-3 An arrangement for generating a traveling electromagnetic wave in the shortwave radio region of the spectrum: an *LC* oscillator produces a sinusoidal current in the antenna, which generates the wave. P is a distant point at which a detector can monitor the wave traveling past it.

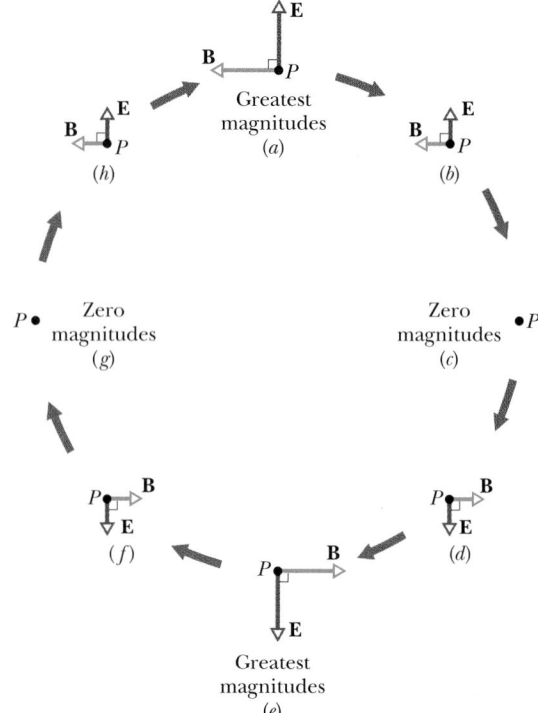

FIGURE 34-4 (a)–(h) The variation in the electric field **E** and the magnetic field **B** at the distant point P of Fig. 34-3 as one wavelength of the electromagnetic wave travels past it. In this perspective, the wave is traveling directly out of the page. The two fields vary sinusoidally in magnitude and direction. Note that they are always perpendicular to each other and to the direction of travel of the wave.

1. The electric and magnetic fields **E** and **B** are always perpendicular to the direction of travel of the wave. Thus the wave is a *transverse wave,* as discussed in Chapter 17.

2. The electric field is always perpendicular to the magnetic field.

3. The cross product **E** × **B** always gives the direction of travel of the wave.

4. The fields always vary sinusoidally, just like the transverse waves discussed in Chapter 17. Moreover, the fields vary with the same frequency and *in phase* (in step) with each other.

In keeping with these features, we can assume that the electromagnetic wave is traveling toward P in the positive direction of an x axis, that the electric field in Fig. 34-4 is oscillating parallel to the y axis, and that the magnetic field oscillates parallel to the z axis (using a right-handed coordinate system, of course). Then we can write the electric and magnetic fields as sinusoidal functions of position x

and time t:

$$E = E_m \sin(kx - \omega t), \qquad (34\text{-}1)$$

$$B = B_m \sin(kx - \omega t), \qquad (34\text{-}2)$$

in which E_m and B_m are the amplitudes of the fields and, as in Chapter 17, ω and k are the angular frequency and angular wave number of the wave, respectively. From these equations, note that not only do the two fields form the electromagnetic wave but each forms its own wave. Equation 34-1 gives the *electric wave component* of the electromagnetic wave, and Eq. 34-2 gives the *magnetic wave component.* As we shall discuss below, these two wave components cannot exist independently.

From Eq. 17-12, we know that the speed of the wave is ω/k. However, since this is an electromagnetic wave, its speed (in vacuum) is given the symbol c rather than v. In the next section you will see that c has the value

$$c = \frac{1}{\sqrt{\mu_0 \epsilon_0}} \qquad \text{(wave speed),} \qquad (34\text{-}3)$$

which is about 3.0×10^8 m/s. In other words:

All electromagnetic waves, including visible light, have the same speed c in vacuum.

You will also see that the wave speed c and the amplitudes of the electric and magnetic fields are related by

$$\frac{E_m}{B_m} = c \qquad \text{(amplitude ratio).} \qquad (34\text{-}4)$$

If we divide Eq. 34-1 by Eq. 34-2 and then substitute with Eq. 34-4, we find that the magnitudes of the fields at every instant are related by

$$\frac{E}{B} = c \qquad \text{(magnitude ratio).} \qquad (34\text{-}5)$$

We can represent the electromagnetic wave as in Fig. 34-5a, with a *ray* (a directed line showing the wave's direction of travel) and *wavefronts* (imaginary surfaces over which the wave has the same magnitude of electric field). The two wavefronts shown in Fig. 34-5a are separated by one wavelength $\lambda (= 2\pi/k)$ of the wave. (Waves traveling in approximately the same direction form a *beam,* such as a laser beam, which can be represented with a ray.)

We can also represent the wave as in Fig. 34-5b, which shows the electric and magnetic field vectors in a "snapshot" of the wave at a certain instant. The curves through the tips of the arrows represent the sinusoidal

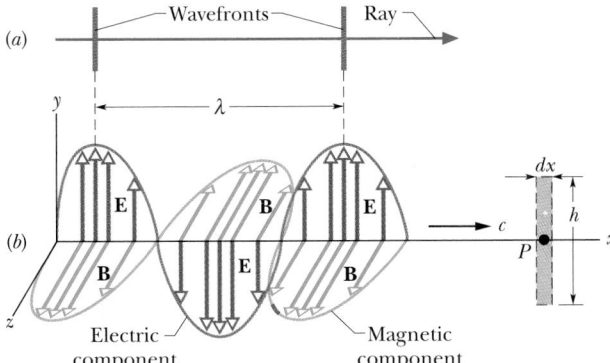

FIGURE 34-5 (*a*) An electromagnetic wave represented with a ray and two wavefronts; the wavefronts are separated by one wavelength λ. (*b*) The same wave represented in a "snapshot" of its electric field **E** and magnetic field **B** at points on the x axis, along which the wave travels at speed c. As it travels past point P, the fields vary as shown in Fig. 34-4. The electric component of the wave consists of only the electric fields; the magnetic component consists of only the magnetic fields. The dashed rectangle at P is used in Fig. 34-6.

oscillations given by Eqs. 34-1 and 34-2; the wave components **E** and **B** are in phase, perpendicular to each other, and perpendicular to the wave's direction of travel.

Interpretation of Fig. 34-5*b* requires some care. The similar drawings for a transverse wave on a taut string that we discussed in Chapter 17 represented the up and down displacement of sections of the string as the wave passed (*something actually moved*). Figure 34-5*b* is more abstract. At the instant shown, the electric and magnetic fields have certain magnitudes and directions (but are always perpendicular to the x axis) at each point along the x axis. We choose to represent these vector quantities with arrows, so we must draw arrows of different lengths, all pointing away from the x axis, like thorns on a rose stem. But the arrows represent only field values for points that are on the x axis. Neither the arrows nor the sinusoidal curves represent a sideways motion of anything, nor do the arrows connect points on the x axis with points off the axis.

Drawings like Fig. 34-5 help us visualize what is actually a very complicated situation. First consider the magnetic field. Because it varies sinusoidally, it induces (via Faraday's law of induction) a perpendicular electric field that also varies sinusoidally. But because that electric field is varying sinusoidally, it induces (via Maxwell's law of induction) a perpendicular magnetic field that also varies sinusoidally. And so on. The two fields continuously create each other via induction, and the resulting sinusoidal variations in the fields travel as a wave, the electromagnetic wave. Without this amazing result, we could not see; indeed, because we need electromagnetic waves from the

Sun to maintain Earth's temperature, we could not even exist without this result.

A Most Curious Wave

The waves we discussed in Chapters 17 and 18 require a *medium* (some material) through which or along which to travel. We had waves traveling along a string, through Earth, and through the air. But an electromagnetic wave (let's use the term *light wave* or *light*) is curiously different in that it requires no medium for its travel. It can, indeed, travel through a medium such as air or glass, but it can also travel through the vacuum of space between a star and us.

Once the special theory of relativity became accepted, long after Einstein published it in 1905, the speed of light waves was realized to be special. One reason was that light has the same speed regardless of the frame of reference from which it is measured. If you send a beam of light along an axis and ask several observers to measure its speed while they move at different speeds along that axis, either in the direction of the light or opposite it, they will all measure the *same speed* for the light. This result is an amazing one and quite different from what would have been found if those observers had measured the speed of any other type of wave; for other waves, the speed of the observers relative to the wave would have affected their measurements.

The meter has now been defined so that the speed of light (any electromagnetic wave) in vacuum has the exact value

$$c = 299{,}792{,}458 \text{ m/s,}$$

which can be used as a standard. In fact, if you now measure the travel time of a pulse of light from one point to another, you are not really measuring the speed of the light but rather the distance between those two points.

34-3 THE TRAVELING ELECTROMAGNETIC WAVE, QUANTITATIVELY

We shall now derive Eqs. 34-3 and 34-4 and, even more important, explore the dual induction of electric and magnetic fields that gives us light.

Equation 34-4 and the Induced Electric Field

The dashed rectangle of dimensions dx and h in Fig. 34-6 is fixed at point P on the x axis and in the xy plane (it is shown at the right in Fig. 34-5*b*). As the electromagnetic wave moves rightward past the rectangle, the magnetic flux Φ_B through the rectangle changes and—according to

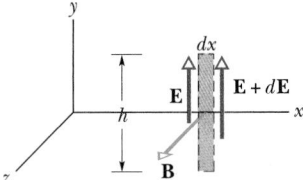

FIGURE 34-6 As the electromagnetic wave travels rightward past point P in Fig. 34-5, the sinusoidal variation of the magnetic field **B** through a rectangle centered at P induces electric fields along the rectangle. At the instant shown, **B** is decreasing in magnitude and the induced electric field is therefore greater in magnitude on the right side of the rectangle than on the left.

Faraday's law of induction—induced electric fields appear throughout the region of the rectangle. We take **E** and **E** + d**E** to be the induced fields along the two long sides of the rectangle. These induced electric fields are, in fact, the electric component of the electromagnetic wave.

Let us consider these fields at the instant when the magnetic wave component passing through the rectangle is the small section marked with red in Fig. 34-5b. Just then, the magnetic field through the rectangle points in the positive z direction and is decreasing in magnitude (the magnitude was greater just before the red section arrived). Because the magnetic field is decreasing, the magnetic flux Φ_B through the rectangle is also decreasing. According to Faraday's law, this change in flux is opposed by induced electric fields, which produce a magnetic field **B** in the positive z direction.

According to Lenz's law, this in turn means that if we imagine the boundary of the rectangle to be a conducting loop, a counterclockwise induced current would have to appear in it. There is, of course, no conducting loop; but this analysis shows that the induced electric field vectors **E** and **E** + d**E** are indeed oriented as shown in Fig. 34-6, with the magnitude of **E** + d**E** greater than that of **E**. Otherwise, the net induced electric field would not act counterclockwise around the rectangle.

Let us now apply Faraday's law of induction,

$$\oint \mathbf{E} \cdot d\mathbf{s} = -\frac{d\Phi_B}{dt}, \qquad (34\text{-}6)$$

counterclockwise around the rectangle of Fig. 34-6. There is no contribution to the integral from the top or bottom of the rectangle because **E** and d**s** are perpendicular there. The integral then has the value

$$\oint \mathbf{E} \cdot d\mathbf{s} = (E + dE)h - Eh = h\, dE. \qquad (34\text{-}7)$$

The flux Φ_B through this rectangle is

$$\Phi_B = (B)(h\, dx), \qquad (34\text{-}8)$$

where B is the magnitude of **B** within the rectangle and $h\, dx$ is the area of the rectangle. Differentiating Eq. 34-8 with respect to t gives

$$\frac{d\Phi_B}{dt} = h\, dx\, \frac{dB}{dt}. \qquad (34\text{-}9)$$

If we substitute Eqs. 34-7 and 34-9 into Eq. 34-6, we find

$$h\, dE = -h\, dx\, \frac{dB}{dt}$$

or

$$\frac{dE}{dx} = -\frac{dB}{dt}. \qquad (34\text{-}10)$$

Actually, both B and E are functions of *two* variables, x and t, as Eqs. 34-1 and 34-2 imply. However, in evaluating dE/dx, we must assume that t is constant because Fig. 34-6 is an "instantaneous snapshot." Also, in evaluating dB/dt we must assume that x is constant because we are dealing with the time rate of change of B at a particular place, the point P in Fig. 34-5b. The derivatives under these circumstances are *partial derivatives*, and Eq. 34-10 must be written

$$\frac{\partial E}{\partial x} = -\frac{\partial B}{\partial t}. \qquad (34\text{-}11)$$

The minus sign in this equation is appropriate and necessary because, although E is increasing with x at the site of the rectangle in Fig. 34-6, B is decreasing with t.

From Eq. 34-1, we have

$$\frac{\partial E}{\partial x} = kE_m \cos(kx - \omega t)$$

and from Eq. 34-2,

$$\frac{\partial B}{\partial t} = -\omega B_m \cos(kx - \omega t).$$

Then Eq. 34-11 reduces to

$$kE_m \cos(kx - \omega t) = \omega B_m \cos(kx - \omega t). \qquad (34\text{-}12)$$

The ratio ω/k for a traveling wave is its speed, which we are calling c. Equation 34-12 then becomes

$$\frac{E_m}{B_m} = c \quad \text{(amplitude ratio)}, \qquad (34\text{-}13)$$

which is just Eq. 34-4.

Equation 34-3 and the Induced Magnetic Field

Figure 34-7 shows another dashed rectangle at point P of Fig. 34-5; this one is in the xz plane. As the electromagnetic wave moves rightward past this new rectangle, the

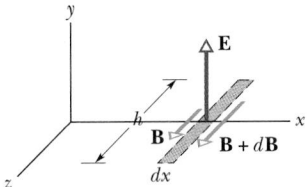

FIGURE 34-7 The sinusoidal variation of the electric field through this rectangle, located (but not shown) at point P in Fig. 34-5, induces magnetic fields along the rectangle. The instant shown is that of Fig. 34-6: **E** is decreasing in magnitude and the induced magnetic field is greater in magnitude on the right side of the rectangle than on the left.

electric flux Φ_E through the rectangle changes and—according to Maxwell's law of induction—induced magnetic fields appear throughout the region of the rectangle. These induced magnetic fields are, in fact, the magnetic component of the electromagnetic wave.

We see from Fig. 34-5 that at the instant chosen for the magnetic field in Fig. 34-6, the electric field through the rectangle of Fig. 34-7 is directed as shown. Recall that at the chosen instant, the magnetic field in Fig. 34-6 is decreasing. Because the two fields are in phase, the electric field in Fig. 34-7 must also be decreasing and so must the electric flux Φ_E through the rectangle. By applying the same reasoning we applied to Fig. 34-6, we see that the changing flux Φ_E will induce a magnetic field with vectors **B** and **B** + d**B** oriented as shown in Fig. 34-7, where **B** + d**B** is greater than **B**.

Let us apply Maxwell's law of induction,

$$\oint \mathbf{B} \cdot d\mathbf{s} = \mu_0 \epsilon_0 \frac{d\Phi_E}{dt}, \qquad (34\text{-}14)$$

by proceeding counterclockwise around the dashed rectangle of Fig. 34-7. Only the long sides of the rectangle contribute to the integral, whose value is

$$\oint \mathbf{B} \cdot d\mathbf{s} = -(B + dB)h + Bh = -h \, dB. \quad (34\text{-}15)$$

The flux Φ_E through the rectangle is

$$\Phi_E = (E)(h \, dx), \qquad (34\text{-}16)$$

where E is the average magnitude of **E** within the rectangle. Differentiating Eq. 34-16 with respect to t gives

$$\frac{d\Phi_E}{dt} = h \, dx \frac{dE}{dt}. \qquad (34\text{-}17)$$

If we substitute Eqs. 34-15 and 34-17 into Eq. 34-14, we find

$$-h \, dB = \mu_0 \epsilon_0 \left(h \, dx \frac{dE}{dt} \right)$$

or, changing to partial-derivative notation as we did before (Eq. 34-11),

$$-\frac{\partial B}{\partial x} = \mu_0 \epsilon_0 \frac{\partial E}{\partial t}. \qquad (34\text{-}18)$$

Again, the minus sign in this equation is necessary because, although B is increasing with x at point P in the rectangle in Fig. 34-7, E is decreasing with t.

Evaluating Eq. 34-18 by using Eqs. 34-1 and 34-2 leads to

$$-kB_m \cos(kx - \omega t) = -\mu_0 \epsilon_0 \omega E_m \cos(kx - \omega t),$$

which we can write as

$$\frac{E_m}{B_m} = \frac{1}{\mu_0 \epsilon_0 (\omega/k)} = \frac{1}{\mu_0 \epsilon_0 c}. \qquad (34\text{-}19)$$

Combining Eqs. 34-13 and 34-19 leads at once to

$$c = \frac{1}{\sqrt{\mu_0 \epsilon_0}} \qquad \text{(wave speed),} \qquad (34\text{-}20)$$

which is exactly Eq. 34-3.

CHECKPOINT **1:** The magnetic field **B** through the rectangle of Fig. 34-6 is shown at a different instant in part 1 of the accompanying figure; **B** is directed in the xz plane, parallel to the z axis, and its magnitude is increasing. (a) Complete part 1 by drawing the induced electric fields, indicating both directions and relative magnitudes (as in Fig. 34-6). (b) For the same instant, complete part 2 of the figure by drawing the electric field of the electromagnetic wave. Also draw the induced magnetic fields, indicating both directions and relative magnitudes (as in Fig. 34-7).

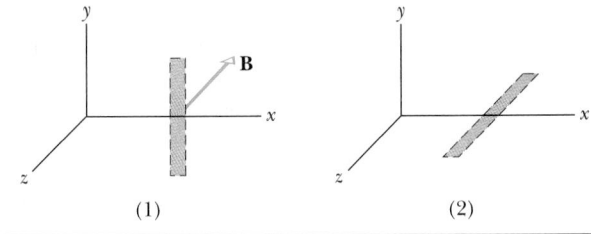

34-4 ENERGY TRANSPORT AND THE POYNTING VECTOR

All sunbathers know that an electromagnetic wave can transport energy and deliver it to a body on which it falls. The rate of energy transport per unit area in such a wave is described by a vector **S**, called the **Poynting vector** after John Henry Poynting (1852–1914), who first discussed its properties. **S** is defined as

$$\mathbf{S} = \frac{1}{\mu_0} \mathbf{E} \times \mathbf{B} \quad \text{(Poynting vector)} \quad (34\text{-}21)$$

and has the SI unit watt per square meter (W/m^2).

The direction of the Poynting vector **S** of an electromagnetic wave at any point gives the wave's direction of travel and the direction of energy transport at that point.

Because **E** and **B** are perpendicular to each other in an electromagnetic wave, the magnitude of $\mathbf{E} \times \mathbf{B}$ is EB. Then the magnitude of **S** is

$$S = \frac{1}{\mu_0} EB, \quad (34\text{-}22)$$

in which S, E, and B are instantaneous values. E and B are so closely coupled to each other that we need to deal with only one of them; we choose E, largely because most instruments for detecting electromagnetic waves deal with the electric component of the wave rather than the magnetic component. So, using $B = E/c$ from Eq. 34-5, we can rewrite Eq. 34-22 as

$$S = \frac{1}{c\mu_0} E^2 \quad \text{(instantaneous energy flow rate).} \quad (34\text{-}23)$$

By substituting $E = E_m \sin(kx - \omega t)$ into Eq. 34-23, we could obtain an equation for the energy transport rate as a function of time. More useful in practice, however, is the average energy transported over time; for that, we need to find the time-averaged value of S, written $\overline{S}$ and also called the **intensity** I of the wave. That is,

$$I = \overline{S} = \frac{1}{c\mu_0} \overline{E^2} = \frac{1}{c\mu_0} \overline{E_m^2 \sin^2(kx - \omega t)}, \quad (34\text{-}24)$$

where in each term the bar means "average value of." The average value of $\sin^2 \theta$, for any angular variable θ, is $\frac{1}{2}$ (see Fig. 33-14). In addition, we define a new quantity E_{rms}, the *root-mean-square* value of the electric field, as

$$E_{\text{rms}} = \frac{E_m}{\sqrt{2}}. \quad (34\text{-}25)$$

We can then rewrite Eq. 34-24 as

$$I = \frac{1}{c\mu_0} E_{\text{rms}}^2. \quad (34\text{-}26)$$

Because $E = cB$ and c is such a very large number, you might conclude that the energy associated with the electric field is much larger than that associated with the magnetic field. That conclusion is incorrect; the two energies are exactly equal. To show this, we start with Eq. 26-23, which gives the energy density $u \, (= \frac{1}{2}\epsilon_0 E^2)$ within an electric field, and substitute cB for E; then we can write

$$u_E = \tfrac{1}{2}\epsilon_0 E^2 = \tfrac{1}{2}\epsilon_0 (cB)^2.$$

If we now substitute for c with Eq. 34-3, we get

$$u_E = \tfrac{1}{2}\epsilon_0 \frac{1}{\mu_0 \epsilon_0} B^2 = \frac{B^2}{2\mu_0}.$$

But Eq. 31-56 tells us that $B^2/2\mu_0$ is the energy density u_B of the magnetic field, so we see that $u_E = u_B$.

Variation of Intensity with Distance

How intensity varies with distance from a real source of electromagnetic radiation is often complex—especially when the source (like a searchlight at a movie premier) beams the radiation in a particular direction. However, in some situations we can assume that the source is a *point source* that emits the light *isotropically,* that is, with equal intensity in all directions. The spherical wavefronts spreading from such an isotropic point source S at a particular instant are shown in cross section in Fig. 34-8.

Let us assume that the energy of the waves is conserved as they spread from this source. Let us also center an imaginary sphere of radius r on the source, as shown in Fig. 34-8. All the energy emitted by the source must pass through the sphere. Thus, the rate at which energy is transferred through the sphere by the radiation must equal the rate at which energy is emitted by the source, that is, the power P_s of the source. The intensity I at the sphere must

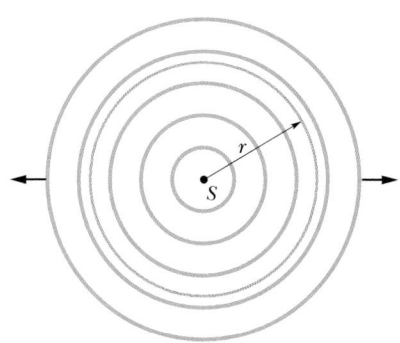

FIGURE 34-8 A point source S emits electromagnetic waves uniformly in all directions. The spherical wavefronts pass through an imaginary sphere of radius r that is centered on S.

then be

$$I = \frac{P_s}{4\pi r^2}, \qquad (34\text{-}27)$$

where $4\pi r^2$ is the area of the sphere. Equation 34-27 tells us that the intensity of the electromagnetic radiation from an isotropic point source decreases with the square of the distance r from the source.

CHECKPOINT **2:** The figure gives the electric field of an electromagnetic wave at a certain point and a certain instant. The wave is transporting energy in the negative z direction. What is the direction of the magnetic field of the wave at that point and instant?

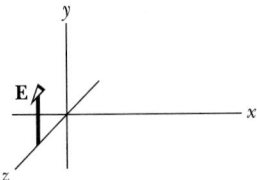

SAMPLE PROBLEM 34-1

An observer is 1.8 m from a point light source whose power P_s is 250 W. Calculate the rms values of the electric and magnetic fields due to the source at the position of the observer.

SOLUTION: Combining Eqs. 34-27 and Eq. 34-26 gives us

$$I = \frac{P_s}{4\pi r^2} = \frac{1}{c\mu_0} E_{\text{rms}}^2,$$

where $4\pi r^2$ is the area of a sphere of radius r centered on the source. The rms value of the electric field is then

$$E_{\text{rms}} = \sqrt{\frac{P_s c \mu_0}{4\pi r^2}}$$

$$= \sqrt{\frac{(250 \text{ W})(3.00 \times 10^8 \text{ m/s})(4\pi \times 10^{-7} \text{ H/m})}{(4\pi)(1.8 \text{ m})^2}}$$

$$= 48.1 \text{ V/m} \approx 48 \text{ V/m}. \qquad \text{(Answer)}$$

The rms value of the magnetic field follows from Eq. 34-5 and is

$$B_{\text{rms}} = \frac{E_{\text{rms}}}{c} = \frac{48.1 \text{ V/m}}{3.00 \times 10^8 \text{ m/s}}$$

$$= 1.6 \times 10^{-7} \text{ T}. \qquad \text{(Answer)}$$

Note that E_{rms} ($= 48$ V/m) is appreciable as judged by ordinary laboratory standards, but B_{rms} ($= 1.6 \times 10^{-7}$ T) is quite small. This discrepancy helps to explain why most instruments used for the detection and measurement of electromagnetic waves are designed to respond to the electric component of the wave. It is wrong, however, to say that the electric component of an electromagnetic wave is ''stronger'' than the magnetic component. You cannot compare quantities that are measured in different units. As we have seen, the electric and magnetic components are on an absolutely equal basis as far as the propagation of the wave is concerned, because their average energies, which *can* be compared, are exactly equal.

34-5 RADIATION PRESSURE

Electromagnetic waves have linear momentum as well as energy. This means that we can exert a pressure—a **radiation pressure**—on an object by shining light on it. However, the pressure must be very small because, for example, you do not feel it when a camera flash is used to take your photograph.

To find an expression for the pressure, let us shine a beam of electromagnetic radiation—light, for example—on an object for a time interval Δt. Further, let us assume that the object is free to move and that the radiation is entirely **absorbed** (taken up) by the object. This means that during the interval Δt, the object gains an energy ΔU from the radiation. Maxwell showed that the object also gains linear momentum. The magnitude Δp of the momentum change of the object is related to the energy change ΔU by

$$\Delta p = \frac{\Delta U}{c} \qquad \text{(total absorption),} \qquad (34\text{-}28)$$

where c is the speed of light. The direction of the momentum change of the object is the direction of the *incident* (incoming) beam that the object absorbs.

Instead of being absorbed, the radiation can be **reflected** by the object; that is, the radiation can be sent off in a new direction as if it bounced off the object. If the radiation is entirely reflected back along its original path, the magnitude of the momentum change of the object is twice that given above, or

$$\Delta p = \frac{2\,\Delta U}{c} \qquad \begin{array}{l}\text{(total reflection} \\ \text{back along path).}\end{array} \qquad (34\text{-}29)$$

In the same way, an object undergoes twice as much momentum change when a perfectly elastic tennis ball is bounced from it as when it is struck by a perfectly inelastic ball (a lump of wet putty, say) of the same mass and velocity. If the incident radiation is partly absorbed and partly reflected, the momentum change of the object is between $\Delta U/c$ and $2\,\Delta U/c$.

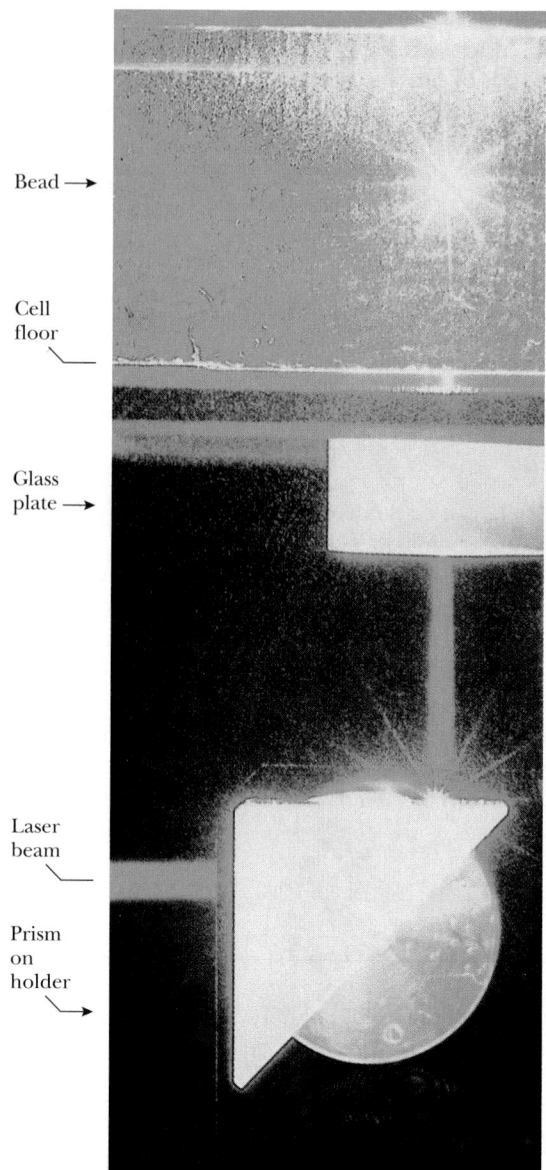

Bead →

Cell floor

Glass plate →

Laser beam

Prism on holder

An initially horizontal laser beam of green light is sent upward by a glass prism into an evacuated transparent cell and onto a glass sphere 20 μm in diameter. The sphere scatters the light, giving the starlike appearance in the upper part of the photograph. Before the laser was turned on, the glass sphere was on the floor of the cell. But the radiation pressure of the laser light has lifted the sphere by about 1 cm.

To find expressions for the force exerted by radiation in terms of the intensity I of the radiation, suppose that a flat surface of area A, perpendicular to the path of the radiation, intercepts the radiation. In time interval Δt, the energy intercepted by area A is

$$\Delta U = IA \, \Delta t. \qquad (34\text{-}31)$$

If the energy is completely absorbed, then Eq. 34-28 tells us that $\Delta p = IA \, \Delta t/c$ and, from Eq. 34-30, the magnitude of the force on the area A is

$$F = \frac{IA}{c} \quad \text{(total absorption).} \qquad (34\text{-}32)$$

Similarly, if the radiation is totally reflected back along its original path, Eq. 34-29 tells us that $\Delta p = 2IA \, \Delta t/c$ and, from Eq. 34-30,

$$F = \frac{2IA}{c} \quad \begin{array}{l}\text{(total reflection}\\ \text{back along path).}\end{array} \qquad (34\text{-}33)$$

If the radiation is partly absorbed and partly reflected, the magnitude of the force on area A is between the values of IA/c and $2IA/c$.

The force per unit area on an object due to radiation is the radiation pressure p_r. We can find it for the situations of Eqs. 34-32 and 34-33 by dividing both sides of each equation by A. We obtain

$$p_r = \frac{I}{c} \quad \text{(total absorption)} \qquad (34\text{-}34)$$

and

$$p_r = \frac{2I}{c} \quad \begin{array}{l}\text{(total reflection}\\ \text{back along path).}\end{array} \qquad (34\text{-}35)$$

Be careful not to confuse the symbol p_r for radiation pressure with the symbol p for momentum.

The development of laser technology has permitted researchers to achieve radiation pressures much greater than, say, that due to a camera flashlamp. This comes about because a beam of laser light—unlike a beam of light from a small lamp filament—can be focused to a tiny spot only a few wavelengths in diameter. This permits the delivery of very large energy to small objects placed at that spot.

From Newton's second law, we know that a change in momentum is related to a force by

$$F = \frac{\Delta p}{\Delta t}. \qquad (34\text{-}30)$$

CHECKPOINT 3: Light of uniform intensity shines perpendicularly on a totally absorbing surface, fully illuminating the surface. If the area of the surface is decreased, do (a) the radiation pressure and (b) the radiation force on the surface increase, decrease, or stay the same?

SAMPLE PROBLEM 34-2

When dust is released by a comet, it does not continue along the comet's orbit because radiation pressure from sunlight pushes it radially outward from the Sun. Assume that a dust particle is spherical with radius R, has density $\rho = 3.5 \times 10^3$ kg/m³, and totally absorbs the sunlight it intercepts. For what value of R does the gravitational force F_g on the dust particle due to the Sun just balance the radiation force F_r on it from the sunlight?

SOLUTION: From Eq. 34-27, the intensity of sunlight on a dust particle (or anything else) at distance r from the Sun is

$$I = \frac{P_S}{4\pi r^2}, \qquad (34\text{-}36)$$

where $4\pi r^2$ is the surface area of a sphere of radius r centered on the sun, and P_S ($= 3.9 \times 10^{26}$ W) is the average power radiated by the Sun. From Eq. 34-32,

$$F_r = \frac{IA}{c} = \frac{I\pi R^2}{c}, \qquad (34\text{-}37)$$

where the area A of the particle that intercepts the sunlight is the particle's cross-sectional area πR^2 (and *not* half its surface area). Substituting Eq. 34-36 into Eq. 34-37 yields

$$F_r = \frac{P_S R^2}{4cr^2}. \qquad (34\text{-}38)$$

From Eq. 14-1 we can write the gravitational force F_g on the particle as

$$F_g = \frac{GM_S m}{r^2} = \frac{4GM_S \rho \pi R^3}{3r^2}, \qquad (34\text{-}39)$$

where M_S ($= 1.99 \times 10^{30}$ kg) is the Sun's mass and we have replaced the dust particle's mass m with $\rho(4/3)\pi R^3$. Setting $F_r = F_g$, and solving for R, we find

$$R = \frac{3P_S}{16\pi c\rho GM_S}.$$

The denominator is

$$(16\pi)(3 \times 10^8 \text{ m/s})(3.5 \times 10^3 \text{ kg/m}^3)$$
$$\times (6.67 \times 10^{-11} \text{ N}\cdot\text{m}^2/\text{kg}^2)(1.99 \times 10^{30} \text{ kg})$$
$$= 7.0 \times 10^{33} \text{ N/s}.$$

We then have

$$R = \frac{(3)(3.9 \times 10^{26} \text{ W})}{7.0 \times 10^{33} \text{ N/s}} = 1.7 \times 10^{-7} \text{ m}. \quad \text{(Answer)}$$

Note that this result is independent of the particle's distance r from the Sun.

Dust particles with $R \approx 1.7 \times 10^{-7}$ m follow an approximately straight path like path b in Fig. 34-9. For larger values of R, comparison of Eqs. 34-38 and 34-39 shows that,

because F_g varies with R^3 and F_r varies with R^2, the gravitational force F_g dominates the radiation force F_r. Thus such particles follow a path that is curved toward the Sun like path c in Fig. 34-9. Similarly, for smaller values of R, the radiation force dominates, and the dust follows a path that is curved away from the Sun like path a. The composite of these dust particles is the dust tail of the comet.

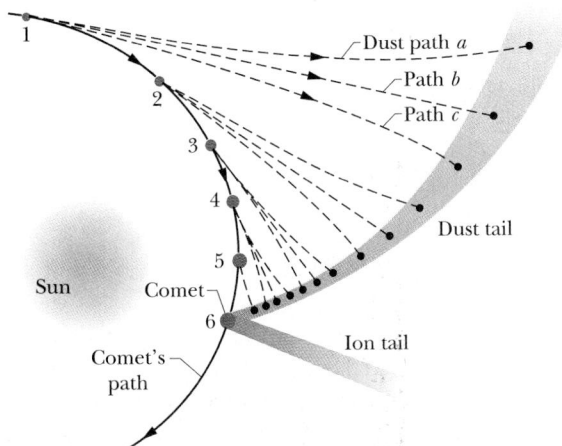

FIGURE 34-9 Sample Problem 34-2. A comet is now at position 6. Dust it has released at five previous positions has been pushed outward by radiation pressure from sunlight, has taken the dashed paths, and now forms the comet's curved dust tail.

34-6 POLARIZATION

VHF (very high frequency) television antennas in England are oriented vertically, but those in North America are horizontal. The difference is due to the direction of oscillation of the electromagnetic waves carrying the TV signal. In England, the transmitting equipment is designed to produce waves that are **polarized** vertically; that is, their electric field oscillates vertically. So, for the electric field of the incident television waves to drive a current along an antenna (and thus provide a signal to a television set), the antenna must be vertical. In North America, the waves are horizontally polarized.

Figure 34-10a shows an electromagnetic wave with its electric field oscillating parallel to the vertical y axis. The plane containing the **E** vectors is called the **plane of oscillation** of the wave (hence, the wave is said to be *plane-polarized* in the y direction). We can represent the wave's **polarization** (state of being polarized) by showing the extent of the electric field oscillations in a "head-on" view of the plane of oscillation, as in Fig. 34-10b.

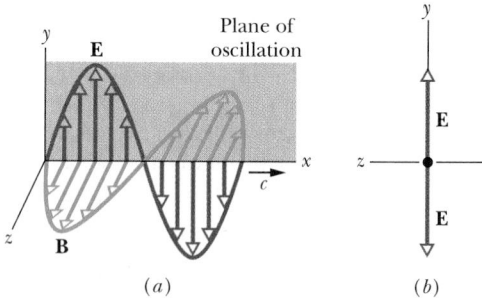

FIGURE 34-10 (a) The plane of oscillation of a polarized electromagnetic wave. (b) To represent the polarization, we view the plane of oscillation "head-on" and indicate the amplitude of the oscillating electric field.

Polarized Light

The electromagnetic waves emitted by a television station all have the same polarization, but the electromagnetic waves emitted by any common source of light (such as the Sun or a bulb) are **polarized randomly** or **unpolarized;** that is, the electric field at any given point is always perpendicular to the direction of travel of the waves but changes directions randomly. So if we try to represent a head-on view of the oscillations over some time period, we do not have a simple drawing like that of Fig. 34-10b; instead we have a mess like that in Fig. 34-11a.

In principle, we can simplify the mess by resolving each electric field of Fig. 34-11a into y and z components and then finding the net fields along the y axis and the z axis separately, as shown in Fig. 34-11b. In doing so, we mathematically change unpolarized light into the superposition of two polarized waves whose planes of oscillation are perpendicular to each other. The result is the double-

arrow representation of Fig. 34-11b, which simplifies drawings of unpolarized light. Similarly we can represent light that is **partially polarized** (its field oscillations are not completely random as in Fig. 34-11a nor are they parallel to a single axis as in Fig. 34-10b). For this situation, we can draw one of the arrows of the double-arrow representation longer than the other arrow.

We can actually transform unpolarized visible light into polarized light by sending it through a *polarizing sheet,* as is shown in Fig. 34-12. Such sheets, commercially known as Polaroids or Polaroid filters, were invented in 1932 by Edwin Land while he was an undergraduate student. A sheet consists of certain long molecules embedded in plastic. When the sheet is manufactured, it is stretched to align the molecules in parallel rows, like rows in a plowed field. When light is then sent through the sheet, electric field components along one direction pass through the sheet, while components perpendicular to that direction are absorbed by the molecules and disappear.

We shall not dwell on the molecules but, instead, shall assign to the sheet a *polarizing direction,* along which electric field components are passed:

An electric field component parallel to the polarizing direction is passed *(transmitted)* by a polarizing sheet; a component perpendicular to it is absorbed.

Thus the light emerging from the sheet consists of only the components that are parallel to the polarizing direction of the sheet; hence the light must be polarized in that direction. In Fig. 34-12, the vertical components are transmitted by the sheet; the horizontal components are absorbed. The transmitted waves are then vertically polarized.

We next consider the intensity of the transmitted light. We start with unpolarized light, whose electric field oscillations we can resolve into y and z components as in Fig.

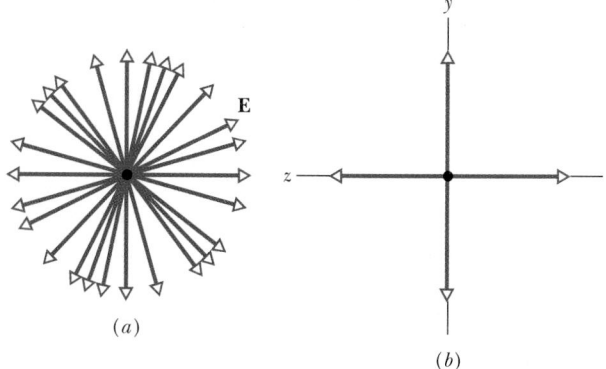

FIGURE 34-11 (a) Unpolarized light consists of waves with randomly directed electric fields. Here the waves are all traveling along the same axis, directly out of the page, and all have the same amplitude E. (b) A second way of representing unpolarized light: the light is the superposition of two polarized waves whose planes of oscillation are perpendicular to each other.

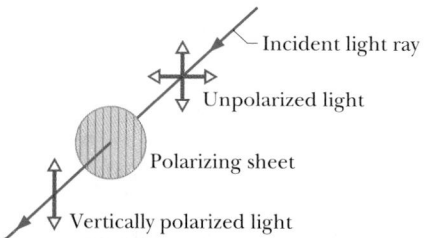

FIGURE 34-12 Unpolarized light becomes polarized when it is sent through a polarizing sheet. Its direction of polarization is then parallel to the polarizing direction of the sheet, which is represented here by the vertical lines drawn in the sheet.

34-11b. Further, we can arrange for the y axis to be parallel to the polarizing direction of the sheet. Then only the y components of the light are passed by the sheet; the z components are absorbed. As suggested by Fig. 34-11b, if the original waves are randomly oriented, the sum of the y components and the sum of the z components are equal. When the z components are absorbed, half the intensity I_0 of the original light is lost. The intensity I of the emerging polarized light is then

$$I = \tfrac{1}{2}I_0. \qquad (34\text{-}40)$$

Let us call this the *one-half rule;* we can use it *only* when the light reaching a polarizing sheet is unpolarized.

Suppose now that the light reaching a polarizing sheet is already polarized. Figure 34-13 shows a polarizing sheet in the plane of the page and the electric field **E** of such a polarized light wave traveling toward the sheet (and thus prior to any absorption). We can resolve **E** into two components relative to the polarizing direction of the sheet: parallel component E_y is transmitted by the sheet, and perpendicular component E_z is absorbed. Since θ is the angle between **E** and the polarizing direction of the sheet, the transmitted parallel component is

$$E_y = E \cos \theta. \qquad (34\text{-}41)$$

Recall that the intensity of an electromagnetic wave (such as our light wave) is proportional to the square of the electric field's magnitude (Eq. 34-26). In our present case then, the intensity I of the emerging wave is proportional to E_y^2 and the intensity I_0 of the original wave is proportional to E^2. Hence, from Eq. 34-41 we can write $I/I_0 = \cos^2 \theta$, or

$$I = I_0 \cos^2 \theta. \qquad (34\text{-}42)$$

Let us call this the *cosine-squared rule;* we can use it *only* when the light reaching a polarizing sheet is already polarized. The transmitted intensity I is a maximum and is equal to the original intensity I_0 when the original wave is polarized parallel to the polarizing direction of the sheet (when θ in Eq. 34-42 is 0° or 180°). And I is zero when the original wave is polarized perpendicular to the polarizing direction of the sheet (when θ is 90°).

Figure 34-14 shows an arrangement in which initially unpolarized light is sent through two polarizing sheets P_1 and P_2. (Often, the first sheet is called the *polarizer,* and the second the *analyzer.*) Because the polarizing direction of P_1 is vertical, the light transmitted by P_1 to P_2 is polarized vertically. If the polarizing direction of P_2 is also vertical, then all the light transmitted by P_1 is transmitted by P_2. If the polarizing direction of P_2 is horizontal, none of the light transmitted by P_1 is transmitted by P_2. We reach the same conclusions by considering only the *relative* orientations of the two sheets: if their polarizing directions are parallel, all the light passed by the first sheet is passed by the second sheet. If those directions are perpendicular (the sheets are said to be *crossed*), no light is passed by the second sheet. These two extremes are displayed with polarized sunglasses in Fig. 34-15.

Finally, if the two polarizing directions of Fig. 34-14 make an angle between 0° and 90°, some of the light transmitted by P_1 will be transmitted by P_2. The intensity of that light is determined by Eq. 34-42.

Light can be polarized by means other than polarizing sheets, such as by reflection (discussed in Section 34-9) and by scattering from atoms or molecules. In *scattering,* light that is intercepted by an object, such as a molecule, is sent off in many, perhaps random, directions. An example is the scattering of sunlight by molecules in the atmosphere, which gives the sky its general glow.

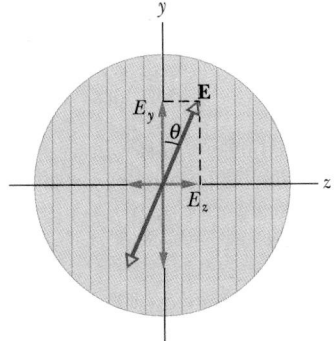

FIGURE 34-13 Polarized light approaching a polarizing sheet. The electric field **E** of the light can be resolved into components E_y (parallel to the polarizing direction of the sheet) and E_z (perpendicular to that direction). Component E_y will be transmitted by the sheet; component E_z will be absorbed.

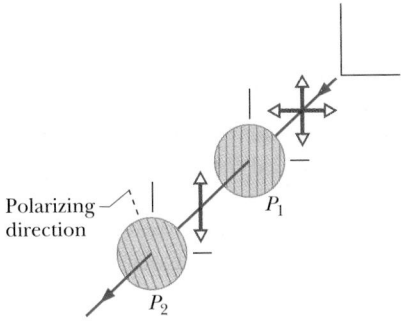

FIGURE 34-14 The light transmitted by polarizing sheet P_1 is vertically polarized, as represented by the vertical arrow. The amount of that light that is then transmitted by polarizing sheet P_2 depends on the angle between the polarization direction of that light and the polarizing direction of P_2 (indicated by the lines drawn in the sheet and by the dashed line).

(a)

(b)

FIGURE 34-15 Polarizing sunglasses consist of sheets whose polarizing directions are vertical when the sunglasses are worn. (a) Overlapping sunglasses transmit light fairly well when their polarizing directions have the same orientation, but (b) they block most of the light when they are crossed.

Although direct sunlight is unpolarized, light from much of the sky is at least partially polarized by such scattering. Bees use the polarization of sky light in navigating to and from their hives. Similarly, the Vikings used it to navigate across the North Sea when the daytime Sun was below the horizon (because of the high latitude of the North Sea). These early seafarers had discovered certain crystals (now called cordierite) that changed color when rotated in polarized light. By looking at the sky through such a crystal while rotating it about their line of sight, they could locate the hidden Sun and thus determine which way was south.

SAMPLE PROBLEM 34-3

Figure 34-16a shows a system of three polarizing sheets in the path of initially unpolarized light. The polarizing direction of the first sheet is parallel to the y axis, that of the second sheet is 60° counterclockwise from the y axis, and that of the third sheet is parallel to the x axis. What fraction of the initial intensity I_0 of the light emerges from the system, and how is that light polarized?

SOLUTION: We work the problem in steps, sheet by sheet. The original light wave is represented in Fig. 34-16b, using the head-on double-arrow representation of Fig. 34-11b. Because the light is initially unpolarized, the intensity I_1 of the light transmitted by the first sheet is given by the one-half rule (Eq. 34-40):

$$I_1 = \tfrac{1}{2}I_0.$$

The polarization of this transmitted light is (as always) parallel to the polarizing direction of the sheet transmitting it; here it is parallel to the y axis, as shown in the head-on view of Fig. 34-16c.

Because the light reaching the second sheet is polarized, the intensity I_2 of the light transmitted by that sheet is given by the cosine-squared rule (Eq. 34-42). The angle θ in the rule is the angle between the polarization direction of the entering light (parallel to the y axis) and the polarizing direction of the second sheet (60° counterclockwise from the y axis), and so θ is 60°. Then

$$I_2 = I_1 \cos^2 60°.$$

The polarization of this transmitted light is parallel to the polarizing direction of the sheet transmitting it, that is, 60° counterclockwise from the y axis, as shown in the head-on view of Fig. 34-16d.

Because this light is polarized, the intensity I_3 of the light transmitted by the third sheet is given by the cosine-squared rule. The angle θ is now the angle between the polarization direction of the entering light (Fig. 34-16d) and the polarizing direction of the third sheet (parallel to the x axis), and so $\theta = 30°$. Thus,

$$I_3 = I_2 \cos^2 30°.$$

This final transmitted light is polarized parallel to the x axis (Fig. 34-16e). We find its intensity by substituting first for I_2 and then for I_1 in the equation above:

$$I_3 = I_2 \cos^2 30° = (I_1 \cos^2 60°) \cos^2 30°$$
$$= (\tfrac{1}{2}I_0) \cos^2 60° \cos^2 30° = 0.094 I_0.$$

Thus $$\frac{I_3}{I_0} = 0.094. \qquad \text{(Answer)}$$

That is, 9.4% of the initial intensity emerges from the three-sheet system. (If we now remove the second sheet, what fraction of the initial intensity emerges from the system?)

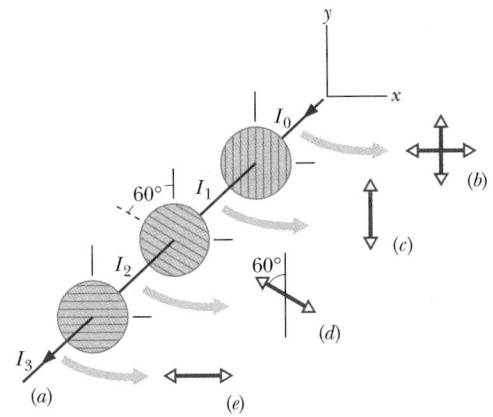

FIGURE 34-16 Sample Problem 34-3. (a) Initially unpolarized light of intensity I_0 is sent into a system of three polarizing sheets. The intensities I_1, I_2, and I_3 of the light transmitted by the sheets are indicated. Shown also are the polarizations, from head-on views, of (b) the initial light, as well as the light transmitted by (c) the first sheet, (d) the second sheet, and (e) the third sheet.

CHECKPOINT **4:** The figure shows four pairs of polarizing sheets, seen face-on. Each pair is mounted in the path of initially unpolarized light (like the three sheets in Fig. 34-16a). The polarizing direction of each sheet (indicated by the dashed line) is referenced to either a horizontal x axis or a vertical y axis. Rank the pairs according to the fraction of the initial intensity that they pass, greatest first.

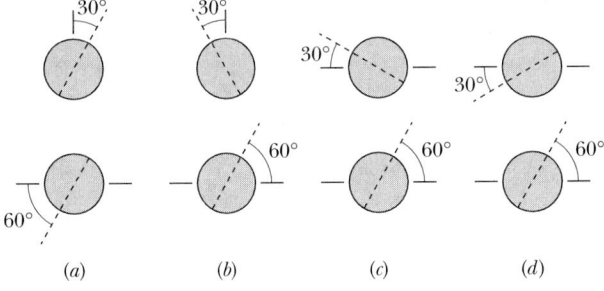

34-7 REFLECTION AND REFRACTION

Although a light wave spreads as it moves away from its source, we can often approximate its travel as being in a straight line; we did so for the light wave in Fig. 34-5a. The study of the properties of light waves under that approximation is called *geometrical optics*. For the rest of this chapter and all of Chapter 35, we shall discuss the geometrical optics of visible light.

The black-and-white photograph in Fig. 34-17a shows an example of light waves traveling in approximately

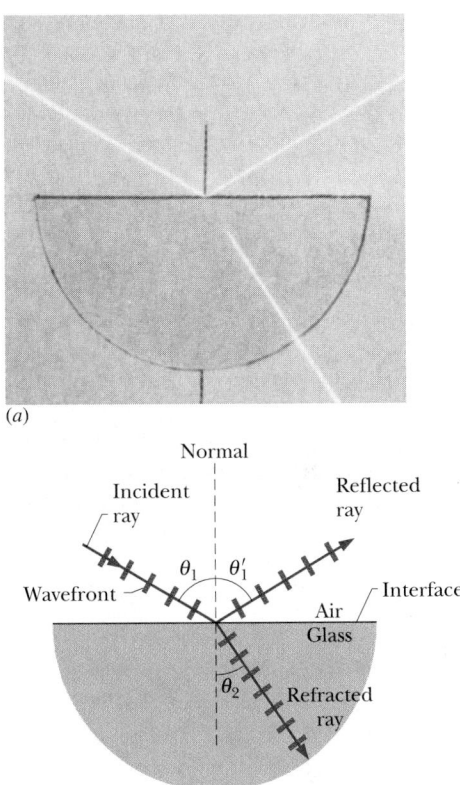

FIGURE 34-17 (a) A black-and-white photograph showing the reflection and refraction of an incident beam of light by a horizontal plane glass surface. (A portion of the refracted beam within the glass was not well photographed.) At the bottom surface, which is curved, the beam is perpendicular to the surface; so the refraction there does not bend the beam. (b) A representation of (a) using rays. The angles of incidence (θ_1), of reflection (θ_1'), and of refraction (θ_2) are marked.

straight lines. A narrow beam of light, angled downward from the left and traveling through air, encounters a *plane* (flat) glass surface. Part of the light is **reflected** by the surface, forming a beam directed upward toward the right, traveling as if the original beam had bounced from the surface. The rest of the light travels through the surface and into the glass, forming a beam directed downward to the right. Because light can travel through the glass like this, the glass is said to be *transparent*; that is, we can see through it. (In this chapter we shall consider only transparent materials.)

The travel of light through a surface (or *interface*) that separates two media is called **refraction,** and the light is said to be *refracted*. Unless an incident beam of light is perpendicular to a surface, refraction by the surface changes the light's direction of travel. For this reason, the beam is said to be ''bent'' by the refraction. Note in Fig.

The stealth aircraft F-117A is virtually invisible to radar largely because of the flat panels that are angled so as to reflect incident radar signals up or down, rather than back to the radar station.

34-17a that the bending occurs only at the surface; within the glass, the light travels in a straight line.

In Figure 34-17b, the beams of light in the photograph are represented with an *incident ray*, a *reflected ray*, and a *refracted ray*. Each ray is oriented with respect to a line, called the *normal*, that is perpendicular to the surface at the point of reflection and refraction. In Fig. 34-17b, the **angle of incidence** is θ_1, the **angle of reflection** is θ_1', and the **angle of refraction** is θ_2, all measured *relative to the normal* as shown. The plane containing the incident ray and the normal is the *plane of incidence*, which is in the plane of the page in Fig. 34-17b.

Experiment shows that reflection and refraction are governed by two laws:

Law of reflection: A reflected ray lies in the plane of incidence and has an angle of reflection equal to the angle of incidence. In Fig. 34-17b, this means that

$$\theta_1' = \theta_1 \quad \text{(reflection).} \qquad (34\text{-}43)$$

(We shall now drop the prime on the angle of reflection.)
Law of refraction: A refracted ray lies in the plane of incidence and has an angle of refraction that is related to the angle of incidence by

$$n_2 \sin \theta_2 = n_1 \sin \theta_1 \quad \text{(refraction).} \qquad (34\text{-}44)$$

Here each of the symbols n_1 and n_2 is a dimensionless constant called the **index of refraction** that is associated with a medium involved in the refraction. We derive this equation, called Snell's law, in Chapter 36. As we shall discuss there, the index of refraction of a medium is equal to c/v, where v is the speed of light in that medium and c is its speed in vacuum.

Table 34-1 gives the indices of refraction of vacuum and some common substances. For vacuum, n is defined to be exactly 1; for air, n is very close to 1.0 (an approximation we shall often make). Nothing has an index of refraction below 1.

We can rearrange Eq. 34-44 as

$$\sin \theta_2 = \frac{n_1}{n_2} \sin \theta_1 \qquad (34\text{-}45)$$

to compare the angle of refraction θ_2 with the angle of incidence θ_1. We can then see that the relative value of θ_2 depends on the relative values of n_2 and n_1. In fact, we can have three basic results:

1. If n_2 is equal to n_1, then θ_2 is equal to θ_1. In this case, refraction does not bend the light beam, which continues in an *undeflected direction*, as in Fig. 34-18a.

2. If n_2 is greater than n_1, then θ_2 is less than θ_1. In this

TABLE 34-1 SOME INDICES OF REFRACTION[a]

MEDIUM	INDEX	MEDIUM	INDEX
Vacuum	exactly 1	Typical crown glass	1.52
Air (STP)[b]	1.00029	Sodium chloride	1.54
Water (20°C)	1.33	Polystyrene	1.55
Acetone	1.36	Carbon disulfide	1.63
Ethyl alcohol	1.36	Heavy flint glass	1.65
Sugar solution (30%)	1.38	Sapphire	1.77
Fused quartz	1.46	Heaviest flint glass	1.89
Sugar solution (80%)	1.49	Diamond	2.42

[a]For a wavelength of 589 nm (yellow sodium light).

[b]STP means "standard temperature (0°C) and pressure (1 atm)."

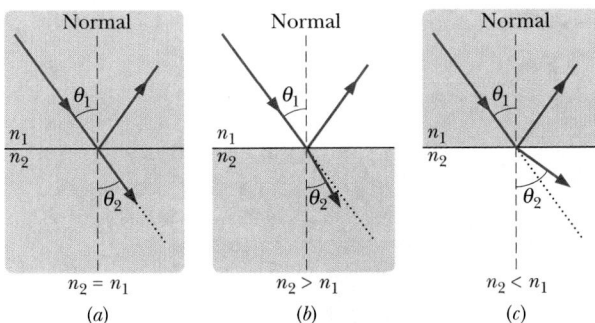

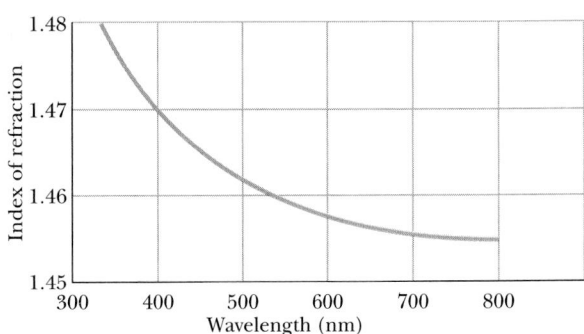

FIGURE 34-18 Light refracting from a medium with an index of refraction n_1 and into a medium with an index of refraction n_2. (a) The beam does not bend when $n_2 = n_1$; the refracted light then travels in the *undeflected direction* (the dotted line), which is the same as the direction of the incident beam. The beam bends (b) toward the normal when $n_2 > n_1$ and (c) away from the normal when $n_2 < n_1$.

FIGURE 34-19 The index of refraction as a function of wavelength for fused quartz. The graph indicates that a beam of short-wavelength light, for which the index of refraction is higher, is bent more upon entering or leaving quartz than a beam of long-wavelength light.

case, refraction bends the light beam away from the undeflected direction and toward the normal, as in Fig. 34-18b.

3. If n_2 is less than n_1, then θ_2 is greater than θ_1. In this case, refraction bends the light beam away from the undeflected direction and away from the normal, as in Fig. 34-18c.

Refraction cannot bend a beam so much that the refracted ray is on the same side of the normal as the incident ray.

Chromatic Dispersion

The index of refraction n encountered by light in any medium except vacuum depends on the wavelength of the light. The dependence of n on wavelength implies that when a light beam consists of rays of different wavelengths, the rays will be refracted at different angles by a surface. That is, the light will be spread out by the refraction. This spreading of light is called **chromatic dispersion,** in which "chromatic" refers to the colors associated with the individual wavelengths and "dispersion" refers to the spreading of the light according to its wavelengths or colors. The refractions of Figs. 34-17 and 34-18 do not show chromatic dispersion because the beams are *monochromatic* (of a single wavelength or color).

Generally, the index of refraction in a given medium is *greater* for a shorter wavelength (corresponding to, say, blue light) than for a longer wavelength (say, red light). As an example, Fig. 34-19 shows how the index of refraction for fused quartz depends on the wavelength of light. Such dependence means that when a beam with waves of both blue and red light is refracted through a surface, such as from air into quartz or vice versa, the blue *component* (the ray corresponding to the wave of blue light) bends more than the red component.

A beam of *white light* consists of components of all (or nearly all) the colors in the visible spectrum with approximately uniform intensities. When you see such a beam, you perceive white rather than the individual colors. In Fig. 34-20a, a beam of white light in air is incident on a glass surface. (Because the pages of this book are white, a beam of white light is represented with a gray ray here. Also, a

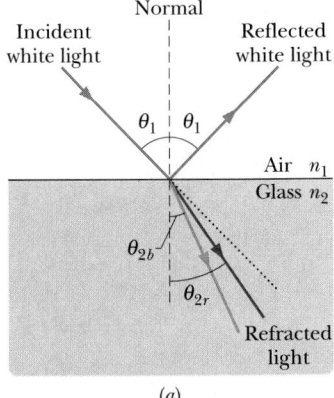

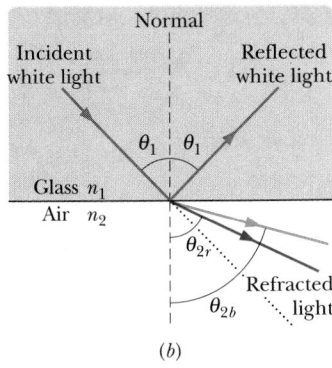

FIGURE 34-20 Chromatic dispersion of white light. The blue component is bent more than the red component. (a) Passing from air to glass, the blue component ends up with the smaller angle of refraction. (b) Passing from glass to air, the blue component ends up with the larger angle of refraction.

beam of monochromatic light is generally represented with a red ray.) Of the refracted light in Fig. 34-20a, only the red and blue components are shown. Because the blue component is bent more than the red component, the angle of refraction θ_{2b} for the blue component is *smaller* than the angle of refraction θ_{2r} for the red component. (Remember, angles are measured relative to the normal.) In Fig. 34-20b, a ray of white light in glass is incident on a glass–air interface. Again, the blue component is bent more than the red component, but now $\theta_{2b} > \theta_{2r}$.

To increase the color separation, we can use a solid glass prism with a triangular cross section, as in Fig. 34-21a. The dispersion at the first surface (on the left in Fig. 34-21a,b) is then enhanced by that at the second surface.

The most charming example of chromatic dispersion is a rainbow. When white sunlight is intercepted by a falling raindrop, some of the light refracts into the drop, reflects from the drop's inner surface, and then refracts out of the drop (Fig. 34-22). As with a prism, the first refraction separates the sunlight into its component colors, and the second refraction increases the separation.

The rainbow you see is formed by light refracted by many such drops: the red comes from drops angled slightly higher in the sky, the blue from drops angled slightly lower, and the intermediate colors from drops at interme-

(a)

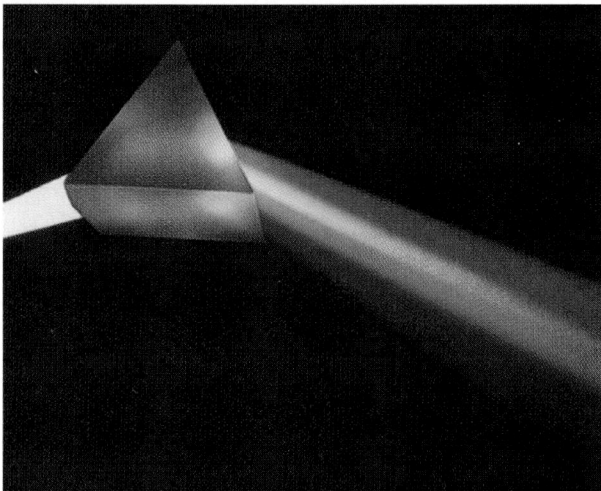

(a)

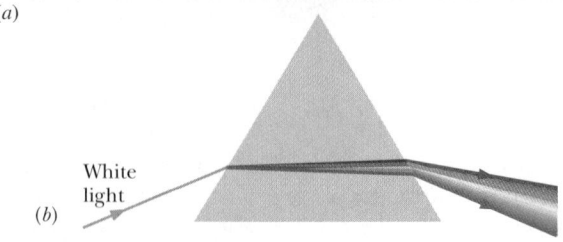

White light

(b)

FIGURE 34-21 (a) A triangular prism separating white light into its component colors. (b) Chromatic dispersion occurs at the first surface and is increased at the second surface.

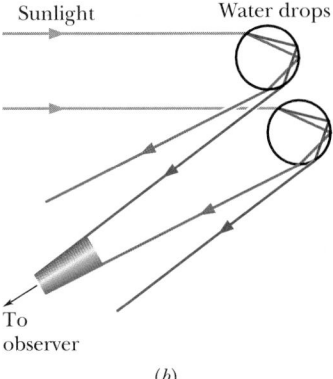

Sunlight   Water drops

To observer

(b)

FIGURE 34-22 (a) A rainbow is always a circular arc that is centered on the direction you would look if you looked directly away from the Sun. Under normal conditions, you are lucky if you see a long arc, but if you are looking downward from an elevated position, you might actually see a full circle. (b) The separation of colors when sunlight refracts into and out of falling raindrops leads to a rainbow. The figure shows the situation for the Sun on the horizon (the rays of sunlight are then horizontal). The paths of red and blue rays from two drops are indicated. Many other drops also contribute red and blue rays, as well as the intermediate colors of the visible spectrum.

diate angles. All the drops sending separated colors to you are angled at about 42° from a point that is directly opposite the Sun in your view. If the rainfall is extensive and brightly lit, you see a circular arc of color, with red on top and blue on bottom. Your rainbow is a personal one, because another observer intercepts light from other drops.

$\mathbb{C}$HECKPOINT **5:** Which of the three drawings (if any) show physically possible refraction?

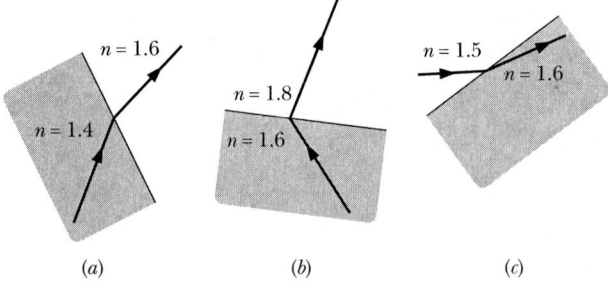

(a)   (b)   (c)

SAMPLE PROBLEM 34-4

A submerged swimmer is looking directly upward through the air–water interface in a pool.

(a) Over what range of angles do rays reach the swimmer's eyes from light sources external to the water? Assume that the light is monochromatic and that the index of refraction of water is 1.33.

SOLUTION: Light reaches the swimmer's eyes from external sources by refracting through the air–water interface, in accordance with Eq. 34-44. Let us associate subscript 1 in that equation with air and subscript 2 with water. Then, with $n_1 = 1.00$ and $n_2 = 1.33$, we have $n_2 > n_1$. So, the refraction of light rays at the air–water interface bends the rays *toward* the normal, as in Fig. 34-18b. The bending of an arbitrary light ray, with angle of incidence θ_1 and angle of refraction θ_2, is shown in Fig. 34-23a; the swimmer's eyes are located at point E. Note that the refracted ray also makes an angle θ_2 with the vertical at E.

From Eq. 34-44 we know that the angle of refraction θ_2 depends on the angle of incidence θ_1 and is given by

$$\sin \theta_2 = \frac{n_1}{n_2} \sin \theta_1. \qquad (34\text{-}46)$$

So, to find the angles at which rays reach point E from external sources, we must consider the range of values of θ_1. That will give us the range of values of θ_2.

The least value of θ_1 is 0°, which is the value for an incident ray that is perpendicular to the air–water interface. For that value Eq. 34-46 gives us

$$\sin \theta_2 = \frac{1.00}{1.33} \sin 0° = 0,$$

or $\theta_2 = 0°.$

Incident ray A in Fig. 34-23b shows this refraction situation: the incident ray is not bent; it reaches E along the vertical through E.

The maximum value of θ_1 is approximately 90°, which is the value for an incident ray that is almost parallel to the air–water interface. Equation 34-46 now gives us

$$\sin \theta_2 = \frac{1.00}{1.33} \sin 90° = 0.752,$$

or $\theta_2 = 48.8°.$

In Fig. 34-23b, the two incident rays B show this refraction situation: these two rays are incident on the interface at the greatest possible angle (90°) but are 48.8° to the vertical when they reach the swimmer.

Figure 34-23b shows only one plane of the swimmer's field of view. But if we rotate that plane about the vertical, we

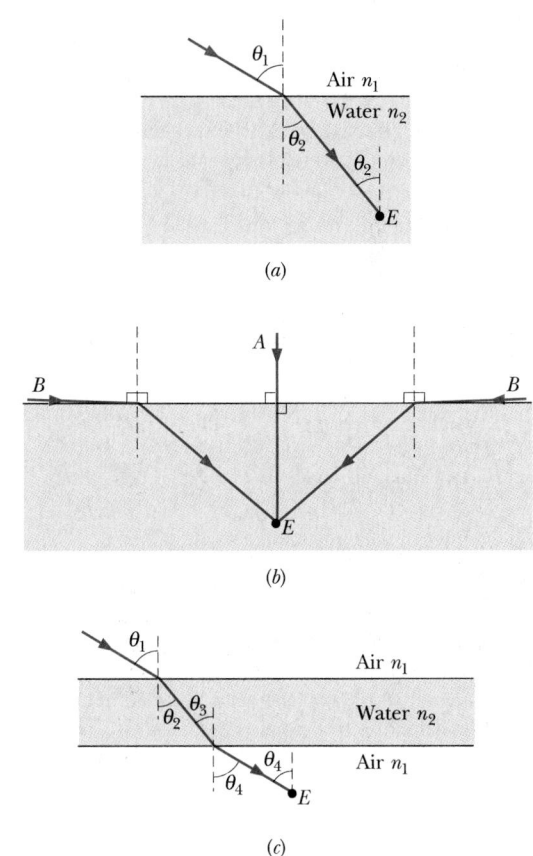

FIGURE 34-23 Sample Problem 34-4. (a) A ray of light, refracted into water, reaches a swimmer's eyes located at point E. The angle of the ray at E is measured relative to the vertical through E. (b) Ray A, perpendicular to the air–water interface, and rays B, almost parallel to the interface, reach point E. (c) The swimmer's eyes at point E are now in air that is trapped by a mask of transparent plastic. A ray reaching E refracts at two air–water interfaces.

see that it then gives the swimmer's entire field of view. Thus, all the refracted rays reaching the swimmer from external sources are contained within a vertical cone that has its apex located at E and intersects the air–water interface in a circle directly above E. Moreover, the apex angle is

$$2\theta_2 = 97.6° \approx 100°. \qquad \text{(Answer)}$$

The swimmer's entire view of the external world comes through the overhead circle, which acts as a personal window for the swimmer.

(b) The swimmer now wears a swimming mask over his eyes; the thin flat layer of transparent plastic through which the swimmer looks is horizontal; and air fills the interior of the mask. Over what range of angles do rays now reach the swimmer's eyes from light sources external to the water? (Neglect refraction by the plastic faceplate of the mask; its inclusion would not alter the result.)

SOLUTION: To answer, we examine the refraction of an arbitrary light ray as shown in Figure 34-23c. As before, the ray refracts from air into water, but now, to reach point E, it must refract a second time, from the water into the air held by the mask. Let the angle of incidence at the second refraction be θ_3 and the angle of refraction be θ_4. We seek the angle at which the ray reaches point E, which is equal to θ_4.

Because the two air–water interfaces in Fig. 34-23c are parallel, $\theta_3 = \theta_2$. From Eq. 34-44, the angle of refraction θ_4 is given by

$$\sin \theta_4 = \frac{n_2}{n_1} \sin \theta_3.$$

Substituting θ_2 for θ_3 and then substituting for $\sin \theta_2$ from Eq. 34-46, we find

$$\sin \theta_4 = \frac{n_2}{n_1} \sin \theta_2 = \frac{n_2 n_1}{n_1 n_2} \sin \theta_1 = \sin \theta_1,$$

which (in this situation) gives us

$$\theta_4 = \theta_1.$$

In words, the arbitrary ray in Fig. 34-23c reaches point E traveling parallel to its original direction. So do all other rays reaching E from external sources. Thus the range of those rays is from about 90° on one side of the vertical to about 90° on the other side. This means that with the mask, the swimmer sees the external world as if the water were not present, and not compressed into a 100° cone.

34-8 TOTAL INTERNAL REFLECTION

Figure 34-24 shows rays of monochromatic light from a point source S in glass incident on the interface between the glass and air. For ray a, which is perpendicular to the interface, part of the light reflects at the interface and the rest travels through it with no change in direction.

For rays b through e, which have progressively larger angles of incidence at the interface, there are also both

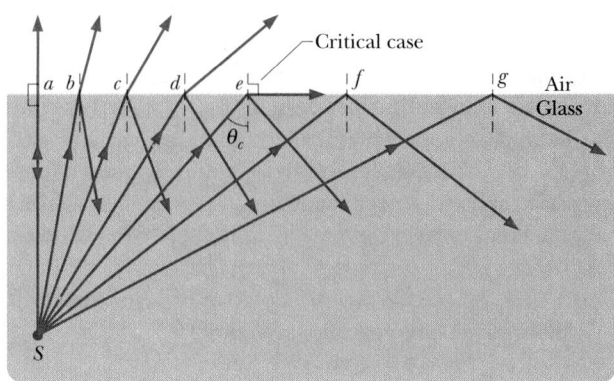

FIGURE 34-24 Total internal reflection of light from a point source S in glass occurs for all angles of incidence greater than the critical angle θ_c. At the critical angle, the refracted ray points along the air–glass interface.

reflection and refraction at the interface. As the angle of incidence increases, the angle of refraction increases; for ray e it is 90°, which means that the refracted ray points directly along the interface. The angle of incidence giving this situation is called the **critical angle** θ_c. For angles of incidence larger than θ_c, such as for rays f and g, there is no refracted ray and *all* the light is reflected; this effect is called **total internal reflection.**

To find θ_c, we use Eq. 34-44: we arbitrarily associate subscript 1 with the glass and subscript 2 with the air, and then we substitute θ_c for θ_1 and 90° for θ_2, finding

$$n_1 \sin \theta_c = n_2 \sin 90°,$$

which gives us

$$\theta_c = \sin^{-1} \frac{n_2}{n_1} \qquad \text{(critical angle)}. \qquad (34\text{-}47)$$

Because the sine of an angle cannot exceed unity, n_2 cannot exceed n_1 in this equation. This restriction tells us that total internal reflection cannot occur when the incident light is in the medium of lower index of refraction. If source S were in the air in Fig. 34-24, all its rays that are incident on the air–glass interface (including f and g) would be both reflected *and* refracted at the interface.

Total internal reflection has found many applications in medical technology. For example, a physician can search for an ulcer in the stomach of a patient by running two thin bundles of *optical fibers* (Fig. 34-25) down the patient's throat. Light introduced at the outer end of one bundle undergoes repeated total internal reflection within the fibers so that, even though the bundle provides a curved path, the light ends up illuminating the interior of the stomach. Some of the light reflected from the interior then comes back up the second bundle in a similar way, to be detected and converted to an image on a monitor's screen for the physician to view.

FIGURE 34-25 Light sent into one end of an optical fiber is transmitted to the opposite end with little loss of the light through the sides of the fiber, because most of the light undergoes repeated total internal reflection along those sides.

SAMPLE PROBLEM 34-5

Figure 34-26 shows a triangular prism of glass in air; a ray incident perpendicularly to one face is totally reflected at the glass–air interface indicated. If θ_1 is 45°, what can you say about the index of refraction n of the glass?

SOLUTION: Using Eq. 34-47, approximating the index of refraction n_2 of air as unity, and substituting the index of refraction n of the glass for n_1, we find for the critical angle θ_c:

$$\theta_c = \sin^{-1} \frac{n_2}{n_1} = \sin^{-1} \frac{1}{n}.$$

Since total internal reflection occurs, θ_c must be less than θ_1, which is 45°. Then

$$\sin^{-1} \frac{1}{n} < 45°,$$

which gives us

$$\frac{1}{n} < \sin 45°,$$

or

$$n > \frac{1}{\sin 45°} = 1.4. \qquad \text{(Answer)}$$

The index of refraction of the glass must be greater than 1.4; otherwise total internal reflection would not occur for the incident ray shown.

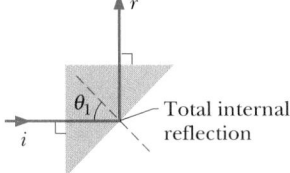

FIGURE 34-26 Sample Problem 34-5. The incident ray i is totally internally reflected at the glass–air interface, becoming the reflected ray r.

CHECKPOINT **6:** Suppose the prism in Sample Problem 34-5 has the index of refraction $n = 1.4$. Does the light still totally internally reflect if we keep the incident ray horizontal but rotate the prism (a) 10° clockwise and (b) 10° counterclockwise in Fig. 34-26?

34-9 POLARIZATION BY REFLECTION

You can increase and decrease the glare you see in sunlight that has been reflected from, say, water by looking through a polarizing sheet (such as a polarizing sunglass lens) and then rotating the sheet's polarizing axis around your line of sight. You can do so because reflected light is fully or partially polarized by the reflection from a surface.

Figure 34-27 shows a ray of unpolarized light incident on a glass surface. Let us resolve the electric field vectors of the light into two components. The *perpendicular components* are perpendicular to the plane of incidence and thus also to the page in Fig. 34-27; these components are represented with dots (as if we see the tips of the vectors). The *parallel components* are parallel to the plane of incidence and the page; they are represented with double-headed arrows. Because the light is unpolarized, these two components are of equal magnitude.

In general, the reflected light also has both components but with unequal magnitudes. This means that the reflected light is partially polarized—the electric fields

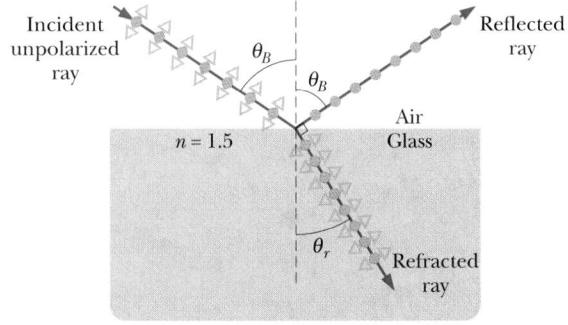

    •   Component perpendicular to page
    ◁→   Component parallel to page

FIGURE 34-27 A ray of unpolarized light in air is incident on a glass surface at the Brewster angle θ_B. The electric field vectors are resolved into components perpendicular to the page (the plane of incidence) and components parallel to the page. The reflected light consists only of components perpendicular to the page and is thus polarized in that direction. The refracted light consists of the original components parallel to the page and weaker components perpendicular to the page; this light is partially polarized.

oscillating along one direction have greater amplitudes than those oscillating along other directions. However, when the light is incident at a particular incident angle, called the *Brewster angle* θ_B, the reflected light has only perpendicular components, as shown in Fig. 34-27. The reflected light is then fully polarized perpendicular to the plane of incidence. The parallel components of the incident light do not disappear; they and perpendicular components form the light that is refracted through the glass surface.

Glass, water, and the other dielectric materials discussed in Section 26-7 can partially and fully polarize light by reflection. When you intercept sunlight reflected from such a surface, you see a bright spot (the glare) on the surface where the reflection takes place. If the surface is horizontal as in Fig. 34-27, the reflected light is partially or fully polarized horizontally. To eliminate such glare from horizontal surfaces, the lenses in polarizing sunglasses are mounted with their polarizing direction vertical.

Brewster's Law

For light incident at the Brewster angle θ_B, we find experimentally that the reflected and refracted rays are perpendicular to each other. Because the reflected ray is reflected at the angle θ_B in Fig. 34-27 and the refracted ray is at angle θ_r, we have

$$\theta_B + \theta_r = 90°.$$

These two angles can also be related with Eq. 34-44. Arbitrarily assigning subscript 1 in Eq. 34-44 to the material through which the incident and reflected rays travel, we have, from that equation,

$$n_1 \sin \theta_B = n_2 \sin \theta_r.$$

Combining these equations leads to

$$n_1 \sin \theta_B = n_2 \sin(90° - \theta_B) = n_2 \cos \theta_B,$$

which gives us

$$\theta_B = \tan^{-1} \frac{n_2}{n_1} \quad \text{(Brewster angle).} \quad (34\text{-}48)$$

(Note carefully that the subscripts in Eq. 34-48 are *not* arbitrary because of our decision as to their meanings.) If the incident and reflected rays travel *in air,* we can approximate n_1 as unity and let n represent n_2 in order to write Eq. 34-48 as

$$\theta_B = \tan^{-1} n \quad \text{(Brewster's law).} \quad (34\text{-}49)$$

This simplified version of Eq. 34-48 is known as **Brewster's law.** Like θ_B, it is named after Sir David Brewster, who found both experimentally in 1812.

SAMPLE PROBLEM 34-6

We wish to use a glass plate with index of refraction $n = 1.57$ to polarize light in air.

(a) At what angle of incidence is the light reflected by the glass fully polarized?

SOLUTION: Because the glass is in air, we can use Eq. 34-49 to find the Brewster angle:

$$\theta_B = \tan^{-1} n = \tan^{-1} 1.57 = 57.5°. \quad \text{(Answer)}$$

(b) What is the corresponding angle of refraction?

SOLUTION: Since $\theta_B + \theta_r = 90°$, we have

$$\theta_r = 90° - \theta_B = 90° - 57.5° = 32.5°. \quad \text{(Answer)}$$

REVIEW & SUMMARY

Electromagnetic Waves

An electromagnetic wave consists of oscillating electric and magnetic fields. The various possible frequencies of electromagnetic waves form a *spectrum*, a small part of which is visible light. An electromagnetic wave traveling along an x axis has an electric field **E** and a magnetic field **B** with magnitudes that depend on x and t:

$$E = E_m \sin(kx - \omega t)$$

and $\qquad\qquad B = B_m \sin(kx - \omega t), \qquad (34\text{-}1, 34\text{-}2)$

where E_m and B_m are the amplitudes of **E** and **B**. The electric field induces the magnetic field and vice versa. The speed of any electromagnetic wave in vacuum is c, which can be written as

$$c = \frac{E}{B} = \frac{1}{\sqrt{\mu_0 \epsilon_0}}, \qquad (34\text{-}5, 34\text{-}3)$$

where E and B are the simultaneous magnitudes of the fields.

Energy Flow

The rate per unit area at which energy is transported via an electromagnetic wave is given by the Poynting vector **S**:

$$\mathbf{S} = \frac{1}{\mu_0} \mathbf{E} \times \mathbf{B}. \qquad (34\text{-}21)$$

The direction of **S** (and thus of the wave's travel and the energy transport) is perpendicular to the directions of both **E** and **B**. The time-averaged rate per unit area at which energy is transported is $\bar{S}$, which is called the *intensity I* of the wave:

$$I = \frac{1}{c\mu_0} E_{rms}^2, \qquad (34\text{-}26)$$

in which $E_{rms} = E_m/\sqrt{2}$. A *point source* of electromagnetic waves emits the waves *isotropically,* that is, with equal intensity in all directions. The intensity of the waves at distance r from a point source of power P_s is

$$I = \frac{P_s}{4\pi r^2}. \qquad (34\text{-}27)$$

Radiation Pressure

When a surface intercepts electromagnetic radiation, a force and a pressure are exerted on the surface. If the radiation is totally absorbed by the surface, the force is

$$F = \frac{IA}{c} \qquad \text{(total absorption)}, \qquad (34\text{-}32)$$

in which I is the intensity of the radiation and A is the area of the surface perpendicular to the path of the radiation. If the radiation is totally reflected back along its original path, the force is

$$F = \frac{2IA}{c} \qquad \text{(total reflection back along path)}. \qquad (34\text{-}33)$$

The radiation pressure p_r is the force per unit area:

$$p_r = \frac{I}{c} \qquad \text{(total absorption)} \qquad (34\text{-}34)$$

and

$$p_r = \frac{2I}{c} \qquad \begin{array}{l}\text{(total reflection)} \\ \text{back along path).}\end{array} \qquad (34\text{-}35)$$

Polarization

Electromagnetic waves are **polarized** if their electric field vectors are all in a single plane, called the *plane of oscillation*. Light waves from common sources are not polarized, that is, they are **unpolarized** or **randomly polarized.**

Polarizing Sheets

When a polarizing sheet is placed in the path of light, only electric field components of the light parallel to the sheet's **polarizing direction** are *transmitted* by the sheet; components perpendicular to the polarizing direction are absorbed. The light that emerges from a polarizing sheet is polarized parallel to the polarizing direction of the sheet.

If the original light is initially unpolarized, the transmitted intensity I is half the original intensity I_0:

$$I = \tfrac{1}{2}I_0. \qquad (34\text{-}40)$$

If the original light is initially polarized, the transmitted intensity depends on the angle θ between the polarization direction of the original light and the polarizing direction of the sheet:

$$I = I_0 \cos^2 \theta. \qquad (34\text{-}42)$$

Geometrical Optics

Geometrical optics is an approximate treatment in which light waves are represented as straight-line rays.

Reflection and Refraction

When a light ray encounters a boundary between two transparent media, a **reflected** ray and a **refracted** ray generally appear. Both rays remain in the plane of incidence. The **angle of reflection** is equal to the angle of incidence, and the **angle of refraction** is related to the angle of incidence by

$$n_1 \sin \theta_1 = n_2 \sin \theta_2 \qquad \text{(refraction)}, \qquad (34\text{-}44)$$

where n_1 and n_2 are the indices of refraction of the media in which the incident and refracted rays travel.

Total Internal Reflection

A wave encountering a boundary across which the index of refraction decreases will experience **total internal reflection** if the angle of incidence exceeds a **critical angle** θ_c, where

$$\theta_c = \sin^{-1} \frac{n_2}{n_1} \qquad \text{(critical angle)}. \qquad (34\text{-}47)$$

Polarization by Reflection

A reflected wave will be fully **polarized,** with its **E** vectors perpendicular to the plane of incidence, if it strikes a boundary at the **Brewster angle** θ_B, where

$$\theta_B = \tan^{-1} \frac{n_2}{n_1} \qquad \text{(Brewster angle)}. \qquad (34\text{-}48)$$

QUESTIONS

1. If the magnetic field of a light wave oscillates parallel to a y axis and is given by $B_y = B_m \sin(kz - \omega t)$, (a) in what direction does the wave travel and (b) parallel to which axis does the associated electric field oscillate?

2. Figure 34-28 shows the electric and magnetic fields of an electromagnetic wave at a certain instant. Is the wave traveling into the page or out of it?

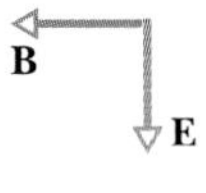

B

E

FIGURE 34-28 Question 2.

3. (a) Fig. 34-29 shows light reaching a polarizing sheet whose polarizing direction is parallel to a y axis. We shall rotate the

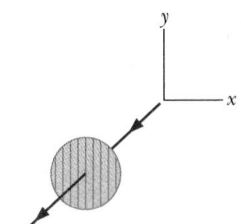

FIGURE 34-29 Question 3.

sheet 40° clockwise about the light's indicated line of travel. During this rotation, does the fraction of the initial light intensity passed by the sheet increase, decrease, or remain the same if the light is (a) initially unpolarized, (b) initially polarized parallel to the x axis, and (c) initially polarized parallel to the y axis?

4. Light is sent through the two-sheet polarizing system of Fig. 34-30. If the ratio of the emerging intensity to the initial intensity is 0.7, is the light initially polarized or unpolarized?

5. Initially unpolarized light is sent through the two-sheet polarizing system of Fig. 34-30. The emerging light is polarized 20° clockwise from the y axis and has half of the initial intensity. What are the polarizing directions of the sheets?

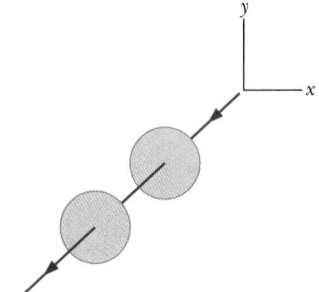

FIGURE 34-30 Questions 4, 5, and 7.

6. In Fig. 34-16a, start with light that is initially polarized parallel to the x axis, and write the ratio of its final intensity I_3 to its initial intensity I_0 as $I_3/I_0 = A \cos^n \theta$. What are A, n, and θ if we rotate the polarizing direction of the first sheet (a) 60° counterclockwise and (b) 90° clockwise from what is shown?

7. Three polarizing sheets are positioned like the two sheets in Fig. 34-30, and initially unpolarized light is sent into the system. How many different final intensities can be produced if the polarizing directions of the sheets are the following: one is parallel to the y axis, one is rotated 20° clockwise from the y axis about the light's line of travel, and one is rotated 20° in the opposite direction about that line of travel.

8. Suppose we rotate the second sheet in Fig. 34-16a, starting with its polarization direction aligned with the y axis ($\theta = 0$) and ending with its polarization direction aligned with the x axis ($\theta = 90°$). Which of the curves in Fig. 34-31 best shows the intensity of the light through the three-sheet system during this 90° rotation?

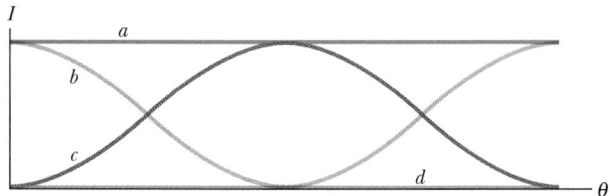

FIGURE 34-31 Question 8.

9. Figure 34-32 shows the multiple reflections of a light ray along a glass corridor where the walls are either parallel or perpendicular to one another. If the angle of incidence at point a is 30°, what are the angles of reflection at points b, c, d, e, and f?

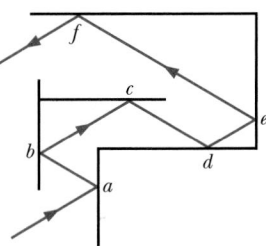

FIGURE 34-32 Question 9.

10. Figure 34-33 shows rays of monochromatic light passing through three materials a, b, and c. Rank the materials according to their indices of refraction, greatest first.

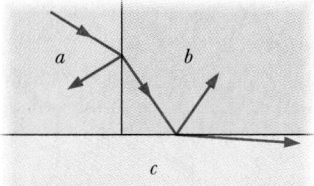

FIGURE 34-33 Question 10.

11. In Fig. 34-34, light travels from material a, through three layers of other materials with surfaces parallel to one another, and then back into another layer of material a. The refractions (but not the associated reflections) at the surfaces are shown. Rank the materials according to their indices of refraction, greatest first.

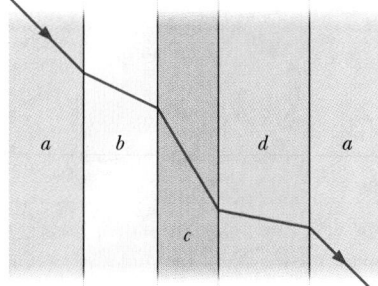

FIGURE 34-34 Question 11.

12. Which of the three parts of Fig. 34-35 show physically possible refraction?

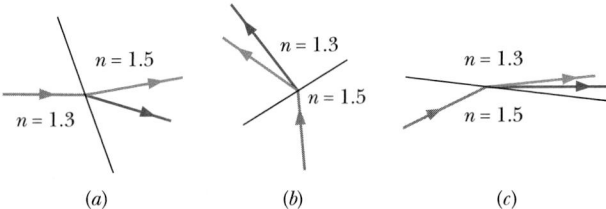

FIGURE 34-35 Question 12.

13. (a) Figure 34-36a shows a ray of sunlight that just barely passes a vertical stick in a pool of water. Does that ray end in the general region of point a or point b? (b) Does the red or the blue component of the ray end up closer to the stick? (c) Figure 34-36b shows a flat object (such as a double-edged razor blade) floating in shallow water and illuminated vertically. The object's weight

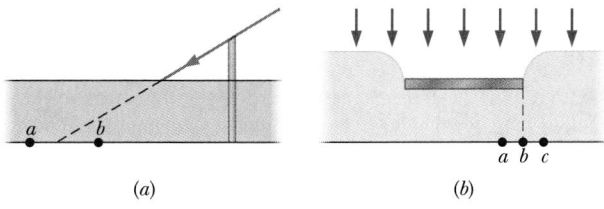

FIGURE 34-36 Question 13.

causes the water surface to curve as shown. In which general region (*a*, *b*, or *c*) is the edge of the object's shadow? (To the right of the edge of the shadow, many rays of sunlight are concentrated and produce an especially bright region, said to be a *caustic.*)

14. Figure 34-22 shows some of the rays of sunlight responsible for the *primary rainbow* (which involves one reflection inside each water drop). A fainter, less frequent *secondary rainbow* (involving two reflections inside each water drop) can appear above a primary rainbow, formed by rays entering and exiting water drops as shown in Fig. 34-37 (without color indicated). Which ray, *a* or *b*, corresponds to red light?

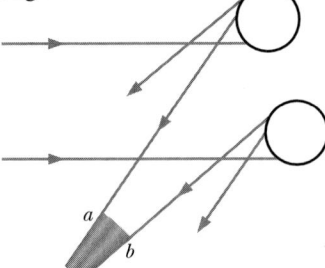

FIGURE 34-37 Question 14.

15. Figure 34-38 shows four long horizontal layers of different materials, with air above and below them. The index of refraction of each material is given. Rays of light are sent into the left ends of each layer as shown. In which layer (give the index of refrac-

tion) is there the possibility of totally trapping the light in that layer so that, after many reflections, all the light reaches the right end of the layer?

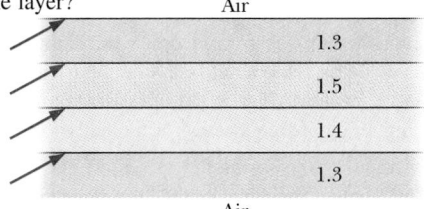

FIGURE 34-38 Question 15.

16. Figure 34-39*a* shows an overhead view of a rectangular room with fully reflecting walls. The room length *L* and width *W* are both integer numbers of units. You fire a laser from corner *a* at 45° to the walls. Small clay armadillos are located in the other corners. You are concerned as to whether your shot will hit one of those targets or yourself. Figure 34-39*b* shows a way to find out by drawing repeated reflections of the room and extending a straight light path through them. Which corner is hit depends on the ratio *L/W*, once that ratio is reduced to lowest terms. (For example, 4/2 reduces to 2/1.) Figure 34-39*b* is for *L/W* = 2/1. We see that the target in corner *d* is hit after one reflection. Determine which corner is hit for any (reduced) *L/W* in the form of (a) even number/odd number, (b) odd number/even number, and (c) odd number/odd number.

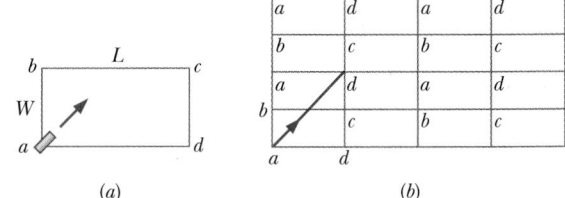

FIGURE 34-39 Question 16.

EXERCISES & PROBLEMS

SECTION 34-1 Maxwell's Rainbow

1E. Project Seafarer was an ambitious program to construct an enormous antenna, buried underground on a site about 4000 square miles in area. Its purpose was to transmit signals to submarines while they were deeply submerged. If the effective wavelength was 1.0×10^4 Earth radii, what would be (a) the frequency and (b) the period of the radiations emitted? Ordinarily electromagnetic radiations do not penetrate very far into conductors such as seawater.

2E. (a) How long does it take a radio signal to travel 150 km from a transmitter to a receiving antenna? (b) We see a full Moon by reflected sunlight. How much earlier did the light that enters our eye leave the Sun? The Earth–Moon and Earth–Sun distances are 3.8×10^5 km and 1.5×10^8 km. (c) What is the round-trip travel time for light between Earth and a spaceship

orbiting Saturn, 1.3×10^9 km distant? (d) The Crab nebula, which is about 6500 light-years (ly) distant, is thought to be the result of a supernova explosion recorded by Chinese astronomers in A.D. 1054. In approximately what year did the explosion actually occur?

3E. (a) The wavelength of the most energetic x rays produced when electrons are accelerated to a kinetic energy of 18 GeV in the Stanford Linear Accelerator and then slam into a solid target is 0.067 fm. What is the frequency of these x rays? (b) A VLF (very low frequency) radio wave has a frequency of only 30 Hz. What is its wavelength?

4E. (a) At what wavelengths does the eye of a standard observer have half its maximum sensitivity? (b) What are the wavelength, the frequency, and the period of the light for which the eye is the most sensitive?

5E. In Fig. 34-1, verify that the uniform spaces between successive powers of 10 must be the same on the wavelength scale and on the frequency scale.

6E. A certain helium–neon laser emits red light in a narrow band of wavelengths centered at 632.8 nm and with a "width" (such as on the scale of Fig. 34-1) of 0.0100 nm. What is the corresponding range of frequencies for the emission?

7P. One method for measuring the speed of light, based on observations by Roemer in 1676, consisted in observing the apparent times of revolution of one of the moons of Jupiter. The true period of revolution is 42.5 h. (a) Taking into account the finite speed of light, how would you expect the apparent time for one revolution to change as Earth moves in its orbit from point x to point y in Fig. 34-40? (b) What observations would be needed to compute the speed of light? Neglect the motion of Jupiter in its orbit. Figure 34-40 is not drawn to scale.

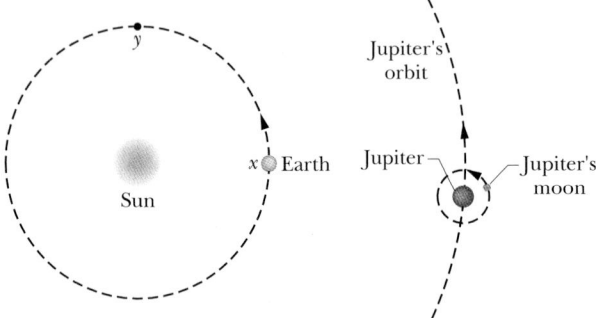

FIGURE 34-40 Problem 7.

SECTION 34-2 The Traveling Electromagnetic Wave, Qualitatively

8E. What is the wavelength of the electromagnetic wave emitted by the oscillator–antenna system of Fig. 34-3 if $L = 0.253$ μH and $C = 25.0$ pF?

9E. What inductance must be connected to a 17 pF capacitor in an oscillator capable of generating 550 nm (i.e., visible) electromagnetic waves? Comment on your answer.

10P. Figure 34-41 shows an LC oscillator connected by a transmission line to an antenna of a so-called *magnetic* dipole type. Compare it with Fig. 34-3, which shows a similar arrangement but with an *electric* dipole type of antenna. (a) What is the basis for the names of these two antenna types? (b) Draw a figure like Fig. 34-4 to describe the electromagnetic wave that sweeps past an observer at point P in Fig. 34-41.

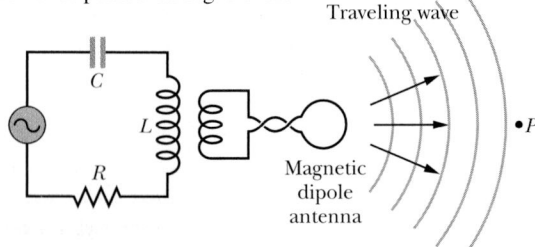

FIGURE 34-41 Problem 10.

SECTION 34-3 The Traveling Electromagnetic Wave, Quantitatively

11E. A plane electromagnetic wave has a maximum electric field of 3.20×10^{-4} V/m. Find the maximum magnetic field.

12E. The electric field of a certain plane electromagnetic wave is given by $E_x = 0$; $E_y = 0$; $E_z = 2.0 \cos[\pi \times 10^{15}(t - x/c)]$, with $c = 3.0 \times 10^8$ m/s and all quantities in SI units. The wave is propagating in the positive x direction. Write expressions for the components of the magnetic field of the wave.

13P. Start from Eqs. 34-11 and 34-18 and show that $E(x, t)$ and $B(x, t)$, the electric and magnetic field components of a plane traveling electromagnetic wave, must satisfy the "wave equations"

$$\frac{\partial^2 E}{\partial t^2} = c^2 \frac{\partial^2 E}{\partial x^2} \quad \text{and} \quad \frac{\partial^2 B}{\partial t^2} = c^2 \frac{\partial^2 B}{\partial x^2}.$$

14P. (a) Show that Eqs. 34-1 and 34-2 satisfy the wave equations displayed in Problem 13. (b) Show that any expressions of the form

$$E = E_m f(kx \pm \omega t) \quad \text{and} \quad B = B_m f(kx \pm \omega t),$$

where $f(kx \pm \omega t)$ denotes an arbitrary function, also satisfy these wave equations.

SECTION 34-4 Energy Transport and the Poynting Vector

15E. Show, by finding the direction of the Poynting vector **S**, that the directions of the electric and magnetic fields at all points in Figs. 34-4 to 34-7 are consistent at all times with the assumed directions of propagation.

16E. Currently operating neodymium–glass lasers can provide 100 TW of power in 1.0 ns pulses at a wavelength of 0.26 μm. How much energy is contained in a single pulse?

17E. Our closest stellar neighbor, Proxima Centauri, is 4.3 ly away. It has been suggested that TV programs from our planet have reached this star and may have been viewed by the hypothetical inhabitants of a hypothetical planet orbiting it. Suppose a television station on Earth has a power of 1.0 MW. What is the intensity of its signal at Proxima Centauri?

18E. An electromagnetic wave is traveling in the negative y direction. At a particular position and time, the electric field is along the positive z axis and has a magnitude of 100 V/m. What are the direction and magnitude of the magnetic field at that position and at that time?

19E. Earth's mean radius is 6.37×10^6 m and the mean Earth–Sun distance is 1.50×10^8 km. What fraction of the radiation emitted by the Sun is intercepted by the disk of Earth?

20E. The radiation emitted by a laser spreads out in the form of a narrow cone with circular cross section. The angle θ of the cone (see Fig. 34-42) is called the *full-angle beam divergence*. An argon laser, radiating at 514.5 nm, is aimed at the Moon in a ranging experiment. If the beam has a full-angle beam divergence of 0.880 μrad, what area on the Moon's surface is illuminated by the laser?

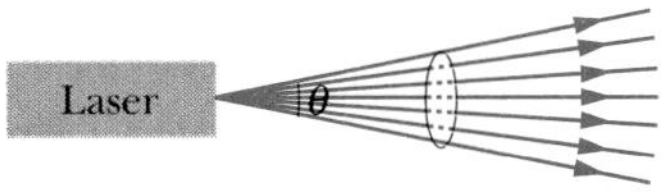

FIGURE 34-42
Exercise 20.

21E. The intensity of direct solar radiation that is not absorbed by the atmosphere on a particular summer day is 100 W/m². How close would you have to stand to a 1.0 kW electric heater to feel the same intensity? Assume that the heater radiates uniformly in all directions.

22E. Show that in a plane traveling electromagnetic wave the intensity, that is, the average rate of energy transport per unit area, is given by

$$\bar{S} = \frac{E_m^2}{2\mu_0 c} = \frac{cB_m^2}{2\mu_0}.$$

23E. What is the intensity of a plane traveling electromagnetic wave if B_m is 1.0×10^{-4} T?

24E. In a plane radio wave the maximum value of the electric field component is 5.00 V/m. Calculate (a) the maximum value of the magnetic field component and (b) the wave intensity.

25P. You walk 150 m directly toward a street lamp and find that the intensity increases to 1.5 times the intensity at your original position. How far from the lamp were you first standing? (Assume that the lamp is an isotropic point source of light.)

26P. Prove that the intensity of an electromagnetic wave is the product of the wave's energy density and its speed.

27P. Sunlight just outside Earth's atmosphere has an intensity of 1.40 kW/m². Calculate E_m and B_m for sunlight, assuming it to be a plane wave.

28P. The maximum electric field at a distance of 10 m from a point light source is 2.0 V/m. What are (a) the maximum value of the magnetic field and (b) the average intensity of the light there? (c) What is the power of the source?

29P. Frank D. Drake, an active investigator in the SETI (Search for Extra-Terrestrial Intelligence) program, has said that the large radio telescope in Arecibo, Puerto Rico, "can detect a signal which lays down on the entire surface of the earth a power of only one picowatt." See Fig. 34-43. (a) What is the power actually received by the Arecibo antenna for such a signal? The antenna diameter is 1000 ft. (b) What would be the power of a source at the center of our galaxy that could provide such a signal? The galactic center is 2.2×10^4 ly away. Take the source as radiating uniformly in all directions.

30P. A helium–neon laser, radiating at 632.8 nm, has a power output of 3.0 mW and a full-angle beam divergence (see Exercise 20) of 0.17 mrad. (a) What is the intensity of the beam 40 m from the laser? (b) What is the power of a point source that provides this same intensity at the same distance?

31P. An airplane flying at a distance of 10 km from a radio transmitter receives a signal of power 10 μW/m². Calculate (a) the amplitude of the electric field at the airplane due to this signal; (b) the amplitude of the magnetic field at the airplane; and (c) the total power of the transmitter, assuming the transmitter to radiate uniformly in all directions.

32P. During a test, a NATO surveillance radar system, operating at 12 GHz at 180 kW of power, attempts to detect an incoming stealth aircraft at 90 km. Assume that the radar beam is emitted uniformly over a hemisphere. (a) What is the intensity of the beam at the aircraft's location? The aircraft reflects radar waves as though it has a cross-sectional area of only 0.22 m². (b) What is the power of the aircraft's reflection? Assume that the beam is reflected uniformly over a hemisphere. Back at the radar site, what are (c) the intensity, (d) the maximum value of the electric field vector, and (e) the rms value of the magnetic field of the reflected radar beam?

SECTION 34-5 Radiation Pressure

33E. A black, totally absorbing piece of cardboard of area $A = 2.0$ cm² intercepts light with an intensity of 10 W/m² from a camera strobe light. What radiation pressure is produced on the cardboard by the light?

34E. High-power lasers are used to compress a plasma (a gas of charged particles) by radiation pressure. A laser generating pulses of radiation of peak power 1.5×10^3 MW is focused onto 1.0 mm² of high-electron-density plasma. Find the pressure exerted on the plasma if the plasma reflects perfectly.

35E. The average intensity of the solar radiation that falls normally on a surface just outside Earth's atmosphere is 1.4 kW/m². (a) What radiation pressure is exerted on this surface, assuming complete absorption? (b) How does this pressure compare with Earth's sea-level atmospheric pressure, which is 1.0×10^5 Pa?

36E. Radiation from the Sun reaching Earth (just outside the atmosphere) has an intensity of 1.4 kW/m². (a) Assuming that Earth (and its atmosphere) behaves like a flat disk perpendicular to the Sun's rays and that all the incident energy is absorbed, calculate the force on Earth due to radiation pressure. (b) Compare it with the force due to the Sun's gravitational attraction.

37E. What is the radiation pressure 1.5 m away from a 500 W lightbulb? Assume that the surface on which the pressure is exerted faces the bulb and is perfectly absorbing and that the bulb radiates uniformly in all directions.

FIGURE 34-43 Problem 29.

38P. A plane electromagnetic wave, with wavelength 3.0 m, travels in vacuum in the positive x direction with its electric vector **E**, of amplitude 300 V/m, directed along the y axis. (a) What is the frequency f of the wave? (b) What are the direction and amplitude of the magnetic field associated with the wave? (c) If $E = E_m \sin(kx - \omega t)$, what are the values of k and ω? (d) What is the time-averaged rate of energy flow in watts per square meter associated with this wave? (e) If the wave falls on a perfectly absorbing sheet of area 2.0 m², at what rate would momentum be delivered to the sheet and what is the radiation pressure exerted on the sheet?

39P. A helium–neon laser of the type often found in physics laboratories has a beam power of 5.00 mW at a wavelength of 633 nm. The beam is focused by a lens to a circular spot whose effective diameter may be taken to be equal to 2.00 wavelengths. Calculate (a) the intensity of the focused beam, (b) the radiation pressure exerted on a tiny perfectly absorbing sphere whose diameter is that of the focal spot, (c) the force exerted on this sphere, and (d) the acceleration imparted to it. Assume a sphere density of 5.00×10^3 kg/m³.

40P. In Fig. 34-44, a laser beam of power 4.60 W and diameter 2.60 mm is directed upward at one circular face (of diameter $d < 2.60$ mm) of a perfectly reflecting cylinder, which is made to "hover" by the beam's radiation pressure. The cylinder's density is 1.20 g/cm³. What is the cylinder's height H?

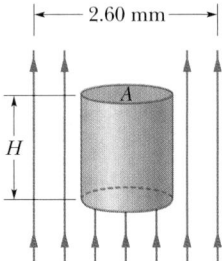

FIGURE 34-44 Problem 40.

41P. Radiation of intensity I is normally incident on an object that absorbs a fraction *frac* of it and reflects the rest back along the original path. What is the radiation pressure on the object?

42P. Prove, for a plane wave that is normally incident on a plane surface, that the radiation pressure on the surface is equal to the energy density in the beam outside the surface. (This relation holds no matter what fraction of the incident energy is reflected.)

43P. A laser beam of intensity I reflects from a flat, totally reflecting surface of area A whose normal makes an angle θ with the direction of the beam. Write an expression for the radiation pressure $p_r(\theta)$ exerted on the surface, in terms of the pressure $p_{r\perp}$ that would be exerted if the beam were perpendicular to the surface.

44P. Prove that the average pressure of a stream of bullets striking a plane surface perpendicularly is twice the kinetic energy density in the stream above the surface. Assume that the bullets are completely absorbed by the surface. Contrast this with the behavior of light in Problem 42.

45P. A small spaceship whose mass is 1.5×10^3 kg (including an astronaut) is drifting in outer space with negligible gravitational forces acting on it. If the astronaut turns on a 10 kW laser beam, what speed will the ship attain in 1.0 day because of the momentum carried away by the beam?

46P. It has been proposed that a spaceship might be propelled in the solar system by radiation pressure, using a large sail made of foil. How large must the sail be if the radiation force is to be equal in magnitude in the Sun's gravitational attraction? Assume that the mass of the ship + sail is 1500 kg, that the sail is perfectly reflecting, and that the sail is oriented perpendicularly to the Sun's rays. See Appendix C for needed data. (With a larger sail, the ship is continually driven away from the Sun.)

47P. A particle in the solar system is under the combined influence of the Sun's gravitational attraction and the radiation force due to the Sun's rays. Assume that the particle is a sphere of density 1.0×10^3 kg/m³ and that all the incident light is absorbed. (a) Show that, if its radius is less than some critical radius R, the particle will be blown out of the solar system. (b) Calculate the critical radius.

SECTION 34-6 Polarization

48E. The magnetic field equations for an electromagnetic wave in vacuum are $B_x = B \sin(ky + \omega t)$, $B_y = B_z = 0$. (a) What is the direction of propagation? (b) Write the electric field equations. (c) Is the wave polarized? If so, in what direction?

49E. A beam of unpolarized light of intensity 10 mW/m² is sent through a polarizing sheet as in Fig. 34-12. (a) Find the maximum value of the electric field of the transmitted beam. (b) What radiation pressure is exerted on the polarizing sheet?

50E. A beam of unpolarized light is sent through two polarizing sheets placed one on top of the other. What must be the angle between the polarizing directions of the sheets if the intensity of the transmitted light is to be one-third the incident intensity?

51E. Three polarizing plates are stacked. The first and third are crossed; the one between has its polarizing direction at 45° to the polarizing directions of the other two. What fraction of the intensity of an originally unpolarized beam is transmitted by the stack?

52E. In Fig. 34-45, initially unpolarized light is sent through three polarizing sheets whose polarizing directions make angles of $\theta_1 = \theta_2 = \theta_3 = 50°$ with the direction of the y axis. What percentage of the initial intensity is transmitted by the system of the three sheets?

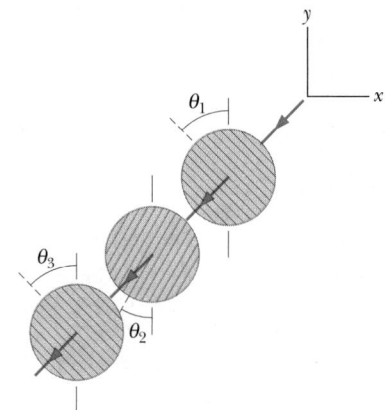

FIGURE 34-45 Exercise 52 and Problem 53.

53P. In Fig. 34-45, initially unpolarized light is sent through three polarizing sheets whose polarizing directions make angles of $\theta_1 = 40°$, $\theta_2 = 20°$, and $\theta_3 = 40°$ with the direction of the y axis. What percentage of the light's initial intensity is transmitted by the system?

54P. An unpolarized beam of light is sent through a stack of four polarizing sheets, oriented so that the angle between the polarizing directions of adjacent sheets is 30°. What fraction of the incident intensity is transmitted by the system?

55P. A beam of polarized light is sent through a system of two polarizing sheets. Relative to the polarization direction of that incident light, the polarizing directions of the sheets are at angles θ for the first sheet and 90° for the second sheet. If 0.10 of the incident intensity is transmitted by the two sheets, what is θ?

56P. A horizontal beam of vertically polarized light of intensity 43 W/m² is sent through two polarizing sheets. The polarizing direction of the first is at 70° to the vertical, and that of the second is horizontal. What is the intensity of the light transmitted by the pair of sheets?

57P. Suppose that in Problem 56 the initial beam is unpolarized. What then is the intensity of the transmitted light?

58P. A beam of partially polarized light can be considered to be a mixture of polarized and unpolarized light. Suppose we send such a beam through a polarizing filter and then rotate the filter through 360° while keeping it perpendicular to the beam. If the transmitted intensity varies by a factor of 5.0 during the rotation, what fraction of the intensity of the original beam is associated with the beam's polarized light?

59P. We want to rotate the direction of polarization of a beam of polarized light through 90° by sending the beam through one or more polarizing sheets. (a) What is the minimum number of sheets required? (b) What is the minimum number of sheets required if the transmitted intensity is to be more than 60% of the original intensity?

60P. At a beach the light is generally partially polarized owing to reflections off sand and water. At a particular beach on a particular day near sundown, the horizontal component of the electric field vector is 2.3 times the vertical component. A standing sunbather puts on polarizing sunglasses; the glasses eliminate the horizontal field component. (a) What fraction of the light intensity received before the glasses were put on now reaches the sunbather's eyes? (b) The sunbather, still wearing the glasses, lies on his side. What fraction of the light intensity received before the glasses were put on now reaches his eyes?

SECTION 34-7 Reflection and Refraction

61E. Figure 34-46 shows light reflecting from two perpendicular reflecting surfaces A and B. Find the angle between the incoming ray i and the outgoing ray r'.

62E. Light in vacuum is incident on the surface of a glass slab. In the vacuum the beam makes an angle of 32.0° with the normal to the surface, while in the glass it makes an angle of 21.0° with the normal. What is the index of refraction of the glass?

63E. When the rectangular metal tank in Fig. 34-47 is filled to

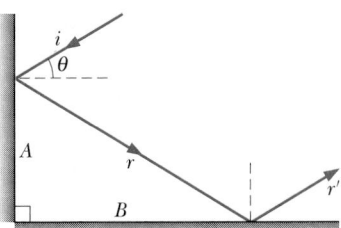

FIGURE 34-46
Exercise 61.

the top with an unknown liquid, an observer with eyes level with the top of the tank can just see the corner E; a ray that refracts toward the observer at the top surface of the liquid is shown. Find the index of refraction of the liquid.

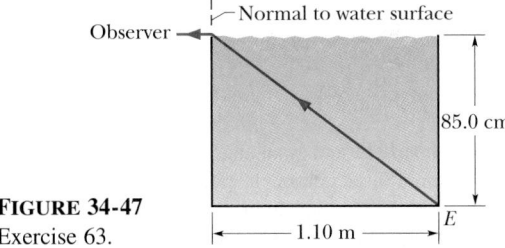

FIGURE 34-47
Exercise 63.

64E. In about A.D. 150, Claudius Ptolemy gave the following measured values for the angle of incidence θ_1 and the angle of refraction θ_2 for a light beam passing from air to water:

θ_1	θ_2	θ_1	θ_2
10°	8°	50°	35°
20°	15°30′	60°	40°30′
30°	22°30′	70°	45°30′
40°	29°	80°	50°

(a) Are these data consistent with the law of refraction? (b) If so, what index of refraction results? These data are interesting as perhaps the oldest recorded physical measurements.

65P. In Fig. 34-48, a 2.00 m long vertical pole extends from the bottom of a swimming pool to a point 50.0 cm above the water. Sunlight is incident at 55.0° above the horizon. What is the length of the shadow of the pole on the level bottom of the pool?

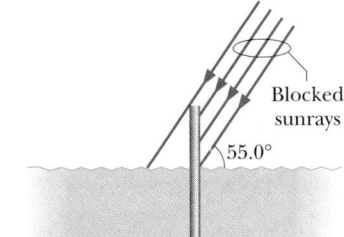

FIGURE 34-48
Problem 65.

66P. A catfish is 2.00 m below the surface of a smooth lake. (a) What is the diameter of the circle on the surface through which the fish can see the world outside the water? (b) If the fish descends, does the diameter of the circle increase, decrease, or remain the same?

67P. Prove that a ray of light incident on the surface of a sheet of plate glass of thickness t emerges from the opposite face parallel to its initial direction but displaced sideways, as in Fig. 34-49. Show that, for small angles of incidence θ, this displacement is given by

$$x = t\theta \frac{n-1}{n},$$

where n is the index of refraction of the glass and θ is measured in radians.

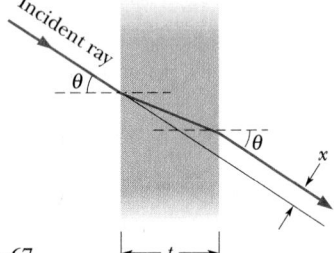

FIGURE 34-49 Problem 67.

68P. A ray of white light makes an angle of incidence of 35° on one face of a prism of fused quartz; the prism's cross section is an equilateral triangle. Sketch the light as it passes through the prism, showing the paths traveled by rays representing (a) blue light, (b) yellow-green light, and (c) red light.

69P. In Figure 34-50, two light rays pass from air through five transparent layers of plastic whose boundaries are parallel, whose indexes of refraction are as given, and whose thickness are unknown. The rays emerge back into air at the right. With respect to a normal taken on the last boundary, what is the angle of (a) emerging ray a and (b) emerging ray b? (c) What are your answers if there is glass, with $n = 1.5$, instead of air on the left and right sides of the plastic layers? (*Hint:* Save yourself much time by first solving the problems algebraically.)

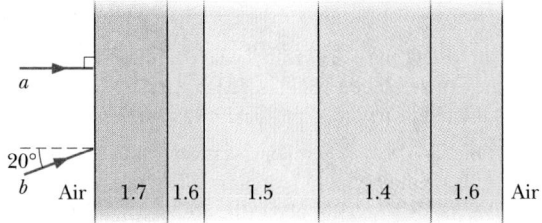

FIGURE 34-50 Problem 69.

70P. In Fig. 34-51, two perpendicular mirrors form the sides of a vessel filled with water. (a) A light ray is incident from above, normal to the water surface. Show that the emerging ray is parallel to the incident ray. Assume that there are two reflections at the mirror surfaces. (b) Repeat the analysis for the case of oblique incidence, with the ray lying in the plane of the figure.

FIGURE 34-51 Problem 70.

71P. In Fig. 34-52, a ray is incident on one face of a triangular glass prism in air. The angle of incidence θ is chosen so that the emerging ray also makes the same angle θ with the normal to the

other face. Show that the index of refraction n of the glass prism is given by

$$n = \frac{\sin \frac{1}{2}(\psi + \phi)}{\sin \frac{1}{2}\phi},$$

where ϕ is the vertex angle of the prism and ψ is the *deviation angle,* the total angle through which the beam is turned in passing through the prism. (Under these conditions the deviation angle ψ has the smallest possible value, which is called the *angle of minimum deviation.*)

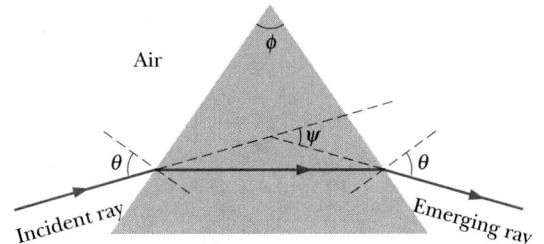

FIGURE 34-52 Problems 71 and 81.

72P. When the atmosphere is cold, moisture can form ice crystals of various shapes. If the atmosphere in the direction of the Sun happens to contain a sufficient number of ice crystals in the shape of flat hexagonal plates, a bright (perhaps colorful) region, called a *sun dog,* appears to the left or right of the Sun. A sun dog is formed by rays of sunlight that pass through the ice plates. These rays are parallel with one another when they arrive at Earth. The rays that pass through an ice plate are redirected by refraction, and those that pass through at the angle of minimum deviation ψ (shown from overhead in Fig. 34-53; see Problem 71) can form a sun dog. The sun dog can then be seen at an angle ψ away from the Sun. If the index of refraction of ice is 1.31, what is ψ?

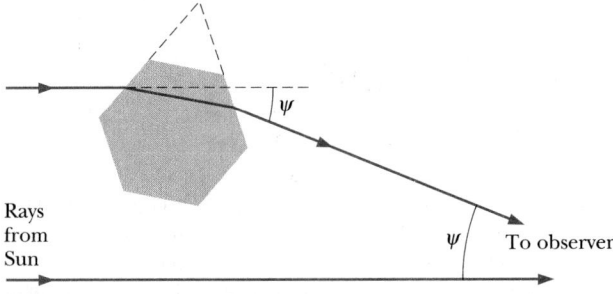

FIGURE 34-53 Problem 72.

73P. A ray of light goes through an equilateral triangular prism that is in the orientation for minimum deviation (see Problem 71). The total deviation is $\psi = 30.0°$. What is the index of refraction of the prism?

SECTION 34-8 Total Internal Reflection

74E. The refractive index of benzene is 1.8. What is the critical angle for a light ray traveling in benzene toward a plane layer of air above the benzene?

75E. In Fig. 34-54, a light ray enters a glass slab at point A and then undergoes total internal reflection at point B. What minimum value for the index of refraction of the glass can be inferred from this information?

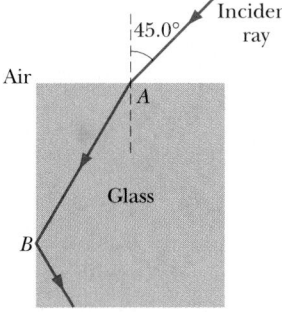

FIGURE 34-54 Exercise 75.

76E. In Fig. 34-55, a ray of light is perpendicular to the face ab of a glass prism ($n = 1.52$). Find the largest value for the angle ϕ so that the ray is totally reflected at face ac if the prism is immersed (a) in air and (b) in water.

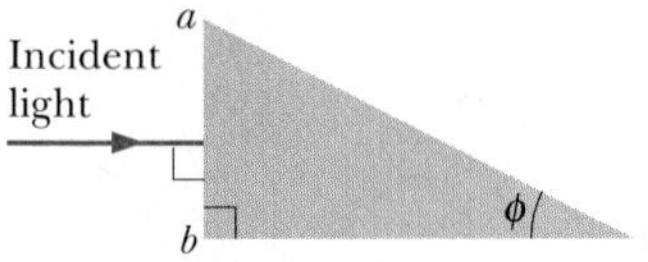

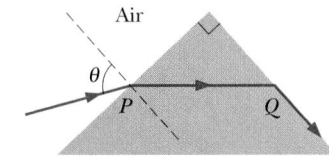

FIGURE 34-55
Exercise 76.

77E. A point source of light is 80.0 cm below the surface of a body of water. Find the diameter of the circle at the surface through which light emerges from the water.

78P. A solid glass cube, of edge length 10 mm and index of refraction 1.5, has a small spot at its center. (a) What parts of the cube face must be covered to prevent the spot from being seen, no matter what the direction of viewing? (Neglect the subsequent behavior of internally reflected rays.) (b) What fraction of the cube surface must be so covered?

79P. A ray of white light traveling in fused quartz strikes a plane surface of the quartz with an angle of incidence θ. Is it possible for the internally reflected beam to appear (a) bluish or (b) reddish? (c) If so, what value of θ is required? (*Hint:* White light will appear bluish if wavelengths corresponding to red are removed from the spectrum, and vice versa.)

80P. In Fig. 34-56, light enters a 90° triangular prism at point P with incident angle θ and then some of it refracts at point Q with an angle of refraction of 90°. (a) What is the index of refraction of the prism in terms of θ? (b) What, numerically, is the maximum value that the index of refraction can have? Explain what happens to the light at Q if the incident angle at Q is (c) increased slightly and (d) decreased slightly.

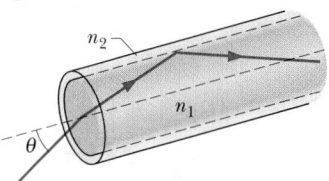

FIGURE 34-56
Problem 80.

81P. Suppose the prism of Fig. 34-52 has apex angle $\phi = 60.0°$ and index of refraction $n = 1.60$. (a) What is the smallest angle of incidence θ for which a ray can enter the left face of the prism and

exit the right face? (b) What angle of incidence θ is required for the ray to exit the prism with an identical angle θ for its refraction, as it does in Fig. 34-52? (See Problem 71.)

82P. A point source of light is placed a distance h below the surface of a large deep lake. (a) Neglecting reflection at the surface except where it is total, show that the fraction *frac* of the light energy that escapes directly from the water surface is independent of h and is given by

$$frac = \tfrac{1}{2}(1 - \sqrt{1 - 1/n^2}),$$

where n is the index of refraction of the water. (b) Evaluate this fraction for $n = 1.33$.

83P. An optical fiber consists of a glass core (index of refraction n_1) surrounded by a coating (index of refraction $n_2 < n_1$). Suppose a beam of light enters the fiber from air at an angle θ with the fiber axis as shown in Fig. 34-57. (a) Show that the greatest possible value of θ for which a ray can travel down the fiber is given by $\theta = \sin^{-1} \sqrt{n_1^2 - n_2^2}$. (b) If the indices of refraction of the glass and coating are 1.58 and 1.53, respectively, what is this value of the incident angle θ?

FIGURE 34-57
Problems 83 and 84.

84P. In an optical fiber (see Problem 83), different rays travel different paths along the fiber, leading to different travel times. This causes a light pulse to spread out as it travels along the fiber, resulting in information loss. The delay time should be minimized by the design of the fiber. Consider a ray that travels a distance L directly along a fiber axis and another that is repeatedly reflected, at the critical angle, as it travels to the same point as the first ray. (a) Show that the difference Δt in the times of arrival is given by

$$\Delta t = \frac{L}{c} \frac{n_1}{n_2} (n_1 - n_2),$$

where n_1 is the index of refraction of the glass core and n_2 is the index of refraction of the fiber coating. (b) Evaluate Δt for the fiber of Problem 83, with $L = 300$ m.

SECTION 34-9 Polarization by Reflection

85E. (a) At what angle of incidence will the light reflected from water be completely polarized? (b) Does this angle depend on the wavelength of the light?

86E. Light traveling in water of refractive index 1.33 is incident on a plate of glass of refractive index 1.53. At what angle of incidence will the reflected light be fully polarized?

87E. Calculate the upper and lower limits of the Brewster angles for white light incident on fused quartz. Assume that the wavelength limits of the light are 400 and 700 nm.

88P. When red light in vacuum is incident at the Brewster angle on a certain glass slab, the angle of refraction is 32.0°. What are (a) the index of refraction of the glass and (b) the Brewster angle?

35
Images

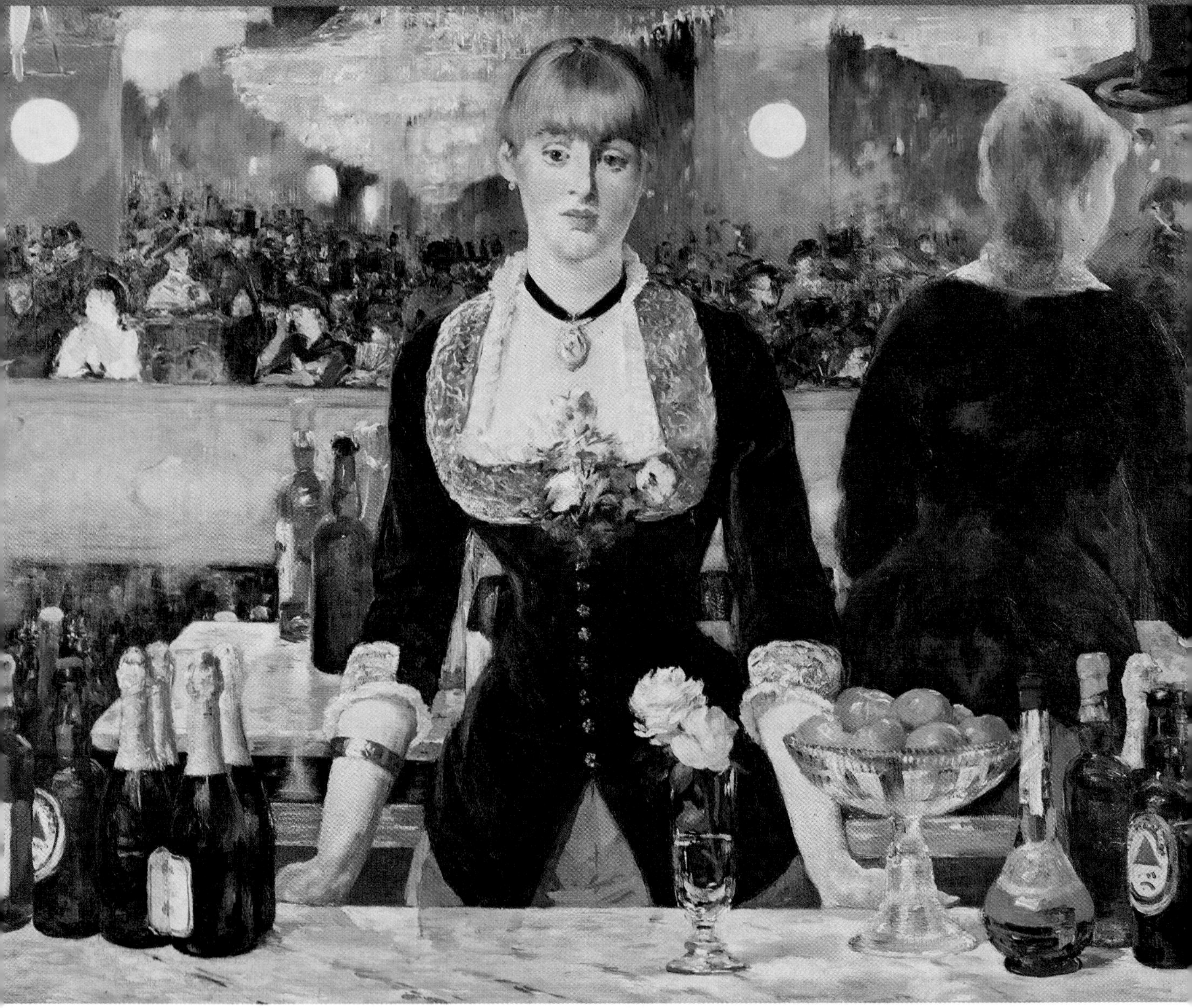

Edouard Manet's A Bar at the Folies-Bergère *has enchanted viewers ever since it was painted in 1882. Part of its appeal lies in the contrast between an audience ready for entertainment and a bartender whose eyes betray her fatigue. But its appeal also depends on a subtle distortion of reality that Manet hid in the painting—a distortion that gives an eerie feel to the scene even before you recognize what is "wrong." Can you find it?*

35-1 TWO TYPES OF IMAGE

For you to see, say, a penguin, your eye must intercept some of the light rays spreading from the penguin and then redirect them onto the retina at the rear of the eye. Your visual system, starting with the retina and ending with the visual cortex at the rear of your brain, automatically and subconsciously processes the information provided by the light. That system identifies edges, orientations, textures, shapes, and colors and then rapidly brings to your consciousness an **image** (a reproduction derived from light) of the penguin: you perceive and recognize the penguin as being in the direction from which the light rays came and at the proper distance.

Your visual system goes through this processing and recognition even if the light rays do not come directly from the penguin, but instead reflect toward you from a mirror or refract through the lenses in a pair of binoculars. However, you now see the penguin in the direction from which the light rays came after they reflected or refracted, and the distance you perceive may be quite different from the penguin's true distance.

For example, if the light rays have been reflected toward you from a standard flat mirror, the penguin appears to be behind the mirror because the rays you intercept come from that direction. Of course, the penguin is not back there. This type of image, which is called a **virtual image,** truly exists only within the brain but nevertheless is *said* to exist at the perceived location.

A **real image** differs in that it can be formed on a surface, such as a card or a movie screen. You can see a real image (otherwise movie theaters would be empty), but the existence of the image does not depend on your seeing it and it is present even if you are not.

In this chapter we explore several ways in which virtual and real images are formed by reflection (as with mirrors) and refraction (as with lenses). We also distinguish between the two types of image more clearly, but here first is an example of a natural virtual image.

A Common Mirage

A common example of a virtual image is a pool of water that appears to lie on the road some distance ahead of you on a sunny day but which you can never reach. The pool is a *mirage* (a type of illusion), formed by light rays coming from the low section of the sky in front of you (Fig. 35-1a). As the rays approach the road, they travel through progressively warmer air that has been heated by the road, which is usually relatively warm. With an increase in air temperature, the speed of light in air increases slightly and, correspondingly, the index of refraction of the air decreases slightly. So, as the rays descend, encountering progres-

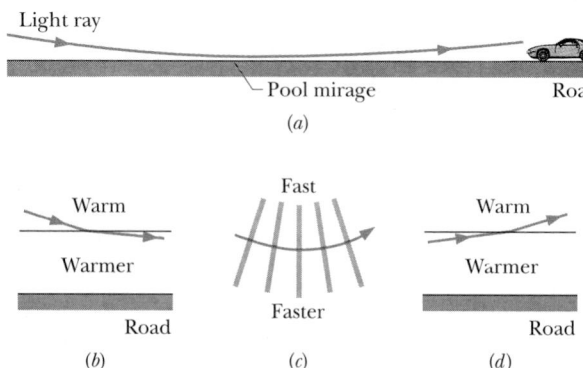

FIGURE 35-1 (*a*) A ray from a low section of the sky refracts through air that is heated by a road (without reaching the road). An observer who intercepts the light perceives it to be from a pool of water on the road. (*b*) Bending (exaggerated) of a light ray descending across an imaginary boundary from warm air to warmer air. (*c*) Bending of a ray when the ray is horizontal, which occurs because the lower ends of wavefronts move faster in warmer air. (*d*) Bending of a ray ascending across an imaginary boundary to warm air from warmer air.

sively smaller indices of refraction, they continuously bend toward the horizontal (Fig. 35-1b).

Once a ray is horizontal, somewhat above the road's surface, it still bends because the lower portion of each associated wavefront is in slightly warmer air and is moving slightly faster than the upper portion of the wavefront (Fig. 35-1c). This nonuniform motion of the wavefronts bends the ray upward. As the ray then ascends, it continues to bend upward through progressively larger indices of refraction (Fig. 35-1d).

If you intercept some of this light, your visual system automatically infers that it originated along a backward extension of the rays you have intercepted and, to make sense of the light, assumes that it came from the road surface. If the light happens to be bluish from blue sky, the mirage appears bluish, like water. Because the air is probably turbulent due to the heating, the mirage shimmies, as if water waves were present. The bluish coloring and the shimmy enhance the illusion of a pool of water, but you are actually seeing a virtual image of a low section of the sky.

35-2 PLANE MIRRORS

A **mirror** is a surface that can reflect a beam of light in one direction instead of either scattering it widely into many directions or absorbing it. A shiny metal surface acts as a mirror; a concrete wall does not. In this section we examine the images that a **plane mirror** (a flat reflecting surface) can produce.

Figure 35-2 shows a point source of light *O*, which we shall call the *object*, at a perpendicular distance *p* in front of a plane mirror. The light that is incident on the mirror is

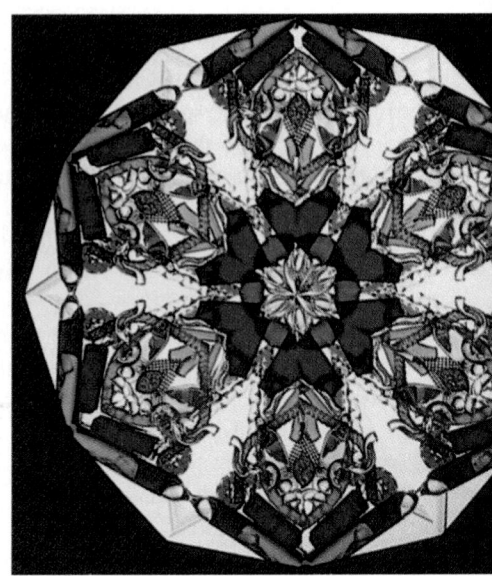

What clues tell you whether this photograph is upside down? There are several.

A section of what you see in a kaleidoscope is a direct view of what lies at the far end of the kaleidoscope tube; the rest consists of images of the direct view that are produced by mirrors extending along the tube. How many mirrors are in this kaleidoscope, and how are they arranged? (The answer is given with the Checkpoint answers.)

represented with rays spreading from O. The reflection of that light is represented with reflected rays spreading from the mirror. If we extend the reflected rays backward (behind the mirror), we find that the extensions intersect at a point that is a perpendicular distance i behind the mirror.

If you look into the mirror of Fig. 35-2, your eyes intercept some of the reflected light. To make sense of what you see, you perceive a point source of light located at the point of intersection of the extensions. This point source is the image I of object O. It is called a *point image* because it is a point, and it is a virtual image because the rays do not actually pass through it. (As you will see, rays *do* pass through a point of intersection for a real image.)

Figure 35-3 shows two rays selected from the many rays in Fig. 35-2. One reaches the mirror at point b, perpendicularly. The other reaches it at an arbitrary point a, with an angle of incidence θ. The extensions of the two reflected rays are also shown. The right triangles $aOba$ and

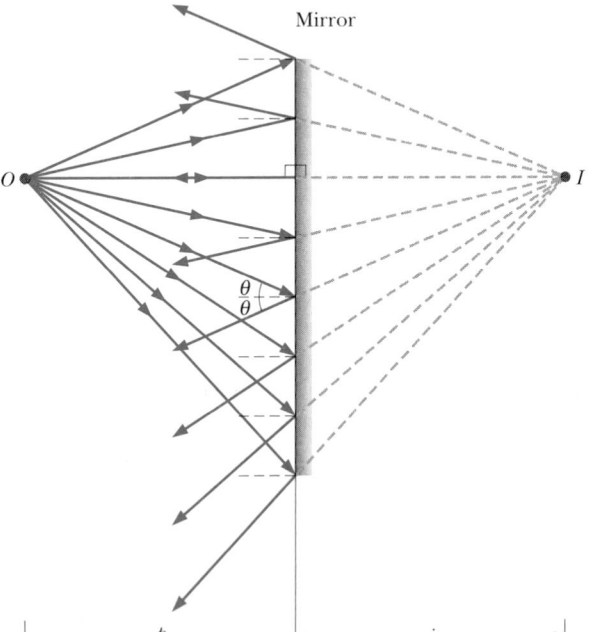

FIGURE 35-2 A point source of light O, called the *object*, is a perpendicular distance p in front of a plane mirror. Light rays reaching the mirror from O reflect from the mirror. If your eye intercepts some of the reflected rays, you perceive a point source of light I to be behind the mirror, at a perpendicular distance i. The perceived source I is a virtual image of object O.

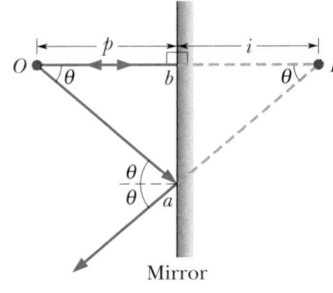

FIGURE 35-3 Two rays from Fig. 35-2. Ray Oa makes an arbitrary angle θ with the normal to the mirror surface. Ray Ob is perpendicular to the mirror.

alba have a common side and three equal angles and are thus congruent. So their horizontal sides are congruent. That is,

$$Ib = Ob, \qquad (35\text{-}1)$$

where *Ib* and *Ob* are the distances from the mirror to the image and the object, respectively. Equation 35-1 tells us that the image is as far behind the mirror as the object is in front of it. By convention (that is, to get our equations to work out), *object distances p* are taken to be positive quantities, and *image distances i* for virtual images (as here) are taken to be negative quantities. Thus Eq. 35-1 can be written as $|i| = p$, or as

$$i = -p \quad \text{(plane mirror).} \qquad (35\text{-}2)$$

Only rays that are fairly close together can enter the eye after reflection at a mirror. For the eye position shown in Fig. 35-4, only a small portion of the mirror near point *a* (a portion smaller than the pupil of the eye) is useful in forming the image. To find this portion, close one eye and look at the mirror image of a small object such as the tip of a pencil. Then move your fingertip over the mirror surface until you cannot see the image. Only that small portion of the mirror under your fingertip produced the image.

Extended Objects

In Fig. 35-5, an extended object *O*, represented by an upright arrow, is at perpendicular distance *p* in front of a plane mirror. Each small portion of the object that faces the mirror acts like the point source *O* of Figs. 35-2 and 35-3. If you intercept the light reflected by the mirror, you perceive a virtual image *I* that is a composite of the virtual point images of all those portions of the object and seems to be at distance *i* behind the mirror. Distances *i* and *p* are related by Eq. 35-2.

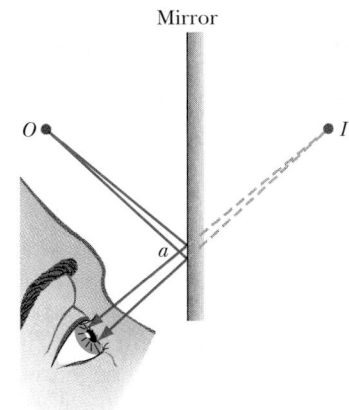

FIGURE 35-4 A "pencil" of rays from *O* enters the eye after reflection at the mirror. Only a small portion of the mirror near *a* is involved in this reflection. The light appears to originate at point *I* behind the mirror.

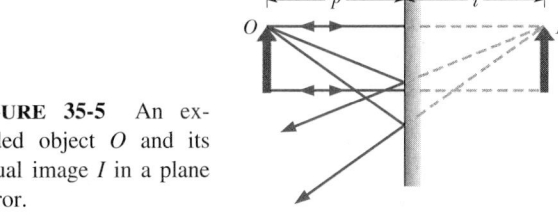

FIGURE 35-5 An extended object *O* and its virtual image *I* in a plane mirror.

We can also locate the image of an extended object as we did for a point object in Fig. 35-2: we draw some of the rays that reach the mirror from the top of the object, draw the corresponding reflected rays, and then extend those reflected rays behind the mirror until they intersect to form an image of the top of the object. We then do the same for rays from the bottom of the object. As shown in Fig. 35-5, we find that virtual image *I* has the same orientation and *height* (measured parallel to the mirror) as object *O*.

Manet's "Folies-Bergère"

In *A Bar at the Folies-Bergère* you see the barroom via reflection by a large mirror on the wall behind the woman tending bar, but the reflection is subtly wrong in three ways. First note the bottles at the left. Manet painted their reflections in the mirror but misplaced them, painting them farther toward the front of the bar than they really were.

Now note the reflection of the woman. Since your view is from directly in front of the woman, her reflection should be behind her, with only a little of it (if any) visible to you; yet Manet painted her reflection well off to the right. Finally, note the reflection of the man facing her. He must be you, because the reflection shows that he is directly in front of the woman, and thus he must be the viewer of the painting. You are looking into Manet's work and seeing your reflection well off to your right. The effect is eerie because it is not what we expect from a painting or from a mirror.

CHECKPOINT **1:** In the figure you look into a system of two vertical parallel mirrors *A* and *B* separated by distance *d*. A grinning gargoyle is perched at point *O*, a distance 0.2*d* from mirror *A*. Each mirror produces a *first* (least deep) image of the gargoyle. Then each mirror produces a *second* image with the object being the first image in the opposite mirror. Then each mirror produces a *third* image with the object being

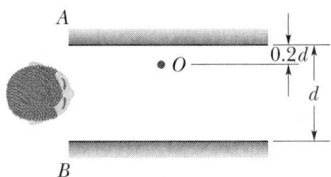

the second image in the opposite mirror. And so on—you might see hundreds of grinning gargoyle images. How deep behind mirror A are the first, second, and third images in mirror A?

SAMPLE PROBLEM 35-1

Charles Barkley is 198 cm tall. How tall must a vertical mirror be if he is to be able to see his entire length in it?

SOLUTION: In Fig. 35-6, the heights of the top of Barkley's head (h), his eyes (e), and the bottoms of his feet (f) are marked by dots. (Dot h has been drawn slightly too high for clarity.) The figure shows the paths followed by rays that leave his head and his feet and enter his eyes, reflecting from the mirror at points a and c, respectively. The mirror need occupy only the vertical distance H between those points.

From the geometry and Eq. 34-43,

$$ab = \tfrac{1}{2}he \quad \text{and} \quad bc = \tfrac{1}{2}ef.$$

So the required height is

$$H = ab + bc = \tfrac{1}{2}(he + ef)$$
$$= (\tfrac{1}{2})(198 \text{ cm}) = 99 \text{ cm.} \qquad \text{(Answer)}$$

Thus the mirror need be no taller than half the athlete's height. And this result is independent of his distance from the mirror. (If you have a full-length mirror available, you might experiment by taping newspaper over the portions of the mirror that do not contribute to your image. You will find that what you have left is just half your height. Mirrors that extend below point c just allow you to look at an image of the floor.)

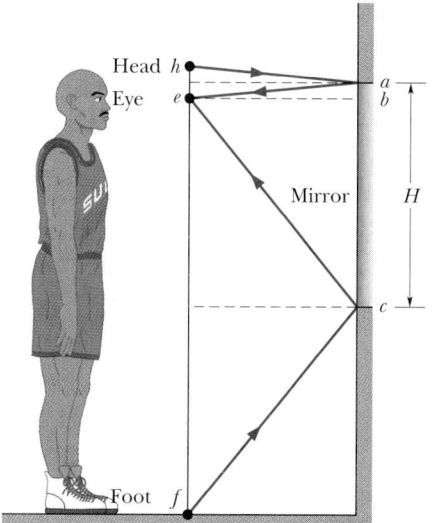

FIGURE 35-6 Sample Problem 35-1. A "full-length mirror" need be only half a person's height.

35-3 SPHERICAL MIRRORS

We turn now from images produced by plane mirrors to images produced by mirrors with curved surfaces. In particular, we shall consider a spherical mirror, which is simply a mirror in the shape of a small section of the surface of a sphere. A plane mirror is in fact a spherical mirror with an infinitely large *radius of curvature.*

Making a Spherical Mirror

We start with the plane mirror of Fig. 35-7a, which faces leftward toward an object O that is shown and an observer that is not shown. We make a **concave mirror** by curving the mirror's surface so it is *concave* ("caved in") as in Fig. 35-7b. Curving the surface in this way changes several characteristics of the mirror and the image it produces of the object:

1. The *center of curvature C* (the center of the sphere of which the mirror's surface is part) was infinitely far from the plane mirror; it is now closer but still in front of the concave mirror.

2. The *field of view*—the extent of the scene that is reflected to the observer—was wide; it is now smaller.

3. The image of the object was as far behind the plane mirror as the object was in front; the image is farther behind the concave mirror; that is, $|i|$ is greater.

4. The height of the image was equal to the height of the object; the height of the image is now greater. This feature is why many makeup mirrors and shaving mirrors are concave—they produce a larger image of a face.

We can make a **convex mirror** by curving a plane mirror so its surface is *convex* ("flexed out") as in Fig. 35-7c. Curving the surface in this way moves the center of curvature C to behind the mirror and increases the field of view. It also moves the image of the object closer to the mirror and shrinks it. Store surveillance mirrors are usually convex to take advantage of the increase in the field of view—more of the store can then be monitored.

Focal Points of Spherical Mirrors

For a plane mirror, the magnitude of the image distance i is always equal to the object distance p. Before we can determine how these two distances are related for a spherical mirror, we must consider the reflection of light from an object O located an effectively infinite distance in front of a spherical mirror, on the mirror's *central axis*. That axis extends through the center of curvature C and the center c

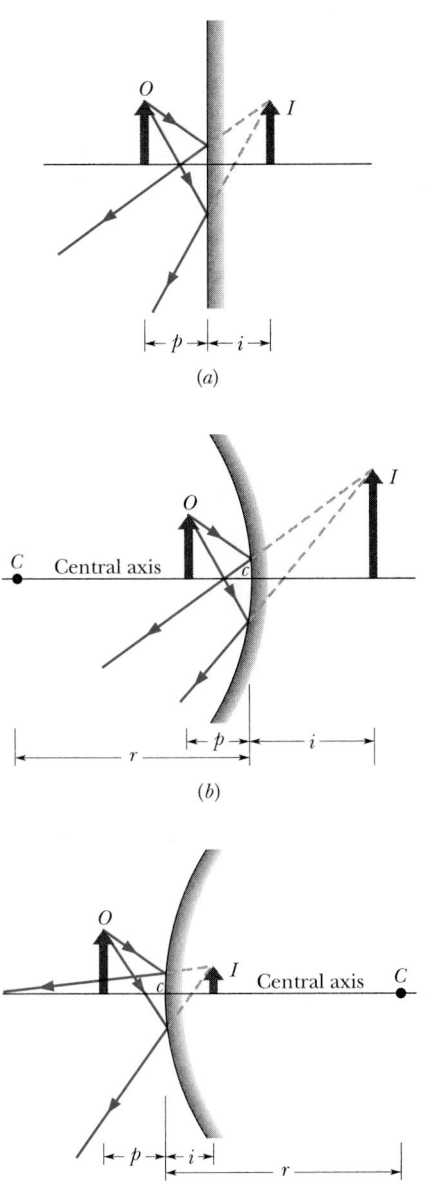

FIGURE 35-7 (*a*) An object *O* forms a virtual image *I* in a plane mirror. (*b*) If the mirror is bent so that it becomes *concave,* the image moves farther away and becomes larger. (*c*) If the plane mirror is bent so that it becomes *convex,* the image moves closer and becomes smaller.

of the mirror. Because of the great distance between the object and the mirror, the light waves spreading from the object are plane waves when they reach the mirror along the central axis. This means that the rays representing the light waves are all parallel to the central axis when they reach the mirror.

When these parallel rays reach a concave mirror like that of Fig. 35-8*a*, those close to the central axis are re-

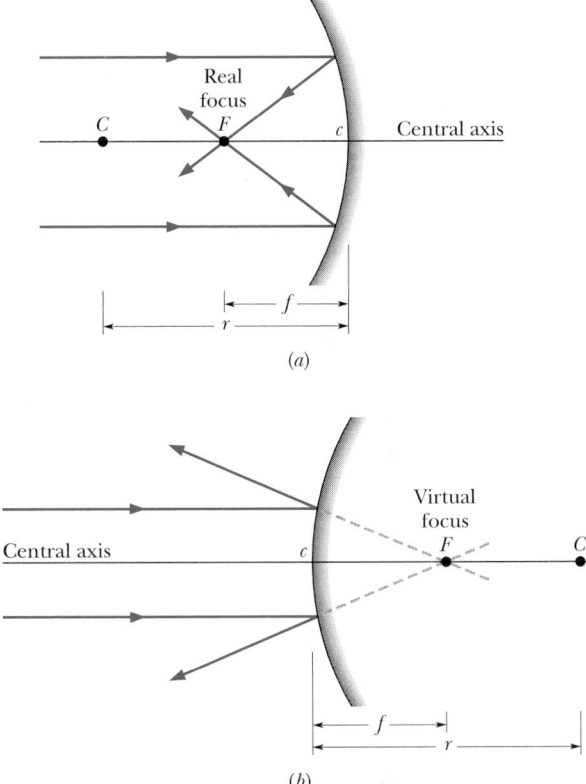

FIGURE 35-8 (*a*) In a concave mirror, incident parallel light rays are brought to a real focus at *F*, on the same side of the mirror as the light rays. (*b*) In a convex mirror, incident parallel light rays seem to diverge from a virtual focus at *F*, on the side of the mirror opposite the light rays.

flected through a common point *F*; two of these reflected rays are shown in the figure. If we placed a card at *F*, a point image of the infinitely distant object *O* would appear on the card. (This would occur for any infinitely distant object.) Point *F* is called the **focal point** (or **focus**) of the mirror, and its distance from the center of the mirror is the **focal length** *f* of the mirror.

If we now substitute a convex mirror for the concave mirror, we find that the parallel rays are no longer reflected through a common point. Instead, they diverge as shown in Fig. 35-8*b*. However, if your eye intercepts some of the reflected light, you perceive the light as originating from a point source behind the mirror. This perceived source is located where extensions of the reflected rays pass through a common point (*F* in Fig. 35-8*b*). That point is the focal point (or focus) *F* of the convex mirror, and its distance from the mirror surface is the focal length *f* of the mirror. If we placed a card at this focal point, an image of object *O* would *not* appear on the card. So, this focal point is not like that of a concave mirror.

To distinguish the actual focal point of a concave

mirror from the perceived focal point of a convex mirror, the former is said to be a *real focal point* and the latter is said to be a *virtual focal point*. Moreover, the focal length f of a concave mirror is taken to be a positive quantity, and that of a convex mirror a negative quantity. For mirrors of both types, the focal length f is related to the radius of curvature r of the mirror by

$$f = \tfrac{1}{2}r \quad \text{(spherical mirror),} \qquad (35\text{-}3)$$

where, consistent with the signs for the focal length, r is a positive quantity for a concave mirror and a negative quantity for a convex mirror.

35-4 IMAGES FROM SPHERICAL MIRRORS

With the focal point of a spherical mirror defined, we can find the relation between image distance i and object distance p for concave and convex spherical mirrors. We begin by placing the object O *inside the focal point* of the concave mirror, that is, between the mirror and its focal point F (Fig. 35-9a). An observer can then see a virtual image of O in the mirror: The image appears to be behind the mirror, and it has the same orientation as the object.

If we now move the object away from the mirror until it is at the focal point, the image moves farther back from the mirror until it is at infinity (Fig. 35-9b). The image is then ambiguous and imperceptible because neither the rays reflected by the mirror nor the ray extensions behind the mirror cross to form an image of O.

If we next move the object *outside the focal point*, that is, farther away from the mirror than the focal point, the rays reflected by the mirror converge to form an *inverted* image of object O (Fig. 35-9c) in front of the mirror. That image moves in from infinity as we move the object farther outside F. If you were to hold a card at the position of the image, the image would show up on the card—the image is said to be *focused* on the card by the mirror. (The verb ''focus,'' which in this context means to produce an image, differs from the noun ''focus,'' which is another name for the focal point.) Because this image can actually appear on a surface, it is a real image—the rays actually intersect to create the image, regardless of whether an observer is present. The image distance i of a real image is a positive quantity, in contrast to that for a virtual image. We also see that:

Real images form on the same side of a mirror as where the object is, and virtual images form on the opposite side.

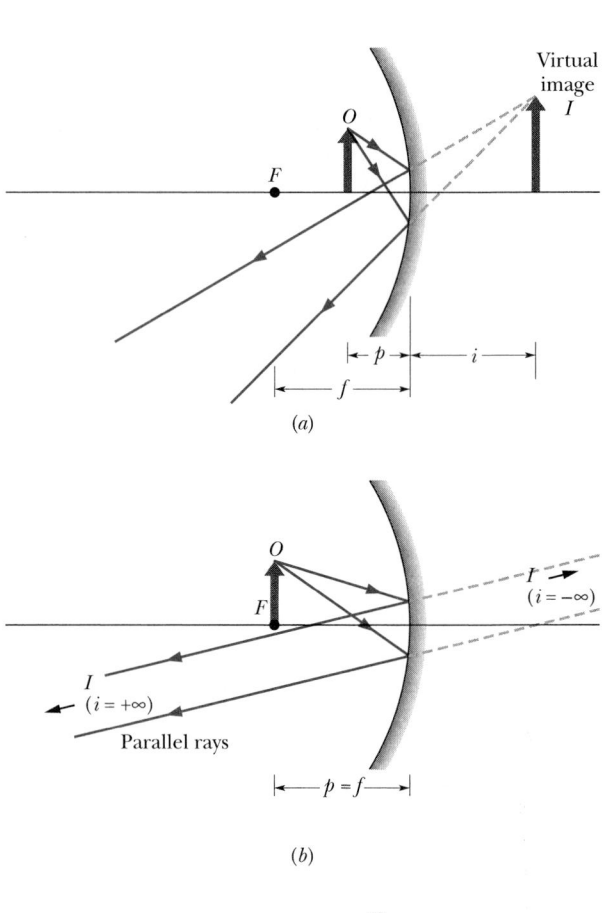

(a)

(b)

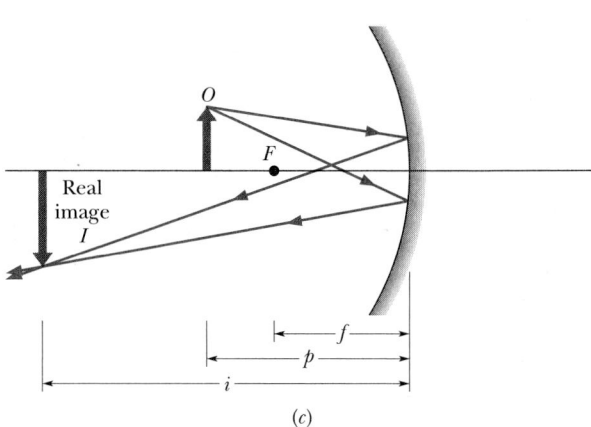

(c)

FIGURE 35-9 (a) An object O inside the focal point of a concave mirror, and its virtual image I. (b) The object at the focal point F. (c) The object outside the focal point, and its real image I.

As we shall prove in Section 35-8, when light rays from an object make only small angles with the central axis of a spherical mirror, a simple equation relates the object distance p, the image distance i, and the focal length f:

$$\frac{1}{p} + \frac{1}{i} = \frac{1}{f} \quad \text{(spherical mirror).} \qquad (35\text{-}4)$$

(For clarity in figures such as Fig. 35-9, the rays are drawn with exaggerated angles.)

The size of an object or image, as measured *perpendicular* to the mirror's central axis, is called the object or image *height*. Let h represent the height of the object, and h' the height of the image. Then the ratio h'/h is called the **lateral magnification** m produced by the mirror. However, by convention, the lateral magnification always includes a plus sign when the image orientation is that of the object and a minus sign when the image orientation is opposite that of the object. For this reason, we write the formula for m as

$$|m| = \frac{h'}{h} \qquad \text{(lateral magnification).} \quad (35\text{-}5)$$

We shall soon prove that the lateral magnification can also be written as

$$m = -\frac{i}{p} \qquad \text{(lateral magnification).} \quad (35\text{-}6)$$

For a plane mirror, for which $i = -p$, we have $m = +1$. The magnification of 1 means that the image is the same size as the object. The plus sign means that the image and the object have the same orientation. For the concave mirror of Fig. 35-9c, $m \approx -1.5$.

Equations 35-3 through 35-6 hold for all plane mirrors, concave spherical mirrors, and convex spherical mirrors. In addition to those equations, you have been asked to absorb a lot of information about these mirrors, and you should organize it for yourself by filling in Table 35-1. Under Image Location, decide if the image is on the *same* side of the mirror as the object or on the *opposite* side. Under Image Type, decide if the image is *real* or *virtual*. Under Image Orientation, decide if the image has the *same* orientation as the object or is *inverted*. Under Sign, give the sign of the quantity or fill in $\pm$ if the sign is ambiguous. You will need this organization to tackle homework or a test.

Locating Images by Drawing Rays

Figures 35-10a and b show an object O in front of a concave mirror. We can graphically locate the image of any off-axis point of the object by drawing a *ray diagram* with any two of four special rays through the point:

1. A ray that is initially parallel to the central axis reflects through the focal point F (ray 1 in Fig. 35-10a).

2. A ray that reflects from the mirror after passing through the focal point emerges parallel to the central axis (ray 2 in Fig. 35-10a).

3. A ray that reflects from the mirror after passing through the center of curvature C returns along itself (ray 3 in Fig. 35-10b).

4. A ray that reflects from the mirror at its intersection c with the central axis is reflected symmetrically about that axis (ray 4 in Fig. 35-10b).

The image of the point is at the intersection of the two special rays you choose. The image of the object can then be found by locating the images of two or more of its off-axis points. You need to modify the descriptions of the rays slightly to apply them to convex mirrors, as in Figs. 35-10c and d.

Proof of Equation 35-6

We are now in a position to derive Eq. 35-6 ($m = -i/p$), the expression for the lateral magnification of an object reflected in a mirror. Consider ray 4 in Fig. 35-10b. It is reflected at point c so that the incident and reflected rays make equal angles with the axis of the mirror at that point.

The two right triangles abc and edc in the figure are similar, so we can write

$$\frac{de}{ab} = \frac{cd}{ca}.$$

TABLE 35-1 **YOUR ORGANIZING TABLE FOR MIRRORS**

MIRROR TYPE	OBJECT LOCATION	IMAGE			SIGN			
		LOCATION	TYPE	ORIENTATION	OF f	OF r	OF i	OF m
Plane	Anywhere							
Concave	Inside F							
	Outside F							
Convex	Anywhere							

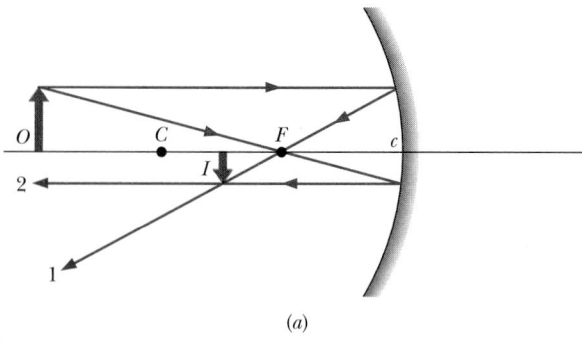

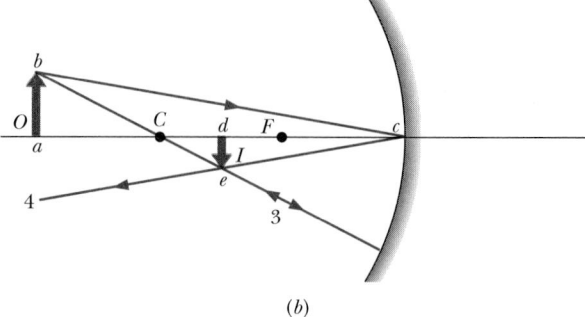

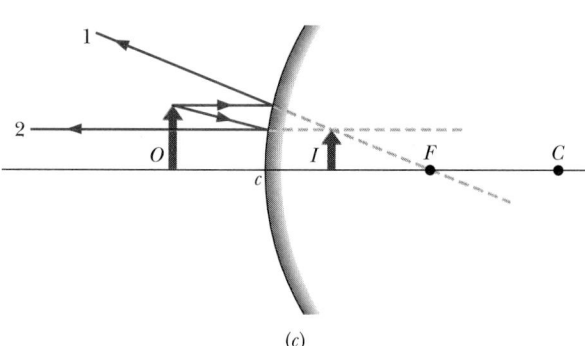

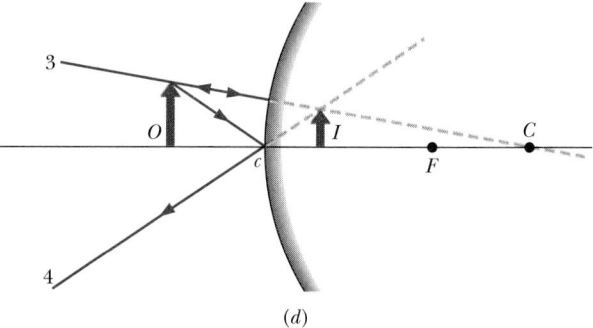

(a)   (b)

(c)   (d)

FIGURE 35-10 (a, b) Four rays that may be drawn to find the image of an object in a concave mirror. For the object position shown, the image is real, inverted, and smaller than the object. (c, d) Four similar rays for the case of a convex mirror. For a convex mirror, the image is always virtual, oriented like the object, and smaller than the object. [In (c), ray 2 is initially directed toward focal point F.]

The quantity on the left (apart from the question of sign) is the lateral magnification m produced by the mirror. Since we indicate an inverted image as a *negative* magnification, we symbolize this as $-m$. But $cd = i$ and $ca = p$, so we have at once

$$m = -\frac{i}{p} \quad \text{(magnification)}, \quad (35\text{-}7)$$

which is the relation we set out to prove.

SAMPLE PROBLEM 35-2

A tarantula of height h sits cautiously before a spherical mirror whose focal length has absolute value $|f| = 40$ cm. The image of the tarantula produced by the mirror has the same orientation as the tarantula and has height $h' = 0.20h$.

(a) Is the image real or virtual, and is it on the same side of the mirror as the tarantula or the opposite side?

SOLUTION: Because the image has the same orientation as the tarantula (the object), it must be virtual and on the opposite side of the mirror. (You can easily see this result if you have filled out Table 35-1.)

(b) Is the mirror concave or convex, and what is its focal length f, sign included?

SOLUTION: Can we tell the type of mirror by the type of image it produces? No, because both types of mirror can produce a virtual image. Can we tell the type of mirror by finding the sign of f with our only two equations that involve f (Eqs. 35-3 and 35-4)? No, we don't have enough information. The only approach left is to consider the magnification information. We know that the ratio of image height h' to object height h is 0.20. So, from Eq. 35-5 we have

$$|m| = \frac{h'}{h} = 0.20.$$

Because the object and image have the same orientation, we know that m must be positive: $m = +0.20$. Substituting this into Eq. 35-6 and solving for, say, i gives us

$$i = -0.20p,$$

which does not appear to be of help in finding f. But it is helpful if we substitute it into Eq. 35-4. That equation gives us

$$\frac{1}{f} = \frac{1}{i} + \frac{1}{p} = \frac{1}{-0.20p} + \frac{1}{p} = \frac{1}{p}(-5 + 1),$$

from which we find

$$f = -p/4.$$

Now we have it: because p is positive, f must be negative, which means that the mirror is convex with

$$f = -40 \text{ cm}. \quad \text{(Answer)}$$

CHECKPOINT 2: A Central American vampire bat, dozing on the central axis of a spherical mirror, is magnified by $m = -4$. Is its image (a) real or virtual, (b) inverted or of the same orientation as the bat, and (c) on the same side of the mirror as the bat or on the opposite side?

35-5 SPHERICAL REFRACTING SURFACES

We now turn from images formed by reflections to images formed by refraction through surfaces of transparent materials, such as glass. We shall consider only spherical surfaces, with radius of curvature r and center of curvature C. The light will be emitted by a point object O in a medium with index of refraction n_1; it will refract through a spherical surface into a medium of index of refraction n_2.

Our concern is whether the light rays, after refracting through the surface, form a real image (no observer necessary) or a virtual image (assuming that an observer intercepts the rays). The answer depends on the relative values of n_1 and n_2 and on the geometry of the situation.

Six possible results are shown in Fig. 35-11. In each part of the figure, the medium with the greater index of refraction is shaded, and object O is always in the medium with index of refraction n_1, to the left of the refracting surface. And in each part, a representative ray is shown refracting through the surface. (That ray and a ray along the central axis suffice to determine the position of the image in each case.)

At the point of refraction of the representative ray, the normal to the refracting surface is a radial line through the center of curvature C. Because of the refraction, the ray bends toward the normal if it is entering a medium of larger index of refraction, and away from the normal if it is entering a medium of smaller index of refraction. If the refracted ray is then directed toward the central axis, it and other (undrawn) rays will form a real image on that axis. If it is directed away from the central axis, it cannot form a real image; however, backward extensions of it and other refracted rays can form a virtual image, provided (as with mirrors) some of those rays are intercepted by an observer.

Real images I are formed (at image distance i) in parts a and b of Fig. 35-11, where the refraction directs the ray *toward* the central axis. Virtual images are formed in parts c and d, where the refraction directs the ray *away* from the central axis. Note, in these four parts, that real images are formed when the object is relatively far from the refracting surface, and virtual images are formed when the object is nearer the refracting surface. In the final situations (Fig.

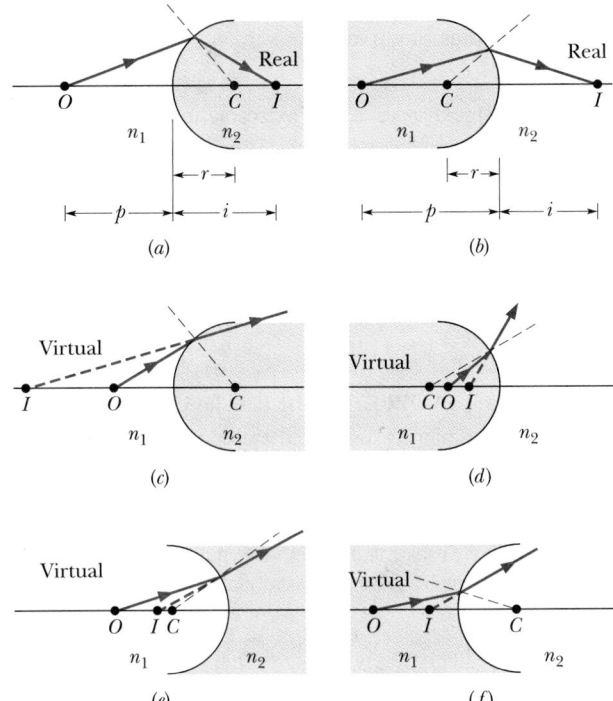

FIGURE 35-11 Six possible ways in which an image can be formed by refraction through a spherical surface of radius r and center of curvature C. The surface separates a medium with index of refraction n_1 from a medium with index of refraction n_2. The point object O is always in the medium with n_1, to the left of the surface. The material with the smaller index of refraction is unshaded (think of it as being air, and the other material as being glass). Real images are formed in (a) and (b); virtual images are formed in the other four situations.

35-11e and f), refraction always directs the ray away from the central axis and virtual images are always formed, regardless of the object distance.

Note the following major difference from reflected images:

Real images form on the side of a refracting surface that is opposite the object, and virtual images form on the same side as the object.

In Section 35-8, we shall show that (for light rays making only small angles with the central axis)

$$\frac{n_1}{p} + \frac{n_2}{i} = \frac{n_2 - n_1}{r}. \qquad (35\text{-}8)$$

Just as with mirrors, the object distance p is positive, and the image distance i is positive for a real image and negative for a virtual image. However, to keep all the signs correct in Eq. 35-8, we must use the following rule for the sign of the radius of curvature r:

> When the object faces a convex refracting surface, the radius of curvature r is positive. When it faces a concave surface, r is negative.

Be careful: this is just the reverse of the sign convention we have for mirrors.

SAMPLE PROBLEM 35-3

A Jurassic mosquito is discovered embedded in a chunk of amber, which has index of refraction 1.6. One surface of the amber is spherically convex with radius of curvature 3.0 mm (Fig. 35-12). The mosquito head happens to be along the central axis of that surface and, when viewed along the axis, appears to be buried 5.0 mm into the amber. How deep is it really?

SOLUTION: First we must realize what is meant by "appears": the observer (in air) sees an image of the mosquito head in the amber and 5.0 mm from the spherical surface of the amber. Because the object (the head) and its image are on the same side of the refracting surface, the image must be virtual and so $i = -5.0$ mm. Also, because the object is always taken to be in the medium of index of refraction n_1, we must have $n_1 = 1.6$ and $n_2 = 1.0$. Finally, because the object faces a concave refracting surface, the radius of curvature r is negative and so $r = -3.0$ mm. What we seek is the object distance p. Substituting the data into Eq. 35-8,

$$\frac{n_1}{p} + \frac{n_2}{i} = \frac{n_2 - n_1}{r},$$

so

$$\frac{1.6}{p} + \frac{1.0}{-5.0 \text{ mm}} = \frac{1.0 - 1.6}{-3.0 \text{ mm}}$$

and

$$p = 4.0 \text{ mm.} \qquad \text{(Answer)}$$

FIGURE 35-12 Sample Problem 35-3. A piece of amber with a mosquito from the Jurassic period, with the head buried at point O. The spherical refracting surface at the right end, with center of curvature C, provides an image I to an observer intercepting rays from the object at O.

This insect has been entombed in amber for about 25 million years. Because we view the insect through a curved refracting surface, the image we see does not coincide with the insect.

CHECKPOINT **3:** A bee is hovering in front of the concave spherical refracting surface of a glass sculpture. (a) Which of the general situations of Fig. 35-11 is like this situation? (b) Is the image produced by the surface real or virtual, and is it on the same side as the bee or the opposite side?

35-6 THIN LENSES

A **lens** is a transparent object with two refracting surfaces whose central axes coincide. The common central axis is the central axis of the lens. When a lens is surrounded by air, light refracts from the air into the lens, crosses through the lens, and then refracts back into the air. Each refraction can change the direction of travel of the light.

A lens that causes light rays initially parallel to the central axis to converge is (reasonably) called a **converging lens.** If, instead, it causes such rays to diverge, the lens is a **diverging lens.** When an object is placed in front of a lens of either type, refraction by the lens of light rays from the object can produce an image of the object.

We shall consider only the special case of a **thin lens,** that is, a lens in which the thickest part is thin compared to the object distance p, the image distance i, and the radii of curvature r_1 and r_2 of the two surfaces of the lens. We shall also consider only light rays that make small angles with the central axis (they are exaggerated in the figures here). In Section 35-8 we shall prove that in such light, a thin lens has a focal length f. Moreover, i and p are related to each other by

$$\frac{1}{f} = \frac{1}{p} + \frac{1}{i} \quad \text{(thin lens),} \qquad (35\text{-}9)$$

which is the same form of equation we had for mirrors. We shall also prove that when a thin lens with index of refraction n is surrounded by air, this focal length f is given by

$$\frac{1}{f} = (n-1)\left(\frac{1}{r_1} - \frac{1}{r_2}\right) \quad \text{(thin lens in air),} \quad (35\text{-}10)$$

which is often called the *lens maker's equation.* Here r_1 is the radius of curvature of the lens surface nearer the object, and r_2 is that of the other surface. The signs of these radii are found with the rules in Section 35-5 for the radii of spherical refracting surfaces. If the lens is surrounded by some medium other than air (say, corn oil) with index of refraction n_{medium}, we replace n in Eq. 35-10 with n/n_{medium}. Keep in mind the basis of Eqs. 35-9 and 35-10:

A lens can produce an image of an object only because it can bend light rays; but it can bend light rays only if its index of refraction differs from that of the surrounding medium.

Figure 35-13a shows a thin lens with convex refracting surfaces, or *sides.* When rays that are parallel to the central axis of the lens are sent through the lens, they refract twice, as is shown enlarged in Fig. 35-13b. This double refraction causes the rays to converge and pass through a common point F_2 at a distance f from the center of the lens. Hence this lens is a converging lens; further, a *real* focal point (or focus) exists at F_2 (because the rays really do pass through it), and the associated focal length is f. When rays parallel to the central axis are sent in the opposite direction through the lens, we find another real focal point at F_1 on the other side of the lens. For a thin lens, these two focal points are equidistant from the lens.

Because the focal points of a converging lens are real, we take the associated focal lengths f to be positive, just as we do with a real focus of a concave mirror. But signs in optics can be tricky, so we had better check this in Eq. 35-10. The left side of that equation is positive if f is positive; how about the right side? We examine it term by term. Because the index of refraction n of glass or any other material is greater than 1, the term $(n-1)$ must be positive. Because the source of the light (which is the object) is at the left and faces the convex left side of the lens, the radius of curvature r_1 of that side must be positive according to the sign rule for refracting surfaces. Similarly, because the object faces a concave right side of the lens, the radius of curvature r_2 of that side must be negative. Thus, the term $(1/r_1 - 1/r_2)$ is positive, the whole right side of Eq. 35-10 is positive, and all the signs are consistent.

Figure 35-13c shows a thin lens with concave sides. When rays that are parallel to the central axis of the lens are sent through this lens, they refract twice, as is shown enlarged in Fig. 35-13d; these rays *diverge,* never passing

FIGURE 35-13 (*a*) Rays initially parallel to the central axis of a converging lens are made to converge to a real focal point F_2 by the lens. The lens is thinner than drawn, with a width like the vertical line through it, where we consider the bending of rays to occur. (*b*) An enlargement of the top part of the lens of (*a*); normals to the surfaces are shown dashed. Note that both refractions of the ray at the surfaces bend the ray downward, toward the central axis. (*c*) The same initially parallel rays are made to diverge by a diverging lens. Extensions of the diverging rays pass through a virtual focal point F_2. (*d*) An enlargement of the top part of the lens of (*c*). Note that both refractions of the ray at the surfaces bend the ray upward, away from the central axis.

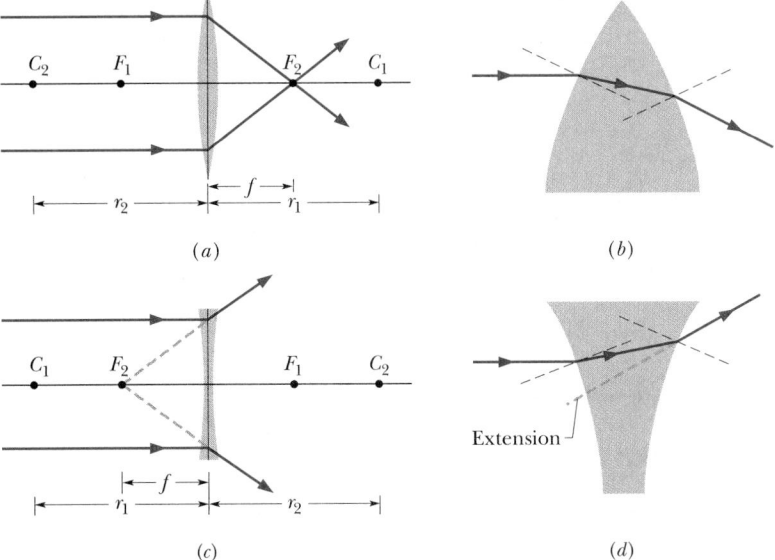

A fire is being started by focusing sunlight onto newspaper by means of a converging lens made of clear ice. The lens was made by freezing water in the shallow vessel (which has a curved bottom).

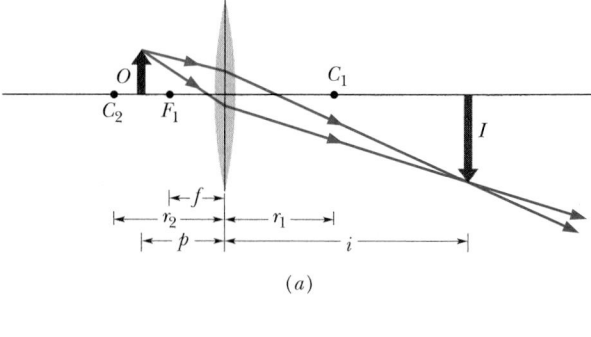

(a)

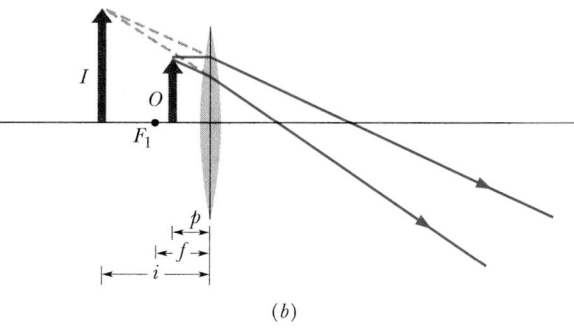

(b)

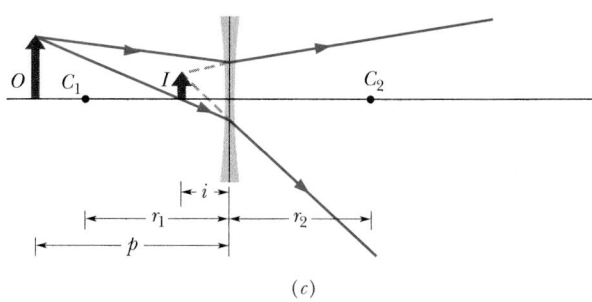

(c)

FIGURE 35-14 (a) A real, inverted image I is formed by a converging lens when the object O is outside the focal point F_1. (b) The image I is virtual and has the same orientation as O when O is inside the focal point. (c) A diverging lens forms a virtual image I, with the same orientation as the object O, whether O is inside or outside the focal point of the lens.

through any common point, and so this lens is a diverging lens. However, extensions of the rays do pass through a common point F_2 at a distance f from the center of the lens. Hence the lens has a *virtual* focal point at F_2. (If your eye intercepts some of the diverging rays, you perceive a bright spot to be at F_2, as if it is the source of the light.) Another virtual focus exists on the opposite side of the lens at F_1, symmetrically placed if the lens is thin. Because the focal points of a diverging lens are virtual, we take the focal length f to be negative.

Images from Lenses

We now consider the types of image formed by converging and diverging lenses. Figure 35-14a shows an object O outside the focal point F_1 of a converging lens. The two rays drawn in the figure show that the lens forms a real, inverted image I of the object on the side of the lens opposite the object.

When the object is placed inside the focal point F_1, as in Fig. 35-14b, the lens forms a virtual image I on the same side of the lens as the object and with the same orientation. Hence, a converging lens can form either a real image or a virtual image, depending on whether the object is outside or inside the focal point, respectively.

Figure 35-14c shows an object O in front of a diverging lens. Regardless of the object distance (regardless of whether O is inside or outside the virtual focal point), this lens produces a virtual image that is on the same side of the lens as the object and has the same orientation.

As with mirrors, we take the image distance i to be positive when the image is real and negative when the image is virtual. However, the locations of real and virtual images from lenses are the reverse of those from mirrors:

Real images form on the side of a lens that is opposite the object, and virtual images form on the same side as the object.

The lateral magnification m produced by converging and diverging lenses is given by Eqs. 35-5 and 35-6, the same as for mirrors.

You have been asked to absorb a lot of information in this section, and you should organize it for yourself by filling in Table 35-2 for thin symmetric lenses. Under Image Location decide whether the image is on the *same* side of the lens as the object or on the *opposite* side. Under Image Type decide whether the image is *real* or *virtual*. Under Image Orientation decide whether the image has the *same* orientation as the object or is *inverted*.

PROBLEM SOLVING TACTICS

TACTIC 1: *Signs of Trouble with Mirrors and Lenses*

Be careful: a mirror with a convex surface has a negative focal length f, just the opposite of a lens with convex surfaces. And a mirror with a concave surface has a positive focal length f, just the opposite of a lens with concave surfaces. Confusing lens properties with mirror properties is a common mistake.

Locating Images of Extended Objects by Drawing Rays

Figure 35-15a shows an object O outside focal point F_1 of a converging lens. We can graphically locate the image of any off-axis point on such an object (such as the tip of the arrow in Fig. 35-15a) by drawing a ray diagram with any two of three special rays through the point. These special rays, chosen from all those that pass through the lens to form the image, are the following:

1. A ray that is initially parallel to the central axis of the lens will pass through focal point F_2 (ray 1 in Fig. 35-15a).

2. A ray that initially passes through focal point F_1 will emerge from the lens parallel to the central axis (ray 2 in Fig. 35-15a).

3. A ray that is initially directed toward the center of the lens will emerge from the lens with no change in its direction (ray 3 in Fig. 35-15a) because the ray encounters the two sides of the lens where they are almost parallel.

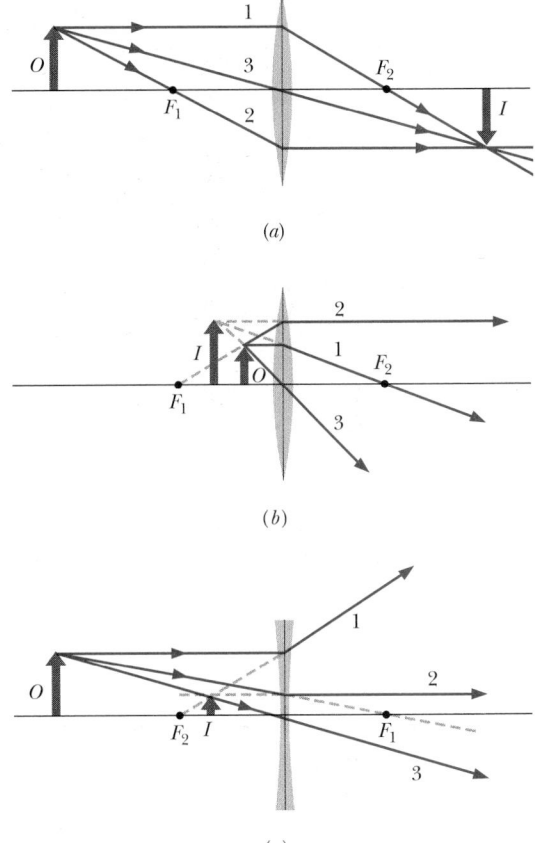

FIGURE 35-15 Three special rays allow us to locate an image formed by a thin lens whether the object O is (*a*) outside or (*b*) inside the focal point of a converging lens, or (*c*) anywhere in front of a diverging lens.

The image of the point is located where the rays intersect on the far side of the lens. The image of the object is found by locating the images of two or more of its points.

Figure 35-15b shows how the extensions of the special three rays can be used to locate the image of an object placed inside focal point F_1 of a converging lens. Note that the description of ray 2 requires modification (it is now a ray whose backward extension passes through F_1). You need to modify the descriptions of rays 1 and 2 to use them to

TABLE 35-2 YOUR ORGANIZING TABLE FOR LENSES

LENS TYPE	OBJECT LOCATION	IMAGE			SIGN		
		LOCATION	TYPE	ORIENTATION	OF f	OF i	OF m
Converging	Inside F						
	Outside F						
Diverging	Anywhere						

locate an image placed (anywhere) in front of a diverging lens (Fig. 35-15c).

Two-Lens Systems

When an object O is placed in front of a system of two lenses whose central axes coincide, we can locate the final image of the system (that is, the image produced by the lens farther from the object) by working in steps. Let lens 1 be the nearer lens and lens 2 the farther lens.

STEP 1. We let p_1 represent the distance of object O from lens 1. We then find the distance i_1 of the image produced by lens 1, either by use of Eq. 35-9 or by drawing rays.

STEP 2. Now, ignoring the presence of lens 1, we treat the image found in step 1 *as the object* for lens 2. If this new object is located beyond lens 2, the object distance p_2 for lens 2 is taken to be negative. (Note this exception to the rule that says the object distance is positive; the exception occurs because the object here is on the side opposite the source of light.) Otherwise, p_2 is taken to be positive as usual. We then find the distance i_2 of the (final) image produced by lens 2 by use of Eq. 35-9 or by drawing rays.

A similar step-by-step solution can be used for any number of lenses or if a mirror is substituted for lens 2.

The overall lateral magnification M produced by a system of two lenses is the product of the lateral magnifications m_1 and m_2 produced by the two lenses:

$$M = m_1 m_2. \tag{35-11}$$

SAMPLE PROBLEM 35-4

A praying mantis preys along the central axis of a thin symmetrical lens, 20 cm from the lens. The lateral magnification of the mantis provided by the lens is $m = -0.25$, and the index of refraction of the lens material is 1.65.

(a) Determine the type of image produced by the lens; the type of lens; whether the object (mantis) is inside or outside the focal point; on which side of the lens the image appears; and whether the image is inverted.

SOLUTION: From Eq. 35-6 ($m = -i/p$) and the given value for m, we see that

$$i = -mp = 0.25p.$$

Even without finishing the calculation, we can answer the questions. Because p is positive, i here must be positive. That means we have a real image, which means we have a converging lens (the only lens that can produce a real image). The object must be outside the focal point (the only way a real image can be produced). And the image is inverted and on the side of the lens opposite the object. (That is how a converging lens makes a real image.)

(b) What is the magnitude r of the two radii of curvature of the lens?

SOLUTION: Because the lens is symmetric, r_1 (for the surface nearer the object) and r_2 have the same magnitude r. Because the lens is a converging lens, $r_1 = +r$ and $r_2 = -r$. We have only one equation (Eq. 35-10) that includes the radius of curvature of a lens, but we lack a value for f to insert in that equation. We can get f from Eq. 35-9 if we first find i. So, we must finish the calculation for i, inserting the given value of p and obtaining

$$i = (0.25)(20 \text{ cm}) = 5.0 \text{ cm}.$$

Now Eq. 35-9 give us

$$\frac{1}{f} = \frac{1}{p} + \frac{1}{i} = \frac{1}{20 \text{ cm}} + \frac{1}{5.0 \text{ cm}},$$

from which we find $f = 4.0$ cm.

Equation 35-10 then gives us

$$\frac{1}{f} = (n-1)\left(\frac{1}{r_1} - \frac{1}{r_2}\right) = (n-1)\left(\frac{1}{+r} - \frac{1}{-r}\right)$$

or, with known values inserted,

$$\frac{1}{4.0 \text{ cm}} = (1.65 - 1)\frac{2}{r},$$

which yields

$$r = (2)(0.65)(4.0 \text{ cm}) = 5.2 \text{ cm}. \quad \text{(Answer)}$$

SAMPLE PROBLEM 35-5

In Fig. 35-16a, a jalapeño seed O_1 is placed in front of two thin symmetrical coaxial lenses 1 and 2, with focal lengths $f_1 = +24$ cm and $f_2 = +9.0$ cm, respectively, and with lens separation $L = 10$ cm. The seed is 6.0 cm from lens 1. Where is its final image?

SOLUTION: We work this problem in steps by first ignoring lens 2 and finding the image of O_1 produced by lens 1 alone (Fig. 35-16b). Equation 35-9, written for lens 1, is

$$\frac{1}{p_1} + \frac{1}{i_1} = \frac{1}{f_1}.$$

Inserting the given data, we have

$$\frac{1}{+6.0 \text{ cm}} + \frac{1}{i_1} = \frac{1}{+24 \text{ cm}},$$

which yields $i_1 = -8.0$ cm.

This tells us that the image I_1 is 8.0 cm from lens 1 and virtual. (We could have guessed that it is virtual by noting that the seed is inside the focal point of lens 1.) Since I_1 is virtual, it is on the same side of the lens as object O_1 and has the same orientation as the seed, as shown in Fig. 35-16b.

In the second step of our solution, we treat image I_1 as an object O_2 for the second lens alone, ignoring lens 1. Because this object O_2 is outside the focal point of lens 2, we can guess that the image I_2 produced by lens 2 is real, inverted, and on the side of the lens opposite O_2. Let us see.

The distance p_2 between object O_2 and lens 2 is, from Fig. 35-16c,

$$p_2 = L + |i_1| = 10 \text{ cm} + 8.0 \text{ cm} = 18 \text{ cm}.$$

Then Eq. 35-9, now written for lens 2, yields

$$\frac{1}{+18 \text{ cm}} + \frac{1}{i_2} = \frac{1}{+9.0 \text{ cm}},$$

hence $\qquad\qquad i_2 = +18 \text{ cm}.$ (Answer)

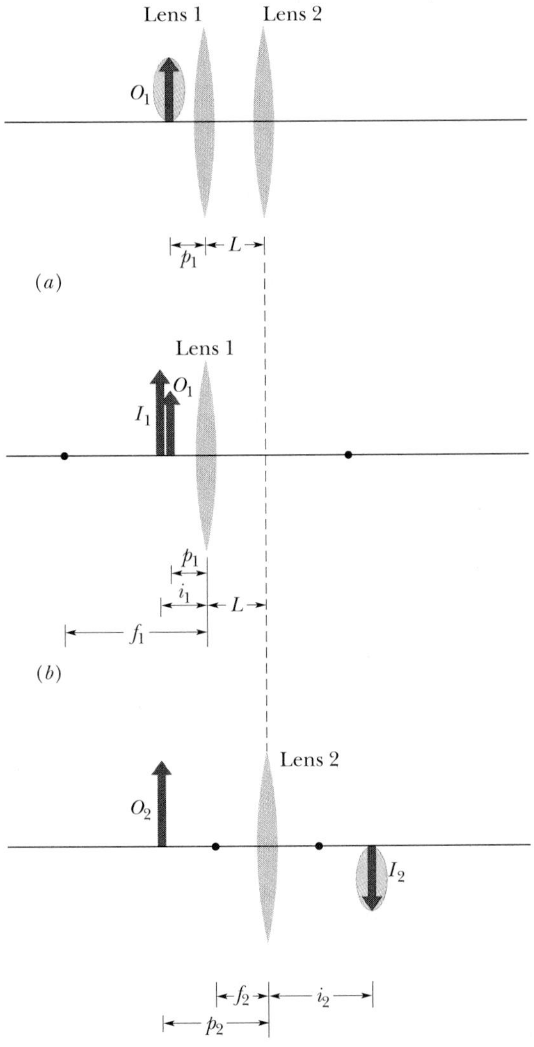

(a)

(b)

(c)

FIGURE 35-16 Sample Problem 35-5. (a) Seed O_1 is distance p_1 from a two-lens system with lens separation L. We use the arrow to orient the seed. (b) The image I_1 produced by lens 1 alone. (c) Image I_1 acts as object O_2 for lens 2 alone, which produces the final image I_2.

The plus sign confirms our guess: image I_2 produced by lens 2 is real, inverted, and on the side of lens 2 opposite O_2, as shown in Fig. 35-16c.

$\mathbb{C}$HECKPOINT **4:** A thin symmetric lens provides an image of a fingerprint with a magnification of $+0.2$ when the fingerprint is 1.0 cm farther from the lens than the focal point of the lens. What are the type and orientation of the image, and what is the type of lens?

35-7 OPTICAL INSTRUMENTS

The human eye is a remarkably effective organ, but its range can be extended in many ways by optical instruments such as eyeglasses, simple magnifying lenses, motion picture projectors, cameras (including TV cameras), microscopes, and telescopes. Many such devices extend the scope of our vision beyond the visible range; satellite-borne infrared cameras and x-ray microscopes are just two examples.

The mirror and thin-lens formulas can be applied only as approximations to most sophisticated optical instruments. The lenses in typical laboratory microscopes are by no means "thin." In most optical instruments the lenses are compound lenses; that is, they are made of several components, the interfaces rarely being exactly spherical. Now we discuss three optical instruments, assuming, for simplicity, that the thin-lens formulas apply.

Simple Magnifying Lens

The normal human eye can focus a sharp image of an object on the retina (at the rear of the eye) if the object is located anywhere from infinity to a certain point called the *near point* P_n. If you move the object closer to the eye than the near point, the perceived retinal image becomes fuzzy. The location of the near point normally varies with age. We have all heard about people who claim not to need glasses but read their newspapers at arm's length; their near points are receding. To find your own near point, remove your glasses or contacts, close one eye, and then bring this page closer to your open eye until it becomes indistinct. In what follows, we take the near point to be 25 cm from the eye, a bit more than the typical value for 20-year-olds.

Figure 35-17a shows an object O placed at the near point P_n of an eye. The size of the image of the object produced on the retina depends on the angle θ that the object occupies in the field of view from that eye. By moving the object closer to the eye, as in Fig. 35-17b, you can

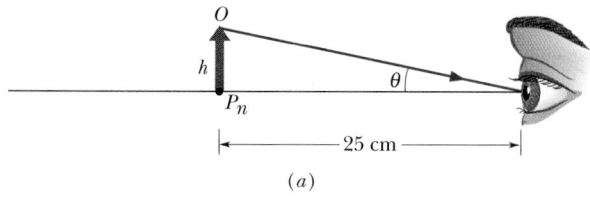

(a)

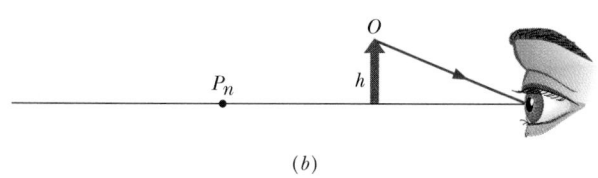

(b)

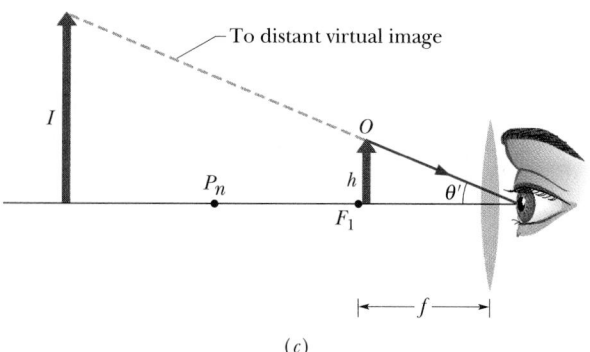

(c)

FIGURE 35-17 (a) An object O of height h, placed at the near point of a human eye, occupies angle θ in the eye's view. (b) The object is moved closer to increase the angle, but now the observer cannot bring the object into focus. (c) A converging lens is placed between the object and the eye, with the object just inside the focal point F_1 of the lens. The image produced by the lens is then far enough away to be focused by the eye, and the image occupies a larger angle θ' than object O does in (a).

increase the angle, hence the possibility of distinguishing details of the object. However, because the object is then closer than the near point, it is no longer *in focus*; that is, the image is no longer clear.

You can restore the clarity by looking at O through a converging lens, placed so that O is just inside the focal point F_1 of the lens, which is at focal length f (Fig. 35-17c). What you then see is the virtual image of O produced by the lens. That image is farther away than the near point; thus the eye can see it clearly.

Moreover, the angle θ' occupied by the virtual image is larger than the largest angle θ that the object alone can occupy and still be seen clearly. The *angular magnification* m_θ (not to be confused with lateral magnification m) of what is seen is

$$m_\theta = \theta'/\theta.$$

In words, the angular magnification of a simple magnifying lens is a comparison of the angle occupied by the image the lens produces with the angle occupied by the

object when the object is moved to the near point of the viewer.

From Fig. 35-17, assuming that O is at the focal point of the lens, and approximating $\tan \theta$ as θ and $\tan \theta'$ as θ', for small angles, we have

$$\theta \approx h/25 \text{ cm} \quad \text{and} \quad \theta' \approx h/f.$$

We then find that

$$m_\theta \approx \frac{25 \text{ cm}}{f} \quad \text{(simple magnifier).} \quad (35\text{-}12)$$

Compound Microscope

Figure 35-18 shows a thin-lens version of a compound microscope. The instrument consists of an *objective* (the front lens) of focal length f_{ob} and an *eyepiece* (the lens near the eye) of focal length f_{ey}. It is used for viewing small objects that are very close to the objective.

The object O to be viewed is placed just outside the first focal point F_1 of the objective, close enough to F_1 that we can approximate its distance p from the lens as being f_{ob}. The separation between the lenses is then adjusted so that the enlarged, inverted, real image I produced by the objective is located just inside the first focal point F_1' of the eyepiece. The *tube length s* shown in Fig. 35-18 is actually large relative to f_{ob}, and we can approximate the distance i between the objective and the image I as being length s.

From Eq. 35-6, and using our approximations for p and i, we can write the lateral magnification produced by the objective as

$$m = -\frac{i}{p} = -\frac{s}{f_{ob}}. \quad (35\text{-}13)$$

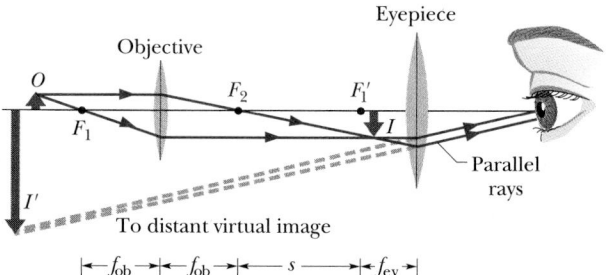

FIGURE 35-18 A thin-lens representation of a compound microscope (not to scale). The objective produces a real image I of object O just inside the focal point F_1' of the eyepiece. Image I then acts as an object for the eyepiece, which produces a virtual final image I' that is seen by the observer. The objective has focal length f_{ob}; the eyepiece has focal length f_{ey}; and s is the tube length.

Since the image I is located just inside the focal point F_1' of the eyepiece, the eyepiece acts as a simple magnifying lens, and an observer sees a final (virtual, inverted) image I' through it. The overall magnification of the instrument is the product of the lateral magnification m produced by the objective, given by Eq. 35-13, and the angular magnification m_θ produced by the eyepiece, given by Eq. 35-12. That is,

$$M = mm_\theta = -\frac{s}{f_{ob}}\frac{25\ \text{cm}}{f_{ey}} \quad \text{(microscope).} \quad (35\text{-}14)$$

Refracting Telescope

Telescopes come in a variety of forms. The form we describe here is the simple refracting telescope that consists of an objective and an eyepiece, both represented in Fig. 35-19 with simple lenses, although in practice, as is also true for most microscopes, each lens is actually a compound lens system.

The lens arrangements for telescopes and for microscopes are similar, but telescopes are designed to view large objects, such as galaxies, stars, and planets, at large distances, whereas microscopes are designed for just the opposite purpose. This difference requires that in the telescope of Fig. 35-19 the second focal point of the objective F_2 coincide with the first focal point of the eyepiece F_1', whereas in the microscope of Fig. 35-18 these points are separated by the tube length s.

In Fig. 35-19a parallel rays from a distant object strike the objective, making an angle θ_{ob} with the telescope axis and forming a real, inverted image at the common focal point F_2, F_1'. This image acts as an object I for the eyepiece, and an observer sees a distant (still inverted) virtual image I' through it. The rays defining the image make an angle θ_{ey} with the telescope axis.

The angular magnification m_θ of the telescope is θ_{ey}/θ_{ob}. From Fig. 35-19b, for rays close to the central axis, we can write $\theta_{ob} = h'/f_{ob}$ and $\theta_{ey} = h'/f_{ey}$, which gives us

$$m_\theta = -\frac{f_{ob}}{f_{ey}} \quad \text{(telescope),} \quad (35\text{-}15)$$

where the minus sign indicates that I' is inverted. In words, the angular magnification of a telescope is a comparison of the angle occupied by the image the telescope produces with the angle occupied by the distant object as seen without the telescope.

Magnification is only one of the design factors for an astronomical telescope and is indeed easily achieved. A good telescope needs *light-gathering power*, which determines how bright the image is. This is important for viewing faint objects such as distant galaxies and is accomplished by making the objective diameter as large as possible. A telescope also needs *resolving power*, which is the ability to distinguish between two distant objects (stars, say) whose angular separation is small. Field of view is another important parameter. A telescope designed to look at galaxies (which occupy a tiny field of view) is much different from one designed to track meteors (which move over a wide field of view).

The telescope designer must also take into account the difference between real lenses and the ideal thin lenses we have discussed. A real lens with spherical surfaces does not form sharp images, a flaw called *spherical aberration*. And because refraction by the two surfaces of a real lens depends on wavelength, a real lens does not focus light of different wavelengths to the same point, a flaw called *chromatic aberration*.

This brief discussion by no means exhausts the design parameters of astronomical telescopes—many others are involved. And we could make a similar listing for any other high-performance optical instrument.

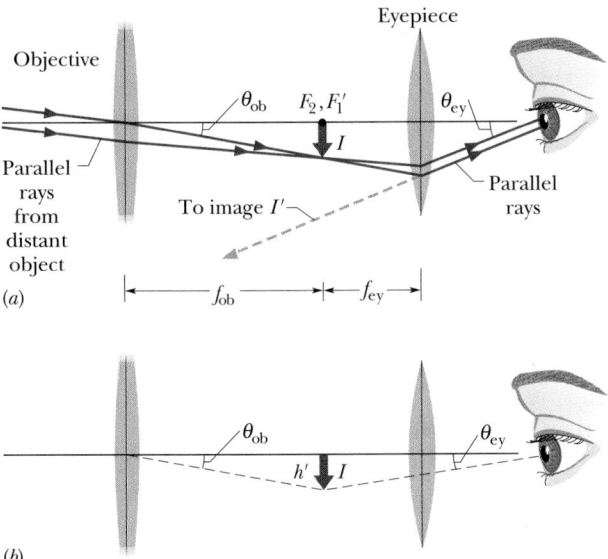

FIGURE 35-19 (*a*) A thin-lens representation of a refracting telescope. The objective produces a real image I of a distant source of light (the object), with approximately parallel light rays at the objective. (One end of the object is assumed to lie on the central axis.) Image I, formed at the common focal points F_2 and F_1', acts as an object for the eyepiece, which produces a virtual final image I' at a great distance from the observer. The objective has focal length f_{ob}; the eyepiece has focal length f_{ey}. (*b*) Image I has height h' and takes up angle θ_{ob} measured from the objective and angle θ_{ey} measured from the eyepiece. [Angles θ_{ey} in (*a*) and (*b*) are equal for rays closer to the central axis than what is drawn here.]

35-8 THREE PROOFS (OPTIONAL)

The Spherical Mirror Formula (Eq. 35-4)

Figure 35-20 shows a point object O placed on the central axis of a concave spherical mirror, outside its center of curvature C. A ray from O that makes an angle α with the axis intersects the axis at I after reflection from the mirror at a. A ray that leaves O along the axis is reflected back along itself at c and also passes through I. Thus I is the image of O; it is a *real* image because light actually passes through it. Let us find the image distance i.

A theorem that is useful here tells us that an exterior angle of a triangle is equal to the sum of the two opposite interior angles. Applying this to triangles OaC and OaI in Fig. 35-20 yields

$$\beta = \alpha + \theta \quad \text{and} \quad \gamma = \alpha + 2\theta.$$

If we eliminate θ between these two equations, we find

$$\alpha + \gamma = 2\beta. \tag{35-16}$$

We can write angles α, β, and γ, in radian measure, as

$$\alpha \approx \frac{\widehat{ac}}{cO} = \frac{\widehat{ac}}{p}, \qquad \beta = \frac{\widehat{ac}}{cC} = \frac{\widehat{ac}}{r},$$

and

$$\gamma \approx \frac{\widehat{ac}}{cI} = \frac{\widehat{ac}}{i}. \tag{35-17}$$

Only the equation for β is exact, because the center of curvature of arc ac is at C. However, the equations for α and γ are approximately correct if these angles are small enough (that is, for rays close to the central axis). Substituting Eqs. 35-17 into Eq. 35-16, using Eq. 35-3 to replace r with $2f$, and canceling $\widehat{ac}$ lead exactly to Eq. 35-4, the relation that we set out to prove.

The Refracting Surface Formula (Eq. 35-8)

The incident ray from point object O in Fig. 35-21 that falls on point a is refracted there according to Eq. 34-44,

$$n_1 \sin \theta_1 = n_2 \sin \theta_2.$$

If α is small, θ_1 and θ_2 will also be small and we can replace the sines of these angles with the angles themselves. Thus the equation above becomes

$$n_1\theta_1 \approx n_2\theta_2. \tag{35-18}$$

We again use the fact that an exterior angle of a triangle is equal to the sum of the two opposite interior angles. Applying this to triangles COa and ICa yields

$$\theta_1 = \alpha + \beta \quad \text{and} \quad \beta = \theta_2 + \gamma. \tag{35-19}$$

If we use Eqs. 35-19 to eliminate θ_1 and θ_2 from Eq. 35-18, we find

$$n_1\alpha + n_2\gamma = (n_2 - n_1)\beta. \tag{35-20}$$

In radian measure the angles α, β, and γ are

$$\alpha \approx \frac{\widehat{ac}}{p}; \quad \beta = \frac{\widehat{ac}}{r}; \quad \gamma \approx \frac{\widehat{ac}}{i}. \tag{35-21}$$

Only the second of these equations is exact. The other two are approximate because I and O are not the centers of circles of which $\widehat{ac}$ is a part. However, for α small enough (for rays close to the axis), the inaccuracies in Eqs. 35-21 are small. Substituting Eqs. 35-21 into Eq. 35-20 leads directly to Eq. 35-8, the relation we set out to prove.

The Thin-Lens Formulas (Eqs. 35-9 and 35-10)

Our plan is to consider each lens surface as a separate refracting surface, and to use the image formed by the first surface as the object for the second.

We start with the thick glass "lens" of length L in Fig. 35-22a whose left and right refracting surfaces are ground to radii r' and r''. A point object O' is placed near the left surface as shown. A ray leaving O' along the central axis is not deflected on entering or leaving the lens.

A second ray leaving O' at an angle α with the central axis intersects the left surface at point a', is refracted, and intersects the second (right) surface at point a''. The ray is again refracted and crosses the axis at I'', which, being the intersection of two rays from O', is the image of point O', formed after refraction at two surfaces.

Figure 35-22b shows that the first (left) surface also forms a virtual image of O' at I'. To locate I', we use Eq. 35-8,

$$\frac{n_1}{p} + \frac{n_2}{i} = \frac{n_2 - n_1}{r}.$$

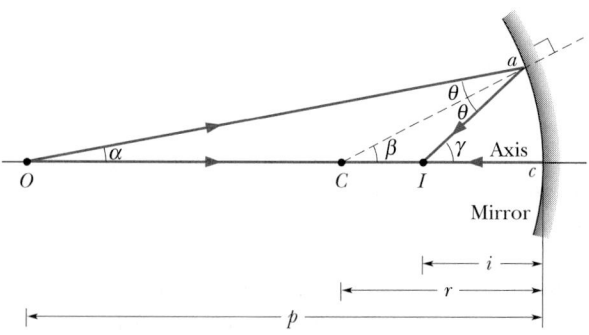

FIGURE 35-20 A concave spherical mirror forms a real point image I by reflecting light rays from a point object O.

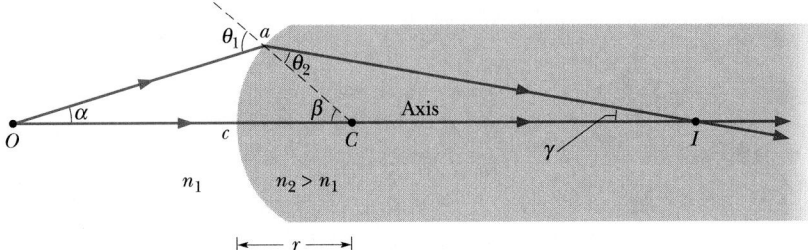

FIGURE 35-21 A real point image I of a point object O is formed by refraction at a spherical convex surface between two media.

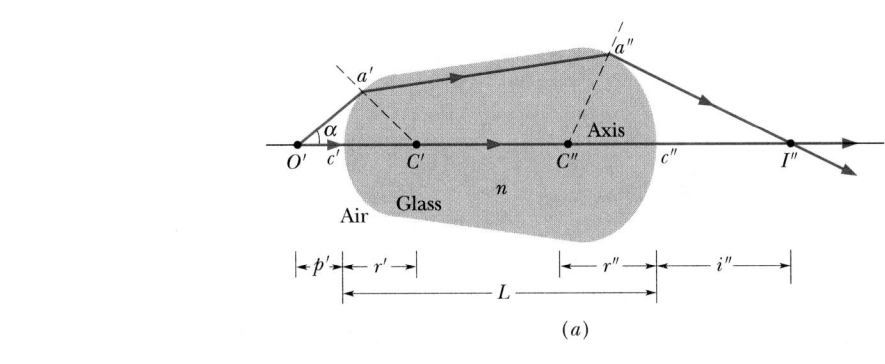

(a)

(b)

FIGURE 35-22 (a) Two rays from point object O' form a real image I'' after refracting through two spherical surfaces of a "lens." The object faces a convex surface at the left side of the lens and a concave surface at the right side. The ray traveling through points a' and a'' is actually close to the central axis through the lens. (b) The left side and (c) the right side, shown separately.

(c)

Putting $n_1 = 1$ and $n_2 = n$ and bearing in mind that the image distance is negative (that is, $i = -i'$ in Fig. 35-22b), we obtain

$$\frac{1}{p'} - \frac{n}{i'} = \frac{n-1}{r'}. \qquad (35\text{-}22)$$

In this equation i' will be a positive number because we have already introduced the minus sign appropriate to a virtual image.

Figure 35-22c shows the second surface again. Unless an observer at point a'' were aware of the existence of the first surface, the observer would think that the light striking that point originated at point I' in Fig. 35-22b and that the region to the left of the surface was filled with glass as indicated. Thus the (virtual) image I' formed by the first surface serves as a real object O'' for the second surface. The distance of this object from the second surface is

$$p'' = i' + L. \qquad (35\text{-}23)$$

To apply Eq. 35-8 to the second surface, we must insert $n_1 = n$ and $n_2 = 1$ because the object now is effectively imbedded in glass. If we substitute with Eq. 35-23, then Eq. 35-8 becomes

$$\frac{n}{i' + L} + \frac{1}{i''} = \frac{1 - n}{r''}. \qquad (35\text{-}24)$$

Let us now assume that the thickness L of the ''lens'' in Fig. 35-22a is so small that we can neglect it in comparison with our other linear quantities (such as p', i', p'', i'', r', and r''). In all that follows we make this *thin-lens approximation*. Putting $L = 0$ in Eq. 35-24 and rearranging the right side lead to

$$\frac{n}{i'} + \frac{1}{i''} = -\frac{n - 1}{r''}. \qquad (35\text{-}25)$$

Adding Eqs. 35-22 and 35-25 leads to

$$\frac{1}{p'} + \frac{1}{i''} = (n - 1)\left(\frac{1}{r'} - \frac{1}{r''}\right).$$

Finally, calling the original object distance simply p and the final image distance simply i leads to

$$\frac{1}{p} + \frac{1}{i} = (n - 1)\left(\frac{1}{r'} - \frac{1}{r''}\right), \qquad (35\text{-}26)$$

which, with a small change in notation, is Eqs. 35-9 and 35-10, the relations we set out to prove.

REVIEW & SUMMARY

Real and Virtual Images

An *image* is a reproduction of an object via light. If the image can form on a surface, it is a *real image* and can exist even if no observer is present. If the image requires the visual system of an observer, it is a *virtual image.*

Image Formation

Spherical mirrors, spherical refracting surfaces, and *thin lenses* can form images of a source of light—the object—by redirecting rays emerging from the source. The image occurs where the redirected rays cross (forming a real image) or where backward extensions of those rays cross (forming a virtual image). If the rays are sufficiently close to the *central axis* through the spherical mirror, refracting surface, or thin lens, we have the following relations between the *object distance p* (which is positive) and the *image distance i* (which is positive for real images and negative for virtual images):

1. Spherical Mirror:

$$\frac{1}{p} + \frac{1}{i} = \frac{1}{f} = \frac{2}{r}, \qquad (35\text{-}4, 35\text{-}3)$$

where f is the mirror's focal length and r is the mirror's radius of curvature. A *plane mirror* is a special case for which $r \rightarrow \infty$, so that $p = -i$. Real images form on the side of a mirror where the object is located, and virtual images form on the opposite side.

2. Spherical Refracting Surface:

$$\frac{n_1}{p} + \frac{n_2}{i} = \frac{n_2 - n_1}{r} \quad \text{(single surface),} \qquad (35\text{-}8)$$

where n_1 is the index of refraction of the material where the object is located, n_2 is the index of refraction of the material on the other side of the refracting surface, and r is the radius of curvature of the surface. When the object faces a convex refracting surface, the radius r is positive. When it faces a concave surface, r is negative. Real images form on the side of a refracting surface that is opposite the object, and virtual images form on the same side as the object.

3. Thin Lens:

$$\frac{1}{p} + \frac{1}{i} = \frac{1}{f} = (n - 1)\left(\frac{1}{r_1} - \frac{1}{r_2}\right), \qquad (35\text{-}9, 35\text{-}10)$$

where f is the lens's focal length, n is the index of refraction of the lens material, and r_1 and r_2 are the radii of curvature of the two sides of the lens, which are spherical surfaces. A convex lens surface that faces the object has a positive radius of curvature; a concave lens surface that faces the object has a negative radius of curvature. Real images form on the side of a lens that is opposite the object, and virtual images form on the same side as the object.

Lateral Magnification

The *lateral magnification m* produced by a spherical mirror or a thin lens is

$$m = -\frac{i}{p}. \qquad (35\text{-}6)$$

The magnitude of m is given by

$$|m| = \frac{h'}{h}, \qquad (35\text{-}5)$$

where h and h' are the heights (measured perpendicular to the central axis) of the object and image, respectively.

Optical Instruments

Three optical instruments that extend human vision are:

1. The *simple magnifying lens,* which produces an *angular magnification m_θ* given by

$$m_\theta = \frac{25 \text{ cm}}{f}, \qquad (35\text{-}12)$$

where f is the focal length of the magnifying lens.

2. The *compound microscope,* which produces an *overall magnification M* given by

$$M = mm_\theta = -\frac{s}{f_{ob}} \frac{25 \text{ cm}}{f_{ey}}, \qquad (35\text{-}14)$$

where m is the lateral magnification produced by the objective, m_θ is the angular magnification produced by the eyepiece, s is the tube length, and f_{ob} and f_{ey} are the focal lengths of the objective and eyepiece, respectively.

3. The *refracting telescope,* which produces an *angular magnification m_θ* given by

$$m_\theta = -\frac{f_{ob}}{f_{ey}}. \qquad (35\text{-}15)$$

QUESTIONS

1. Lake monsters, mermen, and mermaids have long been "sighted" by observers located either on a shore or on a low deck of a ship. From such a low point, an observer can intercept rays of light that leave a floating object (say, a log or a porpoise) and bend slightly back downward toward the observer (one is shown in Fig. 35-23a). The observer then perceives the object as being elongated upward from the water (and probably oscillating because of air turbulence) in a mirage that might easily resemble one of the fabled creatures. Figure 35-23b gives several plots of height from the water surface versus air temperature. Which one best illustrates the conditions giving rise to this mirage?

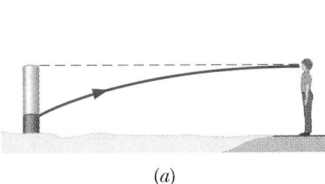

(a)   (b)

FIGURE 35-23 Questions 1 and 2.

2. When Erik the Red was exiled from Iceland by the other Vikings, he headed directly toward the nearest part of Greenland, apparently knowing where that undiscovered land was because of an occasional mirage that brought a virtual image of Greenland around the curve of Earth (Fig. 35-24). Figure 35-23b gives plots of height from Earth's surface versus air temperature. Which one best shows the conditions giving rise to that mirage?

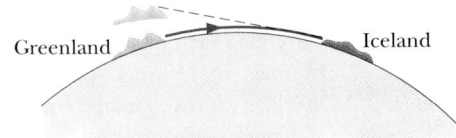

FIGURE 35-24 Question 2.

3. Figure 35-25 shows a fish and a fish stalker in water. (a) Does the stalker see the fish in the general region of point a or point b? (b) Does the fish see the (wild) eyes of the stalker in the general region of point c or point d?

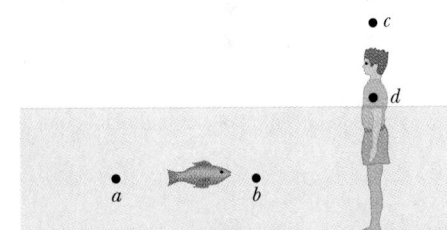

FIGURE 35-25
Question 3.

4. Figure 35-26 shows a coordinate system in front of a flat mirror, with the x axis perpendicular to the mirror. Draw the image of the system in the mirror. (a) Which axis is reversed by the reflection? (b) If you face a mirror, is your image inverted (top for bottom)? Are your left and right reversed (as in common belief)? (c) What then is reversed?

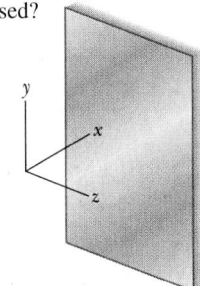

FIGURE 35-26 Question 4.

5. *Putt-putt optics:* Figure 35-27 is an overhead view of a room whose walls are covered with flat mirrors, showing also a sketch of the path of a ray of light from a source S to a target T. The light reflects three times between S and T. The path that produces a "hit" with three reflections must reflect off mirrors ab, bc, and cd; it can be found in the following way. Draw the virtual image I_1 of the target in mirror cd. Then draw the virtual image I_2 of I_1 in the dotted extension of mirror bc shown. Then draw the virtual image I_3 of I_2 in the dotted extension of mirror ab. Now aim the ray toward I_3 and reflect it off the three mirrors. Is there a way to hit the target with (a) two reflections and (b) four reflections?

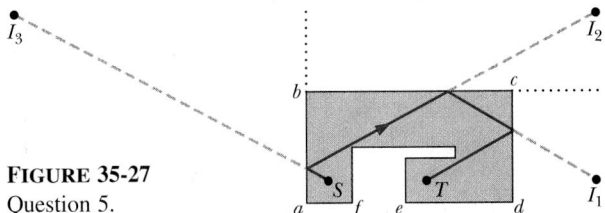

FIGURE 35-27
Question 5.

(a)

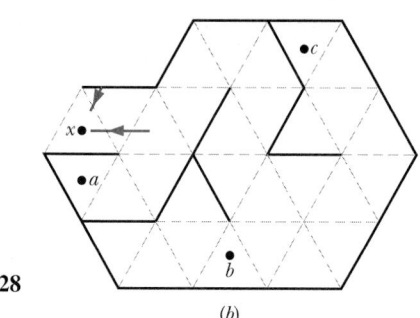

FIGURE 35-28
Question 6.

(b)

6. In the mirror maze of Fig. 35-28a, many "virtual hallways" seem to extend away from you because you see multiple reflections from the mirrors that form the walls of the maze. Those mirrors are placed along some sides of repeated equilateral triangles on the floor. The floor plan for a similar but different maze is shown in Fig. 35-28b; every wall section within this maze is mirrored. If you stand at entrance x, (a) which of the maze monsters a, b, and c hiding in the maze can you see along the virtual hallways extending from entrance x; (b) how many times does each visible monster appear in a hallway; and (c) what is at the far end of a hallway? (*Hint:* The two rays shown are coming down virtual hallways; follow them back into the maze. Do they pass through a triangle with a monster? If so, how many times? For additional analysis, see Jearl Walker, "The Amateur Scientist," *Scientific American,* Vol. 254, pages 120–126, June 1986.

7. A penguin waddles along the central axis of a concave mirror, from the focal point to an effectively infinite distance. (a) How does its image move? (b) Does the height of its image increase continually, decrease continually, or change in some more complicated manner?

8. When a *T. rex* pursues a jeep in the movie *Jurassic Park,* we see a reflected image of the *T. rex* via a side-view mirror, on which is printed the (then darkly humorous) warning: "Objects in mirror are closer than they appear." Is the mirror flat, convex, or concave?

9. Figure 35-29 shows four thin lenses, all of the same material, with sides that either are flat or have a radius of curvature of magnitude 10 cm. Without written calculation, rank the lenses according to the magnitude of the focal length, greatest first.

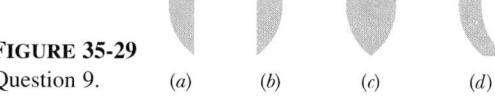

FIGURE 35-29
Question 9.   (a)   (b)   (c)   (d)

10. An object lies before a thin symmetric converging lens. Does the image distance increase, decrease, or remain the same if we increase (a) the index of refraction n of the lens, (b) the magnitude of the radius of curvature of the two sides, and (c) the index of refraction n_{med} of the surrounding medium, keeping n_{med} less than n?

11. A concave mirror and a converging lens (glass with $n = 1.5$) both have a focal length of 3 cm when in air. When they are in water ($n = 1.33$), are their focal lengths greater than, less than, or equal to 3 cm?

12. A converging lens with index of refraction 1.5 is submerged in three separate liquids with indices of refraction 1.3, 1.5, and 1.7. (a) Rank the liquids (by giving their indices of refraction) according to the magnitude of the lens's focal length f in them, greatest first. (b) What is the sign of f in each liquid?

13. The table details six variations of the basic arrangement of two thin lenses represented in Fig. 35-30. (The points labeled F_1 and F_2 are the focal points of lenses 1 and 2.) An object is distance p_1 to the left of lens 1, as in Fig. 35-16. (a) For which variations can we tell, *without calculation,* whether the final image (that due to lens 2) is to the left or right of lens 2 and whether it has the same orientation as the object? (b) For those "easy" variations, give the image location as "left" or "right" and the orientation as "same" or "inverted."

VARIATION	LENS 1	LENS 2	
1	Converging	Converging	$p_1 < f_1$
2	Converging	Converging	$p_1 > f_1$
3	Diverging	Converging	$p_1 < f_1$
4	Diverging	Converging	$p_1 > f_1$
5	Diverging	Diverging	$p_1 < f_1$
6	Diverging	Diverging	$p_1 > f_1$

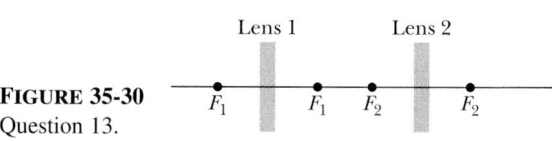

FIGURE 35-30
Question 13.

14. Figure 35-31 shows two situations in which an object sits before two thin lenses whose focal lengths have the same magnitude (one focal point is shown). In each situation, does the final image (due to lens 2) move leftward, move rightward, or remain in place if we move lens 2 toward lens 1?

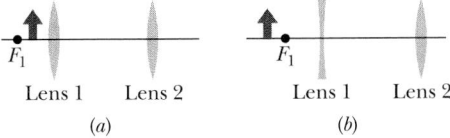

(a)   (b)

FIGURE 35-31 Question 14.

15. Much of the bending of light rays necessary for human vision occurs at the cornea (at the air–eye interface). The cornea has an index of refraction somewhat greater than that of water. (a) When your eye is submerged in a swimming pool, is the bending of light rays at the cornea greater than, less than, or the same as in air? (b) The Central American fish *Anableps anableps* can see simultaneously above and below water because it swims with its eyes partially extending above the water surface. To provide clear sight in both media, is the radius of curvature of the submerged portion of the cornea greater than, less than, or equal to that of the exposed portion?

EXERCISES & PROBLEMS

SECTION 35-2 Plane Mirrors

1E. Figure 35-32 shows an idealized submarine periscope (without submarine) that consists of two parallel plane mirrors set at 45° to the vertical periscope axis and separated by distance L. A penguin is sighted at a distance D from the top mirror. (a) Is the image seen by a submarine officer peering into the periscope real or virtual? (b) Does it have the same or the opposite orientation as the penguin? (c) Is the size (height) of the image greater than, less than, or the same size as that of the penguin? (d) What is the distance of the image from the bottom mirror?

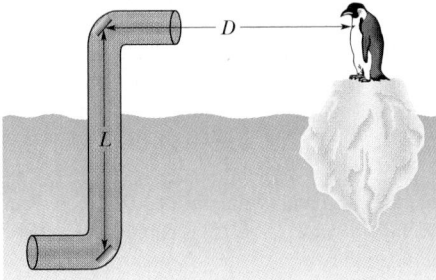

FIGURE 35-32 Exercise 1.

2E. If you move directly toward a plane mirror at speed v, at what speed does your image move toward you (a) in your reference frame and (b) in the reference frame of the mirror?

3E. A moth at about eye level is 10 cm in front of a plane mirror; you are behind the moth, 30 cm from the mirror. For what distance must you focus your eyes to see the image of the moth in the mirror; that is, what is the distance between your eyes and the apparent position of the image?

4E. You look through a camera toward an image of a hummingbird in a plane mirror. The camera is 4.30 m in front of the mirror. The bird is at camera level, 5.00 m to your right and 3.30 m from the mirror. For what distance must you focus your camera lens to get a clear photo of the image; that is, what is the distance between the lens and the apparent position of the image?

5E. Light travels from point A to point B via reflection at point O on the surface of a mirror. Without using calculus, show that length AOB is a minimum when the angle of incidence θ is equal to the angle of reflection ϕ. (*Hint:* Consider the virtual image of A in the mirror.)

6E. Figure 35-33a is an overhead view of two vertical plane mirrors with an object O placed between them. If you look into the mirrors, you see multiple images of O. You can find them by drawing the reflection in each mirror of the angular region between the mirrors, as is done for the left-hand mirror in Fig. 35-33b. Then draw the reflection of the reflection. Continue this on the left and on the right until the reflections meet or overlap at the rear of the mirrors. Then you can count the number of images of O. (a) If $\theta = 90°$, how many images of O would you see? (b) Draw their locations and orientations (as in Fig. 35-33b).

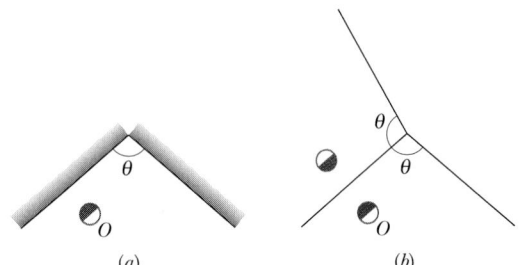

(a)   (b)

FIGURE 35-33 Exercise 6 and Problem 7.

7P. Repeat Exercise 6 for the mirror angle θ equal to (a) 45°, (b) 60°, and (c) 120°. (d) Explain why there are several possible answers for (c).

8P. Figure 35-34 shows an overhead view of a corridor with a plane mirror M mounted at one end. A burglar B sneaks along the corridor directly toward the center of the mirror. If $d = 3.0$ m, how far from the mirror will she be when the security guard S can first see her in the mirror?

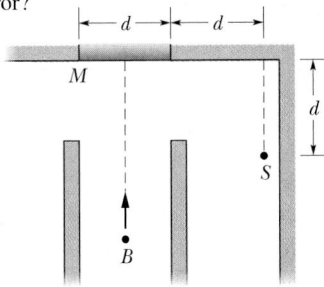

FIGURE 35-34 Problem 8.

9P. Prove that if a plane mirror is rotated through an angle α, the reflected beam is rotated through an angle 2α. Show that this result is reasonable for $\alpha = 45°$.

10P. A point object is 10 cm away from a plane mirror while the eye of an observer (with pupil diameter 5.0 mm) is 20 cm away. Assuming both the eye and the point to be on the same line perpendicular to the mirror surface, find the area of the mirror used in observing the reflection of the point. (*Hint:* Adapt Fig. 35-4.)

11P. You put a point source of light S a distance d in front of a screen A. How is the light intensity at the center of the screen changed if you put a completely reflecting mirror M a distance d behind the source, as in Fig. 35-35? (*Hint:* Use Eq. 34-27.)

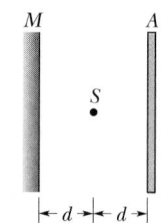

FIGURE 35-35
Problem 11.

12P. Figure 35-36 shows a small lightbulb suspended 250 cm above the surface of the water in a swimming pool. The water is 200 cm deep, and the bottom of the pool is a large mirror. How far below the mirror's surface is the image of the bulb? (*Hint:* Construct a diagram of two rays like that of Fig. 35-3, but take into account the bending of light rays by refraction. Assume that the rays are close to a vertical axis through the bulb, and use the small-angle approximation that $\sin\theta \approx \tan\theta \approx \theta$.)

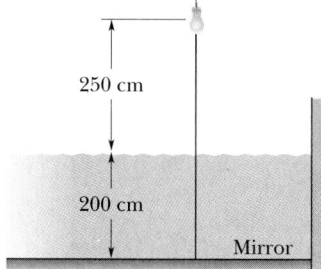

FIGURE 35-36
Problem 12.

13P*. A *corner reflector,* much used in optical, microwave, and other applications, consists of three plane mirrors fastened together to form the corner of a cube. The device has the following property: after three reflections, an incident ray is returned with its direction exactly reversed. Prove this result.

SECTION 35-4   Images from Spherical Mirrors

14E. Equation 35-4 is accurate only if we restrict our attention to rays reflected nearly along the central axis of a mirror, unlike what is drawn (for clarity) in Figs. 35-7b and c. With a ruler, measure r and p in those two parts of Fig. 35-7 and calculate, with Eq. 35-4, the predicted value of i. Then measure i and compare the predicted and measured values.

15E. A concave shaving mirror has a radius of curvature of 35.0 cm. It is positioned so that the (upright) image of a man's

face is 2.50 times the size of the face. How far is the mirror from the face?

16P. Fill in Table 35-3, each row of which refers to a different combination of an object and either a plane mirror, a spherical convex mirror, or a spherical concave mirror. Distances are in centimeters. If a number lacks a sign, find the sign. Sketch each combination and draw in enough rays to locate the object and its image.

TABLE 35-3   PROBLEM 16: MIRRORS

TYPE	f	r	i	p	m	REAL IMAGE?	INVERTED IMAGE?
(a) Concave	20			+10			
(b)				+10	+1.0	No	
(c)		+20		+30			
(d)				+60	−0.50		
(e)		−40	−10				
(f)	20				+0.10		
(g) Convex		40	4.0				
(h)				+24	0.50		Yes

17P. A short straight object of length L lies along the central axis of a spherical mirror, a distance p from the mirror. (a) Show that its image in the mirror will have a length L' where

$$L' = L\left(\frac{f}{p-f}\right)^2.$$

(*Hint:* Locate the two ends of the object.) (b) Show that the *longitudinal magnification* m' $(= L'/L)$ is equal to m^2, where m is the lateral magnification.

18P. (a) A luminous point is moving at speed v_O toward a spherical mirror, along the central axis of the mirror. Show that the image of this point is moving at speed

$$v_I = -\left(\frac{r}{2p-r}\right)^2 v_O,$$

where p is the distance of the luminous point from the mirror at any given time. (*Hint:* Start with Eq. 35-4.) Now assume that the mirror is concave, with $r = 15$ cm, and let $v_O = 5.0$ cm/s. Find the speed of the image when (b) $p = 30$ cm (far outside the focal point), (c) $p = 8.0$ cm (just outside the focal point), and (d) $p = 10$ mm (very near the mirror).

SECTION 35-5   Spherical Refracting Surfaces

19P. A beam of parallel light rays from a laser is incident on a solid transparent sphere of index of refraction n (Fig. 35-37). (a) If a point image is produced at the back of the sphere, what is the index of refraction of the sphere? (b) What index of refraction, if any, will produce a point image at the center of the sphere?

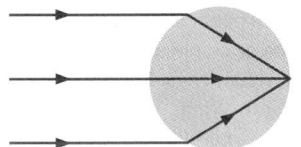

FIGURE 35-37
Problem 19.

20P. Fill in Table 35-4, each row of which refers to a different combination of a point object and a spherical refracting surface separating two media with different indices of refraction. Distances are in centimeters. If a number lacks a sign, find the sign. Sketch each combination and draw in enough rays to locate the object and image.

TABLE 35-4 PROBLEM 20: SPHERICAL REFRACTING SURFACES

	n_1	n_2	p	i	r	INVERTED IMAGE?
(a)	1.0	1.5	+10		+30	
(b)	1.0	1.5	+10	−13		
(c)	1.0	1.5		+600	+30	
(d)	1.0		+20	−20	−20	
(e)	1.5	1.0	+10	−6.0		
(f)	1.5	1.0		−7.5	−30	
(g)	1.5	1.0	+70		+30	
(h)		1.5	+100	+600	−30	

21P. You look downward at a penny that lies at the bottom of a pool of liquid with depth d and index of refraction n (Fig. 35-38). Because you view with two eyes, which intercept different rays of light from the penny, you perceive the penny to be where extensions of the intercepted rays cross, at depth d_a instead of d. Assuming that the intercepted rays in Fig. 35-38 are close to a vertical axis through the penny, show that $d_a = d/n$. (*Hint:* Use the small-angle approximation that $\sin \theta \approx \tan \theta \approx \theta$.)

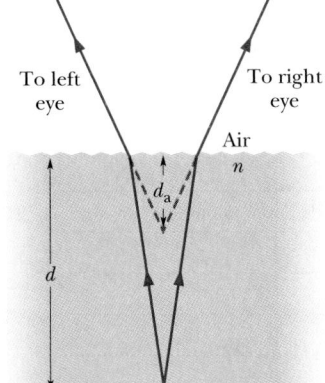

FIGURE 35-38
Problem 21.

22P. A 20-mm-thick layer of water ($n = 1.33$) floats on a 40-mm-thick layer of carbon tetrachloride ($n = 1.46$) in a tank. A penny lies at the bottom of the tank. At what depth below the top

water surface do you perceive the penny? (*Hint:* Use the result and assumptions of Problem 21 and work with a ray diagram of the situation.)

23P*. A goldfish in a spherical fish bowl of radius R is at the level of the center of the bowl and at distance $R/2$ from the glass. What magnification of the fish is produced by the water of the bowl for a viewer looking along a line that includes the fish and the center, from the fish's side of the center? The index of refraction of the water in the bowl is 1.33. Neglect the glass wall of the bowl. Assume the viewer looks with one eye. (*Hint:* Equation 35-5 holds, but Eq. 35-6 does not. You need to work with a ray diagram of the situation and assume that the rays are close to the observer's line of sight.)

SECTION 35-6 Thin Lenses

24E. An object is 20 cm to the left of a thin diverging lens having a 30 cm focal length. What is the image distance i? Find the image position with a ray diagram.

25E. Two coaxial converging lenses, with focal lengths f_1 and f_2, are positioned a distance $f_1 + f_2$ apart, as shown in Fig. 35-39. Arrangements like this are called *beam expanders* and are often used to increase the diameter of a light beam from a laser. (a) If W_1 is the incident beam width, show that the width of the emerging beam is $W_2 = (f_2/f_1)W_1$. (b) Show how a combination of one diverging and one converging lens can also be arranged as a beam expander. Incident rays parallel to the lens axis should exit parallel to the axis.

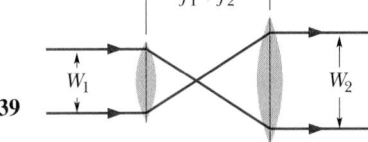

FIGURE 35-39
Exercise 25.

26E. Calculate the ratio of the intensity of the beam emerging from the beam expander of Exercise 25 to the intensity of the incident beam.

27E. A double-convex lens is to be made of glass with an index of refraction of 1.5. One surface is to have twice the radius of curvature of the other and the focal length is to be 60 mm. What are the radii?

28E. You produce an image of the Sun on a screen, using a thin lens whose focal length is 20.0 cm. What is the diameter of the image? (See Appendix C for needed data on the Sun.)

29E. A lens is made of glass having an index of refraction of 1.5. One side of the lens is flat, and the other convex with a radius of curvature of 20 cm. (a) Find the focal length of the lens. (b) If an object is placed 40 cm in front of the lens, where will the image be located?

30E. Using the lens maker's formula (Eq. 35-10), decide which of the thin lenses in Fig. 35-40 are converging and which are diverging for incident light rays that are parallel to the central axis of the lens.

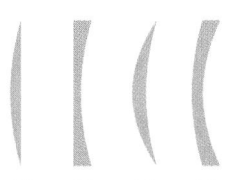

FIGURE 35-40
Exercise 30.   (a)   (b)   (c)   (d)

31E. The formula

$$\frac{1}{p} + \frac{1}{i} = \frac{1}{f}$$

is called the *Gaussian* form of the thin-lens formula. Another form of this formula, the *Newtonian* form, is obtained by considering the distance x from the object to the first focal point and the distance x' from the second focal point to the image. Show that

$$xx' = f^2.$$

32E. A movie camera with a (single) lens of focal length 75 mm takes a picture of a 180 cm high person standing 27 m away. What is the height of the image of the person on the film?

33P. You have a supply of flat glass disks ($n = 1.5$) and a lens-grinding machine that can be set to grind radii of curvature of either 40 or 60 cm. You are asked to prepare a set of six lenses like those shown in Fig. 35-41. What will be the focal length of each lens? Will the lens form a real or a virtual image of the Sun? (*Note:* Where you have a choice of radii of curvature, select the smaller one.)

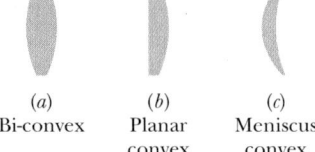

(a)   (b)   (c)
Bi-convex   Planar   Meniscus
             convex   convex

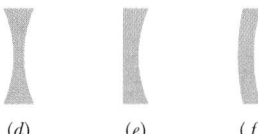

(d)   (e)   (f)
Bi-concave   Planar   Meniscus
             concave   concave

FIGURE 35-41
Problem 33.

34P. To the extent possible, fill in Table 35-5, each row of which refers to a different combination of an object and a thin lens. Distances are in centimeters. For the type of lens, use C for converging and D for diverging. If a number (except for the index of refraction) lacks a sign, find the sign. Sketch each combination and draw in enough rays to locate the object and image.

35P. A converging lens with a focal length of $+20$ cm is located 10 cm to the left of a diverging lens having a focal length of -15 cm. If an object is located 40 cm to the left of the converging lens, locate and describe completely the final image formed by the diverging lens.

36P. An object is placed 1.0 m in front of a converging lens, of focal length 0.50 m, which is 2.0 m in front of a plane mirror. (a) Where is the final image, measured from the lens, that would be seen by an eye looking toward the mirror through the lens (and just past the object)? (b) Is the final image real or virtual? (c) Is the orientation of the final image the same as the object or inverted? (d) What is the lateral magnification?

37P. In Fig. 35-42, an object is placed a distance in front of a converging lens equal to twice the focal length f_1 of the lens. On the other side of the lens is a concave mirror of focal length f_2 separated from the lens by a distance $2(f_1 + f_2)$. (a) Find the location, type, orientation, and lateral magnification of the final image, as seen by an eye looking toward the mirror through the lens and just past (to one side of) the object. (b) Draw a ray diagram to locate the image.

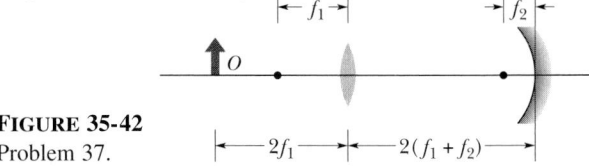

FIGURE 35-42
Problem 37.

38P. In Fig. 35-43, a real inverted image I of an object O is formed by a certain lens (not shown); the object–image separation is $d = 40.0$ cm, measured along the central axis of the lens.

TABLE 35-5 PROBLEM 34: THIN LENSES

	TYPE	f	r_1	r_2	i	p	n	m	REAL IMAGE?	INVERTED IMAGE?
(a)	C	10				$+20$				
(b)		$+10$				$+5.0$				
(c)		10				$+5.0$		>1.0		
(d)		10				$+5.0$		<1.0		
(e)			$+30$	-30		$+10$	1.5			
(f)			-30	$+30$		$+10$	1.5			
(g)			-30	-60		$+10$	1.5			
(h)						$+10$		0.50		No
(i)						$+10$		-0.50		

The image is just half the size of the object. (a) What kind of lens must be used to produce this image? (b) How far from the object must the lens be placed? (c) What is the focal length of the lens?

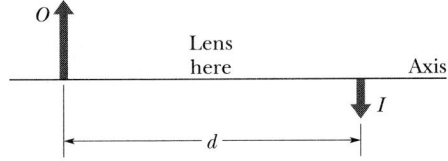

FIGURE 35-43 Problem 38.

39P. An object is 20 cm to the left of a lens with a focal length of +10 cm. A second lens of focal length +12.5 cm is 30 cm to the right of the first lens. (a) Find the location and relative size of the final image. (b) Verify your conclusions by drawing the lens system to scale and constructing a ray diagram. (c) Is the final image real or virtual? (d) Is it inverted?

40P. Two thin lenses of focal lengths f_1 and f_2 are in contact. Show that they are equivalent to a single thin lens with a focal length given by

$$f = \frac{f_1 f_2}{f_1 + f_2}.$$

41P. The *power P* of a lens is defined as $P = 1/f$, where f is the focal length. The unit of power is the *diopter,* where 1 diopter = 1 m^{-1}. (a) Why is this a reasonable definition to use for lens power? (b) Show that the net power of two lenses in contact is given by $P = P_1 + P_2$, where P_1 and P_2 are the powers of the two lenses. (*Hint:* See Problem 40.)

42P. An illuminated slide is held 44 cm from a screen. How far from the slide must a lens of focal length 11 cm be placed to form an image of the slide's picture on the screen?

43P. Show that the distance between an object and its real image formed by a thin converging lens is always greater than or equal to four times the focal length of the lens.

44P. A luminous object and a screen are a fixed distance D apart. (a) Show that a converging lens of focal length f, placed between object and screen, will form a real image on the screen for two lens positions that are separated by a distance

$$d = \sqrt{D(D - 4f)}.$$

(b) Show that the ratio of the two image sizes for these two lens positions is

$$\left(\frac{D - d}{D + d}\right)^2.$$

45P. A narrow beam of parallel light rays is incident on a glass sphere from the left, directed toward the center of the sphere. (The sphere is a lens but certainly not a *thin* lens.) Approximate the angle of incidence of the rays as 0°, and assume that the index of refraction of the glass is $n < 2.0$. Find the image distance i (from the right side of the sphere) in terms of n and the radius r of the sphere. (*Hint:* Apply Eq. 35-8 to locate the image produced by refraction at the left side of the sphere; then use that image as the object for refraction at the right side of the sphere to locate the final image. In the second refraction, is the object distance p positive or negative?)

SECTION 35-7 Optical Instruments

46E. In a microscope of the type shown in Fig. 35-18, the focal length of the objective is 4.00 cm, and that of the eyepiece is 8.00 cm. The distance between the lenses is 25.0 cm. (a) What is the tube length s? (b) If image I in Fig. 35-18 is to be just inside focal point F_1', how far from the objective should the object be? (c) What then is the lateral magnification m of the objective? (d) What then is the angular magnification m_θ of the eyepiece? (e) What then is the overall magnification M of the microscope?

47E. If the angular magnification of an astronomical telescope is 36 and the diameter of the objective is 75 mm, what is the minimum diameter of the eyepiece required to collect all the light entering the objective from a distant point source located on the axis of the instrument?

48P. A simple magnifying lens of focal length f is placed near the eye of someone whose near point P_n is 25 cm from the eye. An object is positioned so that its image in the magnifying lens appears at P_n. (a) What is the angular magnification of the lens? (b) What is the angular magnification if the object is moved so that its image appears at infinity? (c) Evaluate the angular magnifications of (a) and (b) for $f = 10$ cm. (Viewing an image at P_n requires effort by muscles in the eye, whereas for many people viewing an image at infinity requires no effort.)

49P. (a) Show that if the object O in Fig. 35-17c is moved from focal point F_1 toward the eye, the image moves in from infinity and the angle θ' (and thus the angular magnification m_θ) increases. (b) If you continue this process, at what image location will m_θ have its maximum usable value? (You can then still increase m_θ, but the image will no longer be clear.) (c) Show that the maximum usable value of m_θ is 1 + (25 cm)/f. (d) Show that in this situation the angular magnification is equal to the lateral magnification.

50P. Figure 35-44a shows the basic structure of a human eye. Light refracts into the eye through the cornea and is then further redirected by a lens whose shape (and thus ability to focus the light) is controlled by muscles. We can treat the cornea and eye lens as a single effective thin lens (Fig. 35-44b). If the muscles are relaxed, a ''normal'' eye focuses parallel light rays from a distant object O to a point on the retina at the back of the eye, where processing of the visual information begins. As an object is brought close to the eye, the muscles change the shape of the lens so that rays form an inverted real image on the retina (Fig. 35-44c). (a) Suppose the ''relaxed'' focal length f of the effective thin lens of the eye is 2.50 cm. If an object is located at a distance $p = 40.0$ cm, what focal length f' of the effective thin lens is required for the object to be seen clearly? (b) Do the eye muscles increase or decrease the radii of curvature of the eye lens to produce focal length f'?

51P. In an eye that is *farsighted,* the eye focuses parallel rays so that the image would form behind the retina, as in Fig. 35-45a. In an eye that is *nearsighted,* the image is formed in front of the retina, as in Fig. 35-45b. (a) How would you design a corrective lens for each eye defect? Make a ray diagram for each case. (b) If you need eyeglasses only for reading, are you nearsighted or far-

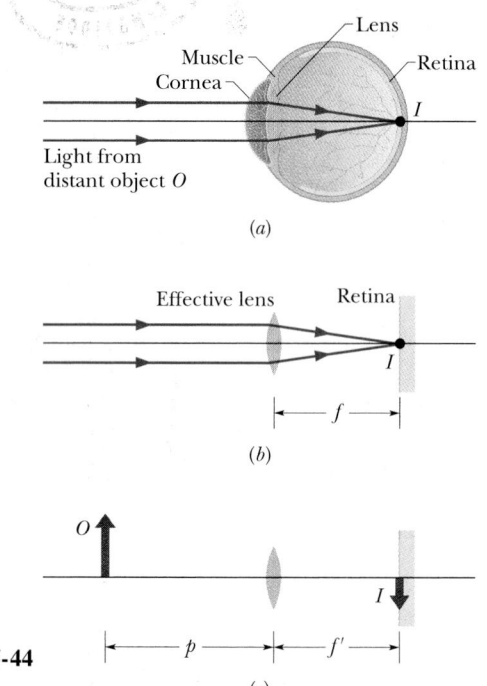

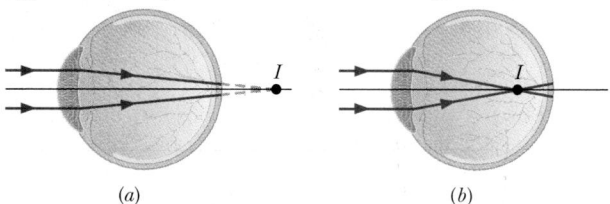

FIGURE 35-44
Problem 50.

(c)

sighted? (c) What is the function of bifocal glasses, in which the upper and lower parts have different focal lengths?

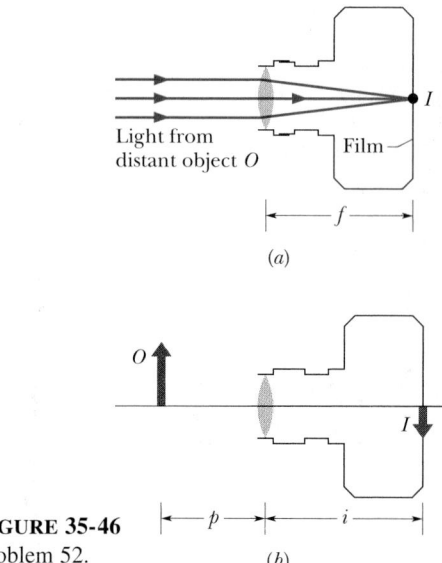

FIGURE 35-45 Problem 51.

52P. Figure 35-46a shows the basic structure of a camera. A lens can be moved forward or back to produce an image on film at the back of the camera. For a certain camera, with the distance between the lens and the film set at $f = 5.0$ cm, parallel light rays from a distant object O converge to a point image on the film. The object is now brought closer, to a distance of $p = 100$ cm, and the lens–film distance is adjusted so that an inverted real image forms on the film (Fig. 35-46b). (a) What is the lens–film distance (the image distance i) now? (b) By how much was the lens–film distance changed?

FIGURE 35-46
Problem 52.

(b)

53P. In a certain compound microscope, the object is 10.0 mm from the objective. The lenses are 300 mm apart and the intermediate image is 50.0 mm from the eyepiece. What overall magnification is produced?

Electronic Computation

54. The equation $1/p + 1/i = 2/r$ for spherical mirrors is an approximation that is valid if the image is formed by rays that make only small angles with the central axis. In reality, many of the angles are large, which smears out the image a little. You can use a computer to find out how much. Refer to Fig. 35-20 and consider a ray that leaves a point source (the object) on the central axis and that makes an angle α with that axis.

First find the point of intersection of the ray with the mirror. If the coordinates of this point are x and y and the origin is placed at the center of curvature, then $y = (x + p - r)\tan \alpha$ and $x^2 + y^2 = r^2$, where p is the object distance and r is the mirror's radius of curvature. Use $\tan \beta = y/x$ to find the angle β at the point of intersection, and then use $\alpha + \gamma = 2\beta$ to find the value of γ. Finally use $\tan \gamma = y/(x + i - r)$ to find the image distance i.

(a) Suppose $r = 12$ cm and $p = 20$ cm. For each of the following values of α, find the position of the image, that is, the position of the point where the reflected ray crosses the central axis: 0.500, 0.100, 0.0100 rad. Compare the results with that obtained with the equation $1/p + 1/i = 2/r$. (b) Repeat the calculations for $p = 4.00$ cm.

36
Interference

*At first glance, the top surface of the <u>Morpho</u> butterfly's wing is simply a
beautiful blue-green. But there is something strange about the color, for it
almost glimmers, unlike the colors of most objects. And if you change your
perspective, or if the wing moves, the tint of the color changes. The wing is said
to be iridescent, and the blue-green we see hides the wing's "true" dull brown
color that appears on the bottom surface. What, then, is so different about the
top surface that gives us this arresting display?*

36-1 INTERFERENCE

Sunlight, as the rainbow shows us, is a composite of all the colors of the visible spectrum. The colors reveal themselves in the rainbow because the incident wavelengths are bent through different angles as they pass through raindrops that produce the bow. However, soap bubbles and oil slicks can also show striking colors, produced not by refraction but by constructive and destructive **interference** of light. The interfering waves combine either to enhance or to suppress certain colors in the spectrum of the incident sunlight. Interference of light waves is thus a superposition phenomenon like those we discussed in Chapter 17.

This selective enhancement or suppression of wavelengths has many applications. When light encounters an ordinary glass surface, for example, about 4% of the incident energy is reflected, thus weakening the transmitted beam by that amount. This unwanted loss of light can be a real problem in optical systems with many components. A thin, transparent ''interference film,'' deposited on the glass surface, can reduce the amount of reflected light (and thus enhance the transmitted light) by destructive interference. The bluish cast of a camera lens reveals the presence of such a coating. Interference coatings can also be used to enhance—rather than reduce—the ability of a surface to reflect light.

To understand interference, we must go beyond the restrictions of geometrical optics and employ the full power of wave optics. In fact, as you will see, the existence of interference phenomena is perhaps our most convincing evidence that light is a wave—because interference cannot be explained other than with waves.

36-2 LIGHT AS A WAVE

The first person to advance a convincing wave theory for light was Dutch physicist Christian Huygens, in 1678. While much less comprehensive than the later electromagnetic theory of Maxwell, Huygens' theory was simpler mathematically and remains useful today. Its great advantages are that it accounts for the laws of reflection and refraction in terms of waves and gives physical meaning to the index of refraction.

Huygens' wave theory is based on a geometrical construction that allows us to tell where a given wavefront will be at any time in the future if we know its present position. This construction is based on **Huygens' principle,** which is:

All points on a wavefront serve as point sources of spherical secondary wavelets. After a time t, the new position of the wavefront will be that of a surface tangent to these secondary wavelets.

Here is a simple example. At the left in Fig. 36-1, the present location of a wavefront of a plane wave traveling to the right in vacuum is represented by plane ab, perpendicular to the page. Where will the wavefront be at time Δt later? We let several points on plane ab (the dots) serve as sources of spherical secondary wavelets that are emitted at $t = 0$. At time Δt, the radius of all these spherical wavelets will have grown to $c\,\Delta t$, where c is the speed of light in vacuum. We draw plane de tangent to these wavelets at time Δt. This plane represents the wavefront of the plane wave at time Δt; it is parallel to plane ab and a perpendicular distance $c\,\Delta t$ from it.

The Law of Refraction

We now use Huygens' principle to derive the law of refraction, Eq. 34-44 (Snell's law). Figure 36-2 shows three stages in the refraction of several wavefronts at a plane interface between air (medium 1) and glass (medium 2). We arbitrarily choose the wavefronts in the incident beam to be separated by λ_1, the wavelength in medium 1. Let the speed of light in air be v_1 and that in glass be v_2. We assume that $v_2 < v_1$, which happens to be true.

Angle θ_1 in Fig. 36-2a is the angle between the wavefront and the interface; this is the same as the angle between the *normal* to the wavefront (that is, the incident ray) and the *normal* to the interface; thus θ_1 is the angle of incidence. As the wave moves into the glass (Fig. 36-2b), the time ($= \lambda_1/v_1$) for a Huygens wavelet to expand from

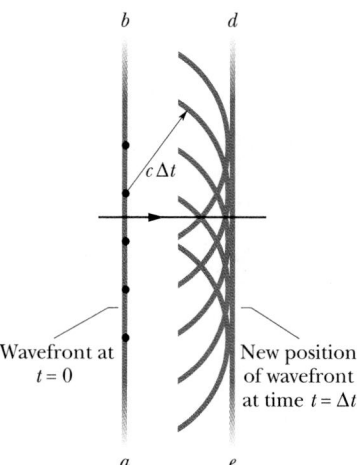

FIGURE 36-1 The propagation of a plane wave in vacuum, as portrayed by Huygens' principle.

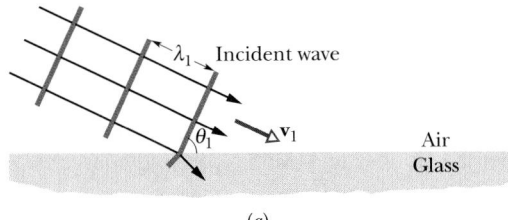

(a)

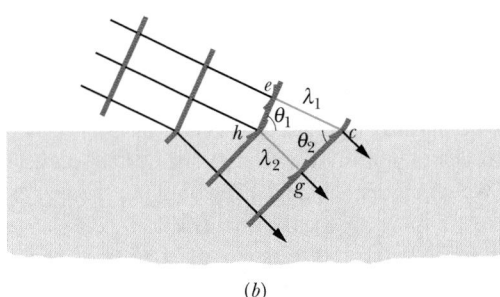

(b)

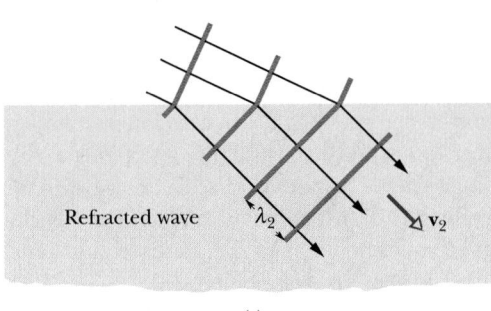

(c)

FIGURE 36-2 The refraction of a plane wave at an air–glass interface, as portrayed by Huygens' principle. The wavelength in glass is smaller than that in air. For simplicity, the reflected wave is not shown.

point e to include point c will equal the time $(= \lambda_2/v_2)$ for a wavelet in the glass to expand at the reduced speed v_2 from h to include g. By equating these times, we obtain the relation

$$\frac{\lambda_1}{\lambda_2} = \frac{v_1}{v_2}, \quad (36\text{-}1)$$

which shows that the wavelengths of light in two media are proportional to the speeds of light in those media.

By Huygens' principle, the refracted wavefront must be tangent to an arc of radius λ_2 centered on h, say at point g. The refracted wavefront must also be tangent to an arc of radius λ_1 centered on e, say at c. Then the refracted wavefront must be oriented as shown. Note that θ_2, the angle between the refracted wavefront and the interface, is actually the angle of refraction.

For the right triangles hce and hcg in Fig. 36-2b we

may write

$$\sin \theta_1 = \frac{\lambda_1}{hc} \quad \text{(for triangle } hce\text{)}$$

and

$$\sin \theta_2 = \frac{\lambda_2}{hc} \quad \text{(for triangle } hcg\text{)}.$$

Dividing the first of these two equations by the second and using Eq. 36-1, we find

$$\frac{\sin \theta_1}{\sin \theta_2} = \frac{\lambda_1}{\lambda_2} = \frac{v_1}{v_2}. \quad (36\text{-}2)$$

We can define an **index of refraction** for each medium as the ratio of the speed of light in vacuum to the speed of light v in the medium. Thus

$$n = \frac{c}{v} \quad \text{(index of refraction).} \quad (36\text{-}3)$$

In particular, for our two media, we have

$$n_1 = \frac{c}{v_1} \quad \text{and} \quad n_2 = \frac{c}{v_2}. \quad (36\text{-}4)$$

If we combine Eqs. 36-2 and 36-4 we find

$$\frac{\sin \theta_1}{\sin \theta_2} = \frac{c/n_1}{c/n_2} = \frac{n_2}{n_1} \quad (36\text{-}5)$$

or

$$n_1 \sin \theta_1 = n_2 \sin \theta_2 \quad \begin{array}{c}\text{(law of}\\ \text{refraction),}\end{array} \quad (36\text{-}6)$$

as introduced in Chapter 34.

CHECKPOINT 1: The figure shows a monochromatic ray of light traveling across parallel interfaces, from an original material a, through layers of material b and c, and then back into material a. Rank the materials according to the speed of light in them, greatest first.

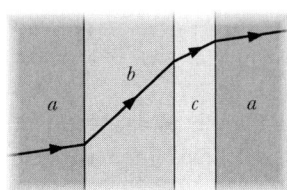

Wavelength and Index of Refraction

We have now seen that the wavelength of light changes when the speed of the light changes, as happens when light crosses an interface from one medium into another. Further, the speed of light in any medium depends on the index of refraction of the medium, according to Eq. 36-3. Thus the wavelength of light in any medium depends on the index of refraction of the medium. Let a certain monochromatic light have wavelength λ and speed c in vacuum and wavelength λ_n and speed v in a medium with an index of refraction n. Then we can rewrite Eq. 36-1 as

$$\lambda_n = \lambda \frac{v}{c}. \tag{36-7}$$

Using Eq. 36-3 to substitute $1/n$ for v/c then yields

$$\lambda_n = \frac{\lambda}{n}. \tag{36-8}$$

This equation relates the wavelength of light in any medium to its wavelength in vacuum. It tells us that the larger the index of refraction of a medium, the smaller the wavelength of light in that medium.

This fact is important in certain situations involving the interference of light waves. For example, in Fig. 36-3, the *waves of the rays* (that is, the waves represented by the rays) have identical wavelengths λ and are initially in phase in air ($n \approx 1$). One of the waves travels through medium 1 of index of refraction n_1 and length L. The other travels through medium 2 of index of refraction n_2 and the same length L. Because the wavelength of the light differs in the two media, the two waves may no longer be in phase when they leave these media.

The phase difference between two light waves can change if the waves travel through different materials having different indices of refraction.

As we shall discuss soon, this change in the phase difference can determine the interference of the light waves if they reach some common point. To find their new phase difference in terms of wavelengths, we first count the number N_1 of wavelengths there are in the length L of medium 1. From Eq. 36-8, the wavelength in medium 1 is $\lambda_{n1} = \lambda/n_1$. So

$$N_1 = \frac{L}{\lambda_{n1}} = \frac{Ln_1}{\lambda}. \tag{36-9}$$

Similarly, we count the number N_2 of wavelengths there are in the length L of medium 2, where the wavelength is

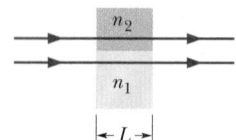

FIGURE 36-3 Two light rays travel through two media having different indices of refraction.

$\lambda_{n2} = \lambda/n_2$:

$$N_2 = \frac{L}{\lambda_{n2}} = \frac{Ln_2}{\lambda}. \tag{36-10}$$

To find the new phase difference between the waves, we subtract the smaller of N_1 and N_2 from the larger. Assuming $n_2 > n_1$, we would obtain

$$N_2 - N_1 = \frac{Ln_2}{\lambda} - \frac{Ln_1}{\lambda} = \frac{L}{\lambda}(n_2 - n_1). \tag{36-11}$$

Suppose Eq. 36-11 tells us that the waves now have a phase difference of 45.6 wavelengths. That is equivalent to taking the initially in-phase waves and shifting one of them by 45.6 wavelengths. However, a shift of an integer number of wavelengths (such as 45) would put the waves back in phase. So it is only the decimal fraction (here, 0.6) that is important. A phase difference of 45.6 wavelengths is equivalent to a phase difference of 0.6 wavelength.

A phase difference of 0.5 wavelength puts the waves exactly out of phase. If the waves were to reach some common point, they would then undergo fully destructive interference, producing darkness at that point. With a phase difference of 0.0 or 1.0 wavelength, they would, instead, undergo fully constructive interference, resulting in brightness at the common point. Our phase difference of 0.6 wavelength is an intermediate situation, but closer to destructive interference, and the waves would produce a dimly illuminated crossing point.

We can also express phase difference in terms of radians and degrees, as we have done already. A phase difference of one wavelength is equivalent to phase differences of 2π rad and $360°$.

SAMPLE PROBLEM 36-1

In Fig. 36-3, the two light waves that are represented by the rays have wavelength 550.0 nm before entering media 1 and 2. Medium 1 is now just air, and medium 2 is a transparent plastic layer of index of refraction 1.600 and thickness 2.600 μm.

(a) What is the phase difference of the emerging waves, in wavelengths?

SOLUTION: From Eq. 36-11, with $n_1 = 1.000$, $n_2 = 1.600$, $L = 2.600$ μm, and $\lambda = 550.0$ nm, we have

$$N_2 - N_1 = \frac{L}{\lambda}(n_2 - n_1)$$

$$= \frac{2.600 \times 10^{-6} \text{ m}}{5.500 \times 10^{-7} \text{ m}}(1.600 - 1.000)$$

$$= 2.84, \tag{Answer}$$

which is equivalent to a phase difference of 0.84 wavelength.

(b) If the rays of the waves were angled slightly so that the waves reached the same point on a distant viewing screen, what type of interference would the waves produce at that point?

SOLUTION: The effective phase difference of 0.84 wavelength is an intermediate situation, but closer to fully constructive interference (1.0) than to fully destructive interference (0.5).

(c) What is the phase difference in radians and in degrees?

SOLUTION: In radians,

$$(0.84)(2\pi \text{ rad}) = 5.3 \text{ rad.} \qquad \text{(Answer)}$$

In degrees,

$$(0.84)(360°) = 302° \approx 300°. \qquad \text{(Answer)}$$

CHECKPOINT 2: The light waves of the rays in Fig. 36-3 have the same wavelength and are initially in phase. (a) If 7.60 wavelengths fit within the length of the top layer and 5.50 wavelengths fit within that of the bottom layer, which layer has the greater index of refraction? (b) If the rays are angled slightly so that they meet at the same point on a distant screen, will the interference there result in brightness, bright intermediate illumination, dark intermediate illumination, or darkness?

36-3 DIFFRACTION

In the next section we shall discuss the experiment that first proved that light is a wave. To prepare for that discussion, we must introduce the idea of **diffraction** of waves, a phenomenon that we explore much more fully in Chapter 37. Its essence is this: if a wave encounters a barrier that has an opening of dimensions similar to the wavelength, the part of the wave that passes through the opening will flare out —will *diffract*—into the region beyond the barrier. The flaring out is consistent with the spreading of the wavelets in the Huygens construction of Fig. 36-1. Diffraction occurs for waves of all types, not just light waves; Fig. 36-4 shows the diffraction of water waves traveling across the surface of water in a shallow tank.

Figure 36-5a shows the situation schematically for an incident plane wave of wavelength λ encountering a slit that has width $a = 6.0\lambda$ and extends into and out of the page. The wave flares out on the far side of the slit. Figures 36-5b (with $a = 3.0\lambda$) and 36-5c ($a = 1.5\lambda$) illustrate the main feature of diffraction: the narrower the slit, the greater the diffraction.

Diffraction limits geometrical optics, in which we represent an electromagnetic wave with a ray. If we actu-

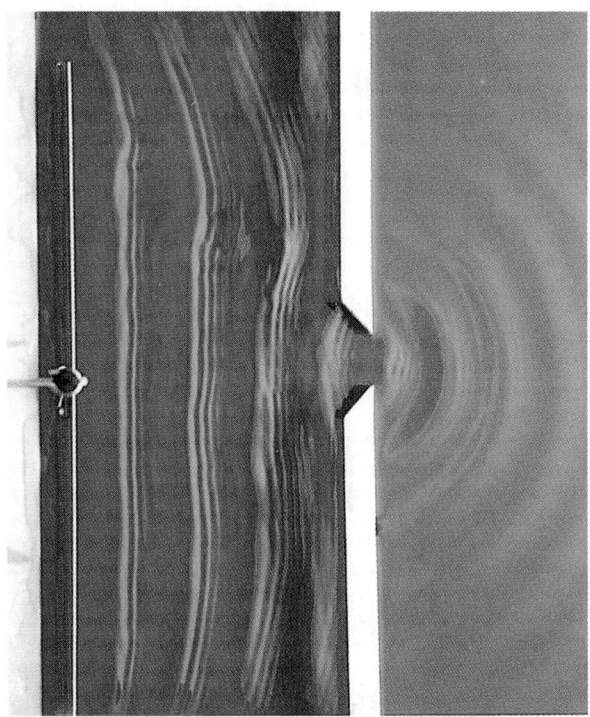

FIGURE 36-4 The diffraction of water waves in a ripple tank. Waves moving from left to right flare out through an opening in a barrier along the water surface.

ally try to form a ray by sending light through a narrow slit, or through a series of narrow slits, diffraction will always defeat our effort because it always causes the light to spread. Indeed, the narrower we make the slits (in the hope of producing a narrower beam), the greater the spreading is. Thus, geometrical optics holds only when slits or other apertures that might be located in the path of light do not have dimensions comparable to or smaller than the wavelength of the light.

36-4 YOUNG'S INTERFERENCE EXPERIMENT

In 1801 Thomas Young experimentally proved that light is a wave, contrary to what most other scientists then thought. He did so by demonstrating that light undergoes interference, as do water waves, sound waves, and waves of all other types. In addition, he was able to measure the average wavelength of sunlight; his value, 570 nm, is impressively close to the modern accepted value of 555 nm. We shall here examine Young's historic experiment as an example of the interference of light waves.

Figure 36-6 gives the basic arrangement of Young's experiment. Light from a distant monochromatic source illuminates slit S_0 in screen A. The emerging light then

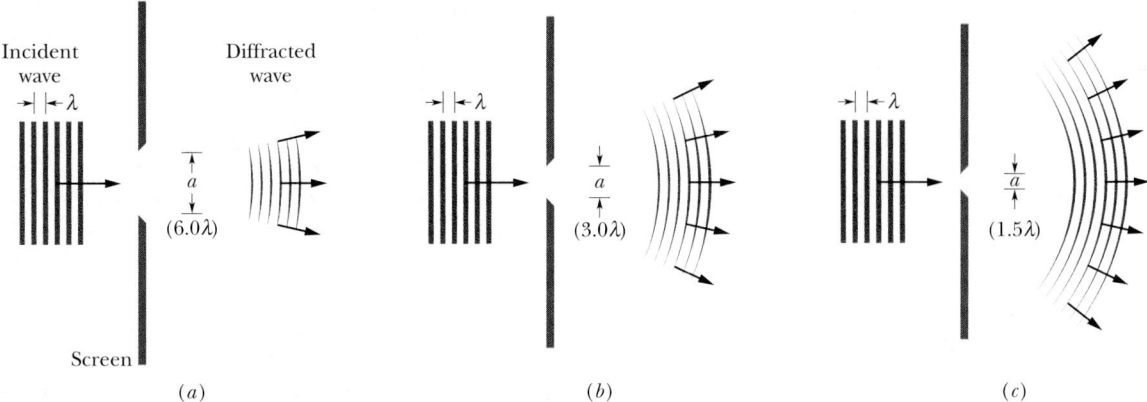

FIGURE 36-5 Diffraction represented schematically. For a given wavelength λ, the diffraction is more pronounced the smaller the slit width a. The figures show the cases for (*a*) slit width $a = 6.0\lambda$, (*b*) slit width $a = 3.0\lambda$, and (*c*) slit width $a = 1.5\lambda$. In all three cases, the screen and the length of the slit extend well into and out of the page, perpendicular to it.

spreads via diffraction to illuminate two slits S_1 and S_2 in screen B. Diffraction of the light by the two slits sends overlapping circular waves into the region beyond screen B, where the waves from one slit interfere with the waves from the other slit.

In the "snapshot" of Fig. 36-6, points at which the interference is fully constructive (interference maxima) are marked with dots. We cannot see such points except where a viewing screen C intercepts the light. On the screen, points of interference maxima form visible bright rows — called *bright bands, bright fringes,* or (loosely speaking) *maxima* — that extend across the screen (into and out of the page in Fig. 36-6). Dark regions — called *dark bands, dark fringes,* or (loosely speaking) *minima* — result from fully destructive interference and are visible between adjacent pairs of bright fringes. (*Maxima* and *minima* more properly refer to the center of a band.) The pattern of bright and dark fringes on the screen is called an **interference pattern.** Figure 36-7 is a photograph of the interference pattern; the photograph has been rotated by 90° to save space.

Locating the Fringes

Waves produce fringes in a *Young's double-slit interference experiment,* as it is called, but what exactly determines the locations of the fringes? To answer, we shall use the arrangement in Fig. 36-8a. There, a plane wave of monochromatic light is incident on two slits S_1 and S_2 in screen B; the light diffracts through the slits and produces an interference pattern on screen C. We draw a central axis from the point halfway between the slits to screen C as a

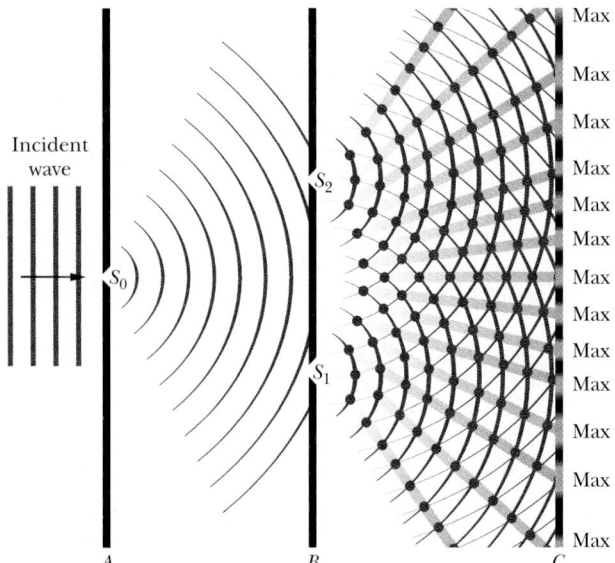

FIGURE 36-6 In Young's interference experiment, incident monochromatic light is diffracted by slit S_0, which then acts as a point source of light that emits semicircular wavefronts. As that light reaches screen B, it is diffracted by slits S_1 and S_2, which then act as two point sources of light. The light waves traveling from slits S_1 and S_2 overlap and undergo interference, forming an interference pattern of maxima and minima on viewing screen C. This figure is a cross section; the screens, slits, and interference pattern extend into and out of the page.

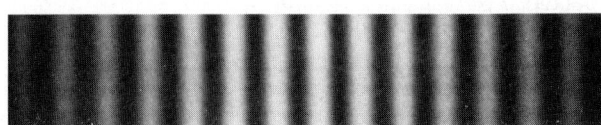

FIGURE 36-7 A photograph of an interference pattern produced by the arrangement shown in Fig. 36-6. (The photograph is a front view of part of screen C and has been rotated clockwise by 90°.) The alternating maxima and minima are called *interference fringes* (because they resemble the decorative fringe sometimes used on clothing and drapery).

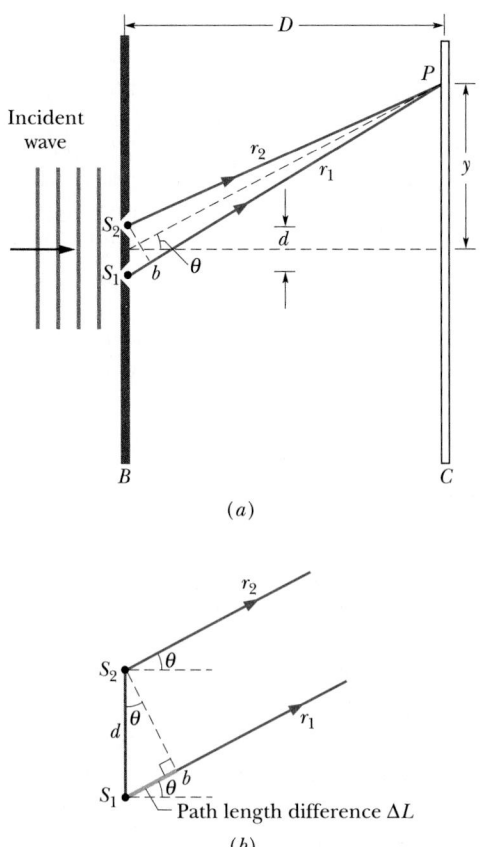

(a)

(b)

FIGURE 36-8 (a) Waves from slits S_1 and S_2 (into and out of the page) combine at P, an arbitrary point on screen C at distance y from the central axis. The angle θ serves as a convenient locator for P. (b) For $D \gg d$, we can approximate rays r_1 and r_2 as being parallel, at angle θ to the central axis.

reference. We then pick, for discussion, an arbitrary point P on the screen, at angle θ to the central axis. This point intercepts the wave of ray r_1 from the bottom slit and the wave of ray r_2 from the top slit.

These waves are in phase when they pass through the two slits because there they are just portions of the same incident wave. But once they have passed the slits, the two waves must travel different distances to reach P. We saw a similar situation in Section 18-4 with sound waves and concluded that

The phase difference between two waves can change if the waves travel paths of different lengths.

The change in phase difference is due to the *path length difference* ΔL in the paths taken by the waves. Consider two waves initially exactly in phase, traveling along paths with a path length difference ΔL, and then passing through

some common point. When ΔL is zero or an integer number of wavelengths, the waves arrive at the common point exactly in phase and they interfere fully constructively there. If that is true for the waves of rays r_1 and r_2 in Fig. 36-8, then point P is part of a bright fringe. When, instead, ΔL is an odd multiple of half a wavelength, the waves arrive at the common point exactly out of phase and they interfere fully destructively there. If that is true for the waves of rays r_1 and r_2, then point P is part of a dark fringe. (And, of course, we can have intermediate situations of interference and thus intermediate illumination at P.) Thus:

What appears at each point on the viewing screen in a Young's interference experiment is determined by the path length difference ΔL of the rays reaching that point.

We can specify where each bright or dark fringe is located on the screen by giving the angle θ from the central axis to that fringe. To find θ, we must relate it to ΔL. We start with Fig. 36-8a by finding a point b along ray r_1 such that the path length from b to P equals the path length from S_2 to P. Then the path length difference ΔL between the two rays is the distance from S_1 to b.

The relation between this S_1-to-b distance and θ is complicated, but we can simplify it considerably if we arrange for the distance D from the slits to the screen to be much greater than the slit separation d. Then we can approximate rays r_1 and r_2 as being parallel to each other and at angle θ to the central axis (Fig. 36-8b). We can also approximate the triangle formed by points S_1, S_2, and b as being a right triangle, and the angle inside that triangle at S_2 as being θ. Then, for that triangle, $\sin \theta = \Delta L / d$ and thus

$$\Delta L = d \sin \theta \quad \text{(path length difference)}. \quad (36\text{-}12)$$

For a bright fringe, we saw that ΔL must be zero or an integer number of wavelengths. Using Eq. 36-12, we can write this requirement as

$$\Delta L = d \sin \theta = (\text{integer})(\lambda), \quad (36\text{-}13)$$

or as

$$d \sin \theta = m\lambda, \quad \text{for } m = 0, 1, 2, \ldots$$
$$\text{(maxima—bright fringes)}. \quad (36\text{-}14)$$

For a dark fringe, ΔL must be an odd multiple of half a wavelength. Again using Eq. 36-12, we can write this requirement as

$$\Delta L = d \sin \theta = (\text{odd number})(\tfrac{1}{2}\lambda), \quad (36\text{-}15)$$

or as

$$d \sin \theta = (m + \tfrac{1}{2})\lambda, \qquad \text{for } m = 0, 1, 2, \ldots$$
$$\text{(minima—dark fringes).} \quad (36\text{-}16)$$

With Eqs. 36-14 and 36-16, we can find the angle θ to any fringe and thus locate that fringe; further, we can use the values of m to label the fringes. For $m = 0$, Eq. 36-14 tells us that a bright fringe is at $\theta = 0$, that is, on the central axis. This *central maximum* is the point at which waves arriving from the two slits have a path length difference $\Delta L = 0$, hence zero phase difference.

For, say, $m = 2$, Eq. 36-14 tells us that *bright* fringes are at

$$\theta = \sin^{-1}\left(\frac{2\lambda}{d}\right)$$

above and below the central axis. Waves from the two slits arrive at these two fringes with $\Delta L = 2\lambda$ and with a phase difference of two wavelengths. These fringes are said to be the *second-order fringes* (meaning $m = 2$) or the *second side maxima* (the second maxima to the side of the central maximum), or they are described as being the second fringes from the central maximum.

For $m = 1$, Eq. 36-16 tells us that *dark* fringes are at

$$\theta = \sin^{-1}\left(\frac{1.5\lambda}{d}\right)$$

above and below the central axis. Waves from the two slits arrive at these two fringes with $\Delta L = 1.5\lambda$ and with a phase difference, in wavelengths, of 1.5. These fringes are called the *second dark fringes* or *second minima* because they are the second dark fringes from the central axis. (The first dark fringes, or first minima, are at locations for which $m = 0$ in Eq. 36-16.)

Equations 36-14 and 36-16 are derived for the situation of $D \gg d$. However, they also apply if we place a converging lens between the slits and the viewing screen and then move the viewing screen to the focal point of the lens. (The screen is then said to be in the *focal plane* of the lens; that is, it is in the plane perpendicular to the central axis at the focal point.) The rays that now arrive at any point on the screen must have been exactly parallel (rather than approximately) when they left the slits—they are like the initially parallel rays in Fig. 35-13a that are directed to a point by a lens.

CHECKPOINT **3:** In Fig. 36-8a, what are ΔL (as a multiple of the wavelength) and the phase difference (in wavelengths) for the two rays if point P is (a) a third side maximum and (b) a third minimum?

SAMPLE PROBLEM 36-2

What is the distance on screen C in Fig. 36-8a between adjacent maxima near the center of the interference pattern? The wavelength λ of the light is 546 nm, the slit separation d is 0.12 mm, and the slit–screen separation D is 55 cm. Assume that the angle θ in Fig. 36-8 is small enough to permit use of the approximations $\sin \theta \approx \tan \theta \approx \theta$, in which θ is expressed in radian measure.

SOLUTION: From Fig. 36-8 we see that, for some value of m (a low value, to ensure that the corresponding maximum will be near the center of the pattern as required),

$$\tan \theta \approx \theta = \frac{y_m}{D},$$

where y_m is the distance to the mth maximum. From Eq. 36-14 we have, for the same value of m,

$$\sin \theta \approx \theta = \frac{m\lambda}{d}.$$

If we equate these two expressions for θ and solve for y_m, we find

$$y_m = \frac{m\lambda D}{d}. \qquad (36\text{-}17)$$

For the adjacent maximum that is farther out, we have

$$y_{m+1} = \frac{(m + 1)\lambda D}{d}. \qquad (36\text{-}18)$$

We find the distance between adjacent maxima by subtracting Eq. 37-17 from Eq. 36-18:

$$\Delta y = y_{m+1} - y_m = \frac{\lambda D}{d}$$
$$= \frac{(546 \times 10^{-9} \text{ m})(55 \times 10^{-2} \text{ m})}{0.12 \times 10^{-3} \text{ m}}$$
$$= 2.50 \times 10^{-3} \text{ m} \approx 2.5 \text{ mm}. \qquad \text{(Answer)}$$

As long as d and θ in Fig. 36-8a are small, the separation of the interference fringes is independent of m; that is, the fringes are evenly spaced.

36-5 COHERENCE

For the interference pattern to appear on viewing screen C in Fig. 36-6, the light waves reaching any point P on the screen must have a phase difference that does not vary in time. That is the case in Fig. 36-6, because the waves passing through slits S_1 and S_2 are portions of the single light wave that illuminates the slits. Because the phase difference remains constant, the light from slits S_1 and S_2 is said to be completely **coherent.**

Direct sunlight is partially coherent; that is, sunlight waves intercepted at two points have a constant phase difference only if the points are very close. If you look closely at your fingernail in bright sunlight, you can see a faint interference pattern called *speckle* that causes the nail to appear covered with specks. You see this effect because light waves scattering from very close points on the nail are sufficiently coherent to interfere with one another at your eye. The slits in a double-slit experiment, however, are not close enough, and in direct sunlight, the light at the slits is **incoherent.** To get coherent light, we have to send the sunlight through a single slit; because that single slit is small, the light that passes through it is coherent. In addition, the smallness of the slit causes the light to spread sufficiently via diffraction to illuminate both slits in the double-slit experiment with that coherent light.

If we replace the double slits with two similar but independent monochromatic light sources, such as two fine incandescent wires, the phase difference between the waves emitted by the sources varies rapidly and randomly. This occurs because the light is emitted by vast numbers of atoms in the wires, acting randomly and independently for extremely short times (of the order of nanoseconds). As a result, at any given point on the viewing screen, the interference between the waves from the two sources varies rapidly and randomly between fully constructive and fully destructive. The eye (and most common optical detectors) cannot follow such changes, and no interference pattern can be seen. The fringes disappear, and the screen is seen as being uniformly illuminated. Such light is said to be completely incoherent.

A *laser* differs from common light sources in that its atoms emit light in a cooperative manner, thereby making the light coherent. Moreover, the light is almost monochromatic, is emitted in a thin beam with little spreading, and can be focused to a width that almost matches the wavelength of the light.

36-6 INTENSITY IN DOUBLE-SLIT INTERFERENCE

Equations 36-14 and 36-16 tell us how to locate the maxima and minima of the double-slit interference pattern on screen C of Fig. 36-8 as a function of the angle θ in that figure. Here we wish to derive an expression for the intensity I of the fringes as a function of θ.

The light leaving the slits is in phase. However, let us assume that the electric field components of the light waves arriving at point P in Fig. 36-8 from the two slits are not in phase and vary with time as

$$E_1 = E_0 \sin \omega t \qquad (36\text{-}19)$$

and

$$E_2 = E_0 \sin (\omega t + \phi), \qquad (36\text{-}20)$$

where ω is the angular frequency of the waves and ϕ is the phase constant of wave E_2. Note that the two waves have the same amplitude E_0 and a phase difference of ϕ. Because that phase difference does not vary, the waves are coherent. We shall show that these two waves will combine at P to produce an illumination of intensity I given by

$$I = 4I_0 \cos^2 \tfrac{1}{2}\phi, \qquad (36\text{-}21)$$

and that

$$\phi = \frac{2\pi d}{\lambda} \sin \theta. \qquad (36\text{-}22)$$

In Eq. 36-21, I_0 is the intensity of the light that arrives on the screen from one slit when the other slit is temporarily covered. We assume that the slits are so narrow in comparison to the wavelength that this single-slit intensity is essentially uniform over the region of the screen in which we wish to examine the fringes.

Equations 36-21 and 36-22, which together tell us how the intensity I of the fringe pattern varies with the angle θ in Fig. 36-8, necessarily contain information about the location of the maxima and minima. Let us see if we can extract it.

Study of Eq. 36-21 shows that intensity maxima will occur when

$$\tfrac{1}{2}\phi = m\pi, \quad \text{for } m = 0, 1, 2, \ldots . \quad (36\text{-}23)$$

If we put this result into Eq. 36-22, we find

$$2m\pi = \frac{2\pi d}{\lambda} \sin \theta, \quad \text{for } m = 0, 1, 2, \ldots$$

or

$$d \sin \theta = m\lambda, \quad \text{for } m = 0, 1, 2, \ldots$$
$$\text{(maxima)}, \quad (36\text{-}24)$$

which is exactly Eq. 36-14, the expression that we derived earlier for the locations of the maxima.

The minima in the fringe pattern occur when

$$\tfrac{1}{2}\phi = (m + \tfrac{1}{2})\pi, \quad \text{for } m = 0, 1, 2, \ldots .$$

If we combine this relation with Eq. 36-22 we are led at once to

$$d \sin \theta = (m + \tfrac{1}{2})\lambda \quad \text{for } m = 0, 1, 2, \ldots$$
$$\text{(minima)}, \quad (36\text{-}25)$$

FIGURE 36-9 A plot of Eq. 36-21, showing the intensity of a double-slit interference pattern as a function of the phase difference between the waves from the two slits. I_0 is the (uniform) intensity that would appear on the screen if one slit were covered. The average intensity of the fringe pattern is $2I_0$, and the *maximum* intensity (for coherent light) is $4I_0$.

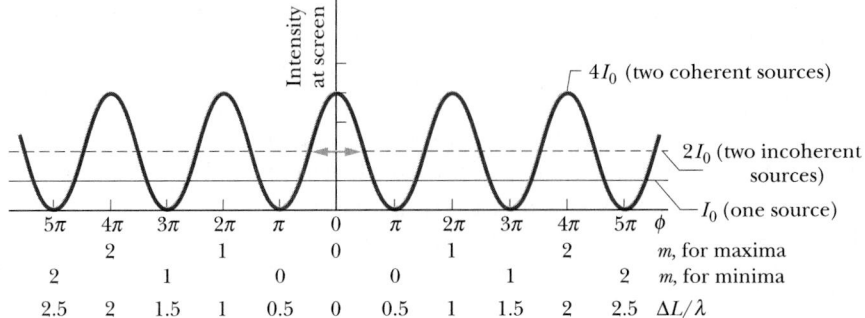

which is just Eq. 36-16, the expression derived earlier for the locations of the fringe minima.

Figure 36-9, which is a plot of Eq. 36-21, shows the intensity pattern for double-slit interference as a function of the phase difference ϕ at the screen. The horizontal solid line is I_0, the (uniform) intensity on the screen when one of the slits is covered up. Note in Eq. 36-21 and the graph that the intensity I (which is always positive) varies from zero at the fringe minima to $4I_0$ at the fringe maxima.

If the waves from the two sources (slits) were *incoherent,* so that no enduring phase relation existed between them, there would be no fringe pattern and the intensity would have the uniform value $2I_0$ for all points on the screen; the horizontal dashed line in Fig. 36-9 shows this uniform value.

Interference cannot create or destroy energy but merely redistributes it over the screen. Thus the *average* intensity on the screen must be the same $2I_0$ regardless of whether the sources are coherent. This follows at once from Eq. 36-21; if we substitute $\frac{1}{2}$, the average value of the cosine-squared function, this equation reduces to $\bar{I} = 2I_0$.

Proof of Eqs. 36-21 and 36-22

We shall combine the electric field components E_1 and E_2, given by Eqs. 36-19 and 36-20, respectively, by the method of phasors discussed in Section 17-10. In Fig. 36-10a, the waves with components E_1 and E_2 are represented by phasors of magnitude E_0 that rotate around the origin at angular speed ω. The values of E_1 and E_2 at any time are the projections of the corresponding phasors onto the vertical axis. Figure 36-10a shows the phasors and their projections at an arbitrary time t. Consistent with Eqs. 36-19 and 36-20, the phasor for E_1 has a rotation angle ωt and the phasor for E_2 has a rotation angle $\omega t + \phi$.

To combine the field components E_1 and E_2 on a phasor diagram, we add them vectorially, as shown in Fig. 36-10b. The magnitude of the vector sum is the amplitude E of the resultant wave, and that wave has a certain phase constant β. To find the amplitude E in Fig. 36-10b, we first

note that the two angles marked β are equal because they are opposite equal-length sides of a triangle. From the theorem (for triangles) that an exterior angle (ϕ) is equal to the sum of the two opposite interior angles ($\beta + \beta$), we see that $\beta = \frac{1}{2}\phi$. Thus we have

$$E = 2(E_0 \cos \beta) = 2E_0 \cos \tfrac{1}{2}\phi. \quad (36\text{-}26)$$

If we square each side of this relation we obtain

$$E^2 = 4E_0^2 \cos^2 \tfrac{1}{2}\phi. \quad (36\text{-}27)$$

From Eq. 34-24, we know that the intensity of an electromagnetic wave is proportional to the square of its amplitude. So the waves we are combining in Fig. 36-10b, whose amplitudes are E_0, have an intensity I_0 that is proportional to E_0^2. And the resultant wave, with amplitude E, has an intensity I that is proportional to E^2. Thus,

$$\frac{I}{I_0} = \frac{E^2}{E_0^2}.$$

Substituting Eq. 36-27 into this and rearranging then yield

$$I = 4I_0 \cos^2 \tfrac{1}{2}\phi,$$

which is Eq. 36-21, which we set out to prove.

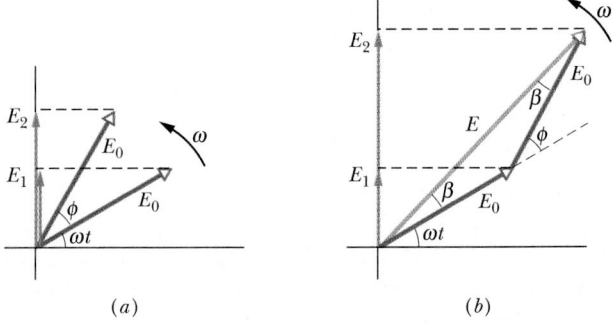

FIGURE 36-10 (a) Phasors representing the electric field components of the waves given by Eqs. 36-19 and 36-20. They both have magnitude E_0 and rotate with angular speed ω. (b) Vector addition of the two phasors gives the phasor representing the resultant wave, with magnitude E and phase constant β.

It remains to prove Eq. 36-22, which relates the phase difference ϕ between the waves arriving at any point P on the screen of Fig. 36-8 to the angle θ that serves as a locator of that point.

The phase difference ϕ in Eq. 36-20 is associated with the path difference S_1b in Fig. 36-8. If S_1b is $\frac{1}{2}\lambda$, then ϕ is π; if S_1b is λ, then ϕ is 2π, and so on. This suggests

$$\begin{pmatrix}\text{phase}\\\text{difference}\end{pmatrix} = \frac{2\pi}{\lambda}\begin{pmatrix}\text{path length}\\\text{difference}\end{pmatrix}. \quad (36\text{-}28)$$

The path difference S_1b in Fig. 36-8b is just $d\sin\theta$, so Eq. 36-28 becomes

$$\phi = \frac{2\pi d}{\lambda}\sin\theta,$$

which is just Eq. 36-22, the other equation that we set out to prove.

Combining More than Two Waves

In a more general case, we might want to find the resultant of more than two sinusoidally varying waves. The general procedure is this:

1. Construct a series of phasors representing the functions to be added. Draw them end to end, maintaining the proper phase relations between adjacent phasors.

2. Construct the vector sum of this array. The length of this vector sum gives the amplitude of the resultant phasor. The angle between the vector sum and the first phasor is the phase of the resultant with respect to this first phasor. The projection of this vector-sum phasor on the vertical axis gives the time variation of the resultant wave.

SAMPLE PROBLEM 36-3

Find the resultant wave $E(t)$ of the following waves:

$$E_1 = E_0 \sin \omega t,$$

$$E_2 = E_0 \sin(\omega t + 60°),$$

$$E_3 = E_0 \sin(\omega t - 30°).$$

SOLUTION: The resultant wave is

$$E(t) = E_1(t) + E_2(t) + E_3(t).$$

In using the method of phasors to find this sum, we are free to evaluate the phasors at any time t. To simplify the problem we choose $t = 0$, for which the phasors representing the three waves are shown in Fig. 36-11. We now treat the addition of the phasors as we would any other addition of vectors. The sum of the horizontal components of E_1, E_2, and E_3 is

$$\sum E_h = E_0 \cos 0 + E_0 \cos 60° + E_0 \cos(-30°)$$
$$= E_0 + 0.500E_0 + 0.866E_0 = 2.37E_0.$$

The sum of the vertical components, which is the value of E at $t = 0$, is

$$\sum E_v = E_0 \sin 0 + E_0 \sin 60° + E_0 \sin(-30°)$$
$$= 0 + 0.866E_0 - 0.500E_0 = 0.366E_0.$$

The resultant wave $E(t)$ has an amplitude E_R of

$$E_R = \sqrt{(2.37E_0)^2 + (0.366E_0)^2} = 2.4E_0,$$

and a phase angle β relative to phasor E_1 of

$$\beta = \tan^{-1}\left(\frac{0.366E_0}{2.37E_0}\right) = 8.8°.$$

We can now write, for the resultant wave $E(t)$,

$$E = E_R \sin(\omega t + \beta)$$
$$= 2.4E_0 \sin(\omega t + 8.8°). \quad \text{(Answer)}$$

Be careful to interpret the angle β correctly in Fig. 36-11: it is the constant angle between E_R and E_1 as the four phasors rotate as a single unit around the origin. The angle between E_R and the horizontal axis does not remain equal to β.

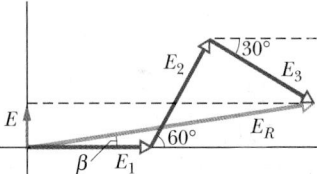

FIGURE 36-11 Sample Problem 36-3. Three phasors E_1, E_2, and E_3, shown at time $t = 0$, combine to give resultant phasor E_R.

CHECKPOINT 4: Each of four pairs of light waves arrives at a certain point on a screen. The waves have the same wavelength. At the arrival point, their amplitudes and phase differences are (a) $2E_0$, $6E_0$, and π rad; (b) $3E_0$, $5E_0$, and π rad; (c) $9E_0$, $7E_0$, and 3π rad; (d) $2E_0$, $2E_0$, and 0 rad. Rank the four pairs according to the intensity of the light at those points, greatest first. (*Hint:* Draw phasors.)

36-7 INTERFERENCE FROM THIN FILMS

The colors we see when sunlight illuminates a soap bubble or an oil slick are caused by the interference of light waves reflected from the front and back surfaces of a thin transparent film. The thickness of the soap or oil film is typically of the order of magnitude of the wavelength of the (visible) light involved. (We shall not consider greater thicknesses, which spoil the coherence of the light needed to produce colors by interference; we shall discuss lesser thicknesses shortly.)

Figure 36-12 shows a thin transparent film of uniform thickness L and index of refraction n_2, illuminated by bright light of wavelength λ from a distant point source. For now, we assume that air lies on both sides of the film and thus that $n_1 = n_3$ in Fig. 36-12. For simplicity, we also assume that the light rays are almost perpendicular to the film ($\theta \approx 0$). We are interested in whether the film is bright or dark to an observer viewing it almost perpendicularly. (Since the film is brightly illuminated, how could it possibly be dark? You will see.)

The incident light, represented by ray i, intercepts the front (left) surface of the film at point a and undergoes both reflection and refraction there. The reflected ray r_1 is intercepted by the observer's eye. The refracted light crosses the film to point b on the back surface, where it undergoes both reflection and refraction. The light reflected at b crosses back through the film to point c, where it undergoes both reflection and refraction. The light refracted at c, represented by ray r_2, is intercepted by the observer's eye.

If the light waves of rays r_1 and r_2 are exactly in phase at the eye, they produce an interference maximum, and region ac on the film is bright to the observer. If they are exactly out of phase, they produce an interference minimum, and region ac is dark to the observer, *even though it is illuminated*. And if there is some intermediate phase difference, there are intermediate interference and intermediate brightness.

So the key to what the observer sees is the phase difference between the waves of rays r_1 and r_2. Both rays are

derived from the same ray i, but the path involved in producing r_2 involves light traveling twice across the film (a to b, and then b to c), whereas the path involved in producing r_1 involves no travel through the film. Because θ is about zero, we approximate the path length difference between the waves of r_1 and r_2 as $2L$. However, to find the phase difference between the waves, we cannot just find the number of wavelengths λ that is equivalent to a path length difference of $2L$. This simple approach is impossible for two reasons: (1) the path length difference occurs in a medium other than air, and (2) reflections are involved, which can change the phase.

The phase difference between two waves can change if one or both are reflected.

Before we continue our discussion of interference from thin films, we must discuss changes in phase that are caused by reflections.

Reflection Phase Shifts

Refraction at an interface never causes a phase change. But reflection can, depending on the indices of refraction on the two sides of the interface. Figure 36-13 shows what happens when reflection causes a phase change, using pulses on a denser string (along which pulse travel is relatively slow) and a lighter string (along which pulse travel is relatively fast).

When a pulse traveling slowly along the denser string in Fig. 36-13a reaches the interface with the lighter string, the pulse is partially transmitted and partially reflected, with no change in orientation. For light, this situation corresponds to the incident wave traveling in the medium of greater index of refraction n (recall that greater n means slower speed). In that case, the wave that is reflected at the interface does not undergo a change in phase; that is, the *reflection phase shift* is zero.

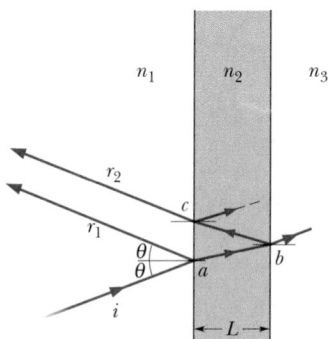

FIGURE 36-12 Light waves, represented with ray i, are incident on a thin film of thickness L and index of refraction n_2. Rays r_1 and r_2 represent light waves reflected by the front and back surfaces of the film. (All three rays are actually nearly perpendicular to the film.) The interference of the waves of r_1 and r_2 with each other depends on their phase difference. The index of refraction n_1 of the medium at the left can differ from the index of refraction n_3 of the medium at the right, but for now we assume that both media are air, with $n_1 = n_3 = 1.0$, which is less than n_2.

FIGURE 36-13 Phase changes when a pulse is reflected at the interface between two stretched strings of different linear densities. The wave speed is greater in the lighter string. (a) The incident pulse is in the denser string. (b) The incident pulse is in the lighter string. Only here is there a phase change.

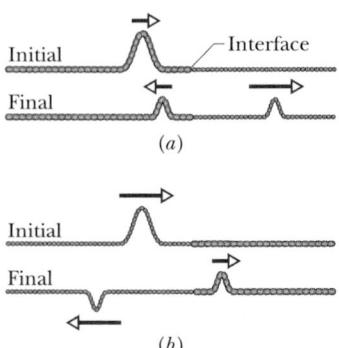

When a pulse traveling more quickly along the lighter string in Fig. 36-13b reaches the interface with the denser string, the transmitted pulse has the same orientation as the incident pulse, but the reflected pulse is inverted. For a sinusoidal wave, such an inversion involves a phase change of π rad, or half a wavelength. For light, this situation corresponds to the incident wave traveling in the medium of lesser index of refraction (with greater speed). In that case, the wave that is reflected at the interface undergoes a phase shift of π rad, or half a wavelength. We can summarize these results for light in terms of the index of refraction of the medium off which (or from which) the light reflects:

Reflection	Reflection phase shift
Off lower index	0
Off higher index	0.5 wavelength

This might be remembered as "higher means half."

Equations for Thin-Film Interference

In this chapter we have now seen three ways in which the phase difference between two waves can change:

1. by reflection
2. by the waves traveling along paths of different lengths
3. by the waves traveling through media of different indices of refraction

When light reflects from a thin film, producing the waves of rays r_1 and r_2 in Fig. 36-12, all three ways are involved. Let us consider them one by one.

We first reexamine the two reflections in Fig. 36-12. At point a on the front interface, the incident wave (in air) reflects from the medium having the higher of the two indices of refraction, so the wave of reflected ray r_1 has its phase shifted by 0.5 wavelength. At point b on the back interface, the incident wave reflects from the medium (air) having the lower of the two indices of refraction, so the wave reflected there is not shifted in phase by the reflection, and thus neither is the portion of it that exits the film as ray r_2. We can organize this information with the first line in Table 36-1. It tells us that, so far, as a result of the reflection phase shifts, the waves of r_1 and r_2 have a phase difference of 0.5 wavelength and thus are exactly out of phase.

Now we must consider the path length difference $2L$ that occurs because the wave of ray r_2 crosses the film twice. (This difference $2L$ is shown on the second line in Table 36-1.) If the waves of r_1 and r_2 are to be exactly in phase so that they produce fully constructive interference,

TABLE 36-1 AN ORGANIZING TABLE FOR THIN-FILM INTERFERENCE IN AIR[a]

	r_1	r_2
Reflection phase shifts	0.5 wavelength	0
Path length difference	2L	
Index in which path length difference occurs	n_2	
In phase[a]:	$2L = \dfrac{\text{odd number}}{2} \times \dfrac{\lambda}{n_2}$	
Out of phase[a]:	$2L = \text{integer} \times \dfrac{\lambda}{n_2}$	

[a]Valid for $n_2 > n_1$ and $n_2 > n_3$.

the path length $2L$ must cause an additional phase difference of 0.5, 1.5, 2.5, . . . wavelengths. Only then will the net phase difference be an integer number of wavelengths. Thus, for a bright film, we must have

$$2L = \frac{\text{odd number}}{2} \times \text{wavelength}$$
$$\text{(in-phase waves).} \quad (36\text{-}29)$$

The wavelength we need here is the wavelength λ_{n2} of the light in the medium containing path length $2L$, that is, in the medium with index of refraction n_2. So, we can rewrite Eq. 36-29 as

$$2L = \frac{\text{odd number}}{2} \times \lambda_{n2} \quad \text{(in-phase waves).} \quad (36\text{-}30)$$

If, instead, the waves are to be exactly out of phase so that there is fully destructive interference, the path length $2L$ must cause either no additional phase difference or a phase difference of 1, 2, 3, . . . wavelengths. Only then will the net phase difference be an odd number of half-wavelengths. So, for a dark film, we must have

$$2L = \text{integer} \times \text{wavelength}, \quad (36\text{-}31)$$

where, again, the wavelength is the wavelength λ_{n2} in the medium containing $2L$. So, this time we have

$$2L = \text{integer} \times \lambda_{n2} \quad \text{(out-of-phase waves).} \quad (36\text{-}32)$$

Now recalling that the wave of ray r_2 traveled through a medium of index of refraction n_2 whereas the wave of ray r_1 did not, we can use Eq. 36-8 ($\lambda_n = \lambda/n$) to write the wavelength of the wave inside the film as

$$\lambda_{n2} = \frac{\lambda}{n_2}, \quad (36\text{-}33)$$

where λ is the wavelength of the incident light in vacuum (and approximately also in air). Substituting Eq. 36-33 into

Eq. 36-30 and replacing "odd number/2" with $(m + \frac{1}{2})$ give us

$$2L = (m + \tfrac{1}{2}) \frac{\lambda}{n_2}, \quad \text{for } m = 0, 1, 2, \ldots$$
(maxima—bright film in air). (36-34)

Similarly, with m replacing "integer," Eq. 36-32 yields

$$2L = m \frac{\lambda}{n_2}, \quad \text{for } m = 0, 1, 2, \ldots$$
(minima—dark film in air). (36-35)

For a given film thickness L, Eqs. 36-34 and 36-35 tell us the wavelengths of light for which the film appears bright and dark, respectively, one wavelength for each value of m. Intermediate wavelengths give intermediate brightnesses. For a given wavelength λ, Eqs. 36-34 and 36-35 tell us the thicknesses of the films that appear bright and dark in that light, respectively, one thickness for each value of m. Intermediate thicknesses give intermediate brightnesses.

A special situation arises when a film is so thin that L is much less than λ, say, $L < 0.1\lambda$. Then the path length difference $2L$ can be neglected, and the phase difference between r_1 and r_2 is due *only* to reflection phase shifts. If the

FIGURE 36-14 The reflection of light from a soapy water film spanning a vertical loop. The top portion is so thin that the light reflected there undergoes destructive interference, making that portion dark. Colored interference fringes, or bands, decorate the rest of the film but are marred by circulation of liquid within the film as the liquid is gradually pulled downward by gravitation.

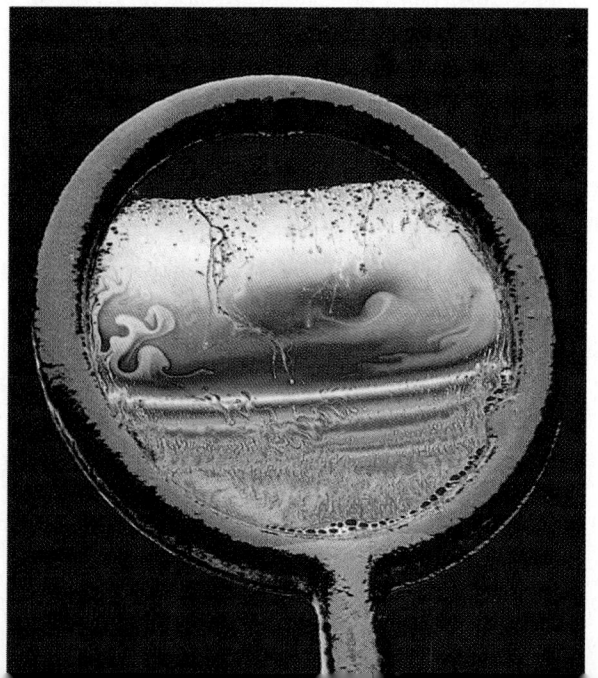

film of Fig. 36-12, where the reflections cause a phase difference of 0.5 wavelength, has thickness $L < 0.1\lambda$, then r_1 and r_2 are exactly out of phase, and thus the film is dark, regardless of the wavelength and even the intensity of the light that illuminates it. This special situation corresponds to $m = 0$ in Eq. 36-35. We shall count any $L < 0.1\lambda$ as being the least thickness that makes the film of Fig. 36-12 dark. The next greater thickness that makes the film dark is that corresponding to $m = 1$.

Figure 36-14 shows a vertical soap film whose thickness increases from top to bottom because the weight of the film has caused it to slump. Bright white light illuminates the film. However, the top portion is so thin that it is dark. In the (somewhat thicker) middle we see fringes, or bands, whose color depends primarily on the wavelength at which reflected light undergoes fully constructive interference for a particular thickness. Toward the (thickest) bottom of the film the fringes become progressively narrower and the colors begin to overlap and fade.

PROBLEM SOLVING TACTICS

TACTIC 1: *Thin-Film Equations*
Some students believe that Eq. 36-34 gives the maxima and Eq. 36-35 gives the minima for *all* thin-film situations. This is not true. These relations were derived only for the situation in which $n_2 > n_1$ and $n_2 > n_3$ in Fig. 36-12.

The appropriate equations for other relative values of the indices of refraction can be derived by following the reasoning of this section and constructing new versions of Table 36-1. In each case you will end up with Eqs. 36-34 and 36-35, but sometimes Eq. 36-34 will give the minima and Eq. 36-35 will give the maxima—the opposite of what we found here. Which equation gives which depends on whether the reflections at the two interfaces give the same reflection phase shift.

$\mathbf{C}$HECKPOINT **5:** The figure shows four situations in which light reflects perpendicularly from a thin film (as in Fig. 36-12), with the indices of refraction as given. (a) For which situations does reflection cause a zero phase difference for the two reflected rays? (b) For which situations will the film be dark if the path length difference $2L$ causes a phase difference of 0.5 wavelength?

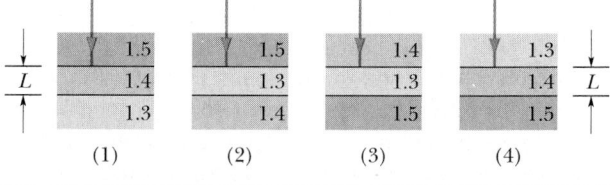

SAMPLE PROBLEM 36-4

White light, with a uniform intensity across the visible wavelength range of 400–690 nm, is perpendicularly incident on a water film, of index of refraction $n_2 = 1.33$ and thickness $L = 320$ nm, that is suspended in air. At what wavelength λ is the light reflected by the film brightest to an observer?

SOLUTION: This situation is like that of Fig. 36-12, for which Eq. 36-34 gives the interference maxima. Solving for λ and inserting the given data, we obtain

$$\lambda = \frac{2n_2 L}{m + \frac{1}{2}} = \frac{(2)(1.33)(320 \text{ nm})}{m + \frac{1}{2}} = \frac{851 \text{ nm}}{m + \frac{1}{2}}.$$

For $m = 0$, this gives us $\lambda = 1700$ nm, which is in the infrared region. For $m = 1$, we find $\lambda = 567$ nm, which is yellow-green light, near the middle of the visible spectrum. For $m = 2$, $\lambda = 340$ nm, which is in the ultraviolet region. So the wavelength at which the light seen by the observer is brightest is

$$\lambda = 567 \text{ nm.} \qquad \text{(Answer)}$$

Solving for L and inserting the given data, we obtain

$$L = \frac{\lambda}{4n_2} = \frac{550 \text{ nm}}{(4)(1.38)} = 99.6 \text{ nm.} \qquad \text{(Answer)}$$

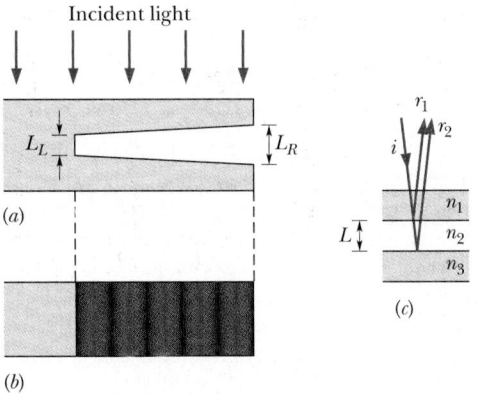

FIGURE 36-15 Sample Problem 36-5. Unwanted reflections from glass can be suppressed (at a chosen wavelength) by coating the glass with a thin transparent film of magnesium fluoride of a properly chosen thickness.

SAMPLE PROBLEM 36-5

A glass lens is coated on one side with a thin film of magnesium fluoride (MgF_2) to reduce reflection from the lens surface (Fig. 36-15). The index of refraction of MgF_2 is 1.38; that of the glass is 1.50. What is the least coating thickness that eliminates (via interference) the reflections at the middle of the visible spectrum ($\lambda = 550$ nm)? Assume that the light is approximately perpendicular to the lens surface.

SOLUTION: Figure 36-15 differs from Fig. 36-12 in that now $n_3 > n_2 > n_1$. This means there is now a reflection phase shift of 0.5 wavelength associated with the reflections at *both* front and back interfaces of the thin film. Constructing a table like Table 36-1, we fill in 0.5 and 0.5 for the first line. For the second and third lines, the path length difference is still $2L$ and it still occurs in a medium (here MgF_2) having index of refraction n_2.

The reflections alone tend to put the waves of r_1 and r_2 in phase. For these rays to be out of phase so that the reflections from the lens are eliminated, the path length difference $2L$ within the film must be

$$2L = \frac{\text{odd number}}{2} \times \text{wavelength}$$

$$= (m + \tfrac{1}{2})\lambda_{n2}, \qquad \text{for } m = 0, 1, 2, \ldots .$$

Substituting λ/n_2 for λ_{n2} yields

$$2L = (m + \tfrac{1}{2})\frac{\lambda}{n_2}, \qquad \text{for } m = 0, 1, 2, \ldots .$$

We want the least thickness for the coating, that is, the smallest L. Thus we choose $m = 0$, the smallest value of m.

SAMPLE PROBLEM 36-6

Figure 36-16a shows a transparent plastic block with a thin wedge of the plastic removed at the right. A broad beam of red light, with wavelength $\lambda = 632.8$ nm, is directed directly downward through the top of the block (at an incidence angle of 0°). Some of the light is reflected back up from the top and bottom surfaces of the wedge, which acts as a thin film (of air) with a thickness that varies uniformly and gradually from L_L at the left-hand end to L_R at the right-hand end. An observer

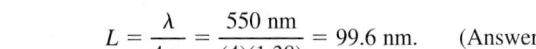

FIGURE 36-16 Sample Problem 36-6. (a) Red light is incident on a thin air-filled wedge in the side of a transparent plastic block. The thickness of the wedge is L_L at the left end and L_R at the right end. (b) The view from above the block: an interference pattern of six dark fringes and five bright red fringes lies over the region of the wedge. (c) A representation of the incident ray i, reflected rays r_1 and r_2, and thickness L of the wedge anywhere along the length of the wedge.

looking down on the block sees an interference pattern consisting of six dark fringes and five bright red fringes along the wedge. What is the change in thickness $\Delta L\,(= L_R - L_L)$ along the wedge?

SOLUTION: This thin-film problem differs from preceding problems because the thickness of the film varies. It is that variation that produces the observed variation between dark and bright fringes along the wedge. Because the observer sees more dark fringes than bright fringes, we can assume that a dark fringe is produced at both the left and right ends of the film. Thus the interference pattern is that shown in Fig. 36-16b, which we can use to determine the change in thickness ΔL of the wedge.

We can represent the reflection of light by the top and bottom surfaces of the wedge anywhere along its length with Fig. 36-16c, at a place where we assume that the wedge has thickness L. From what we know about phase shifts by reflection, we see that the reflection phase shift for ray r_1 is zero and that for ray r_2 is 0.5 wavelength. Constructing a table like Table 36-1, we fill in 0 and 0.5 for the first line. For the second and third lines, the path length difference is still $2L$ and it still occurs in a medium (here air) with index of refraction n_2. Thus, for fully destructive interference we find that

$$2L = \text{integer} \times \frac{\lambda}{n_2} = m\,\frac{\lambda}{n_2}. \qquad (36\text{-}36)$$

We can apply this equation at any point along the wedge where a dark fringe is observed. The least value of the integer m is associated with the least thickness of the wedge where a dark fringe is observed. And progressively greater values of m are associated with progressively greater thicknesses of the wedge where a dark fringe is observed.

A dark fringe happens to be observed at the left end of the wedge, where the thickness is least. Applying Eq. 36-36 to that end, substituting L_L for L, and then solving for L_L, we have

$$L_L = \frac{m_L \lambda}{2n_2}, \qquad (36\text{-}37)$$

where m_L is the integer associated with the dark fringe at the left end and n_2 is the index of refraction of the material inside the wedge (air).

We can also apply Eq. 36-36 to the right end of the wedge, where another dark fringe is seen. There the thickness is L_R, and the integer associated with L_R and this dark fringe is $m_L + 5$ (because the fringe is the fifth one from the fringe at the left-hand end). Substituting L_R for L and $m_L + 5$ for m into Eq. 36-36 and solving for L_R yield

$$L_R = \frac{(m_L + 5)\lambda}{2n_2}. \qquad (36\text{-}38)$$

Subtracting Eq. 36-37 from Eq. 36-38 then gives us the change in thickness ΔL of the wedge:

$$\Delta L = L_R - L_L = \frac{(m_L + 5)\lambda}{2n_2} - \frac{m_L \lambda}{2n_2} = \frac{5}{2}\frac{\lambda}{n_2}.$$

Substituting 632.8×10^{-9} m for λ and 1.00 for n_2 into this equation, we find

$$\Delta L = \frac{5}{2}\frac{632.8 \times 10^{-9}\text{ m}}{1.00}$$

$$= 1.58 \times 10^{-6}\text{ m}. \qquad \text{(Answer)}$$

SAMPLE PROBLEM 36-7

The iridescence seen in the top surface of *Morpho* butterfly wings is due to constructive interference of the light reflected by thin terraces of transparent cuticle-like material. The terraces extend outward, parallel to the wings, from a central structure that is approximately perpendicular to the wing. Cross sections of the central structure and terraces are shown in the electron micrograph of Fig. 36-17a. The terraces have index of refraction $n = 1.53$ and thickness $D_t = 63.5$ nm; they are separated (by air) by $D_a = 127$ nm. If the incident light is perpendicular to the terraces (see Fig. 36-17b, where the angle of the incident light is exaggerated), at what wavelength of visible light do the reflections from the terraces have an interference maximum?

SOLUTION: Let us first consider rays r_1 and r_2 in Fig. 36-17b, which involve reflections at points a and b. This situation is just like that of Fig. 36-12, and so Eq. 36-34 gives the interference maxima. Solving Eq. 36-34 for λ gives us

$$\lambda = \frac{2n_2 L}{m + \frac{1}{2}}.$$

Substituting D_t ($= 63.5$ nm) for L and n ($= 1.53$) for n_2, we have

$$\lambda = \frac{2nD_t}{m + \frac{1}{2}} = \frac{(2)(1.53)(63.5\text{ nm})}{m + \frac{1}{2}} = \frac{194\text{ nm}}{m + \frac{1}{2}}.$$

For $m = 0$, we find an interference maximum at $\lambda = 388$ nm, which is in the ultraviolet region. For all larger values of m, λ is even smaller, farther into the ultraviolet. So rays r_1 and r_2 do not produce the bright blue-green color of the *Morpho*.

Let us next consider rays r_1 and r_3 in Fig. 36-17b. The wave producing the latter passes through a terrace and then through air to the next terrace, where it reflects at point d. Then it travels upward, resulting in ray r_3. The path length difference between the waves leading to rays r_1 and r_3 is $2D_t + 2D_a$. This situation differs considerably from that of Fig. 36-12, and Eq. 36-34 does not apply. To find a new equation for interference maxima for this new situation, we first consider the reflections involved and then count the wavelengths along path length difference $2D_t + 2D_a$.

The reflections at points a and d both introduce a phase change of half a wavelength. So the reflections alone tend to put the waves of rays r_1 and r_3 in phase. Thus for these waves actually to end up in phase, the number of wavelengths along the path length difference $2D_t + 2D_a$ must be an integer. The wavelength within the terrace is $\lambda_n = \lambda/n$. So the number of

incident light is not exactly perpendicular to the terraces but travels along a slanted path, the paths taken by the waves represented by r_1 and r_3 change, and so does the wavelength of maximum interference. Thus as the wing moves in your view, the wavelength at which the wing is brightest changes slightly, producing iridescence of the wing.

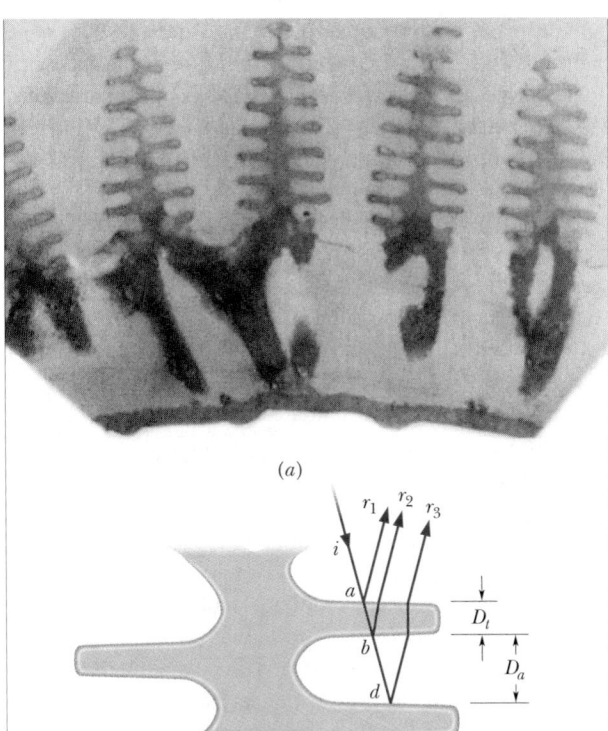

FIGURE 36-17 Sample Problem 36-7. (a) An electron micrograph shows the cross section of terrace structures of cuticle material that stick up from the top surface of a *Morpho* wing. (b) Light waves reflecting at points a and b on a terrace, represented by rays r_1 and r_2, interfere at the eye of an observer. The wave of ray r_1 also interferes with the wave that reflects at point d and is represented by ray r_3.

wavelengths in length $2D_t$ is

$$N_t = \frac{2D_t}{\lambda_n} = \frac{2D_t n}{\lambda}.$$

Similarly, the number of wavelengths in length $2D_a$ is

$$N_a = \frac{2D_a}{\lambda}.$$

For the waves of rays r_1 and r_3 to be in phase, we need $N_t + N_a$ to be equal to an integer m. Thus for an interference maximum,

$$\frac{2D_t n}{\lambda} + \frac{2D_a}{\lambda} = m, \qquad \text{for } m = 1, 2, 3, \ldots.$$

Solving for λ and substituting the given data, we obtain

$$\lambda = \frac{(2)(63.5 \text{ nm})(1.53) + (2)(127 \text{ nm})}{m} = \frac{448 \text{ nm}}{m}.$$

For $m = 1$, we find

$$\lambda = 448 \text{ nm}. \qquad \text{(Answer)}$$

This wavelength corresponds to the bright blue-green color of the top surface of a *Morpho* wing. Further, when the

36-8 MICHELSON'S INTERFEROMETER

An **interferometer** is a device that can be used to measure lengths or changes in length with great accuracy by means of interference fringes. We describe the form originally devised and built by A. A. Michelson in 1881. Consider light that leaves point P on extended source S (Fig. 36-18) and encounters a *beam splitter M*. This is a mirror with the following property: it transmits half the incident light, reflecting the rest. In the figure we have assumed, for convenience, that this mirror possesses negligible thickness. At M the light thus divides into two waves. One proceeds by transmission toward mirror M_1; the other proceeds by reflection toward M_2. The waves are reflected at each of these mirrors and are sent back along their directions of incidence, each wave eventually entering the telescope T. What the observer sees is a pattern of curved or approximately straight interference fringes; the latter resemble the stripes on a zebra.

The path length difference for the two waves when they recombine is $2d_2 - 2d_1$, and anything that changes this path difference will cause a change in the phase between these two waves at the eye. As an example, if mirror M_2 is moved by a distance $\frac{1}{2}\lambda$, the path length difference is changed by λ and the fringe pattern is shifted by one fringe (as if each dark stripe on a zebra had moved to where the

FIGURE 36-18 Michelson's interferometer, showing the path of light originating at point P of an extended source S. Mirror M splits the light into two beams, which reflect from mirrors M_1 and M_2 back to M and then to telescope T. In the telescope an observer sees a pattern of interference fringes.

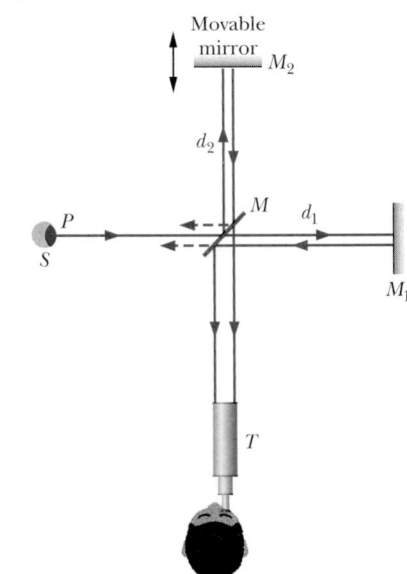

adjacent dark stripe had been). Similarly, moving mirror M_2 by $\frac{1}{4}\lambda$ causes a shift by half a fringe (each zebra stripe shifts by one stripe width).

A shift in the fringe pattern can also be caused by the insertion of a thin transparent material into the optical path of one of the mirrors, say, M_1. If the material has thickness L and index of refraction n, then the number of wavelengths along the light's doubled-back path through the material is

$$N_m = \frac{2L}{\lambda_n} = \frac{2Ln}{\lambda}. \qquad (36\text{-}39)$$

The number of wavelengths in the same thickness $2L$ of air before the insertion of the material is

$$N_a = \frac{2L}{\lambda}. \qquad (36\text{-}40)$$

So when the material is inserted, the light returned by mirror M_1 undergoes a phase change (in terms of wavelengths) of

$$N_m - N_a = \frac{2Ln}{\lambda} - \frac{2L}{\lambda} = \frac{2L}{\lambda}(n-1). \qquad (36\text{-}41)$$

For each phase change of one wavelength, the fringe pattern is shifted by one fringe. Thus by counting the number of fringes through which the material causes the pattern to shift, and substituting that number for $N_m - N_a$ in Eq. 36-41, you can determine the thickness L of the material in terms of λ.

By such techniques the lengths of objects can be expressed in terms of the wavelengths of light. In Michelson's day, the standard of length—the meter—was chosen by international agreement to be the distance between two fine scratches on a certain metal bar preserved at Sèvres, near Paris. Michelson was able to show, using his interferometer, that the standard meter was equivalent to 1,553,163.5 wavelengths of a certain monochromatic red light emitted from a light source containing cadmium. For this careful measurement, Michelson received the 1907 Nobel prize in physics. His work laid the foundation for the eventual abandonment (in 1961) of the meter bar as a standard of length and for the redefinition of the meter in terms of the wavelength of light. By 1983, as we have seen, even this wavelength standard was not precise enough to meet the growing requirements of science and technology, and it was replaced with a new standard based on a defined value for the speed of light.

REVIEW & SUMMARY

Huygens' Principle

The three-dimensional transmission of waves, including light, may often be predicted by *Huygens' principle*, which states that all points on a wavefront serve as point sources of spherical secondary wavelets. After a time t, the new position of the wavefront will be that of a surface tangent to these secondary wavelets.

The law of refraction can be derived from Huygens' principle by assuming that the index of refraction of any medium is $n = c/v$, in which v is the speed of light in the medium and c is the speed of light in vacuum.

Wavelength and Index of Refraction

The wavelength λ_n of light in a medium depends on the index of refraction n of the medium:

$$\lambda_n = \frac{\lambda}{n}, \qquad (36\text{-}8)$$

in which λ is the wavelength of the light in vacuum. Because of this dependency, the phase difference between two waves can change if they pass through different materials with different indices of refraction.

Geometrical Optics and Diffraction

Attempts to isolate a ray by forcing light through a narrow slit fail because of **diffraction**, the flaring out of the light into the geometrical shadow of the slit. If such slits are present, the approxi-

mations of geometrical optics (Chapters 34 and 35) fail, and the full treatment of wave optics must be used.

Young's Experiment

In **Young's interference experiment**, light passing through a single slit falls on two slits in a screen. The light leaving these slits flares out (by diffraction), and interference occurs in the region beyond the screen. A fringe pattern, due to the interference, forms on a viewing screen.

The light intensity at any point on the viewing screen depends in part on the difference in the path lengths from the slits to that point. If this difference is an integer number of wavelengths, the waves interfere constructively and an intensity maximum results. If it is an odd number of half-wavelengths, there is destructive interference and an intensity minimum occurs. The conditions for maximum and minimum intensity are

$$d \sin\theta = m\lambda, \qquad \text{for } m = 0, 1, 2, \ldots$$
$$\text{(maxima—bright fringes)}, \qquad (36\text{-}14)$$

$$d \sin\theta = (m + \tfrac{1}{2})\lambda, \qquad \text{for } m = 0, 1, 2, \ldots$$
$$\text{(minima—dark fringes)}, \qquad (36\text{-}16)$$

where θ is the angle the light path makes with a central axis and d is the slit separation.

Coherence

If two overlapping light waves are to interfere perceptibly, the phase difference between them must remain constant with time; that is, the waves must be **coherent.** When two coherent waves overlap, the resulting intensity may be found by using phasors.

Intensity in Two-Slit Interference

In Young's interference experiment, two waves, each with intensity I_0, yield a resultant wave of intensity I at the viewing screen:

$$I = 4I_0 \cos^2(\tfrac{1}{2}\phi), \qquad \text{where } \phi = \frac{2\pi d}{\lambda} \sin \theta.$$

$$(36\text{-}21, 36\text{-}22)$$

Equations 36-14 and 36-16, which identify the positions of the fringe maxima and minima, are contained within this relation.

Thin-Film Interference

When light is incident on a thin transparent film, the light waves reflected from the front and back surfaces interfere. For near-normal incidence the wavelength conditions for maximum and minimum intensity of the light reflected from a *film in air* are

$$2L = (m + \tfrac{1}{2}) \frac{\lambda}{n_2}, \qquad \text{for } m = 0, 1, 2 \ldots$$

$$\text{(maxima—bright film in air)}, \quad (36\text{-}34)$$

$$2L = m \frac{\lambda}{n_2}, \qquad \text{for } m = 0, 1, 2 \ldots$$

$$\text{(minima—dark film in air)}, \quad (36\text{-}35)$$

where n_2 is the index of refraction of the film, L is its thickness, and λ is the wavelength of the light in air.

If the light incident at an interface between media with different indices of refraction is in the medium with the smaller index of refraction, the reflection causes a phase change of π rad, or half a wavelength, in the reflected wave. Otherwise, there is no phase change due to the reflection. Refraction at an interface does not cause a phase shift.

The Michelson Interferometer

In *Michelson's interferometer* a light wave is split into two beams, which, after traversing paths of different lengths, are recombined so that they interfere and form a fringe pattern. Varying the path length of one of the beams allows distances to be accurately expressed in terms of wavelengths of light, by counting the number of fringes through which the fringe pattern shifts.

QUESTIONS

1. In Fig. 36-19, three pulses of light—*a*, *b*, and *c*—of the same wavelength are sent through layers of plastic whose indices of refraction are given. Rank the pulses according to their travel time through the plastic, greatest first.

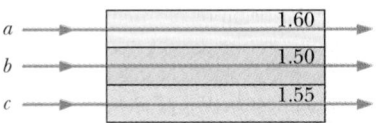

FIGURE 36-19
Question 1.

2. Light travels along the length of a 1500 nm long nanostructure. When a peak of the wave is at one end of the nanostructure, is there a peak or a valley at the other end if the wavelength is (a) 500 nm and (b) 1000 nm?

3. Figure 36-20 shows two rays of light, of wavelength 600 nm, that reflect from glass surfaces separated by 150 nm. The rays are initially in phase. (a) What is the path length difference of the rays? (b) When they have cleared the reflection region, are the rays exactly in phase, exactly out of phase, or in some intermediate state?

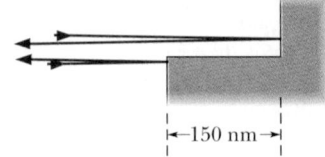

FIGURE 36-20
Question 3.

4. Figure 36-21 shows two light rays that are initially exactly in phase and reflect from several glass surfaces. Neglect the slight slant in the path of the light in the second arrangement. (a) What is the path length difference of the rays? In wavelengths λ, (b) what should that path length difference equal if the rays are to be exactly out of phase when they emerge, and (c) what is the smallest value of *d* that will allow that final phase difference?

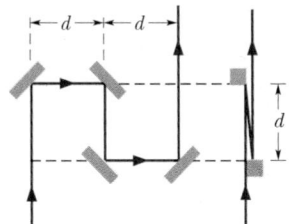

FIGURE 36-21
Question 4.

5. Figure 36-22 shows three situations in which two rays of sunlight penetrate slightly into and then scatter out of lunar soil. Assume that the rays are initially in phase. In which situation are the associated waves most likely to end up in phase? (Just as the Moon becomes full, its brightness suddenly peaks, becoming 25% greater than its brightness on the nights before and after, because at full Moon we intercept light waves that are scattered by lunar soil back toward the Sun and undergo constructive interference at our eyes.)

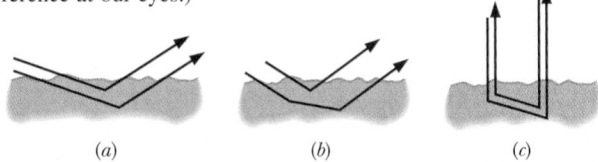

FIGURE 36-22 Question 5.

6. Is there an interference maximum, a minimum, an intermediate state closer to a maximum, or an intermediate state closer to a minimum at point P in Fig. 36-8 if the path length difference of the two rays is (a) 2.2λ, (b) 3.5λ, (c) 1.8λ, and (d) 1.0λ? For each situation, give the value of m associated with the maximum or minimum involved.

7. (a) If you move from one bright fringe in a two-slit interference pattern to the next one farther out, (a) does the path length difference ΔL increase or decrease and (b) by how much does it change, in wavelengths λ?

8. Does the spacing between fringes in a two-slit interference pattern increase, decrease, or stay the same if (a) the slit separation is increased, (b) the color of the light is switched from red to blue, and (c) the whole apparatus is submerged in cooking sherry? (d) If the slits are illuminated with white light, then at any side maximum, does the blue component or the red component peak closer to the central maximum?

9. In Fig. 36-23, a thin, transparent plastic layer has been placed over the lower slit in a double-slit experiment. Does this cause the central maximum (the fringe where waves arrive with a phase difference of zero wavelengths) to move up or down the screen? (*Hint*: Is the wavelength in the plastic greater than or less than that in air?)

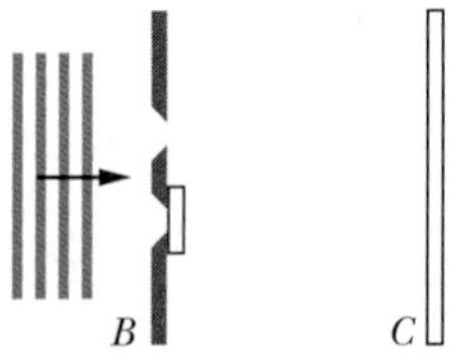

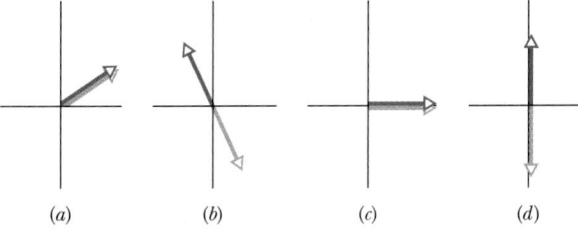

FIGURE 36-23   Question 9.

10. Figure 36-24 shows, at different times, the phasors representing the two light waves arriving at four different points on the viewing screen in a double-slit interference experiment. Assuming all eight phasors have the same length, rank the points according to the intensity of the light there, greatest first.

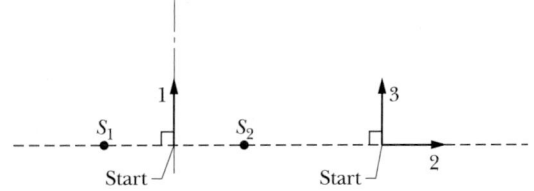

(a)   (b)   (c)   (d)

FIGURE 36-24   Question 10.

11. Figure 36-25 shows two sources S_1 and S_2 that emit radio waves of wavelength λ in all directions. The sources are exactly in phase and are separated by a distance equal to 1.5λ. The vertical broken line is the perpendicular bisector of the distance be-

tween the sources. (a) If we start at the indicated start point and travel along path 1, does the interference produce a maximum all along the path, a minimum all along the path, or alternating maxima and minima? Repeat for (b) path 2 and (c) path 3.

12. Whole milk is a liquid suspension of fat and other particles. If you hold a spoon partially filled with milk in bright sunlight, you will see fleeting points of color near the perimeter of the milk. What causes them?

13. Figure 36-26 shows two rays of light encountering interfaces, where they reflect and refract. Which of the resulting waves are shifted in phase at the interface?

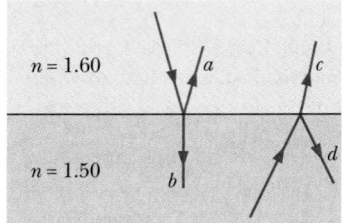

FIGURE 36-26   Question 13.

14. Suppose that the equation $2L = (m + \frac{1}{2})\lambda/n_2$ gives the maxima for interference by a certain thin film. (a) For a given film thickness, does $m = 2$ correspond to the maximum due to the second longest wavelength, the second shortest wavelength, the third longest wavelength, or the third shortest wavelength? (b) For a given wavelength, what value of m corresponds to the third least thickness giving a maximum?

15. Figure 36-27a shows the cross section of a vertical thin film whose width increases downward owing to its weight. Figure 36-27b is a face-on view of the film, showing four bright interference fringes that result when the film is illuminated with a perpendicular beam of red light. Points in the cross section corresponding to the bright fringes are labeled. In terms of the wavelength of the light inside the film, what is the difference in film thickness between (a) points a and b and (b) points b and d?

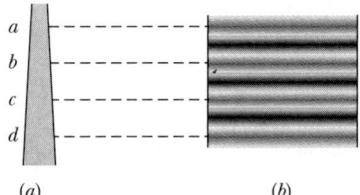

(a)   (b)

FIGURE 36-27   Question 15.

16. Figure 36-28 shows the transmission of light through a thin film in air by a perpendicular beam (tilted in the figure for clarity). (a) Did ray r_3 undergo a phase shift due to reflection? (b) In wavelengths, what is the reflection phase shift for ray r_4? (c) If the film thickness is L, what is the path length difference between rays r_3 and r_4?

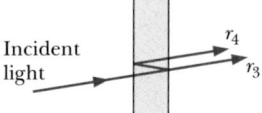

FIGURE 36-28   Question 16.

FIGURE 36-25   Question 11.

17. Sunlight illuminates a thin film of oil that floats on water, which has a greater index of refraction than the oil. The edge of the film has thickness $L < 0.1\lambda$. Is the edge dark (like the corresponding thin region of the soap film in Fig. 36-14) or bright?

18. The eyes of some animals contain reflectors that send light to receptors where the light is absorbed. In the scallop, the reflector consists of many thin transparent layers alternating between high and low indices of refraction. With the proper layer thicknesses, the combined reflections from the interfaces end up in phase with one another, thereby giving a much brighter reflection than a single biological surface or layer could give. Figure 36-29 shows such an arrangement of alternating layers, along with the reflec-

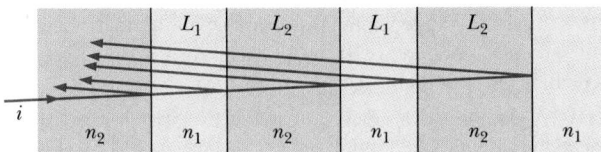

FIGURE 36-29 Question 18.

tions due to a single perpendicularly incident ray i. In terms of the indices of refraction n_1 and n_2 and the wavelength λ of visible light, should the thicknesses be (a) $L_1 = \lambda/4n_1$ and $L_2 = \lambda/4n_2$ or (b) $L_1 = \lambda/2n_1$ and $\lambda/2n_2$?

19. Figure 36-30 shows four situations in which light of wavelength λ is incident perpendicularly on a very thin layer. The indicated indices of refraction are $n_1 = 1.33$ and $n_2 = 1.50$. In each situation the thin layer has thickness $L < 0.1\lambda$. In which situations will the light reflected by the thin layer be approximately eliminated by interference?

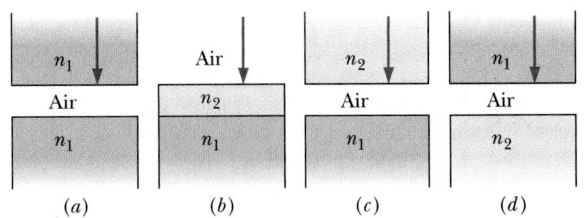

FIGURE 36-30 Question 19.

EXERCISES & PROBLEMS

SECTION 36-2 Light as a Wave

1E. The wavelength of yellow sodium light in air is 589 nm. (a) What is its frequency? (b) What is its wavelength in glass whose index of refraction is 1.52? (c) From the results of (a) and (b) find its speed in this glass.

2E. How much faster, in meters per second, does light travel in sapphire than in diamond? See Table 34-1.

3E. Derive the law of reflection using Huygens' principle.

4E. The speed of yellow light (from a sodium lamp) in a certain liquid is measured to be 1.92×10^8 m/s. What is the index of refraction of this liquid for the light?

5E. What is the speed in fused quartz of light of wavelength 550 nm? (See Fig. 34-19.)

6E. When an electron moves through a medium at a speed exceeding the speed of light in that medium, the electron radiates electromagnetic energy (the *Cerenkov effect*). What minimum speed must an electron have in a liquid of refractive index 1.54 in order to radiate?

7E. A laser beam travels along the axis of a straight section of pipeline, 1 mi long. The pipe normally contains air at standard temperature and pressure (see Table 34-1), but it may also be evacuated. In which case would the travel time for the beam be greater, and by how much?

8P. One end of a stick is pushed through water at speed v, which is greater than the speed u of water waves. Applying Huygens' construction to the water waves produced by the stick, show that a conical wavefront is set up and that its half-angle θ (see Fig. 18-22) is given by

$$\sin \theta = u/v.$$

This is familiar as the bow wave of a ship and the shock wave caused by an object moving through air with a speed exceeding that of sound.

9P. Ocean waves moving at a speed of 4.0 m/s are approaching a beach at an angle of 30° to the normal, as shown in Fig. 36-31. Suppose the water depth changes abruptly at a certain distance from the beach and the wave speed there drops to 3.0 m/s. Close to the beach, what is the angle θ between the direction of wave motion and the normal? (Assume the same law of refraction as for light.) Explain why most waves come in normal to a shore even though at large distances they approach at a variety of angles.

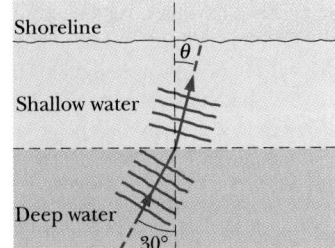

FIGURE 36-31
Problem 9.

10P. In Fig. 36-32, light travels from point A to point B, through two regions having indices of refraction n_1 and n_2. Show that the path that requires the least travel time from A to B is the path for which θ_1 and θ_2 in the figure satisfy Eq. 36-6.

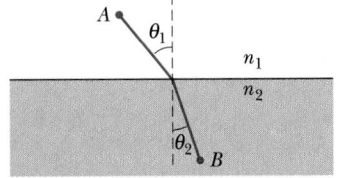

FIGURE 36-32
Problem 10.

11P. In Fig. 36-33, two pulses of light are sent through layers of plastic with the indices of refraction indicated and with thicknesses of either L or $2L$ as shown. (a) Which pulse travels through the plastic in less time? (b) In terms of L/c, what is the difference in the traversal times of the pulses?

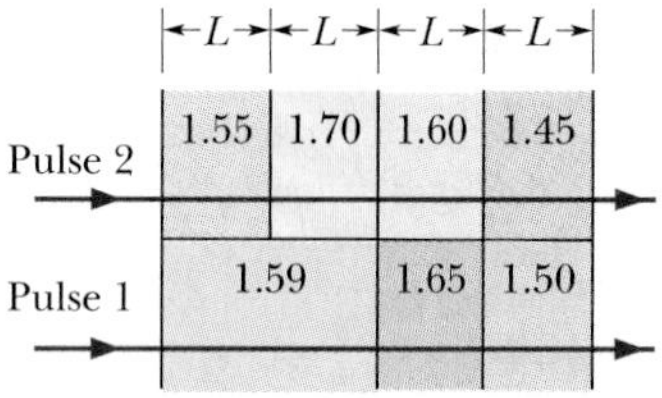

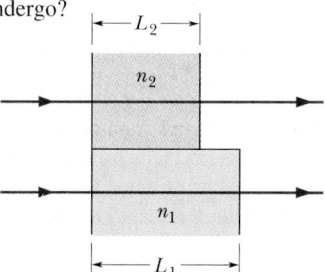

FIGURE 36-33
Problem 11.

12P. In Fig. 36-3, assume two waves of light in air, of wavelength 400 nm, are initially in phase. One travels through a glass layer of index of refraction $n_1 = 1.60$ and thickness L. The other travels through an equally thick plastic layer of index of refraction $n_2 = 1.50$. (a) What is the (least) value of L if the waves are to end up with a phase difference of 5.65 rad? (b) If the waves arrive at some common point after emerging, what type of interference do they undergo?

13P. Suppose the two waves in Fig. 36-3 have wavelength 500 nm in air. In wavelengths, what is their phase difference after traversing media 1 and 2 if (a) $n_1 = 1.50$, $n_2 = 1.60$, and $L = 8.50 \ \mu m$; (b) $n_1 = 1.62$, $n_2 = 1.72$, and $L = 8.50 \ \mu m$; and (c) $n_1 = 1.59$, $n_2 = 1.79$, and $L = 3.25 \ \mu m$? (d) Suppose that in each of these three situations the waves arrive at a common point after emerging. Rank the situations according to the brightness the waves produce at the common point.

14P. In Fig. 36-3, assume the two light waves, of wavelength 620 nm in air, are initially out of phase by π rad. The indices of refraction of the media are $n_1 = 1.45$ and $n_2 = 1.65$. (a) What is the least thickness L that will put the waves exactly in phase once they pass through the two media? (b) What is the next greater L that will do this?

15P. Two waves of light in air, of wavelength 600.0 nm, are initially in phase. They then travel through plastic layers as shown in Fig. 36-34, with $L_1 = 4.00 \ \mu m$, $L_2 = 3.50 \ \mu m$, $n_1 = 1.40$, and $n_2 = 1.60$. (a) In wavelengths, what is their phase difference after they both have emerged from the layers? (b) If the waves later arrive at some common point, what type of interference do they undergo?

FIGURE 36-34 Problem 15.

SECTION 36-4 Young's Interference Experiment

16E. Monochromatic green light, of wavelength 550 nm, illuminates two parallel narrow slits 7.70 μm apart. Calculate the angular deviation (θ in Fig. 36-8) of the third-order (for $m = 3$) bright fringe (a) in radians and (b) in degrees.

17E. What is the phase difference between the waves from the two slits arriving at the mth dark fringe in a Young's double-slit experiment?

18E. If the slit separation d in Young's experiment is doubled, how must the distance D of the viewing screen be changed to maintain the same fringe spacing?

19E. Suppose Young's experiment is performed with blue-green light of wavelength 500 nm. The slits are 1.20 mm apart, and the viewing screen is 5.40 m from the slits. How far apart are the bright fringes?

20E. Find the slit separation of a double-slit arrangement that will produce interference fringes 0.018 rad apart on a distant screen. Assume sodium light ($\lambda = 589$ nm).

21E. A double-slit arrangement produces interference fringes for sodium light ($\lambda = 589$ nm) that have an angular separation of 3.50×10^{-3} rad. For what wavelength would the angular separation be 10.0% greater?

22E. In a double-slit arrangement the slits are separated by a distance equal to 100 times the wavelength of the light passing through the slits. (a) What is the angular separation in radians between the central maximum and an adjacent maximum? (b) What is the distance between these maxima on a screen 50.0 cm from the slits?

23E. In a double-slit experiment (Fig. 36-8), $\lambda = 546$ nm, $d = 0.10$ mm, and $D = 20$ cm. On a viewing screen, what is the distance between the fifth maximum and the seventh minimum from the central maximum?

24E. A double-slit arrangement produces interference fringes for sodium light ($\lambda = 589$ nm) that are 0.20° apart. What is the angular fringe separation if the entire arrangement is immersed in water ($n = 1.33$)?

25E. Two radio-frequency point sources separated by 2.0 m are radiating in phase with $\lambda = 0.50$ m. A detector moves in a circular path around the two sources in a plane containing them. Without written calculation, find how many maxima it detects.

26E. Source A and B emit long-range radio waves of wavelength 400 m, with the phase of the emission from A ahead of that from source B by 90°. The distance r_A from A to a detector is greater than the corresponding distance r_B by 100 m. What is the phase difference at the detector?

27P. In a double-slit experiment the distance between slits is 5.0 mm and the slits are 1.0 m from the screen. Two interference patterns can be seen on the screen: one due to light with wavelength 480 nm, and the other due to light with wavelength 600 nm. What is the separation on the screen between the third-order ($m = 3$) bright fringes of the two different patterns?

28P. If the distance between the first and tenth minima of a double-slit pattern is 18 mm and the slits are separated by 10.15 mm with the screen 50 cm from the slits, what is the wavelength of the light used?

29P. In Fig. 36-35, A and B are identical radiators of waves that are in phase and of the same wavelength λ. The radiators are separated by distance $d = 3.00\lambda$. Find the greatest distance from A, along the x axis, for which fully destructive interference occurs. Express this distance in wavelengths.

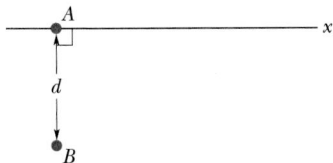

FIGURE 36-35 Problems 29 and 39.

30P. Laser light of wavelength 632.8 nm passes through a double-slit arrangement at the front of a lecture room, reflects off a mirror 20.0 m away at the back of the room, and then produces an interference pattern on a screen at the front of the room. The distance between adjacent bright fringes is 10.0 cm. (a) What is the slit separation? (b) What happens to the pattern when the lecturer places a thin cellophane sheet over one slit, thereby increasing by 2.50 the number of wavelengths along the path that includes the cellophane?

31P. Sodium light ($\lambda = 589$ nm) illuminates two slits separated by $d = 2.0$ mm. The slit–screen distance D is 40 mm. What percentage error is made by using Eq. 36-14 to locate the $m = 10$ bright fringe on the screen rather than using the exact path length difference?

32P. Two point sources, S_1 and S_2 in Fig. 36-36, emit waves in phase and at the same frequency. Show that all curves (such as that given) over which the phase difference for rays r_1 and r_2 is a constant are hyperbolas. (*Hint:* A constant phase difference implies a constant difference in length between r_1 and r_2.)

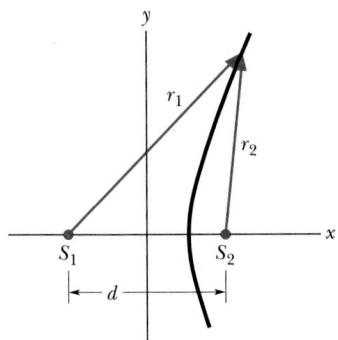

FIGURE 36-36 Problem 32.

33P. A thin flake of mica ($n = 1.58$) is used to cover one slit of a double-slit arrangement. The central point on the screen is now occupied by what had been the seventh bright side fringe ($m = 7$) before the mica was used. If $\lambda = 550$ nm, what is the thickness of the mica? (*Hint:* Consider the wavelength within the mica.)

34P. One slit of a double-slit arrangement is covered by a thin glass plate of refractive index 1.4, and the other by a thin glass plate of refractive index 1.7. The point on the screen at which the central maximum fell before the glass plates were inserted is now occupied by what had been the $m = 5$ bright fringe. Assuming that $\lambda = 480$ nm and that the plates have the same thickness t, find t.

SECTION 36-6 Intensity in Double-Slit Interference

35E. Find the sum y of the following quantities:

$$y_1 = 10 \sin \omega t \quad \text{and} \quad y_2 = 8.0 \sin(\omega t + 30°).$$

36E. Two waves of the same frequency have amplitudes 1.00 and 2.00. They interfere at a point where their phase difference is 60.0°. What is the resultant amplitude?

37E. Light of wavelength 600 nm is incident normally on two parallel narrow slits separated by 0.60 mm. Sketch the intensity pattern observed on a distant screen as a function of angle θ for the range of values $0 \leq \theta \leq 0.0040$ rad.

38E. Add the following quantities using the phasor method:

$$y_1 = 10 \sin \omega t$$

$$y_2 = 15 \sin(\omega t + 30°)$$

$$y_3 = 5.0 \sin(\omega t - 45°)$$

39P. A and B in Fig. 36-35 are point sources of electromagnetic waves of wavelength 1.00 m. They are in phase and separated by $d = 4.00$ m, and they emit at the same power. (a) If a detector is moved to the right along the x axis from point A, at what distances from A are the first three interference maxima detected? (b) Is the intensity of the nearest minimum exactly zero? (*Hint:* Does the intensity of a wave from a point source remain constant with an increase in distance from the source?)

40P. The double horizontal arrow in Fig. 36-9 marks the points on the intensity curve where the intensity of the central fringe is half the maximum intensity. Show that the angular separation $\Delta\theta$ between the corresponding points on the screen is

$$\Delta\theta = \frac{\lambda}{2d}$$

if θ in Fig. 36-8 is small enough so that $\sin \theta \approx \theta$.

41P*. Suppose that one of the slits of a double-slit arrangement is wider than the other, so that the amplitude of the light reaching the central part of the screen from one slit, acting alone, is twice that from the other slit, acting alone. Derive an expression for the light intensity I at the screen in terms of θ, corresponding to Eqs. 36-21 and 36-22.

SECTION 36-7 Interference from Thin Films

42E. In Fig. 36-37, light wave W_1 reflects once from a reflecting surface while light wave W_2 reflects twice from that surface and once from a reflecting sliver at distance L from the mirror. The waves are initially in phase and have a wavelength of 620 nm. Neglect the slight tilt of the rays. (a) For what least value of L are the reflected waves exactly out of phase? (b) How far must the sliver be moved to put the waves exactly out of phase again?

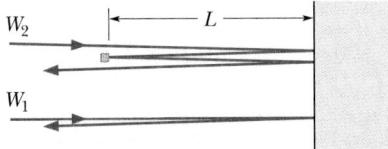

FIGURE 36-37
Exercise 42.

43E. Suppose the light waves of Exercise 42 are initially exactly out of phase. Find an expression for the values of L in terms of the wavelength λ for the situations in which the reflected waves are exactly in phase.

44E. Bright light of wavelength 585 nm is incident perpendicularly on a soap film ($n = 1.33$) of thickness 1.21 μm, suspended in air. Is the light reflected by the two surfaces of the film closer to interfering fully destructively or fully constructively?

45E. Light of wavelength 624 nm is incident perpendicularly on a soap film (with $n = 1.33$) suspended in air. What are the smallest two thicknesses of the film for which the reflections from the film undergo fully constructive interference?

46E. A lens with index of refraction greater than 1.30 is coated with a thin transparent film of index of refraction 1.30 to eliminate by interference the reflection of red light at wavelength 680 nm that is incident perpendicularly on the lens. What minimum film thickness is needed?

47E. A camera lens with index of refraction greater than 1.30 is coated with a thin transparent film of index of refraction 1.25 to eliminate by interference the reflection of light at wavelength λ that is incident perpendicularly on the lens. In terms of λ, what minimum film thickness is needed?

48E. A thin film suspended in air is 0.410 μm thick and illuminated with white light that is incident perpendicularly on its surface. The index of refraction of the film is 1.50. At what wavelengths will visible light reflected from the two surfaces of the film undergo fully constructive interference?

49E. The rhinestones in costume jewelry are glass with index of refraction 1.50. To make them more reflective, they are often coated with a layer of silicon monoxide of index of refraction 2.00. What is the minimum coating thickness needed to ensure that light of wavelength 560 nm and of perpendicular incidence will be reflected from the two surfaces of the coating with fully constructive interference?

50E. We wish to coat flat glass ($n = 1.50$) with a transparent material ($n = 1.25$) so that reflection of light at wavelength 600 nm is eliminated by interference. What minimum thickness can the coating have to do this?

51P. In Fig. 36-38, light of wavelength 600 nm is incident perpendicularly on five sections of a transparent structure suspended in air. The structure has index of refraction 1.50. The thickness of each section is given in terms of $L = 4.00$ μm. For which sections will the light that is reflected from the top and bottom surfaces of that section undergo fully constructive interference?

52P. In Fig. 36-39, light is incident perpendicularly on four thin layers of thickness L. The indices of refraction of the thin layers and of the media above and below these layers are given. Let λ represent the wavelength of the light in air, and n_2 represent the index of refraction of the thin layer in each situation. Consider only the transmission of light that undergoes no reflection or two reflections, as in Fig. 36-39a. For which of the situations does the expression

$$\lambda = \frac{2Ln_2}{m}, \qquad \text{for } m = 0, 1, 2, \ldots ,$$

give the wavelengths of the transmitted light that undergoes fully constructive interference?

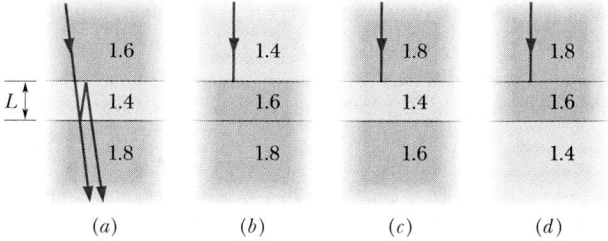

FIGURE 36-39 Problems 52, 53, and 56.

53P. A disabled tanker leaks kerosene ($n = 1.20$) into the Persian Gulf, creating a large slick on top of the water ($n = 1.30$). (a) If you are looking straight down from an airplane, while the Sun is overhead, at a region of the slick where its thickness is 460 nm, for which wavelength(s) of visible light is the reflection brightest because of constructive interference? (b) If you are scuba diving directly under this same region of the slick, for which wavelength(s) of visible light is the transmitted intensity strongest? (*Hint:* Use Fig. 36-39a with appropriate indices of refraction.)

54P. A plane wave of monochromatic light is incident normally on a uniformly thin film of oil that covers a glass plate. The wavelength of the source can be varied continuously. Fully destructive interference of the reflected light is observed for wavelengths of 500 and 700 nm and for no wavelengths in between. If the index of refraction of the oil is 1.30 and that of the glass is 1.50, find the thickness of the oil film.

55P. The reflection of perpendicularly incident white light by a soap film in air has an interference maximum at 600 nm and a minimum at 450 nm, with no minimum in between. If $n = 1.33$ for the film, what is the film thickness, assumed uniform?

56P. A sheet of glass having an index of refraction of 1.40 is to be coated with a film of material having a refractive index of 1.55 such that green light with a wavelength of 525 nm is preferentially transmitted via constructive interference. (a) What is the minimum thickness of the film that will achieve the result? (*Hint:* Use Fig. 36-39a with appropriate indices of refraction.) (b) Why are other parts of the visible spectrum not also preferentially transmitted? (c) Will the transmission of any colors be sharply reduced? If so, which colors?

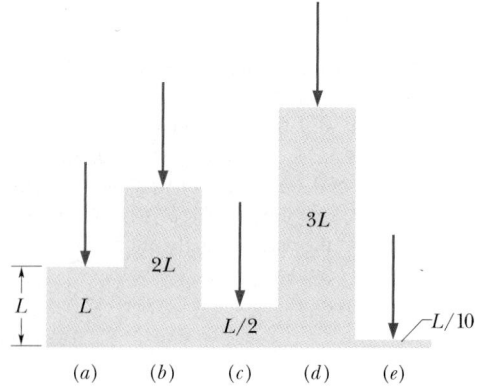

FIGURE 36-38 Problem 51.

57P. A plane monochromatic light wave in air is perpendicularly incident on a thin film of oil that covers a glass plate. The wavelength of the source may be varied continuously. Fully destructive interference of the reflected light is observed for wavelengths of 500 and 700 nm and for no wavelength in between. The index of refraction of glass is 1.50. Show that the index of refraction of the oil must be less than 1.50.

58P. A thin film of acetone ($n = 1.25$) coats a thick glass plate ($n = 1.50$). White light is incident normal to the film. In the reflections, fully destructive interference occurs at 600 nm and fully constructive interference at 700 nm. Calculate the thickness of the acetone film.

59P. Suppose that in Fig. 36-12 the light is not incident perpendicularly on the thin film but at an angle $\theta_i > 0$. Find an equation like Eqs. 36-34 and 36-35 that gives the interference maxima for the waves of rays r_1 and r_2. The wavelength is λ, the film thickness is L, and $n_2 > n_1 = n_3 = 1.0$.

60P. From a medium of index of refraction n_1, monochromatic light of wavelength λ is incident normally on a thin film of uniform thickness L (where $L > 0.1\lambda$) and index of refraction n_2. The light transmitted by the film travels into a medium with index of refraction n_3. Find expressions for the minimum film thickness (in terms of λ and the indices of refraction) for the following cases: (a) minimum light is reflected (hence maximum light is transmitted) with $n_1 < n_2 > n_3$; (b) minimum light is reflected (hence maximum light is transmitted) with $n_1 < n_2 < n_3$; and (c) maximum light is reflected (hence minimum light is transmitted) with $n_1 < n_2 < n_3$.

61P. In Sample Problem 36-5 assume that the coating eliminates the reflection of light of wavelength 550 nm at normal incidence. Calculate the factor by which reflection is diminished by the coating at 450 and 650 nm.

62P. In Fig. 36-40, a broad beam of light of wavelength 683 nm is sent directly downward through the top plate of a pair of glass plates. The plates are 120 mm long, touch at the left end, and are separated by a wire of diameter 0.048 mm at the right end. The air between the plates acts as a thin film. How many bright fringes will be seen by an observer looking down through the top plate?

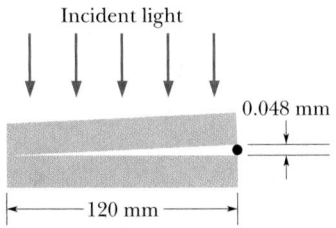

Incident light

0.048 mm

← 120 mm →

FIGURE 36-40 Problems 62 and 63.

63P. In Fig. 36-40, white light is sent directly downward through the top plate of a pair of glass plates. The plates touch at the left end and are separated by wire (of diameter 0.048 mm) at the right end; the air between the plates acts as a thin film. An observer looking down through the top plate sees bright and dark fringes due to that film. (a) Is a dark fringe or a bright fringe seen at the left end? (b) To the right of that end, fully destructive interference occurs at different locations for different wave-

lengths of the light. Does it occur first for the red end or the blue end of the visible spectrum?

64P. In Fig. 36-41a, a broad beam of light of wavelength 600 nm is sent directly downward through a glass plate ($n = 1.5$) that, with a plastic plate ($n = 1.2$), forms a thin wedge of air which acts as a thin film. An observer looking down through the top plate sees the fringe pattern shown in Fig. 36-41b, with dark fringes centered on ends A and B. (a) What is the thickness of the wedge at B? (b) How many dark fringes will the observer see if the air between the plates is replaced with water ($n = 1.33$)?

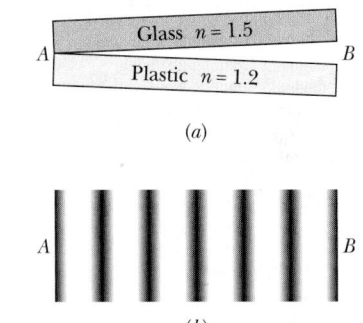

Glass $n = 1.5$

Plastic $n = 1.2$

(a)

A ‖‖‖‖‖‖‖ B

(b)

FIGURE 36-41 Problem 64.

65P. A broad beam of light of wavelength 630 nm is incident at 90° on a thin, wedge-shaped film with index of refraction 1.50. An observer intercepting the light transmitted by the film sees 10 bright and 9 dark fringes along the length of the film. By how much does the film thickness change over this length?

66P. Two glass plates are held together at one end to form a wedge of air that acts as a thin film. A broad beam of light of wavelength 480 nm is directed through the plates, perpendicular to the first plate. An observer intercepting light reflected from the plates sees on the plates an interference pattern that is due to the wedge of air. How much thicker is the wedge at the sixteenth bright fringe than it is at the sixth bright fringe, counting from where the plates touch?

67P. A broad beam of monochromatic light is directed perpendicularly through two glass plates that are held together at one end, creating a wedge of air between them. An observer intercepting light reflected from the wedge of air, which acts as a thin film, sees 4001 dark fringes along the length of the wedge. When the air between the plates is evacuated, only 4000 dark fringes are seen. Calculate the index of refraction of air from these data.

68P. Figure 36-42a shows a lens with radius of curvature R lying on a plane glass plate and illuminated from above by light with wavelength λ. Figure 36-42b shows that circular interference fringes (called *Newton's rings*) appear, associated with the variable thickness d of the air film between the lens and the plate. Find the radii r of the interference maxima assuming $r/R \ll 1$.

69P. In a Newton's rings experiment (see Problem 68), the radius of curvature R of the lens is 5.0 m and its diameter is 20 mm. (a) How many bright rings are produced? Assume that $\lambda = 589$ nm. (b) How many bright rings would be produced if the arrangement were immersed in water ($n = 1.33$)?

70P. A Newton's rings apparatus is to be used to determine the radius of curvature of a lens (see Fig. 36-42 and Problem 68). The radii of the nth and $(n + 20)$th bright rings are measured and found to be 0.162 and 0.368 cm, respectively, in light of wavelength 546 nm. Calculate the radius of curvature of the lower surface of the lens.

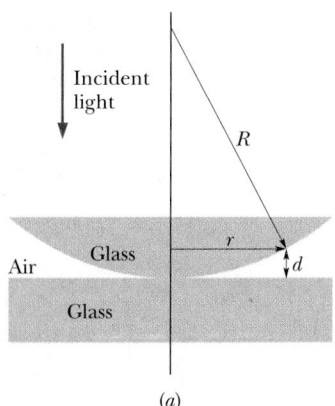

(a)

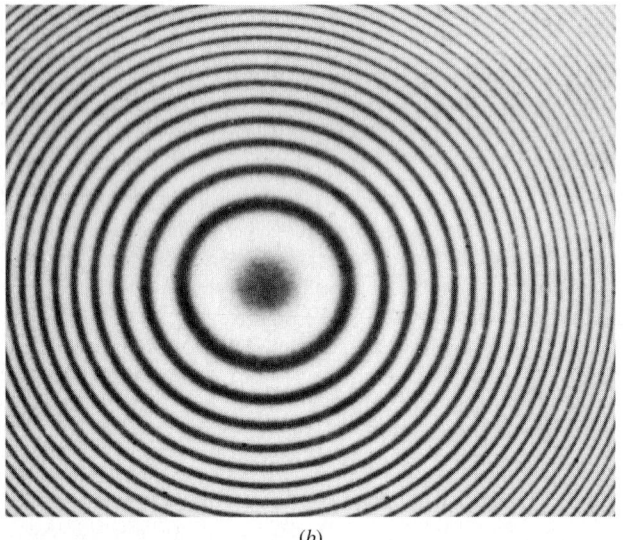

(b)

FIGURE 36-42   Problems 68 through 71.

71P.   (a) Use the result of Problem 68 to show that, in a Newton's rings experiment, the difference in radius between adjacent bright rings (maxima) is given by

$$\Delta r = r_{m+1} - r_m \approx \tfrac{1}{2}\sqrt{\lambda R/m},$$

assuming $m \gg 1$. (b) Now show that the *area* between adjacent bright rings is given by

$$A = \pi \lambda R,$$

assuming $m \gg 1$. Note that this area is independent of m.

72P.   In Fig. 36-43, a microwave transmitter at height a above the water level of a wide lake transmits microwaves of wavelength λ toward a receiver on the opposite shore, a distance x above the water level. The microwaves reflecting from the water interfere with the microwaves arriving directly from the transmitter. Assuming that the lake width D is much larger than a and x, and that $\lambda \geq a$, at what values of x is the signal at the receiver maximum? (*Hint:* Does the reflection cause a phase change?)

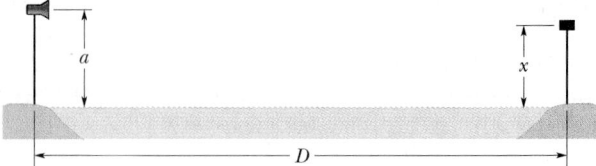

FIGURE 36-43   Problem 72.

SECTION 36-8   Michelson's Interferometer

73E.   If mirror M_2 in a Michelson interferometer is moved through 0.233 mm, a shift of 792 fringes occurs. What is the wavelength of the light producing the fringe pattern?

74E.   A thin film with index of refraction $n = 1.40$ is placed in one arm of a Michelson interferometer, perpendicular to the optical path. If this causes a shift of 7.0 fringes of the pattern produced by light of wavelength 589 nm, what is the film thickness?

75P.   An airtight chamber 5.0 cm long with glass windows is placed in one arm of a Michelson interferometer as indicated in Fig. 36-44. Light of wavelength $\lambda = 500$ nm is used. When the air has been completely evacuated from the chamber, there has been a shift of 60 fringes. From these data, find the index of refraction of air at atmospheric pressure.

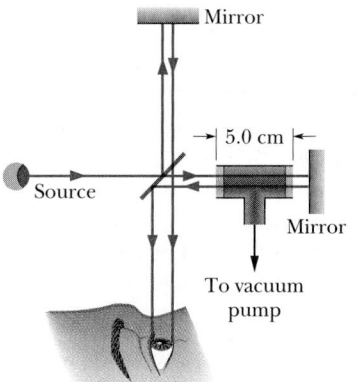

FIGURE 36-44   Problem 75.

76P.   The element sodium can emit light at two wavelengths, $\lambda_1 = 589.10$ nm and $\lambda_2 = 589.59$ nm. If light from sodium is used in a Michelson interferometer, through what distance must one mirror be moved to cause the fringe pattern for one wavelength to shift 1.00 fringe more than the pattern for the other wavelength?

77P.   Write an expression for the intensity observed in a Michelson interferometer (Fig. 36-18) as a function of the position of the movable mirror. Measure the position of the mirror from the point at which $d_1 = d_2$.

78P. By the late 1800s, most scientists believed that light (any electromagnetic wave) required a medium in which to travel, that it could not travel through vacuum. One reason for this belief was that any other type of wave known to the scientists requires a medium. For example, sound waves can travel through air, water, or ground but not through vacuum. Thus, reasoned the scientists, when light travels from the Sun or any other star to Earth, it cannot be traveling through vacuum; instead, it must be traveling through a medium that fills all of space and through which Earth slips. Presumably, light has a certain speed c through this medium, which was called *aether* (or *ether*).

In 1887 Michelson and Edward Morely used a version of Michelson's interferometer to test for the effects of aether on the travel of light within the device. Specifically, the motion of the device through aether as Earth moves around the Sun should affect the interference pattern produced by the device. Scientists assumed that the Sun is approximately stationary in aether; hence the speed of the interferometer through aether should be Earth's speed v about the Sun.

Figure 36-45a shows the basic arrangement of mirrors in the 1887 experiment. The mirrors were mounted on a heavy slab that was suspended on a pool of mercury so that the slab could be rotated smoothly about a vertical axis. Michelson and Morely wanted to monitor the interference pattern as they rotated the slab, changing the orientation of the interferometer arms relative to the motion through aether. A fringe shift in the interference pattern during the rotation would clearly signal the presence of aether.

Figure 36-45b, an overhead view of the equipment, shows the path of the light. To improve the possibility of fringe shift, the light was reflected several times along the arms of the interferometer, instead of only once along each arm as indicated in the basic interferometer of Fig. 36-18. This repeated reflection increased the effective length of each arm to about 10 m. In spite of the added complexity, the interferometer of Figs. 36-45a and b functions just like the simpler interferometer of Fig. 36-18; so we can use Fig. 36-18 in our discussion here by merely taking the arm lengths d_1 and d_2 to be 10 m each.

Let us assume that there is aether through which light has speed c. Figure 36-45c shows a side view of the arm of length d_1 from the aether reference frame as the interferometer moves rightward through it with velocity **v**. (For simplicity, the beam splitter M of Fig. 36-18 is drawn parallel to the mirror M_1 at the far end of the arm.) Figure 36-45d shows the arm just as a particular portion of the light (represented by a dot) begins its travel along the arm. We shall follow this light to find the path length along the arm.

As the light moves at speed c rightward through aether and toward mirror M_1, that mirror moves rightward at speed v. Figure 36-45e shows the positions of M and M_1 when the light reaches M_1, reflecting there. The light now moves leftward through aether at speed c while M moves rightward. Figure 36-45f shows the positions of M and M_1 when the light has returned to M. (a) Show that the total time of travel for this light, from M to M_1 and then back to M, is

$$t_1 = \frac{2cd_1}{c^2 - v^2}$$

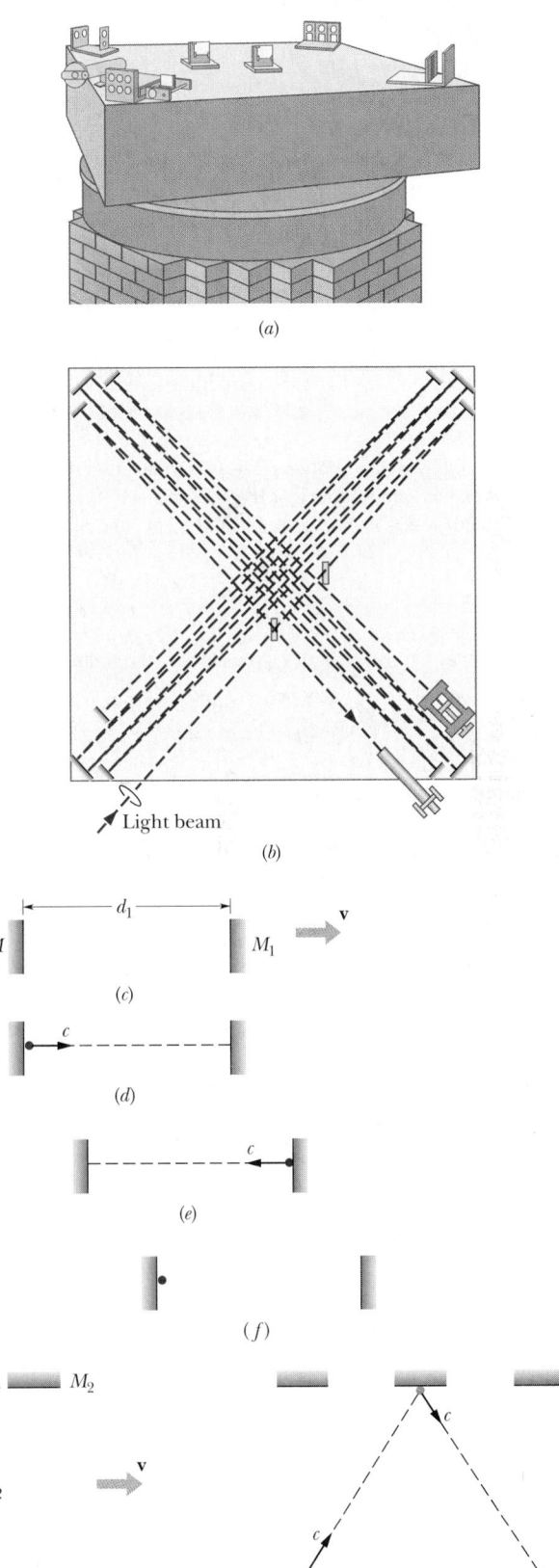

FIGURE 36-45 Problem 78.

and thus that the path length L_1 traveled by the light along this arm is

$$L_1 = ct_1 = \frac{2c^2 d_1}{c^2 - v^2}.$$

Figure 36-45g shows a view of the arm of length d_2; that arm also moves rightward with velocity **v** through the aether. For simplicity, the beam splitter M of Fig. 36-18 is now drawn parallel to the mirror M_2 at the far end of this arm. Figure 36-45h shows the arm just as a particular portion of the light (the dot) begins its travel along the arm. Because the arm moves rightward during the flight of the light, the path of the light is angled rightward toward the position that M_2 will have when the light reaches that mirror (Fig. 36-45i). The reflection of the light from M_2 sends the light angled rightward toward the position that M will have when the light returns to it (Fig. 36-45j). (b) Show that the total time of travel for the light, from M to M_2 and then back to M, is

$$t_2 = \frac{2d_2}{\sqrt{c^2 - v^2}}$$

and thus that the path length L_2 traveled by the light along this arm is

$$L_2 = ct_2 = \frac{2cd_2}{\sqrt{c^2 - v^2}}.$$

Substitute d for d_1 and d_2 in the expressions for L_1 and L_2. Then expand the two expressions by using the binomial expansion (given in Appendix E and explained in the Problem Solving Tactic on page 145); retain the first two terms in each expansion.

(c) Show that path length L_1 is greater than path length L_2 and that their difference ΔL is

$$\Delta L = \frac{dv^2}{c^2}.$$

(d) Next show that the phase difference between the light traveling along L_1 and that along L_2 is

$$\frac{\Delta L}{\lambda} = \frac{dv^2}{\lambda c^2},$$

where λ is the wavelength of the light. This phase difference determines the fringe pattern produced by the light arriving at the telescope in the interferometer.

Now rotate the interferometer by 90° so that the arm of length d_2 is along the direction of motion through the aether and the arm of length d_1 is perpendicular to that direction. (e) Show that the shift in the fringe pattern due to the rotation is

$$\text{shift} = \frac{2dv^2}{\lambda c^2}.$$

(f) Evaluate the shift, setting $c = 3.0 \times 10^8$ m/s, $d = 10$ m, and $\lambda = 500$ nm and using data about Earth given in Appendix C.

This expected fringe shift would have been easily observable. However, Michelson and Morely observed no fringe shift, which cast grave doubt on the existence of aether. In fact, the idea of aether soon disappeared. Moreover, the null result of Michelson and Morely led, at least indirectly, to Einstein's special theory of relativity.

37
Diffraction

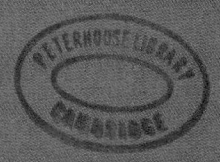

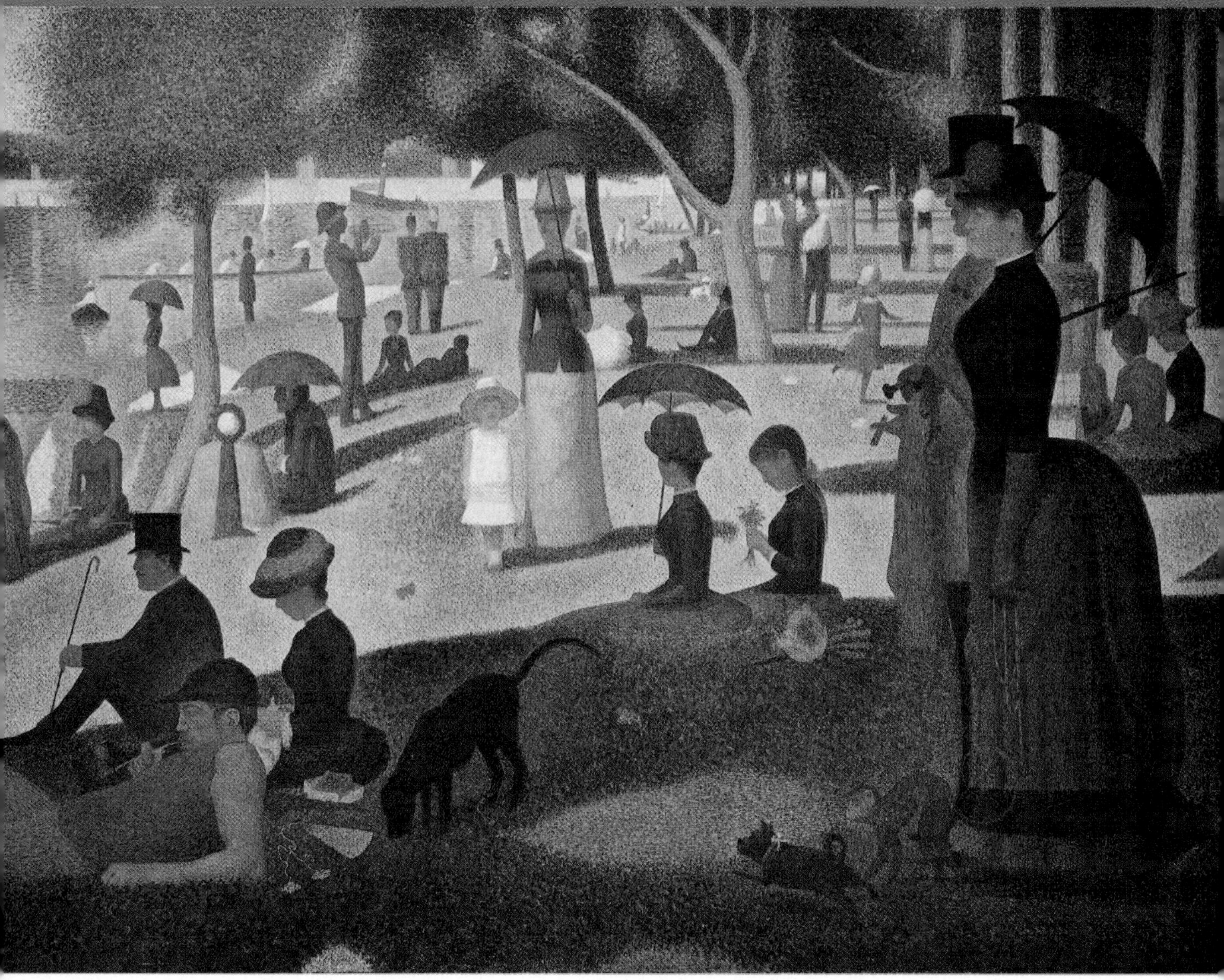

Georges Seurat painted Sunday Afternoon on the Island of La Grande Jatte using not brush strokes in the usual sense, but rather a myriad of small colored dots, in a style of painting now known as pointillism. You can see the dots if you stand close enough to the painting, but as you move away from it, they eventually blend and cannot be distinguished. Moreover, the color that you see at any given place on the painting changes as you move away—which is why Seurat painted with the dots. What causes this change in color?

37-1 DIFFRACTION AND THE WAVE THEORY OF LIGHT

In Chapter 36 we defined diffraction rather loosely as the flaring of light as it emerges from a narrow slit. More than just flaring occurs, however, because the light produces an interference pattern called a **diffraction pattern.** For example, when monochromatic light from a distant source (or a laser) passes through a narrow slit and is then intercepted by a viewing screen, the light produces on the screen a diffraction pattern like that in Fig. 37-1. This pattern consists of a broad and intense (very bright) central maximum and a number of narrower and less intense maxima (called **secondary** or **side** maxima) to both sides. In between the maxima are minima.

Such a pattern would be totally unexpected in geometrical optics: if light traveled in straight lines as rays, then the slit would merely allow some of those rays through and they would form a sharp, bright rendition of the slit on the viewing screen. As in Chapter 36, we again must conclude that geometrical optics is only an approximation.

Diffraction of light is not limited to situations of light passing through a narrow opening (such as a slit or pinhole). It also occurs when light passes an edge, such as the edges of the razor blade in Fig. 37-2. Note the lines of maxima and minima that run approximately parallel to the edges, both on the inside of the blade and on the outside. As the light passes, say, the vertical edge at the left, it flares left and right and undergoes interference, producing the pattern along the left edge. The right-most portion of that pattern actually lies within what would have been the shadow of the blade if geometrical optics prevailed.

You encounter a common example of diffraction when you look at a clear blue sky and see tiny specks and hairlike structures floating in your view. These *floaters,* as they are called, are produced when light passes the edges of tiny bits of vitreous humor (the transparent material fill-

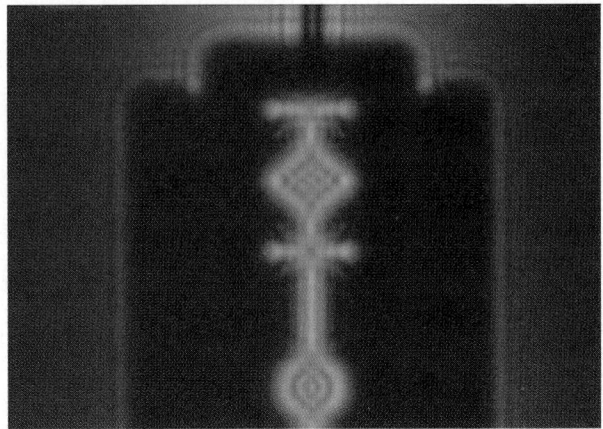

FIGURE 37-2 The diffraction pattern of a razor blade in monochromatic light. Note the lines of alternating maximum and minimum intensity.

ing most of the eyeball). These bits have broken off from the main section and now float in a water layer just in front of the retina where light is detected. What you are seeing when a floater is in your field of vision is the diffraction pattern produced by one of these floating bits. If you sight through a pinhole in an otherwise opaque sheet so as to make the light entering your eye approximately a plane wave, you might be able to distinguish individual maxima and minima in the patterns.

The Fresnel Bright Spot

Diffraction finds a ready explanation in the wave theory of light. However, this theory, originally advanced by Huygens and used 123 years later by Young to explain double-slit interference, was very slow in being adopted, largely because it ran counter to Newton's theory that light was a stream of particles.

Newton's view was the prevailing view in French scientific circles of the early nineteenth century, when Augustin Fresnel was a young military engineer. Fresnel, who believed in the wave theory of light, submitted a paper to the French Academy of Sciences describing his experiments and his wave-theory explanations of them.

In 1819, the Academy, dominated by the supporters of Newton and thinking to challenge the wave point of view, organized a prize competition for an essay on the subject of diffraction. Fresnel won. The Newtonians, however, were neither converted nor silenced. One of them, S. D. Poisson, pointed out the "strange result" that if Fresnel's theories were correct, then light waves should flare into the shadow region of a sphere as they pass the edge of the sphere, producing a bright spot at the center of the shadow. The prize committee arranged a test of the famous mathemati-

FIGURE 37-1 This diffraction pattern appeared on a viewing screen when light that had passed through a narrow horizontal slit reached the screen. The diffraction process causes light to flare out perpendicular to the long sides of the slit. The process also produces an interference pattern consisting of a broad central maximum, less intense and narrower secondary (or side) maxima, and minima.

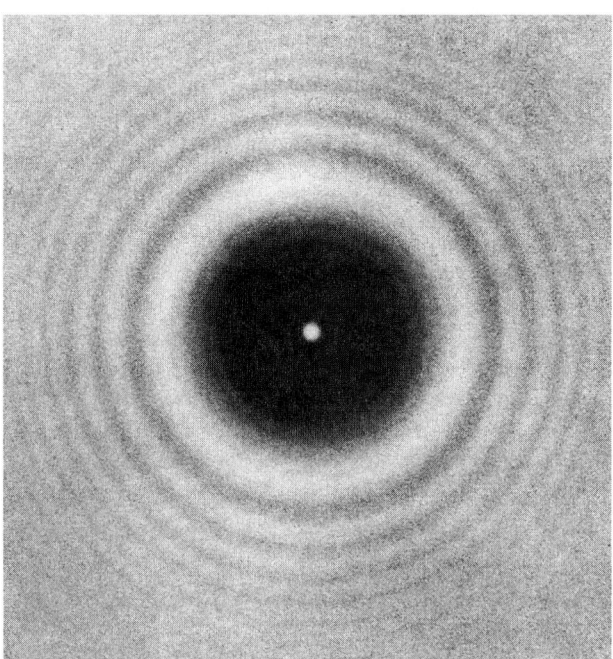

FIGURE 37-3 The diffraction pattern of a disk. Note the concentric diffraction rings and the Fresnel bright spot at the center of the pattern. This experiment is essentially identical to that arranged by the committee testing Fresnel's theories, because both the sphere they used and the disk used here have a cross section with a circular edge.

cian's prediction and discovered (see Fig. 37-3) that the predicted *Fresnel bright spot,* as we call it today, was indeed there! Nothing builds confidence in a theory so much as having one of its unexpected and counterintuitive predictions verified by experiment.

37-2 DIFFRACTION BY A SINGLE SLIT: LOCATING THE MINIMA

Let us now consider how plane waves of light of wavelength λ are diffracted by a single long narrow slit of width a in an otherwise opaque screen B, as shown in cross section in Fig. 37-4a. (In that figure, the slit's length extends into and out of the page.) When the diffracted light reaches viewing screen C, waves from different points within the slit undergo interference and produce a diffraction pattern of bright and dark fringes (interference maxima and minima) on the screen. To locate the fringes, we shall use a procedure somewhat similar to the one we used to locate the fringes in a two-slit interference pattern. However, diffraction is more mathematically challenging, and here we shall be able to find equations for only the dark fringes.

Before we do that, however, we can justify the central

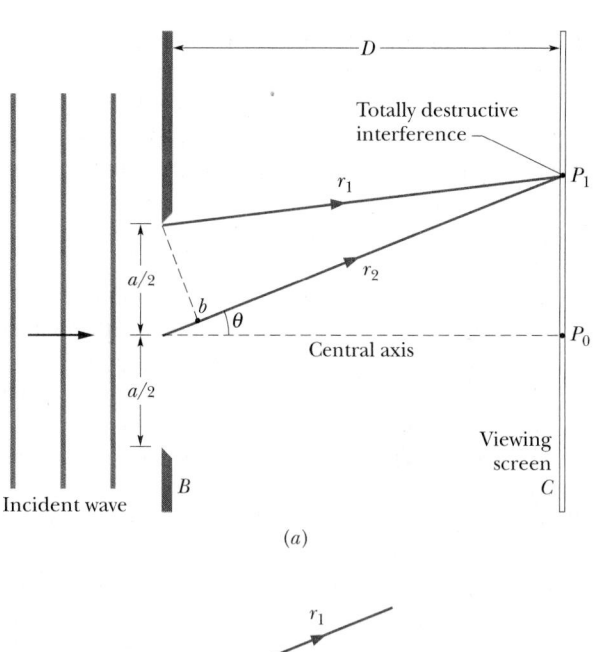

FIGURE 37-4 (a) Waves from the top points of two zones of width $a/2$ undergo totally destructive interference at point P_1 on viewing screen C. (b) For $D \gg a$, we can approximate rays r_1 and r_2 as being parallel, at angle θ to the central axis.

bright fringe seen in Fig. 37-1 by noting that the waves from all points in the slit travel about the same distance to reach the center of the pattern and thus are in phase there. As for the other bright fringes, we can say only that they are approximately halfway between adjacent dark fringes.

To find the dark fringes, we shall use a clever (and simplifying) strategy that involves pairing up all the rays coming through the slit and then finding what conditions cause the waves of the rays in each pair to cancel each other. Figure 37-4a shows how we apply this strategy to locate the first dark fringe, at point P_1. First, we mentally divide the slit into two *zones* of equal widths $a/2$. Then we extend to P_1 a light ray r_1 from the top point of the top zone and a light ray r_2 from the top point of the bottom zone. A central axis is drawn from the center of the slit to screen C, and P_1 is located at an angle θ to that axis.

The waves of the pair of rays r_1 and r_2 are in phase within the slit because they originate from the same wavefront passing through the slit. However, to produce the first dark fringe they must be out of phase by $\lambda/2$ when they reach P_1; this phase difference is due to their path length difference, with the wave of r_2 traveling a longer path to reach P_1 than the wave of r_1. To display this path length difference, we find a point b on ray r_2 such that the path length from b to P_1 matches the path length of ray r_1. Then the path length difference between the two rays is the distance from the center of the slit to b.

When viewing screen C is near screen B, as in Fig. 37-4a, the diffraction pattern on C is difficult to describe mathematically. However, we can simplify the mathematics considerably if we arrange for the screen separation D to be much larger than the slit width a. Then we can approximate rays r_1 and r_2 as being parallel, at angle θ to the central axis (Fig. 37-4b). We can also approximate the triangle formed by point b, the top point of the slit, and the center of the slit as being a right triangle, and one of the angles inside that triangle as being θ. The path length difference between rays r_1 and r_2 (which is still the distance from the center of the slit to point b) is then equal to $(a/2) \sin \theta$.

We can repeat this analysis for any other pair of rays originating at corresponding points in the two zones (say, at the midpoints of the zones) and extending to point P_1. Each such pair of rays has the same path length difference $(a/2) \sin \theta$. Setting this common path length difference equal to $\lambda/2$, we have

$$\frac{a}{2} \sin \theta = \frac{\lambda}{2},$$

which gives us

$$a \sin \theta = \lambda \quad \text{(first minimum).} \quad (37\text{-}1)$$

Given slit width a and wavelength λ, Eq. 37-1 tells us the angle θ of the first dark fringe above and (by symmetry) below the central axis.

Note that if we begin with $a > \lambda$ and then narrow the slit while holding the wavelength constant, the angle for the first dark fringe increases; that is, the extent of the diffraction (the extent of the flaring) is *greater* for a *narrower* slit. For $a = \lambda$, the angle of the first dark fringes is 90°. Since these dark fringes mark the two edges of the central bright fringe, that bright fringe must cover the entire viewing screen.

We find the second dark fringes above and below the central axis as we found the first dark fringes, except that we now divide the slit into *four* zones of equal widths $a/4$, as shown in Fig. 37-5a. We then extend rays r_1, r_2, r_3, and r_4 from the top points of the zones to point P_2, the location of the second dark fringe above the central axis. To produce that fringe, the path length difference between r_1 and r_2, that between r_2 and r_3, and that between r_3 and r_4 must each be equal to $\lambda/2$.

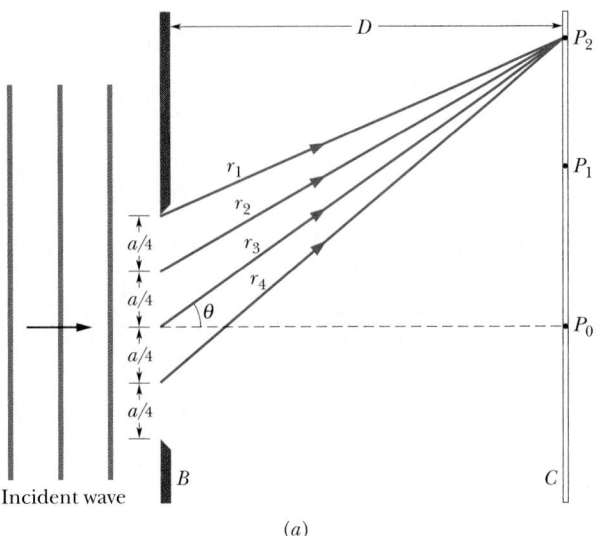

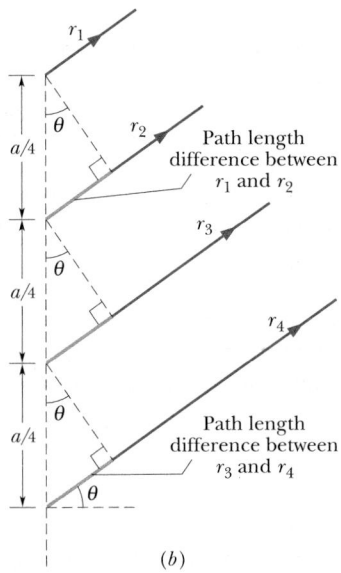

FIGURE 37-5 (*a*) Waves from the top points of four zones of width $a/4$ undergo totally destructive interference at point P_2. (*b*) For $D \gg a$, we can approximate rays r_1, r_2, r_3, and r_4 as being parallel, at angle θ to the central axis.

For $D \gg a$, we can approximate these four rays as being parallel, at angle θ to the central axis. To display their path length differences, we extend a perpendicular line through each adjacent pair of rays, as shown in Fig. 37-5b, to form a series of right triangles, each of which has a path length difference as one side. We see from the top triangle that the path length difference between r_1 and r_2 is $(a/4) \sin \theta$. Similarly, from the bottom triangle, the path length difference between r_3 and r_4 is also $(a/4) \sin \theta$. In fact, the path length difference between the members of any pair of rays that originate at corresponding points in two adjacent zones is $(a/4) \sin \theta$. Since in each such case the path length difference is equal to $\lambda/2$, we have

$$\frac{a}{4} \sin \theta = \frac{\lambda}{2},$$

which gives us

$$a \sin \theta = 2\lambda \quad \text{(second minimum).} \quad (37\text{-}2)$$

We could now continue to locate dark fringes in the diffraction pattern by splitting up the slit into more zones of equal width. We would always choose an even number of zones so that the zones (and their waves) could be paired as we have been doing. We would find that the dark fringes can be located with the following general equation:

$$a \sin \theta = m\lambda, \quad \text{for } m = 1, 2, 3, \ldots$$
$$\text{(minima—dark fringes).} \quad (37\text{-}3)$$

You can remember this result in the following way. Draw a triangle like the one in Fig. 37-4b, but for the full slit width a, and note that the path length difference between the top and bottom rays from the slit equals $a \sin \theta$. So, Equation 37-3 says:

In a single-slit diffraction experiment, dark fringes are produced where the path length differences ($a \sin \theta$) between the top and bottom rays are equal to λ, 2λ, 3λ, ...

This may seem to be wrong, because the waves of those two particular rays will be exactly in phase with each other. However, they each will still be part of a pair of waves that are exactly out of phase with each other; thus, *each* will be canceled by some other wave.

Equations 37-1, 37-2, and 37-3 are derived for the case of $D \gg a$. However, they also apply if we place a converging lens between the slit and the viewing screen and then move the screen in so that it coincides with the focal plane of the lens. The rays that now arrive at any point on the screen are *exactly* parallel (rather than approximately) when they leave the slit—they are like the initially parallel rays of Fig. 35-13a that are directed to a point by a lens.

CHECKPOINT 1: We produce a diffraction pattern on a viewing screen by using a long narrow slit illuminated by blue light. Does the pattern expand away from the bright center or contract toward it if we (a) switch to yellow light or (b) decrease the slit width?

SAMPLE PROBLEM 37-1

A slit of width a is illuminated by white light.

(a) For what value of a will the first minimum for red light of $\lambda = 650$ nm be at $\theta = 15°$?

SOLUTION: At the first minimum, $m = 1$ in Eq. 37-3. Solving for a, we then find

$$a = \frac{m\lambda}{\sin \theta} = \frac{(1)(650 \text{ nm})}{\sin 15°}$$
$$= 2511 \text{ nm} \approx 2.5 \text{ μm.} \quad \text{(Answer)}$$

For the incident light to flare out that much ($\pm 15°$) the slit has to be very fine indeed, amounting to about four times the wavelength. Note that a fine human hair may be about 100 μm in diameter.

(b) What is the wavelength λ' of the light whose first side diffraction maximum is at $15°$, thus coinciding with the first minimum for the red light?

SOLUTION: This maximum is about halfway between the first and second minima produced with wavelength λ'. We can find it without too much error by putting $m = 1.5$ in Eq. 37-3, obtaining

$$a \sin \theta = 1.5\lambda'.$$

Solving for λ' and substituting known data yield

$$\lambda' = \frac{a \sin \theta}{1.5} = \frac{(2511 \text{ nm})(\sin 15°)}{1.5}$$
$$= 430 \text{ nm.} \quad \text{(Answer)}$$

Light of this wavelength is violet. The first side maximum for light of wavelength 430 nm will always coincide with the first minimum for light of wavelength 650 nm, no matter what the slit width. If the slit is relatively narrow, the angle θ at which this overlap occurs will be relatively large, and conversely.

37-3 INTENSITY IN SINGLE-SLIT DIFFRACTION, QUALITATIVELY

In Section 37-2 we saw how to find the positions of the maxima and the minima in a single-slit diffraction pattern. Now we turn to a more general problem: find an expression for the intensity I of the pattern as a function of θ, the angular position of a point on a viewing screen.

To do this, we divide the slit of Fig. 37-4a into N zones of equal widths Δx so small that we can assume that each zone acts as a source of Huygens' wavelets. We wish to superimpose the wavelets arriving at an arbitrary point P on the viewing screen, at angle θ to the central axis, so that we can determine the amplitude E_θ of the resultant wave at P. The intensity of the light at P is then proportional to the square of the amplitude.

To find E_θ, we need the phase relationships among the arriving wavelets. The phase difference between wavelets from adjacent zones is given by

$$\begin{pmatrix}\text{phase} \\ \text{difference}\end{pmatrix} = \left(\frac{2\pi}{\lambda}\right)\begin{pmatrix}\text{path length} \\ \text{difference}\end{pmatrix}.$$

For point P at angle θ, the path length difference between wavelets from adjacent zones is $\Delta x \sin \theta$. So the phase difference $\Delta\phi$ between wavelets from adjacent zones is

$$\Delta\phi = \left(\frac{2\pi}{\lambda}\right)(\Delta x \sin \theta). \qquad (37\text{-}4)$$

We assume that the wavelets arriving at P all have the same amplitude ΔE. To find the amplitude E_θ of the resultant wave at P, we add the amplitudes ΔE via phasors. To do this, we construct a diagram of N phasors, one corresponding to the wavelet from each zone in the slit.

For point P_0 at $\theta = 0$ on the central axis of Fig. 37-4a, Eq. 37-4 tells us that the phase difference $\Delta\phi$ between the wavelets is zero. That is, the wavelets all arrive in phase. Figure 37-6a is the corresponding phasor diagram; adjacent phasors represent wavelets from adjacent zones and are arranged head to tail. Because there is zero phase difference between the wavelets, there is zero angle between each pair of adjacent phasors. The amplitude E_θ of the net wave at P_0 is the vector sum of these phasors. This arrangement of the phasors turns out to be the one that gives the greatest value for the wave amplitude E_θ. We call this value E_m; that is, E_m is the value of E_θ for $\theta = 0$.

We next consider a point P that is at a small angle θ to the central axis. Equation 37-4 now tells us that the phase difference $\Delta\phi$ between wavelets from adjacent zones is no longer zero. Figure 37-6b shows the corresponding phasor diagram; as before, the phasors are arranged head to tail, but now there is an angle $\Delta\phi$ between adjacent phasors. The amplitude E_θ at this new point is still the vector sum of the phasors, but it is smaller than that in Fig. 37-6a, which means that the intensity of the light is less at this new point P than at P_0.

If we continue to increase θ, the angle $\Delta\phi$ between adjacent phasors increases, and eventually the chain of phasors curls completely around so that the head of the last phasor reaches the tail of the first phasor (Fig. 37-6c). The amplitude E_θ is now zero, which means that the intensity of the light is also zero. We have reached the first minimum, or dark fringe, in the diffraction pattern. The first and last phasors now have a phase difference of 2π rad, which means that the path length difference between the top and bottom rays through the slit equals one wavelength. Recall that this is the condition we determined for the first diffraction minimum.

As we continue to increase θ, the angle $\Delta\phi$ between adjacent phasors increases, the chain of phasors begins to wrap back on itself, and the resulting coil begins to shrink. Amplitude E_θ now grows larger until it reaches a maximum value in the arrangement shown in Fig. 37-6d. This arrangement corresponds to the first side maximum in the diffraction pattern.

If we increase θ a bit more, the resulting shrinkage of the coil decreases E_θ, which means that the intensity also decreases. When θ is increased enough, the head of the last phasor again meets the tail of the first phasor. We have then reached the second minimum.

We could continue this qualitative method of determining the maxima and minima of the diffraction pattern but, instead, we shall now turn to a quantitative method.

CHECKPOINT **2:** The figures represent, in smoother form (with more phasors) than Fig. 37-6, the phasor diagrams for points on opposite sides of a certain diffraction maximum. (a) Which maximum is it? (b) What is the approximate value of m (in Eq. 37-3) that corresponds to this maximum?

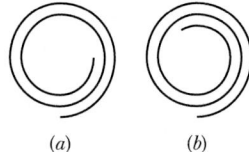

(a)   (b)

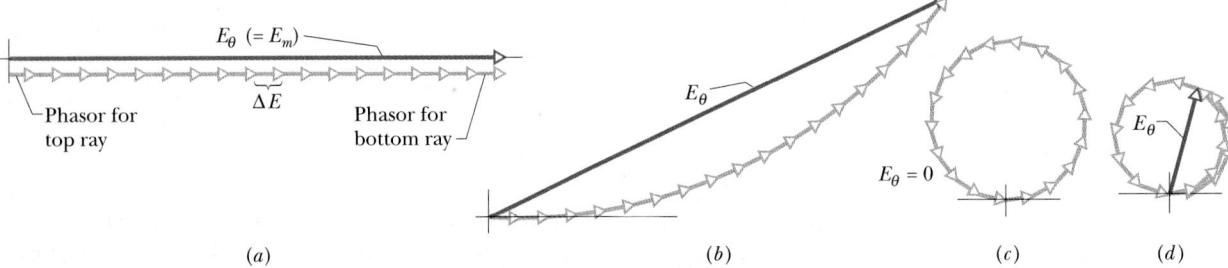

(a)

(b)   (c)   (d)

FIGURE 37-6 Phasor diagrams for $N = 18$ phasors, corresponding to 18 zones in a single slit. Resultant amplitudes E_θ are shown for (a) the central maximum at $\theta = 0$, (b) a point on the screen lying at a small angle θ to the central axis, (c) the first minimum, and (d) the first side maximum.

37-4 INTENSITY IN SINGLE-SLIT DIFFRACTION, QUANTITATIVELY

Equation 37-3 tells us how to locate the minima of the single-slit diffraction pattern on screen C of Fig. 37-4a as a function of the angle θ in that figure. Here we wish to derive an expression for the intensity I of the pattern as a function of θ. We state, and shall prove below, that the intensity is given by

$$I = I_m \left(\frac{\sin \alpha}{\alpha} \right)^2, \qquad (37\text{-}5)$$

where

$$\alpha = \tfrac{1}{2}\phi = \frac{\pi a}{\lambda} \sin \theta. \qquad (37\text{-}6)$$

The symbol α is just a convenient connection between the angle θ that locates a point on the viewing screen and the light intensity I at that point. I_m is the greatest value of the intensities I_θ in the pattern, and it occurs at the central maximum (where $\theta = 0$). And ϕ is the phase difference (in radians) between the top and bottom rays from the slit.

Study of Eq. 37-5 shows that intensity minima will occur when

$$\alpha = m\pi, \qquad \text{for } m = 1, 2, 3, \ldots . \qquad (37\text{-}7)$$

If we put this result into Eq. 37-6 we find

$$m\pi = \frac{\pi a}{\lambda} \sin \theta, \qquad \text{for } m = 1, 2, 3, \ldots$$

or $\qquad a \sin \theta = m\lambda, \qquad \text{for } m = 1, 2, 3, \ldots$
$$\text{(minima—dark fringes),} \qquad (37\text{-}8)$$

which is exactly Eq. 37-3, the expression that we derived earlier for the location of the minima.

Figure 37-7 shows plots of the intensity of a single-slit diffraction pattern, calculated with Eqs. 37-5 and 37-6 for three slit widths: $a = \lambda$, $a = 5\lambda$, and $a = 10\lambda$. Note that as the slit width increases (relative to the wavelength), the width of the *central diffraction maximum* (the central hill-like region of the graphs) decreases; that is, the light undergoes less flaring by the slit. The secondary maxima also decrease in width (and become weaker). In the limit of slit width a being much greater than wavelength λ, the secondary maxima due to the slit disappear; we then no longer have single-slit diffraction (but we still have diffraction due to the edges of the wide slit, like that produced by the edges of the razor blade in Fig. 37-2).

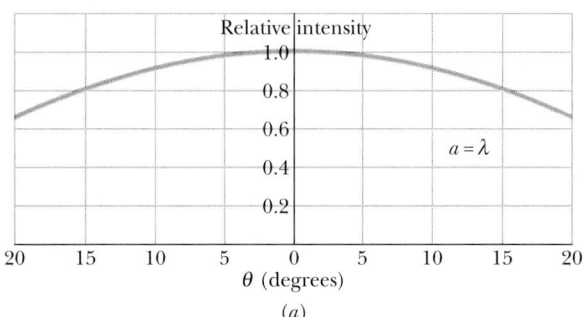

(a)

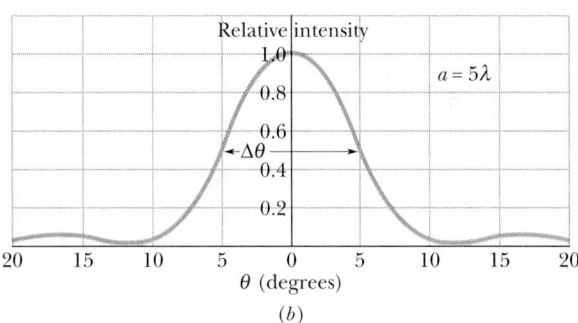

(b)

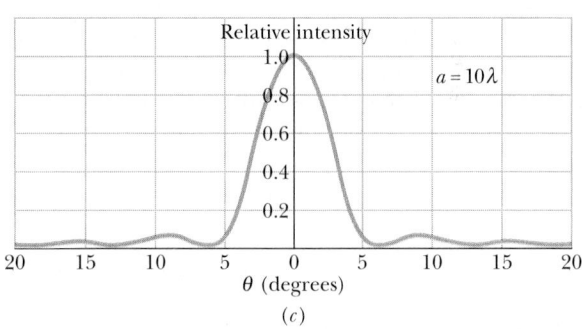

(c)

FIGURE 37-7 The relative intensity in single-slit diffraction for three values of the ratio a/λ. The wider the slit, the narrower is the central diffraction maximum.

Proof of Eqs. 37-5 and 37-6

The arc of phasors in Fig. 37-8 represents the wavelets that reach an arbitrary point P on the viewing screen of Fig. 37-4, corresponding to a particular small angle θ. The amplitude E_θ of the resultant wave at P is the vector sum of these phasors. If we divide the slit of Fig. 37-4 into infinitesimal zones of width Δx, the arc of phasors in Fig. 37-8 approaches the arc of a circle; we call its radius R as indicated in that figure. The length of the arc must be E_m, the amplitude at the center of the diffraction pattern, because if we straightened out the arc we would have the phasor arrangement of Fig. 37-6a (shown lightly in Fig. 37-8).

The angle ϕ in the lower part of Fig. 37-8 is the difference in phase between the infinitesimal vectors at the left

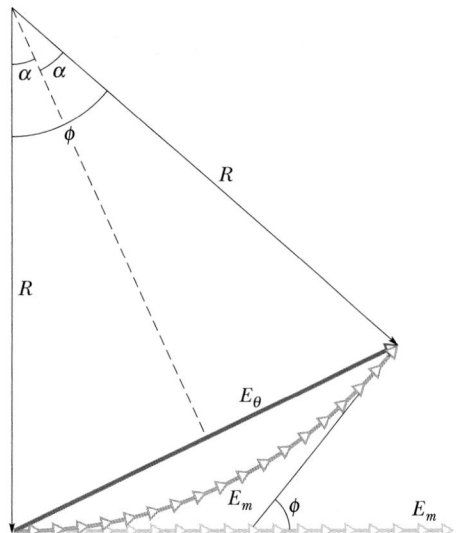

FIGURE 37-8 A construction used to calculate the intensity in single-slit diffraction. The situation shown corresponds to that of Fig. 37-6*b*.

and right ends of arc E_m. From the geometry, ϕ is also the angle between the two radii marked R in Fig. 37-8. The dashed line in that figure then forms two congruent triangles with angle $\frac{1}{2}\phi$. From either triangle we can write

$$\sin \tfrac{1}{2}\phi = \frac{E_\theta}{2R}. \qquad (37\text{-}9)$$

In radian measure, ϕ is (with E_m considered to be a circular arc)

$$\phi = \frac{E_m}{R}.$$

Solving this equation for R and substituting in Eq. 37-9 yield, after some manipulation,

$$E_\theta = \frac{E_m}{\frac{1}{2}\phi} \sin \tfrac{1}{2}\phi. \qquad (37\text{-}10)$$

In Section 34-4 we saw that the intensity of an electromagnetic wave is proportional to the square of the amplitude of its electric field. Here, this means that the maximum intensity I_m (at the center of the diffraction pattern) is proportional to E_m^2 and the intensity I at angle θ is proportional to E_θ^2. Thus, we may write

$$\frac{I}{I_m} = \frac{E_\theta^2}{E_m^2}. \qquad (37\text{-}11)$$

Substituting for E_θ with Eq. 37-10 and then substituting $\alpha = \frac{1}{2}\phi$, we are led to the following expression for the

intensity as a function of θ:

$$I = I_m \left(\frac{\sin \alpha}{\alpha} \right)^2.$$

This is exactly Eq. 37-5, one of the two equations we set out to prove.

The second equation we wish to prove relates α to θ. The phase difference ϕ between the rays from the top and bottom of the entire slit may be related to a path length difference with Eq. 37-4; it tells us that

$$\phi = \left(\frac{2\pi}{\lambda} \right) (a \sin \theta),$$

where a is the sum of the widths Δx of the infinitesimal strips. But $\phi = 2\alpha$, so this equation reduces to Eq. 37-6.

SAMPLE PROBLEM 37-2

Find the intensities of the first three secondary maxima (side maxima) in the single-slit diffraction pattern of Fig. 37-1, measured relative to the intensity of the central maximum.

SOLUTION: The secondary maxima lie approximately halfway between the minima, which are given by Eq. 37-7 ($\alpha = m\pi$). The secondary maxima are then given (approximately) by

$$\alpha = (m + \tfrac{1}{2})\pi, \qquad \text{for } m = 1, 2, 3, \ldots,$$

with α in radian measure. If we substitute this result into Eq. 37-5 we obtain

$$\frac{I}{I_m} = \left(\frac{\sin \alpha}{\alpha} \right)^2 = \left(\frac{\sin(m + \tfrac{1}{2})\pi}{(m + \tfrac{1}{2})\pi} \right)^2,$$

$$\text{for } m = 1, 2, 3, \ldots.$$

The first of the secondary maxima occurs for $m = 1$, its relative intensity being

$$\frac{I_1}{I_m} = \left(\frac{\sin(1 + \tfrac{1}{2})\pi}{(1 + \tfrac{1}{2})\pi} \right)^2 = \left(\frac{\sin 1.5\pi}{1.5\pi} \right)^2$$

$$= 4.50 \times 10^{-2} \approx 4.5\%. \qquad \text{(Answer)}$$

For $m = 2$ and $m = 3$ we find that

$$\frac{I_2}{I_m} = 1.6\% \quad \text{and} \quad \frac{I_3}{I_m} = 0.83\%. \qquad \text{(Answer)}$$

Successive secondary maxima decrease rapidly in intensity. The pattern of Fig. 37-1 was deliberately overexposed to reveal them.

CHECKPOINT **3:** Two wavelengths, 650 and 430 nm, are used separately in a single-slit diffraction experiment. The figure shows the results as graphs of intensity I versus angle θ for the two diffraction patterns. If both wavelengths are then used simultaneously, what color will be seen in the combined diffraction pattern at (a) angle *A* and (c) angle *B*?

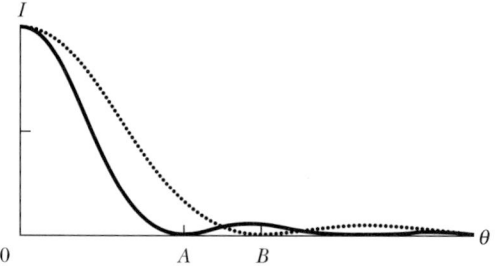

37-5 DIFFRACTION BY A CIRCULAR APERTURE

Here we consider diffraction by a circular aperture, that is, a circular opening such as a circular lens, through which light can pass. Figure 37-9 shows the image of a distant point source of light (a star, for instance) formed on photographic film placed in the focal plane of a converging lens. This image is not a point, as the geometrical optics treatment would suggest, but a circular disk surrounded by several progressively fainter secondary rings. Comparison

with Fig. 37-1 leaves little doubt that we are dealing with a diffraction phenomenon. Here, however, the aperture is a circle of diameter *d* rather than a rectangular slit.

The analysis of such patterns is complex. It shows, however, that the first minimum for the diffraction pattern of a circular aperture of diameter *d* is given by

$$\sin \theta = 1.22 \frac{\lambda}{d} \qquad \text{(first minimum; circular aperture).} \qquad (37\text{-}12)$$

Compare this with Eq. 37-1,

$$\sin \theta = \frac{\lambda}{a} \qquad \text{(first minimum; single slit),} \qquad (37\text{-}13)$$

which locates the first minimum for a long narrow slit of width *a*. The main difference is the factor 1.22, which enters because of the circular shape of the aperture.

Resolvability

The fact that lens images are diffraction patterns is important when we wish to *resolve* (distinguish) two distant point objects whose angular separation is small. Figure 37-10 shows the visual appearances and corresponding intensity patterns for two distant point objects (stars, say) with small angular separations. In Figure 37-10a, the objects are not resolved because of diffraction; that is, their diffraction patterns overlap so much that the two objects cannot be distinguished from a single point object. In Fig. 37-10b the objects are barely resolved, and in Fig. 37-10c they are fully resolved.

In Fig. 37-10b the angular separation of the two point sources is such that the central maximum of the diffraction

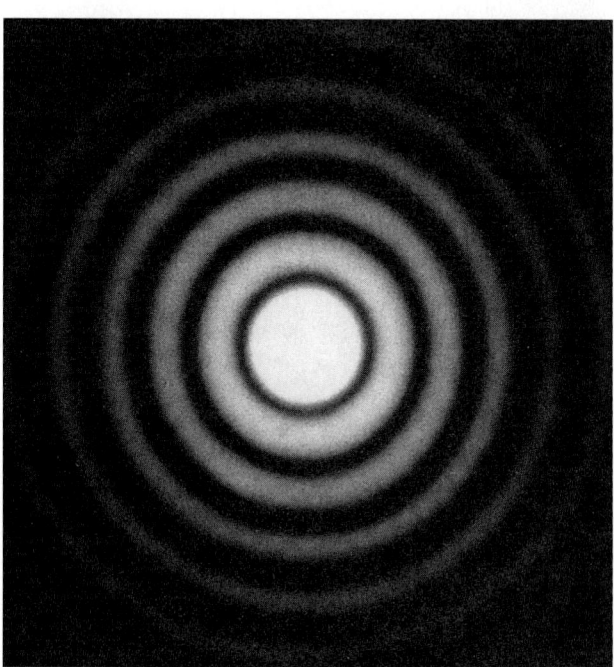

FIGURE 37-9 The diffraction pattern of a circular aperture. Note the central maximum and the circular secondary maxima. The figure has been overexposed to bring out these secondary maxima, which are much less intense than the central maximum.

The construction of a Soviet aircraft carrier can be seen in this image made by a spy satellite and published in 1984. The image has been "cleaned" by a computer to remove diffraction effects and to improve resolution. Today, images from spy satellites can resolve much smaller details than shown here.

FIGURE 37-10 Above, the images of two point sources (stars), formed by a converging lens. Below, representations of the image intensities. In (a) the angular separation of the sources is too small for them to be distinguished; in (b) they can be marginally distinguished, and in (c) they are clearly distinguished. Rayleigh's criterion is just satisfied in (b), with the central maximum of one diffraction pattern coinciding with the first minimum of the other.

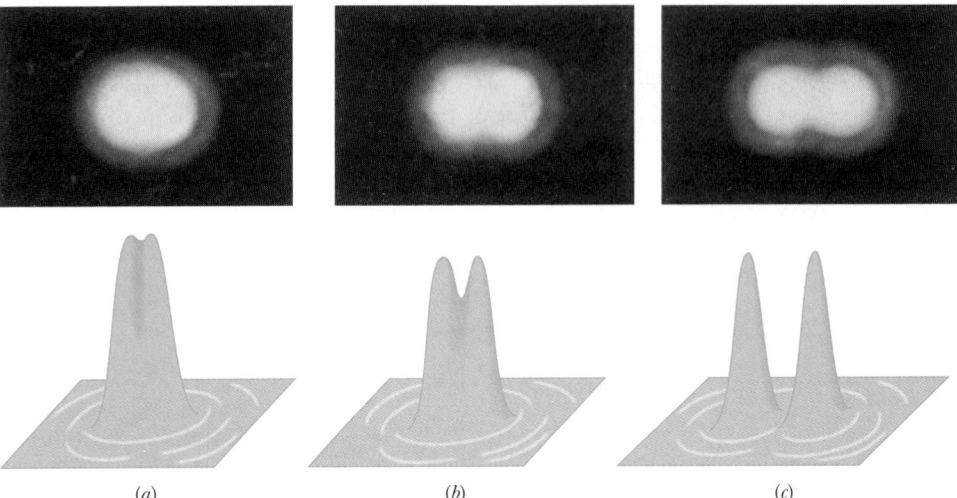

(a)   (b)   (c)

pattern of one source is centered on the first minimum of the diffraction pattern of the other, a condition called **Rayleigh's criterion** for resolvability. From Eq. 37-12, two objects that are barely resolvable by this criterion must have an angular separation θ_R of

$$\theta_R = \sin^{-1} \frac{1.22\lambda}{d}.$$

Since the angles involved are small, we can replace $\sin \theta_R$ with θ_R expressed in radians:

$$\theta_R = 1.22 \frac{\lambda}{d} \quad \text{(Rayleigh's criterion)}. \quad (37\text{-}14)$$

Rayleigh's criterion for resolvability is only an approximation, because resolvability depends on many factors, such as the relative brightness of the sources and their surroundings, turbulence in the air between the sources and the observer, and the functioning of the observer's visual system. However, for the sake of calculations here, we shall take Eq. 37-14 as being a precise criterion: if the angular separation θ between the sources is greater than θ_R, we can resolve the sources; if it is less, we cannot.

When we wish to use a lens to resolve objects of small angular separation, it is desirable to make the diffraction pattern as small as possible. According to Eq. 37-14, this can be done either by increasing the lens diameter or by using light of a shorter wavelength.

For this reason ultraviolet light is often used with microscopes; because of its shorter wavelength, it permits finer detail to be examined than would be possible for the same microscope operated with visible light. In Chapter 40 of the extended version of this text, we show that beams of electrons behave like waves under some circumstances. In an *electron microscope* such beams may have an effective wavelength that is 10^{-5} of the wavelength of visible light. They permit the detailed examination of tiny structures, like that in Fig. 37-11, that would be blurred by diffraction if viewed with an optical microscope.

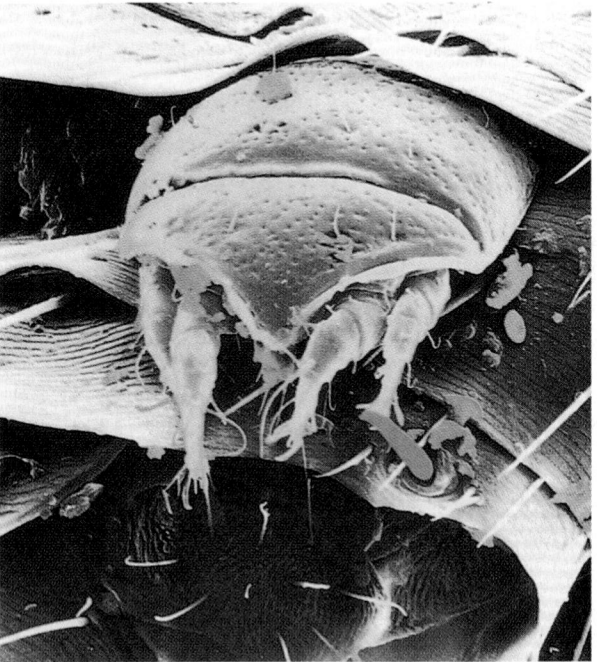

FIGURE 37-11 A false-color scanning electron micrograph of a mite that is on the back of a hedgehog flea.

SAMPLE PROBLEM 37-3

A circular converging lens, with diameter $d = 32$ mm and focal length $f = 24$ cm, forms images of distant point objects in the focal plane of the lens. Light of wavelength $\lambda = 550$ nm is used.

(a) Considering diffraction by the lens, what angular separation must two such objects have to satisfy Rayleigh's criterion for resolvability?

SOLUTION: Figure 37-12 shows two distant point objects P_1 and P_2, the lens, and a viewing screen in the focal plane of the lens. It also shows, at the right, plots of the light intensity I versus position on the screen for the central maxima of the images formed by the lens. From the perspective of the lens, the angular separation θ_o of the objects equals the angular separation θ_i of the images. So, if the images are to satisfy Rayleigh's criterion for resolvability, the angular separations on both sides of the lens must be given by Eq. 37-14 (assuming small angles). Substituting the given data, we obtain from Eq. 37-14

$$\theta_o = \theta_i = \theta_R = 1.22 \frac{\lambda}{d}$$

$$= \frac{(1.22)(550 \times 10^{-9} \text{ m})}{32 \times 10^{-3} \text{ m}} = 2.1 \times 10^{-5} \text{ rad.} \quad \text{(Answer)}$$

At this angular separation, each central maximum in the two intensity curves of Fig. 37-12 is centered on the first minimum of the other curve.

(b) What is the separation Δx of the centers of the *images* in the focal plane? (That is, what is the separation of the *central* peaks in the two curves?)

SOLUTION: From either triangle between the lens and the screen in Fig. 37-12, we see that $\tan \theta_i/2 = \Delta x/2f$. Rearranging this and making the approximation $\tan \theta \approx \theta$, we find

$$\Delta x = f\theta_i, \quad (37\text{-}15)$$

where θ_i is in radian measure. Substituting known data then yields

$$\Delta x = (0.24 \text{ m})(2.1 \times 10^{-5} \text{ rad}) = 5.0 \ \mu\text{m}. \quad \text{(Answer)}$$

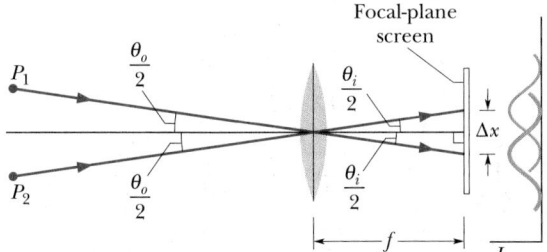

FIGURE 37-12 Sample Problem 37-3. Light from two distant point objects P_1 and P_2 passes through a converging lens and forms images on a viewing screen in the focal plane of the lens. Only one representative ray from each object is shown. The images are not points but diffraction patterns, with intensities approximately as plotted at the right. The angular separation of the objects is θ_o and that of the images is θ_i; the central maxima of the images have a separation Δx.

SAMPLE PROBLEM 37-4

Approximate the colored dots in Seurat's *Sunday Afternoon on the Island of La Grande Jatte* as closely spaced circles with center-to-center separations $D = 2.0$ mm (Fig. 37-13). If the diameter of the pupil of your eye is $d = 1.5$ mm, what is the minimum viewing distance from which you cannot distinguish any dots?

SOLUTION: Consider any two adjacent dots that you can distinguish when you are close to the painting. As you move away, you can distinguish the dots until their angular separation θ (in your view) decreases to that given by Rayleigh's criterion (Eq. 37-14):

$$\theta_R = 1.22 \frac{\lambda}{d}. \quad (37\text{-}16)$$

Because the angular separation is then small, we can approximate $\sin \theta$ as θ and then write

$$\theta = \frac{D}{L}, \quad (37\text{-}17)$$

in which L is your distance from the dots.

Setting θ of Eq. 37-17 equal to θ_R of Eq. 37-16 and solving for L, we obtain

$$L = \frac{Dd}{1.22\lambda}. \quad (37\text{-}18)$$

Equation 37-18 tells us that L is larger for smaller λ. Thus, as you move away from the painting, adjacent red dots (corresponding to a long wavelength) become indistinguishable before adjacent blue dots do. So to find the least distance L at which *no* colored dots are distinguishable, we substitute $\lambda = 400$ nm (blue or violet light) and the given data into Eq. 37-18, finding

$$L = \frac{(2.0 \times 10^{-3} \text{ m})(1.5 \times 10^{-3} \text{ m})}{(1.22)(400 \times 10^{-9} \text{ m})} = 6.1 \text{ m.} \quad \text{(Answer)}$$

At this or a greater distance, the colors of all adjacent dots blend together. The color you then perceive at any given spot on the painting is a blended color that may not actually exist there. In other words, Seurat uses the viewer's eyes to create the colors of his art.

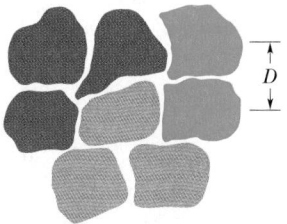

FIGURE 37-13 Sample Problem 37-4. Representation of dots on a Seurat painting.

CHECKPOINT 4: Suppose that you can barely resolve two red dots, owing to diffraction by the pupil of your eye. If we increase the general illumination around you so that the pupil decreases in diameter, does the resolvability of the dots improve or diminish? Consider only diffraction. (You might experiment to check your answer.)

37-6 DIFFRACTION BY A DOUBLE SLIT

In the double-slit experiments of Chapter 36, we implicitly assumed that the slits were narrow compared to the wavelength of the light illuminating them; that is, $a \ll \lambda$. For such narrow slits, the central maximum of the diffraction pattern of either slit covers the entire viewing screen. Moreover, the interference of light from the two slits produces bright fringes that all have approximately the same intensity (Fig. 36-9).

In practice with visible light, however, the condition $a \ll \lambda$ is often not met. For relatively wide slits, the interference of light from two slits produces bright fringes that do not all have the same intensity. In fact, their intensity is modified by the diffraction of the light through each slit.

As an example, the intensity plot of Fig. 37-14a suggests the double-slit fringe pattern that would occur if the slits were infinitely narrow (and thus $a \ll \lambda$); all the bright interference fringes would have the same intensity. The intensity plot of Fig. 37-14b is that of a single actual slit; the diffraction pattern has a broad central maximum and weaker secondary maxima at $\pm 17°$. The plot of Fig. 37-14c suggests the resulting interference pattern for two actual slits. The plot was constructed by using the curve of Fig. 37-14b as an envelope on the intensity plot in Fig. 37-14a. The positions of the fringes are not changed; only the intensity is affected.

Figure 37-15a shows an actual pattern in which both double-slit interference and diffraction are evident. If one slit is covered, the single-slit diffraction pattern of Fig.

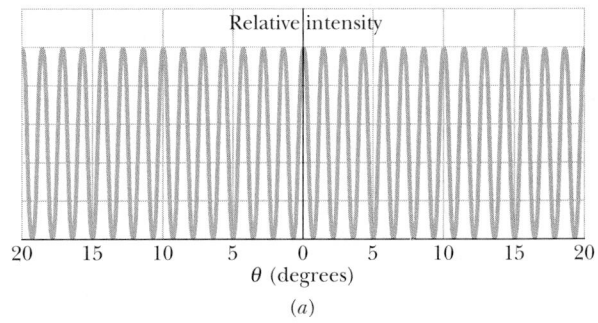

(a)

FIGURE 37-14 (a) The intensity pattern to be expected in a double-slit interference experiment with vanishingly narrow slits. (b) The intensity pattern in diffraction by a typical slit of width a (not vanishingly narrow). (c) The intensity pattern to be expected for two slits of width a. The curve of (b) acts as an envelope, limiting the intensity of the double-slit fringes in (a). Note that the first minima of the diffraction pattern of (b) eliminate the double-slit fringes that would occur near 12° in (c).

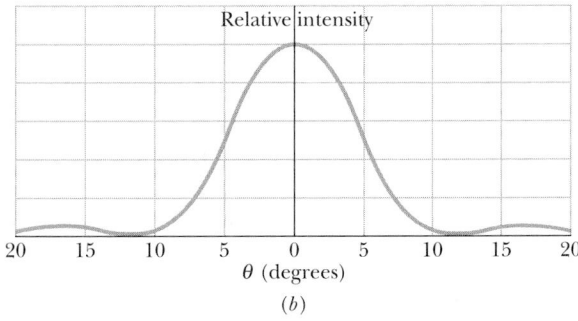

(b)

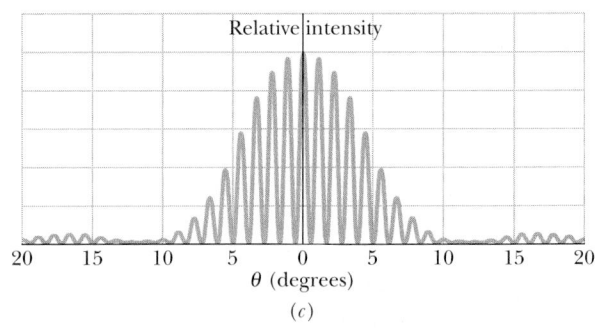

(c)

(a)

(b)

FIGURE 37-15 (a) Interference fringes for a double-slit system; compare with Fig. 37-14c. (b) The diffraction pattern of a single slit; compare with Fig. 37-14b.

37-15*b* results. Note the correspondence between Figs. 37-15*a* and 37-14*c*, and between Figs. 37-15*b* and 37-14*b*. In comparing these figures bear in mind that Fig. 37-15 has been deliberately overexposed to bring out the faint secondary maxima and that two secondary maxima (rather than one) are shown.

With diffraction effects taken into account, the intensity of a double-slit interference pattern is given by

$$I = I_m \, (\cos^2 \beta) \left(\frac{\sin \alpha}{\alpha} \right)^2 \quad \text{(double slit),} \quad (37\text{-}19)$$

in which

$$\beta = \left(\frac{\pi d}{\lambda} \right) \sin \theta \quad (37\text{-}20)$$

and

$$\alpha = \left(\frac{\pi a}{\lambda} \right) \sin \theta. \quad (37\text{-}21)$$

Here *d* is the distance between the centers of the slits, and *a* is the slit width. Note carefully that the right side of Eq. 37-19 is the product of I_m and two factors. (1) The *interference factor* $\cos^2 \beta$ is due to the interference between two slits with slit separation *d* (as given by Eqs. 36-21 and 36-22). (2) The *diffraction factor* $[(\sin \alpha)/\alpha]^2$ is due to diffraction by a single slit of width *a* (as given by Eqs. 37-5 and 37-6).

Let us check these factors. If we let $a \to 0$ in Eq. 37-21, for example, then $\alpha \to 0$ and $(\sin \alpha)/\alpha \to 1$. Equation 37-19 then reduces, as it must, to an equation describing the interference pattern for a pair of vanishingly narrow slits with slit separation *d*. Similarly, putting $d = 0$ in Eq. 37-20 is equivalent physically to causing the two slits to merge into a single slit of width *a*. Then Eq. 37-20 yields $\beta = 0$ and $\cos^2 \beta = 1$. In this case Eq. 37-19 reduces, as it must, to an equation describing the diffraction pattern for a single slit of width *a*.

The double-slit pattern described by Eq. 37-19 and displayed in Fig. 37-15*a* combines interference and diffraction in an intimate way. Both are superposition effects, in that they result from the combining of waves with different phases at a given point. If the combining waves originate from a finite (and usually small) number of elementary coherent sources—as in a double-slit experiment with $a \ll \lambda$—we call the process *interference*. If the combining waves originate in a single wavefront—as in a single-slit experiment—we call the process *diffraction*. This distinction between interference and diffraction (which is somewhat arbitrary and not always adhered to) is a convenient one, but we should not forget that both are superposition effects and usually both are present simultaneously (as in Fig. 37-15*a*).

In a double-slit experiment, the wavelength λ of the light source is 405 nm, the slit separation *d* is 19.44 μm, and the slit width *a* is 4.050 μm.

(a) How many bright fringes are within the central peak of the diffraction envelope?

SOLUTION: The limits of the central diffraction peak are the first diffraction minima, each of which is located at the angle θ given by Eq. 37-3 with $m = 1$:

$$a \sin \theta = \lambda. \quad (37\text{-}22)$$

The locations of the bright fringes of the double-slit interference pattern are given by Eq. 36-14:

$$d \sin \theta = m\lambda, \quad \text{for } m = 0, 1, 2, \ldots. \quad (37\text{-}23)$$

We can locate the first diffraction minimum within the double-slit fringe pattern by dividing Eq. 37-23 by Eq. 37-22 and solving for *m*. By doing so and then substituting the given data, we obtain

$$m = \frac{d}{a} = \frac{19.44 \; \mu\text{m}}{4.050 \; \mu\text{m}} = 4.8.$$

This tells us that the first diffraction minimum occurs just before the bright fringe for $m = 5$ in Eq. 37-23. So, within the central diffraction peak we have the central bright fringe ($m = 0$) and four bright fringes (up to $m = 4$) on each side of it. Thus a total of nine bright fringes of the double-slit interference pattern are within the central peak of the diffraction envelope. The bright fringes to one side of the central bright fringe are shown in Fig. 37-16.

(b) How many bright fringes are within either of the first side peaks of the diffraction envelope?

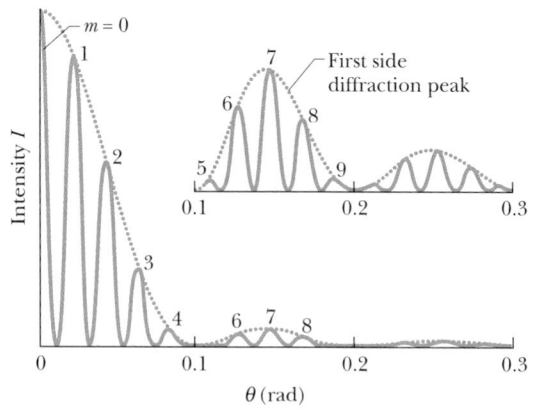

FIGURE 37-16 Sample Problem 37-5. One side of the intensity pattern for a two-slit interference experiment; the diffraction envelope is indicated by dots. The insert shows (vertically expanded) the intensity pattern for the first and second side diffraction peaks.

SOLUTION: The outer limits of the first side diffraction peaks are the second diffraction minima, each of which is at the angle θ given by Eq. 37-3 with $m = 2$:

$$a \sin \theta = 2\lambda. \qquad (37\text{-}24)$$

Dividing Eq. 37-23 by Eq. 37-24, we find

$$m = \frac{2d}{a} = \frac{(2)(19.44 \ \mu m)}{4.050 \ \mu m} = 9.6.$$

This tells us that the second diffraction minimum occurs just before the bright fringe for $m = 10$ in Eq. 37-23. So, within the first side diffraction peak we have the fringes from $m = 5$ to $m = 9$ and thus a total of five bright fringes of the double-slit interference pattern (shown in the insert of Fig. 37-16). However, if the $m = 5$ bright fringe, which is almost eliminated by the first diffraction minimum, is considered too dim to count, then only four bright fringes are in the first side diffraction peak.

CHECKPOINT 5: If we increase the wavelength of the light source in Sample Problem 37-5 to 550 nm, do (a) the width of the central diffraction peak and (b) the number of bright fringes within that peak increase, decrease, or remain the same?

37-7 DIFFRACTION GRATINGS

One of the most useful tools in the study of light and of objects that emit and absorb light is the **diffraction grating**. Somewhat like the double-slit arrangement of Fig. 36-8, this device has a much greater number N of slits, often called *rulings*, perhaps as many as several thousand per millimeter. An idealized grating consisting of only five slits is represented in Fig. 37-17. When monochromatic light is sent through the slits, it forms narrow interference fringes that can be analyzed to determine the wavelength of the light. (Diffraction gratings can also be opaque surfaces with narrow parallel grooves arranged like the slits in Fig. 37-17. Light then scatters back from the grooves to form interference fringes rather than being transmitted through open slits.)

With monochromatic light incident on a diffraction grating, if we gradually increase the number of slits from two to a large number N, the intensity pattern changes from the typical double-slit pattern of Fig. 37-14c to a much more complicated pattern and then eventually to a simple pattern like that shown in Fig. 37-18a. The maxima are now very narrow (and so are called *lines*); they are separated by relatively wide dark regions. What you would see on a viewing screen using monochromatic red light from, say, a helium–neon laser, is shown in Fig. 37-18b.

We use a familiar procedure to find the locations of the bright lines on the viewing screen. We first assume that the screen is far enough from the grating that the rays reaching a particular point P on the screen are approximately parallel when they leave the grating (Fig. 37-19). Then we apply to each pair of adjacent rulings the same reasoning we used for double-slit interference. The separation d between rulings is called the *grating spacing*. (If N rulings occupy a total width w, then $d = w/N$.) The path length difference between adjacent rays is again $d \sin \theta$ (Fig. 37-19), where θ is the angle from the central axis of the grating (and of the pattern) to point P. A line is located at P if the path length difference between adjacent rays is an integer number of wavelengths, that is, if

$$d \sin \theta = m\lambda, \qquad \text{for } m = 0, 1, 2, \ldots$$
$$\text{(maxima—lines)}, \qquad (37\text{-}25)$$

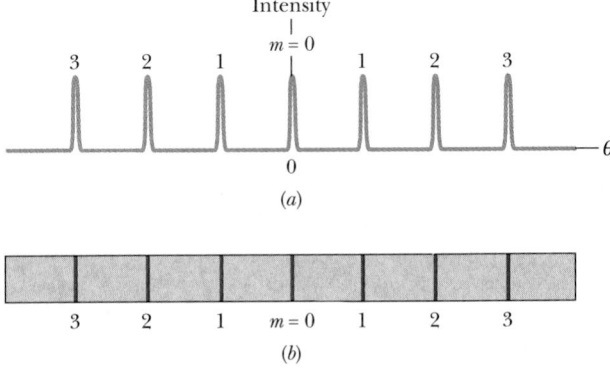

FIGURE 37-18   (a) The intensity pattern produced by a diffraction grating with a great many rulings consists of narrow peaks that are labeled with an order number m. (b) The corresponding bright fringes seen on the screen are called lines and are also labeled with m. Lines of the zeroth, first, second, and third orders are shown.

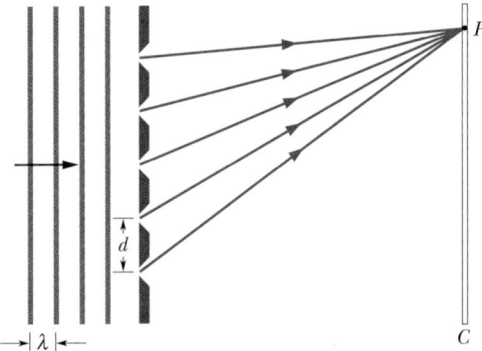

FIGURE 37-17   An idealized diffraction grating, consisting of only five rulings, that produces an interference pattern on a distant viewing screen C.

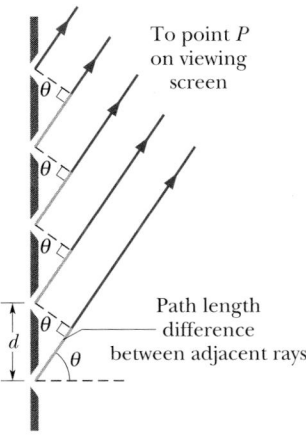

FIGURE 37-19 The rays from the rulings in a diffraction grating to a distant point P are approximately parallel. The path length difference between each two adjacent rays is $d \sin \theta$, where θ is measured as shown. (The rulings extend into and out of the page.)

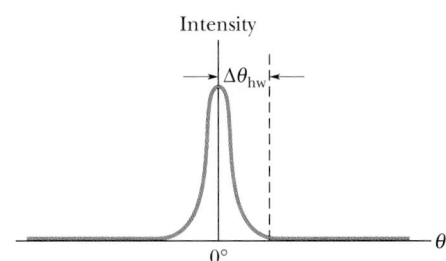

FIGURE 37-20 The half-width $\Delta\theta_{hw}$ of the central line is measured from the center of that line to the adjacent minimum on a plot of I versus θ like Fig. 37-18a.

where λ is the wavelength of the light. Each integer m represents a different line; hence these integers can be used to label the lines, as in Fig. 37-18. The integers are then called the *order numbers,* and the lines are called the *zeroth-order* line (the central line, with $m = 0$), the *first-order* line, the *second-order* line, and so on.

If we rewrite Eq. 37-25 as $\theta = \sin^{-1}(m\lambda/d)$ we see that, for a given diffraction grating, the angle from the central axis to any line (say, the third-order line) depends on the wavelength of the light being used. Thus, when light of an unknown wavelength is sent through a diffraction grating, measurements of the angles to the higher order lines can be used in Eq. 37-25 to determine the wavelength. Even light of several unknown wavelengths can be distinguished and identified in this way. We cannot do that with the double-slit arrangement of Section 36-4, even though the same equation and wavelength dependence apply there. In double-slit interference, the bright fringes due to different wavelengths overlap too much to be distinguished.

Width of the Lines

A grating's ability to resolve (separate) lines of different wavelengths depends on the width of the lines. We shall here derive an expression for the *half-width* of the central line (the line for which $m = 0$) and then state an expression for the half-widths of the higher order lines. We measure the half-width of the central line as the angle $\Delta\theta_{hw}$ from the center of the line at $\theta = 0$ outward to where the line effectively ends and darkness effectively begins with the first minimum (Fig. 37-20). At such a minimum, the N rays from the N slits of the grating cancel one another. (The actual width of the central line is, of course, $2\Delta\theta_{hw}$, but line widths are usually compared via half-widths.)

In Section 37-2 we were also concerned with the cancellation of a great many rays, there due to diffraction through a single slit. We obtained Eq. 37-3 which, owing to the similarity of the two situations, we can use to find the first minimum here. It tells us that the first minimum

occurs where the path length difference between the top and bottom rays equals λ. For single-slit diffraction, this difference is $a \sin \theta$. For a grating of N rulings, each separated from the next by distance d, the distance between the top and bottom rulings is Nd (Fig. 37-21). So, the path length difference between the top and bottom rays here is $Nd \sin \Delta\theta_{hw}$. Thus, the first minimum occurs where

$$Nd \sin \Delta\theta_{hw} = \lambda. \qquad (37\text{-}26)$$

Because $\Delta\theta_{hw}$ is small, $\sin \Delta\theta_{hw} = \Delta\theta_{hw}$ (in radian measure). Substituting this in Eq. 37-26 gives the half-width of the central line as

$$\Delta\theta_{hw} = \frac{\lambda}{Nd} \qquad \text{(half-width of central line).} \quad (37\text{-}27)$$

We state without proof that the half-width of any other line depends on its location relative to the central axis and is

$$\Delta\theta_{hw} = \frac{\lambda}{Nd \cos \theta} \qquad \text{(half-width of line at } \theta\text{).} \quad (37\text{-}28)$$

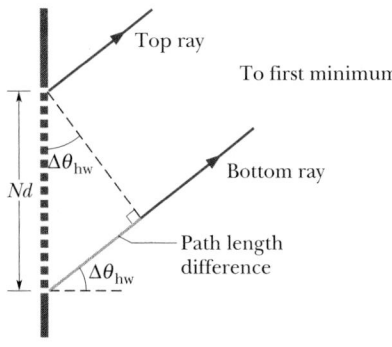

FIGURE 37-21 The top and bottom rulings of a diffraction grating of N rulings are separated by distance Nd. The top and bottom rays passing through these rulings have a path length difference of $Nd \sin \Delta\theta_{hw}$, if $\Delta\theta_{hw}$ is measured as shown. (The angle here is greatly exaggerated for clarity.)

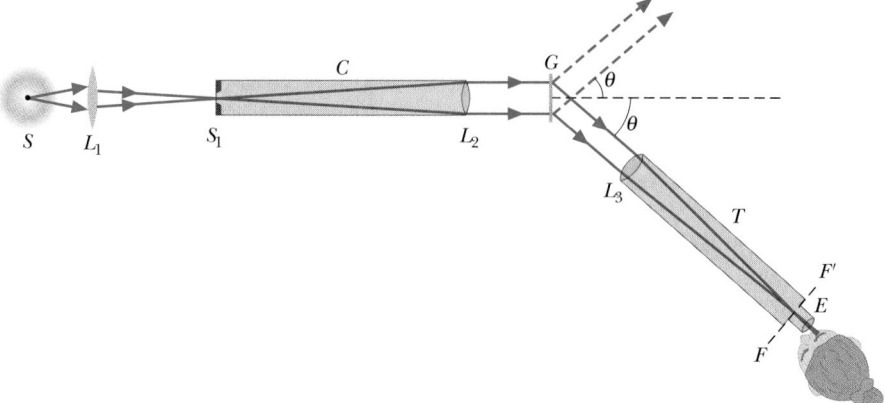

FIGURE 37-22 A simple type of grating spectroscope used to analyze the wavelengths of the light emitted by source S.

Note that for light of a given wavelength λ and a given ruling separation d, the widths of the lines decrease with an increase in the number N of rulings. Thus, of two diffraction gratings, the grating with the larger value of N is better able to distinguish between wavelengths because its diffraction lines are narrower and so produce less overlap.

An Application of Diffraction Gratings

Diffraction gratings are widely used to determine the wavelengths that are emitted by sources of light ranging from lamps to stars. Figure 37-22 shows a simple *grating spectroscope* in which a grating is used for this purpose. Light from source S is focused by lens L_1 on a slit S_1 placed in the focal plane of lens L_2. The light emerging from tube C (called a *collimator*) is a plane wave and is incident perpendicularly on grating G, where it is diffracted into a diffraction pattern, with the $m = 0$ order diffracted at angle $\theta = 0$ along the central axis of the grating.

We can view the diffraction pattern that would appear on a viewing screen at any angle θ simply by orienting telescope T in Fig. 37-22 to that angle. Lens L_3 of the telescope then focuses the light diffracted at angle θ (and at slightly smaller and larger angles) onto a focal plane FF' within the telescope. When we look through eyepiece E, we see a magnified view of this focused image.

By changing the angle θ of the telescope, we can examine the entire diffraction pattern. For any order number other than $m = 0$, the original light is spread out according to wavelength (or color) so that we can determine, with Eq. 37-25, just what wavelengths are being emitted by the source. If the source emits a broad band of wavelengths, what we see as we rotate the telescope through the angles corresponding to an order m is a broad band of color, with the shorter wavelength end at a smaller angle θ than the longer wavelength end. If the source emits discrete wavelengths, what we see are discrete vertical lines of color corresponding to those wavelengths.

For example, the light emitted by a hydrogen lamp, which contains hydrogen gas, has four discrete wavelengths in the visible range. If our eyes intercept this light directly, it appears to be white. If, instead, we view it through a grating spectroscope, we can distinguish, in several orders, the lines of the four colors corresponding to these visible wavelengths. (Such lines are called *emission lines*.) Four orders are represented in Fig. 37-23. In the central order ($m = 0$), the lines corresponding to all four wavelengths are superimposed, giving a single white line at $\theta = 0$. The colors are separated in the higher orders.

The third order is not shown in Fig. 37-23 for the sake of clarity; it actually overlaps the second and fourth orders. The fourth-order red line is missing because it is not formed by the grating used here (which is the grating of Sample Problem 37-6). That is, when we attempt to solve Eq. 37-25 for the angle θ for the red wavelength when $m = 4$, we find that $\sin \theta$ is greater than unity, which is not possible. The fourth order is then said to be *incomplete* for this grating; it might not be incomplete for a grating with greater spacing d, which will spread the lines less than in Fig. 37-23. Figure 37-24 is a photograph of the visible emission lines produced by cadmium.

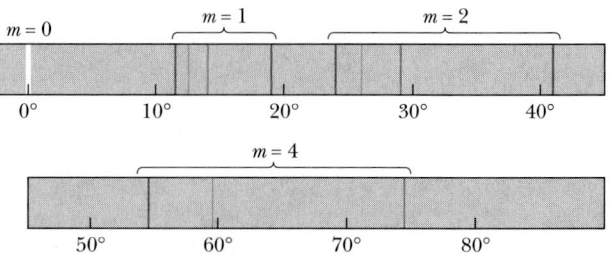

FIGURE 37-23 The zeroth, first, second, and fourth orders of the visible emission lines from hydrogen. Note that the lines are farther apart at greater angles. (They are also dimmer and wider, although that is not shown here.)

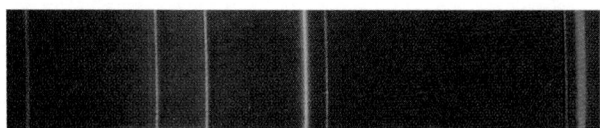

FIGURE 37-24 The visible emission lines of cadmium, as seen through a grating spectroscope.

SAMPLE PROBLEM 37-6

A diffraction grating has 1.26×10^4 rulings uniformly spaced over width $w = 25.4$ mm. It is illuminated at normal incidence by blue light of wavelength 450 nm.

(a) At what angles to the central axis do the second-order maxima occur?

SOLUTION: The grating spacing d is

$$d = \frac{w}{N} = \frac{25.4 \times 10^{-3} \text{ m}}{1.26 \times 10^4}$$

$$= 2.016 \times 10^{-6} \text{ m} = 2016 \text{ nm}.$$

The second-order maxima correspond to $m = 2$ in Eq. 37-25. For $\lambda = 450$ nm, we thus have

$$\theta = \sin^{-1}\frac{m\lambda}{d} = \sin^{-1}\frac{(2)(450 \text{ nm})}{2016 \text{ nm}}$$

$$= 26.51° \approx 26.5°. \qquad \text{(Answer)}$$

(b) What is the half-width of the second-order line?

SOLUTION: From Eq. 37-28,

$$\Delta\theta_{hw} = \frac{\lambda}{Nd \cos \theta} = \frac{450 \text{ nm}}{(1.26 \times 10^4)(2016 \text{ nm})(\cos 26.51°)}$$

$$= 1.98 \times 10^{-5} \text{ rad}. \qquad \text{(Answer)}$$

CHECKPOINT **6:** The figure shows lines of different orders produced by a diffraction grating in monochromatic red light. (a) Is the center of the pattern to the left or right? (b) If we switch to monochromatic green light, will the half-widths of the lines then produced in the same orders be greater than, less than, or the same as the half-widths of the lines shown?

37-8 GRATINGS: DISPERSION AND RESOLVING POWER (OPTIONAL)

Dispersion

To be useful in distinguishing wavelengths that are close to each other (as in a grating spectroscope), a grating must spread apart the diffraction lines associated with the

The fine rulings, each 0.5 μm wide, on a compact disc function as a diffraction grating. When a small source of white light illuminates a disc, the diffracted light forms colored "lanes" that are the composite of the diffraction patterns from the rulings.

various wavelengths. This spreading, called **dispersion,** is defined as

$$D = \frac{\Delta\theta}{\Delta\lambda} \qquad \text{(dispersion defined).} \qquad (37\text{-}29)$$

Here $\Delta\theta$ is the angular separation of two lines whose wavelengths differ by $\Delta\lambda$. The greater D is, the greater is the distance between two emission lines whose wavelengths differ by $\Delta\lambda$. We show below that the dispersion of a grating at angle θ is given by

$$D = \frac{m}{d \cos \theta} \qquad \begin{array}{l}\text{(dispersion} \\ \text{of a grating).}\end{array} \qquad (37\text{-}30)$$

Thus to achieve high dispersion, we must use a grating of small grating spacing (small d) and work in high orders (large m). Note that the dispersion does not depend on the number of rulings. The SI unit for D is the degree per meter or the radian per meter.

Resolving Power

To distinguish lines whose wavelengths are close together, the line widths should also be as narrow as possible. Expressed otherwise, the grating should have a high **resolving power** R, defined as

$$R = \frac{\lambda_{av}}{\Delta\lambda} \qquad \begin{array}{l}\text{(resolving power} \\ \text{defined).}\end{array} \qquad (37\text{-}31)$$

Here λ_{av} is the mean wavelength of two spectrum lines that can barely be recognized as separate, and $\Delta\lambda$ is the wavelength difference between them. The greater R is, the closer two emission lines can be and still be resolved. We shall show below that the resolving power of a grating is given by the simple expression

$$R = Nm \qquad \text{(resolving power of a grating).} \qquad (37\text{-}32)$$

To achieve high resolving power, we must use many rulings (large N in Eq. 37-32).

Proof of Eq. 37-30

Let us start with Eq. 37-25, the expression for the locations of the lines in the diffraction pattern of a grating:

$$d \sin \theta = m\lambda.$$

Let us regard θ and λ as variables and take differentials of this equation. We find

$$d \cos \theta \, d\theta = m \, d\lambda.$$

For small enough angles, we can write these differentials as small differences; thus

$$d \cos \theta \, \Delta\theta = m \, \Delta\lambda \qquad (37\text{-}33)$$

or

$$\frac{\Delta\theta}{\Delta\lambda} = \frac{m}{d \cos \theta}.$$

The ratio on the left is simply D (see Eq. 37-29), so we have indeed derived Eq. 37-30.

Proof of Eq. 37-32

We start with Eq. 37-33, which was derived from Eq. 37-25, the expression for the locations of the lines in the diffraction pattern formed by a grating. Here $\Delta\lambda$ is the small wavelength difference between two waves that are diffracted by the grating, and $\Delta\theta$ is the angular separation between them in the diffraction pattern. If $\Delta\theta$ is to be the smallest angle that will permit the two lines to be resolved, it must (by Rayleigh's criterion) be equal to the half-width of each line, which is given by Eq. 37-28:

$$\Delta\theta_{hw} = \frac{\lambda}{Nd \cos \theta}.$$

If we substitute $\Delta\theta_{hw}$ as given here for $\Delta\theta$ in Eq. 37-33, we find that

$$\frac{\lambda}{N} = m \, \Delta\lambda,$$

from which it readily follows that

$$R = \frac{\lambda}{\Delta\lambda} = Nm.$$

This is Eq. 37-32, which we set out to derive.

Dispersion and Resolving Power Compared

The resolving power of a grating must not be confused with its dispersion. Table 37-1 shows the characteristics of three gratings, all illuminated with light of wavelength $\lambda = 589$ nm, whose diffracted light is viewed in the first order ($m = 1$ in Eq. 37-25). You should verify that the values of D and R as given in the table can be calculated with Eqs. 37-30 and 37-32, respectively. (In the calculations for D, you will need to convert radians per meter to degrees per micrometer.)

For the conditions noted in Table 37-1, gratings A and B have the same *dispersion* and A and C have the same *resolving power*.

Figure 37-25 shows the intensity patterns (also called *line shapes*) that would be produced by these gratings for two lines of wavelengths λ_1 and λ_2, in the vicinity of $\lambda = 589$ nm. Grating B, with the higher resolving power, produces narrower lines and thus is capable of distinguishing

TABLE 37-1   THREE GRATINGS[a]

GRATING	N	d (nm)	θ	D ($°/\mu m$)	R
A	10,000	2540	13.4°	23.2	10,000
B	20,000	2540	13.4°	23.2	20,000
C	10,000	1370	25.5°	46.3	10,000

[a]Data are for $\lambda = 589$ nm and $m = 1$.

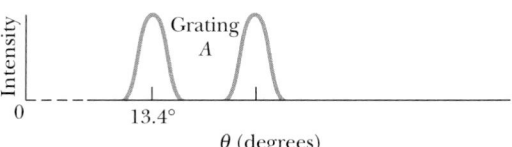

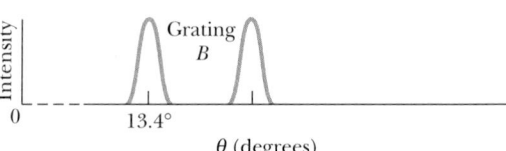

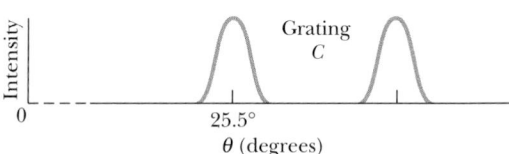

FIGURE 37-25 The intensity patterns for light of two wavelengths sent through the gratings of Table 37-1. Grating B has the highest resolving power, and grating C the highest dispersion.

lines that are much closer together in wavelength than those in the figure. Grating *C*, with the higher dispersion, produces the greater angular separation between the lines.

SAMPLE PROBLEM 37-7

The diffraction grating of Sample Problem 37-6 is illuminated at normal incidence by yellow light from a sodium vapor lamp. This light contains two closely spaced emission lines (known as the sodium doublet) of wavelengths 589.00 nm and 589.59 nm.

(a) At what angle does the first-order maximum occur for the first of these wavelengths?

SOLUTION: The first-order maximum corresponds to *m* = 1 in Eq. 37-25. From Sample Problem 37-6a, we know that the grating spacing *d* is 2016 nm. We thus have

$$\theta = \sin^{-1}\frac{m\lambda}{d} = \sin^{-1}\frac{(1)(589.00 \text{ nm})}{2016 \text{ nm}}$$
$$= 16.99° \approx 17.0°. \qquad \text{(Answer)}$$

(b) In the first order, what is the angular separation between the two lines?

SOLUTION: Here the *dispersion* of the grating comes into play. From Eq. 37-30, the dispersion is

$$D = \frac{m}{d\cos\theta} = \frac{1}{(2016 \text{ nm})(\cos 16.99°)}$$
$$= 5.187 \times 10^{-4} \text{ rad/nm}.$$

From Eq. 37-29, the defining equation for dispersion, we have

$$\Delta\theta = D\,\Delta\lambda$$
$$= (5.187 \times 10^{-4} \text{ rad/nm})$$
$$\times (589.59 \text{ nm} - 589.00 \text{ nm})$$
$$= 3.06 \times 10^{-4} \text{ rad} = 0.0175°. \quad \text{(Answer)}$$

This result depends on the grating spacing *d* but not on the number of rulings there are in the grating.

(c) How close in wavelength can two lines be and still be separated by this grating in the first order?

SOLUTION: Here the *resolving power* of the grating comes into play. From Eq. 37-32, the resolving power is

$$R = Nm = (1.26 \times 10^4)(1) = 1.26 \times 10^4.$$

From Eq. 37-31, the defining equation for resolving power, we have

$$\Delta\lambda = \frac{\lambda}{R} = \frac{589 \text{ nm}}{1.26 \times 10^4} = 0.0467 \text{ nm}. \quad \text{(Answer)}$$

Thus this grating can easily resolve the two sodium lines, which have a wavelength separation of 0.59 nm. Note that this

---

result depends only on the number of grating rulings and is independent of *d*, the spacing between adjacent rulings.

(d) How many rulings must a grating have to just resolve the sodium doublet lines?

SOLUTION: From Eq. 37-31, the defining equation for *R*, the grating must have a resolving power of

$$R = \frac{\lambda}{\Delta\lambda} = \frac{589 \text{ nm}}{0.59 \text{ nm}} = 998.$$

From Eq. 37-32, the number of rulings needed to achieve this resolving power (in the first order) is

$$N = \frac{R}{m} = \frac{998}{1} = 998 \text{ rulings.} \qquad \text{(Answer)}$$

Since our grating has about 13 times as many rulings as this, it can easily resolve the sodium doublet lines, as we have already shown in (c).

37-9 X-RAY DIFFRACTION

X rays are electromagnetic radiation whose wavelengths are of the order of 1 Å (= 10^{-10} m). Compare this with a wavelength of 550 nm (= 5.5×10^{-7} m) at the center of the visible spectrum. Figure 37-26 shows that x rays are produced when electrons escaping from a heated filament *F* are accelerated by a potential difference *V* and strike a metal target *T*.

A standard optical diffraction grating cannot be used to discriminate between different wavelengths in the x-ray wavelength range. For $\lambda = 1$ Å (= 0.1 nm) and *d* = 3000 nm, for example, Eq. 37-25 shows that the first-order maximum occurs at

$$\theta = \sin^{-1}\frac{m\lambda}{d} = \sin^{-1}\frac{(1)(0.1 \text{ nm})}{3000 \text{ nm}} = 0.0019°.$$

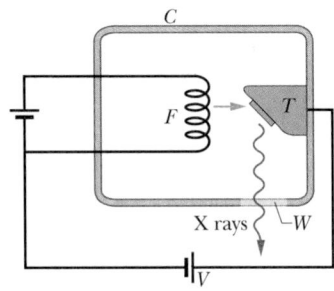

FIGURE 37-26 X rays are generated when electrons from heated filament *F*, accelerated through a potential difference *V*, strike a metal target *T*. The "window" *W* in the evacuated chamber *C* is transparent to x rays.

This is too close to the central maximum to be practical. A grating with $d \approx \lambda$ is desirable, but, since x-ray wavelengths are about equal to atomic diameters, such gratings cannot be constructed mechanically.

In 1912 it occurred to German physicist Max von Laue that a crystalline solid, which consists of a regular array of atoms, might form a natural three-dimensional "diffraction grating" for x rays. The idea is that in a crystal such as sodium chloride (NaCl) a basic unit of atoms (called the *unit cell*) repeats itself throughout the array. In NaCl four sodium ions and four chlorine ions are associated with each unit cell. Figure 37-27a represents a section through a crystal of NaCl and identifies this basic unit. The unit cell is a cube measuring a_0 on each side.

When an x-ray beam enters a crystal such as NaCl, x rays are *scattered*, that is, redirected, in all directions by the crystal structure. In some directions the scattered waves undergo destructive interference, resulting in intensity minima; in other directions the interference is constructive, resulting in intensity maxima. This process of scattering and interference is a form of diffraction, although it is unlike the diffraction of light traveling through a slit or past an edge as we discussed earlier.

Although the process of diffraction of x rays by a crystal is complicated, the maxima turn out to be in directions as *if* the x rays were reflected by a family of parallel *reflecting planes* (or *crystal planes*) that extend through the atoms within the crystal and that contain regular arrays of the atoms. (The x rays are not actually reflected; we use these fictional planes only to simplify the analysis of the actual diffraction process.)

Figure 37-27b shows three of the family of planes, with *interplanar spacing d*, from which the incident rays shown are said to reflect. Rays 1, 2, and 3 reflect from the first, second, and third planes, respectively. At each reflection the angle of incidence and the angle of reflection are represented with θ. Contrary to the custom in optics, these angles are defined relative to the *surface* of the reflecting plane rather than a normal to it. For the situation of Fig. 37-27b, the interplanar spacing happens to be equal to the unit cell dimension a_0.

Figure 37-27c shows an edge-on view of reflection from an adjacent pair of planes. The waves of rays 1 and 2 arrive at the crystal in phase. After they are reflected, they must again be in phase, because the reflections and the reflecting planes have been defined solely to explain the intensity maxima in the diffraction of x rays by a crystal. Unlike light rays, the x rays do not refract upon entering the crystal; moreover, we do not define an index of refraction for this situation. So the relative phase between the waves of rays 1 and 2 as they leave the crystal is set solely by their path length difference. For these rays to be in phase, the path length difference must be equal to an integer multiple of the wavelength λ of the x rays.

By drawing the dashed perpendiculars in Fig. 37-27c, we find that the path length difference is $2d \sin \theta$. In fact, this is true for any pair of adjacent planes in the family of planes represented in Fig. 37-27b. Thus we have, as the criterion for intensity maxima for x-ray diffraction,

$$2d \sin \theta = m\lambda, \quad \text{for } m = 1, 2, 3, \dots$$
$$\text{(Bragg's law),} \quad (37\text{-}34)$$

FIGURE 37-27 (a) The cubic structure of NaCl, showing the sodium and chlorine ions and a unit cell (shaded). (b) Incident x rays undergo diffraction by the structure of (a). The x rays are diffracted as if they were reflected by a family of parallel planes, with the angle of reflection equal to the angle of incidence, both angles measured relative to the planes (not relative to a normal as in optics). (c) The path length difference between waves effectively reflected by two adjacent planes is $2d \sin \theta$. (d) A different orientation of the x rays relative to the structure. A different family of parallel planes now effectively reflects the x rays.

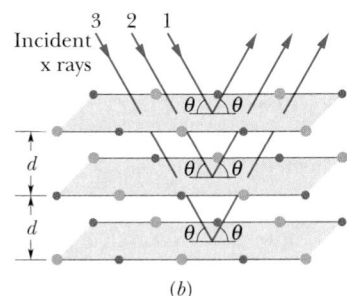

(b)

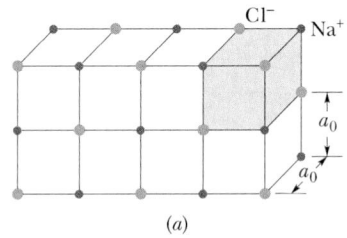

(c)

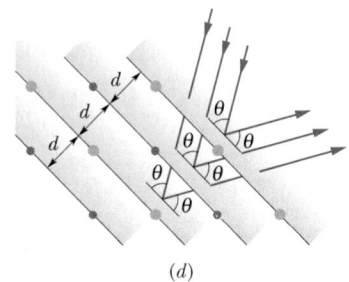

(d)

where m is the order number of an intensity maximum. Equation 37-34 is called **Bragg's law** after British physicist W. L. Bragg, who first derived it. (He and his father shared the 1915 Nobel prize for their use of x rays to study the structures of crystals.) The angle of incidence and reflection in Eq. 37-34 is called a *Bragg angle*.

Regardless of the angle at which x rays enter a crystal, there is always a family of planes from which they can be said to reflect so that we can apply Bragg's law. In Fig. 37-27*d*, the crystal structure has the same orientation as it does in Fig. 37-27*a* but the angle at which the beam enters the structure differs from that shown in Fig. 37-27*b*. This new angle requires a new family of planes, with a different interplanar spacing d and different Bragg angle θ, in order to explain the x-ray diffraction via Bragg's law.

Figure 37-28 shows how the interplanar spacing d can be related to the unit cell dimension a_0. For the particular family of planes shown there,

$$5d = \sqrt{5}a_0,$$

or
$$d = \frac{a_0}{\sqrt{5}}. \qquad (37\text{-}35)$$

Figure 37-28 suggests how the dimensions of the unit cell can be found once the interplanar spacing has been measured by means of x-ray diffraction.

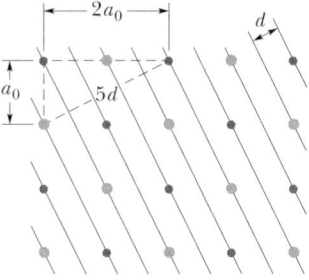

FIGURE 37-28 A family of planes through the structure of Fig. 37-27*a*, and a way to relate the edge length a_0 of a unit cell to the interplanar spacing d.

X-ray diffraction is a powerful tool for studying both x-ray spectra and the arrangement of atoms in crystals. To study spectra, a particular set of crystal planes, having a known spacing d, is chosen. These planes effectively reflect different wavelengths at different angles. A detector that can discriminate one angle from another can then be used to determine the wavelength of radiation reaching it. The crystal itself can be studied with a monochromatic x-ray beam, to determine not only the spacing of various crystal planes but also the structure of the unit cell.

SAMPLE PROBLEM 37-8

At what Bragg angles must x rays with $\lambda = 1.10$ Å be incident on the family of planes represented in Fig. 37-28 if effective reflections from the planes are to result in diffraction intensity maxima? Assume the material to be sodium chloride ($a_0 = 5.63$ Å).

SOLUTION: The interplanar spacing d for these planes is given by Eq. 37-35 as

$$d = \frac{a_0}{\sqrt{5}} = \frac{5.63 \text{ Å}}{\sqrt{5}} = 2.518 \text{ Å}.$$

Equation 37-34 then gives, for the Bragg angles,

$$\theta = \sin^{-1}\frac{m\lambda}{2d} = \sin^{-1}\left(\frac{(m)(1.10 \text{ Å})}{(2)(2.518 \text{ Å})}\right)$$
$$= \sin^{-1}(0.2184m).$$

Maxima are possible for $\theta = 12.6°$ ($m = 1$), $\theta = 25.9°$ ($m = 2$), $\theta = 40.9°$ ($m = 3$), and $\theta = 60.9°$ ($m = 4$). Higher order maxima cannot exist because they require that $\sin \theta$ be greater than 1.

Actually, the unit cell in cubic crystals such as NaCl has diffraction properties such that the intensity of diffracted x-ray beams corresponding to odd values of m is zero. Thus beams are expected only for $\theta = 25.9°$ and $\theta = 60.9°$.

REVIEW & SUMMARY

Diffraction

When waves encounter an edge or an obstacle or aperture with a size comparable to the wavelength of the waves, those waves spread in their direction of travel and undergo interference. This is called **diffraction.**

Single-Slit Diffraction

Waves passing through a long narrow slit of width a produce a **single-slit diffraction pattern** that includes a central maximum and other maxima, separated by minima located at angles θ to the central axis that satisfy

$$a \sin \theta = m\lambda, \qquad \text{for } m = 1, 2, 3, \ldots$$
$$\text{(minima).} \quad (37\text{-}3)$$

The intensity of the diffraction pattern at any given angle θ is

$$I = I_m \left(\frac{\sin \alpha}{\alpha}\right)^2, \qquad \text{where} \qquad \alpha = \frac{\pi a}{\lambda} \sin \theta \quad (37\text{-}5, 37\text{-}6)$$

and I_m is the intensity at the center of the pattern.

Circular Aperture Diffraction

Diffraction by a circular aperture or a lens with diameter d produces a central maximum and concentric maxima and minima, with the first minimum at an angle θ given by

$$\sin \theta = 1.22 \frac{\lambda}{d} \quad \begin{array}{l}\text{(first minimum;}\\ \text{circular aperture).}\end{array} \quad (37\text{-}12)$$

Rayleigh's Criterion

Rayleigh's criterion suggests that two objects are on the verge of resolvability if the central diffraction maximum of one is at the first minimum of the other. Their angular separation must then be at least

$$\theta_R = 1.22 \frac{\lambda}{d} \quad \text{(Rayleigh's criterion),} \quad (37\text{-}14)$$

in which d is the diameter of the aperture.

Double-Slit Diffraction

Waves passing through two slits, each of width a, whose centers are a distance d apart, display diffraction patterns whose intensity I at various diffraction angles θ is given by

$$I = I_m (\cos^2 \beta) \left(\frac{\sin \alpha}{\alpha} \right)^2 \quad \text{(double slit),} \quad (37\text{-}19)$$

with $\beta = (\pi d/\lambda) \sin \theta$ and α the same as for the case of single-slit diffraction.

Multiple-Slit Diffraction

Diffraction by N (multiple) slits results in maxima (lines) at

angles θ such that

$$d \sin \theta = m\lambda, \quad \text{for } m = 0, 1, 2 \dots$$
$$\text{(maxima),} \quad (37\text{-}25)$$

with the half-widths of the lines given by

$$\Delta\theta_{hw} = \frac{\lambda}{Nd \cos \theta} \quad \text{(half-widths).} \quad (37\text{-}28)$$

Diffraction Gratings

A *diffraction grating* is a series of "slits" used to separate an incident wave into its component wavelengths by separating and displaying their diffraction maxima. A grating is characterized by its dispersion D and resolving power R:

$$D = \frac{\Delta\theta}{\Delta\lambda} = \frac{m}{d \cos \theta}$$

$$R = \frac{\lambda_{av}}{\Delta\lambda} = Nm. \quad (37\text{-}29 \text{ to } 37\text{-}32)$$

X-Ray Diffraction

The regular array of atoms in a crystal is a three-dimensional diffraction grating for short-wavelength waves such as x rays. For analysis purposes, the atoms can be visualized as being arranged in planes with characteristic interplanar spacing d. Diffraction maxima (due to constructive interference) occur if the incident direction of the wave, measured from the surfaces of these planes, and the wavelength λ of the radiation satisfy **Bragg's law:**

$$2d \sin \theta = m\lambda, \quad \text{for } m = 1, 2, 3 \dots$$
$$\text{(Bragg's law).} \quad (37\text{-}34)$$

QUESTIONS

1. Light of frequency f illuminating a long narrow slit produces a diffraction pattern. (a) If we switch to light of frequency $1.3f$, does the pattern expand away from the center or contract toward it? (b) Does the pattern expand or contract if, instead, we submerge the equipment in clear corn syrup?

2. You are conducting a single-slit diffraction experiment with light of wavelength λ. What appears, on a distant viewing screen, at a point at which the top and bottom rays through the slit have a path length difference equal to (a) 5λ and (b) 4.5λ?

3. If you speak with the same intensity with and without a megaphone in front of your mouth, in which situation do you sound louder to someone directly in front of you?

4. Figure 37-29 shows four choices for the rectangular opening of a source of either sound waves or light waves. The sides have lengths of either L or $2L$, with L being 3.0 times the wavelength of the waves. Rank the openings according to the extent of (a)

left–right spreading and (b) up–down spreading of the waves due to diffraction, greatest first.

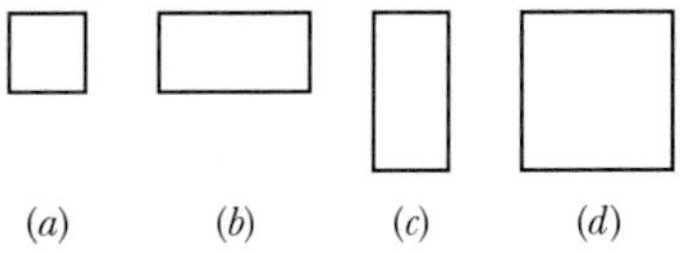

5. In a single-slit diffraction experiment, the top and bottom rays through the slit arrive at a certain point on the viewing screen with a path length difference of 4.0 wavelengths. In a phasor representation like those in Fig 37-6, how many overlapping circles does the chain of phasors make?

6. A vertical spider thread lies between you and the early morning Sun. As you move perpendicular to a line that extends through the thread and the Sun, toward that line, you begin to see the diffraction pattern of sunlight produced by the thread. As your eyes move through the first side maximum of the pattern, what color, red or blue, do you see first? (That is, in a diffraction pattern of white light, is red or blue diffracted at a greater angle?)

7. At night many people see rings (called *entoptic halos*) surrounding bright outdoor lamps in otherwise dark surroundings. The rings are the first of the side maxima in diffraction patterns produced by structures that are thought to be within the cornea (or possibly the lens) of the observer's eye. (The central maxima of such patterns overlap the lamp.) (a) Would a particular ring become smaller or larger if the lamp were switched from

FIGURE 37-29
Question 4.   (a)   (b)   (c)   (d)

blue to red light? (b) If a lamp emits white light, is blue or red on the outside edge of the ring?

8. Figure 37-30 shows the bright fringes that lie within the central diffraction envelope in two double-slit diffraction experiments using the same wavelength of light. Are (a) the slit width a (b) the slit separation d, and (c) the ratio d/a in experiment B greater than, less than, or the same as those in experiment A?

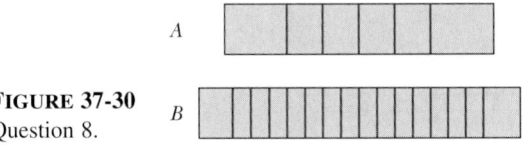

FIGURE 37-30
Question 8.

9. Figure 37-31 shows a red line and a green line of the same order in the pattern produced by a diffraction grating. If we increased the number of rulings in the grating, say, by removing tape that had covered half the rulings, would (a) the half-widths of the lines and (b) the separation of the lines increase, decrease, or remain the same? (c) Would the lines shift to the right, shift to the left, or remain in place?

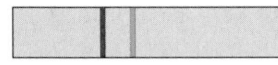

FIGURE 37-31   Questions 9 and 10.

10. For the situation of Question 9 and Fig. 37-31, if instead we

increased the grating spacing, would (a) the half-widths of the lines and (b) the separation of the lines increase, decrease, or remain the same? (c) Would the lines shift to the right, shift to the left, or remain in place?

11. (a) Figure 37-32a shows the lines produced by diffraction gratings A and B using light of the same wavelength; the lines are of the same order and at the same angles θ. Which grating has the greater number of rulings? (b) Figure 37-32b shows lines of two orders produced by a single diffraction grating using light of two wavelengths, both in the red region of the spectrum. Which lines, the left pair or right pair, are in the order with greater m? Is the center of the diffraction pattern to the left or right in (c) Fig. 37-32a and (d) Fig. 37-32b?

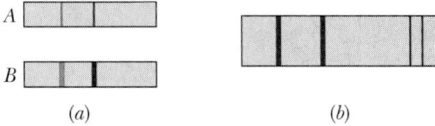

FIGURE 37-32   Question 11.

12. (a) For a given diffraction grating, does the least difference $\Delta\lambda$ in two wavelengths that can be resolved increase, decrease, or remain the same as the wavelength increases? (b) For a given wavelength region (say, around 500 nm), is $\Delta\lambda$ greater in the first order or in the third order?

EXERCISES & PROBLEMS

SECTION 37-2   Diffraction by a Single Slit: Locating the Minima

1E. When monochromatic light is incident on a slit 0.022 mm wide, the first diffraction minimum is observed at an angle of 1.8° from the direction of the incident light. What is the wavelength?

2E. Monochromatic light of wavelength 441 nm is incident on a narrow slit. On a screen 2.00 m away, the distance between the second diffraction minimum and the central maximum is 1.50 cm. (a) Calculate the angle of diffraction θ of the second minimum. (b) Find the width of the slit.

3E. Light of wavelength 633 nm is incident on a narrow slit. The angle between the first diffraction minimum on one side of the central maximum and the first minimum on the other side is 1.20°. What is the width of the slit?

4E. A single slit is illuminated by light of wavelengths λ_a and λ_b, so chosen that the first diffraction minimum of the λ_a component coincides with the second minimum of the λ_b component. (a) What relationship exists between the two wavelengths? (b) Do any other minima in the two diffraction patterns coincide?

5E. The distance between the first and fifth minima of a single-slit diffraction pattern is 0.35 mm with the screen 40 cm away from the slit, using light of wavelength 550 nm. (a) Find the slit width. (b) Calculate the angle θ of the first diffraction minimum.

6E. What must be the ratio of the slit width to the wavelength for a single slit to have the first diffraction minimum at $\theta = 45.0°$?

7E. A plane wave of wavelength 590 nm is incident on a slit with $a = 0.40$ mm. A thin converging lens of focal length +70 cm is placed between the slit and a viewing screen and focuses the light on the screen. (a) How far is the screen from the lens? (b) What is the distance on the screen from the center of the diffraction pattern to the first minimum?

8P. A slit 1.00 mm wide is illuminated by light of wavelength 589 nm. We see a diffraction pattern on a screen 3.00 m away. What is the distance between the first two diffraction minima on the same side of the central diffraction maximum?

9P. Sound waves with frequency 3000 Hz and speed 343 m/s diffract through the rectangular opening of a speaker cabinet and into a large auditorium. The opening, which has a horizontal width of 30.0 cm, faces a wall 100 m away (Fig. 37-33). Where

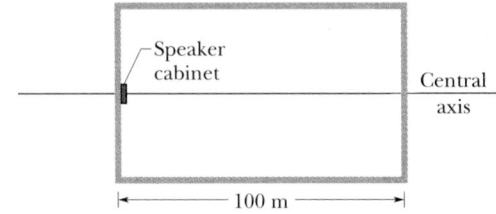

FIGURE 37-33   Problem 9.

along that wall will a listener be at the first diffraction minimum and thus have difficulty hearing the sound? (Neglect reflections from the walls.)

10P. Manufacturers of wire (and other objects of small dimensions) sometimes use a laser to continually monitor the thickness of the product. The wire intercepts the laser beam, producing a diffraction pattern like that of a single slit of the same width as the wire diameter (see Fig. 37-34). Suppose a helium–neon laser, of wavelength 632.8 nm, illuminates a wire, and the diffraction pattern appears on a screen 2.60 m away. If the desired wire diameter is 1.37 mm, what is the observed distance between the two tenth-order minima (one on each side of the central maximum)?

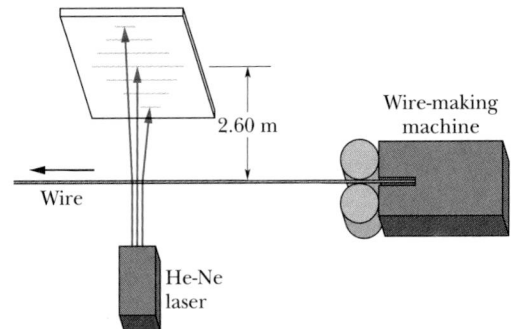

FIGURE 37-34 Problem 10.

SECTION 37-4 Intensity in Single-Slit Diffraction, Quantitatively

11E. A 0.10-mm-wide slit is illuminated by light of wavelength 589 nm. Consider rays that are diffracted at $\theta = 30°$ and calculate the phase difference at the screen of Huygens' wavelets from the top and midpoint of the slit. (*Hint:* See Eq. 37-4.)

12E. Monochromatic light with wavelength 538 nm is incident on a slit with width 0.025 mm. The distance from the slit to a screen is 3.5 m. Consider a point on the screen 1.1 cm from the central maximum. (a) Calculate θ for that point. (b) Calculate α. (c) Calculate the ratio of the intensity at this point to the intensity at the central maximum.

13P. If you double the width of a single slit, the intensity of the central maximum of the diffraction pattern increases by a factor of 4, even though the energy passing through the slit only doubles. Explain this quantitatively.

14P. *Babinet's Principle.* A monochromatic beam of parallel light is incident on a "collimating" hole of diameter $x \gg \lambda$. Point P lies in the geometrical shadow region on a *distant* screen (Fig. 37-35). Two obstacles, shown in Fig. 37-35b, are placed in turn over the collimating hole. A is an opaque circle with a hole in it and B is the "photographic negative" of A. Using superposition concepts, show that the intensity at P is identical for the two diffracting objects A and B.

15P. The full width at half-maximum (FWHM) of the central diffraction maximum is defined as the angle between the two points in the pattern where the intensity is one-half that at the center of the pattern. (See Fig. 37-7b.) (a) Show that the intensity

drops to one-half the maximum value when $\sin^2 \alpha = \alpha^2/2$. (b) Verify that $\alpha = 1.39$ radians (about 80°) is a solution to the transcendental equation of (a). (c) Show that the FWHM is $\Delta\theta = 2 \sin^{-1}(0.443\lambda/a)$. (d) Calculate the FWHM of the central maximum for slits whose widths are 1.0, 5.0, and 10 wavelengths.

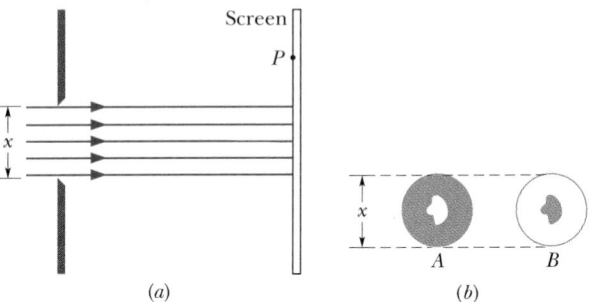

FIGURE 37-35 Problem 14.

16P. (a) Show that the values of α at which intensity maxima for single-slit diffraction occur can be found exactly by differentiating Eq. 37-5 with respect to α and equating the result to zero, obtaining the condition $\tan \alpha = \alpha$. (b) Find the values of α satisfying this relation by plotting the curve $y = \tan \alpha$ and the straight line $y = \alpha$ and finding their intersections or by using a pocket calculator to find an appropriate value of α by trial and error. (c) Find the (noninteger) values of m corresponding to successive maxima in the single-slit pattern. Note that the secondary maxima do not lie exactly halfway between minima.

17P*. Derive this expression for the intensity pattern for a three-slit "grating":

$$I = \tfrac{1}{9}I_m(1 + 4 \cos \phi + 4 \cos^2 \phi),$$

where $\phi = (2\pi d \sin \theta)/\lambda$. Assume that $a \ll \lambda$; be guided by the derivation of the corresponding double-slit formula (Eq. 36-21).

SECTION 37-5 Diffraction by a Circular Aperture

18E. Assume that the lamp in Question 7 emits light at wavelength 550 nm. If a ring has an angular diameter of 2.5°, approximately what is the (linear) diameter of the structure in the eye that causes the ring?

19E. The two headlights of an approaching automobile are 1.4 m apart. At what (a) angular separation and (b) maximum distance will the eye resolve them? Assume that the pupil diameter is 5.0 mm, and use a wavelength of 550 nm. Also assume that diffraction effects alone limit the resolution.

20E. An astronaut in a space shuttle claims she can just barely resolve two point sources on Earth's surface, 160 km below. Calculate their (a) angular and (b) linear separation, assuming ideal conditions. Take $\lambda = 540$ nm and the pupil diameter of the astronaut's eye to be 5.0 mm.

21E. Find the separation of two points on the Moon's surface that can just be resolved by the 200 in. (= 5.1 m) telescope at Mount Palomar, assuming that this separation is determined by diffraction effects. The distance from Earth to the Moon is 3.8 × 10^5 km. Assume a wavelength of 550 nm.

22E. The wall of a large room is covered with acoustic tile in which small holes are drilled 5.0 mm from center to center. How far can a person be from such a tile and still distinguish the individual holes, assuming ideal conditions, the pupil diameter of the observer's eye to be 4.0 mm, and the wavelength of the room light to be 550 nm?

23E. The pupil of a person's eye has a diameter of 5.00 mm. What distance apart must two small objects be if their images are just resolved when they are 250 mm from the eye and illuminated with light of wavelength 500 nm?

24E. Under ideal conditions, estimate the linear separation of two objects on the planet Mars that can just be resolved by an observer on Earth (a) using the naked eye and (b) using the 200 in. (= 5.1 m) Mount Palomar telescope. Use the following data: distance to Mars = 8.0×10^7 km; diameter of pupil = 5.0 mm; wavelength of light = 550 nm.

25E. If Superman really had x-ray vision at 0.10 nm wavelength and a 4.0 mm pupil diameter, at what maximum altitude could he distinguish villains from heroes, assuming that he needs to resolve points separated by 5.0 cm to do this?

26E. A navy cruiser employs radar with a wavelength of 1.6 cm. The circular antenna has a diameter of 2.3 m. At a range of 6.2 km, what is the smallest distance that two speedboats can be from each other and still be resolved as two separate objects by the radar system?

27P. Nuclear-pumped x-ray lasers are seen as a possible weapon to destroy ICBM booster rockets at ranges up to 2000 km. One limitation on such a device is the spreading of the beam due to diffraction, with resulting dilution of beam intensity. Consider such a laser operating at a wavelength of 1.40 nm. The element that emits light is the end of a wire with diameter 0.200 mm. (a) Calculate the diameter of the central beam at a target 2000 km away from the beam source. (b) By what factor is the beam intensity reduced in transit to the target? (The laser is fired from space, so that atmospheric absorption can be ignored.)

28P. (a) How far from grains of red sand must you be to position yourself just at the limit of resolving the grains if your pupil diameter is 1.5 mm, the grains are spherical with radius 50 μm, and the light from the grains has wavelength 650 nm? (b) If the grains were blue and the light from them had wavelength 400 nm, would the answer to (a) be larger or smaller?

29P. The wings of tiger beetles (Fig. 37-36) are colored by interference due to thin cuticle-like layers. In addition, these layers are arranged in patches that are 60 μm across and produce different colors. The color you see is a pointillistic mixture of thin-film interference colors that varies with perspective. Approximately what viewing distance from a wing puts you at the limit of resolving the different colored patches according to Rayleigh's criterion? Use 550 nm as the wavelength of light and 3.00 mm as the diameter of your pupil.

30P. (a) A circular diaphragm 60 cm in diameter oscillates at a frequency of 25 kHz as an underwater source of sound used for submarine detection. Far from the source the sound intensity is distributed as the diffraction pattern of a circular hole whose diameter equals that of the diaphragm. Take the speed of sound in

water to be 1450 m/s and find the angle between the normal to the diaphragm and the direction of the first minimum. (b) Repeat for a source having an (audible) frequency of 1.0 kHz.

FIGURE 37-36 Problem 29. Tiger beetles are colored by pointillistic mixtures of thin-film interference colors.

31P. In June 1985 a laser beam was fired from the Air Force Optical Station on Maui, Hawaii, and reflected back from the shuttle *Discovery* as it sped by, 220 mi overhead. The diameter of the central maximum of the beam at the shuttle position was said to be 30 ft, and the beam wavelength was 500 nm. What is the effective diameter of the laser aperture at the Maui ground station? (*Hint:* A laser beam spreads only because of diffraction; assume a circular exit aperture.)

32P. A spy satellite orbiting at 160 km above Earth's surface has a lens with a focal length of 3.6 m and can resolve objects on the ground as small as 30 cm; it can easily measure the size of an aircraft's air intake. What is the effective lens diameter, determined by diffraction consideration alone? Assume λ = 550 nm.

33P. Millimeter-wave radar generates a narrower beam than conventional microwave radar, making it less vulnerable to antiradar missiles. (a) Calculate the angular width of the central maximum, from first minimum to first minimum, produced by a 220 GHz radar beam emitted by a 55.0-cm-diameter circular antenna. (The frequency is chosen to coincide with a low-absorption atmospheric "window.") (b) Calculate the same quantity for the ship's radar described in Exercise 26.

34P. (a) How small is the angular separation of two stars if their images are barely resolved by the Thaw refracting telescope at the Allegheny Observatory in Pittsburgh? The lens diameter is 76 cm and its focal length is 14 m. Assume λ = 550 nm. (b) Find the distance between these barely resolved stars if each of them is 10 light-years distant from Earth. (c) For the image of a single star in this telescope, find the diameter of the first dark ring in the diffraction pattern, as measured on a photographic plate placed at the focal plane of the telescope lens. Assume that the structure of the image is associated entirely with diffraction at the lens aperture and not with lens "errors."

35P. A circular obstacle produces the same diffraction pattern as a circular hole of the same diameter (except very near $\theta = 0$). Airborne water drops are examples of such obstacles. When you see the Moon through suspended water drops, such as in a fog, you intercept the diffraction pattern from many drops; the composite is a bright circular pattern surrounding the Moon (Fig. 37-37). Next to the Moon, the pattern is white. (a) What color, red or blue, outlines that white pattern? (b) Suppose the outlining ring has an angular diameter that is 1.5 times the angular diameter of the Moon, which is 0.50°. Suppose also that the drops all have about the same diameter; approximately what is that diameter?

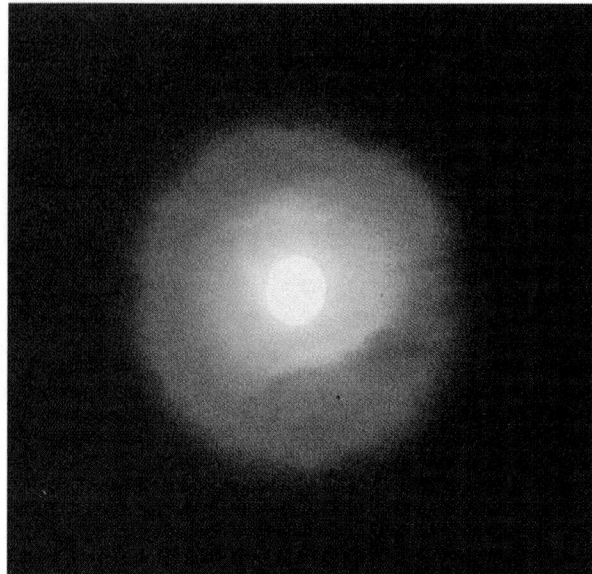

FIGURE 37-37 Problem 35. The corona around the Moon is a composite of the diffraction patterns of airborne water drops.

36P. In a joint Soviet–French experiment to monitor the Moon's surface with a light beam, pulsed radiation from a ruby laser ($\lambda = 0.69$ μm) was directed to the Moon through a reflecting telescope with a mirror radius of 1.3 m. A reflector on the Moon behaved like a circular plane mirror with radius 10 cm, reflecting the light directly back toward the telescope on Earth. The reflected light was then detected after being brought to a focus by this telescope. What fraction of the original light energy was picked up by the detector? Assume that for each direction of travel all the energy is in the central diffraction peak.

SECTION 37-6 Diffraction by a Double Slit

37E. Suppose that the central diffraction envelope of a double-slit diffraction pattern contains 11 bright fringes and the first diffraction minima eliminate (are coincident with) bright fringes. How many bright fringes lie between the first and second minima of the diffraction envelope?

38E. For $d = 2a$ in Fig. 37-38, how many bright interference fringes lie in the central diffraction envelope?

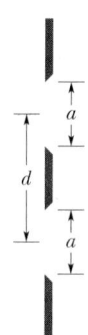

FIGURE 37-38 Exercise 38 and Problem 39.

39P. If we put $d = a$ in Fig. 37-38, the two slits coalesce into a single slit of width $2a$. Show that Eq. 37-19 reduces to the diffraction pattern for such a slit.

40P. (a) In a double-slit system, what ratio of d to a causes diffraction to eliminate the fourth bright side fringe? (b) What other bright fringes are also eliminated?

41P. Two slits of width a and separation d are illuminated by a coherent beam of light of wavelength λ. What is the linear separation of the bright interference fringes observed on a screen that is at a distance D away?

42P. (a) How many fringes appear between the first diffraction-envelope minima to either side of the central maximum for a double-slit pattern if $\lambda = 550$ nm, $d = 0.150$ mm, and $a = 30.0$ μm? (b) What is the ratio of the intensity of the third bright fringe to the intensity of the central fringe?

43P. Light of wavelength 440 nm passes through a double slit, yielding a diffraction pattern whose graph of intensity I versus deflection angle θ is shown in Fig. 37-39. Calculate (a) the slit width and (b) the slit separation. (c) Verify the displayed intensities of the $m = 1$ and $m = 2$ interference fringes.

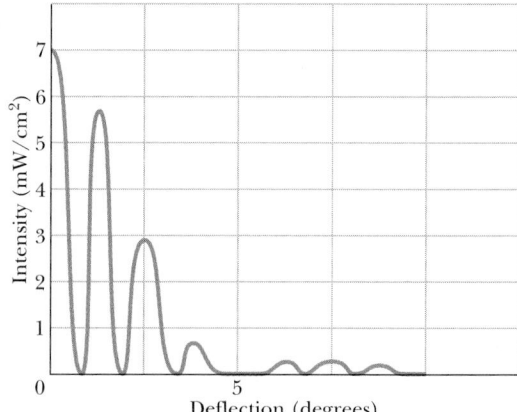

FIGURE 37-39 Problem 43.

44P. An acoustic double-slit system (of slit separation d and slit width a) is driven by two loudspeakers as shown in Fig. 37-40. By use of a variable delay line, the phase of one of the speakers may be varied. Describe in detail what changes occur in the double-slit diffraction pattern at large distances as the phase difference between the speakers is varied from zero to 2π. Take both interference and diffraction effects into account.

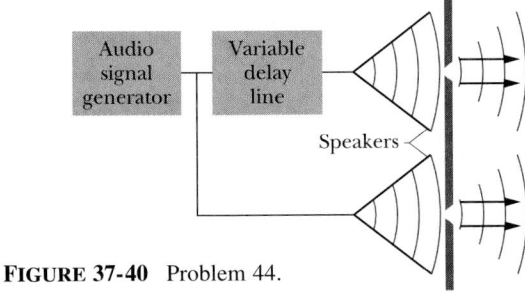

FIGURE 37-40 Problem 44.

SECTION 37-7 Diffraction Gratings

45E. A diffraction grating 20.0 mm wide has 6000 rulings. (a) Calculate the distance d between adjacent rulings. (b) At what angles will intensity maxima occur if the incident radiation has a wavelength of 589 nm?

46E. A diffraction grating has 200 rulings/mm, and it produces an intensity maximum at $\theta = 30.0°$. (a) What are the possible wavelengths of the incident visible light? (b) To what colors do they correspond?

47E. A grating has 315 rulings/mm. For what wavelengths in the visible spectrum can fifth-order diffraction be observed?

48E. Given a grating with 400 lines/mm, how many orders of the entire visible spectrum (400–700 nm) can it produce in addition to the $m = 0$ order?

49E. A diffraction grating 3.00 cm wide produces the second order at 33.0° with light of wavelength 600 nm. What is the total number of lines on the grating?

50E. Some tropical gyrinid beetles (whirligig beetles) are colored by optical interference that is due to scales whose alignment forms a diffraction grating (which uses scattered instead of transmitted light). If the incident light is perpendicular on the grating, the angle between the first-order maxima (on opposite sides of the zeroth-order maximum) is about 26°. What is the grating spacing of the beetle? Use 550 nm as the wavelength of light.

51E. A diffraction grating 1.0 cm wide has 10,000 parallel slits. Monochromatic light that is incident normally is deviated through 30° in the first order. What is the wavelength of the light?

52P. Light of wavelength 600 nm is incident normally on a diffraction grating. Two adjacent maxima occur at angles given by $\sin \theta = 0.2$ and $\sin \theta = 0.3$, respectively. The fourth-order maxima are missing. (a) What is the separation between adjacent slits? (b) What is the smallest possible individual slit width? (c) Which orders of intensity maxima are produced by the grating, assuming the values derived in (a) and (b)?

53P. A diffraction grating is made up of slits of width 300 nm with separation 900 nm. The grating is illuminated by monochromatic plane waves of wavelength $\lambda = 600$ nm at normal incidence. (a) How many diffraction maxima are there in the full pattern? (b) What is the width of a spectral line observed in the first order if the grating has 1000 slits?

54P. Assume that the limits of the visible spectrum are arbitrarily chosen as 430 and 680 nm. Calculate the number of rulings per millimeter of a grating that will spread the first-order spectrum through an angle of 20°.

55P. With light from a gaseous discharge tube incident normally on a grating with slit separation 1.73 μm, sharp maxima of green light are produced at angles $\theta = \pm17.6°$, 37.3°, $-37.1°$, 65.2°, and $-65.0°$. Compute the wavelength of the green light that best fits these data.

56P. Light is incident on a grating at an angle ψ as shown in Fig. 37-41. Show that bright fringes occur at angles θ that satisfy the equation

$$d(\sin \psi + \sin \theta) = m\lambda, \qquad \text{for } m = 0, 1, 2, \ldots .$$

(Compare this equation with Eq. 37-25.) Only the special case $\psi = 0$ has been treated in this chapter.

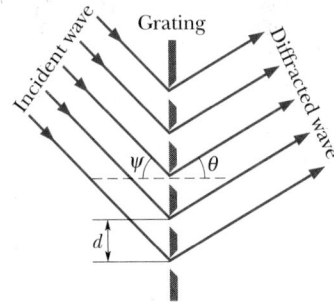

FIGURE 37-41 Problem 56.

57P. A grating with $d = 1.50$ μm is illuminated at various angles of incidence by light of wavelength 600 nm. Plot, as a function of the angle of incidence (0 to 90°), the angular deviation of the first-order maximum from the incident direction. (See Problem 56.)

58P. Two emission lines have wavelengths λ and $\lambda + \Delta\lambda$, respectively, where $\Delta\lambda \ll \lambda$. Show that their angular separation $\Delta\theta$ in a grating spectrometer is given approximately by

$$\Delta\theta = \frac{\Delta\lambda}{\sqrt{(d/m)^2 - \lambda^2}},$$

where d is the slit separation and m is the order at which the lines are observed. Note that the angular separation is greater in the higher orders than in lower orders.

59P. White light (consisting of wavelengths from 400 nm to 700 nm) is normally incident on a grating. Show that, no matter what the value of the grating spacing d, the second order and third order overlap.

60P. Show that a grating made up of alternately transparent and opaque strips of equal width eliminates all the even orders of maxima (except $m = 0$).

61P. A grating has 350 rulings per millimeter and is illuminated at normal incidence by white light. A spectrum is formed on a screen 30 cm from the grating. If a hole 10 mm square is cut in the screen, its inner edge being 50 mm from the central maximum and parallel to it, what range of wavelengths passes through the hole?

62P. Derive Eq. 37-28, the expression for the line widths.

SECTION 37-8 Gratings: Dispersion and Resolving Power

63E. The D line in the spectrum of sodium is a doublet with wavelengths 589.0 and 589.6 nm. Calculate the minimum number of lines needed in a grating that will resolve this doublet in the second-order spectrum. See Sample Problem 37-7.

64E. A grating has 600 rulings/mm and is 5.0 mm wide. (a) What is the smallest wavelength interval it can resolve in the third order at $\lambda = 500$ nm? (b) How many higher orders of maxima can be seen?

65E. A source containing a mixture of hydrogen and deuterium atoms emits red light at two wavelengths whose mean is 656.3 nm and whose separation is 0.180 nm. Find the minimum number of lines needed in a diffraction grating that can resolve these lines in the first order.

66E. (a) How many rulings must a 4.00-cm-wide diffraction grating have to resolve the wavelengths 415.496 and 415.487 nm in the second order? (b) At what angle are the maxima found?

67E. With a particular grating the sodium doublet (see Sample Problem 37-7) is viewed in the third order at 10° to the normal and is barely resolved. Find (a) the grating spacing and (b) the total width of the rulings.

68E. Show that the dispersion of a grating is $D = (\tan \theta)/\lambda$.

69E. A grating has 40,000 rulings spread over 76 mm. (a) What is its expected dispersion D for sodium light ($\lambda = 589$ nm) in the first three orders? (b) What is the grating's resolving power in these orders?

70P. Light containing a mixture of two wavelengths, 500 and 600 nm, is incident normally on a diffraction grating. It is desired (1) that the first and second maxima for each wavelength appear at $\theta \le 30°$, (2) that the dispersion be as high as possible, and (3) that the third order for 600 nm be a missing order. (a) What should be the slit separation? (b) What is the smallest possible individual slit width? (c) For the 600 nm wavelength, which orders of intensity maxima are produced by the grating, assuming the values derived in (a) and (b)?

71P. (a) In terms of the angle θ locating a line produced by a grating, find the product of that line's half-width and the resolving power of the grating. (b) Evaluate that product for the grating of Problem 53, for first order.

72P. A diffraction grating has resolving power $R = \lambda_{av}/\Delta\lambda = Nm$. (a) Show that the corresponding frequency range Δf that can just be resolved is given by $\Delta f = c/Nm\lambda$. (b) From Fig. 37-17, show that the times required for light to travel along the two extreme rays differ by an amount $\Delta t = (Nd/c) \sin \theta$. (c) Show that $(\Delta f)(\Delta t) = 1$, this relation being independent of the various grating parameters. Assume $N \gg 1$.

SECTION 37-9 X-Ray Diffraction

73E. X rays of wavelength 0.12 nm are found to undergo second-order reflection at a Bragg angle of 28° from a lithium fluoride crystal. What is the interplanar spacing of the reflecting planes in the crystal?

74E. What is the smallest Bragg angle for x rays of wavelength 30 pm to undergo reflection from reflecting planes of spacing 0.30 nm in a calcite crystal?

75E. If first-order reflection occurs in a crystal at Bragg angle 3.4°, at what Bragg angle does second-order reflection occur from the same family of reflecting planes?

76E. Figure 37-42 is a graph of intensity versus diffraction angle for the diffraction of an x-ray beam by a crystal. The beam consists of two wavelengths, and the spacing between the reflecting planes is 0.94 nm. What are the two wavelengths?

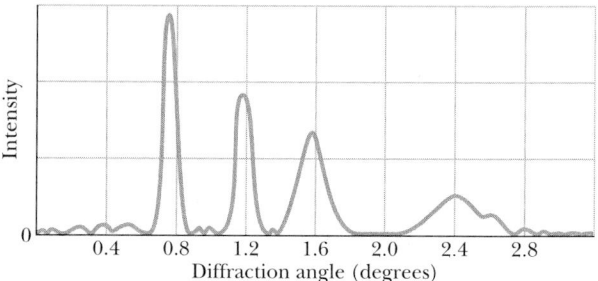

FIGURE 37-42   Exercise 76.

77E. An x-ray beam of wavelength A undergoes a first-order reflection from a crystal when its angle of incidence to a crystal face is 23°, and an x-ray beam of wavelength 97 pm undergoes third-order reflection when its angle of incidence to that face is 60°. Assuming that the two beams reflect from the same family of reflecting planes, find (a) the interplanar spacing and (b) the wavelength A.

78E. An x-ray beam of a certain wavelength is incident on a NaCl crystal, at 30.0° to a certain family of reflecting planes of spacing 39.8 pm. If the reflection from those planes is of the first order, what is the wavelength of the x rays?

79P. Prove that it is not possible to determine both wavelength of radiation and spacing of reflecting planes in a crystal by measuring the Bragg angles in several orders.

80P. In Fig. 37-43, an x-ray beam of wavelengths from 95.0 pm to 140 pm is incident on a family of reflecting planes with spacing $d = 275$ pm. At which wavelengths will these planes produce intensity maxima in their reflections?

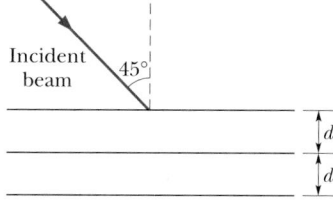

FIGURE 37-43   Problems 80 and 83.

81P. In Fig. 37-44, first-order reflection from the reflection planes shown occurs when an x-ray beam of wavelength 0.260 nm makes an angle of 63.8° with the top face of the crystal. What is the unit cell size a_0?

82P. Consider a two-dimensional square crystal structure, such as one side of the structure shown in Fig. 37-27a. One interplanar spacing of reflecting planes is the unit cell size a_0. Calculate and sketch the next five smaller interplanar spacings. (b) Show that your results in (a) obey the general formula

$$d = \frac{a_0}{\sqrt{h^2 + k^2}},$$

where h and k are relatively prime integers (they have no common factor other than unity).

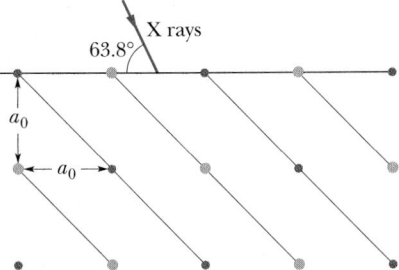

FIGURE 37-44 Problem 81.

83P. In Fig. 37-43, let a beam of x rays of wavelength 0.125 nm be incident on an NaCl crystal at an angle of 45.0° to the top face of the crystal. Let the reflecting planes have separation $d = 0.252$ nm. Through what angles must the crystal be turned about an axis that is perpendicular to the plane of the page for these reflecting planes to give intensity maxima in their reflections?

Electronic Computation

84. A computer can be used to sum the phasors corresponding to Huygens' wavelets and so find a diffraction pattern. Suppose light with a wavelength of 500 nm is incident normally on a single slit with a width of 5.00×10^{-6} m. To approximate the diffraction pattern, sum the phasors corresponding to $N = 200$ wavelets spreading from uniformly distributed sources within the slit. The horizontal and vertical components of the resultant are proportional to

$$E_h = \sum_{i=1}^{N} \cos \phi_i \quad \text{and} \quad E_v = \sum_{i=1}^{N} \sin \phi_i,$$

respectively, where ϕ_i is the phase of wavelet i. The intensity ratio is $I/I_m = (E_h^2 + E_v^2)/N^2$. The factor $1/N^2$ assures that $I/I_m = 1$ when all the wavelets have the same phase. If you consider light that is diffracted at the angle θ to the straight-ahead direction, then you may take the phase of the first wavelet to be zero and the phase of each successive wavelet to be $(2\pi/\lambda)\Delta x \sin \theta$ greater than that of the preceding wavelet. Here Δx is the distance between wavelet sources; that is, $\Delta x = a/(N - 1)$, where a is the slit width. Use this technique to search for the diffraction angles corresponding to the first three secondary maxima and find the intensity ratios for those maxima.

38
Relativity

In modern long-range navigation, the precise location and speed of moving craft are continuously monitored and updated. A system of navigation satellites called NAVSTAR permits locations and speeds anywhere on Earth to be determined to within about 16 m and 2 cm/s. However, if relativity effects were not taken into account, speeds could not be determined any closer than about 20 cm/s, which is unacceptable for modern navigation systems. How can something as abstract as Einstein's special theory of relativity be involved in something as practical as navigation?

38-1 WHAT IS RELATIVITY ALL ABOUT?

Relativity has to do with measurements of events (things that happen): where and when they happen, and by how much any two events are separated in space and in time. In addition, relativity has to do with transforming such measurements and others between reference frames that move relative to each other. (Hence the name *relativity*.) We discussed such matters in Sections 4-8 and 4-9.

Transformations and moving reference frames were well understood and quite routine to physicists in 1905. Then Albert Einstein (Fig. 38-1) published his **special theory of relativity.** The adjective *special* means that the theory deals only with **inertial reference frames,** which are frames that move at constant velocities relative to one another. (Einstein's *general theory of relativity* treats the more challenging situation in which reference frames accelerate; in this chapter the term *relativity* implies only inertial reference frames.)

Starting with two deceivingly simple postulates, Einstein stunned the scientific world by showing that the old ideas about relativity were wrong, even though everyone was so accustomed to them that they seemed to be unquestionable common sense. This supposed common sense, however, was derived from experience only with things that move rather slowly. Einstein's relativity, which turns out to be correct for all possible speeds, predicted many effects that were, at first study, bizarre because no one had experienced them.

In particular, Einstein demonstrated that space and time are entangled; that is, the time between two events depends on how far apart they occur, and vice versa. And the entanglement is different for observers who move relative to each other. One result is that time does not pass at a fixed rate, as if it were ticked off with mechanical regularity on some master grandfather clock that controls the universe. Rather, that rate is adjustable: relative motion can change the rate at which time passes. Prior to 1905, no one but a few daydreamers would have thought that. Now, engineers and scientists take it for granted because their experience with special relativity has reshaped their common sense.

Special relativity has the reputation of being difficult. It is not difficult mathematically, at least not here. But it is difficult in that we must be very careful about *who* measures *what* about an event and just *how* that measurement is made. And it can be difficult because it can contradict experience. Before you read further, you might want to review some of the special relativity discussed earlier in this book (Table 38-1 is a guide).

38-2 THE POSTULATES

We now examine the two postulates of relativity, on which Einstein's theory is based:

---

1. The Relativity Postulate: The laws of physics are the same for observers in all inertial reference frames. No frame is preferred.

Galileo assumed that the laws of *mechanics* were the same in all inertial reference frames. (Newton's first law of motion is one important consequence.) Einstein extended that idea to include *all* the laws of physics, especially electromagnetism and optics. This postulate does *not* say that the

FIGURE 38-1 Einstein in the early 1900s, at his desk at the patent office in Bern, Switzerland, where he was employed when he published his special theory of relativity.

TABLE 38-1 **EARLIER SECTIONS ON RELATIVITY**

SECTION	TITLE
4-10	Relative Motion at High Speeds
7-8	Kinetic Energy at High Speeds
8-8	Mass and Energy
10-6	Reactions and Decay Processes

measured values of all physical quantities are the same for all inertial observers; most are not the same. It is the *laws of physics,* which relate these measurements to each other, that are the same.

---

2. The Speed of Light Postulate: The speed of light in vacuum has the same value *c* in all directions and in all inertial reference frames.

We can also phrase this postulate to say that there is in nature an *ultimate speed c,* the same in all directions and in all inertial reference frames. Light happens to travel at this ultimate speed, as do any massless particles (neutrinos might be an example). But any entity that carries energy or information cannot exceed this limit. Moreover, any particle that does have mass cannot actually reach speed *c*, no matter how much or how long it is accelerated.

Both postulates have been exhaustively tested, and no exceptions have ever been found.

The Ultimate Speed

The reality of the existence of a limit to the speed of accelerated electrons was shown in a 1964 experiment of W. Bertozzi. He accelerated electrons to various measured speeds (see Fig. 38-2) and—by an independent method—also measured their kinetic energies. He found that as the force that acts on a very fast electron is increased, the electron's measured kinetic energy increases toward very large values but its speed does not increase appre-

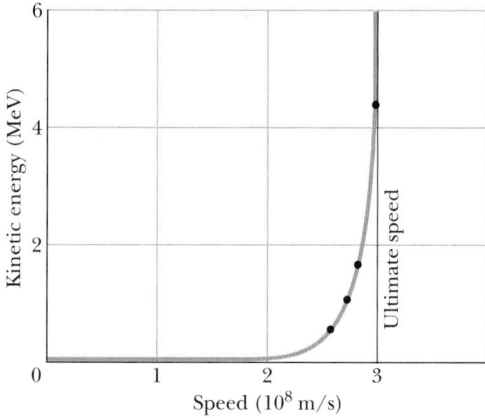

FIGURE 38-2 The dots show measured values of the kinetic energy of an electron plotted against its measured speed. No matter how much energy is given to an electron (or to any other particle having mass), its speed can never equal or exceed the ultimate limiting speed *c*. (The plotted curve through the dots shows the predictions of Einstein's special theory of relativity.)

ciably. Electrons have been accelerated to at least 0.999 999 999 95 times the speed of light but—close though it may be—that speed is still less than the ultimate speed *c*.

Testing the Speed of Light Postulate

If the speed of light is the same in all inertial reference frames, then the speed of light that is emitted by a moving source should be the same as the speed of light that is emitted by a source at rest in the laboratory. This claim has been tested directly, in an experiment of high precision. The "light source" was the *neutral pion* (symbol π^0), an unstable, short-lived particle that can be produced by collisions in a particle accelerator. It decays into two gamma rays by the process

$$\pi^0 \rightarrow \gamma + \gamma. \tag{38-1}$$

Gamma rays are part of the electromagnetic spectrum and obey the speed of light postulate, just as visible light does.

In a 1964 experiment, physicists at CERN, the European particle-physics laboratory near Geneva, generated a beam of pions moving at a speed of 0.999 75*c* with respect to the laboratory. The experimenters then measured the speed of the gamma rays emitted from these very rapidly moving sources. They found that the speed of the light emitted by the pions was the same as would have been measured if the pions had been at rest in the laboratory.

SAMPLE PROBLEM 38-1

An electron with a kinetic energy of 20 GeV (which is said to be a 20 GeV electron) can be shown to have a speed $v = 0.999\ 999\ 999\ 67c$. If such an electron raced a light pulse to the nearest star outside the solar system (Proxima Centauri, 4.3 light-years, or 4.0×10^{16} m, distant), by how much time would the light pulse win the race?

SOLUTION: If *L* is the distance to the star, the difference in travel times is

$$\Delta t = \frac{L}{v} - \frac{L}{c} = L\frac{c - v}{vc}.$$

Now *v* is so close to *c* that we can put $v = c$ in the denominator of this expression (but not in the numerator!). If we do so, we find

$$\Delta t = \frac{L}{c}\left(1 - \frac{v}{c}\right)$$

$$= \frac{(4.0 \times 10^{16}\ \text{m})(1 - 0.999\ 999\ 999\ 67)}{3.00 \times 10^8\ \text{m/s}}$$

$$= 0.044\ \text{s} = 44\ \text{ms}. \qquad \text{(Answer)}$$

38-3 MEASURING AN EVENT

An **event** is something that happens, to which an observer can assign three space coordinates and one time coordinate. Among many possible events are (1) the turning on or off of a tiny lightbulb, (2) the collision of two particles, (3) the passage of a pulse of light through a specified point, (4) an explosion, and (5) the coincidence of the hand of a clock with a marker on the rim of the clock. An observer, fixed in a certain inertial reference frame, may assign to an event A the following coordinates:

RECORD OF EVENT A	
COORDINATE	VALUE
x	3.58 m
y	1.29 m
z	0 m
t	34.5 s

Because in relativity space and time are entangled with each other, we can describe these coordinates collectively as *spacetime* coordinates. The coordinate system itself is part of the reference frame of the observer.

A given event may be recorded by any number of observers, each in a different inertial reference frame. In general, different observers will assign different spacetime coordinates to the same event. Note that an event does not, in any sense, "belong" to a particular inertial reference frame. An event is just something that happens, and anyone in any reference frame may look at it and assign spacetime coordinates to it.

Making such an assignment can be complicated by a practical problem. For example, suppose a balloon bursts 1 km to your right while a firecracker pops 2 km to your left, both at 9:00 A.M. But you do not detect either event precisely at 9:00 A.M. because light from the events has not yet reached you. Because light from the pop has farther to go, it arrives at your eyes later than does light from the balloon burst and thus the pop will seem to have occurred later than the burst. To sort out the actual times and to assign 9:00 A.M. to both events, you must calculate the travel times of the light and then subtract them from the arrival times.

This procedure can be very messy in more challenging situations, and we need an easier procedure that automatically eliminates any concern about the travel time from an event to an observer. To set up such a procedure, we shall construct an imaginary array of measuring rods and clocks throughout the observer's inertial frame (the array moves rigidly with the observer). This construction may seem contrived, but it spares us much confusion and calculation and allows us to find the space coordinates, the time coordinate, and the spacetime coordinates, as follows.

1. **The Space Coordinates.** We imagine the observer's coordinate system fitted with a close-packed, three-dimensional array of measuring rods, one set of rods parallel to each of the three coordinate axes. These rods provide a way to determine coordinates along the axes. Thus if the event is, say, the turning on of a small lightbulb, the observer, in order to locate the position of the event, need read only the three space coordinates at the bulb's location.

2. **The Time Coordinate.** For the time coordinate, we imagine that every point of intersection in the array of measuring rods has a tiny clock, which the observer can read by the light generated by the event. Figure 38-3 suggests the "jungle gym" of clocks and measuring rods that we have described.

The array of clocks must be synchronized properly. It is not enough to assemble a set of identical clocks, set them all to the same time, and then move them to their assigned positions. We do not know, for example, whether moving the clocks will change their rates. (Actually, it will.) We must put the clocks in place and *then* synchronize them.

If we had a method of transmitting signals at infinite speed, synchronization would be a simple matter. However, no known signal has this property. We therefore choose light (interpreted broadly to include the entire electromagnetic spectrum) to send out our synchronizing signals because, in vacuum, light travels at the highest possible speed, the limiting speed c.

Here is one of many ways that we might synchronize an array of clocks with the help of light signals. The observer enlists the help of a large number of temporary helpers, one for each clock. The observer then stands at a point selected as the origin and sends out a pulse of light

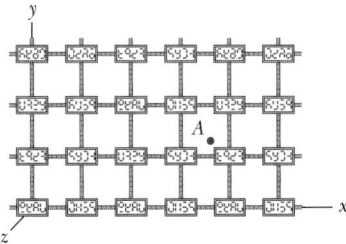

FIGURE 38-3 One section of a three-dimensional array of clocks and measuring rods by which an observer can assign spacetime coordinates to an event, such as a flash of light at point A. The space coordinates are approximately $x = 3.7$ rod lengths, $y = 1.2$ rod lengths, and $z = 0$. The time coordinate is whatever time appears on the clock closest to A at the instant of the flash.

when the origin clock reads $t = 0$. When the light pulse reaches each helper, that helper sets his or her clock to read $t = r/c$, where r is the distance between the helper and the origin. The clocks are then synchronized.

3. The Spacetime Coordinates. The observer can now assign spacetime coordinates to an event by simply recording the time on a clock at the event and the position as measured on the nearest measuring rods. If there are two events, the observer computes their separation in time as the difference of the times on clocks near each, and their separation in space from the differences of coordinates on rods near each. We thus avoid the practical problem of waiting for signals to reach the observer from events and then calculating the travel times of those signals.

38-4 THE RELATIVITY OF SIMULTANEITY

Suppose that one observer (Sam) notes that two independent events (event Red and event Blue) occur at the same time. Suppose also that another observer (Sally), who is moving at a constant velocity **v** with respect to Sam, also records these same two events. Will Sally also find that they occur at the same time?

The answer is that in general she will not:

If two observers are in relative motion, they will not, in general, agree as to whether two events are simultaneous. If one observer finds them to be simultaneous, the other generally will not, and conversely.

We cannot say that one observer is right and the other wrong. Their observations are equally valid, and there is no reason to favor one over the other.

The realization that two contradictory statements about the same natural event can be correct is a seemingly strange outcome of Einstein's theory. However, in Chapter 18 we discussed another way in which motion can affect measurement, without balking at the contradictory results: in the Doppler effect, the frequency an observer measures for a sound wave depends on the relative motion of the observer and the source. So two observers moving relative to one another can measure different frequencies for the same wave. And both measurements are correct.

We conclude the following:

Simultaneity is not an absolute concept but a relative one, depending on the motion of the observer.

If the relative speed of the observers is very much less than the speed of light, then measured departures from simultaneity are so small that they are not noticeable. Such is the case for all our experiences of daily living; this is why the relativity of simultaneity is unfamiliar.

A Closer Look at Simultaneity

Let us clarify the relativity of simultaneity with an example based on the postulates of relativity, no clocks or measuring rods being directly involved. Figure 38-4 shows two long spaceships (the SS *Sally* and the SS *Sam*), which can serve as inertial reference frames for observers Sally and Sam. The two observers are stationed at the midpoints of their ships. The ships are separating along a common x axis, the relative velocity of *Sally* with respect to *Sam* being **v**. Figure 38-4a shows the ships with the two observer stations momentarily aligned opposite each other.

Two large meteorites strike the ships, one setting off a red flare (event Red) and the other a blue flare (event Blue), not necessarily simultaneously. Each event leaves a permanent mark on each ship, at positions R, R' and B, B'.

Let us suppose that the expanding wavefronts from the two events happen to reach Sam at the same time, as Fig. 38-4c shows. Let us further suppose that, after the episode, Sam finds, by measurement, that he was indeed stationed exactly halfway between the markers B and R on his ship when the two events occurred. He will say:

SAM: Light from event Red and light from event Blue reached me at the same time. From the marks on my spaceship, I find that I was standing halfway between the two sources when the light from them reached me. Therefore event Red and event Blue are simultaneous events.

As study of Fig. 38-4 shows, however, the expanding wavefront from event Red will reach Sally *before* the expanding wavefront from event Blue does. She will say:

SALLY: Light from event Red reached me before light from event Blue did. From the marks on my spaceship, I found that I too was standing halfway between the two sources. Therefore the events were *not* simultaneous; event Red occurred first, followed by event Blue.

These reports do not agree. Nevertheless, *both* observers are correct.

Note carefully that there is only one wavefront expanding from the site of each event and that *this wavefront travels with the same speed c in both reference frames*, exactly as the speed of light postulate requires.

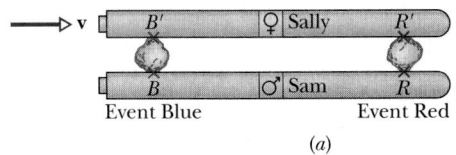

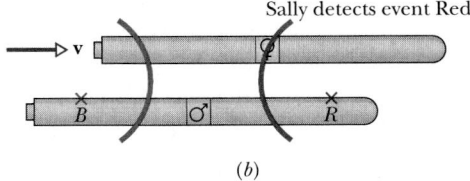

Sally detects event Red

(b)

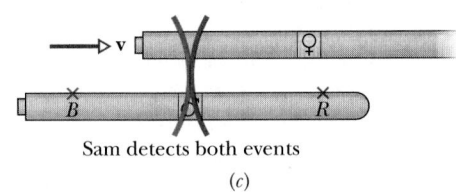

Sam detects both events

(c)

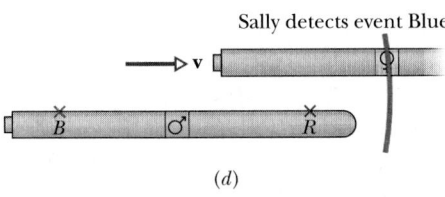

Sally detects event Blue

(d)

FIGURE 38-4 The spaceships of Sally and Sam and the occurrences of events from Sam's view. Sally's ship moves rightward with velocity **v**. (a) Event Red occurs at positions R, R' and event Blue occurs at positions B, B'; each event sends out a wave of light. (b) Sally detects the wave from event Red. (c) Sam simultaneously detects the waves from event Red and event Blue. (d) Sally detects the wave from event Blue.

It *might* have happened that the meteorites struck the ships in such a way that the two hits appeared to Sally to be simultaneous. If that had been the case, then Sam would have declared them not to be simultaneous. The experiences of the two observers are exactly symmetrical.

38-5 THE RELATIVITY OF TIME

If observers who move relative to each other measure the time interval (or *temporal separation*) between two events, they generally will find different results. Why? Because the spatial separation of the events can affect the time intervals measured by the observers.

> The time interval between two events depends on how far apart they are; that is, their spatial and temporal separations are entangled.

In this section we discuss an example of this entanglement by means of an example; however, the example is restricted in a crucial way: *to one of two observers, the two events occur at the same location.* We shall not get to more general examples until Section 38-7.

Figure 38-5a shows the basics of an experiment Sally conducts while she and her equipment ride in a train moving with constant velocity **v** relative to a station. A pulse of light leaves a light source B (event 1), travels vertically upward, is reflected vertically downward by a mirror, and

FIGURE 38-5 (a) Sally, on the train, measures the time interval Δt_0 between events 1 and 2 using a single clock C on the train. That clock is shown twice: first for event 1 and then for event 2. (b) Sam, watching from the station as the events occur, requires two synchronized clocks, C_1 at event 1 and C_2 at event 2, to measure the time interval between the two events; his measured time interval is Δt.

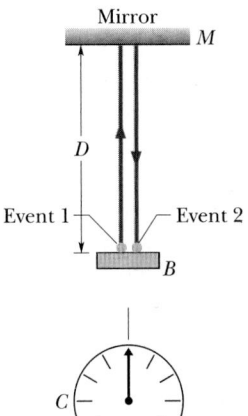

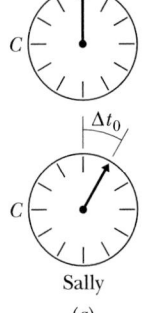

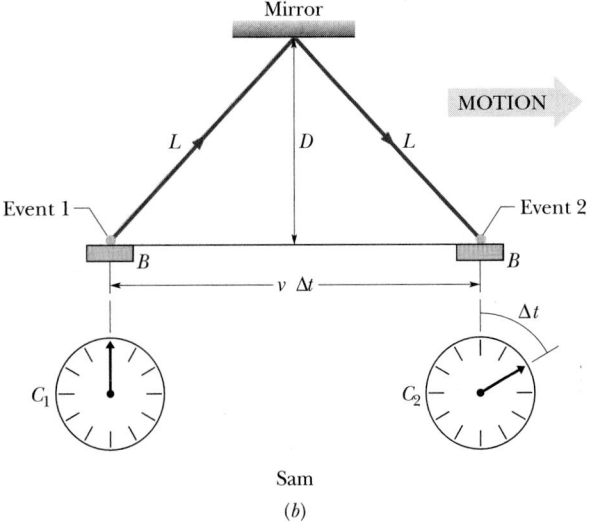

then is detected back at the source (event 2). Sally measures a certain time interval Δt_0 between the two events, related to the distance D from source to mirror by

$$\Delta t_0 = \frac{2D}{c} \quad \text{(Sally).} \quad (38\text{-}2)$$

The two events occur at the same location in Sally's reference frame, and she needs only one clock C at that location to measure the time interval. Clock C is shown twice in Fig. 38-5a, at the beginning and end of the interval.

Consider now how these same two events are measured by Sam, who is standing on the station platform as the train passes. Because the equipment moves with the train during the travel time of the light, Sam sees the path of the light as shown in Fig. 38-5b. For him, the two events occur at different places in his reference frame. So to measure the time interval between events, Sam must use *two* synchronized clocks, C_1 and C_2, one at each event. According to Einstein's speed of light postulate, the light travels at the same speed c for Sam as for Sally. But now, the light travels distance $2L$ between events 1 and 2. The time interval measured by Sam between the two events is

$$\Delta t = \frac{2L}{c} \quad \text{(Sam),} \quad (38\text{-}3)$$

in which

$$L = \sqrt{(\tfrac{1}{2}v\,\Delta t)^2 + D^2}. \quad (38\text{-}4)$$

From Eq. 38-2, we can write this as

$$L = \sqrt{(\tfrac{1}{2}v\,\Delta t)^2 + (\tfrac{1}{2}c\,\Delta t_0)^2}. \quad (38\text{-}5)$$

If we eliminate L between Eqs. 38-3 and 38-5 and solve for Δt, we find

$$\Delta t = \frac{\Delta t_0}{\sqrt{1 - (v/c)^2}}. \quad (38\text{-}6)$$

Equation 38-6 tells us how Sam's measured interval Δt compares with Sally's interval Δt_0. Because v must be less than c, the denominator in Eq. 38-6 must be less than unity. Thus Δt must be greater than Δt_0: Sam measures a *greater* time interval between the two events than does Sally. Sam and Sally have measured the time interval between the *same* two events, but the relative motion between Sam and Sally made their measurements *different*. We conclude that relative motion can change the *rate* at which time passes between two events; the key to this effect is the fact that the speed of light is the same for both observers.

We distinguish between the measurements of Sam and Sally with the following terminology:

When two events occur at the same location in an inertial reference frame, the time interval between them, measured in that frame, is called the **proper time interval** or the **proper time.** Measurements of the same time interval from any other inertial reference frame are always greater.

So Sally measures a proper time interval, and Sam measures a greater time interval. (The term *proper* is unfortunate in that it implies that any other measurement is improper or nonreal. That is just not so.) The increase in the time interval between two events from the proper time interval is called **time dilation.** (To dilate is to expand or stretch; here the time interval is expanded or stretched.)

Often the dimensionless ratio v/c in Eq. 38-6 is replaced with β, called the **speed parameter.** And the dimensionless inverse square root in Eq. 38-6 is often replaced with γ, called the **Lorentz factor:**

$$\gamma = \frac{1}{\sqrt{1 - \beta^2}}. \quad (38\text{-}7)$$

With these replacements, we can rewrite Eq. 38-6 as

$$\Delta t = \gamma\,\Delta t_0 \quad \text{(time dilation).} \quad (38\text{-}8)$$

The speed parameter β is always less than unity and, provided v is not zero, γ is always greater than unity. However, the difference between γ and 1 is not significant unless $v > 0.1c$. Thus, in general, "old relativity" works well enough for $v < 0.1c$, but we must use special relativity for greater values of v. As shown in Fig. 38-6, γ increases rapidly in magnitude as β approaches 1 (as v approaches c). So the greater the relative speed between Sally and Sam, the greater will be the time interval measured by Sam, until at a great enough speed, the interval takes effectively forever.

You might wonder what Sally says about Sam's having measured a greater time interval than she did. His measurement comes as no surprise to her, because to her, he failed to synchronize his clocks C_1 and C_2 in spite of his insistence that he did. Recall that observers in relative motion generally do not agree about simultaneity. Here, Sam insists that his two clocks simultaneously read the same time when event 1 occurred. To Sally, however, Sam's clock C_2 was erroneously set ahead. So when Sam read the time of event 2 on it, to Sally he was reading off a time that

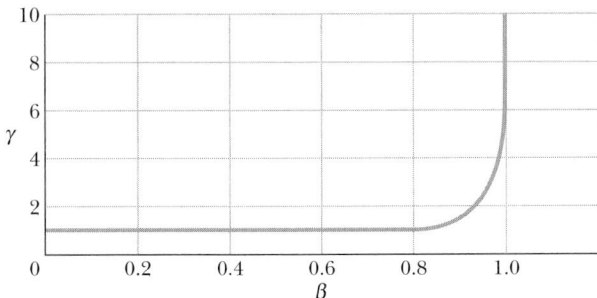

FIGURE 38-6 A plot of the Lorentz factor γ as a function of the speed parameter β ($= v/c$).

was too large, and that is why the time interval he measured between the two events was greater than the interval she measured.

Two Tests of Time Dilation

1. Microscopic Clocks. Subatomic particles called *muons* are unstable; that is, when a muon is produced, it lasts for only a short time before it *decays* (transforms into particles of other types). The *lifetime* of a muon is the time interval between its production (event 1) and its decay (event 2). When muons are stationary and their lifetimes are measured with stationary clocks (say, in a laboratory), their average lifetime is 2.200 μs. This is a proper time interval because, for each muon, events 1 and 2 occur at the same location in the reference frame of the muon. We can represent this proper time interval with Δt_0; moreover, we can call the reference frame in which it is measured the *rest frame* of the muon.

If, instead, the muons are moving, say, through a laboratory, then measurements of their lifetimes made with the laboratory clocks should yield a greater average lifetime (a dilated average lifetime). To check this conclusion, measurements were made of the average lifetime of muons moving with a speed of 0.9994c relative to laboratory clocks. From Eq. 38-7, with $\beta = 0.9994$, the Lorentz factor for this speed is

$$\gamma = \frac{1}{\sqrt{1 - \beta^2}} = \frac{1}{\sqrt{1 - (0.9994)^2}} = 28.87,$$

which is substantially greater than unity. Equation 38-8 then yields, for the average dilated lifetime,

$$\Delta t = \gamma \, \Delta t_0 = (28.87)(2.200 \ \mu s) = 63.5 \ \mu s.$$

The actual measured value matched this result within experimental error.

2. Macroscopic Clocks. In October 1977, Joseph Hafele and Richard Keating carried out what must have been a grueling experiment. They flew four portable atomic clocks twice around the world on commercial airlines, once in each direction. Their purpose was "to test Einstein's theory of relativity with macroscopic clocks." As we have just seen, the time dilation predictions of Einstein's theory have been confirmed on a microscopic scale, but there is great comfort in seeing a confirmation made with an actual clock. Such macroscopic measurements became possible only because of the very high precision of modern atomic clocks. Hafele and Keating verified the predictions of the theory to within 10%. (Einstein's *general* theory of relativity, which predicts that the rate of a clock is influenced by gravitation, also plays a role in this experiment.)

A few years later, physicists at the University of Maryland carried out a similar experiment with improved precision. They flew an atomic clock round and round over Chesapeake Bay for flights lasting 15 h and succeeded in checking the time dilation prediction to better than 1%. Today, when atomic clocks are transported from one place to another for calibration or other purposes, the time dilation caused by their motion is always taken into account.

CHECKPOINT 1: Standing beside railroad tracks, we are suddenly startled by a relativistic boxcar traveling past us as shown in the figure. Inside, a well-equipped hobo fires a laser pulse from the front of the boxcar to its rear. (a) Is our measurement of the speed of the pulse greater than, less than, or the same as that measured by the hobo? (b) Is his measurement of the flight time of the pulse a proper time? (c) Are his measurement and our measurement of the flight time related by Eq. 38-8?

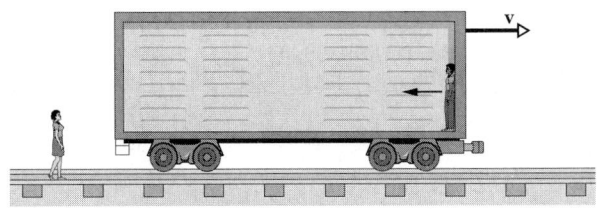

SAMPLE PROBLEM 38-2

Your starship passes Earth with a relative speed of 0.9990c. After traveling 10.0 y (your time), you stop at lookout post LP13, turn, and then travel back to Earth with the same relative speed. The trip back takes another 10.0 y (your time). How long does the round trip take according to measurements made on Earth? (Neglect any effects due to the accelerations involved with stopping and turning.)

SOLUTION: On the journey out, the start and end of the journey occur at the same location in your reference frame, namely on your ship. Hence, you measure proper time Δt_0 for the trip, which is the given 10.0 y. Equation 38-6 gives us the corresponding time Δt as measured in the Earth reference frame:

$$\Delta t = \frac{\Delta t_0}{\sqrt{1 - (v/c)^2}}$$

$$= \frac{10.0 \text{ y}}{\sqrt{1 - (0.9990c/c)^2}} = (22.37)(10.0 \text{ y}) = 224 \text{ y}.$$

On the journey back, we have the same situation and the same data. Thus the round trip requires 20 y of your time but

$$\Delta t_{\text{total}} = (2)(224 \text{ y}) = 448 \text{ y} \qquad \text{(Answer)}$$

of Earth time. In other words, you have aged 20 y while the Earth has aged 448 y. Although you cannot travel into the past (as far as we know), you can travel into the future of, say, Earth, by using high-speed relative motion to adjust the rate at which time passes.

SAMPLE PROBLEM 38-3

The elementary particle known as the *positive kaon* (K$^+$) has, on average, a lifetime of 0.1237 μs when stationary, that is, when the lifetime is measured in the rest frame of the kaon. If positive kaons with a speed of 0.990c relative to a laboratory reference frame are produced, how far can they travel in that frame during their lifetime?

SOLUTION: In the laboratory frame, the distance d traveled by a kaon is related to its speed v (= 0.990c) and its travel time Δt_k by $d = v \Delta t_k$. (This statement does not involve relativity because all quantities are measured in the same reference frame.) If special relativity did not apply, the travel time would be just the 0.1237 μs lifetime of the particle, and thus the travel distance would be just

$$d = v \Delta t_k = (0.990)(3.00 \times 10^8 \text{ m/s})(1.237 \times 10^{-7} \text{ s})$$

$$= 36.7 \text{ m.} \qquad \text{(Wrong Answer)}$$

However, special relativity does apply and the travel time of the kaon in the laboratory frame is its dilated lifetime Δt. With Eq. 38-6, we can find Δt from the kaon's proper lifetime Δt_0 (= 0.1237 μs), as measured in its rest frame:

$$\Delta t = \frac{\Delta t_0}{\sqrt{1 - (v/c)^2}}$$

$$= \frac{0.1237 \times 10^{-6} \text{ s}}{\sqrt{1 - (0.990c/c)^2}} = 8.769 \times 10^{-7} \text{ s.}$$

This is about seven times longer than the kaon's proper lifetime. (This calculation does involve relativity because we must transform data from the particle's rest frame to the laboratory frame.) We can now find the travel distance in the laboratory frame as

$$d = v \Delta t_k = v \Delta t$$

$$= (0.990)(3.00 \times 10^8 \text{ m/s})(8.769 \times 10^{-7} \text{ s})$$

$$= 260 \text{ m.} \qquad \text{(Answer)}$$

This is about seven times our first (wrong) answer. Experiments like the one outlined here, which verify special relativity, became routine in physics laboratories decades ago.

38-6 THE RELATIVITY OF LENGTH

If you want to measure the length of a rod that is at rest with respect to you, you can—at your leisure—note the positions of its end points on a long stationary scale and subtract the two readings. If the rod is moving, however, you must note the positions of the end points *simultaneously* (in your reference frame) or your measurement cannot be called a length. Figure 38-7 suggests the difficulty of trying to measure the length of a moving penguin by locating its front and back at different times. Because simultaneity is relative and it enters into length measurements, length should also be a relative quantity.

Let L_0 be the length of a rod that you measure when the rod is stationary (meaning you and it are in the same

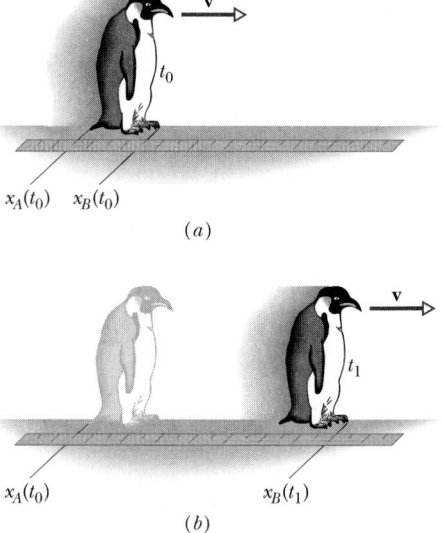

(a)

(b)

FIGURE 38-7 If you want to measure the front-to-back length of a penguin while it is moving, you must mark the positions of its front and back simultaneously (in your reference frame), as in (a), rather than at different times, as in (b).

reference frame, the rod's rest frame). If, instead, there is relative motion at speed v between you and the rod *along the length of the rod,* you measure a length L given by

$$L = L_0 \sqrt{1 - \beta^2} = \frac{L_0}{\gamma} \qquad \text{(length contraction).} \qquad (38\text{-}9)$$

Because the Lorentz factor γ is always greater than unity if there is relative motion, L is less than L_0. The relative motion causes a *length contraction,* and L is called a *contracted length.* Because γ increases with speed v, the length contraction also increases with the relative speed.

The length L_0 of an object measured in the rest frame of the object is its **proper length** or **rest length**. Measurements of the length from any reference frame that is in relative motion parallel to that length are always less than the proper length.

Be careful: length contraction occurs only along the direction of relative motion. Also, the length that is measured does not have to be that of an object like a rod or a circle. Instead, it can be the length (or distance) between two objects in the same rest frame—for example, the Sun and a nearby star (which are, at least approximately, at rest relative to each other).

Does the object *really* shrink? Reality is based on observations and measurements; if the results are always consistent and if no error can be determined, then what is observed and measured is real. In that sense, the object really does shrink. However, a more precise statement is that the object *is really measured* to shrink—motion affects that measurement and thus reality.

Can a high-speed photograph show a contracted object? No, because what it records is not limited to the light emitted by an object at one specific instant (simultaneously). Instead it records all light from the object that happens to arrive at the camera at the instant of exposure, regardless of when the light was emitted.

When you measure a contracted length for, say, a rod, what does an observer moving with the rod say of your measurement? To that observer, you did not locate the two ends of the rod simultaneously. (Recall that observers in motion relative to each other do not agree about simultaneity.) To the observer, you first located the rod's front end and then, slightly later, its rear end, and that is why you measured a length that is less than the proper length.

Proof of Eq. 38-9

Length contraction is a direct consequence of time dilation. Consider once more our two observers. Both Sally, seated on a train moving through a station, and Sam, again on the station platform, want to measure the length of the platform. Sam, using a tape measure, finds the length to be L_0, a proper length because the platform is at rest with respect to him. Sam also notes that Sally, on the train, moves through this length in a time $\Delta t = L_0/v$, where v is the speed of the train. That is,

$$L_0 = v\,\Delta t \qquad \text{(Sam).} \qquad (38\text{-}10)$$

This time interval Δt is not a proper time interval because the two events that define it (Sally passes back of platform and Sally passes front of platform) occur at two different places and Sam must use two synchronized clocks to measure the time interval Δt.

For Sally, however, the platform is moving. She finds that the two events measured by Sam occur *at the same place* in her reference frame. She can time them with a single stationary clock, so the interval Δt_0 that she measures is a proper time interval. To her, the length L of the platform is given by

$$L = v\,\Delta t_0 \qquad \text{(Sally).} \qquad (38\text{-}11)$$

If we divide Eq. 38-11 by Eq. 38-10 and apply Eq. 38-8, the time dilation equation, we have

$$\frac{L}{L_0} = \frac{v\,\Delta t_0}{v\,\Delta t} = \frac{1}{\gamma},$$

or

$$L = \frac{L_0}{\gamma}, \qquad (38\text{-}12)$$

which is Eq. 38-9, the length contraction equation.

SAMPLE PROBLEM 38-4

In Fig. 38-8, Sally (at point A) and Sam's spaceship (of proper length $L_0 = 230$ m) pass each other with constant relative speed v. Sally measures a time interval of 3.57 μs for the ship to pass her (from the passage of point B to the passage of point C). What is the speed parameter β between Sally and the ship?

SOLUTION: If the relative speed v between Sally and Sam were, say, less than $0.1c$, we might have seen this situation in Chapter 2, where we would have said that a ship of length L and speed v passes Sally in a time interval

$$\Delta t = \frac{L}{v}.$$

(No relativity is involved in this statement.)

But here we probably have a relativistic problem, with $v > 0.1c$. In that case, we know that the length L that Sally would measure is not the proper length L_0 of the ship, but a contracted length, given by Eq. 38-9:

$$L = L_0 \sqrt{1 - \beta^2}.$$

(This statement involves relativity, because we are transforming data between Sam's frame and Sally's frame.) According to Sally, the time required for the passage is now written as

$$\Delta t = \frac{\text{contracted length } L}{v} = \frac{L_0 \sqrt{(1 - \beta^2)}}{\beta c}.$$

Solving for β and then substituting the given data, we find, after a little algebra,

$$\beta = \frac{L_0}{\sqrt{(c \, \Delta t)^2 + L_0^2}}$$

$$= \frac{230 \text{ m}}{\sqrt{(3.00 \times 10^8 \text{ m/s})^2 (3.57 \times 10^{-6} \text{ s})^2 + (230 \text{ m})^2}}$$

$$= 0.210. \qquad \text{(Answer)}$$

Thus the relative speed between Sally and the ship is 21% of the speed of light. Note that only the relative motion of Sally and Sam matters here; whether either is stationary relative to, say, a space station is irrelevant. In Fig. 38-8 we took Sally to be stationary, but we could instead have taken the ship to be stationary, with Sally flying past it. Nothing would have changed in our results.

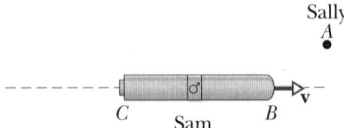

FIGURE 38-8 Sample Problem 38-4. Sally measures how long a spaceship takes to pass her at point A.

CHECKPOINT **2:** In Sample Problem 38-4, Sally measures the passage time of the ship. If Sam does also, (a) which measurement, if either, is a proper time and (b) which measurement is smaller?

SAMPLE PROBLEM 38-5

Caught by surprise near a supernova, you race away from the explosion in your spaceship, hoping to outrun the high-speed material ejected toward you. Your Lorentz factor relative to the inertial reference frame of the local stars is 22.4.

(a) To reach a safe distance, you figure you need to cover 9.00×10^{16} m as measured in the reference frame of the local stars. How long will the flight take, as measured in that frame?

SOLUTION: The length $L_0 = 9.00 \times 10^{16}$ m is a proper length in the reference frame of the local stars, because its two ends are at rest in that frame. Figure 38-6 tells us that with such a large Lorentz factor, your speed relative to the local stars is $v \approx c$. So, with that approximation, to move through

length L_0 requires the time

$$\Delta t = \frac{L_0}{v} = \frac{L_0}{c}$$

$$= \frac{9.00 \times 10^{16} \text{ m}}{3.00 \times 10^8 \text{ m/s}} = 3.00 \times 10^8 \text{ s} = 9.49 \text{ y.} \quad \text{(Answer)}$$

(b) How long does that trip take according to you (in your reference frame)?

SOLUTION: From your reference frame, the distance you cover is a contracted length L that races past you at relative speed $v \approx c$. Equation 38-9 tells us that $L = L_0/\gamma$. So, the time you measure for the passage of that contracted length is

$$\Delta t_0 = \frac{L}{v} = \frac{L_0/\gamma}{v} = \frac{L_0}{c\gamma}$$

$$= \frac{9 \times 10^{16} \text{ m}}{(3.00 \times 10^8 \text{ m/s})(22.4)}$$

$$= 1.339 \times 10^7 \text{ s} = 0.424 \text{ y.} \quad \text{(Answer)}$$

This is a proper time, because the start and end of the passage occur at the same point in your reference frame (at your ship). You can check the validity of the two answers by substituting them into Eq. 38-8 (for time dilation) and solving for γ.

38-7 THE LORENTZ TRANSFORMATION

As Fig. 38-9 shows, inertial reference frame S' is moving with speed v relative to frame S, in the common positive direction of their horizontal axes (marked x and x'). An observer in S reports spacetime coordinates x, y, z, t for an event, and an observer in S' reports x', y', z', t' for the same event. How are these sets of numbers related?

We claim at once (although it requires proof) that the y and z coordinates, which are perpendicular to the motion, are not affected by the motion. That is, $y = y'$ and $z = z'$. Our interest then reduces to the relation between x and x' and between t and t'.

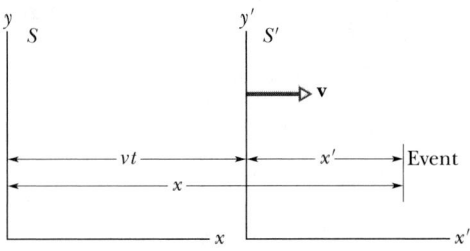

FIGURE 38-9 Two inertial reference frames: frame S' has velocity **v** relative to frame S.

The Galilean Transformation Equations

Prior to Einstein's publication of his special theory of relativity, the four coordinates of interest were assumed to be related by the *Galilean transformation equations:*

$$x' = x - vt$$

$$t' = t$$

(Galilean transformation equations; approximately valid at low speeds). (38-13)

These equations are written with the assumption that $t = t' = 0$ when the origins of S and S' coincide. You can verify the first equation with Fig. 38-9. The second equation effectively claims that time passes at the same rate for observers in both reference frames. That would have been so obviously true to a scientist prior to Einstein that it would not even have been mentioned. When speed v is small compared to c, Eqs. 38-13 generally work well.

The Lorentz Transformation Equations

We state without proof that the correct transformation equations, which remain valid for all speeds up to the speed of light, can be derived from the postulates of relativity. The results, called the **Lorentz transformation equations***, are

$$x' = \gamma(x - vt),$$

$$y' = y,$$

$$z' = z,$$

$$t' = \gamma(t - vx/c^2)$$

(Lorentz transformation equations; valid at all speeds). (38-14)

Note the spatial values x and the temporal values t are bound together in the first and last equations. This entanglement of space and time was a prime message of Einstein's theory, a message that was long rejected by many of his contemporaries.

It is a formal requirement of relativistic equations that they should reduce to familiar classical equations if we let c approach infinity. That is, if the speed of light were infinitely great, *all* finite speeds would be "low" and classical equations would never fail. If we let $c \to \infty$ in Eqs. 38-14, $\gamma \to 1$ and these equations reduce—as we expect—to the Galilean equations (Eqs. 38-13). You should check this.

Equations 38-14 are written in a form that is useful if we are given x and t and wish to find x' and t'. We may

---

*You may wonder why we do not call these the *Einstein transformation equations* (and why not the *Einstein factor* for γ). H. A. Lorentz actually derived these equations before Einstein did, but as the great Dutch physicist graciously conceded, he did not take the further bold step of interpreting these equations as describing the true nature of space and time. It is this interpretation, first made by Einstein, that is at the heart of relativity.

TABLE 38-2 THE LORENTZ TRANSFORMATION EQUATIONS FOR PAIRS OF EVENTS

1. $\Delta x = \gamma(\Delta x' + v\,\Delta t')$	1ʹ. $\Delta x' = \gamma(\Delta x - v\,\Delta t)$
2. $\Delta t = \gamma(\Delta t' + v\,\Delta x'/c^2)$	2ʹ. $\Delta t' = \gamma(\Delta t - v\,\Delta x/c^2)$

$$\gamma = \frac{1}{\sqrt{1 - (v/c)^2}} = \frac{1}{\sqrt{1 - \beta^2}}$$

wish to go the other way, however. In that case we simply solve Eqs. 38-14 for x and t, obtaining

$$x = \gamma(x' + vt'),$$

$$t = \gamma(t' + vx'/c^2).$$

(38-15)

Comparison shows that, starting from either Eqs. 38-14 or Eqs. 38-15, you can find the other set by interchanging primed and unprimed quantities and reversing the sign of the relative velocity v.

Equations 38-14 and 38-15 relate the coordinates of a single event as seen by two observers. Sometimes we want to know not the coordinates of a single event but the differences between coordinates for a pair of events. That is, if we label our events 1 and 2, we may want to relate

$$\Delta x = x_2 - x_1 \quad \text{and} \quad \Delta t = t_2 - t_1,$$

as measured by an observer in S, and

$$\Delta x' = x'_2 - x'_1 \quad \text{and} \quad \Delta t' = t'_2 - t'_1,$$

as measured by an observer in S'.

Table 38-2 displays the Lorentz equations in difference form, suitable for analyzing pairs of events. The equations in the table were derived by simply substituting differences (such as Δx and $\Delta x'$) for the four variables in Eqs. 38-14 and 38-15.

Be careful: when substituting values for these differences, you must be consistent and not mix the values for the first event with those for the second event. And if, say, Δx is a negative quantity, you must be certain to include the minus sign.

CHECKPOINT 3: The following figure shows three situations in which a blue reference frame and a green reference frame are in relative motion along the common direction of their x and x' axes, as indicated by the velocity vector attached to one of the frames. For each situation, if we choose the blue frame to be stationary, then is v in the equations of Table 38-2 a positive or negative quantity?

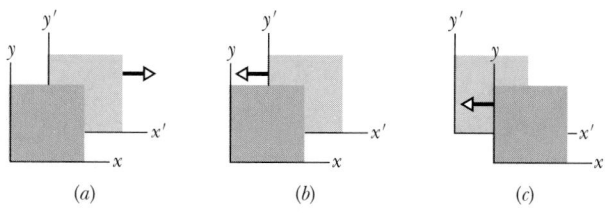

(a)   (b)   (c)

38-8 SOME CONSEQUENCES OF THE LORENTZ EQUATIONS

Here we use the transformation equations of Table 38-2 to affirm some of the conclusions that we reached earlier by arguments based directly on the postulates.

Simultaneity

Consider Eq. 2 of Table 38-2,

$$\Delta t = \gamma\left(\Delta t' + \frac{v\,\Delta x'}{c^2}\right). \qquad (38\text{-}16)$$

If two events occur at different places in reference frame S' of Fig. 38-9, then $\Delta x'$ in this equation is not zero. It follows that even if the events are simultaneous in S' (so $\Delta t' = 0$), they will not be simultaneous in frame S. The time interval between them in S will be

$$\Delta t = \gamma\frac{v\,\Delta x'}{c^2} \qquad \text{(simultaneous events in } S').$$

This is in accord with our conclusion in Section 38-4.

Time Dilation

Suppose now that two events occur at the same place in S' (so $\Delta x' = 0$) but at different times (so $\Delta t' \neq 0$). Equation 38-16 then reduces to

$$\Delta t = \gamma\,\Delta t' \qquad \begin{array}{l}\text{(events in same}\\ \text{place in } S').\end{array} \qquad (38\text{-}17)$$

This confirms time dilation. Because the two events occur at the same place in S', the time interval $\Delta t'$ between them can be measured with a single clock, located at that place. Under these conditions, the measured interval is a proper time interval, and we can label it Δt_0. Thus Eq. 38-17 becomes

$$\Delta t = \gamma\,\Delta t_0 \qquad \text{(time dilation)},$$

which is exactly Eq. 38-8, the time dilation equation.

Length Contraction

Consider Eq. 1' of Table 38-2,

$$\Delta x' = \gamma(\Delta x - v\,\Delta t). \qquad (38\text{-}18)$$

If a rod lies parallel to the x,x' axes of Fig. 38-9 and is at rest in reference frame S', an observer in S' can measure its length at leisure. The value $\Delta x'$ that is obtained by subtracting the coordinates of the end points of the rod will be its proper length L_0.

Suppose the rod is moving in frame S. This means that Δx can be identified as the length L of the rod only if the coordinates of the end points are measured *simultaneously*, that is, if $\Delta t = 0$. If we put $\Delta x' = L_0$, $\Delta x = L$, and $\Delta t = 0$ in Eq. 38-18, we find

$$L = \frac{L_0}{\gamma} \qquad \text{(length contraction)}, \qquad (38\text{-}19)$$

which is exactly Eq. 38-9, the length contraction equation.

SAMPLE PROBLEM 38-6

An Earth starship has been sent to check an Earth outpost on the planet P1407, whose moon houses a battle group of the often hostile Reptulians. As the ship follows a straight-line course first past the planet and then past the moon, it detects a high-energy microwave burst at the Reptulian moon base and then, 1.10 s later, an explosion at the Earth outpost, which is 4.00×10^8 m from the Reptulian base as measured from the ship's reference frame. The Reptulians have obviously attacked the Earth outpost; so the starship begins to prepare for a confrontation with them.

(a) The speed of the ship relative to the planet and its moon is $0.980c$. What are the distance between the burst and the explosion and the time interval between them as measured in the planet–moon inertial frame (and thus according to the occupants of the stations)?

SOLUTION: The situation is shown in Fig. 38-10, where the ship's frame S is chosen to be stationary and the planet–moon frame S' is chosen to be moving with positive velocity (rightward). (This is an arbitrary choice: we could, instead, choose the planet–moon frame to be stationary. Then we would redraw $\mathbf{v}$ in Fig. 38-10 as attached to the S frame and being leftward; v would then be a negative quantity. The result would be the same.) Let subscripts e and b represent the explosion and burst, respectively. Then the given data, all in the unprimed reference frame, are

$$\Delta x = x_e - x_b = +4.00 \times 10^8 \text{ m}$$

and

$$\Delta t = t_e - t_b = +1.10 \text{ s}.$$

Here Δx is a positive quantity because in Fig. 38-10, the coordinate x_e for the explosion is greater than the coordinate x_b for the burst; Δt is also a positive quantity because the time t_e of the explosion is greater (later) than the time t_b of the burst.

We seek $\Delta x'$ and $\Delta t'$, which we shall get by transforming the given S-frame data to the planet–moon frame S'. Because we are considering a pair of events, we choose transformation equations from Table 38-2, namely Eqs. 1′ and 2′:

$$\Delta x' = \gamma(\Delta x - v\,\Delta t) \qquad (38\text{-}20)$$

and

$$\Delta t' = \gamma\left(\Delta t - \frac{v\,\Delta x}{c^2}\right). \qquad (38\text{-}21)$$

Here, $v = +0.980c$, and the Lorentz factor is

$$\gamma = \frac{1}{\sqrt{1-(v/c)^2}} = \frac{1}{\sqrt{1-(+0.980c/c)^2}} = 5.0252.$$

So, Eq. 38-20 becomes

$$\Delta x' = (5.0252)$$
$$\times\,[4.00\times10^8\text{ m} - (+0.980)(3.00\times10^8\text{ m/s})(1.10\text{ s})]$$
$$= 3.85\times10^8\text{ m}, \qquad \text{(Answer)}$$

and Eq. 38-21 becomes

$$\Delta t' = (5.0252)$$
$$\times\left[(1.10\text{ s}) - \frac{(+0.980)(3.00\times10^8\text{ m/s})(4.00\times10^8\text{ m})}{(3.00\times10^8\text{ m/s})^2}\right]$$
$$= -1.04\text{ s}. \qquad \text{(Answer)}$$

(b) What is the meaning of the minus sign in the computed value for $\Delta t'$?

SOLUTION: Recall how we originally defined the time interval between burst and explosion: $\Delta t = t_e - t_b = +1.10$ s. To be consistent with that choice of notation, our definition of $\Delta t'$ must be $t'_e - t'_b$; thus, we have found that

$$\Delta t' = t'_e - t'_b = -1.04\text{ s}.$$

This tells us that $t'_b > t'_e$; that is, in the planet–moon reference frame, the burst occurs 1.04 s *after* the explosion, not 1.10 s *before* the explosion as detected in the ship frame.

(c) Did the burst cause the explosion, or did the explosion cause the burst?

SOLUTION: The sequence of events measured in the planet–moon reference frame is the reverse of that measured in the ship frame. In either situation, if there is a causal relationship between the two events, information must travel from one event to cause the other. Let us check the required speed of the information. In the ship frame, this speed is

$$v_{\text{info}} = \frac{\Delta x}{\Delta t} = \frac{4.00\times10^8\text{ m}}{1.10\text{ s}} = 3.64\times10^8\text{ m/s},$$

but that speed is impossible because it exceeds c. In the planet–moon frame, the speed comes out to be 3.70×10^8 m/s, also impossible. So, neither event could possibly cause the other event; that is, they are *unrelated* events. Thus the starship should not confront the Reptulians.

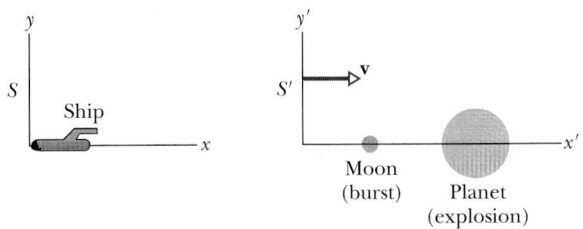

FIGURE 38-10 Sample Problem 38-6. A planet and its moon in reference frame S' move with speed v relative to a starship in reference frame S.

SAMPLE PROBLEM 38-7

Figure 38-11 shows an inertial reference frame S in which event 1 (a rock is kicked up by a truck at coordinates x_1 and t_1) causes event 2 (the rock hits you at coordinates x_2 and t_2). Is there another inertial reference frame S' from which those events can be measured to be reversed in sequence, so that the effect occurs before its cause? (Can you be injured now as a result of a future event?)

SOLUTION: To find the temporal separation $\Delta t'$ of a pair of events in frame S' when we have data for frame S, we use Eq. 2′ of Table 38-2:

$$\Delta t' = \gamma\left(\Delta t - \frac{v\,\Delta x}{c^2}\right). \qquad (38\text{-}22)$$

Recall that v is the relative velocity between S and S'. We take frame S to be stationary; frame S' then has velocity v.

Let $\Delta t = t_2 - t_1$. Then Δt is a positive quantity and, to be consistent with this notation, we must have $\Delta x = x_2 - x_1$ and $\Delta t' = t'_2 - t'_1$. As Fig. 38-11 is drawn, Δx is a positive quantity because $x_2 > x_1$.

We are interested in the possibility that $\Delta t'$ is a negative quantity, which would mean that time t'_1 of event 1 is later (and thus greater) than time t'_2 of event 2. From Eq. 38-22, we see that $\Delta t'$ can be negative only if

$$\frac{v\,\Delta x}{c^2} > \Delta t.$$

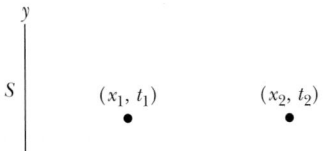

FIGURE 38-11 Sample Problem 38-7. Event 1 at spacetime coordinates (x_1, t_1) causes event 2 at spacetime coordinates (x_2, t_2). Can the sequence of cause and effect be reversed in some other reference frame?

This condition can be rearranged to produce the equivalent condition

$$\frac{\Delta x/\Delta t}{c}\frac{v}{c} > 1.$$

The ratio $\Delta x/\Delta t$ is just the speed at which information (here via a rock) travels from event 1 to produce event 2. That speed cannot exceed c. (Information could travel at c if it comes via light; rocks travel more slowly, of course.) So $(\Delta x/\Delta t)/c$ must be at most 1. And v/c cannot equal or exceed 1. Thus the left side of the last inequality must be less than 1, and the inequality cannot be satisfied.

So there is no frame S' in which event 2 occurs before its cause, event 1. More generally, although the sequence of unrelated events can sometimes be reversed in relativity (as in Sample Problem 38-6), events involving cause and effect can never be reversed.

38-9 THE RELATIVITY OF VELOCITIES

Here we wish to use the Lorentz transformation equations to compare the velocities that two observers in different inertial reference frames S and S' would measure for the same moving particle. We assume again that S' moves with velocity v relative to S.

Suppose that the particle, moving with constant velocity parallel to the x, x' axes in Fig. 38-12, sends out two signals as it moves. Each observer measures the space interval and the time interval between these two events. These four measurements are related by Eqs. 1 and 2 of Table 38-2,

$$\Delta x = \gamma(\Delta x' + v\,\Delta t')$$

and

$$\Delta t = \gamma\left(\Delta t' + \frac{v\,\Delta x'}{c^2}\right).$$

If we divide the first of these equations by the second, we find

$$\frac{\Delta x}{\Delta t} = \frac{\Delta x' + v\,\Delta t'}{\Delta t' + v\,\Delta x'/c^2}.$$

Dividing the numerator and denominator of the right side

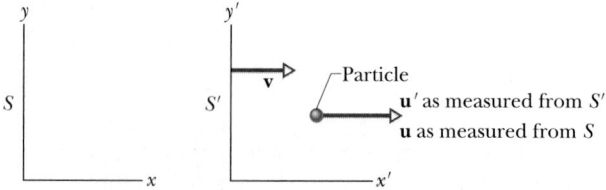

FIGURE 38-12 Reference frame S' moves with velocity $\mathbf{v}$ relative to frame S. A particle has velocity $\mathbf{u}'$ relative to reference frame S' and velocity $\mathbf{u}$ relative to reference frame S.

by $\Delta t'$, we find

$$\frac{\Delta x}{\Delta t} = \frac{\Delta x'/\Delta t' + v}{1 + v(\Delta x'/\Delta t')/c^2}.$$

But, in the differential limit, $\Delta x/\Delta t$ is u, the velocity of the particle as measured in S, and $\Delta x'/\Delta t'$ is u', the velocity of the particle as measured in S'. Then we have, finally,

$$u = \frac{u' + v}{1 + u'v/c^2} \qquad \begin{array}{l}\text{(velocity}\\ \text{transformation)}\end{array} \qquad (38\text{-}23)$$

as the relativistic velocity transformation equation. We discussed this equation, using a different notation, in Section 4-10. You may wish to reread that section and, in particular, to study Sample Problems 4-15 and 4-16. Equation 38-23 reduces to the classical, or Galilean, velocity transformation equation,

$$u = u' + v \qquad \begin{array}{l}\text{(classical velocity}\\ \text{transformation),}\end{array} \qquad (38\text{-}24)$$

when we apply the formal test of letting $c \to \infty$.

38-10 DOPPLER EFFECT FOR LIGHT

In Section 18-8 we discussed the Doppler effect (a shift in detected frequency) for sound waves traveling in air. For such waves, the Doppler effect depends on two velocities, namely, the velocities with respect to the air, which is the medium that transmits the waves, of the source and the detector.

That is not the situation with light waves, for they (and other electromagnetic waves) require no medium, being able to travel even through vacuum. The Doppler effect for light waves depends on only one velocity, the relative velocity $\mathbf{v}$ between source and detector, as measured from the reference frame of either. Let f_0 represent the **proper frequency** of the source, that is, the frequency that is measured by an observer in the rest frame of the source. Let f represent the frequency detected by an observer moving with velocity $\mathbf{v}$ relative to that rest frame. Then, when the direction of $\mathbf{v}$ is directly away from the source,

$$f = f_0\sqrt{\frac{1 - \beta}{1 + \beta}} \qquad \begin{array}{l}\text{(source and}\\ \text{detector separating),}\end{array} \qquad (38\text{-}25)$$

where $\beta = v/c$. When the direction of $\mathbf{v}$ is directly toward the source, we must change the signs in front of both β symbols in Eq. 38-25.

According to Eq. 38-25, when the separation between source and detector is increasing, the detected frequency f is less than the proper frequency f_0. Recalling that $f = c/\lambda$, where λ is the wavelength of the light, we see that the decrease in frequency corresponds to an increase in wavelength. In Section 18-9 we called such an increase in wavelength a *red shift* (because the red portion of the visible spectrum has the longest wavelengths). Similarly, when the source–detector separation is decreasing, f is greater than f_0; this corresponds to a decrease in wavelength, that is, a *blue shift*.

For low speeds ($\beta \ll 1$), Eq. 38-25 can be expanded in a power series in β and approximated as

$$f = f_0(1 - \beta + \tfrac{1}{2}\beta^2) \qquad \text{(low speeds)}. \quad (38\text{-}26)$$

The corresponding low-speed equation for sound waves (or any waves except light waves) has the same first two terms but a different coefficient in the third term. Thus the relativistic effect for low-speed light sources and detectors shows up only with the β^2 term.

As discussed briefly in Chapter 18, police radar units employ the Doppler effect with microwaves. A source in the radar unit emits a microwave signal at a certain frequency f_0 along the road. A car that is moving toward the unit intercepts a microwave signal whose frequency is shifted up to the frequency f in Eq. 38-25 (with the signs of β changed) by the Doppler effect. The car reflects that wave back toward the radar unit. Because the car is moving toward the radar unit, the detector in the unit intercepts a reflected signal that is further shifted up in frequency. The unit compares that detected frequency with f_0 and computes the speed of the car.

CHECKPOINT **4:** The figure shows a source that emits light of proper frequency f_0 while moving directly toward the right with speed $c/4$ as measured from reference frame S. The figure also shows a light detector, which measures a frequency $f > f_0$ for the emitted light. (a) Is the detector moving toward the left or the right? (b) Is the speed of the detector as measured from reference frame S more than $c/4$, less than $c/4$, or equal to $c/4$?

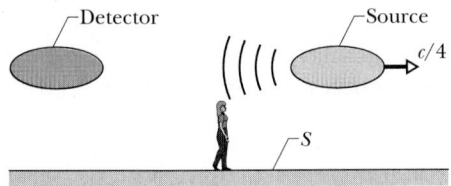

Transverse Doppler Effect

So far, we have discussed the Doppler effect, here and in Chapter 18, only for situations in which the source and the detector move either directly toward or directly away from each other. Figure 38-13 shows a different arrangement, in

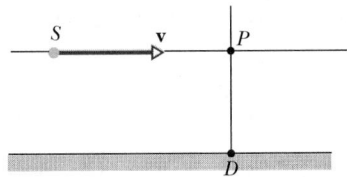

FIGURE 38-13 A light source S travels with velocity **v** past a detector at D. The special theory of relativity predicts a transverse Doppler effect as the source passes through point P, where the direction of travel is then perpendicular to the line extending through D. Classical theory predicts no such effect.

which a source S moves past a detector D. When S reaches point P, its velocity is perpendicular to the line joining S and D and, at that instant, it is moving neither toward nor away from D. If the source is emitting sound waves of frequency f, D detects that frequency (with no Doppler effect) when it intercepts the waves that were emitted at point P. However, if the source is emitting light waves, there is still a Doppler effect, called the **transverse Doppler effect.** In this situation, the detected frequency of light emitted by the source at point P is

$$f = f_0\sqrt{1 - \beta^2} \qquad \text{(transverse Doppler effect)}. \quad (38\text{-}27)$$

For low speeds ($\beta \ll 1$), Eq. 38-27 can be expanded in a power series in β and approximated as

$$f = f_0(1 - \tfrac{1}{2}\beta^2) \qquad \text{(low speeds)}. \quad (38\text{-}28)$$

Here the first term is what we would expect for sound waves and, again, the relativistic effect for low-speed light sources and detectors appears with the β^2 term.

In principle, a police radar unit can determine the speed of a car even when the path of the radar pulse is perpendicular (transverse) to the path of the car. However, Eq. 38-28 tells us that because β is small even for a fast car, the relativistic term $\beta^2/2$ in the transverse Doppler effect is extremely small. So $f \approx f_0$ and the radar unit computes a speed of zero. For this reason, police officers always try to direct the radar pulse along the car's path to get a Doppler shift that gives the car's actual speed. Any deviation from that alignment works in favor of the motorist, because it reduces the measured speed.

The transverse Doppler effect is really another test of time dilation. If we rewrite Eq. 38-27 in terms of the period T of oscillation of the emitted light wave instead of the frequency, we have, since $T = 1/f$,

$$T = \frac{T_0}{\sqrt{1 - \beta^2}} = \gamma T_0, \quad (38\text{-}29)$$

in which $T_0 \, (= 1/f_0)$ is the **proper period** of the source. As comparison with Eq. 38-8 shows, Eq. 38-29 is simply the time dilation formula, since a period is a time interval.

The NAVSTAR Navigation System

Each NAVSTAR satellite continually broadcasts radio signals giving its location, at a frequency that is set and controlled by precise atomic clocks. When the signal is sensed by the detector on, say, an aircraft, the frequency has been Doppler-shifted. By detecting the signals from several NAVSTAR satellites simultaneously, the detector can determine the direction to any one of them and the direction of the velocity of that satellite. From the Doppler shift of the signal, the detector then determines the speed of the aircraft.

Let us use some rough numbers to see how well this can be done. The speed of a NAVSTAR satellite relative to the center of Earth is about 1.0×10^4 m/s. The associated β is about 3.0×10^{-5}. Thus the term $\beta^2/2$ in Eqs. 38-26 and 38-28 (that is, the relativity term) is about 4.5×10^{-10}. In other words, relativity changes the Doppler shift of the detected signal by about 4.5 parts in 10^{10}, which seems hardly worth considering.

However, it is indeed important. The atomic clocks in the satellites are so precise that the variation in the frequency of the satellite signal is only 2 parts in 10^{12}. From Eq. 38-28, we see that β (hence v) depends on the square root of f/f_0. Thus the clock's frequency variation of 2×10^{-12} causes a variation of

$$\sqrt{2 \times 10^{-12}} = 1.4 \times 10^{-6}$$

in the measured value of the relative speed v between satellite and aircraft.

Since v is due primarily to the satellite's great speed, 1.0×10^4 m/s, this means that v (hence the aircraft's speed) can be determined to an accuracy of about

$$(1.4 \times 10^{-6})(1.0 \times 10^4 \text{ m/s}) = 1.4 \text{ cm/s}.$$

Suppose the aircraft flies for 1 h (3600 s). Knowing the speed to about 1.4 cm/s allows the location at the end of that hour to be predicted to about

$$(0.014 \text{ m/s})(3600 \text{ s}) = 50 \text{ m},$$

which is acceptable in modern navigation.

If relativity effects were not taken into account, the speed of the aircraft could not be known any closer than 21 cm/s, and its location after an hour's flight could not be predicted any better than within 760 m.

38-11 A NEW LOOK AT MOMENTUM

Suppose that a number of observers, each in a different inertial reference frame, watch an isolated collision between two particles. In classical mechanics, we have seen that—even though the observers measure different velocities for the colliding particles—they all find that the law of conservation of momentum holds. That is, they find that the momentum of the system of particles after the collision is the same as it was before the collision.

How is this situation affected by relativity? We find that if we continue to define the momentum **p** of a particle as $m\mathbf{v}$, the product of its mass and its velocity, momentum is *not* conserved for all inertial observers. We have two choices: (1) give up the law of conservation of momentum or (2) see if we can redefine the momentum of a particle in some new way so that the law of conservation of momentum still holds. We choose the second route.

Consider a particle moving with constant speed v in the x direction. Classically, its momentum has magnitude

$$p = mv = m \frac{\Delta x}{\Delta t} \qquad \text{(classical momentum),} \qquad (38\text{-}30)$$

in which Δx is the distance covered in time Δt. To find a relativistic expression for momentum, we start with the new definition

$$p = m \frac{\Delta x}{\Delta t_0}.$$

Here, as before, Δx is the distance covered by a moving particle as viewed by an observer watching that particle. However, Δt_0 is the time required to cover that distance, measured not by the observer watching the moving particle but by an observer moving with the particle. The particle is at rest with respect to this second observer, with the result that the time this observer measures is a proper time Δt_0.

Using the time dilation formula (Eq. 38-8), we can then write

$$p = m \frac{\Delta x}{\Delta t_0} = m \frac{\Delta x}{\Delta t} \frac{\Delta t}{\Delta t_0} = m \frac{\Delta x}{\Delta t} \gamma.$$

But since $\Delta x/\Delta t$ is just the particle velocity v,

$$p = \gamma m v \qquad \text{(momentum).} \qquad (38\text{-}31)$$

Note that this differs from the classical definition of Eq. 38-30 only by the Lorentz factor γ. However, that difference is important: unlike classical momentum, relativistic momentum approaches infinite values as v approaches c.

We can generalize the definition of Eq. 38-31 to vector form as

$$\mathbf{p} = \gamma m \mathbf{v} \qquad \text{(momentum).} \qquad (38\text{-}32)$$

We introduced this definition without elaboration in Section 9-4 as a foretaste of things to come (see Eq. 9-24). We state without further proof that, if we adopt the definition of momentum presented in Eq. 38-32, we can continue to apply the principle of conservation of momentum up to the very highest particle speeds.

38-12 A NEW LOOK AT ENERGY

In Section 7-8 we introduced, without elaboration, a relativistic expression for the kinetic energy of a particle:

$$K = mc^2 \left(\frac{1}{\sqrt{1 - (v/c)^2}} - 1 \right).$$

We can now write this equation as

$$K = mc^2(\gamma - 1) \quad \text{(kinetic energy).} \quad (38\text{-}33)$$

We showed in Section 7-8 that—unlikely as it may seem—this expression reduces to the familiar classical $K = \frac{1}{2}mv^2$ at low speeds. Furthermore, Eq. 38-33 can be derived in exactly the same way as the classical kinetic energy expression: by setting the kinetic energy K equal to the work required to accelerate the particle from rest to its observed speed. Let us point out some of the consequences of Eq. 38-33.

Total Energy

We start by defining the *total energy* E of a particle as γmc^2. With the help of Eq. 38-33 we can then write

$$\begin{aligned} E &= \gamma mc^2 \\ &= mc^2 + K \end{aligned} \quad \begin{array}{l}\text{(total energy;}\\ \text{single particle).}\end{array} \quad (38\text{-}34)$$

We interpret Eq. 38-34 as implying that the total energy E of a moving particle is made up of mc^2, which we call the **mass energy** or the **rest energy** of the particle, and K, its kinetic energy. Table 8-1 lists the rest energies of a few particles and other objects. The rest energy of an electron, for example, is 0.511 MeV, and for a proton it is 938 MeV.

The total energy of a system of n particles is

$$E = \sum_{j=1}^{n} E_j = \sum_{j=1}^{n} (\gamma_j m_j c^2) = \sum_{j=1}^{n} m_j c^2 + \sum_{j=1}^{n} K_j$$
$$\text{(total energy; system of particles).} \quad (38\text{-}35)$$

In relativity, the principle of *conservation of energy* is stated as follows:

For an isolated system of particles, the total energy E of the system, which is defined by Eq. 38-35, remains constant, no matter what interactions may occur among the particles.

Thus in any isolated reaction or decay process involving two or more particles, the total energy of the system after the process must be equal to the total energy before the process. During the process, the total rest energy of the interacting particles may change but then the total kinetic energy must also change by an equal amount in the opposite direction to compensate.

Considerations of this sort are at the root of Einstein's well-known relation $E = mc^2$, which asserts that rest energy can be converted to other forms. All reactions—whether chemical or nuclear—in which energy is released or absorbed involve a corresponding change in the rest energy of the reactants. We discussed the relation $E = mc^2$ in detail in Section 8-8.

Momentum and Kinetic Energy

In classical mechanics, the momentum p of a particle is mv and its kinetic energy K is $\frac{1}{2}mv^2$. If we eliminate v between these two expressions, we find a direct relation between momentum and kinetic energy:

$$p^2 = 2Km \quad \text{(classical).} \quad (38\text{-}36)$$

We can find a similar connection in relativity by eliminating v between the relativistic definition of momentum (Eq. 38-31) and the relativistic definition of kinetic energy (Eq. 38-33). Doing so leads, after some algebra, to

$$(pc)^2 = K^2 + 2Kmc^2. \quad (38\text{-}37)$$

With the aid of Eq. 38-34, we can transform Eq. 38-37 into a relation between the momentum p and the total energy E of a particle:

$$E^2 = (pc)^2 + (mc^2)^2. \quad (38\text{-}38)$$

The right triangle of Fig. 38-14 helps to keep these useful relations in mind. You can also show that, in that triangle,

$$\sin \theta = \beta \quad \text{and} \quad \sin \phi = 1/\gamma. \quad (38\text{-}39)$$

With Eq. 38-38 we can see that the product pc must have the same unit as energy E; thus we can express the unit of momentum p as an energy unit divided by c. In fact, momentum in particle physics is often reported in the units MeV/c or GeV/c.

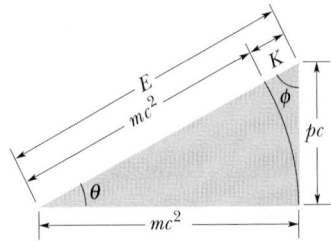

FIGURE 38-14 A useful mnemonic device for remembering the relativistic relations among the total energy E, the rest energy or mass energy mc^2, the kinetic energy K, and the momentum p.

CHECKPOINT 5: Are (a) the kinetic energy and (b) the total energy of a 1 GeV electron more than, less than, or equal to those of a 1 GeV proton?

SAMPLE PROBLEM 38-8

(a) What is the total energy E of a 2.53 MeV electron?

SOLUTION: From Eq. 38-34 we have

$$E = mc^2 + K.$$

From Table 8-1, mc^2 for an electron is 0.511 MeV; so

$$E = 0.511 \text{ MeV} + 2.53 \text{ MeV} = 3.04 \text{ MeV}. \quad \text{(Answer)}$$

(b) What is its momentum p?

SOLUTION: From Eq. 38-38,

$$E^2 = (pc)^2 + (mc^2)^2,$$

we can write

$$(3.04 \text{ MeV})^2 = (pc)^2 + (0.511 \text{ MeV})^2.$$

Then

$$pc = \sqrt{(3.04 \text{ MeV})^2 - (0.511 \text{ MeV})^2} = 3.00 \text{ MeV},$$

and, giving the momentum in units of energy divided by c, we have

$$p = 3.00 \text{ MeV}/c. \quad \text{(Answer)}$$

(c) What is the Lorentz factor γ for the electron?

SOLUTION: From Eq. 38-34, we have

$$E = \gamma mc^2.$$

With $E = 3.04$ MeV and $m = 9.11 \times 10^{-31}$ kg, we then have

$$\gamma = \frac{E}{mc^2} = \frac{(3.04 \times 10^6 \text{ eV})(1.6 \times 10^{-19} \text{ J/eV})}{(9.11 \times 10^{-31} \text{ kg})(3.0 \times 10^8 \text{ m/s})^2}$$

$$= 5.93. \quad \text{(Answer)}$$

SAMPLE PROBLEM 38-9

The most energetic proton ever detected in the cosmic rays coming in from space had an astounding kinetic energy of 3.0×10^{20} eV (enough energy to warm a teaspoon of water by a few degrees).

(a) Calculate the proton's Lorentz factor γ and speed v.

SOLUTION: Solving Eq. 38-33 for γ, we find

$$\gamma = \frac{K + mc^2}{mc^2} = \frac{K}{mc^2} + 1 = \frac{3.0 \times 10^{20} \text{ eV}}{938 \times 10^6 \text{ eV}} + 1$$

$$= 3.198 \times 10^{11} \approx 3.2 \times 10^{11}. \quad \text{(Answer)}$$

Here we used 938 MeV for a proton's rest energy.

This computed value for γ is so large that we cannot use the definition of γ (Eq. 38-7) to find v. Try it; your calculator will tell you that β is effectively equal to 1 and thus that v is effectively equal to c. Actually, v is almost c, but we want a more accurate answer, which we can obtain by first solving Eq. 38-7 for $1 - \beta$. To begin we write

$$\gamma = \frac{1}{\sqrt{1 - \beta^2}} = \frac{1}{\sqrt{(1 - \beta)(1 + \beta)}} \approx \frac{1}{\sqrt{2(1 - \beta)}},$$

where we have used the fact that β is so close to unity that $1 + \beta$ is very close to 2. Solving for $1 - \beta$ then yields

$$1 - \beta = \frac{1}{2\gamma^2} = \frac{1}{(2)(3.198 \times 10^{11})^2}$$

$$= 4.9 \times 10^{-24} \approx 5 \times 10^{-24}.$$

So

$$\beta = 1 - 5 \times 10^{-24},$$

and since $v = \beta c$,

$$v \approx 0.999\,999\,999\,999\,999\,999\,999\,995c. \quad \text{(Answer)}$$

(b) Suppose that the proton travels along a diameter (9.8×10^4 ly) of the Milky Way galaxy. Approximately how long does the proton take to travel that diameter as measured from the common reference frame of Earth and the galaxy?

SOLUTION: We just saw that this *ultrarelativistic* proton is traveling at a speed barely less than c. By the definition of light-year, light takes 9.8×10^4 y to travel 9.8×10^4 ly, and this proton should take almost the same time. Thus, from our Earth–Milky Way reference frame, the trip takes

$$\Delta t = 9.8 \times 10^4 \text{ y.} \quad \text{(Answer)}$$

(c) How long does the trip take as measured in the rest frame of the proton?

SOLUTION: Because the start of the trip and the end of the trip occur at the same location in the proton's rest frame, namely, coincident with the proton itself, what we seek is the proper time of the trip. We can use the time dilation equation (Eq. 38-8) to transform Δt from the Earth–Milky Way frame to the proton rest frame:

$$\Delta t_0 = \frac{\Delta t}{\gamma} = \frac{9.8 \times 10^4 \text{ y}}{3.198 \times 10^{11}}$$

$$= 3.06 \times 10^{-7} \text{ y} = 9.7 \text{ s.} \quad \text{(Answer)}$$

In our frame, the trip takes 98,000 y. In the proton's frame, it takes 9.7 s! As promised at the start of this chapter, relative motion can alter the rate at which time passes, and we have here an extreme example.

REVIEW & SUMMARY

The Postulates

Special theory of relativity is based on two postulates:

1. The laws of physics are the same for observers in all inertial reference frames. No frame is preferred.

2. The speed of light in vacuum has the same value c in all directions and in all inertial reference frames.

The speed of light c in vacuum is an ultimate speed that cannot be exceeded by any entity carrying either energy or information.

Coordinates of an Event

Three space coordinates and one time coordinate specify an **event.** One task of special relativity is to relate these coordinates as assigned by two observers who are in uniform motion with respect to each other.

Simultaneous Events

If two observers are in relative motion, they will not, in general, agree as to whether two events are simultaneous. If one observer finds two events at different locations to be simultaneous, the other will not, and conversely. Simultaneity is *not* an absolute concept but a relative one, depending on the motion of the observer. The relativity of simultaneity is a direct consequence of the finite ultimate speed c.

Time Dilation

If two successive events occur at the same place in an inertial reference frame, the time interval Δt_0 between them, measured on a single clock where they occur, is the **proper time** between the events. *Observers in frames moving relative to that frame will measure a larger value for this interval.* For an observer moving with relative speed v, the measured time interval is

$$\Delta t = \frac{\Delta t_0}{\sqrt{1 - (v/c)^2}} = \frac{\Delta t_0}{\sqrt{1 - \beta^2}} = \gamma \, \Delta t_0$$

$$\text{(time dilation).} \quad \text{(38-6 to 38-8)}$$

Here $\beta = v/c$ is the **speed parameter** and $\gamma = 1/\sqrt{1 - \beta^2}$ is the **Lorentz factor.** An important consequence of time dilation is that moving clocks run slow as measured by an observer at rest.

Length Contraction

The length L_0 of an object measured by an observer in an inertial reference frame in which the object is at rest is called its **proper length.** *Observers in frames mvoing relative to that frame and parallel to that length will measure a shorter length.* For an observer moving with relative speed v, the measured length is

$$L = L_0\sqrt{1 - \beta^2} = \frac{L_0}{\gamma} \qquad \text{(length contraction).} \quad \text{(38-9)}$$

The Lorentz Transformation

The *Lorentz transformation* equations relate the spacetime coordinates of a single event as seen by observers in two inertial frames, S and S', where S' is moving relative to S with velocity v in the positive x, x' direction. The four coordinates are related by

$$x' = \gamma(x - vt),$$
$$y' = y,$$
$$z' = z,$$
$$t' = \gamma(t - vx/c^2)$$

(Lorentz transformation equations; valid at all speeds). (38-14)

Relativity of Velocities

When a particle is moving with speed u' in the positive x' direction in an inertial reference frame S' that itself is moving with speed v parallel to the x direction of a second inertial frame S, the speed u of the particle as measured in S is

$$u = \frac{u' + v}{1 + u'v/c^2} \qquad \text{(relativistic velocity).} \quad \text{(38-23)}$$

Relativistic Doppler Effect

If a source emitting light waves of frequency f_0 moves directly away from a detector with relative velocity **v**, the frequency f measured by the detector is

$$f = f_0 \sqrt{\frac{1 - \beta}{1 + \beta}}. \quad \text{(38-25)}$$

Transverse Doppler Effect

If the relative motion of the source is perpendicular to the source–detector line, the Doppler formula is

$$f = f_0 \sqrt{1 - \beta^2}. \quad \text{(38-27)}$$

This **transverse Doppler effect** is due to time dilation.

Momentum and Energy

The definitions of linear momentum **p**, kinetic energy K, and total energy E that are valid at any possible speed are

$$\mathbf{p} = \gamma m\mathbf{v} \qquad \text{(momentum),} \quad \text{(38-32)}$$
$$K = mc^2(\gamma - 1) \qquad \text{(kinetic energy),} \quad \text{(38-33)}$$
$$E = \gamma mc^2 = mc^2 + K \qquad \text{(total energy, single particle).} \quad \text{(38-34)}$$

With these definitions, the principle of conservation of total energy for a system of particles takes the form

$$E = \sum_{j=1}^{n} (\gamma_j m_j c^2) = \sum_{j=1}^{n} m_j c^2 + \sum_{j=1}^{n} K_j$$

(total energy, system of particles). (38-35)

Two additional energy relationships, derivable from Eqs. 38-22, 38-33, and 38-34, are often useful:

$$(pc)^2 = K^2 + 2Kmc^2, \quad \text{(38-37)}$$
$$E^2 = (pc)^2 + (mc^2)^2. \quad \text{(38-38)}$$

QUESTIONS

1. In Fig. 38-15, ship A sends a laser pulse to an oncoming ship B, while scout ship C races away. The indicated speeds of the ships are all measured from the same reference frame. Rank the ships according to the speed of the pulse as measured from each ship, greatest first.

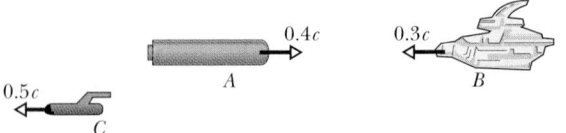

FIGURE 38-15   Questions 1 and 10.

2. Figure 38-16a shows two clocks in stationary frame S (they are synchronized in that frame) and one clock in moving frame S'. Clocks C_1 and C_1' read zero when they pass each other. When clocks C_1' and C_2 pass each other, (a) which clock has the smaller reading and (b) which clock measures a proper time?

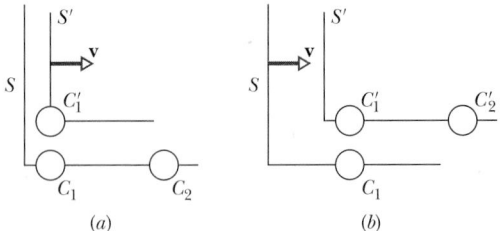

FIGURE 38-16   Questions 2 and 3.

3. Figure 38-16b shows two clocks in stationary frame S' (they are synchronized in that frame) and one clock in moving frame S. Clocks C_1 and C_1' read zero when they pass each other. When clocks C_1 and C_2' pass each other, (a) which clock has the smaller reading and (b) which clock measures a proper time?

4. Sam leaves Venus in a spaceship to Mars and passes Sally, who is on Earth, with a relative speed of $0.5c$. (a) Each measures the Venus–Mars voyage time. Who measures a proper time: Sam, Sally, or neither? (b) On the way, Sam sends a pulse of light to Mars. Each measures the travel time of the pulse. Who measures a proper time?

5. Figure 38-17 shows three situations in which a starship passes Earth (the dot) and then makes a round trip that brings it back past Earth, each at the given Lorentz factor. As measured in the rest frame of Earth, the round-trip distances are as follows: trip 1, $2D$; trip 2, $4D$; trip 3, $6D$. Without written calculation and neglecting any time needed for accelerations, rank the situations according to the travel times of the trips, greatest first, as measured from (a) the rest frame of Earth and (b) the rest frame of the starship. (*Hint:* See Sample Problem 38-5.)

FIGURE 38-17   Question 5.

6. Figure 38-18 is a map of the travel lanes allowed through a stellar region by the indigenous alien government. Each lane (between two lettered junction points) is labeled with the maximum value of γ allowed for that lane. In the rest frame of the junctions, successive junctions are separated by a distance of L or $2L$. (a) Beginning at Home Port, pick the route to Far Base that minimizes your travel time, neglecting the time needed for acceleration when γ changes. (*Hint:* See Sample Problem 38-5.) Who measures (b) less time and (c) less distance for that route, you or someone at rest relative to the junctions?

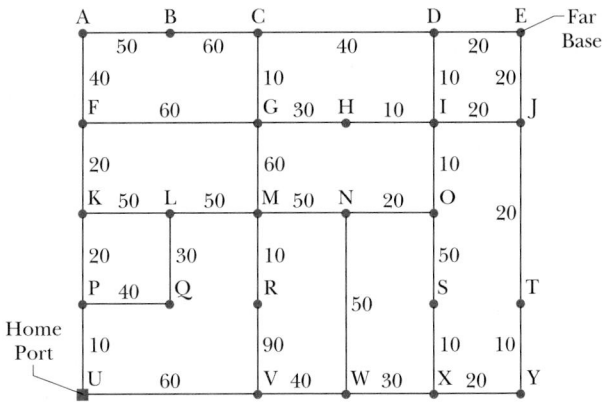

FIGURE 38-18   Question 6.

7. Figure 38-19 shows a ship (with on-board reference frame S') passing us (with reference frame S). A proton is fired at nearly the speed of light along the length of the ship, from the front to the rear. (a) Is the spatial separation $\Delta x'$ between the firing of the proton and its impact a positive or negative quantity? (b) Is the temporal separation $\Delta t'$ between those events a positive or negative quantity?

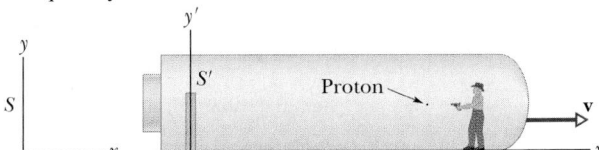

FIGURE 38-19   Question 7.

8. (a) In Fig. 38-9, suppose an observer in frame S' measures two events to be at the same location (say, at x') but not at the same time. Can an observer in frame S possibly measure them to be at the same location? (b) If two events occur simultaneously at the same place for one observer, will they be simultaneous for all other observers? (c) Will they occur at the same place for all other observers?

9. Figure 38-20 shows a starship and an asteroid. In four situations, the velocity of the starship relative to us (on a scout ship) and the velocity of the asteroid relative to the starship are, in that order, (a) $+0.4c$, $+0.4c$; (b) $+0.5c$, $+0.3c$; (c) $+0.9c$, $-0.1c$; and (d) $+0.3c$, $+0.5c$. Without written calculation, rank the situations according to the magnitude of the velocity of the asteroid relative to us, greatest first.

FIGURE 38-20 Question 9.

10. Ships A and B in Fig. 38-15 are moving directly toward each other; the velocities indicated are all measured from the same reference frame. Is the speed of ship A relative to ship B more than $0.7c$, less than $0.7c$, or equal to $0.7c$?

11. Figure 38-21 shows one of four star cruisers that are in a race. As each cruiser passes the starting line, a shuttle craft leaves the cruiser and races toward the finish line. You, judging the race, are stationary relative to the start and finish lines. The speeds v_c of the cruisers relative to you and the speeds v_s of the shuttle craft relative to their starships are, in that order, (1) $0.70c$, $0.40c$; (2) $0.40c$, $0.70c$; (3) $0.20c$, $0.90c$; (4) $0.50c$, $0.60c$. (a) Without written calculation, rank the shuttle craft according to their speeds relative to you, greatest first. (b) Still without written calculation, rank the shuttle craft according to the distances their pilots measure from the starting line to the finish line, greatest first. (c) Each starship sends a signal to its shuttle craft at a certain frequency f_0 as measured on board the starship. Again without written calculation, rank the shuttle craft according to the frequencies they detect, greatest first.

12. While on board a starship, you intercept signals from four shuttle craft that are moving either directly toward or directly away from you. The signals have the same proper frequency f_0. The speed and direction (both relative to you) of the shuttle craft are (a) $0.3c$ toward, (b) $0.6c$ toward, (c) $0.3c$ away, and (d) $0.6c$ away. Rank the shuttle craft according to the frequency you receive, greatest first.

13. Figure 38-22 shows three starships that move either left or right along the axis shown. All emit microwave signals of the same proper frequency f_0. Ship C detects the signal from ship A with a frequency $f_1 > f_0$. Ship A detects the signal from ship B with a frequency $f_2 < f_0$. Is the signal from ship B that is detected by ship C less than f_0, greater than f_1, or between f_0 and f_1?

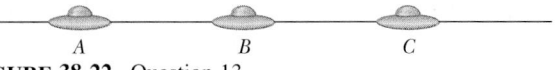

$$\hspace{3em} A \hspace{5em} B \hspace{5em} C$$

FIGURE 38-22 Question 13.

14. The rest energy and total energy, respectively, of three particles, expressed in terms of a basic amount A are (1) A, $2A$; (2) A, $3A$; (3) $3A$, $4A$. Without written calculation, rank the particles according to (a) their mass, (b) their kinetic energy, (c) their Lorentz factor, and (d) their speed, greatest first.

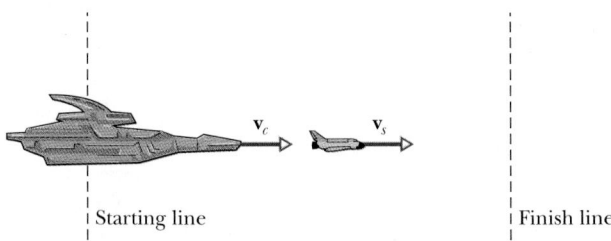

FIGURE 38-21 Question 11.

EXERCISES & PROBLEMS

SECTION 38-2 The Postulates

1E. What fraction of the speed of light does each of the following speeds represent; that is, what is the speed parameter β? (a) A typical rate of continental drift (1 in./y). (b) A highway speed limit of 55 mi/h. (c) A supersonic plane flying at Mach 2.5 (1200 km/h). (d) The escape speed of a projectile from the surface of Earth. (e) A typical recession speed of a distant quasar (3.0×10^4 km/s).

2E. Quite apart from effects due to Earth's rotational and orbital motions, a laboratory reference frame is not strictly an inertial frame because a particle placed at rest there will not, in general, remain at rest; it will fall. Often, however, events happen so quickly that we can ignore the gravitational acceleration and treat the frame as inertial. Consider, for example, an electron of speed $v = 0.992c$, projected horizontally into a laboratory test chamber and moving through a distance of 20 cm. (a) How long would that journey take, and (b) how far would the electron fall during this interval? What can you conclude about the suitability of the laboratory as an inertial frame in this case?

3P. Find the speed of a particle that takes 2.0 y longer than light to travel a distance of 6.0 ly.

SECTION 38-5 The Relativity of Time

4E. What must be the speed parameter β if the Lorentz factor γ is (a) 1.01, (b) 10.0, (c) 100, and (d) 1000?

5E. The mean lifetime of stationary muons is measured to be 2.2 μs. The mean lifetime of high-speed muons in a burst of cosmic rays observed from Earth is measured to be 16 μs. Find the speed of these cosmic-ray muons relative to Earth.

6P. An unstable high-energy particle enters a detector and leaves a track 1.05 mm long before it decays. Its speed relative to the detector was $0.992c$. What is its proper lifetime? That is, how long would the particle have lasted before decay had it been at rest with respect to the detector?

7P. A pion is created in the higher reaches of Earth's atmosphere when an incoming high-energy cosmic-ray particle collides with an atomic nucleus. A pion so formed descends toward Earth with a speed of $0.99c$. In a reference frame in which they are at rest, pions decay with an average life of 26 ns. As measured in a frame fixed with respect to Earth, how far (on the average) will such a pion move through the atmosphere before it decays?

8P. You wish to make a round trip from Earth in a spaceship, traveling at constant speed in a straight line for 6 months and then returning at the same constant speed. You wish further, on your return, to find Earth as it will be a thousand years in the future. (a) How fast must you travel? (b) Does it matter whether you travel in a straight line on your journey? If, for example, you traveled in a circle for 1 year, would you still find that 1000 years had elapsed by Earth clocks when you returned?

SECTION 38-6 The Relativity of Length

9E. A rod lies parallel to the x axis of reference frame S, moving along this axis at a speed of $0.630c$. Its rest length is 1.70 m. What will be its measured length in frame S?

10E. The length of a spaceship is measured to be exactly half its rest length. (a) What is the speed of the spaceship relative to the observer's frame? (b) By what factor do the spaceship's clocks run slow, compared to clocks in the observer's frame?

11E. A meter stick in frame S' makes an angle of 30° with the x' axis. If that frame moves parallel to the x axis with speed $0.90c$ relative to frame S, what is the length of the stick as measured from S?

12E. An electron of $\beta = 0.999\,987$ moves along the axis of an evacuated tube that has a length of 3.00 m as measured by a laboratory observer S at rest relative to the tube. An observer S' at rest relative to the electron, however, would see this tube moving with speed $v\ (= \beta c)$. What length would observer S' measure for the tube?

13E. The rest radius of Earth is 6370 km, and its orbital speed about the Sun is 30 km/s. Suppose Earth moves past an observer at this speed. To the observer, by how much would Earth's diameter be contracted along the direction of motion?

14E. A spaceship of rest length 130 m races past a timing station at a speed of $0.740c$. (a) What is the length of the spaceship as measured by the timing station? (b) What time interval will the station clock record between the passage of the front and back ends of the ship?

15P. A space traveler takes off from Earth and moves at speed $0.99c$ toward the star Vega, which is 26 ly distant. How much time will have elapsed by Earth clocks (a) when the traveler reaches Vega and (b) when Earth observers receive word from the traveler that she has arrived? (c) How much older will Earth observers calculate the traveler to be (according to her) when she reaches Vega than she was when she started the trip?

16P. An airplane whose rest length is 40.0 m is moving at uniform velocity with respect to Earth, at a speed of 630 m/s. (a) By what fraction of its rest length is it shortened to an observer on Earth? (b) How long would it take, according to Earth clocks, for the airplane's clock to fall behind by 1.00 μs? (Use special relativity in your calculations.)

17P. (a) Can a person, in principle, travel from Earth to the galactic center (which is about 23,000 ly distant) in a normal lifetime? Explain, using either time-dilation or length-contraction arguments. (b) What constant speed would be needed to make the trip in 30 y (proper time)?

SECTION 38-8 Some Consequences of the Lorentz Equations

18E. Observer S assigns the spacetime coordinates

$$x = 100 \text{ km} \quad \text{and} \quad t = 200 \text{ μs}$$

to an event. What are the coordinates of this event in frame S', which moves in the direction of increasing x with speed $0.950c$ relative to S? Assume $x = x' = 0$ at $t = t' = 0$.

19E. Observer S reports that an event occurred on his x axis at $x = 3.00 \times 10^8$ m at time $t = 2.50$ s. (a) Observer S' is moving in the direction of increasing x at a speed of $0.400c$. Further, $x = x' = 0$ at $t = t' = 0$. What coordinates does observer S' report for the event? (b) What coordinates would S' report if she were moving in the direction of *decreasing* x at this same speed?

20E. Inertial frame S' moves at a speed of $0.60c$ with respect to frame S (Fig. 38-9). Further, $x = x' = 0$ at $t = t' = 0$. Two events are recorded. In frame S, event 1 occurs at the origin at $t = 0$ and event 2 occurs on the x axis at $x = 3.0$ km at $t = 4.0$ μs. What times of occurrence does observer S' record for these same events? Explain the difference in the time order.

21E. An experimenter arranges to trigger two flashbulbs simultaneously, producing a big flash located at the origin of his reference frame and a small flash at $x = 30.0$ km. An observer, moving at a speed of $0.250c$ in the direction of increasing x, also views the flashes. (a) What time interval between them does she find? (b) Which flash does she say occurs first?

22E. In Table 38-2 the Lorentz transformation equations in the right-hand column can be derived from those in the left-hand column simply by (1) exchanging primed and unprimed quantities and (2) changing the sign of v. Verify this procedure by deriving one set of equations directly from the other by algebraic manipulation.

23P. A clock moves along the x axis at a speed of $0.600c$ and reads zero as it passes the origin. (a) Calculate the Lorentz factor. (b) What time does the clock read as it passes $x = 180$ m?

24P. An observer S sees a big flash of light 1200 m from his position and a small flash of light 720 m closer to him directly in line with the big flash. He measures the time interval between the flashes to be 5.00 μs, the big flash occurring first. (a) What is the relative velocity **v** (give both magnitude and direction) of a second observer S' who records these flashes as occurring at the same place? (b) From the point of view of S', which flash occurs first? (c) What time interval between them does S' measure?

25P. In Problem 24, observer S sees the two flashes in the same positions as before, but they now occur closer together in time. How close together in time can they be in the frame of S and still allow the possibility of finding a frame S' in which they occur at the same place?

SECTION 38-9 The Relativity of Velocities

26E. A particle moves along the x' axis of frame S' with a speed of $0.40c$. Frame S' moves with a speed of $0.60c$ with respect to frame S. What is the measured speed of the particle in frame S?

27E. Frame S' moves relative to frame S at $0.62c$ in the direction of increasing x. In frame S' a particle is measured to have a velocity of $0.47c$ in the direction of increasing x'. (a) What is the velocity of the particle with respect to frame S? (b) What would be the velocity of the particle with respect to S if the particle moved (at $0.47c$) in the direction of *decreasing* x' in the S' frame? In each case, compare your answers with the predictions of the classical velocity transformation equation.

28E. One cosmic-ray particle approaches Earth along Earth's north-south axis with a velocity of $0.80c$ toward the geographic north pole, and another approaches with a velocity $0.60c$ toward the geographic south pole (See Fig. 38-23). What is the relative speed of approach of one particle with respect to the other? (*Hint:* It is useful to consider Earth and one of the particles as the two inertial reference frames.)

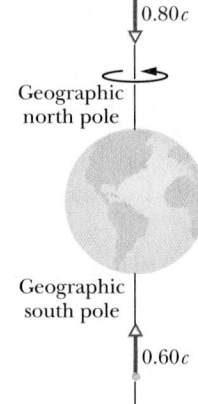

FIGURE 38-23
Exercise 28.

29E. Galaxy A is reported to be receding from us with a speed of $0.35c$. Galaxy B, located in precisely the opposite direction, is also found to be receding from us at this same speed. What recessional speed would an observer on Galaxy A find (a) for our galaxy and (b) for Galaxy B?

30E. It is concluded from measurements of the red shift of the emitted light that quasar Q_1 is moving away from us at a speed of $0.800c$. Quasar Q_2, which lies in the same direction in space but is closer to us, is moving away from us at a speed $0.400c$. What velocity for Q_2 would be measured by an observer on Q_1?

31P. A spaceship whose rest length is 350 m has a speed of $0.82c$ with respect to a certain reference frame. A micrometeorite, also with a speed of $0.82c$ in this frame, passes the spaceship on an antiparallel track. How long does it take this object to pass the spaceship as measured on the ship?

32P. To circle Earth in low orbit, a satellite must have a speed of about 17,000 mi/h. Suppose that two such satellites orbit Earth in opposite directions. (a) What is their relative speed as they pass, according to the classical Galilean velocity transformation equation? (b) What fractional error do you make in (a) by not using the (correct) relativistic transformation equation?

33P. A spaceship, at rest in a certain reference frame S, is given a speed increment of $0.50c$. Relative to its new rest frame, it is then given a further $0.50c$ increment. This process is continued until its speed with respect to its original frame S exceeds $0.999c$. How many increments does this process require?

34P. An armada of spaceships that is 1.00 ly long (in its rest system) moves with speed $0.800c$ relative to ground station S. A messenger travels from the rear of the armada to the front with a speed of $0.950c$ relative to S. How long does the trip take as measured (a) in the messenger's rest system, (b) in the armada's rest system, and (c) by an observer in system S?

SECTION 38-10 Doppler Effect for Light

35E. A spaceship, moving away from Earth at a speed of $0.900c$, reports back by transmitting on a frequency (measured in the spaceship frame) of 100 MHz. To what frequency must Earth receivers be tuned to receive the report?

36E. Some of the familiar hydrogen lines appear in the spectrum of quasar 3C9, but they are shifted so far toward the red that their wavelengths are observed to be three times longer than those observed for hydrogen atoms that are stationary in a laboratory. (a) Show that the classical Doppler equation gives a relative velocity of recession greater than c for this situation. (b) Assuming that the relative motion of 3C9 and Earth is due entirely to recession, find the recession speed that is predicted by the relativistic Doppler equation.

37E. Give the Doppler wavelength shift $\lambda - \lambda_0$, if any, for the sodium D_2 line (589.00 nm) emitted by a source moving in a circle with constant speed ($= 0.100c$) as measured by an observer fixed at the center of the circle.

38P. A spaceship is receding from Earth at a speed of $0.20c$. A light source on the rear of the ship appears blue ($\lambda = 450$ nm) to passengers on the ship. What color would the source appear to an observer on Earth monitoring the receding spaceship?

39P. A radar transmitter T is fixed to a reference frame S' that is moving to the right with speed v relative to reference frame S (see Fig. 38-24). A mechanical timer (essentially a clock) in frame S', having a period τ_0 (measured in S'), causes transmitter T to emit timed radar pulses, which travel at the speed of light and are received by R, a receiver fixed in frame S. (a) What is the period τ of the timer as detected by observer A, who is fixed in frame S? (b) Show that at the receiver R the time interval between pulses arriving from T is not τ or τ_0, but

$$\tau_R = \tau_0 \sqrt{\frac{c + v}{c - v}}.$$

(c) Explain why the receiver R and observer A, who are in the same reference frame, measure a different period for the transmitter. (*Hint:* A clock and a radar pulse are not the same thing.)

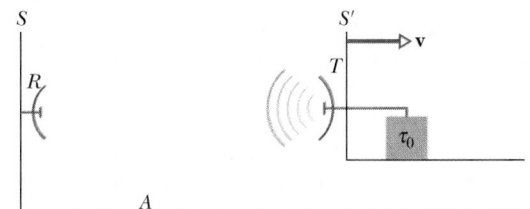

FIGURE 38-24 Problem 39.

SECTION 38-12 A New Look at Energy

40E. How much work must be done to increase the speed of an electron from rest to (a) $0.50c$, (b) $0.990c$, and (c) $0.9990c$?

41E. An electron is moving at a speed such that it could circumnavigate Earth at the equator in 1.00 s. (a) What is its speed, in terms of the speed of light? (b) What is its kinetic energy K? (c) What percent error do you make if you use the classical formula to calculate K?

42E. Find the speed parameter β and Lorentz factor γ for an electron whose kinetic energy is (a) 1.00 keV, (b) 1.00 MeV, and (c) 1.00 GeV.

43E. Find the speed parameter β and Lorentz factor γ for a particle whose kinetic energy is 10.0 MeV if the particle is (a) an electron, (b) a proton, and (c) an alpha particle.

44E. What is the speed of an electron whose kinetic energy is 100 MeV?

45E. A particle has a speed of $0.990c$ in a laboratory reference frame. What are its kinetic energy, its total energy, and its momentum if the particle is (a) a proton and (b) an electron?

46E. In 1979 the United States consumption of electrical energy was about 2.2×10^{12} kW·h. How much mass is equivalent to the consumed energy in that year? Does it make any difference to your answer if this energy is generated in oil-burning, nuclear, or hydroelectric plants?

47E. Quasars are thought to be the nuclei of active galaxies in the early stages of their formation. A typical quasar radiates energy at the rate of 10^{41} W. At what rate is the mass of this quasar being reduced to supply this energy? Express your answer in solar mass units per year, where one solar mass unit (1 smu = 2.0×10^{30} kg) is the mass of our Sun.

48P. How much work must be done to increase the speed of an electron from (a) $0.18c$ to $0.19c$ and (b) $0.98c$ to $0.99c$? Note that the speed increase is $0.01c$ in both cases.

49P. What is the speed of a particle (a) whose kinetic energy is equal to twice its rest energy and (b) whose total energy is equal to twice its rest energy?

50P. A particle with mass m has speed $c/2$ relative to inertial frame S. The particle collides with an identical particle at rest relative to frame S. What is the speed of a frame S' relative to S in which the total momentum of these particles is zero? This frame is called the *center of momentum frame*.

51P. (a) What potential difference would accelerate an electron to speed c, according to classical physics? (b) With this potential difference, what speed would the electron actually attain?

52P. A particle of mass m has a momentum equal to mc. What are (a) its Lorentz factor, (b) its speed, and (c) its kinetic energy?

53P. What must be the momentum of a particle with mass m so that the total energy of the particle is 3 times its rest energy?

54P. Consider the following, all moving in free space: a 2.0 eV photon, a 0.40 MeV electron, and a 10 MeV proton. (a) Which is moving the fastest? (b) The slowest? (c) Which has the greatest momentum? (d) The least? (*Note:* A photon, which is a particle of light, has zero mass.)

55P. A 5.00 grain aspirin tablet has a mass of 320 mg. For how many miles would the energy equivalent of this mass power an automobile? Assume 30.0 mi/gal and a heat of combustion of 1.30×10^8 J/gal for the gasoline used in the automobile.

56P. (a) If the kinetic energy K and the momentum p of a particle can be measured, it should be possible to find its mass m and thus identify the particle. Show that

$$m = \frac{(pc)^2 - K^2}{2Kc^2}.$$

(b) Show that this expression reduces to an expected result as $u/c \to 0$, in which u is the speed of the particle. (c) Find the mass of a particle whose kinetic energy is 55.0 MeV and whose momentum is 121 MeV/c. Express your answer in terms of the mass of the electron.

57P. In a high-energy collision between a cosmic-ray particle and a particle near the top of Earth's atmosphere, 120 km above sea level, a pion is created. The pion has a total energy E of 1.35×10^5 MeV and is traveling vertically downward. In the pion's rest frame, the pion decays 35.0 ns after its creation. At what altitude above sea level, as measured from Earth's reference frame, does the decay occur? The rest energy of a pion is 139.6 MeV.

58P. The average lifetime of muons at rest is 2.20 μs. A laboratory measurement on muons traveling in a beam emerging from a particle accelerator yields an average muon lifetime of 6.90 μs. What are (a) the speed of these muons in the laboratory, (b) their kinetic energy, and (c) their momentum? The mass of a muon is 207 times that of an electron.

59P. (a) How much energy is released in the explosion of a fission bomb containing 3.0 kg of fissionable material? Assume that 0.10% of the mass is converted to released energy. (b) What mass of TNT would have to explode to provide the same energy release? Assume that each mole of TNT liberates 3.4 MJ of energy on exploding. The molecular mass of TNT is 0.227 kg/mol. (c) For the same mass of explosive, how much more effective are nuclear explosions than TNT explosions? That is, compare the fractions of the mass that are converted to energy in each case.

60P. In Section 29-5 we showed that a particle of charge q and mass m moving with speed v perpendicular to a uniform magnetic field B moves in a circle of radius r given by Eq. 29-16:

$$r = \frac{mv}{qB}.$$

Also, it was demonstrated that the period T of the circular motion is independent of the speed of the particle. These results hold only if $v \ll c$. For particles moving faster, the radius of the circular path must be obtained with

$$r = \frac{p}{qB} = \frac{m(\gamma v)}{qB} = \frac{mv}{qB\sqrt{1 - \beta^2}}.$$

This equation is valid at all speeds. Compute the radius of the path of a 10.0 MeV electron moving perpendicular to a uniform 2.20 T magnetic field, using the (a) classical and (b) relativistic formulas. (c) Calculate the period $T = 2\pi r/v$ of the circular motion. Is the result independent of the speed of the electron?

61P. Ionization measurements show that a certain low-mass nuclear particle carries a double charge ($= 2e$) and is moving with a speed of $0.710c$. The radius of curvature of its path in a magnetic field of 1.00 T is 6.28 m. (The path is a circle whose plane is perpendicular to the magnetic field.) Find the mass of the particle and identify it. [*Hint:* Low-mass nuclear particles are made up of neutrons (which have no charge) and protons (charge $= +e$), in roughly equal numbers. Take the mass of each of these particles to be 1.00 u. Also, see Problem 60.]

62P. A 10 GeV proton in cosmic radiation approaches Earth with its velocity **v** perpendicular to Earth's magnetic field **B**, in a region over which Earth's average magnetic field is 55 μT. What is the radius of the proton's curved path in that region? (See Problem 60.)

63P. A 2.50 MeV electron moves perpendicular to a magnetic field in a path whose radius of curvature is 3.0 cm. What is the magnetic field B? (See Problem 60.)

64P. The proton synchrotron at Fermilab accelerates protons to a kinetic energy of 500 GeV. At such a large energy, relativistic effects are important; in particular, as the speed of the proton increases, the time the proton takes to make a trip around its circular orbit in the synchrotron also increases. In a cyclotron, where the magnetic field and the oscillator are at fixed values, this effect of time dilation will put the proton's circling out of synchronization with the oscillator. That eliminates repeated acceleration; hence, the proton will not reach an energy as great as 500 GeV. But in a synchrotron, both the magnitude of the magnetic field and the oscillation frequency are varied to allow for the increase in the time dilation.

At the energy of 500 GeV, calculate (a) the Lorentz factor, (b) the speed parameter, and (c) the magnetic field at the proton orbit, which has a radius of curvature of 750 m. (See Problem 60; use 938.3 MeV as the proton's rest energy.)

Electronic Computation

65. A space probe leaves Earth and travels at $0.97c$. Clocks on Earth and in the probe are all set to zero at launch. Every 6.0 h, mission control sends a signal (traveling at the speed of light) to the probe requesting a status report. The probe immediately sends back a reply that includes the reading on the probe clock at the time the signal was received. For each of the first five signals, calculate the time, according to Earth clocks, at which the reply is received on Earth and the time reported in the reply. Also compute the distance of the probe from Earth when each signal is received by the probe.

Additional Problems

66P. *The Car-in-the-Garage Problem.* Carman has just purchased the world's longest stretch limo, whose proper length is

$L_c = 30.5$ m. In Fig. 38-25*a*, it is shown parked in front of a garage, whose proper length is $L_g = 6.00$ m. The garage has a front door (shown open) and a back door (shown closed). The limo is obviously longer than the garage. Still, Garageman, who owns the garage and knows something about relativistic length contraction, makes a bet with Carman that the limo can fit in the garage with both doors closed. Carman, who dropped the physics course before reaching special relativity, says such a thing, even in principle, is impossible.

To analyze Garageman's scheme, an x_c axis is attached to the limo, with $x_c = 0$ at the rear bumper, and an x_g axis is attached to the garage, with $x_g = 0$ at the (now open) front door. Then Carman is to drive the limo directly toward the front door at a velocity of $0.9980c$ (which is, of course, both technically and financially impossible). Carman is stationary in the x_c reference frame; Garageman is stationary in the x_g reference frame.

There are two events to consider. *Event 1:* When the rear bumper clears the front door, the front door is closed. Let the time of this event be zero to both Carman and Garageman: $t_{g1} = t_{c1} = 0$. The event occurs at $x_c = x_g = 0$. Figure 38-25*b* shows event 1 according to the x_g reference frame. *Event 2:* When the front bumper reaches the back door, that door opens. Figure 38-25*c* shows event 2 according to the x_g reference frame.

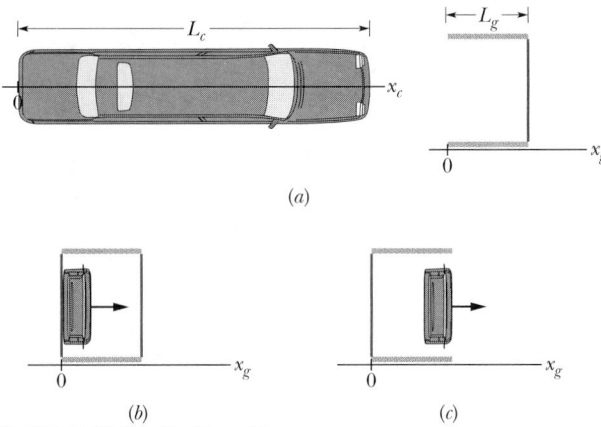

(a)

(b)   (c)

FIGURE 38-25 Problem 66.

According to Garageman, (a) what is the length of the limo and (b) what are the spacetime coordinates x_{g2} and t_{g2} of event 2? (c) For how long is the limo temporarily "trapped" inside the garage, with both doors shut?

Now consider the situation from the x_c reference frame, in which the garage comes racing past the limo at a velocity of $-0.9980c$. According to Carman, (d) what is the length of the passing garage, (e) what are the spacetime coordinates x_{c2} and t_{c2} of event 2, (f) is the limo ever in the garage with both doors shut, and (g) which event occurs first? (h) Sketch events 1 and 2 as seen by Carman. (Are the events causally related; that is, does one of them cause the other?) (i) Finally, who wins the bet?

67P. *Superluminal Jets.* Figure 38-26a shows the path taken by a knot in a jet of ionized gas that has been expelled from a galaxy.

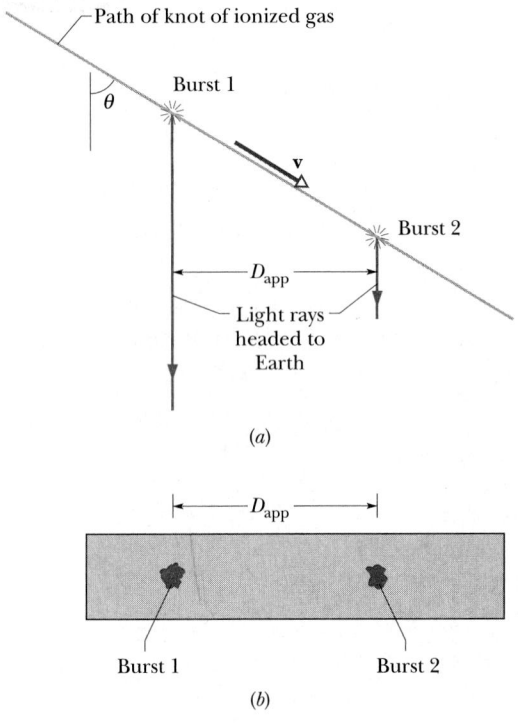

Path of knot of ionized gas

Burst 1

θ

v

Burst 2

D_{app}

Light rays
headed to
Earth

(a)

D_{app}

Burst 1   Burst 2

(b)

FIGURE 38-26 Problem 67.

The knot travels at a constant velocity **v** at an angle θ from the direction of Earth. The knot occasionally emits a burst of light, which is eventually detected on Earth. Two bursts are indicated in Fig. 38-26a, separated in time by t as measured in a stationary frame near the bursts. The bursts are shown in Fig. 38-26b as if they were photographed on the same film, first when the light from burst 1 arrived on Earth and then later when the light from burst 2 arrived. The apparent distance D_{app} traveled by the knot between the two bursts is the distance across an Earth-observer's view of the knot's path. The apparent time T_{app} between the bursts is the difference in the arrival times of the light from them. The apparent speed of the knot is then $V_{app} = D_{app}/T_{app}$. In terms of v, t, and θ, what are (a) D_{app} and (b) T_{app}? (c) Evaluate V_{app} for $v = 0.980c$ and $\theta = 30.0°$.

Tracks of tiny vapor bubbles in this bubble-chamber image reveal where electrons (tracks color-coded green) and positrons (red) moved. A gamma ray (which left no track when it entered at the top) kicked an electron out of one of the hydrogen atoms filling the chamber and then converted to an electron–positron pair. Another gamma ray underwent another pair production farther down. These tracks (curved because of a magnetic field) clearly show that electrons and positrons are particles that move along narrow paths. Yet, those particles can also be interpreted in terms of waves. Can a particle be a wave?

39-1 A NEW DIRECTION

Our discussion of Einstein's theory of relativity took us into a world far beyond that of ordinary experience — the world of objects moving at speeds close to the speed of light. Among other surprises, Einstein's theory predicts that the rate at which a clock runs depends on how fast the clock is moving relative to the observer: the faster the motion, the slower the clock rate. This and other predictions of the theory have passed every experimental test devised thus far, and relativity theory has led us to a deeper and more satisfying view of the nature of space and time.

Now you are about to explore a second world that is outside ordinary experience — the subatomic world. You will encounter a new set of surprises that, though they may sometimes seem bizarre, have led physicists step by step to a deeper view of the nature of reality.

Quantum mechanics, as our new subject is called, answers such questions as: Why do the stars shine? Why do the elements exhibit the order that is so apparent in the periodic table? How do transistors and other microelectronic devices work? Why does copper conduct electricity but glass does not? Because quantum mechanics accounts for all of chemistry, including biochemistry, we need to understand it if we are to understand life itself.

Some of the predictions of quantum mechanics seem strange even to the physicists and philosophers who study its foundations. Nevertheless, this theory too has passed every experimental test — and there have been many — with flying colors. Its predictions have never failed.

39-2 LIGHT WAVES AND PHOTONS

We have described light as a wave, having a wavelength λ, a frequency f, and a speed c, such that

$$c = \lambda f. \qquad (39\text{-}1)$$

In Chapter 34 we used Maxwell's equations to show that a light wave is an interdependent combination of electric and magnetic fields, each alternating at frequency f. We showed further that visible light is part of an electromagnetic spectrum that extends, in a continuous range of wavelengths, from gamma rays to long radio waves.

In 1905 Einstein proposed a property of visible light and other electromagnetic radiation that is *not* predicted by Maxwell's equations and gives us the first of our quantum surprises: when an atom emits or absorbs light, energy is transferred not in a smooth continuous fashion but in small, discrete "lumps" of energy. We now call such lumps of energy **photons** (the word was not introduced until 1926).

According to Einstein's proposal, the energy E transferred by a single photon associated with a light wave of frequency f is

$$E = hf \qquad \text{(photon energy).} \qquad (39\text{-}2)$$

Here h is the **Planck constant**, which has the value

$$h = 6.63 \times 10^{-34} \text{ J} \cdot \text{s} = 4.14 \times 10^{-15} \text{ eV} \cdot \text{s}. \qquad (39\text{-}3)$$

The Planck constant is the basic constant of quantum mechanics, much as the speed of light c is the basic constant of relativity. If c were infinite (it is "large" but it isn't infinite), there would be no special relativity; if h were zero (it is "small" but it isn't zero), there would be no quantum mechanics.

CHECKPOINT 1: Rank the following radiations according to their associated photon energies, greatest first: (a) yellow light from a sodium vapor lamp, (b) a gamma ray emitted by a radioactive nucleus, (c) a radio wave emitted by the antenna of a commercial radio station, (d) a microwave beam emitted by an airport traffic control radar.

SAMPLE PROBLEM 39-1

A 100 W sodium vapor lamp is placed at the center of a large sphere, which absorbs all the sodium light that falls on it. At what rate are photons delivered to the sphere? The wavelength of sodium light is 590 nm.

SOLUTION: From Eq. 39-2 the energy per photon for sodium light is

$$E = hf = \frac{hc}{\lambda}$$

$$= \frac{(6.63 \times 10^{-34} \text{ J} \cdot \text{s})(3.00 \times 10^{8} \text{ m/s})}{590 \times 10^{-9} \text{ m}}$$

$$= 3.37 \times 10^{-19} \text{ J}.$$

Thus, every time a sodium atom emits a photon, the atom loses 3.37×10^{-19} J, or 2.1 eV, of energy. When sodium light is absorbed by the sphere, energy is transferred to the sphere in "lumps" of this same size.

To find the rate R at which photons are absorbed by the sphere, we divide E into the rate (the power P) at which the lamp emits energy:

$$R = \frac{P}{E} = \frac{100 \text{ W}}{3.37 \times 10^{-19} \text{ J/photon}}$$

$$= 3.0 \times 10^{20} \text{ photons/s.} \qquad \text{(Answer)}$$

That's a lot of photons! If you could read words ("photons" of information?) at this rate, you could read every book in the Library of Congress in about one nanosecond.

39-3 THE PHOTOELECTRIC EFFECT

If you shine a beam of light of short enough wavelength onto a clean metal surface, the light will knock electrons out of that surface. This **photoelectric effect** is used in many devices, including TV cameras, camcorders, and night vision viewers. Einstein supported his photon concept by using it to explain this effect, which simply cannot be understood in terms of classical physics.

Let us analyze two photoelectric experiments, each using the apparatus of Fig. 39-1 in which light of frequency f falls on target T and knocks electrons out of it. A potential difference V is maintained between target T and collector cup C to sweep up these electrons, said to be **photoelectrons**. This collection produces a **photoelectric current** i in meter A.

First Photoelectric Experiment

We adjust the potential difference V by moving the sliding contact in Fig. 39-1 so that collector C is negative with respect to target T. This potential difference acts to slow down the ejected electrons. We then vary V until it reaches a certain value, called the **stopping potential** V_{stop}, at which the reading of meter A has just dropped to zero. When $V = V_{stop}$, the most energetic ejected electrons are turned back just before reaching the collector. Then K_{max}, the kinetic energy of these most energetic electrons, is

$$K_{max} = eV_{stop}, \tag{39-4}$$

where e is the elementary charge.

Measurements show that for light of a given frequency, K_{max} *does not depend on the intensity of the light source.* Whether the source is dazzling bright or so feeble that you can scarcely detect it (or has some intermediate brightness), the maximum kinetic energy of the ejected electrons always has the same value.

This is a puzzle for classical physics. If you view the incident light as a classical electromagnetic wave, you have in mind the image of an electron in the target oscillating back and forth under the influence of the alternating electric field of the incident light wave. Under certain conditions, the oscillating electron will pick up enough energy to break through the surface of the target. If you increase the intensity of the incident light beam, you increase the amplitude of the alternating electric field, and it seems reasonable to suppose that this stronger alternating field will give a more energetic "kick" to the ejected electron. *That is not what happens.* For a given frequency, intense light beams and feeble light beams give exactly the same maximum kick to breakaway electrons.

The actual result follows naturally if we think in terms of photons. Now the maximum energy that an electron of

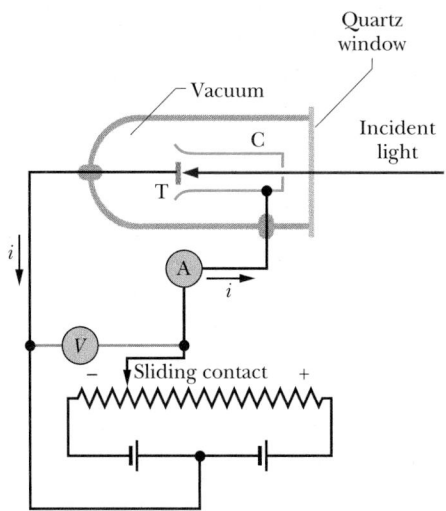

FIGURE 39-1 An apparatus used to study the photoelectric effect. The incident light falls on target T, ejecting electrons, which are collected by collector cup C. The electrons move in the circuit in a direction opposite the conventional current arrows. The batteries and the variable resistor are used to produce and adjust the electric potential difference between T and C.

target T in Fig. 39-1 can pick up from the incident light is only that of a single photon. Increasing the light intensity increases the *number* of photons at the target surface, but the *energy per photon*, given by Eq. 39-2, remains unchanged. Hence the maximum kinetic energy imparted to an electron does not change.

Second Photoelectric Experiment

Now we vary the frequency f of the incident light and measure the associated stopping potential V_{stop}. Figure 39-2 is a plot of V_{stop} against f. Note that the photoelectric effect does not occur if the frequency is less than a certain **cutoff frequency** f_0; it turns out that this is so *no matter how intense the incident light is.*

This is another puzzle for classical physics. If you view light as an electromagnetic wave, you must expect that no matter how low the frequency, electrons can always be ejected if you supply them with enough energy—that is, if you use a bright enough light source. *That is not what happens.* For light below a certain frequency, the photoelectric effect does not occur, no matter how bright the light source.

The existence of a cutoff frequency, however, is just what we should expect if the energy is transferred via photons. The electrons within the target are held there by electric forces. (If they weren't, they would drip out of the target under the influence of gravity!) To just escape from the target, an electron must pick up a certain minimum

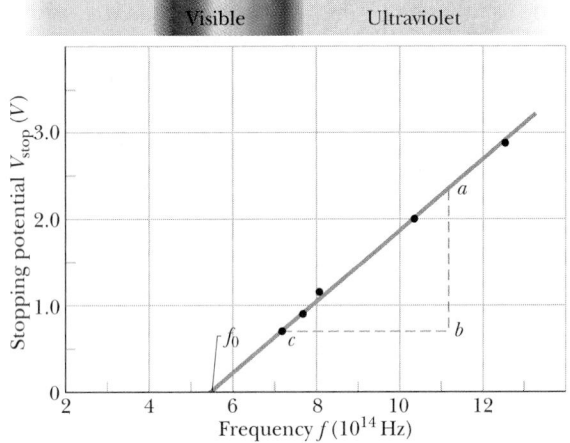

FIGURE 39-2 The stopping potential V_{stop} as a function of the frequency f of the incident light for a sodium target T in the apparatus of Fig. 39-1. (Data reported by R. A. Millikan in 1916.)

energy Φ, where Φ is a property of the target material called its **work function**. If the energy hf transferred to an electron by a photon exceeds the work function of the material (that is, if $hf > \Phi$), the electron can escape through the target surface. If the energy transferred does not exceed the work function (that is, if $hf < \Phi$), the electron cannot escape. This is just what Fig. 39-2 shows.

The Photoelectric Equation

Einstein summed up the results of our two photoelectric experiments in the equation

$$hf = K_{max} + \Phi \qquad \text{(photoelectric equation)}, \qquad (39\text{-}5)$$

which is a statement of the conservation of energy for a single interaction between a photon of frequency f and an electron in a target made of a material with work function Φ. If the electron is to escape from the target, it must pick up energy at least equal to Φ. Any remaining energy $(hf - \Phi)$ that the electron acquires from the interaction appears as kinetic energy of the electron. In the most favorable circumstance, the electron can escape through the surface without losing any of this kinetic energy in the process; it then appears outside the target with the maximum possible kinetic energy K_{max}.

Let us rewrite Eq. 39-5 by substituting for K_{max} from Eq. 39-4. After a little rearranging we get

$$V_{stop} = \left(\frac{h}{e}\right)f - \frac{\Phi}{e}. \qquad (39\text{-}6)$$

The ratios h/e and Φ/e are constants, so we expect a plot of the measured stopping potential V_{stop} against the frequency f to be a straight line, as in Fig. 39-2. Further, the slope of

that straight line should be h/e. As a check, we measure ab and bc in Fig. 39-2 and write

$$\frac{h}{e} = \frac{ab}{bc} = \frac{2.35 \text{ V} - 0.72 \text{ V}}{(11.2 \times 10^{14} - 7.2 \times 10^{14}) \text{ Hz}}$$
$$= 4.1 \times 10^{-15} \text{ V} \cdot \text{s}.$$

Multiplying this result by the elementary charge e, we find

$$h = (4.1 \times 10^{-15} \text{ V} \cdot \text{s})(1.6 \times 10^{-19} \text{ C})$$
$$= 6.6 \times 10^{-34} \text{ J} \cdot \text{s}.$$

This value of the Planck constant agrees with values measured by many other methods.

$\mathbb{C}$HECKPOINT 2: The figure shows data like those of Fig. 39-2 for targets of cesium, potassium, sodium, and lithium. The plots are parallel. (a) Rank the targets according to their work functions, greatest first. (b) Rank the plots according to the value of h they yield, greatest first.

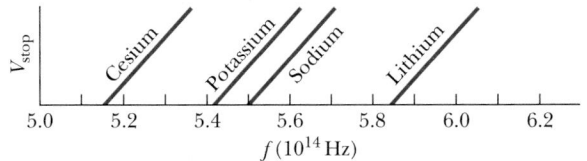

SAMPLE PROBLEM 39-2

A potassium foil is a distance $r = 3.5$ m from an isotropic light source that emits energy at the rate $P = 1.5$ W. The work function Φ of potassium is 2.2 eV. Suppose that the energy transported by the incident light were transferred to the target foil continuously and smoothly (that is, if classical physics prevailed and there were no photons). How long would it take for the foil to soak up enough energy to eject an electron? Assume that the foil absorbs all the energy that falls on it and that the to-be-ejected electron collects energy from a circular patch of the foil whose radius is 5.0×10^{-11} m, about that of a typical atom.

SOLUTION: Let us calculate the rate R at which light energy falls on the target patch. Knowing this, we can easily find how long it takes for the target patch to absorb 2.2 eV.

We assume that the energy emitted by the light source is spread uniformly over expanding spherical wavefronts centered on the source. From Eq. 34-27, the light intensity at the patch is

$$I = \frac{P}{4\pi r^2}$$
$$= \frac{1.5 \text{ W}}{4\pi(3.5 \text{ m})^2} = 9.74 \times 10^{-3} \text{ W/m}^2.$$

The patch has area $A = \pi(5.0 \times 10^{-11}\text{ m})^2$, which is 7.85×10^{-21} m². The rate at which it absorbs energy is then

$$R = IA = (9.74 \times 10^{-3}\text{ W/m}^2)(7.85 \times 10^{-21}\text{ m}^2)$$
$$= 7.65 \times 10^{-23}\text{ W}.$$

If all this energy were delivered to a single electron, the time required for the electron to soak up 2.2 eV would be

$$t = \frac{2.2\text{ eV}}{R} = \left(\frac{2.2\text{ eV}}{7.65 \times 10^{-23}\text{ J/s}}\right)\left(\frac{1.60 \times 10^{-19}\text{ J}}{1\text{ eV}}\right)$$
$$= 4600\text{ s} \approx 1.3\text{ h}. \qquad \text{(Answer)}$$

Thus, you would have to wait more than an hour after turning on the light source for a photoelectron to be ejected. The actual waiting time is less than 10^{-9} s. Apparently, then, the electron does *not* have to "soak up" energy from an incoming wave; it absorbs that energy *all at once* in a single photon–electron interaction.

SAMPLE PROBLEM 39-3

Find the work function Φ of sodium from the data plotted in Fig. 39-2.

SOLUTION: The cutoff frequency f_0 at which the graph in Fig. 39-2 intercepts the frequency axis appears to be about 5.5×10^{14} Hz. Photons at the cutoff frequency have energy just equal to the work function, so we may write Eq. 39-2 as

$$E = hf_0 = \Phi,$$

which gives us

$$\Phi = hf_0 = (6.63 \times 10^{-34}\text{ J·s})(5.5 \times 10^{14}\text{ Hz})$$
$$= 3.6 \times 10^{-19}\text{ J} = 2.3\text{ eV}. \qquad \text{(Answer)}$$

39-4 PHOTONS HAVE MOMENTUM

In 1916 Einstein extended his photon concept by asserting that when light interacts with matter, not only energy but also linear momentum is transferred via photons. Like energy, momentum is transferred in discrete amounts and at pointlike locations instead of broad regions.

The magnitude p of the momentum of a photon associated with a wave of frequency f is

$$p = \frac{hf}{c} = \frac{h}{\lambda} \qquad \text{(photon momentum).} \quad (39\text{-}7)$$

Equation 39-2 ($E = hf = hc/\lambda$) and Eq. 39-7 tell us, for example, that photons associated with a beam of x rays ($\lambda \approx 50$ pm) have much greater values of both energy and momentum than do photons associated with a beam of visible light ($\lambda \approx 500$ nm $= 5 \times 10^5$ pm.)

Some years later, in 1923, Arthur Compton at Washington University in St. Louis carried out an experiment that gave solid support to the view that both momentum and energy are transferred via photons. He arranged for a beam of x rays of wavelength λ to fall on a target made of carbon, as shown in Fig. 39-3. Compton measured the wavelengths and intensities of the x rays scattered in various directions from this target.

Figure 39-4 shows his results. Although there is only a single wavelength ($\lambda = 71.1$ pm) in the incident x-ray beam, we see that the scattered x rays contain a range of wavelengths with two prominent intensity peaks. One peak is centered about the incident wavelength λ, the other about a wavelength λ' that is larger than λ by an amount $\Delta\lambda$, which is called the **Compton shift**. The value of the Compton shift varies with the angle at which the scattered x rays are detected.

Figure 39-4 is still another puzzle for classical physics. If you think of the incident x-ray beam as an electromagnetic wave, you must imagine an electron in the carbon target oscillating back and forth under the influence of the alternating electric field of the incident wave. The electron will oscillate at the frequency of the alternating electric field and—like a tiny radio transmitting antenna—it will radiate *at this same frequency*. The scattered x rays should have the same frequency, and thus the same wavelength, as the incident beam. But they don't.

Compton interpreted the scattering of x rays from carbon in terms of energy and momentum transfers, via photons, between the incident x-ray beam and loosely bound electrons in the carbon target. Let us see, first conceptually and then quantitatively, how this quantum mechanical picture leads to an understanding of Compton's results.

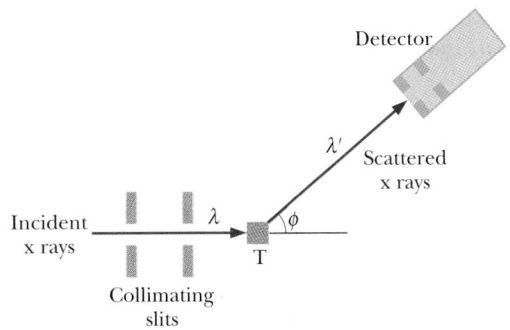

FIGURE 39-3 Compton's apparatus. A beam of x rays of wavelength $\lambda = 71.1$ pm falls on a carbon target T. The x rays scattered from the target are observed at various angles ϕ to the direction of the incident beam. The detector measures both the intensity of the scattered x rays and their wavelength.

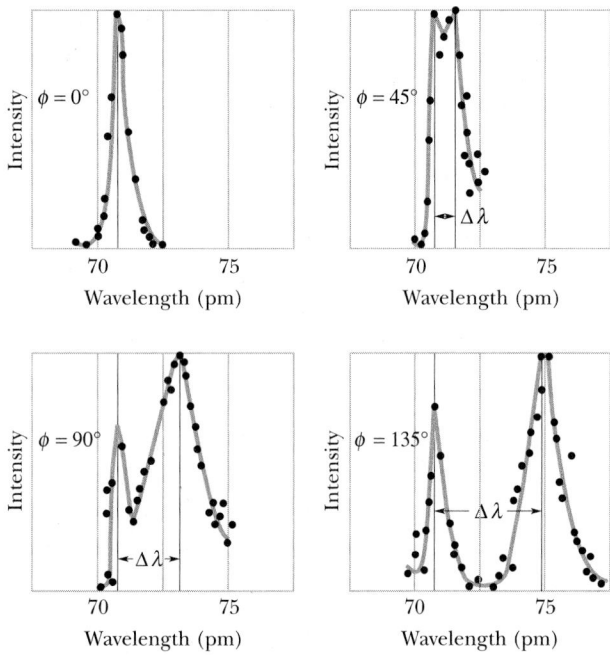

FIGURE 39-4 Compton's results for four values of the scattering angle ϕ. Note that the Compton shift $\Delta\lambda$ increases as the scattering angle increases.

Suppose a single photon (of energy $E = hf$) is associated with the interaction between the incident x-ray beam and a stationary electron. In general, the x-ray direction will change (an x-ray photon is said to be scattered) and the electron will recoil. Thus the electron will pick up some kinetic energy. Because energy is conserved in the interaction, the energy of the scattered photon ($E' = hf'$) must be less than that of the incident photon. The scattered x rays must then have a lower frequency f' and thus a longer wavelength λ' than the incident x rays, just as Compton's experimental results in Fig. 39-4 show.

For the quantitative part, we first apply the law of conservation of energy. Figure 39-5 suggests a "collision" between an x ray and an initially stationary free electron in the target. As a result of the collision, an x ray of wave-

length λ' moves off at an angle ϕ and the electron moves off at an angle θ, as shown. The conservation of energy then gives us

$$hf = hf' + K,$$

in which hf is the energy of the incident x-ray photon, hf' is the energy of the scattered x-ray photon, and K is the kinetic energy of the recoiling electron. Because the electron may recoil with a speed comparable to that of light, we must use the relativistic expression of Eq. 38-33,

$$K = mc^2(\gamma - 1),$$

for its kinetic energy. Here m is the electron's mass and γ is the Lorentz factor

$$\gamma = \sqrt{1 - (v/c)^2}.$$

Substituting for K in the conservation of energy equation yields

$$hf = hf' + mc^2(\gamma - 1).$$

Substituting c/λ for f and c/λ' for f' then leads to the energy conservation equation

$$\frac{h}{\lambda} = \frac{h}{\lambda'} + mc(\gamma - 1). \tag{39-8}$$

Now we need to apply the (vector) law of conservation of momentum to the x-ray–electron collision of Fig. 39-5. The momentum of the incident and scattered photons is given by Eq. 39-7 ($p = h/\lambda$) and that of the scattered electron by Eq. 38-31 ($p = \gamma mv$). By writing separate equations for the conservation of momentum for the x and y directions, we get

$$\frac{h}{\lambda} = \frac{h}{\lambda'} \cos\phi + \gamma mv \cos\theta \qquad (x \text{ direction}) \tag{39-9}$$

and

$$0 = \frac{h}{\lambda'} \sin\phi - \gamma mv \sin\theta \qquad (y \text{ direction}). \tag{39-10}$$

We want to find $\Delta\lambda$ ($= \lambda' - \lambda$), the Compton shift of the scattered x rays. Of the five collision variables (λ, λ', v, ϕ, and θ) that appear in Eqs. 39-8, 39-9, and 39-10, we choose to eliminate v and θ, which deal only with the recoiling electron. Carrying out the algebra (it is somewhat complicated) leads to an equation for the Compton shift as a function of the scattering angle ϕ:

$$\Delta\lambda = \frac{h}{mc}(1 - \cos\phi) \qquad (\text{Compton shift}). \tag{39-11}$$

Here the quantity h/mc, called the **Compton wavelength** of the electron, is a constant. Equation 39-11 agrees exactly with Compton's experimental results.

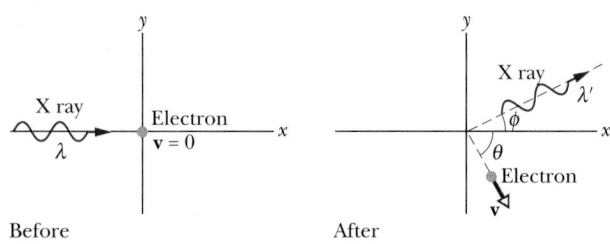

FIGURE 39-5 An x ray of wavelength λ interacts with a stationary electron. The x ray is scattered at angle ϕ, with an increased wavelength λ'. The electron moves off with speed v at angle θ.

A Loose End

It remains to explain the peak at the incident wavelength λ (= 71.1 pm) in Fig. 39-4. This peak arises not from encounters between x rays and the very loosely bound electrons in the target but from encounters between x rays and the electrons that are *tightly* bound to the carbon atoms making up the target. Effectively, each of these latter collisions is between an incident x ray and an entire carbon atom. If we substitute for m in Eq. 39-11 the mass of a carbon atom (which is about 22,000 times that of an electron), we see that $\Delta\lambda$ is about 22,000 times smaller than the Compton shift for an electron—too small to detect. Thus the x rays scattered in these collisions have a detected wavelength equal to that of the incident x rays.

SAMPLE PROBLEM 39-4

X rays of wavelength $\lambda = 22$ pm (photon energy = 56 keV) are scattered from a carbon target, and the scattered rays are detected at 85° to the incident beam.

(a) What is the Compton shift of the scattered rays?

SOLUTION: From Eq. 39-11 we have

$$\Delta\lambda = \frac{h}{mc}(1 - \cos\phi)$$
$$= \frac{(6.63 \times 10^{-34} \text{ J·s})(1 - \cos 85°)}{(9.11 \times 10^{-31} \text{ kg})(3.00 \times 10^8 \text{ m/s})}$$
$$= 2.21 \times 10^{-12} \text{ m} \approx 2.2 \text{ pm.} \quad \text{(Answer)}$$

(b) What percentage of the initial x-ray photon energy is transferred to an electron in such scattering?

SOLUTION: The fractional energy loss *frac* is

$$frac = \frac{E - E'}{E} = \frac{hf - hf'}{hf} = \frac{c/\lambda - c/\lambda'}{c/\lambda} = \frac{\lambda' - \lambda}{\lambda'}$$
$$= \frac{\Delta\lambda}{\lambda + \Delta\lambda}. \quad (39\text{-}12)$$

Substitution yields

$$frac = \frac{2.21 \text{ pm}}{22 \text{ pm} + 2.21 \text{ pm}} = 0.091 \quad \text{or} \quad 9.1\%. \quad \text{(Answer)}$$

Although the Compton shift $\Delta\lambda$ is independent of the wavelength λ of the incident x rays (see Eq. 39-11), the *fractional* photon energy loss of the x rays does depend on λ, increasing as the wavelength of the incident radiation decreases, as indicated by Eq. 39-12.

CHECKPOINT **3:** Compare Compton scattering for x rays ($\lambda \approx 20$ pm) and visible light ($\lambda \approx 500$ nm) at a particular angle of scattering. Which has the greater (a) Compton shift, (b) fractional wavelength shift, (c) fractional photon energy change, and (d) energy imparted to the electron?

39-5 LIGHT AS A PROBABILITY WAVE

You may wonder how light can be a wave and still be generated and absorbed as fixed amounts of energy—photons—in interactions. A new look at the double-slit experiment of Section 36-4 provides some insights. We consider three versions of this important experiment.

The Standard Version

Figure 39-6 reminds us of the original experiment carried out by Thomas Young in 1801, which we first saw in Fig. 36-6. Light falls on screen B, which contains two narrow parallel slits. The light waves emerging from each slit spread out by diffraction and overlap on screen C where, by interference, they form a pattern of alternating intensity maxima and minima. In Section 36-4 we took the existence of these interference fringes as evidence for the wave nature of light.

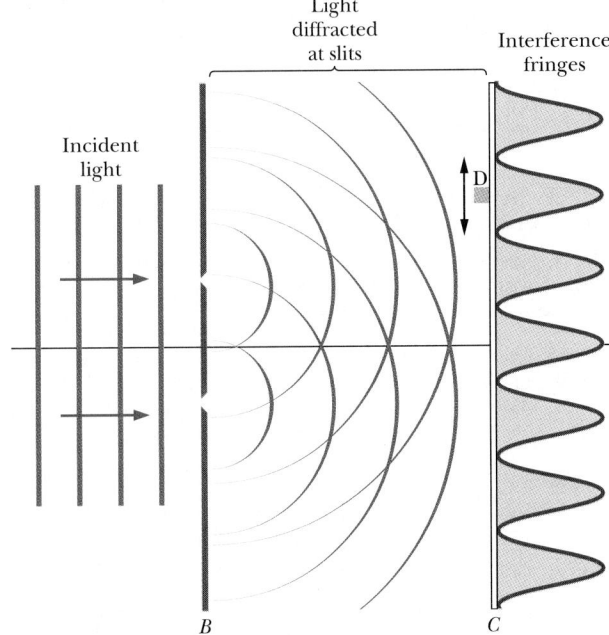

FIGURE 39-6 Light falls on screen B, which contains two parallel slits. Light emerging from these slits spreads out by diffraction. The two diffracted waves overlap at screen C and form a pattern of interference fringes. A small photon detector D in the plane of screen C generates a sharp click for each photon that it absorbs.

Let us place a tiny photon detector D—perhaps a photoelectric device that clicks when it absorbs a photon—at one point in the plane of screen C. We find that it produces a series of clicks, randomly spaced in time, each click signaling the transfer of energy from the light wave to the screen in a single photon-sized lump.

If we move the detector very slowly up or down as indicated in Fig. 39-6, we find that the click rate increases and decreases, passing through alternate maxima and minima that correspond exactly to the maxima and the minima of the interference fringes.

The point of this thought experiment is as follows. We cannot predict when a photon will be detected at any particular point on screen C; photons are detected at individual points at random intervals. We can, however, predict that the relative *probability* that a single photon will be detected at a particular point in a specified time interval is proportional to the intensity of the incident light at that point on screen C.

We saw in Section 34-4 that the intensity I of a light wave at any point is proportional to the square of E_m, the amplitude of the alternating electric field vector, at that point. Thus,

---

The probability (per unit time interval) that a photon will be detected in any small volume centered on a given point in a light wave is proportional to the square of the amplitude of the wave's electric field vector at that point.

We now have a probabilistic description of a light wave, hence another way to view light. It is not only an electromagnetic wave but it is also a **probability wave**. That is, to every point in a light wave we can attach a numerical probability (per unit time interval) that a photon can be detected in any small volume centered on that point.

The Single-Photon Version

A single-photon version of the double-slit experiment was first carried out by G. I. Taylor in 1909 and has been repeated many times since. It differs from the standard version in that the light source is so extremely feeble that it emits only one photon at a time, at random intervals. Astonishingly, interference fringes still build up on screen C if the experiment runs long enough (several months for Taylor's early experiment).

What explanation can we offer for this single-photon, double-slit experiment? Before we can even consider the results, we are compelled to ask questions like these: If the photons move through the apparatus one at a time, through

which of the two slits in screen B does a given photon pass? And how does a given photon even "know" that there is another slit present so that interference is a possibility? And can a single photon somehow pass through both slits and interfere with itself?

Bear in mind that photons manifest themselves only when light interacts with matter. Thus we know that photons *originate* in the source that generates the incident light of Fig. 39-6. And photons *vanish* in screen C, where light interacts with the solid matter that makes up the screen. Between source and detector, however, we *postulate* that light travels *not* as a stream of photons but as a probability wave. Such a wave, no matter how feeble the light source, can be diffracted at each slit. The two diffracted probability subwaves (one from each slit) can then interfere with each other when they meet at any point on screen C, producing a pattern of maximum and minimum "probability fringes" on that screen. Photons will tend to occur in regions of maximum probability and to not occur in regions of minimum probability.

That seems to be a satisfactory explanation of the single-photon experiment, since classical physics offers no explanation at all. According to physicist Richard Feynman,

[the single-photon, double-slit experiment is] a phenomenon which is impossible, *absolutely* impossible, to explain in any classical way, and which has in it the heart of quantum mechanics.

The Single-Photon, Wide-Angle Version

Figure 39-7 shows the arrangement used in another version of the two-slit experiment, reported in 1992 by Ming Lai and Jean-Claude Diels of the University of New Mexico. Source S contains molecules that emit photons that are well

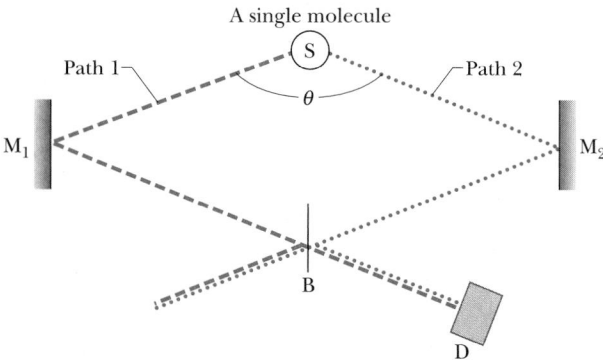

FIGURE 39-7 The light from a single photon emission in source S travels over two widely separated paths and interferes with itself at detector D after being recombined by the beam splitter B. (After Ming Lai and Jean-Claude Diels, *Journal of the Optical Society of America B*, **9**, 2290–2294, December 1992.)

separated in time. Mirrors M_1 and M_2 are positioned to reflect light that the source emits along two distinct paths, 1 and 2, that are separated by an angle θ, which is close to 180°. This arrangement differs from the standard two-slit experiment, in which the angle between the paths of the light waves falling on two slits is very small.

After reflection from mirrors M_1 and M_2, the light waves traveling along paths 1 and 2 meet at beam splitter B. (A beam splitter is an optical device that transmits half the light incident upon it and reflects the other half.) On the right side of the beam splitter in Fig. 39-7 the light wave traveling along path 2 and reflected by B combines with the light wave traveling along path 1 and transmitted by B. These two waves then interfere with each other as they arrive at detector D (a *photomultiplier tube* that can detect individual photons).

The output of the detector is a randomly spaced series of electronic pulses, one for each detected photon. In the experiment, the beam splitter is moved slowly in a horizontal direction (in the reported experiment, only about 50 μm maximum), and the detector output is recorded on a chart recorder. Moving the beam splitter changes the lengths of paths 1 and 2, producing a phase shift between the light waves arriving at detector D. Interference maxima and minima appear in the detector's output signal.

This experiment is difficult to understand in traditional terms. For example, when a molecule in the source emits a single photon, does that photon travel along path 1 or path 2 in Fig. 39-7 (or along any other path)? How can it move in both directions at once? To answer, we assume that when the emitting molecule makes a quantum transition to a lower energy level, a probability wave radiates in all directions from it. The experiment samples this wave in two of those directions, chosen to be nearly opposite each other.

We see that we can interpret all three versions of the double-slit experiment if we assume that (1) light is generated in the source as photons, (2) light is absorbed in the detector as photons, and (3) light travels between source and detector as a probability wave.

39-6 ELECTRONS AND MATTER WAVES

Physicists have rarely gone wrong by assuming the symmetry of nature. Thus, when you learn that a changing magnetic field produces an electric field, you might guess —as both Faraday and Maxwell did—that a changing electric field produces a magnetic field. And it does.

In 1924 French physicist Louis de Broglie made the following appeal to symmetry. A beam of light is a wave, but it transfers energy and momentum to matter in photon-sized lumps. Why can't a beam of particles have the same properties? That is, why can't we think of a moving electron—or any other particle, for that matter—as a **matter wave**?

In particular, de Broglie suggested that Eq. 39-7 ($p = h/\lambda$) might apply not only to photons but also to particles of matter, the electron being a convenient prototype. We used that equation in Section 39-4 to assign a momentum p to a photon, knowing the wavelength λ of its associated wave. We now use it, in the form

$$\lambda = \frac{h}{p} \quad \text{(de Broglie wavelength)} \quad (39\text{-}13)$$

to assign a wavelength λ to a particle whose momentum is p. The wavelength calculated from Eq. 39-13 is called the **de Broglie wavelength** of the moving particle.

De Broglie's prediction of the existence of matter waves was first verified experimentally in 1927, by C. J. Davisson and L. H. Germer of the Bell Telephone Laboratories and by George P. Thomson of the University of Aberdeen in Scotland. More recently, the wave nature of a beam of electrons was demonstrated in a 1989 double-slit experiment like that used to demonstrate the wave nature of light. Figure 39-8 suggests how the fringe pattern builds up with time in this experiment as individual electrons strike the detecting screen. And in 1994 interference fringes were generated with beams of iodine molecules, which are about 500,000 times more massive than electrons.

Figure 39-9a suggests another experiment in which a beam, of either x rays or electrons, is allowed to fall on a target consisting of a powder of tiny aluminum crystals. Scattered by the crystals, the beam emerges from the target with circular symmetry about its initial direction and, because of Bragg reflections on the atomic planes of the aluminum crystals (see Section 37-9), forms concentric rings on a sheet of photographic film placed as indicated. Figure 39-9b shows the result for an x-ray beam; Fig. 39-9c shows the result for an electron beam. The geometries of the rings are identical, showing that both x rays and electrons behave like waves in this experiment. (The energy of the x-ray photons and the momentum of the electrons were chosen so that both beams had the same wavelength.)

We now take the wave nature of matter for granted. Diffraction studies involving beams of electrons or neutrons are used routinely to study the atomic structures of solids and liquids. Matter waves are a valuable supplement to x rays in such studies. Electrons, for example, are less penetrating than x rays and so are particularly useful in studying surface features. Also, x rays interact largely with electrons in a target and for that reason are not effective in locating low-mass atoms—particularly hydrogen—that have few electrons. Neutrons, on the other hand, interact primarily with the nuclei of the target atoms and so are useful where x rays are not.

FIGURE 39-8 The buildup of an interference pattern by a beam of electrons in a two-slit interference experiment like that of Fig. 39-6. Matter waves, like light waves, are *probability waves*. From top to bottom the approximate numbers of electrons involved are 7, 100, 3000, 20,000, and 70,000.

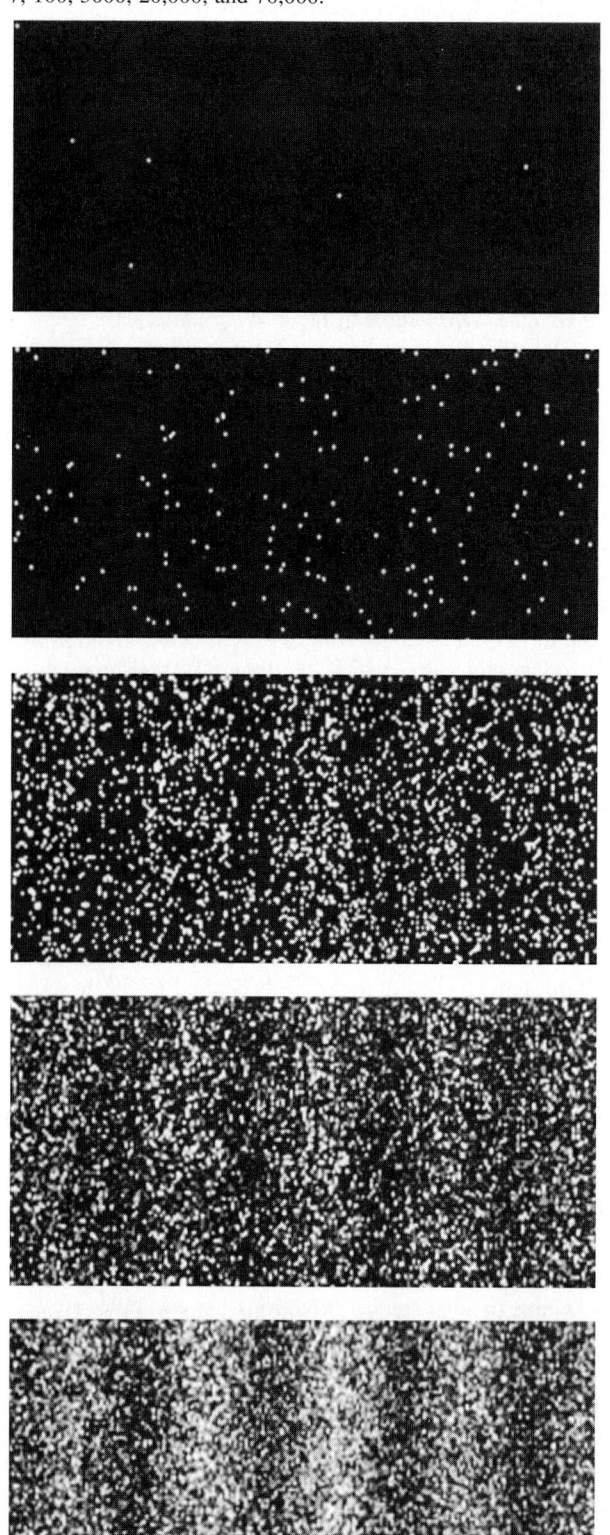

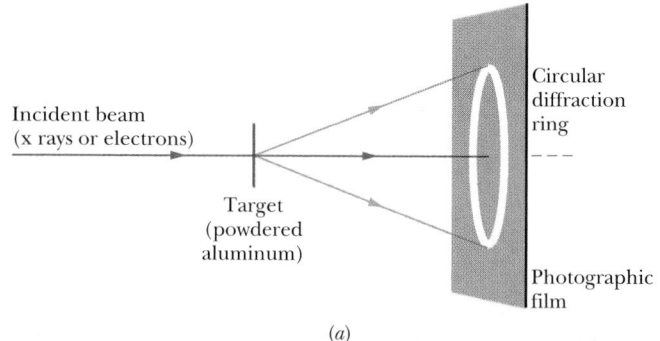

(a)

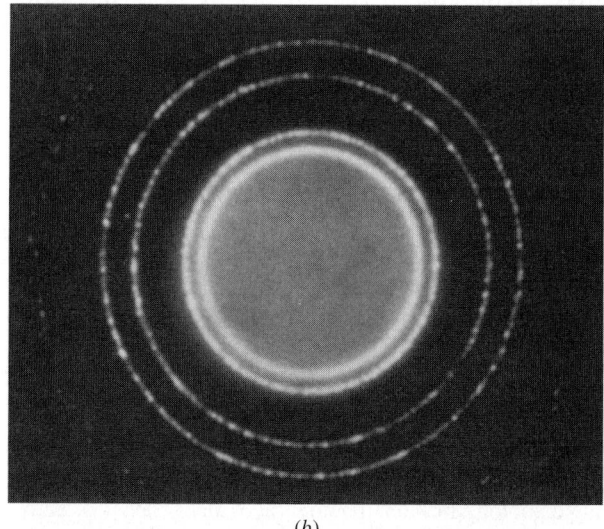

(b)

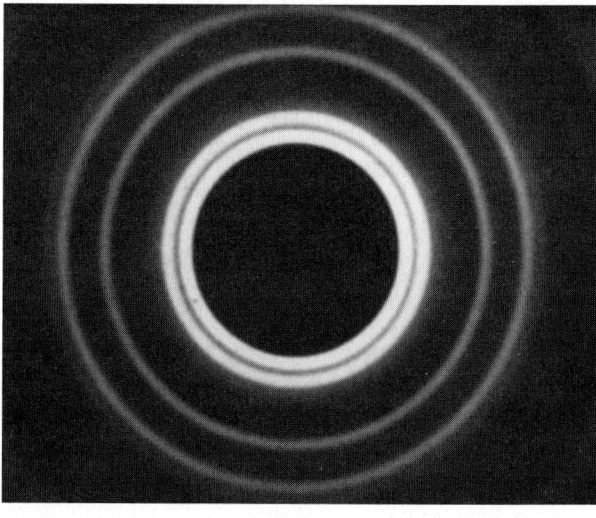

(c)

FIGURE 39-9 (a) An experimental arrangement used to demonstrate, by diffraction techniques, the wavelike character of the incident beam. (b) The diffraction pattern when the incident beam is an x-ray beam (light wave). (c) The diffraction pattern when the incident beam is an electron beam (matter wave). Note the basic geometrical identity of the patterns.

Figure 39-10 shows the structure of solid benzene as deduced from neutron diffraction studies. Each set of concentric blue circles shows the location of one of the six carbon atoms that form the familiar benzene ring. Each set of red circles shows the location of a hydrogen atom that is coupled to a carbon atom.

Waves and Particles

Figures 39-9 and 39-10 are convincing evidence of the *wave* nature of matter. But we have at least as many experiments that suggest the *particle* nature of matter. Consider the tracks generated by electrons and displayed in the opening photo of this chapter. Surely these tracks—which are strings of bubbles left in the liquid hydrogen that fills the bubble chamber—strongly suggest the passage of a particle. Where is the wave?

To simplify the situation, let us turn off the magnetic field so that the strings of bubbles will then be straight. We can view each bubble as a detection point for the electron. Matter waves traveling between detection points such as I and F in Fig. 39-11 explore all possible paths, a few of which are shown.

In general, for every path connecting I and F there will be a neighboring path such that matter waves following the two paths cancel each other by interference. This is not true, however, for the straight-line path joining I and F; in this case, matter waves traversing all neighboring paths reinforce the wave following the direct path. You can think of the bubbles that form the track as a series of detection points at which the matter wave undergoes constructive interference.

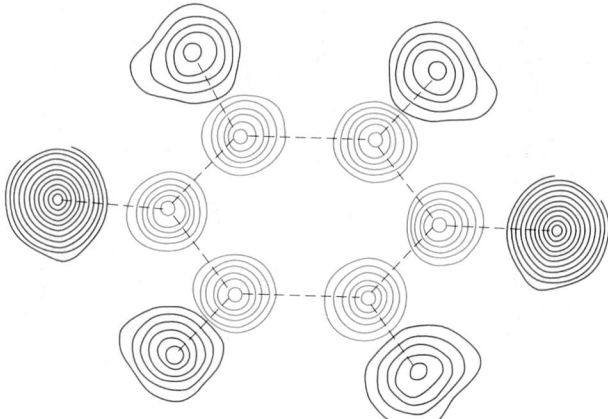

FIGURE 39-10 The atomic structure of solid benzene as revealed using neutron diffraction. The closed curves suggest the patterns of electron density in the solid target. You can easily see the familiar benzene ring of six carbon atoms (blue) and the hydrogen atoms (red) that are coupled to them.

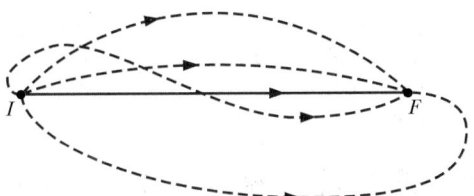

FIGURE 39-11 A few of the many paths that connect two particle detection points I and F. Only matter waves that follow paths close to the straight line between these points interfere constructively. For all other paths, the waves following neighboring paths interfere destructively. Thus a matter wave leaves a straight track.

SAMPLE PROBLEM 39-5

What is the de Broglie wavelength of an electron whose kinetic energy is 120 eV?

SOLUTION: We can find the de Broglie wavelength from Eq. 39-13 if we know the momentum of the electron. From the (nonrelativistic) relation $K = \frac{1}{2}mv^2$ you can show that, for $K = 120$ eV, $v = 6.5 \times 10^6$ m/s, which is well below the speed of light. This low value for v justifies our use of the nonrelativistic expressions for both kinetic energy and momentum ($p = mv$). Eliminating the particle speed v between these two equations yields

$$p = \sqrt{2mK}$$
$$= \sqrt{(2)(9.11 \times 10^{-31} \text{ kg})(120 \text{ eV})(1.60 \times 10^{-19} \text{ J/eV})}$$
$$= 5.91 \times 10^{-24} \text{ kg} \cdot \text{m/s}.$$

From Eq. 39-13 then

$$\lambda = \frac{h}{p} = \frac{6.63 \times 10^{-34} \text{ J} \cdot \text{s}}{5.91 \times 10^{-24} \text{ kg} \cdot \text{m/s}}$$
$$= 1.12 \times 10^{-10} \text{ m} = 112 \text{ pm}. \qquad \text{(Answer)}$$

This is about the size of a typical atom.

CHECKPOINT **4:** An electron and a proton have the same (a) kinetic energy, (b) momentum, (c) speed. In each case, which particle has the shorter de Broglie wavelength?

39-7 SCHRÖDINGER'S EQUATION

A simple traveling wave of any kind, be it a wave on a string, a sound wave, or a light wave, is described in terms of some quantity that varies in a wavelike fashion. For light waves, for example, this quantity is $\mathbf{E}$ (x, y, z, t), the electric field component of the wave. Its observed value at any point depends on the location of that point and on the time at which the observation is made.

What varying quantity should we use to describe a matter wave? We should expect this quantity, which we call the **wave function** $\Psi (x, y, z, t)$, to be more complicated than the corresponding quantity for a light wave because a matter wave, in addition to energy and momentum, transports mass and (often) electric charge. It turns out that Ψ, the uppercase Greek letter psi, usually represents a function that is complex in the mathematical sense; that is, we can always write its values in the form $a + ib$, in which a and b are real numbers and $i^2 = -1$.

In all the situations you will meet here, the space and time variables can be grouped separately and Ψ can be written in the form

$$\Psi (x, y, z, t) = \psi (x, y, z)\, e^{-i\omega t}, \qquad (39\text{-}14)$$

where $\omega \,(= 2\pi f)$ is the angular frequency of the matter wave. Note that ψ, the lowercase Greek letter psi, represents only the space-dependent part of the complete, time-dependent wave function Ψ. We shall deal almost exclusively with ψ. Two questions arise: What is meant by the wave function, and how do we find it?

What does the wave function mean? A matter wave, like a light wave, is a probability wave. Suppose, for example, that a matter wave falls on a particle detector that is small; then the probability that a particle will be detected in a specified time interval is proportional to $|\psi|^2$, where $|\psi|$ is the absolute value of the wave function at the location of the detector. Although ψ is usually a complex quantity, $|\psi|^2$ is always both real and positive. It is, then, $|\psi|^2$, which we call the **probability density**, and not ψ, that has *physical* meaning. Speaking loosely, the meaning is this:

The probability (per unit time) of detecting a particle in a small volume centered on a given point in a matter wave is proportional to the value of $|\psi|^2$ at that point.

Because ψ is usually a complex quantity, we find the square of its absolute value by multiplying ψ by ψ^*, the *complex conjugate* of ψ. (To find ψ^* we replace the imaginary number i in ψ with $-i$, wherever it occurs.)

How do we find the wave function? Sound waves and waves in strings are described by the equations of Newtonian mechanics. Light waves are described by Maxwell's equations. Matter waves are described by **Schrödinger's equation**, advanced in 1926 by Austrian physicist Erwin Schrödinger.

Many of the situations that we shall discuss involve a particle traveling in the x direction through a region in which forces acting on the particle cause it to have a potential energy $E_{pot}(x)$. In this special case, Schrödinger's equation reduces to

$$\frac{d^2\psi}{dx^2} + \frac{8\pi^2 m}{h^2}[E - E_{pot}(x)]\psi = 0$$

(Schrödinger's equation, (39-15)
one-dimensional motion),

in which E is the total mechanical energy (potential energy plus kinetic energy) of the moving particle. We cannot derive Schrödinger's equation from more basic principles; it *is* the basic principle.

If $E_{pot}(x)$ in Eq. 39-15 is zero, that equation describes a **free particle**, that is, a moving particle on which no net force acts. The particle's total energy in this case is all kinetic, and thus E in Eq. 39-15 is $\frac{1}{2}mv^2$. That equation then becomes

$$\frac{d^2\psi}{dx^2} + \frac{8\pi^2 m}{h^2}\left(\frac{mv^2}{2}\right)\psi = 0,$$

which we can recast as

$$\frac{d^2\psi}{dx^2} + \left(2\pi\frac{p}{h}\right)^2\psi = 0.$$

To obtain this equation, we replaced mv with the momentum p and regrouped terms.

From Eq. 39-13 we recognize p/h in the equation above as $1/\lambda$, where λ is the de Broglie wavelength of the moving particle. We further recognize $2\pi/\lambda$ as the *angular wave number* k, which we defined in Eq. 17-5. With this substitution, the equation above becomes

$$\frac{d^2\psi}{dx^2} + k^2\psi = 0 \qquad \begin{array}{c}\text{(Schrödinger's equation,} \\ \text{free particle).}\end{array} \qquad (39\text{-}16)$$

The most general solution of Eq. 39-16 is

$$\psi(x) = Ae^{ikx} + Be^{-ikx}, \qquad (39\text{-}17)$$

in which A and B are arbitrary constants. You can show that this equation is indeed a solution of Eq. 39-16 by substituting $\psi(x)$ and its second derivative into that equation and noting that an identity results.

If we combine Eqs. 39-14 and 39-17 we find, for the time-dependent wave function Ψ of a free particle traveling in the x direction,

$$\Psi(x, t) = \psi(x)e^{-i\omega t}$$
$$= (Ae^{ikx} + Be^{-ikx})e^{-i\omega t}$$
$$= Ae^{i(kx-\omega t)} + Be^{-i(kx+\omega t)}. \qquad (39\text{-}18)$$

Finding the Probability Density $|\psi|^2$

In Section 17-5 we saw that *any function F of the form* $F(kx \pm \omega t)$ represents a traveling wave. This applies to exponential functions like those in Eq. 39-18 as well as to the sinusoidal functions we have used to describe waves on strings. In fact, these two representations of functions are related by

$$e^{i\theta} = \cos\theta + i\sin\theta \quad \text{and} \quad e^{-i\theta} = \cos\theta - i\sin\theta,$$

where θ is any angle.

The first term on the right in Eq. 39-18 thus represents a wave traveling in the direction of increasing x and the second a wave traveling in the direction of decreasing x. However, we have assumed that the free particle we are considering travels only in the direction of *increasing x*. To reduce the general solution (Eq. 39-18) to our case of interest, we choose the arbitrary constant B in Eqs. 39-18 and 39-17 to be zero. At the same time, we relabel the constant A as ψ_0. Equation 39-17 then becomes

$$\psi(x) = \psi_0\, e^{ikx}. \tag{39-19}$$

To calculate the probability density, we take the square of the absolute value of $\psi(x)$. We get

$$|\psi|^2 = |\psi_0\, e^{ikx}|^2 = (\psi_0^2)\,|e^{ikx}|^2.$$

Now, because

$$|e^{ikx}|^2 = (e^{ikx})(e^{ikx})^* = e^{ikx}\, e^{-ikx} = e^{ikx-ikx} = e^0 = 1,$$

we get

$$|\psi|^2 = (\psi_0^2)(1)^2 = \psi_0^2 \qquad \text{(a constant)}.$$

Figure 39-12 is a plot of the probability density $|\psi|^2$ versus x for a free particle—a straight line parallel to the x axis from $-\infty$ to $+\infty$. We see that the probability density $|\psi|^2$ is the same for all values of x, which means that the particle has equal probabilities of being *everywhere* along the x axis. There is no distinguishing feature by which we can predict a most likely position for the particle.

We'll see what this means in the next section.

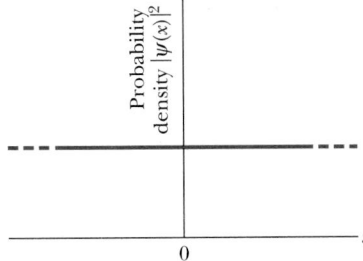

FIGURE 39-12 A plot of the probability density $|\psi|^2$ for a free particle moving in the positive x direction. Since $|\psi|^2$ has the same constant value for all values of x, the particle has the same probability of detection at all points.

39-8 HEISENBERG'S UNCERTAINTY PRINCIPLE

Our inability to predict the position of the free particle in Fig. 39-12 is our first example of **Heisenberg's uncertainty principle**, proposed in 1927 by German physicist Werner Heisenberg. It states that measured values cannot be assigned to the position **r** and the momentum **p** of a particle simultaneously with unlimited precision.

For the components of **r** and **p**, Heisenberg's principle gives the following limits in terms of $\hbar = h/2\pi$ (called "h-bar"):

$$
\begin{aligned}
\Delta x \cdot \Delta p_x &\geq \hbar \\
\Delta y \cdot \Delta p_y &\geq \hbar \\
\Delta z \cdot \Delta p_z &\geq \hbar
\end{aligned}
\qquad
\begin{aligned}
&\text{(Heisenberg's} \\
&\text{uncertainty} \\
&\text{principle).}
\end{aligned}
\qquad (39\text{-}20)
$$

Here Δx and Δp_x, as examples, represent the intrinsic uncertainties in the measurements of the x components of **r** and **p**. Even with the best measuring instruments that modern technology can provide, each product of a position uncertainty and a momentum uncertainty in Eq. 39-20 will be greater than $\hbar$; it can *never* be less.

The particle whose probability density is plotted in Fig. 39-12 is a free particle. That is, no force acts on it, so its momentum **p** must be constant. We implied—without making a point of it—that we can determine **p** with absolute precision. That is, we assumed that $\Delta p_x = \Delta p_y = \Delta p_z = 0$ in Eq. 39-20. That assumption then requires $\Delta x \to \infty$, $\Delta y \to \infty$, and $\Delta z \to \infty$. With such infinitely great uncertainties, the position of the particle is completely unspecified, just as Fig. 39-12 shows.

Do not think that the particle *really has* a sharply defined position that is, for some reason, hidden from us. If its momentum can be specified with absolute precision, the words "position of the particle" simply lose all meaning. The particle in Fig. 39-12 can be found *with equal probability* anywhere along the x axis.

SAMPLE PROBLEM 39-6

An electron of kinetic energy 12.0 eV can be shown to have a speed of 2.05×10^6 m/s. Assume that the electron is moving in the x direction and that you can measure its speed with a precision of 0.50%. What is the minimum uncertainty (that is, the uncertainty due to Heisenberg's uncertainty principle) with which you can simultaneously measure the position of the electron along the x axis?

SOLUTION: The electron's speed is well below the speed of light, so we can find its momentum from the nonrelativistic formula:

$$
\begin{aligned}
p = p_x = mv &= (9.11 \times 10^{-31}\ \text{kg})(2.05 \times 10^6\ \text{m/s}) \\
&= 1.87 \times 10^{-24}\ \text{kg} \cdot \text{m/s}.
\end{aligned}
$$

The uncertainty Δp_x in the momentum measurement is 0.50% of this, or 9.35×10^{-27} kg·m/s. From Heisenberg's principle (Eq. 39-20), the minimum uncertainty in the position measurement is then

$$\Delta x \approx \frac{\hbar}{\Delta p_x} = \frac{(6.63 \times 10^{-34} \text{ J} \cdot \text{s})/2\pi}{9.35 \times 10^{-27} \text{ kg} \cdot \text{m/s}}$$

$$= 1.13 \times 10^{-8} \text{ m} \approx 11 \text{ nm}, \quad \text{(Answer)}$$

which is about 100 atomic diameters. Given your measurement of the electron's momentum, it makes no sense to try to pin down the electron's position to any greater precision.

39-9 BARRIER TUNNELING

Suppose you repeatedly flip a jelly bean along a tabletop on which a book is positioned somewhere along the jelly bean's path. You would be very surprised to see the jelly bean appear on the other side of the book instead of bouncing from it. Don't expect this to happen for jelly beans. However, something very much like it, called **barrier tunneling**, *does* happen for electrons and other particles with small masses.

Figure 39-13*a* shows an electron of total energy E moving parallel to the x axis. Forces act on it such that its potential energy is zero except when it is in the region

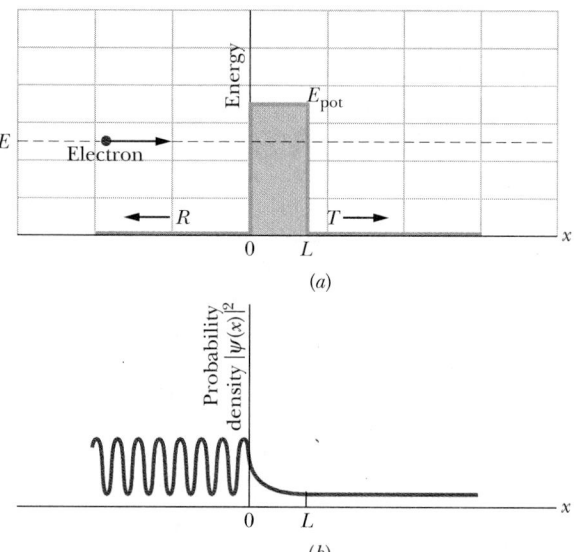

(a)

(b)

FIGURE 39-13 (*a*) An energy diagram showing a potential energy barrier of height E_{pot} and thickness L. An electron with total energy E approaches the barrier from the left. (*b*) The probability density $|\psi|^2$ of the matter wave representing the electron, showing the tunneling of the electron through the barrier. The pattern to the left of the barrier is a standing matter wave due to the superposition of the incident and reflected matter waves.

$0 < x < L$, where its potential energy has the constant value E_{pot}. We define this region as a **potential energy barrier** (often loosely called a **potential barrier**) of height E_{pot} and thickness L.

Classically, because $E < E_{\text{pot}}$, an electron approaching the barrier from the left would be reflected from the barrier and would move back in the direction from which it came. In quantum mechanics, however, there is a finite chance that the matter wave associated with the electron will "leak through" the barrier and appear on the other side. This means that there is a finite probability that the electron will be found on the far side of the barrier, moving to the right.

The wave function $\psi(x)$ describing the electron can be found by solving Schrödinger's equation (Eq. 39-15) separately for the three regions shown in Fig. 39-13*a*: (1) to the left of the barrier, (2) within the barrier, and (3) to the right of the barrier. The arbitrary constants that appear in the solutions are then chosen so that the values of $\psi(x)$ join smoothly (no jumps, no kinks) at $x = 0$ and at $x = L$. Squaring the absolute value of $\psi(x)$ then yields the probability density.

Figure 39-13*b* shows a plot of the result. The oscillating curve to the left of the barrier (for $x < 0$) in Fig. 39-13*a* is a combination of the incident matter wave and the reflected matter wave (which has a smaller amplitude than the incident wave). The oscillations occur because these two waves, traveling in opposite directions, interfere with each other, setting up a standing wave pattern.

Within the barrier (for $0 < x < L$) the probability density decreases exponentially with x. However, provided L is small, the probability density is not quite zero at $x = L$.

To the right of the barrier of Fig. 39-13 (for $x > L$), the probability density plot describes a transmitted wave with low but constant amplitude. Thus the electron can be found in this region with an equal but relatively small probability anywhere along the x axis if it happens to tunnel through the barrier. (Compare this part of the figure with Fig. 39-12 for a free particle.)

We can assign a *transmission coefficient* T to the incident matter wave and the barrier in Fig. 39-13*a*. This coefficient gives the probability with which an approaching electron will be transmitted through the barrier, that is, that tunneling will occur. As an example, if $T = 0.02$, then of every 1000 electrons fired at the barrier, 20 (on average) will tunnel through it and 980 will be reflected.

The **transmission coefficient** T is approximately

$$T \approx e^{-2kL}, \tag{39-21}$$

in which

$$k = \sqrt{\frac{8\pi^2 m(E_{\text{pot}} - E)}{h^2}}. \tag{39-22}$$

Because of the exponential form of Eq. 39-21, the value of T is very sensitive to the three variables on which it depends: particle mass m, barrier thickness L, and energy difference $E_{pot} - E$.

Barrier tunneling finds many applications in technology, among them the tunnel diode, in which the flow of electrons (by tunneling through a device) can be rapidly turned on or off by controlling the barrier height. This can be done very quickly (within 5 ps), so the device is suitable for applications demanding a high-speed response. The 1973 Nobel prize was shared by three "tunnelers," Leo Esaki (for tunneling in semiconductors), Ivar Giaever (for tunneling in superconductors), and Brian Josephson (for the Josephson junction, a rapid quantum switching device based on tunneling). The 1986 Nobel prize was awarded to Gerd Binnig and Heinrich Rohrer to recognize their development of another useful device based on tunneling, the scanning tunneling microscope.

CHECKPOINT 5: Is the wavelength of the transmitted wave in Fig. 39-13b larger than, smaller than, or the same as that of the incident wave?

The Scanning Tunneling Microscope (STM)

A device based on tunneling, the STM allows one to make detailed maps of surfaces, revealing features on the atomic scale with a resolution much greater than can be obtained with an optical or electron microscope. Figure 39-14 shows an example, the individual atoms of the surface being readily apparent.

Figure 39-15 shows the heart of the scanning tunneling microscope. A fine metallic tip, mounted at the intersection of three mutually perpendicular quartz rods, is

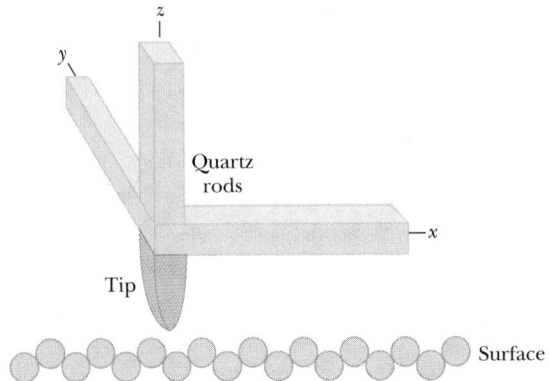

FIGURE 39-15 The essence of a scanning tunneling microscope (STM). Three quartz rods are used to scan a sharply pointed conducting tip across the surface of interest and to maintain a constant separation between tip and surface. The tip thus moves up and down to match the contours of the surface, and a record of its movement is a map like that of Fig. 39-14.

placed close to the surface to be examined. A small potential difference, perhaps only 10 mV, is applied between tip and surface.

Crystalline quartz has an interesting property called *piezoelectricity*: when an electric potential difference is applied across a sample of crystalline quartz, the dimensions of the sample will change slightly. This property is used to change the length of each of the three rods in Fig. 39-15, smoothly and by tiny amounts, so that the tip can be scanned across the surface (in the x and y directions) and also lowered or raised with respect to the surface (in the z direction).

The space between the surface and the tip forms a potential energy barrier, much like that of Fig. 39-13a. If the tip is close enough to the surface, electrons from the sample can tunnel through this barrier from the surface to the tip, forming a tunneling current.

In operation, an electronic feedback arrangement adjusts the vertical position of the tip to keep the tunneling current constant as the tip is scanned back and forth over the surface. This means that the tip–surface separation also remains constant during the scan. The output of the device—for example, Fig. 39-14—is a video display of the varying vertical position of the tip, hence of the surface contour, as a function of the tip position in the xy plane.

Scanning tunneling microscopes are available commercially and are used in laboratories all over the world.

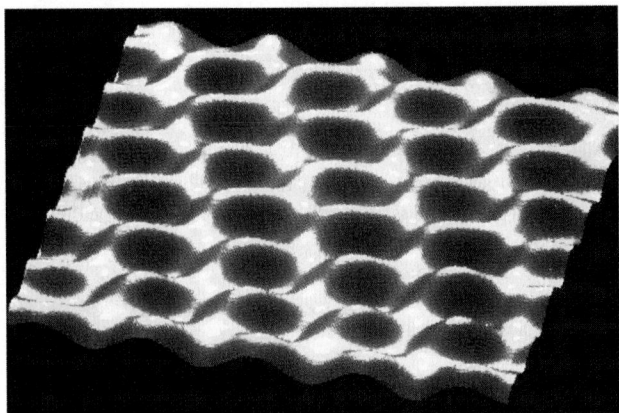

FIGURE 39-14 The contour of a graphite surface as revealed by a scanning tunneling microscope. The carbon atoms and the hexagonal patterns they form are visible.

SAMPLE PROBLEM 39-7

Suppose that the electron in Fig. 39-13a, having a total energy E of 5.1 eV, approaches a barrier of height $E_{pot} = 6.8$ eV and thickness $L = 750$ pm.

(a) What is the approximate transmission coefficient T for this electron–barrier combination?

SOLUTION: To compute T with Eq. 39-21, we first evaluate Eq. 39-22:

$$k = \sqrt{\frac{8\pi^2 m(E_{pot} - E)}{h^2}}.$$

The numerator of the fraction under the square-root sign is

$(8\pi^2)(9.11 \times 10^{-31}$ kg$)(6.8$ eV $- 5.1$ eV$)$

$\times (1.60 \times 10^{-19}$ J/eV$) = 1.956 \times 10^{-47}$ J·kg.

Thus

$$k = \sqrt{\frac{1.956 \times 10^{-47} \text{ J·kg}}{(6.63 \times 10^{-34} \text{ J·s})^2}} = 6.67 \times 10^9 \text{ m}^{-1}.$$

The (dimensionless) quantity $2kL$ is then

$$2kL = (2)(6.67 \times 10^9 \text{ m}^{-1})(750 \times 10^{-12} \text{ m}) = 10.0$$

and, from Eq. 39-21, the transmission coefficient is

$$T \approx e^{-2kL} = e^{-10.0} = 45 \times 10^{-6}. \quad \text{(Answer)}$$

Thus, of every million electrons that strike the barrier, about 45 will tunnel through it.

(b) What would be the transmission coefficient if the incident particle were a proton?

SOLUTION: Carrying out the calculation once more but with the proton mass (1.67×10^{-27} kg) substituted for the electron mass yields $T \approx 10^{-186}$. The transmission coefficient is enormously reduced for this more massive particle. *Imagine how small it would be for a jelly bean!*

REVIEW & SUMMARY

Photons

When light and matter interact, energy and momentum are transferred at pointlike locations in discrete amounts via "bundles" of energy called **photons**. The energy and momentum of a photon are

$$E = hf \quad \text{(photon energy)} \quad (39\text{-}2)$$

$$p = \frac{hf}{c} = \frac{h}{\lambda} \quad \text{(photon momentum)}, \quad (39\text{-}7)$$

in which f and λ are, respectively, the frequency and wavelength of the associated light wave.

Photoelectric Effect

When light of high enough frequency falls on a clean metal surface, electrons are emitted from the surface by photon–electron interactions within the metal. The governing relation is

$$hf = K_{max} + \Phi, \quad (39\text{-}5)$$

in which hf is the photon energy, K_{max} is the kinetic energy of the most energetic emitted electrons, and Φ is the **work function** of the target material, that is, the minimum energy an electron must have if it is to emerge from the surface of the target. If hf is less than Φ, the photoelectric effect does not occur.

Compton Shift

When x rays are scattered by loosely bound electrons in a target, some of the scattered x rays have a longer wavelength than do the incident x rays. This **Compton shift** (in wavelength) is given by

$$\Delta\lambda = \frac{h}{mc}(1 - \cos\phi), \quad (39\text{-}11)$$

in which ϕ is the angle at which the x rays are scattered.

Light Waves and Photons

When light interacts with matter, energy and momentum are transferred via photons. When light is in transit, however, we interpret the light wave as a **probability wave**, in which the probability (per unit time) that a photon can be detected is proportional to E_m^2, where E_m is the amplitude of the oscillating electric field associated with the light wave at the detector.

Matter Waves

A moving particle such as an electron or a proton can be described as a **matter wave**; its wavelength (called the **de Broglie wavelength**) is given by $\lambda = h/p$, where p is the momentum of the particle.

The Wave Function

The displacement of a matter wave is given by its **wave function** $\Psi(x, y, z, t)$, which can be separated into a space-dependent part $\psi(x, y, z)$ and a time-dependent part $e^{-i\omega t}$. For a particle of mass m moving in the x direction with constant total energy E through a region in which its potential energy is $E_{pot}(x)$, $\psi(x)$ can be found by solving the simplified **Schrödinger equation**:

$$\frac{d^2\psi}{dx^2} + \frac{8\pi^2 m}{h^2}[E - E_{pot}(x)]\psi = 0. \quad (39\text{-}15)$$

A matter wave, like a light wave, is a probability wave in the sense that if a particle detector is inserted into the wave, the probability that the detector will register a particle during any specified time interval is proportional to $|\psi|^2$, a quantity called the **probability density**.

For a free particle—that is, a particle for which $E_{pot}(x) = 0$—moving in the x direction, $|\psi|^2$ has a constant value for all positions along the x axis.

Heisenberg's Uncertainty Principle

The probabilistic nature of quantum mechanics places an important limitation on detecting a particle's position and momentum. That is, it is not possible to measure the position **r** and the momentum **p** of a particle simultaneously with unlimited precision.

The uncertainties in the components of these quantities are given by

$$\Delta x \cdot \Delta p_x \geq \hbar$$
$$\Delta y \cdot \Delta p_y \geq \hbar \qquad (39\text{-}20)$$
$$\Delta z \cdot \Delta p_z \geq \hbar.$$

Barrier Tunneling

According to classical physics, an incident particle will be reflected from a potential energy barrier whose height is greater than the particle's kinetic energy. According to quantum mechanics, however, the probability wave associated with a particle has a finite probability of tunneling through such a barrier.

The probability that a given particle of mass m and energy E will tunnel through a barrier of height E_{pot} and thickness L is given by the transmission coefficient T:

$$T \approx e^{-2kL}, \qquad (39\text{-}21)$$

in which

$$k = \sqrt{\frac{8\pi^2 m(E_{pot} - E)}{h^2}}. \qquad (39\text{-}22)$$

QUESTIONS

1. Of the electromagnetic waves generated in a microwave oven and those generated in your dentist's x-ray machine, which has (a) the greater wavelength, (b) the greater frequency, and (c) the greater photon energy?

2. Of the following statements about the photoelectric effect, which are true and which are false? (a) The greater the frequency of the incident light, the greater the stopping potential. (b) The greater the intensity of the incident light, the greater the cutoff frequency. (c) The greater the work function of the target material, the greater the stopping potential. (d) The greater the work function of the target material, the greater the cutoff frequency. (e) The greater the frequency of the incident light, the greater the maximum kinetic energy of the ejected electrons. (f) The greater the energy of the photons, the smaller the stopping potential.

3. According to the figure for Checkpoint 2, is the maximum kinetic energy of the ejected electrons greater for a target made of sodium or of potassium for a given frequency of incident light?

4. In the photoelectric effect (for a given target and a given frequency of the incident light), which of these quantities, if any, depend on the intensity of the incident light beam: (a) the maximum kinetic energy of the electrons, (b) the maximum photoelectric current, (c) the stopping potential, (d) the cutoff frequency?

5. If you shine ultraviolet light on an isolated metal plate, the plate emits electrons for a while. Why does it eventually stop?

6. A metal plate is illuminated with light of a certain frequency. Which of the following determine whether or not electrons are emitted: (a) the intensity of the light; (b) the length of time of exposure to the light; (c) the thermal conductivity of the plate; (d) the area of the plate; (e) the material of the plate?

7. In a Compton-shift experiment, an x-ray photon is scattered in the forward direction, at $\phi = 0$ in Fig. 39-3. How much energy does the electron acquire during this encounter?

8. According to Eq. 39-11 the Compton shift is the same for x rays and for visible light. Why is it that the Compton shift for x rays can be measured readily but that for visible light cannot?

9. Photon A has twice the energy of photon B. (a) Is the momentum of A less than, equal to, or greater than that of B? (b) Is the wavelength of A less than, equal to, or greater than that of B?

10. Compare a photon from a dental x-ray unit (photon A) and one from a microwave oven (photon B). Which has the greater (a) wavelength, (b) energy, (c) frequency, and (d) momentum?

11. The data shown in Fig. 39-4 were taken by allowing x rays to strike a carbon target. In what essential way, if any, would these data differ if the target were sulfur instead of carbon?

12. An electron and a proton have the same kinetic energy. Which has the greater de Broglie wavelength?

13. (a) If you double the kinetic energy of a nonrelativistic particle, how does its de Broglie wavelength change? (b) What if you double the speed of the particle?

14. The following nonrelativistic particles all have the same kinetic energy. Rank them in order of their de Broglie wavelengths, greatest first: electron, alpha particle, neutron.

15. Is the de Broglie wavelength of a speeding bullet extremely large or extremely small?

16. Compare the Compton wavelength of an electron to its de Broglie wavelength. Which of these statements (if any) is true? (a) The Compton wavelength is always larger. (b) The Compton wavelength is always smaller. (c) The two wavelengths are always the same. (d) They are independent of each other.

17. Figure 39-16 shows four situations in which an electron is moving through a field. It is moving (a) opposite an electric field, (b) in the same direction as an electric field, (c) in the same direction as a magnetic field, (d) perpendicular to a magnetic field. For each situation, is the de Broglie wavelength of the electron increasing, decreasing, or remaining the same?

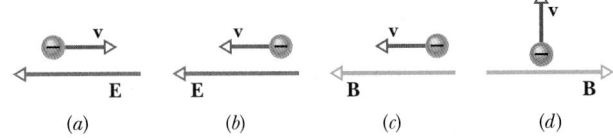

FIGURE 39-16 Question 17.

18. A proton and a deuteron, each having a kinetic energy of 3 MeV, approach a potential energy barrier whose height E_{pot} is 10 MeV. Which particle has the greater chance of tunneling through the barrier? (A deuteron is twice as massive as a proton.)

19. Which has the greater effect on the transmission coefficient T of a potential energy barrier: (a) raising the barrier height E_{pot} by 1% or (b) lowering the kinetic energy E of the incident electron by 1%?

20. At the left in Fig. 39-13b, why are the minima in the values of $|\psi|^2$ greater than zero?

21. Suppose that the height of the potential energy barrier in Fig. 39-13a is infinite. (a) What value would you expect for the transmission coefficient of electrons approaching the barrier? (b) Does Eq. 39-21 predict your expected result?

22. The table gives relative values for three situations for the barrier tunneling experiment of Fig. 39-13. Rank the situations according to the probability of the electron tunneling through the barrier, greatest first.

	ELECTRON ENERGY	BARRIER HEIGHT	BARRIER THICKNESS
(a)	E	$5E$	L
(b)	E	$17E$	$L/2$
(c)	E	$2E$	$2L$

EXERCISES & PROBLEMS

SECTION 39-2 Light Waves and Photons

1E. Show that the energy E of a photon, in electron-volts, is related to its wavelength λ, in nanometers, by $E = 1240/\lambda$.

2E. The orange-colored light from a highway sodium lamp has a wavelength of 589 nm. What is the energy of the photons associated with this light?

3E. Express the Planck constant h in terms of the unit electron-volt–femtoseconds.

4E. Monochromatic light falls on a sheet of photographic film. Individual photons will be recorded if they have enough energy to dissociate an AgBr molecule in the film, and the minimum energy required to do this is about 0.6 eV. What is the greatest wavelength of light that will be recorded? In what region of the spectrum does this light fall?

5E. A spectral emission line that is important in astronomy has a wavelength of 21 cm. What is its corresponding photon energy?

6E. How fast must an electron move to have a kinetic energy equal to the energy of a photon of sodium light ($\lambda = 590$ nm)?

7E. At what rate does the Sun emit photons? Assume for simplicity that the Sun's light is monochromatic, with $\lambda = 550$ nm. The Sun emits energy at the rate of 3.9×10^{26} W.

8E. A helium–neon laser emits a beam of red light ($\lambda = 633$ nm) that is about 3.5 mm in diameter. If the emitted power is 5.0 mW, at what rate per unit area are photons transferred to a detector placed across the beam path? Assume that the detector absorbs the beam completely.

9E. At one time the meter was defined as 1,650,763.73 wavelengths of the orange light emitted by a light source containing krypton-86 atoms. What is the corresponding photon energy of this radiation?

10P. Under ideal conditions, the human eye will record a visual sensation for light whose wavelength is 550 nm if energy is trans-ferred to the eye at a rate as low as 100 photons per second. To what power does this correspond?

11P. An ultraviolet lightbulb emitting light of wavelength 400 nm and an infrared lightbulb emitting light of wavelength 700 nm are both rated at 400 W. (a) Which bulb generates photons at the greater rate and (b) what is that rate?

12P. A satellite in Earth orbit maintains a panel of solar cells of 2.60 m² area perpendicular to the direction of the Sun's rays. Solar energy arrives at the rate of 1.39 kW/m². (a) At what rate does solar energy strike the panel? (b) At what rate are solar photons absorbed by the panel? Assume that the solar radiation is monochromatic, with a wavelength of 550 nm, and that all the solar radiation striking the panel is absorbed. (c) How long would it take for a "mole of photons" to be absorbed by the panel?

13P. A special kind of lightbulb emits monochromatic light of wavelength 630 nm. It is rated at 60 W and is 93% efficient in converting electric energy to light. How many photons will the bulb generate over its 730 h lifetime?

14P. The emerging beam from a 1.5 W argon laser ($\lambda = 515$ nm) has a diameter d of 3.0 mm. The beam is focused by a lens system whose effective focal length f_L is 2.5 mm. The focused beam falls on a totally absorbing screen, where it forms a circular diffraction pattern whose central disk has a radius R given by $1.22 f_L \lambda/d$. It can be shown that 84% of the incident energy falls within this central disk, the rest falling in the fainter, concentric diffraction rings that surround the central disk. At what rate are photons absorbed by the screen in the central disk of the diffraction pattern?

15P. A 100 W sodium lamp ($\lambda = 589$ nm) radiates energy uniformly in all directions. (a) At what rate are photons generated in the lamp? (b) At what distance from the lamp will a totally absorbing screen absorb photons at the rate of 1.00 photon/cm²·s? (c) What is the photon flux (photons per unit area per unit time) on a small screen 2.00 m from the lamp?

SECTION 39-3 The Photoelectric Effect

16E. The work functions for potassium and cesium are 2.25 and 2.14 eV, respectively. (a) Will the photoelectric effect occur for either of these elements with incident light of wavelength 565 nm? (b) With light of wavelength 518 nm?

17E. You wish to pick a substance for a photocell that will operate via the photoelectric effect with visible light. Which of the following will do (work functions are in parentheses): tantalum (4.2 eV), tungsten (4.5 eV), aluminum (4.2 eV), barium (2.5 eV), lithium (2.3 eV)?

18E. (a) The energy needed to remove an electron from metallic sodium is 2.28 eV. Does sodium show a photoelectric effect for red light, with $\lambda = 680$ nm? (b) What is the cutoff wavelength for photoelectric emission from sodium? To what color does that correspond?

19E. Find the maximum kinetic energy of electrons emitted from a certain material if the material's work function is 2.3 eV and the frequency of the incident radiation is 3.0×10^{15} Hz.

20E. Light strikes a sodium surface, causing photoelectric emission. The stopping potential for the emitted electrons is 5.0 V, and the work function of sodium is 2.2 eV. What is the wavelength of the incident light?

21E. The work function of tungsten is 4.50 eV. Calculate the speed of the fastest electrons emitted when light whose photon energy is 5.80 eV falls on a tungsten surface.

22P. Light of wavelength 200 nm falls on an aluminum surface. In aluminum, 4.20 eV is required to remove an electron. What is the kinetic energy of (a) the fastest and (b) the slowest emitted electrons? (c) What is the stopping potential for this situation? (d) What is the cutoff wavelength for aluminum?

23P. (a) If the work function for a certain metal is 1.8 eV, what is its stopping potential for light of wavelength 400 nm? (b) What is the maximum speed of electrons emitted via the photoelectric effect as they leave the metal surface?

24P. The wavelength associated with the cutoff frequency for silver is 325 nm. Find the maximum kinetic energy of electrons ejected from a silver surface by ultraviolet light of wavelength 254 nm.

25P. An orbiting satellite can become charged by the photoelectric effect when sunlight ejects electrons from the vehicle's outer surface. Satellites must be designed to minimize such charging. Suppose a satellite is coated with platinum, a metal with a very large work function ($\Phi = 5.32$ eV). Find the longest wavelength of incident sunlight that can eject an electron from the platinum.

26P. The stopping potential for electrons emitted from a surface illuminated by light of wavelength 491 nm is 0.710 V. When the incident wavelength is changed to a new value, the stopping potential is found to be 1.43 V. (a) What is this new wavelength? (b) What is the work function for the surface?

27P. In a photoelectric experiment using a sodium surface, you find a stopping potential of 1.85 V for a wavelength of 300 nm and a stopping potential of 0.820 V for a wavelength of 400 nm. From these data find (a) a value for the Planck constant, (b) the

work function Φ for sodium, and (c) the cutoff wavelength λ_0 (the wavelength corresponding to the cutoff frequency) for sodium.

28P. In about 1916, R. A. Millikan found the following stopping-potential data for lithium in his photoelectric experiments:

Wavelength (nm)	433.9	404.7	365.0	312.5	253.5
Stopping potential (V)	0.55	0.73	1.09	1.67	2.57

Use these data to make a plot like Fig. 39-2 (which is for sodium) and then use the plot to find (a) the Planck constant and (b) the work function for lithium.

29P. Suppose the *fractional efficiency* of a cesium surface (with work function 1.80 eV) is 1.0×10^{-16}; that is, on average one electron is emitted for every 10^{16} photons that fall on the surface. What would be the current of electrons emitted from such a surface if it were illuminated with 600 nm light from a 2.00 mW laser and all the emitted electrons took part in the charge flow?

30P. X rays with a wavelength of 71 pm eject from a gold foil electrons originating deep within the gold atoms. The ejected electrons move in circular paths of radius r in a region of uniform magnetic field **B**, with $Br = 1.88 \times 10^{-4}$ T·m. Find (a) the maximum kinetic energy of the emitted electrons and (b) the work done in removing them from the gold atoms.

SECTION 39-4 Photons Have Momentum

31E. A certain x-ray beam has a wavelength of 35.0 pm. (a) What is the corresponding frequency? Calculate the corresponding (b) photon energy and (c) photon momentum.

32E. (a) What is the momentum of a photon whose energy equals the rest energy of an electron? What are (b) the wavelength and (c) the frequency of the corresponding radiation?

33E. Light of wavelength 2.4 pm falls on a target containing free electrons. (a) Find the wavelength of light scattered at 30° from the incident direction. (b) Do the same for a scattering angle of 120°.

34P. Gamma rays of photon energy 0.511 MeV fall on free electrons in an aluminum target and are scattered in various directions. (a) What is the wavelength of the incident gamma rays? (b) What is the wavelength of gamma rays scattered at 90.0° to the incident beam? (c) What is the photon energy of the rays scattered in this direction?

35P. Show, by analyzing a collision between a photon and a free electron (using relativistic mechanics), that it is impossible for a photon to give all its energy to a free electron.

36P. An x-ray beam of wavelength 0.01 nm strikes a target containing free electrons. Consider x rays scattered from the target at an angle of 180°. Determine (a) the change in wavelength of the scattered x rays, (b) the change in photon energy between the incident and scattered beams, (c) the kinetic energy transferred to an electron, and (d) the electron's direction of motion.

37P. Calculate the Compton wavelength for (a) an electron and (b) a proton. What is the energy of a photon whose associated wavelength is equal to the Compton wavelength of (c) the electron and (d) the proton?

38P. Calculate the percentage change in photon energy during a collision like that in Fig. 39-5 for $\phi = 90°$ and for radiation in (a) the microwave range, with $\lambda = 3.0$ cm, (b) the visible range, with $\lambda = 500$ nm, (c) the x-ray range, with $\lambda = 25$ pm, and (d) the gamma-ray range, the gamma photon energy being 1.0 MeV. (e) What are your conclusions about the feasibility of detecting the Compton shift in these various regions of the electromagnetic spectrum, judging solely by the criterion of energy loss in a single photon–electron encounter?

39P. What percentage increase in wavelength leads to a 75% loss of photon energy in a photon–free electron collision?

40P. What is the maximum wavelength shift for a Compton collision between a photon and a free *proton*?

41P. An electron of mass m and speed v undergoes a head-on collision with a gamma-ray photon of energy hf_0, scattering the gamma-ray photon back in the direction of incidence. Verify that the energy of the scattered gamma-ray photon, as measured in the laboratory system, is

$$E = hf_0 \left(1 + \frac{2hf_0}{mc^2} \sqrt{\frac{1 + v/c}{1 - v/c}}\right)^{-1}.$$

42P. What would be (a) the Compton shift, (b) the fractional Compton shift, and (c) the change in photon energy for light of wavelength 590 nm scattering from a free, initially stationary electron if the scattering is at 90° to the direction of the incident beam? (d) Calculate the same quantities for x rays whose photon energy is 50.0 keV.

43P. Consider a collision between an x-ray photon of initial energy 50.0 keV and an electron at rest, in which the photon is scattered backward and the electron is knocked forward. (a) What is the energy of the back-scattered photon? (b) What is the kinetic energy of the electron?

44P. Show that $\Delta E/E$, the fractional loss of energy of a photon during a collision with a particle of mass m, is given by

$$\frac{\Delta E}{E} = \frac{hf'}{mc^2}(1 - \cos \phi),$$

where E is the energy of the incident photon, f' is the frequency of the scattered photon, and ϕ is defined as in Fig. 39-5.

45P. Through what angle must a 200 keV photon be scattered by a free electron so that the photon loses 10% of its energy?

46P. Show that when a photon of energy E is scattered from a free electron, the maximum kinetic energy transferred to the electron is given by

$$K_{max} = \frac{E^2}{E + mc^2/2}.$$

47P. What is the maximum kinetic energy of electrons knocked out of a thin copper foil by an incident beam of 17.5 keV x rays?

48P. Derive Eq. 39-11, the equation for the Compton shift, from Eqs. 39-8, 39-9, and 39-10 by eliminating v and θ.

SECTION 39-6 Electrons and Matter Waves

49E. A bullet of mass 40 g travels at 1000 m/s. (a) What wavelength can we associate with it? (b) Why is the wave nature of the bullet not revealed through diffraction effects?

50E. Using the classical relation between momentum and kinetic energy, shows that an electron's de Broglie wavelength in nanometers can be written as $\lambda = 1.226/\sqrt{K}$, in which K is the electron's kinetic energy in electron-volts.

51E. In an ordinary television set, electrons are accelerated through a potential difference of 25.0 kV. What is the de Broglie wavelength of such electrons? (Ignore relativistic effects.)

52E. Calculate the de Broglie wavelength of (a) a 1.00 keV electron, (b) a 1.00 keV photon, and (c) a 1.00 keV neutron.

53P. The wavelength of the yellow spectral emission line of sodium is 590 nm. At what kinetic energy would an electron have the same de Broglie wavelength?

54P. An electron and a photon each have a wavelength of 0.20 nm. Calculate (a) their momenta and (b) their energies.

55P. Neutrons in thermal equilibrium with matter have an average kinetic energy of $(3/2)kT$, where k is the Boltzmann constant and T, which may be taken to be 300 K, is the temperature of the environment of the neutrons. (a) What is the average kinetic energy of such a neutron? (b) What is the corresponding de Broglie wavelength?

56P. If the de Broglie wavelength of a proton is 100 fm, (a) what is the speed of the proton and (b) through what electric potential would the proton have to be accelerated to acquire this speed?

57P. Consider a balloon filled with helium gas at room temperature and pressure. (a) Calculate the average de Broglie wavelength of the helium atoms and the average distance between atoms under these conditions. The average kinetic energy of an atom is equal to $(3/2)kT$, where k is the Boltzmann constant. (b) Can the atoms be treated as particles under these conditions?

58P. (a) A photon has an energy of 1.00 eV, and an electron has a kinetic energy of that same amount. What are their wavelengths? (b) Repeat for an energy of 1.00 GeV.

59P. (a) A photon and an electron both have a wavelength of 1.00 nm. Give the energy of the photon and the kinetic energy of the electron. (b) Repeat for a wavelength of 1.00 fm.

60P. Singly charged sodium ions are accelerated through a potential difference of 300 V. (a) What is the momentum acquired by such an ion? (b) What is its de Broglie wavelength?

61P. The large electron accelerator at Stanford University provides a beam of electrons with kinetic energies of 50 GeV. Electrons with this energy have small wavelengths, suitable for probing the fine details of nuclear structure via scattering. What is the de Broglie wavelength of a 50 GeV electron? How does this wavelength compare with the radius of an average nucleus, taken to be about 5.0 fm? (At this energy it is sufficient to use the extreme relativistic relationship between momentum and energy, namely, $p = E/c$. This is the same relationship used for light and is justified when the kinetic energy of the particle is much greater than its rest energy, as in this case.)

62P. The existence of the atomic nucleus was discovered in 1911 by Ernest Rutherford, who properly interpreted some experiments in which a beam of alpha particles was scattered from a metal foil of atoms such as gold. (a) If the alpha particles had a kinetic energy of 7.5 MeV, what was their de Broglie wavelength? (b) Should the wave nature of the incident alpha particles have been taken into account in interpreting these experiments? The mass of an alpha particle is 4.00 u (atomic mass units), and its distance of closest approach to the nuclear center in these experiments was about 30 fm. (The wave nature of matter was not postulated until more than a decade after these crucial experiments were first performed.)

63P. A nonrelativistic particle is moving three times as fast as an electron. The ratio of the de Broglie wavelength of the particle to that of the electron is 1.813×10^{-4}. By calculating its mass, identify the particle.

64P. The highest achievable resolving power of a microscope is limited only by the wavelength used; that is, the smallest detail that can be separated has dimensions about equal to the wavelength. Suppose one wishes to "see" inside an atom. Assuming the atom to have a diameter of 100 pm, this means that one must resolve detail of separation of, say, 10 pm. (a) If an electron microscope is used, what minimum electron energy is required? (b) If a light microscope is used, what minimum photon energy is required? (c) Which microscope seems more practical? Why?

65P. What accelerating voltage for the electrons would be required if an electron microscope is to obtain the same ultimate resolving power as could be obtained using 100 keV gamma rays? (See Problem 64.)

SECTION 39-7 Schrödinger's Equation

66E. (a) Let $n = a + ib$ be a complex number, where a and b are real (positive or negative) numbers. Show that the product nn^* is always a positive real number. (b) Let $m = c + id$ be another complex number. Show that $|nm| = |n|\,|m|$.

67P. Show that Eq. 39-17 is indeed a solution of Eq. 39-16 by substituting $\psi(x)$ and its second derivative into Eq. 39-16 and noting that an identity results.

68P. (a) Write the wave function $\psi(x)$ displayed in Eq. 39-19 in the form $\psi(x) = a + ib$, where a and b are real quantities. (Assume that ψ_0 is real.) (b) Write the time-dependent wave function $\Psi(x, t)$ that corresponds to $\psi(x)$.

69P. Show that the angular wave number k for a free particle of mass m can be written as

$$k = \frac{2\pi\sqrt{2\,mK}}{h},$$

in which K is the particle's kinetic energy.

70P. The function $\psi(x)$ displayed in Eq. 39-19 describes a free particle, for which we assumed that $E_{pot}(x) = 0$ in Schrödinger's equation (Eq. 39-15). Assume now that $E_{pot}(x) = E_0$ a constant in that equation. Show that Eq. 39-19 is still a solution of Schrödinger's equation, with the wave number k of the particle now given by

$$k = \frac{2\pi}{h}\sqrt{2m(E - E_0)}.$$

71P. Show that $|\psi|^2 = |\Psi|^2$, with ψ and Ψ defined as in Eq. 39-14. That is, show that the probability density does not depend on the time variable.

72P. Suppose that we had put $A = 0$ in Eq. 39-17 and relabeled B as ψ_0. What would the resulting wave function then describe? How, if at all, would Fig. 39-12 be altered?

73P. In Eq. 39-18 keep both terms, putting $A = B = \psi_0$. The equation then describes the superposition of two matter waves of equal amplitude, traveling in opposite directions. (Recall that this is the condition for a standing wave.) (a) Show that $|\Psi(x, t)|^2$ is then given by

$$|\Psi(x, t)|^2 = 2\psi_0^2[1 + \cos 2kx].$$

(b) Plot this function, and demonstrate that it describes the square of the amplitude of a standing matter wave. (c) Show that the nodes of this standing wave are located at

$$x = (2n + 1)(\tfrac{1}{4}\lambda), \qquad \text{where } n = 0, 1, 2, 3 \ldots$$

and λ is the de Broglie wavelength of the particle. (d) Write an expression for the locations of the most probable positions for finding the particle.

SECTION 39-8 Heisenberg's Uncertainty Principle

74E. Figure 39-12 shows that because of Heisenberg's uncertainty principle, it is not possible to assign an x coordinate to the position of the electron. (a) Can you assign a y or a z coordinate? (*Hint:* The momentum of the electron has no y or z component.) (b) Describe the extent of the matter wave in three dimensions.

75E. The uncertainty in the position of an electron is given as 50 pm, which is about equal to the radius of a hydrogen atom. What is the least uncertainty in any simultaneous measurement of the momentum of this electron?

76E. Imagine playing baseball in a universe (not ours!) where the Planck constant is 0.60 J·s. What would be the uncertainty in the position of a 0.50 kg baseball that is moving at 20 m/s along an axis if the uncertainty in the speed is 1.0 m/s?

77P. Figure 39-12 shows a case in which the momentum p_x of a particle is fixed so that $\Delta p_x = 0$; then, from Heisenberg's uncertainty principle (Eq. 39-20), the position x of the particle is completely unknown. From the same principle it follows that the opposite is also true. That is, if the position of a particle is exactly known ($\Delta x = 0$), the uncertainty in its momentum is infinite.

Consider an intermediate case, in which the position of a particle is measured, not to infinite precision, but to within a distance of $\lambda/2\pi$, where λ is the particle's de Broglie wavelength. Show that the uncertainty in the (simultaneously measured) momentum is then equal to the momentum itself; that is, $\Delta p_x = p$. Under these circumstances, would a measured momentum of zero surprise you? What about a measured momentum of $0.5p$? Of $2p$? Of $12p$?

78P. You will find in Chapter 40 that we no longer imagine electrons to move in definite orbits within atoms, like the planets in our solar system. To see why, let us try to "observe" such an orbiting electron by using a light microscope to measure the electron's presumed orbital position with a precision of, say 10 pm (a typical atom has a radius of about 100 pm). The wavelength of the light used in the microscope must then be about 10 pm. (a) What would be the photon energy of this light? (b) How much energy would such a photon impart to an electron in a head-on collision? (c) What do these results tell you about the possibility of "viewing" an atomic electron at two or more points along its presumed orbital path? (*Hint:* The outer electrons of atoms are bound to the atom by energies of only a few electron-volts.)

SECTION 39-9 Barrier Tunneling

79P. A proton and a deuteron (the latter has the same charge as a proton but twice the mass) strike a potential energy barrier that is 10 fm thick and 10 MeV high. Each particle has a kinetic energy of 3.0 MeV before it strikes the barrier. (a) What is the transmission probability for each? (b) What are their respective kinetic energies after they pass through the barrier (assuming that they do so)? (c) What are their respective kinetic energies if they are reflected from the barrier?

80P. Consider a potential energy barrier like that of Fig. 39-13a but whose height E_{pot} is 6.0 eV and whose thickness L is 0.70 nm. What is the energy of an incident electron whose transmission probability is 0.0010?

81P. (a) Suppose a beam of 5.0 eV protons strikes a potential energy barrier of height 6.0 eV and thickness 0.70 nm, at a rate equivalent to a current of 1000 A. How long would you have to wait—on average—for one proton to be transmitted? (b) How long would you have to wait if the particle was an electron rather than a proton?

82P. Consider the barrier-tunneling situation in Sample Problem 39-7. What percentage change in the transmission coefficient T occurs for a 1.0% change in (a) the barrier height, (b) the barrier thickness, and (c) the kinetic energy of the incident electron?

83P. A 1500 kg car moving at 20 m/s approaches a hill that is 24 m high and 30 m long. What is the probability that the car will tunnel quantum mechanically through the hill, appearing on the other side? That is, what is the car's transmission coefficient for this hill? (*Hint:* The potential energy is gravitational in this case.)

84P. A particle of momentum p approaches the barrier of Fig. 39-13a from the left. In the region to the left of the barrier, $E_{pot} = 0$ and Schrödinger's equation takes the form of Eq. 39-16. (a) Show that, in this region,

$$\psi(x) = Ae^{ikx} + Be^{-ikx}$$

is a solution of Schrödinger's equation. Here A and B are real but arbitrary constants and $k = 2\pi p/h$. (b) Show further that

$$|\psi|^2 = A^2 + B^2 + 2AB \cos 2kx.$$

(c) Show that $|\psi|^2$ oscillates between the limits of $(A + B)^2$ and $(A - B)^2$. Note that if $A \neq B$, $|\psi|^2$ is always positive and never zero. Check these results against the plot of $|\psi|^2$ in Fig. 39-13b.

85P. A particle of momentum p and kinetic energy E approaches the barrier of Fig. 39-13a from the left. In the region within the barrier, $E < E_{pot}$ and Schrödinger's equation, in the form of Eq. 39-16, still holds. (a) Show that in this region

$$\psi(x) = Ce^{-kx},$$

in which

$$k = \sqrt{\frac{8\pi^2 m(E_{pot} - E)}{h^2}}$$

and C is an arbitrary constant. (b) Show further that

$$|\psi|^2 = C^2 e^{-2kx}.$$

This function describes the exponential decay of the probability density within the barrier, as displayed in Fig. 39-13b.

40
More About Matter Waves

This spectacular computer image was produced in 1993 at IBM's Almaden Research Center in California. The 48 peaks forming the circle mark the positions of individual atoms of iron on a specially prepared copper surface. The circle, which is about 14 nm in diameter, is called a <u>quantum corral</u>. How do these atoms come to be arranged in a circle? And what is the significance of the ripples that are visible within the corral''?

40-1 ATOM BUILDING

Early in the twentieth century nobody knew how the electrons in an atom are arranged, what their motions are, how atoms emit or absorb light, or even why atoms are stable. Without this knowledge it is not possible to understand how atoms combine to form molecules or stack up to form solids. As a consequence, the foundations of chemistry—including biochemistry, which underlies the nature of life itself—were more or less a mystery.

In 1926 all these questions and many others were answered with the development of **quantum mechanics.** Its basic premise is that moving electrons, protons, and particles of any kind are best viewed as matter waves, whose motions are governed by Schrödinger's equation. Although quantum mechanics also applies to massive particles, there is no point in treating baseballs, automobiles, planets, and so on with quantum mechanics. For such massive, slow-moving objects, Newtonian mechanics and quantum mechanics yield the same answers.

Before we can apply quantum mechanics to the problem of atomic structure, we need to develop some insights by applying quantum ideas in a few simpler situations. These ''practice problems'' may seem artificial but, as you will see, they provide a firm foundation for understanding a very real problem that we shall analyze in Section 40-6 —the structure of the hydrogen atom.

40-2 WAVES ON STRINGS AND MATTER WAVES

In Chapter 17 we saw that waves of two kinds can be set up on a stretched string. If the string is so long that we can take it to be infinitely long, we can set up a *traveling wave* of essentially any frequency. However, if the stretched string has only a finite length, perhaps because it is rigidly clamped at both ends, we can set up only *standing waves* on it; further, these standing waves can have only discrete frequencies. In other words, confining the wave to a finite region of space leads to *quantization* of the motion—to the existence of discrete *states* for the wave, each state with a sharply defined frequency.

This observation applies to waves of all kinds, including matter waves. For matter waves, however, it is more convenient to deal with the energy E of the associated particle than with the frequency f of the wave. In all that follows we shall focus on the matter wave associated with a moving electron, which we choose as a prototype particle for study.

Consider the matter wave associated with an electron moving in the x direction and subject to no net force—a so-called *free particle.* The energy of such an electron can

have any reasonable value, just as a wave traveling along a stretched string of infinite length can have any reasonable frequency.

Consider next the matter wave associated with an atomic electron, perhaps the *valence* (least tightly bound) electron in a sodium atom. Such an electron—held within the atom by the attractive Coulomb force due to the positively charged nucleus—is *not* a free particle. It can exist only in a set of discrete states, each having a discrete energy E. This sounds much like the discrete states and quantized frequencies that are available to a stretched string of finite length. For matter waves, then, as for waves of all kinds, we may state a **confinement principle:**

Confinement of a wave leads to quantization, that is, to the existence of discrete states with discrete energies.

40-3 TRAPPING AN ELECTRON

Here we examine the matter wave associated with an electron confined to a limited region of space. We do so by analogy with standing waves on a string of finite length, stretched along an x axis and confined between rigid supports. Because the supports are rigid, the two ends of the string are nodes, or points at which the string is always at rest. There may be other nodes along the string, but these two must always be present, as Fig. 17-17 shows.

The states, or discrete standing wave patterns in which the string can oscillate, are those for which the length L of the string is equal to an integer number of half-wavelengths. That is, the string can occupy only states for which

$$L = \frac{n\lambda}{2}, \qquad \text{for } n = 1, 2, 3, \cdots . \qquad (40\text{-}1)$$

Each value of n identifies a state of the oscillating string; using the language of quantum mechanics, we can call the integer n a **quantum number.**

For each state of the string permitted by Eq. 40-1, the transverse displacement of the string at various positions along the string is given by

$$y_n(x) = A \sin\left(\frac{n\pi}{L}\right)x, \quad \text{for } n = 1, 2, 3, \cdots , \qquad (40\text{-}2)$$

in which the quantum number n identifies the oscillation pattern and the amplitude A depends on the time at which you inspect the string. (Equation 40-2 is a short version of Eq. 17-51.) We see that for all values of n and for all times, there is a point of zero displacement (a node) at $x = 0$ and at $x = L$, as there must be. Figure 17-17 shows a time exposure of such a stretched string for $n = 2, 3,$ and 4.

Now let us turn our attention to matter waves. Our first problem is to physically confine an electron that is moving along the x axis so that it remains within a finite segment of that axis. Figure 40-1 shows a conceivable "electron trap." It consists of two semi-infinitely long cylinders, each of which has an electric potential approaching $-\infty$; between them is a hollow cylinder of length L, which has an electric potential of zero. We introduce a single electron into this central cylinder, setting the electron in motion parallel to the x axis.

The trap of Fig. 40-1 is easy to analyze but is not very practical. Single electrons *can*, however, be trapped in the laboratory with traps that are more complex in design but similar in concept. At the University of Washington, for example, a single electron has been held in a trap for months on end, permitting scientists to make extremely precise measurements of its properties.

Finding the Quantized Energies

Figure 40-2 shows the potential energy of the electron as a function of its position along the x axis of the idealized trap of Fig. 40-1. When the electron is in the central cylinder, its potential energy E_{pot} $(= -eV)$ is zero because there the potential V is zero. If the electron could get outside this region, its potential energy would be positive and of infinite magnitude, because there $V \rightarrow -\infty$. We call the potential energy pattern of Fig. 40-2 an **infinitely deep potential energy well** or, for short, an *infinite potential well*. It is a "well" because an electron placed in the central cylinder of Fig. 40-1 cannot escape from it. As the electron approaches either end of the cylinder, a force of essentially infinite magnitude reverses the electron's motion.

Just like the standing wave in a length of stretched string, the matter wave describing the confined electron must have nodes at $x = 0$ and $x = L$. Moreover, Eq. 40-1 applies to such a matter wave if we interpret λ in that equation as the de Broglie wavelength associated with the moving electron.

The de Broglie wavelength λ is defined in Eq. 39-13 as $\lambda = h/p$, where p is the magnitude of the electron's momentum. This magnitude p is related to the kinetic energy K by $p = \sqrt{2mK}$, where m is the mass of the electron.

FIGURE 40-2 The electric potential energy $E_{pot}(x)$ of an electron confined to the central cylinder of the idealized trap of Fig. 40-1. We see that $E_{pot} = 0$ for $0 < x < L$, and $E_{pot} \rightarrow \infty$ for $x < 0$ and $x > L$.

For an electron moving within the central cylinder of Fig. 40-1, where $E_{pot} = 0$, the total energy E is equal to the kinetic energy. Hence we can write the de Broglie wavelength of this electron as

$$\lambda = \frac{h}{p} = \frac{h}{\sqrt{2mE}}. \qquad (40\text{-}3)$$

If we substitute Eq. 40-3 into Eq. 40-1 and solve for the energy E, we find that

$$E_n = \left(\frac{h^2}{8mL^2}\right) n^2, \qquad \text{for } n = 1, 2, 3, \cdots. \qquad (40\text{-}4)$$

Because the electron is confined, its energy can have only the values given by Eq. 40-4. The number n, which identifies the quantum state of the electron, is a quantum number. Figure 40-3 shows some of the discrete energy values (or *energy levels*) for an electron in an infinite well with $L = 100$ pm, according to Eq. 40-4. (This value for L was

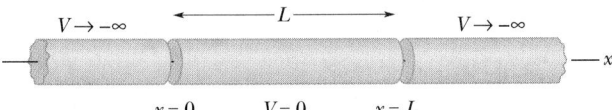

FIGURE 40-1 The elements of an idealized "trap" designed to confine an electron to the central cylinder. We take the semi-infinitely long end cylinders to be at an infinitely great negative potential and the central cylinder to be at zero potential.

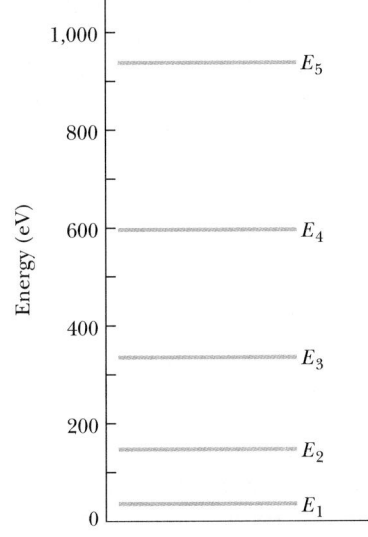

FIGURE 40-3 Several of the allowed energies, given by Eq. 40-4, for an electron confined to the infinite well of Fig. 40-2, with width $L = 100$ pm. Such a plot is called an *energy level diagram*.

chosen because it corresponds roughly to the dimensions of a typical atom.)

The quantum state of a confined electron having the lowest possible energy is called the *ground state*; in Fig. 40-3, this state has energy E_1. The quantum states having greater energies are called *excited states*. In Fig. 40-3, the state with energy E_2 is the first excited state, the state with energy E_3 is the second excited state, and so on.

Energy Changes

To have the lowest allowed energy, the electron tends to occupy the ground state; it can move to an excited state (in which it has greater energy), only if an external source provides energy equal to the energy difference between the ground state and the excited state. An electron that receives such energy is said to make a *quantum jump* (or transition) to the excited state.

One way an electron can gain energy to make a quantum jump up to a greater energy level is to absorb a photon. However, this absorption and quantum jump can occur only if the following condition is met:

If a confined electron is to absorb a photon, the energy hf of the photon must equal the energy difference between the initial energy level of the electron and a higher level.

When an electron reaches an excited state, it does not stay there but quickly *de-excites* by decreasing its energy. One way it can decrease its energy is by emitting a photon under the following condition:

If a confined electron emits a photon, the energy hf of that photon equals the energy difference between the initial energy level of the electron and a lower level.

SAMPLE PROBLEM 40-1

An electron is confined to an infinitely deep potential energy well of width $L = 100$ pm.

(a) What is the energy difference between two adjacent energy levels, with quantum numbers n and $n + 1$? What is the energy difference if $n = 1$?

SOLUTION: From Eq. 40-4 we have

$$\Delta E = E_{n+1} - E_n$$
$$= \frac{h^2}{8mL^2}[(n + 1)^2 - n^2]$$

$$= \frac{h^2}{8mL^2}(2n + 1) \tag{40-5}$$

$$= \frac{(6.63 \times 10^{-34} \text{ J}\cdot\text{s})^2}{(8)(9.11 \times 10^{-31} \text{ kg})(100 \times 10^{-12} \text{ m})^2}(2n + 1)$$

$$= (6.03 \times 10^{-18} \text{ J})(1 \text{ eV}/1.6 \times 10^{-19} \text{ J})(2n + 1)$$

$$= (37.7 \text{ eV})(2n + 1). \tag{Answer}$$

Thus, the energy difference between adjacent levels becomes larger as the quantum number n increases.

The energy difference between the state with $n = 1$ and that with $n = 2$ is found by putting $n = 1$ in the preceding equation, obtaining

$$\Delta E = (37.7 \text{ eV})(3) = 113 \text{ eV}, \tag{Answer}$$

which is consistent with Fig. 40-3.

(b) What is the energy difference between adjacent energy levels for an electron confined to an evacuated tube 3.0 m long?

SOLUTION: We proceed exactly as in (a) except that we put $L = 3.0$ m in place of $L = 100$ pm. The result is

$$\Delta E = (4.19 \times 10^{-20} \text{ eV})(2n + 1). \tag{Answer}$$

For $n = 1$ this yields $\Delta E = 1.3 \times 10^{-19}$ eV, an energy too small to offer any hope of measurement. It is about equal to the energy required to lift a single electron two nanometers vertically against Earth's gravitational force on the electron. When an electron is "confined" to such a large region of space, the energies of its allowed states are so close together that they cannot be identified experimentally as discrete states. For all practical purposes, the quantization of energy and the existence of discrete states are then not detectable.

SAMPLE PROBLEM 40-2

An electron is in the initial state $n_i = 3$ of a infinite potential well of width 100 pm. If it is to make a quantum jump to the state $n_f = 6$ by absorbing a photon, what must be the energy hf of the photon and the wavelength λ associated with it?

SOLUTION: The electron is to jump from an energy level E_3 with $n_i = 3$ to an energy level E_6 with $n_f = 6$; this jump requires an increase in the electron's energy, which is to be provided by the photon absorption. From Eq. 40-4, the change ΔE in the energy of the electron is

$$\Delta E = E_6 - E_3 = \frac{h^2}{8mL^2}n_f^2 - \frac{h^2}{8mL^2}n_i^2$$

$$= \frac{h^2}{8mL^2}(n_f^2 - n_i^2).$$

Equating the photon energy hf to ΔE and then substituting known data, we find

$$hf = \frac{h^2}{8mL^2}(n_f^2 - n_i^2)$$

$$= \frac{(6.63 \times 10^{-34} \text{ J·s})^2}{(8)(9.11 \times 10^{-31} \text{ kg})(100 \times 10^{-12} \text{ m})^2}(6^2 - 3^2)$$

$$= 1.628 \times 10^{-16} \text{ J} \approx 1.63 \times 10^{-16} \text{ J.} \qquad \text{(Answer)}$$

To find the wavelength λ associated with the photon, we substitute c/λ for the frequency f associated with the photon. We then have

$$h\frac{c}{\lambda} = 1.628 \times 10^{-16} \text{ J,}$$

which yields

$$\lambda = \frac{(6.63 \times 10^{-34} \text{ J·s})(3.00 \times 10^8 \text{ m/s})}{1.628 \times 10^{-16} \text{ J}}$$

$$= 1.22 \times 10^{-9} \text{ m.} \qquad \text{(Answer)}$$

This wavelength is in the x-ray region of the electromagnetic spectrum.

CHECKPOINT **1:** Rank the following pairs of quantum states for an electron confined to an infinite well according to the energy differences between the states, greatest first: (a) $n = 3$ to $n = 1$, (b) $n = 5$ to $n = 4$, (c) $n = 4$ to $n = 3$.

Finding the Wave Functions

It turns out that the wave functions for the allowed states of the electron in an infinite potential well are given by

$$\psi_n(x) = A \sin\left(\frac{n\pi}{L}\right)x, \text{ for } n = 1, 2, 3, \cdots, \quad (40\text{-}6)$$

for the range $0 \le x \le L$ (the wave function is zero outside that range). In Eq. 40-6, A is an arbitrary amplitude constant; you will see shortly how to evaluate A.

We could have derived Eq. 40-6 by solving Schrödinger's equation, given the potential energy function plotted in Fig. 40-2. It is far simpler, however, to assume — correctly, as it turns out — that the wave function $\psi_n(x)$ has the same form as the displacement function $y_n(x)$ for a standing wave on a string stretched between rigid supports (see Eq. 40-2).

We are more interested in the *probability density* $\psi_n^2(x)$ than in $\psi_n(x)$ because it is the probability density that has physical meaning. (Although wave functions are usually complex quantities, in Eq. 40-6 the wave function is a real quantity, so we do not have to be concerned about taking its absolute value before squaring.) Recall from Section 39-7 that the value of $\psi_n^2(x)$ at any point measures the probability that the electron will be found near that

point. Specifically, the probability that an electron in an infinite well will be found to lie between the points x and $x + dx$ is $\psi_n^2(x)\,dx$. Thus $\psi_n^2(x)$ is a probability per unit length. From Eq. 40-6 we see that the probability density for an electron in an infinite well is

$$\psi_n^2(x) = A^2 \sin^2\left(\frac{n\pi}{L}\right)x,$$

$$\text{for } n = 1, 2, 3, \cdots, \qquad (40\text{-}7)$$

for the range $0 \le x \le L$ (the probability density is zero outside that range). Figure 40-4 shows $\psi_n^2(x)$ for $n = 1, 2, 3,$ and 15 for an electron in an infinite well whose width L is 100 pm.

If classical physics prevailed, we would expect the trapped electron to appear with equal probability in all parts of the well. From Fig. 40-4 we see that it does not. For example, inspection of that figure and of Eq. 40-7 shows that for the state with $n = 2$, the electron is most likely to be found near $x = 25$ pm and $x = 75$ pm. It will be found with near-zero probability near $x = 0$, $x = 50$ pm, and $x = 100$ pm.

The case of $n = 15$ in Fig. 40-4 suggests that as n increases, it becomes more and more likely that the electron *will* be found with equal probability in all parts of the well. This result is an instance of a general principle called the **correspondence principle:**

At large enough quantum numbers, the predictions of quantum mechanics merge smoothly with those of classical physics.

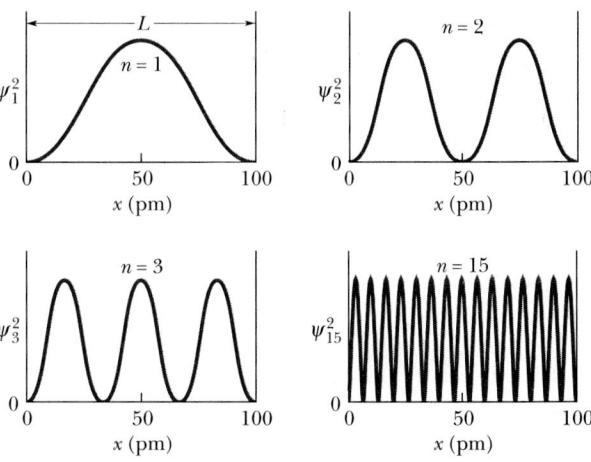

FIGURE 40-4 The probability density $\psi_n^2(x)$ for four states of an electron trapped in an infinite well; their quantum numbers are $n = 1, 2, 3,$ and 15. The electron is most likely to be found where $\psi_n^2(x)$ is high, and least likely to be found where it is low.

This principle, first advanced by Danish physicist Niels Bohr, holds for all quantum predictions. It should remind you of a similar principle concerning the theory of relativity, namely, that at low-enough particle speeds, the predictions of special relativity merge smoothly with those of classical physics.

CHECKPOINT 2: The figure shows three infinite potential wells of widths L, $2L$, and $3L$; each contains an electron in the state for $n = 10$. Rank the wells according to (a) the number of maxima for the probability density of the electron and (b) the energy of the electron, greatest first.

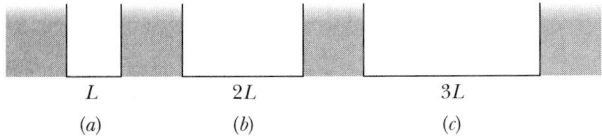

| L | $2L$ | $3L$ |
| (a) | (b) | (c) |

Normalization

The product $\psi_n^2(x)\,dx$ gives the probability that an electron in an infinite well will be found in the interval of the x axis that lies between x and $x + dx$. We know that the electron must be *somewhere* in the infinite well, so it must be that

$$\int_{-\infty}^{+\infty} \psi_n^2(x)\,dx = 1 \quad \text{(normalization equation)}, \quad (40\text{-}8)$$

since the probability 1 corresponds to certainty. Although the integral is taken over the entire x axis, only the region from $x = 0$ to $x = L$ makes any contribution to the probability. Graphically, the integral in Eq. 40-8 represents the area under each of the plots of Fig. 40-4.

In Sample Problem 40-3 we shall show that if you substitute $\psi_n^2(x)$ from Eq. 40-7 into Eq. 40-8, it is possible to assign a specific value to the arbitrary amplitude constant A that appears in Eq. 40-7, namely, $A = \sqrt{2/L}$. This process of using Eq. 40-8 to evaluate the amplitude of a wave function is called **normalizing** the wave function. The process applies to *all* one-dimensional wave functions.

Zero-Point Energy

Substituting $n = 1$ in Eq. 40-4 defines the state of lowest energy for an electron in the infinite well, the ground state. It is the state that the confined electron will occupy unless energy is supplied to it to raise it to an excited state.

The question arises: Why can't we include $n = 0$ among the possibilities listed for n in Eq. 40-4? Putting $n = 0$ in this equation would indeed yield a ground-state

energy of zero. However, putting $n = 0$ in Eq. 40-7 would also yield $\psi_n^2(x) = 0$ for all x, which we can interpret only to mean that there is no electron in the well. But there is, so $n = 0$ is not a possible quantum number.

It is an important conclusion of quantum mechanics that confined systems cannot exist in states with zero energy. They must always have a certain minimum energy called the **zero-point energy**.

We can make the zero-point energy as small as we like by making the infinite well wider, that is, by increasing L in Eq. 40-4 for $n = 1$. In the limit as $L \rightarrow \infty$, the zero-point energy E_1 approaches zero. In this limit, however, with an infinitely wide well, the electron is a free particle, no longer confined in the x direction. And because the energy of a free particle is not quantized, that energy can have any value, including zero. Only a confined particle must have a finite zero-point energy and can never be at rest.

CHECKPOINT 3: Each of the following particles is confined to an infinite well of the same width: (a) an electron, (b) a proton, (c) a deuteron, and (d) an alpha particle. Rank their zero-point energies, greatest first. The particles are listed in order of increasing mass.

SAMPLE PROBLEM 40-3

Use Eq. 40-8, the normalization equation, to evaluate the arbitrary amplitude constant A in Eq. 40-6 for an infinite potential well extending from $x = 0$ to $x = L$.

SOLUTION: Substituting Eq. 40-7 into Eq. 40-8 and taking the constant outside the integral yield

$$A^2 \int_0^L \sin^2\left[\left(\frac{n\pi}{L}\right)x\right] dx = 1. \quad (40\text{-}9)$$

We have changed the limits of the integral from $-\infty$ and $+\infty$ to 0 and L because the wave function is zero outside these new limits (so there's no need to integrate out there).

We can further simplify this equation by changing the variable from x to the dimensionless variable y, where

$$y = \left(\frac{n\pi}{L}\right)x, \quad (40\text{-}10)$$

hence

$$dx = \left(\frac{L}{n\pi}\right)dy.$$

When we change the variable, we must also change the integration limits (again). Equation 40-10 tells us that $y = 0$ when $x = 0$ and that $y = n\pi$ when $x = L$. So 0 and $n\pi$ are our new limits. With all these substitutions, Eq. 40-9 becomes

$$A^2 \frac{L}{n\pi} \int_0^{n\pi} \sin^2 y\,dy = 1.$$

We can use integral 11 in Appendix E to evaluate the integral, obtaining the equation

$$\frac{A^2 L}{n\pi} \left[\frac{y}{2} - \frac{\sin 2y}{4} \right]_0^{n\pi} = 1.$$

Evaluating the limits yields

$$\frac{A^2 L}{n\pi} \frac{n\pi}{2} = 1,$$

so $$A = \sqrt{\frac{2}{L}}.\qquad \text{(Answer)}\quad (4\text{-}11)$$

This result tells us that the dimension for A^2, and thus for $\psi_n^2(x)$, is an inverse length. This is appropriate because a probability density is a probability *per unit length*.

40-4 AN ELECTRON IN A FINITE WELL

A potential energy well of infinite depth is an idealization. Figure 40-5 shows a realizable potential energy well—one in which the potential energy of an electron outside the well is not infinitely great but has a finite positive value E_{pot}^*, called the **well depth.** The analogy between waves on a stretched string and matter waves fails us for wells of finite depth because we can no longer be sure that matter wave nodes exist at $x = 0$ and at $x = L$. (As we shall see, they don't.)

To find the wave functions describing the quantum states of an electron in the finite well of Fig. 40-5, we *must* resort to Schrödinger's equation, the basic equation of quantum mechanics. From Section 39-7 recall that, for motion in one dimension, we use Schrödinger's equation in the form of Eq. 39-15:

$$\frac{d^2\psi}{dx^2} + \frac{8\pi^2 m}{h^2}[E - E_{pot}(x)]\psi = 0. \quad (40\text{-}12)$$

Rather than attempting to solve this equation for the finite well, we simply state the results for particular numerical values of E_{pot}^* and L. Figure 40-6 shows these results as

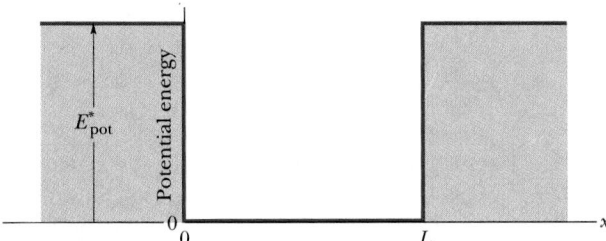

FIGURE 40-5 A *finite* potential energy well. The depth of the well is E_{pot}^* and its width is L. As in the infinite well of Fig. 40-2, the motion of the trapped electron is restricted to the x direction.

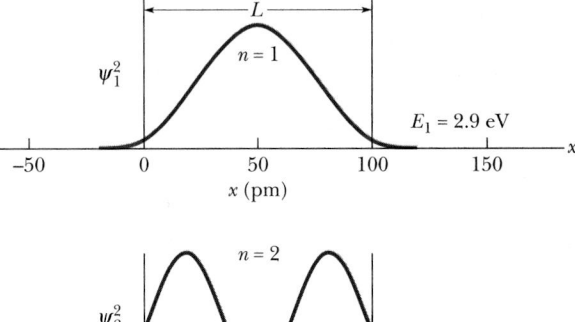

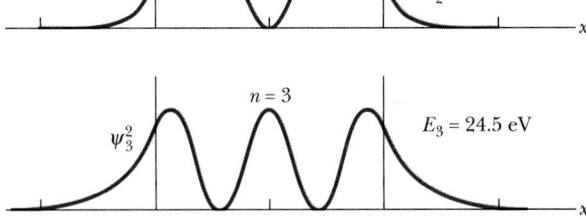

FIGURE 40-6 The probability densities $\psi_n^2(x)$ for an electron confined to the finite well of Fig. 40-5 for the states with $n = 1, 2,$ and 3. The depth of the well is 30 eV, and its width is 100 pm. No other quantum states exist for this trapped electron.

graphs of $\psi_n^2(x)$, the probability density, for a well with $E_{pot}^* = 30$ eV and $L = 100$ pm.

The probability density $\psi_n^2(x)$ for each graph in Fig. 40-6 satisfies Eq. 40-8, the normalization equation. So we know that the areas under all three probability density plots are numerically equal to 1.

If you compare Fig. 40-6, for a finite well, with Fig. 40-4, for an infinite well, you will see one striking difference: for a finite well, there is a finite probability that the electron matter wave (and thus the electron itself) can penetrate the walls of the well—in a region in which Newtonian mechanics says the electron cannot exist. This possibility should not be surprising, because we saw in Section 39-9 that an electron can tunnel through a potential energy barrier. "Leaking" into the walls of a finite potential energy well is a similar phenomenon.

Although you can't tell from Fig. 40-6, the three states shown are the *only* quantum states that can exist with energies less than the well depth, which is 30 eV for this particular finite well. Electrons with $E > 30$ eV are not confined to the well and have energies that are not quantized. Figure 40-7 shows the energy levels for an electron trapped in this well.

SAMPLE PROBLEM 40-4

Suppose a finite well with $E_{pot}^* = 30.0$ eV and $L = 100$ pm confines a single electron in its ground state.

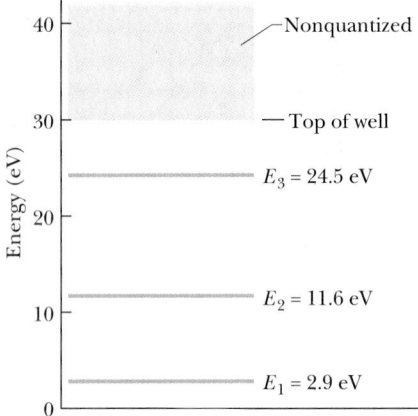

FIGURE 40-7 The energy level diagram for an electron confined to a finite well 30 eV deep and 100 pm wide. Only three discrete quantum states are possible for an electron in this well. States with $E > 30$ eV have a continuous range of energies.

(a) The electron can be raised to quantum states of higher energy by illuminating the well with light of the appropriate wavelength. What discrete wavelengths of incident light would the electron absorb? In what region of the electromagnetic spectrum would these wavelengths (said to be *absorption spectrum lines*) lie?

SOLUTION: The energy imparted by the incident light could raise the electron from its ground state ($n = 1$) to the state with $n = 2$ or to the state with $n = 3$. No other discrete states are possible. The energy difference for the smaller of these energy jumps is $\Delta E = E_2 - E_1$. This energy difference must be contributed by an absorbed photon, of energy $hf = hc/\lambda$. Equating the two energies, we get

$$E_2 - E_1 = \frac{hc}{\lambda}.$$

Solving for λ and substituting energy values from Fig. 40-7 yield

$$\lambda = \frac{hc}{E_2 - E_1}$$
$$= \frac{(6.63 \times 10^{-34} \text{ J} \cdot \text{s})(3.00 \times 10^8 \text{ m/s})}{(11.6 \text{ eV} - 2.9 \text{ eV})(1.60 \times 10^{-19} \text{ J/eV})}$$
$$= 1.43 \times 10^{-7} \text{ m} = 143 \text{ nm}. \quad \text{(Answer)}$$

This wavelength lies in the ultraviolet region of the spectrum.

The jump from the electron's ground state to the state with $n = 3$ requires an energy difference of

$$\Delta E = E_3 - E_1 = 24.5 \text{ eV} - 2.9 \text{ eV} = 21.6 \text{ eV}.$$

Repeating the calculation above with this energy difference leads to

$$\lambda = 57.6 \text{ nm}, \quad \text{(Answer)}$$

which is also in the ultraviolet region of the spectrum.

If the electron is initially in its ground state, as we have assumed, no other discrete spectral absorption lines will occur

because the three discrete states shown in Fig. 40-7 are the only ones that exist in this particular finite well.

(b) Can the electron, initially in the ground state, absorb a photon with a wavelength $\lambda = 100$ nm?

SOLUTION: This wavelength is intermediate between the wavelength of 143 nm (required for the jump to the first excited state) and the wavelength of 57.6 nm (required for the jump to the second excited state). To absorb a photon at the intermediate wavelength 100 nm would require that the electron jump to an intermediate quantum state, but there is none. Thus, the electron cannot absorb light at this wavelength (or any other intermediate wavelength).

(c) With the electron in the ground state, what wavelength of light is needed to barely free the electron from the potential well by a single photon absorption?

SOLUTION: To be free of the potential well, the electron must receive an energy that puts it into the nonquantized energy region of Fig. 40-7. So, it must then have an energy of at least E_{pot}^* ($= 30.0$ eV). As in (a), we can write

$$E_{\text{pot}}^* - E_1 = \frac{hc}{\lambda},$$

from which we find

$$\lambda = \frac{hc}{E_{\text{pot}}^* - E_1}$$
$$= \frac{(6.63 \times 10^{-34} \text{ J} \cdot \text{s})(3.00 \times 10^8 \text{ m/s})}{(30.0 \text{ eV} - 2.9 \text{ eV})(1.60 \times 10^{-19} \text{ J/eV})}$$
$$= 4.59 \times 10^{-8} \text{ m} = 45.9 \text{ nm}. \quad \text{(Answer)}$$

(d) Can the electron, initially in the ground state, absorb a photon with an associated wavelength of 20.2 nm? If so, in what state is the electron after the absorption?

SOLUTION: The photon energy hf at this wavelength is

$$hf = h\frac{c}{\lambda} = \frac{(6.63 \times 10^{-34} \text{ J} \cdot \text{s})(3.00 \times 10^8 \text{ m/s})}{20.2 \times 10^{-9} \text{ m}}$$
$$= 9.847 \times 10^{-18} \text{ J} = 61.5 \text{ eV}.$$

This energy exceeds the 30.0 eV depth of the potential well. Thus, the electron can absorb a photon of this energy; the absorption allows the electron to escape from the well. It is then a free particle with a kinetic energy of

$$K = hf - E_{\text{pot}}^* = 61.5 \text{ eV} - 30.0 \text{ eV} = 31.5 \text{ eV}$$

and is no longer in a quantum state.

CHECKPOINT **4:** Figure 40-6 shows the three quantum states of an electron trapped in a 30 eV finite well. (a) In which state is the electron most likely to be found near the midpoint of the well? (b) Rank the three states according to the probability of the electron being outside the well, greatest first.

40-5 MORE ELECTRON TRAPS

Here we discuss three types of artificial electron traps.

Nanocrystallites

Perhaps the most direct way to construct a potential energy well in the laboratory is to prepare a sample of a semiconducting material in the form of a powder whose granules are small—in the nanometer range—and of uniform size. Each such **nanocrystallite** acts as a potential well for the electrons trapped within it.

Equation 40-4 shows that we can increase the energy of the least energetic quantum state of an electron trapped in an infinite well by reducing the width L of that well. This is also true for the wells formed by individual nanocrystallites. Thus, the smaller the nanocrystallite, the higher its lowest available level, that is, the higher the threshold energy for the photons of light that it can absorb.

If we shine sunlight on a powder of nanocrsytallites, the crystallites can absorb all photons with energies above a certain threshold energy E_t ($= hf_t$). That is, they can absorb all light whose wavelength is *below* a certain threshold λ_t, where

$$\lambda_t = \frac{c}{f_t} = \frac{ch}{E_t}. \qquad (40\text{-}13)$$

Since light not absorbed is scattered, our powder of nanocrystallites will scatter all wavelengths above λ_t.

We see the powder sample by the light it scatters back to our eyes. Thus, by controlling the size of the nanocrystallites in a sample, we can control the wavelengths of the light scattered by the sample, hence the sample's color.

Figure 40-8 shows two samples of the semiconductor cadmium selenide, each consisting of a powder of nanocrystallites of uniform size. The upper sample scatters light at the red end of the spectrum. The lower sample differs from the upper sample *only* in that the lower sample is composed of smaller nanocrystallites. For this reason its threshold energy E_t is larger and, from Eq. 40-13, its threshold wavelength λ_t is smaller. The sample takes on a color of lower wavelength—in this case yellow.

The striking contrast in color between the two samples is compelling evidence of the quantization of the energies of trapped electrons and the dependence of these energies on the size of the electron trap. We remark again that the two samples in Fig. 40-8 are chemically identical; they differ only in the size of the nanocrystallites of which they are composed.

Quantum Dots

The highly developed techniques used to fabricate computer chips can be used to construct, atom by atom, indi-

FIGURE 40-8 Two samples of powdered cadmium selenide, a semiconductor, differing only in the size of their granules. Each granule serves as an electron trap. The upper sample has the larger granules and consequently the smaller spacing between energy levels and the smaller photon energy threshold for the absorption of light. Light not absorbed is scattered, causing the sample to appear red. The lower sample, because of its smaller granules, and consequently its larger level spacing and its larger energy threshold for absorption, appears yellow.

vidual potential energy wells that behave, in many respects, like artificial atoms. These **quantum dots,** as they are usually called, have promising applications in electron optics and computer technology.

In one such arrangement, a "sandwich" is fabricated in which a thin layer of a semiconducting material, shown in purple in Fig. 40-9a, is deposited between two insulating layers, one of which is much thinner than the other. Metal end caps with conducting leads are added at both ends. The materials are chosen to ensure that the potential energy of an electron in the central layer is less than it is in the two insulating layers, causing the central layer to act as a potential energy well. Figure 40-9b is a photograph of an actual quantum dot; the well in which individual electrons can be trapped is the purple region.

The lower (but not the upper) insulating layer in Fig. 40-9a is thin enough to permit electrons to tunnel through it if an appropriate potential difference is applied between the leads. In this way the number of electrons confined to the well can be controlled. The arrangement does indeed behave like an artificial atom with the property that the number of electrons it contains can be controlled. Quantum dots can be constructed in two-dimensional arrays that could well form the basis for computing systems of great speed and storage capacity.

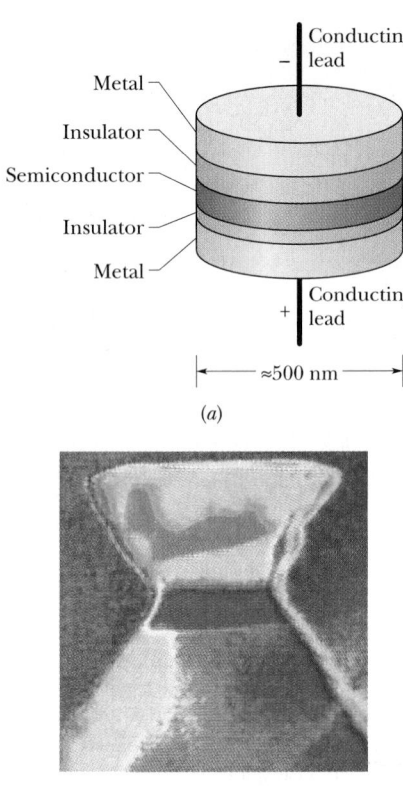

(a)

(b)

FIGURE 40-9 A quantum dot, or "artificial atom." (a) A central semiconducting layer forms a potential energy well in which electrons are trapped. The lower insulating layer is thin enough to allow electrons to be added to or removed from the central layer by barrier tunneling if an appropriate voltage is applied between the leads. (b) A photograph of an actual quantum dot. The central purple band is the electron confinement region.

Quantum Corrals

When a scanning tunneling microscope (described in Section 39-9 and Fig. 39-15) is in operation, its tip exerts a small force on isolated atoms that may be located on an otherwise smooth surface. By careful manipulation of the position of the tip, such isolated atoms can be "dragged" across the surface and deposited at another location. Using this technique, scientists at IBM's Almaden Research Center moved iron atoms across a carefully prepared copper surface, forming the atoms into a circle, which they named a **quantum corral.** The result is shown in the photograph that opens this chapter. Each iron atom in the circle is nestled in a hollow in the copper surface, equidistant from three nearest-neighbor copper atoms. The corral was fabricated at a low temperature (about 4 K) to minimize the tendency of the iron atoms to move randomly about on the surface because of their thermal energies.

The ripples within the corral are due to matter waves associated with electrons that can move over the copper surface but are largely trapped in the potential well of the corral. The dimensions of the ripples are in excellent agreement with the predictions of quantum mechanics.

40-6 THE HYDROGEN ATOM

We now move from artificial atoms to real ones, using the simplest atom—hydrogen—as our example. This atom consists of a single electron (charge $-e$) bound to its central nucleus, a single proton (charge $+e$), by the attractive Coulomb force that acts between them. The hydrogen atom, like all atoms, is an electron trap; it confines its single electron to a region of space. From the confinement principle, we then expect that the electron can exist only in a discrete set of quantum states, each with a well-defined energy. We wish to identify the energies and the wave functions of these states.

The Energies of the Hydrogen Atom States

In Chapter 25 we wrote Eq. 25-43 for the (electric) potential energy of a two-particle system with charges q_1 and q_2:

$$U = \frac{1}{4\pi\epsilon_0}\frac{q_1 q_2}{r},$$

where r is the distance between the particles. For the two-particle system of a hydrogen atom, we change notation a little and write the potential energy as

$$E_{\text{pot}} = \frac{1}{4\pi\epsilon_0}\frac{(e)(-e)}{r} = -\frac{1}{4\pi\epsilon_0}\frac{e^2}{r}. \quad (40\text{-}14)$$

The plot of Fig. 40-10 suggests the three-dimensional potential well in which the hydrogen atom's electron is trapped. This well differs from the finite potential well of Fig. 40-5 in that, for the hydrogen atom, E_{pot} is negative for all values of r because we have (arbitrarily) chosen our zero of potential energy to correspond to $r = \infty$. For the finite well of Fig. 40-5, however, we (equally arbitrarily) chose to assign the zero of potential energy to the region inside the well.

To find the wave functions that define the discrete quantum states of the hydrogen atom and the quantized energies of those states, we must solve Schrödinger's equation, with Eq. 40-14 substituted for E_{pot} in that equation. However, we cannot use the form of Schrödinger's equation given by Eq. 40-12 because that equation holds only for an electron moving in one dimension. In the hydrogen atom, the electron is free to move in three dimensions, so we must use the three-dimensional form of Schrödinger's equation.

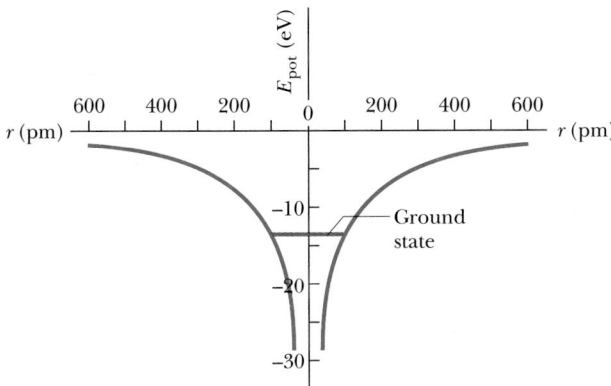

FIGURE 40-10 A plot of Eq. 40-14, which gives the potential energy E_{pot} of a hydrogen atom as a function of the separation r between the electron and the central proton. The plot is shown twice (on the left and on the right) to suggest the three-dimensional spherically symmetric trap to which the electron is confined. Note that r is always a positive quantity.

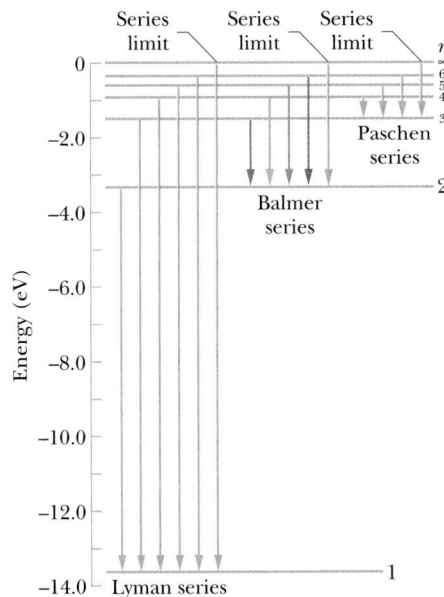

FIGURE 40-11 A plot of Eq. 40-15, showing a few of the energy levels of the hydrogen atom. The downward transitions between these levels, corresponding to the emission of light from the atom, are also shown. The transitions are grouped into series, each labeled with the name of a person associated with the study of the series spectrum.

Solving that equation reveals that the energies of the allowed states are given by

$$E_n = -\frac{me^4}{8\epsilon_0^2 h^2}\frac{1}{n^2} = -\frac{13.6\text{ eV}}{n^2}, \quad (40\text{-}15)$$

$$\text{for } n = 1, 2, 3, \cdots,$$

where n is a quantum number and m is the mass of an electron. The lowest energy, which is for the ground state with $n = 1$, is indicated in Fig. 40-10. Figure 40-11 shows the energy levels of the ground state and five excited states, each labeled with its quantum number n. It also shows the energy level for the greatest value of n, namely, $n = \infty$, for which $E_n = 0$. For any greater energy, the electron and proton are not bound together (there is no hydrogen atom), and the corresponding region in Fig. 40-11 is like the non-quantized region for the finite well of Fig. 40-7.

The quantized energy values given by Eq. 40-15 are actually those of the hydrogen atom, that is, of the *electron + proton* system. However, we can usually attribute the energy to the electron alone because its mass is much less than that of the proton. (Similarly, we can attribute the energy of a *ball + Earth* system to the ball alone.) Thus, we can say that when an electron is trapped in a hydrogen atom, the *electron* can have only energy values given by Eq. 40-15.

As with the other potential wells we have discussed, the electron tends to be in the ground state but can jump to an excited state if given the proper amount of energy. One way the electron can gain the energy to make an upward jump between energy levels is to absorb a photon. The energy hf of the photon must then equal the energy differ-

ence between the initial energy level of the electron and the higher energy level. When the electron reaches a higher energy level, it does not stay there but quickly de-excites to a lower energy level. One way it can de-excite is by emitting a photon. In that case the energy hf of the photon equals the energy difference between the initial higher level and the lower level.

All the possible jumps can be grouped into *series*, each series consisting of the upward jumps that start on, or downward jumps that end on, a particular (home-base) level. Figure 40-11 shows some of the downward jumps for three series. The *Lyman series*, for example, has the ground state as the home-base level. Each series has a *series limit* corresponding to a jump between the home-base level and $n = \infty$. This is the greatest possible jump between quantized levels and thus corresponds to the greatest change in energy of the atom for the given home base.

Figure 40-12 shows the spectrum of the *Balmer series* of the hydrogen atom, as photographed (in one order) with a spectroscope. (The spectral lines are like those shown in Figs. 37-23 and 37-24.) The Balmer series, whose home-base level is $n = 2$, has four spectral lines that are in the visible range, as indicated in Fig. 40-12 and also by the colors used in the Balmer series of Fig. 40-11. The small triangle shown at $\lambda = 364.6$ nm in Fig. 40-12 marks the series limit.

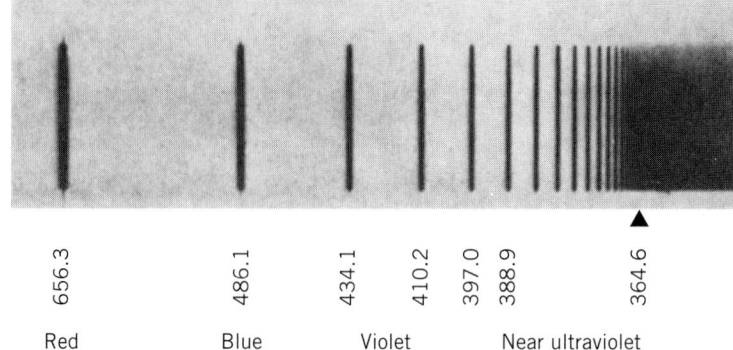

FIGURE 40-12 The spectrum lines of the Balmer series of the hydrogen atom. Whereas Fig. 40-11 shows four transitions of this series, along with the series limit, this figure shows about a dozen lines of this series; note how they are progressively closer toward the series limit, which is marked with a triangle.

λ(nm) 656.3 486.1 434.1 410.2 397.0 388.9 364.6

Red   Blue   Violet   Near ultraviolet

Bohr's Theory of the Hydrogen Atom

In 1913, some 13 years before the formulation of Schrödinger's equation, Bohr proposed a model of the hydrogen atom based on a clever combination of classical and early quantum concepts. His basic assumption—that atoms exist in discrete quantum states of well-defined energy—was a bold break with classical ideas; it carries over today as an indispensable concept in modern quantum mechanics. With this assumption, Bohr made skillful use of the correspondence principle (see Section 40-3), not only to derive Eq. 40-15 for the energies of the quantum states of the hydrogen atom but also to derive a numerical value (the *Bohr radius*) for the effective radius of that atom. In spite of its successes, Bohr's specific model of the hydrogen atom, based on the assumption that the electron moves in planet-like orbits around the nucleus, is inconsistent with the uncertainty principle and has been replaced by the modern probability density model. For his brilliant achievements relating to atomic structure, which greatly stimulated progress toward the modern quantum theory, Bohr was awarded the Nobel prize in 1922.

Quantum Numbers for the Hydrogen Atom

Although the energies of the hydrogen atom states can be described by the single quantum number n, the wave functions describing these states require three quantum numbers, corresponding to the three dimensions in which the electron can move. The three quantum numbers, along with their names and the values that they may have, are shown in Table 40-1.

Each set of quantum numbers (n, l, m_l) identifies the wave function of a particular quantum state. The quantum number n, called the **principal quantum number,** appears in Eq. 40-15 and describes the energy of the state. We state without proof that the **orbital quantum number** l is a measure of the magnitude of the angular momentum associated with the quantum state. The **orbital magnetic quantum number** m_l is related to the orientation in space

of this angular momentum vector. The restrictions on the values of the quantum numbers for the hydrogen atom, as listed in Table 40-1, are not arbitrary but emerge naturally in the process of solving Schrödinger's equation. Note that for the ground state ($n = 1$), the restrictions require that $l = 0$ and $m_l = 0$. That is, the hydrogen atom in its ground state has zero angular momentum.

CHECKPOINT 5: (a) A group of quantum states of the hydrogen atom has $n = 5$. How many values of l are possible for states within this group? (b) A subgroup of hydrogen atom states within the $n = 5$ group has $l = 3$. How many values of m_l are possible for states within this subgroup?

The Wave Function of the Hydrogen Atom Ground State

The wave function for the ground state of the hydrogen atom, as obtained by solving the three-dimensional Schrödinger equation and normalizing the result, is

$$\psi(r) = \frac{1}{\sqrt{\pi} a^{3/2}} e^{-r/a}. \tag{40-16}$$

Here a is the **Bohr radius**, a constant with the dimension *length*. This radius is loosely taken to be the effective radius of a hydrogen atom and turns out to be a convenient unit of length for other situations involving atomic dimen-

TABLE 40-1 QUANTUM NUMBERS FOR THE HYDROGEN ATOM

SYMBOL	NAME	ALLOWED VALUES
n	Principal quantum number	1, 2, 3, . . .
l	Orbital quantum number	0, 1, 2, . . . , $n-1$
m_l	Orbital magnetic quantum number	$-l$, $-(l-1)$, . . . , $+(l-1)$, $+l$

sions. Its value is

$$a = \frac{h^2 \epsilon_0}{\pi m e^2} = 5.29 \times 10^{-11} \text{ m} = 52.9 \text{ pm.} \quad (40\text{-}17)$$

As with other wave functions, ψ in Eq. 40-16 does not have physical meaning but ψ^2 does. Specifically, $\psi^2(r) \, dV$ is the probability that the electron will be found in any given (infinitesimal) volume element dV. Because $\psi^2(r)$ here depends only on r, it makes sense to choose, as a volume element dV, the volume between two concentric spherical shells whose radii are r and $r + dr$. That is, we take the volume element dV to be

$$dV = (4\pi r^2) \, dr, \quad (40\text{-}18)$$

in which $4\pi r^2$ is the area of the inner shell and dr is the radial distance between the two shells. Then

$$\psi^2(r) \, dV = \frac{4}{a^3} e^{-2r/a} r^2 \, dr. \quad (40\text{-}19)$$

We now define a **radial probability density** $P(r)$ such that $P(r) \, dr$ gives the probability that the electron will be found in the volume element defined by Eq. 40-18. In other words, we define $P(r)$ so that $P(r) \, dr = \psi^2(r) \, dV$. Thus, from Eq. 40-19,

$$P(r) = \frac{4}{a^3} r^2 e^{-2r/a} \quad \begin{array}{l}\text{(radial probability density,}\\ \text{hydrogen atom ground state).}\end{array} \quad (40\text{-}20)$$

Figure 40-13 is a plot of Eq. 40-20. The area under the plot is unity; that is,

$$\int_0^\infty P(r) \, dr = 1. \quad (40\text{-}21)$$

This equation simply states that in a normal hydrogen atom, the electron must be *somewhere* in the space surrounding the nucleus.

The triangular marker on the horizontal axis of Fig. 40-13 is located one Bohr radius from the origin. The graph tells us that in the ground state of the hydrogen atom, the electron is most likely to be found near this radius.

Figure 40-13 conflicts sharply with the popular view that electrons in atoms follow well-defined orbits like planets moving around the Sun. *This popular view, however familiar, is incorrect.* Figure 40-13 shows us all that we can ever know about the location of the electron in the ground state of the hydrogen atom. The appropriate question is not ''When will the electron arrive at such-and-such a point?'' but ''What are the odds that the electron will be found in a small volume centered on such-and-such a point?'' Figure 40-14, which we call a dot plot, suggests the probabilistic nature of the wave function and provides a useful mental model of the hydrogen atom in its ground state. Think of the atom in this state as a fuzzy ball with no sharply defined boundary and no hint of orbits.

It is not easy for a beginner to envision subatomic particles in this probabilistic way. The difficulty is our natural impulse to regard an electron as something like a tiny jelly bean, located at certain places at certain times and following a well-defined path. Electrons and other subatomic particles simply do not behave in this way. Do your best to avoid this *jelly bean fallacy*, as we may call it.

The energy of the ground state, found by putting $n = 1$ in Eq. 40-15, is $E_1 = -13.6$ eV. The wave function of Eq. 40-16 results if you solve Schrödinger's equation with this value of the energy. Actually, you can find a solution of this equation for *any* value of the energy, say $E = -11.6$ eV or -14.3 eV. This may suggest that the energies

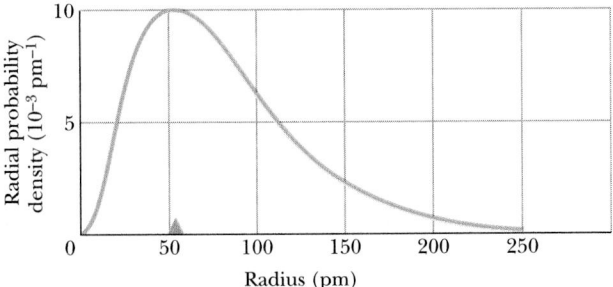

FIGURE 40-13 A plot of the radial probability density $P(r)$ for the ground state of the hydrogen atom. The triangular marker is located at one Bohr radius from the origin, and the origin represents the center of the atom.

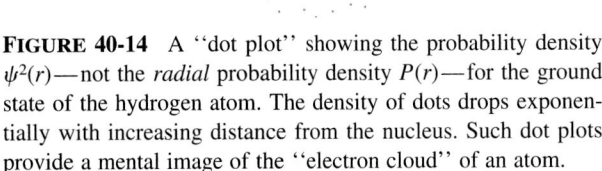

FIGURE 40-14 A ''dot plot'' showing the probability density $\psi^2(r)$ — not the *radial* probability density $P(r)$ — for the ground state of the hydrogen atom. The density of dots drops exponentially with increasing distance from the nucleus. Such dot plots provide a mental image of the ''electron cloud'' of an atom.

of the hydrogen atom states are not quantized. But we know that they are.

The puzzle is solved when we realize that such solutions of Schrödinger's equation are not physically acceptable because they have increasingly large values as $r \rightarrow \infty$. Such ''wave functions'' tell us that the electron is more likely to be found very far from the nucleus than closer to it, which makes no sense. We get rid of these unwanted solutions by imposing a so-called **boundary condition,** in which we agree to accept only solutions of Schrödinger's equation for which $\psi(r) \rightarrow 0$ as $r \rightarrow \infty$. That is, we agree to deal only with *confined* electrons. With this restriction, the solutions of this equation now form a discrete set, with quantized energies given by Eq. 40-15.

SAMPLE PROBLEM 40-5

(a) What is the wavelength of the least energetic photon emitted in the Lyman series of the hydrogen atom spectrum lines?

SOLUTION: For any series, the transition that produces the least energetic photon is the transition between the home-base level that defines the series and the level immediately above it. Figure 40-11 shows that for the Lyman series, the transition with the least energetic photon is the transition from the level with $n = 2$ to that with $n = 1$. From Eq. 40-15 the energy difference for this transition is

$$\Delta E = E_2 - E_1 = -(13.6 \text{ eV}) \left(\frac{1}{2^2} - \frac{1}{1^2} \right) = 10.2 \text{ eV}.$$

The corresponding wavelength is found from Eq. 39-2 ($E = hf$), written in the form

$$\Delta E = hf = \frac{hc}{\lambda},$$

where hf is the energy of the emitted photon. Solving for the wavelength λ yields

$$\lambda = \frac{hc}{\Delta E} = \frac{(6.63 \times 10^{-34} \text{ J} \cdot \text{s})(3.00 \times 10^8 \text{ m/s})}{(10.2 \text{ eV})(1.60 \times 10^{-19} \text{ J/eV})}$$

$$= 1.22 \times 10^{-7} \text{ m} = 122 \text{ nm}. \quad \text{(Answer)}$$

Light with this wavelength is in the ultraviolet region of the electromagnetic spectrum.

(b) What is the wavelength corresponding to the series limit for the Lyman series?

SOLUTION: Figure 40-11 shows that the series limit corresponds to a transition from the level with $n = \infty$ to that with $n = 1$, the home base for this series. From Eq. 40-15, the energy difference for this transition is

$$\Delta E = E_\infty - E_1 = -(13.6 \text{ eV}) \left(\frac{1}{\infty^2} - \frac{1}{1^2} \right)$$

$$= -(13.6 \text{ eV})(0 - 1) = 13.6 \text{ eV}.$$

The corresponding wavelength is found as in (a) and is

$$\lambda = \frac{hc}{\Delta E} = \frac{(6.63 \times 10^{-34} \text{ J} \cdot \text{s})(3.00 \times 10^8 \text{ m/s})}{(13.6 \text{ eV})(1.60 \times 10^{-19} \text{ J/eV})}$$

$$= 9.14 \times 10^{-8} \text{ m} = 91.4 \text{ nm}. \quad \text{(Answer)}$$

Light with this wavelength is also in the ultraviolet region of the electromagnetic spectrum.

SAMPLE PROBLEM 40-6

Show that the radial probability density for the ground state of the hydrogen atom has a maximum at $r = a$.

SOLUTION: The radial probability density we want is given by Eq. 40-20,

$$P(r) = \frac{4}{a^3} r^2 e^{-2r/a}.$$

To find the maximum of any function, we must differentiate it and set the result equal to zero. If we differentiate $P(r)$ with respect to r, using derivative 7 of Appendix E and the chain rule for differentiating products, we get

$$\frac{dP}{dr} = \frac{4}{a^3} r^2 \left(\frac{-2}{a} \right) e^{-2r/a} + \frac{4}{a^3} 2r \, e^{-2r/a}$$

$$= \frac{8r}{a^3} e^{-2r/a} - \frac{8r^2}{a^4} e^{-2r/a}$$

$$= \frac{8}{a^4} r(a - r) e^{-2r/a}.$$

If we set the right side equal to zero, the resulting equation is true if $r = a$. In other words, dP/dr is equal to zero when $r = a$. (Note that we also have $dP/dr = 0$ at $r = 0$ and at $r = \infty$. However, these conditions correspond to a *minimum* in $P(r)$, as you can see in Fig. 40-13.)

SAMPLE PROBLEM 40-7

It can be shown that the probability $p(r)$ that the electron in the ground state of the hydrogen atom will be found inside a sphere of radius r is given by

$$p(r) = 1 - e^{-2x}(1 + 2x + 2x^2),$$

in which x, a dimensionless quantity, is equal to r/a. Find r for $p(r) = 0.90$.

SOLUTION: We seek the radius of a sphere for which $p(r) = 0.90$. Substituting that value in the expression above for $p(r)$, we have

$$0.90 = 1 - e^{-2x}(1 + 2x + 2x^2)$$

or $\qquad 10e^{-2x}(1 + 2x + 2x^2) = 1.$

We must find the value of x that satisfies this equality. It is not possible to solve explicitly for x, but a little trial and error with a pocket calculator (write a small program for it) yields $x = 2.67$. This means that the radius of a sphere such that the electron will be detected inside it 90% of the time is $2.67a$. Mark this position on the horizontal axis of Fig. 40-13 and ask yourself whether it is a reasonable answer.

Hydrogen Atom States with $n = 2$

According to rules given in Table 40-1 there are four states of the hydrogen atom with $n = 2$; their quantum numbers are listed in Table 40-2. Consider first the state with $n = 2$ and $l = m_l = 0$; its probability density is represented by the dot plot of Fig. 40-15. Note that this plot, like the plot for the ground state shown in Fig. 40-14, is spherically symmetric. That is, the probability density is a function of the radial coordinate r only and is independent of the angular coordinates θ and ϕ of Fig. 40-16.

It turns out that all quantum states with $l = 0$ have spherically symmetric wave functions. This is reasonable because the quantum number l is a measure of the angular

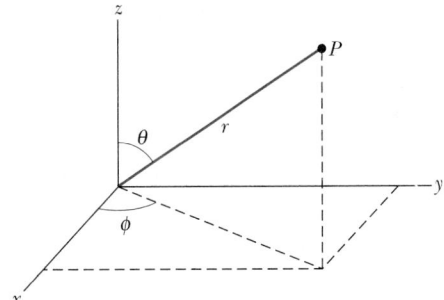

FIGURE 40-16 The relationship between the coordinates x, y, and z of the rectangular coordinate system and the coordinates r, θ, and ϕ of the spherical coordinate system. The latter are more appropriate for analyzing situations involving spherical symmetry, such as the hydrogen atom.

momentum associated with a given state. If $l = 0$, the angular momentum is also zero, which requires that the probability density representing the state have no preferred axis of symmetry.

Dot plots of ψ^2 for the three states with $n = 2$ and $l = 1$ are shown in Fig. 40-17. The probability densities for the states with $m_l = +1$ and $m_l = -1$ are identical. Although these plots are symmetric about the z axis, they are *not* spherically symmetric. That is, the probability densities for these three states are functions of both r and the angular coordinate θ.

Here is a puzzle: What is there about the hydrogen atom that establishes the axis of symmetry that is so obvious in Fig. 40-17? The answer: *absolutely nothing*.

The solution to this puzzle comes about when we realize that all three states shown in Fig. 40-17 have the same

TABLE 40-2 QUANTUM NUMBERS FOR HYDROGEN ATOM STATES WITH $n = 2$

n	l	m_l
2	0	0
2	1	+1
2	1	0
2	1	-1

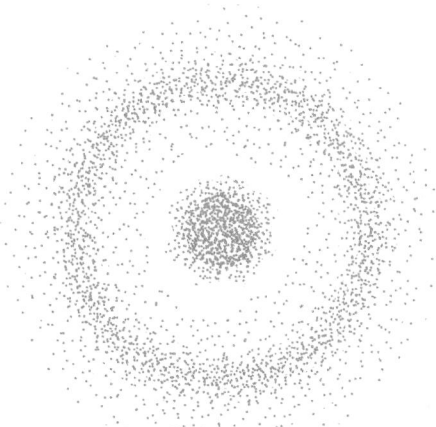

FIGURE 40-15 A dot plot showing the probability density $\psi^2(r)$ for the hydrogen atom in the quantum state with $n = 2$, $l = 0$, and $m_l = 0$. The plot has spherical symmetry about the central nucleus. The gap in the dot density pattern marks a spherical surface over which $\psi^2(r) = 0$.

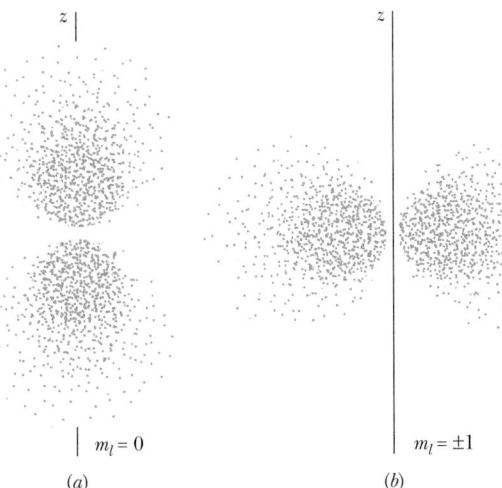

FIGURE 40-17 Dot plots of the probability density $\psi^2(r, \theta)$ for the hydrogen atom in states with $n = 2$ and $l = 1$. (a) Plot for $m_l = 0$. (b) Plot for $m_l = +1$ and $m_l = -1$. Both plots show that the probability density is symmetric about the z axis.

energy. Recall that the energy of a state, given by Eq. 40-15, depends only on the principal quantum number n and is independent of l and m_l. In fact, for an *isolated* hydrogen atom there is no way to differentiate experimentally among the three states of Fig. 40-17.

If we add the probability densities for these three states, the combined probability density turns out to be spherically symmetrical, with no unique axis. One can, then, think of the electron as spending one-third of its time in each of the three states of Fig. 40-17, and one can think of the weighted sum of the three independent wave functions as defining a spherically symmetric **subshell,** specified by the quantum numbers $n = 2$, $l = 1$. The individual states will display their separate existence only if we place the hydrogen atom in an external electric or magnetic field. The three states of the $n = 2$, $l = 1$ subshell will then have different energies, and the field direction will establish the necessary symmetry axis.

The $n = 2$, $l = 0$ state, whose probability density is shown in Fig. 40-15, *also* has the same energy as each of the three states of Fig. 40-17. We can view all four states whose quantum numbers are listed in Table 40-2 as forming a spherically symmetric **shell,** specified by the single quantum number n. The importance of shells and subshells will become evident in Chapter 41, where we discuss atoms having more than one electron.

To round out our picture of the hydrogen atom, we display in Fig. 40-18 a dot plot of the probability density for a hydrogen atom state with a relatively large quantum number ($n = 45$) and the largest orbital quantum number that the restrictions of Table 40-1 permit ($l = n - 1 = 44$). The probability density forms a ring that is symmetrical about the z axis and lies very close to the xy plane. The

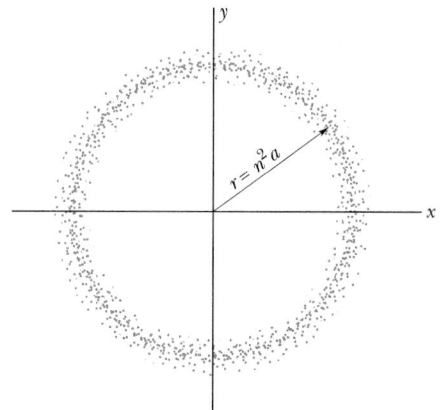

FIGURE 40-18   A dot plot of the probability density $P(r)$ for the hydrogen atom in a quantum state with a relatively large principal quantum number, namely, $n = 45$, and angular momentum quantum number $l = n - 1 = 44$. The dots lie close to the xy plane, the ring of dots suggesting a classical electron orbit.

mean radius of the ring is n^2a, where a is the Bohr radius. This mean radius is more than 2000 times the effective radius of the hydrogen atom in its ground state.

Figure 40-18 suggests the electron orbit of classical physics. Thus we have another illustration of Bohr's correspondence principle, namely, that at large quantum numbers the predictions of quantum mechanics merge smoothly with those of classical physics. Imagine what a dot plot like that of Figure 40-18 would look like for *really* large values of n and l, say $n = 1000$ and $l = 999$.

40-7 QUANTUM WEIRDNESS: AN EXAMPLE

Bohr used to say that if you aren't confused by quantum mechanics, then you don't really understand it. We cannot leave our discussion of quantum matters without examining a prediction of quantum mechanics that seems weird but, as experiment has shown, is undoubtedly correct.

In 1935 Einstein, together with his colleagues Boris Podolsky and Nathan Rosen, explored the quantum mechanics of a two-particle system. They proposed a "thought experiment" (now called the EPR experiment, after their initials) that predicted results so strange that Einstein rejected them, concluding that such predictions indicated a deep flaw in quantum mechanics.

The EPR experiment was carried out in the 1980s and the weird results predicted by quantum mechanics were in fact observed. We shall discuss the EPR experiment in broad outline and then illustrate it by analogy.

In Fig. 40-19 a source S emits two photons, labeled A and B, simultaneously and in opposite directions. Each photon has a certain property X that may have two values, say X_1 and X_2. (The property is actually the polarization direction of the quantum probability wave associated with the photon, but that detail need not concern us.) Because of the way the two photons were generated (simultaneously, in a coordinated emission), it is always true that if photon A

FIGURE 40-19   A source S emits two photons, simultaneously and in opposite directions. An experimenter A may choose, arbitrarily, to reveal either of two possible properties of photon A. Similarly, a second experimenter B may choose, arbitrarily, to reveal either of the same two possible properties of photon B. However, once experimenter A has made her choice, all choice is removed from experimenter B. The result of his measurement is quite predictable, even though the photons may be very far apart and no information has passed between the experimenters.

has value X_1 then photon B will have value X_2, and conversely. There is nothing weird about that.

These two photons, taken together, constitute a single quantum system that can exist in two states; we can call them state (AX_1, BX_2) and state (AX_2, BX_1). Before any measurement is made, quantum mechanics predicts that the *actual* state of this two-photon system is an intimate equal-parts mixture of *both* states. You can imagine the two-particle system oscillating between the states, spending equal time in both.

By making a suitable measurement on photon A, an experimenter can choose to reveal *either* value X_1 or value X_2 for this photon. Let us say that the experimenter chooses to reveal X_1. It then follows that the two-particle system is no longer a mixture of states (AX_1, BX_2) and (AX_2, BX_1). The act of measurement has caused the system to "collapse" into state (AX_1, BX_2) alone. A measurement made on photon B must reveal only the value X_2. In short, the kind of measurement one conducts on A (a matter of arbitrary choice) automatically removes any choice for the state of photon B. Quantum mechanics predicts this to be true even if the photons are far apart (even kilometers apart) when the first measurement is made; no wonder Einstein called this prediction "spooky actions at a distance." Nevertheless, the 1980s experiments show that that is exactly what happens. Most physicists accept the results of these experiments as an impressive endorsement of the validity of quantum mechanics.

Now for a loose analogy. Suppose that a jelly bean can exist in either of two states, red or green. Let Sally and Sam meet in Chicago. Then let Sally move to Boston, taking with her a jelly bean of each color. Sam moves to Los Angeles, with a jelly bean of each color in his pocket. At a certain time let Sally, without communicating in any way with Sam, decide to eat one of her jelly beans and let her deliberately choose the red one. After this time let Sam, without looking, pull a jelly bean from his pocket; *he will always find it to be green*. Furthermore, Sally's green jelly bean and Sam's red jelly bean will simply disappear: the system has collapsed into its Sally-red, Sam-green state.

If Sally had chosen to eat her green jelly bean, the two-bean system would have collapsed into its Sally-green, Sam-red state, and the two other jelly beans would have vanished. Thus, in our analogy, Sally's arbitrary choice in Boston determines the color of the jelly bean Sam pulls out of his pocket in Los Angeles. Spooky indeed!

If you actually tried the Sally–Sam experiment, it would of course not turn out as we have discussed; our story is just an analogy. To make the analogy exact, we would have to give both Sally and Sam single "quantum jelly beans" that were each red and green at the same time, each jelly bean alternating rapidly between the two states in a correlated way. Such quantum behavior is so small for objects as big as jelly beans that it is hopeless to try to detect such quantum behavior. At the quantum level, however, such events really occur. It may seem weird, but that is the way the world is!

REVIEW & SUMMARY

The Confinement Principle

The **confinement principle** applies to waves of all kinds, including waves on a string and the matter of waves of quantum mechanics. It states that confinement leads to quantization, that is, to the existence of discrete states with discrete energies.

An Electron in an Infinite Well

An infinite well is a device for confining an electron. From the confinement principle we expect that the matter wave representing a trapped electron can exist only in a set of discrete states. The energies associated with these states are

$$E_n = \left(\frac{h^2}{8mL^2} \right) n^2, \quad \text{for } n = 1, 2, 3, \ldots, \quad (40\text{-}4)$$

in which L is the width of the well and n is a **quantum number.** The **wave functions** associated with these states are

$$\psi_n(x) = A \sin \left(\frac{n\pi}{L} \right) x, \quad \text{for } n = 1, 2, 3, \ldots. \quad (40\text{-}6)$$

The **probability density** $\psi_n^2(x)$ for an allowed state has the physical meaning that $\psi_n^2(x) \, dx$ is the probability that the electron will be found in the interval between x and $x + dx$. For an electron in an infinite well, the probability densities are

$$\psi_n^2(x) = A^2 \sin^2 \left(\frac{n\pi}{L} \right) x, \quad \text{for } n = 1, 2, 3, \ldots. \quad (40\text{-}7)$$

At high quantum numbers n, the electron tends toward classical behavior in that it tends to occupy all parts of the well with equal probability. This fact leads to the **correspondence principle**: At large enough quantum numbers, the predictions of quantum mechanics merge smoothly with those of classical physics.

Normalization and Zero-Point Energy

The amplitude A^2 in Eq. 40-7 can be found from the **normalizing equation,**

$$\int_{-\infty}^{+\infty} \psi_n^2(x) \, dx = 1, \quad (40\text{-}8)$$

which asserts that the electron must be *somewhere* within the well, because the probability 1 implies certainty.

From Eq. 40-4 we see that the lowest permitted energy for the electron is not zero but the energy that corresponds to $n = 1$. This lowest energy is called the **zero-point energy** of the electron–well system.

An Electron in a Finite Well

A finite potential energy well is one for which the potential energy of an electron inside the well is less than that for one outside the well by a finite amount E_{pot}^*. There is a finite probability that an electron trapped in such a well can, nevertheless, be outside the well (in the wall).

The Hydrogen Atom

The potential energy function for the hydrogen atom is

$$E_{pot} = -\frac{1}{4\pi\epsilon_0}\frac{e^2}{r}. \tag{40-14}$$

The energies of the quantum states of the hydrogen atom are found from the three-dimensional form of Schrödinger's equation to be

$$E_n = -\frac{me^4}{8\epsilon_0^2 h^2}\frac{1}{n^2} = -\frac{13.6\text{ eV}}{n^2} \quad n = 1, 2, 3, \ldots, \tag{40-15}$$

in which n is the **principal quantum number**. The hydrogen atom requires three quantum numbers for its complete description; their names and allowed values are shown in Table 40-1.

The **radial probability density** $P(r)$ for a hydrogen atom state is defined so that $P(r)\,dr$ is the probability that the electron will be found between two concentric shells, centered on the atom's nucleus, whose radii are r and $r + dr$. For the hydrogen atom's ground state,

$$P(r) = \frac{4}{a^3} r^2 e^{-2r/a}, \tag{40-20}$$

in which a, the **Bohr radius**, is a length unit whose value is 52.9 pm. Figure 40-13 is a plot of $P(r)$ for the ground state.

Figures 40-15 and 40-17 represent the probability densities (not the *radial* probability densities) for the four hydrogen atom states with $n = 2$. The plot of Fig. 40-15 ($n = 2, l = 0, m_l = 0$) is spherically symmetric. The plots of Fig. 40-17 ($n = 2, l = 1$, $m_l = 0, +1, -1$) are symmetric about the z axis but, when added together, are also spherically symmetric.

All four states with $n = 2$ have the same energy and may be usefully regarded as a **shell,** identified as the $n = 2$ shell. The three states of Fig. 40-17, taken together, may be regarded as the $n = 2, l = 1$ **subshell.** It is not possible to separate the four $n = 2$ states experimentally unless the hydrogen atom is placed in an electric or a magnetic field, to permit the establishment of a definite symmetry axis.

Quantum Weirdness

In a two-particle quantum mechanical system, an arbitrary choice of the kind of measurement an experimenter makes on one of the particles can completely fix the results that a second experimenter finds for the other particle. That is, the first experimenter to make a measurement has an arbitrary choice; the second has no choice.

QUESTIONS

1. If you double the width of an infinite potential well, (a) is the energy of the ground state of the trapped electron multiplied by 4, 2, $\frac{1}{2}$, $\frac{1}{4}$, or some other number? (b) Are the energies of the higher energy states multiplied by this factor or by some other factor, depending on their quantum number?

2. Three electrons are trapped in three different infinite potential wells of widths (a) 50 pm, (b) 200 pm, and (c) 100 pm. Rank the electrons according to their zero-point energies, greatest first.

3. If you wanted to use the idealized trap of Fig. 40-1 to trap a positron, would you need to change (a) the geometry of the trap, (b) the electric potential of the central cylinder, or (c) the electric potentials of the two semi-infinite end cylinders? (A positron has the same mass as an electron but is positively charged.)

4. An electron is trapped in an infinite potential well in a state with $n = 17$. How many (a) nodes and (b) probability maxima does its matter wave have?

5. Figure 40-20 shows three infinite potential wells, each on an x axis. Without written calculation, determine the wave function ψ for a ground-state electron trapped in each well.

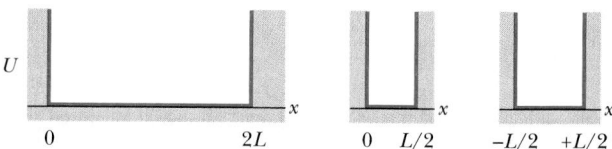

FIGURE 40-20   Question 5.

6. Figure 40-21 indicates the lowest energy levels (in electron-volts) for five situations in which an electron is trapped in an infinite potential well. In wells B, C, D, and E, the electron is in the ground state. We shall excite the electron in well A to the fourth excited state (at 25 eV). The electron can then de-excite to the ground state by emitting one or more photons, corresponding to one long jump or several short jumps. What photon *emission* energies of this de-excitation match a photon *absorption* energy (from the ground state) of the other four wells? Give the corresponding quantum numbers.

7. Is the zero-point energy of a proton trapped in an infinite potential well greater than, less than, or equal to that of an electron trapped in the same potential well?

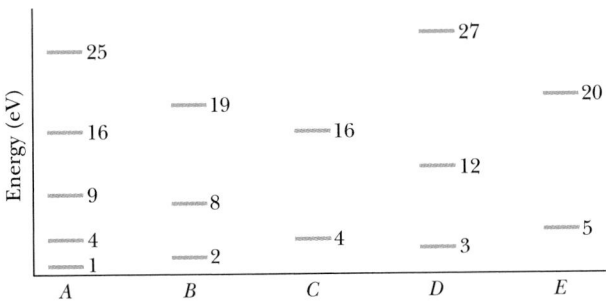

FIGURE 40-21 Question 6.

8. A proton and an electron are trapped in identical infinite potential wells; both particles are in the ground state. At the center of the wells, is the probability density for the proton greater than, less than, or equal to that of the electron?

9. You want to modify the finite potential well of Fig. 40-5 to allow its trapped electron to exist in more than three quantum states. Could you do so by making the well (a) wider or narrower, (b) deeper or shallower?

10. An electron is trapped in a finite potential well that is deep enough to allow the electron to exist in a state with $n = 4$. How many (a) nodes and (b) probability maxima does its associated matter wave have (within the well)?

11. From a visual inspection of Fig. 40-6, rank the quantum numbers of the three quantum states according to the de Broglie wavelength of the electron, greatest first.

12. From a visual inspection of Fig. 40-6, rank the quantum numbers of the three quantum states according to the probability that the electron will be found outside the region given by $0 < x < L$, greatest first.

13. An electron, trapped in a finite potential energy well such as that of Fig. 40-5, is in its state of lowest energy. Are (a) its de Broglie wavelength, (b) the magnitude of its momentum, and (c)

its energy greater than, the same as, or less than they would be if the potential well were infinite, as in Fig. 40-2?

14. The table lists the quantum numbers for five proposed hydrogen atom states. Which of them are not possible?

	n	l	m_l
(a)	3	2	0
(b)	2	3	1
(c)	4	3	-4
(d)	5	5	0
(e)	5	3	-2

15. In 1996 physicists working at an accelerator laboratory succeeded in producing atoms of antihydrogen. Such atoms consist of a positron moving in the electric field of an antiproton. A positron has the same mass but opposite charge of an electron. An antiproton has the same mass but opposite charge of a proton. Would you expect the spectrum of antihydrogen be the same as that of normal hydrogen or different?

16. (a) From Fig. 40-11, the energy level diagram for the hydrogen atom, you can show that the photon energy of the second spectral line of the Lyman series is equal to the sum of the photon energies of two other lines. What are those lines? (b) The photon energy of the second spectral line of the Lyman series is also equal to the *difference* between the photon energies of two other lines. What are *those* lines?

17. A hydrogen atom is in the third excited state. To what state (give the quantum number n) should it jump to (a) emit light with the longest possible wavelength, (b) emit light with the shortest possible wavelength, and (c) absorb light with the longest possible wavelength?

EXERCISES & PROBLEMS

SECTION 40-3 Trapping an Electron

1E. You wish to reduce by one-half the zero-point energy of an electron trapped in an infinite potential well. By what factor must you change the width of the potential well?

2E. What is the ground state energy of (a) an electron and (b) a proton if each is trapped in an infinite potential well that is 100 pm wide?

3E. What must be the width of an infinite potential well if an electron trapped in it in the $n = 3$ state is to have an energy of 4.7 eV?

4E. Consider an atomic nucleus to be equivalent to an infinite potential well with $L = 1.4 \times 10^{-14}$ m, a typical nuclear diameter. What would be the ground-state energy of an electron if it

were trapped in such a potential well? (*Note:* Nuclei do not contain electrons.)

5E. The ground-state energy of an electron trapped in an infinite potential well is 2.6 eV. What will this quantity be if the width of the potential well is doubled?

6E. An electron, trapped in an infinite potential well 250 pm wide, is in its ground state. How much energy must it absorb if it is to jump up to the state with $n = 4$?

7E. What is the SI unit for the probability density of an electron trapped in an infinite potential well?

8E. A proton is confined to an infinite potential well 100 pm wide. What is its zero-point energy?

9E. Show that $\Delta E/E$ for an electron in an infinite potential well approaches the value $2/n$ at large quantum numbers, where ΔE is defined in Sample Problem 40-1. (Note that although ΔE does not approach zero at large quantum numbers, $\Delta E/E$ does, in accordance with the correspondence principle.)

10P. An electron is trapped in an infinite potential well. (a) What pair of adjacent energy levels (if any) will have three times the energy difference that exists between levels $n = 3$ and $n = 4$? (b) What pair (if any) will have twice that energy difference?

11P. An electron is trapped in an infinite potential well. Show that the energy difference ΔE between its quantum levels n and $n + 2$ is $(h^2/4mL^2)(n + 2)$.

12P. An electron is trapped in an infinite potential well. (a) What pair of adjacent energy levels (if any) has an energy difference equal to the energy of the electron in the state with $n = 5$? (b) With $n = 6$?

13P. An electron is trapped in an infinite potential well that is 100 pm wide; the electron is in ground state. What is the probability that you can detect the electron in an interval of width $\Delta x = 5.0$ pm centered at $x = $ (a) 25 pm, (b) 50 pm, and (c) 90 pm? (*Hint:* The interval Δx is so narrow that you can take the probability density to be constant within it.)

14P. Sample Problem 40-1b deals with an electron confined to move parallel to the axis of an evacuated cylinder 3.0 m long. (a) At what quantum number would the energy between adjacent levels be 1.0 eV, a measurable quantity? (b) What would be the energy of the electron at this quantum number? (c) Is this energy in the relativistic range?

15P. A particle is confined to an infinite potential well of width L. If the particle is in its ground state, what is the probability that it will be found between (a) $x = 0$ and $x = L/3$, (b) $x = L/3$ and $x = 2L/3$, and (c) $x = 2L/3$ and $x = L$?

SECTION 40-4 An Electron in a Finite Well

16E. Figure 40-7 gives the energy levels for an electron trapped in a finite potential energy well 30 eV deep. If the electron is in the $n = 3$ state, what is its kinetic energy?

17E. An electron in the $n = 2$ state in the finite potential well of Fig. 40-5 absorbs 31.7 eV of energy from an external source. What is its kinetic energy after this absorption, assuming that the electron moves to a position for which $x > L$?

18E. (a) Show that each term in Schrödinger's equation (Eq. 40-12) has the same dimensions. (b) What is the common SI unit for each of these terms?

19P. As Fig. 40-6 suggests, the probability density for the region $x > L$ in the finite potential well of Fig. 40-5 drops off exponentially according to

$$\psi^2(x) = Ce^{-2kx},$$

where C is a constant. (a) Show that the wave function $\psi(x)$ that may be found from the equation above is a solution of Schrödinger's equation in its one-dimensional form. (b) What must be the value of k for this to be true?

20P. Show that for the region $x > L$ in the finite potential well of Fig. 40-5, $\psi(x) = De^{2kx}$ is a solution of Schrödinger's equation in its one-dimensional form, where D is a constant. On what basis do we find this mathematically acceptable solution to be physically unacceptable?

21P. As Fig. 40-6 suggests, the probability density for an electron in the region $0 < x < L$ for the finite potential well of Fig. 40-5 is sinusoidal, being given by

$$\psi^2(x) = B \sin^2 kx,$$

in which B is a constant. (a) Show that the wave function $\psi(x)$ that may be found from the equation above is a solution of Schrödinger's equation in its one-dimensional form. (b) What must be the value of k for this to be true?

SECTION 40-6 The Hydrogen Atom

22E. Verify that the constant appearing in Eq. 40-15 is 13.6 eV.

23E. An atom (not a hydrogen atom) absorbs a photon whose associated frequency is 6.2×10^{14} Hz. By what amount does the energy of the atom increase?

24E. An atom (not a hydrogen atom) absorbs a photon whose associated wavelength is 375 nm and then immediately emits photon whose associated wavelength is 580 nm. How much net energy is absorbed by the atom in this process?

25E. Repeat Sample Problem 40-5 for the Balmer series of the hydrogen atom.

26E. (a) What is the energy of the hydrogen atom electron whose probability density is represented by the dot plot of Fig. 40-15? (b) What minimum energy is needed to remove this electron from the atom?

27E. What are (a) the energy, (b) the magnitude of the momentum, and (c) the wavelength of a photon emitted when a hydrogen atom undergoes a transition from a state with $n = 3$ to a state with $n = 1$?

28E. What is the ratio of the shortest wavelength of the Balmer series to the shortest wavelength of the Lyman series?

29E. A neutron, with a kinetic energy of 6.0 eV, collides with a stationary hydrogen atom in its ground state. Explain why the collision must be elastic—that is, why kinetic energy must be conserved. (*Hint:* Show that the hydrogen atom cannot be raised to a higher excitation state as a result of the collision.)

30E. A hydrogen atom is excited from its ground state to the state with $n = 4$. (a) How much energy must be absorbed by the atom? (b) Calculate and display on an energy level diagram the different photon energies that may be emitted as the atom returns to its ground state.

31E. Calculate the radial probability density $P(r)$ for the hydrogen atom in its ground state at (a) $r = 0$, (b) $r = a$, and (c) $r = 2a$, where a is the Bohr radius.

32E. For the hydrogen atom in its ground state, calculate (a) the probability density $\psi^2(r)$ and (b) the radial probability density $P(r)$ for $r = a$, where a is the Bohr radius.

33P. How much work must be done to pull apart the electron and the proton that make up the hydrogen atom if the atom is initially in (a) its ground state and (b) the state with $n = 2$?

34P. What are the widths of the wavelength intervals over which (a) the Lyman series and (b) the Balmer series extend? (Each width begins at the longest wavelength and ends at the series limit.) (c) What are the widths of the corresponding frequency intervals? Express the frequency intervals in terahertz (1 THz = 10^{12} Hz).

35P. In the ground state of the hydrogen atom, the electron has a total energy of -13.6 eV. What are (a) its kinetic energy and (b) its potential energy if the electron is one Bohr radius from the central nucleus?

36P. A hydrogen atom, initially at rest in the $n = 4$ quantum state, undergoes a transition to the ground state, emitting a photon in the process. What is the speed of the recoiling hydrogen atom?

37P. Light of wavelength 486.1 nm is emitted by a hydrogen atom. (a) What transition of the atom is responsible for this radiation? (b) To what series does this transition belong?

38P. (a) Find, using the energy level diagram of Fig. 40-11, the quantum numbers corresponding to a transition in which the wavelength of the emitted radiation is 121.6 nm. (b) To what series does this transmission belong?

39P. A hydrogen atom in a state having a *binding energy* (the energy required to remove an electron) of 0.85 eV makes a transition to a state with an *excitation energy* (the difference between the energy of the state and that of the ground state) of 10.2 eV. (a) What is the energy of the photon emitted as a result of the transition? (b) Identify this transition, using the energy level diagram of Fig. 40-11.

40P. Verify the wavelengths given in Fig. 40-12 for the visible spectral lines of the Balmer series.

41P. A hydrogen atom emits light of wavelength 102.6 nm. What are the initial and final quantum numbers for this transition?

42P. What is the probability that in the ground state of the hydrogen atom, the electron will be found at a radius greater than the Bohr radius? (*Hint:* See Sample Problem 40-7.)

43P. Calculate the probability that the electron in the hydrogen atom, in its ground state, will be found between spherical shells whose radii are a and $2a$, where a is the Bohr radius. (*Hint:* See Sample Problem 40-7.)

44P. Schrödinger's equation for states of the hydrogen atom for which the orbital quantum number l is zero is

$$\frac{1}{r^2}\frac{d}{dr}\left(r^2\frac{d\psi}{dr}\right) + \frac{8\pi^2 m}{h^2}[E - E_{\text{pot}}]\psi = 0.$$

Verify that Eq. 40-16, which describes the ground state of the hydrogen atom, is a solution of this equation.

45P. Verify that Eq. 40-20, the radial probability density for the ground state of the hydrogen atom, is normalized. That is, verify that

$$\int_0^\infty P(r)\,dr = 1.$$

46P. (a) For a given value of the principal quantum number n, how many values of the orbital quantum number l are possible? (b) For a given value of l, how many values of the orbital magnetic quantum number m_l are possible? (c) For a given value of n, how many values of m_l are possible?

47P. For what value of the principal quantum number n would the effective radius of the probability density dot plot for the electron in the hydrogen atom be 1.0 mm? Assume that l has its maximum value of $n - 1$. (*Hint:* Be guided by Fig. 40-18.)

48P. What is the probability that an electron in the ground state of the hydrogen atom will be found between two spherical shells whose radii are r and $r + \Delta r$, (a) if $r = 0.500a$ and $\Delta r = 0.010a$ and (b) if $r = 1.00a$ and $\Delta r = 0.01a$, where a is the Bohr radius? (*Hint:* Δr is small enough to permit the radial probability density to be taken to be constant between r and $r + dr$.)

49P*. In Sample Problem 40-6 we showed that the radial probability density for the ground state of the hydrogen atom is a maximum when $r = a$, where a is the Bohr radius. Show that the *average* value of r, defined as

$$\bar{r} = \int P(r)\,r\,dr,$$

has the value 1.5a. In this expression for $\bar{r}$, each value of $P(r)$ is weighted with the value of r at which it occurs. Note that the average value of r is greater than the value of r for which $P(r)$ is a maximum.

50P*. The wave function for the hydrogen atom quantum state shown in Fig. 40-15, which has $n = 2$ and $l = m_l = 0$, is

$$\psi_{200}(r) = \frac{1}{4\sqrt{2\pi}}\,a^{-3/2}\left(2 - \frac{r}{a}\right)e^{-r/2a},$$

in which a is the Bohr radius and the subscript on $\psi(r)$ gives the values of the quantum numbers n, l, m_l. (a) Plot $\psi_{200}^2(r)$ and show that your plot is consistent with the dot plot of Fig. 40-15. (b) Show analytically that $\psi_{200}^2(r)$ has a maximum at $r = 4a$. (c) Find the radial probability density $P_{200}(r)$ for this state. (d) Show that

$$\int_0^\infty P_{200}(r)\,dr = 1,$$

and thus that the expression above for the wave function $\psi_{200}(r)$ has been properly normalized.

51P. The wave functions for the three states shown in Fig. 40-17, which have $n = 2$, $l = 1$ and $m_l = 0$, $+1$, and -1, are

$$\psi_{210}(r, \theta) = (1/4\sqrt{2\pi})(a^{-3/2})(r/a)e^{-r/2a}\cos\theta,$$
$$\psi_{21+1}(r, \theta) = (1/8\sqrt{\pi})(a^{-3/2})(r/a)e^{-r/2a}(\sin\theta)\,e^{+i\phi},$$
$$\psi_{21-1}(r, \theta) = (1/8\sqrt{\pi})(a^{-3/2})(r/a)e^{-r/2a}(\sin\theta)\,e^{-i\phi},$$

in which the subscripts on $\psi(r, \theta)$ give the values of the quantum numbers n, l, m_l and the angles θ and ϕ are defined in Fig. 40-16. Note that the first wave function is real but the others, which involve the imaginary number i, are complex. (a) Find the probability density for each wave function and show that each is consistent with its dot plot in Fig. 40-17. (b) Add the three probability densities derived in (a) and show that their sum is spherically symmetric, depending only on the radial coordinate r.

41
All About Atoms

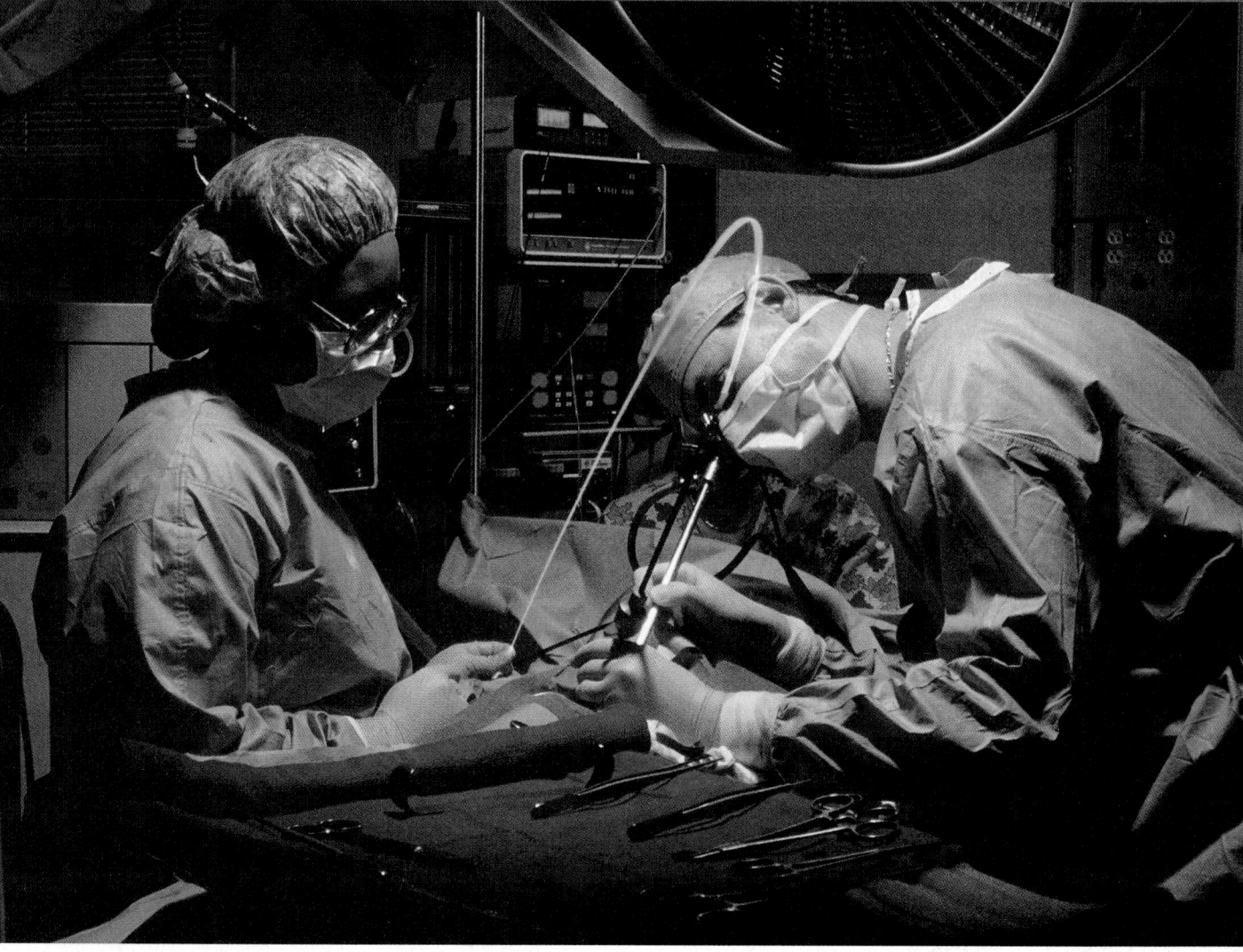

Soon after lasers were invented in the 1960s, they became novel sources of light in uncountable research laboratories. But today, lasers are ubiquitous, being found in such diverse applications as voice and data transmission, surveying, welding, and grocery-store price scanning. The photograph shows surgery being performed with laser light transmitted via optical fibers. Light from a laser and light from any other source are both due to emissions by atoms. What, then, is so different about the light from a laser?

41-1 ATOMS AND THE WORLD AROUND US

In the early years of this century quite a few prominent scientists doubted the very existence of atoms. Today, however, every well-informed person believes that atoms exist and are the building blocks of the material world. Today, we can even pick up individual atoms and move them around. That's how the quantum corral on the opening page of Chapter 40 was formed. You can easily count the 48 iron atoms that make up the circle in that image. We can even photograph single atoms by the light they emit. For example, the faint blue dot at the center of Figure 41-1 is due to light emitted by a single barium atom (actually, an ion) held in a trap at the University of Washington.

41-2 SOME PROPERTIES OF ATOMS

You may think the details of atomic physics are remote from your daily life. However, consider how the following properties of atoms—so basic that we rarely think about them—affect the way we live in our world.

Atoms are stable. Essentially all the atoms that form our tangible world have existed without change for billions of years. What would the world be like if all atoms changed into other forms, perhaps after a few weeks or a few years?

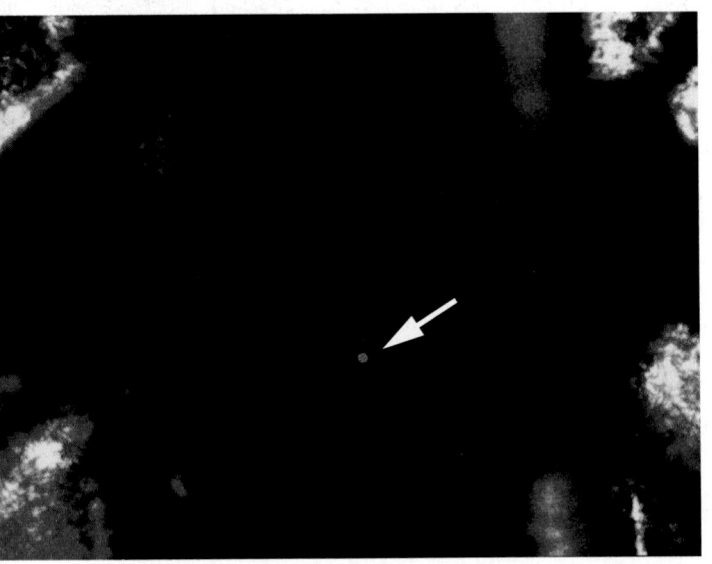

FIGURE 41-1 The blue dot is a photograph of the light emitted from a single barium atom (actually, an ion) held for a long time in a trap at the University of Washington. Special techniques caused the ion to emit light over and over again as it underwent transitions between the same pair of energy levels. The dot represents the cumulative emission of many photons.

Atoms combine with each other. They stick together to form stable molecules and stack up to form rigid solids. An atom is mostly empty space, but you can stand on a floor—made up of atoms—without falling through it.

These basic properties of atoms can be explained by quantum mechanics, as can the three less apparent properties that follow.

Atoms Are Put Together Systematically

Figure 41-2 shows an example of a repetitive property of the elements as a function of their position in the periodic table (Appendix G). The figure is a plot of the **ionization energy** of the elements: the energy required to remove the most loosely bound electron from a neutral atom is plotted as a function of the position in the periodic table of the element to which the atom belongs. The remarkable similarities in the chemical and physical properties of the elements in each vertical column of the periodic table are evidence enough that the atoms are constructed according to systematic rules.

The elements are arranged in the periodic table in six horizontal **periods**; except for the first, each period starts at the left with a highly reactive alkali metal (lithium, sodium, potassium, and so on) and ends at the right with a chemically inert noble gas (neon, argon, krypton, and so on). Quantum mechanics accounts for the chemical properties of these elements. The numbers of elements in the six periods are

$$2, 8, 8, 18, 18, \text{ and } 32.$$

Quantum mechanics predicts these numbers.

Atoms Emit and Absorb Light

We have already seen that atoms can exist only in discrete quantum states, each state having a certain energy. An atom can make a transition from one state to another by emitting light (to jump to a lower energy state) or by absorbing light (to jump to a higher energy state). The frequency f of the light is given by the **Bohr frequency condition**, so called because it was postulated by Bohr well before the advent of modern quantum mechanics:

$$hf = E_{\text{high}} - E_{\text{low}} \quad \text{(Bohr frequency condition)}. \quad (41\text{-}1)$$

Here E_{high} is the higher energy and E_{low} is the lower energy of the pair of quantum states involved in the transition, and hf is the photon energy of the emitted or absorbed light.

Thus the problem of finding the frequencies of the light emitted or absorbed by an atom reduces to the prob-

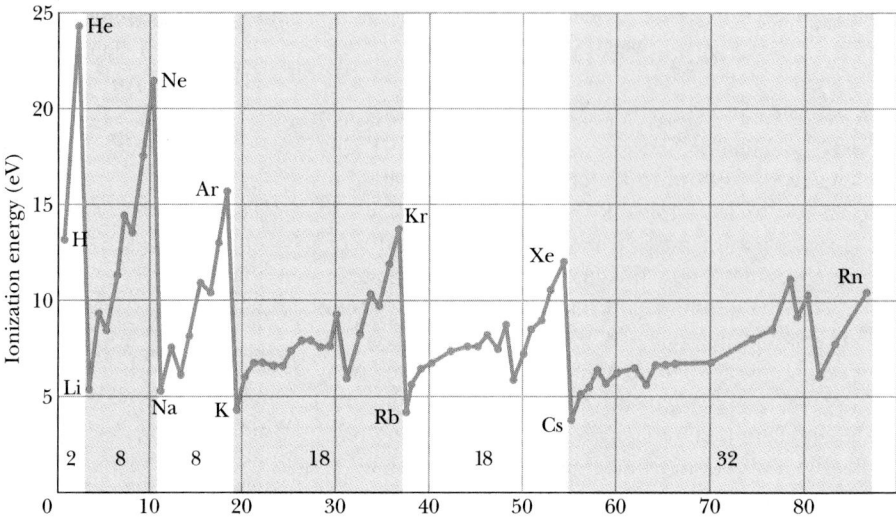

FIGURE 41-2 A plot of the ionization energies of the elements as a function of atomic number, showing the periodic repetition of properties through the six complete horizontal periods of the periodic table. The number of elements in each of these periods is indicated.

lem of finding the energies of the quantum states of that atom. Quantum mechanics allows us—in principle at least—to calculate these energies.

Atoms Have Angular Momentum and Magnetism

Figure 41-3 shows a negatively charged particle moving in a circular orbit around a fixed center. As we discussed in Section 32-4, the orbiting particle has both an angular momentum **L** and (since it is equivalent to a tiny current loop) a magnetic dipole moment **μ**. (Here, for brevity, we drop the subscript orb that we used in Chapter 32.) As Fig. 41-3 shows, **L** and **μ** are both perpendicular to the plane of the orbit but, because of the negative sign of the charge, they point in opposite directions.

The model of Fig. 41-3 is strictly classical and does not accurately represent an electron in an atom. In quantum mechanics, the rigid orbit model has been replaced by the probability density model, best visualized as a dot plot. In quantum mechanics, however, it is still true that in general,

FIGURE 41-3 A classical model showing a particle of mass m and charge $-e$ moving with speed v in a circle of radius r. The moving particle has an angular momentum **L** given by $\mathbf{r} \times \mathbf{p}$, where **p** is its linear momentum $m\mathbf{v}$. The particle's motion is equivalent to a current loop that has an associated magnetic momentum **μ** which is directed opposite **L**.

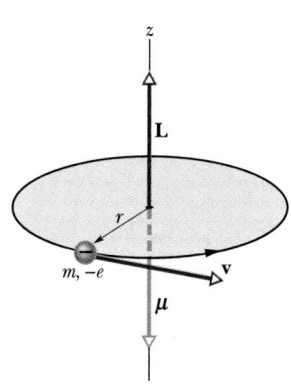

each quantum state of an electron in an atom involves an angular momentum **L** and a magnetic dipole moment **μ** that point in opposite directions.

The Einstein–de Haas Experiment

In 1915, well before the discovery of quantum mechanics, Albert Einstein and Dutch physicist W. J. de Haas carried out a clever experiment designed to verify the coupling of the angular momentum and the magnetic moment of individual atoms.

Einstein and de Haas suspended an iron cylinder from a thin fiber, as shown in Fig. 41-4a. A solenoid was placed around the cylinder but not touching it. Initially, the magnetic dipole moments **μ** of the atoms of the cylinder point in random directions, so their external magnetic effects cancel (Fig. 41-4a). However, when a current is switched on in the solenoid (Fig. 41-4b) so that a magnetic field **B** is set up parallel to the axis of the cylinder, the magnetic dipole moments of the atoms of the cylinder reorient themselves, lining up with that field. If the angular momentum **L** of each atom is coupled to its magnetic moment **μ**, then this alignment of the atomic magnetic moments must cause an alignment of the atomic angular momenta opposite the magnetic field.

No external torques initially act on the cylinder; thus its angular momentum must remain at its initial zero value. However, when **B** is turned on and the atomic angular momenta line up antiparallel to **B**, they tend to give a net angular momentum to the cylinder as a whole (directed downward in Fig. 41-4b). To maintain zero angular momentum, the cylinder begins to rotate around its central axis to produce an angular momentum in the opposite direction (upward in Fig. 41-4b).

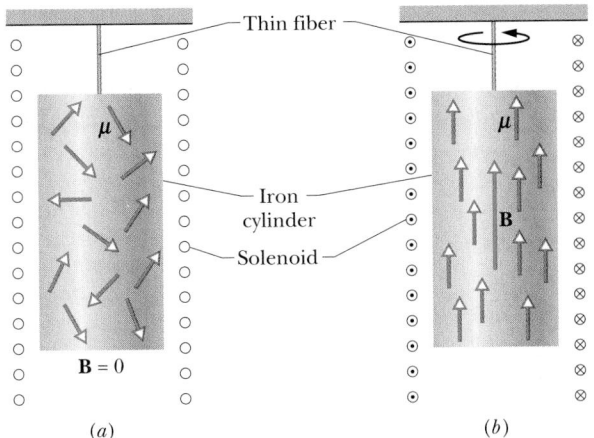

FIGURE 41-4 The Einstein–de Haas experimental setup. (*a*) Initially, the magnetic field in the iron cylinder is zero and the magnetic dipole moment vectors $\boldsymbol{\mu}$ of its atoms are randomly oriented. The atomic angular momentum vectors (not shown) are directed opposite the magnetic dipole moment vectors and thus are also randomly oriented. (*b*) When a magnetic field **B** is applied along the cylinder's axis, the magnetic dipole moment vectors line up parallel to **B**, which means that the angular momentum vectors line up opposite **B**. Because the cylinder is initially isolated from external torques, its angular momentum is conserved and the cylinder as a whole must begin to rotate as shown.

Were it not for the fiber, the cylinder would continue to rotate for as long as the magnetic field is present. However, the twisting of the fiber quickly produces a torque that momentarily stops the cylinder's rotation and then sends the cylinder rotating in the opposite direction as the twisting is undone. Thereafter, the fiber will twist and untwist as the cylinder oscillates about its initial orientation in angular simple harmonic motion.

The observation of the cylinder's rotation verified that the angular momentum and the magnetic dipole moment of an atom are coupled in opposite directions. Moreover, it

demonstrated that the angular momentum associated with the microscopic systems of atoms can result in visible rotation of an object of everyday size.

41-3 ELECTRON SPIN

As we discussed in Section 32-4, whether an electron is trapped in an atom or is free, it has an intrinsic **spin angular momentum S**, often called simply **spin**. (Recall that *intrinsic* means that **S** is a basic characteristic of an electron, like its mass and electric charge.) As we shall discuss in the next section, the magnitude of **S** is quantized and depends on a **spin quantum number** s, which is always $\frac{1}{2}$ for electrons (and also for protons and neutrons). In addition, the component of **S** measured along any axis is quantized and depends on a **spin magnetic quantum number** m_s, which can have only the value $+\frac{1}{2}$ or $-\frac{1}{2}$.

The existence of electron spin was postulated on an empirical basis by two Dutch graduate students, George Uhlenbeck and Samuel Goudsmit, from their studies of atomic spectra. The quantum mechanical basis for electron spin was provided a few years later, by British physicist P. A. M. Dirac, who developed (in 1929) a relativistic quantum theory of the electron.

It is tempting to account for electron spin by thinking of the electron as a tiny sphere spinning about an axis. However, that classical model, like the classical model of orbits, does not hold up. In quantum mechanics, spin angular momentum is best thought of as a measurable intrinsic property of the electron; you simply can't visualize it by a mechanical model.

Table 41-1 shows the four quantum numbers n, l, m_l, and m_s that completely specify the quantum states of the electron in a hydrogen atom. The same quantum numbers also specify the allowed states of any single electron in a multielectron atom.

TABLE 41-1 ELECTRON STATES FOR AN ATOM

QUANTUM NUMBER	SYMBOL	ALLOWED VALUES	RELATED TO
Principal	n	1, 2, 3, . . .	Distance from the nucleus
Orbital	l	$0, 1, 2, \ldots, (n-1)$	Orbital angular momentum
Orbital magnetic	m_l	$0, \pm 1, \pm 2, \ldots, \pm l$	Orbital angular momentum (z component)
Spin magnetic	m_s	$\pm 1/2$	Spin angular momentum (z component)

All states with the same value of n form a **shell**.
There are $2n^2$ states in a shell.

All states with the same values of n and l form a **subshell**.
All states in a subshell have the same energy.
There are $2(2l + 1)$ states in a subshell.

41-4 ANGULAR MOMENTA AND MAGNETIC DIPOLE MOMENTS

Every quantum state of an electron in an atom has an associated orbital angular momentum and a corresponding orbital magnetic dipole moment. Every electron, whether trapped in an atom or free, has a spin angular momentum and a corresponding spin magnetic dipole moment. We discuss these quantities separately first, and then discuss their combination.

Orbital Angular Momentum and Magnetism

The magnitude L of the orbital angular momentum **L** of an electron in an atom is quantized; that is, it can take on only values that belong to a discrete set. These values are

$$L = \sqrt{l(l+1)}\hbar, \qquad (41\text{-}2)$$

in which l is the orbital quantum number and $\hbar$ is $h/2\pi$. According to Table 41-1, l must be either zero or a positive integer no greater than $n-1$. For a state with $n=3$, for example, only $l=2$, $l=1$, and $l=0$ are permitted.

In an isolated atom, there is no preferred direction with respect to which we can discuss the orientation in space of the vector **L**; all directions are equivalent. However, magnets line up with magnetic fields, and the orbital angular momentum of a quantum state is tightly coupled to its orbital magnetic dipole moment. Thus, if we immerse the atom in a uniform magnetic field we can now use the direction of the field—which we can take as a z axis—to discuss the orientations of the electron's orbital magnetic dipole moment $\boldsymbol{\mu}_{\text{orb}}$ and its orbital angular momentum **L**.

As we discussed in Section 32-4, $\boldsymbol{\mu}_{\text{orb}}$ cannot itself be measured; only a projection (a component) can be measured, and that projection is quantized. The quantized values are given by Eq. 32-11 as

$$\mu_{\text{orb},z} = -m_l \mu_B. \qquad (41\text{-}3)$$

Here m_l is the orbital magnetic quantum number of Table 41-1 and μ_B is the *Bohr magneton* (Eq. 32-5):

$$\mu_B = \frac{eh}{4\pi m} = \frac{e\hbar}{2m} = 9.274 \times 10^{-24} \text{ J/T}$$

$$\text{(Bohr magneton)}, \qquad (41\text{-}4)$$

where m is the electron mass.

In Section 32-4, we also discussed that the orbital angular momentum **L** itself cannot be measured; as above, only a projection can be measured, and that projection is quantized. From Eq. 32-9, the quantized values are given by

$$L_z = m_l \hbar. \qquad (41\text{-}5)$$

Figure 41-5 shows the five quantized projections L_z of the orbital angular momentum for an electron with $l=2$ and also the associated orientations of the angular momentum **L**. (Do not take the figure literally—we cannot measure or detect **L**.) Note that for a given value of l, there are $2l+1$ different values of m_l. The restriction imposed by quantum mechanics on the direction of the projections L_z is called **space quantization**.

Note that Eq. 41-3 has a minus sign and Eq. 41-5 does not, reflecting the fact that the orbital angular momentum and the orbital magnetic dipole moment of an electron in an atom point in opposite directions.

Spin Angular Momentum and Spin Magnetic Dipole Moment

The magnitude S of the spin angular momentum **S** of any electron, whether free or trapped, has the single value given by

$$S = \sqrt{s(s+1)}\hbar$$
$$= \sqrt{(\tfrac{1}{2})(\tfrac{1}{2}+1)}\hbar = 0.866\hbar, \qquad (41\text{-}6)$$

where $s\,(=\tfrac{1}{2})$ is the spin quantum number of the electron. Spin angular momentum **S** cannot itself be measured; again, only a projection can be measured, and that projection is quantized. From Eq. 32-3, the quantized values are given by

$$S_z = m_s \hbar, \qquad (41\text{-}7)$$

in which m_s, the spin magnetic quantum number of Table 41-1, can have only two values: $+\tfrac{1}{2}$ and $-\tfrac{1}{2}$.

Correspondingly, the spin magnetic dipole moment $\boldsymbol{\mu}_s$ of an electron cannot be directly measured, but its projection can be measured and is quantized. From Eq. 32-4, the quantized values are given by

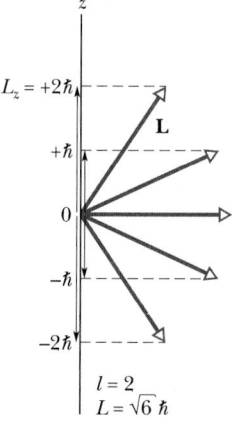

FIGURE 41-5 The allowed values of L_z for an electron in a quantum state with $l=2$. For every orbital angular momentum vector **L** in the figure, there is a vector pointing in the opposite direction, representing the magnitude and direction of the orbital magnetic dipole moment.

$$\mu_{s,z} = -2m_s\mu_B. \qquad (41\text{-}8)$$

Figure 41-6 shows the two quantized projections on the z axis of the spin angular momentum and the spin magnetic dipole moment.

Note that Eq. 41-8, which refers to the spin magnetic dipole moment, differs from Eq. 41-3, which refers to the orbital magnetic dipole moment, by a factor of 2.*

CHECKPOINT **1:** An electron is in a quantum state for which the magnitude of the electron's orbital angular momentum **L** is $2\sqrt{3}\hbar$. How many projections of the electron's orbital magnetic dipole moment on a z axis are allowed?

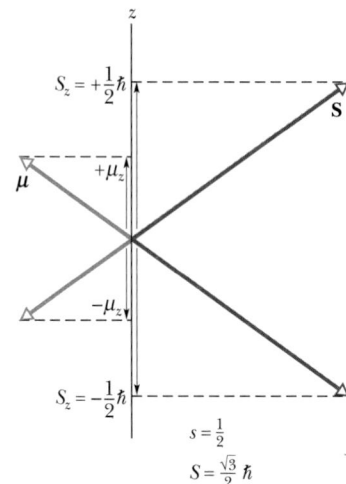

FIGURE 41-6 The allowed values of S_z and μ_z for an electron.

Orbital and Spin Angular Momenta Combined

For an atom that contains more than one electron, we define a total angular momentum **J**, which is the vector sum of the angular momenta of the individual electrons—both their orbital and their spin angular momenta. The number of electrons (and the number of protons) in a neutral atom is the **atomic number** (or **charge number**) Z. Thus for a neutral atom,

$$\mathbf{J} = (\mathbf{L}_1 + \mathbf{L}_2 + \mathbf{L}_3 + \cdots + \mathbf{L}_Z) + \\ (\mathbf{S}_1 + \mathbf{S}_2 + \mathbf{S}_3 + \cdots + \mathbf{S}_Z). \quad (41\text{-}9)$$

Similarly, the total magnetic dipole moment of the multielectron atom is obtained by vectorially adding the magnetic dipole moments (both orbital and spin) of its individual electrons. However, because of the factor of 2 in Eq. 41-8, the resultant magnetic dipole moment for the atom will not point in the direction of the vector $-\mathbf{J}$; instead, it will make a certain angle with that vector. The **effective magnetic dipole moment** $\boldsymbol{\mu}_{\text{eff}}$ for the atom is the (vector) component of the (vector) sum of the individual magnetic dipole moments in the direction of $-\mathbf{J}$.

As you will see in the next section, in typical atoms the orbital angular momenta and the spin angular momenta of most of the electrons add vectorially to zero. Then **J** and $\boldsymbol{\mu}_{\text{eff}}$ turn out to be associated with a relatively small number of electrons, often with a single valence electron.

Figure 41-7 suggests a classical model that helps us to visualize the space quantization of the total angular mo-

mentum vector **J** and the effective magnetic moment vector $\boldsymbol{\mu}_{\text{eff}}$. It shows these coupled vectors rotating about the z axis, with the vectors tracing out cones; the motion is called *precession*. The projections of **J** and $\boldsymbol{\mu}_{\text{eff}}$ on the z axis remain constant during the precession.

The more we learn about physics, the more we tend to look at the world differently. Edward Purcell, a Nobel laureate, said in his Nobel prize lecture that his investigations of atomic angular momenta and magnetic dipole moments caused him to look at snow in a new way. He viewed the snow lying on his doorstep as "full of protons quietly precessing in the Earth's magnetic field."

Precession and the Uncertainty Principle

Heisenberg's uncertainty principle suggests a limitation of the classical model of Fig. 41-7. In its angular form, and for components in the z direction, the uncertainty principle is

$$\Delta J_z \, \Delta\phi \approx \hbar \quad (z\text{ component}), \qquad (41\text{-}10)$$

in which ϕ is the angle of rotation about the z axis in Fig. 41-7. The projection J_z remains constant throughout the

---

*An advanced formulation of quantum mechanics, called **quantum electrodynamics**, predicts that the factor "2" in Eq. 41-8 is actually 2.00231930476. This quantity has also been measured experimentally. Within the precision of the experiment, the result agrees with the theoretical prediction.

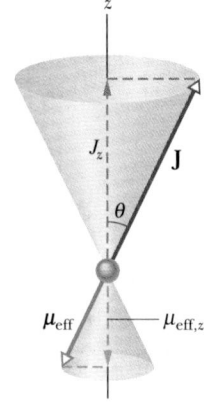

FIGURE 41-7 A classical model showing the total angular momentum vector **J** and the effective magnetic moment vector $\boldsymbol{\mu}_{\text{eff}}$ precessing about a z axis that is defined by imposing a weak magnetic field in that direction. Both vectors maintain angle θ with the z axis; hence the projections of **J** and $\boldsymbol{\mu}_{\text{eff}}$ on the z axis remain constant during the motion.

motion so that $\Delta J_z = 0$. Equation 41-10 then requires that $\Delta\phi$ be infinitely great. This means that, although $J_x^2 + J_y^2$ remains constant, the separate values of J_x and J_y are not measurable quantities. We conclude:

It is the projections of **J** and $\boldsymbol{\mu}_{\text{eff}}$ in the direction of an imposed magnetic field that are the important measurable quantities.

41-5 THE STERN–GERLACH EXPERIMENT

In 1922 Otto Stern and Walther Gerlach at the University of Hamburg in Germany verified space quantization experimentally. At that early date quantum mechanics had not been developed, and the concept of electron spin had not been established. It was known, however, that the atoms of many elements have an angular momentum and a magnetic dipole moment and the possibility of space quantization had been proposed.

Figure 41-8 shows the Stern–Gerlach apparatus. Silver is vaporized in an electrically heated oven; the resulting silver atoms escape through a narrow slit in the oven wall into the rest of the apparatus, from which the air has been pumped. Some of the atoms (which are electrically neutral but have a magnetic moment) pass through a slit in a screen (called a *collimator*), forming a narrow beam. The beam passes between the poles of an electromagnet, finally forming a silver deposit on a glass plate.

A Magnetic Dipole in a Nonuniform Magnetic Field

Let us now digress long enough to find out what force acts on a silver atom as it passes between the pole faces of the

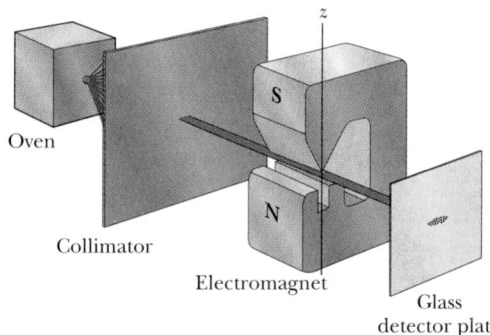

FIGURE 41-8 The apparatus used by Stern and Gerlach in 1922 to demonstrate space quantization. The entire apparatus is contained in an evacuated enclosure.

electromagnet of Fig. 41-8, which are shaped to make the magnetic field as *nonuniform* as possible.

Figure 41-9a shows a dipole of magnetic moment $\boldsymbol{\mu}$, making an angle θ with a *uniform* magnetic field. We can imagine the dipole as having north and south poles, with its magnetic dipole moment vector pointing (by convention) from the south pole to the north pole. For a uniform field, there is no net force on the dipole. The oppositely directed forces $\mathbf{F}_N$ and $\mathbf{F}_S$ on the poles have the same magnitude, and they cancel no matter what the orientation of the dipole. (A magnetic *torque* acts on the dipole of Fig. 41-9a, but that does not concern us here.)

Figures 41-9b and 41-9c show the situation in a nonuniform field. Here the forces $\mathbf{F}_N$ and $\mathbf{F}_S$ do *not* have the same magnitude because the two poles are immersed in fields of different strengths. In this case, there *is* a net force $\mathbf{F}_{\text{net}}$ whose magnitude and direction depend on the orientation of the dipole, that is, on the value of θ. In Fig. 41-9b the net force is directed upward, and in Fig. 41-9c it is directed downward. This tells us that a silver atom in the beam of Fig. 41-8 will be deflected as it passes between the pole faces of the electromagnet, the direction and the magnitude of its deflection depending on the orientation of its magnetic dipole moment.

Now let us calculate the deflecting force, that is, the force in the direction of **B**, which is our z axis. First we note that the magnetic potential energy of a magnetic dipole in a magnetic field **B** is, from Eq. 29-36,

$$E_{\text{pot}} = -\boldsymbol{\mu} \cdot \mathbf{B} = -(\mu \cos \theta)\, B, \qquad (41\text{-}11)$$

in which θ is the angle between the directions of $\boldsymbol{\mu}$ and **B**, as in Fig. 41-9. Then, from Eq. 8-19, the z component F_z of the net force acting on the atom is $-dE_{\text{pot}}/dz$; so, from Eq. 41-11,

$$F_z = -\frac{dE_{\text{pot}}}{dz} = (\mu \cos \theta)\frac{dB}{dz}. \qquad (41\text{-}12)$$

In Figs. 41-9b and c, B increases as z increases, so the magnetic field *gradient* dB/dz (the field's rate of change) is positive. Thus the sign of the deflecting force in Eq. 41-12 is determined by the angle θ. If $\theta < 90°$ (as in Fig. 41-9b), the atom will be deflected upward; if $\theta > 90°$ (as in Fig. 41-9c), the deflection will be downward. Equation 41-12 also shows why the magnetic field in the Stern–Gerlach apparatus is made as nonuniform as possible; the deflecting force on the silver atoms is directly proportional not to B but to dB/dz.

The Experimental Results

Upon examining the silver deposit on the glass plate, Stern and Gerlach deduced the following: when the electromag-

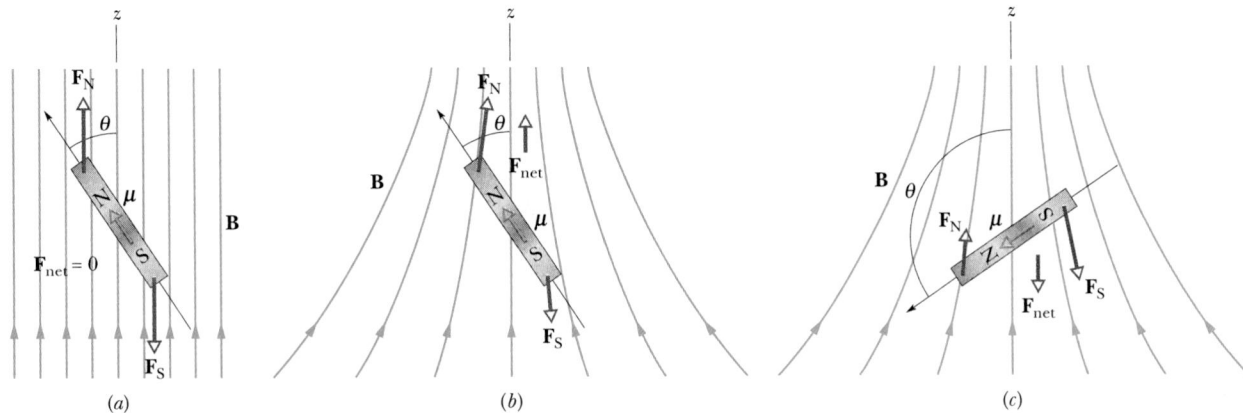

FIGURE 41-9 A magnetic dipole, represented as a bar magnet with two poles, in (a) a uniform magnetic field and (b, c) a nonuniform magnetic field. The net force $\mathbf{F}_{net}$ acting on the magnet is zero in (a), is directed upward in (b), and is directed downward in (c).

net was turned off, the beam passed through to the glass plate undeflected; when the electromagnet was turned on, the beam of silver atoms was split by the magnetic field into two subbeams, each subbeam corresponding to a different orientation of the magnetic moment of the silver atom. We know now (but it was not known then) that all the spin and orbital magnetic moments of the electrons in a silver atom cancel except for the spin magnetic dipole moment of the atom's single valence electron. From Eq. 41-8 and Fig. 41-6 we expect two subbeams, and not some other number of subbeams, in exact agreement with this experiment. Stern and Gerlach ended the published report of their work with the words: "We view these results as direct experimental evidence of space quantization in a magnetic field." Physicists everywhere agreed.

Figure 41-10 shows, as a graph of beam intensity versus detector position, the results of a more recent repetition of the Stern–Gerlach experiment. Here the experimenters used cesium atoms and a different detection scheme; in all other respects the arrangement was the same

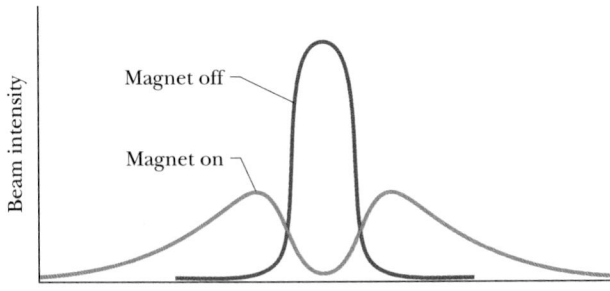

FIGURE 41-10 Results of a modern repetition of the Stern–Gerlach experiment. With the electromagnet turned off, there is only a single beam; with the electromagnet turned on, the original beam splits into two subbeams. The two subbeams correspond to parallel and antiparallel alignment of the magnetic moments of cesium atoms with the external magnetic field.

as that shown in Fig. 41-8. The separation of the original beam into two subbeams when the magnetic field is turned on is especially clear.

SAMPLE PROBLEM 41-1

In the magnet in a Stern–Gerlach experiment, the magnetic field gradient dB/dz through which the beam passes is 1.4 T/mm, and the length w of the beam path through the magnet is 3.5 cm. The temperature of the oven in which the silver is evaporated is adjusted so that the most probable speed v for the atoms in the beam is 750 m/s. Find the vertical deflection d of either subbeam as it emerges from the magnet. (The mass M of a silver atom is 1.8×10^{-25} kg and its effective magnetic moment is 1.0 Bohr magneton, or 9.27×10^{-24} J/T.)

SOLUTION: The vertical acceleration of a silver atom as it passes through the magnet is, from Newton's second law and Eq. 41-12,

$$a = \frac{F_z}{M} = \frac{(\mu \cos \theta)(dB/dz)}{M}.$$

Moving horizontally at speed v, each silver atom passes through the length w of the magnet in a time $t = w/v$. The vertical deflection of any atom as it clears the magnet is then

$$d = \tfrac{1}{2}at^2 = (\tfrac{1}{2})\frac{(\mu \cos \theta)(dB/dz)}{M}\left(\frac{w}{v}\right)^2.$$

Setting $\theta = 0°$ for maximum deflection and inserting given data then yield

$$d = \frac{(\mu \cos \theta)(dB/dz)w^2}{2Mv^2}$$

$$= (9.27 \times 10^{-24} \text{ J/T})(1)(1.4 \times 10^3 \text{ T/m})$$

$$\times \frac{(3.5 \times 10^{-2} \text{ m})^2}{(2)(1.8 \times 10^{-25} \text{ kg})(750 \text{ m/s})^2}$$

$$= 7.85 \times 10^{-5} \text{ m} \approx 0.08 \text{ mm}. \quad \text{(Answer)}$$

The separation between the two subbeams is twice this, or 0.16 mm. This separation is not large but it is easily measured.

41-6 MAGNETIC RESONANCE

As we discussed briefly in Section 32-4, a proton has an intrinsic spin angular momentum **S** and an associated spin magnetic dipole moment $\boldsymbol{\mu}$ that (because the proton is positively charged) is always in the same direction as **S**. If a proton is located in a uniform magnetic field **B** directed along a z axis, the z component μ_z of the spin magnetic dipole moment can have only two quantized orientations: either parallel to **B** or antiparallel to **B**, as shown in Fig. 41-11a. From Eq. 29-37, we know that these two orientations differ in energy by $2\mu_z B$, which is the energy involved in reversing a magnetic dipole in a uniform magnetic field. The lower energy state is with μ_z parallel to **B**, and the higher energy state is with μ_z antiparallel to **B**.

Let us place a drop of water in a uniform magnetic field **B**; then the protons in the hydrogen of the water molecules each have μ_z either parallel or antiparallel to **B**. If we next apply to the drop an alternating electromagnetic field of a certain frequency f, the protons in the lower energy state can undergo reversal in their orientation of μ_z. This process of reversal is called *spin flipping* (because the reversal of a proton's magnetic dipole moment requires a reversal of the proton's spin). The frequency f required for the spin flipping is given by

$$hf = 2\mu_z B, \qquad (41\text{-}13)$$

a condition called **magnetic resonance** (or, as originally, **nuclear magnetic resonance**). In words, if the alternating electromagnetic field is to cause the protons to spin-flip in the magnetic field, the photons associated with that field must have an energy hf equal to the energy difference $2\mu_z B$ between the two possible orientations of μ_z (and thus proton spin) in that field.

Once a proton is spin-flipped to the higher energy state, it can drop back to the lower energy state by emitting a photon of the same energy hf given by Eq. 41-13. Normally more protons are in the lower state than in the higher energy state, as Fig. 41-11b suggests. This means that there will be a detectable net *absorption* of energy from the alternating electromagnetic field.

The constant field **B** in Eq. 41-13 is actually *not* the imposed external field $\mathbf{B}_{\text{ext}}$ in which the water drop is placed; rather, it is that field as modified by the small, local, internal magnetic field $\mathbf{B}_{\text{local}}$ due to the magnetic moments of the atoms and nuclei near a given proton. Thus we can rewrite Eq. 41-13 as

$$hf = 2\mu_z(B_{\text{ext}} + B_{\text{local}}). \qquad (41\text{-}14)$$

To achieve magnetic resonance, it is customary to leave the frequency f of the electromagnetic oscillations fixed and to vary B_{ext} until Eq. 41-14 is satisfied and an absorption peak is recorded.

Nuclear magnetic resonance is a property that is the basis for a valuable analytical tool, particularly for the identification of unknown compounds. Figure 41-12 shows a **nuclear magnetic resonance spectrum**, as it is called, for ethanol, whose formula we may write as $CH_3\text{-}CH_2\text{-}OH$. The various resonance peaks all represent spin flips of protons. They occur at different values of B_{ext}, however, because the local environments of the six protons within the ethanol molecule differ from one another. The spectrum of Fig. 41-12 is a unique signature for ethanol.

Spin technology, here called **magnetic resonance imaging** (MRI), has been applied to medical diagnostics with great success. The protons in the various tissues of the human body find themselves in different local magnetic environments. When the body, or part of it, is immersed in a strong external magnetic field, these environmental differences can be detected by spin-flip techniques and translated by computer processing into an image resembling those produced by x rays. Figure 41-13, for example,

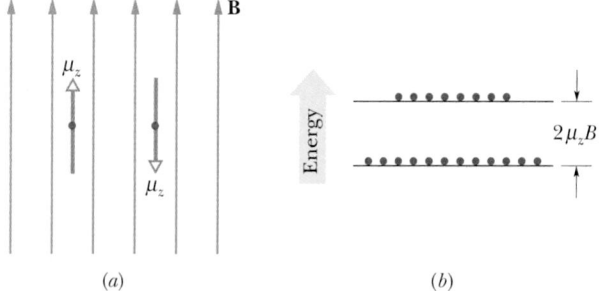

(a)   (b)

FIGURE 41-11 (*a*) A proton, whose spin component in the direction of an applied magnetic field is $\frac{1}{2}\hbar$, can occupy either of two quantized orientations in an external magnetic field. If Eq. 41-13 is satisfied, the protons in the sample can be induced to flip from one orientation to the other. (*b*) Normally, there are more protons in the lower energy state than in the higher energy state.

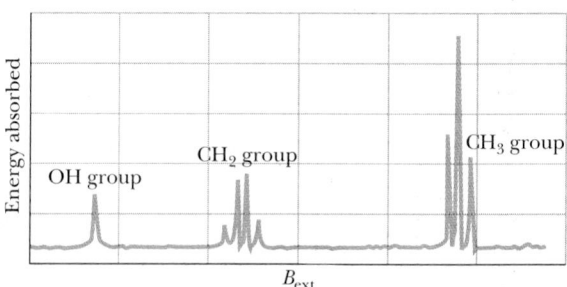

FIGURE 41-12 A nuclear magnetic resonance spectrum for ethanol. The spectral lines represent the absorption of energy associated with spin flips of protons. The three groups of lines correspond, as indicated, to protons in the OH group, the CH_2 group, and the CH_3 group of the ethanol molecule. Note that the two protons in the CH_2 group occupy four different local environments. The entire horizontal axis covers less than 10^{-4} T.

shows a cross section of a human head imaged by this method.

SAMPLE PROBLEM 41-2

A drop of water is suspended in a magnetic field **B** of magnitude 1.80 T and an alternating electromagnetic field is applied, its frequency adjusted to produce spin flips of the protons in the water. The component μ_z of the magnetic dipole moment of a proton, measured along the direction of **B**, is 1.41×10^{-26} J/T. Assume that the local magnetic fields are negligible compared to **B**. What are the frequency f and wavelength λ of the alternating field?

SOLUTION: From Eq. 41-13, we have

$$f = \frac{2\mu_z B}{h} = \frac{(2)(1.41 \times 10^{-26} \text{ J/T})(1.80 \text{ T})}{6.63 \times 10^{-34} \text{ J} \cdot \text{s}}$$

$$= 7.66 \times 10^7 \text{ Hz} = 76.6 \text{ MHz}. \qquad \text{(Answer)}$$

The corresponding wavelength is

$$\lambda = \frac{c}{f} = \frac{3.00 \times 10^8 \text{ m/s}}{7.66 \times 10^7 \text{ Hz}} = 3.92 \text{ m}. \quad \text{(Answer)}$$

This frequency and wavelength are in the short radio wave region of the electromagnetic spectrum.

41-7 BUILDING THE PERIODIC TABLE

The four quantum numbers of Table 41-1 identify the quantum states of individual electrons in a multielectron

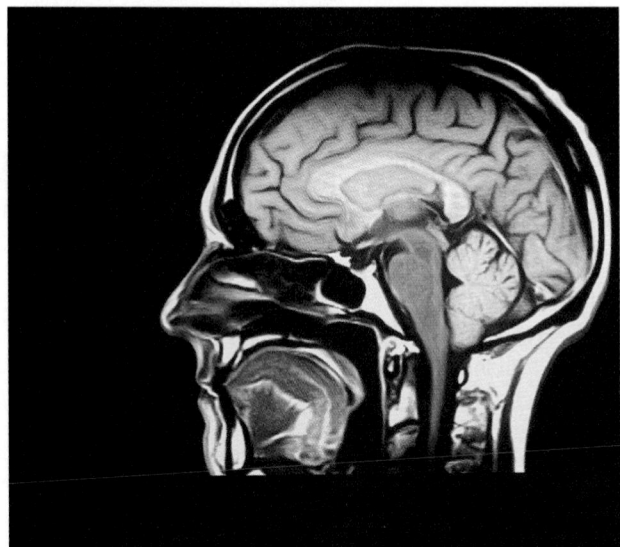

FIGURE 41-13 A cross-sectional view of a human head produced by magnetic resonance imaging. Some of the details visible here would not show up on an x-ray image, even with a modern computerized axial tomography scanner (CAT scanner).

atom. The wave functions for these states, however, are not the same as the wave functions for the corresponding states of the hydrogen atom because, in multielectron atoms, the potential energy associated with a given electron is determined not only by the charge and position of the atom's nucleus but also by the charges and positions of all the other electrons in the atom. Solutions of Schrödinger's equation for multielectron atoms can be carried out numerically—in principle at least—using a computer.

As we discussed in Section 40-6, all states with the same values of the quantum numbers n and l form a subshell. For a given value of l, there are $2l + 1$ possible values of the magnetic quantum number m_l and, for each m_l, there are two possible values for the spin quantum number m_s. Thus, there are $2(2l + 1)$ states in a subshell. It turns out that *all states in a given subshell have the same energy,* its value being determined primarily by the value of n and to a lesser extent by the value of l.

For the purpose of labeling subshells, the values of l are represented by letters:

$$l = 0 \quad 1 \quad 2 \quad 3 \quad 4 \quad 5 \quad \cdots$$
$$s \quad p \quad d \quad f \quad g \quad h \quad \cdots$$

In this notation, for example, the $n = 3$, $l = 2$ subshell would be labeled the $3d$ subshell.

When we assign electrons to states in a multielectron atom, we must be guided by the **Pauli exclusion principle**:

No two electrons in an atom can have the same set of the quantum numbers n, l, m_l, and m_s.

(The principle also holds for protons and neutrons but not for all types of particles.) If this important principle did not hold, all the electrons in an atom would pile up in the state of lowest energy, just as a dozen marbles in a bowl all end up at the bottom of the bowl. Let us examine the atoms of a few elements, to see how the Pauli principle operates in the building up of the periodic table.

Neon

The neon atom has 10 electrons. Only two of them fit into the lowest energy subshell, the $1s$ subshell. These two electrons both have $n = 1$, $l = 0$, and $m_l = 0$, but one has $m_s = +\frac{1}{2}$ and the other has $m_s = -\frac{1}{2}$. The $1s$ subshell, according to Table 41-1, contains $2(2l + 1) = 2$ states. Because this subshell then contains all the electrons permitted by the Pauli principle, it is said to be **closed**.

Two of the remaining eight electrons fill the next lowest energy subshell, the $2s$ subshell. The last six electrons just fill the $2p$ subshell which, with $l = 1$, holds $2(2l + 1) = 6$ states.

In a closed subshell, all allowed z projections of the orbital angular momentum vector **L** are present and, as you can verify from Fig. 41-5, these projections cancel for the subshell as a whole: for every positive projection there is a corresponding negative projection of the same magnitude. Similarly, the z projections of the spin angular momenta also cancel. Thus a closed subshell has no angular momentum and no magnetic moment of any kind. Furthermore, its probability density is spherically symmetric. So neon with its three closed subshells (1s, 2s, and 2p) has no "loosely dangling electrons" to encourage chemical interaction with other atoms. Neon, like the other **noble gases** that form the right-hand column of the periodic table, is chemically inert.

Sodium

Next after neon in the periodic table comes sodium, with 11 electrons. Ten of them form a closed neonlike core, which, as we have seen, has zero angular momentum. The remaining electron is largely outside this inert core, in the 3s subshell—the next lowest energy subshell. Because this **valence electron** of sodium is in a state with $l = 0$ (that is, an s state), the sodium atom's angular momentum and magnetic dipole moment must be due entirely to the spin of this single electron.

Sodium readily combined with other atoms that have a "vacancy" into which sodium's loosely bound valence electron can fit. Sodium, like the other **alkali metals** that form the left-hand column of the periodic table, is chemically active.

Chlorine

The chlorine atom, which has 17 electrons, has a closed 10-electron, neonlike core, with 7 electrons left over. Two of them fill the 3s subshell, leaving five to be assigned to the 3p subshell, which is the subshell next lowest in energy. This subshell, which has $l = 1$, can hold $2(2l + 1) = 6$ electrons, so there is a vacancy, or a "hole," in this subshell.

Chlorine is receptive to interacting with other atoms that have a valence electron that might fill this hole. Sodium chloride (NaCl), for example, is a very stable compound. Chlorine, like the other **halogens** that form column VIIA of the periodic table, is chemically active.

Iron

The arrangement of the 26 electrons of the iron atom can be represented as follows:

$$\underbrace{1s^2 \quad 2s^2\, 2p^6 \quad 3s^2\, 3p^6} 3d^6 \quad 4s^2$$

The subshells are listed in numerical order and, following convention, a superscript gives the number of electrons in each subshell. From Table 41-1 we can see that an s-subshell can hold 2 electrons, a p-subshell 6, and a d-subshell 10. Thus iron's first 18 electrons form the five filled subshells that are marked off by the bracket, leaving 8 electrons to be accounted for. Six of the eight go into the 3d subshell and the remaining two go into the 4s subshell.

The last two electrons do not also go into the 3d subshell (which can hold 10 electrons) because the $3d^6\, 4s^2$ configuration results in a lower energy state for the atom as a whole than would the $3d^8$ configuration. An iron atom with 8 electrons (rather than 6) in the 3d subshell would quickly make a transition to the $3d^6\, 4s^2$ configuration, emitting electromagnetic radiation in the process. The lesson here is that except for the simplest elements, the states may not be filled in what one we might think of as their "logical" sequence.

SAMPLE PROBLEM 41-3

Account for the number of elements in the six horizontal periods of the periodic table in terms of the populations of the subshells.

SOLUTION: As Appendix G shows, the numbers of elements in the six horizontal rows are 2, 8, 8, 18, 18, and 32. The population of a subshell depends only on the quantum number l and is $2(2l + 1)$. Thus

ORBITAL QUANTUM NUMBER l	SUBSHELL POPULATION $2(2l + 1)$
0	2
1	6
2	10
3	14

We can account for each horizontal period in terms of closed subshells in this way:

PERIOD NUMBERS	ELEMENTS IN THE PERIOD	SUMS OF SUBSHELL POPULATIONS
1	2	2
2, 3	8	$2 + 6 = 8$
4, 5	18	$2 + 6 + 10 = 18$
6	32	$2 + 6 + 10 + 14 = 32$

41-8 X RAYS AND THE NUMBERING OF THE ELEMENTS

When a solid target, such as solid copper or tungsten, is bombarded with electrons whose kinetic energies are in the kiloelectron-volt range, electromagnetic radiations called **x rays** are emitted. Our concern here is what these rays— whose medical, dental, and industrial usefulness is so well known and widespread—can teach us about the atoms that absorb or emit them.

Figure 41-14 shows the wavelength spectrum of the x rays produced when a beam of 35 keV electrons falls on a molybdenum target. We see a broad, continuous spectrum of radiation on which are superimposed two peaks of sharply defined wavelengths. The continuous spectrum and the peaks arise in different ways, which we next discuss separately.

The Continuous X-Ray Spectrum

Here we examine the continuous x-ray spectrum of Fig. 41-14, ignoring for the time being the two prominent peaks that rise from it. Consider an electron of initial kinetic energy K_0 that happens to collide (interact) with one of the target atoms, as in Fig. 41-15. The electron may well lose an amount of energy ΔK, which will appear as the energy of an x-ray photon that is radiated away from the site of the collision. (The energy transferred to the recoiling atom is small because of the relatively large mass of the atom; here we neglect that transfer.)

The scattered electron in Fig. 41-15, whose energy is now less than K_0, may have a second collision with a target atom, generating a second photon, whose energy will in general be different from the energy of the photon pro-

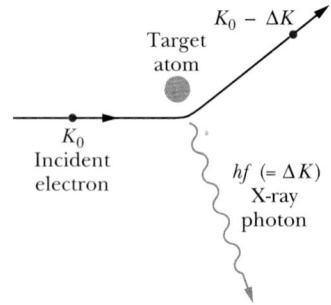

FIGURE 41-15 An electron of kinetic energy K_0 passing near an atom in the target may generate an x-ray photon, the electron losing part of its energy in the process. The continuous x-ray spectrum arises in this way.

duced in the first collision. This electron-scattering process can continue until the electron is approximately stationary. All the photons generated by these collisions form part of the continuous x-ray spectrum.

The prominent feature of that spectrum of Fig. 41-14 is the sharply defined **cutoff wavelength** λ_{min}, below which the continuous spectrum does not exist. This minimum wavelength corresponds to a collision in which an incident electron loses *all* its initial kinetic energy K_0 in a single head-on collision with a target atom. Essentially all this energy appears as the energy of a single photon, whose associated wavelength—the minimum possible x-ray wavelength—is found from

$$K_0 = hf = \frac{hc}{\lambda_{min}},$$

which yields

$$\lambda_{min} = \frac{hc}{K_0} \quad \text{(cutoff wavelength).} \quad (41\text{-}15)$$

The cutoff wavelength is totally independent of the target material. If we were to switch from a molybdenum target to a copper target, for example, all features of the x-ray spectrum of Fig. 41-14 would change *except* the cutoff wavelength.

CHECKPOINT 2: Does the cutoff wavelength λ_{min} of the continuous x-ray spectrum increase, decrease, or remain the same if you (a) increase the kinetic energy of the electrons that strike the x-ray target, (b) allow the electrons to strike a thin foil rather than a thick block of the target material, (c) change the target to an element of higher atomic number?

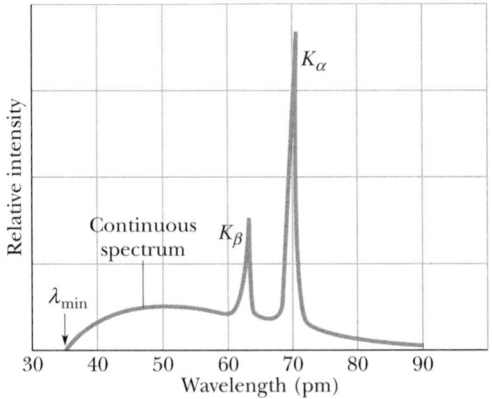

FIGURE 41-14 The distribution by wavelength of the x rays produced when 35 keV electrons strike a molybdenum target. The sharp peaks and the continuous spectrum from which they rise are produced by different mechanisms.

SAMPLE PROBLEM 41-4

A beam of 35.0 keV electrons strikes a molybdenum target, generating the x rays whose spectrum is shown in Fig. 41-14.

(a) What is the cutoff wavelength?

SOLUTION: From Eq. 41-15 we have

$$\lambda_{min} = \frac{hc}{K_0} = \frac{(4.14 \times 10^{-15} \text{ eV} \cdot \text{s})(3.00 \times 10^8 \text{ m/s})}{35.0 \times 10^3 \text{ eV}}$$

$$= 3.55 \times 10^{-11} \text{ m} = 35.5 \text{ pm}. \quad \text{(Answer)}$$

(b) Suppose that one of the incident electrons loses kinetic energy by small amounts until its energy is reduced from 35.0 keV to 20.0 keV. What is the associated wavelength λ of a photon that it would then generate if it lost all this remaining kinetic energy in a single head-on collision with an atom?

SOLUTION: We carry out the calculation as in (a), substituting 20.0 keV for 35.0 keV. The result is

$$\lambda_{min} = 62.1 \text{ pm}. \quad \text{(Answer)}$$

This wavelength is larger than the minimum wavelength calculated in (a) because less energy is involved.

The Characteristic X-Ray Spectrum

We now turn our attention to the two peaks of Fig. 41-14, labeled K_α and K_β. These peaks, together with other peaks that appear at wavelengths beyond the wavelength range displayed in Fig. 41-14, form the **characteristic x-ray spectrum** of the target material.

The peaks arise in a two-part process. (1) An energetic electron strikes an atom in the target and, while it is being scattered, the incident electron knocks out one of the atom's deep-lying (low n value) electrons. If the deep-lying electron is in the shell defined by $n = 1$ (called, for historical reasons, the K shell), there remains a vacancy, or hole, in this shell. (2) An electron in one of the shells located farther from the nucleus transfers to the K shell, filling the hole in this shell. During this transfer, the atom emits a characteristic x-ray photon. If the electron that fills the K-shell vacancy transfers from the shell with $n = 2$ (called the L shell), the emitted radiation is the K_α line of Fig. 41-14; if it transfers from the shell with $n = 3$ (called the M shell), it produces the K_β line, and so on. The hole left in either the L or M shell will be filled by an electron from still farther out in the atom.

In studying x rays, it is more convenient to keep track of the hole created deep in the atom's "electron cloud" rather than recording the changes in the quantum state of the electrons that transfer to fill that hole. Figure 41-16 does exactly that; it is an energy level diagram for molybdenum, the element to which Fig. 41-14 refers. The baseline ($E = 0$) represents the neutral atom in its ground state. The level marked K (at $E = 20$ keV) represents the energy of the molybdenum atom with a hole in its K shell. Similarly, the level marked L (at $E = 2.7$ keV) represents the atom with a hole in its L shell, and so on.

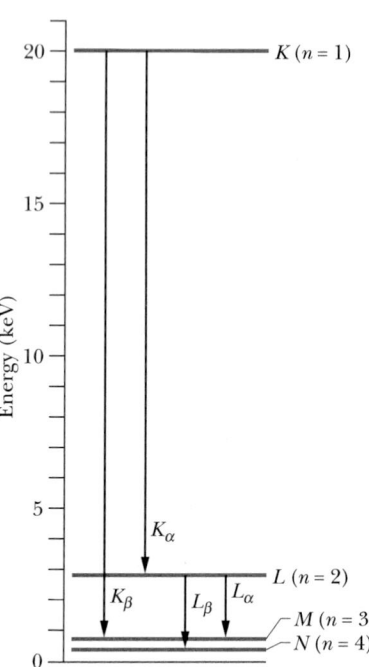

FIGURE 41-16 A simplified atomic energy level diagram for molybdenum, showing the transitions (of holes rather than electrons) that give rise to some of the characteristic x rays of that element. Each horizontal line represents the energy of the atom with a hole (a missing electron) in the shell indicated.

The transitions marked K_α and K_β in Fig. 41-16 are the ones that produce the two x-ray peaks in Fig. 41-14. The K_α spectral line, for example, originates when an electron from the L shell fills a hole in the K shell. This activity corresponds to a downward transition on the energy level diagram of Fig. 41-16 from the K level to the L level.

Numbering the Elements

In 1913 British physicist H. G. J. Moseley generated characteristic x rays for as many elements as he could find—he found 38—by using them as targets for electron bombardment in an evacuated tube of his own design. By means of a trolley manipulated by strings, Moseley was able to move the individual targets into the path of an electron beam. He measured the wavelengths of the x rays by the crystal diffraction method described in Section 37-9.

Moseley then sought (and found) regularities in these spectra as he moved from element to element in the periodic table. In particular, he noted that if, for a given spectral line such as K_α, he plotted for each element the square root of the line frequency f against the position of the element in the periodic table, a straight line resulted. Figure 47-17 shows a portion of his extensive data. Moseley's conclusion was:

> We have here a proof that there is in the atom a fundamental quantity, which increases by regular steps as we pass from one element to the next. This quantity can only be the charge on the central nucleus.

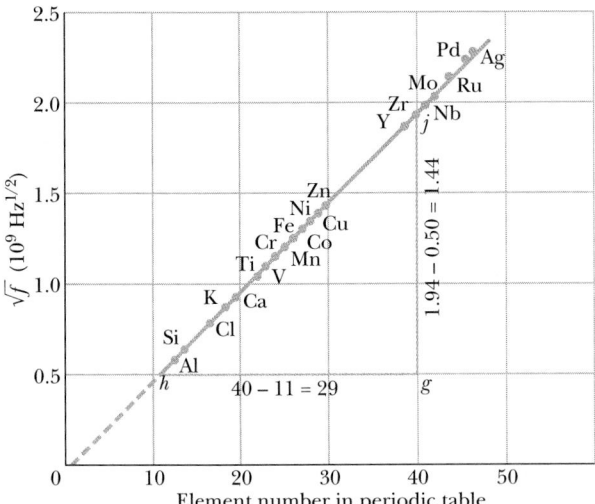

FIGURE 41-17 A Moseley plot of the K_α line of the characteristic x-ray spectra of 21 elements. The frequency is calculated from the measured wavelength. Measurements indicated along the hgj triangle are used in Sample Problem 41-6 to determine the slope of the straight-line plot.

Owing to Moseley's work, the characteristic x-ray spectrum became the universally accepted signature of an element, permitting the solution of a number of periodic table puzzles. Prior to that time (1913), the position of an element in the table was assigned in order of atomic *weight*, although there were several pairs of elements in which it was necessary to invert this order because of compelling chemical evidence; Moseley showed that it is the nuclear charge (that is, the atomic *number Z*) that is the real basis for numbering the elements.

In 1913 the periodic table had several empty squares, and a surprising number of claims for new elements had been advanced. The x-ray spectrum provided a conclusive test of such claims. The lanthanide elements, often called the rare earth elements, had been sorted out only imperfectly because of their similar chemical properties. Once Moseley's work was reported, these elements were properly organized in short order. In more recent times, the identities of elements beyond uranium are pinned down beyond dispute when the elements are available in quantities large enough to permit a study of their individual x-ray spectra.

It is not hard to see why the characteristic x-ray spectrum shows such impressive regularities from element to element while the optical spectrum in the visible and near-visible region does not: the key to the identity of an element is the charge on its nucleus. Gold, for example, is what it is because its atoms have a nuclear charge of $+79e$ (that is, $Z = 79$). An atom with one more elementary charge on its nucleus is mercury; one less is platinum. The

K electrons, which play such a large role in the production of the x-ray spectrum, lie very close to the nucleus and are thus sensitive probes of its charge. The optical spectrum, on the other hand, involves transitions of the outermost electrons, which are heavily screened from the nucleus by the remaining electrons of the atom and are thus *not* sensitive probes of nuclear charge.

Accounting for the Moseley Plot

Moseley's experimental data, of which the Moseley plot of Fig. 41-17 is but a part, can be used directly to assign the elements to their proper squares in the periodic table. This can be done even if no theoretical basis for Moseley's results can be established. However, there is such a basis.

According to Eq. 40-15 the energy of the hydrogen atom is

$$E_n = -\frac{me^4}{8\epsilon_0^2 h^2}\frac{1}{n^2} = -\frac{13.6 \text{ eV}}{n^2},$$
$$\text{for } n = 1, 2, 3, \cdots . \quad (41\text{-}16)$$

Consider now one of the two innermost electrons in the K shell of a multielectron atom. Because of the presence of the other K-shell electron, our electron "sees" an effective nuclear charge of approximately $(Z - 1)e$, where e is the elementary charge and Z is the atomic number of the element. The factor e^4 in Eq. 41-16 is the product of e^2—the square of hydrogen's nuclear charge—and $(-e)^2$—the square of an electron's charge. So for a multielectron atom, we can approximate the effective energy of the atom by replacing the factor e^4 in Eq. 41-16 with $(Z - 1)^2 e^2 \times (-e)^2$, or $e^4(Z - 1)^2$. We find

$$E_n = -\frac{(13.6 \text{ eV})(Z - 1)^2}{n^2}. \quad (41\text{-}17)$$

We saw that the K_α x-ray photon (of energy hf) arises when an electron makes a transition from the L shell (with $n = 2$ and energy E_2) to the K shell (with $n = 1$ and energy E_1). Thus, using Eq. 41-17, we may write the energy of the emitted photon as

$$hf = E_2 - E_1$$
$$= \frac{-(13.6 \text{ eV})(Z - 1)^2}{2^2} - \frac{-(13.6 \text{ eV})(Z - 1)^2}{1^2}$$
$$= (10.2 \text{ eV})(Z - 1)^2.$$

Then the frequency f of the K_α line is

$$f = \frac{hf}{h} = \frac{(10.2 \text{ eV})(Z - 1)^2}{(4.14 \times 10^{-15} \text{ eV} \cdot \text{s})}$$
$$= (2.46 \times 10^{15} \text{ Hz})(Z - 1)^2. \quad (41\text{-}18)$$

Taking the square root of both sides yields

$$\sqrt{f} = CZ - C, \tag{41-19}$$

in which C is a constant. Equation 41-19 is the equation of a straight line. It shows that if we plot the square root of the frequency of the K_α x-ray spectral line against the atomic number Z, we should obtain a straight line. As Fig. 41-17 shows, that is exactly what Moseley found.

CHECKPOINT 3: The K_α x rays arising from a cobalt ($Z = 27$) target have a wavelength of about 179 pm. Is the wavelength of the K_α x rays arising from a nickel ($Z = 28$) target greater than or less than 179 pm?

SAMPLE PROBLEM 41-5

A cobalt target is bombarded with electrons, and the wavelengths of its characteristic x-ray spectrum are measured. There is also a second, fainter characteristic spectrum, which is due to an impurity in the cobalt. The wavelengths of the K_α lines are 178.9 pm (cobalt) and 143.5 pm (impurity). What is the impurity?

SOLUTION: Let us apply Eq. 41-19 to both the cobalt (Co) and the impurity (X). Substituting c/λ for f, we obtain

$$\sqrt{\frac{c}{\lambda_{Co}}} = CZ_{Co} - C \quad \text{and} \quad \sqrt{\frac{c}{\lambda_X}} = CZ_X - C.$$

Dividing the second equation by the first yields

$$\sqrt{\frac{\lambda_{Co}}{\lambda_X}} = \frac{Z_X - 1}{Z_{Co} - 1}.$$

Substituting the given data, and $Z_{Co} = 27$, yields

$$\sqrt{\frac{178.9 \text{ pm}}{143.5 \text{ pm}}} = \frac{Z_X - 1}{27 - 1}.$$

Solving for the unknown, we find that

$$Z_X = 30.0. \tag{Answer}$$

A glance at the periodic table identifies the impurity as zinc.

SAMPLE PROBLEM 41-6

(a) Evaluate the constant C in Eq. 41-19.

SOLUTION: Comparing Eq. 41-18 and Eq. 41-19 reveals that

$$C = \sqrt{2.46 \times 10^{15} \text{ Hz}} = 4.96 \times 10^7 \text{ Hz}^{1/2}. \tag{Answer}$$

(b) Verify from the Moseley plot of Fig. 41-17 that C is the slope of the straight line in that figure.

SOLUTION: If we measure the lines hg and gj in Fig. 41-17, we find that

$$C = \frac{gj}{hg} = \frac{(1.94 - 0.50) \times 10^9 \text{ Hz}^{1/2}}{40 - 11}$$

$$= 4.96 \times 10^7 \text{ Hz}^{1/2}. \tag{Answer}$$

These two results are in full agreement. The agreement is not nearly as good for lines other than K_α in the x-ray spectrum; for them one must make more careful calculations of the effects of the surrounding electrons on the electron producing a line.

41-9 LASERS AND LASER LIGHT

In the late 1940s and again in the early 1960s, quantum mechanics made two enormous contributions to technology: the **transistor**, which ushered in the computer revolution, and the **laser**. Laser light, like the light from an ordinary lightbulb, is emitted when atoms make a transition from one quantum state to a quantum state of lower energy. In a laser, however—but not in other light sources—the atoms act together to produce light with special characteristics, some of which we now describe.

1. *Laser light is highly monochromatic.* Light from an ordinary incandescent lightbulb, being spread over a continuous range of wavelengths, cannot even be discussed in terms of monochromaticity. The spectral lines from a fluorescent neon sign *are* monochromatic, to about 1 part in about 10^6. However, the sharpness of definition of laser light can be many times greater, as much as 1 part in 10^{15}.

2. *Laser light is highly coherent.* Individual long waves (*wave trains*) for laser light can be several hundred kilometers long. When two separated beams that have traveled such distances over separate paths are recombined, they "remember" their common origin and are able to form a pattern of interference fringes. The corresponding *coherence length* for wave trains emitted by a lightbulb is typically less than a meter.

3. *Laser light is highly directional.* A laser beam spreads very little; it departs from strict parallelism only because of diffraction at the exit aperture of the laser. For example, a laser pulse used to measure the distance to the Moon generates a spot on the Moon's surface whose diameter is only one-millionth of the Moon's diameter. Light from an ordinary bulb can be made into an approximately parallel beam by a lens, but the beam divergence is much greater than for laser light. Each point on a lightbulb's filament forms its own separate beam, and the angular divergence of the overall composite beam is set by the size of the filament.

4. *Laser light can be sharply focused.* If two light beams transport the same amount of energy, the beam that can be focused to the smaller spot will have the greater intensity at that spot. For laser light, the focused spot can be so small that an intensity of 10^{17} W/cm² is readily obtained. An oxyacetylene flame, by contrast, has an intensity of only about 10^3 W/cm².

Lasers Have Many Uses

The smallest lasers, used for voice and data transmission over optical fibers, have as their active medium a semiconducting crystal about the size of a pinhead. Small as they are, such lasers can generate about 200 mW of power. The largest lasers, used for nuclear fusion research and for astronomical and military applications, fill a large building. The largest such laser can generate brief pulses of laser light with a power level, during the pulse, of about 10^{14} W. This is a few hundred times greater than the total electric power generating capacity of the United States. To avoid a brief national power blackout during a pulse, the energy required for each pulse is stored up at a steady rate during the relatively long interpulse interval.

Among the many uses of lasers are reading bar codes, manufacturing and reading compact disks, performing surgery of many kinds (see the opening photo of this chapter and Fig. 41-18), surveying, cutting cloth in the garment industry (several hundred layers at a time), welding auto bodies, and generating holograms.

41-10 HOW LASERS WORK

The word "laser" is an acronym for "light amplification by the stimulated emission of radiation," so you should not be surprised that **stimulated emission** is the key to laser operation. Einstein introduced this concept in 1917. Although the world had to wait until 1960 to see an operating laser, the groundwork for its development was put in decades earlier.

Consider an isolated atom that can exist either in its state of lowest energy (its ground state), whose energy is E_0, or in a state of higher energy (an excited state), whose energy is E_x. Here are three processes by which the atom can move from one of these states to the other:

1. ***Absorption.*** Figure 41-19*a* shows the atom initially in its ground state. If the atom is placed in an electromagnetic field that is alternating at frequency *f*, the atom can absorb an amount of energy *hf* from that field and move to the higher energy state. From conservation of energy we have

$$hf = E_x - E_0. \qquad (41\text{-}20)$$

We call this process **absorption.**

2. ***Spontaneous emission.*** In Fig. 41-19*b* the atom is in its excited state and no external radiation is present. After a certain time, the atom will move *of its own accord* to

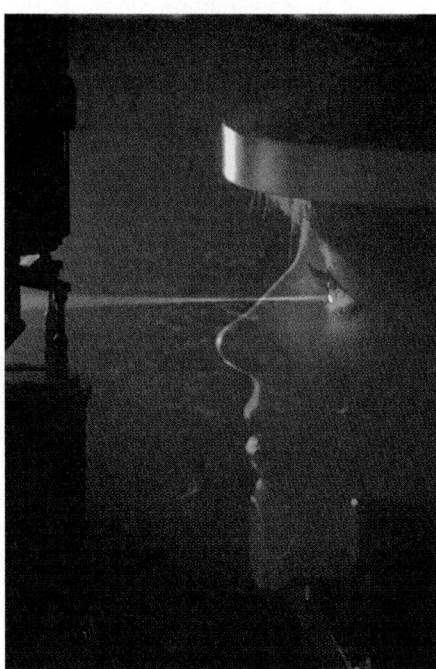

FIGURE 41-18 A laser beam is sent into the eye of a diabetes patient to seal blood vessels in her retina.

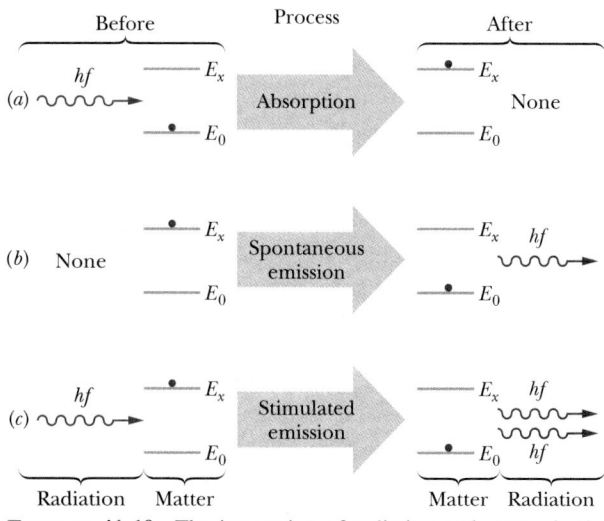

FIGURE 41-19 The interaction of radiation and matter in the processes of (*a*) absorption, (*b*) spontaneous emission, and (*c*) stimulated emission. An atom (matter) is represented by the red dot; the atom is in either a lower quantum state with energy E_0 or a higher quantum state with energy E_x. In (*a*) the atom absorbs a photon of energy *hf* from a passing light wave. In (*b*) it emits a light wave by emitting a photon of energy *hf*. In (*c*) a passing light wave with photon energy *hf* causes the atom to emit a photon of the same energy, increasing the energy of the light wave.

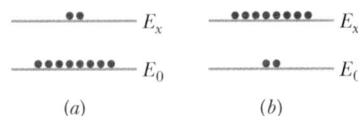

FIGURE 41-20 (a) The equilibrium distribution of atoms between the ground state E_0 and excited state E_x, accounted for by thermal agitation. (b) An inverted population, obtained by special methods. Such an inverted population is essential for laser action.

its ground state, emitting a photon of energy hf in the process. We call this process **spontaneous emission**—*spontaneous* because the event was not triggered by any outside influence. The light from the filament of an ordinary lightbulb is generated in this way.

Normally, the mean life of excited atoms before spontaneous emission occurs is about 10^{-8} s. However, for some excited states, this mean life is perhaps as much as 10^5 times longer. We call such long-lived states **metastable**; they play an important role in laser operation.

3. Stimulated emission. In Fig. 41-19c the atom is again in its excited state, but this time radiation with a frequency given by Eq. 41-20 is present. A photon of energy hf can stimulate the atom to move to its ground state, during which process the atom emits an additional photon, whose energy is also hf. We call this process **stimulated emission**—*stimulated* because the event is triggered by the external photon.

The emitted photon in Fig. 41-19c is in every way identical to the stimulating photon. Thus the waves associated with the photons have the same energy, phase, polarization, and direction of travel. Under proper conditions, a chain reaction of similar stimulated emission processes can be triggered by a single initial photon of the correct frequency. Laser light is generated in this way.

Note, parenthetically, that the photons associated with a beam of laser light are all in the same quantum state. Photons, unlike electrons in atoms, do *not* obey the Pauli exclusion principle. Photons "like" to pile up in the same quantum state; electrons are forbidden to do so.

Figure 41-19c describes stimulated emission for a single atom. Suppose now that a sample contains a large number of atoms in thermal equilibrium at temperature T. Before any radiation is directed at the sample, a number N_0 of these atoms are in their ground state with energy E_0, and a number N_x are in a state of higher energy E_x. Ludwig Boltzmann showed that N_x is given in terms of N_0 by

$$N_x = N_0\, e^{-(E_x - E_0)/kT}, \qquad (41\text{-}21)$$

in which k is Boltzmann's constant. This equation seems reasonable. The quantity kT is the mean kinetic energy of an atom at temperature T. The higher the temperature, the more atoms—on average—will have been "bumped up" by thermal agitation (that is, by atom–atom collisions) to the higher energy state E_x. Also, because $E_x > E_0$, Eq. 41-21 requires that $N_x < N_0$. That is, there will always be fewer atoms in the excited state than in the ground state. This is what we would expect if the level populations are determined only by the action of thermal agitation. Figure 41-20a illustrates this situation.

If we now flood the atoms of Fig. 41-20a with photons

of energy $E_x - E_0$, photons will disappear via absorption by ground-state atoms, and photons will be generated largely via stimulated emission of excited-state atoms. Einstein showed that the probabilities per atom for these two processes are identical. Thus, because there are more atoms in the ground state, the *net* effect will be the absorption of photons.

To produce laser light, we must have more photons emitted than absorbed. That is, we must have a situation in which stimulated emission dominates. The direct way to bring this about is to start with more atoms in the excited state than in the ground state, as in Fig. 41-20b. Since, however, such a **population inversion** is not consistent with thermal equilibrium, we must think up clever ways to set up and maintain one.

The Helium–Neon Gas Laser

Figure 41-21 shows a type of laser commonly found in student laboratories. It was developed in 1961 by Ali Javan and his coworkers. The glass discharge tube is filled with a 20:80 mixture of helium and neon gases, neon being the medium in which laser action occurs.

Figure 41-22 shows simplified energy level diagrams for the two atoms. An electric current passed through the helium–neon gas mixture serves—through collisions between helium atoms and electrons of the current—to raise many helium atoms to state E_3, which is metastable.

The energy of helium state E_3 (20.61 eV) is very close to the energy of neon state E_2 (20.66 eV). Thus when a metastable (E_3) helium atom and a ground-state (E_0) neon atom collide, the excitation energy of the helium atom is

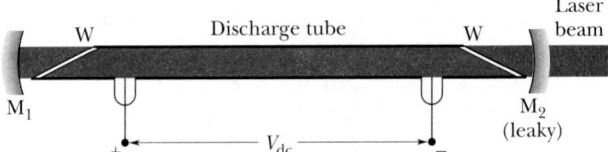

FIGURE 41-21 The elements of a helium–neon gas laser. An applied potential V_{dc} sends electrons through a discharge tube containing a mixture of helium gas and neon gas. Electrons collide with helium atoms, which then collide with neon atoms, which emit light along the length of the tube. The light passes through transparent windows W and reflects back and forth through the tube from mirrors M_1 and M_2 to cause more neon atom emissions. Some of the light leaks through mirror M_2 to form the laser beam.

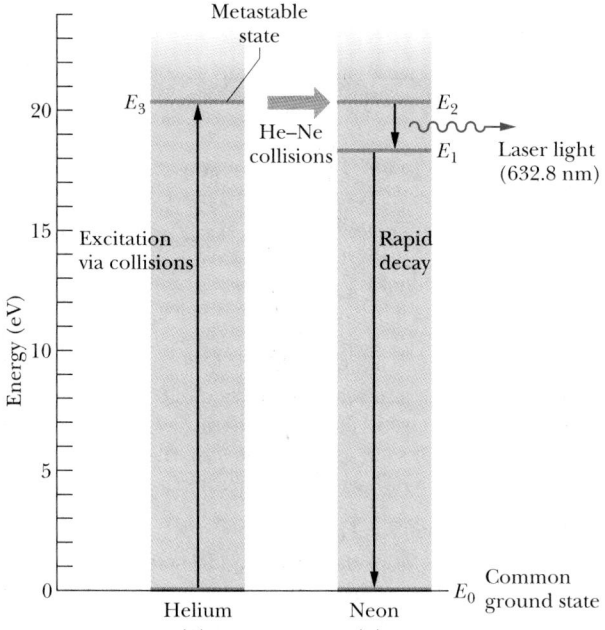

FIGURE 41-22 Four essential energy levels for helium and neon atoms in a helium–neon gas laser. Laser action occurs between levels E_2 and E_1 of neon when more atoms are in the E_2 level than in the E_1 level.

often transferred to the neon atom, which then moves to state E_2. In this way, neon level E_2 in Fig. 41-22 can become more heavily populated than neon level E_1.

This population inversion is relatively easy to set up and maintain because (1) initially there are essentially no neon atoms in state E_1, (2) the metastability of helium level E_3 ensures a ready supply of neon atoms in level E_2, and (3) atoms in level E_1 decay rapidly (through intermediate levels not shown) to the neon ground state E_0.

Suppose now that a single photon is spontaneously emitted as a neon atom transfers from state E_2 to state E_1. Such a photon can trigger a stimulated emission event which, in turn, can trigger other stimulated emission events. Through such a chain reaction, a coherent beam of red laser light, moving parallel to the tube axis, can build up rapidly. This light, of wavelength 632.8 nm, can move back and forth through the discharge tube many times by successive reflections from mirrors M_1 and M_2 (Fig. 41-21), accumulating additional stimulated emission photons with each passage.

Mirror M_1 is coated with an almost totally reflecting film. Mirror M_2, on the other hand, is coated so as to be slightly "leaky," permitting a small fraction of the laser light to escape at each reflection and form a useful external beam of laser light.

CHECKPOINT 4: The wavelength of light from laser A (a helium–neon gas laser) is 632.8 nm; that from laser

B (a carbon dioxide gas laser) is 10.6 μm. That from laser C (a gallium arsenide semiconductor laser) is 840 nm. Rank these lasers according to the energy interval between the two quantum states responsible for laser action, greatest first.

SAMPLE PROBLEM 41-7

In the helium–neon laser of Fig. 41-21, laser action occurs between two excited states of the neon atom. However, in many lasers, laser action (*lasing*) occurs between the ground state and an excited state, as suggested in Fig. 41-20.

(a) Consider such a laser that emits at a wavelength $\lambda = 550$ nm. If a population inversion had not been generated, what is the ratio of the population of atoms in state E_x to that in the ground state E_0?

SOLUTION: From Eq. 41-21, the ratio is

$$N_x/N_0 = e^{-(E_x - E_0)/kT}. \qquad (41-22)$$

The separation between the two energy levels must be

$$\begin{aligned}
E_x - E_0 = hf &= \frac{hc}{\lambda} \\
&= \frac{(6.63 \times 10^{-34}\ \text{J·s})(3.00 \times 10^8\ \text{m/s})}{(550 \times 10^{-9}\ \text{m})(1.60 \times 10^{-19}\ \text{J/eV})} \\
&= 2.26\ \text{eV}.
\end{aligned}$$

The mean energy of thermal agitation kT for an atom at room temperature (300 K) is

$$kT = (8.62 \times 10^{-5}\ \text{eV/K})\,(300\ \text{K}) = 0.0259\ \text{eV}.$$

Substituting the last two results into Eq. 41-22 yields

$$\begin{aligned}
N_x/N_0 &= e^{-(2.26\ \text{eV}/0.0259\ \text{eV})} \\
&= e^{-87.26} \approx 1.3 \times 10^{-38}. \qquad \text{(Answer)}
\end{aligned}$$

This is an extremely small number. It is not unreasonable, however. An atom whose mean thermal agitation energy is only 0.0259 eV will not often impart an energy of 2.26 eV to another atom in a collision.

(b) For the conditions of (a), at what temperature would the ratio N_x/N_0 be 1/2?

SOLUTION: Making this substitution into Eq. 41-22, taking the natural logarithm of both sides, and solving for T yield

$$\begin{aligned}
T = \frac{E_x - E_0}{k(\ln 2)} &= \frac{2.26\ \text{eV}}{(8.62 \times 10^{-5}\ \text{eV/K})\,(\ln 2)} \\
&= 38{,}000\ \text{K}. \qquad \text{(Answer)}
\end{aligned}$$

This is much hotter than the surface of the Sun. It is clear that if we are to invert the populations of these two levels, some specific mechanism for bringing this about is needed. No temperature, however high, will generate a population inversion.

REVIEW & SUMMARY

Some Properties of Atoms

The energies of atoms are quantized; that is, the atoms have only certain specific values of energy associated with different quantum states. Atoms can make transitions between different quantum states by emitting or absorbing a photon; the frequency f associated with that light is given by the *Bohr frequency condition*:

$$hf = E_{\text{high}} - E_{\text{low}}, \qquad (41\text{-}1)$$

where E_{high} is the higher energy and E_{low} is the lower energy of the pair of quantum states involved in the transition. Atoms also have quantized angular momenta and magnetic dipole moments.

Angular Momenta and Magnetic Dipole Moments

An electron trapped in an atom has an *orbital angular momentum* **L** and an associated *orbital magnetic dipole moment* $\boldsymbol{\mu}_{\text{orb}}$, which are always in opposite directions. The magnitude of **L** is given by

$$L = \sqrt{l(l+1)}\hbar, \qquad (41\text{-}2)$$

where l is the *angular momentum quantum number* (which can have the values given by Table 41-1) and $\hbar = h/2\pi$. The projection L_z of **L** on an arbitrary z axis is quantized and measurable and can have the values

$$L_z = m_l \hbar, \qquad (41\text{-}5)$$

where m_l is the *orbital magnetic quantum number* (which can have the values given by Table 41-1).

The projection $\mu_{\text{orb},z}$ of the orbital magnetic dipole moment on the z axis is quantized and measurable and can have the values

$$\mu_{\text{orb},z} = -m_l \mu_B, \qquad (41\text{-}3)$$

where m_l is the *orbital magnetic quantum number* and μ_B is the *Bohr magneton*:

$$\mu_B = \frac{eh}{4\pi m} = 9.274 \times 10^{-24} \text{ J/T}. \qquad (41\text{-}4)$$

An electron, whether trapped or free, has an intrinsic *spin angular momentum* (or just *spin*) **S** and an intrinsic *spin magnetic dipole moment* $\boldsymbol{\mu}_s$, which are always in opposite directions. The magnitude of **S** is given by

$$S = \sqrt{s(s+1)}\,\hbar, \qquad (41\text{-}6)$$

where s is the *spin quantum number* of the electron, which is always $\frac{1}{2}$. The projection S_z of **S** on an arbitrary z axis is quantized and measurable and can have the values

$$S_z = m_s \hbar, \qquad (41\text{-}7)$$

where m_s is the *spin magnetic quantum number* of the electron, which can be $+\frac{1}{2}$ or $-\frac{1}{2}$. The projection $\mu_{s,z}$ of the spin magnetic dipole moment $\boldsymbol{\mu}_s$ on the arbitrary z axis is quantized and measurable and can have the values

$$\mu_{s,z} = -2m_s \mu_B. \qquad (41\text{-}8)$$

Spin and Magnetic Resonance

A proton has an intrinsic spin angular momentum **S** and an associated spin magnetic dipole moment $\boldsymbol{\mu}$ that is always in the *same* direction as **S**. If a proton is located in an external magnetic field **B**, the projection μ_z of $\boldsymbol{\mu}$ on an axis z (defined to be along the direction of **B**) can have only two quantized orientations: parallel to **B** or antiparallel to **B**. The associated energy difference between these orientations is $2\mu_z B$. The energy required of a photon to *spin-flip* the proton between the two orientations is

$$hf = 2\mu_z(B_{\text{ext}} + B_{\text{local}}), \qquad (41\text{-}14)$$

where B_{ext} now represents the external field and B_{local} is the local magnetic field set up by the atoms and nuclei surrounding the proton. Detection of such spin flips can lead to *nuclear magnetic resonance spectra* by which specific substances can be identified.

Building the Periodic Table

Electrons in atoms obey the **Pauli exclusion principle**, which states that *no two electrons in the same atom can have the same set of the quantum numbers* n, l, m_l, and m_s. The elements are listed in the periodic table in order of increasing atomic number Z; the nuclear charge is Ze, and Z is both the number of protons in the nucleus and the number of electrons in the neutral atom.

States with the same value of n form a **shell**, and those with the same values of both n and l form a **subshell**. In *closed* shells and subshells, which are those that contain the maximum number of electrons, the angular momenta and the magnetic moments of the individual electrons add to zero.

X Rays and the Numbering of the Elements

A **continuous spectrum** of x rays arises when high-energy electrons lose some of their energy in a collision with an atomic nucleus. The **cutoff wavelength** λ_{min} results when such electrons lose *all* their initial energy in a single such encounter and is given by

$$\lambda_{\text{min}} = \frac{hc}{K_0}, \qquad (41\text{-}15)$$

in which K_0 is the initial kinetic energy of the electrons that strike the target.

Characteristic x rays arise when high-energy electrons eject electrons from deep within the atom; when the resulting "hole" is filled by an electron from farther out in the atom, a photon of the characteristic x-ray spectrum is generated.

In 1913 the British physicist H. G. J. Moseley measured the frequencies of the characteristic x rays from a number of elements. He noted that when the square root of the frequency is plotted against the position of the element in the periodic table, a straight line results, as in the **Moseley plot** of Fig. 41-17. This allowed Moseley to conclude that the property of the atom that determines the position of an element in the periodic table is not its atomic mass but its **atomic number** Z, that is, the number of protons in its nucleus.

Lasers and Laser Light

Laser light arises by **stimulated emission**. That is, when radiation of frequency given by

$$hf = E_x - E_0 \qquad (41\text{-}20)$$

is present, a transition from an upper energy level to a lower energy level of an atom can occur, a photon of frequency f being emitted. The stimulating photon and the emitted photon are identical in every respect and combine to form laser light.

For the emission process to predominate, there must normally be a **population inversion**; that is, there must be more atoms in the upper energy level than in the lower.

QUESTIONS

1. An electron in an atom of gold is in a state with $n = 4$. Which of these values of l are possible for it: $-3, 0, 2, 3, 4, 5$?

2. An atom of silver has closed $3d$ and $4d$ subshells. Which subshell has the greater number of electrons, or do they have the same number?

3. An atom of uranium has closed $6p$ and $7s$ subshells. Which subshell has the greater number of electrons?

4. An electron in a mercury atom is in the $3d$ subshell. Which values of m_l are possible for it: $-3, -1, 0, 1, 2$?

5. (a) How many subshells are there in the $n = 2$ shell? How many electron states? (b) Repeat (a) for the $n = 5$ shell.

6. From which atom of each of the following pairs is it easier to remove an electron? (a) Krypton or bromine? (b) Rubidium or cerium? (c) Helium or hydrogen?

7. On what quantum numbers does the energy of an electron depend in (a) a hydrogen atom and (b) a vanadium atom?

8. Label these statements as true or false: (a) One (and only one) of these subshells cannot exist: $2p, 4f, 3d, 1p$. (b) The number of values of m_l that are allowed depends only on l and not on n. (c) There are four subshells with $n = 4$. (d) The least value of n that can be associated with a given value of l is $l + 1$. (e) All states with $l = 0$ also have $m_l = 0$. (f) There are n subshells for each value of n.

9. Which (if any) of these statements about the Einstein–de Haas experiment or its results are true? (a) Atoms have angular momentum. (b) The angular momentum of atoms is quantized. (c) Atoms have magnetic moments. (d) The magnetic moments of atoms are quantized. (e) The angular momentum of an atom is strongly coupled to its magnetic moment. (f) The experiment relies on the conservation of angular momentum.

10. Consider the elements krypton and rubidium. (a) Which is more suitable for use in a Stern–Gerlach experiment of the kind described in connection with Fig. 41-8? (b) Which, if either, would not work at all?

11. The x-ray spectrum of Fig. 41-14 is for 35.0 keV electrons striking a molybdenum ($Z = 42$) target. If you substitute a silver ($Z = 47$) target for the molybdenum target, will (a) λ_{min}, (b) the wavelength for the K_α line, (c) the wavelength for the K_β line increase, decrease, or remain unchanged?

12. The K_α x-ray line for any element arises because of a transition between the K shell ($n = 1$) and the L shell ($n = 2$). Figure 41-14 shows this line (for a molybdenum target) occurring at a single wavelength. With higher resolution, however, the line splits into several wavelength components because the L shell does not have a unique energy. (a) How many components does the K_α line have? (b) Similarly, how many components does the K_β line have?

13. Which (if any) of the following is essential for laser action to occur between two energy levels of an atom? (a) There are more atoms in the upper level than in the lower. (b) The upper level is metastable. (c) The lower level is metastable. (d) The lower level is the ground state of the atom. (e) The lasing medium is a gas.

14. Figure 41-22 shows partial energy level diagrams for the helium and neon atoms that are involved in the operation of a helium–neon laser. It is said that a helium atom in state E_3 can collide with a neon atom in its ground state and raise the neon atom to state E_2. The energy of helium state E_3 (20.61 eV) is close to, but not exactly equal to, the energy of neon state E_2 (20.66 eV). How can the energy transfer take place if these energies are not *exactly* equal?

EXERCISES & PROBLEMS

SECTION 41-4 Angular Momenta and Magnetic Dipole Moments

1E. Show that $\hbar = 1.06 \times 10^{-34}$ J·s $= 6.59 \times 10^{-16}$ eV·s.

2E. How many electron states are there in the following subshells: (a) $n = 4, l = 3$; (b) $n = 3, l = 1$; (c) $n = 4, l = 1$; (d) $n = 2, l = 0$?

3E. How many electron states are there in the following shells: (a) $n = 4$, (b) $n = 1$, (c) $n = 3$, (d) $n = 2$?

4E. (a) What is the magnitude of the orbital angular momentum in a state with $l = 3$? (b) What is the magnitude of its largest projection on an imposed z axis?

5E. (a) What number of possible l values are associated with $n = 3$? (b) What number of possible m_l values are associated with $l = 1$?

6E. An electron in a hydrogen atom is in a state with $l = 5$. What is the minimum possible angle between $\mathbf{L}$ and L_z?

7E. Write down all the quantum numbers for states that form the shell with $n = 4$ and $l = 3$.

8E. An electron in a multielectron atom is known to have the quantum number $l = 3$. What are the possible n, m_l, and m_s quantum numbers?

9E. An electron in a multielectron atom has a maximum m_l value of $+4$. What can you say about the rest of its quantum numbers?

10E. How many electron states are there in a shell defined by the quantum number $n = 5$?

11E. An electron is in a state with $n = 3$. What are (a) the number of possible values of l, (b) the number of possible values of m_l, (c) the number of possible values of m_s, (d) the number of states in the $n = 3$ shell, and (e) the number of subshells in the $n = 3$ shell?

12P. An electron is in a state with $l = 3$. Calculate and tabulate the allowed values of L_z, μ_z, and θ, where θ is the angle made by the corresponding vector with the positive direction of the z axis. Find also the magnitudes of $\mathbf{L}$ and $\boldsymbol{\mu}$.

13P. (A correspondence principle problem.) Estimate (a) the quantum number l for the orbital motion of Earth around the Sun and (b) the number of allowed orientations of the plane of Earth's orbit, according to the rules of space quantization. (c) Find θ_{min}, the half-angle of the smallest cone that can be swept out by a perpendicular to Earth's orbit as Earth revolves around the Sun.

14P. If $\mathbf{L}$ is measured along, say, the z axis to give a value for L_z, show that the most that can be said about the other two components of $\mathbf{L}$ is

$$(L_x^2 + L_y^2)^{1/2} = [l(l + 1) - m_l^2]^{1/2}\,\hbar.$$

SECTION 41-5 The Stern–Gerlach Experiment

15E. Calculate the two possible angles between the electron spin angular momentum vector and the magnetic field in Sample Problem 41-1. Bear in mind that the orbital angular momentum of the valence electron in the silver atom is zero.

16E. What is the acceleration of the silver atom as it passes through the deflecting magnet in the Stern–Gerlach experiment of Sample Problem 41-1?

17E. Assume that in the Stern–Gerlach experiment described for neutral silver atoms, the magnetic field **B** has a magnitude of 0.50 T. (a) What is the energy difference between the orientations of the silver atoms in the two subbeams? (b) What is the frequency of the radiation that would induce a transition between these two states? (c) What is its wavelength, and to what part of the electromagnetic spectrum does it belong? The magnetic moment of a neutral silver atom is 1 Bohr magneton.

18P. Suppose that a hydrogen atom in its ground state moves 80 cm through and perpendicular to a vertical magnetic field that has a magnetic field gradient, dB/dz of 1.6×10^2 T/m. (a) What magnitude of force does the field gradient exert on the atom due to the magnetic moment of its electron, which we take to be 1 Bohr magneton? (b) What is the vertical displacement of the atom in the 80 cm of travel if its speed is 1.2×10^5 m/s?

SECTION 41-6 Magnetic Resonance

19E. What is the wavelength of a photon that will induce a transition of an electron spin from parallel to antiparallel orientation in a magnetic field of magnitude 0.200 T? Assume that $l = 0$.

20E. The proton, like the electron, has a spin quantum number s of $\frac{1}{2}$. In the hydrogen atom in its ground state ($n = 1$ and $l = 0$), there are two energy levels, depending on whether the electron and proton spins are parallel or antiparallel. If an atom has a spin flip from the state of higher energy to that of lower energy, a photon of wavelength 21 cm is emitted. Radio astronomers observe this 21 cm radiation coming from deep space. What is the effective magnetic field (due to the magnetic dipole moment of the proton) experienced by the electron emitting this radiation?

21E. An external oscillating magnetic field of frequency 34 MHz is applied to a sample that contains hydrogen atoms. Resonance is observed when the strength of the constant external magnetic field equals 0.78 T. Calculate the strength of the local magnetic field at the site of the protons that are undergoing spin flips, assuming the external and local fields are parallel there.

22E. Excited sodium atoms emit two closely spaced spectrum lines (the sodium doublet; see Fig. 41-23) with wavelengths 588.995 and 589.592 nm. (a) What is the difference in energy between the two upper energy levels? (b) This energy difference occurs because the electron's spin magnetic moment (= 1 Bohr magneton) can be oriented either parallel or antiparallel to the internal magnetic field associated with the electron's orbital motion. Use your result in (a) to find the strength of this internal magnetic field.

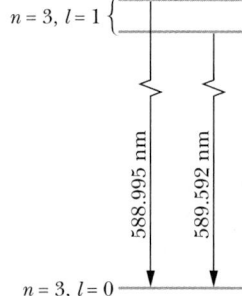

FIGURE 41-23
Exercise 22.

SECTION 41-7 Building the Periodic Table

23P. Show that if the 63 electrons in an atom of europium were assigned to shells according to the "logical" sequence of quantum numbers, this element would have chemical properties similar to those of sodium.

24P. Consider the elements selenium ($Z = 34$), bromine ($Z = 35$), and krypton ($Z = 36$). In their part of the periodic table, the subshells of the electronic states are filled in the sequence

$$1s \quad 2s \quad 2p \quad 3s \quad 3p \quad 3d \quad 4s \quad 4p \cdots$$

For each element, identify the highest occupied subshell and state how many electrons are in it.

25P. Suppose that the electron had no spin and that the Pauli exclusion principle still held. Which, if any, of the present noble gases would remain in that category?

26P. What are the four quantum numbers for the two electrons of the helium atom in its ground state?

27P. Two electrons in lithium ($Z = 3$) have for their quantum numbers $n = 1$, $l = 0$, $m_l = 0$, and $m_s = \pm\frac{1}{2}$. What quantum numbers can the third electron have if the atom is to be in (a) its ground state and (b) its first excited state?

28P. Suppose there are two electrons in the same atom, both of which have $n = 2$ and $l = 1$. (a) If the exclusion principle did not apply, how many combinations of states would conceivably be possible? (b) How many states does the exclusion principle forbid? Which are they?

29P. Show that the number of states with the same quantum number n is given by $2n^2$.

SECTION 41-8 X Rays and the Numbering of the Elements

30E. Show that the cutoff wavelength (in picometers) in the continuous x-ray spectrum from any target is given by $\lambda_{\min} = 1240/V$, where V is the potential difference (in kilovolts) through which the electrons are accelerated before they strike the target.

31E. Knowing that the minimum x-ray wavelength produced by 40.0 keV electrons striking a target is 31.1 pm, determine the Planck constant h.

32E. What is the minimum potential difference across an x-ray tube that will produce x rays with a wavelength of 0.100 nm?

33P. A 20 keV electron is brought to rest by undergoing two successive nuclear encounters such as that of Fig. 41-15, thus transferring its kinetic energy to the energy of two photons. The wavelength associated with the second photon is 130 pm greater than the wavelength of the first photon. (a) Find the kinetic energy of the electron after its first encounter. (b) What are the associated wavelengths and energies of the two photons?

34P. X rays are produced in an x-ray tube by a target potential of 50.0 kV. An electron makes three collisions in the target before coming to rest and loses half its remaining kinetic energy in each of the first two collisions. Determine the wavelengths of the resulting photons. (Neglect the recoil of the heavy target atoms.)

35P. Show that a moving electron cannot spontaneously change into an x-ray photon in free space. A third body (atom or nucleus) must be present. Why is it needed? (*Hint:* Examine the conservation of energy and momentum.)

36E. When electrons bombard a molybdenum target, they produce both continuous and characteristic x rays as shown in Fig. 41-14. In that figure the kinetic energy of the incident electrons is 35.0 keV. If the accelerating potential is increased to 50.0 keV, what mean values of (a) $\lambda_{\min}$, (b) the wavelength of the K_α line, and (c) the wavelength of the K_β line result?

37E. In Fig. 41-14, the x rays shown are produced when 35.0 keV electrons strike a molybdenum ($Z = 42$) target. If the accelerating potential is maintained at this value but a silver ($Z = 47$) target is used instead, what values of (a) $\lambda_{\min}$, (b) the wavelength of the K_α, and (c) the wavelength of the K_β line result? The K, L, and M atomic x-ray levels for silver (compare Fig. 41-16) are 25.51, 3.56, and 0.53 keV.

38E. The wavelength of the K_α line from iron is 193 pm. What is the energy difference between the two states of the iron atom that give rise to this transition?

39P. From Fig. 41-14, calculate approximately the energy difference $E_L - E_M$ for molybdenum. Compare it with the value that may be obtained from Fig. 41-16.

40E. Calculate the ratio of the wavelength of the K_α line for niobium (Nb) to that for gallium (Ga). Take needed data from the periodic table of Appendix G.

41P. Here are the K_α wavelengths (pm) of a few elements:

T	275	Co	179
V	250	Ni	166
Cr	229	Cu	154
Mn	210	Zn	143
Fe	193	Ga	134

Make a Moseley plot (like that in Fig. 41-17) from these data and verify that its slope agrees with the value calculated in Sample Problem 41-6.

42P. A tungsten ($Z = 74$) target is bombarded by electrons in an x-ray tube. (a) What is the minimum value of the accelerating potential that will permit the production of the characteristic K_α and K_β lines of tungsten? (b) For this same accelerating potential, what is $\lambda_{\min}$? (c) What are the K_α and K_β wavelengths? The K, L, and M energy levels for tungsten (see Fig. 41-16) have energies 69.5, 11.3, and 2.30 keV, respectively.

43P. A molybdenum ($Z = 42$) target is bombarded with 35.0 keV electrons and the x-ray spectrum of Fig. 41-14 results. Here the K_β and the K_α wavelengths are 63.0 and 71.0 pm, respectively. (a) What are the corresponding photon energies? (b) It is desired to filter these radiations through a material that will absorb the K_β line much more strongly than it will absorb the K_α line. What substance would you use? The K ionization energies for molybdenum and for four neighboring elements are:

	Zr	Nb	Mo	Tc	Ru
Z	40	41	42	43	44
E_K (keV)	18.00	18.99	20.00	21.04	22.12

(*Hint:* A substance will absorb one x radiation more strongly than another if the photons of the first have enough energy to eject a K electron from the atom of the substance but the photons of the second do not.)

44P. The binding energies of K-shell and L-shell electrons in copper are 8.979 and 0.951 keV, respectively. If a K_α x ray from copper is incident on a sodium chloride crystal and gives a first-order Bragg reflection at an angle of 74.1° measured relative to parallel planes of sodium atoms, what is the spacing between these parallel planes?

45P. (a) Using Eq. 41-18, estimate the ratios of photon energies due to K_α transitions in two atoms whose atomic numbers are Z and Z'. (b) What is this ratio for uranium and aluminum? (c) For uranium and lithium?

46P. Determine how close the theoretical K_α x-ray photon energies, as obtained from Eq. 41-19, are to the measured energies of the low-mass elements from lithium to magnesium. To do this, (a) first determine the constant C in Eq. 41-19 to five significant figures by finding C in terms of the fundamental constants in Eq. 41-16 and then using data from Appendix B to evaluate those constants. (b) Next, calculate the percentage deviations of the theoretical from the measured energies. (c) Finally, plot the deviations and comment on the trend. The measured energies (eV) of the K_α photons for these elements are

Li	54.3	O	524.9
Be	108.5	F	676.8
B	183.3	Ne	848.6
C	277	Na	1041
N	392.4	Mg	1254

(There is actually more than one K_α ray because of the splitting of the L energy level, but that effect is negligible for the elements listed here.)

SECTION 41-9 Lasers and Laser Light

47E. Lasers can be used to generate pulses of light whose durations are as short as 10 fs. (a) How many wavelengths of light ($\lambda = 500$ nm) are contained in such a pulse? (b) Supply the missing quantity (in years):

$$\frac{10 \text{ fs}}{1 \text{ s}} = \frac{1 \text{ s}}{X}.$$

48E. For the conditions of Sample Problem 41-7, how many moles of lasing material are needed to put 10 atoms in the excited state E_x?

49E. By measuring the go-and-return time for a laser pulse to travel from an Earth-bound observatory to a reflector on the Moon, it is possible to measure the separation between these bodies. (a) What is the predicted value of this time? (b) The separation can be measured to a precision of about 15 cm. To what uncertainty in travel time does this correspond? (c) The laser beam forms a spot on the Moon 3 km in diameter. What is the angular divergence of the beam?

50E. A hypothetical atom has energy levels evenly separated by 1.2 eV. For a temperature of 2000 K, what is the ratio of the number of atoms in the 13th excited state to the number in the 11th excited state?

51E. A hypothetical atom has only two atomic energy levels, separated by 3.2 eV. In the atmosphere of a star there are $6.1 \times 10^{13}/\text{cm}^3$ of these atoms in the higher energy state and $2.5 \times 10^{15}/\text{cm}^3$ in the lower energy state. What is the temperature of the star's atmosphere?

52E. A population inversion for two energy levels is often de-scribed by assigning a negative Kelvin temperature to the system. What negative temperature would describe a system in which the population of the upper energy level exceeds that of the lower level by 10% and the energy difference between the two levels is 2.1 eV?

53E. A helium–neon laser emits laser light at a wavelength of 632.8 nm and a power of 2.3 mW. At what rate are photons emitted by this device?

54E. A pulsed laser emits light at a wavelength of 694.4 nm. The pulse duration is 12 ps and the energy per pulse is 0.150 J. (a) What is the length of the pulse? (b) How many photons are emitted in each pulse?

55E. The active volume of a laser constructed of the semicon-ductor GaAlAs is only 200 μm^3 (smaller than a grain of sand) and yet the laser can continuously deliver 5.0 mW of power at a wavelength of 0.80 μm. At what rate does it generate photons?

56E. Assume that lasers are available whose wavelengths can be precisely "tuned" to anywhere in the visible range, that is, in the range 450 nm $< \lambda <$ 650 nm. If every television channel occu-pies a bandwidth of 10 MHz, how many channels could be ac-commodated within this wavelength range?

57E. A high-powered laser beam ($\lambda = 600$ nm) with a beam diameter of 12 cm is aimed at the Moon, 3.8×10^5 km distant. The beam spreads only because of diffraction. The angular loca-tion of the edge of the central diffraction disk (see Eq. 37-12) is given by

$$\sin \theta = \frac{1.22\lambda}{d},$$

where d is the diameter of the beam aperture. What is the diame-ter of the central diffraction disk on the Moon's surface?

58P. The active medium in a particular laser that generates laser light at a wavelength of 694 nm is 6.00 cm long and 1.00 cm in diameter. (a) Treat the medium as an optical resonance cavity analogous to a closed organ pipe. How many standing wave nodes are there along the laser axis? (b) By what amount Δf would the beam frequency have to shift to increase this number by one? (c) Show that Δf is just the inverse of the travel time of laser light for one round trip back and forth along the laser axis. (d) What is the corresponding fractional frequency shift $\Delta f/f$? The appropriate index of refraction of the lasing medium (a ruby crystal) is 1.75.

59P. The mirrors in the laser of Fig. 41-21, which are separated by 8.0 cm, form an optical cavity in which standing waves of laser light can be set up. In the vicinity of $\lambda = 533$ nm, how far apart in wavelength are the adjacent allowed operating modes?

60P. A hypothetical atom has two energy levels, with a transi-tion wavelength between them of 580 nm. In a particular sample at 300 K, 4.0×10^{20} such atoms are in the state of lower energy. (a) How many atoms are in the upper state, assuming conditions of thermal equilibrium? (b) Suppose, instead, that 3.0×10^{20} of these atoms are "pumped" into the upper state by an external process, with 1.0×10^{20} atoms remaining in the lower state. What is the maximum energy that could be released by the atoms in a single laser pulse if each is affected once?

61P. The beam from an argon laser (of wavelength 515 nm) has a diameter d of 3.00 mm and a continuous wave power output of 5.00 W. The beam is focused onto a diffuse surface by a lens whose focal length f is 3.50 cm. A diffraction pattern such as that of Fig. 37-9 is formed, the radius of the central disk being given by

$$R = \frac{1.22 f \lambda}{d}$$

(see Eq. 37-12 and Sample Problem 37-3). The central disk can be shown to contain 84% of the incident power. (a) What is the radius of the central disk? (b) What is the average power flux density in the incident beam? (c) What is the average power flux density in the central disk?

62P. Can an incoming intercontinental ballistic missile be destroyed by an intense laser beam? A beam of intensity 10^8 W/m² would probably burn into and destroy a hardened (nonspinning) missile in 1 s. (a) If the laser had 5.0 MW power, 3.0 μm wavelength, and a 4.0 m beam diameter (a very powerful laser indeed), would it destroy a missile at a distance of 3000 km? (b) If the wavelength could be changed, what maximum value would work? Use the equation for the central disk given in Exercise 57, and take the focal length to be the distance to the target.

Additional Problems

63. *Martian CO_2 laser.* Where sunlight shines on the atmosphere of Mars, carbon dioxide molecules at an altitude of about 75 km undergo naturally occurring laser action. The energy levels involved in the action are shown in Fig. 41-24; population inversion occurs between energy levels E_2 and E_1. (a) What wavelength of sunlight excites the molecules in the lasing action? (b) At what wavelength does lasing occur? (c) In what region of the electromagnetic spectrum do the excitation and lasing wavelengths lie?

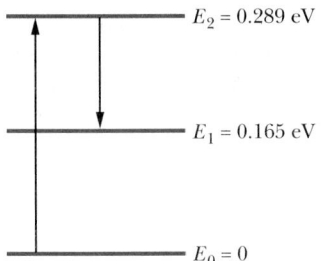

$E_2 = 0.289$ eV

$E_1 = 0.165$ eV

$E_0 = 0$

FIGURE 41-24 Problem 63.

64. *Comet stimulated emission.* When a comet approaches the Sun, the increased warmth evaporates water from the frozen ice on the surface of the comet nucleus, producing a thin atmosphere of water vapor around the nucleus. Sunlight can then dissociate the water vapor into H and OH. The sunlight can also excite the OH molecules into higher energy levels, two of which are represented in Fig. 41-25.

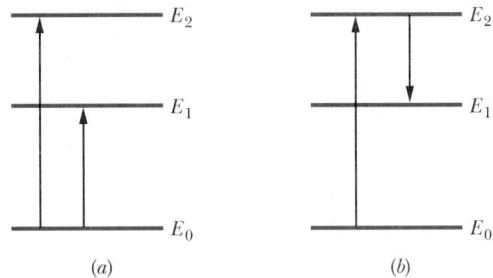

E_2

E_1

E_0

(a)

E_2

E_1

E_0

(b)

FIGURE 41-25 Problem 64.

When the comet is still relatively far from the Sun, the sunlight causes equal excitation to the E_2 and E_1 levels (Fig. 41-25a). Hence, there is no population inversion between the two levels. However, as the comet approaches the Sun, the excitation to the E_1 level decreases and population inversion occurs. The reason has to do with one of the many wavelengths—said to be *Fraunhofer lines*—that are missing in sunlight because, as the light travels outward through the Sun's atmosphere, those particular wavelengths are absorbed by the atmosphere.

As a comet approaches the Sun, the Doppler effect due to the comet's speed relative to the Sun shifts the Fraunhofer lines in wavelength, apparently overlapping one of them with the wavelength required for the excitation to the E_1 level in the OH molecules. Population inversion then occurs in those molecules, and they radiate stimulated emission (Fig. 41-25b). For example, as comet Kohoutek approached the Sun in December 1973 and January 1974, it radiated stimulated emission at about 1666 MHz during mid-January. (a) What was the energy difference $E_2 - E_1$ for that emission? (b) In what region of the electromagnetic spectrum was the emission?

A few of the workers at the Fab 11 factory at Rio Rancho, New Mexico. The plant, which represents an investment of $2.5 billion, has floor space equivalent to about two dozen football fields. According to the New York Times, the plant, "on a high desert mesa in New Mexico is probably the most productive factory in the world, in terms of the value of the goods that it makes." But what do these workers manufacture? Why are they suited up like astronauts? And why is the floor they are standing on perforated?

42-1 SOLIDS

You have seen how well quantum mechanics works when we apply it to questions involving individual atoms. In this chapter we hope to show, by a single broad example, that this theory works just as well when we apply it to questions involving assemblies of atoms in the form of solids.

Every solid has an enormous range of properties that we can choose to examine. Is it transparent? Can it be hammered out into a thin sheet? At what speeds do sound waves travel through it? Is it magnetic? Is it a good heat conductor? . . . The list goes on and on. However, we choose to focus this entire chapter on a single question: *What are the mechanisms by which a solid conducts, or does not conduct, electricity?* As you will see, quantum mechanics provides the answer.

42-2 THE ELECTRICAL PROPERTIES OF SOLIDS

We shall examine only **crystalline solids**, that is, solids whose atoms are arranged in a repetitive three-dimensional structure called a **lattice**. We shall not consider such solids as wood, plastic, glass, or rubber, whose atoms are not arranged in such repetitive patterns. Figure 42-1 shows the basic repetitive units (the **unit cells**) of the lattice structures of copper, our prototype of a metal, and silicon and diamond, our prototypes of a semiconductor and an insulator, respectively.

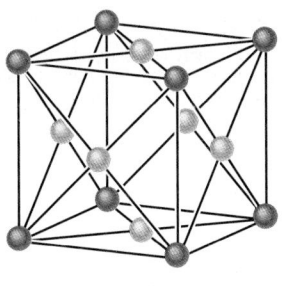 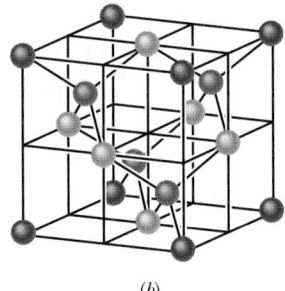

(a)   *(b)*

FIGURE 42-1 (*a*) The unit cell for copper is a cube. There is one copper atom (darker) at each corner of the cube and one copper atom (lighter) at the center of each face of the cube. The arrangement is called *face-centered cubic*. (*b*) The unit cell for silicon and diamond is also a cube, the atoms being arranged in a so-called *diamond lattice*. There is one atom (darkest) at each corner of the cube and one atom (lightest) at the center of each cube face; in addition, four atoms (medium color) lie within the cube. Every atom is bonded to its four nearest neighbors by a two-electron covalent bond (only the four atoms within the cube show all four *nearest* neighbors).

We can classify solids electrically according to three basic properties:

1. Their **resistivity** ρ at room temperature, with the SI unit ohm-meter ($\Omega \cdot$ m); resistivity is defined in Section 27-4.

2. Their **temperature coefficient of resistivity** α, defined as $\alpha = (1/\rho)(d\rho/dT)$ and having the SI unit inverse kelvin (K^{-1}). We can evaluate α for any solid by measuring ρ over a range of temperatures.

3. Their **number density of charge carriers** n. This quantity, the number of charge carriers per unit volume, can be found from measurements of the Hall effect, as discussed in Section 29-4, and from other measurements. It has the SI unit inverse cubic meter (m^{-3}).

From measurements of room-temperature resistivity alone, we discover that there are some materials—we call them **insulators**—that for all practical purposes do not conduct electricity at all. These are materials with very high resistivity. Diamond, an excellent example, has a resistivity greater than that of copper by the enormous factor of about 10^{24}. Thus we may immediately classify crystalline solids into insulators and noninsulators.

We can then use measurements of ρ, α, and n to divide most noninsulators, at least at low temperatures, into two major categories: **metals** and **semiconductors**.

Semiconductors have a considerably larger resistivity ρ than metals.

Semiconductors have a temperature coefficient of resistivity α that is both large and negative. That is, the resistivity of a semiconductor *decreases* with temperature, whereas that of a metal *increases*.

Semiconductors have a considerably smaller number density of charge carriers n than metals.

Table 42-1 shows values of these quantities for copper, our prototype metal, and silicon, our prototype semiconductor.

TABLE 42-1 SOME ELECTRIC PROPERTIES OF TWO MATERIALS[a]

	UNIT	COPPER	SILICON
Type of conductor		Metal	Semiconductor
Number density of charge carriers, n	m^{-3}	9×10^{28}	1×10^{16}
Resistivity, ρ	$\Omega \cdot$ m	2×10^{-8}	3×10^{3}
Temperature coefficient of resistivity, α	K^{-1}	$+4 \times 10^{-3}$	-70×10^{-3}

[a]All values are for room temperature.

Now, with measurements of ρ, α, and n in hand, we have an experimental basis for refining our central question about the conduction of electricity in solids: *What features make diamond an insulator, copper a metal, and silicon a semiconductor?* Again, quantum mechanics provides the answers.

42-3 ENERGY LEVELS IN A CRYSTALLINE SOLID

The distance between adjacent copper atoms in solid copper is 260 pm. Figure 42-2a shows two isolated copper atoms separated by a distance r that is much greater than that. As Fig. 42-2b shows, each of these isolated neutral atoms stacks up its 29 electrons in an array of discrete subshells, as follows:

$$1s^2 \, 2s^2 \, 2p^6 \, 3s^2 \, 3p^6 \, 3d^{10} \, 4s^1.$$

Here we use the shorthand notation of Section 41-7 to identify the subshells. Recall, for example, that the subshell with principal quantum number $n = 3$ and orbital quantum number $l = 1$ is called the $3p$ subshell; it can hold up to $2(2l + 1) = 6$ electrons; the number it actually contains is indicated by a numerical superscript. The first six subshells in copper are filled, but the (outermost) $4s$ subshell, which can hold 2 electrons, holds only one.

If we bring the atoms of Fig. 42-2a closer together, they will — speaking loosely — begin to sense each other's presence. In the language of quantum mechanics, their wave functions will start to overlap, beginning with those of the outermost electrons.

With wave functions that overlap, we speak not of two independent atoms but of a single two-atom system containing $2 \times 29 = 58$ electrons. The Pauli exclusion principle also applies to this larger system and requires that each of these 58 electrons occupy a different quantum state. In

fact, 58 quantum states are available because each energy level of the isolated atom splits into *two* levels for the two-atom system.

If we bring up more atoms, we gradually assemble a lattice of solid copper. If, say, our lattice contains N atoms, then each level of an isolated copper atom must split into N levels in the solid. Thus, the individual energy levels of the solid form energy **bands**, adjacent bands being separated by an energy **gap**, which represents a range of energies that no electron can possess. A typical band is only a few electron-volts wide. Since N may be of the order of 10^{24}, we see that the individual levels within a band are very close together indeed, and there are a vast number of levels.

Figure 42-3 suggests the band–gap structure of the energy levels in a generalized crystalline solid. Note that bands of lower energy are narrower than those of higher energy. This occurs because electrons that occupy the lower energy bands spend most of their time deep within the atom's electron cloud. The wave functions of these core electrons do not overlap as much as the wave functions of the outer electrons. Hence the splitting of these levels is not as great as it is for the higher energy levels normally occupied by the outer electrons.

42-4 INSULATORS

A solid is said to be an insulator if no current exists when we apply a potential difference across it. For a current to exist, the kinetic energy of the average electron must increase. In other words, some electrons in the solid must move to a higher energy level. But as Fig. 42-4a shows, in an insulator the highest band containing any electrons is fully occupied, and the Pauli exclusion principle keeps electrons from moving to occupied levels.

So the electrons in the filled band of an insulator have no place to go; they are in gridlock. It is as if a child tries to

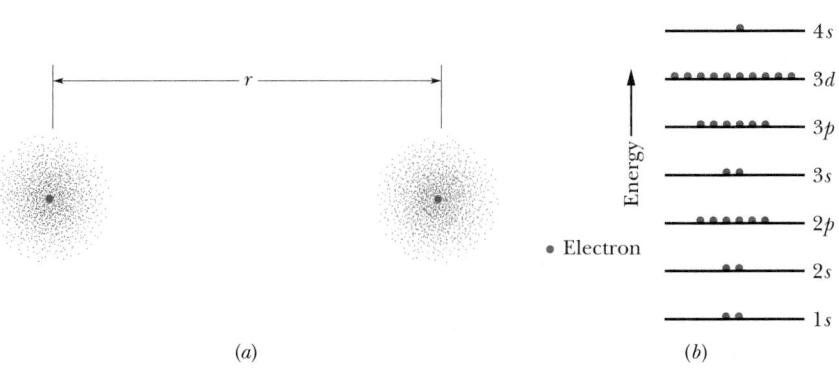

FIGURE 42-2 (*a*) Two copper atoms separated by a large distance; their electron distributions are represented by dot plots. (*b*) Each copper atom has 29 electrons distributed among a set of subshells. In the neutral atom in its ground state, all subshells up through the $3d$ level are filled, the $4s$ subshell contains one electron (it can hold two), and higher subshells are empty. For simplicity, the subshells are shown as being evenly spaced in energy.

(a) (b)

FIGURE 42-3 The band-gap pattern of energy levels for an idealized crystalline solid. As the magnified view suggests, each band consists of a very large number of very closely spaced energy levels. (In many solids, adjacent bands may overlap; for clarity, we have not shown this condition.)

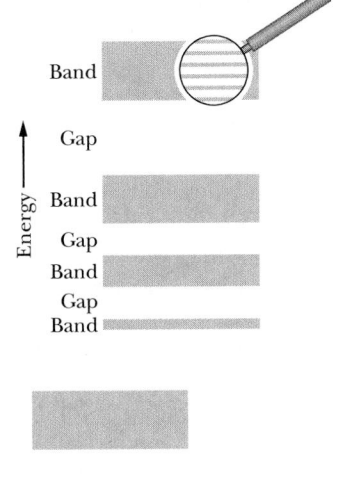

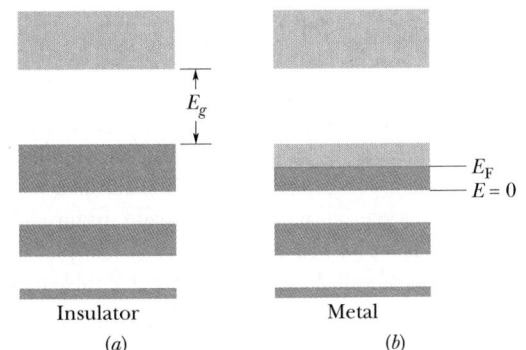

Insulator   Metal
(a)   (b)

FIGURE 42-4 (a) The band-gap pattern for an insulator; filled levels are shown in red. Note that the highest filled level lies at the top of a band and the next highest vacant level is separated from it by a relatively large energy gap E_g. (b) The band-gap pattern for a metal. The highest filled level, called the Fermi level, lies near the middle of a band. Since vacant levels are available within that band, electrons in the band can easily change levels, and conduction can take place.

climb a ladder that already has a child standing on each rung; since there are no vacant rungs, no one can move.

There are plenty of vacant levels in the band above the filled band in Fig. 42-4a. However, if an electron is to occupy one of those levels, it must acquire enough energy to bridge the substantial gap that separates the two bands. In diamond, this gap is so wide (the energy needed to cross it is 5.5 eV, about 140 times the average thermal energy of a free particle at room temperature) that essentially no electron can jump it. Diamond is thus an insulator, and a very good one.

SAMPLE PROBLEM 42-1

In Chapter 41 we used Eq. 41-21,

$$\frac{N_x}{N_0} = e^{-(E_x - E_0)/kT}, \qquad (42\text{-}1)$$

to relate the population N_x of the atoms at energy level E_x to the population N_0 at energy level E_0, where the atoms are part

of a system at temperature T (in kelvins). The constant k is the Boltzmann constant (8.62×10^{-5} eV/K, from Eq. 20-21).

We can use the same equation to find the likelihood that an electron in an insulator will jump the energy gap E_g in Fig. 42-4a. To do so, we set $E_x - E_0 = E_g$; then N_x/N_0 is the ratio of the number of electrons just above the energy gap to the number of electrons just below the energy gap.

What is the probability that, at room temperature (300 K), an electron at the top of the valence band in diamond will jump the gap E_g which, for diamond, is 5.5 eV?

SOLUTION: For diamond, the exponent in Eq. 42-1 is

$$-\frac{E_g}{kT} = -\frac{5.5 \text{ eV}}{(8.62 \times 10^{-5} \text{ eV/K})(300 \text{ K})} = -213.$$

The required probability is then

$$\frac{N_x}{N_0} = e^{-(E_g/kT)} = e^{-213} \approx 3 \times 10^{-93}. \qquad \text{(Answer)}$$

No wonder diamond is such a good insulator. Even in a diamond as large as Earth, the chance for a single electron to jump the gap at 300 K would be vanishingly small!

42-5 METALS

The feature that defines a metal is that, as Fig. 42-4b shows, the highest occupied energy level falls somewhere near the middle of an energy band. If we apply a potential difference across a sample of such a solid, a current can exist because there are plenty of vacant levels at higher energies into which electrons can be raised. So a metal can conduct electricity because electrons in its highest occupied band can easily move into higher energy levels within that band.

In Section 27-6 we introduced the **free-electron model** of a metal, in which the conduction electrons are free to move throughout the volume of the sample like the molecules of a gas in a closed container. We used this model to derive an expression for the resistivity of a metal, assuming that the electrons follow the laws of Newtonian mechanics. Here we use that same model to explain the behavior of the electrons—called the **conduction electrons**—in the partially filled band of Fig. 42-4b. However, we follow the laws of quantum mechanics by assuming the energies of these electrons to be quantized and the Pauli exclusion principle to hold.

We assume too that the electric potential energy of a conduction electron has the same constant value at all points within the lattice. If we choose this value of the potential energy to be zero, as we are free to do, then the energy E of the conduction electrons is entirely kinetic.

The level at the bottom of the partially filled band of Fig. 42-4b corresponds to $E = 0$. The highest occupied level in this band (at absolute zero, $T = 0$ K) is called the **Fermi level**, and the energy corresponding to it is called the **Fermi energy** E_F; for copper, $E_F = 7.0$ eV.

The electron speed corresponding to the Fermi energy is called the **Fermi speed** v_F. For copper the Fermi speed is 1.6×10^6 m/s. This fact should be enough to shatter the popular misconception that all motion ceases at absolute zero; at that temperature — and solely because of the Pauli exclusion principle — the conduction electrons are stacked up in the partially filled band of Fig. 42-4b with energies that range from zero to the Fermi energy.

Conductivity at $T > 0$

Our practical interest in the conduction of electricity in metals is at temperatures above absolute zero. What happens to the electron distribution of Fig. 42-4b at such higher temperatures? As we shall see, surprisingly little.

Of the electrons in the partially filled band of Fig. 42-4b, only those that are close to the Fermi energy find vacant levels above them, and only those electrons are free to be boosted to these higher levels by thermal agitation. Even at $T = 1000$ K, a temperature at which copper would glow brightly in a dark room, the distribution of electrons among the available levels does not differ much from the distribution at $T = 0$ K.

Let us see why. The quantity kT, where k is the Boltzmann constant, is a convenient measure of the energy that may be given to a conduction electron by the random thermal motions of the lattice. At $T = 1000$ K, we have $kT = 0.086$ eV. No electron can hope to have its energy changed by more than a few times this relatively small amount by thermal agitation alone. So at best, only those few conduction electrons whose energies are close to the Fermi energy are likely to be boosted to higher energy levels by thermal agitation. Poetically stated, thermal agitation normally causes only ripples on the surface of the Fermi sea of electrons; the vast depths of that sea lie undisturbed.

How Many Quantum States Are There?

The ability of a metal to conduct electricity depends on how many quantum states are available to its electrons and what the energies of these states are. Thus a question arises: What are the energies of the individual states in the partially filled band of Fig. 42-4b? This question is too difficult to answer because we cannot possibly list the energies of so many states individually. We ask instead: How many states have energies in the energy range E to $E + dE$? We write this number as $N(E)\, dE$, where $N(E)$ is called the **density of states** at energy E. The conventional unit for $N(E)\, dE$ is states per cubic meter (states/m³, or

simply m⁻³); the corresponding unit for $N(E)$ is states per cubic meter per electron-volt (m⁻³ eV⁻¹).

We can find an expression for the density of states by counting the number of standing electron matter waves that can fit into a box the size of the metal sample we are considering. This is analogous to counting the number of standing waves of sound that can exist in a closed organ pipe. The differences are that our problem is three-dimensional (the organ pipe problem is one-dimensional) and the waves are quantum-mechanical matter waves (the organ-pipe waves are sound waves). The result of such counting can be shown to be

$$N(E) = \frac{8\sqrt{2}\pi m^{3/2}}{h^3} E^{1/2} \quad \text{(density of states)}, \quad (42\text{-}2)$$

where m is the mass of the electron and E is the kinetic energy at which $N(E)$ is to be evaluated. Note that nothing in this equation involves the shape of the sample, its temperature, or the material of which it is made. Figure 42-5, a half-parabola, is a plot of Eq. 42-2. As an example, it tells us that there are about 2×10^{28} states per cubic meter of sample whose energies lie in the 1.0 eV energy range centered at 8 eV.

CHECKPOINT **1:** (a) Is the spacing between adjacent energy levels at $E = 4$ eV in copper larger than, the same as, or smaller than the spacing at $E = 6$ eV? (b) Is the spacing between adjacent energy levels at $E = 4$ eV in copper larger than, the same as, or smaller than the spacing for an identical volume of aluminum at that same energy?

The Occupancy Probability $P(E)$

The ability of a metal to conduct electricity depends on the probability that available vacant levels will actually be occupied. Thus another question arises: If an energy level is

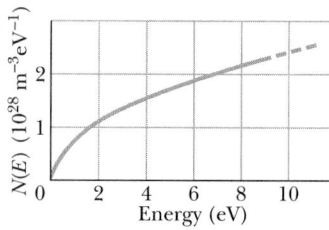

FIGURE 42-5 The density of states $N(E)$, that is, the number of electron energy levels per unit energy interval and per unit volume, plotted as a function of electron energy. The density of states function simply counts the available states; it says nothing about whether these states are or are not occupied by electrons.

available at energy E, what is the probability $P(E)$ that it is actually occupied by an electron? At $T = 0$ K, we know that for all levels with energies below the Fermi energy, $P(E) = 1$, corresponding to a certainty that the level is occupied. We also know that, at $T = 0$ K, for all levels with energies above the Fermi energy, $P(E) = 0$, corresponding to a certainty that the level is *not* occupied. Figure 42-6a illustrates this situation.

To find $P(E)$ at temperatures above absolute zero, we must use a set of quantum counting rules called **Fermi–Dirac statistics**, named for the physicists who introduced them. Using these rules, it is possible to show that the **occupancy probability** $P(E)$ is

$$P(E) = \frac{1}{e^{(E-E_F)/kT} + 1} \qquad \begin{array}{l}\text{(occupancy}\\\text{probability),}\end{array} \quad (42\text{-}3)$$

in which E_F is the Fermi energy. Note that $P(E)$ depends not on the energy E of the level but only on the difference $E - E_F$, which may be positive or negative.

To see whether Eq. 42-3 describes Fig. 42-6a, we substitute $T = 0$ K in it. Then,

For $E < E_F$, the exponential term in Eq. 42-3 is $e^{-\infty}$, or zero, so $P(E) = 1$, in agreement with Fig. 42-6a.

For $E > E_F$, the exponential term is $e^{+\infty}$, so $P(E) = 0$, again in agreement with Fig. 42-6a.

Figure 42-6b is a plot of $P(E)$ for $T = 1000$ K. It shows that, as stated above, changes in the distribution of electrons among the available states involve only states whose energies are near the Fermi energy E_F. Note that if $E = E_F$ (no matter what the temperature T), the exponential term in Eq. 42-3 is $e^0 = 1$ and $P(E) = 0.5$. This leads us to a more useful definition of the Fermi energy:

> The Fermi energy of a given material is the energy of a quantum state that has the probability 0.5 of being occupied by an electron.

In a given sample, however, there may not be a quantum state available at that energy.

How Many *Occupied* States Are There?

Equation 42-2 and Fig. 42-5 tell us how the available states are distributed in energy. The occupancy probability of Eq. 42-3 gives us the probability that any given state will actually be occupied by an electron. To find $N_o(E)$, the density of *occupied* states, we must weight each available state by the appropriate value of the occupancy probability; that is,

$$N_o(E) = N(E)\,P(E) \qquad \begin{array}{l}\text{(density of}\\\text{occupied states).}\end{array} \quad (42\text{-}4)$$

Figure 42-7a is a plot of Eq. 42-4 for copper at $T = 0$ K. It is found by multiplying at each energy, the value of the density of states function (Fig. 42-5) by the value of the occupancy probability for absolute zero (Fig. 42-6a). Fig-

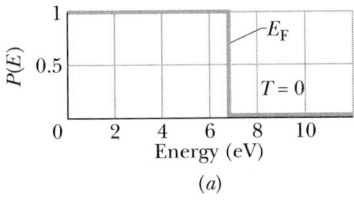

(a)

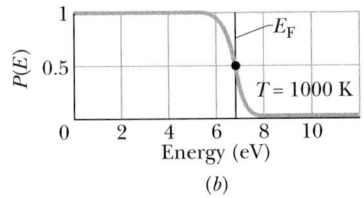

(b)

FIGURE 42-6 The occupancy probability $P(E)$ gives the probability that an energy level will be occupied by an electron. (*a*) At $T = 0$ K, $P(E)$ is unity for levels with energies E up to the Fermi energy and zero for levels with higher energies. (*b*) At $T = 1000$ K, a few electrons whose energies were slightly less than the Fermi energy at $T = 0$ K move up to states with energies slightly greater than the Fermi energy. The dot on the curve shows that, for $E = E_F$, $P(E) = 0.5$.

FIGURE 42-7 (*a*) The density of occupied states $N_o(E)$ for copper at absolute zero. The area under the curve is the number density of electrons n. Note that all states with energies up to the Fermi energy are occupied, and all those with energies above the Fermi energy are vacant. (*b*) The same for copper at $T = 1000$ K. Note that only electrons whose energies are near the Fermi energy have been affected by thermal agitation and redistributed.

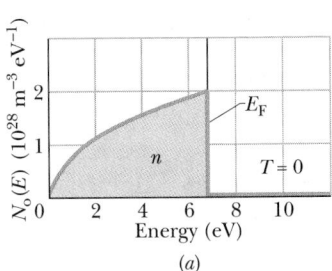

(a)

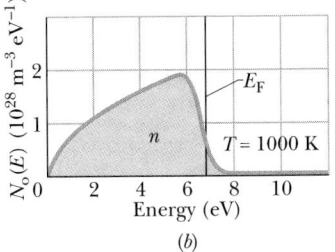

(b)

ure 42-7b, calculated similarly, shows the density of occupied states for copper at $T = 1000$ K.

Calculating the Fermi Energy

Suppose we add up (via integration) the number of occupied states in Fig. 42-7a at all energies between $E = 0$ and $E = E_F$. The result must equal n, the number of conduction electrons per unit volume for the metal. In equation form, we have

$$n = \int_0^{E_F} N_0(E)\, dE. \qquad (42\text{-}5)$$

(Graphically, the integral here represents the area under the distribution curve of Fig. 42-7a.) Because $P(E) = 1$ for all energies below the Fermi energy, we can replace $N_0(E)$ in Eq. 42-5 by $N(E)$ and then use Eq. 42-5 to find the Fermi energy E_F. If we substitute Eq. 42-2 into Eq. 42-5, we find that

$$n = \frac{8\sqrt{2}\pi m^{3/2}}{h^3} \int_0^{E_F} E^{1/2}\, dE = \frac{8\sqrt{2}\pi m^{3/2}}{h^3} \frac{2E_F^{3/2}}{3}.$$

Solving for E_F now leads to

$$E_F = \left(\frac{3}{16\sqrt{2}\pi}\right)^{2/3} \frac{h^2}{m} n^{2/3} = \frac{0.121 h^2}{m} n^{2/3}. \quad (42\text{-}6)$$

Thus when we know n, the number of conduction electrons per unit volume, we can find the Fermi energy for a metal.

SAMPLE PROBLEM 42-2

A cube of copper is 1.00 cm on edge. In the partially filled band of Fig. 42-4b, what is the number N of quantum states in the energy range from $E = 5.000$ eV to $E = 5.010$ eV? (These energy values are so close that we can assume the density of states $N(E)$ is constant over the interval.)

SOLUTION: We can find the number of states N from

$$\begin{pmatrix} \text{number of states} \\ \text{in sample} \end{pmatrix}$$

$$= \begin{pmatrix} \text{density of} \\ \text{states, m}^{-3}\text{eV}^{-1} \end{pmatrix} \begin{pmatrix} \text{energy} \\ \text{range, eV} \end{pmatrix} \begin{pmatrix} \text{volume of} \\ \text{sample, m}^3 \end{pmatrix},$$

or $\qquad\qquad N = N(E)\, \Delta E\, V, \qquad (42\text{-}7)$

where $\Delta E = 0.010$ eV and V is the volume of the cubical sample. From Eq. 42-2 with $E = 5.000$ eV, we get

$$N(E) = \frac{8\sqrt{2}\pi m^{3/2}}{h^3} E^{1/2}$$

$$= (8\sqrt{2}\pi)(9.11 \times 10^{-31}\text{ kg})^{3/2}$$

$$\times \frac{(5.000\text{ eV})^{1/2}(1.60 \times 10^{-19}\text{ J/eV})^{1/2}}{(6.63 \times 10^{-34}\text{ J·s})^3}$$

$= 9.48 \times 10^{46}\text{ m}^{-3}\text{ J}^{-1} = 1.52 \times 10^{28}\text{ m}^{-3}\text{eV}^{-1}$.

Since $V = a^3$, where a is the length of the cube edge, Eq. 42-7 gives us

$$N = N(E)\, \Delta E\, a^3$$

$$= (1.52 \times 10^{28}\text{ m}^{-3}\text{eV}^{-1})(0.010\text{ eV})(1 \times 10^{-2}\text{ m})^3$$

$$= 1.52 \times 10^{20}. \qquad \text{(Answer)}$$

This is an enormous number of states, but expectedly so. Even though all these states fall into an energy range that is only 0.01 eV wide, they originate from the enormous number of atoms that forms our sample.

SAMPLE PROBLEM 42-3

(a) What is the probability that a quantum state whose energy is 0.10 eV above the Fermi energy will be occupied? Assume a sample temperature of 800 K.

SOLUTION: We can find $P(E)$ from Eq. 42-3. But let us first calculate the (dimensionless) exponent in that equation:

$$\frac{E - E_F}{kT} = \frac{0.10\text{ eV}}{(8.62 \times 10^{-5}\text{ eV/K})(800\text{ K})} = 1.45.$$

Inserting this exponent into Eq. 42-3 yields

$$P(E) = \frac{1}{e^{1.45} + 1} = 0.19 \text{ or } 19\%. \qquad \text{(Answer)}$$

(b) What is the probability of occupancy for a state that is 0.10 eV *below* the Fermi energy?

SOLUTION: The exponent in Eq. 42-3 has the same absolute value as in (a) but is now negative. Thus from this equation

$$P(E) = \frac{1}{e^{-1.45} + 1} = 0.81 \text{ or } 81\%. \qquad \text{(Answer)}$$

For states below the Fermi energy, we are often more interested in the probability that the state is *not* occupied. This is just $1 - P(E)$, or 19%. Note that it is the same as the probability of occupancy in (a).

42-6 SEMICONDUCTORS

If you compare Fig. 42-8a with Fig. 42-4a, you can see that the band structure of a semiconductor is like that of an insulator. The main difference is that the semiconductor has a much smaller energy gap E_g between the top of the highest filled band (called the **valence band**) and the bottom of the vacant band just above it (called the **conduction band**). Thus there is no doubt that silicon ($E_g = 1.1$ eV) is a semiconductor and diamond ($E_g = 5.5$ eV) is an insulator. In silicon—but not in diamond—there is a real possi-

FIGURE 42-8 (*a*) The band–gap pattern for a semiconductor. It resembles that of an insulator (see Fig. 42-4*a*) except that here the energy gap E_g is much smaller; thus electrons, because of their thermal agitation, have some reasonable probability of being able to jump the gap. (*b*) Thermal agitation has caused a few electrons to jump the gap from the valence band to the conduction band, leaving an equal number of holes in the valence band.

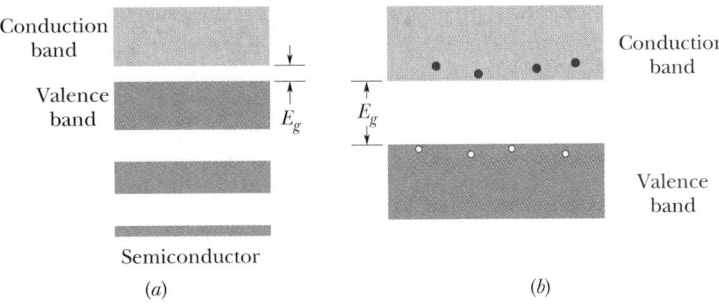

bility that thermal agitation at room temperature will cause electrons to jump the gap from the valence band to the conduction band.

In Table 42-1 we compared three basic electrical properties of copper, our prototype metallic conductor, and silicon, our prototype semiconductor. Let us look again at that table, one row at a time, to see how a semiconductor differs from a metal.

Number Density of Charge Carriers n

The second row of Table 42-1 shows that copper has far more charge carriers per unit volume than silicon, by a factor of about 10^{13}. For copper, each atom contributes one electron, its single valence electron, to the conduction process. Charge carriers in silicon arise only because, at thermal equilibrium, thermal agitation causes a certain (very small) number of valence band electrons to jump the energy gap into the conduction band, leaving an equal number of vacant energy states, called **holes**, in the valence band. Figure 42-8*b* shows the situation.

Both the electrons in the conduction band and the holes in the valence band serve as charge carriers. The holes do so by permitting a certain freedom of movement to electrons in the valence band that, in the absence of holes, would be gridlocked. If an electric field **E** is set up in a semiconductor, the electrons in the valence band, being negatively charged, tend to drift in the direction opposite **E**. This causes the positions of the holes to drift in the direction of **E**. In effect, the holes behave like moving particles of charge $+e$.

It may help to think of a row of cars parked bumper to bumper, the leading car being one car's length from a barrier. If the leading car moves forward to the barrier, it opens up a car's length space behind it. The second car can then move up to fill that space, allowing the third car to move up, and so on. The motions of the many cars toward the barrier are most simply analyzed by focusing attention

on the drift of the single "hole" (parking space) away from the barrier.

In semiconductors, conduction by holes is just as important as conduction by electrons. In thinking about hole conduction, it is well to imagine that all unoccupied states in the valence band are occupied by particles of charge $+e$, and that all electrons in the valence band have been removed, so that these positive charge carriers can move freely throughout the band.

The Resistivity ρ

From Chapter 27 recall that the resistivity ρ of a material is $m/e^2n\tau$, where m is the electron mass, e is the fundamental charge, n is the number of charge carriers per unit volume, and τ is the mean time between collisions of the charge carriers. Table 42-1 shows that, at room temperature, the resistivity of silicon is higher than that of copper, by a factor of about 10^{11}. This vast difference can be accounted for by the vast difference in n. Other factors enter, but their effect on the resistivity is swamped by the enormous difference in n.

The Temperature Coefficient of Resistivity α

Recall that α (see Eq. 27-17) is the fractional change in resistivity per unit change in temperature:

$$\alpha = \frac{1}{\rho}\frac{d\rho}{dT}. \qquad (42\text{-}8)$$

The resistivity of copper *increases* with temperature (that is, $d\rho/dT > 0$) because collisions of copper's charge carriers occur more frequently at higher temperatures. Thus α is *positive* for copper.

The collision frequency also increases with temperature for silicon. However, the resistivity of silicon actually *decreases* with temperature ($d\rho/dT < 0$) because the number of charge carriers (electrons in the conduction band and holes in the valence band) increases so rapidly with temperature. (More electrons jump the gap from the valence band to the conduction band.) Thus the fractional change α is *negative* for silicon.

CHECKPOINT 2: The research laboratory of a large corporation developed three new solid materials whose electrical properties are shown below. Anticipating patent applications, the laboratory identified these materials with code names. Classify each material as a metal, an insulator, a semiconductor, or none of the above:

MATERIAL (CODE NAME)	n (m^{-3})	ρ $(\Omega \cdot m)$	α (K^{-1})
Cleveland	10^{29}	10^{-8}	$+10^{-3}$
Troy	10^{28}	10^{-9}	-10^{-3}
Seattle	10^{15}	10^{3}	-10^{-2}

42-7 DOPED SEMICONDUCTORS

The usefulness of semiconductors in technology can be greatly improved by introducing a small number of suitable replacement atoms (it seems pejorative to call them impurities) into the semiconductor lattice—a process called **doping**. Typically, only about 1 silicon atom in 10^{7} is replaced by a dopant atom. Essentially all modern semiconducting devices are based on doped material. Such materials are of two types, called ***n*-type** and ***p*-type**; we discuss each in turn.

n-Type Semiconductors

The electrons in an isolated silicon atom are arranged in subshells according to the scheme

$$1s^2\, 2s^2\, 2p^6\, 3s^2\, 3p^2,$$

in which, as usual, the superscripts (which add to 14, the atomic number of silicon) represent the numbers of electrons in the specified subshell.

Figure 42-9a is a flattened out representation of a portion of the lattice of pure silicon in which the portion has been projected onto a plane; compare the figure with Fig. 42-1b, which represents the unit cell of the lattice in three dimensions. Each silicon atom contributes its pair of $3s$ electrons and its pair of $3p$ electrons to form a rigid two-electron covalent bond with each of its four nearest neighbors. (A covalent bond is a link between two atoms in which the atoms share a pair of electrons.) The four atoms that lie within the unit cell in Fig. 42-1b show these bonds.

The electrons that form the silicon–silicon bonds constitute the valence band of the silicon sample. If an electron is torn from one of these bonds so that it becomes free to wander throughout the lattice, we say that the electron has been raised from the valence band to the conduction band. The minimum energy required to do this is the gap energy E_g.

Because four of its electrons are involved in bonds, each silicon ''atom'' is actually an ion consisting of an inert neonlike electron cloud (containing 10 electrons) surrounding a nucleus whose charge is $+14e$, where 14 is the atomic number of silicon. The net charge of these ions is thus $+4e$, and the ions are said to have a *valence number* of 4.

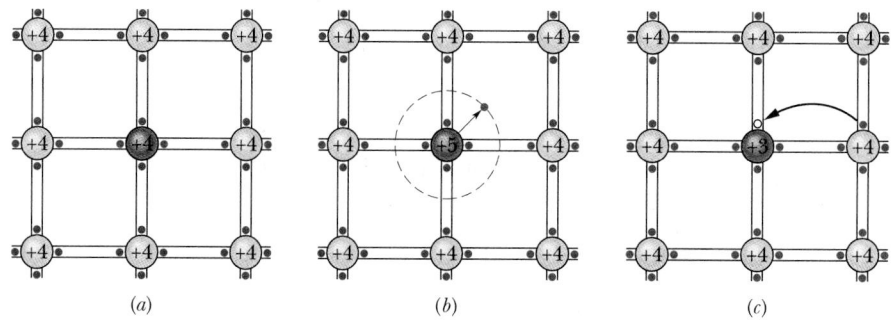

(a)   *(b)*   *(c)*

FIGURE 42-9 (a) A flattened-out representation of the lattice structure of pure silicon. Each silicon ion is coupled to its four nearest neighbors by a two-electron covalent bond (represented by a pair of red dots between two parallel black lines). The electrons belong to the bond—not to the individual atoms—and form the valence band of the sample. (b) One silicon atom is replaced by a phosphorus atom (valence = 5). The ''extra'' electron is only loosely bound to its ion core and may easily be ele- vated to the conduction band, where it is free to wander through the volume of the lattice. (c) One silicon atom is replaced by an aluminum atom (valence = 3). There is now a hole in one of the covalent bonds and thus in the valence band of the sample. The hole can easily migrate through the lattice as electrons from neighboring bonds move in to fill it. In the move shown, the hole migrates rightward.

In Fig. 42-9*b* the central silicon ion has been replaced by an atom of phosphorus (valence = 5). Four of the valence electrons of the phosphorus form bonds with the four surrounding silicon ions. The fifth ("extra") electron is only loosely bound to the phosphorus ion core. On an energy band diagram, we usually say that such an electron occupies a localized energy state that lies within the energy gap, at an average energy interval E_d below the bottom of the conduction band; this is indicated in Fig. 42-10*a*. Because $E_d \ll E_g$, the energy required to excite electrons from *these* levels into the conduction band is much less than that required to excite silicon valence electrons into the conduction band.

The phosphorus atom is called a **donor** atom because it readily *donates* an electron to the conduction band. In fact, at room temperature virtually *all* the electrons contributed by the donor atoms are in the conduction band. By adding donor atoms, it is possible to increase greatly the number of electrons in the conduction band, by a factor very much larger than Fig. 42-10*a* suggests.

Semiconductors doped with donor atoms are called ***n*-type semiconductors**; the "*n*" stands for "negative," to imply that the negative charge carriers introduced into the conduction band greatly outnumber the positive charge carriers, which are the holes in the valence band. In *n*-type semiconductors, the electrons are called the **majority carriers**, and the holes the **minority carriers**.

p-Type Semiconductors

Now consider Fig. 42-9*c*, in which one of the silicon atoms (valence = 4) has been replaced by an atom of aluminum (valence = 3). The aluminum atom can bond covalently with only three silicon atoms, so there is now a "missing" electron (a hole) in one aluminum–silicon bond. With a small expenditure of energy, an electron can be torn from a neighboring silicon–silicon bond to fill this hole, thereby creating a hole in *that* bond. And, similarly, an electron from some other bond can be moved to fill the second hole. In this way, the hole can migrate through the lattice.

The aluminum atom is called an **acceptor** atom because it readily *accepts* an electron from a neighboring bond, that is, from the valence band of silicon. As Fig. 42-10*b* suggests, this electron occupies a localized acceptor state that lies within the energy gap, at an average energy interval E_a above the top of the valence band. By adding acceptor atoms, it is possible to increase very greatly the number of holes in the valence band, by a factor much larger than Fig. 42-10*b* suggests. In silicon at room temperature, virtually *all* the acceptor levels are occupied by electrons.

Semiconductors doped with acceptor atoms are called ***p*-type semiconductors**; the "*p*" stands for "positive" to imply that the holes introduced into the valence band, which behave like positive charge carriers, greatly outnumber the electrons in the conduction band. In *p*-type semiconductors, holes are the majority carriers and electrons are the minority carriers.

Table 42-2 summarizes the properties of a typical *n*-type and a typical *p*-type semiconductor. Note particularly that the donor and acceptor ion cores, although they are charged, are not charge *carriers* because at normal temperatures they remain fixed in their lattice sites.

SAMPLE PROBLEM 42-4

The number density n_0 of conduction electrons in pure silicon at room temperature is about 10^{16} m^{-3}. Assume that, by doping the silicon lattice with phosphorus, we want to increase this number by a factor of a million (10^6). What fraction of silicon atoms must we replace with phosphorus atoms? (Recall that at room temperature, thermal agitation is so effective that essentially every phosphorus atom donates its "extra" electron to the conduction band.)

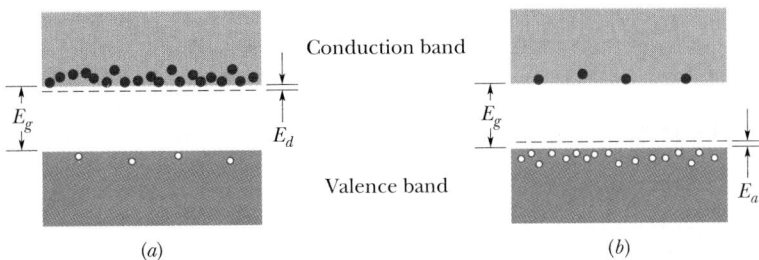

(a)   (b)

FIGURE 42-10 (*a*) In a doped *n*-type semiconductor, the energy levels of donor electrons lie a small interval E_d below the bottom of the conduction band. Because donor electrons can be easily excited to the conduction band, there are now many more electrons in that band. The valence band contains the same small number of holes as before. (*b*) In a doped *p*-type semiconductor, the acceptor levels lie a small interval E_a above the top of the valence band. There are now relatively many more holes in the valence band. The conduction band contains the same small number of electrons as before. The ratio of majority carriers to minority carriers in both (*a*) and (*b*) is very much greater than is suggested by these diagrams.

TABLE 42-2 PROPERTIES OF TWO DOPED SEMICONDUCTORS

| PROPERTY | TYPE OF SEMICONDUCTOR | |
	n	p
Matrix material	Silicon	Silicon
Matrix nuclear charge	$+14e$	$+14e$
Matrix energy gap	1.2 eV	1.2 eV
Dopant	Phosphorus	Aluminum
Type of dopant	Donor	Acceptor
Majority carriers	Electrons	Holes
Minority carriers	Holes	Electrons
Dopant energy gap	0.045 eV	0.067 eV
Dopant valence	5	3
Dopant nuclear charge	$+15e$	$+13e$
Dopant net ion charge	$+e$	$-e$

SOLUTION: The number density of conduction electrons added by doping will be equal to n_P, the number density of phosphorus atoms added. We want the total number density of electrons in the conduction band after doping, original plus added electrons, to be $10^6 n_0$, so

$$10^6 n_0 = n_0 + n_P.$$

Then

$$n_P = 10^6 n_0 - n_0 \approx 10^6 n_0$$
$$= (10^6)(10^{16} \text{ m}^{-3}) = 10^{22} \text{ m}^{-3}.$$

This tells us that we must add 10^{22} atoms of phosphorus to each cubic meter of silicon.

The number density of silicon atoms in a pure silicon lattice may be found from

$$n_{Si} = \frac{N_A d}{A},$$

in which N_A is the Avogadro constant (6.02×10^{23} mol^{-1}), d is the density of silicon (2330 kg/m^3), and A is the molar mass of silicon (28.1 g/mol, or 0.0281 kg/mol). Substitution yields

$$n_{Si} = \frac{(6.02 \times 10^{23} \text{ mol}^{-1})(2330 \text{ kg/m}^3)}{0.0281 \text{ kg/mol}}$$
$$= 5 \times 10^{28} \text{ m}^{-3}.$$

The fraction we seek is approximately

$$\frac{n_P}{n_{Si}} = \frac{10^{22} \text{ m}^{-3}}{5 \times 10^{28} \text{ m}^{-3}} = \frac{1}{5 \times 10^6}. \quad \text{(Answer)}$$

If we replace only *one silicon atom in five million* with a phosphorus atom, the number of electrons in the conduction band will be increased by a factor of a million.

How can such a tiny admixture of phosphorus have what seems to be such a big effect? The answer is that, although the effect is very significant, it is not "big." The number density of conduction electrons was 10^{16} m^{-3} before doping and

10^{22} m^{-3} after doping. For copper, however, the conduction-electron number density (given in Table 42-1) is about 10^{29} m^{-3}. Thus, even after doping, the number density of conduction electrons in silicon remains much less than that of a typical metal, such as copper, by a factor of about 10^7.

42-8 THE p-n JUNCTION

A ***p-n* junction**, as Fig. 42-11a shows, is a single semiconductor crystal that has been selectively doped so that one region is n-type material and the adjacent region is p-type

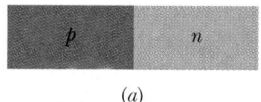

(a)

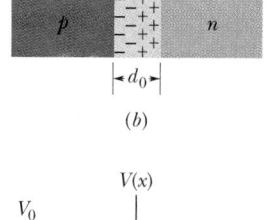

(b)

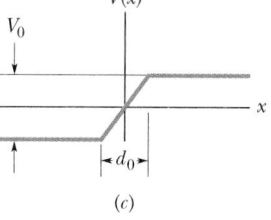

(c)

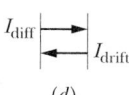

(d)

FIGURE 42-11 (a) A sample of n-type material and a sample of p-type material are intimately joined to form a p-n junction. (b) Motions of the majority charge carriers across the junction plane uncover a space charge associated with uncompensated donor ions (to the right of the plane) and acceptor ions (to the left). This depletion zone contains no free charge carriers; the donor and acceptor ions are fixed in their lattice sites. (c) Associated with the space charge is a contact potential difference V_0, which acts to limit the flow of majority carriers. (d) The diffusion of majority carriers (both electrons and holes) across the junction plane produces a diffusion current I_{diff}; in an isolated p-n junction, that current is just balanced by the drift current I_{drift} produced by minority carriers, with the result that the net current through the junction plane is zero. (In a real p-n junction, the boundaries of the depletion zone would not be sharp, as shown here, and the contact potential curve (c) would be smooth, with no sharp corners.)

material. Such junctions are at the heart of essentially all semiconductor devices.

We assume, for simplicity, that the junction has been formed mechanically, by butting together a bar of *n*-type semiconductor with a bar of *p*-type semiconductor. Thus the transition from one region to the other is perfectly sharp, occurring at a single **junction plane**.

Let us discuss the motions of electrons and holes just after the *n*-type bar and the *p*-type bar, both electrically neutral, have been jammed together to form the junction. We first examine the majority carriers, which are electrons in the *n*-type material and holes in the *p*-type material.

Motions of the Majority Carriers

If you burst a helium-filled balloon, helium atoms will diffuse (spread) outward into the surrounding air. This happens because there are very few helium atoms in normal air. In more formal language, there is a helium *density gradient* at the balloon–air interface (the number density of helium atoms varies across the interface); the helium atoms move so as to reduce the gradient.

In the same way, electrons on the *n* side of Fig. 42-11*a* that are close to the junction plane tend to diffuse across it (from right to left in the figure) and into the *p* side, where there are very few free electrons. Similarly, holes on the *p* side that are close to the junction plane tend to diffuse across that plane (from left to right) and into the *n* side, where there are very few holes. The motions of both the electrons and the holes contribute to a **diffusion current** I_{diff}, conventionally directed from left to right as indicated in Fig. 42-11*d*.

Recall that the *n*-side is studded throughout with positively charged donor ions, fixed firmly in their lattice sites. Normally, the excess positive charge of each of these ions is compensated electrically by one of the conduction-band electrons. When an *n*-side electron diffuses across the junction plane, however, the diffusion "uncovers" one of these donor ions, thus introducing a fixed positive charge near the junction plane on the *n* side.

When the diffusing electron arrives on the *p* side, it quickly combines with an acceptor ion (which lacks one electron), thus introducing a fixed negative charge near the junction plane on the *p* side.

In this way electrons diffusing through the junction plane from right to left in Fig. 42-11*a* result in a buildup of **space charge** on each side of the junction plane, as indicated in Fig. 42-11*b*. Holes diffusing through the junction plane from left to right have exactly the same effect. (Take the time now to convince yourself of that.) The motions of both majority carriers—electrons and holes—contribute to the buildup of these two space charge regions, one posi-

tive and one negative. These two regions form a **depletion zone**, so named because it is relatively free of *mobile* charge carriers; its width is shown as d_0 in Fig. 42-11*b*.

The buildup of space charge generates an associated **contact potential difference** V_0 across the depletion zone, as Fig. 42-11*c* shows. This potential difference serves to limit further diffusion of electrons and holes across the junction plane. Negative charges tend to avoid regions of low potential. Thus, an electron approaching the junction plane from the right in Fig. 42-11*b* is moving toward a region of low potential and would tend to turn back into the *n* side. Similarly, a positive charge (a hole) approaching the junction plane from the left is moving toward a region of high potential and would tend to turn back into the *p* side.

Motions of the Minority Carriers

As Fig. 42-10*a* shows, although the majority carriers in *n*-type material are electrons, there are nevertheless a few holes. Likewise in *p*-type material (Fig. 42-10*b*), although the majority carriers are holes, there are also a few electrons. These few holes and electrons are the minority carriers in the corresponding materials.

Although the potential difference V_0 in Fig. 42-11*c* acts as a barrier for the majority carriers, it is a downhill trip for the minority carriers, be they electrons on the *p*-side or holes on the *n*-side. Positive charges (holes) tend to seek regions of low potential; negative charges (electrons) tend to seek regions of high potential. Thus both types of carriers are *swept across* the junction plane by the contact potential difference and, together, constitute a **drift current** I_{drift} across the junction plane from right to left, as Fig. 42-11*d* indicates.

Thus an isolated *p-n* junction is in an equilibrium state in which a contact potential difference V_0 exists between its ends. At equilibrium, the average diffusion current I_{diff} that moves through the junction plane from the *p* side to the *n* side is just balanced by an average drift current I_{drift} that moves in the opposite direction. These two currents cancel because the net current through the junction plane must be zero; otherwise charge would be transferred without limit from one end of the junction to the other.

CHECKPOINT 3: Which of the following five currents across the junction plane of Fig. 42-11*a* must be zero?
(a) the net current due to holes, both majority and minority carriers included
(b) the net current due to electrons, both majority and minority carriers included
(c) the net current due to both holes and electrons, both majority and minority carriers included

(d) the net current due to majority carriers, both holes and electrons included

(e) the net current due to minority carriers, both holes and electrons included

42-9 THE JUNCTION RECTIFIER

Look now at Fig. 42-12. It shows that, if we place a potential difference across a *p-n* junction in one direction (here labeled + and ''Forward bias''), there will be a current through the junction. However, if we reverse the direction of the potential difference, there will be approximately zero current through the junction.

One application of this property is the **junction rectifier**, whose symbol is shown in Fig. 42-13*b*: the arrowhead corresponds to the *p*-type terminal of the device and points in the allowed direction of conventional current. A sine wave input potential to the device (Fig. 42-13*a*) is transformed to a half-wave output potential (Fig. 42-13*c*) by the junction rectifier; that is, the rectifier acts as essentially a closed switch (zero resistance) for one polarity of the input potential and as essentially an open switch (infinite resistance) for the other.

The average value of the input voltage in Fig. 42-13*a* is zero, but that of the output voltage in Fig. 42-13*c* is not. Thus a junction rectifier can be used, with appropriate electronic filtering that is not shown in the figure, to convert an alternating potential difference into a constant potential difference, as for an electronic power supply.

Figure 42-14 shows why a *p-n* junction operates as a junction rectifier. In Fig. 42-14*a*, a battery is connected across the junction with its positive terminal connected at the *p* side. In this **forward-bias connection**, the *p* side becomes more positive than it was before the connection and the *n* side becomes more negative, thus *decreasing* the height of the potential barrier V_0 of Fig. 42-11*c*. More of the majority carriers can now surmount this smaller barrier;

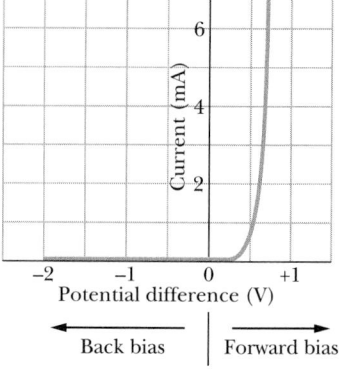

FIGURE 42-12 A current–voltage plot for a *p-n* junction, showing that the junction is highly conducting when forward-biased and essentially nonconducting when back-biased.

hence the diffusion current I_{diff} increases markedly.

The minority carriers that form the drift current, however, sense no barrier, so the drift current I_{drift} is not affected by the external battery. The nice current balance that existed at zero bias (see Fig. 42-11*d*) is thus upset and, as shown in Fig. 42-14*a*, a large net forward current I_F appears in the circuit.

Another effect of forward bias is to narrow the depletion zone, as a comparison of Figs. 42-11*b* and Fig. 42-14*a* shows. The depletion zone narrows because the reduced potential barrier associated with forward bias must be associated with a smaller space charge. Because the ions producing the space charge are fixed in their lattice sites, a reduction in their number can come about only through a reduction in the width of the depletion zone.

Because the depletion zone normally contains very few charge carriers, it is normally a region of high resistivity. But when its width is substantially reduced by a forward bias, its resistance is also reduced substantially, as is consistent with the large forward current.

Figure 42-14*b* shows the **back-bias** connection, in which the negative terminal of the battery is connected at the *p*-type end of the *p-n* junction. Now the applied emf *increases* the contact potential difference, the diffusion

FIGURE 42-13 A *p-n* junction connected as a junction rectifier. The action of the circuit in (*b*) is to pass the positive half of the input wave form (*a*) but to suppress the negative half. The average potential of the input wave form is zero; that of the output wave form (*c*) has a positive value V_{av}.

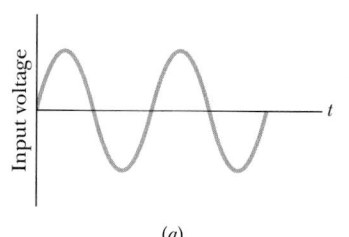

(*a*)

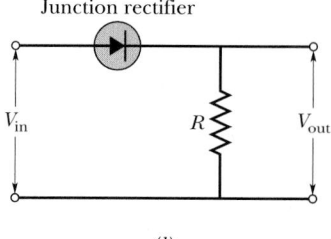

Junction rectifier

(*b*)

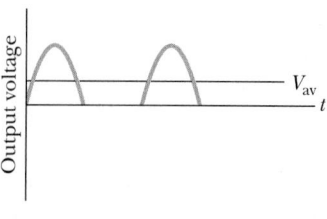

(*c*)

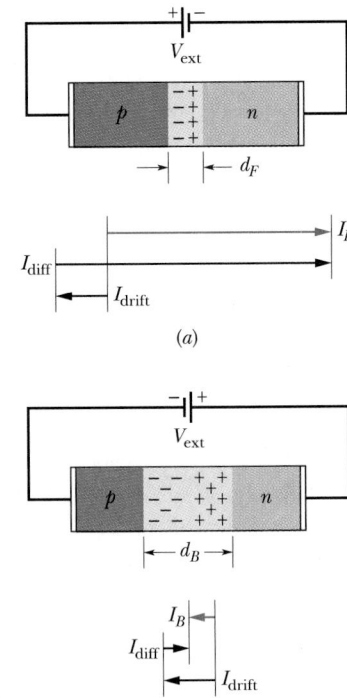

FIGURE 42-14 (*a*) The forward-bias connection of a *p-n* junction, showing the narrowed depletion zone and the large forward current I_F. (*b*) The back-bias connection, showing the widened depletion zone and the small back current I_B.

current *decreases* substantially while the drift current remains unchanged, and a relatively *small* back current I_B results. The depletion zone *widens*, its *high* resistance being consistent with the *small* back current I_B.

42-10 THE LIGHT-EMITTING DIODE (LED)

Nowadays, we can hardly avoid the brightly colored ''electronic'' numbers that glow at us from cash registers and gasoline pumps, microwave ovens and alarm clocks. And we cannot seem to do without the invisible infrared beams that control elevator doors and operate television sets via remote control. In nearly all cases this light is emitted from a *p-n* junction operating as a **light-emitting diode (LED)**. How can a *p-n* junction generate light?

Consider first a simple semiconductor. When an electron from the bottom of the conduction band falls into a hole at the top of the valence band, an energy E_g equal to the gap width is released. In silicon, germanium, and many other semiconductors, this energy is largely transformed into thermal energy of the vibrating lattice, and as a result, no light is emitted.

In some semiconductors, however, including gallium arsenide, the energy can be emitted as a photon of energy

hf and at wavelength

$$\lambda = \frac{c}{f} = \frac{c}{E_g/h} = \frac{hc}{E_g}. \qquad (42\text{-}9)$$

To emit enough light to be useful as an LED, the material must have a suitably large number of electron–hole transitions. This condition is *not* satisfied by a pure semiconductor because, at room temperature, there are simply not enough electron–hole pairs. As Fig. 42-10 suggests, doping will not help. In doped *n*-type material the number of conduction electrons is greatly increased, but there are not enough holes for them to combine with; in doped *p*-type material there are plenty of holes but not enough electrons to combine with them. Thus neither a pure semiconductor nor a doped semiconductor can provide enough electron–hole transitions to serve as a practical LED.

What we need is a semiconductor material with a very large number of electrons in the conduction band *and* a correspondingly large number of holes in the valence band. A device with this property can be fabricated by placing a strong forward bias on a heavily doped *p-n* junction, as in Fig. 42-15. In such an arrangement the current I through the device serves to inject electrons into the *n*-type material and to inject holes into the *p*-type material. If the doping is heavy enough and the current is great enough, the depletion zone can become very narrow, perhaps only a few micrometers wide. The result is a great number density of electrons in the *n*-type material facing a correspondingly great number density of holes in the *p*-type material, across the narrow depletion zone. Many electron–hole combinations occur, causing light to be emitted from that zone. Figure 42-16 shows the construction of an actual LED.

Commercial LEDs designed for the visible region are commonly based on gallium, suitably doped with arsenic

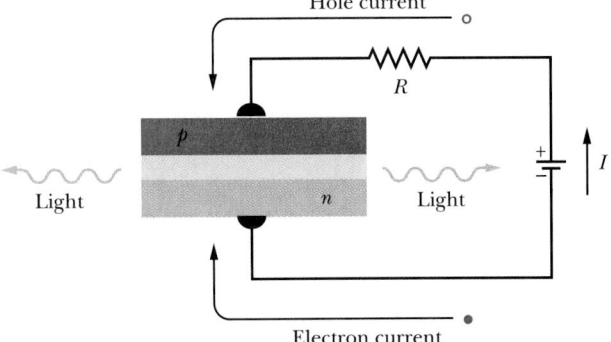

FIGURE 42-15 A forward-biased *p-n* junction, showing electrons being injected into the *n*-type material and holes into the *p*-type material. (Holes move in the conventional direction of the current I, equivalent to electrons moving in the opposite direction.) Light is emitted from the narrow depletion zone each time an electron and a hole combine across that zone.

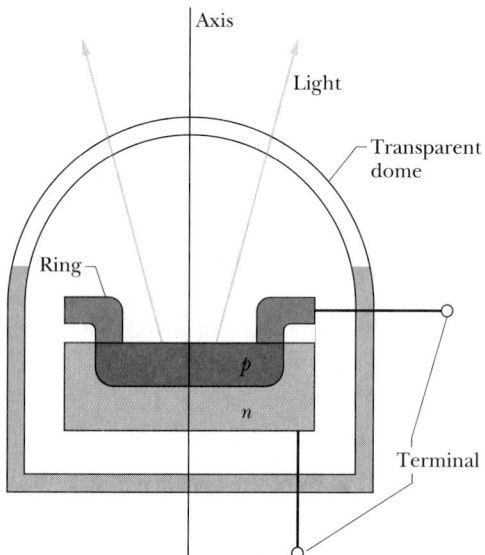

FIGURE 42-16   Cross section of an LED (the device has rotational symmetry about the central axis). The p-type material, which is thin enough to transmit light, is in the form of a circular disk. A connection is made to the p-type material through a circular metal ring that touches the disk at its periphery. The depletion zone between the n-type material and the p-type material is not shown.

and phosphorus atoms. An arrangement in which 60% of the nongallium sites are occupied by arsenic ions and 40% by phosphorus ions results in a gap width E_g of about 1.8 eV, corresponding to red light. Other doping and transition level arrangements make it possible to construct LEDs that emit light in essentially any desired region of the visible and near-visible spectra.

The Photo-Diode

Passing a current through a suitably arranged p-n junction can generate light. The reverse is also true. That is, shining light on a suitably arranged p-n junction can produce a current in a circuit that includes the junction. This is the basis for the **photo-diode**.

When you click your remote control, an LED in the device sends out a coded sequence of pulses of infrared light. The receiving device in your television set is an elaboration of the simple (two-terminal) photo-diode that not only detects the infrared signals but also amplifies them and transforms them into electrical signals that change the channel or adjust the volume, among other tasks.

The Junction Laser

In the arrangement of Fig. 42-15 there are many electrons in the conduction band of the n-type material and many holes in the valence band of the p-type material. Thus there is a **population inversion** for the electrons; that is, there are more electrons in higher energy levels than in lower energy levels. As we discussed in Section 41-10, this is normally a necessary—but not a sufficient—condition for laser action.

When a single electron moves from the conduction band to the valence band, it can release its energy as a photon. This photon can stimulate a second electron to fall into the valence band, producing a second photon by stimulated emission. In this way, if the current through the junction is great enough, a chain reaction of stimulated emission events can occur and laser light can be generated. To bring this about, opposite faces of the p-n junction crystal must be flat and parallel, so that light can be reflected back and forth within the crystal. (Recall that in the helium–neon laser of Fig. 41-21, a pair of mirrors served this purpose.) Thus a p-n junction can act as a **junction laser**, its light output being highly coherent and much more sharply defined in wavelength than light from an LED.

Junction lasers are built into compact disk (CD) players, where, by detecting reflections from the rotating disk, they are used to translate microscopic pits in the disk into sound. They are also much used in optical communication systems based on optical fibers. Figure 42-17 suggests their tiny scale. They are usually designed to operate in the infrared region of the electromagnetic spectrum because optical fibers have two "windows" in that region (at $\lambda = 1.31$ and $1.55 \ \mu$m) for which the absorption per unit length of the fiber is a minimum.

FIGURE 42-17   A semiconducting laser developed at the AT&T Bell Laboratories. The cube at the right is a grain of salt.

SAMPLE PROBLEM 42-5

An LED is constructed from a p-n junction based on a certain Ga-As-P semiconducting material, whose energy gap is 1.9 eV. What is the wavelength of the emitted light?

SOLUTION: If we assume that the transitions are from the bottom of the conduction band to the top of the valence band, Eq. 42-9 holds. From this equation

$$\lambda = \frac{hc}{E_g} = \frac{(6.63 \times 10^{-34} \text{ J} \cdot \text{s})(3.00 \times 10^8 \text{ m/s})}{(1.9 \text{ eV})(1.60 \times 10^{-19} \text{ J/eV})}$$

$$= 6.5 \times 10^{-7} \text{ m} = 650 \text{ nm}. \qquad \text{(Answer)}$$

Light of this wavelength is red.

CHECKPOINT 4: In Sample Problem 42-5 we calculated the wavelength of the light emitted from a certain LED to be 650 nm. Is this (a) the only possible wavelength to be emitted, (b) the maximum emitted wavelength, (c) the minimum emitted wavelength, or (d) the average emitted wavelength?

42-11 THE TRANSISTOR

A **transistor** is a three-terminal semiconducting device that can be used to amplify input signals. Figure 42-18 shows a generalized **f**ield-**e**ffect **t**ransistor (FET); in it, the flow of electrons from terminal S (the **source**) to terminal D (the **drain**) can be controlled by an electric field (hence **field effect**) set up within the device by a suitable electric potential applied to terminal G (the **gate**). Transistors are available in many types; we shall discuss only a particular FET called a **MOSFET**, or **m**etal-**o**xide-**s**emiconductor-**f**ield-**e**ffect **t**ransistor. The MOSFET has been described as the workhorse of the modern electronics industry.

For many applications the MOSFET is operated in only two states: with the drain-to-source current I_{DS} ON ("gate open") or with it OFF ("gate closed"). The first of these can represent a "1" and the other a "0" in the binary arithmetic on which digital logic is based, and therefore MOSFETs can be used in digital logic circuits. Switching between the ON and OFF states can occur at high speed, so that binary logic data can be moved through MOSFET-based circuits very rapidly. As of 1996, MOSFETs about 500 nm in length—about the same as the wavelength of yellow light—were routinely being fabricated for use in electronic devices of all kinds.

Figure 42-19 shows the basic structure of a MOSFET. A single crystal of silicon or other semiconductor is lightly doped to form p-type material. Embedded in this substrate, by heavily "overdoping" with n-type dopants, are two "islands" of n-type material, forming the drain D and the source S. The drain and source are connected by a thin channel of n-type material, called the **n channel**. A thin insulating layer of silicon dioxide (hence the "O" in MOSFET) is deposited on the crystal and penetrated by two metallic terminals (hence the "M") at D and S, so that electrical contact can be made with the drain and the source. A thin metallic layer—the gate G—is deposited facing the n channel. Note that the gate makes no electrical contact with the transistor proper, being separated from it by the insulating oxide layer.

Consider first that the source and p-type substrate are grounded (at zero potential) and the gate is "floating"; that is, the gate is not connected to an external source of emf. Let a potential V_{DS} be applied between the drain and the source, such that the drain is positive. Electrons will then flow through the n channel from source to drain, and the conventional current I_{DS}, as shown in Fig. 42-19, will be from drain to source.

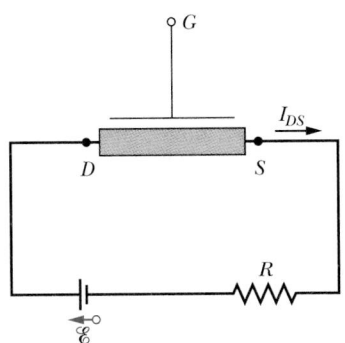

FIGURE 42-18 A representation of a generalized field effect transistor, in which electrons flow through the device from the source terminal S to the drain terminal D. (The conventional current I_{DS} is in the opposite direction.) The magnitude of I_{DS} is controlled by the electric field set up within the body of the device by a potential applied to G, the gate terminal.

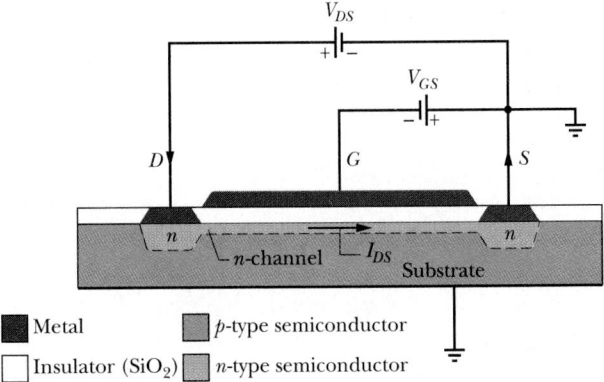

FIGURE 42-19 Representation of a particular type of field-effect transistor known as a MOSFET. The magnitude of the drain–source current through the n channel is controlled by the potential difference V_{GS} applied between the source S and the gate G. A depletion zone that exists between the n-type material and the p-type material is not shown.

Now let a potential V_{GS} be applied to the gate, making it negative with respect to the source. The negative gate sets up within the device an electric field (hence the "field effect") that tends to repel electrons from the *n* channel into the substrate. This electron movement widens the (naturally occurring) depletion zone between the *n* channel and the substrate, at the expense of the *n* channel. The reduced width of the *n* channel, coupled with a reduction in the number of charge carriers in that channel, increases the resistance of that channel and thus decreases the current I_{DS}. With the proper value of V_{GS}, this current can be shut off completely; hence, by controlling V_{GS}, the MOSFET can be switched between its ON and OFF modes.

Charge carriers do not flow through the *substrate* because the substrate (1) is lightly doped, (2) is not a good conductor, and (3) is separated from the *n*-channel and the two *n*-type islands by an insulating depletion zone, not specifically shown in Fig. 42-19. Such a depletion zone always exists at a boundary between *n*-type material and *p*-type material, as Fig. 42-11*b* shows.

Integrated Circuits

Computers and other electronic devices employ thousands (if not millions) of transistors and other electronic components such as capacitors and resistors. These are not assembled as separate units but are crafted into a single semiconducting **chip**, forming an **integrated circuit**.

Figure 42-20 shows a Power PC 620 microprocessor chip, manufactured by Motorola. It incorporates almost 7

FIGURE 42-21 Enlarged photograph of the layout of the chip shown in Fig. 42-20.

million transistors, along with many other electronic components. Figure 42-21 shows a greatly enlarged view of part of the layout of that chip, the different colors identifying different layers of the chip.

At Intel's Rio Rancho plant, chips are fabricated in a 140 step process on 8 inch silicon wafers, each wafer holding about 300 chips. The individual electronic chip components are so small that the tiniest speck of dust can ruin a chip. Precautions are taken to maintain a dust-free atmosphere in the plant's clean rooms, which are thousands of times more pristine than a hospital operating room. That is the reason for the workers' protective clothing, shown in the photograph that opens this chapter. As part of the cleaning process, highly filtered air circulates through the perforated floor at about 100 ft/min. There are also air showers and wipedown stations for removing cosmetics from employees.

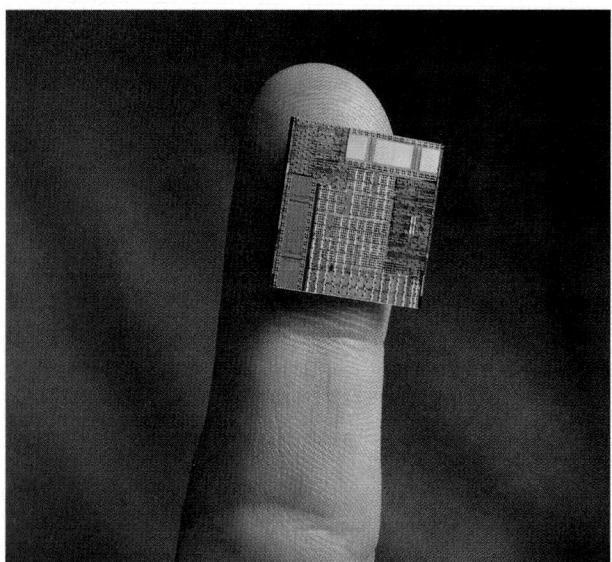

FIGURE 42-20 An integrated circuit for the Motorola Power PC 620 chip, for use mainly in computer workstations and file servers. It will be encapsulated in a ceramic coating for installation and use.

$\mathbf{C}$HECKPOINT **5:** In the MOSFET of Fig. 42-19 the gate potential V_{GS} is increased in magnitude until the *n* channel is blocked off so that I_{DS} falls to zero. Does the blocking (a) start near the source end of the *n* channel and progress along its length, (b) start near the drain end and progress along the channel, or (c) occur simultaneously at all points along the channel?

REVIEW & SUMMARY

Conductors, Semiconductors, and Insulators

Three electrical properties that can be used to distinguish among crystalline solids are the **resistivity** ρ, the **temperature coefficient of resistivity** α, and the **number density of charge carriers** n. Solids can be broadly divided into **conductors** (with small ρ) and **insulators** (with large ρ). Conductors can be further divided into **metals** (with small ρ, positive α, large n) and **semiconductors** (with larger ρ, negative α, and smaller n).

Energy Levels and Gaps in a Crystalline Solid

An isolated atom can exist in only a discrete set of energy levels. As atoms come together to form a solid, the levels of the individual atoms merge to form the discrete energy **bands** of the solid. These energy bands are separated by energy **gaps**, each of which corresponds to a range of energies that no electron may possess.

Any energy band is made up of an enormous number of very closely spaced levels. The Pauli exclusion principle asserts that only one electron may occupy each of these levels.

Insulators

In an insulator, the highest band containing electrons is completely filled and is separated from the vacant band above it by an energy gap so large that electrons can essentially never become thermally agitated enough to jump across the gap.

Metals

In a **metal**, the highest band that contains any electrons is only partially filled. The energy of the highest filled level at a temperature of 0 K is called the **Fermi energy** E_F for the metal; for copper, $E_F = 7.0$ eV.

The **density of states** function $N(E)$ is the number of available energy levels per unit volume of the sample and per unit energy interval and is given by

$$N(E) = \frac{8\sqrt{2}\,\pi m^{3/2}}{h^3} E^{1/2} \quad \text{(density of states).} \quad (42\text{-}2)$$

The **occupancy probability** $P(E)$ (the probability that a given available state will be occupied by an electron) is given by

$$P(E) = \frac{1}{e^{(E - E_F)/kT} + 1} \quad \begin{array}{l}\text{(occupancy}\\\text{probability).}\end{array} \quad (42\text{-}3)$$

The **density of occupied states** $N_o(E)$ is given by the product of the two quantities above:

$$N_o(E) = N(E)\, P(E) \quad \begin{array}{l}\text{(density of}\\\text{occupied states).}\end{array} \quad (42\text{-}4)$$

The Fermi energy for a metal can be found by integrating $N_o(E)$ for $T = 0$ from $E = 0$ to $E = E_F$. The result is

$$E_F = \left(\frac{3}{16\sqrt{2}\,\pi}\right)^{2/3} \frac{h^2}{m}\, n^{2/3} = \frac{0.121h^2}{m}\, n^{2/3}. \quad (42\text{-}6)$$

Semiconductors

The band structure of a **semiconductor** is like that of an insulator except that the gap width E_g is much smaller in the semiconductor. For silicon (a semiconductor) at room temperature, thermal agitation raises a few electrons to the **conduction band**, leaving an equal number of **holes** in the **valence band**. Both electrons and holes serve as charge carriers.

The number of electrons in the conduction band of silicon can be increased greatly by doping with small amounts of phosphorus, thus forming **n-type material**. The number of holes in the valence band can be greatly increased by doping with aluminum, thus forming **p-type material**.

The p-n Junction

A **p-n junction** is a single semiconducting crystal with one end doped to form p-type material and the other end doped to form n-type material, the two types meeting at a **junction plane**. At thermal equilibrium, the following occurs at that plane:

The **majority carriers** (electrons on the n side and holes on the p side) diffuse across the junction plane, producing a **diffusion current** I_{diff}.

The **minority carriers** (holes on the n side and electrons on the p side) are swept across the junction plane, forming a **drift current** I_{drift}. These two currents are equal in magnitude, so the net current is zero.

A **depletion zone**, consisting largely of space-charged donor and acceptor ions, forms across the junction plane.

A **contact potential difference** of height V_0 develops across the depletion zone.

Applications of the p-n Junction

When a potential difference is applied across a p-n junction, the device conducts electricity more readily for one polarity of the applied potential difference than for the other. Thus a p-n junction can serve as a **junction rectifier**.

When a p-n junction is forward biased, it can emit light, hence can serve as a **light-emitting diode** (LED). The wavelength of the emitted light is given by

$$\lambda = \frac{c}{f} = \frac{hc}{E_g}. \quad (42\text{-}9)$$

A strongly forward-biased p-n junction with parallel end faces can operate as a **junction laser**, emitting light of a sharply defined wavelength.

MOSFETS

In a MOSFET, a type of three-terminal transistor, a potential applied to the **gate** terminal G controls the internal flow of electrons from the **source** terminal S to the **drain** terminal D. Commonly, a MOSFET is operated only in its ON (conducting) or its OFF (not conducting) conditions. Installed by the thousands and millions on silicon wafers (**chips**) to form **integrated circuits**, MOSFETs form the basis for computer hardware.

QUESTIONS

1. Figure 42-1*a* shows 14 atoms that represent the unit cell of copper. Since, however, each of these atoms is shared with one or more adjoining unit cells, only a fraction of each atom belongs to the unit cell shown. What is the number of atoms per unit cell for copper? (To answer, count up the fractional atoms belonging to a single unit cell.)

2. Figure 42-1*b* shows 18 atoms that represent the unit cell of silicon. Fourteen of these atoms, however, are shared with one or more adjoining unit cells. What is the number of atoms per unit cell for silicon? (See Question 1.)

3. Does the interval between adjacent energy levels in the highest occupied band of a metal depend on (a) the material of which the sample is made, (b) the size of the sample, (c) the position of the level in the band, (d) the temperature of the sample, or (e) the Fermi energy of the metal?

4. Compare the drift speed v_d of the conduction electrons in a current-carrying copper wire with the Fermi speed v_F for copper? Is v_d (a) about equal to v_F, (b) much greater than v_F, or (c) much less than v_F?

5. In a silicon lattice, where should you look if you want to find (a) a conduction electron, (b) a valence electron, and (c) an electron associated with the $2p$ subshell of the isolated silicon atom?

6. Which of the following statements, if any, are true? (a) At low enough temperatures, silicon behaves like an insulator. (b) At high enough temperatures, silicon becomes a good conductor. (c) At high enough temperatures, silicon behaves like a metal.

7. The energy gaps E_g for the semiconductors silicon and germanium are, respectively, 1.12 and 0.67 eV. Which of the following statements, if any, are true? (a) Both substances have the same number density of charge carriers at room temperature. (b) At room temperature, germanium has a greater number density of charge carriers than silicon. (c) Both substances have a greater number density of conduction electrons than holes. (d) For each substance, the number density of electrons equals that of holes.

8. An isolated atom of germanium has 32 electrons, arranged in subshells according to this scheme:

$$1s^2\ 2s^2\ 2p^6\ 3s^2\ 3p^6\ 3d^{10}\ 4s^2\ 4p^2.$$

This element has the same crystal structure as silicon and, like silicon, is a semiconductor. Which of these electrons form the valence band of crystalline germanium?

9. Germanium ($Z = 32$) has the same crystal structure and the same bonding pattern as silicon. Is the net charge on a germanium ion within its lattice $+e$, $+2e$, $+4e$, $+28e$, or $+32e$?

10. (a) Of the elements arsenic, indium, tin, gallium, antimony, and boron, which would produce *n*-type material if used as a dopant in silicon? (b) Which would produce *p*-type material? (c) Which would be unsuitable as a dopant? (*Hint*: Consult the periodic table in Appendix G.)

11. A sample of silicon is doped with phosphorus. Which of the following statements, if any, are true? (a) The number of holes in the sample is slightly increased. (b) The resistivity is increased. (c) The sample becomes positively charged. (d) The sample becomes negatively charged. (e) The gap between the valence band and the conduction band decreases slightly.

12. To fabricate an *n*-type semiconductor, would you use (a) silicon doped with arsenic or (b) germanium doped with indium? (*Hint*: Consult the periodic table.)

13. In the biased *p-n* junctions shown in Fig. 42-14, there is an electric field **E** in the two depletion zones, associated with the potential difference that exists across the zone in each case. (a) Does **E** point from left to right or from right to left? (b) Is its magnitude greater for forward bias or for back bias?

14. A certain isolated *p-n* junction develops a contact potential difference V_0 across its depletion zone of 0.78 V. A voltmeter is connected across the terminals of the junction, the positive terminal of the meter being connected to the *p* side of the junction. Will the meter read (a) +0.78 V, (b) −0.78 V, (c) zero, or (d) something else? (*Hint*: Contact potentials appear at the connections between the *p-n* junction and the voltmeter leads.)

15. Which of the following obey Ohm's law: (a) a bar of pure silicon, (b) a bar of *n*-type silicon, (c) a bar of *p*-type silicon, (d) a *p-n* junction?

16. An LED based on a gallium–arsenic–phosphorus semiconducting crystal emits red light. If you look at a white surface through such a crystal, will you see (a) red, (b) blue, (c) nothing, because the crystal is opaque, or (d) white?

EXERCISES & PROBLEMS

SECTION 42-5 Metals

1E. Copper is a monovalent metal with a molar mass of 63.5 g/mol and a density of 8.96 g/cm³. Show that the number density of conduction electrons in copper is 8.43×10^{28} m⁻³.

2E. At what pressure, in atmospheres, would an ideal gas have a number density of molecules equal to the number density of the conduction electrons in copper, with both gas and copper at temperature $T = 300$ K?

3E. Verify the numerical factor 0.121 in Eq. 42-6.

4E. What is the number density of conduction electrons in gold, which is a monovalent metal? Use the molar mass and density provided in Appendix F.

5E. Calculate $d\rho/dT$ at room temperature for (a) copper and (b) silicon, using data from Table 42-1.

6E. Use Eq. 42-6 to verify that the Fermi energy of copper is 7.0 eV.

7E. The Fermi energy of copper is 7.0 eV. Verify that the corresponding Fermi speed is 1600 km/s.

8E. (a) Show that Eq. 42-2 can be written as $N(E) = CE^{1/2}$, where $C = 6.78 \times 10^{27}$ m^{-3}eV$^{-3/2}$. (b) Calculate $N(E)$ for $E = 5.00$ eV.

9E. What is the probability that a state 0.062 eV above the Fermi energy will be occupied at (a) $T = 0$ K and (b) $T = 320$ K?

10E. Calculate the density of states $N(E)$ for a metal at energy $E = 8.0$ eV and show that your result is consistent with the curve of Fig. 42-5.

11E. Show that Eq. 42-6 can be written as $E_F = An^{2/3}$, where the constant A has the value 3.65×10^{-19} m$^2 \cdot$eV.

12E. The density of gold is 19.3 g/cm^3, and each gold atom contributes one electron to the conduction band. Use the result of Exercise 4 to calculate the Fermi energy of gold.

13E. A state 63 meV above the Fermi level has a probability of occupancy of 0.090. What is the probability of occupancy for a state 63 meV *below* the Fermi level?

14P. The Fermi energy for copper is 7.0 eV. For copper at 1000 K, (a) find the energy of the energy level whose probability of being occupied by an electron is 0.90. For this energy, evaluate (b) the density of states and (c) the density of occupied states.

15P. The density and molar mass of sodium (a metal) are 971 kg/m^3 and 23.0 g/mol, respectively; the radius of the Na$^+$ ion is 98 pm. (a) What percent of metallic sodium is available to its conduction electrons? (b) Carry out the same calculation for copper, which has density, molar mass, and ionic radius of 8960 kg/m^3, 63.5 g/mol, and 135 pm, respectively. (c) For which of these metals do you think the conduction electrons behave more like a free-electron gas?

16P. Show that $P(E)$, the occupancy probability in Eq. 42-3, is symmetrical about the value of the Fermi energy. That is, show that

$$P(E_F + \Delta E) + P(E_F - \Delta E) = 1.$$

17P. In Eq. 42-3 let $E - E_F = \Delta E = 1.00$ eV. (a) At what temperature does the result of using this quantum equation differ by 1.0% from the result of using the classical Boltzmann equation $P(E) = e^{-\Delta E/kT}$? (b) At what temperature do these results differ by 10%?

18P. What is the probability that an electron will jump from the valence band to the conduction band in a diamond whose mass is equal to the mass of Earth? Use the result of Sample Problem 42-1 and the molar mass of carbon in Appendix F; assume that in diamond there is one valence electron per carbon atom.

19P. Calculate the number density for (a) molecules of oxygen gas at 0°C and 1.0 atm pressure and (b) conduction electrons in copper. (c) What is the ratio of the latter to the former? (d) What is the average distance between particles in each case? Assume this distance is the edge length of a cube whose volume is equal to the available volume per particle.

20P. Calculate $N_o(E)$, the density of occupied states, for copper at $T = 1000$ K for the energies $E = 4.00$, 6.75, 7.00, 7.25, and 9.00 eV. Compare your results with the graph of Fig. 42-7b. The Fermi energy for copper is 7.00 eV.

21P. The Fermi energy for silver is 5.5 eV. (a) At $T = 0$°C, what are the probabilities that states with the following energies are occupied: 4.4, 5.4, 5.5, 5.6, and 6.4 eV? (b) At what temperature is the probability 0.16 that a state with energy $E = 5.6$ eV is occupied?

22P. The Fermi energy of aluminum is 11.6 eV; its density and molar mass are 2.70 g/cm^3 and 27.0 g/mol, respectively. From these data, determine the number of free electrons per atom.

23P. Show that the probability $P_h(E)$ that a hole exists at energy E (that is, that an energy level at energy E is not occupied) is

$$P_h(E) = \frac{1}{e^{-\Delta E/kT} + 1},$$

where $\Delta E = E - E_F$.

24P. Zinc is a bivalent metal. Calculate (a) the number density of conduction electrons, (b) the Fermi energy, (c) the Fermi speed, and (d) the de Broglie wavelength corresponding to this electron speed. See Appendix F for the needed data on zinc.

25P. Silver is a monovalent metal. Calculate (a) the number density of conduction electrons, (b) the Fermi energy, (c) the Fermi speed, and (d) the de Broglie wavelength corresponding to this electron speed. See Appendix F for the needed data on silver.

26P. At $T = 300$ K, how close to the Fermi energy will we find a state whose probability of occupation by a conduction electron is 0.10?

27P. (a) Show that the density of states at the Fermi energy is given by

$$N(E_F) = \frac{(4)(3^{1/3})(\pi^{2/3})mn^{1/3}}{h^2}$$
$$= (4.11 \times 10^{18} \text{ m}^{-2}\text{eV}^{-1})n^{1/3},$$

in which n is the number density of conduction electrons. (b) Calculate $N(E_F)$ for copper using the result of Exercise 1, and verify your calculation with the curve of Fig. 42-5, recalling that $E_F = 7.0$ eV for copper.

28P. (a) Show that the slope dP/dE of Eq. 42-3 at $E = E_F$ is $-1/4kT$. (b) Show that the tangent line to the curve of Fig. 42-6b at $E = E_F$ intercepts the horizontal axis at $E = E_F + 2kT$.

29P. Show that, at $T = 0$ K, the average energy E_{av} of the conduction electrons in a metal is equal to $\frac{3}{5}E_F$. (*Hint*: By definition of average, $E_{av} = (1/n)\int E \, N_o(E) \, dE$, where n is the number density of charge carriers.)

30P. Use the result of Problem 29 to calculate the total translational kinetic energy of the conduction electrons in 1.0 cm^3 of copper at $T = 0$ K.

31P. (a) Using the result of Problem 29, estimate how much energy would be released by the conduction electrons in a penny (assumed all copper and of mass 3.1 g) if we could suddenly turn off the Pauli exclusion principle. (b) For how long would this amount of energy light a 100 W lamp? (*Note*: There is no way to turn off the Pauli principle!)

32P. At 1000 K, the fraction of the conduction electrons in a metal that have energies greater than the Fermi energy is equal to the area under the curve of Fig. 42-7b beyond E_F divided by the area under the entire curve. It is difficult to find these areas by direct integration. However, an approximation to this fraction at any temperature T is

$$frac = \frac{3kT}{2E_F}.$$

Note that *frac* = 0 for $T = 0$ K, just as we would expect. What is this fraction for copper at (a) 300 K and at (b) 1000 K? For copper, $E_F = 7.0$ eV. (c) If you can, check your answers by numerical integration using Eq. 42-4.

33P. At what temperature do 1.3% of the conduction electrons in lithium (a metal) have energies greater than the Fermi energy E_F, which is 4.7 eV? (See Problem 32.)

34P. Silver melts at 961°C. At the melting point, what fraction of the conduction electrons are in states with energies greater than the Fermi energy of 5.5 eV? (See Problem 32.)

SECTION 42-6 Semiconductors

35P. (a) Find the angle θ between adjacent nearest-neighbor bonds in the silicon lattice. Recall that each silicon atom is bonded to four of its nearest neighbors. The four neighbors form a regular tetrahedron: a three-sided pyramid whose sides and base are equilateral triangles. (b) Find the bond length, given that the atoms at the corners of the tetrahedron are 388 pm apart.

36P. The compound gallium arsenide is a commonly used semiconductor, having an energy gap E_g of 1.43 eV. Its crystal structure is like that of silicon, except that half the silicon atoms are replaced by gallium atoms and half by arsenic atoms. Draw a flattened-out sketch of the gallium arsenide lattice, following the pattern of Fig. 42-9a. (a) What are the net charges of the gallium and the arsenic ion cores? (b) How many electrons per bond are there? (*Hint*: Consult the periodic table in Appendix G.)

37P. (a) What is the maximum wavelength of the light that will excite an electron in the valence band of diamond to the conduction band? The energy gap is 5.5 eV. (b) In what part of the electromagnetic spectrum does this wavelength lie?

38P. The occupancy probability function (Eq. 42-3) can be applied to semiconductors as well as to metals. In semiconductors the Fermi energy is close to the midpoint of the gap between the valence band and the conduction band; see Problem 39. For germanium, the gap width is 0.67 eV. What is the probability that (a) a state at the bottom of the conduction band is occupied and (b) a state at the top of the valence band is not occupied. Assume that $T = 290$ K. (*Note*: Figure 42-4b shows that, in a metal, the Fermi

energy lies symmetrically between the population of conduction electrons and the population of holes. To match this scheme in a semiconductor, the Fermi energy must lie near the center of the gap. There need not be an available state at the location of the Fermi energy.)

39P. In a simplified model of an undoped semiconductor, the actual distribution of energy states may be replaced by one in which there are N_v states in the valence band, all these states having the same energy E_v, and N_c states in the conduction band, all these states having the same energy E_c. The number of electrons in the conduction band equals the number of holes in the valence band. (a) Show that this last condition implies that

$$\frac{N_c}{\exp(\Delta E_c/kT) + 1} = \frac{N_v}{\exp(\Delta E_v/kT) + 1},$$

in which

$$\Delta E_c = E_c - E_F \quad \text{and} \quad \Delta E_v = -(E_v - E_F).$$

(*Hint*: See Problem 23.) (b) If the Fermi level is in the gap between the two bands and is far from both bands compared with kT, then the exponentials dominate in the denominators. Under these conditions show that

$$E_F = \frac{(E_c + E_v)}{2} + \frac{kT \ln(N_v/N_c)}{2}$$

and that, if $N_v \approx N_c$, the Fermi level for the undoped semiconductor is close to the gap's center, as stated in Problem 38.

SECTION 42-7 Doped Semiconductors

40P. Pure silicon at room temperature has an electron number density in the conduction band of about 5×10^{15} m^{-3} and an equal density of holes in the valence band. Suppose that one of every 10^7 silicon atoms is replaced by a phosphorus atom. (a) Which type will the doped semiconductor be, n or p? (b) What charge carrier number density will the phosphorus add? (c) What is the ratio of the charge carrier number density (electrons in the conduction band and holes in the valence band) in the doped silicon to that in pure silicon?

41P. What mass of phosphorus is needed to dope 1.0 g of silicon to the extent described in Sample Problem 42-4?

42P. Doping changes the Fermi energy of a semiconductor. Consider silicon, with a gap of 1.11 eV between the top of the valence band and the bottom of the conduction band. At 300 K the Fermi level of the pure material is nearly at the midpoint of the gap. Suppose that silicon is doped with donor atoms, each of which has a state 0.15 eV below the bottom of the conduction band, and suppose further that doping raises the Fermi level to 0.11 eV below the bottom of that band (Fig. 42-22). (a) For both pure and doped silicon, calculate the probability that a state at the bottom of the conduction band is occupied. (b) Calculate the probability that a donor state in the doped material is occupied.

43P. A silicon sample is doped with atoms having a donor state 0.110 eV below the bottom of the conduction band. (The energy

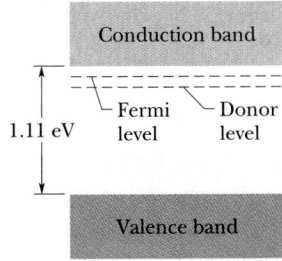

FIGURE 42-22 Problem 42.

gap in silicon is 1.11 eV.) (a) If each of these donor states is occupied with a probability of 5.00×10^{-5} at $T = 300$ K, where is the Fermi level with respect to the top of the valence band? (b) What then is the probability that a state at the bottom of the conduction band is occupied?

SECTION 42-9 The Junction Rectifier

44P. When a photon enters the depletion zone of a *p-n* junction, electron–hole pairs can be created as electrons absorb part of the photon's energy and are excited from the valence band to the conduction band. These junctions are thus often used as detectors for photons, especially in the x-ray and the gamma-ray regions of the electromagnetic spectrum. When a single 662 keV gamma-ray photon is totally absorbed by a semiconductor with an energy gap of 1.1 eV, what is the average number of electron–hole pairs created?

45P. For an ideal *p-n* junction rectifier, with a sharp boundary between its two semiconducting sides, the current I is related to the potential difference V across the rectifier by

$$I = I_0(e^{eV/kT} - 1),$$

where I_0, which depends on the materials but not on the current or the potential difference, is called the *reverse saturation current*.

V is positive if the rectifier is forward-biased and negative if it is back-biased. (a) Verify that this expression predicts the behavior of a junction rectifier by graphing I versus V over the range -0.12 V to $+0.12$ V. Take $T = 300$ K and $I_0 = 5.0$ nA. (b) For the same temperature, calculate the ratio of the current for a 0.50 V forward-bias to the current for a 0.50 V back-bias.

SECTION 42-10 The Light-Emitting Diode (LED)

46P. (a) In a particular crystal, the highest occupied band is full. The crystal is transparent to light of wavelengths longer than 295 nm but opaque at shorter wavelengths. Calculate, in electron-volts, the gap between the highest occupied band and the next higher (empty) band for this material.

47P. A potassium chloride crystal has an energy band gap of 7.6 eV above the topmost occupied band, which is full. Is this crystal opaque or transparent to light of wavelength 140 nm?

SECTION 42-11 The Transistor

48P. A Pentium computer chip, which is about the size of a postage stamp (1.0 in. × 0.875 in.), contains about 3.5 million transistors. If the transistors are square, what must be their *maximum* dimension? (*Note*: Devices other than transistors are also on the chip, and there must be room for the interconnections among the circuit elements. Transistors as small as 0.7 μm are commonly fabricated.)

49P. A silicon-based MOSFET has a square gate 0.50 μm on edge. The insulating silicon oxide layer that separates it from the *p*-type substrate is 0.20 μm thick and has a dielectric constant of 4.5. (a) What is the equivalent gate–substrate capacitance? (b) How many elementary charges e appear in the gate when there is a gate–source potential difference of 1.0 V?

Radioactive nuclei that are injected into a patient collect at certain sites within the patient's body, undergo radioactive decay, and emit gamma rays. These gamma rays are recorded by a detector, and a color-coded image of the patient's body is produced on a video monitor. In the images reproduced here (the left one is a front view of a patient and the right one is a back view), you can tell just where the radioactive nuclei have collected (spine, pelvis, and ribs) by the color-coding of brown and orange. What happens to radioactive nuclei when they undergo decay, and what exactly does "decay" mean?

43-1 DISCOVERING THE NUCLEUS

In the first years of the twentieth century not much was known about the structure of atoms beyond the fact that they contain electrons. The electron had been discovered (by J. J. Thomson) in 1897, and its mass was unknown in those early days. Thus it was not possible even to say how many negatively charged electrons a given atom contained. Atoms were electrically neutral so they must also contain some positive charge, but nobody knew what form this compensating positive charge took.

In 1911 Ernest Rutherford proposed that the positive charge of the atom is densely concentrated at the center of the atom, forming its **nucleus**, and that, furthermore, the nucleus is responsible for most of the mass of the atom. Rutherford's proposal was no mere conjecture but was based firmly on the results of an experiment suggested by him and carried out by his collaborators, Hans Geiger (of Geiger counter fame) and Ernest Marsden, a 20-year-old student who had not yet earned his bachelor's degree.

In Rutherford's day it was known that certain elements, called **radioactive**, transform themselves into other elements spontaneously, emitting particles in the process. One such element is radon, which emits alpha (α) particles with energies of about 5.5 MeV. We now know that these useful particles are the nuclei of the atoms of helium.

Rutherford's idea was to direct energetic alpha particles at a thin target foil and measure the extent to which they were deflected as they passed through the foil. Alpha particles, which are about 7300 times more massive than electrons, have a charge of $+2e$.

Figure 43-1 shows the experimental arrangement of Geiger and Marsden. Their alpha source was a thin-walled glass tube of radon gas. The experiment involves counting the number of alpha particles that are deflected through various scattering angles ϕ.

Figure 43-2 shows their results. Note especially that the vertical scale is logarithmic. We see that most of the particles are scattered through rather small angles but—

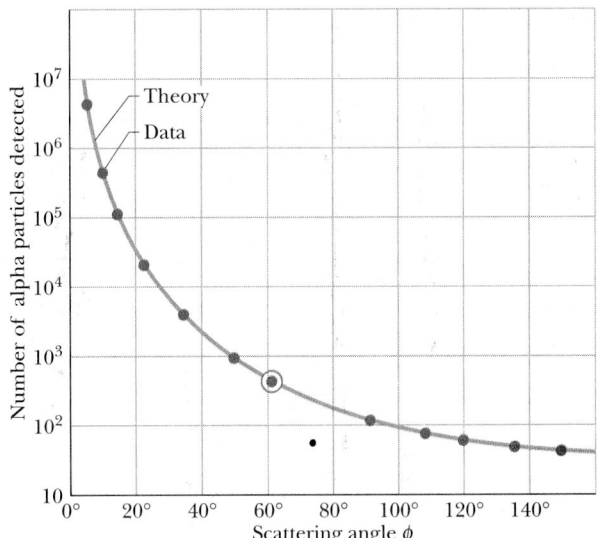

FIGURE 43-2 The dots are alpha-particle scattering data for a gold foil, obtained by Geiger and Marsden using the apparatus of Fig. 43-1. The solid curve is the theoretical prediction, based on the assumption that the atom has a small, massive, positively charged nucleus. Note that the vertical scale is logarithmic, covering six orders of magnitude. The data have been adjusted to fit the theoretical curve at the experimental point that is enclosed in a circle.

and this was the big surprise—a very small fraction of them are scattered through very large angles, approaching 180°. In Rutherford's words: "It was quite the most incredible event that ever happened to me in my life. It was almost as incredible as if you had fired a 15-inch shell at a piece of tissue paper and it came back and hit you."

Why was Rutherford so surprised? At the time of these experiments, most physicists believed in the so-called plum pudding model of the atom, which had been advanced by J. J. Thomson. In this view the positive charge of the atom was thought to be spread out through the entire volume of the atom. The electrons (the "plums") were thought to vibrate about fixed points within this sphere of positive charge (the "pudding").

The maximum deflecting force that could act on an alpha particle as it passed through such a large positive sphere of charge would be far too small to deflect the alpha particle by even as much as 1°. (The expected deflection has been compared to what you would observe if you fired a bullet through a sack of snowballs.) The electrons in the atom would also have very little effect on the massive, energetic alpha particle. They would, in fact, be themselves strongly deflected, much as a swarm of gnats would be brushed aside by a stone thrown through them.

Rutherford saw that, to deflect the alpha particle backward, there must be a large force; this force could be

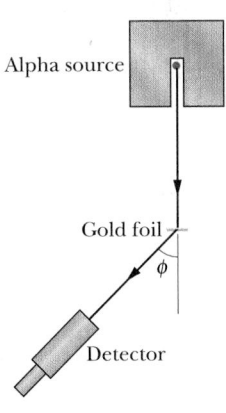

FIGURE 43-1 An arrangement (top view) used in Rutherford's laboratory in 1911–1913 to study the scattering of α particles by thin metal foils. The detector can be rotated to various values of the scattering angle ϕ. The alpha source was radon gas, a decay product of radium. With this simple "tabletop" apparatus, the nucleus was discovered.

provided if the positive charge, instead of being spread throughout the atom, were concentrated tightly at its center. Then the incoming alpha particle could get very close to the center of the positive charge without penetrating it; such a close encounter would result in a large deflecting force.

Figure 43-3 shows possible paths taken by typical alpha particles as they pass through the atoms of the target foil. As we see, most are either undeflected or only slightly deflected, but a few (those whose incoming paths pass, by chance, very close to a nucleus) are deflected through large angles. From an analysis of the data, Rutherford concluded that the radius of the nucleus must be smaller than the radius of an atom by a factor of about 10^4. In other words, the atom is mostly empty space!

SAMPLE PROBLEM 43-1

A 5.30 MeV alpha particle happens, by chance, to be headed directly toward the nucleus of an atom of gold ($Z = 79$). How close does the alpha particle get to the center of the gold nucleus before it comes momentarily to rest and reverses its course? Neglect the recoil of the relatively massive nucleus.

SOLUTION: Initially the total mechanical energy of these two interacting bodies is just equal to K_α (= 5.30 MeV), the initial kinetic energy of the alpha particle. At the moment the alpha particle comes to rest, the total energy is the electric potential energy of the two-body system. Because energy must be conserved, the magnitudes of these two energies must be equal; so, from Eq. 25-43

$$K_\alpha = \frac{1}{4\pi\epsilon_0} \frac{q_\alpha q_{Au}}{d},$$

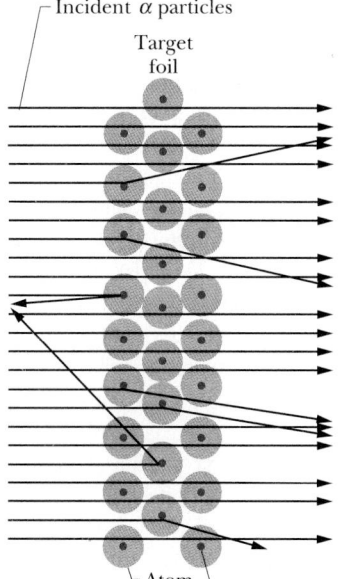

Incident α particles

Target foil

FIGURE 43-3 The angle through which an alpha particle is scattered depends on how close the particle's incident path lies to an atomic nucleus. Large deflections result only from very close encounters.

Atom Nucleus

in which q_α (= 2e) is the charge of the alpha particle, q_{Au} (= 79e) is the charge of the gold nucleus, and d is the distance between the centers of these two bodies.

Substituting for the charges and solving for d yield

$$d = \frac{(2e)(79e)}{4\pi\epsilon_0 K_\alpha}$$

$$= \frac{(2 \times 79)(1.60 \times 10^{-19}\ C)^2}{(4\pi)(8.85 \times 10^{-12}\ F/m)(5.30\ MeV)}$$

$$\times \frac{1\ MeV}{1.60 \times 10^{-13}\ J}$$

$$= 4.29 \times 10^{-14}\ m = 42.9\ fm. \qquad (Answer)$$

This is a small distance by atomic standards but not by nuclear standards. It is, in fact, considerably larger than the sum of the radii of the gold nucleus and the alpha particle. Thus the alpha particle reverses its course without ever actually "touching" the gold nucleus.

43-2 SOME NUCLEAR PROPERTIES

Table 43-1 shows some properties of a few atomic nuclei. When we are interested primarily in their properties as specific nuclear species (rather than as parts of atoms), we call these particles **nuclides**.

Some Nuclear Terminology

Nuclei are made up of protons and neutrons. The number of protons in a nucleus (called the **atomic number** or **proton number** of the nucleus) is represented by the symbol Z; the number of neutrons (the **neutron number**) is represented by the symbol N. The total number of neutrons and protons in a nucleus is called its **mass number** A, so

$$A = Z + N. \qquad (43-1)$$

Neutrons and protons, when considered collectively, are called **nucleons**.

We represent nuclides with symbols such as those displayed in the first column of Table 43-1. Consider ^{197}Au, for example. The superscript (197) is the mass number A. The chemical symbol tells us that this element is gold, whose atomic number is 79. From Eq. 43-1 we see that the neutron number of this nuclide is $197 - 79$, or 118.

Nuclides with the same atomic number Z but different neutron numbers N are called **isotopes** of each other. As it happens, the element gold has 32 isotopes, ranging from ^{173}Au to ^{204}Au. Only one of them (^{197}Au) is stable, the remaining 31 being radioactive. Such **radionuclides** undergo **decay** (or **disintegration**) by emitting a particle and thereby transforming to another type of nuclide.

TABLE 43-1 SOME PROPERTIES OF SELECTED NUCLIDES

NUCLIDE	Z	N	A	STABILITY[a]	MASS[b] (u)	SPIN[c]	BINDING ENERGY (MeV/nucleon)
^{1}H	1	0	1	99.985%	1.007 825	$\frac{1}{2}$	—
^{7}Li	3	4	7	92.5%	7.016 003	$\frac{3}{2}$	5.60
^{31}P	15	16	31	100%	30.973 762	$\frac{1}{2}$	8.48
^{84}Kr	36	48	84	57.0%	83.911 507	0	8.72
^{120}Sn	50	70	120	32.4%	119.902 199	0	8.51
^{157}Gd	64	93	157	15.7%	156.923 956	$\frac{3}{2}$	8.21
^{197}Au	79	18	197	100%	196.966 543	$\frac{3}{2}$	7.91
^{227}Ac	89	138	227	21.8 y	227.027 750	$\frac{3}{2}$	7.65
^{239}Pu	94	145	239	24,100 y	239.052 158	$\frac{1}{2}$	7.56

[a]For stable nuclides, the **isotopic abundance** is given; this is the fraction of atoms of this type found in a typical sample of the element. For radioactive nuclides, the half-life is given.
[b]Following standard practice, the reported mass is that of the neutral atom, not that of the bare nucleus.
[c]Spin angular momentum in units of $\hbar$.

Organizing the Nuclides

The neutral atoms of all isotopes of an element (all with the same Z) have the same number of electrons and the same chemical properties, and they fit into the same box in the chemist's periodic table of the elements. The *nuclear* properties of the various isotopes of a given element, however, are very different. Thus the periodic table is of limited use to the nuclear physicist, the nuclear chemist, or the nuclear engineer.

We organize the nuclides on a **nuclidic chart** like that in Fig. 43-4, in which a nuclide is represented by plotting its proton number against its neutron number. The stable nuclides in this figure are represented by the green, the radionuclides by the beige. As you can see, the radionuclides tend to lie on either side of—and at the upper end of—a well-defined band of stable nuclides. Note too that light stable nuclides tend to lie close to the line $N = Z$, which means that they have the same numbers of neutrons and protons. Heavier nuclides, however, tend to have many more neutrons than protons. For example, we saw that ^{197}Au has 118 neutrons and only 79 protons, a *neutron excess* of 39.

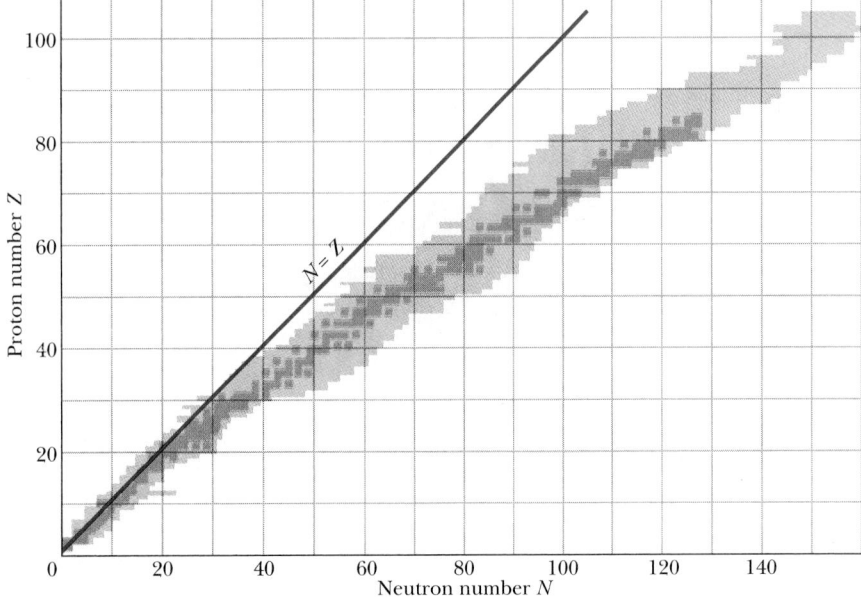

FIGURE 43-4 A plot of the known nuclides. The green shading identifies the band of stable nuclides, the beige shading the radionuclides. Low-mass, stable nuclides have essentially equal numbers of neutrons and protons, but more massive nuclides have an increasing excess of neutrons. The figure shows that there are no stable nuclides with $Z > 83$ (bismuth).

Nuclidic charts are available as wall charts, in which each small box on the chart is filled with data about the nuclide it represents. Figure 43-5 shows a section of such a chart, centered on ¹⁹⁷Au. Relative abundances are shown for stable nuclides, and half-lives (a measure of decay rate) are shown for radionuclides. The sloping line represents a line of **isobars**—nuclides of the same mass number, $A = 198$ in this case.

As of 1996, nuclides with atomic numbers as high as $Z = 112$ ($A = 278$) had been found. Such high-Z nuclides are very unstable and are identified by the products of their radioactive decay. These nuclides are created in accelerator laboratories, one nucleus at a time; in one case, the production rate was so small that a gong was arranged to sound every time one of these nuclei was produced!

CHECKPOINT 1: Based on Fig. 43-4, which of the following nuclides do you conclude are not likely to be detected: ⁵²Fe ($Z = 26$), ⁹⁰As ($Z = 33$), ¹⁵⁸Nd ($Z = 60$), ¹⁷⁵Lu ($Z = 71$), ²⁰⁸Pb ($Z = 82$)?

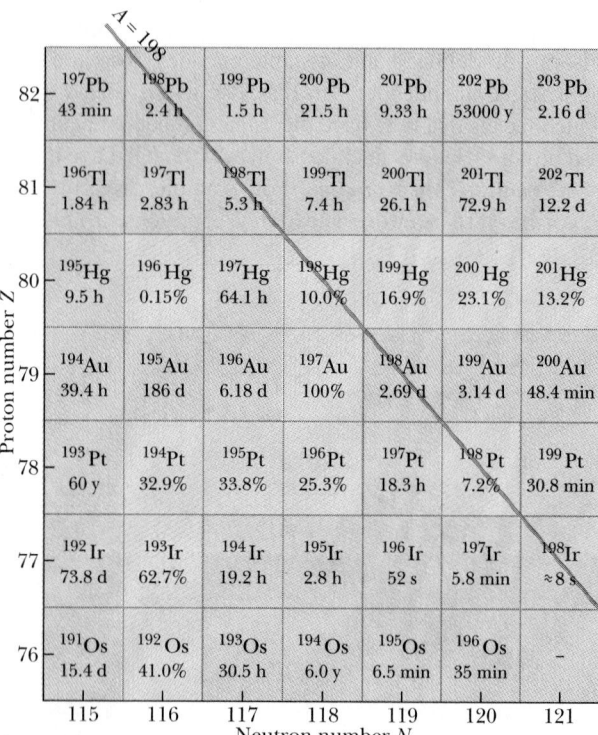

FIGURE 43-5 An enlarged and detailed section of the nuclidic chart of Fig. 43-4, centered on ¹⁹⁷Au. Green squares represent stable nuclides, for which relative isotopic abundances are given. Beige squares represent radionuclides, for which half-lives are given. Isobaric lines of constant mass number A slope as shown by the example line for $A = 198$.

Nuclear Radii

A convenient unit for measuring distances on the scale of nuclei is the *femtometer* ($= 10^{-15}$ m). This unit is often called the *fermi*; the two names share the same abbreviation. Thus

$$1 \text{ femtometer} = 1 \text{ fermi} = 1 \text{ fm} = 10^{-15} \text{ m.} \quad (43\text{-}2)$$

We can learn about the size and structure of nuclei by bombarding them with a beam of high-energy electrons and observing how the nuclei deflect the incident electrons. The electrons must be energetic enough (at least 200 MeV) to have de Broglie wavelengths that are smaller than the nuclear structures they are to probe.

The nucleus, like the atom, is not a solid object with a well-defined surface. Furthermore, although most nuclides are spherical, some are notably ellipsoidal. Nevertheless electron-scattering experiments (as well as experiments of other kinds) allow us to assign to each nuclide an effective radius given by

$$R = R_0 A^{1/3}, \quad (43\text{-}3)$$

in which A is the mass number and $R_0 \approx 1.2$ fm. We see that the volume of a nucleus, which is proportional to R^3, is directly proportional to the mass number A, being independent of the separate values of Z and N.

Nuclear Masses

Atomic masses can be measured with great precision using modern mass spectrometer and nuclear reaction techniques. Recall from Section 1-6 that such masses are reported in atomic mass units u, chosen so that the atomic mass (not the nuclear mass) of ¹²C is exactly 12 u. The relation of this unit to the SI mass unit is, approximately,

$$1 \text{ u} = 1.661 \times 10^{-27} \text{ kg.} \quad (43\text{-}4)$$

The mass number A of a nuclide is so named because the number represents the mass of the nuclide, expressed in atomic mass units and rounded off to the nearest integer. Thus the atomic mass of ¹⁹⁷Au is 196.966573 u, which we round to 197 u.

In nuclear reactions, Einstein's relation $Q = \Delta m \, c^2$ (Eq. 8-40) is an indispensable workaday tool. As we saw in Section 8-8, Q in this equation is the energy released (or absorbed) when the mass of a closed interacting system of particles decreases (or increases) by the magnitude Δm.

The energy equivalent of 1 u can easily be shown to be 931.5 MeV. Thus c^2 can be written as 931.5 MeV/u, and we can use this value to find the energy equivalent (in million electron-volts) of any mass or mass difference (in atomic mass units), or conversely.

Nuclear Binding Energies

The total energy required to tear a nucleus apart into its constituent protons and neutrons can be calculated from $Q = \Delta m\, c^2$ and is called the **nuclear binding energy**. If we divide the binding energy of a nucleus by its mass number, we get the *binding energy per nucleon*. Figure 43-6 plots this quantity as a function of mass number. The "drooping" of this binding energy curve at both high and low mass numbers has practical consequences of the greatest importance.

The drooping of the binding energy curve at high mass numbers tells us that nucleons are more tightly bound when they are assembled into two middle-mass nuclides than into a single high-mass nuclide. In other words, energy can be released by the **nuclear fission**, or splitting, of a single massive nucleus into two smaller fragments.

The drooping of the binding energy curve at low mass numbers, on the other hand, tells us that energy will be released if two nuclides of low mass number combine to form a single middle-mass nuclide. This process, the reverse of fission, is called **nuclear fusion**. It occurs inside our Sun and other stars and in thermonuclear explosions. Controlled nuclear fusion as a practical energy source is the subject of much current attention.

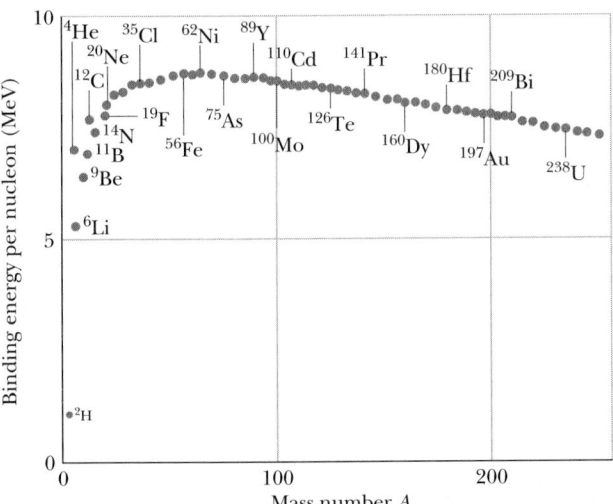

FIGURE 43-6 The binding energy per nucleon for some representative nuclides. The nuclide ^{62}Ni has the highest binding energy per nucleon (8.794 60 ± 0.000 03 MeV/nucleon) of any known stable nuclide. Note that the alpha particle (^{4}He) has a higher binding energy per nucleon than its neighbors in the periodic table and is thus also particularly stable.

Nuclear Energy Levels

The energy of nuclei, like that of atoms, is quantized. That is, nuclei can exist only in discrete quantum states, each with a well-defined energy. Figure 43-7 shows some of these energy levels for ^{28}Al, a typical low-mass nuclide. Note that the energy scale is in millions of electron-volts, rather than the electron-volts used for atoms. When a nucleus makes a transition from one level to a level of lower energy, the emitted photon is typically in the gamma-ray region of the electromagnetic spectrum.

Nuclear Spin and Magnetism

Many nuclides have an intrinsic *nuclear angular momentum*, or spin, and an associated intrinsic *nuclear magnetic moment*. Although nuclear angular momenta are roughly of the same magnitude as the angular momenta of atomic electrons, nuclear magnetic moments are much smaller than typical atomic magnetic moments, by a factor of about 1000.

The Nuclear Force

The force that controls the motions of the atomic electrons is the familiar electromagnetic force. To bind the nucleus together, however, there must be a strong attractive nuclear force of a totally different kind, strong enough to overcome the repulsive force between the (positively charged) nuclear protons and to bind both protons and neutrons into the tiny nuclear volume. The nuclear force must also be of short range because its influence does not extend very far beyond the nuclear "surface."

The present view is that the nuclear force that binds neutrons and protons in the nucleus is not a fundamental force of nature but is a secondary, or "spillover," effect of

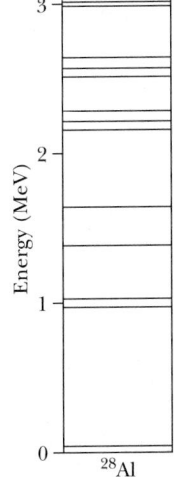

FIGURE 43-7 Energy levels for the nuclide ^{28}Al, deduced from nuclear reaction experiments.

the **strong force** that binds quarks together to form neutrons and protons (see Section 2-9). In much the same way, the attractive force between certain neutral molecules is a spillover effect of the Coulomb electric force that acts within each molecule to bind it together.

SAMPLE PROBLEM 43-2

We can think of all nuclides as made up of a neutron-proton mixture that we can call *nuclear matter*. What is its density?

SOLUTION: We know that this density is high because virtually all the mass of the atom is found in its tiny central nucleus. The volume of a nucleus (assumed spherical) of mass number A and radius R is

$$V = \tfrac{4}{3}\pi R^3 = \tfrac{4}{3}\pi (R_0 A^{1/3})^3 = \tfrac{4}{3}\pi R_0^3 A,$$

where we have used Eq. 43-3 to obtain the third expression. Such a nucleus contains A nucleons so that its nucleon number density ρ_n, expressed in nucleons per unit volume, is

$$\rho_n = \frac{A}{V} = \frac{A}{\tfrac{4}{3}\pi R_0^3 A}$$

$$= \frac{3}{(4\pi)(1.2 \text{ fm})^3} = 0.138 \text{ nucleon/fm}^3.$$

We can consider nuclear matter as having a single density for all nuclides only because A cancels in the equation above.

The mass of a nucleon (neutron *or* proton) is about 1.67×10^{-27} kg. The mass density of nuclear matter in SI units is then

$$\rho = \left(0.138 \frac{\text{nucleons}}{\text{fm}^3}\right)\left(1.67 \times 10^{-27} \frac{\text{kg}}{\text{nucleon}}\right)$$

$$\times \left(10^{15} \frac{\text{fm}}{\text{m}}\right)^3$$

$$\approx 2 \times 10^{17} \text{ kg/m}^3. \qquad \text{(Answer)}$$

This is about 2×10^{14} times the density of water.

SAMPLE PROBLEM 43-3

(a) How much energy is required to separate the typical middle-mass nucleus ^{120}Sn into its constituent nucleons?

SOLUTION: We can find this energy from $Q = \Delta m\, c^2$. Following standard practice, we carry out such calculations in terms of the masses of the neutral atoms involved, not those of the bare nuclei. As Table 43-1 shows, one *atom* of ^{120}Sn (nucleus plus 50 electrons) has a mass of 119.902 199 u. This atom can be separated into 50 hydrogen atoms (50 protons, each with one of the 50 electrons) and 70 neutrons. Each hydrogen atom has a mass of 1.007 825 u, and each neutron a mass of 1.008 665 u. The combined mass of the constituent

particles is

$$m = 50 \times 1.007\,825 \text{ u} + 70 \times 1.008\,665 \text{ u}$$

$$= 120.997\,80 \text{ u}.$$

This exceeds the atomic mass of ^{120}Sn by

$$\Delta m = 120.997\,80 \text{ u} - 119.902\,199 \text{ u}$$

$$= 1.095\,601 \text{ u} \approx 1.096 \text{ u}.$$

Note that since the masses of the 50 electrons cancel in the subtraction, this is also the mass difference that is obtained when a bare ^{120}Sn nucleus is separated into 50 (bare) protons and 70 neutrons. In energy terms this mass difference becomes

$$Q = \Delta m\, c^2 = (1.096 \text{ u})(931.5 \text{ MeV/u})$$

$$= 1021 \text{ MeV}. \qquad \text{(Answer)}$$

(b) What is the binding energy per nucleon for this nuclide?

SOLUTION: The total binding energy Q is the total energy that must be supplied to dismantle the nucleus. The binding energy per nucleon E_n is then

$$E_n = \frac{Q}{A} = \frac{1021 \text{ MeV}}{120} = 8.51 \text{ MeV/nucleon}, \qquad \text{(Answer)}$$

in agreement with the value shown in Table 43-1.

43-3 RADIOACTIVE DECAY

As Fig. 43-4 shows, most of the nuclides that have been identified are radioactive. Such a nuclide spontaneously emits a particle, transforming itself in the process into a different nuclide, occupying a different square on the nuclidic chart.

Radioactive decay provided the first evidence that the laws that govern the subatomic world are statistical. Consider, for example, a 1 mg sample of uranium metal. It contains 2.5×10^{18} atoms of the very long-lived radionuclide ^{238}U. The nuclei of these particular atoms have existed without decaying since they were created—well before the formation of our solar system. During any given second only about 12 of the nuclei in our sample will happen to decay by emitting an alpha particle, transforming themselves into nuclei of ^{234}Th. However,

---

There is absolutely no way to predict whether any given nucleus in the sample will be among the small number of nuclei that decay during the next second. All have an equal chance.

We can express the statistical nature of the decay process by saying that if a sample contains N radioactive nuclei, then the rate ($= -dN/dt$) at which nuclei decay is proportional to N:

$$-\frac{dN}{dt} = \lambda N, \qquad (43\text{-}5)$$

in which λ, the **disintegration constant**, has a characteristic value for every radionuclide. Its SI unit is the inverse second (s^{-1}). Equation 43-5 may be integrated to yield

$$N = N_0 e^{-\lambda t} \quad \text{(radioactive decay)}, \qquad (43\text{-}6)$$

in which N_0 is the number of radioactive nuclei in the sample at $t = 0$ and N is the number remaining at any subsequent time t. Note that lightbulbs (for one example) follow no such exponential decay law. If we life-test 1000 bulbs, we expect that they will all "decay" (that is, burn out) at more or less the same time. The decay of radionuclides follows quite a different law.

We are often more interested in the decay rate R ($= -dN/dt$) than in N itself. Differentiating Eq. 43-6, we find

$$R = -\frac{dN}{dt} = \lambda N_0 e^{-\lambda t},$$

or

$$R = R_0 e^{-\lambda t} \quad \text{(radioactive decay)}, \qquad (43\text{-}7)$$

an alternative form of the law of radioactive decay (Eq. 43-6). Here R_0 ($= \lambda N_0$) is the decay rate at $t = 0$, and R is the rate at any subsequent time t.

The total decay rate R of a sample of a radionuclide is called the **activity** of that sample. The SI unit for activity is the **becquerel**, named for Henri Becquerel, the discoverer of radioactivity:

$$1 \text{ becquerel} = 1 \text{ Bq} = 1 \text{ decay per second}.$$

An older unit, the **curie**, is still in common use:

$$1 \text{ curie} = 1 \text{ Ci} = 3.7 \times 10^{10} \text{ Bq}.$$

An example of the use of these units is the following statement: "The activity of spent reactor fuel rod #5658 on January 15, 1997, was 3.5×10^{15} Bq ($= 9.5 \times 10^4$ Ci)." Thus, on that day 3.5×10^{15} radioactive nuclei in the rod decayed each second. The identities of the radionuclides in the fuel rod, their disintegration constants λ, and the types of radiation they emit have no bearing on this measure of activity.

Often a radioactive sample will be placed near a detector that, for reasons of geometry or detector inefficiency, does not record all the disintegrations that occur in the sample. The reading of the detector under these circumstances is proportional to (and smaller than) the true activity of the sample. Such proportional activity measurements are reported not in becquerel units but simply in counts per unit time.

A quantity of special interest is the **half-life** τ, defined as the time after which both N and R are reduced to one-half their initial values. Putting $R = \frac{1}{2}R_0$ in Eq. 43-7 and substituting τ for t, we have

$$\tfrac{1}{2}R_0 = R_0 e^{-\lambda \tau}.$$

Solving for τ yields

$$\tau = \frac{\ln 2}{\lambda}, \qquad (43\text{-}8)$$

a relation between the half-life τ and the disintegration constant λ.

CHECKPOINT **2:** The nuclide ^{131}I is radioactive, with a half-life of 8.04 days. At noon on January 1, the activity of a certain sample is 600 Bq. Using the concept of half-life, without written calculation, determine whether the activity on January 24 will be a little less than 200 Bq, a little more than 200 Bq, a little less than 75 Bq, or a little more than 75 Bq.

SAMPLE PROBLEM 43-4

The table that follows shows some measurements of the decay rate of a sample of ^{128}I, a radionuclide often used medically as a tracer to measure the rate at which iodine is absorbed by the thyroid gland.

TIME (MIN)	R (COUNTS/S)	TIME (MIN)	R (COUNTS/S)
4	392.2	132	10.9
36	161.4	164	4.56
68	65.5	196	1.86
100	26.8	218	1.00

Find the disintegration constant and the half-life for this radionuclide.

SOLUTION: If we take the natural logarithm of each side of Eq. 43-7, we find

$$\ln R = \ln R_0 - \lambda t.$$

Thus if we plot $\ln R$ against t, we should obtain a straight line whose slope is $-\lambda$. This is done in Fig. 43-8, from which we find

$$-\lambda = -\frac{6.2 - 0}{225 \text{ min } - 0}$$

or $\qquad \lambda = 0.0275 \text{ min}^{-1} \approx 1.7 \text{ h}^{-1}. \qquad$ (Answer)

We find the half-life readily from Eq. 43-8, obtaining

$$\tau = \frac{\ln 2}{\lambda} = \frac{\ln 2}{0.0275 \text{ min}^{-1}} \approx 25 \text{ min}. \qquad \text{(Answer)}$$

The activity of a given sample of ^{128}I will drop to half its initial value in 25 min, no matter what the initial value was. And the number of ^{128}I nuclei in the sample will drop to half the initial number in the same 25 min, no matter what that initial number was.

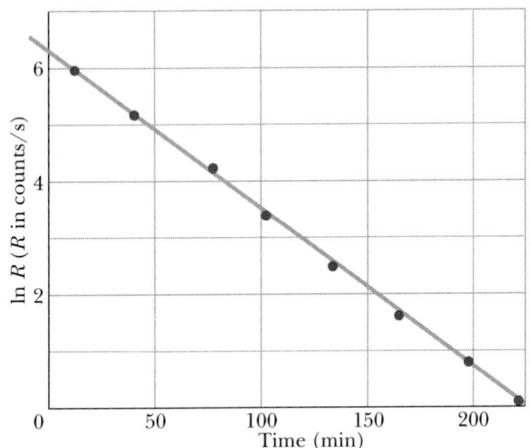

FIGURE 43-8 Sample Problem 43-4. A semilogarithmic plot of the decay of a sample of ^{128}I, based on the data in the table. The half-life of this radionuclide (25 min) can be found from the slope of this curve.

SAMPLE PROBLEM 43-5

A 2.71 g sample of KCl from the chemistry stockroom is found to be radioactive, and it is decaying at a constant rate of 4490 Bq. The decays are traced to the element potassium and in particular to the isotope ^{40}K, which constitutes 1.17% of normal potassium. Calculate the half-life of this nuclide.

SOLUTION: We can find the half-life from Eq. 43-8. Since the activity appears to be constant, the half-life must be very long and we cannot calculate λ by the method of Sample Problem 43-4. We must find it by substituting for N and dN/dt in Eq. 43-5.

From Appendix F, the molar mass of KCl is 74.6 g/mol, so the number of potassium atoms in the sample is

$$N_K = \frac{(6.02 \times 10^{23}\ \text{mol}^{-1})(2.71\ \text{g})}{74.6\ \text{g/mol}} = 2.187 \times 10^{22}.$$

Of these, the number of ^{40}K atoms is

$$N_{40} = (2.187 \times 10^{22})(0.0117) = 2.559 \times 10^{20}.$$

From Eq. 43-5, we then have

$$\lambda = \frac{-dN/dt}{N} = \frac{R_{40}}{N_{40}} = \frac{4490\ \text{s}^{-1}}{2.559 \times 10^{20}}$$

$$= 1.755 \times 10^{-17}\ \text{s}^{-1}.$$

The half-life follows from Eq. 43-8:

$$\tau = \frac{\ln 2}{\lambda} = \frac{(\ln 2)(1\ \text{y}/3.16 \times 10^7\ \text{s})}{1.755 \times 10^{-17}\ \text{s}^{-1}}$$

$$= 1.25 \times 10^9\ \text{y.} \qquad \text{(Answer)}$$

This is of the order of magnitude of the age of the universe! No wonder we cannot measure the half-life of this radionuclide by waiting around for its activity to decrease. Interestingly, the potassium in our own bodies has its normal share of this radioisotope; we are all slightly radioactive.

43-4 ALPHA DECAY

The radionuclide ^{238}U decays by emitting an alpha particle (a helium nucleus) according to the scheme

$$^{238}\text{U} \rightarrow {}^{234}\text{Th} + {}^{4}\text{He}, \qquad Q = 4.25\ \text{MeV.} \quad (43\text{-}9)$$

Here Th is the symbol for the element thorium ($Z = 90$). The half-life for this decay is 4.47×10^9 y. Q is the disintegration energy of the process, that is, the amount of energy released during a single decay. We may well ask: If energy is released in every such decay event, why did the ^{238}U nuclei not decay shortly after they were created? Why did they wait so long? To answer, we must examine the mechanism of alpha decay.

We choose a model in which the alpha particle is imagined to exist (already formed) inside the nucleus before it escapes from the nucleus. Figure 43-9 shows the approximate potential energy $U(r)$ for the alpha particle and the residual ^{234}Th nucleus as a function of their separation r. This energy is due to the combination of (1) a potential well associated with the (attractive) strong nuclear force that acts in the nuclear interior and (2) a Coulomb potential associated with the (repulsive) electric force that acts between the two particles before and after the decay has occurred.

The horizontal black line marked $Q = 4.25$ MeV shows the disintegration energy for the process. If we assume that this represents the total energy of the alpha particle during the decay process, then the part of the $U(r)$ curve above this line constitutes a potential energy barrier like that in Fig. 39-13. This barrier cannot be surmounted. If the alpha particle were able to be at some separation r within the barrier, its potential energy U would exceed its total energy E. This would mean, classically, that its kinetic energy K (which equals $E - U$) would be negative, an impossible situation.

We can see now why the alpha particle is not immediately emitted from the ^{238}U nucleus! That nucleus is surrounded by an impressive potential barrier, occupying —if you think of it in three dimensions—the volume lying

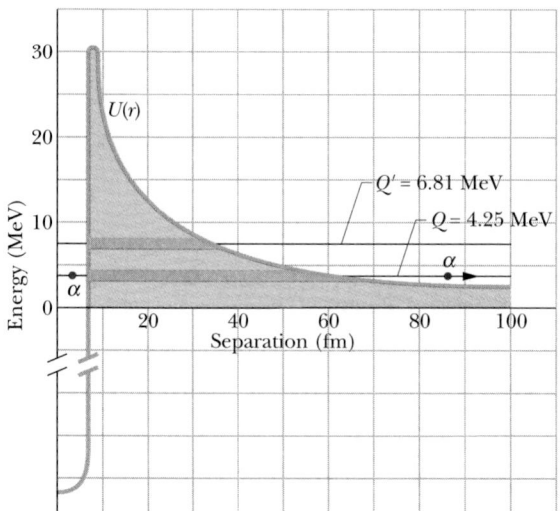

FIGURE 43-9 A potential energy function for the emission of an alpha particle by ^{238}U. The horizontal black line marked $Q = 4.25$ MeV shows the disintegration energy for the process. The thick gray portion of this line represents separations r that are classically forbidden to the alpha particle. The alpha particle is represented by a dot, both inside the barrier (at the left) and outside it (at the right), after the particle has tunneled through. The horizontal black line marked $Q' = 6.81$ MeV shows the disintegration energy for the alpha decay of ^{228}U. (Both isotopes have the same potential energy function because they have the same nuclear charge.)

between two spherical shells (of radii about 8 and 60 fm). This argument is so convincing that we now change our question and ask: How, since the particle seems permanently trapped inside the nucleus by the barrier, can the ^{238}U nucleus *ever* emit an alpha particle? The answer is that, as you learned in Section 39-9, there is a finite probability that a particle can tunnel through an energy barrier that is classically insurmountable. And, in fact, alpha decay occurs as a result of barrier tunneling.

Since the half-life of ^{238}U is very long, the barrier must not be very "leaky." The alpha particle, presumed to be rattling back and forth within the nucleus, must arrive at the inner surface of the barrier about 10^{38} times before it succeeds in tunneling through the barrier. This is about 10^{21} times per second for about 4×10^9 years (the age of Earth)! We, of course, are waiting on the outside, able to count only the alpha particles that *do* manage to escape.

We can test this explanation of alpha decay by examining other alpha emitters. For an extreme contrast, consider the alpha decay of another uranium isotope, ^{228}U, which has a disintegration energy Q' of 6.81 MeV, about 60% higher than that of ^{238}U. (The value of Q' is also shown as a horizontal black line in Fig. 43-9.) Recall from

Section 39-9 that the transmission coefficient of a barrier is very sensitive to small changes in the total energy of the particle seeking to penetrate it. Thus we expect alpha decay to occur more readily for this nuclide than for ^{238}U. Indeed it does. As Table 43-2 shows, its half-life is only 9.1 min! An increase in Q by a factor of only 1.6 produces a decrease in half-life (that is, in the effectiveness of the barrier) by a factor of 3×10^{14}. This is sensitivity indeed.

TABLE 43-2 TWO ALPHA EMITTERS COMPARED

RADIONUCLIDE	Q	HALF-LIFE
^{238}U	4.25 MeV	4.5×10^9 y
^{228}U	6.81 MeV	9.1 min

SAMPLE PROBLEM 43-6

We are given the following atomic masses:

^{238}U	238.05079 u	^{4}He	4.00260 u
^{234}Th	234.04363 u	^{1}H	1.00783 u
^{237}Pa	237.05121 u		

Here Pa is the symbol for the element protactinium ($Z = 91$).

(a) Calculate the energy released during the alpha decay of ^{238}U. The decay process is

$$^{238}\text{U} \rightarrow {}^{234}\text{Th} + {}^4\text{He}.$$

Note, incidentally, how nuclear charge is conserved in this equation: the atomic numbers of thorium (90) and helium (2) add up to the atomic number of uranium (92). The number of nucleons is also conserved: $238 = 234 + 4$.

SOLUTION: The total atomic mass of the decay products in the foregoing process (234.04363 u + 4.00260 u) is less than the atomic mass of ^{238}U by $\Delta m = 0.00456$ u, whose energy equivalent is

$$Q = \Delta m\, c^2 = (0.00456 \text{ u})(931.5 \text{ MeV/u})$$
$$= 4.25 \text{ MeV}. \qquad \text{(Answer)}$$

This disintegration energy appears as the kinetic energies of the alpha particle and the recoiling ^{234}Th atom.

Note again that, following standard practice, we work with the masses of the neutral atoms rather than those of the bare nuclei; when we calculate the mass difference Δm, the masses of the electrons cancel out.

(b) Show that ^{238}U cannot decay spontaneously by emitting a proton.

SOLUTION: If this happened, the decay process would be

$$^{238}\text{U} \rightarrow {}^{237}\text{Pa} + {}^1\text{H}.$$

(You should verify that both nuclear charge and the number of nucleons are conserved in this process.) In this situation, the mass of the two decay products (= 237.05121 u + 1.00783 u) would *exceed* the mass of ^{238}U by $\Delta m =$ 0.00825 u, with energy equivalence $Q = -7.68$ MeV. A minus sign is included here to indicate that we must *add* energy in the amount of 7.68 MeV to a ^{238}U nucleus before it will emit a proton; it will certainly not do so spontaneously.

43-5 BETA DECAY

A nucleus that decays spontaneously by emitting an electron or a positron (a positively charged particle with the mass of an electron) is said to undergo **beta decay**. This, like alpha decay, is a spontaneous process, with a definite disintegration energy and half-life. Again like alpha decay, beta decay is a statistical process, governed by Eqs. 43-6 and 43-7. Here are two examples:

$$^{32}\text{P} \rightarrow {}^{32}\text{S} + \text{e}^- + \nu \quad (\tau = 14.3 \text{ d}) \quad (43\text{-}10)$$

and

$$^{64}\text{Cu} \rightarrow {}^{64}\text{Ni} + \text{e}^+ + \nu \quad (\tau = 12.7 \text{ h}). \quad (43\text{-}11)$$

The symbol ν represents a **neutrino**, a virtually (if not completely) massless, neutral particle that is emitted from the nucleus along with the electron or positron during the decay process. Neutrinos interact only very weakly with matter and—for that reason—are so extremely difficult to detect that their presence long went unnoticed.*

Both charge and nucleon number are conserved in the above two processes. In the decay of Eq. 43-10, for example, we can write for charge conservation

$$(+15e) = (+16e) + (-e) + (0)$$

and for nucleon conservation

$$(32) = (32) + (0) + (0),$$

where we have recognized that neither the electron nor the neutrino is a nucleon and the neutrino has no charge.

It may seem surprising that nuclei can emit electrons, positrons, and neutrinos, since we have said that nuclei are made up of neutrons and protons only. However, we saw earlier that atoms emit photons, and we certainly do not say that atoms "contain" photons. We say that the photons are created during the emission process.

So it is with the electrons, positrons, and neutrinos emitted from nuclei during beta decay. They are created during the emission process. A neutron transforms into a proton within the nucleus according to

$$\text{n} \rightarrow \text{p} + \text{e}^- + \nu \quad (43\text{-}12)$$

or a proton transforms into a neutron via

$$\text{p} \rightarrow \text{n} + \text{e}^+ + \nu. \quad (43\text{-}13)$$

Both of these beta-decay processes provide evidence that—as was pointed out—neutrons and protons are not truly fundamental particles. Note (as in Eqs. 43-10 and 43-11) that the mass number A of a nuclide undergoing beta decay does not change; one of its constituent nucleons simply changes its character according to Eq. 43-12 or 43-13, the total number of nucleons remaining fixed.

In both alpha decay and beta decay, the same amount of energy is released in every individual decay event. In the alpha decay of a particular radionuclide, every emitted alpha particle has the same sharply defined kinetic energy. (In some cases, a radionuclide may emit more than one group of alpha particles, each group having a sharply defined kinetic energy.) However, in the beta decay of Eq. 43-12 for electron emission, the disintegration energy Q is shared—in varying proportions—between the electron and the neutrino. Sometimes the electron gets nearly all the energy, sometimes the neutrino does. In every case, however, the sum of the electron's energy and the neutrino's energy gives a constant value Q. Such sharing of energy, with a sum equal to Q, is also true of the other type of beta decay (Eq. 43-13) that involves a positron emission.

Thus in beta decay the energy of the emitted electrons or positrons may range from zero up to a certain maximum K_{max}. Figure 43-10 shows the distribution of positron energies for the beta decay of ^{64}Cu (see Eq. 43-11). The maximum positron energy K_{max} must equal the disintegration energy Q because the neutrino carries away no energy

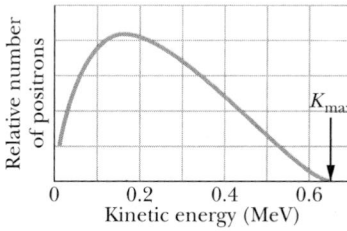

FIGURE 43-10 The distribution of the kinetic energies of positrons emitted in the beta decay of ^{64}Cu. The maximum kinetic energy of the distribution (K_{max}) is 0.653 MeV. In all ^{64}Cu decay events, this energy is shared between the positron and the neutrino, in varying proportions. The *most probable* energy for an emitted positron is about 0.15 MeV.

---

*Beta decay also includes *electron capture*, in which a nucleus decays by absorbing one of its atomic electrons, emitting a neutrino in the process. We do not consider that process here. Also, the neutral particle emitted in the decay process of Eq. 43-10 is actually an *antineutrino*, a distinction we shall not make in this introductory treatment.

when the positron carries away K_{max}. That is,

$$Q = K_{max}. \qquad (43\text{-}14)$$

The Neutrino

Wolfgang Pauli first suggested the existence of neutrinos in 1930. His neutrino hypothesis not only permitted an understanding of the energy distribution of electrons or positrons in beta decay but also solved another early beta-decay puzzle involving "missing" angular momentum.

The neutrino is a truly elusive particle; the mean free path of an energetic neutrino in water has been calculated as no less than several thousand light-years. At the same time, neutrinos left over from the Big Bang that presumably marked the creation of the universe are the most abundant particles of physics. Billions of them pass through our bodies every second, leaving no trace.

In spite of their elusive character, neutrinos have been detected in the laboratory. This was first done in 1953 by F. Reines and C. L. Cowan, using neutrinos generated in a high-power nuclear reactor. (In 1995 Reines, the surviving member of the pair, received a Nobel Prize for this work.) In spite of the difficulties of detection, experimental neutrino physics is now a well-developed branch of experimental physics, with avid practitioners at several laboratories throughout the world.

The Sun emits neutrinos copiously from the nuclear furnace at its core and, at night, these messengers from the center of the Sun come up at us from below, Earth being almost totally transparent to them. In February 1987, light from an exploding star in the Large Magellanic Cloud (a nearby galaxy) reached Earth after having traveled for 170,000 years. Enormous numbers of neutrinos were generated in this explosion, and about 10 of them were picked up by a sensitive neutrino detector in Japan; Fig. 43-11 shows a record of their passage.

Radioactivity and the Nuclidic Chart

Study of alpha and beta decay permits us to look at the nuclidic chart of Fig. 43-4 in a new way. Let us add a third dimension to that chart by plotting the **mass excess** of each nuclide in a direction perpendicular to the *N-Z* plane. The mass excess of a nuclide is (in spite of its name) an energy that approximates the nuclide's *total* binding energy. It is defined as $(m - A)c^2$, where m is the atomic mass of the nuclide and A is its mass number, both expressed in atomic mass units, and c^2 is 931.5 MeV/u.

The surface so formed gives a graphic representation of nuclear stability. As Fig. 43-12 shows (for the low mass nuclides), this surface describes a "valley of the nu-

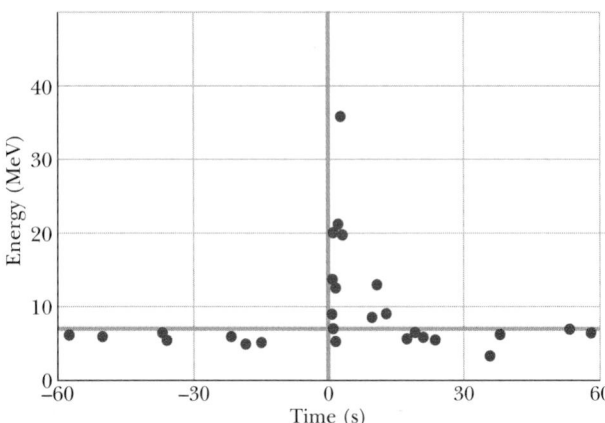

FIGURE 43-11 A burst of neutrinos from the supernova SN 1987A, which occurred at (relative) time 0, stands out from the usual reception of neutrinos. (For neutrinos, 10 is a "burst"!) The particles were detected by an elaborate detector housed deep in a mine in Japan. The supernova was visible only in the Southern Hemisphere, so the neutrinos had to penetrate Earth (a trifling barrier for them) to reach the detector.

clides," the stability band of Fig. 43-4 running along its bottom. Nuclides on the proton-rich side of the valley decay into it by emitting positrons, and those on the neutron-rich side do so by emitting electrons.

$\mathbb{C}$HECKPOINT **3:** ^{238}U decays into ^{234}Th by the emission of an alpha particle. There follows a chain of further radioactive decays, either by alpha decay or by beta decay. Eventually a stable nuclide is reached and after that, no further radioactive decay is possible. Which of the following stable nuclides is the end product of the ^{238}U radioactive decay chain: ^{206}Pb, ^{207}Pb, ^{208}Pb, or ^{209}Pb? (*Hint:* You can decide by considering the changes in mass number A for the two types of decay.)

SAMPLE PROBLEM 43-7

Calculate the disintegration energy Q for the beta decay of ^{32}P, as described by Eq. 43-10. The needed atomic masses are 31.97391 u for ^{32}P and 31.97207 u for ^{32}S.

SOLUTION: Because of the presence of the emitted electron, we must be careful to distinguish between nuclear and atomic masses. Let the boldface symbols $\mathbf{m}_P$ and $\mathbf{m}_S$ represent the nuclear masses of ^{32}P and ^{32}S and let the italic symbols m_P and m_S represent their atomic masses. We take the disintegration energy Q to be $\Delta m\, c^2$, where, from Eq. 43-10,

$$\Delta m = \mathbf{m}_P - (\mathbf{m}_S + m_e),$$

FIGURE 43-12 A portion of the valley of the nuclides, showing only the nuclides of low mass. Deuterium, tritium, and helium lie at the nearest end of the plot, with helium at the high point. The valley stretches away from us, with the plot stopping at about $Z = 22$ and $N = 35$. Nuclides with large values of A, which would be plotted much beyond the valley, can decay into the valley by repeated alpha emissions and by fission (splitting of a nuclide).

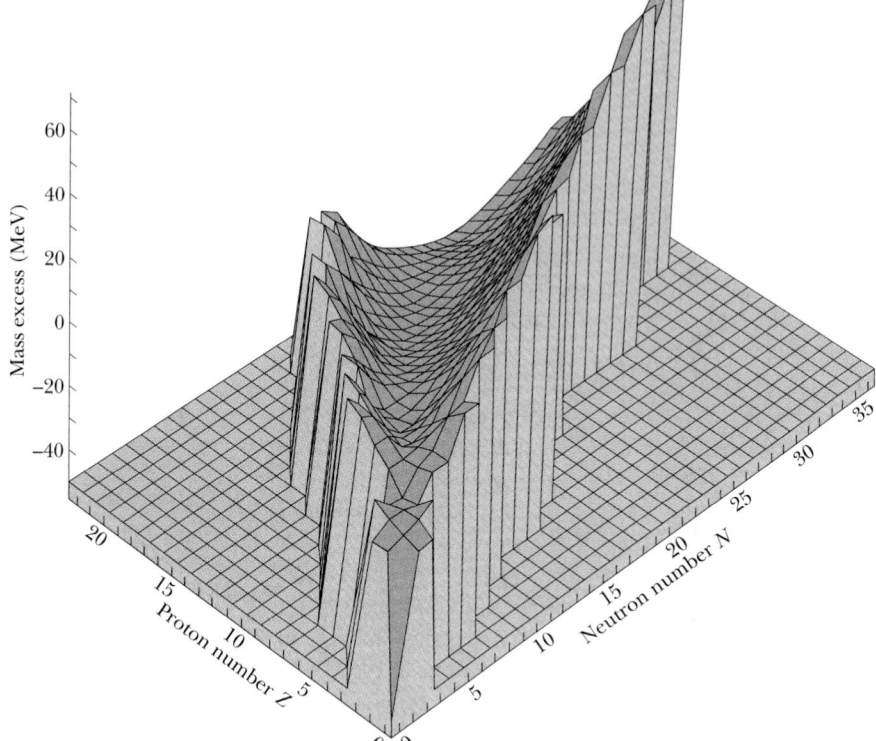

in which m_e is the mass of the electron. If we add and subtract $15m_e$ on the right side of this equation, we obtain

$$\Delta m = (\mathbf{m}_P + 15m_e) - (\mathbf{m}_S + 16m_e).$$

The quantities in parentheses are the atomic masses of ^{32}P and ^{32}S, so

$$\Delta m = m_P - m_S.$$

We thus see that if we subtract the atomic masses in this way, the mass of the emitted electron is automatically taken into account. (This will not work for positron emission.)

The disintegration energy for the ^{32}P decay is then

$$Q = \Delta m \, c^2$$
$$= (31.97391 \text{ u} - 31.97207 \text{ u})(931.5 \text{ MeV/u})$$
$$= 1.71 \text{ MeV}. \qquad \text{(Answer)}$$

Experimentally, this calculated quantity proves to be equal (as Eq. 43-14 requires) to K_{max}, the maximum energy of the emitted electrons. Although 1.71 MeV is released every time a ^{32}P nucleus decays, in essentially every case the electron carries away less energy than this. The neutrino gets essentially all the rest, carrying it stealthily out of the laboratory.

43-6 RADIOACTIVE DATING

If you know the half-life of a given radionuclide, you can in principle use the decay of that radionuclide as a clock to measure time intervals. The decay of very long-lived nuclides, for example, can be used to measure the age of rocks, that is, the time that has elapsed since they were formed. Measurements for rocks from Earth and the Moon, and for meteorites, yield a consistent maximum age of about 4.5×10^9 y for these bodies.

The radionuclide ^{40}K, for example, decays to ^{40}Ar, a stable isotope of the noble gas argon. The half-life for this decay is 1.25×10^9 y. By measuring the ratio of ^{40}K to ^{40}Ar found in the rock in question, one can then calculate the age of that rock. Other long-lived decays, such as that of ^{235}U to ^{207}Pb (involving a number of intermediate stages), can be used to verify these measurements.

For measuring shorter time intervals, in the range of historical interest, radiocarbon dating has proved invaluable. The radionuclide ^{14}C (with $\tau = 5730$ y) is produced at a constant rate in the upper atmosphere as atmospheric nitrogen is bombarded by cosmic rays. This radiocarbon mixes with the carbon that is normally present in the atmosphere (as CO_2) so that there is about one atom of ^{14}C for every 10^{13} atoms of ordinary stable ^{12}C. Because of biological activity such as photosynthesis and breathing, the atoms of atmospheric carbon trade places randomly, one atom at a time, with the atoms of carbon in every living thing, including broccoli, mushrooms, penguins, and humans. Eventually an exchange equilibrium is reached at which the carbon atoms of every living thing contain a fixed small fraction of the radioactive nuclide ^{14}C.

This exchange persists as long as the organism is alive. When the organism dies, the exchange with the atmosphere stops and the amount of radiocarbon trapped in the organism, since it is no longer being replenished, dwindles

A fragment of the Dead Sea scrolls and the caves from which the scrolls were recovered. The age of the scrolls was determined by radiocarbon dating a sample of the cloth used to plug the jars in which the scrolls were sealed.

away with a half-life of 5730 y. By measuring the amount of radiocarbon per gram of organic matter, it is possible to measure the time that has elapsed since the organism died. Charcoal from ancient campfires, the Dead Sea scrolls, and many prehistoric artifacts have been dated in this way.

SAMPLE PROBLEM 43-8

Mass spectrometric analysis of potassium and argon atoms in a Moon rock sample shows that the ratio of the number of (stable) ^{40}Ar atoms present to the number of (radioactive) ^{40}K atoms is 10.3. Assume that all the argon atoms were produced by the decay of potassium atoms, with a half-life of 1.25×10^9 y. How old is the rock?

SOLUTION: If N_0 potassium atoms were present at the time the rock was formed by solidification from a molten form, the number of potassium atoms remaining at the time of analysis is, from Eq. 43-6,

$$N_K = N_0 e^{-\lambda t}, \qquad (43\text{-}15)$$

in which t is the age of the rock. For every potassium atom that decays, an argon atom is produced. Thus the number of argon atoms present at the time of the analysis is

$$N_{Ar} = N_0 - N_K. \qquad (43\text{-}16)$$

We cannot measure N_0; so let's eliminate it from Eqs. 43-15 and 43-16. We find, after some algebra,

$$\lambda t = \ln\left(1 + \frac{N_{Ar}}{N_K}\right), \qquad (43\text{-}17)$$

---

in which N_{Ar}/N_K *can* be measured. Solving for t and using Eq. 43-8 to replace λ with $(\ln 2)/\tau$ yield

$$t = \frac{\tau \ln(1 + N_{Ar}/N_K)}{\ln 2}$$

$$= \frac{(1.25 \times 10^9 \text{ y})[\ln(1 + 10.3)]}{\ln 2}$$

$$= 4.37 \times 10^9 \text{ y}. \qquad \text{(Answer)}$$

Lesser ages may be found for other lunar or terrestrial rock samples, but no substantially greater ones. Thus the solar system must be about 4 billion years old.

43-7 MEASURING RADIATION DOSAGE

The effect of radiation such as gamma rays, electrons, and alpha particles on living tissue (particularly our own) is a matter of public interest. Such radiation is found in nature in cosmic rays and arises from radioactive elements in Earth's crust. Radiation associated with some human activities, such as using x rays and radionuclides in medicine and in industry, also contributes. The disposal of radioactive waste and the evaluation of the probability of nuclear accidents are subjects that attract national and international concern.

It is not our task here to explore the various sources of radiation but simply to describe the units in which the properties and effects of such radiations are expressed. We have already discussed the *activity* of a radioactive source. There are two remaining quantities of interest.

1. **Absorbed Dose.** This is a measure of the radiation dose (as energy per unit mass) actually absorbed by a specific object, such as a patient's hand or chest. Its SI unit is the **gray** (Gy). An older unit, the **rad** (from **r**adiation **a**bsorbed **d**ose) is still in common use. The terms are defined and related as follows:

$$1 \text{ Gy} = 1 \text{ J/kg} = 100 \text{ rad}.$$

A typical statement is: "A whole-body, short-term gamma-ray dose of 3 Gy (= 300 rad) will cause death in 50% of the population exposed to it." By way of comfort, we note that the present average absorbed dose per year, from sources of both natural and human origin, is about 2 mGy (= 0.2 rad).

2. **Dose Equivalent.** Although different types of radiation (gamma rays and neutrons, say) may deliver the same amount of energy to the body, they do not have the same biological effect. The dose equivalent allows us to express the biological effect by multiplying the absorbed dose (in grays or rads) by a numerical **RBE** factor (from **r**elative

biological **e**ffectiveness). For x rays and electrons, for example, RBE = 1; for slow neutrons RBE = 5; for alpha particles RBE = 10, and so on. Personnel-monitoring devices such as film badges register the dose equivalent.

The SI unit of dose equivalent is the **sievert** (Sv). An earlier unit, the **rem**, is still in common use. Their relationship is

$$1 \text{ Sv} = 100 \text{ rem.}$$

An example of the correct use of these terms is: "The recommendation of the National Council on Radiation Protection is that no individual who is (nonoccupationally) exposed to radiation should receive a dose equivalent greater than 5 mSv (= 0.5 rem) in any one year." This includes radiations of all kinds; of course the appropriate RBE factor must be used for each kind.

SAMPLE PROBLEM 43-9

We have seen that a gamma-ray dose of 3 Gy is lethal to half of those people exposed to it. If the equivalent energy were absorbed as heat, what rise in body temperature would result?

SOLUTION: An absorbed dose of 3 Gy corresponds to an absorbed energy per unit mass of 3 J/kg. Assume that c, the specific heat of the human body, is the same as that of water, 4180 J/kg·K. From Eq. 19-15, the temperature rise is

$$\Delta T = \frac{Q/m}{c} = \frac{3 \text{ J/kg}}{4180 \text{ J/kg·K}}$$

$$= 7.2 \times 10^{-4} \text{ K} \approx 700 \text{ } \mu\text{K}. \qquad \text{(Answer)}$$

Obviously the damage done by ionizing radiation has nothing to do with thermal heating. The harmful effects arise because the radiation succeeds in breaking molecular bonds and thus interfering with the normal functioning of the tissues in which it is absorbed.

43-8 NUCLEAR MODELS

Nuclei are more complicated than atoms. For atoms, the basic force law (Coulomb's law) is simple in form and there is a natural force center, the nucleus. For nuclei the force law is complicated and cannot, in fact, be written down explicitly in full detail. Furthermore, the nucleus—a jumble of protons and neutrons—has no natural force center to simplify the calculations.

In the absence of a comprehensive nuclear *theory*, we turn to the construction of nuclear *models*. A nuclear model is simply a way of looking at the nucleus that gives a physical insight into as wide a range of its properties as possible. The usefulness of a model is tested by its ability to provide predictions that can be verified experimentally in the laboratory.

Two models of the nucleus have proved useful. Although based on assumptions that seem flatly to exclude each other, each accounts very well for a selected group of nuclear properties. After describing them separately, we shall see how these two models may be combined to form a single coherent picture of the atomic nucleus.

The Collective Model

In the *collective model*, formulated by Niels Bohr, the nucleons, moving around at random, are imagined to interact strongly with each other, like the molecules in a drop of liquid. A given nucleon collides frequently with other nucleons in the nuclear interior, its mean free path as it moves about being substantially less than the nuclear radius. This constant "jiggling around" reminds us of the thermal agitation of the molecules in a drop of liquid.

The collective model permits us to correlate many facts about nuclear masses and binding energies; it is useful (as you will see later) in explaining nuclear fission. It is also useful for understanding a large class of nuclear reactions.

Consider, for example, a generalized reaction of the form

$$X + a \rightarrow C \rightarrow Y + b. \qquad (43\text{-}18)$$

We imagine that projectile a enters target nucleus X, forming a **compound nucleus** C and conveying to it a certain amount of excitation energy. The projectile, perhaps a neutron, is at once caught up by the random motions that characterize the nuclear interior. It quickly loses its identity—so to speak—and the excitation energy it carried into the nucleus is quickly shared with all the other nucleons in C.

The quasi-stable state represented by C in Eq. 43-18 may endure for as long as 10^{-16} s before it decays to Y and b. By nuclear standards, this is a very long time, being about one million times longer than the time required for a nucleon with a few million electron-volts of energy to travel across a nucleus. The central feature of this compound-nucleus concept is that the formation of the compound nucleus and its eventual decay are totally independent events. At the time of its decay, the compound nucleus has "forgotten" how it was formed. Hence its mode of decay is not influenced by its mode of formation. As an example, Fig. 43-13 shows three possible ways in which the compound nucleus ^{20}Ne might be formed and three in which it might decay. Any of the three formation modes can lead to any of the three decay modes.

The Independent Particle Model

In the collective model, we assume that the nucleons move around at random and bump into each other frequently.

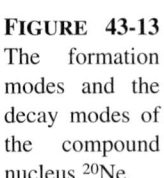

FIGURE 43-13 The formation modes and the decay modes of the compound nucleus ^{20}Ne.

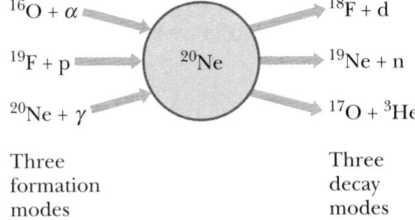

^{16}O + α ^{18}F + d
^{19}F + p ^{20}Ne ^{19}Ne + n
^{20}Ne + γ ^{17}O + ^{3}He

Three formation modes   Three decay modes

The *independent particle model*, however, is based on just the opposite assumption, namely, that each nucleon remains in a well-defined quantum state within the nucleus and hardly makes any collisions at all! The nucleus, unlike the atom, has no fixed center of charge; we assume in this model that each nucleon moves in a potential well that is determined by the smeared-out (time-averaged) motions of all the other nucleons.

A nucleon in a nucleus, like an electron in an atom, has a set of quantum numbers that defines its state of motion. Also, nucleons obey the Pauli exclusion principle, just as electrons do. That is, no two nucleons may occupy the same state at the same time. In this regard, the neutrons and the protons are treated separately, each particle type having its own array of available quantized states.

The fact that nucleons obey the Pauli exclusion principle helps us to understand the relative stability of nucleon states. If two nucleons within the nucleus are to collide, the energy of each of them after the collision must correspond to the energy of an *unoccupied* state. If these states happen to be filled by a nucleon of the same type, a collision simply cannot occur. Thus any given nucleon experiencing repeated "frustrated collision opportunities" will maintain its state of motion long enough to give meaning to the statement that it exists in a quantum state with a well-defined energy.

In the atomic realm, the repetitions of physical and chemical properties that we find in the periodic table are associated with a property of atomic electrons, namely, that they arrange themselves in shells that have a special stability when fully occupied. We can take the atomic numbers of the noble gases,

$$2, 10, 18, 36, 54, 86, \ldots ,$$

as *magic electron numbers* that mark the completion (or closure) of such shells.

Nuclei also show such closed-shell effects, associated with certain **magic nucleon numbers**:

$$2, 8, 20, 28, 50, 82, 126, \ldots .$$

Any nuclide whose proton number Z or neutron number N has one of these values turns out to have a special stability that may be made apparent in a variety of ways.

Examples of "magic" nuclides are ^{18}O ($Z = 8$), ^{40}Ca ($Z = 20$, $N = 20$), ^{92}Mo ($N = 50$), and ^{208}Pb ($Z = 82$, $N = 126$). Both ^{40}Ca and ^{208}Pb are said to be "doubly magic" because they contain both filled shells of protons *and* filled shells of neutrons.

The magic number 2 shows up in the exceptional stability of the alpha particle (^{4}He), which, with $Z = N = 2$, is doubly magic. For example, on the binding energy curve of Fig. 43-6, the binding energy per nucleon for this nuclide stands well above that of its periodic-table neighbors hydrogen, lithium, and beryllium. The alpha particle is so tightly bound, in fact, that it is impossible to add another particle to it; there is no stable nuclide with $A = 5$.

The central idea of a closed shell is that a single particle outside a closed shell can be relatively easily removed, but considerably more energy must be expended to remove a particle from the shell itself. The sodium atom, for example, has one (valence) electron outside a closed electron shell. Only about 5 eV is required to strip the valence electron away from a sodium atom; to remove a *second* electron, however, (which must be plucked out of a closed shell) requires 22 eV. As a nuclear case, consider ^{121}Sb ($Z = 51$), which contains a single proton outside a closed shell of 50 protons. To remove this lone proton requires 5.8 MeV; to remove a *second* proton, however, requires an energy of 11 MeV. There is much additional experimental evidence that the nucleons in a nucleus form closed shells and that these shells exhibit stable properties.

We have seen that quantum mechanics can account beautifully for the magic electron numbers, that is, for the populations of the subshells into which atomic electrons are grouped. It turns out that, by making certain assumptions, quantum mechanics can account equally well for the magic nucleon numbers! The 1963 Nobel prize was, in fact, awarded to Maria Mayer and Hans Jensen "for their discoveries concerning nuclear shell structure."

A Combined Model

Consider a nucleus in which a small number of neutrons (or protons) exist outside a core of closed shells that contains a magic number of neutrons (or protons). The outside nucleons occupy quantized states in a potential well established by the central core, thus preserving the central feature of the independent-particle model. These outside nucleons also interact with the core, deforming it and setting up "tidal wave" motions of rotation or vibration within it. These collective motions of the core preserve the central feature of that model. Such a model of nuclear structure thus succeeds in combining the seemingly irreconcilable points of view of the collective and independent-particle models. It has been remarkably successful.

SAMPLE PROBLEM 43-10

Consider the neutron capture reaction

$$^{109}\text{Ag} + \text{n} \rightarrow {}^{110}\text{Ag} \rightarrow {}^{110}\text{Ag} + \gamma, \quad (43\text{-}19)$$

in which a compound nucleus (^{110}Ag) is formed. Figure 43-14 shows the relative rate at which such events take place, plotted against the energy of the incoming neutron. Find the mean lifetime of this compound nucleus by using the uncertainty principle written in the form

$$\Delta E \cdot \Delta t \approx \hbar. \quad (43\text{-}20)$$

Here ΔE is a measure of the uncertainty with which the energy of an atomic or nuclear state can be defined. The quantity Δt is a measure of the time available to measure this energy. In fact, Δt is just $\bar{t}$, the mean life of the compound nucleus before it decays to its ground state.

SOLUTION: We see that the relative reaction rate is sharply peaked at a neutron energy of about 5.2 eV. This suggests that we are dealing with a single excited level of the compound nucleus ^{110}Ag. When the available energy just matches the energy of this level above the ^{110}Ag ground state, we have "resonance" and the reaction of Eq. 43-19 really "goes."

However, the resonance peak is not infinitely sharp, having an approximate half-width (see ΔE in the figure) of about

0.20 eV. We account for this by saying that the excited level is not sharply defined in energy, having an energy uncertainty ΔE of about 0.20 eV. Thus we have

$$\Delta t = \bar{t} \approx \frac{\hbar}{\Delta E} = \frac{(4.14 \times 10^{-15}\ \text{eV}\cdot\text{s})/2\pi}{0.20\ \text{eV}}$$

$$\approx 3 \times 10^{-15}\ \text{s}. \quad \text{(Answer)}$$

This is several hundred times greater than the time a 0.20 eV neutron takes to cross the diameter of a ^{109}Ag nucleus. So, the neutron is spending this time of 3×10^{-15} s *as part of* the nucleus.

FIGURE 43-14 Sample Problem 43-10. A plot of the relative number of reaction events of the type described by Eq. 43-19, as a function of the energy of the incident neutron. The half-width ΔE of this resonance peak is about 0.20 eV.

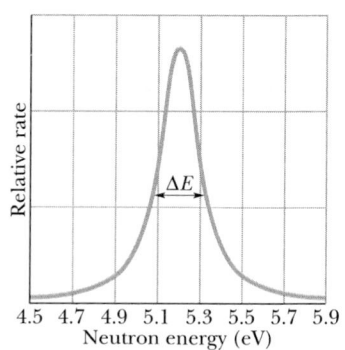

REVIEW & SUMMARY

The Nuclides

Approximately 2000 **nuclides** are known to exist. Each is characterized by an **atomic number** Z (the number of protons), a **neutron number** N, and a **mass number** A (the total number of **nucleons**—protons and neutrons). Thus $A = Z + N$. Nuclides with the same atomic number but different neutron numbers are **isotopes** of each other. Nuclei have a mean radius R given by

$$R = R_0 A^{1/3}, \quad (43\text{-}3)$$

where $R_0 \approx 1.2$ fm.

Mass–Energy Exchanges

The energy equivalent of one mass unit (u) is 931.5 MeV. The binding energy curve shows that middle-mass nuclides are the most stable and that energy can be released both by fission of high-mass nuclei and by fusion of low-mass nuclei.

The Nuclear Force

Nuclei are held together by an attractive force acting among the nucleons. It is thought to be a secondary effect of the **strong force** acting between the quarks that make up the nucleons. Nuclei can exist in a number of discrete energy states, each with a characteristic intrinsic angular momentum and magnetic moment.

Radioactive Decay

Most known nuclides are radioactive; they spontaneously decay

at a rate $R\ (= -dN/dt)$ that is proportional to the number N of radioactive atoms present, the proportionality constant being the **disintegration constant** λ. This leads to the law of exponential decay:

$$N = N_0 e^{-\lambda t}, \quad R = \lambda N = R_0 e^{-\lambda t}$$

$$\text{(radioactive decay).} \quad (43\text{-}6,\ 43\text{-}7)$$

The half-life $\tau = (\ln 2)/\lambda$ of a radioactive nuclide is the time required for the decay rate R (or the number N) in a sample to drop to half its initial value.

Alpha Decay

Some nuclides decay by emitting an alpha particle. Such decay is inhibited by a potential energy barrier that cannot be penetrated according to classical mechanics but is subject to tunneling according to quantum mechanics. The barrier penetrability, and thus the half-life for alpha decay, is very sensitive to the energy of the alpha particle.

Beta Decay

In **beta decay** either an electron or a positron is emitted by a nucleus, along with a neutrino. The emitted particles share the available disintegration energy. The electrons and positrons emitted in beta decay have a continuous spectrum of energies up to a limit $K_{\text{max}}\ (= Q = \Delta m\,c^2)$.

Radioactive Dating

Naturally occurring radioactive nuclides provide a means for estimating the dates of historic and prehistoric events. For example, the ages of organic materials can often be found by measuring their ^{14}C content; rock samples can be dated using the radioactive isotope ^{40}K.

Radiation Dosage

Three units are used to describe exposure to ionizing radiation. The **becquerel** (1 Bq = 1 decay per second) measures the **activity** of a source. The amount of energy actually absorbed is measured in **grays**, with 1 Gy corresponding to 1 J/kg. The estimated biological effect of the absorbed energy is measured in **sieverts**; a dose equivalent of 1 Sv causes the same biological effect regardless of the radiation type by which it was acquired.

Nuclear Models

The **collective** model of nuclear structure assumes that nucleons collide constantly and that relatively long-lived **compound nuclei** are formed when a projectile is captured. The formation of a compound nucleus and the eventual decay of that nucleus are totally independent events.

The **independent particle** model of nuclear structure assumes that each nucleon moves, essentially without collisions, in a quantized state within the nucleus. The model predicts nucleon levels and **magic numbers** of nucleons (2, 8, 20, 28, 50, 82, and 126) associated with closed shells of nucleons; nuclides with any of these numbers of neutrons or protons are particularly stable.

The **combined** model, in which extra nucleons occupy quantized states outside a central core of closed shells, is highly successful in predicting many nuclear properties.

QUESTIONS

1. Suppose the alpha particle of Sample Problem 43-1 is replaced with a proton of the same initial kinetic energy and also headed directly toward the nucleus of the gold atom. Will the distance from the center of the nucleus at which the proton stops be greater than, less than, or the same as that of the alpha particle?

2. In your body are there more protons than neutrons, more neutrons than protons, or about the same number of each?

3. The nuclide ^{244}Pu ($Z = 94$) is an alpha-particle emitter. Into which of the following nuclides does it decay: ^{240}Np ($Z = 93$), ^{240}U ($Z = 92$), ^{248}Cm ($Z = 96$), or ^{244}Am ($Z = 95$)?

4. A certain nuclide is said to be particularly stable. Does its binding energy per nucleon lie slightly above or slightly below the binding energy curve of Fig. 43-6?

5. Is the mass excess of an alpha particle (use a straight edge on Fig. 43-12) greater than or less than the particle's total binding energy (use the binding energy per nucleon from Fig. 43-6)?

6. The radionuclide ^{196}Ir decays by emitting electrons. (a) Into which square in Fig. 43-5 is it transformed? (b) Do further decays then occur?

7. A lead nuclide contains 82 protons. (a) If it also contained 82 neutrons, where would it be located on the plot of Fig. 43-4? (b) If such a nucleus could be formed, would it emit positrons, emit electrons, or be stable? (c) From Fig. 43-4, about how many neutrons do you expect to find in a stable lead nuclide?

8. The nuclide ^{238}U ($Z = 92$) can fission into two parts that have identical atomic numbers and mass numbers. (a) Is the nuclide ^{238}U above or below the $N = Z$ line of Fig. 43-4? (b) Are the two fragments above or below this line? (c) Are these fragments stable or radioactive?

9. Radionuclides decay exponentially, as in Eq. 43-7. Batteries, stars, and even students also decay, where ''decay'' stands for ''burn out.'' Do these items decay exponentially?

10. At $t = 0$, a sample of radionuclide A has the same decay rate as a sample of radionuclide B has at $t = 30$ min. The disintegration constants are λ_A and λ_B, with $\lambda_A < \lambda_B$. Will the two samples ever have (simultaneously) the same decay rate? (*Hint*: Sketch a graph of their activities.)

11. At $t = 0$, a sample of radionuclide A has twice the decay rate as a sample of radionuclide B. The disintegration constants are λ_A and λ_B, with $\lambda_A > \lambda_B$. Will the two samples ever have (simultaneously) the same decay rate?

12. Figure 43-15 gives the activities of three radioactive samples versus time. Rank the samples according to their (a) half-life and (b) disintegration constant, greatest first. (*Hint*: For (a), use a straight-edge on the graph.)

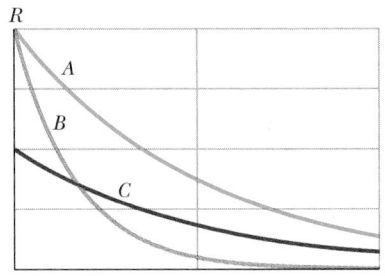

FIGURE 43-15 Question 12.

13. If the mass of a radioactive sample is doubled, do (a) the activity of the sample and (b) the disintegration constant of the sample increase, decrease, or remain the same?

14. At $t = 0$ we begin to observe two identical radioactive nuclei with a half-life of 5 min. At $t = 1$ min, one of the nuclei decays. Does that event increase or decrease the chance of the second nucleus decaying in the next 4 min, or is there no effect on the second nucleus?

15. The radionuclide ^{49}Sc has a half-life of 57.0 min. The counting rate of a sample of this nuclide at $t = 0$ is 6000 counts/min above the general background activity, which is 30 counts/min. Without actual computation, determine whether the counting rate of the sample will be about equal to the background rate in about 3 h, 7 h, 10 h, or a time much longer than 10 h.

16. The radionuclides ^{209}At and ^{209}Po emit alpha particles with energies of 5.65 and 4.88 MeV, respectively. Which nuclide has the longer half-life?

17. The magic numbers for nuclei are given in Section 43-8 as 2, 8, 20, 28, 50, 82, and 126. Are nuclides magic (that is, especially stable) when (a) only the mass number A, (b) only the atomic number Z, (c) only the neutron number N, or (d) either Z or N is equal to one of these numbers? Pick the correct statements.

18. (a) Which of the following nuclides are magic: (a) ^{122}Sn, (b) ^{132}Sn, (c) ^{98}Cd, (d) ^{198}Au, (e) ^{208}Pb? (b) Which, if any, are doubly magic?

EXERCISES & PROBLEMS

SECTION 43-1 Discovering the Nucleus

1E. Calculate the distance of closest approach for a head-on collision between a 5.30 MeV alpha particle and the nucleus of a copper atom.

2E. Assume that a gold nucleus has a radius of 6.23 fm and an alpha particle has a radius of 1.80 fm. What minimum energy must an incident alpha particle have in order to penetrate the gold nucleus?

3P. When an alpha particle collides elastically with a nucleus, the nucleus recoils. Suppose a 5.00 MeV alpha particle has a head-on elastic collision with a gold nucleus that is initially at rest. Give the kinetic energy of (a) the recoiling nucleus and (b) the rebounding alpha particle.

SECTION 43-2 Some Nuclear Properties

4E. A neutron star is a stellar object whose density is about that of nuclear matter, as calculated in Sample Problem 43-2. Suppose that the Sun were to collapse and become such a star without losing any of its present mass. What would be its radius?

5E. The nuclide ^{14}C contains (a) how many protons and (b) how many neutrons?

6E. The radius of a nucleus is measured, by electron-scattering methods, to be 3.6 fm. What is the likely mass number of the nucleus?

7E. Locate the nuclides displayed in Table 43-1 on the nuclidic chart of Fig. 43-4. Verify that they lie in the stability zone.

8E. Using a nuclidic chart, write the symbols for (a) all stable isotopes with $Z = 60$, (b) all radioactive nuclides with $N = 60$, and (c) all nuclides with $A = 60$.

9E. The electric potential energy of a uniform sphere of charge q and radius R is

$$U = \frac{3q^2}{20\pi\epsilon_0 R}.$$

(a) Find the electric potential energy for the nuclide ^{239}Pu, assumed to be spherical with radius 6.64 fm. (b) For this nuclide, compare the electric potential energy per nucleon, and also per proton, with the binding energy per nucleon of 7.56 MeV. (c) What do you conclude?

10E. The strong neutron excess of high-mass nuclei is illustrated by noting that most high-mass nuclides could never fission into two stable nuclei without neutrons being left over. For example, consider the spontaneous fission of a ^{235}U nucleus into two stable daughter nuclei with atomic numbers 39 and 53. By referring to Fig. 43-4, determine what the daughter nuclides are and the number of neutrons left over.

11E. Arrange the 25 nuclides $^{118-122}$Te, $^{117-121}$Sb, $^{116-120}$Sn, $^{115-119}$In, and $^{114-118}$Cd in squares in a nuclidic chart similar to Fig. 43-5. Draw in and label (a) all isobaric (constant A) lines and (b) all lines of constant neutron excess, defined as $N - Z$.

12E. Calculate and compare (a) the nuclear mass density ρ_m and (b) the nuclear charge density ρ_q for the fairly low-mass nuclide ^{55}Mn and for the fairly high-mass nuclide ^{209}Bi. (c) Are the differences what you would expect? Explain.

13E. Verify the binding energy per nucleon given in Table 43-1 for ^{239}Pu, 7.56 MeV/nucleon. The needed atomic masses are 239.05216 u (^{239}Pu), 1.00783 u (^{1}H), and 1.00867 u (neutron).

14E. (a) Show that an approximate formula for the mass M of an atom is $M = Am_p$, where A is the mass number and m_p is the proton mass. (b) What percent error is committed in using this formula to calculate the masses of the atoms in Table 43-1? The mass of the bare proton is 1.007276 u. (c) Is this formula accurate enough to be used for calculations of nuclear binding energy?

15E. The characteristic nuclear time is a useful but loosely defined quantity, taken to be the time required for a nucleon with a few million electron-volts of kinetic energy to travel a distance equal to the diameter of a middle-mass nuclide. What is the order of magnitude of this quantity? Consider 5 MeV neutrons traversing a nuclear diameter of ^{197}Au; use Eq. 43-3.

16E. Nuclear radii may be measured by scattering high-energy electrons from nuclei. (a) What is the de Broglie wavelength for 200 MeV electrons? (c) Are these electrons suitable probes for this purpose?

17E. Because a nucleon is confined to a nucleus, we can take the uncertainty in its position to be approximately the nuclear radius R. What does the uncertainty principle say about the kinetic energy of a nucleon in a nucleus with, say, $A = 100$? (*Hint*: Take the uncertainty in momentum Δp to be the actual momentum p.)

18E. The atomic masses of ^{1}H, ^{12}C, and ^{238}U are 1.007825 u, 12.000000 u (this one is exact by definition), and 238.050785 u, respectively. (a) What would these masses be if the mass unit were defined to give the mass of ^{1}H as (exactly) 1.000000 u? (b)

Use your result to suggest why this perhaps obvious choice was not made.

19P. (a) Show that the energy tied up in nuclear, or strong force, bonds is proportional to A, the mass number of the nucleus in question. (b) Show that the energy tied up in Coulomb force bonds between the protons is proportional to $Z(Z - 1)$. (c) Show that, as we move to larger and larger nuclei (see Fig. 43-4), the importance of the Coulomb force increases more rapidly than does that of the strong force.

20P. In the periodic table, the entry for magnesium is

$$\boxed{\begin{array}{c} 12 \\ Mg \\ 24.312 \end{array}}$$

There are three isotopes:

^{24}Mg, atomic mass = 23.98504 u.

^{25}Mg, atomic mass = 24.98584 u.

^{26}Mg, atomic mass = 25.98259 u.

The abundance of ^{24}Mg is 78.99% by weight. Calculate the abundances of the other two isotopes.

21P. You are asked to pick apart an alpha particle (^{4}He) by removing, in sequence, a proton, a neutron, and a proton. Calculate (a) the work required for each step, (b) the total binding energy of the alpha particle, and (c) the binding energy per nucleon. Some needed atomic masses are

^{4}He	4.00260 u	^{2}H	2.01410 u
^{3}H	3.01605 u	^{1}H	1.00783 u
n	1.00867 u		

22P. Because the neutron has no charge, its mass must be found in some way other than by using a mass spectrometer. When a neutron and a proton meet (assume both to be almost stationary), they combine and form a deuteron, emitting a gamma ray whose energy is 2.2233 MeV. The masses of the proton and the deuteron are 1.007 825 035 and 2.014 101 9 u, respectively. Find the mass of the neutron from these data, to as many significant figures as the data warrant. (A value of the mass–energy conversion factor that is more precise than the one presented in the text is 931.502 MeV/u.)

23P. A penny has a mass of 3.0 g. Calculate the nuclear energy that would be required to separate all the neutrons and protons in this coin. Ignore the binding energy of the electrons, and for simplicity assume that the penny is made entirely of ^{63}Cu atoms (of mass 62.92960 u). The masses of the proton and the neutron are 1.00783 and 1.00867 u, respectively.

24P. To simplify calculations, atomic masses are sometimes tabulated not as the actual atomic mass m but as $(m - A)c^2$, where A is the mass number expressed in atomic mass units. This quantity, usually reported in millions of electron-volts, is called the *mass excess*, represented with symbol Δ. Using data from Sample Problem 43-3, find the mass excesses for (a) ^{1}H, (b) the neutron, and (c) ^{120}Sn.

25P. (See Problem 24.) Show that the total binding energy of a nuclide can be written as

$$E = Z\Delta_H + N\Delta_n - \Delta,$$

where Δ_H, Δ_n, and Δ are the appropriate mass excesses. Using this method calculate the binding energy per nucleon for ^{197}Au. Compare your result with the value listed in Table 43-1. The needed mass excesses, rounded to three significant figures, are $\Delta_H = +7.29$ MeV, $\Delta_n = +8.07$ MeV, and $\Delta_{197} = -31.2$ MeV. Note the economy of calculation that results when mass excesses are used in place of the actual masses.

SECTION 43-3 Radioactive Decay

26E. The half-life of a particular radioactive isotope is 6.5 h. If there are initially 48×10^{19} atoms of this isotope, how many remain after 26 h?

27E. The half-life of a radioactive isotope is 140 d. How many days would it take for the decay rate of a sample of this isotope to fall to one-fourth of its initial value?

28E. A radioactive nuclide has a half-life of 30 y. What fraction of an initially pure sample of this nuclide will remain undecayed after (a) 60 y and (b) 90 y?

29E. Consider an initially pure 3.4 g sample of ^{67}Ga, an isotope that has a half-life of 78 h. (a) What is its initial decay rate? (b) What is its decay rate 48 h later?

30E. A radioactive isotope of mercury, ^{197}Hg, decays into gold, ^{197}Au, with a disintegration constant of 0.0108 h^{-1}. (a) Calculate its half-life. What fraction of a sample will remain (b) after three half-lives and (c) after 10.0 days?

31E. From data presented in the first few paragraphs of Section 43-3, deduce (a) the disintegration constant λ and (b) the half-life of ^{238}U.

32E. The plutonium isotope ^{239}Pu is produced as a by-product in nuclear reactors, hence is accumulating in our environment. It is radioactive, decaying by alpha decay with a half-life of 2.41×10^4 y. (a) How many nuclei of Pu constitute a chemically lethal dose of 2 mg? (b) What is the decay rate of this amount?

33E. Cancer cells are more vulnerable to x and gamma radiation than are healthy cells. In the past, the standard source for radiation therapy was radioactive ^{60}Co, which decays into an excited nuclear state of ^{60}Ni, which immediately drops into the ground state, emitting two gamma-ray photons, each with an approximate energy of 1.2 MeV. The controlling beta-decay half-life is 5.27 y. How many radioactive ^{60}Co nuclei were present in a 6000 Ci source of the type used in hospitals? (Energetic particles from linear accelerators are now used in radiation therapy.)

34P. After long effort, in 1902 Marie and Pierre Curie succeeded in separating from uranium ore the first substantial quantity of radium, one decigram of pure RaCl$_2$. The radium was the radioactive isotope ^{226}Ra, which has a decay half-life of 1600 y. (a) How many radium nuclei had the Curies isolated? (b) What was the decay rate of their sample, in disintegrations per second?

35P. The radionuclide ^{64}Cu has a half-life of 12.7 h. How much of an initially pure 5.50 g sample of ^{64}Cu will decay during the 2.00 h period beginning 14.0 h later?

36P. The radionuclide ^{32}P ($\tau = 14.28$ d) is often used as a tracer to follow the course of biochemical reactions involving phosphorus. (a) If the counting rate in a particular experimental setup is initially 3050 counts/s, after what time will it fall to 170 counts/s? (b) A solution containing ^{32}P is fed to the root system of an experimental tomato plant and the ^{32}P activity in a leaf is measured 3.48 days later. By what factor must this reading be multiplied to correct for the decay that has occurred since the experiment began?

37P. A source contains two phosphorus radionuclides, ^{32}P ($\tau = 14.3$ d) and ^{33}P ($\tau = 25.3$ d). Initially 10.0% of the decays come from ^{33}P. How long must one wait until 90.0% do so?

38P. A 1.00 g sample of samarium emits alpha particles at a rate of 120 particles/s. The responsible isotope is ^{147}Sm, whose natural abundance in bulk samarium is 15.0%. Calculate the half-life for the decay process.

39P. Plutonium isotope ^{239}Pu decays by alpha decay with a half-life of 24,100 y. How many grams of helium are produced by an initially pure 12.0 g sample of ^{239}Pu after 20,000 y? (Consider only the helium produced directly by the plutonium and not by any by-products of the decay process.)

40P. Calculate the mass of a sample of (initially pure) ^{40}K with an initial decay rate of 1.70×10^5 disintegrations/s. The isotope has a half-life of 1.28×10^9 y.

41P. One of the dangers of radioactive fallout from a nuclear bomb is its ^{90}Sr, which decays with a 29 year half-life. Because it has chemical properties much like those of calcium, the strontium, if ingested by a cow, becomes concentrated in the cow's milk. Some of the ^{90}Sr ends up in the bones of whoever drinks the milk. The energetic electrons emitted in the beta decay of ^{90}Sr damage the bone marrow and thus impair the production of red blood cells. A 1 megaton bomb produces approximately 400 g of ^{90}Sr. If the fallout spreads uniformly over a 2000 km^2 area, what ground area would receive radioactivity equal to the allowed bone burden for one person, which is 74,000 counts/s?

42P. After a brief neutron irradiation of silver, two isotopes are present: ^{108}Ag ($\tau = 2.42$ min) with an initial decay rate of 3.1×10^5/s, and ^{110}Ag ($\tau = 24.6$ s) with an initial decay rate of 4.1×10^6/s. Make a semilog plot similar to Fig. 43-8 showing the total combined decay rate of the two isotopes as a function of time from $t = 0$ until $t = 10$ min. We used Fig. 43-8 to illustrate the extraction of the half-life for simple (one isotope) decays. Given only your plot of total decay rate, suggest a way to analyze it in order to find the half-lives of both isotopes.

43P. A certain radionuclide is being manufactured, say, in a cyclotron, at a constant rate R. It is also decaying, with a disintegration constant λ. Assume that the production process has been going on for a time that is long compared to the half-life of the radionuclide. Show that the number of radioactive nuclei present after such time remains constant and is given by $N = R/\lambda$. Now show that this result holds no matter how many radioactive nuclei

were present initially. The nuclide is said to be in *secular equilibrium* with its source; in this state its decay rate is just equal to its production rate.

44P. (See Problem 43.) The radionuclide ^{56}Mn has a half-life of 2.58 h and is produced in a cyclotron by bombarding a manganese target with deuterons. The target contains only the stable manganese isotope ^{55}Mn and the reaction that produces ^{56}Mn is

$$^{55}\text{Mn} + \text{d} \rightarrow {}^{56}\text{Mn} + \text{p}.$$

After bombardment for a time much longer than 2.58 h, the activity of the target, due to ^{56}Mn, is 8.88×10^{10} s^{-1}. (a) At what constant rate R are ^{56}Mn nuclei being produced in the cyclotron during the bombardment? (b) At what rate are they decaying (also during the bombardment)? (c) How many ^{56}Mn nuclei are present at the end of the bombardment? (d) What is their total mass?

45P. (See Problems 43 and 44.) A radium source contains 1.00 mg of ^{226}Ra, which decays with a half-life of 1600 y to produce ^{222}Rn, a noble gas. This radon isotope in turn decays by alpha emission with a half-life of 3.82 d. (a) What is the rate of disintegration of ^{226}Ra in the source? (b) How long does it take for the radon to come to secular equilibrium with its radium parent? (c) At what rate is the radon then decaying? (d) How much radon is in equilibrium with its radium parent?

SECTION 43-4 Alpha Decay

46E. Consider a ^{238}U nucleus to be made up of an alpha particle (^{4}He) and a residual nucleus (^{234}Th). Plot the electrostatic potential energy $U(r)$, where r is the distance between these particles. Cover the approximate range 10 fm $< r <$ 100 fm and compare your plot with that of Fig 43-9.

47E. Generally, more massive nuclides tend to be more unstable to alpha decay. For example, the most stable isotope of uranium, ^{238}U, has an alpha decay half-life of 4.5×10^9 y. The most stable isotope of plutonium is ^{244}Pu with an 8.0×10^7 y half-life, and for curium we have ^{248}Cm and 3.4×10^5 y. When half of an original sample of ^{238}U has decayed, what fractions of the original samples of these isotopes of plutonium and curium are left?

48P. A ^{238}U nucleus emits a 4.196 MeV alpha particle. Calculate the disintegration energy Q for this process, taking the recoil energy of the residual ^{234}Th nucleus into account.

49P. Consider that a ^{238}U nucleus emits (a) an alpha particle or (b) a sequence of neutron, proton, neutron, proton. Calculate the energy released in each case. (c) Convince yourself both by reasoned argument and by direct calculation that the difference between these two numbers is just the total binding energy of the alpha particle. Find that binding energy. Some needed atomic and particle masses are

^{238}U	238.050 79 u	^{234}Th	234.043 63 u
^{237}U	237.048 73 u	^{4}He	4.002 60 u
^{236}Pa	236.048 91 u	^{1}H	1.007 83 u
^{235}Pa	235.045 44 u	n	1.008 67 u

50P. Under certain circumstances, a nucleus can decay by emitting a particle more massive than an alpha particle. Such decays are very rare and have only recently been observed. Consider the decays

$$^{223}\text{Ra} \rightarrow {}^{209}\text{Pb} + {}^{14}\text{C}$$

and

$$^{223}\text{Ra} \rightarrow {}^{219}\text{Rn} + {}^{4}\text{He}.$$

(a) Calculate the Q values for these decays and determine that both are energetically possible. (b) The Coulomb barrier height for alpha particles in this decay is 30.0 MeV. What is the barrier height for ^{14}C decay? The needed atomic masses are

^{223}Ra	223.018 50 u	^{14}C	14.003 24 u
^{209}Pb	208.981 07 u	^{4}He	4.002 60 u
^{219}Rn	219.009 48 u		

51P. Heavy radionuclides emit an alpha particle rather than other combinations of nucleons because the alpha particle is such a stable, tightly bound structure. To confirm this statement, calculate the disintegration energies for these hypothetical decay processes and discuss the meaning of your findings:

$$^{235}\text{U} \rightarrow {}^{232}\text{Th} + {}^{3}\text{He}, \qquad Q_3;$$

$$^{235}\text{U} \rightarrow {}^{231}\text{Th} + {}^{4}\text{He}, \qquad Q_4;$$

$$^{235}\text{U} \rightarrow {}^{230}\text{Th} + {}^{5}\text{He}, \qquad Q_5.$$

The needed atomic masses are

^{232}Th	232.0381 u	^{3}He	3.0160 u
^{231}Th	231.0363 u	^{4}He	4.0026 u
^{230}Th	230.0331 u	^{5}He	5.0122 u
^{235}U	235.0439 u		

SECTION 43-5 Beta Decay

52E. A certain stable nuclide, after absorbing a neutron, emits an electron, and the new nuclide splits spontaneously into two alpha particles. Identify the nuclide.

53E. The cesium isotope ^{137}Cs is present in the fallout from aboveground detonations of nuclear bombs. Because it decays with a slow (30.2 y) half-life into ^{137}Ba, releasing considerable energy in the process, it is of environmental concern. The atomic masses of the Cs and Ba are 136.9071 and 136.9058 u, respectively; calculate the total energy released in such a decay.

54E. Large mass radionuclides, which may be either alpha or beta emitters, belong to one of four decay chains, depending on whether their mass numbers A are of the form $4n$, $4n + 1$, $4n + 2$, or $4n + 3$, where n is a positive integer. (a) Justify this statement and show that if a nuclide belongs to one of these families, all its decay products belong to the same family. (b) Classify these nuclides as to family: ^{235}U, ^{236}U, ^{238}U, ^{239}Pu, ^{240}Pu, ^{245}Cm, ^{246}Cm, ^{249}Cf, and ^{253}Fm.

55E. A free neutron decays according to Eq. 43-12. If the neutron–hydrogen atom mass difference is 840 μu, what is the maximum kinetic energy K_{max} of the electron energy spectrum?

56E. An electron is emitted from a middle-mass nuclide ($A = 150$, say) with a kinetic energy of 1.0 MeV. (a) What is its de Broglie wavelength? (b) Calculate the radius of the emitting nucleus. (c) Can such an electron be confined as a standing wave in a ''box'' of such dimensions? (d) Can you use these numbers to disprove the argument (long since abandoned) that electrons actually exist in nuclei?

57P. Some radionuclides decay by capturing one of their own atomic electrons, a K-shell electron, say. An example is

$$^{49}\text{V} + e^- \rightarrow {}^{49}\text{Ti} + \nu, \qquad \tau = 331 \text{ d}.$$

Show that the disintegration energy Q for this process is given by

$$Q = (m_{\text{V}} - m_{\text{Ti}})c^2 - E_K,$$

where m_{V} and m_{Ti} are the atomic masses of ^{49}V and ^{49}Ti, respectively, and E_K is the binding energy of the vanadium K electron. (*Hint:* Put $\mathbf{m}_{\text{V}}$ and $\mathbf{m}_{\text{Ti}}$ as the corresponding nuclear masses and proceed as in Sample Problem 43-7.)

58P. Find the disintegration energy Q for the decay of ^{49}V by K-electron capture, as described in Problem 57. The needed data are $m_{\text{V}} = 48.94852$ u, $m_{\text{Ti}} = 48.94787$ u, and $E_K = 5.47$ keV.

59P. The radionuclide ^{11}C decays according to

$$^{11}\text{C} \rightarrow {}^{11}\text{B} + e^+ + \nu, \qquad \tau = 20.3 \text{ min}.$$

The maximum energy of the emitted positrons is 0.960 MeV. (a) Show that the disintegration energy Q for this process is given by

$$Q = (m_{\text{C}} - m_{\text{B}} - 2m_e)c^2,$$

where m_{C} and m_{B} are the atomic masses of ^{11}C and ^{11}B, respectively, and m_e is the mass of a positron. (b) Given that $m_{\text{C}} = 11.011434$ u, $m_{\text{B}} = 11.009305$ u, and $m_e = 0.0005486$ u, calculate Q and compare it with the maximum energy of the emitted positron given above. (*Hint:* Let $\mathbf{m}_{\text{C}}$ and $\mathbf{m}_{\text{B}}$ be the nuclear masses and proceed as in Sample Problem 43-7 for beta decay. Note that positron decay is an exception to the general rule that if atomic masses are used in nuclear decay calculations, the mass of the emitted electron is automatically taken care of.)

60P. Two radioactive materials that are unstable with regard to alpha decay, ^{238}U and ^{232}Th, and one that is unstable with regard to beta decay, ^{40}K, are sufficiently abundant in granite to contribute significantly to the heating of Earth through the decay energy produced. The alpha-unstable isotopes give rise to decay chains that stop when stable lead isotopes are formed. The isotope ^{40}K has a single beta decay. Here is the information:

PARENT	DECAY MODE	HALF-LIFE (y)	STABLE END POINT	Q (MeV)	f (ppm)
^{238}U	α	4.47×10^9	^{206}Pb	51.7	4
^{232}Th	α	1.41×10^{10}	^{208}Pb	42.7	13
^{40}K	β	1.28×10^9	^{40}Ca	1.31	4

In the table Q is the total energy released in the decay of one parent nucleus to the final stable endpoint and f is the abundance of the isotope in kilograms per kilogram of granite; ppm means parts per million. (a) Show that these materials give rise to a total heat production of 1.0×10^{-9} W for each kilogram of granite. (b) Assuming that there is 2.7×10^{22} kg of granite in a 20 km thick spherical shell at the surface of Earth, estimate the power of this decay process over all of Earth. Compare this power with the total solar power intercepted by Earth, 1.7×10^{17} W.

61P*. The radionuclide ^{32}P decays to ^{32}S as described by Eq. 43-10. In a particular decay event, a 1.71 MeV electron is emitted, the maximum possible value. What is the kinetic energy of the recoiling ^{32}S atom in this event? (*Hint:* For the electron it is necessary to use the relativistic expressions for kinetic energy and linear momentum. Newtonian mechanics may safely be used for the relatively slow-moving ^{32}S atom.)

SECTION 43-6 Radioactive Dating

62E. ^{238}U decays to ^{206}Pb with a half-life of 4.47×10^9 y. Although the decay occurs in many individual steps, the first step has by far the longest half-life; therefore one can often consider the decay to go directly to lead. That is,

$$^{238}\text{U} \rightarrow {}^{206}\text{Pb} + \text{various decay products.}$$

A rock is found to contain 4.20 mg of ^{238}U and 2.135 mg of ^{206}Pb. Assume that the rock contained no lead at formation, so all the lead now present arose from the decay of uranium. (a) How many atoms of ^{238}U and ^{206}Pb does the rock now contain? (b) How many atoms of ^{238}U did the rock contain at formation? (c) What is the age of the rock?

63E. A 5.00 g charcoal sample from an ancient fire pit has a ^{14}C activity of 63.0 disintegrations/min. A living tree has a ^{14}C activity of 15.3 disintegrations/min per 1.00 g. The half-life of ^{14}C is 5730 y. How old is the charcoal sample?

64P. A particular rock is thought to be 260 million years old. If it contains 3.70 mg of ^{238}U, how much ^{206}Pb should it contain? See Exercise 62.

65P. A rock, recovered from far underground, is found to contain 0.86 mg of ^{238}U, 0.15 mg of ^{206}Pb, and 1.6 mg of ^{40}Ar. How much ^{40}K will it likely contain? Needed half-lives are listed in Problem 60.

SECTION 43-7 Measuring Radiation Dosage

66E. A Geiger counter records 8700 counts in 1 min. Calculate the activity of the source in becquerels and in curies, assuming that the counter records all decays.

67E. The nuclide ^{198}Au, with a half-life of 2.70 d, is used in cancer therapy. What mass of this nuclide is required to produce an activity of 250 Ci?

68E. An airline pilot spends an average of 20 h per week flying at 35,000 ft, at which altitude the dose equivalent due to cosmic radiation is 7.0 μSv/h. What is the annual (52 week) dose equivalent from this source alone? Note that the maximum permitted yearly dose equivalent (from all sources) for the general population is 5 mSv, and for radiation workers it is 50 mSv.

69E. A 75 kg person receives a whole-body radiation dose of 2.4×10^{-4} Gy, delivered by alpha particles for which the RBE factor is 12. Calculate (a) the absorbed energy in joules and (b) the dose equivalent in sieverts and in rem.

70P. A typical chest x-ray radiation dose is 250 μSv, delivered by x rays with an RBE factor of 0.85. Assuming that the mass of the exposed tissue is one-half the patient's mass of 88 kg, calculate the energy absorbed in joules.

71P. An 85 kg worker at a breeder reactor plant accidentally ingests 2.5 mg of ^{239}Pu dust. ^{239}Pu has a half-life of 24,100 y, decaying by alpha decay. The energy of the emitted alpha particles is 5.2 MeV, with an RBE factor of 13. Assume that the plutonium resides in the worker's body for 12 h and that 95% of the emitted alpha particles are stopped within the body. Calculate (a) the number of plutonium atoms ingested, (b) the number that decay during the 12 h, (c) the energy absorbed by the body, (d) the resulting physical dose in grays, and (e) the dose equivalent in sieverts.

SECTION 43-8 Nuclear Models

72E. An intermediate nucleus in a particular nuclear reaction decays within 10^{-22} s of its formation. (a) What is the uncertainty ΔE in our knowledge of this intermediate state? (b) Can this state be called a compound nucleus? (See Sample Problem 43-10.)

73E. A typical kinetic energy for a nucleon in a middle-mass nucleus may be taken as 5.00 MeV. To what effective nuclear temperature does this correspond, using the assumptions of the collective model of nuclear structure?

74E. In the following list of nuclides, identify (a) those with filled nucleon shells, (b) those with one nucleon outside a filled shell, and (c) those with one vacancy in an otherwise filled shell: ^{13}C, ^{18}O, ^{40}K, ^{49}Ti, ^{60}Ni, ^{91}Zr, ^{92}Mo, ^{121}Sb, ^{143}Nd, ^{144}Sm, ^{205}Tl, and ^{207}Pb.

75P. Consider the three formation processes shown for the compound nucleus ^{20}Ne in Fig. 43-13. What energy must (a) the alpha particle, (b) the proton, and (c) the γ-ray photon have to provide 25.0 MeV of excitation energy to the compound nucleus? Some needed atomic and particle masses are

^{20}Ne	19.992 44 u	α	4.002 60 u
^{19}F	18.998 40 u	p	1.007 83 u
^{16}O	15.994 91 u		

76P. Consider the three decay processes shown for the compound nucleus ^{20}Ne in Fig. 43-13. If the compound nucleus is initially at rest and has an excitation energy of 25.0 MeV, what kinetic energy, measured in the laboratory, will (a) the deuteron, (b) the neutron, and (c) the ^{3}He nucleus have when the nucleus decays? Some needed atomic and particle masses are

^{20}Ne	19.992 44 u	d	2.014 10 u
^{19}Ne	19.001 88 u	n	1.008 67 u
^{18}F	18.000 94 u	^{3}He	3.016 03 u
^{17}O	16.999 13 u		

77P. The nuclide ^{208}Pb is doubly magic in that its proton number Z ($= 82$) and its neutron number N ($= 126$) represent filled nucleon shells. An additional proton would yield ^{209}Bi, and an additional neutron ^{209}Pb. These extra nucleons should be easier to remove than a proton or a neutron from the filled shells of ^{208}Pb. (a) Calculate the energy required to remove the extra proton from ^{209}Bi and compare it with the energy required to remove a proton from the filled proton shell of ^{208}Pb. (b) Calculate the energy required to remove the extra neutron from ^{209}Pb and compare it with the energy required to remove a neutron from the filled neutron shell of ^{208}Pb. Do your results agree with expectation? Use these atomic mass data:

NUCLIDE	Z	N	ATOMIC MASS (u)
^{209}Bi	$82 + 1$	126	208.9804
^{208}Pb	82	126	207.9767
^{207}Tl	$82 - 1$	126	206.9774
^{209}Pb	82	$126 + 1$	208.9811
^{207}Pb	82	$126 - 1$	206.9759

The masses of the proton and the neutron are 1.00783 and 1.00867 u, respectively.

78P. The nucleus ^{91}Zr ($Z = 40$, $N = 51$) has a single neutron outside a filled 50 neutron core. Because 50 is a magic number, this single neutron should perhaps be especially loosely bound. (a) What is its binding energy? (b) What is the binding energy of the next neutron, which would have to be extracted from the filled core? (c) What is the binding energy per nucleon for the entire nucleus? Compare these three numbers and discuss. Some needed atomic masses are

$$
\begin{array}{llll}
^{91}\text{Zr} & 90.905\ 64\ \text{u} & \text{n} & 1.008\ 67\ \text{u} \\
^{90}\text{Zr} & 89.904\ 71\ \text{u} & \text{p} & 1.007\ 83\ \text{u} \\
^{89}\text{Zr} & 88.908\ 90\ \text{u} & &
\end{array}
$$

79P. Calculate (a) the energy needed to remove a proton from a ^{121}Sb nucleus, and (b) the energy needed to remove a proton from the resulting ^{120}Sn nucleus. Needed atomic masses are

$$
\begin{array}{ll}
^{121}\text{Sb} & 120.9038\ \text{u} \\
^{120}\text{Sn} & 119.9022\ \text{u} \\
^{119}\text{In} & 118.9058\ \text{u}
\end{array}
$$

Additional Problem

80. At the end of World War II, Dutch authorities arrested Dutch artist Hans van Meegeren for treason because, during the war, he sold a masterpiece painting to the infamous Nazi Hermann Goering. The painting, *Christ and His Disciples at Emmaus* by the Dutch master Johannes Vermeer (1632–1675), had been discovered in 1937 by van Meegeren, having been lost for almost 300 years. Soon after the discovery, art experts proclaimed that

Emmaus was possibly the best Vermeer ever seen. Selling such a Dutch national treasure to the enemy was unthinkable treason.

However, shortly after being imprisoned, van Meegeren suddenly announced that he, not Vermeer, had painted *Emmaus*. He explained that he had carefully mimicked Vermeer's style, using a 300-y-old canvas and Vermeer's choice of pigments; he then signed Vermeer's name to the work and baked the painting to give it an authentically old look.

Was van Meegeren lying to avoid a conviction of treason, hoping to be convicted of only the lesser crime of fraud? To art experts, *Emmaus* certainly looked like a Vermeer but, at the time of van Meegeren's trial in 1947, there was no scientific way to answer the question. However, in 1968 Bernard Keisch of Carnegie-Mellon University was able to answer the question with newly developed techniques of radioactive analysis.

Specifically, he analyzed a small sample of white lead-bearing pigment removed from *Emmaus*. This pigment is refined from lead ore, in which the lead is produced by a long radioactive decay series that starts with unstable ^{238}U and ends with stable ^{206}Pb. To follow the spirit of Keisch's analysis, focus on the following, abbreviated portion of that decay series,

$$
^{230}\text{Th} \xrightarrow[75.4\ \text{ky}]{} {}^{226}\text{Ra} \xrightarrow[1.60\ \text{ky}]{} {}^{210}\text{Pb} \xrightarrow[22.6\ \text{y}]{} {}^{206}\text{Pb},
$$

in which intermediate, relatively short-lived radionuclides have been omitted. The longer and more important half-lives in this portion of the decay series are indicated.

(a) Show that in a unit sample of lead ore, the rate at which the number of ^{210}Pb changes is given by

$$
\frac{dN_{210}}{dt} = \lambda_{226} N_{226} - \lambda_{210} N_{210},
$$

where N_{210} and N_{226} are the numbers of ^{210}Pb and ^{226}Ra in the unit sample and λ_{210} and λ_{226} are the corresponding disintegration constants.

Because the decay series has been active for billions of years and because the half-life of ^{210}Pb is much less than that of ^{226}Ra, the nuclides ^{226}Ra and ^{210}Pb are in *equilibrium*; that is, their numbers, or concentrations, in the sample do not change. (b) What is the ratio R_{226}/R_{210} of the activities of these nuclides in the unit sample of lead ore? (c) What is the ratio N_{226}/N_{210} of their numbers?

When lead pigment is refined from the ore, most of the ^{226}Ra is eliminated. Assume that only 1.00% remains. Just after the pigment is produced, what are the ratios (d) R_{226}/R_{210} and (e) N_{226}/N_{210}?

Keisch realized that with time the ratio R_{226}/R_{210} of the pigment would gradually change from the value of freshly refined pigment back to the value of the ore as equilibrium between the ^{210}Pb and the remaining ^{226}Ra is established in the pigment. If *Emmaus* was painted by Vermeer and the sample of pigment taken from it was 300 years old when examined in 1968, the ratio would be close to the answer of (b). If *Emmaus* was painted by van Meegeren in the 1930s and the sample was only about 30 years old, the ratio would be close to the answer of (d). Keisch found a ratio of 0.09. (f) Is *Emmaus* a Vermeer?

44
Energy from the Nucleus

The image that has transfixed the world since World War II. When Robert Oppenheimer, the head of the scientific team that developed the atomic bomb, witnessed the first atomic explosion, he quoted from a sacred Hindu text: ''Now I am become Death, the destroyer of worlds.'' What is the physics behind this image that has so horrified the world?

44-1 THE ATOM AND ITS NUCLEUS

When we get energy from coal by burning the fuel in a furnace, we are tinkering with atoms of carbon and oxygen, rearranging their outer *electrons* into more stable combinations. When we get energy from uranium in a nuclear reactor, we are again burning a fuel, but then we are tinkering with its nucleus, rearranging its *nucleons* into more stable combinations.

Electrons are held in atoms by the electromagnetic Coulomb force, and it takes only a few electron-volts to pull one of them out. On the other hand, nucleons are held in nuclei by the strong force, and it takes a few *million* electron-volts to pull one of *them* out. This factor of a few million is reflected in the fact that we can extract about that much more energy from a kilogram of uranium than we can from a kilogram of coal.

In both atomic and nuclear burning, the release of energy is accompanied by a decrease in mass, according to Einstein's equation $Q = \Delta m \, c^2$. The central difference between burning uranium and burning coal is that, in the former case, a much larger fraction of the available mass (again, by a factor of a few million) is consumed.

The different processes that can be used for atomic or nuclear burning do provide different levels of power, or rates at which the energy is delivered. In the nuclear case we can burn a kilogram of uranium explosively in a bomb or slowly in a power reactor. In the atomic case, we might consider exploding a stick of dynamite or digesting a jelly doughnut. (Surprisingly, the total energy release is greater in the second case than in the first!)

Table 44-1 shows how much energy can be extracted from 1 kg of matter by doing various things to it. Instead of reporting the energy directly, the table shows how long the extracted energy could operate a 100 W lightbulb. Only processes in the first three rows of the table have actually been carried out; the remaining three represent theoretical

limits that may not be attainable in practice. The bottom row, the total mutual annihilation of matter and antimatter, is an ultimate energy production goal. When you have converted all the available mass, you can do no more.

Keep in mind that the comparisons of Table 44-1 are computed on a per-unit-mass basis. Kilogram for kilogram, you get several million times more energy from uranium than you do from coal or from falling water. On the other hand, there is a lot of coal in Earth's crust, and water is easily backed up behind a dam.

44-2 NUCLEAR FISSION: THE BASIC PROCESS

In 1932 English physicist James Chadwick discovered the neutron. A few years later Enrico Fermi and his co-workers in Rome found that when various elements are bombarded by neutrons, new radioactive elements are produced. Fermi had predicted that the neutron, being uncharged, would be a useful nuclear projectile; unlike the proton or the alpha particle, it experiences no repulsive Coulomb force when it approaches a nuclear surface. Even *thermal neutrons,* which are slowly moving neutrons in thermal equilibrium with the surrounding matter at room temperature, with a mean kinetic energy of only about 0.04 eV, are useful projectiles in nuclear studies.

In the late 1930s physicist Lise Meitner and chemists Otto Hahn and Fritz Strassmann, working in Berlin and following up the work of Fermi and co-workers, bombarded solutions of uranium salts with such thermal neutrons. They found that after the bombardment a number of new radionuclides were present. In 1939 (after Meitner had fled to neutral Sweden), one of the radionuclides produced in this way was positively identified, by repeated tests, as barium. But how, Hahn and Strassmann wondered, could this middle-mass element ($Z = 56$) be produced by bombarding uranium ($Z = 92$) with neutrons?

The puzzle was solved within a few weeks by Meitner and her nephew Otto Frisch. They suggested the mechanism by which a uranium nucleus, having absorbed a thermal neutron, could split, with the release of energy, into two roughly equal parts, one of which might well be barium. Frisch named the process **fission**.

Meitner's central role in the discovery of fission was not fully recognized until recent historical research brought it to light. She did not share in the Nobel prize in chemistry that was awarded to Otto Hahn in 1944. However, both Hahn and Meitner have been honored by having elements named after them: hahnium (symbol Ha, $Z = 105$) and meitnerium (symbol Mt, $Z = 109$.)

TABLE 44-1 **ENERGY RELEASED BY 1 KG OF MATTER**

FORM OF MATTER	PROCESS	TIME[a]
Water	A 50 m waterfall	5 s
Coal	Burning	8 h
Enriched UO_2	Fission in a reactor	690 y
^{235}U	Complete fission	3×10^4 y
Hot deuterium gas	Complete fusion	3×10^4 y
Matter and antimatter	Complete annihilation	3×10^7 y

[a] This column shows how long the energy generated could power a 100 W lightbulb.

Fission, a Closer Look

Figure 44-1 shows the distribution by mass number of the fragments produced when ^{235}U is bombarded with thermal neutrons. The most probable mass numbers, occurring in about 7% of the events, are centered around $A \approx 95$ and $A \approx 140$. Curiously, the "double-peaked" character of Fig. 44-1 is still not understood.

In a typical ^{235}U fission event, a ^{235}U nucleus absorbs a thermal neutron, producing a compound nucleus ^{236}U in a highly excited state. It is *this* nucleus that actually undergoes fission, splitting into two fragments. These fragments —between them—rapidly emit two neutrons, leaving (in a typical case) ^{140}Xe ($Z = 54$) and ^{94}Sr ($Z = 38$) as fission fragments. Thus the overall fission equation for this event is

$$^{235}\text{U} + \text{n} \rightarrow {}^{236}\text{U} \rightarrow {}^{140}\text{Xe} + {}^{94}\text{Sr} + 2\text{n}. \quad (44\text{-}1)$$

Note that during the formation and fission of the compound nucleus, there is conservation of the number of protons and of the number of neutrons involved in the process (and thus their total number and the net charge).

In Eq. 44-1, the fragments ^{140}Xe and ^{94}Sr are both highly unstable, undergoing beta decay (with the conversion of a neutron to a proton and the emission of an electron and a neutrino) until each reaches a stable end product. For xenon, the decay chain is

$$^{140}\text{Xe} \rightarrow {}^{140}\text{Cs} \rightarrow {}^{140}\text{Ba} \rightarrow {}^{140}\text{La} \rightarrow {}^{140}\text{Ce}$$

τ	14 s	64 s	13 d	40 h	Stable
Z	54	55	56	57	58

$$(44\text{-}2)$$

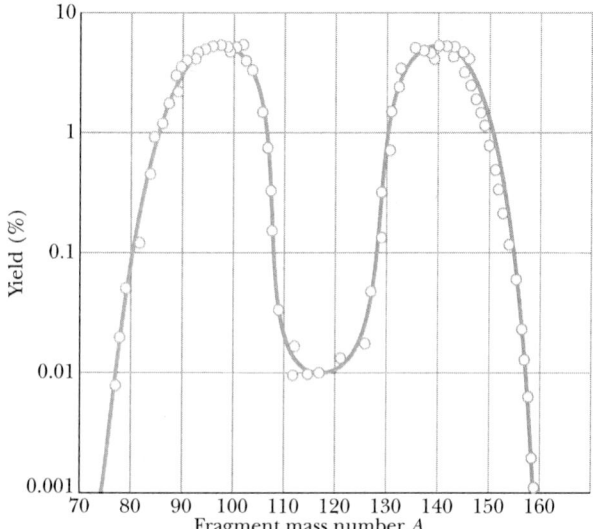

FIGURE 44-1 The distribution by mass number of the fragments that are found when many fission events of ^{235}U are examined. Note that the vertical scale is logarithmic.

For strontium, it is

$$^{94}\text{Sr} \rightarrow {}^{94}\text{Y} \rightarrow {}^{94}\text{Zr}$$

τ	75 s	19 min	Stable
Z	38	39	40

$$(44\text{-}3)$$

As we should expect from Section 43-5, the mass numbers (140 and 94) of the fragments remain unchanged during these beta-decay processes, and the atomic numbers (initially 54 and 38) increase by unity at each step.

Inspection of the stability band on the nuclidic chart of Fig. 43-4 can show us why the fission fragments are unstable. The nuclide ^{236}U, which is the fissioning nucleus in the reaction of Eq. 44-1, has 92 protons and $236 - 92$, or 144 neutrons, for a neutron/proton ratio of about 1.6. The primary fragments formed immediately after the fission reaction have about this same neutron/proton ratio. However, stable nuclides in the middle-mass region have smaller neutron/proton ratios, in the range 1.3–1.4. The primary fragments will thus be *neutron rich* (they have too many neutrons) and will "boil off" a few neutrons, two in the case of the reaction of Eq. 44-1. The fragments that remain are still too neutron rich to be stable. Beta decay offers a mechanism for getting rid of the excess neutrons, namely, by changing them into protons within the nucleus.

We can use the binding energy curve of Fig. 43-6 to estimate the energy released in fission. From this curve, we see that for high-mass nuclides ($A \approx 240$) the mean binding energy per nucleon is about 7.6 MeV. For middle-mass nuclides ($A \approx 120$), it is about 8.5 MeV. The difference in total binding energy between a single large nucleus ($A = 240$) and two fragments (assumed equal in nucleons) into which it may split is then

$$Q = 2(8.5 \text{ MeV})(120) - (7.6 \text{ MeV})(240)$$
$$\approx 200 \text{ MeV}. \quad (44\text{-}4)$$

The more careful calculation of Sample Problem 44-1 agrees remarkably well with this rough estimate.

CHECKPOINT **1:** A generic fission event is

$$^{235}\text{U} + \text{n} \rightarrow X + Y + 2\text{n}.$$

Which of the following pairs *cannot* represent X and Y: (a) ^{141}Xe and ^{93}Sr; (b) ^{139}Cs and ^{95}Rb; (c) ^{156}Nd and ^{79}Ge; (d) ^{121}In and ^{113}Ru?

SAMPLE PROBLEM 44-1

Calculate the disintegration energy Q for the fission event of Eq. 44-1, taking into account the decay of the fission fragments as displayed in Eqs. 44-2 and 44-3.

SOLUTION: We can calculate the disintegration energy from $Q = \Delta m\, c^2$. Some atomic and particle masses that we will need are

$$^{235}\text{U} \quad 235.0439 \text{ u} \qquad ^{140}\text{Ce} \quad 139.9054 \text{ u}$$
$$\text{n} \quad 1.00867 \text{ u} \qquad ^{94}\text{Zr} \quad 93.9063 \text{ u}$$

If we combine Eq. 44-1 with Eqs. 44-2 and 44-3, we see that the overall transformation is

$$^{235}\text{U} \rightarrow {}^{140}\text{Ce} + {}^{94}\text{Zr} + \text{n}. \tag{44-5}$$

Only the single neutron appears here because the initiating neutron on the left side of Eq. 44-1 cancels one of the two neutrons on the right of that equation. The mass difference for the reaction of Eq. 44-5 is

$$\Delta m = (235.0439 \text{ u})$$
$$\quad - (139.9054 \text{ u} + 93.9063 \text{ u} + 1.00867 \text{ u})$$
$$\quad = 0.22353 \text{ u},$$

and the corresponding disintegration energy is

$$Q = \Delta m\, c^2 = (0.22353 \text{ u})(931.5 \text{ MeV/u})$$
$$\quad = 208 \text{ MeV}, \qquad \text{(Answer)}$$

in good agreement with our estimate of Eq. 44-4.

If the fission event takes place in a bulk solid, most of this disintegration energy appears eventually as an increase in the internal energy of that body, revealing itself as a rise in temperature. Five or six percent or so of the disintegration energy, however, is associated with neutrinos that are emitted during the beta decay of the primary fission fragments. This energy is carried out of the system and is lost.

44-3 A MODEL FOR NUCLEAR FISSION

Soon after the discovery of fission, Niels Bohr and John Wheeler used the collective model, based on the analogy between a nucleus and a charged liquid drop, to explain its main features. Figure 44-2 suggests how the fission process proceeds from this point of view. When a heavy nucleus—let us say ^{235}U—absorbs a slow (thermal) neutron, as in Fig. 44-2a, that neutron falls into the potential well associated with the strong forces that act in the nuclear interior. The neutron's potential energy is then transformed into internal excitation energy of the nucleus, as Fig. 44-2b suggests. The amount of excitation energy that a slow neutron carries into the nucleus is equal to the work required to pull a neutron out of the nucleus, that is, to the binding energy E_n of the neutron.

Figures 44-2c and d show that the nucleus, behaving like an energetically oscillating charged liquid drop, will sooner or later develop a short "neck" and will begin to separate into two charged "globs." If the electric repulsion between these two globs forces them far enough apart to

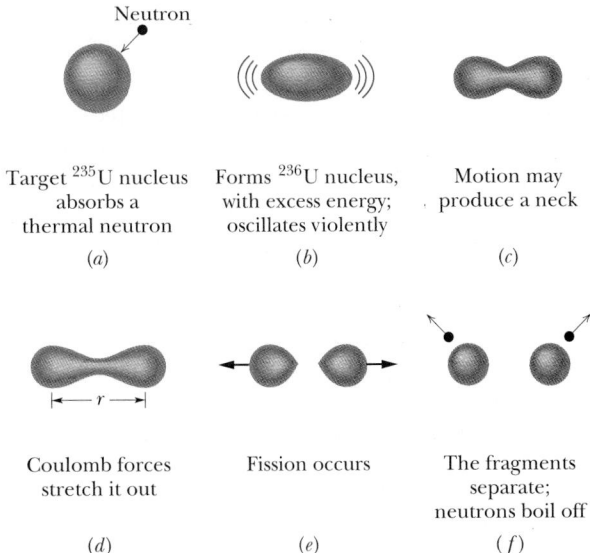

FIGURE 44-2 The stages of a typical fission process, according to the collective model of Bohr and Wheeler.

break the neck, the two fragments, each still carrying some residual excitation energy, will fly apart (Figs. 44-2e and f). Fission has occurred.

Thus far this model gave a good qualitative picture of the fission process. It remained to be seen, however, whether it could answer a hard question: Why are some heavy nuclides (^{235}U and ^{239}Pu, say) readily fissionable by thermal neutrons when other, equally massive nuclides (^{238}U and ^{243}Am, say) are not?

Bohr and Wheeler were able to answer this question. Figure 44-3 shows a graph of the potential energy at various stages of the fissioning nucleus, derived from their model for the fission process. This energy is plotted against the *distortion parameter r*, which is a rough measure of the extent to which the oscillating nucleus departs from a spherical shape. Figure 44-2d suggests how this parameter is defined just before fission occurs. When the fragments

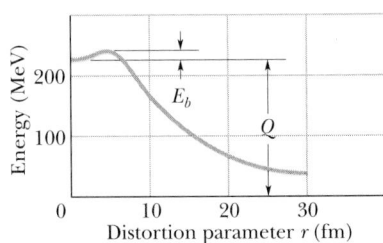

FIGURE 44-3 The potential energy at various stages in the fission process, as predicted from the collective model of Bohr and Wheeler. The Q of the reaction (about 200 MeV) and the fission barrier height E_b are both indicated.

are far apart, this parameter is simply the distance between their centers.

The energy difference between the initial state and the final state of the fissioning nucleus—that is, the disintegration energy Q—is labeled in Fig. 44-3. The central feature of that figure, however, is that the potential energy curve passes through a maximum at a certain value of r. Thus there is a *potential barrier* of height E_b that must be surmounted (or tunneled through) before fission can occur. This reminds us of alpha decay (Fig. 43-9), which is also a process that is inhibited by a potential barrier.

We see then that fission will occur only if the absorbed neutron provides an excitation energy E_n great enough to overcome the barrier. This energy E_n need not be *quite* as great as the barrier height E_b because of the possibility of quantum-mechanical tunneling.

Table 44-2 shows a test of fissionability by thermal neutrons applied to four high-mass nuclides, chosen from dozens of possible candidates. For each nuclide both the barrier height E_b and the excitation energy E_n are given. The former was calculated from the theory of Bohr and Wheeler; the latter was computed from the known masses, using $Q = \Delta m \, c^2$.

For ^{235}U and ^{239}Pu we see that $E_n > E_b$. This means that fission by absorption of a thermal neutron is predicted to occur for these nuclides. For the other two nuclides (^{238}U and ^{243}Am), we have $E_n < E_b$; thus there is not enough energy from a thermal neutron for the excited nucleus to surmount the barrier or to tunnel through it effectively. Instead of fissioning, the nucleus gets rid of its excitation energy by emitting a gamma ray photon.

^{238}U and ^{243}Am *can* be made to fission, however, if they absorb a substantially energetic (rather than a thermal) neutron. For ^{238}U, for example, the absorbed neutron must have at least 1.3 MeV of energy for this *fast fission* process to have a good chance of occurring.

44-4 THE NUCLEAR REACTOR

For large-scale energy release due to fission, one fission event must trigger others, so that the process spreads throughout the nuclear fuel like flame through a log. The fact that more neutrons are produced in fission than are consumed raises the possibility of just such a **chain reaction**, with each neutron produced potentially triggering another fission. The reaction can be either rapid (as in a nuclear bomb) or controlled (as in a nuclear reactor).

Suppose that we wish to design a reactor based on the fission of ^{235}U by thermal neutrons. Natural uranium contains 0.7% of this isotope, the remaining 99.3% being ^{238}U, which is not fissionable by thermal neutrons. Let us give ourselves an edge by artificially enriching the uranium fuel so that it contains perhaps 3% ^{235}U. Three difficulties still stand in the way of a working reactor.

1. *The Neutron Leakage Problem.* Some of the neutrons produced by fission will leak out of the reactor and so not be part of the chain reaction. Leakage is a surface effect, its magnitude being proportional to the square of a typical reactor dimension (the surface area of a cube of edge length a is $6a^2$). Neutron production, however, occurs throughout the volume of the fuel and is thus proportional to the cube of a typical dimension (the volume of a cube is a^3). We can make the fraction of neutrons lost by leakage as small as we wish by making the reactor core large enough, thereby reducing the surface-to-volume ratio ($= 6/a$ for a cube).

2. *The Neutron Energy Problem.* The neutrons produced by fission are fast, with kinetic energies of about 2 MeV. However, fission is induced most effectively by thermal neutrons. The fast neutrons can be slowed down by mixing the uranium fuel with a substance—called a **moderator**—that has two properties: it is effective in slowing down neutrons by elastic collisions, and it does not remove neutrons from the core by absorbing them in ways that do not result in fission. Most power reactors in North America use water as a moderator; the hydrogen nuclei (protons) in the water are the effective component. We saw in Chapter 10 that, if a moving particle has a head-on elastic collision with a stationary particle, the moving particle loses *all* its kinetic energy if the two particles have the same mass. Thus protons form an effective moderator because they have essentially the same mass as the fast neutrons whose speeds we wish to reduce.

3. *The Neutron Capture Problem.* As the fast (2 MeV) neutrons generated by fission are slowed down in the moderator to thermal energies (about 0.04 eV), they must pass through a critical energy interval (from 1 to 100 eV) in which they are particularly susceptible to nonfission capture by ^{238}U nuclei. Such *resonance capture*, which results in the emission of a gamma ray, removes the neutron from the fission chain. To minimize such nonfission capture, the

TABLE 44-2 TEST OF THE FISSIONABILITY OF FOUR NUCLIDES

TARGET NUCLIDE	NUCLIDE BEING FISSIONED	E_n (MeV)	E_b (MeV)	FISSION BY THERMAL NEUTRONS?
^{235}U	^{236}U	6.5	5.2	Yes
^{238}U	^{239}U	4.8	5.7	No
^{239}Pu	^{240}Pu	6.4	4.8	Yes
^{243}Am	^{244}Am	5.5	5.8	No

uranium fuel and the moderator are not intimately mixed but are "clumped together", occupying different regions of the reactor volume.

In a typical reactor, the uranium fuel is in the form of uranium oxide pellets, which are inserted end to end into long hollow metal tubes. The liquid moderator surrounds bundles of these **fuel rods**, forming the reactor core. This geometric arrangement increases the probability that a fast neutron, produced in a fuel rod, will find itself in the moderator when it passes through the critical energy interval. Once the neutron has reached thermal energies, it may *still* be captured in ways that do not result in fission (called *thermal capture*). However, it is much more likely that the thermal neutron will wander back into a fuel rod and produce a fission event.

Figure 44-4 shows the neutron balance in a typical power reactor operating at constant power. Let us trace a sample of 1000 thermal neutrons through one complete cycle, or *generation*, in the reactor core. They produce 1330 neutrons by fission in the ^{235}U fuel and 40 neutrons by fast fission in ^{238}U, which gives 370 neutrons more than the original 1000, all of them fast. When the reactor is operating at a steady power level, exactly the same number of neutrons (370) is then lost by leakage from the core and by nonfission capture, leaving 1000 thermal neutrons to continue the chain reaction. In this cycle, of course, each of the 370 neutrons produced by fission events represents a deposit of energy in the reactor core, heating it the core.

The *multiplication factor k*—an important reactor parameter—is the ratio of the number of neutrons present at the beginning of a particular generation to the number present at the beginning of the next generation. In Fig. 44-4, the multiplication factor is 1000/1000, or exactly unity. For $k = 1$, the operation of the reactor is said to be exactly *critical*, which is what we wish it to be for steady-power operation. Reactors are actually designed so that they are inherently *supercritical* ($k > 1$); the multiplication factor is then adjusted to critical operation ($k = 1$) by inserting **control rods** into the reactor core. These rods, containing a material, such as cadmium, that absorbs neutrons readily, can be inserted farther to reduce the operating power level and withdrawn to increase the power level or to compensate for the tendency of reactors to go *subcritical* as (neutron-absorbing) fission products build up in the core during continued operation.

If you pulled out one of the control rods rapidly, how fast would the reactor power level increase? This *response time* is controlled by the fascinating circumstance that a small fraction of the neutrons generated by fission do not escape promptly from the newly formed fission fragments but are emitted from these fragments later, as the fragments decay by beta emission. Of the 370 "new" neutrons produced in Fig. 44-4, for example, perhaps 16 are delayed, being emitted from fragments following beta decays whose half-lives range from 0.2 to 55 s. These delayed neutrons are few in number, but they serve the essential purpose of slowing the reactor response time to match practical mechanical reaction times.

Figure 44-5 shows the broad outlines of an electric power plant based on a *pressurized-water reactor* (PWR), a type in common use in North America. In such a reactor, water is used both as the moderator and as the heat transfer

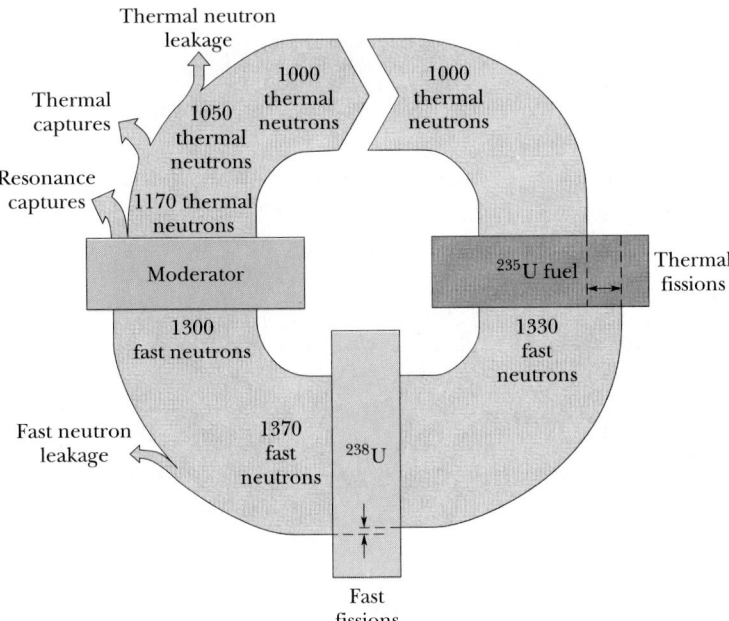

FIGURE 44-4 Neutron bookkeeping in a reactor. A generation of 1000 thermal neutrons interacts with the ^{235}U fuel, the ^{238}U matrix, and the moderator. They produce 1370 neutrons by fission; 370 of these are lost by nonfission capture or by leakage; so 1000 thermal neutrons are left to form the next generation. The figure is drawn for a reactor running at a steady power level.

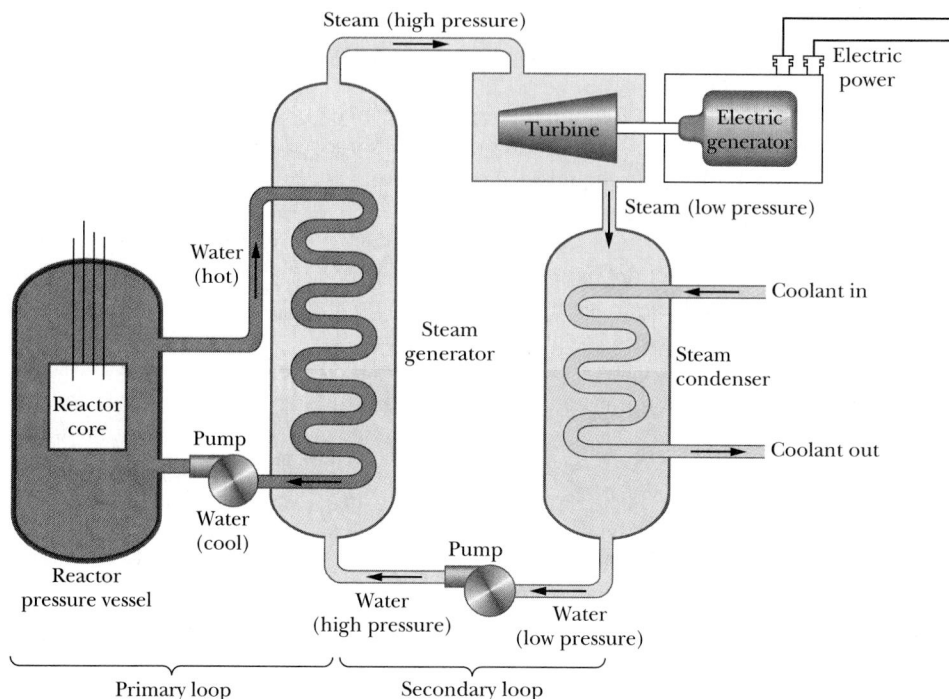

FIGURE 44-5 A simplified layout of a nuclear power plant, based on a pressurized-water reactor. Many features are omitted —among them the arrangement for cooling the reactor core in case of an emergency.

medium. In the *primary loop,* water at high temperature and pressure (possibly 600 K and 150 atm) circulates through the reactor vessel and transfers energy from the hot reactor core to the steam generator, which is part of the *secondary loop.* In the steam generator, evaporation provides high-pressure steam to operate the turbine that drives the electric generator. To complete the secondary loop, low-pressure steam from the turbine is cooled and con-

densed to water and forced back into the steam generator by a pump. To give some idea of scale, a typical reactor vessel for a 1000 MW (electric) plant may be 40 ft high and weigh 450 tons. Water flows through the primary loop at a rate of about 300,000 gal/min.

An unavoidable feature of reactor operation is the accumulation of radioactive wastes, including both fission products and heavy *transuranic* nuclides such as plutonium and americium. One measure of their radioactivity is the rate at which they release energy in thermal form. Figure 44-6 shows the thermal power generated by such wastes from one year's operation of a typical large nuclear plant. Note that both scales are logarithmic. Most "spent"

The scene 20 m from the Chernobyl reactor unit 4 (near Kiev), after it exploded in April 1986. Nearly all the volatile radionuclides inside the reactor were released into the air.

FIGURE 44-6 The thermal power released by the radioactive wastes from one year's operation of a typical large nuclear power plant, shown as a function of time. The curve is the superposition of the effects of many radionuclides, with a wide variety of half-lives. Note that both scales are logarithmic.

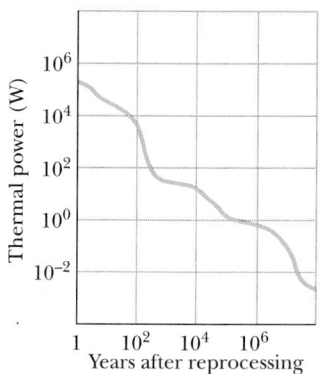

fuel rods from power reactor operation are stored on site, immersed in water; permanent secure storage facilities have yet to be completed.

Much weapons-derived radioactive waste accumulated during World War II and in subsequent years is also still in on-site storage. Figure 44-7, for example, shows an underground storage tank farm under construction at the Hanford Site in Washington State; each large tank holds 250,000 gallons of high-level radioactive liquid waste. There are now 152 such tanks at the site. In addition, much solid waste, both low-level radioactive waste (contaminated clothing, for example) and high-level waste (reactor cores from decommissioned nuclear submarines, for example) is buried in trenches.

SAMPLE PROBLEM 44-2

A large electric generating station is powered by a pressurized-water nuclear reactor. The thermal power in the reactor core is 3400 MW, and 1100 MW of electricity is generated. The fuel charge is 86,000 kg of uranium, in the form of 110 tons of uranium oxide, distributed among 57,000 fuel rods. The uranium is enriched to 3.0% ^{235}U.

(a) What is the plant efficiency?

SOLUTION: From the definition of efficiency (Eq. 21-5), we have

$$\text{eff} = \frac{\text{useful output}}{\text{energy input}} = \frac{1100 \text{ MW (electric)}}{3400 \text{ MW (thermal)}}$$

$$= 0.32, \text{ or } 32\%. \qquad \text{(Answer)}$$

FIGURE 44-7 An underground tank farm under construction during World War II at the Hanford Site in Washington State. Note the trucks and the workers. Each large tank holds 250,000 gallons of high-level radioactive waste.

The efficiency—as for all power plants, whether based on fossil fuel or nuclear fuel—is controlled by the second law of thermodynamics. To run this plant, energy at the rate of 3400 MW − 1100 MW, or 2300 MW, must be discharged as thermal energy to the environment.

(b) At what rate R do fission events occur in the reactor core?

SOLUTION: If $P = 3400$ MW is the thermal power in the core and $Q = 200$ MeV is the average energy released per fission event, then, in steady-state operation,

$$R = \frac{P}{Q}$$

$$= \left(\frac{3.4 \times 10^9 \text{ W}}{200 \text{ MeV/fission}}\right)\left(\frac{1 \text{ MeV}}{1.60 \times 10^{-13} \text{ J}}\right)\left(\frac{1 \text{ J/s}}{1 \text{ W}}\right)$$

$$= 1.06 \times 10^{20} \text{ fissions/s}$$

$$\approx 1.1 \times 10^{20} \text{ fissions/s}. \qquad \text{(Answer)}$$

(c) At what rate is the ^{235}U fuel disappearing? Assume conditions at start-up.

SOLUTION: ^{235}U disappears by fission at the rate calculated in (b). It is also consumed by (nonfission) neutron capture at a rate about one-fourth as large. The total ^{235}U consumption rate is then $(1.25)(1.06 \times 10^{20} \text{ atoms/s})$, or 1.33×10^{20} atoms/s. We recast this as a mass rate as follows, using the molar mass of 0.235 kg/mol and the Avogadro constant:

$$\frac{dM}{dt} = (1.33 \times 10^{20} \text{ atoms/s})\left(\frac{0.235 \text{ kg/mol}}{6.02 \times 10^{23} \text{ atoms/mol}}\right)$$

$$= 5.19 \times 10^{-5} \text{ kg/s} \approx 4.5 \text{ kg/d}. \qquad \text{(Answer)}$$

(d) At this rate of fuel consumption, how long would the fuel supply last?

SOLUTION: From the data given, we can calculate that, at start-up, about $(0.030)(86,000 \text{ kg})$, or 2580 kg, of ^{235}U was present. Thus a somewhat simplistic answer would be

$$T = \frac{2580 \text{ kg}}{4.5 \text{ kg/d}} \approx 570 \text{ d}. \qquad \text{(Answer)}$$

In practice, the fuel rods must be replaced (usually in batches) before their ^{235}U content is entirely consumed.

(e) At what rate is mass being converted to other forms of energy in the reactor core?

SOLUTION: From Einstein's relation $E = mc^2$, we can write

$$\frac{dM}{dt} = \frac{dE/dt}{c^2} = \frac{3.4 \times 10^9 \text{ W}}{(3.00 \times 10^8 \text{ m/s})^2}$$

$$= 3.8 \times 10^{-8} \text{ kg/s} = 3.3 \text{ g/d}. \qquad \text{(Answer)}$$

We see that the mass conversion rate is about the mass of one penny every day! This rate of conversion of mass to other forms of energy is quite a different quantity from the fuel consumption rate (loss of ^{235}U) calculated in (c).

C HECKPOINT **2:** In Sample Problem 44-2 we saw that the generated power of the nuclear power plant (P_{gen} = 1100 MW) was less than the power discharged to the environment (P_{dis} = 2300 MW). Does the second law of thermodynamics: (a) Require that P_{gen} always be less than P_{dis}? (b) Permit P_{gen} to be greater than P_{dis}? (c) Permit P_{dis} to be zero, assuming optimum reactor design?

44-5 A NATURAL NUCLEAR REACTOR (OPTIONAL)

On December 2, 1942, when the reactor assembled by Enrico Fermi and his associates first went critical (Fig. 44-8), the physicists had every right to assume that they had put into operation the first fission reactor that had ever existed on this planet. About 30 years later it was discovered that, if they did in fact think that, they were wrong.

Some two billion years ago, in a uranium deposit now being mined in Gabon, West Africa, a natural fission reactor apparently went into operation and ran for perhaps several hundred thousand years before shutting down. We can analyze this conjecture by considering two questions:

1. *Was There Enough Fuel?* The fuel for a uranium-based fission reactor must be the easily fissionable isotope ^{235}U, which constitutes only 0.72% of natural uranium. This isotopic ratio has been measured for terrestrial samples, in Moon rocks, and in meteorites; in all cases the abundance values are the same. The clue to the discovery

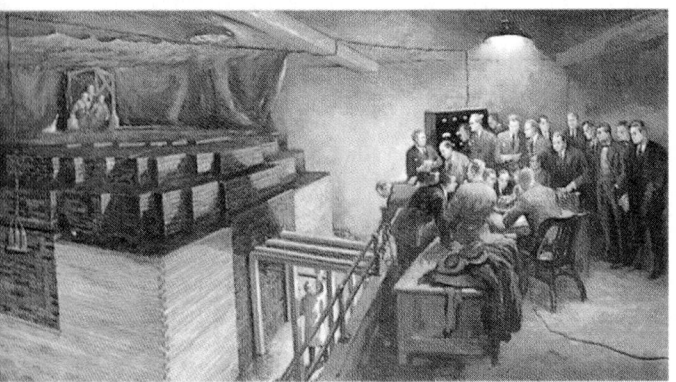

FIGURE 44-8 A painting of the first nuclear reactor, assembled during World War II on a squash court at the University of Chicago, by a team headed by Enrico Fermi. This reactor, which went critical on December 2, 1942, was built of lumps of uranium embedded in blocks of graphite. It served as a prototype for later reactors whose purpose was to manufacture plutonium for the construction of nuclear weapons.

in West Africa was that the uranium in that deposit was deficient in ^{235}U, some samples having abundances as low as 0.44%. Investigation led to the speculation that this deficit in ^{235}U could be accounted for if, at some time in the past, the isotope was partially consumed by the operation of a natural fission reactor.

The serious problem remains that, with an isotopic abundance of only 0.72%, a reactor can be assembled (as Fermi and his team learned) only after thoughtful design and with scrupulous attention to detail. There seems no chance that a nuclear reactor could go critical "naturally."

However, things were different in the distant past. Both ^{235}U and ^{238}U are radioactive, with half-lives of 7.04×10^8 y and 44.7×10^8 y, respectively. Thus the half-life of the readily fissionable ^{235}U is about 6.4 times shorter than that of ^{238}U. Because ^{235}U decays faster, there was more of it, relative to ^{238}U, in the past. Two billion years ago, in fact, this abundance was not 0.72%, as it is now, but 3.8%. This abundance happens to be just about the abundance to which natural uranium is artificially enriched to serve as fuel in modern power reactors.

With this readily fissionable fuel available, the presence of a natural reactor (provided certain other conditions are met) is less surprising. The fuel was there. Two billion years ago, incidentally, the highest order of life-forms to have evolved were the blue-green algae.

2. *What Is the Evidence?* The mere depletion of ^{235}U in an ore deposit does not prove the existence of a natural fission reactor. One looks for more convincing evidence.

If there were a reactor, there must be fission products. Of the 30 or so elements whose stable isotopes are produced in this way, some must still remain. Study of their isotopic abundances could provide the convincing evidence we need.

Of the several elements investigated, the case of neodymium is spectacularly convincing. Figure 44-9*a* shows the isotopic abundances of the seven stable neodymium isotopes as they are normally found in nature. Figure 44-9*b* shows these abundances as they appear among the ultimate stable fission products of the fission of ^{235}U. The clear differences are not surprising, considering the totally different origins of the two sets of isotopes. Note particularly that ^{142}Nd, the dominant isotope in the natural element, is totally absent from the fission products.

The big question is: What do the neodymium isotopes found in the uranium ore body in West Africa look like? If a natural reactor operated there, we would expect to find isotopes from *both* sources (that is, natural isotopes as well as fission-produced isotopes). Figure 44-9*c* shows the results after this and other corrections have been made to the raw data. Comparison of Figs. 44-9*b* and *c* indicates that there was indeed a natural fission reactor at work.

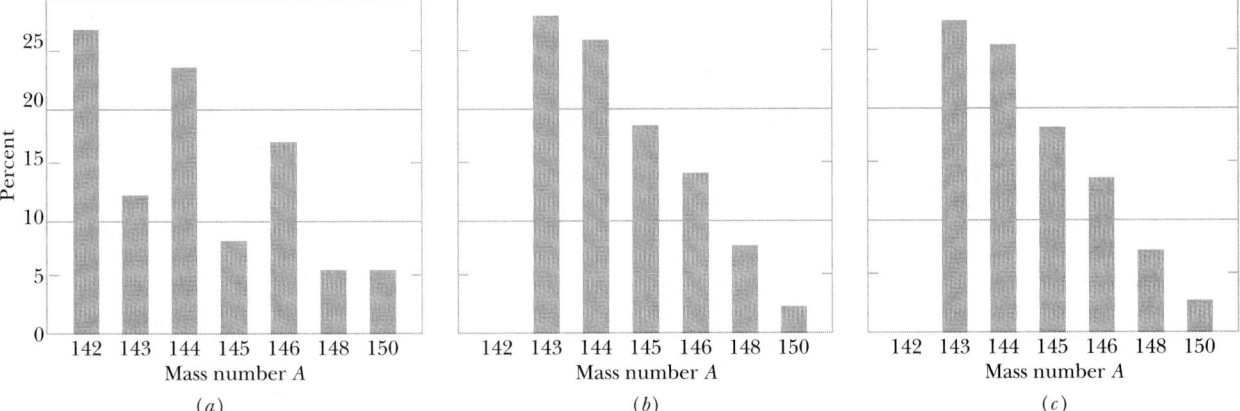

FIGURE 44-9 The distribution by mass number of the isotopes of neodymium as they occur in (a) natural terrestrial deposits of the ores of this element and (b) the spent fuel of a power reactor. (c) The distribution (after several corrections) found for neodymium from the uranium mine in Gabon, West Africa. Note that (b) and (c) are virtually identical and are quite different from (a).

An important conclusion can be drawn from the failure of the fission products of the West African natural reactor to migrate far from the region of their production for about 2 billion years. This property may support the feasibility of long-term storage of radioactive waste in suitably chosen geological environments.

SAMPLE PROBLEM 44-3

The ratio of ^{235}U to ^{238}U in natural uranium deposits today is 0.0072. What was this ratio 2.0×10^9 y ago? The half-lives of the two isotopes are 7.04×10^8 y and 44.7×10^8 y, respectively.

SOLUTION: Consider two samples that, at a time t in the past, contained $N_5(0)$ and $N_8(0)$ atoms of ^{235}U and ^{238}U, respectively. The numbers of atoms remaining at the present time t are

$$N_5(t) = N_5(0)e^{-\lambda_5 t} \quad \text{and} \quad N_8(t) = N_8(0)e^{-\lambda_8 t},$$

respectively, in which λ_5 and λ_8 are the corresponding disintegration constants. Dividing gives

$$\frac{N_5(t)}{N_8(t)} = \frac{N_5(0)}{N_8(0)} e^{-(\lambda_5 - \lambda_8)t}.$$

Expressed in terms of isotopic ratios $r = N_5/N_8$, this becomes

$$r(0) = r(t)e^{(\lambda_5 - \lambda_8)t}.$$

The disintegration constants are related to the half-lives by Eq. 43-8, which yields

$$\lambda_5 = \frac{\ln 2}{\tau_5} = \frac{\ln 2}{7.04 \times 10^8 \text{ y}} = 9.85 \times 10^{-10} \text{ y}^{-1}$$

and

$$\lambda_8 = \frac{\ln 2}{\tau_8} = \frac{\ln 2}{44.7 \times 10^8 \text{ y}} = 1.55 \times 10^{-10} \text{ y}^{-1}.$$

The exponent in the expression for $r(0)$ above is then

$$(\lambda_5 - \lambda_8)t = [(9.85 - 1.55) \times 10^{-10} \text{ y}^{-1}](2 \times 10^9 \text{ y})$$
$$= 1.66.$$

The isotopic ratio is then

$$r(0) = r(t)e^{(\lambda_5 - \lambda_8)t} = (0.0072)(e^{1.66})$$
$$= 0.0379 \approx 3.8\%. \qquad \text{(Answer)}$$

Two billion years ago, the ratio of ^{235}U to ^{238}U in natural uranium deposits was much higher than it is today. You should be able to show that when Earth was formed (4.5 billion years ago), this ratio was 30%.

44-6   THERMONUCLEAR FUSION: THE BASIC PROCESS

The binding energy curve of Fig. 43-6 shows that energy can be released if two light nuclei combine to form a single larger nucleus, a process called nuclear **fusion.** The process is hindered by the Coulomb repulsion that acts to prevent the two positively charged particles from getting close enough to be within range of their attractive nuclear forces and thus "fusing." The height of this *Coulomb barrier* depends on the charges and the radii of the two interacting nuclei. We show in Sample Problem 44-4 that, for two protons ($Z = 1$), the barrier height is 400 keV. For more highly charged particles, of course, the barrier is correspondingly higher.

To generate useful amounts of power, nuclear fusion must occur in bulk matter. The best hope for bringing this about is to raise the temperature of the material until the particles have enough energy—due to their thermal motions alone—to penetrate the Coulomb barrier. We call this process **thermonuclear fusion**.

In thermonuclear studies, temperatures are reported in terms of the kinetic energy K of interacting particles via the relation

$$K = kT, \tag{44-6}$$

in which K is the kinetic energy corresponding to the *most probable speed* of the interacting particles, k is the Boltzmann constant, and T is in kelvins. Thus rather than saying, "The temperature at the center of the Sun is 1.5×10^7 K," it is more common to say, "The temperature at the center of the Sun is 1.3 keV."

Room temperature corresponds to $K \approx 0.03$ eV; a particle with only this amount of energy could not hope to overcome a barrier as high as, say, 400 keV. Even at the center of the Sun, where $kT = 1.3$ keV, the outlook for thermonuclear fusion does not seem promising at first glance. Yet we know that thermonuclear fusion not only occurs in the core of the Sun but is the dominant feature of that body and of all other stars.

The puzzle is solved when we realize two facts: (1) The energy calculated with Eq. 44-6 is that of the particles with the *most probable* speed, as defined in Section 20-7; there is a long tail of particles with much higher speeds and, correspondingly, much higher energies. (2) The barrier heights that we have calculated represent the *peaks* of the barriers. Barrier tunneling can occur at energies considerably lower than these peaks, as we saw in the case of alpha decay in Section 43-4.

Figure 44-10 sums things up. The curve marked $n(K)$ in this figure is a Maxwell distribution curve for the protons in the Sun's core, drawn to correspond to the Sun's central temperature. This curve differs from the Maxwell distribution curve of Fig. 20-7 in that here the curve is drawn in terms of energy and not of speed. Specifically, for any kinetic energy K, the expression $n(K)\,dK$ gives the probability that a proton will have a kinetic energy lying between K and $K + dK$. The value of kT in the core of the Sun is indicated by the vertical line in the figure; note that many of the Sun's core protons have energies greater than this value.

The curve marked $p(K)$ in Fig. 44-10 is the probability of barrier penetration for two colliding protons. The two curves in Fig. 44-10 suggest that there is a particular proton energy at which proton–proton fusion events occur at a maximum rate. At energies much above this value, the barrier is transparent enough but too few protons have these energies, and the reaction cannot be sustained. At

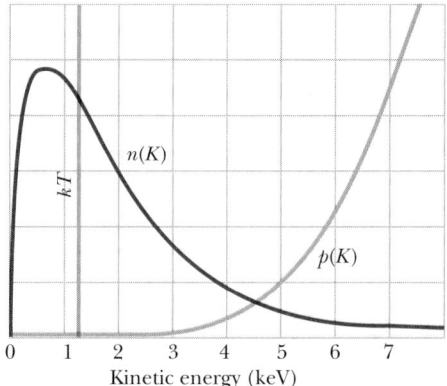

FIGURE 44-10 The curve marked $n(K)$ gives the number density per unit energy for protons at the center of the Sun. The curve marked $p(K)$ gives the probability of barrier penetration for proton–proton collisions at the Sun's central temperature. The vertical line marks the value of kT at this temperature. Note that the two curves are drawn to (separate) arbitrary vertical scales.

energies much below this value, plenty of protons have these energies but the barrier is too formidable.

CHECKPOINT **3:** Which of these potential fusion reactions will *not* result in the net release of energy: (a) ^{6}Li + ^{6}Li, (b) ^{4}He + ^{4}He, (c) ^{12}C + ^{12}C, (d) ^{20}Ne + ^{20}Ne, (e) ^{35}Cl + ^{35}Cl, and (f) ^{14}N + ^{35}Cl? (*Hint:* Consult the binding energy curve of Fig. 43-6.)

SAMPLE PROBLEM 44-4

Assume a proton is a sphere of radius $R \approx 1$ fm. Two protons are fired at each other with the same kinetic energy K.

(a) What must K be if the particles are brought to rest by their mutual Coulomb repulsion when they are just "touching" each other? We can take this value of K as a representative measure of the height of the Coulomb barrier.

SOLUTION: Because the two protons are momentarily at rest when they just touch, all their initial kinetic energy has been transformed into electric potential energy. Their centers are separated by a distance $2R$ and we have, from Eq. 25-43,

$$2K = \frac{1}{4\pi\epsilon_0}\frac{q_1 q_2}{r} = \frac{1}{4\pi\epsilon_0}\frac{e^2}{2R}.$$

This yields, with known values,

$$K = \frac{e^2}{16\pi\epsilon_0 R}$$

$$= \frac{(1.60 \times 10^{-19}\ \text{C})^2}{(16\pi)(8.85 \times 10^{-12}\ \text{F/m})(1 \times 10^{-15}\ \text{m})}$$

$$= 5.75 \times 10^{-14}\ \text{J} = 360\ \text{keV} \approx 400\ \text{keV}. \quad \text{(Answer)}$$

(b) At what temperature would a proton in a gas of protons have the average kinetic energy calculated in (a), and thus have energy equal to the height of the Coulomb barrier?

SOLUTION: Treating a proton gas as an ideal gas, we take the average energy of the protons to be $K_{av} = \frac{3}{2}kT$ (Eq. 20-20), where k is the Boltzmann constant. Solving this equation for T and using the result of (a) yield

$$T = \frac{2K_{av}}{3k} = \frac{(2)(5.75 \times 10^{-14} \text{ J})}{(3)(1.38 \times 10^{-23} \text{ J/K})} \approx 3 \times 10^9 \text{ K}.$$

The temperature of the core of the Sun is only about 1.5×10^7 K, so it is clear that fusion in the Sun's core must involve protons whose energies are *far* above the average energy.

44-7 THERMONUCLEAR FUSION IN THE SUN AND OTHER STARS

The Sun radiates energy at the rate of 3.9×10^{26} W and has been doing so for several billion years. Where does all this energy come from? Chemical burning is ruled out; if the Sun had been made of coal and oxygen—in the right proportions for combustion—it would have lasted for only about 1000 y. Another possibility is that the Sun is slowly shrinking, under the action of its own gravitational forces. By transferring gravitational potential energy to thermal energy, the Sun might maintain its temperature and continue to radiate. Calculation, however, shows that this mechanism also fails; it produces a solar lifetime that is too short by a factor of at least 500. That leaves only thermonuclear fusion. The Sun, as you will see, burns not coal but hydrogen, and in a nuclear furnace, not an atomic or chemical one.

The fusion reaction in the Sun is a multistep process in which hydrogen is burned into helium, hydrogen being the "fuel" and helium the "ashes." Figure 44-11 shows the **proton–proton** (p–p) **cycle** by which this occurs.

The p–p cycle starts with the collision of two protons (^{1}H + ^{1}H) to form a deuteron (^{2}H), with the simultaneous creation of a positron (e^+) and a neutrino (ν). The positron very quickly encounters a free electron (e^-) in the Sun and both particles annihilate (see Section 22-6), their mass energy appearing as two gamma-ray photons (γ).

A pair of such events is shown in the top row of Fig. 44-11. These events are actually extremely rare. In fact, only once in about 10^{26} proton–proton collisions is a deuteron formed; in the vast majority of cases, the two protons simply rebound elastically from each other. It is the slowness of this "bottleneck" process that regulates the rate of energy production and keeps the Sun from exploding. In spite of this slowness, there are so very many protons in the huge and dense volume of the Sun's core that deuterium is produced in just this way at the rate of 10^{12} kg/s.

Once a deuteron has been produced, it quickly collides with another proton and forms a ^{3}He nucleus, as the middle row of Fig. 44-11 shows. Two such ^{3}He nuclei may eventually (within 10^5 y; there is plenty of time) find each other, forming an alpha particle (^{4}He) and two protons, as the bottom row in the figure shows.

Overall, we see from Fig. 44-11 that the p–p cycle amounts to the combination of four protons and two electrons to form an alpha particle, two neutrinos, and six gamma ray photons. Thus

$$4\,^1\text{H} + 2e^- \rightarrow {^4}\text{He} + 2\nu + 6\gamma. \quad (44\text{-}7)$$

Now let us add two electrons to each side of Eq. 44-7, obtaining

$$(4\,^1\text{H} + 4e^-) \rightarrow ({^4}\text{He} + 2e^-) + 2\nu + 6\gamma. \quad (44\text{-}8)$$

The quantities in the two sets of parentheses then represent *atoms* (not bare nuclei) of hydrogen and of helium.

The energy release in the reaction of Eq. 44-8 is

$$\begin{aligned} Q &= \Delta m\, c^2 \\ &= [(4)(1.007825 \text{ u}) - 4.002603 \text{ u}][931.5 \text{ MeV/u}] \\ &= 26.7 \text{ MeV}, \end{aligned}$$

in which 1.007825 u is the mass of a hydrogen atom and 4.002603 u is the mass of a helium atom. Neutrinos have at most a negligibly small mass, and gamma-ray photons

FIGURE 44-11 The proton–proton mechanism that accounts for energy production in the Sun. In this process, protons fuse to form an alpha particle (^{4}He), with a net energy release of 26.7 MeV for each event.

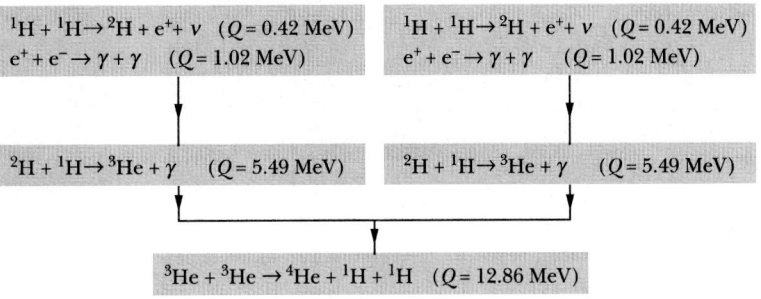

$$^1\text{H} + {}^1\text{H} \rightarrow {}^2\text{H} + e^+ + \nu \quad (Q = 0.42 \text{ MeV})$$
$$e^+ + e^- \rightarrow \gamma + \gamma \quad (Q = 1.02 \text{ MeV})$$

$$^1\text{H} + {}^1\text{H} \rightarrow {}^2\text{H} + e^+ + \nu \quad (Q = 0.42 \text{ MeV})$$
$$e^+ + e^- \rightarrow \gamma + \gamma \quad (Q = 1.02 \text{ MeV})$$

$$^2\text{H} + {}^1\text{H} \rightarrow {}^3\text{He} + \gamma \quad (Q = 5.49 \text{ MeV})$$

$$^2\text{H} + {}^1\text{H} \rightarrow {}^3\text{He} + \gamma \quad (Q = 5.49 \text{ MeV})$$

$$^3\text{He} + {}^3\text{He} \rightarrow {}^4\text{He} + {}^1\text{H} + {}^1\text{H} \quad (Q = 12.86 \text{ MeV})$$

have no mass; thus they do not enter into the calculation of the disintegration energy.

This same value of Q follows (as it must) from adding up the Q values for the separate steps of the proton–proton cycle in Fig. 44-11. Thus

$$Q = (2)(0.42 \text{ MeV}) + (2)(1.02 \text{ MeV})$$
$$+ (2)(5.49 \text{ MeV}) + 12.86 \text{ MeV}$$
$$= 26.7 \text{ MeV}.$$

About 0.5 MeV of this energy is carried out of the Sun by the two neutrinos in Eq. 44-8; the rest (= 26.2 MeV) is deposited in the core of the Sun as thermal energy.

The burning of hydrogen in the Sun's core is alchemy on a grand scale in the sense that one element is turned into another. The medieval alchemists, however, were more interested in changing lead into gold than in changing hydrogen into helium. In a sense, they were on the right track, except that their furnaces were not hot enough. Instead of being at a temperature of, say, 600 K, the ovens should have been at least as hot as 10^8 K.

Hydrogen burning has been going on in the Sun for about 5×10^9 y, and calculations show that there is enough hydrogen left to keep the Sun going for about the same length of time into the future. In 5 billion years, however, the Sun's core, which by that time will be largely helium, will begin to cool and the Sun will start to collapse under its own gravity. This will raise the core temperature and cause the outer envelope to expand, turning the Sun into what astronomers call a *red giant*.

If the core temperature increases to about 10^8 K again, energy can be produced once more by burning helium to make carbon. As a star evolves and becomes still hotter, other elements can be formed by other fusion reactions. However, elements more massive than those with $A \approx 56$ cannot be manufactured by further fusion processes. Mass number $A = 56$ marks the approximate peak of the binding energy curve of Fig. 43-6, and fusion producing nuclides beyond this point involves the consumption—not the production—of energy.

Elements with mass numbers beyond $A \approx 56$ are thought to be formed by neutron capture during cataclysmic stellar explosions that we call *supernovas* (Fig. 44-12). In such an event the outer shell of the star is blown outward into space, where it mixes with—and becomes part of—the tenuous medium that fills the space between the stars. It is from this medium, continually enriched by debris from stellar explosions, that new stars form, by condensation under the influence of the gravitational force.

The abundance on Earth of elements heavier than hydrogen and helium suggests that our solar system has condensed out of interstellar material that contained the remnants of such explosions. Thus all the elements around us—including those in our own bodies—were manufactured in the interiors of stars that no longer exist. As one scientist put it: "In truth, we are the children of the stars."

SAMPLE PROBLEM 44-5

At what rate is hydrogen being consumed in the core of the Sun by the p–p cycle of Fig. 44-11?

SOLUTION: As we discussed above, 26.2 MeV appears as thermal energy in the Sun for every four protons that are consumed, a rate of 6.6 MeV/proton. We can express this energy transfer rate as

(a)

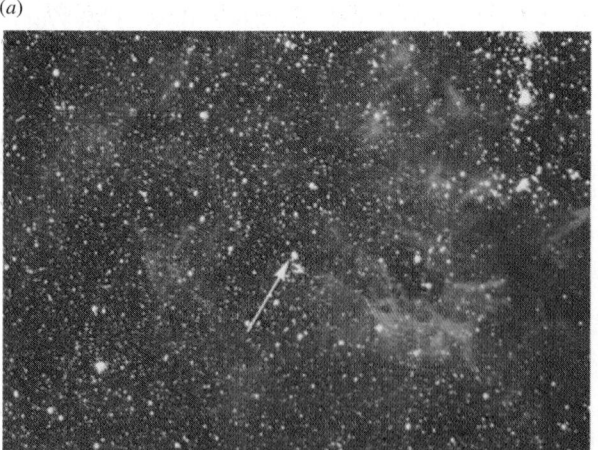

(b)

FIGURE 44-12 (*a*) The star known as Sanduleak, as it appeared until 1987. (*b*) We then began to intercept light from the supernova of that star; the explosion was 100 million times brighter than our Sun and could be seen with the unaided eye. The explosion took place 155,000 light-years away and thus actually occurred 155,000 years ago.

$$\frac{dE}{dm} = (6.6 \text{ MeV/proton}) \left(\frac{1 \text{ proton}}{1.67 \times 10^{-27} \text{ kg}} \right)$$

$$\times \left(\frac{1.60 \times 10^{-13} \text{ J}}{1 \text{ MeV}} \right)$$

$$= 6.3 \times 10^{14} \text{ J/kg}.$$

This tells us that the Sun radiates away 6.3×10^{14} J of energy for every kilogram of protons consumed. The hydrogen consumption rate is then the Sun's power (= 3.9×10^{26} W) divided by the quantity above; thus

$$\text{rate} = \frac{3.9 \times 10^{26} \text{ W}}{6.3 \times 10^{14} \text{ J/kg}} = 6.2 \times 10^{11} \text{ kg/s}. \quad \text{(Answer)}$$

This seems like a tremendous mass loss every second but—to keep things in perspective—it is small compared to the Sun's mass, which is 2×10^{30} kg.

44-8 CONTROLLED THERMONUCLEAR FUSION

The first thermonuclear reaction on Earth occurred at Eniwetok Atoll on November 1, 1952, when the United States exploded a fusion device, generating an energy release equivalent to 10 million tons of TNT. The high temperatures and densities needed to initiate the reaction were provided by using a fission bomb as a trigger.

A sustained and controllable source of fusion power —a fusion reactor—is considerably more difficult to achieve. The goal, however, is being pursued vigorously in many countries around the world because many people look to the fusion reactor as the power source of the future, at least as far as the generation of electricity is concerned.

The p–p scheme displayed in Fig. 44-11 is not suitable for an Earth-bound fusion reactor because it is hopelessly slow. The process succeeds in the Sun only because of the enormous density of protons in the center of the Sun. The most attractive reactions for terrestrial use appear to be two deuteron–deuteron (d–d) reactions,

$$^2\text{H} + {}^2\text{H} \rightarrow {}^3\text{He} + \text{n} \qquad Q = +3.27 \text{ MeV}, \qquad (44\text{-}9)$$

$$^2\text{H} + {}^2\text{H} \rightarrow {}^3\text{H} + {}^1\text{H} \qquad Q = +4.03 \text{ MeV}, \qquad (44\text{-}10)$$

and the deuteron–triton (d–t) reaction*

$$^2\text{H} + {}^3\text{H} \rightarrow {}^4\text{He} + \text{n} \qquad Q = +17.59 \text{ MeV}. \qquad (44\text{-}11)$$

Deuterium, the source of deuterons for these reactions, has an isotopic abundance of only 1 part in 6700, but is available in unlimited quantities as a component of seawater. Proponents of power from the nucleus have described our

---

*The nucleus of the hydrogen isotope ^{3}H (tritium) is called the *triton*. It is a radionuclide, with a half-life of 12.3 y.

ultimate power choice—when we have burned up all our fossil fuels—as either "burning rocks" (fission of uranium extracted from ores) or "burning water" (fusion of deuterium extracted from water).

There are three requirements for a successful thermonuclear reactor:

1. *A High Particle Density n.* The (number) density of interacting particles (the number of, say, deuterons per unit volume) must be great enough to ensure that the d–d collision rate is high enough. At the high temperatures required, the deuterium would be completely ionized, forming a neutral **plasma** (ionized gas) of deuterons and electrons.

2. *A High Plasma Temperature T.* The plasma must be hot. Otherwise the colliding deuterons will not be energetic enough to penetrate the Coulomb barrier that tends to keep them apart. A plasma ion temperature of 35 keV, corresponding to 4×10^8 K, has been achieved in the laboratory. This is about 30 times higher than the Sun's central temperature.

3. *A Long Confinement Time τ.* A major problem is containing the hot plasma long enough to maintain it at a density and a temperature sufficiently high to ensure the fusion of enough of the fuel. It is clear that no solid container can withstand the high temperatures that are necessary, so clever confining techniques are called for; we shall shortly discuss two techniques.

It can be shown that, for the successful operation of a thermonuclear reactor using the d–t reaction, it is necessary to have

$$n\tau > 10^{20} \text{ s/m}^3. \qquad (44\text{-}12)$$

This condition, known as **Lawson's criterion**, tells us that we have a choice between confining a lot of particles for a short time or fewer particles for a longer time. Beyond meeting this criterion, it is still necessary that the plasma temperature be high enough.

Two approaches to controlled nuclear power generation are currently under study. Although neither approach has yet been successful, both are being pursued because of their promise and because of the potential importance of controlled fusion to the world's energy problems.

Magnetic Confinement

In one version of this approach, a suitably shaped magnetic field is used to confine a hot plasma in an evacuated doughnut-shaped chamber called a **tokamak** (the name is an abbreviation consisting of parts of three Russian words). The magnetic forces acting on the charged particles that make up the hot plasma keep the plasma from touching the walls of the chamber. Figure 44-13 shows such a device at the Plasma Physics Laboratory of Princeton University.

The plasma is heated by inducing a current to flow in it and also by bombarding the plasma with an externally accelerated beam of particles. The first goal is to achieve **breakeven**, which occurs when the Lawson criterion is met or exceeded. The ultimate goal is **ignition**, which corresponds to a self-sustaining thermonuclear reaction, with a net generation of energy. As of 1996, ignition has not been achieved, either in tokamaks or in other magnetic confinement devices.

Inertial Confinement

This technique for confining and heating fusion fuel so that a thermonuclear reaction can occur involves ''zapping'' a solid fuel pellet from all sides with intense laser beams, evaporating some material from the surface of the pellet. This boiled-off material causes an inward-moving shock wave that compresses the core of the pellet, increasing both its particle density and its temperature. The process is called *inertial confinement* because (a) the fuel is *confined* to the pellet and (b) the particles do not escape from the heated pellet during the very short zapping interval because of their *inertia* (their mass).

Laser fusion, using the confinement technique, is being investigated in many laboratories in the United States and elsewhere. At the Lawrence Livermore Laboratory, for example, deuterium–tritium fuel pellets, each smaller than a grain of sand (Fig. 44-14), are to be zapped by 10 synchronized high-power laser pulses symmetrically arranged around the pellet. The laser pulses are designed to deliver, in total, some 200 kJ of energy to each fuel pellet in less than a nanosecond. This is a delivered

FIGURE 44-14   The small spheres on the quarter are deuterium–tritium fuel pellets, designed to be used in a laser fusion chamber.

power of about 2×10^{14} W during the pulse, which is roughly 100 times the total installed (sustained) electric power generating capacity of the world!

In an operating thermonuclear reactor of the laser-fusion type, fuel pellets are to be exploded—like miniature hydrogen bombs—at a rate of perhaps 10–100 per second. The feasibility of laser fusion as the basis of a thermonuclear power reactor has not been demonstrated as of 1996, but research is continuing at a vigorous pace.

FIGURE 44-13   The Tokamak Fusion Test Reactor at Princeton University.

SAMPLE PROBLEM 44-6

Suppose that a fuel pellet in a laser fusion device contains equal numbers of deuterium and tritium atoms. The density $d = 200$ kg/m³ of the pellet is increased by a factor of 10^3 by the action of the laser pulses.

(a) How many particles per unit volume (both deuterons and tritons) does the pellet contain in its compressed state?

SOLUTION: We can write, for the density d^* of the compressed pellet,

$$d^* = 10^3 d = m_d \left(\frac{n}{2} \right) + m_t \left(\frac{n}{2} \right),$$

in which n is the total number of fusible particles per unit volume (both deuterons and tritons) in the compressed pellet, m_d is the mass of a deuterium *atom,* and m_t is the mass of a tritium *atom.* These atomic masses are related to the Avogadro constant N_A and to the corresponding molar masses ($M_d = 2.0 \times 10^{-3}$ kg/mole and $M_t = 3.0 \times 10^{-3}$ kg/mole) by

$$m_d = \frac{M_d}{N_A} \quad \text{and} \quad m_t = \frac{M_t}{N_A}.$$

Combining the foregoing equations and solving for n lead to

$$n = \frac{2000 d N_A}{M_d + M_t},$$

which gives us

$$n = \frac{(2000)(200 \text{ kg/m}^3)(6.02 \times 10^{23} \text{ mol}^{-1})}{2.0 \times 10^{-3} \text{ kg/mol} + 3.0 \times 10^{-3} \text{ kg/mol}}$$

$$= 4.8 \times 10^{31} \text{ m}^{-3}. \qquad \text{(Answer)}$$

(b) According to Lawson's criterion, how long must the pellet maintain this particle density if break-even operation is to take place?

SOLUTION: From Lawson's criterion (Eq. 44-12) we have

$$\tau > \frac{10^{20} \text{ s/m}^3}{4.8 \times 10^{31} \text{ m}^{-3}} \approx 10^{-12} \text{ s}. \qquad \text{(Answer)}$$

The pellet must remain compressed for at least 10^{-12} s if break-even operation is to occur. (The plasma temperature must also be suitably high.)

REVIEW & SUMMARY

Energy from the Nucleus

Nuclear processes are about a million times more effective, per unit mass, than chemical processes in transforming mass into other forms of energy.

Nuclear Fission

Equation 44-1 shows a **fission** of ^{236}U induced by thermal neutrons bombarding ^{235}U. Equations 44-2 and 44-3 show the beta-decay chains of the primary fragments. The energy released in such a fission event is $Q \approx 200$ MeV.

Fission can be understood in terms of the collective model, in which a nucleus is likened to a charged liquid drop carrying a certain excitation energy. A potential barrier must be tunneled if fission is to occur. Fissionability depends on the relationship between the barrier height E_b and the excitation energy E_n.

The neutrons released during fission make possible a fission **chain reaction**. Figure 44-4 shows the neutron balance for one cycle of a typical reactor. Figure 44-5 suggests the outlines of a complete nuclear power plant.

Nuclear Fusion

The release of energy by the **fusion** of two light nuclei is inhibited by their mutual Coulomb barrier. Fusion can occur in bulk matter only if the temperature is high enough (that is, if the particle energy is high enough) for appreciable barrier tunneling to occur.

The Sun's energy arises mainly from the thermonuclear burning of hydrogen to form helium by the **proton–proton cycle** outlined in Fig. 44-11. Elements up to $A \approx 56$ (the peak of the binding energy curve) can be built up by other fusion processes once the hydrogen fuel supply of a star has been exhausted.

Controlled Fusion

Controlled **thermonuclear fusion** for power generation has not yet been achieved, even on a laboratory scale. The d–d and the d–t reactions are the most promising mechanisms. A successful fusion reactor must satisfy **Lawson's criterion,**

$$n\tau > 10^{20} \text{ s/m}^3, \qquad (44-12)$$

and must have a suitably high plasma temperature T.

In a **tokamak** the plasma is confined by a magnetic field. In **laser fusion** inertial confinement is used.

QUESTIONS

1. In Table 44-1, does the relation $Q = \Delta m c^2$ apply to (a) all the processes, (b) all the processes except the waterfall, (c) the fission processes only, (d) the fission and fusion processes only?

2. According to Fig. 44-1, fission of ^{235}U by thermal neutrons into two equally massive fragments occurs in about one event in (a) 10,000, (b) 1000, (c) 100, (c) 10.

3. Do the initial fragments formed by fission have (a) more protons than neutrons, (b) more neutrons than protons, or (c) about the same number of each?

4. Consider the fission reaction

$$^{235}\text{U} + \text{n} \rightarrow \text{X} + \text{Y} + 2\text{n}.$$

Rank the following possible nuclide values for X (or Y), most likely first: (a) ^{152}Nd, (b) ^{140}I, (c) ^{128}In, (d) ^{115}Pd, (e) ^{105}Mo. (*Hint:* See Fig. 44-1.)

5. Pick the most likely member of each of these pairs to be one of the initial fragments formed by a fission event: (a) ^{93}Sr or ^{93}Ru, (b) ^{140}Gd or ^{140}I, (c) ^{155}Nd or ^{155}Lu. (*Hint:* See Fig. 43-4 and the periodic table.)

6. Suppose that a ^{238}U nucleus ''swallows'' a neutron and then decays not by fission but by beta decay, emitting an electron and a neutrino. Which nuclide remains after this decay: (a) ^{239}Pu, (b) ^{238}Np, (c) ^{239}Np, or (d) ^{238}Pa?

7. A nuclear reactor is operating at a certain power level, with its multiplication factor k adjusted to unity. If the control rods are used to reduce the power output of the reactor to 25% of its former value, is the multiplication factor now (a) a little less than unity, (b) substantially less than unity, or (c) still equal to unity?

8. A nuclear reactor core should have the smallest possible surface-to-volume ratio. Order the following solids according to their surface-to-volume ratios, greatest first: (a) a cube of edge a, (b) a sphere of radius a, (c) a cone of height a and base radius a, and (d) a cylinder of radius a and height a. (The area of the *curved* surface of the cone is $\sqrt{2}\,\pi a^2$ and its volume is $\pi a^3/3$.)

9. Figure 44-6 shows how the heat generated by the nuclear waste from one year's operation of a large nuclear power plant decays over time. By what approximate factor has this thermal energy output decreased at the end of 100 years: (a) 20, (b) 200, (c) 2000, (d) more than 2000?

10. Which of these elements is *not* ''cooked up'' by thermonuclear fusion processes in stellar interiors: (a) carbon, (b) silicon, (c) chromium, (d) bromine?

11. About 2% of the energy generated in the Sun's core by the p–p reaction is carried out of the Sun by neutrinos. Is the energy associated with this neutrino flux (a) equal to, (b) greater than, or (c) less than the energy radiated from the Sun's surface as electromagnetic radiation?

12. Lawson's criterion for the d–t reaction (Eq. 44-12) is $n\tau > 10^{20}$ s/m^3. For the d–d reaction, do you expect the number on the right-hand side to be (a) the same, (b) smaller, or (c) larger?

EXERCISES & PROBLEMS

SECTION 44-2 Nuclear Fission: The Basic Process

1E. (a) How many atoms are contained in 1.0 kg of pure ^{235}U? (b) How much energy, in joules, is released by the complete fissioning of 1.0 kg of ^{235}U? Assume $Q = 200$ MeV. (c) For how long would this energy light a 100 W lamp?

2E. The fission properties of the plutonium isotope ^{239}Pu are very similar to those of ^{235}U. The average energy released per fission is 180 MeV. How much energy, in MeV, is liberated if all the atoms in 1.00 kg of pure ^{239}Pu undergo fission?

3E. At what rate must ^{235}U nuclei undergo fission by neutrons to generate energy at the rate of 1.0 W? Assume that $Q = 200$ MeV.

4E. Fill in the following table, which refers to the generalized fission reaction

$$^{235}\text{U} + \text{n} \rightarrow \text{X} + \text{Y} + b\text{n}.$$

X	Y	b
^{140}Xe	—	1
^{139}I	—	2
—	^{100}Zr	2
^{141}Cs	^{92}Rb	—

5E. Verify that, as stated in Section 44-2, neutrons in equilibrium with matter at room temperature, 300 K, have an average kinetic energy of about 0.04 eV.

6E. Calculate the disintegration energy Q for the fission of ^{52}Cr into two equal fragments. The masses you will need are 51.94051 u for ^{52}Cr and 25.98259 u for ^{26}Mg.

7E. Calculate the disintegration energy Q for the fission of ^{98}Mo into two equal parts. The masses you will need are 97.90541 u for ^{98}Mo and 48.95002 u for ^{49}Sc. If Q turns out to be positive, discuss why this process does not occur spontaneously.

8E. Calculate the energy released in the fission reaction

$$^{235}\text{U} + \text{n} \rightarrow {}^{141}\text{Cs} + {}^{93}\text{Rb} + 2\text{n}.$$

Needed atomic and particle masses are

^{235}U	235.04392 u	^{93}Rb	92.92157 u
^{141}Cs	140.91963 u	n	1.00867 u

9E. ^{235}U decays by alpha emission with a half-life of 7.0×10^8 y. It also decays (rarely) by spontaneous fission, and if the alpha decay did not occur, its half-life due to this process alone would be 3.0×10^{17} y. (a) At what rate do spontaneous fission decays occur in 1.0 g of ^{235}U? (b) How many ^{235}U alpha-decay events are there for every spontaneous fission event?

10P. Verify that, as reported in Table 44-1, fissioning of the ^{235}U in 1.0 kg of UO$_2$ (enriched so that ^{235}U is 3.0% of the total uranium) could keep a 100 W lamp burning for 690 y.

11P. Consider the fission of ^{238}U by fast neutrons. In one fission event no neutrons were emitted and the final stable end products, after the beta decay of the primary fission fragments, were ^{140}Ce and ^{99}Ru. (a) How many beta-decay events were there in the two beta-decay chains, considered together? (b) Calculate Q. The relevant atomic and particle masses are

^{238}U	238.05079 u	^{140}Ce	139.90543 u
n	1.00867 u	^{99}Ru	98.90594 u

12P. In a particular fission event in which ^{235}U is fissioned by slow neutrons, no neutron is emitted and one of the primary fission fragments is ^{83}Ge. (a) What is the other fragment? (b) How is the disintegration energy $Q = 170$ MeV split between the two fragments? (c) Calculate the initial speed of each fragment.

13P. Assume that just after the fission of ^{236}U according to Eq. 44-1, the resulting ^{140}Xe and ^{94}Sr nuclei are just touching at their surfaces. (a) Assuming the nuclei to be spherical, calculate the electric potential energy (in MeV) associated with the repulsion between the two fragments. (*Hint:* Use Eq. 43-3 to calculate the radii of the fragments.) (b) Compare this energy with the energy released in a typical fission event.

14P. A ^{236}U nucleus undergoes fission and breaks up into two middle-mass fragments, ^{140}Xe and ^{96}Sr. (a) By what percentage does the surface area of the ^{236}U nucleus change during this process? (b) By what percentage does its volume change? (c) By what percentage does its electric potential energy change? The potential energy of a uniformly charged sphere of radius r and charge Q is given by

$$U = \frac{3}{5}\left(\frac{Q^2}{4\pi\epsilon_0 r}\right).$$

SECTION 44-4 The Nuclear Reactor

15E. A 200 MW fission reactor consumes half its fuel in 3.00 y. How much ^{235}U did it contain initially? Assume that all the energy generated arises from the fission of ^{235}U and that this nuclide is consumed only by the fission process.

16E. Repeat Exercise 15 taking into account nonfission neutron capture by the ^{235}U.

17E. ^{238}Np requires 4.2 MeV for fission. To remove a neutron from this nuclide requires an energy expenditure of 5.0 MeV. Is ^{237}Np fissionable by thermal neutrons?

18P. The thermal energy generated when radiations from radionuclides are absorbed in matter can serve as the basis for a small power source for use in satellites, remote weather stations, and so on. Such radionuclides are manufactured in abundance in nuclear reactors and may be separated chemically from the spent fuel. One suitable radionuclide is ^{238}Pu ($\tau = 87.7$ y), which is an alpha emitter with $Q = 5.50$ MeV. At what rate is thermal energy generated in 1.00 kg of this material?

19P. (See Problem 18.) Among the many fission products that may be extracted chemically from the spent fuel of a nuclear reactor is ^{90}Sr ($\tau = 29$ y). This isotope is produced in typical large reactors at the rate of about 18 kg/y. By its radioactivity it generates thermal energy at the rate of 0.93 W/g. (a) Calculate the effective disintegration energy Q_{eff} associated with the decay of a ^{90}Sr nucleus. (Q_{eff} includes contributions from the decay of the ^{90}Sr daughter products in its decay chain but not from neutrinos, which escape totally from the sample.) (b) It is desired to construct a power source generating 150 W (electric) to use in operating electronic equipment in an underwater acoustic beacon. If the power source is based on the thermal energy generated by ^{90}Sr and if the efficiency of the thermal–electric conversion process is 5.0%, how much ^{90}Sr is needed?

20P. Many fear that helping additional nations develop nuclear power reactor technology will increase the likelihood of nuclear war because reactors can be used not only to produce electrical energy but, as a by-product through neutron capture with inexpensive ^{238}U, to make ^{239}Pu, which is a "fuel" for nuclear bombs. What simple series of reactions involving neutron capture and beta decay would yield this plutonium isotope?

21P. In an atomic bomb, energy release is due to the uncontrolled fission of plutonium ^{239}Pu (or ^{235}U). The bomb's rating is the magnitude of the released energy, specified in terms of the mass of TNT required to produce the same energy release. One megaton (10^6 tons) of TNT releases 2.6×10^{28} MeV of energy. (a) Calculate the rating, in tons of TNT, of an atomic bomb containing 95 kg of ^{239}Pu, of which 2.5 kg actually undergoes fission. (See Exercise 2.) (b) Why is the other 92.5 kg of ^{239}Pu needed if it does not fission?

22P. A 66 kiloton atomic bomb (see Problem 21) is fueled with pure ^{235}U (Fig. 44-15), 4.0% of which actually undergoes fission. (a) How much uranium is in the bomb? (b) How many primary fission fragments are produced? (c) How many neutrons generated in the fissions are released to the environment? (On the average, each fission produces 2.5 neutrons.)

FIGURE 44-15 Problem 22. A "button" of ^{235}U, ready to be recast and machined for a warhead.

23P. The neutron generation time t_{gen} in a reactor is the average time needed for a fast neutron emitted in one fission to be slowed to thermal energies by the moderator and to initiate another fission. Suppose that the power output of a reactor at time $t = 0$ is P_0. Show that the power output a time t later is $P(t)$, where

$$P(t) = P_0 k^{t/t_{gen}},$$

where k is the multiplication factor. For constant power output, $k = 1$.

24P. The neutron generation time (see Problem 23) of a particular power reactor is 1.3 ms. It is generating energy at the rate of 1200 MW. To perform certain maintenance checks, the power level must temporarily be reduced to 350 MW. It is desired that the transition to the reduced power level take 2.6 s. To what (constant) value should the multiplication factor be set to effect the transition in the desired time?

25P. The neutron generation time t_{gen} (see Problem 23) in a particular reactor is 1.0 ms. If the reactor is operating at a power level of 500 MW, about how many free neutrons are present in the reactor at any moment?

26P. A reactor operates at 400 MW with a neutron generation time (see Problem 23) of 30.0 ms. If its power increases for 5.00 min with a multiplication factor of 1.0003, what is the power output at the end of the 5.00 min?

27P. (a) A neutron of mass m_n and kinetic energy K makes a head-on elastic collision with a stationary atom of mass m. Show that the fractional kinetic energy loss of the neutron is given by

$$\frac{\Delta K}{K} = \frac{4m_n m}{(m + m_n)^2},$$

in which m_n is the neutron mass. (b) Find $\Delta K/K$ for each of the following as the stationary atom: hydrogen, deuterium, carbon, and lead. (c) If $K = 1.00$ MeV initially, how many such collisions would it take to reduce the neutron's kinetic energy to a thermal value (0.025 eV) if the stationary atoms it collides with are deuterium, a commonly used moderator? (*Note:* In actual moderators, most collisions are not head-on.)

SECTION 44-5 A Natural Nuclear Reactor

28E. How long ago was the ratio $^{235}U/^{238}U$ in natural uranium deposits equal to 0.15?

29E. The natural fission reactor discussed in Section 44-5 is estimated to have generated 15 gigawatt-years of energy during its lifetime. (a) If the reactor lasted for 200,000 y, at what average power level did it operate? (b) How much ^{235}U did it consume during its lifetime?

30P. Mixed in with ^{238}U, uranium mined today contains 0.72% of fissionable ^{235}U, too little to make reactor fuel for thermal-neutron fission. For this reason, the natural uranium must be enriched with ^{235}U. Both ^{235}U ($\tau = 7.0 \times 10^8$ y) and ^{238}U ($\tau = 4.5 \times 10^9$ y) are radioactive. How far back in time would natural uranium have been a practical reactor fuel, with a $^{235}U/^{238}U$ ratio of 3.0%?

31P. Some uranium samples from the natural reactor site described in Section 44-5 were found to be slightly *enriched* in ^{235}U, rather than depleted. Account for this in terms of neutron absorption by the abundant isotope ^{238}U and the subsequent beta and alpha decay of its products.

SECTION 44-6 Thermonuclear Fusion: The Basic Process

32E. Calculate the height of the Coulomb barrier for the head-on collision of two deuterons. The effective radius of a deuteron may be taken to be 2.1 fm.

33E. From information given in the text, collect and write down the approximate heights of the Coulomb barriers for (a) the alpha decay of ^{238}U and (b) the fission of ^{235}U by thermal neutrons.

34E. Verify that the fusion of 1.0 kg of deuterium by the reaction

$$^2H + {}^2H \rightarrow {}^3He + n \quad (Q = +3.27 \text{ MeV})$$

could keep a 100 W lamp burning for 3×10^4 y.

35E. Methods other than heating the material have been suggested for overcoming the Coulomb barrier for fusion. For example, one might consider particle accelerators. If you were to use two of them to accelerate two beams of deuterons directly toward each other so as to collide head-on, (a) what voltage would each accelerator require for the colliding deuterons to overcome the Coulomb barrier? (b) Why do you suppose this method is not presently used?

36P. In Fig. 44-10, the equation for $n(K)$, the number density per unit energy for particles, is

$$n(K) = 1.13n \frac{K^{1/2}}{(kT)^{3/2}} e^{-K/kT},$$

where n is the total number density of particles. At the center of the Sun the temperature is 1.50×10^7 K, and the mean proton energy $\overline{K}$ is 1.94 keV. Find the ratio of the number density of protons at 5.00 keV to that at the mean proton energy.

37P. Calculate the Coulomb barrier height for two ^7Li nuclei that are fired at each other with the same initial kinetic energy K. (*Hint:* Use Eq. 43-3 to calculate the radii of the nuclei.)

38P. Expressions for the Maxwell speed and energy distributions for the molecules in a gas are given in Chapter 20. (a) Show that the *most probable energy* is given by

$$K_p = \tfrac{1}{2}kT.$$

Verify this result with the energy distribution curve of Fig. 44-10, for which $T = 1.5 \times 10^7$ K. (b) Show that the *most probable speed* is given by

$$v_p = \sqrt{\frac{2kT}{m}}.$$

Find its value for protons at $T = 1.5 \times 10^7$ K. (c) Show that *the energy corresponding to the most probable speed* (which is not the same as the most probable energy) is

$$K_{v,p} = kT.$$

Locate this quantity on the curve of Fig. 44-10.

SECTION 44-7 Thermonuclear Fusion in the Sun and Other Stars

39E. We have seen that Q for the overall proton–proton cycle is 26.7 MeV. How can you relate this number to the Q values for the reactions that make up this cycle, as displayed in Fig. 44-11?

40E. Show that the energy released when three alpha particles fuse to form ^{12}C is 7.27 MeV. The atomic mass of ^4He is 4.0026 u, and that of ^{12}C is 12.0000 u.

41E. At the center of the Sun the density is 1.5×10^5 kg/m^3 and the composition is essentially 35% hydrogen by mass and 65% helium. (a) What is the density of protons at the center of the Sun? (b) How much greater is this than the density of particles in an ideal gas at standard conditions of temperature (0°C) and pressure $(1.01 \times 10^5 \text{ Pa})$?

42P. Verify the three Q values reported for the reactions in Fig. 44-11. The needed atomic and particle masses are

^{1}H	1.007 825 u	^{4}He	4.002 603 u
^{2}H	2.014 102 u	$e^{\pm}$	0.000 5486 u
^{3}He	3.016 029 u		

(*Hint:* Distinguish carefully between atomic and nuclear masses, and take the positrons properly into account.)

43P. Calculate and compare the energy released by (a) the fusion of 1.0 kg of hydrogen deep within the Sun and (b) the fission of 1.0 kg of ^{235}U in a fission reactor.

44P. The Sun has a mass of 2.0×10^{30} kg and radiates energy at the rate of 3.9×10^{26} W. (a) At what rate does the Sun transfer its mass to other forms of energy? (b) What fraction of its original mass has the Sun lost in this way since it began to burn hydrogen, about 4.5×10^9 y ago?

45P. (a) Calculate the rate at which the Sun generates neutrinos. Assume that energy production is entirely by the proton–proton cycle. (b) At what rate do solar neutrinos reach Earth?

46P. Coal burns according to the reaction $C + O_2 \rightarrow CO_2$. The heat of combustion is 3.3×10^7 J/kg of atomic carbon consumed. (a) Express this in terms of energy per carbon atom. (b) Express it in terms of energy per kilogram of the initial reactants, carbon and oxygen. (c) Suppose that the Sun (mass = 2.0×10^{30} kg) were made of carbon and oxygen in combustible proportions and that it continued to radiate energy at its present rate of 3.9×10^{26} W. How long would it last?

47P. In certain stars the *carbon cycle* is more likely than the proton–proton cycle to be effective in generating energy. This cycle is

$$^{12}\text{C} + {}^1\text{H} \rightarrow {}^{13}\text{N} + \gamma, \qquad Q_1 = 1.95 \text{ MeV},$$
$$^{13}\text{N} \rightarrow {}^{13}\text{C} + e^+ + \nu, \qquad Q_2 = 1.19,$$
$$^{13}\text{C} + {}^1\text{H} \rightarrow {}^{14}\text{N} + \gamma, \qquad Q_3 = 7.55,$$
$$^{14}\text{N} + {}^1\text{H} \rightarrow {}^{15}\text{O} + \gamma, \qquad Q_4 = 7.30,$$
$$^{15}\text{O} \rightarrow {}^{15}\text{N} + e^+ + \nu, \qquad Q_5 = 1.73,$$
$$^{15}\text{N} + {}^1\text{H} \rightarrow {}^{12}\text{C} + {}^4\text{He}, \qquad Q_6 = 4.97.$$

(a) Show that this cycle of reactions is exactly equivalent in its overall effects to the proton–proton cycle of Fig. 44-11. (b) Verify that the two cycles, as expected, have the same Q value.

48P. Assume that the core of the Sun has one-eighth of the Sun's mass and is compressed within a sphere whose radius is one-fourth of the solar radius. Assume further that the composition of the core is 35% hydrogen by mass and that essentially all the Sun's energy is generated there. If the Sun continues to burn hydrogen at the rate calculated in Sample Problem 44-5, how long will it be before the hydrogen is entirely consumed? The Sun's mass is 2.0×10^{30} kg.

49P. The effective Q for the proton–proton cycle of Fig. 44-11 is 26.2 MeV. (a) Express this as energy per kilogram of hydrogen consumed. (b) The power of the Sun is 3.9×10^{26} W. If its energy derives from the proton–proton cycle, at what rate is it losing hydrogen? (c) At what rate is it losing mass? Account for the difference in the results for (b) and (c). (d) The Sun's mass is 2.0×10^{30} kg. If it loses mass at the constant rate calculated in (c), how long will it take to lose 0.10% of its mass?

50P. A star converts all its hydrogen to helium, achieving a 100% helium composition. It now proceeds to convert the helium to carbon via the triple-alpha process,

$$^4\text{He} + {}^4\text{He} + {}^4\text{He} \rightarrow {}^{12}\text{C} + 7.27 \text{ MeV}.$$

The mass of the star is 4.6×10^{32} kg, and it generates energy at the rate of 5.3×10^{30} W. How long will it take to convert all the helium to carbon at this rate?

51P. Figure 44-16 shows an early proposal for a hydrogen bomb. The fusion fuel is deuterium, ^{2}H. The high temperature and particle density needed for fusion are provided by an atomic bomb "trigger," which involves a ^{235}U or ^{239}Pu fission fuel that is arranged to impress an imploding, compressive shock wave on the deuterium. The operative fusion reaction is

$$5 \, {}^2\text{H} \rightarrow {}^3\text{He} + {}^4\text{He} + {}^1\text{H} + 2\text{n}.$$

(a) Calculate Q for the fusion reaction. For needed atomic masses, see Problem 42. (b) Calculate the rating (see Problem 21) of the fusion part of the bomb if it contains 500 kg of deuterium, 30.0% of which undergoes fusion.

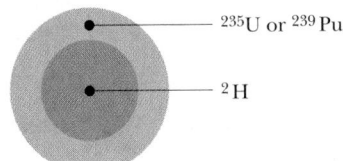

FIGURE 44-16 Problem 51.

SECTION 44-8 Controlled Thermonuclear Fusion

52E. Verify the Q values reported in Eqs. 44-9, 44-10, and 44-11. The needed masses are

^{1}H	1.007 825 u	^{4}He	4.002 603 u
^{2}H	2.014 102 u	n	1.008 665 u
^{3}H	3.016 049 u		

53E. In the deuteron–triton fusion reaction of Eq. 44-11, how is the reaction energy Q shared between the alpha particle and the neutron? Neglect the relatively small kinetic energies of the two combining particles.

54P. Ordinary water consists of roughly 0.0150% by mass of "heavy water," in which one of the two hydrogens is replaced with deuterium, ^{2}H. How much average fusion power could be obtained if we "burned" all the ^{2}H in 1 liter of water in 1 day through the reaction $^2\text{H} + {}^2\text{H} \rightarrow {}^3\text{He} + \text{n}$?

45
Quarks, Leptons, and the Big Bang

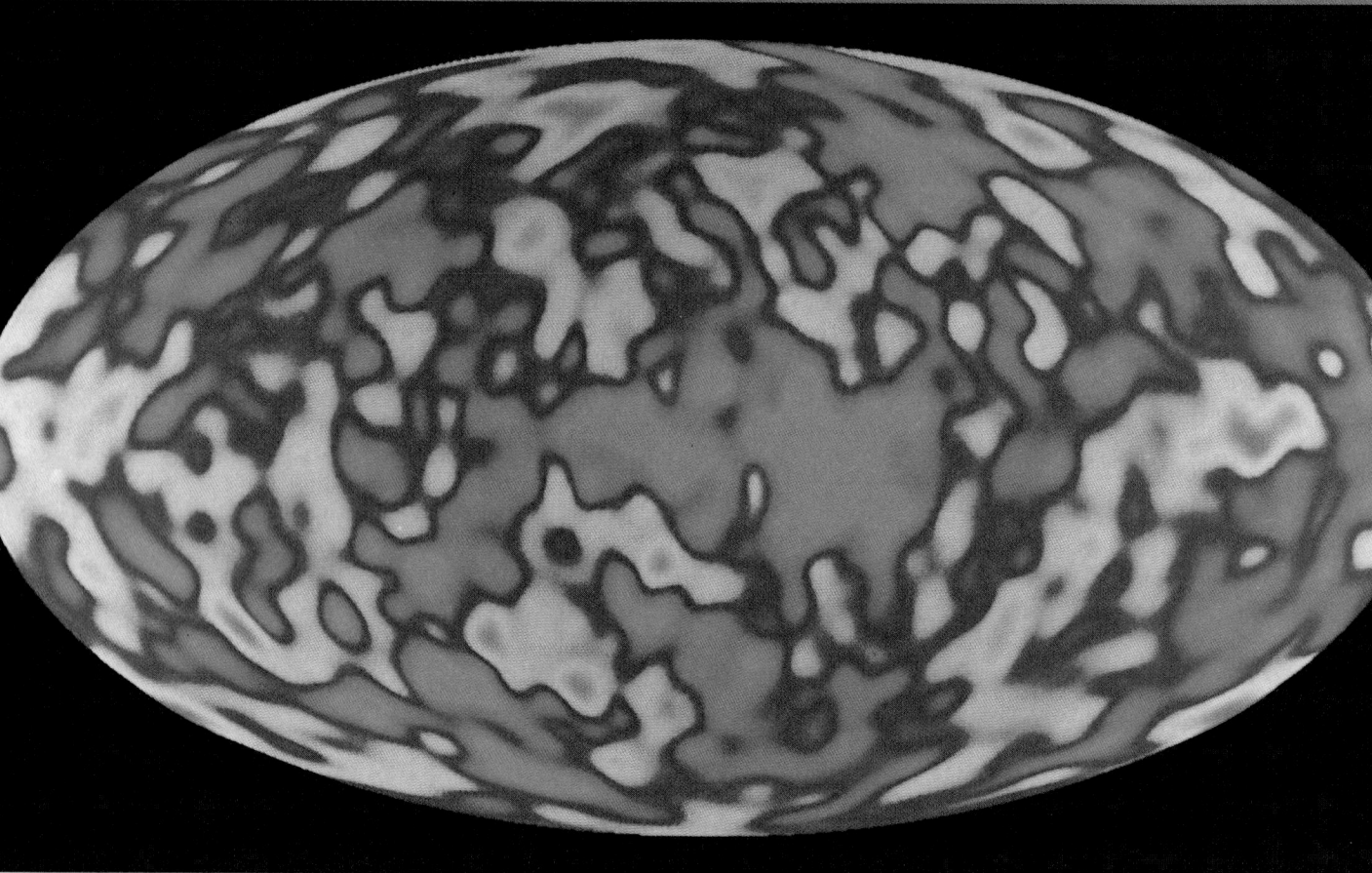

This color-coded image is effectively a photograph of the universe when it was only 300,000 y old, which was about 15×10^9 y ago. This is what you would have seen then as you looked away in all directions (the view has been condensed to this oval). Patches of light from atoms stretch across the "sky," but galaxies, stars, and planets have not yet formed. How can such a photograph of the early universe be taken?

45-1 LIFE AT THE CUTTING EDGE

Physicists often refer to the theories of relativity and quantum mechanics as "modern physics," to distinguish them from the theories of Newtonian mechanics and Maxwellian electromagnetism, which are lumped together as "classical physics." As the years go by, the word "modern" seems less and less appropriate for theories whose foundations were laid down in the opening years of the twentieth century. Nevertheless, the label hangs on.

In this closing chapter we consider two lines of investigation that are truly "modern" but at the same time have the most ancient of roots. They center around two deceptively simple questions:

What is the universe made of?
How did the universe come to be the way it is?

Progress in answering these questions has been rapid in the last few decades.

Many new insights are based on experiments carried out with large particle accelerators. But as physicists bang particles together at higher and higher energies, using larger and larger accelerators, they come to realize that no conceivable Earth-bound accelerator can generate particles with energies great enough to test their theories. There has been only one source of particles with these energies, and that was the universe itself within the first few minutes of its existence. The "quark soup" that constituted the universe in the first few seconds on the cosmic clock is the ultimate testing ground for the theories of particle physics!

In this chapter you will encounter a host of new terms and a veritable flood of particles with names that you should not try to remember. If you are temporarily bewildered, you are sharing the bewilderment of the physicists who lived through these developments and who at times saw nothing but increasing complexity with little hope of understanding. If you stick with it, however, you will come to share the excitement physicists felt as marvelous new accelerators poured out new results, as the theorists put forth ideas each more daring than the last, and as clarity finally sprang from obscurity.

You may wish to begin by rereading Section 2-9, in which we first discussed the basic particles of physics.

45-2 PARTICLES, PARTICLES, PARTICLES

In the 1930s, there were many scientists who thought that the problem of the ultimate structure of matter was well on the way to being solved. The atom could be understood in terms of only three particles—the electron, the proton, and the neutron. Quantum theory accounted well for the structure of the atom and for radioactive alpha decay. The neutrino had been postulated and, although not yet observed, had been incorporated by Enrico Fermi into a successful theory of beta decay. There was hope that quantum theory, applied to protons and neutrons, would soon account for the structure of the nucleus. What else was there?

The euphoria did not last. The end of that same decade saw the beginning of a period of discovery of new particles that continues to this day. The new particles have names and symbols such as *muon* (μ), *pion* (π), *kaon* (K), and *sigma* (Σ). All the new particles are unstable, their half-lives ranging from about 10^{-6} to 10^{-23} s. This last value is so small that the very existence of such particles can be established only by indirect methods.

These new particles are commonly produced in head-on collisions between protons or electrons accelerated to high energies in accelerators at places like Fermilab (near Chicago), CERN (near Geneva), SLAC (at Stanford), and DESY (near Hamburg, Germany). They are discovered with particle detectors that have grown in sophistication until (see Fig. 45-1) they rival the size and complexity of entire accelerators of only a decade or so ago.

FIGURE 45-1 The OPAL (omni-purpose apparatus) particle detector at CERN, the European high-energy particle physics laboratory near Geneva, Switzerland. OPAL is designed to measure the energies of particles produced in electron−positron collisions at energies of 50 GeV. Although the detector is huge (weighing over 3000 tons), it is small compared with the accelerator itself, which is a ring with a circumference of 27 km.

Today there are several hundred known particles. Naming them has strained the resources of the Greek alphabet, and most are known only by an assigned number in a periodically issued compilation. To make sense of this array of particles, we look for simple physical criteria, each of which will allow us to place each particle into either of two categories. We can make a first rough cut among the particles in at least the following three ways.

Fermion or Boson?

All particles have an intrinsic angular momentum called **spin**, as we discussed for electrons, protons, and neutrons in Section 32-4. Generalizing the notation of that section, we can write the component of spin **S** in any direction (assume it to be along a z axis) as

$$S_z = m_s \hbar \qquad \text{for} \qquad m_s = s, s - 1, \ldots, -s, \qquad (45\text{-}1)$$

in which $\hbar$ is $h/2\pi$, m_s is the *spin magnetic quantum number*, and s is the *spin quantum number*. The latter can have either half-integer values ($\frac{1}{2}$, $\frac{3}{2}$, . . .) or integer values (0, 1, 2, . . .). For example, an electron has the value $s = \frac{1}{2}$. Hence the spin of an electron (measured along any direction) can have the values

$$s_z = \tfrac{1}{2}\hbar \qquad \text{(spin up)}$$

or
$$s_z = -\tfrac{1}{2}\hbar \qquad \text{(spin down)}.$$

Confusingly, the term *spin* is actually used in two ways: it properly means a particle's intrinsic angular momentum **S** but it is often used loosely to mean the particle's spin quantum number s. In the latter case, for example, an electron is said to be a spin-$\frac{1}{2}$ particle.

Particles with half-integer spin quantum numbers (like electrons) are called **fermions**, after Fermi, who (simultaneously with Paul Dirac) discovered the statistical laws that govern their behavior. Like electrons, protons and neutrons also have $s = \frac{1}{2}$ and are fermions.

Particles with integer (or zero) spin quantum numbers are called **bosons**, after Indian physicist Satyendra Nath Bose, who (simultaneously with Albert Einstein) discovered the governing statistical laws for *these* particles. Photons, which have $s = 1$, are bosons; you will soon meet other particles in this class.

This may seem a trivial way to classify particles but it is very important for this reason:

Fermions obey the Pauli exclusion principle, which asserts that only a single particle can be assigned to a given quantum state. Bosons *do not* obey this princi-

ple. Any number of bosons can be assigned to a given quantum state.

We have seen how important the Pauli exclusion principle is in assigning (spin-$\frac{1}{2}$) electrons in an atom to individual quantum states. Doing so leads to a full accounting of the structure and properties of atoms of different types and of solids such as metals and semiconductors.

Because bosons do *not* obey the Pauli principle, these particles tend to pile up in the quantum state of lowest energy. In 1995 a group in Boulder, Colorado, succeeded in producing a condensate of about 2000 rubidium-87 atoms—they are bosons—in a single quantum state of approximately zero energy.

For this to happen, the rubidium has to be a vapor with a temperature so low and a density so great that the de Broglie wavelengths of the individual atoms are greater than the average separation between the atoms. When this condition is met, the wave functions of the individual atoms overlap and the entire assembly becomes a single quantum system, called a *Bose–Einstein condensate*. Figure 45-2 shows that, as the temperature of the rubidium vapor is lowered to about 1.70×10^{-7} K, this system does indeed "collapse" into a single sharply defined state corresponding to approximately zero speed for its atoms.

Hadron or Lepton?

We can also classify particles in terms of the forces that act on them. In Section 6-5 (which you may wish to reread) we outlined the four known fundamental forces. The *gravitational force* acts on *all* particles, but its effects at the level of subatomic particles are so weak that we need not consider them (at least not in today's research). The *electromagnetic force* acts on all *charged* particles; its effects are well known and we can take them into account when we need to; we largely ignore this force in this chapter.

We are left with the *strong force*, which is the force that binds nucleons together, and the *weak force*, which is involved in beta decay and similar processes. The weak force acts on all particles, the strong force only on some.

We can roughly classify particles on the basis of whether the strong force acts on them. Particles on which the *strong force* acts are called **hadrons**. Particles on which the strong force does *not* act, leaving the weak force as the dominant force, are called **leptons**. Protons, neutrons, and pions are hadrons; electrons and neutrinos are leptons. You will soon meet other members of each class.

We can make a further distinction among the hadrons because some of them are bosons (we call them **mesons**);

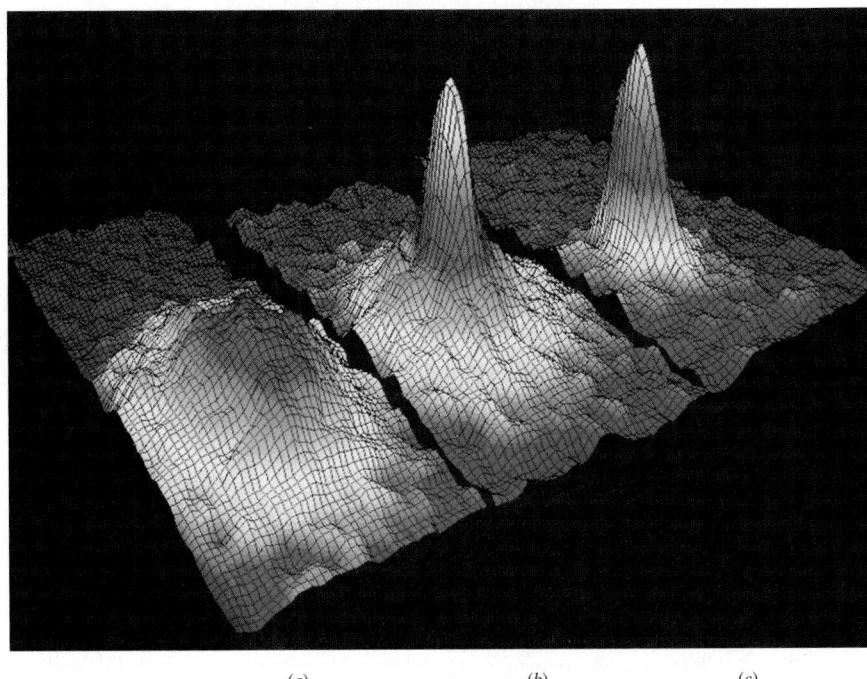

(a)   (b)   (c)

FIGURE 45-2 Three computer-generated plots of the particle speed distribution in a vapor of rubidium-87 atoms. The temperature of the vapor is successively reduced from plot (a) to plot (c). Plot (c) shows a sharp peak centered around zero speed. That is, all the atoms are in the same quantum state. The achievement of such a Bose–Einstein condensate, often called the Holy Grail of atomic physics, was finally recorded in 1995.

the pion is an example. The other hadrons are fermions (we call them **baryons**); the proton is an example.

Particle or Antiparticle?

In 1928 Dirac predicted that the electron e^- should have a positively charged counterpart of the same mass and spin. The counterpart, the *positron* e^+, was discovered in cosmic radiation in 1932 by Carl Anderson. Physicists then gradually realized that *every* particle has a corresponding **antiparticle**. The members of such pairs have the same mass and spin but opposite signs of charge (if they are charged) and opposite signs of other quantum numbers that we have not yet discussed.

At first, *particle* referred to the common particles such as electrons, protons, and neutrons, and *antiparticle* referred to their rarely detected counterparts. For the less common particles, the assignment of *particle* and *antiparticle* is made so as to be consistent with certain conservation laws that we discuss later in this chapter. We often, but not always, represent an antiparticle by putting a bar over the symbol for the particle. Thus p is the symbol for the proton, and $\bar{p}$ (pronounced "p bar") is the symbol for the antiproton.

When a particle meets its antiparticle, the two can annihilate each other. That is, the two (types of) particles disappear and their combined energies reappear in other

forms. For an electron annihilating with a positron, this energy reappears as two gamma-ray photons:

$$e^- + e^+ \rightarrow \gamma + \gamma \qquad (Q = 1.02 \text{ MeV}). \quad (45\text{-}2)$$

If the two particles are stationary when they annihilate, the photons share the energy equally between them and—to conserve momentum and because photons cannot be stationary—they fly off in opposite directions.

In 1996 physicists at CERN succeeded in producing, for a few fleeting nanoseconds, a handful of antihydrogen atoms, each consisting of a positron and an antiproton bound together (presumably just as the electron and proton of a hydrogen atom are bound together). Such an assembly of antiparticles is called *antimatter* to distinguish it from an assembly of particles (*matter*).

One can speculate that there are galaxies of antimatter, complete with atoms, molecules, and even physicists. One can even contemplate the disaster that would occur if, say, an asteroid freed from such a galaxy collided with (hence annihilating) a section of Earth. Thankfully, however, the present view is that not only our galaxy but the universe as a whole consists largely of matter rather than antimatter. (This lack of symmetry is disturbing to a physicist, who normally expects to find symmetry in nature.)

We sum up this section by saying that if you find a new particle, you should ask three questions:

• Is it a fermion or a boson?

• Is it a lepton or a hadron? If the latter, is it a meson or a baryon?

• Is it a particle or an antiparticle?

45-3 AN INTERLUDE

Before pressing on with the task of classifying the particles, let us step aside for a moment and capture some of the spirit of particle research by analyzing a typical particle event, namely that shown in the bubble-chamber photograph of Fig. 45-3a.

The tracks in this figure are streams of bubbles formed in the wake of energetic charged particles as they move through a chamber filled with liquid hydrogen. We can identify the particle that leaves a particular track by—among other ways—measuring the relative spacing between the bubbles. A magnetic field permeates the chamber, deflecting the tracks of positively charged particles counterclockwise and those of negatively charged particles clockwise. By measuring the radius of curvature of a track, we can calculate the momentum of the particle that made it. Table 45-1 shows some properties of the particles and antiparticles that participated in the event of Fig. 45-3a. Following common practice, we express the masses of the particles listed in Table 45-1—and in all other tables in this chapter—in the unit MeV/c^2. The reason for this notation is that the rest energy of a particle is more often needed than its mass. Thus, the mass of a proton is shown in Table 45-1 to be 938.3 MeV/c^2. To find the proton's rest energy, multiply this mass by c^2, obtaining 938.3 MeV.

Our tools for analysis are the laws of conservation of energy, linear momentum, angular momentum, and charge, along with other conservation laws that we have not yet discussed. Figure 45-3a is one member of a stereo pair of photographs so that, in practice, these analyses are carried out in three dimensions.

The event of Fig. 45-3a is triggered by an energetic antiproton ($\overline{p}$) that, generated in an accelerator at the Lawrence Berkeley Laboratory, enters the chamber from

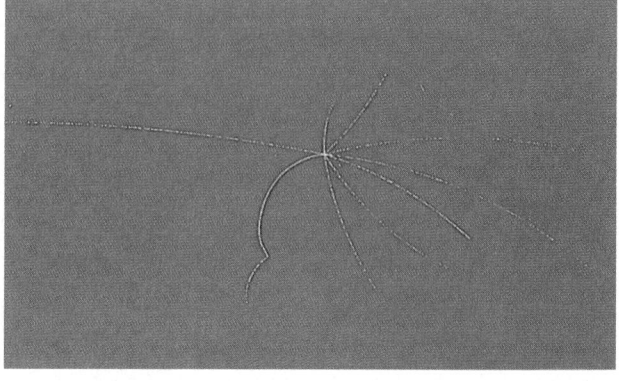

(a)

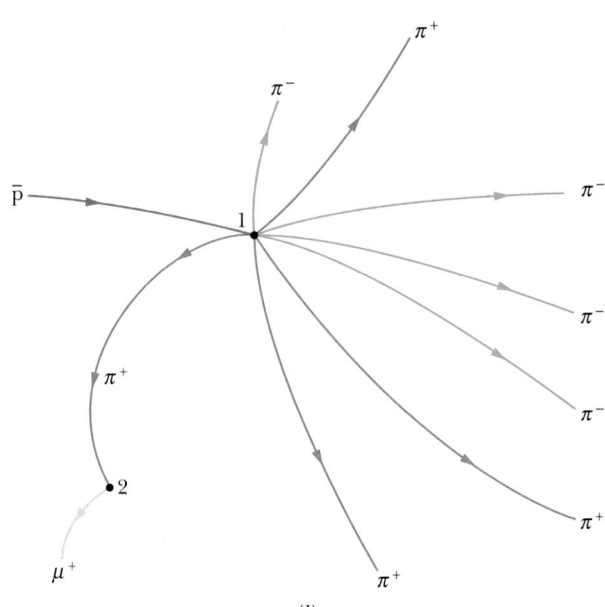

(b)

FIGURE 45-3 (a) A bubble-chamber photograph of a series of events initiated by an antiproton that enters the chamber from the left. (b) The tracks redrawn and labeled for clarity. The dots at points 1 and 2 indicate the sites of specific secondary events that are described in the text. The tracks are curved because a magnetic field present in the chamber exerts a deflecting force on each moving charged particle.

TABLE 45-1 THE PARTICLES OR ANTIPARTICLES INVOLVED IN THE EVENT OF FIG. 45-3

PARTICLE	SYMBOL	CHARGE	MASS (MeV/c^2)	SPIN	IDENTITY	MEAN LIFE[a] (s)	ANTIPARTICLE
Neutrino	ν	0	0	$\frac{1}{2}$	Lepton	Stable	$\overline{\nu}$
Electron	e^-	-1	0.511	$\frac{1}{2}$	Lepton	Stable	e^+
Muon	μ^-	-1	105.7	$\frac{1}{2}$	Lepton	2.2×10^{-6}	μ^+
Pion	π^+	$+1$	139.6	0	Meson	2.6×10^{-8}	π^-
Proton	p	$+1$	938.3	$\frac{1}{2}$	Baryon	Stable	$\overline{p}$

[a]The *mean life* ($= 1/\lambda$) differs from the *half-life* [$= (\ln 2)/\lambda$]; see Eq. 43-8.

the left. There are three separate subevents; two occur at points 1 and 2 in Fig. 45-3b, and the third occurs out of the frame of the figure.

1. *Proton–Antiproton Annihilation.* At point 1 in Fig. 45-3b, the initiating antiproton (blue track) slams into a proton of the chamber fluid, and the result is mutual annihilation. We can tell that annihilation occurred while the incoming antiproton was in flight because most of the particles generated in the encounter move in the forward direction, that is, toward the right in Fig. 45-3. From the principle of conservation of linear momentum then, the incoming antiproton must have had a forward momentum when it underwent annihilation.

The total energy involved in the collision of the antiproton and the proton is the sum of the antiproton's kinetic energy and the two (identical) rest energies of the particles (2 × 938.3 MeV, or 1876.6 MeV). This is enough energy to create a number of lighter particles and give them kinetic energy. In this case, the annihilation produces four positive pions (red tracks) and four negative pions (green tracks). (For simplicity, we assume that no gamma-ray photons, which would leave no tracks because they lack electric charge, are produced.) The process is

$$p + \bar{p} \rightarrow 4\pi^+ + 4\pi^-. \quad (45\text{-}3)$$

We see from Table 45-1 that the positive pions (π^+) are *particles* and the negative pions (π^-) are *antiparticles*. The reaction of Eq. 45-3 is a *strong interaction* (it involves the strong force), because all the particles involved in the reaction are hadrons.

Note that charge is conserved. We can write the charge of a particle as Qe, in which Q is a *charge quantum number*. (Don't confuse this Q with that representing energy in reactions like Eq. 45-2.) The Q values for the interaction of Eq. 45-3 are

$$(+1) + (-1) = 4 \times (+1) + 4 \times (-1),$$

which tells us that the net charge is zero before the interaction and zero afterward.

For the energy balance, note from above that the energy available from the p–$\bar{p}$ annihilation process is at least the sum of the proton and antiproton rest energies, 1876.6 MeV. The rest energy of a pion is 139.6 MeV, so the rest energies of the eight pions amount to 8 × 139.6 MeV, or 1116.8 MeV. This leaves a substantial amount of energy (at least about 760 MeV) to distribute among the eight pions as kinetic energy. Thus, the requirement of energy conservation is easily met.

2. *Pion Decay.* Pions are unstable particles; charged pions decay with a mean life of 2.6×10^{-8} s. At point 2 in Fig. 45-3b one of the positive pions comes to rest in the

chamber and decays spontaneously into an antimuon μ^+ (yellow track) and a neutrino ν:

$$\pi^+ \rightarrow \mu^+ + \nu. \quad (45\text{-}4)$$

The neutrino, being uncharged, leaves no track. Both the antimuon and the neutrino are leptons; that is, they are particles on which the strong force does not act. Thus the decay process of Eq. 45-4, which is governed by the weak force, is described as a *weak interaction*. The rest energy of an antimuon is 105.7 MeV, so an energy of 139.6 MeV − 105.7 MeV, or 33.9 MeV, is available to share between the antimuon and the neutrino as kinetic energy.

The spin quantum number of the pion is zero and the spin quantum numbers of the antimuon and the neutrino are both $\frac{1}{2}$; therefore angular momentum is conserved in Eq. 45-4 if the spins of the antimuon and the neutrino are opposite each other (one must be spin up and the other spin down):

$$0\hbar = \tfrac{1}{2}\hbar + (-\tfrac{1}{2}\hbar).$$

Equation 45-4 also shows that charge is conserved.

3. *Muon Decay.* Muons (whether μ^- or μ^+) are also unstable, decaying with a mean life of 2.2×10^{-6} s. Not shown in the frame of Fig. 45-3b, the antimuon produced in the reaction of Eq. 45-4 comes to rest and decays spontaneously according to

$$\mu^+ \rightarrow e^+ + \nu + \bar{\nu}. \quad (45\text{-}5)$$

The rest energy of the muon is 105.7 MeV and that of the positron is only 0.511 MeV, leaving 105.2 MeV to be shared as kinetic energy among the three particles produced in Eq. 45-5.

You may wonder: Why *two* neutrinos in Eq. 45-5? Why not just one, as in the pion decay in Eq. 45-4? One answer is that the spin quantum numbers of the antimuon, the positron, and the neutrino are each $\frac{1}{2}$; with only one neutrino, spin angular momentum could not be conserved in the antimuon decay of Eq. 45-5. In Section 45-4 we shall discuss another reason.

SAMPLE PROBLEM 45-1

In 1964 some experiments at the Brookhaven National Laboratory employed a focused beam of kaons (K^-). The kaons, whose kinetic energy was 5000 MeV and which were generated in the synchrotron located there, traveled at a relativistic speed through 140 m of an evacuated beam tube to reach a bubble chamber, where the experiments took place.

The rest energy mc^2 of a kaon is 494 MeV and its half-life τ_0 for decay is 8.6×10^{-9} s. By what factor had the intensity of the kaon beam diminished while the particles were traveling from the synchrotron to the bubble chamber?

SOLUTION: The kinetic energy of a kaon is related to its rest energy mc^2 by Eq. 38-33:

$$K = mc^2(\gamma - 1).$$

Hence the Lorentz factor γ for a 5000 MeV kaon is

$$\gamma = \frac{K}{mc^2} + 1 = \frac{5000 \text{ MeV}}{494 \text{ MeV}} + 1 = 11.1.$$

The half-life τ of these moving kaons in the reference frame of the laboratory is related to their half-life at rest by the time dilation factor (Eq. 38-8):

$$\tau = \gamma\tau_0 = (11.1)(8.6 \times 10^{-9} \text{ s}) = 9.55 \times 10^{-8} \text{ s}.$$

Traveling at approximately the speed of light during time τ, a kaon in the beam could cover a distance, as measured in the laboratory frame, of

$$L = c\tau = (3.00 \times 10^8 \text{ m/s})(9.55 \times 10^{-8} \text{ s}) = 28.7 \text{ m}.$$

Thus, the distance $L = 28.7$ m in the laboratory frame corresponds to the half-life of the kaons, and we expect the number of kaons in the beam to be halved with each such length. So, at the end of the full travel distance of 140 m, the number of kaons in the beam (hence the beam intensity) drops to

$$\left(\frac{1}{2}\right)^{(140/28.7)} = 0.034, \text{ or } 3.4\% \qquad \text{(Answer)}$$

of the initial value due to particle decay alone.

Such a beam loss—though unwelcome—is acceptable. Note, however, that *if not for the time dilation effect,* the beam would have weakened to

$$\left(\frac{1}{2}\right)^{(140/28.7)(11.1)} \approx 5 \times 10^{-17}$$

of its initial value. (To see, retrace our reasoning above.) Thus time dilation increased the beam intensity by a factor of nearly a million billion.

SAMPLE PROBLEM 45-2

A stationary positive pion decays as described by Eq. 45-4:

$$\pi^+ \rightarrow \mu^+ + \nu.$$

What is the kinetic energy of the antimuon μ^+? What is the kinetic energy of the neutrino?

SOLUTION: From Table 45-1 the rest energies of the pion and the antimuon are 139.6 MeV and 105.7 MeV, respectively. The difference between these quantities must appear as kinetic energy of the antimuon and the neutrino; that is,

$$139.6 \text{ MeV} - 105.7 \text{ MeV} = 33.9 \text{ MeV} = K_\mu + K_\nu. \qquad (45\text{-}6)$$

Because the pion was stationary when it decayed, the principle of conservation of linear momentum requires that

$$p_\mu = p_\nu,$$

in which p_μ is the *magnitude* of the linear momentum of the antimuon and p_ν that of the neutrino. For convenience, we cast this in the form

$$(p_\mu c)^2 = (p_\nu c)^2. \qquad (45\text{-}7)$$

Equation 38-37,

$$(pc)^2 = K^2 + 2Kmc^2, \qquad (45\text{-}8)$$

gives the relativistic relation between the kinetic energy K of a particle and its momentum p. If we apply this relation to Eq. 45-7, we find that

$$K_\mu^2 + 2K_\mu m_\mu c^2 = K_\nu^2, \qquad (45\text{-}9)$$

in which we assume $mc^2 = 0$ for the neutrino. Solving Eq. 45-6 for K_ν and then substituting this value in Eq. 45-9 and solving for K_μ, we find

$$K_\mu = \frac{(33.9 \text{ MeV})^2}{(2)(33.9 \text{ MeV} + m_\mu c^2)} = \frac{(33.9 \text{ MeV})^2}{(2)(33.9 \text{ MeV} + 105.7 \text{ MeV})} = 4.12 \text{ MeV}. \qquad \text{(Answer)}$$

The kinetic energy of the neutrino is then, from Eq. 45-6,

$$K_\nu = 33.9 \text{ MeV} - K_\mu = 33.9 \text{ MeV} - 4.12 \text{ MeV} = 29.8 \text{ MeV}. \qquad \text{(Answer)}$$

We see that, although the magnitudes of the momenta of the two recoiling particles are the same, the neutrino gets the larger share (88%) of the kinetic energy.

SAMPLE PROBLEM 45-3

Protons in a bubble chamber are bombarded by energetic negative pions, and the following reaction occurs:

$$\pi^- + \text{p} \rightarrow \text{K}^- + \Sigma^+.$$

The rest energies of these particles are

π^-	139.6 MeV	K^-	493.7 MeV
p	938.3 MeV	Σ^+	1189.4 MeV

What is the disintegration energy of the reaction?

SOLUTION: The disintegration energy is given by

$$Q = (m_\pi c^2 + m_\text{p} c^2) - (m_\text{K} c^2 + m_\Sigma c^2) = (139.6 \text{ MeV} + 938.3 \text{ MeV}) - (493.7 \text{ MeV} + 1189.4 \text{ MeV}) = -605 \text{ MeV}. \qquad \text{(Answer)}$$

The minus sign means that the reaction is *endothermic.* That is, if the proton is at rest, the incoming pion (π^-) must have a kinetic energy greater than a certain threshold value to make the reaction go. The threshold energy is greater than 605 MeV because linear momentum must be conserved, which means that the kaon (K^-) and the sigma (Σ^+) must not only be created but also must be given some kinetic energy. A relativistic calculation whose details are beyond our scope shows that the threshold energy for the incident pion is 907 MeV.

45-4   THE LEPTONS

Now let us press on with our classification program for the particles. We turn first to leptons, those particles on which the strong force does *not* act.

So far, we have encountered, as leptons, the familiar electron and the neutrino that accompanies it in beta decay. The muon, whose decay is described in Eq. 45-5, is another member of this family. Physicists gradually learned that the neutrino that appears in Eq. 45-4, associated with the production of a muon, is *not the same particle* as the neutrino produced in beta decay, associated with the appearance of an electron. We call the former the **muon neutrino** (symbol ν_μ) and the latter the **electron neutrino** (symbol ν_e) when it is necessary to distinguish between them.

These two types of neutrino are known to be different particles because, if a beam of muon neutrinos (produced from pion decay as in Eq. 45-4) is allowed to strike a solid target, *only muons*—and never electrons—are produced. On the other hand, if electron neutrinos (produced by the beta decay of fission products in a nuclear reactor) are allowed to strike a solid target, *only electrons*—and never muons—are produced.

Another lepton, the **tau**, was discovered at SLAC in 1975; its discoverer, Martin Perl, shared the 1995 Nobel prize in physics. It has its own associated neutrino, different still from the other two. Table 45-2 lists the known leptons. There are reasons for dividing the leptons into three families, each consisting of a particle (electron, muon, or tau) and its associated neutrino. Furthermore,

there are reasons to believe that there are *only* the three families of leptons shown in Table 45-2. Leptons have no discernible internal structure and no measurable dimensions; they are believed to be truly pointlike fundamental particles when they interact with other particles or with electromagnetic waves.

The Conservation of Lepton Number

In each of the three lepton families of Table 45-2, we can assign a quantum number, the **lepton number**: we assign $L = +1$ to each particle and $L = -1$ to each antiparticle. To particles of other types (such as a proton), we assign $L = 0$. It is an experimental fact that, in all particle interactions, the lepton number *for each lepton family* is separately conserved. Thus there are three lepton numbers, L_e, L_μ, and L_τ, *each of which* must remain unchanged during any particle interaction.

We illustrate this by reconsidering the antimuon decay process of Eq. 45-5, which we now rewrite more fully as

$$\mu^+ \rightarrow e^+ + \nu_e + \bar{\nu}_\mu. \tag{45-10}$$

Consider this first in terms of the muon family of leptons. The μ^+ is an antiparticle (see Table 45-2) and thus has a muon lepton number $L_\mu = -1$. The two particles e^+ and ν_e do not belong to the muon family and thus have $L_\mu = 0$. This leaves $\bar{\nu}_\mu$ on the right which, being an antiparticle, also has the muon lepton number of $L_\mu = -1$. Thus both sides of Eq. 45-10 have the same muon lepton number, namely, $L_\mu = -1$; if they didn't, the μ^+ would not decay by this process.

No members of the electron family appear on the left in Eq. 45-10, so there the electron lepton number must be $L_e = 0$. On the right side of Eq. 45-10 the positron, being an antiparticle (again see Table 45-2), has an electron lepton number $l = -1$. The electron neutrino ν_e, being a particle, has an electron number of $L_e = +1$. Thus the net

---

TABLE 45-2   THE LEPTONS[a]

FAMILY	PARTICLE	SYMBOL	MASS (MeV/c^2)	CHARGE	ANTIPARTICLE
ELECTRON	Electron	e^-	0.511	-1	e^+
	Electron neutrino[b]	ν_e	0	0	$\bar{\nu}_e$
MUON	Muon	μ^-	105.7	-1	μ^+
	Muon neutrino[b]	ν_μ	0	0	$\bar{\nu}_\mu$
TAU	Tau	τ^-	1777	-1	τ^+
	Tau neutrino[b]	ν_τ	0	0	$\bar{\nu}_\tau$

[a] All leptons (including particles and antiparticles) have spin $\frac{1}{2}$ and are thus fermions.

[b] If the neutrino masses are not zero, they are very small. As of 1996, this is an open question.

electron lepton number for these two particles on the right in Eq. 45-10 is also zero; the electron lepton number is also conserved.

No members of the tau family appear on either side of Eq. 45-10, so we must have $L_\tau = 0$ on each side. Thus each of the lepton quantum numbers L_μ, L_e, and L_τ remains unchanged during the decay process of Eq. 45-10, their constant values being -1, 0, and 0, respectively. This example is but one illustration of a general law called the **conservation of lepton number**; the law holds for all particle interactions.

CHECKPOINT 1: (a) The π^+ meson decays by the process $\pi^+ \rightarrow \mu^+ + \nu$. To what lepton family does this neutrino belong? (b) Is this neutrino a particle or an antiparticle? (c) What is its lepton number?

45-5 ANOTHER CONSERVATION LAW

We are now ready to consider hadrons (baryons and mesons), those particles whose interactions are governed by the strong force. We start by adding another conservation law to our list: **conservation of baryon number.**

To develop this conservation law, let us consider the proton decay process

$$p \rightarrow e^+ + \nu_e \qquad (Q = 937.8 \text{ MeV}). \quad (45\text{-}11)$$

This process *never* happens. We should be glad that it does not because otherwise all protons in the universe would gradually change into positrons, with disastrous consequences. Yet this decay process violates none of the conservation laws we have so far discussed, including the conservation of lepton number.

We account for the apparent stability of the proton—and for the absence of many other processes that might otherwise occur—by introducing a new quantum number, the **baryon number** B, and a new conservation law, the conservation of baryon number.

To every baryon we assign $B = +1$. To every antibaryon we assign $B = -1$. To all particles of other types we assign $B = 0$. A particle interaction cannot occur if it changes the net baryon number.

A proton is a baryon, whereas a positron and a neutrino are not. Thus the process of Eq. 45-11 cannot occur because it violates the law of conservation of baryon

number; that is,

$$(+1) \neq (0) + (0).$$

Baryon number conservation is useful in accounting for the many particle decays and reactions that—although not otherwise forbidden—simply do not occur.

CHECKPOINT 2: This mode of decay for a neutron is *not* observed:

$$n \rightarrow p + e^-.$$

Which of the following conservation laws does this process violate: (a) energy, (b) angular momentum, (c) linear momentum, (d) charge, (e) lepton number, (f) baryon number? The masses are the following: $m_n = 939.6 \text{ MeV}/c^2$, $m_p = 938.3 \text{ MeV}/c^2$, and $m_e = 0.511 \text{ MeV}/c^2$.

SAMPLE PROBLEM 45-4

Analyze the proposed decay of a stationary proton according to the scheme

$$p \rightarrow \pi^0 + \pi^+ \qquad \text{(doesn't happen!)}$$

by testing it against the various conservation laws. (Both pions are mesons, with spin and baryon number both equal to zero. The rest energy of the π^0 meson is 135.0 MeV.)

SOLUTION: We see at once that charge is conserved and that linear momentum can also be readily conserved. For the latter, all that is necessary is that the two pions move in opposite directions from the site of the stationary proton, with momenta of equal magnitude. In addition, the lepton number is easily conserved because the lepton number for each of the three particles in the decay is zero.

The disintegration energy is found by subtracting the rest energies of the particles. Thus, using Table 45-1, we have

$$Q = (m_p c^2) - (m_0 c^2 + m_+ c^2)$$
$$= (938.3 \text{ MeV}) - (135.0 \text{ MeV} + 139.6 \text{ MeV})$$
$$= 663.7 \text{ MeV}.$$

The fact that Q is positive shows that the process cannot be ruled out on energy conservation grounds; the energy is there.

We have noted that both pions have zero spin. The proton, however, has a spin quantum number of $\frac{1}{2}$. Thus angular momentum is *not* conserved and—this violation is reason enough—the process cannot occur.

Moreover, baryon number is not conserved. For the proton we have $B = +1$, and for the two pions we have $B = 0$. The process is thus doubly forbidden, violating two of our five conservation laws.

SAMPLE PROBLEM 45-5

A particle called xi-minus and having the symbol Ξ^- decays as follows:

$$\Xi^- \rightarrow \Lambda^0 + \pi^-.$$

The Λ^0 particle (called lambda-zero) and the π^- particle are both unstable. The following decay processes occur in cascade until only stable products remain:

$$\Lambda^0 \rightarrow p + \pi^- \quad \pi^- \rightarrow \mu^- + \bar{\nu}_\mu \quad \mu^- \rightarrow e^- + \nu_\mu + \bar{\nu}_e.$$

(a) Write the overall decay scheme for the Ξ^- particle.

SOLUTION: Study of the decay equations shows that the overall decay scheme is

$$\Xi^- \rightarrow p + 2(e^- + \bar{\nu}_e) + 2(\nu_\mu + \bar{\nu}_\mu). \quad \text{(Answer)}$$

All the products on the right side of this overall decay scheme are stable. Note that charge is conserved, the net charge quantum number being -1 on each side.

(b) Is the Ξ^- particle a meson or a baryon?

SOLUTION: The proton in the overall equation is a baryon. All the other particles on the right side of the equation have $B = 0$. Thus, for conservation of baryon number, the baryon number of the Ξ^- particle must be $+1$. Hence the particle is a *baryon*. If it were a meson, its baryon number would be zero.

(c) Are lepton numbers conserved in the overall decay scheme?

SOLUTION: The Ξ^- particle on the left of the overall decay equation does not appear in Table 45-2, which lists all the leptons; so it must have lepton number zero. The right side of the equation contains eight leptons, four in the electron family and four in the muon family. Within each family, these leptons occur in particle–antiparticle pairs with lepton numbers of opposite signs. Thus both the net electron lepton number L_e and the net muon lepton number L_μ are zero, and so lepton numbers are conserved.

(d) What can you say about the spin of the Ξ^- particle?

SOLUTION: All nine particles on the right side of the overall decay equation are spin-$\frac{1}{2}$ particles. Nine values of $m_s = \pm\frac{1}{2}$ will always combine to form a net *half-integer* value, no matter how you align the individual S_z components parallel or antiparallel to a given axis. Thus the resultant spin quantum number of the Ξ^- particle must be half-integer. (Actually, the quantum number is $\frac{1}{2}$; the Ξ^- particle is listed with other spin-$\frac{1}{2}$ baryons in Table 45-3.)

45-6 STILL ANOTHER CONSERVATION LAW

Particles have intrinsic properties in addition to the ones we have listed so far: mass, charge, spin, lepton number,

and baryon number. The first of these additional properties emerged when researchers observed that certain new particles, such as the kaon (K) and the sigma (Σ), always seemed to be produced in pairs. It seemed impossible to produce only one of them at a time. Thus if a beam of energetic pions interacts with the protons in a bubble chamber, the reaction

$$\pi^+ + p \rightarrow K^+ + \Sigma^+ \tag{45-12}$$

often occurs. The reaction

$$\pi^+ + p \rightarrow \pi^+ + \Sigma^+, \tag{45-13}$$

which violates no conservation law known in the early days of particle physics, never occurs.

It was eventually proposed (by Murray Gell-Mann in the United States and independently by K. Nishijima in Japan) that certain particles possess a new property, called **strangeness**, with its own quantum number S and its own conservation law. (Be careful not to confuse the symbol S here with spin.) The name *strangeness* arises from the fact that, before the identities of these particles were pinned down, they were known as "strange particles," and the label stuck.

The proton, neutron, and pion have $S = 0$; that is, they are not "strange." It was proposed, however, that the K^+ particle has strangeness $S = +1$ and that Σ^+ has $S = -1$. Thus strangeness is conserved in Eq. 45-12,

$$(0) + (0) = (+1) + (-1) \quad \text{(values of } S\text{)},$$

but is *not* conserved in Eq. 45-13,

$$(0) + (0) \neq (0) + (-1) \quad \text{(values of } S\text{)}.$$

The reaction of Eq. 45-13 does not occur because it violates the law of **conservation of strangeness**:

> Strangeness is conserved in interactions involving the strong force.

It may seem heavy-handed to invent a new property of particles just to account for a little puzzle like that posed by Eqs. 45-12 and 45-13. However, strangeness and its quantum number soon revealed themselves in many other areas in particle physics, and strangeness is now fully accepted as a legitimate particle attribute, on a par with charge and spin. To those who know and love particles, strangeness is no longer strange.

Do not be misled by the whimsical character of the name. Strangeness is no more mysterious a property of particles than is charge. Both are properties that particles may (or may not) have; each is described by an appropriate quantum number. Each obeys a conservation law. Still other properties of particles have been discovered and given even more whimsical names, such as *charm* and *bottomness,* but all are perfectly legitimate properties. Let us see, as an example, how the new property of strangeness ''earns its keep'' by leading us to uncover important regularities in the properties of the particles.

45-7 THE EIGHTFOLD WAY

There are eight baryons—the neutron and the proton among them—that have a spin quantum number of $\frac{1}{2}$. Table 45-3 shows some of their other properties. Figure 45-4a shows the fascinating pattern that emerges if we plot the strangeness of these baryons against their charge quantum number, using a sloping axis for the charge quantum numbers. Six of the eight form a hexagon with the two remaining baryons at its center.

Let us turn now from the hadrons called baryons to the hadrons called mesons. Nine with a spin of zero are listed in Table 45-4. If we plot them on a sloping strangeness–charge diagram, as in Fig. 45-4b, the same fascinating pattern emerges! These and related plots, called the *Eightfold Way* patterns,* were proposed independently in 1961 by Murray Gell-Mann at the California Institute of Technology and by Yuval Ne'eman at Imperial College, London. The two patterns of Fig. 45-4 are representative of a larger number of symmetrical patterns in which groups of baryons and mesons can be displayed.

The symmetry of the Eightfold Way pattern for the spin-$\frac{3}{2}$ baryons (not shown here) calls for *ten* particles arranged in a pattern like that of the tenpins in a bowling alley. However, when the pattern was first proposed, only *nine* such particles were known; the ''headpin'' was missing. In 1962, guided by theory and the symmetry of the pattern, Gell-Mann made a prediction in which he essentially said:

> There exists a spin-$\frac{3}{2}$ baryon with a charge of −1, a strangeness of −3, and a rest energy of about 1680 MeV. If you look for this *omega minus* particle (as I propose to call it), I think you will find it.

A team of physicists headed by Nicholas Samios of the Brookhaven National Laboratory took up the challenge and found the ''missing'' particle, confirming all of its

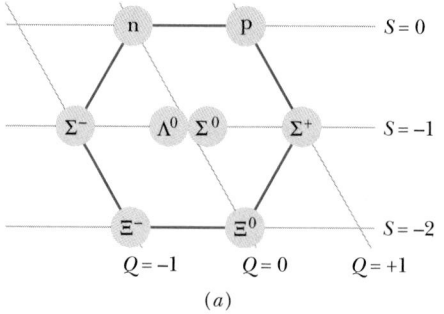

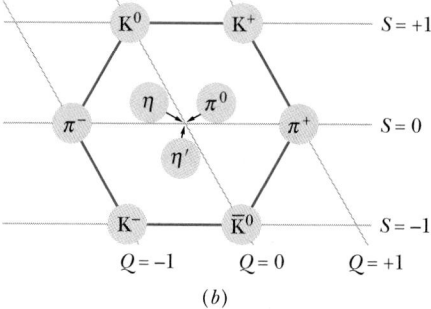

FIGURE 45-4 (a) The Eightfold Way pattern for the eight spin-$\frac{1}{2}$ baryons listed in Table 45-3. The particles are represented as disks on a strangeness–charge plot, using a sloping axis for the charge quantum number. (b) A similar pattern for the nine spin-zero mesons listed in Table 45-4.

predicted properties. Nothing beats prompt experimental confirmation for building confidence in a theory!

The Eightfold Way patterns bear the same relationship to particle physics that the periodic table does to chemistry. In each case, there is a pattern of organization in which vacancies (missing particles or missing elements) stick out like sore thumbs, guiding experimenters in their searches. In the case of the periodic table, its very existence strongly suggests that the atoms of the elements are not fundamental particles but have an underlying structure. Similarly, the Eightfold Way patterns strongly suggest that the mesons and the baryons must have an underlying structure, in terms of which their properties can be understood. That structure is the *quark model,* which we now discuss.

45-8 THE QUARK MODEL

In 1964 Gell-Mann and George Zweig independently pointed out that the Eightfold Way patterns can be understood in a simple way if the mesons and the baryons are built up out of subunits that Gell-Mann called **quarks.** We deal first with three of them, called the *up quark* (symbol u), the *down quark* (symbol d), and the *strange quark*

---

*A borrowing from Eastern mysticism. The ''Eight'' refers to the eight quantum numbers (only a few of which we have defined here) that are involved in the symmetry-based theory that predicts the existence of the patterns.

TABLE 45-3 **THE EIGHT SPIN-$\frac{1}{2}$ BARYONS**

| | | | QUANTUM NUMBERS | |
PARTICLE	SYMBOL	MASS (MeV/c^2)	CHARGE	STRANGENESS
Proton	p	938.3	+1	0
Neutron	n	939.6	0	0
Lambda	Λ^0	1115.6	0	−1
Sigma	Σ^+	1189.4	+1	−1
Sigma	Σ^0	1192.5	0	−1
Sigma	Σ^-	1197.3	−1	−1
Xi	Ξ^0	1314.9	0	−2
Xi	Ξ^-	1321.3	−1	−2

(symbol s), and we assign to them the properties displayed in Table 45-5. (The names of the quarks, along with those assigned to three other quarks that we shall meet later, have no meanings other than as convenient labels. Collectively, these names are called the *quark flavors*. We could have just as well called them vanilla, chocolate, and strawberry instead of up, down, and strange.)

The fractional charge quantum numbers of the quarks may jar you a little. However, withhold judgment until you see how neatly these fractional charges account for the observed integral charges of the mesons and the baryons. Quarks have not (yet) been convincingly observed in the laboratory as free particles, and theorists have put forward plausible reasons why this should be the case. In any event, the quark model is so useful that the failure to see free quarks has not proved a barrier to universal acceptance by physicists of the quark model.

We have seen how we can put atoms together by combining electrons and nuclei. Now let us see how we can put mesons and baryons together by combining quarks. We state in advance that success will be complete. That is, for particles formed from the up, down, and strange quarks,

There is no known meson or baryon whose properties cannot be understood in terms of an appropriate combination of quarks. Conversely, there is no possible quark combination that does not correspond to an observed meson or baryon.

Let us look first at the baryons.

Quarks and Baryons

Each baryon is a combination of three quarks; the combinations are given in Fig. 45-5a. With regard to baryon number, we see that any three quarks (each with $B = +\frac{1}{3}$) yield a proper baryon (with $B = +1$). The spins work out also. With three spin quantum numbers of $\frac{1}{2}$ to work with, we can arrange them so that two spins are parallel and one

TABLE 45-4 **NINE SPIN-ZERO MESONS**[a]

| | | | QUANTUM NUMBERS | |
PARTICLE	SYMBOL	MASS (MeV/c^2)	CHARGE	STRANGENESS
Pion	π^0	135.0	0	0
Pion	π^+	139.6	+1	0
Pion	π^-	139.6	−1	0
Kaon	K^+	493.7	+1	+1
Kaon	K^-	493.7	−1	−1
Kaon	K^0	497.7	0	+1
Kaon	$\overline{K}^0$	497.7	0	−1
Eta	η	547.5	0	0
Eta prime	η'	957.8	0	0

[a] All mesons are bosons, having spins of 0, 1, 2, ...

TABLE 45-5   **THE QUARKS**[a]

			QUANTUM NUMBERS			
					BARYON	
PARTICLE	SYMBOL	MASS (MeV/c^2)	CHARGE Q	STRANGENESS S	NUMBER B	ANTIPARTICLE
Up	u	5	$+\frac{2}{3}$	0	$+\frac{1}{3}$	$\bar{u}$
Down	d	10	$-\frac{1}{3}$	0	$+\frac{1}{3}$	$\bar{d}$
Charm	c	1500	$+\frac{2}{3}$	0	$+\frac{1}{3}$	$\bar{c}$
Strange	s	200	$-\frac{1}{3}$	-1	$+\frac{1}{3}$	$\bar{s}$
Top	t	$\approx 180{,}000$	$+\frac{2}{3}$	0	$+\frac{1}{3}$	$\bar{t}$
Bottom	b	4300	$-\frac{1}{3}$	0	$+\frac{1}{3}$	$\bar{b}$

[a] All quarks have spin $\frac{1}{2}$ and thus are fermions. The quantum numbers Q, S, and B for an antiquark are opposite that for the corresponding quark.

antiparallel. This leads to a net spin of $s = \frac{1}{2}$, which is the spin quantum number of all the baryons displayed in Table 45-3 and Fig. 45-4a.

Charges also work out, as we can see from three examples. The proton has a quark composition of uud, so its charge quantum number is

$$Q(\text{uud}) = (+\tfrac{2}{3}) + (+\tfrac{2}{3}) + (-\tfrac{1}{3}) = +1.$$

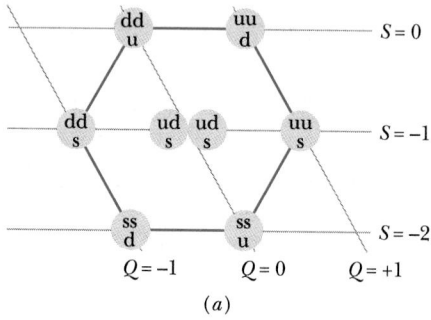

(a)

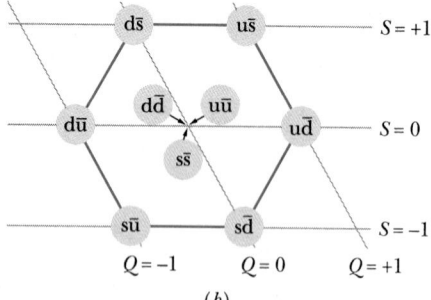

(b)

FIGURE 45-5 (a) The quark compositions of the eight spin-$\frac{1}{2}$ baryons plotted in Fig. 45-4a. (Although the two central baryons share the same quark structure, the sigma is an excited state of the lambda, decaying into the lambda by emission of a gamma-ray photon.) (b) The quark compositions of the nine spin-zero mesons plotted in Fig. 45-4b.

The neutron has a quark composition of udd and its charge quantum number is

$$Q(\text{udd}) = (+\tfrac{2}{3}) + (-\tfrac{1}{3}) + (-\tfrac{1}{3}) = 0.$$

The Σ^- (sigma-minus) particle has a quark composition of dds and its charge quantum number is

$$Q(\text{dds}) = (-\tfrac{1}{3}) + (-\tfrac{1}{3}) + (-\tfrac{1}{3}) = -1.$$

The strangeness quantum numbers work out as well. Check this by using Table 45-3 for the Σ^- and Table 45-5 for the quarks.

Quarks and Mesons

Mesons are quark–antiquark pairs; their compositions are given in Fig. 45-5b and are consistent with the fact that the spins of all the mesons displayed in Fig. 45-4b and Table 45-4 are zero. Both quarks and antiquarks have $s = \frac{1}{2}$, so the two particles that make up a meson must have opposite spins to give spin-zero for the meson.

The quark–antiquark model is also consistent with the fact that mesons are not baryons; that is, mesons have a baryon number $B = 0$. The baryon number for a quark is $+\frac{1}{3}$ and for an antiquark is $-\frac{1}{3}$; thus the combination of baryon numbers in a meson is zero.

Consider the meson π^+, which is made up of an up quark u and an antidown quark $\bar{d}$. We see from Table 45-5 that the charge quantum number of the up quark is $+\frac{2}{3}$ and that of the antidown quark is $+\frac{1}{3}$ (the sign is opposite that of the down quark). This adds nicely to a charge quantum number of $+1$ for the π^+ meson. Thus

$$Q(\text{u}\bar{d}) = (+\tfrac{2}{3}) + (+\tfrac{1}{3}) = +1.$$

All the charge and strangeness quantum numbers of Fig. 45-5b agree with those of Table 45-4 and Fig. 45-4b. Convince yourself that all possible up, down, and strange quark–antiquark combinations are used and that all known spin-zero mesons are accounted for. Everything fits.

CHECKPOINT **3:** A particle is a combination of a down quark (d) and an antiup quark ($\bar{u}$). Is the particle (a) a π^0 meson, (b) a proton, (c) a π^- meson, (d) a π^+ meson, or (e) a neutron?

A New Look at Beta Decay

Let us see how beta decay appears from the quark point of view. In Eq. 43-10, we presented a typical example of this process:

$$^{32}\text{P} \rightarrow {}^{32}\text{S} + e^- + \nu.$$

After the neutron was discovered and Fermi had worked out his theory of beta decay, physicists came to view the fundamental beta-decay process as the changing of a neutron into a proton inside the nucleus:

$$\text{n} \rightarrow \text{p} + e^- + \bar{\nu}_e,$$

in which the neutrino is identified more completely. Today we look deeper and see that a neutron (udd) can change into a proton (uud) by changing a down quark into an up quark. We now view the fundamental beta-decay process as

$$\text{d} \rightarrow \text{u} + e^- + \bar{\nu}_e.$$

Thus as we come to know more and more about the fundamental nature of matter, we can examine familiar processes at deeper and deeper levels. We see too that the quark model not only helps us to understand the structure of particles but also throws light on their interactions.

Still More Quarks

There are other particles and other Eightfold Way patterns that we have not discussed. To account for them, it turns out that we need to postulate three more quarks, the *charm quark* c, the *top quark* t, and the *bottom quark* b.

We note in Table 45-5 that three quarks are exceptionally massive, the most massive of them (top) being almost 170 times more massive than a proton. To generate particles that contain such quarks, we must go to higher and higher energies, which is the reason that these three quarks were not discovered earlier.

The first observed particle that contains a charm quark was the J/Ψ meson, whose quark structure is $c\bar{c}$). It was discovered simultaneously and independently in 1974 by groups headed by Samuel Ting at the Brookhaven National Laboratory and by Burton Richter at Stanford University.

The top quark defied all efforts to generate it in the laboratory until 1995, when its existence was finally demonstrated in the Tevatron, a large particle accelerator at Fermilab. In this accelerator, protons and antiprotons, each with an energy of 0.9 TeV (= 9×10^{11} eV), are made to collide at the centers of two large particle detectors. In a very few cases, the colliding protons generate a top–antitop ($t\bar{t}$) quark pair, which quickly decays into particles of lower energy. Figure 45-6 shows a computer-generated plot of the tracks of these decay products. The existence of the top–antitop quark pair is deduced by careful analysis of this ''debris.'' The discovery of the much-sought-after top quark is regarded as a triumph for the entire quark–lepton scheme of particle physics.

Look back for a moment at Table 45-5 (the quark family) and Table 45-2 (the lepton family) and notice the neat symmetry of these two ''six-packs'' of particles, each dividing naturally into three corresponding two-particle families. In terms of what we know today, the quarks and the leptons seem to be truly fundamental particles with no internal structure.

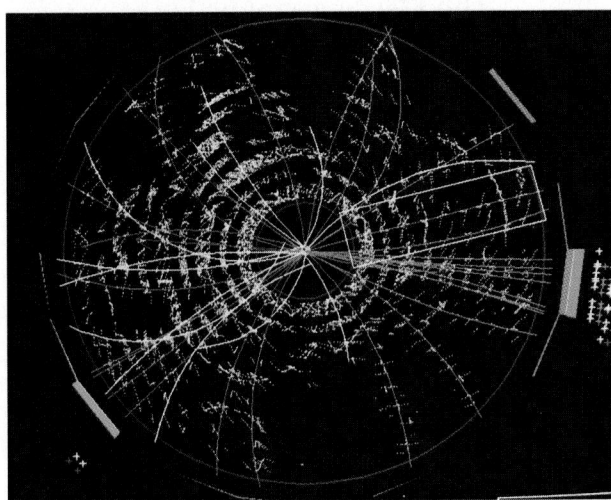

FIGURE 45-6 A computer-generated display of the tracks formed by particles produced in the decay of a top–antitop pair. The pair was formed in the Tevatron accelerator at Fermilab, by the head-on collision of a high-energy proton and antiproton.

SAMPLE PROBLEM 45-6

The Ξ^- particle has a spin quantum number of $\frac{1}{2}$ and quantum numbers $Q = -1$ and $S = -2$. It is known to be a three-quark combination involving only up, down, and strange quarks. What must this combination be?

SOLUTION: Because its strangeness is -2, this particle must contain two strange quarks, each of which (see Table 45-5) has $S = -1$. The third quark must then be either an up quark or a down quark (both having $S = 0$). The two strange quarks have a combined charge of $(-\frac{1}{3}) + (-\frac{1}{3})$, or $-\frac{2}{3}$. Since we require a charge of -1 for the Ξ^- particle, we must add a third quark whose charge is $-\frac{1}{3}$; that is the down quark. Thus the quark composition of Ξ^- is dss.

As a final check, note that Table 45-5 shows that the baryon number for both the down and the strange quark is $+\frac{1}{3}$. Thus the baryon number for the Ξ^- particle (dss) is

$$B = (\tfrac{1}{3}) + (\tfrac{1}{3}) + (\tfrac{1}{3}) = +1.$$

This is what we expect because Ξ^-, being made up of quarks (not antiquarks), is a *particle* (not an antiparticle). If it were an antiparticle, it would have $B = -1$.

45-9 THE BASIC FORCES AND MESSENGER PARTICLES (OPTIONAL)

We turn now from cataloguing the particles to considering the forces that act between them.

The Electromagnetic Force

At the atomic level, we say that two electrons exert electromagnetic forces on each other according to Coulomb's law. At a deeper level, this interaction is described by a highly successful theory called **quantum electrodynamics** (QED). From this point of view we say that each electron senses the presence of the other by exchanging photons with it.

We cannot detect these photons because they are emitted by one electron and absorbed by the other a very short time later. Because of their transitory existence, we call them **virtual photons**. Because of their role in communicating between the two interacting charges, we sometimes call these photons *messenger particles*.

If a stationary electron emits a photon and remains itself unchanged, energy is not conserved. The principle of conservation of energy is saved, however, by the uncertainty principle, written in the form

$$\Delta E \cdot \Delta t \approx \hbar. \tag{45-14}$$

We interpret this relation to mean that you can "overdraw" an amount of energy ΔE, violating conservation of energy, *provided* you "return" it within an interval Δt given by $\hbar/\Delta E$. The virtual photons do just that. Consider a pair of interacting electrons. When, say, an electron A emits a virtual photon, the overdraw in energy is quickly set right when that electron receives a virtual photon from electron B, and the violation of the principle of conservation of energy for the electron pair is hidden by the inherent uncertainty.

The Weak Force

A theory of the weak force, which acts on all particles, was developed by analogy with the theory of the electromagnetic force. The messenger particles that transmit the weak force between particles, however, are not (massless) photons but massive particles, identified by the symbols W and Z. The theory was so successful that it revealed the electromagnetic force and the weak force as being different aspects of a single **electroweak force**. This accomplishment is a logical extension of the work of Maxwell, who revealed the electric and magnetic forces as being different aspects of a single *electromagnetic* force.

The electroweak theory was specific in predicting the properties of the messenger particles. Their charges and masses, for example, were predicted to be

PARTICLE	CHARGE	MASS
W	$\pm e$	80.6 GeV/c^2
Z	0	91.2 GeV/c^2

Recall that the proton mass is only 0.938 GeV/c^2; these are massive particles! The Nobel physics prize for 1979 was awarded to Sheldon Glashow, Steven Weinberg, and Abdus Salam for their development of the electroweak theory.

The theory was confirmed in 1983 by Carlo Rubbia and his group at CERN. Both messenger particles were observed, their masses agreeing with the predicted values. The Nobel physics prize for 1984 went to Rubbia and Simon van der Meer for this brilliant experimental work.

Some notion of the complexity of particle physics in this day and age can be found by looking at an earlier Nobel prize particle physics experiment—the discovery of the neutron. This vitally important discovery was a "tabletop" experiment, employing particles emitted by naturally occurring radioactive materials as projectiles; it was reported in 1932 under the title "Possible Existence of a Neutron," the single author being James Chadwick.

The discovery of the W and Z messenger particles in 1983, by contrast, was carried out at a large particle accelerator, about 7 km in circumference and operating in the range of several hundred billion electron-volts. The principal particle detector alone weighed 2000 tons. The experiment employed more than 130 physicists from 12 institutions in 8 countries, along with a large support staff.

The Strong Force

A theory of the strong force—that is, the force that acts between quarks to bind hadrons together—has also been developed. The messenger particles in this case are called

gluons and, like the photon, they are predicted to be mass-less. The theory assumes that each "flavor" of quark comes in three varieties that, for convenience, have been labeled *red, yellow,* and *blue.* Thus there are three up quarks, one of each color, and so on. The antiquarks also come in three colors, which we call *antired, antiyellow,* and *antiblue.* You must not think that quarks are actually colored, like tiny jelly beans. The names are labels of convenience but (for once) they do have a certain formal justification, as you shall see.

The force acting between quarks is called a **color force** and the underlying theory, by analogy with quantum electrodynamics (QED), is called **quantum chromodynamics** (QCD). An important prediction of the theory is that quarks can be assembled only in combinations that are *color-neutral.*

There are two ways to bring about color neutrality. In the theory of actual colors, red + yellow + blue yields white, which is color-neutral; thus we can assemble three quarks to form a baryon. Antired + antiyellow + antiblue is also white, so that we can assemble three antiquarks to form an antibaryon. Finally, red + antired, or yellow + antiyellow, or blue + antiblue also yields white. Thus we can assemble a quark–antiquark combination to form a meson. The color-neutral rule does not permit any other combination of quarks, and none are observed.

The color force not only acts to bind together quarks as baryons and mesons, but it also acts between such particles, in which case it has traditionally been called the strong force. Hence, not only does the color force bind together quarks to form protons and neutrons, but it also binds together the protons and neutrons to form nuclei.

Einstein's Dream

The attempt to unify the fundamental forces of nature—which occupied Einstein's attention for much of his later life—is very much a current problem. Table 6-2 summarizes the current status. We have seen that the weak force has been successfully combined with electromagnetism so that they may be jointly viewed as aspects of a single *electroweak force.* Theories that attempt to add the strong force to this combination—called *grand unification theories* (GUTs)—are being pursued actively. Theories that seek to complete the job by adding gravity—sometimes called *theories of everything* (TOE)—are at an encouraging but speculative stage at this time.

45-10 A PAUSE FOR REFLECTION

Let us put what you have just learned in perspective. If all we are interested in is the structure of the world around us, we can get along nicely with the electron, the neutrino, the neutron, and the proton. As someone has said, we can operate "Spaceship Earth" quite well with just these particles. We can see a few of the more exotic particles by looking for them in the cosmic rays but, to see most of them, we must build massive accelerators and look for them carefully at great effort and expense.

The reason we must go to such lengths is that—measured in energy terms—we live in a world of very low temperatures. Even at the center of the Sun, the value of kT is only about 1 keV. To produce the exotic particles, we must be able to accelerate protons or electrons to energies in the GeV and TeV range and higher. Once upon a time, however, the temperature (everywhere) *was* high enough to provide just such energies, and far greater. That was when the universe began. So let us now turn our attention to that time.

When astronomers look out in space, they are also looking back in time. Thus we can study the universe at earlier times by examining distant celestial objects. The most distant objects we can "see" are **quasars** (*quasi-stellar* objects), which are thought to be the extremely luminous centers of galaxies in the process of formation. As of January 1996, the most distant quasar, found by astronomers at the California Institute of Technology, is 14×10^9 light-years distant from Earth. The light from this object that is now entering our telescopes left it about 14 billion years ago, so we see the object as it was then.

45-11 THE UNIVERSE IS EXPANDING

As we saw in Section 18-9, it is possible to measure the relative speeds at which galaxies are approaching us or receding from us by measuring the Doppler shift of the light that they emit. If we look only at distant galaxies, beyond our immediate galactic neighbors, we find an astonishing fact. They are all moving away from us!

In 1929 Edwin P. Hubble established a connection between the apparent speed of recession v of a galaxy and its distance r from us, namely, that they are directly proportional. That is,

$$v = Hr \qquad \text{(Hubble's law)}, \qquad (45\text{-}15)$$

in which H, the proportionality constant, is called the **Hubble constant**. Its value is somewhat uncertain because of the difficulty of measuring the distances to remote galaxies. One value, based on observations made in 1994 with the Hubble Space Telescope, is

$$H = 80 \pm 17 \text{ km/s} \cdot \text{Mpc}, \qquad (45\text{-}16)$$

in which Mpc stands for the length unit megaparsec:

$$1 \text{ Mpc} = 3.084 \times 10^{19} \text{ km} = 3.260 \times 10^6 \text{ ly}. \qquad (45\text{-}17)$$

Another value, based on a long series of measurements and announced in 1996, is 57 km/s · Mpc. These two measurements suggest the uncertainty that surrounds the value of this important parameter. In calculations here, we shall use the value $H = 80$ km/s · Mpc, which is equivalent to 24.5 mm/s · ly.

We interpret Hubble's law to mean that the universe is expanding, much as the raisins in what is to be a loaf of raisin bread grow farther apart as the dough rises. Observers on all other galaxies would find that distant galaxies were rushing away from them also, in accordance with Hubble's law. In keeping with our analogy, no raisin (galaxy) has a unique or preferred view.

Hubble's law fits in well with the hypothesis that the universe originated in a powerful explosion (the Big Bang) some billions of years ago. The receding galaxies are the fragments of this primordial explosion, modified in structure with the passage of time and expanding away from one another because of the expansion of the universe itself.

SAMPLE PROBLEM 45-7

If Hubble's law holds, how far away is a quasar with an apparent recession of 2.8×10^8 m/s. (Note that this is 93% of the speed of light.)

SOLUTION: From Hubble's law (Eq. 45-15),

$$r = \frac{v}{H}$$

$$= \frac{2.8 \times 10^8 \text{ m/s}}{80 \text{ km/s} \cdot \text{Mpc}} \left(\frac{3.084 \times 10^{19} \text{ km}}{1 \text{ Mpc}} \right) \left(\frac{1 \text{ ly}}{9.46 \times 10^{15} \text{ m}} \right)$$

$$= 11 \times 10^9 \text{ ly.} \qquad \text{(Answer)}$$

The result is only approximate because the quasar has not always been receding from us at the same apparent speed.

SAMPLE PROBLEM 45-8

Assume that the quasar in Sample Problem 45-7 has been moving at its calculated speed with respect to us ever since the Big Bang. What minimum limit does this impose on how long ago the Big Bang occurred? That is, what is the minimum age of the universe based on that speed?

SOLUTION: We can find the time from

$$t = \frac{r}{v} = \frac{r}{rH} = \frac{1}{H}$$

$$= \frac{1}{80 \text{ km/s} \cdot \text{Mpc}} \left(\frac{3.084 \times 10^{19} \text{ km}}{1 \text{ Mpc}} \right) \left(\frac{1 \text{ y}}{3.16 \times 10^7 \text{ s}} \right)$$

$$= 12 \times 10^9 \text{ y.} \qquad \text{(Answer)}$$

The result depends on the value of the Hubble constant. For example, a smaller value for H would lead to a greater age for the universe.

45-12 THE COSMIC BACKGROUND RADIATION

In 1965 Arno Penzias and Robert Wilson, of what was then the Bell Telephone Laboratories, were testing a sensitive microwave receiver used for communications research. They discovered a faint background "hiss" that remained unchanged in intensity no matter where their antenna was pointed. It soon became clear that Penzias and Wilson were observing a **cosmic background radiation**, generated in the early universe and filling all space almost uniformly. This radiation, whose maximum intensity occurs at a wavelength of 1.1 mm, has the same distribution in wavelength as does radiation in a cavity whose walls are held at a temperature of 2.7 K, the "cavity" in this situation being the entire universe. Penzias and Wilson were awarded the 1978 Nobel prize in physics for their discovery.

This radiation originated about 300,000 years after the Big Bang, when the universe suddenly became transparent to electromagnetic waves. The radiation at that time corresponded to cavity radiation at a temperature of perhaps 10^5 K. As the universe expanded, however, the temperature dropped to its present value of 2.7 K, much as the temperature of a gas expanding under adiabatic conditions will fall.

45-13 DARK MATTER

At the Kitt Peak National Observatory in Arizona, Vera Rubin and her co-worker Kent Ford measured the rotational rates of a number of distant galaxies. They did so by measuring the Doppler shifts of bright clusters of stars located within each galaxy at various distances from the galactic center. As Fig. 45-7 shows, their results were surprising: the orbital speed of stars at the outer visible edge of the galaxy is about the same as that of stars close to the galactic center.

As the solid curve in Fig. 45-7 attests, that is not what we would expect to find if all the mass of the galaxy were represented by visible light. Nor is the pattern found by Rubin and Ford what we find in the solar system. For example, the orbital speed of Pluto (the planet most distant from the Sun) is only about one tenth that of Mercury (the planet closest to the Sun).

The only explanation for the findings of Rubin and Ford that is consistent with Newtonian mechanics is that a

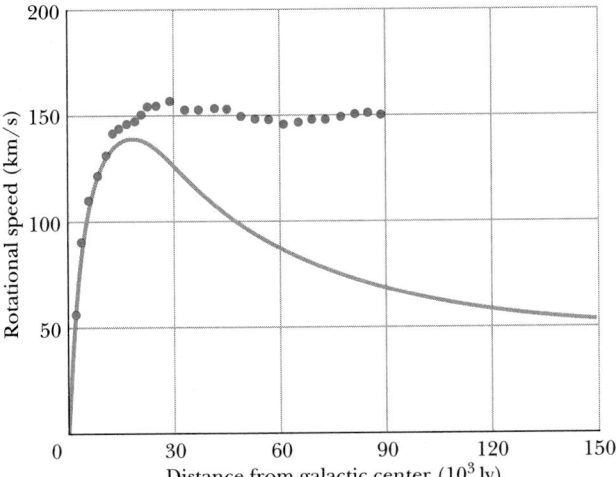

FIGURE 45-7 The rotational speed of stars in a typical galaxy as a function of their distance from the galactic center. The solid curve shows that if all the mass in the galaxy is visible, the rotational speed should drop off with distance at large distances. The dots are the experimental data, which show that the rotational speed is approximately constant at large distances.

typical galaxy contains much more matter than what we can actually see. In fact, the visible portion of a galaxy represents only about 5 to 10% of the total mass of the galaxy. In addition to these studies of galactic rotation, many other observations lead to the conclusion that the universe abounds in matter that we cannot see.

What then is this **dark matter** that permeates and surrounds a typical galaxy like a gigantic halo whose diameter is perhaps 30 times the diameter of the visible galaxy? Dark matter candidates fall into two classes, whimsically called WIMPs (*w*eakly *i*nteracting *m*assive *p*articles) and MACHOs (*m*assive *c*ompact *h*alo *o*bjects). If neutrinos have mass, they are possible WIMP candidates. MACHOs can include objects such as black holes, white dwarf stars, and brown dwarf stars; the latter are Jupiter-size objects that are not massive enough to become actual stars, which shine because of fusion.

As of 1996, there is convincing evidence that MACHOs do indeed exist in our own galaxy. Assume that an (invisible) MACHO in our galaxy passes, by chance, in front of a star in a nearby galaxy. Einstein, in his general theory of relativity, predicted that light rays passing near any massive object will be deflected by the mass of that object (see Section 14-9). Thus, if star, MACHO, and Earth are aligned, the MACHO will act as a *gravitational lens,* focusing the light rays from the star that pass near it and causing the image of the star to brighten while the MACHO is eclipsing it.

Enough such events have been observed to convince astronomers that MACHOs can account for a substantial fraction (some say 50%) of the dark matter in our own galaxy. Observations are ongoing.

45-14   THE BIG BANG

In 1985, a physicist remarked at a scientific meeting:

> It is as certain that the universe started with a Big Bang about 15 billion years ago as it is that the Earth goes around the Sun.

This strong statement suggests the level of confidence in which the Big Bang theory, first advanced by Belgian physicist Georges Lemaître, is held by those who study these matters.

You must not imagine that the Big Bang was like the explosion of some gigantic firecracker and that, in principle at least, you could have stood to one side and watched. There was no "one side" because the Big Bang represents the beginning of spacetime itself. From the point of view of our present universe, there is no position in space to which you can point and say, "The Big Bang happened there." It happened everywhere.

Moreover, there was no "before the Big Bang," because time *began* with that creation event. In this context, the word "before" loses its meaning. We can, however, conjecture about what went on during succeeding intervals of time after the Big Bang.

$t \approx 10^{-43}$ *s.* This is the earliest time at which we can say anything meaningful about the development of the universe. It is at this moment that the concepts of space and time come to have their present meanings and the laws of physics as we know them become applicable. At this instant, the entire universe is much smaller than a proton, say, in the tip of your nose, and its temperature is $\approx 10^{32}$ K.

$t \approx 10^{-34}$ *s.* By this moment the universe has undergone a tremendously rapid inflation, increasing its size by a factor of about 10^{30}. It has become a hot soup of quarks and leptons, at a temperature of $\approx 10^{27}$ K.

$t \approx 10^{-4}$ *s.* Quarks can now combine to form protons and neutrons and their antiparticles. The universe has now cooled to such an extent by continued (but much slower) expansion that photons lack the energy needed to break up these new particles. Particles of matter and antimatter collide and annihilate each other. There is a slight excess of matter, which, failing to find annihilation partners, survives to form the world of matter that we know today.

t ≈ 1 min. The universe has now cooled enough so that protons and neutrons, in colliding, can stick together to form the nuclei of the low-mass elements ^{2}H, ^{3}He, ^{4}He, and ^{7}Li. The predicted relative abundances of these nuclides are just what we observe in the universe today. There is plenty of radiation present, but light cannot travel far before it interacts with a nucleus. The universe is opaque to its own radiation.

t ≈ 300,000 y. The temperature has now fallen to ≈10^4 K, and electrons can stick to bare nuclei when they collide, forming atoms. Because light does not interact appreciably with (uncharged) particles such as neutral atoms, the light is now free to travel great distances. This radiation forms the cosmic *background radiation* discussed in Section 45-12. Atoms of hydrogen and helium, under the influence of gravity, begin to clump up, starting the formation of galaxies and stars.

Early measurements suggested that the cosmic background radiation is uniform in all directions, implying that all matter in the universe 300,000 y after the Big Bang was uniformly distributed. This finding was most puzzling because matter in the present universe is not uniformly distributed, but instead is collected in galaxies, clusters of galaxies, and superclusters of galactic clusters. There are also vast *voids* in which there is relatively little matter, and there are regions so crowded with matter that they are called *walls*. If the Big Bang theory of the beginning of the universe is even approximately correct, the seeds for this nonuniform distribution of matter must have been in place before the universe was 300,000 y old and now should show up as a nonuniform distribution of the microwave background radiation.

In 1992, measurements made by NASA's Cosmic Background Explorer (COBE) satellite revealed that the background radiation is, in fact, not perfectly uniform. The image shown in this chapter's opening photograph was made from those measurements and shows us the universe when it was only 300,000 y old. As you can see, large-scale collecting of matter had already begun; thus, the Big Bang theory is, in principle, on the right track.

45-15 A SUMMING UP

Let us, in these closing paragraphs, consider where our rapidly accumulating store of knowledge about the universe is leading us. That it provides satisfaction to a host of curiosity-motivated physicists and astronomers is beyond dispute. However, some view it as a humbling experience in that each increase in knowledge seems to reveal more clearly our own relative insignificance in the grand scheme of things. Thus in roughly chronological order, we humans have come to realize that

> Our Earth is not the center of the solar system.
>
> Our Sun is but one star among many.
>
> Our galaxy is but one of many, and our Sun is an insignificant star near its outer edge.
>
> Our Earth has existed for perhaps only a third of the age of the universe and will surely disappear when our Sun burns up its fuel and becomes a red giant.
>
> Our species has inhabited Earth for less than a million years, a blink in cosmological time.
>
> The last crushing blow: The neutrons and protons of which we are made are not the predominant form of matter in the universe.

On the bright side, however, it is we ourselves who discovered all these facts. Although our position in the universe may be insignificant, the laws of physics that we have discovered (uncovered?) seem to hold throughout the universe and—as far as we know—for all past and future time. At least, there is no evidence that other laws hold in other parts of the universe. Thus until someone complains, we are entitled to stamp the laws of physics "Discovered on Earth." There remains much more to be discovered, and so we close this text by quoting these forward-looking words: *The universe is full of magical things, patiently waiting for our wits to grow sharper.*

REVIEW & SUMMARY

Leptons and Quarks

Current research supports the view that all matter is made of six kinds of **leptons** (Table 45-2) and six kinds of **quarks** (Table 45-5). All these particles have spin quantum numbers equal to $\frac{1}{2}$ and are thus **fermions** (particles with half-integer spin quantum

numbers). There are also 12 **antiparticles**, one corresponding to each of the leptons and quarks.

The Interactions

Particles with electric charge interact by the electromagnetic force by exchanging **virtual photons**. Leptons can interact with each other and with quarks through the **weak force**, via massive W and Z particles as messengers. In addition, quarks interact with each other by the **color force**. The electromagnetic and weak forces are different manifestations of the same force, called the **electroweak force**.

Leptons

Three of the leptons (the **electron**, **muon**, and **tau**) have electric charge equal to $-1e$; these also have nonzero mass. There are also three uncharged **neutrinos** (also leptons), one corresponding to each of the charged leptons. The neutrinos have very small, possibly zero, mass. The antiparticles for the charged leptons have positive charge.

Quarks

The six quarks (up, down, strange, charm, bottom, and top, in order of increasing mass) each have baryon number $+\frac{1}{3}$ and charge equal to either $+(\frac{2}{3})e$ or $-(\frac{1}{3})e$. The strange quark has strangeness -1 while the others all have strangeness 0. These four algebraic signs are reversed for the antiquarks.

Hadrons: Baryons and Mesons

Quarks combine into strongly interacting particles called **hadrons**. **Baryons** are hadrons with half-integer spin quantum numbers ($\frac{1}{2}$ or $\frac{3}{2}$). **Mesons** are hadrons with integer spin quantum numbers (0 or 1). Baryons are fermions, and mesons are bosons. Mesons have baryon number equal to zero; baryons have baryon number equal to $+1$ or -1. **Quantum chromodynamics** predicts that the possible combinations of quarks are either a quark with an antiquark, three quarks, or three antiquarks (this prediction is consistent with experiment). All hadrons, except for protons, are unstable.

Expansion of the Universe

Current evidence strongly suggests that the universe is expanding, with the distant galaxies moving away from us at a rate v given by **Hubble's law**:

$$v = Hr \qquad \text{(Hubble's law).} \qquad (45\text{-}15)$$

Here we take H, the **Hubble constant**, as having the value

$$H = 80 \text{ km/s} \cdot \text{Mpc} = 24.5 \text{ mm/s} \cdot \text{ly}. \qquad (45\text{-}16)$$

History of the Universe

The expansion described by Hubble's law and the presence of ubiquitous background microwave radiation suggest that the universe began in a "big bang" about 15 billion years ago.

QUESTIONS

1. Not only particles such as electrons and protons but also entire atoms can be classified as fermions or bosons, depending whether their overall spin quantum numbers are, respectively, half-integral or integral. Consider the helium isotopes ^3He and ^4He. Which of the following statements is correct? (a) Both are fermions. (b) Both are bosons. (c) ^4He is a fermion and ^3He a boson. (d) ^3He is a fermion and ^4He a boson. (The two helium electrons form a closed shell and play no role in this determination.)

2. Is the direction of the magnetic field in Fig. 45-3b out of the page or into the page?

3. Which of the eight pions in Fig. 45-3b has the smallest kinetic energy?

4. An electron cannot decay into two neutrinos. Which of the following conservation laws would necessarily be violated if it did: (a) energy, (b) angular momentum, (c) charge, (d) lepton number, (e) linear momentum, (f) baryon number?

5. A proton cannot decay into a neutron and a neutrino. Which of the following conservation laws would necessarily be violated if it did: (a) energy (assume the proton is stationary), (b) angular momentum, (c) charge, (d) lepton number, (e) linear momentum, (f) baryon number?

6. A proton has enough energy to decay into a shower made up of electrons, neutrinos, and their antiparticles. Which of the following conservation laws would necessarily be violated if it did: (a) energy, (b) angular momentum, (c) charge, (d) lepton number, (e) linear momentum, (f) baryon number?

7. As we have seen, the π^- meson has the quark structure $d\bar{u}$. Which of the following conservation laws would necessarily be violated if we considered forming a π^- with a d quark and a u quark: (a) energy, (b) angular momentum, (c) charge, (d) lepton number, (e) linear momentum, (f) baryon number?

8. A Σ^+ particle has these quantum numbers: strangeness $S = -1$, charge $Q = +1$, and spin $s = \frac{1}{2}$. To which of the following quark combinations does it belong: (a) dds, (b) $s\bar{s}$, (c) uus, (d) ssu, or (e) $uu\bar{s}$?

9. The left column deals with atomic physics and the right column with particle physics. Match the items in the two columns.

1. chemistry	a. Eightfold Way patterns
2. electrons	b. missing hadrons
3. periodic table	c. quantum chromodynamics
4. missing elements	d. particle physics
5. quantum mechanics	e. quarks

10. Consider the neutrino whose symbol is $\bar{\nu}_\tau$. (a) Is it a quark, a lepton, a meson, or a baryon? (b) Is it a particle or an antiparticle? (c) Is it a boson or a fermion? (d) Is it stable against spontaneous decay?

11. Match the items in these two columns:

1.	tau	a.	quark
2.	pion	b.	lepton
3.	proton	c.	meson
4.	positron	d.	baryon
5.	charm	e.	antiparticle

12. List these particles in order of mass, least first: (a) proton, (b) neutrino, (c) π^+ meson, (d) strange quark, (e) tau, (f) electron, and (g) Σ^-. (*Hint:* Consult the tables in this chapter.)

13. What are the lepton numbers of these particles: (a) π^-, (b) e^-, (c) μ^+, (d) τ^-, (e) $\bar{\nu}_\mu$?

EXERCISES & PROBLEMS

SECTION 45-3 An Interlude

1E. Calculate the difference in mass, in kilograms, between the muon and pion of Sample Problem 45-2.

2E. A neutral pion decays into two gamma rays: $\pi^0 \rightarrow \gamma + \gamma$. Calculate the wavelengths of the gamma rays produced by the decay of a neutral pion at rest.

3E. An electron and a positron are separated by a distance r. Find the ratio of the gravitational force to the electric force between them. From the result, what can you conclude concerning the forces acting between particles detected in a bubble chamber?

4E. The positively charged pion decays by Eq. 45-4: $\pi^+ \rightarrow \mu^+ + \nu$. What then must be the decay scheme of the negatively charged pion? (*Hint:* The π^- is the antiparticle of the π^+.)

5E. How much energy would be released if Earth were annihilated by collision with an anti-Earth?

6P. A neutral pion has a rest energy of 135 MeV and a mean life of 8.3×10^{-17} s. If it is produced with an initial kinetic energy of 80 MeV and decays after one mean lifetime, what is the longest possible track this particle could leave in a bubble chamber? Use relativistic time dilation. (*Hint:* See Sample Problem 45-1.)

7P. Observations of neutrinos emitted by the supernova SN1987a (Fig. 44-12) in the Large Magellanic Cloud place an upper limit of 20 eV on the rest energy of the electron neutrino. Suppose that the rest energy of this neutrino, rather than being zero, is in fact equal to 20 eV. How much slower than the speed of light would a 1.5 MeV neutrino emitted in a beta decay move?

8P. Certain theories predict that the proton is unstable, with a half-life of about 10^{32} years. Assuming that this is true, calculate the number of proton decays you would expect to occur in one year in the water of an Olympic-sized swimming pool holding 114,000 gallons of water.

9P. A positive tau (τ^+, rest energy = 1777 MeV) is moving with 2200 MeV of kinetic energy in a circular path perpendicular to a uniform 1.20 T magnetic field. (a) Calculate the momentum of the tau in kilogram-meters per second. Relativistic effects must be considered. (b) Find the radius of the circular path.

10P. The rest energies of many short-lived particles cannot be measured directly but must be inferred from the measured momenta and known rest energies of the decay products. Consider the ρ^0 meson, which decays by the reaction $\rho^0 \rightarrow \pi^+ + \pi^-$. Calculate the rest energy of the ρ^0 meson given that the oppositely directed momenta of the created pions each have magnitude 358.3 MeV/c. See Table 45-4 for the rest energies of the pions.

11P. (a) A stationary particle m_0 decays into two particles m_1 and m_2, which move off with equal but oppositely directed momenta. Show that the kinetic energy K_1 of m_1 is given by

$$K_1 = \frac{1}{2E_0}[(E_0 - E_1)^2 - E_2^2],$$

where m_0, m_1, and m_2 are masses and E_0, E_1, and E_2 are the corresponding rest energies. (*Hint:* Follow the arguments of Sample Problem 45-2 except that, in this case, neither of the created particles has zero mass.) (b) Show that the result in (a) yields the kinetic energy of the muon as calculated in Sample Problem 45-2.

SECTION 45-5 Another Conservation Law

12E. Verify that the hypothetical proton decay scheme given in Eq. 45-11 does not violate the conservation laws of (a) charge, (b) energy, (c) linear momentum, and (d) angular momentum.

13E. Which conservation law is violated in each of these proposed decays? Assume that the initial particle is stationary and the decay products have zero orbital angular momentum. (a) $\mu^- \rightarrow e^- + \nu_\mu$; (b) $\mu^- \rightarrow e^+ + \nu_e + \bar{\nu}_\mu$; (c) $\mu^+ \rightarrow \pi^+ + \nu_\mu$.

14P. The A_2^+ particle and its products decay according to the following schemes:

$$A_2^+ \rightarrow \rho^0 + \pi^+, \quad \mu^+ \rightarrow e^+ + \nu + \bar{\nu},$$
$$\rho^0 \rightarrow \pi^+ + \pi^-, \quad \pi^- \rightarrow \mu^- + \bar{\nu},$$
$$\pi^+ \rightarrow \mu^+ + \nu, \quad \mu^- \rightarrow e^- + \nu + \bar{\nu}.$$

(a) What are the final stable decay products? (b) From the evidence, is the A_2^+ particle a fermion or a boson? Is that particle a meson or a baryon? What is its baryon number? (*Hint:* See Sample Problem 45-5.)

SECTION 45-7 The Eightfold Way

15E. The reaction $\pi^+ + p \rightarrow p + p + \bar{n}$ proceeds by the strong interaction. By applying the conservation laws, deduce the charge, baryon number, and strangeness of the antineutron.

16E. By examining strangeness, determine which of the following decays or reactions proceed via the strong interaction: (a) $K^0 \rightarrow \pi^+ + \pi^-$; (b) $\Lambda^0 + p \rightarrow \Sigma^+ + n$; (c) $\Lambda^0 \rightarrow p + \pi^-$; (d) $K^- + p \rightarrow \Lambda^0 + \pi^0$.

17E. Which conservation law is violated in each of these proposed reactions and decays? (Assume that the products have zero orbital angular momentum.) (a) $\Lambda^0 \rightarrow p + K^-$; (b) $\Omega^- \rightarrow \Sigma^- + \pi^0$ ($S = -3$, $Q = -1$ for Ω^-); (c) $K^- + p \rightarrow \Lambda^0 + \pi^+$.

18E. Calculate the disintegration energy of the reactions (a) $\pi^+ + p \rightarrow \Sigma^+ + K^+$ and (b) $K^- + p \rightarrow \Lambda^0 + \pi^0$.

19E. A Σ^- particle moving with 220 MeV of kinetic energy decays according to $\Sigma^- \rightarrow \pi^- + n$. Calculate the total kinetic energy of the decay products.

20P. Use the conservation laws to identify the particle labeled x in each of the following reactions, which proceed by means of the strong interaction: (a) $p + p \rightarrow p + \Lambda^0 + x$; (b) $p + \bar{p} \rightarrow n + x$; (c) $\pi^- + p \rightarrow \Xi^0 + K^0 + x$.

21P. Show that if, instead of plotting S versus Q for the spin-$\frac{1}{2}$ baryons in Fig. 45-4a and for the spin-zero mesons in Fig. 45-4b, the quantity $Y = B + S$ is plotted against the quantity $T_z = Q - \frac{1}{2}B$, then the hexagonal patterns emerge with the use of nonsloping (perpendicular) axes. (The quantity Y is called *hypercharge* and T_z is related to a quantity called *isospin*.)

22P. Consider the decay $\Lambda^0 \rightarrow p + \pi^-$ with the Λ^0 at rest. (a) Calculate the disintegration energy. (b) Find the kinetic energy of the proton. (c) What is the kinetic energy of the pion? (*Hint:* See Problem 11.)

SECTION 45-8 The Quark Model

23E. The quark makeups of the proton and neutron are uud and udd, respectively. What are the quark makeups of (a) the antiproton and (b) the antineutron?

24E. From Tables 45-3 and 45-5, determine the identities of the baryons formed from the following combinations of quarks. Check your answers with the baryon octet shown in Fig. 45-4a: (a) ddu; (b) uus; (c) ssd.

25E. What quark combinations are needed to form (a) a Λ^0 and (b) a Ξ^0?

26E. Using the up, down, and strange quarks only, construct, if possible, a baryon (a) with $Q = +1$ and $S = -2$ and (b) with $Q = +2$ and $S = 0$.

27E. There are 10 baryons with spin $\frac{3}{2}$. Their symbols and quantum numbers are as follows:

	Q	S		Q	S
Δ^-	-1	0	Σ^{*0}	0	-1
Δ^0	0	0	Σ^{*+}	$+1$	-1
Δ^+	$+1$	0	Ξ^{*-}	-1	-2
Δ^{++}	$+2$	0	Ξ^{*0}	0	-2
Σ^{*-}	-1	-1	Ω^-	-1	-3

Make a charge–strangeness plot for these baryons, using the sloping coordinate system of Fig. 45-4. Compare your plot with this figure.

28P. There is no known meson with $Q = +1$ and $S = -1$ or with $Q = -1$ and $S = +1$. Explain why, in terms of the quark model.

29P. The spin-$\frac{3}{2}$ Σ^{*0} baryon (see Exercise 27) has a rest energy of 1385 MeV (with an intrinsic uncertainty ignored here); the spin-$\frac{1}{2}$ Σ^0 baryon has a rest energy of 1192.5 MeV. If each of these particles has a kinetic energy of 1000 MeV, which, if either, is moving faster and by how much?

SECTION 45-11 The Universe Is Expanding

30E. If Hubble's law can be extrapolated to very large distances, at what distance would the apparent recessional speed become equal to the speed of light?

31E. What is the observed wavelength of the 656.3 nm H_α line of hydrogen emitted by a galaxy at a distance of 2.40×10^8 ly?

32E. In the laboratory, one of the lines of sodium is emitted at a wavelength of 590.0 nm. In the light from a particular galaxy, however, this line is seen at a wavelength of 602.0 nm. Calculate the distance to the galaxy, assuming that Hubble's law holds.

33P. The apparent recessional speeds of galaxies and quasars at great distances are close to the speed of light, so the relativistic Doppler shift formula (Eq. 38-25) must be used. The red shift is reported as fractional red shift $z = \Delta\lambda/\lambda_0$. (a) Show that, in terms of z, the recessional speed parameter $\beta = v/c$ is given by

$$\beta = \frac{z^2 + 2z}{z^2 + 2z + 2}.$$

(b) A quasar detected in 1987 has $z = 4.43$. Calculate its speed parameter. (c) Find the distance to the quasar, assuming that Hubble's law is valid to these distances.

34P. Will the universe continue to expand forever? To attack this question, make the (reasonable?) assumption that the recessional speed v of a galaxy a distance r from us is determined only by the matter that lies inside a sphere of radius r centered on us. If the total mass inside this sphere is M, the escape speed v_e from the sphere is given by $v_e = \sqrt{2GM/r}$ (Eq. 14-26). (a) Show that to prevent unlimited expansion, the average density ρ inside the sphere must be at least equal to

$$\rho = \frac{3H^2}{8\pi G}.$$

(b) Evaluate this "critical density" numerically; express your answer in terms of hydrogen atoms per cubic meter. Measurements of the actual density are difficult and are complicated by the presence of dark matter.

SECTION 45-12 The Cosmic Background Radiation

35P. Due to the presence everywhere of the microwave background radiation, the minimum possible temperature of a gas in interstellar or intergalactic space is not 0 K but 2.7 K. This implies that a significant fraction of the molecules in space that can occupy excited states of low excitation energy may, in fact, be in those excited states. Subsequent de-excitation would lead to the emission of radiation that could be detected. Consider a (hypothetical) molecule with just one excited state. (a) What would the excitation energy have to be for 25% of the molecules to be in the excited state? (*Hint:* See Eq. 41-21.) (b) What would be the wavelength of the photon emitted in a transition back to the ground state?

SECTION 45-13 Dark Matter

36E. What would the mass of the Sun have to be if Pluto (the outermost planet most of the time) were to have the same orbital speed that Mercury (the innermost planet) has now? Use data from Appendix C, and express your answer in terms of the Sun's current mass M. (Assume circular orbits.)

37P. Suppose the radius of the Sun were increased to 5.90×10^{12} m (the average radius of the orbit of the planet Pluto, the outermost planet), that the density of this expanded Sun were uniform, and that the planets revolved within this tenuous object. (a) Calculate Earth's orbital speed in this new configuration and compare this with its present orbital speed of 29.8 km/s. Assume that the radius of Earth's orbit remains unchanged. (b) What would be the new period of revolution of Earth? (The Sun's mass remains unchanged.)

38P. Suppose that the matter (stars, gas, dust) of a particular galaxy, of total mass M, is distributed uniformly throughout a sphere of radius R. A star of mass m is revolving about the center of the galaxy in a circular orbit of radius $r < R$. (a) Show that the orbital speed v of the star is given by

$$v = r\sqrt{GM/R^3},$$

and therefore that the period T of revolution is

$$T = 2\pi\sqrt{R^3/GM},$$

independent of r. Ignore any resistive forces. (b) What is the corresponding formula for the orbital period, assuming that the mass of the galaxy is strongly concentrated toward the center of the galaxy, so that essentially all the mass is at distances from the center less than r?

SECTION 45-14 The Big Bang

39E. It is possible to derive the following relation between the temperature T of a cavity radiator and the wavelength λ_{max} at which it radiates most strongly:

$$\lambda_{max}T = 2898 \ \mu\text{m} \cdot \text{K}.$$

(This is Wien's law.) (a) The microwave background radiation peaks in intensity at a wavelength of 1.1 mm. To what temperature does this correspond? (b) About 300,000 y after the Big Bang, the universe became transparent to electromagnetic radiation. Its temperature then was about 10^5 K. What was the wavelength at which the background radiation was then most intense?

40E. The wavelength of the photons at which a radiation field of temperature T radiates most intensely is given by $\lambda_{max} = (2898 \ \mu\text{m} \cdot \text{K})/T$ (see Exercise 39). (a) Show that the energy E of such a photon can be computed from

$$E = (4.28 \times 10^{-10} \ \text{MeV/K})T.$$

(b) At what minimum temperature can this photon create an electron–positron pair in a pair production process (as discussed in Section 22-6)?

Additional Problems

41. Figure 45-8 shows part of the experimental arrangement in which antiprotons were discovered in the 1950s. A beam of 6.2 GeV protons emerged from a particle accelerator and collided with nuclei in a copper target. According to theoretical predictions at the time, the collisions with the protons and neutrons in those nuclei should produce antiprotons via the reactions

$$p + p \rightarrow p + p + p + \bar{p}$$

and

$$p + n \rightarrow p + n + p + \bar{p}.$$

However, even if these reactions did occur, they would be rare compared to the reactions

$$p + p \rightarrow p + p + \pi^+ + \pi^-$$

and

$$p + n \rightarrow p + n + \pi^+ + \pi^-.$$

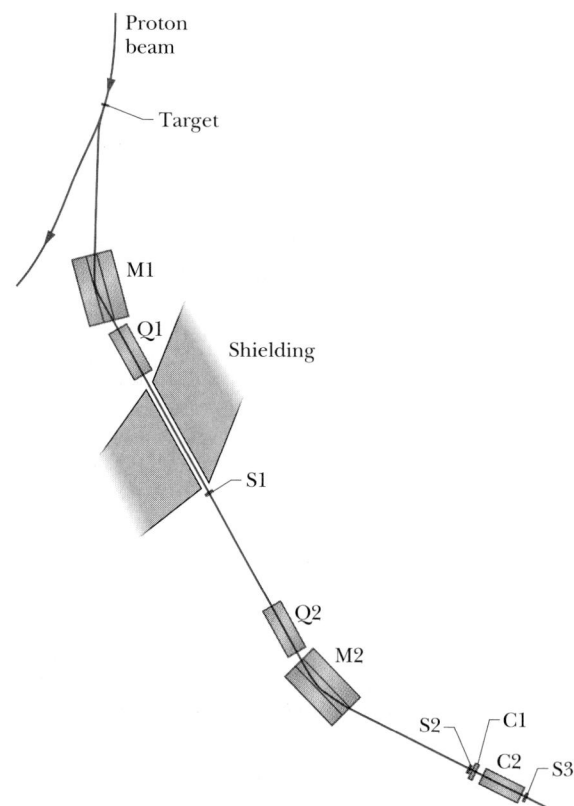

FIGURE 45-8   Problem 41.

Thus, most of the particles produced by the collisions between the 6.2 GeV protons and the copper target were pions.

To prove that antiprotons existed and were also produced by the collisions, particles leaving the target were sent into a series of magnetic fields and detectors as shown in Fig. 45-8. The first magnetic field M1 curved the paths of any charged particles passing through it; moreover, the field was arranged so that the only particles that emerged from it to reach the second magnetic field (Q1) had to be negatively charged (either a $\bar{p}$ or a π^-) and have a momentum of 1.19 GeV/c. Q1 was a special type of magnetic field (a *quadrapole field*) that focused the particles reaching it into a beam, allowing them to pass through a hole through thick shielding to a *scintillation counter* S1. The travel of a charged particle through such a counter triggers a signal (much like a conventional television screen emits a pulse of light when struck by an electron). Thus, each signal indicated the passage of either a 1.19 GeV/c π^- or (presumably) a 1.19 GeV/c $\bar{p}$.

After being refocused by magnetic field Q2, the particles were directed by magnetic field M2 through a second scintillation counter S2 and then through two *Cerenkov counters* C1 and C2. These latter detectors can be manufactured so that they send a signal only when a particle passes through them with a speed in a certain range. In the experiment, a particle with a speed greater than 0.79c would trigger C1 and a particle with a speed between 0.75c and 0.78c would trigger C2.

There were then two ways to distinguish the predicted rare antiprotons from the abundant negative pions. Both ways involved the fact that the speed of a 1.19 GeV/c $\bar{p}$ would differ from that of a 1.19 GeV/c π^-: (1) According to calculations, a $\bar{p}$ would trigger one of the Cerenkov counters and a π^- would trigger the other. (2) Also, the time interval Δt between signals from S1 and S2, which were separated by 12 m, would be one value for a $\bar{p}$ and another value for a π^-. Thus, if the correct Cerenkov counter is triggered and the time interval Δt has the correct value, the experiment would prove the existence of antiprotons.

What is the speed of (a) an antiproton with a momentum of 1.19 GeV/c and (b) a negative pion with a momentum of 1.19 GeV/c? The speed of an antiproton through the Cerenkov detectors would actually be slightly slower than calculated here because an antiproton would lose a little energy within the detectors. Which Cerenkov detector was triggered by (c) an antiproton and (d) a negative pion? What time interval Δt indicated the passage of (e) an antiproton and (f) a negative pion? [Problem adapted from O. Chamberlain, E. Segrè, C. Wiegand, and T. Ypsilantis, "Observation of Antiprotons," *Physical Review*, Vol. 100, pages 947–950 (1955).]

42. Cosmological red-shift. The expansion of the universe is often represented with a drawing like Fig. 45-9a. In that figure, we are located at the symbol labelled MW (for the Milky Way gal-

axy), at the origin of the *r* axis that extends radially away from us in any direction. Other, very distant galaxies are also represented. Superimposed on their symbols are their velocity vectors as inferred from the red shift of the light reaching us from the galaxies. In accord with Hubble's law, the velocity of each galaxy is proportional to its distance from us. Such drawings can be misleading because they imply (1) that the red shifts are due to the motions of galaxies relative to us, as they rush away from us through static (stationary) space, and (2) that we are at the center of all this motion.

Actually, the expansion of the universe and the increased separation of the galaxies are due not to an outward rush of the galaxies into pre-existing space but to an expansion of space itself throughout the universe. *Space is dynamic, not static.*

Figures 45-9b, c, and d show a different way of representing the universe and its expansion. Each part of the figure gives part of a one-dimensional section of the universe (along an *r* axis); the other two spatial dimensions of the universe are not shown. Each of the three parts of the figure show the Milky Way and six other galaxies; the parts are positioned along a time axis, with time increasing upward. In part b, at the earliest time of the three parts, the Milky Way and the six other galaxies are represented as being relatively close. As time progresses upward in the figures, space expands, causing the galaxies to move apart. Note that relative to the Milky Way, all the other galaxies move away from it because of the expansion. However, there is nothing special about the Milky Way—the galaxies move away from any other choice of observation point.

Figures 45-9e and f focuses on just the Milky Way galaxy and one of the other galaxies, galaxy A, at two particular times during the expansion. In part e, galaxy A is a distance r from us in the Milky Way and is emitting a light wave of wavelength λ. In part f, after a time interval Δt, that light wave is being detected at Earth. Let us represent the expansion rate per unit length of space with α, which we assume to be constant during time interval Δt.

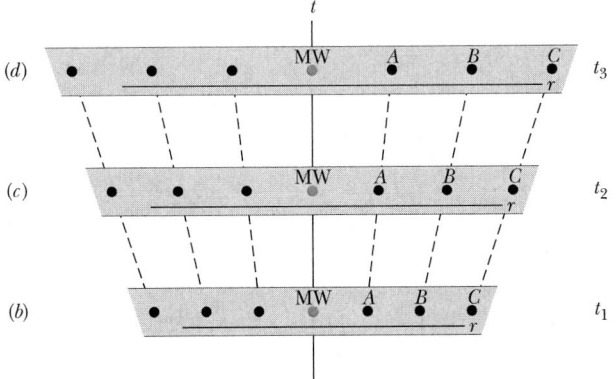

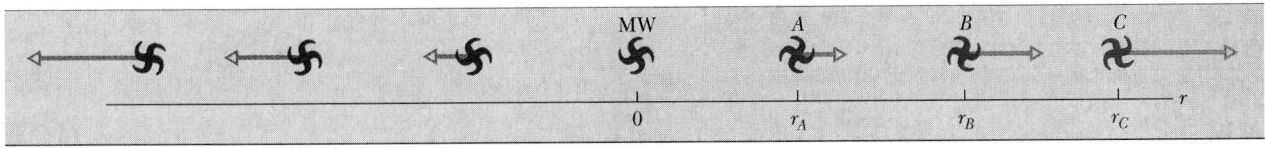

(a)

FIGURE 45-9 Problem 42.

Then during Δt, every unit length of space (say, every meter) expands by $\alpha \Delta t$; hence a distance r expands by $r\alpha \Delta t$. The light wave of Figs. 45-9e and f travels at speed c from galaxy A to Earth. (a) Show that

$$\Delta t = \frac{r}{c - r\alpha}.$$

The detected wavelength λ' of the light is greater than the emitted wavelength λ because space expanded during time interval Δt. This increase in wavelength is called the **cosmological red-shift**; it is not a Doppler effect. (b) Show that the change in wavelength $\Delta\lambda \, (= \lambda' - \lambda)$ is given by

$$\frac{\Delta\lambda}{\lambda} = \frac{r\alpha}{c - r\alpha}.$$

(c) Expand the right side of this equation, using the binomial expansion (given in Appendix E and explained in the Problem Solving Tactic on page 145). (d) Retain the first term in the expansion. What is the resulting expression for $\Delta\lambda/\lambda$?

If, instead, we assume that Fig. 45-9a applies and that $\Delta\lambda$ is due to a Doppler effect, then from Eq. 18-57 we have

$$\frac{\Delta\lambda}{\lambda} = \frac{v}{c},$$

where v is radial velocity of Galaxy A relative to Earth. (e) Using Hubble's law, compare this Doppler-effect result with the cosmological-expansion result of (d) and find a value for α. From this analysis you can see that the two results, derived with very different models about the red-shift of the light we detect from distant galaxies, are compatible.

Suppose that the light we detect from Galaxy A has a red-shift of $\Delta\lambda/\lambda = 0.050$ and that the expansion rate of the universe has been constant at the current value given in the chapter. (f) Using the result of (b), find the distance between the galaxy and Earth when the light was emitted. Find how long ago the light was emitted by the galaxy (g) by using the result of (a) and (h) by assuming that the red-shift is a Doppler effect. (*Hint*: For (h), the time is just the distance at the time of emission divided by the speed of light, because if the red-shift is just a Doppler effect, the distance does not change during the light's travel to us. Here the two models about the red-shift of the light differ in their results.) (i) At the time of detection, what is the distance between Earth and Galaxy A? (We make the assumption that Galaxy A still exists; if it ceased to exist, humans would not know about its death until the last light emitted by the galaxy reached Earth.)

Now suppose that the light we detect from Galaxy B (Fig. 45-9g) has a red-shift of $\Delta\lambda/\lambda = 0.080$. (j) Using the result of (b), find the distance between Galaxy B and Earth when the light was emitted. (k) Using the result of (a), find how long ago the light was emitted by Galaxy B. (l) When the light that we detect from Galaxy A was emitted, what was the distance between Galaxy A and Galaxy B?

43. *Cosmic background radiation.* Figure 45-10a represents a section through the universe when the universe was 300,000 years old and Figure 45-10b represents the universe now. (See Problem 42 for an explanation of this type of drawing.) At time $t = 300{,}000$ y, free electrons and protons combined to form hydrogen, emitting light in every direction (in Fig. 45-10a only light traveling along the length of the section is represented). The emitted light is now being detected on Earth as part of the cosmic background radiation.

Assume that the universe is 14 billion y old and the expansion rate is constant throughout the expansion of the universe. (a) By how much have the wavelengths emitted at $t = 300{,}000$ y been expanded by the expansion of the universe when they are now detected? (b) Assuming that hydrogen atoms formed at $t = 300{,}000$ y are like hydrogen atoms now, what detected wavelength corresponds to the Lyman series limit in the emitted light? (c) The cosmic background radiation has a peak in intensity at a detected wavelength of 1.1 mm. What is the corresponding emitted wavelength?

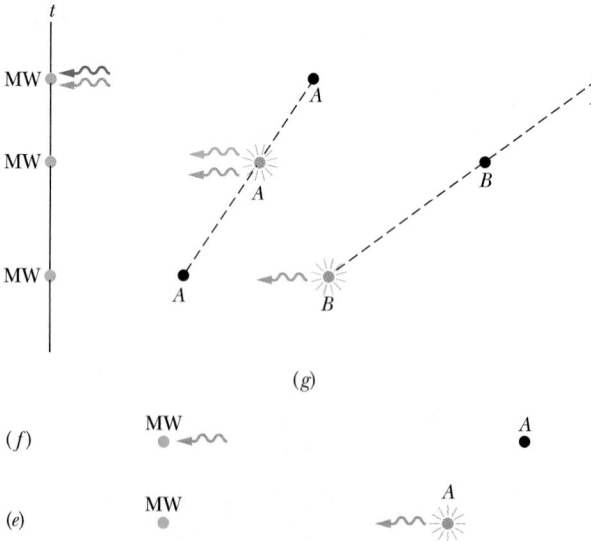

(g)

(f)

(e)

FIGURE 45-9 Problem 42 (continued).

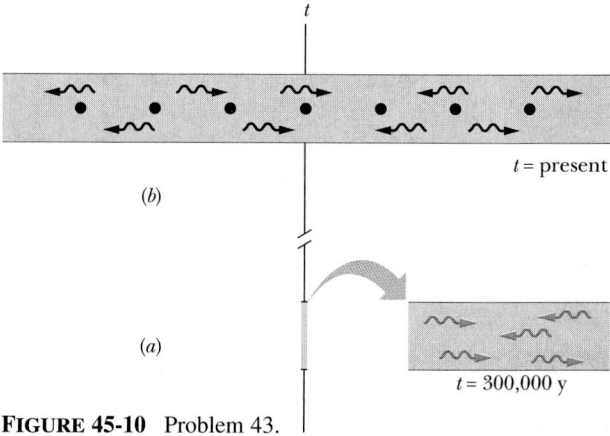

(b)

$t = \text{present}$

(a)

$t = 300{,}000$ y

FIGURE 45-10 Problem 43.

Appendix A
The International System of Units (SI)*

1. THE SI BASE UNITS

QUANTITY	NAME	SYMBOL	DEFINITION
length	meter	m	". . . the length of the path traveled by light in vacuum in 1/299,792,458 of a second." (1983)
mass	kilogram	kg	". . . this prototype [a certain platinum–iridium cylinder] shall henceforth be considered to be the unit of mass." (1889)
time	second	s	". . . the duration of 9,192,631,770 periods of the radiation corresponding to the transition between the two hyperfine levels of the ground state of the cesium-133 atom." (1967)
electric current	ampere	A	". . . that constant current which, if maintained in two straight parallel conductors of infinite length, of negligible circular cross section, and placed 1 meter apart in vacuum, would produce between these conductors a force equal to 2×10^{-7} newton per meter of length." (1946)
thermodynamic temperature	kelvin	K	". . . the fraction 1/273.16 of the thermodynamic temperature of the triple point of water." (1967)
amount of substance	mole	mol	". . . the amount of substance of a system which contains as many elementary entities as there are atoms in 0.012 kilogram of carbon-12." (1971)
luminous intensity	candela	cd	". . . the luminous intensity, in the perpendicular direction, of a surface of 1/600,000 square meter of a blackbody at the temperature of freezing platinum under a pressure of 101.325 newtons per square meter." (1967)

---

*Adapted from "The International System of Units (SI)," National Bureau of Standards Special Publication 330, 1972 edition. The definitions above were adopted by the General Conference of Weights and Measures, an international body, on the dates shown. In this book we do not use the candela.

2. SOME SI DERIVED UNITS

QUANTITY	NAME OF UNIT	SYMBOL	
area	square meter	m^2	
volume	cubic meter	m^3	
frequency	hertz	Hz	s^{-1}
mass density (density)	kilogram per cubic meter	kg/m^3	
speed, velocity	meter per second	m/s	
angular velocity	radian per second	rad/s	
acceleration	meter per second per second	m/s^2	
angular acceleration	radian per second per second	rad/s^2	
force	newton	N	$kg \cdot m/s^2$
pressure	pascal	Pa	N/m^2
work, energy, quantity of heat	joule	J	$N \cdot m$
power	watt	W	J/s
quantity of electric charge	coulomb	C	$A \cdot s$
potential difference, electromotive force	volt	V	W/A
electric field strength	volt per meter (or newton per coulomb)	V/m	N/C
electric resistance	ohm	Ω	V/A
capacitance	farad	F	$A \cdot s/V$
magnetic flux	weber	Wb	$V \cdot s$
inductance	henry	H	$V \cdot s/A$
magnetic flux density	tesla	T	Wb/m^2
magnetic field strength	ampere per meter	A/m	
entropy	joule per kelvin	J/K	
specific heat	joule per kilogram kelvin	$J/(kg \cdot K)$	
thermal conductivity	watt per meter kelvin	$W/(m \cdot K)$	
radiant intensity	watt per steradian	W/sr	

3. THE SI SUPPLEMENTARY UNITS

QUANTITY	NAME OF UNIT	SYMBOL
plane angle	radian	rad
solid angle	steradian	sr

Appendix B
*Some Fundamental Constants of Physics**

CONSTANT	SYMBOL	COMPUTATIONAL VALUE	BEST (1986) VALUE VALUE[a]	BEST (1986) VALUE UNCERTAINTY[b]
Speed of light in a vacuum	c	3.00×10^8 m/s	2.99792458	exact
Elementary charge	e	1.60×10^{-19} C	1.60217733	0.30
Gravitational constant	G	6.67×10^{-11} m³/s²·kg	6.67259	128
Universal gas constant	R	8.31 J/mol·K	8.314510	8.4
Avogadro constant	N_A	6.02×10^{23} mol⁻¹	6.0221367	0.59
Boltzmann constant	k	1.38×10^{-23} J/K	1.380658	8.5
Stefan-Boltzmann constant	σ	5.67×10^{-8} W/m²·K⁴	5.67051	34
Molar volume of ideal gas at STP[d]	V_m	2.24×10^{-2} m³/mol	2.241409	8.4
Permittivity constant	ϵ_0	8.85×10^{-12} F/m	8.85418781762	exact
Permeability constant	μ_0	1.26×10^{-6} H/m	1.25663706143	exact
Planck constant	h	6.63×10^{-34} J·s	6.6260755	0.60
Electron mass[c]	m_e	9.11×10^{-31} kg	9.1093897	0.59
		5.49×10^{-4} u	5.48579903	0.023
Proton mass[c]	m_p	1.67×10^{-27} kg	1.6726231	0.59
		1.0073 u	1.0072764660	0.005
Ratio of proton mass to electron mass	m_p/m_e	1840	1836.152701	0.020
Electron charge-to-mass ratio	e/m_e	1.76×10^{11} C/kg	1.75881961	0.30
Neutron mass[c]	m_n	1.68×10^{-27} kg	1.6749286	0.59
		1.0087 u	1.0086649235	0.0023
Hydrogen atom mass[c]	m_{1_H}	1.0078 u	1.0078250316	0.0005
Deuterium atom mass[c]	m_{2_H}	2.0141 u	2.0141017779	0.0005
Helium atom mass[c]	$m_{4_{He}}$	4.0026 u	4.0026032	0.067
Muon mass	m_μ	1.88×10^{-28} kg	1.8835326	0.61
Electron magnetic moment	μ_e	9.28×10^{-24} J/T	9.2847701	0.34
Proton magnetic moment	μ_p	1.41×10^{-26} J/T	1.41060761	0.34
Bohr magneton	μ_B	9.27×10^{-24} J/T	9.2740154	0.34
Nuclear magneton	μ_N	5.05×10^{-27} J/T	5.0507866	0.34
Bohr radius	r_B	5.29×10^{-11} m	5.29177249	0.045
Rydberg constant	R	1.10×10^7 m⁻¹	1.0973731534	0.0012
Electron Compton wavelength	λ_C	2.43×10^{-12} m	2.42631058	0.089

[a]Values given in this column should be given the same unit and power of 10 as the computational value. [b]Parts per million. [c]Masses given in u are in unified atomic mass units, where 1 u = $1.6605402 \times 10^{-27}$ kg. [d]STP means standard temperature and pressure: 0°C and 1.0 atm (0.1 MPa).

*The values in this table were largely selected from a longer list in *Symbols, Units and Nomenclature in Physics* (IUPAP), prepared by E. Richard Cohen and Pierre Giacomo, 1986.

Appendix C
Some Astronomical Data

SOME DISTANCES FROM THE EARTH

To the moon*	3.82×10^8 m
To the sun*	1.50×10^{11} m
To the nearest star (Proxima Centauri)	4.04×10^{16} m
To the center of our galaxy	2.2×10^{20} m
To the Andromeda Galaxy	2.1×10^{22} m
To the edge of the observable universe	$\sim 10^{26}$ m

*Mean distance.

THE SUN, THE EARTH, AND THE MOON

PROPERTY	UNIT	SUN	EARTH	MOON
Mass	kg	1.99×10^{30}	5.98×10^{24}	7.36×10^{22}
Mean radius	m	6.96×10^8	6.37×10^6	1.74×10^6
Mean density	kg/m^3	1410	5520	3340
Free-fall acceleration at the surface	m/s^2	274	9.81	1.67
Escape velocity	km/s	618	11.2	2.38
Period of rotation[a]	—	37 d at poles[b] 26 d at equator[b]	23 h 56 min	27.3 d
Radiation power[c]	W	3.90×10^{26}		

[a]Measured with respect to the distant stars.

[b]The sun, a ball of gas, does not rotate as a rigid body.

[c]Just outside the Earth's atmosphere solar energy is received, assuming normal incidence, at the rate of 1340 W/m^2.

SOME PROPERTIES OF THE PLANETS

	MERCURY	VENUS	EARTH	MARS	JUPITER	SATURN	URANUS	NEPTUNE	PLUTO
Mean distance from sun, 10^6 km	57.9	108	150	228	778	1430	2870	4500	5900
Period of revolution, y	0.241	0.615	1.00	1.88	11.9	29.5	84.0	165	248
Period of rotation,[a] d	58.7	-243[b]	0.997	1.03	0.409	0.426	-0.451[b]	0.658	6.39
Orbital speed, km/s	47.9	35.0	29.8	24.1	13.1	9.64	6.81	5.43	4.74
Inclination of axis to orbit	$<28°$	$\approx 3°$	23.4°	25.0°	3.08°	26.7°	97.9°	29.6°	57.5°
Inclination of orbit to Earth's orbit	7.00°	3.39°		1.85°	1.30°	2.49°	0.77°	1.77°	17.2°
Eccentricity of orbit	0.206	0.0068	0.0167	0.0934	0.0485	0.0556	0.0472	0.0086	0.250
Equatorial diameter, km	4880	12,100	12,800	6790	143,000	120,000	51,800	49,500	2300
Mass (Earth = 1)	0.0558	0.815	1.000	0.107	318	95.1	14.5	17.2	0.002
Density (water = 1)	5.60	5.20	5.52	3.95	1.31	0.704	1.21	1.67	2.03
Surface value of g,[c] m/s^2	3.78	8.60	9.78	3.72	22.9	9.05	7.77	11.0	0.5
Escape velocity,[c] km/s	4.3	10.3	11.2	5.0	59.5	35.6	21.2	23.6	1.1
Known satellites	0	0	1	2	16 + ring	18 + rings	15 + rings	8 + rings	1

[a]Measured with respect to the distant stars.

[b]Venus and Uranus rotate opposite their orbital motion.

[c]Gravitational acceleration measured at the planet's equator.

Appendix **D**
Conversion Factors

Conversion factors may be read directly from these tables. For example, 1 degree = 2.778×10^{-3} revolutions, so $16.7° = 16.7 \times 2.778 \times 10^{-3}$ rev. The SI quantities are fully capitalized.

Adapted in part from G. Shortley and D. Williams, *Elements of Physics,* Prentice-Hall, Englewood Cliffs, NJ, 1971.

PLANE ANGLE

	°	′	″	RADIAN	rev
1 degree =	1	60	3600	1.745×10^{-2}	2.778×10^{-3}
1 minute =	1.667×10^{-2}	1	60	2.909×10^{-4}	4.630×10^{-5}
1 second =	2.778×10^{-4}	1.667×10^{-2}	1	4.848×10^{-6}	7.716×10^{-7}
1 RADIAN =	57.30	3438	2.063×10^{5}	1	0.1592
1 revolution =	360	2.16×10^{4}	1.296×10^{6}	6.283	1

SOLID ANGLE

1 sphere = 4π steradians = 12.57 steradians

LENGTH

	cm	METER	km	in.	ft	mi
1 centimeter =	1	10^{-2}	10^{-5}	0.3937	3.281×10^{-2}	6.214×10^{-6}
1 METER =	100	1	10^{-3}	39.37	3.281	6.214×10^{-4}
1 kilometer =	10^{5}	1000	1	3.937×10^{4}	3281	0.6214
1 inch =	2.540	2.540×10^{-2}	2.540×10^{-5}	1	8.333×10^{-2}	1.578×10^{-5}
1 foot =	30.48	0.3048	3.048×10^{-4}	12	1	1.894×10^{-4}
1 mile =	1.609×10^{5}	1609	1.609	6.336×10^{4}	5280	1

1 angström = 10^{-10} m
1 nautical mile = 1852 m
    = 1.151 miles = 6076 ft

1 fermi = 10^{-15} m
1 light-year = 9.460×10^{12} km
1 parsec = 3.084×10^{13} km

1 fathom = 6 ft
1 Bohr radius = 5.292×10^{-11} m
1 yard = 3 ft

1 rod = 16.5 ft
1 mil = 10^{-3} in.
1 nm = 10^{-9} m

AREA

	METER^2	cm^2	ft^2	in.^{2}
1 SQUARE METER =	1	10^{4}	10.76	1550
1 square centimeter =	10^{-4}	1	1.076×10^{-3}	0.1550
1 square foot =	9.290×10^{-2}	929.0	1	144
1 square inch =	6.452×10^{-4}	6.452	6.944×10^{-3}	1

1 square mile = 2.788×10^{7} ft^2
    = 640 acres
1 barn = 10^{-28} m^2

1 acre = 43,560 ft^2
1 hectare = 10^{4} m^2 = 2.471 acres

VOLUME

	METER^3	cm^3	L	ft^3	in.^{3}
1 CUBIC METER = 1		10^6	1000	35.31	6.102×10^4
1 cubic centimeter = 10^{-6}		1	1.000×10^{-3}	3.531×10^{-5}	6.102×10^{-2}
1 liter = 1.000×10^{-3}		1000	1	3.531×10^{-2}	61.02
1 cubic foot = 2.832×10^{-2}		2.832×10^4	28.32	1	1728
1 cubic inch = 1.639×10^{-5}		16.39	1.639×10^{-2}	5.787×10^{-4}	1

1 U.S. fluid gallon = 4 U.S. fluid quarts = 8 U.S. pints = 128 U.S. fluid ounces = 231 in.^{3}

1 British imperial gallon = 277.4 in.^{3} = 1.201 U.S. fluid gallons

MASS

Quantities in the colored areas are not mass units but are often used as such. When we write, for example, 1 kg ''='' 2.205 lb, this means that a kilogram is a *mass* that *weighs* 2.205 pounds at a location where *g* has the standard value of 9.80665 m/s^2.

	g	KILOGRAM	slug	u	oz	lb	ton
1 gram = 1		0.001	6.852×10^{-5}	6.022×10^{23}	3.527×10^{-2}	2.205×10^{-3}	1.102×10^{-6}
1 KILOGRAM = 1000		1	6.852×10^{-2}	6.022×10^{26}	35.27	2.205	1.102×10^{-3}
1 slug = 1.459×10^4		14.59	1	8.786×10^{27}	514.8	32.17	1.609×10^{-2}
1 atomic mass unit = 1.661×10^{-24}		1.661×10^{-27}	1.138×10^{-28}	1	5.857×10^{-26}	3.662×10^{-27}	1.830×10^{-30}
1 ounce = 28.35		2.835×10^{-2}	1.943×10^{-3}	1.718×10^{25}	1	6.250×10^{-2}	3.125×10^{-5}
1 pound = 453.6		0.4536	3.108×10^{-2}	2.732×10^{26}	16	1	0.0005
1 ton = 9.072×10^5		907.2	62.16	5.463×10^{29}	3.2×10^4	2000	1

1 metric ton = 1000 kg

DENSITY

Quantities in the colored areas are weight densities and, as such, are dimensionally different from mass densities. See note for mass table.

	slug/ft^3	KILOGRAM/ METER^3	g/cm^3	lb/ft^3	lb/in.^{3}
1 slug per foot^3 = 1		515.4	0.5154	32.17	1.862×10^{-2}
1 KILOGRAM per METER^3 = 1.940×10^{-3}		1	0.001	6.243×10^{-2}	3.613×10^{-5}
1 gram per centimeter^3 = 1.940		1000	1	62.43	3.613×10^{-2}
1 pound per foot^3 = 3.108×10^{-2}		16.02	1.602×10^{-2}	1	5.787×10^{-4}
1 pound per inch^3 = 53.71		2.768×10^4	27.68	1728	1

TIME

	y	d	h	min	SECOND
1 year = 1		365.25	8.766×10^3	5.259×10^5	3.156×10^7
1 day = 2.738×10^{-3}		1	24	1440	8.640×10^4
1 hour = 1.141×10^{-4}		4.167×10^{-2}	1	60	3600
1 minute = 1.901×10^{-6}		6.944×10^{-4}	1.667×10^{-2}	1	60
1 SECOND = 3.169×10^{-8}		1.157×10^{-5}	2.778×10^{-4}	1.667×10^{-2}	1

SPEED

	ft/s	km/h	METER/ SECOND	mi/h	cm/s
1 foot per second = 1		1.097	0.3048	0.6818	30.48
1 kilometer per hour = 0.9113		1	0.2778	0.6214	27.78
1 METER per SECOND = 3.281		3.6	1	2.237	100
1 mile per hour = 1.467		1.609	0.4470	1	44.70
1 centimeter per second = 3.281×10^{-2}		3.6×10^{-2}	0.01	2.237×10^{-2}	1

1 knot = 1 nautical mi/h = 1.688 ft/s   1 mi/min = 88.00 ft/s = 60.00 mi/h

FORCE

Force units in the colored areas are now little used. To clarify: 1 gram-force (= 1 gf) is the force of gravity that would act on an object whose mass is 1 gram at a location where g has the standard value of 9.80665 m/s².

	dyne	NEWTON	lb	pdl	gf	kgf
1 dyne = 1		10^{-5}	2.248×10^{-6}	7.233×10^{-5}	1.020×10^{-3}	1.020×10^{-6}
1 NEWTON = 10^5		1	0.2248	7.233	102.0	0.1020
1 pound = 4.448×10^5		4.448	1	32.17	453.6	0.4536
1 poundal = 1.383×10^4		0.1383	3.108×10^{-2}	1	14.10	1.410×10^{-2}
1 gram-force = 980.7		9.807×10^{-3}	2.205×10^{-3}	7.093×10^{-2}	1	0.001
1 kilogram-force = 9.807×10^5		9.807	2.205	70.93	1000	1

PRESSURE

	atm	dyne/cm²	inch of water	cm Hg	PASCAL	lb/in.²	lb/ft²
1 atmosphere = 1		1.013×10^6	406.8	76	1.013×10^5	14.70	2116
1 dyne per centimeter² = 9.869×10^{-7}		1	4.015×10^{-4}	7.501×10^{-5}	0.1	1.405×10^{-5}	2.089×10^{-3}
1 inch of water[a] at 4°C = 2.458×10^{-3}		2491	1	0.1868	249.1	3.613×10^{-2}	5.202
1 centimeter of mercury[a] at 0°C = 1.316×10^{-2}		1.333×10^4	5.353	1	1333	0.1934	27.85
1 PASCAL = 9.869×10^{-6}		10	4.015×10^{-3}	7.501×10^{-4}	1	1.450×10^{-4}	2.089×10^{-2}
1 pound per inch² = 6.805×10^{-2}		6.895×10^4	27.68	5.171	6.895×10^3	1	144
1 pound per foot² = 4.725×10^{-4}		478.8	0.1922	3.591×10^{-2}	47.88	6.944×10^{-3}	1

[a] Where the acceleration of gravity has the standard value of 9.80665 m/s².

1 bar = 10^6 dyne/cm² = 0.1 MPa   1 millibar = 10^3 dyne/cm² = 10^2 Pa   1 torr = 1 mm Hg

ENERGY, WORK, HEAT

Quantities in the colored areas are not properly energy units but are included for convenience. They arise from the relativistic mass–energy equivalence formula $E = mc^2$ and represent the energy released if a kilogram or unified atomic mass unit (u) is completely converted to energy (bottom two rows) or the mass that would be completely converted to one unit of energy (rightmost two columns).

	Btu	erg	ft·lb	hp·h	JOULE	cal	kW·h	eV	MeV	kg	u
1 British thermal unit =	1	1.055×10^{10}	777.9	3.929×10^{-4}	1055	252.0	2.930×10^{-4}	6.585×10^{21}	6.585×10^{15}	1.174×10^{-14}	7.070×10^{12}
1 erg =	9.481×10^{-11}	1	7.376×10^{-8}	3.725×10^{-14}	10^{-7}	2.389×10^{-8}	2.778×10^{-14}	6.242×10^{11}	6.242×10^{5}	1.113×10^{-24}	670.2
1 foot-pound =	1.285×10^{-3}	1.356×10^{7}	1	5.051×10^{-7}	1.356	0.3238	3.766×10^{-7}	8.464×10^{18}	8.464×10^{12}	1.509×10^{-17}	9.037×10^{9}
1 horsepower-hour =	2545	2.685×10^{13}	1.980×10^{6}	1	2.685×10^{6}	6.413×10^{5}	0.7457	1.676×10^{25}	1.676×10^{19}	2.988×10^{-11}	1.799×10^{16}
1 JOULE =	9.481×10^{-4}	10^{7}	0.7376	3.725×10^{-7}	1	0.2389	2.778×10^{-7}	6.242×10^{18}	6.242×10^{12}	1.113×10^{-17}	6.702×10^{9}
1 calorie =	3.969×10^{-3}	4.186×10^{7}	3.088	1.560×10^{-6}	4.186	1	1.163×10^{-6}	2.613×10^{19}	2.613×10^{13}	4.660×10^{-17}	2.806×10^{10}
1 kilowatt-hour =	3413	3.600×10^{13}	2.655×10^{6}	1.341	3.600×10^{6}	8.600×10^{5}	1	2.247×10^{25}	2.247×10^{19}	4.007×10^{-11}	2.413×10^{16}
1 electron-volt =	1.519×10^{-22}	1.602×10^{-12}	1.182×10^{-19}	5.967×10^{-26}	1.602×10^{-19}	3.827×10^{-20}	4.450×10^{-26}	1	10^{-6}	1.783×10^{-36}	1.074×10^{-9}
1 million electron-volts =	1.519×10^{-16}	1.602×10^{-6}	1.182×10^{-13}	5.967×10^{-20}	1.602×10^{-13}	3.827×10^{-14}	4.450×10^{-20}	10^{-6}	1	1.783×10^{-30}	1.074×10^{-3}
1 kilogram =	8.521×10^{13}	8.987×10^{23}	6.629×10^{16}	3.348×10^{10}	8.987×10^{16}	2.146×10^{16}	2.497×10^{10}	5.610×10^{35}	5.610×10^{29}	1	6.022×10^{26}
1 unified atomic mass unit =	1.415×10^{-13}	1.492×10^{-3}	1.101×10^{-10}	5.559×10^{-17}	1.492×10^{-10}	3.564×10^{-11}	4.146×10^{-17}	9.320×10^{8}	932.0	1.661×10^{-27}	1

POWER

	Btu/h	ft·lb/s	hp	cal/s	kW	WATT
1 British thermal unit per hour =	1	0.2161	3.929×10^{-4}	6.998×10^{-2}	2.930×10^{-4}	0.2930
1 foot-pound per second =	4.628	1	1.818×10^{-3}	0.3239	1.356×10^{-3}	1.356
1 horsepower =	2545	550	1	178.1	0.7457	745.7
1 calorie per second =	14.29	3.088	5.615×10^{-3}	1	4.186×10^{-3}	4.186
1 kilowatt =	3413	737.6	1.341	238.9	1	1000
1 WATT =	3.413	0.7376	1.341×10^{-3}	0.2389	0.001	1

MAGNETIC FIELD

	gauss	TESLA	milligauss
1 gauss =	1	10^{-4}	1000
1 TESLA =	10^{4}	1	10^{7}
1 milligauss =	0.001	10^{-7}	1

MAGNETIC FLUX

	maxwell	WEBER
1 maxwell =	1	10^{-8}
1 WEBER =	10^{8}	1

1 tesla = 1 weber/meter^2

Appendix E
Mathematical Formulas

GEOMETRY

Circle of radius r: circumference $= 2\pi r$; area $= \pi r^2$.

Sphere of radius r: area $= 4\pi r^2$; volume $= \frac{4}{3}\pi r^3$.

Right circular cylinder of radius r and height h:
area $= 2\pi r^2 + 2\pi rh$; volume $= \pi r^2 h$.

Triangle of base a and altitude h: area $= \frac{1}{2}ah$.

QUADRATIC FORMULA

If $ax^2 + bx + c = 0$, then $x = \dfrac{-b \pm \sqrt{b^2 - 4ac}}{2a}$.

TRIGONOMETRIC FUNCTIONS OF ANGLE θ

$\sin \theta = \dfrac{y}{r}$ $\cos \theta = \dfrac{x}{r}$

$\tan \theta = \dfrac{y}{x}$ $\cot \theta = \dfrac{x}{y}$

$\sec \theta = \dfrac{r}{x}$ $\csc \theta = \dfrac{r}{y}$

PYTHAGOREAN THEOREM

In this right triangle,
$$a^2 + b^2 = c^2$$

TRIANGLES

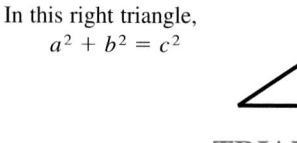

Angles are A, B, C

Opposite sides are a, b, c

Angles $A + B + C = 180°$

$\dfrac{\sin A}{a} = \dfrac{\sin B}{b} = \dfrac{\sin C}{c}$

$c^2 = a^2 + b^2 - 2ab \cos C$

Exterior angle $D = A + C$

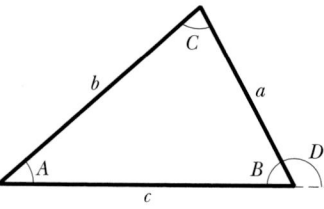

MATHEMATICAL SIGNS AND SYMBOLS

$=$ equals

$\approx$ equals approximately

$\sim$ is the order of magnitude of

$\neq$ is not equal to

$\equiv$ is identical to, is defined as

$>$ is greater than ($\gg$ is much greater than)

$<$ is less than ($\ll$ is much less than)

$\geq$ is greater than or equal to (or, is no less than)

$\leq$ is less than or equal to (or, is no more than)

$\pm$ plus or minus

$\propto$ is proportional to

Σ the sum of

$\bar{x}$ the average value of x

TRIGONOMETRIC IDENTITIES

$\sin(90° - \theta) = \cos \theta$

$\cos(90° - \theta) = \sin \theta$

$\sin \theta / \cos \theta = \tan \theta$

$\sin^2 \theta + \cos^2 \theta = 1$

$\sec^2 \theta - \tan^2 \theta = 1$

$\csc^2 \theta - \cot^2 \theta = 1$

$\sin 2\theta = 2 \sin \theta \cos \theta$

$\cos 2\theta = \cos^2 \theta - \sin^2 \theta = 2 \cos^2 \theta - 1 = 1 - 2 \sin^2 \theta$

$\sin(\alpha \pm \beta) = \sin \alpha \cos \beta \pm \cos \alpha \sin \beta$

$\cos(\alpha \pm \beta) = \cos \alpha \cos \beta \mp \sin \alpha \sin \beta$

$\tan(\alpha \pm \beta) = \dfrac{\tan \alpha \pm \tan \beta}{1 \mp \tan \alpha \tan \beta}$

$\sin \alpha \pm \sin \beta = 2 \sin \frac{1}{2}(\alpha \pm \beta) \cos \frac{1}{2}(\alpha \mp \beta)$

$\cos \alpha + \cos \beta = 2 \cos \frac{1}{2}(\alpha + \beta) \cos \frac{1}{2}(\alpha - \beta)$

$\cos \alpha - \cos \beta = -2 \sin \frac{1}{2}(\alpha + \beta) \sin \frac{1}{2}(\alpha - \beta)$

BINOMIAL THEOREM

$$(1 + x)^n = 1 + \frac{nx}{1!} + \frac{n(n-1)x^2}{2!} + \cdots \qquad (x^2 < 1)$$

EXPONENTIAL EXPANSION

$$e^x = 1 + x + \frac{x^2}{2!} + \frac{x^3}{3!} + \cdots$$

LOGARITHMIC EXPANSION

$$\ln(1 + x) = x - \tfrac{1}{2}x^2 + \tfrac{1}{3}x^3 - \cdots \qquad (|x| < 1)$$

TRIGONOMETRIC EXPANSIONS
(θ in radians)

$$\sin \theta = \theta - \frac{\theta^3}{3!} + \frac{\theta^5}{5!} - \cdots$$

$$\cos \theta = 1 - \frac{\theta^2}{2!} + \frac{\theta^4}{4!} - \cdots$$

$$\tan \theta = \theta + \frac{\theta^3}{3} + \frac{2\theta^5}{15} + \cdots$$

CRAMER'S RULE

Two simultaneous equations in unknowns x and y,

$$a_1 x + b_1 y = c_1 \qquad \text{and} \qquad a_2 x + b_2 y = c_2,$$

have the solutions

$$x = \frac{\begin{vmatrix} c_1 & b_1 \\ c_2 & b_2 \end{vmatrix}}{\begin{vmatrix} a_1 & b_1 \\ a_2 & b_2 \end{vmatrix}} = \frac{c_1 b_2 - c_2 b_1}{a_1 b_2 - a_2 b_1}$$

and

$$y = \frac{\begin{vmatrix} a_1 & c_1 \\ a_2 & c_2 \end{vmatrix}}{\begin{vmatrix} a_1 & b_1 \\ a_2 & b_2 \end{vmatrix}} = \frac{a_1 c_2 - a_2 c_1}{a_1 b_2 - a_2 b_1}.$$

PRODUCTS OF VECTORS

Let $\mathbf{i}$, $\mathbf{j}$, and $\mathbf{k}$ be unit vectors in the x, y, and z directions. Then

$$\mathbf{i} \cdot \mathbf{i} = \mathbf{j} \cdot \mathbf{j} = \mathbf{k} \cdot \mathbf{k} = 1, \qquad \mathbf{i} \cdot \mathbf{j} = \mathbf{j} \cdot \mathbf{k} = \mathbf{k} \cdot \mathbf{i} = 0,$$

$$\mathbf{i} \times \mathbf{i} = \mathbf{j} \times \mathbf{j} = \mathbf{k} \times \mathbf{k} = 0,$$

$$\mathbf{i} \times \mathbf{j} = \mathbf{k}, \qquad \mathbf{j} \times \mathbf{k} = \mathbf{i}, \qquad \mathbf{k} \times \mathbf{i} = \mathbf{j},$$

Any vector $\mathbf{a}$ with components a_x, a_y, and a_z along the x, y, and z axes can be written

$$\mathbf{a} = a_x \mathbf{i} + a_y \mathbf{j} + a_z \mathbf{k}.$$

Let $\mathbf{a}$, $\mathbf{b}$, and $\mathbf{c}$ be arbitrary vectors with magnitudes a, b, and c. Then

$$\mathbf{a} \times (\mathbf{b} + \mathbf{c}) = (\mathbf{a} \times \mathbf{b}) + (\mathbf{a} \times \mathbf{c})$$

$$(s\mathbf{a}) \times \mathbf{b} = \mathbf{a} \times (s\mathbf{b}) = s(\mathbf{a} \times \mathbf{b}) \qquad (s = \text{a scalar}).$$

Let θ be the smaller of the two angles between $\mathbf{a}$ and $\mathbf{b}$. Then

$$\mathbf{a} \cdot \mathbf{b} = \mathbf{b} \cdot \mathbf{a} = a_x b_x + a_y b_y + a_z b_z = ab \cos \theta$$

$$\mathbf{a} \times \mathbf{b} = -\mathbf{b} \times \mathbf{a} = \begin{vmatrix} \mathbf{i} & \mathbf{j} & \mathbf{k} \\ a_x & a_y & a_z \\ b_x & b_y & b_z \end{vmatrix}$$

$$= \mathbf{i} \begin{vmatrix} a_y & a_z \\ b_y & b_z \end{vmatrix} - \mathbf{j} \begin{vmatrix} a_x & a_z \\ b_x & b_z \end{vmatrix} + \mathbf{k} \begin{vmatrix} a_x & a_y \\ b_x & b_y \end{vmatrix}$$

$$= (a_y b_z - b_y a_z)\mathbf{i}$$

$$+ (a_z b_x - b_z a_x)\mathbf{j} + (a_x b_y - b_x a_y)\mathbf{k}$$

$$|\mathbf{a} \times \mathbf{b}| = ab \sin \theta$$

$$\mathbf{a} \cdot (\mathbf{b} \times \mathbf{c}) = \mathbf{b} \cdot (\mathbf{c} \times \mathbf{a}) = \mathbf{c} \cdot (\mathbf{a} \times \mathbf{b})$$

$$\mathbf{a} \times (\mathbf{b} \times \mathbf{c}) = (\mathbf{a} \cdot \mathbf{c})\mathbf{b} - (\mathbf{a} \cdot \mathbf{b})\mathbf{c}$$

DERIVATIVES AND INTEGRALS

In what follows, the letters u and v stand for any functions of x, and a and m are constants. To each of the indefinite integrals should be added an arbitrary constant of integration. The *Handbook of Chemistry and Physics* (CRC Press Inc.) gives a more extensive tabulation.

1. $\dfrac{dx}{dx} = 1$

2. $\dfrac{d}{dx}(au) = a\dfrac{du}{dx}$

3. $\dfrac{d}{dx}(u + v) = \dfrac{du}{dx} + \dfrac{dv}{dx}$

4. $\dfrac{d}{dx}x^m = mx^{m-1}$

5. $\dfrac{d}{dx}\ln x = \dfrac{1}{x}$

6. $\dfrac{d}{dx}(uv) = u\dfrac{dv}{dx} + v\dfrac{du}{dx}$

7. $\dfrac{d}{dx}e^x = e^x$

8. $\dfrac{d}{dx}\sin x = \cos x$

9. $\dfrac{d}{dx}\cos x = -\sin x$

10. $\dfrac{d}{dx}\tan x = \sec^2 x$

11. $\dfrac{d}{dx}\cot x = -\csc^2 x$

12. $\dfrac{d}{dx}\sec x = \tan x \sec x$

13. $\dfrac{d}{dx}\csc x = -\cot x \csc x$

14. $\dfrac{d}{dx}e^u = e^u\dfrac{du}{dx}$

15. $\dfrac{d}{dx}\sin u = \cos u\dfrac{du}{dx}$

16. $\dfrac{d}{dx}\cos u = -\sin u\dfrac{du}{dx}$

1. $\displaystyle\int dx = x$

2. $\displaystyle\int au\, dx = a\int u\, dx$

3. $\displaystyle\int (u + v)\, dx = \int u\, dx + \int v\, dx$

4. $\displaystyle\int x^m\, dx = \dfrac{x^{m+1}}{m+1} \quad (m \neq -1)$

5. $\displaystyle\int \dfrac{dx}{x} = \ln |x|$

6. $\displaystyle\int u\dfrac{dv}{dx}\, dx = uv - \int v\dfrac{du}{dx}\, dx$

7. $\displaystyle\int e^x\, dx = e^x$

8. $\displaystyle\int \sin x\, dx = -\cos x$

9. $\displaystyle\int \cos x\, dx = \sin x$

10. $\displaystyle\int \tan x\, dx = \ln |\sec x|$

11. $\displaystyle\int \sin^2 x\, dx = \tfrac{1}{2}x - \tfrac{1}{4}\sin 2x$

12. $\displaystyle\int e^{-ax}\, dx = -\dfrac{1}{a}e^{-ax}$

13. $\displaystyle\int xe^{-ax}\, dx = -\dfrac{1}{a^2}(ax + 1)e^{-ax}$

14. $\displaystyle\int x^2 e^{-ax}\, dx = -\dfrac{1}{a^3}(a^2x^2 + 2ax + 2)e^{-ax}$

15. $\displaystyle\int_0^\infty x^n e^{-ax}\, dx = \dfrac{n!}{a^{n+1}}$

16. $\displaystyle\int_0^\infty x^{2n} e^{-ax^2}\, dx = \dfrac{1 \cdot 3 \cdot 5 \cdots (2n-1)}{2^{n+1}a^n}\sqrt{\dfrac{\pi}{a}}$

17. $\displaystyle\int \dfrac{dx}{\sqrt{x^2 + a^2}} = \ln(x + \sqrt{x^2 + a^2})$

18. $\displaystyle\int \dfrac{x\, dx}{(x^2 + a^2)^{3/2}} = -\dfrac{1}{(x^2 + a^2)^{1/2}}$

19. $\displaystyle\int \dfrac{dx}{(x^2 + a^2)^{3/2}} = \dfrac{x}{a^2(x^2 + a^2)^{1/2}}$

Appendix F
Properties of the Elements

All physical properties are for a pressure of 1 atm unless otherwise specified.

ELEMENT	SYMBOL	ATOMIC NUMBER, Z	MOLAR MASS, g/mol	DENSITY, g/cm³ AT 20°C	MELTING POINT, °C	BOILING POINT, °C	SPECIFIC HEAT, J/(g·°C) AT 25°C
Actinium	Ac	89	(227)	10.06	1323	(3473)	0.092
Aluminum	Al	13	26.9815	2.699	660	2450	0.900
Americium	Am	95	(243)	13.67	1541	—	—
Antimony	Sb	51	121.75	6.691	630.5	1380	0.205
Argon	Ar	18	39.948	1.6626×10^{-3}	−189.4	−185.8	0.523
Arsenic	As	33	74.9216	5.78	817 (28 atm)	613	0.331
Astatine	At	85	(210)	—	(302)	—	—
Barium	Ba	56	137.34	3.594	729	1640	0.205
Berkelium	Bk	97	(247)	14.79	—	—	—
Beryllium	Be	4	9.0122	1.848	1287	2770	1.83
Bismuth	Bi	83	208.980	9.747	271.37	1560	0.122
Boron	B	5	10.811	2.34	2030	—	1.11
Bromine	Br	35	79.909	3.12 (liquid)	−7.2	58	0.293
Cadmium	Cd	48	112.40	8.65	321.03	765	0.226
Calcium	Ca	20	40.08	1.55	838	1440	0.624
Californium	Cf	98	(251)	—	—	—	—
Carbon	C	6	12.01115	2.26	3727	4830	0.691
Cerium	Ce	58	140.12	6.768	804	3470	0.188
Cesium	Cs	55	132.905	1.873	28.40	690	0.243
Chlorine	Cl	17	35.453	3.214×10^{-3} (0°C)	−101	−34.7	0.486
Chromium	Cr	24	51.996	7.19	1857	2665	0.448
Cobalt	Co	27	58.9332	8.85	1495	2900	0.423
Copper	Cu	29	63.54	8.96	1083.40	2595	0.385
Curium	Cm	96	(247)	13.3	—	—	—
Dysprosium	Dy	66	162.50	8.55	1409	2330	0.172
Einsteinium	Es	99	(254)	—	—	—	—
Erbium	Er	68	167.26	9.15	1522	2630	0.167
Europium	Eu	63	151.96	5.243	817	1490	0.163
Fermium	Fm	100	(237)	—	—	—	—
Fluorine	F	9	18.9984	1.696×10^{-3} (0°C)	−219.6	−188.2	0.753
Francium	Fr	87	(223)	—	(27)	—	—
Gadolinium	Gd	64	157.25	7.90	1312	2730	0.234
Gallium	Ga	31	69.72	5.907	29.75	2237	0.377
Germanium	Ge	32	72.59	5.323	937.25	2830	0.322
Gold	Au	79	196.967	19.32	1064.43	2970	0.131
Hafnium	Hf	72	178.49	13.31	2227	5400	0.144
Hahnium	Ha	105	—	—	—	—	—
Hassium	Hs	108	—	—	—	—	—

continued on next page

ELEMENT	SYMBOL	ATOMIC NUMBER, Z	MOLAR MASS, g/mol	DENSITY, g/cm³ AT 20°C	MELTING POINT, °C	BOILING POINT, °C	SPECIFIC HEAT, J/(g·°C) AT 25°C
Helium	He	2	4.0026	0.1664×10^{-3}	−269.7	−268.9	5.23
Holmium	Ho	67	164.930	8.79	1470	2330	0.165
Hydrogen	H	1	1.00797	0.08375×10^{-3}	−259.19	−252.7	14.4
Indium	In	49	114.82	7.31	156.634	2000	0.233
Iodine	I	53	126.9044	4.93	113.7	183	0.218
Iridium	Ir	77	192.2	22.5	2447	(5300)	0.130
Iron	Fe	26	55.847	7.874	1536.5	3000	0.447
Krypton	Kr	36	83.80	3.488×10^{-3}	−157.37	−152	0.247
Lanthanum	La	57	138.91	6.189	920	3470	0.195
Lawrencium	Lr	103	(257)		—	—	
Lead	Pb	82	207.19	11.35	327.45	1725	0.129
Lithium	Li	3	6.939	0.534	180.55	1300	3.58
Lutetium	Lu	71	174.97	9.849	1663	1930	0.155
Magnesium	Mg	12	24.312	1.738	650	1107	1.03
Manganese	Mn	25	54.9380	7.44	1244	2150	0.481
Meitnerium	Mt	109	—	—	—	—	—
Mendelevium	Md	101	(256)	—	—	—	—
Mercury	Hg	80	200.59	13.55	−38.87	357	0.138
Molybdenum	Mo	42	95.94	10.22	2617	5560	0.251
Neodymium	Nd	60	144.24	7.007	1016	3180	0.188
Neon	Ne	10	20.183	0.8387×10^{-3}	−248.597	−246.0	1.03
Neptunium	Np	93	(237)	20.25	637	—	1.26
Nickel	Ni	28	58.71	8.902	1453	2730	0.444
Nielsbohrium	Ns	107	—	—	—	—	—
Niobium	Nb	41	92.906	8.57	2468	4927	0.264
Nitrogen	N	7	14.0067	1.1649×10^{-3}	−210	−195.8	1.03
Nobelium	No	102	(255)	—	—	—	—
Osmium	Os	76	190.2	22.59	3027	5500	0.130
Oxygen	O	8	15.9994	1.3318×10^{-3}	−218.80	−183.0	0.913
Palladium	Pd	46	106.4	12.02	1552	3980	0.243
Phosphorus	P	15	30.9738	1.83	44.25	280	0.741
Platinum	Pt	78	195.09	21.45	1769	4530	0.134
Plutonium	Pu	94	(244)	19.8	640	3235	0.130
Polonium	Po	84	(210)	9.32	254	—	—
Potassium	K	19	39.102	0.862	63.20	760	0.758
Praseodymium	Pr	59	140.907	6.773	931	3020	0.197
Promethium	Pm	61	(145)	7.22	(1027)	—	—
Protactinium	Pa	91	(231)	15.37 (estimated)	(1230)	—	—
Radium	Ra	88	(226)	5.0	700	—	—
Radon	Rn	86	(222)	9.96×10^{-3} (0°C)	(−71)	−61.8	0.092
Rhenium	Re	75	186.2	21.02	3180	5900	0.134
Rhodium	Rh	45	102.905	12.41	1963	4500	0.243
Rubidium	Rb	37	85.47	1.532	39.49	688	0.364
Ruthenium	Ru	44	101.107	12.37	2250	4900	0.239
Rutherfordium	Rf	104	—	—	—	—	—
Samarium	Sm	62	150.35	7.52	1072	1630	0.197

continued on next page

ELEMENT	SYMBOL	ATOMIC NUMBER, Z	MOLAR MASS, g/mol	DENSITY, g/cm³ AT 20°C	MELTING POINT, °C	BOILING POINT, °C	SPECIFIC HEAT, J/(g·°C) AT 25°C
Scandium	Sc	21	44.956	2.99	1539	2730	0.569
Seaborgium	Sg	106	—	—	—	—	—
Selenium	Se	34	78.96	4.79	221	685	0.318
Silicon	Si	14	28.086	2.33	1412	2680	0.712
Silver	Ag	47	107.870	10.49	960.8	2210	0.234
Sodium	Na	11	22.9898	0.9712	97.85	892	1.23
Strontium	Sr	38	87.62	2.54	768	1380	0.737
Sulfur	S	16	32.064	2.07	119.0	444.6	0.707
Tantalum	Ta	73	180.948	16.6	3014	5425	0.138
Technetium	Tc	43	(99)	11.46	2200	—	0.209
Tellurium	Te	52	127.60	6.24	449.5	990	0.201
Terbium	Tb	65	158.924	8.229	1357	2530	0.180
Thallium	Tl	81	204.37	11.85	304	1457	0.130
Thorium	Th	90	(232)	11.72	1755	(3850)	0.117
Thulium	Tm	69	168.934	9.32	1545	1720	0.159
Tin	Sn	50	118.69	7.2984	231.868	2270	0.226
Titanium	Ti	22	47.90	4.54	1670	3260	0.523
Tungsten	W	74	183.85	19.3	3380	5930	0.134
Uranium	U	92	(238)	18.95	1132	3818	0.117
Vanadium	V	23	50.942	6.11	1902	3400	0.490
Xenon	Xe	54	131.30	5.495×10^{-3}	-111.79	-108	0.159
Ytterbium	Yb	70	173.04	6.965	824	1530	0.155
Yttrium	Y	39	88.905	4.469	1526	3030	0.297
Zinc	Zn	30	65.37	7.133	419.58	906	0.389
Zirconium	Zr	40	91.22	6.506	1852	3580	0.276

The values in parentheses in the column of molar masses are the mass numbers of the longest-lived isotopes of those elements that are radioactive. Melting points and boiling points in parentheses are uncertain.

The data for gases are valid only when these are in their usual molecular state, such as H_2, He, O_2, Ne, etc. The specific heats of the gases are the values at constant pressure.

Source: Adapted from Wehr, Richards, Adair, *Physics of the Atom,* 4th ed., Addison-Wesley, R ding, MA, 1984, and from J. Emsley, *The Elements,* 2nd ed., Clarendon Press, Oxford, 1991.

Appendix G
Periodic Table of the Elements

Metals

Metalloids

Nonmetals

Alkali metals
IA

Noble gases
0

1	IIA												IIIA	IVA	VA	VIA	VIIA	2
H																		**He**

Transition metals

VIIIB

THE HORIZONTAL PERIODS

Period																		
1	1 H	IIA											IIIA	IVA	VA	VIA	VIIA	2 He
2	3 Li	4 Be											5 B	6 C	7 N	8 O	9 F	10 Ne
3	11 Na	12 Mg	IIIB	IVB	VB	VIB	VIIB		VIIIB		IB	IIB	13 Al	14 Si	15 P	16 S	17 Cl	18 Ar
4	19 K	20 Ca	21 Sc	22 Ti	23 V	24 Cr	25 Mn	26 Fe	27 Co	28 Ni	29 Cu	30 Zn	31 Ga	32 Ge	33 As	34 Se	35 Br	36 Kr
5	37 Rb	38 Sr	39 Y ·	40 Zr	41 Nb	42 Mo	43 Tc	44 Ru	45 Rh	46 Pd	47 Ag	48 Cd	49 In	50 Sn	51 Sb	52 Te	53 I	54 Xe
6	55 Cs	56 Ba	57-71 *	72 Hf	73 Ta	74 W	75 Re	76 Os	77 Ir	78 Pt	79 Au	80 Hg	81 Tl	82 Pb	83 Bi	84 Po	85 At	86 Rn
7	87 Fr	88 Ra	89-103 †	104 Rf	105 Ha	106 Sg	107 Ns	108 Hs	109 Mt	110	111	112						

Inner transition metals

Lanthanide series *	57 La	58 Ce	59 Pr	60 Nd	61 Pm	62 Sm	63 Eu	64 Gd	65 Tb	66 Dy	67 Ho	68 Er	69 Tm	70 Yb	71 Lu
Actinide series †	89 Ac	90 Th	91 Pa	92 U	93 Np	94 Pu	95 Am	96 Cm	97 Bk	98 Cf	99 Es	100 Fm	101 Md	102 No	103 Lr

The names for elements 104–109 (Rutherfordium, Hahnium, Seaborgium, Nielsbohrium, Hassium, and Meitnerium, respectively) are those recommended by the American Chemical Society Nomenclature Committee. As of 1996, the names and symbols for elements 104–108 have not yet been approved by the appropriate international body. Elements 110, 111 and 112 have been discovered but, as of 1996, have not been provisionally named.

Appendix H
Winners of the Nobel Prize in Physics*

1901 Wilhelm Konrad Röntgen *(1845–1923)* for the discovery of x rays

1902 Hendrik Antoon Lorentz *(1853–1928)* and Pieter Zeeman *(1865–1943)* for their researches into the influence of magnetism upon radiation phenomena

1903 Antoine Henri Becquerel *(1852–1908)* for his discovery of spontaneous radioactivity

Pierre Curie *(1859–1906)* and Marie Sklowdowska-Curie *(1867–1934)* for their joint researches on the radiation phenomena discovered by Becquerel

1904 Lord Rayleigh (John William Strutt) *(1842–1919)* for his investigations of the densities of the most important gases and for his discovery of argon

1905 Philipp Eduard Anton von Lenard *(1862–1947)* for his work on cathode rays

1906 Joseph John Thomson *(1856–1940)* for his theoretical and experimental investigations on the conduction of electricity by gases

1907 Albert Abraham Michelson *(1852–1931)* for his optical precision instruments and metrological investigations carried out with their aid

1908 Gabriel Lippmann *(1845–1921)* for his method of reproducing colors photographically based on the phenomena of interference

1909 Guglielmo Marconi *(1874–1937)* and Carl Ferdinand Braun *(1850–1918)* for their contributions to the development of wireless telegraphy

1910 Johannes Diderik van der Waals *(1837–1932)* for his work on the equation of state for gases and liquids

1911 Wilhelm Wien *(1864–1928)* for his discoveries regarding the laws governing the radiation of heat

1912 Nils Gustaf Dalén *(1869–1937)* for his invention of automatic regulators for use in conjunction with gas accumulators for illuminating lighthouses and buoys

1913 Heike Kamerlingh Onnes *(1853–1926)* for his investigations of the properties of matter at low temperatures which led, among other things, to the production of liquid helium

1914 Max von Laue *(1879–1960)* for his discovery of the diffraction of Röntgen rays by crystals

1915 William Henry Bragg *(1862–1942)* and William Lawrence Bragg *(1890–1971)* for their services in the analysis of crystal structure by means of x rays

1917 Charles Glover Barkla *(1877–1944)* for his discovery of the characteristic x rays of the elements

1918 Max Planck *(1858–1947)* for his discovery of energy quanta

1919 Johannes Stark *(1874–1957)* for his discovery of the Doppler effect in canal rays and the splitting of spectral lines in electric fields

1920 Charles-Édouard Guillaume *(1861–1938)* for the service he rendered to precision measurements in physics by his discovery of anomalies in nickel steel alloys

1921 Albert Einstein *(1879–1955)* for his services to theoretical physics, and especially for his discovery of the law of the photoelectric effect

1922 Niels Bohr *(1885–1962)* for the investigation of the structure of atoms, and of the radiation emanating from them

1923 Robert Andrews Millikan *(1868–1953)* for his work on the elementary charge of electricity and on the photoelectric effect

1924 Karl Manne Georg Siegbahn *(1888–1979)* for his discoveries and research in the field of x-ray spectroscopy

1925 James Franck *(1882–1964)* and Gustav Hertz *(1887–1975)* for their discovery of the laws governing the impact of an electron upon an atom

1926 Jean Baptiste Perrin *(1870–1942)* for his work on the discontinuous structure of matter, and especially for his discovery of sedimentation equilibrium

1927 Arthur Holly Compton *(1892–1962)* for his discovery of the effect named after him

Charles Thomson Rees Wilson (1869–1959) for his method of making the paths of electrically charged particles visible by condensation of vapor

1928 Owen Willans Richardson *(1879–1959)* for his work on the thermionic phenomenon and especially for the discovery of the law named after him

1929 Prince Louis Victor de Broglie *(1892–1987)* for his discovery of the wave nature of electrons

*See *Nobel Lectures, Physics,* 1901–1970, Elsevier Publishing Company, for biographies of the awardees and for lectures given by them on receiving the prize.

1930 Sir Chandrasekhara Venkata Raman *(1888–1970)* for his work on the scattering of light and for the discovery of the effect named after him

1932 Werner Heisenberg *(1901–1976)* for the creation of quantum mechanics, the application of which has, among other things, led to the discovery of the allotropic forms of hydrogen

1933 Erwin Schrödinger *(1887–1961)* and Paul Adrien Maurice Dirac *(1902–1984)* for the discovery of new productive forms of atomic theory

1935 James Chadwick *(1891–1974)* for his discovery of the neutron

1936 Victor Franz Hess *(1883–1964)* for the discovery of cosmic radiation

Carl David Anderson *(1905–1991)* for his discovery of the positron

1937 Clinton Joseph Davisson *(1881–1958)* and George Paget Thomson *(1892–1975)* for their experimental discovery of the diffraction of electrons by crystals

1938 Enrico Fermi *(1901–1954)* for his demonstrations of the existence of new radioactive elements produced by neutron irradiation, and for his related discovery of nuclear reactions brought about by slow neutrons

1939 Ernest Orlando Lawrence *(1901–1958)* for the invention and development of the cyclotron and for results obtained with it, especially for artificial radioactive elements

1943 Otto Stern *(1888–1969)* for his contribution to the development of the molecular-ray method and his discovery of the magnetic moment of the proton

1944 Isidor Isaac Rabi *(1898–1988)* for his resonance method for recording the magnetic properties of atomic nuclei

1945 Wolfgang Pauli *(1900–1958)* for the discovery of the Exclusion Principle (also called Pauli Principle)

1946 Percy Williams Bridgman *(1882–1961)* for the invention of an apparatus to produce extremely high pressures and for the discoveries he made therewith in the field of high-pressure physics

1947 Sir Edward Victor Appleton *(1892–1965)* for his investigations of the physics of the upper atmosphere, especially for the discovery of the so-called Appleton layer

1948 Patrick Maynard Stuart Blackett *(1897–1974)* for his development of the Wilson cloud-chamber method, and his discoveries therewith in nuclear physics and cosmic radiation

1949 Hideki Yukawa *(1907–1981)* for his prediction of the existence of mesons on the basis of theoretical work on nuclear forces

1950 Cecil Frank Powell *(1903–1969)* for his development of the photographic method of studying nuclear processes and his discoveries regarding mesons made with this method

1951 Sir John Douglas Cockcroft *(1897–1967)* and Ernest Thomas Sinton Walton *(1903–   )* for their pioneer work on the transmutation of atomic nuclei by artificially accelerated atomic particles

1952 Felix Bloch *(1905–1983)* and Edward Mills Purcell *(1912–   )* for their development of new nuclear-magnetic precision methods and discoveries in connection therewith

1953 Frits Zernike *(1888–1966)* for his demonstration of the phase-contrast method, especially for his invention of the phase-contrast microscope

1954 Max Born *(1882–1970)* for his fundamental research in quantum mechanics, especially for his statistical interpretation of the wave function

Walther Bothe *(1891–1957)* for the coincidence method and his discoveries made therewith

1955 Willis Eugene Lamb *(1913–   )* for his discoveries concerning the fine structure of the hydrogen spectrum

Polykarp Kusch *(1911–1993)* for his precision determination of the magnetic moment of the electron

1956 William Shockley *(1910–1989)*, John Bardeen *(1908–1991)* and Walter Houser Brattain *(1902–1987)* for their researches on semiconductors and their discovery of the transistor effect

1957 Chen Ning Yang *(1922–   )* and Tsung Dao Lee *(1926–   )* for their penetrating investigation of the parity laws which has led to important discoveries regarding the elementary particles

1958 Pavel Aleksejevič Čerenkov *(1904–   )*, Il' ja Michajlovič Frank *(1908–1990)* and Igor' Evgen' evič Tamm *(1895–1971)* for the discovery and interpretation of the Cerenkov effect

1959 Emilio Gino Segrè *(1905–1989)* and Owen Chamberlain *(1920–   )* for their discovery of the antiproton

1960 Donald Arthur Glaser *(1926–   )* for the invention of the bubble chamber

1961 Robert Hofstadter *(1915–1990)* for his pioneering studies of electron scattering in atomic nuclei and for his thereby achieved discoveries concerning the structure of the nucleons

Rudolf Ludwig Mössbauer *(1929–   )* for his researches concerning the resonance absorption of γ rays and his discovery in this connection of the effect which bears his name

1962 Lev Davidovič Landau *(1908–1968)* for his pioneering theories of condensed matter, especially liquid helium

1963 Eugene P. Wigner *(1902–1995)* for his contributions to the theory of the atomic nucleus and the elementary particles, particularly through the discovery and application of fundamental symmetry principles

Maria Goeppert Mayer *(1906–1972)* and J. Hans D. Jensen *(1907–1973)* for their discoveries concerning nuclear shell structure

1964 Charles H. Townes *(1915–   )*, Nikolai G. Basov *(1922–   )* and Alexander M. Prochorov *(1916–   )* for fundamental work in the field of quantum electronics which has led to the construction of oscillators and amplifiers based on the maser–laser principle

1965 Sin-itiro Tomonaga *(1906–1979)*, Julian Schwinger *(1918–1994)* and Richard P. Feynman *(1918–1988)* for their fundamental work in quantum electrodynamics, with deep-ploughing consequences for the physics of elementary particles

1966 Alfred Kastler *(1902–1984)* for the discovery and development of optical methods for studying Hertzian resonance in atoms

1967 Hans Albrecht Bethe *(1906–   )* for his contributions to the theory of nuclear reactions, especially his discoveries concerning the energy production in stars

1968 Luis W. Alvarez *(1911–1988)* for his decisive contribution to elementary particle physics, in particular the discovery of a large number of resonance states, made possible through his development of the techniques of using the hydrogen bubble chamber and its data analysis

1969 Murray Gell-Mann *(1929–   )* for his contributions and discoveries concerning the classification of elementary particles and their interactions

1970 Hannes Alfvén *(1908–1995)* for fundamental work and discoveries in magneto-hydrodynamics with fruitful applications in different parts of plasma physics

Louis Néel *(1904–   )* for fundamental work and discoveries concerning antiferromagnetism and ferrimagnetism which have led to important applications in solid state physics

1971 Dennis Gabor *(1900–1979)* for his discovery of the principles of holography

1972 John Bardeen *(1908–1991)*, Leon N. Cooper *(1930–   )* and J. Robert Schrieffer *(1931–   )* for their development of a theory of superconductivity

1973 Leo Esaki *(1925–   )* for his discovery of tunneling in semiconductors

Ivar Giaever *(1929–   )* for his discovery of tunneling in superconductors

Brian D. Josephson *(1940–   )* for his theoretical prediction of the properties of a supercurrent through a tunnel barrier

1974 Antony Hewish *(1924–   )* for the discovery of pulsars

Sir Martin Ryle *(1918–1984)* for his pioneering work in radioastronomy

1975 Aage Bohr *(1922–   )*, Ben Mottelson *(1926–   )* and James Rainwater *(1917–1986)* for the discovery of the connection between collective motion and particle motion and the development of the theory of the structure of the atomic nucleus based on this connection

1976 Burton Richter *(1931–   )* and Samuel Chao Chung Ting *(1936–   )* for their (independent) discovery of an important fundamental particle

1977 Philip Warren Anderson *(1923–   )*, Nevill Francis Mott *(1905–   )* and John Hasbrouck Van Vleck *(1899–1980)* for their fundamental theoretical investigations of the electronic structure of magnetic and disordered systems

1978 Peter L. Kapitza *(1894–1984)* for his basic inventions and discoveries in low-temperature physics

Arno A. Penzias *(1933–   )* and Robert Woodrow Wilson *(1936–   )* for their discovery of cosmic microwave background radiation

1979 Sheldon Lee Glashow *(1932–   )*, Abdus Salam *(1926–   )*, and Steven Weinberg *(1933–   )* for their unified model of the action of the weak and electromagnetic forces and for their prediction of the existence of neutral currents

1980 James W. Cronin *(1931–   )* and Val L. Fitch *(1923–   )* for the discovery of violations of fundamental symmetry principles in the decay of neutral K mesons

1981 Nicolaas Bloembergen *(1920–   )* and Arthur Leonard Schawlow *(1921–   )* for their contribution to the development of laser spectroscopy

Kai M. Siegbahn *(1918–   )* for his contribution to high-resolution electron spectroscopy

1982 Kenneth Geddes Wilson *(1936–   )* for his method of analyzing the critical phenomena inherent in the changes of matter under the influence of pressure and temperature

1983 Subrehmanyan Chandrasekhar *(1910–1995)* for his theoretical studies of the structure and evolution of stars

William A. Fowler *(1911–1995)* for his studies of the formation of the chemical elements in the universe

1984 Carlo Rubbia *(1934–   )* and Simon van der Meer *(1925–   )* for their decisive contributions to the Large Project, which led to the discovery of the field particles W and Z, communicators of the weak interaction

1985 Klaus von Klitzing *(1943–   )* for his discovery of the quantized Hall resistance

1986 Ernst Ruska *(1906–1988)* for his invention of the electron microscope

Gerd Binnig *(1947–   )*, Heinrich Rohrer *(1933–   )* for their invention of the scanning tunneling microscope

1987 Karl Alex Müller *(1927–   )* and J. George Bednorz *(1950–   )* for their discovery of a new class of superconductors

1988 Leon M. Lederman *(1922–   )*, Melvin Schwartz *(1932–   )* and Jack Steinberger *(1921–   )* for the first use of a neutrino beam and the discovery of the muon neutrino

1989 Norman Ramsey *(1915–   )*, Hans Dehmelt *(1922–   )* and Wolfgang Paul *(1913–1993)* for their work that led to the development of atomic clocks and precision timing

1990 Jerome I. Friedman *(1930–   )*, Henry W. Kendall *(1926–   )* and Richard E. Taylor *(1929–   )* for demonstrating that protons and neutrons consist of quarks

1991 Pierre de Gennes *(1932–   )* for studies of order phenomena, such as in liquid crystals and polymers

1992 George Charpak *(1924–   )* for his invention of fast electronic detectors for high-energy particles

1993 Joseph H. Taylor *(1941–)* and Russell A. Hulse *(1950–)* for the discovery and interpretation of the first binary pulsar.

1994 Bertram N. Brockhouse *(1918–)* and Clifford G. Shull *(1915–)* for the development of neutron scattering techniques

1995 Martin L. Perl *(1927–)* for the discovery of the tau lepton

Frederick Reines *(1918–)* for the detection of the neutrino

Answers to Checkpoints, Odd-Numbered Questions, Exercises, and Problems

Chapter 1

EP **3.** (a) 186 mi; (b) 3.0×10^8 mm **5.** (a) 10^9; (b) 10^{-4}; (c) 9.1×10^5 **7.** 32.2 km **9.** 0.020 km³ **11.** (a) 250 ft²; (b) 23.3 m²; (c) 3060 ft³; (d) 86.6 m³ **13.** 8×10^2 km **15.** (a) 11.3 m²/L; (b) 1.13×10^4 m⁻¹; (c) 2.17×10^{-3} gal/ft² **17.** (a) $d_{Sun}/d_{Moon} = 400$; (b) $V_{Sun}/V_{Moon} = 6.4 \times 10^7$; (c) 3.5×10^3 km **19.** (a) 0.98 ft/ns; (b) 0.30 mm/ps **21.** 3.156×10^7 s **23.** 5.79×10^{12} days **25.** (a) 0.013; (b) 0.54; (c) 10.3; (d) 31 m/s **27.** 15° **29.** 3.3 ft **31.** 2 days 5 hours **33.** (a) 2.99×10^{-26} kg; (b) 4.68×10^{46} **35.** 1.3×10^9 kg **37.** (a) 10^3 kg/m³; (b) 158 kg/s **39.** (a) 1.18×10^{-29} m³; (b) 0.282 nm

Chapter 2

CP **1.** b and c **2.** zero **3.** (a) 1 and 4; (b) 2 and 3; (c) 3 **4.** (a) plus; (b) minus; (c) minus; (d) plus **5.** 1 and 4 **6.** (a) plus; (b) minus; (c) $a = -g = -9.8$ m/s² **Q** **1.** (a) yes; (b) no; (c) yes; (d) yes **3.** (a) 2, 3; (b) 1, 3; (c) 4 **5.** all tie (see Eq. 2-16) **7.** (a) $-g$; (b) 2 m/s upward **9.** same **11.** $x = t^2$ and $x = 8(t - 2) + (1.5)(t - 2)^2$ **13.** increase **EP** **1.** (a) Lewis: 10.0 m/s, Rodgers: 5.41 m/s; (b) 1 h 10 min **3.** 309 ft **5.** 2 cm/y **7.** 6.71×10^8 mi/h, 9.84×10^8 ft/s, 1.00 ly/y **9.** (a) 5.7 ft/s; (b) 7.0 ft/s **11.** (a) 45 mi/h (72 km/h); (b) 43 mi/h (69 km/h); (c) 44 mi/h (71 km/h); (d) 0 **13.** (a) 28.5 cm/s; (b) 18.0 cm/s; (c) 40.5 cm/s; (d) 28.1 cm/s; (e) 30.3 cm/s **15.** (a) mathematically, an infinite number; (b) 60 km **17.** (a) 4 s > t > 2 s; (b) 3 s > t > 0; (c) 7 s > t > 3 s; (d) $t = 3$ s **19.** 100 m **23.** (a) The signs of v and a are: AB: +, −; BC: 0, 0; CD: +, +; DE: +, 0; (b) no; (c) no **25.** (e) situations (a), (b), and (d) **27.** (a) 80 m/s; (b) 110 m/s; (c) 20 m/s² **29.** (a) 1.10 m/s, 6.11 mm/s²; (b) 1.47 m/s, 6.11 mm/s² **31.** (a) 2.00 s; (b) 12 cm from left edge of screen; (c) 9.00 cm/s², to the left; (d) to the right; (e) to the left; (f) 3.46 s **33.** 0.556 s **35.** each, 0.28 m/s² **37.** 2.8 m/s² **39.** 1.62×10^{15} m/s² **41.** $21g$ **43.** (a) $25g$; (b) 400 m **45.** 90 m **47.** (a) 5.0 m/s²; (b) 4.0 s; (c) 6.0 s; (d) 90 m **49.** (a) 5.00 m/s; (b) 1.67 m/s²; (c) 7.50 m **51.** (a) 0.74 s; (b) −20 ft/s² **53.** (a) 0.75 s; (b) 50 m **55.** (a) 34.7 ft; (b) 41.6 s **57.** (a) 3.26 ft/s² **61.** (a) 31 m/s; (b) 6.4 s **63.** (a) 48.5 m/s; (b) 4.95 s; (c) 34.3 m/s; (d) 3.50 s **65.** (a) 5.44 m/s; (b) 53.3 m/s; (d) 5.80 m **67.** (a) 3.2 s; (b) 1.3 s **69.** 4.0 m/s **71.** (a) 350 ms; (b) 82 ms (each is for ascent and descent through the 15 cm) **73.** 857 m/s², upward **75.** (a) 1.23 cm; (b) 4 times, 9 times, 16 times, 25 times **77.** (a) 8.85 m/s; (b) 1.00 m **79.** 22 cm and 89 cm below the nozzle **81.** (a) 3.41 s; (b) 57 m **83.** (a) 40.0 ft/s **85.** 1.5 s **87.** (a) 5.4 s; (b) 41 m/s **89.** 20.4 m

91. (a) $d = v_i^2/2a' + T_R v_i$; (b) 9.0 m/s²; (c) 0.66 s. **93.** (a) $v_j^2 = 2a' d_0(j - 1) + v_1^2$; (c) 7.0 m/s²; (d) 14 m.

Chapter 3

CP **1.** (a) 7 m; (b) 1 m **2.** c, d, f **3.** (a) +, +; (b) +, −; (c) +, + **4.** (a) 90°; (b) 0 (vectors are parallel); (c) 180° (vectors are antiparallel) **5.** (a) 0° or 180°; (b) 90° **Q** **1.** **A** and **B** **3.** No, but **a** and $-\mathbf{b}$ are commutative: $\mathbf{a} + (-\mathbf{b}) = (-\mathbf{b}) + \mathbf{a}$. **5.** (a) **a** and **b** are parallel; (b) $\mathbf{b} = 0$; (c) **a** and **b** are perpendicular **7.** (a)–(c) yes (example: 5**i** and −2**i**) **9.** all but e **11.** (a) minus, minus; (b) minus, minus **13.** (a) **B** and **C**, **D** and **E**; (b) **D** and **E** **15.** no (their orientations can differ) **17.** (a) 0 (vectors are parallel); (b) 0 (vectors are antiparallel) **EP** **1.** The displacements should be (a) parallel, (b) antiparallel, (c) perpendicular **3.** (b) 3.2 km, 41° south of west **5.** **a** + **b**: 4.2, 40° east of north; **b** − **a**: 8.0, 24° north of west **7.** (a) 38 units at 320°; (b) 130 units at 1.2°; (c) 62 units at 130° **9.** $a_x = -2.5$, $a_y = -6.9$ **11.** $r_x = 13$ m, $r_y = 7.5$ m **13.** (a) 14 cm, 45° left of straight down; (b) 20 cm, vertically up; (c) zero **15.** 4.74 km **17.** 168 cm, 32.5° above the floor **19.** $r_x = 12$, $r_y = -5.8$, $r_z = -2.8$ **21.** (a) 8**i** + 2**j**, 8.2, 14°; (b) 2**i** − 6**j**, 6.3, −72° relative to **i** **23.** (a) 5.0, −37°; (b) 10, 53°; (c) 11, 27°; (d) 11, 80°; (e) 11, 260°; the angles are relative to +x, the last two vectors are in opposite directions **25.** 4.1 **27.** (a) $r_x = 1.59$, $r_y = 12.1$; (b) 12.2; (c) 82.5° **29.** 3390 ft, horizontally **31.** (a) −2.83 m, −2.83 m, +5.00 m, 0 m, 3.00 m, 5.20 m; (b) 5.17 m, 2.37 m; (c) 5.69 m, 24.6° north of east; (d) 5.69 m, 24.6° south of west **35.** (a) $a_x = 9.51$ m, $a_y = 14.1$ m; (b) $a_x' = 13.4$ m, $a_y' = 10.5$ m **37.** (a) +y; (b) −y; (c) 0; (d) 0; (e) +z; (f) −z; (g) ab, both; (h) ab/d, +z **39.** yes **41.** (a) up, unit magnitude; (b) zero; (c) south, unit magnitude; (d) 1.00; (e) 0 **43.** (a) −18.8; (b) 26.9, +z direction **45.** (a) 12, out of page; (b) 12, into page; (c) 12, out of page **47.** (a) 11**i** + 5**j** − 7**k**; (b) 120° **51.** (a) 57°; (b) $c_x = \pm 2.2$, $c_y = \mp 4.5$ **53.** (a) −21; (b) −9; (c) 5**i** − 11**j** − 9**k**

Chapter 4

CP **1.** (a) $(8\mathbf{i} - 6\mathbf{j})$ m; (b) yes, the xy plane **2.** (a) first; (b) third **3.** (1) and (3) a_x and a_y are both constant and thus **a** is constant; (2) and (4) a_y is constant but a_x is not, thus **a** is not **4.** 4 m/s³, −2 m/s, 3 m **5.** (a) v_x constant; (b) v_y initially positive, decreases to zero, and then becomes progressively more negative; (c) $a_x = 0$ throughout; (d) $a_y = -g$ throughout **6.** (a) −(4 m/s)**i**; (b) −(8 m/s²)**j** **7.** (1) 0, distance not changing; (2) +70 km/h, distance increasing; (3) +80 km/h, distance decreasing **Q** **1.** (1) and (3) a_y is constant but a_x is not and thus **a** is not; (2) a_x is constant but a_y

is not and thus **a** is not; (4) a_x and a_y are both constant and thus **a** is constant; -2 m/s², 3 m/s **3.** (a) highest point; (b) lowest point **5.** (a) all tie; (b) 1 and 2 tie (the rocket is shot upward), then 3 and 4 tie (it is shot into the ground!) **7.** $(2\mathbf{i} - 4\mathbf{j})$ m/s **9.** (a) all tie; (b) all tie; (c) c, b, a; (d) c, b, a **11.** (a) no; (b) same **13.** (a) in your hands; (b) behind you; (c) in front of you **15.** (a) straight down; (b) curved; (c) more curved **17.** (a) 3; (b) 4. **EP 1.** (a) $(-5.0\mathbf{i} + 8.0\mathbf{j})$ m; (b) 9.4 m, 122° from $+x$; (d) $(8\mathbf{i} - 8\mathbf{j})$ m; (e) 11 m, $-45°$ from $+x$ **3.** (a) $(-7.0\mathbf{i} + 12\mathbf{j})$ m; (b) xy plane **5.** (a) 671 mi, 63.4° south of east; (b) 298 mi/h, 63.4° south of east; (c) 400 mi/h **7.** (a) 6.79 km/h; (b) 6.96° **9.** (a) $(3\mathbf{i} - 8t\mathbf{j})$ m/s; (b) $(3\mathbf{i} - 16\mathbf{j})$ m/s; (c) 16 m/s, $-79°$ to $+x$ **11.** (a) $(8t\mathbf{j} + \mathbf{k})$ m/s; (b) $8\mathbf{j}$ m/s² **13.** $(-2.10\mathbf{i} + 2.81\mathbf{j})$ m/s² **15.** (a) $-1.5\mathbf{j}$ m/s; (b) $(4.5\mathbf{i} - 2.25\mathbf{j})$ m **17.** 60.0° **19.** (a) 63 ms; (b) 1.6×10^3 ft/s **21.** (a) 2.0 ns; (b) 2.0 mm; (c) $(1.0 \times 10^9 \mathbf{i} - 2.0 \times 10^8 \mathbf{j})$ cm/s **23.** (a) 3.03 s; (b) 758 m; (c) 29.7 m/s **25.** (a) 16 m/s, 23° above the horizontal; (b) 27 m/s, 57° below the horizontal **27.** (a) 32.4 m; (b) -37.7 m **29.** (b) 76° **31.** (a) 51.8 m; (b) 27.4 m/s; (c) 67.5 m **33.** (a) 194 m/s; (b) 38° **35.** 1.9 in. **37.** (a) 11 m; (b) 23 m; (c) 17 m/s, 63° below horizontal **41.** (a) 73 ft; (b) 7.6°; (c) 1.0 s **43.** 23 ft/s **45.** (a) 11 m; (b) 45 m/s **47.** 30 m above the release point **49.** 19 ft/s **51.** (a) 202 m/s; (b) 806 m; (c) 161 m/s, -171 m/s **53.** (a) 20 cm; (b) no, the ball hits the net only 4.4 cm above the ground **55.** yes; its center passes about 4.1 ft above the fence **57.** (a) 9.00×10^{22} m/s², toward the center; (b) 1.52×10^{-16} s **59.** (a) 6.7×10^6 m/s; (b) 1.4×10^{-7} s **61.** (a) 7.49 km/s; (b) 8.00 m/s² **63.** (a) 0.94 m; (b) 19 m/s; (c) 2400 m/s², toward center; (d) 0.05 s **65.** (a) 1.3×10^5 m/s; (b) 7.9×10^5 m/s² or $(8.0 \times 10^4)g$, toward the center; (c) both answers increase **67.** (a) 0.034 m/s²; (b) 84 min **69.** 2.58 cm/s² **71.** 160 m/s² **73.** 36 s, no **75.** 0.018 mi/s² from either frame **77.** 130° **79.** 60° **81.** (a) 5.8 m/s; (b) 16.7 m; (c) 67° **83.** 185 km/h, 22° south of west **85.** (a) from 75° east of south; (b) 30° east of north; substitute west for east to get second solution **87.** (a) 30° upstream; (b) 69 min; (c) 80 min; (d) 80 min; (e) perpendicular to the current, the shortest possible time is 60 min **89.** $0.83c$ **91.** (a) $0.35c$; (b) $0.62c$ **93.** For launch angles from 5° to 70°, it always moves away from the launch site. For a 75° launch angle, it moves toward the site from 11.5 s to 18.5 s after launch. For an 80° launch angle, it moves toward the site from 10.5 s to 20.5 s after launch. For an 85° launch angle, it moves toward the site from 10.5 s to 20.5 s after launch. For a 90° launch angle, it moves toward the site from 10 s to 20.5 s after launch. **95.** (a) 1.6 s; (b) no; (c) 14 m/s; (d) yes **97.** (a) $\Delta \mathbf{D} = (1.0$ m$)\mathbf{i} - (2.0$ m$)\mathbf{j} + (1.0$ m$)\mathbf{k}$; (b) 2.4 m; (c) $\bar{\mathbf{v}} = (0.025$ m/s$)\mathbf{i} - (0.050$ m/s$)\mathbf{j} + (0.025$ m/s$)\mathbf{k}$ (d) cannot be determined without additional information

Chapter 5

CP 1. c, d, and e **2.** (a) and (b) 2 N, leftward (acceleration is zero in each situation) **3.** (a) and (b) 1, 4, 3, 2 **4.** (a) equal; (b) greater (acceleration is upward, thus net force on body must be upward) **5.** (a) equal; (b) greater; (c) less **6.** (a) increase; (b) yes; (c) same; (d) yes **7.** (a) $F \sin \theta$; (b) increase **8.** 0 **Q 1.** (a) yes; (b) yes; (c) yes; (d) yes **3.** (a) 2 and 4; (b) 2 and 4 **5.** (a) 50 N, upward; (b) 150 N, upward **7.** (a) less; (b) greater **9.** (a) no; (b) no; (c) no **11.** (a) increases; (b) increases; (c) decreases; (d) decreases **13.** (a) 20 kg; (b) 18 kg; (c) 10 kg; (d) all tie; (e) 3, 2, 1 **15.** d, c, a, b **EP 1.** (a) $F_x = 1.88$ N, $F_y = 0.684$ N; (b) $(1.88\mathbf{i} + 0.684\mathbf{j})$ N **3.** (a) $(-6.26\mathbf{i} - 3.23\mathbf{j})$ N; (b) 7.0 N, 207° relative to $+x$ **5.** $(-2\mathbf{i} + 6\mathbf{j})$ N **7.** (a) 0; (b) $+20$ N; (c) -20 N; (d) -40 N; (e) -60 N **9.** (a) $(1\mathbf{i} - 1.3\mathbf{j})$ m/s²; (b) 1.6 m/s² at $-50°$ from $+x$ **11.** (a) $\mathbf{F}_2$ and $\mathbf{F}_3$ are in the $-x$ direction, $\mathbf{a} = 0$; (b) $\mathbf{F}_2$ and $\mathbf{F}_3$ are in the $-x$ direction, $\mathbf{a}$ is on the x axis, $a = 0.83$ m/s²; (c) $\mathbf{F}_2$ and $\mathbf{F}_3$ are at 34° from $-x$ direction; $\mathbf{a} = 0$ **13.** (a) 22 N, 2.3 kg; (b) 1100 N, 110 kg; (c) 1.6×10^4 N, 1.6×10^3 kg **15.** (a) 11 N, 2.2 kg; (b) 0, 2.2 kg **17.** (a) 44 N; (b) 78 N; (c) 54 N; (d) 152 N **19.** 1.18×10^4 N **21.** 1.2×10^5 N **23.** 16 N **25.** (a) 13 ft/s²; (b) 190 lb **27.** (a) 42 N; (b) 72 N; (c) 4.9 m/s² **29.** (a) 0.02 m/s²; (b) 8×10^4 km; (c) 2×10^3 m/s **31.** (a) 1.1×10^{-15} N; (b) 8.9×10^{-30} N **33.** (a) 5500 N; (b) 2.7 s; (c) 4 times as far; (d) twice the time **35.** (a) 4.9×10^5 N; (b) 1.5×10^6 N **37.** (a) 110 lb, up; (b) 110 lb, down **39.** (a) 0.74 m/s²; (b) 7.3 m/s² **41.** (a) $\cos \theta$; (b) $\sqrt{\cos \theta}$ **43.** 1.8×10^4 N **45.** (a) 4.6×10^3 N; (b) 5.8×10^3 N **47.** (a) 250 m/s²; (b) 2.0×10^4 N **49.** 23 kg **51.** (a) 620 N; (b) 580 N **53.** 1.9×10^5 lb **55.** (a) rope breaks; (b) 1.6 m/s² **57.** 4.6 N **59.** (a) allow a downward acceleration with magnitude ≥ 4.2 ft/s²; (b) 13 ft/s or greater **61.** 195 N, up **63.** (a) 566 N; (b) 1130 N **65.** 18,000 N **67.** (a) 1.4×10^4 N; (b) 1.1×10^4 N; (c) 2700 N, toward the counterweight **69.** 6800 N, at 21° to the line of motion of the barge **71.** (a) 4.6 m/s²; (b) 2.6 m/s² **73.** (b) $F/(m + M)$; (c) $MF/(m + M)$; (d) $F(m + 2M)/2(m + M)$ **75.** $T_1 = 13$ N, $T_2 = 20$ N, $a = 3.2$ m/s²

Chapter 6

CP 1. (a) zero (because there is no attempt at sliding); (b) 5 N; (c) no; (d) yes **2.** (a) same (10 N); (b) decreases; (c) decreases **3.** greater **4.** (a) **a** downward; **N** upward; (b) **a** and **N** upward **5.** (a) $4R_1$; (b) $4R_1$ **6.** (a) same; (b) increases; (c) increases **Q 1.** They slide at the same angle for all orders. **3.** (a) upward; (b) horizontal, toward you; (c) no change; (d) increases; (e) increases **5.** The frictional force $\mathbf{f}_s$ is initially directed up the ramp, decreases in magnitude to zero, and then is directed down the ramp, increasing in magnitude until the magnitude reaches $f_{s,max}$; thereafter, the magnitude of the frictional force is f_k, which is a constant smaller value. **7.** (a) decreases; (b) decreases; (c) increases; (d) increases **9.** (a) zero; (b) infinite **11.** 4, 3; then 1, 2, and 5 tie **13.** (a) less; (b) greater **EP 1.** (a) 200 N; (b) 120 N **3.** 2° **5.** 440 N

7. (a) 110 N; (b) 130 N; (c) no; (d) 46 N; (e) 17 N
9. (a) 90 N; (b) 70 N; (c) 0.89 m/s^2 **11.** (a) no; (b) $(-12\mathbf{i} +$
$5\mathbf{j})$ N **13.** 20° **15.** (a) 0.13 N; (b) 0.12 **17.** $\mu_s = 0.58$,
$\mu_k = 0.54$ **19.** (a) 0.11 m/s^2, 0.23 m/s^2; (b) 0.041, 0.029
21. 36 m **23.** (a) 300 N; (b) 1.3 m/s^2 **25.** (a) 66 N; (b) 2.3
m/s^2 **27.** (a) $\mu_k mg/(\sin\theta - \mu_k \cos\theta)$; (b) $\theta_0 = \tan^{-1}\mu_s$
29. (b) 3.0×10^7 N **31.** 100 N **33.** 3.3 kg **35.** (a) 11
ft/s^2; (b) 0.46 lb; (c) blocks move independently
37. (a) 27 N; (b) 3.0 m/s^2 **39.** (a) 6.1 m/s^2, leftward;
(b) 0.98 m/s^2, leftward **41.** (a) 3.0×10^5 N; (b) 1.2°
43. 9.9 s **45.** 3.75 **47.** 12 cm **49.** 68 ft
51. (a) 3210 N; (b) yes **53.** 0.078 **55.** (a) 0.72 m/s;
(b) 2.1 m/s^2; (c) 0.50 N **57.** $\sqrt{Mgr/m}$ **59.** (a) 30 cm/s;
(b) 180 cm/s^2, radially inward; (c) 3.6×10^{-3} N, radially
inward; (d) 0.37 **61.** (a) 275 N; (b) 877 N **63.** 874 N
65. (a) at the bottom of the circle; (b) 31 ft/s **67.** (a) 9.5
m/s; (b) 20 m **69.** 13° **71.** (a) 0.0338 N; (b) 9.77 N

Chapter 7

CP **1.** (a) decrease; (b) same; (c) negative, zero **2.** d, c,
b, a **3.** (a) same; (b) smaller **4.** (a) positive; (b) negative;
(c) zero **5.** zero **Q** **1.** all tie **3.** (a) increasing;
(b) same; (c) same; (d) increasing **5.** (a) positive;
(b) negative; (c) negative **7.** (a) positive; (b) zero;
(c) negative; (d) negative; (e) zero; (f) positive **9.** all tie
11. c, d, a and b tie; then f, e. **13.** (a) 3 m; (b) 3 m; (c) 0
and 6 m; (d) negative direction of x **15.** (a) A; (b) B
17. twice **EP** **1.** 1.8×10^{13} J **3.** (a) 3610 J;
(b) 1900 J; (c) 1.1×10^{10} J **5.** (a) 1×10^5 megatons TNT;
(b) 1×10^7 bombs **7.** father, 2.4 m/s; son, 4.8 m/s
9. (a) 200 N; (b) 700 m; (c) -1.4×10^5 J; (d) 400 N, 350 m,
-1.4×10^5 J **11.** 5000 J **13.** 47 keV **15.** 7.9 J
17. 530 J **19.** -37 J **21.** (a) 314 J; (b) -155 J; (c) 0;
(d) 158 J **23.** (a) 98 N; (b) 4.0 cm; (c) 3.9 J; (d) -3.9 J
25. (a) $-3Mgd/4$; (b) Mgd; (c) $Mgd/4$ (d) $\sqrt{gd/2}$ **27.** 25 J
31. -6 J **33.** (a) 12 J; (b) 4.0 m; (c) 18 J
35. (a) -0.043 J; (b) -0.13 J **37.** (a) 6.6 m/s; (b) 4.7 m
39. (a) up; (b) 5.0 cm; (c) 5.0 J **41.** 270 kW
43. 235 kW **45.** 490 W **47.** (a) 100 J; (b) 67 W;
(c) 33 W **49.** 0.99 hp **51.** (a) 0; (b) -350 W
53. (a) 79.4 keV; (b) 3.12 MeV; (c) 10.9 MeV **55.** (a) 32 J;
(b) 8 W; (c) 78°

Chapter 8

CP **1.** no **2.** 3, 1, 2 **3.** (a) all tie; (b) all tie **4.** (a) CD,
AB, BC (zero); (b) positive direction of x **5.** 2, 1, 3
6. decrease **7.** (a) seventh excited state, with energy E_7;
(b) 1.3 eV **Q** **1.** -40 J **3.** (c) and (d) tie; then (a) and
(b) tie **5.** (a) all tie; (b) all tie **7.** (a) 3, 2, 1; (b) 1, 2, 3
9. less than (smaller decrease in potential energy)
11. (a) $E < 3$ J, $K < 2$ J; (b) $E < 5$ J, $K < 4$ J
13. (a) increasing; (b) decreasing; (c) decreasing; (d) constant
in AB and BC, decreasing in CD **EP** **1.** 15 J
3. (a) 167 J; (b) -167 J; (c) 196 J; (d) 29 J **5.** (a) 0;
(b) $mgh/2$; (c) mgh; (d) $mgh/2$; (e) mgh **7.** (a) -0.80 J;
(b) -0.80 J; (c) $+1.1$ J **9.** (a) $mgL(1 - \cos\theta)$;
(b) $-mgL(1 - \cos\theta)$; (c) $mgL(1 - \cos\theta)$ **11.** (a) 18 J;

(b) 0; (c) 30 J; (d) 0; (e) parts b and d **13.** (a) 2.08 m/s;
(b) 2.08 m/s **15.** (a) $\sqrt{2gL}$; (b) $2\sqrt{gL}$; (c) $\sqrt{2gL}$
17. 830 ft **19.** (a) 6.75 J; (b) -6.75 J; (c) 6.75 J; (d) 6.75 J;
(e) -6.75 J; (f) 0.459 m **21.** (a) 21.0 m/s; (b) 21.0 m/s
23. (a) 0.98 J; (b) -0.98 J; (c) 3.1 N/cm **25.** (a) 39.2 J;
(b) 39.2 J; (c) 4.00 m **27.** (a) 54 m/s; (b) 52 m/s; (c) 76 m,
below **29.** (a) 39 ft/s; (b) 4.3 in. **31.** (a) 300 J; (b) 93.8 J;
(c) 6.38 m **33.** (a) 4.8 m/s; (b) 2.4 m/s
35. (a) $[v_0^2 + 2gL(1 - \cos\theta_0)]^{1/2}$; (b) $(2gL\cos\theta_0)^{1/2}$;
(c) $[gL(3 + 2\cos\theta_0)]^{1/2}$ **37.** (a) $U(x) = -Gm_1m_2/x$;
(b) $Gm_1m_2d/x_1(x_1 + d)$ **39.** (a) $8mg$ leftward and mg
downward; (b) $2.5R$ **43.** $mgL/32$ **47.** (a) $1.12(A/B)^{1/6}$;
(b) repulsive; (c) attractive **49.** (a) turning point on left, none
on right; molecule breaks apart; (b) turning points on both left
and right; molecule does not break apart; (c) -1.2×10^{-19} J;
(d) 2.2×10^{-19} J; (e) $\approx 1 \times 10^{-9}$ on each, directed toward the
other; (f) $r < 0.2$ nm; (g) $r > 0.2$ nm; (h) $r = 0.2$ nm
51. -25 J **53.** (a) 2200 J; (b) -1500 J; (c) 700 J
55. 17 kW **57.** (a) -0.74 J; (b) -0.53 J **59.** -12 J
61. 54% **63.** 880 MW **65.** (a) 39 kW; (b) 39 kW
67. (a) 1.5 MJ; (b) 0.51 MJ; (c) 1.0 MJ; (d) 63 m/s
69. (a) 67 J; (b) 67 J; (c) 46 cm **71.** Your force on the
cabbage does work. **73.** (a) -0.90 J; (b) 0.46 J; (c) 1.0 m/s
75. (a) 18 ft/s; (b) 18 ft **77.** 4.3 m **79.** (a) 31.0 J;
(b) 5.35 m/s; (c) conservative **81.** 1.2 m **85.** in the center
of the flat part **87.** (a) 24 ft/s; (b) 3.0 ft; (c) 9.0 ft;
(d) 49 ft **89.** (a) 216 J; (b) 1180 N; (c) 432 J; (d) motor also
supplies thermal energy to crate and belt **91.** (a) $1.1 \times$
10^{17} J; (b) 1.2 kg **93.** 7.28 MeV **95.** (a) release;
(b) 17.6 MeV **97.** (a) 5.3 eV; (b) 0.9 eV **99.** (a) 7.2 J;
(b) -7.2 J; (c) 86 cm; (d) 26 cm

Chapter 9

CP **1.** (a) origin; (b) fourth quadrant; (c) on y axis below
origin; (d) origin; (e) third quadrant; (f) origin **2.** (a) to (c)
at the center of mass, still at the origin (their forces are internal
to the system and cannot move the center of mass) **3.** (a) 1,
3, and then 2 and 4 tie (zero force); (b) 3 **4.** (a) 0; (b) no;
(c) negative x **5.** (a) 500 km/h; (b) 2600 km/h; (c) 1600
km/h **6.** (a) yes; (b) no **Q** **1.** point 4 **3.** (a) at the
center of the sled; (b) $L/4$, to the right; (c) not at all (no net
external force); (d) $L/4$, to the left; (e) L; (f) $L/2$; (g) $L/2$
5. (a) ac, cd, and bc; (b) bc; (c) bd and ad **7.** (a) 2 N,
rightward; (b) 2 N, rightward; (c) greater than 2 N, rightward
9. b, c, a **11.** (a) yes; (b) 6 kg·m/s in $-x$ direction; (c) can't
tell **EP** **1.** (a) 4600 km; (b) $0.73R_e$ **3.** (a) $x_{cm} = 1.1$ m,
$y_{cm} = 1.3$ m; (b) shifts toward topmost particle **5.** $x_{cm} =$
-0.25 m, $y_{cm} = 0$ **7.** in the iron, at midheight and midwidth,
2.7 cm from midlength **9.** $x_{cm} = y_{cm} = 20$ cm, $z_{cm} =$
16 cm **11.** (a) $H/2$; (b) $H/2$; (c) descends to lowest point
and then ascends to $H/2$; (d) $(HM/m)(\sqrt{1 + m/M} - 1)$
13. 72 km/h **15.** (a) center of mass does not move;
(b) 0.75 m **17.** 4.8 m/s **19.** (a) 22 m; (b) 9.3 m/s
21. 53 m **23.** 13.6 ft **25.** (a) 52.0 km/h; (b) 28.8 km/h
27. a proton **29.** (a) 30°; (b) $-0.572\mathbf{j}$ kg·m/s
31. (a) $(-4.0 \times 10^4 \mathbf{i})$ kg·m/s; (b) west; (c) 0 **33.** $0.707c$
35. 0.57 m/s, toward center of mass **37.** it increases by

4.4 m/s **39.** (a) rocket case: 7290 m/s, payload: 8200 m/s; (b) before: 1.271×10^{10} J, after: 1.275×10^{10} J **41.** (a) -1; (b) 1830; (c) 1830; (d) same **43.** 14 m/s, 135° from the other pieces **45.** 190 m/s **47.** (a) $0.200v_{rel}$; (b) $0.210v_{rel}$; (c) $0.209v_{rel}$ **49.** (a) 1.57×10^6 N; (b) 1.35×10^5 kg; (c) 2.08 km/s **51.** 108 m/s **53.** 2.2×10^{-3} **57.** fast barge: 46 N more; slow barge: no change **59.** (a) 7.8 MJ; (b) 6.2 **61.** 690 W **63.** 5.5×10^6 N **65.** 24 W **67.** 100 m **69.** (a) 860 N; (b) 2.4 m/s **71.** (a) 3.0×10^5 J; (b) 10 kW; (c) 20 kW **73.** (a) 2.1×10^6 kg; (b) $\sqrt{100 + 1.5t}$ m/s; (c) $(1.5 \times 10^6)/\sqrt{100 + 1.5t}$ N; (d) 6.7 km **75.** $t = (3d/2)^{2/3}(m/2P)^{1/3}$

Chapter 10

CP **1.** (a) unchanged; (b) unchanged; (c) decreased **2.** (a) zero; (b) positive; (c) positive direction of y **3.** (a) 4 kg·m/s; (b) 8 kg·m/s; (c) 3 J **4.** (a) 0; (b) 4 kg·m/s **5.** (a) 10 kg·m/s; (b) 14 kg·m/s; (c) 6 kg·m/s **6.** (a) 2 kg·m/s; (b) 3 kg·m/s **7.** (a) increases; (b) increases **Q** **1.** all tie **3.** b and c **5.** (a) one stationary; (b) 2; (c) 5; (d) equal (pool player's result) **7.** (a) 1 and 4 tie; then 2 and 3 tie; (b) 1; 3 and 4 tie; then 2 **9.** (a) rightward; (b) rightward; (c) smaller **11.** positive direction of x axis **EP** **1.** (a) 750 N; (b) 6.0 m/s **3.** 6.2×10^4 N **5.** 3000 N ($= 660$ lb) **7.** 1.1 m **9.** (a) 42 N·s; (b) 2100 N **11.** (a) $(7.4 \times 10^3 \, \mathbf{i} - 7.4 \times 10^3 \, \mathbf{j})$ N·s; (b) $(-7.4 \times 10^3 \, \mathbf{i})$ N·s; (c) 2.3×10^3 N; (d) 2.1×10^4 N; (e) $-45°$ **13.** (a) 1.0 kg·m/s; (b) 250 J; (c) 10 N; (d) 1700 N **15.** 5 N **17.** $2\mu v$ **19.** 990 N **21.** (a) 1.8 N·s, to the left; (b) 180 N, to the right **25.** 8 m/s **27.** 38 km/s **29.** 4.2 m/s **31.** (a) 99 g; (b) 1.9 m/s; (c) 0.93 m/s **33.** (a) 1.2 kg; (b) 2.5 m/s **35.** 7.8 kg **37.** (a) 1/3; (b) $4h$ **39.** 35 cm **41.** 3.0 m/s **43.** (a) $(10\mathbf{i} + 15\mathbf{j})$ m/s; (b) 500 J lost **45.** (a) 2.7 m/s; (b) 1400 m/s **47.** (a) A: 4.6 m/s, B: 3.9 m/s; (b) 7.5 m/s **49.** 20 J for the heavy particle, 40 J for the light particle **51.** $mv^2/6$ **53.** 13 tons **55.** 25 cm **57.** 0.975 m/s, 0.841 m/s **59.** (a) 4.1 ft/s; (b) 1700 ft·lb; (c) $v_{24} = 5.3$ ft/s, $v_{32} = 3.3$ ft/s **61.** (a) 30° from the incoming proton's direction; (b) 250 m/s and 430 m/s **63.** (a) 41°; (b) 4.76 m/s; (c) no **65.** $v = V/4$ **67.** (a) 117° from the final direction of B; (b) no **69.** 120° **71.** (a) 1.9 m/s, 30° to initial direction; (b) no **73.** (a) 3.4 m/s, deflected by 17° to the right; (b) 0.95 MJ **75.** (a) 117 MeV; (b) equal and opposite momenta; (c) π^- **77.** (a) 4.94 MeV; (b) 0; (c) 4.85 MeV; (d) 0.09 MeV

Chapter 11

CP **1.** (b) and (c) **2.** (a) and (d) **3.** (a) yes; (b) no; (c) yes; (d) yes **4.** all tie **5.** 1, 2, 4, 3 **6.** (a) 1 and 3 tie, 4; then 2 and 5 tie (zero) **7.** (a) downward in the figure; (b) less **Q** **1.** (a) positive; (b) zero; (c) negative; (d) negative **3.** (a) 2 and 3; (b) 1 and 3; (c) 4 **5.** (a) and (c) **7.** (a) all tie; (b) 2, 3; then 1 and 4 tie **9.** b, c, a **11.** less **13.** 90°; then 70° and 110° tie **15.** Finite angular

displacements are not commutative. **EP** **1.** (a) 1.50 rad; (b) 85.9°; (c) 1.49 m **3.** (a) 0.105 rad/s; (b) 1.75×10^{-3} rad/s; (c) 1.45×10^{-4} rad/s **5.** (a) $\omega(2) = 4.0$ rad/s, $\omega(4) = 28$ rad/s; (b) 12 rad/s^2; (c) $\alpha(2) = 6.0$ rad/s^2, $\alpha(4) = 18$ rad/s^2 **7.** (a) $\omega_0 + at^4 - bt^3$; (b) $\theta_0 + \omega_0 t + at^5/5 - bt^4/4$ **9.** 11 rad/s **11.** (a) 9000 rev/min^2; (b) 420 rev **13.** (a) 30 s; (b) 1800 rad **15.** 200 rev/min **17.** (a) 2.0 rad/s^2; (b) 5.0 rad/s; (c) 10 rad/s; (d) 75 rad **19.** (a) 13.5 s; (b) 27.0 rad/s **21.** (a) 340 s; (b) -4.5×10^{-3} rad/s^2; (c) 98 s **23.** (a) 1.0 rev/s^2; (b) 4.8 s; (c) 9.6 s; (d) 48 rev **25.** 6.1 ft/s^2 (1.8 m/s^2), toward the center **27.** 0.13 rad/s **29.** 5.6 rad/s^2 **31.** (a) 5.1 h; (b) 8.1 h **33.** (a) 2.50×10^{-3} rad/s; (b) 20.2 m/s^2; (c) 0 **35.** (a) -1.1 rev/min^2; (b) 9900 rev; (c) -0.99 mm/s^2; (d) 31 m/s^2 **37.** (a) 310 m/s; (b) 340 m/s **39.** (a) 1.94 m/s^2; (b) 75.1°, toward the center of the track **41.** 16 s **43.** (a) 73 cm/s^2; (b) 0.075; (c) 0.11 **45.** 12.3 kg·m^2 **47.** first cylinder: 1100 J; second cylinder: 9700 J **49.** (a) 221 kg·m^2; (b) 1.10×10^4 J **51.** (a) 6490 kg·m^2; (b) 4.36 MJ **53.** 0.097 kg·m^2 **57.** (a) 1300 g·cm^2; (b) 550 g·cm^2; (c) 1900 g·cm^2; (d) $A + B$ **59.** (a) 49 MJ; (b) 100 min **61.** 4.6 N·m **63.** (a) $r_1 F_1 \sin \theta_1 - r_2 F_2 \sin \theta_2$; (b) -3.8 N·m **65.** 1.28 kg·m^2 **67.** 9.7 rad/s^2, counterclockwise **69.** (a) 155 kg·m^2; (b) 64.4 kg **71.** (a) 420 rad/s^2; (b) 500 rad/s **73.** small sphere: (a) 0.689 N·m and (b) 3.05 N; large sphere: (a) 9.84 N·m and (b) 11.5 N **75.** 1.73 m/s^2; 6.92 m/s^2 **77.** (a) 1.4 m/s; (b) 1.4 m/s **79.** (a) 19.8 kJ; (b) 1.32 kW **81.** (a) 8.2×10^{28} N·m; (b) 2.6×10^{29} J; (c) 3.0×10^{21} kW **83.** $\sqrt{9g/4\ell}$ **85.** (a) 4.8×10^5 N; (b) 1.1×10^4 N·m; (c) 1.3×10^6 J **87.** (a) $3g(1 - \cos \theta)$; (b) $\frac{3}{2}g \sin \theta$; (c) 41.8° **89.** (a) 5.6 rad/s^2; (b) 3.1 rad/s **91.** (a) 42.1 km/h; (b) 3.09 rad/s^2; (c) 7.57 kW **93.** (a) 3.4×10^5 g·cm^2; (b) 2.9×10^5 g·cm^2; (c) 6.3×10^5 g·cm^2; (d) (1.2 cm) $\mathbf{i}$ + (5.9 cm) $\mathbf{j}$

Chapter 12

CP **1.** (a) same; (b) less **2.** less **3.** (a) $\pm z$; (b) $+y$; (c) $-x$ **4.** (a) 1 and 3 tie, then 2 and 4 tie, then 5 (zero); (b) 2 and 3 **5.** (a) 3, 1; then 2 and 4 tie (zero); (b) 3 **6.** (a) all tie (same τ, same t, thus same ΔL); (b) sphere, disk, hoop (reverse order of I) **7.** (a) decreases; (b) same; (c) increases **Q** **1.** (a) same; (b) block; (c) block **3.** (a) greater; (b) same **5.** (a) L; (b) 1.5L **7.** b, then c and d tie; then a and e tie (zero) **9.** a, then b and c tie; then e, d (zero) **11.** (a) same; (b) increases, because of decrease in rotational inertia **13.** (a) 30 units clockwise; (b) 2 then 4, then the others; or 4 then 2, then the others **15.** (a) spins in place; (b) rolls toward you; (c) rolls away from you **EP** **1.** 1.00 **3.** (a) 59.3 rad/s; (b) -9.31 rad/s^2; (c) 70.7 m **5.** (a) -4.11 m/s^2; (b) -16.4 rad/s^2; (c) -2.54 N·m **7.** (a) 8.0°; (b) 0.14g **9.** (a) 4.0 N, to the left; (b) 0.60 kg·m^2 **11.** (a) $\frac{1}{2}mR^2$; (b) a solid circular cylinder **13.** (a) $mg(R - r)$; (b) 2/7; (c) $(17/7)mg$ **15.** (a) 2.7R; (b) $(50/7)mg$ **17.** (a) 13 cm/s^2; (b) 4.4 s; (c) 55 cm/s; (d) 1.8×10^{-2} J; (e) 1.4 J; (f) 27 rev/s **21.** (a) 24 N·m, in $+y$ direction; (b) 24 N·m, $-y$; (c) 12 N·m, $+y$;

(d) 12 N·m, $-y$ **23.** (a) $(-1.5\mathbf{i} - 4.0\mathbf{j} - \mathbf{k})$ N·m;
(b) $(-1.5\mathbf{i} - 4.0\mathbf{j} - \mathbf{k})$ N·m **25.** $-2.0\mathbf{i}$ N·m **27.** 9.8
kg·m²/s **29.** (a) 12 kg·m²/s, out of page; (b) 3.0 N·m,
out of page **31.** (a) 0; (b) $(8.0\mathbf{i} + 8.0\mathbf{k})$ N·m **33.** (a) mvd;
(b) no; (c) 0, yes **35.** (a) 3.15×10^{43} kg·m²/s; (b) 0.616
37. 4.5 N·m, parallel to xy plane at $-63°$ from $+x$
39. (a) 0; (b) 0; (c) $30t^3$ kg·m²/s, $90t^2$ N·m, both in $-z$
direction; (d) $30t^3$ kg·m²/s, $90t^2$ N·m, both in $+z$ direction
41. (a) $\frac{1}{2}mgt^2v_0 \cos\theta_0$; (b) $mgtv_0 \cos\theta_0$; (c) $mgtv_0 \cos\theta_0$
43. (a) -1.47 N·m; (b) 20.4 rad; (c) -29.9 J; (d) 19.9 W
45. (a) 12.2 kg·m²; (b) 308 kg·m²/s, down **47.** (a) 1/3;
(b) 1/9 **49.** $\omega_0 R_1 R_2 I_1/(I_1 R_2^2 + I_2 R_1^2)$ **51.** (a) 3.6 rev/s;
(b) 3.0; (c) work done by man in moving weights inward
53. (a) 267 rev/min; (b) 2/3 **55.** 3.0 min **57.** 2.6 rad/s
59. (a) they revolve in a circle of 1.5 m radius at 0.93 rad/s;
(b) 8.4 rad/s; (c) $K_a = 98$ J, $K_b = 880$ J; (d) from the work
done in pulling inward **61.** $m/(M + m)(v/R)$
63. (a) $mvR/(I + MR^2)$; (b) $mvR^2/(I + MR^2)$ **65.** 1300 m/s
67. (a) 18 rad/s; (b) 0.92
69. $\theta = \cos^{-1}\left[1 - \dfrac{6m^2h}{\ell(2m + M)(3m + M)}\right]$
71. 5.28×10^{-35} J·s **73.** Any three are spin up; the other is
spin down. **75.** (a) The magnitude of the angular momentum
increases in proportion to t^2 and the magnitude of the torque
increases in proportion to t, in agreement with the second law
for rotation. (b) The magnitudes of the angular momentum and
torque again increase with time. But the change in the
magnitude of the angular momentum in any interval is less than
is predicted by proportionality to t^2 law and the change in the
torque is less than is predicted by proportionality to t. At any
position of the projectile the torque is less when drag is present
than when it is not.

Chapter 13

CP **1.** c, e, f **2.** (a) no; (b) at site of $\mathbf{F}_1$, perpendicular to
plane of figure; (c) 45 N **3.** (a) at C (to eliminate forces there
from a torque equation); (b) plus; (c) minus; (d) equal **4.** d
5. (a) equal; (b) B; (c) B Q **1.** (a) yes; (b) yes; (c) yes;
(d) no **3.** b **5.** (a) yes; (b) no; (c) no (it could balance the
torques but the forces would then be unbalanced) **7.** (a) a,
then b and c tie, then d **9.** (a) 20 N (the key is the pulley
with the 20 N weight); (b) 25 N **11.** (a) $\sin\theta$; (b) same;
(c) larger **13.** tie of A and B, then C EP **1.** (a) two;
(b) seven **3.** (a) 2.5 m; (b) 7.3° **5.** 120° **7.** 7920 N
9. (a) 840 N; (b) 530 N **11.** 0.536 m **13.** (a) 2770 N;
(b) 3890 N **15.** (a) 1160 N, down; (b) 1740 N, up; (c) left,
stretched; (d) right, compressed **17.** (a) 280 N; (b) 880 N,
71° above the horizontal **19.** bars BC, CD, and DA are under
tension due to forces T, diagonals AC and BD are compressed
due to forces $\sqrt{2}T$ **21.** (a) 1800 lb; (b) 822 lb; (c) 1270 lb
23. (a) 49 N; (b) 28 N; (c) 57 N; (d) 29° **25.** (a) 1900 N, up;
(b) 2100 N, down **27.** (a) 340 N; (b) 0.88 m; (c) increases,
decreases **29.** $W\sqrt{2rh - h^2}/(r - h)$ **31.** (a) $L/2$; (b) $L/4$;
(c) $L/6$; (d) $L/8$; (e) $25L/24$ **33.** (a) 6630 N; (b) $F_h = 5740$ N;
(c) $F_v = 5960$ N **35.** 2.20 m **37.** (a) 1.50 m; (b) 433 N;

(c) 250 N **39.** (a) $a_1 = L/2$, $a_2 = 5L/8$, $h = 9L/8$;
(b) $b_1 = 2L/3$, $b_2 = L/2$, $h = 7L/6$ **41.** (a) 47 lb; (b) 120 lb;
(c) 72 lb **43.** (a) 445 N; (b) 0.50; (c) 315 N
45. (a) 3.9 m/s²; (b) 2000 N on each rear wheel, 3500 N on
each front wheel; (c) 790 N on each rear wheel, 1410 N on
each front wheel **47.** (a) 1.9×10^{-3}; (b) 1.3×10^7 N/m²;
(c) 6.9×10^9 N/m² **49.** 3.1 cm **51.** 2.4×10^9 N/m²
53. (a) 1.8×10^7 N; (b) 1.4×10^7 N; (c) 16 **55.** (a) 867 N;
(b) 143 N; (c) 0.165

Chapter 14

CP **1.** all tie **2.** (a) 1, tie of 2 and 4, then 3; (b) line d
3. negative y direction **4.** (a) increase; (b) negative
5. (a) 2; (b) 1 **6.** (a) path 1 (decreased E (more negative)
gives decreased a); (b) less than (decreased a gives decreased T)
Q **1.** (a) between, closer to less massive particle; (b) no;
(c) no (other than infinity) **3.** $3GM^2/d^2$, leftward **5.** b, tie
of a and c, then d **7.** b, a, c **9.** (a) negative; (b) negative;
(c) postive; (d) all tie **11.** (a) all tie; (b) all tie
13. (a) same; (b) greater EP **1.** 19 m **3.** 2.16
5. 1/2 **7.** 3.4×10^5 km **9.** (a) 3.7×10^{-5} N, increasing y **11.** $M = m$ **13.** 3.2×10^{-7} N **15.** $(GmM/d^2) \times$
$\left[1 - \dfrac{1}{8(1 - R/2d)^2}\right]$ **17.** 2.6×10^6 m **19.** (a) $1.3 \times$
10^{12} m/s²; (b) 1.6×10^6 m/s **21.** (a) 17 N; (b) 2.5
23. (b) 1.9 h **27.** (a) $a_g = (3.03 \times 10^{43}$ kg·m/s²)/M_h;
(b) decrease; (c) 9.82 m/s²; (d) 7.30×10^{-15} m/s²; (e) no
29. 7.91 km/s **31.** (a) $(3.0 \times 10^{-7}m)$ N; (b) $(3.3 \times 10^{-7}m)$
N; (c) $(6.7 \times 10^{-7}mr)$ N **33.** (a) 9.83 m/s²; (b) 9.84 m/s²;
(c) 9.79 m/s² **35.** (a) -1.4×10^{-4} J; (b) less; (c) positive;
(d) negative **37.** (a) 0.74; (b) 3.7 m/s²; (c) 5.0 km/s
39. (a) 0.0451; (b) 28.5 **41.** $-Gm(M_E/R + M_M/r)$
43. (a) 5.0×10^{-11} J; (b) -5.0×10^{-11} J **45.** (a) 1700 m/s;
(b) 250 km; (c) 1400 m/s **47.** (a) 2.2×10^{-7} rad/s;
(b) 90 km/s **51.** (a) -1.67×10^{-8} J; (b) 0.56×10^{-8} J
55. 6.5×10^{23} kg **57.** 5×10^{10} **59.** (a) 7.82 km/s;
(b) 87.5 min **61.** (a) 6640 km; (b) 0.0136 **63.** (a) 39.5
AU³/M_S·y²; (b) $T^2 = r^3/M$ **65.** (a) 1.9×10^{13} m;
(b) $3.5R_P$ **67.** south, at 35.4° above the horizon
71. $2\pi r^{3/2}/\sqrt{G(M + m/4)}$ **73.** $\sqrt{GM/L}$ **75.** (a) 2.8 y;
(b) 1.0×10^{-4} **77.** (a) 1/2; (b) 1/2; (c) B, by 1.1×10^8 J
79. (a) 54 km/s; (b) 960 m/s; (c) $R_p/R_a = v_a/v_p$ **81.** (a) $4.6 \times$
10^5 J; (b) 260 **83.** (a) 7.5 km/s; (b) 97 min; (c) 410 km;
(d) 7.7 km/s; (e) 92 min; (f) 3.2×10^{-3} N; (g) if the satellite–
Earth system is considered isolated, its $\mathbf{L}$ is conserved
85. (a) 5540 s; (b) 7.68 km/s; (c) 7.60 km/s; (d) 5.78×10^{10} J;
(e) -11.8×10^{10} J; (f) -6.02×10^{10} J; (g) $6.63 \times$
10^6 m; (h) 170 s, new orbit **87.** (a) $(-7.0$ mm)$\mathbf{i} +$
$(3.0$ cm)$\mathbf{j}$; (b) $(-0.19$ m/s)$\mathbf{i} + (0.40$ m/s)$\mathbf{j}$ **89.** (a) $1.98 \times$
10^{30} kg; (b) 1.96×10^{30} kg

Chapter 15

CP **1.** all tie **2.** (a) all tie; (b) $0.95\rho_0$, ρ_0, $1.1\rho_0$
3. 13 cm³/s, outward **4.** (a) all tie; (b) 1, then 2 and 3 tie, 4;

(c) 4, 3, 2, 1   **Q** **1.** e, then b and d tie, then a and c tie
3. (a) 1, 3, 2; (b) all tie; (c) no (you must consider the weight
exerted on the scale via the walls)   **5.** 3, 4, 1, 2
7. (a) downward; (b) downward; (c) same   **9.** (a) same;
(b) same; (c) lower; (d) higher   **11.** (a) block 1,
counterclockwise; block 2, clockwise; (b) block 1, tip more;
block 2, right itself   **EP** **1.** 1000 kg/m³   **3.** 1.1×10^5 Pa
or 1.1 atm   **5.** 2.9×10^4 N   **7.** 6.0 lb/in.²   **9.** 1.90×10^4
Pa   **11.** 5.4×10^4 Pa   **13.** 0.52 m   **15.** (a) 6.06×10^9 N;
(b) 20 atm   **17.** 0.412 cm   **19.** $\frac{1}{4}\rho g A (h_2 - h_1)^2$
21. 44 km   **23.** (a) $\rho g W D^2/2$; (b) $\rho g W D^3/6$; (c) $D/3$
25. (a) 2.2; (b) 2.4   **27.** -3.9×10^{-3} atm   **29.** (a) fA/a;
(b) 20 lb   **31.** 1070 g   **33.** 1.5 g/cm³   **35.** 600 kg/m³
37. (a) 670 kg/m³; (b) 740 kg/m³   **39.** 390 kg
41. (a) 1.2 kg; (b) 1300 kg/m³   **43.** 0.126 m³   **45.** five
47. (a) 1.80 m³; (b) 4.75 m³   **49.** 2.79 g/cm³
51. (a) 9.4 N; (b) 1.6 N   **53.** 4.0 m   **55.** 28 ft/s   **57.** 43
cm/s   **59.** (a) 2.40 m/s; (b) 245 Pa   **61.** (a) 12 ft/s; (b) 13
lb/in.²   **63.** 0.72 ft·lb/ft³   **65.** (a) 2; (b) $R_1/R_2 = \frac{1}{2}$; (c) drain
it until $h_2 = h_1/4$ **67.** 116 m/s   **69.** (a) 6.4 m³; (b) 5.4 m/s;
(c) 9.8×10^4 Pa   **71.** (a) 560 Pa; (b) 5.0×10^4 N
73. 40 m/s   **75.** (b) $H - h$; (c) $H/2$ **77.** (b) 5.4 ft³/s
79. (b) 63.3 m/s

Chapter 16

CP **1.** (a) $-x_m$; (b) $+x_m$; (c) 0   **2.** a   **3.** (a) 5 J; (b) 2 J;
(c) 5 J   **4.** all tie (in Eq. 16-32, m is included in I)
5. 1, 2, 3 (the ratio m/b matters; k does not)   **Q** **1.** c
3. (a) 0; (b) between 0 and $+x_m$; (c) between $-x_m$ and 0;
(d) between $-x_m$ and 0   **5.** (a) toward $-x_m$; (b) toward $+x_m$;
(c) between $-x_m$ and 0; (d) between $-x_m$ and 0; (e) decreasing;
(f) increasing   **7.** (a) 3, 2, 1; (b) all tie   **9.** 3, 2, 1
11. system with spring A **13.** b (infinite period; does not
oscillate), c, a **15.** (a) same; (b) same; (c) same; (d) smaller;
(e) smaller; (f) and (g) larger ($T = \infty$)   **EP** **1.** (a) 0.50 s;
(b) 2.0 Hz; (c) 18 cm   **3.** (a) 245 N/m; (b) 0.284 s
5. 708 N/m   **7.** $f > 500$ Hz   **9.** (a) 100 N/m; (b) 0.45 s
11. (a) 6.28×10^5 rad/s; (b) 1.59 mm   **13.** (a) 1.0 mm;
(b) 0.75 m/s; (c) 570 m/s²   **15.** (a) 1.29×10^5 N/m;
(b) 2.68 Hz   **17.** (a) 4.0 s; (b) $\pi/2$ rad/s; (c) 0.37 cm;
(d) (0.37 cm) cos $\frac{\pi}{2}t$; (e) $(-0.58$ cm/s) sin $\frac{\pi}{2}t$; (f) 0.58 cm/s;
(g) 0.91 cm/s²; (h) 0; (i) 0.58 cm/s   **19.** (b) 12.47 kg;
(c) 54.43 kg   **21.** 1.6 kg   **23.** (a) 1.6 Hz; (b) 1.0 m/s, 0;
(c) 10 m/s², ± 10 cm; (d) $(-10$ N/m)x **25.** 22 cm
27. (a) 25 cm; (b) 2.2 Hz   **29.** (a) 0.500 m; (b) -0.251 m;
(c) 3.06 m/s   **31.** (a) $0.183A$; (b) same direction
37. (a) $k_1 = (n + 1)k/n$, $k_2 = (n + 1)k$; (b) $f_1 = \sqrt{(n + 1)/n}f$,
$f_2 = \sqrt{n + 1}f$ **39.** (b) 42 min   **41.** (a) 200 N/m;
(b) 1.39 kg; (c) 1.91 Hz   **43.** (a) 130 N/m; (b) 0.62 s; (c) 1.6
Hz; (d) 5.0 cm; (e) 0.51 m/s   **45.** (a) 3/4; (b) 1/4; (c) $x_m/\sqrt{2}$
47. (a) 3.5 m; (b) 0.75 s   **49.** (a) 0.21 m; (b) 1.6 Hz;
(c) 0.10 m   **51.** (a) 0.0625 J; (b) 0.03125 J   **53.** 12 s
55. (a) 39.5 rad/s; (b) 34.2 rad/s; (c) 124 rad/s²   **57.** (a) 8.3 s;
(b) no   **59.** 9.47 m/s²   **61.** 8.77 s   **63.** 5.6 cm
65. $2\pi\sqrt{(R^2 + 2d^2)/2gd}$ **67.** (a) 0.205 kg·m²; (b) 47.7 cm;
(c) 1.50 s   **71.** (a) $2\pi\sqrt{(L^2 + 12x^2)/12gx}$; (b) 0.289 m

73. 9.78 m/s²   **75.** $2\pi\sqrt{m/3k}$ **77.** $(1/2\pi)(\sqrt{g^2 + v^4/R^2}/L)^{1/2}$
79. (b) smaller   **81.** (a) 2.0 s; (b) 18.5 N·m/rad
83. $0.29L$ **85.** 0.39   **87.** (a) 0.102 kg/s; (b) 0.137 J
89. $k = 490$ N/cm, $b = 1100$ kg/s   **91.** 1.9 in.
93. (a) $y_m = 8.8 \times 10^{-4}$ m, $T = 0.18$ s, $\omega = 35$ rad/s;
(b) $y_m = 5.6 \times 10^{-2}$ m, $T = 0.48$ s, $\omega = 13$ rad/s;
(c) $y_m = 3.3 \times 10^{-2}$ m, $T = 0.31$ s, $\omega = 20$ rad/s

Chapter 17

CP **1.** a, 2; b, 3; c, 1   **2.** (a) 2, 3, 1; (b) 3, then 1 and 2 tie
3. a **4.** 0.20 and 0.80 tie, then 0.60, 0.45   **5.** (a) 1; (b) 3;
(c) 2   **6.** (a) 75 Hz; (b) 525 Hz   **Q** **1.** $7d$ **3.** tie of A
and B, then C, D **5.** intermediate (closer to fully destructive
interference)   **7.** a and d tie, then b and c tie   **9.** (a) 8;
(b) antinode; (c) longer; (d) lower   **11.** (a) integer multiples
of 3; (b) node; (c) node   **13.** string A **15.** decrease
EP **1.** (a) 75 Hz; (b) 13 ms   **3.** (a) 7.5×10^{14} to 4.3×10^{14}
Hz; (b) 1.0 to 200 m; (c) 6.0×10^{16} to 3.0×10^{19} Hz
5. $y = 0.010$ sin $\pi(3.33x + 1100t)$, with x and y in m and t
in s   **11.** (a) $z = 3.0$ sin($60y - 10\pi t$), with z in mm, y in cm,
and t in s; (b) 9.4 cm/s   **13.** (a) $y = 2.0$ sin $2\pi(0.10x -$
$400t)$, with x and y in cm and t in s; (b) 50 m/s; (c) 40 m/s
15. (b) 2.0 cm/s; (c) $y = (4.0$ cm) sin($\pi x/10 - \pi t/5 + \pi$),
where x is in cm and t is in s; (d) -2.5 cm/s   **17.** 129 m/s
19. 135 N   **23.** (a) 15 m/s; (b) 0.036 N   **25.** $y =$
0.12 sin($141x + 628t$), with y in mm, x in m, and t in s
27. (a) 5.0 cm; (b) 40 cm; (c) 12 m/s; (d) 0.033 s; (e) 9.4 m/s;
(f) 5.0 sin($16x + 190t + 0.79$), with x in m, y in cm, and t in s
29. (a) $v_1 = 28.6$ m/s, $v_2 = 22.1$ m/s; (b) $M_1 = 188$ g, $M_2 =$
313 g   **31.** (a) $\sqrt{k(\Delta l)(l + \Delta l)/m}$ **33.** (a) $P_2 = 2P_1$;
(b) $P_2 = P_1/4$ **35.** (a) 3.77 m/s; (b) 12.3 N; (c) zero; (d) 46.3
W; (e) zero; (f) zero; (g) ± 0.50 cm   **37.** 82.8°, 1.45 rad,
0.23 wavelength   **39.** 5.0 cm   **41.** (a) 4.4 mm; (b) 112°
43. (a) $0.83y_1$; (b) 37°   **45.** (a) $2f_3$; (b) λ_3 **47.** 10 cm
49. (a) 82.0 m/s; (b) 16.8 m; (c) 4.88 Hz   **51.** 240 cm, 120 cm,
80 cm   **53.** 7.91 Hz, 15.8 Hz, 23.7 Hz   **55.** $f_{1A} = f_{4B}$,
$f_{2A} = f_{8B}$ **57.** (a) 2.0 Hz, 200 cm, 400 cm/s; (b) $x = 50$ cm,
150 cm, 250 cm, etc.; (c) $x = 0$, 100 cm, 200 cm, etc.
63. (a) 1.3 m; (b) $y' = 0.002$ sin($9.4x$) cos($3800t$), with x and
y in m and t in s   **67.** (b) in the positive x direction;
interchange the amplitudes of the original two traveling waves;
(c) largest at $x = \lambda/4 = 6.26$ cm; smallest at $x = 0$ and
$x = \lambda/2 = 12.5$ cm; (d) the largest amplitude is 4.0 mm, which
is the sum of the amplitudes of the original traveling waves; the
smallest amplitude is 1.0 mm, which is the difference of the
amplitudes of the original traveling waves

Chapter 18

CP **1.** beginning to decrease (example: mentally move the
curves of Fig. 18-6 rightward past the point at $x = 42$ m)
2. (a) fully constructive, 0; (b) fully constructive, 4   **3.** (a) 1
and 2 tie, then 3; (b) 3, then 1 and 2 tie   **4.** second
5. loosen   **6.** a, greater; b, less; c, can't tell; d, can't tell;
e, greater; f, less   **7.** (a) 222 m/s; (b) 262 m/s   **Q** **1.** pulse

along path 2 **3.** (a) 2.0 wavelengths; (b) 1.5 wavelengths; (c) fully constructive, fully destructive **5.** (a) exactly out of phase; (b) exactly out of phase **7.** 70 dB **9.** (a) two; (b) antinode **11.** all odd harmonics **13.** 501, 503, and 508 Hz; or 505, 507, and 508 Hz **EP 1.** (a) $\approx 6\%$ **3.** the radio listener by about 0.85 s **5.** 7.9×10^{10} Pa **7.** If only the length is uncertain, it must be known to within 10^{-4} m. If only the time is imprecise, the uncertainty must be no more than one part in 10^8. **9.** 43.5 m **11.** 40.7 m **13.** 100 kHz **15.** (a) 2.29, 0.229, 22.9 kHz; (b) 1.14, 0.114, 11.4 kHz **17.** (a) 6.0 m/s; (b) $y = 0.30 \sin(\pi x/12 + 50\pi t)$, with x and y in cm and t in s **19.** 4.12 rad **21.** (a) $343 \times (1 + 2m)$ Hz, with m being an integer from 0 to 28; (b) $686m$ Hz, with m being an integer from 1 to 29 **23.** (a) eight; (b) eight **25.** 64.7 Hz, 129 Hz **27.** (a) 0.080 W/m²; (b) 0.013 W/m² **29.** 36.8 nm **31.** (a) 1000; (b) 32 **33.** (a) 39.7 μW/m²; (b) 171 nm; (c) 0.893 Pa **35.** (a) 59.7; (b) 2.81×10^{-4} **37.** $s_m \propto r^{-1/2}$ **39.** (a) 5000; (b) 71; (c) 71 **41.** 171 m **43.** 3.16 km **45.** (a) 5200 Hz; (b) amplitude$_{SAD}$/amplitude$_{SBD}$ = 2 **47.** 20 kHz **49.** by a factor of 4 **51.** water filled to a height of $\frac{7}{8}, \frac{5}{8}, \frac{3}{8}, \frac{1}{8}$ m **53.** (a) 5.0 cm from one end; (b) 1.2; (c) 1.2 **55.** (a) 1130, 1500, and 1880 Hz **57.** (a) 230 Hz; (b) higher **59.** (a) node; (c) 22 s **61.** 387 Hz **63.** 0.02 **65.** 3.8 Hz **67.** (a) 380 mi/h, away from owner; (b) 77 mi/h, away from owner **69.** 15.1 ft/s **71.** 2.6×10^8 m/s **73.** (a) 77.6 Hz; (b) 77.0 Hz **75.** 33.0 km **79.** (a) 970 Hz; (b) 1030 Hz; (c) 60 Hz, no **81.** (a) 1.02 kHz; (b) 1.04 kHz **83.** 1540 m/s **85.** 41 kHz **87.** (a) 2.0 kHz; (b) 2.0 kHz **89.** (a) 485.8 Hz; (b) 500.0 Hz; (c) 486.2 Hz; (d) 500.0 Hz **91.** 1×10^6 m/s, receding **93.** $0.13c$

Chapter 19

CP 1. (a) all tie; (b) 50°X, 50°Y, 50°W **2.** (a) 2 and 3 tie, then 1, then 4; (b) 3, 2, then 1 and 4 tie **3.** A **4.** c and e **5.** (a) zero; (b) positive **6.** b and d tie, then a, c **Q 1.** 25 S°, 25 U°, 25 R° **3.** c, then the rest tie **5.** B, then A and C tie **7.** (a) both clockwise; (b) both clockwise **9.** c, a, b **11.** upward (with liquid water on the exterior and at the bottom, $\Delta T = 0$ horizontally and downward) **13.** at the temperature of your fingers **15.** 3, 2, 1 **EP 1.** 2.71 K **3.** 0.05 kPa, nitrogen **5.** (a) 320°F; (b) -12.3°F **7.** (a) -96°F; (b) 56.7°C **9.** (a) -40°; (b) 575°; (c) Celsius and Kelvin cannot give the same reading **11.** (a) Dimensions are inverse time **13.** 4.4×10^{-3} cm **15.** 0.038 in. **17.** (a) 9.996 cm; (b) 68°C **19.** 170 km **21.** 0.32 cm² **23.** 29 cm³ **25.** 0.432 cm³ **27.** -157°C **29.** 360°C **35.** +0.68 s/h **37.** (b) use 39.3 cm of steel and 13.1 cm of brass **39.** (a) 523 J/kg·K; (b) 0.600; (c) 26.2 J/mol·K **41.** 94.6 L **43.** 109 g **45.** 1.30 MJ **47.** 1.9 times as great **49.** (a) 33.9 Btu; (b) 172 F° **51.** (a) 52 MJ; (b) 0°C **53.** (a) 411 g; (b) 3.1¢ **55.** 0.41 kJ/kg·K **57.** 3.0 min **59.** 73 kW **61.** 2.17 g **63.** 33 m² **65.** 33 g **67.** (a) 0°C; (b) 2.5°C **69.** 2500 J/kg·K **71.** A: 120 J, B: 75 J, C: 30 J **73.** (a) -200 J; (b) -293 J;

(c) -93 J **75.** -5.0 J **77.** 33.3 kJ **79.** 766°C **81.** (a) 1.2 W/m·K, 0.70 Btu/ft·F°·h; (b) 0.030 ft²·F°·h/Btu **83.** 1660 J/s **87.** arrangement b **89.** (a) 2.0 MW; (b) 220 W **91.** (a) 17 kW/m²; (b) 18 W/m² **93.** -6.1 nW **95.** 0.40 cm/h **97.** Cu-Al, 84.3°C; Al-brass, 57.6°C

Chapter 20

CP 1. all but c **2.** (a) all tie; (b) 3, 2, 1 **3.** gas A **4.** 5 (greatest change in T), then tie of 1, 2, 3, and 4 **5.** 1, 2, 3 ($Q_3 = 0$, Q_2 goes into work W_2, but Q_1 goes into greater work W_1 and increases gas temperature) **Q 1.** increased but less than doubled **3.** a, c, b **5.** 1180 J **7.** d, tie of a and b, then c **9.** constant-volume process **11.** (a) same; (b) increases; (c) decreases; (d) increases **13.** -4 J **15.** (a) 1, polyatomic; 2, diatomic; 3, monatomic; (b) more **EP 1.** (a) 0.0127; (b) 7.65×10^{21} **3.** 6560 **5.** number of molecules in the ink $\approx 3 \times 10^{16}$; number of people $\approx 5 \times 10^{20}$; statement is wrong, by a factor of about 20,000 **7.** (a) 5.47×10^{-8} mol; (b) 3.29×10^{16} **9.** (a) 106; (b) 0.892 m³ **11.** 27.0 lb/in.² **13.** (a) 2.5×10^{25}; (b) 1.2 kg **15.** 5600 J **17.** 1/5 **19.** (a) -45 J; (b) 180 K **21.** 100 cm³ **23.** 198°F **25.** 2.0×10^5 Pa **27.** 180 m/s **29.** 9.53×10^6 m/s **31.** 313°C **33.** 1.9 kPa **35.** (a) 0.0353 eV, 0.0483 eV; (b) 3400 J, 4650 J **37.** 9.1×10^{-6} **39.** (a) 6.75×10^{-20} J; (b) 10.7 **41.** 0.32 nm **43.** 15 cm **45.** (a) 3.27×10^{10}; (b) 172 m **47.** (a) 22.5 L; (b) 2.25; (c) 8.4×10^{-5} cm; (d) same as (c) **51.** (a) 3.2 cm/s; (b) 3.4 cm/s; (c) 4.0 cm/s **53.** (a) v_P, v_{rms}, $\overline{v}$ (b) reverse ranking **55.** (a) 1.0×10^4 K, 1.6×10^5 K; (b) 440 K, 7000 K **57.** 4.7 **59.** (a) $2N/3v_0$; (b) $N/3$; (c) $1.22v_0$; (d) $1.31v_0$ **61.** $RT \ln(V_f/V_i)$ **63.** (a) 15.9 J; (b) 34.4 J/mol·K; (c) 26.1 J/mol·K **65.** $(n_1C_1 + n_2C_2 + n_3C_3)/(n_1 + n_2 + n_3)$ **67.** (a) -5.0 kJ; (b) 2.0 kJ; (c) 5.0 kJ **69.** (a) 0.375 mol; (b) 1090 J; (c) 0.714 **71.** (a) 14 atm; (b) 620 K **79.** 0.63 **81.** (a) monatomic; (b) 2.7×10^4 K; (c) 4.5×10^4 mol; (d) 3.4 kJ, 340 kJ; (e) 0.01 **83.** 5 m³ **85.** (a) in joules, in the order Q, ΔE_{int}, W: $1 \rightarrow 2$: 3740, 3740, 0; $2 \rightarrow 3$: 0, -1810, 1810; $3 \rightarrow 1$: -3220, -1930, -1290; cycle: 520, 0, 520; (b) $V_2 = 0.0246$ m³, $p_2 = 2.00$ atm, $V_3 = 0.0373$ m³, $p_3 = 1.00$ atm

Chapter 21

CP 1. a, b, c **2.** smaller **3.** c, b, a **4.** a, d, c, b **5.** b **Q 1.** increase **3.** (a) increase; (b) same **5.** equal **7.** lower the temperature of the low temperature reservoir **9.** (a) same; (b) increase; (c) decrease **11.** (a) same; (b) increase; (c) decrease **13.** more than the age of the universe **EP 1.** 1.86×10^4 J **3.** 2.75 mol **7.** (a) 5.79×10^4 J; (b) 173 J/K **9.** $+3.59$ J/K **11.** (a) 14.6 J/K; (b) 30.2 J/K **13.** (a) 4.45 J/K; (b) no **15.** (a) 4500 J; (b) -5000 J; (c) 9500 J **17.** (a) 57.0°C; (b) -22.1 J/K; (c) $+24.9$ J/K; (d) $+2.8$ J/K **19.** (a) -710 mJ/K; (b) $+710$ mJ/K; (c) $+723$ mJ/K;

(d) -723 mJ/K; (e) $+13$ mJ/K; (f) 0   **23.** (a) (I) constant T, $Q = pV \ln 2$; constant V, $Q = 4.5pV$; (II) constant T, $Q = -pV \ln 2$; constant p, $Q = 7.5pV$; (b) (I) constant T, $W = pV \ln 2$; constant V, $W = 0$; (II) constant T, $W = -pV \ln 2$; constant p, $W = 3pV$; (c) $4.5pV$ for either case; (d) $4R \ln 2$ for either case   **25.** 0.75 J/K   **27.** (a) -943 J/K; (b) $+943$ J/K; (c) yes   **29.** (a) $3p_0V_0$; (b) $6RT_0$, $(3/2)R \ln 2$; (c) both are zero   **33.** (a) 31%; (b) 16 kJ   **35.** engine A, first; engine B, first and second; engine C, second; engine D, neither   **37.** 97 K   **39.** 99.99995%   **41.** 7.2 J/cycle   **43.** (a) 7200 J; (b) 960 J; (c) 13%   **45.** (a) 2270 J; (b) 14,800 J; (c) 15.4%; (d) 75.0%, greater   **49.** (a) 78%; (b) 81 kg/s   **55.** (a) 49 kJ; (b) 7.4 kJ   **57.** 21 J   **59.** (a) 0.071 J; (b) 0.50 J; (c) 2.0 J; (d) 5.0 J   **61.** 1.08 MJ   **63.** $[1 - (T_2/T_1)] \div [1 - (T_4/T_3)]$ **67.** (a) 1.27×10^{30}; (b) 7.9%; (c) 7.3%; (d) 7.3%; (e) 1.1%; (f) 0.0023%   **69.** (a) $W = N!/(n_1! \, n_2! \, n_3!)$; (b) $[(N/2)! \, (N/2)!]/[(N/3)! \, (N/3)! \, (N/3)!]$; (c) 4.2×10^{16}

Chapter 22

CP **1.** C and D attract; B and D attract   **2.** (a) leftward; (b) leftward; (c) leftward   **3.** (a) a, c, b; (b) less than   **4.** $-15e$ (net charge of $-30e$ is equally shared)   **Q** **1.** No, only for charged particles, charged particle-like objects, and spherical shells (including solid spheres) of uniform charge   **3.** a and b **5.** two points: one to the left of the particles and one between the protons   **7.** $6q^2/4\pi\epsilon_0 d^2$, leftward   **9.** (a) same; (b) less than; (c) cancel; (d) add; (e) the adding components; (f) positive direction of y; (g) negative direction of y; (h) positive direction of x; (i) negative direction of x **11.** (a) A, B, and D; (b) all four; (c) Connect A and D; disconnect them; then connect one of them to B. (There are two more solutions.)   **13.** (a) possibly; (b) definitely   **15.** same   **17.** D **EP** **1.** 0.50 C   **3.** 2.81 N on each   **5.** (a) 4.9×10^{-7} kg; (b) 7.1×10^{-11} C   **7.** $3F/8$ **9.** (a) 1.60 N; (b) 2.77 N   **11.** (a) $q_1 = 9q_2$; (b) $q_1 = -25q_2$ **13.** either -1.00 μC and $+3.00$ μC or $+1.00$ μC and -3.00 μC   **15.** (a) 36 N, $-10°$ from the x axis; (b) $x = -8.3$ cm, $y = +2.7$ cm   **17.** (a) 5.7×10^{13} C, no; (b) 6.0×10^5 kg   **19.** (a) $Q = -2\sqrt{2}q$; (b) no   **21.** 3.1 cm   **23.** 2.89×10^{-9} N   **25.** -1.32×10^{13} C   **27.** (a) 3.2×10^{-19} C; (b) two   **29.** (a) 8.99×10^{-19} N; (b) 625   **31.** 5.1 m below the electron   **33.** 1.3 days   **35.** (a) 0; (b) 1.9×10^{-9} N   **37.** 10^{18} N   **39.** (a) ^{9}B; (b) ^{13}N; (c) ^{12}C   **41.** (a) $F = (Q^2/4\pi\epsilon_0 d^2)\alpha(1 - \alpha)$; (c) 0.5; (d) 0.15 and 0.85

Chapter 23

CP **1.** (a) rightward; (b) leftward; (c) leftward; (d) rightward (p and e have same charge magnitude, and p is farther)   **2.** all tie   **3.** (a) toward positive y; (b) toward positive x; (c) toward negative y **4.** (a) leftward; (b) leftward; (c) decrease   **5.** (a) all tie; (b) 1 and 3 tie, then 2 and 4 tie   **Q** **1.** (a) toward positive x; (b) downward and to the right;

(c) A **3.** two points: one to the left of the particles, the other between the protons   **5.** (a) yes; (b) toward; (c) no (the field vectors are not along the same line); (d) cancel; (e) add; (f) adding components; (g) toward negative y **7.** (a) 3, then 1 and 2 tie (zero); (b) all tie; (c) 1 and 2 tie, then 3   **9.** (a) rightward; (b) $+q_1$ and $-q_3$, increase; q_2, decrease; n, same   **11.** a, b, c **13.** (a) 4, 3, 1, 2; (b) 3, then 1 and 4 tie, then 2   **EP** **1.** (a) 6.4×10^{-18} N; (b) 20 N/C   **3.** to the right in the figure   **7.** 56 pC   **9.** 3.07×10^{21} N/C, radially outward   **13.** (a) $q/8\pi\epsilon_0 d^2$, to the left; $3q/\pi\epsilon_0 d^2$, to the right; $7q/16\pi\epsilon_0 d^2$, to the left   **15.** 0   **17.** 9:30   **19.** $E = q/\pi\epsilon_0 a^2$, along bisector, away from triangle   **21.** $7.4q/4\pi\epsilon_0 d^2$, 28° counterclockwise to $+x$ **23.** 6.88×10^{-28} C·m   **25.** $(1/4\pi\epsilon_0)(p/r^3)$, antiparallel to $\mathbf{p}$ **29.** $R/\sqrt{2}$ **31.** $(1/4\pi\epsilon_0)(4q/\pi R^2)$, toward increasing y **37.** (a) 0.10 μC; (b) 1.3×10^{17}; (c) 5.0×10^{-6} **39.** 3.51×10^{15} m/s^2 **41.** 6.6×10^{-15} N   **43.** 2.03×10^{-7} N/C, up   **45.** (a) -0.029 C; (b) repulsive forces would explode the sphere   **47.** (a) 1.92×10^{12} m/s^2; (b) 1.96×10^5 m/s   **49.** (a) 8.87×10^{-15} N; (b) 120   **51.** 1.64×10^{-19} C ($\approx$3% high)   **53.** (a) 0.245 N, 11.3° clockwise from the $+x$ axis; (b) $x = 108$ m, $y = -21.6$ m   **55.** 27μm   **57.** (a) yes; (b) upper plate, 2.73 cm   **59.** (a) 0; (b) 8.5×10^{-22} N·m; (c) 0   **61.** $(1/2\pi)\sqrt{pE/I}$ **63.** (a) $E = (2q/4\pi\epsilon_0 d^2)(\alpha/(1 + \alpha^2)^{3/2})$; (c) 0.707; (d) 0.21 and 1.9

Chapter 24

CP **1.** (a) $+EA$; (b) $-EA$; (c) 0; (d) 0   **2.** (a) 2; (b) 3; (c) 1   **3.** (a) equal; (b) equal; (c) equal   **4.** (a) $+50e$; (b) $-150e$ **5.** 3 and 4 tie, then 2, 1   **Q** **1.** (a) 8 N·m^2/C; (b) 0   **3.** (a) all tie (zero); (b) all tie   **5.** $+13q/\epsilon_0$ **7.** all tie   **9.** all tie   **11.** 2σ, σ, 3σ; or 3σ, σ, 2σ **13.** (a) all tie ($E = 0$); (b) all tie   **15.** (a) same ($E = 0$); (b) decrease; (c) decrease (to zero); (d) same   **EP** **1.** (a) 693 kg/s; (b) 693 kg/s; (c) 347 kg/s; (d) 347 kg/s; (e) 575 kg/s   **3.** (a) 0; (b) -3.92 N·m^2/C; (c) 0; (d) 0 for each field   **5.** (a) enclose $2q$ and $-2q$, or enclose all four charges; (b) enclose $2q$ and q; (c) not possible   **7.** 2.0×10^5 N·m^2/C   **9.** $q/6\epsilon_0$ **11.** (a) $-\pi R^2 E$; (b) $\pi R^2 E$ **13.** -4.2×10^{-10} C   **15.** 0 through each of the three faces meeting at q, $q/24\epsilon_0$ through each of the other faces   **17.** 2.0 μC/m^2 **19.** (a) 4.5×10^{-7} C/m^2; (b) 5.1×10^4 N/C   **21.** (a) -3.0×10^{-6} C; (b) $+1.3 \times 10^{-5}$ C   **23.** (a) 0.32 μC; (b) 0.14 μC   **27.** (a) $E = q/2\pi\epsilon_0 Lr$, radially inward; (b) $-q$ on both inner and outer surfaces; (c) $E = q/2\pi\epsilon_0 Lr$, radially outward   **29.** 3.6 nC   **31.** (b) $\rho R^2/2\epsilon_0 r$ **33.** (a) 5.3×10^7 N/C; (b) 60 N/C   **35.** 5.0 nC/m^2 **37.** 0.44 mm   **39.** (a) 4.9×10^{-22} C/m^2; (b) downward   **41.** (a) $\rho x/\epsilon_0$; (b) $\rho d/2\epsilon_0$ **43.** (a) -750 N·m^2/C; (b) -6.64 nC   **45.** (a) 4.0×10^6 N/C; (b) 0   **47.** (a) 0; (b) $q_a/4\pi\epsilon_0 r^2$; (c) $(q_a + q_b)/4\pi\epsilon_0 r^2$ **51.** (a) $-q$; (b) $+q$; (c) $E = q/4\pi\epsilon_0 r^2$, radially outward; (d) $E = 0$; (e) $E = q/4\pi\epsilon_0 r^2$, radially outward; (f) 0; (g) $E = q/4\pi\epsilon_0 r^2$, radially outward; (h) yes, charge is induced; (i) no; (j) yes; (k) no; (l) no   **53.** (a) $E = (q/4\pi\epsilon_0 a^3)r$; (b) $E = q/4\pi\epsilon_0 r^2$; (c) 0; (d) 0; (e) inner, $-q$; outer, 0   **55.** $q/2\pi a^2$ **59.** $\alpha = 0.80$

Chapter 25

CP **1.** (a) negative; (b) increase **2.** (a) positive; (b) higher **3.** (a) rightward; (b) 1, 2, 3, 5: positive; 4, negative; (c) 3, then 1, 2, and 5 tie, then 4 **4.** all tie **5.** a, c (zero), b **6.** (a) 2, then 1 and 3 tie; (b) 3; (c) accelerate leftward **7.** closer (half of 9.23 fm) Q **1.** (a) higher; (b) positive; (c) negative; (d) all tie **3.** (a) 1 and 2; (b) none; (c) no; (d) 1 and 2 yes, 3 and 4 no **5.** b, then a, c, and d tie **7.** (a) negative; (b) zero **9.** (a) 1, then 2 and 3 tie; (b) 3 **11.** left **13.** a, b, c **15.** (a) c, b, a; (b) zero **17.** (a) positive; (b) positive; (c) negative; (d) all tie **19.** (a) no; (b) yes **21.** no (a particle at the intersection would have two different potential energies) **23.** (a)–(b) all tie; (c) C, B, A; (d) all tie EP **1.** 1.2 GeV **3.** (a) 3.0×10^{10} J; (b) 7.7 km/s; (c) 9.0×10^4 kg **7.** 2.90 kV **9.** 8.8 mm **11.** (a) 136 MV/m; (b) 8.82 kV/m **13.** (b) because $V = 0$ point is chosen differently; (c) $q/(8\pi\epsilon_0 R)$; (d) potential differences are independent of the choice for the $V = 0$ point **15.** (a) -4500 V; (b) -4500 V **17.** 843 V **19.** 2.8×10^5 **21.** $x = d/4$ and $x = -d/2$ **23.** none **25.** (a) 3.3 nC; (b) 12 nC/m^2 **27.** 6.4×10^8 V **29.** 190 MV **31.** (a) -4.8 nm; (b) 8.1 nm; (c) no **33.** 16.3 μV **35.** (a) $\dfrac{2\lambda}{4\pi\epsilon_0} \ln\left[\dfrac{L/2 + (L^2/4 + d^2)^{1/2}}{d}\right]$; (b) 0 **37.** (a) $-5Q/4\pi\epsilon_0 R$; (b) $-5Q/4\pi\epsilon_0(z^2 + R^2)^{1/2}$ **39.** $0.113\sigma R/\epsilon_0$ **41.** $(Q/4\pi\epsilon_0 L)\ln(1 + L/d)$ **43.** 670 V/m **45.** $p/2\pi\epsilon_0 r^3$ **47.** 39 V/m, $-x$ direction **51.** (a) $\dfrac{c}{4\pi\epsilon_0}[\sqrt{L^2 + y^2} - y]$; (b) $\dfrac{c}{4\pi\epsilon_0}\left[1 - \dfrac{y}{\sqrt{L^2 + y^2}}\right]$ **53.** (a) 2.5 MV; (b) 5.1 J; (c) 6.9 J **55.** (a) 2.72×10^{-14} J; (b) 3.02×10^{-31} kg, about $\frac{1}{3}$ of accepted value **57.** (a) 0.484 MeV; (b) 0 **59.** 2.1 d **61.** 0 **63.** (a) 27.2 V; (b) -27.2 eV; (c) 13.6 eV; (d) 13.6 eV **65.** 1.8×10^{-10} J **67.** 1.48×10^7 m/s **69.** $qQ/4\pi\epsilon_0 K$ **71.** 0.32 km/s **73.** 1.6×10^{-9} m **77.** (a) $V_1 = V_2$; (b) $q_1 = q/3$, $q_2 = 2q/3$; (c) 2 **79.** (a) -0.12 V; (b) 1.8×10^{-8} N/C, radially inward **81.** (a) 12,000 N/C; (b) 1800 V; (c) 5.8 cm **83.** (c) 4.24 V

Chapter 26

CP **1.** (a) same; (b) same **2.** (a) decreases; (b) increases; (c) decreases **3.** (a) V, $q/2$; (b) $V/2$, q **4.** (a) $q_0 = q_1 + q_{34}$; (b) equal (C_3 and C_4 are in series) **5.** (a) same; (b)–(d) increase; (e) same (same potential difference across same plate separation) **6.** (a) same; (b) decrease; (c) increase Q **1.** a, 2; b, 1; c, 3 **3.** (a) increase; (b) same **5.** (a) parallel; (b) series **7.** (a) $C/3$; (b) $3C$; (c) parallel **9.** (a) equal; (b) less **11.** (a)–(d) less **13.** (a) 2; (b) 3; (c) 1 **15.** Increase plate separation d, but also plate area A, keeping A/d constant. EP **1.** 7.5 pC **3.** 3.0 mC **5.** (a) 140 pF; (b) 17 nC **7.** (a) 84.5 pF; (b) 191 cm^2 **9.** (a) 11 cm^2; (b) 11 pF; (c) 1.2 V **13.** (b) 4.6×10^{-5}/K **15.** 7.33 μF **17.** 315 mC **19.** (a) 10.0 μF; (b) $q_2 = 0.800$ mC, $q_1 = 1.20$ mC; (c) 200 V for both **21.** (a) $d/3$; (b) $3d$ **25.** (a) five in series; (b) three

arrays as in (a) in parallel (and other possibilities) **27.** 43 pF **29.** (a) 50 V; (b) 5.0×10^{-5} C; (c) 1.5×10^{-4} C **31.** (a) $q_1 = 9.0$ μC, $q_2 = 16$ μC, $q_3 = 9.0$ μC, $q_4 = 16$ μC; (b) $q_1 = 8.4$ μC, $q_2 = 17$ μC, $q_3 = 11$ μC, $q_4 = 14$ μC **33.** 99.6 nJ **35.** 72 F **37.** 4.9% **39.** 0.27 J **41.** 0.11 J/m^3 **43.** (a) 2.0 J **45.** (a) $q_1 = 0.21$ mC, $q_2 = 0.11$ mC, $q_3 = 0.32$ mC; (b) $V_1 = V_2 = 21$ V, $V_3 = 79$ V; (c) $U_1 = 2.2$ mJ, $U_2 = 1.1$ mJ, $U_3 = 13$ mJ **47.** (a) $q_1 = q_2 = 0.33$ mC, $q_3 = 0.40$ mC; (b) $V_1 = 33$ V, $V_2 = 67$ V, $V_3 = 100$ V; (c) $U_1 = 5.6$ mJ, $U_2 = 11$ mJ, $U_3 = 20$ mJ **53.** Pyrex **55.** (a) 6.2 cm; (b) 280 pF **57.** 0.63 m^2 **59.** (a) 2.85 m^3; (b) 1.01×10^4 **61.** (a) $\epsilon_0 A/(d-b)$; (b) $d/(d-b)$; (c) $-q^2 b/2\epsilon_0 A$, sucked in **65.** $\dfrac{\epsilon_0 A}{4d}\left(\kappa_1 + \dfrac{2\kappa_2\kappa_3}{\kappa_2 + \kappa_3}\right)$ **67.** (a) 13.4 pF; (b) 1.15 nC; (c) 1.13×10^4 N/C; (d) 4.33×10^3 N/C **69.** (a) 7.1; (b) 0.77 μC **71.** (a) 0.606; (b) 0.394

Chapter 27

CP **1.** 8 A, rightward **2.** (a)–(c) rightward **3.** a and c tie, then b **4.** Device 2 **5.** (a) and (b) tie, then (d), then (c) Q **1.** a, b, and c tie, then d (zero) **3.** b, a, c **5.** tie of A, B, and C, then a tie of $A + B$ and $B + C$, then $A + B + C$ **7.** (a)–(c) 1 and 2 tie, then 3 **9.** C, A, B **11.** b, a, c **13.** (a) conductors: 1 and 4; semiconductors: 2 and 3; (b) 2 and 3; (c) all four EP **1.** 1.25×10^{15} **3.** 6.7 μC/m^2 **5.** 14-gauge **7.** (a) 2.4×10^{-5} A/m^2; (b) 1.8×10^{-15} m/s **9.** 0.67 A, toward the negative terminal **11.** (a) 0.654 μA/m^2; (b) 83.4 MA **13.** 13 min **15.** (a) $J_0 A/3$; (b) $2J_0 A/3$ **17.** 2.0×10^{-8} $\Omega\cdot$m **19.** 100 V **21.** (a) 1.53 kA; (b) 54.1 MA/m^2; (c) 10.6×10^{-8} $\Omega\cdot$m, platinum **23.** (a) 253°C; (b) yes **25.** (a) 0.38 mV; (b) negative; (c) 3 min 58 s **27.** 54 Ω **29.** 3.0 **31.** (a) 1.3 mΩ; (b) 4.6 mm **33.** (a) 6.00 mA; (b) 1.59×10^{-8} V; (c) 21.2 nΩ **35.** 2000 K **37.** (a) copper: 5.32×10^5 A/m^2, aluminum: 3.27×10^5 A/m^2; (b) copper: 1.01 kg/m, aluminum: 0.495 kg/m **39.** 0.40 Ω **41.** (a) $R = \rho L/\pi ab$ **43.** 14 kC **45.** 11.1 Ω **47.** (a) 1.0 kW; (b) 25¢ **49.** 0.135 W **51.** (a) 1.74 A; (b) 2.15 MA/m^2; (c) 36.3 mV/m; (d) 2.09 W **53.** (a) 1.3×10^5 A/m^2; (b) 94 mV **55.** (a) \$4.46 for a 31-day month; (b) 144 Ω; (c) 0.833 A **57.** 660 W **59.** (a) 3.1×10^{11}; (b) 25 μA; (c) 1300 W, 25 MW **61.** 27 cm/s **63.** (a) 120 Ω; (b) 107 Ω; (c) 5.3×10^{-3}/C°; (d) 5.9×10^{-3}/C°; (e) 276 Ω

Chapter 28

CP **1.** (a) rightward; (b) all tie; (c) b, then a and c tie; (d) b, then a and c tie **2.** (a) all tie; (b) R_1, R_2, R_3 **3.** (a) less; (b) greater; (c) equal **4.** (a) $V/2$, i; (b) V, $i/2$ **5.** (a) 1, 2, 4, 3; (b) 4, tie of 1 and 2; then 3 Q **1.** 3, 4, 1, 2 **3.** (a) no; (b) yes; (c) all tie (the circuits are the same) **5.** parallel, R_2, R_1, series **7.** (a) equal; (b) more **9.** (a) less; (b) less; (c) more **11.** C_1, 15 V; C_2, 35 V; C_3, 20 V; C_4, 20 V; C_5, 30 V **13.** 60 μC **15.** c, b, a **17.** (a) all tie; (b) 1, 3, 2 **19.** 1, 3, and 4 tie (8 V on each resistor), then 2 and 5 tie (4 V on each resistor) EP **1.** (a) \$320; (b) 4.8¢ **3.** 11 kJ

5. (a) counterclockwise; (b) battery 1; (c) B **7.** (a) 80 J;
(b) 67 J; (c) 13 J converted to thermal energy within battery
9. (a) 14 V; (b) 100 W; (c) 600 W; (d) 10 V, 100 W
11. (a) 50 V; (b) 48 V; (c) B is connected to the negative
terminal **13.** 2.5 V **15.** (a) 6.9 km; (b) 20 Ω
17. 8.0 Ω **19.** 10^{-6} **21.** the cable **23.** (a) 1000 Ω;
(b) 300 mV; (c) 2.3×10^{-3} **25.** (a) 3.41 A or 0.586 A;
(b) 0.293 V or 1.71 V **27.** 5.56 A **29.** 4.0 Ω and 12 Ω
31. 4.50 Ω **33.** 0.00, 2.00, 2.40, 2.86, 3.00, 3.60, 3.75,
3.94 A **35.** $V_d - V_c = +0.25$ V, by all paths **37.** three
39. (a) 2.50 Ω; (b) 3.13 Ω **41.** nine **43.** (a) left branch:
0.67 A down; center branch: 0.33 A up; right branch: 0.33 A
up; (b) 3.3 V **47.** (a) 120 Ω; (b) $i_1 = 51$ mA, $i_2 = i_3 =$
19 mA, $i_4 = 13$ mA **49.** (a) 19.5 Ω; (b) 0; (c) ∞; (d) 82.3 W,
57.6 W **51.** (a) Cu: 1.11 A, Al: 0.893 A; (b) 126 m
53. (a) 13.5 kΩ; (b) 1500 Ω; (c) 167 Ω; (d) 1480 Ω
55. 0.45 A **57.** (a) 12.5 V; (b) 50 A **59.** -0.9%
65. (a) 0.41τ; (b) 1.1τ **67.** 4.6 **69.** (a) 0.955 μC/s;
(b) 1.08 μW; (c) 2.74 μW; (d) 3.82 μW **71.** 2.35 MΩ
73. 0.72 MΩ **75.** 24.8 Ω to 14.9 kΩ **77.** (a) at $t = 0$,
$i_1 = 1.1$ mA, $i_2 = i_3 = 0.55$ mA; at $t = \infty$, $i_1 = i_2 = 0.82$ mA,
$i_3 = 0$; (c) at $t = 0$, $V_2 = 400$ V; at $t = \infty$, $V_2 = 600$ V;
(d) after several time constants ($\tau = 7.1$ s) have elapsed
79. (a) $V_T = -ir + \mathcal{E}$; (b) 13.6 V; (c) 0.060 Ω
81. (a) 6.4 V; (b) 3.6 W; (c) 17 W; (d) -5.6 W; (e) a

Chapter 29

CP **1.** a, $+z$; b, $-x$; c, $F_B = 0$ **2.** 2, then tie of 1 and 3
(zero); (b) 4 **3.** (a) $+z$ and $-z$ tie, then $+y$ and $-y$ tie, then
$+x$ and $-x$ tie (zero); (b) $+y$ **4.** (a) electron;
(b) clockwise **5.** $-y$ **6.** (a) all tie; (b) 1 and 4 tie, then 2
and 3 tie **Q** **1.** (a) all tie; (b) 1 and 2 (charge is negative)
3. a, no, $\mathbf{v}$ and $\mathbf{F}_B$ must be perpendicular; b, yes; c, no, $\mathbf{B}$ and
$\mathbf{F}_B$ must be perpendicular **5.** (a) $\mathbf{F}_E$; (b) $\mathbf{F}_B$ **7.** (a) right;
(b) right **9.** (a) negative; (b) equal; (c) equal; (d) half a
circle **11.** (a) $\mathbf{B}_1$; (b) $\mathbf{B}_1$ into page; $\mathbf{B}_2$ out of page; (c) less
13. all **15.** all tie **17.** (a) positive; (b) (1) and (2) tie, then
(3) which is zero **EP** **1.** M/QT **3.** (a) 9.56×10^{-14} N, 0;
(b) $0.267°$ **5.** (a) east; (b) 6.28×10^{14} m/s^2; (c) 2.98 mm
7. $0.75\mathbf{k}$ T **9.** (a) 3.4×10^{-4} T, horizontal and to the left
as viewed along $\mathbf{v}_0$; (b) yes, if velocity is the same as the
electron's velocity **11.** $(-11.4\mathbf{i} - 6.00\mathbf{j} + 4.80\mathbf{k})$ V/m
13. 680 kV/m **17.** (b) 2.84×10^{-3} **19.** (a) 1.11×10^7 m/s;
(b) 0.316 mm **21.** (a) 0.34 mm; (b) 2.6 keV
23. (a) 2.05×10^7 m/s; (b) 467 μT; (c) 13.1 MHz; (d) 76.3 ns
25. (a) 2.60×10^6 m/s; (b) 0.109 μs; (c) 0.140 MeV;
(d) 70 kV **29.** (a) 1.0 MeV; (b) 0.5 MeV **31.** $R_d = \sqrt{2}R_p$;
$R_\alpha = R_p$. **33.** (a) $B\sqrt{mq/2V}\,\Delta x$; (b) 8.2 mm **37.** (a) $-q$;
(b) $\pi m/qB$ **39.** $B_{min} = \sqrt{mV/2ed^2}$ **41.** (a) 22 cm;
(b) 21 MHz **43.** neutron moves tangent to original path,
proton moves in a circular orbit of radius 25 cm **45.** 28.2 N,
horizontally west **47.** 20.1 N **49.** $Bitd/m$, away from
generator **51.** $-0.35\mathbf{k}$ N **53.** 0.10 T, at 31° from the
vertical **55.** 4.3×10^{-3} N$\cdot$m, negative y **59.** $qvaB/2$
61. (a) 540 Ω, in series; (b) 2.52 Ω, in parallel
63. (a) 12.7 A; (b) 0.0805 N$\cdot$m **65.** (a) 0.184 A$\cdot$m^2;

(b) 1.45 N$\cdot$m **67.** (a) 20 min; (b) 5.9×10^{-2} N$\cdot$m
69. (a) $(8.0 \times 10^{-4}$ N$\cdot$m$)(-1.2\mathbf{i} - 0.90\mathbf{j} + 1.0\mathbf{k})$;
(b) -6.0×10^{-4} J

Chapter 30

CP **1.** a, c, b **2.** b, c, a **3.** d, tie of a and c, then b
4. d, a, tie of b and c (zero) **Q** **1.** c, d, then a and b tie
3. 2 and 4 **5.** a, b, c **7.** b, d, c, a (zero) **9.** (a) 1, $+x$;
2, $-y$; (b) 1, $+y$; 2, $+x$ **11.** outward **13.** c and
d tie, then b, a **15.** d, then tie of a and e, then b, c
17. 0 (dot product is zero) **EP** **1.** 32.1 A
3. (a) 3.3 μT; (b) yes **5.** (a) $(0.24\mathbf{i})$ nT; (b) 0; (c) $(-43\mathbf{k})$ pT;
(d) $(0.14\mathbf{k})$ nT **7.** (a) 16 A; (b) west to east **9.** 0
11. (a) 0; (b) $\mu_0 i/4R$, into the page; (c) same as (b)
13. $\mu_0 i\theta\,(1/b - 1/a)/4\pi$, out of page **15.** (a) 1.0 mT, out of
the figure; (b) 0.80 mT, out of the figure **25.** 200 μT, into
page **27.** (a) it is impossible to have other than $B = 0$
midway between them; (b) 30 A **29.** 4.3 A, out of page
35. $0.338\mu_0 i^2/a$, toward the center of the square
37. (b) to the right **39.** (b) 2.3 km/s **41.** $+5\mu_0 i$

47. (a) $\mu_0 ir/2\pi c^2$; (b) $\mu_0 i/2\pi r$; (c) $\dfrac{\mu_0 i}{2\pi(a^2 - b^2)}\,\dfrac{a^2 - r^2}{r}$;

(d) 0 **49.** $3i/8$, into page **53.** 0.30 mT **55.** 108 m
61. 0.272 A **63.** (a) 4; (b) 1/2 **65.** (a) 2.4 A$\cdot$m^2;
(b) 46 cm **67.** (a) $\mu_0 i(1/a + 1/b)/4$, into page; (b) $\frac{1}{2}i\pi(a^2 + b^2)$,
into page **69.** (a) 79 μT; (b) 1.1×10^{-6} N$\cdot$m
71. (b) $(0.060$ $\mathbf{j})$ A$\cdot$m^2; (c) $(9.6 \times 10^{-11}\mathbf{j})$ T, $(-4.8 \times
10^{-11}\mathbf{j})$ T **73.** (a) B from sum: 7.069×10^{-5} T; $\mu_0 in =
5.027 \times 10^{-5}$ T; 40% difference; (b) B from sum: $1.043 \times
10^{-4}$ T; $\mu_0 in = 1.005 \times 10^{-4}$ T; 4% difference; (c) B from
sum: 2.506×10^{-4} T; $\mu_0 in = 2.513 \times 10^{-4}$ T; 0.3%
difference **75.** (a) $\mathbf{B} = (\mu_0/2\pi)\,[i_1/(x - a) + i_2/x]\mathbf{j}$;
(b) $\mathbf{B} = (\mu_0/2\pi)\,(i_1/a)\,(1 + b/2)\mathbf{j})$

Chapter 31

CP **1.** b, then d and e tie, and then a and c tie (zero) **2.** a
and b tie, then c (zero) **3.** c and d tie, then a and b tie
4. b, out; c, out; d, into; e, into **5.** d and e **6.** (a) 2, 3, 1
(zero); (b) 2, 3, 1 **7.** a and b tie, then c **Q** **1.** (a) all tie
(zero); (b) all tie (nonzero); (c) 3, then tie of 1 and 2 (zero)
3. out **5.** (a) into; (b) counterclockwise; (c) larger
7. (a) leftward; (b) rightward **9.** c, a, b **11.** (a) 1, 3, 2;
(b) 1 and 3 tie, then 2 **13.** a, tie of b and c **15.** (a) more;
(b) same; (c) same; (d) same (zero) **17.** a, 2; b, 4; c, 1; d, 3
EP **1.** 57 μWb **3.** 1.5 mV **5.** (a) 31 mV; (b) right to
left **7.** (a) 0.40 V; (b) 20 A **9.** (b) 58 mA **11.** 1.2 mV
13. 1.15 μWb **15.** 51 mV, clockwise when viewed along the
direction of $\mathbf{B}$ **17.** (a) 1.26×10^{-4} T, 0, -1.26×10^{-4} T;
(b) 5.04×10^{-8} V **19.** (b) no **21.** 15.5 μC
23. (a) 24 μV; (b) from c to b **25.** (b) design it so that
$Nab = (5/2\pi)$ m^2 **27.** (a) 0.598 μV; (b) counterclockwise

29. (a) $\dfrac{\mu_0 ia}{2\pi}\left(\dfrac{2r + b}{2r - b}\right)$; (b) $2\mu_0 iabv/\pi R(4r^2 - b^2)$

31. $A^2 B^2/R\Delta t$ **33.** (a) 48.1 mV; (b) 2.67 mA; (c) 0.128 mW
35. $v_t = mgR/B^2L^2$ **37.** 268 W **39.** (a) 240 μV; (b) 0.600

mA; (c) 0.144 μW; (d) 2.88 $\times$ 10^{-8} N; (e) same as (c)
41. 1, -1.07 mV; 2, -2.40 mV; 3, 1.33 mV **43.** at a:
4.4×10^7 m/s^2, to the right; at b: 0; at c: 4.4×10^7 m/s^2, to the
left **45.** 0.10 μWb **47.** (a) 800; (b) 2.5×10^{-4} H/m
49. (a) $\mu_0 i/W$; (b) $\pi\mu_0 R^2/W$ **51.** (a) decreasing;
(b) 0.68 mH **53.** (a) 0.10 H/m; (b) 1.3 V/m **55.** (a) 16 kV;
(b) 3.1 kV; (c) 23 kV **57.** (b) so that the changing magnetic
field of one does not induce current in the other;
(c) $1/L_{eq} = \sum_{j=1}^{N} (1/L_j)$ **59.** 6.91 **61.** 1.54 s **63.** (a) 8.45
ns; (b) 7.37 mA **65.** $(42 + 20t)$ V **67.** 12.0 A/s
69. (a) $i_1 = i_2 = 3.33$ A; (b) $i_1 = 4.55$ A, $i_2 = 2.73$ A;
(c) $i_1 = 0$, $i_2 = 1.82$ A; (d) $i_1 = i_2 = 0$ **71.** $\mathcal{E}L_1/R(L_1 + L_2)$
73. (a) $i(1 - e^{-Rt/L})$ **75.** $1.23\tau_L$ **77.** (a) 240 W;
(b) 150 W; (c) 390 W **79.** (a) 97.9 H; (b) 0.196 mJ
81. (a) 10.5 mJ; (b) 14.1 mJ **83.** (a) 34.2 J/m^3; (b) 49.4
mJ **85.** 1.5×10^8 V/m **87.** $(\mu_0 l/2\pi)\ln(b/a)$ **89.** (a) 1.3
mT; (b) 0.63 J/m^3 **91.** (a) 1.0 J/m^3; (b) 4.8×10^{-15} J/m^3
93. (a) 1.67 mH; (b) 6.00 mWb **95.** 13 H **99.** magnetic
field exists only within the cross section of solenoid 1

Chapter 32

CP **1.** d, b, c, a (zero) **2.** (a) 2; (b) 1 **3.** (a) away;
(b) away; (c) less **4.** (a) toward; (b) toward; (c) less
5. a, c, b, d (zero) **6.** tie of b, c, and d, then a
Q **1.** (a) a, c, f; (b) bar gh **3.** supplied **5.** (a) all down;
(b) 1 up, 2 down, 3 zero **7.** (a) 1 up, 2 up, 3 down;
(b) 1 down, 2 up, 3 zero **9.** (a) rightward; (b) leftward
11. (a) decreasing; (b) decreasing **13.** (a) tie of a and b, then
c, d; (b) none (plate lacks circular symmetry, so **B** is not
tangent to a circular loop); (c) none **15.** 1/4 **EP** **1.** (b)
sign is minus; (c) no, compensating positive flux through open
end near magnet **3.** 47 μWb, inward **5.** 55 μT **7.** (a)
600 MA; (b) yes; (c) no **9.** (a) 31.0 μT, 0°; (b) 55.9 μT,
73.9°; (c) 62.0 μT, 90° **11.** 4.6×10^{-24} J **13.** (a) $5.3 \times$
10^{11} V/m; (b) 20 mT; (c) 660 **15.** (a) 7; (b) 7; (c) $3h/2\pi$, 0;
(d) $3eh/4\pi m$, 0; (e) $3.5h/2\pi$; (f) 8 **17.** (b) in the direction
of the angular momentum vector **19.** $\Delta\mu = e^2 r^2 B/4m$
21. 20.8 mJ/T **23.** yes **25.** (a) 4 K; (b) 1 K
29. (a) 3.0 μT; (b) 5.6×10^{-10} eV **31.** (a) 8.9 A·m^2;
(b) 13 N·m **35.** (a) 0.14 A; (b) 79 μC **37.** $2.4 \times$
10^{13} V/m·s **39.** 1.9 pT **41.** 7.5×10^5 V/s
43. 7.2×10^{12} V/m·s **45.** (a) 2.1×10^{-8} A, downward;
(b) clockwise **47.** (a) 0.63 μT; (b) 2.3×10^{12} V/m·s
49. (a) 2.0 A; (b) 2.3×10^{11} V/m·s; (c) 0.50 A;
(d) 0.63 μT·m **51.** (a) 7.60 μA; (b) 859 kV·m/s;
(c) 3.39 mm; (d) 5.16 pT

Chapter 33

CP **1.** (a) $T/2$, (b) T, (c) $T/2$, (d) $T/4$ **2.** (a) 5 V;
(b) 150 μJ **3.** (a) 1; (b) 2 **4.** (a) C, B, A; (b) 1, A;
2, B; 3, S; 4, C; (c) A **5.** (a) increases; (b) decreases
6. (a) 1, lags; 2, leads; 3, in phase; (b) 3 ($\omega_d = \omega$ when
$X_L = X_C$) **7.** (a) increase (circuit is mainly capacitive;
increase C to decrease X_C to be closer to resonance for

maximum P_{av}); (b) closer **8.** step-up **Q** **1.** (a) $T/4$,
(b) $T/4$, (c) $T/2$ (see Fig. 33-2), (d) $T/2$ (see Eq. 31-40)
3. b, a, c **5.** (a) 3, 1, 2; (b) 2, tie of 1 and 3
7. slower **9.** (a) 1 and 4; (b) 2 and 3 **11.** (a) 3,
then 1 and 2 tie; (b) 2, 1, 3 **13.** (a) negative; (b) lead
15. (a)–(c) rightward, increase **EP** **1.** 9.14 nF
3. 45.2 mA **5.** (a) 6.00 μs; (b) 167 kHz; (c) 3.00 μs
7. (a) 89 rad/s; (b) 70 ms; (c) 25 μF **9.** 38 μH
11. 7.0×10^{-4} s **15.** (a) 3.0 nC; (b) 1.7 mA; (c) 4.5 nJ
17. (a) 3.60 mH; (b) 1.33 kHz; (c) 0.188 ms **19.** 600, 710,
1100, 1300 Hz **21.** (a) $Q/\sqrt{3}$; (b) 0.152 **25.** (a) 1.98 μJ;
(b) 5.56 μC; (c) 12.6 mA; (d) $-46.9°$; (e) $+46.9°$ **27.** (a) 0;
(b) $2i(t)$ **29.** (a) 356 μs; (b) 2.50 mH; (c) 3.20 mJ
31. 8.66 mΩ **33.** $(L/R)\ln 2$ **35.** (a) $\pi/2$ rad; (b) $q =$
$(I/\omega)\, e^{-Rt/2L} \sin \omega' t$ **39.** (a) 0.0955 A; (b) 0.0119 A
41. (a) 4.60 kHz; (b) 26.6 nF; (c) $X_L = 2.60$ kΩ, $X_C =$
0.650 kΩ **43.** (a) 0.65 kHz; (b) 24 Ω **45.** (a) 39.1 mA;
(b) 0; (c) 33.9 mA **47.** (a) 6.73 ms; (b) 2.24 ms;
(c) capacitor; (d) 59.0 μF **49.** (a) $X_C = 0$, $X_L = 86.7$ Ω,
$Z = 182$ Ω, $I = 198$ mA, $\phi = 28.5°$ **51.** (a) $X_C = 37.9$ Ω,
$X_L = 86.7$ Ω, $Z = 167$ Ω, $I = 216$ mA, $\phi = 17.1°$
53. (a) 2.35 mH; (b) they move away from 1.40 kHz
55. 1000V **57.** (a) 36.0 V; (b) 27.3 V; (c) 17.0 V;
(d) -8.34 V **59.** (a) 224 rad/s; (b) 6.00 A; (c) 228 rad/s,
219 rad/s; (d) 0.040 **61.** (a) 707 Ω; (b) 32.2 mH;
(c) 21.9 nF **63.** (a) resonance at $f = 1/2\pi\sqrt{LC} = 85.7$ Hz;
(b) 15.6 μF; (c) 225 mA **65.** (a) 796 Hz; (b) no change;
(c) decreased; (d) increased **69.** 141 V **71.** (a) taking;
(b) supplying **73.** 0, 9.00 W, 3.14 W, 1.82 W
75. 177 Ω, no **77.** 7.61 A **83.** (a) 117 μF; (b) 0;
(c) 90.0 W, 0; (d) 0°, 90°; (e) 1, 0 **85.** (a) 2.59 A;
(b) 38.8 V, 159 V, 224 V, 64.2 V, 75.0 V; (c) 100 W for R,
0 for L and C. **87.** (a) 2.4 V; (b) 3.2 mA, 0.16 A
89. (a) 1.9 V, 5.9 W; (b) 19 V, 590 W; (c) 0.19 kV, 59 kW
91. (a) $X_C = [(2\pi)(45 \times 10^{-6}\text{ F})f]^{-1}$; (c) 17.7 Hz
93. (a) $X_L = (2\pi)(40 \times 10^{-3}\text{ H})f$; (c) 796 Hz
95. (b) 61 Hz; (c) 90 Ω and 61 Hz

Chapter 34

CP **1.** (a) (Use Fig. 34-5.) On right side of rectangle, **E** is in
negative y direction; on left side, $\mathbf{E} + d\mathbf{E}$ is greater and in
same direction; (b) **E** is downward. On right side, **B** is in
negative z direction; on left side, $\mathbf{B} + d\mathbf{B}$ is greater and in
same direction. **2.** positive direction of x **3.** (a) same;
(b) decrease **4.** a, d, b, c (zero) **5.** a **6.** (a) yes; (b) no
Q **1.** (a) positive direction of z; (b) x **3.** (a) same;
(b) increase; (c) decrease **5.** both 20° clockwise from the y
axis **7.** two **9.** b, 30°; c, 60°; d, 60°; e, 30°; f, 60°
11. d, b, a, c **13.** (a) b; (b) blue; (c) c **15.** 1.5
EP **1.** (a) 4.7×10^{-3} Hz; (b) 3 min 32 s **3.** (a) 4.5×10^{24}
Hz; (b) 1.0×10^4 km or 1.6 Earth radii **7.** it would steadily
increase; (b) the summed discrepancies between the apparent
time of eclipse and those observed from x; the radius of Earth's
orbit **9.** 5.0×10^{-21} H **11.** 1.07 pT **17.** 4.8×10^{-29}
W/m^2 **19.** 4.51×10^{-10} **21.** 89 cm **23.** 1.2 MW/m^2
25. 820 m **27.** (a) 1.03 kV/m; 3.43 μT **29.** (a) 1.4 $\times$

10^{-22} W; (b) 1.1×10^{15} W **31.** (a) 87 mV/m; (b) 0.30 nT; (c) 13 kW **33.** 3.3×10^{-8} Pa **35.** (a) 4.7×10^{-6} Pa; (b) 2.1×10^{10} times smaller **37.** 5.9×10^{-8} Pa **39.** (a) 3.97 GW/m²; (b) 13.2 Pa; (c) 1.67×10^{-11} N; (d) 3.14×10^{3} m/s² **41.** $I(2 - frac)/c$ **43.** $p_{r\perp} \cos^2 \theta$ **45.** 1.9 mm/s **47.** (b) 580 nm **49.** (a) 1.9 V/m; (b) 1.7×10^{-11} Pa **51.** 1/8 **53.** 3.1% **55.** 20° or 70° **57.** 19 W/m² **59.** (a) 2 sheets; (b) 5 sheets **61.** 180° **63.** 1.26 **65.** 1.07 m **69.** (a) 0; (b) 20°; (c) still 0 and 20° **73.** 1.41 **75.** 1.22 **77.** 182 cm **79.** (a) no; (b) yes; (c) about 43° **81.** (a) 35.6°; (b) 53.1° **83.** (b) 23.2° **85.** (a) 53°; (b) yes **87.** 55.5°; 55.8°

Chapter 35

CP Kaleidoscope answer: two mirrors that form a V with an angle of 60° **1.** $0.2d$, $1.8d$, $2.2d$ **2.** (a) real; (b) inverted; (c) same **3.** (a) e; (b) virtual, same **4.** virtual, same as object, diverging **Q** **1.** c **3.** (a) a; (b) c **5.** (a) no; (b) yes (fourth is off mirror ed) **7.** (a) from infinity to the focal point; (b) decrease continually **9.** d (infinite), tie of a and b, then c **11.** mirror, equal; lens, greater **13.** (a) all but variation 2; (b) for 1, 3, and 4: right, inverted; for 5 and 6: left, same **15.** (a) less; (b) less **EP** **1.** (a) virtual; (b) same; (c) same; (d) $D + L$ **3.** 40 cm **7.** (a) 7; (b) 5; (c) 1 to 3; (d) depends on the position of O and your perspective **11.** new illumination is 10/9 of the old **15.** 10.5 cm **19.** (a) 2.00; (b) none **23.** 1.14 **25.** (b) separate the lenses by a distance $f_2 - |f_1|$, where f_2 is the focal length of the converging lens **27.** 45 mm, 90 mm **29.** (a) +40 cm; (b) at infinity **33.** (a) 40 cm, real; (b) 80 cm, real; (c) 240 cm, real; (d) −40 cm, virtual; (e) −80 cm, virtual; (f) −240 cm, virtual **35.** same orientation, virtual, 30 cm to left of second lens, $m = 1$ **37.** (a) final image coincides in location with the object; it is real, inverted, and $m = -1.0$ **39.** (a) coincides in location with the original object and is enlarged 5.0 times; (c) virtual; (d) yes **45.** $i = \dfrac{(2 - n)r}{2(n - 1)}$, to the right of the right side of the sphere **47.** 2.1 mm **49.** (b) when image is at near point **51.** (b) farsighted **53.** −125

Chapter 36

CP **1.** b (least n), c, a **2.** (a) top; (b) bright intermediate illumination (phase difference is 2.1 wavelengths) **3.** (a) 3λ, 3; (b) 2.5λ, 2.5 **4.** a and d tie (amplitude of resultant wave is $4E_0$), then b and c tie (amplitude of resultant wave is $2E_0$) **5.** (a) 1 and 4; (b) 1 and 4 **Q** **1.** a, c, b **3.** (a) 300 nm; (b) exactly out of phase **5.** c **7.** (a) increase; (b) 1λ **9.** down **11.** (a) maximum; (b) minimum; (c) alternates **13.** d **15.** (a) 0.5 wavelength; (b) 1 wavelength **17.** bright **19.** all **EP** **1.** (a) 5.09×10^{14} Hz; (b) 388 nm; (c) 1.97×10^8 m/s **5.** 2.1×10^8 m/s **7.** the time is longer for the pipeline containing air, by about 1.55 ns **9.** 22°, refraction reduces θ **11.** (a) pulse 2; (b) $0.03L/c$ **13.** (a) 1.70 (or 0.70); (b) 1.70 (or 0.70); (c) 1.30

(or 0.30); (d) brightness is identical, close to fully destructive interference **15.** (a) 0.833; (b) intermediate, closer to fully constructive interference **17.** $(2m + 1)\pi$ **19.** 2.25 mm **21.** 648 nm **23.** 1.6 mm **25.** 16 **27.** 0.072 mm **29.** 8.75λ **31.** 0.03% **33.** 6.64 μm **35.** $y = 17 \sin(\omega t + 13°)$ **39.** (a) 1.17 m, 3.00 m, 7.50 m; (b) no **41.** $I = \frac{1}{9}I_m[1 + 8\cos^2(\pi d \sin \theta/\lambda)]$, I_m = intensity of central maximum **43.** $L = (m + \frac{1}{2})\lambda/2$, for $m = 0, 1, 2, \ldots$ **45.** 0.117 μm, 0.352 μm **47.** $\lambda/5$ **49.** 70.0 nm **51.** none **53.** (a) 552 nm; (b) 442 nm **55.** 338 nm **59.** $2n_2L \cos \theta_r = (m + \frac{1}{2})\lambda$, for $m = 0, 1, 2, \ldots$, where $\theta_r = \sin^{-1}[(\sin \theta_i)/n_2]$ **61.** intensity is diminished by 88% at 450 nm and by 94% at 650 nm **63.** (a) dark; (b) blue end **65.** 1.89 μm **67.** 1.00025 **69.** (a) 34; (b) 46 **73.** 588 nm **75.** 1.00030 **77.** $I = I_m \cos^2(2\pi x/\lambda)$

Chapter 37

CP **1.** (a) expand; (b) expand **2.** (a) second side maximum; (b) 2.5 **3.** (a) red; (b) violet **4.** diminish **5.** (a) increase; (b) same **6.** (a) left; (b) less **Q** **1.** (a) contract; (b) contract **3.** with megaphone (larger opening, less diffraction) **5.** four **7.** (a) larger; (b) red **9.** (a) decrease; (b) same; (c) in place **11.** (a) A; (b) left; (c) left; (d) right **EP** **1.** 690 nm **3.** 60.4 μm **5.** (a) 2.5 mm; (b) 2.2×10^{-4} rad **7.** (a) 70 cm; (b) 1.0 mm **9.** 41.2 m from the central axis **11.** 160° **15.** (d) 53°, 10°, 5.1° **19.** (a) 1.3×10^{-4} rad; (b) 10 km **21.** 50 m **23.** 30.5 μm **25.** 1600 km **27.** (a) 17.1 m; (b) 1.37×10^{-10} **29.** 27 cm **31.** 4.7 cm **33.** (a) 0.347°; (b) 0.97° **35.** (a) red; (b) 130 μm **37.** five **41.** $\lambda D/d$ **43.** (a) 5.05 μm; (b) 20.2 μm **45.** (a) 3.33 μm; (b) 0, $\pm 10.2°$, $\pm 20.7°$, $\pm 32.0°$, $\pm 45.0°$, $\pm 62.2°$ **47.** all wavelengths shorter than 635 nm **49.** 13,600 **51.** 500 nm **53.** (a) three; (b) 0.051° **55.** 523 nm **61.** 470 nm to 560 nm **63.** 491 **65.** 3650 **67.** (a) 1.0×10^4 nm; (b) 3.3 mm **69.** (a) 0.032°/nm, 0.076°/nm, 0.24°/nm; (b) 40,000, 80,000, 120,000 **71.** (a) tan θ; (b) 0.89 **73.** 0.26 nm **75.** 6.8° **77.** (a) 170 pm; (b) 130 pm **81.** 0.570 nm **83.** 30.6°, 15.3° (clockwise); 3.08°, 37.8° (counterclockwise)

Chapter 38

CP **1.** (a) same (speed of light postulate); (b) no (the start and end of the flight are spatially separated); (c) no (again, because of the spatial separation) **2.** (a) Sally's; (b) Sally's **3.** a, negative; b, positive; c, negative **4.** (a) right; (b) more **5.** (a) equal; (b) less **Q** **1.** all tie (pulse speed is c) **3.** (a) C_1; (b) C_1 **5.** (a) 3, 2, 1; (b) 1 and 3 tie, then 2 **7.** (a) negative; (b) positive **9.** c, then b and d tie, then a **11.** (a) 3, tie of 1 and 2, then 4; (b) 4, tie of 1 and 2, then 3; (c) 1, 4, 2, 3 **13.** greater than f_1 **EP** **1.** (a) 3×10^{-18}; (b) 8.2×10^{-8}; (c) 1.1×10^{-6}; (d) 3.7×10^{-5}; (e) 0.10 **3.** $0.75c$ **5.** $0.99c$ **7.** 55 m **9.** 1.32 m **11.** 0.63 m **13.** 6.4 cm **15.** (a) 26 y; (b) 52 y; (c) 3.7 y **17.** (b) 0.999 999 15c **19.** (a) $x' = 0$, $t' = 2.29$ s; (b) $x' =$

6.55×10^8 m, $t' = 3.16$ s **21.** (a) 25.8 μs; (b) small flash
23. (a) 1.25; (b) 0.800 μs **25.** 2.40 μs **27.** (a) 0.84c, in the direction of increasing x; (b) 0.21c, in the direction of increasing x; the classical predictions are 1.1c and 0.15c
29. (a) 0.35c; (b) 0.62c **31.** 1.2 μs **33.** seven
35. 22.9 MHz **37.** +2.97 nm **39.** (a) $\tau_0/\sqrt{1 - v^2/c^2}$
41. (a) 0.134c; (b) 4.65 keV; (c) 1.1% **43.** (a) 0.9988, 20.6; (b) 0.145, 1.01; (c) 0.073, 1.0027 **45.** (a) 5.71 GeV, 6.65 GeV, 6.58 GeV/c; (b) 3.11 MeV, 3.62 MeV, 3.59 MeV/c **47.** 18 smu/y **49.** (a) 0.943c; (b) 0.866c
51. (a) 256 kV; (b) 0.746c **53.** $\sqrt{8}mc$ **55.** 6.65×10^6 mi, or 270 earth circumferences **57.** 110 km **59.** (a) 2.7×10^{14} J; (b) 1.8×10^7 kg; (c) 6.0×10^6 **61.** 4.00 u, probably a helium nucleus **63.** 330 mT
65.

SIGNAL	TIME SENT (h)	TIME REPLY RECEIVED (h)	TIME REPORTED	DISTANCE (m)
1	6.0	400	11.8	2.10×10^{14}
2	12.0	800	23.6	4.19×10^{14}
3	18.0	1200	35.5	6.29×10^{14}
4	24.0	1600	47.3	8.38×10^{14}
5	30.0	2000	59.1	1.05×10^{15}

67. (a) $vt \sin \theta$; (b) $t[1 - (v/c) \cos \theta]$; (c) 3.24c

Chapter 39

CP **1.** b, a, d, c **2.** (a) lithium, sodium, potassium, cesium; (b) all tie **3.** (a) same; (b)–(d) x rays **4.** (a) proton; (b) same; (c) proton **5.** same Q **1.** (a) microwave; (b) x ray; (c) x ray **3.** potassium **5.** Positive charge builds up on the plate, inhibiting further electron emission.
7. none **9.** (a) greater; (b) less **11.** no essential change **13.** (a) decreases by a factor of $1/\sqrt{2}$; (b) decreases by a factor of 1/2 **15.** extremely small **17.** (a) increasing; (b) decreasing; (c) same; (d) same **19.** a **21.** (a) zero; (b) yes EP **3.** 4.14 eV·fs **5.** 5.9 μeV **7.** 1.0×10^{45} photons/s **9.** 2.047 eV **11.** (a) infrared bulb; (b) 1.4×10^{21} photons/s **13.** 4.7×10^{26} photons **15.** (a) 2.96×10^{20} photons/s; (b) 48,600 km; (c) 5.89×10^{18} photons/m$^2 \cdot$s
17. barium and lithium **19.** 10 eV **21.** 676 km/s
23. (a) 1.3 V; (b) 680 km/s **25.** 233 nm **27.** (a) 6.60×10^{-34} J·s; (b) 2.27 eV; (c) 545 nm **29.** 9.68×10^{-20} A
31. (a) 8.57×10^{18} Hz; (b) 35.4 keV; (c) 1.89×10^{-23} kg·m/s = 35.4 keV/c **33.** (a) 2.7 pm; (b) 6.05 pm
37. (a) 2.43 pm; (b) 1.32 fm; (c) 0.511 MeV; (d) 938 MeV
39. 300% **43.** (a) 41.8 keV; (b) 8.2 keV **45.** 44°
47. 1.12 keV **49.** (a) 1.7×10^{-35} m; (b) de Broglie wavelength too small **51.** 7.77 pm **53.** 4.3 μeV
55. (a) 38.8 meV; (b) 146 pm **57.** (a) 73 pm, 3.4 nm; (b) yes, their average de Broglie wavelength is much smaller than their average separation **59.** (a) 1.24 keV, 1.50 eV; (b) 1.24 GeV, 1.24 GeV **61.** 0.025 fm, about 200 times smaller than a nuclear radius **63.** neutron **65.** 9.70 kV (relativistic calculation), 9.79 kV (classical calculation)

73. (d) $x = n(\lambda/2)$, where $n = 0, 1, 2, \ldots$ **75.** 2.1×10^{-24} kg·m/s **79.** (a) proton: 9.2×10^{-6}; deuteron, 7.6×10^{-8}; (b) 3.0 MeV for each; (c) 3.0 MeV for each
81. (a) 10^{104} years (don't hold your breath); (b) 2×10^{-19} s (the smaller mass of the electron makes an enormous difference) **83.** $T = 10^{-x}$, where $x = 3.1 \times 10^{39}$, a *very* small number

Chapter 40

CP **1.** b, a, c **2.** (a) all tie; (b) a, b, c **3.** a, b, c, d
4. (a) $n = 1$; (b) $n = 3, n = 2, n = 1$ **5.** (a) 5; (b) 7
Q **1.** (a) 1/4; (b) same factor **3.** c
5. (a) $(\sqrt{1/L})\sin(\pi/2L)x$; (b) $(\sqrt{4/L})\sin(2\pi/L)x$; (c) $(\sqrt{2/L})\sin(\pi/L)x$ **7.** less **9.** (a) wider; (b) deeper
11. $n = 1, n = 2, n = 3$ **13.** (a) greater; (b) less; (c) less
15. same **17.** (a) $n = 3$; (b) $n = 1$; (c) $n = 5$
EP **1.** multiply it by $\sqrt{2}$ **3.** 850 pm **5.** 0.65 eV
7. meter^{-1} **13.** (a) 5.0%; (b) 10%; (c) 0.95%
15. (a) 19.6%; (b) 60.8%; (c) 19.6% **17.** 13.3 eV
19. (b) $k = (2\pi/h)[2m(E_{pot} - E)]^{1/2}$ **21.** (b) $k = (2\pi/h)(2mE)^{1/2}$ **23.** 2.6 eV **25.** (a) 658 nm; (b) 366 nm **27.** (a) 12 eV; (b) 6.5×10^{-27} kg·m/s; (c) 103 nm **31.** (a) 0; (b) 10.2 nm^{-1}; (c) 5.54 nm^{-1}
33. (a) 13.6 eV; (b) 3.40 eV **35.** (a) 13.6 eV; (b) -27.2 eV **37.** (a) $n = 4$ to $n = 2$; (b) Balmer series
39. (a) 2.6 eV; (b) $n = 4$ to $n = 2$ **41.** $n = 3$ to $n = 1$
43. 43.9% **47.** $n \approx 4348$ **51.** (a) $P_{210} = (r^4/8a^5)e^{-r/a}\cos^2 \theta$; $P_{21+1} = P_{21-1} = (r^4/16a^5)e^{-r/a}\sin^2 \theta$

Chapter 41

CP **1.** 7 **2.** (a) decrease; (b)–(c) same **3.** less
4. A, C, B Q **1.** 0, 2, and 3 **3.** 6p **5.** (a) 2, 8; (b) 5, 50 **7.** (a) n; (b) n and l **9.** a, c, e, f
11. (a) unchanged; (b) decrease; (c) decrease **13.** a and b
EP **3.** (a) 32; (b) 2; (c) 18; (d) 8 **5.** (a) 3; (b) 3 **7.** $n = 3$; for $l = 0$, $m_l = 0$; for $l = 1$, $m_l = 0, \pm 1$; for $l = 2$, $m_l = 0$, $\pm 1, \pm 2$ **9.** $l = 4$; $n \geq 5$; $m_s = \pm 1/2$ **11.** (a) 3; (b) 9; (c) 2; (d) 18; (e) 3 **13.** (a) 3×10^{74}; (b) 6×10^{74}; (c) 6×10^{-38} rad **15.** 54.7° and 125° **17.** (a) 58 μeV; (b) 14 GHz; (c) 2.1 cm; short radio wave region **19.** 5.35 cm
21. 19 mT **25.** argon **27.** (a) 2, 0, 0, $\pm\frac{1}{2}$; (b) $n = 2, l = 1$, $m_l = 1, 0,$ or -1, $m_s = \pm\frac{1}{2}$ **33.** (a) 5.7 keV; (b) 87 pm, 14 keV; 220 pm, 5.7 keV **37.** (a) 35.4 pm, as for molybdenum; (b) 57 pm; (c) 50 pm **43.** (a) 19.7 keV, 17.5 keV; (b) Zr or Nb (Zr better) **45.** (a) $(Z - 1)^2/(Z' - 1)^2$; (b) 57.5; (c) 2070
47. (a) 6; (b) 3.2×10^6 years **49.** (a) 2.55 s; (b) 500 ps; (c) $(4.5 \times 10^{-4})°$ or 1.6″ of arc **51.** 10,000 K
53. 7.3×10^{15} per second **55.** 2.0×10^{16} per second
57. 4.6 km **59.** 1.8 pm **61.** (a) 7.33 μm; (b) 7.07×10^5 W/m^2; (c) 2.49×10^{10} W/m^2 **63.** (a) 4.3 μm; (b) 10 μm; (c) infrared

Chapter 42

CP **1.** (a) larger; (b) same **2.** Cleveland, metal; Troy, none; Seattle, semiconductor **3.** a, b, and c **4.** b **5.** b
Q **1.** 4 **3.** b and c **5.** (a) anywhere in the lattice; (b) in

any silicon–silicon bond; (c) in a silicon ion core, at a lattice site **7.** b and d **9.** $+4e$ **11.** none **13.** (a) right to left; (b) back bias **15.** a, b, and c **EP** **1.** 8.43×10^{28} m^{-3} **5.** (a) $+8.0 \times 10^{-11}$ $\Omega \cdot$m/K; (b) -210 $\Omega \cdot$m/K **9.** (a) 0; (b) 0.096 **13.** 0.91 **15.** (a) 90%; (b) 12%; (c) sodium **17.** (a) 2500 K; (b) 5300 K **19.** (a) 2.7×10^{25} m^{-3}; (b) 8.4×10^{28} m^{-3}; (c) 3100; (d) molecules: 3.3 nm; electrons: 0.23 nm **21.** (a) 1.0, 0.99, 0.50, 0.014, 2.5×10^{-17}; (b) 700 K **25.** (a) 5.9×10^{28} m^{-3}; (b) 5.5 eV; (c) 1390 km/s; (d) 0.52 nm **27.** (b) 1.80×10^{28} m^{-3}eV^{-1} **31.** (a) 19.7 kJ; (b) 197 s **33.** 200°C **35.** (a) 109.5°; (b) 235 pm **37.** (a) 225 nm; (b) ultraviolet **41.** 0.22 μg **43.** (a) 0.744 eV above; (b) 7.13×10^{-7} **45.** (b) 2.5×10^{8} **47.** opaque **49.** (a) 5.0×10^{-17} F; (b) about $300e$

Chapter 43

CP **1.** ^{90}As and ^{158}Nd **2.** a little more than 75 Bq (elapsed time is a little less than three half-lives) **3.** ^{206}Pb **Q** **1.** less **3.** ^{240}U **5.** less **7.** (a) on the $N = Z$ line; (b) positrons; (c) about 120 **9.** no **11.** yes **13.** (a) increase; (b) same **15.** 7 h **17.** d **EP** **1.** 15.8 fm **3.** (a) 0.390 MeV; (b) 4.61 MeV **5.** (a) six; (b) eight **9.** (a) 1150 MeV; (b) 4.81 MeV/nucleon, 12.2 MeV/proton **15.** 4×10^{-22} s **17.** $K \approx 30$ MeV **21.** (a) 19.8 MeV, 6.26 MeV, 2.22 MeV; (b) 28.3 MeV; (c) 7.07 MeV **23.** 1.6×10^{25} MeV **25.** 7.92 MeV **27.** 280 d **29.** (a) 7.6×10^{16} s^{-1}; (b) 4.9×10^{16} s^{-1} **31.** (a) 4.8×10^{-18} s^{-1}; (b) 4.6×10^{9} y **33.** 5.3×10^{22} **35.** 265 mg **37.** 209 d **39.** 87.8 mg **41.** 730 cm^2 **45.** (a) 3.66×10^{7} s^{-1}; (b) $t \gg 3.82$ d; (c) 3.66×10^{7} s^{-1}; (d) 6.42 ng **47.** Pu: 1.2×10^{-17}, Cm: $e^{-9173} \approx 0$ **49.** (a) 4.25 MeV; (b) -24.1 MeV; (c) 28.3 MeV **51.** $Q_3 = -9.50$ MeV, $Q_4 = 4.66$ MeV, $Q_5 = -1.30$ MeV **53.** 1.21 MeV **55.** 0.782 MeV **59.** (b) 0.961 MeV **61.** 78.4 eV **63.** 1600 y **65.** 1.8 mg **67.** 1.02 mg **69.** (a) 18 mJ; (b) 2.9 mSv = 0.29 rem **71.** (a) 6.3×10^{18}; (b) 2.5×10^{11}; (c) 0.20 J; (d) 2.3 mGy; (e) 30 mSv **73.** 3.87×10^{10} K **75.** (a) 25.4 MeV; (b) 12.8 MeV; (c) 25.0 MeV

77. (a) 3.85 MeV, 7.95 MeV; (b) 3.98 MeV, 7.33 MeV **79.** (a) 5.8 MeV; (b) 11 MeV

Chapter 44

CP **1.** c and d **2.** (a) no; (b) yes; (c) no **3.** e **Q** **1.** a **3.** b **5.** (a) ^{93}Sr; (b) ^{140}I; (c) ^{155}Nd **7.** c **9.** a **11.** c **EP** **1.** (a) 2.6×10^{24}; (b) 8.2×10^{13} J; (c) 2.6×10^{4} y **3.** 3.1×10^{10} s^{-1} **7.** $+5.00$ MeV **9.** (a) 16 fissions/d; (b) 4.3×10^{8} **11.** (a) 10; (b) 226 MeV **13.** (a) 252 MeV; (b) typical fission energy is 200 MeV **15.** 461 kg **17.** yes **19.** (a) 1.2 MeV; (b) 3.2 kg **21.** (a) 44 kton **25.** 1.6×10^{16} **27.** (b) 1.0, 0.89, 0.28, 0.019; (c) 8 **29.** (a) 75 kW; (b) 5800 kg **33.** (a) 30 MeV; (b) 6 MeV **35.** (a) 170 kV **37.** 1.41 MeV **41.** (a) 3.1×10^{31} protons/m^3; (b) 1.2×10^{6} times **43.** (a) 4.0×10^{27} MeV; (b) 5.1×10^{26} MeV **45.** (a) 1.83×10^{38} s^{-1}; (b) 8.25×10^{28} s^{-1} **49.** (a) 6.3×10^{14} J/kg; (b) 6.2×10^{11} kg/s; (c) 4.3×10^{9} kg/s; (d) 15×10^{9} y **51.** (a) 24.9 MeV; (b) 8.65 megatons TNT **53.** $K_\alpha = 3.52$ MeV, $K_n = 14.1$ MeV

Chapter 45

CP **1.** (a) the muon family; (b) a particle; (c) $L_\mu = +1$ **2.** b and e **3.** c **Q** **1.** d **3.** the π^+ pion whose track curves downward at the left **5.** a, b, c, d **7.** c, f **9.** 1d, 2e, 3a, 4b, 5c **11.** 1b, 2c, 3d, 4e, 5a **13.** (a) 0; (b) $+1$; (c) -1; (d) $+1$; (e) -1 **EP** **1.** 6.03×10^{-29} kg **3.** 2.4×10^{-43} **5.** 1.08×10^{42} J **7.** 2.7 cm/s **9.** (a) 1.90×10^{-18} kg $\cdot$ m/s; (b) 9.90 m **13.** (a) L_μ, L_e, angular momentum; (b) charge; (c) energy, L_μ **15.** $Q = 0$, $B = -1$, $S = 0$ **17.** (a) energy; (b) strangeness; (c) charge **19.** 338 MeV **23.** (a) $u\bar{u}d$; (b) $\bar{u}dd$ **25.** (a) sud; (b) uss **29.** Σ^0, 7530 km/s **31.** 669 nm **33.** (b) 0.934; (c) 1.15×10^{10} ly **35.** (a) 256 μeV; (b) 4.84 mm **37.** (a) 122 m/s; (b) 246 y **39.** (a) 2.6 K; (b) 29 nm **41.** (a) $0.785c$; (b) $0.993c$; (c) C2; (d) C1; (e) 51 ns; (f) 40 ns **43.** (a) 4.7×10^{4}; (b) 4.3 mm; (c) 24 nm

Index

Page references followed by lowercase roman t indicate material in tables. References followed by lowercase italic *n* indicate material in footnotes.

Photo Credits

Chapter 16
Page 372: Tom van Dyke/Sygma. Page 373: Kent Knudson/FPG International. Page 387: Bettmann Archive. Page 392: Courtesy NASA.

Chapter 17
Page 399: John Visser/Bruce Coleman, Inc. Page 415: Richard Megna/Fundamental Photographs. Page 416: Courtesy T.D. Rossing, Northern Illinois University.

Chapter 18
Page 425: Stephen Dalton/Animals Animals. Page 426: Howard Sochurak/The Stock Market. Page 433: Ben Rose/The Image Bank. Page 434: Bob Gruen/Star File. Page 435: John Eastcott/Yva Momativk/DRK Photo. Page 441: Philippe Plailly/Science Photo Library/Photo Researchers. Page 444: Courtesy NASA.

Chapter 19
Page 453: Tom Owen Edmunds/The Image Bank. Page 459: AP/Wide World Photos. Page 471: Peter Arnold/Peter Arnold, Inc. Page 472: Courtesy Daedalus Enterprises, Inc.

Chapter 20
Page 484: Bryan and Cherry Alexander Photography.

Chapter 21
Page 509: Steven Dalton/Photo Researchers. Page 517: Richard Ustinich/The Image Bank. Page 523 (left): Cary Wolinski/Stock, Boston. Page 523 (right): Courtesy of Professor Bernard Hallet, Quaternary Research Center, University of Washington, Seattle. Page 534: Jeff Werner.

Chapter 22
Page 537: Michael Watson. Page 538: Fundamental Photographs. Page 539: Courtesy Xerox. Page 540: Johann Gabriel Doppelmayr, Neuentdeckte Phaenomena von Bewünderswurdigen Würckungen der Natur, Nuremberg, 1744. Page 547: Courtesy Lawrence Berkeley Laboratory.

Chapter 23
Page 554: Quesada/Burke Studios. Page 565: Russ Kinne/Comstock, Inc. Page 566: Courtesy Environmental Elements Corporation.

Chapter 24
Page 579: E.R. Degginger/Bruce Coleman, Inc. Page 589 (top): Courtesy E. Philip Krider, Institute for Atmospheric Physics, University of Arizona, Tucson. Page 589 (bottom): C. Johnny Autery.

Chapter 25
Pages 601 and 605: Courtesy NOAA. Page 617: Courtesy Westinghouse Corporation.

Chapter 26
Page 628: Goivaux Communication/Phototake. Page 629: Paul Silvermann/Fundamental Photographs. Page 638: ©The Harold E. Edgerton 1992 Trust/Courtesy Palm Press, Inc. Page 639: Courtesy The Royal Institute, England.

Chapter 27
Page 651: UPI/Corbis-Bettmann. Page 657: The Image Works. Page 663: Laurie Rubie/Tony Stone Images/New York, Inc. Page 666 (left): Courtesy AT&T Bell Laboratories. Page 666 (right): Courtesy Shoji Tonaka/International Superconductivity Technology Center, Tokyo, Japan.

Chapter 28
Page 673: Norbert Wu. Page 674: Courtesy Southern California Edison Company.

Chapter 29
Page 700: Johnny Johnson/Earth Scenes/Animals Animals. Page 701: Schneps/The Image Bank. Page 703: Courtesy Lawrence Berkeley Laboratory, University of California. Page 704: Courtesy Dr. Richard Cannon, Southeast Missouri State University, Cape Girardeau. Page 708: Courtesy John Le P. Webb, Sussex University, England. Page 710: Courtesy Dr. L.A. Frank, University of Iowa. Page 712: Courtesy Fermi National Accelerator Laboratory.

Chapter 30
Page 728: Michael Brown/Florida Today/Gamma Liaison. Page 730: Courtesy Education Development Center.

Chapter 31
Page 752: Dan McCoy/Black Star. Page 756: Courtesy Fender Musical Instruments Corporation. Page 760: Courtesy Jenn-Air Co. Page 765: Courtesy The Royal Institute, England.

Chapter 32
Page 786: Bob Zehring. Page 787: Runk/Schoenberger/Grant Heilman Photography. Page 794: Peter Lerman. Page 797 (top): Courtesy Ralph W. DeBlois. Page 797 (bottom): R.E. Rosenweig, Research and Science Laboratory, courtesy Exxon Co. USA.

Chapter 33
Page 811: Courtesy Haverfield Helicopter Co. Page 814: Courtesy Hewlett Packard. Page 827 (left): Steve Kagan/Gamma Liaison. Page 827 (right): Ted Cowell/Black Star.

Chapter 34
Page 841: Courtesy Hansen Publications. Page 850: ©1992 Ben and Miriam Rose, from the collection of the Center for Creative Photography, Tucson. Page 854: Diane Schiumo/Fundamental Photographs. Page 855: PSSC Physics, 2nd edition; ©1975 D.C. Heath and Co. with Education Development Center, Newton, MA. Page 856: Courtesy Lockheed Advanced Development Company. Page 858 (top right): Tony Stone Images/New York, Inc. Page 858 (bottom left): Courtesy Bausch & Lomb. Page 861: Will and Deni McIntyre/Photo Researchers. Page 867: Cornell University.

Chapter 35
Page 872: Courtesy Courtauld Institute Galleries, London. Page 874 (left): Frans Lanting/Minden Pictures, Inc. Page 874 (right): Wayne Sorce. Page 882: Dr. Paul A. Zahl/Photo Researchers.

Page 884: Courtesy Matthew J. Wheeler. Page 894: Piergiorgio Scharandis/Black Star.

Chapter 36

Page 901: E.R. Degginger. Page 905: Runk Schoenberger/Grant Heilman Photography. Page 906: From *Atlas of Optical Phenomena* by M. Cagnet et al., Springer-Verlag, Prentice Hall, 1962. Page 914: Richard Megna/Fundamental Photographs. Page 917: Courtesy Dr. Helen Ghiradella, Department of Biological Sciences, SUNY, Albany. Page 926: Courtesy Bausch & Lomb.

Chapter 37

Page 929: Georges Seurat, French, 1859–1891, *A Sunday on La Grande Jatte,* 1884. Oil on canvas; 1884–86, 207.5 × 308 cm; Helen Birch Bartlett Memorial Collection, 1926. Photograph ©1996, The Art Institute of Chicago. All Rights Reserved. Page 930: Ken Kay/Fundamental Photographs. Pages 931, 937 (bottom left) and 940: From *Atlas of Optical Phenomena* by Cagnet, Francon, Thierr, Springer-Verlag, Berlin, 1962. Page 937 (bottom right): AP/Wide World Photos, Inc. Page 938: Cath Ellis/Science Photo Library/Photo Researchers. Page 945 (left): Department of Physics, Imperial College/Science Photo Library/Photo Researchers. Page 945 (top right): Peter L. Chapman/Stock, Boston. Page 953: Kjell B. Sandved/Bruce Coleman, Inc. Page 954: Courtesy Professor Robert Greenler, Physics Department, University of Wisconsin.

Chapter 38

Page 958: T. Tracy/FPG International. Page 959: Courtesy Hebrew University of Jerusalem, Israel.

Chapter 39

Page 985: Lawrence Berkeley Laboratory/Science Photo Library/Photo Researchers. Page 994 (left): Courtesy A. Tonomura, J. Endo, T. Matsuda, and T. Kawasaki/Advanced Research Laboratory, Hitachi, Ltd., Kokubinju, Tokyo; H. Ezawa, Department of Physics, Gakushuin University, Mejiro, Tokyo. Page 994 (top right): Courtesy Riber Division of Instruments, Inc. Page 994 (center right): From PSSC film ''Matter Waves,'' courtesy Education Development Center, Newton, Massachusetts. Page 999: Philippe Plailly/Science Photo Library/Photo Researchers.

Chapter 40

Page 1007: Courtesy International Business Machines Corporation, Almaden Research Center, CA. Page 1015: From ''Scientific American'', January 1993, page 122. Reproduced with permission of Michael Steigerwald, Bell Labs-Lucent Technologies. Page 1016: From ''Scientific American'', September 1995, page 67. Image reproduced with permission of H. Temkin, Texas Tech University. Page 1018: From W. Finkelnburg, Structure of Matter, Springer-Verlag, 1964. Reproduced with permission.

Chapter 41

Page 1028: Kurt Coste/Fertilim Institute of New Orleans/Tony Stone Image/New York, Inc. Page 1029: Courtesy Warren Nagourney. Page 1037: SBHA/Tony Stone Images/New York, Inc. Page 1043: Will & Deni McIntyre/Photo Researchers.

Chapter 42

Page 1052: Steven Northrup/New York Times Pictures. Page 1066: Courtesy AT&T Bell Laboratories. Page 1068: Scot Hill Photography.

Chapter 43

Page 1074: Elscint/Science Photo Library/Photo Researchers. Page 1087: George Rockwin/Bruce Coleman, Inc. Page 1087 (inset): R. Perry/Sygma.

Chapter 44

Page 1098: Courtesy U.S. Department of Energy. Page 1104: Ivleva/Magnum Photos, Inc. Page 1105: Courtesy U.S. Department of Energy. Page 1106: Courtesy Chicago Historical Society. Page 1110: Courtesy Anglo-Australian Telescope Board. Page 1112 (top right): Courtesy Los Alamos National Laboratory, New Mexico. Page 1112: (bottom left): Courtesy Princeton Plasma Physics Laboratory. Page 1115: Courtesy Martrin Marietta Energy Systems and U.S. Department of Energy.

Chapter 45

Page 1118: Courtesy NASA. Page 1119: David Parker/Science Photo Library/Photo Researchers. Page 1121: Courtesy Michael Mathews. Page 1122: Courtesy Lawrence Berkeley Laboratory. Page 1131: Courtesy Fermilab Visual Media Services.

PROBLEM SOLVING TACTICS